# AutoCAD® 2027 INSTRUCTOR

A Student Guide for
In-Depth Coverage of AutoCAD's
Commands and Features

James A. Leach
Shawna Lockhart

SDC
PUBLICATIONS

SDC Publications
P.O. Box 1334
Mission, KS 66222
913-262-2664
www.SDCpublications.com
Publisher: Stephen Schroff



ISBN 13: 978-1-63057-807-7
ISBN 10: 1-63057-807-X

Printed and bound in the United States of America.

# Table of Contents

# Preface

## ABOUT THIS BOOK

### This Book Is Your AutoCAD 2027 Instructor

The objective of *AutoCAD 2027 Instructor* is to provide you with extensive knowledge of AutoCAD, whether you are taking an instructor-led course or learning AutoCAD on your own.

### In-Depth Coverage

*AutoCAD 2027 Instructor* is written to instruct you in a broad and deep range of AutoCAD *2027* features. Commands, options, system variables, and features are explained fully with many examples and illustrations. This text can be used for a two- or three-course sequence.

### Graphically Oriented

Because *AutoCAD 2027 Instructor* discusses concepts that are graphical by nature, many illustrations (approximately 2000) are used to communicate the concepts, commands, and applications.

### Pedagogical Progression

*AutoCAD 2027 Instructor* is presented in a pedagogical format by delivering the fundamental concepts first, then moving toward the more advanced and specialized features of AutoCAD. The book begins with small pieces of information explained in a simple form and then builds on that experience to deliver more complex ideas requiring a synthesis of earlier concepts. The chapter exercises follow the same progression, beginning with a simple tutorial approach and ending with more challenging problems requiring a synthesis of earlier exercises.

### Multi-chapter "Reuse" Exercises

About half of the Chapter Exercises are used again later in subsequent chapters. This concept emphasizes the natural pedagogical progression of the text, creates connections between concepts, and maximizes the student's efforts. Reuse Exercises are denoted with a "Reuse" icon and the related drawing file names are printed in a reversed font.

### Important "Tips"

Tips, reminders, notes, and cautions are given in the book and denoted by a "TIP" marker in the margin. This feature helps you identify and remember important concepts, commands, procedures, and tricks used by professionals that would otherwise be discovered only after much experience.

### Markers

Older editions of this text denoted added commands by a "2024", "2023", or "2022" and so on marker in the margin. The commands are up-to-date for the software version of the text in this case 2027. To find which commands are added and deleted in various software versions, you can use the handy New Selection from the user Help, if you so desire.

### Valuable Reference Guide

*AutoCAD 2027 Instructor* is structured to be used as a reference guide to AutoCAD. Several important tables and lists are "tabbed" on the edge of the page for easy access. Every command throughout the book is given with a "command table" listing the possible methods of invoking the command.
A comprehensive index gives an alphabetical listing of AutoCAD commands, command options, and concepts discussed.

### For Professionals and Students in Diverse Areas

*AutoCAD 2027 Instructor* is written for professionals and students in the fields of engineering, architecture, design, construction, manufacturing, and any other field that has a use for AutoCAD. Applications and examples from many fields are given throughout the text. The applications and examples are not intended to have an inclination toward one particular field. Instead, applications to a particular field are used when they best explain an idea or use of a command.

### Online Resources

The web page for this book provides students and instructors with access to additional resources to accompany *AutoCAD 2027 Instructor.* Sample drawings for the chapter exercises are available for students. Instructor resources include solution drawings to the chapter exercises and 20 test/review questions and answers for each chapter in Word® format. Learn more at:

**www.sdcpublications.com/Book/978-1-63057-493-2**

### Have Fun

We predict you will have a positive experience learning AutoCAD. Although learning AutoCAD is not a trivial endeavor, you will have fun learning this exciting technology. In fact, we predict that more than once in your learning experience you will say to yourself, "Sweet!" (or something to that effect).

*Shawna Lockhart and James A. Leach*

## ABOUT THE AUTHORS

**James A. Leach** (B.I.D., M.Ed.) is Professor Emeritus, currently teaching part-time at Auburn University for the Industrial and Systems Engineering Department. Jim taught Engineering Graphics for 25 years at University of Louisville and 15 years at Auburn University. He began teaching AutoCAD early in 1984 using Version 1.4, the first version of AutoCAD to operate on IBM personal computers. For 18 years, Jim was Director of the AutoCAD Training Center established at the University of Louisville in 1985, one of the first fifteen centers to be authorized by Autodesk.

**Shawna Lockhart** combines her university teaching experience at Montana State University and Embry Riddle Aeronautical University with wide ranging industrial experience to write engineering design communication and CAD software texts. An early adopter of 2D and 3D CAD technology, she is the author of multiple AutoCAD texts, from AutoCAD Release 10 in 1992 to the current 2027 release. Shawna also writes industrial manuals on a variety of topics including robotics, dynos, mineral processing and lasers.

## CONTRIBUTING AUTHORS

Thomas D. Bledsaw (B.S., M.S. Industrial Technology) is a CAD consultant and former National Chair for the School of Drafting and Design for ITT Educational Services. He began teaching AutoCAD Version 2.5 in 1991. Prior to this, Tom was employed in the defense industry as an engineering designer using CATIA and CADAM. In his 22 years of teaching AutoCAD, Tom has worked with several authors and completed technical reviews on numerous textbooks and has developed curriculum and courseware taught in over 100 schools. Tom was coauthor for *AutoCAD 2012 Instructor* and was a consultant for *AutoCAD 2017 Instructor*.

Michael E. Beall is owner of CAD Trainer Guy, LLC in Shelbyville, Kentucky, near Louisville. His book, *The AutoCAD Workbench, 2nd Edition,* is a compilation of over 300 AutoCAD tips and insights from *Michael's Corner* and is available from his website, www.cadtrainerguy.com. Michael has been presenting CAD training seminars domestically and internationally to the A&D community since 1982, including presentations at Autodesk University where he won the 2014 Speaker Award for Hands-on Labs. Michael received Autodesk's Instructional Quality Award twice while teaching at the University of Louisville's Authorized Autodesk Training center. Michael assisted with several topics in *AutoCAD 2017 Instructor*.

Steven H. Baldock (B.S., B.Che., M.Eng.) is an engineer at a consulting firm in Louisville, Kentucky, and offers CADD consulting and training as well. Steve is an Autodesk Certified Instructor and teaches courses at the University of Louisville's AutoCAD Training Center and at J.B. Speed School of Engineering. He has also instructed and developed course materials for various community colleges. He has 25 years experience using AutoCAD in architectural, civil, structural, and process design applications. Steve prepared material for several sections of *AutoCAD 2010 Instructor*, such as xrefs, groups, text, multilines, and dimensioning, and created several hundred figures used in *AutoCAD 2017 Instructor*.

Tom Heffner has been the Coordinator and AutoCAD Instructor at Winnipeg Technical College, Continuing Education since 1989. Tom prepares the curriculum and teaches the Continuing Education AutoCAD Program to Level I, II, and III (3D) students. Over the past 34 years, Tom has worked for several engineering companies in Canada, spending 20 years with The City of Winnipeg, Public Works Department, Engineering Division as Supervisor of Drafting and Graphic Services. Tom was instrumental in introducing and training City of Winnipeg Engineering employees AutoCAD starting with Version 2.6 up to the current Autodesk release. Tom has been a contributing author for McGraw-Hill for *AutoCAD 2008 Instructor, AutoCAD 2007 Instructor,* and *AutoCAD 2006 Companion* chapter exercises and chapter test/review questions.

Bruce Duffy has been an AutoCAD trainer and consultant for 29 years. He has aided many Fortune 500 companies with training, curriculum development, and productivity assessment/training. Bruce worked as an instructor at the University of Louisville and the Authorized Autodesk Training Center for seven years. Since 2002, Bruce has been working with Florida International University in Miami, Florida, as an Instructional Design and Online Learning consultant. He still consults and provides AutoCAD training for national and international clients. Bruce Duffy contributed much of the material for Chapter 31 of *AutoCAD 2017 Instructor*.

Patrick McCuistion, Ph.D., has taught in the Department of Industrial Technology at Ohio University since 1989. Dr. McCuistion taught three years at Texas A&M University and previously worked in various engineering design, drafting, and checking positions at several manufacturing industries. He has provided instruction in geometric dimensioning and tolerancing to many industry, military, and educational institutions and has written one book, several articles and given many presentations on the topic. Dr. McCuistion is an active chairman or member of the following ANSI/ASME subcommittees: Y14.2 Line Conventions & Lettering, Y14.3 Multiview and Section View Drawings, Y14.5 Dimensioning & Tolerancing, Y14.8 Castings, Forging & Molded Part Drawings, Y14.5.2 Geometric Dimensioning Certification, Y14.35 Revisions, Y14.36 Surface Texture, Y14.43 Functional Gages, and the Y14 Main committee. Dr. McCuistion contributed the material on geometric dimensioning and tolerancing for *AutoCAD 2017 Instructor*.

## ACKNOWLEDGMENTS

I want to thank all of the contributing authors for their assistance in writing *AutoCAD 2027 Instructor*. Without their help, this text could not have been as application-specific nor could it have been completed in the short time frame.

I also want to thank all of the readers that have contacted me with comments and suggestions on specific sections of this text. Your comments help me improve this book, assist me in developing new ideas, and keep me abreast of ways AutoCAD and this text are used in industrial and educational settings.

*James A. Leach and Shawna Lockhart*

## TRADEMARK AND COPYRIGHT ACKNOWLEDGMENTS

The following drawings used for Chapter Exercises appear courtesy of James A. Leach, *Problems in Engineering Graphics Fundamentals, Series A* and *Series B,* ©1984 and 1985 by Contemporary Publishing Company of Raleigh, Inc.: Gasket A, Gasket B, Pulley, Holder, Angle Brace, Saddle, V-Block, Bar Guide, Cam Shaft, Bearing, Cylinder, Support Bracket, Corner Brace, and Adjustable Mount. Reprinted or redrawn by permission of the publisher.

The following are trademarks of Autodesk, Inc. in the United States. For additional trademark information, please contact the Autodesk, Inc., Legal Department: 415-507-6744. 3DEC (design/logo)®, 3December®, 3December.com®, 3ds Max®, ActiveShapes®, ActiveShapes (logo)®, Actrix®, ADI®, AEC Authority (logo)®, Alias®, AliasStudio™, [Alias] "Swirl Design" (design/logo)®, Alias|Wavefront (design/logo)®, ATC®, AUGI® [Note: AUGI is a registered trademark of Autodesk, Inc., licensed exclusively to the Autodesk User Group International.], AutoCAD®, AutoCAD Learning Assistance™, AutoCAD LT®, AutoCAD Simulator™, AutoCAD SQL Extension™, AutoCAD SQL Interface™, Autodesk®, Autodesk Envision®, Autodesk Insight™, Autodesk Intent™, Autodesk Inventor®, Autodesk Map®, Autodesk MapGuide®, Autodesk Streamline®, AutoLISP®, AutoSketch®, AutoSnap™, AutoTrack™, Backburner™, Backdraft®, Built with ObjectARX (+product) (logos)™, Burn™, Buzzsaw®, CAiCE™, Can You Imagine®, Character Studio®, Cinestream™, Civil 3D®, Cleaner®, Cleaner Central™, ClearScale™, Colour Warper™, Combustion®, Communication Specification™, Constructware®, Content Explorer™, Create>what's>Next> (design/logo)®, Dancing Baby, The (image)™, DesignCenter™, Design Doctor™, Designer's Toolkit™, DesignKids™, DesignProf™, Design Server™, DesignStudio®, Design|Studio (design/logo)®, Design Your World®, Design Your World (design/logo)®, Discreet®, DWF™, DWG™, DWG (design/logo)™, DWG TrueConvert™, DWG TrueView™, DXF™, EditDV®, Education by Design®, Extending the Design Team™, FBX®, Field to Field®, Filmbox®, FMDesktop™, GDX Driver™, Gmax®, Heads-up Design™, Heidi®, HOOPS®, HumanIK®, i-drop®, iMOUT™, Incinerator®, IntroDV®, Inventor™, Kaydara®, Kaydara (design/logo)®, LocationLogic™, Lustre®, Maya®, Mechanical Desktop®, MotionBuilder™, ObjectARX®, ObjectDBX™, Open Reality®, PolarSnap™, PortfolioWall®, Powered with Autodesk Technology™, Productstream®, ProjectPoint®, Reactor®, RealDWG™, Real-time Roto™, Render Queue™, Revit®, Showcase™, SketchBook®, Topobase™, Toxik™, Visual®, Visual Bridge™, Visual Construction®, Visual Drainage®, Visual Hydro®, Visual Landscape®, Visual LISP®,Visual Roads®, Visual Survey®, Visual Syllabus™, Visual Toolbox®, Visual Tugboat®, Voice Reality®, Volo®, Wiretap™

The following, and all other brand names, product names, or trademarks belong to their respective holders.

CATIA® is a trademark of Dassault Systems
MicroStation® is a trademark of Bentley Systems
Pro/Engineer® is a trademark of PTC Systems
Rhinoceros is a trademark of Robert McNeel and Associates.
SolidEdge® is a trademark of Siemens PLM
SOLIDWORKS® is a trademark of Dassault Systems

## LEGEND

The following special treatment of characters and fonts in the textual content is intended to assist you in translating the meaning of words or sentences in *AutoCAD 2017 Instructor.*

| | |
|---|---|
| Underline | Emphasis of a word or an idea. |
| **Helvetica font** | An AutoCAD prompt appearing on the screen at the command line or in a text window. |
| *Italic (Upper and Lower)* | An AutoCAD command, option, menu, toolbar, or dialog box name. |
| UPPER CASE | A file name. |
| *UPPER CASE ITALIC* | An AutoCAD system variable. |

Anything in **Bold** represents user input:

| | |
|---|---|
| **Bold** | What you should type or press on the keyboard. |
| ***Bold Italic*** | An AutoCAD command that you should type or menu item that you should select. |
| **BOLD UPPER CASE** | A file name that you should type. |
| ***BOLD UPPER CASE ITALIC*** | A system variable that you should type. |
| **PICK** | Move the cursor to the indicated position on the screen and press the select button (button #1 or left mouse button). |

# Introduction

## WHAT IS CAD?

CAD is an acronym for Computer-Aided Design. CAD allows you to accomplish design and drafting activities using a computer. A CAD software package, such as AutoCAD, enables you to create designs and generate drawings to document those designs.

Design is a broad field involving the process of making an idea into a real product or system. The design process requires repeated refinement of an idea or ideas until a solution results—a manufactured product or constructed system. Traditionally, design involves the use of sketches, drawings, renderings, two-dimensional (2D) and three-dimensional (3D) models, prototypes, testing, analysis, and documentation. Drafting is generally known as the production of drawings that are used to document a design for manufacturing or construction or to archive the design.

CAD is a tool that can be used for design and drafting activities. CAD can be used to make "rough" idea drawings, although it is more suited to creating accurate finished drawings and renderings. CAD can be used to create a 2D or 3D computer model of the product or system for further analysis and testing by other computer programs. In addition, CAD can be used to supply manufacturing equipment such as lathes, mills, laser cutters, or 3D printers with numerical data to manufacture a product. CAD is also used to create 2D and 3D documentation for communicating and archiving the design.

The tangible result of CAD activity is often a drawing that can be printed but can be a PDF file, shared drawing file, rendering or numerical data for use with another software package or manufacturing device. Regardless of the purpose for using CAD, the resulting drawing or model is stored in a CAD file. The file consists of numeric data in binary form usually saved to a storage drive, cloud, or intranet.

## WHY SHOULD YOU USE CAD?

Although there are other methods used for design and drafting activities, CAD offers the following advantages over other methods in many cases:

1. Accuracy
2. Productivity for repetitive operations
3. Collaborating and sharing the CAD file with other people and software

### Accuracy

Since CAD technology is based on computer technology, it offers great accuracy. When you draw with a CAD system, the graphical elements, such as lines, arcs, and circles, are stored in the CAD file as numeric data. CAD systems store that numeric data with great precision. For example, AutoCAD stores values with fourteen significant digits. The value 1, for example, is stored digitally as the equivalent of 1.0000000000000. This precision provides you with the ability to create designs and drawings that are 100% accurate for almost every case.

### Productivity for Repetitive Operations

It may be faster to create a simple "rough" drawing, such as a sketch by hand (pencil and paper), than it would by using a CAD system. However, for larger and more complex drawings, particularly those involving similar shapes or repetitive operations, CAD methods are very efficient. Any kind of shape or operation accomplished with the CAD system can be easily duplicated since it is stored in a CAD file.

In short, it may take some time to set up the first drawing and create some of the initial geometry, but any of the existing geometry or drawing setups can be easily duplicated in the current drawing or for new drawings.

Likewise, making changes to a CAD file is generally much faster than creating the original geometry. Since all the graphical elements in a CAD drawing are stored, only the affected components of the design or drawing need to be altered, and the drawing can be printed again or converted to other formats.

As CAD and the associated technology advance and software becomes more interconnected, more productive developments are available. For example, it is possible to make a change to a 3D model that automatically causes a related change in the linked 2D engineering drawing. One of the main advantages of these technological advances is productivity.

## Collaborating and Sharing The CAD File With Other People and Software

Of course, CAD is not the only form of industrial activity that is making technological advances. Most industries use computer software to increase capability and productivity. Since software is written using digital information and may be written for the same or similar computer operating systems, it is possible and desirable to share data, work interactively with a team from a single set of files, or interconnect CAD data with other applications, even appearing simultaneously on one screen.

For example, word processing programs can generate text that can be imported into a drawing file, or a drawing can be created and imported into a text file as an illustration. A drawing created with a CAD system such as AutoCAD can be exported to a finite element analysis program that can read the computer model and compute and analyze stresses. 3D CAD models can be exported to create finished parts using additive manufacturing devices such as 3D printers. CAD files can be dynamically "linked" to spreadsheets or databases in such a way that changing a value in a spreadsheet or text in a database can automatically make the related change in the drawing, or vice versa.

Another advance in CAD technology is the automatic creation and interconnectivity of a 2D drawing and a 3D model in one CAD file. With this tool, you can design a 3D model and have the 2D drawings automatically generated, or even document the design within the 3D drawing. The resulting set has bi-directional associativity; that is, a change in either the 2D drawings or the 3D model is automatically updated in the other.

With the introduction of the new Web technologies, designers and related professionals can easily collaborate by viewing, editing, marking up, and transferring drawings over the Internet. CAD drawings can contain Internet links to other drawings, text information, or other related Web sites. Multiple CAD users can share a single CAD session from remote locations over the Internet.

CAD, however, may not be the best tool for every design related activity. For example, CAD may help develop ideas but probably won't replace the idea sketch, at least not with present technology. A 3D CAD model can save much time and expense for some analysis and testing but may not replace the "feel" of an actual model, unless expensive virtual reality technology is used.

With everything considered, CAD offers many opportunities for increased accuracy, productivity, and interconnectivity. Considering the speed at which this technology is advancing, many more opportunities are rapidly obtainable. However, we need to start with the basics. Beginning by learning to create an AutoCAD drawing is a good start.

## WHY USE AutoCAD?

CAD systems are available for a number of computer platforms: laptops, personal computers (PCs), and workstations. AutoCAD, offered to the public in late 1982, was one of the first PC-based CAD software products. Since that time, it grew to be the world leader in market share for all CAD products. Autodesk, the manufacturer of AutoCAD, is one of the largest software producers in the world and has millions of customers in more than 150 countries.

Learning AutoCAD as a first CAD system gives you a good foundation for learning other CAD packages because many concepts and commands introduced by AutoCAD are utilized by other systems. In some cases, AutoCAD features become industry standards. The .DXF file format, for example, was introduced by Autodesk and has become an industry standard for CAD file conversion between systems.

Learning AutoCAD offers a number of advantages for you. As a student, learning AutoCAD gives you a high probability of using your skills in industry. Because AutoCAD is such a widely used CAD software program, using it gives you a great probability of being able to share CAD files and related data and information with your colleagues, vendors, and clients.

Compatibility of hardware and software is an important issue in industry. Maintaining compatible hardware and software allows you the highest probability of sharing data and information with others as well as offering you flexibility in experimenting with and utilizing the latest technological advancements. AutoCAD provides you with broad compatibilities in the CAD domain.

This introduction is not intended as a selling point but to remind you of the importance and potential of the task you are about to undertake. If you are a professional or a student, you have most likely already made up your mind that you want to learn to use AutoCAD as a design or drafting tool. If you have made up your mind, then you can accomplish anything. Let's begin.

CHAPTER

# 1

# Getting Started

**Chapter Objectives**

After completing this chapter you should:

1. understand how the X, Y, Z coordinate system is used to define the location of drawing elements in digital format in a CAD drawing file;

2. understand why you should create drawings full size in the actual units with CAD;

3. be able to start AutoCAD to begin drawing;

4. recognize the areas of the AutoCAD Drawing Editor and know the function of each;

5. be able to use the many methods of entering commands;

6. be able to turn on and off the *Snap*, *Grid*, *Ortho*, *Polar*, and *Dynamic Input* drawing aids;

7. know how to customize the AutoCAD for Windows screen to your preferences.

## CONCEPTS

### Coordinate Systems

Any location in a drawing, such as the endpoint of a line, can be described in X, Y, and Z coordinate values (Cartesian coordinates). If a line is drawn on a sheet of paper, for example, its endpoints can be charted by giving the distance over and up from the lower-left corner of the sheet (Fig. 1-1).

These distances, or values, can be expressed as X and Y coordinates; X is the horizontal distance from the lower-left corner (origin) and Y is the vertical distance from that origin. In a three-dimensional coordinate system, the third dimension, Z, is measured from the origin in a direction perpendicular to the plane defined by X and Y.

Two-dimensional (2D) and three-dimensional (3D) CAD systems use coordinate values to define the location of drawing elements such as lines and circles (called <u>objects</u> in AutoCAD).

In a 2D drawing, a line is defined by the X and Y coordinate values for its two endpoints (Fig. 1-2).

In a 3D drawing, a line can be created and defined by specifying X, Y, and Z coordinate values (Fig. 1-3). Coordinate values are always expressed by the X value first separated by a comma, then Y, then Z.

FIGURE 1-1

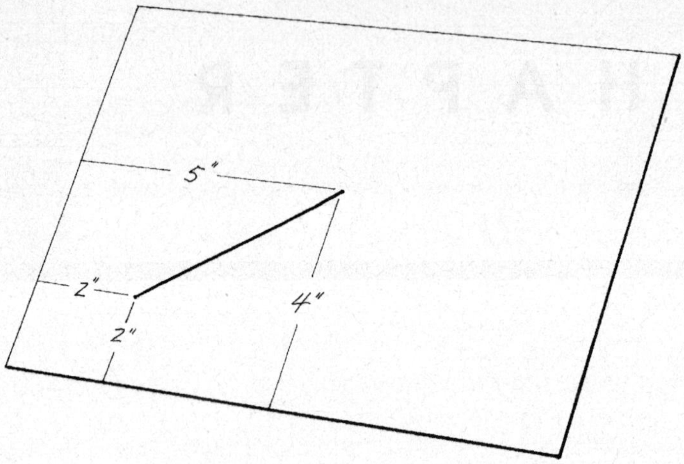


FIGURE 1-2

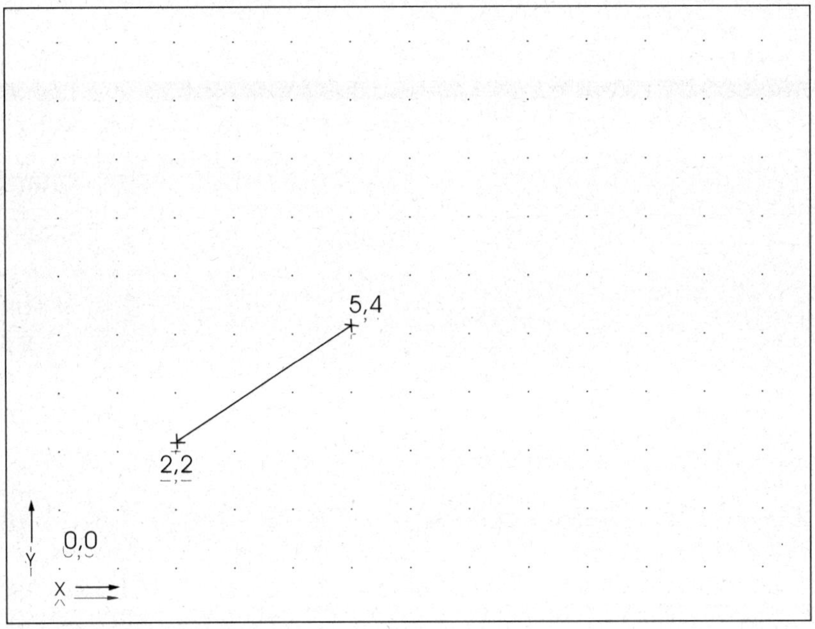


FIGURE 1-3

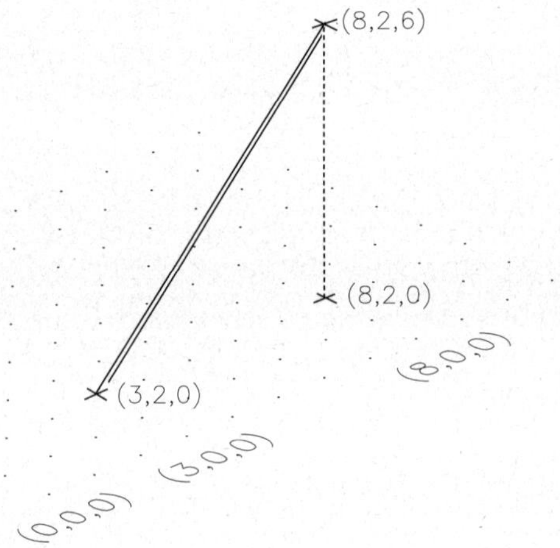

## The CAD Database

A CAD (Computer-Aided Design) file, which is the electronically stored version of the drawing, keeps data in binary digital form. These digits describe coordinate values for all of the endpoints, center points, radii, vertices, etc. for all the objects composing the drawing, along with another code that describes the kinds of objects (line, circle, arc, ellipse, etc.). Figure 1-4 shows part of an AutoCAD DXF (Drawing Interchange Format) file giving numeric data defining lines and other objects. Knowing that a CAD system stores drawings by keeping coordinate data helps you understand the input that is required to create objects and how to translate the meaning of prompts on the screen.

**FIGURE 1-4**

```
LINE
  8
0
 62
      8
 10
15.0
 20
8.105789
 30
0.0
 11
15.644291
 21
8.75
 31
0.0
  0
LINE
  8
0
 62
     8
 18
```

## Angles in AutoCAD

Angles in AutoCAD are measured in a counterclockwise direction. Angle 0 is positioned in a positive X direction, that is, horizontally from left to right. Therefore, 90 degrees is in a positive Y direction, or straight up; 180 degrees is in a negative X direction, or to the left; and 270 degrees is in a negative Y direction, or straight down (Fig. 1-5).

The position and direction of measuring angles in AutoCAD can be changed; however, the defaults listed here are used in most cases.

**FIGURE 1-5**

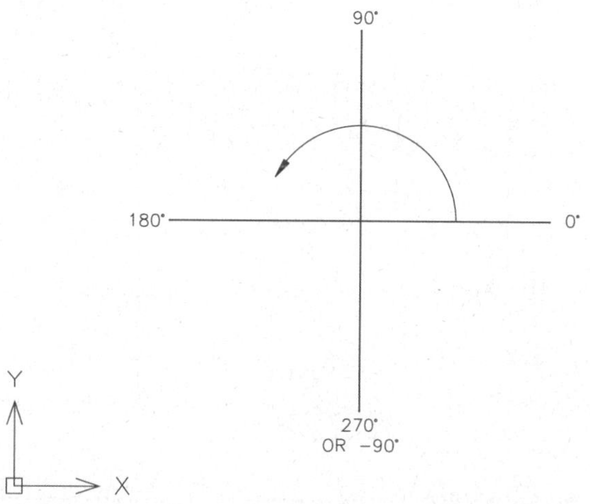


## Draw True Size

When creating a drawing with pencil and paper tools, you must first determine a scale to use so the drawing will be proportional to the actual object and will fit on the sheet (Fig. 1-6). However, when creating a drawing on a CAD system, there is no fixed size drawing area. The number of drawing units that appear on the screen is variable and is assigned to fit the application.

**FIGURE 1-6**

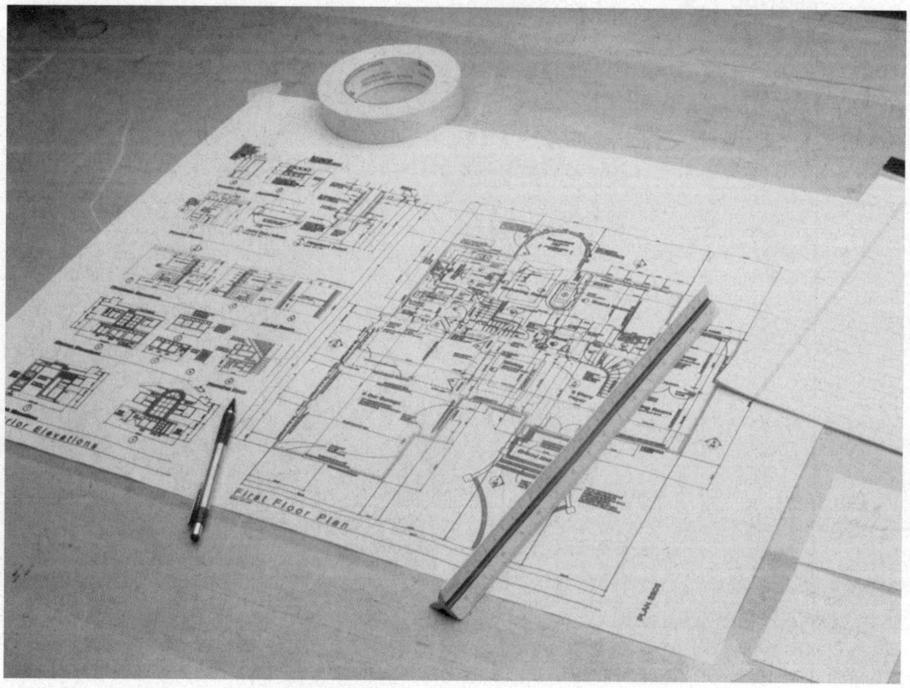

TIP

Create CAD drawings true size using real-world units so the values will be useful for analysis and measurements. The user specifies what units are to be used (architectural, engineering, etc.) and then specifies what size drawing area is needed (in X and Y values) to draw the necessary geometry (Fig. 1-7).

**FIGURE 1-7**

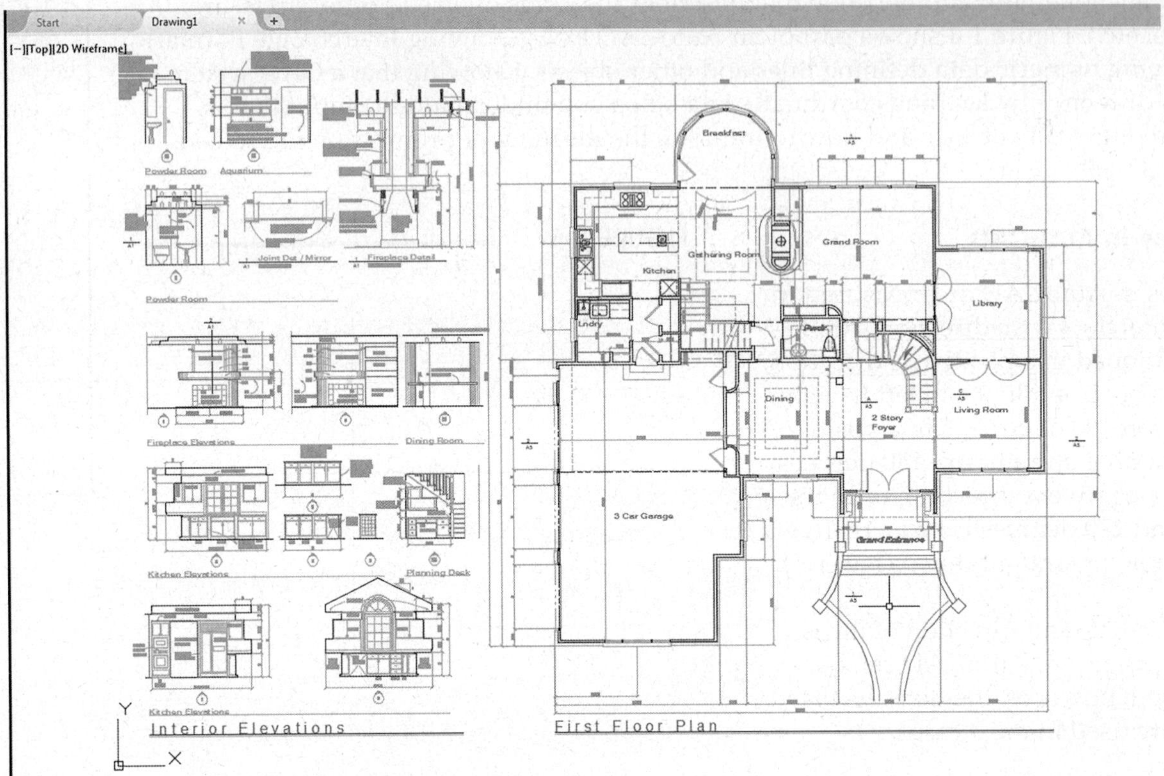


Whatever the specified size of the drawing area, it can be displayed on the screen in its entirety (Fig. 1-7) or as only a portion of the drawing area (Fig. 1-8).

**FIGURE 1-8**

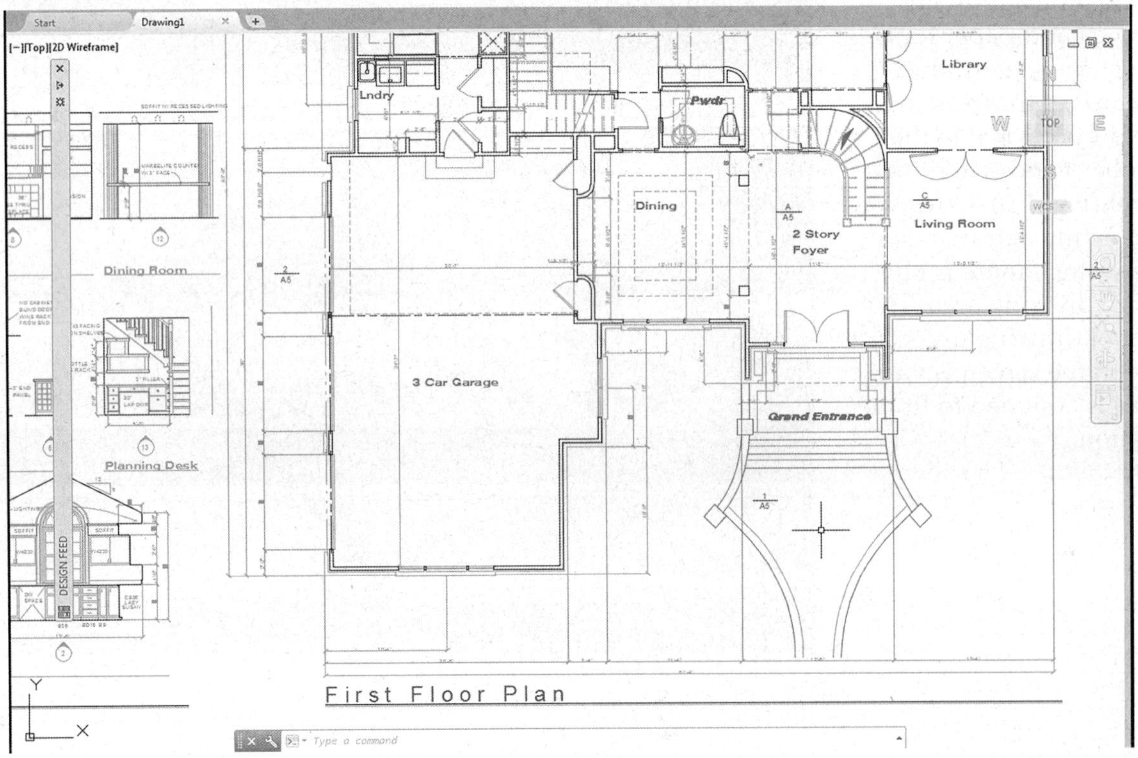

### Plot to Scale

As long as a drawing exists as a CAD file or is visible on the screen, it is considered a virtual, full-sized object. Only when the CAD drawing is transferred to paper by a plotter or printer is it converted (usually reduced) to a size that will fit on a sheet. A CAD drawing can be automatically scaled to fit on the sheet regardless of sheet size; however, this action results in a plotted drawing that is not to a typical scale (not to a whole number proportion of the real object). Usually it is desirable to plot a drawing so that the resulting drawing is a proportion of the actual object size. It is preferred to use whole numbers for the scale proportions, for example 4 drawing units equals 1 unit on the paper and not proportions such as 3.5 drawing units equals 1 unit on the paper. The scale to enter as the plot scale (Fig. 1-9) is simply the proportion of the plotted drawing size to the actual object.

FIGURE 1-9

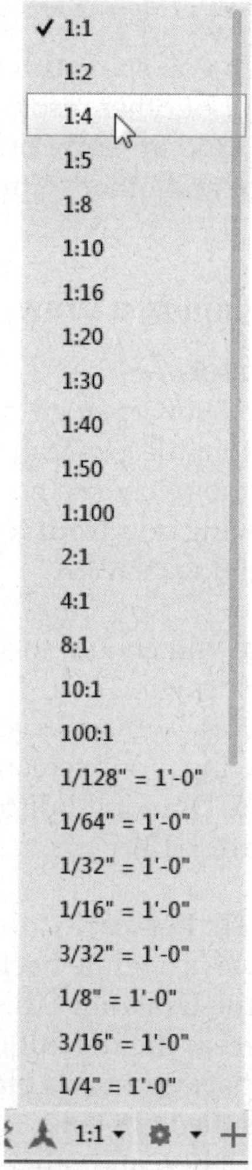


## STARTING AutoCAD

Assuming that AutoCAD has been installed and configured properly for your system, you are ready to begin using AutoCAD.

To start AutoCAD 2027, locate the "AutoCAD 2027" shortcut icon on the desktop (Fig. 1-10). Double-clicking on the icon launches AutoCAD 2027.

If you cannot locate the AutoCAD 2027 shortcut icon on the desktop, select the "Windows" button and search for "AutoCAD 2027" in the Programs list (Fig. 1-11).

FIGURE 1-10

FIGURE 1-11

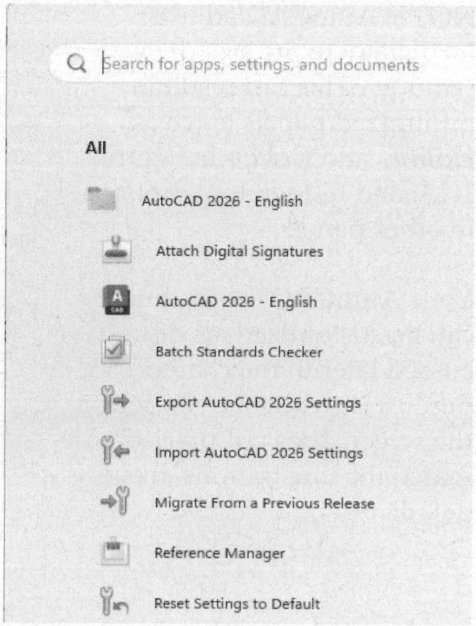

## THE AutoCAD DRAWING EDITOR

When you start AutoCAD, you see the Start Screen (Figure 1-12) that allows you to launch the Drawing Editor or other resources. The Drawing Editor is composed of a central drawing area, called the "graphics" area, and an array of toolbars, menus, a command line, and other elements, depending on the workspace and other settings that you can change.

### Beginning a Drawing

#### *Start* tab

When you open an AutoCAD session, the *Start* tab appears for you to choose what drawing you want to work with (Fig. 1-12).

Once you choose to start a new drawing (Fig. 1-12 inset) or work with an existing drawing, your screen changes to the Drawing Editor (see Figure 1-13).

NOTE: For clarity, this book displays a white background for the Drawing Editor, whereas the default color for the background is black. It also uses the light color scheme for the ribbon and other tools.

NOTE: AutoCAD adjusts its display to the size of its window. This can result in minimized panels (such as *Utilities* and *Groups* in Figure 1-13) and hidden tool captions in other panels.

Your AutoCAD environment can be customized, as discussed later in the chapter. So, don't worry if what you see on the screen does not precisely match the illustrations in every detail.

FIGURE 1-12

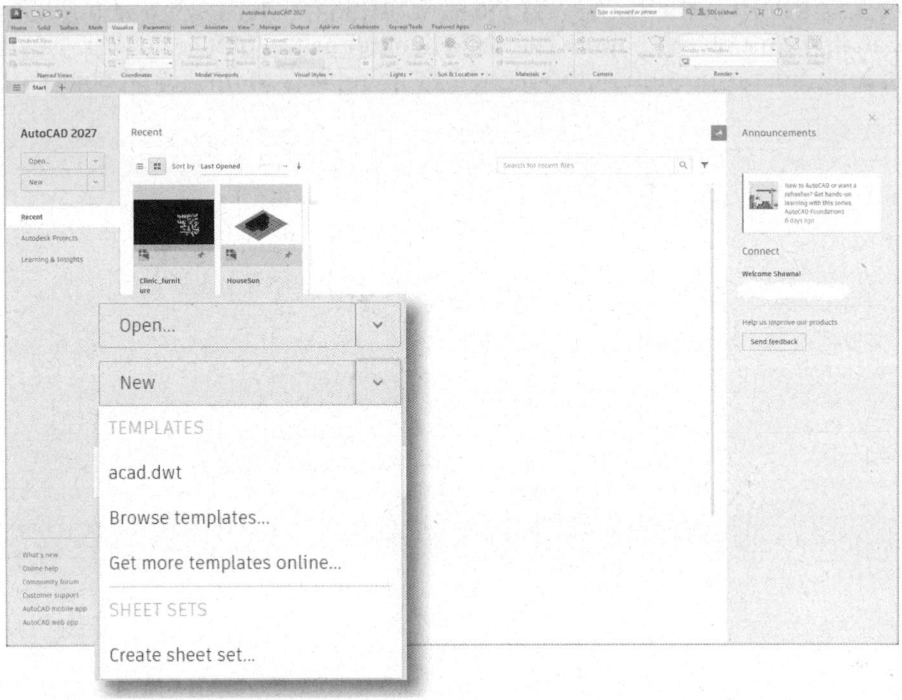


FIGURE 1-13

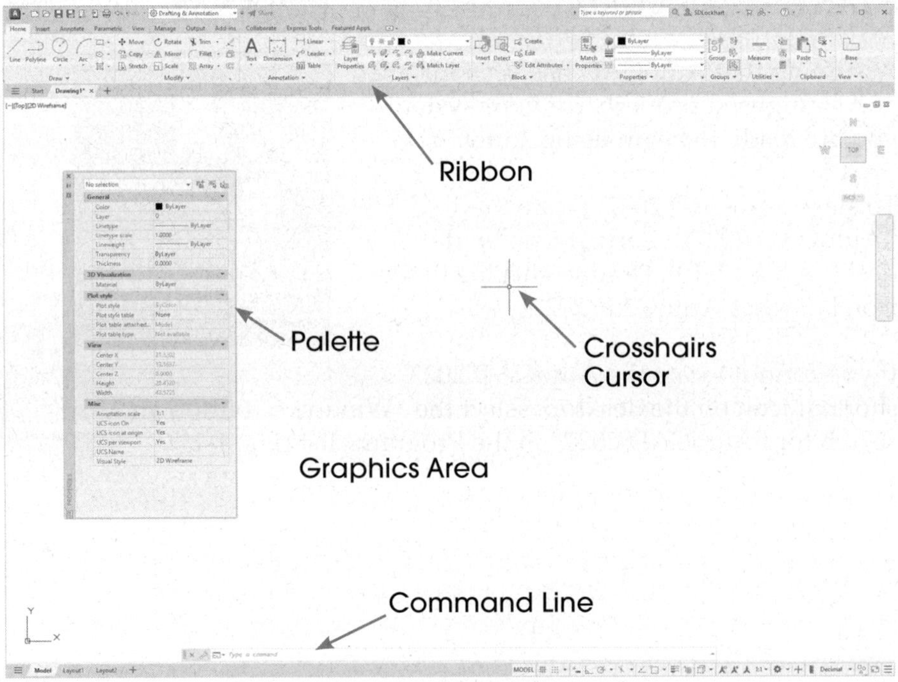

In the *Start* tab, you can create a new drawing by selecting the *New* option. The new drawing is based on the most recently used template. You could also select from other templates using the drop-down list using the small arrow at the right of the New selection. Normally a template drawing contains no geometry (lines, arcs, circles, etc.) but has specific settings based on what and how you intend to draw. For example, if you intend to use inch units, use the ACAD.DWT template drawing; if you intend to use metric units, use the ACADISO.DWT.

Alternately, you can select *Open...* to locate and open an existing drawing, *Open a Sheet Set* or *Explore Sample Drawings* that are installed when you install AutoCAD. You can also select from *Recent Documents* (drawings) shown in the list in the center of the *Create* screen.

The center section of the opening screen (*Start* tab) lists the *Recent Documents* that were opened (Fig. 1-12, center). Here you can select from a list rather than browsing for files using the *Open Files* option on the left. The *Recent Documents* list can be changed to large thumbnail images (as shown in Figure 1-12), or to details with no image (file name, date, time).

## Drafting & Annotation Workspace

The workspace that typically appears in the Drawing Editor when you begin drawing for the first time is the *Drafting & Annotation* workspace. This workspace offers functions for creating 2-dimensional (2D) drawings. Two other workspaces are available for working with 3-dimensional (3D) drawings (see Chapter 33). The basic organization of all workspaces is the same—changing the workspace simply provides different command options from the "Ribbon" near the top of the screen.

### Ribbon

The Ribbon is located along the top of the screen (Fig. 1-13) and consists of <u>tabs</u>, <u>panels</u> and <u>tools</u>. Panels are the groups of related tools (commands) between the vertical separators with titles such as *Draw, Modify,* and *Layers*. The panels that appear are based on the active tab that is selected (above the panels) such as *Home, Insert, Annotate,* and so on. For example, selecting the *Home* tab produces the *Draw, Modify, Annotation, Layers, Block* and other panels.

### Graphics Area

The large central area of the screen is the Graphics Area. It displays the lines, circles, and other objects you create to make up the drawing. The cursor is the intersection of the crosshairs cursor (crossing vertical and horizontal lines that follow the mouse movements). The default size of the graphics area for English settings is 12 units (X or horizontal) by 9 units (Y or vertical). This usable drawing area (12 x 9) is called the drawing Limits and can be changed to any size to fit the application. To see the display of coordinates in the AutoCAD screen, turn on the *Coordinate Display* on the Status Bar. See "Status Bar, Coordinates," in the next section.

### Palettes

When you activate the *Drafting & Annotation* workspace, one or more palettes can be activated as needed on your screen. Figure 1-13 displays the *Properties* palette. Each palette serves a particular function as explained in later chapters. You can close the palettes by clicking on the "X" in the upper corner of the palette.

### Command Line

The Command line consists of the one to three text lines (by default) at the bottom of the screen (see Fig. 1-13). Any command that is entered as well as the prompts that AutoCAD issues normally appear here. The Command line gives the current state of drawing activity. It is very important to keep aware of the Command line as you draw while you are learning AutoCAD. You should develop the habit of glancing at the Command line while you work in AutoCAD. The Command line can be set to display any number of lines and it can be moved to another location (see "Customizing the AutoCAD Screen" later in this chapter).

### Application Menu and Quick Access Toolbar

All workspaces provide the Application Menu and Quick Access toolbar (Fig. 1-14). The Application Menu is accessed by selecting the large, red letter "A" in the upper-left corner of the AutoCAD window. Selecting this button produces a list of options that allow you to manage drawing files. For example, you can create new drawings, open existing drawings, or save, export, and print drawings. You can also use the search box above the list to search for commands. The Quick Access toolbar, just to the right of the letter "A", contains frequently used commands such as *Qnew*, *Open*, *Save*, *Undo*, and *Plot*.

**FIGURE 1-14**

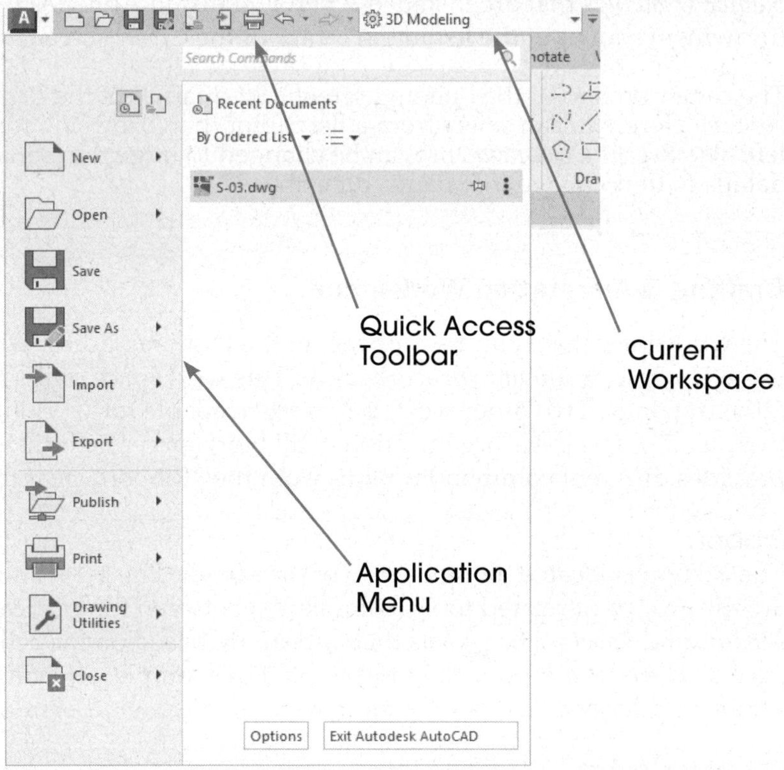


**FIGURE 1-15**

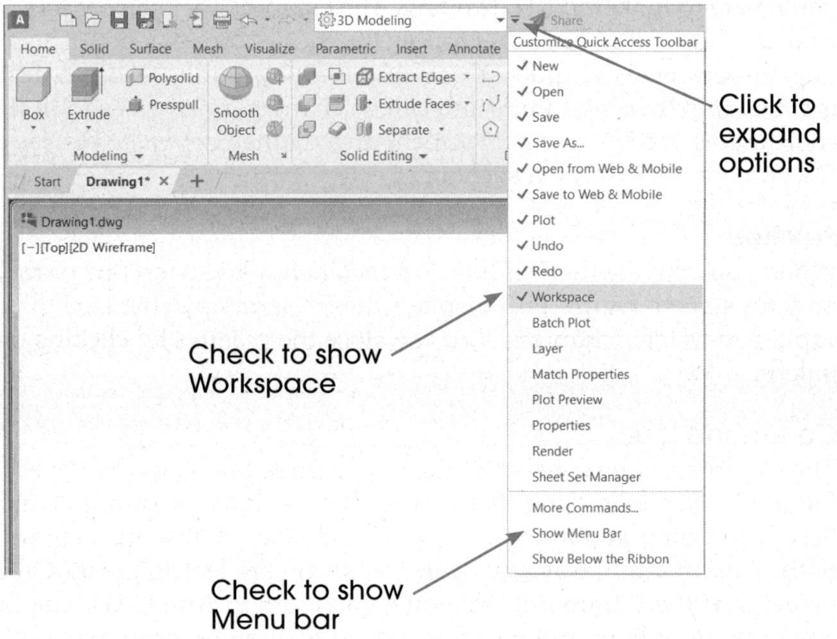


**TIP**

### Menu Bar

A Menu Bar is available that provides a series of pull-down menus along the top of the Drawing Editor just above the Ribbon. The Menu Bar is useful for learning AutoCAD because the command names are listed beside each icon. If you are looking for a particular command, searching the pull-down menus is faster and easier than searching the Ribbon. You can enable the Menu Bar by selecting the *Show Menu Bar* option from the Quick Access toolbar drop-down arrow (Fig. 1-15).

Selecting any of the words in the Menu Bar activates, or pulls down, the respective menu (Fig. 1-16). Selecting a word appearing with an arrow activates a cascading menu with other options. Selecting a word with an ellipsis (. . .) activates a dialog box (see "Dialog Boxes and Palettes"). Words in the pull-down menus are not necessarily the same as the formal command names used when typing commands. Menus can be canceled by pressing Escape or picking in the graphics area.

**FIGURE 1-16**

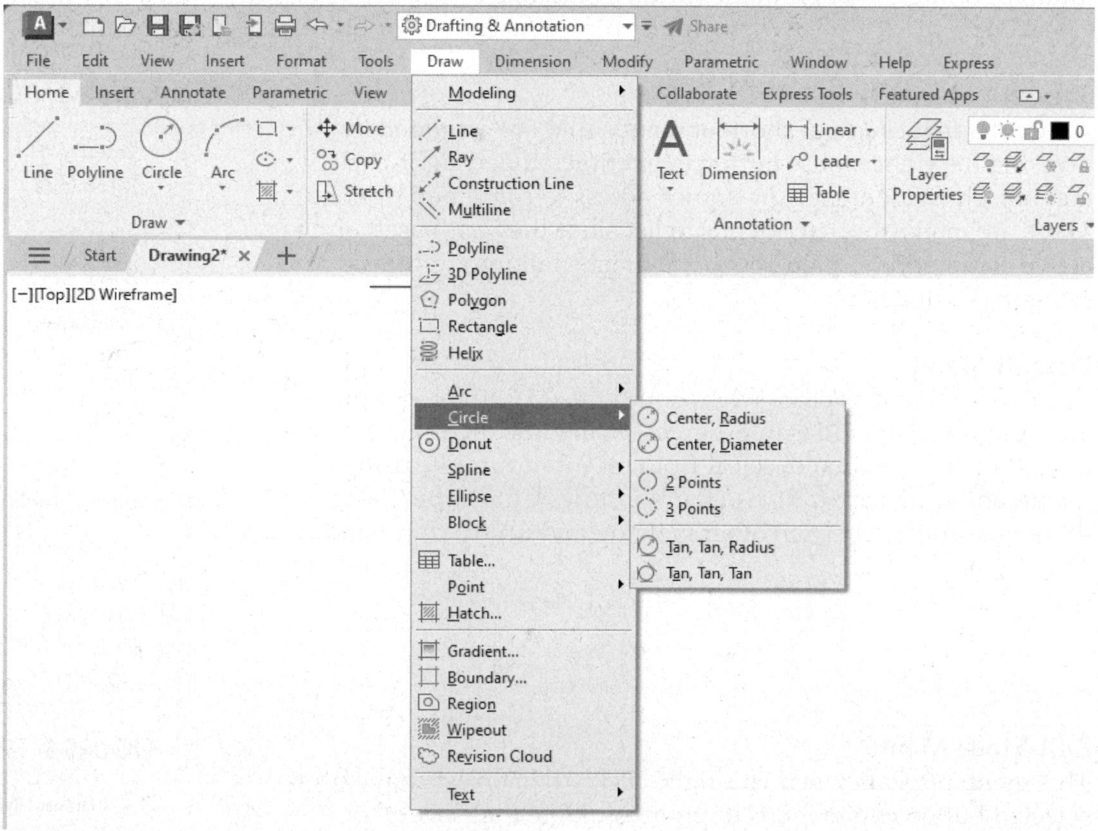


## Dialog Boxes and Palettes

Dialog boxes and Palettes provide interfaces for controlling complex commands or a group of related commands. Depending on the command, dialog boxes and palettes allow you to select among multiple options and sometimes give a preview of the effect of selections. The *Layer Properties Manager* palette (Fig. 1-17) gives complete control of layer colors, linetypes, and visibility.

**FIGURE 1-17**

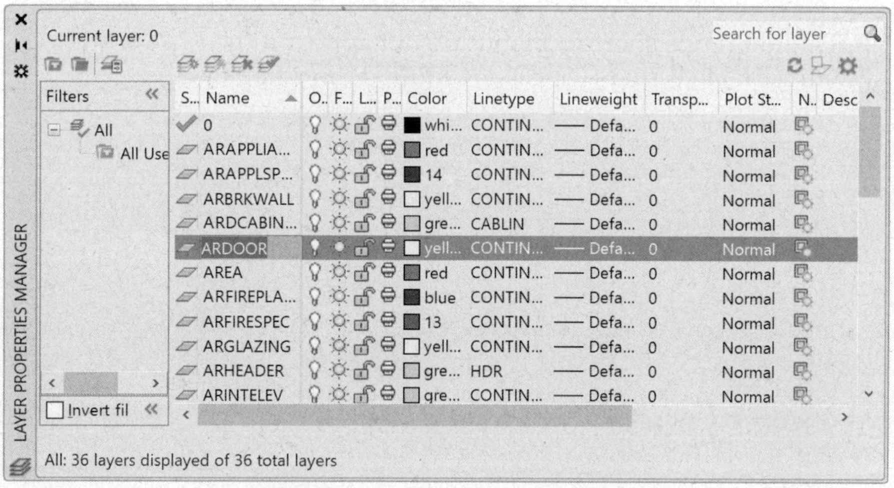


Dialog boxes and palettes can be invoked by typing a command, selecting an icon button, or PICKing from the menus. For example, typing *Layer*, PICKing the *Layer Properties Manager* button, or selecting *Layer* from the *Format* pull-down menu causes the *Layer Properties Manager* to appear. Palettes and dialog boxes can be resized to display more or less information.

The basic elements of a dialog box are:

| | |
|---|---|
| Button | Resembles a push button and triggers some type of action. |
| Edit box | Allows typing or editing of a single line of text. |
| Image tile | A button that displays a graphical image. |
| List box | A list of text strings from which one or more can be selected. |
| Drop-down list | A text string that drops down to display a list of selections. |
| Radio button | A group of buttons, only one of which can be turned on at a time. |
| Checkbox | A checkbox for turning a feature on or off (displays a check mark when on). |

## Shortcut Menus

AutoCAD makes use of shortcut menus that are activated by pressing the right mouse button (sometimes called right-click menus). Shortcut menus give quick access to command options. There are many shortcut menus to list since they are based on the active command or dialog box. The menus fall into five basic categories listed here.

FIGURE 1-18

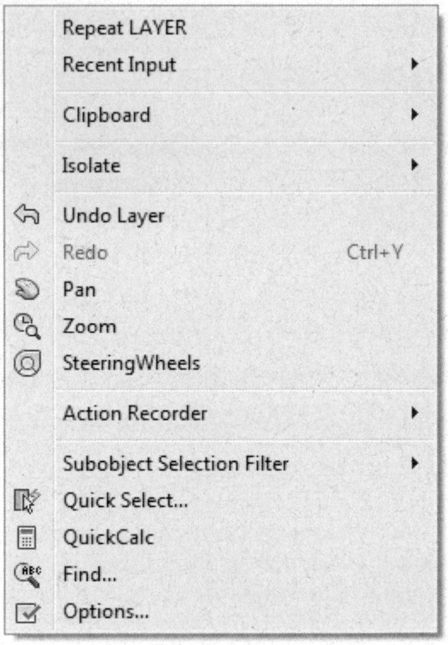


### Default Menu

The default menu appears when you right-click in the drawing area and no command is in progress. Using this menu, you can repeat the last command, select from recent input (like using the up and down arrows), use the Windows Cut, Copy, Paste functions, and select from other viewing and utility commands (Fig. 1-18).

### Edit-Mode Menu

This menu appears when you right-click when objects have been selected but no command is in progress. Note that several of AutoCAD's *Modify* commands are available on the menu such as *Erase, Move, Copy, Scale,* and *Rotate* (Fig. 1-19).

NOTE: Edit mode shortcut menus do not appear if the *PICKFIRST* system variable is set to 0 (see "*PICKFIRST*" in Chapter 20).

FIGURE 1-19

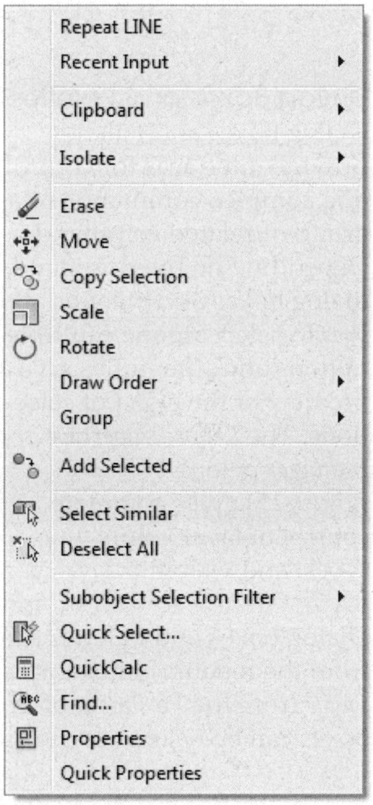

**Command-Mode Menu**

These menus appear when you right-click when a command is in progress. This menu changes since the options are specific to the command (Fig. 1-20).

FIGURE 1-20

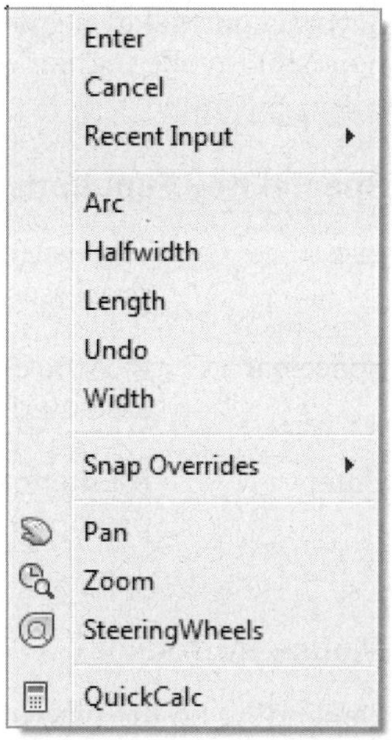


**Dialog-Mode Menu**

When the pointer is in a dialog box or tab and you right-click, this menu appears. The options on this menu can change based on the current dialog box (Fig. 1-21).

FIGURE 1-21

### Other Menus

There are other menus that can be invoked. For example, a menu appears if you right-click in the Command line area.

Because there are so many shortcut menus, don't be too concerned about learning these until you have had some experience. The best advice at this time is just remember to experiment by right-clicking often to display the possible options.

## Command Entry Using the Keyboard

Although commands are accessible using toolbars, the dashboard, pull-down menus, and shortcut menus, you can enter the command using the keyboard. As you type the letters, they appear on the Command line (near the bottom of the screen) or in the *Dynamic Input* box near the cursor (see "Drawing Aids," "*Dynamic Input*"). Using the keyboard to type in commands offers several features:

| | |
|---|---|
| Command name | Type in the full command name. |
| Command alias | Type the one- or two-letter shortcut for the command (see App. A). |
| Accelerator keys | Type a Ctrl key plus another key to invoke the command (see App. B). |
| Up and down arrows | Use the up and down arrows to cycle through the most recent input. |
| Tab key | Type in a few letters and press the Tab key to "AutoComplete" the command. |

### Accelerator Keys (Control Key Sequences)

Several control key sequences (holding down the Ctrl key or Alt key and pressing another key simultaneously) invoke regular AutoCAD commands or produce special functions (see App. B).

### Special Key Functions

| | |
|---|---|
| Esc | The Esc (escape) key cancels a command, menu, or dialog box or interrupts some processes. |
| Space bar | In AutoCAD, the space bar performs the same action as the Enter key. Only when you are entering text into a drawing does the space bar create a space. |
| Enter | If Enter or the space bar is pressed when no command is in use (the open Command: prompt is visible), the last command used is invoked again. |

### Mouse Buttons

Depending on the type of mouse used for cursor control, a different number of buttons is available. In any case, the buttons normally perform the following tasks:

| | | |
|---|---|---|
| left button | PICK | Used to select commands or pick locations on the screen. |
| right button | Enter or shortcut menu | Depending on the status of the drawing or command, this button either performs the same function as the Enter key or produces a shortcut menu. |
| wheel | *Pan* | If you press and drag, you can pan the drawing about on the screen. |
| | *Zoom* | If you turn the wheel, you can zoom in and out centered on the location of the cursor. |

## COMMAND ENTRY

### Methods for Entering Commands

There are many possible methods for entering commands in AutoCAD depending on your system configuration. Generally, most of the methods can be used to invoke a particular commonly used command or dialog box.

| | |
|---|---|
| 1. Keyboard | Type the command name, command alias, or accelerator keys at the keyboard. |
| 2. Menu Bar | Select the command or dialog box from a pull-down menu. |
| 3. Tools (icon buttons) | Select the command or dialog box by PICKing a tool (icon button) from a toolbar, palette, or Ribbon. |
| 4. Shortcut menus | Select the command from the right-click shortcut menu. Right-clicking produces a shortcut menu depending on whether a command is active, objects are selected, or the pointer is in a dialog box. |

## Understanding the "Command Tables" in this Book

Command tables, like the one below, are used throughout this book to show the possible methods for entering a particular command. The table shows the icon used in the toolbars, the palettes, and the Ribbon, indicates the tab and panel on the Ribbon for locating the icon, gives the selections to make for the pull-down menu, gives the correct spelling for entering commands and command aliases at the keyboard, and gives the shortcut menu and option. This example uses the *Line* command.

### *Line*

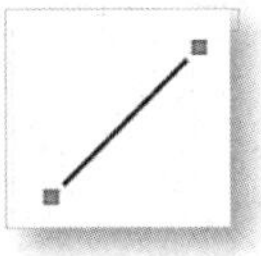

| Ribbon | Menu Bar | Command (Type) | Alias (Type) | Shortcut |
|---|---|---|---|---|
| *Home*<br>*Draw*<br>*Line* | *Draw*<br>*Line* | *Line* | *L* | ... |

## Command Entry Methods Practice

Start AutoCAD. Use the **ACAD.DWT** template. Invoke the *Line* command using each of the command entry methods as follows.

**1. Type the command**

| Steps | Command Prompt | Perform Action | Comments |
|---|---|---|---|
| 1. | | press **Escape** if another command is in use | |
| 2. | Command: | type *Line* and press **Enter** | |
| 3. | LINE Specify first point: | **PICK** any point | a "rubberband" line appears |
| 4. | Specify next point or [Undo]: | **PICK** any point | another "rubberband" line appears |
| 5. | Specify next point or [Undo]: | press **Enter** | to complete command |

**2. Type the command alias**

| Steps | Command Prompt | Perform Action | Comments |
|---|---|---|---|
| 1. | | press **Escape** if another command is in use | |
| 2. | Command: | type *L* and press **Enter** | |
| 3. | LINE Specify first point: | **PICK** any point | a "rubberband" line appears |
| 4. | Specify next point or [Undo]: | **PICK** any point | another "rubberband" line appears |
| 5. | Specify next point or [Undo]: | press **Enter** | to complete command |

**3. Menu Bar**

| Steps | Command Prompt | Perform Action | Comments |
|---|---|---|---|
| 1. | Command: | select the *Draw* menu from the menu bar | |
| 2. | Command: | select *Line* | menu disappears |
| 3. | LINE Specify first point: | **PICK** any point | a "rubberband" line appears |
| 4. | Specify next point or [Undo]: | **PICK** any point | another "rubberband" line appears |
| 5. | Specify next point or [Undo]: | press **Enter** | to complete command |

**4. Ribbon (*Drafting & Annotation* workspace)**

| Steps | Command Prompt | Perform Action | Comments |
|---|---|---|---|
| 1. | Command: | select the *Line* icon from the *Draw* panel on the *Home* tab | the *Line* tool is the first icon in the *Draw* panel |
| 2. | LINE Specify first point: | **PICK** any point | a "rubberband" line appears |
| 3. | Specify next point or [Undo]: | **PICK** any point | another "rubberband" line appears |
| 4. | Specify next point or [Undo]: | press **Enter** | to complete command |

**5. Shortcut menu**

| Steps | Command Prompt | Perform Action | Comments |
|---|---|---|---|
| 1. | Command: | right-click in the drawing area and select *Repeat LINE* from menu | *Repeat LINE* should be first in the list |
| 2. | LINE Specify first point: | **PICK** any point | a "rubberband" line appears |
| 3. | Specify next point or [Undo]: | **PICK** any point | another "rubberband" line appears |
| 4. | Specify next point or [Undo]: | press **Enter** | to complete command |

**6. Click an Option from the Command Prompt**

1. Select the *Circle* icon from the *Draw* panel on the *Home* tab.
2. **PICK** on the 2P option from the command prompt.
3. press **Esc** to cancel the command.

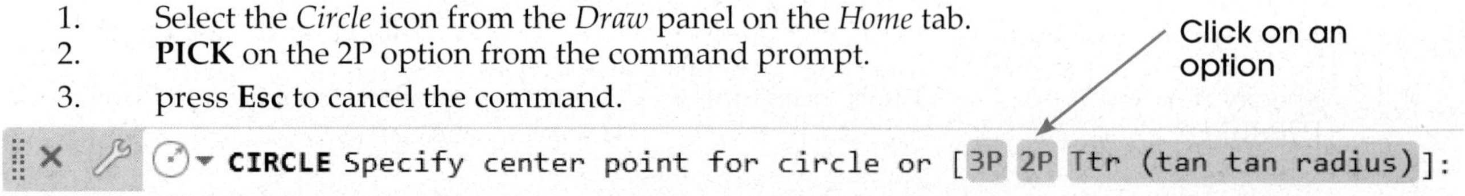


When you are finished practicing, use the Application Menu or the *File* menu and select *Exit* to exit AutoCAD. You do not have to "Save Changes."

## DRAWING AIDS

This section explains several features that appear near the bottom of the Drawing Editor including the Status Bar drawing aids, Model and Layout tabs, and the Text window.

### Status Bar

The Status Bar appears at the bottom of the AutoCAD window and contains a set of icons that allow you to turn on or off drawing aids such as *Grid*, *Snap*, *Dynamic Input*, and *Polar Tracking* (Fig. 1-22, bottom). Gray icons indicate the feature is off and blue icons indicate the feature is on. You can turn some features on or off using a function key as an alternative to clicking the icons, such as F7 to turn *Grid* on or off (see "Function Keys" later in this chapter).
You can determine which drawing aid icons appear on the Status Bar by using the *Customization* button (Fig. 1-22, bottom-right corner). This button produces a list of drawing aids that you check or uncheck (Fig. 1-22, right). Therefore, checking or unchecking a word in the list determines if that icon appears on the Status Bar.

**FIGURE 1-22**

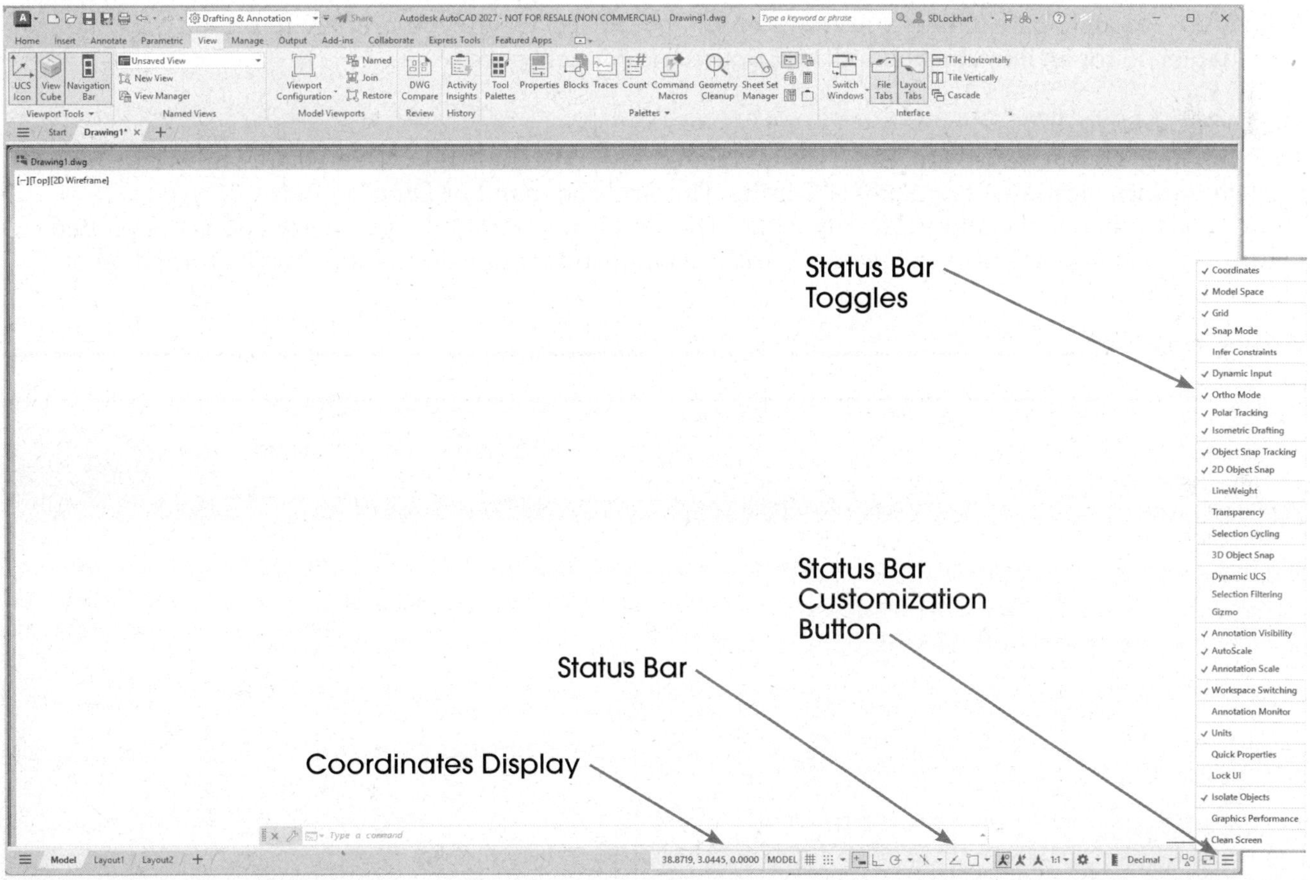


*Coordinates*

The Coordinate display (sometimes referred to as "Coords") displays the position of the cursor (cross-hairs) in X, Y, and Z coordinates. The coordinate display is located at the left end of the Status Bar (see Fig. 1-22). This can be very helpful when you draw because the current coordinate position of the cursor is displayed as the cursor moves, sometimes called "cursor tracking." It is recommended to turn the coordinates on while you begin learning AutoCAD. You can click on the numbers to turn this feature on or off. The default setting (in the acad.dwt template) displays decimal format with four places to the right of the decimal. Although this format can be changed, AutoCAD always records 14 significant places for every coordinate!

TIP

When the coordinate display is turned on, there are four possible formats: *Absolute, Relative, Geographic,* and *Specific*. Access these options by right-clicking on the coordinate display.

| | |
|---|---|
| *Absolute* | This option tracks the current cursor location in absolute coordinates (relative to 0,0,0). |
| *Relative* | This option tracks the current cursor location in coordinates relative to the last point established but in a polar format (distance and angle). This option appears only if a point has been established, such as when the first point of a Line has been designated. |
| *Geographic* | Use this option when you have set a geographic location in the drawing (*Geographiclocation* command). |
| *Specific* | This option displays the coordinates only when you click to select a point rather than continuously tracking the current location of the cursor. |

### *MODEL/PAPER* (*Model Space* or *Paper Space*)

Next to the Coordinate display the word *PAPER* or *MODEL* appears by default. There are two spaces in AutoCAD—Model Space, where you draw, and Paper Space, where you prepare to print a drawing. See Chapter 13 for an introduction to Paper Space and Model Space.

### *Grid* (*Grid Mode*, F7)

This drawing aid can be used to give a visual reference of units of length. The *Grid* default value for English settings (acad.dwt template) is .5 units. The *Grid* command or *Drawing Aids* dialog box allows you to change the interval to any value. The *Grid* is not part of the geometry and is not plotted. Figure 1-23 displays a *Grid* of .5. *Snap* and *Grid* are independent functions—they can be turned on or off independently.

**FIGURE 1-23**

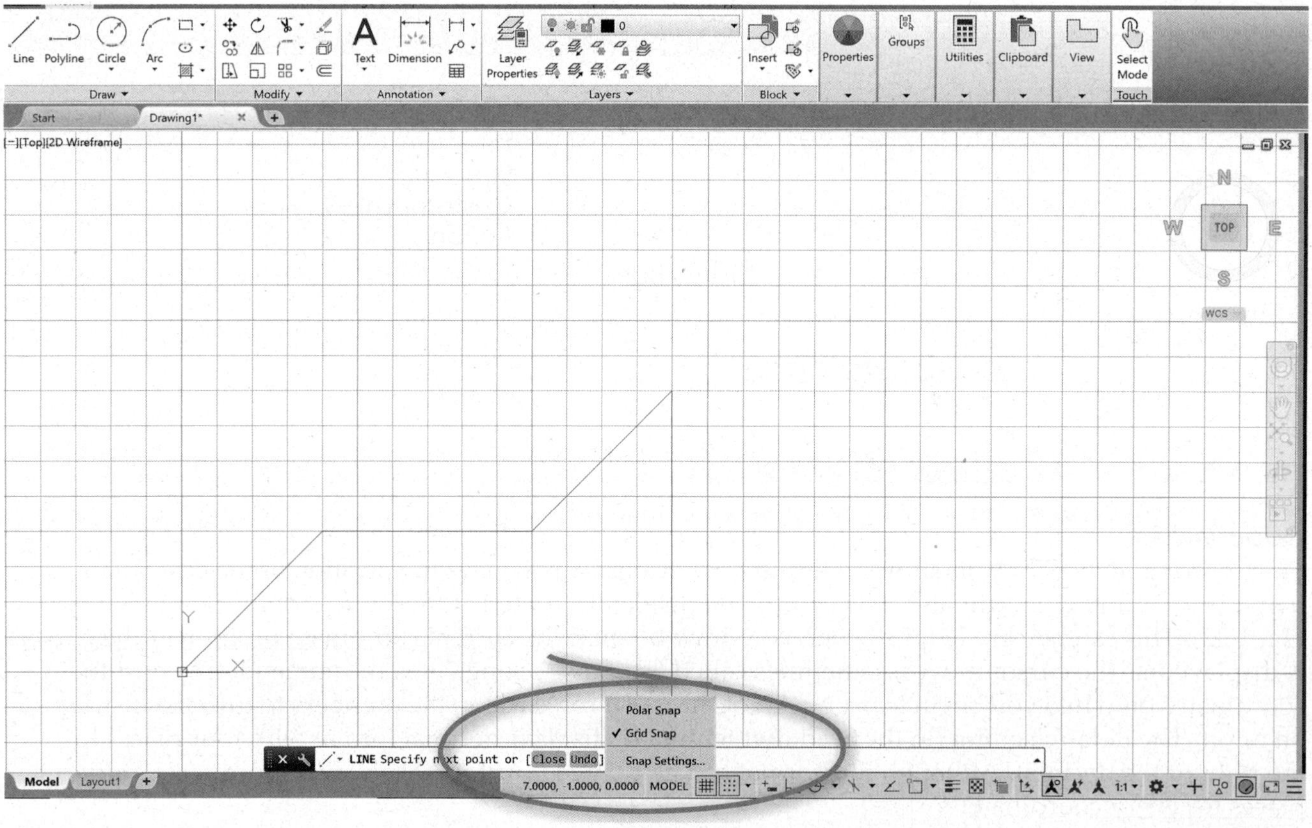

### *Snap Mode* (F9)

*Snap* has two modes in AutoCAD: *Grid Snap* and *Polar Snap*. Only one of the two modes can be active at one time. *Grid Snap* is a function that forces the cursor to "snap" to regular intervals (.5 units is the default setting), which aids in creating geometry accurate to interval lengths. You can use the *Snap* command or the *Drafting Settings* dialog box to specify any value for the *Snap* increment. Figure 1-23 displays a *Snap* setting of .5 (note the values in the coordinate display).

The other mode of *Snap* is *Polar Snap*. *Polar Snap* forces the cursor to snap to regular intervals along angular lines. *Polar Snap* is functional only when *Polar Tracking* is also toggled on since it works in conjunction with *Polar Tracking*. The *Polar Snap* interval uses the *Grid Snap* setting by default but can be changed to any value using the *Snap* command or the *Drafting Settings* dialog box. *Polar Snap* is discussed in detail in Chapter 3.

Since you can have only one *Snap* mode on at a time (*Grid Snap* or *Polar Snap*), you can select which of the two is on by right-clicking on the icon on the Status Bar. You can also access the *Drafting Settings* dialog box by selecting *Snap Settings...* from the pop-up menu (see Figure 1-23).

### *Infer Constraints*

Enabling *Infer Constraints* mode automatically applies constraints between the object you are creating or editing and the object or points associated with object snaps. For most drawings, turn this feature off.

### *Dynamic Input* (F12)

*Dynamic Input* is a feature that helps you visualize and specify coordinate values and angular values when drawing lines, arcs, circles, etc. *Dynamic Input* may display absolute Cartesian coordinates (X and Y values) or relative polar coordinates (distance and angle) depending on the current command prompt and the settings you prefer (Fig. 1-24).

**FIGURE 1-24**

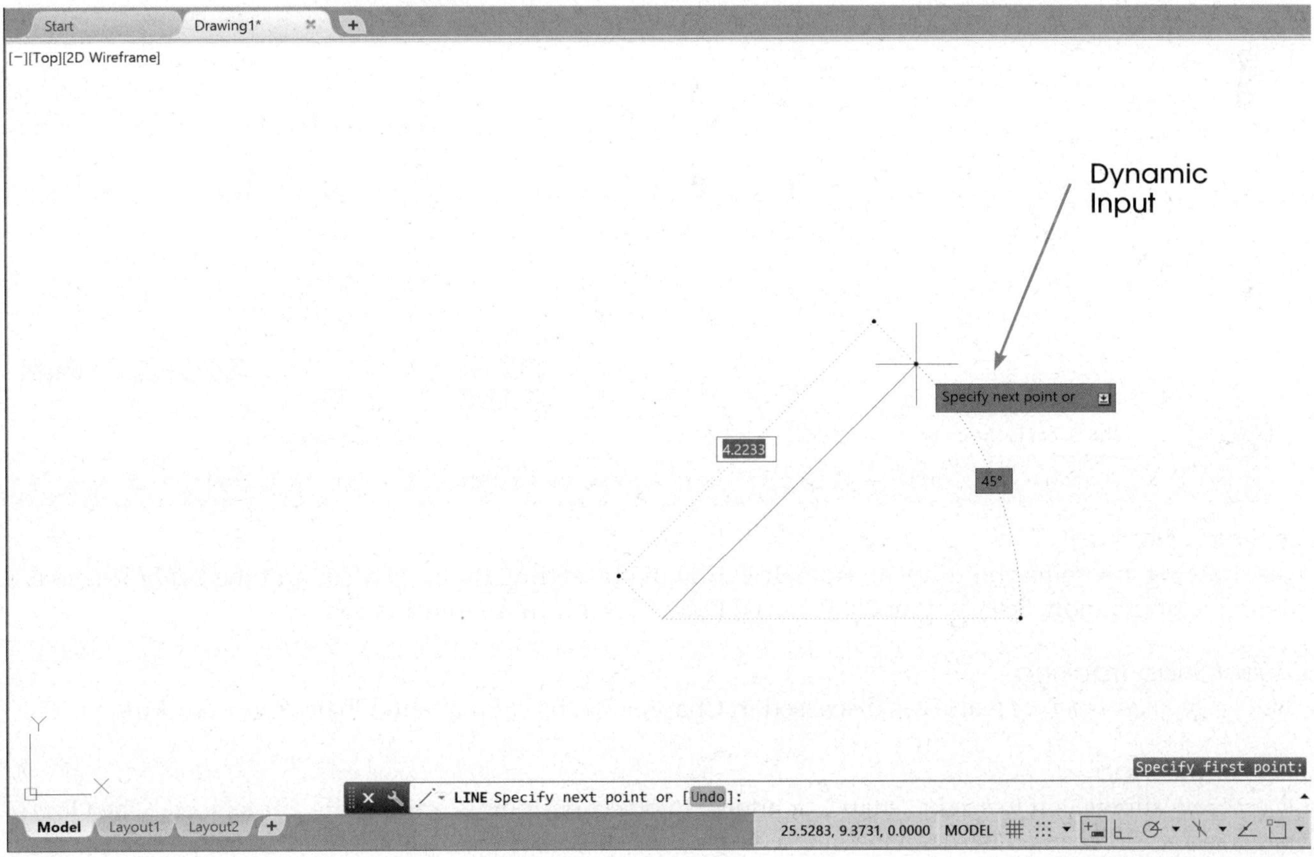

*Dynamic Input* is explained further in Chapter 3. You can toggle *Dynamic Input* on or off by using the icon or the F12 key. You can change the format of the display during input by right-clicking on the icon and selecting *Dynamic Input Settings* from the menu. For example, you can set *Dynamic Input* to display a "tool tip" that gives the current coordinate location of the cursor (cursor tracking mode) when no commands are in use.

### *Ortho Mode* (F8)

If *Ortho Mode* is on, lines are forced to an orthogonal alignment (horizontal or vertical) when drawing. *Ortho* is often helpful since so many drawings are composed mainly of horizontal and vertical lines.

### *Polar Tracking* (F10)

*Polar Tracking* makes it easy to draw lines at regular angular increments, such as 30, 45, or 90 degrees. The F10 key or the icon button toggles *Polar Tracking* on or off. When *Polar Tracking* is on, a polar tracking vector (a faint dotted line) appears when the "rubber band" line approaches the desired angular increment as shown in Figure 1-25. By default, *Polar Tracking* is set to 90 degrees, but can be set to any angular increment by right-clicking on the icon and selecting an angle option (bottom-right, Figure 1-25).

**FIGURE 1-25**

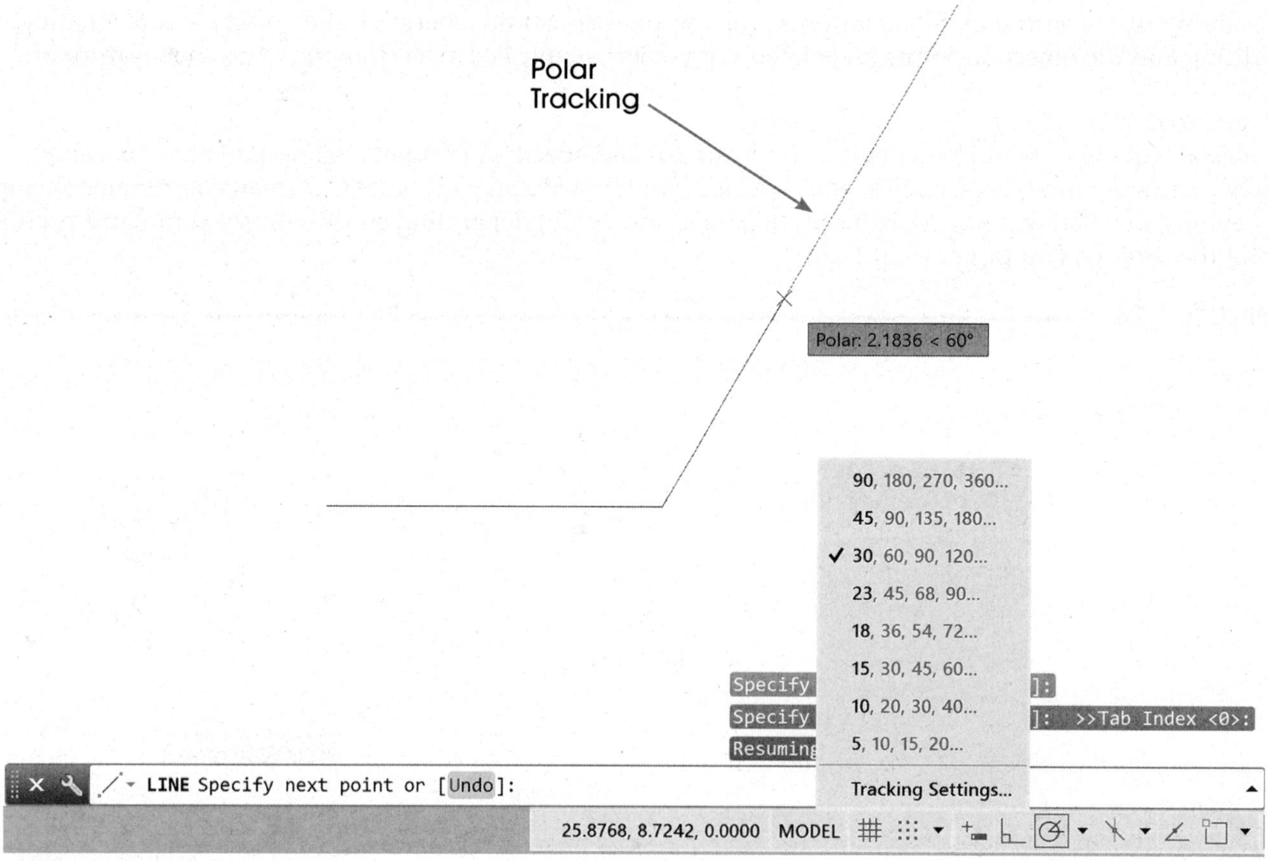


### *Isometric Drafting*

This drawing aid helps you draw an isometric drawing by setting the *Snap Mode* and the *Grid Mode* to a 30-degree orientation. See Chapter 25, Pictorial Drawings, for more information.

### *Object Snap Tracking*

The *Object Snap Tracking* feature is discussed in Chapter 7, Object Snap and Object Snap Tracking.

### *2D Object Snap*

*Object Snap* allows you to easily "snap," or attach, to object endpoints, midpoints, centers, etc. The *Object Snap* feature is discussed in detail in Chapter 7.

### *LineWeight*
Use this aid to display line thicknesses that may be applied to objects. See Chapter 11.

### *Transparency*
Imported images can have a transparency quality applied. Use this icon to display or not display transparency. See Chapter 11 for a full explanation.

### *Selection Cycling*
Chapter 4 explains how to use *Selection Cycling*.

### *3D Object Snap*
You can snap to vertices, midpoints, faces, etc. of 3D objects. This icon is used to turn this feature on or off.

### *Dynamic UCS*
You can use *Dynamic User Coordinate Systems* when creating or editing 3D objects. See Chapter 35 for more information.

### *Selection Filtering*
Discussed in Chapter 20, Advanced Selection Sets, this feature assists when selecting objects for editing.

### *Gizmo*
A *Gizmo* can be used to create and edit 3D objects. This feature is discussed with the solid modeling information.

### *Annotation Visibility*
Dimensions and text objects can be made to change size automatically. This option allows you to see changes made by scale changes to the current viewport.

### *AutoScale*
This feature toggles the automatic scaling of annotation objects on or off.

### *Annotation Scale*
This pop-up list of standard scales allows you to set the desired scale for annotation in the current viewport.

### *Workspace Switching*
Use this list to select from the three workspaces (*Drafting & Annotation, 3D Basics,* and *3D Modeling*) as well as control the customization of your workspaces.

### *Annotation Monitor*
Dimensions are associated (attached) to objects. You can use the *Annotation Monitor* to flag dimensions that lose associativity when objects are moved or updated.

### *Units*
When turned on, the current units for the drawing are displayed such as decimal or feet and inches.

### *Quick Properties*
When you select objects to edit, a *Quick Properties* panel (giving properties of the selected objects) may appear based on your setting with this icon. See Chapter 16.

### *Lock UI*
Use *Lock UI* to "lock" in place user interface elements such as windows, palettes, and toolbars from being accidentally or otherwise moved or changed.

### *Isolate Objects*

This icon provides a quick method for using the *Hideobjects, Isolateobjects,* and *Unisolateobjects* commands. See Chapter 20 for more information on these features.

### *Graphics Performance*

This icon provides access to the *Graphics Performance* dialog box where you can tune the performance of your graphics display.

### *Clean Screen*

*Clean Screen* removes the Ribbon so you have more screen area to draw. See the "Customizing the AutoCAD Screen" section next in this chapter.

## Function Keys

Several function keys are usable with AutoCAD. They offer a quick method of turning on or off (toggling) drawing aids and other features.

| | | |
|---|---|---|
| **F1** | *Help* | Opens a help window providing written explanations on commands and variables. |
| **F2** | *Text window* | Activates the text window showing the previous command line activity (command history). |
| **F3** | *Object Snap* | Turns Running *Object Snaps* on or off. If no Running *Object Snaps* are set, F3 produces the *Object Snap Settings* dialog box (discussed in Chapter 7). |
| **F4** | *3D Object Snap* | Turns *3D Object Snap* on or off. |
| **F5** | *Isoplane* | When using an *Isometric* style *Snap* and *Grid* setting, toggles the cursor (with *Ortho* on) to draw on one of three isometric planes. |
| **F6** | *Dynamic UCS* | Turns *Dynamic UCS* on or off for 3D modeling (see Chapters 35 and 36). |
| **F7** | *Grid* | Turns the *Grid* on or off. |
| **F8** | *Ortho* | Turns *Ortho* on or off. |
| **F9** | *Snap* | Turns *Snap* on or off. |
| **F10** | *Polar Tracking* | Turns *Polar Tracking* on or off. |
| **F11** | *Object Snap Tracking* | Turns *Object Snap Tracking* on or off. |
| **F12** | *Dynamic Input* | Turns *Dynamic Input* on or off. |

## AutoCAD Text Window

FIGURE 1-26

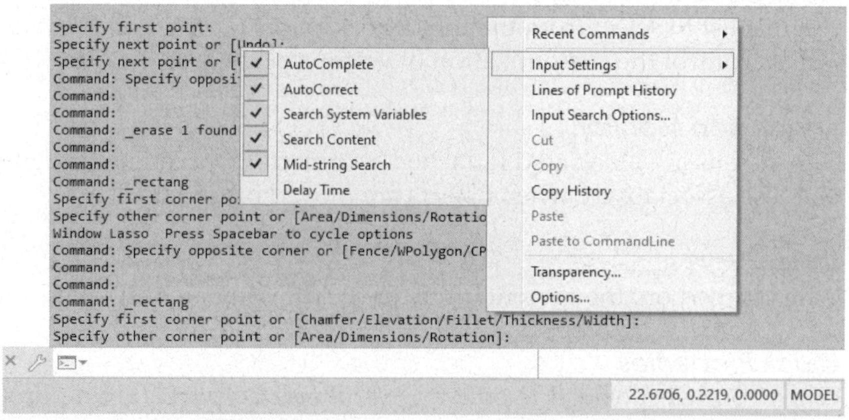


Pressing the F2 key activates the *Text Window,* sometimes called the Command History (Fig. 1-26). Here you can see the text activity that occurred at the Command line—kind of an "expanded" Command line. Press F2 again to close the text window. The right-click menu in this text window provides several options. If you highlight text in the window, you can then *Paste to Command line, Copy* it to another program such as a word processor, *Copy History* (entire command history) to another program, or *Paste* text into the window. The *Options* choice invokes the *Options* dialog box (discussed later).

Several options in the right-click menu of the text window allow you to customize how the Command line operates. For example, selecting *Input Settings* provides options for how AutoCAD responds as you type words at the Command line. Here you can toggle *AutoComplete* and *AutoCorrect* as well as what content in the drawing is searched when you begin typing.

Selecting *Input Search Options* from the text window right-click menu invokes the dialog box shown in Figure 1-27. This box provides complete control of how AutoCAD completes, corrects, and searches as you type words into the Command line.

**FIGURE 1-27**

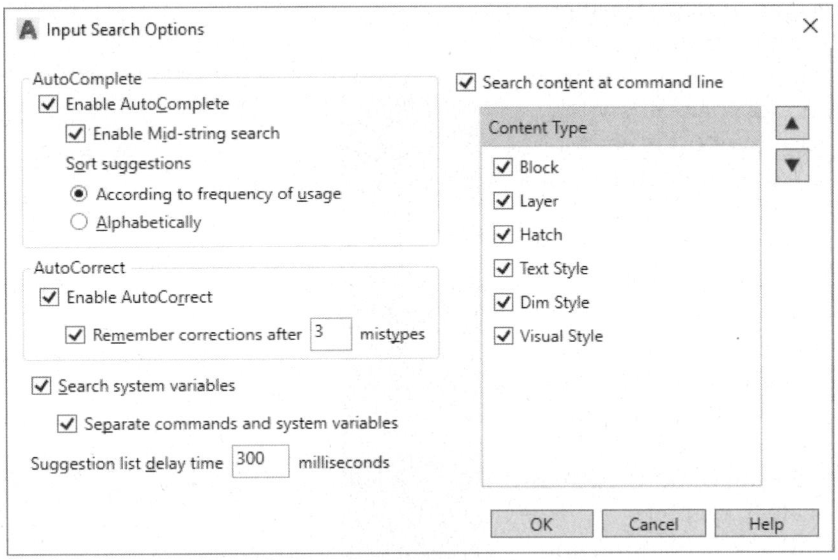


Selecting the *Transparency* option produces a window that controls the percentage of transparency you prefer for the Command line. Use the *Lines of Prompt History* to specify the maximum number of text lines that appear when the Command line automatically expands as you work. The setting is saved in the *CLIPROMPTLINES* system variable and is set to 3 by default.

## Model Space and Layouts

### Model Space

When you begin a drawing, model space is the current drawing space. This area is also known as the *Model* tab (Fig. 1-28, lower-left). In this area you should create the geometry representing the subject of your drawing, such as a floor plan, a mechanical part, or an electrical schematic. Dimensions are usually created and attached to your objects in model space.

**FIGURE 1-28**

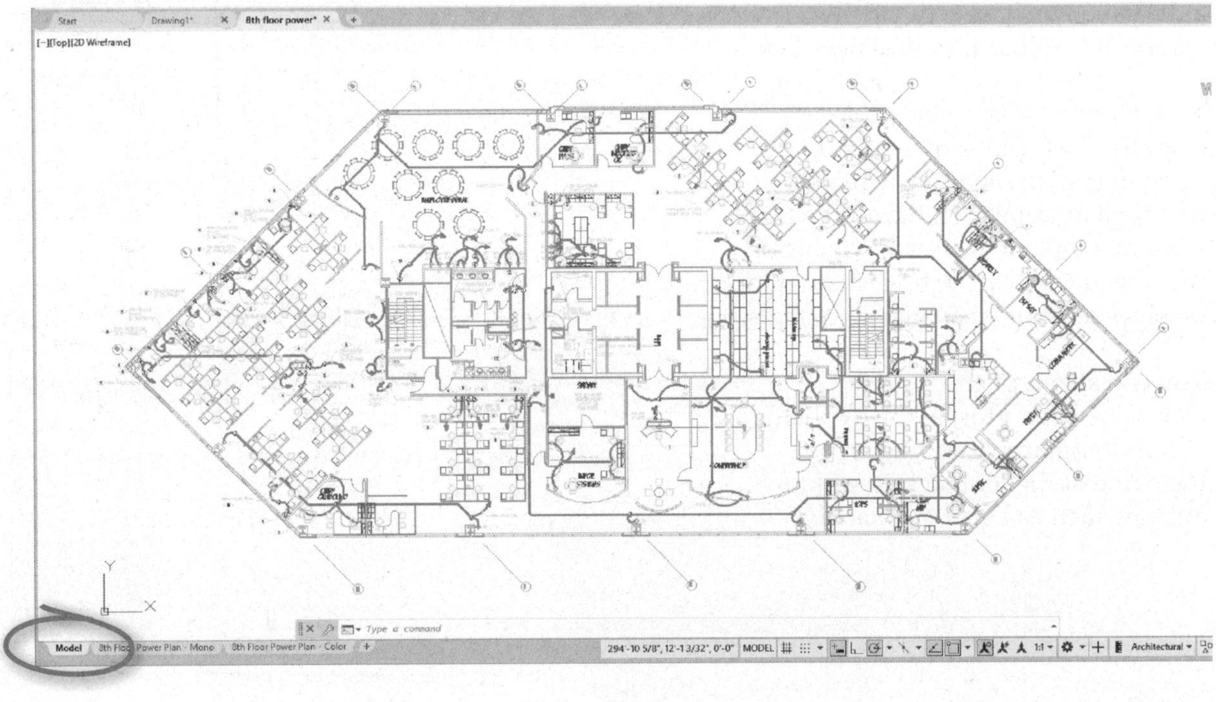

### Layouts

When you are finished with your drawing, you can print it directly from model space or switch to a layout (Fig. 1-29). Layouts, sometimes known as paper space, represent sheets of paper that you print on.

You must use several commands to set up the layout to display the geometry and set all the printing options such as scale, paper size, device, and so on. Printing and layouts are discussed in detail in Chapters 13, 14, and 32.

FIGURE 1-29

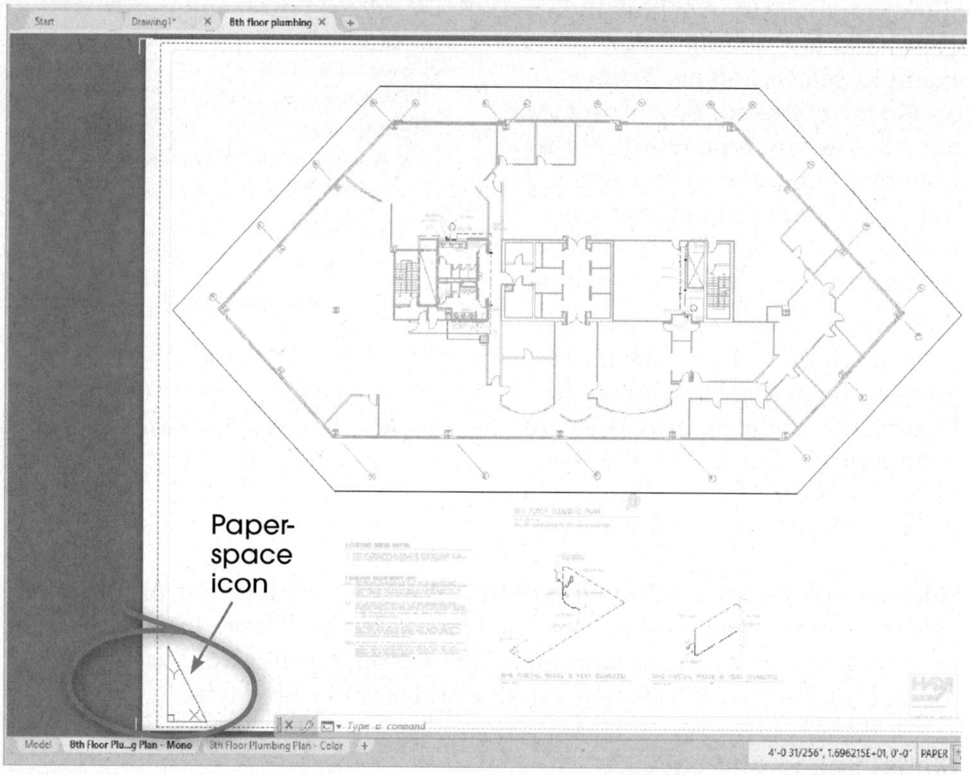


## CUSTOMIZING THE AutoCAD SCREEN

### Toolbars

Many toolbars are available, each with a group of related commands for specialized functions. For example, when you are ready to dimension a drawing, you can activate the *Dimension* toolbar for efficiency.

To activate toolbars, select *Toolbars* from the *Tools* pull-down menu. Selecting any toolbar name makes that toolbar appear on the screen (Fig. 1-30). If a toolbar is already visible on the screen, you can right-click on it to produce a list of all available toolbars.

You can also type the *-Toolbar* command (don't forget the hyphen) at the Command line. Next, type the name of the toolbar you want to appear, such as *Draw* or *Dimension*.

FIGURE 1-30

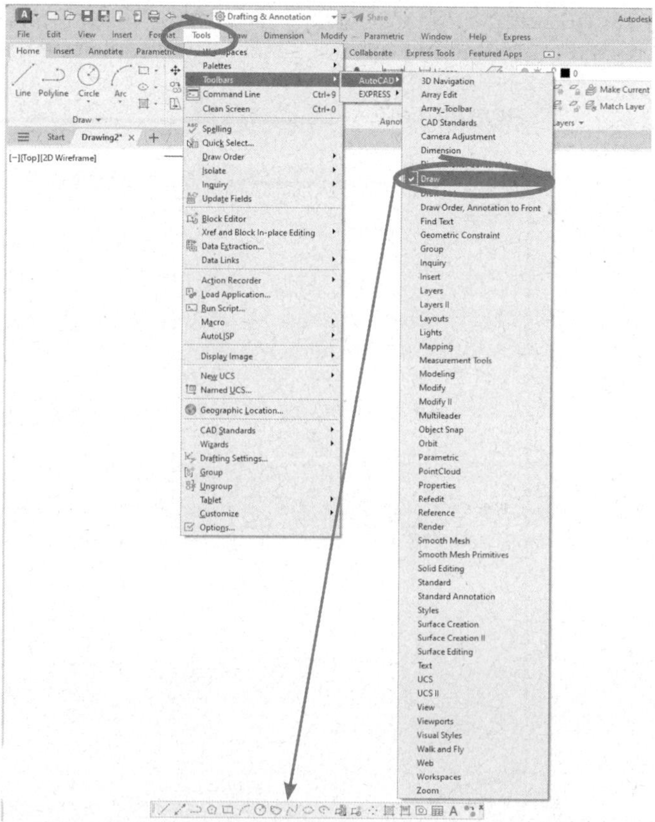

Use the *Show* option to make the toolbar appear on the screen. (The *Toolbar* command [without the hyphen] produces the *Customize User Interface* dialog box that is used for customizing the Ribbon, menus, toolbars, palettes, and workspaces.)

By default, most toolbars that are newly activated are floating (see Fig. 1-30). A floating toolbar can be easily moved to any location on the screen if it obstructs an important area of a drawing. Placing the pointer in the title background allows you to move the toolbar by holding down the left button and dragging it to a new location (Fig. 1-31). Floating toolbars can also be resized by placing the pointer on the narrow border until a two-way arrow appears, then dragging left, right, up, or down.

**FIGURE 1-31**

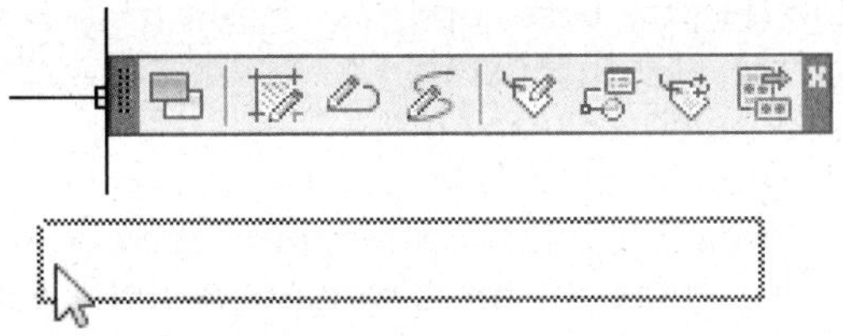

A floating toolbar can be docked against any border (right, left, top, bottom) by dragging it to the desired location (Fig. 1-32). Several toolbars can be stacked in a docked position. By the same method, docked toolbars can be dragged back onto the graphics area.

Toolbars can be removed from the screen by clicking once on the "X" symbol in the upper right of a floating toolbar.

**FIGURE 1-32**

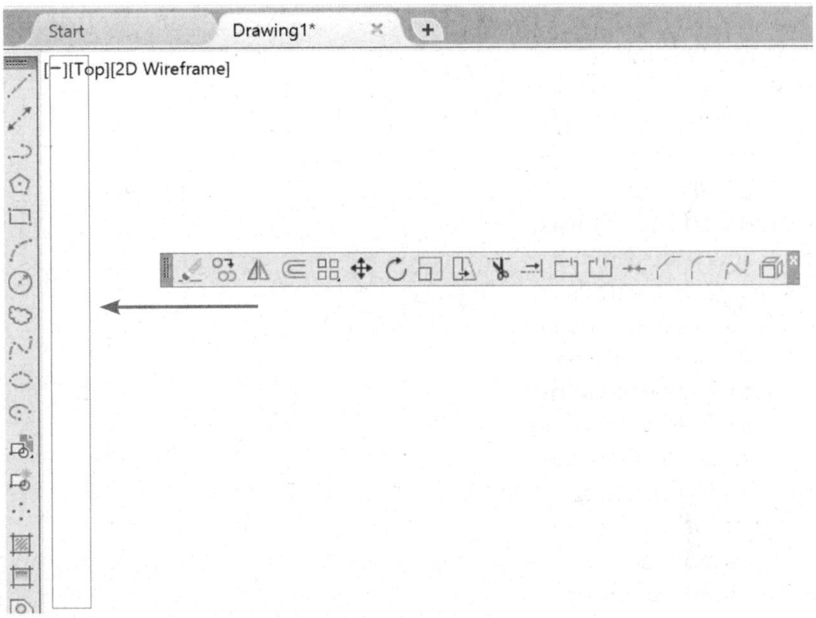


## Ribbon

You can customize the Ribbon to your liking. Right-click on any tab to produce the shortcut menu shown in Figure 1-33. You can select which tabs and panels you want to appear. For example, if you want only certain panels to appear for a specific tab, first click on the tab, then use this menu to produce a list of toggles for all the panels that can appear for that tab, as shown in the figure. If you choose *Undock* from the shortcut menu the Ribbon is converted to a palette. You can drag the Ribbon palette to the right or left side of the screen or back to the top of the screen. You can further customize how

**FIGURE 1-33**

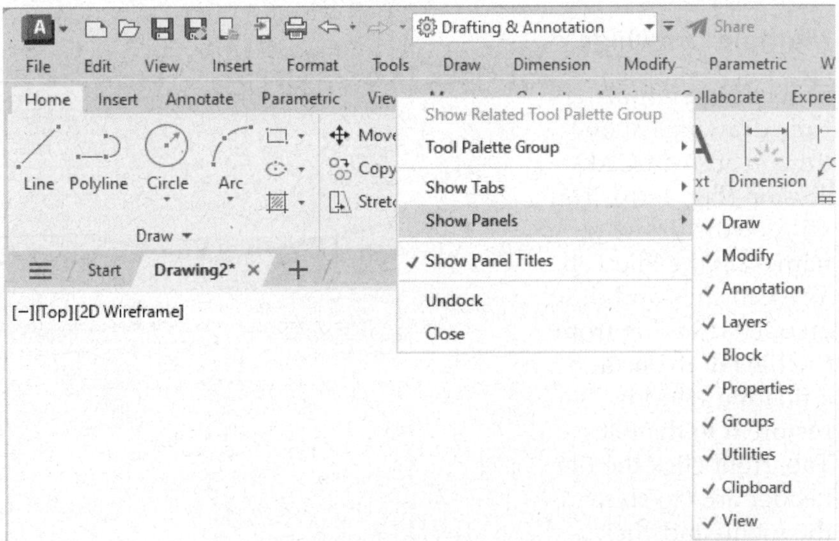

much of the panels appear in the Ribbon. There is a very small drop-down arrow just to the right of the last tab (Figure 1-34, top right). Use this list to choose how you want the panels in each tab to appear. You can choose *Minimize to Panel Titles* to increase space in the graphics area. With this option, only the panel titles appear without the tools (Fig. 1-34). To activate a panel and display the tools contained in the panel, momentarily rest the pointer on the panel title until the full panel pops down to display the full set of tools (Fig. 1-34, top left).

**FIGURE 1-34**

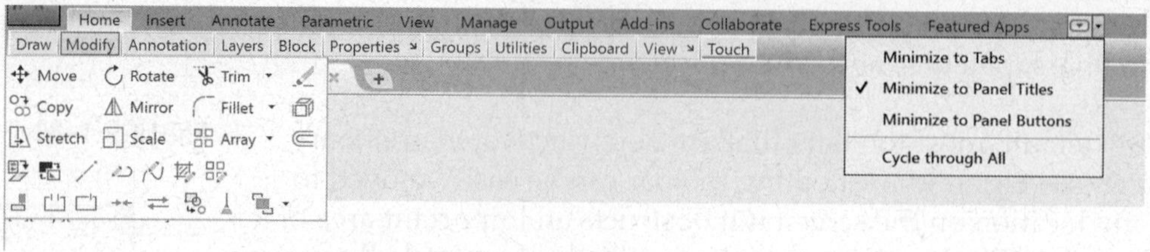


## Clean Screen

You can change the AutoCAD window from a normal window to a window with no toolbars or Ribbon (Fig. 1-35). Notice that the Command line, Status Bar, and pull-down menus (if activated) are still visible. This is useful if you want to maximize the drawing area to view a drawing or to make a presentation. Activate *Clean Screen* by toggling the small square in the extreme lower-right corner of the screen, by toggling the Ctrl+0 (Ctrl and zero) key sequence, or by entering the *Cleanscreenon* and *Cleanscreenoff* commands.

**FIGURE 1-35**

## Multiple Drawings

You can open multiple drawings at one time in an AutoCAD session (Fig. 1-36). You will see the drawing names on the File Tab. You can click and drag an open drawing from the file tab to become a floating window. To restore it to the File Tab, right click the file header area to show the menu and click Move to File Tab.

**FIGURE 1-36**

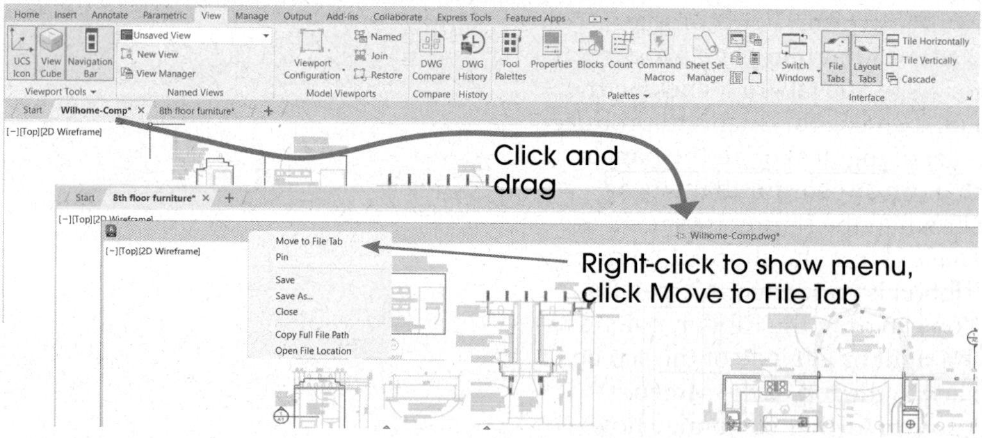

## Options

Fonts, colors, and other features of the AutoCAD drawing editor can be customized to your liking by using the *Options* command and selecting the *Display* tab. Typing *Options* or selecting *Options...* from the *Tools* pull-down menu or default shortcut menu activates the dialog box shown in Figure 1-37. Selecting the *Color...* tile provides a dialog box for customizing the screen colors.

NOTE: All changes made to the Windows screen by any of the options discussed in this section are automatically saved for the next drawing session. The changes are saved as the current profile. However, if you are working in a school laboratory, it is possible that the computer systems are set up to present the same screen defaults each time you start AutoCAD.

**FIGURE 1-37**

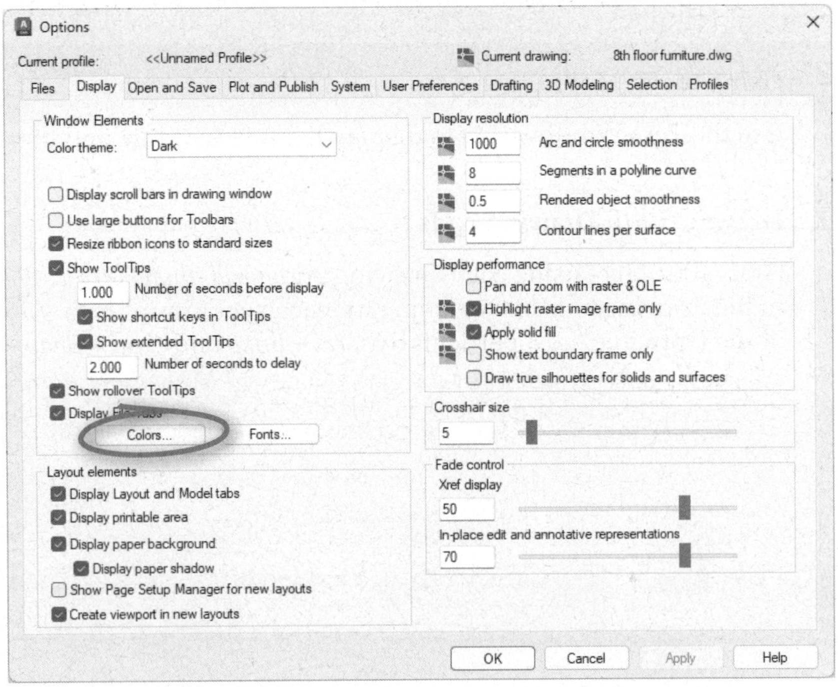


## CHAPTER EXERCISES

1. **Starting and Exiting AutoCAD**

   Start AutoCAD by double-clicking the "**AutoCAD**" shortcut icon or selecting "**AutoCAD**" from the Start menu. Select *Start Drawing* in the opening AutoCAD screen. Draw a *Line*. Exit AutoCAD by selecting the *Exit Autodesk AutoCAD* option from the Application Menu ("A" in the upper-left corner). Answer *No* to the "Save changes to Drawing1.dwg?" prompt. Repeat these steps until you are confident with the procedure.

2. **Using Drawing Aids**

   Start AutoCAD. Turn on and off each of the following modes:

   *Snap, Grid, Ortho, Polar Tracking*

3. **Understanding Coordinates**

   Use the Status Bar *Customize* button (lower-right corner) to turn on the display of *Coordinates* and the *Dynamic Input* icon. Start the *Line* command. Begin drawing a *Line* by PICKing a "Specify first point:". Toggle *Dynamic Input* on and then toggle *Coordinates* to display the *Relative* and *Absolute* formats (HINT: right-click on the coordinate display to access the format options).

PICK several other points at the "Specify next point or [Undo]:" prompt. Pay particular attention to the coordinate values displayed for each point and visualize the relationship between that point and coordinate 0,0 (absolute Cartesian value) or the last point established and the distance and angle (relative polar value). Finish the command by pressing Enter.

4. **Using the Text Window**

   Use the Text Window (**F2**) to display the text window and the graphics screen.

5. **Drawing with Drawing Aids**

   Draw four *Lines* using each Drawing Aid: *Grid, Snap, Ortho, Polar Tracking*. Toggle on and off each of the drawing aids one at a time for each set of four Lines. Next, draw *Lines* using combinations of the Drawing Aids, particularly *Grid* + *Snap* and *Grid* + *Snap* + *Polar Tracking*.

CHAPTER

# 2

# Working with Files

## Chapter Objectives

After completing this chapter you should:

1. be able to name drawing files;
2. be able to use file-related dialog boxes;
3. be able to use the Windows right-click shortcut menus in file dialog boxes;
4. be able to create *New* drawings;
5. be able to *Open* and *Close* existing drawings;
6. be able to *Save* drawings;
7. be able to use *Saveas* to save a drawing under a different name, path, and/or format;
8. understand how *Partialopen* and *Partiaload* can open part of a drawing;
9. be able to practice good file management techniques.

# AutoCAD DRAWING FILES

## Naming Drawing Files

What is a drawing file? A CAD drawing file is the electronically stored data form of a drawing. The computer's hard disk is the principal magnetic storage device used for saving and restoring CAD drawing files. Flash drives, compact disks (CDs), networks, and the Internet (including the "cloud") are used to transport files from one computer to another, as in the case of transferring CAD files among clients, consultants, or vendors in industry. The AutoCAD commands used for saving drawings to and restoring drawings from files are explained in this chapter.

An AutoCAD drawing file has a name that you assign and a file extension of ".DWG." An example of an AutoCAD drawing file is:

PART-024.DWG

file name extension

The file name you assign must be compliant with the Windows file name conventions; that is, it can have a maximum of 256 alphanumeric characters. File names and directory names (folders) can be in UPPERCASE, Title Case, or lowercase letters. Characters such as _ - $ # ( ) ^ and spaces can be used in names, but other characters such as \ / : * ? < > | are not allowed. AutoCAD automatically appends the extension of .DWG to all AutoCAD-created drawing files.

NOTE: The chapter exercises and other examples in this book generally list file names in UPPERCASE letters for easy recognition. The file names and directory (folder) names on your system may appear as UPPERCASE, Title Case, or lowercase.

## Beginning and Saving a Drawing

TIP

When you start AutoCAD, the Drawing Editor appears and allows you to begin drawing even before using any file commands. As you draw, you should develop the habit of saving the drawing periodically (about every 15 or 20 minutes) using *Save*. *Save* stores the drawing in its most current state.

The typical drawing session would involve using *New* to begin a new drawing or using *Open* to open an existing drawing. *Save* would be used periodically, and *Exit* would be used for the final save and to end the session.

## Accessing File Commands

Proper use of the file-related commands covered in this chapter allows you to manage your AutoCAD drawing files in a safe and efficient manner. Although the file-related commands can be invoked by several methods, they are easily accessible via the Application Menu (Fig. 2-1). Most of the selections from this menu invoke dialog boxes for selection or specification of file names.

FIGURE 2-1

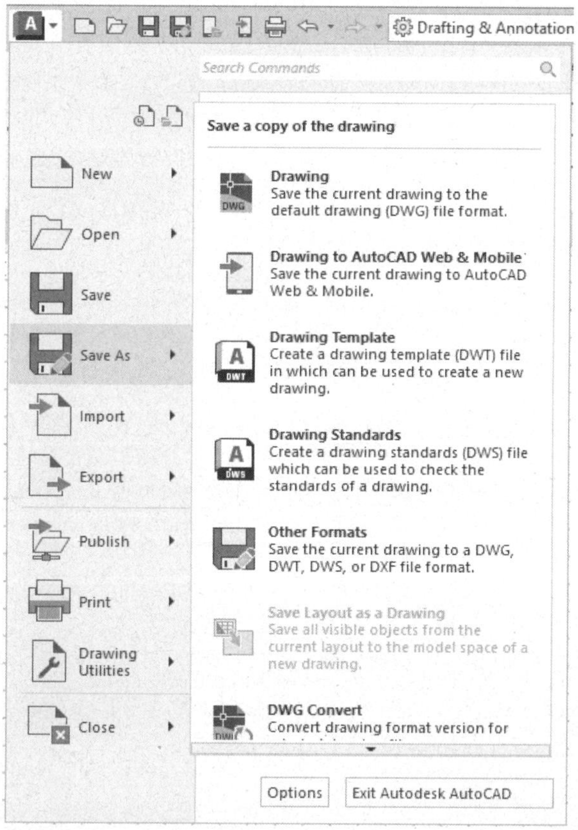

The Application Menu (select the "A" at the top left of the AutoCAD screen) has tools (icon buttons) for *New*, *Open*, and *Save*. File commands can also be entered at the keyboard by typing the formal command name, the command alias, or using the Ctrl keys sequences. File commands and related dialog boxes are also available from the *File* pull-down menu, if activated.

## File Navigation Dialog Box Functions

FIGURE 2-2

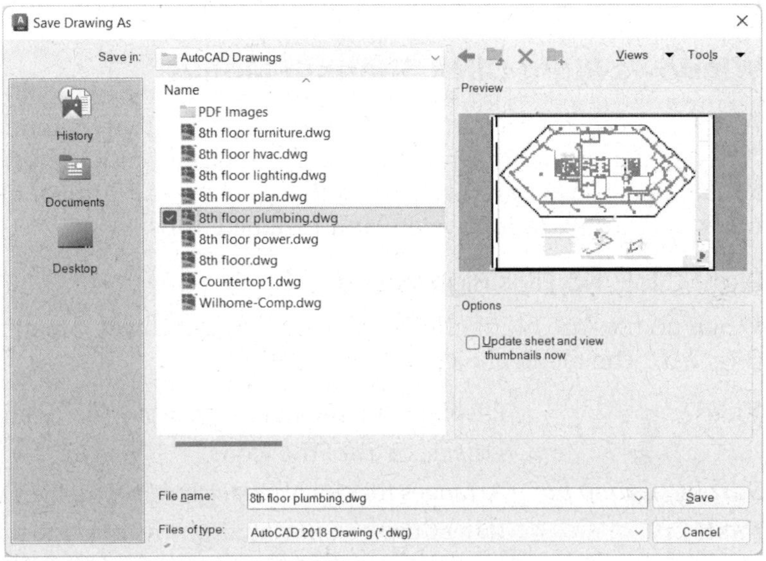


There are many dialog boxes that appear in AutoCAD to help you manage files. All of these dialog boxes operate in a similar manner. A few guidelines will help you use them. The *Save Drawing As* dialog box (Fig. 2-2) is used as an example.

- The top of the box gives the title describing the action to be performed. It is very important to glance at the title before acting, especially when saving or deleting files.
- The desired file can be selected by PICKing it, then PICKing *OK*. Double-clicking on the selection accomplishes the same action. File names can also be typed in the *File name:* edit box.
- Every file name has an extension (three letters following the period) called the type. File types can be selected from the *Files of Type:* section of the dialog boxes, or the desired file extension can be entered in the *File name:* edit box.
- The current folder (directory) is listed in the drop-down box near the top of the dialog box. You can select another folder (directory) or drive by using the drop-down list displaying the current path, by selecting the *Back* arrow, or by selecting the *Up One Level* icon to the right of the list. (Rest your pointer on an icon momentarily to make the tool tip appear.)
- Selecting one of the folders displayed on the left side of the file dialog boxes allows you to navigate to the following locations. The files or locations found appear in the list in the central area.

  *History:* Lists a history of files you have opened from most to least recent.

  *Documents:* Lists all files and folders saved in the Documents folder.

  *Desktop:* Shows files, folders, and shortcuts on your desktop.

  *OneDrive:* Shows files and folders for your Microsoft cloud services, if set up.
- You can use the *Search the Web* icon to display the *Browse the Web* dialog box (not shown). The default site is http://www.autodesk.com/.
- Any highlighted file(s) can be deleted using the "X" (*Delete*) button.
- A new folder (subdirectory) can be created within the current folder (directory) by selecting the *New Folder* icon.
- The *Views* drop-down list allows you to toggle the listing of files to a *List* or to show *Details*. The *List* option displays only file names, whereas the *Details* option gives file-related information such as file size, file type, time, and date last modified (see Fig. 2-2).

- The detailed list can be sorted alphabetically by name, by file size, alphabetically by file type, or chronologically by time and date. Do this by clicking the Name, Type, Size, or Date Modified tiles immediately above the list. Double-clicking on one of the tiles reverses the order of the list. You can use the Preview option to show a preview of the highlighted file (see Fig. 2-2) or to disable the preview.
- Use the *Tools* drop-down to add files or folders to *Favorites*. *Places* adds files or folders to the folder list on the left side of the dialog box. *Options* allows you to specify formats for saving DWG and DXF files. *Digital Signatures* allows you to apply a digital signature to the file (see "*Saveas*").

NOTE: Most dialog boxes can be resized by positioning the pointer on the border and dragging.

## Windows Right-Click Shortcut Menus

AutoCAD utilizes the right-click shortcut menus that operate with the Windows operating systems. Activate the shortcut menus by pressing the right mouse button (right-clicking) inside the file list area of a dialog box. There are two menus that give you additional file management capabilities based on features of your operating system and installed apps.

TIP

### Right-Click, No Files Highlighted

When no files are highlighted, right-clicking produces a menu for file list and file management options (Fig. 2-3). The menu choices are as follows:

FIGURE 2-3

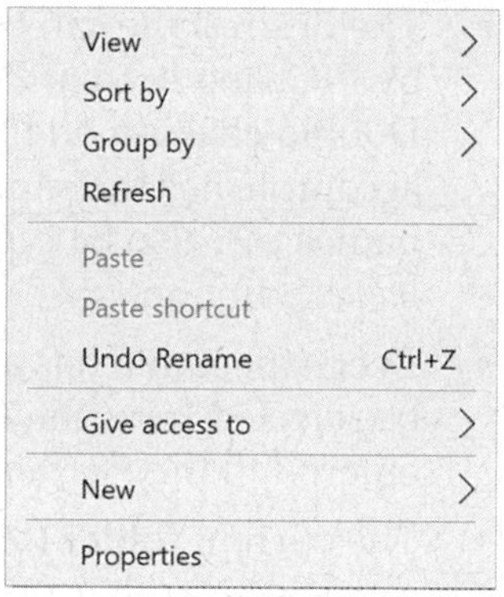


*View* — Displays the files as *Icons*, a *List* (no details), a list with *Details*, or *Tiles* the icons.

*Sort by, Group by* — Arranges files by *Name*, *Date Modified*, file *Type*, *Size*, or *Tags*.

*Paste* — If the *Copy* or *Cut* function was previously used (see "Right-Click, File Highlighted," Fig. 2-4), *Paste* can be used to place the copied file into the displayed folder.

*Paste Shortcut* — Use this option to place a *Shortcut* (to open a file) into the current folder.

*Undo* — Can be used to undo the last selection. You can use Ctrl+Z also.

*Give access to* — Lets you control permission and sharing for the file.

*New* — Creates a new *Folder*, *Shortcut*, or document.

*Properties* — Displays a dialog box listing properties of the current folder.

### Right-Click, File Highlighted

When a file is selected, right-clicking produces a menu with options for the selected file (Fig. 2-4). The menu choices are among the following:

FIGURE 2-4

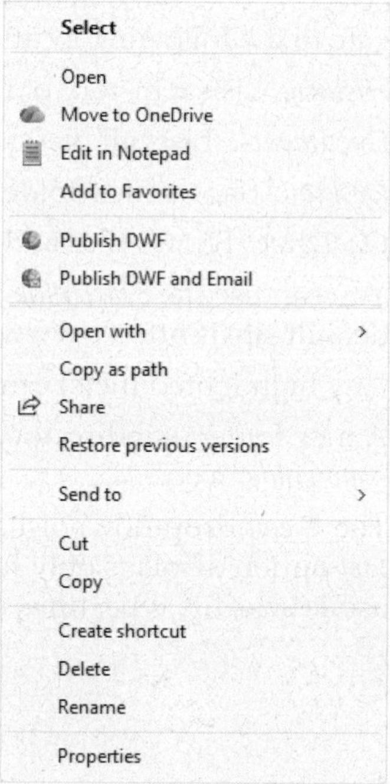


*Select* — Processes the file (like selecting the *OK* button) according to the dialog box function. For example, if the *SaveAs* dialog box is open, *Select* saves the highlighted file; or if the *Select File* dialog box is active, *Select* opens the drawing.

*Open* — Opens the application (AutoCAD or other program) and loads the selected file. Since AutoCAD can have multiple drawings open, you can select several drawings to open with this feature.

*Move to OneDrive* — Select this option to move the drawing to your Microsoft cloud storage area, if set up.

*Add to Favorites* — Creates a shortcut to the file in the Windows 11 File Explorer "Home" section for quick access

| | |
|---|---|
| ***Publish DWF*** | Select this option to open the drawing and produce the *Specify DWF File* dialog box. |
| ***Publish and Email*** | Publishes a DWF, then opens the system's email program with the DWF attachment. |
| ***Convert to Adobe PDF*** | Opens the Save Adobe PDF file dialog to for creating a PDF from the selected file. You may have various options if Adobe Acrobat is installed on your system. |
| ***Open with*** | Expand the options to show installed software which may open the file. |
| ***Copy as path*** | Copies the directory location along with the file name. This is very useful for emailing a link to the file on your network to a colleague, among other uses. |
| ***Scan*** | Scans the selected file with the system's default virus protection software. |
| ***Share*** | Provides sharing capabilities for the file based on the network options. |
| ***Restore Previous...*** | You can restore previous versions of the file if a Windows backup was used. |
| ***Send To*** | Copies the selected file to the selected device. This is an easy way to copy a file from your hard drive to another drive. |
| ***Cut*** | In conjunction with *Paste,* allows you to move a file from one location to another. |
| ***Copy*** | In conjunction with *Paste,* allows you to copy the selected file to another location. You can copy the file to the same folder, but Windows renames the file to "Copy of . . .". |
| ***Create Shortcut*** | Use this option to create a *Shortcut* (to open the highlighted file) in the current folder. The shortcut can be moved to another location using drag and drop. |
| ***Delete*** | Sends the selected file to the Recycle Bin. |
| ***Rename*** | Allows you to rename the selected file. Move your cursor to the highlighted file name, click near the letters you want to change, then type or use the backspace, delete, space, or arrow keys. |
| ***Properties*** | Shows file properties if available. |

## AutoCAD FILE COMMANDS

### *New*

| Ribbon | Menu Bar | Command (Type) | Alias (Type) | Shortcut |
|---|---|---|---|---|
| ... | *File* *New...* | *New* | ... | *Ctrl+N* |

Type *New* at the command prompt, use the *Ctrl+N* key sequence or select *New* from the Application Menu or the *File* pull-down menu. The *New* command begins a new drawing. Use this option if you want to create a new drawing based on an existing template drawing. A template drawing is one that may have some of the setup steps performed but usually contains no geometry (graphical objects).

The *Select Template* dialog box (Fig. 2-5) appears when you use the *New* command (see NOTE on the next page). Here you can select from all AutoCAD-supplied templates. Selecting the ACAD.DWT template begins a new drawing using the Imperial default settings with a drawing area of 12 x 9 units.

**FIGURE 2-5**

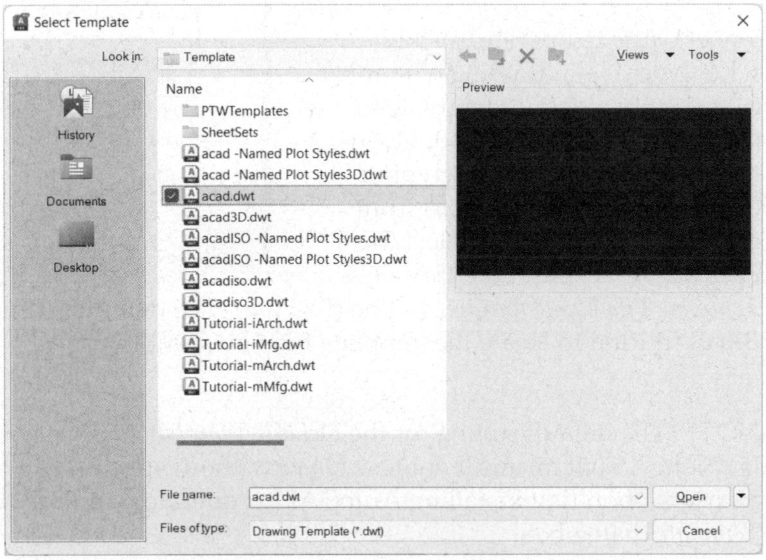

Selecting the ACADISO.DWT template begins a new drawing using the metric settings having a drawing area of 420 x 279 units. Details of these and other template drawings are discussed in Chapter 12, Advanced Drawing Setup.

For the purposes of learning AutoCAD starting with the basic principles and commands discussed in this text, it is helpful to use the ACAD.DWT (Imperial inch settings) when beginning a new drawing. Until you read Chapters 6 and 12, you can begin new drawings for completing the exercises and practicing the examples in Chapters 1 through 5 using the ACAD.DWT template.

NOTE: The setting for the *STARTUP* system variable may affect what happens when you use the *New* command. The default setting is 3. (Also see Chapter 6, Startup Options.)

0 Starts a drawing without defined settings.
1 Displays the *Startup* or the *Create New Drawing* dialog box.
2 A new tab is displayed. If available in the application, a custom dialog box is displayed.
3 A new tab is displayed and the ribbon is pre-loaded when you open or create a new drawing.

### Qnew

| Ribbon | Menu Bar | Command (Type) | Alias (Type) | Shortcut |
|---|---|---|---|---|
| ... | ... | *Qnew* | ... | ... |

Type *Qnew* at the command line or select *Qnew* from the Quick Access toolbar. *Qnew* (Quick New) is intended to immediately begin a new drawing using the template file specified in the *Files* tab of the *Options* dialog box. If the proper settings are made, *Qnew* is faster to use than the *New* command since it does not force the *Create New Drawing* or the *Select Template* dialog box to open, as the *New* command does.

**FIGURE 2-6**

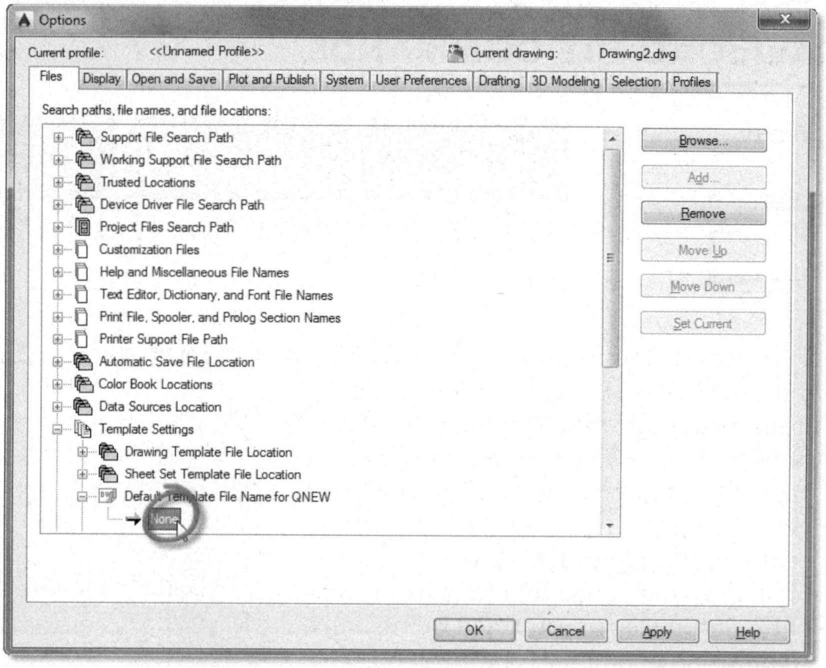


To set *Qnew* to operate quickly when opening a drawing, you must first specify a default template to use when *Qnew* is invoked. Open the *Options* dialog box by typing *Options* or selecting *Options* from the Application Menu. Select the *Files* tab and expand the *Default Template File Name for QNEW* line (Fig. 2-6). The default setting is *None*. Highlight *None*, then pick the *Browse* button to locate the template file you want to use.

NOTE: The default setting for the *Default Template File Name for QNEW* line (in the *Options* dialog box) is "None." That means if you use *Qnew* without specifying a template drawing, as will be the case for all users upon first installing AutoCAD, it operates similarly to the *New* command by opening the *Select Template* dialog box.

## *Open*

| Ribbon | Menu Bar | Command (Type) | Alias (Type) | Shortcut |
|---|---|---|---|---|
| ... | *File Open* | *Open* | ... | *Ctrl+O* |

Use *Open* to select an existing drawing to be loaded into AutoCAD. Normally you would open an existing drawing (one that is completed or partially completed) so you can continue drawing or to make a print or plot. You can *Open* multiple drawings at one time in AutoCAD.

If you select the *Open* command from the Application Menu (Fig. 2-7), you can open drawings on your local computer or network, open a *Drawing from the Cloud*, open a *Sheet Set*, open a *DGN* file (MicroStation drawing) into AutoCAD, or open sample drawings provided by Autodesk.

**FIGURE 2-7**

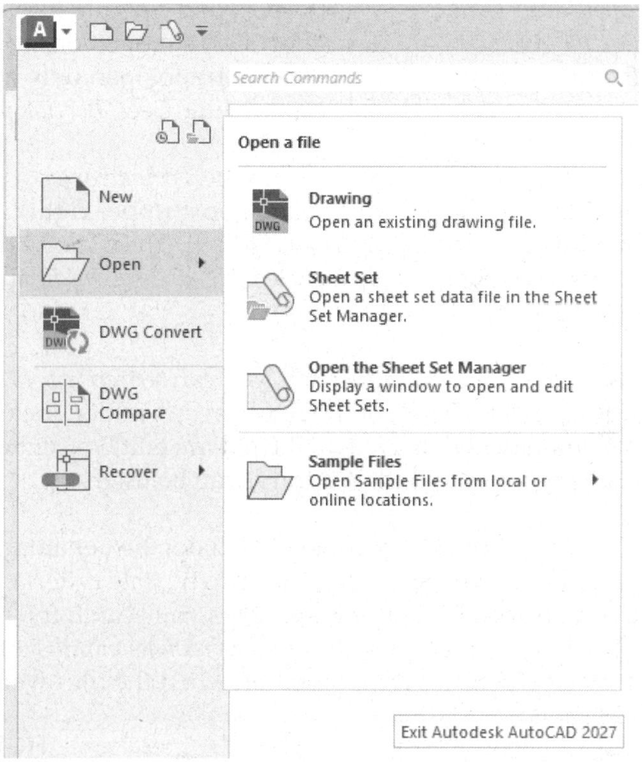


The *Open* command produces the *Select File* dialog box (Fig. 2-8). In this dialog box you can select any drawing from the current directory list. PICKing a drawing name from the list displays a small bitmap image of the drawing in the *Preview* tile. Select the *Open* button or double-click on the file name to open the highlighted drawing. You could instead type the file name (and path) of the desired drawing in the *File name:* edit box and press Enter, but a preview of the typed entry will not appear.

**FIGURE 2-8**

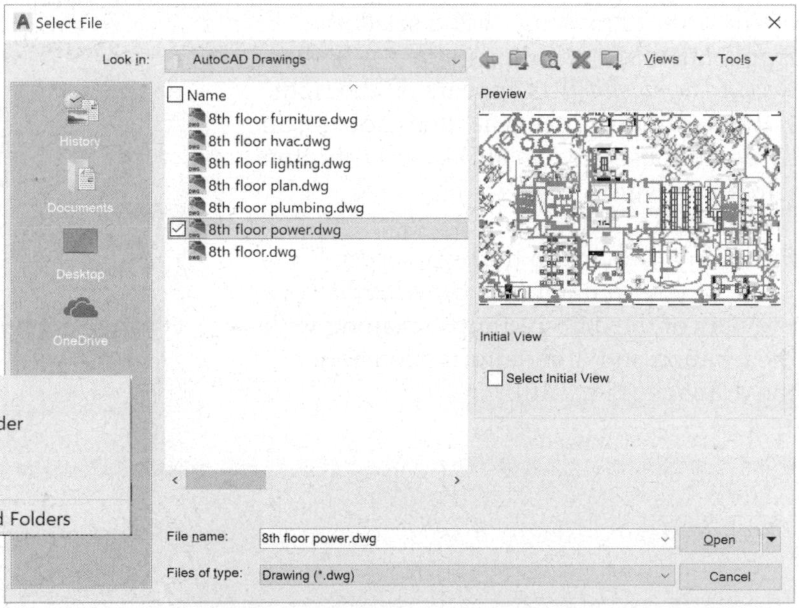

Right-click in the left panel to show the menu and use *Add Current Folder* to make a shortcut to your project folder.

**TIP** You can open multiple drawings at one time by holding down the Shift key to select a range (all files between and including the two selected) or holding down the Ctrl key to select multiple drawings not in a range. You can open a .DWG (drawing), .DWS (drawing standards), .DXF (drawing interchange format), or .DWT (drawing template) file by selecting from the *Files of type:* drop-down list.

Using the *Open* drop-down list (lower right) you can select from these options:

| | |
|---|---|
| *Open* | Opens a drawing file and allows you to edit the file. |
| *Open Read-Only* | Opens a drawing file for viewing, but you cannot edit the file. |
| *Partial Open* | Allows you to open only a part of a drawing for editing. See "*Partialopen.*" |
| *Partial Open Read-Only* | Allows you to open only a part of a drawing for viewing, but you cannot edit the drawing. See "*Partialopen.*" |

Selecting the *Find...* option from the *Tools* drop-down list in the *Select File* dialog box (upper right) invokes the *Find* dialog box (Fig. 2-9). The dialog box offers two criteria to locate files: *Name & Location* and *Date Modified*.

FIGURE 2-9

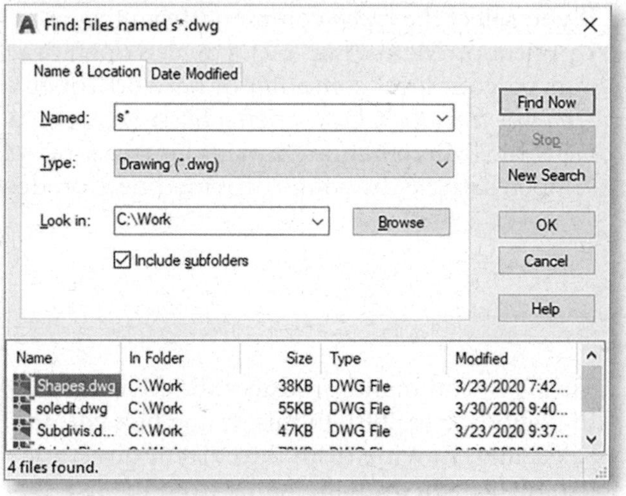


Using the *Name & Location* tab, you can search for .DWG, .DWS, .DWT, or .DXF files in any folder, including subfolders if desired. In the *Named*: edit box, enter a drawing name to find. Wildcards can be used.

The *Date Modified* tab of the *Find* dialog box enables you to search for files meeting specific date criteria that you specify. This feature helps you search for files *between* two dates, *during the previous months*, or *during the previous days*, based on when the files were last saved.

### *Recent Documents*

You can use the *Recent Documents* list as an alternative to the *Open* command to open current drawings. If you select the Application Menu button (letter "A"), the *Recent Documents* lists appears on the right. If the recent documents list does not appear by default, use the *Recent Documents* button (circled in Figure 2-10).

Select any drawing in the list to open. If you rest your pointer on a drawing name, a preview of the drawing appears along with the location and other information about the drawing (Fig. 2-10).

FIGURE 2-10

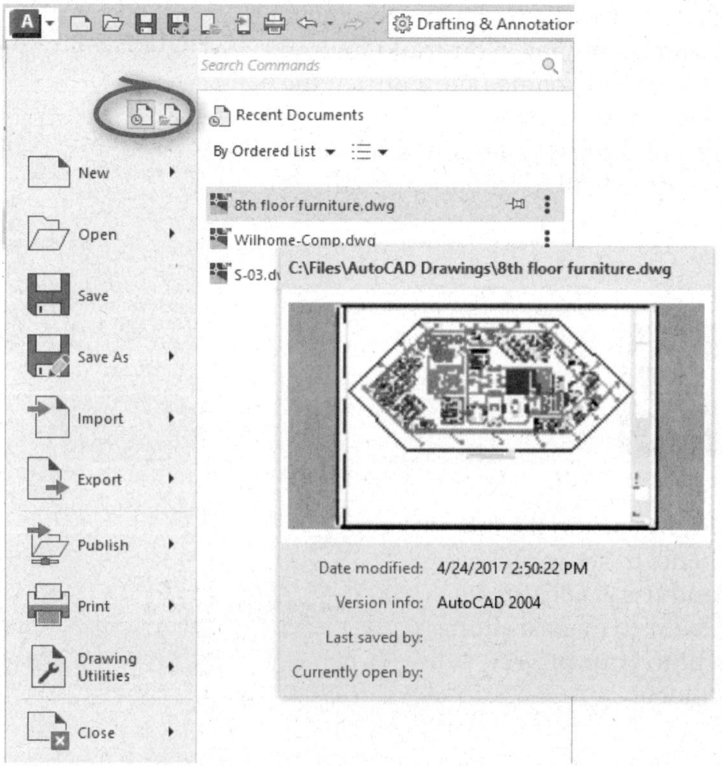

## *Save*

| Ribbon | Menu Bar | Command (Type) | Alias (Type) | Shortcut |
|---|---|---|---|---|
| ... | *File*<br>*Save* | *Save* | ... | *Ctrl+S* |

The *Save* command is intended to be used periodically during a drawing session (every 15 to 20 minutes is recommended). When *Save* is selected from a menu, the current version of the drawing is saved without interruption of the drawing session. The first time an unnamed drawing is saved, the *Save Drawing As* dialog box (see Fig. 2-2) appears, which prompts you for a drawing name. Typically, however, the drawing already has an assigned name, in which case *Save* actually performs a *Qsave* (quick save). A *Qsave* gives no prompts or options, nor does it display a dialog box.

When the file has an assigned name, using *Save* by selecting the command from a menu or icon button automatically performs a quick save (*Qsave*). Therefore, *Qsave* automatically saves the drawing in the same drive and directory from which it was opened or where it was first saved. In contrast, typing *Save* always produces the *Save Drawing As* dialog box, where you can enter a new name and/or path to save the drawing or press Enter to keep the same name and path (see *SaveAs*).

NOTE: If you want to save a drawing directly to another drive (such as a flash drive), first *Save* the drawing to the local hard drive, then use the *Send To* option in the right-click menu in the *Save*, *Save Drawing As*, or *Select File* dialog box (see "Windows Right-Click Shortcut Menus" and Fig. 2-4).

## *Qsave*

| Ribbon | Menu Bar | Command (Type) | Alias (Type) | Shortcut |
|---|---|---|---|---|
| ... | *File*<br>*Save* | *Qsave* | ... | *Ctrl+S* |

*Qsave* (quick save) is normally invoked automatically when *Save* is used (see *Save*) but can also be typed. *Qsave* saves the drawing under the previously assigned file name. No dialog boxes appear nor are any other inputs required. This is the same as using *Save* (from a menu), assuming the drawing name has been assigned. However, if the drawing has not been named when *Qsave* is invoked, the *Save Drawing As* dialog box appears.

## *Saveas*

| Ribbon | Menu Bar | Command (Type) | Alias (Type) | Shortcut |
|---|---|---|---|---|
| ... | *File*<br>*Save As* | *Saveas* | ... | ... |

The *Saveas* command can fulfill four functions:

1. Save the drawing file under a new name if desired.
2. Save the drawing file to a new path (drive and directory location) if desired.
3. In the case of either 1 or 2, assign the new file name and/or path to the current drawing (change the name and path of the current drawing).
4. Save the drawing in a format other than the default AutoCAD 2018 format or as a different file type (.DWT, .DWS, or .DXF).

Therefore, assuming a name has previously been assigned, *Saveas* allows you to save the current drawing under a different name and/or path; but beware, because *Saveas* sets the current drawing name and/or path to the last one entered. This dialog box is shown in Figure 2-2.

TIP

*Saveas* can be a benefit when creating two similar drawings. A typical scenario follows. A design engineer wants to make two similar but slightly different design drawings. During construction of the first drawing, the engineer periodically saves under the name DESIGN1 using *Save*. The first drawing is then completed and *Saved*. Instead of starting a *New* drawing, *Saveas* is used to save the current drawing under the name DESIGN2. *Saveas* also resets the current drawing name to DESIGN2. The designer then has two separate but identical drawing files on disk which can be further edited to complete the specialized differences. The engineer continues to work on the current drawing DESIGN2.

TIP

You can save an AutoCAD drawing in several formats other than the default format (*AutoCAD 2018 Drawing *.dwg*). Use the drop-down list at the bottom of the *Save* (or *Save Drawing As*) dialog box (Fig. 2-11) to save the current drawing as an earlier version drawing (.DWG) of AutoCAD or LT, a template file (.DWT), or a .DWS or a .DXF file. As of AutoCAD 2017, a drawing can be saved to an earlier release version and later *Opened* in the current AutoCAD release and all newer features are retained during the "round trip."

**FIGURE 2-11**

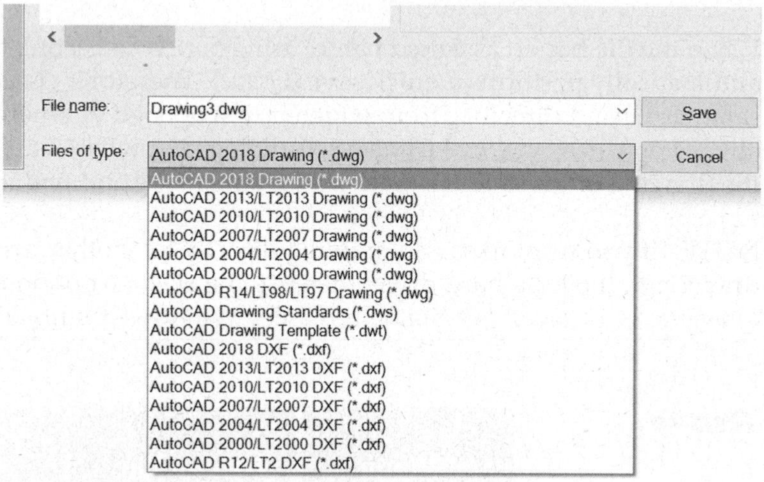


### *Tools*

You can access the *Saveas Options* dialog box and attach *Digital Signatures* by selecting the down arrow (circled) in the upper-right corner of the *Save* or *Save As* dialog box (Fig. 2-12).

**FIGURE 2-12**

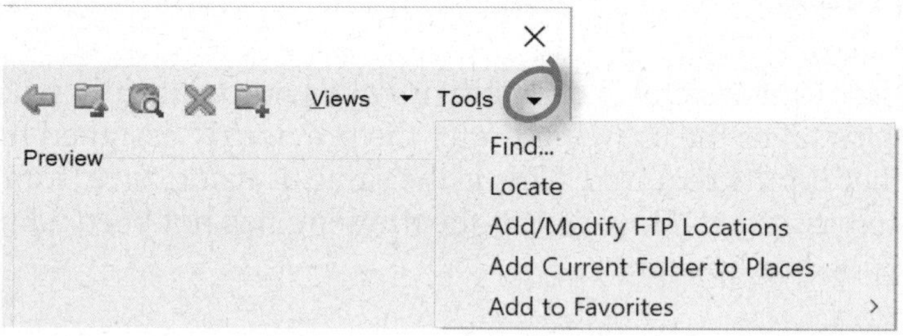


*Options...*
This choice produces the *Saveas Options* dialog box (not shown). Options include saving drawings as DXF files, saving drawings as earlier versions of AutoCAD, and specifying the Index type. Layer and spacial indexing are useful with the *Partialopen* and *Partiaload* commands (discussed in this chapter), and with externally referenced drawings (see Chapter 30, Xreferences).

*Digital Signatures...*
You can attach digital signatures to drawings. In order to use this feature, you must aquire a digital signature through a third-party certificate authority. If a drawing has a digital signature assigned, it is designated as an original signed and dated drawing. Users are assured that the drawing is original and has not been changed in any way. If such a drawing is later modified, the digital signature is then invalidated, and anyone opening the drawing is notified that the drawing has been changed and the digital signature is invalid. You can also use the *Digitalsign* command to access this feature.

### *Close*

| Ribbon | Menu Bar | Command (Type) | Alias (Type) | Shortcut |
|---|---|---|---|---|
| ... | *File*<br>*Close* | *Close* | ... | ... |

Use the *Close* command to close the current drawing. Because AutoCAD allows you to have several drawings open, *Close* gives you control to close one drawing while leaving others open. If the drawing has been changed but not saved, AutoCAD prompts you to save or discard the changes. In this case, a warning box appears (Fig. 2-13). *Yes* causes AutoCAD to *Save* then close the drawing; *No* closes the drawing without saving; and *Cancel* aborts the close operation so the drawing stays open.

**FIGURE 2-13**

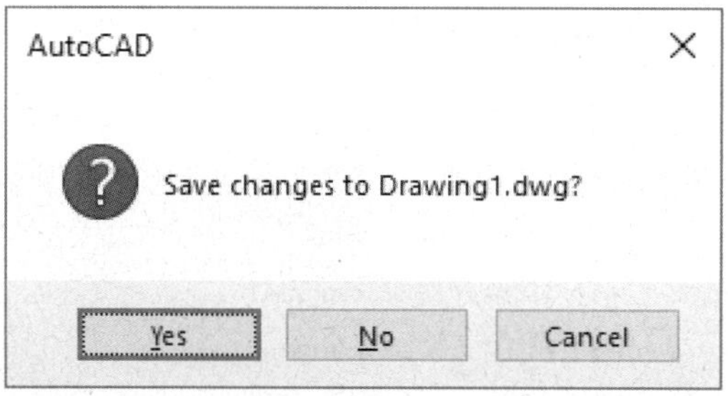


### *Closeall*

| Ribbon | Menu Bar | Command (Type) | Alias (Type) | Shortcut |
|---|---|---|---|---|
| ... | *Window*<br>*Closeall* | *Closeall* | ... | ... |

The *Closeall* command closes all drawings currently open in your AutoCAD session. If any of the drawings have been changed but not saved, you are prompted to save or discard the changes for each drawing.

### *Closeallother*

| Ribbon | Menu Bar | Command (Type) | Alias (Type) | Shortcut |
|---|---|---|---|---|
| ... | ... | *Closeallother* | ... | ... |

Type *Closeallother* at the Command prompt to close all open drawings other than the current drawing.

### *Exit*

| Ribbon | Menu Bar | Command (Type) | Alias (Type) | Shortcut |
|---|---|---|---|---|
| ... | *File*<br>*Exit AutoCAD* | *Exit* | ... | *Ctrl+Q* |

This is the simplest method to use when you want to exit AutoCAD. If any changes have been made to the drawings since the last *Save, Exit* invokes a warning box asking if you want to *Save changes to . . .?* before ending AutoCAD (see Fig. 2-13). An alternative to using the *Exit* option is to use the standard Windows methods for exiting an application. That is, Select the "**X**" in the extreme upper-right corner of the AutoCAD window. Using this option is the same as using *Exit*.

### *Quit*

| Ribbon | Menu Bar | Command (Type) | Alias (Type) | Shortcut |
|---|---|---|---|---|
| ... | ... | *Quit* | *Exit* | ... |

The *Quit* command accomplishes the same action as *Exit*. *Quit* discontinues (exits) the AutoCAD session and usually produces the dialog box shown in Figure 2-13 if changes have been made since the last *Save*.

### *Partialopen*

| Ribbon | Menu Bar | Command (Type) | Alias (Type) | Shortcut |
|---|---|---|---|---|
| ... | *File*<br>*Open*<br>*Partial Open* | *Partialopen* | ... | ... |

This feature in AutoCAD speeds up and simplifies loading, viewing, and working with large drawings. With *Partialopen* you can select which views and layers, and therefore which geometry, you want to load. This feature frees up system memory, speeds up regenerations, and prevents having to view and manipulate unneeded layers and geometry.

There are two ways you can use *Partialopen*: entering the command at the Command line or using the *Open* command to access the *Partial Open* dialog box interface. Generally you should use the *Open* command to access the *Select File* dialog box, then the *Partial Open* dialog box. Use the *Open* drop-down list in the lower-right corner of the *Select File* dialog box to access *Partial Open* (Fig. 2-14).

FIGURE 2-14

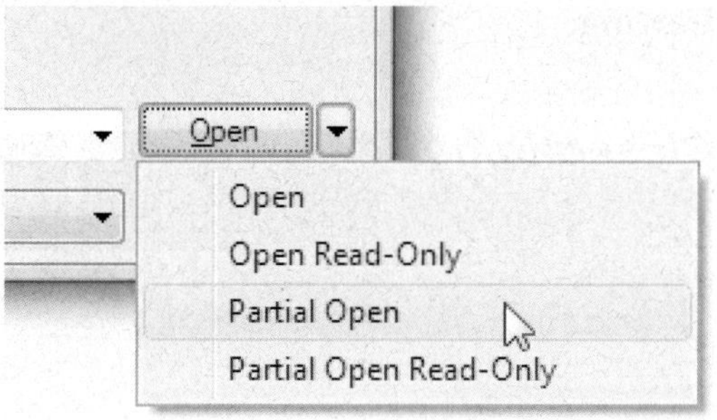


Choosing the *Partial Open* option from the *Select File* dialog box produces the *Partial Open* dialog box (Fig. 2-15). The dialog box allows you to select a *View* (if a named *View* has previously been saved in the drawing) and select which layers you want to open. (See Chapter 10 and Chapter 11 for information on views and layers.) You can select a range of layers by holding down Shift when selecting a layer at both ends of the range.

FIGURE 2-15

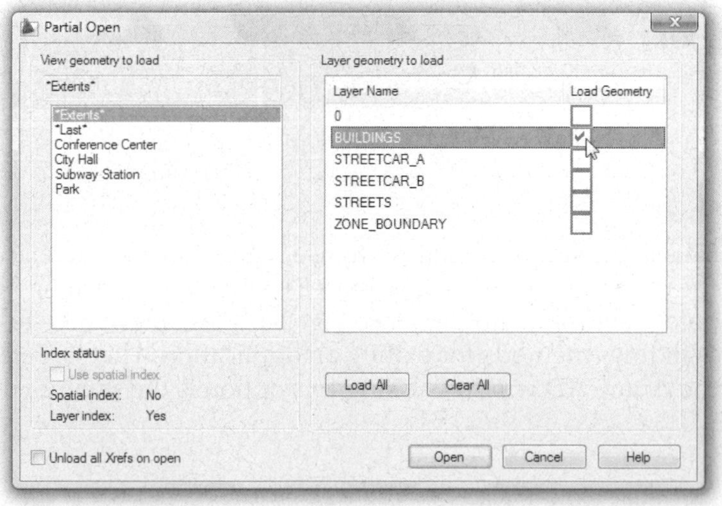

When a drawing is partially open, only the geometry on the selected layers is visible and can be edited. Although all of the layer names will appear in any list of layers in the drawing, only the geometry on the selected layers is actually loaded. In addition, all other named objects in the drawing (blocks, dimension styles, layouts, linetypes, text styles, UCSs, views, and viewport configurations) are loaded into AutoCAD.

For example, assume a drawing of the downtown area was partially opened. Selecting only the BUILDINGS layer (see Fig. 2-15) may yield a display as shown in Figure 2-16 revealing only the objects on the BUILDINGS layer.

**FIGURE 2-16**

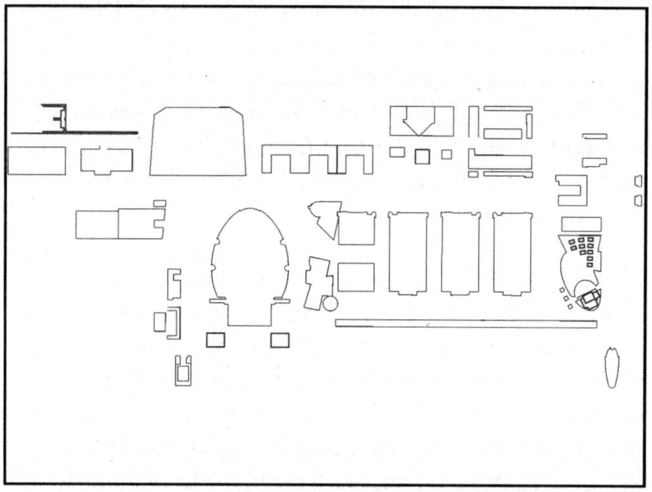

You can also use Spatial and Layer indexing (see Fig. 2-15, lower-left corner) if the drawing's *INDEXCTL* variable is set to allow "demand loading." This feature speeds loading and saves system memory for large drawings. Also, any attached *Xref* drawings can be unloaded if desired (see Chapter 30, Xreferences).

If you choose to enter the *Partialopen* command at the Command line, a text interface appears instead of the dialog box.

## Partiaload

| Ribbon | Menu Bar | Command (Type) | Alias (Type) | Shortcut |
|---|---|---|---|---|
| ... | *File*<br>*Partial Load* | *Partiaload* | ... | ... |

You must have a partially opened drawing for *Partiaload* to operate (see *Partialopen* described previously). *Partiaload* loads additional geometry into a partially opened drawing.

Accessing this command by a menu or by typing *Partiaload* produces the *Partial Load* dialog box. Here you select the additional layers or another area of the drawing you want to open. For example, Figure 2-17 displays the *Partial Load* dialog box displaying *Views* and *Layers* from the same drawing shown in Figure 2-16.

**FIGURE 2-17**

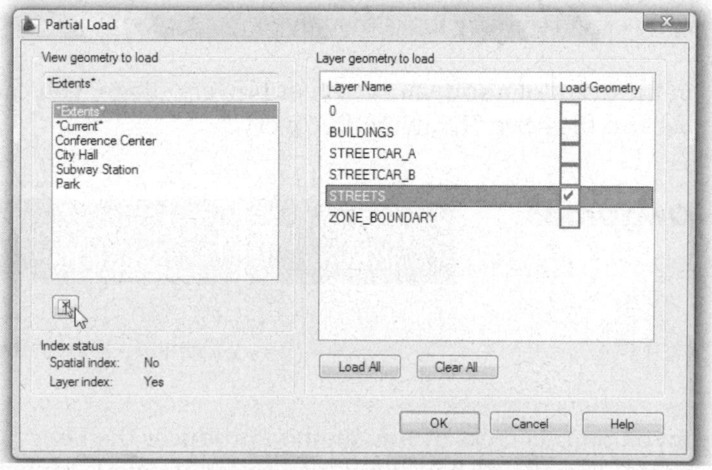

In this case, the STREETS layer is loaded to reveal the additional geometry (see Fig. 2-18). Any information that is loaded into the file using *Partiaload* cannot be unloaded, not even with *Undo*. The *Partial Load* dialog box is essentially the same as the *Partial Open* dialog box (see Fig. 2-15) with the exclusion of the *Index* and *Xref* options in the lower-left corner.

FIGURE 2-18

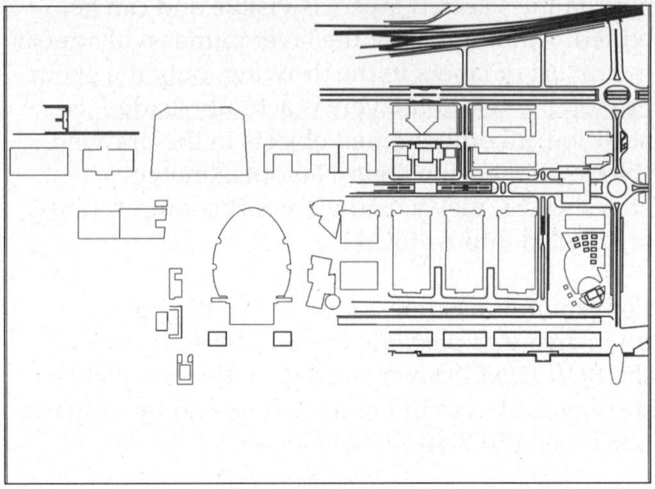

Notice the *Pick a Window* button in the lower-left corner of the dialog box. This feature allows you to return to the drawing and define an area (window) of the drawing to view. All geometry within and crossing the windowed area for the defined layers appears in the drawing. For example, the *Pick a Window* feature was used to select the right side of the drawing shown in Figure 2-18. Notice how the STREETS layer is displayed only on the right side of the drawing. This feature allows you to view two or more areas of the same drawing, each displaying a different set of layers.

If you enter *-Partiaload* at the Command line (with the hyphen prefix), the text interface appears at the Command line.

## Recover

| Ribbon | Menu Bar | Command (Type) | Alias (Type) | Shortcut |
|---|---|---|---|---|
| ... | *File*<br>*Drawing Utilities >*<br>*Recover...* | *Recover* | ... | ... |

This option is used only in the case that *Open* does not operate on a particular drawing because of damage to the file. Damage to a drawing file can occur from improper exiting of AutoCAD (such as power failure) or from damage to a drive. *Recover* can usually reassemble the file to a usable state and reload the drawing. The *Recover* command produces the *Select File* dialog box. Select the drawing to recover. A recovery log is displayed in a text window reporting results of the recover operation.

In the event of a software crash or power outage, you can use the Drawing Recovery Manager to recover backup files. See "Drawing Recovery."

## Dwgprops

| Ribbon | Menu Bar | Command (Type) | Alias (Type) | Shortcut |
|---|---|---|---|---|
| ... | *File*<br>*Drawing Properties* | *Dwgprops* | ... | ... |

Invoking *Dwgprops* by any method produces the *Drawing Properties* dialog box (Fig. 2-19, on the next page). This dialog box allows you to input details for better management and tracking of drawings such as title, author, subject, and other keywords. The information stored in the *Drawing Properties* dialog box should be specified by you while other information, such as time and date of last save, is read from the system. You must *Open* the drawing to input data into the user fields and *Save* the drawing to save the properties you enter.

The drawing properties can be read from Windows Explorer by right-clicking a .DWG file and selecting *Properties*. Using an AutoCAD feature called DesignCenter, unopened drawings can be located by finding keywords stored in the drawing properties (see Chapter 21). The four tabs in the *Drawing Properties* dialog box are explained here.

### *General* Tab

This tab (Fig. 2-19) displays the drawing type, location, size, and other information. All the information here is supplied by the operating system; therefore, all fields are read-only.

FIGURE 2-19

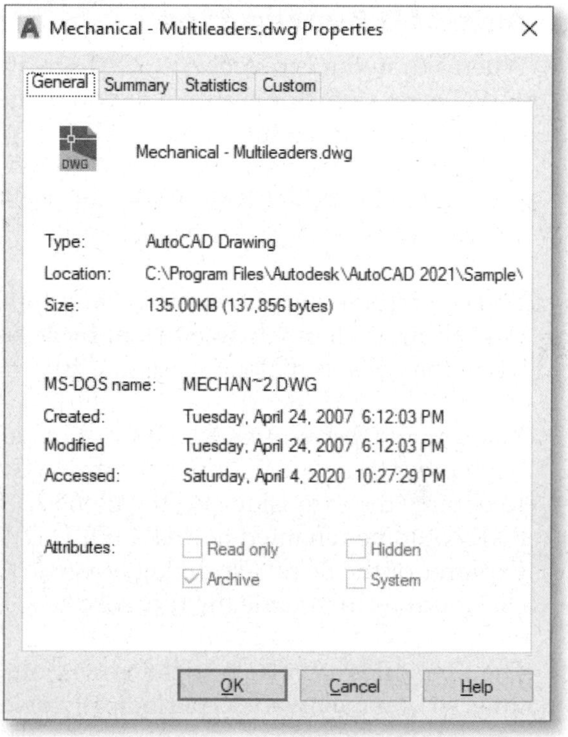


### *Summary* Tab

This tab is used to input information you want to use to track the drawing. Data you enter is viewable through AutoCAD's DesignCenter and Windows Explorer *Properties*. You can enter the drawing title, subject, author, keywords, comments, and a hyperlink base. The *Hyperlink Base* is a base address for all relative links you insert within the drawing using the *Hyperlink* command (see Chapter 23). You can enter a URL (an Internet address) or a path to a folder on a network drive in the *Hyperlink Base* field.

### *Statistics* Tab

This tab shows read-only data such as file size and the dates that files were created and last modified (not shown). You can search for all files created at a certain time or modified on a certain date. The *Created*, *Modified*, and *Total Editing Time* values are stored in the *TDCREATE*, *TDUPDATE*, and *TDINDWG* read-only system variables.

### *Custom* Tab

The custom fields can also be used in searches to help locate the drawing using the *Find* dialog box of AutoCAD DesignCenter (not shown). AutoCAD also provides access to these properties using its native programming language, AutoLISP. Enter any data relevant to the drawing that may be useful in a search.

## ***SAVETIME***

*SAVETIME* is actually a system variable that controls AutoCAD's temporary save feature. AutoCAD automatically saves a temporary drawing file for you at time intervals that you specify. The default time interval is 10 (minutes). A value of 0 disables this feature. The value for *SAVETIME* can also be set in the *Open and Save* tab of the *Options* dialog box (see Fig. 2-20, on the next page).

When automatic saving occurs, the drawing is saved under the current name with a randomly generated suffix and a .SV$ extension is assigned. For example, if the Electrical.DWG drawing were automatically saved, the assigned file might be Electrical_1_1_9912.SV$. In this way the current drawing name is not overwritten. The path (location) of the automatically saved file can be specified using the *SAVEFILEPATH* system variable or the *Files* tab of the *Options* dialog box.

NOTE: The automatically saved file is only a temporary file. That is, once you save or close the drawing, the temporary save file is automatically deleted. Therefore, this feature is useful when you experience an improper shutdown of AutoCAD (such as a power outage). In this case, the temporary file can be retrieved and renamed so all of your work is not lost. See "Drawing Recovery."

## AutoCAD Backup Files

When a drawing is saved, AutoCAD creates a file with a .DWG extension. For example, if you name the drawing PART1, using *Save* creates a file named PART1.DWG. The next time you save, AutoCAD makes a new PART1.DWG and renames the old version to PART1.BAK. Only one .BAK (backup) file is kept automatically by AutoCAD by default.

You can disable the automatic backup function by changing the *ISAVEBAK* system variable to 0 or by accessing the *Open and Save* tab in the *Options* dialog box (Fig. 2-20).

You cannot *Open* a .BAK file. It must be renamed to a .DWG file. Remember that you already have a .DWG file by the same name, so rename the extension and the filename. For example, PART1. BAK could be renamed to PART1OLD.DWG. Use Windows Explorer or the *Select File* dialog box (use *Open*) with the right-click options to rename the file. Also see "Drawing Recovery."

**FIGURE 2-20**

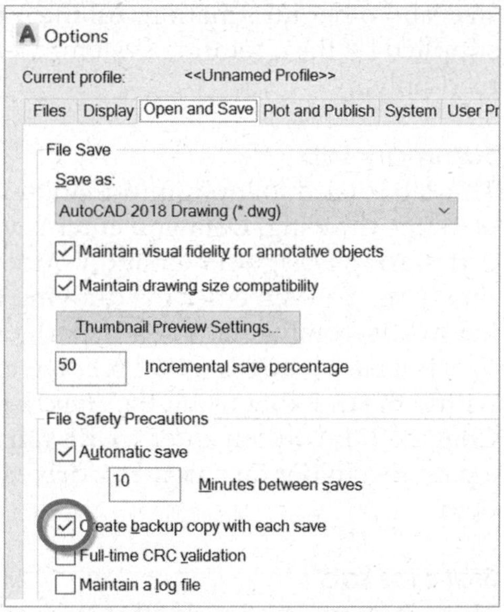


The .BAK files can also be deleted without affecting the .DWG files. The .BAK files accumulate after time, so you may want to periodically delete the unneeded ones to conserve disk space.

## Drawing Recovery

In the case of an unexpected or accidental shutdown, such as a power outage, drawings that were open when the shutdown occurred can be recovered. If a shutdown does occur, restart your computer system and begin AutoCAD. A *Drawing Recovery* message automatically appears giving instructions on recovering the drawings that were open before the system shutdown. Click *OK* to bypass the message and begin recovery.

The *Drawing Recovery Manager* also automatically appears and allows you to recover any or all drawings that were open when the shutdown occurred (Fig. 2-21). Highlight each of the .DWG (drawing), .BAK (backup), and SV$ (temporary backup) files that appear in the *Backup Files* list to determine which has the most current *Last Saved* data. Then simply double-click on the most current file to *Open* the drawing.

**FIGURE 2-21**

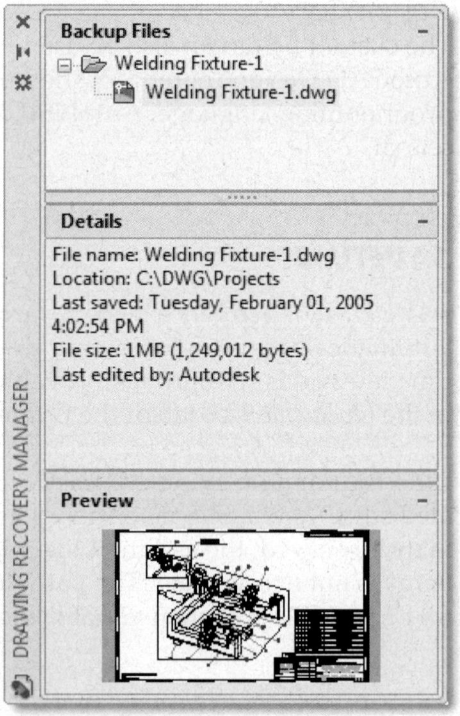

If you select a .BAK or .SV$ file, AutoCAD automatically locates the file from its previously stored location, renames it with a .DWG file extension, and relocates the file in the folder where the last .DWG was saved. In this case, you should then use *Saveas* to rename the file.

Although the *Drawing Recovery Manager* is available at any time by using the *Drawingrecovery* command or by selecting it from the *Drawing Utilities* Application Menu, the manager has no use until an accidental shutdown occurs. Invoke the command anytime after the shutdown to recover the drawings.

## AutoCAD Drawing File Management

TIP

AutoCAD drawing files should be stored in folders (directories) used exclusively for that purpose. For example, you may use Windows Explorer to create a new folder called "Drawings" or "DWG." This could be a folder in your Documents library or a folder in another appropriate location.

You can create folders for different types of drawings or different projects. For example, you may have a folder named "DWG" with several subfolders for each project, drawing type, or client, such as "COMPONENTS" and "ASSEMBLIES," or "In Progress" and "Completed," or "TWA," "Ford," and "Dupont." If you are working in an office or school laboratory, a directory structure most likely already has been created for you to use.

Safety and organization are important considerations for storage of AutoCAD drawing files. AutoCAD drawings should be saved to designated folders to prevent accidental deletion of system or other important files. If you work with AutoCAD long enough or if you work in an office or laboratory, most likely many drawings are saved on the computer hard drive or network drives. It is imperative that a logical directory structure be maintained so important drawings can be located easily and saved safely.

## Using Other File Formats

See Chapter 23 for file commands related to importing and exporting images and exchanging files in other than *.DWG format.

# CHAPTER EXERCISES

Start AutoCAD. Use the ACAD.DWT template. Type *Zoom*, then type *All*. NOTE: The chapter exercises in this book list file names in UPPERCASE letters for easy recognition. The file names and directory (folder) names on your system may appear as UPPERCASE, Title Case, or lower case.

1. **The Working Directory**

   In this exercise you will create or determine the name of the folder ("working" directory) for opening and saving AutoCAD files on your computer system. If you are working in an office or laboratory, a folder (directory) has most likely been created for saving your files. If you have installed AutoCAD yourself and are learning on your home or office system, you should create a folder for saving AutoCAD drawings. It should have a name like "C:\ACAD\DWG," "C:\Documents\Dwgs," or "C:\Acad\Drawing Files." (HINT: Use the ***Saveas*** command. The name of the folder last used for saving files appears at the top of the *Save Drawing As* dialog box.)

2. ***Save* and name a drawing file**

   Draw 2 vertical ***Lines***. Select ***Save*** from the Application Menu or by another method. (The *Save Drawing As* dialog box appears since a name has not yet been assigned.) Name the drawing "**CH2 VERTICAL.**"

3. **Using *Qsave***

   Draw 2 more vertical ***Lines***. Select ***Save*** again from the menu or by any other method except typing. (Notice that the *Qsave* command appears at the Command line since the drawing has already been named.) Draw 2 more vertical ***Lines*** (a total of six lines now). Select ***Save*** again. Do not *Close.*

4. **Start a *New* drawing**

   Invoke ***New*** from the Application Menu or by any other method. Use the ACAD.DWT template. Draw 2 horizontal ***Lines.*** Use ***Save***. Enter "**CH2 HORIZONTAL**" as the name for the drawing. Draw 2 more horizontal ***Lines,*** but <u>do not</u> *Save*. Continue to exercise 5.

5. ***Close* the current drawing**

   Use ***Close*** to close CH2 HORIZONTAL. Notice that AutoCAD first forces you to answer ***Yes*** or ***No*** to *Save changes to CH2 HORIZONTAL.DWG?* PICK ***Yes*** to save the changes.

6. **Using *Saveas***

   The CH2 VERTICAL drawing should now be the current drawing. Draw 2 inclined (angled) ***Lines*** in the CH2 VERTICAL DRAWING. Invoke ***Saveas*** to save the drawing under a new name. Enter "**CH2 INCLINED**" as the new name. Notice the current drawing name displayed in the AutoCAD title bar (at the top of the screen) is reset to the new name. Draw 2 more inclined ***Lines*** and ***Save.***

7. ***Open* an AutoCAD sample drawing**

   ***Open*** a drawing named "**HOUSE DESIGNER.DWG**" usually located in the "C:\Program Files\Autodesk\AutoCAD 2021\Sample\en-us\DesignCenter" folder. Use the *Look in:* section of the dialog box to change drive and folder if necessary. Pick ***Yes*** to open the file read-only, which does not let you change or save the drawing. Practice the ***Open*** command by looking at other sample drawings in the Sample folder. Finally, ***Close*** all the sample drawings.

8. **Check the *SAVETIME* setting**

   At the command prompt, enter ***SAVETIME***. The reported value is the interval (in minutes) between automatic saves. Your system may be set to 10 (default setting). If you are using your own system, change the setting to **5**.

9. **Find and rename a backup file**

   ***Close*** all open drawings. Use the ***Open*** command. When the *Select File* dialog box appears, locate the folder where your drawing files are saved, then enter "***.BAK**" in the *File name:* edit box, without the quotes ( * is a wildcard that means "all files"). All of the .BAK files should appear. Search for a backup file that was created the last time you saved **CH2 VERTICAL.DWG,** named **CH2 VERTICAL.BAK.** **Right-click** on the file name. Select ***Rename*** from the shortcut menu and rename the file "**CH2 VERT 2.DWG.**" Select ***Yes*** in the rename warning dialog box. Next, enter ***.DWG** in the *File name:* edit box to make all the .DWG files reappear. Highlight **CH2 VERT 2.DWG** and the thumbnail image should appear in the preview display. You can now ***Open*** the file if you wish.

CHAPTER 3

# Draw Command Concepts

## Chapter Objectives

After completing this chapter you should:

1. be able to understand the formats and types of coordinates;
2. be able to draw lines and circles using mouse input with drawing aids such as *Snap* and *Grid*;
3. be able to draw lines and circles using *Dynamic Input*;
4. be able to draw lines and circles using Command Line Input;
5. be able to use *Polar Tracking* with the three coordinate entry methods.

## AutoCAD OBJECTS

The smallest component of a drawing in AutoCAD is called an object (sometimes referred to as an entity). An example of an object is a *Line*, an *Arc*, or a *Circle* (Fig. 3-1). A rectangle created with the *Line* command contains four line objects.

Draw commands create objects. The draw command names are the same as the object names.

Simple objects are *Point, Line, Arc,* and *Circle.*

Complex objects are shapes such as *Polygon, Rectangle, Ellipse, Polyline,* and *Spline* which are created with one command (Fig. 3-2). Even though they appear to have several segments, they are treated by AutoCAD as one object.

**FIGURE 3-1**

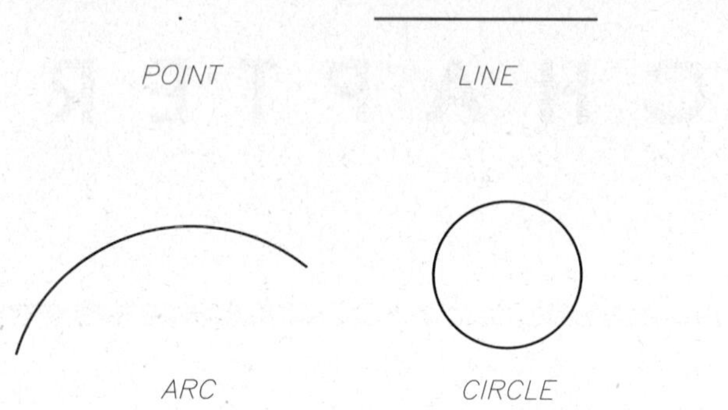


**FIGURE 3-2**

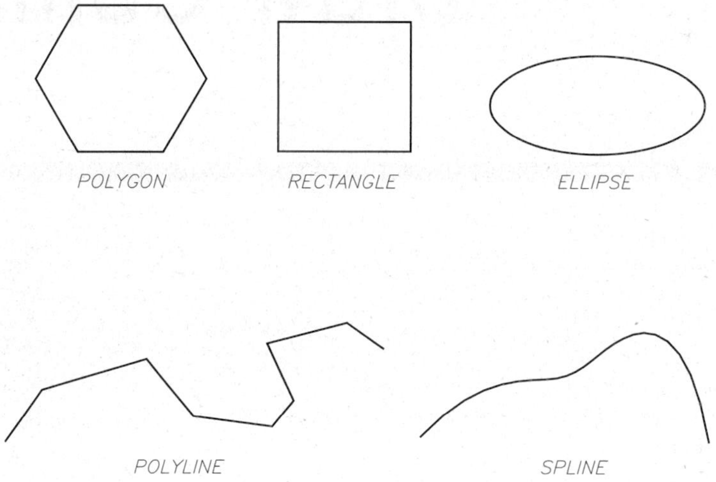


**TIP**

It is not always apparent whether a shape is composed of one or more objects. However, if you "hover over" an object with the cursor, an object is "highlighted," or shown in a shaded line pattern (Fig. 3-3). This highlighting reveals whether the shape is composed of one or several objects.

**FIGURE 3-3**

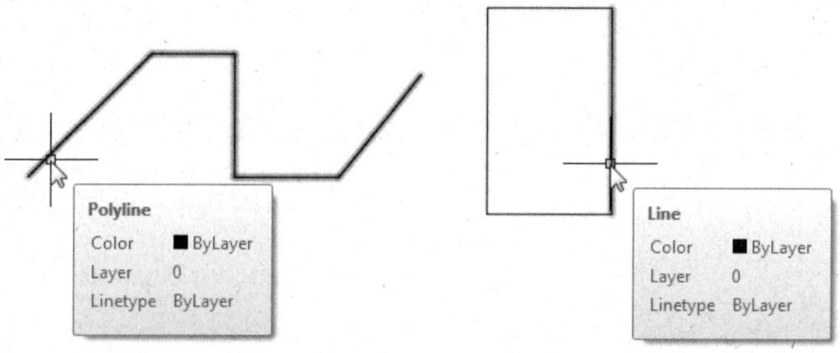


## LOCATING THE DRAW COMMANDS

To invoke *Draw* commands, any of these command entry methods can be used depending on your computer setup.

| | | |
|---|---|---|
| 1. | Keyboard | Type the command name or command alias using the keyboard. |
| 2. | Menu Bar | Select the command or dialog box from a pull-down menu. |
| 3. | Tools (icon buttons) | Select the command or dialog box by PICKing a tool (icon button) from a toolbar, palette, or panel on the Ribbon. |

To start a draw command, you can select the tool from the Ribbon in the *Home* tab, *Draw* panel (Fig. 3-4). Note that some commands or command options may be located on a drop-down menu. For example, Figure 3-4 shows the options for *Circle* on a drop-down menu.

Alternately, you can type in the command or command alias at the Command line. For example, type *C* for the *Circle* command.

**FIGURE 3-4**

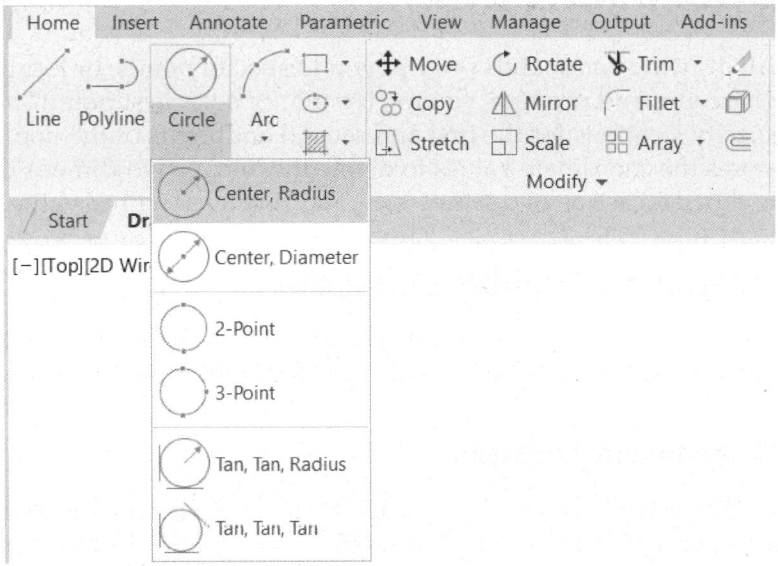


If the Menu Bar is activated, you can select draw commands from the *Draw* pull-down menu (Fig. 3-5). Note that command options are located on fly-out menus.

**FIGURE 3-5**

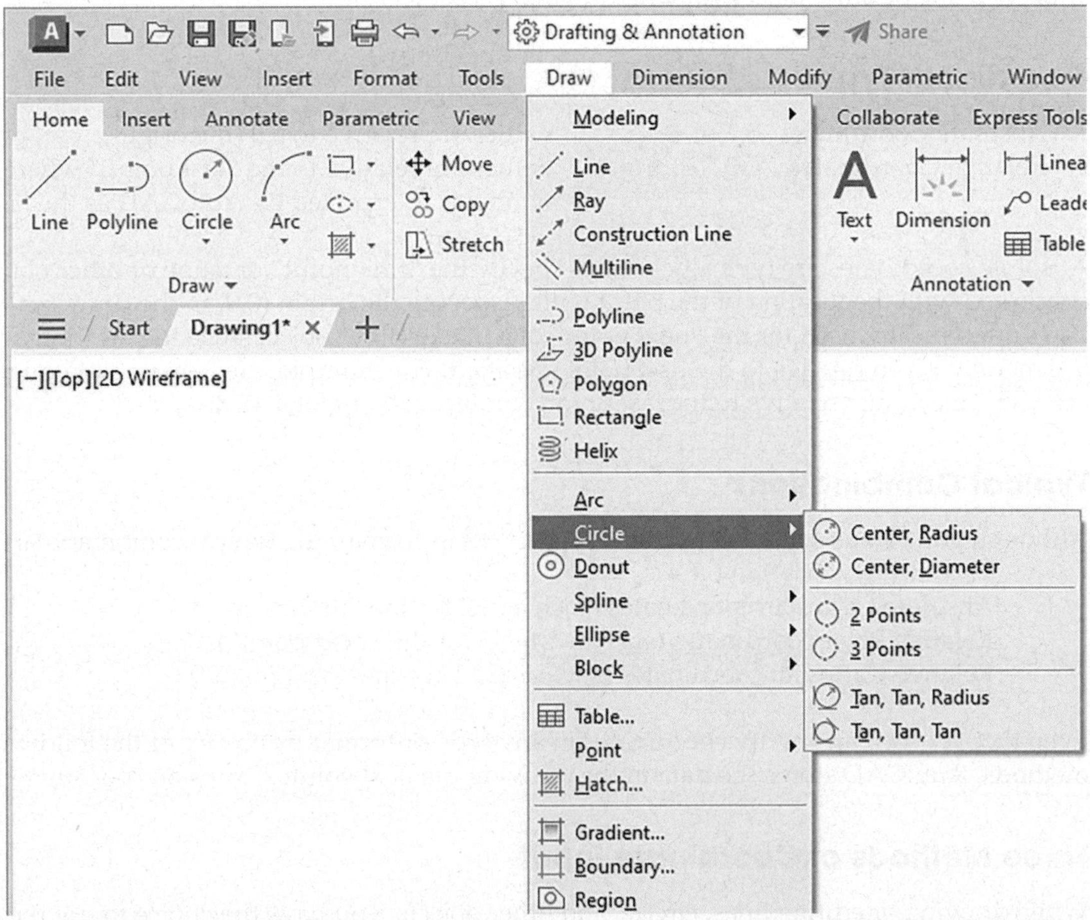


You can also select tools from the *Draw* toolbar if activated (not shown). See Chapter 1, Getting Started, for information on activating the Menu Bar and toolbars.

# COORDINATE ENTRY

All drawing commands prompt you to specify points, or locations, in the drawing. For example, the *Line* command prompts you to give the "Specify first point:" and "Specify next point:," expecting you to specify locations for the first and second endpoints of the line. After you specify those points, AutoCAD stores the coordinate values to define the line. A two-dimensional line in AutoCAD is defined and stored in the database as two sets of X, Y, and Z values (with Z values of 0), one for each endpoint.

## Coordinate Formats and Types

There are many ways you can specify coordinates, or tell AutoCAD the location of points, when you draw or edit objects. You can also use different types and formats for the coordinates that you enter.

### Coordinate Formats

| | | |
|---|---|---|
| 1. Cartesian Format | 3,7 | Specifies an X and Y value. |
| 2. Polar Format | 6<45 | Specifies a distance and angle, where 6 is the distance value, the "less than" symbol indicates "angle of," and 45 is the angular value in degrees. |

The Cartesian format is useful when AutoCAD prompts you for a point and you know the exact X,Y coordinates of the desired location. The Polar format is useful when you know the specific length and angle of a line or other object you are drawing.

### Coordinate Types

| | | |
|---|---|---|
| 1. Absolute Coordinates | 3,7 or 6<45 | Values are relative to 0,0 (the origin). |
| 2. Relative Coordinates | @3,7 or @6<45 | Values are relative to the "last point," where the "@" (at) symbol is interpreted by AutoCAD as the last point specified. |

Absolute coordinates are typically used to specify the "first point" of a line or other object since you are concerned with the location of the point with respect to the origin (0,0) in the drawing. Relative coordinates are typically used for the "next point" of a line or other object since the location of the next point is generally given <u>relative</u> to the first point specified. For example, the "next point" might be given as "@6<45", meaning "relative to the last point, 6 units at an angle of 45 degrees."

### Typical Combinations

Although you can use either coordinate type in either format, the typical combinations are:

Absolute Cartesian coordinates (such as 3,7 for the "first point").
Relative Polar coordinates (such as @6<45 for the "next point").
Relative Cartesian coordinates (such as @3,7 for the "next point").

Even though you can specify coordinates in any type or format using any of the following three methods, AutoCAD stores the data in the drawing file in absolute Cartesian coordinate format.

### Three Methods of Coordinate Input

While drawing or editing lines, circles, and other objects, you have the choice to use the following three methods to tell AutoCAD the location of points. Even though you may be simply picking points on the screen, technically you are entering coordinates.

1. Mouse Input — PICK (press the left mouse button to specify a point on the screen).
2. Dynamic Input — When Dynamic Input is on, you can key values into the edit boxes near the cursor and press Tab to change boxes.
3. Command Line Input — You can key in values in any format at the Command prompt.

## Mouse Input

To use mouse input simply move the cursor and press the left mouse button to pick points. You can use the mouse to input points at any time regardless of the Status Bar settings; however, you should always use a drawing aid with mouse input, such as *Snap* and *Grid* or *Object Snap*. With *Snap* and *Grid* on, you can watch the coordinate display in the bottom center of the screen to locate points (see Fig. 3-6).

**FIGURE 3-6**

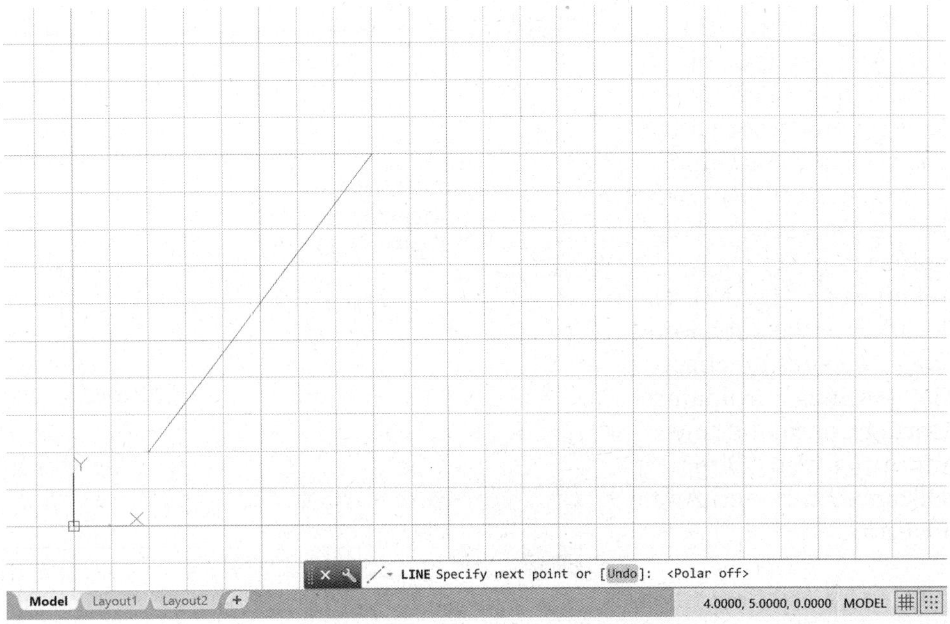


For example, to draw a Line from 1,1 to 4,5, first turn on Snap and Grid. Then invoke the Line command and at the "Specify first point:" prompt, watch the coordinate display while moving the cursor to location 1,1, then press the left mouse button to "pick" that point. At the "Specify next point:" prompt, locate position 4,5 and pick (see Fig. 3-6).

**FIGURE 3-7**

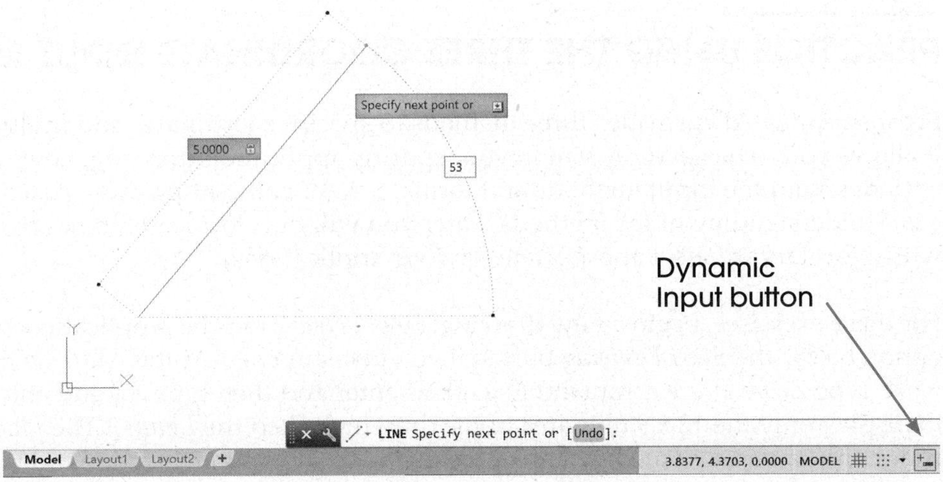


### *Dynamic Input*

Turn on *Dynamic Input* by clicking the *Dynamic Input* button on the Status Bar (bottom-right, Fig. 3-7). You can then type values into the input boxes that appear in the drawing area. Press the Tab key to change edit boxes for input, then press Enter to complete the entries. After pressing Enter, the values appear on the Command line. By default, *Dynamic Input* is in an absolute Cartesian format for the "first point" and a relative polar format when entering the "next point" (see Fig. 3-7). With *Dynamic Input*, you do not use the "@" (at) symbol to specify relative coordinates. You can change *Dynamic Input* to Cartesian or relative format if desired. The prompt line near the cursor (for example "Specify next point or") can also be turned off (see "*Dynamic Input* Settings").

For example, using *Dynamic Input* to draw a *Line* starting at 1,1 and having a length of 5 units and an angle of 53 degrees, first invoke the *Line* command. Then, at the "Specify first point:" prompt, key

"1" into the first edit box, press Tab, then key "1" into the second edit box, then Enter. Finally, at the "Specify next point:" prompt, key "5" into the first edit box, press Tab, then key "53" into the second edit box, then Enter. For some applications you may prefer instead to key a value into only one edit box, press Enter, and specify the other value (length or angle) using mouse input. See also "Direct Distance Entry and Angle Override."

### Command Line Input

You can enter coordinate values in any type or format at the Command line. If you enter relative coordinates, you must key in the "@" (at) symbol before Cartesian or polar values to make them relative (see Fig. 3-8). The location of the cursor on the screen has no effect on entering coordinates during Command line input. See also "Direct Distance Entry and Angle Override."

**FIGURE 3-8**

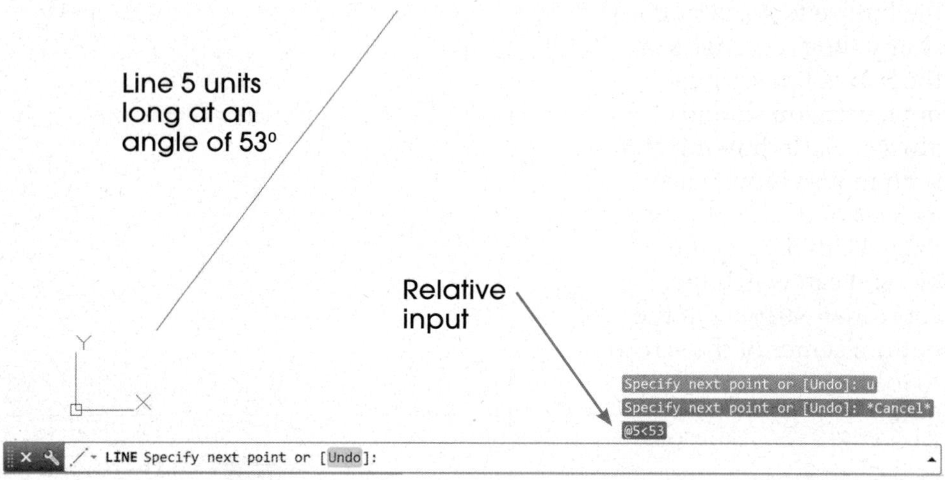


NOTE: To use Command line input, *Dynamic Input* can be on or off. If *Dynamic Input* is off, values are entered directly at the Command line. However, if *Dynamic Input* is on, the values entered appear in the *Dynamic Input* edit boxes but do not actually appear on the Command line until after pressing Enter.

## PRACTICE USING THE THREE COORDINATE INPUT METHODS

Because AutoCAD provides three methods to specify coordinates and multiple formats for coordinates, it allows you to be very efficient in your specific application drawing; however, it will take some practice to understand the input methods and formats. After completing these practice exercises, you will have a basic understanding of the methods. Later you will develop a preference for which input methods and which formats are most appropriate for your applications.

For these exercises, begin a new drawing. Select *New* from the Application Menu or use the *Templates* option below the *Start Drawing* button if you just launched AutoCAD. Use the ACAD.DWT template. Next, type *Zoom* at the command line, press enter and then type *All* and enter. (This action zooms in to display only the 12 x 9 unit area of the drawing called the *Limits*.) The *Line* command and the *Circle* command can be activated by any of the methods shown in the command tables.

### *Line*

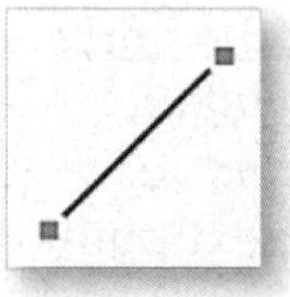

| Ribbon | Menu Bar | Command (Type) | Alias (Type) | Shortcut |
|---|---|---|---|---|
| *Home*<br>*Draw*<br>*Line* | *Draw*<br>*Line* | *Line* | L | ... |

**Circle**

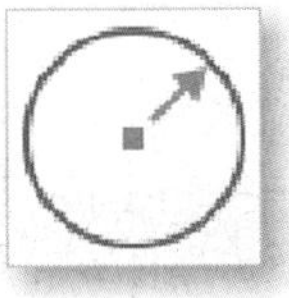

| Ribbon | Menu Bar | Command (Type) | Alias (Type) | Shortcut |
|---|---|---|---|---|
| *Home Draw Circle* | *Draw Circle >* | *Circle* | C | ... |

## Using Mouse Input to Draw Lines and Circles

You can use the mouse to pick points at any time, no matter if *Dynamic Input* or any other drawing aids appearing on the Status Bar are on or off. However, for clarity it is best to use the settings given on this page for this exercise.

**FIGURE 3-9**

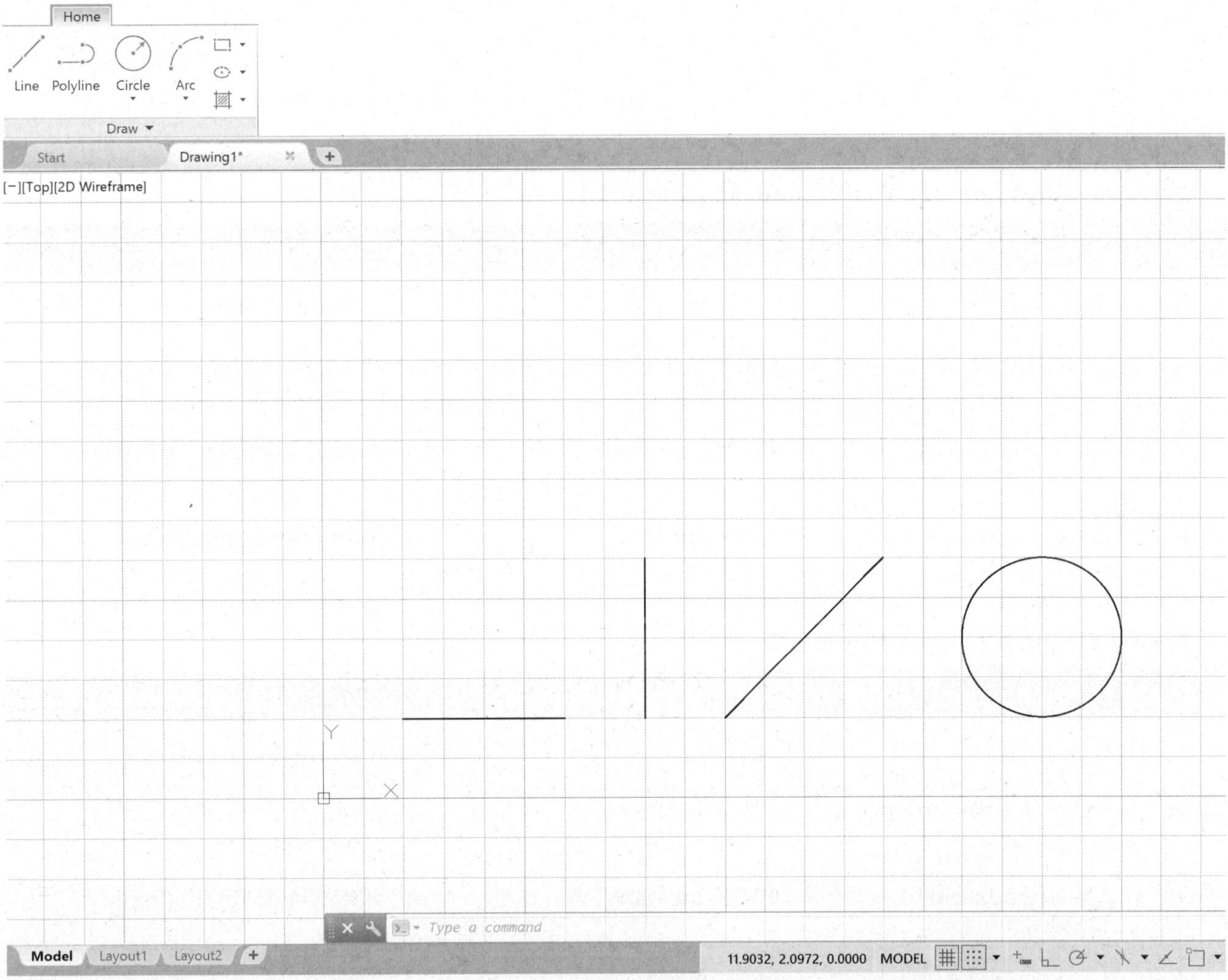


Status Bar settings to begin this exercise: *Model* tab should be active, *Snap* and *Grid* should be on (the buttons appear blue when using default colors). All other settings (*Ortho, Polar Tracking, Object Snap, Object Snap Tracking, Dynamic Input)*, and so forth should be off (see Fig. 3-9). Also ensure *Grid Snap* is on, not *Polar Snap*, by right-clicking on the *Snap* icon and selecting *Grid Snap*. Remember that AutoCAD stores coordinate values to an accuracy of 14 places, so it is important to use *Snap* and *Grid* or some other appropriate drawing aid with mouse input to assist you in picking exact points.

In the following 4 exercises you will draw three *Lines* and one *Circle* in specific locations as shown in Figure 3-9. Begin by following the steps below to construct the first horizontal *Line*.

**1. Draw a horizontal *Line* of 2 units length starting at 1,1.**

| Steps | Command Prompt | Perform Action | Comments |
|---|---|---|---|
| 1. | Command: | select or type ***Line*** | use any method |
| 2. | LINE Specify first point: | **PICK** location 1,1 | watch coordinate display, lower left |
| 3. | Specify next point or [Undo]: | **PICK** location 3,1 | watch coordinate display |
| 4. | Specify next point or [Undo]: | press **Enter** | completes command |

Check your work with the first *Line* as shown in Figure 3-9. Then proceed to construct the other two *Lines* and the *Circle*.

**2. Draw a vertical *Line* of 2 units length starting at 4,1.**

| Steps | Command Prompt | Perform Action | Comments |
|---|---|---|---|
| 1. | Command: | select or type ***Line*** | use any method |
| 2. | LINE Specify first point: | **PICK** location 4,1 | watch coordinate display, lower left |
| 3. | Specify next point or [Undo]: | **PICK** location 4,3 | watch coordinate display |
| 4. | Specify next point or [Undo]: | press **Enter** | completes command |

**3. Draw a diagonal *Line* starting at 5,1.**

| Steps | Command Prompt | Perform Action | Comments |
|---|---|---|---|
| 1. | Command: | select or type ***Line*** | use any method |
| 2. | LINE Specify first point: | **PICK** location 5,1 | watch coordinate display, lower left |
| 3. | Specify next point or [Undo]: | **PICK** location 7,3 | watch coordinate display |
| 4. | Specify next point or [Undo]: | press **Enter** | completes command |

**4. Draw a *Circle* of 1 unit radius with the center at 9,2.**

| Steps | Command Prompt | Perform Action | Comments |
|---|---|---|---|
| 1. | Command: | select or type ***Circle*** | use any method |
| 2. | CIRCLE Specify center point: | **PICK** location 9,2 | watch coordinate display |
| 3. | Specify radius of circle... | **PICK** location 10,2 | completes command |

Check your work with the objects shown in Figure 3-9. Remember that when you use mouse input you should use *Snap* and *Grid*, *Object Snap*, or other appropriate drawing aids to ensure you pick exact points.

## Using *Dynamic Input* to Draw Lines and Circles

Turn on *Dynamic Input* by selecting the *Dynamic Input* button on the Status Bar (bottom of AutoCAD screen). For this exercise *Dynamic Input* should be on and all other settings (*Snap*, *Grid*, *Ortho*, *Polar Tracking*, etc.) should be off. Ensure the *Model* tab is active.

Using the default setting for *Dynamic Input*, the "first point" specified is absolute Cartesian format and the "next point:" is relative polar format.

In the next 4 exercises you will draw another set of three *Lines* and one *Circle* located directly above the previous set as shown in Figure 3-10.

**FIGURE 3-10**

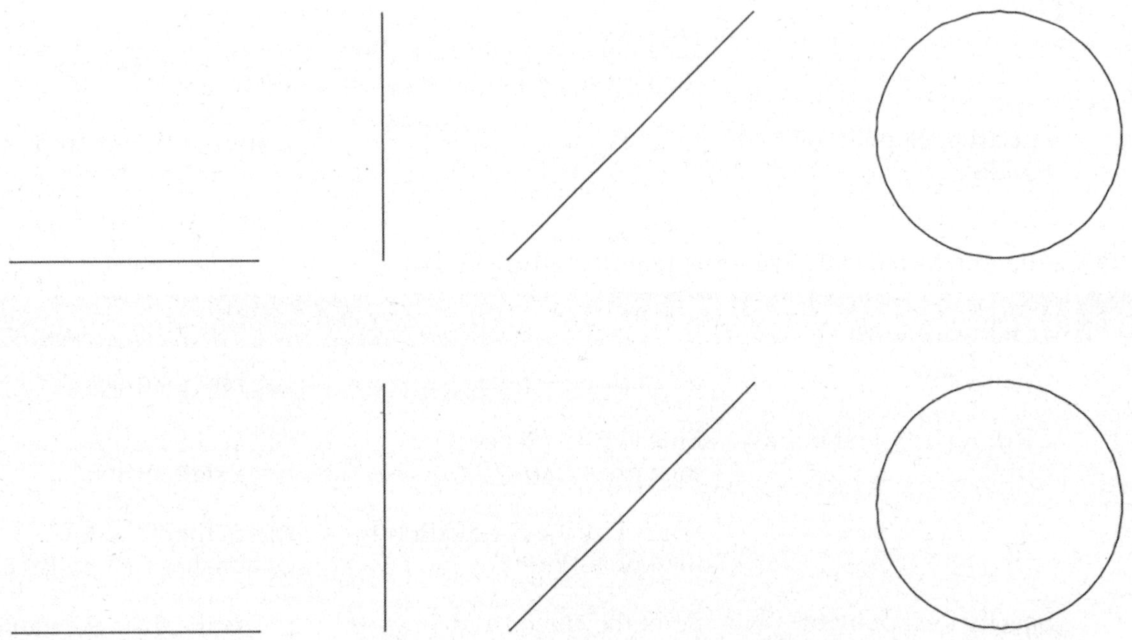

5. **Draw a horizontal *Line* of 2 units length starting at 1,4.**

| Steps | Command Prompt | Perform Action | Comments |
|---|---|---|---|
| 1. | Command: | select or type ***Line*** | use any method |
| 2. | LINE Specify first point: | Enter **1** into first edit box, then press **Tab** | this is the X value for Cartesian format |
| 3. | | Enter **4** into second edit box, then press **Enter** | this is the Y value; establishes first endpoint |
| 4. | Specify next point or [Undo]: | Enter **2** into first edit box, then press **Tab** | this is the distance value for relative polar format |
| 5. | | Enter **0** into second edit box, then press **Enter** | this is the angular value; establishes second endpoint |
| 6. | Specify next point or [Undo]: | press **Enter** | completes command |

6. **Draw a vertical *Line* of 2 units length starting at 4,4.**

| Steps | Command Prompt | Perform Action | Comments |
|---|---|---|---|
| 1. | Command: | select or type ***Line*** | use any method |
| 2. | LINE Specify first point: | Enter **4** into first edit box, then press **Tab** | this is the X value for Cartesian format |
| 3. | | Enter **4** into second edit box, then press **Enter** | this is the Y value; establishes first endpoint |
| 4. | Specify next point or [Undo]: | Enter **2** into first edit box, then press **Tab** | this is the distance value for relative polar format |
| 5. | | Enter **90** into second edit box, then press **Enter** | this is the angular value; establishes second endpoint |
| 6. | Specify next point or [Undo]: | press **Enter** | completes command |

7. **Draw a diagonal *Line* of 2.8284 units length starting at 5,4.**

| Steps | Command Prompt | Perform Action | Comments |
|---|---|---|---|
| 1. | Command: | select or type ***Line*** | use any method |
| 2. | LINE Specify first point: | Enter **5** into first edit box, then press **Tab** | this is the X value for Cartesian format |
| 3. | | Enter **4** into second edit box, then press **Enter** | this is the Y value; establishes first endpoint |
| 4. | Specify next point or [Undo]: | Point the cursor to the upper-right, then enter **2.8284** into first edit box, then press **Tab** | this is the distance value for relative polar format |

| | | | |
|---|---|---|---|
| 5. | | Enter 45 into second edit box, then press **Enter** | this is the angular value; establishes second endpoint |
| 6. | Specify next point or [Undo]: | press **Enter** | completes command |

See Figure 3-10 for an illustration of this exercise.

**8. Draw a *Circle* of 1 unit radius with the center at 9,5.**

| Steps | Command Prompt | Perform Action | Comments |
|---|---|---|---|
| 1. | Command: | select or type *Circle* | use any method |
| 2. | CIRCLE Specify first point... | Enter **9** into first edit box, then press **Tab** | this is the X value for Cartesian format |
| 3. | | Enter **5** into second edit box, then press **Enter** | this is the Y value; establishes circle center |
| 4. | Specify radius of circle... | Enter **1** into edit box, then press **Enter** | this is the distance value; completes command |

## Using Command Line Input to Draw Lines and Circles

You can use Command line input when *Dynamic Input* is on or off; however, for clarity in this exercise (Fig. 3-11) it is best to toggle *Dynamic Input* off. Although the settings for the other drawing aids do not affect Command line input, in this exercise it is best to set only the *Model* tab active and all other settings should be off. With Command line input you can key in coordinates at the Command line in any type or format. When using relative format, you must key in the "@" (at) symbol.

**FIGURE 3-11**

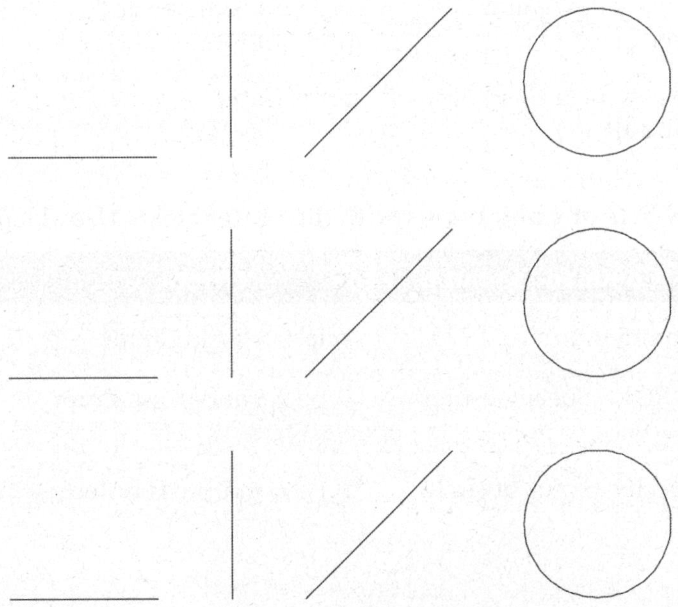

**9. Draw a horizontal *Line* of 2 units length starting at 1,7. Use absolute Cartesian coordinates.**

| Steps | Command Prompt | Perform Action | Comments |
|---|---|---|---|
| 1. | Command: | select or type ***Line*** | use any method |
| 2. | LINE Specify first point: | type **1,7** and press **Enter** | establishes first endpoint |
| 3. | Specify next point or [Undo]: | type **3,7** and press **Enter** | establishes second endpoint |
| 4. | Specify next point or [Undo]: | press **Enter** | completes command |

**10. Draw a vertical *Line* of 2 units length. Use relative Cartesian coordinates.**

| Steps | Command Prompt | Perform Action | Comments |
|---|---|---|---|
| 1. | Command: | select or type ***Line*** | use any method |
| 2. | LINE Specify first point: | type **@1,0** and press **Enter** | @ means "last point" |
| 3. | Specify next point or [Undo]: | type **@0,2** and press **Enter** | establishes second endpoint |
| 4. | Specify next point or [Undo]: | press **Enter** | completes command |

**11. Draw a diagonal *Line* of 2.8284 units length. Use absolute Cartesian and relative polar coordinates.**

| Steps | Command Prompt | Perform Action | Comments |
|---|---|---|---|
| 1. | Command: | select or type ***Line*** | use any method |
| 2. | LINE Specify first point: | type **5,7** and press **Enter** | establishes first endpoint |
| 3. | Specify next point or [Undo]: | type **@2.8284<45** and press **Enter** | < means "angle of"; establishes second endpoint |
| 4. | Specify next point or [Undo]: | press **Enter** | completes command |

**12. Draw a *Circle* of 1 unit radius with the center at 9,8. Use absolute Cartesian coordinates.**

| Steps | Command Prompt | Perform Action | Comments |
|---|---|---|---|
| 1. | Command: | select or type ***Circle*** | use any method |
| 2. | CIRCLE Specify center point... | type **9,8** and press **Enter** | establishes center |
| 3. | Specify radius of circle... | type **1** and press **Enter** | completes command |

## Direct Distance Entry and Angle Override

When prompted for a "next point," instead of using relative polar values preceded by the "@" (at) symbol to specify a distance and an angle, you can key in a distance-only value (called Direct Distance Entry) or key in an angle-only value (Angle Override). With these features, you can key in one value (either distance or angle) and specify the other value using mouse input. This capability represents an older, less graphic version of *Dynamic Input*. Both Direct Distance Entry and Angle Override can be used with Command Line Input and with *Dynamic Input*.

### Direct Distance Entry

Direct Distance Entry allows you to directly key in a distance value. Use Direct Distance Entry in response to the "next point:" prompt. With Direct Distance Entry, you point the mouse so the "rubberband" line indicates the angle of the desired line, then simply key in the desired distance value (length) at the Command line in response to the "next point:" prompt. For example, if you wanted to draw a horizontal line of 7.5 units, you would first turn on *Polar Tracking*, point the cursor in a horizontal direction, key in "7.5," and press Enter. Direct Distance Entry can be used with *Dynamic Input* on or off.

TIP

Direct Distance Entry is best used in conjunction with *Polar Tracking* (turn on *Polar Tracking* on the Status Bar). *Polar Tracking* assists you in creating a "rubberband" line that is horizontal, vertical, or at a regular angular increment you specify such as 30 or 45 degrees. Indicating the angle of the line with the mouse without using *Polar Tracking* or *Ortho* will not likely result in a precise angle. See "*Polar Tracking* and *Polar Snap*."

### Angle Override

If you want to specify that a line is drawn at a specific angle, but specify the line length using mouse input, use Angle Override. At the "next point" prompt, key in the "<" (less than) symbol and an angular value, such as "<30", then press Enter. AutoCAD responds with "Angle Override: 30." The rubberband line is then constrained to 30 degrees but allows you to pick a point to specify the length. Angle Override can be used with Command Line Input and with *Dynamic Input*.

## *Dynamic Input* Settings

The default setting for *Dynamic Input* is absolute Cartesian coordinate input for the "first point" and relative polar coordinates for the "second point." To change these settings, right-click on the icon on the Status Bar and select *Dynamic Input Settings*.... This action produces the *Dynamic Input* tab of the *Drafting Settings* dialog box (Fig. 3-12). You can also select *Drafting Settings* from the *Tools* pull-down menu or type the command alias *DS* to produce this dialog box.

The *Pointer Input* section allows you to change settings for the type and format of coordinates that are used during *Dynamic Input*. The *Dimension Input* section controls the appearance of the dimension lines that are shown around the edit boxes during the "next point" specification.

FIGURE 3-12

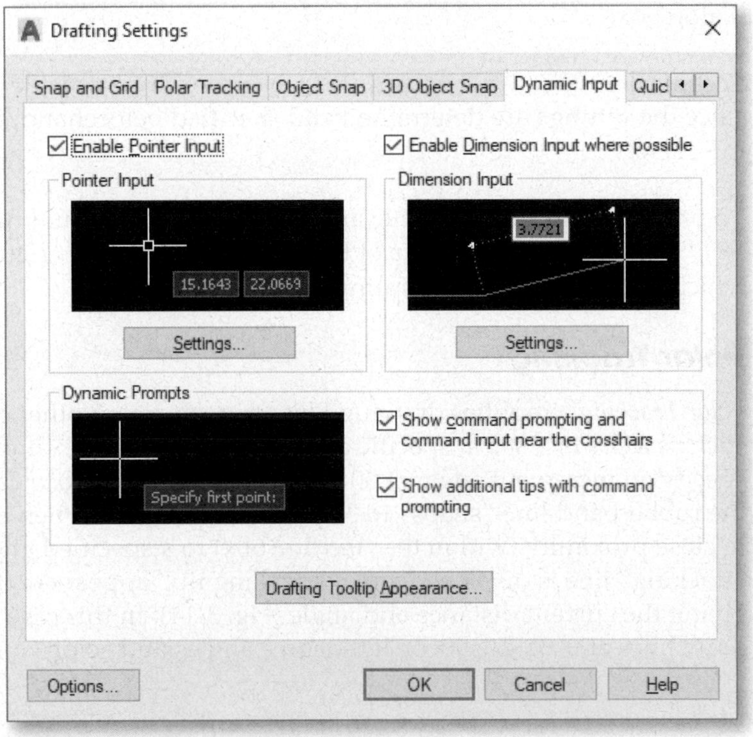

Select *Settings...* in the *Pointer Input* section to produce the *Polar Input Settings* dialog box (Fig. 3-13). Note that *Polar format* and *Relative coordinates* are the default settings for the "next point" specification. You cannot change the format for the "first point" specification, which is always absolute Cartesian format.

FIGURE 3-13

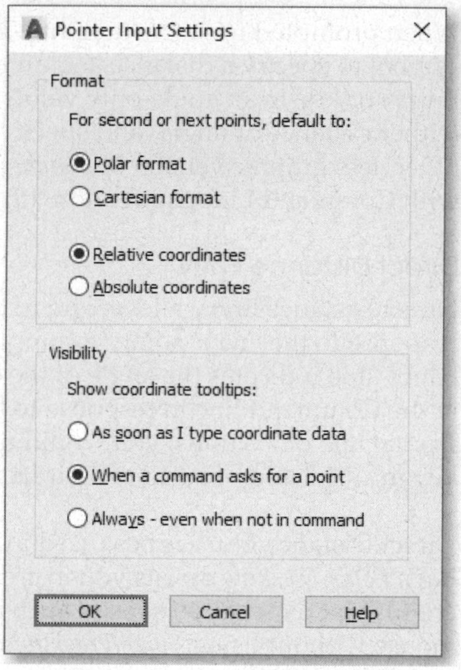


## POLAR TRACKING AND POLAR SNAP

Remember that all draw commands prompt you to specify locations, or coordinates, in the drawing. You can indicate these coordinates interactively (using the cursor), entering values (at the keyboard), or using a combination of both. Features in AutoCAD that provide an easy method for specifying coordinate locations interactively are *Polar Tracking* and *Polar Snap*. Another feature, *Object Snap Tracking*, is discussed in Chapter 7. These features help you draw objects at specific angles and in specific relationships to other objects.

*Polar Tracking* helps the rubberband line snap to angular increments such as 45, 30, or 15 degrees. *Polar Snap* can be used in conjunction to make the rubberband line snap to incremental lengths such as .5 or 1.0. You can toggle these features on and off with the *Snap* and *Polar Tracking* buttons on the Status Bar or by toggling F9 and F10, respectively. First, you should specify the settings in the *Drafting Settings* dialog box.

Technically, *Polar Tracking* and *Polar Snap* would be considered a variation of the Mouse Input method since the settings are determined and specified beforehand, then points on the screen are PICKed using the cursor.

To practice with *Polar Tracking* and *Polar Snap*, you should turn *Dynamic Input*, *Object Snap*, and *Object Snap Tracking* off. Since some of these features generate alignment vectors, it can be difficult to determine which of the features is operating when they are all on.

### Polar Tracking

*Polar Tracking* simplifies drawing *Lines* or performing other operations such as *Move* or *Copy* at specific angle increments. For example, you can specify an increment angle of 30 degrees. Then when *Polar Tracking* is on, the rubberband line "snaps" to 30-degree increments when the cursor is in close proximity (within the *Aperture* box) to a specified angle. A dotted "tracking" line is displayed and a "tracking tip" appears at the cursor giving the current distance and angle (Fig. 3-14). In this case, it is simple to draw lines at 0, 30, 60, 90, or 120-degree angles and so on.

FIGURE 3-14

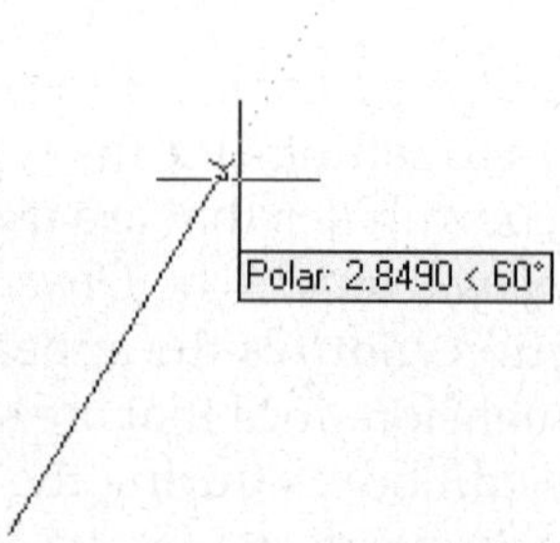

Figure 3-15 displays possible positions for *Polar Tracking* when a 30-degree angle is specified. Available angle options are 90, 45, 30, 22.5, 18, 15, 10, and 5 degrees, or you can specify any user-defined angle. This is a tremendous aid for drawing lines at typical angles.

FIGURE 3-15

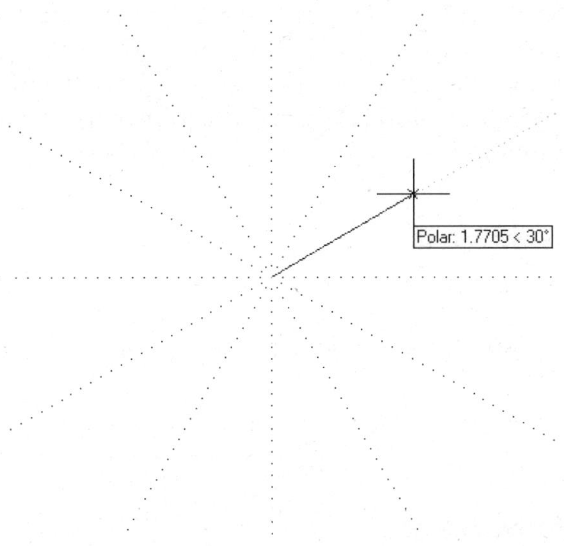


TIP

You can access the settings using the *Polar Tracking* tab of the *Drafting Settings* dialog box (Fig. 3-16). Invoke the dialog box by the following methods:

1. Enter *Dsettings* or *DS* at the Command line, then select the *Polar Tracking* tab.

2. Right-click on the icon on the Status Bar, then choose *Tracking Settings...* from the menu.

3. Select *Drafting Settings* from the *Tools* pull-down menu, then select the *Polar Tracking* tab.

In this dialog box, you can select the *Increment Angle* from the drop-down list (as shown) or specify *Additional Angles*. Use the *New* button to create user-defined angles. Highlighting a user-defined angle and selecting the *Delete* button removes the angle from the list.

The *Object Snap Tracking Settings* are used only with Object Snap (discussed in Chapter 7). When the *Polar Angle Measurement* is set to *Absolute*, the angle reported on the "tracking tip" is an absolute angle (relative to angle 0 of the current coordinate system).

FIGURE 3-16

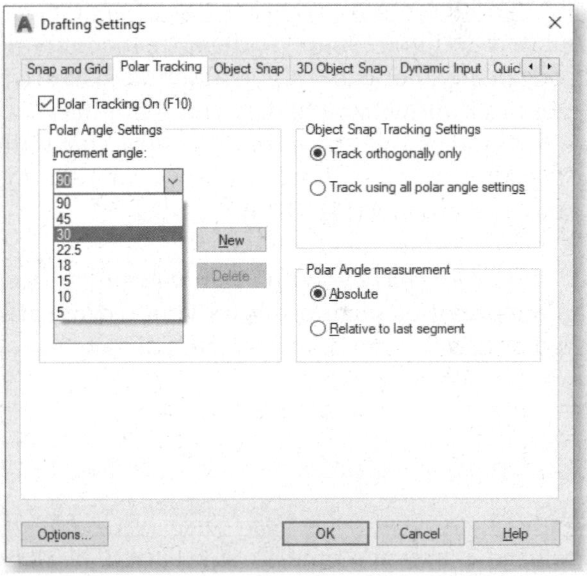


## *Polar Tracking* with Direct Distance Entry

With normal *Polar Tracking* (*Polar Snap* is off), the line "snaps" to the set angles but can be drawn to any length (distance). This option is particularly useful in conjunction with Direct Distance Entry since you specify the angular increment in the *Polar Tracking* tab of the *Drafting Settings* dialog box, but indicate the distance by entering values at the Command line. In other words, when drawing a *Line*, move the cursor so it "snaps" to the desired angle, then enter the desired distance at the keyboard (Fig. 3-17). See previous discussion, "Direct Distance Entry and Angle Override."

FIGURE 3-17

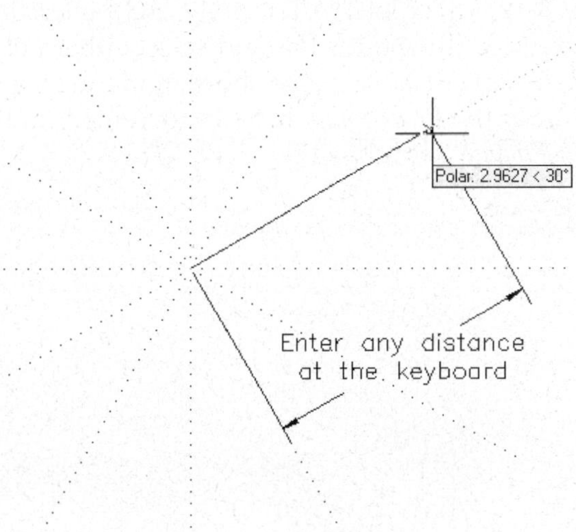


TIP

## ***Polar Tracking*** **Override**

When *Polar Tracking* is on, you can override the previously specified angle increment and draw to another specific angle. The new angle is valid only for one point specification. To enter a polar override angle, enter the left angle bracket (<) and an angle value whenever a command asks you to specify a point. The following command prompt sequence shows a 12-degree override entered during a *Line* command.

```
Command: LINE
Specify first point: PICK
Specify next point or [Undo]: <12
Angle Override: 12
Specify next point or [Undo]: PICK
```

*Polar Tracking* Override can be used with Command Line Input or *Dynamic Input*.

## ***Polar Snap***

*Polar Tracking* makes the rubberband line "snap" to angular increments, whereas *Polar Snap* makes the rubberband line "snap" to distance increments that you specify. For example, setting the distance increment to 2 allows you to draw lines at intervals of 2, 4, 6, 8, and so on. Therefore, *Polar Tracking* with *Polar Snap* allows you to draw at specific angular and distance increments (Fig. 3-18).

Using *Polar Tracking* with *Polar Snap* off, as described in the previous section, allows you to draw at specified angles but not at any specific distance intervals.

FIGURE 3-18

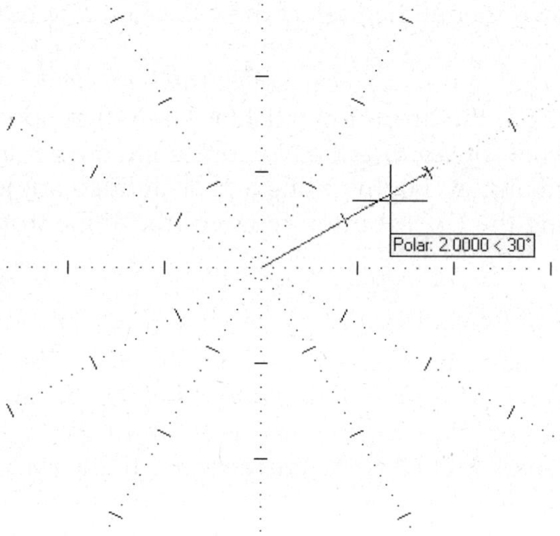


Only one type of *Snap*, *Polar Snap*, or *Grid Snap* can be used at any one time. Toggle either *Polar Snap* or *Grid Snap* on by setting the radio button in the *Snap and Grid* tab of the *Drafting Settings* dialog box (Fig. 3-19, lower left). Alternately, right-click on the *Snap* icon at the Status Bar and select either *Polar Snap* or *Grid Snap* from the shortcut menu (see Fig. 3-20). Set the *Polar Snap* distance increment in the *Polar Distance* edit box (Fig. 3-19, left-center).

FIGURE 3-19

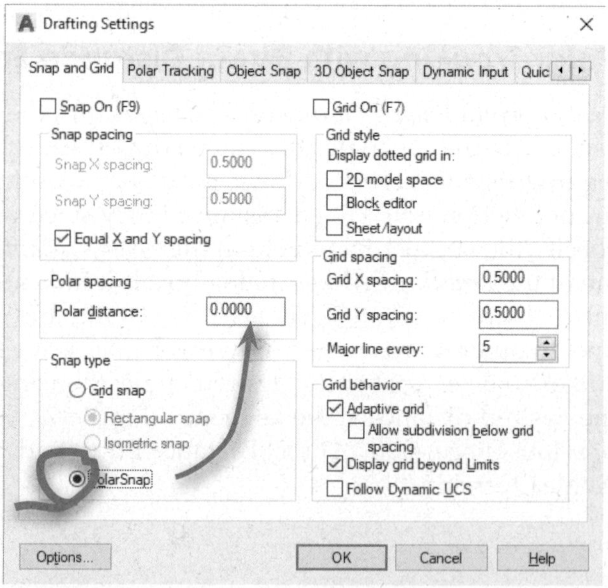

To utilize *Polar Snap*, both *Snap* (F9) and *Polar Tracking* (F10) must be turned on, *Polar Snap* must be on, and a *Polar Distance* value must be specified in the *Drafting Settings* dialog box. As a check, settings should be as shown in Figure 3-20 (*Snap* and *Polar Tracking* are on), with the addition of some *Polar Distance* setting in the dialog box. *Snap* turns on or off whichever of the two snap types is active, *Grid Snap* or *Polar Snap*. For example, if *Polar Snap* is the active snap type, toggling F9 turns only *Polar Snap* off and on.

**FIGURE 3-20**

## DRAWING LINES USING *POLAR TRACKING*

For these exercises you will use combinations of *Polar Tracking* with *Polar Snap*, *Grid Snap*, and *Dynamic Input*. Also, *Object Snap* and *Object Snap Tracking* should be turned off. Turn off *Object Snap* and *Object Snap Tracking* by toggling the *Object Snap* and *Object Snap Tracking* buttons on the Status Bar or using F3 and F11 to do so.

### Using *Polar Tracking* and *Grid Snap*

Draw the shape in Figure 3-21 using *Polar Tracking* in conjunction with *Grid Snap*. Follow the steps below.

1. Begin a *New* drawing. Use the **ACAD.DWT** template. Next, type **Z** at the command line (command alias for *Zoom*) and enter. Then enter ***A*** for the *All* option.

2. Toggle on *Grid Snap* (rather than *Polar Snap*) by right-clicking on the *Snap* icon on the Status Bar, then selecting ***Grid Snap*** from the menu.

3. Turn on *Polar Tracking* and make the appropriate settings. Do this by right-clicking on the *Polar Tracking* icon on the Status Bar (Fig. 3-22). Set the *Increment Angle* to **45**.

**FIGURE 3-21**

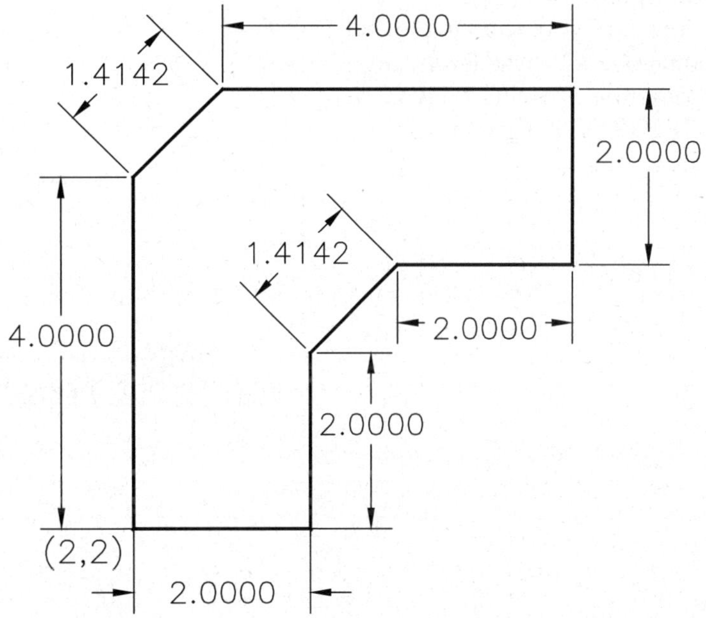


**FIGURE 3-22**

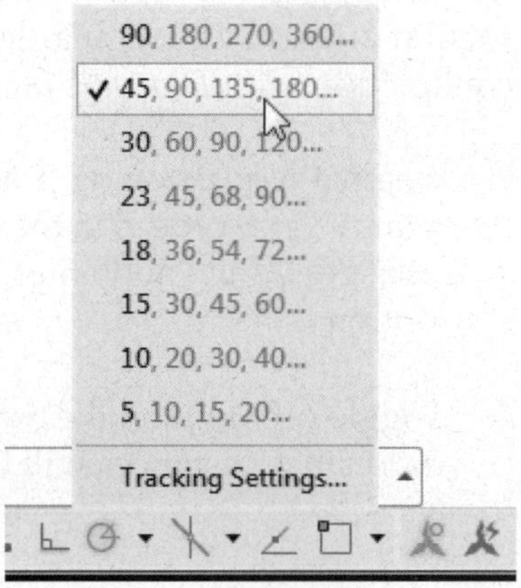

4. Make these other settings by single-clicking the icons on the Status Bar or using the Function keys. The Status Bar indicates the on/off status of these drawing aids such that a light blue button means on and a gray button means off. The command prompt also indicates the on or off status when they are changed.

    ***Grid*** (**F9**) is on
    ***Snap*** (**F7**) is on and set to ***Grid Snap***
    ***Polar Tracking*** (**F10**) is on
    All other drawing aids are off

5. Use the ***Line*** command. Starting at location 2.000, 2.000, 0.000 (watch the *Coordinate* display), begin drawing the shape shown in Figure 3-21. The tracking tip should indicate the current length and angle of the lines as you draw (Fig. 3-23).

6. Complete the shape. Compare your drawing to that in Figure 3-21. Use ***Save*** and name the drawing **PTRACK-GRIDSNAP**. ***Close*** the drawing.

**FIGURE 3-23**

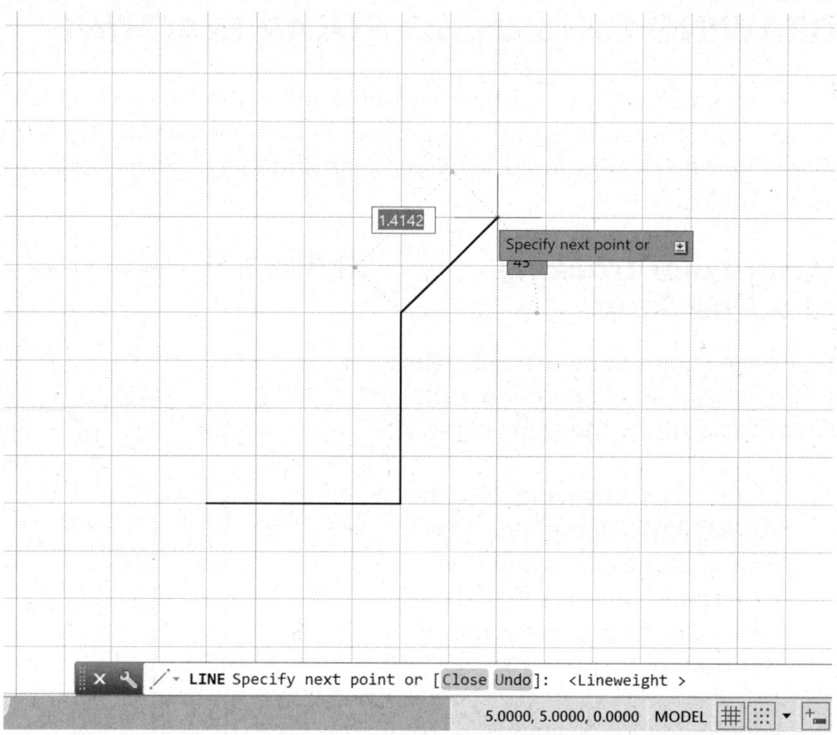


## Using *Polar Tracking* and *Dynamic Input*

Draw the shape in Figure 3-24. Since the lines are at regular angles but irregular lengths, use *Polar Tracking* in conjunction with *Dynamic Input*. Follow these steps.

1. Begin a *New* drawing. Use the **ACAD.DWT** template. Next, type **Z** at the command line (command alias for *Zoom*) and enter. Then enter ***A*** for the *All* option.

2. Toggle off ***Snap*** on the Status Bar. No *Snap* type is used since distances will be entered at the keyboard.

**FIGURE 3-24**

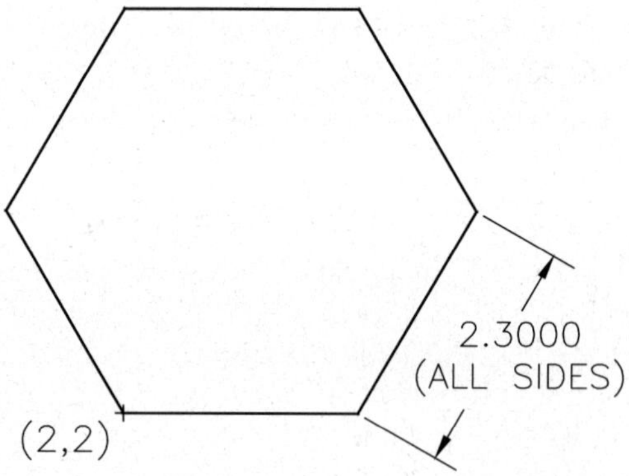

3. Turn on *Polar Tracking* and make the appropriate settings. Do this by right-clicking on the icon on the Status Bar. Ensure the *Increment Angle* is set to **30**, and ***Polar Tracking*** is ***On***.

4. Ensure these other settings are correct using the buttons on the Status Bar or using the Function keys.

   ***Dynamic Input*** (**F12**) is on
   ***Polar Tracking*** (**F10**) is on
   All other drawing aids are off

5. Use the ***Line*** command. Starting at location 2.000, 2.000, 0.000 (enter absolute coordinates of **2,2**), begin drawing the shape shown in Figure 3-24. Use *Polar Tracking* by moving the mouse in the desired direction (Fig. 3-25). When the angle edit box indicates the correct angle for each line, enter the distance for each line (**2.3**) into the distance edit box and press Enter.

6. Complete the shape. Compare your drawing to that in Figure 3-24. Use ***Save*** and name the drawing **PTRACK-DYN**. ***Close*** the drawing and ***Exit*** AutoCAD.

**FIGURE 3-25**

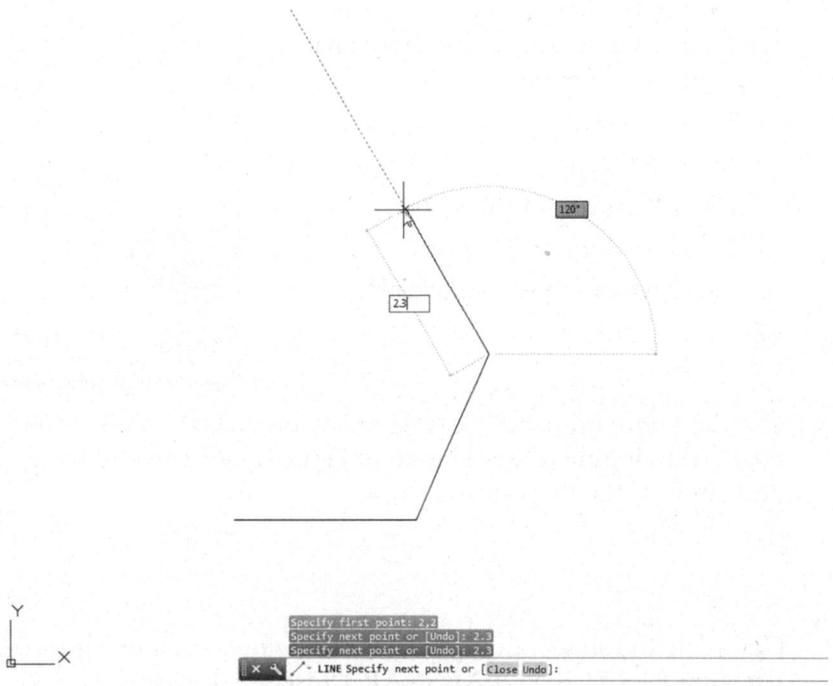


## Using *Polar Tracking, Polar Snap*, and *Dynamic Input*

Draw the shape in Figure 3-26. Since the lines are regular interval lengths, use *Polar Tracking* in conjunction with *Polar Snap*. Follow these steps.

1. Begin a *New* drawing. Use the **ACAD.DWT** template. Next, type ***Z*** at the command line (command alias for *Zoom*) and enter. Then enter ***A*** for the *All* option.

2. Toggle on *Polar Snap* (rather than *Grid Snap*) by right-clicking on the icon on the Status Bar, then checking ***Polar Snap***.

**FIGURE 3-26**

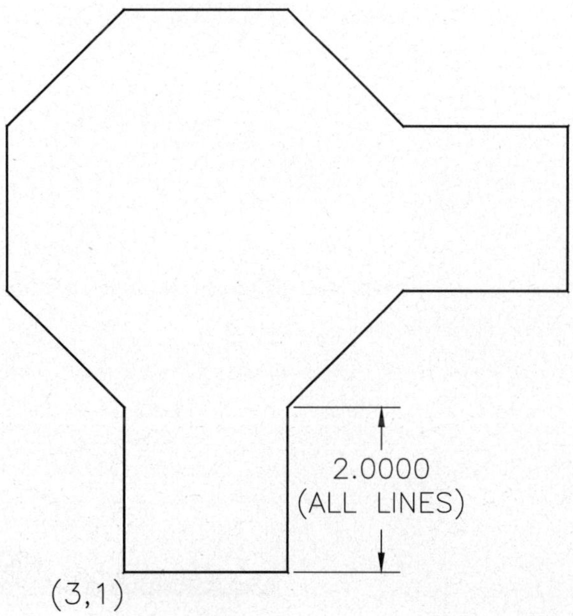

3. Make the appropriate *Polar Snap* settings. Do this by right-clicking on the icon on the Status Bar and selecting ***Snap Settings*** from the menu. In the ***Snap and Grid*** tab of the *Drafting Settings* dialog box that appears, set the *Polar Distance* to **1** (see Fig. 3-27). Next, access the ***Polar Tracking*** tab and ensure the *Increment Angle* is set to **45**. Select the ***OK*** button.

4. Ensure these other settings are correct by clicking the words on the Status Bar or using the Function keys.

   ***Grid*** (**F9**) is on

   ***Snap*** (**F7**) is on

   ***Polar Tracking*** (**F10**) is on

   ***Dynamic Input*** (**F12**) is on

   All other drawing aids are off

**FIGURE 3-27**

5. Use the ***Line*** command. Starting at location 3.000, 1.000, 0.000 (enter absolute coordinates of **3,1**), begin drawing the shape shown in Figure 3-26. The edit boxes should indicate the current length and angle of the lines as you draw (Fig. 3-28).

6. Complete the shape. Compare your drawing to that in Figure 3-26. Use ***Save*** and name the drawing **PTRACK-POLARSNAP**. ***Close*** the drawing.

**FIGURE 3-28**

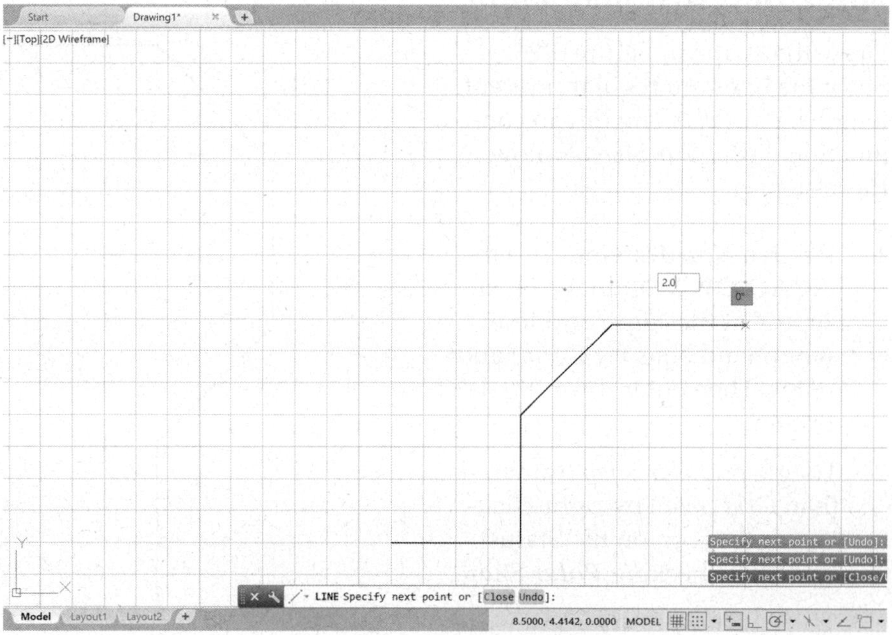

## CHAPTER EXERCISES

1. **Start a *New* Drawing**

   Start a new drawing from the **ACAD.DWT** template. Type ***Z*** at the command line (command alias for *Zoom*) and enter. Then use ***A*** for the zoom *All* option. Next, use the ***Save*** command and save the drawing as **CH3EX1.** Remember to ***Save*** often as you complete the following exercises.

**FIGURE 3-29**

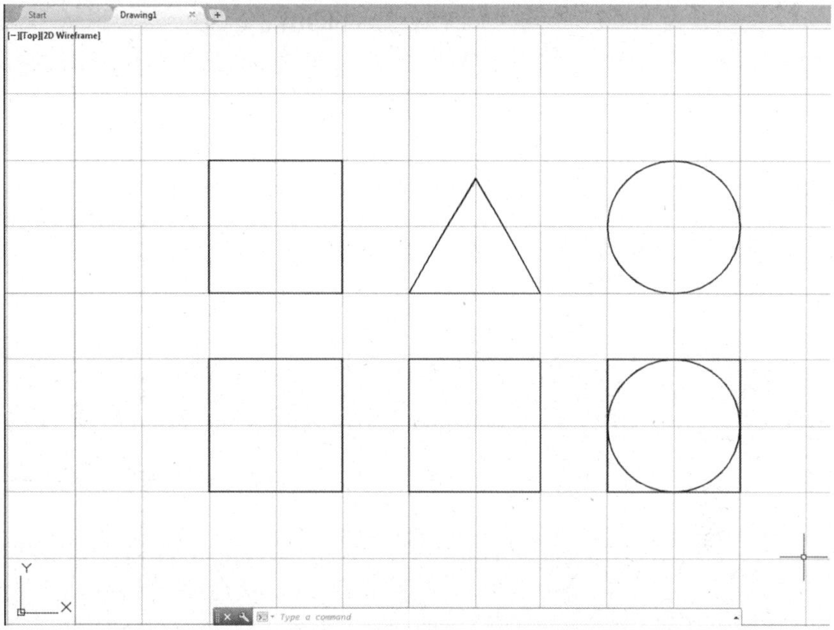


2. **Use Mouse Input**

   Draw a square with sides of 2 unit length. Locate the lower-left corner of the square at **2,2**. Use the ***Line*** command with mouse input. (HINT: Turn on ***Grid Snap***, ***Grid***, and ***Polar Tracking***.)

3. **Use Dynamic Input**

   Draw another square with 2 unit sides using the ***Line*** command. Turn on ***Dynamic Input***. Enter the correct values into the edit boxes.

4. **Use Command Line Input**

   Draw a third square (with 2 unit sides) using the ***Line*** command. Turn off ***Dynamic Input***. Enter coordinates in any format at the Command line.

5. **Use Direct Distance Entry**

   Draw a fourth square (with 2 unit sides) beginning at the lower-left corner of **2,5**. Ensure that ***Dynamic Input*** is off. Complete the square drawing *Lines* using direct distance coordinate entry. Don't forget to turn on ***Ortho*** or ***Polar Tracking***.

6. **Use Dynamic Input and Polar Tracking**

   Draw an equilateral triangle with sides of 2 units. Turn on ***Polar Tracking*** and ***Dynamic Input***. Make the appropriate angular increment settings in *Drafting Settings* dialog box. Begin by locating the lower-left corner at **5,5**. (HINT: An equilateral triangle has interior angles of 60 degrees.)

7. **Use Mouse Input**

   Turn on ***Snap*** and ***Grid***. Use mouse input to draw a ***Circle*** with a 1 unit radius. Locate the center at **9,6**.

8. **Use any method**

   Draw another ***Circle*** with a 2 unit diameter. Using any method, locate the center 3 units below the previous *Circle*.

9. ***Save* your drawing**

   Use ***Save***. Compare your results with Figure 3-29. When you are finished, ***Close*** the drawing.

10. In the next series of steps, you will create the Stamped Plate shown in Figure 3-30. Begin a ***New*** drawing. Use the **ACAD.DWT** template. Next, type ***Z*** at the command line (command alias for *Zoom*) and enter. Then enter ***A*** for the *All* option. Next, use the ***Save*** command and assign the name **CH3EX2**. Remember to ***Save*** often as you work.

**FIGURE 3-30**

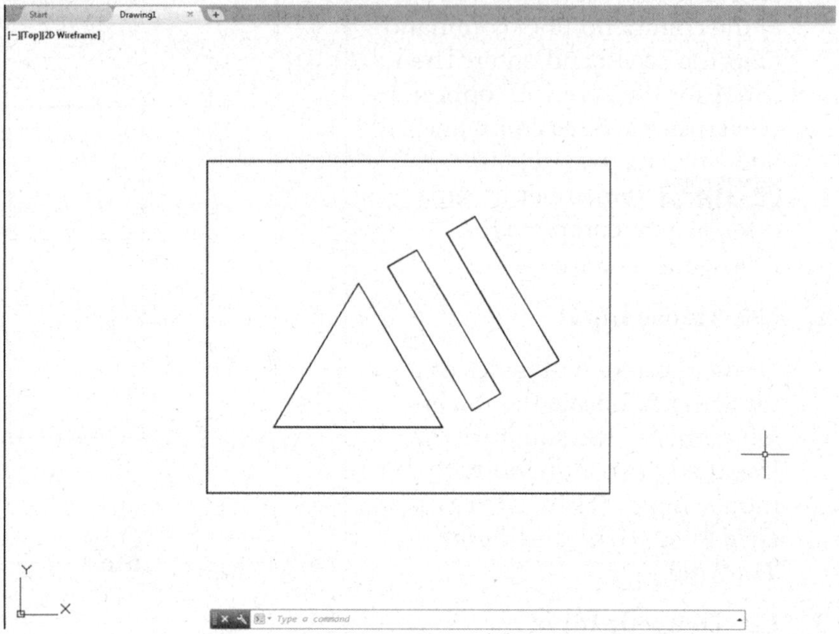


11. First, you will create the equilateral triangle as shown in Figure 3-31. All sides are equal and all angles are equal. To create the shape easily, you should use *Polar Tracking* with *Dynamic Input*. Access the *Polar Tracking* tab of the ***Drafting Settings*** dialog box and set the ***Increment Angle*** to **30**. (Or, you can right-click on the *Polar Tracking* icon and select ***30*** from the shortcut menu.)

**FIGURE 3-31**

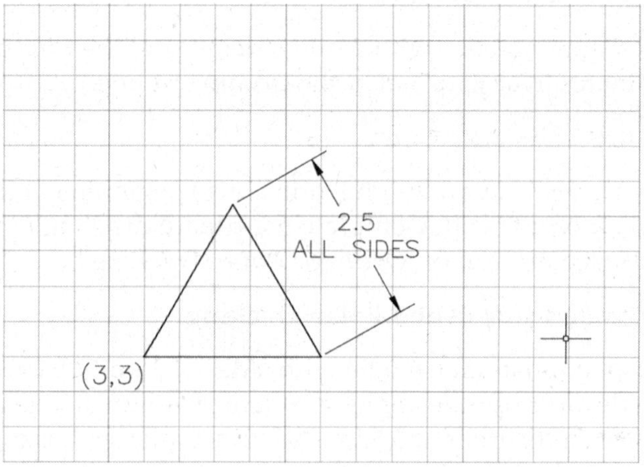


On the Status Bar, make sure ***Polar Tracking*** and ***Dynamic Input*** are on, but not *Snap, Ortho, Object Snap, Object Snap Tracking* or *Dynamic UCS*. *Grid* is optional.

12. Use the ***Line*** command to create the equilateral triangle. Start at position **3,3** by entering absolute coordinates. All sides should be **2.5** units, so enter the distance values at the keyboard as you position the mouse in the desired direction for each line. The drawing at this point should look like that in Figure 3-31 (not including the notation). ***Save*** the drawing but do not *Close* it.

13. Next, create the outside shape (Fig. 3-32). Since the dimensions are all at even unit intervals, use *Grid Snap* to create the rectangle. (HINT: Right-click on the icon and check ***Grid Snap***.) At the Status Bar, make sure that ***Snap*** and ***Polar*** are on. Use the ***Line*** command to create the rectangular shape starting at point **2,2** (enter absolute coordinates).

**FIGURE 3-32**

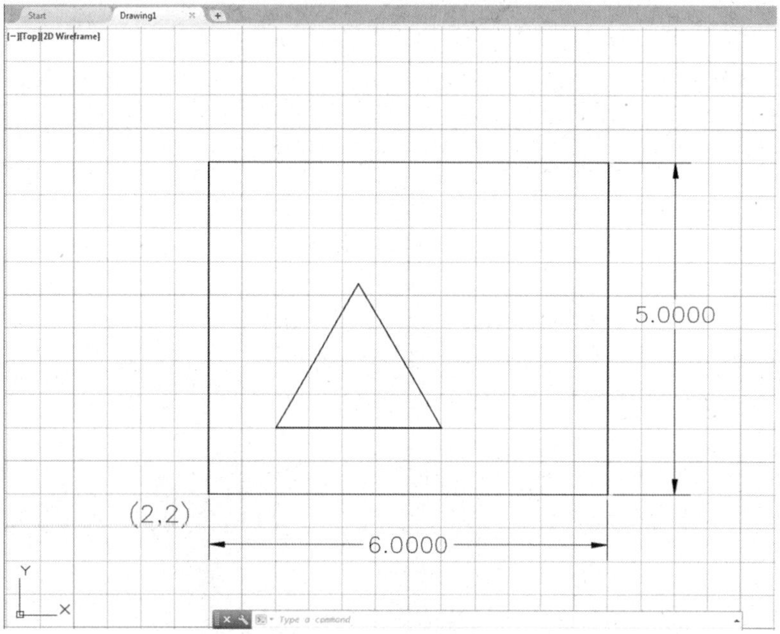


14. The two inside shapes are easily created using *Polar Tracking* in combination with *Polar Snap*. Access the ***Grid and Snap*** tab of the *Drafting Settings* dialog box. (HINT: Right-click on the ***Snap*** icon and select ***Snap Settings***.) Set the *Snap Type* to ***Polar Snap*** and set the *Polar Distance* to **.5**. At the Status Bar, make sure that ***Snap, Polar Tracking,*** and ***Dynamic Input*** are on. Next, use the ***Line*** command and create the two shapes as shown in Figure 3-33. Specify the starting positions for each shape (**5.93,3.25** and **6.80,3.75**) by entering absolute coordinates. Each rectangular shape is .5 units x 2.5 units.

**FIGURE 3-33**

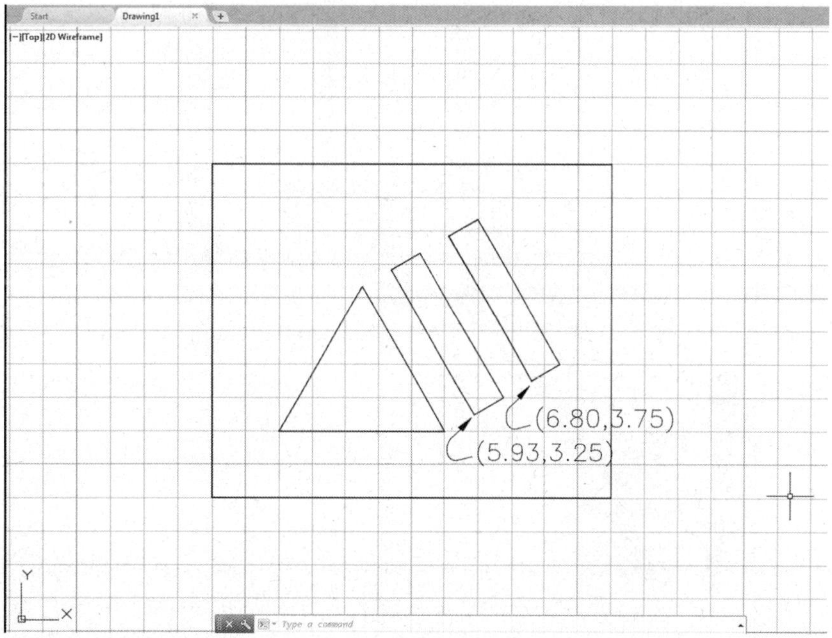


15. When you are finished, ***Save*** the drawing and ***Exit*** AutoCAD.

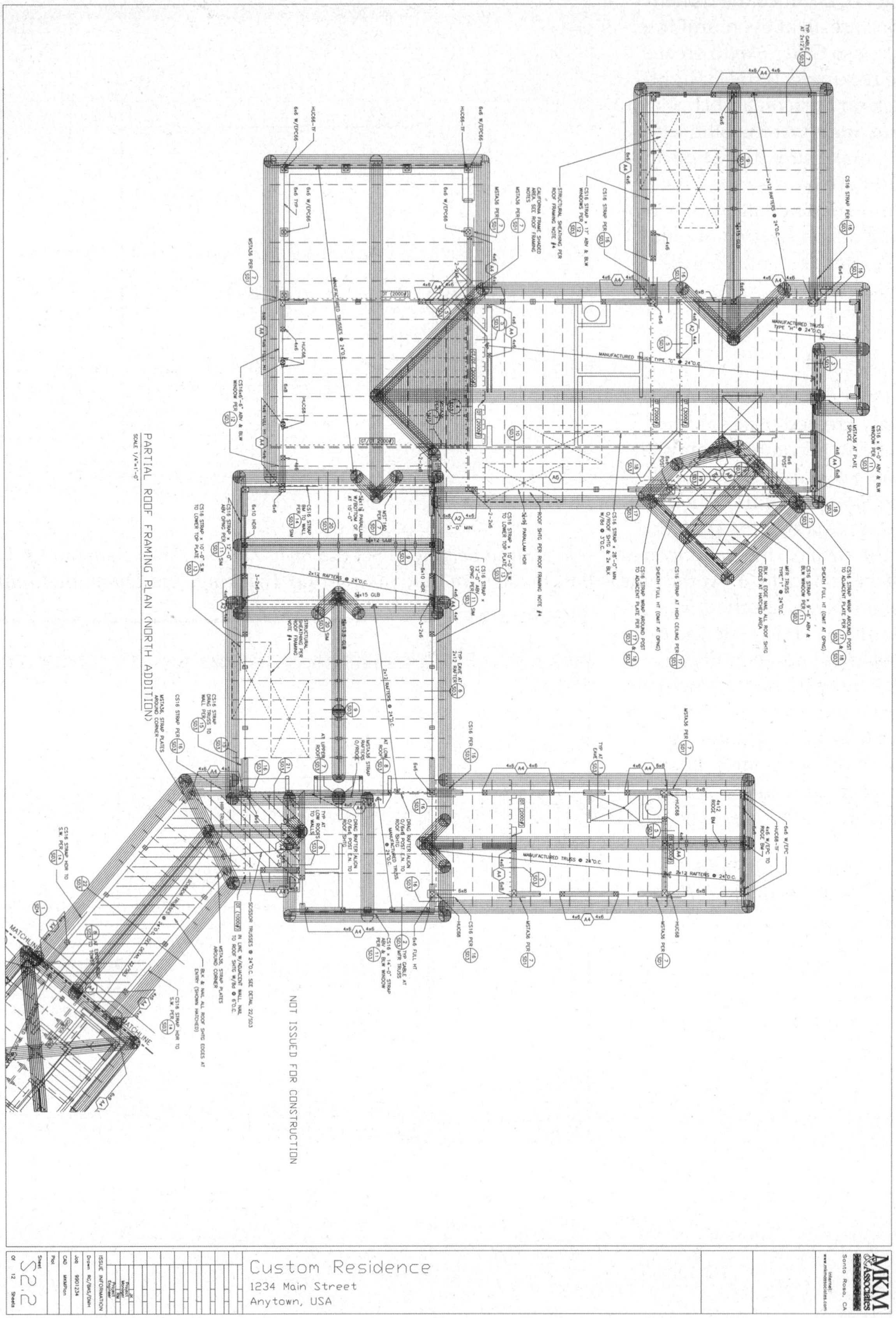


**MKM PLAN.DWG, Courtesy of Autodesk, Inc.**

CHAPTER

# 4

# Selection Sets

## Chapter Objectives

After completing this chapter you should:

1. know that *Modify* commands and many other commands require you to select objects;
2. be able to create a selection set using each of the specification methods;
3. be able to *Erase* objects from the drawing;
4. be able to *Move* and *Copy* objects from one location to another;
5. understand Noun/Verb and Verb/Noun order of command syntax.

## MODIFY COMMAND CONCEPTS

*Draw* commands create objects. *Modify* commands change existing objects or use existing objects to create new ones. Examples are *Copy* an existing *Circle*, *Move* a *Line*, or *Erase* a *Circle*. Since all of the Modify commands use or modify existing objects, you must first select the objects that you want to act on. The process of selecting the objects you want to use is called building a selection set. For example, if you want to *Copy*, *Erase*, or *Move* several objects in the drawing, you must first select the set of objects that you want to act on.

Remember that any of these three command entry methods (depending on your setup) can be used to invoke Modify commands.

| | | |
|---|---|---|
| 1. | Keyboard | Type the command name or command alias at the keyboard. |
| 2. | Menu bar | Select from the *Modify* pull-down menu. |
| 3. | Tools (icon buttons) | Select from the *Modify* or *Modify II* toolbar, *Modify* tool palette, or *Modify* panel on the Ribbon. |

All of the *Modify* commands will be discussed in detail later, but for now we will focus on how to build selection sets.

## SELECTION SETS

Two *Circles* and five *Lines*, as shown in Figure 4-1, will be used to illustrate the selection process in this chapter. In Figure 4-1, no objects are selected.

FIGURE 4-1

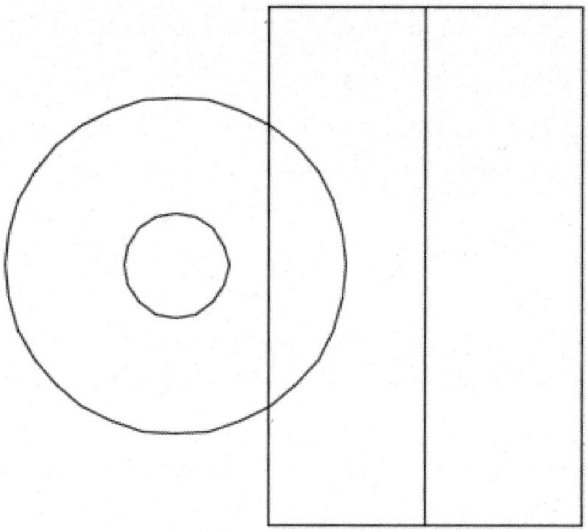

### Selection Preview

When you pass the pickbox over an object such as a *Line* or *Circle*, the object becomes thick and gray as shown by the vertical *Line* in Figure 4-2. This feature, called selection preview, or sometimes "rollover highlighting," assists you by indicating the objects you are about to select before you actually PICK them. Selection preview may occur both during a modify command at the "Select objects:" prompt (when only the pickbox appears) and also when no commands are in use (when the "cross-hairs" appear). This feature, turned on with the default settings, can be turned off or changed if desired.

FIGURE 4-2

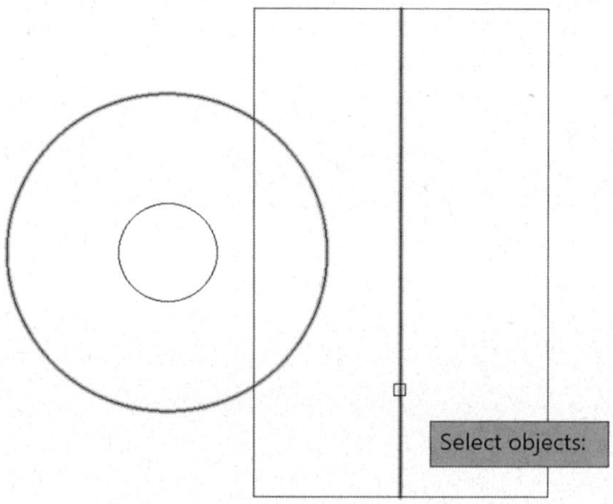


### Highlighting

After you PICK an object using the pickbox or any of the methods explained next, the objects become "highlighted," or displayed with light blue shading, as shown by the large *Circle* in Figure 4-2. Highlighted (selected) objects become part of the selection set.

NOTE: The color of the selected objects can be changed from the default (blue) color to another color using the *Options* dialog box, *Selection* tab, *Selection effect color* drop-down list (*SELECTIONEFFECTCOLOR* system variable).

### Verb/Noun Selection

There are two times you can select objects. You can select objects before using a command, called Noun/Verb (object/command) order, or you can select objects during a Modify command when the "Select objects:" prompt appears, called Verb/Noun (command/object) order. It is recommended that you begin learning the selection options by using the Modify command first (command/object), then select objects during the "Select objects:" prompt. (See "Noun/Verb Syntax" for more information.)

### Grips

When you select objects before using a command (Noun/Verb selection), blue squares called grips may appear on the selected objects. Grips offer advanced editing features that are discussed in Chapter 19. When you begin AutoCAD, grips are most likely turned on by default. It is recommended to turn off grips until you learn the basic selection options and Modify commands. Turn grips off by typing in *GRIPS* at the Command line and changing the setting to ***0*** (zero). (*GRIPS* are set to 2 by default.)

TIP

### Selection

No matter which of the methods you use to invoke a Modify command, you must specify a selection set during the command operation. For example, examine the command syntax that would appear at the Command line when the *Erase* command is used.

Command: ***ERASE***
Select objects:

The "Select objects:" prompt is your cue to use any of several methods to PICK the objects to erase. As a matter of fact, most Modify commands begin with the same "Select objects:" prompt (unless you selected immediately before invoking the command). When the "Select objects:" prompt appears, the "crosshairs" disappear and only a small, square pickbox appears at the cursor.

Use the pickbox to select one object at a time. Once an object is selected, it becomes highlighted. The Command line also verifies the selection by responding with "1 found." You can select as many additional objects as you want, and you can use many selection methods as described on the next few pages. Therefore, object selection is a cumulative process in that selected objects remain highlighted while you pick additional objects. AutoCAD continually responds with the "Select objects:" prompt, so press Enter to indicate when you are finished selecting objects and are ready to proceed with the command.

## Selection Set Options

The pickbox is the default option, which can automatically be changed to a window or lasso by PICKing in an open area (PICKing no objects). Since the pickbox, windows, and lassos (sometimes known as the *AUto* options) are the default selection methods, no action is needed on your part to activate these methods. The other methods can be invoked by typing the capitalized letters shown in the option names following.

All the possible selection set options are listed if you enter a "?" (question mark symbol) at the "Select objects:" prompt.

TIP

Command: ***MOVE***
Select objects: ***?***
*Invalid selection*
Expects a point or
Window/Last/Crossing/BOX/ALL/Fence/WPolygon/CPolygon/Group/Add/Remove/Multiple/Previous/Undo/AUto/SIngle/SUbobject/Object
Select objects:

Use any option by typing the indicated uppercase letters and pressing Enter.

### pickbox

This default option is used for selecting one object at a time. Locate the pickbox so that an object crosses through it and PICK (Fig. 4-3). You do not have to type or select anything to use this option.

FIGURE 4-3

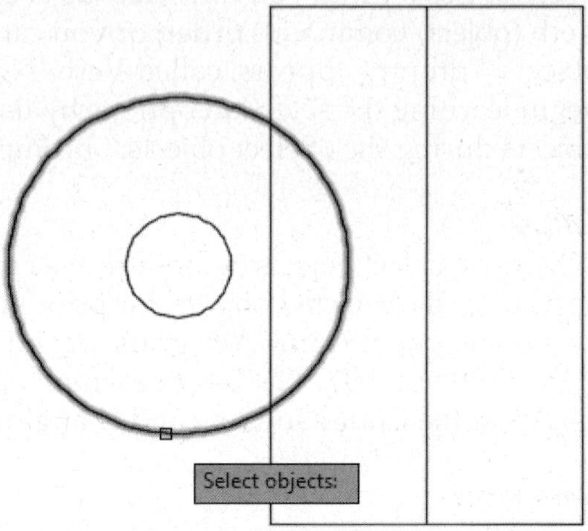


### window (*AUto* option)

Only objects completely within the window are selected, not objects that cross through or are outside the window (Fig. 4-4). The window border is a solid linetype rectangular box filled with a transparent blue color.

To use this option, you do not have to type or select anything from a menu. Instead, when the pickbox appears, position it in an open area (so that no objects cross through it); then PICK to start a window. If you move the cursor to the right, a window is created. Next, PICK the other diagonal corner.

Therefore, you must PICK each corner of the window rather than click-and-drag. If you click-and-drag, the *Lasso* options may begin based on your system settings (see *Window Lasso, Crossing Lasso,* and *Fence Lasso*).

FIGURE 4-4

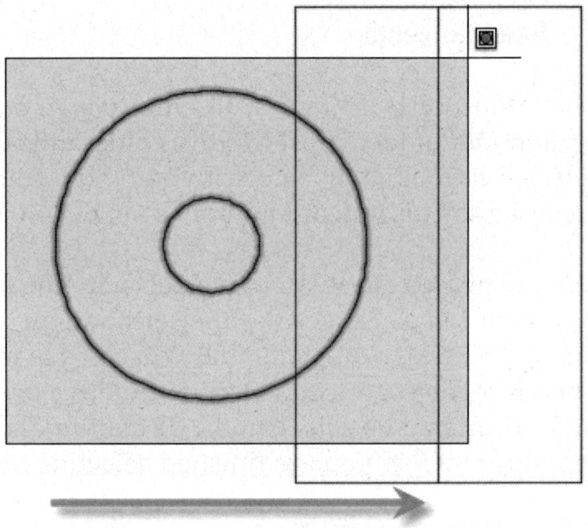

#### *Window*

You can also type the letter ***W*** at the "Select objects:" prompt to create a selection window as shown in Figure 4-4. In this case, no pickbox appears, and you can PICK the diagonal corners of the window by PICKing corners in either direction.

### crossing window (*AUto* option)

Note that all objects within and crossing through a crossing window are selected (compare Fig. 4-5 with Fig. 4-4). The crossing window is displayed as a broken linetype rectangular box filled with a transparent green color.

This option is similar to using a window in that when the pickbox appears you can position it in an open area and PICK to start the window. However, move the cursor to the left to form a crossing window. Moving to the right creates a window instead. Pick again for the second corner.

FIGURE 4-5

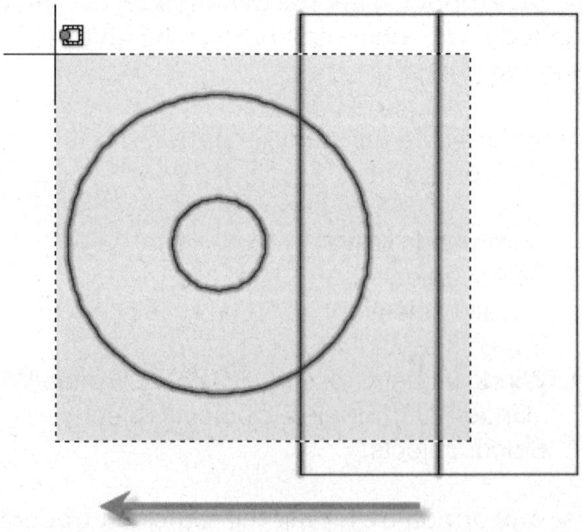

#### *Crossing Window*

You can also type the letters ***CW*** at the "Select objects:" prompt to create a crossing window as shown in Figure 4-5. With this option, no pickbox appears, and you can PICK the diagonal corners by moving in either direction to form the crossing window.

### *Window Polygon*

Type the letters ***WP*** at the "Select objects:" prompt to use this option. Note that the *Window Polygon* operates like a *Window* in that only objects completely within the window are selected, but the box can be *any* irregular polygonal shape (Fig. 4-6). You can pick any number of corners to form any shape rather than picking just two diagonal corners to form a regular rectangular *Window*. Polygon segments cannot intersect (the window cannot cross over itself).

**FIGURE 4-6**

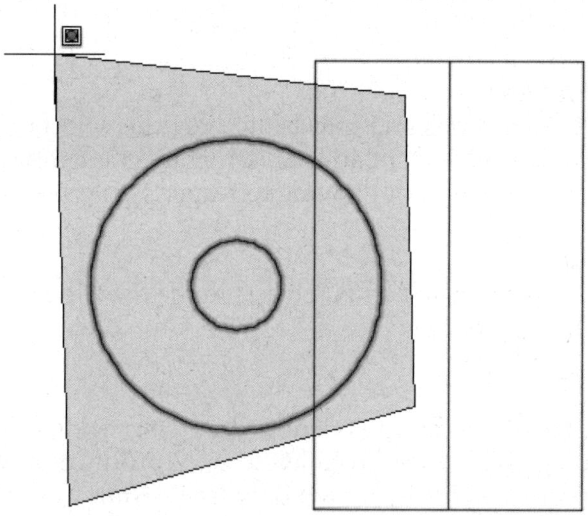

### *Crossing Polygon*

Type the letters ***CP*** at the "Select objects:" prompt to use this option. The *Crossing Polygon* operates like a *Crossing Window* in that all objects within and crossing through the *Crossing Polygon* are selected; however, the polygon can have any number of corners (Fig. 4-7). Polygon segments cannot intersect (the window cannot cross over itself).

**FIGURE 4-7**

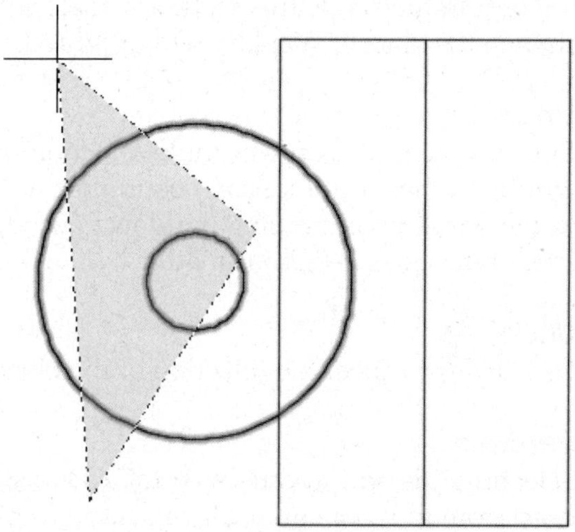

### *Fence*

To invoke this option, type the letter ***F*** at the "Select objects:" prompt. A *Fence* operates like a crossing line. Any objects crossing the *Fence* are selected. The *Fence* can have any number of segments (Fig. 4-8).

**FIGURE 4-8**

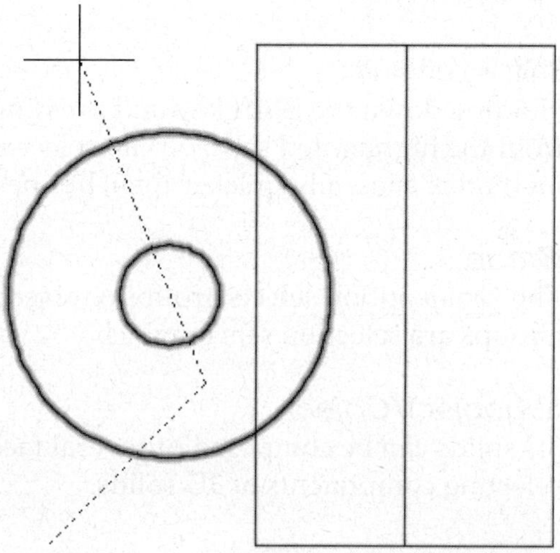

### *Last*

This option automatically finds and selects only the last object created. *Last* does not find the last object modified (with *Move, Stretch,* etc.).

### *Previous*

*Previous* finds and selects the previous selection set, that is, whatever was selected during the previous command (except after *Erase*). This option allows you to use several editing commands on the same set of objects without having to respecify the set.

### *ALL*

This option selects all objects in the drawing except those on *Frozen* or *Locked* layers (*Layers* are covered in Chapter 11).

### *BOX*

This option is equivalent to the *AUto* window/crossing window option without the pickbox. PICKing diagonal corners from left to right produces a window and PICKing diagonal corners from right to left produces a crossing window (see *AUto*).

### *Multiple*

The *Multiple* option allows selection of objects with the pickbox only; however, the selected objects are not highlighted. Use this method to select very complex objects to save computing time required to change the objects' display to highlighted.

### *SIngle*

This option allows only a single selection using one of the *AUto* methods (pickbox, window, or crossing window), then automatically continues with the command. Therefore, you can select multiple objects (if the window or crossing window is used), but only one selection is allowed. You do not have to press Enter after the selection is made.

### *Undo*

Use *Undo* to cancel the selection of the object(s) most recently added to the selection set.

### *Remove*

Selecting this option causes AutoCAD to switch to the "Remove objects:" mode. Any selection options used from this time on remove objects from the highlighted selection set.

### *Add*

The *Add* option switches back to the default "Select objects:" mode so additional objects can be added to the selection set.

TIP

### Shift + Left Button

Holding down the **Shift** key and pressing the left mouse button simultaneously removes objects selected from the highlighted set. You can select with any of the *AUto* options (pickbox, windows, or lassos). This method is generally quicker than, but performs the same action as, *Remove*.

### *Group*

The *Group* option selects groups of objects that were previously specified using the *Group* command. Groups are selection sets to which you can assign a name (see Chapter 20).

### *SUbobject/Object*

3D solids can be composed of several faces, edges, and primitive solids. These options are used for selecting components of 3D solids.

TIP

Lasso options work automatically (you do not have to type in an option), similar to the window and crossing window *AUto* options. To use the lasso options, you PICK, drag while holding down the button, then release to form the lasso, whereas with the *AUto* window and crossing window, you must PICK both corners of the window (PICK, move, PICK).

There are three lasso options: Window Lasso, Crossing Lasso, and Fence Lasso. You do not type anything at the Command line to use these options. You can cycle through the options by pressing the space bar after starting a lasso. (See "Lasso Settings" if lassos do not operate by default on your system.)

**FIGURE 4-9**

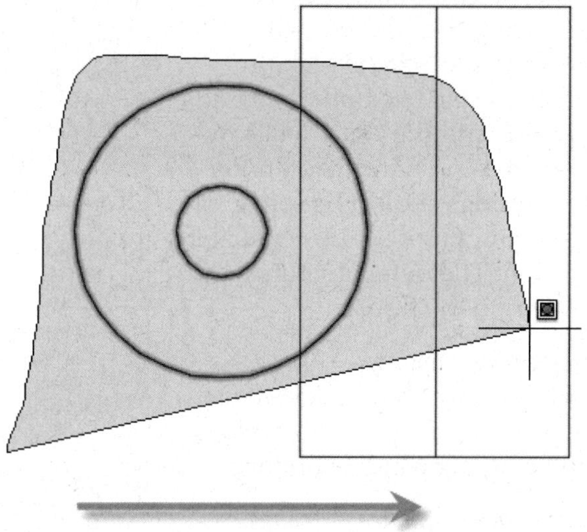

### Window Lasso

A Window Lasso selects only objects that are completely within the lasso. To use this option, PICK in an open area (so no objects are selected) and while holding down the mouse button, drag to the right to form the lasso (Fig. 4-9). Release the mouse button to complete the selection. If you want to switch to another lasso option, press space bar before releasing as indicated by the Command prompt.

Window Lasso
Press Spacebar to cycle options

**FIGURE 4-10**

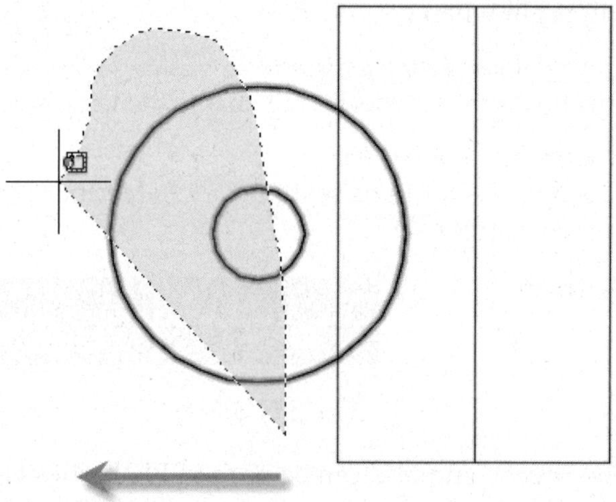

### Crossing Lasso

Similar to a crossing window, a Crossing Lasso selects all objects within and crossing the lasso. PICK and drag the lasso to the left while holding down the mouse button to collect the desired objects (Fig. 4-10). Release the mouse button to complete the selection.

**FIGURE 4-11**

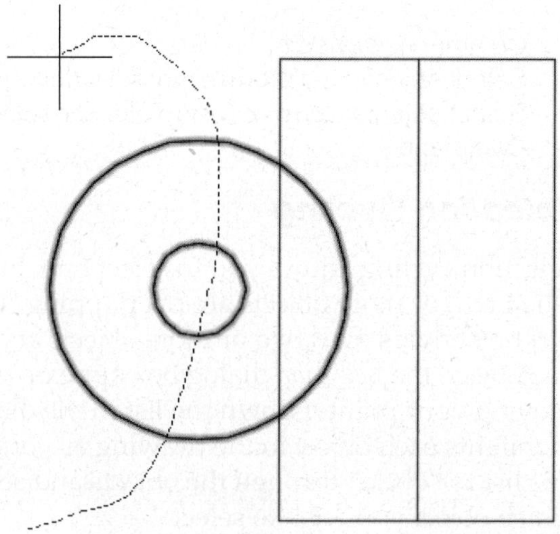

### Fence Lasso

To use the Fence Lasso you must first begin a Window Lasso or Crossing Lasso, and then press the space bar while holding down the mouse button and cycle to the Fence Lasso option. The Fence Lasso operates similarly to the *Fence* selection option, in that all objects that cross the fence are selected. Drag the fence to highlight the desired objects and then release the mouse button to complete the selection (Fig. 4-11).

### Lasso Settings

Control the lasso settings by accessing the *Options* dialog box and the *Selection* tab. In order to best use both the lasso options and the *AUto* window and crossing window options, the settings should be set as shown in Figure 4-12, center-left section. The default settings are shown in the figure.

FIGURE 4-12

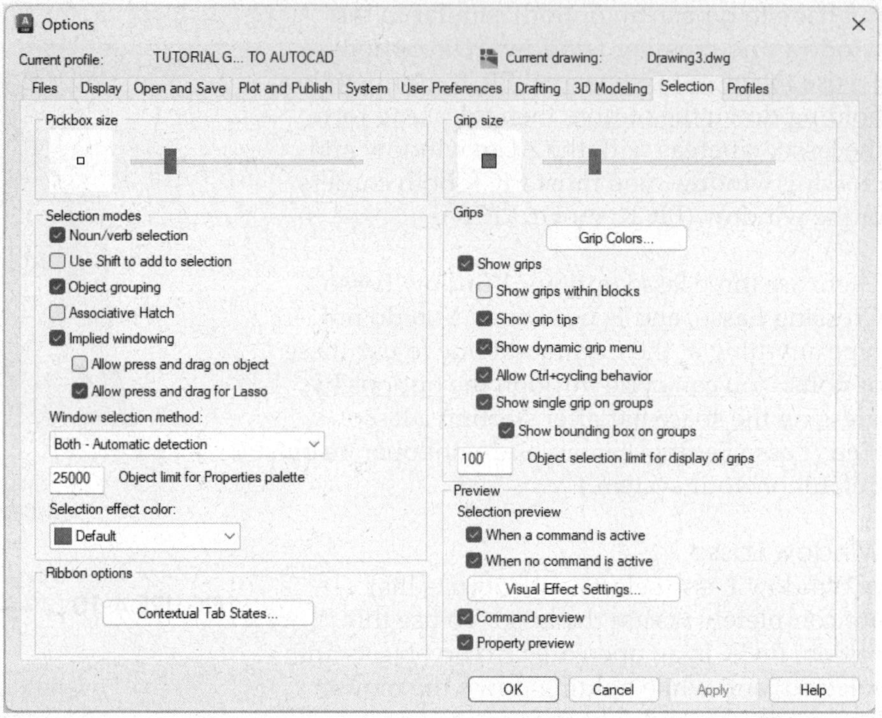


*Implied windowing*
This box must be checked to use the lasso options and the window and crossing window options.

*Allow press and drag on object*
If this box is checked, you cannot use the pickbox to select objects—some type of window will begin when you PICK.

*Allow press and drag for lasso*
This box must be checked to enable the lasso selection options.

*Window selection method*
This option should be set to *Both – Automatic detection* to use the *AUto* options (pickbox, windows and the lasso options).

## Select

| Ribbon | Menu Bar | Command (Type) | Alias (Type) | Shortcut |
|---|---|---|---|---|
| ... | ... | *Select* | ... | ... |

The *Select* command can be used to PICK objects to be saved in the selection set buffer for subsequent use with the *Previous* option. Any of the selection methods can be used to PICK the objects. The selected objects become unhighlighted when you complete the command by pressing Enter. The objects become highlighted again and are used as the selection set if you use the *Previous* selection option in the next editing command.

Command: ***SELECT***
Select objects: **PICK** (use any selection option)
Select objects: **Enter** (completes the selection)
Command:

## Selection Cycling

Selection cycling allows you to select only one object when two or more objects are overlapping. When the pickbox hovers over two or more objects and you PICK one object, the *Selection* dialog box appears (Fig. 4-13). Moving your pointer down the list in the dialog box highlights each object in the drawing as you point. Use this list to "cycle" through the objects and select exactly which object you wish to select.

FIGURE 4-13

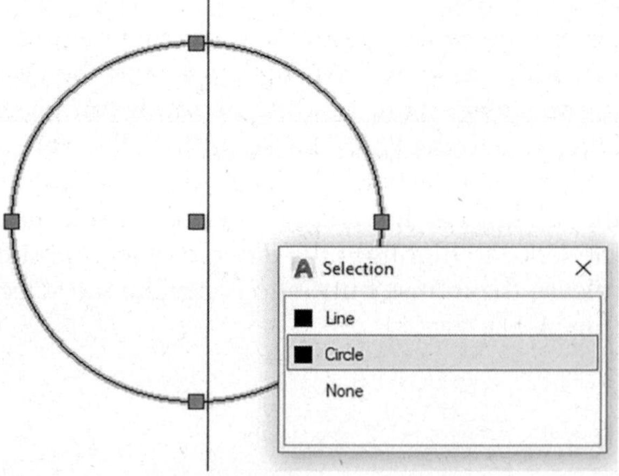

*Selection Cycling* can be toggled on and off using the icon on the Status Bar, using the Ctrl+W key sequence, or setting the *SELECTIONCYCLING* system variable to *0* (off) or *2* (*Selection* dialog box appears). If the icon does not appear on your Status Bar, activate it using the *Customization* button in the lower-right corner of the Status Bar.

You can also use the *Selection Cycling* tab in the *Drafting Settings* dialog box to turn selection cycling on or off as shown in Fig. 4-14. Check or uncheck *Allow selection cycling*.

**FIGURE 4-14**

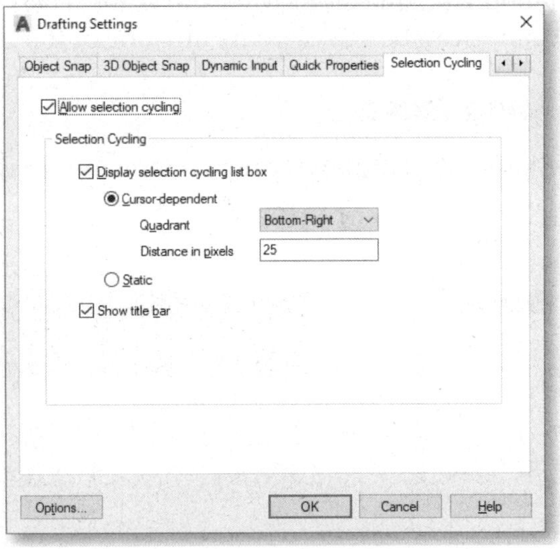


## NOUN/VERB SYNTAX

An object is the <u>noun</u> and a command is the <u>verb</u>. Noun/Verb syntax order means to pick objects (nouns) first, then use an editing command (verb) second. If you select objects first (at the open Command: prompt) and then immediately choose a Modify command, AutoCAD recognizes the selection set and passes through the "Select objects:" prompt to the next step in the command.

Verb/Noun means to invoke a command and then select objects within the command. For example, if the *Erase* command (verb) is invoked first, AutoCAD then issues the prompt to "Select objects:"; therefore, objects (nouns) are PICKed second. In the previous examples, and with much older versions of AutoCAD, only Verb/Noun syntax order was used.

You can use <u>either</u> order you want (Noun/Verb or Verb/Noun) and AutoCAD automatically understands. If objects are selected first, the selection set is passed to the next editing command used, but if no objects are selected first, the editing command automatically prompts you to "Select objects:".

If you use Noun/Verb order, you are limited to using only the *AUto* options for object selection (pickbox, windows, and lassos). You can only use the other options (e.g., *Crossing Polygon, Fence, Previous*) if you invoke the desired Modify command first, then select objects when the "Select objects:" prompt appears.

TIP

The *PICKFIRST* variable (a very descriptive name) enables Noun/Verb syntax. The default setting is 1 (On). If *PICKFIRST* is set to 0 (Off), Noun/Verb syntax is disabled and the selection set can be specified only <u>within</u> the editing commands (Verb/Noun).

Setting *PICKFIRST* to 1 provides two options: Noun/Verb and Verb/Noun. You can use <u>either</u> order you want. If objects are selected first, the selection set is passed to the next editing command, but if no objects are selected first, the editing command prompts you to select objects. See Chapter 20 for a complete explanation of *PICKFIRST* and advanced selection set features.

## SELECTION SETS PRACTICE

TIP

NOTE: While learning and practicing with the editing commands, it is suggested that *GRIPS* be turned off. This can be accomplished by typing in **GRIPS** and setting the *GRIPS* variable to a value of **0**. The AutoCAD default has *GRIPS* on (set to 2). *GRIPS* are covered in Chapter 19.

Begin a *New* drawing to complete the selection set practice exercises. Next, type **Z** at the command line (command alias for *Zoom*) and enter. Then enter ***A*** for the *All* option. The *Erase, Move,* and *Copy* commands can be activated by any one of the methods shown in the command tables that follow.

## Using *Erase*

*Erase* is the simplest editing command. *Erase* removes objects from the drawing. The only action required is the selection of objects to be erased.

### *Erase*

| Ribbon | Menu Bar | Command (Type) | Alias (Type) | Shortcut |
|---|---|---|---|---|
| *Home* *Modify* (icon) | *Modify* *Erase* | *Erase* | E | (Edit Mode) *Erase* |

1. Draw several *Lines* and *Circles*. Practice using the object selection options with the *Erase* command. The following sequence uses the pickbox, *Window*, and *Crossing Window*.

| Steps | Command Prompt | Perform Action | Comments |
|---|---|---|---|
| 1. | Command: | type ***E*** and press **Spacebar** | *E* is the alias for *Erase*; Spacebar can be used like Enter |
| 2. | ERASE Select objects: | use pickbox to select one or two objects | objects are highlighted |
| 3. | Select objects: | type **W** to use a ***Window***, then select more objects | objects are highlighted |
| 4. | Select objects: | type **C** to use a ***Crossing Window***, then select objects | objects are highlighted |
| 5. | Select objects: | press **Enter** | objects are erased |

2. Draw several more *Lines* and *Circles*. Practice using the *Erase* command with the *AUto Window* and *AUto Crossing Window* options as indicated below.

| Steps | Command Prompt | Perform Action | Comments |
|---|---|---|---|
| 1. | Command: | select *Erase* from the *Modify* panel | |
| 2. | ERASE Select objects: | use pickbox to **PICK** an open area, create *Window* to the right to select objects | objects inside *Window* are highlighted |
| 3. | Select objects: | **PICK** an open area, create *Crossing Window* to the left to select objects | objects inside and crossing through *Window* are highlighted |
| 4. | Select objects: | press **Enter** | objects are erased |

## Using *Move*

The *Move* command specifically prompts you to (1) select objects, (2) specify a "base point," a point to move from, and (3) specify a "second point of displacement," a point to move to.

### *Move*

| Ribbon | Menu Bar | Command (Type) | Alias (Type) | Shortcut |
|---|---|---|---|---|
| *Home* *Modify* (icon) | *Modify* *Move* | *Move* | *M* | (Edit Mode) *Move* |

1. Draw a *Circle* and two *Lines*. Use the *Move* command to practice selecting objects and to move one *Line* and the *Circle* as indicated in the following table.

| Steps | Command Prompt | Perform Action | Comments |
|---|---|---|---|
| 1. | Command: | type ***M*** and press **Spacebar** | *M* is the command alias for *Move* |
| 2. | MOVE Select objects: | use pickbox to select one *Line* and the *Circle* | objects are highlighted |
| 3. | Select objects: | press **Spacebar** or **Enter** | |
| 4. | Select base point or displacement: | **PICK** near the *Circle* center | base point is the handle, or where to move from |
| 5. | Specify second point: | **PICK** near the other *Line* | second point is where to move to |

2. Use *Move* again to move the *Circle* back to its original position. Select the *Circle* with the window lasso option.

| Steps | Command Prompt | Perform Action | Comments |
|---|---|---|---|
| 1. | Command: | select *Move* from the *Modify* panel | |
| 2. | MOVE Select objects: | **press** on the left and **drag** up and to the right to create a window lasso around the *Circle* | lasso only the circle; object is highlighted |
| 3. | Select objects: | press **Spacebar** or **Enter** | |
| 4. | Select base point or displacement: | **PICK** near the *Circle* center | base point is the handle, or where to move from |
| 5. | Specify second point: | **PICK** near the original location | second point is where to move to |

## Using *Copy*

The *Copy* command is similar to *Move* because you are prompted to (1) select objects, (2) specify a "base point," a point to copy from, and (3) specify a "second point of displacement," a point to copy to.

### *Copy*

| Ribbon | Menu Bar | Command (Type) | Alias (Type) | Shortcut |
|---|---|---|---|---|
| *Home* *Modify* (icon) | *Modify* *Copy* | *Copy* | *CO or CP* | (Edit Mode) *Copy Selection* |

1. Using the *Circle* and 2 *Lines* from the previous exercise, use the *Copy* command to practice selecting objects and to make copies of the objects as indicated in the following table.

| Steps | Command Prompt | Perform Action | Comments |
|---|---|---|---|
| 1. | Command: | type ***CO*** and press **Enter** or **Spacebar** | *CO* is the command alias for *Copy* |
| 2. | COPY Select objects: | type ***F*** and press **Enter** | *F* invokes the *Fence* selection option |
| 3. | Specify first fence point: | **PICK** a point near two *Lines* | Starts the "fence" |
| 4. | Specify the next fence point: | **PICK** a second point across the two *Lines* | lines are highlighted |
| 5. | Specify next fence point: | Press **Enter** | completes *Fence* option |
| 6. | Select objects: | Press **Enter** | completes object selection |
| 7. | Specify base point or displacement: | **PICK** between the *Lines* | base point is the handle, or where to copy from |
| 8. | Specify second point: | enter **@2<45** and press **Enter** | copies of the lines are created 2 units in distance at 45 degrees from the original location |

2. Practice using the Array option of the *Copy* command by following the steps in the table below.

| Steps | Command Prompt | Perform Action | Comments |
|---|---|---|---|
| 1. | Command: | type ***CO*** and press **Enter** or **Spacebar** | *CO* is the alias for *Copy* |
| 2. | COPY Select objects: | **drag** a crossing box to select a single circle from the drawing. Press Enter to finalize selection. | the circle is selected |
| 3. | Specify base point or displacement: | **PICK** near the circle's center | the basepoint is selected. |

| | | | |
|---|---|---|---|
| 4. | Specify second point or [Array] <use first point as displacement>: | type *A* and press **Enter** or **Spacebar** | the *Array* option of the *Copy* command is selected |
| 5. | Enter number of items to array: | type **4** and press **Enter** or **Spacebar (or whatever number of total copies of the selected items to create including the original)** | copies of the circle rubberband from the selected basepoint. Entering 4 makes 3 additional copies. |
| 6. | Specify second point or [Fit]: | move the cursor and **PICK** to place the pattern of copies. | typing **F** for fit, changes the spacing of the copied items as you move the cursor during rubberbanding |
| 7. | Specify second point or [Array/Exit/Undo] <Exit>: | Press **Enter** | Enter ends the command when you are done placing patterned sets of copies |

3. Practice removing objects from the selection set by following the steps given in the table below.

| Steps | Command Prompt | Perform Action | Comments |
|---|---|---|---|
| 1. | Command: | type *CO* and press **Enter** or **Spacebar** | *CO* is the command alias for *Copy* |
| 2. | COPY Select objects: | **press** in an open area near the right of your drawing, **drag** a crossing lasso to the left to circle and select all objects | all objects within and crossing the lasso are highlighted |
| | Select objects: | hold down **Shift** and **PICK** all highlighted objects except one *Circle* | holding down Shift while PICKing removes objects from the selection set |
| | Select objects: | press **Enter** | only one circle is highlighted |
| | Specify base point or displacement: | **PICK** near the circle's center | the base point is the handle, or where to copy <u>from</u> |
| | Specify second point or [Array] <use first point as displacement>: | turn on ***Ortho***, move the cursor to the right, type **3** and press **Enter** | a copy of the circle is created 3 units to the right of the original circle |

## Noun/Verb Command Syntax Practice

1. Practice using the *Move* command using Noun/Verb syntax order by following the steps in the table below.

| Steps | Command Prompt | Perform Action | Comments |
|---|---|---|---|
| 1. | Command: | **PICK** one *Circle* | circle becomes highlighted (grips may appear if not disabled) |
| 2. | Command: | type ***M*** and press **Enter** or **Spacebar** | *M* is the command alias for *Move* |
| 3. | MOVE 1 found Specify base point or displacement: | **PICK** near the *Circle's* center | command skips the "Select objects:" prompt and proceeds |
| 4. | Specify second point: | enter **@-3,-3** and press **Enter** | circle is moved -3 X units and -3 Y units from the original location |

2. Practice using Noun/Verb syntax order by selecting objects for *Erase*.

| Steps | Command Prompt | Perform Action | Comments |
|---|---|---|---|
| 1. | Command: | **PICK** one *Line* | line becomes highlighted (grips may appear if not disabled) |
| 2. | Command: | type ***E*** and press **Enter** or **Spacebar** | *E* is the command alias for *Erase* |
| 3. | ERASE 1 found: | | command skips the "Select objects:" prompt and erases highlighted line |

## CHAPTER EXERCISES

*Open* drawing **CH3EX1** that you created in Chapter 3 Exercises. Turn off *Snap* (**F9**) to make object selection easier.

REUSE

1. **Use the pickbox to select objects**

   Invoke the ***Erase*** command by any method. Select the lower-left square with the pickbox (Fig. 4-15, highlighted). Each *Line* must be selected individually. Press **Enter** to complete *Erase*. Then use the ***Oops*** command to unerase the square. (Type ***Oops***.)

**FIGURE 4-15**

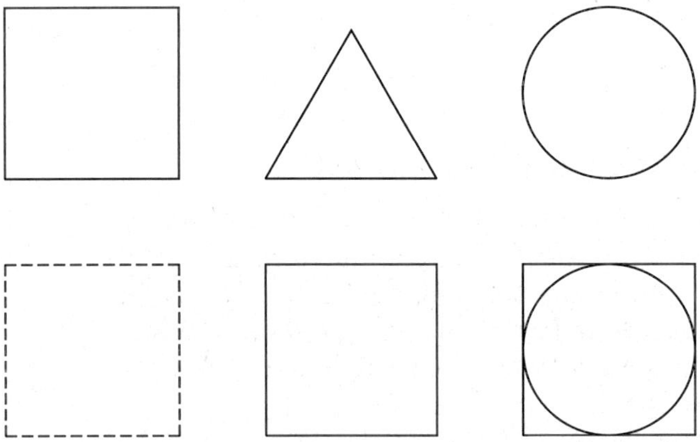

2. **Use the *AUto Window* and *AUto Crossing Window***

   Invoke ***Erase***. Select the center square on the bottom row with the *AUto Window* and select the equilateral triangle with the *AUto Crossing Window*. Press **Enter** to complete the *Erase* as shown in Figure 4-16. Use ***Oops*** to bring back the objects.

**FIGURE 4-16**

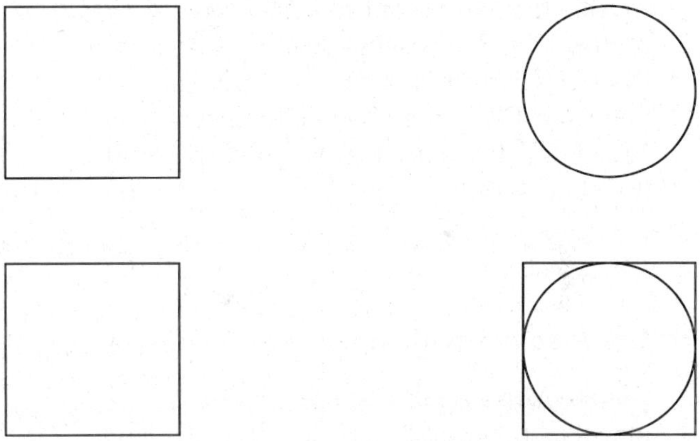

3. **Use the *Fence* selection option**

   Invoke ***Erase*** again. Use the ***Fence*** option to select all the vertical *Lines* and the *Circle* from the squares on the bottom row. Complete the *Erase* (see Fig. 4-17). Use ***Oops*** to unerase.

**FIGURE 4-17**

4. **Use the *ALL* option and deselect**

   Use ***Erase***. Select all the objects with ***ALL***. Remove the four *Lines* (shown highlighted in Fig. 4-18) from the selection set by pressing **Shift** while PICKing. Complete the *Erase* to leave only the four *Lines*. Finally, use ***Oops***.

FIGURE 4-18

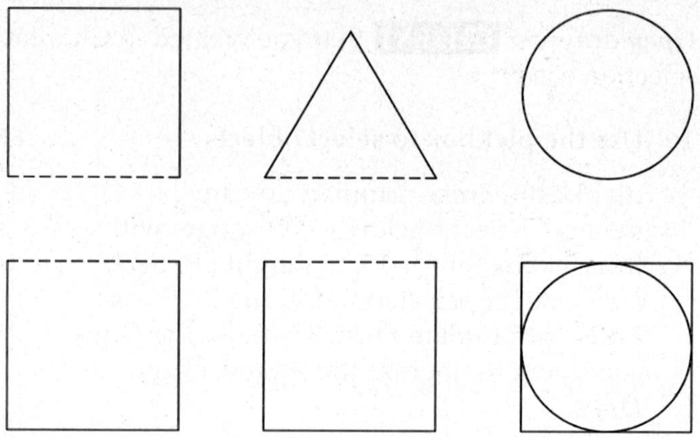

5. **Use Noun/Verb selection**

   Before invoking *Erase,* use the pickbox or *AUto Window* to select the triangle. (Make sure no other commands are in use.) Then invoke ***Erase***. The triangle should disappear. Retrieve the triangle with ***Oops***.

6. **Use *Move* with a crossing lasso**

   Invoke the ***Move*** command by any method. Use a crossing lasso to select only the *Lines* comprising the triangle. Turn on *Snap* and PICK the lower-left corner as the "Base point:". *Move* the triangle up 1 unit. (See Fig. 4-19.)

FIGURE 4-19

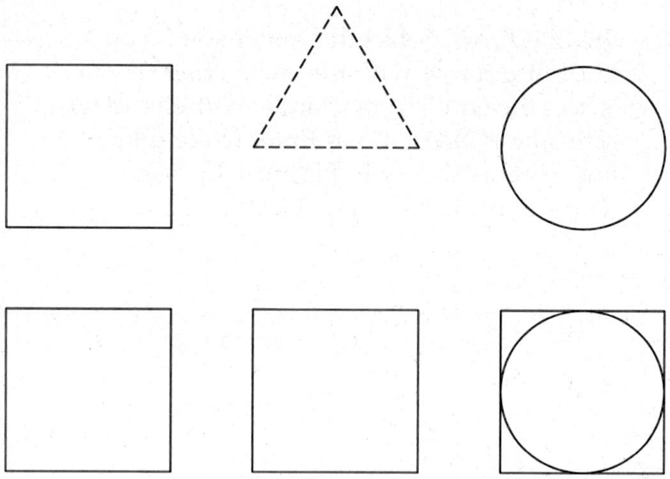

7. **Use *Previous* with *Move***

   Invoke ***Move*** again. At the "Select objects:" prompt, type ***P*** for *Previous*. The triangle should highlight. Using the same base point, move the triangle back to its original position.

8. ***Exit*** AutoCAD and do not save changes.

CHAPTER

# 5

# Helpful Commands

## Chapter Objectives

After completing this chapter you should:

1. be able to use the AutoCAD Assistant;
2. be able to access *Help* for any command or system variable;
3. be able to use *Oops* to unerase objects;
4. be able to use *U* to undo one command or use *Undo* to undo multiple commands;
5. be able to *Redo* commands that were undone;
6. be able to regenerate the drawing with *Regen*.

# CONCEPTS

There are several commands that do not draw or edit objects in AutoCAD but are intended to assist you in using AutoCAD. These commands are used by experienced AutoCAD users and are particularly helpful to the beginner. The commands, as a group, are not located in any one menu but are scattered throughout several menus.

## ASSISTANT

| Ribbon | Menu Bar | Command (Type) | Alias (Type) | Shortcut |
|---|---|---|---|---|
| *InfoCenter Autodesk Assistant icon* | | *Assistantopen* | | |

You can use the Autodesk Assistant's AI-enhanced chat to ask for information using ordinary language. You can also query the drawing database. For example you can have the assistant list specified layers, blocks in use, and compare your drawing to company standards. Some of the Assistant features are in Tech Preview and may not give correct results every time. This is a super useful tool to learn ways to use AutoCAD commands, find information efficiently, and see if your question can quickly be answered by AI tools within AutoCAD. Use the InfoCenter icon to open the Autodesk Assistant palette as shown in Figure 5-1. It is very hard to see the icon in the light user interface. It is located just to the right of the help icon in the Info Center area.

TIP

Try it by entering a question in the prompt window where you see the text "Ask Autodesk Assistant" or use an example provided in the list. Figure 5-2 shows the result for the query, "Help me start a drawing that will be 300 feet by 180 feet."

**FIGURE 5-1**

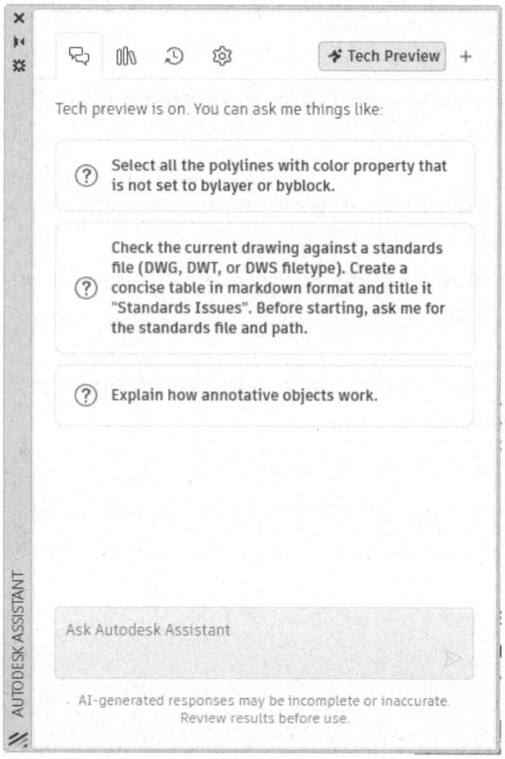


**FIGURE 5-2**

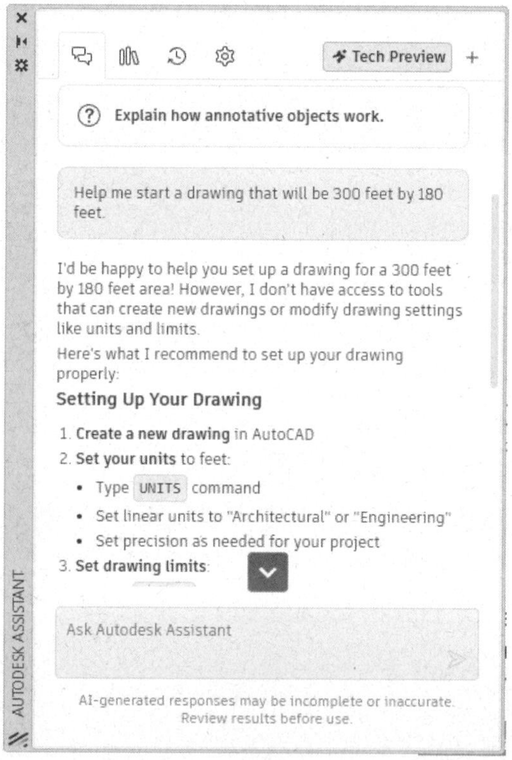

## Writing effective prompts for AutoCAD

### *Use Verb + Object + Context*

Start your prompt with an action, name the thing you want, and add the details.
This structure helps to remove ambiguity and tells the assistant exactly what to do.

*Examples italicized words below are Verbs, Objects, and Context*

- ***Explain*** how ***annotative scaling*** works when ***plotting from a layout***.
- ***Select*** all ***polylines*** whose ***color property*** is anything other than ByLayer or ByBlock.
- ***Teach me*** how to ***display dynamic constraints*** in a drawing.

### *Be Specific About the Output You Want*

Tell the assistant what form the answer should take. For example add an output constraint, like one of the following:

- Use bullet points
- Give step-by-step instructions
- Provide a layer list
- Generate a standards report
- Give troubleshooting steps
- Write a code snippet; AutoLISP, scripts, etc.

### *Add Helpful Details*

Context helps the assistant give accurate, relevant answers. For example, use a prompt like "Describe how to set up a plot style for a mechanical drawing using a specified CTB file. Specify details like the following:

- Drawing purpose (mechanical, architectural, civil)
- Units (imperial vs. metric)
- Overall dimensions
- File standards or templates
- Layer naming conventions

### *Ask the Assistant to improve your prompts*

Asking the Assistant to consolidate or combine your prompts can help turn messy or multi-step conversations into a single, reusable prompt. For example, type "*Combine this chat into one prompt that identifies non-standard object settings*" or "*Improve the above prompt so I can reuse it in the future*" or "*Help me turn this into a clear prompt.*"

### *Tips for Better Results*

Use precise drawing object names (layers, constraints, blocks, Xrefs). Avoid vague words like "simple," "nice," or "clean." Break complex tasks into steps—or ask the assistant to do that for you.
Here are some more helpful example prompts:

- Explain the likely causes of [problem] and give me steps to fix it.
- Check this drawing for [specific standards] and summarize any issues.
- Teach me how to use [feature] with a simple example.
- Write an AutoLISP routine for [task], using [parameters].

### *Save Useful Prompts*

Save prompts that work well by clicking the icon which appears when you hover the mouse in the lower right of each prompt input area (Figure 5-3). Use these prompts as is or as a starting point for the next time you have a question for Autodesk Assistant.

**FIGURE 5-3**

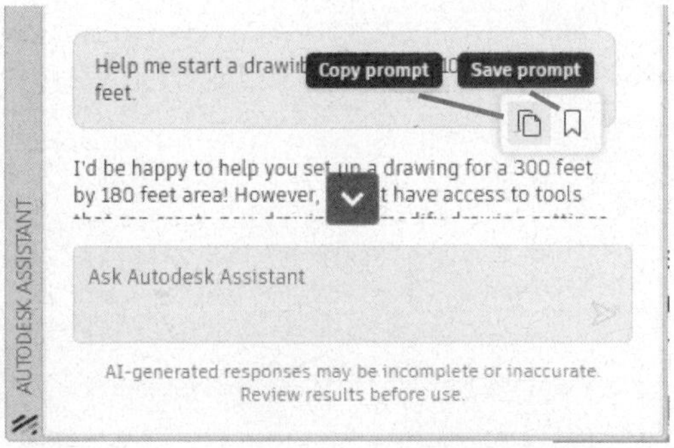

## Help

| Ribbon | Menu Bar | Command (Type) | Alias (Type) | Shortcut |
|---|---|---|---|---|
| ... | *Help*<br>*Help* | *Help* | *?* | *F1* |

*Help* gives you an explanation for any AutoCAD command or system variable as well as help for using the menus and toolbars. *Help* displays a window that accesses the Autodesk web site, which gives a variety of methods for finding the information that you need. There is even help for using *Help*!

*Help* can be used two ways: (1) entered as a command at the open Command: prompt or (2) used transparently while a command is currently in use.

1. If the *Help* command is entered at an open Command: prompt (when no other commands are in use), the *Help* window appears (Fig. 5-4).

2. When *Help* is used transparently (when a command is in use), it is context sensitive; that is, help on the current command is given automatically. For example, Figure 5-5 displays the window that appears if *Help* is invoked during the *Line* command. (If typing a transparent command, an apostrophe (') symbol is typed as a prefix to the command, such as ***'HELP*** or ***'?***. If you press **F1**, it is automatically transparent.)

TIP

**FIGURE 5-4**

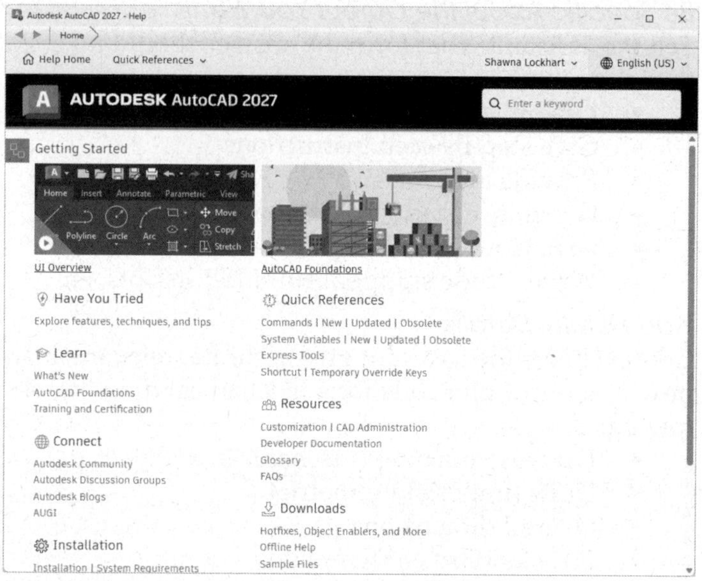


**FIGURE 5-5**

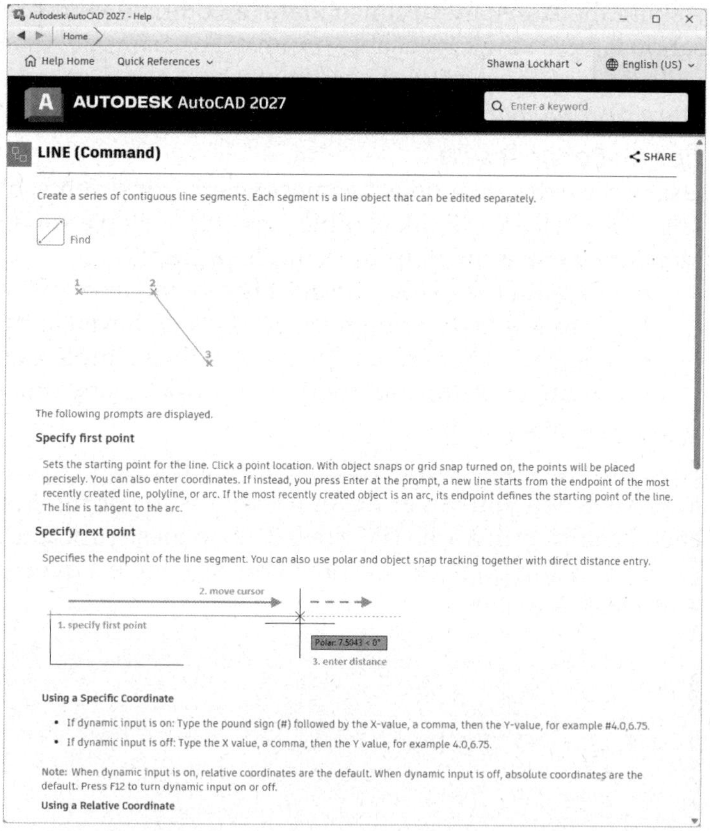

Much of the text that appears in the window can be PICKed to reveal another level of help on that item. This feature, called hypertext, is activated by moving the pointer to a word (usually in a different color) or an icon. When the pointer changes to a small hand, click on the item to activate the new information.

The features in the *Help* window are described next.

FIGURE 5-6

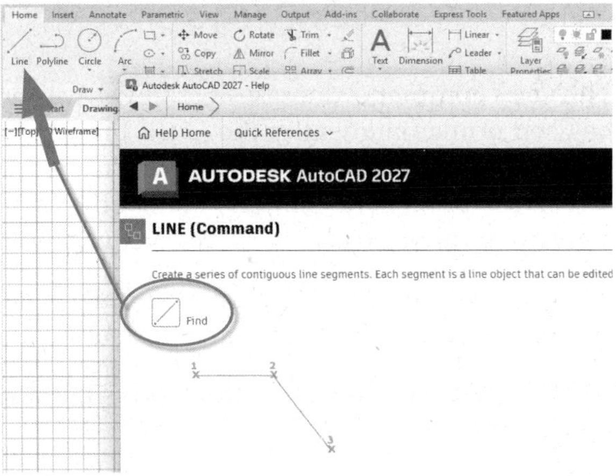


## UI Finder

You can use the *UI Finder* (User Interface Finder) in *Help* to indicate where specific tools (icons) are located in the AutoCAD screen. The *UI Finder* may appear near the Ribbon, Status Bar, or wherever the command or feature is located. For example, Figure 5-6 shows the *UI Finder* indicating the location of the *Line* tool on the Ribbon.

To use this function, open the *Help* window and use *Search* (see "*Search*" below) to display help for the desired command, and then click on the "*Find*" button in the *Help* window.

## Search

*Search* allows you to enter several keywords that describe the topic you want help with. Type in word or words in the edit box (Fig. 5-7) and press **Enter**. The related information appears as a list of topics.

FIGURE 5-7

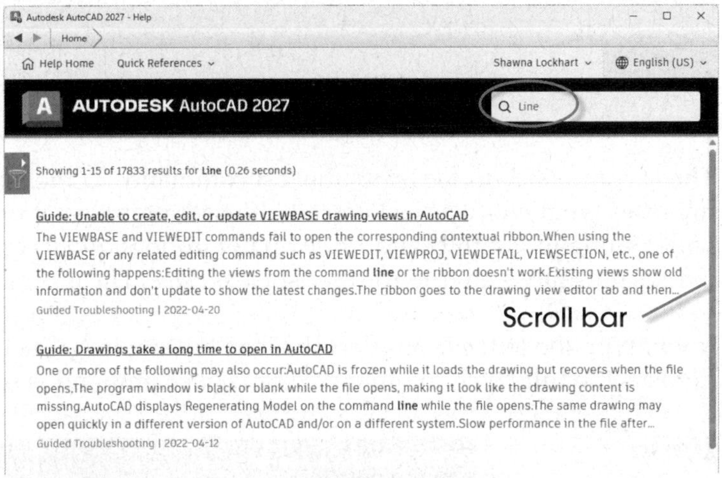


## Refine By

You can click the filter icon to use the *Refine By* options from the left panel to refine the list of help topics.

Several general filter categories are available: *Knowledge Source, Information Category, Product Feature, Content Type, Activity, Content Provider, User Type, Skill Level,* and *Media*. Each category provides several filters. Click the items to apply as many filters from as many categories as you like. Active filters are listed at the top of the results list. To remove a filter, pick the "**X**" icon next to its name. To remove all active filters, use the *Clear All* button. Figure 5-8 shows two active filters.

FIGURE 5-8

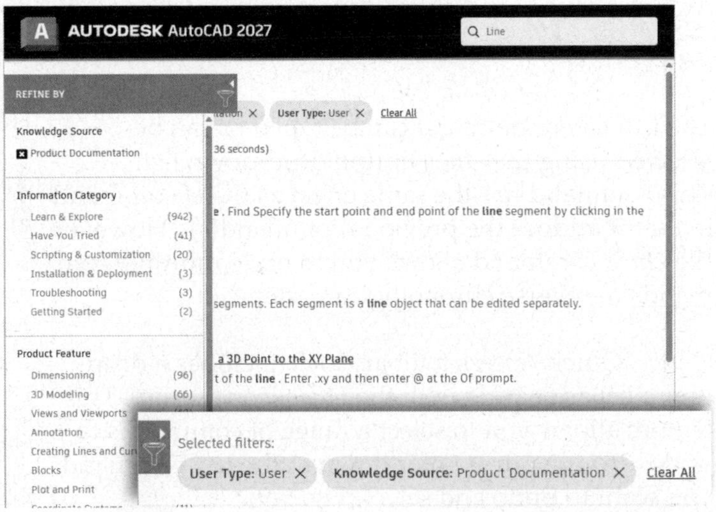

## Oops

| Ribbon | Menu Bar | Command (Type) | Alias (Type) | Shortcut |
|---|---|---|---|---|
| ... | ... | *Oops* | ... | ... |

The *Oops* command unerases whatever was erased with the last *Erase* command. *Oops* does not have to be used immediately after the *Erase*, but can be used at any time after the *Erase*. *Oops* is typically used after an accidental erase. However, *Erase* could be used intentionally to remove something from the screen temporarily to simplify some other action. For example, you can *Erase* a *Line* to simplify PICKing a group of other objects to *Move* or *Copy*, and then use *Oops* to restore the erased *Line*.

*Oops* can be used to restore the original set of objects after the *Block* or *Wblock* command is used to combine many objects into one object (explained in Chapter 21).

TIP

NOTE: *Oops* has no icon button and is not accessible from any menu. *Oops* must be typed at the Command line.

## U

| Ribbon | Menu Bar | Command (Type) | Alias (Type) | Shortcut |
|---|---|---|---|---|
| ... | *Edit*<br>*Undo* | *U* | ... | *Ctrl+Z* or (Default Menu) *Undo* |

The *U* command undoes only the last command. *U* means "undo one command." If used after *Erase,* it unerases whatever was just erased. If used after *Line,* it undoes the group of lines drawn with the last *Line* command. Both *U* and *Undo* do not undo inquiry commands (like *Help*), the *Plot* command, or commands that cause a write-to-disk, such as *Save*. The button is found in the Quick Access Toolbar.

TIP

If you type the letter ***U*** or select the icon button, only the last command is undone. Typing ***Undo*** invokes the full *Undo* command. The *Undo* command is explained next.

## Undo

| Ribbon | Menu Bar | Command (Type) | Alias (Type) | Shortcut |
|---|---|---|---|---|
| ... | ... | *Undo* | ... | ... |

The full *Undo* command can be typed or can be selected using the *Undo* button drop-down list. This command has the same effect as the *U* command in that it undoes the previous command(s). However, the *Undo* command allows you to undo multiple commands in reverse chronological order.

On the Quick Access toolbar, *Undo* includes a drop-down list appearing with the *Undo* icon button. This feature allows you to select a range of commands to undo. Bring your pointer down to the last command you want to undo and select (Fig. 5-9).

FIGURE 5-9

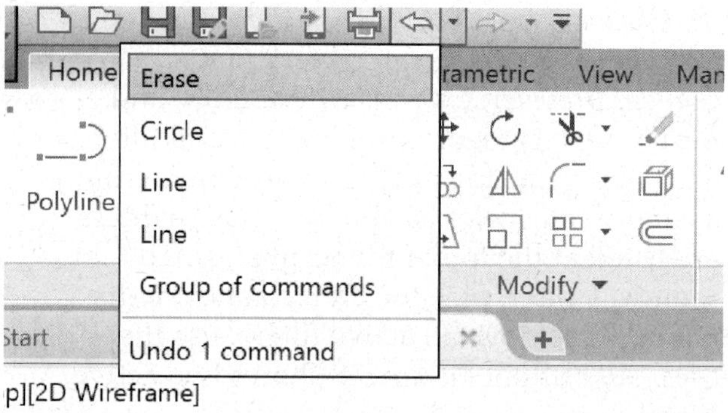

If you type the *Undo* command, the command prompt below appears. The default option of *Undo* is "<1>," which means to undo only the last command (the same action as using the *U* command). Do this by pressing Enter. All of the *Undo* options are listed next.

Command: ***UNDO***
Current settings: Auto = On, Control = All, Combine = Yes; Layer = Yes
Enter the number of operations to undo or [Auto Control BEgin End Mark Back] <1>:

### *<number>*

Enter a value for the number of commands to *Undo*. This is the default option.

### *Mark*

This option sets a marker at that stage of the drawing. The marker is intended to be used by the *Back* option for future *Undo* commands.

### *Back*

This option causes *Undo* to go back to the last marker encountered. Markers are created by the *Mark* option. If a marker is encountered, it is removed. If no marker is encountered, beware, because *Undo* goes back to the beginning of the session. A warning message appears in this case.

### *BEgin*

This option sets the first designator for a group of commands to be treated as one *Undo*.

### *End*

*End* sets the second designator for the end of a group.

### *Auto*

If *On*, *Auto* treats each command as one group; for example, several lines drawn with one *Line* command would all be undone with *U*.

### *Control*

This option gives the following controls.

*All*

*All* turns on the full *Undo* command functions.

*None*

*None* turns off the *U* and *Undo* commands. If you do use *Undo* while it is turned off, only the *All*, *None*, and *One* options are displayed at the command line.

*One*

This option limits *Undo* to a single operation, similar to using the *U* command.

*Combine*

*Combine* controls whether multiple, consecutive zoom and pan commands are combined as a single operation for undo and redo operations (on by default).

*Layer*

This option controls whether multiple actions you perform in the layer dialog box are combined as a single undo operation.

## Redo

| Ribbon | Menu Bar | Command (Type) | Alias (Type) | Shortcut |
|---|---|---|---|---|
| ... | *Edit*<br>*Redo* | *Redo* | ... | *Ctrl+Y* or (Default Menu) *Redo* |

TIP

The *Redo* command undoes an *Undo*. *Redo* must be used as the next command after the *Undo*. The result of *Redo* is as if *Undo* was never used. *Redo* automatically reverses the action of the previous *Undo*, even when multiple commands were undone. Remember, *U* or *Undo* can be used at any time, but *Redo* has no effect unless used immediately after *U* or *Undo*. The *Redo* button is found on the Quick Access Toolbar.

## Mredo

| Ribbon | Menu Bar | Command (Type) | Alias (Type) | Shortcut |
|---|---|---|---|---|
| ... | ... | *Mredo* | ... | ... |

The *Redo* command reverses the total action of the last *Undo* command. In contrast, *Mredo* (multiple redo) allows you to select how many of the commands, in reverse chronological order, you want to redo.

*Mredo* is useful only when the *Undo* command or drop-down list was just used (as the last command) and multiple commands were undone. Then you can immediately use *Mredo* to reverse the action of any number of the previous *Undos*. This feature is available (the icon is enabled) only if *Undo* was just used.

On the Quick Access toolbar, you can achieve a *Mredo* using the *Redo* drop-down list to select from the list of previously undone commands to reverse the action of the *Undo* (Fig. 5-10). Remember, the list includes only commands that were undone during the previous *Undo* command.

**FIGURE 5-10**

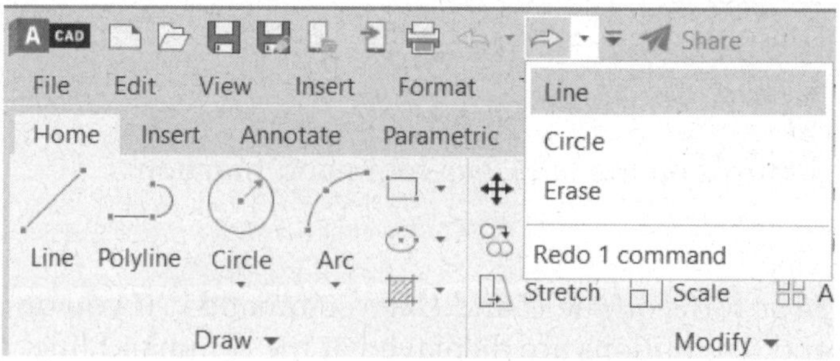


Typing *Mredo* yields the following prompt:

Command: ***MREDO***
Enter number of actions or [All Last]:

Enter a value for the number of commands you want to redo. *All* will redo all of the undos (all the undos shown in the *Redo* drop-down list), the same as using the *Redo* command. The *Last* option generally has the same result as the *All* option; however, when the *U* command was used multiple times in succession, the *Last* option undoes only the last *U*.

### *Regen*

| Ribbon | Menu Bar | Command (Type) | Alias (Type) | Shortcut |
|---|---|---|---|---|
| ... | *View*<br>*Regen* | *Regen* | *RE* | ... |

The *Regen* command reads the database and redisplays the drawing accordingly. Occasionally, the *Regen* command is required to update the drawing to display the latest changes made to some system variables. *Regenall* is used to regenerate all viewports when several viewports are being used.

*Regen* is sometimes needed after you *Zoom* or *Pan* in versions of AutoCAD previous to 2016. Since AutoCAD 2016 you no longer have to perform a manual *Regen* when zooming or panning. AutoCAD automatically executes a *Regen* as needed when zooming or panning.

## CHAPTER EXERCISES

Start AutoCAD. Select the **ACAD.DWT** template, then ***Zoom All***. Complete the following exercises.

1. ***Help***

   Use ***Help*** by any method to find information on the following commands. Use *Search* to locate information on each command. (Use *Help* at the open Command: prompt, not during a command in use.) Read the text screen for each command.

   *New, Open, Save, SaveAs*
   *Oops, Undo, U*

2. **Context-sensitive *Help***

   Invoke each of the commands listed below. When you see the first prompt in each command, enter ***'Help*** or ***'?*** (transparently) or press F1. Read the explanation for each prompt. Select a hypertext item in each screen (underlined).

   *Line, Arc, Circle, Point*

3. ***Oops***

   Draw 3 vertical ***Lines***. ***Erase*** one line then use ***Oops*** to restore it. Next, ***Erase*** two *Lines,* each with a separate use of the *Erase* command. Use ***Oops***. Only the last *Line* is restored. ***Erase*** the remaining two *Lines,* but select both with a *Window*. Now use ***Oops*** to restore both *Lines* (since they were *Erased* at the same time).

4. **Delayed *Oops***

   *Oops* can be used at any time, not only immediately after the *Erase*. Draw several horizontal ***Lines*** near the bottom of the screen. Draw a ***Circle*** on the ***Lines***. Then ***Erase*** the *Circle*. Use ***Move***, select the *Lines* with a *Window*, and displace the *Lines* to another location above. Now use ***Oops*** to make the *Circle* reappear.

5. ***U***

Press the letter ***U*** (make sure no other commands are in use). The *Circle* should disappear (*U* undoes the last command—*Oops*). Do this repeatedly to *Undo* one command at a time until the *Circle* and *Lines* are in their original position (when you first created them).

6. ***Undo***

Use the ***Undo*** command and select the ***Back*** option. Answer ***Yes*** to the warning message. This action should *Undo* everything.

Draw a vertical ***Line***. Next, draw a square with four ***Line*** segments (all drawn in the same *Line* command). Finally, draw a second vertical ***Line***. ***Erase*** the first *Line*.

Now <u>type</u> ***Undo*** and enter a value of **3**. You should have only one (the first) *Line* remaining. *Undo* reversed the following three commands:

| | |
|---|---|
| *Erase* | The first vertical *Line* was unerased. |
| *Line* | The second vertical *Line* was removed. |
| *Line* | The four *Lines* comprising the square were removed. |

7. ***Multiple Undo***

Draw two *Circles* and two more *Lines* so your drawing shows a total of two *Circles* and three *Lines*. Now use the *Multiple Undo* drop-down list. The list should contain *Line, Line, Circle, Circle, Line*. Highlight the first two lines and the two circles in the list, then left-click. Only one *Line* should remain in your drawing.

8. ***Multiple Redo***

Select the *Multiple Redo* drop-down list. The list should contain *Circle, Circle, Line, Line*. Highlight the first two *Circles* in the list, then left-click. The two *Circles* should reappear in your drawing, now showing a total of two *Circles* and one *Line*.

9. ***Exit*** AutoCAD and answer ***No*** to "Save Changes to…?"

# CHAPTER 6

# Basic Drawing Setup

## Chapter Objectives

After completing this chapter you should:

1. know the basic steps for setting up a drawing;
2. know how to start a drawing using a template;
3. be able to specify the desired *Units*, *Angles* format, and *Precision* for the drawing;
4. be able to specify the drawing *Limits*;
5. know how to specify the *Snap* and *Grid* increments;
6. understand the basic function and use of a *Layout*;
7. understand the basic principles for setting the print scale and printing from a *Layout*.

## STEPS FOR BASIC DRAWING SETUP

Assuming the general configuration (dimensions and proportions) of the geometry to be created is known, the following steps are suggested for setting up a drawing:

1. Determine and set the *Units* that are to be used.
2. Determine and set the drawing *Limits;* then *Zoom All.*
3. Set an appropriate *Snap* type and increment.
4. Set an appropriate *Grid* value to be used.

These additional steps for drawing setup are discussed also in Chapter 12, Advanced Drawing Setup.

5. Change the *LTSCALE* value based on the new *Limits.*
6. Create the desired *Layers* and assign appropriate *linetype* and *color* settings.
7. Create desired *Text Styles* (optional).
8. Create desired *Dimension Styles* (optional).
9. Activate a *Layout* tab, set it up for the plot or print device and paper size, and create a viewport (if not already existing).
10. Create or insert a title block and border in the layout.

## STARTUP OPTIONS

As explained in Chapter 1, Getting Started, you can begin a new drawing two ways:

1. When you start AutoCAD, select *Start Drawing* or *Templates* from the *Get Started* button in the *Create* screen.
2. If you are already in the AutoCAD Drawing Editor, select *New* or *Qnew* or select the *Start* tab to go to the *Create* screen.

Also explained in Chapter 1, when you begin a new drawing a template file (.DWT file type) is used as a starting point. Based on your choice for starting a new drawing, you can select from a list of templates or use a previously assigned template as described below.

1. If you use the *New* command or select *Templates* from the *Get Started* button, you can select from any of the supplied templates.
2. If you use the *Qnew* command or select *Start Drawing* from the *Create* screen's *Get Started* button, AutoCAD begins with a previously assigned template. You can specify which template is the automatic starting template in the *Options* dialog box (see Chapter 1). If a template has not yet been specified, the *Qnew* command produces the *Select Template* dialog box, whereas the *Start Drawing* button in the opening screen uses the ACAD.DWT by default.

Chapter 12, Advanced Drawing Setup, explains how to create your own drawing templates to appear in the list of available templates.

NOTE: The *STARTUP* system variable can be used to control the appearance of an older series of dialog boxes for selecting templates or using "wizards" to set up a drawing. These dialog boxes are not especially efficient or effective and simply duplicate the steps for setting up a drawing using the commands explained in the following "Setup Commands" section. In fact, using either of the two wizards (*Quick Setup* wizard or *Advanced Setup* wizard) and selecting the default options or using the *Start from Scratch* dialog box produces a similar result—beginning with the ACAD.DWT (inch) or the ACADISO.DWT (metric) drawing template! The default setting for *STARTUP* is 3 (starting with a preloaded template); a setting of 1 turns on the dialog boxes. (See *New* in Chapter 2 for all *STARTUP* settings.)

## DRAWING TEMPLATES

No matter which method you choose to begin a new drawing, one of the drawing templates is used as a starting point. Template drawings can have some of the setup steps already performed but they generally contain no drawing geometry. The main difference between all of the ACAD.DWT (inch) templates and the ACADISO.DWT (metric) templates is the drawing area provided (set using the *Limits* command) and the sizes of other drawing features related to using inch or metric dimensions.

As explained earlier, you can select a specific template to use or use a previously assigned template. If you wish to select a template when you first launch AutoCAD, you can use the *Template* list in the *New* selection on the *Start* tab area (Fig. 6-1).

**FIGURE 6-1**

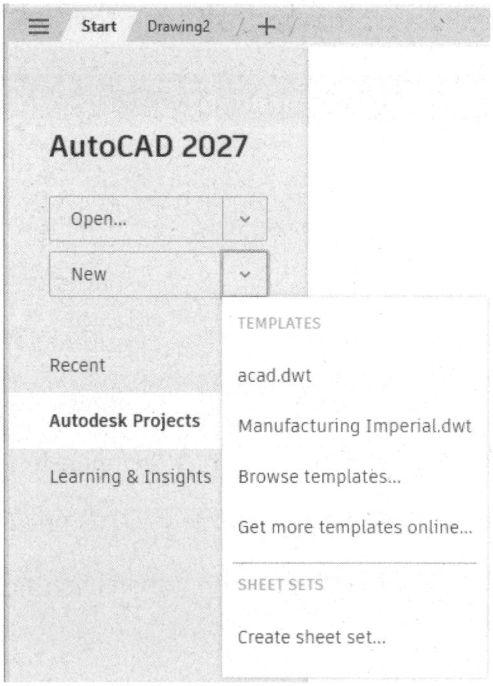


Alternately, if you are already running AutoCAD and see the Drawing Editor, you can use the *New* command to produce the *Select Template* dialog box (Fig. 6-2).

More information on these templates, including creating template drawings and using templates provided by AutoCAD, is given in Chapter 12, Advanced Drawing Setup.

**FIGURE 6-2**

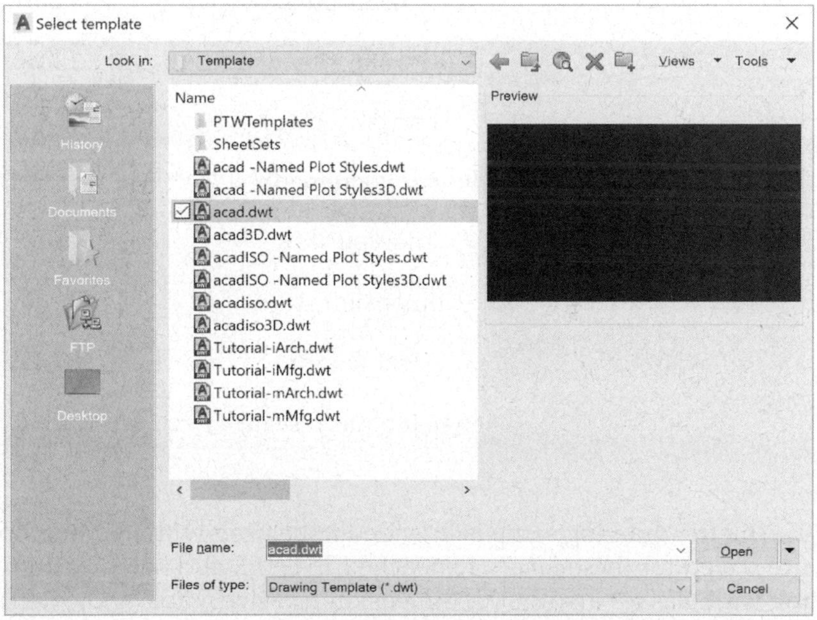


NOTE: After selecting the ACAD.DWT template drawing, it is recommended that you perform a *Zoom* with the *All* option. Do this by selecting *Zoom, All* from the *Navigate* panel in the *View* tab or type *Z*, press Enter, then type *A* and Enter. This action ensures that the drawing displays all of the drawing *Limits*.

TIP

The following two tables list the AutoCAD settings for the two template drawings, ACAD.DWT and ACADISO.DWT. Many of these settings, such as *Units*, *Limits*, *Snap*, and *Grid*, are explained in the following sections.

### Table of Basic Settings for the ACAD (Inch) Templates

| Related Command | Description | System Variable | Default Setting |
|---|---|---|---|
| *Units* | linear units | *LUNITS* | 2 (decimal) |
| *Limits* | drawing area | *LIMMAX* | 12.0000, 9.0000 |
| *Snap* | snap increment | *SNAPUNIT* | .5000, .5000 |
| *Grid* | grid increment | *GRIDUNIT* | .5000, .5000 |
| *LTSCALE* | linetype scale | *LTSCALE* | 1.0000 |
| *DIMSCALE* | dimension scale | *DIMSCALE* | 1.0000 |
| *Text, Mtext* | text height | *TEXTSIZE* | .2000 |
| *Hatch* | hatch pattern scale | *HPSCALE* | 1.0000 |

### Table of Basic Settings for the ACADISO (Metric) Templates

| Related Command | Description | System Variable | Default Setting |
|---|---|---|---|
| *Units* | linear units | *LUNITS* | 2 (decimal) |
| *Limits* | drawing area | *LIMMAX* | 420.0000, 297.0000 |
| *Snap* | snap increment | *SNAPUNIT* | 10.0000, 10.0000 |
| *Grid* | grid increment | *GRIDUNIT* | 10.0000, 10.0000 |
| *LTSCALE* | linetype scale | *LTSCALE* | 1.0000 |
| *DIMSCALE* | dimension scale | *DIMSCALE* | 1.0000 |
| *Text, Mtext* | text height | *TEXTSIZE* | 2.5000 |
| *Hatch* | hatch pattern scale | *HPSCALE* | 1.0000 |

The metric drawing setup is intended to be used with ISO linetypes and ISO hatch patterns, which are pre-scaled for these *Limits*, hence the *LTSCALE* and hatch pattern scale of 1. The individual dimensioning variables for arrow size, dimension text size, gaps, and extensions, etc. are changed so the dimensions are drawn correctly with a *DIMSCALE* of 1 with the ISO-25 dimension style.

# SETUP COMMANDS

Once you have selected a template to start a drawing, and before you actually draw anything such as *Lines* and *Circles*, use the following commands to set up the drawing with the desired units of measure, expected drawing area, and drawing aids values.

## *Units*

| Ribbon | Menu Bar | Command (Type) | Alias (Type) | Shortcut |
|---|---|---|---|---|
| ... | *Format Units...* | *Units* or *-Units* | *UN* or *-UN* | ... |

The *Units* command allows you to specify the type and precision of linear and angular units as well as the direction and orientation of angles to be used in the drawing. The current setting of *Units* determines the display of values by the coordinates display and in some dialog boxes.

You can select *Units ...* from the *Format* pull-down or type *Units* (or command alias *UN*) to invoke the *Drawing Units* dialog box (Fig. 6-3). Type *-Units* (or *-UN*) to produce a text screen (Fig. 6-4).

**FIGURE 6-3**

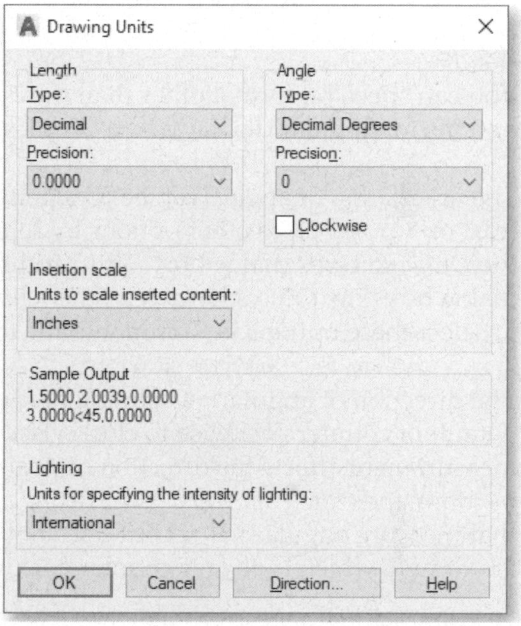


The linear and angular units options are displayed in the dialog box format (Fig. 6-3) and in Command line format (Fig. 6-4). Access the Command line format by typing *-Units* or *-UN*. The choices for both linear and angular *Units* are shown next.

**FIGURE 6-4**

```
UNITS
Command: Specify opposite corner or [Fence/WPolygon/CPolygon]:
Invalid window specification.
Command: Specify opposite corner or [Fence/WPolygon/CPolygon]: *Cancel*
Command: -UN
-UNITS Report formats:      (Examples)
  1.  Scientific      1.55E+01
  2.  Decimal         15.50
  3.  Engineering     1'-3.50"
  4.  Architectural   1'-3 1/2"
  5.  Fractional      15 1/2
With the exception of Engineering and Architectural formats,
these formats can be used with any basic unit of measurement.
For example, Decimal mode is perfect for metric units as well
as decimal English units.
Enter choice, 1 to 5 <2>:
Enter number of digits to right of decimal point (0 to 8) <4>:
Systems of angle measure:       (Examples)
  1.  Decimal degrees             45.0000
  2.  Degrees/minutes/seconds     45d0'0"
  3.  Grads                       50.0000g
  4.  Radians                     0.7854r
  5.  Surveyor's units            N 45d0'0" E
Enter choice, 1 to 5 <1>:
Enter number of fractional places for display of angles (0 to 8) <0>:
Direction for angle 0:
   East    3 o'clock  =  0
   North  12 o'clock  =  90
   West    9 o'clock  =  180
   South   6 o'clock  =  270
Enter direction for angle 0 <0>:
Measure angles clockwise? [Yes/No] <N>
```

| *Units* | Format | |
|---|---|---|
| 1. Scientific | 1.55E + 01 | Generic decimal units with an exponent. |
| 2. Decimal | 15.50 | Generic decimal, usually used for applications in metric or decimal inches. |
| 3. Engineering | 1′-3.50″ | Explicit feet and decimal inches with notation, one unit equals one inch. |
| 4. Architectural | 1′-3 1/2″ | Explicit feet and fractional inches with notation, one unit equals one inch. |
| 5. Fractional | 15 1/2 | Generic fractional units. |

### *Precision*

When setting *Units,* you should also set the precision. *Precision* is the number of places to the right of the decimal or the denominator of the smallest fraction to display. The precision is set by making the desired selection from the *Precision* pop-up list in the *Drawing Units* dialog box (Fig. 6-3) or by keying in the desired selection in Command line format.

TIP

*Precision* controls only the display of values in the coordinate display and in dialog boxes. The actual precision of the drawing database is always the same in AutoCAD: 14 significant digits.

### *Angles*

You can specify a format other than the default (decimal degrees) for expression of angles. Format options for angular display and examples of each are shown in Figure 6-3 (dialog box format).

The orientation of angle 0 can be changed from the default position (east or X positive) to other options by selecting the *Direction* tile in the *Drawing Units* dialog box. This produces the *Direction Control* dialog box (Fig. 6-5). Alternately, the *-Units* command can be typed to select these options in Command line format.

**FIGURE 6-5**

The direction of angular measurement can be changed from its default of counterclockwise to clockwise. The direction of angular measurement affects the direction of positive and negative angles in commands such as *Array Polar, Rotate* and dimension commands that measure angular values but does not change the direction *Arcs* are drawn, which is always counterclockwise.

### *Insertion Scale*

This section in the *Drawing Units* dialog box specifies the units to use when you drag-and-drop *Blocks* from AutoCAD DesignCenter or Tool palettes into the current drawing. There are many choices including *Unitless, Inches, Feet, Millimeters,* and so on. If you intend to insert *Blocks* into the drawing (you are currently setting units for) using drag-and-drop, select a unit from the list that matches the units of the *Blocks* to insert. If you are not sure, select *Unitless*. See Chapter 21, Blocks, DesignCenter, and Tool Palettes, for more information on this subject.

## Keyboard Input of *Units* Values

When AutoCAD prompts for a point or a distance, you can respond by entering values at the keyboard. The values can be in any format—integer, decimal, fractional, or scientific—regardless of the format of *Units* selected.

TIP

You can type in explicit feet and inch values only if *Architectural* or *Engineering* units have been specified as the drawing units. For this reason, specifying *Units* is the first step in setting up a drawing.

Type in explicit feet or inch values by using the apostrophe (′) symbol after values representing feet and the quote (″) symbol after values representing inches. If no symbol is used, the values are understood by AutoCAD to be inches.

TIP

Feet and inches input cannot contain a blank, so a hyphen (-) must be typed between inches and fractions. For example, with *Architectural* units, key in **6′2-1/2″**, which reads "six feet, two and one-half inches." The standard engineering and architectural format for dimensioning, however, places the hyphen between feet and inches (as displayed by the default setting for the coordinate display).

The *UNITMODE* variable set to 1 changes the coordinate display to remind you of the correct format for input of feet and inches (with the hyphen between inches and fractions) rather than displaying the standard format for feet and inch notation (standard format, *UNITMODE* of 0, is the default setting). If options other than *Architectural* or *Engineering* are used, values are read as generic units.

## Limits

| Ribbon | Menu Bar | Command (Type) | Alias (Type) | Shortcut |
|---|---|---|---|---|
| ... | *Format Drawing Limits* | *Limits* | ... | ... |

The *Limits* command allows you to set the size of the drawing area by specifying the lower-left and upper-right corners in X,Y coordinate values.

Command: LIMITS
Reset Model space limits
Specify lower left corner or [ON/OFF] <0.0000,0.0000>: X,Y or Enter (Enter an X,Y value or accept the 0,0 default—normally use 0,0 as lower-left corner.)
Specify upper right corner <12.0000,9.0000>: X,Y (Enter new values to change upper-right corner to allow adequate drawing area.)

The default *Limits* values in the ACAD.DWT are 12 and 9, that is, 12 units in the X direction and 9 units in the Y direction (Fig. 6-6).

Note: To show the grid only in the limits area use the Grid command, Limits option and enter N at the prompt:

Command: GRID
Specify grid spacing(X) or [ON/OFF/Snap/Major/aDaptive/Limits/Follow/Aspect] <1.0000>: **L**
Display grid beyond Limits [Yes/No] <Yes>: n Display grid beyond Limits [Yes/No] <Yes>: **N**

**FIGURE 6-6**

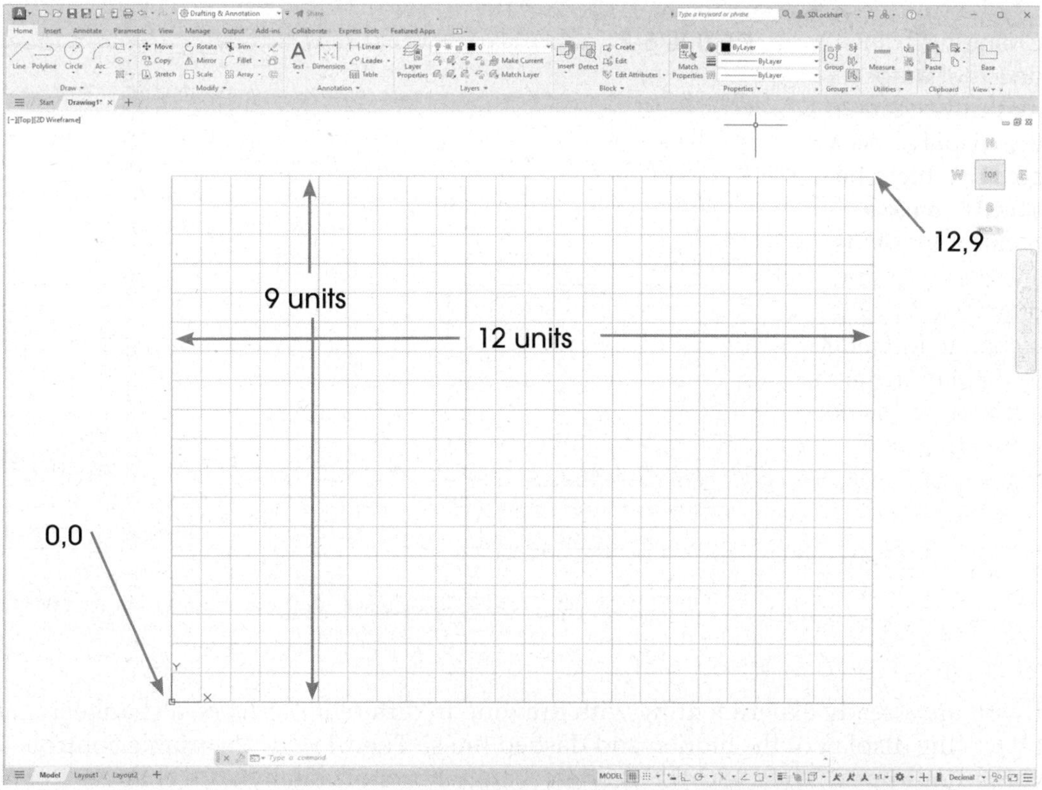

With the default settings for the ACAD.DWT, the *Grid* is displayed past the *Limits*. The grid display can be controlled using the *Limits* option of *Grid* (see "*Grid*"). The units are generic decimal units that can be used to represent inches, feet, millimeters, miles, or whatever is appropriate for the intended drawing. Typically, however, decimal units are used to represent inches or millimeters. If the default units are used to represent inches, the default drawing size would be 12 x 9 inches.

Remember that when a CAD system is used to create a drawing, the geometry should be drawn full size by specifying dimensions of objects in real-world units. A completed CAD drawing or model is virtually an exact dimensional replica of the actual object. Scaling of the drawing occurs only when plotting or printing the file to an actual fixed-size sheet of paper.

TIP

Before beginning to create an AutoCAD drawing, determine the size of the drawing area needed for the intended geometry. After setting *Units*, appropriate *Limits* should be set in order to draw the object or geometry to the real-world size in the actual units. There are no practical maximum or minimum settings for *Limits*.

The X,Y values you enter as *Limits* are understood by AutoCAD as values in the units specified by the *Units* command. For example, if you previously specified *Architectural units*, then the values entered are understood as inches unless the notation for feet (') is given (**240,180** or **20',15'** would define the same coordinate). Remember, you can type in explicit feet and inch values only if *Architectural* or *Engineering* units have been specified as the drawing units.

TIP

If you are planning to plot the drawing to scale, *Limits* should be set to a proportion of the sheet size you plan to plot on. For example, setting limits to 22 x 17 (2 times 11 by 8.5) would allow enough room for drawing an object about 20" x 15" and allow plotting at 1/2 size on the 11" x 8.5" sheet. Simply stated, set *Limits* to a proportion of the paper.

*Limits* also defines the display area when a *Zoom All* is used. *Zoom All* forces the full display of the *Limits*. *Zoom All* can be invoked by typing **Z** (command alias) then **A** for the *All* option.

TIP

Changing *Limits* does not automatically change the display. As a general rule, you should make a habit of invoking a *Zoom All* immediately following a change in *Limits* to display the area defined by the new limits (Fig. 6-7).

**FIGURE 6-7**

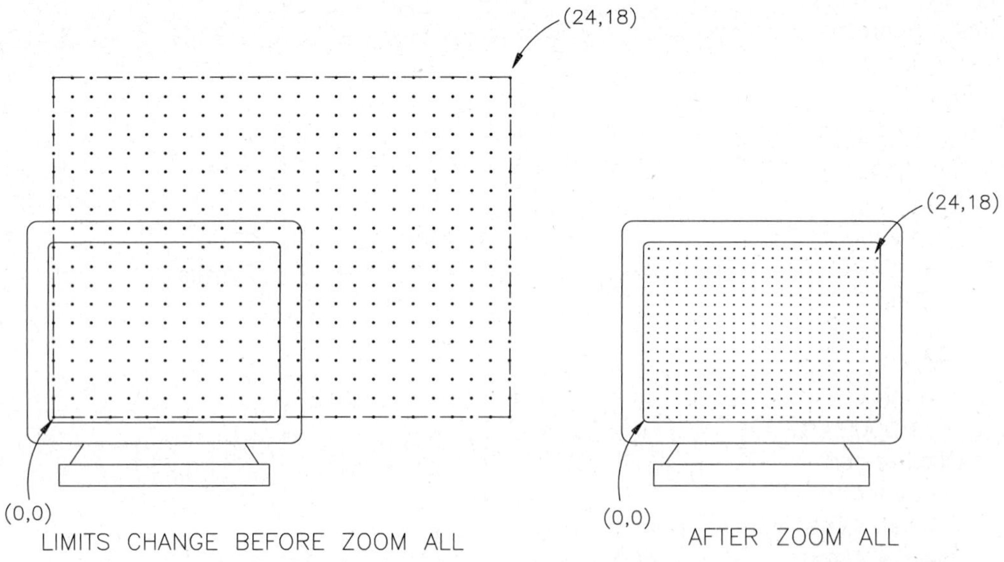


If you are already experimenting with drawing in different *Linetypes*, a change in *Limits* and *Zoom All* affects the display of the hidden and dashed lines. The *LTSCALE* variable controls the spacing of non-continuous lines. The *LTSCALE* is often changed proportionally with changes in *Limits*.

### *ON/OFF*

If the *ON* option of *Limits* is used, limits checking is activated. Limits checking prevents you from drawing objects outside of the limits by issuing an outside-limits error. This is similar to drawing "off the paper." Limits checking is *OFF* by default.

## Snap

| Ribbon | Menu Bar | Command (Type) | Alias (Type) | Shortcut |
|---|---|---|---|---|
| ... | *Tools*<br>*Drafting Settings...*<br>*Snap and Grid* | *Snap* | *SN* | *F9* or *Ctrl+B* |

*Snap* in AutoCAD has two possible types, Grid Snap and Polar Snap. When you are setting up the drawing, you should set the desired *Snap type* and increment. You may decide to use both types of *Snap*, so set both increments initially. You can have only one of the two *Snap types* active at one time. (See Chapter 3, Draw Command Concepts, for further explanation and practice using both *Snap* types.)

*Grid Snap* forces the cursor to preset positions on the screen, similar to the *Grid*. *Grid Snap* is like an invisible grid that the cursor "snaps" to. This function can be of assistance if you are drawing *Lines* and other objects to set positions on the drawing, such as to every .5 unit. The default value for *Grid Snap* is .5, but it can be changed to any value.

*Polar Snap* forces the cursor to move in set intervals from the previously designated point (such as the "first point" selected during the *Line* command) to the next point. *Polar Snap* operates for cursor movement at any previously set *Polar Tracking* angle, whereas *Grid Snap* (since it is rectangular) forces regular intervals only in horizontal or vertical movements. *Polar Tracking* must also be on (F10) for *Polar Snap* to operate.

Set the desired snap type and increments using the *Snap* command or the *Drafting Settings* dialog box. In the *Drafting Settings* dialog box, select the *Snap and Grid* tab (Fig. 6-8).

To set the *Polar Snap* increment, first select *Polar Snap* in the *Snap Type* section (lower left). This action causes the *Polar Spacing* section to be enabled. Next, enter the desired *Polar Distance* in the edit box.

To set the *Grid Snap* increment, first select *Grid Snap* in the *Snap Type* section. This action causes the *Snap* section to be enabled. Next, enter the desired *Snap X Spacing* value in the edit box. If you want a non-square snap grid, remove the check by *Equal X and Y spacing*, then enter the desired values in the *Snap X spacing* and *Snap Y spacing* edit boxes.

**FIGURE 6-8**

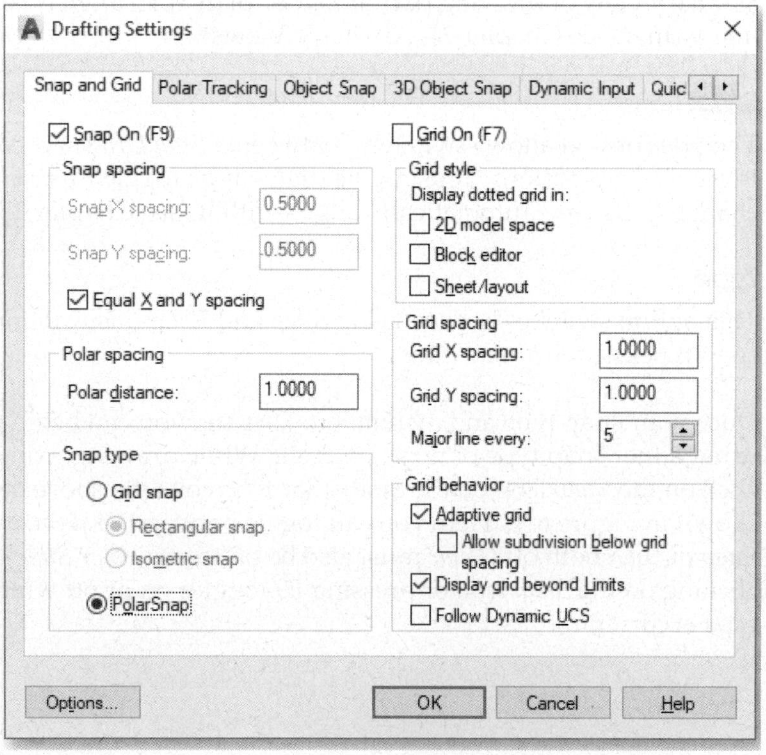

The Command line format of *Snap* is as follows:

```
Command: SNAP
Specify snap spacing or [ON/OFF/Aspect/Legacy/Style/Type] <0.5000>:
```

### Value

Entering a value at the Command line prompt sets the *Grid Snap* increment only. You must use the *Drafting Settings* dialog box to set the *Polar Snap* distance.

### *ON/OFF*

Selecting *ON* or *OFF* accomplishes the same action as toggling the F9 key or selecting the *Snap* icon on the Status Bar.

### *Aspect*

The *Aspect* option allows specification of unequal X and Y spacing for the snap grid. This is identical to entering unequal *Snap X Spacing* and *Snap Y Spacing* in the dialog box.

### *Legacy*

Normally (when *Legacy* is set to *No*), the crosshairs cursor does not "snap" to regular increments unless you are in a drawing command or other command that expects you to specify a point. If you specify a setting of *Yes*, the crosshairs always "snap" to regular increments as specified by the *Value* even when no command is active. The following prompt appears at the Command line for this option.

```
Maintain legacy behavior of always snapping to the grid? [Yes/No] <No>:
```

### *Rotate*

TIP

Although the *Rotate* option is not listed, the *Grid Snap* can be rotated about any point and set to any angle. When *Snap* has been rotated, the *Grid*, *Ortho*, and the "crosshairs" automatically follow this alignment. This action facilitates creating objects oriented at the specified angle, for example, creating an auxiliary view of drawing part or a floor plan at an angle. To do this, use the *Rotate* option in Command line format (see Chapter 27, Auxiliary Views).

### *Style*

The *Style* option allows switching between a *Standard* snap pattern (square or rectangular) and an *Isometric* snap pattern. If using the dialog box, toggle *Isometric* snap. When *Snap Style* or *Rotate* (*Angle*) is changed, the *Grid* automatically aligns with it (see Chapter 25, Pictorial Drawings).

### *Type*

This option switches between *Grid Snap* and *Polar Snap*. Remember, you can have only one of the two snap types active at one time.

Once your snap type and increment(s) are set, you can begin drawing using either snap type or no snap at all. While drawing, you can right-click on the *Snap* icon on the Status Bar to invoke the shortcut menu shown in Figure 6-9. Here you can toggle between *Grid Snap* and *Polar Snap* or turn both off (*Polar* must also be on to use *Polar Snap*). Left-clicking on the *Snap* icon or pressing F9 toggles on or off whichever snap type is current.

**FIGURE 6-9**

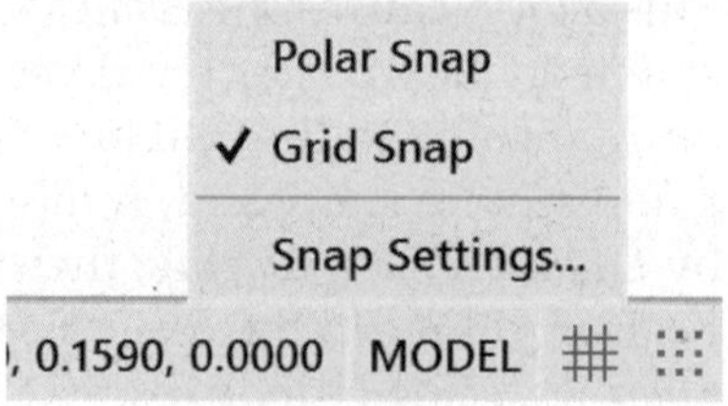

TIP

The process of setting the *Snap Type*, *Grid Snap Spacing*, *Polar Snap Spacing*, and *Polar Tracking Angle* is usually done during the initial stages of drawing setup, although these settings can be changed at any time. The *Grid Snap Spacing* value is stored in the *SNAPUNIT* system variable and saved in the drawing file. The other settings (*Snap Type*, *Polar Snap Spacing*, and *Polar Tracking Angle*) are saved in the system registry (as the *SNAPTYPE*, *POLARDIST*, and *POLARANG* system variables) so that the settings remain in effect for any drawing until changed.

### Grid

| Ribbon | Menu Bar | Command (Type) | Alias (Type) | Shortcut |
|---|---|---|---|---|
| ... | *Tools*<br>*Drafting Settings...*<br>*Snap and Grid* | *Grid* | ... | *F7* or *Ctrl+G* |

The *Grid* is on by default and displays light lines every .5 units and heavier lines every 2.5 units. You can change the display of lines on the grid to dots in the *Grid Style* section of the *Drafting Settings* dialog box (Fig. 6-8). *Grid* is visible on the screen, whereas *Snap* is invisible. *Grid* is only a visible display of some regular interval. *Grid* and *Snap* can be independent of each other. In other words, each can have separate spacing settings and the active state of each (*ON, OFF*) can be controlled independently. The *Grid* follows the *Snap* if *Snap* is rotated or changed to *Isometric Style*. Although the *Grid* spacing can be different than that of *Snap*, it can also be forced to follow *Snap* by using the *Snap* option. The default *Grid* setting is **0.5**.

The *Grid* cannot be plotted. It is not comprised of *Line* or *Point* objects and therefore is not part of the current drawing. *Grid* is only a visual aid.

*Grid* can be accessed by Command line format (shown below) or set via the *Drafting Settings* dialog box (Fig. 6-8). The dialog box can be invoked by menu selection or by typing *Dsettings* or *DS*.

```
Command: GRID
Specify grid spacing(X) or [ON/OFF/Snap/Major/aDaptive/Limits/Follow/Aspect] <0.5000>:
```

#### ON/OFF

The *ON* and *OFF* options simply make the *Grid* visible or not (like toggling the F7 key, pressing Ctrl+G, or clicking the icon on the Status bar).

#### Grid Spacing

If you supply a value in Command line format or in the *Grid X spacing* and *Grid Y spacing* edit boxes, the *Grid* is displayed at the specified spacing regardless of the *Snap* spacing. If you key in an *X* as a suffix to the value (for example, 2X), the *Grid* is displayed at that value times the current *Snap* spacing.

#### Limits (Display grid beyond Limits)

This option determines whether or not the *Grid* is displayed beyond the area set by the *Limits* command. For example, if you chose to display the grid beyond the *Limits*, the grid pattern is displayed to the edge of the drawing area no matter how far you *Zoom* in or out. If you chose not to display the *Grid* beyond the *Limits*, when *Zoomed* out you can "see" the *Limits* by the area filled with the grid pattern. Regardless of this setting, *Zoom All* displays the area defined by the *Limits*.

### *Adaptive grid*

When the grid is not adaptive, lines or dots are displayed at the value specified for *Grid spacing* regardless of how far you *Zoom* in or out. With this setting, at some point the grid becomes too small to display if you *Zoom* out far enough.

If *Adaptive* grid is turned on, the grid adapts to changes in the display due to *Zooming*. For example, if you *Zoom* out, the *Grid* lines or dots may be displayed every 2.5 units instead of every .5 units (the frequency of these grid lines is determined by the setting of the *Major* grid lines). This *Adaptive grid* setting is stored in the *GRIDDISPLAY* system variable.

### *Allow subdivision below grid spacing*

If *Adaptive* grid is turned on, but this option is off, the grid adapts only when you *Zoom* out by increasing the *Grid* spacing. If *Allow subdivision below grid spacing* is turned on, the grid adapts by adding more grid lines or dots when you *Zoom* in. For example, if you *Zoom* in, the *Grid* lines or dots may be displayed every .1 units instead of every .5 units. However, keep in mind that even though the grid lines or dots are more closely spaced, the *Snap* setting does not change. Therefore, the closely spaced lines do not improve the *Snap* accuracy.

### *Major lines*

This value specifies the frequency of grid lines or dots that appear if *Adaptive grid* is turned on. For example, if *Grid spacing* is set to 1 and the *Major lines* setting is 2, grid lines would appear at intervals of 2, 4, 8, and so on, as you continually *Zoom* out. This setting is stored in the *GRIDMAJOR* system variable.

### *Follow*

This option is used only for drawing in 3D. If this option is on, the grid pattern automatically follows the XY plane of the Dynamic UCS. This setting is stored in the *GRIDDISPLAY* system variable.

### *Aspect*

The *Aspect* option of *Grid* allows different X and Y spacing (causing a rectangular rather than a square *Grid*).

## *Dsettings*

| Ribbon | Menu Bar | Command (Type) | Alias (Type) | Shortcut |
|---|---|---|---|---|
| ... | *Tools*<br>*Drafting Settings...* | *Dsettings* | *DS* | *Status Bar*<br>(right-click)<br>*Settings...* |

You can access controls to *Snap* and *Grid* features using the *Dsettings* command. *Dsettings* produces the *Drafting Settings* dialog box described earlier (see Fig. 6-8). There are several tabs in the dialog box. Use the *Snap and Grid* tab to control settings for *Snap* and *Grid* as previously described. Several other tabs are explained in Chapter 7.

## Using *Snap* and *Grid*

Using *Snap* and *Grid* for drawing is a personal preference and should be used whenever appropriate for the drawing. Using *Snap* and *Grid* can be beneficial for some drawings, but may not be useful for others.

Generally, *Snap* and *Grid* can be useful in cases where many of the lines to be drawn or other measurements used are at some regular interval, such as 1 mm or 1/2". Typically, small mechanical drawings and some simple architectural drawings may fall into this category. On the other hand, if you anticipate that few of the measurements in the drawing will be at regular interval lengths, *Snap* and *Grid* may be of little value. Drawings such as civil engineering drawings involving site plans would fall into this category. *Grid* and *Snap* are useful only in cases where the interval is set to a large enough value relative to the screen size (or *Limits*) to facilitate seeing and PICKing points easily.

TIP

## INTRODUCTION TO LAYOUTS AND PRINTING

The last several steps listed in the "Steps for Basic Drawing Setup" on this chapter's first page are also listed below.

5. Change the *LTSCALE* value based on the drawing *Limits*.
6. Create the desired *Layers* and assign appropriate *linetype* and *color* settings.
7. Create desired *Text Styles*.
8. Create desired *Dimension Styles*.
9. Activate a *Layout* tab, set it up for the plot or print device and paper size, and create a viewport (if not already created).
10. Create or insert a title block and border in the layout.

Part of the process of setting the *LTSCALE*, creating *Text*, and creating *Dimension Styles* (steps 5, 7, and 8) is related to the size of the drawing. Steps 9 and 10 prepare the drawing for making a print or plot. Although steps 9 and 10 are often performed after the drawing is complete and just before making a print or plot, it is wise to consider these steps in the drawing setup process. Because you want hidden line dashes, text, dimensions, hatch patterns, etc. to have the correct size in the finished print or plot, it would be sensible to consider the paper size of the print or plot and the drawing scale before you create text, dimensions, and hatch patterns in the drawing. In this way, you can more accurately set the necessary sizes and system variables before you draw, or as you draw, instead of changing multiple settings upon completion of the drawing.

To put it simply, to determine the size for linetypes (*LTSCALE*), text, and dimensions, use the <u>proportion of the drawing to the paper size</u>. In other words, if the size of the drawing area (*Limits*) is 22 x 17 and the paper size you will print on is 11 x 8.5, then the drawing is 2 times the size of the paper. Therefore, set the *LTSCALE* to 2 and create the text and dimensions twice as large as you want them to appear on the print.

TIP

This chapter gives an introduction to layouts and creating paper space viewports. Advanced features of drawing setup, layouts, viewports, and plotting are discussed in Chapter 12, Advanced Drawing Setup, Chapter 13, Layouts and Viewports, and Chapter 14, Printing and Plotting.

### *Model* Tab and *Layout* Tabs

At the bottom left corner of the drawing area you should see one *Model* tab and two *Layout* tabs (Fig. 6-10).

**FIGURE 6-10**

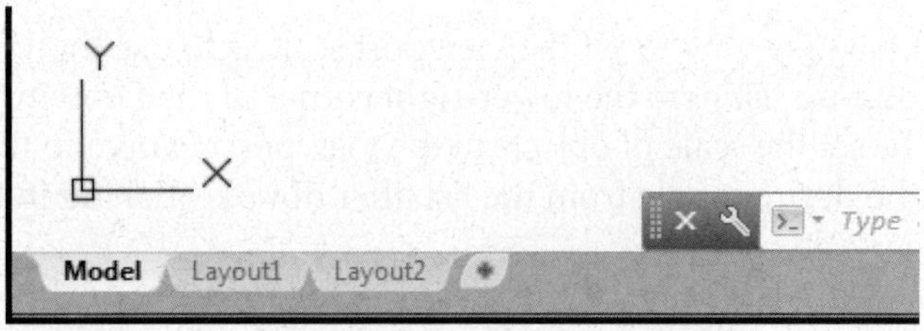

When you start AutoCAD and begin a drawing, the *Model* tab is active by default. Objects that represent the subject of the drawing (model geometry) are normally drawn in the *Model* tab, also known as "model space." Traditionally, dimensions and notes (text) are also created in model space.

A layout is activated by selecting the *Layout1*, *Layout2*, or other layout tab (Fig. 6-11). A layout represents the sheet of paper that you intend to print or plot on (sometimes called paper space). The dashed line around the "paper sheet" in Figure 6-11 represents the maximum printable area for the configured printer or plotter.

**FIGURE 6-11**

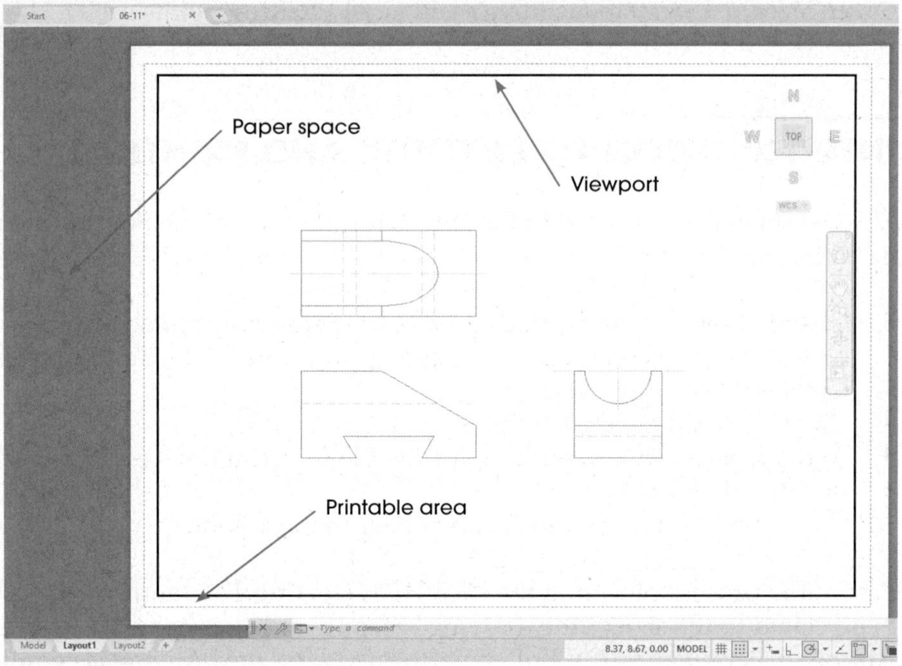


You can have multiple layouts (*Layout1*, *Layout2*, etc.), each representing a different sheet size and/or print or plot device. For example, you may have one layout to print with a laser printer on a 8.5 x 11 inch sheet and another layout set up to plot on an 24 x 18 ("C" size) sheet of paper.

With the default options set when AutoCAD is installed, a viewport is automatically created when you activate a layout. However, you can create layouts using the *Vports* command. A viewport is a "window" that looks into model space. Therefore, you first create the drawing objects in model space (the *Model* tab), then activate a layout and create a viewport to "look into" model space. Typically, only drawing objects such as a title block, border, and some text are created in a layout. Creating layouts and viewports is discussed in detail in Chapter 13, Layouts and Viewports.

**FIGURE 6-12**

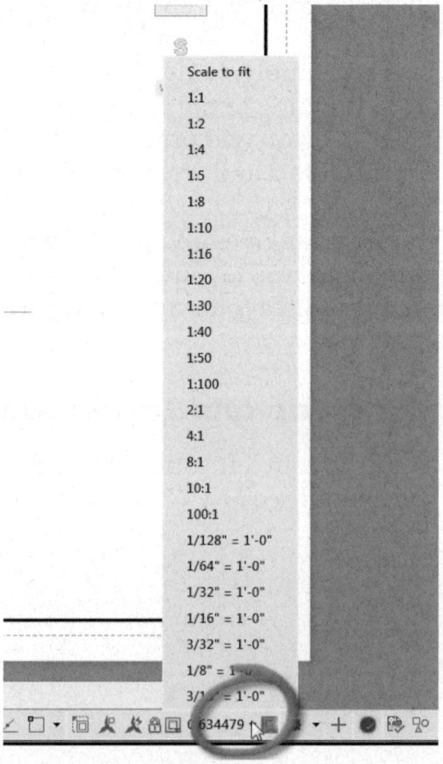


Since a layout represents the actual printed sheet, you normally print the layout full size (1:1). However, the view of the drawing objects appearing in the viewport is scaled to achieve the desired print scale. In other words, you can control the scale of the drawing by setting the "viewport scale"—the proportion of the drawing objects that appear in the viewport relative to the paper size, or simply stated, the proportion of model space to paper space. (This is the same idea as the proportion of the drawing area to the paper size, discussed earlier.)

One easy way to set the viewport scale is to use the *Viewport Scale* pop-up list near the lower-right corner of the Drawing Editor (Fig. 6-12). To set the scale of objects in the viewport relative to the paper size, select the desired scale from the list after double-clicking in the viewport.

Setting the scale of the drawing for printing and plotting using this and other methods is discussed in detail in Chapter 13, Layouts and

Viewports, and Chapter 14, Printing and Plotting.

Setting the *Viewport Scale* changes the size of the geometry in the viewport relative to the paper (*Layout*). Note in Figure 6-13 that selecting *1:1* from the *Viewport Scale* pop-up list changes the size of the drawing in the viewport compared to the drawing shown in Figure 6-11.

Using a *Layout* tab to set up a drawing for printing or plotting is recommended, although you can also make a print or plot directly from the *Model* tab (print from model space). This method is also discussed in Chapter 14, Printing and Plotting.

FIGURE 6-13

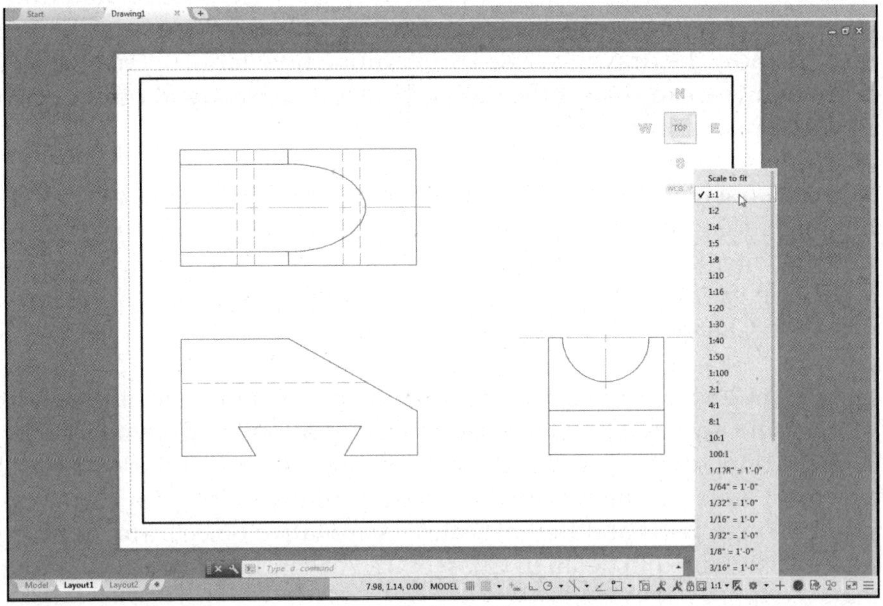


Step 10 in the "Steps for Basic Drawing Setup" is to create a titleblock and border for the layout. The titleblock and text that you want to appear only in the print but not in the drawing (in model space) are <u>typically created in the layout</u> (in paper space). Figure 6-14 shows the drawing after a titleblock has been added.

The advantage of creating a titleblock early in the drawing process is knowing how much space the titleblock occupies so you can plan around it. Note that in Figure 6-14 no drawing objects can be created in the lower-right corner because the titleblock occupies that area.

FIGURE 6-14

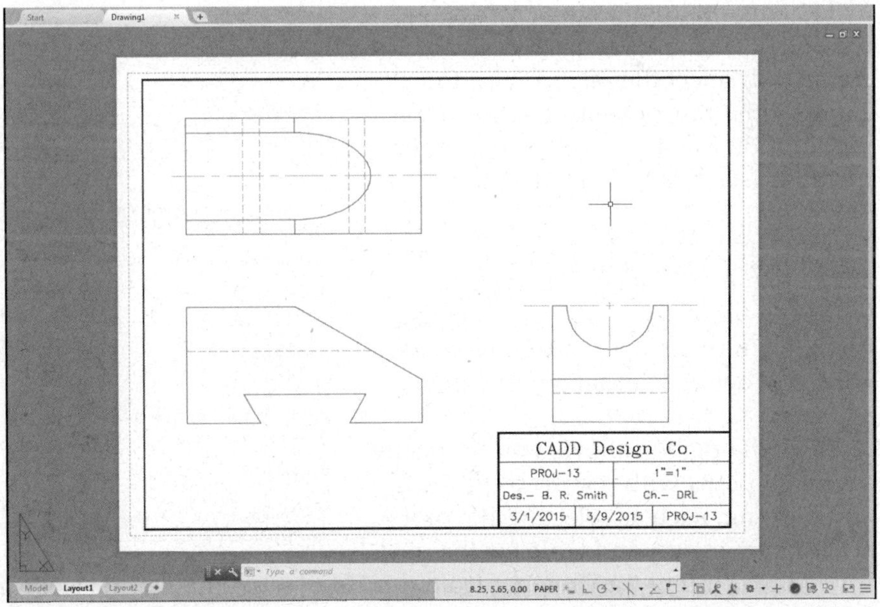


Creating a titleblock is often accomplished by using the *Insert* command to bring a previously created *Block* into the layout. A *Block* in AutoCAD is a group of objects (*Lines*, *Arcs*, *Circles*, *Text*, etc.) that is saved as one object. In this way, you do not have to create the set of objects repeatedly—you need create them only once and save them as a *Block*, then use the *Insert* command to insert the *Block* into your drawing. For example, your school or office may have a standard titleblock saved as a *Block* that you can *Insert*. *Blocks* are covered in Chapter 21, Blocks, DesignCenter, and Tool Palettes.

TIP

## Why Set Up Layouts Before Drawing?

Knowing the intended drawing scale before completing the drawing helps you set the correct size for linetypes, text, dimensions, hatch patterns, and other size-related drawing objects. Since the text and dimensions must be readable in the final printed drawing (usually 1/8" to 1/4" or 3mm-6mm), you should know the drawing scale to determine how large to create the text and dimensions in the drawing. Although you can change the sizes when you are ready to print or plot, knowing the drawing scale early in the drawing process should save you time.

You may not have to go through the steps to create a layout and viewport to determine the drawing scale, although doing so is a very "visible" method to achieve this. The important element is knowing ahead of time the intended drawing scale. If you know the intended drawing scale, you can calculate the correct sizes for creating text and dimensions and setting the scale for linetypes. This topic is a major theme discussed in Chapters 12, 13, and 14.

TIP

AutoCAD 2008 introduced annotative objects. Annotative objects offer an alternative method for determining and specifying a static size for "annotation" in drawings such as text, dimensions, and hatch patterns. Annotative objects are dynamic in that the size of these objects can be easily changed to match the viewport scale. Using this method is particularly helpful for displaying the same text and dimensions in multiple viewports at different scales. Annotative objects are introduced in Chapters 12, 13, 14, and 28 and are explained in detail in Chapter 32, Advanced Layouts, Plotting, and Annotative Objects.

## Printing and Plotting

The *Plot* command allows you to print a drawing (using a printer) or plot a drawing (using a plotter). The *Plot* command invokes the *Plot* dialog box (Fig. 6-15). You can type the *Plot* command, select the *Plot* icon button from *Output* tab, or select *Plot* from the Application Menu, as well as use other methods (see "*Plot*," Chapter 14).

If you want to print the drawing as it appears in model space, invoke *Plot* while the *Model* tab is active. Likewise, if you want to print a layout, invoke *Plot* while the desired *Layout* tab is active.

The *Plot* dialog box allows you to specify several options with respect to printing and plotting, such as selecting the plot device, paper size, and scale for the plot.

FIGURE 6-15

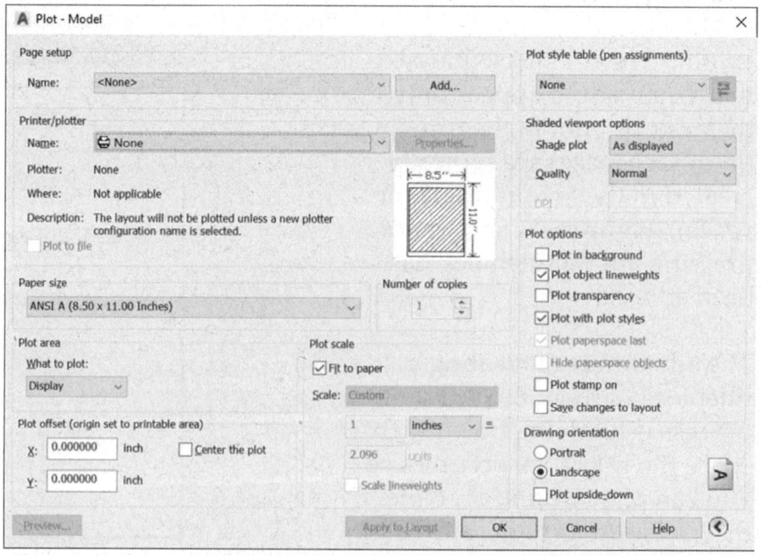


If you are printing the drawing from the *Model* tab, you would select an appropriate scale (such as *1:1*, *1:2*, or *Scaled to Fit*) in the *Scale* drop-down list so the drawing geometry would fit appropriately on the printed sheet (see Fig. 6-15). If you are printing the drawing from a layout, you would select *1:1* in the *Scale* drop-down list. In this case, you want to print the layout (already set to the sheet size for the plot device) at full size.

The drawing geometry that appears in the viewport can be scaled by setting the viewport scale, as described earlier. Any plot specifications you make with the *Plot* dialog box (or the *Page Setup* dialog box) are saved with each *Layout* tab and the *Model* tab; therefore, you can have several layouts, each saved with a particular print or plot setup (scale, device, etc.). Details of printing and plotting, including all the options of the *Plot* dialog box, printing to scale, and configuring printers and plotters, are discussed in Chapter 14.

TIP

## Setting Layout Options and Plot Options

In Chapters 12, 13, and 14 you will learn advanced steps in setting up a drawing, how to create layouts and viewports, and how to print and plot drawings. Until you study those chapters, and for printing drawings before that time, you may need to go through a simple process of configuring your system for a plot device and to automatically create a viewport in layouts. However, if you are at a school or office, some settings may have already been prepared for you, so the following steps may not be needed. Activating a *Layout* tab automatically creates a viewport if you use AutoCAD's default options that appear when it is first installed.

As a check, start AutoCAD and draw a *Circle* in model space. When you activate a *Layout* for the first time in a drawing, you see either a "blank" sheet with no viewport, or a viewport that already exists, or one that is automatically created by AutoCAD. The viewport allows you to view the circle you created in model space. (If the *Page Setup Manager* appears, select *Cancel* this time.) If a viewport exists, but no circle appears, double-click <u>inside</u> the viewport and type *Z* (for *Zoom*), press Enter, then type *A* (for *All*) and press Enter to make the circle appear.

If your system has not previously been configured for you at your school or office, follow the procedure given in "Configuring a Default Output Device" and "Creating Automatic Viewports." After doing so, you should be able to activate a *Layout* tab and AutoCAD will automatically match the layout size to the sheet size of your output device and automatically create a viewport.

TIP

### Configuring a Default Output Device

If no default output device has been specified for your system or to check to see if one has already been specified for you, follow the steps below (Fig. 6-16).

1. Invoke the *Options* dialog box by right-clicking in the drawing area and selecting *Options...* from the bottom of the shortcut menu, or selecting *Options...* from the bottom of the Application Menu.
2. In the *Plot and Publish* tab, locate the *Default plot settings for new drawings* cluster near the top-left corner of the dialog box. Select the desired plot device from the list (such as a laser printer), then select the *Use as default output device* button.
3. Select *OK*.

FIGURE 6-16

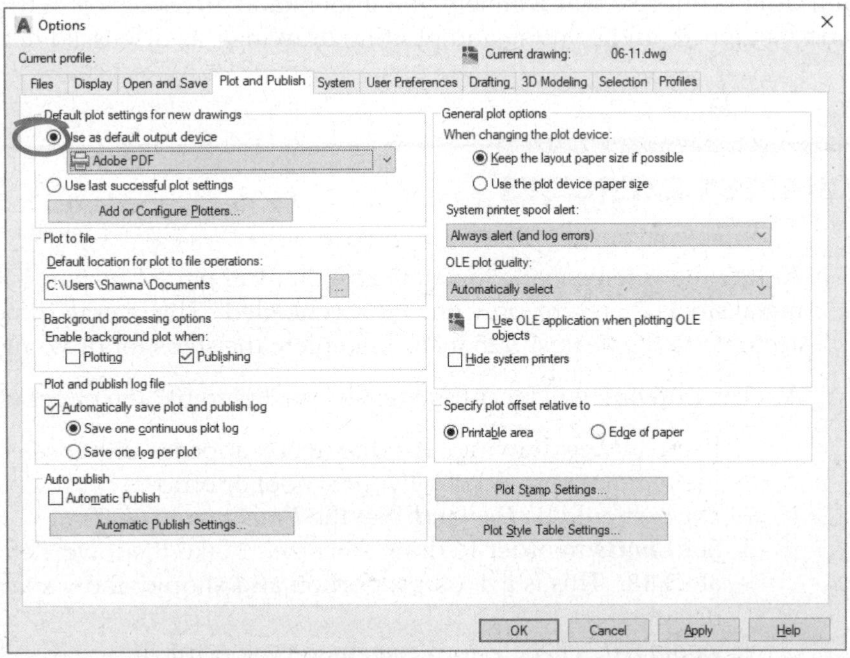


If output devices that are connected to your system do not appear in the list, see "Configuring Plotters and Printers" in Chapter 14.

### Creating Automatic Viewports

If no viewport exists on your screen, configure your system to automatically create a viewport by following these steps (Fig. 6-17).

FIGURE 6-17

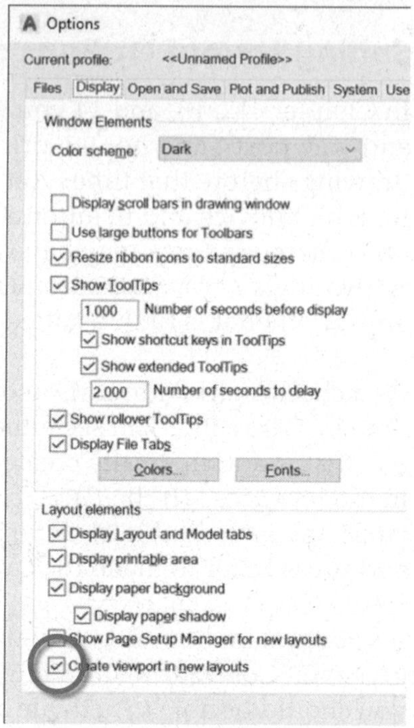


1. Invoke the *Options* dialog box again.
2. In the *Display* tab, find the *Layout elements* cluster near the lower-left corner of the dialog box. Select the *Create viewport in new layouts* checkbox from the bottom of the list.
3. Select *OK*.

### Printing the Drawing

Once an output device has been configured for your system, you are ready to print. For simplicity, you need to perform only three steps to make a print of your drawing, either from the *Model* tab or from a *Layout* tab.

1. Invoke the *Plot* dialog box. Set the desired scale in the *Plot Scale* section, *Scale* drop-down list. If you are printing from a *Layout* tab, select *1:1*. If you are printing from the *Model* tab, select *Scaled to Fit* or other appropriate scale.
2. Select the *Preview* button to ensure the drawing will be printed as you expect.
3. Assuming the preview is as you expect, select the *OK* button to produce the print or plot.

This section, "Introduction to Layouts and Printing," will give you enough information to create prints or plots of drawings for practice and for the next several Chapter Exercises. You also have learned some important concepts that will help you understand all aspects of setting up drawings, creating layouts and viewports, and printing and plotting drawings, as discussed in Chapters 12, 13, and 14.

## CHAPTER EXERCISES

1. A drawing is to be made to detail a mechanical part. The part is to be manufactured from sheet metal stock; therefore, only one view is needed. The overall dimensions are 16 x 10 inches, accurate to the nearest .125 inch. Complete the steps for drawing setup:

   A. The drawing will be automatically "scaled to fit" the paper (no standard scale).

      1. Begin a ***New*** drawing. If a dialog box appears, either select the ***ACAD.DWT*** template or the template provided by your school or office.
      2. ***Units*** should be ***Decimal***. Set the ***Precision*** to **0.000**.
      3. Set ***Limits*** in order to draw full size. Make the lower-left corner **0,0** and the upper right at **24,18**. This is a 4 x 3 proportion and should allow space for the part and dimensions or notes.
      4. ***Zoom All***. (Type **Z** for *Zoom;* then type **A** for *All*.)
      5. Set the ***Grid*** to **1**.
      6. Set the ***Grid Snap*** increment to **.125**. Since the *Polar Snap* increment and *Snap Type* are not saved with the drawing (but are saved in the system registry), it is of no use to set these options at this time.
      7. Save this drawing as **CH6EX1A** (to be used again later). (When plotting at a later time, "Scale to Fit" can be specified.)

B. The drawing will be printed from a layout to scale on engineering A size paper (11" x 8.5").

1. Begin a ***New*** drawing. If a dialog box appears, either select the ***ACAD.DWT*** template or the template provided by your school or office.
2. ***Units*** should be ***Decimal***. Set the ***Precision*** to **0.000**.
3. Set ***Limits*** to a proportion of the paper size, making the lower-left corner **0,0** and the upper-right at **22,17**. This allows space for drawing full size and for dimensions or notes.
4. ***Zoom All***. (Type **Z** for *Zoom;* then type **A** for *All.*)
5. Set the ***Grid*** to **1**.
6. Set the ***Grid Snap*** increment to **.125**. Since the *Polar Snap* increment and *Snap Type* are not saved with the drawing (but are saved in the system registry), it is of no use to set these options at this time.
7. Activate a ***Layout*** tab. Assuming your system is configured for a printer using an 11 x 8.5 sheet and is configured to automatically create a viewport (see "Setting Layout Options and Plot Options" in this chapter), a viewport should appear. If the *Page Setup Manager* appears, select ***Close***.
8. Double-click inside the viewport, then select **1:2** from the ***Viewport Scale*** list (in the lower-right corner). Next, double-click outside the viewport or toggle the ***MODEL*** button on the Status bar to ***PAPER***.
9. Activate the ***Model*** tab. ***Save*** the drawing as **CH6EX1B.**

2. A drawing is to be prepared for a house plan. Set up the drawing for a floor plan that is approximately 50' x 30'. Assume the drawing is to be automatically "Scaled to Fit" the sheet (no standard scale).

REUSE

A. Begin a ***New*** drawing. If a dialog box appears, either select the ***ACAD.DWT*** template or the template provided by your school or office.
B. Set ***Units*** to ***Architectural***. Set the ***Precision*** to **0'-0 1/4"**. Each unit equals 1 inch.
C. Set ***Limits*** to **0,0** and **80',60'**. Use the apostrophe (') symbol to designate feet. Otherwise, enter **0,0** and **960,720** (size in inch units is: 80x12=960 and 60x12=720).
D. ***Zoom All***. (Type **Z** for *Zoom;* then type ***A*** for *All.*)
E. Set ***Grid*** to **24** (2 feet).
F. Set the ***Grid Snap*** increment to **6"**. Since the *Polar Snap* increment and *Snap Type* are not saved with the drawing (but are saved in the system registry), it is of no use to set these options at this time.
G. ***Save*** this drawing as **CH6EX2.**

3. A multiview drawing of a mechanical part is to be made. The part is 125mm in width, 30mm in height, and 60mm in depth. The plot is to be made on an A3 metric sheet size (420mm x 297mm). The drawing will use ISO linetypes and ISO hatch patterns, so AutoCAD's *Metric* default settings can be used.

REUSE

A. Begin a ***New*** drawing using the ***ACADISO.DWT*** template.
B. ***Units*** should be ***Decimal***. Set the ***Precision*** to **0.00**.
C. Calculate the space needed for three views. If ***Limits*** are set to the sheet size, there should be adequate space for the views. Make sure the lower-left corner is at **0,0** and the upper right is at **420,297**. (Since the *Limits* are set to the sheet size, a plot can be made later at 1:1 scale.)
D. Set the ***Grid Snap*** increment to **10**. Make ***Grid Snap*** current.
E. ***Save*** this drawing as **CH6EX3** (to be used again later).

4. Assume you are working in an office that designs many mechanical parts in metric units. However, the office uses a standard laser jet printer for 11″ x 8.5″ sheets. Since AutoCAD does not have a setup for metric drawings on non-metric sheets, it would help to carry out the steps for drawing setup and save the drawing as a template to be used later.

   A. Begin a ***New*** drawing. If a dialog box appears, either select the ***ACAD.DWT*** template or the template provided by your school or office.
   B. Set the ***Units Precision*** to **0.00**.
   C. Change the ***Limits*** to match an 11″ x 8.5″ sheet. Make the lower-left corner **0,0** and the upper right **279,216** (11 x 8.5 times 25.4, approximately). (Since the *Limits* are set to the sheet size, plots can easily be made at 1:1 scale.)
   D. ***Zoom All***. (Type **Z** for *Zoom*, then **A** for *All*).
   E. Change the ***Grid Snap*** increment to **2**. When you begin the drawing in another exercise, you may want to set the *Polar Snap* increment to 2 and make *Polar Snap* current.
   F. Change ***Grid*** to **10**.
   G. At the Command prompt, type ***LTSCALE***. Change the value to **25**.
   H. Activate a ***Layout*** tab. Assuming your system is configured for a printer using an 11 x 8.5 sheet and is configured to automatically create a viewport (see "Setting Layout Options and Plot Options" in this chapter), a viewport should appear. (If the *Page Setup Manager* appears, select ***Close*** and the viewport should appear.)
   I. Produce the ***Page Setup Manager*** by selecting it from the ***Output*** tab or typing ***Pagesetup*** at the Command prompt. Select ***Modify*** so the *Page Setup* dialog box appears. In this dialog box, find the *Printer/Plotter* section and select your configured printer if not already configured. Ensure the *Paper size* is set to *Letter*. In the *Plot Scale* section, use the drop-down list to change *inches* to ***mm***. Then select ***1:1*** in the ***Scale*** drop-down list. This action causes AutoCAD to measure the 11 x 8.5 inch sheet in millimeters (as shown by the image in the *Printer/Plotter* section) and print the sheet at 1:1. Select ***OK*** in the *Page Setup* dialog box and ***Close*** in the *Page Setup Manager*. Using the *Page Setup Manager* saves the print settings with the layout for future use.
   J. Examining the layout, it appears the viewport is extremely small in relation to the printable area. ***Erase*** the viewport.
   K. To make a new viewport, type ***-Vports*** at the Command prompt (don't forget the hyphen) and press **Enter**. Accept the default (***Fit***) option by pressing **Enter**. A viewport appears to fit the printable area.
   L. Double-click inside the viewport, then select ***1:1*** from the ***Viewport Scale*** list (in the lower-right corner).
      Next, double-click outside the viewport or toggle the ***MODEL*** button on the Status bar to ***PAPER***.
   M. Activate the ***Model*** tab. ***Save*** the drawing as **A-METRIC.**

5. Assume you are commissioned by the local parks and recreation department to provide a layout drawing for a major league sized baseball field. Follow these steps to set up the drawing:

   A. Begin a ***New*** drawing. If a dialog box appears, either select the *ACAD.DWT* template or the template provided by your school or office.
   B. Set the ***Units*** to ***Architectural*** and the ***Precision*** to **1/2″**.
   C. Set ***Limits*** to an area of **512′** x **384′** (make sure you key in the apostrophe to designate feet). ***Zoom All***.
   D. Type ***DS*** to invoke the *Drafting Settings* dialog box. Change the ***Grid Snap*** to **10′** (don't forget the apostrophe). When you are ready to draw (at a later time), you may want to set the *Polar Snap* increment to 10′ and make *Polar Snap* the current *Snap Type*.
   E. Use the ***Grid*** command and change the value to **20′**. Ensure *Snap* and *Grid* are on.
   F. Save the drawing and assign the name **BALL FIELD CH6.**

CHAPTER 7

# Object Snap and Object Snap Tracking

## Chapter Objectives

After completing this chapter you should:

1. understand the importance of accuracy in CAD drawings;
2. know the function of each of the Object Snap modes;
3. be able to recognize the Snap Marker symbols;
4. be able to invoke Object Snaps for single point selection;
5. be able to operate Running Object Snap modes;
6. know that you can toggle Running Object Snap off to specify points not at object features;
7. be able to use Object Snap Tracking to create and edit objects that align with existing Object Snap points.

## CAD ACCURACY

Because CAD databases store drawings as digital information with great precision (fourteen numeric places in AutoCAD), it is possible, practical, and desirable to create drawings that are 100% accurate; that is, a CAD drawing should be created as an exact dimensional replica of the actual object. For example, lines that appear to connect should actually connect by having the exact coordinate values for the matching line endpoints.

Only by employing exact precision can dimensions placed in a drawing automatically display the exact intended length, or can a CAD database be used to drive CNC (Computer Numerical Control) machine devices such as milling machines or lathes, or can the CAD database be used for rapid prototyping and manufacturing devices such as Stereo Lithography Apparatus, 3D printers and other Additive Manufacturing devices. With CAD/CAM technology (Computer-Aided Design/Computer-Aided Manufacturing), the CAD database defines the configuration and accuracy of the finished part.

TIP

Accuracy is critical. Therefore, in no case should you create CAD drawings with only visual accuracy such as one might do when sketching using the "eyeball method."

## OBJECT SNAP

AutoCAD provides a capability called "Object Snap" that enables you to "snap" to existing object endpoints, midpoints, centers, intersections, etc. When an Object Snap mode (*Endpoint, Midpoint, Center, Intersection,* etc.) is invoked, you can move the cursor near the desired object feature (endpoint, midpoint, etc.) and AutoCAD locates and calculates the coordinate location of the desired object feature. Available Object Snap modes are:

> *Endpoint, Midpoint, Intersection, Center, Quadrant, Tangent, Perpendicular, Parallel, Extension, Nearest, Insert, Node, Mid Between Two Points, Temporary Tracking, From,* and *Apparent Intersection*

For example, when you want to draw a *Line* and connect its endpoint to an existing *Line,* you can invoke the *Endpoint* Object Snap mode at the "Specify next point or [Undo]:" prompt, then snap exactly to the desired line end by moving the cursor near it and PICKing (Fig. 7-1). Object Snaps can be used for any draw or modify operation—whenever AutoCAD prompts for a point (location).

**FIGURE 7-1**

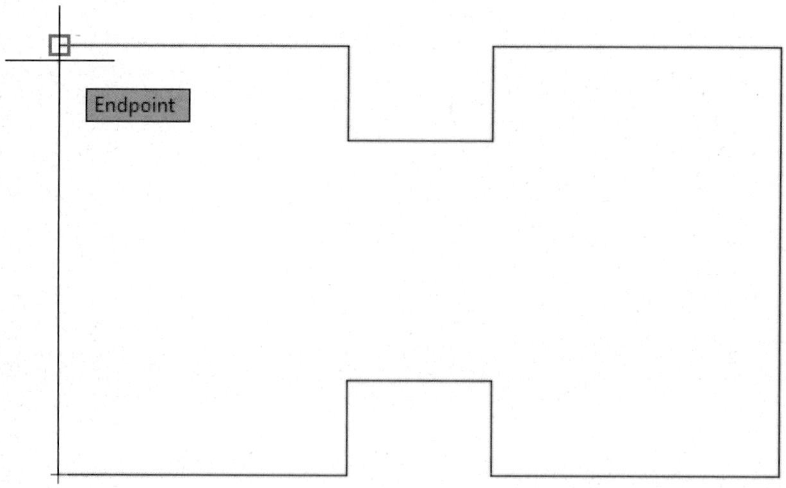

AutoCAD displays a "Snap Marker" indicating the particular object feature (endpoint, midpoint, etc.) when you move the cursor near an object feature. Each Object Snap mode (*Endpoint, Midpoint, Center, Intersection,* etc.) has a distinct symbol (Snap Marker) representing the object feature. This innovation allows you to preview and confirm the snap points before you PICK them (Fig. 7-2).

**FIGURE 7-2**

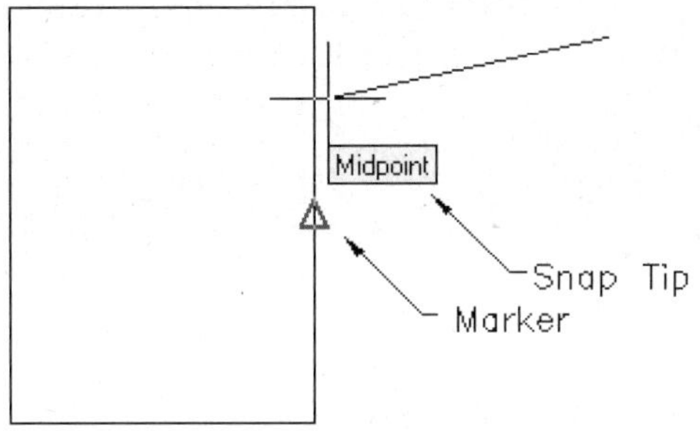


AutoCAD provides two other aids for previewing and confirming Object Snap points before you PICK. A "Snap Tip" appears shortly after the Snap Marker appears (hold the cursor still and wait one second). The Snap Tip gives the name of the found Object Snap point, such as an *Endpoint, Midpoint, Center,* or *Intersection,* (Fig. 7-2). In addition, a "Magnet" draws the cursor to the snap point if the cursor is within the confines of the Snap Marker. This Magnet feature helps confirm that you have the desired snap point before making the PICK.

A visible target box, or "Aperture," can be displayed at the cursor (invisible by default) whenever an Object Snap mode is in effect (Fig. 7-3). The Aperture is a square box larger than the pickbox (default size of 10 pixels square). Technically, this target box (visible or invisible) must be located on an object before a Snap Marker and related Snap Tip appear. The settings for the Aperture, Snap Markers, Snap Tips, and Magnet are controlled in the *Drafting* tab of the *Options* dialog box (discussed later).

**FIGURE 7-3**

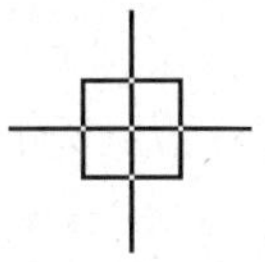

## OBJECT SNAP MODES

The Object Snap modes are explained in this section. Each mode, and its relation to the AutoCAD objects, is illustrated.

Object Snaps must be activated in order for you to "snap" to the desired object features. Two methods for activating Object Snaps, Single Point Selection and Running Object Snaps, are discussed in the sections following Object Snap Modes. The Object Snap modes (*Endpoint, Midpoint, Center, Intersection,* etc.) operate identically for either method.

### *Center*

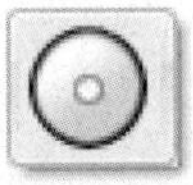

This Object Snap option finds the center of a *Circle* or *Arc* (Fig. 7-4), or a *Donut* or closed *Pline*. You can PICK the *Circle* object or where you think the center is located.

**FIGURE 7-4**

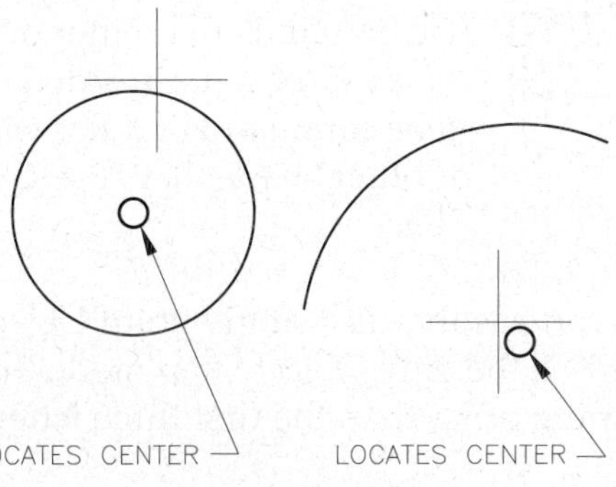

### *Endpoint*

The *Endpoint* option snaps to the endpoint of a *Line, Pline, Spline,* or *Arc* (Fig. 7-5). PICK the object near the desired end.

**FIGURE 7-5**

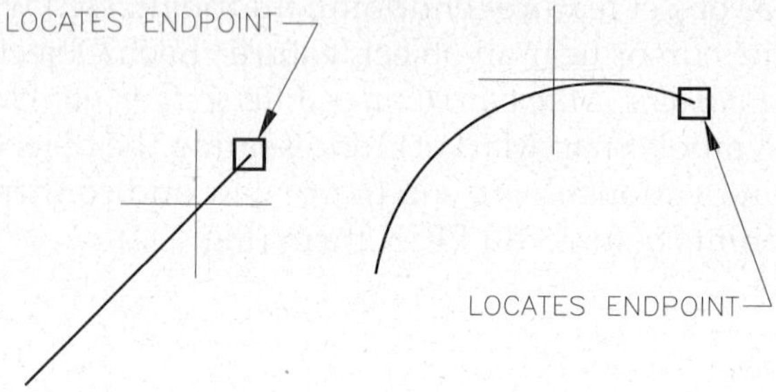


### *Insert*

This option locates the insertion point of *Text* or a *Block* (Fig. 7-6). PICK anywhere on the *Block* or line of *Text*.

**FIGURE 7-6**

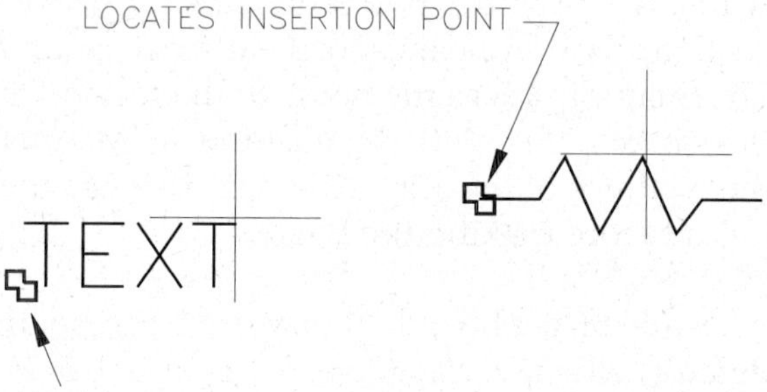


### *Intersection*

Using this option causes AutoCAD to calculate and snap to the intersection of any two objects (Fig. 7-7). You can locate the cursor (Aperture) so that both objects pass near (through) it, or you can PICK each object individually.

Even if the two objects that you PICK do not physically intersect, you can PICK each one individually with the *Intersection* mode and AutoCAD will find the extended intersection.

**FIGURE 7-7**

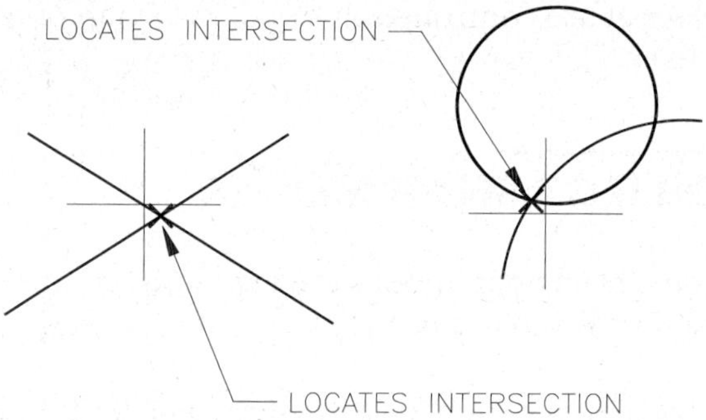


### *Geometric Center*

The *Geometric Center* Object Snap option (*GCEN*) finds the center of a closed *Pline*. The *Pline* can be a multi-sided shape created with the *Pline* command or a *Rectang, Polygon, Boundary,* or other shape that AutoCAD creates using a *Pline*.

If you use this as a single-point Object Snap, type *GCEN*. This is the only Object Snap mode that is accessed by typing other than the first three letters.

**FIGURE 7-8**

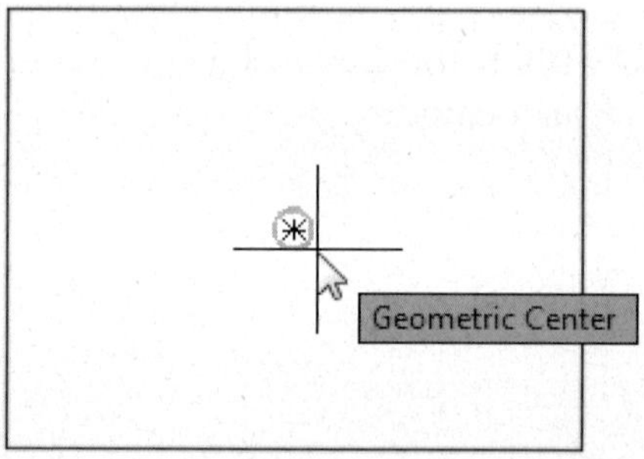

### *Midpoint*

The *Midpoint* option snaps to the point of a *Line* or *Arc* that is halfway between the endpoints. PICK anywhere on the object (Fig. 7-9).

**FIGURE 7-9**

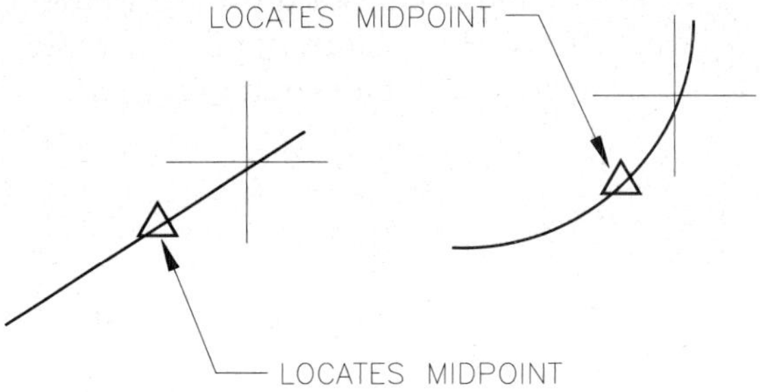


### *Nearest*

The *Nearest* option locates the point on an object nearest to the cursor position (Fig. 7-10). Place the cursor center nearest to the desired location, then PICK.

**FIGURE 7-10**

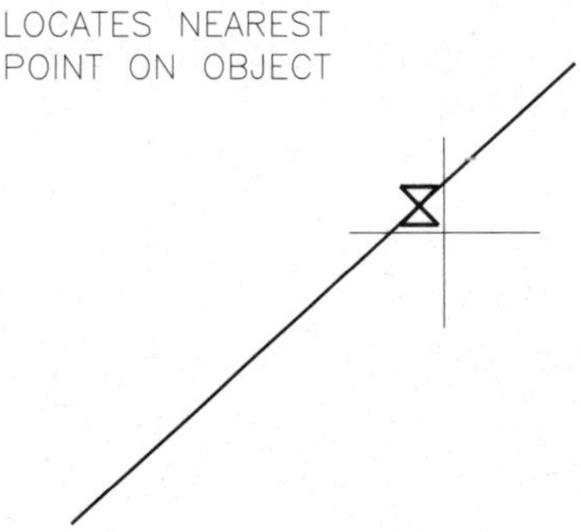


*Nearest* cannot be used effectively with *Ortho* or *Polar Tracking* to draw orthogonal lines because the *Nearest* point takes precedence over *Ortho* and *Polar Tracking*. However, the *Intersection* mode (in Running Object Snap mode only) in combination with *Polar Tracking* does allow you to construct orthogonal lines (or lines at any *Polar Tracking* angle) that intersect with other objects.

TIP

### *Node*

This option snaps to a *Point* object (Fig. 7-11). The *Point* must be within the Aperture (visible or invisible Aperture).

**FIGURE 7-11**

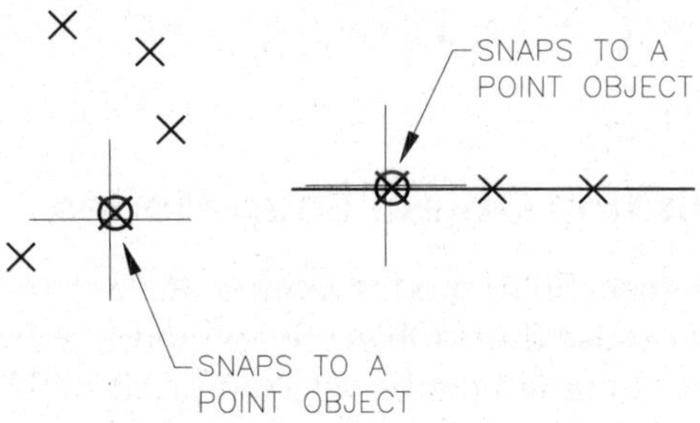


### *Perpendicular*

Use this option to snap perpendicular to the selected object (Fig. 7-12). PICK anywhere on a *Line* or straight *Pline* segment. The *Perpendicular* option is typically used for the second point ("Specify next point:" prompt) of the *Line* command.

**FIGURE 7-12**

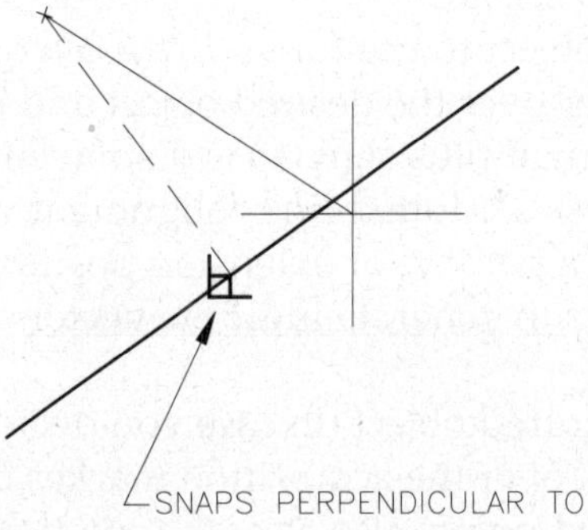

*Quadrant*

The *Quadrant* option snaps to the 0, 90, 180, or 270 degree quadrant of a *Circle* (Fig. 7-13). PICK nearest to the desired *Quadrant*.

**FIGURE 7-13**

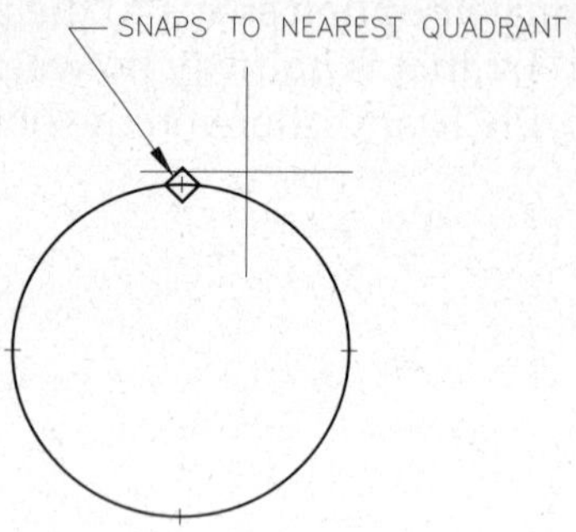


*Tangent*

This option calculates and snaps to a tangent point of an *Arc* or *Circle* (Fig. 7-14). PICK the *Arc* or *Circle* as near as possible to the expected *Tangent* point.

**FIGURE 7-14**

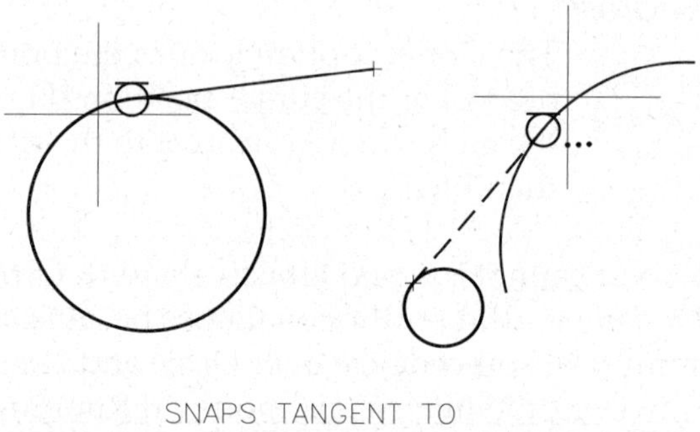


*Apparent intersection*

Use this option when you are working with a 3D drawing and want to snap to a point in space where two objects appear to intersect (from your viewpoint) but do not actually physically intersect.

## Acquisition Object Snap Modes

Three Object Snap modes require an "acquisition" step: *Parallel, Extension*, and *Temporary Tracking*. When you use these Object Snap modes a dotted line appears, called an alignment vector, that indicates a vector along which the selected point will lie. These modes require an additional step—you must "acquire" (select) a point or object. For example, using the *Parallel* mode, you must "acquire" an object to be parallel to. The process used to "acquire" objects is explained here and is similar to that used for Object Snap Tracking (see "Object Snap Tracking" later in this chapter).

### To Acquire an Object

To acquire an object to use for an *Extension* or *Parallel* Object Snap mode, move the cursor over the desired object and pause briefly, but do not pick the object. A small plus sign (+) is displayed when AutoCAD acquires the object (Fig. 7-15). A dotted-line "alignment vector" appears as you move the cursor into a parallel or extension position (Fig. 7-16). You can acquire multiple objects to generate multiple vectors.

**FIGURE 7-15**

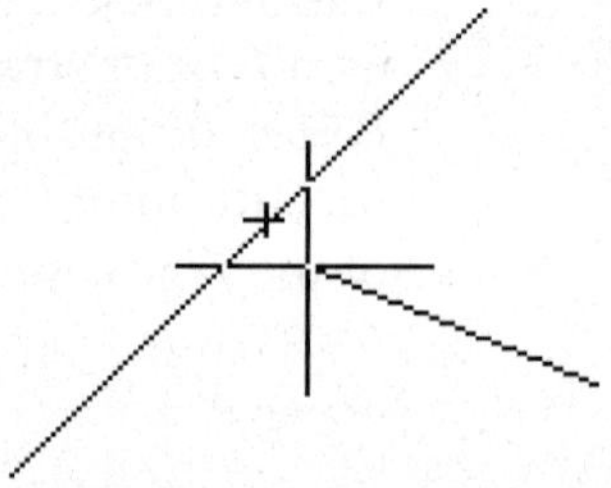

To clear an acquired object (in case you decide not to use that object), move the cursor back over the acquisition marker until the plus sign (+) disappears. Acquired points also clear automatically when another command is issued.

### *Parallel*

The *Parallel* Object Snap option snaps the rubberband line into a parallel relationship with any acquired object (see "To Acquire an Object," earlier this section).

Start a *Line* by picking a start point. When AutoCAD prompts to "Specify next point or [Undo]:," acquire an object to use as the parallel source (a line to draw parallel to). Next, move the cursor to within a reasonably parallel position with the acquired line. The current rubberband line snaps into an exact parallel position (Fig. 7-16). A dotted-line parallel alignment vector appears as well as a tool tip indicating the current *Line* length and angle. The parallel Object Snap symbol appears on the source parallel line. Pick to specify the current *Line* length. Keep in mind multiple lines can be acquired, giving you several parallel options. The acquired source objects lose their acquisition markers when each *Line* segment is completed.

**FIGURE 7-16**

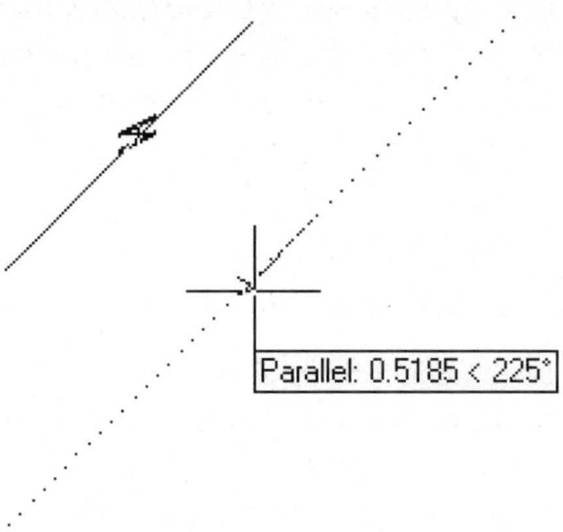


Consider the use of *Parallel* Object Snap with other commands such as *Move* or *Copy*. Figure 7-17 illustrates using *Move* with a *Circle* in a direction *Parallel* to the acquired *Line*.

**FIGURE 7-17**

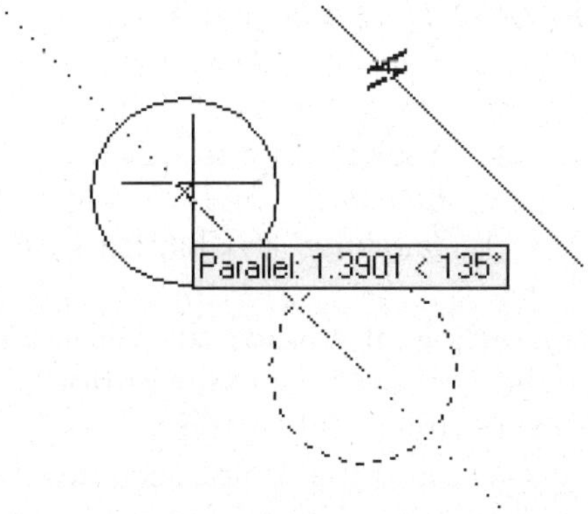


### *Extension*

The *Extension* Object Snap option snaps the rubberband line so that it intersects with an extension of any acquired object (see "To Acquire an Object," earlier this section).

For example, assume you use the *Line* command, then the *Extension* Object Snap mode. When another existing object is acquired, the current *Line* segment intersects with an extension of the acquired object. Figure 7-18 depicts drawing a *Line* segment (upper left) to an *Extension* of an acquired *Line* (lower right). Notice the acquisition marker (plus symbol) on the acquired object.

**FIGURE 7-18**

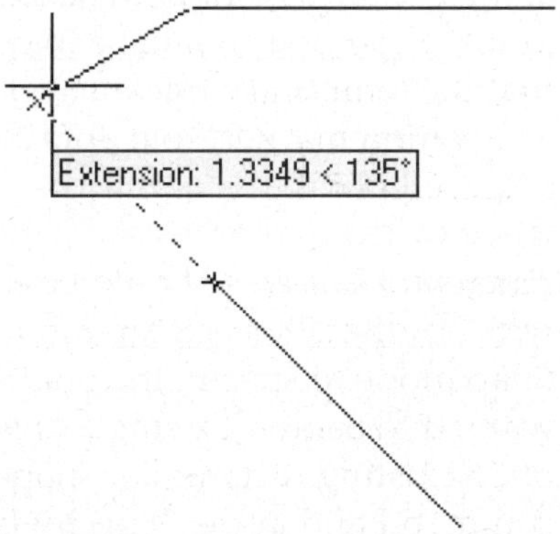


Consider drawing a *Line* to an *Extension* of other objects, such as shown in Figure 7-19. Here a *Line* is drawn to the *Extension* of an *Arc*.

**FIGURE 7-19**

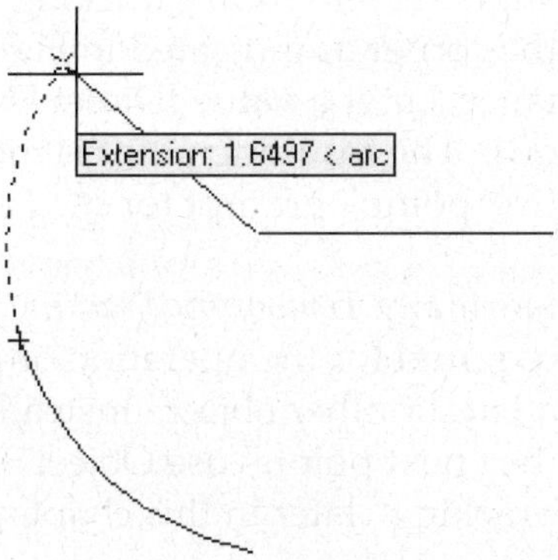

If you set *Extension* as a Running Object Snap mode, each newly created *Line* segment automatically becomes acquired; therefore, you can draw an extension of the previous segment (at the same angle) easily (Fig. 7-20). (See "Object Snap Running Mode" later in this chapter.)

**FIGURE 7-20**

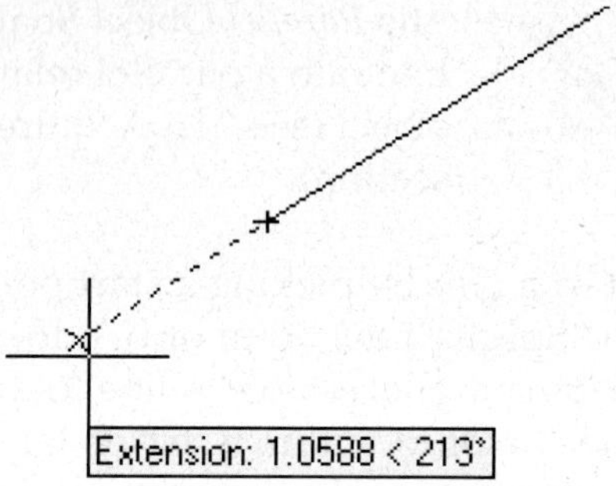


## Object Snap Modifiers

Three Object Snap modes can be thought of as Object Snap modifiers—that is, they are intended to be used with other Object Snap options. They are *Temporary Tracking* or *TT* (also requires an acquisition point), *From*, and *Mid Between Two Points* or *M2p*. To use these options you can select from the *Object Snap* shortcut menu (Shift+right-click) or type the appropriate letters (*TT*, *M2P*, or *From*) at the command line when asked to specify a point.

### *Temporary Tracking (TT)*

*Temporary Tracking* sets up a temporary Polar Tracking point. This option allows you to "track" in a polar direction from any point you select. Tracking is the process of moving from a selected point in a preset angular direction. The preset angles are those specified in the *Increment Angle* section in the *Polar Tracking* tab of the *Drafting Settings* dialog box (see "Polar Tracking" in Chapter 3, Fig. 3-16).

For example, if you use the *Line* command, then use the *Temporary Tracking* button or type ***TT*** at the "Specify first point:" prompt, you can select a point anywhere and that point becomes the temporary tracking point. The cursor moves in a preset angle from that point (Fig. 7-21). The next point selected along the alignment path becomes the *Line's* first point. Polar does not have to be on to use *TT*. Note that the temporary tracking point is indicated by an acquisition marker, so it appears similar to any other acquired point.

**FIGURE 7-21**

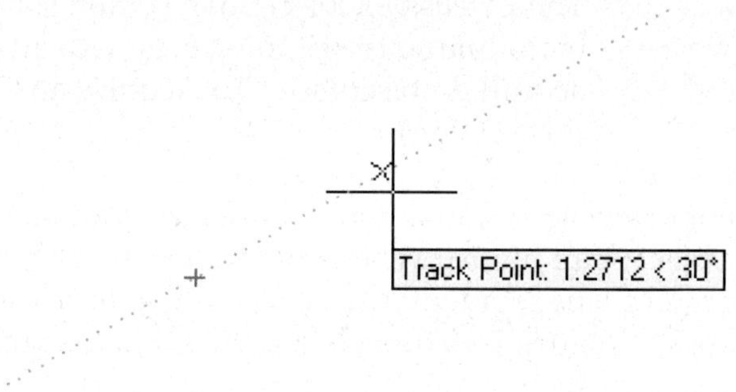


*Temporary Tracking* should be used in conjunction with another Object Snap mode (use *TT*, then another Object Snap mode to specify the point). As an example, assume you wanted to begin drawing a *Line* directly above the corner of an existing rectangular shape. To do this, use the *Line* command and at the "Specify first point:" prompt invoke *TT*. Then use *Endpoint* to acquire the desired corner of the rectangle. A temporary tracking alignment vector appears from the corner as indicated in Figure 7-22. To specify an exact distance, enter a value (Direct Distance Entry with *Dynamic Input* off). The resulting point is the point that satisfies the "Specify first point:" prompt for the *Line*.

**FIGURE 7-22**

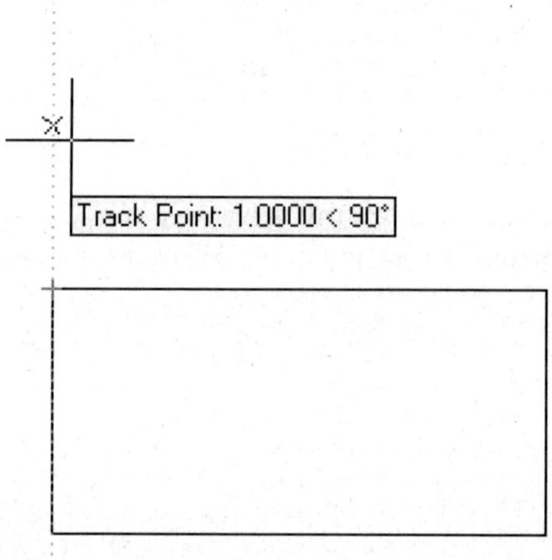


Normally *Temporary Tracking* is used when only one point is required for the operation or possibly for the "first point" of a *Line* or other object. Instead of using *Temporary Tracking* for the "next point," use Object Snap Tracking (see "Object Snap Tracking" later in this chapter).

### *From*

The *From* option is similar to the *Temporary Tracking* mode; however, instead of using a tracking vector to place the desired point, you specify a distance using relative Cartesian or relative polar coordinates. Like *Temporary Tracking*, you use another Object Snap mode to specify the point to track from, except with the *From* option no tracking vector appears. There are two steps: first, select a "base point" (use another Object Snap mode to select this point), then specify an "Offset" (enter relative Cartesian or relative polar coordinates).

Like *Temporary Tracking*, the *From* mode is used for the "first point" or when only one point is required. Use Object Snap Tracking for the "next point" specification rather than *From*.

### *Mid Between 2 Points (M2p)*

*M2p* finds the midpoint between two points. *M2p* is intended to find a point not necessarily on an object—the point can be anywhere between the two selected points. Other Object Snap modes should be used to designate the desired two points.

For example, assume you wanted to draw a *Circle* in the middle of a doorway of a floor plan (Fig. 7-23). Begin the *Circle* command, then at the "Specify center point for circle:" prompt, enter "m2p." Select the two points on each side of the doorway as shown. AutoCAD locates the midpoint between and uses it as the center of the *Circle*.

**FIGURE 7-23**

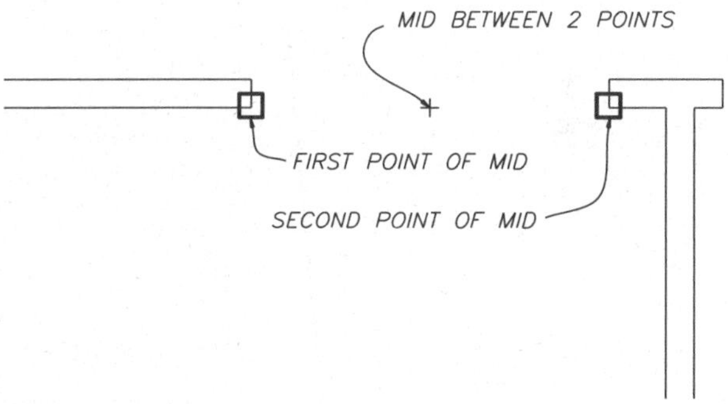


Command: ***CIRCLE***
Specify center point for circle or [3P/2P/Ttr (tan tan radius)]: ***m2p***
First point of mid: **PICK**
Second point of mid: **PICK**
Specify radius of circle or [Diameter]:

*M2p* runs only as a single-point mode and should be typed or selected from the *Object Snap* shortcut menu (Shift+right-click). *M2p* is not available as a "running Object Snap mode" option (not found in the *Object Snap* tab of the *Drafting Settings* dialog box), but is intended to be used with running Object Snaps.

## OBJECT SNAP SINGLE POINT SELECTION

***Object Snaps (Single Point)***

| Ribbon | Menu Bar | Cursor Menu (Shift+button 2) | Command (Type) | Alias (Type) | Shortcut |
|---|---|---|---|---|---|
| ... | ... | *Endpoint* *Midpoint*, etc. | *End*, *Mid*, etc. (first three letters) | ... | ... |

There are two methods for invoking Object Snap modes for single point selection, as shown in the command table. If you prefer to type, enter only the first three letters of the Object Snap mode at the "Specify first point: " prompt, "Specify next point or [Undo]:" prompt, or any time AutoCAD prompts for a point.

In addition to typing the options, a special menu called the cursor menu can be used. The cursor menu pops up at the current location of the cursor and replaces the cursor when invoked (Fig. 7-24). This menu is activated by pressing **Shift+right-click** (hold down the Shift key while clicking the right mouse button).

**FIGURE 7-24**

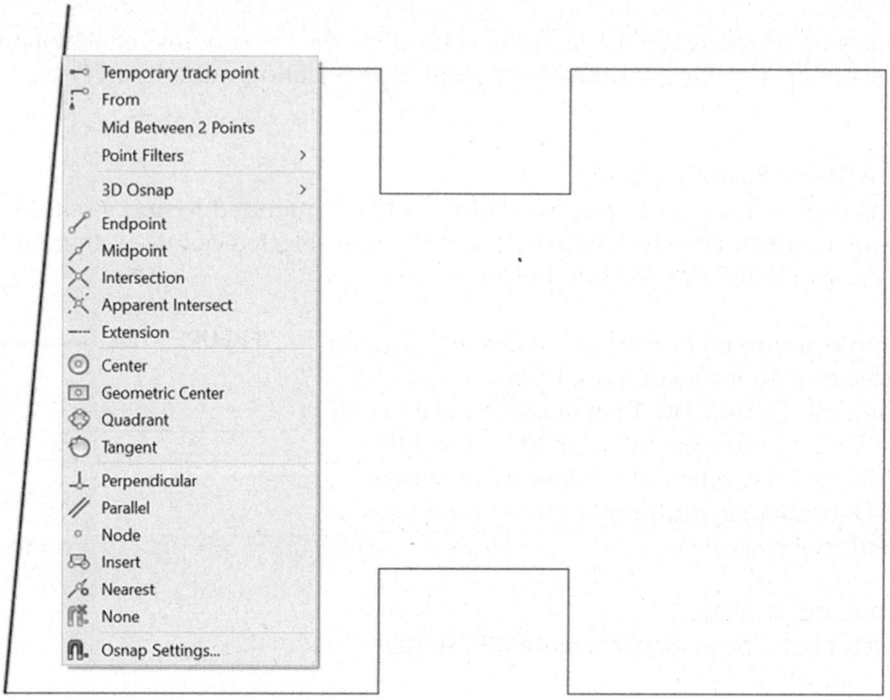


With any of these methods, Object Snap modes are active only for selection of a single point. If you want to *Object Snap* to another point, you must select an Object Snap mode again for the second point. With this method, the desired Object Snap mode is selected transparently (invoked during another command operation) immediately before selecting a point when prompted. In other words, whenever you are prompted for a point (for example, the "Specify first point:" prompt of the *Line* command), select or type an Object Snap option. Then PICK near the desired object feature (endpoint, center, etc.) with the cursor. AutoCAD snaps to the feature of the object and uses it for the point specification. Using Object Snap in this way allows the Object Snap mode to operate only for that single point selection.

For example, when using Object Snap during the *Line* command, the Command line reads as shown:

Command: ***LINE***
Specify first point: ***endp of*** **(PICK)**
Specify next point or [Undo]: ***endp of*** **(PICK)**
Specify next point or [Undo]: **Enter**
Command:

Also see "Running Object Snap Override" for another helpful application of single point selection.

# OBJECT SNAP RUNNING MODE

A more effective method for using Object Snap is called "Running Object Snap" because one or more Object Snap modes (*Endpoint, Center, Midpoint,* etc.) can be turned on and kept running indefinitely. Turn on Running Object Snap by selecting the *Object Snap* button on the Status Bar (Fig. 7-25) or by pressing the F3 key or Ctrl+F key sequence.

FIGURE 7-25

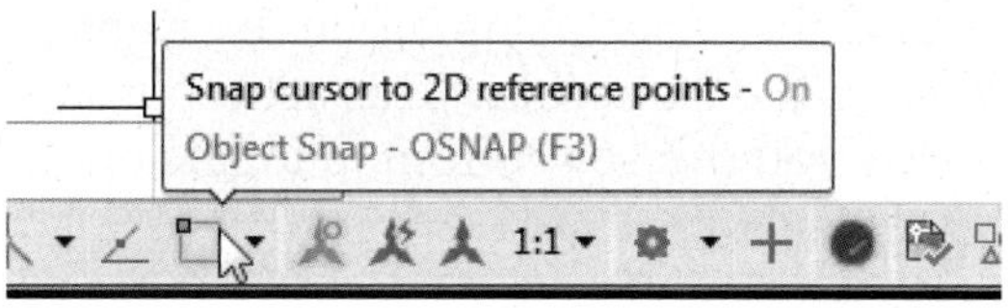


This method can obviously be more productive because you do not have to continually invoke an Object Snap mode each time you need to use one. For example, suppose you have several *Endpoint*s to connect. It would be most efficient to turn on the running *Endpoint* Object Snap mode and leave it running during the multiple selections. This is faster than continually selecting the Object Snap mode each time before you PICK.

TIP

You normally want several Object Snap modes running at the same time. A common practice is to turn on the *Endpoint, Center, Midpoint* modes simultaneously. In that way, if you move your cursor near a *Circle,* the *Center* Marker appears; if you move the cursor near the end or the middle of a *Line,* the *Endpoint,* or the *Midpoint* mode Markers appear.

### *Osnap*

| Ribbon | Menu Bar | Cursor Menu (Shift+button 2) | Command (Type) | Alias (Type) | Shortcut |
|---|---|---|---|---|---|
| ... | *Tools*<br>*Drafting Settings...*<br>*Object Snap* | *Object Snap Settings...* | *Osnap* or *-Osnap* | *OS* or *-OS* | ... |

## Running Object Snap Modes

All features of Running Object Snaps are controlled by the *Drafting Settings* dialog box (Fig. 7-26). This dialog box can be invoked by any of the following methods (see the previous Command Table):

- Type the *Osnap* command.
- Type *OS,* the command alias.
- Select *Drafting Settings…* from the *Tools* pull-down menu.
- Select the *Object Snap Settings* icon button from the *Object Snap* toolbar.
- Right-click on the *Object Snap* icon on the Status Bar, then select *Settings…* from the shortcut menu that appears.

FIGURE 7-26

TIP

Use the *Object Snap* tab to select the desired Object Snap settings. Try using three or four commonly used modes together, such as *Endpoint, Midpoint, Center,* and *Intersection.* The Snap Markers indicate which one of the modes would be used as you move the cursor near different object features.

Access the Object Snap modes by right-clicking on the *Object Snap* icon on the Status bar. This action produces a list of running Object Snaps you can select from to toggle on or off (Fig. 7-27).

**FIGURE 7-27**

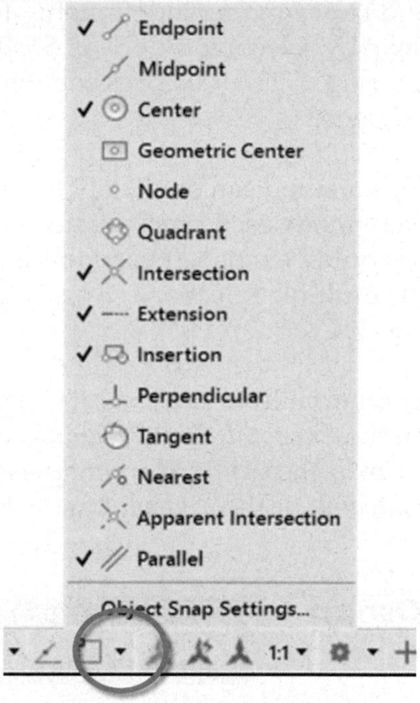


## Running Object Snap Toggle

Another feature that makes Running Object Snap effective is *Object Snap* Toggle. If you need to PICK a point without using Object Snap, use the *Object Snap* Toggle to temporarily turn off the modes. With Running Object Snaps temporarily off, you can PICK any point without AutoCAD forcing your selection to an *Endpoint*, *Center*, or *Midpoint*, etc. When you toggle Running Object Snap on again, AutoCAD remembers which modes were previously set.

The following methods can each be used to toggle Running Object Snaps on and off:

- click the ***Object Snap*** icon that appears on the Status Bar
- press **F3**
- press **Ctrl+F**

### *None*

The *None* Object Snap option is a Running Object Snap effective for only one PICK. None is similar to the Running *Object Snap* Toggle, except it is effective for one PICK, then Running Object Snaps automatically come back on without having to use a toggle. If you have Object Snap modes running but want to deactivate them for a single PICK, use *None* in response to "Specify first point:" or other point selection prompt. In other words, using *None* during a draw or edit command turns off any Running Object Snaps for that single point selection. *None* can be typed at the command prompt (when prompted for a point) or can be selected from the bottom of the *Object Snap* shortcut menu (Fig 7-24).

## Running Object Snap Override

The single point selection mode (using the cursor menu shown in Figure 7-24 or typing in an Object Snap option at the "Specify point" prompt) can be used for another purpose—as an override for Running Object Snap. In other words, if you have a set of Running Object Snaps active but instead want to use a particular Object Snap mode that is not running, type it or use the cursor menu to activate it and override (turn off) the Running Object Snaps for the next PICK. Once the mode is used for a single PICK, the Running Object Snaps become active again.

TIP

This feature is especially helpful for the *Tangent* Object Snap mode. It is not effective to keep both *Tangent* and *Quadrant* running since the *Quadrant* mode often overrides *Tangent*. If you want to use both, set *Quadrant* as a running mode and use *Tangent* as a single point override when needed.

## Object Snap Options

You can set other options for using Object Snap in the *Drafting* tab of the *Options* dialog box (Fig. 7-28, on the next page). Here, more advanced preferences are set such as size and color for the markers and Aperture, and the display of tracking vectors and tool tips.

Access the *Options* dialog box by typing *Options*, selecting the *Options* button from the *Drafting Settings* dialog box, or right-clicking in the drawing area while no commands are in progress, then selecting *Options...* from the shortcut menu. Select the *Drafting* tab. The possible settings are explained next.

**FIGURE 7-28**

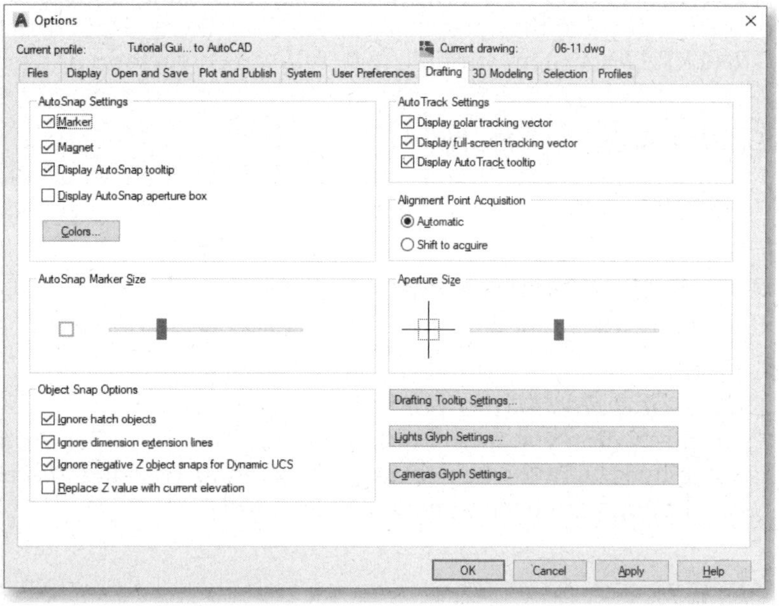


### *Marker*

When this box is checked, a Snap Marker appears when the cursor is moved near an object feature. Each Object Snap mode (*Endpoint, Center, Midpoint,* etc.) has a unique Marker. Normally this box should be checked, especially when using Running Object Snap, because the Markers help confirm when and which Object Snap mode is in effect.

### *Magnet*

The Magnet feature causes the cursor to "lock" onto the object feature (*Endpoint, Center, Midpoint,* etc.) when the cursor is within the confines of the Marker. The Magnet helps confirm the exact location that will be snapped to for the subsequent PICK.

### *Display AutoSnap tooltip*

The Snap Tip (similar to a Tool Tip) is helpful for beginners because it gives a verbal indication of the Object Snap mode in effect (*Endpoint, Center, Midpoint,* etc.) when the cursor is near the object feature. Experienced users may want to turn off the Snap Tips once the Marker symbols are learned.

### *Display AutoSnap aperture box*

This checkbox controls the visibility of the Aperture. Whether the Aperture is visible or not, its size determines how close the cursor must be to an object before a Marker appears and the displayed Object Snap mode takes effect.

### *Colors*

Use this button to select a color for the Markers. You may want to change Marker color if you change the color of the Drawing Editor background. For example, if the background is changed to white, you may want to change the Marker color to dark blue instead of yellow.

### *AutoSnap Marker size*

This control determines the size of the Markers. Remember that if the Magnet is on, the cursor locks to the object feature when the cursor is within the confines of the Magnet. Therefore, the larger the Marker, the greater the Magnet effective area, and the smaller the Marker, the smaller the Magnet effective area.

### *Display polar tracking vector*

Removing the check from this box disables the display of the alignment vector that appears only when you use *Polar Tracking*. This option does not affect the display of Object Snap alignment vectors. You can also change the setting of the *TRACKPATH* system variable to 2.

### *Display full-screen tracking vector*

When the check is removed from this box, tracking vectors that appear are displayed from the acquired point to the cursor only, not across the entire screen. You can also change the setting of the *TRACKPATH* system variable to 0 (full-screen vectors) or 1 (short vectors).

### *Display AutoTrack tooltip*

Removing this check disables the tool tips that appear for Polar Tracking, single point selection Object Snaps such as *Temporary Tracking* or *Extension*, and *Object Snap Tracking*.

### *Automatic (Alignment Point Acquisition)*

Resting the cursor on an existing object for one second automatically acquires that point.

### *Shift to acquire (Alignment Point Acquisition)*

With this option you must hold down the Shift key when resting the cursor on an object to acquire. This is sometimes helpful when you use Object Snap Tracking (discussed later in this chapter) since many unintended points can be automatically acquired.

### *Aperture size*

Notice that the *Aperture Size* can be adjusted through this dialog box or through the use of the *Aperture* command. Whether the Aperture is visible or not (see Fig. 7-29, *Display AutoSnap aperture box*), its size determines how close the cursor must be to an object before the Marker appears and the displayed Object Snap mode takes effect. In other words, the smaller the Aperture, the closer the cursor must be to the object before a Marker appears and the Object Snap mode has an effect, and the larger the Aperture, the farther the cursor can be from the object while still having an effect. The Aperture default size is 10 pixels square. The Aperture size influences Single Point Object Snap selection as well as Running Object Snaps.

### *Ignore hatch objects*

The default setting is to *Ignore Hatch Objects* for Object Snaps. In other words, with this setting you cannot *Object Snap* to a hatch object—the hatch objects are "invisible" to Object Snap modes. If this box is unchecked, an *Endpoint* option, for example, would find endpoints for all hatch pattern lines. You can also use the *OSOPTIONS* system variable. (See Chapter 26, Section Views, for information on *Hatch*.)

### *Ignore dimension extension lines*

This option specifies whether you can snap to extension lines of existing dimensions. The default setting is to *Ignore dimension extension lines* for Object Snaps. With this setting you cannot *Object Snap* to extension lines. You can also use the *OSOPTIONS* system variable.

### *Ignore negative Z object snaps for Dynamic UCS*

Intended for drawing in 3D, this option specifies that object snaps ignore geometry with negative Z values during use of a Dynamic UCS (*OSOPTIONS* system variable).

### *Replace Z value with current elevation*

Normally (with this box unchecked) you can *Object Snap* to any point in 3D space and AutoCAD finds the X, Y, and Z values of the selected point. When this feature is checked and you *Object Snap* to a point in 3D space, AutoCAD uses the selected X and Y coordinates but substitutes the Z value with whatever value you set as the *Elevation*. You could instead use the *OSNAPZ* system variable to turn this feature on and off.

## OBJECT SNAP TRACKING

A feature called Object Snap Tracking helps you draw objects at specific angles and in specific relationships to other objects. Object Snap Tracking works in conjunction with object snaps and displays temporary alignment paths called "tracking vectors" that help you create objects aligned at precise angular positions relative to other objects. You can toggle Object Snap Tracking on and off with the *Object Snap Tracking* button on the Status Bar or by toggling F11.

TIP

To practice with Object Snap Tracking, try turning the *Extension* and *Parallel* Object Snap options off. Since these two new options require point acquisition and generate alignment vectors, it can be difficult to determine which of these features is operating when they are all on.

Object Snap Tracking is similar to Polar Tracking in that it displays and snaps to alignment vectors, but the alignment vectors are generated from <u>other existing objects</u>, not from the current object. These other objects are acquired by Object Snapping to them. Once a point (*Endpoint*, *Midpoint*, etc.) is acquired, alignment vectors generate from it in proximity to the cursor location. This process allows you to construct geometry that has orthogonal or angular relationships to other existing objects.

Using Object Snap Tracking is essentially the same as using the *Temporary Tracking* Object Snap option, then using another Object Snap mode to acquire the tracking point. (See previous Figure 7-22 and related explanation to refresh your memory.) Object Snap Tracking can be used with either Single Point or Running Object Snap mode.

For example, Figure 7-29 displays an alignment vector generated from an acquired *Endpoint* (top of the left *Line*). The current *Line* can then be drawn to a point that is horizontally aligned with the acquired *Endpoint*. Note that in Figure 7-29 and the related figures following, the Object Snap Marker (*Endpoint* marker in this case) is anchored to the acquired point. The alignment vector always rotates about, and passes through, the acquired point (at the *Endpoint* marker). The current *Line* being constructed is at the cursor location.

**FIGURE 7-29**

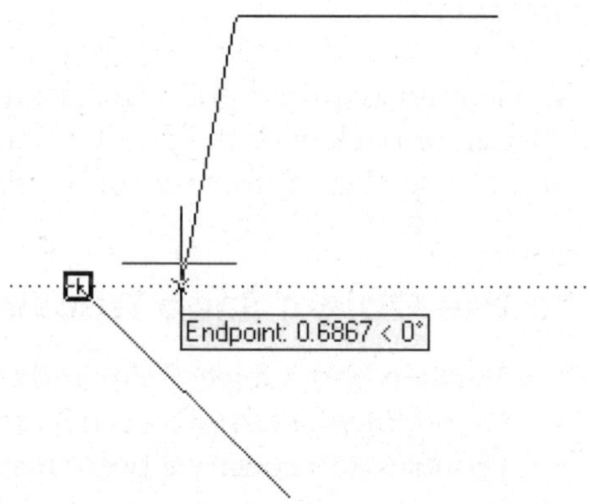


Moving the cursor to another location causes a different alignment vector to appear; this one displays vertical alignment with the same *Endpoint* (Fig. 7-30).

**FIGURE 7-30**

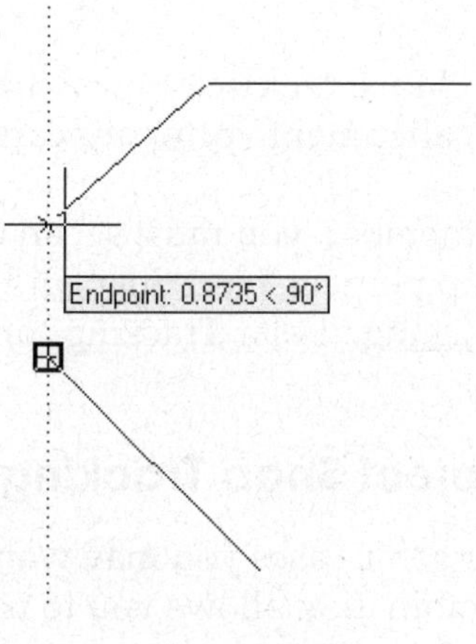

In addition, moving the cursor around an acquired point causes an array of alignment vectors to appear based on the current angular increment set in the *Polar Tracking* tab of the *Drafting Settings* dialog box (see Chapter 3, Fig. 3-16, and related discussion). In Figure 7-31, alignment vectors are generated from the acquired point in 30-degree increments.

**FIGURE 7-31**

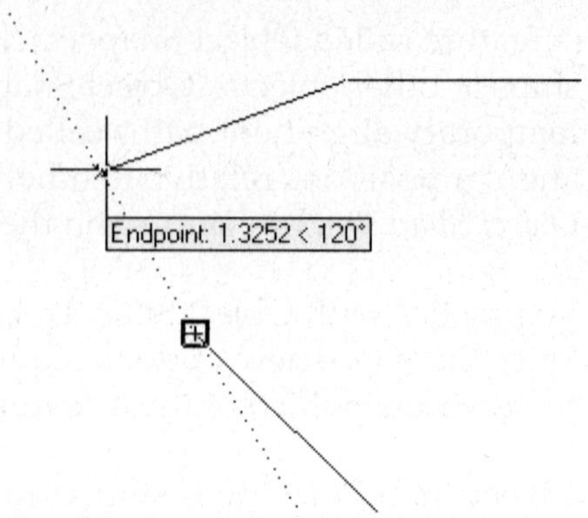


To acquire a point to use for Object Snap Tracking (when AutoCAD prompts to specify a point), move the cursor over the object point and pause briefly when the Object Snap Marker appears, but do not pick the point. A small plus sign (+) is displayed when AutoCAD acquires the point (Fig. 7-32). The alignment vector appears as you move the cursor away from the acquired point. You can acquire multiple points to generate multiple alignment vectors.

**FIGURE 7-32**

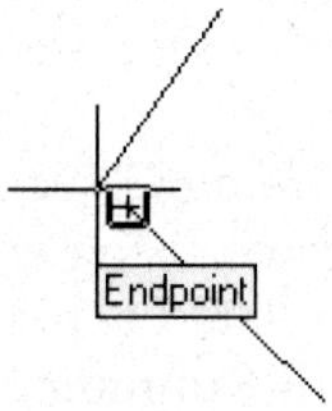


In Figure 7-32, the *Endpoint* object snap is on. Start a *Line* by picking its start point, move the cursor over another line's endpoint to acquire it, and then move the cursor along the horizontal, vertical, or polar alignment vector that appears (not shown in Figure 7-32) to locate the endpoint you want for the line you are drawing.

To clear an acquired point (in case you decide not to use an alignment vector from that point), move the cursor back over the point's acquisition marker until the plus sign (+) disappears. Acquired points also clear automatically when another command is issued.

### To Use Object Snap Tracking:

1. Turn on both Object Snap and Object Snap Tracking (press F3 and F11 or toggle *Object Snap* and *Object Snap Tracking* on the Status Bar). You can use *Object Snap* single point selection when prompted to specify a point instead of using Running Object Snap.
2. Start a *Draw* or *Modify* command that prompts you to specify a point.
3. Move the cursor over an Object Snap point to temporarily acquire it. Do not PICK the point but only pause over the point briefly to acquire it.
4. Move the cursor away from the acquired point until the desired vertical, horizontal, or polar alignment vector appears, then PICK the desired location for the line along the alignment vector.

Remember, you must set an Object Snap (single or running) before you can track from an object's snap point. Object Snap and Object Snap Tracking must both be turned on to use Object Snap Tracking. Polar Tracking can also be turned on, but it is not necessary.

## Object Snap Tracking with Polar Tracking

For some cases you may want to use Object Snap Tracking in conjunction with Polar Tracking. This combination allows you to track from the last point specified (on the current *Line* or other operation) as well as to connect to an alignment vector from an existing object.

Figure 7-33 displays both *Polar Tracking* and *Object Snap Tracking* turned on. The current *Line* (dashed vertical line) is Polar Tracking along a vertical vector from the last point specified (on the horizontal line above). The new *Line's* endpoint falls on the horizontal alignment vector (Object Snap Tracking) acquired from the *Endpoint* of the existing diagonal *Line*. Note that the tool tip displays the Polar Tracking angle ("Polar: <270") and the Object Snap Tracking mode and angle ("Endpoint: <0").

**FIGURE 7-33**

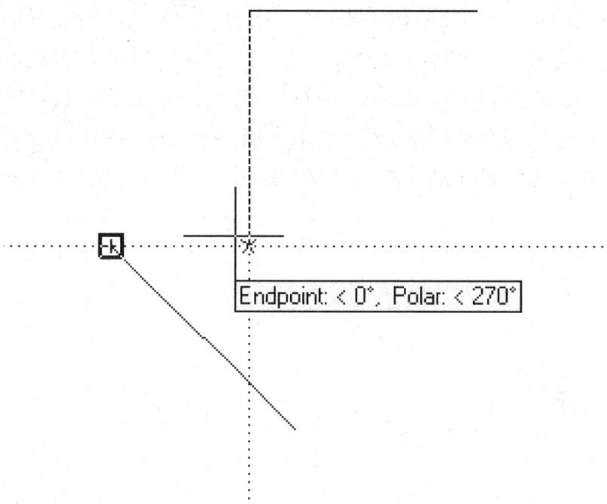


If the cursor is moved from the previous location (in the previous figure), additional Polar Tracking options and Object Snap Tracking options appear. For example, in Figure 7-34, the current *Line* endpoint is tracking at 210 degrees (Polar Tracking) and falls on a vertical alignment vector from the *Endpoint* of the diagonal line (Object Snap Tracking).

**FIGURE 7-34**

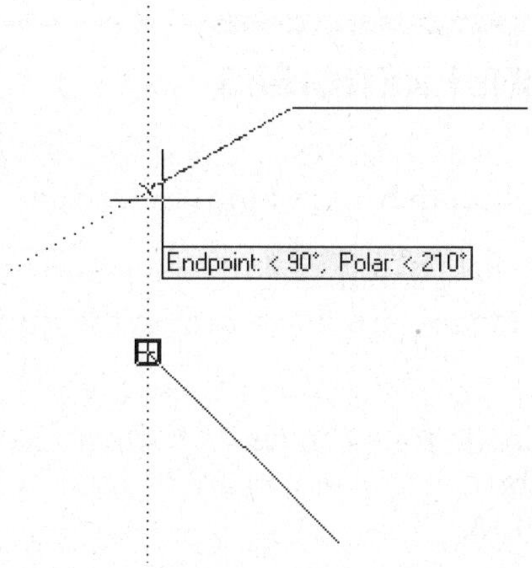


You can accomplish the same capabilities available with the combination of Polar Tracking and Object Snap Tracking by using only Object Snap Tracking and multiple acquired points. For example, Figure 7-35 illustrates the same situation as in Figure 7-33 but only *Object Snap Tracking* is on (*Polar Tracking* is off). Notice that <u>two *Endpoints* have been acquired</u> to generate the desired vertical and horizontal alignment vectors.

**FIGURE 7-35**

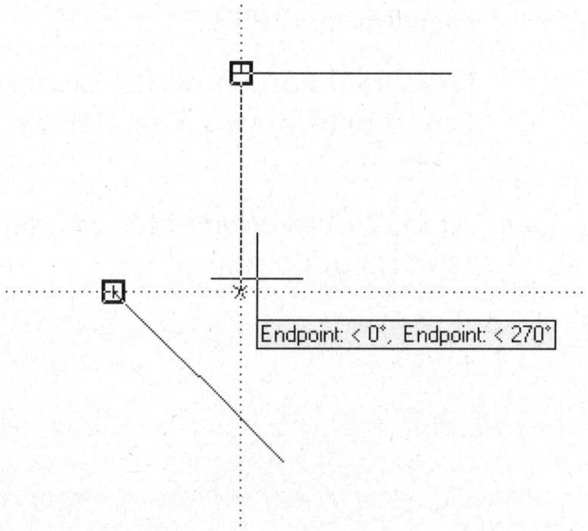

## Object Snap Tracking Settings

You can set the *Object Snap Tracking Settings* in the *Drafting Settings* dialog box (Fig. 7-36). The only options are to *Track orthogonally only* or to *Track using all polar angle settings*. The Object Snap Tracking vectors are determined by the *Increment angle* and *Additional angles* set for Polar Tracking (left side of dialog box). If you want to track using these angles, select *Track using all polar angle settings*.

**FIGURE 7-36**

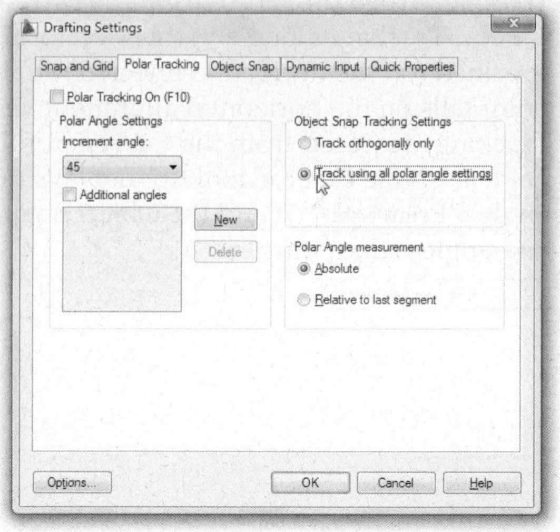


## CHAPTER EXERCISES

1. **Object Snap Single Point Selection**

   ***Open* the CH6EX1A** drawing and begin constructing the sheet metal part. Each unit in the drawing represents one inch.

   A. Create four ***Circles***. All *Circles* have a radius of **1.685**. The *Circles'* centers are located at **5,5**, **5,13**, **19,5**, and **19,13**.

   B. Draw four ***Lines***. The *Lines* (highlighted in Figure 7-37) should be drawn on the outside of the *Circles* by using the ***Quadrant*** Object Snap mode as shown for each *Line* endpoint.

   C. Draw two ***Lines*** from the ***Center*** of the existing *Circles* to form two diagonals as shown in Figure 7-38.

   D. At the ***Intersection*** of the diagonals create a ***Circle*** with a **3** unit radius.

**FIGURE 7-37**

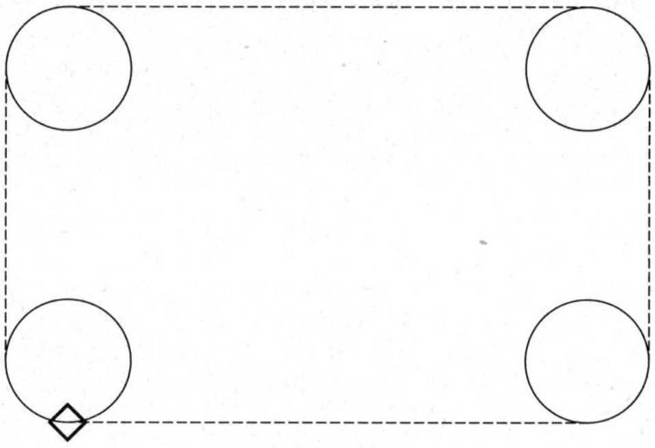

**FIGURE 7-38**

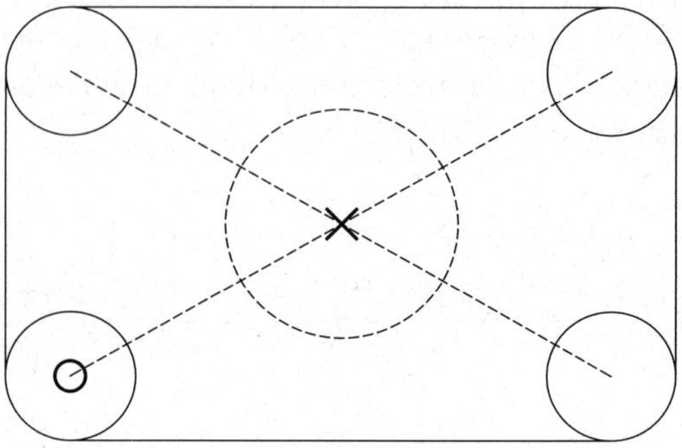

E. Draw two ***Lines***, each from the ***Intersection*** of the diagonals to the ***Midpoint*** of the vertical *Lines* on each side. Finally, construct four new ***Circles*** with a radius of **.25**, each at the ***Center*** of the existing ones (Fig. 7-39).

F. ***SaveAs*** **CH7EX1.**

**FIGURE 7-39**

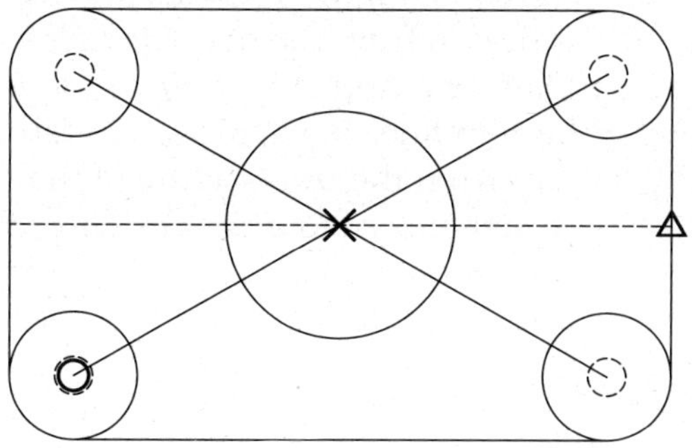

2. **Object Snap Single Point Selection**

A multiview drawing of a mechanical part is to be constructed using the **A-METRIC** drawing. All dimensions are in millimeters, so each unit equals one millimeter.

A. ***Open*** **A-METRIC** from Chapter 6 Exercises. Draw a ***Line*** from **60,140** to **140,140**. Create two ***Circles*** with the centers at the ***Endpoints*** of the *Line,* one *Circle* having a diameter of **60** and the second *Circle* having a diameter of **30**. Draw two ***Lines Tangent*** to the *Circles* as shown in Figure 7-40. ***SaveAs*** **PIVOTARM CH7.**

**FIGURE 7-40**

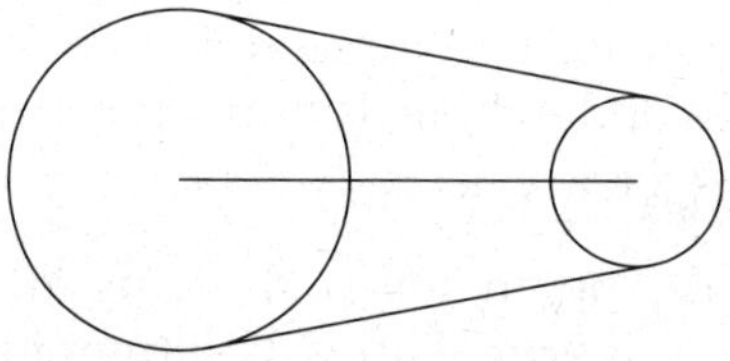

B. Draw a vertical ***Line*** down from the far left ***Quadrant*** of the *Circle* on the left. Specify relative polar coordinates using any input method to make the *Line* 100 units at 270 degrees. Draw a horizontal ***Line*** **125** units from the last ***Endpoint*** using relative polar or direct distance entry coordinates. Draw another ***Line*** between that ***Endpoint*** and the ***Quadrant*** of the *Circle* on the right. Finally, draw a horizontal ***Line*** from point **30,70** and ***Perpendicular*** to the vertical *Line* on the right (Fig. 7-41).

**FIGURE 7-41**

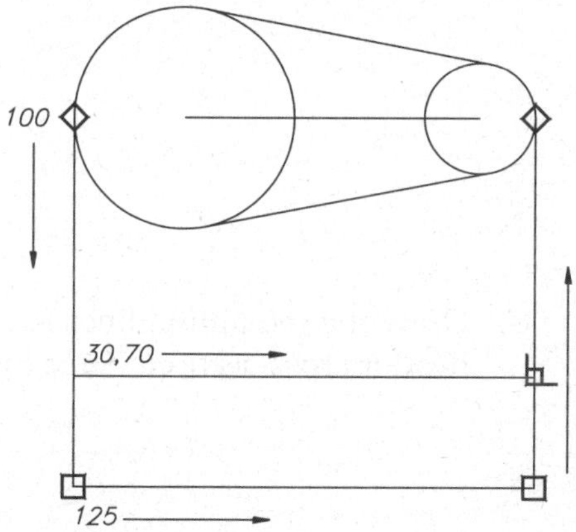


C. Draw two vertical ***Lines*** from the ***Intersections*** of the horizontal *Line* and *Circles* and ***Perpendicular*** to the *Line* at the bottom. Next, draw two ***Circles*** concentric to the previous two and with diameters of **20** and **10** as shown in Figure 7-42.

**FIGURE 7-42**

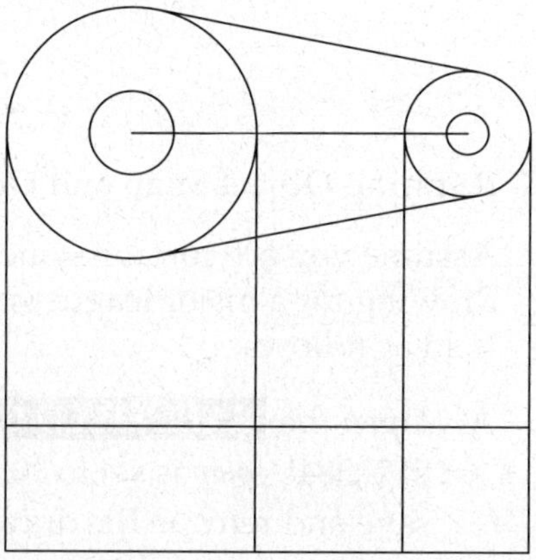

D. Draw four more vertical ***Lines*** as shown in Figure 7-43. Each ***Line*** is drawn from the new *Circles'* ***Quadrant*** and ***Perpendicular*** to the bottom line. Next, draw a miter *Line* from the ***Intersection*** of the corner shown to **@150<45**. ***Save*** the drawing for completion at a later time as another chapter exercise.

FIGURE 7-43

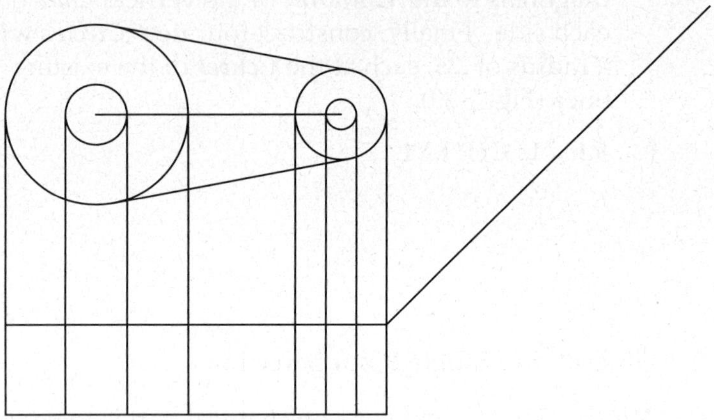

3. **Running Object Snap**

Create a cross-sectional view of a door header composed of two 2 x 6 wooden boards and a piece of 1/2" plywood. (The dimensions of a 2 x 6 are actually 1-1/2" x 5-3/8".)

A. Begin a ***New*** drawing and assign the name **HEADER**. Draw four vertical lines as shown in Figure 7-44.

B. Use the ***OSNAP*** command or select ***Drafting Settings...*** from the ***Tools*** menu and turn on the ***Endpoint*** and ***Intersection*** modes.

FIGURE 7-44

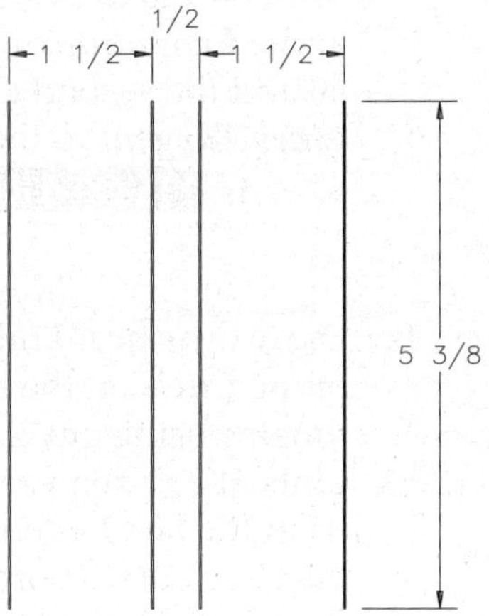


C. Draw the remaining lines as shown in Figure 7-45 to complete the header cross-section. ***Save*** the drawing.

FIGURE 7-45

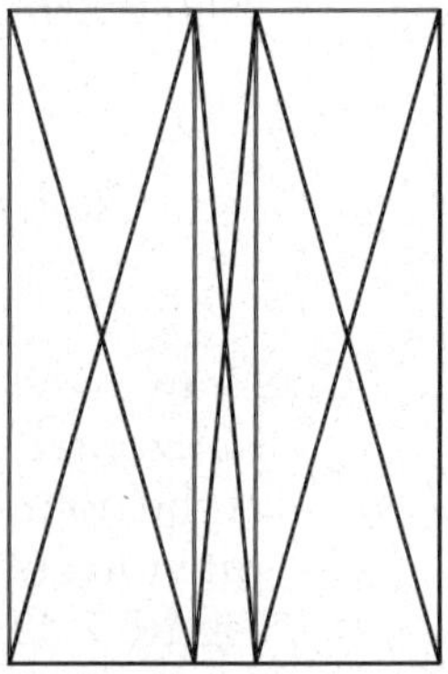

REUSE

4. **Running Object Snap and *Object Snap* Toggle**

Assume you are commissioned by the local parks and recreation department to provide a layout drawing for a major league sized baseball field. Lay out the location of bases, infield, and outfield as follows.

A. ***Open*** the **BALL FIELD CH6.DWG** that you set up in Chapter 6. Make sure *Limits* are set to 512′,384′, *Snap* is set to 10′, and *Grid* is set to 20′. Ensure *Snap* and *Grid* are on. Use ***SaveAs*** to save and rename the drawing to **BALL FIELD CH7**.

B. Begin drawing the baseball diamond by using the ***Line*** command and using ***Dynamic Input***. **PICK** the "Specify first point:" at **20',20'** (watch coordinate display). Draw the foul line to first base by turning on ***Polar Tracking***, move the cursor to the right (along the X direction) and enter a value of **90'** (don't forget the apostrophe to indicate feet). At the "Specify next point or [Undo]:" prompt, continue by drawing a vertical *Line* of **90'**. Continue drawing a square with **90'** between bases (Fig. 7-46).

**FIGURE 7-46**

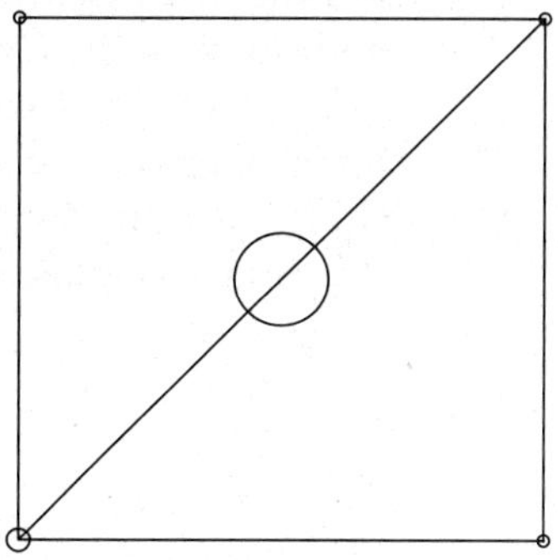

C. Invoke the ***Drafting Settings*** dialog box by any method. Turn on the ***Endpoint***, ***Midpoint***, and ***Center*** object snaps.

D. Use the ***Circle*** command to create a "base" at the lower-right corner of the square. At the "Specify center point for circle or [3P/2P/Ttr]:" prompt, **PICK** the *Line* ***Endpoint*** at the lower-right corner of the square. Enter a *Radius* of **1'** (don't forget the apostrophe). Since Running Object Snaps are on, AutoCAD should display the Marker (square box) at each of the *Line Endpoints* as you move the cursor near; therefore, you can easily "snap" the center of the bases (*Circles*) to the corners of the square. Draw ***Circles*** of the same ***Radius*** at second base (upper-right corner), and third base (upper-left corner of the square). Create home plate with a ***Circle*** of a **2'** ***Radius*** by the same method (see Fig. 7-46).

E. Draw the pitcher's mound by first drawing a ***Line*** between home plate and second base. **PICK** the "Specify first point:" at home plate (***Center*** or ***Endpoint***), then at the "Specify next point or [Undo]:" prompt, PICK the ***Center*** or the ***Endpoint*** at second base. Construct a ***Circle*** of **8'** ***Radius*** at the ***Midpoint*** of the newly constructed diagonal line to represent the pitcher's mound (see Fig. 7-46).

F. ***Erase*** the diagonal *Line* between home plate and second base. Draw the foul lines from first and third base to the outfield. For the first base foul line, construct a ***Line*** with the "Specify first point:" at the ***Endpoint*** of the existing first base line or ***Center*** of the base. Move the cursor (with *Polar Tracking* on) to the right (X positive) and enter a value of **240'** (don't forget the apostrophe). Press **Enter** to complete the *Line* command. Draw the third base foul line at the same length (in the positive Y direction) by the same method (see Fig. 7-47).

**FIGURE 7-47**

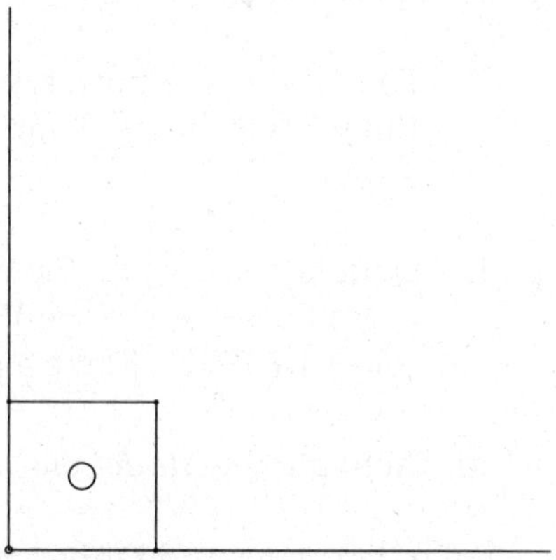

G. Draw the home run fence by using the ***Arc*** command from the ***Draw*** pull-down menu. Select the ***Start, Center, End*** method. **PICK** the end of the first base line (***Endpoint*** object snap) for the ***Start*** of the *Arc* (Fig. 7-48, point 1), **PICK** the pitcher's mound (***Center*** object snap) for the ***Center*** of the *Arc* (point 2), and the end of the third base line (***Endpoint*** object snap) for the ***End*** of the *Arc* (point 3). Don't worry about Polar Tracking in this case because Object Snap overrides Polar Tracking.

**FIGURE 7-48**

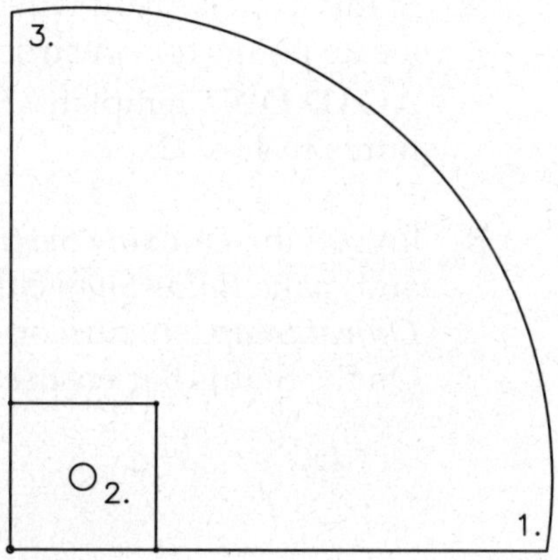

H. Now turn off ***Polar Tracking***. Construct an *Arc* to represent the end of the infield. Select the ***Arc Start, Center, End*** method from the ***Draw*** pull-down menu. For the ***Start*** point of the *Arc,* toggle Running ***Object Snaps*** off by pressing **F3**, **Ctrl+F**, or clicking the icon on the Status Bar and **PICK** location **140′, 20′, 0′** on the first base line (watch the coordinate display and ensure *Snap* and *Grid* are on). (See Fig. 7-49, point 1.) Next, toggle Running ***Object Snaps*** back on, and **PICK** the ***Center*** of the pitcher's mound as the ***Center*** of the *Arc* (point 2). Third, toggle Running ***Object Snaps*** off again and **PICK** the ***End*** point of the *Arc* on the third base line (point 3).

FIGURE 7-49

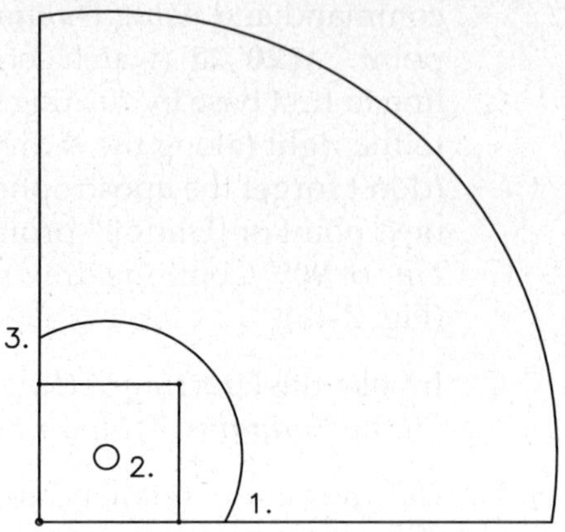


I. Lastly, the pitcher's mound should be moved to the correct distance from home plate. Type ***M*** (the command alias for *Move*) or select ***Move*** from the ***Modify*** pull-down menu. When prompted to "Select objects:" **PICK** the *Circle* representing pitcher's mound. At the "Specify base point or displacement:" prompt, toggle Running ***Object Snaps*** on and **PICK** the ***Center*** of the mound. At the "Specify second point of displacement:" prompt, enter **@3′2″<225**. This should reposition the pitcher's mound to the regulation distance from home plate (60′-6″). Compare your drawing to Figure 7-49. ***Save*** the drawing (as BALL FIELD CH7).

J. Activate a ***Layout*** tab. Assuming your system is configured for a printer using an 11 x 8.5 sheet and is configured to automatically create a viewport (see "Setting Layout Options and Plot Options" in Chapter 6), a viewport should appear. (If the *Page Setup Manager* automatically appears, select ***Close***.) If a viewport appears, proceed to step L. If no viewport appears, complete step K.

K. To make a viewport, type ***-Vports*** at the Command prompt (don't forget the hyphen) and press **Enter**. Accept the default (***Fit***) option by pressing **Enter**. A viewport appears to fit the printable area.

L. Double-click inside the viewport, then type **Z** for *Zoom* and press **Enter**, and type **A** for *All* and press **Enter**. From the ***Viewport Scale*** list (near the lower-right corner of the Drawing Editor), select ***1/64″=1′***. Make a print of the drawing on an 11 x 8.5 inch sheet.

M. Activate the ***Model*** tab. ***Save*** the drawing as **BALL FIELD CH7.**

5. **Polar Tracking and Object Snap Tracking**

A. In this exercise, you will create a table base using Polar Tracking and Object Snap Tracking to locate points for construction of holes and other geometry. Begin a ***New*** drawing and use the **ACAD.DWT** template. ***Save*** the drawing and assign the name **TABLE-BASE.** Set the drawing limits to **48** x **32**.

B. Invoke the ***Drafting Settings*** dialog box. In the ***Snap and Grid*** tab, set ***Polar Distance*** to **1.00** and make ***Polar Snap*** current. In the ***Polar Tracking*** tab, set the ***Increment Angle*** to **45**. In the ***Object Snap*** tab, turn on the ***Endpoint, Midpoint,*** and ***Center*** Object Snap options. Select ***OK***. On the Status Bar ensure ***Snap, Polar Tracking, Object Snap,*** and ***Object Snap Tracking*** are on.

C. Use a ***Line*** and create the three line segments representing the first leg as shown in Figure 7-50. Begin at the indicated location. Use Polar Tracking and Polar Snap to assist drawing the diagonal lines segments. (Do not create the dimensions in your drawing.)

**FIGURE 7-50**

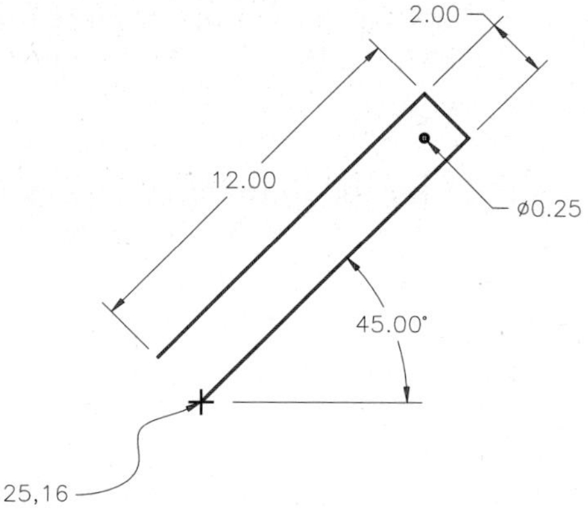


D. Place the first drill hole at the end of the leg using ***Circle*** with the ***Center, Radius*** option. Use ***Object Snap Tracking*** to indicate the center for the hole (*Circle*). Track vertically and horizontally from the indicated corners of the leg in Figure 7-51. Use ***Endpoint*** to snap to the indicated corners. (See Figure 7-50 for hole dimension.)

**FIGURE 7-51**

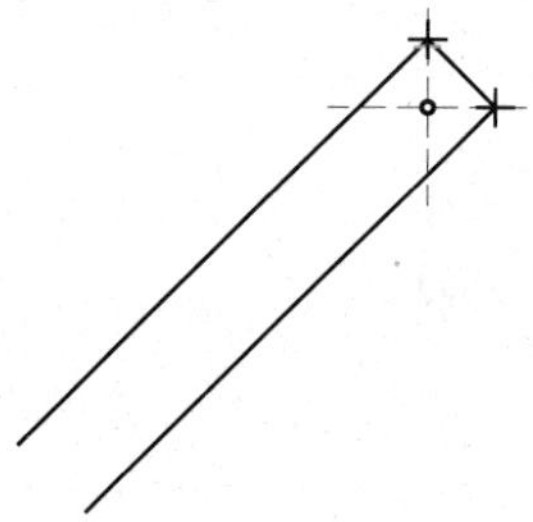

E. Create the other 3 legs in a similar fashion. You can track from the ***Endpoints*** on the bottom of the existing leg to identify the "first point" of the next *Line* segment as shown in Figure 7-52.

**FIGURE 7-52**

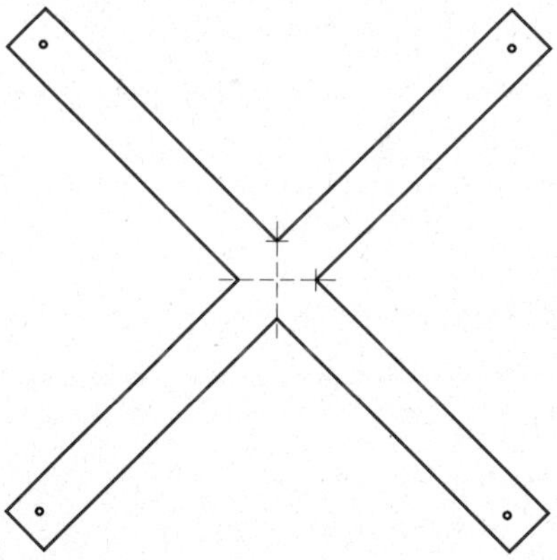

F. Use ***Line*** to create a 24" x 24" square table top on the right side of the drawing as shown in Figure 7-53. Use ***Object Snap Tracking*** and ***Polar Tracking*** with the ***Move*** command to move the square's center point to the center point of the 4 legs (HINT: At the "Specify base point or displacement:" prompt, *Object Snap Track* to the square's ***Midpoints***. At the "Specify second point of displacement:" prompt, use ***Endpoint*** to locate the legs' center.)

**FIGURE 7-53**

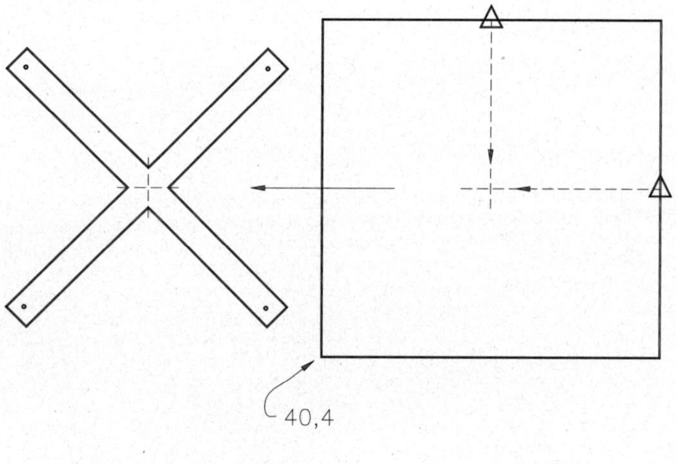

G. Create a smaller square (***Line***) in the center of the table which will act as a support plate for the legs. Track from the ***Midpoint*** of the legs to create the lines as indicated (highlighted) in Figure 7-54.

FIGURE 7-54

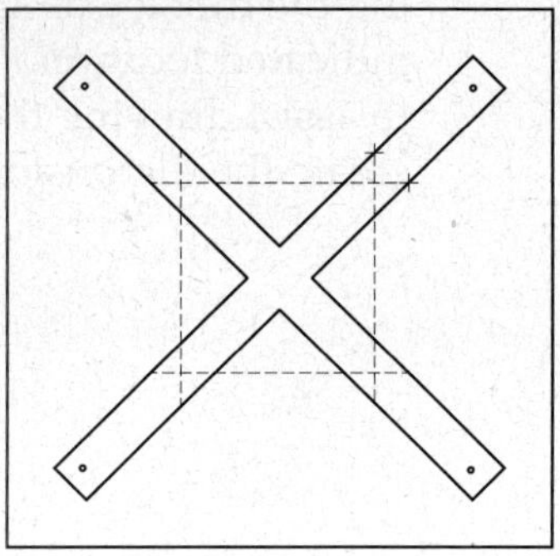

H. ***Save*** the drawing. You will finish the table base in Chapter 9.

CHAPTER 8

# Draw Commands I

## Chapter Objectives

After completing this chapter you should:

1. know where to locate and how to invoke the draw commands;
2. be able to draw *Lines*;
3. be able to draw *Circles* by each of the five options;
4. be able to draw *Arcs* by each of the eleven options;
5. be able to create *Point* objects and specify the *Point Style*;
6. be able to create *Plines* with width and composed of line and arc segments.

# CONCEPTS

## Draw Commands—Simple and Complex

Draw commands create objects. An object is the smallest component of a drawing. The draw commands listed immediately below create simple objects and are discussed in this chapter. Simple objects appear as one entity.

*Line, Circle, Arc,* and *Point*

Other draw commands create more complex shapes. Complex shapes appear to be composed of several components, but each shape is usually one object. An example of an object that is one entity but usually appears as several segments is listed below and is also covered in this chapter:

*Pline*

Other draw commands, listed below, (see Chapter 15) are a combination of simple and complex shapes:

*Xline, Ray, Polygon, Rectangle, Donut, Spline, Ellipse, Divide, Mline, Measure, Sketch, Solid, Region, Boundary, 3DPoly, Helix,* and *Revcould*

FIGURE 8-1

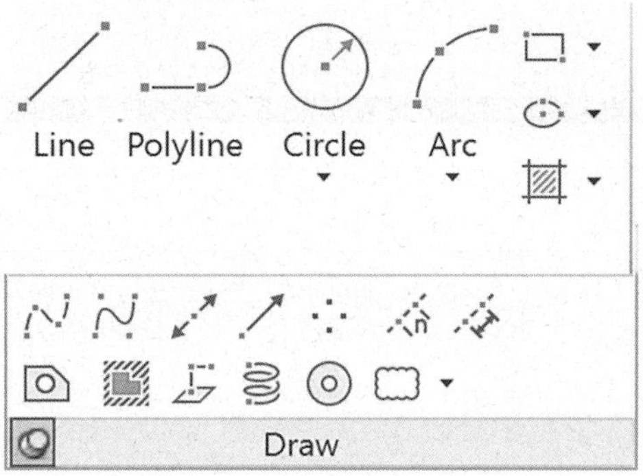


## Draw Command Access

As a review from Chapter 3, Draw Command Concepts, remember that there are many methods you can use to access the draw commands. For example, you can enter the command names or command aliases at the keyboard. You can also select draw commands from the *Draw* panel on the Ribbon in the *Home* tab (Fig. 8-1). Note that many of the draw commands and options are accessible only on drop-down menus.

FIGURE 8-2

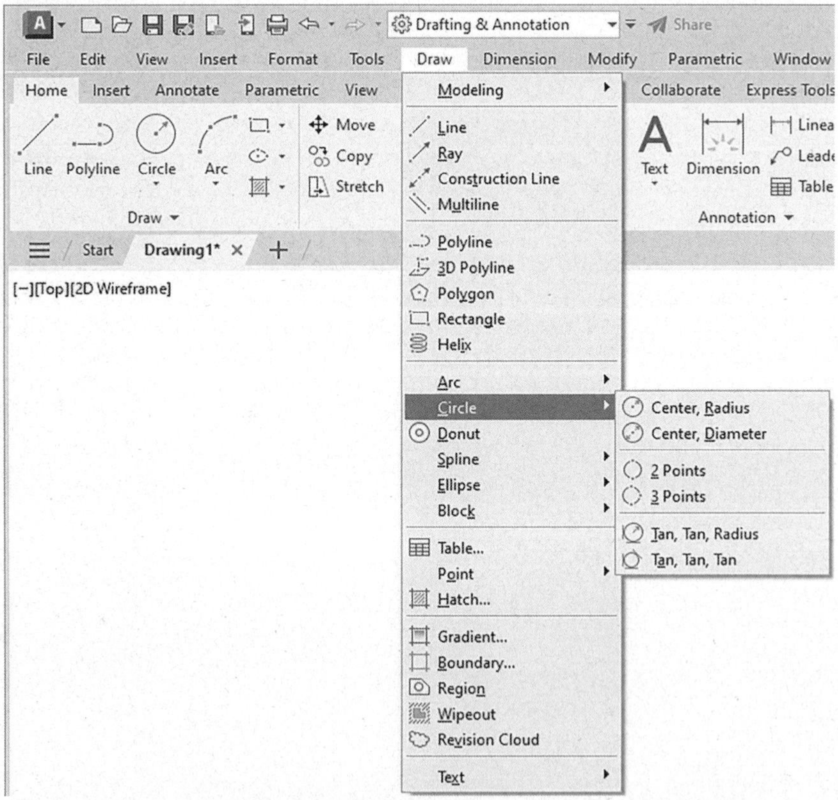


Alternately, you can select commands from the *Draw* pull-down menu if the Menu Bar is activated. Note that on the *Draw* pull-down menu, command options are presented on fly-outs (Fig. 8-2). Additionally, a *Draw* toolbar can be activated (see Chapter 1 for information on activating the Menu Bar and toolbars).

## Coordinate Entry Input Methods

When creating objects with draw commands, AutoCAD always prompts you to indicate points (such as endpoints, centers, radii) to describe the size and location of the objects to be drawn. An example you are familiar with is the *Line* command, where AutoCAD prompts for the "Specify first point:". Indication of these points, called coordinate entry, can be accomplished by three input methods:

| | |
|---|---|
| 1. Mouse Input | PICK (press the left mouse button to specify a point on the screen). |
| 2. Dynamic Input | When *Dynamic Input* is on, you can key values into the edit boxes near the cursor and press Tab to change boxes. |
| 3. Command Line Input | You can key values in any format at the Command prompt. |

Any of these methods can be used whenever AutoCAD prompts you to specify points. (For practice with these methods, see Chapter 3, Draw Command Concepts.)

### Drawing Aids

Also keep in mind that you can specify points interactively using the following AutoCAD features individually or in combination: *Grid, Snap, Ortho, Polar Tracking, Object Snap,* and *Object Snap Tracking*. Use these drawing tools whenever AutoCAD prompts you to select points. (See Chapters 3 and 7 for details on these tools.)

TIP

# COMMANDS

### *Line*

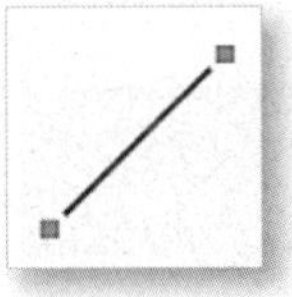

| Ribbon | Menu Bar | Command (Type) | Alias (Type) | Shortcut |
|---|---|---|---|---|
| *Home Draw Line* | *Draw Line...* | *Line* | L | ... |

This is the fundamental drawing command. The *Line* command creates straight line segments; each segment is an object. One or several line segments can be drawn with the *Line* command.

Command: ***LINE***
Specify first point: **PICK** or (**coordinates**) (A point can be designated by interactively selecting with the input device or by entering coordinates. Use any of the drawing tools to assist with interactive entry.)
Specify next point or [Undo]: **PICK** or (**coordinates**) (Again, device input or keyboard input can be used.)
Specify next point or [Undo]: **PICK** or (**coordinates**)
Specify next point or [Close/Undo]: **PICK** or (**coordinates**) or ***C***
Specify next point or [Close/Undo]: press **Enter** to finish command
Command:

**FIGURE 8-3**

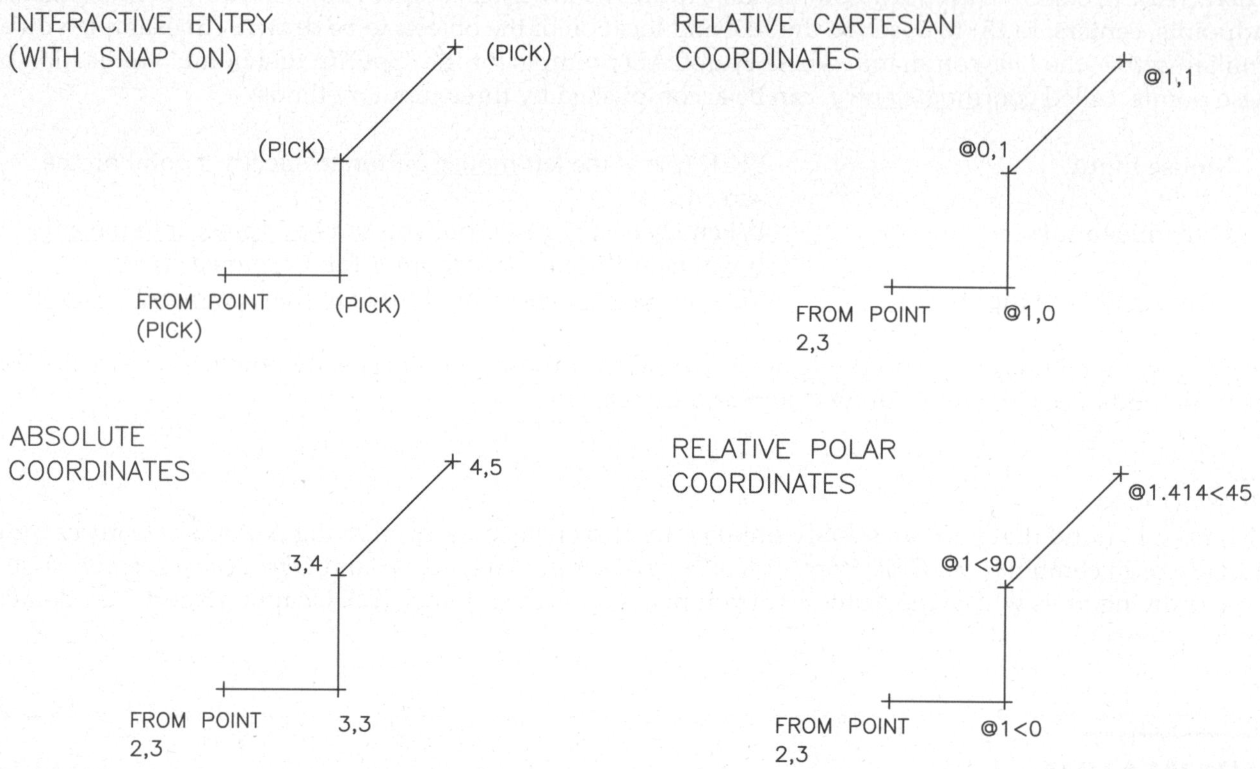


Figure 8-3 shows four examples of creating the same *Line* segments using different formats of coordinate entry. Refer to Chapter 3, Draw Command Concepts, for examples of drawing vertical, horizontal, and inclined lines using the formats for coordinate entry.

## *Circle*

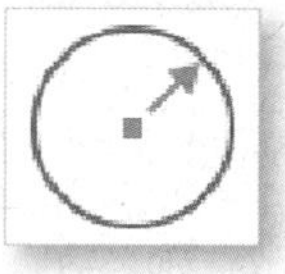

| Ribbon | Menu Bar | Command (Type) | Alias (Type) | Shortcut |
|---|---|---|---|---|
| *Home* *Draw* *Circle* | *Draw* *Circle >* | *Circle* | *C* | ... |

The *Circle* command creates one object. Depending on the option selected, you can provide two or three points to define a *Circle*. As with all commands, the Command line prompt displays the possible options:

Command: **`CIRCLE`**

Specify center point for circle or [3P/2P/Ttr (tan tan radius)]: **`PICK`** or (**`coordinates`**) or (**`option`**). (PICKing or entering coordinates designates the center point for the circle. You can enter *3P*, *2P*, or *T* for another option.)

As with most commands, the default and other options are displayed on the Command line. The default option always appears first. The other options can be invoked by typing the numbers and/or uppercase letter(s) that appear in brackets [ ].

All of the options for creating *Circles* are available from the *Draw* menu and from the *Draw* panel. The tool (icon button) for only the *Center, Radius* method is included in the *Draw* toolbar. Although the *Circle* options are not available on the *Draw* toolbar, you can use the *Circle* button, then right-click for a shortcut menu to select options.

The options, or methods, for drawing *Circles* are listed below. Each figure gives several possibilities for each option, with and without *Object Snaps*.

### *Center, Radius*

Specify a center point, then a radius (Fig. 8-4).

The *Radius* (or *Diameter*) can be specified by entering values or by indicating a length interactively (PICK two points to specify a length when prompted). As always, points can be specified by PICKing or entering coordinates. Watch the Coordinate display for coordinate or distance (polar format) display. *Grid Snap, Polar Snap, Polar Tracking, Object Snap,* and *Object Snap Tracking* can be used for interactive point specification.

**FIGURE 8-4**

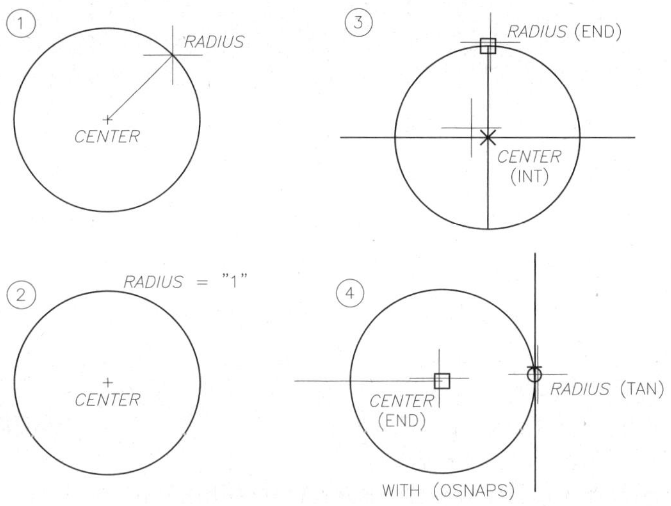


### *Center, Diameter*

Specify the center point, then the diameter (Fig. 8-5).

**FIGURE 8-5**

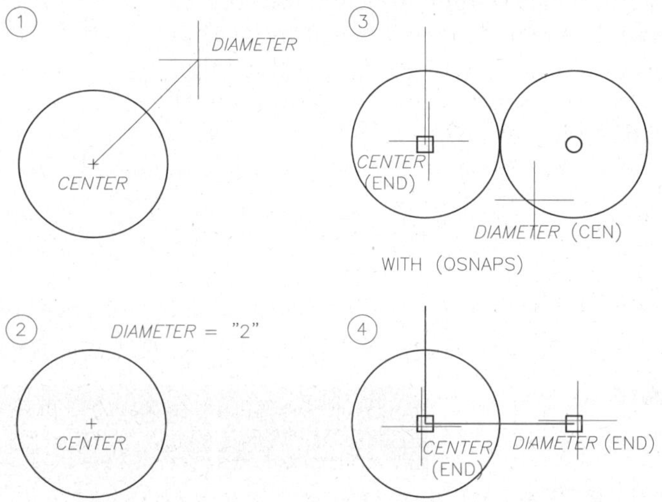


### *2 Points*

The two points specify the location and diameter.

The *Tangent* Object Snaps can be used when selecting points with the *2 Point* and *3 Point* options, as shown in Figures 8-6 and 8-7.

**FIGURE 8-6**

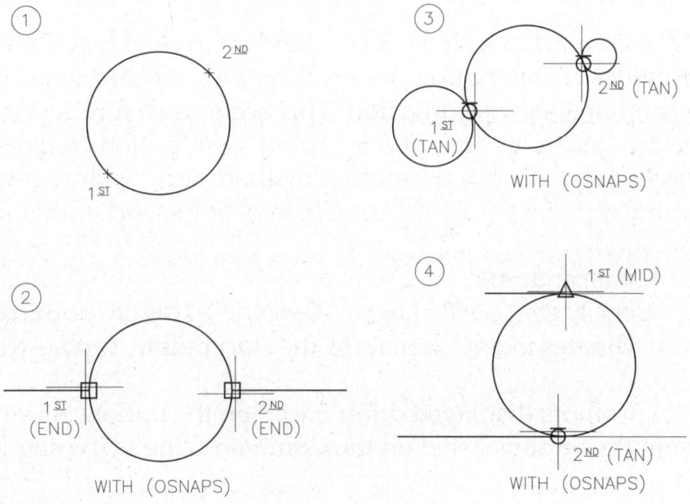

### *3 Points*

The *Circle* passes through all three points specified (Fig. 8-7).

**FIGURE 8-7**

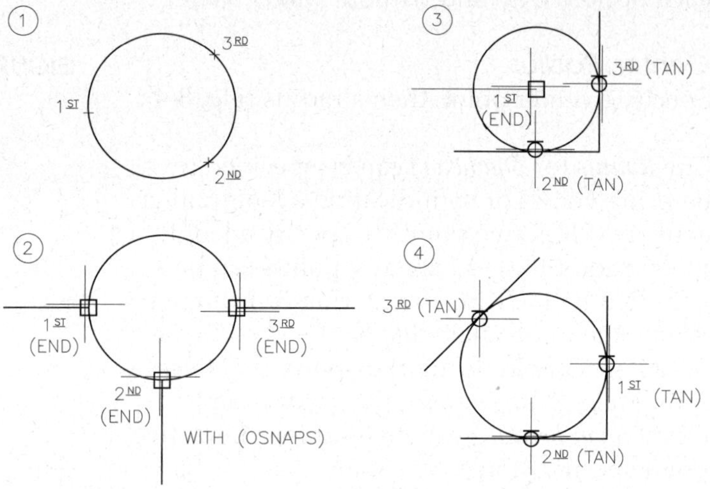


### *Tangent, Tangent, Radius*

Specify two objects for the *Circle* to be tangent to; then specify the radius (Fig. 8-8).

**TIP** The *TTR* (Tangent, Tangent, Radius) method is extremely efficient and productive. The Object Snap *Tangent* modes are automatically invoked. This is the only draw command option that automatically calls Object Snaps.

**FIGURE 8-8**

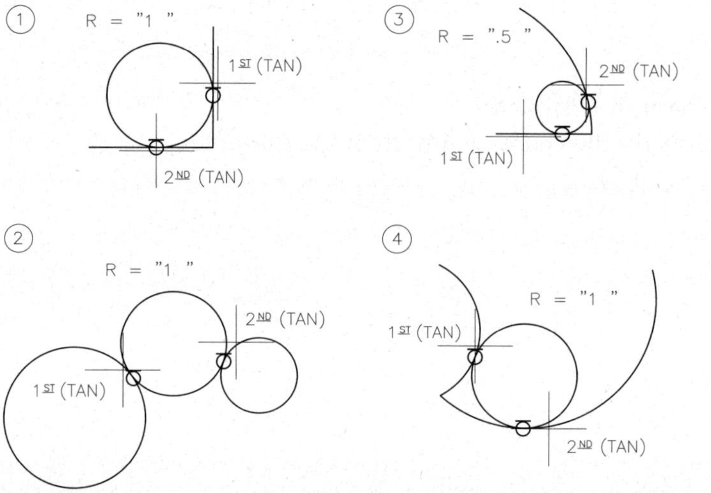


## *Arc*

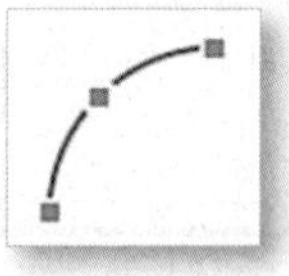

| Ribbon | Menu Bar | Command (Type) | Alias (Type) | Shortcut |
|---|---|---|---|---|
| *Home*<br>*Draw*<br>*Arc* | *Draw*<br>*Arc >* | *Arc* | *A* | ... |

An arc is part of a circle; it is a regular curve of less than 360 degrees. The *Arc* command in AutoCAD provides eleven options for creating arcs. An *Arc* is one object. *Arcs* are always drawn by default in a counterclockwise direction. This occurrence forces you to decide in advance which points should be designated as *Start* and *End* points (for options requesting those points). For this reason, it is often easier to create arcs by another method, such as drawing a *Circle* and then using *Trim* or using the *Fillet* command. (See "Use *Arcs* or *Circles*?" at the end of this section on *Arcs*.) The *Arc* command prompt is:

Command: ***ARC***
Specify start point of arc or [Center]: **`PICK`** or (**`coordinates`**) or ***C*** (Interactively select or enter coordinates in any format for the start point. Type C to use the *Center* option.)

The prompts displayed differ based on the option. At any time while using the command, you can select from the options listed on the Command line by typing in the capitalized letter(s) for the desired option.

Alternately, to use a particular option of the *Arc* command, you can select from the *Draw* menu or from the *Draw* panel. The tool (icon button) for only the *3 Points* method is included in the *Draw* toolbar. However, you can select the *Arc* button, then right-click for a shortcut menu at any time during the *Arc* command to show other possible options at that point. These options require coordinate entry of points in a specific order.

### *3Points*

Specify three points through which the *Arc* passes (Fig. 8-9).

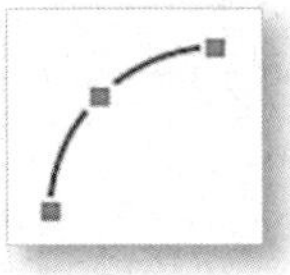

**FIGURE 8-9**

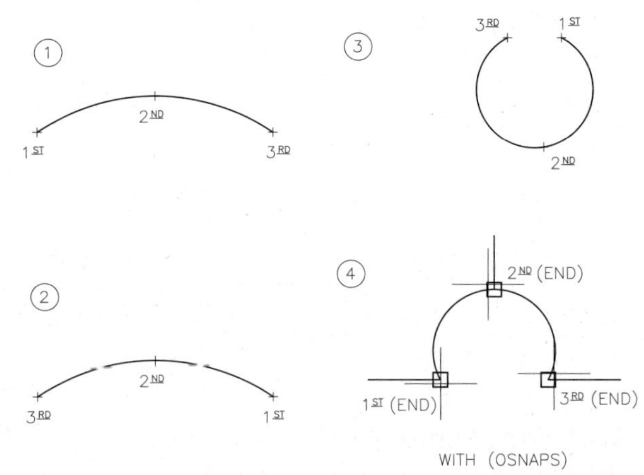


### *Start, Center, End*

The radius is defined by the first two points that you specify (Fig. 8-10).

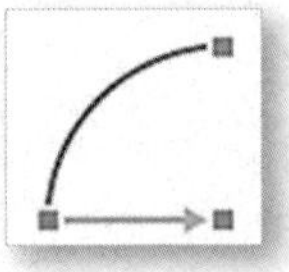

**FIGURE 8-10**

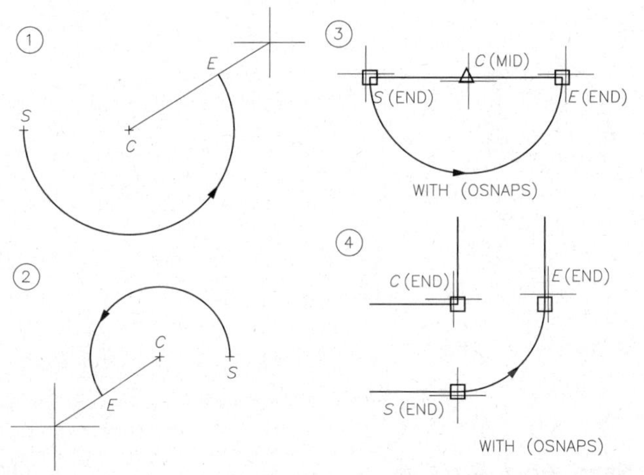


### *Start, Center, Angle*

The angle is the included angle between the sides from the center to the endpoints (Fig. 8-11). A negative angle can be entered to generate an *Arc* in a clockwise direction.

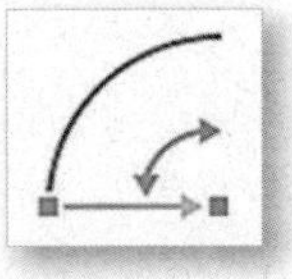

**FIGURE 8-11**

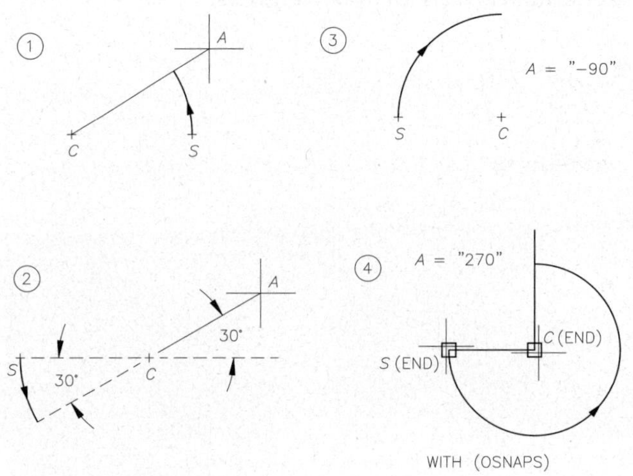

### *Start, Center, Length*

*Length* means length of chord. The length of chord is between the start and the other point specified (Fig. 8-12). A negative chord length can be entered to generate an *Arc* of 180+ degrees.

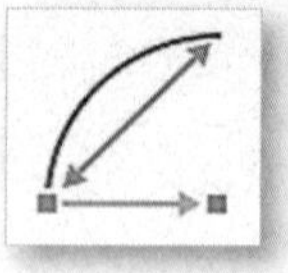

FIGURE 8-12

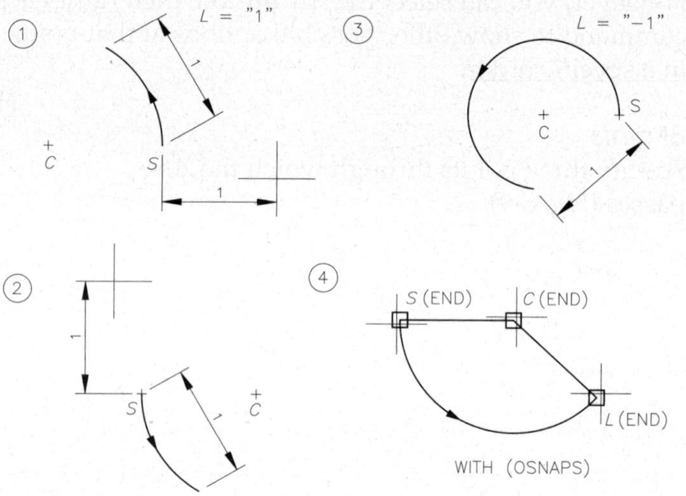


### *Start, End, Angle*

The included angle is between the sides from the center to the endpoints (Fig. 8-13). Negative angles generate clockwise *Arcs*.

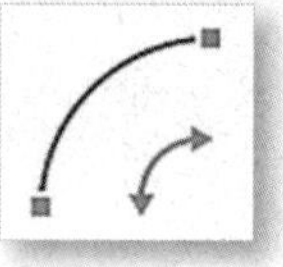

FIGURE 8-13

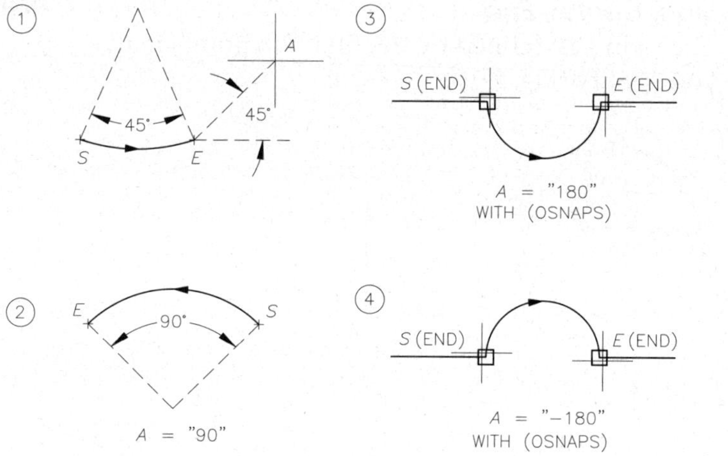


### *Start, End, Radius*

The radius can be PICKed or entered as a value (Fig. 8-14). A negative radius value generates an *Arc* of 180+ degrees.

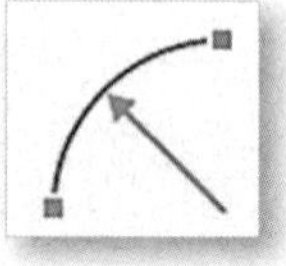

FIGURE 8-14

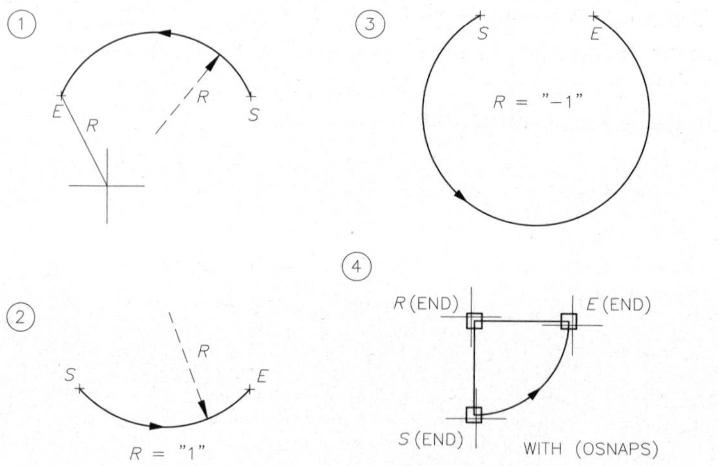

### *Start, End, Direction*

The direction is tangent to the start point (Fig. 8-15).

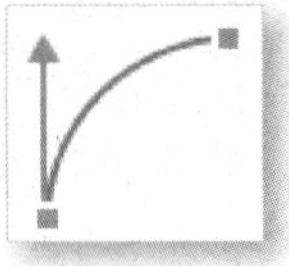

FIGURE 8-15

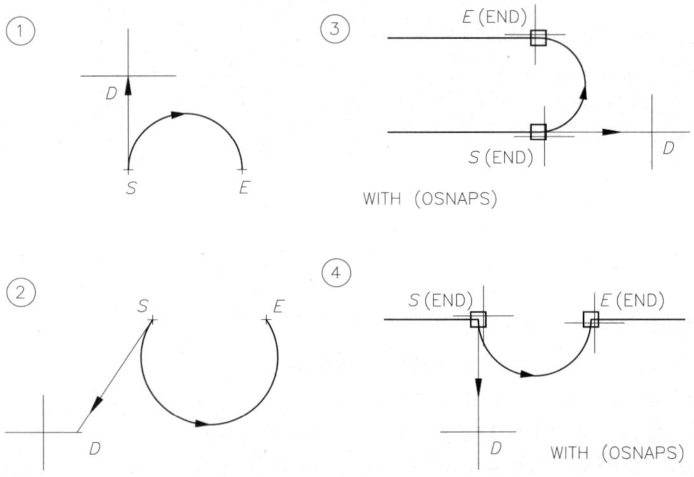


### *Center, Start, End*

This option is like *Start, Center, End* but in a different order (Fig. 8-16).

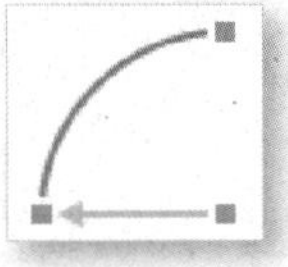

FIGURE 8-16

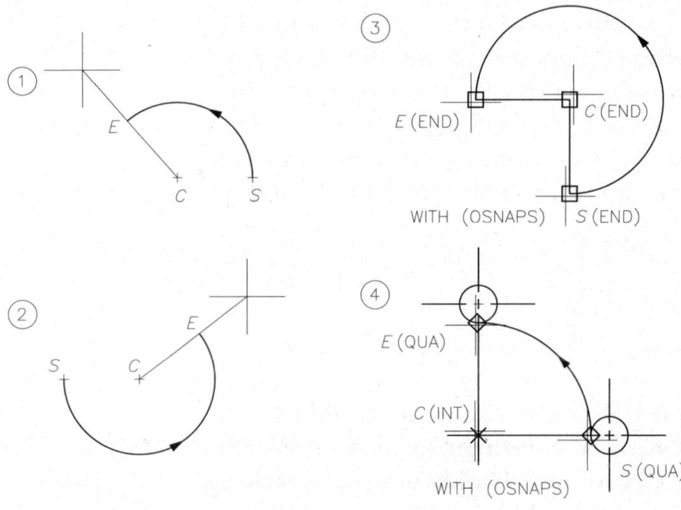


### *Center, Start, Angle*

This option is like *Start, Center, Angle* but in a different order (Fig. 8-17).

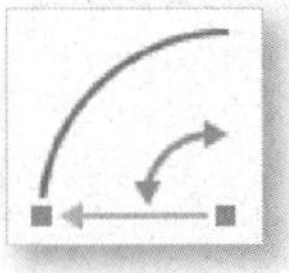

FIGURE 8-17

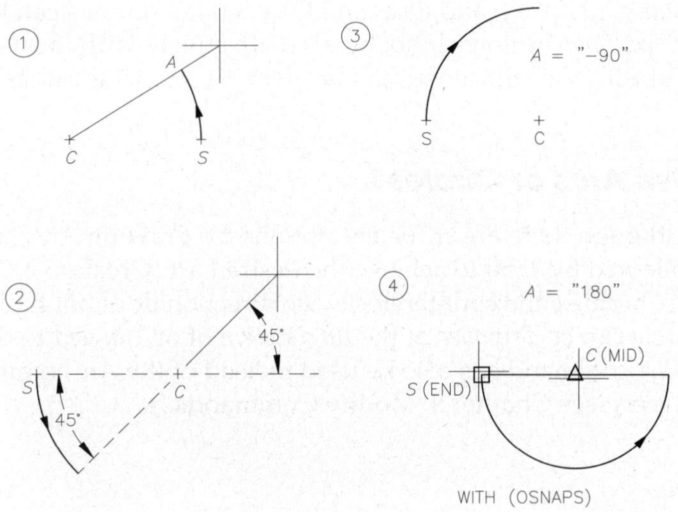

### *Center, Start, Length*

This is similar to the *Start, Center, Length* option but in a different order (Fig. 8-18). *Length* means length of chord.

FIGURE 8-18

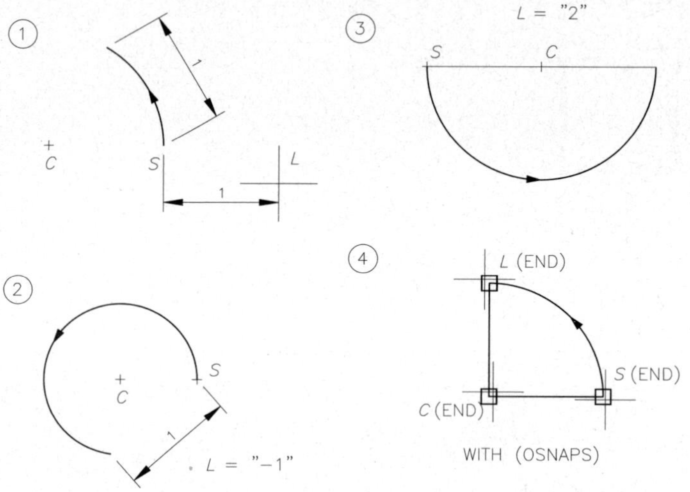


### *Continue*

The new *Arc* continues from and is tangent to the last point (Fig. 8-19). The only other point required is the endpoint of the *Arc*. This method allows drawing *Arcs* tangent to the preceding *Line* or *Arc*.

*Arc*s are always created in a counter-clockwise direction. This fact must be taken into consideration when using any method except the *3-Point*, the *Start, End, Direction*, and the *Continue* options. The direction is explicitly specified with *Start, End, Direction*, and *Continue* methods, and direction is irrelevant for *3-Point* method.

FIGURE 8-19

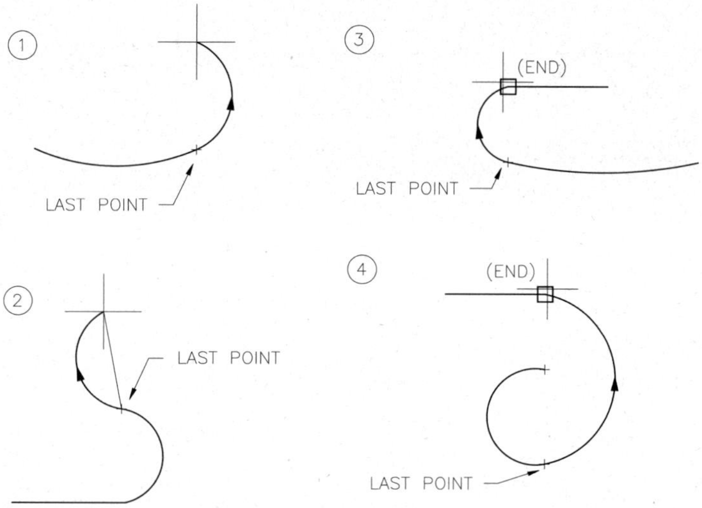


TIP

As usual, points can be specified by PICKing or entering coordinates. Watch the Coordinate display or Polar Tracking tooltip to display coordinate values or distances. *Grid Snap, Polar Snap, Polar Tracking, Object Snap,* and *Object Snap Tracking* can be used when PICKing. The *Endpoint, Intersection, Center, Midpoint,* and *Quadrant* Object Snap options can be used with great effectiveness. The *Tangent Object Snap* option cannot be used effectively with most of the *Arc* options. The *Radius, Direction, Length,* and *Angle* specifications can be given by entering values or by PICKing with or without Object Snaps.

## *Use Arcs or Circles?*

TIP

Although there are sufficient options for drawing *Arcs*, usually it is easier to use the *Circle* command followed by *Trim* to achieve the desired arc. Creating a *Circle* is generally an easier operation than using *Arc* because the counterclockwise direction does not have to be considered. The unwanted portion of the circle can be *Trimmed* at the *Intersection* of or *Tangent* to the connecting objects using Object Snap. The *Fillet* command can also be used instead of the *Arc* command to add a fillet (arc) between two existing objects (see Chapter 9, Modify Commands I).

## Point

| Ribbon | Menu Bar | Command (Type) | Alias (Type) | Shortcut |
|---|---|---|---|---|
| *Home Draw* (icon) | *Draw Point > Single Point* or *Multiple Point* | *Point* | *PO* | ... |

A *Point* is an object that has no dimension; it only has location. A *Point* is specified by giving a coordinate value or by PICKing a location on the screen.

Figure 8-20 compares *Points* to *Line* and *Circle* objects.

Command: ***POINT***
Current point modes: PDMODE=0 PDSIZE=0.0000
Specify a point: **PICK** or (**coordinates**)
(Select a location for the *Point* object.)

**FIGURE 8-20**

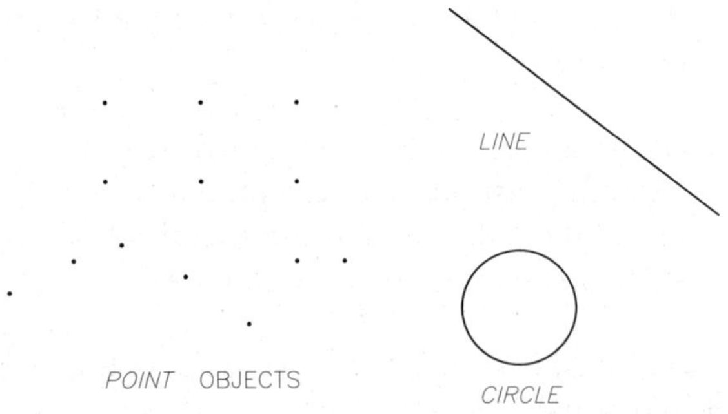


*Points* are useful in construction of drawings to locate points of reference for subsequent construction or locational verification. The *Node Object Snap* option is used to snap to *Point* objects.

Points are drawing objects and therefore appear in prints and plots. The default "style" for points is a tiny dot. The *Point Style* dialog box can be used to define the format you choose for *Point* objects (the *Point* type [*PDMODE*] and size [*PDSIZE*]).

The *Draw* menu offers the *Single Point* and the *Multiple Point* options. The *Single Point* option creates one *Point,* then returns to the command prompt. This option is the same as using the *Point* command by any other method. Selecting *Multiple Point* continues the *Point* command until you press the Escape key.

## Ptype

| Ribbon | Menu Bar | Command (Type) | Alias (Type) | Shortcut |
|---|---|---|---|---|
| *Home Utilities Point Style* | *Format Point Style...* | *Ptype* | ... | ... |

The *Point Style* dialog box (Fig. 8-21) is available only through the methods listed in the command table above. This dialog box allows you to define the format for the display of *Point* objects. The selected style is applied immediately to all newly created *Point* objects. The *Point Style* controls the format of *Points* for printing and plotting as well as for the computer display.

**FIGURE 8-21**

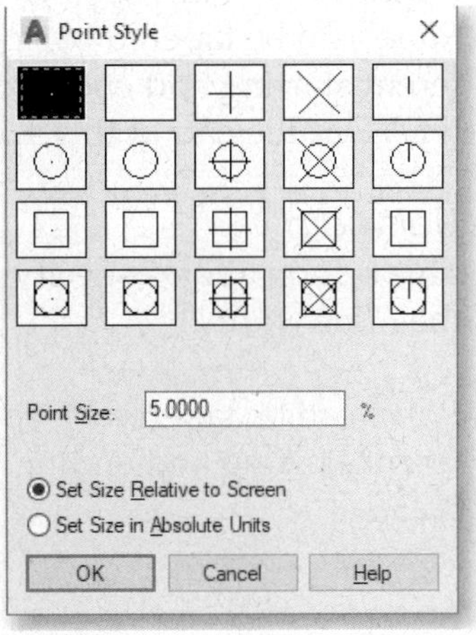


You can set the *Point Size* in *Absolute* units or *Relative to Screen*. The *Relative to Screen* option keeps the *Points* the same size on the display when you *Zoom* in and out, whereas setting *Point Size* in *Absolute Units* gives you control over the size of *Points* for prints and plots. The *Point Size* is stored in the *PDSIZE* system variable. The selected *Point Style* is stored in the *PDMODE* variable.

## Pline

| Ribbon | Menu Bar | Command (Type) | Alias (Type) | Shortcut |
|---|---|---|---|---|
| *Home*<br>*Draw*<br>*Polyline* | *Draw*<br>*Polyline* | *Pline* | *PL* | ... |

A *Pline* (or *Polyline*) has special features that make this object more versatile than a *Line*. Three features are most noticeable when first using *Plines*:

TIP

1. A *Pline* can have a specified *width*, whereas a *Line* has no width.
2. Several *Pline* segments created with one *Pline* command are treated by AutoCAD as one object, whereas individual line segments created with one use of the *Line* command are individual objects.
3. A *Pline* can contain arc segments.

**FIGURE 8-22**

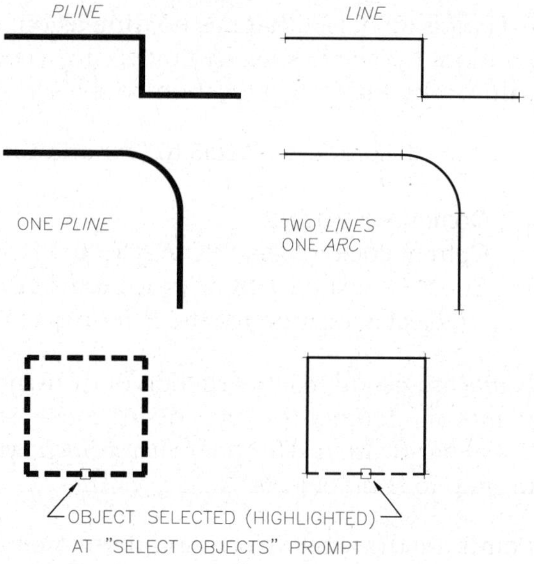


Figure 8-22 illustrates *Pline* versus *Line* and *Arc* comparisons.

The *Pline* command begins with the same prompt as *Line*; however, after the "start point:" is established, the *Pline* options are accessible.

Command: ***PLINE***
Specify start point: **PICK** or (**coordinates**)
Current line-width is 0.0000
Specify next point or [Arc/Close/Halfwidth/Length/Undo/Width]:

Similar to the other drawing commands, you can invoke the *Pline* command by any method, then right-click for a shortcut menu at any time during the command to show other possible options at that point.

The options and descriptions follow.

### *Width*

You can use this option to specify starting and ending widths. Width is measured perpendicular to the centerline of the *Pline* segment (Fig. 8-23). *Plines* can be tapered by specifying different starting and ending widths. See NOTE at the end of this section.

**FIGURE 8-23**

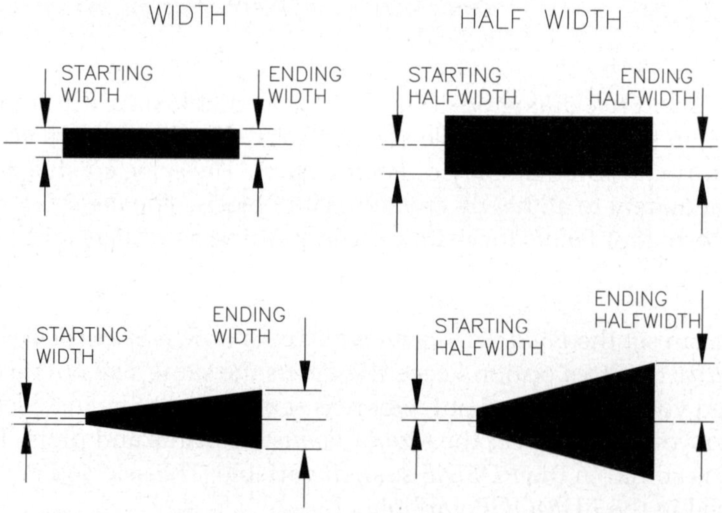


### *Halfwidth*

This option allows specifying half of the *Pline* width.

*Plines* can be tapered by specifying different starting and ending widths (Fig. 8-23).

### *Arc*

This option (by default) creates an arc segment in a manner similar to the *Arc Continue* method (Fig. 8-24). Any of several other methods are possible (see "*Pline Arc* Segments").

FIGURE 8-24

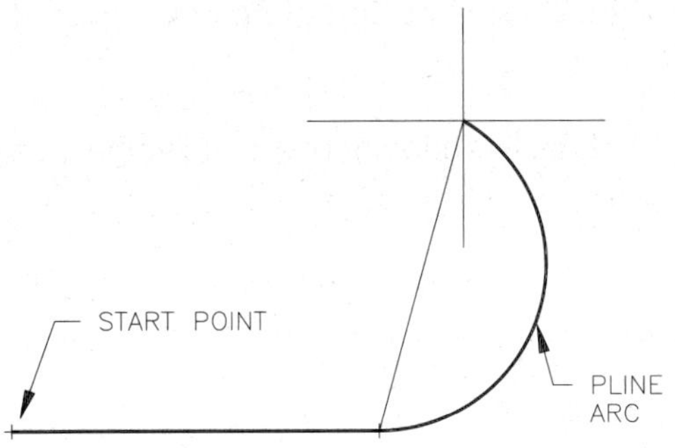


### *Close*

The *Close* option creates the closing segment connecting the first and last points specified with the current *Pline* command as shown in Figure 8-25.

This option can also be used to close a group of connected *Pline* segments into one continuous *Pline*. (A *Pline* closed by PICKing points has a specific start and endpoint.) A *Pline Closed* by this method has special properties if you use *Pedit* for *Pline* editing or if you use the *Fillet* command with the *Pline* option (see *Fillet* in Chapter 9 and *Pedit* in Chapter 16).

FIGURE 8-25

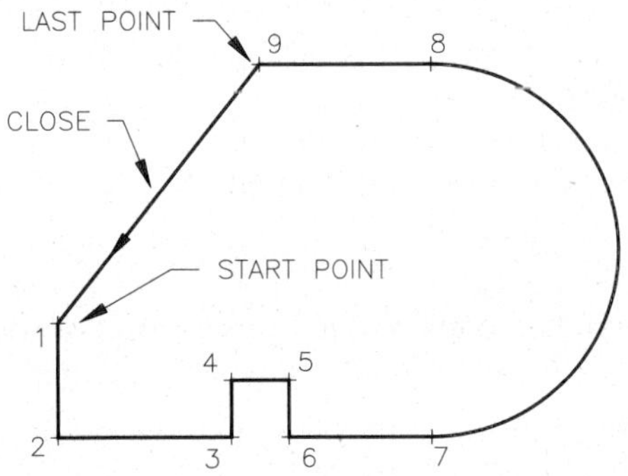


### *Length*

*Length* draws a *Pline* segment at the same angle as and connected to the previous segment and uses a length that you specify. If the previous segment was an arc, *Length* makes the current segment tangent to the ending direction (Fig. 8-26).

FIGURE 8-26

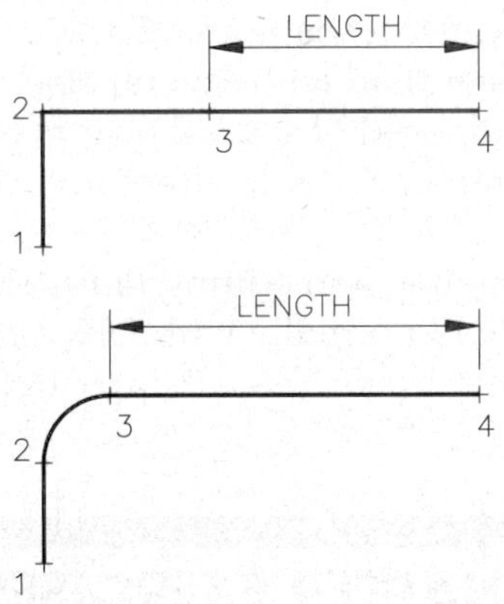


### *Undo*

Use this option to *Undo* the last *Pline* segment. It can be used repeatedly to undo multiple segments.

TIP

NOTE: If you change the *Width* of a *Pline*, be sure to respond to <u>both prompts</u> for width ("Specify starting width:" and "Specify ending width:") before you draw the first *Pline* segment. It is easy to hastily PICK the endpoint of the *Pline* segment after specifying the "Specify starting width:" instead of responding with a value (or Enter) for the "Specify ending width:". In this case (if you PICK at the "Specify ending width:" prompt), AutoCAD understands the line length that you interactively specified to be the ending width you want for the next line segment (you can PICK two points in response to the "Specify starting width:" or "Specify ending width:" prompts).

Command: ***PLINE***
Specify start point: **PICK**
Current line-width is 0.0000
Specify next point or [Arc/Close/Halfwidth/Length/Undo/Width]: ***w***
Specify starting width <0.0000>: **.2**
Specify ending width <0.2000>: Enter a value or press **Enter**—do not PICK the "next point."

### *Pline Arc* Segments

When the *Arc* option of *Pline* is selected, the prompt changes to provide the various methods for construction of arcs:

Specify endpoint of arc or [Angle/CEnter/CLose/Direction/Halfwidth/Line/Radius/Second pt/Undo/Width]:

#### *Angle*

You can draw an arc segment by specifying the included angle (a negative value indicates a clockwise direction for arc generation).

#### *CEnter*

This option allows you to specify a specific center point for the arc segment.

#### *CLose*

This option closes the *Pline* group with an arc segment.

#### *Direction*

*Direction* allows you to specify an explicit starting direction rather than using the ending direction of the previous segment as a default.

#### *Line*

This switches back to the line options of the *Pline* command.

#### *Radius*

You can specify an arc radius using this option.

#### *Second pt*

Using this option allows specification of a 3-point arc.

TIP

Because a shape created with one *Pline* command is one object, manipulation of the shape is generally easier than with several objects. For some applications, one *Pline* shape can have advantages over shapes composed of several objects (see "*Offset*," Chapter 9). Editing *Plines* is accomplished by using the *Pedit* command. As an alternative, *Plines* can be "broken" back down into individual objects with *Explode*.

Drawing and editing *Plines* can be somewhat involved. As an alternative, you can draw a shape as you would normally with *Line, Circle, Arc, Trim*, etc., and then convert the shape to one *Pline* object using *Pedit* or *Join*. (See Chapter 16 for details on converting *Lines* and *Arcs* to *Plines*.)

## CHAPTER EXERCISES

Create a ***New*** drawing. Use the **ACAD.DWT** template, then ***Zoom All***. Set ***Polar Snap On*** and set the ***Polar Distance*** to **.25**. Turn on ***Snap, Polar Tracking, Object Snap, Object Snap Tracking***, and ***Dynamic Input***. ***Save*** the drawing as **CH8EX**. For each of the following problems, ***Open*** **CH8EX**, complete one problem, then use ***SaveAs*** to give the drawing a new name.

1. ***Open*** **CH8EX**. Create the geometry shown in Figure 8-27. Start the first *Circle* center at point **4,4.5** as shown. Do not copy the dimensions. ***SaveAs*** **LINK** (HINT: Locate and draw the two small *Circles* first. Use *Arc, Start, Center, End* or *Center, Start, End* for the rounded ends.)

**FIGURE 8-27**

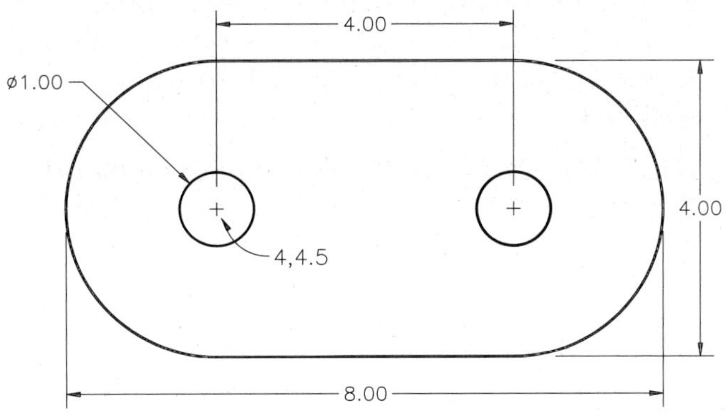


REUSE

2. ***Open*** **CH8EX**. Create the geometry as shown in Figure 8-28. Do not copy the dimensions. Assume symmetry about the vertical axis. ***SaveAs*** **SLOTPLATE CH8.**

**FIGURE 8-28**

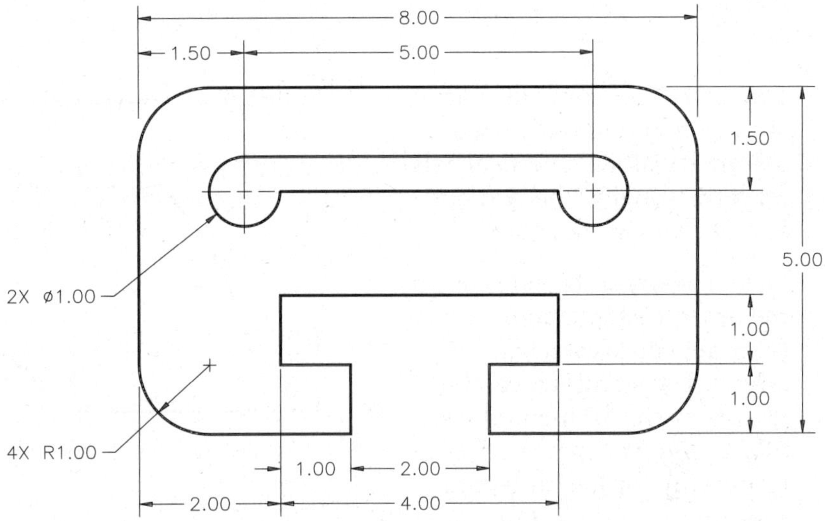


REUSE

3. ***Open*** **CH8EX**. Create the shapes shown in Figure 8-29. Do not copy the dimensions. ***SaveAs*** **CH8EX3**.

   Draw the ***Lines*** at the bottom first, starting at coordinate **1,3**. Then create ***Point*** objects at **5,7**, **5.4,7**, **5.8,7**, etc. Change the ***Point Style*** to an X. Use the ***NODe*** *Object Snap* mode to draw the inclined ***Lines***. Create the ***Arc*** on top with the ***Start, End, Direction*** option.

**FIGURE 8-29**

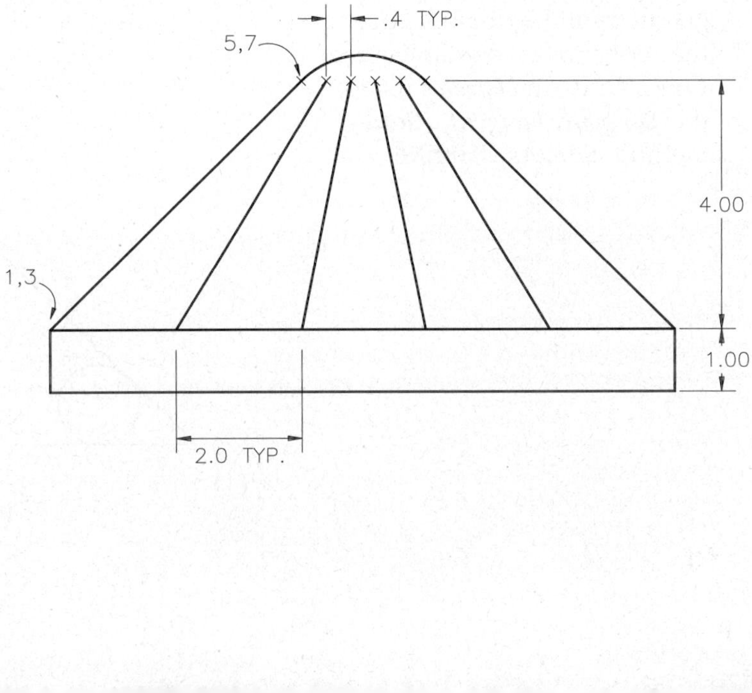

4. *Open* **CH8EX**. Create the shape shown in Figure 8-30. Draw the two horizontal ***Lines*** and the vertical ***Line*** first by specifying the endpoints as given. Then create the ***Circle*** and ***Arcs***. ***SaveAs*** **CH8EX4**.

   HINT: Use the ***Circle 2P*** method with ***Endpoint*** Object Snaps. The two upper ***Arcs*** can be drawn by the ***Start, End, Radius*** method.

**FIGURE 8-30**

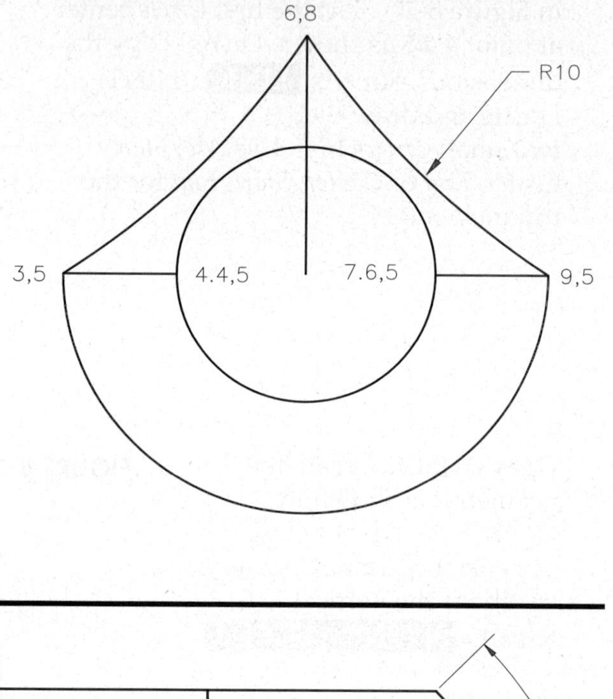


5. *Open* **CH8EX**. Draw the shape shown in Figure 8-31. Assume symmetry along a vertical axis. Start by drawing the two horizontal ***Lines*** at the base.

   Next, construct the side ***Arcs*** by the ***Start, Center, Angle*** method (you can specify a negative angle). The small ***Arc*** can be drawn by the ***3P*** method. Use ***Object Snaps*** when needed (especially for the horizontal ***Line*** on top and the ***Line*** along the vertical axis). ***SaveAs*** **CH8EX5**.

**FIGURE 8-31**

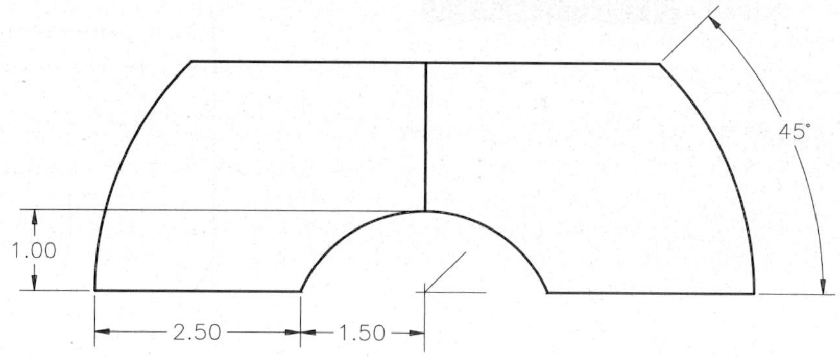


6. *Open* **CH8EX**. Complete the geometry in Figure 8-32. Use the coordinates to establish the ***Lines***. Draw the ***Circles*** using the ***Tangent, Tangent, Radius*** method. ***SaveAs*** **CH8EX6**.

**FIGURE 8-32**

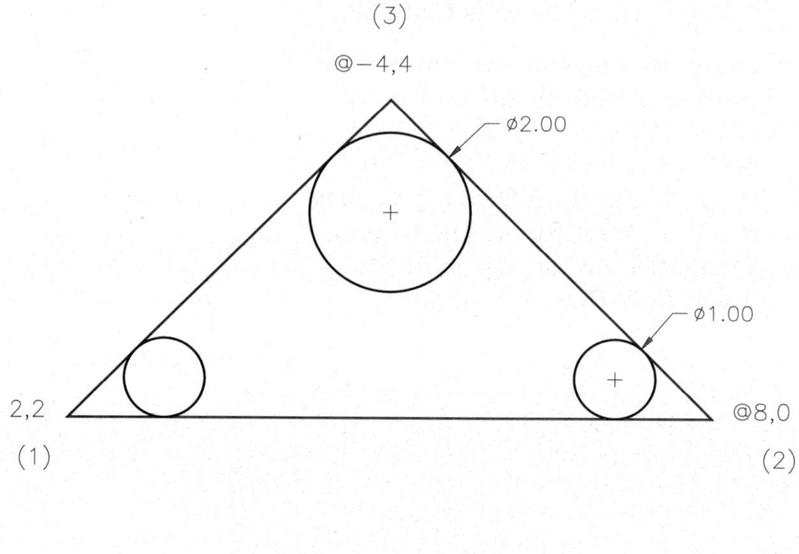

7. **Retaining Wall**

A contractor plans to stake out the edge of a retaining wall located at the bottom of a hill. The following table lists coordinate values based on a survey at the site. Use the drawing you created in Chapter 6 Exercises named **CH6EX2**. ***Save*** the drawing as **RET-WALL.**

REUSE

**FIGURE 8-33**

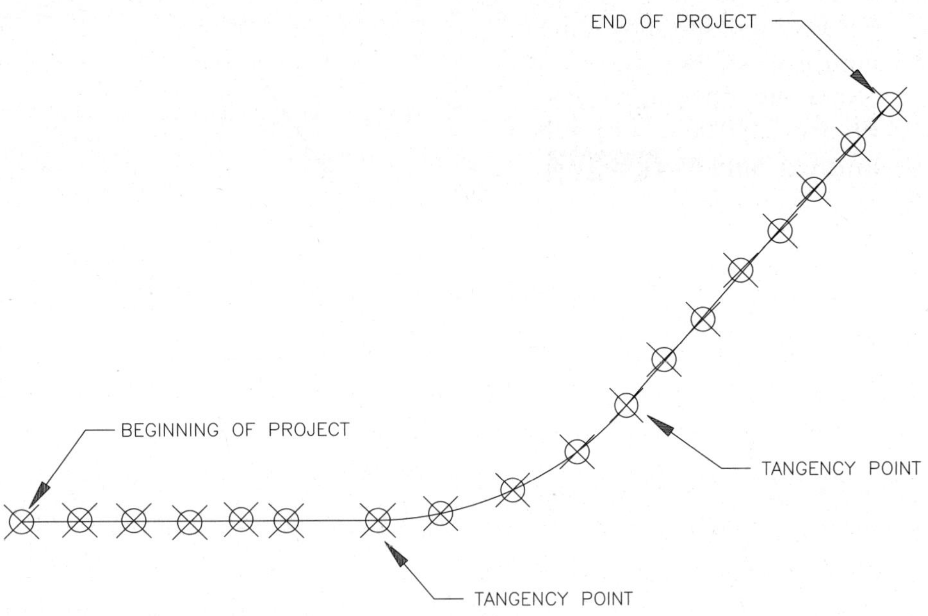


Place ***Points*** at each coordinate value in order to create a set of data points. Determine the location of one ***Arc*** and two ***Lines*** representing the centerline of the retaining wall edge. The centerline of the retaining wall should match the data points as accurately as possible (Fig. 8-33). The data points given in the following table are in inch units.

| Pt # | X | Y | |
|---|---|---|---|
| 1. | 60.0000 | 108.0000 | |
| 2. | 81.5118 | 108.5120 | |
| 3. | 101.4870 | 108.2560 | |
| 4. | 122.2305 | 107.4880 | |
| 5. | 141.4375 | 108.5120 | |
| 6. | 158.1949 | 108.0000 | |
| 7. | 192.0000 | 108.0000 | Recommended tangency point |
| 8. | 215.3914 | 110.5185 | |
| 9. | 242.3905 | 119.2603 | |
| 10. | 266.5006 | 133.6341 | |
| 11. | 285.1499 | 151.0284 | Recommended tangency point |
| 12. | 299.5069 | 168.1924 | |
| 13. | 314.2060 | 183.4111 | |
| 14. | 328.3813 | 201.7784 | |
| 15. | 343.0811 | 216.4723 | |
| 16. | 355.6808 | 232.2157 | |
| 17. | 370.3805 | 249.0087 | |
| 18. | 384.0000 | 264.0000 | |

(Hint: Change the *Point Style* to an easily visible format.)

REUSE

8. ***Pline***

Create the shape shown in Figure 8-34. Draw the outside shape with one continuous ***Pline*** (with 0.00 width). When finished, ***SaveAs*** **PLINE1.**

**FIGURE 8-34**

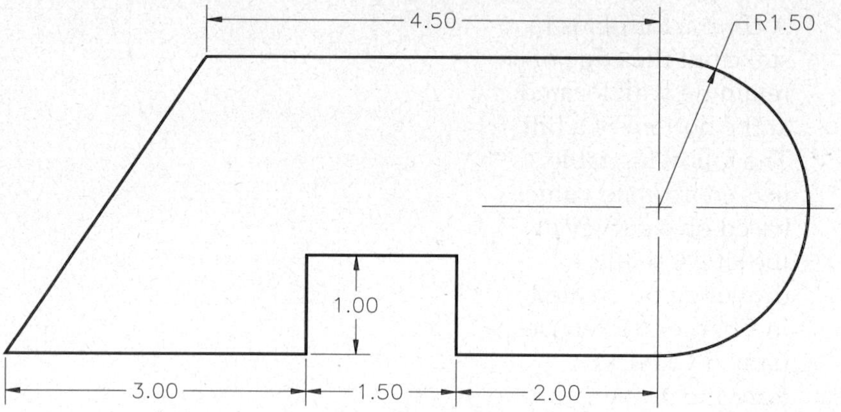

CHAPTER

# 9

# Modify Commands I

## Chapter Objectives

After completing this chapter you should:

1. be able to *Erase* objects;
2. be able to *Move* objects from a base point to a second point;
3. know how to *Rotate* objects about a base point;
4. be able to enlarge or reduce objects with *Scale*;
5. be able to *Stretch* objects and change the length of *Lines* and *Arcs* with *Lengthen*;
6. be able to *Trim* away parts of objects at cutting edges and *Extend* objects to boundary edges;
7. know how to use the *Break* options;
8. be able to *Copy* objects and make *Mirror* images of selected objects;
9. know how to create parallel copies of objects with *Offset*;
10. be able to make *Rectangular* and *Polar Arrays* of objects;
11. be able to create a *Fillet* and a *Chamfer* between two objects.

## CONCEPTS

*Draw* commands are used to create new objects. *Modify* commands are used to change existing objects or to use existing objects to create new and similar objects. The *Modify* commands covered first in this chapter (listed below) only change existing objects.

*Erase, Move, Rotate, Scale, Stretch, Lengthen, Trim, Extend, Break,* and *Join*

The *Modify* commands covered near the end of this chapter (listed below) use existing objects to create new and similar objects. For example, the *Copy* command prompts you to select an object (or set of objects), then creates an identical object (or set).

*Copy, Mirror, Offset, Array* (three types), *Fillet,* and *Chamfer*

Several commands are found in the *Edit* pull-down menu such as *Cut, Copy,* and *Paste.* Although these command names appear similar (or the same in the case of *Copy*) to some of those in the *Modify* menu, these AutoCAD command names are actually *Cutclip, Copyclip,* and *Pasteclip.* These commands are Windows functions for cutting and pasting between two different AutoCAD drawings or other software applications. See Chapter 31, Object Linking and Embedding, for more information.

Remember that there are many methods you can use to access the modify commands. On the Ribbon you can access the commands from the *Modify* panel on the *Home* tab. Note that many of these commands are found on the drop-down section of this panel (Fig. 9-1).

FIGURE 9-1

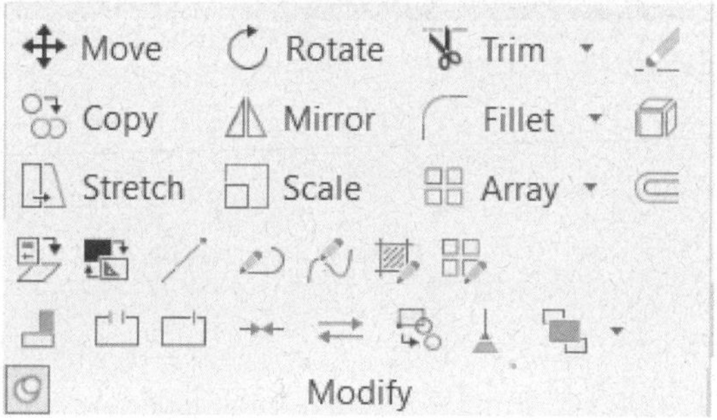


If you have the Menu Bar activated, you can select the *Modify* pull-down menu to access these commands (Fig. 9-2).

FIGURE 9-2

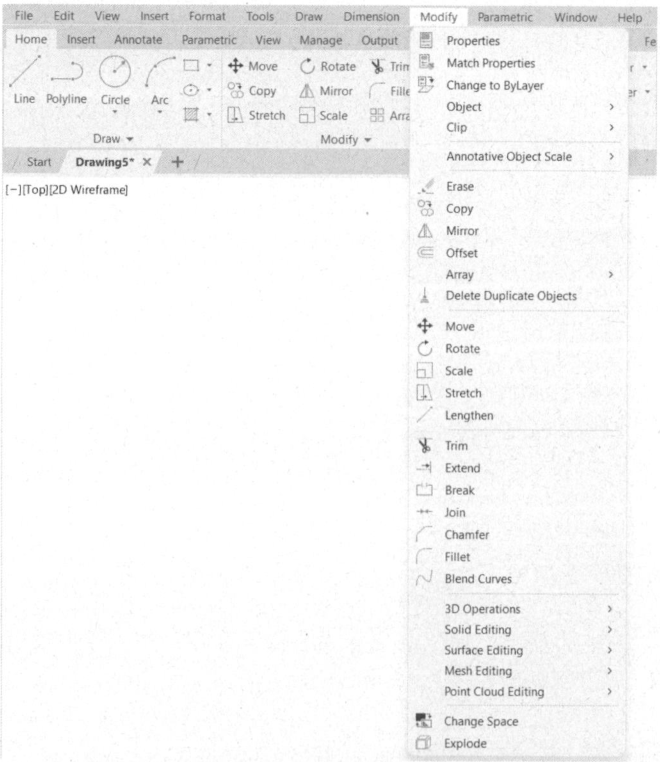


A *Modify* toolbar is also available. Enable the *Modify* toolbar by using the *-Toolbar* command (don't forget the hyphen), then enter *Modify*, then *Show*. If any toolbars are currently visible on your screen, you can right-click on the toolbar title to produce the list of available toolbars.

You can also enter modify commands by typing the command name or command alias at the keyboard.

### Selection

Since all *Modify* commands affect or use existing geometry, the first step in using most *Modify* commands is to construct a selection set (see Chapter 4). This can be done by one of two methods:

1. Invoking the desired command and then creating the selection set in response to the "Select Objects:" prompt (Verb/Noun syntax order) using any of the select object options;
2. Selecting the desired set of objects with the pickbox or *AUto* methods (windows or lassos) before invoking the edit command (Noun/Verb syntax order).

The first method allows use of any of the selection options (*Last, All, WPolygon, Fence,* etc.), while the latter method allows only the use of the pickbox and *AUto* windows and lassos.

### Selection Preview

When you use a *Modify* command you will likely see a preview to assist in your editing process. For example, when using *Trim*, an object's segment that you hover over with the pickbox to trim away (before you actually PICK it) appears as a faded image as shown in Figure 9-3. In addition, many commands display a "cursor badge" indicating the command action. The cursor badges are typically similar to the related icon button, such as the four "arrows" for the *Move* command. Almost all *Modify* commands display a preview or cursor badge.

**FIGURE 9-3**

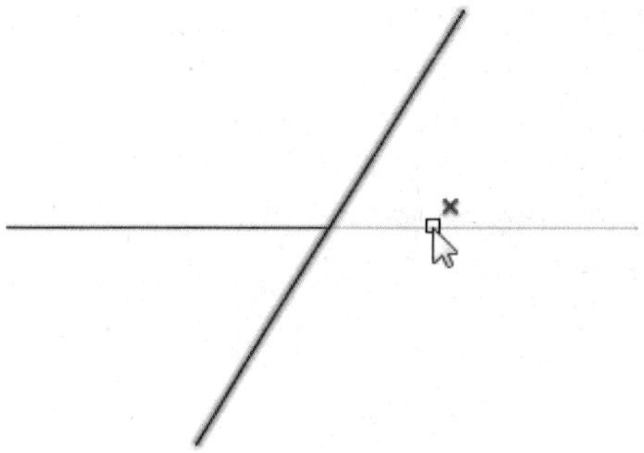

## COMMANDS

### *Erase*

| Ribbon | Menu Bar | Command (Type) | Alias (Type) | Shortcut |
|---|---|---|---|---|
| *Home*<br>*Modify*<br>(icon) | *Modify*<br>*Erase* | *Erase* | *E* | (Edit Mode)<br>*Erase* |

The *Erase* command deletes the objects you select from the drawing. Any of the object selection methods can be used to highlight the objects to *Erase*. The only other required action is for you to press *Enter* to cause the erase to take effect.

Command: ***ERASE***
Select objects: **PICK** (Use any object selection method.)
Select objects: **PICK** (Continue to select desired objects.)
Select objects: **Enter** (Confirms the object selection process and causes *Erase* to take effect.)
Command:

If objects are erased accidentally, *U* can be used immediately following the mistake to undo one step, or *Oops* can be used to bring back into the drawing whatever was *Erased* the last time *Erase* was used (see Chapter 5). If only part of an object should be erased, use *Trim* or *Break*.

## Move

| Ribbon | Menu Bar | Command (Type) | Alias (Type) | Shortcut |
|---|---|---|---|---|
| *Home Modify* (icon) | *Modify Move* | *Move* | *M* | (Edit Mode) *Move* |

*Move* allows you to relocate one or more objects from the existing position in the drawing to any other position you specify. After selecting the objects to *Move*, you must specify the "base point" and "second point of displacement."

You can use any of the coordinate entry methods to specify these points. Examples are shown in Figure 9-4.

**FIGURE 9-4**

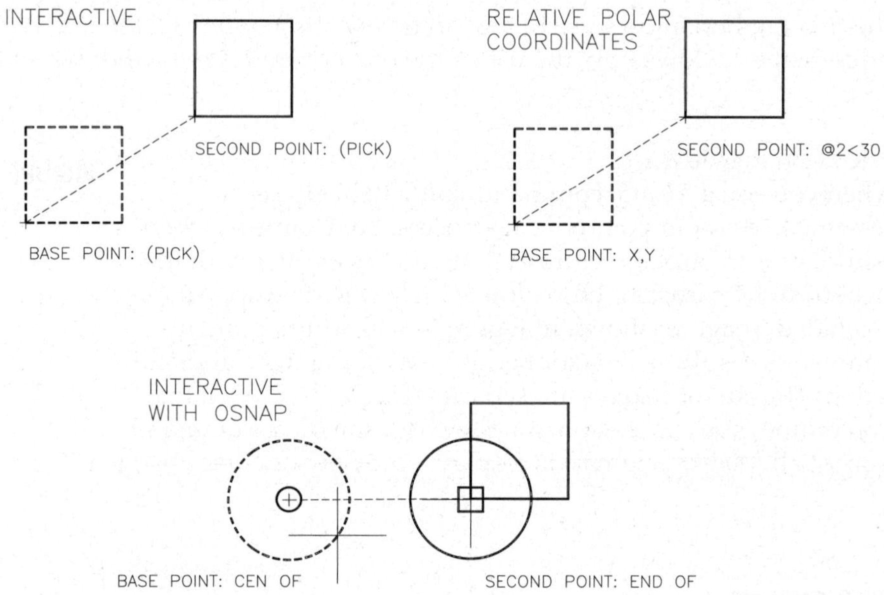


Command: ***MOVE***
Select objects: **PICK** (Use any of the object selection methods.)
Select objects: **PICK** (Continue to select other desired objects.)
Select objects: **Enter** (Press Enter to indicate selection of objects is complete.)
Specify base point or displacement: **PICK** or (**coordinates**) (This is the point to move from. Select a point to use as a "handle." An *Endpoint* or *Center*, etc., can be used.)
Specify second point or <use first point as displacement>: **PICK** or (**coordinates**) (This is the point to move to. Object Snaps can also be used here.)
Command:

Keep in mind that Object Snaps can be used when PICKing any point. It is often helpful to toggle *Ortho* or *Polar Tracking ON* to force the *Move* in a horizontal or vertical direction.

If you know a specific distance and an angle that the set of objects should be moved, *Polar Tracking* or relative polar coordinates can be used. In the following sequence, relative polar coordinates are used to move objects 2 units in a 30 degree direction (see Fig. 9-4, relative polar coordinates).

Command: ***MOVE***
Select objects: **PICK**
Select objects: **PICK**
Select objects: **Enter**
Specify base point or displacement: **X,Y (coordinates)**
Specify second point or <use first point as displacement>: **@2<30**
Command:

*Dynamic Input* and direct distance entry can be used effectively with *Move*. For example, assume you wanted to move the right side view of a multiview drawing 20 units to the right (Fig. 9-5). First, invoke the *Move* command and select all of the objects comprising the right side view in response to the "Select Objects:" prompt. The command sequence is as follows:

**FIGURE 9-5**

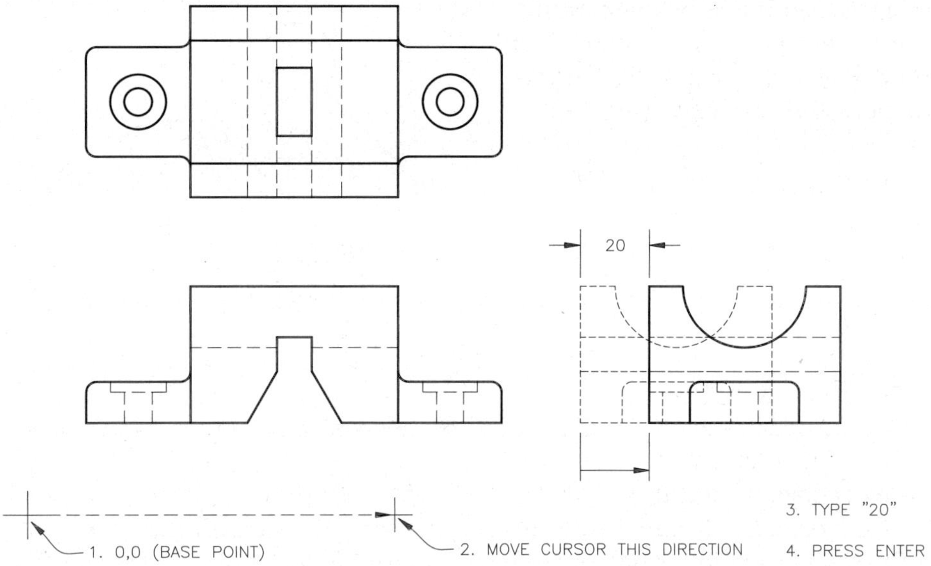


Command: ***MOVE***
Select objects: **PICK**
Select objects: **Enter**
Specify base point or displacement: **0,0** (or any value)
Specify second point or <use first point as displacement>: **20** (with *Ortho* or *Polar Tracking* on, move cursor to right), then press **Enter**

In response to the "Specify base point or displacement:" prompt, PICK a point or enter any coordinate pairs or single value. At the "Specify second point:" prompt, move the cursor to the right any distance (using *Polar Tracking* or *Ortho* on), then type "20" and press Enter. The value specifies the distance, and the cursor location from the last point indicates the direction of movement.

In the previous example, note that any value can be entered in response to the "Specify base point or displacement:" prompt. If a single value is entered ("3," for example), AutoCAD recognizes it as direct distance entry. The point designated is 3 units from the last PICK point in the direction specified by wherever the cursor is at the time of entry.

## Rotate

| Ribbon | Menu Bar | Command (Type) | Alias (Type) | Shortcut |
|---|---|---|---|---|
| *Home Modify* (icon) | *Modify Rotate* | *Rotate* | *RO* | (Edit Mode) *Rotate* |

Selected objects can be rotated to any position with this command. After selecting objects to *Rotate,* you select a "base point" (a point to rotate about) then specify an angle for rotation. AutoCAD rotates the selected objects by the increment specified from the original position.

For example, specifying a value of **45** would *Rotate* the selected objects 45 degrees counterclockwise from their current position; a value of **-45** would *Rotate* the objects 45 degrees in a clockwise direction (Fig. 9-6).

**FIGURE 9-6**

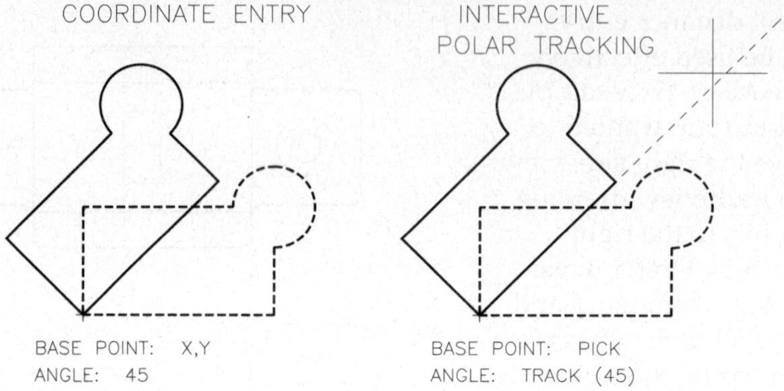


Command: ***ROTATE***
Current positive angle in UCS: ANGDIR=counterclockwise ANGBASE=0
Select objects: **PICK**
Select objects: **PICK**
Select objects: **Enter** (Indicates completion of object selection.)
Specify base point: **PICK** or (**coordinates**) (Select a point to rotate about.)
Specify rotation angle or [Copy/Reference]: **PICK** or (**coordinates**) (Interactively rotate the set or enter a value for the number of degrees to rotate the object set.)
Command:

The base point is often selected interactively with Object Snaps. When specifying an angle for rotation, you can enter an angular value, use *Polar Tracking*, or turn on *Ortho* for 90-degree rotation. Note the status of the two related system variables is given: *ANGDIR* (counterclockwise or clockwise rotation) and *ANGBASE* (base angle used for rotation).

The *Copy* option allows you to make a copy of the selected objects—that is, the original set of objects remains in its original orientation while you create rotated copies. After selecting a "base point," type "C" or right-click and select *Copy* at the following prompt.

Specify rotation angle or [Copy/Reference] <0>: ***C***

Next, enter a rotation angle value or PICK a point. Using this option would result in, using Figure 9-6 for example, a drawing with both sets of objects: the original set (shown with dashed lines) and the new set.

TIP

The *Reference* option can be used to specify a vector as the original angle before rotation (Fig. 9-7). This vector can be indicated interactively (Object Snaps can be used) or entered as an angle using keyboard entry. Angular values that you enter in response to the "new angle:" prompt are understood by AutoCAD as absolute angles for the *Reference* option only.

Specify rotation angle or [Reference]: ***R*** (Indicates the *Reference* option.)
Specify the reference angle <0>: **PICK** or (**value**) (PICK the first point of the vector.)
Specify second point: **PICK** (Indicates the second point of the vector.)
Specify the new angle: (**value**) (Indicates the new angle.)
Command:

**FIGURE 9-7**

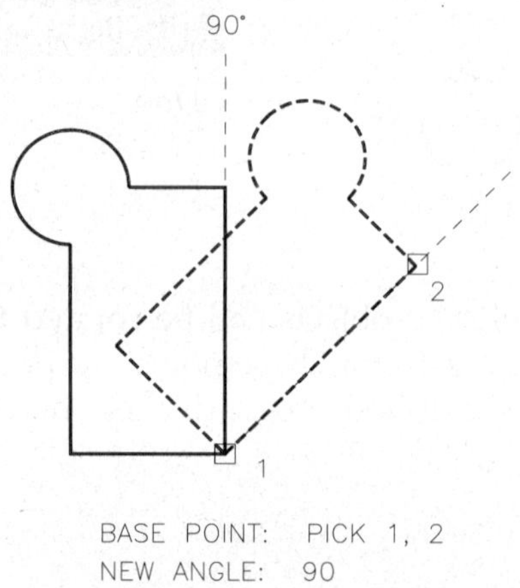

## Scale

| Ribbon | Menu Bar | Command (Type) | Alias (Type) | Shortcut |
|---|---|---|---|---|
| *Home Modify* (icon) | *Modify Scale* | *Scale* | *SC* | (Edit Mode) *Scale* |

The *Scale* command is used to increase or decrease the size of objects in a drawing. The *Scale* command does not normally have any relation to plotting a drawing to scale.

After selecting objects to *Scale,* AutoCAD prompts you to select a "Base point:", which is the stationary point. You can then scale the size of the selected objects interactively or enter a scale factor (Fig. 9-8).

**FIGURE 9-8**

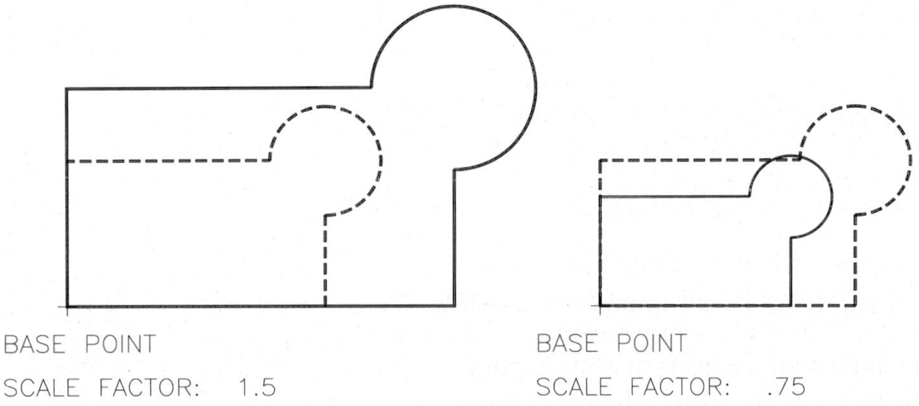


Using interactive input, you are presented with a rubberband line connected to the base point. Making the rubberband line longer or shorter than 1 unit increases or decreases the scale of the selected objects by that proportion; for example, pulling the rubberband line to two units length increases the scale by a factor of two.

Command: ***SCALE***
Select objects: **PICK** or (**coordinates**) (Select the objects to scale.)
Select objects: **Enter** (Indicates completion of the object selection.)
Specify base point: **PICK** or (**coordinates**) (Select the stationary point.)
Specify scale factor or [Copy/Reference]: **PICK** or (**value**) or **R** (Interactively scale the set of objects or enter a value for the scale factor.)
Command:

The *Copy* option allows you to make a copy of the selected objects—that is, the original set of objects remains in its original orientation while you create scaled copies. Type "C" or right-click and select *Copy* in response to "Specify rotation angle or [Copy/Reference] <0>:". Similar to the *Copy* option of the *Rotate* command, using this option would result in two sets of objects: the original selection set and the new set.

It may be desirable in some cases to use the *Reference* option to specify a value or two points to use as the reference length. This length can be indicated interactively (Object Snaps can be used) or entered as a value. This length is used for the subsequent reference length that the rubberband uses when interactively scaling. For example, if the reference distance is 2, then the "rubberband" line must be stretched to a length greater than 2 to increase the scale of the selected objects.

TIP

*Scale* normally should not be used to change the scale of an entire drawing in order to plot on a specific size sheet. CAD drawings should be created full size in actual units.

### Stretch

| Ribbon | Menu Bar | Command (Type) | Alias (Type) | Shortcut |
|---|---|---|---|---|
| *Home*<br>*Modify*<br>(icon) | *Modify*<br>*Stretch* | *Stretch* | *S* | ... |

Objects can be made longer or shorter with *Stretch*. The power of this command lies in the ability to *Stretch* groups of objects while retaining the connectivity of the group (Fig. 9-9). When *Stretched, Lines* and *Plines* become longer or shorter. *Arcs* change radius to become longer or shorter. *Circles* do not stretch; rather, they move if the center is selected within the *Crossing Window*.

Objects to *Stretch* should be selected by a <u>*Crossing Window* or *Crossing Lasso* only</u>. The window or lasso should be created so the objects to *Stretch* <u>cross</u> through the window. *Stretch* actually moves the object endpoints that are located within the window or lasso.

**FIGURE 9-9**

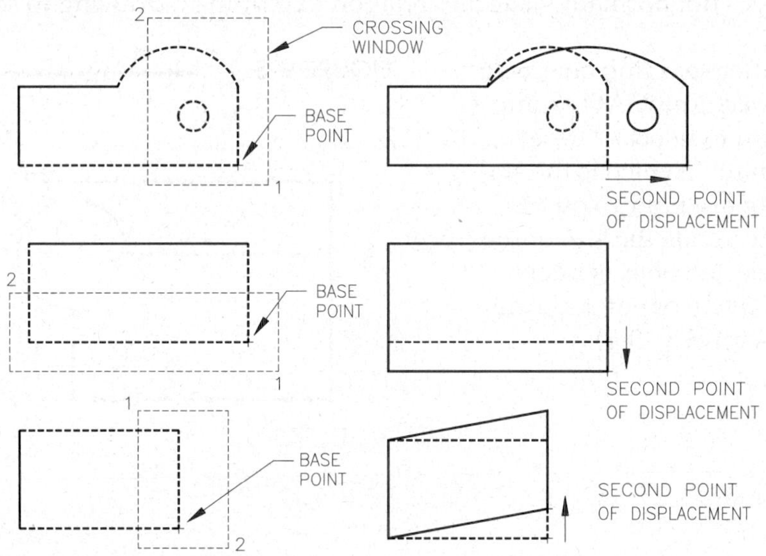


Command: **STRETCH**
Select objects to stretch by crossing-window or crossing-polygon...
Select objects: **PICK**
Specify opposite corner: **PICK**
Select objects: **Enter**
Specify base point or [Displacement] <Displacement>: (Select a point to stretch from.)
Specify second point or <use first point as displacement>: (Select a point to stretch to.)

**TIP**

*Stretch* can be used to lengthen one object while shortening another. Application of this ability would be repositioning a door or window on a wall (Fig. 9-10).

In summary, the *Stretch* command <u>stretches objects that cross</u> the selection window and <u>moves objects that are completely within</u> the selection window, as shown in Figure 9-10.

The *Stretch* command allows you to select objects with other selection methods, such as a pickbox or normal (non-crossing) *Window*. Doing so, however, results in a *Move* since <u>only objects crossing through a crossing window get stretched</u>.

**FIGURE 9-10**

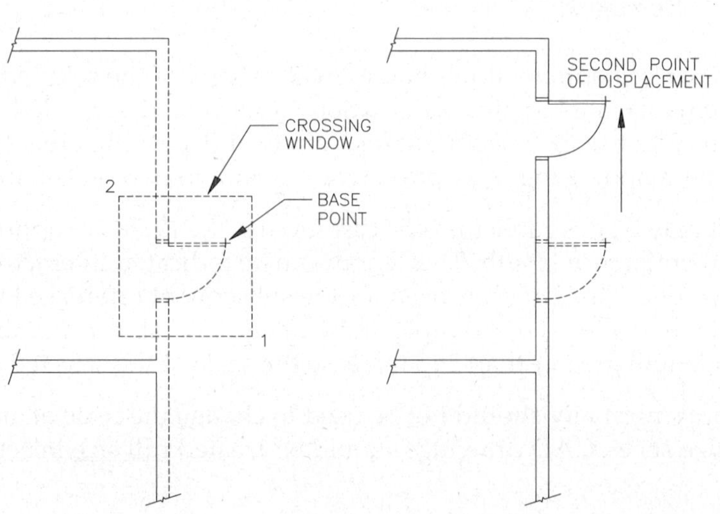

## *Lengthen*

| Ribbon | Menu Bar | Command (Type) | Alias (Type) | Shortcut |
|---|---|---|---|---|
| *Home Modify* (icon) | *Modify Lengthen* | *Lengthen* | *LEN* | ... |

*Lengthen* changes the length (longer or shorter) of linear objects and arcs. No additional objects are required (as with *Trim* and *Extend*) to make the change in length. Many methods are provided as displayed in the command prompt.

Command: ***LENGTHEN***
Select an object or [DElta/Percent/Total/DYnamic]:

### *Select an object*

Selecting an object causes AutoCAD to report the current length of that object. If an *Arc* is selected, the included angle is also given.

### *DElta*

Using this option returns the prompt shown next. You can change the current length of an object (including an arc) by any increment that you specify. Entering a positive value increases the length by that amount, while a negative value decreases the current length. The end of the object that you select changes while the other end retains its current endpoint (Fig. 9-11).

Enter delta length or [Angle] <0.0000>:
(**value**) or ***a***

**FIGURE 9-11**

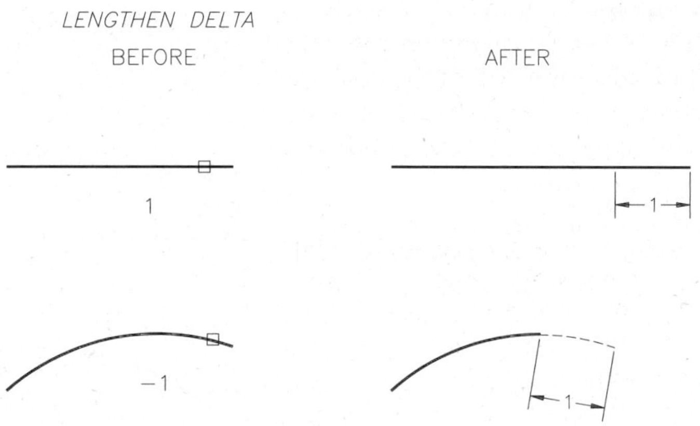


The *Angle* option allows you to change the included angle of an arc (the length along the curvature of the arc can be changed with the *Delta* option). Enter a positive or negative value (degrees) to add or subtract to the current included angle, then select the end of the object to change (Fig. 9-12).

**FIGURE 9-12**

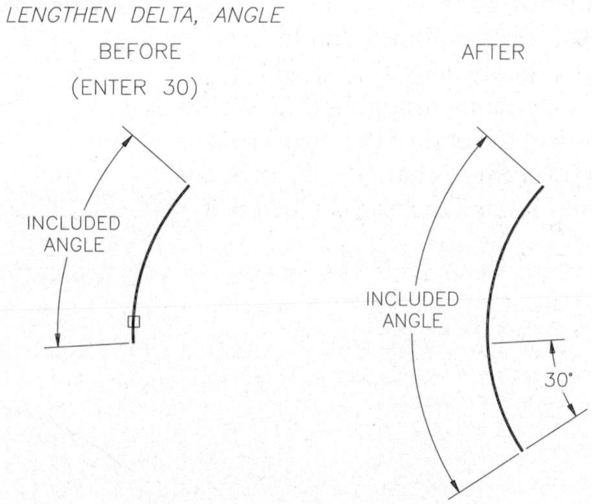

*Percent*
Use this option if you want to change the length by a percentage of the current total length (Fig. 9-13). For arcs, the percentage applied affects the length and the included angle equally, so there is no *Angle* option. A value of greater than 100 increases the current length, and a value of less than 100 decreases the current length. Negative values are not allowed. The end of the object that you select changes.

**FIGURE 9-13**

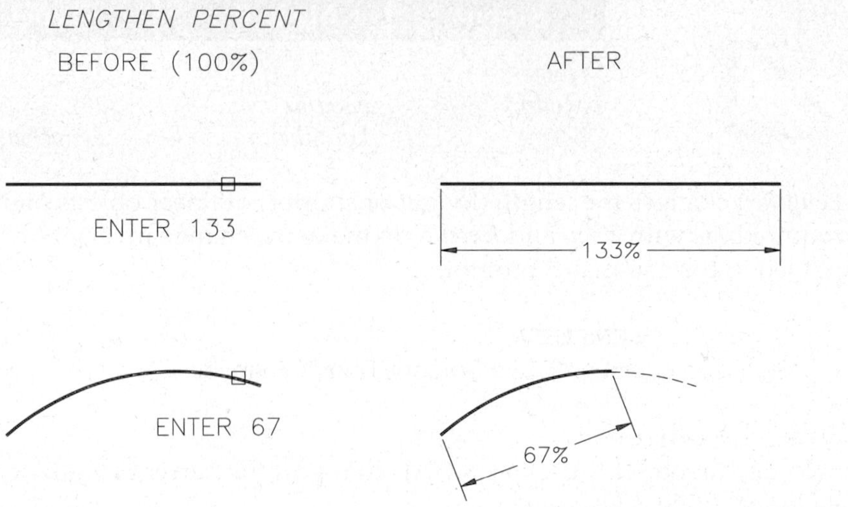


*Total*
This option lets you specify a value for the new total length (Fig. 9-14). Simply enter the value and select the end of the object to change. The angle option is used to change the total included angle of a selected arc.

Specify total length or [Angle] <1.0000>: (**`value`**) or ***A***

**FIGURE 9-14**

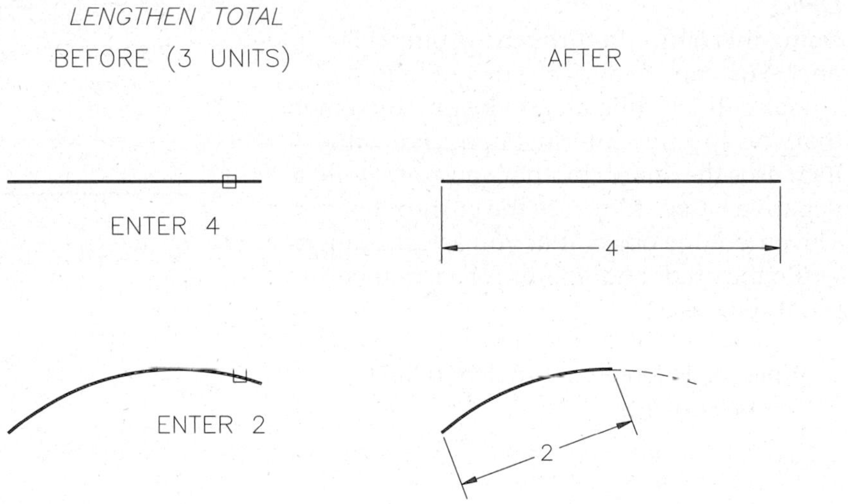


*DYnamic*
This option allows you to change the length of an object by dynamic dragging (Fig. 9-15). Select the end of the object that you want to change. Object Snaps and Polar Tracking can be used.

**FIGURE 9-15**

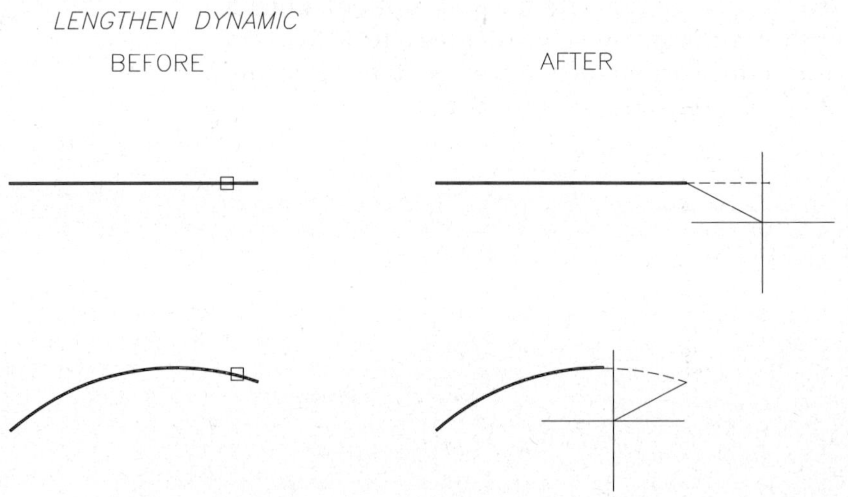

## *Trim*

| Ribbon | Menu Bar | Command (Type) | Alias (Type) | Shortcut |
|---|---|---|---|---|
| *Home Modify* (icon) | *Modify Trim* | *Trim* | *TR* | ... |

The *Trim* command allows you to trim (shorten) the end of an object back to the intersection of another object (Fig. 9-16). The middle section of an object can also be *Trimmed* between two intersecting objects. Objects that do not cross any others can be erased when selected to trim. The mOde option gives you the choice of the following methods:

**FIGURE 9-16**

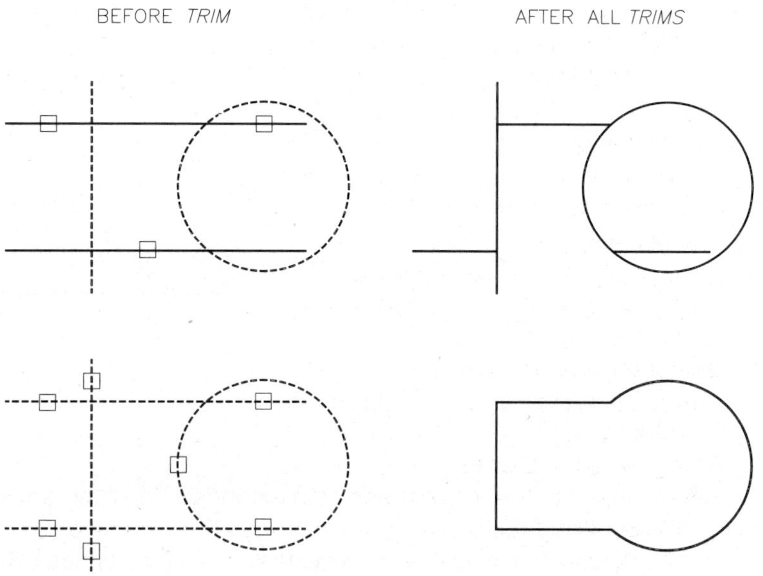


*Quick* (default): during Trim all potential trim boundaries are immediately available. The segments to be trimmed (or extended by holding down Shift during selection) can be selected individually, freehand by holding and dragging with the left mouse button, or by clicking two points to define a "fence" line through the items to trim. Objects in the selection that cannot be trimmed are deleted instead.

*Standard*: first select the boundaries by which the object will be cut and then press Enter. Then select on the portions of the objects to be trimmed off. To use all objects as boundaries, press [Enter] at the first Select Objects prompt.

The TRIMEXTENDMODE system variable controls the default behavior of the Trim command to either Quick (default) or Standard.

### Quick Mode Trim

The default method for the Trim command is now the Quick option. To use this very intuitive method, just click on the portions to be removed, or drag a "freehand" line through them (Fig. 9-17), or click two points to define a line which will trim or erase anything it crosses.

**FIGURE 9-17**

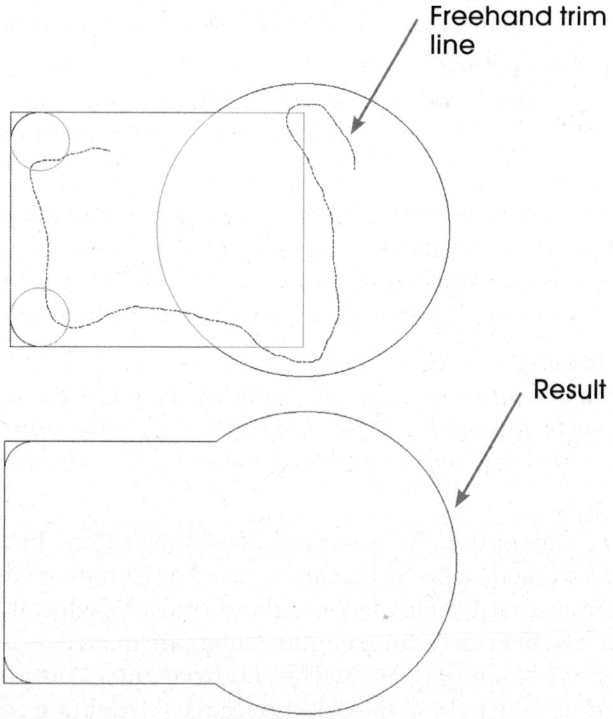


Command: **`TRIM`**
Current settings: Projection=UCS, Edge=None, Mode=Quick
Select object to trim or shift-select to extend or [cuTting edges/Crossing/mOde/Project/eRase]: (Start dragging a line across objects.)
Specify next fence point or [Undo]: (Continue holding down right mouse button and dragging across object to trim. Let up button to trim the highlighted items)
Select object to trim or shift-select to extend or [cuTting edges/Crossing/mOde/Project/eRase/Undo]: **`Enter`** (to end command.)

### Standard Mode Trim

To use the Standard Trim option, first set the mOde to Standard. There are two steps to the Trim command when using the Standard option: first, PICK one or more "cutting edges" (existing objects); then PICK the object or objects to *Trim* (portion to remove). The cutting edges are highlighted after selection. Cutting edges themselves can be trimmed if they intersect other cutting edges, but lose their highlight when trimmed.

Command: ***TRIM***
Select object to trim or shift-select to extend or [cuTting edges/Crossing/mOde/Project/eRase]: **O Enter**
Enter a trim mode option [Quick/Standard] <Quick>: **S Enter**
Select object to trim or shift-select to extend or [cuTting edges/Fence/Crossing/mOde/Project/Edge/eRase/Undo]: **T Enter**
Current settings: Projection=UCS, Edge=None, Mode=Standard
Select cutting edges ...
Select objects or <select all>: **PICK** or press **Enter** to use all objects as cutting edges
Select objects: **Enter**
Select object to trim or shift-select to extend or [cuTting edges/Fence/Crossing/mOde/Project/Edge/eRase]: **PICK** (Select the part of an object to trim away.)
Select object to trim or shift-select to extend or [cuTting edges/Fence/Crossing/mOde/Project/Edge/eRase]: **Enter**

**FIGURE 9-18**

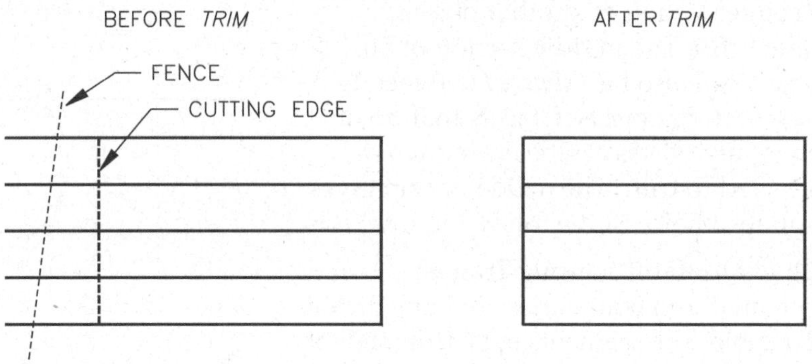


#### *Select all*

If you need to select many objects as <u>cutting edges</u>, you can press Enter at the "Select objects or <select all>:" prompt. In this case, all objects in the drawing become cutting edges; however, the objects are not highlighted.

#### *cuTting edges*

If you want to define specific <u>cutting edges</u>, you can use the T option to allow you to select them. You can use this during Quick mode to briefly use the Standard mode behavior of the command.

#### *Fence*

The *Fence* and *Crossing Window* options are available for selecting the <u>objects to trim</u>. Rather than selecting each object individually with the pickbox, *Fence* allows you to select multiple objects using a crossing line. The *Fence* can have several line segments (see "*Fence*" in Chapter 3). This option is especially useful when you have many objects to trim away in close proximity to each other such as shown in Figure 9-18.

#### *Crossing*

You can also use a *Crossing* window to select the parts of <u>objects you want to trim</u>. This option may be more efficient than using the pickbox to select objects individually.

**FIGURE 9-19**

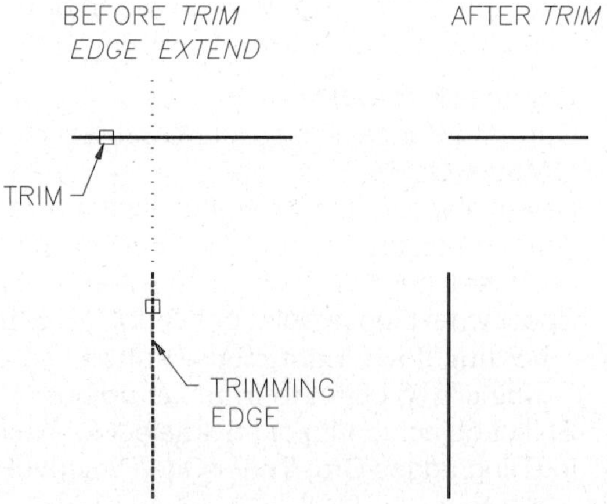


#### *Edge*

The *Edge* option can be set to *Extend* or *No extend*. In the *Extend* mode, objects that are selected as trimming edges will be <u>imaginarily extended</u> to serve as a cutting edge. In other words, lines used for trimming edges are treated as having infinite length (Fig. 9-19). The *No extend* mode considers only the actual length of the object selected as trimming edges.

Select object to trim or shift-select to extend or [Fence/Crossing/Project/Edge/eRase/Undo]: ***Edge***
Enter an implied edge extension mode [Extend/No extend] <No extend>:

### Project

The *Project* switch controls how *Trim* and *Extend* operate in 3D space. *Projection* affects the projection of the cutting edge and boundary edge. The three options are described here:

Select object to trim or shift-select to extend or [Fence/Crossing/Project/Edge/eRase/Undo]: ***project***
Enter a projection option [None/Ucs/View] <Ucs>:

*None*

The *None* option does not project the cutting edge. This mode is used for normal 2D drawing when all objects (cutting edges and objects to trim) lie in the current drawing plane. You can use this mode to *Trim* in 3D space when all objects lie in the plane of the current UCS or in planes parallel to it and the objects physically intersect. In any case, the objects must physically intersect for trimming to occur.

The other two options (*UCS* and *View*) allow you to trim objects that do not truly intersect in 3D space.

*UCS*

This option projects the cutting edge perpendicular to the UCS. Any objects crossing the "projected" cutting edge in 3D space can be trimmed. For example, selecting a *Circle* as the cutting edge creates a projected cylinder used for cutting (Fig. 9-20).

*View*

The *View* option allows you to trim objects that appear to intersect from the current viewpoint. The objects do not have to physically intersect in 3D space. The cutting edge is projected perpendicularly to the screen (parallel to the line of sight). This mode is useful for trimming "hidden edges" of a wireframe model to make the surfaces appear opaque (Fig. 9-21).

**FIGURE 9-20**

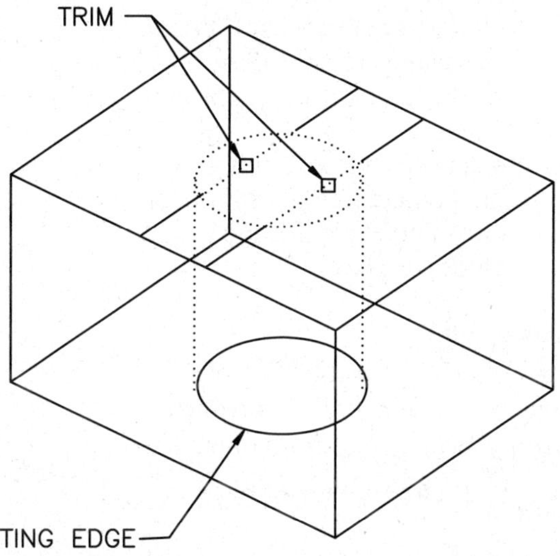


### eRase

The *eRase* option actually allows you to erase entire objects during the *Trim* command. This option operates just like the *Erase* command. When you are finished erasing, you are returned to the "Select object to trim…" prompt.

Select object to trim or shift-select to extend or [Fence/Crossing/Project/Edge/eRase/Undo]: ***r***
Select objects to erase or <exit>: **PICK**
Select objects to erase: **Enter** (Selected objects are erased.)
Select object to trim or shift-select to extend or [Fence/Crossing/Project/Edge/eRase/Undo]:

**FIGURE 9-21**

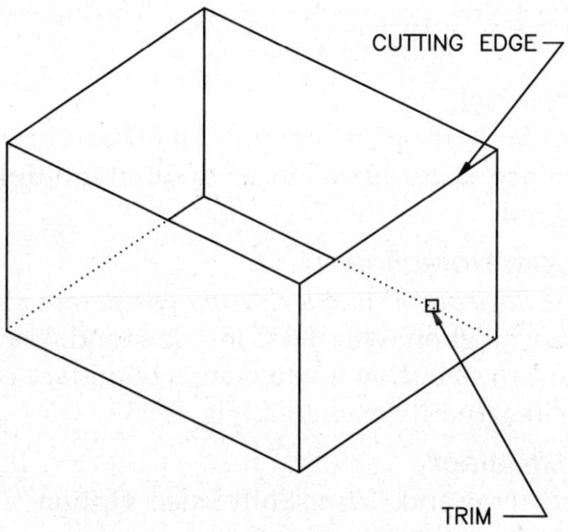


### Undo

*Undo* allows you to undo the last *Trim* in case of an accidental trim.

### Shift-Select

See "*Trim* and *Extend* Shift-Select Option."

## *Extend*

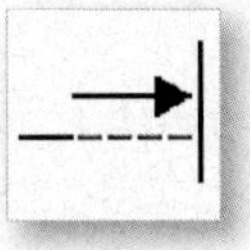

| Ribbon | Menu Bar | Command (Type) | Alias (Type) | Shortcut |
|---|---|---|---|---|
| *Home Modify* (icon) | *Modify Extend* | *Extend* | *EX* | ... |

*Extend* can be thought of as the opposite of *Trim*. Objects such as *Lines, Arcs,* and *Plines* can be *Extended* until intersecting another object called a boundary edge (Fig. 9-22). The command first requires selection of existing objects to serve as boundary edge(s) which become highlighted; then the objects to extend are selected. Objects extend until, and only if, they eventually intersect a boundary edge. An *Extended* object acquires a new endpoint at the boundary edge intersection.

Command: ***EXTEND***
Current settings:
  Projection=UCS, Edge=None
Select boundary edges ...
Select objects or <select all>:
  **PICK** (Select boundary edge.)
Select objects: **Enter**
Select object to extend or shift-select to trim or [Fence/Crossing/Project/Edge/Undo]: **PICK** (Select object to extend.)
Select object to extend or shift-select to trim or [Fence/Crossing/Project/Edge/Undo]: **Enter**

**FIGURE 9-22**

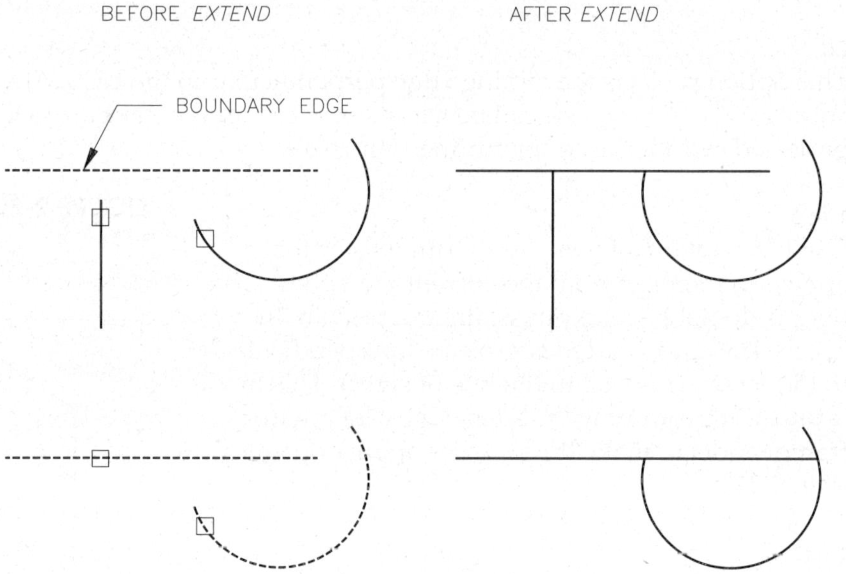


### *Select all*

Pressing Enter at the "Select objects or <select all>:" prompt automatically makes all objects in the drawing boundary edges. The objects are not highlighted.

**FIGURE 9-23**

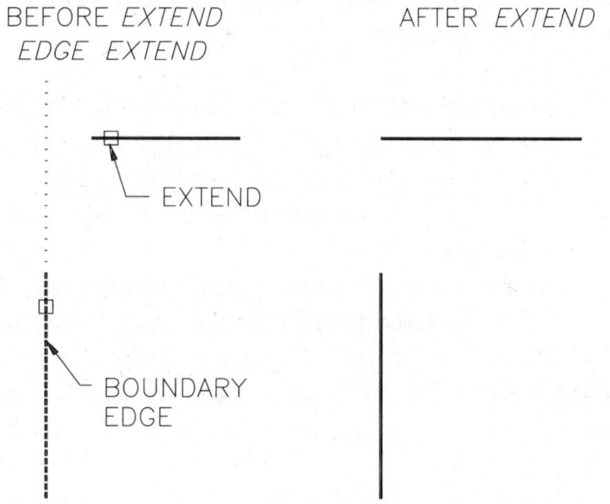


### *Fence*

Use the *Fence* option (a crossing line) to select multiple objects that you want to extend. This operates similar to the *Fence* option of *Trim*.

### *Crossing*

Similar to the same option of the *Trim* command, you can use a *Crossing* window to select multiple objects to extend.

### *Edge/Projection*

The *Edge* and *Projection* switches operate identically to their function with the *Trim* command. Use *Edge* with the *Extend* option if you want a boundary edge object to be imaginarily extended (Fig. 9-23).

### *Shift-Select*

See "*Trim* and *Extend* Shift-Select Option."

## *Trim* and *Extend* Shift-Select Option

TIP

With this feature, you can toggle between *Trim* and *Extend* by holding down the Shift key. For example, if you invoked *Trim* and are in the process of trimming lines but decide to then *Extend* a line, you can simply hold down the Shift key to change to the *Extend* command without leaving *Trim*.

Command: ***TRIM***
Current settings: Projection=UCS, Edge=None
Select cutting edges ...
Select objects or <select all>: **PICK**
Select objects: **Enter**
Select object to trim or shift-select to extend or [Fence/Crossing/Project/Edge/eRase/Undo]: **PICK** (to trim)
Select object to trim or shift-select to extend or [Fence/Crossing/Project/Edge/eRase/Undo]: **Shift**, then **PICK** to extend

TIP

Not only does holding down the Shift key toggle between *Trim* and *Extend*, but objects you selected as cutting edges become boundary edges and vice versa. Therefore, if you want to use this feature effectively, during the first step of either command (*Trim* or *Extend*), you must anticipate and select edges you might potentially use as both cutting edges and boundary edges.

## *Break*

| Ribbon | Menu Bar | Command (Type) | Alias (Type) | Shortcut |
|---|---|---|---|---|
| *Home Modify* (icon) | *Modify Break* | *Break* | *BR* | ... |

*Break* allows you to break a space in an object or break the end off an object. You can think of *Break* as a partial erase. If you choose to break a space in an object, the space is created between two points that you specify (Fig. 9-24). In this case, the Break creates two objects from one.

If *Break*ing a circle (Fig. 9-25), the break is in a counterclockwise direction from the first to the second point specified.

FIGURE 9-24

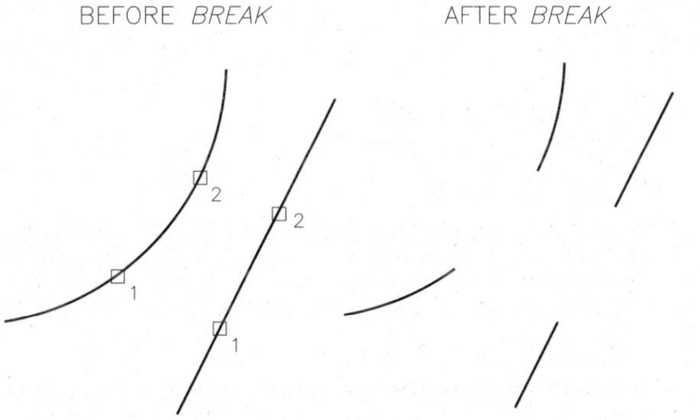


FIGURE 9-25

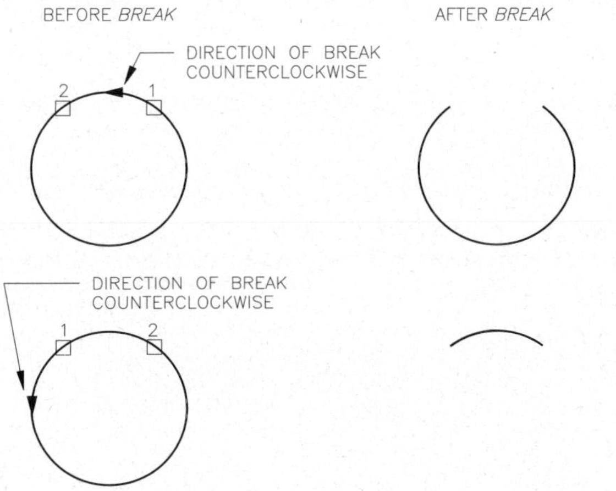

If you want to *Break* the end off a *Line* or *Arc*, the first point should be specified at the point of the break and the second point should be just off the end of the *Line* or *Arc* (Fig. 9-26).

**FIGURE 9-26**

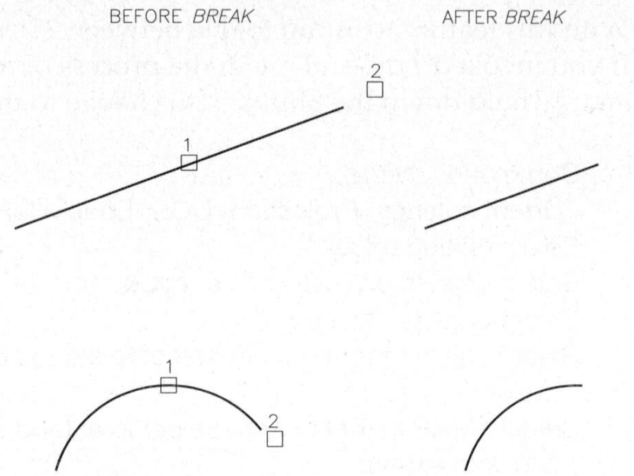


The *Modify* panel includes icons for only the default option (see *Select, Second*) and the *Break at Point* option. Four methods are available only from the screen menu. If you are typing, the desired option is selected by keying the letter ***F*** (for *First point*) or symbol @ (for "last point").

### *Select, Second*

This method (the default method) has two steps: select the object to break; then select the second point of the break. The first point used to select the object is also the first point of the break (see Fig. 9-24, Fig. 9-25, Fig. 9-26).

Command: ***BREAK***
Select object: **PICK** (This is the first point of the break.)
Specify second break point or [First point]: **PICK** (This is the second point of the break.)
Command:

### *Select, 2 Points*

This method uses the first selection only to indicate the object to *Break*. You then specify the point that is the first point of the *Break*, and the next point specified is the second point of the *Break*. This option can be used with Object Snap *Intersection* to achieve the same results as *Trim*. The command sequence is as follows:

TIP

Command: ***BREAK***
Select object: **PICK** (This selects only the object to break.)
Specify second break point or [First point]: ***f*** (Indicates specification of the first point.)
Specify first break point: **PICK** (This is the first point of the break.)
Specify second break point: **PICK** (This is the second point of the break.)
Command:

**FIGURE 9-27**

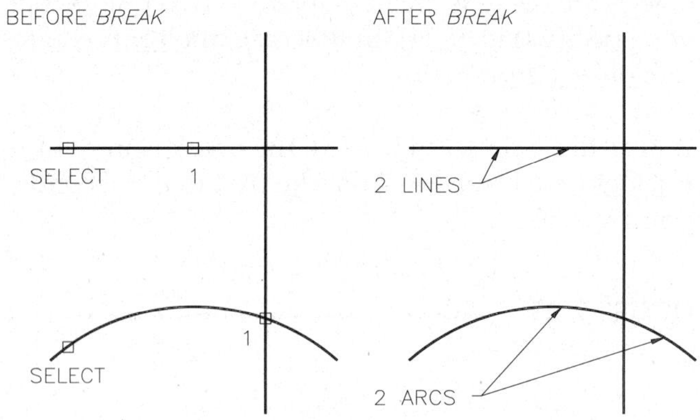


### *Break at Point*

This option creates a break with no space; however, you select the object you want to *Break* first and the point of the *Break* next (Fig. 9-27).

Command: ***BREAK***
Select object: **PICK** (This selects only the object to break.)
Specify break point: **PICK** (This is the first point of the break.)

## Join

| Ribbon | Menu Bar | Command (Type) | Alias (Type) | Shortcut |
|---|---|---|---|---|
| *Home Modify* (icon) | *Modify Join* | *Join* | *J* | ... |

The *Join* command allows you to combine two of the same objects (for example, two collinear *Lines*) into one object (for example, one *Line*). You can also use *Join* to *Close* an *Arc* to form a complete *Circle* or close an elliptical arc to form a complete *Ellipse*. You can *Join* the following objects:

*Arcs, Elliptical* arcs, *Lines, Polylines,* and *Splines*

Prompts and restrictions for each type of object are given next.

### *Lines*

The *Lines* to join must be collinear, but they can have gaps between them (Fig. 9-28).

```
Command: JOIN
Select source object: PICK
Select lines to join to source: PICK
Select lines to join to source: Enter
1 line joined to source
Command:
```

**FIGURE 9-28**

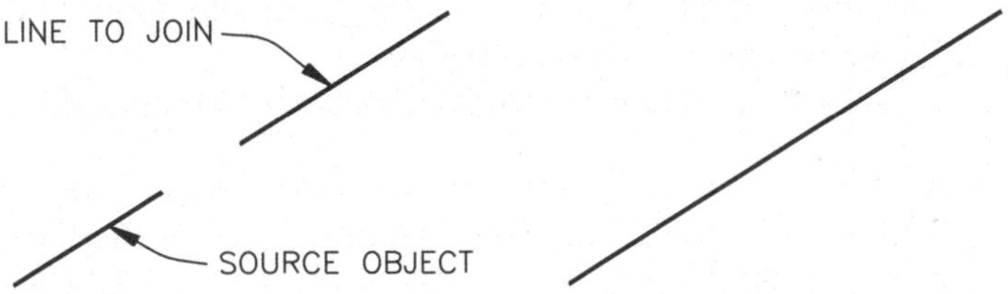


### *Arc*

The *Arc* objects must lie on the same imaginary circle but can have gaps between them. The *Close* option converts the source arc into a *Circle* (Fig. 9-29).

```
Command: JOIN
Select source object: PICK
Select arcs to join to source or [cLose]: l
Arc converted to a circle.
Command:
```

**FIGURE 9-29**

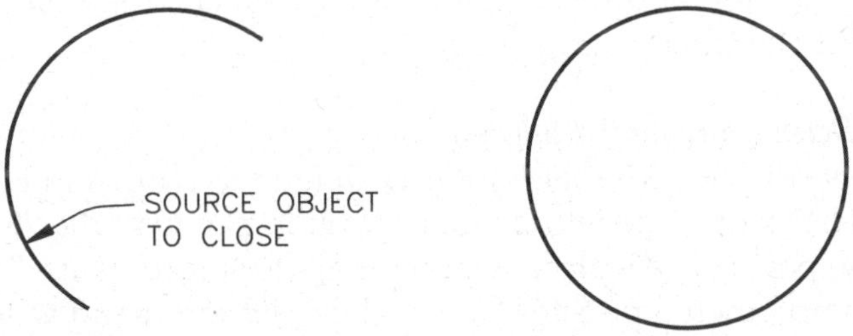


### *Polyline*

Objects to *Join* to *Plines* can be *Plines, Lines*, or *Arcs*. The objects cannot have gaps between them and must lie on the UCS XY plane or on a plane parallel to it. If *Arcs* and *Lines* are joined to the *Pline*, they are converted to one *Pline*.

### *Elliptical Arc*

The elliptical arcs must lie on the same ellipse shape but can have gaps between them. The *Close* option closes the source elliptical arc into a complete *Ellipse*.

### *Spline*

The *Spline* objects must lie in the same plane, and the end points to join must be connected.

## Copy

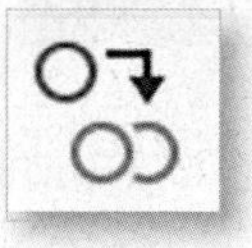

| Ribbon | Menu Bar | Command (Type) | Alias (Type) | Shortcut |
|---|---|---|---|---|
| *Home Modify* (icon) | *Modify Copy* | *Copy* | *CO or CP* | (Edit Mode) *Copy Selection* |

*Copy* creates a duplicate set of the selected objects and allows placement of those copies. The *Copy* operation is like the *Move* command, except with *Copy* the original set of objects remains in its original location. You specify a "base point:" (point to copy from) and a "specify second point of displacement or <use first point as displacement>:" (point to copy to). You can also use the *Array* option to create a pattern of several copies at once. This is different than the Array command. See Figure 9-30.

The command syntax for *Copy* is as follows.

Command: ***COPY***
Select objects: **PICK**
Select objects: **Enter**
Specify base point or [Displacement/mOde] <Displacement>:
**PICK** (This is the point to copy from. Select a point, usually on an object, as a "handle.")
Specify second point or [Array] <use first point as displacement>: **PICK** (This is the point to copy to. Object Snaps or coordinate entry should be used.)
Specify second point or [Array/Exit/Undo] <Exit>: **Enter**

**FIGURE 9-30**

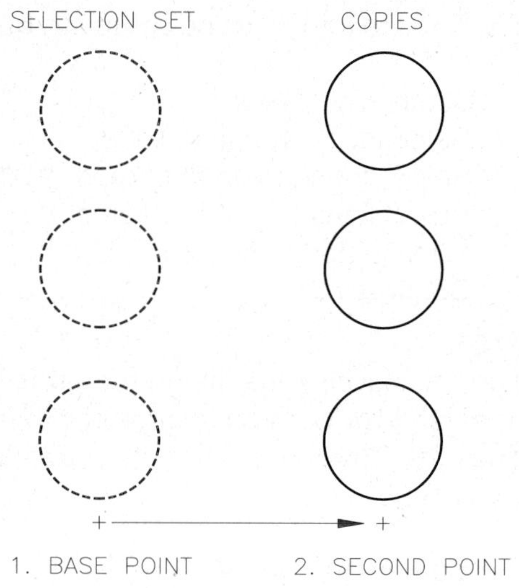


In many applications it is desirable to use Object Snap options to PICK the "base point:" and the "second point of displacement:" (Fig. 9-31, 9-33).

Alternately, you can PICK the "base point:" and use Polar Tracking or enter relative Cartesian coordinates, relative polar coordinates, or direct distance entry to specify the "second point of displacement:" (Fig. 9-32).

**FIGURE 9-31**

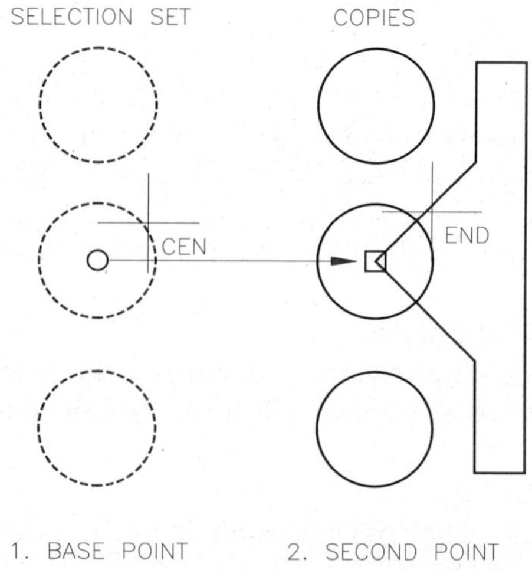


### *mOde: (Single/Multiple)*

*Copy* runs in a *Multiple* mode by default as shown in Figure 9-32. Activate the *Single* mode by typing an ***O*** at the "Specify base point or [Displacement mOde] <Displacement>:" prompt, then type ***S*** for *Single*. The *Copy* command will continue to operate in the *Single* mode until you specify the *Multiple* mode.

Command: ***COPY***
Select objects: **PICK**
Select objects: **Enter**
Current settings: Copy mode = Multiple
Specify base point or [Displacement/mOde]
<Displacement>: ***o*** (for *mOde* option)
Enter a copy mode option [Single/Multiple] <Multiple>: ***s***
(for *Single* option)
Specify base point or [Displacement/mOde/Multiple]
<Displacement>: **PICK**
Specify second point or [Array] <use first point as displacement>: **PICK**
Command:

### *Array*

The *Array* option allows you to make multiple copies in a line, similar to creating a *Path Array* along a straight line (see "*Path Array*"). After specifying the "number of items to array" (total number of items in the line), you can PICK or enter a distance (between objects) or you can select the *Fit* option. Using the *Fit* option, select the end of the line for the items to "fit" along (Fig. 9-34).

Specify second point or [Array] <use first point as displacement>: ***a***
Enter number of items to array: **3**
Specify second point or [Fit]: ***f***
Specify second point or [Array]: **PICK**

**FIGURE 9-32**

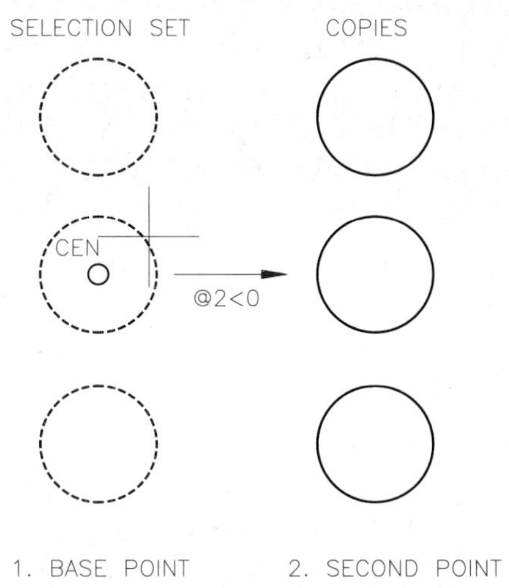


**FIGURE 9-33**

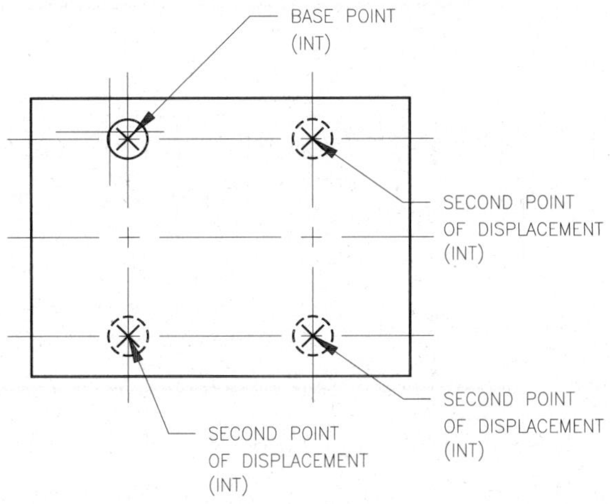


**FIGURE 9-34**

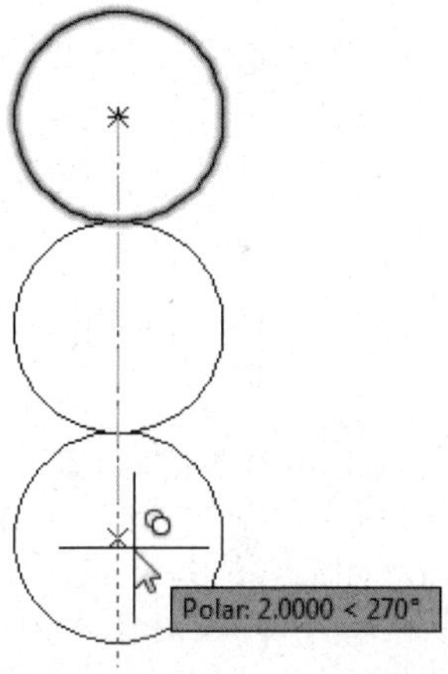


## *Mirror*

| Ribbon | Menu Bar | Command (Type) | Alias (Type) | Shortcut |
|---|---|---|---|---|
| *Home Modify (icon)* | *Modify Mirror* | *Mirror* | *MI* | ... |

This command creates a mirror image of selected existing objects. You can retain or delete the original objects ("source objects"). After selecting objects, you create two points specifying a "rubberband line," or "mirror line," about which to *Mirror*.

Command: ***MIRROR***
Select objects: **PICK** (Select object or group of objects to mirror.)
Select objects: **Enter** (Press Enter to indicate completion of object selection.)
Specify first point of mirror line: **PICK** or (**coordinates**) (PICK first endpoint of mirror axis.)
Specify second point of mirror line: **PICK** or (**coordinates**) (PICK second point of mirror line.)
Erase source objects? [Yes/No] <N>: **Enter** or **Y**
Command:

If you want to *Mirror* only in a vertical or horizontal direction, toggle *Ortho* or *Polar Tracking On* before selecting the "second point of mirror line." The length of the mirror line is unimportant since it represents a vector or axis (Fig. 9-35).

**FIGURE 9-35**

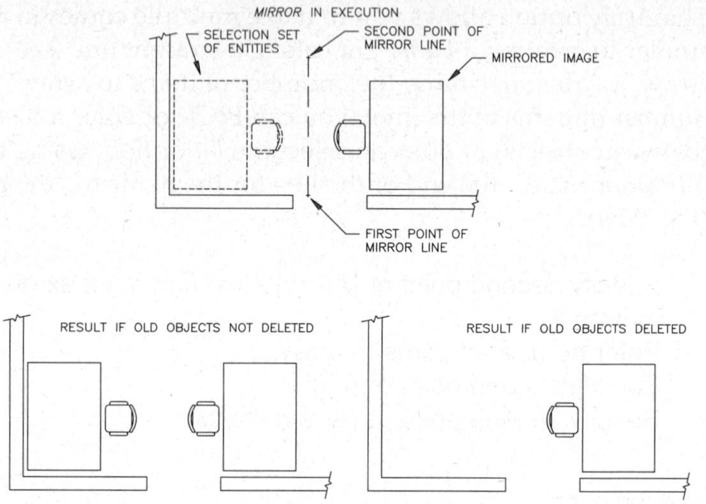


*Mirror* can be used to draw the other half of a symmetrical object, thus saving some drawing time (Fig. 9-36).

**FIGURE 9-36**

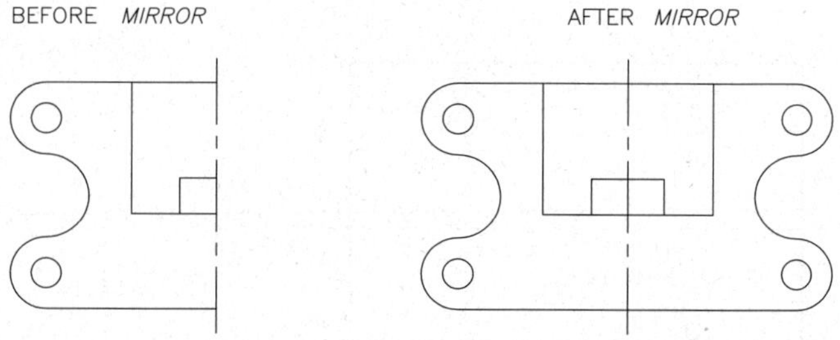


TIP

You can control whether <u>text</u> is mirrored by using the *MIRRTEXT* variable. Type *MIRRTEXT* at the Command prompt and change the value to 1 if you want the text to be reflected; otherwise, the default setting of 0 <u>does not</u> mirror text along with other selected objects (Fig. 9-37). Dimensions are <u>not</u> affected by the *MIRRTEXT* variable and therefore are not reflected (that is, if the dimensions are associative). (See Chapter 28 for details on dimensions.)

**FIGURE 9-37**

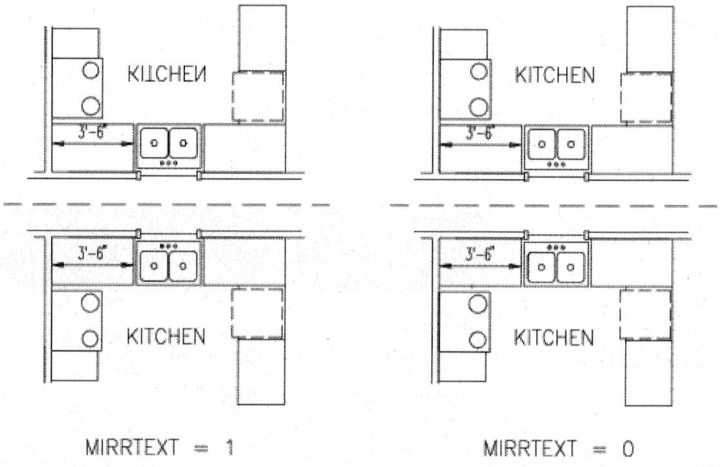


## *Offset*

| Ribbon | Menu Bar | Command (Type) | Alias (Type) | Shortcut |
|---|---|---|---|---|
| *Home Modify* (icon) | *Modify Offset* | *Offset* | *O* | ... |

*Offset* creates a <u>parallel copy</u> of selected objects. Selected objects can be *Lines*, *Arcs*, *Circles*, *Plines*, or other objects. *Offset* is a very useful command that can increase productivity greatly, particularly with *Plines*.

Depending on the object selected, the resulting *Offset* is drawn differently (Fig. 9-38). *Offset* creates a parallel copy of a *Line* equal in length and perpendicular to the original. *Arc*s and *Circle*s have a concentric *Offset*. *Offsetting* closed *Plines* or *Splines* results in a complete parallel shape.

**FIGURE 9-38**

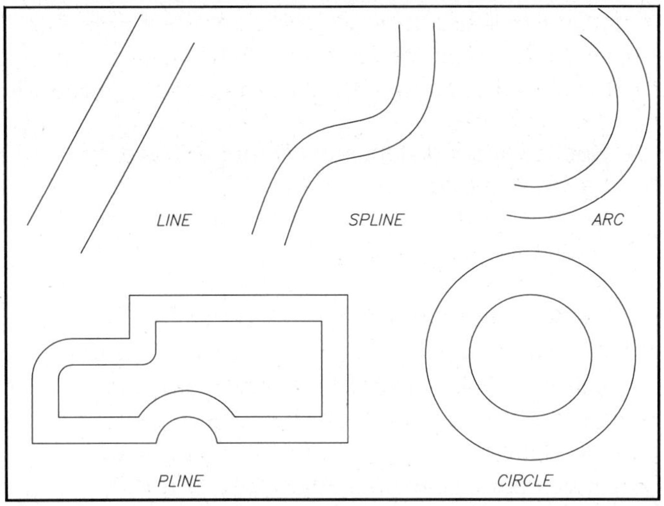


### *Distance*

The default option (*distance*) prompts you to specify a distance between the original object and the newly offset object (Fig. 9-39, left).

> Command: ***OFFSET***
> Current settings: Erase source=No Layer=Source OFFSETGAPTYPE=0
> Specify offset distance or [Through/Erase/Layer] <1.0000>: **(value)** or **PICK** (Specify an offset distance by entering a value or picking two points.)
> Select object to offset or [Exit/Undo] <Exit>: **PICK** (Select the object to offset.)
> Specify point on side to offset or [Exit/Multiple/Undo] <Exit>: **PICK** (Specify which side of original object to create the offset object.)
> Select object to offset or [Exit/Undo] <Exit>: **Enter**

**FIGURE 9-39**

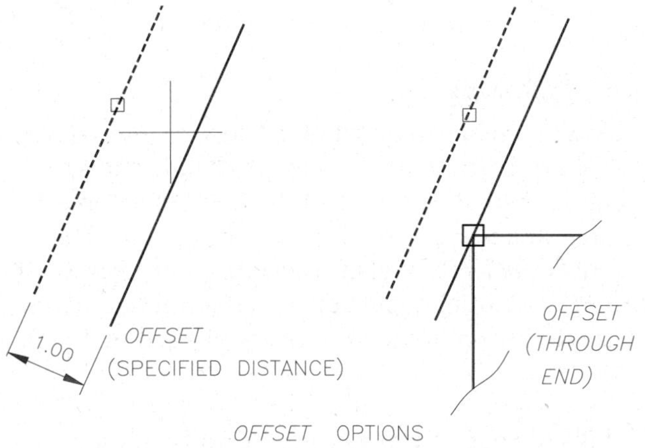


### *Through*

Use the *Through* option to offset an object so it passes through a specified point (Fig. 9-39, right).

> Command: ***OFFSET***
> Current settings: Erase source=No Layer=Source OFFSETGAPTYPE=0
> Specify offset distance or [Through/Erase/Layer] <1.0000>: ***t***
> Select object to offset or [Exit/Undo] <Exit>: **PICK**
> Specify through point or [Exit/Multiple/Undo] <Exit>: **PICK** (Select a point for the object to be offset through. Normally, an Object Snap mode is used to specify the point.)

### *Erase*

Note that the default *Erase source* setting is *No*. With this setting, the original object to offset remains in the drawing. Changing the *Erase* setting to *Yes* causes the selected (source) object to be erased automatically when the offset is created.

> Specify offset distance or [Through/Erase/Layer] <1.0000>: ***e***
> Erase source object after offsetting? [Yes/No] <No>: ***y***

### *Layer*

You can use this setting to create an offset object on a different layer than the original layer of the source object. First, using the *Layer* command (see Chapter 11) change the current layer to the desired layer, next change the *Layer* setting of the *Offset* command to *Current*, then create the offset.

Specify offset distance or [Through/Erase/Layer] <1.0000>: ***1***
Enter layer option for offset objects [Current/Source] <Source>: ***c***

**FIGURE 9-40**

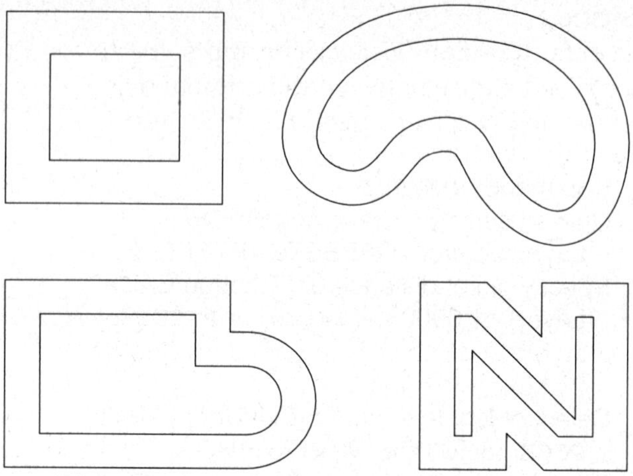

**TIP**

Notice the power of using *Offset* with closed *Pline* or *Spline* shapes (Fig. 9-40). Because a *Pline* or a *Spline* is one object, *Offset* creates one complete smaller or larger "parallel" object. Any closed shape composed of *Line*s and *Arc*s can be converted to one *Pline* object (see "*Pedit*," Chapter 16).

### *OFFSETGAPTYPE*

When you create an offset of a "closed" *Pline* shape with sharp corners, such as the rectangle shown in Figure 9-40, the corners of the new offset object are also sharp (with the default setting of 0). The *OFFSETGAPTYPE* system variable controls how the corners (potential gaps) between the line segments are treated when closed *Plines* are offset (Fig. 9-41).

**FIGURE 9-41**

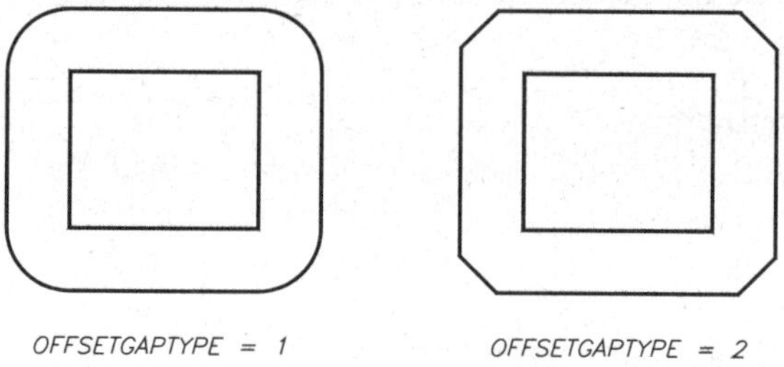


0 Fills the gaps by extending the polyline segments
1 Fills the gaps with filleted arc segments (the radius of each arc segment is equal to the offset distance)
2 Fills the gaps with chamfered line segments (the perpendicular distance to each chamfer is equal to the offset distance)

## *Array*

| Ribbon | Menu Bar | Command (Type) | Alias (Type) | Shortcut |
|---|---|---|---|---|
| ... | ... | *Array* | *AR* | ... |

The *Array* command was used in earlier versions of AutoCAD to create rectangular and circular arrays. Although *Array* still operates, it has been updated by the *Arrayrect*, *Arraypolar*, and *Arraypath* commands (see next sections). Currently *Array* is presented only in command line format to access the newer commands.

Command: ***ARRAY***
Select objects: **PICK**
Select objects: **Enter**
Enter array type [Rectangular/PAth/POlar] <Rectangular>:

## *Arrayrect*

| Ribbon | Menu Bar | Command (Type) | Alias (Type) | Shortcut |
|---|---|---|---|---|
| *Home<br>Modify<br>Rectangular Array* | *Modify<br>Array ><br>Rectangular Array* | *Arrayrect* | ... | ... |

A *Rectangular Array* is a rectangular pattern of objects. One or more objects can be used as the "selection set" to pattern. You specify the number of rows and columns and the distances between the rows and columns. Figure 9-42 illustrates a rectangular array of a set of objects (one table and set of chairs) arrayed to a rectangular pattern of 2 columns and 3 rows.

When you use *Rectangular Array* and "Select objects" to designate your selection set, the *Array Creation* tab appears on the Ribbon that allows you to specify the values that configure your array (Fig. 9-43). Alternately, you can use the Command line format to specify the options. The options are explained next.

**FIGURE 9-42**

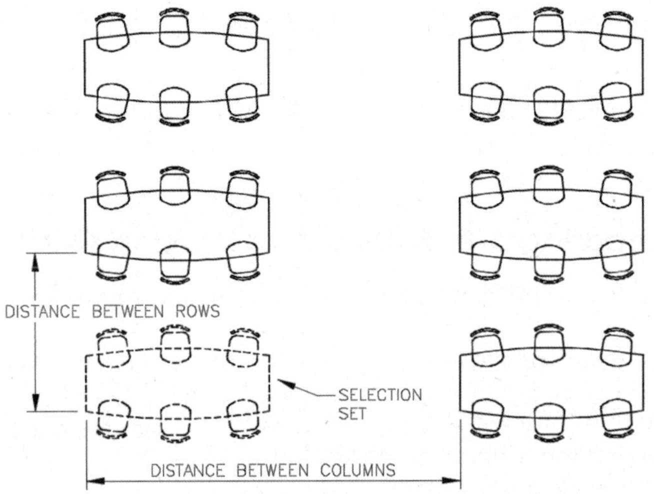


**FIGURE 9-43**

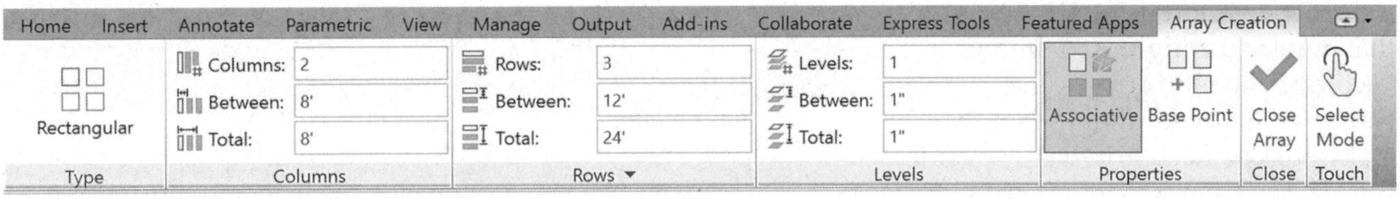


Command: ***ARRAYRECT***
Select objects: **PICK**
Select objects: **Enter**
Type = Rectangular  Associative = Yes
Select grip to edit array or [ASsociative/Basepoint/COUnt/Spacing/COLumns/Rows/Levels/eXit]<eXit>:

### *Columns, Rows, Levels (Count)*

Specify an integer value for the desired pattern. *Columns* are the number of vertical groups (in the X direction), *Rows* are the number of horizontal groups (in the Y direction), and *Levels* are used for a 3-dimensional array (in the Y direction).

### *Between (Spacing)*

These values designate the distances between the rows and columns and levels (see Fig. 9-42). *Between* is the distance between a point on the selection set to the same point on the next set in the adjacent column, row, or level. Enter positive values to create an array to the right (+X) and upward (+Y) from the selected set, or enter negative values to create the array in the –X and –Y directions.

### *Total*

As an alternative to entering values in the *Between* edit boxes, you can specify a *Total* distance for the designated number of *Columns, Rows,* and *Levels.* Entering a value in *Total* changes the *Between* values automatically and vice versa.

### Increment

Use the *Increment* edit box at the bottom of the *Rows* panel to specify an increasing or decreasing elevation for each subsequent row.

### Associative

Arrays can be associative or non-associative. Associative arrays are treated as one complex object rather than individual copies of the selection set. Associative arrays can be easily edited by clicking on an existing associative array and then using the "grips" to change the array parameters or using the same *Array Creation* tab that appears when creating an array. Non-associative arrays are treated as individual copies of the selection set. See "Creating and Editing Arrays Using Grips."

### Base Point

AutoCAD automatically selects an arbitrary point on the selection set to use as the base point to generate the array. You can specify a different base point with this option.

## Creating and Editing Arrays Using Grips

Note that the Command line prompts for you to select a "grip" to edit the array.

```
Select grip to edit array or [ASsociative/Base point/COUnt/Spacing/COLumns/Rows/Levels/eXit]
```

Rather than using the *Array Creation* tab on the Ribbon to specify the parameters of the array (values for rows, columns, between, etc.), you can create the array and manipulate it during creation using the "grips" that appear during the command. Grips are the blue arrows and squares that are located within the array objects. Each grip has a specific purpose as shown in Figure 9-44. For example, to change the distance between rows or columns, click on the appropriate grip and move it. The values in the *Array Creation* tab change automatically as you adjust the grips. Cancel (turn off) the grips by pressing Escape.

**FIGURE 9-44**

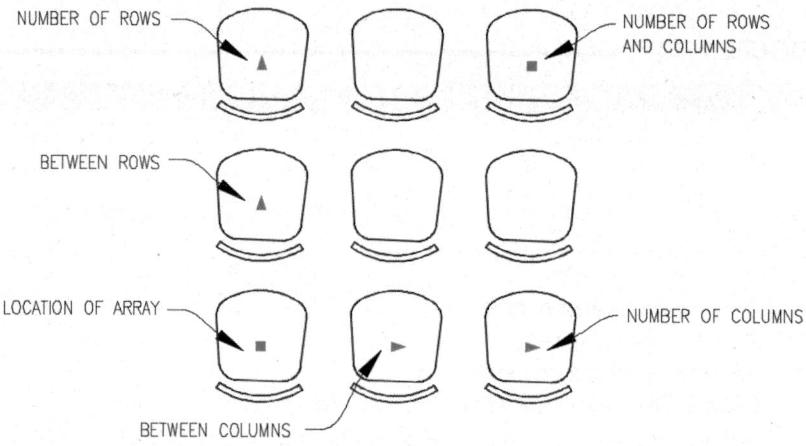


As long as an *Array* was created as an *Associative* array, it can be edited easily at a later time using the grips. Simply click on the array to activate the grips and then manipulate the array as desired by affecting the appropriate grip. See Chapter 19, Grip Editing, for additional information on grips.

### Arraypolar

| Ribbon | Menu Bar | Command (Type) | Alias (Type) | Shortcut |
|---|---|---|---|---|
| *Home*<br>*Modify*<br>*Polar Array* | *Modify*<br>*Array >*<br>*Polar Array* | *Arraypolar* | ... | ... |

The *Polar Array* command generates a circular pattern of one object or a set of objects. For a *Polar Array*, you specify the center point for the array, and then the *Array Creation* tab appears on the Ribbon.

Figure 9-45 illustrates a polar array of the table and set of chairs using 8 as the number of *Items* and 360 as the *Fill angle*. Note also that the items are rotated.

**FIGURE 9-45**

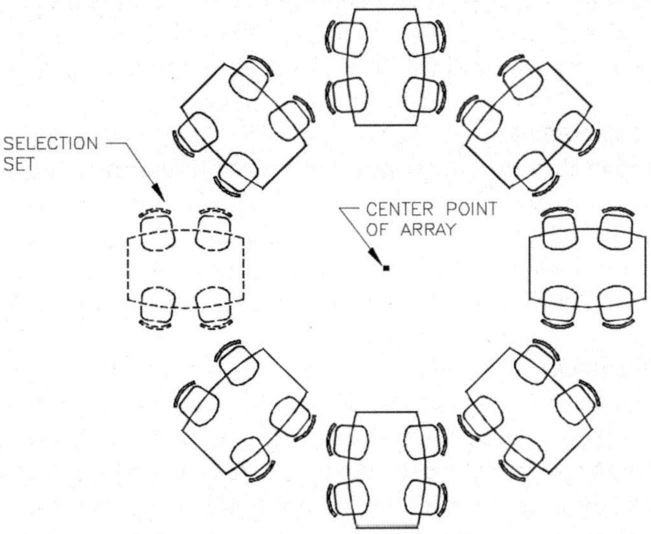


Once you invoke the *Polar Array* command, you must immediately "Select objects" to designate your selection set. Next you must specify the center point for the circular array. A temporary array appears on the screen and the *Array Creation* tab appears on the Ribbon (Fig. 9-46). Similar to creating a *Rectangular Array*, you can either enter values in the *Array Creation* tab to configure your array or edit the array using grips that appear on the array. You can also use the command-line format to specify the options. The options are explained next.

**FIGURE 9-46**

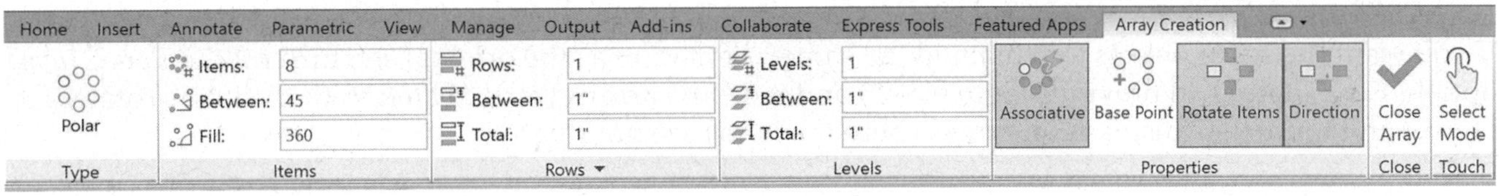


Command: ***ARRAYPOLAR***
Select objects: **PICK**
Select objects: **Enter**
Type = Polar Associative = Yes
Specify center point of array or [Base point/Axis of rotation]:
Select grip to edit array or [ASsociative/Base point/Items/Angle between/Fill Angle/ROWs/Levels/ROTate items/eXit]<eXit>:

### *Items*

Specify an integer value for number of copies of the selection set. This number includes the original selection set.

### *Fill*

This value designates the angle that the array fills. By default, the temporary array will appear as a full 360-degree array. For example, making an array of 5 items (table and set of chairs) through 180 degrees would produce the arrangement shown in Figure 9-47.

**FIGURE 9-47**

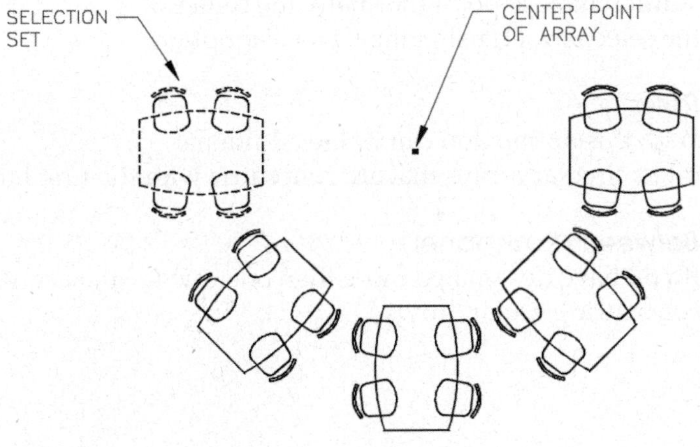

### *Between (Angle between)*

Rather than specify the *Fill* angle, you can specify a value for angle *Between* items. For example, in Figures 9-45 and 9-46, changing this value from 45 to 30 would create an array of 8 sets of tables and chairs that would fill only 210 degrees (there are seven 30-degree segments between the 8 items).

### *Associative*

Remember that arrays can be associative or non-associative. Associative arrays are treated as one complex object rather than individual copies of the selection set and can be easily edited by clicking on an existing associative array and then using the "grips" to change the array parameters or using the same *Array Creation* tab that appears when creating an array. Non-associative arrays are treated as individual copies of the selection set. See previous discussion "Creating and Editing Arrays Using Grips."

### *Direction*

Arrays can be generated in a clockwise or counterclockwise direction. This parameter is important when the array is generated through less than 360 degrees. If the *Direction* button is selected (blue), a counterclockwise array is produced, whereas a clockwise array is produced when the *Direction* button is off (not highlighted in blue). For example, the array in Figure 9-46 was produced with *Direction* on (counterclockwise direction), whereas the array would be generated in an arc above the center if *Direction* were off (clockwise direction).

### *Rotate Items*

If you want the set of objects to remain in the same orientation as the original set, turn off the *Rotate Items* option. For example, in previous Figures 9-45 and 9-47, the arrays were created with the items rotated. However, Figure 9-47 displays an array where the items are not rotated.

### *Base Point*

TIP

AutoCAD automatically selects an arbitrary point on the selection set to use as the base point to generate the array—somewhere on one of the objects. The base point AutoCAD selects is not necessarily in the center of the selection set. This fact is important when you create a circular array when the items are not rotated. For example in Figure 9-48, if AutoCAD set the arbitrary base point as the corner of the table rather than the center of the set, the resulting array would not be generated symmetrically about the designated center point. In such a case, you can specify a different base point (normally the center of the selection set) using the *Base Point* option.

**FIGURE 9-48**

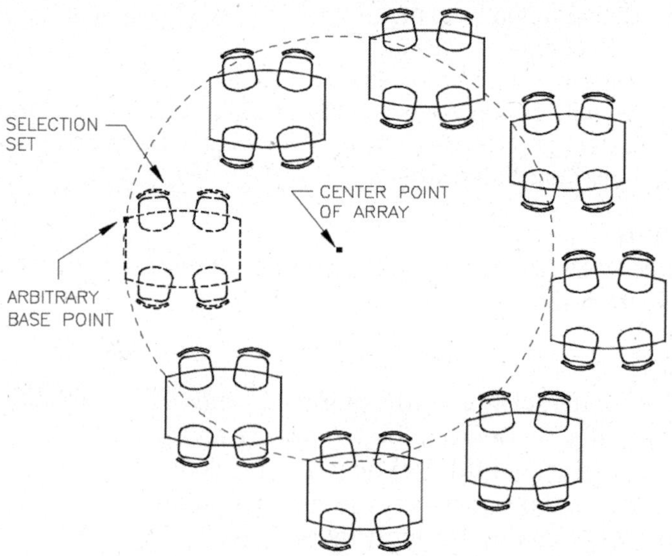


### *Rows*

With this option you can create additional outer circular arrays that are concentric with the first inner array.

### *Between (Row panel)*

If you have designated more than one row for the array, this value specifies the distance between the concentric circular "rows."

### *Total (Row panel)*

If you specified 3 or more "rows," you can specify a distance *Between* each row or specify a *Total* distance from the first to the last concentric circle.

### *Levels, Between, Total*

These options are for generating a 3-dimensional circular array. See Chapter 36, Solid Model Editing, for more information on *Levels*.

### *Close Array*

Use this button to complete the command.

## Arraypath

| Ribbon | Menu Bar | Command (Type) | Alias (Type) | Shortcut |
|---|---|---|---|---|
| *Home* *Modify* *Path Array* | *Modify* *Array >* *Path Array* | *Arraypath* | ... | ... |

A *Path Array* creates a pattern of existing objects around a defined path. The path can be a *Line, Polyline, 3D Polyline, Spline, Helix, Arc, Circle,* or *Ellipse*. After selecting objects to array, you are prompted to select the "path curve," (Fig. 9-49) which can be a straight line or any type of curved object and can be a closed shape (circular or elliptical, for example).

Once the path curve is selected, AutoCAD automatically generates a temporary array along the path and the *Array Creation* panel appears on the Ribbon (see Fig. 9-50). The temporary array that is generated is a "best guess" based on the size of the selection set and length of the path. You can then edit the array using the options on the *Array Creation* panel, the Command line options, or using the grips that appear on the array.

FIGURE 9-49

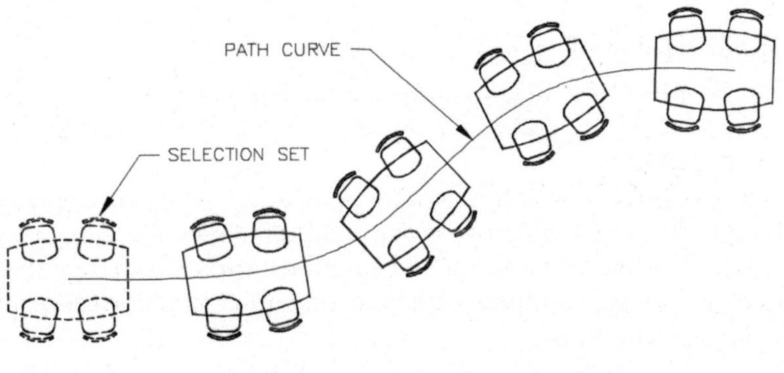


FIGURE 9-50

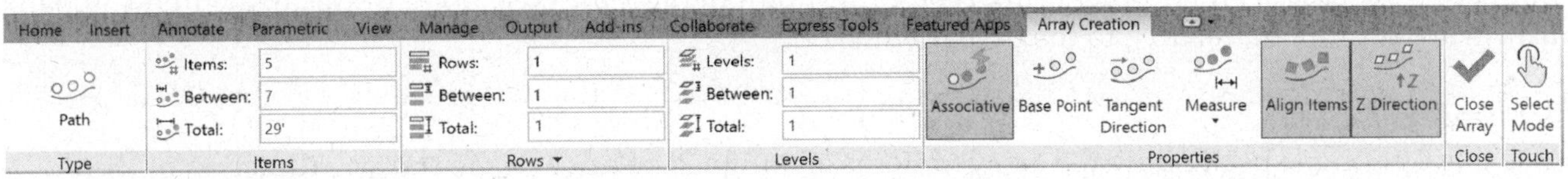


Command: ***ARRAYPATH***
Select objects: **PICK**
Select objects: **Enter**
Type = Path  Associative = Yes
Select path curve: **PICK**
Select grip to edit array or [ASsociative/Method/Base point/Tangent direction/Items/Rows/Levels/Align items/Zdirection/eXit]<eXit>:

Tip: You can also enter a function, such as 32/2

### *Items (Method, Divide)*

Specify an integer value for number of copies of the selection set. This number includes the original selection set. The *Divide* option (Items from the Array Creation Ribbon or Method Divide Command line format) enables this edit box.

### *Between (Method, Measure)*

Rather than specify the number of Items, you can specify a value for the distance *Between* items. For example, in Figure 9-49, changing this value from 7′ to 9′ would create an array of 4 sets of tables and chairs that would fit the given path length. The *Measure* option (on the Ribbon or in Command line format) must be selected to enable this edit box.

### *Rows (Row* panel*)*

With this option you can create additional rows of arrays. The paths of the rows are parallel with the first array; however, the items also align in a parallel manner with the original items, which may result in uneven spacing.

### *Between (Row* panel*)*

If you have designated more than one row for the array, this value specifies the distance between the rows.

### *Total (Row* panel*)*

If you specified 3 or more rows, this value specifies the *Total* distance from the first to the last row.

### *Levels, Between, Total*

These options are to generate a 3-dimensional array.

### *Associative*

Arrays can be associative or non-associative. Associative arrays are treated as one complex object rather than individual copies of the selection set. Associative arrays can be easily edited by clicking on an existing associative array and then using the "grips" to change the array parameters or using the same *Array Creation* tab that appears when creating an array. Non-associative arrays are treated as individual copies of the selection set.

### *Base point*

Similar to the *Base point* for other array types, this point defines the base point of the array. Items in path arrays are positioned relative to the base point.

### *Tangent direction*

Use this option to specify how the arrayed items are aligned relative to the starting direction of the path. This option effectively realigns the orientation of all the items along the path. The default option is *Normal* which orients the direction of the first item with the starting direction of the path curve.

### *Align Items*

This option determines if the items are aligned (rotated) along the path direction relative to the first item's orientation (aligned) or if the items remain in the same orientation as the first item (not aligned). This option is similar to the *Rotate Items* option of a *Polar Array* (see previous Figure 9-47).

### *Z Direction*

In a 3D array, this option controls whether to maintain the items' original Z direction or to naturally bank the items along a 3D path.

## *Fillet*

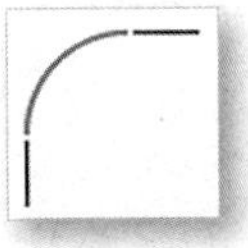

| Ribbon | Menu Bar | Command (Type) | Alias (Type) | Shortcut |
|---|---|---|---|---|
| *Home*<br>*Modify*<br>*Fillet* | *Modify*<br>*Fillet* | *Fillet* | *F* | ... |

The *Fillet* command automatically rounds a sharp corner (intersection of two *Lines, Arcs, Circles,* or *Pline* vertices) with a radius (Fig. 9-51). You specify only the radius and select the objects' ends to be *Filleted.* The objects to fillet do not have to completely intersect or can overlap. You can specify whether or not the objects are automatically extended or trimmed as necessary.

AutoCAD provides a preview when passing the cursor over the second object in the *Fillet* selection.

Command: ***FILLET***
Current settings: Mode = TRIM, Radius = 0.5000
Select first object or [Undo/Polyline/Radius/Trim/Multiple]: **PICK** (Select one *Line, Arc,* or *Circle* near the point where the fillet should be created.)
Select second object or shift-select to apply corner: **PICK** (Select the second object near the desired fillet location.)
Command:

The fillet is created at the corner selected.

Treatment of *Arcs* and *Circles* with *Fillet* is shown in Figure 9-52. Note that the objects to *Fillet* do not have to intersect or can overlap.

**FIGURE 9-51**

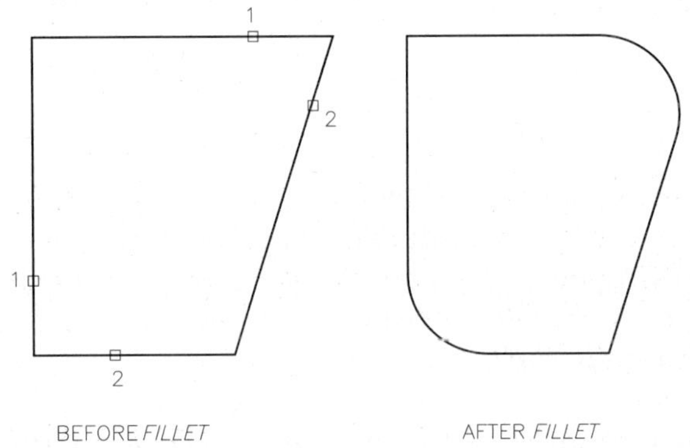


**FIGURE 9-52**

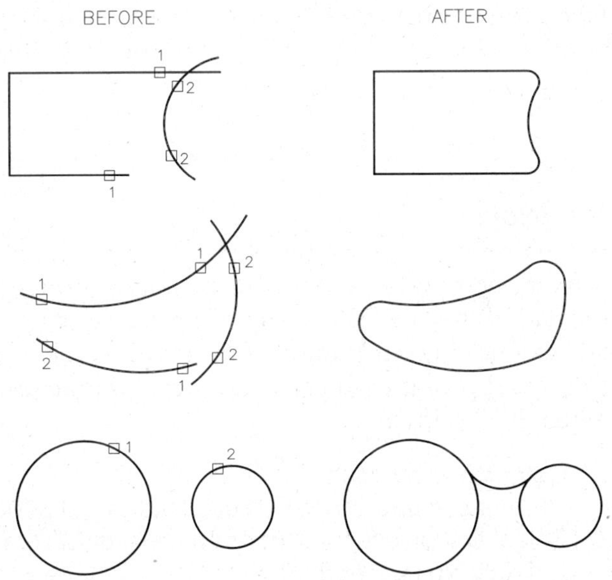


**FIGURE 9-53**

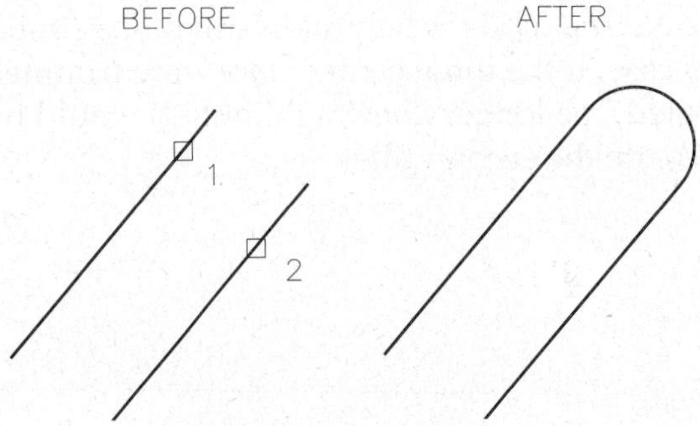


### *Radius*

Use the *Radius* option to specify the desired radius for the fillets. *Fillet* uses the specified radius value for all new fillets until the value is changed.

If parallel objects are selected, the *Fillet* is automatically created to the correct radius (Fig. 9-53). Therefore, parallel objects can be filleted at any time without specifying a radius value.

### *Polyline*

*Fillets* can be created on *Polylines* in the same manner as two *Line* objects. Use *Fillet* as you normally would for *Lines*. However, if you want a fillet equal to the specified radius to be added to each vertex of the *Pline* (except the endpoint vertices), use the *Polyline* option; then select anywhere on the *Pline* (Fig. 9-54).

FIGURE 9-54

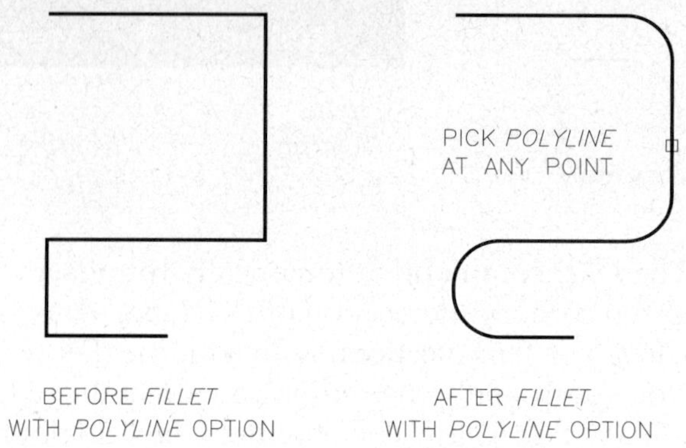


*Closed Plines* created by the *Close* option of *Pedit* react differently with *Fillet* than *Plines* connected by PICKing matching endpoints. Figure 9-55 illustrates the effect of *Fillet Polyline* on a *Closed Pline* and on a connected *Pline*.

When creating a *Fillet* using the *Polylines* option, a preview is displayed for the entire *Polyline*.

FIGURE 9-55

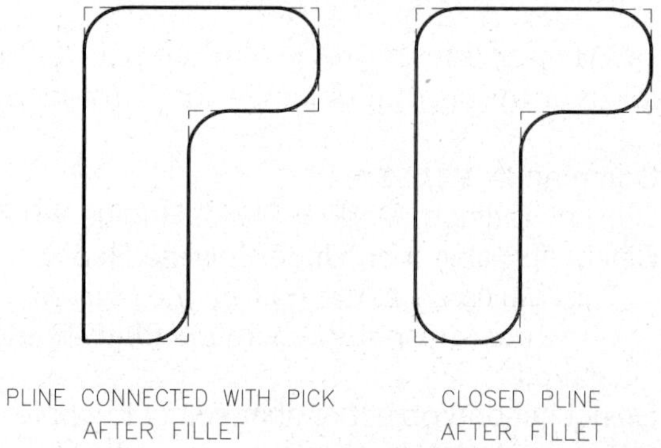


**TIP**

*Fillets* can be created on *Splines* in the same manner as two *Line* objects. Use *Fillet* as you normally would for *Lines*.

### *Trim/Notrim*

In the previous figures, the objects are shown having been filleted with the *Trim* mode—that is, with automatically trimmed or extended objects to meet the end of the new fillet radius. The *Notrim* mode creates the fillet without any extending or trimming of the involved objects (Fig. 9-56). Note that the command prompt indicates the current mode (as well as the current radius) when *Fillet* is invoked.

FIGURE 9-56

> Command: ***FILLET***
> Current settings: Mode = TRIM, Radius = 0.5000
> Select first object or [Undo/Polyline/Radius/Trim/Multiple]: ***t*** (Invokes the Trim/No trim option)
> Enter Trim mode option [Trim/No trim] <Trim>: ***n*** (Sets the option to No trim)
> Select first object or [Undo/Polyline/Radius/Trim/Multiple]:

The *Notrim* mode is helpful in situations similar to that in Figure 9-57. In this case, if the intersecting *Lines* were trimmed when the first fillet was created, the longer *Line* (highlighted) would have to be redrawn in order to create the second fillet.

FIGURE 9-57

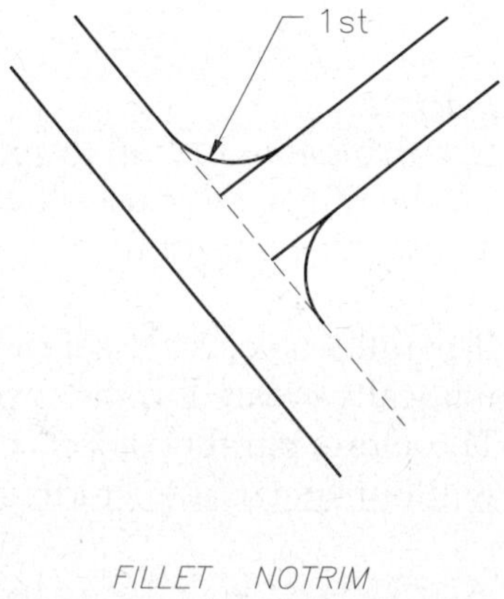

Changing the *Trim/Notrim* option sets the *TRIMMODE* system variable to 0 (*Notrim*) or 1 (*Trim*). This variable controls trimming for both *Fillet* and *Chamfer,* so if you set *Fillet* to *Notrim, Chamfer* is also set to *Notrim.*

### *Multiple*

Normally the *Fillet* command allows you to create one fillet, then the command ends. Use the *Multiple* option if you want to create several filleted corners. For example, if you wanted to *Fillet* the four corners of a rectangular shape you could use the *Multiple* option instead of using the *Fillet* command four separate times.

### *Undo*

The *Undo* option operates only in conjunction with the *Multiple* option. *Undo* will remove the most recent of multiple fillets.

### *Shift-select*

If you want to create a sharp corner instead of a rounded (radiused) corner, you can use the Shift-select option. After selecting the first line, hold down the Shift key and select the second line. Because *Fillet* can automatically trim and extend as needed, a sharp corner is created (Fig. 9-58). This action is identical to setting a *Radius* of 0 and selecting two lines normally. The *Trim* mode setting does not affect using Shift-select to create sharp corners.

**FIGURE 9-58**

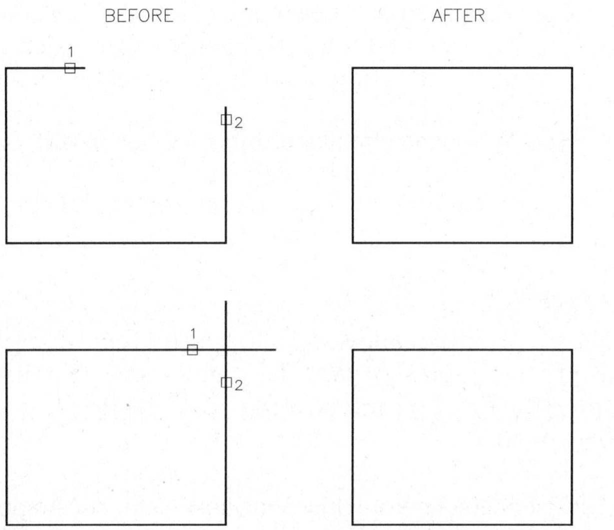


TIP

## *Chamfer*

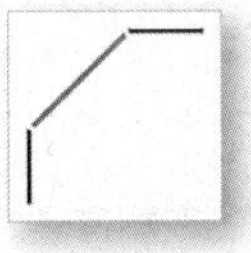

| Ribbon | Menu Bar | Command (Type) | Alias (Type) | Shortcut |
|---|---|---|---|---|
| *Home Modify Chamfer* | *Modify Chamfer* | *Chamfer* | *CHA* | ... |

Chamfering is a manufacturing process used to replace a sharp corner with an angled surface. In AutoCAD, *Chamfer* is commonly used to change the intersection of two *Lines* or *Plines* by adding an angled line. The *Chamfer* command is similar to *Fillet,* but rather than rounding with a radius or "fillet," an angled line is automatically drawn at the distances (from the existing corner) that you specify.

Chamfers can be created by two methods: *Distance* (specify two distances) or *Angle* (specify a distance and an angle). The current method and the previously specified values are displayed at the Command prompt along with the options:

```
Command: CHAMFER
(TRIM mode) Current chamfer Dist1 = 0.0000, Dist2 = 0.0000
Select first line or [Undo/Polyline/Distance/ Angle/Trim/mEthod/Multiple]:
```

### *mEthod*

Use this option to indicate which of the two methods you want to use: *Distance* (specify 2 distances) or *Angle* (specify a distance and an angle).

### *Distance*

The *Distance* option is used to specify the two values applied when the *Distance Method* is used to create the chamfer. The values indicate the distances from the corner (intersection of two lines) to each chamfer endpoint (Fig. 9-59).

**FIGURE 9-59**

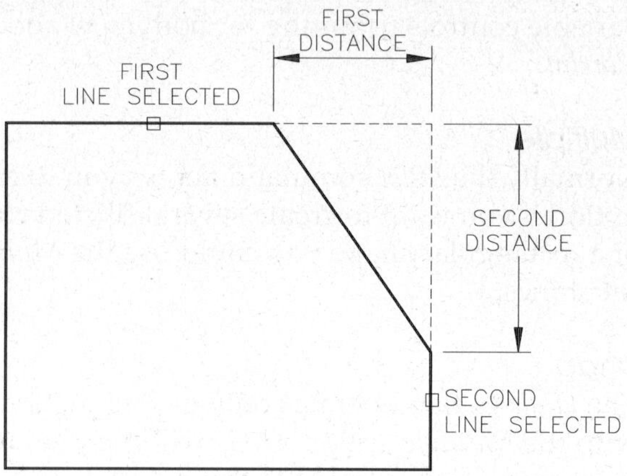


Command: ***CHAMFER***
(TRIM mode) Current chamfer Dist1 = 0.0000,
Dist2 = 0.0000
Select first line or [Undo/Polyline/
Distance/Angle/Trim/mEthod/Multiple]: ***d***
Specify first chamfer distance <0.0000>: **(value)**
(Enter a value for the distance from the existing corner to the endpoint of the chamfer on the first line.)
Specify second chamfer distance <0.7500>: **Enter** or **(value)** (Press Enter to use the same value as the first, or enter another value.)
Select first line or [Undo/Polyline/Distance/Angle/Trim/mEthod/Multiple]:

### *Angle*

The *Angle* option allows you to specify the values that are used for the *Angle Method*. The values specify a distance along the first line and an angle from the first line (Fig. 9-60).

**FIGURE 9-60**

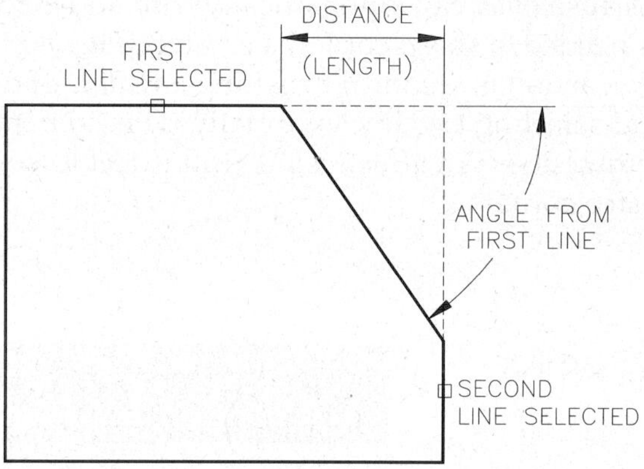


Select first line or [Undo/Polyline/Distance/ Angle/Trim/
mEthod/Multiple]: ***a***
Specify chamfer length on the first line <0.0000>:
**(value)**
Specify chamfer angle from the first line <0>: **(value)**

**FIGURE 9-61**

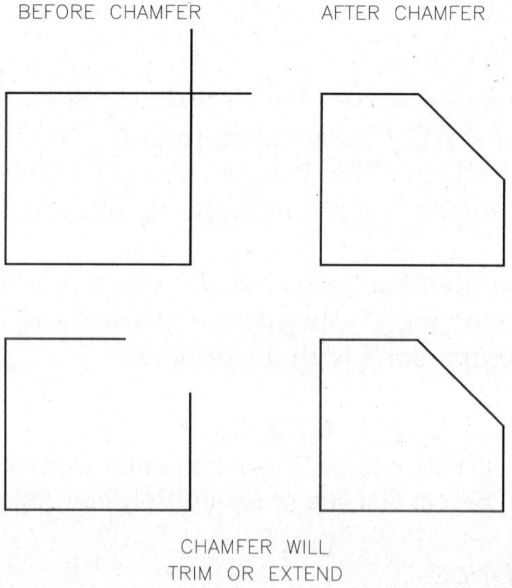


### *Trim/Notrim*

The two lines selected for chamfering do not have to intersect but can overlap or not connect. The *Trim* setting automatically trims or extends the lines selected for the chamfer (Fig. 9-61), while the *Notrim* setting adds the chamfer without changing the length of the selected lines. These two options are the same as those in the *Fillet* command (see "*Fillet*"). Changing these options sets the *TRIMMODE* system variable to 0 (*Notrim*) or 1 (*Trim*). The variable controls trimming for both *Fillet* and *Chamfer*.

AutoCAD provides a preview when passing the cursor over the second object in the *Chamfer* selection.

### *Polyline*

The *Polyline* option of *Chamfer* creates chamfers on all vertices of a *Pline*. All vertices of the *Pline* are chamfered with the supplied distances (Fig. 9-62). The first end of the *Pline* that was drawn takes the first distance. Use *Chamfer* without this option if you want to chamfer only one corner of a *Pline*. This is similar to the same option of *Fillet* (see "*Fillet*").

When creating a chamfer using the *Polylines* option, a preview is displayed for the entire *Polyline*.

**FIGURE 9-62**

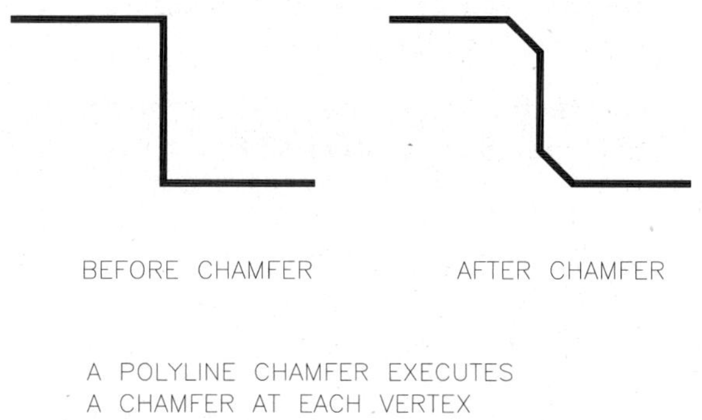


### *Multiple*

Normally the *Chamfer* command ends after you create one chamfer, similar to the operation of the *Fillet* command. Instead of using the *Chamfer* command several times to create several chamfered corners, use the *Multiple* option.

### *Undo*

The *Undo* option operates only in conjunction with the *Multiple* option. *Undo* will remove the most recent of multiple chamfers.

### *Shift-select*

Similar to the same feature of the *Fillet* command, use the Shift-select option if you want to create a sharp corner instead of a chamfered corner. After selecting the first line to chamfer, hold down the Shift key and select the second line.

## CHAPTER EXERCISES

1. ***Move***

   Begin a ***New*** drawing and create the geometry in Figure 9-63A using ***Lines*** and ***Circles***. If desired, set **Polar Distance** to **.25** to make drawing easy and accurate.

   For practice, turn *Snap OFF* (**F9**). Use the ***Move*** command to move the *Circles* and *Lines* into the positions shown in Fig. 9-63B. ***Object Snaps*** are required to *Move* the geometry accurately (since *Snap* is off). ***Save*** the drawing as **MOVE1**.

**FIGURE 9-63**

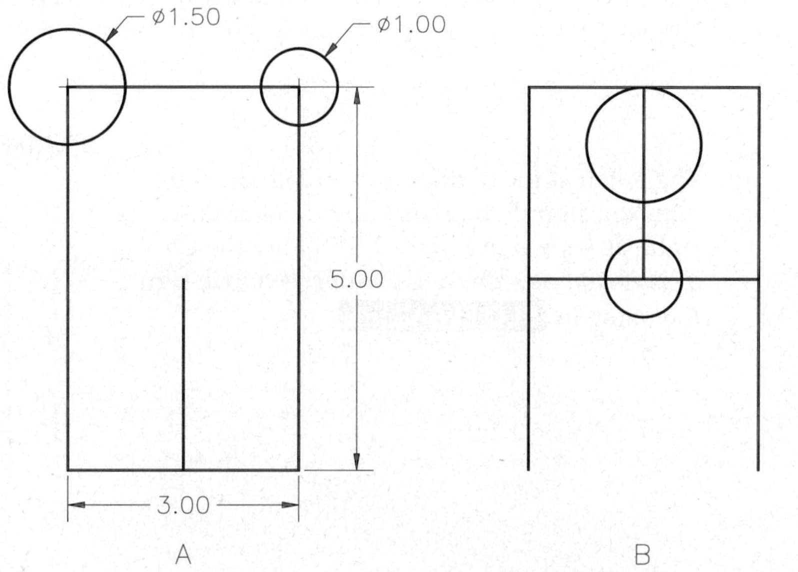

2. ***Rotate***

Begin a ***New*** drawing and create the geometry in Figure 9-64A.

***Rotate*** the shape into position shown in step B. ***SaveAs*** **ROTATE1**.

Use the ***Reference*** option to ***Rotate*** the box to align with the diagonal *Line* as shown in C. ***SaveAs*** **ROTATE2**.

**FIGURE 9-64**

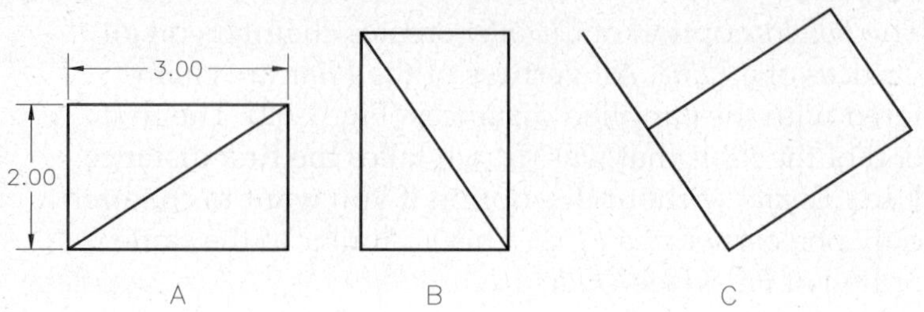


3. ***Scale***

***Open*** **ROTATE1** to again use the shape shown in Figure 9-65B and ***SaveAs*** **SCALE1**. ***Scale*** the shape by a factor of **1.5**.

***Open*** **ROTATE2** to again use the shape shown in C. Use the ***Reference*** option of ***Scale*** to increase the scale of the three other *Lines* to equal the length of the original diagonal *Line* as shown. (HINT: ***Object Snaps*** are required to specify the ***Reference length*** and ***New length***.) ***SaveAs*** **SCALE2**.

**FIGURE 9-65**

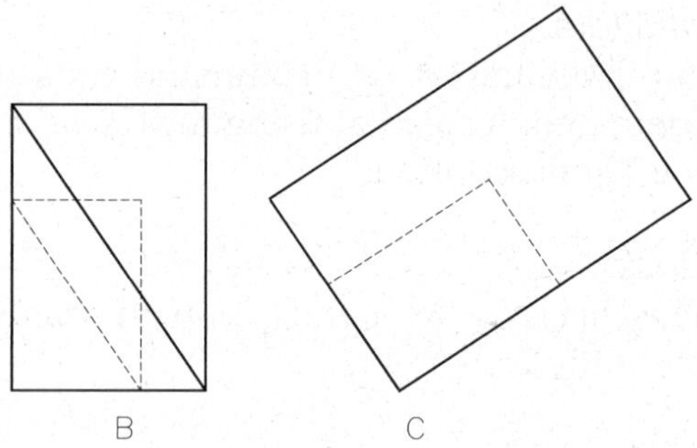


4. ***Stretch***

A design change has been requested. ***Open*** the **SLOTPLATE CH8** drawing and make the following changes.

A. The top of the plate (including the slot) must be moved upward. This design change will add 1″ to the total height of the Slot Plate. Use ***Stretch*** to accomplish the change, as shown in Figure 9-66. Draw the *Crossing Window* as shown.

**FIGURE 9-66**

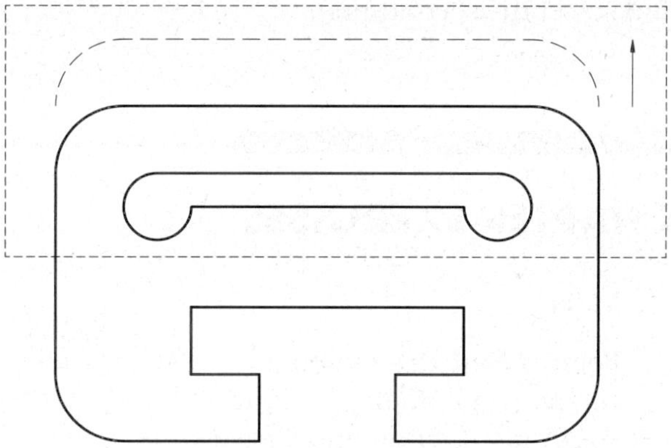

B. The notch at the bottom of the plate must be adjusted slightly by relocating it .50 units to the right, as shown in Figure 9-67. Draw the *Crossing Window* as shown. Use ***SaveAs*** to reassign the name to **SLOTPLATE 2.**

**FIGURE 9-67**

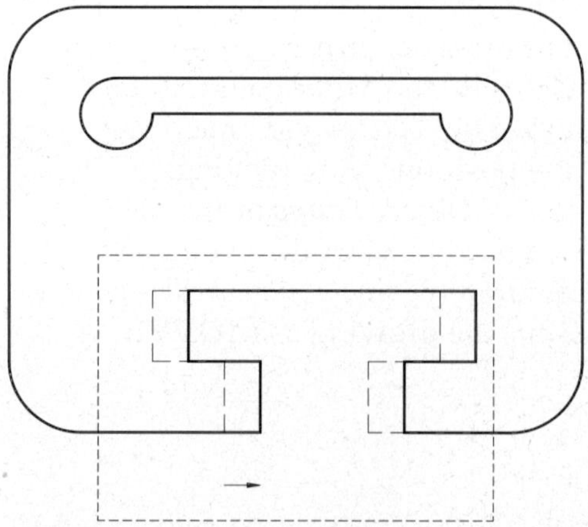

5. *Trim*

   A. Create the shape shown in Figure 9-68A. *SaveAs* **TRIM-EX.**

   B. Use *Trim* to alter the shape as shown in B. *SaveAs* **TRIM1.**

   C. *Open* **TRIM-EX** to create the shapes shown in C and D using *Trim*. *SaveAs* **TRIM2** and **TRIM3.**

6. *Extend*

   *Open* each of the drawings created as solutions for Figure 9-68 (**TRIM1, TRIM2,** and **TRIM3**). Use *Extend* to return each of the drawings to the original form shown in Figure 9-67A. *SaveAs* **EXTEND1, EXTEND2,** and **EXTEND3.**

**FIGURE 9-68**

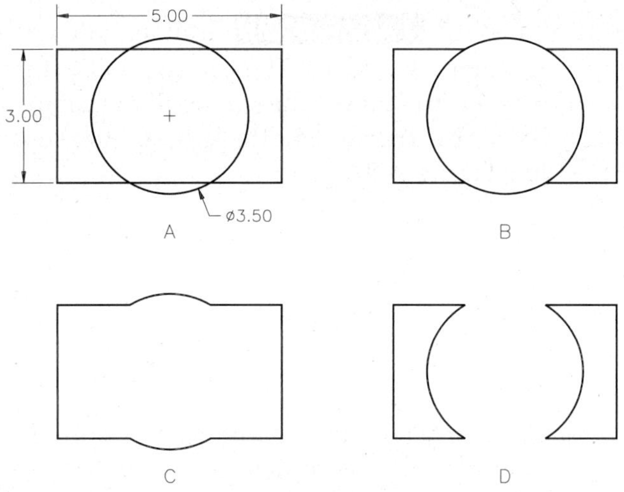


7. *Break*

   A. Create the shape shown in Figure 9-69A. *SaveAs* **BREAK-EX.**

   B. Use *Break* to make the two breaks as shown in B. *SaveAs* **BREAK1.** (HINT: You may have to use *Object Snaps* to create the breaks at the *Intersections* or *Quadrants* as shown.)

   C. Open **BREAK-EX** each time to create the shapes shown in C and D with the *Break* command. *SaveAs* **BREAK2** and **BREAK3.**

**FIGURE 9-69**

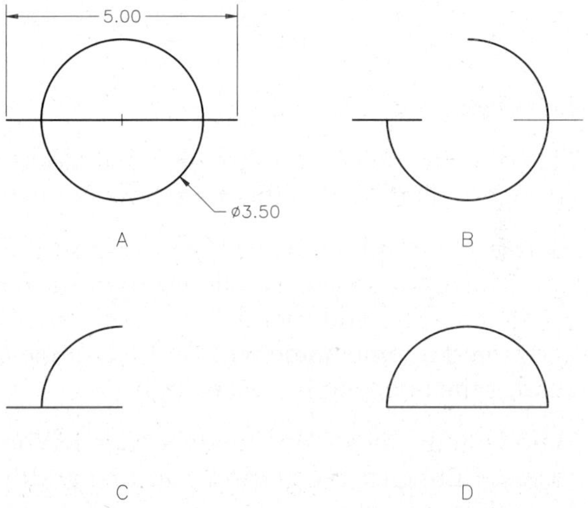

8. **Complete the Table Base**

A. Open the **TABLE-BASE** drawing you created in the Chapter 7 Exercises. The last step in the previous exercise was the creation of the square at the center of the table base. Now, use ***Trim*** to edit the legs and the support plate to look like Figure 9-70.

FIGURE 9-70

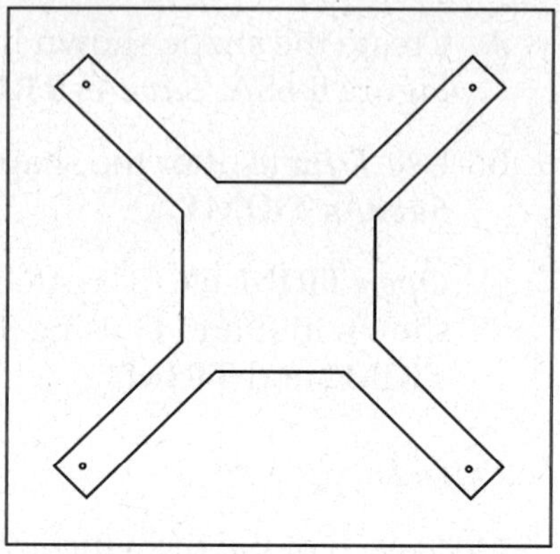

B. Use ***Circle*** with the ***Center, Radius*** option to create a 4″ diameter circle to represent the welded support for the center post for this table. Finally, add the other 4 drill holes (.25″ diameter) on the legs at the intersection of the vertical and horizontal lines representing the edges of the base. Use ***Polar Tracking*** or ***Intersection*** *Object Snap* for this step. The table base should appear as shown in Figure 9-71. *Save* the drawing.

FIGURE 9-71

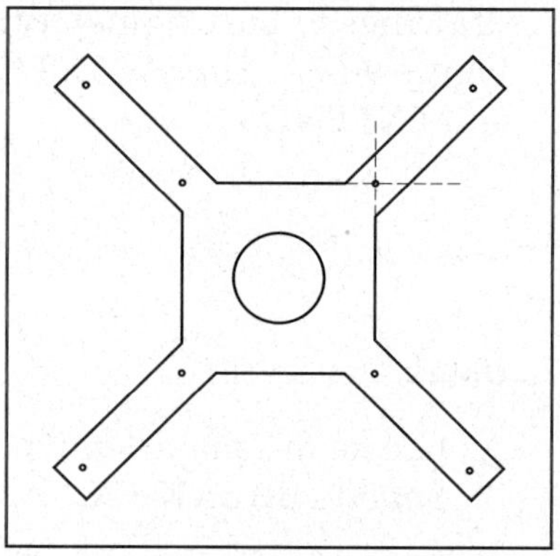

9. ***Lengthen***

Five of your friends went to the horse races, each with $5.00 to bet. Construct a simple bar graph (similar to Fig. 9-72) to illustrate how your friends' wealth compared at the beginning of the day.

Modify the graph with *Lengthen* to report the results of their winnings and losses at the end of the day. The reports were as follows: friend 1 made $1.33 while friend 2 lost $2.40; friend 3 reported a 150% increase and friend 4 brought home 75% of the money; friend 5 ended the day with $7.80 and friend 6 came home with $5.60. Use the *Lengthen* command with the appropriate options to change the line lengths accordingly.

Who won the most? Who lost the most? Who was closest to even? Enhance the graph by adding width to the bars and other improvements as you wish (similar to Fig. 9-73). ***SaveAs*** **FRIENDS GRAPH**.

FIGURE 9-72

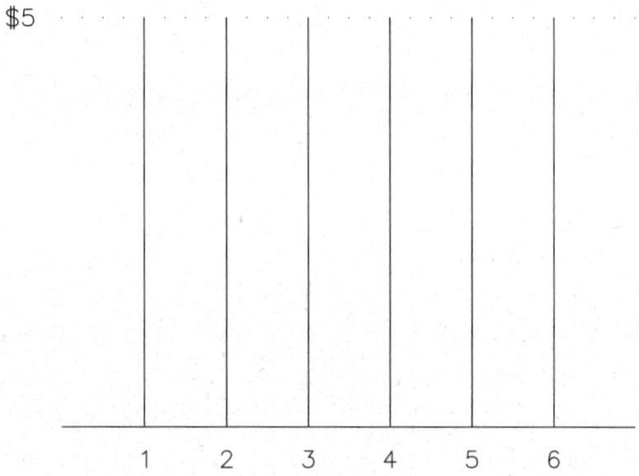


FIGURE 9-73

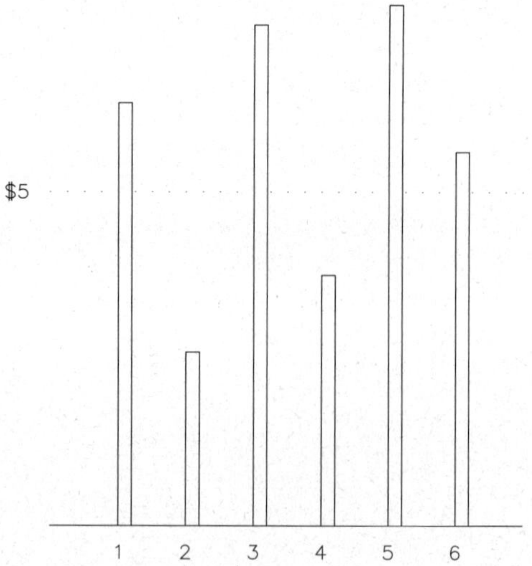

10. ***Trim, Extend***

A. ***Open*** the PIVOTARM CH7 drawing from the Chapter 7 Exercises. Use ***Trim*** to remove the upper sections of the vertical *Lines* connecting the top and front views. Compare your work to Figure 9-74.

**FIGURE 9-74**

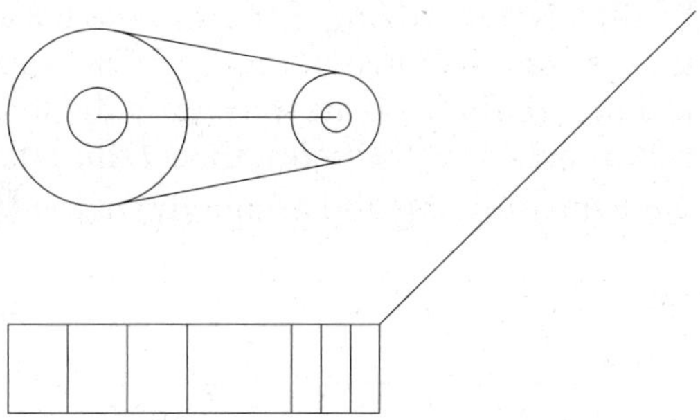

B. Next, draw a horizontal ***Line*** in the front view between the ***Midpoints*** of the vertical *Line* on each end of the view, as shown in Figure 9-75. ***Erase*** the horizontal *Line* in the top view between the *Circle* centers.

**FIGURE 9-75**

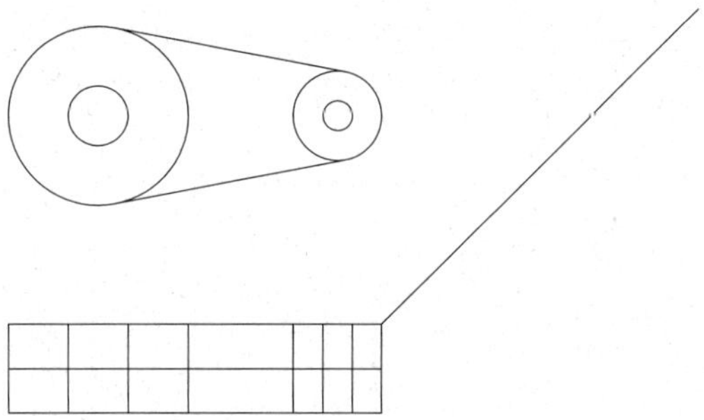

C. Draw vertical ***Lines*** from the ***Endpoints*** of the inclined *Line* (side of the object) in the top view down to the bottom *Line* in the front view, as shown in Figure 9-76. Use the two vertical lines as ***Cutting edges*** for ***Trimming*** the *Line* in the middle of the front view, as shown highlighted.

**FIGURE 9-76**

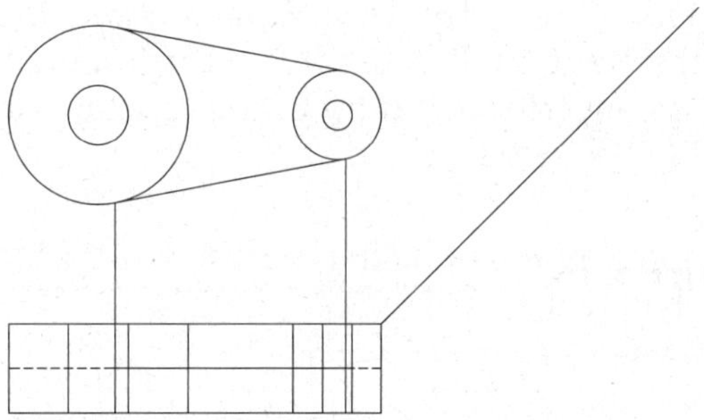

D. Finally, ***Erase*** the vertical lines used for ***Cutting edges*** and then use ***Trim*** to achieve the object as shown in Figure 9-77. Use ***SaveAs*** and name the drawing PIVOTARM CH9.

**FIGURE 9-77**

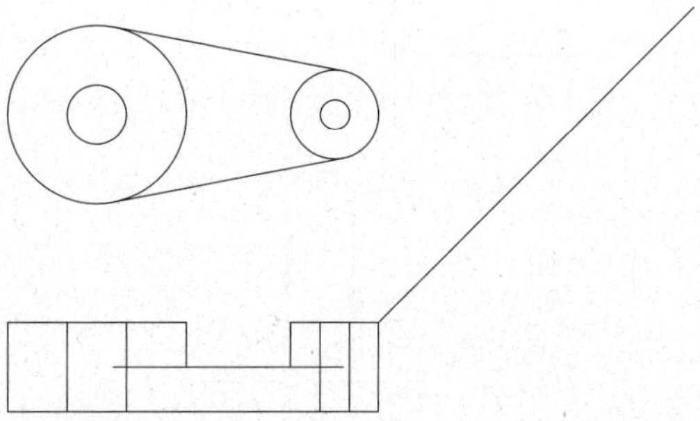

11. **GASKETA**

Begin a *New* drawing. Set *Limits* to **0,0** and **8,6**. Set *Snap* and *Grid* values appropriately. Create the Gasket as shown in Figure 9-78. Construct only the gasket shape, not the dimensions or centerlines. (HINT: Locate and draw the four .5″ diameter *Circles*, then create the concentric .5″ radius arcs as full *Circles*, then *Trim*. Make use of *Object Snaps* and *Trim* whenever applicable.) *Save* the drawing and assign the name **GASKETA.**

**FIGURE 9-78**

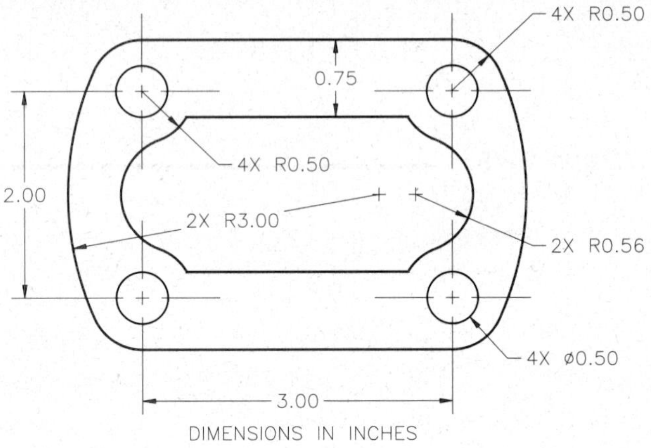


12. **Chemical Process Flow Diagram**

Begin a *New* drawing and re-create the chemical process flow diagram shown in Figure 9-79. Use *Line, Circle, Arc, Trim, Extend, Scale, Break,* and other commands you feel necessary to complete the diagram. Because this is diagrammatic and will not be used for manufacturing, dimensions are not critical, but try to construct the shapes with proportional accuracy. *Save* the drawing as **FLOWDIAG**.

**FIGURE 9-79**

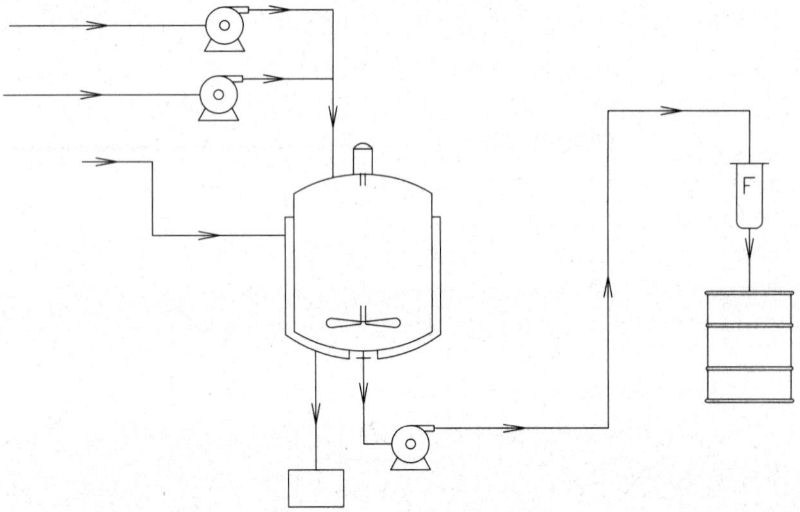

13. *Copy*

Begin a ***New*** drawing. Set the ***Limits*** to **24,18**. Set ***Polar Snap*** to **.5** and set your desired running ***Object Snaps***. Turn on ***Snap, Grid, Object Snap,*** and ***Polar Tracking.*** Create the sheet metal part composed of four ***Lines*** and one ***Circle*** as shown in Figure 9-80. The lower-left corner of the part is at coordinate 2,3.

Use ***Copy*** to create two copies of the rectangle and hole in a side-by-side fashion as shown in Figure 9-81. Allow 2 units between the sheet metal layouts. ***SaveAs*** **PLATES.**

**FIGURE 9-80**

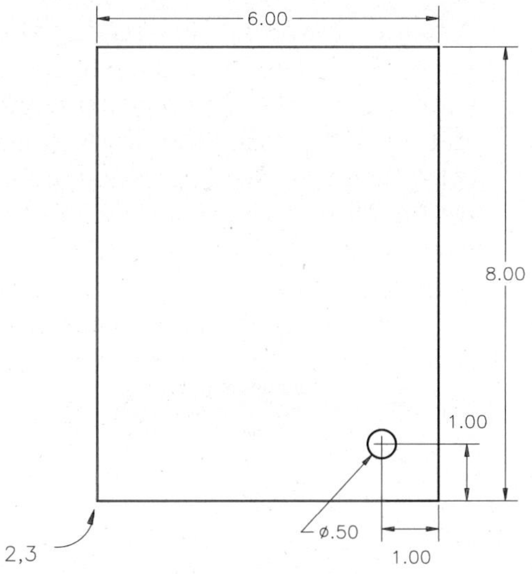


REUSE

A. Use ***Copy*** to create the additional 3 holes equally spaced near the other corners of the part as shown in Figure 9-81A. ***Save*** the drawing.

B. Use ***Copy*** to create the hole configuration as shown in Figure 9-81B. ***Save***.

C. Use ***Copy*** to create the hole placements as shown in C. Each hole center is **2** units at **125** degrees from the previous one (use relative polar coordinates or set an appropriate *Polar Angle*). ***Save*** the drawing.

**FIGURE 9-81**

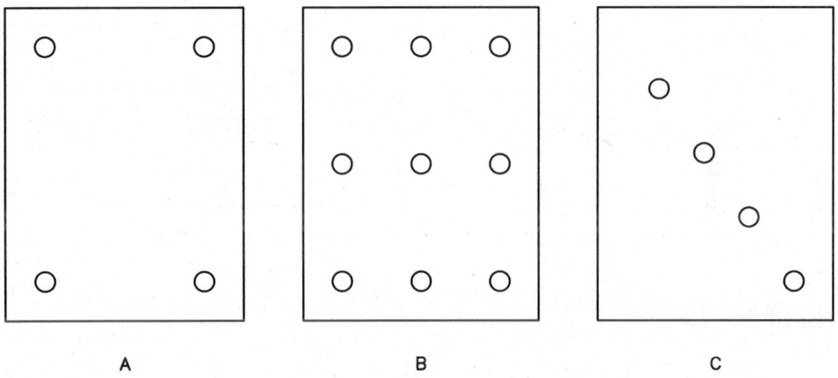

14. *Mirror*

A manufacturing cell is displayed in Figure 9-82. The view is from above, showing a robot centered in a work station. The production line requires 4 cells. Begin by starting a *New* drawing, setting ***Units*** to ***Engineering*** and ***Limits*** to **40′ x 30′**. It may be helpful to set ***Polar Snap*** to **6″**. Draw one cell to the dimensions indicated. Begin at the indicated coordinates of the lower-left corner of the cell. ***SaveAs*** **MANFCELL.**

**FIGURE 9-82**

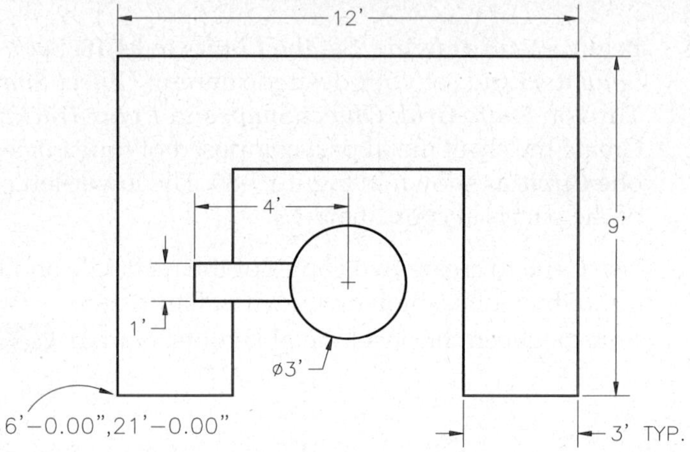


Use ***Mirror*** to create the other three manufacturing cells as shown in Figure 9-83. Ensure that there is sufficient space between the cells as indicated. Draw the two horizontal ***Lines*** representing the walkway as shown. ***Save.***

**FIGURE 9-83**

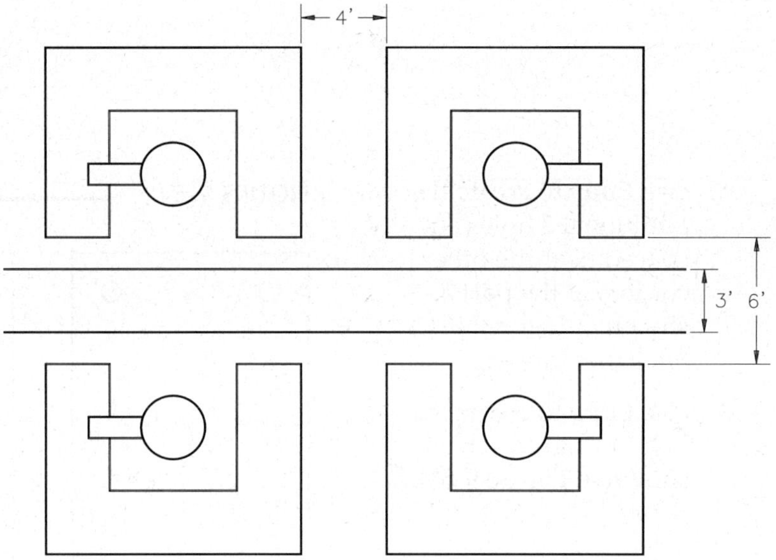


15. *Offset*

Create the electrical schematic as shown in Figure 9-84. Draw ***Lines*** and ***Plines***, then ***Offset*** as needed. Use ***Point*** objects with an appropriate (circular) ***Point Style*** at each of the connections as shown. Check the ***Set Size in Absolute Units*** radio button (in the *Point Style* dialog box) and find an appropriate size for your drawing. Because this is a schematic, you can create the symbols by approximating the dimensions. Omit the text. ***Save*** the drawing as **SCHEMATIC1.** Create a ***Plot*** and check ***Scaled to Fit.***

**FIGURE 9-84**

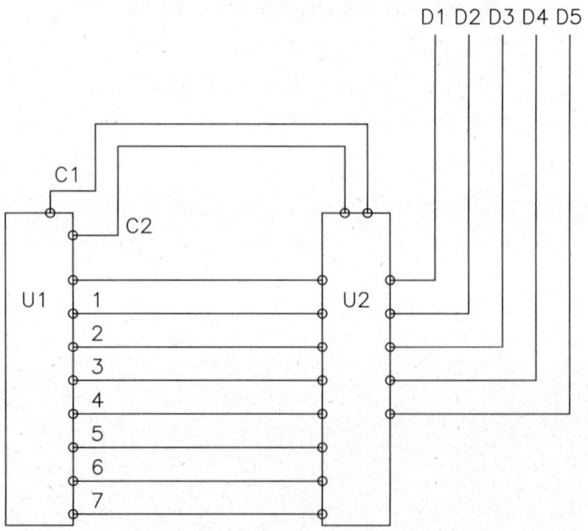

16. ***Array, Polar***

Begin a *New* drawing. Create the starting geometry for a Flange Plate as shown in Figure 9-85. *SaveAs* **ARRAY**.

**FIGURE 9-85**

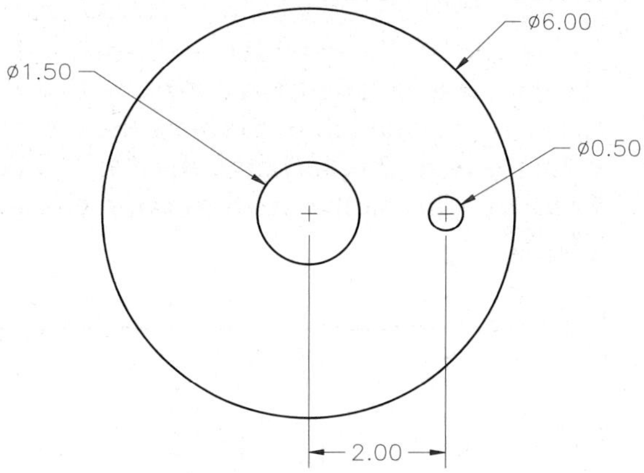


**FIGURE 9-86**

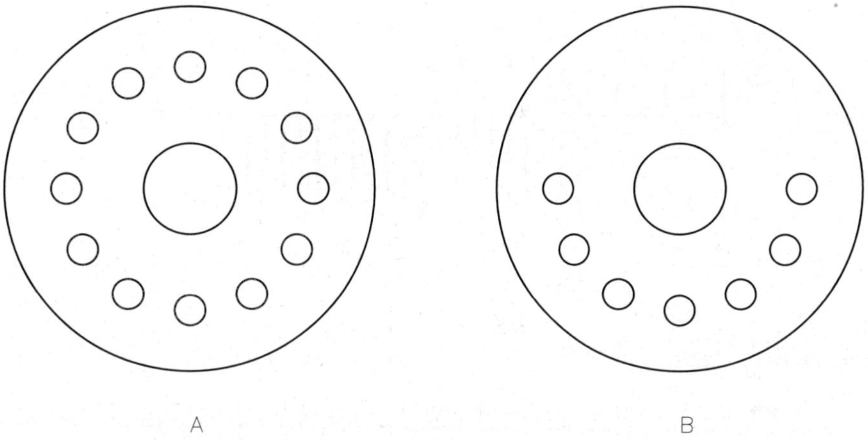


A. Create the ***Polar Array*** as shown in Figure 9-86A. *SaveAs* **ARRAY1.**

B. ***Open*** **ARRAY**. Create the ***Polar Array*** as shown in Figure 9-86B. *SaveAs* **ARRAY2.**

(HINT: Use a negative angle to generate the *Array* in a clockwise direction.)

17. ***Array, Rectangular***

Begin a *New* drawing. Use ***Save*** and assign the name **LIBRARY DESKS**. Create the *Array* of study carrels (desks) for the library as shown in Figure 9-87. The room size is 36′ x 27′ (set ***Units*** and ***Limits*** accordingly). Each carrel is **30″ x 42″**. Design your own chair. Draw the first carrel (highlighted) at the indicated coordinates. Create the ***Rectangular Array*** so that the carrels touch side to side and allow a 6′ aisle for walking between carrels (not including chairs).

**FIGURE 9-87**

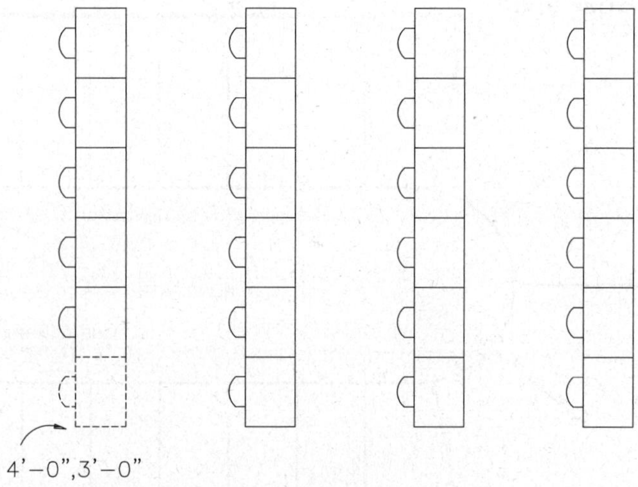

REUSE

18. ***Array, Rectangular***

Create the bolt head with one thread as shown in Figure 9-88A. Create the small line segment at the end (0.40 in length) as a ***Pline*** with a ***Width*** of **.02**. ***Array*** the first thread (both crest and root lines, as indicated in B) to create the schematic thread representation. There is **1** row and **10** columns with **.2** units between each. Add the ***Lines*** around the outside to complete the fastener. ***Erase*** the *Pline* at the small end of the fastener (Fig. 9-88B) and replace it with a ***Line***. ***Save*** as **BOLT.**

**FIGURE 9-88**

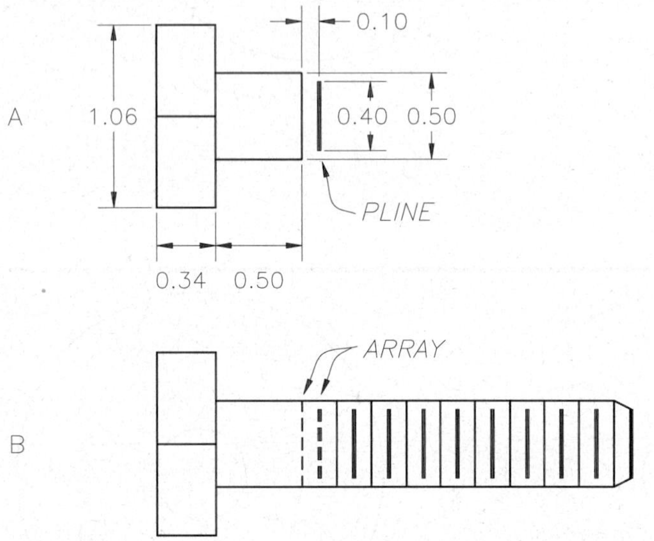


19. ***Array, Path***

Create the parking spaces shown in Figure 9-89. Change ***Units*** to ***Architectural***. Set **Limits** to **80′** x **60′**. Draw a ***Pline*** (path) as shown with a ***Width*** of **3″** composed of two straight line **64′** segments and an arc segment of **12′** radius. Draw another ***Pline*** at **16′** long with a ***Width*** of **3″** at the lower-right end of the path. Make a ***Path Array*** along the path as shown with **21** items at **8′-3 3/8″** between. ***Save*** as **PARK.**

**FIGURE 9-89**

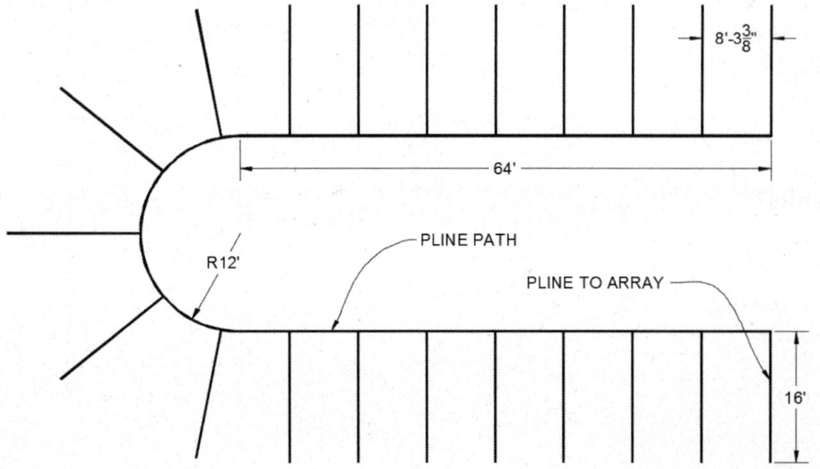

20. ***Fillet***

Create the "T" Plate shown in Figure 9-90. Use ***Fillet*** to create all the fillets and rounds as the last step. When finished, ***Save*** the drawing as **T-PLATE** and make a plot.

**FIGURE 9-90**

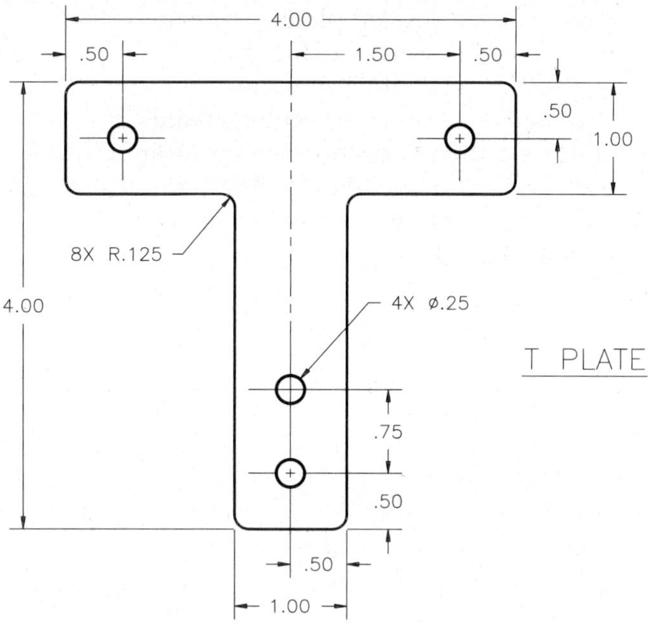


21. ***Copy, Fillet***

Create the Gasket shown in Figure 9-91. Include center-lines in your drawing. ***Save*** the drawing as **GASKETB.**

**FIGURE 9-91**

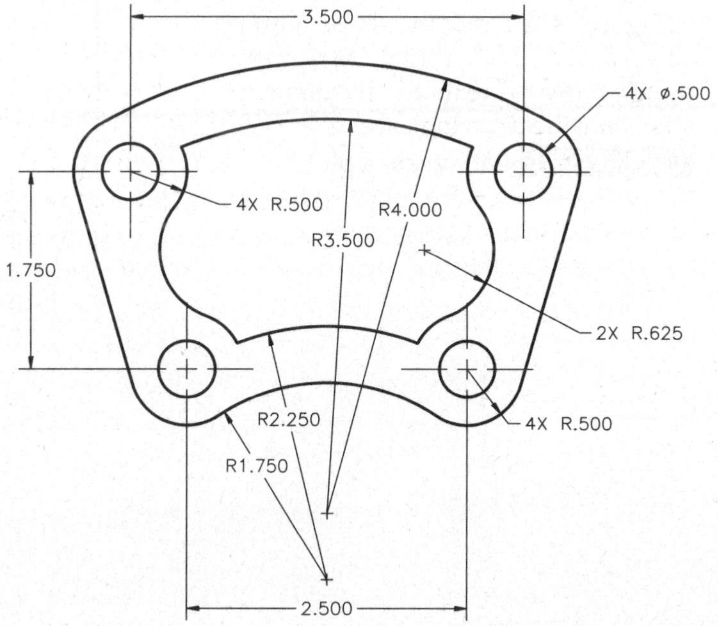

22. *Array*

Create the perforated plate as shown in Figure 9-92. ***Save*** the drawing as **PERFPLAT**. (HINT: For the *Rectangular Array* in the center, create 100 holes, then *Erase* four holes, one at each corner for a total of 96. The two circular hole patterns contain a total of 40 holes each.)

**FIGURE 9-92**

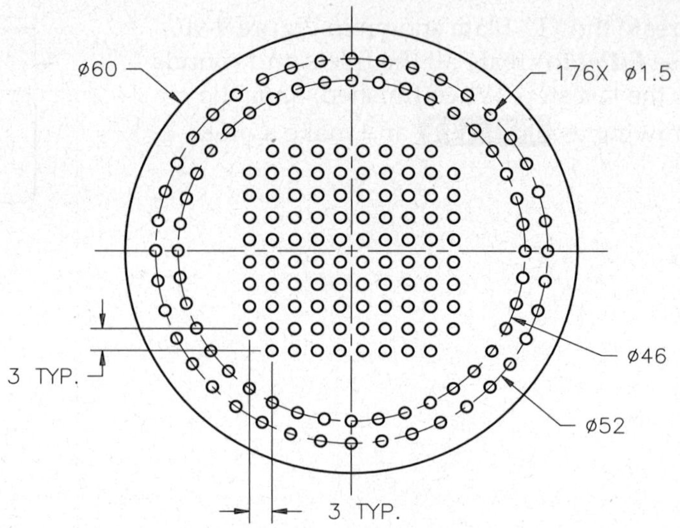


23. *Chamfer*

REUSE

Begin a ***New*** drawing. Set the ***Limits*** to **279,216**. Next, type ***Zoom*** and use the ***All*** option. Set ***Snap*** to **2** and ***Grid*** to **10**. Create the Catch Bracket shown in Figure 9-93. Draw the shape with all vertical and horizontal ***Lines***, then use ***Chamfer*** to create the six chamfers. ***Save*** the drawing as **CBRACKET** and create a plot.

**FIGURE 9-93**

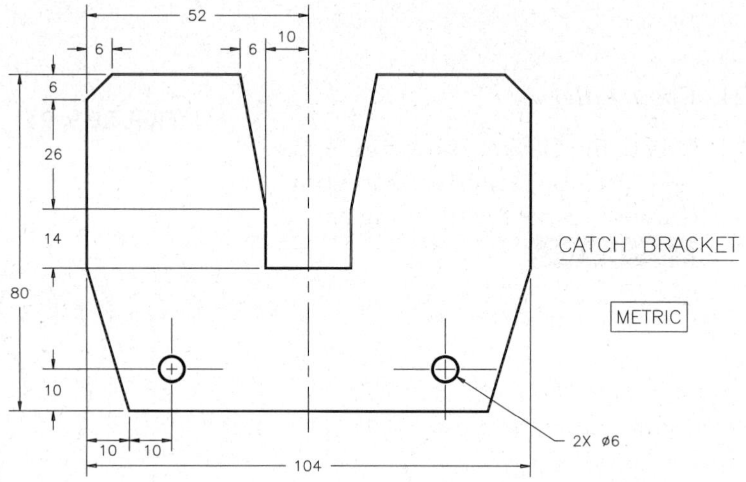

CHAPTER

# 10

# Basic Viewing Commands

## Chapter Objectives

After completing this chapter you should:

1. understand the relationship between the drawing objects and the display of those objects;

2. be able to use all of the *Zoom* options to view areas of a drawing;

3. be able to *Pan* the display about your screen;

4. be able to save and restore *Views*;

5. know how to use *Viewres* to change the display resolution for curved shapes;

6. be able to use *Navswheel* to *Pan* and *Zoom*;

7. be able to turn on and off the *UCS Icon*;

8. be able to create, save, and restore model space viewports using the *Vports* command.

## CONCEPTS

The accepted CAD practice is to draw full size using actual units. Since the drawing is a virtual dimensional replica of the actual object, a drawing could represent a vast area (several hundred feet or even miles) or a small area (only millimeters). The drawing is created full size with the actual units, but any portion can be displayed on the screen. Consider also that CAD systems provide for a very high degree of dimensional precision, which permits the generation of drawings with great detail and accuracy.

TIP

In AutoCAD the commands that facilitate viewing different areas of a drawing are *Zoom, Pan,* and *View*. Surprisingly, the view navigation commands are not available by default on the View tab of the Ribbon in the default AutoCAD *Drafting & Annotation* workspace. There are many quick mouse and keyboard methods you can use to zoom and pan, but when you are learning the software, seeing all of the options on a menu is handy. You can add the *Navigate* panel by right-clicking on the *View* tab, selecting *Show Panels*, and then checking *Navigate* (Fig. 10-1).

**FIGURE 10-1**

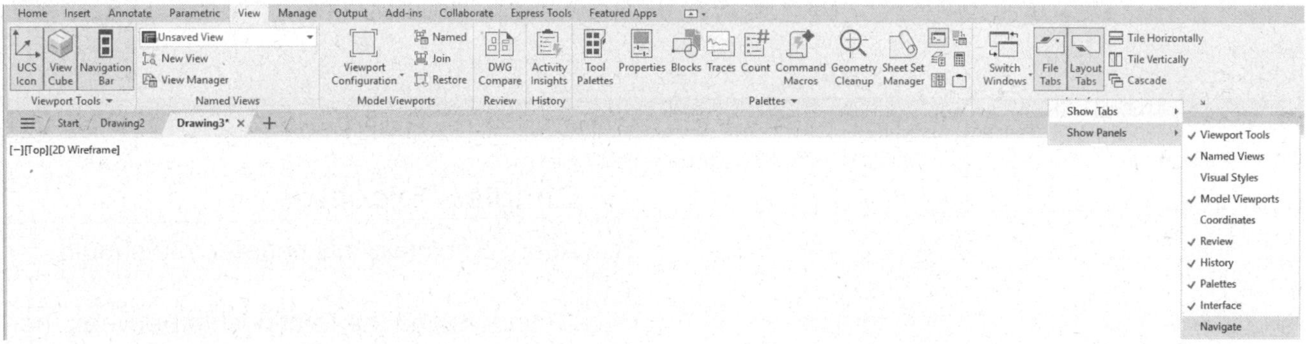


Once the *Navigate* panel is added, the *View* tab shows the Zoom and Pan and other view navigation commands (Fig. 10-2). Note that *Zoom* options are on a drop-down menu.

**FIGURE 10-2**

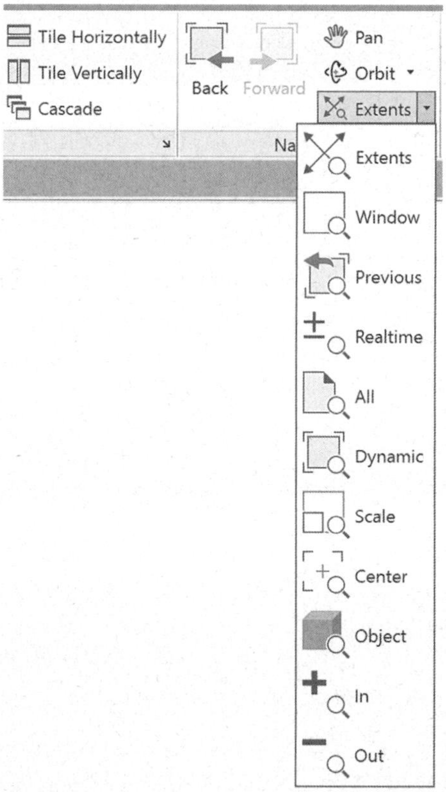


You can also find the viewing commands on the *View* pull-down menu if the Menu Bar is showing. Here the options of the *Zoom* command are located on a fly-out menu. You can show the Menu Bar using the drop-down arrow on the right side of the Quick Access toolbar and selecting *Show Menu Bar* (see Chapter 1 for more information).

Like other commands, the viewing commands can also be invoked by typing the command or command alias.

The commands discussed in this chapter are:

> *Zoom, Pan, View, Dsviewer, Viewres, Ucsicon, Viewgo, Navswheel, Qvdrawing,* and *Vports*

# *ZOOM* AND *PAN* WITH THE MOUSE WHEEL

## *Zoom* and *Pan*

To zoom in means to magnify a small segment of a drawing and to zoom out means to display a larger area of the drawing. Zooming does not change the size of the drawing objects; zooming changes only the display of those objects—the area that is displayed on the screen. All objects in the drawing have the same dimensions before and after zooming. Only your display of the objects changes.

To pan means to move the display area slightly without changing the size of the current view window. Using the pan function, you "drag" the drawing across the screen to display an area outside of the current view in the Drawing Editor.

Although there are many ways to change the area of the drawing you want to view using either the *Zoom* or *Pan* commands, the fastest and easiest method for simple zoom and pan operations is to use the mouse wheel (the small wheel between the two mouse buttons). Using the mouse wheel to zoom or pan does not require you to invoke the *Zoom* or *Pan* commands. Additionally, the mouse wheel zoom and pan functions are transparent, meaning that you can use this method while another command is in use. So, if you have a mouse wheel, you can zoom and pan at any time without using any commands.

## *Zoom* with the Mouse Wheel

If you have a mouse wheel, you can zoom by simply turning the wheel. Turn the wheel forward to zoom in and turn the wheel backward to zoom out. The current location of the cursor is the center for zooming. In other words, if you want to zoom in to an area, simply locate your cursor on the spot and turn the wheel forward. This type of zooming can be done transparently (during another command operation). You can also pan using the wheel (see "*Pan* with the Mouse Wheel or Third Button").

TIP

For example, Figure 10-3 shows a floor plan of a dorm room. Notice the location of the cursor on the left side of the drawing area.

FIGURE 10-3

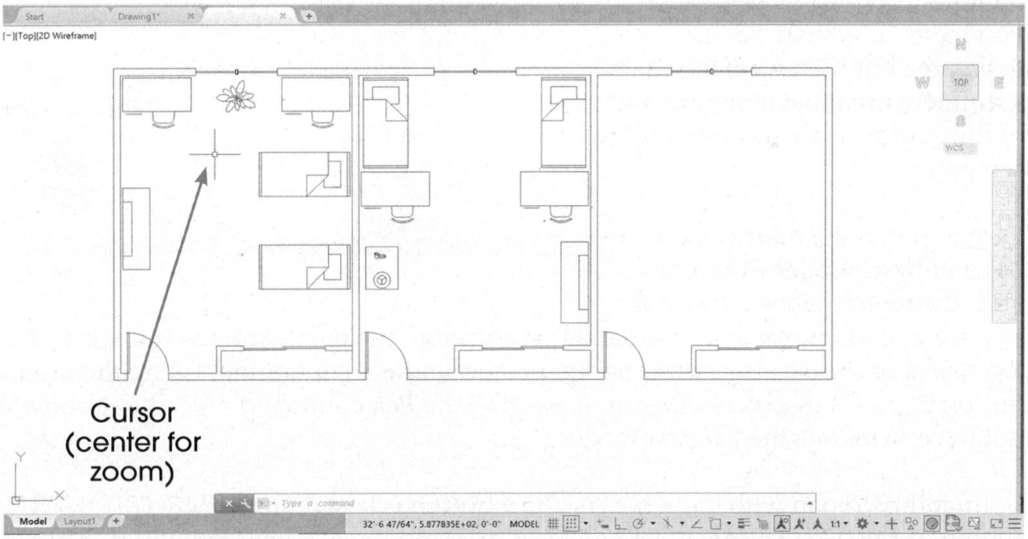

FIGURE 10-4

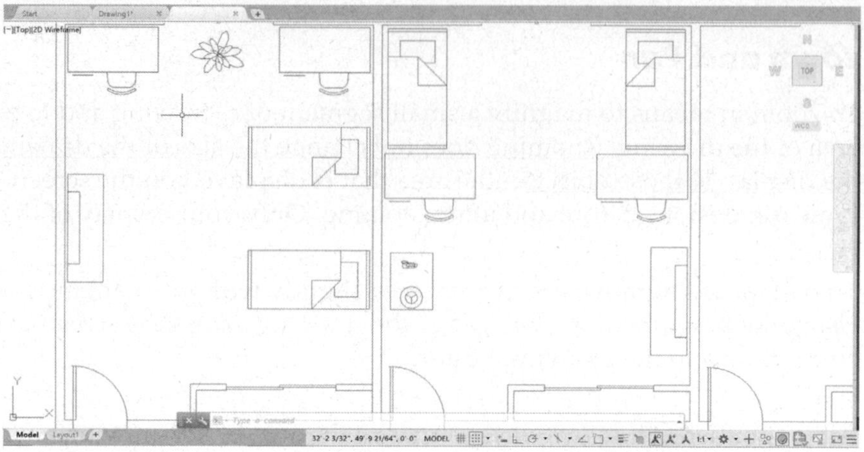

By turning the mouse wheel forward, the display changes by zooming in (enlarging) to the area designated by the cursor location. The resulting display (after turning the mouse wheel forward) is shown in Figure 10-4.

Using the same method, you could zoom out, or change the display from Figure 10-4 to Figure 10-3, by turning the wheel backward. When zooming out, the cursor location also controls the center for zooming.

**TIP**

The amount of zooming in or out that occurs with each turn of the mouse wheel is controlled by the *ZOOMFACTOR* system variable. The default value is 60. Increasing this value causes a greater degree of zooming with each increment of the mouse wheel's forward or backward movement, whereas decreasing the value results in smaller changes in zooming with the wheel movement. The setting for *ZOOMFACTOR* can be between 3 and 100 and is stored with the individual computer (in the Registry).

### ***Pan*** **with the Mouse Wheel**

FIGURE 10-5

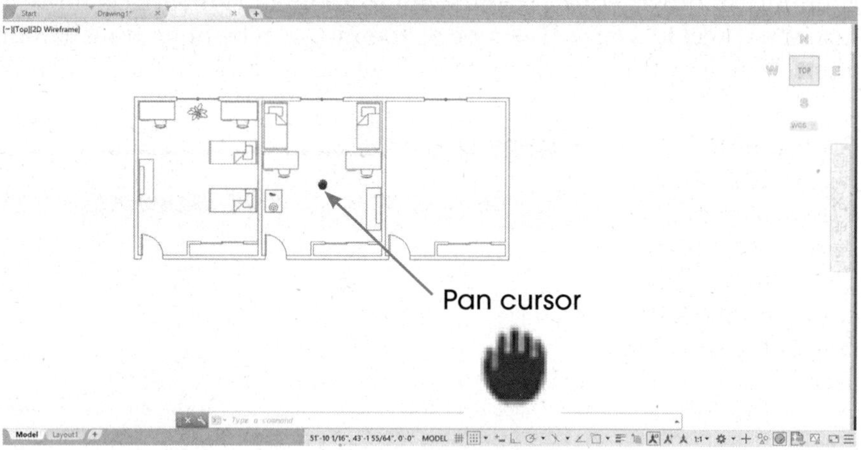


**TIP**

If you have a mouse wheel or third mouse button, you can pan by holding down the wheel or button so the "hand" appears (Fig. 10-5), then move the hand cursor in any direction to "drag" the drawing around on the screen. Panning is typically used when you have previously zoomed in to an area but then want to view a different area that is slightly out of the current viewing area (off the screen).

Panning with the mouse wheel or third button, similar to zooming with the mouse wheel, does not require you to invoke any commands. In addition, the pan feature is transparent, so you can pan with the wheel or third button while another command is in operation. Using the mouse wheel or third button to pan is essentially the same as using the *Pan* command with the *Realtime* option, except you do not have to invoke the *Pan* command.

Your ability to pan with the wheel or third button is based on the *MBUTTONPAN* system variable setting. If *MBUTTONPAN* is set to 1, then pressing the wheel or third button activates the *Realtime Pan* feature. If *MBUTTONPAN* is set to 0, pressing the wheel or button triggers the action defined in the ACAD.CUI file, normally set to activate the *Object Snap* shortcut menu. The *MBUTTONPAN* setting is saved on the individual computer (in the Registry).

## COMMANDS

### *Zoom*

| Ribbon | Menu Bar | Command (Type) | Alias (Type) | Shortcut |
|---|---|---|---|---|
| *View* *Navigate* (option) | *View* *Zoom* > | *Zoom* | Z | (Default Menu) *Zoom* |

The *Zoom* options described next can be typed or selected by any method. Unlike most commands, AutoCAD provides a tool (icon button) for each option of the *Zoom* command. If you type *Zoom,* the options can be invoked by typing the first letter of the desired option or pressing Enter for the *Realtime* option. The Command prompt appears as follows.

Command: ***ZOOM***
Specify corner of window, enter a scale factor (nX or nXP), or [All/Center/Dynamic/Extents/Previous/Scale/Window/Object] <real time>:

#### *Realtime (Rtzoom)*

*Realtime* is the default option of *Zoom*. If you type *Zoom,* just press Enter to activate the *Realtime* option. Alternately, you can type *Rtzoom* to invoke this option directly.

With the *Realtime* option you can interactively zoom in and out with vertical cursor motion. When you activate the *Realtime* option, the cursor changes to a magnifying glass with a plus (+) and a minus (-) symbol. Move the cursor to any location on the drawing area and hold down the PICK (left) mouse button. Move the cursor up to zoom in and down to zoom out (Fig. 10-6). Horizontal movement has no effect. You can zoom in or out repetitively. Pressing Esc or Enter exits *Realtime Zoom* and returns to the Command prompt.

The current drawing window is used to determine the zooming factor. Moving the cursor half the window height (from the center to the top or bottom) zooms in or out to a zoom factor of 100%. Starting at the bottom or top allows zooming in or out (respectively) at a factor of 200%.

If you have zoomed in or out repeatedly, you can reach a zoom limit. In this case the plus (+) or minus (-) symbol disappears when you press the left mouse button. To zoom further with *Rtzoom,* first type *Regen.*

**FIGURE 10-6**

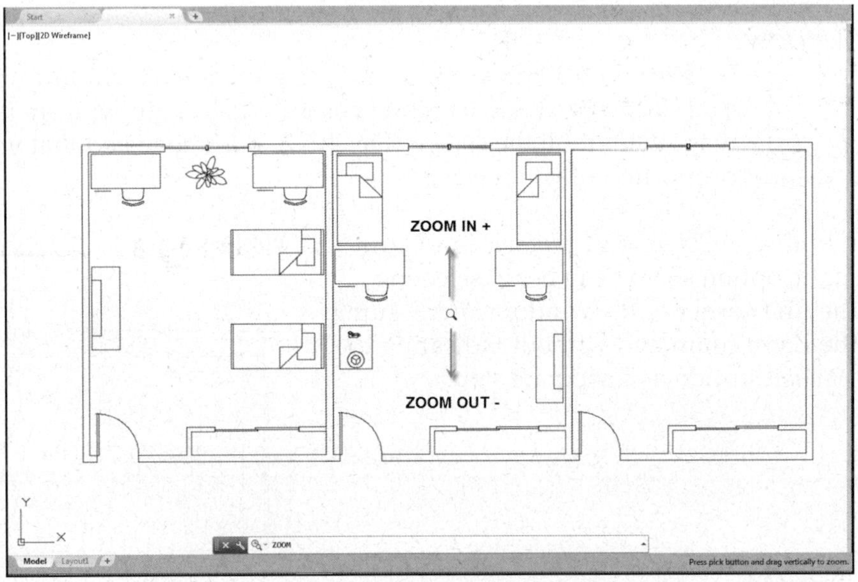

Pressing the right mouse button produces a small cursor menu with other viewing options (Fig. 10-7). This same menu and its options can also be displayed by right-clicking during *Realtime Pan*. The options of the cursor menu are described below.

**FIGURE 10-7**

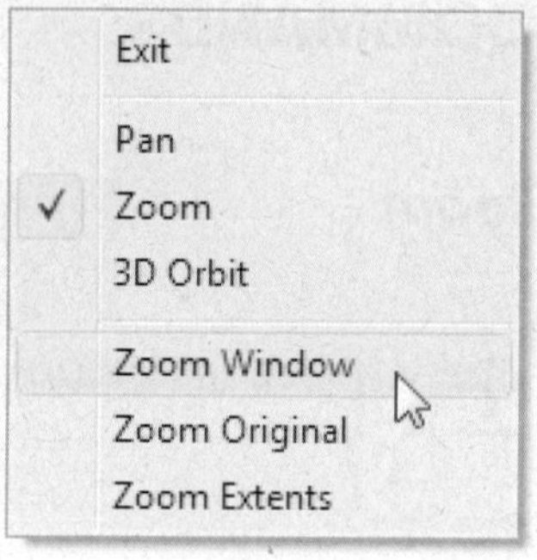


*Exit*
Select *Exit* to exit *Realtime Pan* or *Zoom* and return to the Command prompt. The Escape or Enter key can be used to accomplish the same action.

*Pan*
This selection switches to the *Realtime* option of *Pan*. A check mark appears here if you are currently using *Realtime Pan*. See "*Pan*" next in this chapter.

*Zoom*
A check mark appears here if you are currently using *Realtime Zoom*. If you are using the *Realtime* option of *Pan*, check this option to switch to *Realtime Zoom*.

*Zoom Window*
Select this option if you want to display an area to zoom in to by specifying a *Window*. This feature operates differently but accomplishes the same action as the *Window* option of *Zoom*. With this option, window selection is made with one PICK. In other words, PICK for the first corner, hold down the left button and drag, then release to establish the other corner. With the *Window* option of the *Zoom* command, PICK once for each of two corners. See "*Window*."

*Zoom Original*
Selecting this option automatically displays the area of the drawing that appeared on the screen immediately before using the *Realtime* option of *Zoom* or *Pan*. Using this option successively has no effect on the display. This feature is different from the *Previous* option of the *Zoom* command, in which ten successive previous views can be displayed.

*Zoom Extents*
Use this option to display the entire drawing in its largest possible form in the drawing area. This option is identical to the *Extents* option of the *Zoom* command. See "*Extents*."

### *Window*

To *Zoom* with a *Window* is to draw a rectangular window around the desired viewing area. You PICK a first and a second corner (diagonally) to form the rectangle. The windowed area is magnified to fill the screen (Fig. 10-8). It is suggested that you draw the window to match the proportion of your screen.

If you type *Zoom* or Z, *Window* is an <u>automatic</u> option so you can begin selecting the first corner of the window after issuing the *Zoom* command <u>without</u> indicating the *Window* option as a separate step.

**FIGURE 10-8**

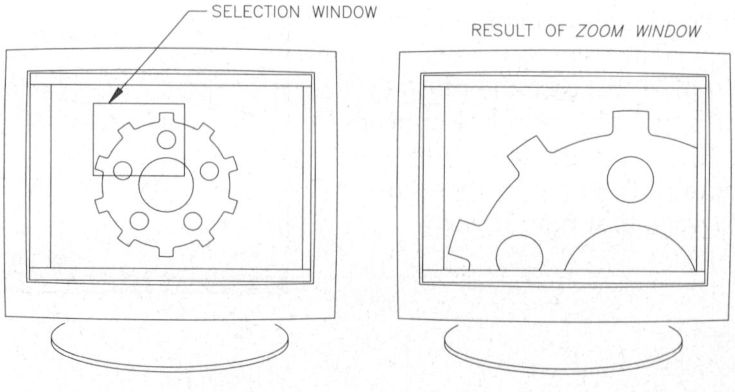

### *All*

This option displays all of the objects in the drawing and all of the *Limits*. In Figure 10-9, notice the effects of *Zoom All*, based on the drawing objects and the drawing *Limits*.

**FIGURE 10-9**

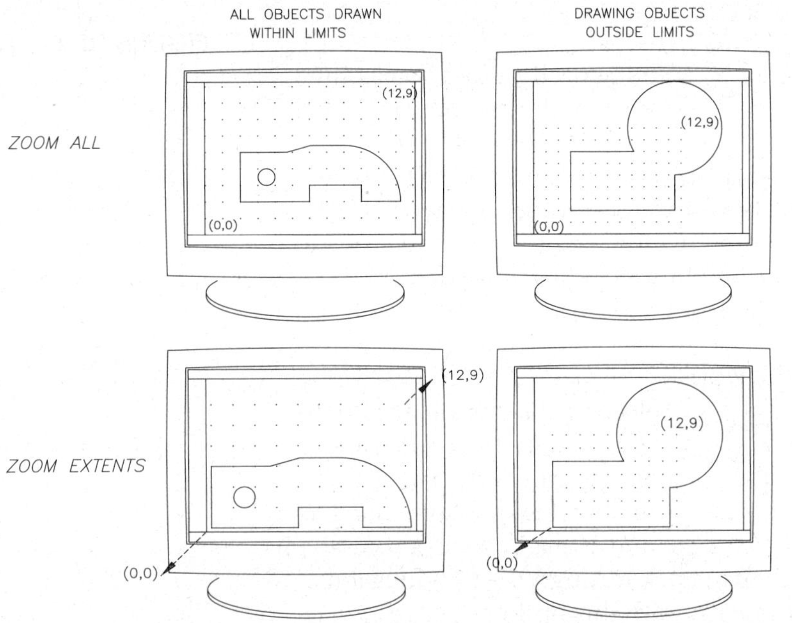


### *Extents*

This option results in the largest possible display of all of the objects, disregarding the *Limits*. (*Zoom All* includes the *Limits*.) See Figure 10-9.

### *Zoom Object*

You can think of this option as "zoom to extents of object." *Zoom Object* allows you to select any object, such as a *Line*, *Circle*, or *Pline*, then AutoCAD automatically zooms to display the entire object as large as possible on the screen.

### *Scale (X/XP)*

This option allows you to enter a scale factor for the desired display.

**FIGURE 10-10**

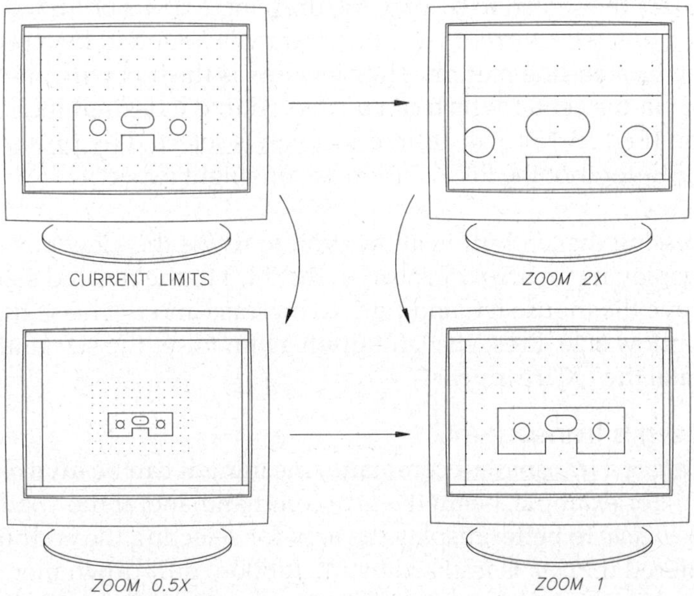


The value entered can be relative to the full view (*Limits*) or to the current display (Fig. 10-10). A value of **1**, **2**, or **.5** causes a display that is 1, 2, or .5 times the size of the *Limits*, centered on the current display. A value of **1X**, **2X**, or **.5X** yields a display 1, 2, or .5 times the size of the current display. If you are in a *Layout* tab (in paper space) and viewing model space, a value of **1XP**, **2XP**, or **.5XP** yields a model space display scaled 1, 2, or .5 times paper space units.

TIP

If you type *Zoom* or *Z*, you can enter a scale factor at the *Zoom* command prompt without having to type the letter *S*.

### *Zoom In*

*Zoom In* magnifies the current display by a factor of 2X (2 times the current display). Using this option is the same as entering a *Zoom Scale* of 2X.

### Zoom Out

*Zoom Out* makes the current display smaller by a factor of .5X (.5 times the current display). Using this option is the same as entering a *Zoom Scale* of .5X.

### Center

First, specify a location as the center of the zoomed area; then specify either a *Magnification factor* (see *Scale X/XP*), a *Height* value for the resulting display, or PICK two points forming a vertical to indicate the height for the resulting display (Fig. 10-11).

**FIGURE 10-11**

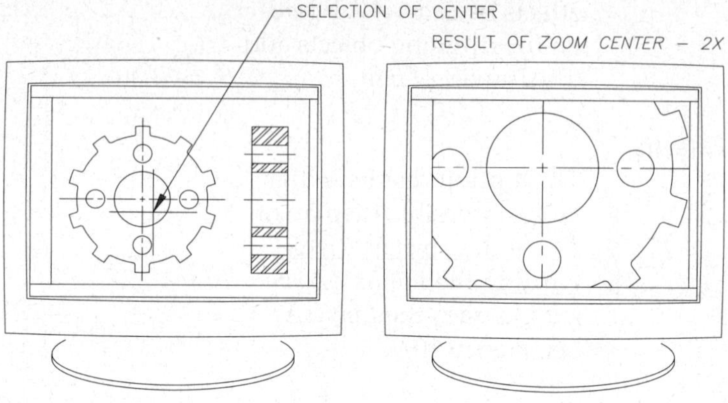


### Previous

Selecting this option automatically changes to the previous display. AutoCAD saves the previous ten displays changed by *Zoom, Pan,* and *View*. You can successively change back through the previous ten displays with this option.

### Dynamic

With this option you can change the display from one windowed area in a drawing to another without using *Zoom All* (to see the entire drawing) as an intermediate step. *Zoom Dynamic* causes the screen to display the drawing *Extents* bounded by a box. The current view or window is bounded by a box in a broken-line pattern. The view box is the box with an "X" in the center which can be moved to the desired location (Fig. 10-12). The desired location is selected by pressing **Enter** (not the PICK button as you might expect).

Pressing the left button allows you to resize the view box (displaying an arrow instead of the "X") to the desired size. Move the mouse left and right to increase and decrease the window size. Press the left button again to set the size and make the "X" reappear.

**FIGURE 10-12**

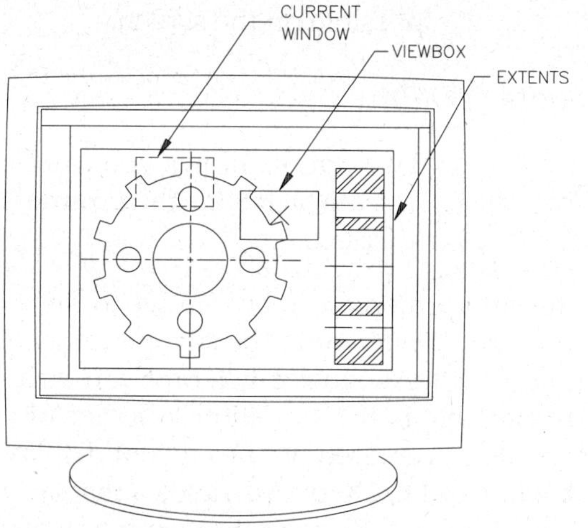


### *Zoom* Is Transparent

*Zoom* is a transparent command, meaning it can be invoked while another command is in operation. You can, for example, begin the *Line* command and at the "Specify next point or [Undo]:" prompt *'Zoom* with a *Window* to better display the area for selecting the endpoint. This transparent feature is automatically entered if *Zoom* is invoked by the Ribbon, pull-down menu, or toolbar icon, but if typed it must be prefixed by the apostrophe (') symbol, e.g., "Specify next point or [Undo]: *'Zoom.*" If *Zoom* has been invoked transparently, the >> symbols appear at the command prompt before the listed options as follows:

```
Command: LINE
Specify first point: 'zoom
>>Specify corner of window, enter a scale factor (nX or nXP), or
  [All/Center/Dynamic/Extents/Previous/Scale/Window/Object] <real time>:
```

## Pan

| Ribbon | Menu Bar | Command (Type) | Alias (Type) | Shortcut |
|---|---|---|---|---|
| *View* *Navigate* *Pan* | *View* *Pan* | *Pan* or *-Pan* | *P* or *-P* | (Default Menu) *Pan* |

Using the *Pan* command by typing or selecting from the tool, screen, or digitizing tablet menu produces the *Realtime* version of *Pan*. The *Point* option is also available from the menus or by typing *-Pan*. The other "automatic" *Pan* options (*Left, Right, Up, Down*) can be selected from the menus.

The *Pan* command is useful if you want to move (pan) the display area slightly without changing the size of the current view window. With *Pan,* you "drag" the drawing across the screen to display an area outside of the current view in the drawing area.

### *Realtime (RTPAN)*

*Realtime* is the default option of *Pan*. It is invoked by selecting *Pan* by any method, including typing *Pan* or *Rtpan*. *Realtime Pan* is essentially the same as using the mouse wheel or third button to pan. That is, *Realtime Pan* allows you to interactively pan by "pulling" the drawing across the screen with the cursor motion. After activating the command, the cursor changes to a hand cursor. Move the hand to any location on the drawing, then hold the PICK (left) mouse button down and drag the drawing around on the screen to achieve the desired view (see Fig. 10-5). When you release the mouse button, panning is discontinued. You can move the hand to another location and pan again without exiting the command.

You must press Escape or Enter or use *Exit* from the shortcut menu to exit *Realtime Pan* and return to the Command prompt. The following Command prompt appears during *Realtime Pan*.

Command: ***PAN***
Press ESC or ENTER to exit, or right-click to display shortcut menu.

If you press the right mouse button, a small cursor menu pops up. This is the same menu that appears during *Realtime Zoom* (see Fig. 10-7). This menu provides options for you to *Exit, Zoom* or *Pan* (realtime), *3Dorbit, Zoom Window, Zoom Original,* or *Zoom Extents*. See "*Zoom*" earlier in this chapter.

### *Point*

The *Point* option is available through the menus or by typing "*-Pan*" (some commands can be typed with a hyphen [-] prefix to invoke the command line version of the command without dialog boxes, etc.). Using this option produces the following Command prompt.

Command: ***-PAN***
Specify base point or displacement: **PICK** or (**value**)
Specify second point: **PICK** or (**value**)
Command:

The "base point:" can be thought of as a "handle," or point to move from, and the "second point:" as the new location to move to (Fig. 10-13). You can PICK each of these points with the cursor.

**FIGURE 10-13**

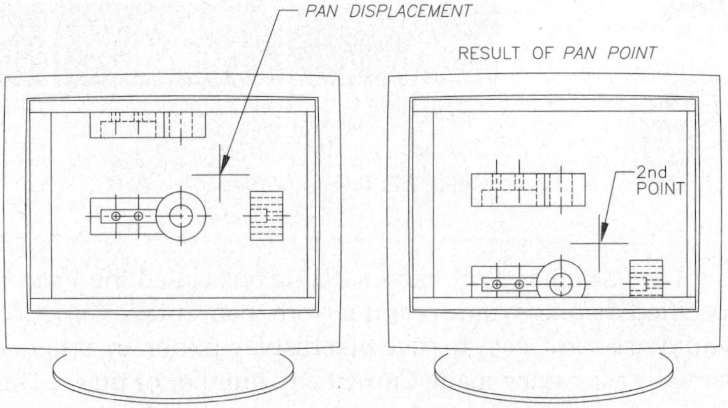

### *Left, Right, Up, Down*

These *Pan* options are available from the pull-down menu only. Each of these options automatically pans to the direction indicated, equal to about 1/8 of the current drawing area width or height.

### *Pan* Is Transparent

*Pan*, like *Zoom*, can be used as a transparent command. The transparent feature is automatically entered if *Pan* is invoked by the screen, pull-down, or tablet menus or icons. However, if you type *Pan* during another command operation, it must be prefixed by the apostrophe (') symbol.

Command: ***LINE***
Specify first point: '***pan***
>>Press ESC or ENTER to exit, or right-click to display shortcut menu. (Pan, then) **Enter**
Resuming LINE command.
Specify next point or [Undo]:

## *Zoom* and *Pan* with *Undo* and *Redo*

By default, consecutive uses of *Zoom* and *Pan* are treated as one command for *Undo* and *Redo* operations. For example, if you *Zoomed* three times consecutively but then wanted to return to the original display, you would have to *Undo* only once. You can change this setting (*Combine zoom and pan commands*) in the lower-right corner of the *User Preferences* tab of the *Options* dialog box.

## Viewback Viewforward

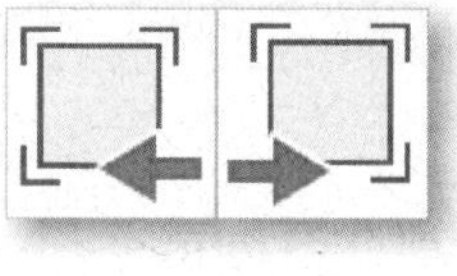

| Ribbon | Menu Bar | Command (Type) | Alias (Type) | Shortcut |
|---|---|---|---|---|
| *View*<br>*Navigate*<br>*Viewback* or<br>*Viewforward* | ... | *Viewback* or<br>*Viewforward* | ... | ... |

The *Viewback* command is similar to *Zoom Previous* in that it displays the previous view (previously *Zoom*ed or *Pan*ned area). Use *Viewforward* to restore the original viewed area (displayed before using *Viewback*). *Zoom Previous* cannot perform the same function as *Viewforward* since it saves ten previous views. Likewise, you cannot always use *Undo* after *Zoom* for this function because multiple *Zooms* and *Pans* are combined into one *Undo* (see previous discussion).

## View

| Ribbon | Menu Bar | Command (Type) | Alias (Type) | Shortcut |
|---|---|---|---|---|
| *View*<br>*Views*<br>*View Manager* | *View*<br>*Named Views...* | *View* or<br>*-View* | *V* or *-V* | ... |

The *View* command provides a dialog box called the *View Manager* for you to create *New* views of a specified display window and restore them (make *Current*) at a later time. For typical applications you would use *View, New* to save that display under an assigned name. Later in the drawing session, the named *View* can be made *Current* any number of times. This method saves time compared to *Zooming* in and out to display the same areas repeatedly. Making a named *View* the *Current* view requires no regeneration time. It also lets you easily return to the exact view when that is helpful.

Using *View* invokes the *View Manager* (Fig. 10-14) and *-View* produces the Command line format. The *View Manager* is available on the Ribbon by right-clicking on the *View* tab, selecting *Show Panels*, and selecting *Views*.

The left side of the View Manager lists the saved views contained in the current drawing. The central area gives several input fields that allow you to change Camera and Target settings, Perspective, Field of view, and Clipping plane settings. These features are used almost exclusively for 3D views (see Chapter 33, 3D Basics, Navigation, and Visual Styles).

**FIGURE 10-14**

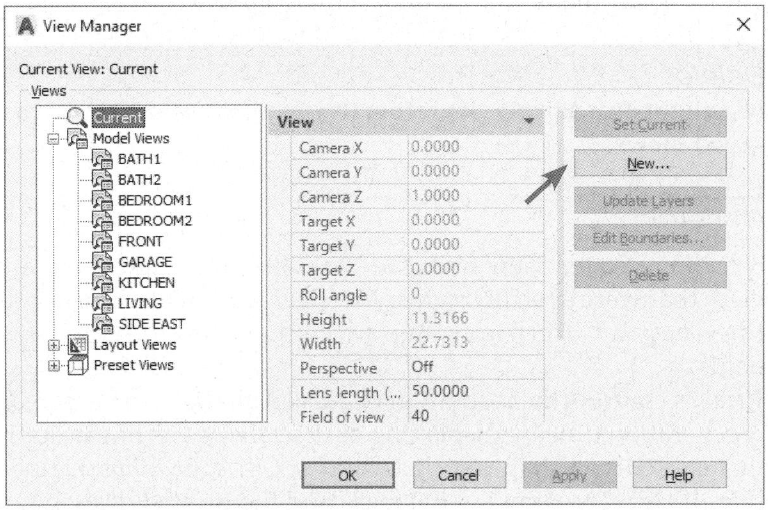


### New

Using the *New* button in the *View Manager* produces the *New View/Shot Properties* dialog box (Fig. 10-15) that you should use to create and specify options for a new named view. Enter a *View Name* and (optional) *View Category*, a description field (such as "Elevation" or "Plan").

The *Boundary* section allows you to specify the area of the drawing you want to display when the view is restored. If already zoomed into the desired view area, select *Current Display*. Otherwise, select *Define window* or press the *Define View Window* button to return to the drawing to specify a window.

**FIGURE 10-15**

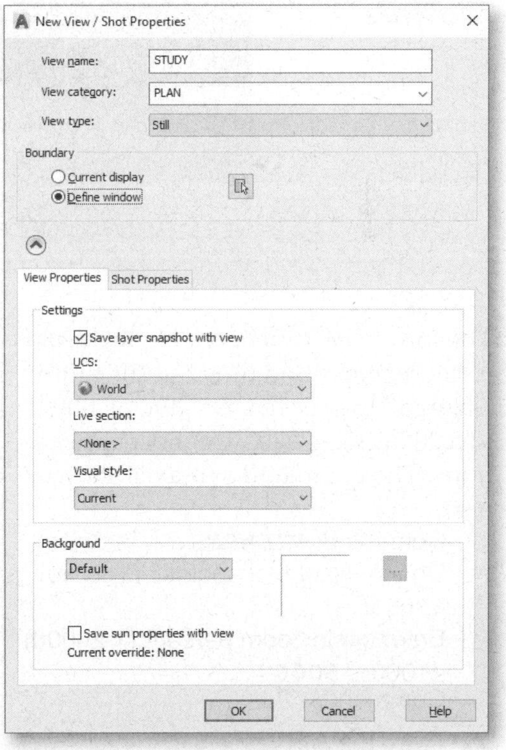


Under *Settings* you can choose to *Save layer snapshot with view* so that when the view is later restored the layer visibility settings return to those when the view was defined or later updated using the *Update Layers* button. In this way, when the view is restored, AutoCAD resets the layer visibility back to when the view was created no matter what the current visibility settings may be at the time you restore the view. (Layers are discussed in detail in Chapter 11.) The *Background* options (*Default*, *Solid*, *Gradient*, *Image*, and *Sun & Sky*) are used primarily for 3D applications (see Chapters 33, 35, and 36).

### Set Current

To *Restore* a named view, double-click on the name from the list or highlight the name and select *Set Current*, then press *OK* (see Figure 10-14). This action makes the selected view current in the drawing area. If the view you saved was in another *Layout tab* or *Model tab*, then using *Set Current* will return to the correct layout and display the desired view.

### Update Layers

If the layer settings were stored when the view was created (the *Save layer snapshot with view* option was used) and you want to reset the layer visibility for the view, use this option. Pressing this button causes the layer visibility information saved with the selected named view to update to the current settings (match the layer visibility in the current model space or layout viewport).

### Edit Boundaries

The *Edit Boundaries* button returns to the drawing with the selected view highlighted. With this unique feature, the screen displays the current view (boundary) in the central area of the screen with the normal background color and the additional area surrounding the old view boundary in a grayed-out representation. Select the new boundary with a window.

### Delete

Highlight any named view from the list, then select *Delete* to remove the named view from the list of saved views.

### General

Selecting a saved view from the *Views* list on the left changes the central area to list the features of the view that were specified when the view was created (using the *New* option), such as *Name, Category,* and *Layer snapshot*. You can change any of these specifications by entering the new information in the input fields.

*Viewres* controls the resolution of curved shapes for the screen display only. Its purpose is to speed regeneration time by displaying curved shapes as linear approximations of curves; that is, a curved shape (such as an *Arc, Circle,* or *Ellipse*) appears as several short, straight line segments (Fig. 10-16). The drawing database and the plotted drawing, however, always define a true curve. The range of *Viewres* is from 1 to 20,000 with 1000 being the default. The higher the value (called "circle zoom percentage"), the more accurate the display of curves and the slower the regeneration time.

## Viewres

| Ribbon | Menu Bar | Command (Type) | Alias (Type) | Shortcut |
|---|---|---|---|---|
| ... | *Tools*<br>*Options...*<br>*Display*<br>*Display Resolution* | *Viewres* | ... | ... |

FIGURE 10-16

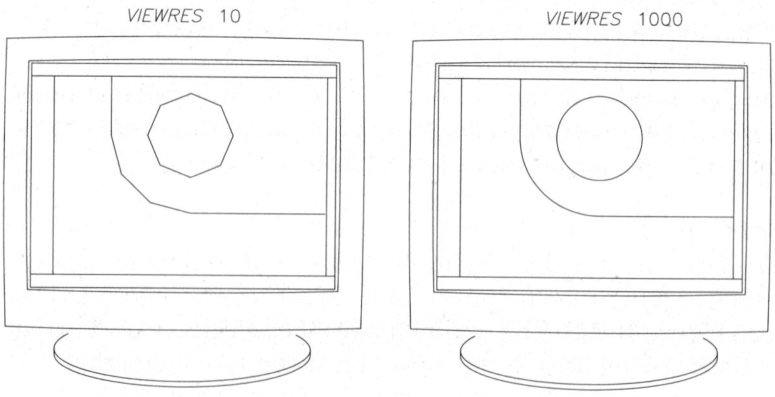


The lower the *Viewres* value, the faster is the regeneration time but the more "jagged" the curves. A value of 1,000 to 5,000 is suggested for most applications. The command syntax is shown next.

Command: ***VIEWRES***
Do you want fast zooms? [Yes/No] <Y>: ***Y***
Enter circle zoom percent (1-20000) <1000>: **5000**
Command:

Fast zooms are no longer a functioning part of this command, so your response to "Do you want fast zooms?" is irrelevant. The prompt is kept only for script compatibility.

## Ucsicon

| Ribbon | Menu Bar | Command (Type) | Alias (Type) | Shortcut |
|---|---|---|---|---|
| *View Viewport Tools* | *View Display > UCS Icon >* | *Ucsicon* | ... | ... |

The icon that may appear in the lower-left corner of the AutoCAD Drawing Editor is the Coordinate System Icon (Fig. 10-17). The icon is sometimes called the "UCS icon" because it automatically orients itself to the new location of a coordinate system that you create, called a "UCS" (User Coordinate System).

The *Ucsicon* command controls the appearance and positioning of the Coordinate System Icon. It is important to display the icon when working with 3D drawings so that you can more easily visualize the orientation of the X, Y, and Z coordinate system. However, when you are working with 2D drawings, it is not necessary to display this icon. (By now you are accustomed to normal orientation of the positive X and Y axes—X positive is to the right and Y positive is up.)

**FIGURE 10-17**

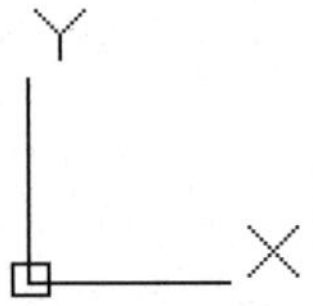


Command: ***UCSICON***
Enter an option [ON/OFF/All/Noorigin/ORigin/Selectable/Properties]
<OFF>: ON

### *ON/OFF*

Use the *ON* and *OFF* options to turn the display of the *Uscicon* on and off. For example, if you draw primarily in 2D, you may want to turn the icon off since it is not necessary to constantly indicate the origin and the XY directions.

NOTE: The setting for the *Display UCS Icon* section of the *Options* dialog box, *3D Modeling* tab, <u>overrides</u> the *ON* and *OFF* options of the *Ucsicon* command. In other words, if the check is removed for *2D Wireframe visual style* in the dialog box (Fig. 10-18), the icon is permanently off and cannot be turned on by using the *Ucsicon* command.

See later chapters for more information and 3D applications of the UCS icon.

**FIGURE 10-18**

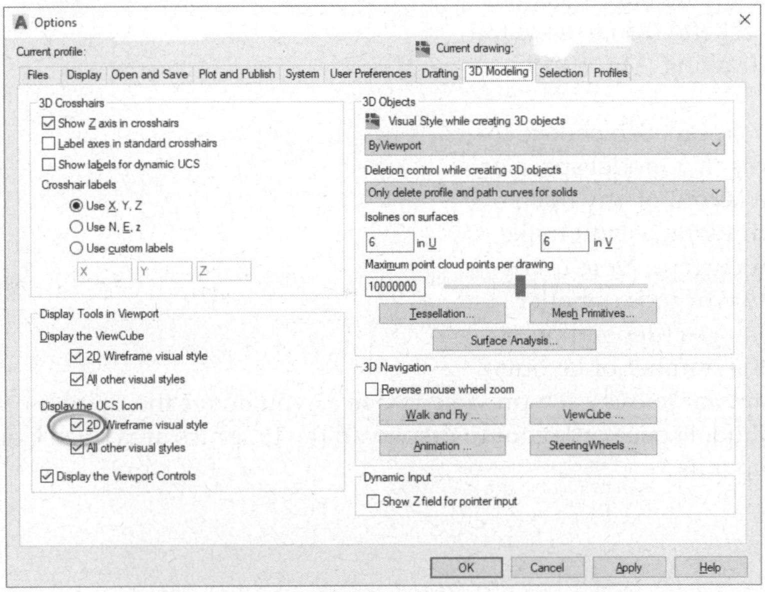

## Viewgo

| Ribbon | Menu Bar | Command (Type) | Alias (Type) | Shortcut |
|---|---|---|---|---|
| ... | ... | *Viewgo* | *VGO* | ... |

*Viewgo* is a command you can use to quickly restore any named view without having to use the *View Manager*. If a layer state, UCS, live section, visual style, or background setting was saved with the view, the settings are also restored.

Command: ***VIEWGO***
Enter view name to go:

## Qvdrawing

| Ribbon | Menu Bar | Command (Type) | Alias (Type) | Shortcut |
|---|---|---|---|---|
| ... | ... | *Qvdrawing* | *QVD* | ... |

*Qvdrawing*, or Quick View Drawings, displays preview images of all drawings currently open in the AutoCAD session. A two-level structure appears at the bottom of the AutoCAD window—the first level displays the images of open drawings and the second level displays the images of model space and layouts for each drawing.

**FIGURE 10-19**

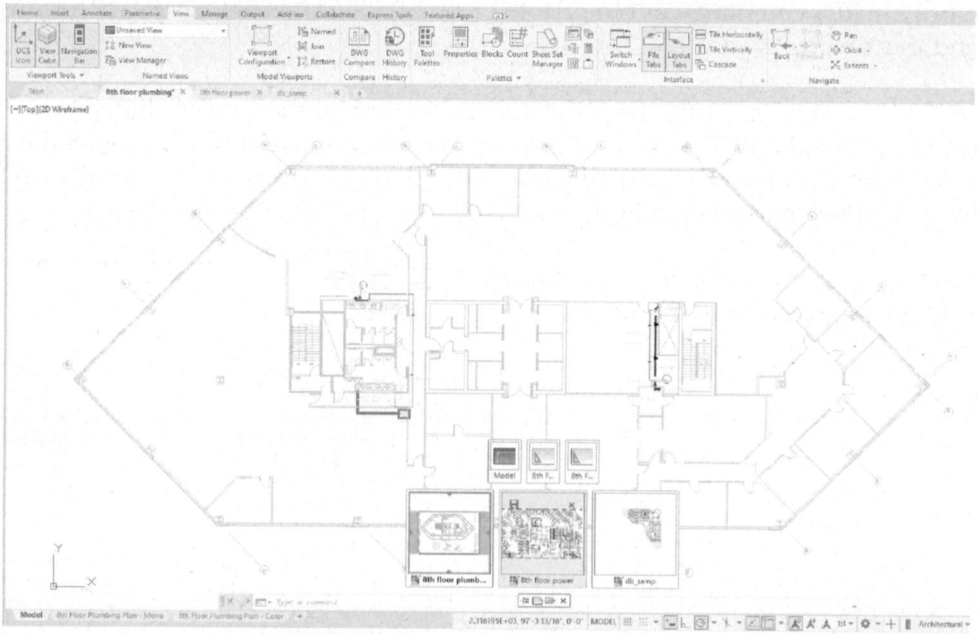

Activate the command by typing *Qvdrawing* at the Command prompt. Select one of the preview images to make the drawing the current drawing displayed in the graphics area. Select the "X" in the preview image to close the drawing and select the diskette icon to save the drawing (Fig. 10-19).

You can also choose to display model space or a layout of any open drawing as you make it current. Note that two or more smaller images (depending on the number of layouts) appear above each preview image as you hover the pointer on the drawing image. Select the desired model space or layout to display in the graphics area. (See Chapters 13 and 32 for more information on layouts.)

## *Vports*

| Ribbon | Menu Bar | Command (Type) | Alias (Type) | Shortcut |
|---|---|---|---|---|
| *View* *Model Viewports* *Named* | *View* *Viewports >* | *Vports* or *-Vports* | ... | ... |

A "viewport" is one of several simultaneous views of a drawing on the screen. The *Vports* command invokes the *Viewports* dialog box. With this dialog box you can create model space viewports or paper space viewports, depending on which space is current when you invoke *Vports*. *Model* tab (model space) viewports are sometimes called "tiled" viewports because they fit together like tiles with no space between.

*Layout* tab (paper space) viewports can be any shape and configuration and are used mainly for setting up several views on a sheet for plotting. Only an introduction to *Model* tab viewports is given in this chapter. Using model space viewports for constructing 3D models is discussed in Chapter 33. *Layout* tab (paper space) viewports are discussed in Chapters 13 and 32, and using paper space viewports for 3D applications is discussed in Chapter 37.

If the *Model* tab is active and you use the *Vports* command, you will create model space viewports. Several viewport configurations are available. *Vports* in this case simply divides the screen into multiple sections, allowing you to view different parts of a drawing. Seeing several views of the same drawing simultaneously can make construction and editing of complex drawings more efficient than repeatedly using other display commands to view detailed areas. Model space *Vports* affect only the screen display. The viewport configuration cannot be plotted. If the *Plot* command is used from Model space, only the current viewport is plotted.

TIP

**FIGURE 10-20**

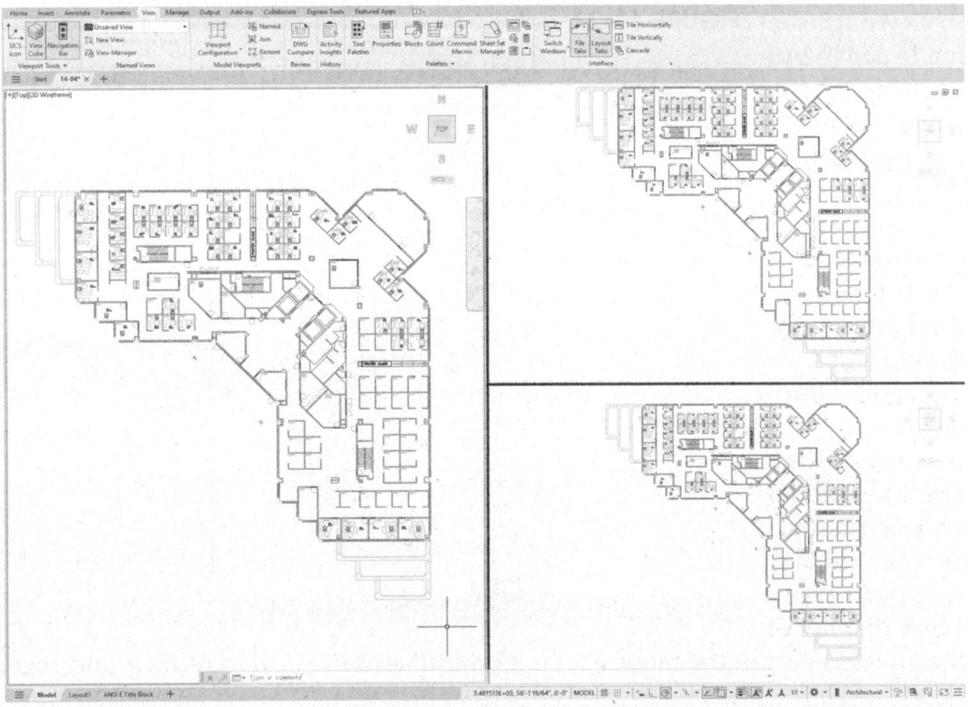

Figure 10-20 displays the AutoCAD drawing editor after the *Vports* command was used to divide the screen into viewports. Tiled viewports always fit together like tiles with no space between.

Once multiple model space viewports have been created, you can adjust the size of the viewports. Simply move the cursor across the viewports until the "crosshairs" change to the "arrow," then use the arrow pointer to click and drag a vertical or horizontal viewport border to change its size. You can eliminate a viewport by dragging its border all the way to the edge of the graphics area, or switch to paperspace and click its border and erase it. If you don't see the viewport border, check that the viewport is not on a frozen layer.

The *View* tab or *View* pull-down menu can be used to display the *Viewports* dialog box (Fig. 10-21). The dialog box allows you to PICK the configuration of the viewports that you want. If you are working on a 2D drawing, make sure you select *2D* in the *Setup* box.

After you select the desired layout, the previous display appears in each of the viewports. For example, if a full view of the office layout is displayed when you use the *Vports* command, the resulting display in each viewport would be the same full view of the office (see Fig. 10-20). It is up to you then to use viewing commands (*Zoom*, *Pan*, etc.) in each viewport to specify what areas of the drawing you want to see in each viewport. There is no automatic viewpoint configuration option for *Vports* for 2D drawings.

**FIGURE 10-21**

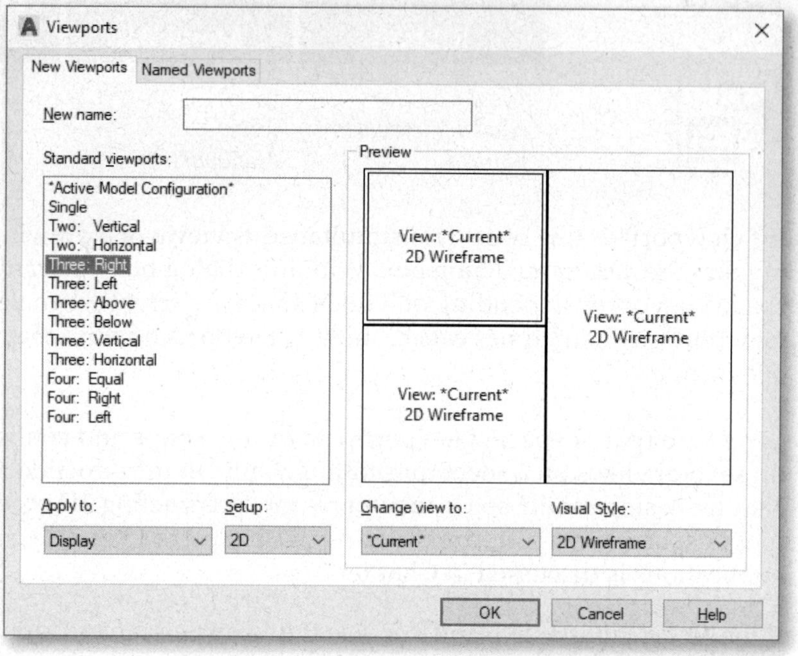


A popular arrangement of viewpoints for construction and editing of 2D drawings is a combination of an overall view and one or two *Zoomed* views (Fig. 10-22). You cannot draw or project from one viewport to another. Keep in mind that there is only one model (drawing) but several views of it on the screen. The active or current viewport displays the cursor, while moving the pointing device to another viewport displays only the pointer in that viewport.

**FIGURE 10-22**

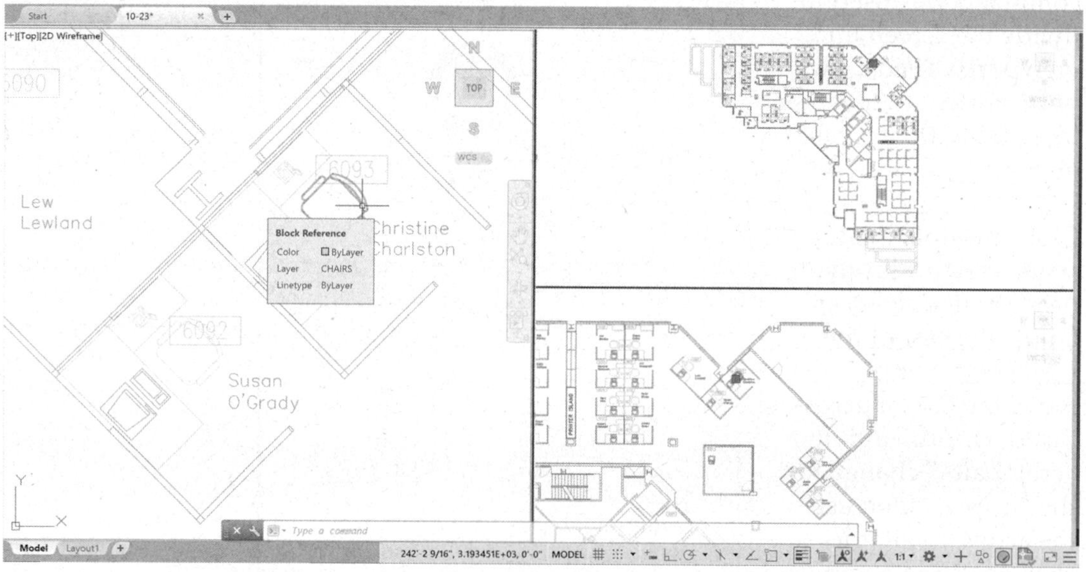


A viewport is made active by PICKing in it. Any display commands (*Zoom*, *Pan*, etc.) and some drawing aids (*Snap*, *Grid*) used affect only the current viewport. Draw and modify commands that affect the model are potentially apparent in all viewports (for every display of the affected part of the model). Use *Redrawall* and *Regenall* to redraw and regenerate all viewports.

You can begin a drawing command in one viewport and finish in another. In other words, you can toggle viewports within a command. For example, you can use the *Line* command to PICK the "first point:" in one viewport, then make another viewport current to PICK the "next point:".

You can save a viewport configuration (including the views you prepare) by entering a name in the *New Name* box. If you save several viewport configurations for a drawing, you can then select from the list in the *Named Viewports* tab. You can type *-Vports* to use the Command line version. The format is as follows:

Command: ***-VPORTS***
Enter an option [Save/Restore/Delete/Join/SIngle/?/2/3/4/Toggle/MOde] <3>:

### *Save*
Allows you to assign a name and save the current viewport configuration.

### *Restore*
Redisplays a previously *Saved* viewport configuration.

### *Delete*
*Deletes* a named viewport configuration.

### *Join*
This option allows you to combine (join) two adjacent viewports. The viewports to join must share a common edge the full length of each viewport. For example, if four equal viewports were displayed, two adjacent viewports could be joined to produce a total of three viewports.

### *SIngle*
Changes back to a single screen display using the current viewport's display.

### *2, 3, 4*
Use these options to create 2, 3, or 4 viewports. You can choose the configuration. The possibilities are illustrated if you use the *Viewports* dialog box.

## CHAPTER EXERCISES

1. ***Zoom Extents, Realtime, Window, Previous, Center***

   Open the **Floor Plan Sample.dwg** drawing from the AutoCAD Sample drawings in the "Database Connectivity" folder. The drawing shows an office building layout (Fig. 10-23). Use *SaveAs* and rename the drawing to **ZOOM TEST** and locate it in your working folder (directory). (When you view sample drawings, it is a good idea to copy them to another name so you do not accidentally change the original drawings.)

**FIGURE 10-23**

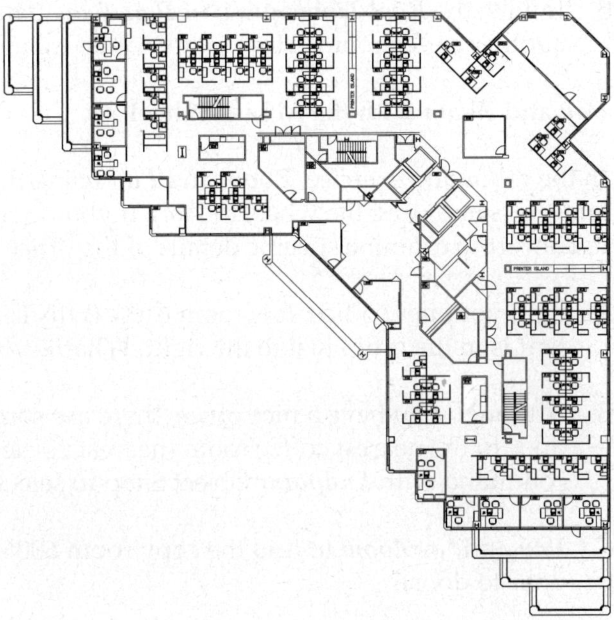

FIGURE 10-24

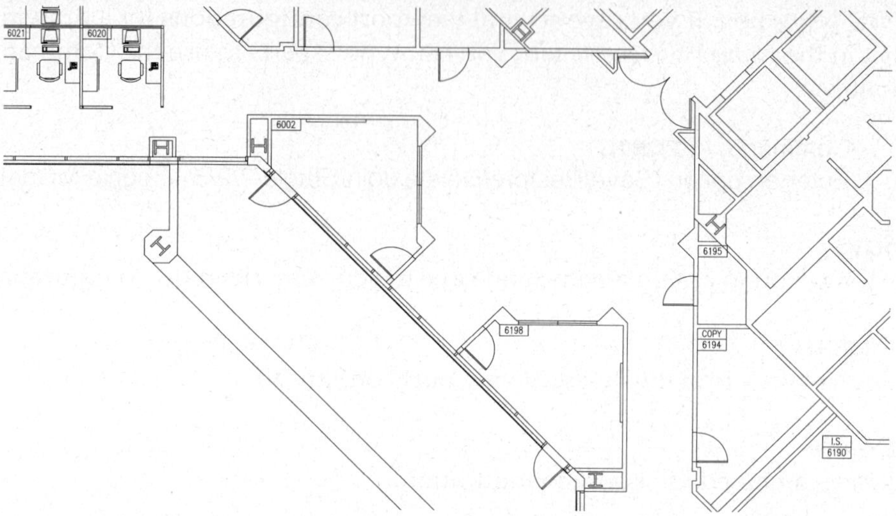


A. First, use *Zoom Extents* to display the office drawing as large as possible on your screen. Next, use *Zoom Realtime* and zoom in to the center of the layout (the center of your screen). You should be able to see rooms 6002 and 6198 located against the exterior wall (Fig. 10-24). Right-click and use *Zoom Extents* from the cursor pop-up menu to display the entire layout again. Press **Esc**, **Enter**, or right-click and select ***Exit*** to exit *Realtime*.

B. Use ***Zoom Window*** to closely examine room 6050 at the top left of the building. Whose office is it? What items on the desk are magenta in color? What item is cyan (light blue)?

C. Use ***Zoom Center***. **PICK** the telephone on the desk next to the computer as the center. Specify a height of **24** (2′). You should see a display showing only the telephone. How many keys are on the telephone's number pad (*Point* objects)?

D. Next, use ***Zoom Previous*** repeatedly until you see the same display of rooms 6002 and 6198 as before (Fig. 10-24). Use ***Zoom Previous*** repeatedly until you see the office drawing *Extents*.

2. ***Pan Realtime***

A. Using the same drawing as exercise 1 (ZOOM TEST.DWG, originally Floor Plan Sample.dwg), use ***Zoom Extents*** to ensure you can view the entire drawing. Next, ***Zoom*** with a ***Window*** to an area about 1/4 the size of the building.

B. Invoke the ***Real Time*** option of ***Pan***. *Pan* about the drawing to find a coffee room. Can you find another coffee room? How many coffee rooms are there?

3. ***Pan* and *Zoom* with the Mouse Wheel**

In the following exercise, ***Zoom*** and ***Pan*** using the wheel on your mouse (turn the wheel to *Zoom* and press and drag the wheel to *Pan*). If you do not have this capability, use ***Realtime Pan*** and ***Zoom*** in concert to examine specific details of the office layout.

A. Find your new office. It is room 6100. (HINT: It is centrally located in the building.) Your assistant is in the office just to the right. What is your assistant's name and room number?

B. Although you have a nice office, there are some disadvantages. How far is it from your office door to the nearest coffee room (nearest corner of the coffee room)? HINT: Use the ***Dist*** command with ***Endpoint*** Object Snap to select the two nearest doors.

C. ***Pan*** and/or ***Zoom*** to find the copy room 6006. How far is it to the copy room (direct distance, door to door)?

D. Perform a ***Zoom Extents***. Use the wheel to zoom in to Kathy Ragerie's office (room 6150) in the lower-right corner of the building. Remember to locate the cursor at the spot in the drawing that you want to zoom in to. If you need to pan, hold down the wheel and "drag" the drawing across the screen until you can view all of room 6150 clearly. How many chairs are in Kathy's office? Finally, use ***Zoom Extents*** to size the drawing to the screen.

4. ***Pan Point***

A. ***Zoom*** in to your new office (room 6100) so that you can see the entire room and room number. Assume you were listening to music on your computer with the door open and calculated it could be heard about 100′ away. Naturally, you are concerned about not bothering the company CEO who has an office in room 6048. Use ***Pan Point*** to see if the CEO's office is within listening range. (HINT: enter **100′<0** at the "Specify base point or displacement:" prompt.) Should you turn down the system?

B. Do not ***Exit*** the **ZOOM TEST** drawing.

5. ***Viewres, Zoom Dynamic***

A. Open the sample drawing named **TABLET.DWG**. This drawing is supplied on the website for this text (the website address for this text is listed on the back cover and in the Preface under "Online Resources"). Use ***SaveAs***, name the new drawing **TABLET TEST**, and locate it in your working folder. Use ***Zoom*** with the ***Window*** option (or you can try *Realtime Zoom*). Zoom in to the first several command icons located in the middle left section of the tablet menu. Your display should reveal icons for *Zoom* and other viewing commands. ***Pan*** or ***Zoom*** in or out until you see the *Zoom* icons clearly.

B. Notice how the *Zoom* command icons are shown as polygons instead of circles. Use the ***Viewres*** command and change the value to **5000**. Do the icons now appear as circles? Use ***Zoom Realtime*** again. Does the new setting change the speed of zooming?

C. Use ***Zoom Dynamic***, then try moving the view box immediately. Can you move the box before the drawing is completely regenerated? Change the view box size so it is approximately equal to the size of one icon. Zoom in to the command in the lower-right corner of the tablet menu. What is the command? What is the command in the lower-left corner? For a drawing of this complexity and file size, which option of *Zoom* is faster and easier for your system—*Realtime* or *Dynamic*?

D. If you wish, you can ***Exit*** the **TABLET TEST** drawing.

6. ***View***

A. Make the **ZOOM TEST** drawing that you used in previous exercises the current drawing. ***Zoom*** in to the office that you will be moving into (room 6100). Make sure you can see the entire office. Use the ***View*** command and create a ***New*** view named **6100**. Next, ***Zoom*** or ***Pan*** to room 6048. ***Save*** the display as a *View* named **6048**.

B. Use the ***View*** dialog box again and make view **6100** ***Current***. In order for the CEO to access information on your new computer, a cable must be stretched from your office to room 6048. Find out what length of cable is needed to connect the upper left corner of office 6100 to the upper right corner of office 6048. (HINT: Use ***Dist*** to determine the distance. Type the ***'-View*** command transparently [prefix with an apostrophe and hyphen] during the *Dist* command and use Running ***Endpoint*** Object Snap to select the two indicated office corners. The *View* dialog box is not transparent.) What length of cable is needed? Do not *Exit* the drawing.

7. *Zoom All, Extents*

A. Begin a ***New*** drawing using the **ACAD.DWT**. ***Zoom All***. Right-click on the ***Grid*** button on the Status bar, select ***Settings***, and remove all checks from the *Grid behavior* section in the *Drafting Settings* dialog box. Turn on the ***Snap*** (**F9**) and ***Grid*** (**F7**). Draw two ***Circles***, each with a **1.5** unit ***radius***. The *Circle* centers are at **3,5** and at **5,5**. See Figure 10-25.

B. Use ***Zoom All***. Does the display change? Now use ***Zoom Extents***. What happens? Now use ***Zoom All*** again. Which option always shows all of the *Limits*?

**FIGURE 10-25**

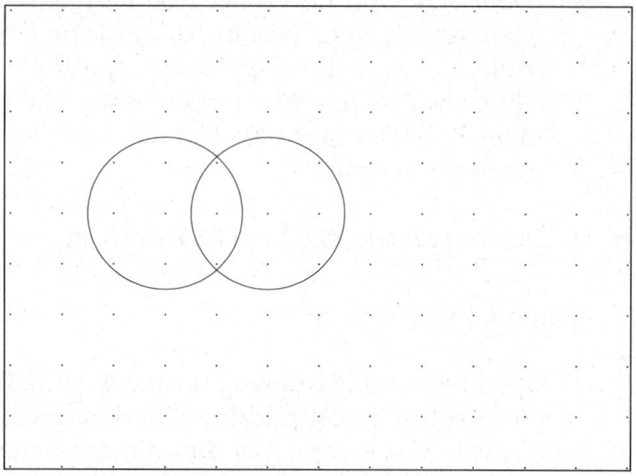

C. Draw a ***Circle*** with the center at **10,10** and with a ***radius*** of **5**. Now use ***Zoom All***. Notice the *Grid* appears only on the area defined by the *Limits*. Can you move the cursor to 0,0? Now use ***Zoom Extents***. What happens? Can you move the cursor to 0,0?

D. ***Erase*** the large ***Circle***. Use ***Zoom All***. Can you move the cursor to 0,0? Use ***Zoom Extents***. Can you find point 0,0?

E. ***Exit*** the drawing and discard changes.

8. *Vports*

A. ***Open*** the **ZOOM TEST** drawing again. Perform a ***Zoom Extents***. Invoke the ***Viewports*** dialog box by any method. When the dialog box appears, choose the ***Three: Right*** option. The resulting viewport configuration should appear as Figure 10-20, shown earlier in the chapter.

B. Using ***Zoom*** and ***Pan*** in the individual viewports, produce a display with an overall view on the top and two detailed views as shown in Figure 10-22.

C. Type the ***Vports*** command. Use the ***New*** option to save the viewport configuration as **3R**.

D. Click in the bottom-left viewport to make it the current viewport. Use ***Vports*** again and change the display to a ***Single*** screen. Now use ***Realtime Zoom*** and ***Pan*** to locate and zoom to room 6100.

E. Use the ***Viewports*** dialog box again. Select the ***Named Viewports*** tab and select **3R** as the viewport configuration to restore. Ensure the **3R** viewport configuration was saved as expected, including the detail views you prepared.

F. Invoke a ***Single*** viewport again. Finally, use the ***Vports*** dialog box to create another viewport configuration of your choosing. ***Save*** the viewport configuration and assign an appropriate name. ***Save*** the **ZOOM TEST** drawing.

CHAPTER 11

# Layers and Object Properties

## Chapter Objectives

After completing this chapter you should:

1. understand the strategy of grouping related geometry with *Layers*;
2. be able to create *Layers*;
3. be able to assign *Color, Linetype,* and *Lineweight* to *Layers*;
4. be able to control a layer's properties and visibility settings (*On, Off, Freeze, Thaw, Lock, Unlock*);
5. be able to manipulate groups of layers by creating *Group Filters* and *Properties Filters*;
6. be able to set *LTSCALE* to adjust the scale of linetypes globally;
7. understand the concept of object properties;
8. be able to change an object's properties (layer, color, linetype, linetype scale, and lineweight) with the Ribbon drop-down lists, the *Properties palette*, and with *Match Properties*;
9. be able to use the *Layer Tools* included with AutoCAD.

## CONCEPTS

In a CAD drawing, layers are used to group related objects in a drawing. Objects (*Lines*, *Circles*, *Arcs*, etc.) that are created to describe one component, function, or process of a drawing are perceived as related information and, therefore, are typically drawn on one layer. A single CAD drawing is generally composed of several components and, therefore, several layers. Use of layers provides you with a method to control visible features of the components of a drawing. For each layer, you can control its color on the screen, the linetype and lineweight it will be displayed with, and its visibility setting (on or off). You can also control if the layer is plotted or not.

FIGURE 11-1

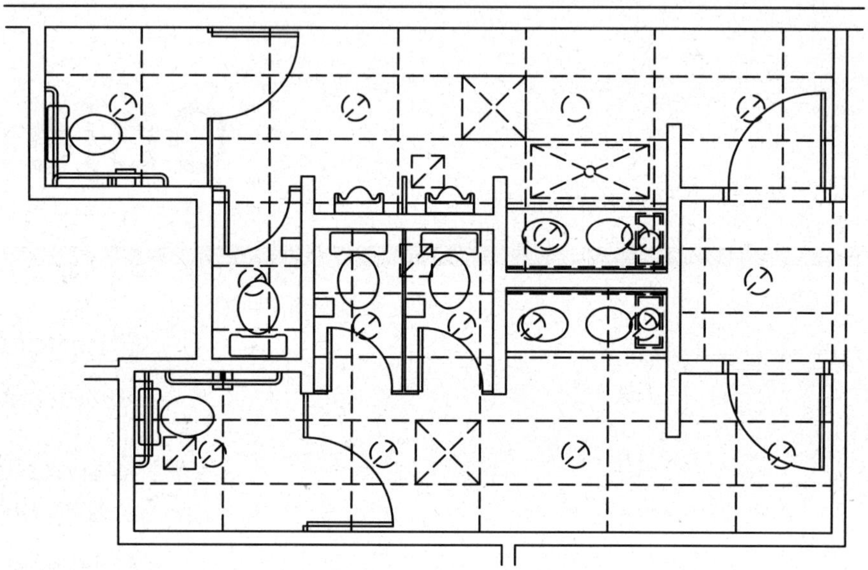

Layers in a CAD drawing can be compared to clear overlay sheets on a manual drawing. For example, in a CAD architectural drawing, the floor plan can be drawn on one layer, electrical layout on another, plumbing on a third layer, and HVAC (heating, ventilating, and air conditioning) on a fourth layer (Fig. 11-1). Each layer of a CAD drawing can be assigned a different color, linetype, lineweight, and visibility setting similar to the way clear overlay sheets on a manual drawing can be used. Layers can be temporarily turned *Off* or *On* to simplify drawing and editing, like overlaying or removing the clear sheets. For example, in the architectural CAD drawing, only the floor plan layer can be made visible while creating the electrical layout, but can later be cross-referenced with the HVAC layout by turning its layer on. Layers can also be made "non-plottable." Final plots or prints can be made of specific layers for the subcontractors and one plot of all layers for the general contractor by controlling the layers' *Plot/No Plot* icon before plotting.

AutoCAD allows you to create a practically unlimited number of layers. You should assign a name to each layer when you create it. The layer names should be descriptive of the information on the layer.

### Assigning Colors, Linetypes, Lineweights, and Transparency

There are two strategies for assigning colors, linetypes, and lineweights in a drawing: assign these properties to layers or assign them to objects.

#### Assign colors, linetypes, lineweights, and transparency to layers

For most drawings, layers are assigned a color, linetype, and lineweight so that all objects drawn on a single layer have the same color, linetype, and lineweight. Assigning colors, linetypes, lineweights, and transparency to layers is called *ByLayer* color, linetype, lineweight, and transparency setting. Using the *ByLayer* method makes it visually apparent which objects are related (on the same layer). All objects on the same layer have the same linetype and color.

### Assign colors, linetypes, lineweights, and transparency to individual objects

Alternately, you can assign colors, linetypes, lineweights, and transparency to specific objects, overriding the layer's color, linetype, and lineweight setting. This method is fast and easy for small drawings and works well when layering schemes are not used or for particular applications. However, using this method makes it difficult to see which layers the objects are located on.

## Object Properties

Another way to describe the assignment of color, linetype, and lineweight properties is to consider the concept of Object Properties. Each object has properties such as a layer, a color, a linetype, and a lineweight. The color, linetype, and lineweight for each object can be designated as *ByLayer* or as a specific color, linetype, or lineweight. Object properties for the two drawing strategies (schemes) are described as follows.

**Object Properties**

| | 1. *ByLayer* Drawing Scheme | 2. Object-Specific Drawing Scheme |
|---|---|---|
| Layer Assignment | Layer name descriptive of geometry on the layer | Layer name descriptive of geometry on the layer |
| Color Assignment | *ByLayer* | *Red*, *Green*, *Blue*, *Yellow*, or other specific color setting |
| Linetype Assignment | *ByLayer* | *Continuous*, *Hidden*, *Center*, or other specific linetype setting |
| Lineweight Assignment | *ByLayer* | 0.05 mm, 0.15 mm, 0.010", or other specific lineweight setting |
| Transparency Assignment | *ByLayer* | *Transparency Value* 0 to 90 |

TIP

It is recommended that beginners use only one method for assigning colors, linetypes, and lineweights. After gaining some experience, it may be desirable to combine the two methods only for specific applications. Usually, the *ByLayer* method is learned first, and the object-specific color, linetype, lineweight and transparency assignment method is used only when layers are not needed or when complex applications are needed. (The *ByBlock* color, linetype, and lineweight assignment has special applications for *Blocks* and is discussed in Chapter 21.)

## Plot Styles

Plot style is an object property that controls how an object appears in a plot. A plot style can be assigned to any object, but can also be assigned to a layer (*ByLayer*). A plot style can control such appearances as (plotted) color, lineweight, screen percentage, line end styles, line join styles, and pen number. A plot style is actually another object property, just like color, linetype, and lineweight, except that assigning it is optional. You should assign plot styles only if you want the plot to have features other than the way the drawing appears on the screen.

There are two types of plot styles: color-dependent and named plot styles. The color-dependent plot style is the simpler of the two because it is based on color. Since colors are usually assigned to layers, each layer's color on the screen controls how the drawing is plotted. Each screen color can be assigned to use a separate pen or print color, screening percentage, line end style, and so on. In this way, all objects that appear in one color will be plotted with the same appearance. On the other hand, named plot styles are not based on color, so they can be assigned to any object. With named plot styles, objects that are drawn in one color can be plotted with different appearances. Plot styles are discussed in detail in Chapter 32, Advanced Layouts, Annotative Objects, and Plotting.

# LAYERS AND LAYER PROPERTIES CONTROLS

TIP

## Layer Drop-Down List

The *Layers* panel (on the Ribbon) contains a drop-down list for making layer control quick and easy (Fig. 11-2). The list normally displays the current layer's name, visibility setting, and properties. When you pull down the list, all the layers (unless otherwise specified in the *Named Layer Filters* dialog box) and their settings are displayed. Selecting any layer name makes it current. Clicking on any of the visibility/properties icons changes that layer's setting as described in the following section. Several layers can be changed in one "drop." You cannot change a layer's linetype or line-weight, nor can you create new layers using this drop-down list. This drop-down list is also available through the Quick Access toolbar, though it is off by default.

FIGURE 11-2

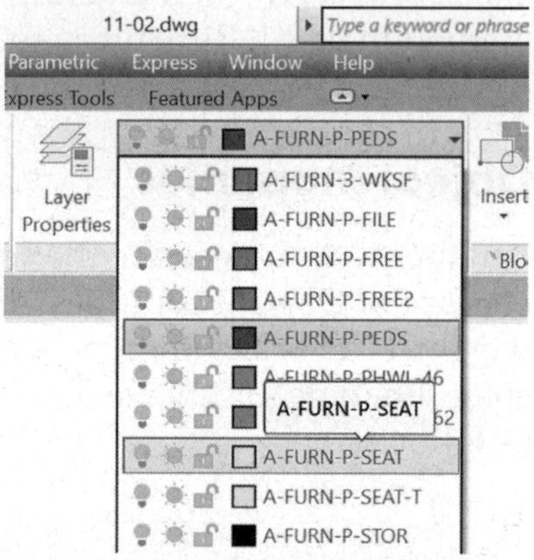


## *Layer*

| Ribbon | Menu Bar | Command (Type) | Alias (Type) | Shortcut |
|---|---|---|---|---|
| *Home*<br>*Layers*<br>*Layer Properties* | *Format*<br>*Layer* | *Layer* or<br>*-Layer* | *LA* | ... |

The *Layer Properties Manager* (Fig. 11-3) lets you create and manage layers. The *Layer Properties Manager* is invoked by using the icon button (shown above), typing the *Layer* command or *LA* command alias, or selecting *Layer* from the *Format* pull-down menu.

FIGURE 11-3

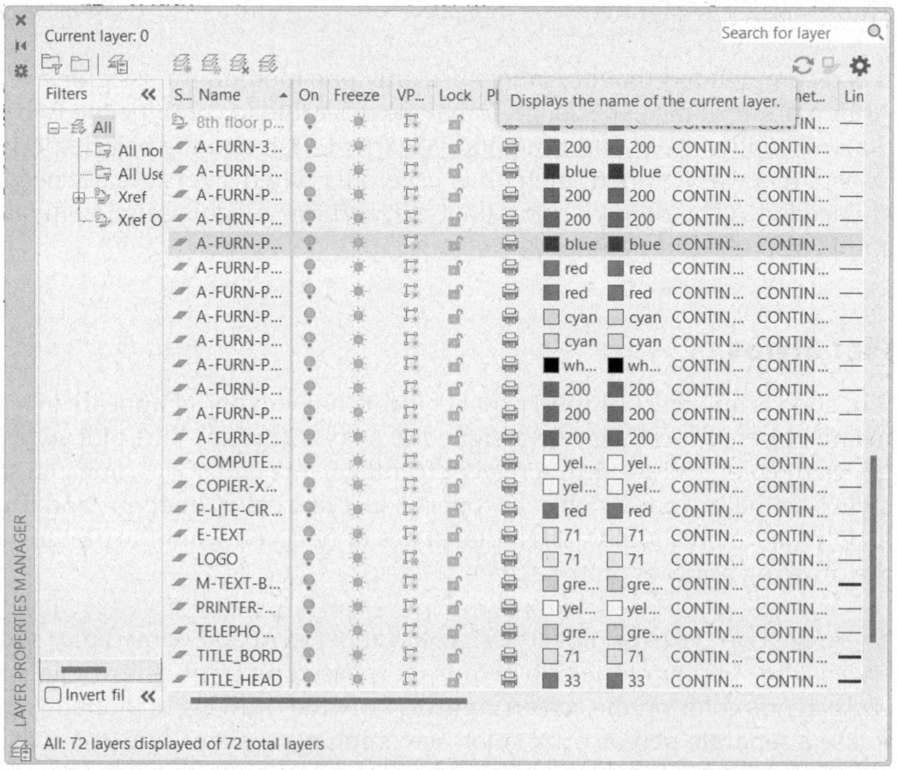


As an alternative to the *Layer Properties Manager*, the *Layer* command can be used in Command line format by typing *Layer* with a "-" (hyphen) prefix. The Command line format shows most of the options available through the *Layer Properties Manager*.

Command: ***-LAYER***
Current layer: "0"
Enter an option
[?/Make/Set/New/Rename/ON/OFF/Color/Ltype/LWeight/TRansparency/MATerial/Plot/Freeze/Thaw/LOck/Unlock/stAte/Description/rEconcile/Xref]:

## *Layer Properties Manager* Layer List

The right-central area of the *Layer Properties Manager* (see Fig. 11-3) is the layer list that displays the list of layers in the drawing. Here you can control the current visibility settings and the properties assigned to each layer. This list can display all the layers in the drawing or a subset of layers if a layer filter is applied. Layer filters are created and applied in the *Filters* pane on the left side of the *Layer Properties Manager* (see "Layer Filters" later in this chapter). For new drawings such as those created from the ACAD.DWT template, only one layer may exist in the drawing—Layer "0" (zero). New layers can be created by selecting the *New Layer* button or the *New Layer* option of the right-click menu (see *New Layer*).

All properties and visibility settings of layers can be controlled by highlighting the layer name and then selecting one of the icons for the layer such as the light bulb icon (*On, Off*), sun/snowflake icon (*Thaw/Freeze*), padlock icon (*Lock/Unlock*), *Color* tile, *Linetype*, *Lineweight*, *Transparency* or *Plot/No plot* icon.

Multiple layers can be highlighted (see Fig. 11-3) by holding down the Ctrl key while PICKing (to highlight one at a time) or holding down the Shift key while PICKing (to select a range of layers between and including two selected names). Right-clicking in the list area displays a cursor menu allowing you to *Select All* or *Clear All* names in the list and other options (see Fig. 11-5). You can rename a layer by clicking twice slowly on the name (this is the same as PICKing an already highlighted name).

FIGURE 11-4

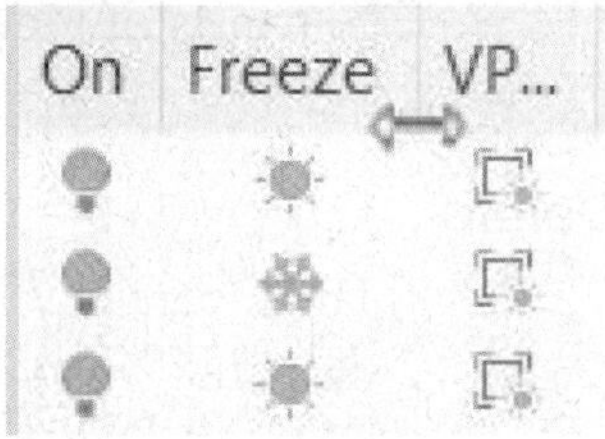


The column widths can be changed by moving the pointer to the separator between column headings until double arrows appear (Fig. 11-4). You can also right-click on a column heading to produce a list of columns to display. Uncheck any columns from the list that you do not want displayed in the *Layer Properties Manager*.

A particularly useful feature is the ability to sort the layers in the list by any one of the headings (*Name, On, Freeze, Linetype*, etc.) by clicking on the heading tile. For example, you can sort the list of names in alphabetical order (or reverse order) by clicking once (or twice) on the *Name* heading above the list of names. Or you may want to sort the *Frozen* and *Thawed* layers or sort layers by *Color* by clicking on the column heading.

FIGURE 11-5

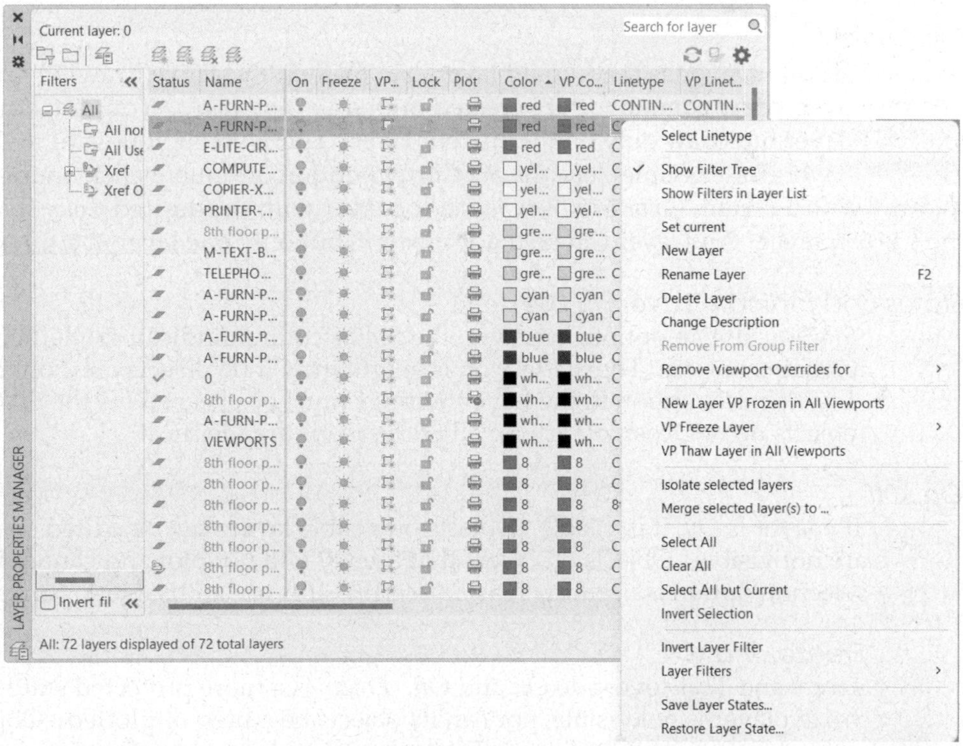


Right-clicking in the layer list of the *Layer Properties Manager* produces a shortcut menu (Fig. 11-5). You have a choice of selecting icon buttons or selecting shortcut menu options for most of the layer controls available in the *Layer Properties Manager*, as noted in the following sections.

There is only one layer in the AutoCAD templates as they are provided to you "out of the box." That layer is Layer 0. Layer 0 is a part of every AutoCAD drawing because it cannot be deleted. You can, however, change the *Color, Linetype, Lineweight,* and *Transparency* of Layer 0 from the defaults (*Continuous* linetype, *Default* lineweight, and color #7 *White*). Layer 0 is generally used as a construction layer or for geometry not intended to be included in the final draft of the drawing. Layer 0 has special properties when creating *Blocks* (see Chapter 21).

### *New Layer*

The *New Layer* option allows you to make new layers. A new layer named "Layer1" (or other number) then appears in the list with the default color (*White*), linetype (*Continuous*), and lineweight (*Default*). If an existing layer name is highlighted when you make a new layer, the highlighted layer's properties (*color, linetype, lineweight,* and *plot style*) are used as a template for the new layer. The new layer name initially appears in the rename mode so you can immediately assign a more appropriate and descriptive name for the layer. You can create many new names quickly by typing (renaming) the first layer name, then typing a comma before other names. A comma forces a new "blank" layer name to appear. Colors, linetypes, and lineweights should be assigned as the next step. New layers for dimensions and hatch patterns can be created using the *DIMLAYER* (see Chapter 28) and *HPLAYER* (see Chapter 26) system variables, respectively.

### *New Layer VP*

Use this option to create a new layer that is frozen in all viewports. See Chapters 13 and 32 for concepts of viewport-specific layer settings.

### *Delete*

The *Delete Layer* option allows you to delete layers. Only layers with no geometry can be deleted. You cannot delete a layer that has objects on it, nor can you delete Layer 0, the current layer, or layers that are part of externally referenced (*Xref*) drawings. If you attempt to *Delete* such a layer, accidentally or intentionally, a warning appears.

### *Set Current*

To *Set* a layer as the *Current* layer is to make it the active drawing layer. Any objects created with draw commands are created on the *Current* layer. You can, however, edit objects on any layer but draw only on the current layer. Therefore, if you want to draw on the FLOORPLAN layer (for example), use the *Set Current* option or double-click on the layer name. If you want to draw with a certain *Color* or *Linetype*, set the layer with the desired *Color, Linetype,* and *Lineweight* as the *Current* layer. Any layer can be made current, but only one layer at a time can be current.

### *Status* and *Indicate Layers in Use*

*Status* is not an option; however, the *Status* column indicates whether or not the related layer contains objects. The *Status* icon is a light gray if no objects exist on the layer. The *Indicate Layers in Use* option in the Layer Settings must be checked for the *Status* icon to indicate if objects do not exist, otherwise all *Status* icons are identical.

### *On, Off*

If a layer is *On*, it is visible. Objects on visible layers can be edited or plotted. Layers that are *Off* are not visible. Objects on layers that are *Off* will not plot and cannot be edited (unless the *ALL* selection option is used, such as *Erase, All*). It is not advisable to turn the current layer *Off*.

### *Freeze, Thaw*

*Freeze* and *Thaw* override *On* and *Off*. *Freeze* is a more protected state than *Off*. Like being *Off*, a frozen layer is not visible, nor can its objects be edited or plotted. Objects on a frozen layer cannot be accidentally *Erased*. *Freezing* unused layers speeds up computing time when working with large and complex drawings. *Thawing* reverses the *Freezing* state. Layers can be *Thawed* and also turned *Off*. Frozen layers are not visible even though the light bulb icon may be on.

### Lock, Unlock

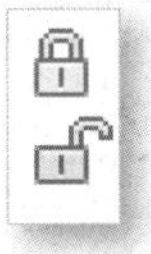

Layers that are *Locked* are protected from being edited but are still visible and can be plotted. *Locking* a layer prevents its objects from being changed even though they are visible. Objects on *Locked* layers cannot be selected with the *ALL* selection option (such as *Erase, All*). Layers can be *Locked* and *Off*.

## Color, Linetype, Lineweight, and Other Properties

Layers have properties of *Color, Linetype,* and *Lineweight* such that (generally) an object that is drawn on, or changed to, a specific layer assumes the layer's linetype and color. Using this scheme (*ByLayer*) enhances your ability to see what geometry is related by layer. It is also possible, however, to assign specific color, linetype, and lineweight to objects which will override the layer's color, linetype, and lineweight (see "*Color, Linetype,* and *Lineweight* Commands" and "Changing Object Properties").

### Color

Selecting one of the small color boxes in the list area of the *Layer Properties Manager* or Layer Control drop-down list causes the *Select Color* dialog box to pop up (Fig. 11-6). The desired color can then be selected or the name or color number (called the ACI—AutoCAD Color Index) can be typed in the edit box. This action retroactively changes the color assigned to a layer. Since the color setting is assigned to the layer, all objects on the layer that have the *ByLayer* setting change to the new layer color. Objects with specific color assigned (not *ByLayer*) are not affected. Alternately, the *Color* option of the *-Layer* command (hyphen prefix) can be typed to enter the color name or ACI number.

The *Select Color* dialog box also contains tabs for the *Index Color* (ACI), *True Color,* and *Color Books.* See "True Color and Color Books" later in this chapter for information.

**FIGURE 11-6**

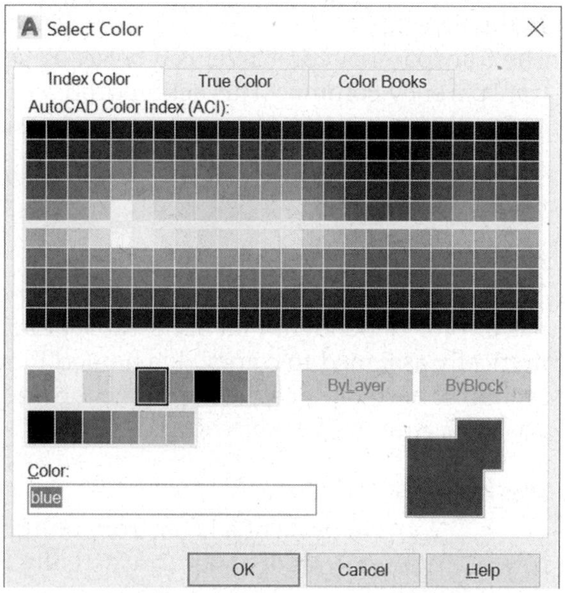


### Linetype

To set a layer's linetype, select the *Linetype* (word such as *Continuous* or *Hidden*) in the layer list, which in turn invokes the *Select Linetype* dialog box (Fig. 11-7). Select the desired linetype from the list. Similar to changing a layer's color, all objects on the layer with *ByLayer* linetype assignment are retroactively displayed in the selected layer linetype while non-*ByLayer* objects remain unchanged.

**FIGURE 11-7**

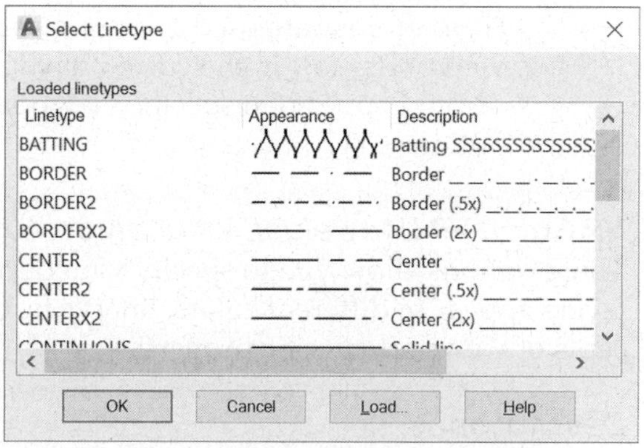


The ACAD.DWT and most other template drawings as supplied by Autodesk have only one linetype available (*Continuous*). Before you can use other linetypes in a drawing, you must load the linetypes by selecting the *Load* tile (see the *Linetype* command) or by using a template drawing that has the desired linetypes already loaded.

### Lineweight

The lineweight for a layer can be set by selecting the *Lineweight* (word such as *Default* or *0.20 mm, 0.50 mm, 0.010"* or *0.020"*, etc.) in the layer list. This action produces the *Lineweight* dialog box (Fig. 11-8). Like the *Color* and *Linetype* properties, all objects on the layer with *ByLayer* lineweight assignment are retroactively displayed in the lineweight assigned to the layer while non-*ByLayer* objects remain unchanged. You can set the units for lineweights (mm or inches) using the *Lineweight* command.

**FIGURE 11-8**

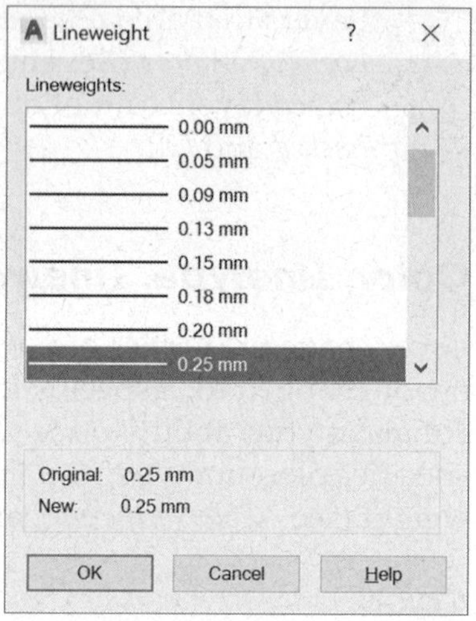


### Transparency

The transparency for a layer can be set by selecting the number in the transparency column. This action produces the *Layer Transparency* dialog box (Fig. 11-9). You can set the *Transparency Value* from 0 to 90.

**FIGURE 11-9**

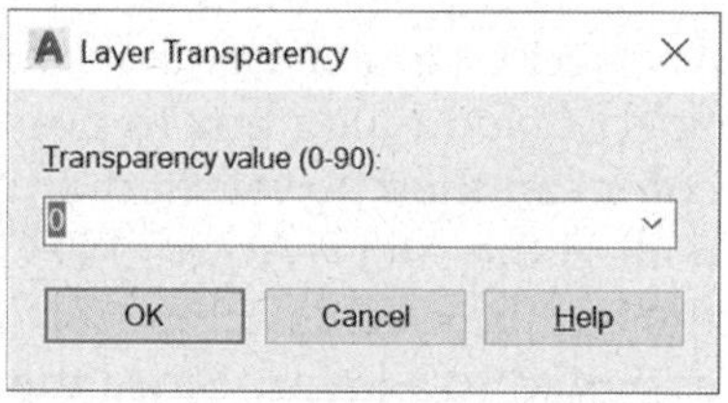


### Plot Style

This column of the *Layer Properties Manager* designates the plot style assigned to the layer. If the drawing has a color-dependent Plot Style Table attached, this section is disabled since the plot styles are automatically assigned to colors. If a named Plot Style Table is attached to the drawing you can select this section to assign plot styles from the table to layers. Plot styles are discussed in Chapter 32, Advanced Layouts, Annotative Objects, and Plotting.

### Plot/No Plot

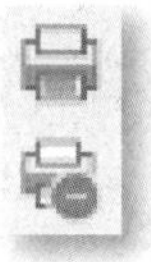

You can prevent a layer from plotting by clicking the printer icon so a red mark appears over the printer symbol. There are actually three methods for preventing a layer from appearing on the plot: *Freeze* it, turn it *Off*, or change the *Plot/No Plot* icon. The *Plot/No Plot* icon is generally preferred since the other two options prevent the layer from appearing in the drawing (screen) as well as in the plot.

### Current VP Freeze, VP Freeze

These options are used and are displayed in the *Layer Properties Manager* only when paper space viewports exist in the current layout tab. Using these options, you can control what geometry (layers) appears in specific viewports. See Chapters 13 and 32 for more information on these options.

### VP Color, VP Lineweight, VP Linetype, VP Plot Style

These options allow you to specify viewport-specific property overrides. That is, you can stipulate that layers appear in different colors, linetypes, etc. for different viewports. See Chapters 13 and 32 for concepts of viewport-specific properties.

### Description

You can add an optional *Description* for each layer that appears in the *Layer Properties Manager*. Select *Change Description* from the right-click menu and add descriptive information about the objects that reside on the layer. For example, layer A-WALL-INT-OPT2 might have the description "Design option 2 for interior walls."

### *Search for layer*

This edit box in the upper-right corner allows you to use wildcards to search for matching names in a long list of layers. For example, entering "A-WALL*" would cause only layers beginning with "A-WALL" to appear in the layer list. Clicking in the edit box makes the "*" (asterisk) wildcard automatically appear. Do not press Enter after typing in your entry.

Note that the *Layer Properties Manager* is in a palette configuration. You can use the *Auto-hide* button (under the "X") to make the palette disappear, leaving only the title bar. Resting the pointer on the title bar restores the palette. Other options for moving, sizing, docking, anchoring and setting the transparency level of the palette are available from the *Properties* button (below *Auto-hide*).

## Layer Filters

When working on drawings with a large number of layers in the *Layer Properties Manager* or in the Layer drop-down list, it is time consuming to scroll through a long list of layer names to change visibility and properties settings. A "layer filter" allows you to reduce the list to a subset of the entire list of layers in a drawing. You can create multiple layer filters, and therefore, multiple "groups" of layers, based on object properties, visibility settings, or objects on the layers. For example, you could make filters to display only layers that are *Thawed*, only layers that are red in color, or only layers that contain furniture.

The Filters area on the left side of the *Layer Properties Manager* allows you to create layer filters (Fig. 11-10). You can even create hierarchical filters, or groups within groups. The groups of layers (filters) that you create are listed in a hierarchical format under the *All* designator. You can create two types of filters: *Group Filters* and *Properties Filters*. If any *Xrefs* are used by the drawing, AutoCAD automatically creates a filter for each *Xref* drawing and displays the list under the *Xref* heading at the bottom of the *Filters* list.

**FIGURE 11-10**

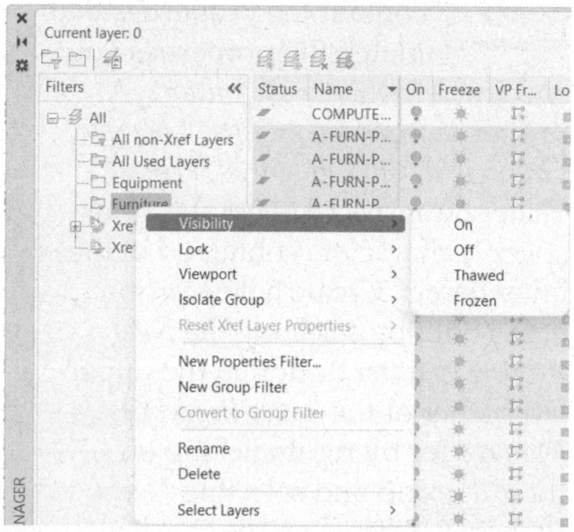


When working with drawings containing a large number of layers, using filters can improve your productivity since filters allow you to reduce the number of names in the layer list. You can click on any group of layers in the Filter Tree View to display its list of layers in the right side of the dialog box. For example, Figure 11-11 displays the list of layers in the "Furniture" group. Checking *Invert filter* causes the list to display all but the filter list. For example, in Figure 11-11, if *Invert filter* was checked, all the layers in the drawing except those beginning with A-FURN would be displayed in the list.

**FIGURE 11-11**

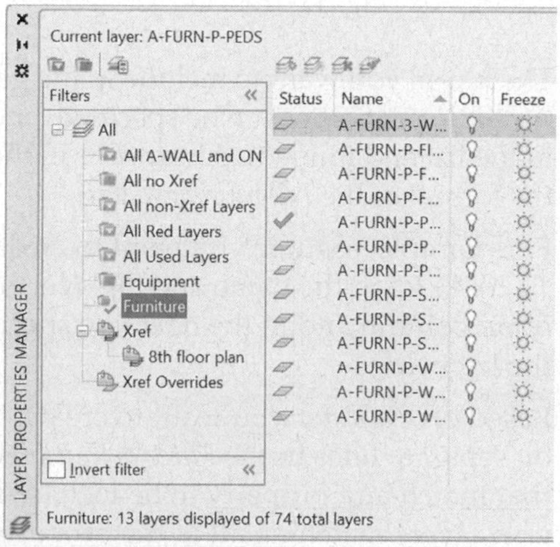


Additionally, you can control visibility and lock settings for an entire group. Right-click on a group (filter) name in the Filter Tree View, then select the *Visibility* option, *On, Off, Thaw, Freeze, Lock* or *Unlock* from the shortcut menu (see Fig. 11-10). You can also *Freeze* or *Thaw* the layers for the current viewport (if applicable). A handy *Isolate Group* option is available. Using this feature freezes all layers except those in the group and displays only the group layers in the drawing. However, beware of this feature since no "unisolate" option is available and *Layerp* (Layer Previous) will not undo the layer isolation (see *Layerp*).

TIP

## Creating a *Group Filter*

Create a group filter by selecting the *New Group Filter* button in the upper-left corner of the *Layer Properties Manager* or by right-clicking on the *All* group and selecting *New Group Filter* from the shortcut menu (see Fig. 11-12). This action causes a new entry named *Group Filter 1* to appear in the filter list under *All*. Assign an appropriate descriptive name for the filter you intend to create, such as "Walls" or "Text." You can then populate the list of layers for the group using two methods. First, you can drag-and-drop layer names from the drawing layer list on the right into the group (layers added to the group are not moved from the drawing layer list, but copied). Second, you can highlight the group name, right-click, and pick *Select Layers* from the shortcut menu. This action returns you to the drawing so you can pick objects on the layers you want to include in the group.

**FIGURE 11-12**

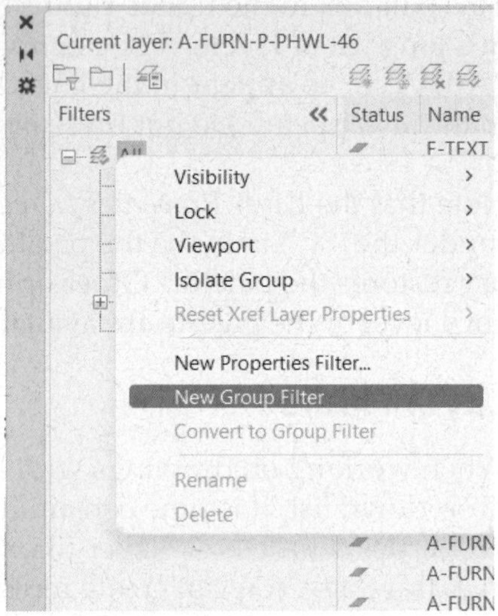


## Creating a *Properties Filter*

A properties filter is a group of layers that contain the properties that match the properties you specify in the *Filter Definition*. A property could be a *Color*, *Linetype*, *Lineweight*, *On/Off* status, *Freeze/Thaw* status, *Lock/Unlock* status, or specific characters contained in the layer name. Create a new properties filter by selecting the *New Properties Filter* button in the upper-left corner of the *Layer Properties Manager* or by right-clicking on the *All* group and selecting *New Properties Filter* from the shortcut menu (see Fig. 11-12). This action invokes the *Layer Filter Properties* dialog box (Fig. 11-13).

**FIGURE 11-13**

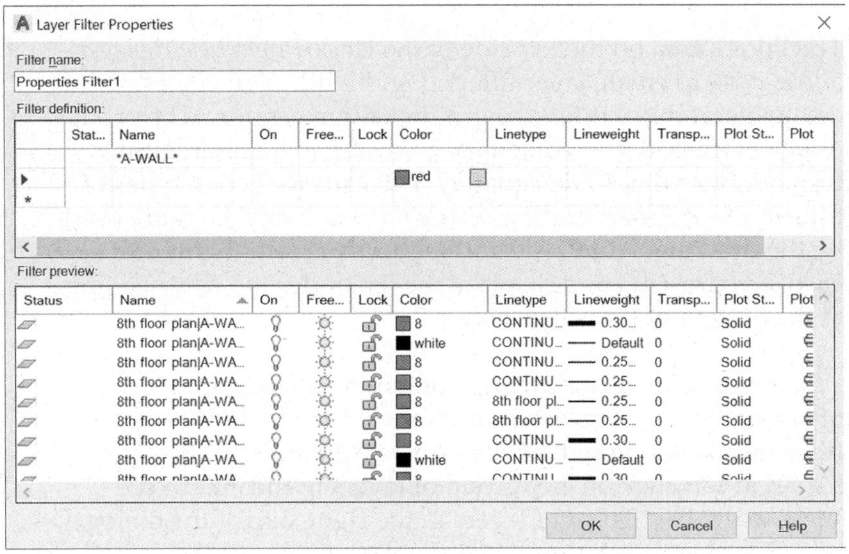


The layers you want to include in the group must match the criteria you specify in the *Filter definition* area. Multiple properties can be specified, and the layers matching these properties are automatically included in the list and immediately appear in the *Filter preview* list below (you cannot directly alter the contents or the format of the *Filter preview* list).

For example, assume you want to create a property filter to include all layers that include the characters "A-WALL" in the name and all layers that are *Red* in color. In the first line, enter "*A-WALL*" in the *Name* column and in the next line specify *Red* in the color column. The matching layers then appear in the list below.

TIP

Based on the filter you want to create, it is important whether you specify the properties on one line or on separate lines in the *Filter definition* box. Specifying properties on separate lines allows any layers that match any property to be included in the list. Specifying properties all on one line requires that only layers matching all of the properties are included in the list. For example, the list in Figure 11-13 includes all layers that have A-WALL in the name or layers that are *Red*, whereas if the two properties were listed on one line, only layers that include A-WALL and are *Red* would be included.

Properties filters are dynamic, whereas group filters are static. For example, if a property filter definition includes an *On* visibility status, all layers that are currently *On* match this definition and therefore are automatically included in the list. If any layer's *On/Off* visibility is then changed, it would cause the layer to move into or out of the group. In contrast, group filters are based solely on the specific layers that are chosen to be in the group, so changing the group is possible only by "manually" moving a layer into or out of the group.

Property filters can be converted to group filters by right-clicking on the filter name and selecting *Convert to Group Filter* from the shortcut menu. Only layers that currently match the properties are included in the new group filter. Keep in mind that the new group is static unless you "manually" change its members. Group filters cannot be converted to property filters.

Consider also that you may want to create subgroups, or filters "nested" within filters, therefore appearing in a hierarchy in the Filter Tree View. Group filters can be created within other group filters. You can likewise create a property filter within another property filter. You can also create a property filter within a group filter, but you cannot create a group filter within a property filter due to the flexibility of the parent property filter.

### *Settings...*

The *Settings* button in the *Layer Properties Manager* displays the *Layer Settings* dialog box. Using this tool, you can specify that a warning appears when new layers are created and saved in the drawing or in an externally referenced (Xref) drawing. This feature is useful when drawings are accessible to multiple users and you need to be notified when new layers are added to the drawing without your knowledge. See Chapter 30, Xreferences.

## Displaying an Object's Properties and Visibility Settings

When the layer list in the Ribbon is in the normal position (not "dropped down"), it can be used to display an object's layer properties and visibility settings. Do this by selecting an object (with the pickbox, window, or crossing window) when no commands are in use.

**TIP**

When an object is selected, the *Properties* panel displays the layer, color, linetype, lineweight, and plot style of the selected object (Fig. 11-14). The layer list displays the selected object's layer name and layer settings rather than that of the current layer. The *Color*, *Linetype*, *Lineweight*, and *Plot Style* boxes also temporarily reflect the color, linetype, lineweight, and plot style properties of the selected object. Pressing the Escape key causes the list boxes to display the current layer name and settings again, as would normally be displayed. If more than one object is selected, and the objects are on different layers or have different color, linetype, and lineweight properties, the list boxes go blank until the objects become unhighlighted or a command is used.

These sections of the Ribbon have another important feature that allows you to change a highlighted object's (or set of objects) properties. See "Changing Object Properties" near the end of this chapter.

**FIGURE 11-14**

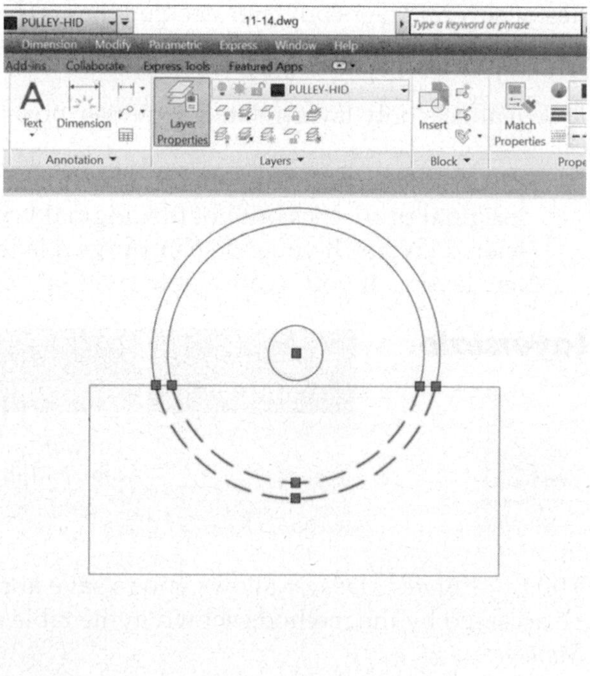

### Laymcur

| Ribbon | Menu Bar | Command (Type) | Alias (Type) | Shortcut |
|---|---|---|---|---|
| *Home*<br>*Layers*<br>(icon) | *Format*<br>*Layer Tools >*<br>*Make Object's*<br>*Layer Current* | *Laymcur* | .... | ... |

TIP

This productive feature can be used to make a desired layer current simply by selecting any object on the layer. This option is generally faster than using the *Layer Properties Manager* or *Layer* drop-down list to set a layer current. The *Make Object's Layer Current* feature is particularly useful when you want to draw objects on the same layer as other objects you see but are not sure of the layer name.

There are two steps in the procedure to make an object's layer current: (1) select the icon and (2) select any object on the desired layer. The selected object's layer becomes current and the new current layer name immediately appears in the Layer Control list box in the Object Properties toolbar.

### Layerp

| Ribbon | Menu Bar | Command (Type) | Alias (Type) | Shortcut |
|---|---|---|---|---|
| *Home*<br>*Layers*<br>(icon) | *Format*<br>*Layer Tools >*<br>*Layer Previous* | *Layerp* | .... | ... |

*Layerp* (*Layer Previous*) is an *Undo* command only for layers. In other words, when you use *Layerp*, all the changes you made the last time you used the *Layer Properties Manager*, the *Layer* drop-down list, or the *-Layer* command are undone. Using *Layerp* does not affect any other activities that occurred to the drawing since the last layer settings, such as creating or editing geometry or viewing controls like *Pan* or *Zoom*.

TIP

Use *Layerp* to return the drawing to the previous layer settings. For example, if you froze several layers and changed some of the geometry in a drawing, but now want to thaw those frozen layers again without affecting the geometry changes, use *Layerp*. Or, if you changed the color and linetype properties of several layers but later decide you prefer the previous property settings, use *Layerp* to undo the changes and restore the original layer settings.

*Layerp* affects only layer-related activities; however, *Layerp* does not undo the following changes:

Renamed layers: If you rename a layer and change its properties, *Layerp* restores the original properties but not the original layer name.
Deleted layers: If you delete or purge a layer, using *Layerp* does not restore it.
New layers: If you create a new layer in a drawing, using *Layerp* does not remove it.

### Layerstate

| Ribbon | Menu Bar | Command (Type) | Alias (Type) | Shortcut |
|---|---|---|---|---|
| ... | *Format*<br>*Layer States*<br>*Manager* | *Layerstate* | *LAS* | ... |

The *Layer States Manager* allows you to save and restore "layer states." The *Layer States Manager* can be accessed by the methods shown in the table above and is also accessible from the *Layer Properties Manager*.

A layer state is a combination of layer settings (*On/Off, Freeze/Thaw*, etc.) that can be saved and later restored. Layer states differ from layer filters in that filters simply allow you to "save" short lists of layers, whereas layer states allow you to save different visibility and properties settings for use at a later time. A layer state saves all the visibility and properties settings for all the layers in the drawing. Saving multiple layer states allows you to quickly reset particular *On/Off, Freeze/Thaw,* etc. combinations to improve your efficiency working with particular groups of layers.

For example, you may require that only a certain combination of layers is *On* for editing a particular aspect of a drawing—say, the floor plan, the HVAC layout, and the HVAC dimensions—while a different set of layers is *On* for working with another aspect of the drawing—say, the floor plan, plumbing layout, and fixtures. Rather than manipulating the list of layers and making the appropriate *On/Off* settings each time you work with the floor plan, the HVAC layout, and the HVAC dimensions and each time you work with floor plan, plumbing layout, and fixtures, you can save these settings as layer states to be *Restored* when needed. A layer state also stores the *Current* layer, so when you *Restore* a layer setting, you are ready to begin drawing on the appropriate layer.

For example, the preceding scenario (saving the HVAC layers settings) would be accomplished by first using the *Layer Properties Manager* as you would normally to make the desired layer visibility settings, that is, turn *On* the floor plan layer, the HVAC layer, and the HVAC dimensions layer, and turn *Off* all other layers. Next, invoke the *Layer States Manager* (Fig. 11-15) to assign a name and save the settings. Then, at any time later, no matter what the current layer settings are, you could use the *Layer States Manager* to *Restore* the visibility settings specified in the HVAC layer state.

**FIGURE 11-15**

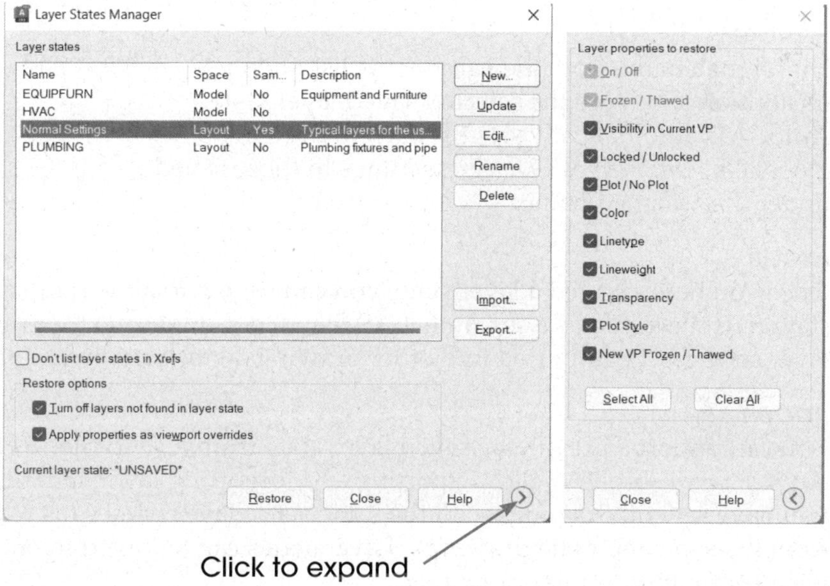


### To Save a Layer State:

The procedure for saving a layer state is as follows:

1. Use the *Layer Properties Manager* as you would normally to make the desired settings (*On/Off, Freeze/Thaw,* etc.) for all layers.
2. Invoke the *Layer States Manager* (icon button or Alt+S) from within the *Layer Properties Manager*.
3. Use the *New* option to produce the *New Layer State to Save* dialog box (not shown). Here, assign an appropriate name and optional description for the current layer settings configuration, then select *OK* to close this dialog box.
4. *Close* the *Layer States Manager*.

TIP

NOTE: It is highly suggested that you first create a "Normal Settings" layer state when the layer settings are in the normal positions and before you begin creating other new layer states. In this way, you can easily restore the layer settings to their normal positions, no matter what changes are made subsequently.

Options in the *Layer State Manager* are described next.

### New

This option produces the *New Layer State to Save* dialog box. The *Description* can contain any characters including spaces; however, the layer state name can contain only alpha-numeric characters and spaces.

### Save

You must use the *Save* option to save any changes you have made in the *Layer States Manager*.

### Edit...

The *Edit Layer State* dialog box appears (Fig. 11-16). Here you can only add or remove layers from the layer state. Highlight the desired layer(s), then select the *Add* or *Remove* button in the lower-left corner.

**FIGURE 11-16**

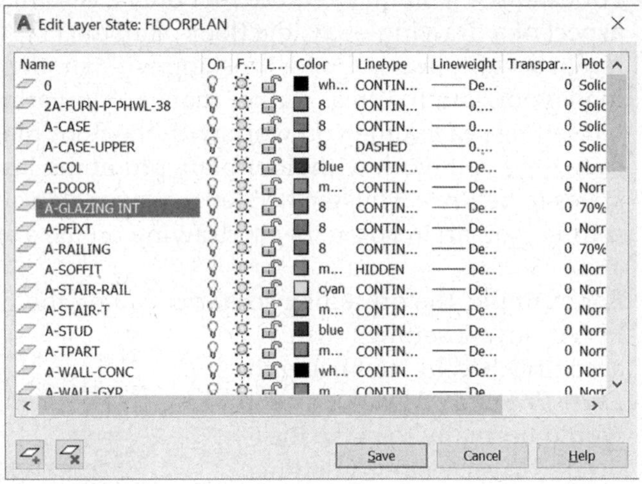


### Rename

Use this option to rename an existing layer state.

### Delete

Delete a layer state by highlighting the name from the list then pressing *Delete*.

### Restore

Highlight a name from the *Layer States* list, then use this option to restore the selected saved layer state. AutoCAD automatically sets the current drawing's global *On, Off, Freeze, Thaw,* etc. settings to those saved under the assigned name.

### Export

Once you have created a layer state, you can save it to a file (with a .LAS extension). Each named layer state must be saved as a separate .LAS file; you cannot save the entire set of layer states to one file. A layer state can be imported into other drawings with the *Import* option.

### Import

You can import a previously saved layer state from a .LAS file. You can also import layer states from a .DWG, .DWT, or .DWS file. Exporting and importing layer states is intended to be used primarily when you have several drawings with the same set of layer names (as when similar drawings are created from the same template drawing). Layer states can be saved in one drawing, then *Exported* to files and *Imported* to other similar drawings.

### Layer Properties to Restore

These checkboxes allow you to specify which settings you want to save with the layer state. In other words, with all the boxes checked you save all the settings as they appear in the *Layer Properties Manager* for each layer. If, on the other hand, only the *On/Off* box is checked, only the *On* or *Off* setting of each layer as it is listed in the *Layer Properties Manager* is saved when the layer state is created. Therefore, if you are interested in saving only visibility settings (*On/Off, Freeze/Thaw*) for layer groups, it is recommended to uncheck the other boxes.

TIP

Although the visibility settings such as *On/Off* and *Freeze/Thaw* are likely to be changed often during the process of working on a drawing, the other properties of layers such as *Color, Linetype, Lineweight,* etc. are not likely to be changed. However, creating layer states using these checkboxes (on the right) can be useful for specific applications. Since properties of layer such as color, lineweight, and plot style usually affect printing and plotting, you can create and restore layer states to prepare for different plot setups. Or, you may prefer a certain color and lineweight "look" on screen for drawing and editing, then restore a layer state that complies with the company standard before saving the drawing or for creating a print or plot. For example, to prepare for printing on a black and white printer, you could create and restore a layer state that quickly sets all layer colors to black. These checkboxes are also helpful if you intend to *Export* a layer state from one drawing, then *Import* it to another.

In such a case, you can clone one drawing's color, linetype, lineweight, and/or plot characteristics to another drawing, assuming each drawing has the same layer names.

#### *Turn off layers not found in layer state*

This checkbox may be useful if you imported a layer state (.LAS file) into your drawing and the current drawing contained more layers than were named in the .LAS file.

#### Apply properties as viewport overrides

Checking this box causes the layer property overrides to be applied to the current viewport. This option is available only when the *Layer States Manager* is accessed when a layout viewport is active. When property overrides are applied to a viewport, affected layers are displayed with a highlighted background in the *Layer Properties Manager* (assuming the *Layer States Manager* is accessed when that specific viewport is active). See Chapters 13 and 32 for concepts on viewport-specific layer properties.

#### Layer States with *Blocks and Xrefs*

Layer states saved with a drawing that is inserted into your current drawing as a *Block* are added to the current drawing. Saved layer states of *Xrefed* drawings are not accessible from the current drawing. (See Chapters 21 and 30 for information on *Blocks* and *Xrefs*.)

## OBJECT-SPECIFIC PROPERTIES CONTROLS

The commands in this section are used to control the color, linetype, and lineweight properties of individual objects. This method of object property assignment is used only in special cases—when you want the objects' color, linetype, and lineweight to override those properties assigned to the layers on which the objects reside. Using this method makes it difficult to see which objects are on which layers. In most cases, the *ByLayer* property is assigned instead to individual objects so the objects assume the properties of their layers. In the case that color, linetype and/or lineweight properties are accidentally set for individual objects or you want to change object properties to a *ByLayer* setting, you can use the *Setbylayer* command (see "*Setbylayer*") to retroactively change object properties to *ByLayer* properties.

TIP

### *Linetype*

| Ribbon | Menu Bar | Command (Type) | Alias (Type) | Shortcut |
|---|---|---|---|---|
| *Home Properties* (drop-down) | *Format Linetype...* | *Linetype* or *-Linetype* | *LT* or *-LT* | ... |

Invoking this command presents the *Linetype Manager* (Fig. 11-17). Even though this looks similar to the dialog box used for assigning linetypes to layers (shown earlier in Fig. 11-7), beware! When linetypes are selected and made *Current* using the *Linetype Manager*, they are assigned to objects—not to layers. That is, selecting a linetype by this manner (making it *Current*) causes all objects from that time on to be drawn using that linetype, regardless of the layer that they are on (unless the *ByLayer* type is selected). In contrast, selecting linetypes during the *Layer* command (using the *Layer Properties Manager*) results in assignment of linetypes to layers.

FIGURE 11-17

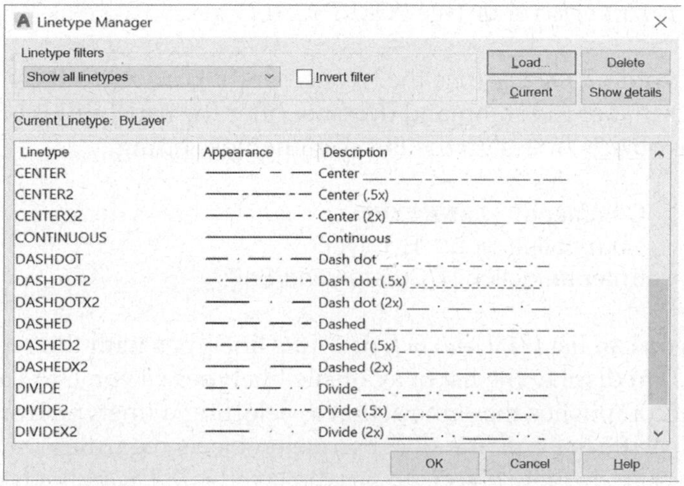


TIP

To draw an object in a specific linetype, simply select the linetype from the *Linetype Manager* list, select *Current* (or double-click the linetype), and select the *OK* tile. That linetype stays in effect (all objects are drawn with that linetype) until another linetype is selected. If you want to draw objects in the layer's assigned linetype, select *ByLayer*. *ByBlock* linetype assignment is discussed in Chapter 21.

### *Load*

The *Linetype Manager* also lets you *Load* linetypes that are not already in the drawing. Just select the *Load* button to view and select from the list of available linetypes in the *Load or Reload Linetypes* dialog box (Fig. 11-18). The linetypes are loaded from the ACAD.LIN or ACADISO. LIN file. If you want to load all linetypes into the drawing, right-click to display a small cursor menu, then choose *Select All* to highlight all linetypes (see Fig. 11-18).

**FIGURE 11-18**

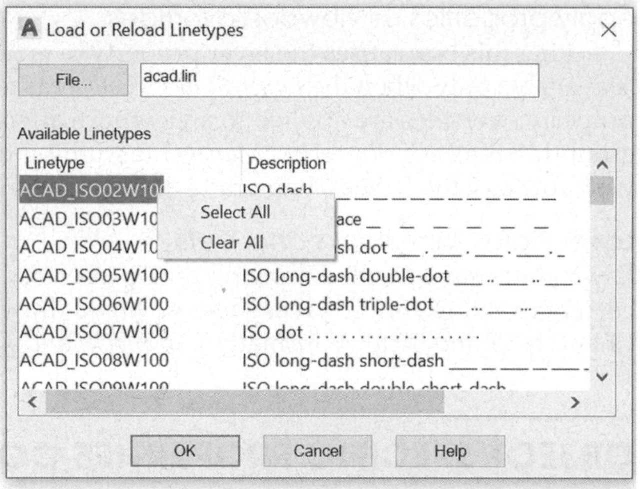


You can also select multiple linetypes by holding down the Ctrl key or select a range by holding down the Shift key. The loaded linetypes are then available for assignment from both the *Linetype Manager* and the *Layer Properties Manager*.

AutoCAD supplies numerous linetypes that can be viewed in the *Load or Reload Linetypes* dialog box (Fig. 11-18). The ACAD_ISO*n*W100 linetypes are intended to be used with metric drawings. For the ACAD_ISO*n*W100 linetypes, *Limits* should be set to metric sheet sizes to accommodate the relatively large linetype spacing. Avoid using both ACAD_ISO*n*W100 linetypes and non-ISO linetypes in one drawing due to the difficulty of managing the two sets of linetype scales.

### *Delete*

The *Delete* button in the *Linetype Manager* is useful for deleting any unused linetypes from the drawing. Unused linetypes are those which have been loaded into the drawing but have not been assigned to layers or to objects. Freeing the drawing of unused linetypes can reduce file size slightly.

The *Linetype Manager* has a *Linetype filters* drop-down list and *Show details/Hide details* button. If details are visible, you can change the *Name* or *Description*, or set the *Global scale factor* (see "*LTSCALE*") or *Current object scale* (see "*CELTSCALE*").

Typing *-Linetype* (use the hyphen prefix) produces the command line format of *Linetype*. Although this format of the command does not offer the capabilities of the dialog box version, you can assign linetypes to objects or assign the *ByLayer* linetype setting.

```
Command: -LINETYPE
Current line type: "ByLayer"
Enter an option [?/Create/Load/Set]:
```

You can list (*?*), *Load*, *Set*, or *Create* linetypes with the typed form of the command. Type a question mark *(?)* to display the list of available linetypes. Type *L* to *Load* any linetypes. The *Set* option of *Linetype* accomplishes the same action as selecting a linetype from the list in the dialog box. That is, you can set a specific linetype for all subsequent objects regardless of the layer's linetype setting, or you can use *Set* to assign the new objects to use the layer's (*ByLayer*) linetype.

You can create your own custom linetypes with the *Create* option or by using a text editor. Both simple linetypes (line, dash, and dot combinations) and complex linetypes (including text or other shapes) are possible.

## Linetype Drop-Down List

The Ribbon contains a drop-down list for selecting linetypes (Fig. 11-19). Although this appears to be quick and easy, you can only assign linetypes to objects by this method unless *ByLayer* is selected to use the layers' assigned linetypes. Any linetype you select from this list becomes the current object linetype. If you want to select linetypes for layers, make sure this list displays the *ByLayer* setting, then use the *Layer Properties Manager* to select linetypes for layers.

FIGURE 11-19

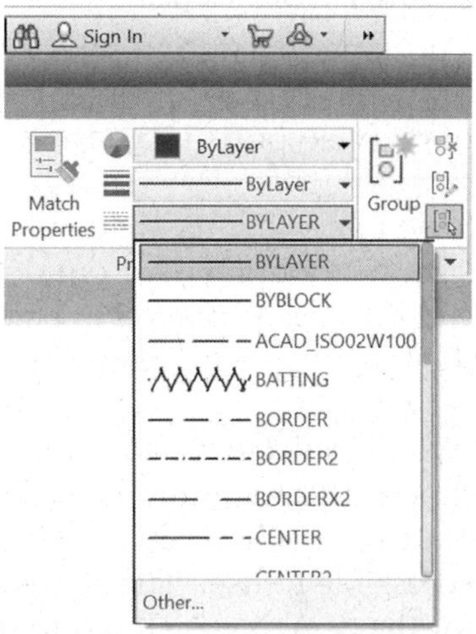


Keep in mind that the *Linetype* drop-down list as well as the *Layer*, *Color*, and *Lineweight* drop-down lists display the current settings for selected objects (objects selected when no commands are in use). When linetypes are assigned to layers rather than objects, the layer drop-down list reports a *ByLayer* setting.

### *Lweight*

| Ribbon | Menu Bar | Command (Type) | Alias (Type) | Shortcut |
|---|---|---|---|---|
| *Home*<br>*Properties*<br>(drop-down) | *Format*<br>*Lineweight...* | *Lweight* | *LW* | ... |

The *Lweight* command (short for lineweight) produces the *Lineweight Settings* dialog box (Fig. 11-20). This dialog box is the lineweight equivalent of the *Linetype Manager*—that is, this dialog box assigns lineweights to objects, not layers (unless the *ByLayer* or *Default* lineweight is selected). See the previous discussion under "*Linetype*." The *ByBlock* setting is discussed in Chapter 21.

FIGURE 11-20

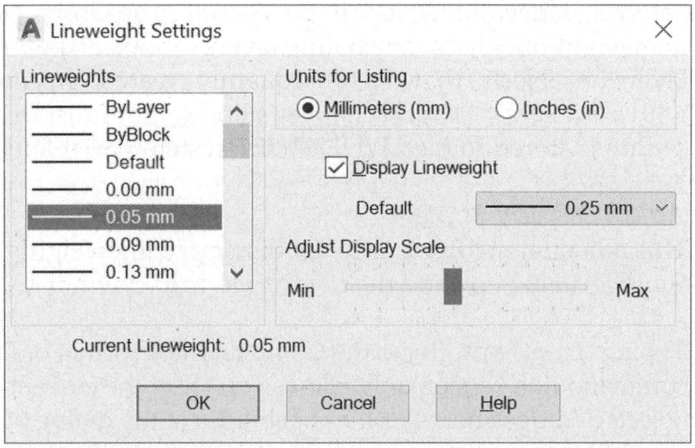


Any lineweight selected in the *Lineweight Settings* dialog box automatically becomes current (without having to select a *Current* button or double-click). The current lineweight is assigned to all subsequently drawn objects. That is, selecting a lineweight by this manner (making it current) causes all objects from that time on to be drawn using that lineweight, regardless of the layer that they are on (unless *ByLayer* or *Default* is selected). In contrast, selecting lineweights during the *Layer* command (using the *Layer Properties Manager*) results in assignment of lineweights to layers.

As a reminder, this type of drawing method (object-specific property assignment) can be difficult for beginning drawings and for complex drawings. If you want to draw objects in the layer's assigned lineweight, select *ByLayer*. If you want all layers to have the same lineweight, select *Default*.

### *Units for Listing*

You can choose *Millimeters (mm)* or *Inches (in)* as the units for the lineweight list. The values reported in the list (0.05 mm, 0.15 mm, 0.010", 0.015", etc.) indicate the thickness of the selected line. The setting is stored in the *LWUNITS* system variable.

### *Display Lineweight*

Checking this option causes the assigned lineweight thickness to be displayed anywhere in the drawing—in model space (the *Model* tab), for model space geometry that appears in paper space viewports (viewports in a *Layout* tab), and in paper space (in a *Layout* tab, but not in a viewport).

When lineweights are displayed in model space (the *Model* tab), the line thickness is relative to the screen size, so zooming in or out does not make the lines appear wider or narrower. However, when lineweights are displayed in paper space (a *Layout* tab), the line thickness is absolute, so zooming changes the lineweight appearance.

The *Display Lineweight* checkbox has the same function as toggling the *Linetype Display* icon on the Status Bar (Fig. 11-21). The toggle is somewhat more accessible to use. The status of the lineweight display for the drawing is stored in the *LWDISPLAY* system variable. Using either the *Display Lineweight* checkbox or the *Lineweight Display* icon changes the setting of the *LWDISPLAY* variable, and vice versa.

**FIGURE 11-21**

### *Default*

Although the *Current* lineweight for new drawings is *ByLayer*, when new layers are created, all layers have the *Default* lineweight. This feature (*Default* rather than *ByLayer*) is exclusive for lineweights. Assigning the *Default* setting ensures that all layers (or objects) with this setting have the same lineweight. The advantage is that the lineweight can be changed globally for all layers (or objects) that have the *Default* setting.

Selecting a new lineweight in the *Default* drop-down list of the *Lineweight Settings* dialog box changes the lineweight thickness for existing layers (or objects) with a *Default* setting and assigns that setting for new layers (or objects) that are subsequently created. In contrast, to change the lineweight for existing layers with a *ByLayer* setting, each layer's lineweight must be changed individually. The *Default* lineweight setting is stored in the *LWDEFAULT* system variable.

### *Adjust Display Scale*

This adjustment affects only the display of lineweights in the *Lineweight Settings* dialog box and the *Lineweight* drop-down list.

Typing *-Lweight* produces the Command line format of *Lweight*. This command line version only allows you to set the current lineweight for new objects and does not offer the capabilities of the dialog box version. Typing *-Lweight* (use the hyphen prefix) produces the following prompt:

Command: **-LWEIGHT**
Current lineweight: ByLayer
Enter default lineweight for new objects or [?]:

**FIGURE 11-22**

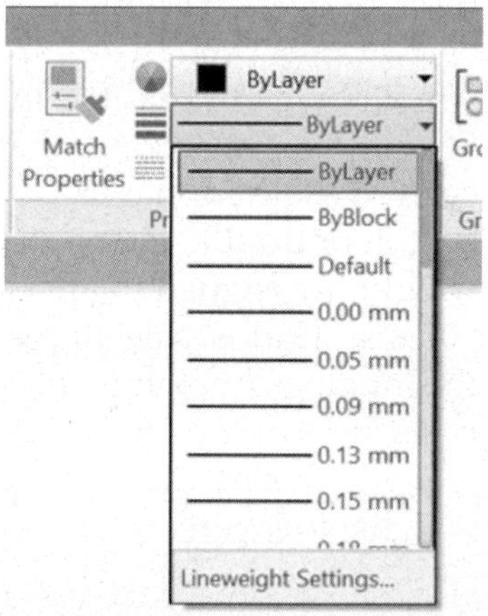


## Lineweight Drop-Down List

The Ribbon contains a drop-down list for selecting lineweights (Fig. 11-22). Similar to the function of the *Linetype* drop-down list, you can only assign lineweights to objects by this method unless *ByLayer* is selected to use the layers' assigned lineweights.

TIP

Any lineweight you select from this list becomes the current object lineweight. If you want to select lineweights for layers, make sure this list displays the *ByLayer* setting, then use the *Layer Properties Manager* to select lineweights for layers.

Remember that the *Lineweight* drop-down list as well as the *Layer*, *Color*, and *Linetype* drop-down lists display the current settings for selected objects (objects selected when no commands are in use). When lineweights are assigned to layers rather than objects, the layer drop-down list reports a *ByLayer* setting.

## Color

| Ribbon | Menu Bar | Command (Type) | Alias (Type) | Shortcut |
|---|---|---|---|---|
| *Home Properties* (drop-down) | *Format Color...* | *Color* | *COL* | ... |

Similar to linetypes and lineweights, colors can be assigned to layers or to objects. Using the *Color* command assigns a color for all newly created objects, regardless of the layer's color designation (unless the *ByLayer* color is selected). This color setting overrides the layer color for any newly created objects so that all new objects are drawn with the specified color no matter what layer they are on. This type of color designation prohibits your ability to see which objects are on which layers by their color; however, for some applications object color setting may be desirable. Use the *Layer Properties Manager* to set colors for layers.

Invoking this command by the menus or by typing *Color* presents the *Select Color* dialog box shown in Figure 11-23. This is essentially the same dialog box used for assigning colors to layers; however, the *ByLayer* and *ByBlock* tiles are accessible. (The buttons are grayed-out when this dialog is invoked from the *Layer Properties Manager* because, in that case, any setting is a *ByLayer* setting.) *ByBlock* color assignment is discussed in Chapter 21.

FIGURE 11-23

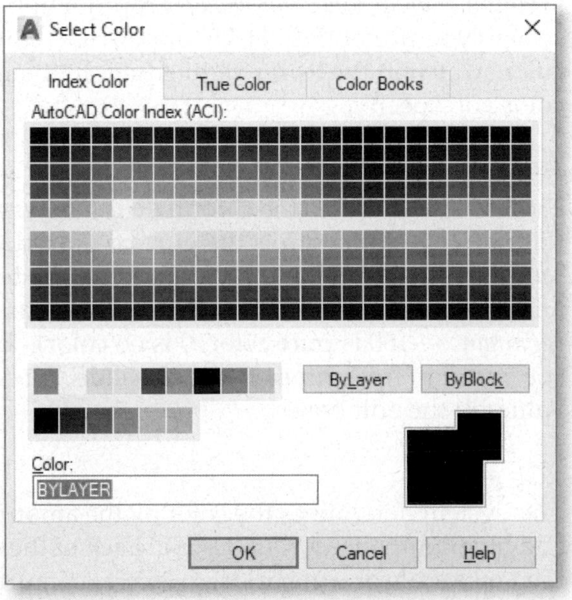


If you type *-Color* (include the hyphen prefix), the Command line format is displayed as follows:

```
Command: -COLOR
Enter default object color [Truecolor/COlorbook] <BYLAYER>:
```

The current color (whether *ByLayer* or specific object color) is saved in the *CECOLOR* (Current Entity Color) variable. The current color can be set by changing the value in the *CECOLOR* variable directly at the command prompt (by typing *CECOLOR*), by using the *Color* command (*Select Color* dialog box), or by using the *Color* drop-down list on the Ribbon. The *CECOLOR* variable accepts a string value (such as "red" or "bylayer") or accepts the ACI number (0 through 255). Valid values for true colors are a string of integers each from 1 to 255 separated by commas and preceded by RGB. A true color setting is entered as follows: RGB:000,000,000.

## Color Drop-Down List

TIP

When an object-specific color has been set, the top item in the *Color* drop-down list displays the current color (Fig. 11-24). Beware—using this list to select a color assigns an object-specific color unless *ByLayer* is selected. Any color you select from this list becomes the current object color. Choosing *More Colors…* from the bottom of the list invokes the *Select Color* dialog box (see Fig. 11-23). If you want to assign colors to layers, make sure this list displays the *ByLayer* setting, then use the *Layer Properties Manager* to select colors for layers.

**FIGURE 11-24**

The *Color* drop-down list as well as the others in the *Properties* panel display the current settings for selected objects (when no commands are in use).

## *True Color* and *Color Books*

AutoCAD offers support for true color and industry-standard color books. Three tabs are available in the *Select Color* dialog box—*Index Color*, *True Color*, and *Color Books*. The *Index Color* tab, discussed earlier in the chapter (see Fig. 11-23), allows you to specify color using the 256-color ACI (AutoCAD Color Index).

### *True Color*

The *True Color* tab in the *Select Color* dialog box provides a tool for you to specify colors by the *HLS* or the *RGB* color systems. Select the *Color Model* drop-down list to select which system you prefer (Fig. 11-25, upper-right corner). Watch the color swatch in the lower-right corner to dynamically display the color you specify.

**FIGURE 11-25**

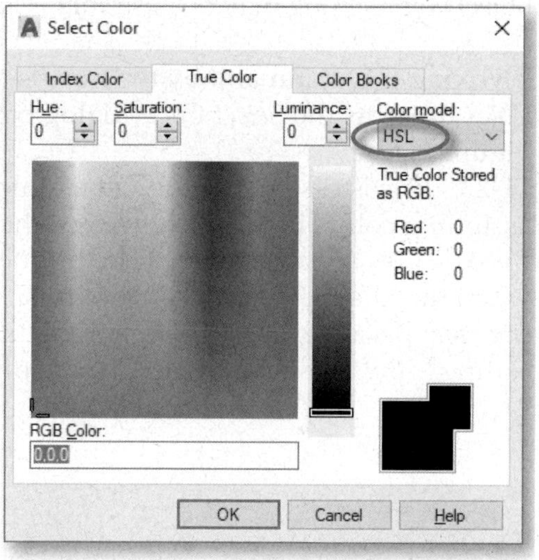


*HLS*

This system specifies the color by *Hue*, *Luminance*, and *Saturation* (Fig. 11-25). *Hue* controls the pure color (red, yellow, blue), *Luminance* controls the color "value" (white to black—100 to 0), and *Saturation* controls the purity of the color (mix of *Hue* and *Luminance*—100 is pure color, 0 is no color). Use the pointer to specify *Hue* and *Saturation* and the slider for *Luminance* or enter values in the edit boxes.

*RGB*

This system determines the color by the amount of red, green, and blue components (Fig. 11-26). Each of these colors is specified on a scale of 0 (no color) to 255 (maximum color). You can use the sliders or enter in the RGB values in the edit box.

**FIGURE 11-26**

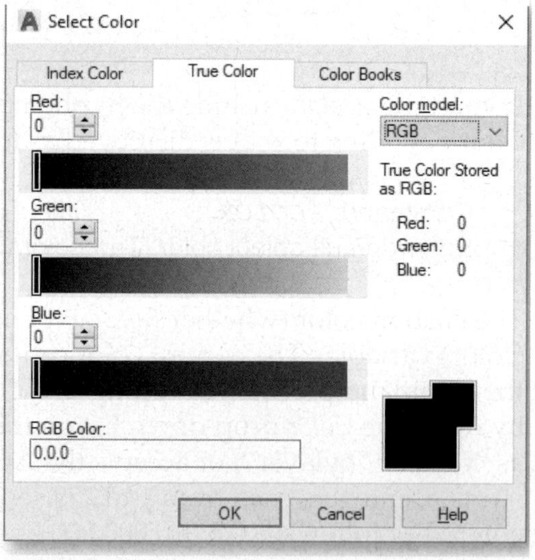


### *Color Books*

The *Color Books* tab in the *Select Color* dialog box allows you to specify colors using the DIC, Pantone, or RAL commercial color systems. These systems are used largely in the architectural and interior design industries as a standard for specifying colors for paint and interior furnishings. These standard colors have been traditionally supplied to professionals in books displaying color samples for design and color matching.

Select the desired DIC, Pantone, or RAL "book" from the *Color book* drop-down list (Fig. 11-27). Use the slider and up/down arrows to display the color samples. Specify a particular color by clicking on it so the color appears in the color swatch (lower-right corner) and the index number appears in the *Color* edit box. If you know the desired index number, you can enter it directly in the edit box.

**FIGURE 11-27**

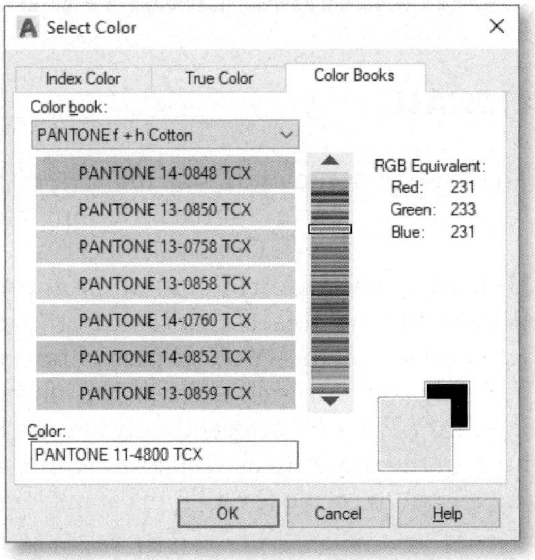


## Setbylayer

| Ribbon | Menu Bar | Command (Type) | Alias (Type) | Shortcut |
|---|---|---|---|---|
| *Home Modify* (icon) | *Modify Change to ByLayer* | *Setbylayer* | ... | ... |

You can change object-specific (*ByBlock*) property settings for selected objects to *ByLayer* settings using this command. *Setbylayer* prompts you to select objects. Properties for the selected objects are automatically changed to the properties of the layer that the objects are on. Properties that can be changed are *Color*, *Linetype*, *Lineweight*, *Material*, and *Plot Style*.

Command: ***SETBYLAYER***
Current active settings: Color Linetype Lineweight Material
Select objects or [Settings]: **PICK**
Select objects or [Settings]: **Enter**
Change ByBlock to ByLayer? [Yes/No] <Yes>: ***Y*** or ***N***
Include blocks? [Yes/No] <Yes>: ***Y*** or ***N***
*nn* objects modified, *nn* objects did not need to be changed.
Command:

If *Settings* is selected, the *SetByLayer Settings* dialog box is displayed (Fig. 11-28). Here you can specify which object properties are set to *ByLayer*.

**FIGURE 11-28**

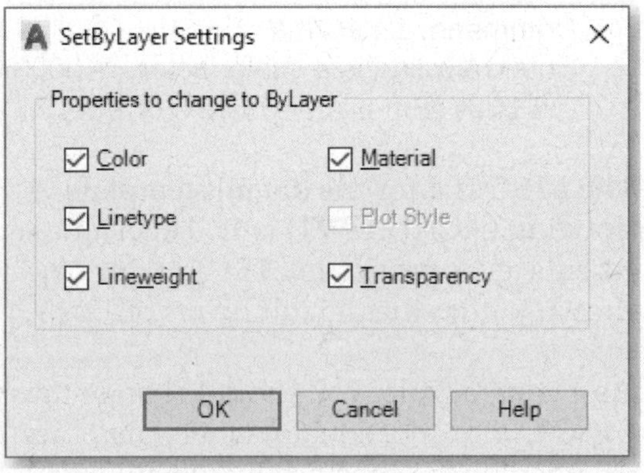


Occasionally you may be required to edit a drawing created by another person or company. This command is especially useful when you want to change such a drawing (that has *ByBlock* properties) to a *ByLayer* setting.

# CONTROLLING LINETYPE SCALE

## *LTSCALE*

The scale of non-continuous linetypes is controlled by the *LTSCALE*. *LTSCALE* is a system variable, and therefore is not found on the Ribbon, menus, or tools.

Hidden, dashed, dotted, and other linetypes that have spaces are called non-continuous linetypes. When drawing objects that have non-continuous linetypes (either *ByLayer* or object-specific linetype designations), the linetype's dashes or dots are automatically created and spaced. The *LTSCALE* (Linetype Scale) system variable controls the length and spacing of the dashes and/or dots. The value that is specified for *LTSCALE* affects the drawing globally and retroactively. That is, all existing non-continuous lines in the drawing as well as new lines are affected by *LTSCALE*. You can therefore adjust the drawing's linetype scale for all lines at any time with this one command.

TIP

If you choose to make the dashes of non-continuous lines smaller and closer together, reduce *LTSCALE*; if you desire larger dashes, increase *LTSCALE*. The *Hidden* linetype is shown in Figure 11-29 at various *LTSCALE* settings. Any positive value can be specified.

**FIGURE 11-29**

*LTSCALE* can be set in the *Details* section of the *Linetype Manager* or in Command line format. In the *Linetype Manager*, select the *Details* button to allow access to the *Global scale factor* edit box (Fig. 11-30). Changing the value in this edit box sets the *LTSCALE* variable. Changing the value in the *Current object scale* edit box sets the linetype scale for the current object only (see "*CELTSCALE*"), but does not affect the global linetype scale (*LTSCALE*).

**FIGURE 11-30**

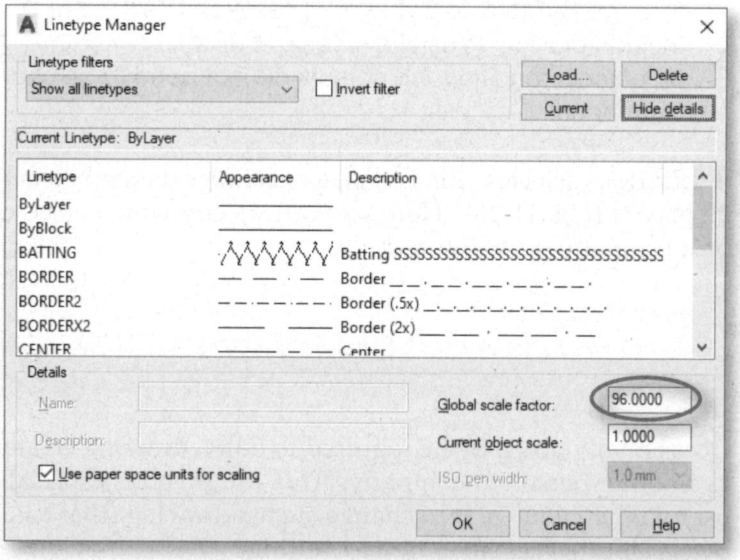


*LTSCALE* can also be used in the Command line format.

Command: ***LTSCALE***
Enter new linetype scale factor <1.0000>:
(**value**) (Enter any positive value.)

The *LTSCALE* for the default template drawing (ACAD.DWT) is 1. This value represents an appropriate *LTSCALE* for objects drawn within the default *Limits* of 12 x 9.

TIP

As a general rule, you should change the *LTSCALE* proportionally when *Limits* are changed (more specifically, when *Limits* are changed to other than the intended plot sheet size).

For example, if you increase the drawing area defined by *Limits* by a factor of 2 (to 24 x 18) from the default (12 x 9), you might also change *LTSCALE* proportionally to a value of 2. Since *LTSCALE* is retroactive, it can be changed at a later time or repeatedly adjusted to display the desired spacing of linetypes.

If you load the *Hidden* linetype, it displays dashes (when plotted 1:1) of 1/4" with a *LTSCALE* of 1. The ANSI and ISO standards state that hidden lines should be displayed on drawings with dashes of approximately 1/8" or 3mm. To accomplish this, you can use the *Hidden* linetype and change *LTSCALE* to .5. It is recommended, however, that you use the *Hidden2* linetype that has dashes of 1/8" when plotted 1:1. Using this strategy, *LTSCALE* remains at a value of 1 to create the standard linetype sizes for *Limits* of 12 x 9 and can easily be changed in proportion to the *Limits*. (For more information on *LTSCALE*, see Chapter 12 and Chapter 13.)

The ACAD_ISO*n*W100 linetypes are intended to be used with metric drawings. Changing *Limits* to metric sheet sizes automatically displays these linetypes with appropriate linetype spacing. For these linetypes only, *LTSCALE* is changed automatically to the value selected in the *ISO Pen Width* box of the *Linetype* tab (see Fig. 11-30). Using both ACAD_ISO*n*W100 linetypes and other linetypes in one drawing is discouraged due to the difficulty managing two sets of linetype scales.

Even though you have some control over the size of spacing for non-continuous lines, you have almost no control over the placement of the dashes for non-continuous lines. For example, the short dashes of center lines cannot always be controlled to intersect at the centers of a series of circles. The spacing can only be adjusted globally (all lines in the drawing) to reach a compromise. You can also adjust individual objects' linetype scale (*CELTSCALE*) to achieve the desired effect. See "*CELTSCALE*" (next) and Chapter 24, Multiview Drawing, for further discussion and suggestions on this subject.

### CELTSCALE

*CELTSCALE* stands for Current Entity Linetype Scale. *CELTSCALE* is actually a system variable for linetype scale stored with each object—an object property. This setting changes the object-specific linetype scale proportional to the *LTSCALE*. The *LTSCALE* value is global and retroactive, whereas *CELTSCALE* sets the linetype scale for all newly created objects and is not retroactive. *CELTSCALE* is object-specific. Using *CELTSCALE* to set an object linetype scale is similar to setting an object color, linetype, and lineweight in that the properties are assigned to the specific object.

For example, if you wanted all non-continuous lines (dashes and spaces) in the drawing to be two times the default size, set *LTSCALE* to 2 (*LTSCALE* is global and retroactive). If you then wanted only two or three lines to have smaller spacing, change *CELTSCALE* to .5, draw the new lines, and then change *CELTSCALE* back to 1.

You can also use the *Linetype Manager* to set the *CELTSCALE* (see Fig. 11-30). Set the desired *CELTSCALE* by listing the *Details* and changing the value in the *Current object scale* edit box. This setting affects all linetypes for all newly created objects. The linetype scale for individual objects can also be changed retroactively using the method explained in the following paragraph.

Using *CELTSCALE* is not the recommended method for adjusting individual lines' linetype scale. Setting this variable each time you wanted to create a new entity with a different linetype scale would be too time consuming and confusing. The recommended method for adjusting linetypes in the drawing to different scales is not to use *CELTSCALE*, but to use the following strategy.

TIP

1. Set *LTSCALE* to an appropriate value for the drawing.
2. Create all objects in the drawing with the desired linetypes.
3. Adjust the *LTSCALE* again if necessary to globally alter the linetype scale.
4. Use *Properties* and *Matchprop* to retroactively change the linetype scale for selected objects (see "*Properties*" and "*Matchprop*").

## CHANGING OBJECT PROPERTIES

Often it is desirable to change the properties of an object after it has been created. For example, an object's *Layer* property could be changed. This can be thought of as "moving" the object from one layer to another. When an object is changed from one layer to another, it assumes the new layer's color, linetype, and lineweight, provided the object was originally created with color, linetype, and lineweight assigned *ByLayer*, as is generally the case. In other words, if an object was created on the wrong layer (possibly with the wrong linetype or color), it could be "moved" to the desired layer, therefore assuming the new layer's linetype and color.

Another example is to change an individual object's linetype scale. In some cases an individual object's linetype scale requires an adjustment to other than the global linetype scale (*LTSCALE*) setting. One of several methods can be used to adjust the individual object's linetype scale to an appropriate value.

Several methods can be used to retroactively change properties of selected objects. The *Properties* panel, the *Properties* palette, and the *Match Properties* command can be used to retroactively change the properties of individual objects. Properties that can be changed by these three methods are *Layer*, *Linetype*, *Lineweight*, *Color*, (object-specific) *Linetype scale*, *Plot Style*, and other properties.

Although some of the commands discussed in this section have additional capabilities, the discussion is limited to changing the object properties covered in this chapter—specifically layer, color, linetype, and lineweight. For this reason, these commands and features are also discussed in Chapter 16, Modify Commands II, and in other chapters.

### *Properties* Panel

The five drop-down lists in the Ribbon in the *Properties* and *Layers* panels (when not "dropped down") generally show the current layer, color, linetype, lineweight, and plot style. However, if an object or set of objects is selected, the information in these lists changes to display the current objects' settings. You can change the selected objects' settings by "dropping down" any of the lists and making another selection.

First, select (highlight) an object when no commands are in use. Use the pickbox (that appears on the cross-hairs), window or crossing window to select the desired object or set of objects. The entries in the five lists (*Layer, Color, Linetype, Lineweight,* and *Plot Style*) then change to display the settings for the selected object or objects. If several objects are selected that have different properties, the boxes display no information (go "blank"). Next, use any of the drop-down lists to make another selection (Fig. 11-31). The highlighted object's properties are changed to those selected in the lists. Press the Escape key to complete the process.

FIGURE 11-31

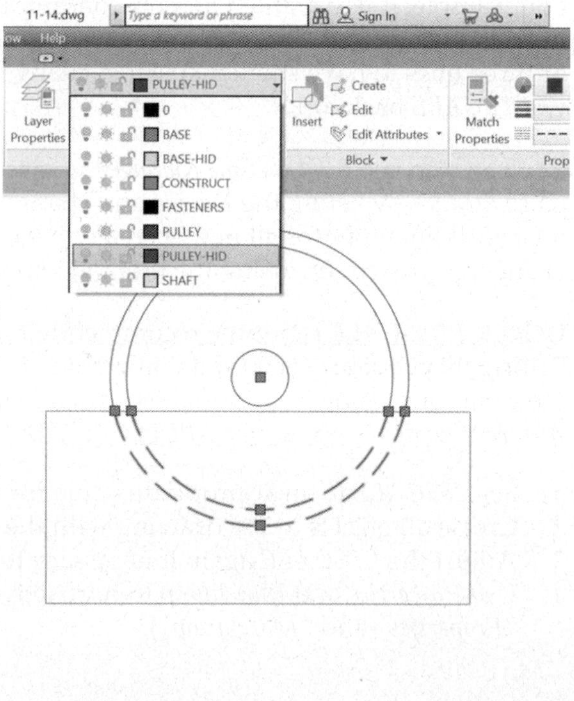


TIP

Remember that in most cases color, linetype, and lineweight settings are assigned *ByLayer*. In this type of drawing scheme, to change the linetype or color properties of an object, you would change the object's layer (see Fig. 11-31). If you are using this type of drawing scheme, refrain from using the *Color, Linetype,* and *Lineweight* drop-down lists for changing properties.

## Properties

| Ribbon | Menu Bar | Command (Type) | Alias (Type) | Shortcut |
|---|---|---|---|---|
| *View Palettes Properties* | *Modify Properties* | *Properties* | *PR* or *CH* | (Edit Mode) *Properties* or *Ctrl+1* |

The *Properties* palette (Fig. 11-32) gives you complete access to one or more objects' properties. The contents of the palette change based on what type and how many objects are selected. You can use the palette two ways: you can invoke the palette, then select (highlight) one or more objects, or you can select objects first and then invoke the palette. Once opened, this palette remains on the screen until dismissed by clicking the "X" in the upper corner. You can even toggle the palette on and off with Ctrl+1 (to appear and disappear). When multiple objects are selected, a drop-down list in the top of the palette appears. Select the object(s) whose properties you want to change.

FIGURE 11-32

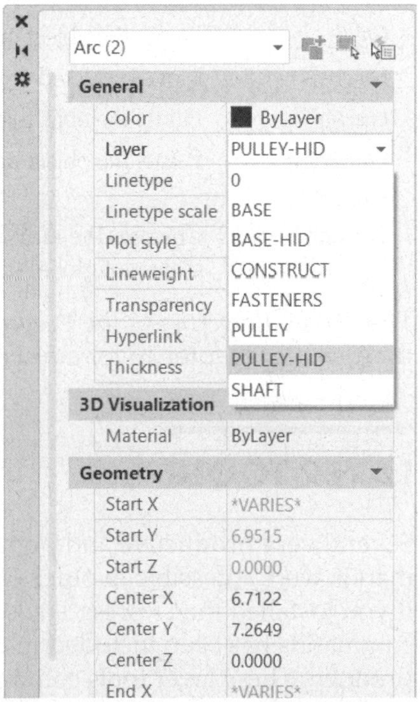


To change an object's layer, linetype, lineweight, or color properties, highlight the desired objects in the drawing and select the properties you want to change from the right side of the palette. In the *General* list, the layer, linetype, lineweight, and color properties are located in the top half of the list.

TIP

If you use the *ByLayer* strategy of linetype, lineweight, and color properties assignment, you can change an object's linetype, lineweight, and color simply by changing its layer (see Fig. 11-32). This method is recommended for most applications.

## Matchprop

| Ribbon | Menu Bar | Command (Type) | Alias (Type) | Shortcut |
|---|---|---|---|---|
| *Home Properties Match Properties* | *Modify Match Properties* | *Matchprop* | *MA* | ... |

*Matchprop* is used to "paint" the properties of one object to another. The process is simple. After invoking the command, select the object that has the desired properties (source object), then select the object you want to "paint" the properties to (destination object). The command prompt is as follows.

Command: ***MATCHPROP***
Select source object: **PICK**
Current active settings: Color Layer Ltype Ltscale Lineweight Transparency Thickness PlotStyle Dim Text Hatch Polyline Viewport Table Material Multileader Center object
Select destination object(s) or [Settings]: **PICK**
Select destination object(s) or [Settings]: **Enter**
Command:

Only one "source object" can be selected, but its properties can be painted to several "destination objects." The "destination object(s)" assume all of the properties of the "source object" (listed as "Current active settings").

Use the *Settings* option to control which of several possible properties and other settings are "painted" to the destination objects. At the "Select destination object(s) or [Settings]:" prompt, type *S* to display the *Property Settings* dialog box (Fig. 11-33). In the dialog box, designate which of the *Basic Properties* or *Special Properties* are to be painted to the "destination objects." The following *Basic Properties* correspond to properties discussed in this chapter.

| | |
|---|---|
| *Color* | Paints the object-specific or *ByLayer* color. |
| *Layer* | Moves selected objects to Source Object layer. |
| *Linetype* | Paints the object-specific or *ByLayer* linetype. |
| *Lineweight* | Paints the object-specific or *ByLayer* lineweight. |
| *Linetype Scale* | Changes the individual object's linetype scale, not global (*LTSCALE*). |

**FIGURE 11-33**

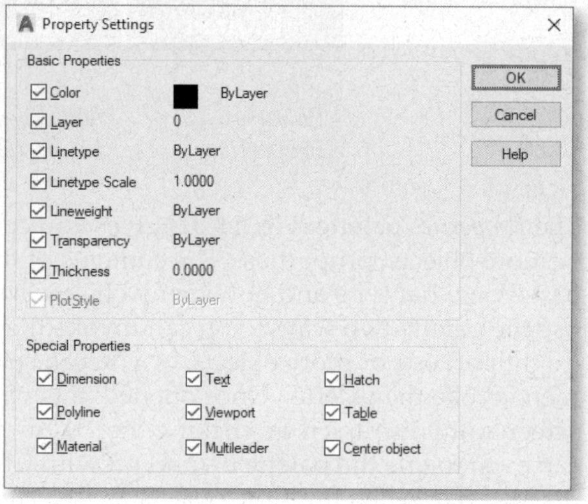


The *Special Properties* of the *Property Settings* dialog box are discussed in Chapter 16, Modify Commands II. Also see Chapter 16 for a full explanation of the *Properties* palette (*Properties*).

## LAYER TOOLS

Several very productive and popular layer-related commands were accessible in older AutoCAD releases only if you installed the Express Tools. This set of useful commands has been included in the AutoCAD core product. These layer tools commands can be found on the Ribbon in the *Home* tab, *Layers* panel. If you are using the Menu Bar, the commands are found in the *Format* pull-down, *Layer Tools* cascading menu. (Fig. 11-34).

**FIGURE 11-34**

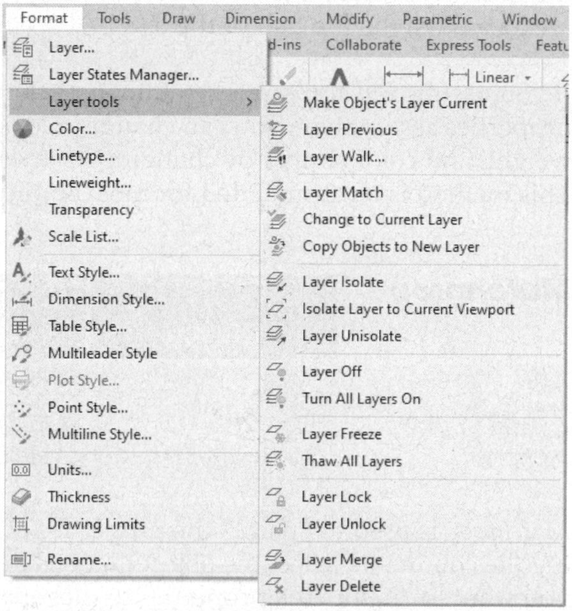


### Laymch

| Ribbon | Menu Bar | Command (Type) | Alias (Type) | Shortcut |
|---|---|---|---|---|
| *Home Layers* (icon) | *Format Layer Tools > Layer Match* | *Laymch* | ... | ... |

*Laymch* allows you to change the layer of selected objects to the layer of the selected destination object. This command performs the same function as the *Matchprop* command (see Chapter 12), but only with respect to layers (*Matchprop* can change other properties such as color, linetype, or linetype scale).

Using *Laymch* produces the following prompt:

Command: ***LAYMCH***
Select objects to be changed:
Select objects: **PICK**
Select objects: **PICK**
Select objects: **Enter**
2 found.
Select object on destination layer or [Name]: **PICK**
2 objects changed to layer 0.
Command:

You can use the *Name* option to select the desired layer to change objects to. The *Change to Layer* dialog box appears (Fig. 11-35). Choose the destination layer for the selected objects.

**FIGURE 11-35**

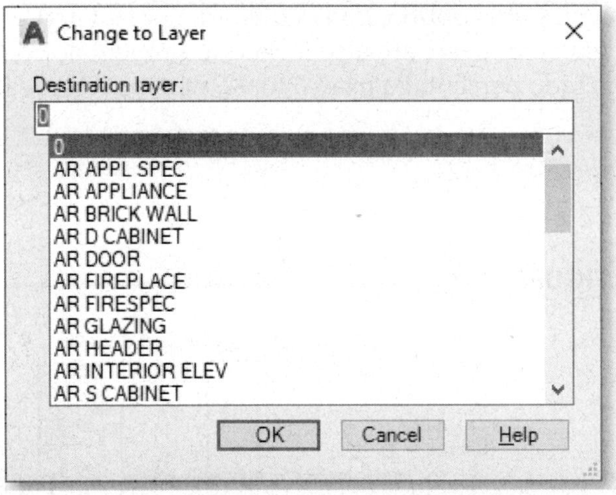


### Laycur

| Ribbon | Menu Bar | Command (Type) | Alias (Type) | Shortcut |
|---|---|---|---|---|
| *Home Layers* (icon) | *Format Layer Tools > Change to Current Layer* | *Laycur* | ... | ... |

*Laycur* changes the layer of selected objects to the current layer. The following prompt is issued:

Command: ***LAYCUR***
Select objects to be changed to the current layer:
Select objects: **PICK**
Select objects: **PICK**
Select objects: **Enter**
2 found. 2 objects changed to layer TEXT (the current layer).
Command:

### Layiso

| Ribbon | Menu Bar | Command (Type) | Alias (Type) | Shortcut |
|---|---|---|---|---|
| *Home Layers* (icon) | *Format Layer Tools > Layer Isolate* | *Layiso* | ... | ... |

*Layiso* allows you to see, or "isolate," specific layers. This is useful if you want to work with one or a few layers in a drawing with many layers. All other layers that are not isolated can be either locked and faded to appear less visible or turned off so they are not visible at all. Select object(s) on the layer(s) to be isolated. The object's layer (or the last selected object's layer) becomes the current layer.

Command: ***LAYISO***
Current setting: Hide layers, Viewports=Off
Select objects on the layer(s) to be isolated or [Settings]: **PICK**
Select objects on the layer(s) to be isolated or [Settings]: **Enter**
Layer A-FURN-P-PHWL-46 has been isolated.
Command:

If *Settings* is selected, the options for visibility and viewports can be specified. The setting you choose persists from session to session.

Enter setting for layers not isolated [Off/Lock and fade] <Lock and fade>:

### *Lock and fade*

Using this option, layers that are not isolated are faded to make the isolated layer(s) more visible. Non-isolated layers are also locked to prevent accidental modification to objects on those layers. You specify a fade percentage from 0 to 90, where 90 represents almost total invisibility.

Enter fade value (0-90) <50>:

**FIGURE 11-36**

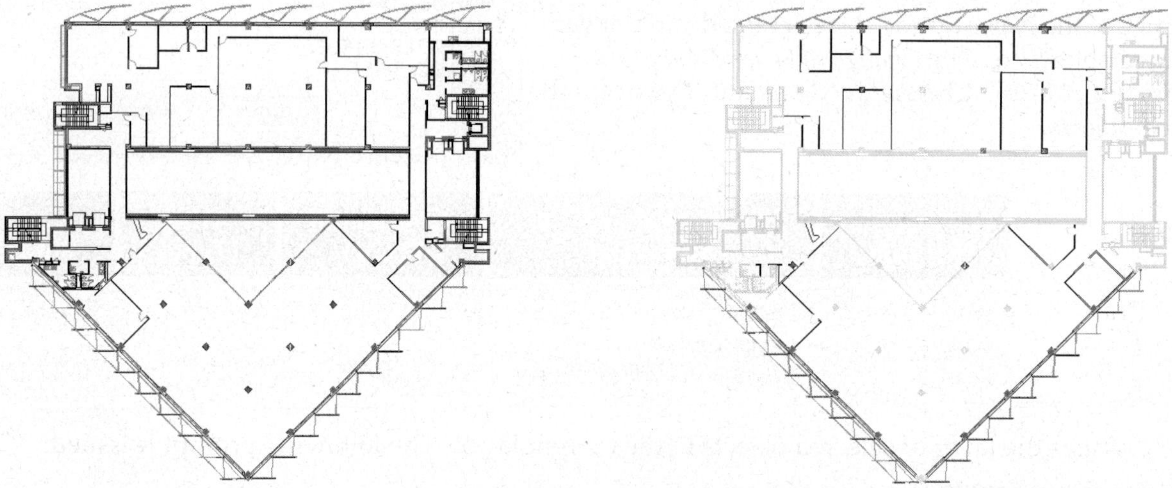

For example, Figure 11-36 illustrates application of the *Lock and fade* option. On the left, no layers are isolated so all layers appear in the same intensity. However, the right figure displays the same drawing using the *Lock and fade* option where the isolated layers appear at full intensity and non-isolated layers are locked and faded.

You can expand the *Layers* panel on the Ribbon to gain acess to some *Layer Tools* commands as well as a *Locked layer fading* slider (Fig. 11-37). The slider determines the fade value and displays in the drawing real-time.

**FIGURE 11-37**

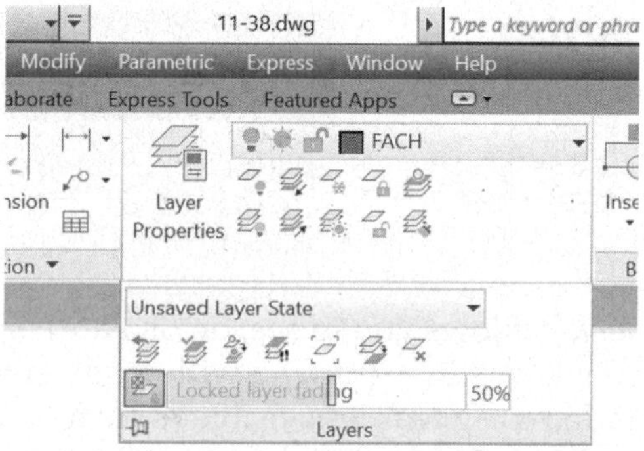


### *Off*

Rather than locking and fading non-isolated layers, you can use the *Off* option to turn off the layer visibility totally for non-isolated layers. You can then determine if you want to turn off all non-isolated layers or only those in the current viewport.

In paper space viewport use [Vpfreeze/Off] <Off>:

*Vpfreeze*

If you are using *Layiso* in paper space, *Vpfreeze* isolates layers only in the current viewport. Objects on the isolated layer(s) are displayed and all other layers in the current viewport are frozen. Other viewports in the drawing are unchanged.

*Off*

This option isolates layers in all viewports—model space and paper space. Objects on the isolated layer are displayed and all other layers are turned off in viewports and in model space.

### *Layuniso*

| Ribbon | Menu Bar | Command (Type) | Alias (Type) | Shortcut |
|---|---|---|---|---|
| *Home*<br>*Layers*<br>(icon) | *Format*<br>*Layer Tools >*<br>*Layer Unisolate* | *Layuniso* | ... | ... |

***Layuniso*** simply thaws all layers that were previously frozen with *Layiso*. All layers are restored to the state they were in just before you used *Layiso*. If *Layiso* was not used, *Layuniso* does not restore any layers.

Command: ***LAYUNISO***
Layers isolated by LAYISO command have been restored.

### *Copytolayer*

| Ribbon | Menu Bar | Command (Type) | Alias (Type) | Shortcut |
|---|---|---|---|---|
| *Home*<br>*Layers*<br>(icon) | *Format*<br>*Layer Tools >*<br>*Copy Objects to New Layer* | *Copytolayer...* | ... | ... |

When you need two different layers that contain similar (or duplicate) objects, such as "Existing Doors" and "Demo Doors," you can use this express feature to select existing object(s), then choose the destination layer for copied objects from a dialog box. The destination layer must exist prior to launching this command.

Command: ***COPYTOLAYER***
Select objects to copy: **PICK**
Select objects to copy: **Enter**
Select object on destination layer or [Name] <Name>: ***n***
1 object(s) copied and placed on layer "CHAIRS".
Specify base point or [Displacement/eXit] <eXit>: **PICK**
Specify second point of displacement or <use first point as displacement>: **PICK**
Command:

When asked for the "destination layer," you can PICK an object on the desired layer or use the *Name* option to select the desired layer to copy the objects to. The *Name* option produces the *Copy to Layer* dialog box (Fig. 11-38) where you select the destination layer for the copied objects.

**FIGURE 11-38**

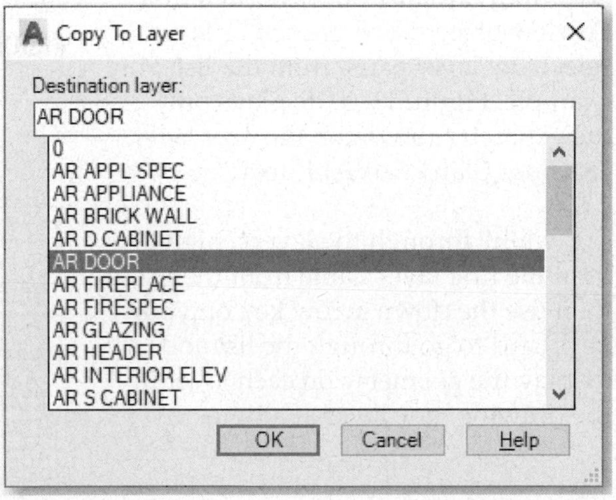


**TIP**

Also note that you can select a new location for the copied objects when prompted for a "base point" and "second point of displacement." In some cases, such as copying objects from an "Existing Doors" layer to a "Demo Doors" layer, you can press Enter to copy the new objects to the same coordinates as the existing objects.

### Laywalk

| Ribbon | Menu Bar | Command (Type) | Alias (Type) | Shortcut |
|---|---|---|---|---|
| *Home Layers* (icon) | *Format Layer Tools > Layer Walk* | *Laywalk* | ... | ... |

TIP

The main feature of this utility enables you to view the objects on any layer(s) in a drawing even if the layers are *Off* or *Frozen*. *Laywalk* enables you to view objects on any specific layer or any set of layers, one layer at a time if desired. *Laywalk* is best utilized to examine the layer contents of unfamiliar drawings, such as those created by an outside source. Although *Laywalk* dynamically changes the *Freeze/Thaw* or *Off/On* state of any layers you select to view, you can choose not to keep the changes made using *Laywalk* and restore the original layer visibility settings when you exit this utility. The *LayerWalk* dialog box provides several options and a shortcut menu with additional features to manipulate the visibility of layers. You also have access to the *Layer Manager* utility from the shortcut menu, enabling you to manipulate and save layer states.

NOTE: Since *Laywalk* shows the contents of each layer, you may want to use *Zoom Extents* just prior to launching *Laywalk* with an unfamiliar drawing.

Invoking the *Laywalk* command by any method produces the *LayerWalk* dialog box. When the *LayerWalk* dialog box first appears, layers that are visible in the drawing are highlighted in the list. For example in Fig. 11-39, the *LayerWalk* dialog box list indicates (highlights) the names of all visible layers in the drawing, but the A-FURN-* layers are not visible in the drawing and are not highlighted in the list.

NOTE: The *LayerWalk* dialog box is sizable, so enlarge the dialog box if needed to see more of the layer names in the list.

You can also easily view objects on each layer individually, one layer at a time. To view objects on a particular layer, select any layer name from the list. For example, Figure 11-40 displays only the geometry drawn on the Xref layer "8th floor plan | A-WALL-INT."

To "walk" through the layers, highlight the first layer name from the list, then use the down arrow key on your keyboard to go through the list and display the geometry on each layer in the drawing, one layer at a time.

FIGURE 11-39

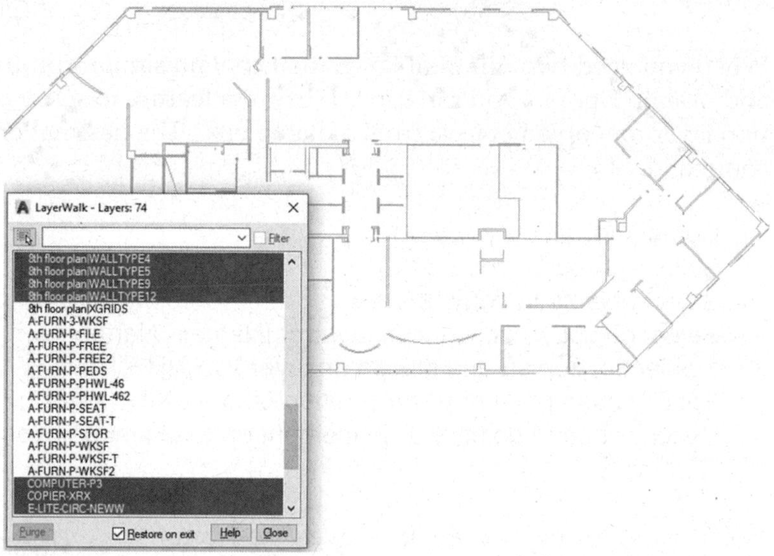


FIGURE 11-40

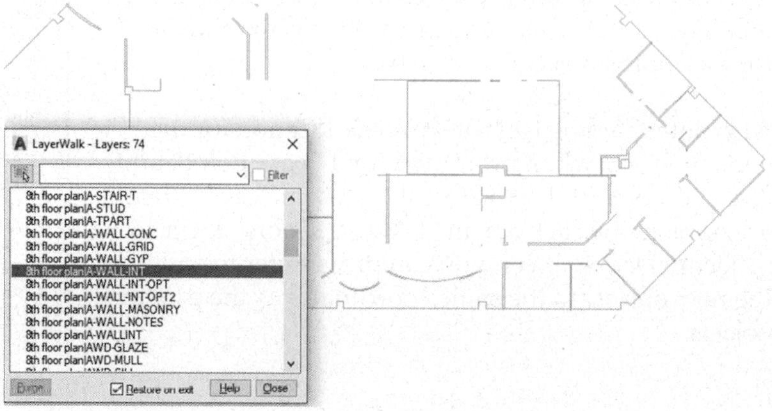

You can also select more than one layer by holding down the left button and dragging your cursor across the layers to view (or use Ctrl to select single layers or Shift to select a range).

### Layer name edit box and *Filter* check box

As an alternative, you can enter individual layer names in the edit box at the top of the *LayerWalk* dialog box to display that layer. Wildcards can be used. For example entering "A-FURN-*" in the edit box would display those layers in the drawing whose names begin with "A-FURN-" (Fig. 11-41).

**FIGURE 11-41**

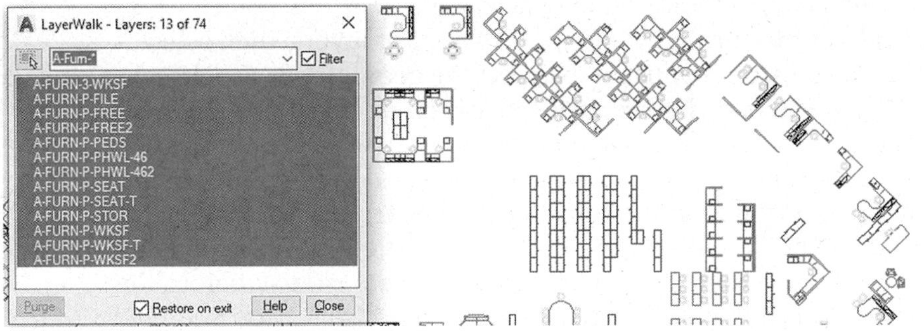


Use a filter to reduce the number of layers appearing in the list and in the drawing. Create a filter by entering layer names with a common prefix, then an asterisk, such as "A-FURN-*" (see Figure 11-41), or enter layer names of an Xref drawing, such as "8th floor plan*." This action creates a temporary filter for the list. When the *Filter* box is checked, only those layers listed in the edit box appear in the list and in the drawing. With a filter in effect, you can still select individual names from the list to view the geometry, etc. Uncheck the *Filter* box to make all layers in the drawing appear in the list. A filter remains in effect until cleared by unchecking the box. Remember that filters are temporary (useful only during the current *LayerWalk* session) unless saved. Use *Save Filter* in the *LayerWalk* dialog box's shortcut menu to save filters for use in subsequent *LayerWalk* sessions.

### *Select Objects* button

This button (upper-left corner of the dialog box) enables you to select one or multiple objects in the drawing, then returns to the *LayerWalk* dialog box with the objects' layer(s) highlighted. This is an excellent way to examine an unfamiliar drawing and find out which layers specific objects were drawn on, especially when selected objects are on more than one layer.

### *Purge*

*Purge* deletes layers from the drawing (even if *Restore on Exit* is checked). The *Purge* button is enabled only if there are one or more unreferenced layers (layers containing no objects) in the drawing and those layers are highlighted in the list. Use the *Select Unreferenced* option in the shortcut menu to locate and highlight any unused layers, then the *Purge* button to delete those layer names from the drawing and from the layer list.

### *Restore on Exit* check box

*Restore on Exit* means to disregard any changes made to the layer status using *LayerWalk* and restore the drawing to the visibility state before using *LayerWalk*. Use this option if you want to examine the geometry on specific layers of a drawing but then return to work on the drawing without saving any of those layer visibility changes. In other words, if *Restore on Exit* is checked when you *Close* the *LayerWalk* dialog box, AutoCAD returns to the layers displayed prior to using *LayerWalk*. If you want to keep the layer display attained using *LayerWalk*, but return to work on the drawing, make sure *Restore on Exit* is not checked before you close the dialog box.

### Shortcut menu

Right-click in the *LayerWalk* dialog to display the shortcut menu (Fig. 11-42).

FIGURE 11-42

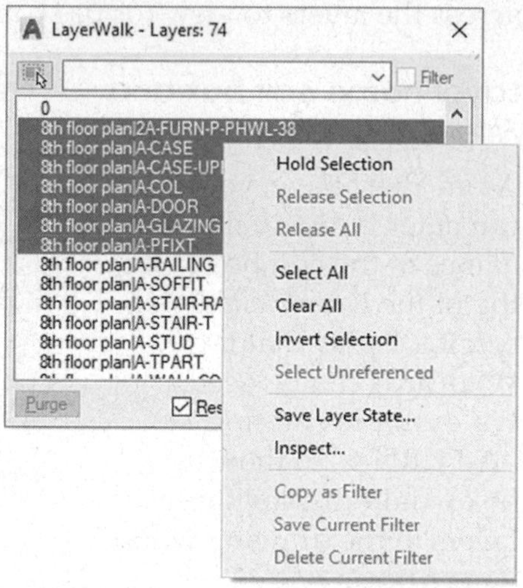


*Hold Selection, Release Selection, Release All*

If you want to turn on the display of a set of layers, then select other layers from the list to examine the geometry in the drawing, use *Hold Selection*. For example, in an architectural setting you could highlight the floorplan layer names, use *Hold Selection* to keep those layers visible, then select other layer names to display that geometry within the context of the floorplan. When you "hold" a list of layers, highlighting is removed from the names, so "held" selections appear in the list with an asterisk before the name. You can "hold" multiple selections. Use *Release Selection* to remove one selection from "hold" status and use *Release All* to remove multiple layer selections.

TIP

*Select All, Clear All*

Use *Select All* to select all names in the list or *Clear All* to clear the list. Use *LayerWalk's Select All* feature to quickly turn on all layers in a drawing, even if the layers selected are *Frozen*!

*Invert Selection*

The *Invert Selection* option highlights (and displays in the drawing) all layers from the list previously not highlighted and clears (and does not display in the drawing) those names previously shown.

*Select Unreferenced*

The *Select Unreferenced* feature highlights any unreferenced layers in the drawing (layers that contain no geometry). Using *Select Unreferenced* is a quick way to locate unused layers so you can then use the *Purge* button to remove those layer names from the drawing.

FIGURE 11-43

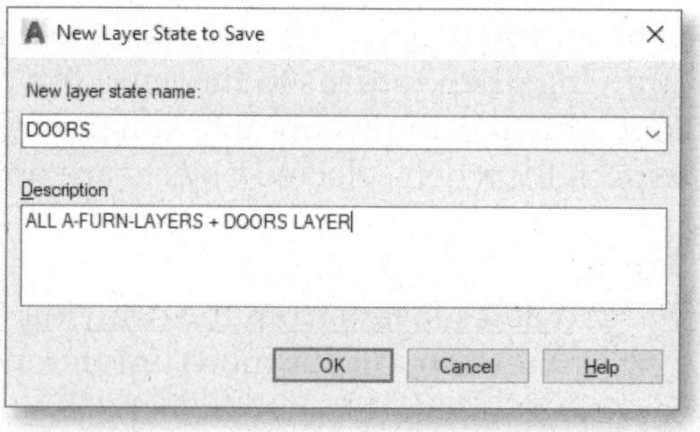


*Save Layer State*

The *Save Layer State* feature enables you to give a name to the highlighted layer group for future use (Fig. 11-43). A *Layer State* is a particular layer visibility setting saved with a name. For example, if you use *LayerWalk* to specify a set of layers for viewing, but expect to view this specific layer set again sometim in the future, *Save* the layer state. In this way you can *Restore* the layer state by name in the future rather than having to select that specific set of layers one by one in the *LayerWalk* dialog box. Named layer states are available in the *Layer States Manager*. See previous discussion, "*Layerstate*," for more information.

*Inspect*

Select *Inspect* to open the *Inspect* dialog box (Fig. 11-44). This utility displays the number of layers in the drawing, the number of layers selected (if any), and the number of objects on the selected layers.

FIGURE 11-44

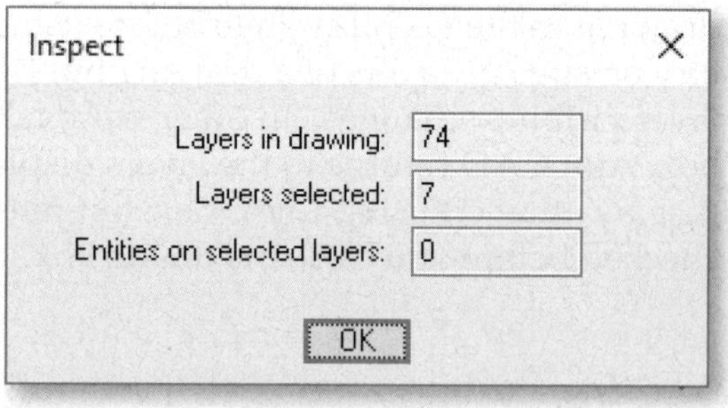


*Copy Filter, Save Current Filter, Delete Current Filter*

Use the *Copy Filter, Save Current Filter, Delete Current Filter* functions to capture and discard layer filter names. Remember that filters are temporary (useful only during the current *LayerWalk* session) unless saved (see previous discussion, "Layer name edit box and *Filter* checkbox").

To save a filter for use in subsequent *LayerWalk* sessions (after closing the *LayerWalk* dialog box), create a filter by entering a layer name string in the edit box (see previous Fig. 11-41), then select *Save Current Filter* from the shortcut menu. Or, use *Copy Filter* if a single layer name has been highlighted and you want to save the name as a filter. The named filter then appears in the drop-down list whenever *LayerWalk* is used again for the drawing. The filter is saved regardless of the setting for *Restore on Exit*. The *LayerWalk* filters are not available as named layer filters from the *Layer Properties Manager*.

## Layfrz

| Ribbon | Menu Bar | Command (Type) | Alias (Type) | Shortcut |
|---|---|---|---|---|
| *Home* *Layers* (icon) | *Format* *Layer Tools >* *Layer Freeze* | *Layfrz* | ... | ... |

This utility *Freezes* the layer(s) of the selected object(s). More than one object (layer) can be selected for *Freezing* with this command, but you select only one object at a time, then the prompt repeats. You cannot *Freeze* the current layer. AutoCAD gives the following prompt:

Command: **LAYFRZ**
Current settings: Viewports=Vpfreeze, Block nesting level=Block
Select an object on the layer to be frozen or [Settings/Undo]: **PICK**
Layer "CHAIRS" has been frozen.
Select an object on the layer to be frozen or [Settings/Undo]: **PICK**
Layer "E-F-DOOR" has been frozen.
Select an object on the layer to be frozen or [Settings/Undo]: **Enter**
Command:

You can use *Undo* to undo the last layer (object) selected. Type *S* for *Settings* to produce the following prompt:

Enter setting type for [Viewports/Block selection]:

### *Viewports*

Using this option you have the choice to freeze layers only for the current viewport (*Viewports*) or freeze the layer globally for the drawing (*Freeze*).

### *Block selection*

This option allows you to control which layers related to the block are frozen when you select a block: *None*, *Block*, or *Entity*.

*None*
If a nested *Block* or an *Xref* is selected, freezes the parent layer containing that *Block* or *Xref*.

*Block*
If a selected object is nested in a *Block*, freezes only the layer of that *Block*. If a selected object is nested in an *Xref*, freezes only the layer of the object.

*Entity*
Freezes the layers of selected individual objects within the *Block* or *Xref* even if they are nested.

## Laythw

| Ribbon | Menu Bar | Command (Type) | Alias (Type) | Shortcut |
|---|---|---|---|---|
| *Home* *Layers* (icon) | *Format* *Layer Tools >* *Thaw All Layers* | *Laythw* | ... | ... |

Use this command to <u>*Thaw* all layers</u> in the drawing. Layers that are *Frozen* <u>and</u> *Off* are only *Thawed* by this command so they still remain invisible until turned *On*:

Command: ***LAYTHW***
All layers have been thawed.
Command:

## Layoff

| Ribbon | Menu Bar | Command (Type) | Alias (Type) | Shortcut |
|---|---|---|---|---|
| *Home* *Layers* (icon) | *Format* *Layer Tools >* *Layer Off* | *Layoff* | ... | ... |

*Layoff* is used to turn *Off* the layer(s) of the selected object(s). Multiple objects can be selected (one at a time) to turn off their layers. The *Options* are the same as those for *Layfrz*:

Command: ***LAYOFF***
Current settings: Viewports=Vpfreeze, Block nesting level=Block
Select an object on the layer to be turned off or [Settings/Undo]: **PICK**
Layer "E-F-DOOR" has been turned off.
Select an object on the layer to be turned off or [Settings/Undo]: **Enter**
Command:

You can use *Undo* to undo the last layer (object) selected. The *Settings* option produces the following prompt:

Enter setting type for [Viewports/Block selection]:

### *Viewports*

Using this option you have the choice to turn off the selected layers only for the current viewport (*Viewport*) or turn off selected layers globally for the drawing (*Off*).

### *Block selection*

This option allows you to control which layers related to the block are frozen when you select a block: *None*, *Block*, or *Entity*. These options are identical to those described for *Layfrz* (see "*Layfrz*").

## Layon

| Ribbon | Menu Bar | Command (Type) | Alias (Type) | Shortcut |
|---|---|---|---|---|
| *Home* *Layers* (icon) | *Format* *Layer Tools >* *Turn All Layers On* | *Layon* | ... | ... |

This command is a quick way to turn <u>all layers on</u>. Layers that are *Frozen* are not affected; only layers that are *Off* can be turned *On* with this command:

Command: ***LAYON***
Warning: layer 0 is frozen. Will not display until thawed.
All layers have been turned on.
Command:

### *Laylck*

| Ribbon | Menu Bar | Command (Type) | Alias (Type) | Shortcut |
|---|---|---|---|---|
| *Home*<br>*Layers*<br>(icon) | *Format*<br>*Layer Tools >*<br>*Layer Lock* | *Laylck* | ... | ... |

*Laylck* locks the layers of selected objects. Simply select the layers to lock. Locked layers are then displayed as faded or dimmed. The amount of fading is determined by the fade value setting of the *Layiso* command or the *Locked layer fading* slider setting in the *Layers* control panel of the dashboard. See "*Layiso*" for information on setting the fade value or using the *Locked layer fading* slider in the *Layers* control panel. There are no options for this command. If objects are selected that are part of an *Xref* or *Block,* the layer that was current when the *Xref* or *Block* was *Attached* or *Inserted* is locked.

Command: ***LAYLCK***
Select an object on the layer to be locked: **PICK**
Layer TEXT has been locked.

### *Layulk*

| Ribbon | Menu Bar | Command (Type) | Alias (Type) | Shortcut |
|---|---|---|---|---|
| *Home*<br>*Layers*<br>(icon) | *Format*<br>*Layer Tools >*<br>*Layer Unlock* | *Layulk* | ... | ... |

*Layulk* has the opposite function of *Laylck;* that is, it unlocks the layers of selected objects. The faded appearance of the locked layers is removed so visibility of the objects returns to full intensity. If the selected objects are part of an *Xref* or *Block,* the layer that was current when the *Xref* or *Block* was *Attached* or *Inserted* is unlocked. There are no options for this command.

Command: ***LAYULK***
Select an object on the layer to be unlocked: **PICK**
Layer TEXT has been unlocked.
Command:

### *Layvpi*

| Ribbon | Menu Bar | Command (Type) | Alias (Type) | Shortcut |
|---|---|---|---|---|
| *Home*<br>*Layers*<br>(icon) | *Format*<br>*Layer Tools >*<br>*Isolate Layer to*<br>*Current Viewport* | *Layvpi* | ... | ... |

This feature is intended to be used to freeze layers within viewports. *Layvpi* freezes specific layers in all viewports <u>except</u> the current one. You identify the layer(s) you want to freeze by selecting any object on the desired layer. Make sure you select the objects from within the layout in which you do <u>not</u> want the layer(s) frozen.

Command: ***LAYVPI***
Select an object on the layer to be Isolated in viewport or [Settings/Undo]: **PICK**
Layer A-FURN-P-SEAT has been frozen in all viewports but the current one.

When you select an object, its layer is immediately frozen in all other viewports. The prompt repeats so you can continually select objects (on layers to freeze). If you type *S* for *Settings*, the following prompt appears:

Select an option [Layouts/Block Selection]:

The *Block Selection* option here is the same as that found in *Layfrz* which allows you to determine the level of freezing for nested *Blocks* or *Xrefs* (see previous discussion under "*Layfrz*").

Normally using *Layvpi* affects viewports only in the current layout. If you want to have layers frozen for all layouts, enter *L* for *Layouts* at the previous prompt to choose from the following:

Isolate layers in all viewports except current for [All layouts/Current layout] <Current layout>:

The *All layouts* option specifies that the selected layer is frozen in all layouts. In other words, with this option, selecting an object on a layer in a viewport not only freezes that layer for all other viewports in that layout, but it also freezes the layer in all other viewports in all the layouts in that drawing.

## *Laymrg*

| Ribbon | Menu Bar | Command (Type) | Alias (Type) | Shortcut |
|---|---|---|---|---|
| *Home Layers* (icon) | *Format Layer Tools > Layer Merge* | *Laymrg* | ... | ... |

*Laymrg* combines selected layers with any other existing layer. Objects on the layers to merge are moved to the target layer, then the merged layers are deleted. For example, merging layers CHAIRS and FILE_CABINETS to layer FURNITURE would move the related objects to layer FURNITURE and delete layers CHAIRS and FILE_CABINETS.

Command: ***LAYMRG***
Select object on layer to merge or [Name]: **PICK**
Selected layers: CHAIRS.
Select object on layer to merge or [Name/Undo]: **PICK**
Selected layers: CHAIRS,FILE_CABINETS.
Select object on layer to merge or [Name/Undo]: **Enter**
Select object on target layer or [Name]: **PICK**
******** WARNING ********
You are about to merge 2 layers into layer "FURNITURE".
Do you wish to continue? [Yes/No] <No>: ***y***
Deleting layer "CHAIRS".
Deleting layer "FILE_CABINETS".
2 layers deleted.
Command:

At the "Select object on layer to merge or [Name]:" prompt you can select an object from the drawing or you can enter *N* for the *Name* option. This option produces the *Merge Layers* dialog box in which you can select the desired layer by name (Fig. 11-45). Once you select objects on the layer(s) to merge, you must press Enter to produce the "target layer" prompt.

**FIGURE 11-45**

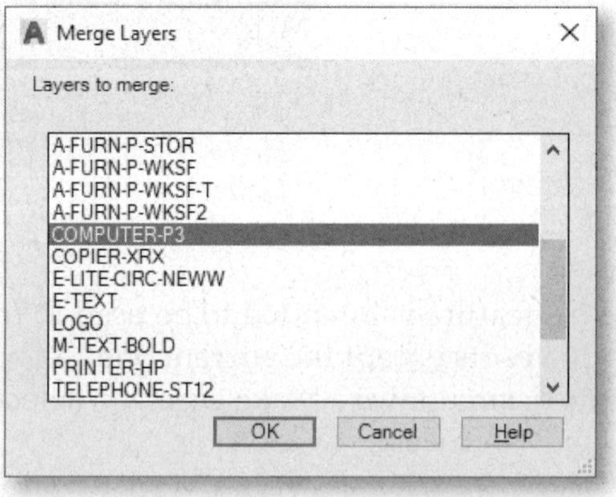

### *Laydel*

| Ribbon | Menu Bar | Command (Type) | Alias (Type) | Shortcut |
|---|---|---|---|---|
| *Home*<br>*Layers*<br>(icon) | *Format*<br>*Layer Tools >*<br>*Layer Delete* | *Laydel* | ... | ... |

*Laydel* deletes all objects on selected layers, then deletes the layers (names). For example, selecting layer E-B-CORE would delete layer E-B-CORE and all objects on that layer.

Command: ***LAYDEL***
Select object on layer to delete or [Name]: **PICK**
Selected layers: E-B-CORE.
Select object on layer to delete or [Name/Undo]: **Enter**
******** WARNING ********
You are about to delete layer "E-B-CORE" from this drawing.
Do you wish to continue? [Yes/No] <No>: ***y***
Deleting layer "E-B-CORE".
1 layer deleted.
Command:

At the "Select object on layer to delete or [Name]:" prompt you can select an object from the drawing or you can enter *N* for the *Name* option. The *Name* option produces the *Delete Layers* dialog box where you can select the desired layer by name. The dialog box is essentially the same as that used for the *Laymrg* command (see Fig. 11-45).

## CHAPTER EXERCISES

1. ***Layer Properties Manager*, Layer drop-down list, *LayerP*, and *Make Object's Layer Current***

   ***Open*** the sample drawing named **FLOOR PLAN SAMPLE.DWG**. Use the ***Explore Sample Drawings*** option in the ***Create*** section of the AutoCAD opening screen or otherwise locate the drawing located in the **AutoCAD 2021/ Sample/Database Connectivity** folder. Next, use ***Saveas*** to save the drawing in your working directory as **DB_SAMP2**.

   A. Invoke the ***Layer*** command in command line format by typing ***-LAYER*** or ***-LA***. Use the ***?*** option to yield the list of layers in the drawing. Notice that all the layers have the same linetype except one. Which layer has a different linetype and what is the linetype?

   B. Use the ***Layer*** drop-down list to indicate the layers of selected items. On the upper-left side of the drawing, **PICK** one or two cyan (light blue) lines. What layer are these lines on? Press the **Esc** key twice to cancel the selection. Next, **PICK** the gray staircase just above the center of the floor plan. What layer is the staircase on? Press the **Esc** key twice to cancel the selection.

   C. Invoke the ***Layer Properties Manager*** by selecting the icon button on the Ribbon or using the ***Format*** pull-down menu. Turn ***Off*** the **CPU** layer (click the light bulb icon). Select ***OK*** to return to the drawing and check the new setting. Are the computers displayed? Next, use the **Layer** drop-down list to turn the **CPU** layer back ***On***.

D. Use the ***Layer Properties Manager*** and determine if any layers are *Frozen*. Which layers are *Frozen*? Next, right-click for the shortcut menu and select the ***Select All*** option. ***Freeze*** all layers. Which layer will not *Freeze*? Select ***OK*** in the warning dialog box, then ***Apply***. Notice the light bulb icon indicates the layers are still *On* although the layers do not appear in the drawing (*Freeze* overrides *On*). Use the same procedure to select all layers and ***Thaw*** them. Then ***Freeze*** the two layers that were previously frozen.

E. Use the ***Layer Properties Manager*** again and select ***All Used Layers*** in the Filter Tree View area. Are there any layers in the drawing that are unused (examine the text just below the Filter Tree View)? How many layers are in the drawing? Select the ***All*** group in the Filter Tree View.

F. Now we want to *Freeze* all the *Cyan* layers. Using the ***Layer Properties Manager*** again, sort the list by color. **PICK** the first layer name in the list of the four cyan ones. Now hold down the **Shift** key and **PICK** the last name in the list. All four layers should be selected. ***Freeze*** all four by selecting any one of the sun/snowflake icons. Select ***OK*** to return to the drawing to examine the changes. Next, use the ***LayerP*** command to undo the last action in the *Layer Properties Manager*.

G. Assuming you wanted to work only with the Electrical layers (those beginning with "E-"). You can create a group filter to display only those layers. First, activate the ***Layer Properties Manager***, then select the ***New Group Filter*** button above the Filter Tree View. Name the filter "**Electrical**." Next, select the ***All*** group to make the list of all layers appear. Then enter "**E-***" (E, hyphen, asterisk) in the ***Search for Layer*** edit box below so only layers beginning with "E-" appear in the list. Select all those layers from the list, then drag and drop them into the new group. Next, in the Filter Tree View select the ***All*** group, then select the ***Electrical*** group again to ensure the group was created correctly. Press ***OK*** to close the ***Layer Properties Manager.*** Now examine the Layer drop-down list. Does it display only layer names beginning with "E-"? Return to the ***Layer Properties Manager*** and ***Freeze*** all layers not beginning with "E-" by checking the ***Invert Filter*** checkbox, then selecting and freezing all the layers that appear in the list. Press ***OK*** to see only the Electrical layers in the drawing. Use ***LayerP*** to undo the last action. ***Save*** the drawing.

H. In the ***Layer Properties Manager,*** remove the check in the ***Invert Filter*** checkbox. Then select the ***All*** group and ensure all layers are ***Thawed***. Sort the list alphabetically by ***Name***. ***Freeze*** all layers beginning with **C** and **E**. Assuming you were working with only the thawed layers, it would be convenient to display only those names in the drop-down list. Make a new Property Filter by selecting the ***New Property Filter*** button. In the *Layer Filter Properties* dialog box that appears, assign the new filter name "**Thawed**," then press the Tab key. In the *Filter Definition* area, select the "sun" icon in the ***Freeze*** column. Select ***OK*** and then examine the layer list. Select the ***All*** layer filter, then the ***Thawed*** filter to display the related names in the layer list. Next, ***Freeze*** the **GRID-08** and **GRIDLN** layers. Note that the list of names in the *Thawed* filter changes since property filters are dynamic. Finally, ***Thaw*** all layers and return to the drawing. **Save** the drawing.

I. Next, you will "move" objects from one layer to another using the *Properties* panel. Make layer **0** (zero) the current layer by using the Layer drop-down list and selecting the name. Then using the ***Layer Properties Manager***, right-click to ***Select All*** names. ***Freeze*** all layers (except the current layer). Then ***Thaw*** layer **CHAIRS**. Select ***OK*** and return to the drawing to examine the chairs (cyan in color). Use a crossing window to select all of the objects in the drawing (the small blue Grips may appear). Next, list the layers using the **Layer** drop-down list and **PICK** layer **0**. Press **Esc** twice to disable the Grips. The objects should be black and on layer 0. You actually "moved" the objects to layer 0. Finally, type ***U*** to undo the last action so the chairs return to layer CHAIRS (cyan). Also, ***Thaw*** all layers.

J. Let's assume you want to draw some additional objects on the GRIDLN layer and the E-F-TERR layer. Select the ***Make Object's Layer Current*** icon button. At the "Select object whose layer will become current" prompt, select one of the horizontal or vertical grid lines (gray color). The **GRIDLN** layer name should appear in the Layer box as the current layer. Draw a ***Line*** and it should be on layer GRIDLN and gray in color. Next, use ***Make Object's Layer Current*** again and select one of the cyan lines at the upper-left corner of the drawing. The **E-F-TERR** layer should appear as the current layer. Draw another ***Line***. It should be on layer E-F-TERR and cyan in color. Keep in mind that this method of setting a layer current works especially well when you do not know the layer names or know what objects are on which layers.

K. ***Close*** the drawing. Do not save the changes.

2. ***Layer Properties Manager*** **and** ***Linetype Manager***

A. Begin a ***New*** drawing and use ***Save*** to assign the name **CH11EX2**. Set ***Limits*** to **11** x **8.5**, then ***Zoom All***. Use the ***Linetype Manager*** to list the loaded linetypes. Are any linetypes already loaded? **PICK** the ***Load*** button then right-click to ***Select All*** linetypes. Select ***OK***, then close the *Linetype Manager*.

B. Use the ***Layer Properties Manager*** to create 3 new layers named **OBJ**, **HID**, and **CEN**. Assign the following colors, linetypes, and lineweights to the layers by clicking on the ***Color***, ***Linetype***, and ***Lineweight*** columns.

| | | | |
|---|---|---|---|
| **OBJ** | *Red* | *Continuous* | *Default* |
| **HID** | *Yellow* | *Hidden2* | *Default* |
| **CEN** | *Green* | *Center2* | *Default* |

C. Make the **OBJ** layer ***Current*** and PICK the ***OK*** tile. Verify the current layer by looking at the ***Layer Control*** drop-down list, then draw the visible object ***Lines*** and the ***Circle*** only (not the dimensions) as shown in Figure 11-46.

**FIGURE 11-46**

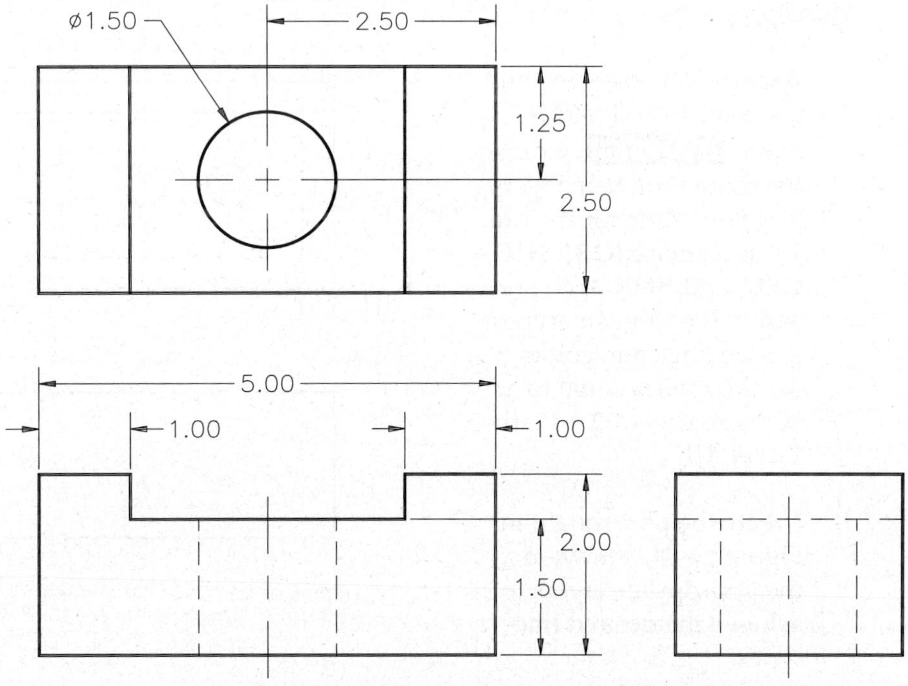

D. When you are finished drawing the visible object lines, create the necessary hidden lines by making layer **HID** the ***Current*** layer, then drawing ***Lines***. Notice that you only specify the *Line* endpoints as usual and AutoCAD creates the dashes.

E. Next, create the center lines for the holes by making layer **CEN** the ***Current*** layer and drawing ***Lines***. Make sure the center lines extend slightly beyond the *Circle* and beyond the horizontal *Lines* defining the hole. ***Save*** the drawing.

F. Now create a ***New*** layer named **BORDER** and draw a border and title block of your design on that layer. The final drawing should appear as that in Figure 11-47.

**FIGURE 11-47**

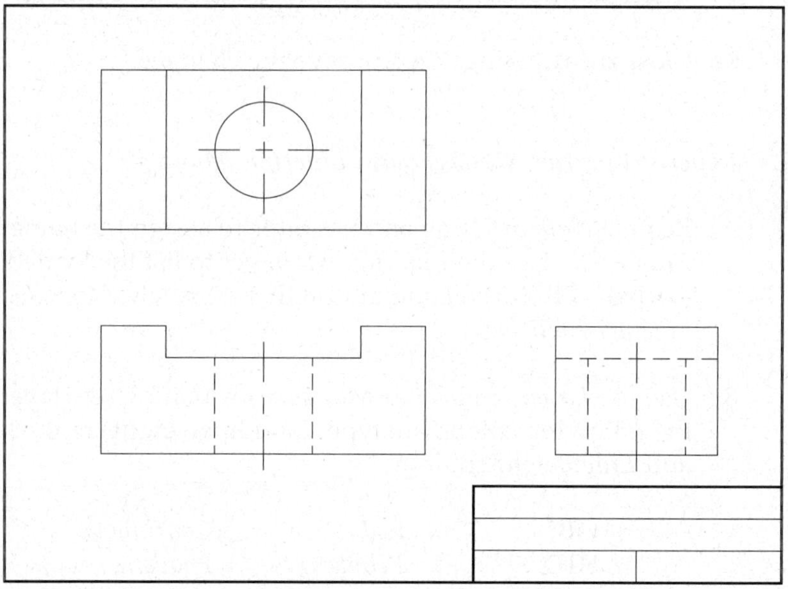

G. Open the ***Linetype Manager*** again and right-click to ***Select All***, then select the ***Delete*** button. All unused linetypes should be removed from the drawing.

H. ***Save*** the drawing, then ***Close*** the drawing.

3. ***LTSCALE, Properties*** **palette,** ***Matchprop***

A. Begin a ***New*** drawing and use ***Save*** to assign the name **CH11EX3.** Create the same four ***New*** layers that you made for the previous exercise (**OBJ, HID, CEN,** and **BORDER**) and assign the same linetypes, lineweights, and colors. Set the ***Limits*** equal to a "C" size sheet, **22** x **17**, then ***Zoom All***.

B. Create the part shown in Figure 11-48. Draw on the appropriate layers to achieve the desired linetypes.

**FIGURE 11-48**

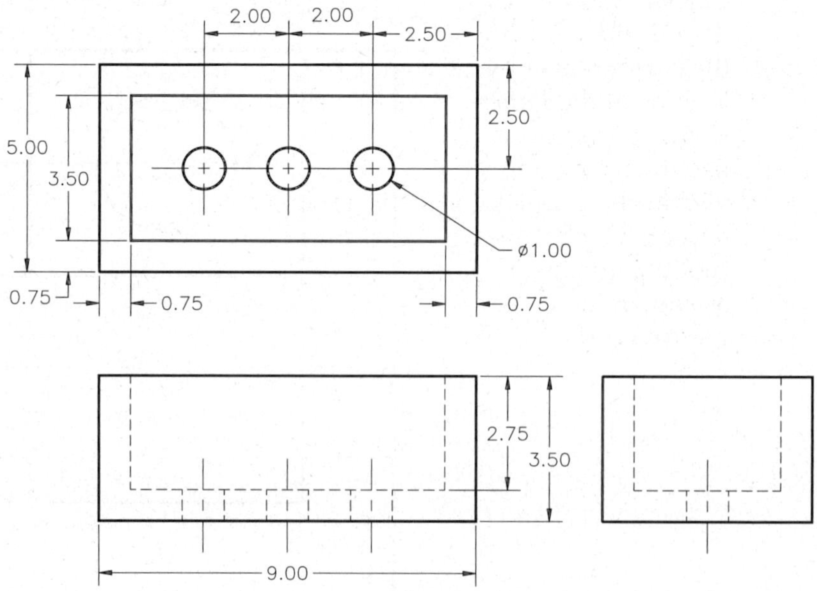

C. Notice that the *Hidden* and *Center* linetype dashes are very small. Use ***LTSCALE*** to adjust the scale of the non-continuous lines. Since you changed the *Limits* by a factor of slightly less than 2, try using **2** as the ***LTSCALE*** factor. Notice that all the lines are affected (globally) and that the new *LTSCALE* is retroactive for existing lines as well as for new lines. If the dashes appear too long or short because of the small hidden line segments, *LTSCALE* can be adjusted. Remember that you cannot control where the multiple short centerline dashes appear for one line segment. Try to reach a compromise between the hidden and center line dashes by adjusting the *LTSCALE*.

D. The six short vertical hidden lines in the front view and two in the side view should be adjusted to a smaller linetype scale. Invoke the *Properties* palette by selecting ***Properties*** from the ***Modify*** pull-down menu or selecting the ***Properties*** icon button on the Ribbon. Select only the two short lines in the side view. Use the palette to retroactively adjust the two individual objects' ***Linetype Scale*** to **0.6000**.

E. When the new *Linetype Scale* for the two selected lines is adjusted correctly, use *Matchprop* to "paint" the *Linetype Scale* to the several other short lines in the front view. Select ***Match Properties*** from the ***Modify*** pull-down menu or select the "paint brush" icon. When prompted for the "source object," select one of the two short vertical lines in the side view. When prompted for the "destination object(s)," select each of the short lines in the front view. This action should "paint" the *Linetype Scale* to the lines in the front view.

F. When you have the desired linetype scales, ***Exit*** AutoCAD and ***Save Changes***. Keep in mind that the drawing size may appear differently on the screen than it will on a print or plot. Both the plot scale and the paper size should be considered when you set *LTSCALE*. This topic will be discussed further in Chapter 12, Advanced Drawing Setup.

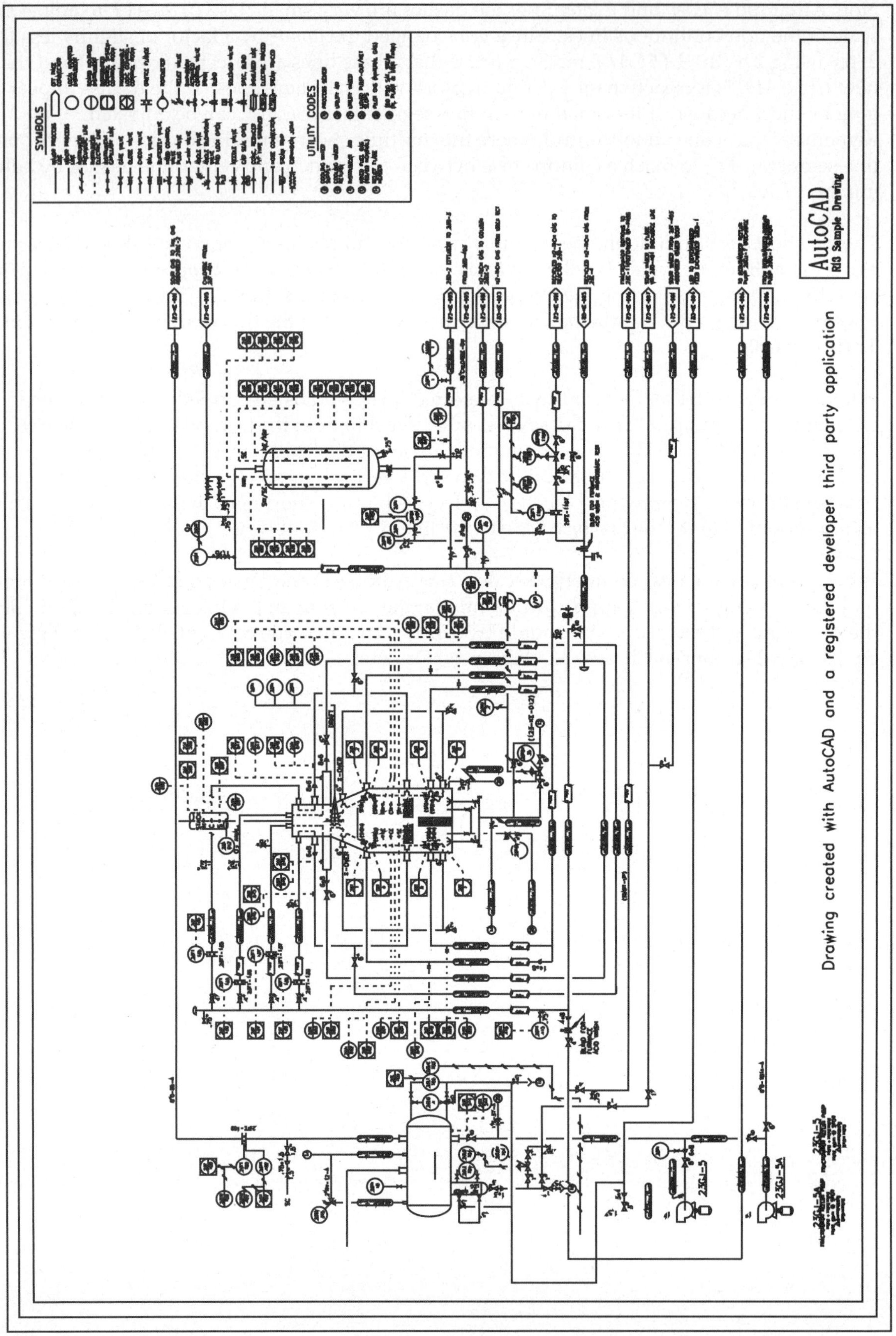


**PNID.DWG, Courtesy of Autodesk, Inc.**

CHAPTER 12

# Advanced Drawing Setup

## Chapter Objectives

After completing this chapter you should:

1. know the steps for setting up a drawing;

2. be able to determine an appropriate Limits setting for the drawing;

3. be able to calculate and apply the "drawing scale factor";

4. be able to access existing and create new template drawings;

5. know what setup steps can be considered for creating template drawings.

## CONCEPTS

When you begin a drawing, there are several steps that are typically performed in preparation for creating geometry, such as setting *Units, Limits,* and creating *Layers* with *linetypes, lineweights,* and *colors*. Some of these basic concepts were discussed in Chapters 6 and 11. This chapter discusses setting *Limits* for correct plotting as well as other procedures, such as layer creation and variables settings, that help prepare a drawing for geometry creation.

To correctly set up a drawing for printing or plotting to scale, <u>any two</u> of the following three variables must be known. The third can be determined from the other two.

Drawing *Limits*
Print or plot sheet (paper) size
Print or plot scale

The method given in this text for setting up a drawing (Chapters 6 and 12) is intended for plotting from both the *Model* tab (model space) and from *Layout* tabs (paper space). The method applies when you plot from the *Model* tab or plot with <u>one viewport</u> in a *Layout* tab. If multiple viewports are created in one layout, the same general steps would be taken; however, the plot scale might vary for each viewport based on the size and number of viewports. (See Chapters 13, 14, and 32 for more information.)

TIP

Rather than performing the steps for drawing setup each time you begin, you can use "template drawings." Template drawings have many setup steps already performed but are otherwise empty. AutoCAD provides several template drawings and you can create your own template drawings. A typical engineering, architectural, design, or construction office generally produces drawings that are similar in format and can benefit from creation and use of individualized template drawings. The drawing similarities may be the subject of the drawing (geometry), plot scale, sheet size, layering schemes, dimensioning styles, and/or text styles. Template drawing creation and use are discussed in this chapter.

## STEPS FOR DRAWING SETUP

Assuming that you have in mind the general dimensions and proportions of the drawing you want to create, and the drawing will involve using layers, dimensions, and text, the following steps are suggested for setting up a drawing:

1. Determine and set the *Units* that are to be used.
2. Determine and set the drawing *Limits;* then *Zoom All.*
3. Set an appropriate *Snap Type* (*Polar Snap* or *Grid Snap*), *Snap* spacing, and *Polar* spacing if useful.
4. Set an appropriate *Grid* value if useful.
5. Change the *LTSCALE* value based on the new *Limits*. Set *PSLTSCALE* and *MSLTSCALE* to 0.
6. Create the desired *Layers* and assign appropriate *linetype, lineweight,* and *color* settings.
7. Create desired *Text Styles* (optional, discussed in Chapter 18).
8. Create desired *Dimension Styles* (optional, discussed in Chapter 29).
9. Activate a *Layout* tab, set it up for the plot or print device and paper size, and create a viewport (if not already existing).
10. Create or insert a title block and border in the layout.

Each of the steps for drawing setup is explained in detail here.

### 1. Set *Units*

This task is accomplished by using the *Units* command. Set the linear units and precision desired. Set angular units and precision if needed. (See Chapter 6 for details on the *Units* command.)

2. **Set *Limits***

Before beginning to create an AutoCAD drawing, determine the size of the drawing area needed for the intended geometry. Using the actual *Units,* appropriate *Limits* should be set in order to draw the object or geometry to the real-world size. *Limits* are set with the *Limits* command by specifying the lower-left and upper-right corners of the drawing area. Always *Zoom All* after changing *Limits.* (See Chapter 6 for details on the *Limits* command.)

If you are planning to plot the drawing to scale, *Limits* should be set to a proportion of the sheet size you plan to plot on. For example, if the sheet size is 11″ x 8.5″, set *Limits* to 11 x 8.5 if you want to plot full size (1″=1″). Setting *Limits* to 22 x 17 (2 times 11 x 8.5) provides 2 times the drawing area and allows plotting at 1/2 size (1/2″=1″) on the 11″ x 8.5″ sheet. Simply stated, set *Limits* to a proportion of the paper size.

Setting *Limits* to the paper size allows plotting at 1=1 scale. Setting *Limits* to a proportion of the sheet size allows plotting at the reciprocal of that proportion. For example, setting *Limits* to 2 times an 11″ x 8.5″ sheet allows you to plot 1/2 size on that sheet. Or setting *Limits* to 4 times an 11″ x 8.5″ sheet allows you to plot 1/4 size on that sheet. (Standard paper sizes are given in Chapter 14.)

Even if you plan to use a layout tab (paper space) and create one viewport, setting model space *Limits* to a proportion of the paper size is recommended. In this way you can easily calculate the viewport scale (the proportion of paper space units to model space units); therefore, the model space geometry can be plotted to a standard scale.

If you plan to plot from a layout, *Limits* in paper space should be set to the actual paper size; however, setting the paper space *Limits* values is automatically done for you when you select a *Plot Device* and *Paper Size* from the *Page Setup* or *Plot* dialog box (see Chapter 13, Layouts and Viewports, and Chapter 14, Printing and Plotting).

### Drawing Scale Factor

The proportion of the *Limits* to the intended print or plot sheet size is the "drawing scale factor." This factor can be used as a general scale factor for other size-related drawing variables such as *LTSCALE,* dimension *Overall Scale* (*DIMSCALE*), and *Hatch* pattern scale. The drawing scale factor can also be used to determine the viewport scales.

Most size-related AutoCAD drawing variables are set to 1 by default. This means that variables (such as *LTSCALE*) that control sizing and spacing of objects are set appropriately for creating a drawing plotted full size (1=1).

**FIGURE 12-1**

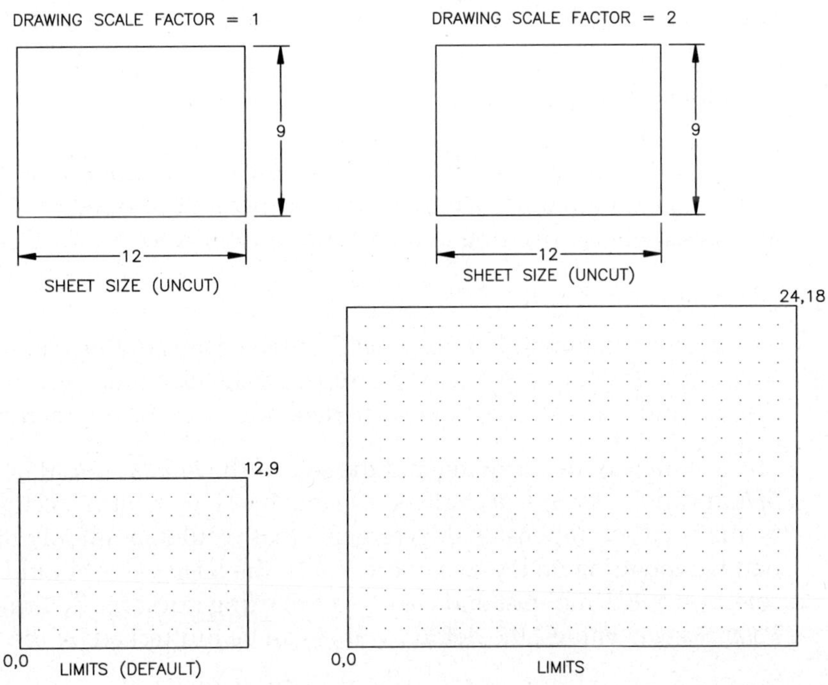

Therefore, when you set *Limits* to the sheet size and print or plot at 1=1, the sizing and spacing of linetypes and other variable-controlled objects are correct. When *Limits* are changed to some proportion of the sheet size, the size-related variables should also be changed proportionally. For example, if you intend to plot on a 12 x 9 sheet and the default *Limits* (12 x 9) are changed by a factor of 2 (to 24 x 18), then 2 becomes the drawing scale factor. Then, as a general rule, the values of variables such as *LTSCALE, DIMSCALE* (*Overall Scale*), and other scales should be multiplied by a factor of 2 (Fig. 12-1, on the previous page). When you then print the 24 x 18 area onto a 12" x 9" sheet, the plot scale would be 1/2, and all sized features would appear correct.

*Limits* should be set to the paper size or to a proportion of the paper size used for plotting or printing. In many cases, "cut" paper sizes are used based on the 11" x 8.5" module (as opposed to the 12" x 9" module called "uncut sizes"). Assume you plan to print on an 11" x 8.5" sheet. Setting *Limits* to the sheet size provides plotting or printing at 1=1 scale. Changing the *Limits* to a proportion of the sheet size by some multiplier makes that value the drawing scale factor. The reciprocal of the drawing scale factor is the scale for plotting on that sheet (Fig. 12-2).

**FIGURE 12-2**

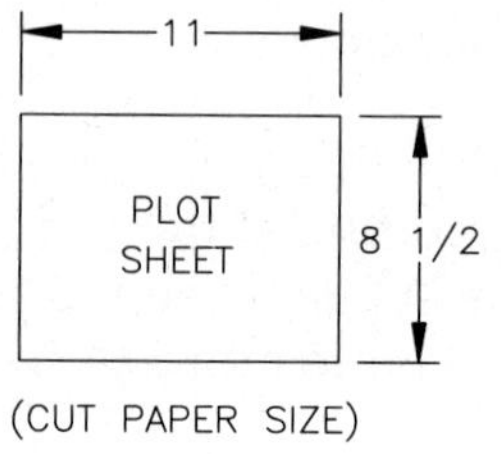


| LIMITS | SCALE FACTOR | PLOT SCALE |
|---|---|---|
| 11 x 8.5 | 1 | 1=1 (FULL SIZE) |
| 22 x 17 | 2 | 1=2 (HALF SIZE) |
| 44 x 34 | 4 | 1=4 (1/4 SIZE) |
| 110 x 85 | 10 | 1=10 (1/10 SIZE) |

Since the drawing scale factor (DSF) is the proportion of *Limits* to the sheet size, you can use this formula:

$$DSF = \frac{\text{Limits}}{\text{Sheet size}}$$

Because plot/print scale is the reciprocal of the drawing scale factor:

$$\text{Plot scale} = \frac{1}{DSF}$$

The term "drawing scale factor" is not a variable or command that can be found in AutoCAD or its official documentation. This concept has been developed by AutoCAD users to set system variables appropriately for printing and plotting drawings to standard scales.

3. Set *Snap*

Use the *Snap* command or *Snap and Grid* tab of the *Drafting Settings* dialog box to set the *Snap* type (*Grid Snap* and/or *Polar Snap*) and appropriate *Snap Spacing* and/or *Polar Spacing* values. The *Snap Spacing* and *Polar Spacing* values are dependent on the interactive drawing accuracy that is desired.

The accuracy of the drawing and the size of the *Limits* should be considered when setting the *Snap Spacing* and *Polar Spacing* values. On one hand, to achieve accuracy and detail, you want the values to be the smallest dimensional increments that would commonly be used in the drawing. On the other hand, depending on the size of the *Limits*, the *Snap* value should be large enough to make interactive selection (PICKing) fast and easy. As a starting point for determining appropriate *Snap Spacing* and *Polar Spacing* values, the default values can be multiplied by the "drawing scale factor."

If you are creating a template drawing (.DWT file), it is of no help to set the *Snap* type (*Grid Snap* or *Polar Snap*), *Polar Spacing* value, or *Polar Tracking* angles since these values are stored in the system Registry, not in the drawing. The *Grid Snap Spacing* value, however, is stored in the drawing file.

4. **Set *Grid***

   The *Grid* value setting is usually set equal to or proportionally larger than that of the *Grid Snap* value. *Grid* should be set to a proportion that is easily visible and to some value representing a regular increment (such as .5, 1, 2, 5, 10, or 20). Setting the value to a proportion of *Grid Snap* gives visual indication of the *Grid Snap* increment. For example, a *Grid* value of 1X, 2X, or 5X would give visual display of every 1, 2, or 5 *Grid Snap* increments, respectively. If you are not using *Snap* because only extremely small or irregular interval lengths are needed, you may want to turn *Grid* off. (See Chapter 6 for details on the *Grid* command.)

5. **Set the *LTSCALE*, *MSLTSCALE*, and *PSLTSCALE***

   A change in *Limits* (then *Zoom All*) affects the display of non-continuous (hidden, dashed, dotted, etc.) lines. The *LTSCALE* variable controls the spacing of non-continuous lines. The drawing scale factor can be used to determine the *LTSCALE* setting. For example, if the default *Limits* (based on sheet size) have been changed by a factor of 2, the drawing scale factor=2; so set the *LTSCALE* to 2. In other words, if the DSF=2 and plot scale=1/2, you would want to double the linetype dashes to appear the correct size in the plot, so *LTSCALE*=2. If you prefer an *LTSCALE* setting of other than 1 for a drawing in the default *Limits*, multiply that value by the drawing scale factor.

   American National Standards Institute (ANSI) requires that hidden lines be plotted with 1/8" dashes. An AutoCAD drawing plotted full size (1 unit=1") with the default *LTSCALE* value of 1 plots the *Hidden* linetype with 1/4" dashes. Therefore, it is recommended for English drawings that you use the *Hidden2*, *Center2*, and other *2 linetypes with dash lengths .5 times as long so an *LTSCALE* of 1 produces the ANSI standard dash lengths. Multiply your *LTSCALE* value by the drawing scale factor when plotting to scale. If individual lines need to be adjusted for linetype scale, use *Properties* when the drawing is nearly complete. (See Chapter 11 for details on *LTSCALE*.)

   Assuming you are planning to create only one viewport in a *Layout* tab, set the *PSLTSCALE* (paper space linetype scale) and *MSLTSCALE* (model space linetype scale) values to 0 (zero). AutoCAD's default setting for both values is 1. The setting of 0 forces the linetype scale to appear the same in *Layout* viewports as in the *Model* tab. Set the *PSLTSCALE* and *MSLTSCALE* values at the command prompt.

6. **Create *Layers*; Assign *Linetypes*, *Lineweights*, and *Colors***

   Use the *Layer Properties Manager* to create the layers that you anticipate needing. Multiple layers can be created with the *New* option. You can type in several new names separated by commas. Assign a descriptive name for each layer, indicating its type of geometry, part name, or function. Include a *linetype* designator in the layer name if appropriate; for example, PART1-H and PART1-V indicate hidden and visible line layers for PART1.

   Once the *Layers* have been created, assign a *Color*, *Linetype*, and *Lineweight* to each layer. *Colors* can be used to give visual relationships to parts of the drawing. Geometry that is intended to be plotted with different pens (pen size or color) or printed with different plotted appearances should be drawn in different screen colors (especially when color-dependent plot styles are used). Use the *Linetype* command to load the desired linetypes. You should also select *Lineweights* for each layer (if desired) using the *Layer Properties Manager*. (See Chapter 11 for details on creating layers and assigning linetypes, lineweights, and colors.)

7. **Create Text Styles**

   AutoCAD has only one text style as part of the standard template drawings (ACAD.DWT and ACADISO.DWT) and one or two text styles for most other template drawings. If you desire other *Text Styles*, they are created using the *Style* command or the *Text Style...* option from the *Format* pull-down menu. (See Chapter 18, Text and Tables.) If you desire engineering standard text, create a *Text Style* using the ROMANS.SHX or .TTF font file.

8. **Create Dimension Styles**

If you plan to dimension your drawing, Dimension Styles can be created at this point; however, they are generally created during the dimensioning process. Dimension Styles are names given to groups of dimension variable settings. (See Chapter 29 for information on creating Dimension Styles.)

Although you do not have to create Dimension Styles until you are ready to dimension the geometry, it is helpful to create Dimension Styles as part of a template drawing. If you produce similar drawings repeatedly, your dimensioning techniques are probably similar. Much time can be saved by using a template drawing with previously created Dimension Styles. Several of the AutoCAD-supplied template drawings have prepared Dimension Styles.

9. **Activate a *Layout*, Set the Plot Device, and Create a Viewport**

This step can be done just before making a print or plot, or you can complete this and the next step before creating your drawing geometry. Completing these steps early allows you to see how your printed drawing might appear and how much area is occupied by the titleblock and border.

Activate a *Layout*. If your *Options* are set as explained in Chapter 6, Basic Drawing Setup, the layout is automatically set up for your default print/plot device and an appropriate-sized viewport is created. Otherwise, use the *Page Setup Manager* (*Pagesetup* command) to set the desired print/plot device and paper size. Then use the *Vports* command to create one or more viewports. This topic is discussed briefly in Chapter 6, Basic Drawing Setup, and is discussed in more detail in Chapter 13, Layouts and Viewports.

10. **Create a Titleblock and Border**

For 2D drawings, it is helpful to insert a titleblock and border early in the drawing process. This action gives a visual drawing boundary as well as reserves the space needed by the titleblock.

Rather than create a new titleblock and border for each new drawing, a common practice is to use the *Insert* command to insert an existing titleblock and border as a *Block*.

You can draw or *Insert* a titleblock and border in the *Model* tab (model space) or into a layout (paper space), depending on how you want to plot the drawing. If you plan to plot from *Model* space, multiply the actual size of the titleblock and border by the drawing scale factor to determine their sizes for the drawing. If you want to plot from a layout, draw or insert the titleblock and border at the actual size (1:1) since a layout represents the plot sheet. (See Chapters 21 and 22 for information on *Block* creation and *Insertion*.)

## USING AND CREATING TEMPLATE DRAWINGS

Instead of going through the steps for setup each time you begin a new drawing, you can create one or more template drawings or use one of the AutoCAD-supplied template drawings. AutoCAD template drawings are saved as a .DWT file format. A template drawing has the initial setup steps (*Units*, *Limits*, *Layers*, *Linetypes*, *Lineweights*, *Colors*, etc.) completed and saved, but no geometry has been created yet. For some templates, *Layout* tabs have been set up with titleblocks, viewports, and plot settings. Template drawings are used as a template or starting point each time you begin a new drawing. AutoCAD actually makes a copy of the template you select to begin the drawing. The creation and use of template drawings can save hours of preparation.

### Using Template Drawings

To use a template drawing, select the *Template* option in the *Create* screen or in the *Select Template* dialog box (Fig. 12-3, on the next page). This option allows you to select the desired template (.DWT) drawing from a list including all templates in the Template folder.

Any template drawings that you create using the *SaveAs* command (as described in the next section) appear in the list of available templates. Select the *Browse...* option to browse other folders.

AutoCAD includes standard templates for Imperial (inch) drawings and ISO (metric) drawings. The basic settings in the standard templates are described in the following table. Templates are also supplied for use with architectural and manufacturing tutorials.

**FIGURE 12-3**

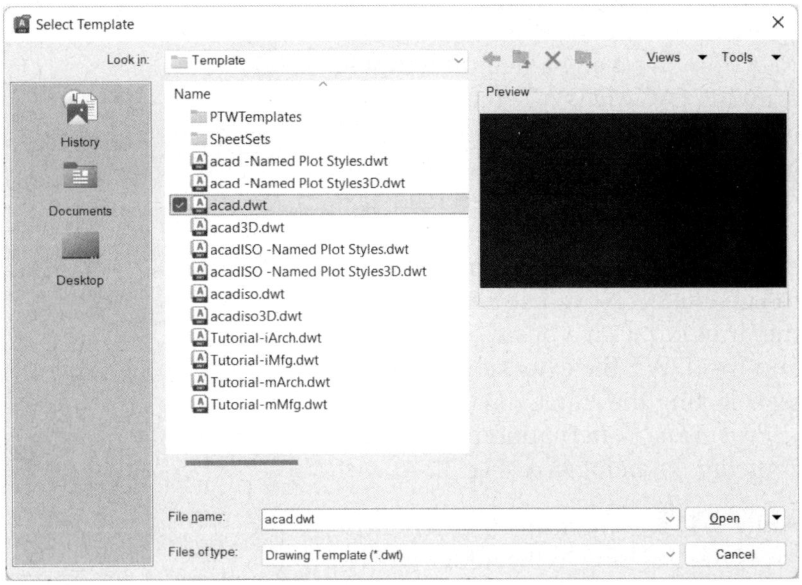


## Table of Standard AutoCAD Templates

| Template | Intended Sheet Size | Limits (Model | Title Block | Orientation | Layers | Dimen. Styles | Plot Style |
|---|---|---|---|---|---|---|---|
| ACAD - Named Plot Styles.dwt | --- | 12,9 | No | Landscape | 0 | Standard, Annotative | Named |
| Acad.dwt | --- | 12,9 | No | Landscape | 0 | Standard, Annotative | Color Dependent |
| ACAD - Named Plot Styles3d.dwt | --- | 12,9 | No | Landscape | 0 | Standard, Annotative | Named |
| ACAD3d.dwt | --- | 12,9 | No | Landscape | 0 | Standard, Annotative | Color Dependent |
| ACADISO - Named Plot Styles.dwt | --- | 420,297 | No | Landscape | 0 | Annotative, ISO-25 | Named |
| Acadiso.dwt | --- | 420,297 | No | Landscape | 0 | Annotative, ISO-25 | Color Dependent |
| ACADISO - Named Plot Styles3d.dwt | --- | 420,297 | No | Landscape | 0 | Annotative, ISO-25 | Named |
| ACADISO3d.dwt | --- | 420,297 | No | Landscape | 0 | Annotative, ISO-25 | Named |

NOTE: The initial *PSLTSCALE* and *MSLTSCALE* values for all templates is 1. (See Chapter 13 for details on *PSLTSCALE* and *MSLTSCALE*.)

See the Tables of *Limits* Settings in Chapter 14 for standard plotting scales and intended sheet sizes for the *Limits* settings in these templates.

## Creating Template Drawings

FIGURE 12-4

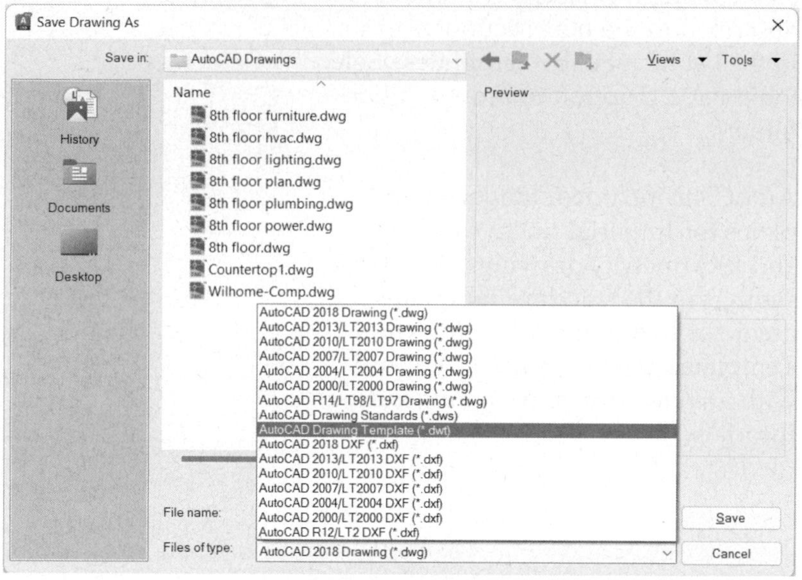


To make a template drawing, begin a *New* drawing using a default template drawing (ACAD.DWT) then do the initial drawing setup. Alternately, *Open* an existing drawing that is already set up (layers created, linetypes loaded, *Limits* and other settings for plotting or printing to scale, etc.), and *Erase* all the geometry. Next, use *SaveAs* to save the drawing with a descriptive name and the .DWT file extension. Do this by selecting the *AutoCAD Drawing Template File (*.dwt)* option in the *Save Drawing As* dialog box (Fig. 12-4).

AutoCAD automatically (by default) saves the drawing in the folder where other template (*DWT) files are found. Use the *Save in*: drop-down list at the top of the *Save Drawing As* dialog box to specify another location for saving the .DWT file.

Location of the Template folder aligns with the Microsoft Windows conventions. That is, the templates are stored by default (for a stand-alone installation) in a folder multiple levels under the user's profile (for example, C:\Users\Username\AppData\Local\Autodesk\AutoCAD 2021\... and so on.) To find the complete path, use the *Options* dialog box, *Files* tab, and expand *Template Settings*.

FIGURE 12-5

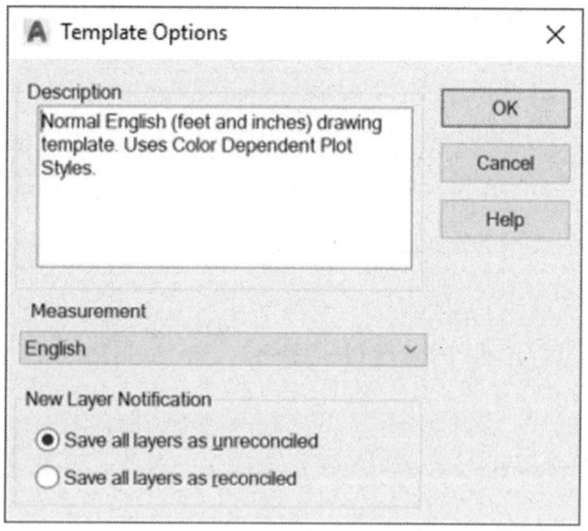


After you assign the file name in the *Save Drawing As* dialog box, you can enter a description of the template drawing in the *Template Description* dialog box that appears (Fig. 12-5). The description is saved with the template drawing and appears in the *Startup* and *Create New Drawing* dialog boxes when you highlight the template drawing name (see Fig. 12-3).

Multiple template drawings can be created, each for a particular situation. For example, you may want to create several templates, each having the setup steps completed but with different *Limits* and with the intention to plot or print each in a different scale or on a different size sheet. Another possibility is to create templates with different layout and layering schemes. There are many possibilities, but the specific settings used depend on your applications (geometry), typical scales, or your plot/print devices.

Typical drawing steps that can be considered for developing template drawings are listed below:

Set *Units*
Set *Limits*
Set *Snap*
Set *Grid*
Set *LTSCALE, MSLTSCALE,* and *PSLTSCALE*
Create *Layers* with color and linetypes assigned
Create *Text Styles* (see Chapter 18)
Create Dimension Styles (see Chapter 29)
Create *Layout* tabs, each with titleblock, border and plot settings saved (plot device, paper size, plot scale, etc.). (See Chapter 13, Layouts and Viewports.)

You can create template drawings for different sheet sizes and different plot scales. If you plot from the *Model* tab, a template drawing should be created for each case (scale and sheet size). When you learn to create *Layouts*, you can save one template drawing that contains several layouts, each layout for a different sheet size and/or scale.

## Additional Advanced Drawing Setup Concepts

The drawing setup described in this chapter is appropriate for learning the basic concepts of setting up a drawing in preparation for creating layouts and viewports and plotting to scale. In Chapter 13, Layouts and Viewports, you will learn the basics of creating layouts and viewports—particularly for drawings that use one large viewport in a layout.

In many cases, you may want to create several viewports in one or more layouts. The primary purpose of this action is to print the same drawing, or parts of the same drawing, in different scales. Annotative objects (such as annotative text, annotative dimensions, and annotative hatch patterns) can be used to facilitate this objective. Annotative objects automatically adjust for size when the *Annotative Scale* for the viewport is changed. That is, annotative objects can be automatically resized to appear "readable" in different viewports at different scales. If you plan to create multiple viewports and intend to use annotative objects, the steps for setting up a drawing are given below.

## Steps for Drawing Setup Using Annotative Objects

The first four steps are the same as those given previously (see "Steps for Drawing Setup"):

1. Determine and set the *Units* that are to be used.
2. Determine and set the drawing *Limits*; then *Zoom All*.
3. Set an appropriate *Snap Type*, *Snap* spacing, and *Polar* spacing if useful.
4. Set an appropriate *Grid* value if useful.

These steps are for using annotative objects specifically:

5. Change *MSLTSCALE* to 0 and *PSLTSCALE* to 1.
6. Set the *Annotation Scale* pop-up list in the *Model* tab (lower-right corner of the Drawing Editor) to the scale you intend to print the drawing for the primary viewport (use the reciprocal of the Drawing Scale Factor). In other words, use the following formula where AS is the *Annotation Scale* and DSF is the Drawing Scale Factor:

   AS = 1/DSF

The remaining steps are the same as previously given (see "Steps for Drawing Setup"), steps 6-10.

Annotative objects, such as annotative text and annotative dimensions, are explained in their respective chapters on text and dimensions. In Chapter 32, several advanced concepts are explained for using multiple viewports, including using annotative objects in multiple viewports. In addition, the "reverse method" for drawing setup is discussed, where the drawing scale factor is based on the viewport scale.

## CHAPTER EXERCISES

For the first three exercises, you will set up several drawings that will be used in other chapters for creating geometry. Follow the typical steps for setting up a drawing given in this chapter.

REUSE

1. **Create a Metric Template Drawing**
   You are to create a template drawing for use with metric units. The drawing will be printed full size on an "A" size sheet (not on an A4 metric sheet), and the dimensions are in millimeters. Start a ***New*** drawing using the **ACAD.DWT** template. Follow these steps.

   A. Set ***Units*** to ***Decimal*** and ***Precision*** to **0.00**.
   B. Set ***Limits*** to a metric sheet size, **279.4** x **215.9**; then ***Zoom All*** (scale factor is approximately 25).
   C. Set ***Snap Spacing*** (for *Grid Snap*) to **1**. (*Polar Tracking* and *Polar Snap* settings do not have to be preset since they are saved in the system Registry.)
   D. Set ***Grid*** to **10**.
   E. Set ***LTSCALE*** to **25** (drawing scale factor of 25.4). Set ***PSLTSCALE*** and ***MSLTSCALE*** to **0**.
   F. ***Load*** the ***Center2*** and ***Hidden2 Linetypes***.
   G. Create the following ***Layers*** and assign the ***Colors***, ***Linetypes***, and ***Lineweights*** as shown:

| | | | |
|---|---|---|---|
| **OBJECT** | ***red*** | ***continuous*** | ***0.40 mm*** |
| **CONSTR** | ***white*** | ***continuous*** | ***0.25 mm*** |
| **CENTER** | ***green*** | ***center2*** | ***0.25 mm*** |
| **HIDDEN** | ***yellow*** | ***hidden2*** | ***0.25 mm*** |
| **DIM** | ***cyan*** | ***continuous*** | ***0.25 mm*** |
| **TITLE** | ***white*** | ***continuous*** | ***0.25 mm*** |
| **VPORTS** | ***white*** | ***continuous*** | ***0.25 mm*** |

   H. Use the ***SaveAs*** command. In the dialog box under *Save as type:*, select ***AutoCAD Drawing Template File (.dwt)***. Assign the name **A-METRIC.** Enter **Metric A Size Drawing** in the *Template Description* dialog box.

REUSE

2. **Barguide Setup**
   A drawing of a mechanical part is to be made and plotted full size (1"=1") on an 11" x 8.5" sheet.

   A. Begin a ***New*** drawing using the **ACAD.DWT** template.
   B. Set the ***Units*** to ***Decimal*** with a ***Precision*** of ***0.000***.
   C. Set the ***Limits*** to **11** x **8.5**.
   D. Set the ***Grid*** to **.500**.
   E. Change the ***Snap*** to **.250**.
   F. Set ***LTSCALE*** to **1**. Set ***PSLTSCALE*** and ***MSLTSCALE*** to **0**.
   G. ***Load*** the ***Center2*** and ***Hidden2 Linetypes***.
   H. Create the following ***Layers*** and assign the given ***Colors***, ***Linetypes***, and ***Lineweights***:

| | | | |
|---|---|---|---|
| **OBJECT** | ***red*** | ***continuous*** | ***0.40 mm*** |
| **CONSTR** | ***white*** | ***continuous*** | ***0.25 mm*** |
| **CENTER** | ***green*** | ***center2*** | ***0.25 mm*** |
| **HIDDEN** | ***yellow*** | ***hidden2*** | ***0.25 mm*** |
| **DIM** | ***cyan*** | ***continuous*** | ***0.25 mm*** |
| **TITLE** | ***white*** | ***continuous*** | ***0.25 mm*** |
| **VPORTS** | ***white*** | ***continuous*** | ***0.25 mm*** |

   I. Save the drawing as **BARGUIDE** for use in another chapter exercise.

3. **Apartment Setup**

A floor plan of an apartment has been requested. Dimensions are in feet and inches. The drawing will be plotted at 1/2″=1′ scale on an 24″ x 18″ sheet.

A. Begin a *New* drawing using the **ACAD.DWT** template.
B. Set the ***Units*** to ***Architectural*** with a ***Precision*** of ***1/8***.
C. Set the ***Limits*** to **48′** x **36′** (make sure you use the apostrophe to designate feet).
D. Set the ***Snap*** value to **1** (inch).
E. Set the ***Grid*** value to **12** (inches).
F. Set the ***LTSCALE*** to **12**. Set ***PSLTSCALE*** and ***MSLTSCALE*** to **0**.
G. Create the following ***Layers*** and assign the ***Colors*** and ***Lineweights:***

| | | |
|---|---|---|
| **FLOORPLN** | *red* | *0.016″* |
| **CONSTR** | *white* | *0.010″* |
| **TEXT** | *green* | *0.010″* |
| **TITLE** | *yellow* | *0.010″* |
| **DIM** | *cyan* | *0.010″* |
| **VPORTS** | *white* | *0.010″* |

H. ***Save*** the drawing and assign the name **APARTMENT** to be used later.

4. **Mechanical Setup**
A drawing is to be created and printed in 1/4″=1″ scale on an A size (11″ x 8.5″) sheet.

A. Begin a *New* drawing using the **ACAD.DWT** template.
B. Set the ***Units*** to ***Decimal*** with a ***Precision*** of ***0.000***.
C. Set the ***Limits*** to **44** x **34**.
D. Set ***Snap*** to **1**.
E. Set ***Grid*** to **2**.
F. Type ***LTS*** at the command prompt and enter **4**.
G. Set ***PSLTSCALE*** and ***MSLTSCALE*** to **0**.
H. Save the drawing and name it **CH12EX4**.

5. **Create a Template Drawing**
In this exercise, you will create a "generic" template drawing that can be used at a later time. Creating templates now will save you time later when you begin new drawings.

A. Create a template drawing for use with decimal dimensions and using standard paper "A" size format. Begin a *New* drawing and use the **ACAD.DWT** template.
B. Set ***Units*** to ***Decimal*** and ***Precision*** to **0.00**.
C. Set ***Limits*** to **11** x **8.5**.
D. Set ***Snap*** to **.25**.
E. Set ***Grid*** to **1**.
F. Set ***LTSCALE*** to **1**. Set ***PSLTSCALE*** and ***MSLTSCALE*** to **0**.
G. Create the following ***Layers***, assign the ***Linetypes*** and ***Lineweights*** as shown, and assign your choice of ***Colors***. Create any other layers you think you may need or any assigned by your instructor.

| | | |
|---|---|---|
| **OBJECT** | *continuous* | *0.016″* |
| **CONSTR** | *continuous* | *0.010″* |
| **TEXT** | *continuous* | *0.010″* |
| **TITLE** | *continuous* | *0.010″* |
| **VPORTS** | *continuous* | *0.010″* |
| **DIM** | *continuous* | *0.010″* |
| **HIDDEN** | *hidden2* | *0.010″* |
| **CENTER** | *center2* | *0.010″* |
| **DASHED** | *dashed2* | *0.010″* |

H. Use ***SaveAs*** and name the template drawing **ASHEET.** Make sure you select ***AutoCAD Drawing Template File (*.dwt)*** from the *Save as Type:* drop-down list in the *Save Drawing As* dialog box. Also, from the ***Save In:*** drop-down list on top, select <u>your working directory</u> as the location to save the template.

# CHAPTER 13

# Layouts and Viewports

## Chapter Objectives

After completing this chapter you should:

1. know the difference between paper space and model space and between tiled viewports and paper space viewports;

2. know that the purpose of a layout is to prepare model space geometry for plotting;

3. know the "Guidelines for Using Layouts and Viewports";

4. be able to create layouts using the *Layout Wizard* and the *Layout* command;

5. know how to set up a layout for plotting using the *Page Setup Manager* and the *Page Setup* dialog box;

6. be able to use the options of *Vports* and *-Vports* to create and control viewports in paper space;

7. know how to scale the model geometry displayed in paper space viewports.

# CONCEPTS

You are already somewhat familiar with model space and paper space. Model space is used for construction of geometry, whereas a paper space layout is used for preparing to print or plot the model geometry. In order to view the model geometry in a layout, one or more viewports must be created. This can be done automatically by AutoCAD when you activate a *Layout* tab or button for the first time, or you can use the *Vports* command to create viewports in a layout. An introduction to these concepts is given in Chapter 6, Basic Drawing Setup, "Introduction to Layouts and Printing." However, this chapter (Chapter 13) gives a full explanation of paper space, layouts, and creating paper space viewports. Advanced features of layouts and plotting are discussed in Chapter 32.

## Paper Space and Model Space

The two drawing spaces that AutoCAD provides are model space and paper space. Model space is activated by selecting the *Model* tab or button and paper space is activated by selecting the *Layout1*, *Layout2*, or other layout tab or button. When you start AutoCAD and begin a drawing, model space is active by default. Objects that represent the subject of the drawing (model geometry) are normally drawn in model space. Dimensioning is traditionally performed in model space because it is associative—directly associated to the model geometry. The model geometry is usually completed before using paper space. It is possible to create dimensions in paper space that are attached to objects in model space; however, this practice is recommended only for certain applications. (See "Dimensioning in Paper Space Layouts" in Chapter 29.)

Paper space represents the paper that you plot or print on. When you enter paper space (by activating a *Layout* tab) for the first time, you may see only a blank "sheet," unless your system is configured to automatically generate a viewport as explained in Chapter 6. Normally the only geometry you would create in paper space is the title block, drawing border, and possibly some other annotation (text). In order to see any model geometry from paper space, viewports must be created (like cutting rectangular holes) so you can "see" into model space. Viewports are created either automatically by AutoCAD or by using the *Vports* command. Any number or size of rectangular or non-rectangular viewports can be created in paper space. Since there is only one model space in a drawing, you see the same model space geometry in each viewport initially. You can, however, control the size and area of the geometry displayed and which layers are *Frozen* and *Thawed* in each viewport. Since paper space represents the actual paper used for plotting, you should plot from paper space at a scale of 1:1.

The *TILEMODE* system variable controls which space is active—paper space (*TILEMODE*=0) or model space (*TILEMODE*=1). *TILEMODE* is automatically set when you activate the *Model* tab or a *Layout* tab.

## Layouts

By default there is a *Layout1* and *Layout2*. You can produce any configuration of layouts by creating new layouts or copying, renaming, deleting, or moving existing layouts. A shortcut menu is available (right-click while your pointer is on a layout tab) that allows you to *Rename* the selected layout as well as create a *New Layout* from scratch or *From Template*, *Delete*, *Move or Copy* a layout, *Select all Layouts*, or set up for plotting. You can rearrange the order of the tabs by simply dragging and dropping the tabs into the sequence you prefer. This is especially helpful since complex drawings can contain many layouts.

The *Qvlayout* (*Quick View Layouts*) command displays a small thumbnail image of model space and of each layout for every open drawing and allows you to activate any of them by selecting its image. The *Quick View Layouts* is also produced by hovering over any drawing tab in the upper-left corner of the graphics area (see Figure 13-1).

The primary function of a layout is to prepare the model geometry for plotting. A layout simulates the sheet of paper you will plot on and allows you to specify plotting parameters and see the changes in settings that you make, similar to a plot preview. Plot specifications are made using the *Page Setup* dialog box to select such parameters as plot device, paper size and orientation, scale, and Plot Style Tables to attach.

FIGURE 13-1

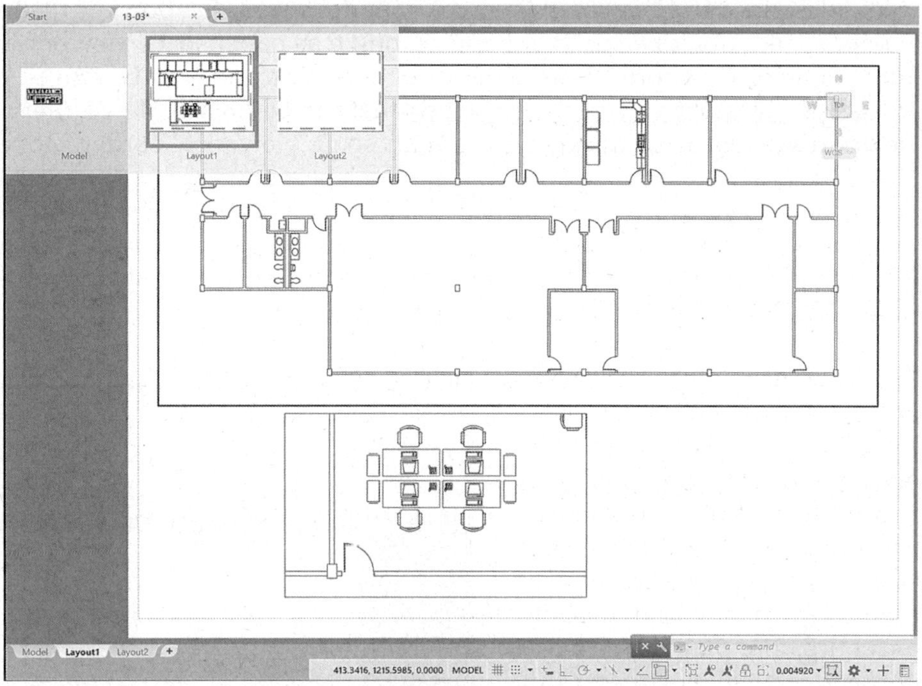

Since you can have multiple layouts in AutoCAD, you can set up multiple plot schemes. For example, assume you have a drawing (in model space) of an office complex. You can set up one layout to plot the office floor plan on a "C" size sheet with an electrostatic plotter in 1/4"=1' scale and set up a second layout to plot the same floor plan on a "A" size sheet with a laser jet printer at 1/8"=1' scale. Since the plotting setup is saved with each layout, you can plot any layout again without additional setup.

NOTE: If the *Model* and *Layout* tabs are not visible at the bottom left of the drawing area, you can make them appear by checking *Display Layout and Model* tabs in the *Display* tab of the *Options* dialog box (see Figure 13-2) or by clicking the *Layout Tabs* toggle in the *View* tab, *Interface* panel, of the Ribbon (see Figure 13-3).

FIGURE 13-2

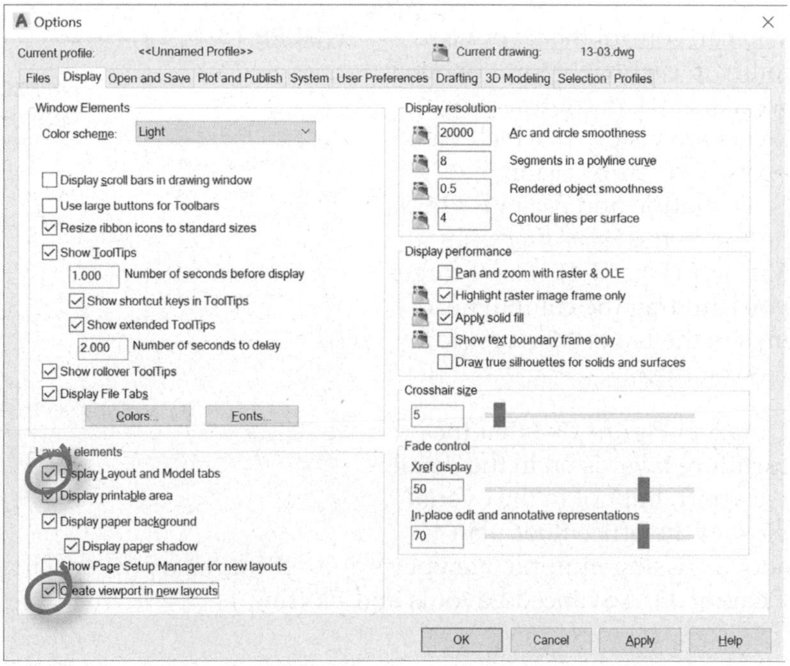


## Viewports

When you activate a layout tab (enter paper space) for the first time, you may see a "blank sheet." However, if *Create Viewport in New Layouts* is checked in the *Display* tab of the *Options* dialog box (see Fig. 13-2, lower-left corner), you may see a viewport that is automatically created. If no viewport exists, you can create one or more with the *Vports* command. A viewport in paper space is a window into model space. Without a viewport, no model geometry is visible in a layout.

There are two types of viewports in AutoCAD, (1) viewports in model space that divide the screen like "tiles" described in Chapter 10 (known as model space viewports or tiled viewports) and (2) viewports in paper space described in this chapter (known as paper space viewports or floating viewports). The term "floating viewports" is used because these viewports can be moved or resized and can take on any shape. Floating viewports are objects that can be *Erased*, *Moved*, *Copied*, stretched with grips, or the viewport's layer can be turned *Off* or *Frozen* so no viewport "border" appears in the plot.

The command generally used to create viewports, *Vports*, can create either tiled viewports or paper space viewports, but only one type at a time depending on the active space—model space or a layout. When model space is active (*TILEMODE*=1), *Vports* creates tiled viewports. When a layout is active (*TILEMODE*=0), *Vports* creates paper space viewports.

Typically there is only one viewport per layout; however, you can create multiple viewports in one layout. This capability gives you the flexibility to display two or more areas of the same model (model space) in one layout. For example, you may want to show the entire floor plan in one viewport and a detail (enlarged view) in a second viewport, both in the same layout (Fig. 13-3).

**FIGURE 13-3**

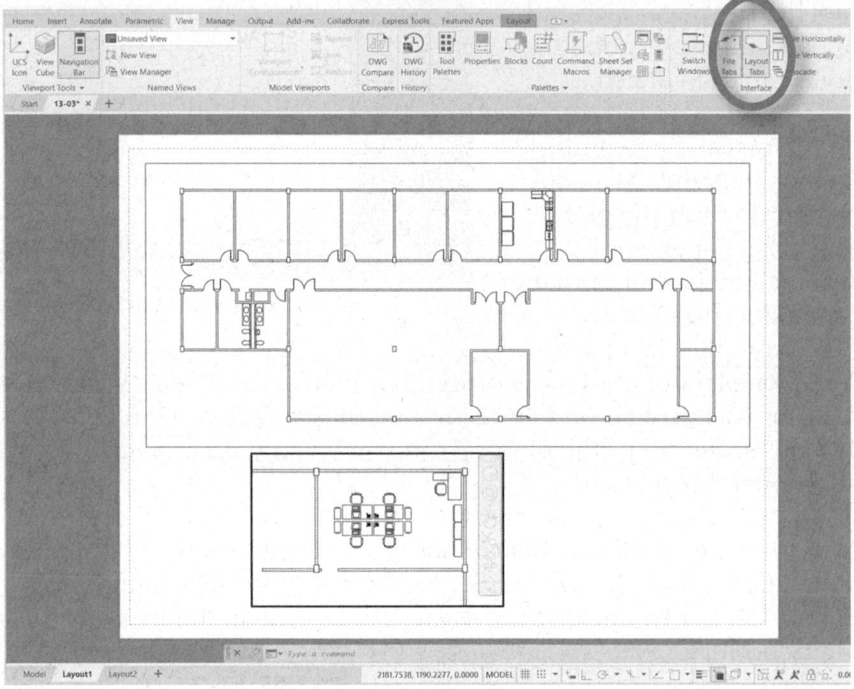

You can control the layer visibility in each viewport. In other words, you can control which layers are visible in which viewports. Do this by using the *VP Freeze* button and *New VP Freeze* button in the *Layer Properties Manager* (Fig. 13-4). Don't forget you can drag the column headings in the Layer Manager to reorder them.

**FIGURE 13-4**

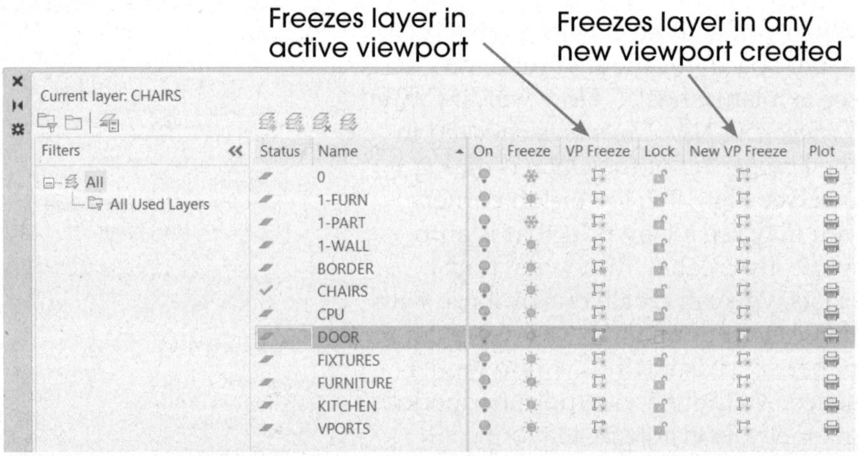


Notice in Figure 13-3 that the furniture layer is on in the detail viewport, but not in the overall view of the office floor plan above. (Using multiple viewports and setting viewport-specific layer visibility is discussed in detail in Chapter 32, Advanced Layouts and Plotting.)

You can also scale the display of the geometry that appears in a viewport. You can set the viewport scale by using the *Viewport Scale* pop-up list (Fig. 13-5), the *Viewports* toolbar drop-down list, the *Properties* palette, or entering a *Zoom XP* factor. This action sets the proportion (or scale) of paper space units to model space units.

**FIGURE 13-5**

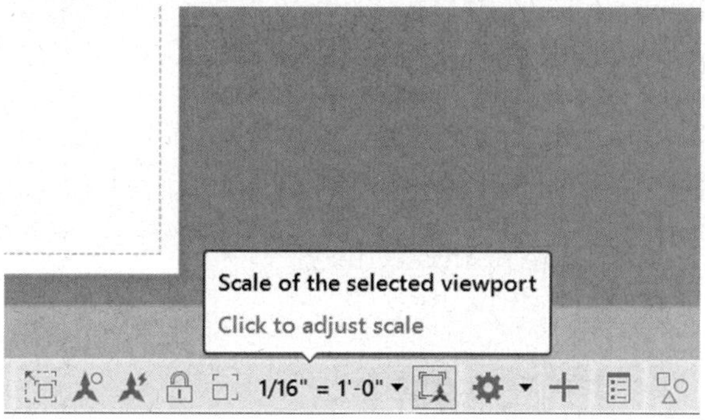


Since a layout represents the plotted sheet, you normally plot the layout at 1:1 scale. However, the geometry in each viewport is scaled to achieve the scale for the drawing objects. For example, you could have two or more viewports in one layout, each displaying the geometry at different scales, as is the case for Figure 13-6. Or, you can set up one layout to display model geometry in one scale and set up a similar layout to display the same geometry at a different scale.

**FIGURE 13-6**

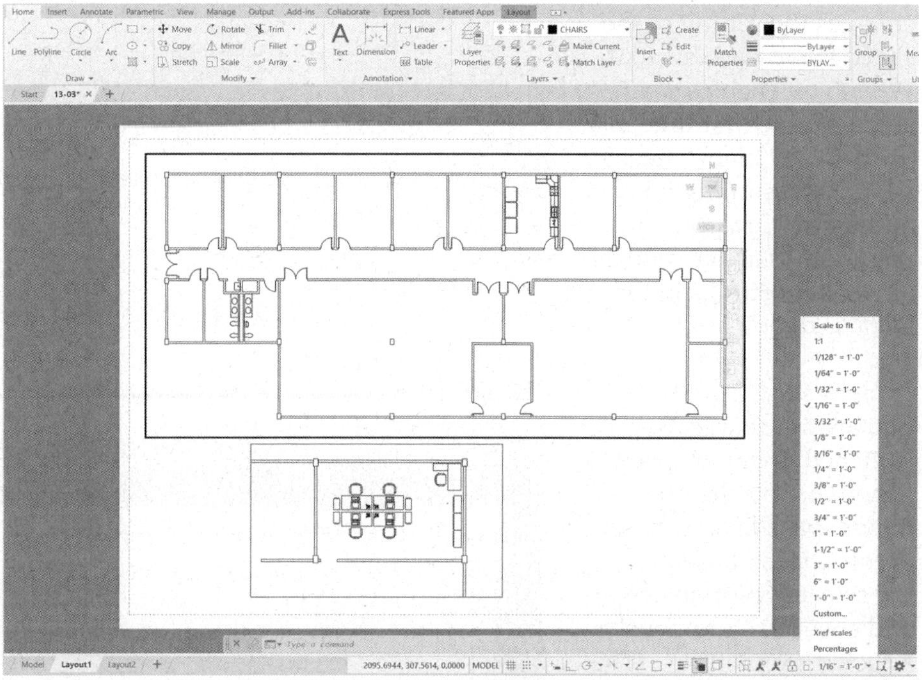

An important concept to remember when using layouts is that the *Plot Scale* that you select in the *Page Setup* and *Plot* dialog boxes should almost always be 1:1. This is because the layout is the actual paper size. Therefore, the geometry displayed in the viewports must be scaled to the desired plot scale (the reciprocal of the "drawing scale factor") which can be set by using the *Viewports Scale* pop-up list, the *Viewports* toolbar, the *Properties* palette for the viewport, or a *Zoom XP* factor.

## Layouts and Viewports Example

Consider this brief example to explain the basics of using layouts and viewports. In order to keep this example simple, only one viewport is created in the layout to set up a drawing for plotting. This example assumes that no automatic viewports or page setups are created. Your system may be configured to automatically create viewports when a *Layout* tab is activated (as recommended in Chapter 6 to simplify viewport creation). Disabling automatic setups is accomplished by ensuring *Show Page Setup Manager for new layouts* and *Create viewport in new layouts* is unchecked in the *Display* tab in the *Options* dialog box.

First, the part geometry is created in model space as usual (Fig. 13-7). Associative dimensions are also normally created in model space. This step is the same method that you would have normally used to create a drawing.

FIGURE 13-7

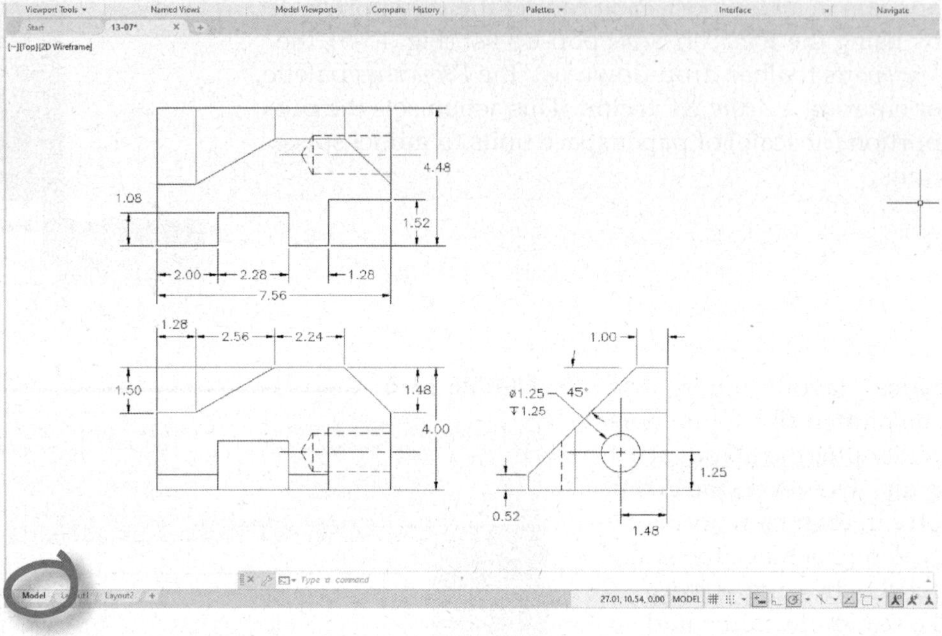

When the part geometry is complete, enable paper space by selecting a layout. AutoCAD automatically changes the setting of the *TILEMODE* variable to 0. When you enable paper space for the first time in a drawing, a "blank sheet" appears (assuming no automatic viewports or page setups are created as a result of settings in the *Options* dialog box).

By activating a layout, you are automatically switched to paper space. When that occurs, the paper space *Limits* are set automatically based on the selected plot device and sheet size previously selected for plotting. (The default plot device is set in the *Plotting* tab of the *Options* dialog box.)

FIGURE 13-8

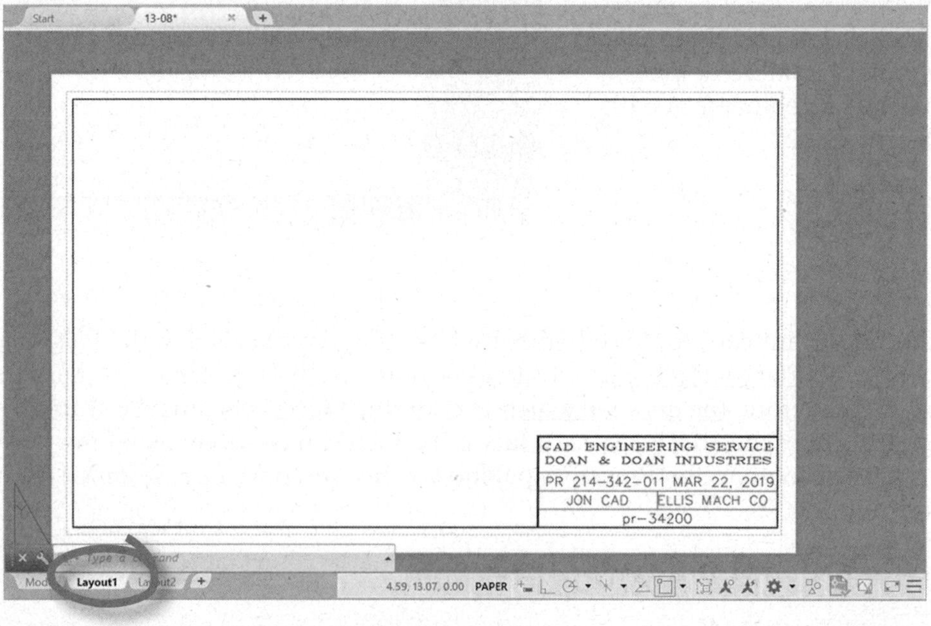


Objects such as a title block and border should be created in the paper space layout (Fig. 13-8). Normally, only objects that are annotations for the drawing (title blocks, tables, border, company logo, etc.) are drawn in the layout.

Next, create a viewport in the "paper" with the *Vports* command in order to "look" into model space. All of the model geometry in the drawing initially appears in the viewport.

At this point, there is no specific scale relation between paper space units and the size of the geometry in the viewport (Fig. 13-9).

You can control what part of model space geometry you see in a viewport and control the scale of model space to paper space. Two methods are used to control the geometry that is visible in a particular viewport.

**FIGURE 13-9**

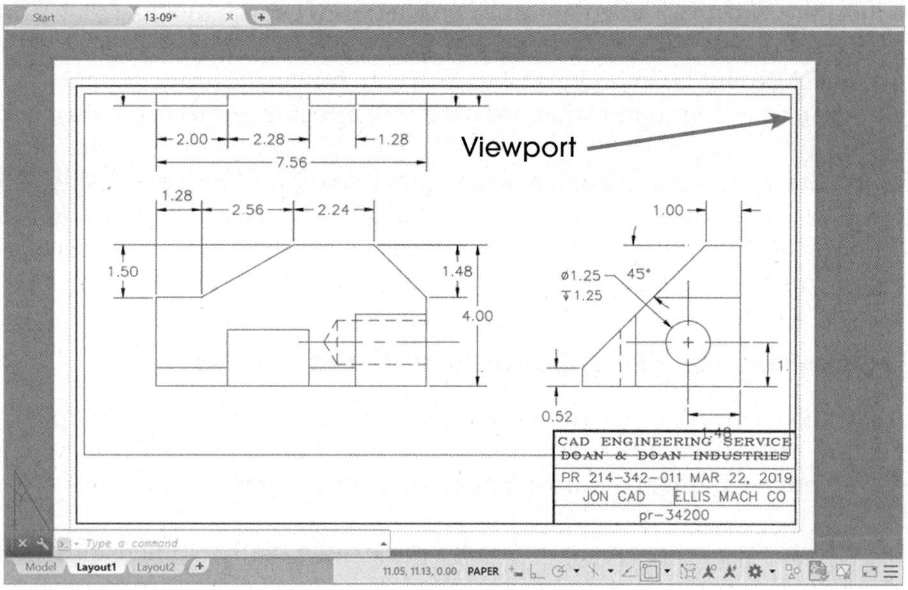


1. **Display commands and Viewport Scale**

   The *Zoom*, *Pan*, *View*, *3Dorbit* and other display commands allow you to specify what area of model space you want to see in a viewport. In addition, the scale of the model geometry in the viewport can be set by using the *Viewport Scale* pop-up list, the *Viewports* toolbar, the *Properties* palette for the viewport, or a *Zoom XP* factor.

2. **Viewport-specific layer visibility control**

   You can control what layers are visible in specific viewports. This function is often used for displaying different model space geometry in separate viewports. You can control which layers appear in which viewports by using the *Layer Properties Manager*. (See Chapter 32 for information on viewport-specific layer visibility control.)

For example, the *Viewport Scale* pop-up list could be used to scale the model geometry in a viewport (Fig. 13-10). In this example, select the viewport (from paper space) and choose *1:2* to scale the model geometry to 1/2 size (model space units equal 1/2 times paper space units).

**FIGURE 13-10**

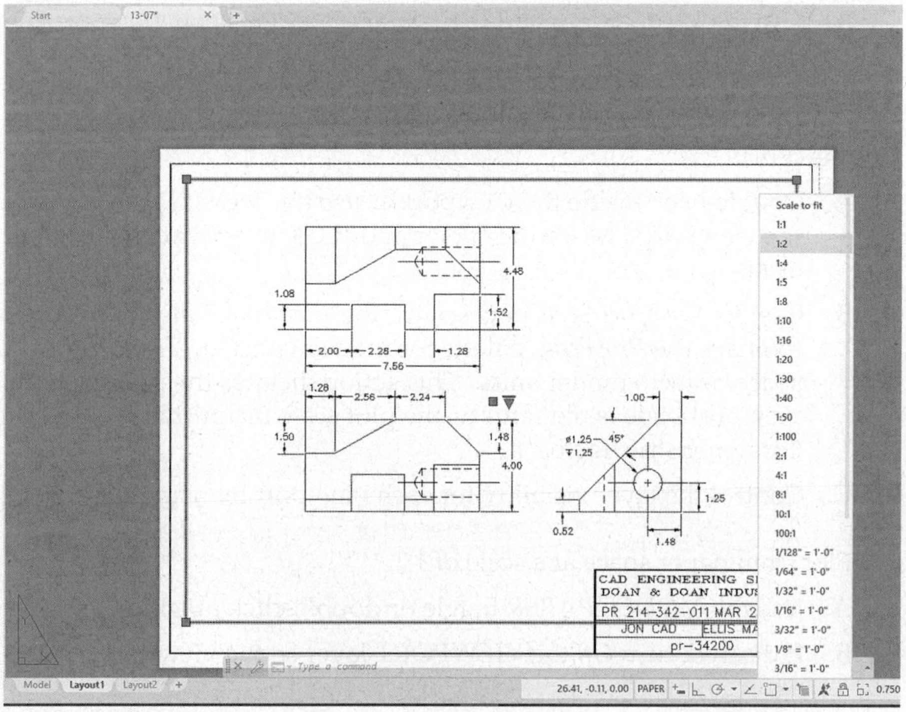


While in a layout with a viewport created to display model space geometry, you can double-click inside or outside the viewport. This action allows you to switch between model space (inside a viewport) and paper space (outside a viewport) so you can draw or edit in either space.

You could instead use the *Mspace* (Model Space) and *Pspace* (Paper Space) commands or single-click the words *MODEL* or *PAPER* on the Status Bar to switch between inside and outside the viewport. The "crosshairs" can move completely across the screen when in paper space, but only within the viewport when in model space (model space inside a viewport).

An object cannot be in both spaces. You can, however, draw in paper space and *Object Snap* to objects in model space. Commands that are used will affect the objects or display of the current space.

When drawing is completed, activate paper space and plot at 1:1 since the layout size (paper space *Limits*) is set to the actual paper size. Typically, the *Page Setup Manager* or the *Plot* dialog box is used to prepare the layout for the plot. The plot scale to use for the layout is almost always 1:1 since the layout represents the plotted sheet and the model geometry is already scaled.

## Guidelines for Using Layouts and Viewports

Although there are other alternatives, the typical steps used to set up a layout and viewports for plotting are listed here. (These guidelines assume that no automatic viewports or page setups are created, as described in Chapter 6, "Introduction to Layouts and Printing.")

1. Create the part geometry in model space. Associative dimensions are typically created in model space.
2. Activate a layout. When this is done, AutoCAD automatically sets up the size of the layout (*Limits*) for the default plot device, paper size, and orientation. Use the *Page Setup Manager* and *Page Setup* dialog box to ensure the correct device, paper size, and orientation are selected. If not, make the desired selections. (If the layout does not automatically reflect the desired settings, check the *Plot and Publish* tab of the *Options* dialog box to ensure that *Use the plot device paper size* is checked.)
3. Set up a title block and border as indicated.
   A. Make a layer named BORDER or TITLE and *Set* that layer as current.
   B. Draw, *Insert*, or *Xref* a border and a title block.
4. Make viewports in paper space.
   A. Make a layer named VIEWPORT (or other descriptive name) and set it as the *Current* layer (it can be turned *Off* later if you do not want the viewport objects to appear in the plot).
   B. Use the *Vports* command to make viewports. Each viewport contains a view of model space geometry.
5. Control the display of model space geometry in each viewport. Complete these steps for each viewport.
   A. Double-click inside the viewport or use the *MODEL/PAPER* toggle to "go into" model space (in the viewport). Move the cursor and PICK to activate the desired viewport if several viewports exist.
   B. Use the *Viewport Scale* pop-up list, the *Viewport Scale Control* drop-down list (on the *Viewports* toolbar), the *Properties* palette for the viewport, or *Zoom XP* to scale the display of the paper space units to model units. This action dictates the plot scale for the model space geometry. The viewport scale is the same as the plot scale factor that would otherwise be used (reciprocal of the "drawing scale factor").
   C. Control the layer visibility for each viewport by using the icons in the *Layer Properties Manager*.
6. Plot from paper space at a scale of 1:1.
   A. Use the *MODEL/PAPER* toggle or double-click outside the viewport to switch to paper space.
   B. If desired, turn *Off* the VIEWPORT layer so the viewport borders do not plot.

C. Invoke the *Plot* dialog box or the *Page Setup* dialog box to set the plot scale and other options. Normally, set the plot scale for the layout to *1:1* since the layout is set to the actual paper size. The paper space geometry will plot full size, and the resulting model space geometry will plot to the scale selected in the *Viewport Scale* pop-up list, *Viewport Scale* (toolbar) drop-down list, *Properties* palette, or by the designated *Zoom XP* factor.

D. Make a *Full Preview* to check that all settings are correct. Make changes if necessary. Finally, make the plot.

## CREATING AND SETTING UP LAYOUTS

Listed below are several ways to create layouts.

| | |
|---|---|
| Shortcut menu | Right-click on a *Layout* tab or the *Quick View Layouts* button and select *New Layout*, *From Template*, or *Move or Copy*. |
| *Layout* command | Type *Layout* at the Command line or select from the *Insert* pull-down menu. Use the *New Layout*, *Layout from Template*, or *Create Layout Wizard* option. |
| *Layout* tab (Ribbon) | Select a *Layout* to activate the *Layout* tab on the Ribbon, then select *New* or *From Template*. |
| *Layoutwizard* command | Type or select from the *Insert* pull-down menu. |

There are several steps involved in correctly setting up a layout, as listed previously in the "Guidelines for Using Layouts and Viewports." The steps involve specifying a plot device, setting paper size and orientation, setting up a new *Layout* tab, and creating the viewports and scaling the model geometry. One method for accomplishing all of these tasks is to use the *Layout Wizard*.
The *Layoutwizard* command produces a wizard to automatically lead you through eight steps to correctly set up a layout tab for plotting. The steps are essentially the same as those listed earlier in the "Guidelines for Using Layouts and Viewports."

### *Layoutwizard*

| Ribbon | Menu Bar | Command (Type) | Alias (Type) | Shortcut |
|---|---|---|---|---|
| ... | *Insert Layout > Create Layout Wizard* | *Layoutwizard* | ... | ... |

#### *Begin*

The first step is intended for you to enter a name for the layout (Fig. 13-11). The default name is whatever layout number is next in the sequence, for example, *Layout3*. Enter any name that has not yet been used in this drawing. You can change the name later if you want using the *Rename* option of the *Layout* command.

**FIGURE 13-11**

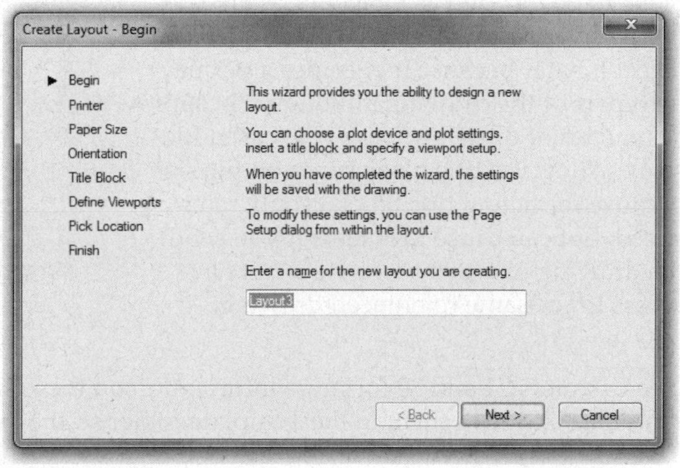

### *Printer*

Because the *Limits* in paper space are automatically set based on the size and orientation of the selected paper, choosing a print or plot device is essential at this step (Fig. 13-12). The list includes all previously configured printer and plotter devices. If you intend to use a different device, it may be wise to *Cancel*, use *Plotter Manager* to configure the new device, then begin *Layoutwizard* again (see Chapter 14, "Configuring Plotters and Printers").

**FIGURE 13-12**

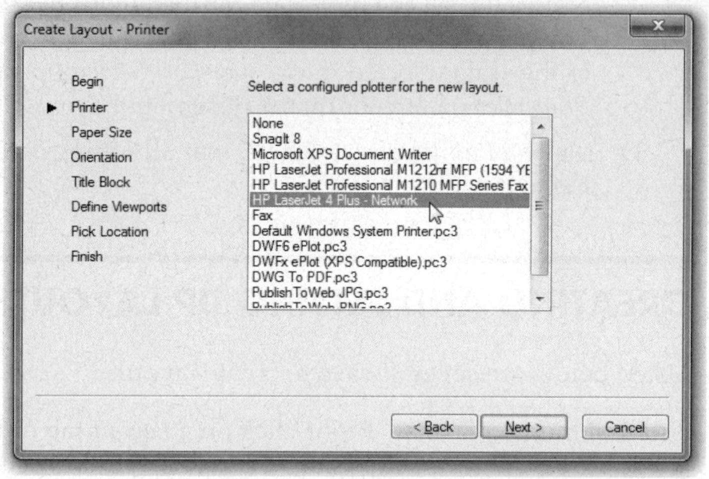


### *Paper Size*

Select the paper size you expect to use for the layout (Fig. 13-13). Make sure you also select the units for the paper (*Drawing Units*) since the layout size (*Limits* in paper space) is automatically set in the selected units.

**FIGURE 13-13**

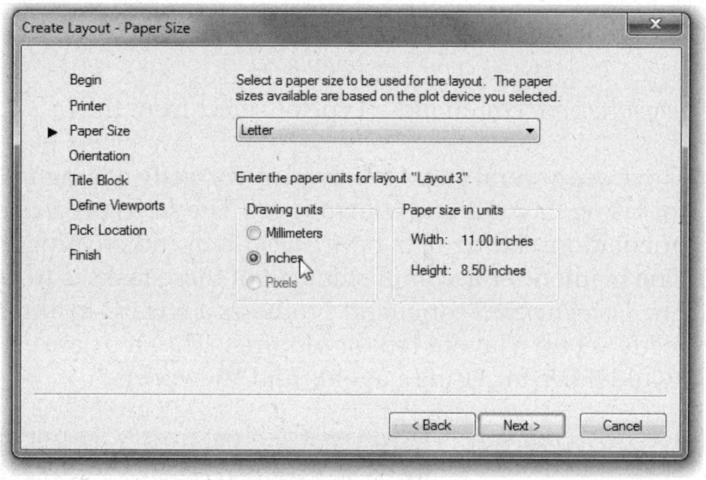


### *Orientation*

Select either *Portrait* or *Landscape* (not shown). *Landscape* plots horizontal lines in the drawing along the long axis of the paper, whereas *Portrait* plots horizontal lines in the drawing along the short axis of the paper.

### *Title Block*

In this step, you can select *None* or select from the AutoCAD-supplied title blocks (Fig. 13-14). The selected title block is inserted into the paper space layout. You can also specify whether you want the title block to be inserted as a *Block* or as an *Xref*. Normally inserting the title block as a *Block* object is safer because it becomes a permanent part of the drawing although it occupies a small bit of drawing space (increased file size). An *Xrefed* title block saves on file size because the actual title block is only referenced, but can cause problems if you send the drawing to a client who does not have access to the same referenced drawing.

**FIGURE 13-14**

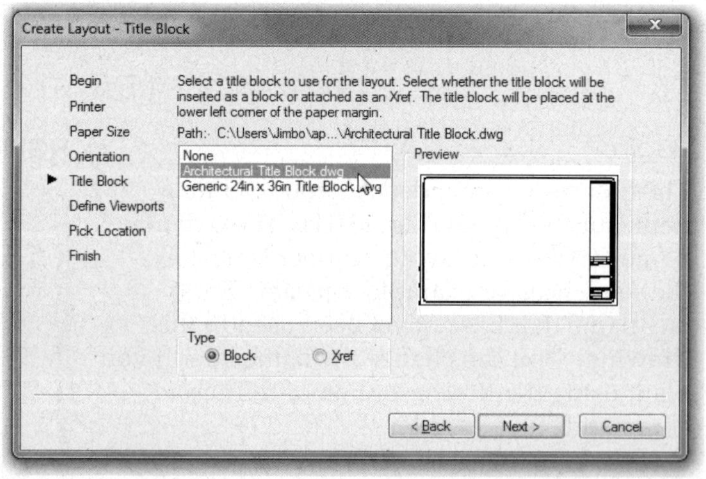


(See Chapters 21 and 30 for more information on these subjects.) You can also create your own title block drawings and save them in the Template folder so they will appear in this list.

### *Define Viewports*

This step has two parts (Fig. 13-15). First select the type and number of viewports to set up. Use *None* if you want to set up viewports at a later time or want to create a non-rectangular viewport. Normally, use *Single* if you need only one rectangular viewport. Use *Std. 3D Engineering Views* if you have a 3D model. The *Array* option creates multiple viewports with the number of *Rows* and *Columns* you enter in the edit boxes below. Second, determine the *Viewport Scale*. This setting determines the scale that the model space geometry will be displayed in the viewport (ratio of paper space units to model space units).

**FIGURE 13-15**

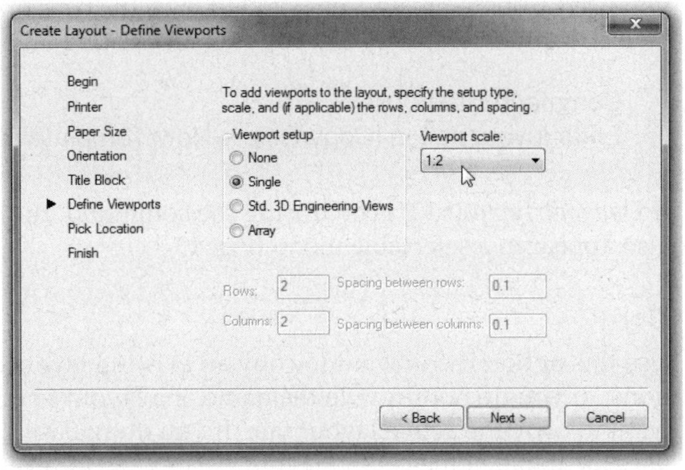


### *Pick Location*

In this step, pick *Select Location* to temporarily exit the wizard and return to the layout to pick diagonal corners for the viewport(s) to fit within (Fig. 13-16). Normally, you do not want the corners of the viewport(s) to overlap the title block or border. You can bypass this step to have AutoCAD automatically draw the viewport border at the extents of the printable area.

**FIGURE 13-16**

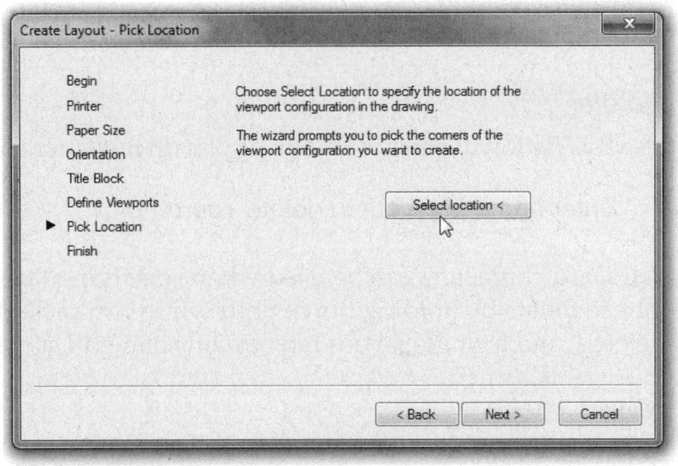


### *Finish*

This is not really a step, rather a confirmation (not shown). In all other steps, you can use *Back* to go backward in the process to change some of the specifications. In this step, *Back* is disabled, so you can only select *Finish* or *Cancel*.

At this time, the new layout is activated so you can see the setup you specified. If you want to make changes to the plot device, paper size, orientation, or viewport scale, you can do so by using the *Page Setup* dialog box. You can also change the viewport sizes or configuration using *Vports* or grips (explained later in this chapter).

## *Layout*

| Ribbon | Menu Bar | Command (Type) | Alias (Type) | Shortcut |
|---|---|---|---|---|
| *Layout* *Layout* *New* | *Insert* *Layout >* *New Layout* | *Layout* | *LO* or *-LO* | ... |

If you prefer to create a layout without using the *Layout Wizard*, type the *Layout* command or select it from the *Layout* tab, *Insert* pull-down menu, or *Layout* toolbar. Using the wizard leads you through all the steps required to create a new layout including selecting plot device, paper size and orientation, and creating viewports. Use the *Layout* command to create viewports "manually" (if your system is not set to automatically create viewports). You will then have to use *Vports* to create viewports in the layout.

It is a good idea to ensure the correct default plot device is specified in the *Options* dialog box before you use *Layout* to create a new layout. See the *New* option for more information. Besides creating a new layout, using a template, and copying an existing layout, you can rename, save, and delete layout.

Command: ***LAYOUT***
Enter layout option [Copy/Delete/New/Template/Rename/SAveas/Set/?] <set>:

TIP

If *Dynamic Input* (F12) is on during the command, the command line options also appear in a selectable menu (Fig. 13-17).

**FIGURE 13-17**

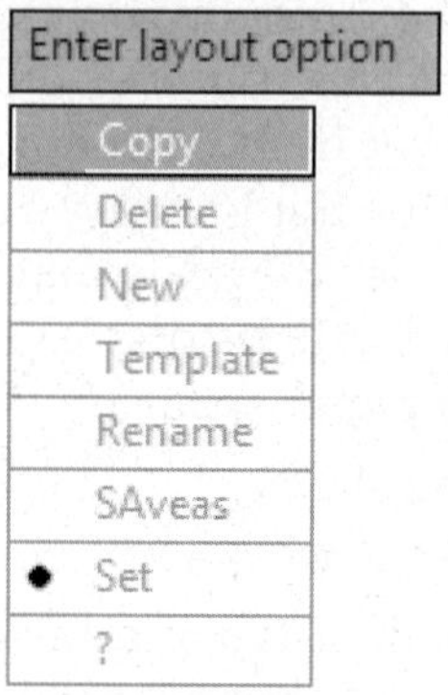


### *Copy*

Use this option if you want to copy an existing layout including its plot specifications. If you do not provide the name of a layout to copy, AutoCAD assumes you want to copy the active layout tab. If you do not assign a new name, AutoCAD uses the name of the copied layout (*Floor Plan*, for example) and adds an incremental number in parentheses, such as *Floor Plan (2)*.

Enter layout to copy <current>:
Enter layout name for copy <default>:

### *Delete*

Use this option to *Delete* a layout. If you do not enter a name, the most current layout is deleted by default.

Enter name of layout to delete <current>:

Wildcard characters can be used when specifying layout names to delete. You can select several *Layout* tabs to delete by holding down Shift while you pick. If you select all layouts to delete, all the layouts are deleted, and a single layout tab remains named *Layout1*. The *Model* tab cannot be deleted.

### *New*

This option creates a new layout tab. AutoCAD creates a new layout from scratch based on settings

defined by the default plot device and paper size (in the *Options* dialog box). If you choose not to assign a new name, the layout name is automatically generated (*Layout3*, for example).

Enter new layout name <Layout#>:

Keep in mind that when the new layout is created, its size (*Limits*) and shape are determined by the *Use as Default Output Device* setting specified in the *Plot and Publish* tab of the *Options* dialog. Therefore, it is a good idea to ensure the correct plot device is specified before you use the *New* option. You can, however, change the layout at a later time using the *Page Setup* dialog box, assuming that *Use the plot device paper size* is checked in the *Plot and Publish* tab of the *Options* dialog box (see *Options* in this chapter).

### *Template*

You can create a new template based on an existing layout in a template file (.DWT) or drawing file (.DWG) with this option. The layout, viewports, associated layers, and all the paper space geometry in the layout (from the specified template or drawing file) are inserted into the current drawing. No dimension styles or other objects are imported. The standard file navigation dialog box (not shown) is displayed for you to select a file to use as a template. Next, the *Insert Layout* dialog box appears for you to select the desired layout from the drawing (Fig. 13-18).

**FIGURE 13-18**

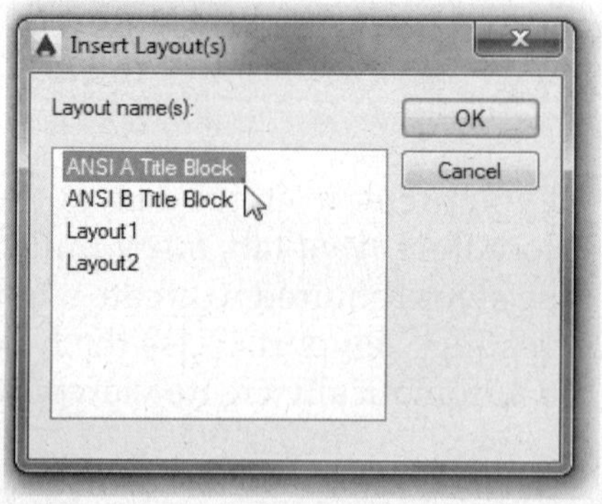

### *Rename*

Use *Rename* to change the name of an existing layout. The last current layout is used as the default for the layout to rename.

```
Enter layout to rename <current>:
Enter new layout name:
```

Layout names can contain up to 255 characters and are not case sensitive. Only the first 32 characters are displayed in the tab.

### *Save*

This option creates a .DWT file. The file contains all of the (paper space) geometry in the layout and all of the plot settings for the layout, but no model space geometry. All *Block* definitions appearing in the layout, such as a titleblock, are also saved to the .DWT file, but not unused *Block* definitions. All layouts are stored in the template folder as defined in the *Options* dialog box. The last current layout is used as the default for the layout to save.

```
Enter layout to save to template <current>:
```

The standard file selection dialog box is displayed in which you can specify a file name for the .DWT file. When you later create new layout from a template (using the *Template* option of *Layout*), all the saved information (plot settings, layout geometry and *Block* definitions) is imported into the new layout.

### *Set*

This option simply makes a layout current. This option has identical results as picking a layout tab.

### ? (List Layouts)

Use this option to list all the layouts when you have turned off the layout tabs in the *Display* tab of the *Options* dialog box. Otherwise, layout tabs are visible at the bottom of the drawing area.

## Inserting Layouts with AutoCAD DesignCenter

A feature in AutoCAD, called DesignCenter, allows you to insert content from any drawing into another drawing. Content that can be inserted includes drawings, *Blocks, Dimstyles, Layers, Layouts, Linetypes, Tablestyles, Textstyles, Xrefs*, raster images, and URLs (Web site addresses). If you have created a layout in a drawing and want to insert it into the current drawing, you can use the *Template* option of *Layout* (as previously explained) or use DesignCenter to drag and drop the layout name into the drawing. With DesignCenter, only layouts from .DWG files (not .DWT files) can be imported. See Chapter 21 Blocks, DesignCenter, and Tool Palettes for more information.

## Setting Up the Layout

When you do not use the *Layout Wizard*, the steps involved in setting up the layout must be done individually. In cases when you need flexibility, such as creating non-rectangular viewports or when some factors are not yet known, it is desirable to take each step individually. These steps are listed in "Guidelines for Using Layouts and Viewports."

### *Options*

| Ribbon | Menu Bar | Command (Type) | Alias (Type) | Shortcut |
|---|---|---|---|---|
| *Output Plot* (arrow) | *Tools Options...* | *Options* | *OP* | (Default Menu) *Options...* |

When you create *New* layouts using the *Layout* command, the size and shape of the layout is determined by the selected plot device and paper size. Default settings are made in the *Plot and Publish* tab of the *Options* dialog box (Fig. 13-19). Although most settings concerned with plotting a layout can later be changed using the *Page Setup* dialog box, it may be efficient to make the settings in the *Options* dialog box before creating layouts from scratch. The following settings in this dialog box affect layouts.

FIGURE 13-19

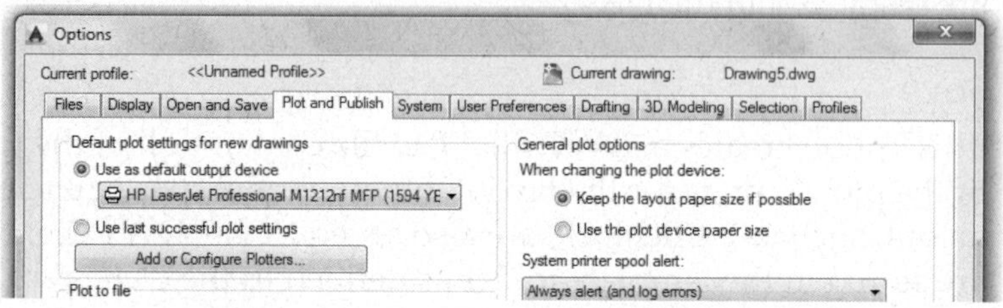


#### *Default plot settings for new drawings*

Your choice in this section determines what plot device is used to automatically set the paper size (paper space *Limits*) when new layouts are created. Your setting here applies to new drawings and the current drawing. (Paper size information for each device is stored either in the plotter configuration file [.PC3] or in the default system settings if the output device is a system printer.)

*Use as default output device*
The selected device is used to determine the paper size (paper space *Limits*) that are set automatically when a new layout is created in the current drawing and for new drawings. This drop-down list contains all configured plot or print devices (any plotter configuration files [PC3] and any system printers that are configured in the system).

*Use last successful plot settings*
This button uses the plotting settings (device and paper size) according to those of the last successful plot that was made on the system (a plot must be made from the layout for this option to be valid) and applies them to new layouts that are created. This setting affects new layouts created for the current drawing as well as new drawings. This option overrides the output device appearing in the *Use as default output device* drop-down list.

#### *General plot options*

These options apply only when plot devices are changed for existing layouts and the layout has a viewport created. Plot devices for layouts can be changed using the *Page Setup* or *Plot* dialog boxes. When you change the plot device for an existing layout, the paper size (*Limits* setting) of the layout is determined by your choice in this section.

*Keep the layout paper size if possible*
If this button is pressed, AutoCAD attempts to use the paper size initially specified for the layout if you decide later to change the plot device for the layout. This option is useful if you have two devices that can use the same size paper as specified in the layout, so you can select either device to use without affecting the layout. However, if the selected output device cannot plot to the previously specified paper size, AutoCAD displays a warning message and uses the paper size specified by the plot device you select in the *Page Setup* or *Plot* dialog box. If no viewports have been created in the layout, you can change the plot device and automatically reset the layout limits without consequence. This button sets the *PAPERUPDATE* system variable to 0 (layout paper size is not updated).

*Use the Plot Device Paper Size*
As the alternative to the previous button, this one sets the paper size (*Limits*) for an existing layout to the paper size of whichever device is selected for that layout in the *Page Setup* or *Plot* dialog box. This option sets *PAPERUPDATE* to 1.

## Pagesetup

| Ribbon | Menu Bar | Command (Type) | Alias (Type) | Shortcut |
|---|---|---|---|---|
| *Output*<br>*Plot*<br>*Page Setup Manager...* | *File*<br>*Page Setup Manager...* | *Pagesetup* | ... | ... |

Plot specifications that you make and save for a layout are called "page setups." There are two separate dialog boxes that are used for saving a page setup. Using the *Pagesetup* command produces the *Page Setup Manager* (Fig. 13-20) which is a "front end" for the *Page Setup* dialog box. The *Page Setup* dialog box (see Fig. 13-21) is used to select the plot specifications for the layout such as plot device, paper size, plot scale, etc. All settings made in the *Page Setup Manager* and related dialog box are saved with the layout. Therefore, the plot settings are prepared when you are ready to make a plot or print.

If you have created a layout using the *Layout* command and the desired settings were made in the *Options* dialog box as explained in the previous section, the next step in setting up a layout is using the *Page Setup Manager* and the related *Page Setup* dialog box. If you used the *Layout Wizard*, you can optionally use the *Page Setup Manager* and related dialog box to alter some of those settings.

To save a page setup for a layout, first activate the desired layout tab, then use *Pagesetup* to produce the *Page Setup Manager*. The active layout tab's name appears highlighted in the *Page setups* list. Next, select the *Modify* option as shown in Figure 13-20 to produce the *Page Setup* dialog box.

When the *Page Setup* dialog box appears (Fig. 13-21), specify the plotting options listed next.

FIGURE 13-20

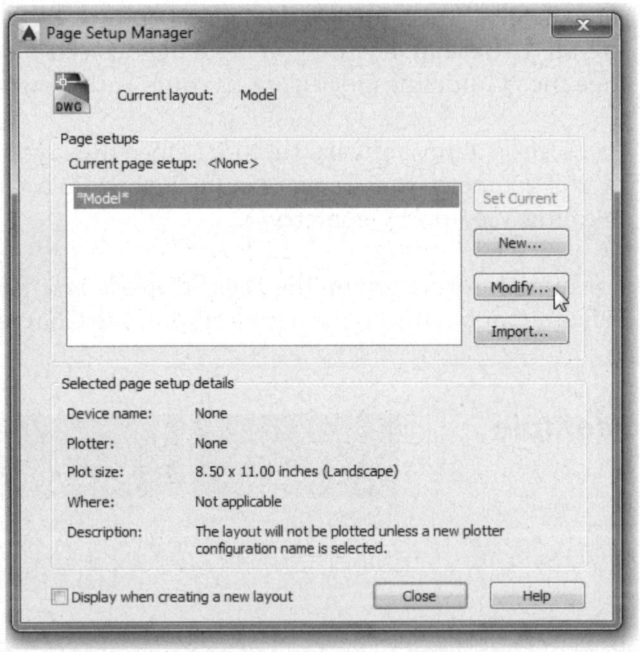


FIGURE 13-21

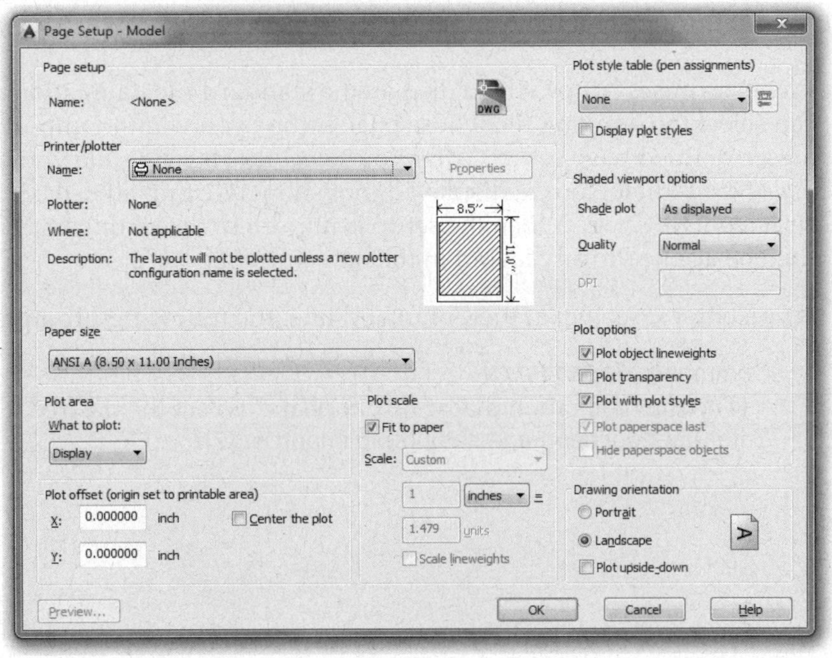

1. Select a plotter or printer in the *Printer/plotter* section.
2. Select the desired *Paper size*.
3. Ensure the *Plot scale* is set to 1:1.
4. Select any other desired options such as *Drawing orientation, Plot options*, etc.
5. Select the *OK* button, then *Close* the *Page Setup Manager* to save the settings for the layout.

Remember that the size of the layout (paper space *Limits*) is automatically set based on *Paper Size* and *Drawing Orientation* you select here. Depending on your selection of *Keep the layout size if possible* or *Use the plot device paper size* in the *Plot and Publish* tab of the *Options* dialog box (see previous section), the layout size and shape may or may not change to the new plot device settings. You can, however, select a new plot device, paper size, or orientation after creating the layout, but before creating viewports.

TIP

Since the layout size and shape may change with a new device, viewports that were previously created may no longer fit on the page. In that case, the viewports must be changed to a new size or shape. Alternately, you can *Erase* the viewports and create new ones. Both of these choices are not recommended. Instead, ensure you have the desired plot device and paper settings before creating viewports. (See the "Guidelines for Using Layouts and Viewports.")

*Plot scale* is almost always set to *1:1* since the layout represents the sheet you will be plotting on. The model geometry that appears in the layout (in a viewport) will be scaled by setting a viewport scale (see "Scaling Viewport Geometry").

For more information on the *Page Setup Manager* and the *Page Setup* dialog box, such as how to apply an existing page setup to the active layout, see Chapter 14.

### Psetupin

| Ribbon | Menu Bar | Command (Type) | Alias (Type) | Shortcut |
|---|---|---|---|---|
| ... | ... | *Psetupin* or *-Psetupin* | ... | ... |

*Psetupin* imports a user-defined page setup into a new drawing layout. This feature provides the ability to import a saved, named page setup from another drawing into the current drawing. User-defined page setups are created using the *Page Setup* dialog box with the *Add* option (see previous section).

*Psetupin* causes AutoCAD to display the standard file navigation dialog box (not shown) in which you can select the drawing file whose page setups you want to import. The selected .DWG file must contain a user-defined page setup. After you select the drawing file you want to use, the *Import User Defined Page Setups* dialog box is displayed (not shown) listing all user-defined page setups contained in the selected .DWG file. When the setup is imported, the settings that are saved in the named page setup can then be applied to any layout in the new drawing.

If you enter *-Psetupin* at the Command line, the following prompts are displayed.

Command: ***-PSETUPIN***
(The file navigation dialog box appears. Select the desired .DWG file.)
Enter user defined page setup(s) to import or [?]:

## USING VIEWPORTS IN PAPER SPACE

If you use the *Layout Wizard*, viewports can be created during that process. If you do not use the wizard, viewports should be created for a layout only after specifying the plot device, paper size, and orientation in the *Page Setup* dialog box or through the default plot options set in the *Options* dialog box. You can create viewports with the *Vport* or *-Vport* commands. The Command line version (*-Vports*) produces several options not available in the dialog box, including options to create non-rectangular viewports.

### *Vports*

| Ribbon | Menu Bar | Command (Type) | Alias (Type) | Shortcut |
|---|---|---|---|---|
| *Layout Layout Viewports* | *View Viewports >* | *Vports* or *-Vports* | ... | ... |

The *Vports* command produces the *Viewports* dialog box. Here you can select from several configurations to create rectangular viewports as shown in Figure 13-22. Making a selection in the *Standard Viewports* list on the left of the dialog box in turn displays the selected *Preview* on the right.

FIGURE 13-22

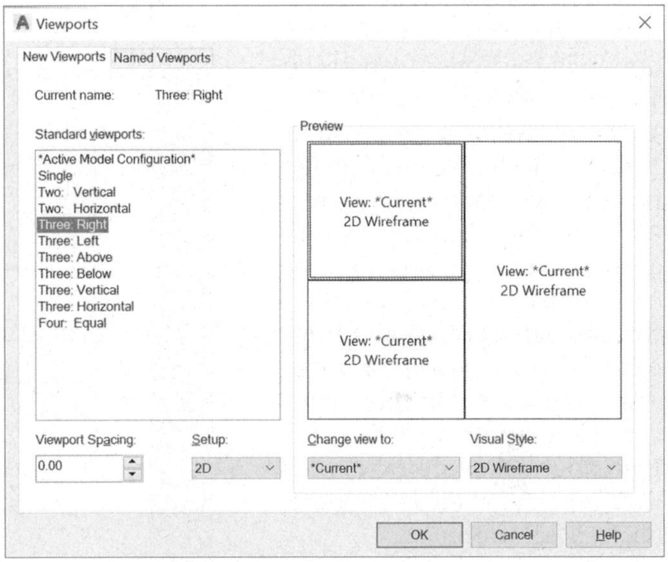


After making the desired selection from the dialog box, AutoCAD issues the following prompt.

Command: ***VPORTS***
Specify first corner or [Fit] <Fit>:

You can pick two diagonal corners to specify the area for the viewports to fill. Generally, pick just inside the titleblock and border if one has been inserted. If you use the *Fit* option, AutoCAD automatically fills the printable area with the specified number of viewports.

FIGURE 13-23

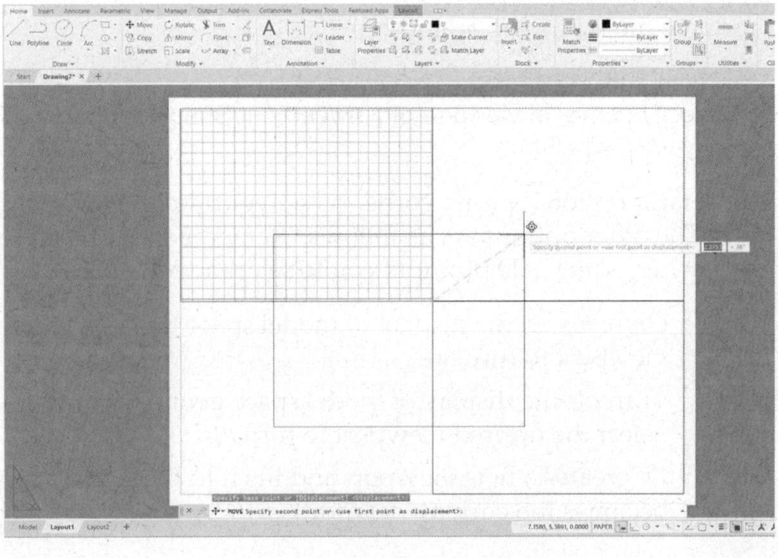

The *Setup: 3D* and *Change view to:* options are intended for use with 3D models.

You may notice that the *Viewports* dialog box is the same dialog box that appears when you use *Vports* while in the *Model* tab. However, when *Vports* dialog is used in the *Model* tab, tiled viewports are created, whereas when the *Vports* is used in a layout (paper space), paper space viewports are created.

Paper space viewports differ from tiled viewports. AutoCAD treats paper space viewports created with *Vports* as objects. Like other objects, paper space viewports can be affected by most editing commands. For example, you could use *Vports* to create one viewport, then use *Copy* or *Array* to create other viewports. You could edit the size of viewports with *Stretch* or *Scale*. You can use Grips to change paper space viewports. Additionally, you can use *Move* (Fig. 13-23, previous page) to relocate the position of viewports. Delete a viewport using *Erase*. To edit a paper space viewport, you must be in paper space to PICK the viewport objects (borders).

Paper Space viewports can also overlap. This feature makes it possible for geometry appearing in different viewports to occupy the same area on the screen or on a plot.

NOTE: Avoid creating one viewport completely within another's border because visibility and selection problems may result.

Another difference between tiled model space viewports and paper space viewports is that viewports in layouts can have space between them. The option in the bottom left of the *Viewports* dialog box called *Viewport Spacing* allows you to create space between the viewports (if you choose any option other than *Single*) by entering a value in the edit box. Figure 13-24 illustrates a layout with three viewports after entering .5 in the *Viewport Spacing* edit box. Also note several of the *-Vports* command options are available in the right-click shortcut menu that appears when a viewport object is selected.

**FIGURE 13-24**

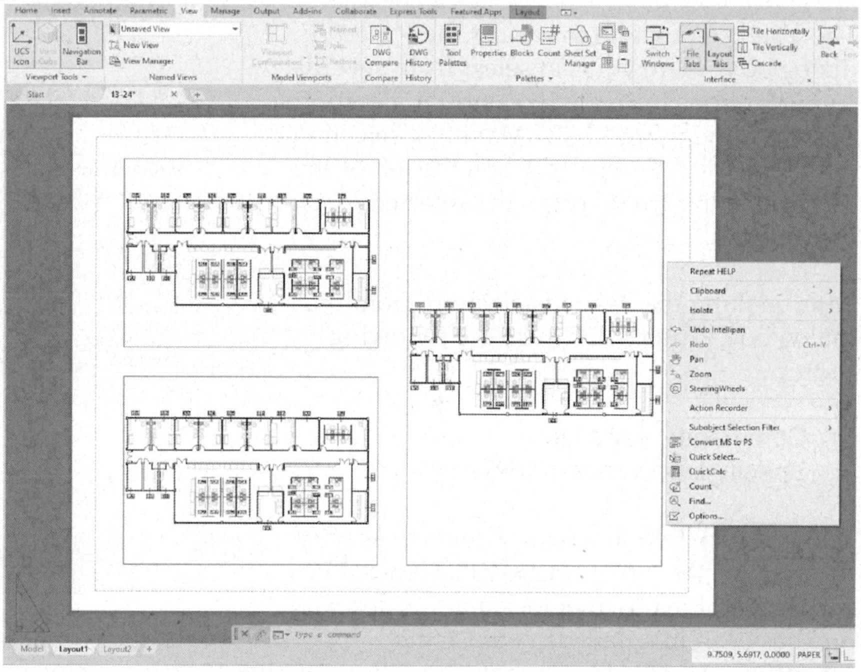


Using the *-Vports* command (with a hyphen prefix) invokes the Command line version and produces the following options.

Command: -***VPORTS***
Specify corner of viewport or [ON/OFF/Fit/Shadeplot/Lock/NEw/NAmed/Object/Polygonal/Restore/LAyer/2/3/4] <Fit>:

The default option (Specify corner of viewport) allows you to pick two diagonal corners for one viewport. The other options are as follows. Remember to select the viewport objects (borders) while in paper space when AutoCAD prompts you to select a viewport.

***ON*** *On* turns on the display of model space geometry in the selected viewport. Select the desired viewport to turn on.

***OFF*** Turn off the display of model space geometry in the selected viewport with this option. Select the desired viewport to turn off.

***Fit*** *Fit* creates a new viewport and fits it to the size of the printable area. The new viewport becomes the current viewport.

### *Shadeplot*

If you are displaying a 3D surface or solid model, *Shadeplot* allows you to select how you want to print the current viewport: *As Displayed*, *Wireframe*, *Hidden*, *Visual Styles*, or *Rendered*. (See Chapter 33, 3D Basics, Navigation, and Visual Styles, for more information.)

### *Lock*

Use this option to lock the scale of the current viewport. Turn viewport locking *On* or *Off*. With paper space viewports, you set the scale of each viewport individually using *Zoom* with an *XP* scale factor or by using the *Viewports* toolbar scale drop-down list (see "Scaling Viewport Geometry"). Double-clicking inside a viewport and then using *Zoom* normally changes the scale of a viewport. If *Lock* is used on the viewport, the specified scale of the viewport cannot be changed (unless viewport locking is turned *Off*). If you attempt to *Zoom* inside a locked viewport, the display of the entire layout (paper space) is zoomed.

TIP

*NEw* Use this option to create a new viewport by specifying its corners.

### *NAmed*

This option allows you to choose a named view for use in the viewport, but you must previously have set that up using the View Manager or similar commands.

### *Object*

You can use *Object* to convert a closed *Polyline*, *Ellipse*, *Spline*, *Region*, or *Circle* into a viewport (border). The polyline you specify must be closed and contain at least three vertices. It can be self-intersecting and can contain arcs as well as line segments. Figure 13-25 displays a closed *Pline* converted to a viewport.

**FIGURE 13-25**

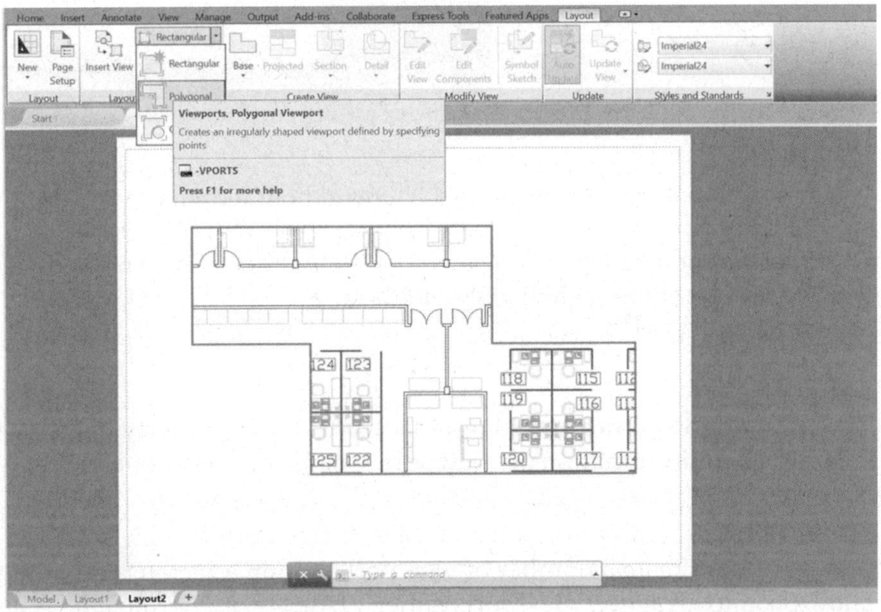


### *Polygonal*

Use this option to create an irregularly shaped viewport defined by specifying points. You can define the viewport by straight line or arc segments.

Command: ***-VPORTS***
Specify corner of viewport or [ON/OFF/Fit/Shadeplot/Lock/Object/Polygonal/Restore/2/3/4] <Fit>: ***p***
Specify start point: **PICK**
Specify next point or [Arc/Close/Length/Undo]: **PICK**
Specify next point or [Arc/Close/Length/Undo]: **PICK**
Specify next point or [Arc/Close/Length/Undo]: ***c***
Regenerating model.

If you choose the *Arc* option, the following prompts allow all arc creation options, identical to those found in the full *Arc* command.

Enter an arc boundary option [Angle/CEnter/CLose/Direction/Line/Radius/Second pt/Undo/Endpoint of arc] <Endpoint>:

Figure 13-26 displays a viewport created with the *Polygonal* option of *-Vports*. Notice the arc segment. Keep in mind both the *Object* and *Polygonal* options can be used to create non-rectangular and curved viewports.

**FIGURE 13-26**

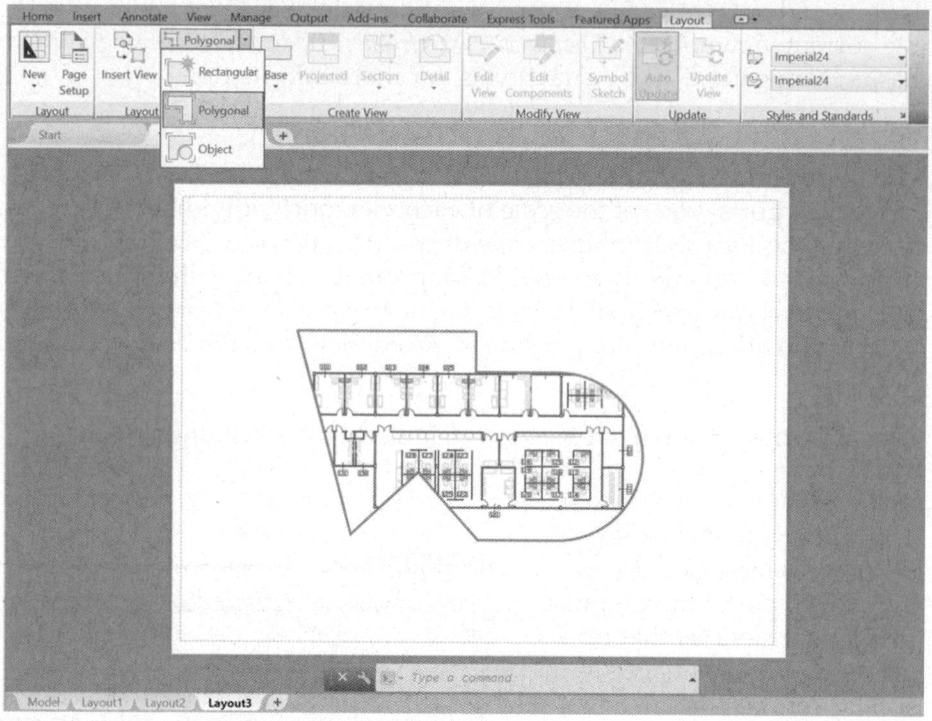


### *Restore*

Use this option to restore a previously created named viewport configuration.

Enter viewport configuration name or [?] <*Active>: Enter named viewport configuration.
Specify first corner or [Fit] <Fit>: **PICK** or ***F***
Regenerating model.

### *Layer*

The *Layer* option allows you to reset viewport layer property overrides back to global properties. See Chapter 32 for information on viewport-specific property overrides.

### *2/3/4*

These options create a number of viewports within a rectangular area that you specify. The *2* option allows you to arrange 2 viewports either *Vertically* or *Horizontally*. The *4* option automatically divides the specified area into 4 equal viewports. The possible configurations using the *3* option are displayed if you use the *Viewports* dialog box. Using the 3 option yields the following prompt:

[Horizontal/Vertical/Above/Below/Left/Right] <Right>:

## *Qvlayout*

| Ribbon | Menu Bar | Command (Type) | Alias (Type) | Shortcut |
|---|---|---|---|---|
| ... | ... | *Qvlayout* | *QVL* | ... |

You can activate the *Qvlayout* (*Quick View Layouts*) command by typing *Qvl* or *Qvlayout* at the Command prompt. *Qvlayout* causes a series of small thumbnail images to be displayed near the bottom of the AutoCAD window (Fig. 13-27, bottom). The *Quick View Layouts* are also produced by hovering over the drawing tab in the upper-left corner of the graphics area (Fig. 13-27, top).

FIGURE 13-27

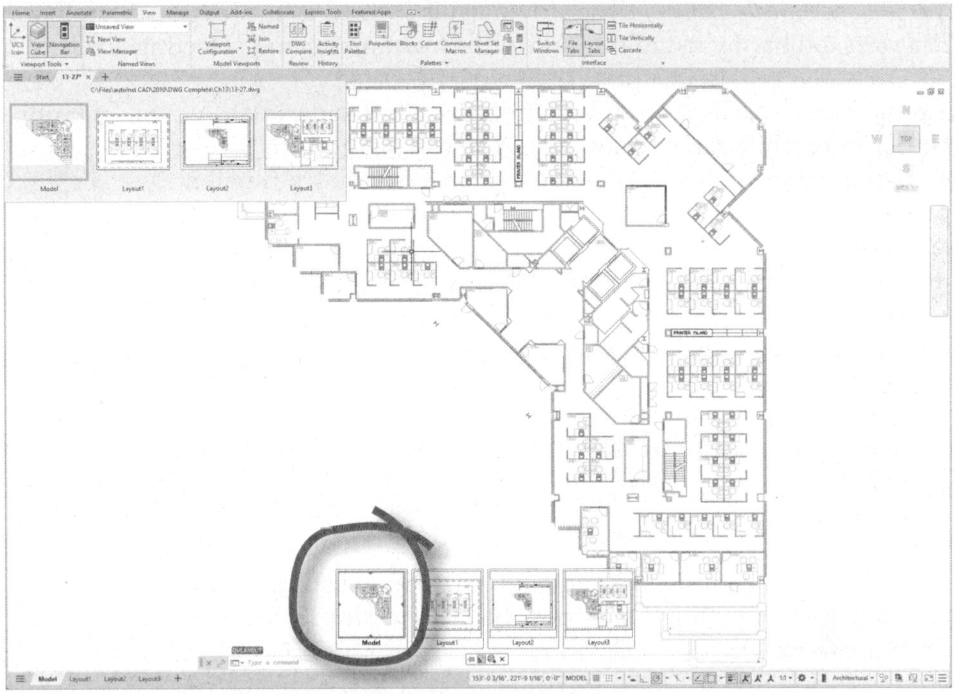

The left-most image represents model space and each layout existing in the drawing is represented by an image. Select an image to make that layout or model space active. Although either of these options can be used as an alternative to selecting the *Model* tab and *Layout* tabs to activate layouts, there are additional functions. You can *Plot* or *Publish* the desired layout without having to activate it first. Do this by selecting the *Plot* or *Publish* button in the upper-left and upper-right corners of the desired image (see Figure 13-27, top left).

## Scaling the Display of Viewport Geometry

Paper space (layout) objects such as title block and border should correspond to the paper at a 1:1 scale. A paper space layout is intended to represent the print or plot sheet. *Limits* in a layout are automatically set to the exact paper size, and the finished drawing is plotted from the layout to a scale of 1:1. The model space geometry, however, should be true scale in real-world units, and *Limits* in model space are generally set to accommodate that geometry.

When model space geometry appears in a viewport in a layout, the size of the displayed geometry can be controlled so that it appears and plots in the correct scale. There are three ways to scale the display of model space geometry in a paper space viewport. You can (1) use the *Viewport Scale* pop-up list, (2) use the *Viewport* toolbar drop-down list, (3) use the *Properties* palette, or (4) use the *Zoom* command and enter an *XP* factor. Once you set the scale for the viewport, use the *Lock* option of *-Vports* or the *Properties* palette to lock the scale of the viewport so it is not accidentally changed when you *Zoom* or *Pan* inside the viewport.

TIP

When you use layouts, the *Plot Scale* specified in the *Plot* dialog box and *Page Setup* dialog box dictates the scale for the layout, which should normally be set to 1:1. Therefore, the viewport scale actually determines the plot scale of the geometry in a finished plot. Set the viewport scale to the reciprocal of the "drawing scale factor," as explained in Chapter 12, Advanced Drawing Setup.

### *Viewport Scale* Pop-up List

The *Viewport Scale* pop-up list is located in the lower-right corner of the screen (see Figs. 13-5 and 13-6). To set the viewport scale (scale of the model geometry in the viewport), first make the viewport current by double-clicking in it or typing *Mspace*. When the crosshairs appear in the viewport and the viewport border is highlighted (a wide border appears), select the desired scale from the list. The entries in the list can be deleted, added, or edited by selecting *Custom...*, which produces the *Edit Scale List* dialog box (see "*Scalelistedit*" near the end of this chapter).

### *Viewport Scale Control* Drop-Down List

The *Viewport Scale Control* drop-down list is located in the *Viewports* toolbar (see Fig. 13-28). Invoke the *Viewports* toolbar by using the *Tools* pull-down menu and selecting *Toolbars, Viewports.*

The *Viewport Scale* lists provide standard scales for decimal, metric, and architectural scales only. You can select from the list or enter a value. Values can be entered as a decimal, a fraction (proper or improper), a proportion using a colon symbol ( : ), or an equation using an equal symbol ( = ). Feet ( ′ ) and inch ( ″ ) unit symbols can also be entered. For example, if you wanted to scale the viewport geometry at 1/2″=1′, you could enter any of the following, as well as other, values.

| | |
|---|---|
| 1/2″=1′ | 1:24 |
| .5″=1′ | 1=24 |
| 1/2=12 | .0417 |
| 1/24 | |

FIGURE 13-28

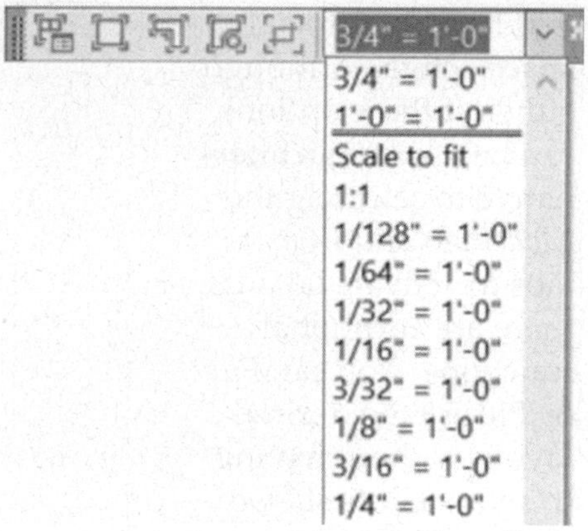


In addition, you can edit the contents of the scale list by adding, deleting, or rearranging the entries. See "*Scalelistedit*" near the end of this chapter.

Remember that the viewport scale is actually the desired plot scale for the geometry in the viewport, which is the reciprocal of the "drawing scale factor" (see Chapter 12).

### *Properties*

The *Properties* palette can also be used to specify the scale for the display of model space geometry in a viewport. Do this by first double-clicking in paper space or issuing the *Pspace* command, then invoking *Properties.* When (or before) the dialog box appears, select the viewport object (border). Ensure the word "Viewport" appears in the top of the box (Fig. 13-29).

FIGURE 13-29

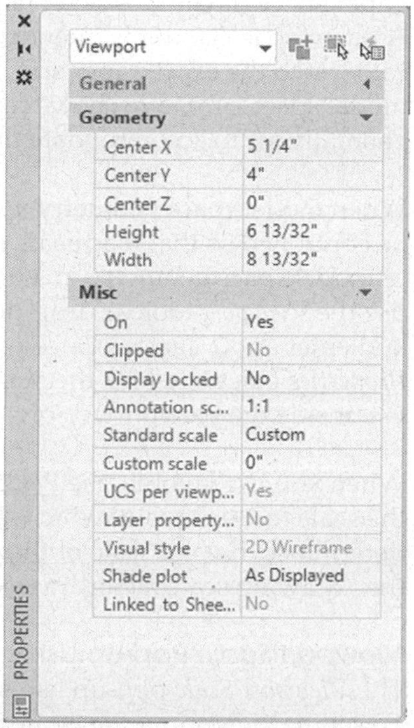


Under *Standard scale,* select the desired scale from the drop-down list. This method offers only standard scales. Alternately, you can enter any values in the *Custom scale* box. (See *Properties*, Chapter 16.) Similar to using the *Viewport Scale Control* edit box, you can enter values as a decimal or fraction and use a colon symbol or an equal symbol. Feet and inch unit symbols can also be entered. You can also *Lock* the viewport scale using the *Display locked* option in the palette. (See *Properties*, Chapter 16.)

### *ZOOM XP Factors*

To scale the display of model space geometry relative to paper space (in a viewport), you can also use the *XP* option of the *Zoom* command. *XP* means "times paper space." Thus, model space geometry is *Zoomed* to some factor "times paper space."

Since paper space is set to the actual size of the paper, the *Zoom XP* factor that should be used for a viewport is equivalent to the plot scale that would otherwise be used for plotting that geometry in model space. The *Zoom XP* factor is the reciprocal of the "drawing scale factor." *Zoom XP* only while you are "in" the desired model space viewport. Fractions or decimals are accepted.

For example, if the model space geometry would normally be plotted at 1/2"=1" or 1:2, the *Zoom* factor would be .5*XP* or 1/2*XP*. If a drawing would normally be plotted at 1/4"=1', the *Zoom* factor would be 1/48*XP*. Other examples are given in the following list.

| | |
|---|---|
| 1:5 | *Zoom* .2*XP* or 1/5 *XP* |
| 1:10 | *Zoom* .1*XP* or 1/10*XP* |
| 1:20 | *Zoom* .05*XP* or 1/20*XP* |
| 1/2"=1" | *Zoom* 1/2*XP* |
| 3/8"=1" | *Zoom* 3/8*XP* |
| 1/4"=1" | *Zoom* 1/4*XP* |
| 1/8"=1" | *Zoom* 1/8*XP* |
| 3"=1' | *Zoom* 1/4*XP* |
| 1"=1' | *Zoom* 1/12*XP* |
| 3/4'=1' | *Zoom* 1/16*XP* |
| 1/2"=1' | *Zoom* 1/24*XP* |
| 3/8"=1' | *Zoom* 1/32*XP* |
| 1/4"=1' | *Zoom* 1/48*XP* |
| 1/8"=1' | *Zoom* 1/96*XP* |

Refer to the "Tables of Limits Settings," Chapter 14, for other plot scale factors.

## Locking Viewport Geometry

Once the viewport scale has been set to yield the correct plot scale, you should lock the scale of the viewport so it is not accidentally changed when you *Zoom* or *Pan* inside the viewport. If viewport lock is on and you use *Zoom* or *Pan* when the viewport is active, AutoCAD automatically switches to paper space (*Pspace*) for zooming and panning.

You can lock the viewport scale by the following methods.

1. Toggle the small *Lock/Unlock Viewport* icon (padlock) located left of the *Viewport Scale* pop-up list (see Fig. 13-5).
2. Use the *Lock* option of the *-Vports* command (see *-Vports*).
3. Use the *Display locked* setting in the *Properties* palette (see Fig. 13-29).
4. Use the *Display locked* option in the right-click shortcut edit menu that appears when a viewport object is selected (see Fig. 13-24).

TIP

## Linetype Scale in Viewports—*PSLTSCALE*

The *LTSCALE* (linetype scale) setting controls how hidden, center, dashed, and other non-continuous lines appear in the *Model* tab. *LTSCALE* can be set to any value. The *PSLTSCALE* (paper space linetype scale) setting determines if those lines appear and plot the same in a layout as they do in model space. *PSLTSCALE* can be set to 1 or 0 (on or off).

If *PSLTSCALE* is set to 0, the non-continuous line scaling that appears in a layout (and in viewports) is the same as that in model space. In this way, the linetype spacing always looks the same relative to model space units, no matter whether you view it from the *Model* tab or from a viewport in a *Layout*. For example, if in model space a particular center line shows one short dash and two long dashes, it will look the same when viewed in a viewport in a layout—one short dash and two long dashes. When *PSLTSCALE* is 0, linetype scaling is controlled exclusively by the *LTSCALE* setting. A *PSLTSCALE* setting of 0 is recommended for most drawings unless they contain multiple viewports or layouts.

TIP

If *PSLTSCALE* is set to 1 (the default setting for all AutoCAD template drawings), the linetype scale for non-continuous lines in viewports is automatically changed relative to the viewport scale; therefore, the scale of those lines in layouts can appear differently than they do in model space. A *PSLTSCALE* setting of 1 is recommended for drawings that contain multiple viewports or layouts, especially when they are at different scales. In such a situation, the line dashes would appear the same size in different viewports relative to paper space, even though the viewport scales were different. Technically speaking, if *PSLTSCALE* is set to 1, *LTSCALE* controls linetype scaling globally for the drawing, but lines that appear in viewports are automatically scaled to the *LTSCALE* times the viewport scale.

Until you gain more experience with layouts and viewports, and in particular creating multiple viewports, a *PSLTSCALE* setting of 0 is the simplest strategy. This setting also agrees with the strategy used in this text discussed earlier (in Chapter 6 and Chapter 12) for drawing setup. You can change the default *PSLTSCALE* setting of 1 to 0 by typing *PSLTSCALE* at the Command prompt.

Chapter 32, Advanced Layouts, Annotative Objects, and Plotting, gives a more detailed discussion of *PSLTSCALE*, creating multiple viewports and layouts, and other drawing setup strategies.

## Vpmax

| Ribbon | Menu Bar | Command (Type) | Alias (Type) | Shortcut |
|---|---|---|---|---|
| ... | ... | *Vpmax* | ... | ... |

You can invoke the *Vpmax* command by typing or using the button located on the right of the Status Bar only when a Layout tab is active (Fig. 13-30). When a *Layout* tab is active, *Vpmax* maximizes the viewport to enable you to draw and edit inside a viewport (in model space). *Vpmax* is intended to be used in conjunction with *Vpmin* (see *Vpmin* next). This enhanced method is used instead of using the *PAPER/MODEL* toggle on the Status Bar when you to want toggle between paper space and model space as well as maximize your drawing area.

**FIGURE 13-30**

For example, assume you are working in a layout (such as in Fig. 13-31) but realize that you need to edit some geometry inside a viewport. Use the *Vpmax* command to best prepare for making the edits in model space. *Vpmax* accomplishes two actions: 1) it activates model space inside the viewport, and 2) it removes all paper space area from the screen to make drawing and editing in model space easier.

**FIGURE 13-31**

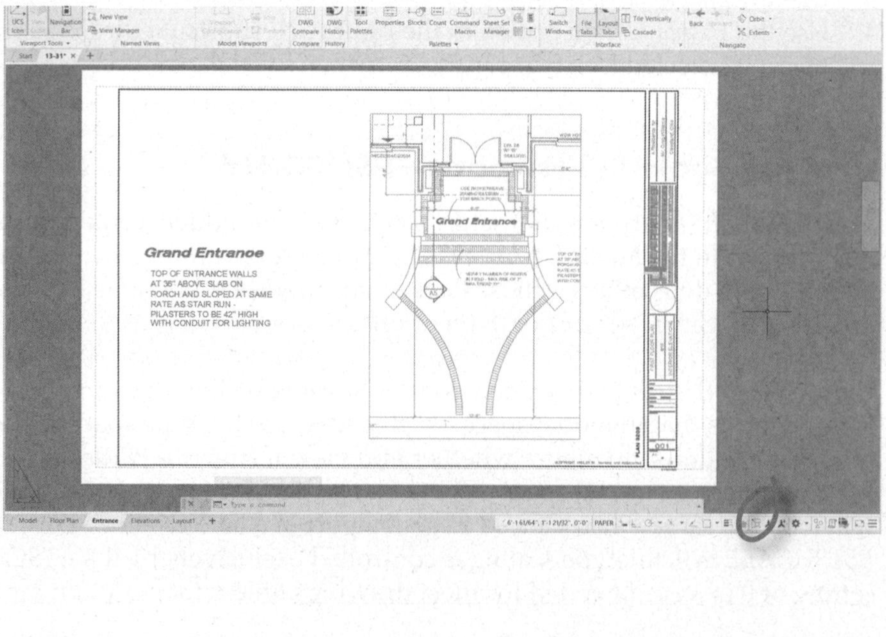

Figure 13-32 shows this example after using *Vpmax*. Model space (inside the viewport) is active and the drawing area has been maximized in the Drawing Editor. Note that *Vpmax* does not *Zoom*, it only displays more area of model space. The size of the model space geometry is the same as before using *Vpmax*. You can, however, use *Zoom* after using *Vpmax* to enlarge the model geometry display. Using *Vpmin* restores the original layout display, even if you used *Zoom* after *Vpmax*.

FIGURE 13-32

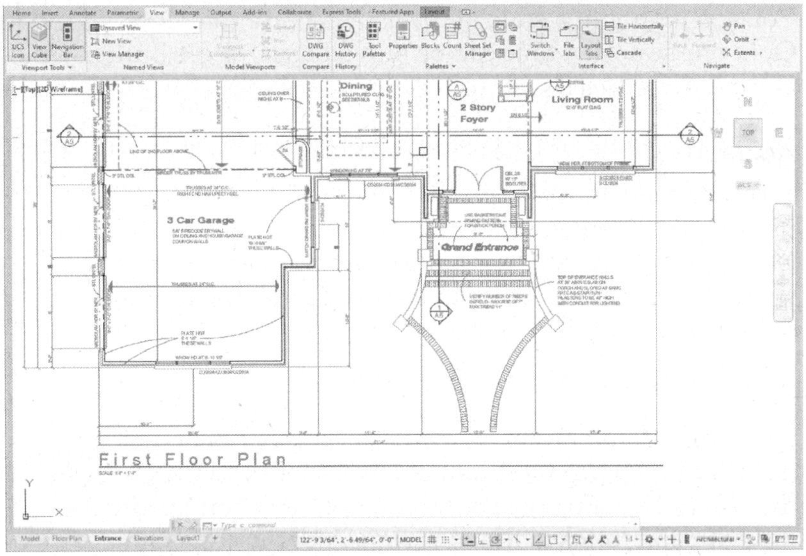


### Vpmin

| Ribbon | Menu Bar | Command (Type) | Alias (Type) | Shortcut |
|---|---|---|---|---|
| ... | ... | *Vpmin* | ... | ... |

The *Vpmin* button appears at the same location as the *Vpmax* button on the Status Bar (Fig. 13-33). *Vpmin* would be used after finishing the needed changes to the drawing to restore the original layout display (to that before using *Vpmax*). For the example used previously, using *Vpmin* would restore the display to that shown in Figure 13-31.

FIGURE 13-33

### Scalelistedit

| Ribbon | Menu Bar | Command (Type) | Alias (Type) | Shortcut |
|---|---|---|---|---|
| *Annotate Annotation Scaling Scale List* | *Format Scale List* | *Scalelistedit* | ... | ... |

To set the scale for the display of geometry in a viewport, use any of the methods described in "Scaling the Display of Viewport Geometry." and illustrated in Figures 13-5, 13-6, 13-28, and 13-29. You can use the *Scalelistedit* command to control the contents of the list of scales. *Scalelistedit* produces the *Edit Drawing Scales* dialog box (Fig. 13-34). You can add new scales, edit existing scales, rearrange the list, and delete unused scales. For example, if you were working in an industry where you typically used only a few scales, you may want to use *Scalelistedit* to move those choices to the top of the list. You can restore the original (default) list at any time.

FIGURE 13-34

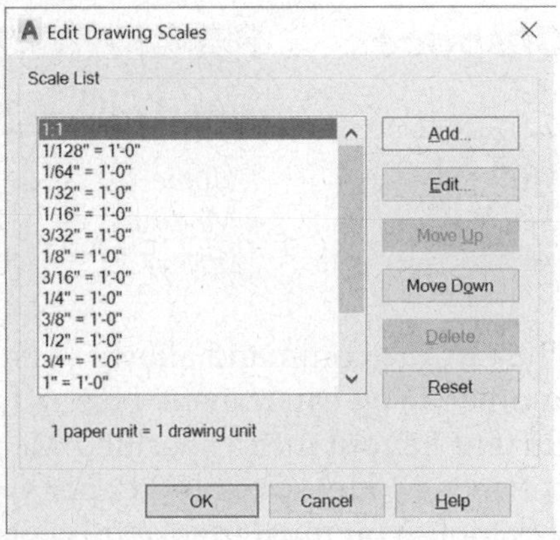

### *Move Up* and *Move Down*

You can highlight a scale in the list, then use the *Move Up* and *Move Down* buttons to rearrange the list. Only one scale at a time can be moved using these buttons; selecting multiple scales disables the *Move Up* and *Move Down* options.

### *Add*

Select the *Add* button to produce the *Add Scale* dialog box (Fig. 13-35) where you can add entries to the list of scales. The *Name appearing in scale list* can contain numbers or letters, as shown in the figure. Enter the desired ratio of paper units to drawing units in the edit boxes below.

**FIGURE 13-35**

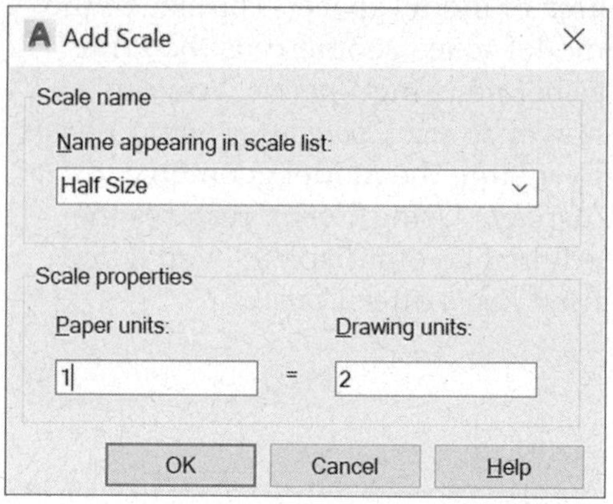


### *Edit*

Selecting the *Edit* button in the *Edit Drawing Scales* dialog box (see Fig. 13-34) produces a dialog box (not shown) similar to the *Add Scale* dialog box; however, you can change only the ratio of paper units to drawing units, not the scale name.

### *Delete*

You can remove scales from the list by highlighting one or more scales (select multiple scales by holding down the Ctrl or Shift keys while selecting), then pressing *Delete*.

### *Reset*

The *Reset* option removes all the customized scales and restores the default list of scales. In this case you will lose all the scales you added, moved, or otherwise edited. A warning appears requiring your confirmation to reset the list to the defaults.

### *Application of the Scale List*

Although this chapter has been primarily concerned with the scale list that appears in the *Viewport* toolbar drop-down list, editing the scale list using the *Scalelistedit* command affects all the scale lists that appear in AutoCAD including:

*Viewport Scale* pop-up list
*Viewport* toolbar drop-down list
*Plot* dialog box
*Page Setup* dialog box
*Properties* palette
*Layout Wizard*
*Sheet Set Manager*

## *Chspace*

| Ribbon | Menu Bar | Command (Type) | Alias (Type) | Shortcut |
|---|---|---|---|---|
| *Home* *Modify* (icon) | *Modify* *Change Space* | *Chspace* | ... | ... |

TIP

The *Chspace* command allows you to transfer selected objects from one space (Paper or Model) to another. You can transfer objects from a Model space viewport to the current Layout or from the current Layout into a specified Model space viewport. For example, if you want to swap objects between Model space and Paper space, such as a sheet border, revision block, or a drawing legend, use this tool. You must invoke this command when in a *Layout*.

Command: ***CHSPACE***
Select objects: **PICK**
Specify opposite corner: **PICK**
Select objects: **Enter**
Set the TARGET viewport active and press ENTER to continue. **Enter**
117 object(s) changed from PAPER space to MODEL space.
Objects were scaled by a factor of 29'-0 7/16" to maintain visual appearance.

When transferred, the objects selected are automatically scaled to appear the same in the destination space. For this reason, when transferring from a Model space viewport to the Layout, it is suggested that you zoom as necessary before invoking *Chspace*.

### Advanced Applications of Paper Space Viewports

In this chapter the basic applications of layouts and viewports are discussed. The basic procedures include creating a layout, setting up the layout for printing or plotting, creating viewports, and scaling the geometry in the viewports to plot to scale.

Layouts can be used for many other advanced applications such as setting up multiple viewports to display views of one drawing at different scales or for different plot devices, controlling what layers appear in each of several viewports, applying Plot Style Tables, displaying multiple drawings (*Xrefs*) in one drawing, and displaying and plotting several views of a 3D model. These topics are discussed in Chapter 30, Xreferences; and Chapter 32, Advanced Layouts, Annotative Objects, and Plotting.

## CHAPTER EXERCISES

1. *Pagesetup, Vports*

   A. ***Open*** the **A-METRIC.DWT** template drawing that you created in the Chapter 12 Exercises. Activate a ***Layout***. If your *Options* are set as explained in Chapter 6, the layout is automatically set up for your default print/plot device. Otherwise, use the ***Page Setup Manager*** and select the ***Modify*** button. Then use the ***Page Setup*** dialog box to set the desired print/plot device and to select an 11 x 8.5 paper size.

   B. If a viewport already exists in the layout, ***Erase*** it. Make **VPORTS** the ***Current*** layer. Then use the ***Vports*** command and select a ***Single*** viewport and accept the default (***Fit***) option to create one viewport at the maximum size for the printable area. Ensure that ***PSLTSCALE*** is set to **0** by typing it at the Command prompt. Finally, make the ***Model*** tab active. ***Save*** and ***Close*** the drawing.

REUSE

2. *Pagesetup, Vports*

   A. ***Open*** the **BARGUIDE** drawing that you set up in the Chapter 12 Exercises. Activate a ***Layout*** tab. If your *Options* are set as explained in Chapter 6, the layout is automatically set up for your default print/plot device. Otherwise, use the ***Page Setup Manager*** and select the ***Modify*** button. Then use the *Page Setup* dialog box that appears to set the desired print/plot device and to select an **11 x 8.5** paper size.

REUSE

B. If a viewport already exists in the layout, ***Erase*** it. Make **VPORTS** the ***Current*** layer. Then use the ***Vports*** command and select a ***Single*** viewport and accept the default (***Fit***) option to create one viewport at the maximum size for the printable area. Type ***PSLTSCALE*** and set the value to **0**. Finally, make the ***Model*** tab active. ***Save*** and ***Close*** the drawing.

REUSE

3. **Create a Layout and Viewport for the ASHEET Template Drawing**

A. ***Open*** the **ASHEET.DWT** template drawing you created in the Chapter 12 Exercises. Activate a ***Layout***. Unless already set up, use ***Pagesetup*** to set the desired print/plot device and to select an **11 x 8.5** paper size.

B. If a viewport already exists in the layout, ***Erase*** it. Make **VPORTS** the ***Current*** layer. Then use the ***Vports*** command and select a ***Single*** viewport and accept the default (***Fit***) option to create one viewport at the maximum size for the printable area. Ensure that ***PSLTSCALE*** is set to **0**. Finally, make the ***Model*** tab active. ***Save*** and ***Close*** the ASHEET template drawing.

REUSE

4. **Create a New BSHEET Template Drawing**

Complete this exercise if your system is configured with a "B" size printer or plotter.

A. Using the template drawing in the previous exercise, create a template for a standard engineering "B" size sheet. First, use the ***New*** command. When the *Select a Template File* dialog box appears, locate and select the **ASHEET.DWT.** When the drawing opens, set the model space ***Limits*** to **17** x **11** and ***Zoom All***.

B. Activate *Layout 1*. ***Erase*** the existing viewport. Activate the ***Page Setup Manager*** and select the ***Modify*** button. In the *Page Setup* dialog box select a print or plot device that can use a "B" size sheet (17 x 11). In the ***Paper Size*** drop-down list, select ***ANSI B (11 x 17 inches)***. Close the dialog box and the *Page Setup Manager*.

C. Make **VPORTS** the ***Current*** layer. Then use the ***Vports*** command and select a ***Single*** viewport and accept the default (***Fit***) option to create one viewport at the maximum size for the printable area. Finally, make the ***Model*** tab active. All other settings and layers are okay as they are. Use ***Saveas*** and save the new drawing as a template (.**DWT**) drawing in your working directory. Assign the name **BSHEET.**

REUSE

5. **Create Multiple Layouts for Plotting on C and D Size Sheets**

Complete this exercise if your system is configured with a "C" and "D" sized plotter.

A. Using the ASHEET.DWT template drawing from an earlier exercise, create a template for standard engineering "C" and "D" size sheets. First, use the ***New*** command and select the ***Template*** option from the ***Create New Drawing*** dialog box. Select the ***Browse*** button. When the *Select a Template File* dialog box appears, select the **ASHEET.DWT.** When the drawing opens, set the model space ***Limits*** to **34** x **22** and ***Zoom All***.

B. Activate *Layout1*. ***Erase*** the existing viewport. Activate the ***Page Setup Manager*** and select the ***Modify*** button. In the *Page Setup* dialog box select a plot device that can use a "C" size sheet (22 x 17). In the ***Paper Size*** drop-down list, select ***ANSI C (22 x 17 inches)***. Close the dialog box and the *Page Setup Manager*.

C. Make **VPORTS** the ***Current*** layer. Then use the ***Vports*** command and select a ***Single*** viewport and use the ***Fit*** option to create one viewport at the maximum size for the printable area. Right-click on the ***Layout 1*** tab and select ***Rename*** from the shortcut menu. Rename the layout to "**C Sheet**."

D. Activate *Layout2*. ***Erase*** the existing viewport if one exists. Activate the ***Page Setup Manager***, select the ***Modify*** button, then select a plot device that can use a "D" size sheet (34 x 22). (This can be the same device you selected in step B, as long as it supports a D size sheet.) In the ***Paper Size*** drop-down list, select ***ANSI D (34 x 22 inches)***. Close this dialog box and the *Page Setup Manager*.

E. Make **VPORTS** the ***Current*** layer. Then use the ***Vports*** command and select ***Single*** to create one viewport. ***Rename*** the layout "**D Sheet**."

F. Finally, make the ***Model*** tab active. All other settings and layers are okay as they are. Use ***Saveas*** and save the new drawing as a template (**.DWT**) drawing in your working directory. Assign the name **C-D-SHEET.**

6. ***Layout Wizard***

In this exercise, you will open an existing drawing and use the ***Layout Wizard*** to set up a layout for plotting.

A. ***Open*** the sample drawing named **FLOOR PLAN SAMPLE.DWG**. Use the ***Explore Sample Drawings*** option in the ***Create*** page of the AutoCAD opening screen or otherwise locate the drawing located in the **AutoCAD 2021/Sample/Database Connectivity** folder. Next, use ***Saveas*** to save the drawing in your working directory as **VB_SAMP**.

B. Create a new layer named **VPORTS** and make it the ***Current*** layer. Invoke the ***Layout Wizard***. In the *Begin* step, enter a name for the new layout, such as **Layout 3**. Select *Next*.

C. In the *Printer* step, the list of available printers in your lab or office appears. The default printer should be highlighted. Choose any device that you know will operate for your lab or office and select ***Next***. If you are unsure of which devices are usable, keep the default selection and select ***Next***.

D. For *Paper Size*, select the largest paper size supported by your device. Ensure ***Inches*** is selected unless you are using a standard metric sheet. Select ***Next***.

E. Select ***Landscape*** for the *Orientation*.

F. Select a *Title Block* that matches the paper size you specified if available. Insert the title block as a ***Block***. Select ***Next***.

G. In the *Define Viewports* step, select a ***Single*** viewport. Under *Viewport Scale*, select ***Scaled to Fit***. Proceed to the next step.

H. Bypass the next step in the wizard, *Pick Location*, by selecting ***Next***. Also, when the *Finish* step appears, select ***Next***. This action causes AutoCAD to create a viewport that fits the extents of the printable area. The resulting layout should look similar to that shown in Figure 13-36. As you can see, the model space geometry extends beyond the title block. Your layout may appear slightly different depending on the sheet size and title block you selected. ***Save*** the drawing.

**FIGURE 13-36**

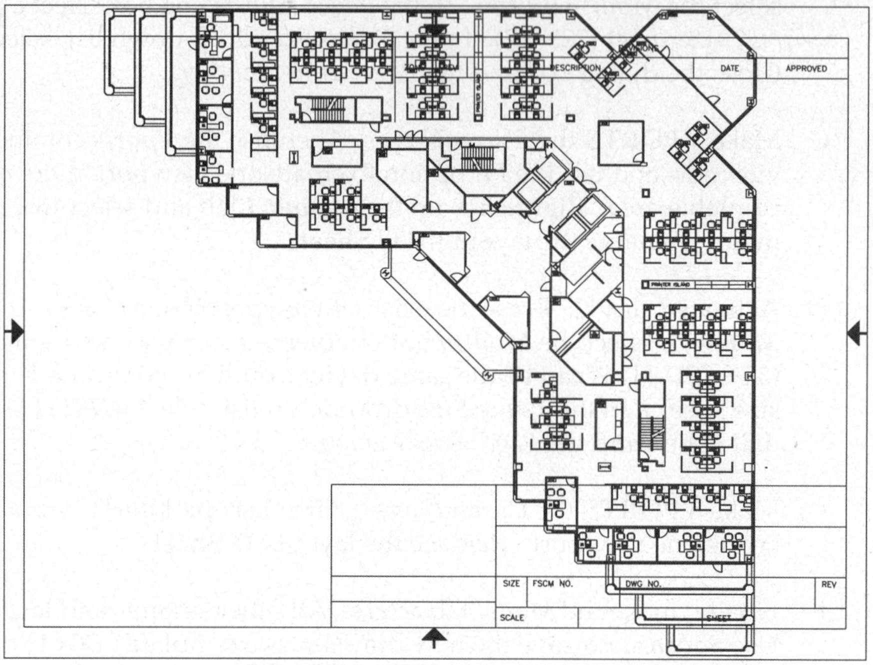


I. Use the ***Mspace*** command or double-click inside the viewport to activate model space. To specify a standard scale for the geometry in the viewport, use the ***Viewport Scale*** pop-up list and select a scale (such as ***1/64″=1′***) until the model geometry appears completely within the title block border (Fig. 13-37).

**FIGURE 13-37**

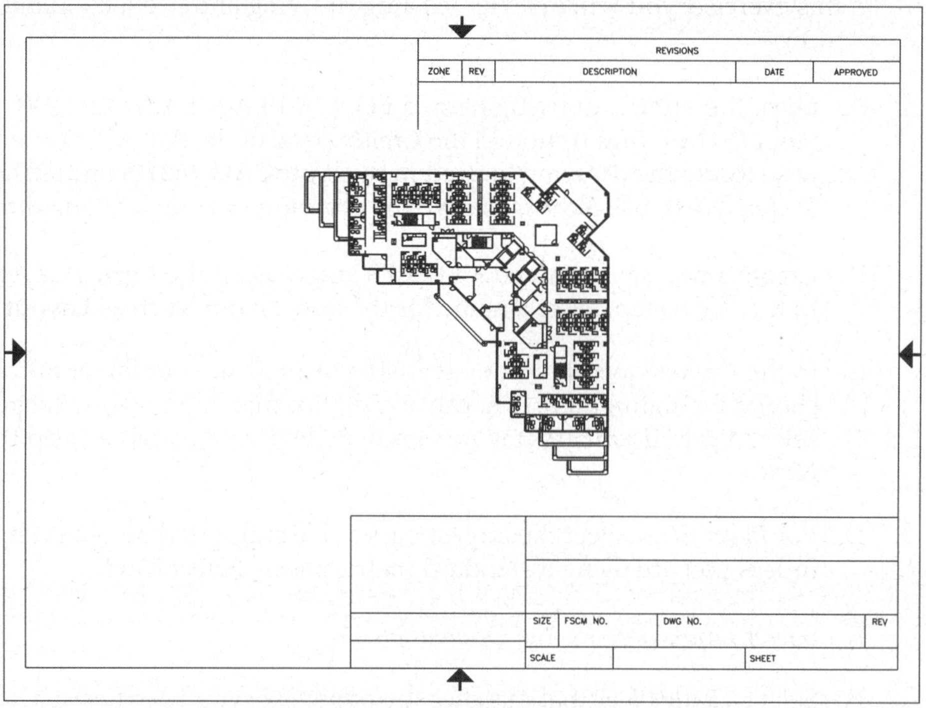


J. ***Freeze*** layer **VPORTS** so the viewport border does not appear. ***Save*** the drawing. With the layout active, access the ***Plot*** dialog box and ensure the *Plot Scale* is set to **1:1**. Select ***Plot***.

7. ***Page Setup*, *Vports*, and *Zoom XP***

In this exercise, you will use an existing metric drawing, access *Layout1*, draw a border and title block, create one viewport, and use *Zoom XP* to scale the model geometry to paper space. Finally, you will *Plot* the drawing to scale on an "A" size sheet.

A. ***Open*** the **CBRACKET** drawing you worked on in Chapter 9 Exercises. If you have already created a title block and border, ***Erase*** them.

B. Invoke the ***Options*** dialog box and access the ***Display*** tab. In the *Layout elements* section (lower left), ensure that *Show Page Setup Manager for new layouts* is checked and *Create viewport in new layouts* is not checked.

C. Activate ***Layout1***. The ***Page Setup Manager*** should appear. Select the ***Modify*** button. In the *Page Setup* dialog box select a print or plot device that will allow you to print or plot on an "A" size sheet (11 x 8.5). In the ***Paper Size*** drop-down list, select the correct sheet size, then select ***Landscape*** orientation. Select ***OK*** in the *Page Setup* dialog box and ***Close*** the *Page Setup Manager*.

D. Set layer **TITLE** ***Current***. Draw a title block and border (in paper space). HINT: Use *Pline* with a *width* of .02, and draw the border with a .5 unit margin within the edge of the *Limits* (border size of 10 x 7.5). Provide spaces in the title block for your school or company name, your name, part name, date, and scale as shown in the following figure. Use ***Saveas*** and assign the name **CBRACKET-PS**.

E. Create layer **VPORTS** and make it ***Current***. Use the ***Vports*** command to create a ***Single*** viewport. Pick diagonal corners for the viewport at **.5,1.5** and **9.5,7**. The CBRACKET drawing should appear in the viewport at no particular scale, similar to Figure 13-38.

**FIGURE 13-38**

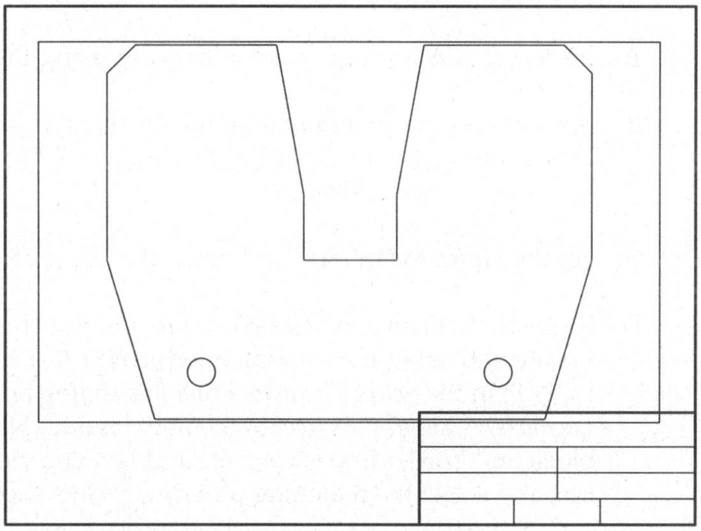

F. Next, use the ***Mspace*** command, the ***MODEL/PAPER*** toggle on the Status Bar, or double-click inside the viewport to bring the cursor into model space. Use ***Zoom*** with a ***.03937XP*** factor to scale the model space geometry to paper space. (The conversion from inches to millimeters is 25.4 and from millimeters to inches is .03937. Model space units are millimeters and paper space units are inches; therefore, enter a *Zoom XP* factor of .03937.)

G. Finally, activate paper space by using the ***Pspace*** command, the ***MODEL/PAPER*** toggle, or double-clicking outside the viewport. Use the ***Layer Properties Manager*** to make layer **VPORTS** ***non-plottable***. Your completed drawing should look like that in Figure 13-39. ***Save*** the drawing. ***Plot*** the drawing from the layout at **1:1** scale.

**FIGURE 13-39**

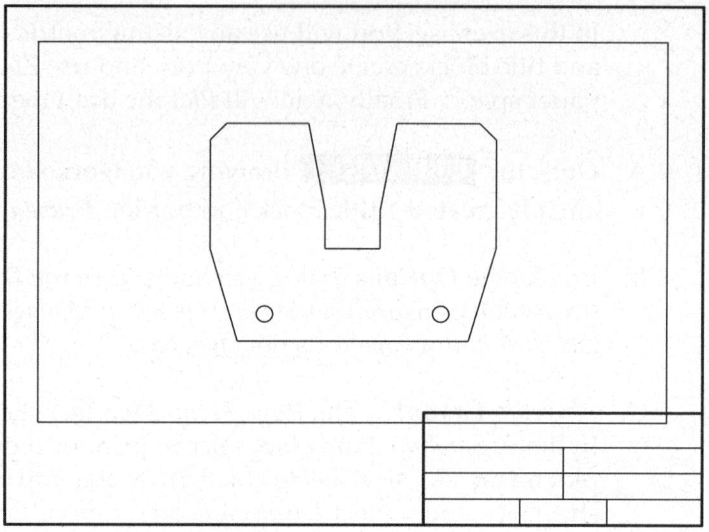

**NOTE:** You will need access to the AutoCAD-supplied ANSI drawing templates (such as ANSI A – Color Dependent Plot Styles.dwt, etc.) for the following exercises. These templates are supplied on the website for this text (the website address for this text is listed on the back cover and in the Preface under "Online Resources").

8. **Using a Layout Template**

This exercise gives you experience using a layout template already set up with a title block and border. You will draw the hammer (see Fig. 13-40) and use a layout template to plot on a B size sheet. It is necessary to have a plot or print device configured for your system that can use a "B" size sheet.

A. Start AutoCAD. Begin a ***New*** drawing using the **ACAD.DWT** drawing template.

B. Now access the ***Display*** tab in the ***Options*** dialog box. In the *Layout elements* section (lower left), remove the check for both *Show Page Setup Manager for new layouts* and *Create viewport in new layouts*. Select ***OK***.

C. Set the ***Units*** to ***Decimal*** and make the ***Precision 0.00***. Set the ***Limits*** to **17 x 11**.

D. Right-click on an existing layout tab and use the ***From Template*** option from the shortcut menu. Locate and select the (downloaded) ***ANSI B -Color Dependent Plot Styles*** template drawing (.DWT) in the *Select Template From File* dialog box, then select the ***ANSI B Title Block*** in the *Insert Layout(s)* dialog box. Access the new layout. Note that the template loads a layout with a title block and border in paper space and has one viewport created. Access the ***Page Setup Manager*** box and select the matching plot device and sheet size for the selected ANSI B title block.

E. Pick the ***Model*** tab. Set up appropriate ***Snap*** and ***Grid*** (for the *Limits* of 17 x 11) if desired. Also set ***LTSCALE*** and any running ***Object Snaps*** you want to use. Type ***PSLTSCALE*** and set the value to **0**.

F. Create the following layers and assign linetypes and line weights. Assign colors of your choice.

**GEOMETRY** ***Continuous***
**CENTER** ***Center2***

Notice the *Title Block* and *Viewport* layers already exist as part of the template drawing.

G. Draw the hammer in the *Model* tab according to the dimensions given in Figure 13-40. ***Save*** the drawing as **HAMMER.**

**FIGURE 13-40**

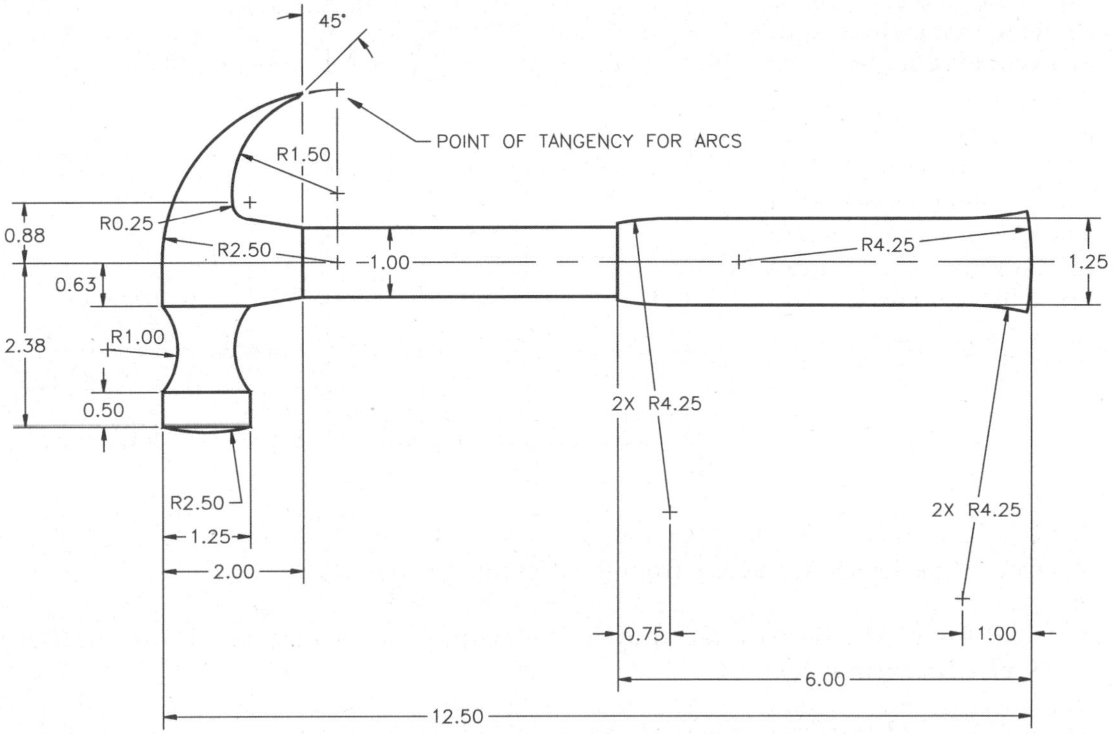


H. When you are finished with the drawing, select the ***ANSI B Title Block*** tab. Now, scale the hammer to plot to a standard scale. You can use either the ***Viewports*** toolbar or the ***Viewport Scale*** pop-up list to set the viewport scale to **1:1**. Alternately, use ***Zoom 1XP*** to correctly size model space units to paper space units.

I. Your completed drawing should look like that in Figure 13-41. Use the ***Layer Properties Manager*** to make layer **Viewports** ***non-plottable. Save*** the drawing. Make a ***Plot*** from the layout at a scale of **1:1**.

**FIGURE 13-41**

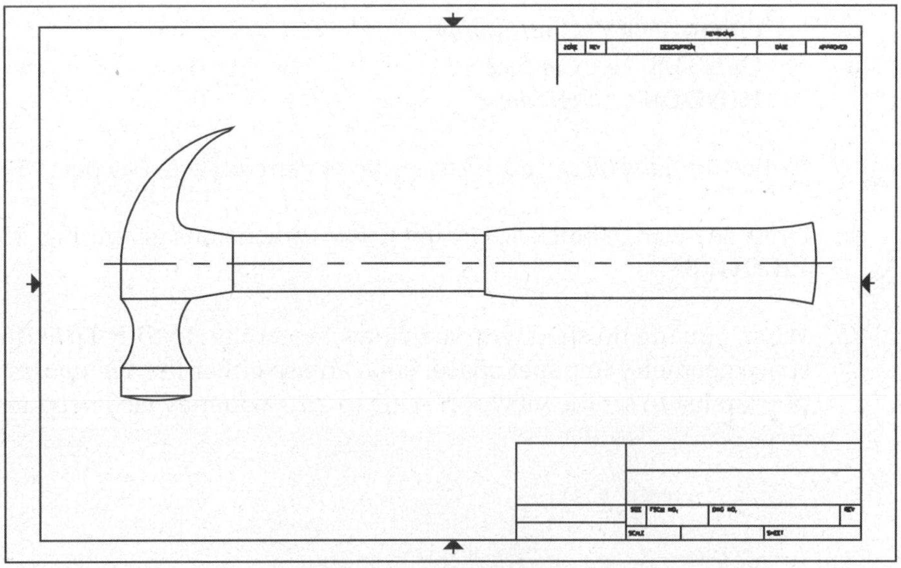

9. **Using a Template Drawing**

This exercise gives you experience using a template drawing that includes a title block and border. You will construct the Wedge Block in Figure 13-42 and plot on a B size sheet. It is necessary to have a plot or print device configured for your system that can use a "B" size sheet.

**FIGURE 13-42**

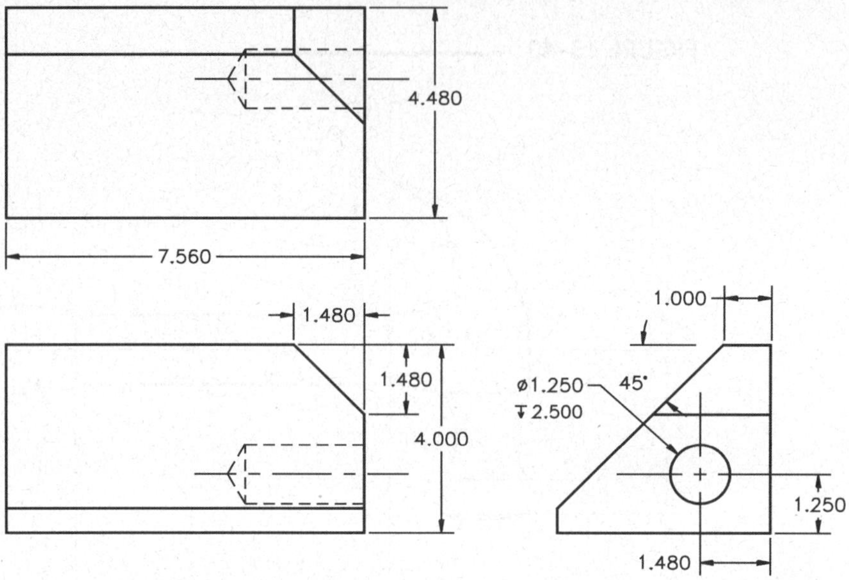


A. Complete steps 8.A. and 8.B. from the previous exercise.

B. Start AutoCAD. Begin a ***New*** drawing using the (downloaded) ***ANSI B-Color Dependent Plot Styles.Dwt*** template.

C. Select the ***ANSI B Title Block*** tab. Then invoke the ***Page Setup Manager*** box and select a matching plot or print device and sheet size for the ANSI B title block.

D. Pick the ***Model*** tab. Set up model space with ***Limits*** of **17** x **11**. Set up appropriate ***Snap*** and ***Grid*** if desired. Also set ***LTSCALE*** and any running ***Object Snaps*** you want to use. Type ***PSLTSCALE*** and set the value to **0**.

E. Create the following layers and assign linetypes and line weights. Assign colors of your choice.

| | |
|---|---|
| **GEOMETRY** | ***Continuous*** |
| **CENTER** | ***Center2*** |
| **HIDDEN** | ***Hidden2*** |

Notice the *Title Block* and *Viewport* layers already exist as part of the template drawing.

F. Draw the wedge block according to the dimensions given (Fig. 13-42). ***Save*** the drawing as **WEDGEBK**.

G. When you are finished with the views, select the ***ANSI B Title Block*** tab. To scale the model space geometry to paper space, you can use either the ***Viewports*** toolbar or the ***Viewport Scale*** pop-up list to set the viewport scale to **1:2.** You may also need to use ***Pan*** to center the geometry.

H. Next, use the ***Layer Properties Manager*** to make layer **Viewports** ***non-plottable***. Your completed drawing should look like that in Figure 13-43. ***Save*** the drawing. From paper space, plot the drawing on a B size sheet and enter a scale of **1:1**.

**FIGURE 13-43**

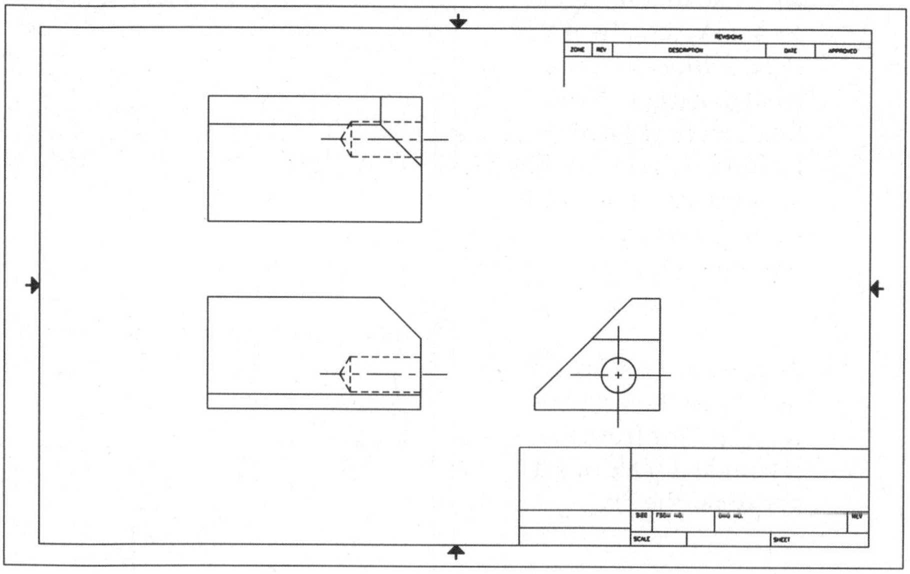

10. **Multiple Layouts**

Assume a client requests that a copy of the HAMMER drawing be faxed to him immediately; however, your fax machine cannot accommodate the B size sheet that contains the plotted drawing. In this exercise you will create a second layout using a layout template to plot the same model geometry on an A size sheet.

A. Open the drawing you created in a previous exercise named **HAMMER**. Access the ***Display*** tab in the ***Options*** dialog box. In the *Layout elements* section (lower left), remove the check for both *Show Page Setup Manager for new layouts* and *Create viewport in new layouts*. Select ***OK***.

B. Right-click on an existing layout tab and use the ***From Template*** option from the shortcut menu. Locate and select the (downloaded) ***ANSI A -Color Dependent Plot Styles*** template drawing (.DWT) to import. When the layout is imported, click the new ***ANSI A Title Block*** tab. Note that the template loads a layout with a title block and border in paper space and has one viewport created. Access the ***Page Setup Manager*** and select the matching print or plot device and sheet size for the selected ANSI A title block.

C. Double-click inside the viewport or use ***Mspace*** or the ***MODEL/PAPER*** toggle. Now scale the hammer to plot to a standard scale by using ***Zoom 3/4XP*** or enter **.75** in the appropriate edit box in the ***Viewports*** toolbar or ***Properties*** palette to correctly scale the model space geometry. Use ***Pan*** in the viewport to center the geometry within the viewport.

D. Your new layout should look like that in Figure 13-44. Access the ***ANSI B Title Block*** tab you created earlier. Note that you now have two layouts, each layout has settings saved to plot the same geometry using different plot devices and at different scales.

E. Access the new layout tab. ***Save*** the drawing. Make a ***Plot*** from the layout at a scale of **1:1** and send the fax.

**FIGURE 13-44**

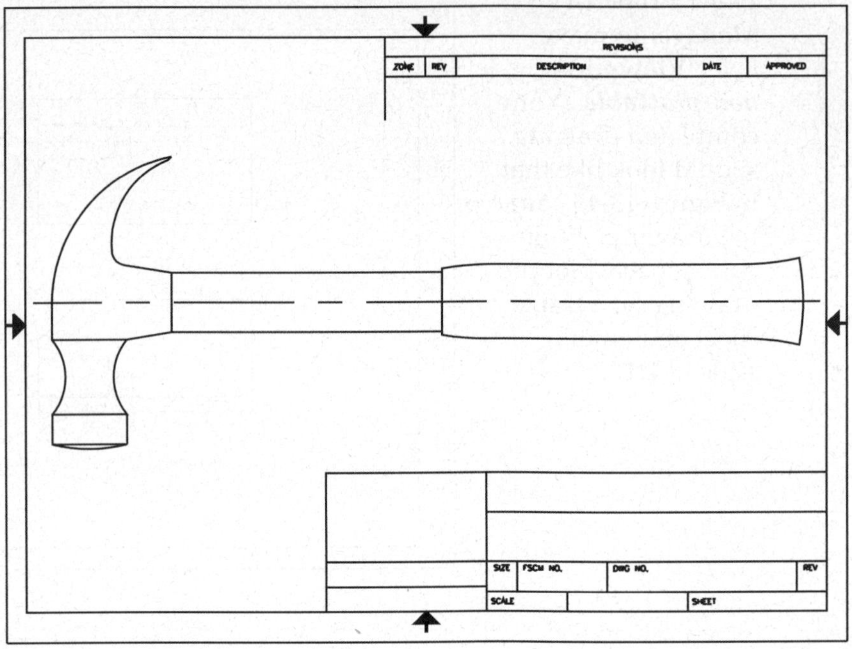

CHAPTER

# 14

# Printing and Plotting

## Chapter Objectives

After completing this chapter you should:

1. know the typical steps for printing or plotting;
2. be able to invoke and use the *Plot* dialog box and *Page Setup Manager*;
3. be able to select from available plotting devices and set the paper size and orientation;
4. be able to specify what area of the drawing you want to plot;
5. be able to preview the plot before creating a plotted drawing;
6. be able to specify a scale for plotting a drawing;
7. know how to set up a drawing for plotting to a standard scale on a standard size sheet;
8. be able to use the Tables of *Limits* Settings to determine *Limits*, scale, and paper size settings;
9. know how to configure plot and print devices in AutoCAD.

## CONCEPTS

Several concepts, procedures, and tools in AutoCAD relate to printing and plotting. Many of these topics have already been discussed, such as setting up a drawing to draw true size, creating layouts and viewports, specifying the viewport scale, and specifying the plot or print options. This chapter explains the typical steps to plotting from either model space or layouts, using the *Plot* dialog box and the *Page Setup Manager*, configuring print and plot devices, and plotting to scale. Other advanced concepts of plotting, such as using plot styles and plot style tables, plot stamping, and advanced features of layouts, are explained in Chapter 32, Advanced Layouts, Annotative Objects, and Plotting.

In AutoCAD, the term "plotting" can refer to plotting on a plot device (such as a pen plotter or electrostatic plotter) or printing with a printer (such as a laser jet printer). The *Plot* command is used to create a plot or print by producing the *Plot* dialog box. You can plot only by using the *Plot* command (Fig. 14-1).

**FIGURE 14-1**

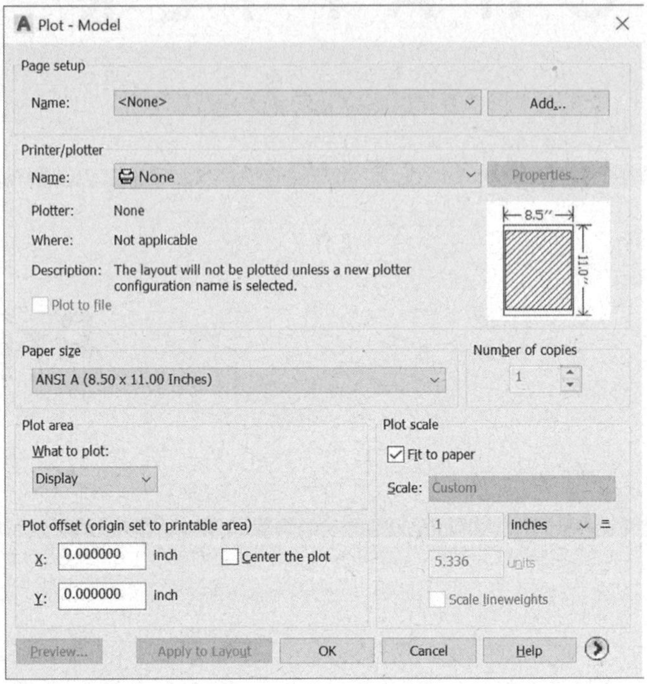


Before making a plot, you specify the parameters of how the drawing will appear on the plot, such as what type of printer or plotter you want to use, the paper size and orientation, what area of the drawing to plot, plot scale, and other options. You can save the plot specifications for each layout. These saved plot specifications are called "page setups." In this way, each layout can have its own page setup—set up to plot or print with a specific device, paper size, orientation, scale, and so on. Therefore, one drawing can be plotted and printed in multiple ways.

Although you can create a plot only by using the *Plot* dialog box, two methods can be used to create and save the plot specifications for a layout:

1. Use the *Plot* dialog box to specify the plot settings and select the *Apply to Layout* button.
2. Use the *Page Setup Manager* and related *Page Setup* dialog box.

These two methods are almost identical since the *Plot* dialog box and the *Page Setup* dialog box have almost the same features and functions except that you cannot plot from the *Page Setup* dialog box. Therefore, for most applications, it is suggested that you use the *Plot* dialog box for saving the page setups and for plotting. This chapter explains the features and distinctions of both methods.

## TYPICAL STEPS TO PLOTTING

Assuming your CAD system and plotting devices have been properly configured, the typical basic steps to plotting model space or a layout are listed below.

1. Use *Save* to ensure the drawing has been saved in its most recent form before plotting (just in case some problem arises while plotting).

2. Make sure the plotter or printer is turned on, has paper (and pens for some devices) loaded, and is ready to accept the plot information from the computer.

3. Invoke the *Plot* dialog box.

4. Select the intended plot device from the *Printer/Plotter* drop-down list. The list includes all devices currently configured for your system.

5. Ensure the desired *Drawing orientation* and *Paper size* are selected.

6. Determine and select the desired *Plot area* for the drawing: *Layout, Limits, Extents, Display, Window,* or *View.*

7. Select the desired scale from the *Scale* drop-down list or enter a *Custom* scale. If no standard scale is needed, select *Fit to paper*. If you are plotting from a *Layout*, normally set the scale to 1:1.

8. If necessary, specify a *Plot offset* or *Center the plot* on the sheet.

9. If necessary, specify the *Plot Options*, such as plotting with lineweights or plot styles.

10. Always preview the plot to ensure the drawing will plot as you expect. Select a *Preview* to view the drawing objects as they will plot. If the preview does not display the plot as you intend, make the appropriate changes. Otherwise, needless time and media could be wasted.

11. If all settings are acceptable and you want to save the page setup (plot specifications) with the layout (recommended), select the *Apply to Layout* button.

12. If you want to make the plot at this time, select the *OK* button. If you do not want to plot, select *Cancel* and the settings will be saved (if you used *Apply to Layout*).

## PRINTING, PLOTTING, AND SAVING PAGE SETUPS

### *Plot*

| Ribbon | Menu Bar | Command (Type) | Alias (Type) | Shortcut |
|---|---|---|---|---|
| *Output Plot Plot* | *File Plot...* | *Plot* or *-Plot* | *PRINT* | *Ctrl+P* |

Using *Plot* invokes the *Plot* dialog box (see Fig. 14-2 on the next page). This dialog box has a collapsed and expanded mode controlled by the arrow button in the lower-right corner. The *Plot* dialog box allows you to set plotting parameters such as plotting device, paper size, orientation, and scale. Settings made to plotting options can be saved for the *Model* tab and each *Layout* tab in the drawing so they do not have to be re-entered the next time you want to make a plot.

#### *Page Setup*

The *Name* drop-down list at the top left corner of the *Plot* dialog box gives the page setups saved in the drawing. You can select a previously saved page setup from the list to apply those settings to the current layout. Although the default page setup is listed as <*None*>, it will be listed as the current tab name (such as *Layout1*) once you save it. You can also assign a specific name using the *Add* button. The list can contain saved page setups (such as *Layout1* or *Layout2*) and "named" (assigned) setups (such as *Setup1*, *Setup2*, or other assigned names). You can also *Import* page setups from other drawings that have assigned names.

### Add

This option produces the *Add Page Setup* dialog box (not shown) where you can assign a name for the current page setup. You can accept the default assigned name (*Setup1*) or enter a name of your choice. You must assign a name if you want to *Import* the page setup into another drawing.

**FIGURE 14-2**

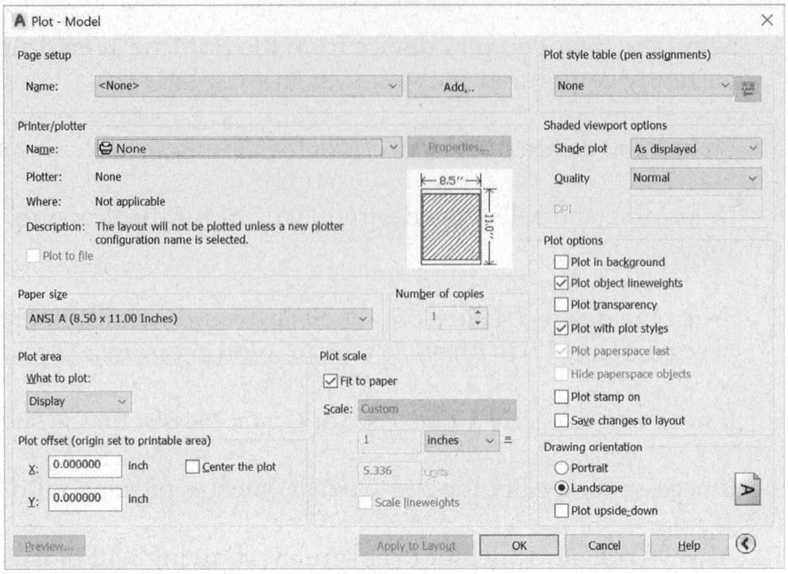


### Printer/plotter

Many devices (printers and plotters) can be configured for use with AutoCAD. Once configured, those devices appear in the *Printer/plotter* list. For example, you may configure an "A" size plotter, a "D" size plotter, and a laser printer. Any one of those devices could then be used to plot the current drawing.

When you install AutoCAD, it automatically configures itself to use the Windows system print devices, two DWF devices, several PDF devices, and two *PublishToWeb* services. Additional devices must be configured within AutoCAD by using the *Plotter Manager* (see "Configuring Plotters and Printers" near the end of this chapter).

*Name:*
Select the desired device from the list. The list is composed of all previously configured devices for your system.

*Properties...*
This button invokes the *Plotter Configuration Editor*. In the editor, you can specify a variety of characteristics for the specific plot device shown in the *Name:* drop-down list. The *Plotter Configuration Editor* is also accessible through the *Plotter Manager* (see *"Plotter Configuration Editor"* near the end of this chapter).

*Plot to file*
This option writes the plot to a file instead of making a plot. This action generally creates a .PLT file type; however, the format of the file (for example, PCL or HP/GL language) depends on the device brand and model that is configured. When you use the *OK* button to create the plot, the *Browse for Plot File* dialog box (not shown) appears for you to specify the name and location for the plot file. The plot file can be printed or plotted later without AutoCAD, assuming the correct interpreter for the file is available.

### *Paper size* and *Number of copies*

The options that appear in this list depend on the selected printer/plotter device. Normally, several paper sizes are available. Select the desired *Paper size* from the drop-down list, then select the *Number of copies* you want to create.

### *Plot area*

In the *What to plot:* list section you can select what aspect of the drawing you want to plot, described next.

*Limits* or *Layout*
This option changes depending on whether you invoke the *Plot* dialog box while the *Model* tab or a *Layout* tab is current. If the model tab is current, this selection plots the area defined by the *Limits* command. If a *Layout* tab is current, the layout is plotted as defined by the selected paper size.

*Extents*
Plotting *Extents* is similar to *Zoom Extents*. This option plots the entire drawing (all objects), disregarding the *Limits*. Use the *Zoom Extents* command in the drawing to make sure the extents have been updated before using this plot option.

*Display*
This option plots the current display on the screen. If using viewports, it plots the current viewport display.

*View*
With this option you can plot a named view previously saved using the *View* command.

*Window*
*Window* allows you to plot any portion of the drawing. You must specify the window by picking or supplying coordinates for the lower-left and upper-right corners.

### Plot offset

Plotting and printing devices cannot plot or print to the edge of the paper. Therefore, the lower left corner of your drawing will not plot at exactly the lower left corner of the paper. You can choose to reposition, or offset, the drawing on the paper by entering positive or negative values in the X and Y edit boxes. Home position for landscape orientation is the lower-left corner and for portrait orientation it is the upper-left corner of the paper. If plotting from the *Model* tab, you can choose to *Center the plot* on the paper.

### Plot scale

You can select a standard scale from the drop-down list or enter a custom scale.

*Fit to paper*
The *Fit to paper* option is disabled if a layout tab is active. If the *Model* tab is active, the *Fit to paper* option is checked and the *Scale* section is disabled. The drawing is automatically sized to fit on the sheet based on the specified area to plot (*Display, Limits, Extents,* etc.). Changing the *inches* or *mm* specification changes only the units of measure for other areas of the *Plot* dialog box.

*Scale*
This section provides the scale drop-down list for you to select a specific scale for the drawing to be plotted. These options are the same as those appearing in all other scale drop-down lists (such as the *Viewport Scale* pop-up list, *Viewport* toolbar, and *Properties* palette). The entries can be edited using the *Scalelistedit* command (see Chapter 13). First ensure the correct paper units are selected (*inches* or *mm*), then select the desired scale from the list.

If you are plotting from a *Layout* tab, select 1:1 from the *Scale* drop-down list since you almost always want to plot paper space at 1:1 scale. (To set the scale for the viewport geometry in a layout, use the *Viewport* toolbar, *Zoom XP* factor, or *Properties* palette as explained in Chapter 13.) If you are plotting the *Model* tab, select an option directly from the *Scale* drop-down list to specify the desired plot scale for the drawing. In either case, the plot scale for the drawing is equal to the reciprocal of the drawing scale factor and can be determined by referencing the Tables of *Limits* Settings. (See "Plotting to Scale.")

Custom Edit Boxes
Instead of selecting from the drop-down list, you can enter the desired ratio in the edit boxes. Decimals or fractions can be entered. The first edit box represents paper units and the second represents drawing units. For example, to prepare for a plot of 3/8 size (3/8"=1"), the following ratios can be entered: 3=8, 3/8=1, or .375=1 (paper units =drawing units). For guidelines on plotting a drawing to scale, see "Plotting to Scale" in this chapter.

*Scale lineweights*
This option scales the lineweights for plots. When checked, lineweights are scaled proportionally with the plot scale, so a 1mm lineweight, for example, would plot at .5mm when the drawing is plotted at 1:2 scale. Normally (no check in this box) lineweights are plotted at their absolute value (0.010", 0.016", 0.25 mm, etc.) at any scale the drawing is plotted.

The following options are available from the expanded mode of the *Plot* dialog box (Fig. 14-3).

**FIGURE 14-3**

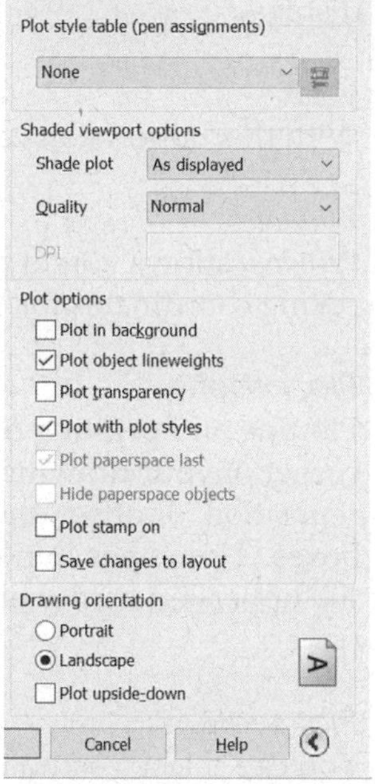


## Plot style table (pen assignments)

If desired, select a plot style table to use with your plot device. A plot style table can contain several plot styles. Each plot style can assign specific line characteristics (lineweight, screening, dithering, end and joint types, fill patterns, etc.). Normally, a plot style table can be assigned to the *Model* tab or to *Layout* tabs. The plot style tables appearing in the list are managed by the *Plot Style Manager*. Selecting the *New* option invokes the *Add Plot Style Table* wizard. (Plot styles are explained in Chapter 32.)

## Shaded viewport options

These options are useful only if you are printing from the *Model* tab and you are printing a 3D surface or solid model. These options do not affect the display of 3D objects in viewports of a *Layout* tab. You should have a laser jet, inkjet, or electrostatic type printer configured as the plot device (one that can print solid areas) since a pen plotter only creates line plots.

*Shade Plot*
Use these options if you are printing from the *Model* tab. You can shade the 3D object on the screen using any Visual Style, then select *As displayed* in the *Shade plot* section to make a shaded print, or you can select any option from the drop-down list for printing the drawing no matter which display is visible on screen.

*Quality* and *DPI*
You can also select from *Draft* (wireframe), *Preview* (shaded at 150 dpi), *Normal* (shaded at 300 dpi), *Presentation* (shaded at 600 dpi), or *Maximum* (for your printer) to specify the resolution quality of the print. Select *Custom* if you want to supply a different value in the *DPI* (dots per inch) edit box.

NOTE: The label for this section is misleading since it cannot be used to control printing for viewports. If you are printing from a *Layout* tab and want to print shaded images of 3D objects in viewports, you must control *Shadeplot* for each viewport using the *-Vports* command or the *Properties* palette. (See "Printing Shaded 3D Models" in Chapter 33.)

## Plot options

These checkboxes control the following options.

*Plot in background*
When this box is not checked, the *Plot Job Progress* window appears while the drawing is processed to the print or plot device so you must wait some time (depending on the complexity of the drawing) before continuing to work. Checking this option allows you to return to the drawing immediately while plotting continues in the background. Plotting in the background generally requires more time to produce the plot.

NOTE: The *Plot in background* setting is valid only for the current plot and is not saved in the page setup. The default setting for this box (checked or unchecked) is controlled by the *Background processing options* section in the *Plot and Publish* tab of the *Options* dialog box (see Fig. 14-7).

*Plot object lineweights*
Checking this option plots the lineweights as they are assigned in the drawing. Lineweights can be assigned to layers or to objects. Checking *Plot with plot styles* disables this option.

*Plot with plot styles*
You can use plot styles to override lineweights assigned in the drawing. You must have a plot style table attached (see *Plot Device Tab, plot style table*) for this option to have effect.

*Plot paperspace last*
Checking this option plots paper space geometry last. Normally, paper space geometry is plotted before model space geometry.

*Hide paperspace objects*
This option is useful only if you have 3D surface or solid objects in paper space and you want to remove hidden lines (edges obscured from view). 3D objects are almost always created in model space. In that case, use the *Shade plot* options (see *Shaded viewport options*) if you are printing from the *Model* tab. If you are printing from a *Layout* tab, you must control *Shadeplot* for each viewport using the *-Vports* command or the *Properties* palette. (See "Printing Shaded 3D Models" in Chapter 33.)

*Plot stamp on*
You have the capability of adding an informational text "stamp" on each plot. For example, the drawing name, date, and time of plot can be added to each drawing that you plot for tracking and verification purposes. To specify the plot stamp, use the *Plotstamp* command. Select this option to apply the stamp you specified to the plotted drawing. (See "*Plotstamp*" in Chapter 32.)

*Save changes to layout*
Check this box if you want settings you make (the page setup) to be saved with the drawing for the selected *Model* tab or *Layout* tab. Selecting this option and making a plot has the same effect as pressing the *Apply to Layout* button. If you want to save changes without plotting, use the *Apply to Layout* button.

### Drawing orientation

*Landscape* is the typical horizontal orientation (similar to the drawing area orientation) and *Portrait* is a vertical orientation. If plotting from a layout tab, this setting affects the orientation of the paper; therefore, set this option before creating your viewport arrangement and title block to ensure they match the paper orientation. If plotting from the *Model* tab, *Portrait* positions horizontal lines in the drawing along the short axis of the paper.

### Preview

Selecting the *Preview* button displays a simulated plot sheet to ensure the drawing will be plotted as you expect (Fig. 14-4). This feature is helpful particularly when plot style tables are attached since the resulting plots may appear differently than the *Model* tab or *Layout* tabs display. The *Zoom* feature is on by default so you can check specific areas of the drawing before making the plot. Right-clicking produces a shortcut menu, allowing you to use *Pan* and other *Zoom* options. Changing the display during a *Preview* does not change the area of the drawing to be plotted or printed.

**FIGURE 14-4**

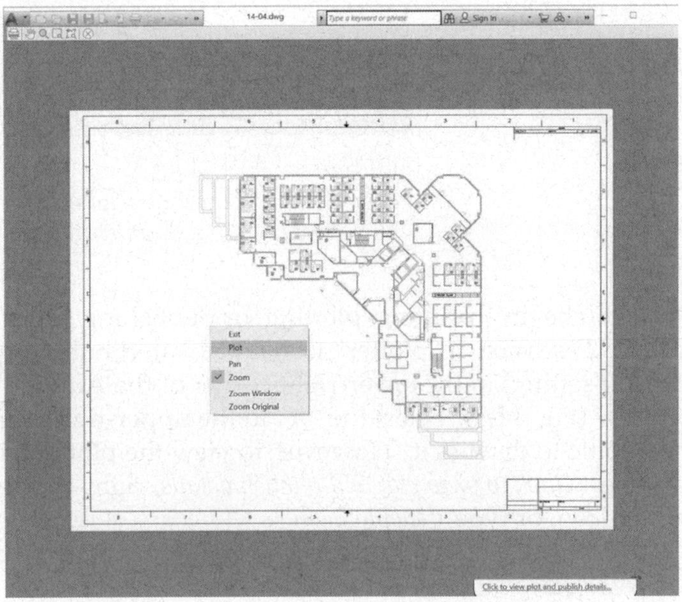

*Apply to Layout*
TIP Using this button saves the current *Plot* dialog box settings to the active layout as a page setup. If no changes have been made to the default settings, this button is disabled. If this button is disabled and you want to save the current settings, assign a specific page setup name using the *Add* button. If you do not want to plot after saving the page setup, then use *Cancel*.

NOTE: Using this option (in the *Plot* dialog box) accomplishes essentially the same results as using the *Page Setup Manager* and *Page Setup* dialog box to save a page setup.

*OK*
Press *OK* to create the print or plot as assigned by the current settings in the *Plot* dialog box. The settings are not saved with the layout as a page setup unless the *Save changes to layout* box is checked or you used the *Apply to Layout* option.

*Cancel*
Cancel the plot and the current settings with this option. If you used the *Apply to Layout* option, *Cancel* will not undo the saved settings.

## -Plot

Entering *-Plot* at the Command line invokes the Command line version of *Plot*. This is an alternate method for creating a plot if you want to use the current settings or if you enter a page setup name to define the plot settings.

```
Command: -PLOT
Detailed plot configuration? [Yes/No] <No>:
Enter a layout name or [?] <Layout 1>:
Enter a page setup name <>:
Enter an output device name or [?] <HP 7586B.pc3>:
Write the plot to a file [Yes/No] <N>:
Save changes to page setup [Yes/No]? <N>
Proceed with plot [Yes/No] <Y>:
Effective plotting area: 33.54 wide by 42.60 high
Plotting viewport 2.
Command:
```

## Viewplotdetails

| Ribbon | Menu Bar | Command (Type) | Alias (Type) | Shortcut |
|---|---|---|---|---|
| *Output Plot View Details* | *File View Plot and Publish Details* | *Viewplotdetails* | ... | ... |

Once you begin a printing, plotting, or publishing job in an AutoCAD session, a "plotter" icon and an alert bubble appear (default setting) in the lower-right corner of the AutoCAD window (Fig. 14-5). Select the "X" in the upper-right corner of the bubble to dismiss it. However, to view the plot details, you can select *Click to view plot and publish details*, right-click on the plotter icon, or type *Viewplotdetails*.

**FIGURE 14-5**

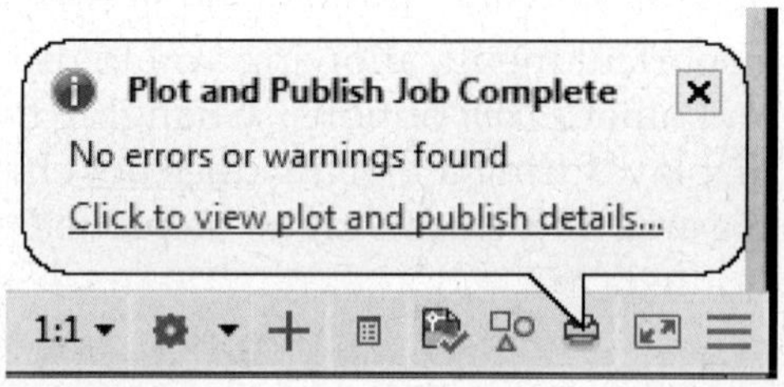

Any one of the three actions described above produces the *Plot and Publish Details* dialog box (Fig. 14-6). Complete details about the plot or print job, including any errors encountered, are displayed in the dialog box.

**FIGURE 14-6**

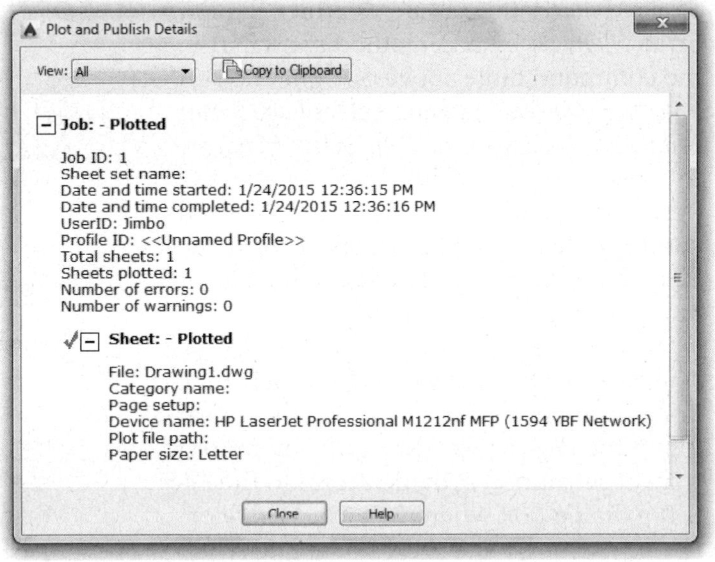


In addition to the plot details, AutoCAD (by default) creates a plot log named *PlotandPublishLog.CSV*. This is a spread sheet format file listing details about the plots such as start time and date, completion time and date, and final status. You can choose to create one continuous or multiple plot logs. The default settings are shown in the lower-left corner of the *Plot and Publish* tab of the *Options* dialog box (Fig. 14-7). You can find or change the location of the plot log file using the *Files* tab of the *Options* dialog box (not shown).

**FIGURE 14-7**

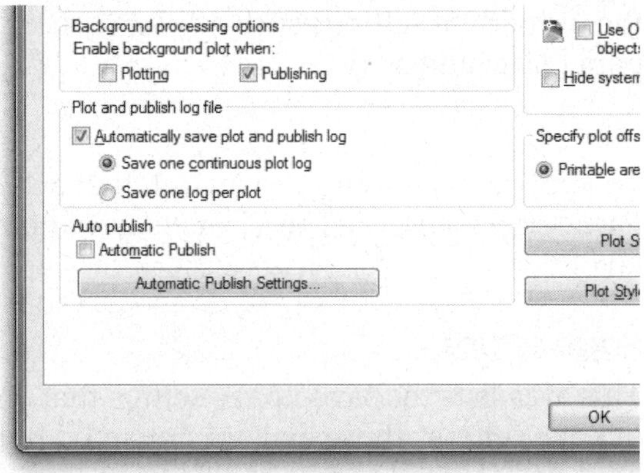


Note that this section of the *Plot and Publish* tab of the *Options* dialog box also contains the control for *Background processing options*. See the *Plot* command, *Plot in background* option in this chapter for more information.

## Pagesetup

| Ribbon | Menu Bar | Command (Type) | Alias (Type) | Shortcut |
|---|---|---|---|---|
| *Output Plot Page Setup Manager* | *File Page Setup Manager...* | *Pagesetup* | ... | ... |

As you already know, AutoCAD allows you to set and save plotting specifications, called page setups, for each layout. In this way, one drawing can contain several layouts, each layout can display the same or different aspects of the drawing, and each layout can be set up to print or plot with the same or different plot devices, paper sizes, scales, plot styles, and so on. Oddly enough, you can create and manage page setups using either the *Plot* dialog box or the *Page Setup Manager* and related *Page Setup* dialog box. The main difference is that you cannot plot from the *Page Setup Manager* or *Page Setup* dialog box. See the following discussion, "Should you use the *Plot* dialog box or the *Page Setup Manager*?"

The *Pagesetup* command produces the *Page Setup Manager* (Fig. 14-8). Produce the *Page Setup Manager* by any method shown in the command table above or right-click on a *Layout* or *Model* tab and select *Page Setup Manager*. To cause the *Page Setup Manager* to appear automatically when you create a new layout, check the box in the lower-left corner of this dialog box or check *Show Page Setup Manager for new layouts* in the *Display* tab of the *Options* dialog box.

The *Page Setup Manager* has two basic functions: 1) apply existing page setups appearing in the *Page Setups* list to any layout or to a sheet set, and 2) provide access to the *Page Setup* dialog box where you can *Modify* or *Import* existing page setups or create *New* page setups. The *Page Setup* dialog box (see Fig. 14-10) allows you to specify and save the individual settings such as plot device, paper size, and so on.

**FIGURE 14-8**

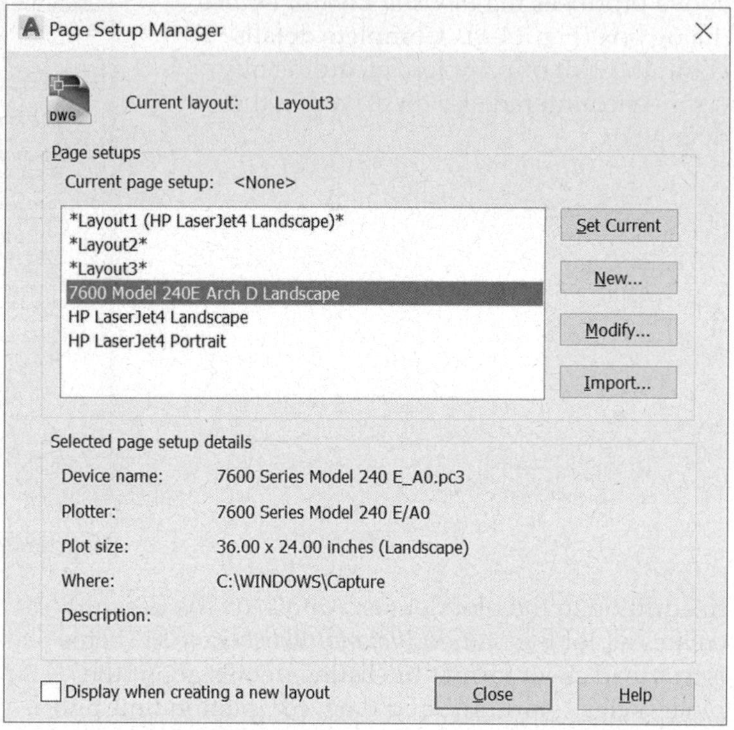


To apply a page setup to a layout, first activate the desired layout tab, then use *Pagesetup* to produce the *Page Setup Manager*. The *Current layout* note at the top of the manager displays the name of the current layout.

### *Page setups*

This area lists the saved page setups that are available in the drawing (or sheet set). There are two types of page setups: those that are unnamed but applied to layouts, such as **Layout1**, and page setups that have an assigned name (named page setups), such as *HP LaserJet4 Landscape*. "Applied" or "attached" page setups are denoted by asterisks.

By default, every initialized layout (initialized by clicking the layout tab) has a default unnamed page setup automatically applied to it, such as **Layout1**. Layouts that have a named page setup applied to them are enclosed in asterisks with the named page setup in parentheses, such as **Layout 2 (HP LaserJet4 Landscape)**. Therefore, the *Page Setups* list can contain 1) unnamed applied page setups such as **Layout1**, 2) named but unapplied page setups such as *HP LaserJet4 Landscape*, and 3) named page setups that have been applied to a layout such as **Layout 2 (HP LaserJet4 Landscape)**.

If the *Page Setup Manager* is opened from the *Sheet Set Manager*, only named page setups in the page setup overrides file (a drawing template [.DWT] file) that have *Plot area* set to *Layout* or *Extents* are listed.

### *Set Current*

To apply or attach a page setup from the list to the current layout, highlight the desired page setup, then use *Set Current*. You can select any named page setup, such as *HP LaserJet4 Landscape*, to apply to the current layout. You can also select and apply another layout's unnamed page setup, such as **Layout3**; however, doing so does not change the name of the current layout's page setup. Once you assign a page setup to the current layout, you cannot "un-apply" or "detach" it. You can only assign another named page setup. *Set Current* is disabled for sheet sets or when you select the current page setup from the list.

### *New*

This button produces the *New Page Setup* dialog box (Fig. 14-9) where you can assign a name for the new page setup. Assign a descriptive name (the default name being "Setup1"). You can select from the existing page setups in the drawing to "start with" as a template. Selecting the *OK* button produces the *Page Setup* dialog box where you specify the individual plotting options. Creating a *New* page setup does not automatically apply the new page setup to the current layout—you must use *Set Current* then select *OK* in the *Page Setup* dialog box to do so.

**FIGURE 14-9**

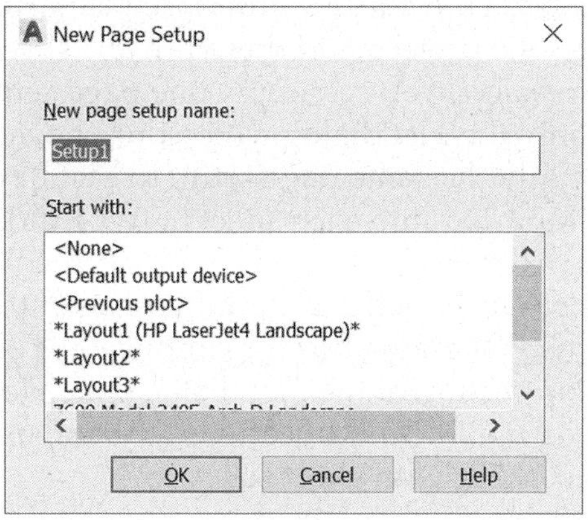


As you can see, the *Page Setup* dialog box (Fig. 14-10) is essentially identical to the *Plot* dialog box, except that you cannot create a plot. A few other small differences are explained in the next section (see "Should You Use the *Plot* Dialog Box or *Page Setup Manager*?").

**FIGURE 14-10**

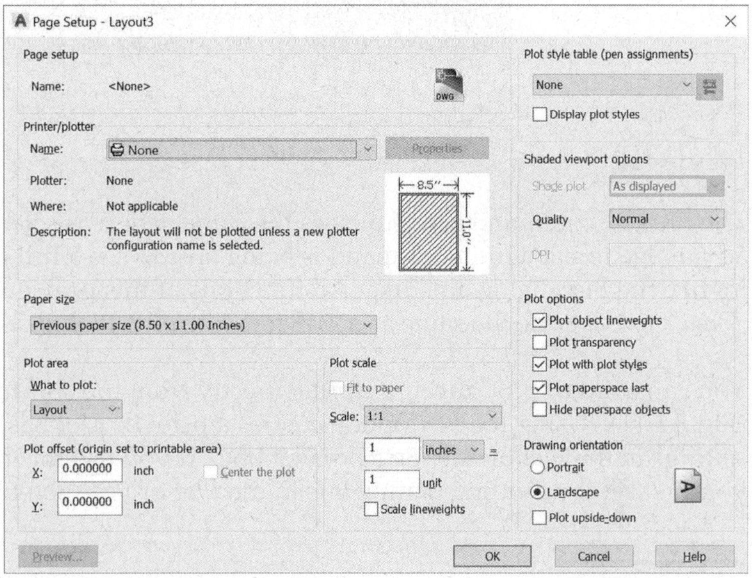


### *Modify*

This button in the *Page Setup Manager* (see previous Fig. 14-8) produces the *Page Setup* dialog box where you can edit the existing settings for the selected page setup. Select any page setup in the *Page Setup* list to modify, including the current layout's page setup.

### *Import*

Selecting this button invokes a standard file selection dialog box (not shown) from which you can import one or more page setups. You can select from any .DWG, .DWT, or .DXF file. Next, the *Import Page Setups* dialog box appears (not shown) for you to select the individual page setups in that specific drawing to import. The imported page setups then appear in the *Page Setup Manager* list, but are not yet applied to any layout.

## Should You Use the *Plot* Dialog Box or the *Page Setup Manager*?

As you probably noticed, the *Page Setup Manager* and related *Page Setup* dialog box are similar and have essentially the same functions as the *Plot* dialog box, including managing page setups, with a few exceptions. The main differences are noted below.

1. You cannot plot from the *Page Setup Manager* or *Page Setup* dialog box.
2. The *Plot* dialog box includes all features in one dialog box compared to two for the *Page Setup Manager* and *Page Setup* dialog box.
3. Only the *Plot* dialog box allows you to set options to *Plot to a file*, *Plot in the background*, and turn *Plot stamp on*.
4. You can modify any page setup from the *Page Setup Manager*, whereas you can modify only the current layout's page setup in the *Plot* dialog box.

So it has probably occurred to you, "Why do I need the *Page Setup Manager*?" Autodesk designed the *Page Setup Manager* interface primarily for use with sheet sets. A "sheet set" is a group of sheets related to one project, and therefore all "sheets" are plotted with the same plot settings. With the *Page Setup Manager*, you can apply one page setup to the entire sheet set. Without the *Page Setup Manager* and only the *Plot* dialog, you would have to open every sheet in the set (therefore every drawing file) and apply the same page setup to each layout in each drawing. Also, the desired page setup may have to be imported into each sheet (.DWG) individually before applying it.

TIP

So which method should you use? If you want to accomplish typical plotting and printing tasks, use the *Plot* dialog box for simplicity. You can also create, import, and assign page setups easily using the top section of the *Plot* dialog box. However, if you want to manage a large group of page setups, including making modifications to them, use the *Page Setup Manager*. Also use the *Page Setup Manager* when working with sheet sets.

### *Preview*

| Ribbon | Menu Bar | Command (Type) | Alias (Type) | Shortcut |
|---|---|---|---|---|
| *Output*<br>*Plot*<br>*Preview* | *File*<br>*Plot Preview* | *Preview* | *PRE* | ... |

The *Preview* command accomplishes the same action as selecting *Preview* in the *Plot* dialog box. The advantage to using this command is being able to see a full print preview directly without having to invoke the *Plot* dialog box first. All functions of this preview (right-click for shortcut menu to *Pan* and *Zoom*, etc.) operate the same as a preview from the *Plot* dialog box excluding one important option: *Plot*.

TIP

You can print or plot during *Preview* directly from the shortcut menu (Fig. 14-11). Using the *Plot* option creates a print or plot based on your settings in the *Plot* or *Page Setup* dialog box. If you use this feature, ensure you first set the plotting parameters in the *Plot* or *Page Setup* dialog box.

**FIGURE 14-11**

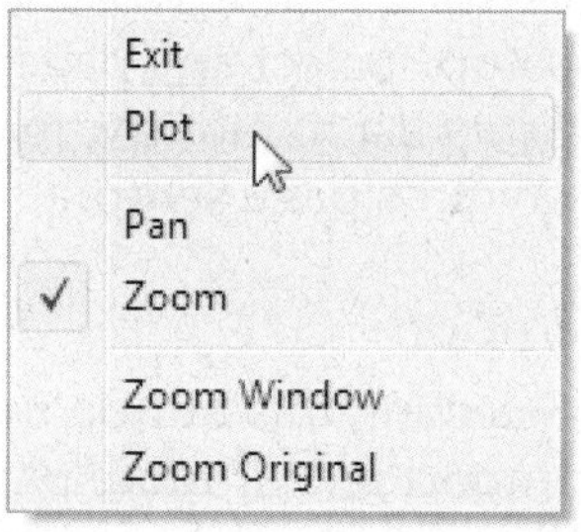


## PLOTTING TO SCALE

When you create a manual drawing, a scale is determined before you can begin drawing. The scale is determined by the proportion between the size of the object on the paper and the actual size of the object. You then complete the drawing in that scale so the actual object is proportionally reduced or enlarged to fit on the paper.

With a CAD drawing you are not restricted to drawing on a sheet of paper, so the geometry can be created full size. Set *Limits* to provide an appropriate amount of drawing space; then the geometry can be drawn using the actual dimensions of the object. The resulting drawing on the CAD system is a virtual full-size replica of the actual object. Not until the CAD drawing is plotted on a fixed size sheet of paper, however, is it scaled to fit on the sheet.

You can specify the plot scale in one of two ways. If you are plotting a *Layout* tab, specify the scale of the geometry that appears in the viewport by using the *Viewport Scale* pop-up list, *Viewport* toolbar, the *Properties* palette, or the *Zoom XP* factor, and then select 1:1 in the *Plot* or *Page Setup* dialog box. If you plot the *Model* tab, select the desired scale directly in the *Plot* or *Page Setup* dialog box. In either case, the value or ratio that you specify is the proportion of paper space units to model space units—the same as the proportion of the paper size to the *Limits*. This ratio is also equal to the reciprocal of the "drawing scale factor," as explained in Chapters 12 and 13. Exceptions to this rule are cases in which multiple viewports are used.

For example, if you want to draw an object 15" long and plot it on an 11" x 8.5" sheet, you can set the model space *Limits* to 22 x 17 (2 x the sheet size). The value of **2** is then the drawing scale factor, and the plot scale is **1:2** (or **1/2**, the reciprocal of the drawing scale factor). If, in another case, the calculated drawing scale factor is **4**, then **1:4** (or **1/4**) would be the plot scale to achieve a drawing plotted at 1/4 actual size. In order to calculate *Limits* and drawing scale factors, it is helpful to know the standard paper sizes.

## Standard Paper Sizes

| Size | Engineering (") (ANSI) | Architectural (") |
|---|---|---|
| A | 8.5 x 11 | 9 x 12 |
| B | 11 x 17 | 12 x 18 |
| C | 17 x 22 | 18 x 24 |
| D | 22 x 34 | 24 x 36 |
| E | 34 x 44 | 36 x 48 |

| Size | Metric (mm) (ISO) |
|---|---|
| A4 | 210 x 297 |
| A3 | 297 x 420 |
| A2 | 420 x 594 |
| A1 | 594 x 841 |
| A0 | 841 x 1189 |

## Calculating the Drawing Scale Factor

TIP

Assuming you are planning to plot from the *Model* tab or from a *Layout* tab using one viewport that fills almost all of the printable area, use the following method to determine the "drawing scale factor" and, therefore, the plot or print scale to specify.

1. In order to provide adequate space to create the drawing geometry full size, it is recommended to set the model space *Limits*, then *Zoom All*. *Limits* should be set to the intended sheet size used for plotting times a factor, if necessary, that provides enough area for drawing. This factor, the proportion of the *Limits* to the sheet size, is the "drawing scale factor."

   $$\text{DSF} = \frac{\textit{Limits}}{\text{sheet size}}$$

   You should also use a proportion that will yield a standard drawing scale (1/2"=1", 1/8"=1', 1:50, etc.), instead of a scale that is not a standard (1/3"=1", 3/5"=1', 1:23, etc.). See the "Tables of *Limits* Settings" for common standard drawing scales.

2. The "drawing scale factor" is used as the value (at least a starting point) for changing all size-related variables (*LTSCALE*, *Overall Scale* for dimensions, text height, *Hatch* pattern scale, etc.). The reciprocal of the drawing scale factor is the plot scale to use for plotting or printing the drawing.

   $$\text{Plot Scale} = \frac{1}{\text{DSF}}$$

3. If you are using millimeter dimensions, set *Units* to *Decimal* and use metric values for the sheet size. In this way, the reciprocal of the drawing scale factor is the plot scale, the same as feet and inch drawings. However, multiply the drawing's scale factor by 25.4 (25.4mm = 1") to determine the factor for changing all size-related variables (*LTSCALE*, *Overall Scale* for dimensions, etc.).

   $$\text{DSF(mm)} = \frac{\textit{Limits}}{\text{sheet size}} \times 25.4$$

If drawing or inserting a border on the sheet, its maximum size cannot exceed the printable area. The printable area is displayed in a layout by the dashed line border just inside the "paper" edge. Since plotters or printers do not draw or print all the way to the edge of the paper, the border should not be drawn outside of the printable area. Generally, approximately 1/4" to 1/2" (6mm to 12mm) offset from each edge of the paper (margin) is required. When plotting a layout, the printable area is denoted by a dashed border around the sheet. If plotting model space, you should multiply the margin (1/2 of the difference between the *Paper Size* and printable area) by the "drawing scale factor" to determine the margin size in model space drawing units.

## Guidelines for Plotting Model Space to Scale

TIP

Even though model space *Limits* can be changed and plot scale can be reset at <u>any time</u> in the drawing process, it is helpful to begin the process during the initial drawing setup (also see Chapter 12, Advanced Drawing Setup).

1. Set *Units* (*Decimal, Architectural, Engineering*, etc.) and *Precision* to be used in the drawing.
2. Set model space *Limits*. Set the *Limits* values to a proportion of the desired sheet size that provides enough area for drawing geometry as described in "Calculating the Drawing Scale Factor." The resulting value is the drawing scale factor.
3. Create the drawing geometry in model space. Use the DSF as the initial value (or multiplier) for linetype scale, text size, dimension scale, hatch scale, etc.
4. Use the DSF to determine the scale factor to create, *Insert* or *Xref* the title block and border in model space, if one is needed.
5. Access the *Plot* dialog box. Select the desired scale in the *Scale* drop-down list or enter a value in the *Custom* edit boxes. The value to enter is 1/DSF. (Values are also given in the "Tables of *Limits* Settings.") Make a plot *Preview*, then make the plot or make the needed adjustments.

## Guidelines for Plotting Layouts to Scale

TIP

Assuming you are using one viewport that fills almost all of the printable area, plot from the *Layout* tab using this method.

1. Set *Units* (*Decimal, Architectural, Engineering*, etc.) and *Precision* to be used in the drawing.
2. Set model space *Limits* and create the geometry in model space. Use the DSF to determine values for linetype scale, text size, dimension scale, hatch scale, etc. Complete the drawing geometry in model space.
3. Activate a layout to begin setting up the layout. You can also use the *Layout Wizard* to create a new layout and complete steps 2 through 5.
4. Use *Page Setup* to select the desired *Plot Device* and *Paper Size*.
5. Create, *Insert*, or *Xref* the title block and border in the layout. If you are using one of AutoCAD's template drawings or another template, the title block and border may already exist.
6. Create a viewport using *Vports*.
7. Click inside the viewport to set the viewport scale. You can set the scale using the *Viewport Scale* pop-up list, *Viewports* toolbar, the *Properties* palette, or use *Zoom* and enter an *XP* factor. Use 1/DSF as the viewport scale.
8. Activate the *Plot* dialog box and ensure the scale in the *Scale* drop-down list is 1:1. Make a plot *Preview*, then make the plot or make the needed adjustments.

If you want to print the same drawing using different devices and different scales, you can create multiple layouts. With each layout you can save the specific plot settings. Beware—a drawing printed at different scales may require changing or setting different values for linetype scale, text size, dimension variables, hatch scales, etc. for each layout. See Chapter 32 for more information on this topic.

### Annotative Objects

Annotative objects (annotative text and dimensions, etc.) automatically change size when you change the viewport scale. However, calculating *Limits* and a drawing scale factor is also a valid strategy for using annotative objects effectively. For example, the drawing scale factor should be used to set the *Annotation Scale* in the *Model* tab. See Chapter 32 for more information on using annotative objects.

### Template Drawings

Simplifying the process of plotting to scale can be accomplished by preparing template drawings. One method is to create a template for each sheet size that is used in the lab or office. In this way, the CAD operator begins the session by selecting the template drawing representing the sheet size and then multiplies those *Limits* by some factor to achieve the desired *Limits* and drawing scale factor. Another method is to create multiple layouts in the template drawing, one for each device and sheet size that you have available. In this way, a template drawing can be selected with the final layouts, title blocks, plot devices, and sheet sizes already specified, then you need only set the model space *Limits* and plot scale(s).

## TABLES OF *LIMITS* SETTINGS

Rather than making calculations of *Limits*, drawing scale factor, and plot scale for each drawing, the Tables of *Limits* Settings on the following pages can be used to make calculating easier. There are five tables, one for each of the following applications:

Mechanical Engineering
Architectural
Metric (using ISO standard metric sheets)
Metric (using ANSI standard engineering sheets)
Civil Engineering

To use the tables correctly, you must know any two of the following three variables. The third variable can be determined from the other two.

Approximate drawing *Limits*
Desired print or plot paper size
Desired print or plot scale

1. If you know the ***Scale*** and **paper size**:
   Assuming you know the scale you want to use, look along the top row to find the desired scale that you eventually want to print or plot. Find the desired paper size by looking down the left column. The intersection of the row and column yields the model space *Limits* settings to use to achieve the desired plot scale.
2. If you know the approximate model space ***Limits*** and **paper size**:
   Calculate how much space (minimum *Limits*) you require to create the drawing actual size. Look down the left column of the appropriate table to find the paper size you want to use for plotting. Look along that row to find the next larger *Limits* settings than your required area. Use these values to set the drawing *Limits*. The scale to use for the plot is located on top of that column.
3. If you know the ***Scale*** and approximate model space ***Limits***:
   Calculate how much space (minimum *Limits*) you need to create the drawing objects actual size. Look along the top row of the appropriate table to find the desired scale that you eventually want to plot or print. Look down that column to find the next larger *Limits* than your required minimum area. Set the model space *Limits* to these values. Look to the left end of that row to give the paper size you need to print or plot the *Limits* in the desired scale.

NOTE: Common scales are given in the Tables of *Limits* Settings. If you need to create a print or plot in a scale that is not listed in the tables, you may be able to use the tables as a guide by finding the nearest table and standard scale to match your needs, then calculating proportional settings.

**MECHANICAL TABLE OF *LIMITS* SETTINGS**
**For ANSI Sheet Sizes**
**(X axis x Y axis)**

| Paper Size (Inches) | Drawing Scale Factor 1<br>Scale 1″ = 1″<br>Proportion 1:1 | 1.33<br>3/4″= 1″<br>3:4 | 2<br>1.2″= 1″<br>1:2 | 2.67<br>3/8″= 1″<br>3:8 | 4<br>1/4″= 1″<br>1:4 | 5.33<br>3/16″= 1″<br>3:16 | 8<br>1/8″= 1″<br>1:8 |
|---|---|---|---|---|---|---|---|
| A. 11 x 8.5 In. | 11.0 x 8.5 | 14.7 x 11.3 | 22.0 x 17.0 | 29.3 x 22.7 | 44.0 x 34.0 | 58.7 x 45.3 | 88.0 x 68.0 |
| B. 17 x 11 In. | 17.0 x 11.0 | 22.7 x 14.7 | 34.0 x 22.0 | 45.3 x 29.3 | 68.0 x 44.0 | 90.7 x 58.7 | 136.0 x 88.0 |
| C. 22 x 17 In. | 22.0 x 17.0 | 29.3 x 22.7 | 44.0 x 34.0 | 58.7 x 45.3 | 88.0 x 68.0 | 117.3 x 90.7 | 176.0 x 136.0 |
| D. 34 x 22 In. | 34.0 x 22.0 | 45.3 x 29.3 | 68.0 x 44.0 | 90.7 x 58.7 | 136.0 x 88.0 | 181.3 x 117.3 | 272.0 x 176.0 |
| E. 44 x 34 In. | 44.0 x 34.0 | 58.7 x 45.3 | 88.0 x 68.0 | 117.3 x 90.7 | 176.0 x 136.0 | 234.7 x 181.3 | 352.0 x 272.0 |

**ARCHITECTURAL TABLE OF *LIMITS* SETTINGS**
**For Architectural Sheet Sizes**
**(X axis x Y axis)**

| Paper Size (Inches) | Drawing Scale Factor 12<br>Scale 1″ = 1′<br>Proportion 1:12 | 16<br>3/4″ x 1′<br>1:16 | 24<br>1/2″ = 1′<br>1:24 | 32<br>3/8″ = 1′<br>1:32 | 48<br>1/4″ = 1′<br>1:48 | 64<br>3/16″ = 1′<br>1:64 | 96<br>1/8″ = 1′<br>1:96 |
|---|---|---|---|---|---|---|---|
| A. 12 x 9 | | | | | | | |
| Ft | 12 x 9 | 16 x 12 | 24 x 18 | 32 x 24 | 48 x 36 | 64 x 48 | 96 x 72 |
| In. | 144 x 108 | 192x 144 | 288 x 216 | 384 x 288 | 576 x 432 | 768 x 576 | 1152 x 864 |
| B. 18 x 12 | | | | | | | |
| Ft | 18 x 12 | 24 x 16 | 36 X 24 | 48 x 32 | 72 x 48 | 96 x 64 | 144 x 96 |
| In. | 216 x 144 | 288 x 192 | 432 x 288 | 576 x 384 | 864 x 576 | 1152 x 768 | 1728 x 1152 |
| C. 24 x 18 | | | | | | | |
| Ft | 24 x 18 | 32 x 24 | 48 x 36 | 64 x 48 | 96 x 72 | 128 x 96 | 192 x 144 |
| In. | 288 x 216 | 384 x 288 | 576 x 432 | 768 x 576 | 1152 x 864 | 1536 x 1152 | 2034 x 1728 |
| D. 36 x 24 | | | | | | | |
| Ft | 36 x 24 | 48 x 32 | 72 x 48 | 96 x 64 | 144 x 96 | 192 x 128 | 288 x 192 |
| In. | 432 x 288 | 576 x 384 | 864 x 576 | 1152 x 768 | 1728 x 1152 | 2304 x 1536 | 3456 x 2304 |
| E. 48 x 36 | | | | | | | |
| Ft | 48 x 36 | 64 x 48 | 96 x 72 | 128 x 96 | 192 x 144 | 256 x 192 | 384 x 288 |
| In. | 576 x 432 | 768 x 576 | 1152 x 864 | 1536 x 1152 | 2304 x 1728 | 3072 x 2304 | 4608 x 3456 |

**METRIC TABLE OF *LIMITS* SETTINGS**
**For Metric (ISO) Sheet Sizes**
**(X axis x Y axis)**

| Paper Size (mm) | Drawing Scale Factor 25.4<br>Scale 1:1<br>Proportion 1:1 | 50:8<br>1:2<br>1:2 | 127<br>1:5<br>1:5 | 254<br>1:10<br>1:10 | 508<br>1:20<br>1:20 | 1270<br>1:50<br>1:50 | 2540<br>1:100<br>1:100 |
|---|---|---|---|---|---|---|---|
| A4 297 x 210 | | | | | | | |
| mm | 297 x 210 | 594 x 420 | 1485 x 1050 | 2970 x 2100 | 5940 x 4200 | 14,850 x 10,500 | 29,700 x 21,000 |
| m | .297 x .210 | .594 x .420 | 1.485 x 1.050 | 2.97 x 2.10 | 5.94 x 4.20 | 14.85 x 10.50 | 29.70 x 21.00 |
| A3 420 x 297 | | | | | | | |
| mm | 420 x 297 | 840 x 594 | 2100 x 1485 | 4200 x 2970 | 8400 x 5940 | 21,000 x 14,850 | 42,000 x 29,700 |
| m | .420 x .297 | .840 x .594 | 2.10 x 1.485 | 4.20 x 2.97 | 8.40 x 5.94 | 21.00 x 14.85 | 42.00 x 29.70 |
| A2 594 x 420 | | | | | | | |
| mm | 594 x 420 | 1188 x 840 | 2970 x 2100 | 5940 x 4200 | 11,880 x 8400 | 29,700 x 21,000 | 59,400 x 42,000 |
| m | .594 x .420 | 1.188 x .840 | 2.97 x 2.10 | 5.94 x 4.20 | 11.88 x 8.40 | 29.70 x 21.00 | 59.40 x 42.00 |
| A1 841 x 594 | | | | | | | |
| mm | 841 x 594 | 1682 x 1188 | 4205 x 2970 | 8410 x 5940 | 16,820 x 11,880 | 42,050 x 29,700 | 84,100 x 59,400 |
| m | .841 x .594 | 1.682 x 1.188 | 4.205 x 2.970 | 8.41 x 5.94 | 16.82 x 11.88 | 42.05 x 29.70 | 84.10 x 59.40 |
| A0 1189 x 841 | | | | | | | |
| mm | 1189 x 841 | 2378 x 1682 | 5945 x 4205 | 11,890 x 8410 | 23,780 x 16,820 | 59,450 x 42,050 | 118,900 x 84,100 |
| m | 1.189 x .841 | 2.378 x 1.682 | 5.945 x 4.205 | 11.89 x 8.41 | 23.78 x 16.82 | 59.45 x 42.05 | 118.90 x 84.10 |

**METRIC TABLE OF *LIMITS* SETTINGS**
**For ANSI Sheet Sizes**
**(X axis x Y axis)**

| Paper Size (mm) | Drawing Scale Factor 25.4<br>Scale TAB 1:1<br>Proportion 1:1 | 33.9<br>1:1.5 (2:3)<br>1.33 | 50.8<br>1:2<br>1:2 | 127<br>1:5<br>1:5 | 254<br>1:10<br>1:10 | 508<br>1:20<br>1:20 | 1270<br>1:50<br>1:50 |
|---|---|---|---|---|---|---|---|
| A 297.4 x 215.9 | | | | | | | |
| mm | 297.4 x 215.9 | 396.5 x 287.9 | 594.8 x 431.8 | 1487.0 x 1079.5 | 2974.0 x 2159.0 | 5948.0 x 4318.0 | 14,870.0 x 10,795.0 |
| m | 0.297 x 0.216 | 0.397 x 0.288 | 0.595 x 0.432 | 1.487 x 1.080 | 2.974 x 2.159 | 5.948 x 4.318 | 14.870 x 10.795 |
| B 431.8 x 297.4 | | | | | | | |
| mm | 431.8 x 297.4 | 575.7 x 396.5 | 863.6 x 594.8 | 2159.0 x 1487.0 | 4318.0 x 2974.0 | 8636.0 x 5948.0 | 21,590.0 x 14,870.0 |
| m | 0.432 x 0.297 | 0.576 x 0.397 | 0.864 x 0.595 | 2.159 x 1.487 | 4.318 x 2.974 | 8.636 x 5.948 | 21.590 x 14.870 |
| C 594.8 x 431.8 | | | | | | | |
| mm | 594.8 x 431.8 | 793.1 x 575.7 | 1189.6 x 863.6 | 2974.0 x 2159.0 | 5948.0 x 4318.0 | 11,896.0 x 8636.0 | 29,740.0 x 21,590.0 |
| m | 0.595 x 0.432 | 0.793 x 0.576 | 1.190 x 0.864 | 2.974 x 2.159 | 5.948 x 4.318 | 11.896 x 8.636 | 29.740 x 21.590 |
| D 863.6 x 594.8 | | | | | | | |
| mm | 863.6 x 594.8 | 1151.5 x 793.1 | 1727.2 x 1189.6 | 4318.0 x 2974.0 | 8636.0 x 5948.0 | 17,272.0 x 11,896.0 | 43,180.0 x 29,740.0 |
| m | 0.864 x 0.595 | 1.151 x 0.793 | 1.727 x 1.190 | 4.318 x 2.974 | 8.636 x 5.948 | 17.272 x 11.896 | 43.180 x 29.740 |
| E 1189.6 x 863.6 | | | | | | | |
| mm | 1190 x 864 | 1586 x 1151 | 2379 x 1727 | 5948 x 4318 | 11,896 x 8636 | 23,792 x 17,272 | 59,480.0 x 43,180 |
| m | 1.190 x 0.864 | 1.586 x 1.151 | 2.379 x 1.727 | 5.948 x 4.318 | 11.896 x 8.636 | 23.792 x 17.272 | 59.480 x 43.180 |

**CIVIL TABLE OF *LIMITS* SETTINGS**
**For ANSI Sheet Sizes**
**(X axis x Y axis)**

| Paper Size (Inches) | Drawing Scale Factor / Scale / Proportion: 120 / 1″ = 10′ / 1:120 | 240 / 1″ = 20′ / 1:240 | 360 / 1″ = 30′ / 1:360 | 480 / 1″ = 40′ / 1:480 | 600 / 1″ = 50′ / 1:600 |
|---|---|---|---|---|---|
| A. 11 x 8.5 In. | 1320 x 1020 | 2640 x 2040 | 3960 x 3060 | 5280 x 4080 | 6600 x 5100 |
| Ft | 110 x 85 | 220 x 170 | 330 x 255 | 440 x 340 | 550 x 425 |
| B. 17 x 11 In. | 2040 x 1320 | 4080 x 2640 | 6120 x 3960 | 8160 x 5280 | 10,200 x 6600 |
| Ft | 170 x 110 | 340 x 220 | 510 x 330 | 680 x 440 | 850 x 550 |
| C. 22 x 17 In. | 2640 x 2040 | 5280 x 4080 | 7920 x 6120 | 10,560 x 8160 | 13,200 x 10,200 |
| Ft | 220 x 170 | 440 x 340 | 660 x 510 | 880 x 680 | 1100 x 850 |
| D. 34 x 22 In. | 4080 x 2640 | 8160 x 5280 | 12,240 x 7920 | 16,320 x 10,560 | 20,400 x 13,200 |
| Ft | 340 x 220 | 680 x 440 | 1020 x 660 | 1360 x 880 | 1700 x 1100 |
| E. 44 x 34 In. | 5280 x 4080 | 10,560 x 8160 | 15, 840 x 12,240 | 21,120 x 16,320 | 26,400 x 20,400 |
| Ft | 440 x 340 | 880 x 680 | 1320 x 1020 | 1760 x 1360 | 2200 x 1700 |

## Examples for Plotting to Scale

Following are several hypothetical examples of drawings that can be created using the "Guidelines for Plotting to Scale." As you read the examples, try to follow the logic and check the "Tables of *Limits* Settings."

A. A one-view drawing of a mechanical part that is 40" in length is to be drawn requiring an area of approximately 40" x 30", and the drawing is to be plotted on an 8.5" x 11" sheet. The template drawing *Limits* are preset to 0,0 and 11,8.5. (The ACAD.DWT template *Limits* are set to 0,0 and 12,9, which represents an uncut sheet size. Template *Limits* of 11 x 8.5 are more practical in this case.) The expected plot scale is 1/4"=1". The following steps are used to calculate the new *Limits*.

1. ***Units*** are set to ***decimal***. Each unit represents 1.00 inches.
2. Multiplying the *Limits* of 11 x 8.5 by a factor of **4**, the new ***Limits*** should be set to **44 x 34**, allowing adequate space for the drawing. The "drawing scale factor" is **4**.
3. All size-related variable default values (***LTSCALE, Overall Scale*** for dimensions, etc.) are multiplied by **4**. The plot scale (for *Layout* tabs, entered in the *Viewport* toolbar, etc., or for the *Model* tab, entered in the *Plot* or *Page Setup* dialog box) is **1:4** to achieve a plotted drawing of 1/4"=1".

B. A floorplan of a residence will occupy 60' x 40'. The drawing is to be plotted on a "D" size architectural sheet (24" x 36"). No prepared template drawing exists, so the ACAD.DWT template drawing (12 x 9) *Limits* are used. The expected plot scale is 1/2"=1'.

1. ***Units*** are set to ***Architectural***. Each unit represents 1".
2. The floorplan size is converted to inches (60' x 40' = 720" x 480"). The sheet size of 36" x 24" (or 3' x 2') is multiplied by **24** to arrive at ***Limits*** of **864" x 576"** (72' x 48'), allowing adequate area for the floor plan. The *Limits* are changed to those values.
3. All default values of size-related variables (***LTSCALE, Overall Scale*** for dimensions, etc.) are multiplied by **24**, the drawing scale factor. The plot scale (for *Layout* tabs, entered in the *Viewport Scale* list, etc., or for the *Model* tab, entered in the *Plot* or *Page Setup* dialog box) is **1/2"=1'** (or 1/24, which is the reciprocal of the drawing scale factor).

C. A roadway cloverleaf is to be laid out to fit in an acquired plot of land measuring 1500' x 1000'. The drawing will be plotted on "D" size engineering sheet (34" x 22"). A template drawing with *Limits* set equal to the sheet size is used. The expected plot scale is 1"=50'.

1. ***Units*** are set to ***Engineering*** (feet and decimal inches). Each unit represents 1.00".
2. The sheet size of 34" x 22" (or 2.833' x 1.833') is multiplied by **600** to arrive at ***Limits*** of **20400" x 13200"** (1700' x 1100'), allowing enough drawing area for the site. The *Limits* are changed to **1700' x 1100'**.
3. All default values of size-related variables (***LTSCALE, Overall Scale*** for dimensions, etc.) are multiplied by **600**, the drawing scale factor. The plot scale (for *Layout* tabs, entered in the *Viewport Scale* list, etc., or for the *Model* tab, entered in the *Plot* or *Page Setup* dialog box) is **1/600** (***inches = drawing units***) to achieve a drawing of 1"=50' scale.

NOTE: Many civil engineering firms use one unit in AutoCAD to represent one foot. This simplifies the problem of using decimal feet (10 parts/foot rather than 12 parts/foot); however, problems occur if architectural layouts are inserted or otherwise combined with civil work.

D. Three views of a small machine part are to be drawn and dimensioned in millimeters. An area of 480mm x 360mm is needed for the views and dimensions. The part is to be plotted on an A4 size sheet. The expected plot scale is 1:2.

1. ***Units*** are set to ***decimal***. Each unit represents 1.00 millimeters.
2. Setting the *Limits* exactly to the sheet size (297 x 210) would not allow enough area for the drawing. The sheet size is multiplied by a factor of **2** to yield ***Limits*** of **594 x 420**, providing the necessary 480 x 360 area.
3. The plot scale (for *Layout* tabs, entered in the *Viewport Scale* list, etc., or for the *Model* tab, entered in the *Plot* or *Page Setup* dialog box) is **1:2** scale. Since this drawing is metric, the drawing scale factor for changing all size-related variable default values (***LTSCALE, Overall Scale*** for dimensions, etc.) is multiplied by 25.4, or 2 x 25.4 = approximately **50**.

# CONFIGURING PLOTTERS AND PRINTERS

## Plotting Components

When a drawing is plotted, three main components are referenced to develop the finished plot.

1. The plot and page settings you set up in the *Plot* dialog box or *Page Setup* dialog box (saved in the .DWG file)
2. The plotter configuration information you specify in the *Add-A-Plotter Wizard* or *Plotter Configuration Editor* (saved in a .PC3 file)
3. A plot style table you create using the *Add-A-Plot Style Table Wizard* or *New* option in the *Plot* dialog box (saved in a .CTB or .STB file)

The three components are stored separately so they can be applied in different combinations. For example, any plot style table can be used with any plot device, or any page setup can use any plot style table.

The *Plot* and *Page Setup* dialog boxes were explained earlier in this chapter. The following section explains how to configure plotters and create .PC3 files using the *Plotter Manager*, *Add-A-Plotter Wizard*, and *Plotter Configuration Editor*. Plot style tables (.CTB and .STB files) are discussed in Chapter 33.

If you are using AutoCAD in a lab or office, the devices used for making plots and prints are most likely already configured for your use. However, no matter if you are a student or a professional, you may be required to configure a new device or change configuration settings on existing devices. This section explains those possibilities. In order to accomplish either of the tasks mentioned here, use the *Plottermanager* command to invoke the AutoCAD *Plotter Manager*.

## *Plottermanager*

| Ribbon | Menu Bar | Command (Type) | Alias (Type) | Shortcut |
|---|---|---|---|---|
| *Output*<br>*Plot*<br>*Plotter Manager* | *File*<br>*Plotter Manager* | *Plottermanager* | ... | ... |

The AutoCAD *Plotter Manager* can be invoked by several methods as shown in the command table above. In addition to those methods, you can select the *Add or Configure Plotters* button in the *Plotting* tab of the *Options* dialog box.

In any case, the *Plotter Manager* is a separate system window which can run as an independent application to AutoCAD (Fig. 14-12). By default, the window displays icons of all plot devices configured specifically for AutoCAD. The default Windows configured device is included in this group. (When you install AutoCAD, the Windows configured system printer is automatically configured if you have a previously installed Windows printer.) You can add or configure a plotter with the *Plotter Manager*.

**FIGURE 14-12**

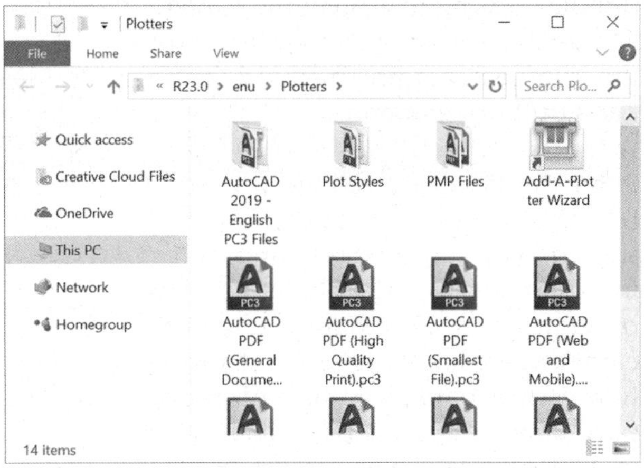


The display of the *Plotter Manager* can also be changed to list *Details* instead of *Icons*, as shown in Figure 14-13 (use the *Views* pull-down menu and select *Details*). Note that the defined plotters are actually saved as .PC3 files. If you want to add a new plotter for use with AutoCAD, select the *Add-A-Plotter Wizard*. To change the settings on an existing plotter using the *Plotter Configuration Editor*, double-click on its icon or file name.

**FIGURE 14-13**

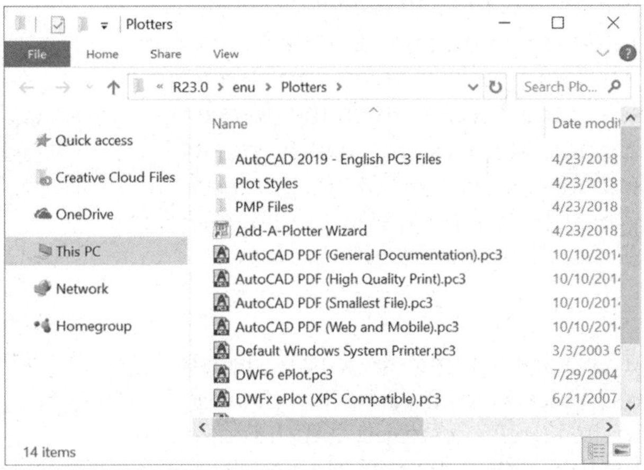


## *Add-A-Plotter Wizard*

The *Add-A-Plotter Wizard* makes the process of configuring a new device very simple. Several choices are available (Fig. 14-14). You can configure a new device to be connected directly to and managed by your system or to be shared and managed by a network server. You can also reconfigure printers that were previously set up as Windows system printers. In this case, icons for the printers will appear in the *Plotter Manager* and you can access and control configuration settings to use specifically for AutoCAD.

**FIGURE 14-14**

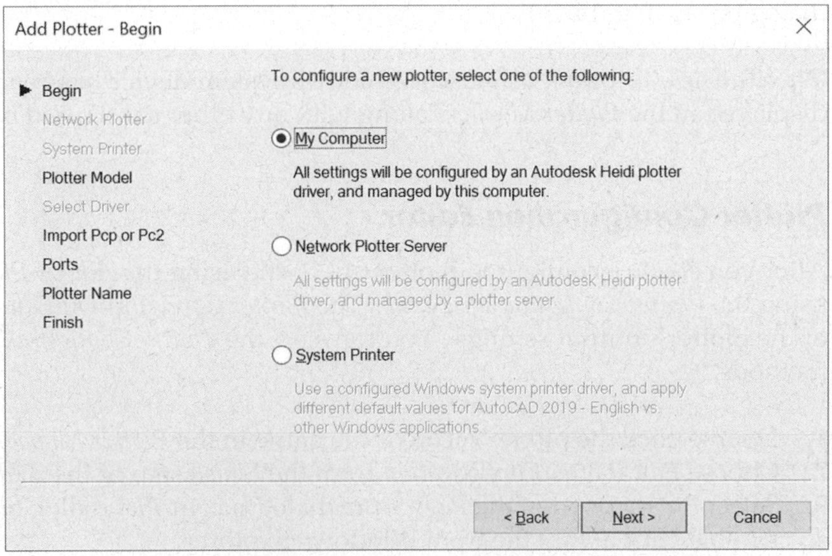

Start the *Add-A-Plotter Wizard* by double-clicking its icon in the *Plotter Manager* or selecting *Add Plotter...* from the *Wizards* cascading menu in the *Tools* pull-down menu.

After you have selected the device (*Plotter Model*) to configure (not shown), you can choose to import information about the plotter if it was previously configured for use with an earlier release of AutoCAD. Information such as paper size, optimization, port names, etc. were previously stored in .PCP and .PC2 files and can be converted to the .PC3 format (Fig. 14-15).

**FIGURE 14-15**

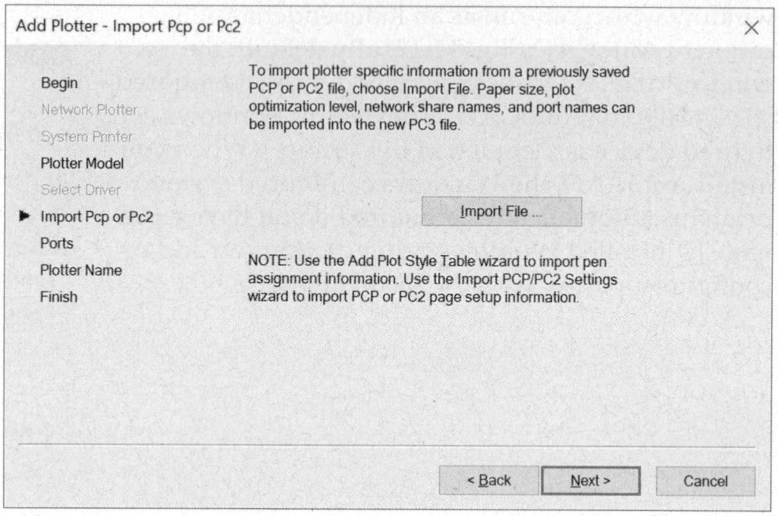


If you selected *My Computer* in the *Begin* step, you are presented with the *Ports* page next (not shown). You can select the port to which the device is connected or specify that the device will always *Plot to File* or *Autospool*. If you specify a port, you can later select *Plot to file* in the *Plot* dialog box.

If you are using a device previously used as a Windows system printer, make sure you supply a different name for the device than the name Windows uses. This action ensures the new name appears in the *Plot* and *Page Setup* dialog boxes (Fig. 14-16).

**FIGURE 14-16**

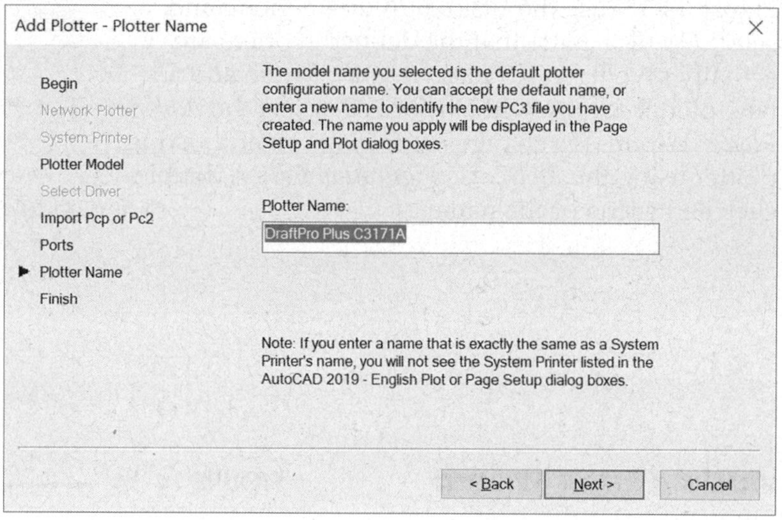


Once the new or previously used Windows system device has been configured for AutoCAD, its icon is displayed in the *Plotter Manager* along with any other configured devices.

## Plotter Configuration Editor

After you create a configured plotter (.PC3) file using the *Add-A-Plotter Wizard*, you can edit the file using the *Plotter Configuration Editor*. The *Plotter Configuration Editor* allows you to specify a number of the plotter's output settings. You can start the *Plotter Configuration Editor* by one of the following methods.

1. Double-click the plotter's icon or file name in the *Plotter Manager*.
2. Choose *Edit Plotter Configuration* from the *Finish* step of the *Add-A-Plotter Wizard*.
3. Select *Properties* from the *Page Setup* dialog box or *Plot* dialog box.
4. Double-click a .PC3 file from Windows Explorer.

The *Plotter Configuration Editor* contains three tabs. The *General* tab contains basic information about the plotter. Only the *Description* area is not read-only and can be edited. Information entered in the *Description* is saved with the .PC3 file.

The *Ports* tab allows you to specify ports to connect with the device. If you are configuring a local, non-Windows system plotter, you can respecify the port to which the device is connected. You can choose a serial (LPT), parallel (COM), or network port. If you are using a serial port, the settings within AutoCAD must match the plotter settings. For additional technical information, check AutoCAD *Help*.

The *Device and Document Settings* tab contains the plotting options. These choices are different depending on your configured plotting device. For example, when you configure a laser printer you have the option to modify the printing characteristics (Fig. 14-17). The *Device and Document Settings* tab can contain some or all of the following general option areas.

**FIGURE 14-17**

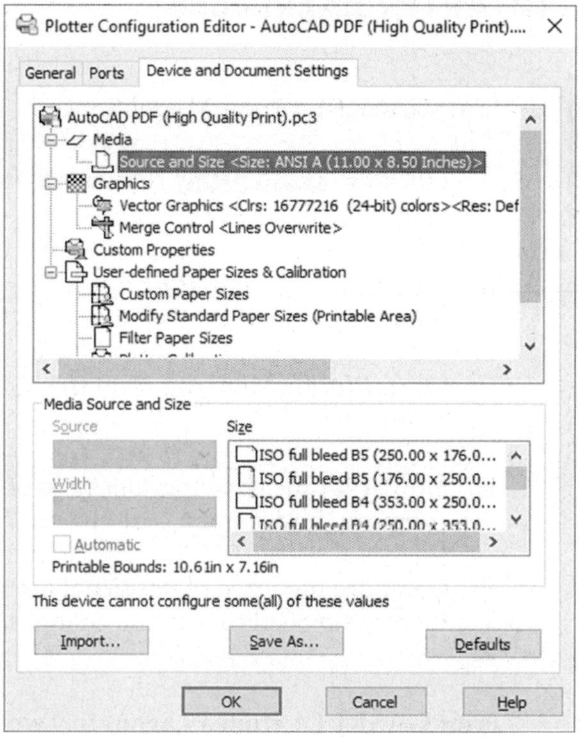


| | |
|---|---|
| *Media* | Specifies a paper source, size, type, and destination. |
| *Physical Pen Configuration* | Specifies settings for pen plotters. |
| *Graphics* | Specifies settings for printing vector and raster graphics and True Type fonts. |
| *Custom Properties* | Displays settings related to the device driver. |
| *Initialization Strings* | Sets pre-initialization, post-initialization, and termination printer strings. |
| *User-Defined Paper Sizes and Calibration* | Attaches a plot model parameter (.PMP) file to the .PC3 file, calibrates plotter, and adds, deletes, or revises custom paper sizes. |

The related .PC3 file contains these same categories of settings. Double-click any of the categories to view and change the specific settings.

You can also filter paper sizes that you do not use so they do not appear as a choice in the *Page Setup* and *Plot* dialog boxes. Expand the *User-defined Paper Sizes & Calibration* section, then select *Filter Paper Sizes*.

Because there are so many possible options for all the possible plotting devices it is not appropriate to include the information in this text. Additional technical information on this subject should be found in AutoCAD *Help*.

## CHAPTER EXERCISES

1. **Print GASKETA from Model Space**

   A. ***Open*** the **GASKETA** drawing that you created in Chapter 9 Exercises. Activate the ***Model*** tab. What are the model space *Limits*?

   B. From the *Model* tab, ***Plot*** the drawing ***Extents*** on an 11″ x 8.5″ sheet. Select the ***Fit to paper*** option from the *Plot scale* section in the *Plot* dialog box.

   C. Next, ***Plot*** the drawing from the *Model* tab again. Plot the ***Limits*** on the same size sheet. ***Scale to Fit***.

   D. Now, ***Plot*** the drawing ***Limits*** as before but plot the drawing at **1:1**. Measure the drawing and compare the accuracy with the dimensions given in the exercise in Chapter 9.

   E. Compare the three plots. What are the differences and why did they occur? (When you finish, there is no need to *Save* the changes.)

2. **Print GASKETA from a Layout in Two Scales**

   A. Using the **GASKETA** drawing again, activate the ***Layout1*** tab. If a viewport exists in the layout, ***Erase*** it. Activate the ***Plot*** or ***Page Setup*** dialog box and configure the layout to use a printer to print on a ***Letter*** (or "A" size) sheet.

   B. Next, make layer **0** ***Current***. Use ***Vports*** to create a ***Single*** viewport using the ***Fit*** option. Double-click inside the viewport and ***Zoom All*** to see all of the model space *Limits*. Next, set the viewport scale to **1:1** using any method (*Zoom XP*, the *Viewport* toolbar, or the *Properties* palette). You may need to ***Pan*** to center the geometry, but do <u>not</u> *Zoom*.

   C. (Optional) Create a title block and border for the drawing in paper space.

   D. Activate the ***Plot*** dialog box. Ensure the *Scale* edit box is set at **1:1**. Plot the ***Layout***. When complete, measure the plot for accuracy by comparing with the dimensions given for the exercise in Chapter 9. ***Save*** the drawing.

   E. Reset the viewport scale to **2:1** using any method (*Zoom XP*, the *Viewport Scale* list, the *Viewport* toolbar, or the *Properties* palette). Activate the ***Plot*** dialog box and make a second plot at the new scale. ***Close*** the drawing, but <u>do not save</u> it.

3. **Print the MANFCELL Drawing in Two Standard Scales**

   A. ***Open*** the **MANFCELL** drawing from Chapter 9. Check to make sure that the ***Limits*** settings are at **0,0** and **40′,30′**.

   B. Activate the ***Layout1*** tab. If a viewport exists, ***Erase*** it. Use ***Pagesetup*** to select a printer to print on a letter ("A") sheet size. Next, create a ***Single*** viewport with the ***Fit*** option on layer **0**.

C. In this step you will make one print of the drawing at the largest possible size using a standard architectural scale. Use any method to set the viewport scale. Use the "Architectural Table of *Limits* Settings" in the chapter for guidance on an appropriate scale to set. (HINT: Look in the "A" size sheet row and find the next larger *Limits* settings that will accommodate your current *Limits*, then use the scale shown above at the top of the column.) ***Save*** the drawing. Make a print using the ***Plot*** dialog box.

D. Now make a plot of the drawing using the next smaller standard architectural scale. This time determine the appropriate scale to set using either the "Architectural Table of *Limits* Settings" or by selecting it from the *Viewport Scale* pop-up list or *Viewport Scale* drop-down list.

4. **Use *Plotter Manager* to Configure a "D" Size Plot Device**
Even if your system currently has a "D" size plot device configured, this exercise will give you practice with *Plotter Manager*. This exercise does not require that the device actually exist in your lab or office or that it be physically attached to your system.

A. Invoke the ***Plotter Manager***. In *Plotter Manager*, double-click on ***Add-A-Plotter Wizard***.

B. Select the ***Next*** button in the *Introduction* page. In the *Begin* page, select ***My Computer***. In the *Plotter Model* page, select any *Manufacturer* and *Model* that can plot on a "D" size sheet, such as ***CalComp 68444 Color Electrostatic***, or ***Hewlett-Packard Draft-Pro EXL (7576A)***.

C. Select ***Next*** in both the *Import PCP or PC2* and *Ports* pages. In the *Plotter Name* step, select the default name but insert the manufacturer's name as a prefix, such as "HP" or "CalComp."

D. Finally, activate the ***Plot*** or ***Page Setup*** dialog box and ensure your new device is listed in the *Name* drop-down list.

5. **Create Multiple Layouts for Different Plot Devices, Set *PSLTSCALE***
In this exercise you will create two layouts—one for plotting on an engineering "A" size sheet and one on a "C" size sheet. Check the table of "Standard Paper Sizes" in the chapter and find the correct size in inches for both engineering "A" and "C" size sheets. You will also adjust *LTSCALE* and *PSLTSCALE* appropriately for the layouts.

REUSE

A. ***Open*** drawing **CH11EX3.** Check to ensure the model space ***Limits*** are set at **22** x **17**. If you plan to print on a "A" size sheet, what is the drawing scale factor? Is the ***LTSCALE*** set appropriately?

B. Activate the ***Layout1*** tab. Right-click and use ***Rename*** to rename the tab to **A Sheet**. If a viewport exists in the layout, ***Erase*** it. Activate the ***Page Setup*** dialog box and configure the layout to use a printer to print on a ***Letter*** ("A" size) sheet. Next, use ***Vports*** to create a viewport on layer **0** using the ***Fit*** option. Double-click inside the viewport and set the viewport scale to **1:2**.

C. Toggle between ***Model*** tab and the ***A Sheet*** tab to examine the hidden and center lines. Do the dashes appear the same in both tabs? If the dashes appear differently in the two tabs, type ***PSLTSCALE*** and ensure the value is set to **0**. (Make sure you always use ***Regenall*** after using ***PSLTSCALE***.)

D. Use the ***Plot*** dialog box to make the plot. Measure the plot for accuracy by comparing it with the dimensions given for the exercise in Chapter 11. To refresh your memory, that exercise involved adjusting the *LTSCALE* factor. Does the *LTSCALE* in the drawing yield hidden line dashes of 1/8" (for the long lines) on the plot? If not, make the ***LTSCALE*** adjustment and plot again. Use *SaveAs* to save and rename the drawing as **CH14EX5**.

E. Next, activate the ***Layout2*** tab. Right-click and use ***Rename*** to rename the tab to **C Sheet**. If a viewport exists in the layout, ***Erase*** it. Activate the ***Page Setup*** dialog box and configure the layout to use a plotter for a ***ANSI C (22 x 17 Inches)*** sheet. Next, on layer **0**, use ***Vports*** to create a viewport using the ***Fit*** option. Double-click inside the viewport and set the viewport scale. Since the model space limits are set to 22 x 17 (same as the sheet size), set the viewport scale to **1:1**.

F. Do the hidden lines appear the same in all tabs? *PSLTSCALE* set to 0 ensures that the hidden and center lines appear with the same *LTSCALE* in the viewports as in the *Model* tab. If everything looks okay, make the plot. Does the plot show hidden line dashes of 1/8″? ***Save*** the drawing.

6. **Create an Architectural Template Drawing**
In this exercise you will create a new template for architectural applications to plot at 1/8″=1′ scale on a "D" size sheet.

A. Begin a ***New*** drawing using the template **ASHEET.DWT** you worked on in Chapter 13 Exercises. Use ***Save*** to create a new template **(.DWT)** named **D-8-AR** in your working folder (not where the AutoCAD-supplied templates are stored).

B. Set ***Units*** to ***Architectural***. Use the "Architectural Table of *Limits* Settings" to determine and set the new model space ***Limits*** for an architectural "**D**" size sheet to plot at **1/8″=1′**. Multiply the existing ***LTSCALE*** (**1.0**) times the drawing scale factor shown in the table. Turn ***Snap*** to ***Polar Snap*** and turn ***Grid*** off.

C. Activate the ***Layout2*** tab. Right-click to produce the shortcut menu and select ***Rename***. Rename the tab to **D Sheet**. ***Erase*** the viewport if one exists.

D. Activate the ***Page Setup*** dialog box. Select a plot device that can use an architectural D size sheet (36 x 24), such as the device you configured for your system in Exercise 4. In the ***Layout Settings*** tab, select ***ARCH D (24 x 36 Inches)*** in the ***Paper Size*** drop-down list. Select ***OK*** to dismiss the *Page Setup* dialog box and save the settings.

E. Make **VPORTS** the current layer. Use the ***Vports*** command to create a ***Single*** viewport using the ***Fit*** option. Double-click inside the viewport and set the viewport scale to **1/8″=1′**.

F. Make the ***Model*** tab active and set the **OBJECT** layer current. Finally, ***Save*** as a template (.DWT) drawing in your working folder.

7. **Create a Civil Engineering Template Drawing**
Create a new template for civil engineering applications to plot at 1″=20′ scale on a "C" size sheet.

A. Begin a ***New*** drawing using the template **C-D-SHEET.** Use ***Save*** to create a new template (.DWT) named **C-CIVIL-20** in your working folder (not where the AutoCAD-supplied templates are stored).

B. Set ***Units*** to ***Engineering***. Use the "Civil Table of *Limits* Settings" to determine and set the new ***Limits*** for a "**C**" size sheet to plot at **1″=20′**. Multiply the existing ***LTSCALE*** (.5) times the scale factor shown. Turn ***Snap*** to ***Polar*** and ***Grid*** Off.

C. Activate the ***D Sheet*** layout tab. Use any method to set the viewport scale to **1:240**. ***Save*** the drawing as a template (.DWT) drawing in your working folder.

CHAPTER

# 15

# Draw Commands II

## Chapter Objectives

After completing this chapter you should:

1. be able to create construction lines using the *Xline* and *Ray* commands;
2. be able to create *Polygons* by the *Circumscribe*, *Inscribe*, and *Edge* methods;
3. be able to create *Rectangles* and *Donuts*;
4. be able to create *Spline* curves passing exactly through the selected points;
5. be able to create *Ellipses* using the *Axis End* method, the *Center* method, and the *Arc* method;
6. be able to use *Divide* to add points at equal parts of an object and *Measure* to add points at specified segment lengths along an object;
7. be able to draw multiple parallel lines with *Mline* and to create multiline styles with *Mstyle*;
8. be able to use the *Sketch* command to create "freehand" sketch lines;
9. be able to create a *Boundary* by PICKing inside a closed area;
10. know that objects forming a closed shape can be combined into one *Region* object;
11. be able to use the *Revcloud* command to draw revision clouds.

## CONCEPTS

Remember that *Draw* commands create AutoCAD objects. The draw commands addressed in this chapter create more complex objects than those discussed in Chapter 8, Draw Commands I. The draw commands covered previously are *Line, Circle, Arc, Point,* and *Pline*. The shapes created by the commands covered in this chapter are more complex. Most of the objects appear to be composed of several simple objects, but each shape is actually treated by AutoCAD as one object. Only the *Divide, Measure* and *Sketch* commands create multiple objects. The following commands are explained in this chapter.

| | | | |
|---|---|---|---|
| *Xline* | *Ray* | *Polygon* | *Rectangle* |
| *Donut* | *Spline* | *Ellipse* | *Divide* |
| *Measure* | *Mline* | *Sketch* | *Solid* |
| *Boundary* | *Region* | *Wipeout* | *Revcloud* |

These draw commands can be accessed using any of the command entry methods, including the menus and icon buttons, as illustrated in Chapter 8. Buttons that appear by the Command tables in this chapter are available in the default *Draw* toolbar or can be activated by creating your own custom toolbar.

## COMMANDS

### Xline

| Ribbon | Menu Bar | Command (Type) | Alias (Type) | Shortcut |
|---|---|---|---|---|
| *Home* *Draw* (icon) | *Draw* *Construction Line* | *Xline* | *XL* | ... |

When you draft with pencil and paper, light "construction" lines are used to lay out a drawing. These construction lines are not intended to be part of the finished object lines but are helpful for preliminary layout such as locating intersections, center points, and projecting between views.

There are two types of construction lines—*Xline* and *Ray*. An *Xline* is a line with infinite length, therefore having no endpoints. A *Ray* has one "anchored" endpoint and the other end extends to infinity. Even though these lines extend to infinity, they do not affect the drawing *Limits* or *Extents* or change the display or plot area in any way.

An *Xline* has no endpoints (*Endpoint Object Snap* cannot be used) but does have a root, which is the theoretical midpoint (*Midpoint Object Snap* can be used). If *Trim* or *Break* is used with *Xlines* or *Rays* such that two endpoints are created, the construction lines become *Line* objects. *Xlines* and *Rays* are drawn on the current layer and assume the current linetype and color (object-specific or *ByLayer*).

There are many ways that you can create *Xlines* as shown by the options appearing at the command prompt. All options of *Xline* automatically repeat so you can easily draw multiple lines.

Command: **XLINE**
Specify a point or [Hor/Ver/Ang/Bisect/Offset]:

### *Specify a point*

The default option only requires that you specify two points to construct the *Xline* (Fig. 15-1). The first point becomes the root and anchors the line for the next point specification. The second point, or "through point," can be PICKed at any location and can pass through any point (*Polar Snap*, *Polar Tracking*, *Object Snaps*, and *Objects Snap Tracking* can be used). If horizontal or vertical *Xlines* are needed, *Ortho* can be used in conjunction with the "Specify a point:" option.

Command: **XLINE**
Specify a point or [Hor/Ver/Ang/Bisect/Offset]:
Specify through point:

**FIGURE 15-1**

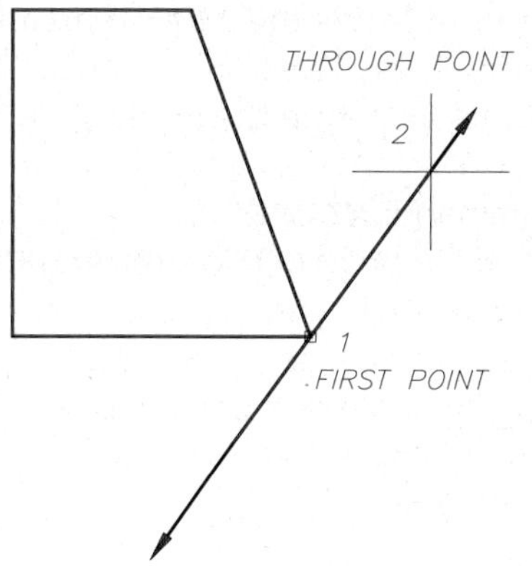


### *Hor*

This option creates a horizontal construction line. Type the letter "H" at the first prompt. You only specify one point, the through point or root (Fig. 15-2).

**FIGURE 15-2**

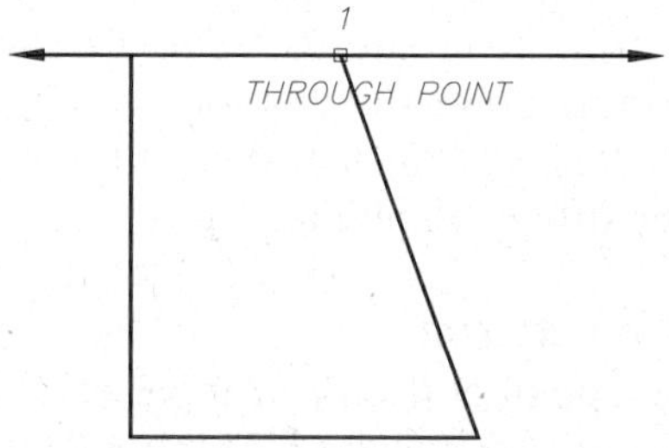


### *Ver*

*Ver* creates a vertical construction line. You only specify one point, the through point (root) (Fig. 15-3).

**FIGURE 15-3**

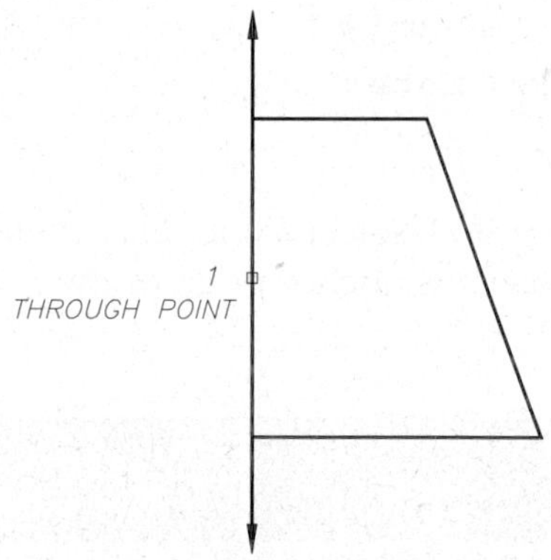


### *Ang*

The *Ang* option provides two ways to specify the desired angle. You can (1) *Enter angle* or (2) select a *Reference* line (*Line*, *Xline*, *Ray*, or *Pline*) as the starting angle, then specify an angle from the selected line (in a counterclockwise direction) for the *Xline* to be drawn (Fig. 15-4).

Command: **XLINE**
Specify a point or [Hor/Ver/Ang/Bisect/Offset]: ***a***
Enter angle of xline (0) or [Reference]:

**FIGURE 15-4**

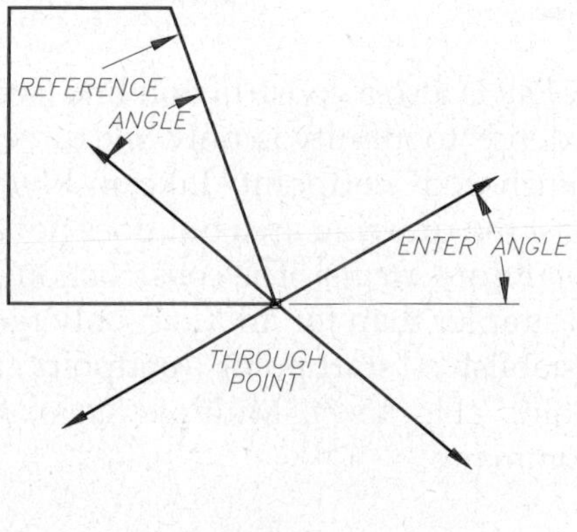

### *Bisect*

This option draws the *Xline* at an angle between two selected points. First, select the angle vertex, then two points to define the angle (Fig. 15-5).

Command: **XLINE**
Specify a point or [Hor/Ver/Ang/Bisect/Offset]: ***b***
Specify angle vertex point: **PICK**
Specify angle start point: **PICK**
Specify angle end point: **PICK**
Specify angle end point: **Enter**
Command:

FIGURE 15-5

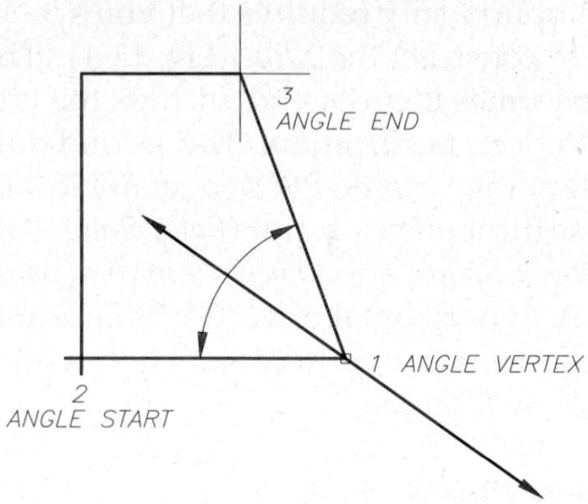


### *Offset*

*Offset* creates an *Xline* parallel to another line. This option operates similarly to the *Offset* command. You can (1) specify a *Distance* from the selected line or (2) PICK a point to create the *Xline Through*. With the *Distance* option, enter the distance value, select a line (*Line, Xline, Ray,* or *Pline*), and specify on which side to create the offset *Xline* (Fig. 15-6).

Command: **XLINE**
Specify a point or [Hor/Ver/Ang/Bisect/Offset]: ***o***
Specify offset distance or [Through] <1.0000>:
  **(value)** or **PICK**
Select a line object: **PICK**
Specify side to offset: **PICK**
Select a line object: **Enter**
Command:

FIGURE 15-6

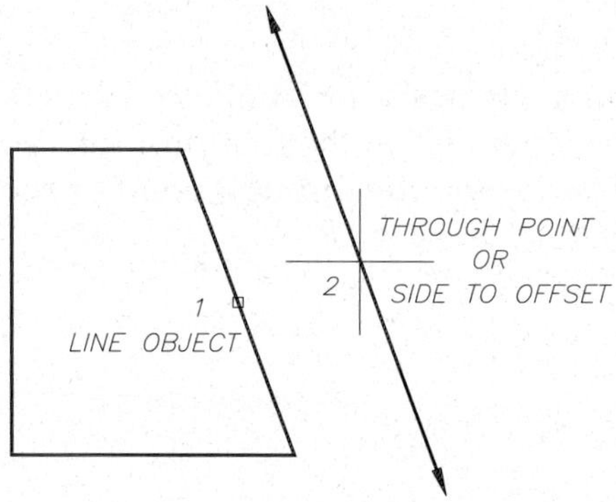


Using the *Through* option, select a line (*Line, Xline, Ray,* or *Pline*); then specify a point for the *Xline* to pass through. In each case, the anchor point of the *Xline* is the "root." (See "*Offset,*" Chapter 9.)

## Ray

| Ribbon | Menu Bar | Command (Type) | Alias (Type) | Shortcut |
|---|---|---|---|---|
| *Home Draw* (icon) | *Draw Ray* | *Ray* | ... | ... |

A *Ray* is also a construction line (see *Xline*), but it extends to infinity in only one direction and has one "anchored" endpoint. Like an *Xline,* a *Ray* extends past the drawing area but does not affect the drawing *Limits* or *Extents*. The construction process for a *Ray* is simpler than for an *Xline,* only requiring you to establish a "start point" (endpoint) and a "through point" (Fig. 15-7). Multiple *Rays* can be created in one command.

FIGURE 15-7

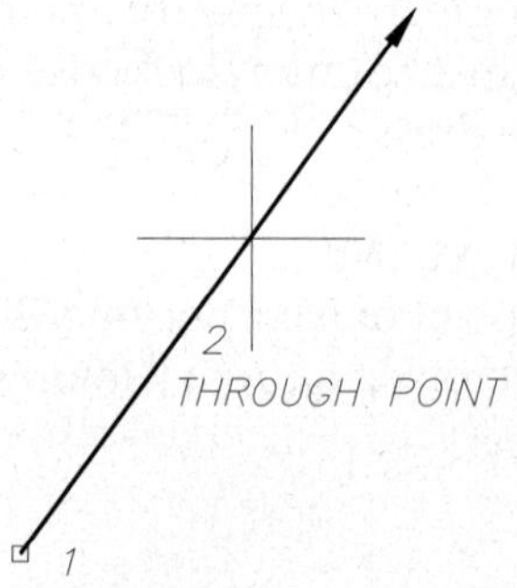

Command: ***RAY***
Specify start point: **PICK** or (**coordinates**)
Specify through point: **PICK** or (**coordinates**)
Specify through point: **PICK** or (**coordinates**)
Specify through point: **Enter**
Command:

*Rays* are especially helpful for construction of geometry about a central reference point or for construction of angular features. In each case, the geometry is usually constructed in only one direction from the center or vertex. If horizontal or vertical *Rays* are needed, just toggle on *Ortho*. To draw a *Ray* at a specific angle, use relative polar coordinates or *Polar Tracking*. *Endpoint* and other appropriate *Object Snaps* can be used with *Rays*. A *Ray* has one *Endpoint* but no *Midpoint*.

## *Polygon*

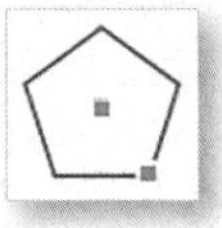

| Ribbon | Menu Bar | Command (Type) | Alias (Type) | Shortcut |
|---|---|---|---|---|
| *Home Draw Polygon* | *Draw Polygon* | *Polygon* | *POL* | ... |

The *Polygon* command creates a regular polygon (all angles are equal and all sides have equal length). A *Polygon* object appears to be several individual objects but, like a *Pline,* is actually one object. In fact, AutoCAD uses *Pline* to create a *Polygon*. There are two basic options for creating *Polygons*: you can specify an *Edge* (length of one side) or specify the size of an imaginary circle for the *Polygon* to *Inscribe* or *Circumscribe*.

### *Inscribe/Circumscribe*

The command sequence for this default method follows:

Command: ***POLYGON***
Enter number of sides <4>: (**value**)
Specify center of polygon or [Edge]: **PICK** or (**coordinates**)
Enter an option [Inscribed in circle/Circumscribed about circle] <I>: ***I*** or ***C***
Specify radius of circle: **PICK** or (**coordinates**)
Command:

**FIGURE 15-8**

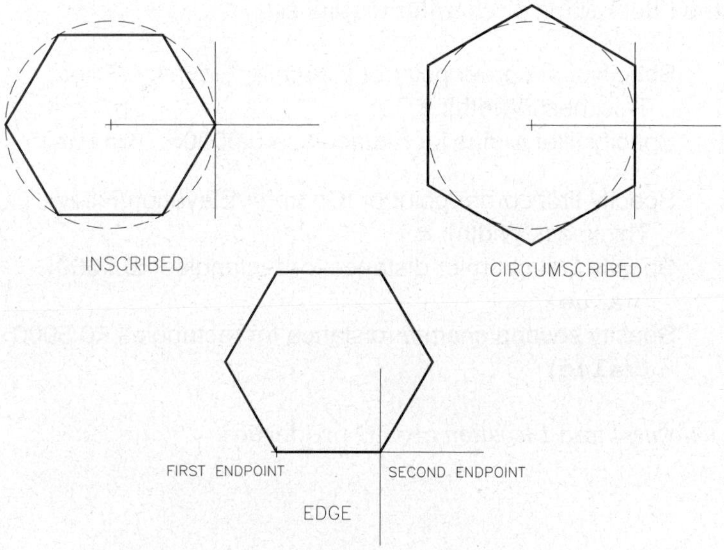


TIP

The orientation of the *Polygon* and the imaginary circle are shown in Figure 15-8. Note that the *Inscribed* option allows control of one-half of the distance across the corners, and the *circumscribed* option allows control of one-half of the distance across the flats.

Using *Ortho ON* with specification of the *radius of circle* forces the *Polygon* to a 90 degree orientation.

### *Edge*

The *Edge* option only requires you to indicate the number of sides desired and to specify the two endpoints of one edge (Fig. 15-8).

Command: ***POLYGON***
Enter number of sides <4>: (**value**)
Specify center of polygon or [Edge]: ***e***
Specify first endpoint of edge: **PICK** or (**coordinates**)
Specify second endpoint of edge: **PICK** or (**coordinates**)
Command:

Because *Polygons* are created as *Plines*, *Pedit* can be used to change the line width or edit the shape in some way (see "*Pedit*," Chapter 16). *Polygons* can also be *Exploded* into individual objects similar to the way other *Plines* can be broken down into component objects (see "*Explode*," Chapter 16).

## Rectang

| Ribbon | Menu Bar | Command (Type) | Alias (Type) | Shortcut |
|---|---|---|---|---|
| *Home*<br>*Draw*<br>*Rectangle* | *Draw*<br>*Rectangle* | *Rectang* | *REC* | ... |

The *Rectang* command only requires the specification of two diagonal corners for construction of a rectangle, identical to making a selection window (Fig. 15-9). The corners can be PICKed or dimensions can be entered. The resulting rectangle is actually a *Pline* object. Therefore, a rectangle is one object, not four separate objects.

Specify first corner point or [Chamfer/Elevation/Fillet/
  Thickness/Width]: **PICK, (coordinates)** or option
Specify other corner point or [Area/Dimensions/Rotation]:
  **PICK, (coordinates)** or option

**FIGURE 15-9**

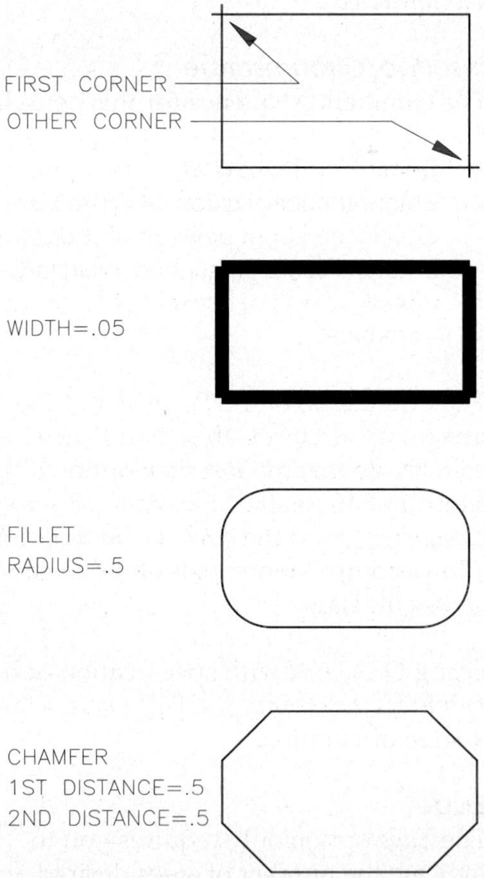


Several options of the *Rectangle* command affect the shape as if it were a *Pline* (see *Pline* and *Pedit*). For example, a *Rectangle* can have *width:*

Specify first corner point or [Chamfer/Elevation/Fillet/
  Thickness/Width]: ***w***
Specify line width for rectangles <0.0000>: (**value**)

The *Fillet* and *Chamfer* options allow you to specify values for the fillet radius or chamfer distances:

Specify first corner point or [Chamfer/Elevation/Fillet/
  Thickness/Width]: ***f***
Specify fillet radius for rectangles <0.0000>: (**value**)

Specify first corner point or [Chamfer/Elevation/Fillet/
  Thickness/Width]: ***c***
Specify first chamfer distance for rectangles <0.0000>:
  (**value**)
Specify second chamfer distance for rectangles <0.5000>:
  (**value**)

*Thickness* and *Elevation* are 3D properties.

After specifying the first corner of the rectangle, you can enter the other corner or use these options (Fig. 15-10). The *Dimensions* option prompts for the length and width rather than picking two diagonal corners.

Specify other corner point or [Area/Dimensions/Rotation]: ***d***
Specify length for rectangles <4.0000>: **(value)**
Specify width for rectangles <2.5000>: **(value)**

The *Area* option prompts for an area value and either the length or width to use for calculating the area.

Enter area of rectangle in current units <100.0000>: **(value)**
Calculate rectangle dimensions based on [Length/Width] <Length>: ***l***
Enter rectangle length <10.0000>: **(value)**

The *Rotation* option allows you to specify a rotational value for the orientation of the *Rectangle*.

Specify other corner point or [Area/Dimensions/Rotation]: **r**
Specify rotation angle or [Pick points] <0>: **(value)**

**FIGURE 15-10**

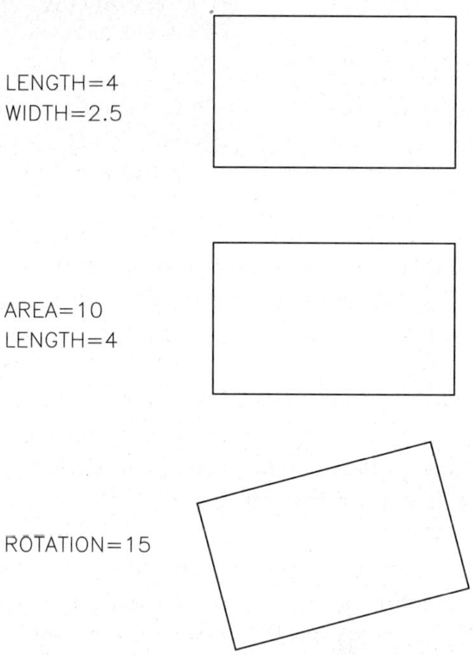


## Donut

| Ribbon | Menu Bar | Command (Type) | Alias (Type) | Shortcut |
|---|---|---|---|---|
| *Home Draw* (icon) | *Draw Donut* | *Donut* | *DO* | ... |

A *Donut* is a circle with width (Fig. 15-11). Invoking the command allows changing the inside and outside diameters and creating multiple *Donuts*:

Command: ***DONUT***
Specify inside diameter of donut <0.5000>: (**value**)
Specify outside diameter of donut <1.0000>: (**value**)
Specify center of donut or <exit>: **PICK**
Specify center of donut or <exit>: **Enter**

**FIGURE 15-11**

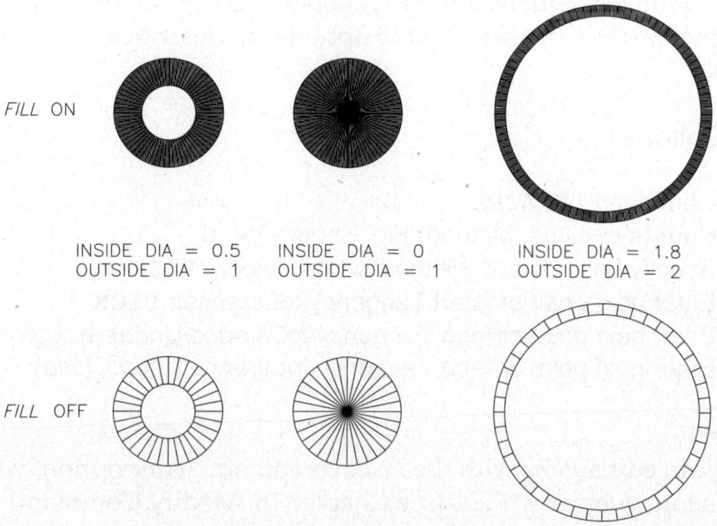


*Donuts* are actually solid filled circular *Plines* with width. The solid fill for *Donuts*, *Plines*, and other "solid" objects can be turned off with the *Fill* command or *FILLMODE* system variable.

## Spline

| Ribbon | Menu Bar | Command (Type) | Alias (Type) | Shortcut |
|---|---|---|---|---|
| *Home Draw* (icon) | *Draw Spline* | *Spline* | *SPL* | ... |

The *Spline* command creates a NURBS (non-uniform rational bezier spline) curve. The non-uniform feature allows irregular spacing of selected points to achieve sharp corners, for example. A *Spline* can also be used to create regular (rational) shapes such as arcs, circles, and ellipses. Irregular shapes can be combined with regular curves, all in one spline curve definition.

All *Splines* have multiple points that control the shape of the curve. These points are used to create a new *Spline* and can be used to edit, or change, the shape of the curve of an existing *Spline*. These points (also called vertices) can pass directly through the *Spline* curve or can fall outside the curve so as to have a "pull" on the curve.

Therefore, there are two types, or "methods," for creating and editing *Splines*: *Fit* and *Control Vertices*. A *Fit Spline* generally passes exactly through the points that are used to create the *Spline* (unless a *Tolerance* value is applied), whereas a *Control Vertices Spline* has points outside the curve that control the shape. Use the *Method* option to specify *Fit* or *CV*.

Enter spline creation method [Fit/CV] <Fit>:

**FIGURE 15-12**

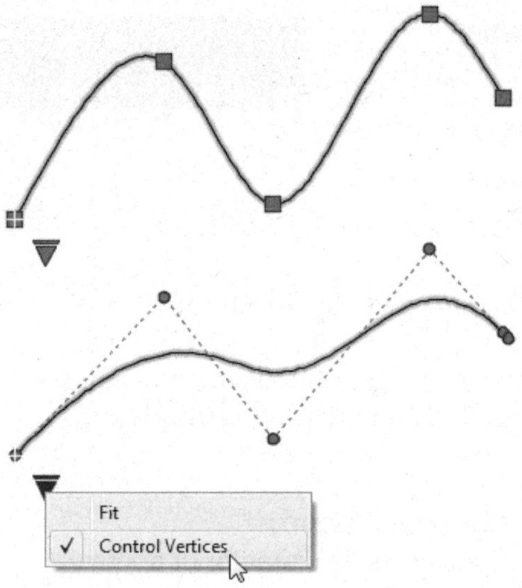


Figure 15-12 shows two *Splines* with grips highlighted—one *Spline* created with the *Fit* method (top) and one *Spline* created with the *Control Vertices* (*CV*) method (bottom). Both *Splines* were created by PICKing the same (relative) location of points (at the location of the grips). Notice how the *Fit* method passes directly through the selected points, whereas the *CV* method falls away from the points.

The command options are different based on which of the two *Methods* are set. All the options are described next.

### *Fit* Options

Command: ***SPLINE***
Current settings: Method=Fit Knots=Chord
Specify first point or [Method/Knots/Object]: **PICK**
Enter next point or [start Tangency/toLerance]: **PICK**
Enter next point or [end Tangency/toLerance/Undo]: **PICK**
Enter next point or [end Tangency/toLerance/Undo/Close]:

*Object*

You can edit a *Pline* with the *Pedit* command, *Spline* option, which changes multiple straight *Pline* sections into one curvy *Pline* (see Chapter 16, Modify Commands II). Use the *Object* option of *Spline* to convert a Spline-fit *Pline* into a true *Spline*.

*Knots*
The *Knots* options specify the computational method used to blend the curve between fit points. The *Square root* method produces a gentler or straighter curve. The *Uniform* method produces the most radical curve, sometimes causing the curves to overshoot the fit points. The *Chord-Length* method creates curve smoothness between the *Square root* and *Uniform* methods.

*Start Tangency/End Tangency*
Use these options if you wish to create a curve with either of the end points tangent to a line segment, similar to adding an additional invisible fit point. Normally (without this option), the blend of the curve is generated based only on the specified points that fall on the curve.

*Tolerance*
A *Tolerance* applied to the *Spline* "loosens" the fit of the curve. A tolerance of 0 (default) causes the *Spline* to pass exactly through the fit points as shown by the dotted *Spline* in Figure 15-13. Entering a positive value allows the curve to fall away from the points to form a smoother curve as shown by the dark *Spline* in Figure 15-13. The tolerance value does not affect the starting and ending points of the curve, which always connect to the fit points. This option for *Fit Splines* is similar to the *Degree* option for *CV Splines*.

**FIGURE 15-13**

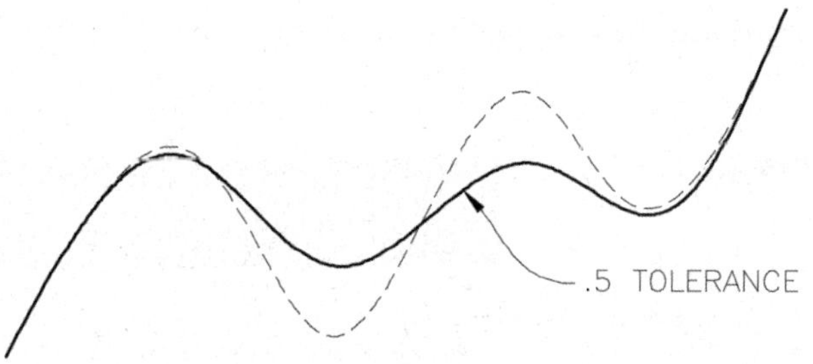


*Close*
The *Close* option allows creation of closed *Splines* where the start and end points are automatically connected. Therefore, these *Splines* can be regular curves if the selected points are symmetrically arranged.

### *Control Vertices* Options

Command: ***SPLINE***
Current settings: Method=CV Degree=4
Specify first point or [Method/Degree/Object]:
Enter next point:

*Degree*
Use this option to create splines of (polynomial) degree 1 (linear), degree 2 (quadratic), degree 3 (cubic), and so on up to degree 10. The higher the value, the smoother (straighter) the curve and the farther it falls from the control vertices. Figure 15-15 illustrates several curves with *Degree* values applied. This option for *CV Splines* is similar to the *Tolerance* option for *Fit Splines*.

**FIGURE 15-14**

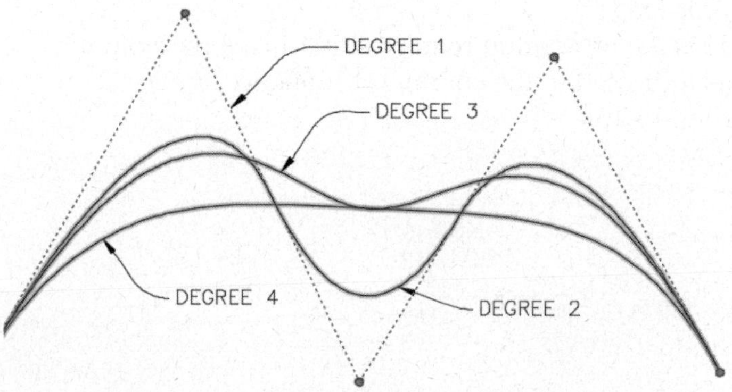

### Editing Splines Using Grips

One of the most valuable characteristics of *Splines* is the ability to easily change the shape of the curve using grips. Note that when a *Spline* is selected, a grip appears at each control point (as shown in Figure 15-15). Therefore, you can select any grip and stretch the location to affect the shape of the curve at that location. Hovering over a grip produces a menu with other options. Using this menu you can convert a *Fit Spline* to a *Control Vertices Spline* and vice versa (see Figure 15-12). Grips are discussed in Chapter 19.

**FIGURE 15-15**

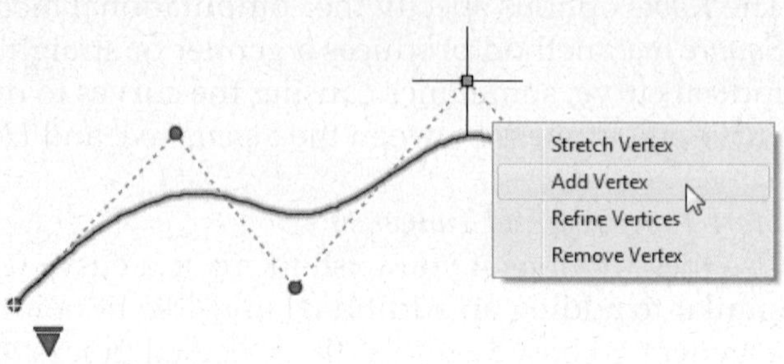


Additionally, you can use the *Splinedit* command to edit the curves. *Splinedit* offers many options for changing the Spline characteristics. See Chapter 16, Modify Commands II.

## *Ellipse*

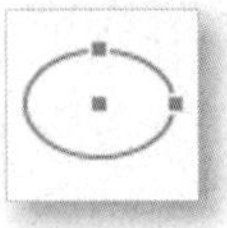

| Ribbon | Menu Bar | Command (Type) | Alias (Type) | Shortcut |
|---|---|---|---|---|
| *Home* *Draw* (icon) | *Draw* *Ellipse* | *Ellipse* | *EL* | ... |

An *Ellipse* is one object. AutoCAD *Ellipses* are (by default) NURBS curves (see *Spline*). There are three methods of creating *Ellipse*s in AutoCAD: (1) specify one axis and the end of the second, (2) specify the center and the ends of each axis, and (3) create an elliptical arc. Each option also permits supplying a rotation angle rather than the second axis length.

Command: ***ELLIPSE***
Specify axis endpoint of ellipse or [Arc/Center]: **PICK** or (**coordinates**)
(This is the first endpoint of either the major or minor axis.)
Specify other endpoint of axis: **PICK** or (**coordinates**)
(Select a point for the other endpoint of the first axis.)
Specify distance to other axis or [Rotation]: **PICK** or (**coordinates**)
(This is the distance measured perpendicularly from the established axis.)
Command:

### *Axis End*

This default option requires PICKing three points as indicated in the command sequence above (Fig. 15-16).

**FIGURE 15-16**

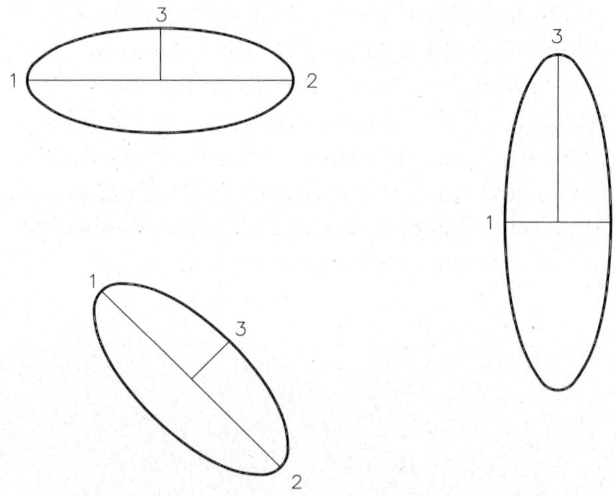

### *Rotation*

If the *Rotation* option is used with the *Axis End* method, the following syntax is used:

> Specify distance to other axis or [Rotation]: ***r***
> Specify rotation around major axis: **PICK** or (**value**)

The specified angle is the number of degrees the shape is rotated <u>from the circular position</u> (Fig. 15-17).

**FIGURE 15-17**

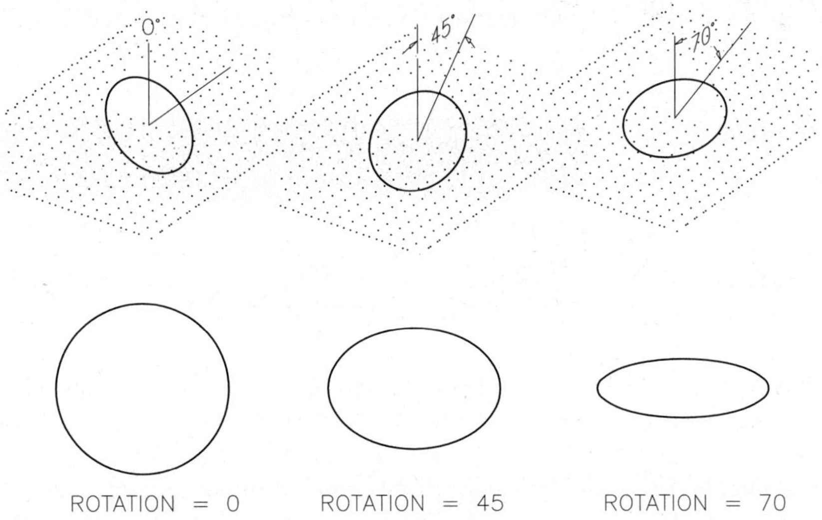


### *Center*

TIP

With many practical applications, the center point of the ellipse is known, and therefore the *Center* option should be used (Fig. 15-18):

> Command: ***ELLIPSE***
> Specify axis endpoint of ellipse or [Arc/Center]: ***c***
> Specify center of ellipse: **PICK** or (**coordinates**)
> Specify endpoint of axis: **PICK** or (**coordinates**)
> (Select a point for the endpoint of the first axis.)
> Specify distance to other axis or [Rotation]: **PICK** or (**coordinates**) (This is the distance measured perpendicularly from the established axis.)
> Command:

**FIGURE 15-18**

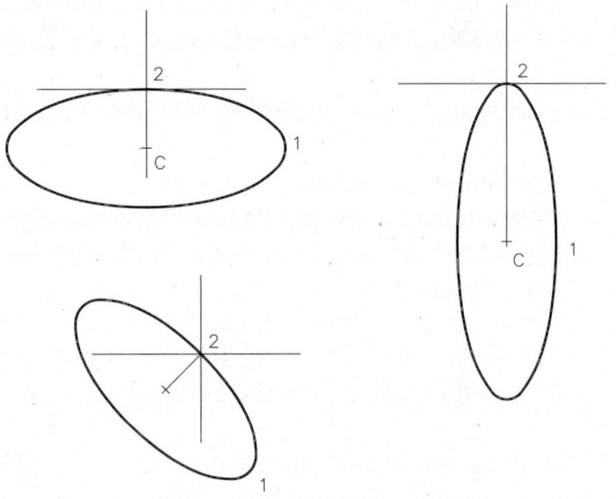


The *Rotation* option appears and can be invoked after specifying the *Center* and first *Axis endpoint*.

### *Arc*

Use this option to construct an elliptical arc (partial ellipse). The procedure is identical to the *Center* option with the addition of specifying the start- and endpoints for the arc (Fig. 15-19):

**FIGURE 15-19**

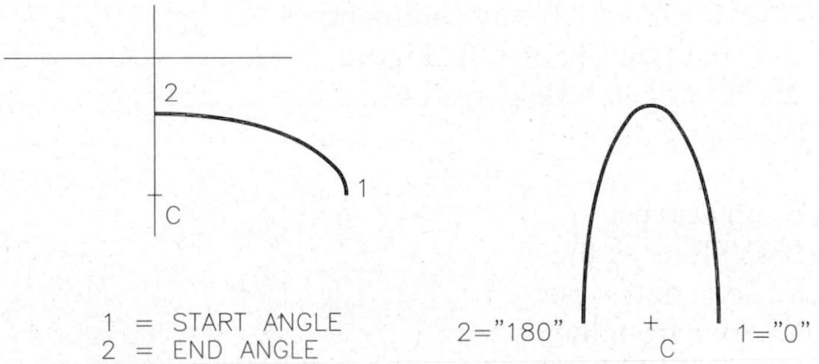


> Command: ***ELLIPSE***
> Specify axis endpoint of ellipse or [Arc/Center]: ***a***
> Specify axis endpoint of elliptical arc or [Center]: **PICK** or (**coordinates**)
> Specify other endpoint of axis: **PICK** or (**coordinates**)
> Specify distance to other axis or [Rotation]: **PICK** or (**coordinates**)
> Specify start angle or [Parameter]: **PICK** or (**angular value**)
> Specify end angle or [Parameter/Included angle]: **PICK** or (**angular value**)
> Command:

The *Parameter* option allows you to specify the start point and endpoint for the elliptical arc. The parameters are based on the parametric vector equation: p(u)=c+a*cos(u)+b*sin(u), where c is the center of the ellipse and a and b are the major and minor axes.

## Divide

| Ribbon | Menu Bar | Command (Type) | Alias (Type) | Shortcut |
|---|---|---|---|---|
| *Home Draw* (icon) | *Draw Point > Divide* | *Divide* | *DIV* | ... |

The *Divide* and *Measure* commands add *Point* objects to existing objects. Both commands are found in the *Draw* pull-down menu under *Point* because they create *Point* objects.

The *Divide* command finds equal intervals along an object such as a *Line, Pline, Spline,* or *Arc* and adds a *Point* object at each interval. The object being divided is not actually broken into parts—it remains as one object. *Point* objects are automatically added to display the "divisions."

The point objects that are added to the object can be used for subsequent construction by allowing you to *Object Snap* to equally spaced intervals (*Nodes*).

The command sequence for the *Divide* command is as follows:

Command: ***DIVIDE***
Select object to divide: **PICK** (Only one object can be selected.)
Enter the number of segments or [Block]: (**value**)
Command:

*Point* objects are added to divide the object selected into the desired number of parts. Therefore, there is one less *Point* added than the number of segments specified.

After using the *Divide* command, the *Point* objects may not be visible unless the point style is changed with the *Point Style* dialog box (*Format* pull-down menu) or by changing the *PDMODE* variable by command line format (see Chapter 8). Figure 15-20 shows *Points* displayed using a *PDMODE* of 3.

**FIGURE 15-20**

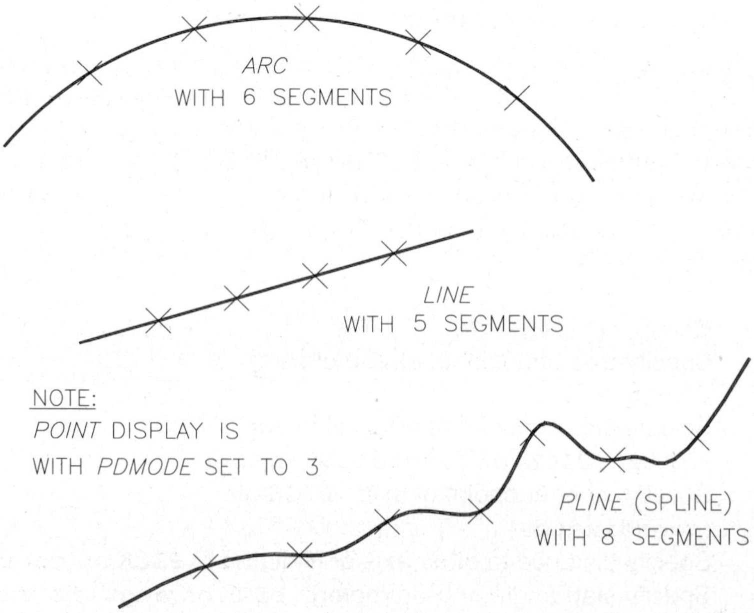


**TIP**

You can also use DDPTYPE to set the point style or choose Point Style from the Home tab, Utilities panel, Measure flyout.

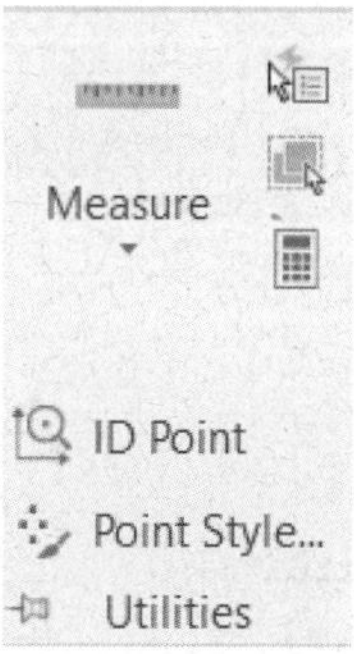

You can set the Divide command to use *Blocks* inserted rather than *Point* objects along equal divisions of the selected object. Figure 15-21 displays a generic rectangular-shaped block inserted with *Divide,* both aligned and not aligned with a *Line, Arc,* and *Pline*. In order to insert a *Block* using the *Divide* command, the name of an existing *Block* must be given. (See Chapter 21.)

**FIGURE 15-21**

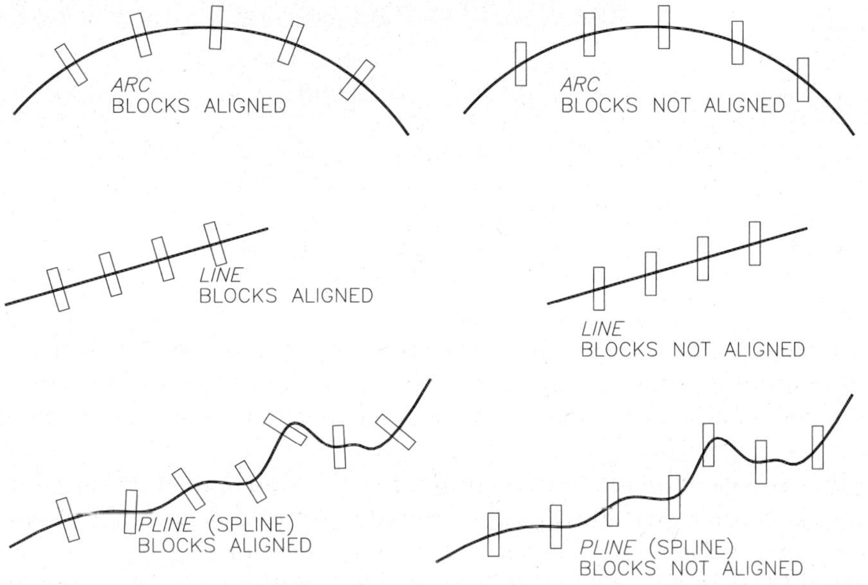


### *Measure*

| Ribbon | Menu Bar | Command (Type) | Alias (Type) | Shortcut |
|---|---|---|---|---|
| *Home Draw* (icon) | *Draw Point > Measure* | *Measure* | *ME* | ... |

The *Measure* command is similar to the *Divide* command in that *Point* objects (or *Blocks*) are inserted along the selected object. The *Measure* command, however, allows you to designate the length of segments rather than the number of segments as with the *Divide* command.

Command: ***MEASURE***
Select object to measure: **PICK** (Only one object can be selected.)
Specify length of segment or [Block]: (**value**) (Enter a value for length of one segment.)
Command:

*Point* objects are added to the selected object at the designated intervals (lengths). One *Point* is added for each interval beginning at the end nearest the end used for object selection (Fig. 15-22). The intervals are of equal length except possibly the last segment, which is whatever length is remaining.

**FIGURE 15-22**

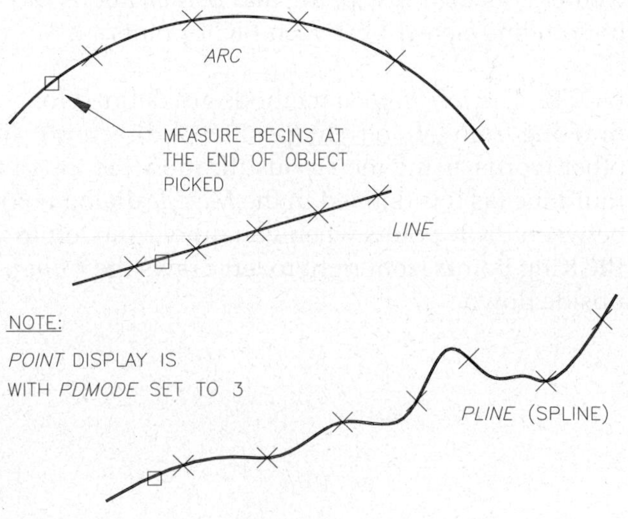


TIP

You can use the Block option so that *Blocks* be inserted rather than *Point* objects at the designated intervals of the selected object. Inserting *Blocks* with *Measure* requires that an existing *Block* be used.

## Mline

| Ribbon | Menu Bar | Command (Type) | Alias (Type) | Shortcut |
|---|---|---|---|---|
| ... | *Draw*<br>*Multiline* | *Mline* | *ML* | ... |

*Mline* is short for multiline. A multiline is a set of parallel lines that behave as one object. The individual lines in the set are called line elements and are defined to your specifications. The set can contain up to 16 individual lines, each having its own offset, linetype, or color.

*Mlines* are useful for creating drawings composed of many parallel lines. An architectural floor plan, for example, contains many parallel lines representing walls. Instead of constructing the parallel lines individually (symbolizing the wall thickness), only one *Mline* needs to be drawn to depict the wall. The thickness and other details of the wall are dictated by the definition of the *Mline*.

*Mline* is the command to draw multilines. To define the individual line elements of a logical set (offset, linetype, color), use the *Mlstyle* command. An *Mlstyle* (multiline style) is a particular combination of defined line elements. Each *Mlstyle* can be saved in an external file with a .MLN file type. In this way, multilines can be used in different drawings without having to redefine the set. See "*Mlstyle*" next. The *Mline* command operates similar to the *Line* command, asking you to specify points:

Command: ***MLINE***
Current settings: Justification = Top, Scale = 1.00, Style = STANDARD
Specify start point or [Justification/Scale/STyle]: **PICK** or (**coordinates**)
Specify next point: **PICK** or (**coordinates**)
Specify next point or [Undo]: **PICK** or (**coordinates**)
Specify next point or [Close/Undo]: **PICK** or (**coordinates**)
Specify next point or [Close/Undo]: **Enter**
Command:

*Mline* provides the following three options for drawing multilines.

### *Justification*

*Justification* determines how the multiline is drawn between the points you specify. The choices are *Top*, *Zero*, or *Bottom* (Fig. 15-23). *Top* aligns the top of the set of line elements along the points you PICK as you draw. A *Zero* setting uses the center of the set of line elements as the PICK point, and *Bottom* aligns the bottom line element between PICK points.

NOTE: The *Justification* methods are defined for drawing from left to right (positive X direction). In other words, using the *Top* justification, the top of the multiline (as it is defined in the *Mlstyle* dialog box) is between PICK points when you draw from left to right. PICKing points from right to left draws the *Mline* upside down.

**FIGURE 15-23**

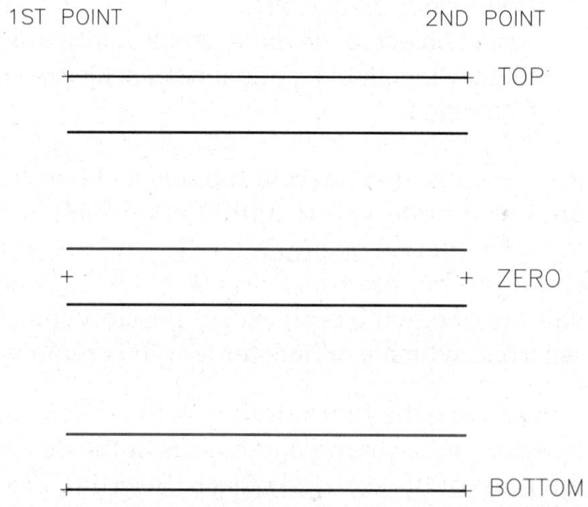

### *Scale*

The *Scale* option controls the overall width of the multiline. The value you specify for the scale factor increases or decreases the defined multiline width proportionately. In Figure 15-24, a scale factor of 2.0 provides a multiple line pattern that is twice as wide as the *Mlstyle* definition; a scale of 3.0 provides a pattern three times the width.

FIGURE 15-24

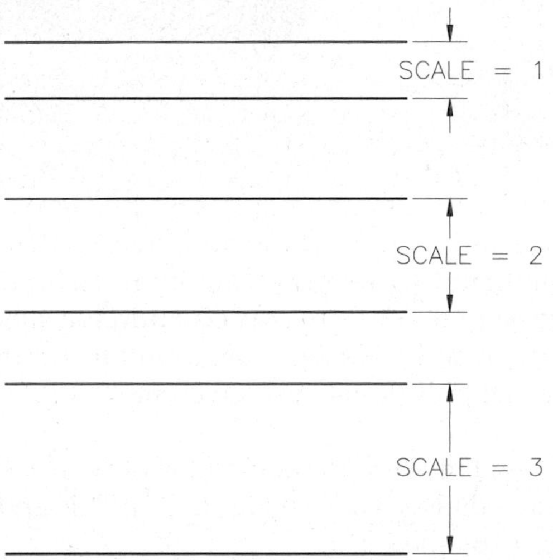


You can also enter negative scale values or a value of 0. A negative scale value flips the order of the multiple line pattern (Fig. 15-25). A value of 0 collapses the multiline into a single line. The utility of the *Scale* option allows you to vary a multiline set proportionally without the need for a complete new multiline definition.

FIGURE 15-25

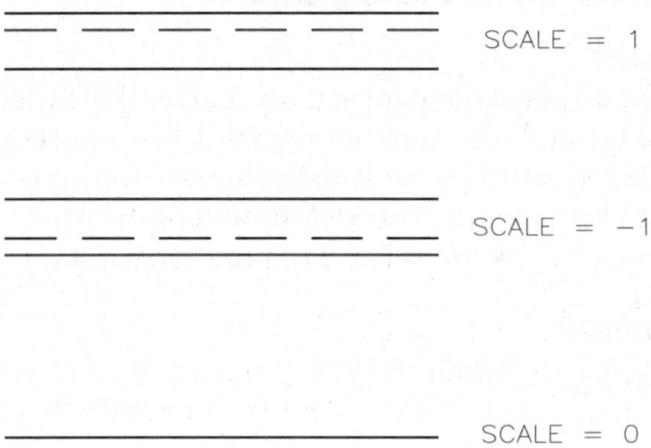


### *Style*

The *Style* option enables you to load different multiline styles provided they have been previously created. Some examples of multiline styles are shown in Figure 15-26. Only one multiline *Style* called STANDARD is provided by AutoCAD. The STANDARD multiline style has two parallel lines defined at 1 unit width.

Using the *Style* option of the *Mline* command provides only the Command line format for setting a current style. Instead, you can use the *Mlstyle* dialog box to select a current multiline style by a more visible method.

FIGURE 15-26

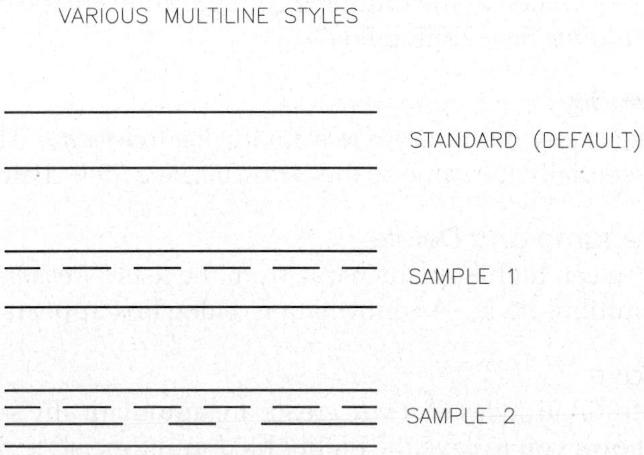

## *Mlstyle*

| Ribbon | Menu Bar | Command (Type) | Alias (Type) | Shortcut |
|---|---|---|---|---|
| ... | *Format* *Multiline Style...* | *Mlstyle* | ... | ... |

The *Mline* command draws multilines, whereas the *Mlstyle* controls the configuration of the multilines that are drawn. Use the *Mlstyle* command to create new multiline definitions or to load, select, and edit previously created multiline styles. *Mlstyle* is not a draw command and is therefore found not in the *Draw* menus but in the *Format* pull-down and screen menus.

Using the *Mlstyle* command by any method invokes the *Multiline Styles* dialog box (see Fig. 15-27). This and the related dialog boxes are explained next.

FIGURE 15-27

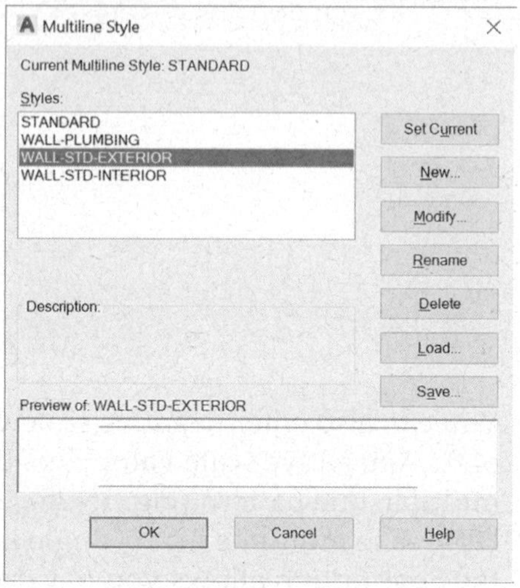


### *Multiline Style* Dialog Box

#### *Styles* List

This list displays the names of the currently loaded multiline styles. The image at the bottom of the dialog box is a representation of the highlighted multiline style definition so you can display each of the loaded multiline styles by selecting the names in the list. The *Style* option of the *Mline* command serves the same purpose, but operates only in command line format and requires you to know the specific *Mlstyle* name and definition.

#### *Set Current*

Highlight a multiline style from the list, then select *Set Current* to make it the current style. Select the *OK* button to return to the drawing to draw with the current style.

#### *New*

Select *New* to create a new multiline style and produce the *Create New Multiline Style* dialog box (not shown). You can select any loaded style to *Start with* as a template. Short names are acceptable since a *Description* of the multiline configuration can be entered in the dialog box that appears next (see "*New Multiline Style* Dialog Box").

#### *Modify*

Select any multiline style from the list to *Modify*. The *Modify Multiline* Style dialog box appears, which is essentially the same as the *New Multiline Style* dialog box (see "*New Multiline Style* Dialog Box").

#### *Rename and Delete*

You can highlight any name from the list to *Rename*. The *Delete* tile allows you to delete the highlighted multiline style. A confirmation dialog box appears.

#### *Save*

Multiline styles that you create are automatically saved with the drawing file. However, using *Save* allows you to save the highlighted multiline style definition to an external (.MLN) file so you can use that multiline style in other drawings without having to recreate the style. The *Save Multiline File* dialog box appears where you can specify the .MLN file name and location. Although multiple multiline styles can be saved to one .MLN file, only one style at a time can be saved. The default multiline definition file is ACAD.MLN, which contains only the STANDARD multiline style. This file can be appended with additional style definitions or a new .MLN file can be created. A .MLN file can contain more than one multiline style.

NOTE: You must choose the *OK* button in the *Multiline Style* dialog box to save any new multiline style definitions to the current drawing file.

### *Load*

The *Load* tile invokes the *Load Multiline Styles* dialog box where you can select one existing multiline style definition from an .MLN file (Fig. 15-28).

**FIGURE 15-28**

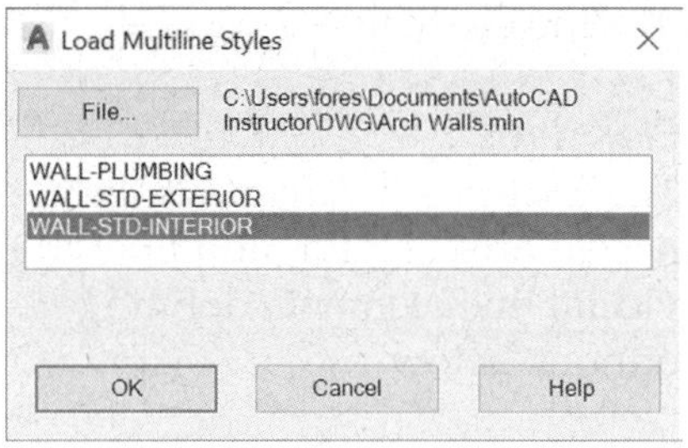


## ***Multiline Style* Dialog Box**

The *New Multiline Style* dialog box (Fig. 15-29) and the identical *Modify Multiline Style* dialog box allow you to specify the multiline style definition—the properties of the line elements contained in the style. Options specified in these dialog boxes are immediately displayed in the *Multiline Style* dialog box preview.

**FIGURE 15-29**

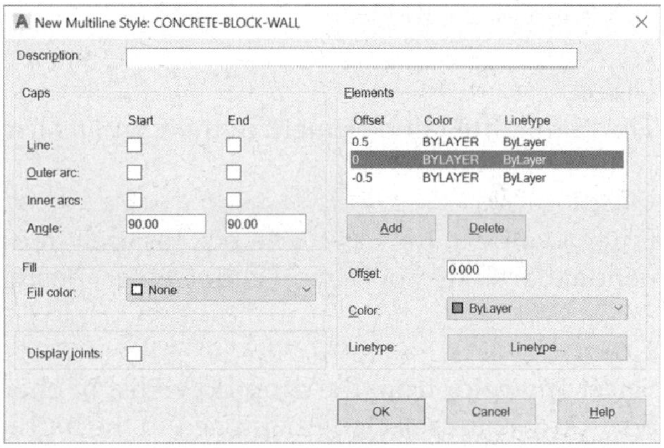


### *Description*

The *Description* edit box allows you to provide an optional description for a new or existing multiline style definition. The description can contain up to 256 characters.

### *Caps*

Use this section to specify how the ends of the multilines are "capped." A cap can be set for the *Start* and/or *End* segments of a multiline series (Fig. 15-30).

**FIGURE 15-30**

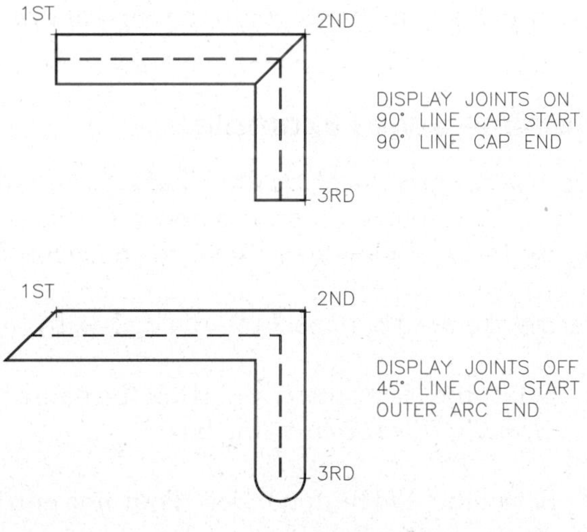


The *Line* checkboxes create straight line caps across either or both of the *Start* or *End* segments of the multiline. The *Line* caps are drawn across all line elements (Fig. 15-30).

The *Outer Arcs* option creates an arc connecting and tangent to the two outermost line elements. You can control both the *Start* and *End* segments.

Selecting *Inner Arcs* creates an arc between interior line segments. This option works with four or more line elements. In the case of an uneven number of line elements (five or more), the middle line is not capped.

The *Angle* option controls the angle of start and end lines or arcs. Valid values range from 10 to 170 degrees.

### *Fill*

Selecting a *Fill Color* produces a multiline with color filled between the outermost lines. Although the image preview does not display the multiline fill, the color fill appears in the drawing.

### *Display Joints*

When several segments of an *Mline* are drawn, AutoCAD automatically miters the "joints" so the segments connect with sharp, even corners. The "joint" is the bisector of the angle created by two adjoining segments. The *Display Joints* checkbox toggles on and off the display of this miter line running across all the line elements.

### *Elements*

The *Elements* list displays the current settings for the multiline style. Each line entry (element) in a style has an offset, color, and linetype associated with it. A multiline definition can have as many as 16 individual line elements defining it. Select any element in the list before changing its properties with the options below.

### *Add*

New line elements are added to the list by selecting the *Add* button; then the *Offset*, *Color*, and *Linetype* properties can be assigned. The new element appears in the list with an offset of 0, color set to *Bylayer*, and linetype set to *Bylayer*.

### *Delete*

The highlighted line element in the *Elements* list is deleted by selecting the *Delete* button.

### *Offset*

Enter a value in the *Offset* edit box for each line element. The offset value is the distance measured perpendicular to the theoretical zero (center) position along the axis of the *Mline*.

### *Color*

Select any color from the drop-down list or choose *Select Color* to produce the standard *Select Color* dialog box. Any color selected is assigned to the highlighted line element in the list.

### *Linetype*

Use this button to assign a linetype to the highlighted line element. The standard *Select Linetype* dialog box appears. Linetypes not in the current drawing file can be *Loaded*.

## *Multiline Styles* Example

Because the process of creating and editing multiline styles is somewhat involved, here is an example that you can follow to create a new multiline style. The specifications for two multilines are given in Figure 15-31. Follow these steps to create SAMPLE1.

1. Invoke the *Multiline Styles* dialog box by any menu or icon or by typing *Mlstyle*.

2. Select the *New* button and enter the name SAMPLE1 in the *Name* edit box. Select *OK* to invoke the *New Multiline Style* dialog box.

3. Select the *Add* button twice. Four line entries should appear in the *Elements* list box. The two new entries have a 0.0 offset.

4. Highlight one of the new entries and change its offset by entering 0.2 in the *Offset* edit box. Next, change the linetype by selecting the *Linetype* button and choosing the *Center* linetype. (If the *Center* linetype is not listed, you should *Load* it.)

5. Highlight the other new entry and change its *Offset* to -0.3. Next, change its *Linetype* to *Hidden*.

6. Select the *OK* button to return to the *Multiline Styles* dialog box. Highlight the new SAMPLE1 style from the list and select *Set Current*. Last, choose the *OK* button to save the multiline style to the drawing.

7. Use *Mline* to draw several segments and to test your new multiline style.

8. Use *Mlstyle* again and select *Modify* to create Line end caps and a *Color Fill* to the style.

Now try to create SAMPLE2 multiline style on your own (see Fig. 15-31).

FIGURE 15-31

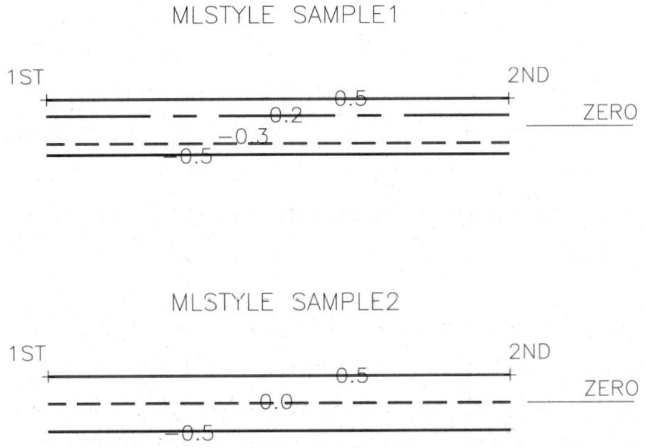


## Editing Multilines

Because a multiline is treated as one object and its complex configuration is defined by the multiline style, a special set of editing functions is designed for *Mlines*. The *Mledit* (multiline edit) command provides the editing functions for existing multilines and is discussed in Chapter 16. Only a few other traditional editing commands such as *Trim* and *Extend* can be used for editing functions. Your usefulness for *Mlines* is limited until you can edit them with *Mledit*.

## *Sketch*

| Ribbon | Menu Bar | Command (Type) | Alias (Type) | Shortcut |
| --- | --- | --- | --- | --- |
| ... | ... | *Sketch* | ... | ... |

The *Sketch* command is unlike other draw commands. *Sketch* quickly creates many short line segments (individual objects) by following the motions of the cursor. *Sketch* is used to give the appearance of a freehand "sketched" line, as used in Figure 15-32 for the tree and bushes. You do not specify individual endpoints, but rather draw a freehand line by placing the "pen" down, moving the cursor, and then picking up the "pen." This action creates a large number of short *Line* segments. *Sketch* line segments are temporary (displayed in a different color) until *Recorded* by exiting the *Sketch* command.

FIGURE 15-32

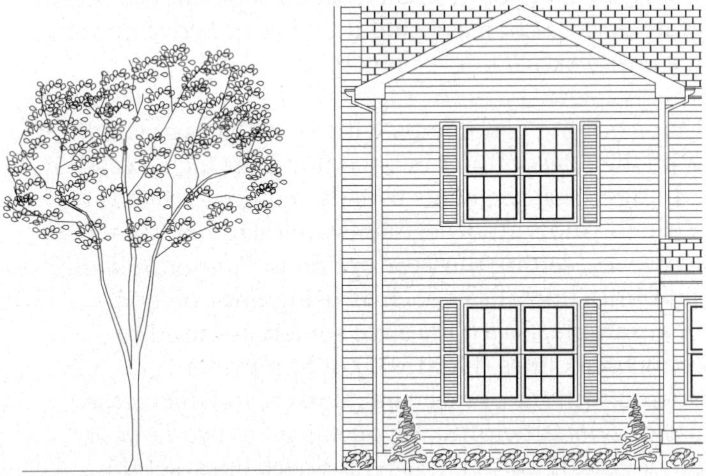

CAUTION: *Sketch* can increase the drawing file size greatly due to the relatively large number of line segment endpoints required to define the *Sketch* line.

TIP

```
Command: SKETCH
Type = Lines  Increment = 0.1000  Tolerance = 0.5000
Specify sketch or [Type/Increment/toLerance]:
Specify sketch: PICK and drag
36 lines recorded.
```

The options for the *Sketch* commands are explained next.

FIGURE 15-33

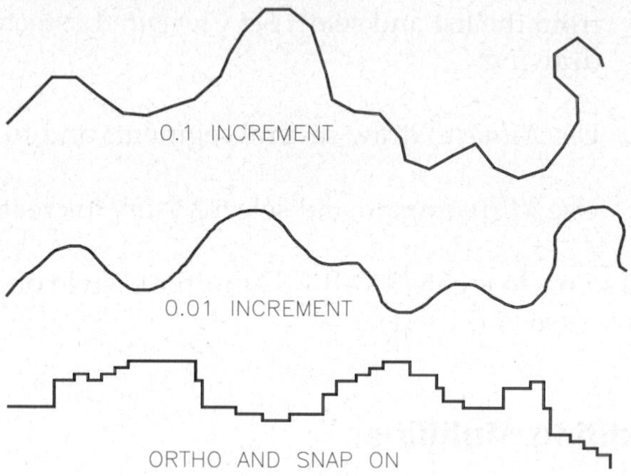


### *Increment*

It is important to specify an *Increment* length for the short line segments that are created. The *Increment* controls the "resolution" of the *Sketch* line (Fig. 15-33). Too large of an *Increment* makes the straight line segments apparent, while too small of an *Increment* unnecessarily increases file size.

The default *Increment* is 0.1 (appropriate for default *Limits* of 12 x 9) and should be changed proportionally with a change in *Limits*. Generally, multiply the default *Increment* of 0.1 times the drawing scale factor. The *Sketch Increment* is stored in the *SKETCHINC* system variable.

Another important rule to consider whenever using *Sketch* is to turn *Snap Off* and *Ortho Off*, unless a "stair-step" effect is desired (see Fig. 15-33). *Polar Snap* and *Polar Tracking* do not affect *Sketch* lines.

FIGURE 15-34

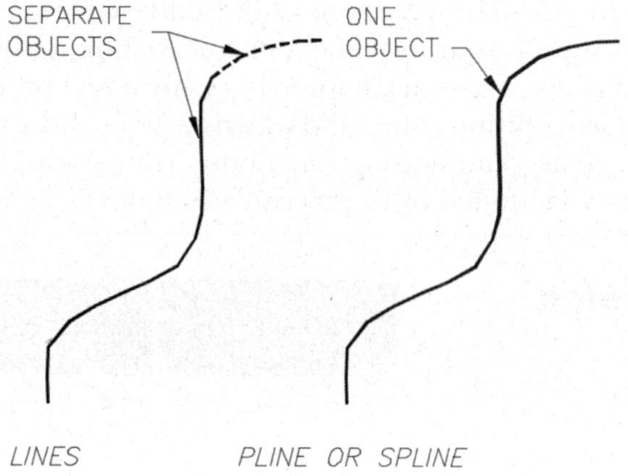


### *Type*

Using the *Type* option, AutoCAD allows you to choose if you want the sketch segments to be *Lines*, *Plines*, or *Splines*.

```
Specify sketch or [Type/Increment/toLerance]: t
Enter sketch type [Lines/Polyline/Spline] <Lines>:
```

Using the *Lines* option, each *Sketch* segment is a separate object, whereas using the *Pline* or *Spline* option, the entire *Sketch* line is one object.

FIGURE 15-35

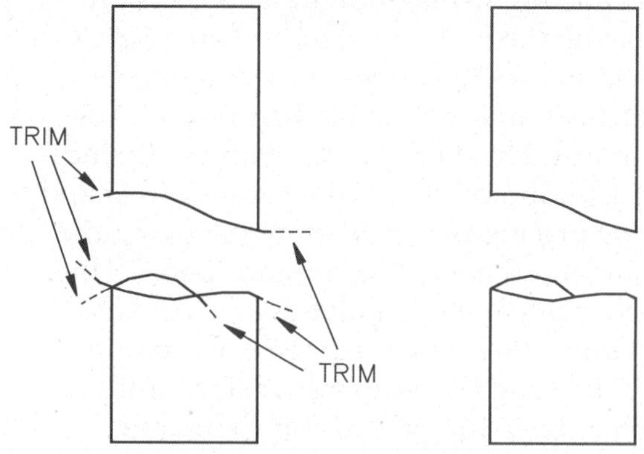


Using editing commands with *Sketch* lines created with the *Line* option can be tedious. In such a case, editing *Sketch* segments usually requires *Zooming* in since the individual segments are relatively small. However, setting the *Type* option to *Pline* or *Spline* simplifies operations such as using *Erase* or *Trim*. For example, *Sketch* lines are sometimes used to draw break lines (Fig. 15-35) or to represent broken sections of mechanical parts, in which case use of *Trim* is helpful. If you expect to use *Trim* or other editing commands with *Sketch* lines, set the *Type* option to *Pline* or *Spline* before creating the *Sketch* lines.

TIP

Using the *Type* option changes the *SKPOLY* system variable. You can type the *SKPOLY* variable at the Command line to change the *Type* of sketch objects. The settings are as follows.

| | |
|---|---|
| *SKPOLY* = 0 | Sketch creates individual *Line* segments |
| *SKPOLY* = 1 | Sketch creates all segments as one *Pline* |
| *SKPOLY* = 2 | Sketch creates all segments as one *Spline* |

*Tolerance*
This setting affects only *Sketch* lines drawn using the *Spline* type. For *Splines*, the setting specifies how closely the spline's curve fits to the freehand sketch. In other words, setting a higher *Tolerance* value has the effect of "smoothing out" the *Sketch* line. For example, Figure 15-36 displays two *Sketch* lines where each one was drawn from one point to the next (designated by Xs); however, one *Sketch* was created with a *Tolerance* of 0 (zero) and the other was created with a *Tolerance* of 1. The valid *Tolerance* range is between 0 and 1. The setting is saved in the *SKTOLERANCE* system variable.

**FIGURE 15-36**

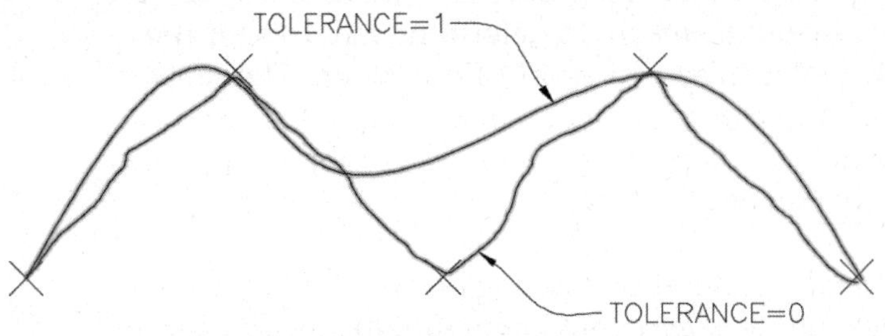


## *Solid*

| Ribbon | Menu Bar | Command (Type) | Alias (Type) | Shortcut |
|---|---|---|---|---|
| ... | ... | *Solid* | *SO* | ... |

The *Solid* command creates a 2D shape that is filled solid with the current color (object or *ByLayer*). The shape is not a 3D solid. Each 2D solid has either three or four straight edges. The construction process is relatively simple, but the order of PICKing corners is critical. In order to construct a square or rectangle, draw a bow-tie configuration (Fig. 15-37). Several *Solids* can be connected to produce more complex shapes.

**FIGURE 15-37**

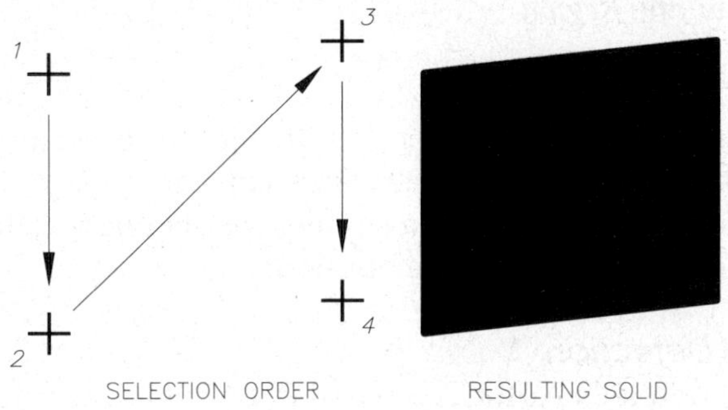


A *Solid* object is useful when you need shapes filled with solid color; for example, three-sided *Solids* can be used for large arrowheads. Keep in mind that solid filled areas may cause problems for some output devices such as plotters with felt pens and thin paper. As an alternative to using *Solid*, enclosed areas can be filled with hatch patterns or solid color (see "*Hatch*," Chapter 26).

## *Boundary*

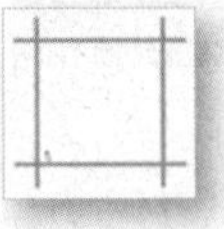

| Ribbon | Menu Bar | Command (Type) | Alias (Type) | Shortcut |
|---|---|---|---|---|
| *Home Draw Boundary* | *Draw Boundary...* | *Boundary* or *-Boundary* | *BO* or *-BO* | ... |

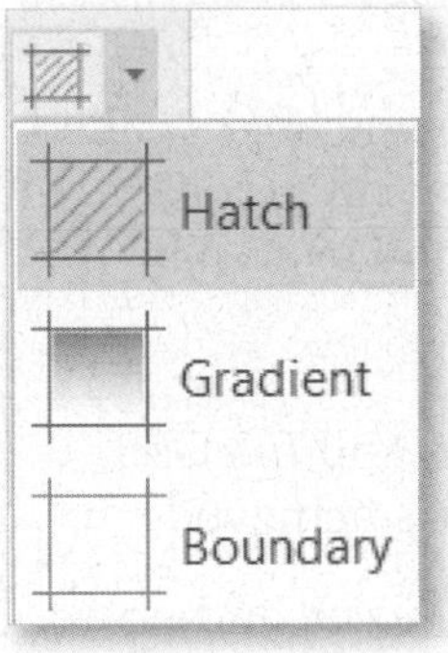


*Boundary* finds and draws a boundary from a group of connected or overlapping shapes forming an enclosed area. The shapes can be any AutoCAD objects in any configuration, as long as they form a totally enclosed area. *Boundary* creates either a *Polyline* or *Region* object as the boundary shape (see *Region* next). The resulting *Boundary* does not affect the existing objects in any way. *Boundary* finds and includes internal closed areas (called "islands") such as circles (holes) and includes them as part of the *Boundary*.

To create a *Boundary*, select an internal point in any enclosed area (Fig. 15-38). *Boundary* finds the boundary (complete enclosed shape) surrounding the internal point selected. You can use the resulting *Boundary* to construct other geometry or use the generated *Boundary* to determine the *Area* for the shape (such as a room in a floor plan). The same technique of selecting an internal point is also used to determine boundaries for sectioning using *Hatch* (Chapter 26).

**FIGURE 15-38**

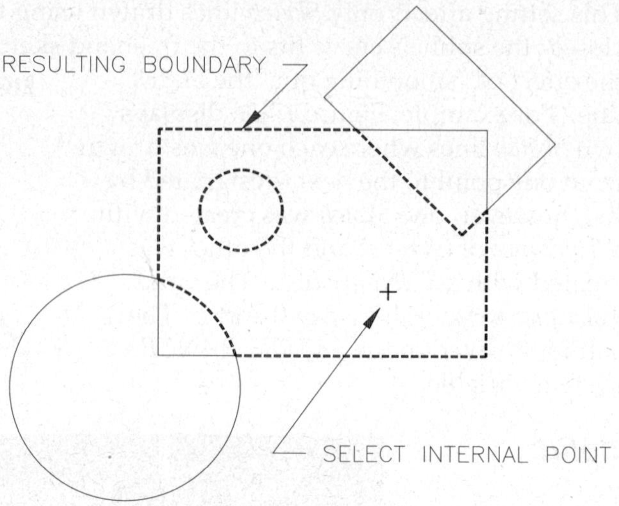


*Boundary* operates in dialog box mode (Fig. 15-39), or type *-Boundary* for Command line mode. After setting the desired options, select the *Pick Points* tile to select the desired internal point (as shown in Fig. 15-38).

### *Object Type*

Construct either a *Region* or *Polyline* boundary (see *Region* next). If islands are found, two or more *Plines* are formed but only one *Region*.

### *Boundary Set*

The *Current Viewport* option is sufficient for most applications; however, for large drawings you can make a new boundary set by selecting a smaller set of objects to be considered for the possible boundary.

### *Island Detection*

This option tells AutoCAD whether or not to include interior enclosed objects in the Boundary.

### *Pick Points*

Use this option to PICK an enclosed area in the drawing that is anywhere <u>inside</u> the desired boundary.

**FIGURE 15-39**

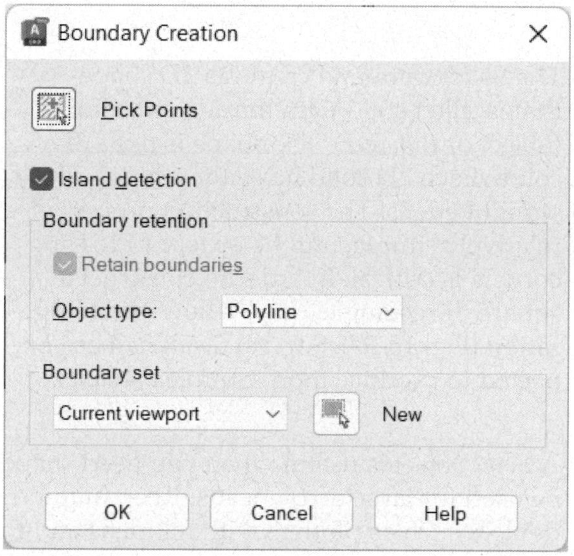


## Region

| Ribbon | Menu Bar | Command (Type) | Alias (Type) | Shortcut |
|---|---|---|---|---|
| *Home Draw* (icon) | *Draw Region* | *Region* | *REG* | ... |

The *Region* command converts one object or a set of objects forming a closed shape into one object called a *Region*. This is similar to the way in which a set of objects (*Lines, Arcs, Plines*, etc.) forming a closed shape and having matching endpoints can be converted into a closed *Pline*. A *Region*, however, has special properties:

1. Several *Regions* can be combined with Boolean operations known as *Union, Subtract*, and *Intersect* to form a "composite" *Region*. This process can be repeated until the desired shape is achieved.
2. A *Region* is considered a planar surface. The surface is defined by the edges of the *Region* and no edges can exist within the *Region* perimeter. *Regions* can be used with surface modeling.

In order to create a *Region*, a closed shape must exist. The shape can be composed of one or more objects such as a *Line, Arc, Pline, Circle, Ellipse,* or anything composed of a *Pline* (*Polygon, Rectangle, Boundary*). If more than one object is involved, the endpoints must match (having no gaps or overlaps). Simply invoke the *Region* command and select all objects to be converted to the *Region*.

Command: ***REGION***
Select objects: **PICK**
Select objects: **PICK**
Select objects: **PICK**
Select objects: **Enter**
1 loop extracted.
1 Region created.
Command:

Consider the shape shown in Figure 15-40 composed of four connecting *Lines*. Using *Region* combines the shape into one object, a *Region*. The appearance of the object does not change after the conversion, even though the resulting shape is one object.

Although the *Region* appears to be no different than a closed *Pline*, (unless you turn on a shaded visual style) it is more powerful because several *Regions* can be combined to form complex shapes (composite *Regions*) using the three Boolean operations explained in Chapter 16.

**FIGURE 15-40**

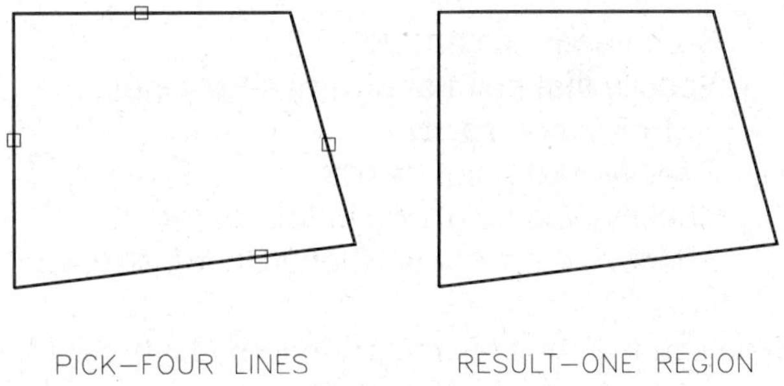


The Boolean operators, *Union, Subtraction,* and *Intersection,* can be used with *Regions* as well as solids. Any number of these commands can be used with *Regions* to form complex geometry (see Chapter 16, Modify Commands II).

**FIGURE 15-41**

As an example, a set of *Regions* (converted *Circles*) can be combined to form the sprocket with only one *Subtract* operation (Fig. 15-41). Compare the simplicity of this operation to the process of using *Trim* to delete each of the unwanted portions of the small circles.

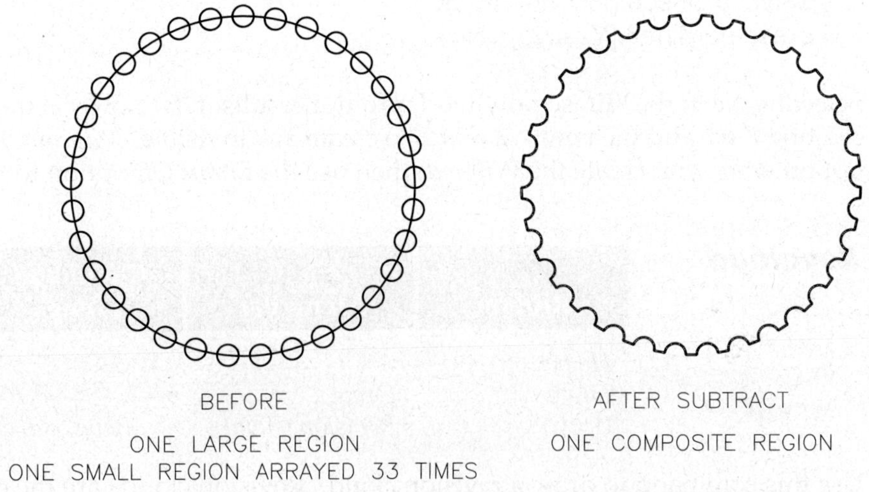

### Wipeout

| Ribbon | Menu Bar | Command (Type) | Alias (Type) | Shortcut |
|---|---|---|---|---|
| *Home*<br>*Draw*<br>(icon) | *Draw*<br>*Wipeout* | *Wipeout* | ... | ... |

FIGURE 15-42

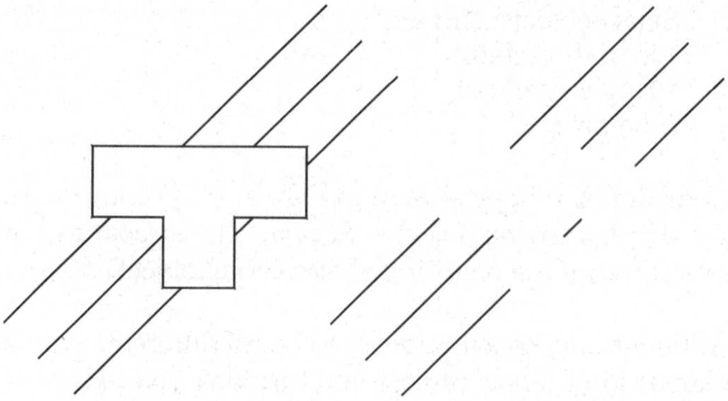

TIP

*Wipeout* creates an "opaque" object that is used to "hide" other objects. The *Wipeout* command creates a closed *Pline* or uses an existing closed *Pline*. All objects behind the *Wipeout* become "invisible." For example, assume a *Pline* shape was created so as to cross other objects. You could use *Wipeout* to make the shape appear to be "on top of" the other objects (Fig. 15-42, left).

Command: ***WIPEOUT***
Specify first point or [Frames/Polyline]
<Polyline>: **PICK**
Specify next point: **PICK**
Specify next point or [Undo]: **PICK**
Specify next point or [Close/Undo]: **Enter**

You can use the *Frames* option to make the *Wipeout* object disappear. Doing so would result in a display similar to Figure 15-42, right.

Specify first point or [Frames/Polyline] <Polyline>: ***f***
Enter mode [ON/OFF] <ON>: ***off***

If you have an existing *Pline* and want to use it to create a *Wipeout* of the same shape, use the *Polyline* option.

Command: ***WIPEOUT***
Specify first point or [Frames/Polyline] <Polyline>: ***p***
Select a closed polyline: **PICK**
Erase polyline? [Yes/No] <No>:

TIP

Selecting *No* to the "Erase polyline?" prompt results in two objects that occupy the same space—the existing *Pline* and the new *Wipeout*. To create an "invisible" *Wipeout* you must specify the *Erase Polyline* option when you create the *Wipeout*, then use the *Frame Off* option to make the *Wipeout* object disappear.

### Revcloud

| Ribbon | Menu Bar | Command (Type) | Alias (Type) | Shortcut |
|---|---|---|---|---|
| *Home*<br>*Draw*<br>(icon) | *Draw*<br>*Revision Cloud* | *Revcloud* | ... | ... |

Use this command to draw a revision cloud. Revision clouds are the customary method of indicating an area of a drawing (usually an architectural application) that contains a revision to the design. Revision clouds should be created on a separate layer so they can be controlled for plotting.

You can draw *Freehand*, *Rectangular*, or *Polygonal* revision clouds. Drawing *Revclouds* is quick and easy because you draw only the basic shape and the arc segments are automatically drawn. *Revclouds* are composed of *Plines*.

Command: **REVCLOUD**
Minimum arc length: 0.5000 Maximum arc length: 0.5000 Style: Normal Type: Freehand
Specify start point or [Arc length/Object/Rectangular/Polygonal/Freehand/Style/Modify] <Object>:

### Freehand

Create a *Freehand Revcloud* by picking a point and moving the cursor in a circular direction (Fig. 15-43). The cloud automatically closes when you approach the start point. You do not have to "pick" an end-point.

Guide crosshairs along cloud path...

### Rectangular

Using the *Rectangular* option, simply pick two diagonal corners for the rectangle (Fig. 15-43).

Specify first corner point or... **PICK**
Specify opposite corner: **PICK**

### Polygonal

You only have to pick "corners" to create a *Polygonal Revcloud*. You can draw three or more corners (Fig 15-43). Press Enter to finish the sequence.

Specify start point or... **PICK**
Specify next point: **PICK**
Specify next point or [Undo]: **PICK**
Specify next point or [Undo]: **PICK** or **Enter**

**FIGURE 15-43**

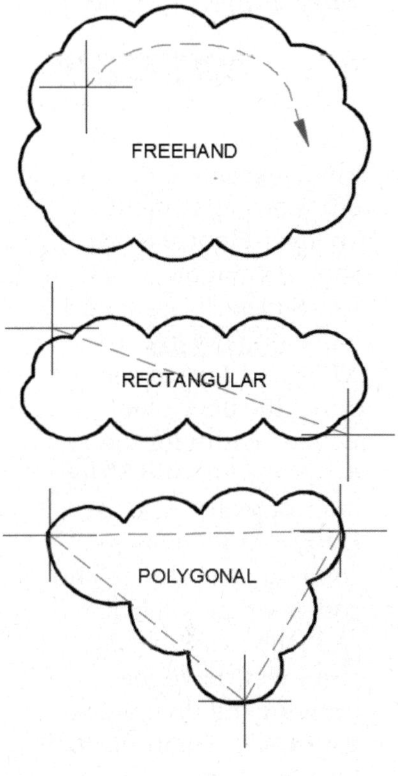


### Arc length

Use the *Arc length* option to specify a value in drawing units for the arc segments. Entering different values for the minimum and maximum arc lengths draws *Revclouds* with different size arcs, similar to those shown in Figure 15-43.

### Object

You can convert some closed objects into *Revclouds* using the *Object* option. *Rectangles*, *Polygons*, *Circles*, and *Ellipses* can be converted to *Revclouds*.

### Style

Use this option to specify the line style that the *Revcloud* is drawn with, either *Normal* or *Calligraphy*. The *Normal* option creates arcs with a constant line width of 0, such as shown in Figure 15-43. The *Calligraphy* option creates a "stroke" similar to a calligraphy pen. With this option each arc is drawn starting with a thin lineweight and ending with a thick lineweight.

### Modify

See Chapter 16 for a full explanation of this option.

### Revcloudvariance Variable

The system variable, Revcloudvariance, allows you to set whether the arcs for the cloud are created with varied (ON/1) or uniform chord (OFF/0) lengths.

## CHAPTER EXERCISES

1. ***Polygon***

   ***Open*** the **PLINE1** drawing you created in Chapter 8 Exercises. Use ***Polygon*** to construct the polygon located at the *Center* of the existing arc, as shown in Figure 15-44. When finished, ***SaveAs*** **POLYGON1.**

FIGURE 15-44

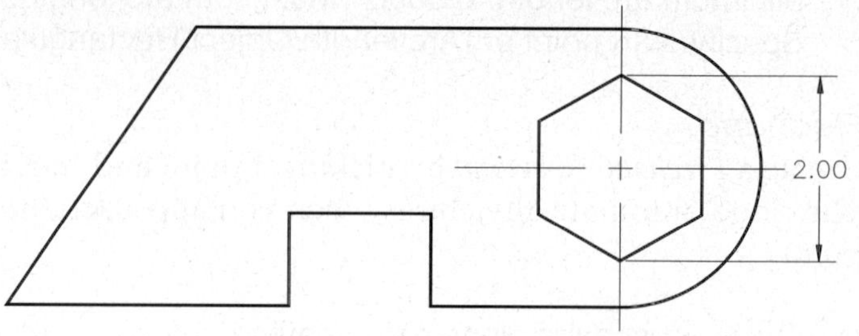


2. ***Xline, Ellipse, Rectangle***

   ***Open*** the **APARTMENT** drawing that you created in Chapter 12. Draw the floor plan of the efficiency apartment on layer **Floorplan** as shown in Figure 15-45. Use ***Saveas*** to assign the name **EFF-APT.** Use ***Xline*** and ***Line*** to construct the floor plan with 8″ width for the exterior walls and 5″ for interior walls. Use the ***Ellipse*** and ***Rectangle*** commands to design and construct the kitchen sink, tub, wash basin, and toilet. Save the drawing but do not exit AutoCAD. Continue to Exercise 3.

FIGURE 15-45

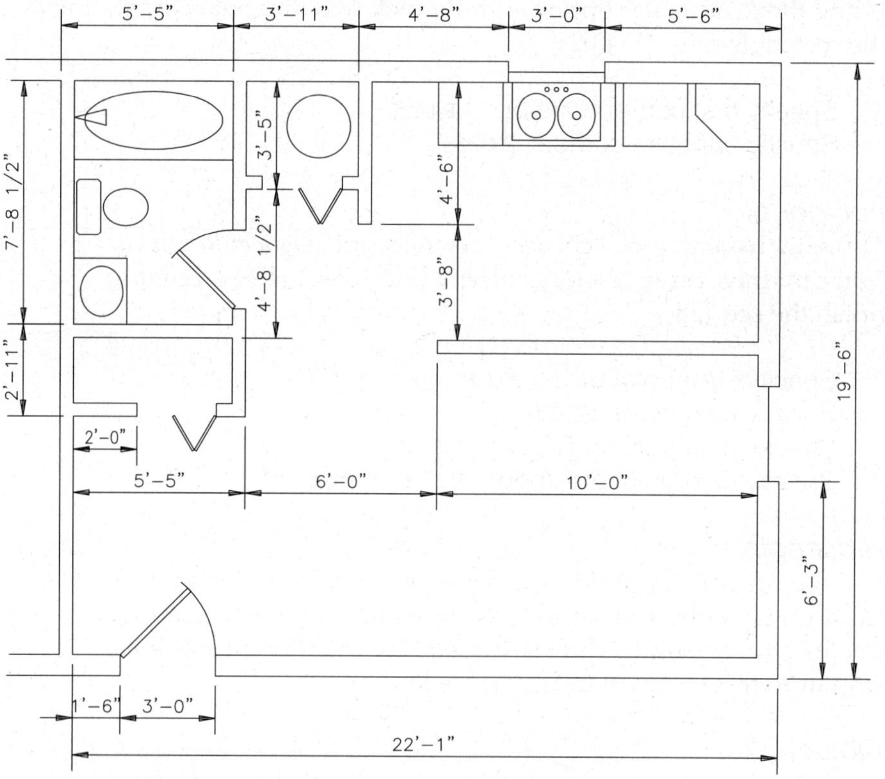


3. ***Polygon, Sketch***

   Create a plant for the efficiency apartment as shown in Figure 15-46. Locate the plant near the entry. The plant is in a hexagonal pot (use ***Polygon***) measuring 18″ across the corners. Use ***Sketch*** to create 2 leaves as shown in Figure A. Create a ***Polar Array*** to develop the other leaves similar to Figure B. ***Save*** the **EFF-APT** drawing.

FIGURE 15-46

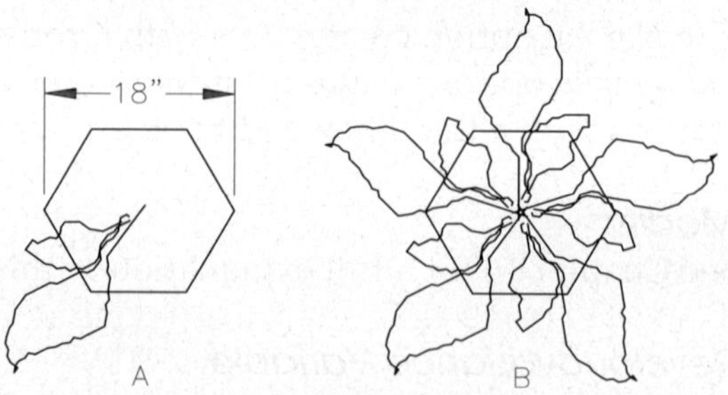

4. ***Pline, Page Setup, Vports,* and *Viewport Scale***

This exercise gives you experience creating a border and plotting the efficiency apartment to scale. To complete the exercise as instructed, you should have a device configured to plot on a "C" size sheet.

A. Invoke the ***Options*** dialog box and access the ***Display*** tab. In the *Layout Elements* section (lower left), ensure that *Show Page Setup Manager for new layouts* and *Create viewport in new layouts* are not checked.

B. Select the ***Layout1*** tab. Activate the ***Page Setup Manager***. Select ***Modify***, then select a plot device from the list that will allow you to plot on an architectural "C" size sheet (**24" x 18"**). Next, select the correct sheet size. Pick ***OK*** to finish the setup.

C. Make layer **TITLE** the ***Current*** layer. Draw a border (in paper space) using a ***Pline*** with a ***width*** of **.04**, and draw the border with a **.75** unit margin within the edge of the *Limits* (border size of 22.5 x 16.5). ***Save*** the drawing as **EFF-APT-PS.**

D. Make layer **VPORTS** ***Current***. Use ***Vports*** to create a ***Single*** viewport. Pick diagonal corners for the viewport at **1,1** and **23,17**. The apartment drawing should appear in the viewport at no particular scale. Use any method to set the viewport scale to ***1/2"=1'***.

E. Next, ***Freeze*** layer **VPORTS**. Your completed drawing should look like that in Figure 15-47. ***Save*** the drawing. ***Plot*** the drawing from the layout at **1:1** scale.

FIGURE 15-47

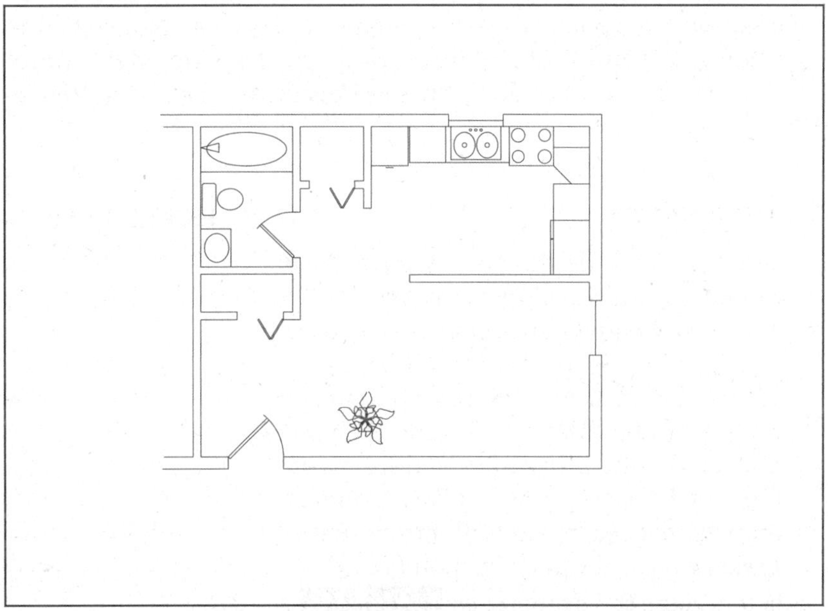

5. ***Donut***

***Open*** **SCHEMATIC1** drawing you created in Chapter 9 Exercises. Use the ***Point Style*** dialog box (***Format*** pull-down menu) to change the style of the *Point* objects to dots instead of small circles. Turn on the ***Node*** Running Object Snap mode. Use the ***Donut*** command and set the ***Inside diameter*** and the ***Outside diameter*** to an appropriate value for your drawing such that the *Inside diameter* is 1/2 the value of the *Outside diameter*. Create a ***Donut*** at each existing ***Node***. ***SaveAs*** **SCHEMATIC1B**. Your finished drawing should look like that in Figure 15-48.

FIGURE 15-48

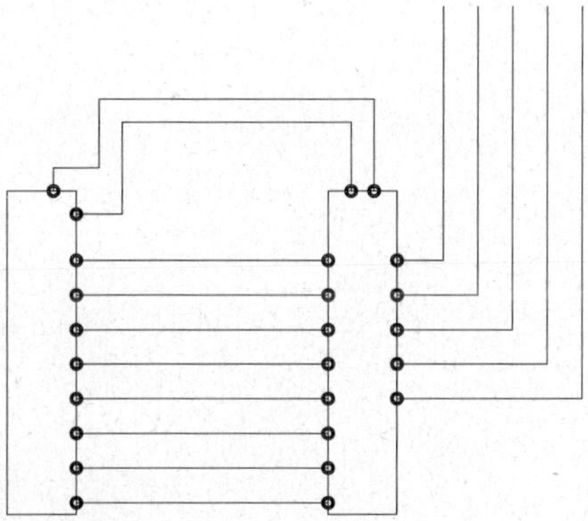

REUSE

6. ***Ellipse, Polygon***

Begin a ***New*** drawing using the template drawing that you created in Chapter 12 named **A-METRIC** Use ***SaveAs*** and assign the name **WRENCH.**

FIGURE 15-49

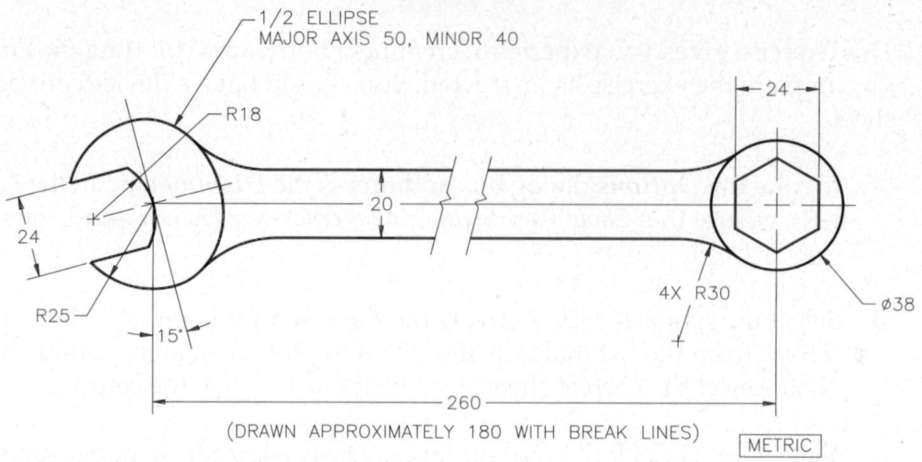


Complete the construction of the wrench shown in Figure 15-49. Center the drawing within the *Limits*. Use ***Line***, ***Circle***, ***Arc***, ***Ellipse***, and ***Polygon*** to create the shape. For the "break lines," create one connected series of jagged ***Lines***, then ***Copy*** for the matching set. Utilize ***Trim***, ***Rotate***, and other edit commands where necessary. HINT: Draw the wrench head in an orthogonal position; then rotate the entire shape 15 degrees. Notice the head is composed of ½ of a *Circle* (radius=25) and ½ of an *Ellipse*. ***Save*** the drawing when completed. Activate a ***Layout*** tab, configure a print or plot device using ***Pagesetup***, make one ***Viewport***, and ***Plot*** the drawing at a standard scale.

REUSE

7. ***Xline, Spline***

Create the handle shown in Figure 15-50. Construct multiple *Horizontal* and *Vertical* ***Xlines*** at the given distances on a separate layer (shown as dashed lines in Figure 15-50). Construct two ***Splines*** that make up the handle sides by PICKing the points indicated at the *Xline* intersections. Note the ***End Tangent*** for one *Spline*. Connect the *Splines* at each end with horizontal ***Lines***. ***Freeze*** the construction (*Xline*) layer. ***Save*** the drawing as **HANDLE.**

FIGURE 15-50

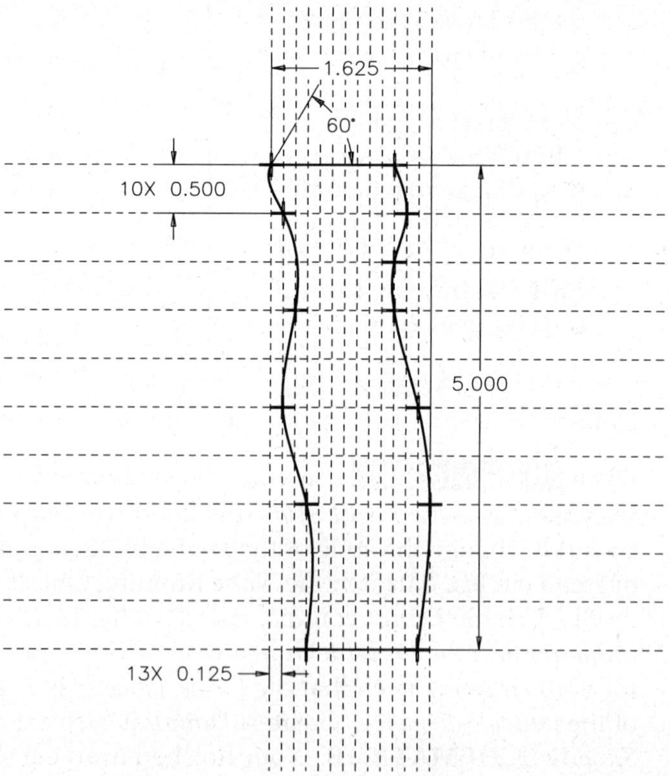

8. *Divide, Measure*

Use the **ASHEET** template drawing, use *SaveAs,* and assign the name **BILLMATL.** Create the table in Figure 15-51 to be used as a bill of materials. Draw the bottom ***Line*** (as dimensioned) and a vertical ***Line***. Use ***Divide*** along the bottom *Line* and ***Measure*** along the vertical *Line* to locate *Points* as desired. Create ***Offsets Through*** the *Points* using ***Node*** Object Snap. (***Polar*** and ***Trim*** may be of help.)

**FIGURE 15-51**

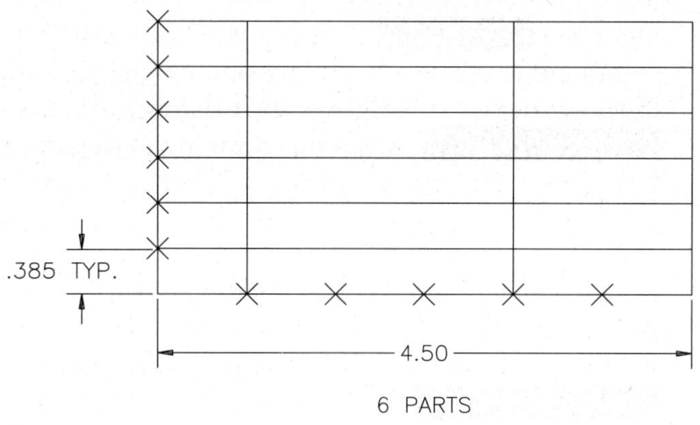


9. *Mline, Mlstyle, Sketch*

Using the drawing that you created in Chapter 8, add the concrete walls, island, and curbs (Fig. 15-52). Use ***Mlstyle*** to create a ***New*** multiline style named **WALL**. The new style should have two line elements with ***Offsets*** of **1** and **-1** and ***Start*** and ***End Caps*** at **90** degrees. Next, use ***Mline*** with the ***Bottom Justification*** and draw one *Mline* along the existing retaining wall centerline. Use ***Endpoint*** to snap to the ends of the existing 2 *Lines* and *Arc*. Construct a second ***Mline*** above using the following coordinates:

**FIGURE 15-52**

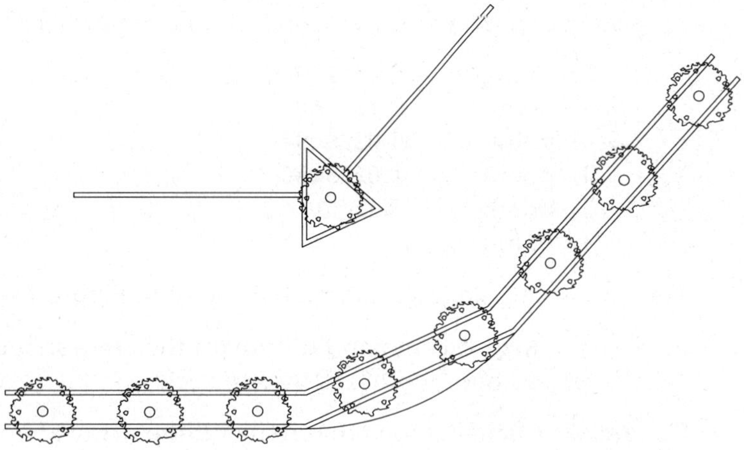

**60,123**
**192,123**
**274,161**
**372.7,274**

Next, construct the ***Closed*** triangular island with an ***Mline*** using ***Top Justification***. The coordinates are:

**192,192**
**222.58,206.26**
**192,233**

For the parking curbs, draw two more ***Mlines***; each *Mline* snaps to the ***Midpoint*** of the island and is **100′** long. The curb on the right has an angle of **49** degrees (use polar coordinate entry). Finally, create trees similar to the ones shown using ***Sketch***. Remember to set the *Sketch* ***Type*** to ***Pline*** to simplify editing at a later time. Create one tree and ***Copy*** it to the other locations. Spacing between trees is approximately **48′**. ***Save*** as **RET-WAL2.**

10. ***Boundary***

*Open* the **GASKETA** drawing that you created in Chapter 9 Exercises. Use the ***Boundary*** command to create a boundary of the outside shape only (no islands). Use ***Move Last*** to displace the new shape to the right of the existing gasket. Use ***SaveAs*** and rename the drawing to **GASKET-AREA.** Keep this drawing to determine the *Area* of the shape after reading Chapter 17.

11. ***Region***

Create a gear, using *Regions*. Begin a ***New*** drawing, using the **ACAD.DWT** template or ***Start from Scratch*** with the ***English*** defaults settings. Use ***Save*** and assign the name **GEAR-REGION.**

A. Set ***Limits*** at **8** x **6** and ***Zoom All***. Create a ***Circle*** of **1** unit ***diameter*** with the center at **4,3**. Create a second concentric ***Circle*** with a ***radius*** of **1.557**. Create a closed ***Pline*** (to represent one gear tooth) by entering the following coordinates:

| | |
|---|---|
| From point: | **5.571,2.9191** |
| To point: | **@.17<160** |
| To point: | **@.0228<94** |
| To point: | **@.0228<86** |
| To point: | **@.17<20** |
| To point: | **c** |

FIGURE 15-53

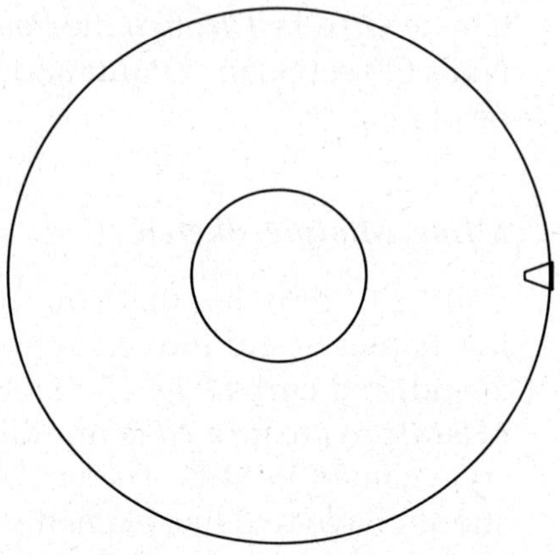

The gear at this stage should look like that in Figure 15-53.

B. Use the ***Region*** command to convert the three shapes (two *Circles* and one "tooth") to three *Regions*.

C. ***Array*** the small *Region* (tooth) in a ***Polar*** array about the center of the gear. There are **40** items that are rotated as they are copied. This action should create all the teeth of the gear (Fig. 15-54).

D. ***Save*** the drawing. The gear will be completed in Chapter 16 Exercises by using the *Subtract* Boolean operation.

FIGURE 15-54

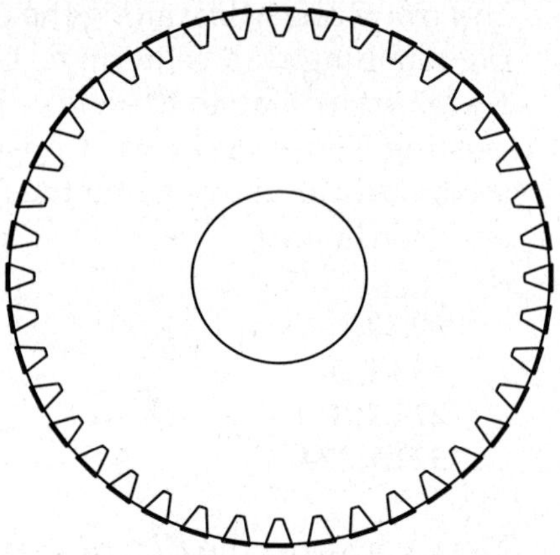

12. ***Revcloud***

Assume you are working with the city engineering office checking the new apartment design submitted to your office for approval. ***Open*** **EFF-APT** drawing from Exercise 2. You notice that the location of the front door has some issues concerning passing the safety codes. Use ***Revcloud*** to denote the area for review. Draw a ***Freehand*** revision cloud similar to that shown in Figure 15-55. Use ***SaveAs*** to save the drawing as **EFF-APT-REV**.

FIGURE 15-55

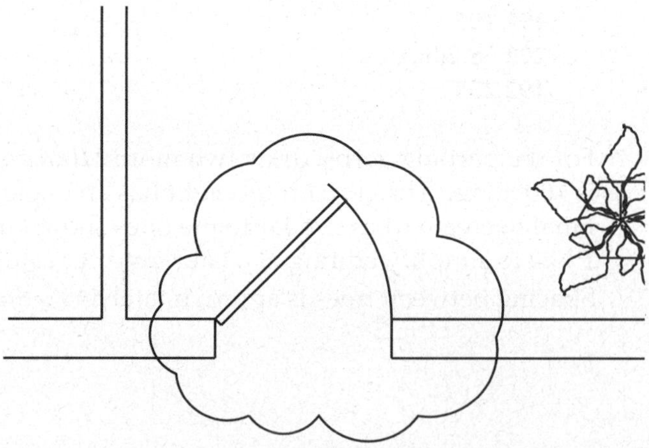

CHAPTER

# 16

# Modify Commands II

## Chapter Objectives

After completing this chapter you should:

1. be able to use *Properties* to modify any type of object;
2. be able to use *Matchprop* and the *Object Properties* lists to change properties of an object;
3. know how to use Quick Properties to change object properties;
4. be able to *Explode a Pline* into its component objects;
5. be able to *Align* objects with other objects;
6. be able to use all the *Pedit* options to modify *Plines* and to convert *Lines* and *Arcs* to *Plines*;
7. be able to modify *Splines* with *Splinedit*;
8. be able to use *Arrayedit* to change existing associative rectangular, polar, and path *Arrays*;
9. be able to edit existing *Mlines* using *Mledit*;
10. know that composite *Regions* can be created with *Union*, *Subtract*, and *Intersect*.

# CONCEPTS

This chapter examines commands that are similar to, but generally more advanced and powerful than, those discussed in Chapter 9, Modify Commands I. Several commands and features mentioned in Chapter 11, Layers and Object Properties, are explained completely in this chapter. Other commands in this chapter are used to modify the properties of objects or convert objects from one type to another. Several commands are used to modify specific types of objects such as *Pedit* (edit *Plines*), *Splinedit* (edits *Splines*), *Arrayedit* (edits *Arrays*), and *Mledit* (edits *Mlines*).

As explained in previous chapters, you can access commands by a variety of methods. For example, you can type in the command names at the Command line or type the command alias (if an alias has been assigned to the command). Most of the commands in this chapter appear in the *Home* tab and *Modify* panel of the Ribbon (Fig. 16-1, top left) or in the *Modify* pull-down menu if the Menu Bar is activated (Fig. 16-1, center). A few commands discussed in this chapter appear in other locations.

**FIGURE 16-1**

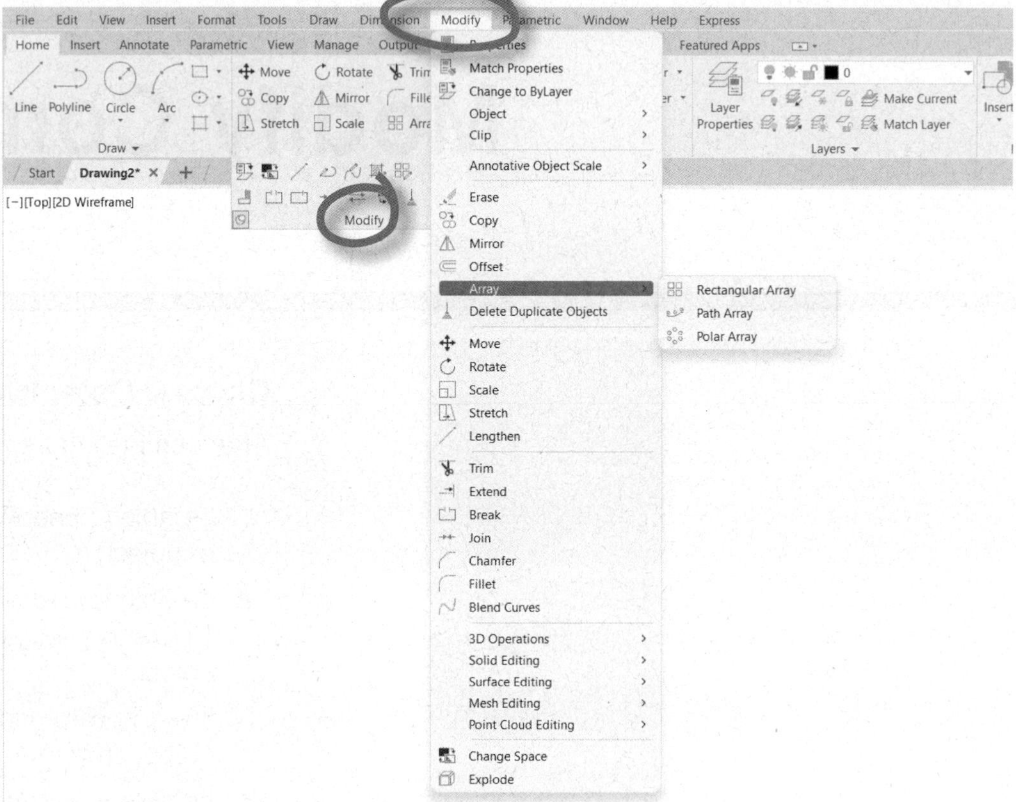


# COMMANDS

You can use several methods to change properties of an object or of several objects. If you want to change an object's layer, color, or linetype only, the quickest method is by using the *Object Properties* toolbar (see "*Object Properties*" toolbar). If you want to change many properties of one or more objects (including layer, color, linetype, linetype scale, text style, dimension style, or hatch style) to match the properties of another existing object, use *Matchprop*. If you want to change any property (including coordinate data) for any object, use *Properties* or Quick Properties.

You can double-click on any object (assuming the *Dblclkedit* command is set to *On*) to produce the *Properties* palette or a more specific editing tool, such as the *Hatch Edit* dialog box or the *Multiline Text Editor*, depending on the type of object you select.

## Considerations for Changing Basic Properties of Objects

The *Properties* palette, the Quick Properties window, the Object Properties drop-down lists (*Color*, *Layer*, *Linetype*, *Lineweight*, and *Plot Style*), and the *Matchprop* and *Chprop* commands can be used to change the basic properties of objects. Here are several basic properties that can be changed and considerations when doing so.

### Layer

By changing an object's *Layer*, the selected object is effectively moved to the designated layer. In doing so, if the object's *Color, Linetype,* and *Lineweight* are set to *ByLayer*, the object assumes the color, linetype, and lineweight of the layer to which it is moved.

### Color

It may be desirable in some cases to assign explicit *Color* to an object or objects independent of the layer on which they are drawn. The properties editing commands allow changing the color of an existing object from one object color to another, or from *ByLayer* assignment to an object color. An object drawn with an object-specific color can also be changed to *ByLayer* with this option.

### Linetype

An object can assume the *Linetype* assigned *ByLayer* or can be assigned an object-specific *Linetype*. The *Linetype* option is used to change an object's *Linetype* to that of its layer (*ByLayer*) or to any object-specific linetype that has been loaded into the current drawing. An object drawn with an object-specific linetype can also be changed to *ByLayer* with this option.

### Linetype Scale

When an individual object is selected, the object's individual linetype scale can be changed with this option, but not the global linetype scale (*LTSCALE*). This is the recommended method to alter an individual object's linetype scale. First, draw all the objects in the current global *LTSCALE*. One of the properties editing commands could then be used with this option to retroactively adjust the linetype scale of specific objects. The result would be similar to setting the *CELTSCALE* before drawing the specific objects. Using this method to adjust a specific object's linetype scale does not reset the global *LTSCALE* or *CELTSCALE* variables.

TIP

### Thickness

An object's *Thickness* can be changed by this option. *Thickness* is a three-dimensional quality (Z dimension) assigned to a two-dimensional object. (Don't confuse this with Polyline Width.)

### Lineweight

An object can assume the *Lineweight* assigned *ByLayer* or can be assigned an object-specific *Linetype*. The *Lineweight* option is used to change an object's *Lineweight* to that of its layer (*ByLayer*) or to any object-specific linetype that has been loaded into the current drawing. An object drawn with an object-specific linetype can also be changed to *ByLayer* with this option.

### Transparency

You can change the *Transparency* of an object by using the *Properties Editing* commands. *Transparency* can be assigned *ByLayer* or can be assigned to an object. *Transparency* works best for fill, gradients, or hatch patterns.

### Plotstyle

Use this option to change an object's *Plotstyle*. Plot styles assigned as *ByLayer* or to individual objects can change the way the objects appear in plots, such as having certain screen patterns, line end joints, plotted colors, plotted lineweights, and so on. (See Chapter 32 for more information on plot styles.)

## Properties

| Ribbon | Menu Bar | Command (Type) | Alias (Type) | Shortcut |
|---|---|---|---|---|
| *View Palettes Properties* | *Modify Properties* | *Properties* | *PR* | (Edit Mode) *Properties* or *Ctrl+1* |

The *Properties* palette (Fig. 16-2) gives you complete access to one or more objects' properties. You can edit the properties simply by changing the entries in the right column.

**FIGURE 16-2**

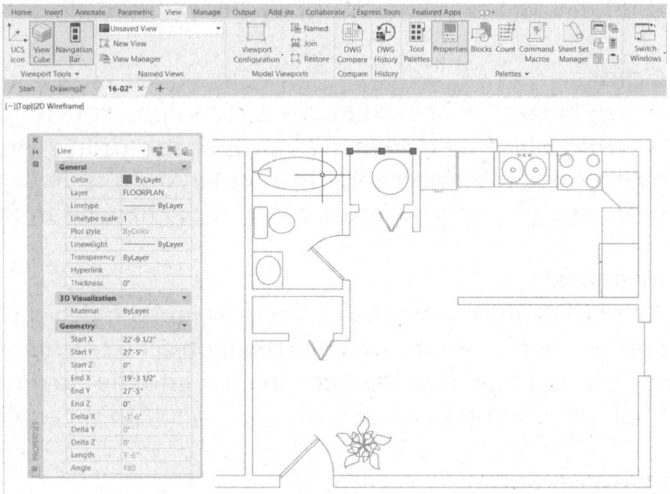

Once opened, this window remains on the screen until dismissed by clicking the "X" in the upper corner. The *Properties* palette can be "docked" on the side of the screen or can be "floating." You can toggle the palette on and off with Ctrl+1 (to appear and disappear).

You can also set the palette to *Auto-hide* so only the title bar appears on your screen until you point to it to produce the full palette (Fig. 16-3). If you click on the bottom icon on the title bar (*Properties*), a shortcut menu displays options to *Move*, *Size*, *Close*, *Allow Docking*, toggle *Auto-hide*, and toggle the *Transparency* of the palette. Changes you make to this palette are "persistent" (remain until changed).

The contents of the palette change based on what type and how many objects are selected. The power of this feature is apparent because the contents of the palette are specific to the type of object that you select. For example, if you select a *Line*, entries specific to that *Line* appear, allowing changes to any properties that the *Line* possesses (see Fig. 16-2). Or, if you select a *Circle* or some *Text*, a window appears specific to the *Circle* or *Text* properties.

**FIGURE 16-3**

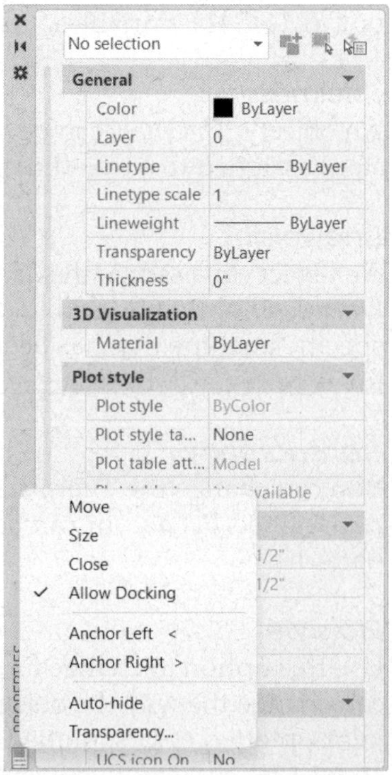


TIP

You can use the palette two ways: you can invoke the palette, then select (highlight) one or more objects, or you can select objects first and then invoke the palette. When multiple objects are selected, a drop-down list in the top of the palette appears. You can select specific object types from the list whose properties you want to change.

If you prefer to leave the palette on your screen, select the objects you want to change when no commands are in use (see Fig. 16-2, highlighted line). After changing the objects' properties, press Escape to deselect the objects. Pressing Escape does not dismiss the *Properties* palette, nor does it undo the changes as long as the cursor is outside the palette.

If no objects are selected, the *Properties* palette displays "No selection" in the drop-down list at the top (Fig. 16-3) and gives the current settings for the drawing. Drawing-wide settings can be changed such as the *LTSCALE* (see Fig. 16-3).

When objects are selected, the dialog box gives access to properties of the selected objects. Figure 16-2 displays the palette after selecting a *Line*. Notice the selection (*Line*) in the drop-down box at the top of the palette. The properties displayed in the central area of the window are specific to the object or group of objects highlighted in the drawing and selected from the drop-down list. In this case (see Fig. 16-2), any aspect of the *Line* can be modified.

If a *Pline* is selected, for example, any property of the *Pline* can be changed. For example, the *Width* of the *Pline* can be changed by entering a new value in the *Properties* palette (Fig. 16-4).

**FIGURE 16-4**

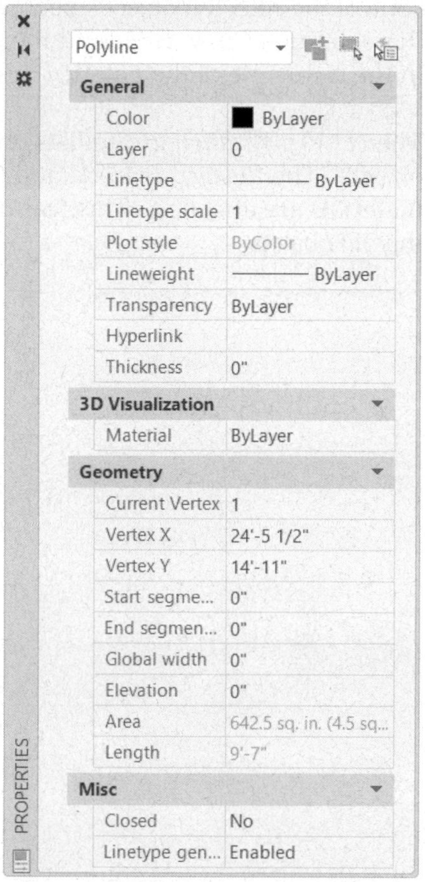


If a dimension is selected, for example, properties specific to a dimension, or to a group of dimensions if selected, can be changed. Note the list of variables that can be changed for a single dimension in Figure 16-5.

The *General* group lists basic properties for an object such as *Layer*, *Color*, or *Linetype*. To change an object's layer, linetype, lineweight, or color properties, highlight the desired objects in the drawing and select the properties you want to change from the right side of the palette. If you use the *ByLayer* strategy of linetype, lineweight, and color properties assignment, you can change an object's linetype, lineweight, and color simply by changing its layer (see Fig. 16-2). This method is recommended for most applications and is described in Chapter 11.

Any "visual" property of selected objects can be previewed using the *Properties* palette. For example, you can highlight objects, then use the drop-down choices for *Color*, *Layer*, *Linetype*, *Lineweight*, *Transparency*, etc. and see a preview in the drawing before making the selection in the *Properties* palette. This feature also applies to the Object Properties drop-down lists on the Ribbon (*Color*, *Layer*, *Linetype*, *Lineweight*, *Transparency*).

**FIGURE 16-5**

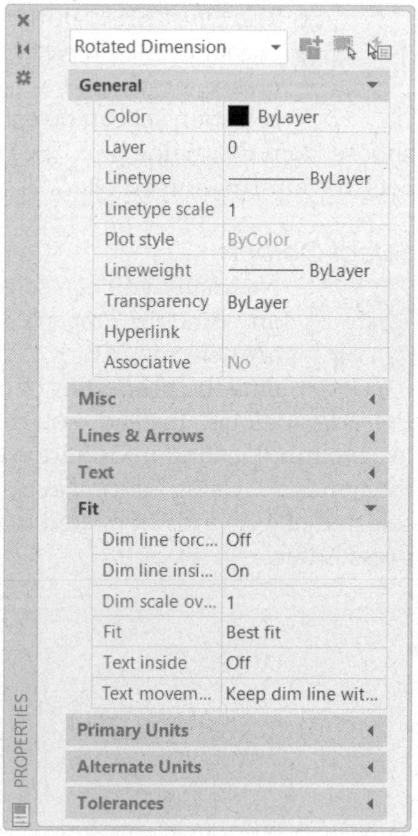

The *Geometry* group lists and allows changing any values that control the selected object's geometry. For example, to change the diameter of a *Circle*, highlight the property and change the value in the right side of the palette (Fig. 16-6).

FIGURE 16-6

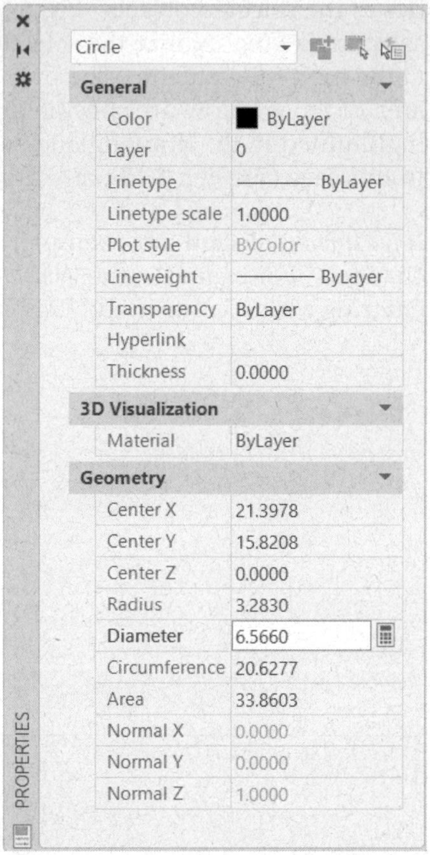


The *3D Visualization* group lists only *Material* by default for most objects. The *Bylayer*, *Byblock*, and *Global* properties determine how materials are attached. This feature is used primarily for rendering 3D objects.

### *Quick Select*

The *Quick Select* button appears near the upper-right corner of the *Properties* palette (Fig. 16-7). Selecting this button produces the *Quick Select* dialog box in which a selection set can be constructed based on criteria you choose from the dialog box. (See Chapter 20, Advanced Selection Sets, for information on *Quick Select*.)

FIGURE 16-7

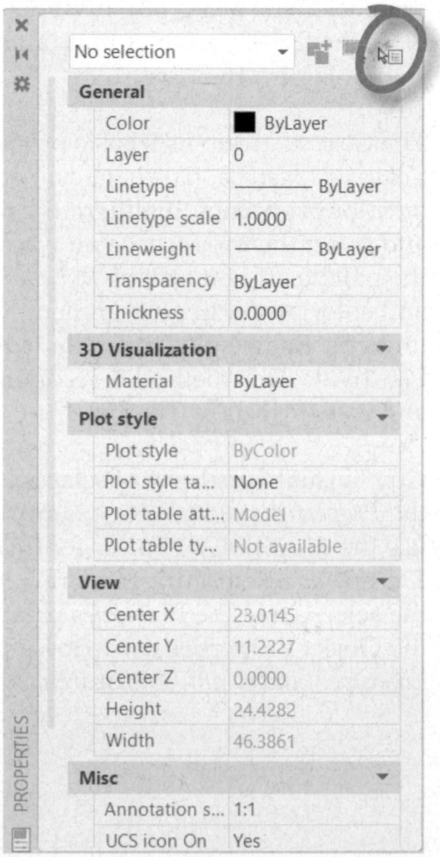


### *Select Objects*

Normally you can select objects by the typical methods any time the *Properties* palette is open. When objects are selected, AutoCAD searches its database and presents the properties of each object, one at a time, in the *Properties* palette. Instead, you can use the *Select Objects* button to save time posting information on each object individually to the window. Using *Select objects*, the properties of the total set of selected objects are not posted to the *Properties* palette until you press Enter.

### *Toggle Value of Pickadd Sysvar*

This button toggles the *PICKADD* system variable on or off (1 or 0). This variable affects object selection at all times, not only for use with the *Properties* palette. *PICKADD* determines whether objects you PICK are added to the current selection set (normal setting, *PICKADD*=1) or replace the previous object or set (*PICKADD*=0).

TIP

The button position of "+" (plus symbol) indicates a setting of 1, or the *PICKADD* system variable is on. In this position, newly selected objects are added to the existing set of selected objects.

If you click the *PICKADD* button, the *PICKADD* system variable changes to a setting of 0 and displays this icon. When *PICKADD* is set to 0, or off, a newly selected object replaces the previous one. This setting works well with the *Properties* palette because selecting a new object replaces the previous set of properties displayed in the *Properties* palette with those of the newly selected object.

NOTE: Since the *PICKADD* setting affects object selection anytime, make sure you set this variable back to your desired setting (usually on, or "+") before dismissing the *Properties* palette. (See Chapter 20, Advanced Selection Sets, for information on *PICKADD*.)

TIP

## ***DBLCLKEDIT***

You can double-click on an object to produce the *Properties* palette or other similar dialog box to edit the object. The *DBLCLKEDIT* variable (double-click edit) controls whether double-clicking an object produces a dialog box. *DBLCLKEDIT* is a system variable; however, you can change the setting simply by typing the variable at the Command line.

```
Command: DBLCLKEDIT
Enter new value for DBLCLKEDIT <ON>:
```

If double-click editing is turned on, the *Properties* palette, Quick Properties window, or other dialog box is displayed when an object is double-clicked. When you double-click most objects, the Quick Properties window is displayed.

However, double-clicking some types of objects displays editing tools that are specific to the type of object. For example, double-clicking a line of *Text* produces the *Text Edit* dialog box. The object types (and resulting editing tool) that do not produce the *Properties* palette or Quick Properties window when the object is double-clicked are listed below. These objects and the related editing tools are discussed fully in upcoming chapters of this text.

| | |
|---|---|
| *Attribute* | Displays the *Edit Attribute Definition* dialog box (*Ddedit*). |
| *Attribute within a block* | Displays the *Enhanced Attribute Editor* (*Eattedit*). |
| *Block* | Displays the *Reference Edit* dialog box (*Refedit*). |
| *Hatch* | Displays the *Hatch Edit* dialog box (*Hatchedit*). |
| *Leader text* | Displays the *Multiline Text Editor* dialog box (*Ddedit*). |
| *Mline* | Displays the *Multiline Edit Tools* dialog box (*Mledit*). |
| *Mtext* | Displays the *Multiline Text Editor* dialog box (*Ddedit*). |
| *Text* | Displays the *Edit Text* dialog box (*Ddedit*). |
| *Xref* | Displays the *Reference Edit* dialog box (*Refedit*). |

NOTE: The *PICKFIRST* system variable must be on (set to 1) for the *Properties* palette or other editing tool to appear when an object is double-clicked. (See Chapter 20, Advanced Selection Sets, for information on the *PICKFIRST* system variable.)

TIP

## Matchprop

| Ribbon | Menu Bar | Command (Type) | Alias (Type) | Shortcut |
|---|---|---|---|---|
| *Home Properties Match Properties* | *Modify Match Properties* | *Matchprop* | *MA* | ... |

*Matchprop* is explained briefly in Chapter 11 but is explained again in this chapter with the full details of the *Special Properties* palette.

*Matchprop* is used to "paint" the properties of one object to another. Simply invoke the command, select the object that has the desired properties ("source object"), then select the object you want to "paint" the properties to ("destination object"). The Command prompt is as follows:

Command: ***MATCHPROP***
Select source object: **PICK**
Current active settings: Color Layer Ltype Ltscale Lineweight Transparency Thickness PlotStyle Dim Text Hatch Polyline Viewport Table Material Multileader Center object
Select destination object(s) or [Settings]: **PICK**
Select destination object(s) or [Settings]: **PICK**
Select destination object(s) or [Settings]: **Enter**
Command:

You can select several destination objects. The destination object(s) assume all of the "Current active settings" of the source object.

**FIGURE 16-8**

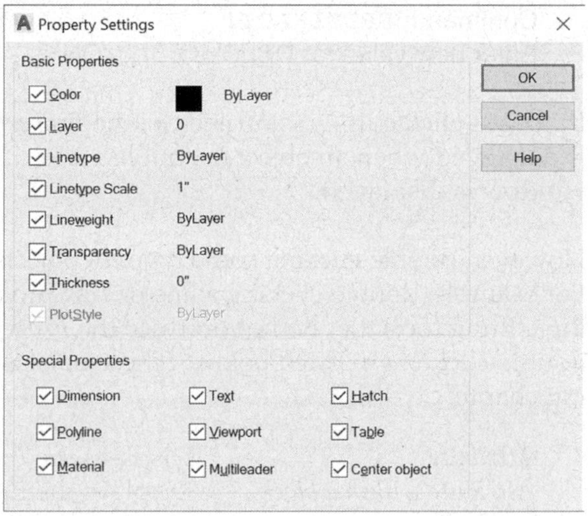


The *Property Settings* dialog box can be used to set which of the *Basic Properties* palette and *Special Properties* palette are to be painted to the destination objects (Fig. 16-8). At the "Select destination object(s) or [Settings]:" prompt, type *S* to display the dialog box. You can control the following *Basic Properties* palette. Only the checked properties are painted to the destination objects.

| | |
|---|---|
| *Color* | This option paints the object-specific or *ByLayer* color. |
| *Layer* | Move selected objects to the source object layer with this option. |
| *Linetype* | Paint the object-specific or *ByLayer* linetype of the source object to the destination object. |
| *Linetype Scale* | This option changes the individual object's linetype scale (*CELTSCALE*) not global linetype scale (*LTSCALE*). |
| *Thickness* | Thickness is a 3-dimensional quality. |
| *Transparency* | This option sets the opacity of an object *ByLayer* or object specific. |
| *Lineweight* | You can paint the object-specific or *ByLayer* lineweight of the source object to the destination object. |
| *PlotStyle* | The plot style of the source object is painted to the destination object when this setting is checked. (See Chapter 32 for information on plot styles.) |

The *Special Properties* palette section allows you to specify features of dimensions, text, hatch patterns, viewports, and polylines to match, as explained next.

*Dimension* This setting paints the *Dimension Style.* A *Dimension Style* defines the appearance of a dimension such as text style, size of arrows and text. (See Chapter 29.)

*Text* This setting paints the source object's text *Style.* The text *Style* defines the text font and many other parameters that affect the appearance of the text. (See Chapter 18.)

*Hatch* Checking this box paints the hatch properties such as *Pattern, Angle, Scale,* and other characteristics. (See Chapter 26.)

*Polyline* If you use *Matchprop* with *Plines*, the *Width* and *Linetype Generation* properties in addition to basic object properties of the first *Pline* are painted to the second. If the source polyline has variable *Width*, it is not transferred.

*Viewport* If you match properties from one paper space viewport to another, the following properties are changed in addition to the basic object properties: *On/Off*, *Display Locking*, standard or custom *Scale*, *Shadeplot*, *Snap*, *Grid*, and *UCSicon* settings.

*Table* Use this option to change the table style of the destination object to that of the source object. (See Chapter 18.)

*Material* This setting changes the material applied to the destination object.

*Multileader* This option paints the *Multileader Style.* A *Multileader Style* determines the appearance of the leaders such as arrowheads, text, and line. (See Chapter 28.)

*Center object* Properties of a *Centerline* or *Centermark* object can be "painted" to another similar object. (See Chapter 24 for information on *Centerline* and *Centermark.*)

## Object Properties Drop-Down Lists

The six drop-down lists in the Ribbon (when not "dropped down") generally show the current layer, color, linetype, lineweight , transparency, and plot style. However, if an object or set of objects is selected, the information in these boxes changes to display the current object's settings. You can change an object's settings by picking an object (when no commands are in use), then dropping down any of the six lists and selecting a different layer, linetype, or color, etc.

Make sure you select (highlight) an object when no commands are in use. Use the pickbox (that appears on the cursor) or *AUto* window or lasso to select the desired object or set of objects. The entries in the six boxes (*Layer*, *Color*, *Linetype*, etc.) then change to display the settings for the selected object or objects. If several objects are selected that have different properties, the boxes display no information (go "blank").

Next, use any of the drop-down lists to make another selection (Fig. 16-9). The highlighted object's properties are changed to those selected in the lists. Press the Escape key to complete the process.

**FIGURE 16-9**

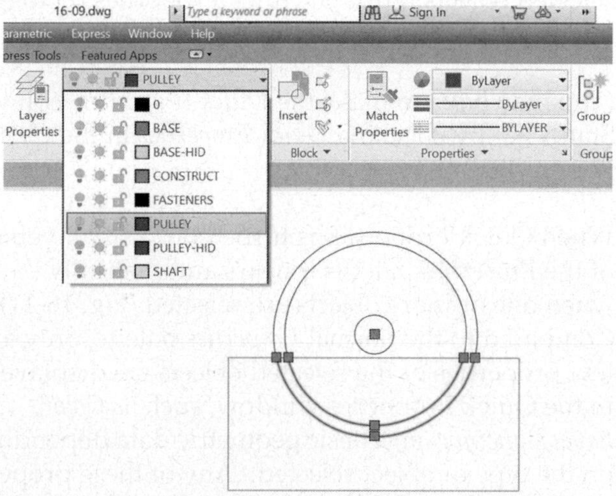


Remember that in most cases color, linetype, lineweight, and transparency settings are assigned *ByLayer*. In this type of drawing scheme, to change the linetype or color properties of an object, you would change only the object's layer (see Fig. 16-9). If you are using this type of drawing scheme, refrain from using the *Color* and *Linetype* drop-down lists for changing properties.

TIP

This method for changing object properties is about as quick and easy as *Matchprop*; however, only the layer, linetype, color, lineweight, transparency, or plot style can be changed with the drop-down lists. This method works well if you do not have other existing objects to match or if you want to change only layer, linetype, color, lineweight, and plot style without matching text, dimension, and hatch styles, and viewport or polyline properties.

### Chprop

| Ribbon | Menu Bar | Command (Type) | Alias (Type) | Shortcut |
|---|---|---|---|---|
| ... | ... | *Chprop* | ... | ... |

*Chprop* allows you to change basic properties of one or more objects using Command line format. If you want to change properties of several objects, pick several objects or select with a window or crossing window at the "Select objects:" prompt.

Command: ***CHPROP***
Select objects: **PICK**
Select objects: **Enter**
Enter property to change [Color/LAyer/LType/ltScale/LWeight/Thickness/TRansparency/Material/Annotative]:

### Change

| Ribbon | Menu Bar | Command (Type) | Alias (Type) | Shortcut |
|---|---|---|---|---|
| ... | ... | *Change* | *-CH* | ... |

*Change* is an old command that allows you to change properties in Command line format. You can change points (such as the end point of a *Line*) or change properties.

Command: ***CHANGE***
Select objects: **PICK**
Specify change point or [Properties]: ***P***
Enter property to change [Color/Elev/LAyer/LType/ltScale/LWeight/Thickness/TRansparency/Material/Annotative]:

## Quick Properties

Quick Properties is not a command, rather it is a feature that allows you to access and change only a few properties of selected objects. Turn on Quick Properties by toggling the button on the Status bar (Fig. 16-10).

(You may have to make the *Quick Properties* icon visible on the Status bar by checking *Quick Properties* in the *Customization* pop-up list.)

FIGURE 16-10

When Quick Properties is turned on, a small version of the *Properties* palette appears automatically when one or more objects are selected (Fig. 16-11). Compared to the normal *Properties* palette, only a few properties of the selected objects are displayed in the Quick Properties window, such as *Color*, *Layer*, *Linetype*, and basic geometric data depending on the type of object selected. Any of these properties can be changed by selecting from a drop-down list or entering new data in an edit box. The Quick Properties window disappears when the object becomes deselected by pressing the Esc key or activating another command.

FIGURE 16-11

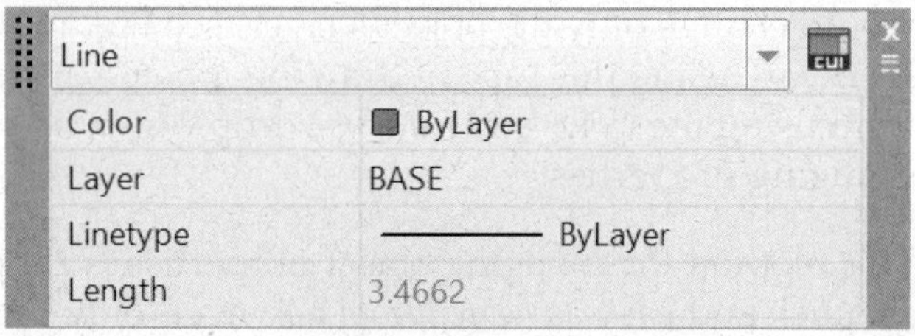

Right-clicking on the border of the Quick Properties window produces a shortcut menu. Select *Settings* to produce the *Quick Properties* tab of the *Drafting Settings* dialog box (Fig. 16-12). You can select to display the Quick Properties panel only for defined objects. Use the *Customize User Interface* window to define the desired objects. Note that you can specify the *Palette Location* to appear at the current cursor position or to repeatedly appear at the same position that you drag it (*QPLOCATION* system variable). *Palette Behavior* defines the number of rows that appear and whether the palette disappears when the objects are deselected.

**FIGURE 16-12**

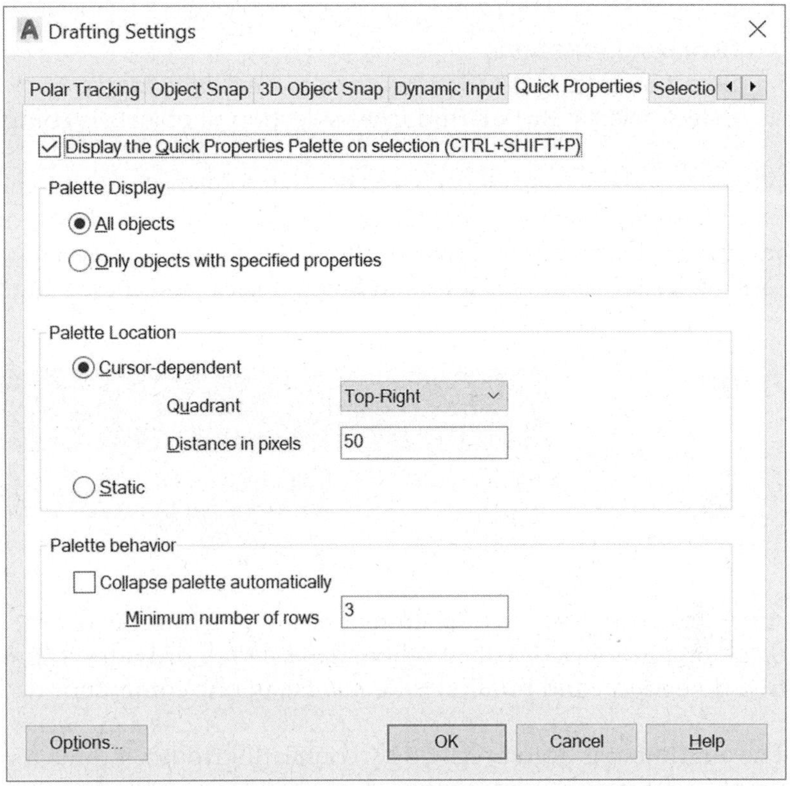


## *Explode*

| Ribbon | Menu Bar | Command (Type) | Alias (Type) | Shortcut |
|---|---|---|---|---|
| *Home* *Modify* (icon) | *Modify* *Explode* | *Explode* | X | ... |

Many graphical shapes can be created in AutoCAD that are made of several elements but are treated as one object, such as *Plines, Polygons, Blocks, Hatch* patterns, and dimensions. The *Explode* command provides you with a means of breaking down or "exploding" the complex shape from one object into its many component segments (Fig. 16-13). Generally, *Explode* is used to allow subsequent editing of one or more of the component objects of a *Pline, Polygon,* or *Block*, etc., which would otherwise be impossible while the complex shape is considered one object.

**FIGURE 16-13**

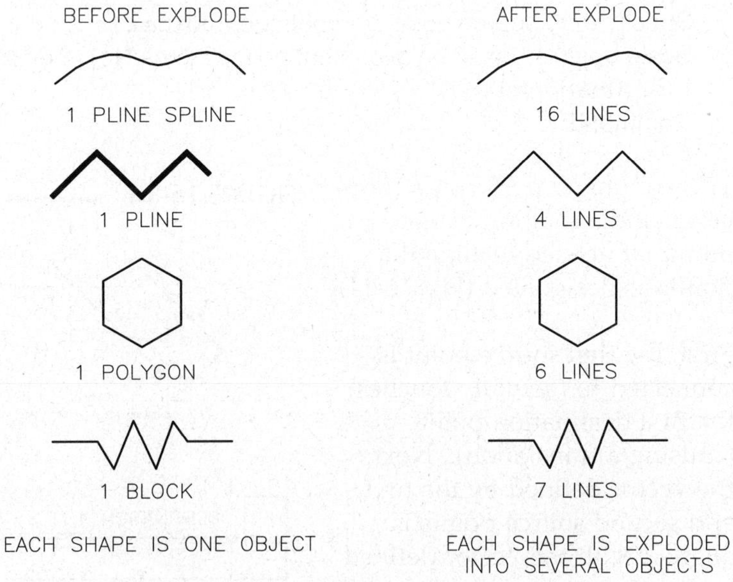

The *Explode* command has no options and is simple to use. You only need to select the objects to *Explode.*

Command: ***EXPLODE***
Select objects: **PICK** (Select one or more *Plines*, *Blocks*, etc.)
Select objects: **Enter** (Indicates selection of objects is complete.)

TIP

When *Plines*, *Polygons*, *Blocks*, or hatch patterns are *Exploded,* they are transformed into *Line*, *Arc*, and *Circle* objects. Beware, *Plines* having *width* lose their width information when *Exploded* since *Line*, *Arc*, and *Circle* objects cannot have width. *Exploding* objects can have other consequences such as losing "associativity" of dimensions and hatch objects and increasing file sizes by *Exploding Blocks*.

## Align

| Ribbon | Menu Bar | Command (Type) | Alias (Type) | Shortcut |
|---|---|---|---|---|
| *Home Modify* (icon) | *Modify 3D Operations > Align* | *Align* | *AL* | ... |

*Align* provides a means of aligning one shape (a simple object, group of objects, *Pline*, *Boundary*, *Region*, *Block*, or a 3D object) with another shape. *Align* provides a complex motion, usually a combined translation (like *Move*) and rotation (like *Rotate*), in one command.

The alignment is accomplished by connecting source points (on the shape to be moved) to destination points (on the stationary shape). You should use Object Snap modes to select the source and destination points to assure accurate alignment. Either a 2D or 3D alignment can be accomplished with this command. The command syntax for alignment in a 2D alignment is as follows:

Command: ***ALIGN***
Select objects: **PICK**
Select objects: **Enter**
Specify first source point: **PICK** (with Object Snap)
Specify first destination point: **PICK** (with Object Snap)
Specify second source point: **PICK** (with Object Snap)
Specify second destination point: **PICK** (with Object Snap)
Specify third source point or <continue>: **Enter**
Scale objects based on alignment points? [Yes/No] <N>: **Enter** (or ***Y*** to scale the source object to match destination object)
Command:

This command performs a translation and a rotation in one motion if needed to align the points as designated (Fig. 16-14).

First, the first source point is connected to (actually touches) the first destination point (causing a translation). Next, the vector defined by the first and second source points is aligned with the vector defined by the first and second destination points (causing rotation).

**FIGURE 16-14**

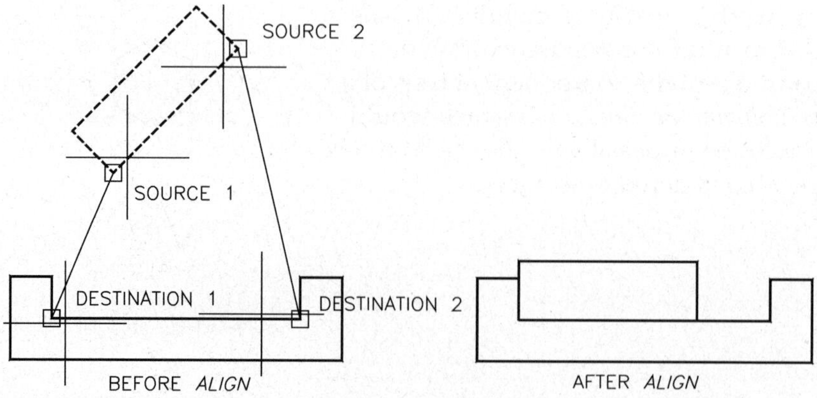

If no third destination point is given (needed only for a 3D alignment), a 2D alignment is assumed and performed on the basis of the two sets of points.

Note that you can scale the source object based on the distance between the source points and the destination points. If you answer "Y" to the "Scale objects based on alignment points?" prompt, the source object is enlarged or reduced so the distance between its alignment points matches that of the destination points.

TIP

## Pedit

| Ribbon | Menu Bar | Command (Type) | Alias (Type) | Shortcut |
|---|---|---|---|---|
| *Home Modify* (icon) | *Modify Object > Polyline* | *Pedit* | *PE* | ... |

This command provides numerous options for editing *Polylines* (*Plines*). As an alternative, *Properties* can be used to change many of the *Pline*'s features in dialog box form (see Fig. 16-4).

The list of options below emphasizes the great flexibility possible with *Polylines*. The first step after invoking *Pedit* is to select the *Pline* to edit.

Command: ***PEDIT***
Select polyline or [Multiple]: **PICK**
Enter an option [Close/Join/Width/Edit vertex/Fit/Spline/Decurve/Ltype gen/Reverse/Undo]:

### *Multiple*

The *Multiple* option allows multiple *Plines* to be edited simultaneously. The selected *Plines* can be totally separate objects and do not have to be connected in any way. Once the *Multiple* option is invoked and the objects are selected, any *Pline* option, such as *Close*, *Open*, *Join*, *Width*, *Fit*, *Spline*, *Decurve*, or *Ltype gen*, operates on the selected *Plines*.

TIP

Command: ***PEDIT***
Select polyline or [Multiple]: ***m***
Select objects: **PICK**
Select objects: **PICK**
Select objects: **Enter**
Enter an option [Close/Open/Join/Edit vertex/Width/Fit/Spline/Decurve/Ltype gen/Reverse/Undo]:

For example, you can change the width of all *Plines* in a drawing simultaneously using the *Multiple* option, selecting all *Plines*, and using the *Width* option.

### *Close*

*Close* connects the last segment with the first segment of an existing "open" *Pline*, resulting in a "closed" *Pline* (Fig. 16-15). A closed *Pline* is one continuous object having no specific start or endpoint, as opposed to one closed by PICKing points. A *Closed Pline* reacts differently to the *Spline* option and to some commands such as *Fillet*, *Pline* option (see "*Fillet*," Chapter 9).

**FIGURE 16-15**

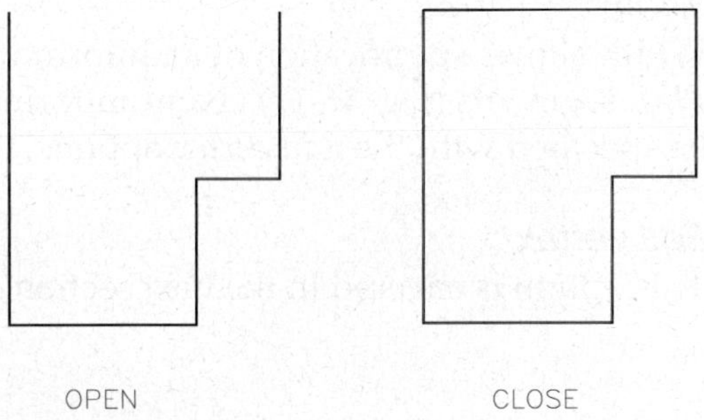


### *Open*

*Open* removes the closing segment if the *Close* option was used previously (Fig. 16-15).

TIP

### Join

The *Join* option combines two or more objects into one *Pline*. The line segments do not have to meet exactly for *Join* to work. You must first use the *Multiple* option, then *Join*.

```
Command: PEDIT
Select polyline or [Multiple]: m
Select objects: PICK
Select objects: Enter
Enter an option [Close/Open/Join/Width/Edit vertex/Fit/Spline/Decurve/Ltype gen/Reverse/Undo]: j
Join Type = Extend
Enter fuzz distance or [Jointype] <0.0000>: .5
1 segments added to polyline
Enter an option [Close/Open/Join/Width/Fit/Spline/Decurve/Ltype gen/Reverse/Undo]:
```

If the ends of the line segments do not touch but are within a distance that you can set, called the *fuzz distance*, the ends can be joined by *Pedit*. *Pedit* handles this automatically by either extending and trimming the line segments or by adding a new line segment based on your setting for *Jointype*.

```
Select objects: PICK
Enter an option [Close/Open/Join/Width/Fit/Spline/Decurve/Ltype gen/Reverse/Undo]: j
Join Type = Extend
Enter fuzz distance or [Jointype] <2.0000>: j
Enter join type [Extend/Add/Both] <Extend>:
```

*Extend*

This option causes AutoCAD to join the selected polylines by extending or trimming the segments to the nearest endpoints (see Fig. 16-16).

*Add*

Use this option to add a straight segment between the nearest endpoints (see Fig. 16-16).

*Both*

If you use this option, the selected polylines are joined by extending or trimming if possible. If not, as in the case of near parallel lines when an extension would be outside the fuzz distance, a straight segment is added between the nearest endpoints.

**FIGURE 16-16**

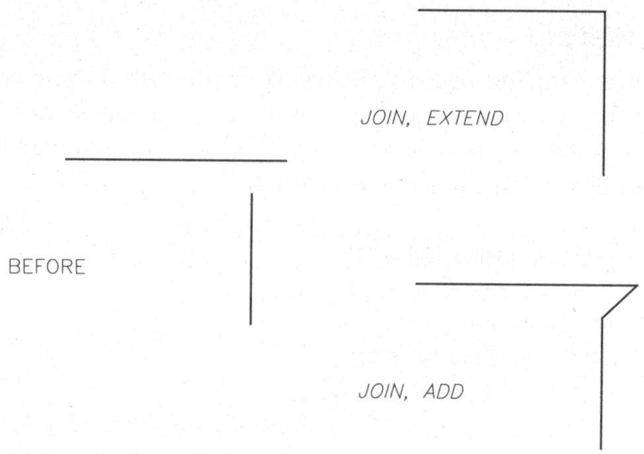


If the properties of several objects being joined into a polyline differ, the resulting polyline inherits the properties of the first object you select.

### Width

Width allows specification of a uniform width for *Pline* segments (Fig. 16-17). Non-uniform width can be specified with the *Edit vertex* option.

### Edit vertex

This option is covered in the next section.

**FIGURE 16-17**

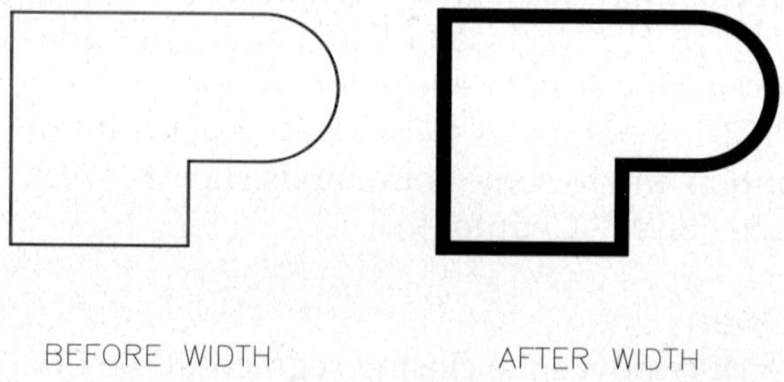

*Fit*

This option converts the *Pline* from straight line segments to arcs. The curve consists of two arcs for each pair of vertices. The resulting curve can be radical if the original *Pline* consists of sharp angles. The resulting curve passes through all vertices (Fig. 16-18).

FIGURE 16-18

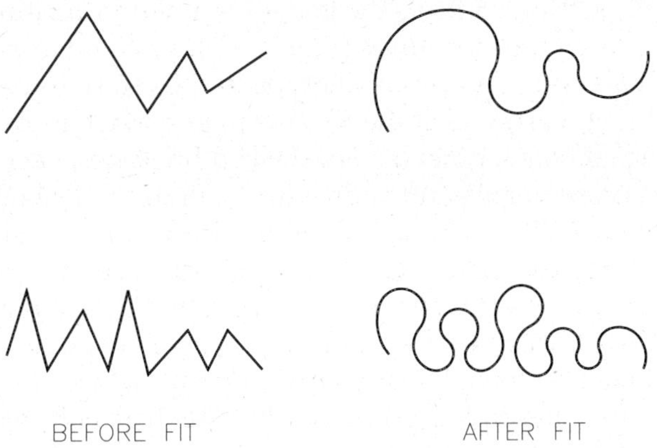


*Spline*

This option converts the *Pline* to a B-spline (Bezier spline) (Fig. 16-19). The *Pline* vertices act as "control points" affecting the shape of the curve. The resulting curve passes through only the end vertices. A *Spline*-fit *Pline* is not the same as a spline curve created with the *Spline* command. This option produces a less versatile version of the newer *Spline* object.

FIGURE 16-19

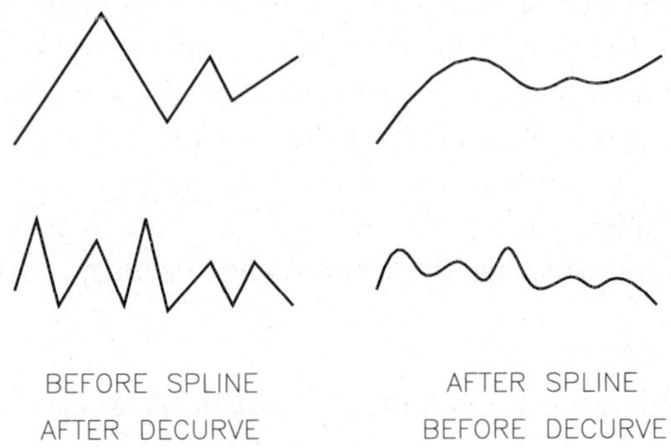


*Decurve*

*Decurve* removes the *Spline* or *Fit* curve and returns the *Pline* to its original straight line segments state (Fig. 16-19).

When you use the *Spline* option of *Pedit,* the amount of "pull" can be affected by setting the *SPLINETYPE* system variable to either 5 or 6 before using the *Spline* option. *SPLINETYPE* applies either a quadratic (5=more pull) or cubic (6=less pull) B-spline function (Fig. 16-20).

FIGURE 16-20

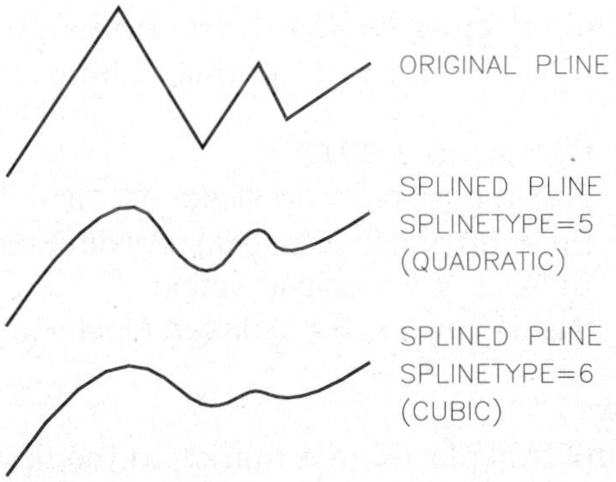


The *SPLINESEGS* system variable controls the number of line segments created when the *Spline* option is used. The variable should be set before using the option to any value (8=default): the higher the value, the more line segments. The actual number of segments in the resulting curve depends on the original number of *Pline* vertices and the value of the *SPLINETYPE* variable (Fig. 16-21).

FIGURE 16-21

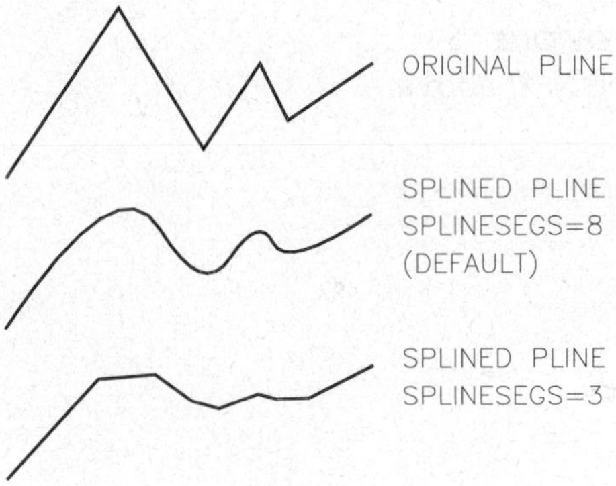

### *Ltype gen*

This setting controls the generation of non-continuous linetypes for *Plines* (*PLINEGEN* system variable). If *Off,* non-continuous linetype dashes start and stop at each vertex, as if the *Pline* segments were individual *Line* segments. For dashed linetypes, each line segment begins and ends with a full dashed segment (Fig. 16-22). If *On,* linetypes are drawn in a consistent pattern, disregarding vertices. In this case, it is possible for a vertex to have a space rather than a dash. Using the *Ltype gen* option retroactively changes *Plines* that have already been drawn. *Ltype gen* affects objects composed of *Plines* such as *Polygons, Rectangles,* and *Boundaries.*

**FIGURE 16-22**

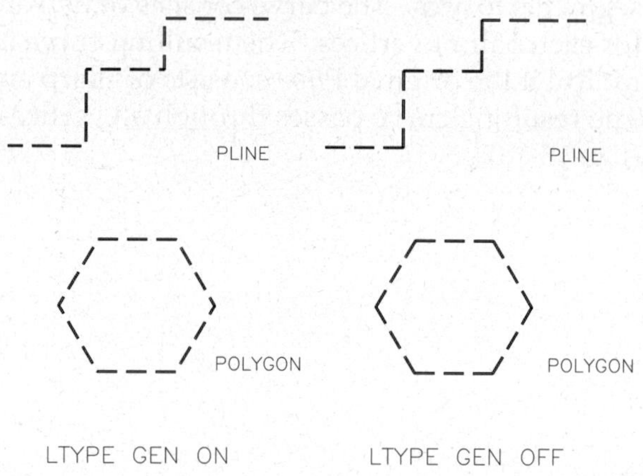


### *Reverse*

This option reverses the order of the first and last vertex of the *Pline.* Using *Reverse* may be helpful before using the *Edit vertex* option since it changes the origin and direction of the vertex marker.

### *Undo*

*Undo* reverses the most recent *Pedit* operation.

### *eXit*

This option exits the *Pedit* options, keeps the changes, and returns to the Command prompt.

## Vertex Editing

Upon selecting the *Edit Vertex* option from the *Pedit* options list, the group of suboptions is displayed on the screen menu and Command line:

Command: ***PEDIT***
Select polyline or [Multiple]: **PICK**
Enter an option [Close/Join/Width/Edit vertex/Fit/Spline/Decurve/Ltype gen/Reverse/Undo]: ***e***
Enter a vertex editing option
[Next/Previous/Break/Insert/Move/Regen/Straighten/Tangent/Width/eXit] <N>:

### *Next*

AutoCAD places an **X** marker at the first endpoint of the *Pline.* The *Next* and *Previous* options allow you to sequence the marker to the desired vertex (Fig. 16-23).

**FIGURE 16-23**

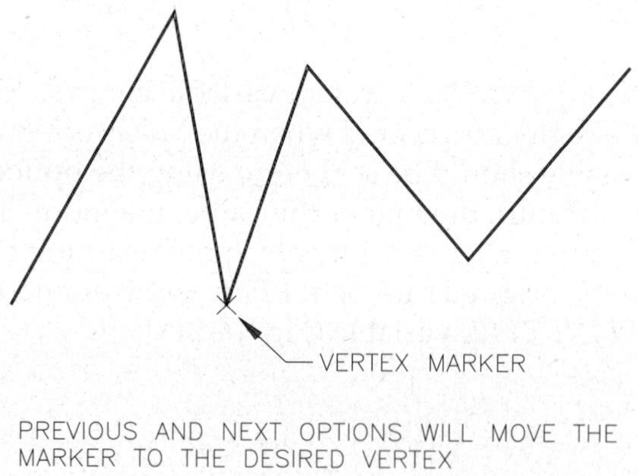


### *Previous*

See *Next* above.

*Break*
This selection causes a break between the marked vertex and another vertex you then select using the *Next* or *Previous* option (Fig. 16-24).

Enter an option [Next/Previous/Go/eXit] <N>:

Selecting *Go* causes the break. An endpoint vertex cannot be selected.

**FIGURE 16-24**

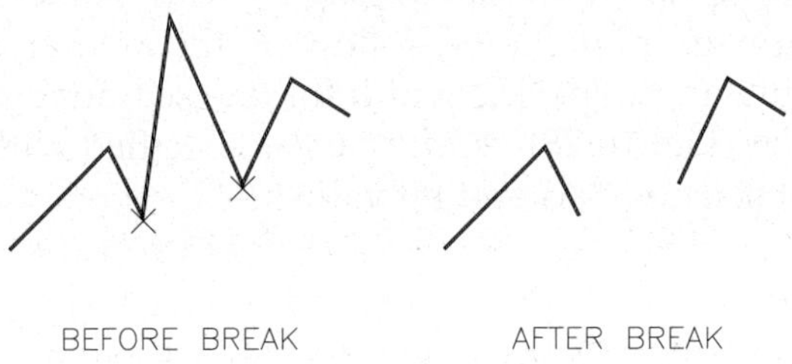


*Insert*
*Insert* allows you to insert a new vertex after the vertex location marked with the **X** (Fig. 16-25). Place the marker at the intended new vertex, use *Insert,* then PICK the new vertex location. If the vertex is the endpoint, a line rubberbands to the cursor location to show as a visual cue.

**FIGURE 16-25**

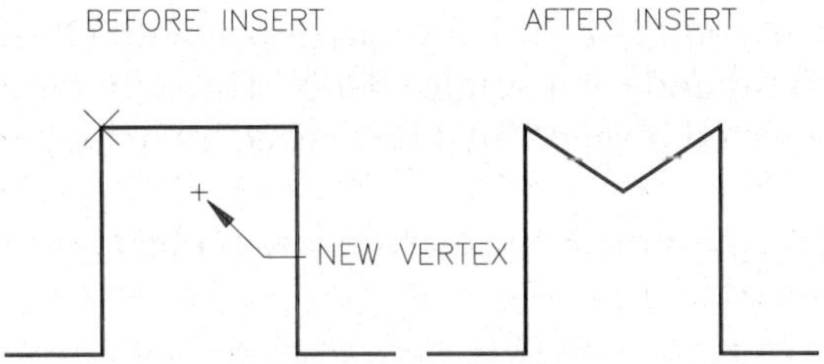


*Move*
You are prompted to indicate a new location to *Move* the marked vertex (Fig. 16-26).

**FIGURE 16-26**

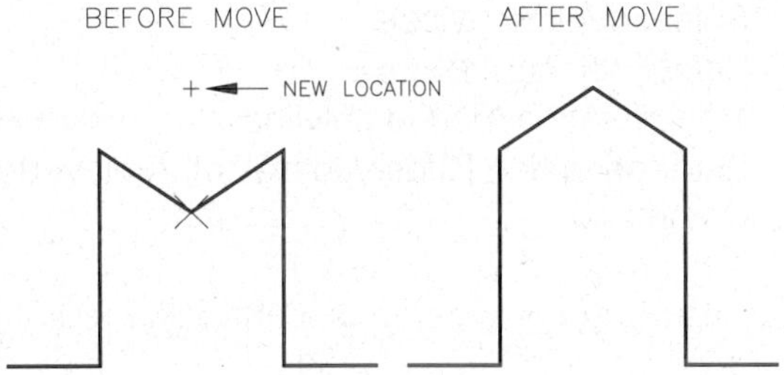


*Regen*
In older releases of AutoCAD, *Regen* should be used after the *Width* option to display the new changes.

*Straighten*
You can *Straighten* the *Pline* segments between the current marker and the other marker that you then place by one of these options:

Enter an option [Next/Previous/Go/eXit] <N>:

Selecting *Go* causes the straightening to occur (Fig. 16-27).

**FIGURE 16-27**

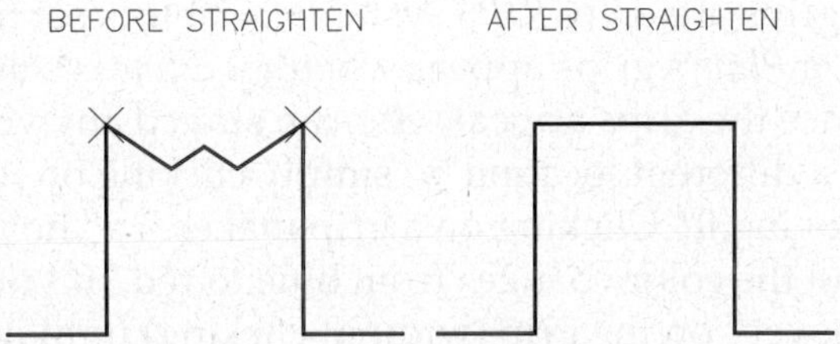


*Tangent*
Tangent allows you to specify the direction of tangency of the current vertex for use with curve *Fitting*.

### *Width*

This option allows changing the *Width* of the *Pline* segment immediately following the marker, thus achieving a specific width for one segment of the *Pline* (Fig. 16-28). *Width* can be specified with different starting and ending values.

FIGURE 16-28

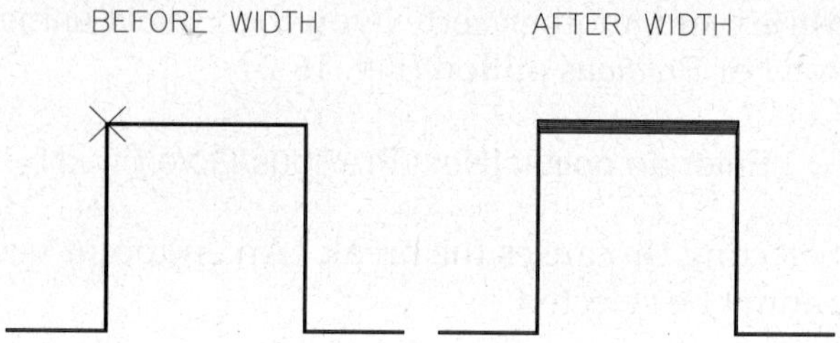


### *eXit*

This option exits from vertex editing, saves changes, and returns to the main *Pedit* prompt.

### Converting *Lines* and *Arcs* to *Plines*

TIP

A very important and productive feature of *Pedit* is the ability to convert *Lines* and *Arcs* to *Plines* and closed *Pline* shapes. Potential uses of this option are converting a series of connected *Lines* and *Arcs* to a closed *Pline* for subsequent use with *Offset* or for inquiry of the area (*Area* command) or length (*List* command) of a single shape. The only requirement for conversion of *Lines* and *Arcs* to *Plines* is that the selected objects must have exact matching endpoints.

To accomplish the conversion of objects to *Plines,* simply select a *Line* or *Arc* object and request to turn it into one:

Command: ***PEDIT***
Select polyline or [Multiple]: **PICK** (Select only one *Line* or *Arc*)
Object selected is not a polyline
Do you want to turn it into one? <Y> **Enter**
Enter an option [Close/Join/Width/Edit vertex/Fit/Spline/Decurve/Ltype gen/Reverse/Undo]: ***j*** (Use the *Join* option)
Select objects: **PICK**
Select objects: **Enter**
1 segments added to polyline
Enter an option [Close/Join/Width/Edit vertex/Fit/Spline/Decurve/Ltype gen/Reverse/Undo]: **Enter**
Command:

The resulting conversion is a one *Polyline* shape.

## Editing *Plines* with Grips

As an alternative to using the *Pedit* command to change existing *Plines*, you can use the *Pline's* grips to change the vertices of a *Pline*. Although grips are discussed in detail in Chapter 19, this section describes only the grip editing features of a *Pline*.

If grips are on (*GRIPS* system variable is set to 1 or 2), a *Pline's* grips appear when the *Pline* is selected. Once the grips appear, you can stretch any vertex to a different location by simply clicking on it and moving it. Clicking on a grip makes it a "hot" grip and the color changes from blue to red. If you "hover" on the grip (without clicking) until it turns pink, a shortcut menu appears with other options (Fig. 16-29). Press Escape to cancel the grips.

FIGURE 16-29

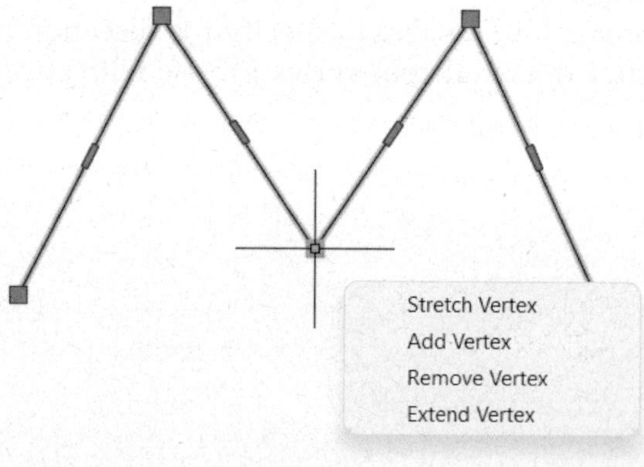


TIP

Tip: You can press *Ctrl* to cycle between the Stretch, Add, Remove, and Extend Vertex options.

Editing *Plines* using grips is generally easier and more intuitive than using the options of the *Pedit* command. Refer to Chapter 19 for more information on grips.

TIP

### *Splinedit*

| Ribbon | Menu Bar | Command (Type) | Alias (Type) | Shortcut |
|---|---|---|---|---|
| *Home Modify* (icon) | *Modify Object > Spline* | *Splinedit* | *SPE* | ... |

*Splinedit* is an extremely powerful command for changing the configuration of existing *Splines*. You can use multiple methods to change *Splines*. All of the *Splinedit* methods fall under two sets of options.

The two groups of options that AutoCAD uses to edit *Splines* are based on two sets of points: fit data points and control vertices. Data points are the points that were specified when the *Spline* was created—the points that the *Spline* actually passes through (Fig. 16-30).

**FIGURE 16-30**

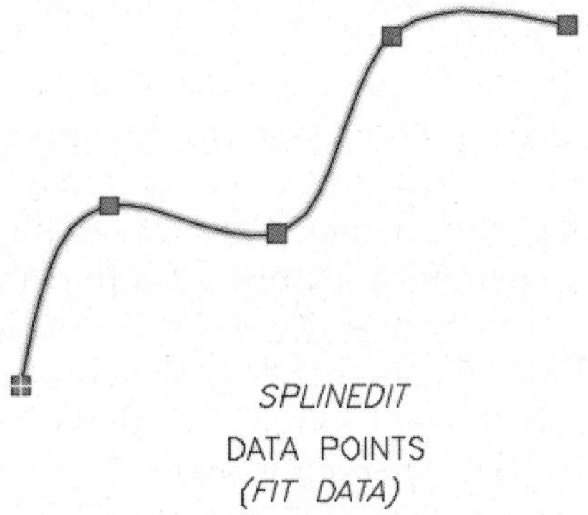


Control vertices are other points outside of the path of the *Spline* that only have a "pull" effect on the curve (Fig. 16-31).

**FIGURE 16-31**

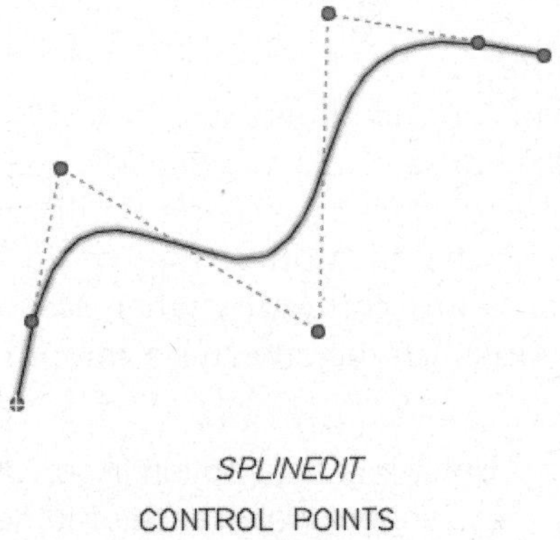


## Editing *Spline* Fit Data Points

The command prompt displays several levels of options. The top level of options uses the control vertices method for editing. Select *Fit Data* to use data points for editing. The *Fit Data* methods are recommended for most applications. Since the curve passes directly through the data points, these options offer direct control of the curve path.

TIP

Command: ***SPLINEDIT***
Select spline: **PICK**
Enter an option [Close/Join/Fit data/Edit vertex/convert to Polyline/Reverse/Undo/eXit] <eXit>: ***f***
Enter a fit data option [Add/Close/Delete/Kink/Move/Purge/Tangents/toLerance/eXit] <eXit>:

### Add

You can add points to the *Spline*. The *Spline* curve changes to pass through the new points. First, PICK an existing point on the curve. That point and the next one in sequence (in the order of creation) become highlighted. The new point will change the curve between those two highlighted data points (Fig. 16-32).

FIGURE 16-32

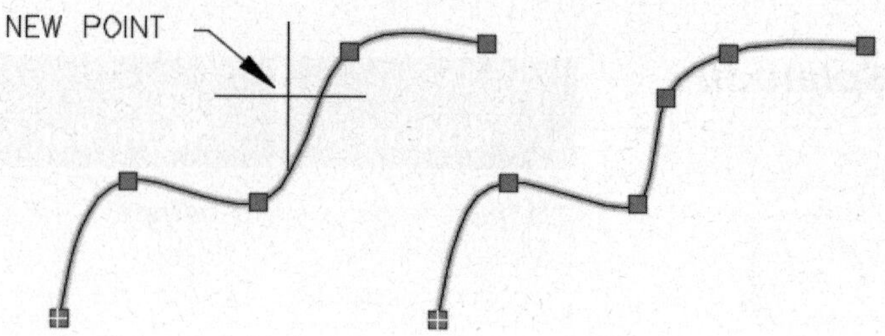


Specify existing fit point on spline <exit>: **PICK**
Specify new fit point to add <exit>: **PICK**
(PICK a new point location between the two marked points.)

### Close/Open

The *Close* option appears only if the existing curve is open, and the *Open* prompt appears only if the curve is closed. Selecting either option automatically forces the opposite change. *Close* causes the two ends to become tangent, forming one smooth curve (Fig. 16-33). This <u>tangent continuity</u> is characteristic of *Closed Splines* only. *Splines* that have matching endpoints do not have tangent continuity unless the *Close* option of *Spline* or *Splinedit* is used.

FIGURE 16-33

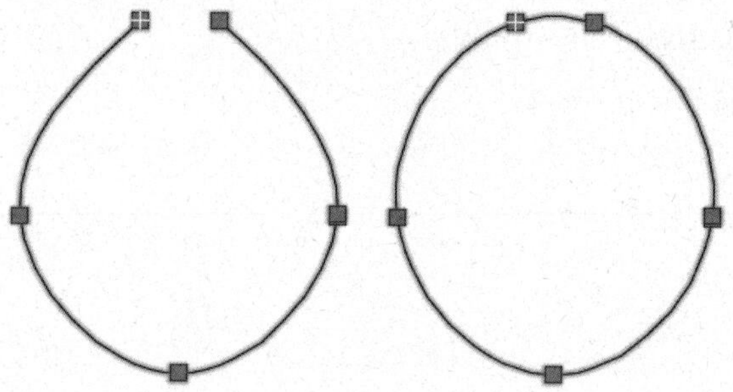


### Delete

Select a data point to delete. The curve then fits to the existing data points.

### Kink

This option allows you to add knots (special fit points) along the curve. Next, use the *Move* option to move the new points to change the curve shape. Unlike adding normal fit data points with the *Add* option, these points do not maintain tangent or curvature continuity when *Moved*. The resulting points are "kinks" in the otherwise smooth curve (Fig 16-34).

FIGURE 16-34

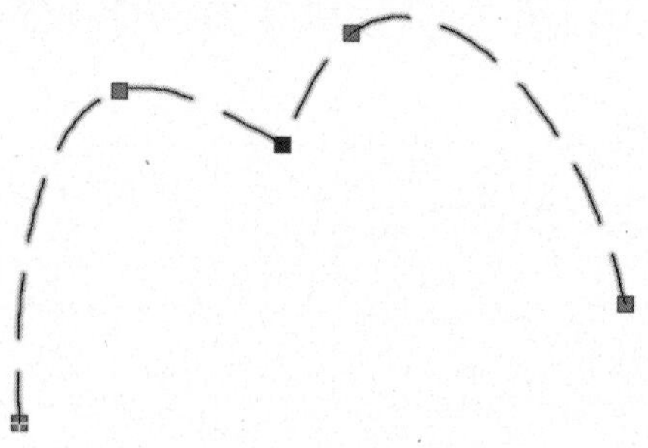

Enter a fit data option [Add/Close/Delete/Kink/Move/Purge/Tangents/toLerance/eXit] <eXit>: ***k***
Specify a point on the spline <exit>:

### Move

You can move any data point to a new location with this option (Fig. 16-35). The beginning endpoint (in the order of creation) becomes highlighted. Type ***N*** for next or ***S*** to select the desired data point to move, then PICK the new location.

FIGURE 16-35

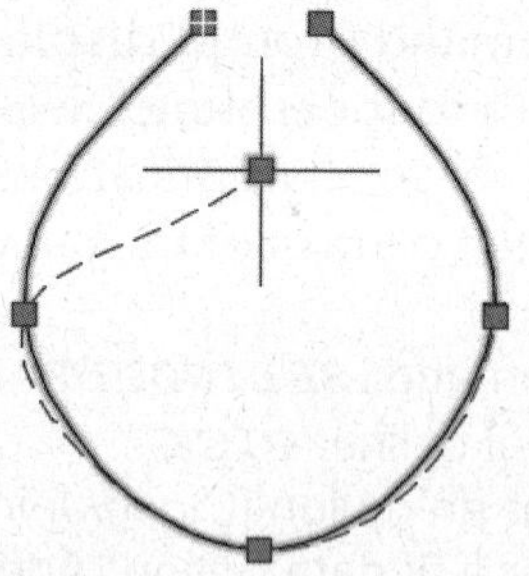


Specify new location or [Next/Previous/Select point/eXit] <N>:

### *Purge*

*Purge* deletes all data points and renders the *Fit Data* set of options unusable. You are returned to the control point options (top level). To reinstate the points, use *Undo*.

### *Tangents*

You can change the directions for the start and endpoint tangents with this option. This action gives the same control that exists with the "Enter start tangent" and "Enter end tangent" prompts of the *Spline* command used when the curves were created (see Fig. 15-13, *Spline, End Tangent*).

### *toLerance*

Use *toLerance* to specify a value, or tolerance, for the curve to "fall" away from the data points. Specifying a tolerance causes the curve to smooth out, or fall, from the data points. The higher the value, the more the curve "loosens." The *toLerance* option of *Splinedit* is identical to the *Fit Tolerance* option available with *Spline* (see Chapter 15, *Spline, Tolerance*).

## Editing *Spline* Control Vertices

Use the top level of command options (except *Fit Data*) to edit the *Spline's* control vertices. These options are similar to those used for editing the data points; however, the results are different since the curve does not pass through the control vertices.

Command: ***SPLINEDIT***
Select spline: **PICK**
Enter an option [Close/Join/Fit data/Edit vertex/convert to Polyline/Reverse/Undo/eXit] <eXit>:

### *Fit Data*

Discussed previously.

### *Close/Open*

These options operate similar to the *Fit Data* equivalents; however, the resulting curve falls away from the control points (Fig. 16-36; see also Fig. 16-33, *Fit Data, Close*).

### *Join*

You can *Join* one *Spline* to another as long as the two curves have shared (connecting) endpoints. The resulting curve may have a "kink" at the joined vertex. You can then use *Edit vertex, Delete* to delete the kink and create a smooth curve.

**FIGURE 16-36**

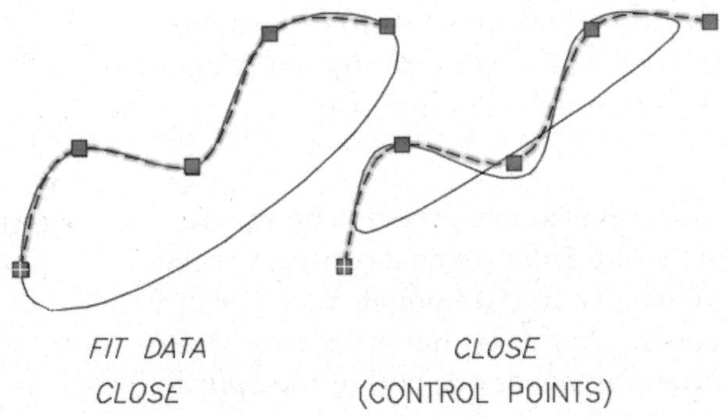


### *Edit Vertex*

*Edit Vertex* allows you to change the location of any control points. This is the control points' equivalent to the *Move* option of *Fit Data* (see Fig. 16-35, *Fit Data, Move*).

*Add control points* is the control points' equivalent to *Fit Data Add* (see Fig. 16-32). At the "Specify a point on the Spline" prompt, simply PICK a point near the desired location for the new point to appear.

*Elevate order* allows you to increase the number of control points uniformly along the length of the *Spline*. Enter a value from *n* to 26, where *n* is the current number of points + one. Once a *Spline* is elevated, it cannot be reduced.

*Weight* is an option that you use to assign a value to the amount of "pull" that a specific control point has on the *Spline* curve. The higher the value, the more "pull," and the closer the curve moves toward the control point. The typical method of selecting points (*Next/Previous/Select point/eXit/*) is used.

### *Reverse*

The *Reverse* option reverses the direction of the *Spline*. The first endpoint (when created) then becomes the last endpoint. Reversing the direction may be helpful for selection during the *Move* option.

### *convert to Polyline*

Use this option to convert the *Spline* to a *Polyline*. You are prompted for a "precision value." The precision value determines how closely the resulting polyline matches the spline or how many polyline segments are created. Valid values are any integer between 0 and 99.

## Editing *Splines* with Grips

TIP

Similar to editing *Plines* with the object's grips, you can edit *Splines* with the grips that appear on *Splines* (see related information in the "Editing *Plines* with Grips" section). Assuming the *GRIPS* system variable is set to 1 or 2 (on), selecting a *Spline* causes the grips to appear at each fit data point or control vertex, depending on the type of *Spline*. Press Escape to cancel grips. Editing *Splines* with grips is generally easier and more intuitive than using the *Splinedit* command.

If you click on a cold grip, it changes from a cold grip (blue) to a hot grip (red) and can be stretched (moved) to any location. For example, selecting the *Spline* in Figure 16-37 causes grips to appear at each fit data point. These vertices can be selected and stretched to any location. If you hover over a grip (the grip turns pink), a menu appears with additional options to add or remove fit points as well as change the tangent direction for endpoint grips.

**FIGURE 16-37**

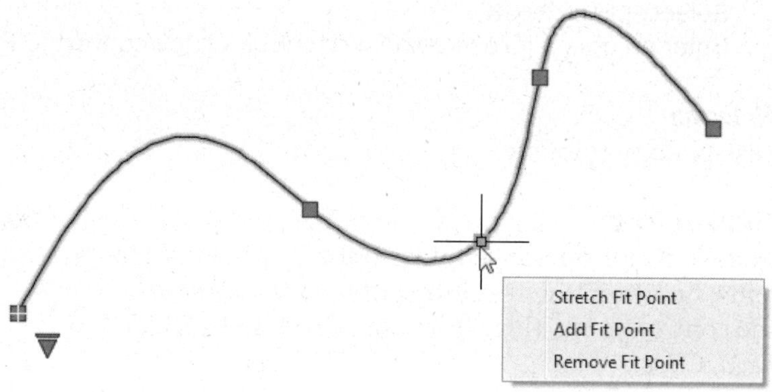


You can also use grips to change the state of a *Spline* from a control vertices *Spline* to a fit data points *Spline* or vice versa. Do this by hovering over the down-arrow at one end of the *Spline* then click so the menu appears (see Figure 16-38).

**FIGURE 16-38**

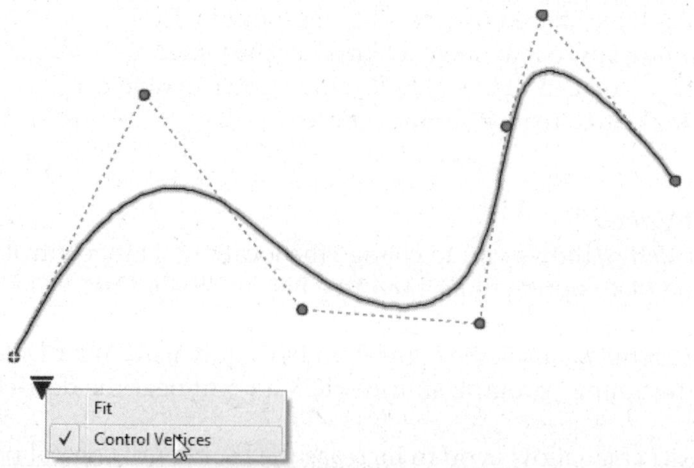

To edit the grips on control vertices, select the object so its grips appear. If you select any grip to make it 'hot" (red), you can then stretch the control vertex to a new location. Hovering over a grip until it turns pink then selecting it produces the menu shown in Figure 16-39. You can add or remove vertices or refine vertices (automatically add a control point that matches the existing curve).

See Chapter 19 for more information on grips.

FIGURE 16-39

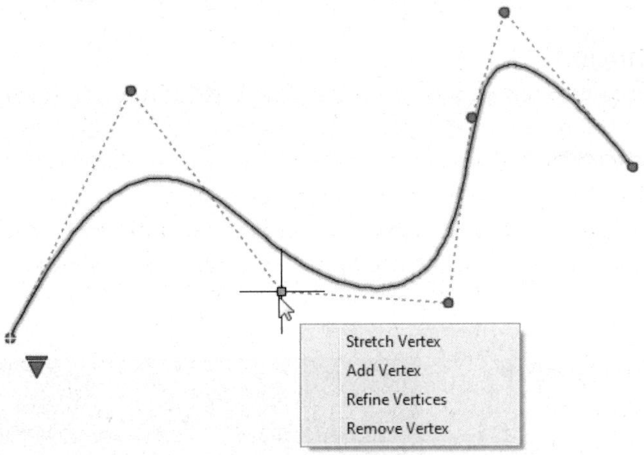


### *Reverse*

| Ribbon | Menu Bar | Command (Type) | Alias (Type) | Shortcut |
|---|---|---|---|---|
| *Home Modify* (icon) | ... | *Reverse* | ... | ... |

Use this command to reverse the order of vertices of selected *Lines*, *Plines*, *Splines*, and *Helixes*. *Reverse* causes the original first point or vertex to become the last, and vice versa. *Reverse* is helpful to reverse the direction of linetypes that include text. For example, a linetype that included text that appears upside down could be reversed to display text right-side up.

### *Blend*

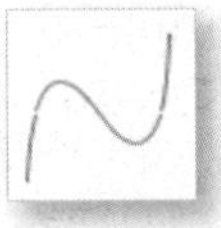

| Ribbon | Menu Bar | Command (Type) | Alias (Type) | Shortcut |
|---|---|---|---|---|
| *Home Modify* (icon) | *Modify Blend Curves* | *Blend* | ... | ... |

This command blends the gap between two selected lines or curves by adding a *Spline*. The resulting *Spline* creates a smooth curve which starts and ends tangent to the two objects that are connected. The shape of the resulting *Spline* depends on the specified *Continuity*. The lengths of the selected objects remain unchanged. You can select *Lines, Arcs, Elliptical* arcs, *Helixes,* open *Polylines*, and open *Splines* to *Blend*.

Command: ***BLEND***
Continuity = Tangent
Select first object or [CONtinuity]:
**PICK**
Select second object: **PICK**

Create the *Blend* by selecting each object near an endpoint. Figure 16-40 shows two *Line* segments connected by *Blend*. Keep in mind that the resulting *Spline* can be further edited by accessing the grips at the *Spline's* control points.

FIGURE 16-40

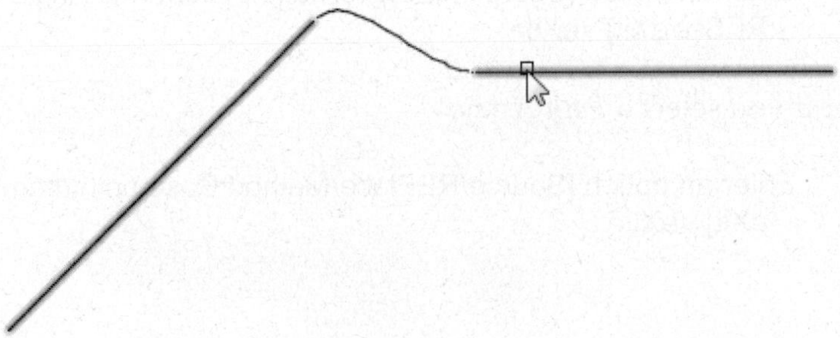

### Continuity

Specify one of two types of blends, *Tangent* or *Smooth*.

### Tangent

*Tangent* creates a degree 3 spline with tangency (G1) continuity to the selected objects at their endpoints.

### Smooth

Use this option to create a degree 5 spline with curvature (G2) continuity to the selected objects at their endpoints. If you use the *Smooth* option, do not switch the display from control vertices to fit points. This action changes the *Spline* to degree 3, which will change the shape of the blended curve.

## Arrayedit

| Ribbon | Menu Bar | Command (Type) | Alias (Type) | Shortcut |
|---|---|---|---|---|
| *Home Modify* (icon) | *Modify Object > Array* | *Arrayedit* | ... | ... |

There are two basic methods for editing arrays: 1) you can <u>select</u> an associative array object, which causes the grips to appear and the *Array* editor panel on the Ribbon to be displayed, or 2) you can use the *Arrayedit* command. Both methods provide the same options.

If you use grips and the *Array* editor panel, most of the tools for making changes to the array are the same as when you created the array using the *Array Creation* panel with the exception of the *Source, Replace,* and *Reset* options which are described below. The array properties available on the *Array* editor panel and the *Arrayedit* command depend on the type of the selected array: rectangular, polar, or path. Refer to Chapter 9, Modify Commands I, for information on using the *Array Creation* panel and grips.

If you use the *Arrayedit* command to edit associative rectangular, polar, or path arrays, the options are in command line format. You can use *Arrayedit* to modify associative arrays by editing the array properties, editing source objects, or replacing items with other objects. Remember that the type of array you select (rectangular, path, or polar) determines the remaining prompts.

Command: ***ARRAYEDIT***
Select array: **PICK**

If you select a *Rectangular Array*:

Enter an option [Source/REPlace/Base point/Rows/Columns/Levels/RESet/eXit] <eXit>:

or if you select a *Polar Array*:

Enter an option [Source/REPlace/Base point/Items/Angle between/Fill angle/Rows/Levels/ROTate items/RESet/eXit] <eXit>:

or if you select a *Path Array*:

Enter an option [Source/REPlace/Method/Base point/Items/Rows/Levels/Align items/Z direction/RESet/eXit] <eXit>:

The options are listed below.

### *Source*

This option activates an editing state in which you can update the array by changing one of its items. Select one of the array items as a source object to modify. Make any type of change such as adding lines, circles, trimming, etc. All changes (including the creation of new objects) you make to the source are applied to all items in the array that reference that source object.

### *Replace*

You can replace one of the arrayed objects with an existing object or set of objects. The other existing items in the array do not change. All items referencing the original source object are also changed to the source. You must select the object(s) to use as the replacement objects, specify a base point for positioning the replacement object, and specify an array item to replace.

### *Base point*

Use this option to redefine the base point of the existing array. Path arrays are repositioned relative to the new base point.

### *Rows*

This option allows you to specify the number and spacing of rows and the incrementing elevation between them. Specify a value for the *Number of rows* in the array or use an *Expression* (mathematical formula or equation). You can also indicate the *Total* distance between the first and last rows. Use *Incrementing elevation* to specify the amount of elevation change between each successive row with this option.

### *Columns* (rectangular arrays)

You can change the number and spacing of columns by indicating the *Number* of columns, the *Distance* between columns, an *Expression*, or the *Total* distance between the first and last columns.

### *Levels*

For 3D arrays you can specify a change in the *Number* of levels, *Distance* between levels, give an *Expression*, or give the *Total* distance between the first and last levels.

### *Method* (path arrays)

This choice controls how to distribute items along the path. Use *Divide* to redistribute items evenly along the length of the path or use *Measure* to maintain current spacing when the path is edited.

### *Items* (path and polar arrays)

You can specify the *Number* of items in the array using a value or expression. For path arrays you can *Fill entire path,* which uses the current distance between items to determine how the path is filled. Use this option if you expect the size of the path to change and want to maintain a specific spacing between array items. For path arrays whose *Method* property is set to *Measure*, you are prompted to redefine the distribution method (*Divide, Total, Expression*).

### *Align items* (path arrays)

This option specifies whether to align each item to be tangent to the path direction. The alignment is relative to the orientation of the first item by answering *Yes* (sets each arrayed item to follow perpendicular to the orientation of the curve) or *No* (sets all arrayed items to the same alignment as the source object in the array).

### *Z direction* (path arrays)

Specify whether to maintain the original Z direction or to naturally bank the items along a 3D path.

*Angle between* (polar arrays)
Specify the angle between items using a value or expression.

*Fill angle* (polar arrays)
Specify the angle between the first and last items in the array using a value or expression.

*Rotate items* (polar arrays)
You can choose to rotate the items as they are arrayed.

*Reset*
This option restores any erased items and removes any item overrides.

## *Mledit*

| Ribbon | Menu Bar | Command (Type) | Alias (Type) | Shortcut |
|---|---|---|---|---|
| ... | *Modify Object > Multiline...* | *Mledit* or *-Mledit* | ... | ... |

The *Mledit* command provides tools for editing multilines created with the *Mline* command. Invoking the *Mledit* command by any method produces the *Multiline Edit Tools* dialog box (Fig. 16-41). Selecting an option closes the dialog box allowing you to select the desired multiline segments for the editing action.

*Mledit* should be used for most *Mline* modifications. Although some typical editing commands such as *Trim*, *Extend*, *Stretch*, and Grips operate with *Mlines*, the *Break*, *Fillet*, and *Chamfer* commands do not operate with *Mlines* unless the *Mlines* are *Exploded*, which is normally not desired. (See "Other Editing Possibilities for *Mlines*" at the end of this section.)

**FIGURE 16-41**

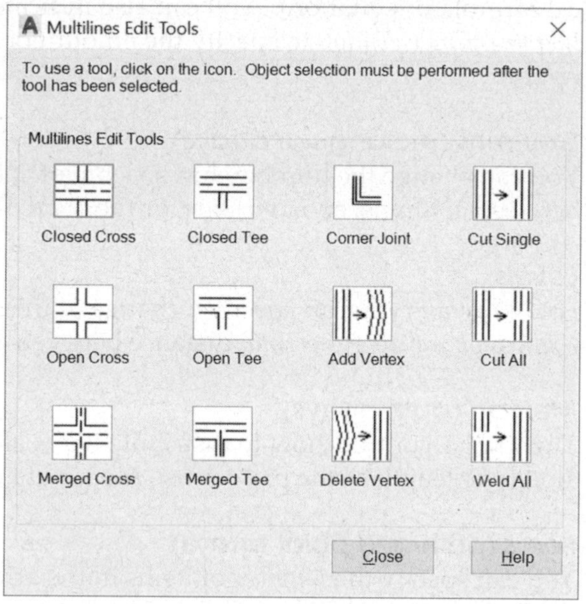


Command: **MLEDIT** (The *Multiline Edit Tools* dialog box appears. Select the desired option.)
Select first mline: **PICK**
Select second mline: **PICK**
Select first mline or [Undo]: **PICK** or **Enter**
Command:

The *Multiline Edit Tools* dialog box is organized in columns as follows:

| Intersection—Cross | Intersection—Tee | Corner, Vertices | Lines |
|---|---|---|---|
| Closed Cross | Closed Tee | Corner Joint | Cut Single |
| Open Cross | Open Tee | Vertex Add | Cut All |
| Merged Cross | Merged Tee | Vertex Delete | Weld |

### *Closed Cross*

Use this option to trim one of two intersecting *Mlines*. The first *Mline* PICKed is trimmed to the outer edges of the second. The second *Mline* is "closed." All line elements in the first multiline are trimmed (Fig. 16-42).

**FIGURE 16-42**

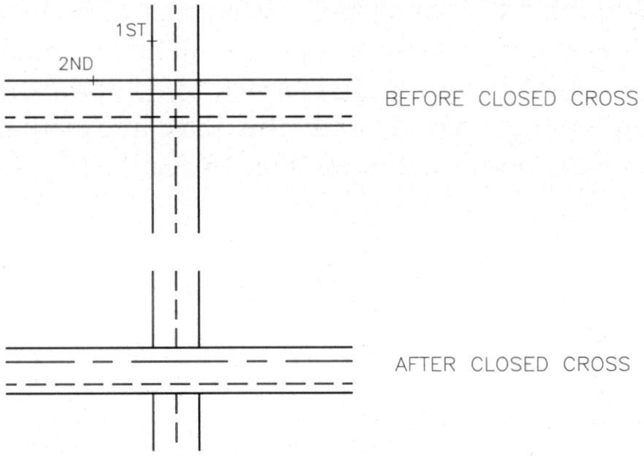


### *Open Cross*

This option trims both intersecting *Mlines*. Both *Mlines* are "open." All line elements of the first *Mline* PICKed are trimmed to the outer edges of the second. Only the outer line elements of the second multiline are trimmed, while the inner lines continue through the intersection (Fig. 16-43).

**FIGURE 16-43**

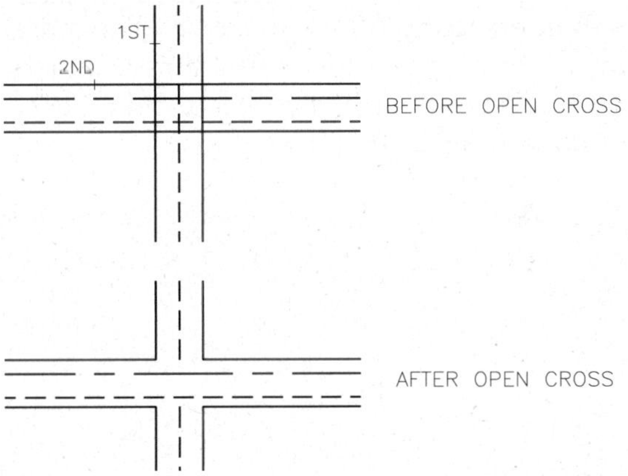


### *Merged Cross*

With this option, the outer line elements of both intersecting *Mlines* are trimmed and the inner line elements merge. The inner line elements merge at the second *Mline's* next set (Fig. 16-44). A full merge (both inner lines continue through) occurs only if the second *Mline* PICKed has no or only one inner line element.

**FIGURE 16-44**

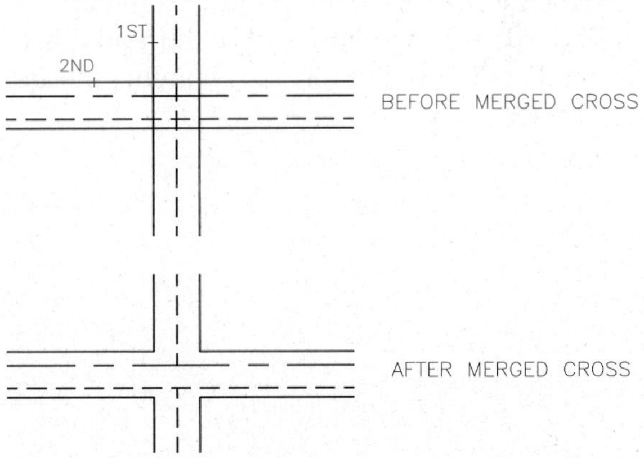


### *Closed Tee*

As indicated by the image tile, a "T" intersection is created rather than a "crossing" intersection. The first *Mline* PICKed is trimmed to the nearest (to the PICK point) outer edge of the second. Only the side of the first *Mline* nearest the PICK point remains. The second *Mline* is not affected (Fig. 16-45).

**FIGURE 16-45**

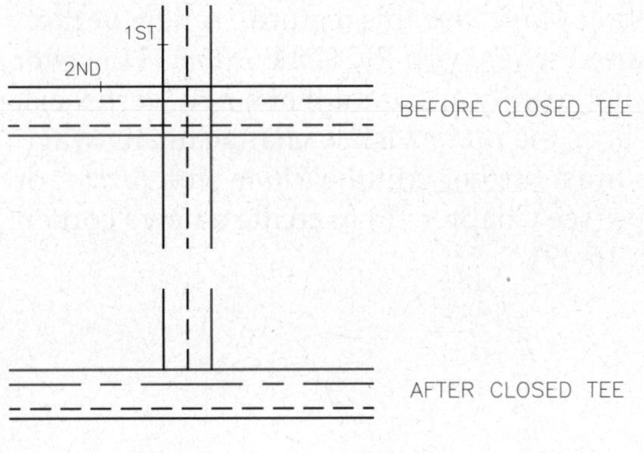

### *Open Tee*

With this "T" intersection, both outer edges of two intersecting *Mlines* are "open." All line elements of the first *Mline* PICKed are trimmed at the outer edges of the second. Only the outer line element of the second is trimmed (Fig. 16-46).

**FIGURE 16-46**

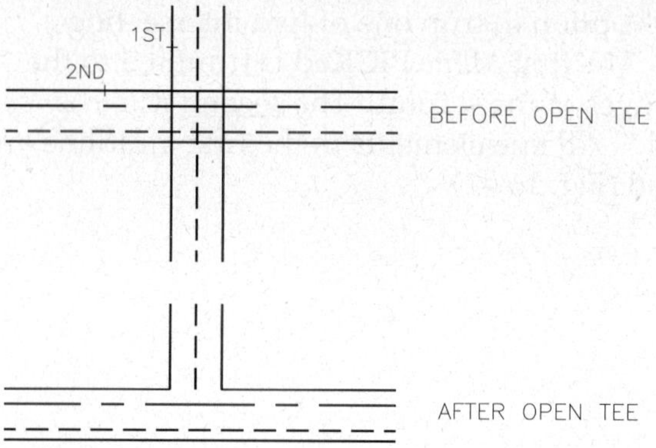


### *Merged Tee*

This "T" option allows the inner line elements of both intersecting *Mlines* to *merge*. The merge occurs at the second *Mline's* first inner element. Only the side of the first line nearest the PICK point remains (Fig. 16-47).

**FIGURE 16-47**

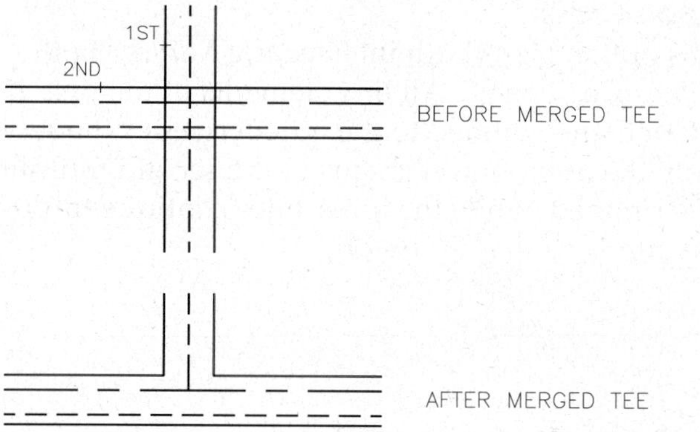


### *Corner Joint*

This option trims both *Mlines* to create a corner. Only the PICKed sides of the *Mlines* remain and the extending portions of both (if any) are trimmed. All of the inner line elements merge (Fig. 16-48).

**FIGURE 16-48**

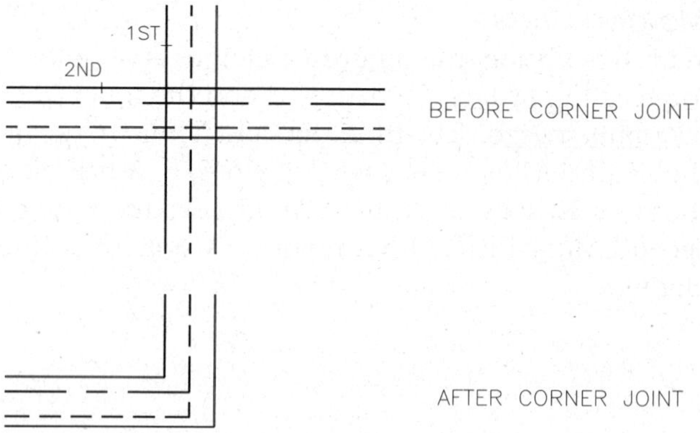


### *Add Vertex*

If you want to add a new corner (vertex) to an existing *Mline*, use this feature. A new vertex is added where you PICK the *Mline*. However, it is not readily apparent that a new vertex exists, nor does the *Mline* visibly change in any way. You must further edit the *Mline* with *Stretch* or Grips (see Chapter 19) to create a new "corner" (Fig. 16-49).

**FIGURE 16-49**

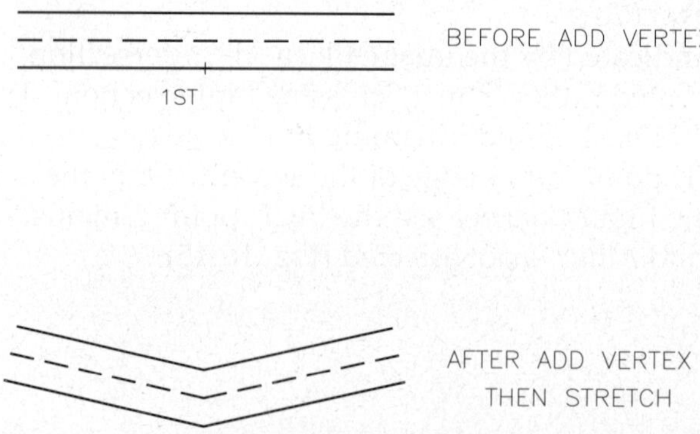

### *Delete Vertex*

Use this feature to remove a corner (vertex) of an *Mline*. The vertex nearest the location you PICK is deleted. The resulting *Mline* contains only one straight segment between the adjacent two vertices. Unlike *Add Vertex,* the *Mline* immediately changes and no further editing is needed to see the effect (Fig. 16-50).

**FIGURE 16-50**

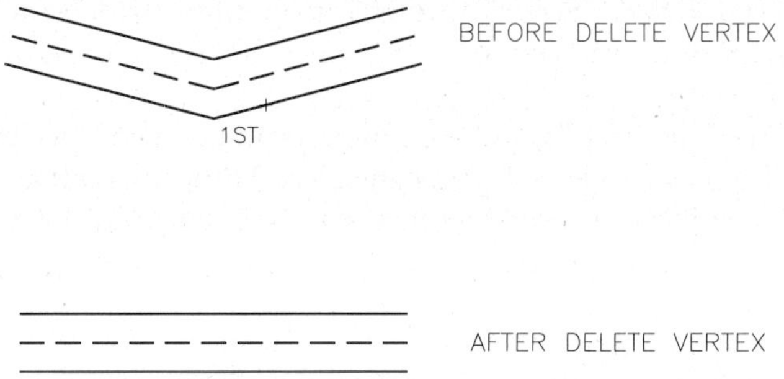


### *Cut Single*

The *Cut* options are used to break (cut) a space in one *Mline*. *Cut Single* breaks any line element that is selected. Similar to *Break 2Points,* the break occurs between the two PICK points (Fig. 16-51). The break points can be on either side of a vertex.

**FIGURE 16-51**

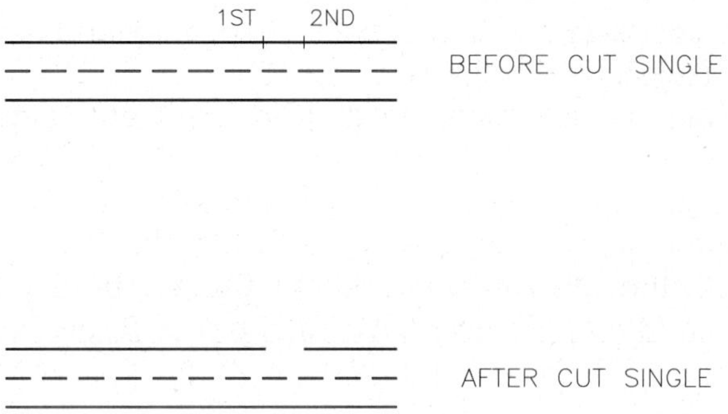


### *Cut All*

This *Cut* option breaks all line elements of the selected *Mline*. Any line elements can be selected. The line elements are cut at the PICK points in a direction perpendicular to the axis of the *Mline* (Fig. 16-52). NOTE: Although the *Mline* appears to be cut into two separate *Mlines,* it remains one single object. Using *Stretch* or Grips to "move" the *Mline* causes the cut to close again.

**FIGURE 16-52**

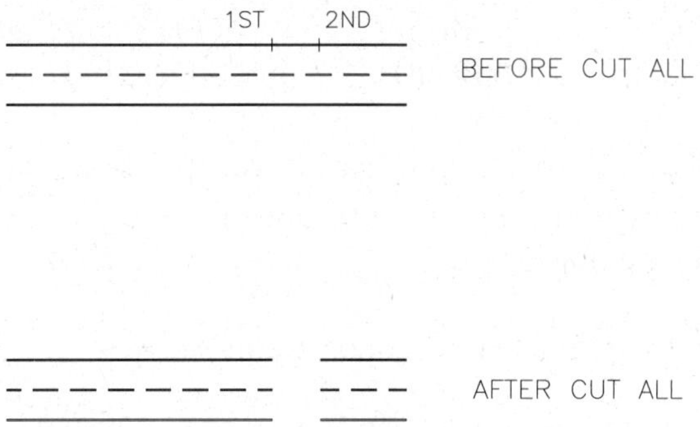


### *Weld*

This option reverses the action of a *Cut*. The *Mline* is restored to its original continuous configuration. PICK on both sides of the "break" (Fig. 16-53).

Like many other dialog box-based commands, *Mledit* can also be used in command line format by using the hyphen (-) symbol as a prefix to the command. Key in "*-MLEDIT*" to force the command line interface. The following prompt appears:

**FIGURE 16-53**

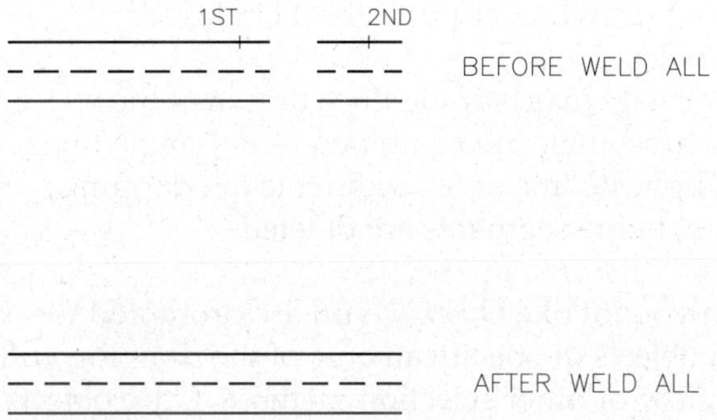


Command: ***-MLEDIT***
Enter mline editing option [CC/OC/MC/CT/OT/MT/CJ/AV/DV/CS/CA/WA]:

### Other Editing Possibilities for *Mlines*

*Trim, Extend, Stretch,* and Grips operate correctly with *Mlines*. *Stretch* and Grips can be used to change endpoints and move vertices; however, doing so may affect intersections with other *Mlines*.

**TIP** The *Trim* and *Extend* commands can be used very effectively with *Mlines*. For the "cutting edges" and "boundary edges," you can select *Mlines* as well as other objects (*Lines, Plines, Arcs,* etc.). If you select an *Mline* as a "cutting edge" or "boundary edge," you can then select from a *Closed, Open,* or *Merged* intersection, producing results identical to using the *Closed Tee, Open Tee,* or *Merged Tee* options of *Mledit*.

Command: **`TRIM`**
Current settings: Projection=UCS, Edge=None
Select cutting edges ...
Select objects or <select all>: **`PICK`**
Select objects: **`Enter`**
Select object to trim or shift-select to extend or [Fence/Crossing/Project/Edge/eRase/Undo]: **`PICK`** (select *Mline* object)
Enter mline junction option [Closed/Open/Merged] <Open>:

You may experience situations where *Mledit, Trim, Extend, Stretch*, or Grips cannot handle your desired editing request. It is possible to *Explode* the *Mline*, which converts each line element to an individual object, allowing you to use other editing options. However, this practice is not recommended because once an *Mline* is *Exploded*, it cannot be converted back to an *Mline*, nor can *Mledit* be used.

## *Overkill*

| Ribbon | Menu Bar | Command (Type) | Alias (Type) | Shortcut |
|---|---|---|---|---|
| *Home Modify* (icon) | *Modify Delete Duplicate Objects* | *Overkill* | ... | ... |

*Overkill* automatically removes duplicate or overlapping lines, arcs, and polylines. In addition, if partially overlapping or contiguous lines, arcs, and polylines are selected, they will be combined to become one object.

*Overkill* treats these particular circumstances:

Duplicate objects from the drawing area or the block editor are deleted.
Duplicate copies of objects are deleted.
Arcs drawn over portions of circles are deleted.
Partially overlapping lines drawn at the same angle are combined into a single line.
Duplicate line or arc segments overlapping polyline segments are deleted.

When you invoke *Overkill*, you are prompted to select objects or specify an area of the drawing with a window or other selection method. Once objects are selected, the *Delete Duplicate Objects* dialog box appears (Fig. 16-54). Here you can specify which object properties are ignored and how colinear objects are treated when the selected objects are compared.

**FIGURE 16-54**

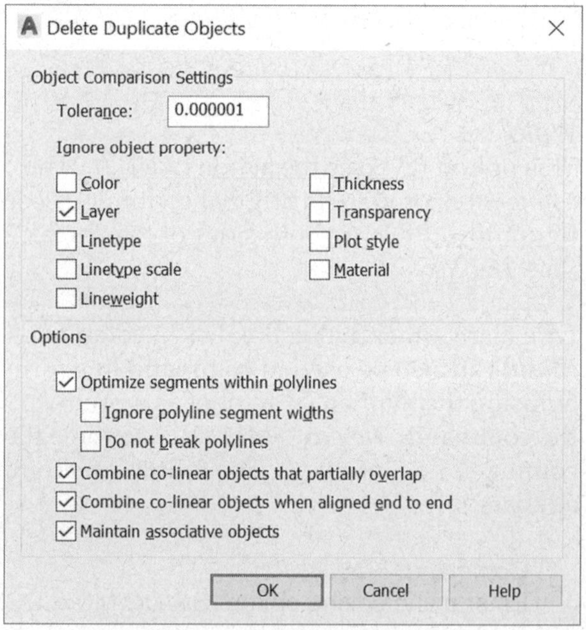

### *Tolerance*

This value controls the precision with which *Overkill* makes numeric distance comparisons as to whether objects actually overlap. If this value is 0, the two objects being compared must match before *Overkill* modifies or deletes one of them.

### *Ignore object property*

In this section of the dialog box you can check which object properties are ignored during comparisons. For example, if *Layer* is checked, one of two objects that overlap will be deleted even if they reside on different layers.

### *Options*

*Optimize segments within polylines*
If this box is checked, individual line and arc segments within selected polylines are compared. Duplicate vertices and segments are removed. *Overkill* also compares individual polyline segments with completely separate line and arc segments. If a polyline segment duplicates a line or arc object, one of them is deleted. If this option is not checked, polylines are compared as discreet objects and the two sub-options are not selectable.

*Ignore polyline segment width*
When checked, different polyline segment widths are ignored while optimizing polyline segments.

*Do not break polylines*
When checked, different polyline objects are unchanged.

*Combine co-linear objects that partially overlap*
If checked, any overlapping objects are combined into single objects.

*Combine co-linear objects when aligned end to end*
Check this option to combine objects that have common endpoints into single objects.

*Maintain associative objects*
Associative objects are not deleted or modified if this box is checked.

## Boolean Commands

*Region* combines one or several objects forming a closed shape into one object, a *Region*. The appearance of the object(s) does not change after the conversion, even though the resulting shape is one object (see "*Region*," Chapter 15).

Although the *Region* appears to be no different than a closed *Pline*, it is more powerful because several *Regions* can be combined to form complex shapes (called "composite *Regions*") using the three Boolean operations explained next. As an example, a set of *Regions* (converted *Circles*) can be combined to form the sprocket with only one *Subtract* operation, as shown in Chapter 15, Figure 15-41.

The Boolean operators, *Union*, *Subtract*, and *Intersect*, can be used with *Regions* as well as solids. Any number of these commands can be used with *Regions* to form complex geometry. To illustrate each of the Boolean commands, consider the shapes shown in Figure 16-55. The *Circle* and the closed *Pline* are first converted to *Regions*, then *Union*, *Subtract*, or *Intersection* can be used.

**FIGURE 16-55**

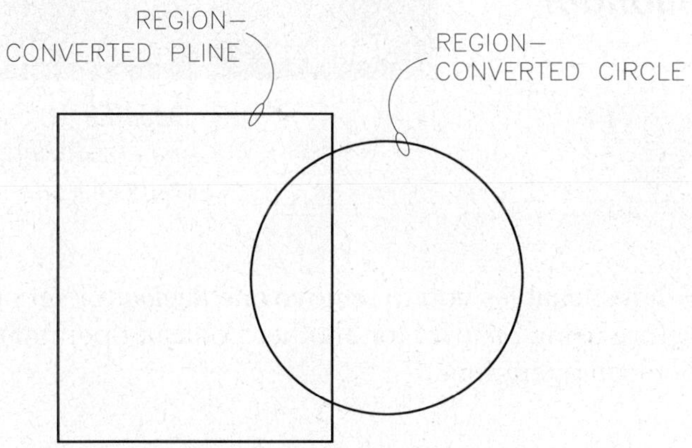

## Union

| Ribbon | Menu Bar | Command (Type) | Alias (Type) | Shortcut |
|---|---|---|---|---|
| ... | *Modify Solid Editing > Union* | *Union* | *UNI* | ... |

*Union* combines two or more *Regions* (or solids) into one *Region* (or solid). The resulting composite *Region* has the encompassing perimeter and area of the original *Regions*. Invoking *Union* causes AutoCAD to prompt you to select objects. You can select only existing *Regions* (or solids).

Command: ***UNION***
Select objects: **PICK** (*Region*)
Select objects: **PICK** (*Region*)
Select objects: **Enter**
Command:

FIGURE 16-56

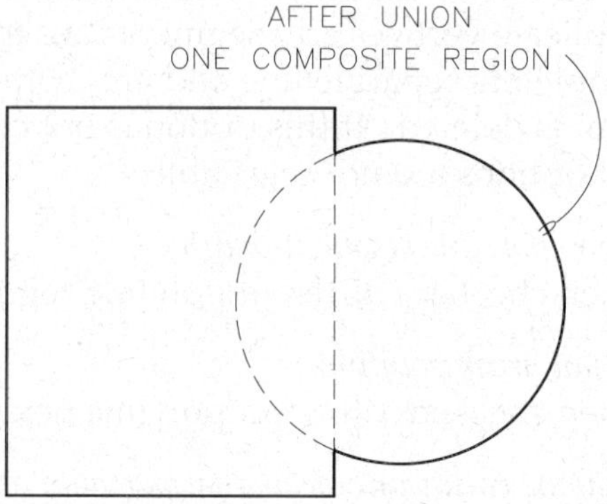


The selected *Regions* are combined into one composite *Region* (Fig. 16-56). Any number of Boolean operations can be performed on the *Region(s)*.

Several *Regions* can be selected in response to the "Select objects:" prompt. For example, a composite *Region* such as that in Figure 16-57 can be created with one *Union*.

Two or more *Regions* can be *Unioned* even if they do not overlap. They are simply combined into one object although they still appear as two.

FIGURE 16-57

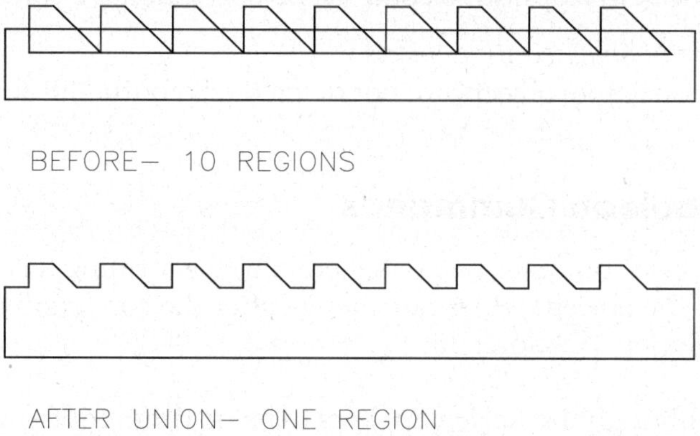


## Subtract

| Ribbon | Menu Bar | Command (Type) | Alias (Type) | Shortcut |
|---|---|---|---|---|
| ... | *Modify Solid Editing > Subtract* | *Subtract* | *SU* | ... |

*Subtract* enables you to remove one *Region* (or set of *Regions*) from another. The *Regions* must be created before using *Subtract* (or another Boolean operator). *Subtract* also works with solids (as do the other Boolean operations).

There are two steps to *Subtract*. First, you are prompted to select the *Region* or set of *Regions* to "subtract from" (those that you wish to keep), then to select the *Regions* "to subtract" (those you want to remove). The resulting shape is one composite *Region* comprising the perimeter of the first set minus the second (sometimes called "difference").

Command: ***SUBTRACT***
Select solids and regions to subtract from...
Select objects: **PICK**
Select objects: **Enter**
Select solids and regions to subtract...
Select objects: **PICK**
Select objects: **Enter**
Command:

Consider the two shapes previously shown in Figure 16-55. The resulting *Region* shown in Figure 16-58 is the result of *Subtracting* the circular *Region* from the rectangular one.

Keep in mind that multiple *Regions* can be selected as the set to keep or as the set to remove. For example, the sprocket illustrated previously (Fig. 15-41) was created by subtracting several circular *Regions* in one operation. Another example is the removal of material to create holes or slots in sheet metal.

**FIGURE 16-58**

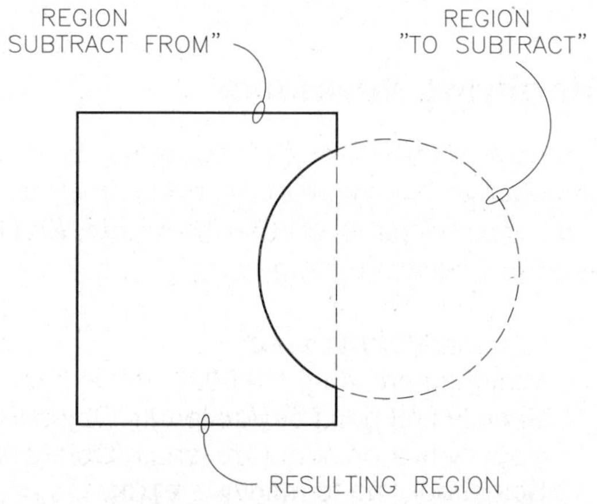


TIP

### *Intersect*

| Ribbon | Menu Bar | Command (Type) | Alias (Type) | Shortcut |
|---|---|---|---|---|
| ... | *Modify Solid Editing > Intersect* | *Intersect* | *IN* | ... |

*Intersect* is the Boolean operator that finds the common area from two or more *Regions*.

Consider the rectangular and circular *Regions* previously shown (Fig. 16-55). Using the *Intersect* command and selecting both shapes results in a *Region* comprising only that area that is shared by both shapes (Fig. 16-59):

Command: ***INTERSECT***
Select objects: **PICK**
Select objects: **PICK**
Select objects: **Enter**
Command:

**FIGURE 16-59**

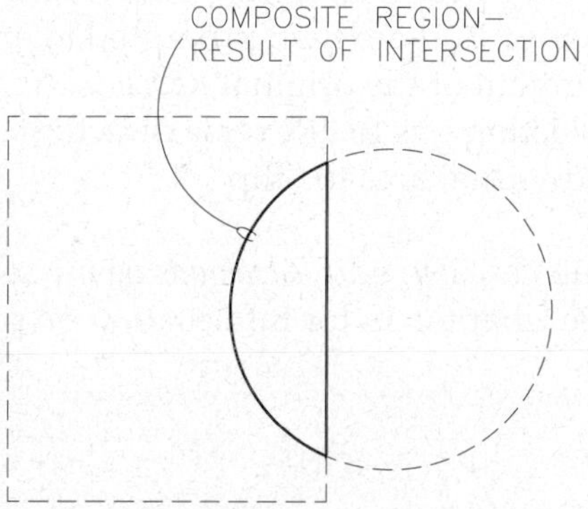

If more than two *Regions* are selected, the resulting *Intersection* is composed of only the common area from all shapes (Fig. 16-60). If all of the shapes selected do not overlap, a null *Region* is created (all shapes disappear because no area is common to all).

FIGURE 16-60

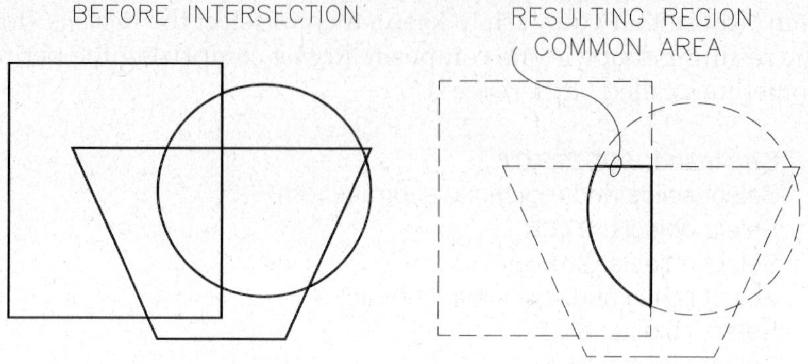


*Intersect* is a powerful operation when used with solid modeling. Techniques for saving time using Boolean operations are discussed in Chapter 35, Solid Modeling Construction.

## Modifying *Revclouds*

You can modify *Revclouds* using the *Revcloud* command with the *Modify* option. With this feature you can change the shape of the revision cloud and reverse the direction of the arcs. You cannot convert one type of *Revcloud* to another (*Freehand, Rectangular*, or *Polygonal*). For example, to edit a *Freehand Revcloud*, use the *Freehand* option.

```
Command: REVCLOUD
Minimum arc length: 0.5000   Maximum arc length: 0.5000   Style: Normal   Type: Freehand
Specify first point or [Arc length/Object/Rectangular/Polygonal/Freehand/Style/Modify] <Object>: F
Specify first point or [Arc length/Object/Rectangular/Polygonal/Freehand/Style/Modify] <Object>: M
Select polyline to modify:  PICK
Specify next point or [First point]:  PICK
Specify next point or [Undo]: PICK
Specify next point or [Undo]: Enter
Pick a side to erase: PICK
Reverse direction [Yes/No] <No>: Enter
```

AutoCAD begins drawing the new shape at the point you select for the "Select polyline to modify:" prompt. (Use the *First point* option to specify a different point.) At the "Specify next point:" prompts, select points to draw the new shape. You are then prompted to erase the old segment of the original *Revcloud* (Fig. 16-61). Selecting Yes to "Reverse direction" causes the individual arcs to "flip."

FIGURE 16-61

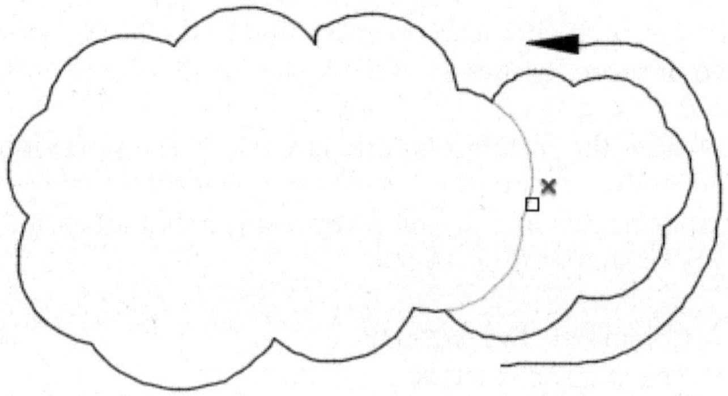

You can also edit *Revclouds* easily using grips. See Chapter 19 for information on grips.

## CHAPTER EXERCISES

1. *Chprop* or *Properties*

   ***Open*** the **PIVOTARM CH9** drawing that you worked on in Chapter 9 Exercises. ***Load*** the ***Hidden2*** and ***Center2 Linetypes***. Make two ***New Layers*** named **HID** and **CEN** and assign the matching linetypes and yellow and green colors, respectively. Check the *Limits* of the drawing, then calculate and set an appropriate ***LTSCALE***. Use ***Chprop*** or ***Properties*** palette to change the *Lines* representing the holes in the front view to the **HID** layer as shown in Figure 16-62. Use ***SaveAs*** and name the drawing **PIVOTARM CH16.**

FIGURE 16-62

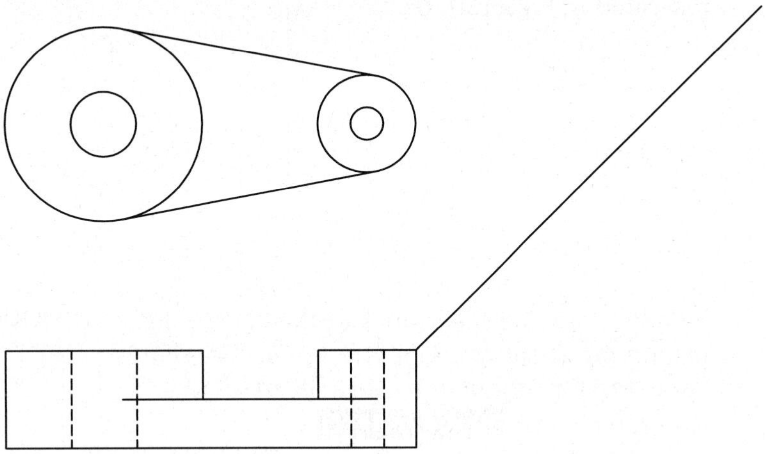

2. *Properties* **Palette**

   A design change is required for the bolt holes in **GASKETA** (from Chapter 9 Exercises). ***Open*** **GASKETA** and invoke the *Properties* palette. Change each of the four bolt holes to **.375** diameter (Fig. 16-63). ***Save*** the drawing.

FIGURE 16-63

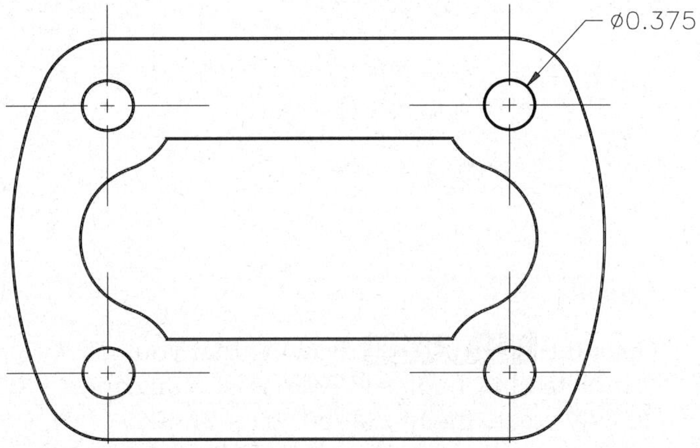


3. *Align*

   ***Open*** the **PLATES** drawing you created in Chapter 9. The three plates are to be stamped at one time on a single sheet of stock measuring 15″ x 12″. Place the three plates together to achieve optimum nesting on the sheet stock.

   A. Use ***Align*** to move the plate in the center (with 9 holes). Select the ***First source*** and ***destination points*** (1S, 1D) and ***Second source*** and ***destination points*** (2S, 2D) as shown in Figure 16-64.

FIGURE 16-64

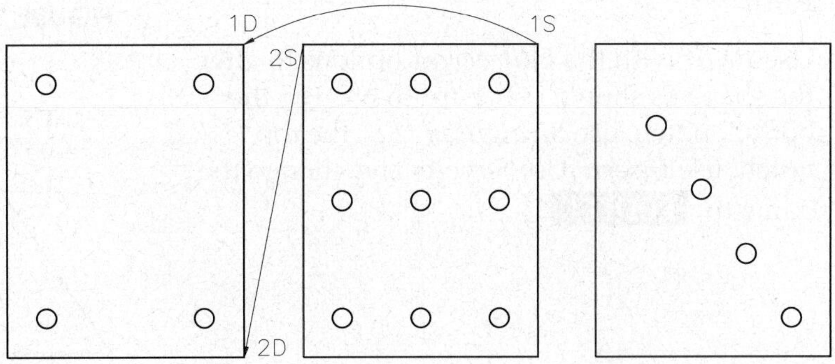

B. After the first alignment is complete, use ***Align*** to move the plate on the right (with 4 diagonal holes). The ***source*** and ***destination points*** are indicated in Figure 16-65.

**FIGURE 16-65**

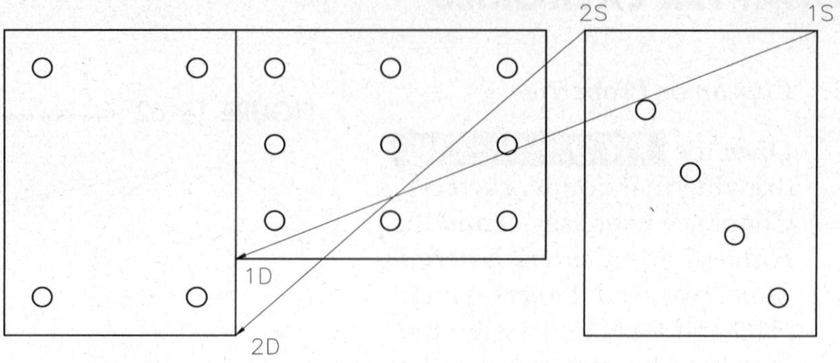


C. Finally, draw the sheet stock outline (15" x 12") using ***Line*** as shown in Figure 16-66. The plates are ready for production. Use ***SaveAs*** and assign the name **PLATENEST.**

**FIGURE 16-66**

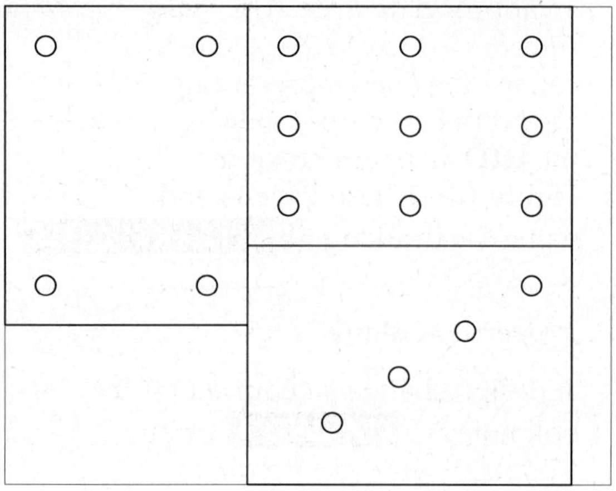

4. ***Explode***

***Open*** the **POLYGON1** drawing that you completed in Chapter 15 Exercises. You can quickly create the five-sided shape shown (continuous lines) in Figure 16-67 by *Exploding* the *Polygon*. First, ***Explode*** the *Polygon* and ***Erase*** the two *Lines* (highlighted). Draw the bottom ***Line*** from the two open ***Endpoints***. Do not exit the drawing.

REUSE

**FIGURE 16-67**

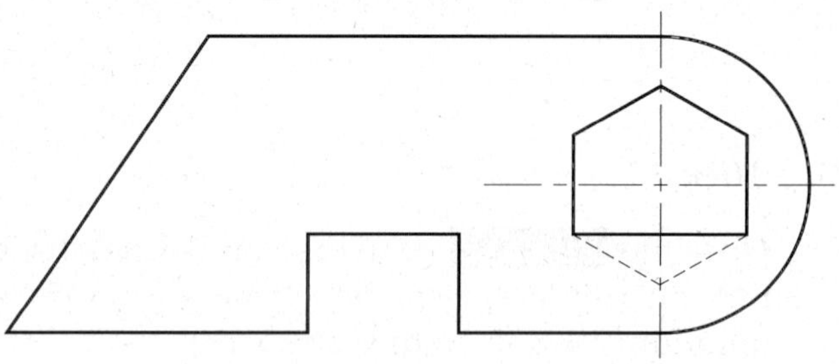

5. ***Pedit***

Use ***Pedit*** with the ***Edit vertex*** options to alter the shape as shown in Figure 16-68. For the bottom notch, use ***Straighten***. For the top notch, use ***Insert***. Use ***SaveAs*** and change the name to **PEDIT1.**

REUSE

**FIGURE 16-68**

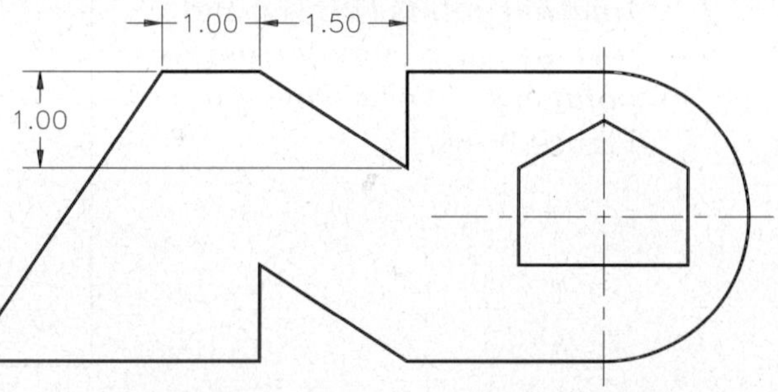

6. ***Pline, Pedit***

A. Create a line graph as shown in Figure 16-69 to illustrate the low temperatures for a week. The temperatures are as follows:

| X axis | Y axis |
|---|---|
| Sunday | 20 |
| Monday | 14 |
| Tuesday | 18 |
| Wednesday | 26 |
| Thursday | 34 |
| Friday | 38 |
| Saturday | 27 |

Use equal intervals along each axis. Use a ***Pline*** for the graph line. ***Save*** the drawing as **TEMP-A**. (You will label the graph at a later time.)

FIGURE 16-69

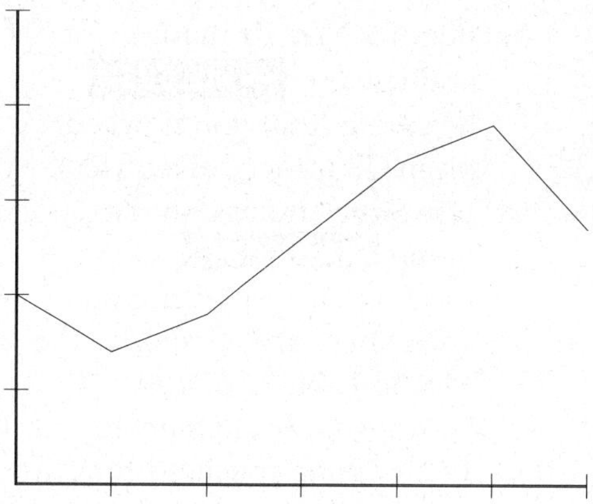

B. Use ***Pedit*** to change the *Pline* to a ***Spline***. Note that the graph line is no longer 100% accurate because it does not pass through the original vertices (see Fig. 16-70). Select the *Spline* to make its original control vertices visible.

Use the ***Properties*** palette and try the ***Cubic*** and ***Quadratic*** options. Which option causes the vertices to have more pull? Find the most accurate option. ***SaveAs*** **TEMP-B**.

FIGURE 16-70

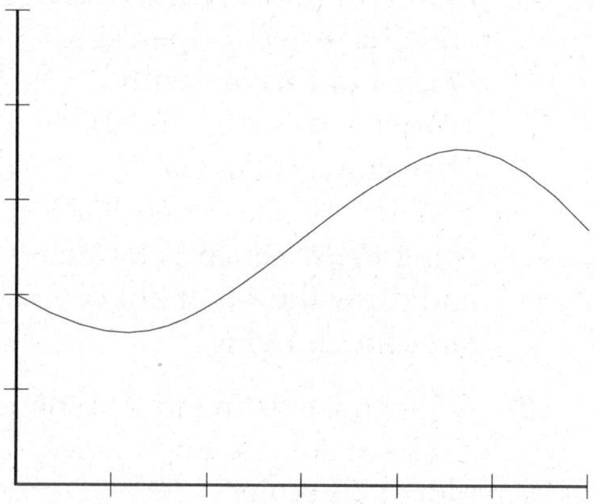

C. Use ***Pedit*** to change the curve from ***Pline*** to ***Fit Curve***. Does the graph line pass through the vertices? ***Saveas*** **TEMP-C**.

D. ***Open*** drawing **TEMP-A**. ***Erase*** the *Splined* ***Pline*** and construct the graph using a ***Spline*** instead (Fig. 16-71, see Exercise 7A for data). ***SaveAs*** **TEMP-D.** Compare the *Spline* with the variations of *Plines*. Which of the four drawings (A, B, C, or D) is smoothest? Which is the most accurate?

FIGURE 16-71

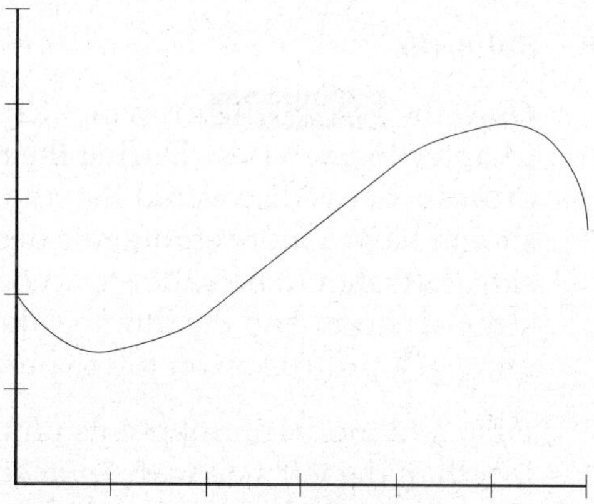

E. It was learned that there was a mistake in reporting the temperatures for that week. Thursday's low must be changed to 38 degrees and Friday's to 34 degrees. ***Open*** **TEMP-D** (if not already open) and use ***Splinedit*** to correct the mistake. Use the ***Move*** option of ***Fit Data*** so that the exact data points can be altered as shown in Figure 16-72. ***SaveAs*** **TEMP-E**.

FIGURE 16-72

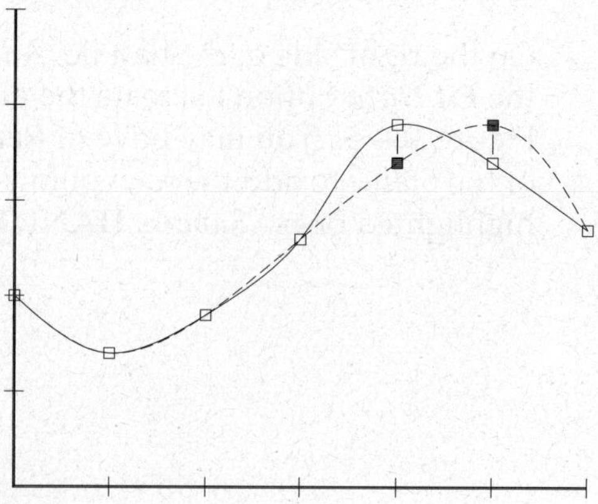

7. **Converting *Lines, Circles,* and *Arcs* to *Plines***

A. Begin a ***New*** drawing and use the **A-METRIC** template (that you worked on in Chapter 13 Exercises). Use ***Save*** and assign the name **GASKETC.** Change the ***Limits*** for plotting on an A sheet at 2:1 (refer to the Metric Table of *Limits* Settings and set *Limits* to 1/2 x *Limits* specified for 1:1 scale). Change the ***LTSCALE*** to **12**. First, draw only the <u>inside</u> shape using ***Lines*** and ***Circles*** (with ***Trim***) or ***Arcs*** (Fig. 16-73). Then convert the *Lines* and *Arcs* to one closed *Pline* using ***Pedit***. Finally, locate and draw the 3 bolt holes. ***Save*** the drawing.

**FIGURE 16-73**

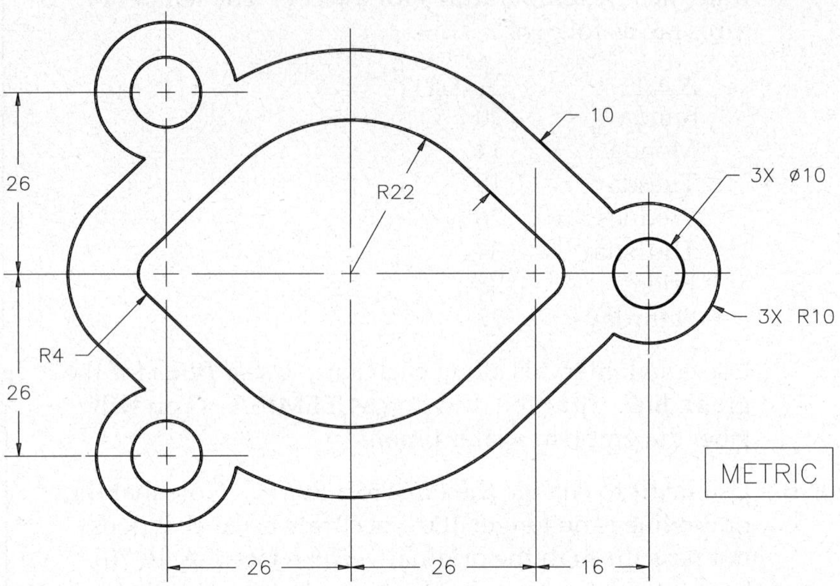


B. ***Offset*** the existing inside shape to create the outside shape. Use ***Offset*** to create concentric circles around the bolt holes. Use ***Trim*** to complete the gasket. ***Save*** the drawing and create a plot at 2:1 scale.

8. ***Splinedit***

***Open*** the **HANDLE** drawing that you created in Chapter 15 Exercises. During the testing and analysis process, it was discovered that the shape of the handle should have a more ergonomic design. The finger side (left) should be flatter to accommodate varying sizes of hands, and the thumb side (right) should have more of a protrusion on top to prevent slippage.

First, add more control points uniformly along the length of the left side with ***Spinedit, Edit vertex. Elevate*** the ***Order*** from 4 to **6**, then use ***Move Vertex*** to align the control points as shown in Figure 16-74.

On the right side of the handle, ***Add*** two points under the ***Fit Data*** option to create the protrusion shown in Figure 16-75. You may have to ***Reverse*** the direction of the *Spline* to add the new points between the two highlighted ones. ***SaveAs*** **HANDLE2**.

**FIGURE 16-74**

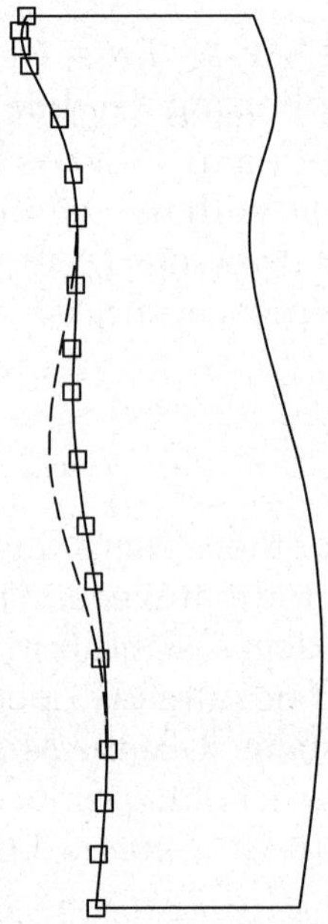

**FIGURE 16-75**

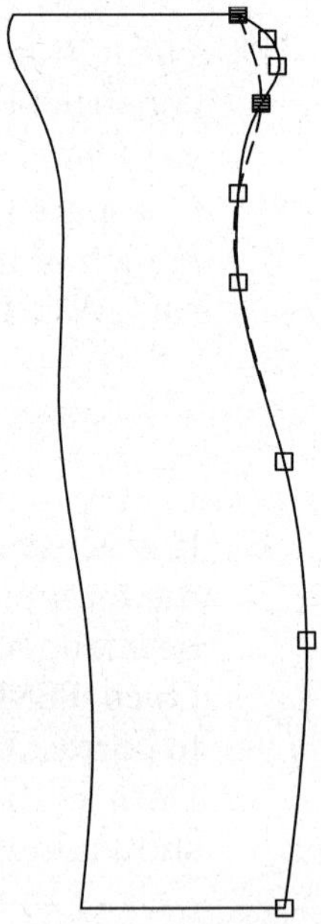

9. ***Mline, Mledit, Stretch***

Draw the floor plan of the storage room shown in Figure 16-76. Use ***Mline*** objects for the walls and ***Plines*** for the windows and doors. Use ***Mledit*** to treat the intersections as shown. When your drawing is complete according to the given specifications, use ***Stretch*** to center the large window along the top wall. Save the drawing as **STORE ROOM.**

REUSE

FIGURE 16-76

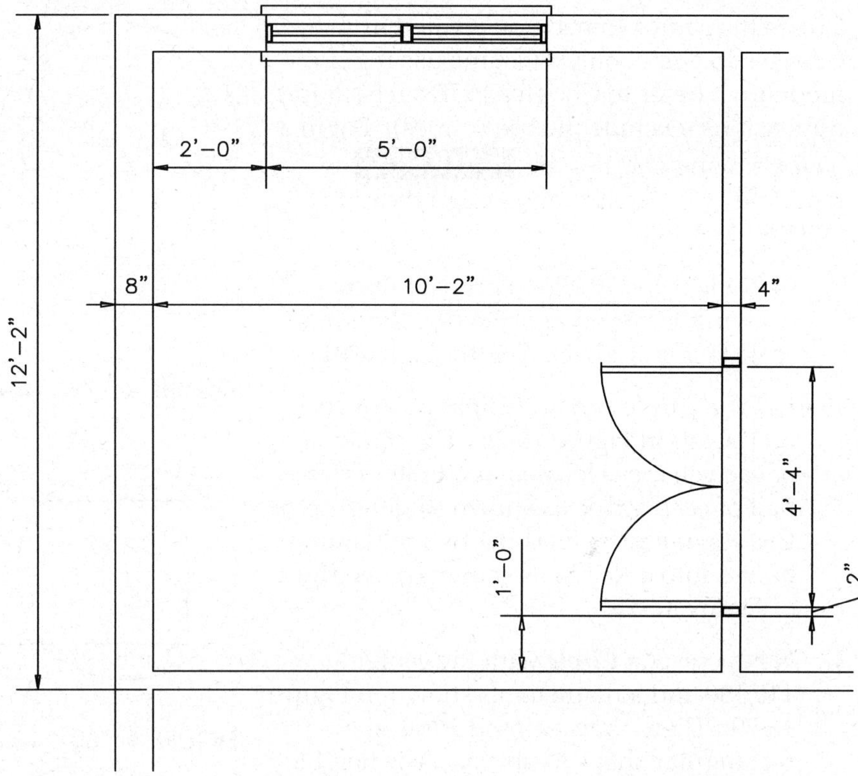


10. **Gear Drawing**

Complete the drawing of the gear you began in Chapter 15 Exercises called **GEAR-REGION.** If you remember, three shapes were created (two *Circles* and one "tooth") and each was converted to a *Region*. Finally, the "tooth" was *Arrayed* to create the total of 40 teeth.

REUSE

A. To complete the gear, *Subtract* the small circular *Region* and all of the teeth from the large circular *Region*. First, use ***Subtract***. At the "Select solids and regions to subtract from…" prompt, PICK the large circular *Region*. At the "Select solids and regions to subtract…" prompt, use a window to select everything (the large circular *Region* is automatically filtered out). The resulting gear should resemble Figure 16-77. ***Save*** the drawing as **GEAR-REGION 2.**

B. Consider the steps involved if you were to create the gear (as an alternative) by using *Trim* to remove 40 small sections of the large *Circle* and all unwanted parts of the teeth. *Regions* are clearly easier in this case.

FIGURE 16-77

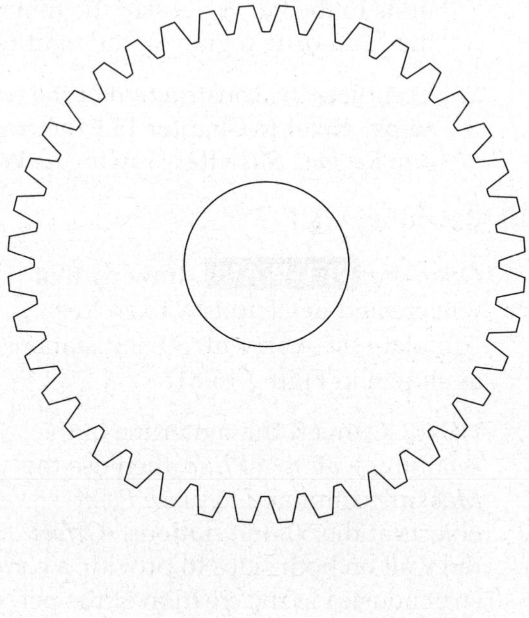

**11. Wrench Drawing**

Create the same wrench you created in Chapter 15 again, only this time use region modeling. Refer to Chapter 15, Exercise 6 for dimensions (exclude the break lines). Begin a *New* drawing and use the **A-METRIC** template. Use ***SaveAs*** and assign the name **WRENCH-REG**.

A. Set ***Limits*** to **372,288** to prepare the drawing for plotting at 3:4 (the drawing scale factor is 33.87). Set the ***Grid*** to **10**.

B. Draw a ***Circle*** and an ***Ellipse*** as shown on the left in Figure 16-78. The center of each shape is located at **60,150**. ***Trim*** half of each shape as shown (highlighted). Use ***Region*** to convert the two remaining halves into a *Region* as shown on the right of Figure 16-78.

C. Next, create a ***Circle*** with the center at **110,150** and a diameter as shown in Figure 16-79. Then draw a closed ***Pline*** in a rectangular shape as shown. The height of the rectangle must be drawn as specified; however, the width of the rectangle can be drawn approximately as shown on the left. Convert each shape to a ***Region***, then use ***Intersect*** to create the region as shown on the right side of the figure.

**FIGURE 16-78**

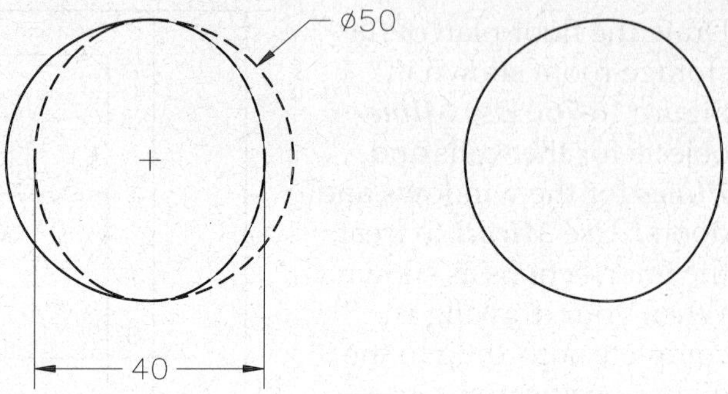


**FIGURE 16-79**

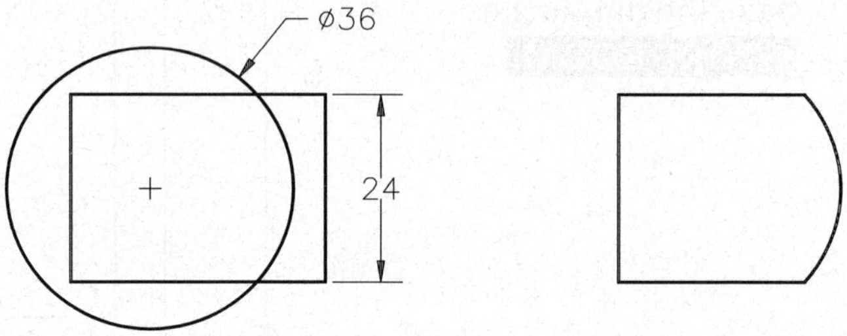


**FIGURE 16-80**

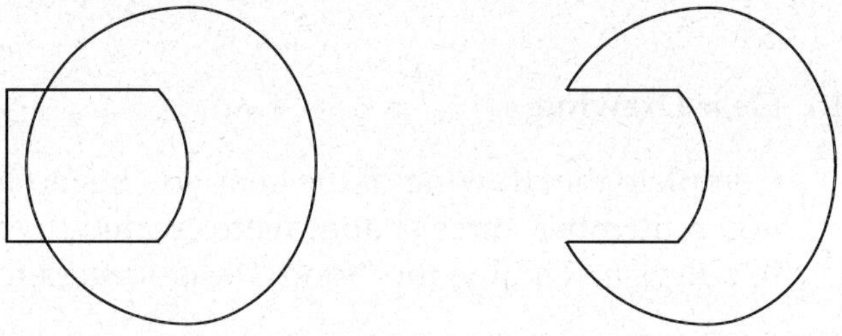

D. ***Move*** the rectangular-shaped region **68** units to the left to overlap the first region as shown in Figure 16-80. Use ***Subtract*** to create the composite region on the right representing the head of the wrench.

E. Complete the construction of the wrench in a manner similar to that used in the previous steps. Refer to Chapter 15 Exercises for dimensions of the wrench. Complete the wrench as one *Region*. ***Save*** the drawing as **WRENCH-REG**.

**12. Retaining Wall**

***Open*** the **RET-WALL** drawing that you created in Chapter 8 Exercises. Annotate the wall with 50 unit stations as shown in Figure 16-81.

HINT: Convert the centerline of the retaining wall to a ***Pline***, then use the ***Measure*** command to place *Point* objects at the 50 unit stations. ***Offset*** the wall on both sides to provide a construction aid in the creation of the perpendicular tick marks. Use ***Node*** and ***Perpendicular*** Object Snaps.
***Save*** the drawing as **RET-WAL3**.

**FIGURE 16-81**

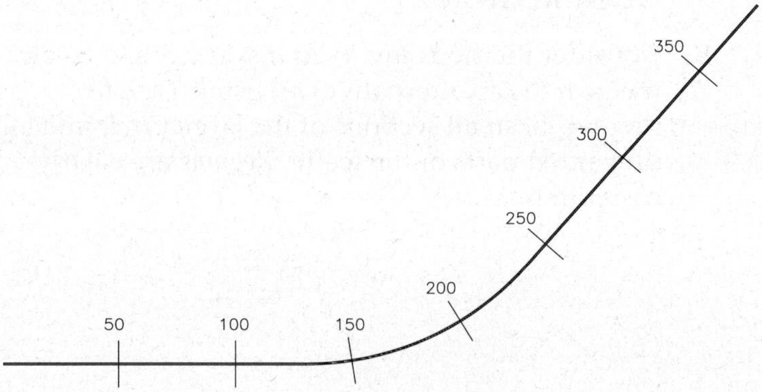

CHAPTER

# 17

# Inquiry Commands

## Chapter Objectives

After completing this chapter you should:

1. be able to list the *Status* of a drawing;
2. be able to *List* the AutoCAD database information about an object;
3. be able to calculate the *Area* of a closed shape with and without "islands";
4. be able to find the distance between two points with *Measure Distance*;
5. be able to report the coordinate value of a selected point using the *ID* command;
6. know how to list the *Time* spent on a drawing or in the current drawing session;
7. be able to use *Setvar* to change system variable settings or list current settings;
8. be able to use the *System Variable Monitor* to track and correct inadvertent changes to your preferred system variable settings

8. be able to use the *Geometry Cleanup* to find and correct drawing issues.

## CONCEPTS

AutoCAD provides several commands that allow you to find out information about the current drawing status and specific objects in a drawing. These commands as a group are known as "*Inquiry* commands" and are grouped together in the menu systems. Most of the *Inquiry* commands are located on the Ribbon's *Home* tab in the *Utilities* panel (Fig. 17-1, upper-right). If you prefer the Menu Bar, the *Inquiry* commands are grouped together on the *Tools* pull-down menu (Fig. 17-1, center). One special command that is included in this section is *Geometry Cleanup*. It is located on the View menu. Geometry Cleanup scans your drawing to find common geometry errors such as gaps, overshoots, undershoots, and angles that don't line up. Drawing errors are marked in the drawing and options to correct them are offered.

Using *Inquiry* commands, you can find out such information as the amount of time spent in the current drawing, the distance between two points, the area of a closed shape, the listing of properties for specific objects (coordinates of endpoints, lengths, angles, etc.), and current settings for system variables as well as other information. The *Inquiry* commands are:

*Status, List, Measuregeom, ID, Time, Setvar, Sysvarmonitor,* and *geometrycleanupopen,* which lets you inquire for drawing errors.

**FIGURE 17-1**

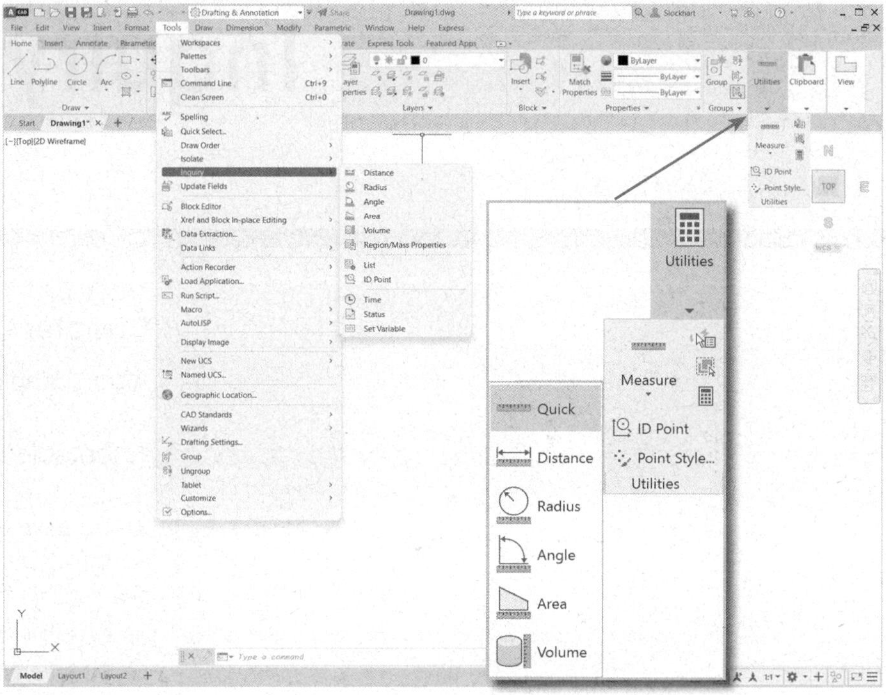


## COMMANDS

### *Status*

| Ribbon | Menu Bar | Command (Type) | Alias (Type) | Shortcut |
|---|---|---|---|---|
| ... | *Tools Inquiry > Status* | *Status* | ... | ... |

The *Status* command gives many pieces of information related to the current drawing. Typing or PICKing the command from the icon or one of the menus causes a text screen to appear similar to that shown in Figure 17-2, on the next page. The information items are:

Total number of objects in the current drawing.
Paper space limits: values set by the *Limits* command in Paper Space. (*Limits* in Paper Space are set by selecting a *Paper size* in the *Page Setup* or *Plot* dialog box.)
Paper space uses: area used by the objects (drawing extents) in Paper Space.
Model space limits: values set by the *Limits* command in Model Space.
Model space uses: area used by the objects (drawing extents) in Model Space.

Display shows: current display or windowed area.
Insertion basepoint: point specified by the *Base* command or default (0,0).
Snap resolution: value specified by the *Snap* command.
Grid spacing: value specified by the *Grid* command.
Current space: Paper Space or Model Space
Current layer: name.
Current color: current color assignment
Current linetype: current linetype assignment
Current lineweight: current lineweight assignment.
Current plot style: current plot style assignment.
Current elevation, thickness: 3D properties—current height above the XY plane and Z dimension.
On or off status: *Fill, Grid, Ortho, Qtext, Snap, Tablet.*
Object Snap Modes: current Running Object Snap modes.
Free dwg disk: space on the current drawing hard disk drive.
Free temp disk: space on the current temporary files hard disk drive.
Free physical memory: amount of free RAM (total RAM).
Free swap file space: amount of free swap file space (total allocated swap file).

**FIGURE 17-2**

```
Command: STATUS
5011 objects in C:\Program Files\Autodesk\AutoCAD 2015\Sample\Database Connectivity
\Floor Plan Sample.dwg
Undo file size:       16630 bytes
Model space limits are X:      0'-0"   Y:      0'-0"  (Off)
                       X:     96'-0"   Y:     72'-0"
Model space uses       X:     70'-3"   Y:     70'-3"
                       X: 285'-11 29/64"   Y:   282'-0" **Over
Display shows          X: -71'-3 7/16"   Y: -258'-6 17/64"
                       X: 372'-7 13/32"   Y: 20'-6 13/32"
Insertion base is      X:      0'-0"   Y:      0'-0"   Z:      0'-0"
Snap resolution is     X:      0'-3"   Y:      0'-3"
Grid spacing is        X:      1'-0"   Y:      1'-0"
Current space:        Model space
Current layout:       Model
Current layer:        "0"
Current color:        BYLAYER -- 7 (white)
Current linetype:     BYLAYER -- "CONTINUOUS"
Current material:     BYLAYER -- "Global"
Current lineweight:   Default
Current elevation:     0'-0"  thickness:     0'-0"
Fill on  Grid off  Ortho off  Qtext off  Snap off  Tablet off
Press ENTER to continue:
Object snap modes:    Center, Endpoint, Intersection, Midpoint, Extension,
Free dwg disk (C:) space: 233927.3 MBytes
Free temp disk (C:) space: 233927.3 MBytes
Free physical memory: 1658.5 Mbytes (out of 3069.9M).
Free swap file space: 4389.6 Mbytes (out of 6138.2M).
```

## *List*

| Ribbon | Menu Bar | Command (Type) | Alias (Type) | Shortcut |
|---|---|---|---|---|
| ... | *Tools Inquiry > List* | *List* | *LS* or *LI* | ... |

The *List* command displays the database list of information in text window format for one or more specified objects. The information displayed depends on the type of object selected. Invoking the *List* command causes a prompt for you to select objects. AutoCAD then displays the list for the selected objects (see Fig. 17-3 and Fig. 17-4, on the next page). A *List* of a *Line* and an *Arc* is given in Figure 17-3.

**FIGURE 17-3**

```
Command: LI
LIST 2 found
                  LINE      Layer: "E-F-TERR"
                            Space: Model space
                   Handle = c9a
              from point, X=281'-2 5/64"  Y=   70'-3"  Z=    0'-0"
                to point, X=254'-5 59/64"  Y=   70'-3"  Z=    0'-0"
          Length =26'-8 5/32",  Angle in XY Plane = 180.00
                  Delta X =-26'-8 5/32", Delta Y =     0'-0", Delta Z =    0'-0"
                  ARC       Layer: "0"
                            Space: Model space
                   LineWeight: Default
                   Handle = b46d
            center point, X=216'-0 21/64"  Y=173'-9 29/32"  Z=    0'-0"
            radius     3'-0"
             start angle   2.40
               end angle  90.02
            length 4'-7 3/64"
Command:
```

For a *Line,* coordinates for the endpoints, line length and angle, current layer, and other information are given.

For an *Arc,* the center coordinate, radius, start and end angles, and length are given.

The *List* for a *Pline* is shown in Figure 17-4. The location of each vertex is given, as well as the length and perimeter of the entire *Pline*.

*Plines* are created and listed as *Lwpolylines*, or "lightweight polylines." The data common to all vertices are stored only once, and only the coordinate data are stored for each vertex. Because this data structure saves file space, the *Plines* are known as lightweight *Plines*.

**FIGURE 17-4**

```
LIST
Select objects: 1 found
Select objects:
                  LWPOLYLINE  Layer: "0"
                            Space: Model space
                   LineWeight: Default
                   Handle = b470
             Open
    Constant width     0'-0"
             area   37922.59 square in. (263.3513 square ft.)
           length   68'-1 13/16"
         at point  X=83'-0 11/16"  Y=168'-3 21/64"  Z=    0'-0"
         at point  X=96'-10 39/64"  Y=168'-3 21/64"  Z=    0'-0"
         at point  X=96'-10 39/64"  Y=178'-4 13/32"  Z=    0'-0"
         at point  X=104'-4 5/32"  Y=178'-4 13/32"  Z=    0'-0"
         at point  X=  123'-3"  Y=178'-4 13/32"  Z=    0'-0"
         at point  X=  123'-3"  Y=  168'-6"  Z=    0'-0"
         at point  X=  131'-3"  Y=  168'-6"  Z=    0'-0"
```

## *Measuregeom*

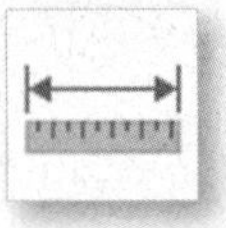

| Ribbon | Menu Bar | Command (Type) | Alias (Type) | Shortcut |
|---|---|---|---|---|
| *Home Utilities Measure* | *Tools Inquiry >* | *Measuregeom* | *MEA* | ... |

The *Measuregeom* (measure geometry) command includes several options for measuring including the older *Distance* and *Area* commands. In addition, *Measuregeom* also includes the *Radius*, *Angle*, and *Volume* options. The options below are explained in the following sections.

Command: ***MEASUREGEOM***
Move cursor or [Distance/Radius/Angle/ARea/Volume/Quick/Mode/eXit] <eXit>:

**TIP** You can turn on *Dynamic Input* in order to present a more graphic display of the determined measurements. Although Object Snap is not automatically turned on, you should use object snaps for most applications of *Measuregeom* to ensure selection of the exact desired points. The *Measuregeom* command will also turn off *dynamic dimensions* when *Dynamic Input* is turned off.

**2020**

### *Quick*

The *Quick* option shows dimensions inferred from objects near the cursor. This is a fast way to get size information about a drawing.

The command is as follows.

Command: ***MEASUREGEOM***
Move cursor or [Distance/Radius/Angle/ARea/Volume/Quick/Mode/eXit] <eXit>: ***Q***
Move cursor or [Distance/Radius/Angle/ARea/Volume/Quick/Mode/eXit] <eXit>: **(move the cursor and note the dimensions that appear on screen Figure 17-5)**
Move cursor or [Distance/Radius/Angle/ARea/Volume/Quick/Mode/eXit] <eXit>:

**FIGURE 17-5**

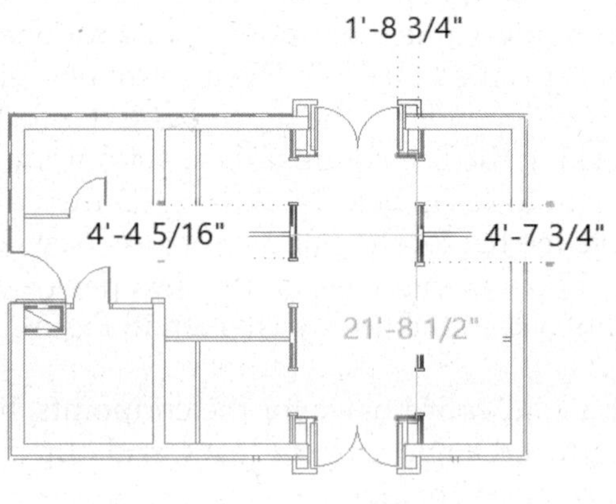


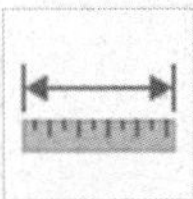

### *Distance*

The *Distance* option reports the distance between any two points. Object Snap can be used to "snap" to the existing points. This option is helpful in many engineering or architectural applications, such as finding the clearance between two mechanical parts, finding the distance between columns in a building, or finding the size of an opening in a doorway.

The command is as follows.

Command: ***MEASUREGEOM***
Move cursor or [Distance/Radius/Angle/ARea/Volume/Quick/Mode/eXit] <eXit>: ***d***
Specify first point: **PICK**
Specify second point or [Multiple points]: **PICK**
Distance = nn.nnn, Angle in XY Plane = nn, Angle from XY Plane = nn
Delta X = nn.nnn, Delta Y = nn.nnn, Delta Z = nn.nnn
Move cursor or [Distance/Radius/Angle/ARea/Volume/Quick/Mode/eXit] <eXit>:

**FIGURE 17-6**

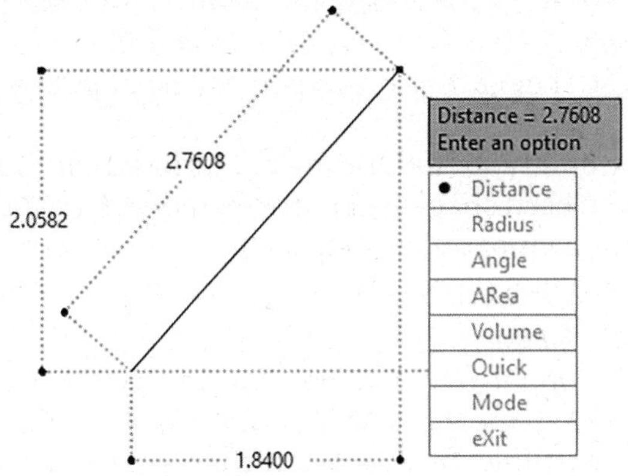


AutoCAD reports the distance in two formats: on the command line and in the drawing (Fig. 17-6). Both formats list the absolute and relative distances as well as the angle of the line between the points. If you turn on *Dynamic Input*, you can then select from the list of options by clicking in the drawing window or typing at the keyboard.

You can use the *Multiple points* option within *Distance* to achieve a sum of distances between multiple points as shown in the following command sequence. The total also appears in the drawing window.

Specify first point: **PICK**
Specify second point or [Multiple points]: ***m***
Specify next point or [Arc/Length/Undo/Total] <Total>: **PICK**
Distance = 6.5009
Specify next point or [Arc/Close/Length/Undo/Total] <Total>: **PICK**
Distance = 13.0089
Specify next point or [Arc/Close/Length/Undo/Total] <Total>: **PICK**
Distance = 16.6668

### *Area*

With the *Area* option AutoCAD calculates the area and the perimeter of any enclosed shape in a matter of milliseconds. You specify the area (shape) to consider for calculation by picking the *Object* (if it is a closed *Pline, Polygon, Circle, Boundary, Region* or other closed object) or by picking points (corners of the outline) to define the shape. The options are given below.

**FIGURE 17-7**

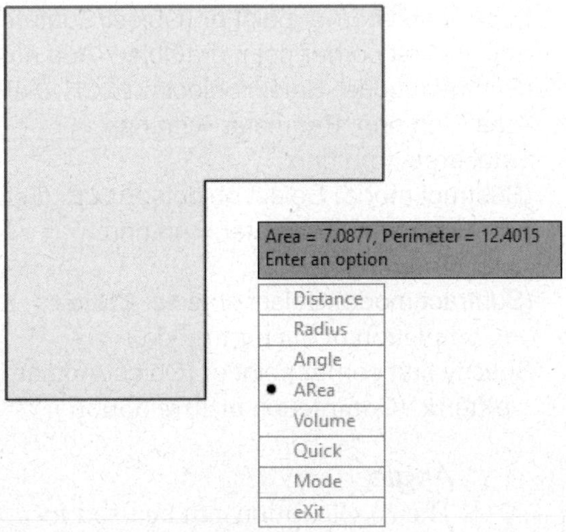


*Specify first corner point*
With this option you specify the area by picking points around the perimeter of any shape. Select points in a sequential order around the perimeter. This method can be used for shapes with straight or radiused sides. An example of the default method (picking points to define the area) is shown in Figure 17-7. Note that AutoCAD shades the selected area and reports the area and perimeter. The command sequence is shown below.

Command: ***MEASUREGEOM***
Move cursor or [Distance/Radius/Angle/ARea/Volume/Quick/Mode/eXit] <eXit>: ***ar***
Specify first corner point or [Object/Add area/Subtract area/eXit] <Object>: **PICK**
Specify next point or [Arc/Length/Undo]: **PICK** or (**option**)
Specify next point or [Arc/Length/Undo]: **PICK** or (**option**)
Area = nn.nnn perimeter = nn.nnn

*Object*
If the shape for which you want to find the area and perimeter is a *Circle*, *Polygon*, *Ellipse*, *Boundary*, *Region*, or closed *Pline*, the *Object* option of *Area* can be used. Select the shape with one PICK (since all of these shapes are considered as one object by AutoCAD).

**TIP** The ability to find the area of a closed *Pline*, *Region*, or *Boundary* is extremely helpful. Remember that any closed shape, even if it includes *Arcs* and other curves, can be converted to a closed *Pline* with the *Pedit* command (as long as there are no gaps or overlaps) or can be used with the *Boundary* command. This method provides you with the ability to easily calculate the area of any shape, curved or straight.

*Add, Subtract*
*Add* and *Subtract* are tools for finding the area of a closed shape that has islands, or negative spaces. For example, finding the surface area of a sheet of material that has punched holes, where the area of the holes is subtracted from the area defined by the perimeter shape. The *Add* and *Subtract* options are used for that purpose. Make sure to press Enter between the *Add* and *Subtract* modes. An example of this command sequence used to find the area of a shape minus the holes is shown in Figure 17-8. Notice that the object selected in the first step is a closed *Pline* shape, including an *Arc*.

**FIGURE 17-8**

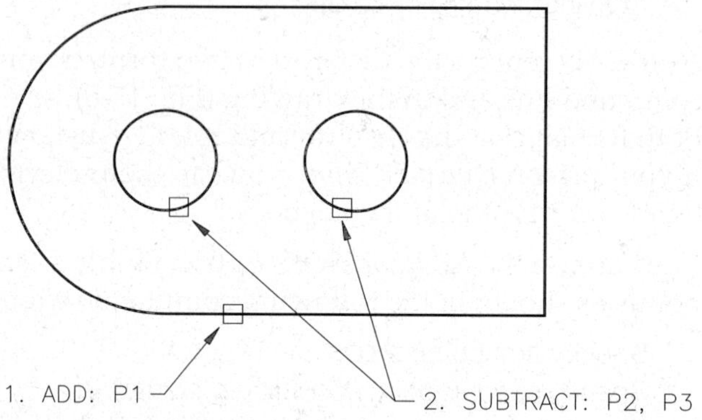


Command: ***MEASUREGEOM***
Move cursor or [Distance/Radius/Angle/ARea/Volume/Quick/Mode/eXit] <eXit>: ***ar***
Specify first corner point or [Object/Add area/Subtract area/eXit] <Object>: ***a*** (Defining the perimeter shape is prefaced by the *Add* option.)
Specify first corner point or [Object/Subtract area/eXit]: ***o*** (for a *Boundary* or closed object.)
(Add mode) Select objects: **PICK** (Select the closed object.)
Area = nn.nnn, Perimeter = nn.nnn
Total area = nn.nnn
(Add mode) Select object: **Enter** (Indicates completion of the *Add* mode.)
Specify first corner point or [Object/Subtract area/eXit]: ***s*** (Change to the *Subtract* mode.)
Specify first corner point or [Object/Add area/eXit]: ***o*** (Assuming the holes are *Circle* objects.)
(Subtract mode) Select objects: **PICK** (Select the first *Circle* to subtract.)
Area = nn.nnn, Perimeter = nn.nnn
Total area = nn.nnn
(Subtract mode) Select objects: **PICK** (Select the second *Circle* to subtract.)
Area = nn.nnn, Perimeter = nn.nnn
Total area = nn.nnn
(Subtract mode) Select objects: **Enter** (Completion of *Subtract* mode.)
Specify first corner point or [Object/Add area/eXit]: ***x*** (Completion of *Area* option.)

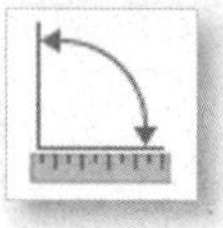

### *Angle*

The *Angle* option can be used to measure the angle of a *Line*, the angle between two *Lines*, the included angle of an *Arc*, or an angle between two points on a *Circle*. The following command sequence relates to measuring the angle between two *Lines* as shown in Figure 17-9.

**FIGURE 17-9**

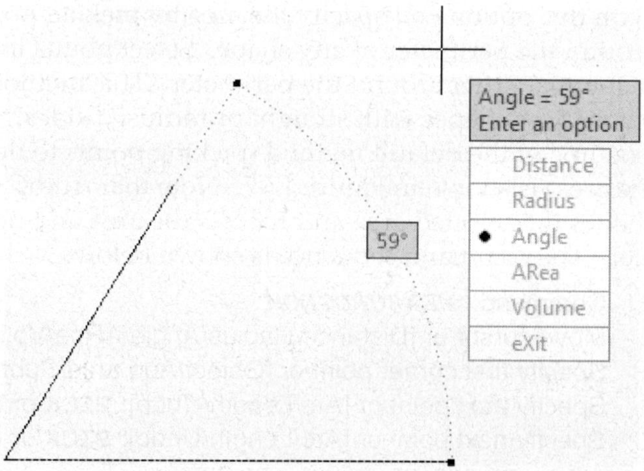

Command: ***MEASUREGEOM***
Move cursor or [Distance/Radius/Angle/ARea/Volume/Quick/Mode/eXit] <eXit>: ***a***
Select arc, circle, line, or <Specify vertex>: **PICK**
Select second line: **PICK**
Angle = 59°

### *Radius*

This option is the simplest option since it requires you to select only one *Arc* or *Circle*. The resulting radius value is displayed on the Command line and in the drawing window.

Command: ***MEASUREGEOM***
Move cursor or [Distance/Radius/Angle/ARea/Volume/Quick/Mode/eXit] <eXit>: ***a***
Select arc or circle: **PICK**
Radius = 0.9329
Diameter = 1.8658

### *Volume*

This option is useful for determining the volume of three-dimensional solid objects. The *Volume* option works similarly to the *Area* option.

## *Id*

| Ribbon | Menu Bar | Command (Type) | Alias (Type) | Shortcut |
|---|---|---|---|---|
| *Home*<br>*Utilities*<br>*ID Point* | *Tools*<br>*Inquiry >*<br>*ID Point* | *Id* | ... | ... |

The *ID* command reports the coordinate value of any point you select with the cursor. If you require the location associated with a specific object, an Object Snap mode (*Endpoint, Midpoint, Center,* etc.) can be used.

Command: ***ID***
Specify point: **PICK** (Use Object Snap if needed.)
X = 7.63 Y = 6.25 Z = 0.00
Command:

TIP

NOTE: *ID* also sets AutoCAD's "last point." The last point can be referenced in commands by using the @ (at) symbol with relative rectangular or relative polar coordinates.

## *Time*

| Ribbon | Menu Bar | Command (Type) | Alias (Type) | Shortcut |
|---|---|---|---|---|
| ... | *Tools*<br>*Inquiry >*<br>*Time* | *Time* | ... | ... |

This command is useful for keeping track of the time spent in the current drawing session or total time spent on a particular drawing. Knowing how much time is spent on a drawing can be useful in an office situation for bidding or billing jobs.

The *Time* command reports the information shown in Figure 17-10. The *Total editing time* is automatically kept, starting from when the drawing was first created until the current time. Plotting and printing time is not included in this total, nor is the time spent in a session when changes are discarded.

**FIGURE 17-10**

```
Command: TIME
Current time:              Friday, May 09, 2014  12:39:52:556 PM
Times for this drawing:
  Created:                 Friday, May 09, 2014  12:29:05:671 PM
  Last updated:            Friday, May 09, 2014  12:29:05:671 PM
  Total editing time:      0 days 00:10:46:893
  Elapsed timer (on):      0 days 00:10:46:892
  Next automatic save in: <no modifications yet>
Enter option [Display/ON/OFF/Reset]:
```

### *Display*

The *Display* option causes *Time* to repeat the display with the updated times.

### *ON/OFF/Reset*

The *Elapsed timer* is a separate compilation of time controlled by the user. The *Elapsed timer* can be turned *ON* or *OFF* or can be *Reset*.

*Time* also reports when the next automatic save will be made. The time interval of the Automatic Save feature is controlled by the *SAVETIME* system variable. To set the interval between automatic saves, type *SAVETIME* at the Command line and specify a value for time (in minutes). See also *SAVETIME* in Chapter 2.

## *Setvar*

| Ribbon | Menu Bar | Command (Type) | Alias (Type) | Shortcut |
|---|---|---|---|---|
| ... | *Tools Inquiry > Set Variable* | *Setvar* | *SET* | ... |

The settings (values or on/off status) that you make for many commands, such as *Limits, Grid, Snap,* Running Object Snap, *Fillet* values, *Pline* width, etc. are saved in system variables. There are over 400 system variables. The variables store the settings that are used to create and edit the drawing. *Setvar* ("set variable") gives you access to the system variables.

*Setvar* allows you to perform two functions: (1) change the setting for any system variable and (2) display the current setting for one or all system variables. To change a setting for a system variable using *Setvar,* just use the command and enter the name of the variable. For example, the following syntax lists values for the *Grid* setting and current *Fillet* radius:

Command: ***SETVAR***
Enter variable name or [?]: ***GRIDUNIT***
Enter new value for GRIDUNIT <1.00,1.00>:

Command: ***SETVAR***
Enter variable name or [?]: ***FILLETRAD***
Enter new value for FILLETRAD <0.50>:

With recent releases of AutoCAD, Setvar is not needed to set system variables. You can enter the variable name directly at the Command prompt without using *Setvar* first.

Command: ***FILLETRAD***
Enter new value for FILLETRAD <0.50>:

TIP

To list the current settings for all system variables, use *Setvar* with the *?* (question mark) option. The complete list of system variables is given in a text window with the current setting for each variable (Fig. 17-11).

You can also use *Help* to list the system variables with a short explanation for each. In the *Help Topics* dialog box, select the *Contents* tab, select *Command Reference*, then *System Variables*.

**FIGURE 17-11**

```
SETVAR
Enter variable name or [?]: ?
Enter variable(s) to list <*>:
3DCONVERSIONMODE  1
3DDWFPREC         2
3DSELECTIONMODE   1
ACADLSPASDOC      0
ACADPREFIX        "C:\Users\Jimbo\appdata\roaming\autodesk\autocad 2015\r20.0
\e..." (read only)
ACADVER           "20.0s (LMS Tech)"                (read only)
ACTPATH           ""
ACTRECORDERSTATE  0                                 (read only)
ACTRECPATH        "C:\Users\Jimbo\appdata\roaming\autodesk\autocad 2015\r20.0\e..."
ACTUI             6
AFLAGS            16
ANGBASE           0
ANGDIR            0
ANNOALLVISIBLE    1
ANNOAUTOSCALE     -4
ANNOTATIVEDWG     0
APBOX             0
APERTURE          10
AREA              0.0000                            (read only)
ATTDIA            1
```

## Sysvarmonitor

| Ribbon | Menu Bar | Command (Type) | Alias (Type) | Shortcut |
|---|---|---|---|---|
| ... | ... | *Sysvarmonitor* | ... | ... |

The *System Variable Monitor* tracks unwanted changes to system variable settings. For example, there may be some system variables that you prefer to always stay at a particular setting, yet the settings may be inadvertently changed occasionally. The *System Variable Monitor* notifies you when specific variables are changed from your preferred settings and allows you to reset them. When a preferred variable setting is changed, two warnings are given—one at the Command line (shown below) and one on the Status Bar (Fig. 17-12).

**FIGURE 17-12**

**** System Variable Changed ****

2 of the monitored system variables have changed from the preferred value. Use SYSVARMONITOR command to view changes.

Use the *Sysvarmonitor* command or click on the *System Variable Monitor* button on the Status Bar to produce the *System Variable Monitor* (Fig. 17-13).

**FIGURE 17-13**

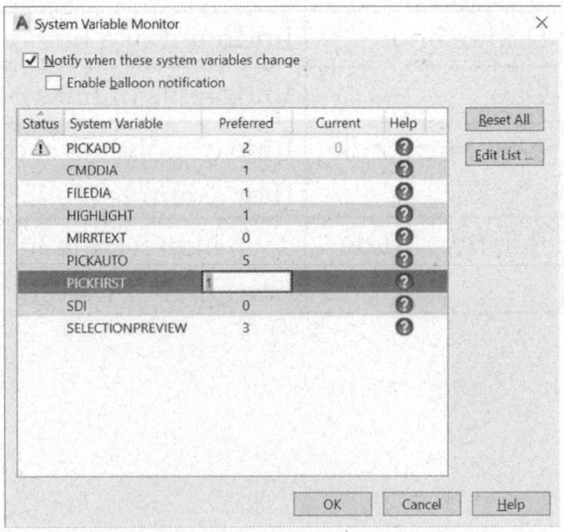


The *System Variable* list is prepopulated with variables that are often inadvertently modified. If any settings are changed from the *Preferred* value, the *Status* displays an "X" and the *Current* setting is given. Using the checkboxes near the top of the dialog box, you can choose whether to *Notify when these system variables change*, or *Enable balloon notification*, or both.

### *Reset All*

Select this button to reset the *Current* setting to the *Preferred* setting for all variables. After you cause a reset, the *Status* column displays green checkmarks and the Command line displays a notice such as that below. *Reset All* is also available in the right-click menu of the System Variable status bar icon.

Command:

2 system variable(s) were reset to preferred value.

FIGURE 17-14

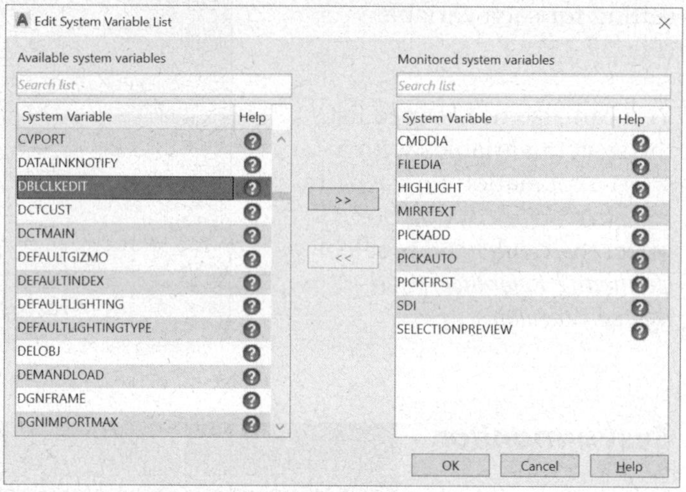


### *Edit List*

Selecting this option in the *System Variable Monitor* produces the *Edit System Variables List* dialog box (Fig. 17-14). Here you can specify which variables you prefer to monitor. Select any variables from the *Available system variables* list to add to the *Monitored system variables* list, and vice versa.

## Massprop

The *Massprop* command is used to give mass and volumetric properties for AutoCAD solids. See Chapter 36, Solid Model Editing, for use of *Massprop*.

## Geometry Cleanup

| Ribbon | Command (Type) | Alias (Type) | Shortcut |
|---|---|---|---|
| *View Palettes Geometry Cleanup* | *Geomcleanupopen* | ... | ... |

*Geometry Cleanup* identifies and can also connect line, polyline, arc, and elliptical arc objects in the drawing. Geometry Cleanup only works on line, polyline, arc, and elliptical arcs in the XY plane of the current UCS. Geometry Cleanup allows you to specify a tolerance range for the following conditions:

| | |
|---|---|
| Overshoot | Endpoint of line or polyline extends past a shape. |
| Undershoot | Endpoint of a line or polyline doesn't meet a shape. |
| Gap | Endpoints of lines or polylines don't coincide. |
| Absolute Angle | Line or polyline is off from desired angle in the UCS (For example 0°, 45°, or 90°, but or any value you specify.) |
| Relative Angle | Two intersecting lines or polylines do not meet at the intended angle relative to one another (not the UCS). |

Select Geometry Cleanup to produce the *Geometry Cleanup* palette (Fig. 17-15).

Click the *Select Objects* button from the panel.

Select objects: **PICK**

Click: *Scan/Rescan* from the Geometry Cleanup Settings panel.

**FIGURE 17-15**

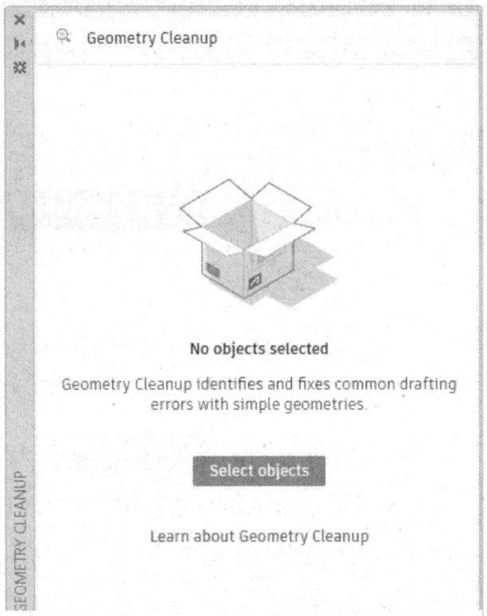


The Geometry Cleanup panel now lists object with unmet tolerances as specified in the Geometry Cleanup Settings (Fig. 17-16). Click to expand items, for example Gap (1).

**FIGURE 17-16**

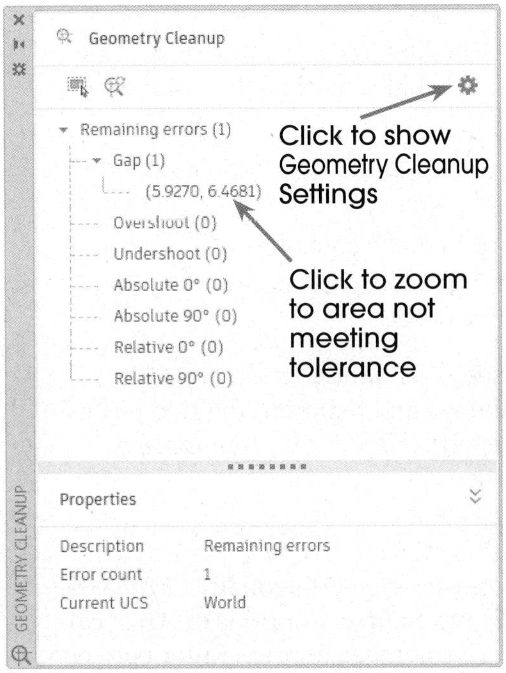


**FIGURE 17-17**

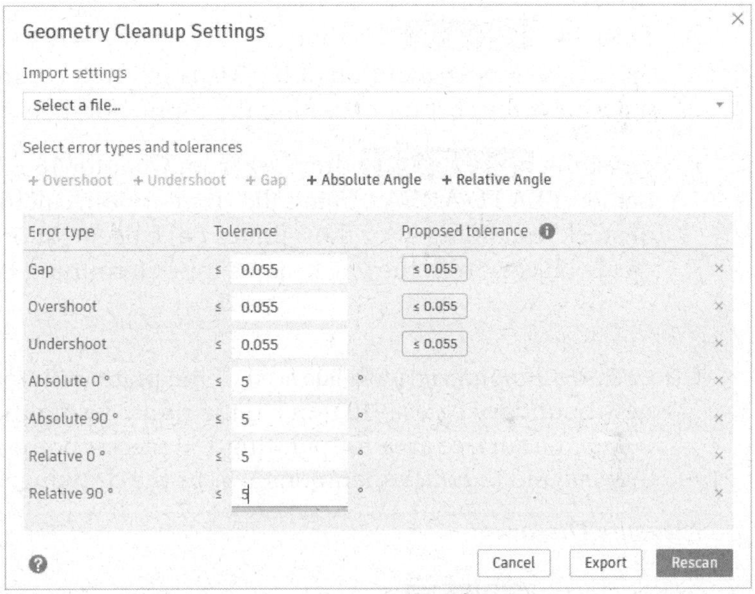


To review and fix the errors, click the numeric value. Your drawing is zoomed to show it in the Geometry Cleanup window.

Right-click to show the pop up menu. As you hover over the menu selections that can be automatically fixed, that repair is shown in the drawing. To apply the repair choice, click the menu selection. You may want to refer to Grip Editing in Chapter 19.

**FIGURE 17-18**

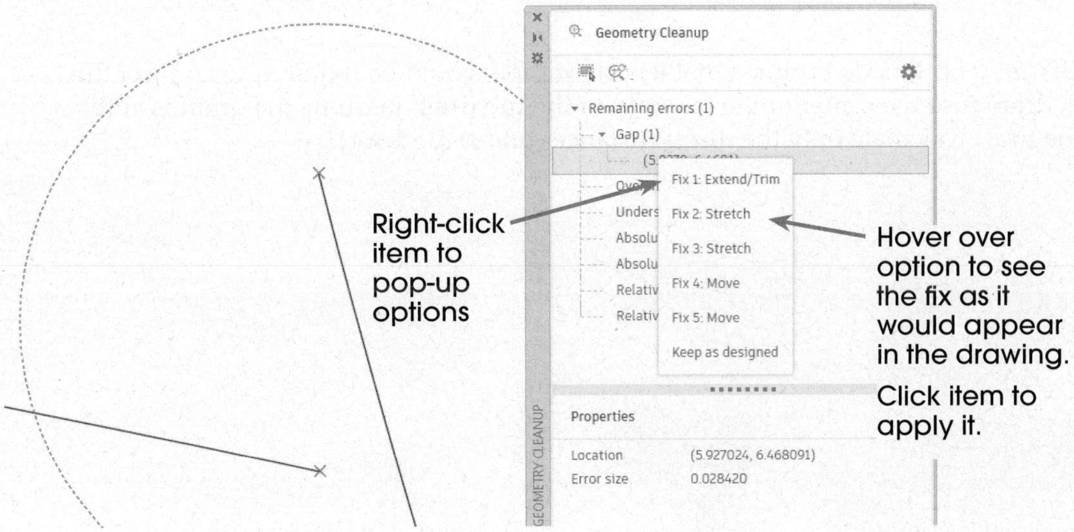

## CHAPTER EXERCISES

1. *List*

A. ***Open*** the **PLATENEST** drawing from Chapter 16 Exercises. Assume that a laser will be used to cut the plates and holes from the stock, and you must program the coordinates. Use the ***List*** command to give information on the ***Line***s and ***Circles*** for the one plate with four holes in a diagonal orientation. Determine and write down the coordinates for the 4 corners of the plate and the centers of the 4 holes.

REUSE

B. ***Open*** the **EFF-APT** drawing from Chapter 15 Exercises. Use ***List*** to determine the area of the inside of the tub. If the tub were filled with 10″ of water, what would be the volume of water in the tub?

REUSE

2. *Measure Area*

A. ***Open*** the **EFF-APT** drawing. The entry room is to be carpeted at a cost of $12.50 per square yard. Use the ***Area*** option of the ***Measuregeom*** command (with the PICK points option) to determine the cost for carpeting the room, not including the closet.

B. ***Open*** the **PLATENEST** drawing from Chapter 16 Exercises. Use ***SaveAs*** to create a file named **PLATE-AREA**. Using the ***Area*** option, calculate the wasted material (the two pieces of stock remaining after the 3 plates have been cut or stamped). HINT: Use ***Boundary*** to create objects from the waste areas for determining the *Area*.

REUSE

C. Create a ***Boundary*** (with islands) of the plate with four holes arranged diagonally. ***Move*** the new boundary objects **10** units to the right. Using the ***Add*** and ***Subtract*** options of ***Area***, calculate the surface area for painting 100 pieces, both sides. (Remember to press Enter between the *Add* and *Subtract* operations.) ***Save*** the drawing.

3. *Measure Distance*

A. ***Open*** the **EFF-APT** drawing. Use the ***Distance*** option of the ***Measuregeom*** command to determine the best location for installing a wall-mounted telephone in the apartment. Where should the telephone be located in order to provide the most equal access from all corners of the apartment? What is the farthest distance that you would have to walk to answer the phone?

B. Using the ***Distance*** option, determine what length of pipe would be required to connect the kitchen sink drain (use the center of the far sink) to the tub drain (assume the drain is at the far end of the tub). Calculate only the direct distance (under the floor).

4. *ID*

*Open* the **PLATENEST** drawing once again. You have now been assigned to program the laser to cut the plate with 4 holes in the corners. Use the ***ID*** command (with *Object Snaps*) to determine the coordinates for the 4 corners and the hole centers.

REUSE

5. ***Time***

Using the ***Time*** command, what is the total amount of editing time you spent with the **PLATENEST** drawing? How much time have you spent in this session? How much time until the next automatic save?

6. ***Setvar***

Using the **PLATENEST** drawing again, use ***Setvar*** to list system variables. What are the current settings for *Fillet* radius (***FILLETRAD***), the last point used (***LASTPOINT***), *Linetype Scale* (***LTSCALE***), automatic file save interval (***SAVETIME***), and text *Style* (***TEXTSTYLE***).

7. ***System Variable Monitor***

Open the ***System Variable Monitor***. Ensure the *PICKFIRST* and *PICKADD* system variables are included in the *System Variable* list, and then close the dialog box. Change either of these variables at the Command line and make sure the notification appears on the Status Bar.

Click on the ***System Variable Monitor*** button to produce the dialog box. Select the ***Reset All*** button and then select ***OK*** to change the variable settings back to the preferred values.

Open the ***System Variable Monitor*** again. Select the **Edit List** button to produce the *Edit System Variables List* dialog box. Add any system variables that you prefer to the *Monitored System Variables* list, and then select ***OK***. (If you are not sure what variables to add, you can try ***GRIDUNIT*** or ***SNAPUNIT***.) Make changes to the system variables at the Command line or by another means. (For example, change the ***Snap*** and ***Grid*** settings in the ***Drafting Settings*** dialog box.) Ensure you receive the proper notifications. If not, make any needed corrections.

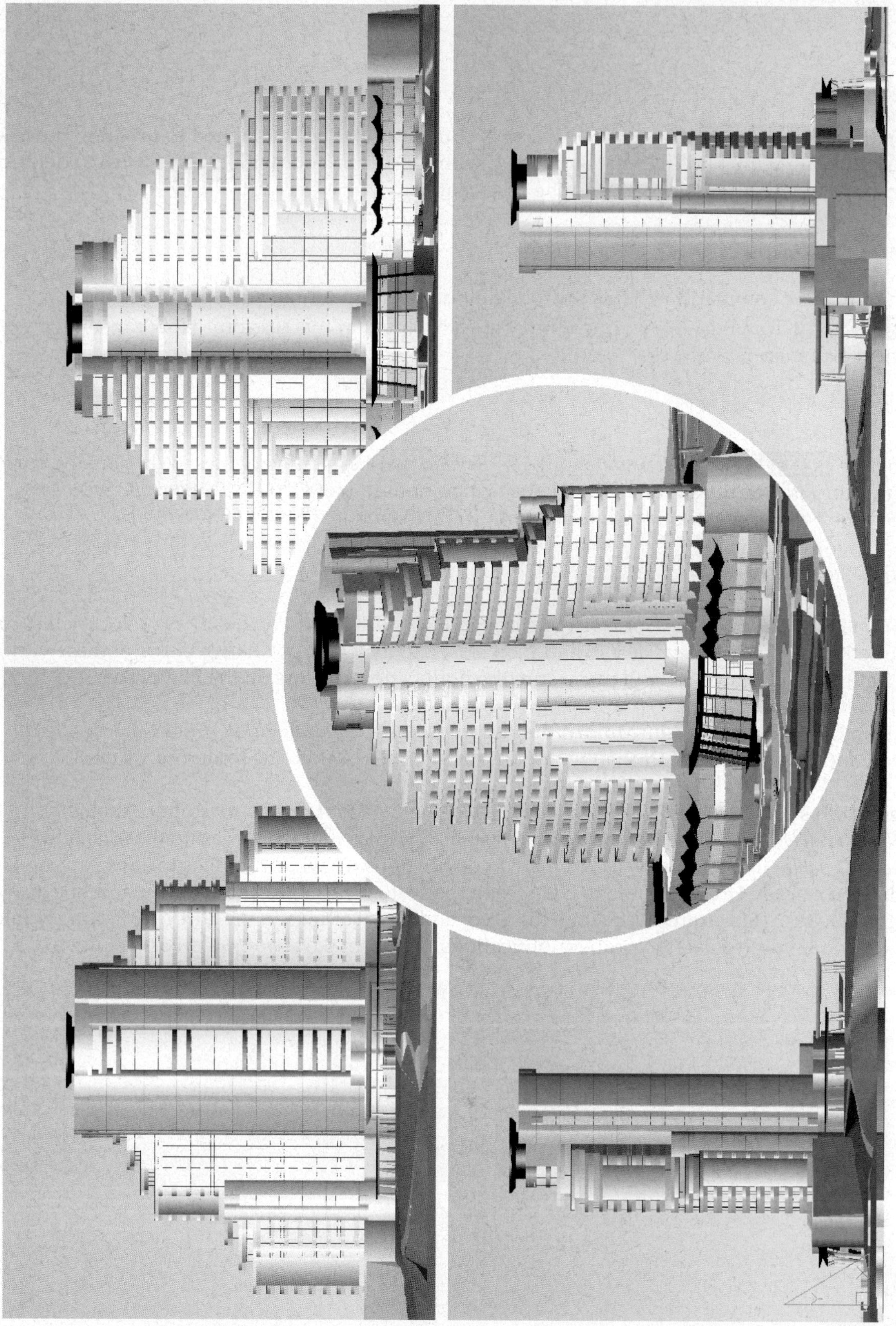

HOTEL MODEL.DWG, Courtesy of Autodesk, Inc.

CHAPTER

# 18

# Text and Tables

## Chapter Objectives

After completing this chapter you should:

1. be able to create lines of text in a drawing using *Text*;
2. be able to create and format paragraph text using *Mtext*;
3. know the options available for creating and editing text using the *Text Editor* tab;
4. be able to create text styles with the *Style* command including annotative styles;
5. know how to edit text using *Textedit*, *Mtedit*, and *Properties*;
6. be able to use *Spell* to check spelling and be able to *Find* and *Replace* text in a drawing;
7. be able to create and update text *Fields*;
8. be able to create *Tables* that contain text, fields, and calculations.

## CONCEPTS

The *Text* and *Mtext* commands provide you with a means of creating text in an AutoCAD drawing. "Text" in CAD drawings usually refers to sentences, words, or notes created from alphabetical or numerical characters that appear in the drawing. The numeric values that are part of specific dimensions are generally <u>not</u> considered "text," since dimensional values are a component of the dimension created automatically with the use of dimensioning commands.

Text in technical drawings is typically in the form of notes concerning information or descriptions of the objects contained in the drawing. For example, an architectural drawing might have written descriptions of rooms or spaces, special instructions for construction, or notes concerning materials or furnishings (Fig. 18-1). An engineering drawing may contain, in addition to the dimensions, manufacturing notes, bill of materials, schedules, or tables (Fig. 18-2). Technical illustrations may contain part numbers or assembly notes. Title blocks also contain text.

**FIGURE 18-1**

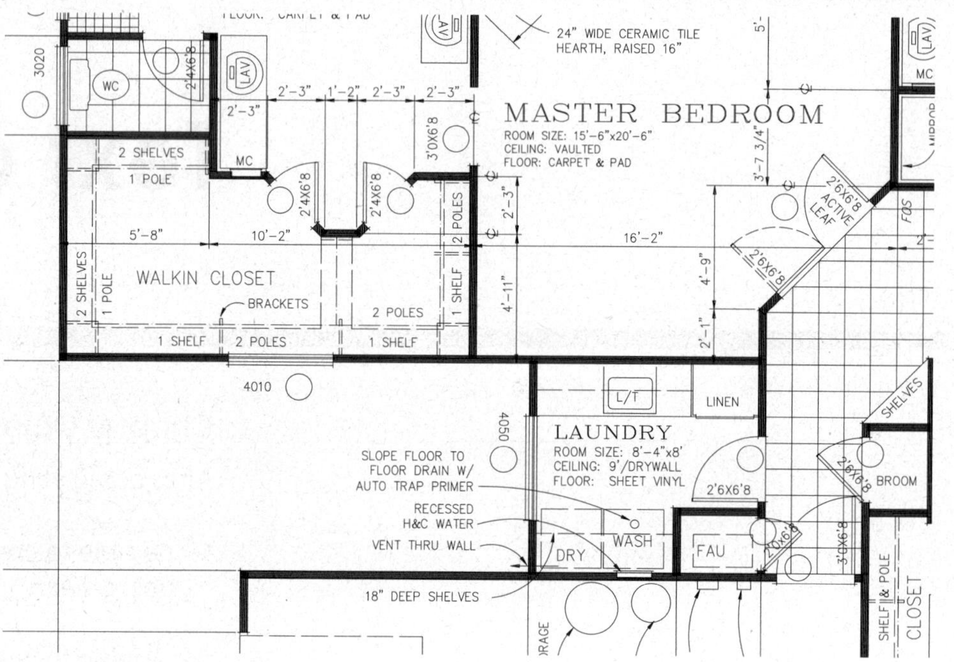


A line of text or paragraph of text in an AutoCAD drawing is treated as an object, just like a *Line* or a *Circle*. Each text object can be *Erased, Moved, Rotated,* or otherwise edited as any other graphical object. The letters themselves can be changed individually with special text editing commands. A spell checker is available by using the *Spell* command. Since text is treated as a graphical element, the use of many lines of text in a drawing can slow regeneration time and increase plotting time significantly.

**FIGURE 18-2**

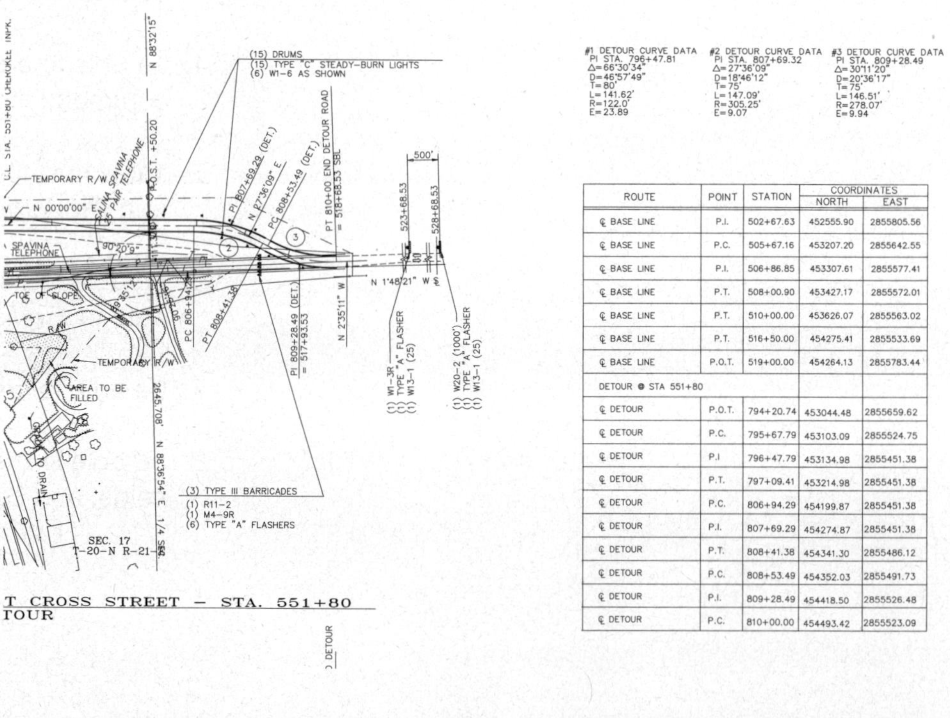

The *Text* and *Mtext* commands perform basically the same function; they create text in a drawing. *Text* creates a single line of text but allows entry of multiple lines with one use of the command. *Mtext* is the more sophisticated method of text entry. With *Mtext* (multiline text) you can create a paragraph of text that "wraps" within a text boundary (rectangle) that you specify. You can also use *Mtext* to import text content from an external file. An *Mtext* paragraph is treated as one AutoCAD object.

The form or shape of the individual letters is partly determined by the text *Style*. Create a style with the *Style* command, select any Windows standard TrueType font (.TTF) or AutoCAD-supplied font file (.TTF or .SHX shape file), then specify other parameters such as width or angle to create a style to suit your needs. When you use the *Text* or *Mtext* commands, you can select any previously created *Style* and apply one of many options for text justification used for aligning multiple lines of text, such as right, center, or left justified. Only two simple text styles, called *Standard* and *Annotative*, have been created as part of the ACAD.DWT and ACADISO.DWT template drawings. The *Standard* text style should be used as the "template" for most text objects. The *Annotative* style allows you to create annotative text, which is useful if you want to display the same text objects in multiple viewports at different scales.

Once the text has been placed in the drawing, many options are available for changing, or "editing," the text. For *Text* objects you can edit the text "in place" (directly in the drawing), and for *Mtext* objects you can use the "text editor" you used to create the text, complete with all the formatting options. Options for editing text include changing the actual content of the text, changing the text style, scale, or justification method, spell checking, and using a find and replace tool. You can also use the *Properties* palette to change any aspect of the text. When text is printed you can specify that the text be unfilled (outline text) or that each line be printed as a rectangular block instead of individual letters.

Additional features described in this chapter allow you to create lettered, numbered, or bulleted lists. You can create text fields that contain strings of text that are variable and display information based on current data. You can even create tables and control the format and content of the tables. Tables can be set up to perform calculations.

Commands related to creating or editing text and tables in an AutoCAD drawing include:

| | |
|---|---|
| *Text* | Places individual lines of text in a drawing. |
| *Mtext* | Places text in paragraph form (with word-wrap) within a text boundary and allows many methods of formatting the appearance of the text using the *Text Editor* tab. |
| *Style* | Creates text styles for use with any of the text creation commands. You can select from font files and specify other parameters to design the appearance of the letters. |
| *Spell* | Checks spelling of existing text in a drawing. |
| *Find* | Used to find or replace a text string globally in the drawing. |
| *Textedit* | Allows you to edit *Text* objects in place. |
| *Mtedit* | Invokes the *Text Editor* tab for editing any aspect of individual *Mtext* characters or the entire *Mtext* paragraph(s). |
| *Spacetrans* | Calculates the size of text in a model space viewport in paper space units. |
| *Qtext, Textfill* | Changes the appearance of text to speed up regenerations and/or plots. |
| *Scaletext* | Allows you to change the text size without changing the insertion point. |
| *Justifytext* | Allows you to change the text justification method without moving the text. |
| *Field* | Creates variable text that changes based on current data in the drawing or system. |
| *Table* | Creates a table based on a table style that contains rows and columns of text. |
| *Tablestyle* | Specifies the properties of a table such as text style, text alignment, and border line properties. |
| *Tabledit* | Edits text content in a table. |
| *Txt2mtxt* | Converts *Text* to *Mtext* or combines multiple Text or Mtext items into one. |

## TEXT CREATION COMMANDS

The commands for creating text are formally named *Text* and *Mtext* (these are the commands used for typing). The *Draw* pull-down menu provides access to the two commonly used text commands, *Multiline Text...* (*Mtext*) and *Single-Line Text* (*Text*) (Fig. 18-3). The *Text* and *Mtext* commands as well as text editing commands are available on the *Text* toolbar (Fig. 18-4).

**FIGURE 18-3**

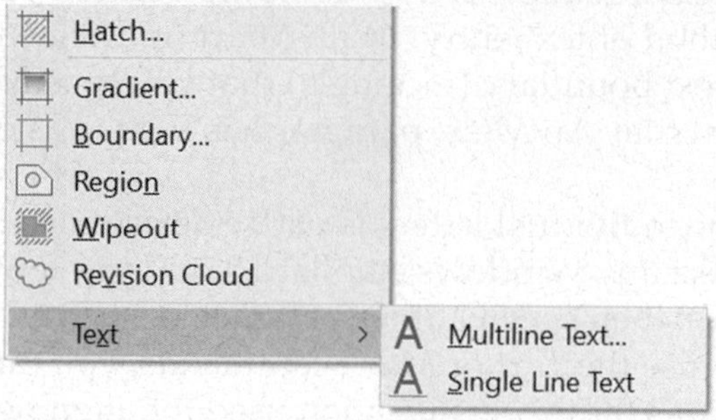


**FIGURE 18-4**

Both the *Text* and *Mtext* commands are available on the Ribbon in two locations: the *Home* tab, *Annotation* panel (Fig. 18-5) and the *Annotate* tab, *Text* panel.

**FIGURE 18-5**

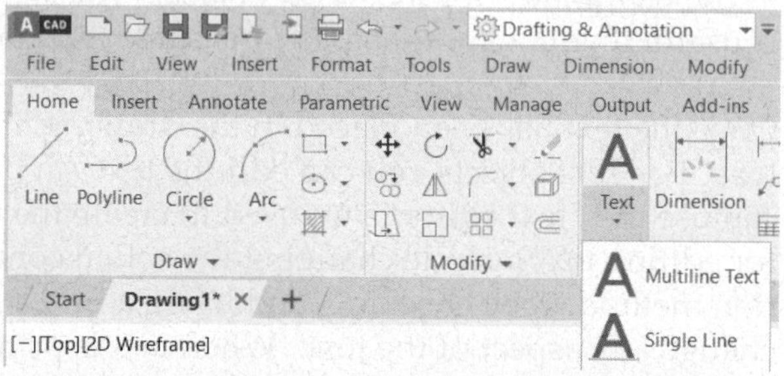


### *Text*

| Ribbon | Menu Bar | Command (Type) | Alias (Type) | Shortcut |
|---|---|---|---|---|
| *Home Annotation Single Line* | *Draw Text > Single Line Text...* | *Text* | *DT* | ... |

*Text* (Single Line Text) lets you insert single lines of text into an AutoCAD drawing. *Text* displays each character in the drawing as it is typed. You can enter multiple lines of text without exiting the *Text* command. The lines of text do not wrap. The options are presented below:

Command: ***TEXT***
Current text style: "Standard" Text height: 0.2000
Annotative: No
Specify start point of text or [Justify/Style]:

**FIGURE 18-6**

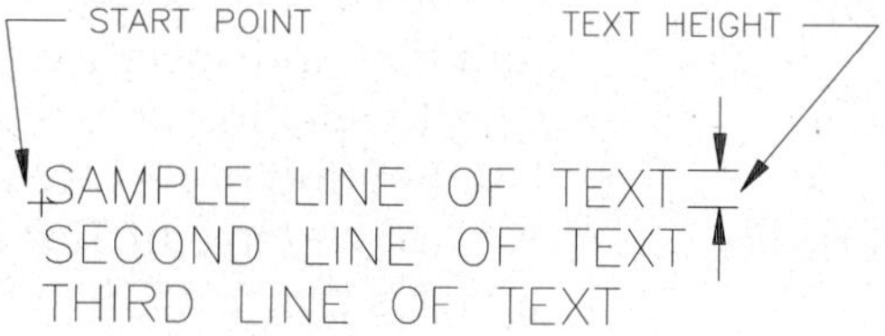


#### *Start Point*

The *Start point* for a line of text is the baseline for the text (Fig. 18-6). *Height* is the distance from the baseline to the top of upper case letters. Additional lines of text are automatically spaced below and justified. The *rotation angle* is the angle of the baseline (Fig. 18-7).

**FIGURE 18-7**

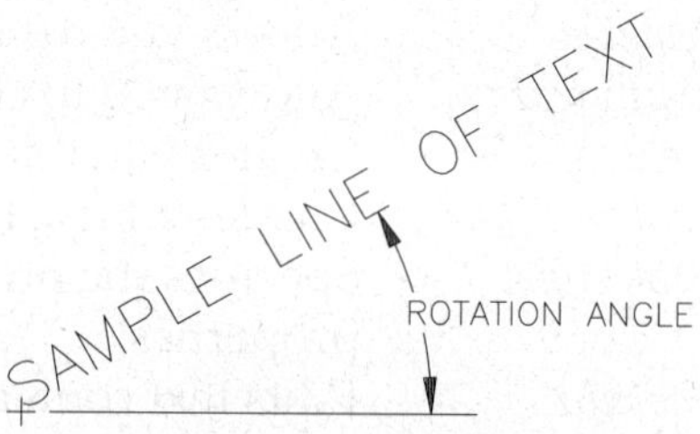


The command sequence for this option is:

Command: ***TEXT***
Current text style: "Standard" Text height: 0.2000
Specify start point of text or [Justify/Style]: **PICK** or (**coordinates**)
Specify height <0.2000>: **Enter** or (**value**)
Specify rotation angle of text <0>: **Enter** or (**value**)
(Type the desired line of text and press **Enter**)
(Press **Enter** once to add an additional line of text. Press **Enter** twice to complete the command.)
Command:

### *Justify*

If you want to use one of the justification methods, invoking this option displays the choices at the prompt:

Command: ***TEXT***
Current text style: "Standard" Text height: 0.2000 Annotative: No
Specify start point of text or [Justify/Style]: ***J*** (*Justify* option)
Enter an option [Left/Center/Right/Align/Middle/Fit/TL/TC/TR/ML/MC/MR/BL/BC/BR]:

*Align*
*Aligns* the line of text between the two points specified (P1, P2). The text height is adjusted automatically (Fig. 18-8).

**FIGURE 18-8**

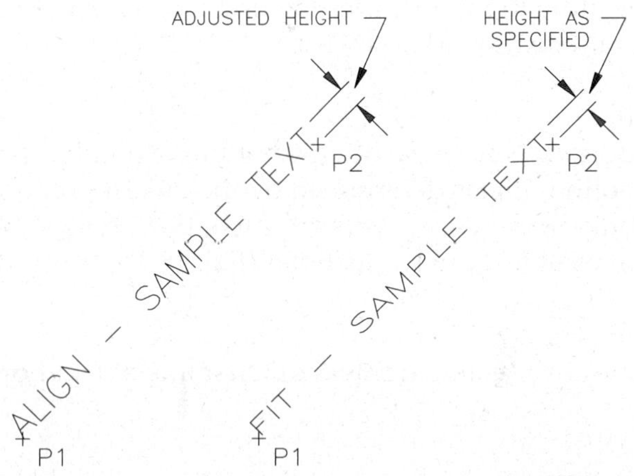


*Fit*
*Fits* (compresses or extends) the line of text between the two points specified (P1, P2). The text height does not change (Fig. 18-8).

*Left*
Creates text left justified at the start point (Fig. 18-9).

*Center*
*Centers* the baseline of the first line of text at the specified point. Additional lines of text are centered below the first (Fig. 18-9).

**FIGURE 18-9**

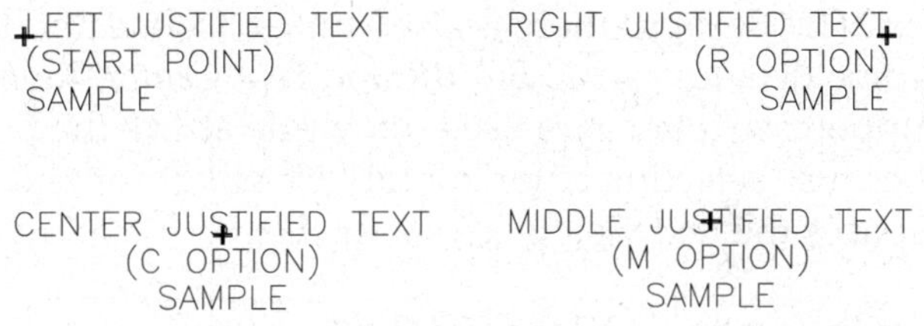


*Middle*
Centers the first line of text both vertically and horizontally about the specified point. Additional lines of text are centered below it (Fig. 18-9).

*Right*
Creates text that is right justified from the specified point (Fig. 18-9).

*TL*
Top Left. Places the text in the drawing so the top line (of the first line of text) is at the point specified and additional lines of text are left justified below the point. The top line is defined by the upper case and tall lower case letters (Fig. 18-10).

**FIGURE 18-10**

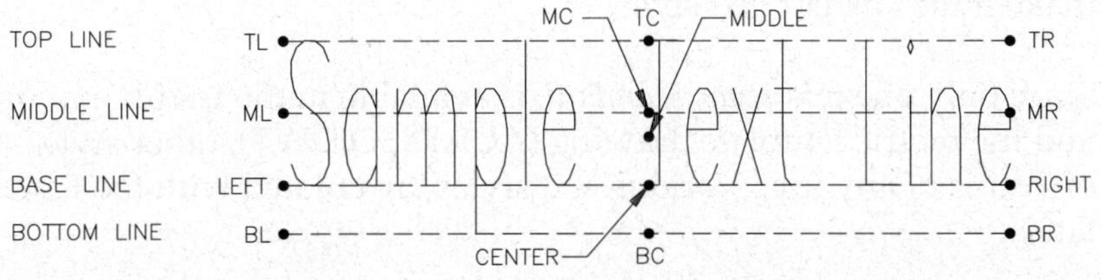


*TC*
Top Center. Places the text so the top line of text is at the point specified and the line(s) of text are centered below the point (Fig. 18-10).

*TR*
Top Right. Places the text so the top right corner of the text is at the point specified and additional lines of text are right justified below that point (Fig. 18-10).

*ML*
Middle Left. Places text so it is left justified and the middle line of the first line of text aligns with the point specified. The middle line is half way between the top line and the baseline, not considering the bottom (extender) line (Fig. 18-10).

*MC*
Middle Center. Centers the first line of text both vertically and horizontally about the midpoint of the middle line. Additional lines of text are centered below that point (Fig. 18-10).

*MR*
Middle Right. Justifies the first line of text at the right end of the middle line. Additional lines of text are right justified (Fig. 18-10).

*BL*
Bottom Left. Attaches the bottom (extender) line of the first line of text to the specified point. The bottom line is determined by the lowest point of lower case extended letters such as y, p, q, j, and g. If only upper-case letters are used, the letters appear to be located above the specified point. Additional lines of text are left justified (Fig. 18-10).

*BC*
Bottom Center. Centers the first line of text horizontally about the bottom (extender) line (Fig. 18-10).

*BR*
Bottom Right. Aligns the bottom (extender) line of the first line of text at the specified point. Additional lines of text are right justified (Fig. 18-10).

NOTE: Because there is a separate baseline and bottom (extender) line, the *MC* and *Middle* points do not coincide and the *BL, BC, BR* and *Left, Center, Right* options differ. Also, because of this feature, when all uppercase letters are used, they ride above the bottom line. This can be helpful for placing text in a table because selecting a horizontal *Line* object for text alignment with *BL, BC,* or *BR* options automatically spaces the text visibly above the *Line*.

### *Style* (option of *Text* or *Mtext*)

The *style* option of the *Text* or *Mtext* command allows you to select from the existing text styles that have been previously created as part of the current drawing. The style selected from the list becomes the current style and is used when placing text with *Text* or *Mtext*. Alternately, before you use *Text* or *Mtext*, you can select the current text style from the *Text Style Control* drop-down list (in the *Annotation* panel) to make it the current text style.

Since only one text *style, Standard,* is available in the traditional (inch) template drawing (ACAD.DWT) and the metric template drawing (ACADISO.DWT), other styles must be created before the *style* option of *Text* is of any use. Various text styles are created with the *Style* command (this topic is discussed later).

Use the *Style* option of *Text* or *Mtext* to list existing styles and specifications. An example listing is shown below:

> Command: ***TEXT***
> Current text style: "Standard" Text height: 0.2000 Annotative: No
> Specify start point of text or [Justify/Style]: ***S*** (*Style* option)
> Enter style name or [?] <STANDARD>: ***?*** (*list* option)
> Enter text style(s) to list <*>: **Enter**
> Text styles:

Style name: "Annotative"   Font files: txt.shx
  Height: 0.0000 Width factor: 1.0000 Obliquing angle: 0
  Generation: Normal

Style name: "ROMANS"   Font files: romans
  Height: 1.50 Width factor: 1.00 Obliquing angle: 0.000
  Generation: Normal

Style name: "STANDARD"   Font files: txt
  Height: 0.00 Width factor: 1.00 Obliquing angle: 0.000
  Generation: Normal

## *Mtext*

| Ribbon | Menu Bar | Command (Type) | Alias (Type) | Shortcut |
|---|---|---|---|---|
| *Home Annotation Multiline Text* | *Draw Text > Multiline Text...* | *Mtext* or *-Mtext* | *T, -T* or *MT* | ... |

*Multiline Text* (*Mtext*) has more editing options than the *Text* command. You can apply underlining, color, bold, italic, font, and height changes to individual characters or words within a paragraph or multiple paragraphs of text.

*Mtext* allows you to create paragraph text defined by a text boundary. The text boundary is a reference rectangle that specifies the paragraph width. The *Mtext* object that you create can be a line, one paragraph, or several paragraphs. AutoCAD references *Mtext* (created with one use of the *Mtext* command) as one object, regardless of the amount of text supplied. Like *Text*, several justification methods are possible.

Command: ***MTEXT***
Current text style: "Standard" Text height: 0.2000 Annotative: No
Specify first corner: **PICK**
Specify opposite corner or [Height/Justify/Line spacing/Rotation/Style/Width/Columns]: **PICK** or
  (**option**)

You can PICK two corners to invoke the *Text Editor* tab, or enter the first letter of one of these options: *Height, Justify, Rotation, Style, Line Spacing,* or *Width.* Other options can also be accessed within the *Text Editor* tab.

Using the default option for the *Mtext* command, you supply a "first corner" and "opposite corner" to define the diagonal corners of the text boundary (like a window). Although this boundary confines the text on two or three sides, one or two arrows indicate the direction text flows if it "spills" out of the boundary (Fig. 18-11). (See "*Justification*.")

**FIGURE 18-11**

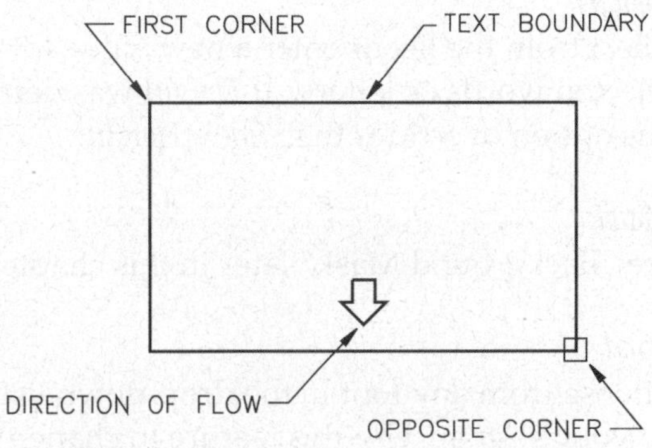

After you PICK the two points defining the text boundary, a text window appears ready for you to enter the desired text (Fig. 18-12). The text wraps based on the width you defined for the text boundary. Select the *OK* button on the *Text Editor* to have the text entered into the drawing.

**FIGURE 18-12**

**FIGURE 18-13**

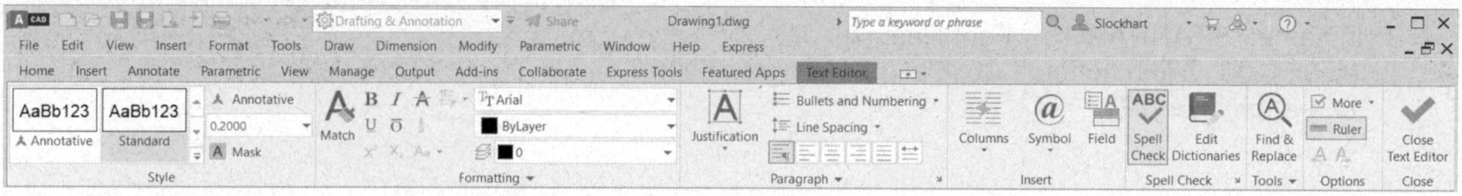


During the *Mtext* command, the Ribbon converts to the *Text Editor* (Fig. 18-13). The *Text Editor* provides many options for entering text including justification, bullets and numbering, columns, tables, symbols, and more. The *Mtext* options are explained next.

### *Ruler*

Use the *Ruler* button or *Show Ruler* option from the shortcut menu to make the ruler on top of the text area appear (Fig. 18-14). Note that you can use the sliders to interactively change both the width and the height of the text boundary.

**FIGURE 18-14**

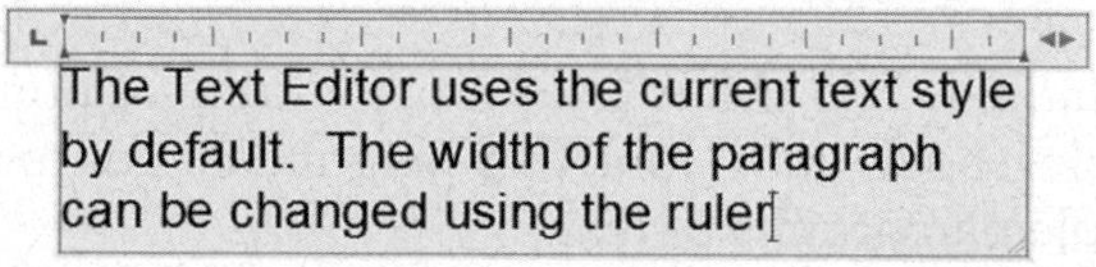


### *Width*

Even though this option does not appear on the *Text Editor* tab, you can interactively change the width of the existing text boundary. Do this by placing your pointer at the right border of the ruler above the text so the double arrows appear, then adjust the paragraph width (Fig. 18-14). The lines of text will automatically "wrap" to adjust to the new text boundary. You can also use the *Width* option from the Command line to set the width to a specific value. Also see "*Justification*" for changing the width of the text boundary with grips.

### *Style*

This option is the first (far left) list in the *Text Editor* tab. The box (not selected) displays the current text *Style* (*Standard* for most template drawings), and the list displays all existing text styles in the drawing. Select the style you want to use (make current) from this list. Alternately, before you use *Text* or *Mtext*, you can select the current text style from the *Text Style* drop-down list (in the *Annotate* tab, *Text* panel) to make it the current text style. (See the *Style* command for information on creating new text styles.)

### *Annotative*

Use this toggle to set the *Annotative* feature of the selected text. See "Annotative Text."

### *Height*

Select from the list or enter a new value for the height to be used globally for the paragraph or for selected words or letters. If *Height* was defined when the selected text *Style* was created for the drawing, this option overrides the *Style's Height*.

### *Mask*

See "Background Mask" later in this chapter.

### *Font*

Choose from any font in the drop-down list in the *Formatting* panel. The list includes all fonts registered on your system. Use this feature to change the appearance of selected words in the paragraph.

Your selection here overrides the font originally assigned to the text *Style* for the selected or newly created words or paragraphs. Even though you can change the font for the entire *Mtext* object (paragraph) using this feature, it is recommended that you set the paragraph to the desired *Style* (in the *Style* panel), rather than change the characters in the *Style* to a different font for the current *Mtext* object. See following NOTE.

### *Text Color*

Select individual text, then use this drop-down list from the *Formatting* panel, to select a color for the selected text. This selection overrides the layer color. See following NOTE.

NOTE: The *Font, Height,* and *Color* options in the *Text Editor* tab override the properties of the text *Style* and layer color. For example, changing the font in the *Font* list overrides the text style's font used for the paragraph so it is possible to have one font used for the paragraph and others for individual characters within the paragraph. This is analogous to object-specific color and linetype assignment in that you can have a layer containing objects with different linetypes and colors than the linetype and color assigned to the layer. To avoid confusion, it is recommended for global formatting to create text *Styles* to be used for the paragraphs, layer color to determine color for the paragraphs, then if needed, use the *Font* and *Color* options in the *Text Editor* tab to change fonts and colors for selected text rather than for the entire paragraph.

### *Bold, Italic, Underline, Overline, Superscript, Subscript, Change Case*

Select (highlight) the desired letters or words then PICK the desired button. Only authentic TrueType fonts (not the AutoCAD-supplied .SHX equivalents) can be bolded or italicized.

### *Match*

*Match* is similar to *Match Properties* but expressly for text formatting. First highlight the source text, then select the *Match* button, and then select the text to apply the formatting.

### *Stack/Unstack*

This option is a toggle to *Stack* or *Unstack* specific text (see "Creating Stacked Text"). Highlight the fraction or text, then use this option to stack or unstack the fraction or text.

### *Clear*

This drop-down list offers three methods to remove formatting. First highlight the source text, then select the desired *Clear* option. *Remove Character Formatting* clears the formatting from the selected text (any options appearing in the *Formatting* panel). *Remove Paragraph Formatting* clears formatting from the entire paragraph (any options appearing in the *Paragraph* panel) even if the entire paragraph is not selected. *Remove All Formatting* clears both character and paragraph formatting.

### *Oblique Angle*

This option as well as *Tracking* and *Width Factor* are available on a drop-down list at the bottom of the *Formatting* panel. *Oblique Angle* allows you to apply an angle (incline) to text. Enter the desired angle (in degrees) in the edit box. Apply an *Oblique Angle* to newly created text or existing text.

### *Tracking*

This parameter, also known as kerning, determines the spacing between letters. This option does not affect the width of the individual letters. The acceptable range is .75 through 4.0, with 1.0 representing normal spacing.

### *Width Factor*

Use this option to increase the width of the individual letters. Apply this characteristic to newly created text or existing text.

### Rotation

The *Rotation* option specifies the rotation angle of the entire *Mtext* object (paragraph) including the text boundary. This option is available only by command line; however, the grips Rotate option can be used after the *Mtext* object has been created. Using the Command line method, you can see the rotated text boundary as you specify the corners (Fig. 18-15).

**FIGURE 18-15**

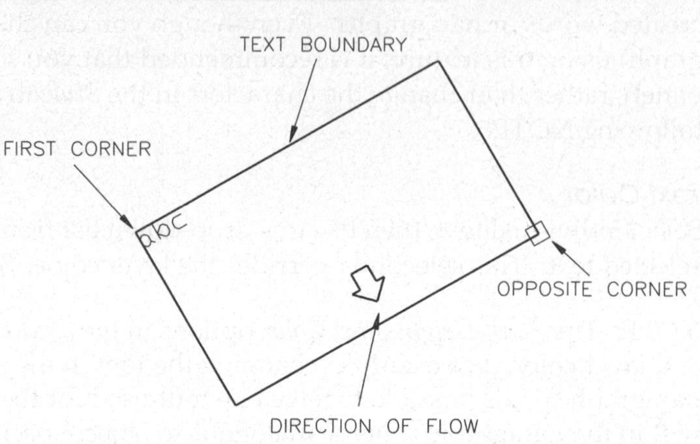


### Cut, Copy, Paste

These selections (only the right-click menu, when text is highlighted) allow you to *Cut* (erase) highlighted text from the paragraph, *Copy* highlighted text, and *Paste* (the *Cut* or *Copied*) text to the current cursor position. Text that you *Cut* or *Copy* is held in the Clipboard until Pasted.

**FIGURE 18-16**

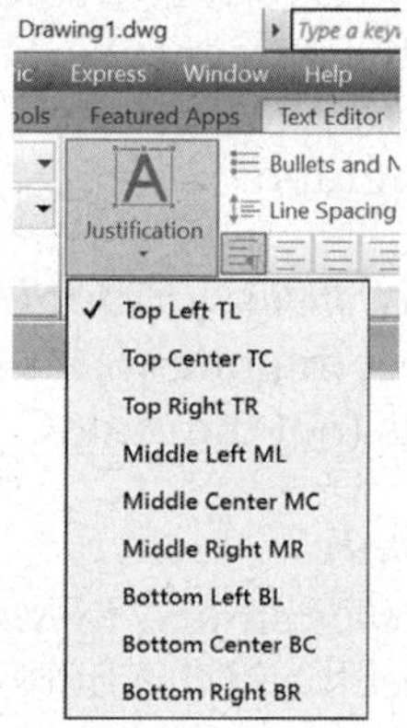


### Justification

The choices (Fig. 18-16) determine how the text in the paragraph is oriented (originally), how the text flows, and what the paragraph *Insertion* point is for Object Snaps. These options are not available in the right-click shortcut menus. These options are <u>not the same</u> as, and can be overridden by, the *Paragraph Alignment* settings (*Left*, *Center*, *Right*, *Justify*, and *Distribute*) available just to the right in the panel or from the right-click shortcut menus.

Justification is the method of aligning the text with respect to the text boundary. The options are *Top Left*, *Top Centered*, *Top Right*, *Middle Left*, *Bottom Left*, etc. The text paragraph (*Mtext* object) is effectively "attached" to the boundary based on the option selected. These options are illustrated in Figure 18-17. The illustration shows the relationship among the selected option, the text boundary, the direction of flow, and the resulting text paragraph.

**FIGURE 18-17**

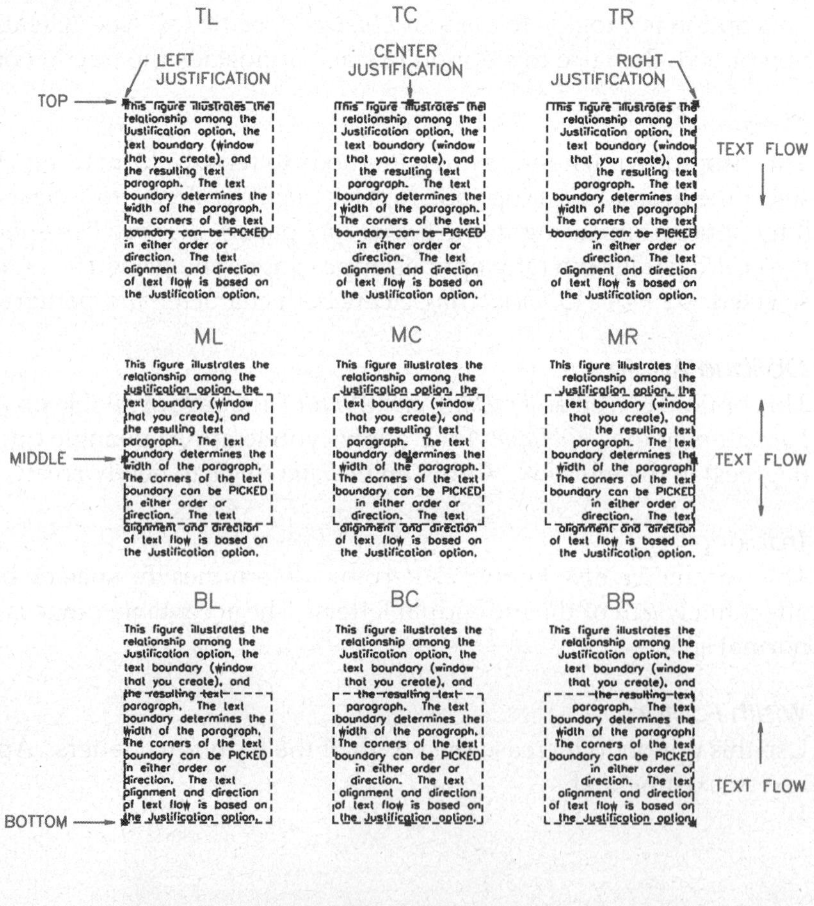

If you want to adjust the text boundary after creating the *Mtext* object, you can do so within the *Text Editor* tab (see previous discussion, "Width") or you can use grips (Chapter 19) to stretch the text boundary. If you activate the grips, two grips appear to change the text boundary and one grip appears at the defined justification point (Fig. 18-18).

NOTE: The *MText Justification* options can be overridden by selecting one of the *Paragraph Alignment* options from the right-click shortcut menus or from the toolbar. If a *Paragraph Alignment* option is used, however, it does not change the paragraph's *Insertion* point.

**FIGURE 18-18**

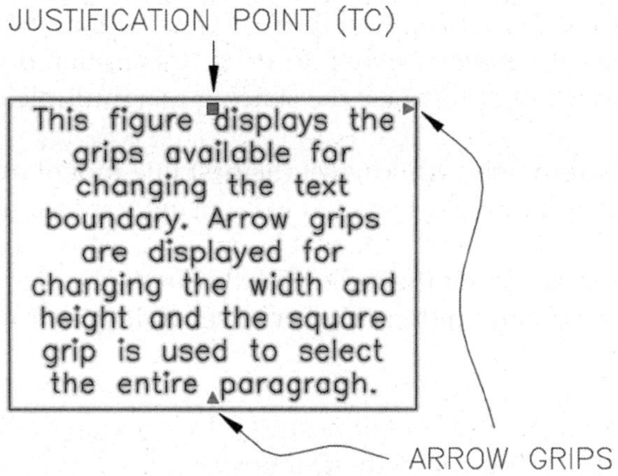


### *Paragraph Alignment*

The *Left*, *Center*, *Right*, *Justify*, and *Distribute* options automatically align the text in the paragraph accordingly and override the *Mtext Justification* setting for the paragraph (see previous NOTE).

*Left, Center, Right*

These options move the text in the paragraph to align flush left, centered, and flush right, respectively. The character spacing is not altered. The text does not have to be selected.

*Justify, Distribute*

Both of these options change the horizontal spacing such that the text in each line is adjusted to begin flush left and end flush right. The *Justify* option changes the spacing between words, whereas the *Distribute* option changes the individual character spacing.

### *Paragraph*

This option, selected from the arrow on the *Paragraph* panel or from shortcut menus, produces the paragraph dialog box (Fig. 18-19). The *Paragraph* dialog box is used to set tabs, indents, paragraph alignment, paragraph spacing, and line spacing.

*Tab*

To designate each tab, specify the increment in units and the tab type (left, center, right, decimal), then select *Add*.

*Left Indent, Right Indent*

Enter desired values for each indent.

*Paragraph Alignment*

These choices are identical to those for *Paragraph Alignment* discussed previously.

**FIGURE 18-19**

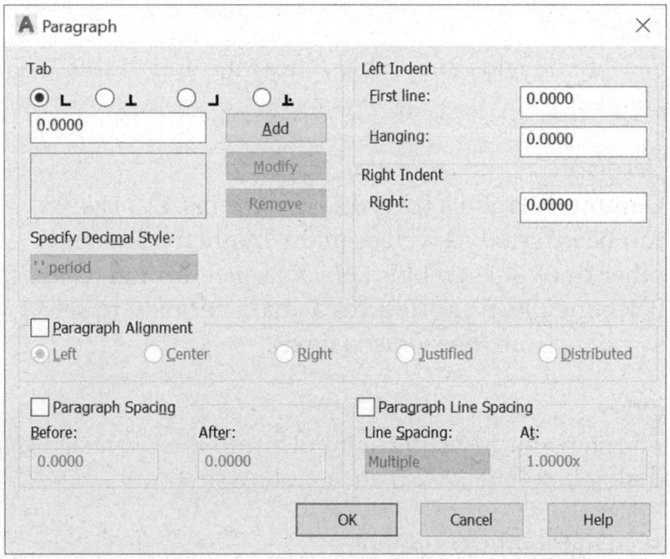


*Paragraph Spacing*

If you have multiple paragraphs created as one *Mtext* object, use the *Before* and *After* edit boxes to specify the space before and after each paragraph.

*Paragraph Line Spacing*
These options affect the vertical spacing between lines in the *Mtext* object. *Exactly* forces the line spacing to be the same for all lines. *At least* automatically adds space between lines based on the height of the largest character; therefore, use this option if you have different sizes of text in the paragraph. The *Multiple* option sets the spacing to a multiple of single line spacing.

Alternately, you can set the first line indent and the paragraph indent interactively by using your pointer on the ruler (just above the text boundary). Move the top indent marker to set the indent for the *First line* (Fig. 18-20) and the bottom indent marker to set an indent for the entire *Paragraph*.

**FIGURE 18-20**

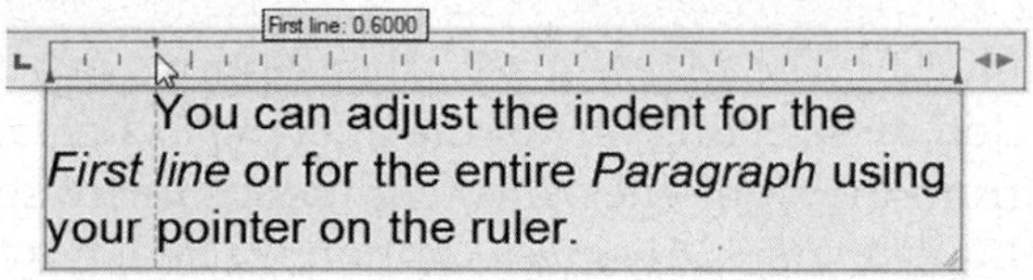


You can also set tab positions by clicking in the ruler appearing above the text box (Fig. 18-21). Pick the desired position to make a tab appear. Drag and drop tab stops out of the ruler to clear them. Select from left, right, center, or decimal tabs as shown by the pointer in the figure.

**FIGURE 18-21**

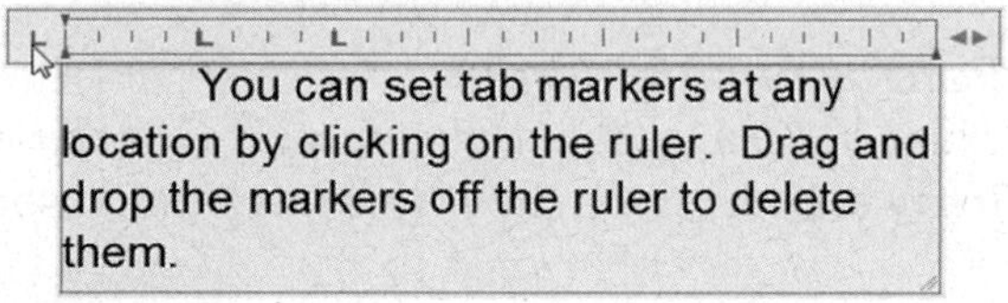


### Bullets and Numbering
You can create a variety of lettered, numbered, or bulleted lists using these options. See "Bullets and Lists" later in this chapter.

### Line Spacing
Place your cursor anywhere in the *Mtext* object, then use this drop-down list to select from the line spacing options. Selecting *More* produces the *Paragraph* dialog box shown in Figure 18-19. *Clear Line Spacing* sets the line spacing back to the default.

### Columns
See "Creating Columns" later in this chapter.

### Field
See "Text Fields and Tables" near the end of this chapter.

### Symbol
Common symbols (*Degrees, Plus/Minus, Diameter*) can be inserted as well as many mathematical and other drawing symbols (*Almost Equal* through *Cubed*). Selecting *Other…* produces a character map to select symbols from (See *Other...* next).

*Other...*
The steps for inserting symbols from the *Character Map* dialog box (Fig. 18-22) are as follows:

**FIGURE 18-22**

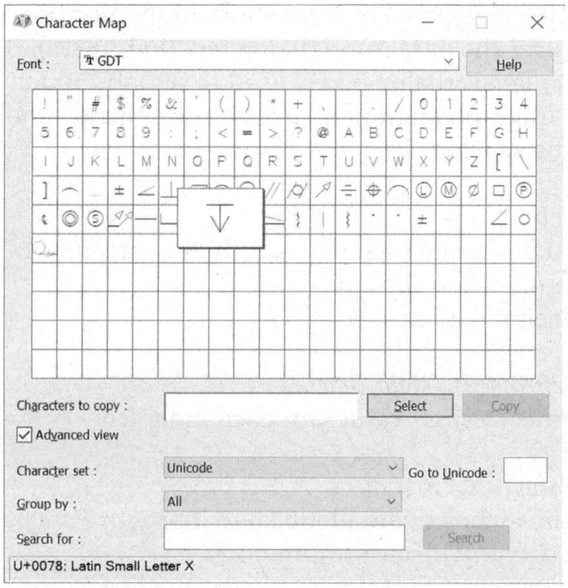


1. Highlight the symbol.
2. Double-click or pick *Select* so the item appears in the *Characters to copy*: edit box.
3. Select the *Copy* button to copy the item(s) to the Windows Clipboard.
4. *Close* the dialog box.
5. In the *Text Editor* tab, move the cursor to the desired location to insert the symbol.
6. Finally, right-click and select *Paste* from the menu.

### *Spell Check*

This toggle turns automatic spell checking on and off. See "Editing Text" later in this chapter for more information on the *Spell* command.

### *Edit Dictionaries*

See *Spell* in the "Editing Text" section later in this chapter.

### *Find & Replace*

See "Editing Text" later in this chapter.

### *Import Text*

See "Importing External Text into AutoCAD" later in this chapter.

### *All CAPS*

This option affects only newly typed and imported text. When checked, the *All CAPS* option is similar to turning on Caps Lock on your keyboard so only upper case letters appear as you type. If needed, use *Change Case* to convert the text back to lower case.

### *Character Set*

This option (available only from the *Options* panel or right-click shortcut menu) displays a menu of code pages. If *Big Fonts* have been used for text styles in the drawing, you can select a text string, then select a code page from the menu to apply it to the text (see "*Big Fonts*").

### *Editor Settings*

This option allows you to choose which features appear in the Text Editor, such as displaying *WYSIWYG* (What You See Is What You Get) or displaying a transparent background in the text window. The *Text Highlight Color* option displays the *Select Color* dialog box which is similar to using the *Color* drop-down list from the *Formatting* panel. You can also select *Show Toolbar* which displays the *Text Formatting* editor that was used in previous versions of AutoCAD (basically a duplication of the newer *Text Editor* tab on the Ribbon). Several other options are available if you select *Editor Settings* from the right-click shortcut menu, such as toggling *Spell Check* and accessing the *Check Spelling Settings* and the *Dictionaries* dialog boxes.

### *Ruler*

Use this toggle to turn the ruler that appears above the text window on and off.

### *Combine Paragraphs*

This option is available from the right-click shortcut menu. If you have more than one paragraph in one *Mtext* object, you can select all the text in the paragraphs and then use this option. The selected paragraphs are converted into a single paragraph and paragraph breaks are replaced with a space.

### *Background Mask*

In some cases, it is necessary to create text "on top of" other objects, such as a hatch pattern (Fig. 18-23). With this option, you can specify that the *Mtext* object you create has a background, or mask, to "cover" the objects "behind" the text. The resulting background is similar to that created by a *Wipeout*, only for text.

**FIGURE 18-23**

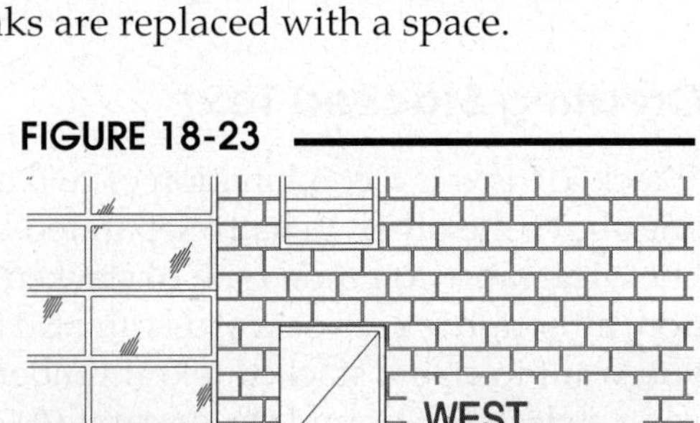

To create a background mask for an *Mtext* object, select *Mask* from the *Text Editor, Style* panel, or select *Background Mask* from the *Options* menu or right-click shortcut menu. This produces the *Background Mask* dialog box (Fig. 18-24) where you specify the *Fill color* for the mask and the *Border offset factor* (distance the mask boundary is offset from the actual text inserted). Selecting the *Use drawing background color* option under *Fill Color* forces AutoCAD to use the screen color as the background, therefore producing a "clear" mask.

**FIGURE 18-24**

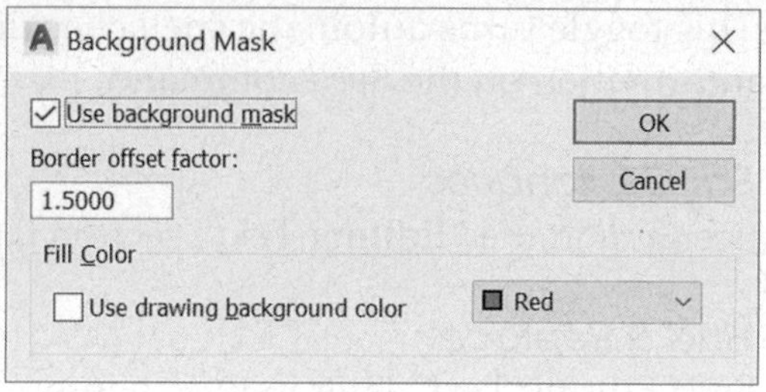


## Bullets and Lists

You can use the *Text Editor* to create a list with numbers, letters, bullets, or other characters as separators (Fig. 18-25). Items in the list are automatically indented based on the tab stops on the ruler. You can add or delete an item or move an item up or down a level and the list numbering automatically adjusts. You can create sub-levels in the list and also remove and reapply list formatting. To turn automatic list creation off completely, remove the check from *Allow Bullets and Lists*.

**FIGURE 18-25**

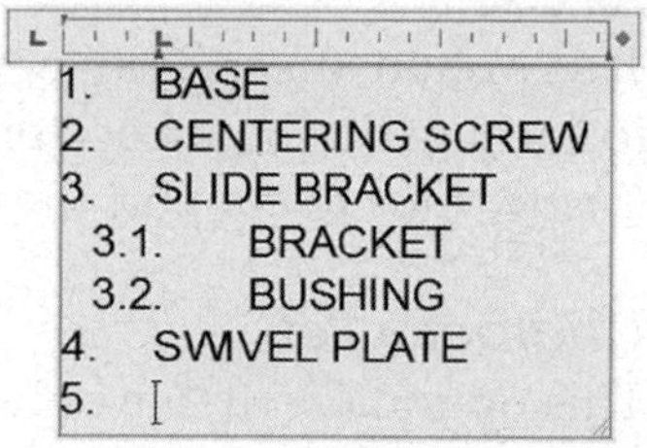


You can create lists in two ways: create the list as you type or apply list formatting to selected text. To create a list as you type, first select the *Lettered, Numbered,* or *Bulleted* option, then begin typing the first item in the list. Pressing Enter adds the next number, letter, or bullet. Press Enter twice to end the list. To create a sub-list (a lower-level list), press the Tab key before the first item in the sub-list. Press Enter to start the next item at the same level, or press Shift+Tab to move the item up a level.

If *Allow Auto Bullets and Numbering* is checked, you can type any letter or number, then a period or another delimiter character, then Tab to create a list. The following characters can be used as delimiters: period (.), colon (:), close parenthesis ()), close angle bracket (>), close square bracket (]), and close curly bracket (}). You can also create a bulleted list (without letters or numbers) using other characters as the "bullet," such as pound (#), asterisk (*), or hyphen (-), then press Tab and enter the first list item.

You can convert existing text to a list by selecting the paragraph, then select the *Numbered, Bulleted,* or *Lettered* option. To remove numbers, bullets, or letters from a list, highlight the list, then select the active list button on the toolbar to make it inactive.

NOTE: Bulleted lists for TrueType fonts use the bullet character that is included in the font. For SHX fonts, the default bullet character is a plus sign.

## Creating Stacked Text

"Stacked" text is a combination of two elements of text, one above the other, usually separated by a horizontal line. The most common type of stacked text is a fraction, although letters or words can also be stacked. If you want to create stacked text (numbers or letters), enter a slash (/), pound (#), or carat (^) as a separator between the upper and lower lines of text. Entering a slash, pound, or carat character between two numbers in the *Text Editor* automatically produces a stacked fraction with the default setting (*Auto Stacking* turned on).

**FIGURE 18-26**

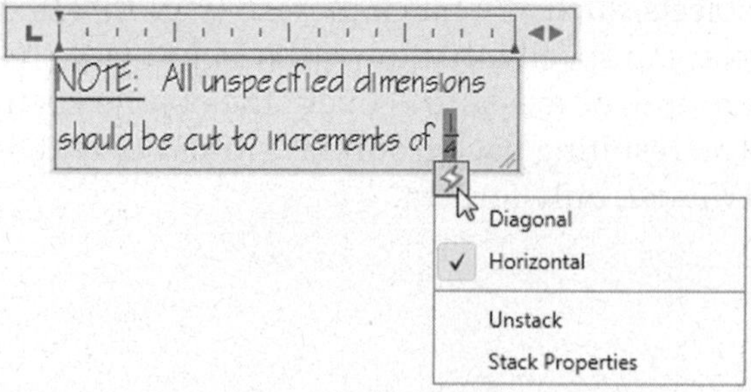

Although the slash, pound, and carat characters each specify a different type of stacked format, these options can be changed if you can click on the small icon that appears (Fig. 18-26). Use this shortcut menu to change the format to *Diagonal* or *Horizontal*. You can also *Unstack* or *Stack* existing fractions.

### *Stack Properties*

This feature is available only if you first highlight stacked text such as a fraction. If you select *Stack Properties* from the shortcut menu the *Stack Properties* dialog box appears (Fig. 18-27). The *Text* section of the *Stack Properties* dialog box allows you to change the text characters (numbers or letters) contained in the stacked set. Although using a slash (/), pound (#), or carat (^) symbol to specify a stack normally creates horizontal, diagonal, and tolerance format, respectively, that format can be changed with the *Style* drop-down list in this dialog box. Use the *Position* option to align the fraction with the *Top*, *Center*, or *Bottom* of the other normal characters in the line of text. You can also specify the *Text size* for the stack (which should generally be smaller than normal characters since the stacked set occupies more vertical space).

**FIGURE 18-27**

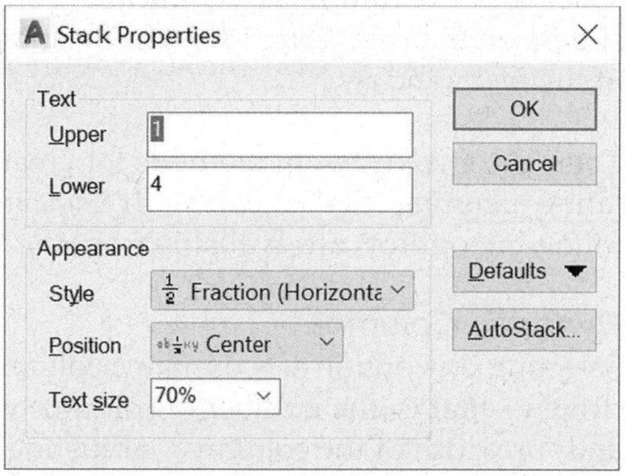


### *AutoStack Properties*

Selecting the *AutoStack...* button from the *Stack Properties* dialog box invokes the *AutoStack Properties* dialog box (Fig. 18-28).

**FIGURE 18-28**

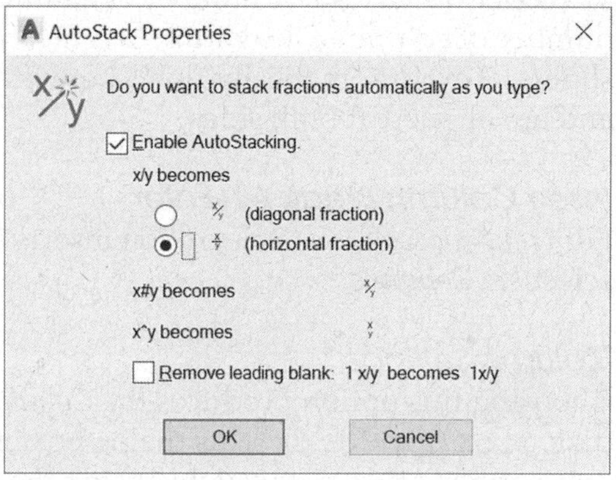


*Enable AutoStacking*

When this box is checked, fractions are automatically stacked as you type. Anytime you enter numeric characters before and after the carat, slash, or pound character, the numbers are automatically stacked. Each of these characters causes the following style of stacking:

| | |
|---|---|
| slash (/) | causes a vertical stack separated by a horizontal line |
| pound (#) | causes a diagonal stack separated by a diagonal line |
| carat (^) | causes a vertical stack without a horizontal separator line |

*Remove Leading Blank*

This option removes blanks between a whole number and a fraction. For example, you would normally type one and one-half with a space (1 1/2), which would convert to a whole number plus a space before the fraction.

## Creating Columns

Creating columns using *Mtext* is relatively simple, and adjusting the columns to your needs is intuitive. The *Columns* button provides several ways to create the columns. To create columns, begin by using *Mtext* and <u>entering the desired text</u>.

Next, either during the process of entering the text or after the paragraph(s) are complete, select the desired column option from the menu. Once the columns have been created, you can easily adjust the height, width, and spacing of the columns dynamically using the multiple sliders on the ruler (Fig. 18-29) or by adjusting the borders of the text window.

**FIGURE 18-29**

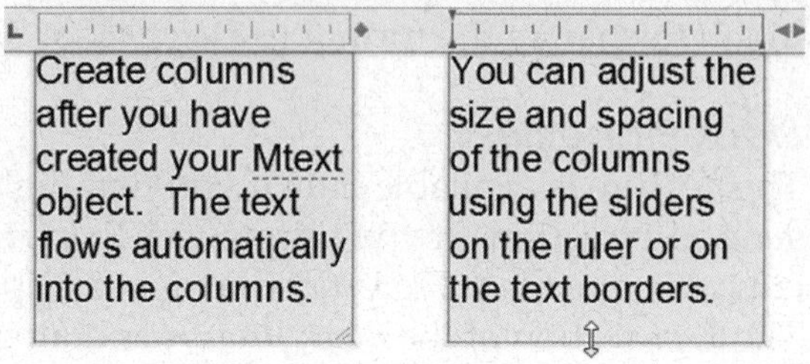


There are two different methods for creating and manipulating columns: *Static Columns* or *Dynamic Columns*. The following options are available from the *Columns* menu.

### *Dynamic Columns*

Selecting this option sets dynamic columns mode to the current *Mtext* object. *Dynamic Columns* are text driven—that is, the number of columns varies depending on how much text is entered and by the height and/or width of the columns. Adjusting the columns affects text flow and text flow causes columns to be added or removed. *Auto height* columns allow you to set the height for all columns with one slider, whereas *Manual height* columns can be individually adjusted for height.

### *Static Columns*

This option sets a static number of columns to the current *Mtext* object. Although you specify a static number of columns, the width, height, and spacing between the columns can be adjusted using the sliders. Text automatically flows between columns as needed. All static columns share the same height and are aligned at both sides.

### *Insert Column Break Alt+Enter*

For *Dynamic Columns*, this option inserts a manual column break. This feature is disabled when you select *No Columns*.

### *Column Settings*

Choosing this option produces the *Column Settings* dialog box (Fig. 18-30). This dialog box offers the same options that are available using the menus and the sliders. The dialog box is useful if you have the need for exact height and width spacing. Even though values can be entered into the edit boxes in the *Height* and *Width* clusters, the columns can be interactively adjusted later using the sliders.

**FIGURE 18-30**

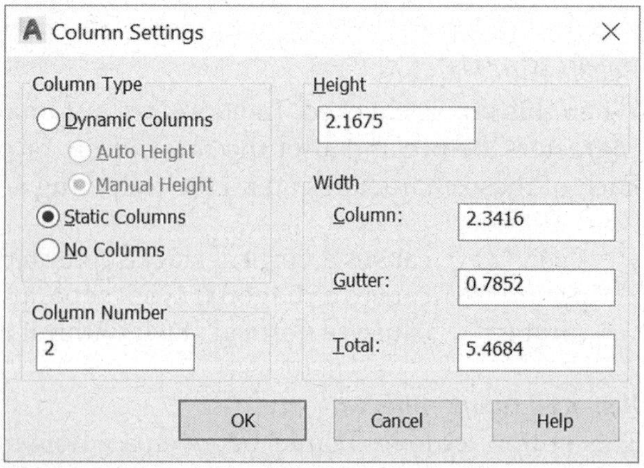


### Grip Editing

Columns can be edited using grips in a fashion similar to using the sliders available within the *Mtext* command. See Chapter 19, Grip Editing.

## Calculating Text Height for Scaled Drawings

To achieve a specific height in a drawing intended to be plotted to scale, multiply the desired (non-annotative) text height for the plot by the drawing scale factor. (See Chapter 14, Printing and Plotting.) For example, if the drawing scale factor is 48 and the desired text height on the plotted drawing is 1/8", enter **6** (1/8 x 48) in response to the "Height:" prompt of the *Text* or *Mtext* command.

If you know the plot scale (for example, 1/4"= 1') but not the drawing scale factor (DSF), calculate the reciprocal of the plot scale to determine the DSF, then multiply the intended text height for the plot by the DSF, and enter the value in response to the "Height:" prompt. For example, if the plot scale is 1/4"=1', then the DSF = 48 (reciprocal of 1/48).

If *Limits* have already been set and you do not know the drawing scale factor, use the following steps to calculate a text height to enter in response to the "Height:" prompt to achieve a specific plotted text height.

1. Determine the sheet size to be used for plotting (for example, 36" x 24").
2. Decide on the text height for the finished plot (for example, .125").
3. Check the *Limits* of the current drawing (for example, 144' x 96' or 1728" x 1152").
4. Divide the *Limits* by the sheet size to determine the drawing scale factor (1728"/36" = 48).
5. Multiply the desired text height by the drawing scale factor (.125 x 48 = 6).

## Spacetrans

| Ribbon | Menu Bar | Command (Type) | Alias (Type) | Shortcut |
|---|---|---|---|---|
| ... | ... | *Spacetrans* | ... | ... |

The *Spacetrans* command is used to convert a distance in one space (model space or paper space) to the equivalent distance in the other space. *Spacetrans* is intended to be used transparently (within a command) to pass the translated value to the current command. For example, using *Spacetrans* while drawing a *Line* in paper space allows you to draw the line length to an equivalent model space distance.

TIP

*Spacetrans* is especially helpful for creating non-annotative text objects in a model space viewport that print at a specific size in paper space. Instead of having to calculate the text height in model space units times the drawing scale factor (text height x DSF) as described previously in "Calculating Text Height for Scaled Drawings," using *Spacetrans* allows you to enter only the desired plotted text height (assuming you are plotting from paper space at 1:1). Using *Spacetrans* in model space (inside a viewport) allows you to create model space text in paper space units.

As an example, assume you are creating non-annotative *Text* inside a viewport (in model space) and want the text height to be equivalent to .125 units (1/8") in paper space. You would use a procedure similar to that shown below—that is, invoking *Spacetrans* transparently during the *Text* command to convert a paper space value (.125) to model space units and pass the value to the *Text* command.

```
Command: TEXT
Current text style: "romans" Text height: 0.2000 Annotative: No
Specify start point of text or [Justify/Style]: PICK
Specify height <0.2000>: 'spacetrans
>>Specify paper space distance <1.0000>: .125
Resuming TEXT command.
Specify height <0.2000>: 0.500000000000000
Specify rotation angle of text <0>: Enter
Enter text:
```

When you are first prompted to "Specify height," invoke *Spacetrans* transparently. Enter the desired height in paper space units at the "Specify paper space distance" prompt. *Spacetrans* then converts the .125 units to the equivalent distance in model space (0.500000000000000) and passes the value to the second "Specify height" prompt, thus creating text in model space to print at .125 in paper space.

Technically, *Spacetrans* computes the entered paper space units times the reciprocal of the viewport scale (or viewport zoom factor). In the previous example, the viewport scale is 1:4; therefore, .125 x 4 = .5.

*Spacetrans* cannot be used in the *Model* tab—you must be in a layout with at least one viewport created, and you must be in the desired space when *Spacetrans* is invoked. If you are in paper space when *Spacetrans* is invoked and there is more than one viewport, you are prompted to "Select a viewport."

NOTE: The *Spacetrans* command cannot be used transparently (during another command) unless it has been invoked previously in the AutoCAD session. Invoke *Spacetrans* at the command prompt, then press Esc (escape). You can then use *Spacetrans* during the *Text* command.

## COMBINING SINGLE AND MULTIPLE-LINE TEXT ELEMENTS

### *Txt2mtxt*

| Ribbon | Menu Bar | Command (Type) | Alias (Type) | Shortcut |
|---|---|---|---|---|
| *Insert*<br>*Import*<br>*Combine Text*<br>(icon) | *Format*<br>*Text Style...* | *Txt2mtxt* | ... | ... |

The TXT2MTXT command converts Text objects to Mtext objects or combines multiple Text and/or Mtext objects into a single Mtext object.

Command: ***TXT2MTXT***
Select objects of [SEttings]: ***PICK***
Select objects or [SEttings]: ***Enter***
Command:

Choosing the Settings option launches the PDF Text Recognition Settings dialog box, which can also be opened by choosing Recognition Settings in the Import section of the Insert tab.

*Combine into a single mtext object*
When selected, this option brings all selected Text or Mtext objects together as a single Mtext object. The order of the source objects in the final Mtext object can be either from the top of the screen down (Sort top-down), or in the order you PICK the selection set (Select order of text).

*Word-wrap text*
When selected, the source objects are combined into a single line that wraps based on the size of the Mtext boundary. This selection is only available when Combine into a single mtext object is selected and Force uniform line spacing is not selected.

FIGURE 18-31

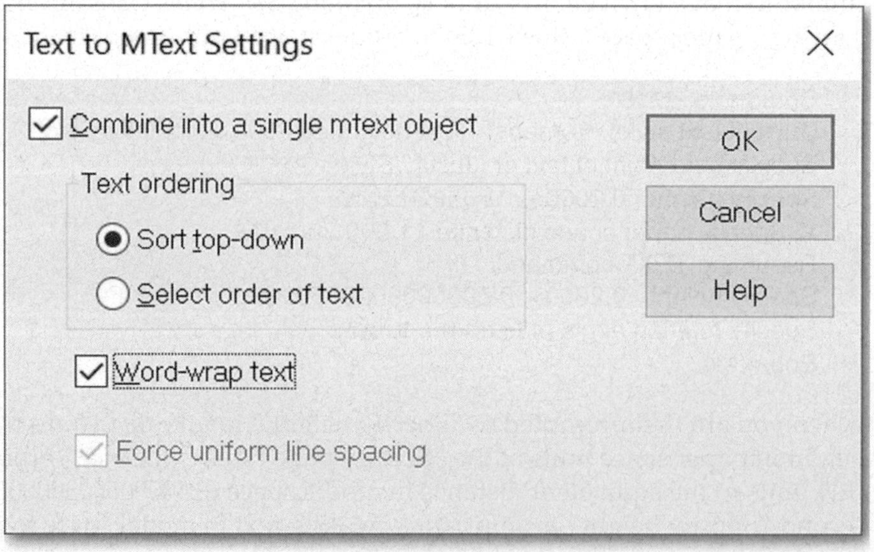

*Force uniform line spacing*

Ensures that all vertical spacing between the selected source lines is equal in the resulting Mtext object. This can move some selected text from its original location. This selection is only available when Combine into a single mtext object is selected and Word-wrap text is not selected.

## TEXT STYLES, FONTS, ANNOTATIVE TEXT, AND EXTERNAL TEXT

### *Style*

| Ribbon | Menu Bar | Command (Type) | Alias (Type) | Shortcut |
|---|---|---|---|---|
| *Home Annotation* (icon) | *Format Text Style...* | *Style* or *- Style* | *ST* | ... |

Text styles can be created by using the *Style* command. A text *Style* is created by selecting a font file as a foundation and then specifying several other parameters to define the configuration of the letters.

The *Style* command invokes the *Text Style* dialog box (Fig. 18-32). Options for creating and modifying text styles are accessible from this device. Selecting the font file is the initial step in creating a style. The font file selected then becomes a foundation for "designing" the new style based on your choices for the other parameters (*Effects*).

**FIGURE 18-32**

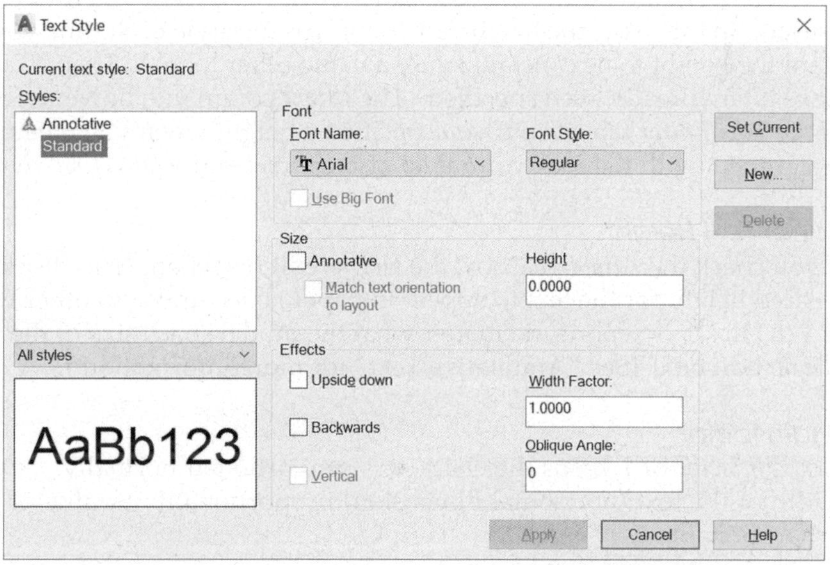


Recommended steps for creating text *Styles* are as follows:

1. Use the *New* button to create a new text *Style*. This button opens the *New Text Style* dialog box (not shown). By default, AutoCAD automatically assigns the name Style*n*, where *n* is a number that starts at 1. Enter a descriptive name into the edit box. Style names can be up to 256 characters long and can contain letters, numbers, and the special characters dollar sign ($), underscore (_), and hyphen (-).
2. Select a font to use for the style from the *Font Name* drop-down list (see Fig. 18-31). Authentic TrueType font files (.TTF) as well as AutoCAD equivalents of the TrueType font files (.SHX) are available.
3. If you want to create annotative text, check the *Annotative* box. Enter the desired *Paper Text Height* in the edit box. (See "Annotative Text" for more information.)
4. Select parameters in the *Effects* section of the dialog box. Changes to the *Upside down, Backwards, Width Factor,* or *Oblique Angle* are displayed immediately in the *Preview* tile.
5. Select *Apply* to save the changes you made to the new style.
6. Create other new *Styles* using the same procedure listed in steps 1 through 5.
7. Select *Close*. The new (or last created) text style that is created automatically becomes the current style inserted when *Text* or *Mtext* is used.

You can modify existing styles using this dialog box by selecting the existing *Style* from the list, making the changes, then selecting *Apply* to save the changes to the existing *Style*.

Alternately, you can enter *-Style* (use the hyphen prefix) at the command prompt to display the Command line format of *Style*.

### Rename

Select an existing *Style* from the list, right click, then select *Rename*. Enter a new name in the edit box.

### Delete

Select an existing *Style* from the list, then select *Delete*. You cannot delete the current *Style* or *Styles* that have been used for creating text in the drawing.

### Annotative

Check the *Annotative* box if you want to create annotative text. (See "Annotative Text" for more information.)

### Height

The height should be 0.000 if you want to be prompted again for height each time the *Text* command is used. In this way, the height is variable for the style each time you create text in the drawing. If you want the height to be constant, enter a value other than 0. Then, *Text* will not prompt you for a height since it has already been specified. The *Mtext* command, however, allows you to change the height in the *Text Editor* tab, even if you specified a height when you created the text style. A specific height assignment with the *Style* command also overrides the *DIMTXT* setting (see Chapter 29).

### Paper Text Height

If you check the *Annotative* box, the *Height* edit box changes to display *Paper Text Height*. Enter the desired height for the text to appear in paper space units. In other words, if you want the text to appear at 1/8" in all viewports, no matter what the model space size of the text is, enter .125 in the *Paper Text Height* edit box. (See "Annotative Text" for more information.)

### Width factor

A *width factor* of 1 keeps the characters proportioned normally. A value less than 1 compresses the width of the text (horizontal dimension) proportionally; a value of greater than 1 extends the text proportionally.

### Obliquing angle

An angle of 0 keeps the font file as vertical characters. Entering an angle of 15, for example, would slant the text forward from the existing position, or entering a negative angle would cause a back-slant on a vertically oriented font (Fig. 18-33).

**FIGURE 18-33**

### Backwards

Backwards characters can be helpful for special applications, such as those in the printing industry.

### Upside-down

Each letter is created upside-down in the order as typed (Fig. 18-34). This is different than entering a rotation angle of 180 in the *Text* command. (Turn this book 180 degrees to read the figure.)

**FIGURE 18-34**

### *Vertical*

Vertical letters are shown in Figure 18-35. The normal rotation angle for vertical text when using *Text* or *Mtext* is 270. Only .SHX fonts can be used for this option. *Vertical* text does not display in the *Preview* image tile.

**FIGURE 18-35**

Since specification of a font file is an initial step in creating a style, it seems logical that different styles could be created using one font file but changing the other parameters. It is possible, and in many cases desirable, to do so. For example, a common practice is to create styles that reference the same font file but have different obliquing angles (often used for lettering on isometric planes). This can be done by using the *Style* command to assign the same font file for each style, but assign different parameters (obliquing angle, width factor, etc.) and a unique name to each style. The relationship between fonts, styles, and resulting text is shown in Figure 18-36.

**FIGURE 18-36**

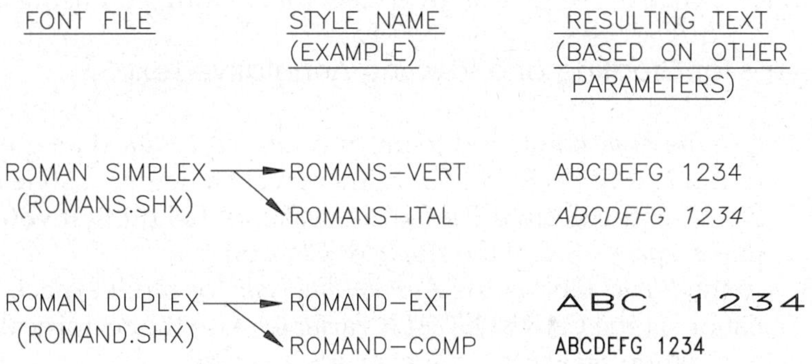


### *.SHX Fonts*

The .SHX fonts are fonts created especially for AutoCAD drawings by Autodesk. They are generally smaller files and are efficient to use for AutoCAD drawings where file size and regeneration time is critical. These fonts, however, are composed of single line segments and are not solid-filled such as most TrueType fonts (use *Zoom* to closely examine and compare .SHX vs. TrueType fonts).

TIP

Because the *Text Editor* tab is a Windows-compliant dialog box, it can display only fonts that are recognized by Windows. Since AutoCAD .SHX fonts are not recognized by Windows, AutoCAD supplies a TrueType equivalent when you select an .SHX or any other non-TrueType font for display in the *Text Editor* tab only. When you then use a text creation command, AutoCAD creates the text in the drawing with the true .SHX font.

All AutoCAD .SHX shape fonts are Unicode fonts. A Unicode font can contain 65,535 characters with shapes for many languages (see *Big Fonts*). Unicode fonts contain many more characters than are shown on your keyboard in your system; therefore, to use a character not directly available from the keyboard, you can enter the escape sequence ***\U+nnnn*** where *nnnn* represents the Unicode hexadecimal value for the character (see "Special Text Characters").

### *Big Fonts*

*Big Fonts* are used for characters not found in the English language. Text files for some alphabets contain thousands of non-ASCII characters. For these applications, AutoCAD provides a special type of shape definition known as a *Big Font* file that contains many other characters. When you check *Use Big Font*, the *Font Style* box changes to a *Big Font* box. You can select *Use Big Font* only when using .SHX fonts. Set a *Style* to use both regular and *Big Font* files for other than English languages. Otherwise, for most applications, select only the regular .SHX font files.

### *TrueType Fonts*

AutoCAD uses the Windows operating system directly to display TrueType text on the screen for most applications. However, when text in AutoCAD has been transformed (mirrored, upside-down, backward, oblique, has a width factor not equal to 1, or is in an orientation that is not co-planar with the screen), AutoCAD must draw the TrueType text. Text that has been transformed might appear slightly more bold in some circumstances, especially at lower resolutions. This difference is only in the screen display of the font and does not affect the printed drawing.

### *Proxy Fonts*

For third-party or custom .SHX fonts that have no TrueType equivalent, AutoCAD supplies up to eight different TrueType fonts called proxy fonts. Proxy fonts appear different in the *Text Editor* tab from the font they represent to indicate that they are substitutions for the fonts used in the drawing.

## Annotative Text

For complex drawings including multiple layouts and viewports, it may be desirable to create annotative objects. Annotative objects, such as annotative text, annotative dimensions, and annotative hatch patterns, automatically adjust for size when the *Annotation Scale* for each viewport is changed. That is, annotative text objects can be automatically resized to appear "readable" in different viewports at different scales. The following section is helpful if you plan to create multiple viewports and want to use annotative text. To give an overview for creating and using annotative text, follow these steps.

### Steps for Creating and Viewing Annotative Text

1. Use the *Style* command to produce the *Text Style* dialog box.
2. In the *Text Style* dialog box, either create a new text style or change an existing style to *Annotative*. Next, set the desired *Paper Text Height*, or height that you want the text to appear in the viewports in paper space. Select the *Apply* or *Close* button.
3. In the *Model* tab, set the *Annotation Scale* for model space. You can use the *Annotation Scale* pop-up list or set the *CANNOSCALE* variable. Use the reciprocal of the "drawing scale factor" (DSF) as the annotation scale (AS) value, or AS = 1/DSF.
4. Create the text in the *Model* tab using *Text* or *Mtext*. Accept the default (model space) height.
5. Create the desired viewports if not already created.
6. Set the desired scale for the current viewport by double-clicking inside the viewport and using the *Viewport Scale* pop-up list or *Viewports* toolbar. Alternately, select the viewport object (border) and use the *Properties* palette for the viewport. Setting the *Viewport Scale* automatically adjusts the annotation scale for the viewport.

To simplify, create annotative text using the following formulas, where DSF represents "drawing scale factor":

*Annotation Scale* in the *Model* tab = 1/DSF
*Paper Text Height* x DSF = model space text height.

For example, assume you are drawing a small apartment. In order to draw the floor plan in the *Model* tab full size, you change the default *Limits* (12, 9) by a factor of 48 to arrive at *Limits* of 48',36' (drawing scale factor = 48). After creating the floor plan, you are ready to label the rooms.

Following steps 1 through 3, create a new *Text Style* as *Annotative*, then set the *Paper Text Height* to 1/8". Next, set the annotative scale for the *Model* tab using the *Annotation Scale* pop-up list in the lower-right corner of the Drawing Editor (see Fig. 18-37). Using the drawing scale factor (DSF) of 48, set the *Annotation Scale* to 1:48 (or AS = 1/DSF). (Also see "Steps for Drawing Setup Using Annotative Objects" in Chapter 12.)

Next, following step 4, use *Text* to create the room labels in the *Model* tab as shown in Figure 18-37. The command prompt appears as shown below. Note that AutoCAD automatically calculates the model space text height (1/8" x 48 = 6", or paper text height x DSF = model text height) so you do not have to specify the height during the *Text* command.

Command: ***TEXT***
Current text style: "ARCH1"
Text height: 0'-6"
Annotative: Yes
Specify start point of text or [Justify/Style]:

**FIGURE 18-37**

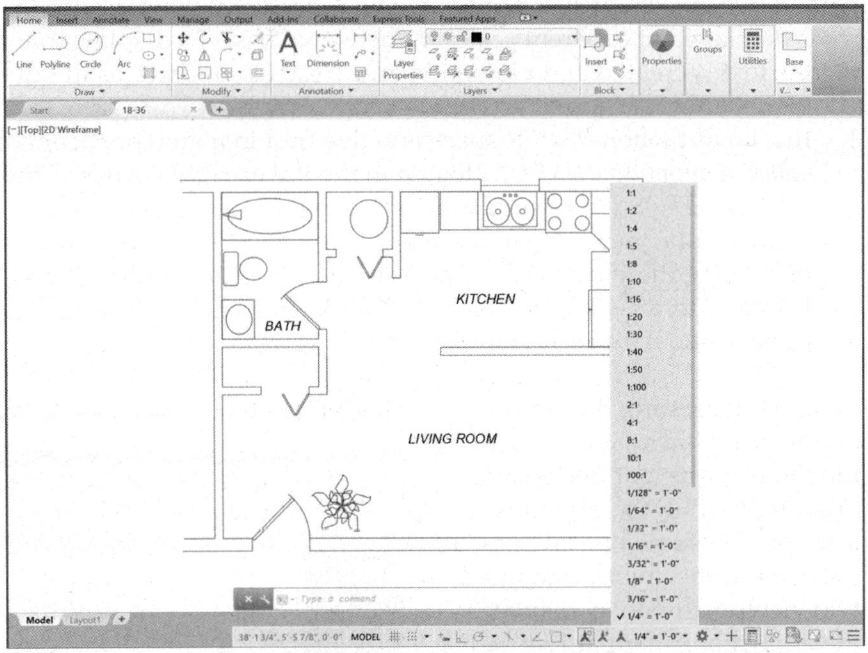


Next, following steps 5 and 6, switch to a *Layout* tab, set it up for the desired plot or print device using the *Page Setup Manager*, and create a viewport. Set the viewport scale for the viewport using the *Viewport Scale* pop-up list as shown in Figure 18-38. The text should appear appropriately in the layout at 1/8" (paper text height). Note that the *Annotation Scale* is locked to the *Viewport Scale* (by default) so its setting changes accordingly as the *Viewport Scale* changes.

**FIGURE 18-38**

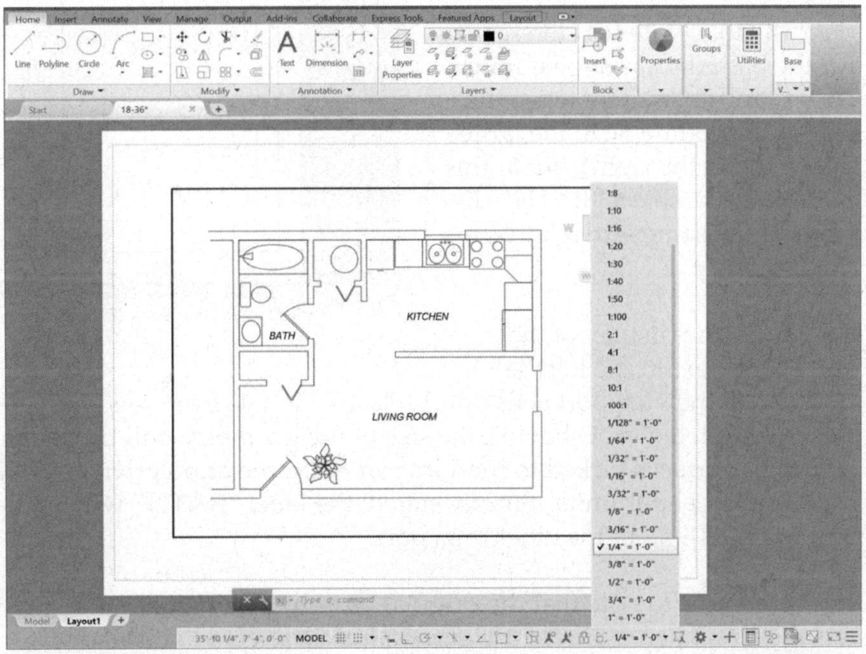


Note that at this point, using only one viewport, there is no real advantage to using annotative text as opposed to non-annotative text. The text appears in the *Model* tab and *Layout* tab at the correct size, just as if non-annotative text were used and set to a (model space) text height of 6".

### Annotative Text for Multiple Viewports

TIP

Annotative objects such as annotative text are useful when multiple viewports are used and the geometry is displayed at different scales in the viewports. To display annotative text in such cases, and assuming the text is created and viewports are already prepared (as in our example), follow these steps:

1. In a layout when *PAPER* space is active (not in a viewport), enable the *Automatically add scales to annotative objects...* toggle in the lower-right corner of the Drawing Editor or set the *ANNOAUTOSCALE* variable to a positive value (1, 2, 3, or 4).
2. Activate one viewport by double-clicking inside the viewport. Set the viewport scale for the viewport using the *Viewport Scale* pop-up list, *Viewports* toolbar, or *Properties* palette. The *Annotation Scale* is locked to the *Viewport Scale* by default, so it changes automatically to match the *Viewport Scale*.
3. Repeat step 2 for each viewport.

To illustrate setting the annotative scale for each viewport, use the same apartment floor plan drawing used in the previous example. Assume one layout was created but the layout contained two viewports, one to display the entire apartment at 1/4″ = 1′ scale and the second viewport to display only the bath at 1/2″ = 1′ scale (Fig. 18-39). Note that the scale for the bath (left) is shown much larger (approximately twice the size of the geometry in the large viewport on the right), but at this point the text object "BATH" displays at the same size relative to the geometry.

FIGURE 18-39

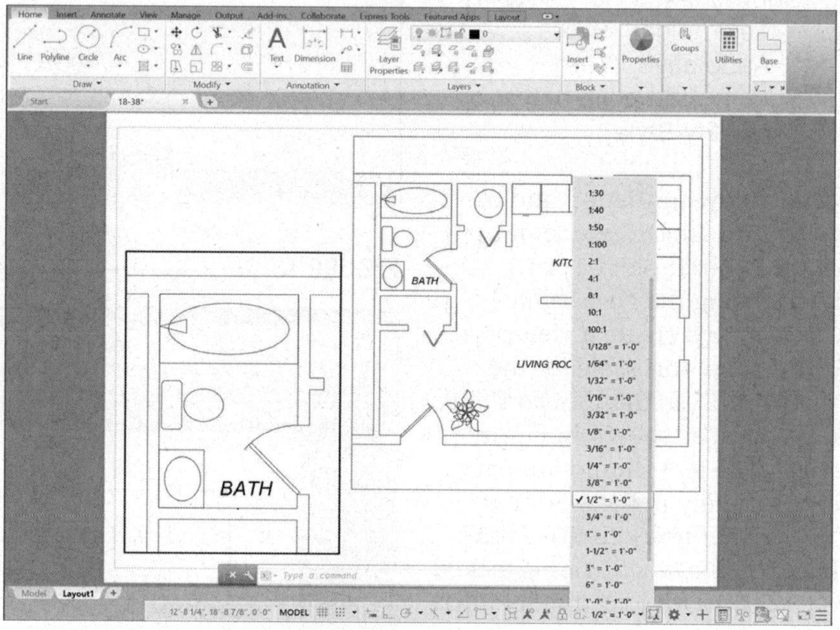


To change the display of the annotative text in the left viewport, make the viewport active and select 1/2″ = 1′ from the *Viewport Scale* pop-up list as shown. Once this is completed (not shown), the size of the geometry will be changed to exactly 1:2, and since the *Annotation Scale* is locked to the *Viewport Scale*, the annotative text object will be displayed at the correct size, 1/8″ in paper units. Simply stated, the label "BATH" will be the same size as "LIVING ROOM" and "KITCHEN" in the other viewport.

TIP

It is important to note that the original text objects do not change size when a different *Viewport Scale* is selected. Instead, a new annotative text object is created as a copy of the original, but in the new scale (Fig. 18-40). When the *Automatically add scales to annotative objects...* toggle (the *ANNOAUTOSCALE* variable) is on, new text objects are created for each new scale selected from the *Viewport Scale* list. Therefore, note that a single annotative object can have multiple *Annotative Scales* (or *Object Scales*), one for each time the object is displayed in a different scale.

FIGURE 18-40

The *SELECTIONANNODISPLAY* system variable, when set to 1, allows you to see each text object when selected, as shown in Figure 18-40. The text objects can be manipulated individually using grips. (See Chapter 32 for more information on using annotative objects in multiple viewports.)

## Importing External Text into AutoCAD

You can import ASCII text files created in other text editors or word processors into an AutoCAD drawing. There are three methods: use the *Import Text* option in the *Text Editor* tab, use *Copy* and *Paste* (using the Windows Clipboard), or "drag and drop" a file icon from the Windows File Explorer.

Importing text in ASCII or RTF files can save you drawing time. For example, you can create a text file of standard notes that you include in many drawings or use information prepared for a report or other document. Instead of entering (typing) this information in the drawing, you can import the file. The imported text becomes an AutoCAD text object, which you can edit as if you created it in AutoCAD. Imported text retains its original formatting properties from the text editor in which it was created.

### *Importing with the Text Editor tab*

To import text into the *Text Editor* tab, first use the *Mtext* command. PICK two points to define the text boundary as usual. When the *Text Editor* tab appears, right-click and select *Import Text* from the shortcut menu. The *Select File* dialog box appears for you to locate and select the desired file. The file to import must be in an ASCII or RTF format. (An ASCII text file often has a .TXT file extension.) The selected text appears in the editor and can be manipulated just as any text you entered.

You can also *Copy* text (to the Windows Clipboard) from any Windows document and *Paste* it into AutoCAD. The best method to use with *Copy* and *Paste* is to use the *Text Editor* tab. Open the editor, then use the Alt+Tab key combination to switch to the other document. Highlight the desired text and select *Copy* from the *Edit* menu or right-click menu. Switch back to the *Text Editor* tab, right-click for the pop-up menu, and select *Paste*. The imported text can be edited like any other AutoCAD *Mtext* object.

### *Using Drag and Drop to Import Text Files*

You can also use the drag-and-drop feature to insert ASCII text into a drawing. Open the Windows File Explorer and size the window so AutoCAD is also visible on the screen. Drag the file name or icon and drop it into an AutoCAD drawing. The pasted text uses the formats and fonts defined by the current AutoCAD text style. The width of the lines or paragraph is determined by line breaks and carriage returns in the original document. You can drag only files with a file extension of .TXT into an AutoCAD drawing.

## Using an External Text Editor for *Multiline Text*

You can specify a different text editor to use instead of the "internal" *Text Editor* tab. For example, you may prefer the Windows Notepad or other editor, although these editors do not offer the text formatting capabilities available in the *Text Editor* tab.

To specify an external text editor, you can use the *Files* tab of the *Options* dialog box from the *Tools* pull-down menu (Fig. 18-41) or use the *MTEXTED* system variable. Enter "internal" to use the internal AutoCAD *Text Editor* tab.

If you use an external text editor (other than the AutoCAD internal editor), you must enter the formatting codes as you enter the text using the text editor. Text features that can be controlled are underline, overline, height, color, spacing, stacked text, and other options. The following table lists the codes and the resulting text.

**FIGURE 18-41**

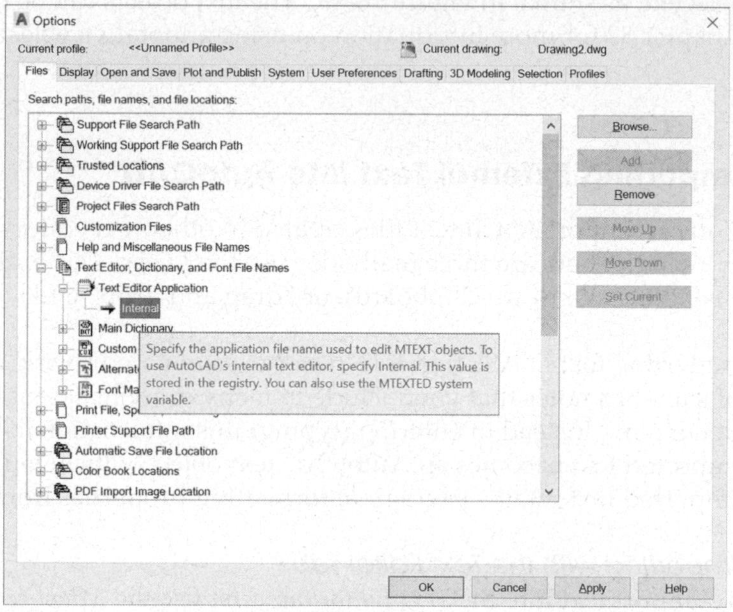


## Format Codes for Creating *Multiline* Text in External Text Editors

| Format code | Purpose | Type this. . . | To produce this |
|---|---|---|---|
| `\O. . .\o` | Turns overline on and off | You have \Omany\o choices | You have many choices |
| `\L. . .\l` | Turns underline on and off | You have \Lmany\l choices | You have <u>many</u> choices |
| `\~` | Inserts a nonbreaking space | Keep these\~words together | Keep these words together |
| `\\` | Inserts a backslash | slash\\backslash | slash\ backslash |
| `\{. . .\}` | Inserts an opening and closing brace | The \{bracketed\} word | The {bracketed} word |
| `\C`*`value;`* | Changes to the specified color | Change \C2;these colors | Change these colors |
| `\F`*`filename;`* | Changes to the specified font file | Change \Farial;these fonts | Change these fonts |
| `\H`*`value;`* | Changes to the specified text height | Change \H2;these sizes | Change these sizes |
| `\S. . .^. . .` | Stacks the subsequent text at the \ or ^ symbol | 1.000\S+0.010^0.000; | $1.000^{+0.010}_{-0.000}$ |
| `\T`*`value;`* | Adjusts the space between characters, from .75 to 4 times | \T2;TRACKING | T R A C K I N G |
| `\Q`*`angle;`* | Changes obliquing angle | \Q20;OBLIQUE | *OBLIQUE* |
| `\W`*`value;`* | Change width factor to produce wide text | \W2;Wide | WIDE |
| `\A` | Sets the alignment value to 0 (bottom), 1 (center), or 2 (top) | \A1;Center\S1/2 | Center |
| `\P` | Ends paragraph | First paragraph\PSecond paragraph | First paragraph<br>Second paragraph |

For example, you may enter the following text in the Windows Notepad editor using the *Mtext* command.

```
\C1;This sample line of text is red in color.\P
\C7;\H.25;This text has a height of 0.25.\P
\H.2;\Fromanc;This line is an example of romanc.shx.\P
\Ftxt;This line contains stacked fractions such as \S1/2
```

Figure 18-42 displays the text as it might appear in the AutoCAD drawing editor.

**FIGURE 18-42**

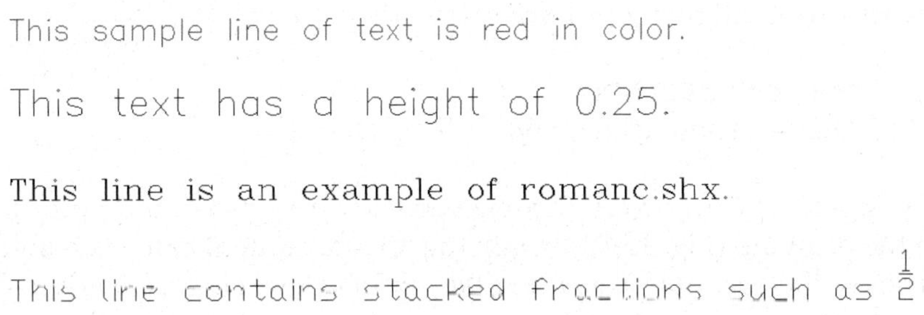


### Special Text Characters

Special characters that are often used in drawings can be entered easily in the internal *Text Editor* tab by selecting *Symbol* from the shortcut menu. In an external editor or with the *Text* command, you can code the text by using the "%%" symbols or by entering the Unicode values for high-ASCII (characters above the 128 range). The %% symbols can be used with the *Text* command, but the Unicode values must be used for creating *Mtext* with an external text editor. The following codes are typed in response to the "Enter text:" prompt in the *Text* command or entered in the external text editor.

| External Editor | *Text* | Result | Description |
|---|---|---|---|
| **\U+2205** | **%%c** | ø | diameter |
| **\U+00b0** | **%%d** | ° | degrees |
| | **%%o** | | overscored text |
| | **%%u** | ___ | underscored text |
| **\U+00b1** | **%%p** | | plus or minus |
| | **%%*nnn*** | varies | ASCII text character number |
| **\U+*nnnn*** | | varies | Unicode text hexadecimal value |

For example, entering "**Chamfer 45%%d**" at the "Enter text:" prompt of the *Text* command draws the following text: **Chamfer 45°**.

## EDITING TEXT

Although several methods are available to edit text, probably the simplest method to use is to double-click on a *Text* or *Mtext* object (assuming *DBLCLKEDIT* is set to *On*). Doing so invokes the appropriate editing method for the specific text. If you double-click on a *Text* object, you can edit the text "in place." No real text editor appears—the text becomes reversed (in color) and you can change the text itself (see "*Textedit*"). If you double-click on an *Mtext* object, the *Text Editor* tab appears (see "*Mtedit*").

You can also edit *Text* and *Mtext* objects using the *Properties* palette. The *Properties* palette is useful for editing all properties of *Text* objects; however, to gain the most control over editing *Mtext* objects, double-click, use *Textedit*, or use *Mtedit*.

### Textedit

| Ribbon | Menu Bar | Command (Type) | Alias (Type) | Shortcut |
|---|---|---|---|---|
| ... | *Modify Object > Text > Edit...* | *Textedit* | *ED* | ... |

*Textedit* can be used for editing either *Text* or *Mtext* objects. The following prompt is issued:

Command: ***TEXTEDIT***
Select an annotation object or [Unit Mode]:

**FIGURE 18-43**

If you select a *Text* object, the text becomes "reversed" and you can edit the contents of the text directly in the drawing (Fig. 18-43). Only the textual content can be changed. If you want to change other properties of the *Text* object such as *Style, Justification, Height, Rotation,* etc., use the *Properties* palette.

The *Mode* can be *Single* or *Multiple*. The *Multiple* mode allows you to edit multiple text objects without having to restart the command each time.

### Mtedit

| Ribbon | Menu Bar | Command (Type) | Alias (Type) | Shortcut |
|---|---|---|---|---|
| ... | ... | *Mtedit* | ... | ... |

The *Mtedit* command is designed specifically for editing *Mtext* objects by producing the *Text Editor* tab. The following prompt appears.

Command: ***MTEDIT***
Select an MTEXT object:

If you select an *Mtext* object, the *Text Editor* tab appears. In the *Text Editor* tab, you can edit the textual contents as well as any properties of the *Mtext* object. The *Text Editor* tab allows you to change any aspect of the text content and all other properties of the *Mtext* objects. (You can also produce the *Text Editor* tab by double-clicking on an *Mtext* object.) *Mtedit* cannot be used to edit *Text* objects. If you select *Mtext* that is too small or too large to read and edit properly, you can *Zoom* with the mouse wheel inside the edit boundary to change the display of the text and text editing boundary.

### Properties

| Ribbon | Menu Bar | Command (Type) | Alias (Type) | Shortcut |
|---|---|---|---|---|
| *View Palettes Properties* | *Modify Properties* | *Properties* | *PR* or *CH* | (Edit Mode) *Find* or *Ctrl+1* |

The *Properties* command invokes the *Properties* palette. As described in Chapter 16, the palette that appears is specific to the type of object that is PICKed—kind of a "smart" properties palette. Remember that you can select objects and then invoke the palette, or you can keep the *Properties* palette on the screen and select objects in the drawing whose properties you want to change. (See Chapter 16 for more information on the *Properties* palette.)

If a line of *Text* is selected, the palette displays properties as shown in Figure 18-44. The *Properties* palette allows you to change almost any properties associated with the text, including the *Contents* (text), *Style, Justification, Height, Rotation, Width, Obliquing,* and *Text alignment.* Simply locate the property in the left column and change the value in the right column.

FIGURE 18-44

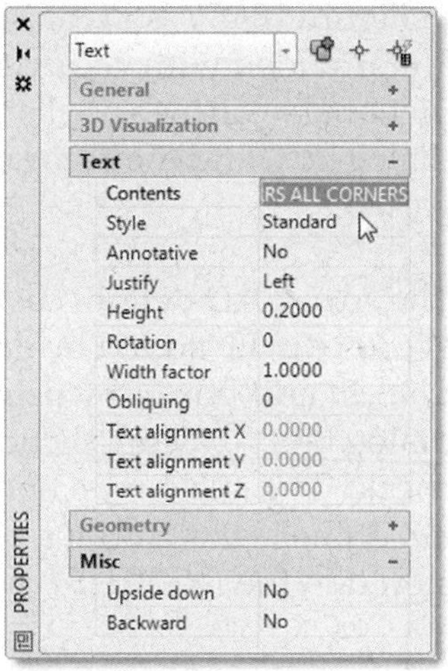


TIP

Note that the palette displays an entry for the *Annotative* property. You can use this section of the *Properties* palette to convert non-annotative objects to annotative by changing the *No* value to *Yes.* In this case, a new entry appears for *Annotative Scale.* A single annotative object can have multiple *Annotative Scales* (or *Object Scales*), one for each time the object is displayed in a different scale (in different viewports).

If you select an *Mtext* paragraph, the *Properties* palette appears allowing you to change the properties of the object (Fig. 18-45). However, if you select the *Contents* section, a small button on the right appears. Pressing this button invokes the *Text Editor* tab with the *Mtext* contents and allows you to edit the text with the full capabilities of the editor.

Note that when using the *Properties* palette, you can create a *Text frame* around the *Mtext* object. This feature basically makes the text boundary visible. To do so, highlight *Text frame* and select *Yes* (Fig. 18-44).

FIGURE 18-45

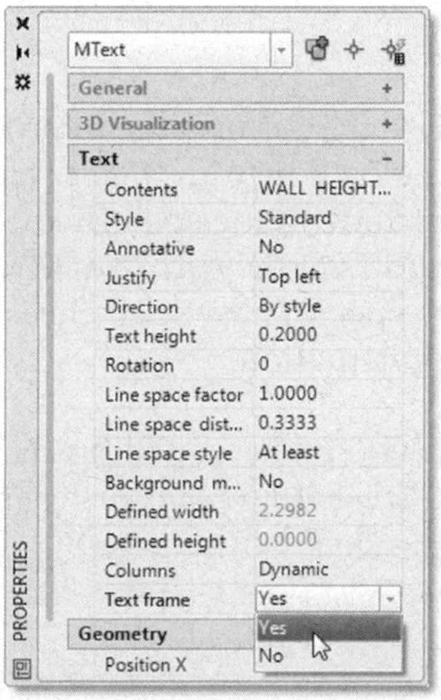


## *Spell*

| Ribbon | Menu Bar | Command (Type) | Alias (Type) | Shortcut |
|---|---|---|---|---|
| *Annotate Text* (icon) | *Tools Spelling* | *Spell* | *SP* | ... |

AutoCAD has an internal spell checker that can be used to spell check and correct existing *Text* or *Mtext.* You can also spell check *Block* attributes and *Blocks* containing text. The *Check Spelling* dialog box (Fig. 18-46) has options to *Ignore* the current word or *Change* to the suggested word. Treat every occurrence of the highlighted word with *Ignore All* and *Change All.*

FIGURE 18-46

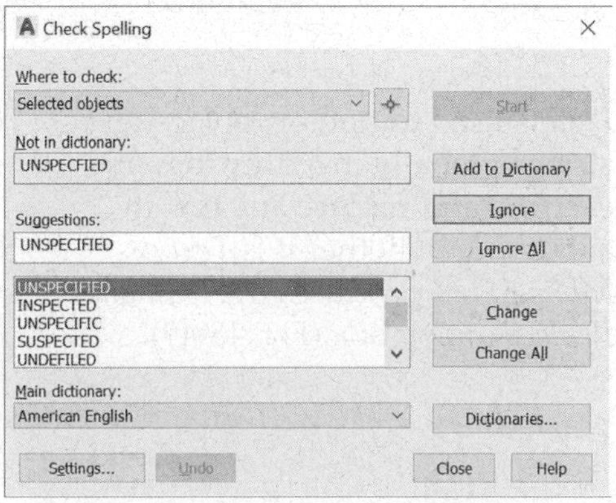


You can check the *Entire drawing, Current space/layout,* or *Select objects.* AutoCAD matches the words to the words in the current dictionary. If the speller indicates that a word is misspelled but it is a proper name or an acronym you use often, it can be added to a custom dictionary. Choose *Add to Dictionary* to leave a word unchanged and add it to the custom dictionary.

Selecting the *Dictionaries...* tile produces the dialog box shown in Figure 18-47. You can select from other *Main Dictionaries* that are provided with your version of AutoCAD. The current main dictionary can also be changed in the *Files* tab of the *Options* dialog box, and the name is stored in the *DCTMAIN* system variable.

The AutoCAD default custom dictionary is SAMPLE.CUS. Words can be added by entering the desired word in the *Content* edit box or by using the *Add* tile in the *Check Spelling* dialog box. Custom dictionaries can be changed during a spell check. The custom dictionary can also be changed in the *Files* tab of the *Options* dialog box (see Fig. 18-41), and its name is stored in the *DCTCUST* system variable.

**FIGURE 18-47**

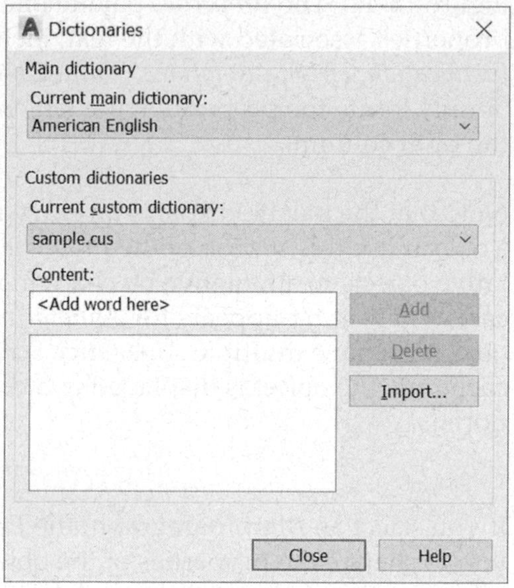


Use the *Settings* tile to produce the *Check Spelling Settings* dialog box (Fig. 18-48). Here you can decide what text to include in the spell check, namely, *Dimensions*, *Block attributes*, and *External references* (Xrefs). The *Options* section allows you to specify how spell checking handles case, numbers, and punctuation.

If you make a spelling change to your drawing using *Change*, *Change All*, *Ignore*, or *Ignore All*, you can undo the change with the *Undo* button (see Fig. 18-46).

**FIGURE 18-48**

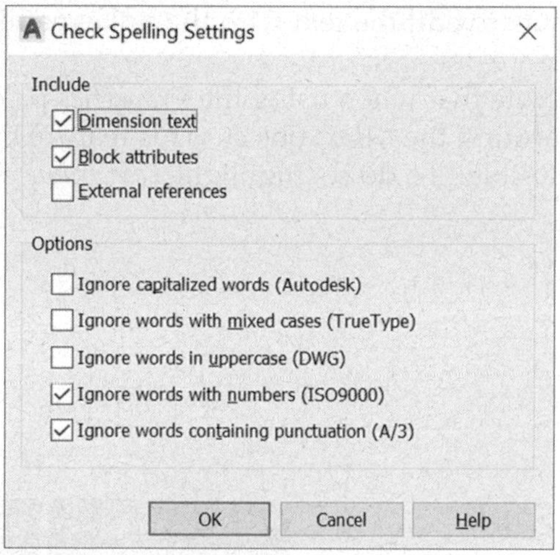


### *Find*

| Ribbon | Menu Bar | Command (Type) | Alias (Type) | Shortcut |
|---|---|---|---|---|
| *Annotate Text* (edit box) | *Edit Find...* | *Find* | ... | (Default menu) *Find...* |

TIP

*Find* is used to find or replace text strings globally in a drawing. It can find and replace any text in a drawing, whether it is *Text* or *Mtext*. *Find* produces the *Find and Replace* dialog box (Fig. 18-49).

**FIGURE 18-49**

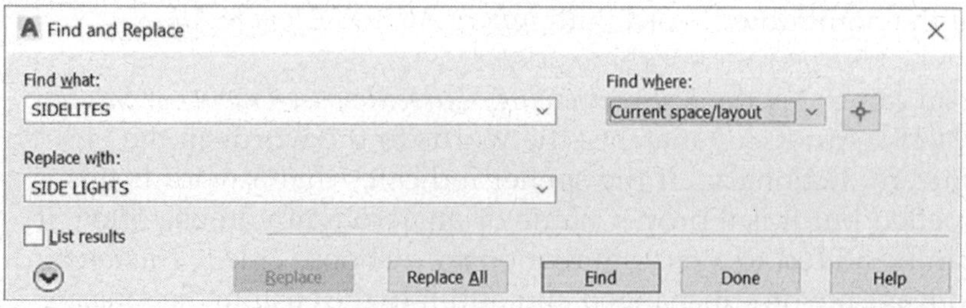

The *Find* command searches a drawing globally and operates with all types of text, including dimensions, tables, and *Block* attributes. In contrast, the *Find/Replace* option in the *Text Editor* tab right-click menu (see "*Find/Replace* in the *Text Editor* tab" next) operates only for *Mtext* and for only one *Mtext* object at a time.

### *Find What, Find/Find Next*

To find text only (not replace), enter the desired string in the *Find What* edit box and press *Find* (Fig. 18-48). When matching text is found, it appears in the context of the sentence or paragraph in the *List Results* area or locates the text in the drawing (see "*List Results*"). When one instance of the text string is located, the *Find* button changes to *Find Next*.

### *Replace With, Replace*

You can find any text string in a drawing, or you can search for text and replace it with other text. If you want to search for a text string and replace it with another, enter the desired text strings in the *Find What* and *Replace With* edit boxes, then press the *Find/Find Next* button. You can verify and replace each instance of the text string found by alternately using *Find Next* and *Replace*, or you can select *Replace All* to globally replace all without verification. The status area confirms the replacements and indicates the number of replacements that were made.

### *Select Objects*

The small button in the upper-right corner of the dialog box allows you to select objects with a pickbox or window to determine your selection set for the search. Press Enter when objects have been selected to return to the dialog box. Once a selection set has been specified, you can search the *Entire drawing* or limit the search to the *Current space/layout* by choosing these options in the *Find where* drop-down list.

### *List Results*

The *List Results* checkbox determines how AutoCAD displays the results of the search. If *List Results* is not checked, AutoCAD locates the text in the drawing and automatically zooms to it as shown in Figure 18-50. If *List Results* is checked, then the results of the search are listed in the display area at the bottom of the *Find and Replace* dialog box listing the *Text Location*, *Object Type*, and *Text String*.

**FIGURE 18-50**

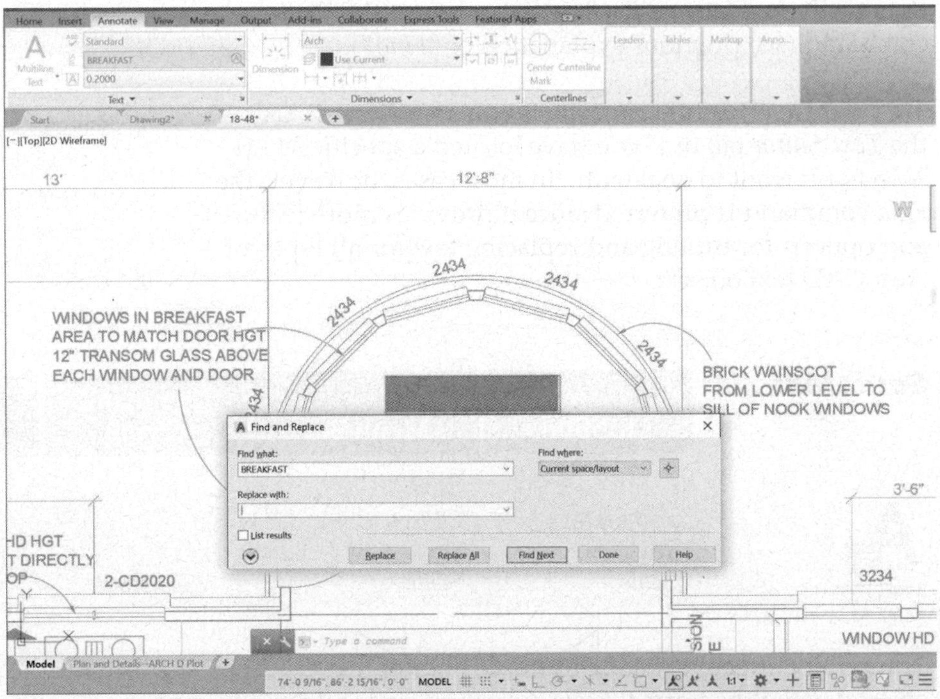

### *More Options*

Selecting the *More Options* button in the *Find and Replace* dialog box produces the *Search Options* section of the dialog box (Fig. 18-51). Note that the *Find* feature can locate and replace *Block Attribute Values, Dimension Text, Text (Mtext, Text), Table Text, Hyperlink Description*, and *Hyperlinks*. Removing any check disables the search for that text type.

**FIGURE 18-51**

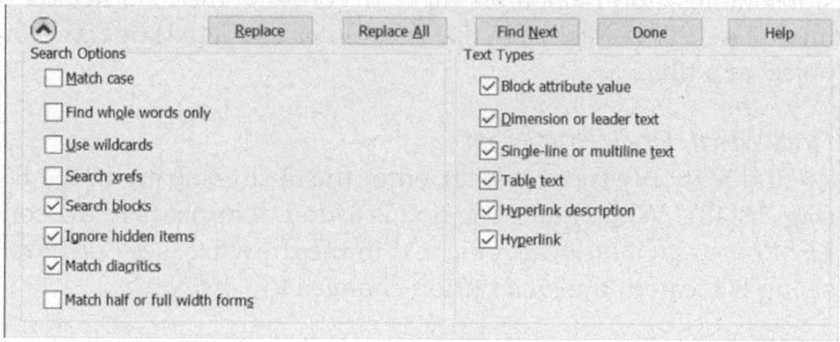


If it is important that the search exactly match the case (upper and lowercase letters) you entered in the *Find What* and *Replace With* edit boxes, check the *Match Case* box. If you want to search for complete words only, check the *Find whole words only* box. For example, without a check in either of these boxes, entering "door" in the *Find What* edit box would yield a find of "Doors" in the *List Results* area. With a check in either *Match Case* or *Find whole words only*, a search for "door" would not find "Doors."

## ***Find/Replace* in the *Text Editor* tab**

Another version of the *Find* command is the *Find and Replace* option available from the *Text Editor* tab menu. With this feature you can find and replace text only within the *Text Editor* tab. *Find and Replace* operates only for *Mtext* objects, not for other text objects.

Selecting *Find and Replace* from the *Text Editor* tab shortcut menu produces the *Find and Replace* dialog box (Fig. 18-52). The options here operate similarly to the same options in the *Find and Replace* dialog box of the *Find* command (see previous explanation of "*Find*").

**FIGURE 18-52**

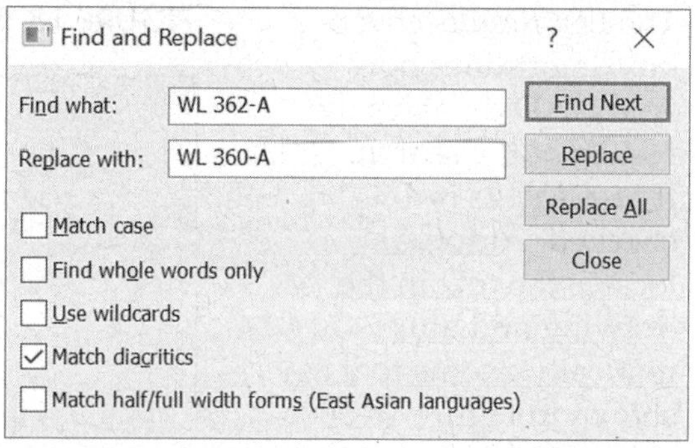


This feature is useful primarily if you are currently using the *Text Editor* tab or if you have located a specific *Mtext* object you want to deal with. In most cases, however, the *Find* command is preferred since it provides more features and options for finding and replacing text for all types of AutoCAD text objects.

## ***Scaletext***

| Ribbon | Menu Bar | Command (Type) | Alias (Type) | Shortcut |
|---|---|---|---|---|
| *Annotate Text Scale* | *Modify Object > Text > Scale* | *Scaletext* | ... | ... |

You cannot effectively scale multiple lines of text (*Text* objects) or multiple paragraphs of text (*Mtext* objects) using the normal *Scale* command since all text objects become scaled with relation to only one base point; therefore, individual lines of text lose their original insertion point.

The text editing command *Scaletext* allows you to scale multiple text objects using this one command, and each line or paragraph is scaled relative to its individual justification point.

For example, consider the three *Text* objects shown in Figure 18-53. Assume each line of text was created using the *Start point* option (left-justified) and with a height of .20. If you wanted to change the height of all three text objects to .25, the following sequence could be used.

**FIGURE 18-53**

SCALETEXT scales each text object individually.
Each line is scaled relative to the justification point.
These sample lines are created by TEXT.

Command: ***SCALETEXT***
Select objects: **PICK**
Select objects: **PICK**
Select objects: **PICK**
Select objects: **Enter**
Enter a base point option for scaling
[Existing/Left/Center/Middle/Right/TL/TC/TR/ML/MC/MR/BL/BC/BR] <Existing>: **Enter**
Specify new model height or [Paper height/Match object/Scale factor] <0.20000>: **.25**
Command:

The resulting three lines of scaled text appear in Figure 18-54. Note that each line increased in height, which is evident since the space between lines has been reduced. Note also that each line of text has retained its insertion point; therefore, the original justification is unchanged.

**FIGURE 18-54**

SCALETEXT scales each text object individually.
Each line is scaled relative to the justification point.
These sample lines are created by TEXT.

*Scaletext* also allows you to specify a new justification point. The justification option you select is applied to each text object individually. For example, selecting the *MC* option would scale each line of text in relation to its own middle center point.

Enter a base point option for scaling
[Existing/Left/Center/Middle/Right/TL/TC/TR/ML/MC/MR/BL/BC/BR] <Existing>: ***mc***
Specify new model height or [Paper height/Match object/Scale factor] <0.20000>: **.25**

### *Scale factor*

This option allows you to specify a relative scale factor rather than a specific height. Entering a value of 1.5, for example, would scale each line of text 1.5 times the current size. This option is particularly helpful when the drawing contains lines of text at different heights and you want to change all lines proportionally.

Specify new model height or [Paper height/Match object/Scale factor] <0.2500>: ***s***
Specify scale factor or [Reference] <0.2000>: **1.5**

### *Paper height*

If you are using annotative text objects, you can scale selected text objects using this option. This action simply changes the size of all instances (all annotation scales) of the text objects, similar to using the *Scale* command in model space; however, you can enter the value in paper space units. This option does not create a new paper space scale copy of the object as if you selected a new *Viewport Scale* for the viewport.

Specify new model height or [Paper height/Match object/Scale factor] <1/8">: ***p***
Specify new paper height <0">: **1/4**

### Match object

You can select an existing text object that you want the text to match. After selecting the matching text, its height is listed and the text (that was selected in the first step) is scaled to the listed height.

```
Specify new model height or [Paper height/Match object/Scale factor] <0.2500>: m
Select a text object with the desired height: PICK
Height=0.2400
```

## Justifytext

| Ribbon | Menu Bar | Command (Type) | Alias (Type) | Shortcut |
|---|---|---|---|---|
| *Annotate Text* (icon) | *Modify Object > Text > Justify* | *Justifytext* | ... | ... |

*Justifytext* changes the justification point of selected text objects without changing the text locations. Technically, this command relocates the insertion point (sometimes called attachment point) for the text object (*Mtext* objects, *Text* objects, leader text objects, and block attributes) and then justifies the text to the new insertion point.

Consider the two *Text* objects and the single *Mtext* object shown in Figure 18-55. In each case, the default justification methods were used when creating the text, resulting in left-justified text and insertion points as shown by the small "blip" (the "blip" is not created as part of the text—it is shown here only for illustration).

**FIGURE 18-55**

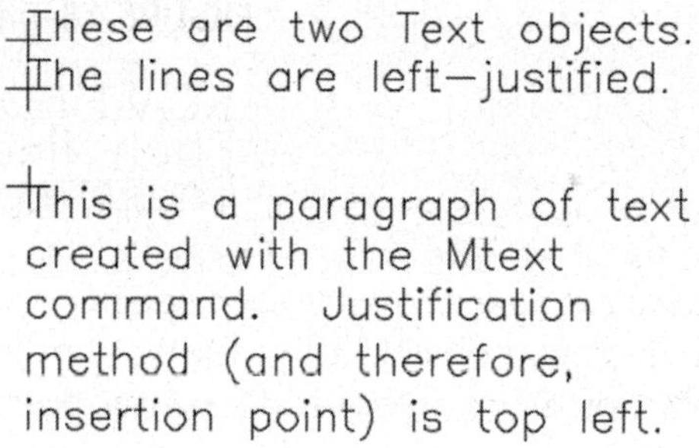


Using previously available methods for changing the text justification yielded different results based on the command used or type of text object selected. For example, using the *Properties* palette to change the *Justify* field and selecting the *Bottom Right* option results in the changes shown in Figure 18-56. Note that for the *Text* objects, the insertion point is not changed, but the text objects are moved to justify according to the original insertion point. On the other hand, the insertion point for the *Mtext* paragraph is changed from the original top-left insertion point (shown as dashed in the figure) to the bottom right and the text is justified to the new insertion point.

**FIGURE 18-56**

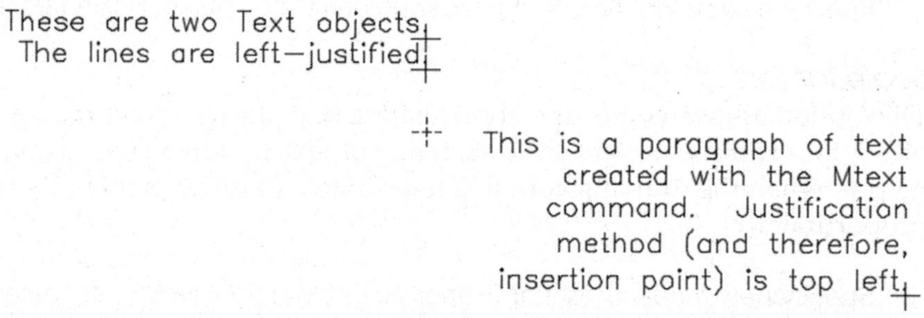


To avoid this problem, use the *Justifytext* command to change multiple text objects to produce similar results for all cases. For example, assume you used the *Justifytext* command to change the three text objects shown previously in Figure 18-54 to a new justification point of *Bottom right*.

Command: ***JUSTIFYTEXT***
Select objects: **PICK**
Select objects: **PICK**
Select objects: **PICK**
Select objects: **Enter**
Enter a justification option
[Left/Align/Fit/Center/Middle/Right/TL/TC/TR/ML/MC/
MR/BL/BC/BR] <TL>: ***br***
Command:

**FIGURE 18-57**

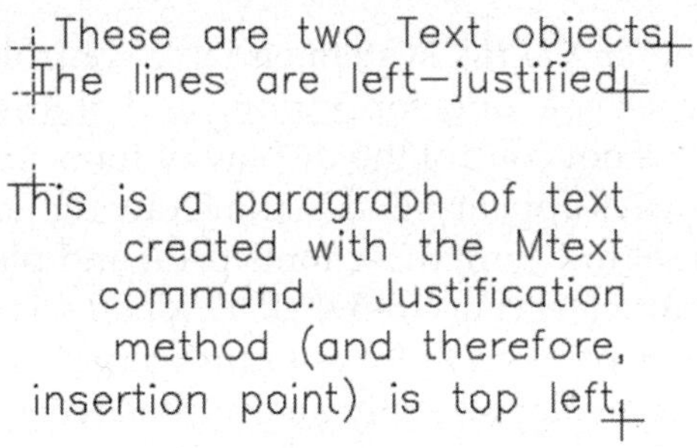


The resulting changes are shown in Figure 18-57. Compare these changes to the original text objects in Figure 18-54. Notice that the *Justifytext* command keeps the text in the same location, although the insertion points have been changed. Each text object has been justified in relation to its own new insertion point.

### Qtext

| Ribbon | Menu Bar | Command (Type) | Alias (Type) | Shortcut |
|---|---|---|---|---|
| ... | ... | *Qtext* | ... | ... |

*Qtext* (quick text) allows you to display a line of text as a box in order to speed up drawing and plotting times. Because text objects are treated as graphical elements, a drawing with much text can be relatively slower to regenerate and take considerably more time plotting than the same drawing with little or no text. When *Qtext* is turned *ON* and the drawing is regenerated (use *Regen*), each text line is displayed as a rectangular box (Fig. 18-58). Each box displayed represents one line of text and is approximately equal in size to the associated line of text.

**FIGURE 18-58**

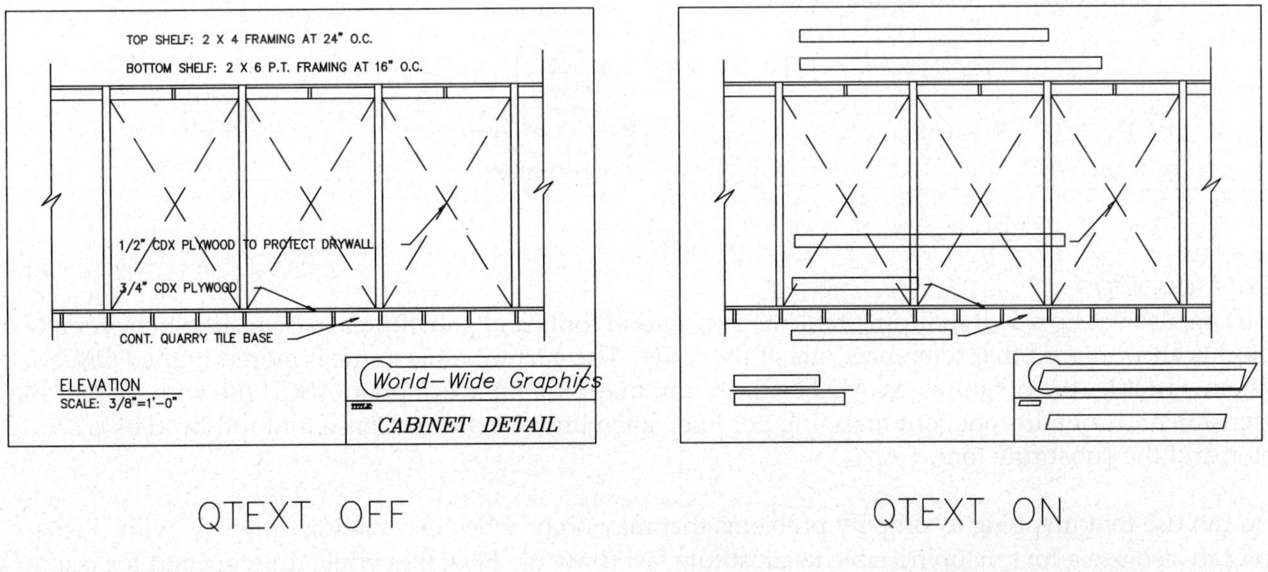


For drawings with considerable amounts of text, *Qtext ON* noticeably reduces regeneration time. For check plots (plots made during the drawing or design process used for checking progress), the drawing can be plotted with *Qtext ON*, requiring considerably less plotting time. *Qtext* is then turned *OFF* and the drawing must be *Regenerated* to make the final plot. When *Qtext* is turned *ON*, the text remains in a readable state until a *Regen* is invoked or caused. When *Qtext* is turned *OFF*, the drawing must be regenerated to read the text again.

TIP

### *TEXTFILL*

The *TEXTFILL* system variable controls the display of TrueType fonts for printing and plotting only. *TEXTFILL* does not control the display of fonts for the screen—fonts always appear filled in the Drawing Editor. If *TEXTFILL* is set to 1 (on), these fonts print and plot with solid-filled characters (Fig. 18-59). If *TEXTFILL* is set to 0 (off), the fonts print and plot as outlined text. The variable controls text display globally and retroactively.

FIGURE 18-59

## Substituting Fonts

### *Alternate Fonts*

Fonts in your drawing are specified by the font file referenced by the current text *Style* and by individual font formats specified for sections of *Mtext*. Text font files are not stored in the drawing; instead, text objects reference the font files located on the current computer system. Occasionally, a drawing is loaded that references a font file not available on your system. This occurrence is more common with the ability to create text styles using TrueType fonts.

By default, AutoCAD substitutes the SIMPLEX.SHX for unreferenced fonts. In other words, when AutoCAD loads a drawing that contains a font not on your system, the SIMPLEX.SHX font is automatically substituted. You can use the *FONTALT* system variable or the *Files* tab of the *Options* dialog box (Fig. 18-60) to specify other font files that you want to use as an alternate font whenever a drawing with unreferenced fonts is loaded. Typically, .SHX fonts are used as alternates.

FIGURE 18-60

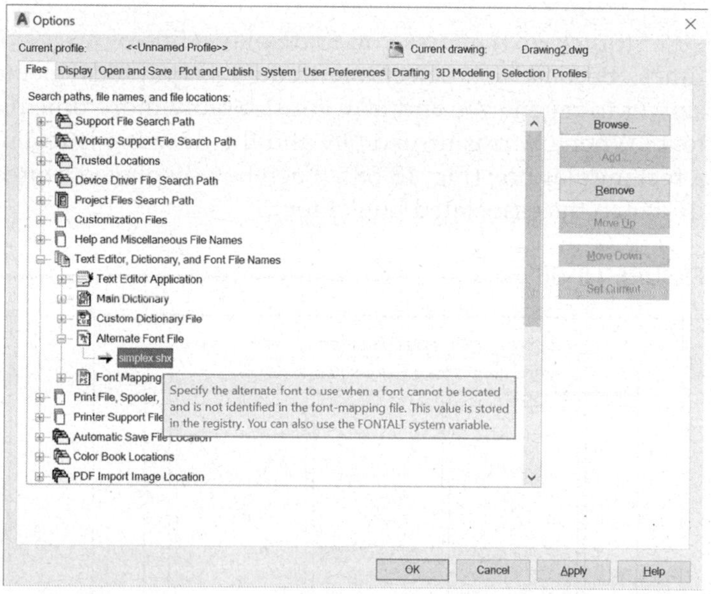


### *Font Mapping*

You can also set up a font mapping table listing several fonts and substitutes to be used whenever a text object is encountered that references one of the fonts. The current table name is stored in the *FONTMAP* system variable (the default is ACAD.FMP). A font mapping table is a plain ASCII file with a .FMP file extension and contains one font mapping per line. Each line contains the base font followed by a semicolon and the substitute font.

TIP

You can use font mapping to simplify problems that may occur when exchanging drawings with clients. You can also use a font mapping table to substitute fast-drawing .SHX files while drawing and for test plots, then use another table to substitute back the more complex fonts for the final draft. For example, you might want to substitute the ROMANS.SHX font for the Times TrueType font and the TXT.SHX font for the Arial TrueType font. (The second font in each line is the new font that you want to appear in the drawing.)

```
times;romans.shx
arial;txt.shx
```

AutoCAD provides a sample font mapping table called ACAD.FMP. The table has 16 substitutions that are used to map the Release 13 PostScript fonts to the newer TrueType fonts. The ACAD.FMP contents are shown here.

```
cibt;CITYB___.TTF
cobt;COUNB___.TTF
eur;EURR____.TTF
euro;EURRO___.TTF
par;PANROMAN.TTF
rom;ROMANTIC.TTF
romb;ROMAB___.TTF
romi;ROMAI___.TTF
sas;SANSS___.TTF
sasb;SANSSB__.TTF
sasbo;SANSSBO_.TTF
saso;SANSSO__.TTF
suf;SUPEF___.TTF
te;TECHNIC_.TTF
teb;TECHB___.TTF
tel;TECHL___.TTF
```

You can edit this table with an ASCII text editor or create a new font mapping table. You can specify a new font mapping file in the *Files* tab of the *Options* dialog box (see Fig. 18-59) or use the *FONTMAP* system variable.

A font mapping table forces the listed substitutions, while an alternate font substitutes the specified font for any font but only when an unreferenced font is encountered.

When a drawing is opened and AutoCAD begins to locate the fonts referenced by text *Styles* in the drawing, AutoCAD follows the following priority:

1. AutoCAD uses the *FONTMAP* value(s) if defined.
2. If not defined, AutoCAD uses the font defined in the text *Style*.
3. If not found, Windows substitutes a similar font (if .TTF) or uses the *FONTALT* specified (if .SHX).
4. If an .SHX is not found, it prompts you for a new font.

## TEXT FIELDS AND TABLES

### Text Fields

You can insert a "text field" as part of a *Text* or *Mtext* object. A text field is a string of text that is variable and displays information based on current data. For example, you might insert text strings in a title block that display the drawing creation date, the current date, the file name, and sheet number. You may also use text fields to display the properties of any object, such as area, color, layer, length, linetype, etc. Other possibilities include using fields to list current system variables settings, plot settings, and settings for sheet sets. The available text fields are predefined in AutoCAD. In other words, you would use a field to display information about a drawing or about objects in a drawing that you might expect to change. Whenever the drawing is updated and then regenerated, saved, or opened, the field automatically displays the current information. Therefore, using fields prevents you from having to make "manual" changes to text as the drawing changes.

### *Field*

| Ribbon | Menu Bar | Command (Type) | Alias (Type) | Shortcut |
|---|---|---|---|---|
| *Insert Data Field* | *Insert Field...* | *Field* | ... | ... |

*Field* is both a command and an option of the *Text* and *Mtext* commands. You can use the *Field* command directly by any method shown in the command table above, you can select *Insert Field…* from the shortcut menu of *Text* (right-click during the "Enter text:" prompt), or you can select *Field…* from the shortcut menu when the *Text Editor* tab is displayed. You can also insert a text field in a *Table* (see "Creating and Editing Tables" next).

**FIGURE 18-61**

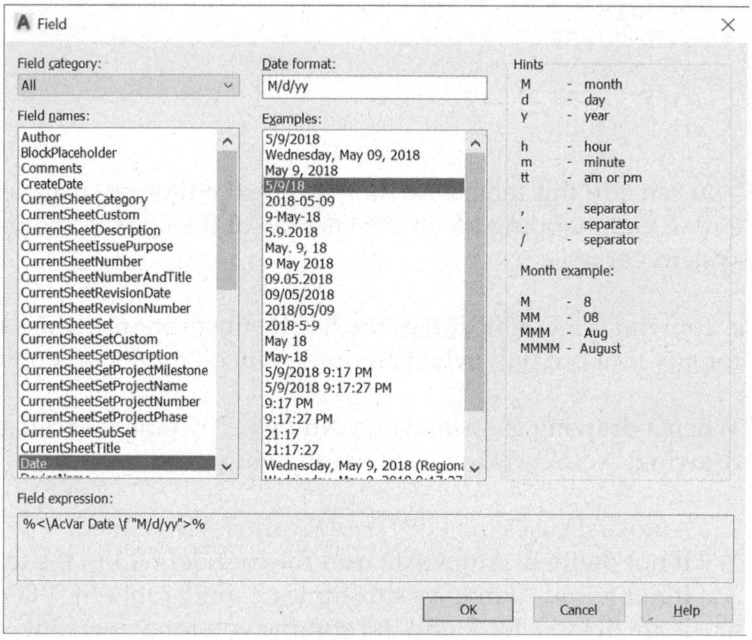


Using the *Field* command or selecting *Insert Field* from a shortcut menu produces the *Field* dialog box (Fig. 18-61). Select the desired information you want displayed from the *Field Names* list, then select the *Format* or other options on the right (based on the selected *Field Name*). This action defines a variable visible in the *Field Expression*: box below. The actual text displayed in the drawing is the current information that satisfies the variable that is taken from the drawing or from your computer system. For example, the *Date, Filesize,* and *Login* variables read and display information from your computer system. The *Title, Author, Subject, Keywords, Hyperlinkbase,* and *Comments* variables read and display information as previously specified in the drawing properties (*Dwgprops*). Other options, such as *Createdate, Filename,* sheet information, plot information, are taken from information contained in the drawing. Selecting some options, such as *Object, NamedObject, Hyperlink, SystemVariable,* and *DieselExpression,* provide edit boxes on the right side of the *Field* dialog box for you to select or otherwise specify the information.

If you want to create a field as a single, independent text string, use the *Field* <u>command</u>. Using the *Field* command, once the desired settings are selected from the *Field* dialog box, an *Mtext* (field) object is placed in your drawing at the location you PICK. The text is entered with the current *text style* and *text height* (previously set using *Mtext* or *Style*).

If you want to insert a field as part of a sentence or paragraph (Fig. 18-62) or use other than the current text style and height, use the *Field* <u>option</u> of the *Text* or *Mtext* command to insert the field. Note that a field object ("222.2 SQ. FT.") is normally displayed with a gray background. Use the *FIELDDISPLAY* system variable to control whether fields are displayed with a gray background (*FIELDDISPLAY*=1) or no background (*FIELDDISPLAY*=0).

**FIGURE 18-62**

Living Room Area = 222.2 SQ. FT.

One example of the power of using *Fields* is to display an object property. As an example, assume you want to insert a text field that displays the area of a closed *Pline* (the *Pline* might represent a floor plan or the outline of a mechanical part). In the *Field* dialog box (Fig. 18-63), first select *Object* from the *Field names*: list. The right side of the dialog box changes to provide appropriate options depending on your choice of field names. Second, select the *Select Objects* button (see pointer in top-center of figure) to return to the drawing and select the desired *Pline* object. Third, select the object *Property:* and *Format:*. Finally, selecting the *OK* button dismisses the dialog box and prompts you for the location of the desired text. The area of the *Pline* is displayed in the text field (see previous Fig. 18-61). If the *Pline* changes, the text field changes (after *Regen* or *Save*) to reflect the new area calculation.

**FIGURE 18-63**

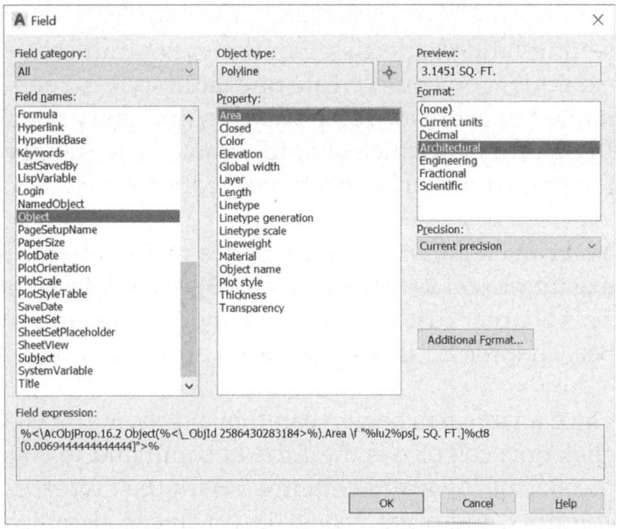


## *Updatefield*

| Ribbon | Menu Bar | Command (Type) | Alias (Type) | Shortcut |
|---|---|---|---|---|
| *Insert Data (icon)* | *Tools Update Fields...* | *Updatefield* | ... | ... |

Most fields automatically update when the related drawing features change and you then use *Regen* or *Save*. However, some fields may not change when you use *Regen* or *Save*, such as fields that display information derived from the computer system. Use the *Updatefield* command to force an update and regeneration of the text contained in <u>any</u> field. You can also enter the *All* option to force all fields in the drawing to update. The following prompt is issued:

Command: ***UPDATEFIELD***
Select objects: **PICK**
Select objects: **Enter**
1 field(s) found
1 field(s) updated
Command:

## Creating and Editing Tables

Many drawings contain tables such as Bills of Materials, Revision Notes, Sheet Lists, Parts Lists, and so on. Rather than drawing *Lines* and inserting the text into cells using *Text*, you can create an AutoCAD *Table*. Use the *Table* command to create *Table* objects. The *Table* objects are variable in that the number and size of rows and columns are flexible. Text or blocks can be easily inserted by right-clicking within the cells and can be easily re-justified for the entire table. You can specify that particular cells contain title headings, column headings, and sub-headings. Properties such as text style, height, color, and fill can be assigned to cells. Border properties can be assigned such as lineweight, color, and grid visibility. You can specify the table direction as up or down. You can even include cells that make calculations in your tables.

Another important feature of tables is they are based on Table Styles, much the same way as text and dimensions are based on Text Styles and Dimension Styles. In other words, you can use the *Tablestyle* command (*Table Style* dialog box) to quickly create a table based on an existing Table Style having specific settings and properties. Similarly, you can create new styles and modify or delete existing styles. For example, you could create one table style named Revision Notes, another named Parts List, and another named Sheet List. Table styles ensure you are using standard fonts, text heights, colors, lineweights, etc. In this way, an office or studio can create and save common Table Styles as part of a template drawing to facilitate quick and easy table creation based on company practices or standards.

You can create a table using three methods: start 1) from an empty table, 2) from a data link such as an existing Excel spreadsheet, and 3) from data extracted from the drawing, specifically, block *Attribute* data. This chapter discusses the first method, creating a table "from scratch" and entering the text into the cells. See Chapter 22, Block Attributes and Data Links, for more information on data links and data extraction.

Once a table has been created in a drawing, it is easy to edit the table and its contents. You can double-click on a cell or use the *Tabledit* command to change the text entries. Using grips, you can click and drag the size of rows and columns. With the *Properties* palette, you change the table style, number of rows and columns, and other properties. You can also highlight the table or cells and access right-click shortcut menus with options specifically for the table and its contents.

### Table

| Ribbon | Menu Bar | Command (Type) | Alias (Type) | Shortcut |
|---|---|---|---|---|
| *Annotate Tables Table* | *Draw Table...* | *Table* | *TB* | ... |

The *Table* command produces the *Insert Table* dialog box (Fig. 18-64). Here you can select from existing Table Styles, specify the column and row settings, and specify how you want to insert the table.

**FIGURE 18-64**

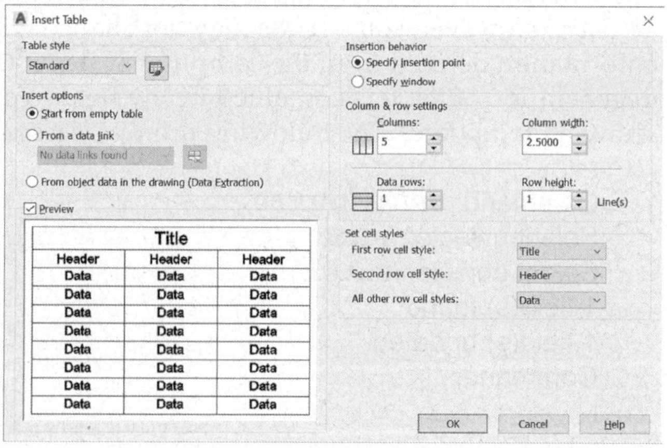


#### *Table Style* Settings

Select the Table Style you want to use from the *Table Style Name* drop-down list (upper-left corner). The table style defines the text, cell, and border properties of the table, such as text height, color, border lineweight, and so on. If this is your first table or you are using the ACAD.DWT or other AutoCAD-supplied template, only the *Standard* style exists. To create a new style, use the button to the right of the drop-down list to produce the *Table Style* dialog box. (See "*Tablestyle*" next.)

#### *Insert options*

You can create a table using one of the three options listed in this section. Only the *Start from empty table* option is described in this chapter. See Chapter 22, Block Attributes and Data Links, for more information on the *From a data link* and *From object data in the drawing (Data Extraction)* options.

#### *Insertion Behavior*

After selecting the table style, specify how you want to insert the table. The *Specify insertion point* option allows you to insert the table in the drawing using its upper-left corner, and the table size is based on the options you specify below in the *Column & Row Settings* section. The *Specify window* option allows you to dynamically determine the size of the table during insertion based on the window you specify.

### *Column & Row Settings*

The available options in the *Column & Row Settings* section are based on which of the *Insertion Behavior* options you select. If you chose to insert the table by the *Specify insertion point* option above (see Fig. 18-64), you can then specify the number and width of columns and number and height of rows before insertion. However, if you had selected the *Specify window* option above, you can specify only one column option and one row option since the other parameters are determined dynamically during insertion.

In other words, if you had selected the *Specify window* option above, then specified the number of columns and rows in this section, the resulting width and height of the cells would be determined dynamically based on the window size you pick during insertion. No matter which of these options you select, you can easily edit the table later to include more or fewer rows or columns and change the size of the rows and columns.

### *Set Cell Styles*

You can specify a particular style for the first row of cells, second row, and other cells. A table style saves the properties (text style, font, height, color, margins, border, etc.) of each of these three types of cells. You can select *Cell Styles* from any *Table Style* saved in the drawing. By default, only three cell styles exist in the AutoCAD-supplied template drawings: *Title*, *Header*, and *Data*.

### Creating the Table Text

Once you have specified the option in the *Insert Table* dialog box, selecting *Close* returns you to the drawing and produces the "Specify insertion point:" prompt. After insertion, the table appears with the first cell highlighted and the *Text Editor* tab (Fig. 18-65). Additionally, the columns and rows appear with temporary labels for columns (A, B, C, etc.) and rows (1, 2, 3, etc.). (The column and row labels do not appear in the completed table.) Simply enter the desired text into each cell. To move around the cells use the arrow keys, or use the Tab key to move across cells and the Enter key to move down the cells. Click inside any cell to enter and edit the text. You can also insert *Blocks* or *Fields* into cells (see *"Tabledit"*).

**FIGURE 18-65**

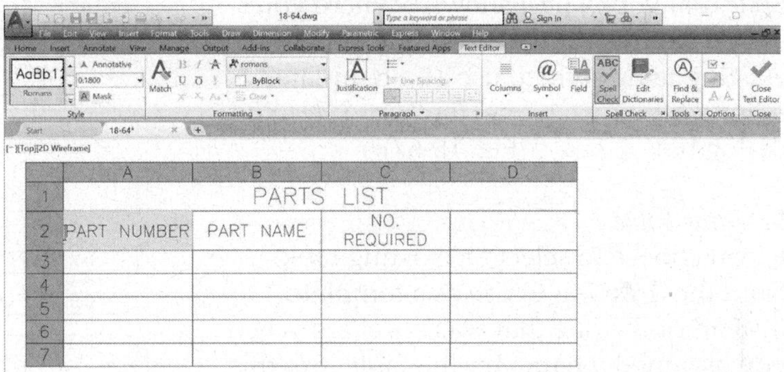


### Editing the Table Text

See *"Tabledit."*

## *Tablestyle*

| Ribbon | Menu Bar | Command (Type) | Alias (Type) | Shortcut |
|---|---|---|---|---|
| *Annotate Tables* (arrow) | *Format Tablestyle...* | *Tablestyle...* | *TS* | ... |

Table Styles are similar to dimension styles in that the table style defines the appearance of the table like a dimension style defines the appearance of a dimension. Use the *Table Style* dialog box to specify the properties of a table, such as text style, alignment, height, color, fill color, border line color, and lineweight. Similar also to the relationship between dimensions and their dimension styles, an existing table in a drawing is automatically updated whenever its style is changed.

The *Table Style* dialog box (Fig. 18-66) can be produced from within the *Insert Table* dialog box or by using the *Tablestyle* command. This dialog box allows you to modify an existing style, make new styles, delete unused styles, or set a style as the current style.

**FIGURE 18-66**

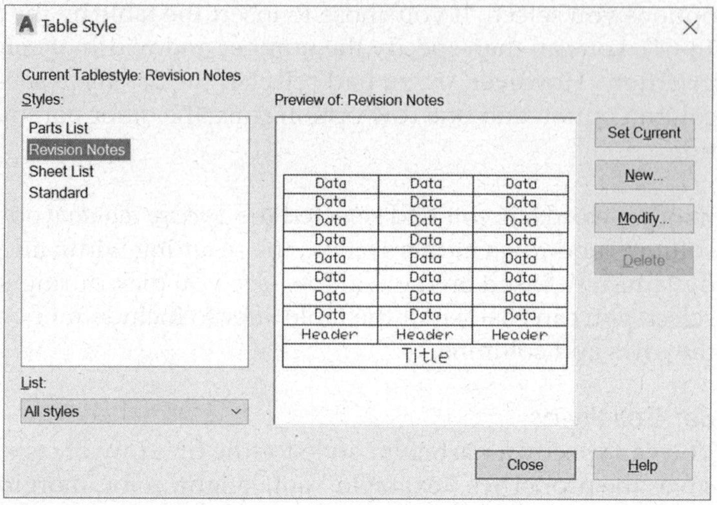


### *Set Current*

Select a style from the *Styles:* list, then select *Set Current*. The current table style is useful only when using the *Table* command. The selected current style appears as the default in the *Insert Table* dialog box (see "Table").

### *New...*

Use this option to create a new table style. This button produces the *Create New Table Style* dialog box (not shown), where you assign a name for the new style and an existing table style to use as a template. After assigning a name, the *New Table Style* dialog box appears (Fig. 18-67).

**FIGURE 18-67**

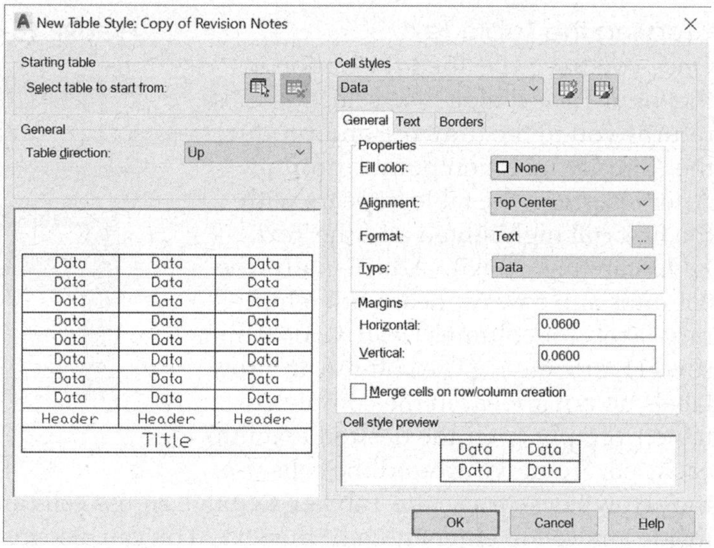


*Starting Table*

If you choose to select an existing table from the drawing to use as a template instead of the one that was specified when you assigned a name for the table, use this option. You are then prompted to PICK a table from the drawing.

*Table direction*

Select *Up* or *Down* from the list. The *Up* option creates a table with the title and headers on the bottom of the table (see the preview area in previous Fig. 18-65).

*Cell Styles*

Select a cell style to set its properties using the tabs and options below. By default, the *Title*, *Header*, and *Data* cell styles exist in all table styles and cannot be renamed or deleted. Other cell styles can be created by selecting the *Create a New Cell Style* and *Manage Cell Styles* options from the drop-down list or from the buttons to the right. The *Manage Cell Styles* dialog box (not shown) allows you to create *New* styles, *Rename*, and *Delete* styles.

*General, Text, Borders*

These three tabs allow you to specify how each cell looks in the finished table. The *General* tab (see Fig. 18-67) offers the options such as *Fill color*, *Alignment* method for the text, *Format* for the text, and *Margins* for the cells. The *Text* tab provides options for the *Text style*, *Text height*, *Text color* and *Text angle*.

Options for the *Borders* tab are shown in Figure 18-68. *Lineweight, Linetype, Color, Double line* and *Spacing* can be assigned. The buttons near the bottom of the section allow you to select which borders appear for the cells.

FIGURE 18-68

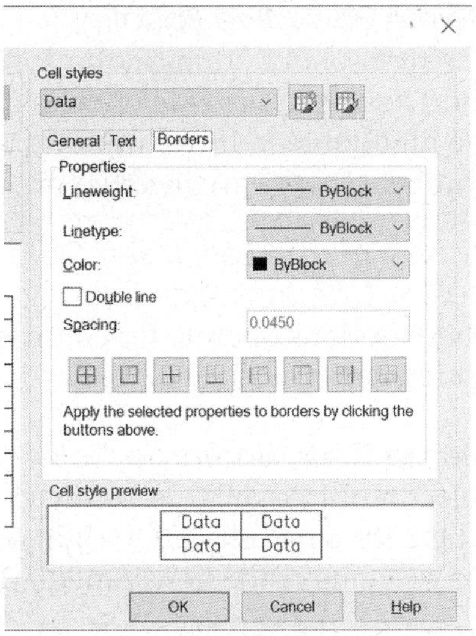


### *Modify...*

The *Modify Table Style* dialog box (not shown) is identical to the *New Table Style* dialog box except for the name. The options for modifying an existing table style are the same as those for creating a new style. However, there is one important feature here: when a table style is modified, all existing tables in the style automatically update to display the new style settings. For example, suppose you have a table in the drawing but want to change the alignment of the cells from middle center to left center. Using the *Modify Table Style* dialog box to change the alignment for the style automatically updates the alignment in the existing table and any other tables created with that style.

TIP

## *Tabledit*

| Ribbon | Menu Bar | Command (Type) | Alias (Type) | Shortcut |
|---|---|---|---|---|
| ... | ... | *Tabledit* | ... | ... |

The *Tabledit* command is used to edit text in a table similar to the way *Textedit* is used to edit *Mtext*—that is, *Tabledit* produces the *Text Editor* tab (see previous Fig. 18-65). As an easy alternative to invoking *Tabledit*, you can simply double-click inside any cell (assuming *Dblclkedit* is on) to produce the *Text Editor* tab. If you invoke the *Tabledit* command otherwise, you are prompted to pick a cell, then the editor appears.

Command: ***TABLEDIT***
Pick a table cell: **PICK**

The options available for editing the text are the same as those used to create the text for the table. Several options are available on the toolbars, the *Options* menu, and the right-click shortcut menu (see *"Mtext"*).

## The *Table Cell* Editor

The *Table Cell* editor appears if you select (left-click) in any cell of a table (Fig. 18-69). You can also highlight one or more cells and right-click to produce a shortcut menu (Fig.18-70). Highlight multiple cells by holding down the Shift key when selecting or use a crossing window to select. Generally, the buttons on the *Table Cell* editor and options in the shortcut menu provide the same options. The options are described on the next page.

FIGURE 18-69

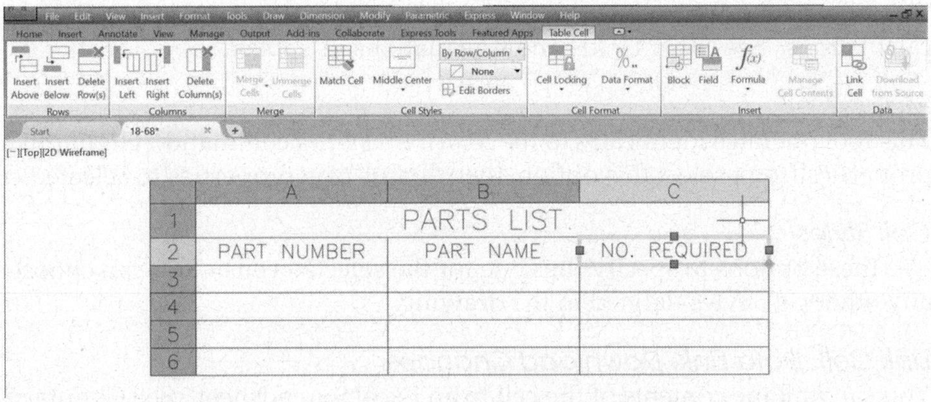

FIGURE 18-70

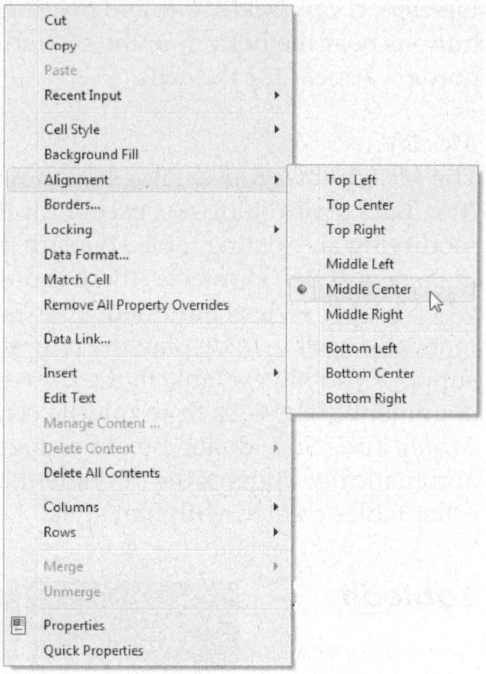


### *Insert Row Above, Insert Row Below, Delete Rows, Size Equally*

The first set of buttons in the *Table Editor* and options from the shortcut menu provide options for changing row properties of the highlighted. If two or more rows are highlighted, the same number of rows are inserted or sized.

### *Insert Column Left, Insert Column Right, Delete Columns, Size Equally*

These options operate for columns identically to the set of options for rows.

### *Merge Cells, Unmerge Cells*

Select multiple cells. You can *Merge Cells* (combine cells) where the first cell (left to right or top to bottom) determines the contents of the newly merged cell.

### *Cell Borders*

This option produces the *Cell Border Properties* dialog box (not shown). Here you can set the *Border Type* (which grid lines appear), *Lineweight*, *Linetype*, *Color*, and *Spacing* value.

### *Alignment*

You can re-justify the contents of selected cells (*Top left*, *Top Center*, etc.) using this drop-down list or cascading menu (see Fig. 18-70).

### *Locking*

The *Locking* options include *Unlocked*, *Content Locked*, *Format Locked*, and both *Content and Format Locked*.

### *Format*

This option produces the *Table Cell Format* dialog box where you can specify the *Data Type*, such as *Angle*, *Currency*, *Date*, etc. For each type, you can specify details of how the data appears in the cell (number of decimals, symbols used, and so on).

### *Insert Block*

You can insert any block into a cell using the provided dialog box *Browse* button. Next, select the desired *Scale*, *Rotation Angle*, and *Cell Alignment* for the block.

### *Insert Text Field*

This powerful option allows you to produce tables that contain variable cell contents, such as totals, or calculations. See "Text Fields and Calculations in Tables."

### *Match Cell*

This tool operates identically to the *Match Properties* command. Highlight any cell you want to copy the properties from, select this option, then "paint" the properties to selected cells.

### *Cell Styles*

Use these options to specify the style for the selected cells. You can choose from *Title*, *Header*, *Data*, or any other cell styles defined in the drawing.

### *Link Cell, Data Link, Download Changes*

You can link the contents of the cell to an Excel Spreadsheet. See Chapter 22, Block Attributes and Data Links, for more information.

### *Remove All Property Overrides*

Using this option generally removes any properties that were changed using the *Properties* palette (see "Editing Tables with the *Properties* Palette") and sets the properties back to those settings defined by the table style. These properties, such as the *Cell Styles, Borders* properties, and *Text* properties, etc., are defined in the *New Table Style* and *Modify Table Style* dialog boxes. However, some properties such as table width or height (even though they may have been set in the *Properties* palette) or properties changed in the *Text Editor* tab may not revert back to the settings defined by the style when *Remove All Property Overrides* is selected (see "Editing Tables with the *Properties* Palette").

### *Properties*

Select *Properties* to produce the *Properties* palette (see "Editing Tables with the *Properties* Palette").

### Shortcut Menu, Table Highlighted

If you select the table on its edge gridline (single-click), the entire table becomes highlighted rather than an individual cell. Right-clicking produces a menu including the portion shown in Figure 18-71. Here you can *Size Columns Equally* and *Size Rows Equally*; however, these options affect all cells in the table. In this case, the overall size of the table remains constant but the individual cells equalize. You can also *Export* tables to a comma-separated file (.CSV) for use in Microsoft Excel or other applications (see "Importing and Exporting Tables"). The *Table Indicator Color* option allows you to change the row (1, 2, 3) and column (A, B, C) labels. You can also create a new table style or update the current style based on the highlighted table by selecting *Table Style*, then *Save as a New Table Style* or *Save as a Table in the Current Table Style*.

FIGURE 18-71

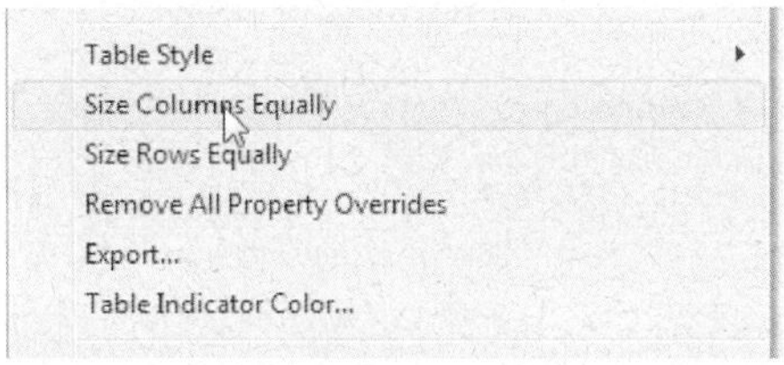


## Editing Tables with Grips

If grips are on, you can select individual cells or select the entire table and dynamically adjust the table size (see Chapter 19). Selecting an individual cell, then a hot grip, you can stretch the size of the column or row of the selected cell. With the entire table selected, use any column grip to stretch the column width. Holding down the Ctrl key while stretching a column keeps the overall table width constant. You can stretch the height of the columns uniformly using the lower-left (arrow-shaped) grip (Fig. 18-72). Use the upper-right arrow grip to stretch all column widths and use the lower-right arrow grip to stretch both rows and columns. Also see "Table Breaks."

FIGURE 18-72

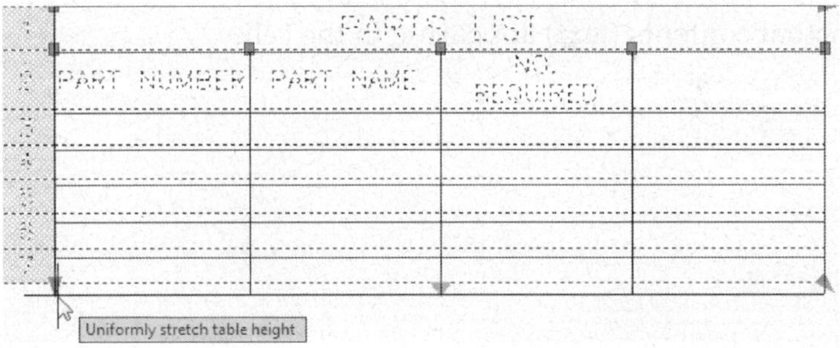


## Editing Tables with the *Properties* Palette

### *Table* Properties

Double-clicking on a cell produces the *Text Editor* tab, while double-clicking on the table edge produces the *Properties* palette (Fig. 18-73). In the *Table* category you cannot change the number of columns or rows, but you can change a table from one table style to another. You can also change the overall width and height.

FIGURE 18-73

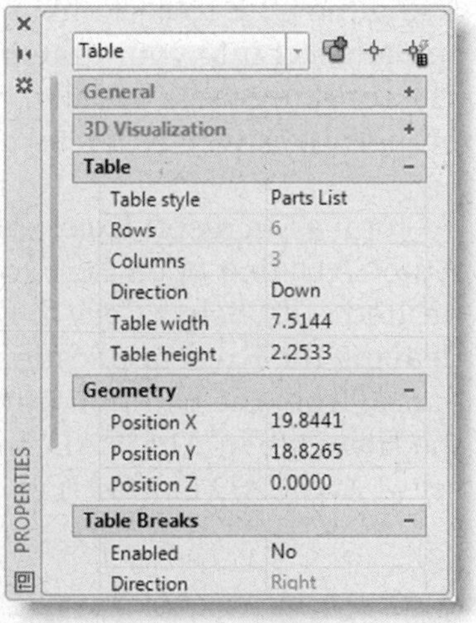

Remember that using the *Properties* palette to edit an individual table or cell usually creates properties overrides for the table. Properties overrides can be removed by right-clicking on the table and selecting *Remove All Property Overrides* from the shortcut menu. However, some properties such as table width or height may not revert back to the settings defined by the style when this option is selected.

### *Cell* Properties

You can also select an individual cell, right-click to invoke the shortcut menu, then select *Properties* to change the properties of the highlighted cell or cells using the *Properties* palette (Fig. 18-74). You can alter the *Cell Style* or the style for the entire row or column. Changing the *Cell width* or *Cell height* changes the entire row or column of the cell. *Alignment* sets the justification for the cell contents. You can set the *Background Fill* and several other border properties, *Vertical* and *Horizontal* cell margins, and *Cell Locking* status.

FIGURE 18-74

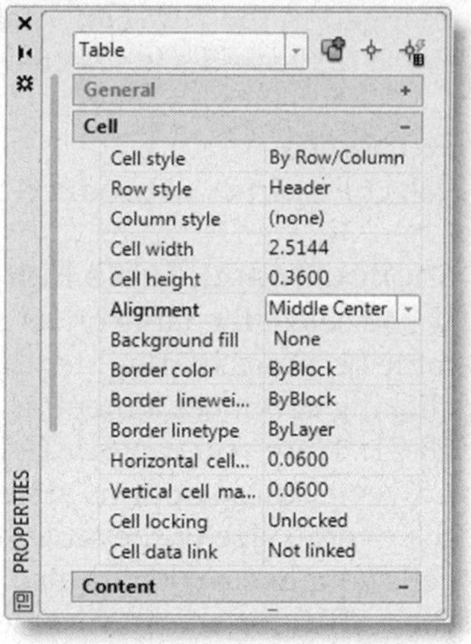


### *Content* Properties

Selecting a cell produces the contents of the *Properties* palette as shown in previous Figure 18-73. However, you can scroll down in the palette to reveal options for the *Content* of the cell (Fig. 18-75). These options affect the appearance of the alpha-numeric contents, such as *Format*, *Data Type*, *Text Style*, *Height*, *Rotation*, *Color*. You can also change the actual contents (text) appearing in the cell.

FIGURE 18-75

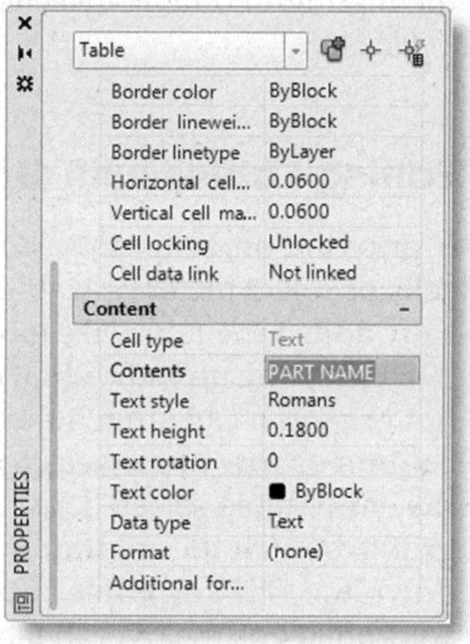


## Importing and Exporting Tables

Using the OLE (object linking and embedding) features in AutoCAD, you can paste a Microsoft Excel spreadsheet into your drawing. The pasted object is converted to a native AutoCAD object—a table. The resulting table has all the features of a normal table including the ability to set and modify the table properties as well as the text content.

FIGURE 18-76

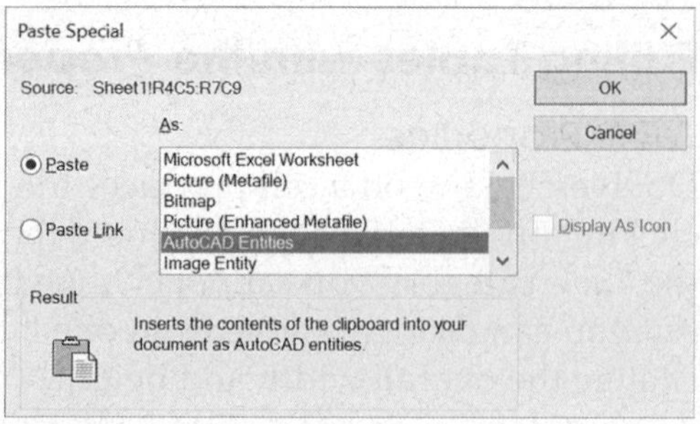


To copy a Microsoft Excel spreadsheet and paste it into AutoCAD, first open the Excel document and highlight the desired cells. Select *Copy* in Excel. Next, return to AutoCAD (leaving the Excel application running) and select *Paste Special* from the *Edit* pull-down menu or *Clipboard* panel in the *Home* tab in AutoCAD. In the *Paste Special* dialog box select *AutoCAD Entities* (Fig. 18-76).

You can also export an AutoCAD table to a comma-separated file (.CSV) for use in Microsoft Excel or other applications. Do this by selecting the entire table and right-clicking, then select *Export* from the shortcut menu (see previous Figure 18-71). The resulting file can be opened directly into Microsoft Excel. Also see Chapter 22, Block Attributes and Data Links.

## Table Breaks

You can break a table into segments using grips. In other words, instead of a table having one column, you can break the table into two or more columns. Multiple columns are called "table breaks." Individual columns, or table breaks, can be "detached" from the original table and moved to anywhere in the drawing if the appropriate settings are made.

FIGURE 18-77

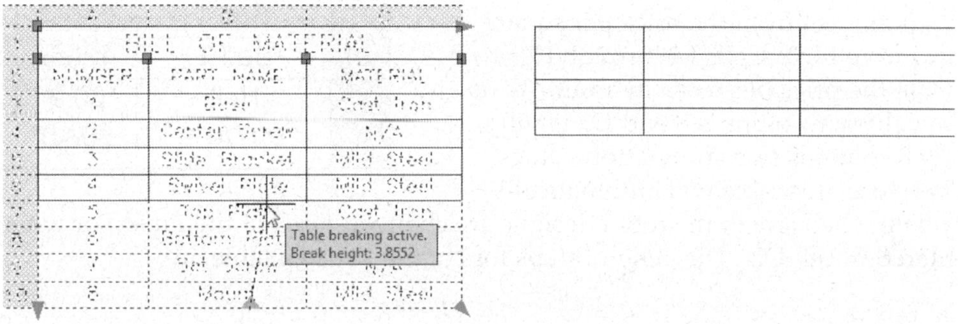


To create table breaks, you must use grips. Turn grips on by changing the *GRIPS* system variable to 1 or 2 (see Chapter 19 for information on grips). Next, select the table on its edge border so the grips appear (Fig. 18-77). Find and select the table breaking (arrow-shaped) grip for the table at the bottom center. Drag the grip up (for top-down tables) into the table so a second column appears. Drag to set the height for the desired columns. Press Esc (Escape) to deactivate the grips.

### Table Fragments

Activating the table breaking grips allows you to create columns for the table. These columns are connected by default, so that if you move the table, both columns move together. However, you can break a table into multiple fragments. Do this by highlighting the table and activating the *Properties* palette. In the *Table Breaks* category, change *Manual positions* to *Yes* (Fig. 18-78). Now you can move the individual columns separately anywhere within your drawing.

FIGURE 18-78

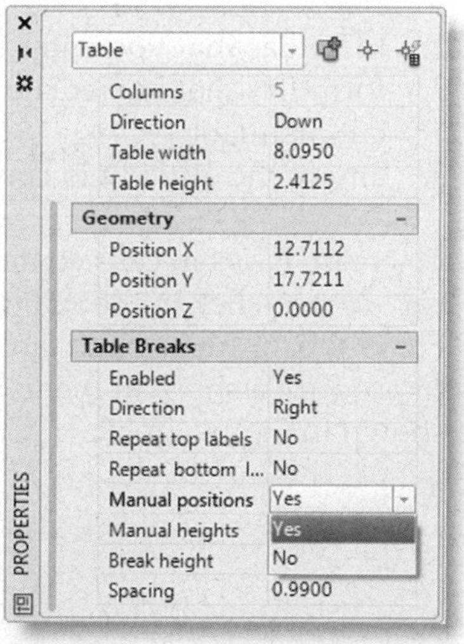


Note also that once table breaks (columns) are created, you can specify that *Labels* are repeated for each column. If table fragments are created, you can set *Manual Heights* in the *Properties* palette so that the fragments can have different heights, in which case you can later edit each table to add or remove rows.

## Text Fields and Calculations in Tables

Traditionally, external spreadsheets have been used to collect information from AutoCAD drawings and to perform calculations with the data. By combining the use of tables and the power of text fields, you can create tables to collect information and perform simple calculations directly in an AutoCAD drawing. This is accomplished by including *Fields* in table cells and composing mathematical expressions within the *Fields*.

Arithmetic calculations that are provided are *Sum, Average,* and *Count*. You can also enter an = (equals) symbol into any table cell, then enter your own formula. You can read values from other cells by specifying the cell labels (such as "B4") or selecting a cell range, then include these cell values in your formulas.

As an example, suppose you want to create a table in a drawing of a house to calculate room square footage and the cost to carpet the rooms (Fig. 18-79). The cells indicated with a gray background contain *Fields* for making the calculations. In this table, cells B4 through B7 calculate the areas for *Plines* that are drawn around each room (see the example in Fig. 18-61). Cells D4 through D7 convert the price in square feet from the price per square yard in cell D2. Cells F4 through F7 calculate the price per room by multiplying the values in columns B and D. Finally, row 8 contains two summation values. Therefore, this table will automatically update when any of the room designs are changed (area of the *Plines*) or when a different carpet cost is entered in cell D2. The general steps for creating such a table are:

**FIGURE 18-79**

| | A | B | C | D | E | F |
|---|---|---|---|---|---|---|
| 1 | CARPET CALCULATION | | | | | |
| 2 | Carpet Cost-> | $/YD | = | 12.50 | | |
| 3 | ROOMS | SQ. FT. | | $/SQ FT | | $/ROOM |
| 4 | Living | 216.0 | x | 1.39 | = | =b4*d4 |
| 5 | Bed 1 | 140.0 | x | 1.39 | = | 194.4 |
| 6 | Bed 2 | 120.0 | x | 1.39 | = | 166.7 |
| 7 | Hall | 30.0 | x | 1.39 | = | 41.7 |
| 8 | TOTALS | 506.0 | | | | $702.78 |

1. Use the *Table* command to create the basic format (rows and columns) of the table.
2. Use the *Tablestyle* command to set the cell properties (text style, height, etc.) and border properties (lineweight, color).
3. Enter in the heading and other static text.
4. Create *Fields* in the desired cells.

*Fields* in table cells can contain:

1. formulas that you enter directly into the cell by first entering an "=" (equals) symbol,
2. formulas that you create using either the *Field* dialog box or by selecting directly from the cell shortcut menu, or
3. any of the other field contents available using the *Field* dialog box (see "*Field*" earlier in the chapter).

To create *Fields* in cells containing formulas, you can double-click in the cell to produce the *Text Editor* tab, or highlight a table cell and right-click to produce a shortcut menu with the options shown in Figure 18-80.

**FIGURE 18-80**

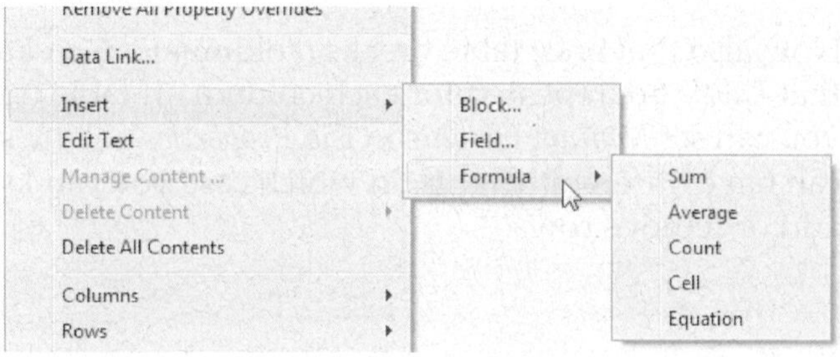


## Entering Formulas Directly Into Cells

Produce the *Text Editor* tab by either double-clicking in a cell or by highlighting a cell, then selecting *Edit Text* from the shortcut menu. Enter an "=" (equals) symbol, then enter the desired arithmetic expression. Typical operators are accepted:

plus (+), minus (-), times (*), divided by (/), exponent (^), parentheses ()

In addition, the following functions are accepted:

sum, average, count

You can also read numerical values from other cells by entering the cell labels (such as "B4" or "D2"). You can specify a range of cells to read using a ":" (colon) symbol (such as "A4:D4").

For example, to sum the contents of cells A1 through A25, enter the following formula: "=sum(a1:a25)." As another example, in Figure 18-79 the highlighted cell (F4) displays the formula ("=b4*d4") for calculating the cost of carpeting the living room. The formula multiplies the values for cell B4 (square footage) times D4 (price).

## Entering Formulas Using the *Field* Dialog Box

Highlight a single cell (not the entire table), right-click, then select *Insert Field* from the cell shortcut menu or *Table Cell* editor. This action produces the *Field* dialog box. If the *Formula* option is not pre-selected, select it (Fig. 18-81). You can enter a formula directly into the *Formula* edit box, or select the *Average, Sum,* or *Count* functions above. Select the *Cell* button to return to the drawing to specify a cell to enter into the *Formula. Evaluate* displays the calculation in the *Preview* box.

**FIGURE 18-81**

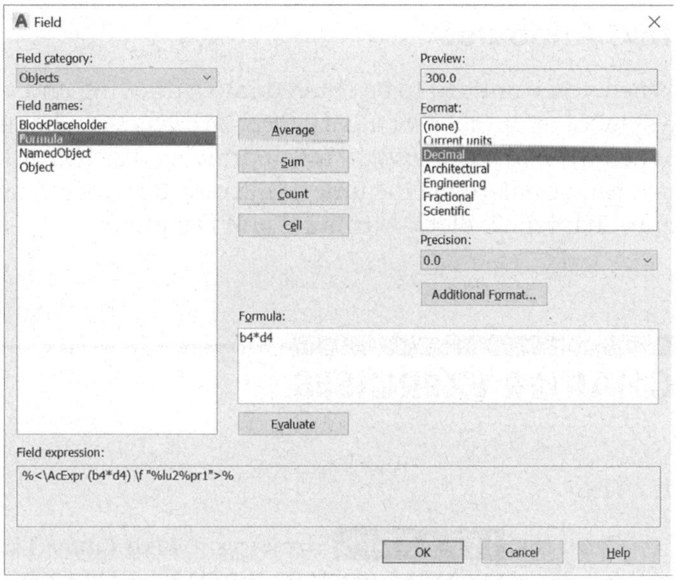


*Additional Format*

The *Additional Format* dialog box provides options for units conversion (Fig. 18-82). If you need to convert the *Current value* to some other unit of length, volume, etc., enter the desired value into the *Conversion factor* input field. You can add a *Prefix* or a *Suffix* (for example, see the "$" [dollar] symbol in cell F8 in Fig. 18-79). Also choose from number separators or zero suppression.

NOTE: You must enter a formula in the *Field* dialog box and press *Evaluate* to populate the *Format* list and activate the *Additional Format* button.

**FIGURE 18-82**

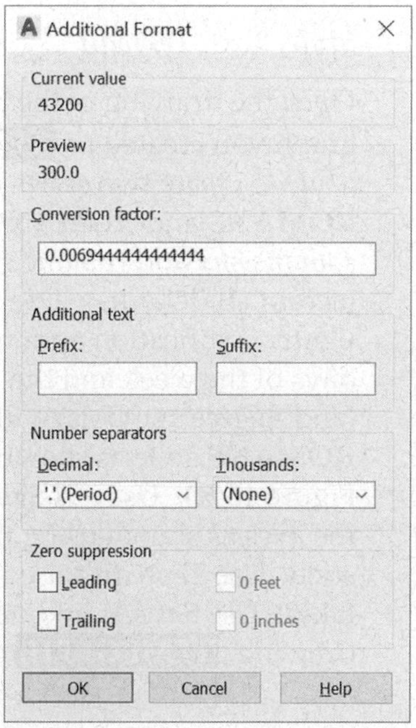

### Entering Formulas Using Cell Shortcut Menu

Highlight a single cell, then right-click and select *Insert Formula* from the shortcut menu (see Fig. 18-80). You can then select from the standard functions, *Sum, Average*, etc. When prompted to "Select first corner of table range:", select the desired cells using a selection window. AutoCAD automatically enters the formula into the cell and produces the *Text Editor* tab so you can further edit the formula if desired.

### Entering Other (Non-formula) Fields Using the *Field* Dialog Box

Highlight a single cell, right-click, then select *Insert Field* from the cell shortcut menu. This action produces the *Field* dialog box (Fig. 18-81). Select the *Formula* option from the list. In the *Field Category* drop-down list, select *All*. Specify any field for the cell. For example in Figure 18-79, the square footage for each room is calculated (column B) by entering *Object* as the field name and specifying that the *Area* for a *Polyline* is displayed in the cell (also see the example in Fig. 18-62).

## Text Attributes

When you want text to be entered into a drawing and associated with some geometry, such as a label for a symbol or number for a part, *Block Attributes* can be used. *Attributes* are text objects attached to *Blocks*. When the *Blocks* are *Inserted* into a drawing, the text *Attributes* are also inserted; however, the content of the text can be entered at the time of the insertion. See Chapter 21, Blocks, DesignCenter, and Tool Palettes, and Chapter 22, Block Attributes and Data Links.

# CHAPTER EXERCISES

1. *Text*

   ***Open*** the **EFF-APT-PS** drawing. Make ***Layer*** **TEXT** ***Current*** and use ***Text*** to label the three rooms: **KITCHEN, LIVING ROOM**, and **BATH**. Use the ***Standard style*** and the ***Start point*** justification option. When prompted for the ***Height:***, enter a value to yield letters of 3/16" on a 1/4"=1' plot (3/16 x the drawing scale factor = text height). ***Save*** the drawing.

2. *Style, Text, Textedit*

   ***Open*** the drawing of the temperature graph you created as **TEMP-D.** Use ***Style*** to create two styles named **ROMANS** and **ROMANC** based on the ***romans.shx*** and ***romanc.shx*** font files (accept all defaults). Use ***Text*** with the ***Center*** justification option to label the days of the week and the temperatures (and degree symbols) with the **ROMANS** style as shown in Figure 18-83. Use a ***Height*** of **1.6**. Label the axes as shown using the **ROMANC** style. Use ***Textedit*** for editing any mistakes. Use ***SaveAs*** and name the drawing **TEMPGRPH.**

**FIGURE 18-83**

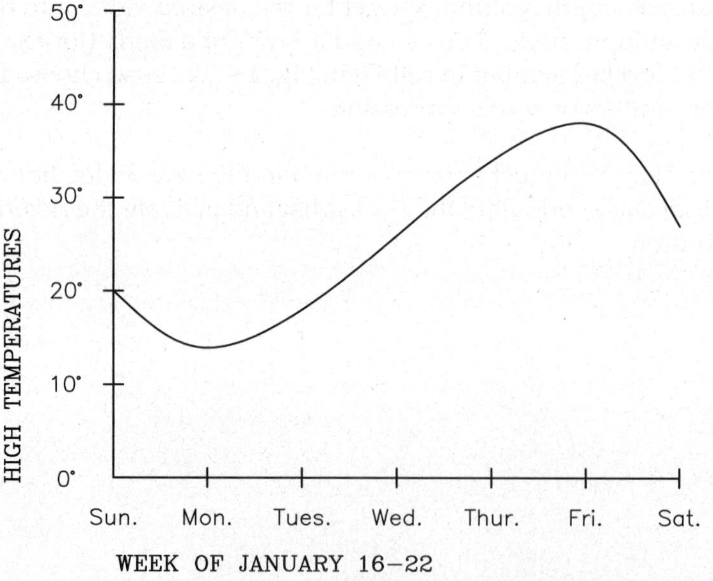

3. ***Style, Text***

*Open* the **BILLMATL** drawing created in the Chapter 15 Exercises. Use ***Style*** to create a new style using the ***romans.shx*** font. Use whatever justification methods you need to align the text information (not the titles) as shown in Figure 18-84. Next, type the ***Style*** command to create a new style that you name as **ROMANS-ITAL**. Use the ***romans.shx*** font file and specify a **15** degree ***obliquing angle***. Use this style for the **NO.**, **PART NAME**, and **MATERIAL**. ***SaveAs*** **BILLMAT2**.

**FIGURE 18-84**

| *NO.* | *PART NAME* | *MATERIAL* |
|---|---|---|
| 1 | Base | Cast Iron |
| 2 | Centering Screw | N/A |
| 3 | Slide Bracket | Mild Steel |
| 4 | Swivel Plate | Mild Steel |
| 5 | Top Plate | Cast Iron |

4. ***Edit Text***

*Open* the **EFF-APT-PS** drawing. Create a new style named **ARCH1** using the ***CityBlueprint*** (.TTF) font file. Next, invoke the ***Properties*** command. Use this dialog box to modify the text style of each of the existing room names to the new style as shown in Figure 18-85. ***SaveAs*** **EFF-APT2.**

**FIGURE 18-85**

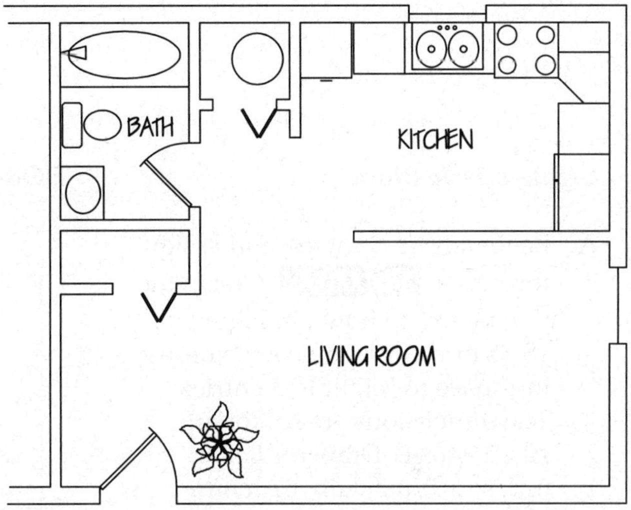


5. ***Text, Mtext***

*Open* the **CBRACKET** drawing from Chapter 9 Exercises. Using ***romans.shx*** font, use ***Text*** to place the part name and METRIC annotation (Fig. 18-86). Use a ***Height*** of **5** and **4**, respectively, and the ***Center Justification*** option. For the notes, use ***Mtext*** to create the boundary as shown. Use the default ***Justify*** method (*TL*)and a ***Height*** of **3**. Use ***Textedit, Mtedit,*** or ***Properties*** if necessary. ***SaveAs*** **CBRACTXT**.

**FIGURE 18-86**

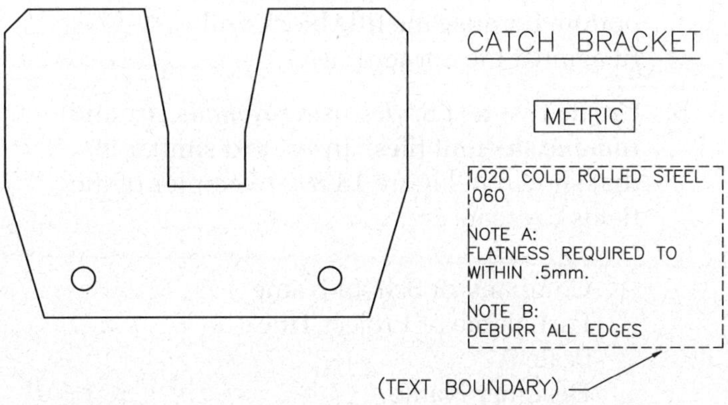


6. ***Style***

Create two new styles for each of your template drawings: **ASHEET**, **BSHEET**, and **C-D-SHEET.** Use the ***romans.shx*** style with the default options for engineering applications or ***CityBlueprint*** (.TTF) for architectural applications. Next, design a style of your choosing to use for larger text as in title blocks or large notes.

7. ***Import Text, Textedit, Properties***

Use a text editor such as Windows Wordpad or Notepad to create a text file containing words similar to "Temperatures were recorded at Sanderson Field by the National Weather Service." Then *Open* the **TEMPGRPH** drawing and use the *Import Text...* option of ***Mtext*** to bring the text into the drawing as a note in the graph as shown in Figure 18-87. Use *Textedit* to edit the text if desired or use *Properties* to change the text style or height.

**FIGURE 18-87**

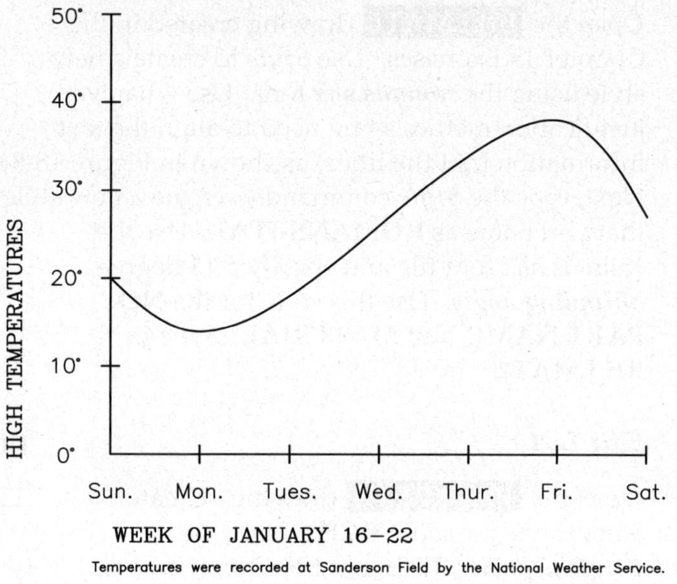


8. ***Create a Title Block***

A. Begin a *New* drawing and assign the name **TBLOCK.** Create the title block as shown in Figure 18-88 or design your own, allowing space for eight text entries. The dimensions are set for an A size sheet. Draw on ***Layer* 0**. Use a ***Pline*** with **.02 *width*** for the boundary and ***Lines*** for the interior divisions. (No *Lines* are needed on the right side and bottom because the title block will fit against the border lines.)

**FIGURE 18-88**

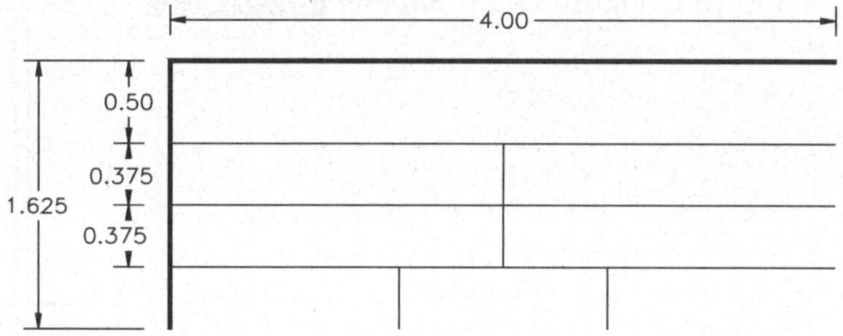


B. Create two text *Styles* using ***romans.shx*** and ***romanc.shx*** font files. Insert text similar to that shown in Figure 18-89. Examples of the fields to create are:

Company or School Name
Part Name or Project Title
Scale
Designer Name
Checker or Instructor Name
Completion Date
Check or Grade Date
Project or Part Number

**FIGURE 18-89**

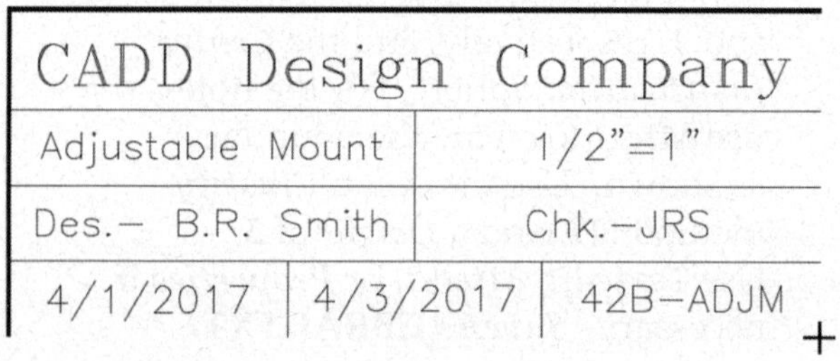


Choose text items relevant to your school or office. *Save* the drawing.

9. *Mtext*

*Open* the **STORE ROOM** drawing that you created in Chapter 16 Exercises. Use *Mtext* to create text paragraphs giving specifications as shown in Figure 18-90. Format the text below as shown in the figure. Use ***CityBlueprint*** (.TTF) as the base font file and specify a base ***Height*** of **3.5″**. Use a ***Color*** of your choice and ***CountryBlueprint*** (.TTF) font to emphasize the first line of the Room paragraph. All paragraphs use the ***TC Justify*** methods except the Contractor Notes paragraph, which is ***TL***. Save the drawing as **STORE ROOM2.**

Room: STORAGE ROOM
11′-2″ x 10′-2″
Cedar Lined - 2 Walls

Doors: 2 - 2268 DOORS
Fire Type A
Andermax

Windows: 2 - 2640 CASEMENT WINDOWS
Triple Pane Argon Filled
Andermax

Notes: Contractor Notes:

Contractor to verify all dimensions in field. Fill door and window rough-outs after door and window placement.

**FIGURE 18-90**

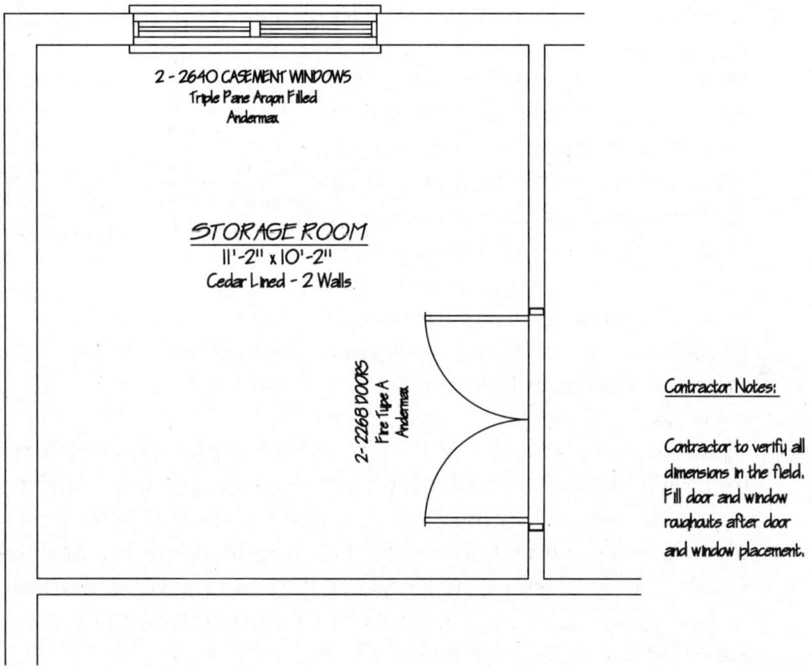


10. ***Tablestyle, Table***

**FIGURE 18-91**

| BILL OF MATERIAL | | |
|---|---|---|
| NUMBER | PART NAME | MATERIAL |
| 1 | Base | Cast Iron |
| 2 | Centering Screw | N/A |
| 3 | Slide Bracket | Mild Steel |
| 4 | Swivel Plate | Mild Steel |
| 5 | Top Plate | Cast Iron |

Begin a *New* drawing and assign the name **MATERIAL TABLE**. Use the ***Tablestyle*** command to create a ***New*** table style named **Bill of Material**. In the ***Text*** tab, assign a new ***Text style*** using the ***Romans.shx*** font file. Accept the default *Text height* and other parameters. Next, use the ***Table*** command to specify a table with ***3 Columns*** and ***5 Data rows***. Select ***Specify window*** under ***Insertion behavior***. Select ***OK***, then PICK the two corners for the table ensuring that you stretch the table wide enough to allow appropriate space in the cells for the intended contents. Enter the text into the cells as shown in Figure 18-91. If you need to make adjustments, highlight the table on an outside edge and use grips to stretch the table. Also, you could right-click while the table is highlighted and select ***Size Columns Equally*** or ***Size Rows Equally***. ***Save*** the drawing.

11. ***Fields***

A. ***Open*** the **EFT-APT2** drawing you worked on in Exercise 4. Use *SaveAs* and assign the name **EFF-APT-AREA**. Make a new ***Layer*** called ***Area*** and make it the ***Current*** layer. In the floor plan, draw a ***Pline*** around the Living Room using ***Object Snap*** to connect to the interior corners of the room as shown in Figure 18-92 by the heavy outline. Make sure you ***Close*** the *Pline*. (You do not have to assign a *Width* for your *Pline*.) Next, draw a second ***Pline*** around the Kitchen using the same method, as shown in the figure. Finally, draw a third ***Pline*** around the other space (bath, closets, and hall) using ***Object Snap*** to connect to the interior walls and two other *Plines* as shown in the figure.

**FIGURE 18-92**

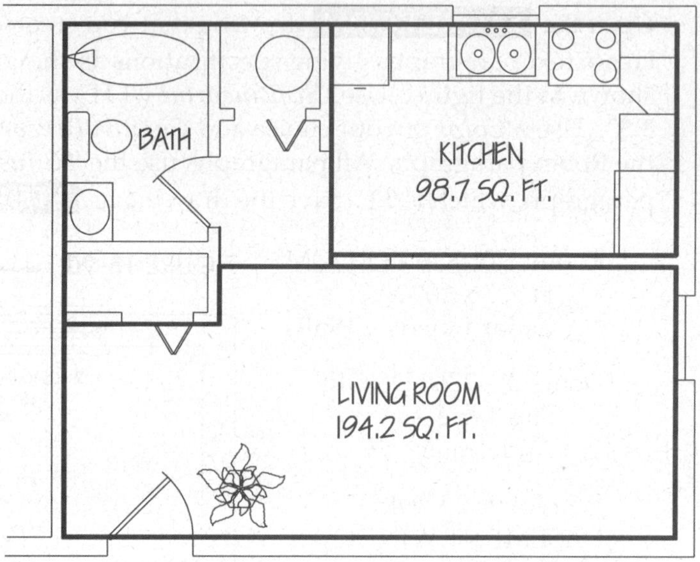


B. Make the text layer ***Current***. Use ***Mtext*** to place text directly below the "LIVING ROOM" text. When the *Text Editor* tab appears, select ***Insert Field*** from the *Options* menu or the right-click menu. In the *Field* dialog box, select ***Object*** from the ***Field names*** list and then use the ***Select Object*** button to return to the drawing and select the ***Pline*** around the Living Room. Select ***Area*** in the ***Property*** list. Set the other ***Format*** options to display text as shown in Figure 18-92. Next, follow the same procedure to create a ***Field*** to display the area of the Kitchen. Do not create a field for the other area. ***Freeze*** the ***Area*** layer. ***Save*** the drawing.

12. ***Fields* and Calculations in a *Table***

A. If not already open, ***Open*** the **EFF-APT-AREA** drawing. ***Thaw*** the ***Area*** layer. Create a new ***Layer*** named **Table** and make it ***Current***.

B. Use ***Tablestyle*** to create a ***New*** table style and assign the name **Area Table**. For the ***Data*** cell style, set the ***Text Style*** to ***Cityblueprint*** and the ***Height*** to **6″**. For the ***Header*** style, set ***Text Style*** to ***Cityblueprint*** and ***Height*** to **8″**. In the ***Title*** style, set ***Text Style*** to ***Cityblueprint*** and ***Height*** to **9″**. Make **Area Table** the ***Current*** table style, then close the dialog box.

C. Use the ***Table*** command to create a table similar to that shown in Figure 18-93. In the *Insert Table* dialog box, choose the following options: ***Specify Insertion Point, 2 Columns, 4 Rows, Column Width*** of **5′**, then select ***OK*** to insert the table just to the right of the floor plan. Create the static text as shown in the figure but do not create text for the four cells listing the square footage calculations. If needed, make any adjustments to the table.

D. To create the area fields for the Living Room, highlight the appropriate cell, then use the Ribbon menu or right-click menu and select ***Insert Field***. In the ***Field*** dialog box select the following options: ***Object*** from ***Field Names*** list, ***Select Objects*** button to select the ***Pline*** around the Living Room, ***Area*** from ***Property*** list, ***Architectural*** from ***Format*** list. Select ***OK*** to close the dialog box and create the field displaying the area for the Living Room. Follow the same procedure in the two cells below for the "Kitchen" area and "Other" area.

E. For the last cell (lower-right corner of the table), use ***Insert Field***, select ***Formula*** from ***Names*** list, ***Sum*** from the formula choices, and ***Decimal*** from ***Format*** list. When prompted to select the corners of the data range, select the three cells displaying the area calculations. When the calculation appears in the last cell, make any adjustments to the table as needed. (Your area calculations may vary slightly from those shown in Figure 18-93 based on where your *Plines* are drawn.) Finally, ***Freeze*** the **Area** layer. ***Save*** the drawing.

**FIGURE 18-93**

| Efficiency Apartment Area | |
|---|---|
| ROOMS | AREA |
| Living Room | 194.2 SQ. FT. |
| Kitchen | 98.7 SQ. FT. |
| Other | 91.2 SQ. FT. |
| TOTAL | 384.0 SQ. FT. |

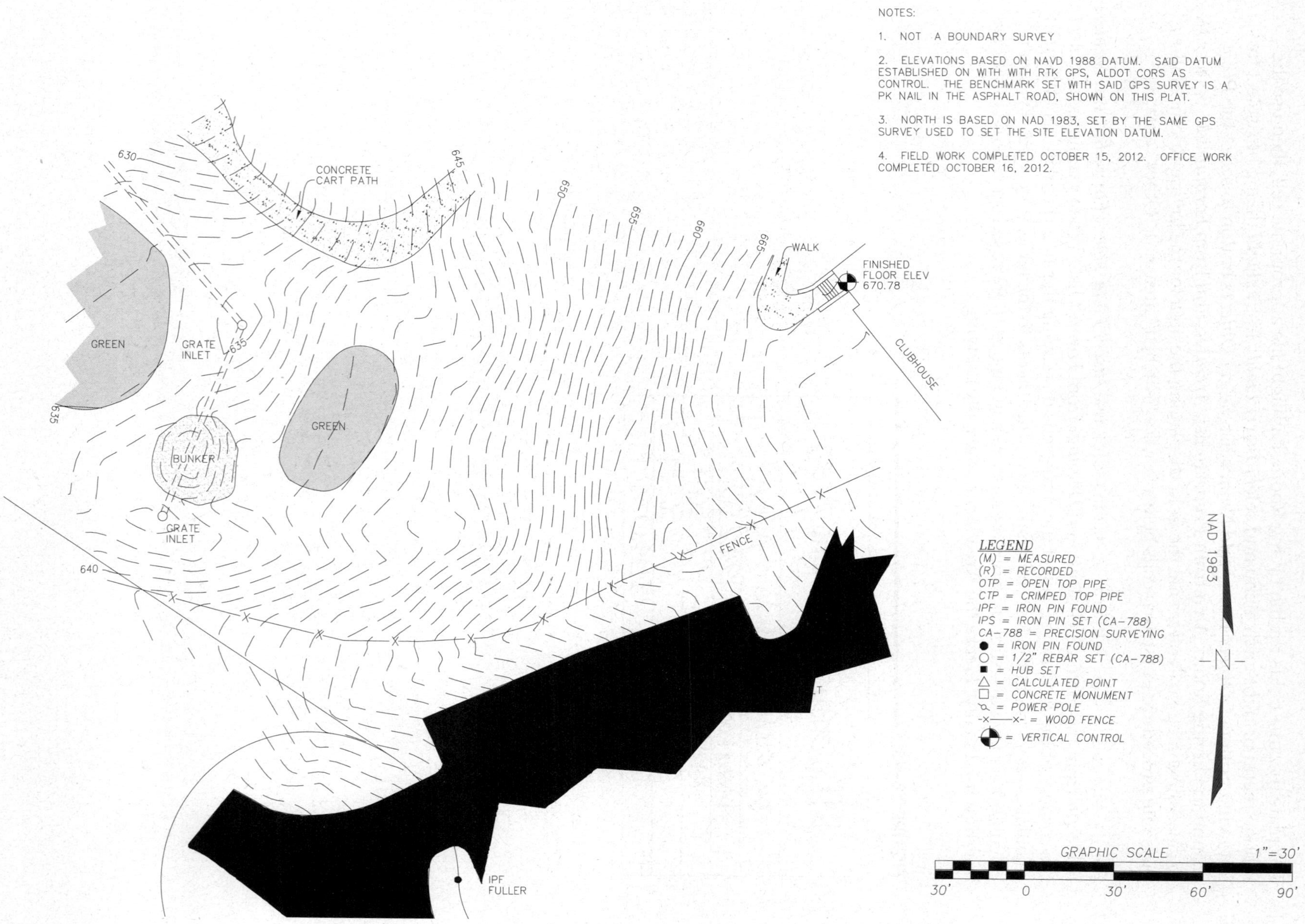
NOTES:
1. NOT A BOUNDARY SURVEY
2. ELEVATIONS BASED ON NAVD 1988 DATUM. SAID DATUM ESTABLISHED ON WITH WITH RTK GPS, ALDOT CORS AS CONTROL. THE BENCHMARK SET WITH SAID GPS SURVEY IS A PK NAIL IN THE ASPHALT ROAD, SHOWN ON THIS PLAT.
3. NORTH IS BASED ON NAD 1983, SET BY THE SAME GPS SURVEY USED TO SET THE SITE ELEVATION DATUM.
4. FIELD WORK COMPLETED OCTOBER 15, 2012. OFFICE WORK COMPLETED OCTOBER 16, 2012.
CONCRETE CART PATH
WALK
FINISHED FLOOR ELEV 670.78
CLUBHOUSE
GREEN
GRATE INLET
GREEN
BUNKER
GRATE INLET
FENCE
IPF FULLER
630
635
640
645
650
655
660
665
LEGEND
(M) = MEASURED
(R) = RECORDED
OTP = OPEN TOP PIPE
CTP = CRIMPED TOP PIPE
IPF = IRON PIN FOUND
IPS = IRON PIN SET (CA-788)
CA-788 = PRECISION SURVEYING
● = IRON PIN FOUND
○ = 1/2" REBAR SET (CA-788)
■ = HUB SET
△ = CALCULATED POINT
□ = CONCRETE MONUMENT
= POWER POLE
-x——x- = WOOD FENCE
= VERTICAL CONTROL
NAD 1983
N
GRAPHIC SCALE
1"=30'
30'
0
30'
60'
90'

CHAPTER

# 19

# Grip Editing

## Chapter Objectives

After completing this chapter you should:

1. be able to use the *GRIPS* variable to enable or disable object grips;
2. be able to activate the grips on any object;
3. be able to make an object's grips hot or cold;
4. be able to use each of the grip editing options, namely, STRETCH, MOVE, ROTATE, SCALE, and MIRROR;
5. be able to use the copy and Base suboptions;
6. be able to use the auxiliary grid that is automatically created when the Copy suboption is used;
7. be able to change grip variable settings using the *Options* dialog box or the Command line format.

## CONCEPTS

Grips provide an alternative method of editing AutoCAD objects. The object grips are available for use by setting the *GRIPS* variable to **1** or **2**. Object grips are small squares appearing on selected objects at endpoints, midpoints, or centers, etc. The object grips are activated (made visible) by **PICK**ing objects with the cursor pickbox only <u>when no commands are in use</u> (at the open Command prompt). Grips are like small, magnetic Object Snaps (*Endpoint*, *Midpoint*, *Center*, *Quadrant*, etc.) that can be used for snapping one object to another, for example. If the cursor is moved within the small square, it is automatically "snapped" to the grip. Grips can replace the use of Object Snap for many applications. The grip option allows you to STRETCH, MOVE, ROTATE, SCALE, MIRROR, or COPY objects without invoking the normal editing commands or Object Snaps.

As an example, the endpoint of a *Line* could be "snapped" to the quadrant of a *Circle* (shown in Fig. 19-1) by the following steps:

1. Activate the grips by selecting both objects. Selection is done when no commands are in use (during the open Command prompt).
2. Select the grip at the endpoint of the *Line* (1). The grip turns **hot** (red).
3. The ** STRETCH ** option appears in place of the Command prompt.
4. STRETCH the *Line* to the quadrant grip on the *Circle* (2). **PICK** when the cursor "snaps" to the grip.
5. The *Line* and the *Circle* should then have connecting points. The Command prompt reappears. Press Escape to cancel (deactivate) the grips.

**FIGURE 19-1**

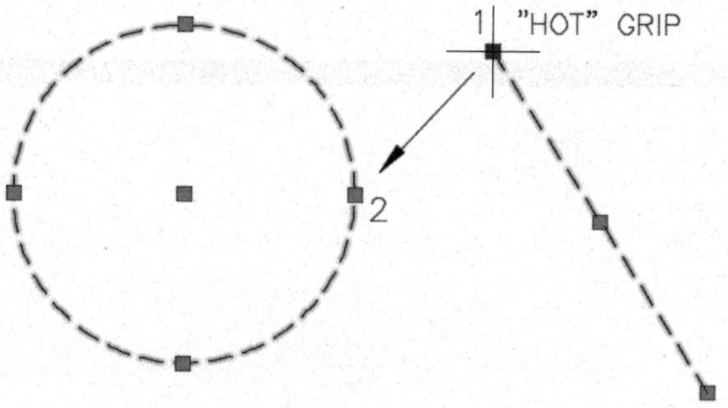


## GRIPS FEATURES

Grips are enabled or disabled by changing the setting of the system variable, *GRIPS*. A setting of 1 or 2 enables or turns *ON GRIPS* and a setting of **0** disables or turns *OFF GRIPS*. This variable can be typed at the Command prompt, or grips can be invoked from the *Selection* tab of the *Options* dialog box, or by typing *Ddgrips* (Fig. 19-2). Using the dialog box, toggle *Show Grips* to turn *GRIPS ON*. The default setting in AutoCAD for the *GRIPS* variable is **2** (*ON*). (See "Grip Settings" near the end of the chapter for explanations of the other options in the *Grips* section of the *Options* dialog box.)

**FIGURE 19-2**

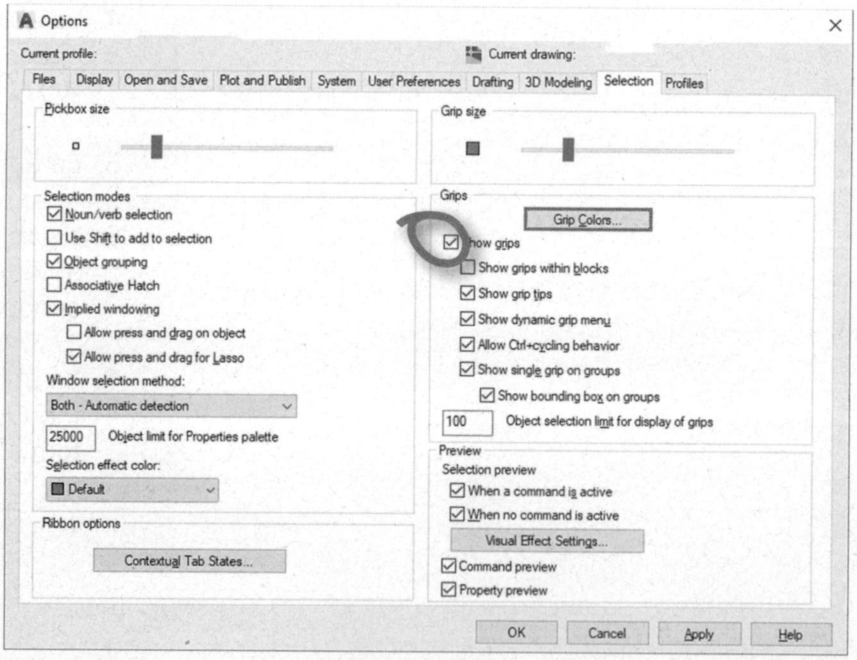

## Activating Grips on Objects

The grips on objects are activated by selecting desired objects with the *AUto* options (pickbox, windows, or lassos). This action is done when no commands are in use (at the open Command prompt). When an object has been selected, two things happen: the grips appear and the object is highlighted. The grips are the small blue (default color) boxes appearing at the endpoints, midpoint, center, quadrants, vertices, insertion point, or other locations depending on the object type (Fig. 19-3). Highlighting indicates that the object is included in the selection set.

**FIGURE 19-3**

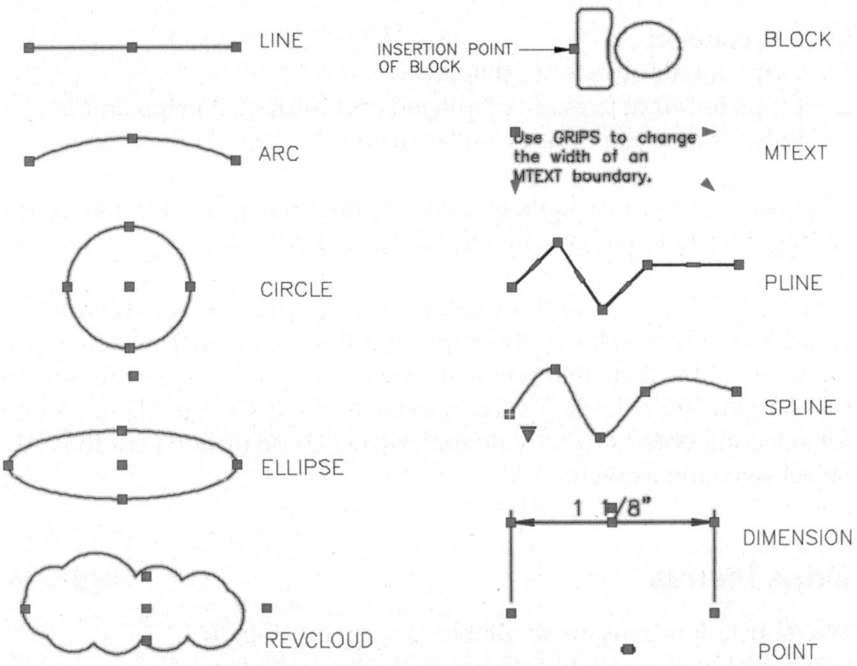


## Cold and Hot Grips

Grips can have only two states: **cold** and **hot**. When grips are turned on, and an object is selected (when no commands are in use), the object's grips become **cold**—its grips are displayed in blue (default color) and the object is highlighted.

One or more grips can then be made **hot**. A hot grip is created by "hovering" the cursor over a cold grip until the grip changes to a pink (default) color, then you PICK the grip so it changes to red (default color). The intermittent pink grip is called a "hover" grip.

A **hot** grip is red (by default) and its object is almost always highlighted. Any grip can be changed to **hot** by selecting the grip itself. A **hot** grip is the default base point used for the editing action such as MOVE, ROTATE, SCALE, or MIRROR, or is the stretch point for STRETCH. When a **hot** grip exists, a new series of prompts appear in place of the Command prompt that displays the various grip editing options. Two or more grips can be made **hot** simultaneously by holding down the Shift key while selecting each grip.

If you have made a grip **hot** and want to deactivate it, possibly to make another grip **hot** instead, press Escape once. This returns the grip to a **cold** state. In effect, this is an undo only for the **hot** grip. Pressing Escape again cancels all grips. Pressing Escape demotes hot grips or cancels all grips:

| Grip State | Press Escape once | Press Escape twice |
|---|---|---|
| only **cold** | grips are deactivated | |
| **hot** and **cold** | **hot** demoted to **cold** | grips are deactivated |

## *GRIPS* System Variable Settings

Settings for the *GRIPS* system variable are as follows:

0 Grips are off.
1 Grips for all objects are displayed.
2 Grips for all objects are displayed and midpoint grips on *Pline* segments are displayed (notice the long rectangular "midpoint" grips on the *Pline* in Figure 19-3).

The *GRIPS* system variable is saved in the Registry rather than in the drawing file. Therefore, the *GRIPS* setting does not change when drawings are edited.

When *GRIPS* are enabled, a small pickbox (5 pixels is the default size) appears at the center of the cursor crosshairs. (The pickbox also appears if the *PICKFIRST* system variable is set to 1 or 2.) This pickbox operates in the same manner as the pickbox appearing during the "Select objects:" prompt. Only the pickbox, *AUTO window*, *AUTO crossing window*, *Crossing Lasso*, and *Window Lasso* methods can be used for selecting objects to activate the grips. (These options are the only options available for Noun/Verb object selection as well.)

## Grips Menus

Two shortcut menus are available for grip editing. If you hover over a grip but do not click, the grip turns pink and a shortcut menu may appear listing the possible grip editing options (Fig. 19-4). Multifunctional grip options can be accessed in this manner. The options appearing in this menu are different depending on the type of object and location of the grip. A menu may not appear if no special options are available for that grip.

FIGURE 19-4

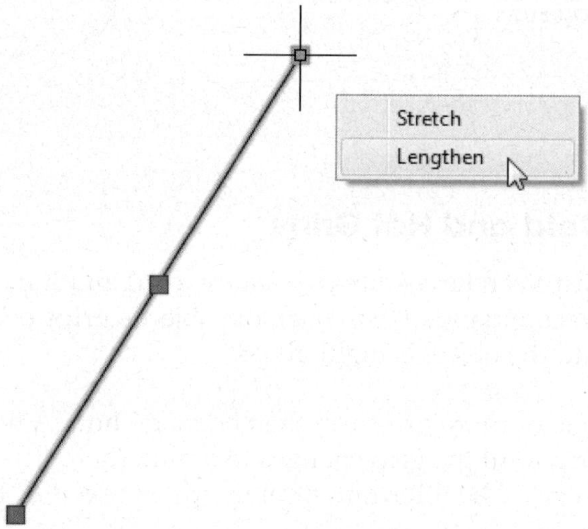


If you first left-click to make a grip hot, then right-click on the hot grip, a menu appears with the standard STRETCH, MOVE, ROTATE, SCALE, MIRROR options (Fig. 19-5). These are the same options listed on the Command line that can also be accessed by pressing the space bar to cycle through the choices (see "Grip Editing Options").

NOTE: Beware that if grips are cold and you right-click in the drawing area (not on a grip) the standard Edit Mode menu appears. The standard Edit Mode menu contains full commands (see "Edit Mode Menu" in Chapter 1).

FIGURE 19-5

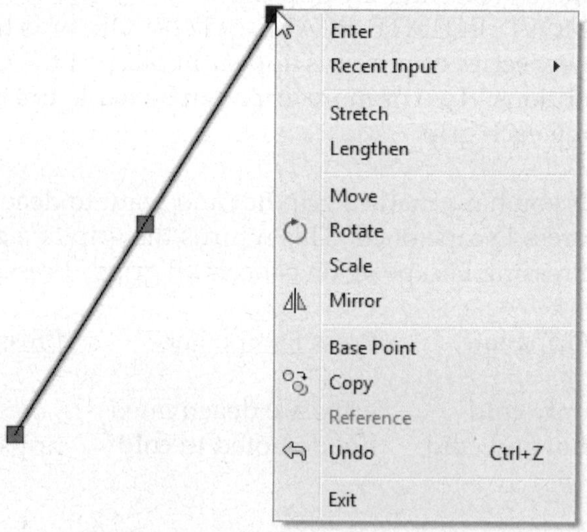

## Grip Editing Options

When a **hot grip** has been activated, the grip editing options are available. The Command prompt is replaced by the STRETCH, MOVE, ROTATE, SCALE, or MIRROR grip editing options. You can sequentially cycle through the options by pressing the Space bar or Enter key. The editing options are displayed in Figures 19-6 through 19-10 below and on the next page.

Alternately, you can right-click when a grip is **hot** to activate the grip menu (see Fig. 19-5). This menu has the same options available in Command line format with the addition of *Reference* and *Lengthen*. The options are described in the following figures.

** STRETCH **
Specify stretch point or [Basepoint/Copy/Undo/eXit]:

Note that for *Arc* objects only, grips appear at the endpoints, midpoint, and center. In addition, arrowshaped grips appear at the endpoints and midpoint (Fig. 19-6). These arrow grips allow you to change the radius (midpoint arrow) or extend the arc's length (endpoint arrows) in the STRETCH mode only.

**FIGURE 19-6**

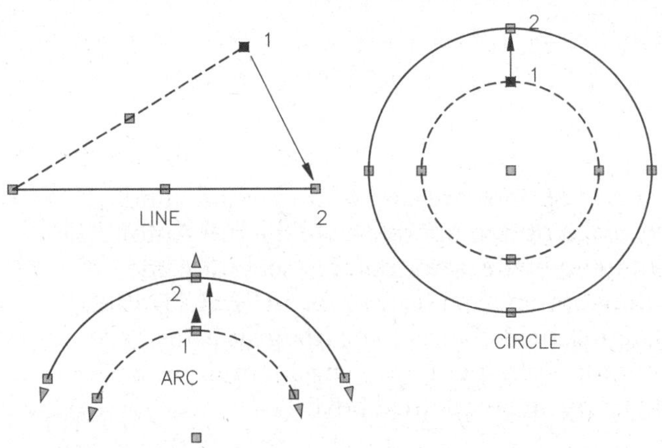


** MOVE **
Specify move point or
[Base point/Copy/Undo/eXit]:
(Fig. 19-7)

**FIGURE 19-7**

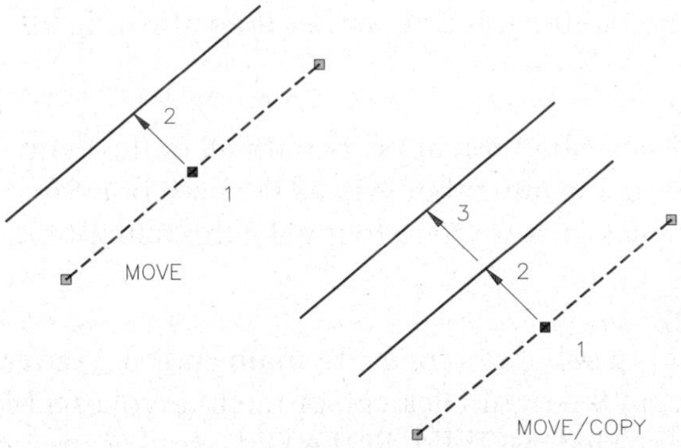


** ROTATE **
Specify rotation angle or
[Base point/Copy/Undo/Reference/eXit]:
(Fig. 19-8)

**FIGURE 19-8**

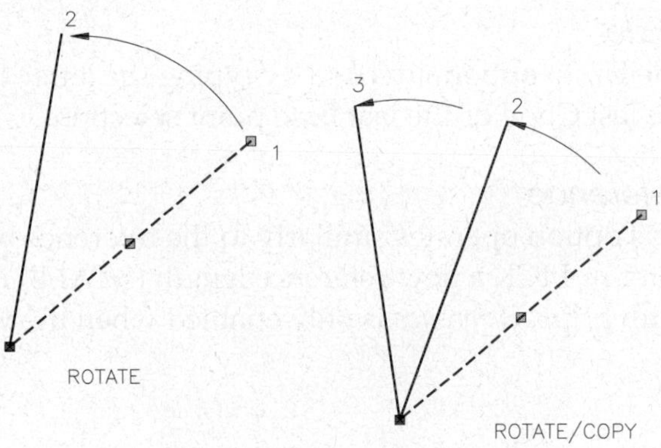

** SCALE **
Specify scale factor or
[Base point/Copy/Undo/Reference/eXit]:
(Fig. 19-9)

**FIGURE 19-9**

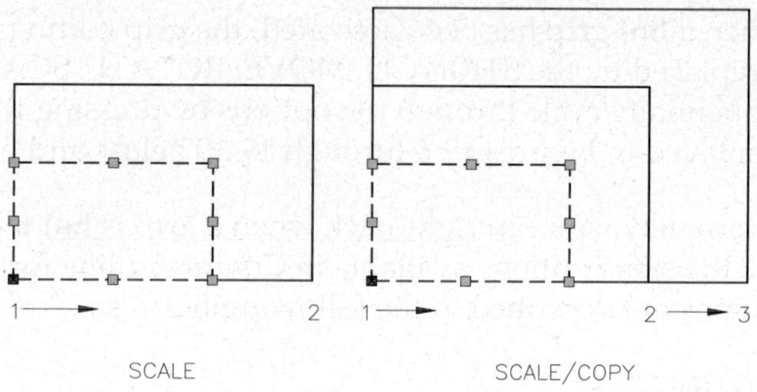


** MIRROR **
Specify second point or
[Base point/Copy/Undo/eXit]:
(Fig. 19-10)

**FIGURE 19-10**

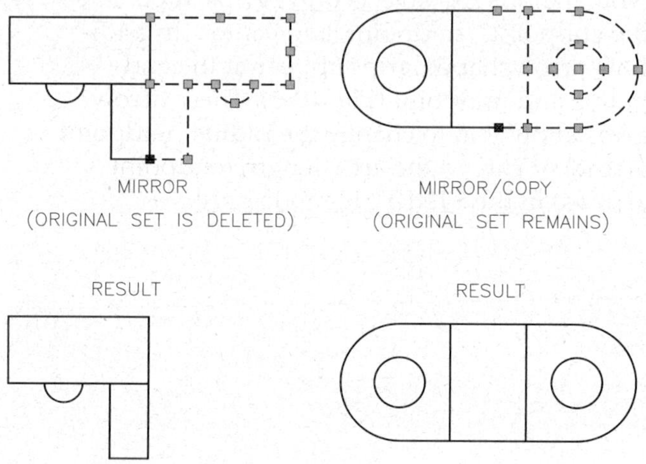


The grip options are easy to understand and use. Each option operates like the full AutoCAD command by the same name. Generally, the editing option used (except for STRETCH) affects all highlighted objects. The **hot** grip is the base point for each operation. The suboptions, Base and Copy, are explained next.

NOTE: The STRETCH option differs from other options in that STRETCH affects only the object that is attached to the **hot** grip, rather than affecting all highlighted (**cold**) objects.

### *Base*

The Base suboption appears with all of the main grip editing options (STRETCH, MOVE, etc.). Base allows using any other grip as the base point instead of the **hot** grip. Type the letter ***B*** or select from the right-click cursor menu to invoke this suboption.

### *Copy*

Copy is a suboption of every main choice. Activating this suboption by typing the letter ***C*** or selecting from the right-click cursor menu invokes a Multiple copy mode, such that whatever set of objects is STRETCHed, MOVEd, ROTATEd, etc., becomes the first of an unlimited number of copies (see the previous five figures). The Multiple mode remains active until exiting back to the Command prompt.

### *Undo*

The Undo option, invoked by typing the letter ***U*** or selecting from the right-click cursor menu, will undo the last Copy or the last Base point selection. Undo functions only after a Base or Copy operation.

### *Reference*

This option operates similarly to the reference option of the *Scale* and *Rotate* commands. Use *Reference* to enter or PICK a new reference length (SCALE) or angle (ROTATE). (See "*Scale* and *Rotate*," Chapter 9.) With grips, *Reference* is only enabled when the SCALE or ROTATE options are active.

### Shift to Turn on *Ortho*

Using the Copy option allows you to make multiple copies, such as MOVE/Copy or ROTATE/Copy. If you hold down the Shift key as you make copies, *Ortho* is turned on as long as Shift is depressed.

### Auxiliary Grid

An auxiliary grid is automatically established on creating the first Copy (Fig. 19-11). The grid is activated by pressing Ctrl while placing the subsequent copies. The subsequent copies are then "snapped" to the grid in the same manner that *Snap* functions. The spacing of this auxiliary grid is determined by the location of the first Copy, that is, the X and Y intervals between the base point and the second point.

**FIGURE 19-11**

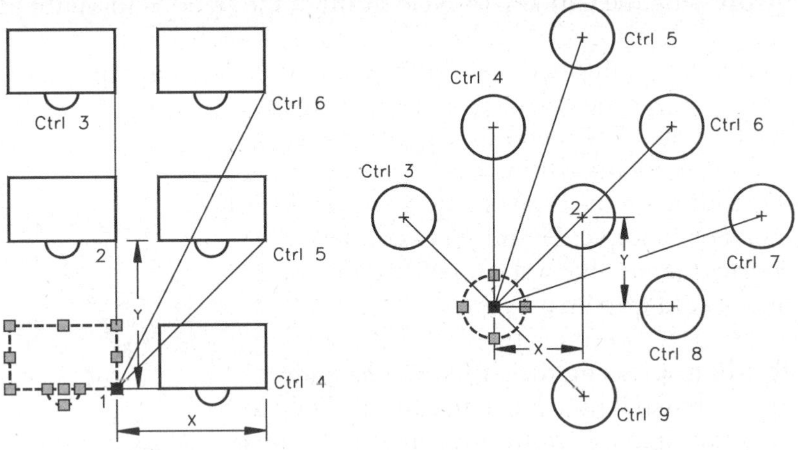


MOVE/COPY
Ctrl: GRID=X,Y

MOVE/COPY
Ctrl: GRID=X,Y

For example, a "polar array" can be simulated with grips by using ROTATE with the Copy suboption (Fig. 19-12). The "array" can be constructed by making one Copy, then using the auxiliary grid to achieve equal angular spacing. The steps for creating a "polar array" are as follows:

1. Select the object(s) to array.
2. Select a grip on the set of objects to use as the center of the array. Cycle to the ROTATE option by pressing Enter or selecting from the right-click cursor menu. Next, invoke the Copy suboption.
3. Make the first copy at any desired location.
4. After making the first copy, activate the auxiliary angular grid by holding down Ctrl while making the other copies.
5. Cancel the grips or select another command from the menus.

**FIGURE 19-12**

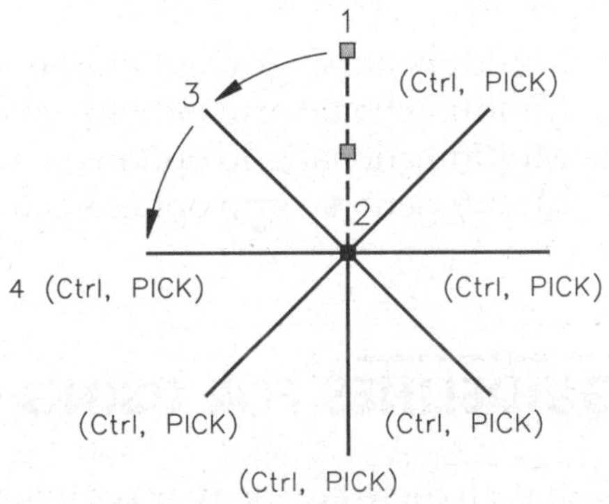


ROTATE/COPY
Ctrl – ANGLES EQUAL

### Editing Dimensions

One of the most effective applications of grips is as a dimension editor. Because grips exist at the dimension's extension line origins, arrowhead endpoints, and dimensional value, a dimension can be changed in many ways and still retain its associativity (Fig. 19-13). See Chapter 28, Dimensioning, for further information about dimensions, associativity, and editing dimensions with grips.

**FIGURE 19-13**

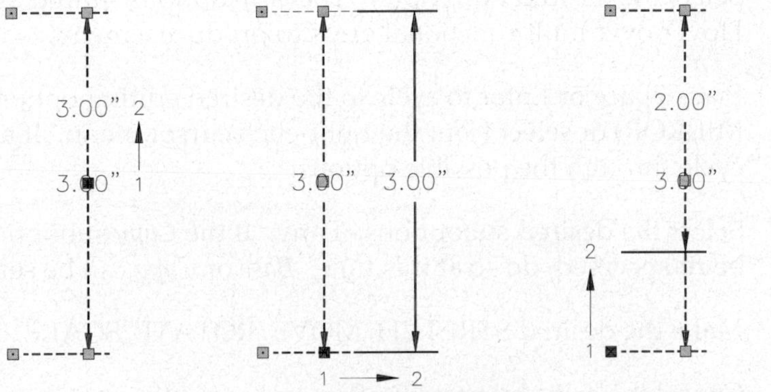


STRETCH TEXT
STRETCH EXTENSION LINE
CHANGE DIMENSION

### Multifunctional Grips

Multifunctional grips provide an enhanced mode of grip editing for specific objects. *Lines, Polylines, Arcs, Elliptical Arcs, Splines, Dimensions, Multileaders, 3D Faces, Edges,* and *Vertices* have multifunctional grip capabilities. "Multifunctional" means multiple special editing options are available on a shortcut menu. You can access these editing options either by hovering on a grip to produce a shortcut menu or by pressing the Ctrl key to cycle through the options for a hot grip.

To use the multifunctional shortcut menu, simply hover the cursor over a cold grip, but do not click. The menu that appears allows you to access the editing options which vary depending on the type of object you are selecting. For example, Figure 19-14 displays the shortcut menu that appears when hovering over a cold grip on a *Spline*.

**FIGURE 19-14**

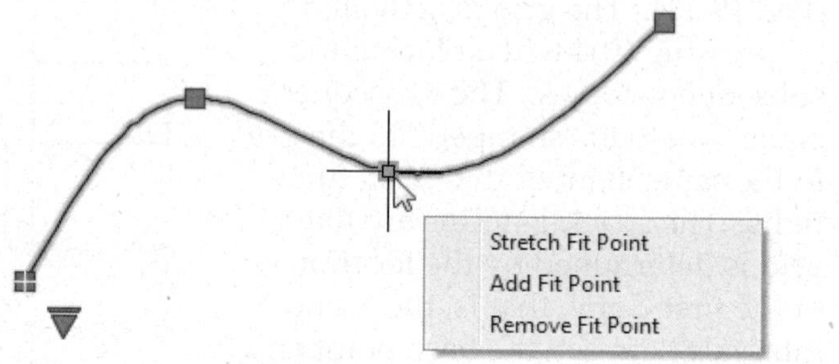


To edit grips using the Ctrl key, select any grip to make it hot. Next, use the Ctrl key to cycle through the multifunction grip options appearing on the Command line. As always, you can also cycle through the typical STRETCH, MOVE, ROTATE, SCALE, MIRROR options by pressing the Space bar.

The system variable that controls multifunctional grips is *GRIPMULTIFUNCTIONAL*. Possible settings are as follows:

0 Multifunctional grip options are not available.
1 Multifunctional grip options can be accessed by pressing Ctrl to change grip behavior (Ctrl-cycling).
2 Multifunctional grip options can be accessed via the menu displayed when you hover over a grip.
3 Multifunctional grip options can be accessed with both Ctrl-cycling and the grip menu.

## GUIDELINES FOR USING GRIPS

Although there are many ways to use grips based on the existing objects and the desired application, a general set of guidelines for using grips is given here:

1. Create the objects to edit.

2. Select objects to make **cold** grips appear.

3. Select the desired **hot** grip(s). The *Grip* options should appear in place of the Command line. Hover over multifunctional grips to produce a menu.

4. Press Space or Enter to cycle to the desired editing option (STRETCH, MOVE, ROTATE, SCALE, MIRROR) or select from the right-click cursor menu. If a multifunctional grip is **hot**, press Ctrl to cycle through the possible options.

5. Select the desired suboptions, if any. If the *Copy* suboption is needed or the base point needs to be re-specified, do so at this time. *Base* or *Copy* can be selected in either order.

6. Make the desired STRETCH, MOVE, ROTATE, SCALE, or MIRROR.

7. Cancel the grips by pressing Escape or selecting a command from a menu.

## GRIPS SETTINGS

FIGURE 19-15

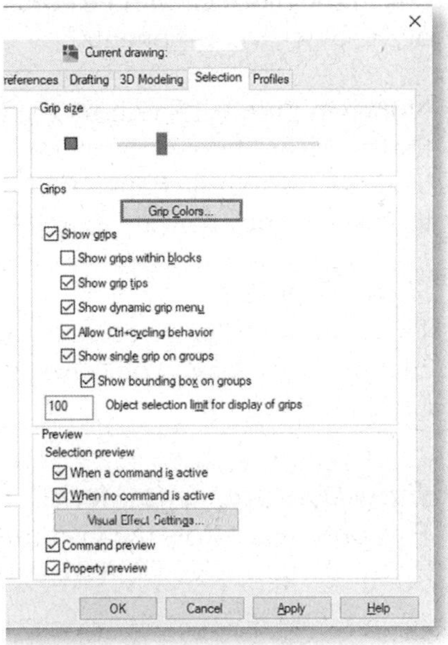


Several settings are available in the *Selection* tab of the *Options* dialog box (Fig. 19-15) that control the way grips appear or operate. The settings can also be changed by typing in the related variable name at the Command prompt.

### *Grip Size*

Use the slider bar to interactively increase or decrease the size of the grip "box." The *GRIPSIZE* variable could alternately be used. The default size is 5 pixels square (*GRIPSIZE*=5).

### *Grip Colors*

Selecting the *Grip Colors* button will open the dialog box shown in Figure 19-16.

*Unselected Grip Color*

This setting enables you to change the color of **cold** grips. Select the desired color from the *Unselected Grip Color* drop-down list. PICKing the *Select Color...* tile produces the *Select Color* dialog box (identical to that used with other color settings).

FIGURE 19-16

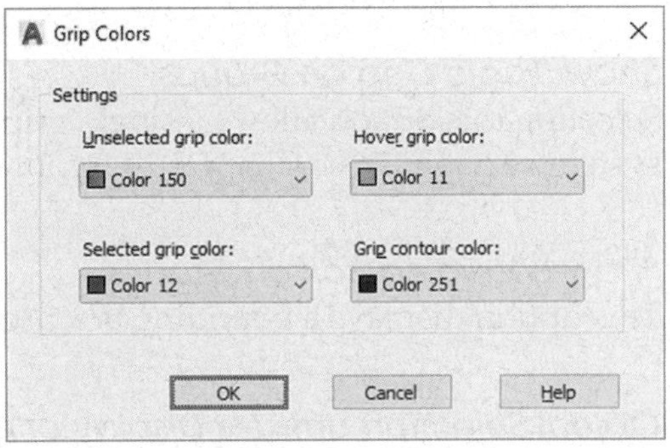


*Selected Grip Color*

The color of hot grips can be specified with this option. Select the desired color from the *Selected Grip Color* drop-down list. The *Select Color...* tile produces the *Select Color* dialog box.

*Hover Grip Color*

This setting specifies the color of grips when you "hover" your cursor over a cold (blue) grip, usually just before you PICK it to make it a hot grip. The drop-down list operates identically to the other grip color options.

*Grip Contour Color*

The contour is the thin outline around the cold grips.

### *Show Grips*

If a check appears in the checkbox, grips are enabled for the workstation. A check sets the *GRIPS* variable to 2 (on). Removing the check disables grips and sets *GRIPS*=0 (off). The default setting is on.

### *Show Grips within Blocks*

When no check appears in this box, only one grip at the block's insertion point is visible (Fig. 19-17). This allows you to work with a block as one object. The related variable is *GRIPBLOCK*, with a setting of 0=off. This is the default setting (disabled).

FIGURE 19-17

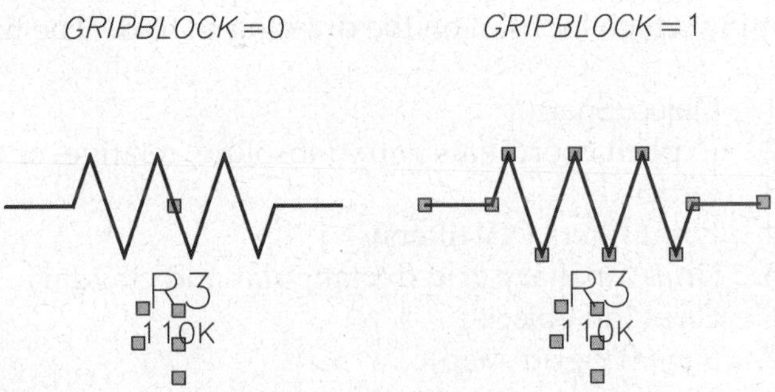

When the box is checked, *GRIPBLOCK* is set to 1. All grips on all objects contained in the block are visible and functional (Fig. 19-18). This allows you to use any grip on any object within the block. This does not mean that the block is *Exploded*—the block retains its single-object integrity and individual entities in the block cannot be edited independently. This feature permits you to use the grips only on each of the block's components.

TIP

Notice in Figure 19-17 how the insertion point is not accessible when *GRIPBLOCK* is set to 1. Notice also that there are two grips on each of the *Normal* attributes and only one on the *Invisible* attribute and that these grips do not change with the two *GRIPBLOCK* settings. Making one of these grips hot allows you to "move" the attribute to any location and the attribute still retains its association with the block.

### *Show Grip Tips*

This option has no effect on standard AutoCAD objects. This option enables the grip tips only when custom (third-party) objects supporting grip tips are included in the drawing. (Alternately, set *GRIPTIPS*=1.)

### *Show Dynamic Grip Menu*

This option controls the visibility of dynamic menu when pausing over a multi-functional grip (*GRIPMULTIFUNCTIONAL* system variable).

### *Allow Ctrl+Cycling Behavior*

Checking this option allows the Ctrl-cycling behavior for multi-functional grips (*GRIPMULTIFUNCTIONAL*).

### *Show Single Grip on Groups*

Selecting this option allows a single grip to be displayed for a *Group* of objects (*GROUPDSIPLAYMODE* system variable). See Chapter 20 for information on *Groups*.

### *Show Bounding Box on Groups*

This option displays a bounding box around the extents of a *Group*.

### *Object Selection Limit for Display of Grips*

If you select a large number of objects, the display of hundreds of grips can be counter-productive. Use this value to suppress the display of grips when the selection set includes more than the specified number of objects. The value is stored in the *GRIPOBJLIMIT* system variable. The valid range is 0 to 32,767. When set to 0, grips are always displayed.

## Point Specification Precedence

When you select a point with the input device, AutoCAD uses a point specification precedence to determine which location on the drawing to find. The hierarchy is listed here:

1. Object Snap
2. Explicit coordinate entry (absolute, relative, or polar coordinates)
3. *Ortho*
4. Point filters (XY filters)
5. *Grips* auxiliary grid (rectangular and circular)
6. *Grips* (on objects)
7. *Snap* (F9 grid snap)
8. Digitizing a point location

Practically, this means that *Object Snap* has priority over any other point selection mode. As far as grips are concerned, *Ortho* overrides grips, so turn off *Ortho* if you want to snap to grips. Although grips override *Snap* (F9), it is suggested that *Snap* be turned *Off* while using grips to simplify PICKing.

## More Grips

Grips have a wide range of editing potential. As presented in this chapter, most of the grip-editing options are alternatives to using the *Stretch, Move, Rotate, Scale, Mirror,* and *Copy* commands. However, grips also provide editing methods for particular objects that are not available by any other means. Editing dimension objects is one example that is discussed in Chapters 28 and 29. Another very powerful application of grips is editing solid models. These techniques are discussed in Chapter 36.

# CHAPTER EXERCISES

1. *Open* the **TEMP-D** drawing from Chapter 16 Exercises. Use the grips on the existing *Spline* to generate a new graph displaying the temperatures for the following week. **PICK** the ***Spline*** to activate grips (Fig. 19-18). Use the **STRETCH** option to stretch the first data point grip to the value indicated below for Sunday's temperature. **Cancel** the grips, then repeat the steps for each data point on the graph. ***Save*** the drawing as **TEMP-F**.

| X axis | Y axis |
|---|---|
| Sunday | 29 |
| Monday | 32 |
| Tuesday | 40 |
| Wednesday | 33 |
| Thursday | 44 |
| Friday | 49 |
| Saturday | 45 |

FIGURE 19-18

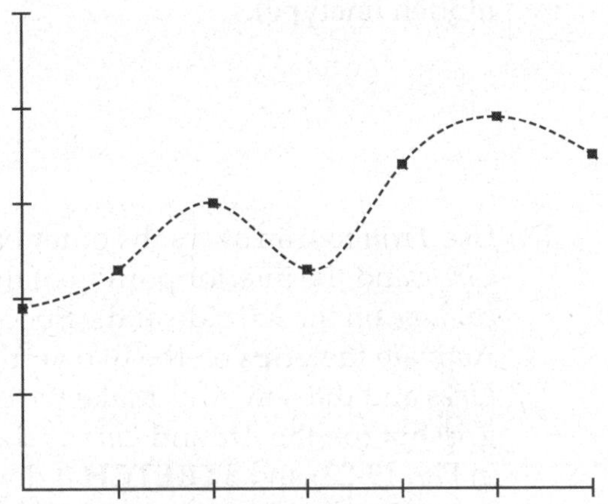

2. This exercise involves all the options of grip editing to create a Space Plate. Begin a *New* drawing or use the **ASHEET *template*** and use ***SaveAs*** to assign the name **SPACEPLT**.

   A. Set the ***Snap*** value to **.125**. Set ***Polar Snap*** to **.125** and turn on ***Polar Tracking***. Draw the geometry shown in Figure 19-19 using the ***Line*** and ***Circle*** commands.

FIGURE 19-19

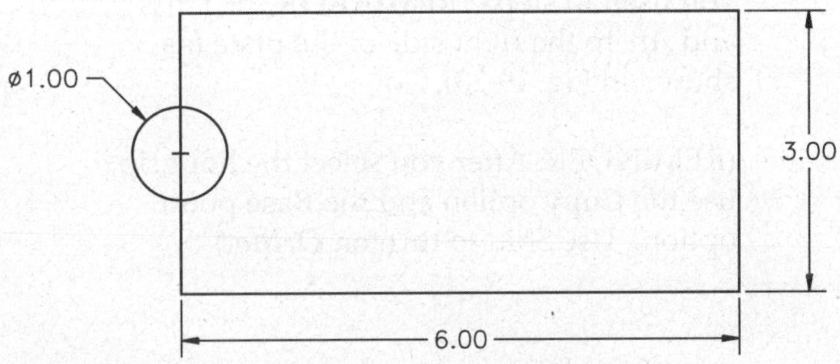

B. Activate the **grips** on the *Circle*. Make the center grip **hot**. Cycle to the **MOVE** option. Enter *C* for the **Copy** option. You should then get the prompt for **MOVE (multiple)**. Make two copies as shown in Figure 19-20. The new *Circles* should be spaced evenly, with the one at the far right in the center of the rectangle. If the spacing is off, use **grips** with the **STRETCH** option to make the correction.

**FIGURE 19-20**

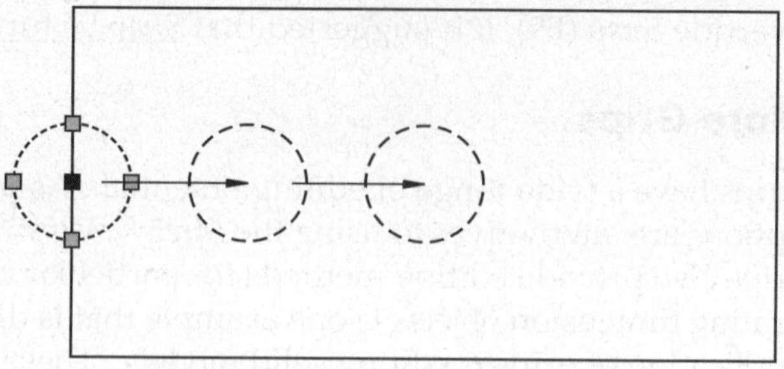

C. Activate the **grips** on the number 2 *Circle* (from the left). Make the center *Circle* grip **hot** and cycle to the **MIRROR** option. Enter *C* for the **Copy** option. (You'll see the **MIRROR (multiple)** prompt.) Then enter *B* to specify a new base point as indicated in Figure 19-21. Press Shift to turn ***On Ortho*** and specify the mirror axis as shown to create the new *Circle* (shown in Fig. 19-21 in hidden linetype).

**FIGURE 19-21**

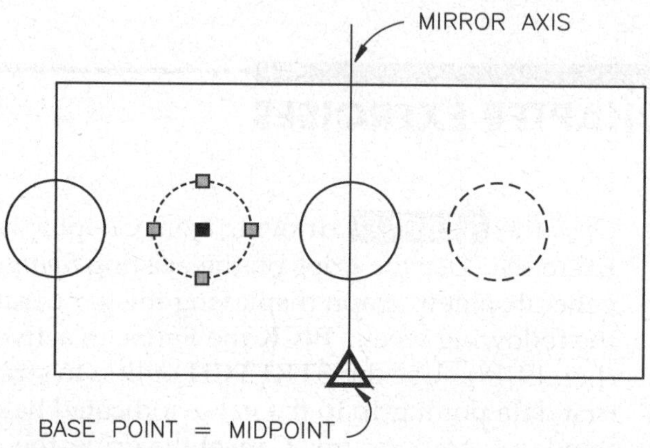


D. Use ***Trim*** to trim away the outer half of the *Circle* and the interior portion of the vertical *Line* on the left side of the Space Plate. Activate the **grips** on the two vertical *Lines* and the new *Arc*. Make the common grip **hot** (on the *Arc* and *Line* as shown in Fig. 19-22) and **STRETCH** it downward .5 units. (Note how you can affect multiple objects by selecting a common grip.) **STRETCH** the upper end of the *Arc* upward .5 units.

**FIGURE 19-22**

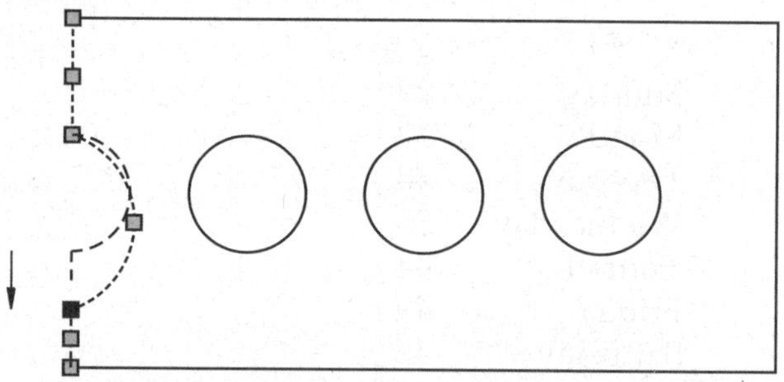

E. ***Erase*** the *Line* on the right side of the Space Plate. Use the same method that you used in step C to **MIRROR** the *Lines* and *Arc* to the right side of the plate (as shown in Fig. 19-23).

(REMINDER: After you select the **hot** grip, use the **Copy** option <u>and</u> the **Base** point option. Use Shift to turn on ***Ortho***.)

**FIGURE 19-23**

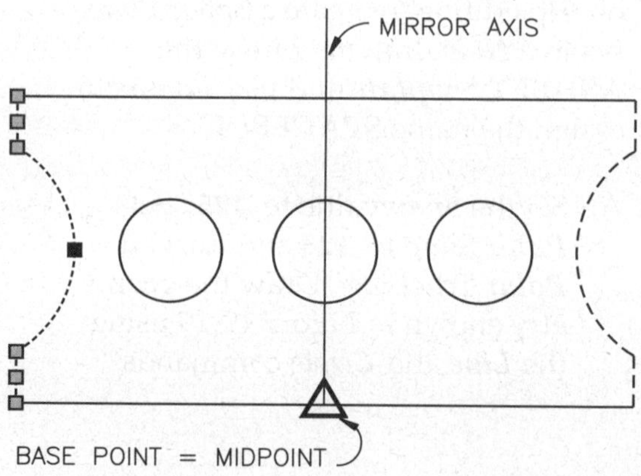

F. In this step, you will **STRETCH** the top edge upward one unit and the bottom edge downward one unit by selecting multiple **hot** grips.

Select the desired horizontal and attached vertical *Lines*. Hold down Shift while selecting each of the endpoint grips, as shown in Figure 19-24. Although they appear **hot**, you must select one of the two again to activate the **STRETCH** option.

**FIGURE 19-24**

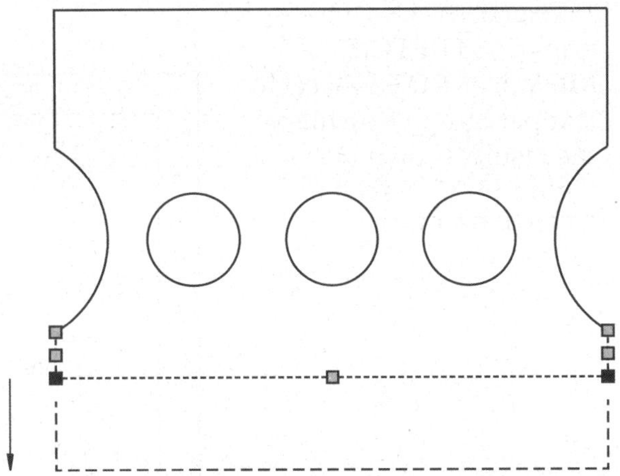

G. In this step, two more *Circles* are created by the **ROTATE** option (see Fig. 19-25). Select the two outside *Circles* to make the grips appear. PICK the center grip to make it **hot**. Cycle to the **ROTATE** option. Enter ***C*** for the **Copy** option. Next, enter ***B*** for the **Base** option and ***Object Snap*** to the ***Center*** as shown. Press Shift to turn on ***Ortho*** and create the new *Circles*.

**FIGURE 19-25**

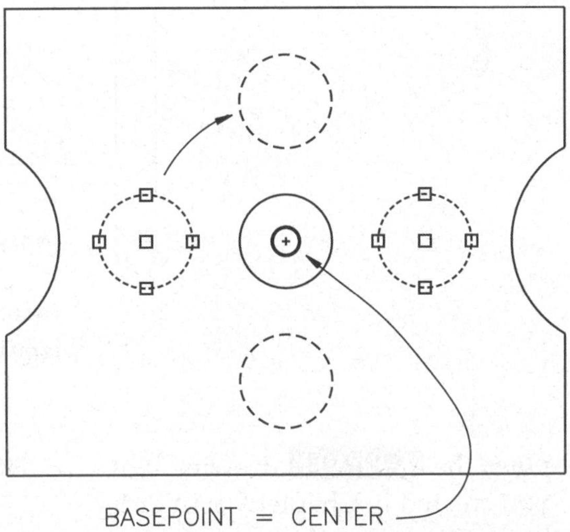


H. Select the center *Circle* and make the center grip **hot** (Fig. 19-26). Cycle to the **SCALE** option. The scale factor is **1.5**. Since the *Circle* is a 1 unit diameter, it can be interactively scaled (watch the *coordinate display* display), or you can enter the value. The drawing is complete. ***Save*** the drawing as **SPACEPLT**.

**FIGURE 19-26**

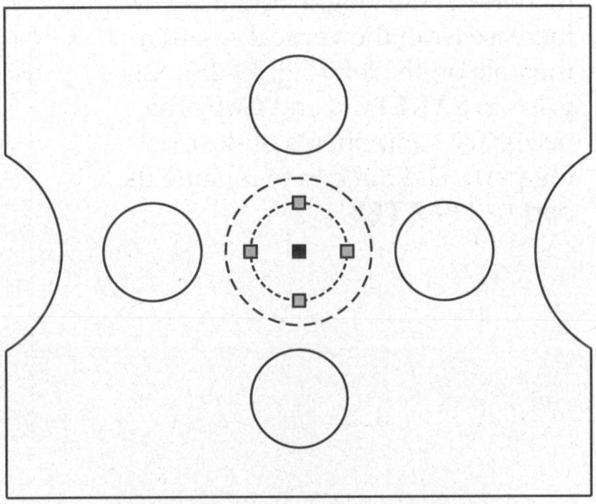

3. *Open* the **STORE ROOM2** drawing from Chapter 18 Exercises. Use the grips to **STRETCH**, **MOVE**, or **ROTATE** each text paragraph to achieve the results shown in Figure 19-27. ***SaveAs*** **STORE ROOM3**.

**FIGURE 19-27**

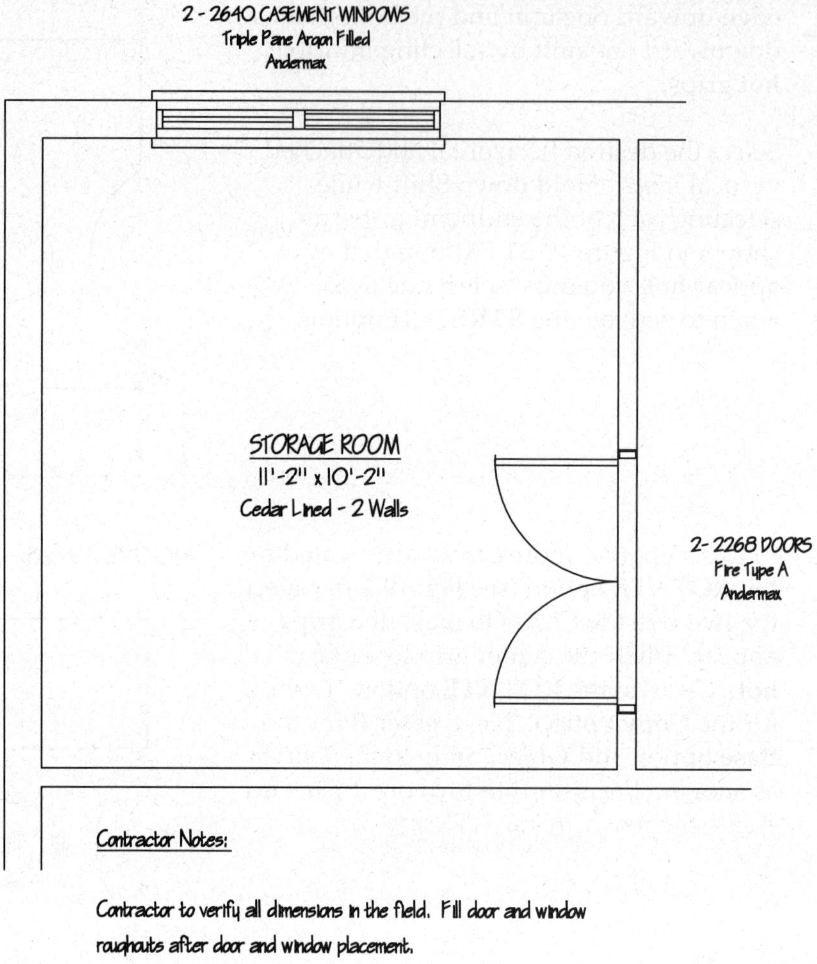


4. *Open* the **T-PLATE** drawing that you created in Chapter 9. An order has arrived for a modified version of the part. The new plate requires two new holes along the top, a 1" increase in the height, and a .5" increase from the vertical center to the hole on the left (Fig. 19-28). Use grips to **STRETCH** and **Copy** the necessary components of the existing part. Use ***SaveAs*** to rename the part to **TPLATEB**.

**FIGURE 19-28**

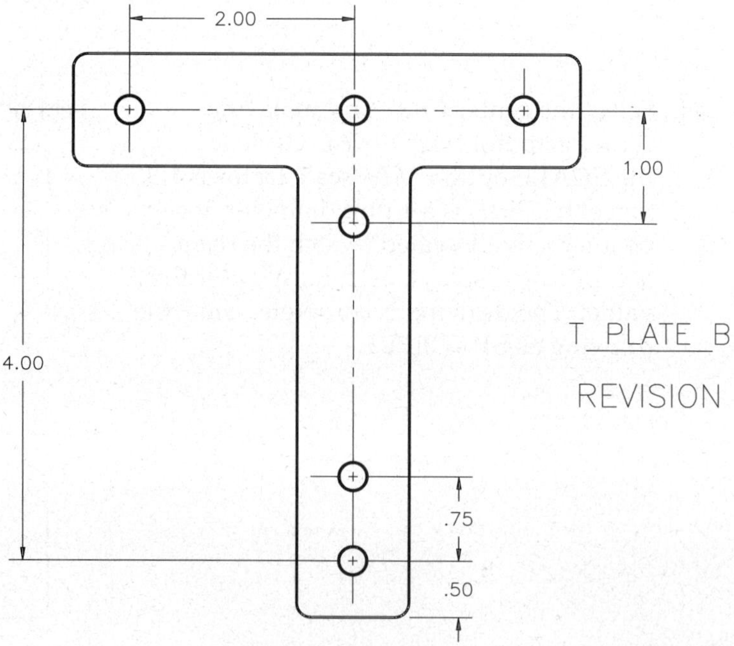

CHAPTER 20

# Advanced Selection Sets

### Chapter Objectives

After completing this chapter you should:

1. be able to control the system variables that affect object selection by using the *Options* dialog box or by changing the variables directly;

2. be able to use tools to enhance object selection, such as *Select Similar, Hide Objects,* and *Isolate Objects*;

3. be able to use *Qselect* to specify criteria for creating a selection set to use with an editing command;

4. be able to use *Object Selection Filters* in a complex drawing to find a selection set for use with an editing command;

5. be able to create named selection sets (Object Groups) with the *Group* command and use options in the *Group Manager* dialog box.

## CONCEPTS

When you "Select Objects:", you determine which objects in the drawing are affected by the subsequent editing action by specifying a selection set. You can select the set of objects in several ways. The fundamental methods of object selection (such as PICKing with the pickbox, a *Window*, a *Crossing Window*, etc.) are explained in Chapter 4, Selection Sets. This chapter deals with advanced methods of specifying selection sets and the variables that control your preferences for how selection methods operate. Specifically, this chapter explains:

Selection Set Variables
Object Filters
Object Groups

## SELECTION SET VARIABLES

Several variables allow you to customize the way you select objects. Keep in mind that selecting objects occurs only for editing commands; therefore, the variables discussed in this chapter affect how you select objects when you edit AutoCAD objects. The variable names and the related action are briefly explained here:

| Variable Name | Default Setting | Related Action |
|---|---|---|
| *PICKBOX* | 3 | Sets the object selection target height in pixels. |
| *PICKFIRST* | 1 | Controls whether you can select objects before you issue a command (noun-verb selection). |
| *PICKADD* | 2 | Determines whether selected objects are added to or replace the current selection set. |
| *PICKSTYLE* | 1 | Determines if all objects in a group are selected when you select one object in that group. Also determines which objects are selected when you select an associative hatch. |
| *PICKAUTO* | 5 | Determines how clicking and/or dragging control the windows and lassos. |
| *PICKDRAG* | 2 | Controls if you can use click-and-click and press-and-drag for selection windows. |

### Changing the Settings

Settings for the selection set variables can be changed in two ways:

1. You can type the variable name at the Command: prompt (just like an AutoCAD command) to change the setting.

2. Use the *Selection* tab of the *Options* dialog box (Fig. 20-1) to change the selection set variables listed above. The variables can be changed in this dialog box, but the syntax in the dialog box is not the same as the variable name. A check in the checkbox by each choice does not necessarily mean a setting of *ON* for the related variable.

When you change any of these variables, the setting is recorded for the user profile, rather than in the current drawing file as with most variable settings. In this way, the change (which is generally the personal preference of the operator) is established at the workstation, not the drawing file.

### *Options*

| Ribbon | Menu Bar | Command (Type) | Alias (Type) | Shortcut |
|---|---|---|---|---|
| ... | *Tools Options...* | *Options* | *OP* | (Default menu) *Options...* |

The *Selection* tab of the *Options* dialog box (Fig. 20-1) can be used to change the selection set variables listed previously. Although the checkboxes in the *Selection* tab do not indicate the formal variable names, they are intended to be more descriptive.

NOTE: It is advisable to turn off grips when you are experimenting with these variables to avoid confusion. Turn off grips by removing the check from *Show Grips* on the right side of the dialog box or by typing *GRIPS* at the command prompt and changing the setting to 0 (off).

**FIGURE 20-1**

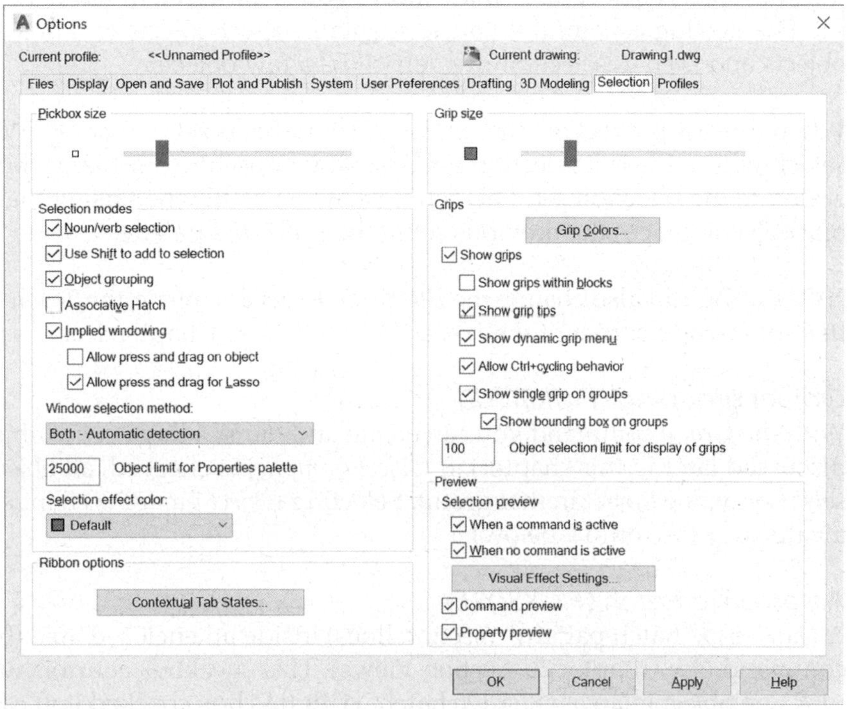


The explanations below describe the options in the *Pickbox size, Selection modes,* and *Preview* sections of the *Selection* tab of the *Options* dialog box. The options in the *Grips* section of the dialog box are explained in Chapter 19.

#### *Pickbox size*

This slider sets the size of the pickbox. The default value is 3 (3 pixels by 3 pixels square).

#### *Noun/Verb selection (PICKFIRST)*

A check in this box sets the *PICKFIRST* system variable to a value of 1, or turns on *Noun/Verb* selection. If *Noun/Verb* selection is on, you can select objects with the pickbox or window selection <u>before</u> using a command. This order of editing is called "Noun/Verb"; the objects are the nouns and the command is the verb. Noun/Verb editing is preferred by some users because you can decide what objects need to be changed, then decide how (what command) you want to change them.

If you select objects before issuing a command, you do not get a chance to "Select objects:" within the command. The selection set PICKed immediately before the command is used for the editing action. The command then passes through the "Select objects:" step to the next prompt in the sequence. Only the pickbox, window, crossing window, and lassos can be used for object selection with Noun/Verb editing. The other object selection methods (*ALL, Last, Previous, Fence, Window Polygon,* and *Crossing Polygon*) are only available when you are presented with the "Select objects:" prompt.

NOTE: You can use the *System Variable Monitor* to track changes to the *PICKADD, PICKAUTO,* and *PICKFIRST* variables since these three variables are already listed in the monitor. You can add other variables to the list. (See *Sysvarmonitor* in Chapter 17.)

NOTE: To disable the cursor pickbox, *PICKFIRST* and *GRIPS* must be Off.

### *Use Shift to add to selection (PICKADD)*

This checkbox controls whether subsequent selections replace the current selection set or add to it. Your choice changes the *PICKADD* system variable setting. *PICKADD* controls whether objects are ADDed to the selection set when they are PICKed or whether selected objects replace the last selection set. When this box is not checked, *PICKADD* is *On* (set to 1 or 2). With the default setting (no check), the selection set is cumulative; that is, each object PICKed is added to the current set. This mode also allows you to use multiple selection methods to build the set. Also with the default setting (when *PICKADD* is *On*), the Shift + left-click combination allows you to deselect, or remove, objects from the current selection set. Deselecting is helpful if you accidentally select objects or if it is easier in some situations to select all objects and then deselect (Shift + left-click) a few objects.

When the box is checked, the *PICKADD* variable is set to 0 or *Off*. With this setting, objects that you select replace the last selection set. Also, if you use the Shift key when selecting objects, those objects are added to the selection set. This setting can be useful when using the *Properties* palette so that selecting one object replaces the previous set in the *Properties* palette.

NOTE: You can also change the *PICKADD* variable using the *Toggle value of PICKADD Sysvar* button in the upper-right corner of the *Properties* palette (see Chapter 16).

### *Object grouping (PICKSTYLE)*

With the *Group* command you can create and name a large set of objects for easier selection (groups are discussed later in this chapter). If *Object grouping* is checked, all objects in a group are selected when you select only one object in that group. Selecting *Object grouping* changes the *PICKSTYLE* system variable to a value of 1 (see NOTE below).

### *Associative Hatch (PICKSTYLE)*

A *Hatch* is a "hatch pattern" (section lines) inside an enclosed area (boundary) created using the *Hatch* command (see Chapter 26, Section Views). This checkbox controls what objects are selected when you pick one object in an associative hatch. With this box checked you can select the entire hatch pattern and the enclosing boundary by picking only one line in the pattern. Checking this box sets the *PICKSTYLE* system variable to 2 (see NOTE below).

NOTE: If both the *Object grouping* and *Associative Hatch* boxes are checked, *PICKSTYLE* is set to 3. If neither box is checked, *PICKSTYLE* is set to 0.

### *Implied windowing (PICKAUTO)*

If this box is not checked, you cannot select objects using any automatic windows or lassos—you have to type in the "W" for *Window* or "CW" for *Crossing Window*, for example, at the "Select objects:" prompt. With this box checked, window and crossing window selection is automatic—drawing the *AUTO* selection window from left to right selects objects that are entirely inside the window's boundaries and drawing from right to left selects objects within and crossing the window's boundaries. This box must also be checked if you want to use lassos. This option affects the *PICKAUTO* system variable setting.

*Allow press and drag on object*

When this box is checked, you can land directly on an object with the pickbox and begin creating a selection window using the press and drag method. If you click (not press) the object is selected. The type of window that is created with the press and drag action (window or lasso) is determined by the *Allow press and drag for Lasso* box setting. This option also affects the *PICKAUTO* system variable setting.

*Allow press and drag for Lasso*

If the *Allow press and drag for Lasso* box is checked, press and drag initiates a lasso window. This option also affects the *PICKAUTO* system variable setting.

*Window selection method (PICKDRAG)*
If you want to use both normal selection windows and lassos, select *Both – Automatic detection* from the drop-down list. This option sets the *PICKDRAG* system variable setting to 2.

## OBJECT SELECTION TOOLS AND FILTERS

Large and complex AutoCAD drawings may require advanced methods of specifying a selection set. For example, consider working with a drawing of a manufacturing plant layout and having to select all of the doors on all floors, or having to select all metal washers of less than 1/2" diameter in a complex mechanical assembly drawing. Rather than spend several minutes PICKing objects, it may be more efficient to use one of the object selection tools or object filters that AutoCAD provides.

If you want to make particular objects you are currently working with "stand out," you can do this by isolating these objects. The *Isolateobjects* command makes all but the selected objects temporarily invisible. Use *Unisolate* to reinstate the visibility. You can also select specific objects to make temporarily invisible with the *Hideobjects* command, then *Unisolate* later.

You can use the *Selall* command to select all the objects in model space or in the layout. However, if you want to easily select objects that are similar, but not have to input specific search criteria, use *Selectsimilar*. *Selectsimilar* automatically finds objects of the same type (line, circle, etc.) with the same properties (color, layer, etc.).

If you want to select different types of objects (lines, circles, and text, for example) that share similar properties like color or layer, AutoCAD provides two robust selection filters: *Quick Select* and *Filter*. Both object selection filters allow you to specify a criteria based on object type and/or object property. You specify what type of object (*Line*, *Arc*, *Mtext*, etc.) and/or what property (*Layer*, *Color*, *Linetype*, etc.) and AutoCAD searches (applies the filter to) the drawing to find objects that match the criteria. Once found, the objects can be used for an editing action, such as *Move*, *Copy*, *Erase*, or change properties.

### Selall

| Ribbon | Menu Bar | Command (Type) | Alias (Type) | Shortcut |
|---|---|---|---|---|
| *Home Utilities* (icon) | *Edit Select All* | *Ai_Selall* | *Selall* | Ctrl+A |

If you want to select all objects in the current space (model or layout tab), use *Select All*. *Select All* does not select objects that are on layers that are *Frozen* but does select objects on layers that are *Off*. This command is not transparent, so you have to use *Select All* first, and then use the desired command to operate on the objects. If you want to use a command, then select all the objects, type *All* at the "Select objects" prompt.

### Selectsimiliar

| Ribbon | Menu Bar | Command (Type) | Alias (Type) | Shortcut |
|---|---|---|---|---|
| ... | ... | *Selectsimilar* | ... | (Edit Mode) *Select Similar* |

*Selectsimilar* allows you to automatically select objects that are similar based on the object type, such as a *Line* or *Circle*, and the same properties such as *Layer* and *Color*. The selection criteria for shared properties are pre-set parameters that you can change using the *Settings* option.

*Selectsimilar* cannot be used transparently (during another command). You can first select an object, then right-click for the Edit Mode shortcut menu and choose *Selectsimilar* and all similar objects are automatically selected and the command ends. Alternately, first invoke *Selectsimilar*, select an object, then press Enter and all similar objects are automatically selected as listed below.

```
Command: SELECTSIMILAR
Select objects or [SEttings]: 1 found
Select objects or [SEttings]: Enter
Command:
```

Only objects of the same type, such as a *Line*, a *Circle*, a *Pline*, etc., are considered similar. You can change other shared properties with the *Settings* option. The *Settings* option produces the *Select Similar Settings* dialog box (Fig. 20-2). Here you can specify which properties are used as criteria in the search. The default settings are *Name* and *Layer*. *Name* considers referenced objects with matching names (such as blocks, xrefs, and images) to be similar. *Object style* considers objects with matching styles (such as text styles, dimension styles, and table styles) to be similar.

**FIGURE 20-2**

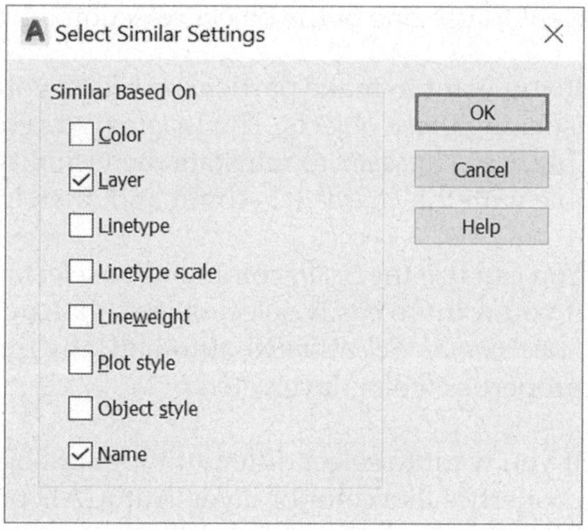


NOTE: You must type the *Selectsimilar* command at the Command line to access the *Settings* option.

TIP

NOTE: You cannot use *Selectsimilar* to select objects of different object types (*Lines, Circles*, etc.). If you want to select different types of objects that share properties like color or linetype, use *Quick Select*.

### Hideobjects

| Ribbon | Menu Bar | Command (Type) | Alias (Type) | Shortcut |
|---|---|---|---|---|
| ... | *Tools Isolate > Hide Objects* | *Hideobjects* | ... | (Edit Mode) *Isolate > Hide Objects* |

*Hideobjects* is useful if you want to make selected objects temporarily invisible. For example, it may be helpful to temporarily "remove" a set of objects to simplify editing other objects in a congested area of the drawing. This idea is similar to *Freezing* layers; however, with *Hideobjects* you can select particular objects to hide regardless of the objects' layers or other properties. Simply invoke the command and select the objects to hide.

```
Command: HIDEOBJECTS
Select objects: PICK
Select objects: Enter
Command:
```

Use *Unisolateobjects* to make all hidden objects in the view visible again.

**FIGURE 20-3**

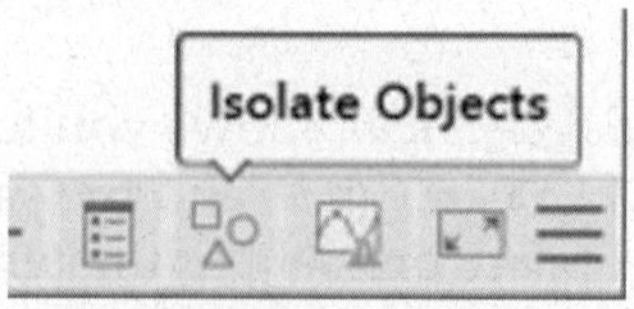


You can use the *Isolate Objects* button on the Status Bar to access both the *Hideobjects* and *Isolateobjects* commands (Fig 20-3). Use the *Customization* button (far right) to activate the *Isolate Objects* button.

### *Unisolateobjects*

| Ribbon | Menu Bar | Command (Type) | Alias (Type) | Shortcut |
|---|---|---|---|---|
| ... | *Tools Isolate > End Object Isolation* | *Unisolateobjects* | *Unhide* | (Edit Mode) *Isolate > End Object Isolation* |

Use *Unisolateobjects* to make all objects visible again that were temporarily hidden from view by the *Hideobjects* or *Isolateobjects* commands. There are no options for this command. Issue the command and all previously hidden or isolated objects in the drawing become visible. By default, when you close and reopen a drawing, all previously hidden objects are displayed (based on the setting of the *OBJECTISOLATIONMODE* system variable).

Command: ***UNISOLATEOBJECTS***
*nn* object(s) unisolated.
Command:

### *Isolateobjects*

| Ribbon | Menu Bar | Command (Type) | Alias (Type) | Shortcut |
|---|---|---|---|---|
| ... | *Tools Isolate > Isolate Objects* | *Isolateobjects* | *Isolate* | (Edit Mode) *Isolate > Isolate Objects* |

*Isolateobjects* has a similar application to the *Hideobjects* command; however, *Isolateobjects* hides all objects in the drawing view except those you select. Invoke the *Isolateobjects* command and select the objects you want to remain visible. All other objects in the view become temporarily invisible until *Unisolateobjects* is issued to reinstate the visibility of all previously hidden or isolated objects.

Command: ***ISOLATEOBJECTS***
Select objects: **PICK**
Select objects: **Enter**
Command:

### *Qselect*

| Ribbon | Menu Bar | Command (Type) | Alias (Type) | Shortcut |
|---|---|---|---|---|
| *Home Utilities* (icon) | *Tools Quick Select...* | *Qselect* | ... | (Default Menu) *Quick Select* |

With *Qselect*, you create a selection set of all objects that either match or do not match the object type and object property criteria that you specify. You can search the entire drawing or an existing selection set. If you use *Qselect* with a partially opened drawing, objects that are not loaded are not considered in the search. When the *OK* button is pressed in the *Quick Select* dialog box, AutoCAD finds and highlights the objects matching the criteria. The objects become the current selection set so any editing command used immediately after *Qselect* (by Noun/Verb selection) automatically finds the matching selection set. *Qselect* is the simpler of the two available object selection filters. Although you cannot name and save selection sets with *Qselect*, you can search and find objects by *Object type* and *Properties* much easier than using the *Object Selection Filters* dialog box (explained in detail in the following section).

*Qselect* can be invoked by several methods listed in the command table above, including choosing *Quick Select* from the shortcut menu that appears when you right-click in the drawing area. The *Qselect* button also appears in the *Properties* palette (see Chapter 16). *Qselect* produces the *Quick Select* dialog box (Fig. 20-4).

**FIGURE 20-4**

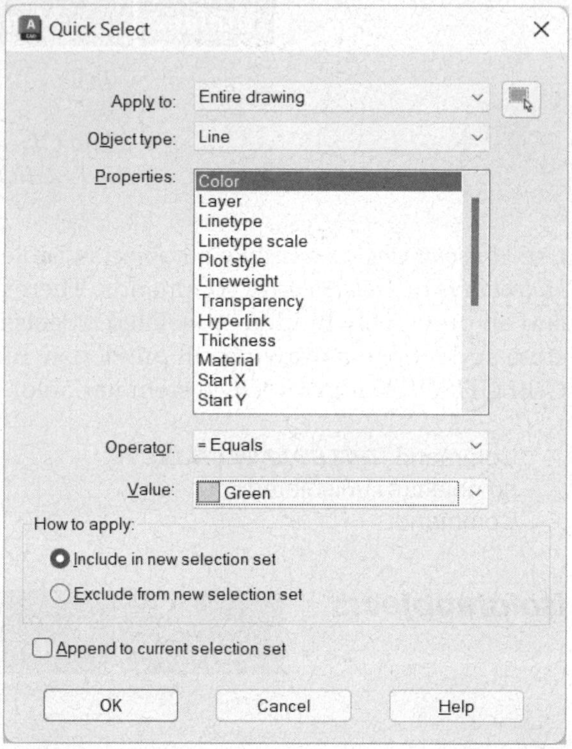


To use the *Quick Select* dialog box, follow these steps:

1. Invoke the *Quick Select* dialog box by any method described above.
2. Specify the *Entire Drawing* to search or use the *Select Object* button to specify a smaller search area.
3. If you want to find one *Object Type*, select it from the drop-down list; otherwise *Multiple* will be used.
4. Select the desired *Property* (only one) to search for (unless you need only to search by *Object Type*).
5. Specify the *Operator* and *Value*.
6. Specify if you want to *Include in New Selection Set* (search for the object type and property you specified) or *Exclude from New Selection Set* (search for everything but the object type and property you specified).
7. Press *OK*. AutoCAD highlights all objects that match the criteria.
8. Use an editing command to act on the highlighted objects (*Move*, *Erase*, change properties, etc.)

For example, assume the criteria shown in Figure 20-4 were applied to a process flow diagram, AutoCAD would find and highlight all *Text* objects in the drawing whose *Color Equals Green* (Fig. 20-5). Note the grips (small squares) that appear on the highlighted objects when grips are enabled.

**FIGURE 20-5**

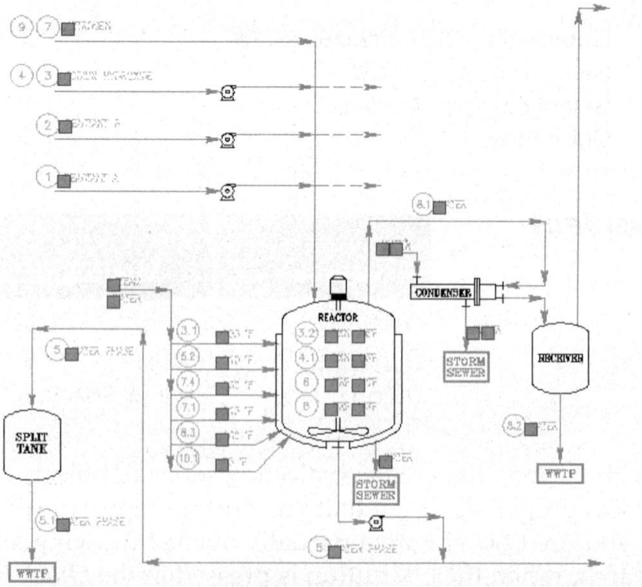


The specific areas of the dialog box are explained as follows.

### *Apply to*

Specify whether you want to apply the filtering criteria to the entire drawing or the current selection set. If there is no current selection set when you invoke *Qselect*, *Entire Drawing* is the default value. Use the *Select Objects* button to return to the drawing and create a selection set.

### *Select Objects*

The *Select Objects* button (upper right corner) temporarily returns to the drawing so you can select the objects on which you want to perform the *Quick Select* operation. AutoCAD uses the new selection set as the area to search for objects matching the specified criteria.

### *Object Type*

Specifies the object type you want to search for (filter). The default is *Multiple*; therefore, if no previous selection set exists, the *Object Type* list includes all and only object types available in the current drawing. For example, if only *Lines* and *Circles* have been created in the drawing, only *Line* and *Circle* object types appear in the list. If a selection set exists, the list includes only the object types of the selected objects.

### *Properties*

This field is specific to the *Object Type* selected in the list above. The *Properties* list includes all searchable properties for the selected object type. For some object types, the list of properties can be very long; for example, dimension objects contain a property field for each dimensioning variable. This feature gives you considerable power. For example, you could use this tool to highlight all *Linear Dimensions* that have a specific type of *Tolerance Display* or find all *Text* that matches a specific *Height* value. The sort order for the properties list (alphabetic or categorized) is based on the current sort order in the *Properties* palette (see "*Properties*" in Chapter 16). The property you select determines the options available in the *Operator* and *Value* boxes.

### *Operator*

The *Operator* that you select controls if the search *Equals*, is *Not Equal To*, *Greater Than*, or *Less Than* the text string or numerical value in the *Value* field. Options in the *Operator* field vary depending on the *Property*. For example, *Greater Than* and *Less Than* are not available for some properties.

### *Value*

Possibilities in this field depend on the selected object and property. If known values for the selected property are stored in the drawing, the *Value* box becomes a drop-down list from which you can choose an available value. For example, if *Color* is selected in the *Properties* list, *ByLayer*, *ByBlock*, *Red*, *Yellow*, and so on appear in the list. If no choices appear in the *Value* field, input a text string or numerical value.

### *How To Apply*

Your choice here specifies whether you want the new selection set to include or exclude objects that match the specified filtering criteria. Choose *Include in new selection set* if you want to create a new selection set composed only of objects that match the filtering criteria. Choose *Exclude from new selection set* to create a new selection set composed of all objects that do not match the filtering criteria.

### *OK*

After specifying the criteria for the objects to search in the fields described above, select *OK*. All objects matching (or not matching) the criteria set in the *Quick Select* dialog box become highlighted in the current drawing. These objects are treated as if you had selected them by the pickbox, window, or other selection method. Assuming Noun/Verb selection is enabled (*PICKFIRST*=1), the next editing command issued automatically finds the *Qselect* objects.

### *Append to Current Selection Set*

Although you can only select one specific object type and one property at a time, you can use *Qselect* repeatedly to search for multiple object types and properties. Do this by specifying one set of criteria, then select *OK*. AutoCAD highlights all objects matching that criteria. Next, while the objects are highlighted, right-click to select *Qselect* again. Specify a second set of criteria and include *Append to current selection set*. When you press *OK*, AutoCAD highlights objects matching the new criteria and "adds" them to the original highlighted set.

## Filter

| Ribbon | Menu Bar | Command (Type) | Alias (Type) | Shortcut |
|---|---|---|---|---|
| ... | ... | *Filter* | *FI* | ... |

The *Object Selection Filters* dialog box is a more advanced version of *Qselect*. Here you can specify multiple selection criteria and save a criteria set under a name that you assign for use at a later time. You can also invoke *Filter* during a command (transparently) to select objects for use with the current command. Using the *Filter* command by any method produces the *Object Selection Filters* dialog box. Figure 20-6 displays the dialog box when it first appears. The *Select Filter* cluster near the left center is where you specify the search criteria. After you designate the filters and values, use *Add to List* to cause your choices to appear in the large area at the top of the dialog box. All filters appearing at the top of the box are applied to the drawing when *Apply* is selected.

**FIGURE 20-6**

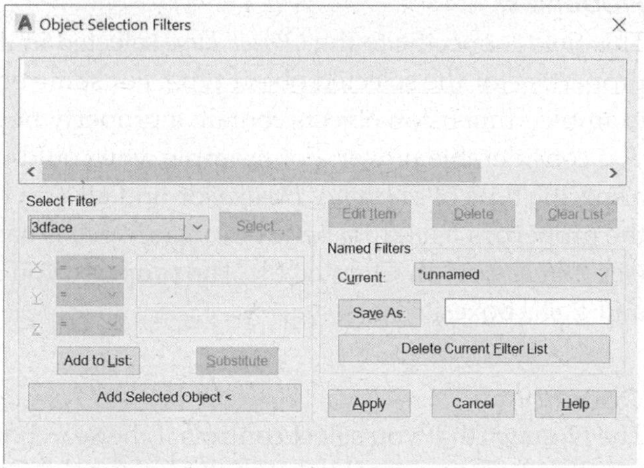


### *Select Filter* (Drop-Down List)

Activating the drop-down list reveals the possible selection criteria (filters) that you can use, such as *Arc* or *Layer*. You can select one or more of the selection filters shown in the following list. You can also specify a set of values that applies to each filter, for example, *Arcs* with a *radius* of less than 1.00" or *Layers* that begin with "AR."

3dface
Arc
Arc Center
Arc Radius
Attribute
Attribute Position
Attribute Tag
Block
Block Name
Block Position
Block Rotation
Body
Camera
Circle
Circle Center
Circle Radius
Color
Dimension
Dimension Style
DWF Underlay
Elevation
Ellipse
Ellipse Center
Hatch
Hatch Pattern Name
Helix
Image
Image Position
Layer
Leader
Light
Line
Line End
Line Start
Linetype
Linetype Scale
Material
Mesh
Multileader
Multiline
Multiline Style
Normal Vector
Point
Point Position
Polyline
Position Marker
Ray
Region
Section Plane
Shape
Shape Name
Shape Position
Solid
Solid Body
Spline
Surface (12 types)
Table
Text
Text Height
Text Position
Text Rotation
Text Style Name
Text Value
Thickness
Tolerance
Trace
Transparency
Viewport
Viewport Center
Xdata ID
Xline

The list also provides a series of typical database grouping operators which include AND, OR, XOR, and NOT. For example, you may want to select all *Text* in a drawing but *NOT* the *Text* in a specific text style.

### *Select*

If you select a filter that has multiple values, such as *Layer*, *Block Name*, and *Text Style Name*, the *Select* button is activated and enables you to select a value from the list (Fig. 20-7). If the *Select* button is grayed out, no existing values are available.

**FIGURE 20-7**

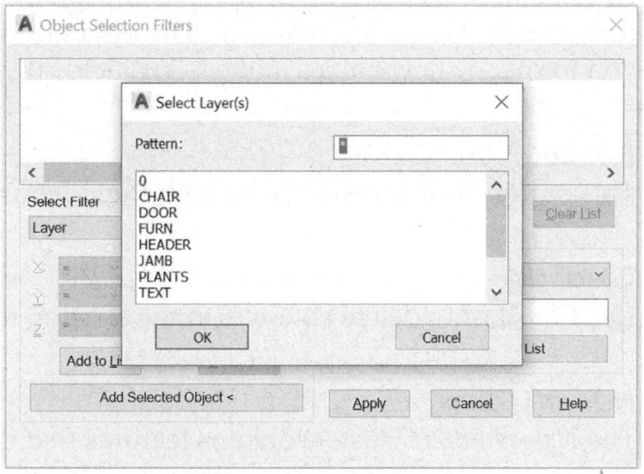


### *X*

If you select a filter that requires alphanumeric input (numbers or text strings), the *X* field is used. For example, if you selected *Arc Radius* as a filter, you would select the operator (=, <, >, etc.) from the first drop-down list and key in a numeric value in the edit box (Fig. 20-8). Other examples requiring numeric input are *Circle Radius*, *Text Rotation*, and *Text Height*.

**FIGURE 20-8**

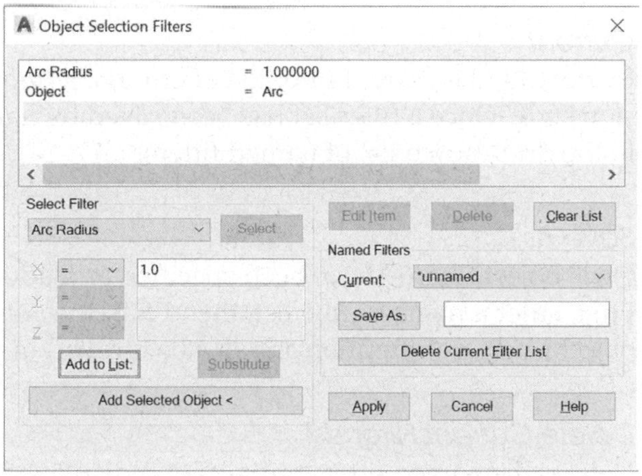


If you select a filter that requires a text string, the string is entered in the *X* field edit box. Examples are *Attribute Tag* and *Text Value*.

The *X* field is also used (in conjunction with the *Y* and *Z* fields) for filters that require coordinate input in an *X,Y,Z* format. Example filters include *Arc Center*, *Block Position*, and *Line Start*.

### *Y,Z*

These fields are used for coordinate entry of the *Y* and *Z* coordinates when needed as one of the criteria for the filter (see "*X*" above). Additionally, you may use standard database relational operators such as equal, less than, or greater than, to more effectively specify the desired value.

### *Add to List*

Use the *Add to List* button after selecting a filter to force the new selection to appear in the list at the top of the dialog box.

### *Substitute*

After choosing a value from the *Select* listing or entering values for *X*, *Y*, and *Z*, *Substitute* replaces the value of the highlighted item in the list (at the top of the dialog box) with the new choice or value.

### *Add Selected Object*

Use this button to select an object from the drawing that you want to include in the filter list.

### Edit Item

Once an item is added to the list, you must use *Edit Item* to change it. First, highlight the item in the filter list (on top), then select *Edit Item*. The filter and values then appear in the *Select Filter* cluster ready for you to specify new values in the *X,Y,Z* fields or you can select a new filter by using the *Select* button.

### Delete

This button simply deletes the highlighted item in the filter list.

### Clear List

Use *Clear List* to delete all filters in the existing filter list.

### Named Filters

The *Named Filters* cluster contains features that enable you to save the current filter configuration to a file for future use with the current drawing. This eliminates the need to rebuild a selection filter used previously.

### Current

By default, the current filter list of criteria is **unnamed*, in much the same fashion as the default dimension style. Once a filter list has been saved by name, it appears as a selectable object filter configuration in the drop-down list of named filters.

### Save As

Once object filter(s) have been specified and added to the list (on top), you can save the list by name. First, enter a name for the list, then PICK the *Save As* button. The list is automatically set as the current object filter configuration and can be recalled in the future for use in the drawing.

### Delete Current Filter List

Deletes the named filter configuration displayed in the *Current* field.

### Apply

TIP

The *Apply* button applies the current filter list to the current AutoCAD drawing. Using *Apply* causes the dialog box to disappear and the "Select objects:" prompt to appear. You must specify a selection set (by any method) to be tested against the filter criteria. If you want the filters to apply to the entire drawing (except *Locked* and *Frozen* layers), enter *All*. If you only need to search a smaller area of the drawing, PICK objects or use a window or other selection method.

TIP

The application of the current filters can be accomplished in two ways:

1. Within a command (transparently)
   At the "Select objects:" prompt, you can invoke *Filter* transparently by selecting from a menu or typing "*'filter*" (prefaced by an apostrophe). Then set the desired filters, select *Apply*, and specify the selection set to test against the criteria. The resulting (filtered) selection set is used for the current command.

2. At the Command prompt
   From the Command: prompt (when no command is in use), type or select *Filter*, specify the filter configuration, and then use *Apply*. You are prompted to "Select objects:". Specify *All* or use a selection method to specify the area of the drawing for the filter to be applied against. Press Enter to complete the operation. The filtered selection set is stored in the selection set buffer and can be recalled by using the *Previous* option in response to the next "Select objects:" prompt.

## OBJECT GROUPS

FIGURE 20-9

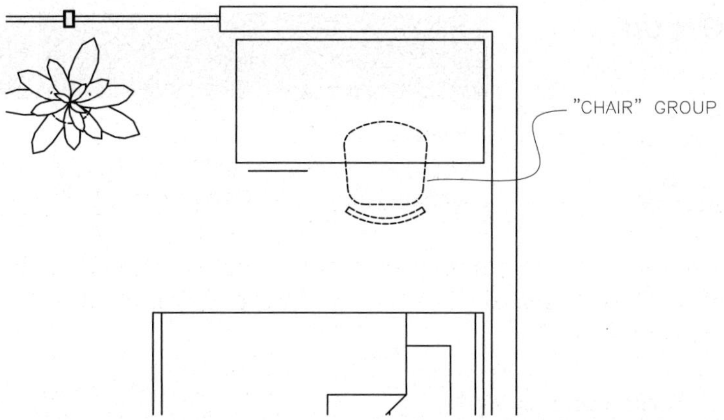


Often, a group of objects that are related in some way in a drawing may require an editing action. For example, all objects (*Lines* and *Arcs*) representing a chair in a floor plan may need to be selected for *Copying* or changing color or changing to another layer (Fig. 20-9). Or you may have to make several manipulations of all the fasteners in a mechanical drawing or all of the data points in a civil engineering drawing. In complex drawings containing a large number of objects, the process of building such a selection set (with the pickbox, window, or other methods) may take considerable effort. One disadvantage of the traditional selection process (without *Groups*) is that selection sets could not be saved and recalled for use at a later time. AutoCAD introduced the *Group* command to identify and organize named selection sets.

A Group is a set of objects that has an assigned name and description. The *Group* command allows you to determine the objects you want to include in the group and to assign a name and description. Usually the objects chosen to participate in a group relate to each other in some manner. Objects can be members of more than one group.

Once a group is created, it is assigned a *Selectable* status by default. If a group is *Selectable*, the entire group can be automatically selected (highlighted) at any "Select objects:" prompt by PICKing any individual member, or by typing the word "group" or the letter "G," then giving the group name. For example, if all the objects in a mechanical drawing representing fasteners were assigned to a group named "fasten," the group could be selected during the *Move* command as follows:

Command: ***MOVE***
Select objects: ***g***
Enter group name: **fasten**
18 found
Select objects: **Enter**
Specify base point or displacement:
etc.

If the group is assigned a nonselectable status, the group cannot be selected as a whole by any method. Nonselectablility prevents accidentally *Copying* the entire group, for example. If a group is assigned a selectable status, the group can be manipulated as a whole or each member can be edited individually. The *PICKSTYLE* variable controls whether PICKing highlights the individual member or the entire group. In other words, you can PICK individual members if *PICKSTYLE*=0 or 2 or an entire selectable group if *PICKSTYLE*=1 or 3. The *Group Selection On/Off* button or the Ctrl+H key sequence toggles the *PICKSTYLE* variable to 2 (*Off*) or 3 (*On*).

A group can be thought of as a set of objects that has a level of distinction beyond that of a typical selection set (specified "on the fly" by the pickbox, window, or other method), but not as formal as a *Block*. A *Block* is a group of objects that is combined into one object (using the *Block* command), but individual entities cannot be edited separately. (Blocks are discussed in Chapter 21.) Using groups for effective drawing organization is similar to good layer control; however, with groups you have the versatility to allow members of a group to reside on different layers.

To summarize, using groups involves two basic activities: (1) use the *Group* command to define the objects and assign a name and (2) at any later time, locate (highlight) the group for editing action at the "Select objects:" prompt by PICKing a member or typing "G" and giving its name.

### Group

| Ribbon | Menu Bar | Command (Type) | Alias (Type) | Shortcut |
|---|---|---|---|---|
| *Home* *Groups* *Group* | *Tools* *Group* | *Group* | G or -G | ... |

Use the *Group* command to create groups. Using *Group* by any method shown in the command table above, you can assign a name and description for the group and specify the members (objects) for the group.

```
Command: GROUP
Select objects or [Name/Description]:
```

You can select the objects you want to comprise the group or enter a name or description in any order. The options are as follows:

#### *Select Objects*

Use any selection method to pick objects you wish to be group members.

#### *Name*

This option is used to enter the name of a new group that you want to create. You can enter the desired name before choosing the set of objects.

In order to better utilize group capabilities and organize your drawing, you should name groups that you create. If you create an unnamed group, AutoCAD automatically assigns a default name such as *A*n*, where *n* is the number of the group created since the beginning of the drawing; however, this name appears only in the *Group Manager*. You can *Rename* groups with the *Groupedit* command or name unnamed groups with the *Group Manager*.

#### *Description*

Including a *Description* for new groups is optional, but this feature is useful in organizing various selection sets in a drawing. A higher level of distinction is added to the groups if a description is included. The description may be up to 256 characters. If a group name is selected from the list in the *Group Manager*, both the group name and description appear in the *Group Identification* cluster.

#### *PICKSTYLE* System Variable and Group Selectability

Newly created groups are *Selectable* by default. A selectable group can be selected as a whole by two methods. When prompted to "Select objects:" during an editing command, you can (1) PICK any individual member of a selectable group or (2) enter "G" and the group name. The *PICKSTYLE* settings are as follows:

| | | |
|---|---|---|
| 0 or 2 | *Off* | The individual members are selectable. If subgroups exist, the subgroups are selectable. |
| 1 or 3 | *On* | The entire group is selectable. |

When *PICKSTYLE* is set to 1 or 3 (the *Group Selectability On/Off* button is *On*), PICKing any member of a selectable group highlights (selects) the entire group. (Settings 2 and 3 affect associative hatches.)

If *PICKSTYLE* is set to 0 or 2 (the *Group Selectability On/Off* button is Off), PICKing any member of a selectable group highlights only that member. You can also use the Ctrl+H key sequence to toggle the *PICKSTYLE* variable setting to 1 or 3 (*Group on*) or 0 or 2 (*Group off*).

*PICKSTYLE* has no effect on nonselectable groups. Existing groups can be made *Unselectable* using the *Group Manager.* A group that is not selectable cannot be selected as a whole.

### Groupedit

| Ribbon | Menu Bar | Command (Type) | Alias (Type) | Shortcut |
|---|---|---|---|---|
| *Home* *Groups* (icon) | ... | *Groupedit* | ... | (Edit Mode) *Group* > *Add* or *Remove* |

Use the *Groupedit* command to add members, remove members, or rename the group.

Command: ***GROUPEDIT***
Select group or [Name]: **PICK** (a group or enter a group name)
Enter an option [Add objects/Remove objects/REName]:

#### *Remove*
The *Remove* option allows you to remove any member of the specified group. You are prompted to "Remove objects." If all members of a group are removed, the object group name still remains defined in the list box of the *Group Manager*.

#### *Add*
Use this option to add new members to the specified group. The "Select objects:" prompt appears.

#### *Rename*
Change the name of the selected group at the following prompt or type "?" to list all the group names.

Enter a new name for group or [?]:

### Ungroup

| Ribbon | Menu Bar | Command (Type) | Alias (Type) | Shortcut |
|---|---|---|---|---|
| *Home* *Groups* (icon) | *Tools* *Ungroup* | *Ungroup* | ... | (Edit Mode) *Group* > *Ungroup* |

The *Ungroup* command reverses the action of the *Group* command. Use *Ungroup* to explode or "break down" the group into its original individual objects.

Command: ***UNGROUP***
Select group or [Name]: **PICK**
Group XXXXX exploded.
Command:

If the object you name or pick is comprised of nested groups, you have the option of accepting or not accepting the "explode." *Accept* explodes the highest level of groups. You can use the *Next* option to have AutoCAD highlight and cycle through the nested groups so you can explode a subgroup.

Command: ***UNGROUP***
Select group or [Name]: **PICK**
Object is a member of more than one group [Accept/Next] <Accept>

### *Group Bounding Box On/Off (GROUPDISPLAYMODE)*

The *Group Bounding Box On/Off* button (on the Ribbon's *Home* tab, *Groups* panel) turns a bounding box on or off for groups. If *On*, a rectangular box appears around the group when the group is selected. If grips are turned on, a single grip appears at the group's center. The button toggles the *GROUPDISPLAYMODE* system variable to a value of either 0 (off) or 2 (on). If *Grips* are turned *On*, the *GROUPDISPLAYMODE* settings are as follows:

0 Bounding box is off, grips are displayed on all objects in the selected group
1 Bounding box is off, a single grip is displayed at the center of the grouped objects
2 Displays the group bounding box with a single grip at the center

## Group Examples

Imagine that you have the task to lay out the facilities for a college dormitory. Using the floor plan provided by the architect, you begin to go through the process of drawing furniture items to be included in a typical room, such as a desk, chair, and accessory furnishings. The drawing may look like that in Figure 20-10.

**FIGURE 20-10**

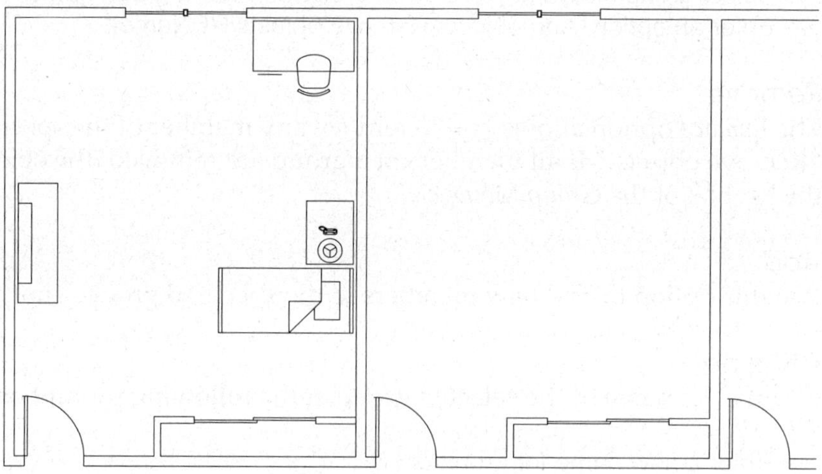

Since each room is to house two students, all objects representing each furniture item must be *Copied*, *Mirrored*, or otherwise manipulated several times to complete the layout for the entire dorm. To *Copy* the chair, for example, by the traditional method (without creating groups), each of the 12 objects (*Lines* and *Arcs*) comprising the chair must be selected each time the chair is manipulated. Instead, combining the individual objects into a group named "CHAIR" would make selection of these items for each operation much easier.

To make the CHAIR group, the *Group* command is used. The CHAIR group is created by selecting all *Lines* and *Arcs* as members (see previous Fig. 20-9).

Now, any time you want to *Copy* the chair, at the "Select objects:" prompt enter the letter "G" and "CHAIR" to cause the entire group to be highlighted. Alternately, the CHAIR group can be selected by PICKing any *Line* or *Arc* on the chair since the group is selectable.

The same procedure is used to combine several objects into a group for each furniture item. When complete, the complete suite of named groups representing each furniture item is BED, CHAIR, DESK, DRESSER, and TABLE. To simplify creating a complete dorm room, the CHAIR and TABLE are combined into one group named CHAIRDESK-SET. Next, the CHAIRDESK-SET is copied since two desks and chairs are needed in the room. This action creates a nested group. Also, the BED is *Copied*. *Copying* a group creates an unnamed group. Therefore, two unnamed groups now exist in the room: the BED copy and the CHAIRDESK-SET copy.

The next step in the drawing is to lay out several other rooms. Assuming that other rooms will have the same arrangement of furniture, another group could be made of the entire room layout. (Objects can be members of more than one group.) A group called ROOM is created in the same manner as before, and all objects that are part of furniture items are selected to be members of the group. If the ROOM group is made selectable, all items can be highlighted with one PICK and copied to the other rooms (Fig. 20-11).

**FIGURE 20-11**

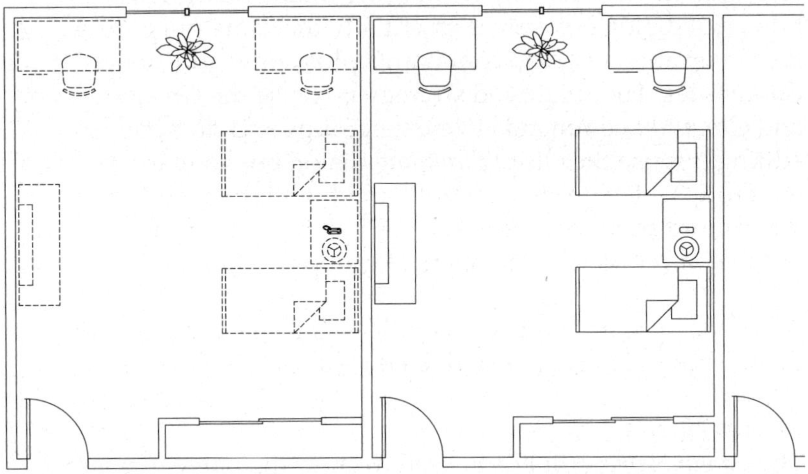

After submitting the layout to the client for review, a request is made for you to present alternative room layouts. Going back to the drawing, you discover that each time you select an individual furniture item to manipulate, the entire ROOM group is highlighted. To remedy the situation, the *Group Selectability On/Off* button is toggled *Off*. Now the individual furniture items can be selected since the original groups (CHAIR, DESK, BED, etc.) are still selectable (Fig. 20-12).

**FIGURE 20-12**

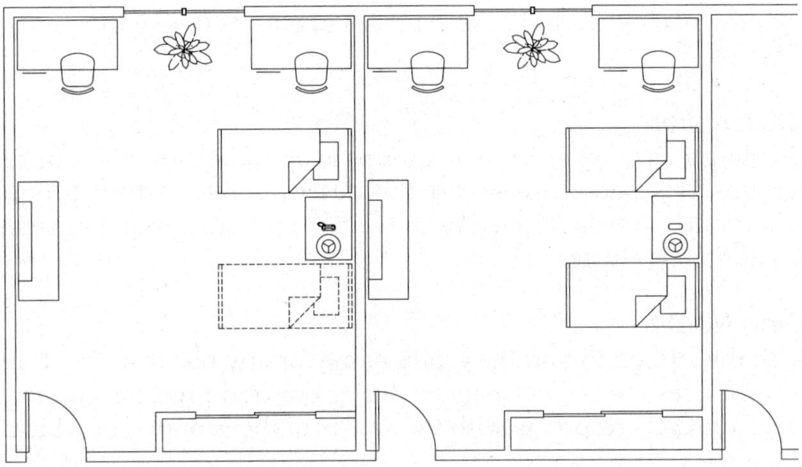

TIP

Another, more typical application is the use of *Groups* in conjunction with *Blocks*. A block is a set of objects combined into one object with the *Block* command. A block always behaves as one object. In our dorm room layout, each furniture item could be *Inserted* as a previously defined block. The entire set of blocks (bed, chair, desk, etc.) could be combined into a group called ROOM. In this case, the ROOM group would be set to *Selectable*. This would allow you to select all furniture items as a group by PICKing only one. However, if you wanted to select only one furniture item (Block), use Ctrl+H or the *Group Selectability* button to change *PICKSTYLE* to 0 or 2. In this way, individual members of any group (blocks in this case) can be PICKed without having to change the *Selectable* status. See Chapter 21 for more information on *Blocks*.

## Classicgroup

| Ribbon | Menu Bar | Command (Type) | Alias (Type) | Shortcut |
|---|---|---|---|---|
| *Home Groups Group Manager* | ... | *Classicgroup* | ... | ... |

Using the *Group Manager* button or typing the *Classicgroup* command produces the *Object Grouping* dialog box (see Figure 20-13, on the next page).

(Typing "*-group*" at the command line produces the command line equivalent to the *Object Grouping* dialog box.) This powerful dialog box serves essentially the same purpose as the *Group* and *Groupedit* commands plus a few additional capabilities. The suggested strategy is to use the *Group* and *Groupedit* commands to create groups and do basic editing but use the *Object Grouping* dialog box to manage the groups—list groups, identify and highlight groups, and other more advanced tasks such as making groups nonselectable or reordering members of groups.

All of the options and details of the *Group Identification* and *Create Group* clusters are explained next.

**FIGURE 20-13**

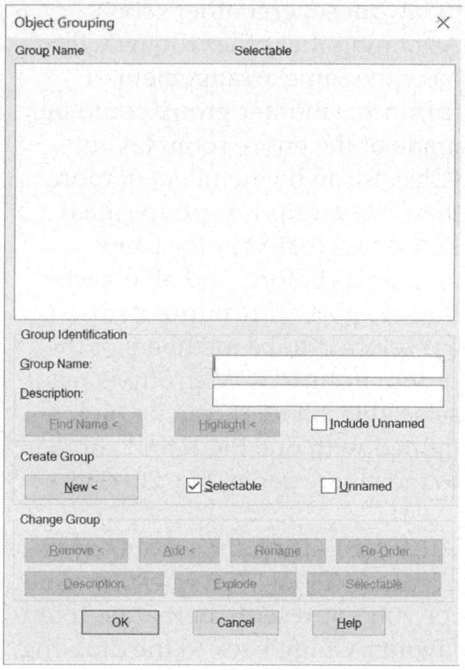


### *Group Name:*

The *Group Name*: edit box is used to enter the name of a new group that you want to create or to display the name of an existing group selected from the list above that requires changing. When creating a new group, enter the desired name before choosing the set of objects (using the *New* tile).

### *Description*

Adding a *Description* for new groups is optional, but this feature may be helpful when using multiple groups. It is a good idea to use this description to identify groups that are copied or otherwise unnamed. If a group name is selected from the list, both the group name and description appear in the *Group Identification* cluster.

### *Find Name*

Use this option to find the group name for any object in the drawing. Using the *Find Name* tile temporarily removes the *Object Grouping* dialog box and prompts you to "Pick a member of a group." After doing so, AutoCAD responds with the names of the group(s) of which the selected object is a member.

### *Highlight*

Choosing this tile highlights all members of the group appearing in the *Group Name*: box. The dialog box temporarily disappears so the drawing can be viewed. Use this feature to verify which objects are members of the current group.

**FIGURE 20-14**

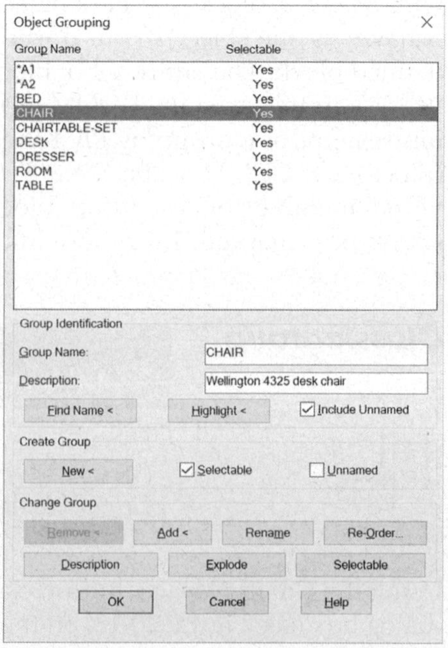


### *Include Unnamed*

When a group is copied (with *Copy, Mirror,* etc.), the objects in the new group have a collective relationship like any other group; however, the group is considered an "unnamed group." Toggling the *Include Unnamed* checkbox forces any unnamed groups to be included in the list above (Fig. 20-14). In this way, you can select and manipulate copied groups like any other group but without having to go through the process to create the group from scratch. A copied group or unnamed new group is given a default name such as *A*n*, where *n* is the number of the group created since the beginning of the drawing. You can create an unnamed group using the *Unnamed* checkbox and the *New* tile. An unnamed group can be *Renamed*.

### *New*

The *New* tile in the *Create Group* cluster is used to select the objects that you want to assign as members of a new group. You are returned to the drawing and prompted to "Select objects:". (See previous discussion under the *Group* command.)

### *Selectable* (checkbox)

Use this toggle to specify the selectable status for the new group to create. If the check appears, new groups that are made with the *Group Manager* are created as *Selectable* groups. Groups created with the *Group* command are selectable by default. (See *Selectable* under the "Changing Properties of a Group" section.)

### *Unnamed*

If you want to create a group with a default name (*A*n*, where *n* is the number of the group created since the beginning of the drawing), use this option. If you keep unnamed groups, toggle on *Include Unnamed* to include these groups in the list. An unnamed group can be *Renamed*.

TIP

When you make a copy of an existing group, a default name (such as *A1) is assigned for the copy. However, if you keep an unnamed group that is an identical copy, a good strategy is to keep the default (*A*n*) name as an identifier and add a description like "copy of fasten group."

## Changing Properties of a Group

Once a group exists, it can be changed in a number of ways. The options for changing a group appear in the *Change Group* cluster of the *Object Grouping* dialog box. To change a group, first select the group to change from the list on top of the *Object Grouping* dialog box (see Figure 20-14). Once a group is selected, the name and description appear in the related edit boxes. A group must be selected before the *Change Group* cluster options are enabled.

### *Remove*

The *Remove* tile allows you to remove any members of the specified group. The dialog box disappears temporarily and you are prompted to "Remove objects." If all members of a group are removed, the object group still remains defined by name in the list box. (Using this option has the same result as using the *Groupedit* command.)

### *Add*

Use this option to add new members to the group appearing in the *Group Name*: edit box. The "Select objects:" prompt appears. (This action is similar to using the *Groupedit* command.)

### *Rename*

First, select the desired group to rename from the list. Next, type over the existing name in the *Group Name*: edit box and enter the desired new name. Then choose the *Rename* button. (You can also use the *Groupedit* command to rename groups.)

### *Re-order*

See "Re-ordering Group Members" at the end of this section.

### *Description*

You can change the description for the specified object group with this option. First, select the desired group from the list. Next, edit the contents of the *Description* edit box. Then, select the *Description* tile.

### *Explode*

Use this option to remove a group from the drawing. The *Explode* option breaks the current group into its original objects. The group definition is removed from the object group list box; however, the individual group members are not affected in any other way. (This option has the same function as the *Ungroup* command.)

### *Selectable (button)*

This option changes the selectable status for a group. This capability is available only through this dialog box. If the group is assigned a nonselectable status, the group cannot be selected or manipulated as a whole by any method. Nonselectability prevents accidentally selecting the entire group or *Copying* the entire group, for example. Highlight a group name from the list, then choose the *Selectable* tile. The *Yes* or *No* indicator in the *Selectable* column of the list toggles as you click the tile.

## Reordering Group Members

When you select objects to be included in new groups, AutoCAD assigns a number to each member. The number corresponds to the sequence that you select the objects. For special applications, the number associated with individual members is critical. For example, a group may be formed to generate a complete tool path where each member of the group represents one motion of the sequence.

The *Re-Order* button (in the *Object Grouping* dialog box) invokes the *Order Group* dialog box (Fig. 20-15). This dialog box allows you to change the order of individual group members.

**FIGURE 20-15**

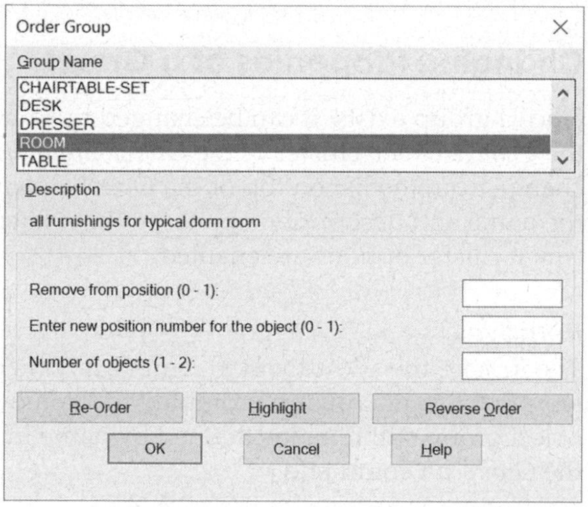


To reverse the order, simply select *Reverse Order*. To reorder members of a group to another sequence, use these steps:

1. In the *Order Group* dialog box, select the group to reorder from the list on top.
2. Select the *Highlight* option to display each group member one at a time. In the small dialog box that appears (not shown), select *Next* repeatedly to go completely through the current sequence of group members, paying attention to the assigned number for each member. The sequence begins with 0, so the first object has position 0. Then select *OK* to return to the *Order Group* dialog box.
3. In the *Number of Objects* edit box, enter the number of objects that you want to reposition.
4. In the *Remove from Position* edit box, enter the current position of the object to reorder.
5. Enter the new position in the *Enter new position number* edit box.
6. Select *Re-order*.
7. Select *Highlight* again to verify the new order.

Operation of this dialog box is difficult, and it is possible to accidentally remove or duplicate members from the group using this method. A more straightforward and stable alternative for reordering is to *Explode* the group and make a new group composed of the same members, but PICK the members in the desired order.

## CHAPTER EXERCISES

1. ***PICKFIRST***

Check to ensure that ***PICKFIRST*** is set to **1** (*ON*). Also make sure *GRIPS* are set to 0 (*OFF*) by typing ***GRIPS*** at the command prompt or invoking the ***Selection*** tab of the ***Options*** dialog box. For each of the following editing commands you use, make sure you PICK the objects FIRST, then invoke the editing command.

A. ***Open*** the **PLATES** drawing that you created as an exercise in Chapter 9 (not the PLATNEST drawing). ***Erase*** 3 of the 4 holes from the plate on the left, leaving only the hole at the lower-left corner (**PICK** the ***Circles***, then invoke ***Erase***). Create a ***Rectangular Array*** with **7** rows and **5** columns and **1** unit between each hole (PICK the *Circle* before invoking *Array*). The new plate should have 35 holes as shown in Figure 20-16, plate A.

FIGURE 20-16

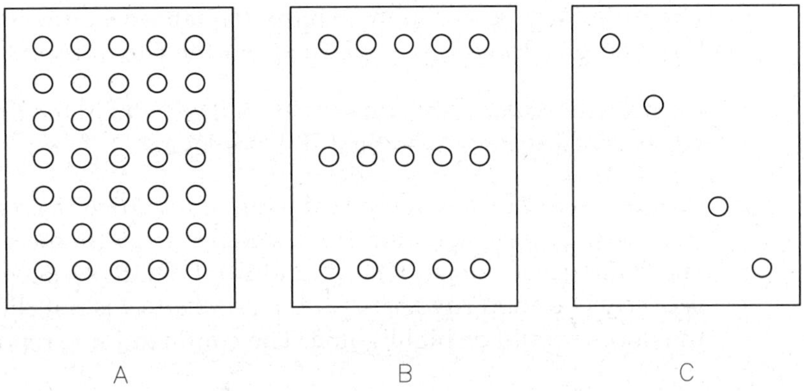


B. For the center plate, **PICK** the 3 holes on the vertical center, then invoke ***Copy***. Create the additional 2 sets of 3 holes on each side of the center column. Your new plate should look like that in Figure 20-16, plate B. Use ***SaveAs*** to assign a new name, **PLATES2**.

C. For the last plate (on the right), **PICK** the two top holes and ***Move*** them upward. Use the ***Center*** of the top hole as the "Specify base point or displacement, or [Multiple]:" and specify a "Specify second point of displacement or <use first point as displacement>:" as **1** unit from the top and left edges. Compare your results to Figure 20-16, plate C. ***Save*** the drawing (as **PLATES2**).

Remember that you can leave the *PICKFIRST* setting *ON* always. This means that you can use both Noun/Verb or Verb/Noun editing at any time. You always have the choice of whether you want to PICK objects first or use the command first.

2. ***PICKADD***

Change ***PICKADD*** to **0** (***OFF***). Remember that if you want to PICK objects and add them to the existing highlighted selection set, hold down the **SHIFT** key, then **PICK**.

FIGURE 20-17

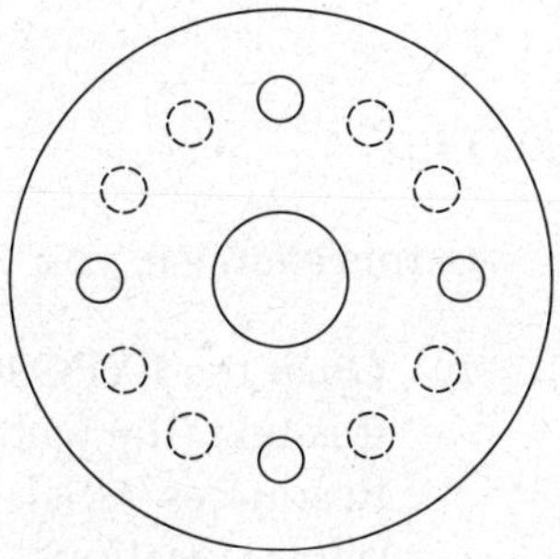

A. ***Open*** the **ARRAY1** drawing from Chapter 9. (The settings for *GRIPS* and *PICKFIRST* have not changed by *Opening* a new drawing.) ***Explode*** the holes (the *Polar Array*). ***Erase*** the holes (highlighted *Circles* shown in Figure 20-17) leaving only 4 holes. Use either Noun/Verb or Verb/Noun editing. ***SaveAs*** **FLANGE1**.

B. ***Open*** drawing **ARRAY2.** Select all the holes and ***Rotate*** them **180** degrees to achieve the arrangement shown in Figure 20-18. Any selection method may be used. Try several selection methods to see how *PICKADD* reacts. (The *Fence* option can be used to select all holes without having to use the **Shift** key.) ***SaveAs*** **FLANGE2**. Change the ***PICKADD*** setting back to **2** (*ON*).

FIGURE 20-18

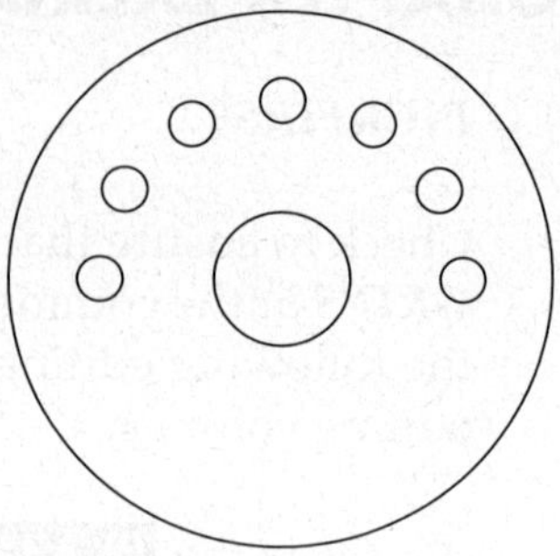

3. ***Qselect***

This exercise involves using object selection filters to change the properties of objects in the drawing. Open the sample drawing named **Floor Plan Sample.dwg** located in the **Sample/Database Connectivity** folder. To find this drawing, use the **Explore Sample Drawings...** option under the ***Get Started*** tile in the opening ***Create*** screen.

A. Activate the ***Model*** tab. Notice that all the doors in the drawing are green in color. Using *List*, verify that any door is on the E-F-Door layer.

B. Invoke ***Qselect*** by any method. Ensure that ***Entire Drawing*** appears in the *Apply to*: section. In the *Object Type* section, select ***Block Reference*** from the drop-down list. In *Properties*, *Operator* and *Value*, select ***Layer***, ***Equals***, and **E-F-DOOR**, respectively. Ensure ***Include in new selection set*** is checked and *Append to current selection set* is not checked. Then select the ***OK*** button. All the doors should be highlighted. The command line reports "39 items selected."

C. With the doors highlighted, use the *Layer* drop-down list and select layer **0**. All the doors are now "moved" to layer 0 and therefore change to black or white (the *ByLayer* color assignment) depending on your background color. Press **Escape** to clear the highlighted items and grips (if enabled).

D. Invoke the ***Properties*** palette (not shown). Use the ***Quick Select*** button to produce the *Quick Select* dialog box. In the dialog box (Fig. 20-19), ensure that ***Entire Drawing*** appears in the *Apply to*: section. In the *Object Type* section, select ***Multiple*** from the drop-down list. In *Properties*, *Operator* and *Value*, select ***Layer***, ***Equals***, and **E-F-STAIR**, respectively. Ensure ***Include in new selection set*** is checked and *Append to current selection set* is not checked. Then select the ***OK*** button. The stairs should be highlighted and the command line reports "217 items selected."

FIGURE 20-19

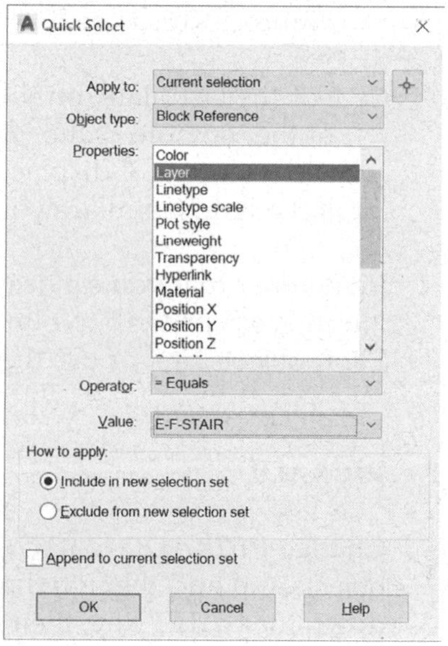


E. In the *Properties* palette, select the ***Color*** property and select ***Green***. Click in the drawing area and press **Escape** to deselect the stair objects. The stairs should now be green.

F. ***Close*** the drawing and do not save the changes.

4. ***Filter***

In this exercise, you will use the *Filter* command for several operations.

A. ***Open*** the **EXPO98 maps.DWG.** The **Expo98** maps drawing is available from the website for this text (the website address is listed on the back cover and in the Preface under "Online Resources"). Select the ***Model*** tab. The map indicates information about services such as telephone locations, transportation, restaurants, and so on.

B. Use the *Object Selection Filters* dialog box to count the number of telephones. Invoke ***Filter***. When the dialog box appears, select ***Block Name*** from the *Select Filter* section drop-down list. Then, click on the ***Select*** button to select from the list of Blocks in the drawing. Select **TELEFONE** from the bottom of the list, then pick ***OK***. In the *Object Selection Filters* dialog box, select ***Add to List***. Two entries in the filter list above should appear (Fig. 20-20).

**FIGURE 20-20**

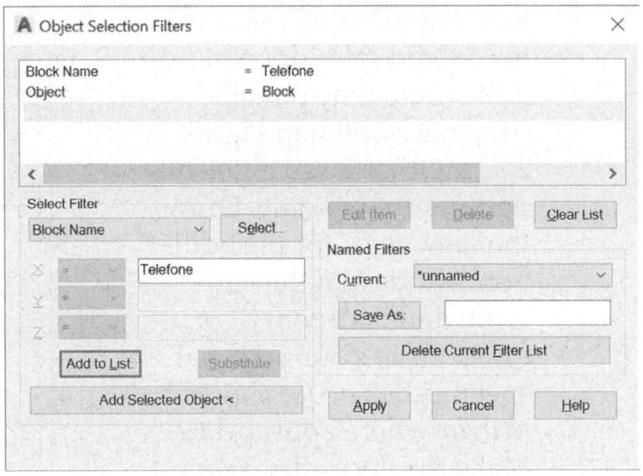


C. Select the ***Apply*** button. The dialog box disappears and the command line prompts to "Select objects:". Enter ***All*** so the filter is applied to the entire drawing. AutoCAD reports the number of (all) objects found and the number that were filtered out (not telephones). At the "Select objects:" prompt, press **Enter**. Although you have exited the *Filter* command, the objects remain highlighted.

D. Use the *Select* command to do the arithmetic. Type ***Select*** at the command prompt. AutoCAD reports "75 found."

E. Next, you will change the cafeteria symbols to layer 0. Do this by typing ***Change*** at the command line. At the "Select objects:" prompt, enter ***'Filter*** (with the apostrophe prefix). When the dialog box appears, highlight the existing items (telephone blocks) from the filter list and select ***Delete*** to remove them. Next, go through the process (as before) to create a filter for ***Block Names*** called **CAFETERIA**. Make sure you ***Add to List***. Finally, select ***Apply*** to exit the dialog box. At the "Select objects:" prompt, type ***All*** so AutoCAD searches the entire drawing.

F. Press **Enter** at the next "Select objects:" prompt. Remember, you are still in the ***Change*** command, so AutoCAD prompts to "Specify change point or [Properties]:". Type ***P*** for the *Properties* option. Enter ***LA*** for *Layer* and specify layer **0** as the new layer name. Complete the *Change* command by pressing **Enter**.

G. Since the CAFETERIA objects have been moved to layer 0, use the ***Layer*** command to make layer **0 *Current***, then right-click to ***Select All*** and ***Freeze*** all layers. AutoCAD reports that the current layer (0) cannot be frozen. Pick ***OK*** to exit the *Layer Properties Manager*. The cafeteria symbols should be visible.

H. ***Close*** the drawing, but do not save the changes.

5. **Object Groups**

A. ***Open*** the **EFF-APT2** drawing that you last worked on in Chapter 18 exercises. Create a ***Group*** for each of the three bathroom fixtures. ***Name*** one group **WASHBASIN**, give it a ***Description***, and include the *Ellipse* and the surrounding *Rectangle* as its members. Make two more groups named **WCLOSET** and **BATHTUB** and select the appropriate members. Next, combine the <u>three</u> fixture groups into <u>one</u> ***Group*** named **BATHROOM** and give it a description. Then, draw a bed of your own design and combine its members into a ***Group*** named **SINGLEBED**. Use ***SaveAs*** to save and rename the drawing to **2BR-APT.**

REUSE

B. A small bedroom and bathroom are to be added to the apartment, but a 10′ maximum interior span is allowed. One possible design is shown in Figure 20-21. Other more efficient designs are possible. Draw the new walls for the addition, but design the bathroom and bedroom door locations to your personal specifications. Use ***Copy*** or ***Mirror*** where appropriate. Make the necessary ***Trims*** and other edits to complete the floor plan.

**FIGURE 20-21**

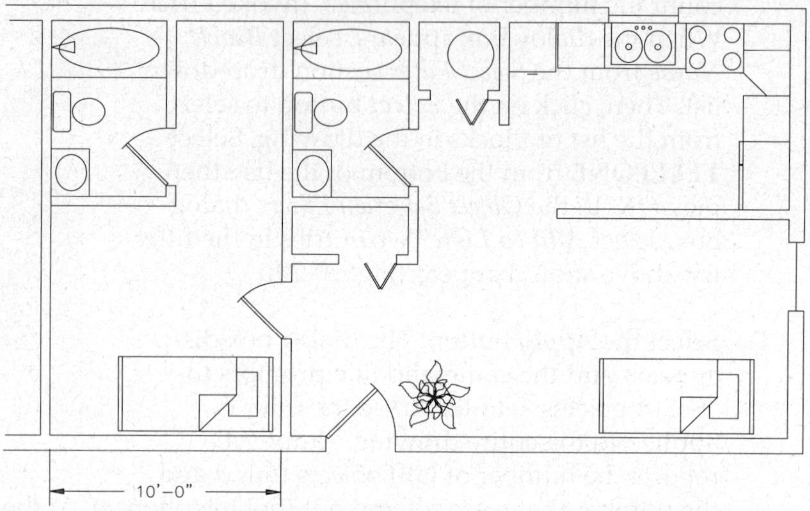


C. Next, ***Copy*** the bathroom fixture groups to the new bathroom. If your design is similar to that in the figure, the BATHROOM group can be copied as a whole. If you need to *Copy* only individual fixtures, take the appropriate action so you don't select the entire BATHROOM group. ***Copy*** or ***Mirror*** the **SINGLEBED**. Once completed, activate the ***Object Grouping*** dialog box (*Group Manager*) and toggle on ***Include Unnamed***. Can you explain what happened? Assign a ***Description*** to each of the unnamed groups and ***Save*** the drawing.

CHAPTER 21

# Blocks, DesignCenter, and Tool Palettes

## Chapter Objectives

After completing this chapter you should:

1. be able to use the *Block* command to transform a group of objects into one object that is stored in the current drawing's block definition table;
2. be able to use the *Insert* and *Minsert* commands to bring *Blocks* into drawings;
3. know that color, linetype and lineweight of *Blocks* are based on conditions when the *Block* is made;
4. be able to convert *Blocks* to individual objects with *Explode*;
5. be able to use *Wblock* to prepare .DWG files for insertion into other drawings;
6. be able to redefine and globally change previously inserted *Blocks*;
7. be able to use DesignCenter™ and Tool Palettes to drag and drop *Blocks* into the current drawing,
8. be able to use Count to find the number of items.

## CONCEPTS

A *Block* is a <u>group</u> of objects that are combined into <u>one</u> object with the *Block* command. The typical application for *Blocks* is in the use of symbols. Many drawings contain symbols, such as doors and windows for architectural drawings, capacitors and resistors for electrical schematics, or pumps and valves for piping and instrumentation drawings. In AutoCAD, symbols are created first by constructing the desired geometry with objects like *Line*, *Arc*, and *Circle*, then transforming the set of objects comprising the symbol into a *Block*. A description of the objects comprising the *Block* is then stored in the drawing's "block definition table." The *Blocks* can then each be *Inserted* into a drawing many times and treated as a single object. Text can be attached to *Blocks* (called *Attributes*) and the text can be modified for each *Block* when inserted.

Figure 21.1 compares a shape composed of a set of objects and the same shape after it has been made into a *Block* and *Inserted* back into the drawing. Notice that the original set of objects is selected (highlighted) individually for editing, whereas the *Block* is only one object.

**FIGURE 21.1**

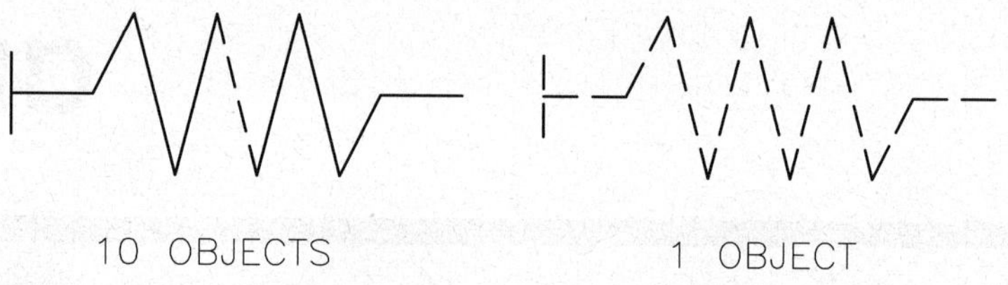


Since an inserted *Block* is one object, it uses less file space than a set of objects that is copied with *Copy*. The *Copy* command creates a duplicate set of objects, so that if the original symbol were created with 10 objects, 3 copies would yield a total of 40 objects. If instead the original set of 10 were made into a *Block* and then *Inserted* 3 times, the total objects would be 13 (the original 10 + 3).

Upon *Inserting* a *Block*, its scale can be changed and rotational orientation specified without having to use the *Scale* or *Rotate* commands (Figure 21.2). If a design change is desired in the *Blocks* that have already been *Inserted*, the original *Block* can be redefined and the previously inserted *Blocks* are automatically updated. *Blocks* can be made to have explicit *Linetype*, *Lineweight* and *Color*, regardless of the layer they are inserted onto, or they can be made to assume the *Color*, *Linetype*, and *Lineweight* of the layer onto which they are *Inserted*.

**FIGURE 21.2**

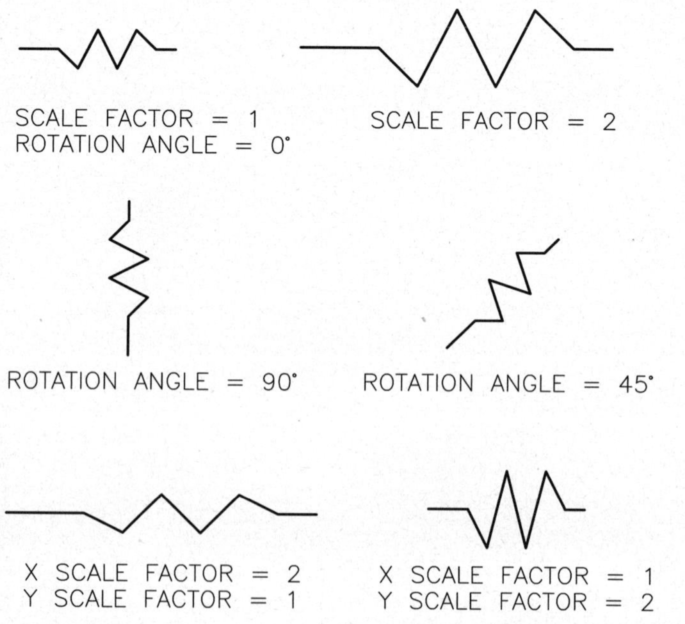


Similar to annotative text objects (see Chapter 18), you can create *Annotative Blocks* that can change size when drawing geometry is displayed in different scales (*Viewport Scale*) for different viewports. However, *Blocks* are generally used to represent parts of the drawing geometry and are normally intended to be displayed in the same scale as other drawing geometry. Therefore, *Annotative Blocks* are best for special applications such as dynamic callouts or view labels in architectural drawings.

*Blocks* can be nested; that is, one *Block* can reference another *Block*. Practically, this means that the definition of *Block* "C" can contain *Block* "A" so that when *Block* "C" is inserted, *Block* "A" is also inserted as part of *Block* "C" (Figure 21.3).

**FIGURE 21.3**

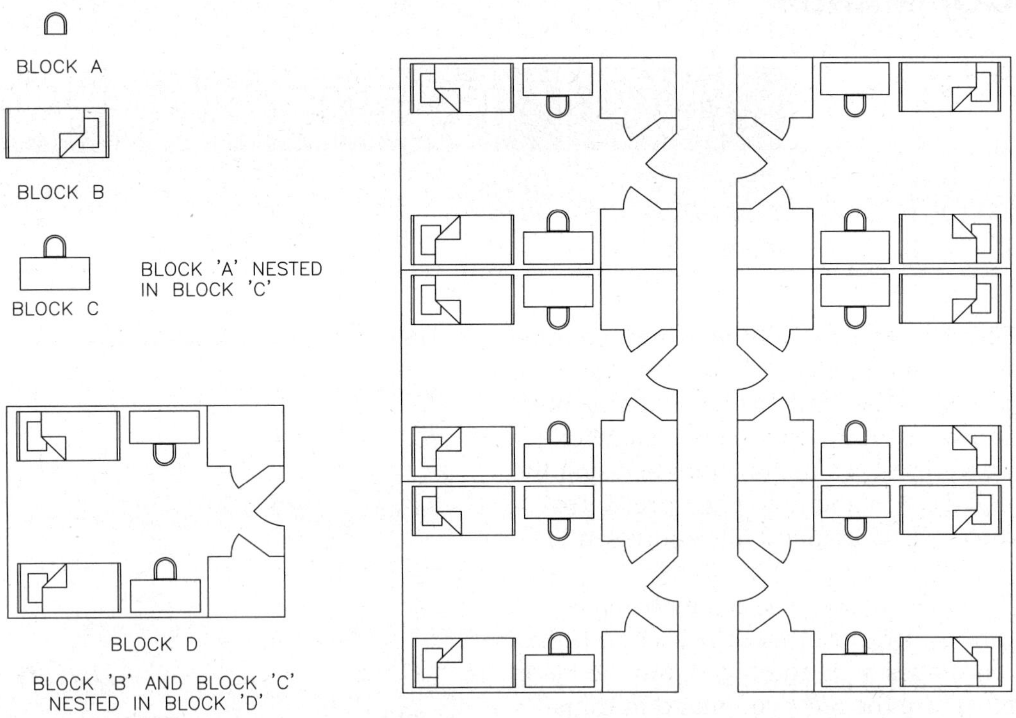


*Blocks* created within the current drawing can be copied to disk as complete and separate drawing files (.DWG file) by using the *Wblock* command (Write Block). This action allows you to *Insert* the *Blocks* into other drawings.

Commands related to using *Blocks* are:

| | |
|---|---|
| *Block* | Creates a *Block* from individual objects |
| *Insert* | Inserts a *Block* into a drawing |
| *Minsert* | Permits a multiple insert in a rectangular pattern |
| *Explode* | Breaks a *Block* into its original set of multiple objects |
| *Wblock* | Writes an existing *Block* or a set of objects to a file on disk |
| *Base* | Allows specification of an insertion base point |
| *Purge* | Deletes uninserted *Blocks* from the block definition table |
| *Rename* | Allows renaming *Blocks* |
| *Bedit* | Invokes the Block Editor for editing blocks |

The following additional tools, discussed in this chapter, provide powerful features for creating, inserting, and editing blocks.

| | |
|---|---|
| Tool palettes | Tool palettes allow you to drag-and-drop *Blocks* and *Hatch* patterns into a drawing. This process is similar to, but easier than, using *Insert*. You can easily add blocks to create your own tool palettes. Tool palettes are discussed also in Chapter 26. |
| DesignCenter | This tool allows you to drag-and-drop *Blocks* from other drawings into the current drawing. You can easily locate content from other drawings such as *Dimension Styles*, *Layers*, *Linetypes*, *Text Styles*, and *Xref* drawings to drag and drop into the current drawing. DesignCenter is discussed also in Chapter 30. |

## COMMANDS

### *Block*

| Ribbon | Menu Bar | Command (Type) | Alias (Type) | Shortcut |
|---|---|---|---|---|
| *Home Create Block* (icon) | *Draw Block > Make...* | *Block*, or *-Block* | *B* or *-B* | ... |

Selecting the icon button, using the pull-down menu, or typing *Block* or *Bmake* produces the *Block Definition* dialog box shown in Figure 21.4. This dialog box provides the same functions as using the *-Block* command (a hyphen prefix produces the Command line equivalent).

FIGURE 21.4

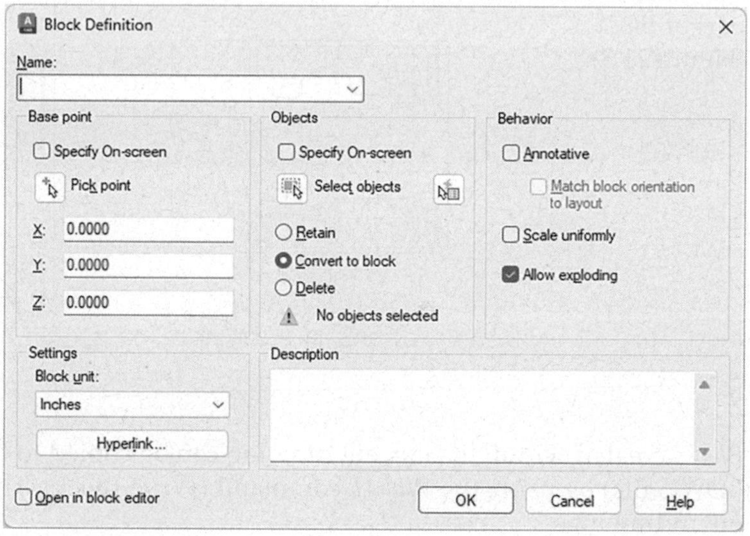


To make a *Block*, first create the *Lines, Circles, Arcs,* or other objects comprising the shapes to be combined into the *Block*. Next, use the *Block* command to transform the objects into one object—a *Block*.

In the *Block Definition* dialog box, enter the desired *Block* name in the *Name* edit box. Then use the *Select objects* button (top center) to return to the drawing temporarily to select the objects you wish to comprise the *Block*. After selection of objects, the dialog box reappears. Use the *Pick Insertion Base Point* button (in the *Base point* section of the dialog box) if you want to use a point other than the default 0,0,0 as the "insertion point" when the *Block* is later inserted. Usually select a point in the corner or center of the set of objects as the base point. When you select *OK*, the new *Block* is defined and stored in the drawing's block definition table awaiting future insertions.

If *Delete* is selected in the *Objects* section of the dialog box, the original set of "template" objects comprising the *Block* disappear even though the definition of the *Block* remains in the table. Checking *Retain* forces AutoCAD to retain the original objects (similar to using *Oops* after the *Block* command), or selecting *Convert to block* keeps the original set of objects visible in the drawing but transforms them into a *Block*.

The *Block Unit* drop-down option and the *Description* edit box are used to specify how *Blocks* are described when DesignCenter or a Ribbon gallery is used to drag and drop the *Blocks* into a drawing instead of using the *Insert* command. The *Block unit* options are described later in this chapter. The *Annotative* checkbox allows you to create annotative blocks. If the *Scale uniformly* checkbox is not checked, you have the option to scale the block non-uniformly when it is inserted into the drawing. If the *Allow exploding* box is checked, blocks can be "broken down" into the individual objects used to create the block (*Line, Arc, Circle,* etc.) using the *Explode* command or the *Explode* option of the *Insert* command. Use the *Hyperlink* button to produce the *Insert Hyperlink* dialog box for attaching a hyperlink to a *Block*. If *Open in block editor* is checked, the Block Editor is opened when you press the *OK* button. Normally, remove this check if you are making a block and do not want to edit it further. The *Names* drop-down list (at the top) is used to select existing *Blocks* if you want to redefine a *Block* (see "Redefining Blocks and the Block Editor" later in this chapter).

If you prefer to type, use *-Block* to produce the Command line equivalent of the *Block Definition* dialog box. The command syntax is as follows:

Command: **-BLOCK**
Block name (or ?): (**name**) (Enter a descriptive name for the *Block* up to 255 characters.)
Specify insertion base point or [Annotative]: **PICK** or (**coordinates**) (Select a point to be used later for insertion.)
Select objects: **PICK**
Select objects: **PICK** (Continue selecting all desired objects.)
Select objects: **Enter**

The *Block* then disappears as it is stored in the current drawing's "block definition table." The *Oops* command can be used to restore the original set of "template" objects (they reappear), but the definition of the *Block* remains in the table. Using the *?* option of the *Block* command lists the *Blocks* stored in the block definition table.

## *Block Color, Linetype,* and *Lineweight* Settings

The color, linetype, and lineweight of the *Block* are determined by one of the following settings when the *Block* is created:

1. When a *Block* is inserted, it is drawn on its original layer with its original *color, linetype,* and *lineweight* (when the objects were created), regardless of the layer or *color, linetype* and *lineweight* settings that are current when the *Block* is inserted (unless conditions 2 or 3 exist).

2. If a *Block* is created on Layer 0 (Layer 0 is current when the original objects comprising the *Block* are created), then the *Block* assumes the *color, linetype,* and *lineweight* of any layer that is current when it is inserted (Figure 21.5).

TIP

**FIGURE 21.5**

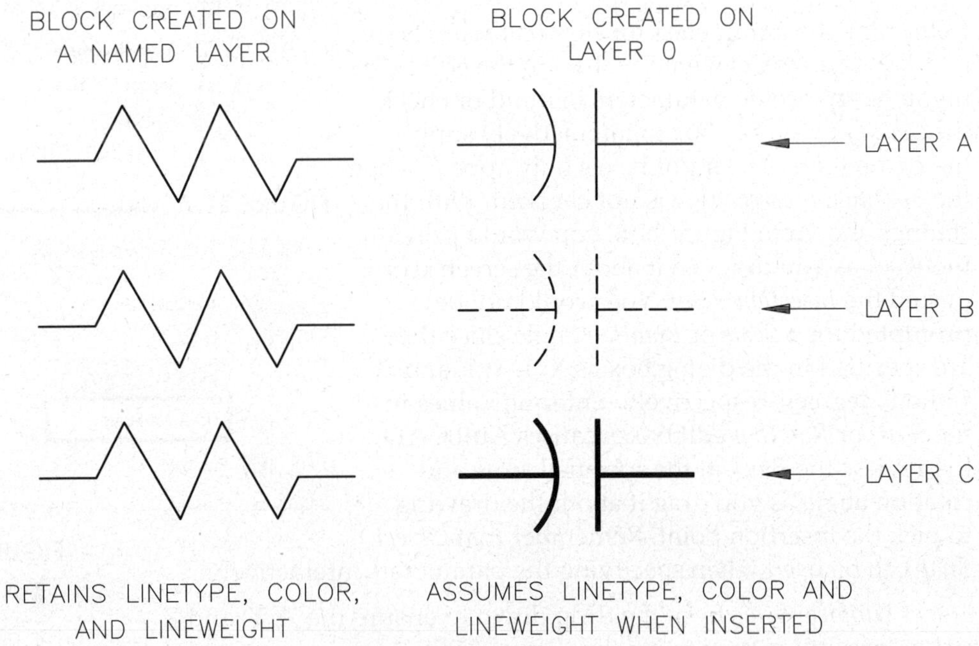


3. If the *Block* is created with the special *ByBlock color, linetype,* and *lineweight* setting, the *Block* is inserted with the *color, linetype,* and *lineweight* settings that are current during insertion, whether the *ByLayer* or explicit object *color, linetype,* and *lineweight* settings are current.

### Insert

| Ribbon | Menu Bar | Command (Type) | Alias (Type) | Shortcut |
|---|---|---|---|---|
| *Home Block, Insert Recent Blocks...* | *Insert Blocks Palette...* | *Insert* or *-Insert* | *I* or *-I* | ... |

Once a *Block* has been created, it can be inserted into the drawing at the desired location(s) using the *Insert* command. The *Insert* command produces the *Blocks palette* (Figure 21.6) which allows you to select which *Block* to insert. The lower section of the Blocks palette has options for specifying the *Insertion point*, *Scale*, *Rotation*, and *Auto-Placement*, either interactively by using the check mark or by entering values into the input boxes that appear when the check box is not selected. You can also choose whether to *Repeat Placement* or *Explode* the inserted block. *Explode* inserts the *Block* as separate individual objects (see "*Explode*" later in this chapter).

FIGURE 21.6

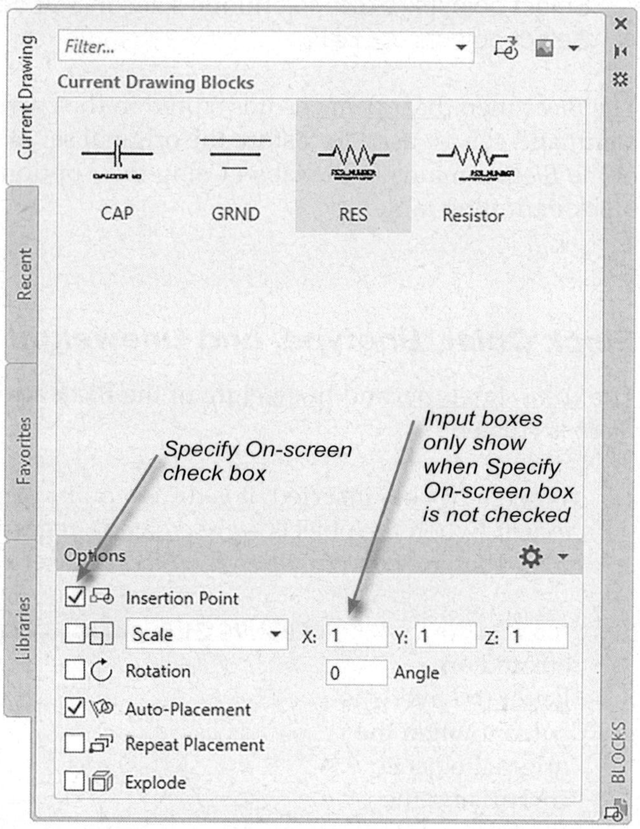


Select the *Block* you want to insert. Use the tabs along the left side of the palette to show blocks in the Current Drawing, Recent, Favorites, or Libraries. All *Blocks* in the drawing's block definition table are listed in the *Current Drawing* tab.

## Insertion Options

Determine the parameters for *Insertion point*, *Scale*, and *Rotation*. You can enter values in the edit boxes if you have specific parameters in mind or check the *Specify On-screen* box to interactively supply the parameters. The input boxes only appear when the *Specify On-screen* box is **not** checked. With the settings shown in Figure 21-6, you would preview the *Block* as you dragged it about the screen and picked the *Insertion Point*. You would not be prompted for a *Scale* or *Rotation* angle since they are specified in the dialog box as X: 1, Y: 1, and Z: 1 and 0 degrees, respectively. Entering values in the *Scale* or *Rotation* edit boxes causes AutoCAD to preview the *Block* at the specified scale and rotation angle as you drag it about the drawing to pick the insertion point. Remember that *Object Snap* can be used when specifying the parameters interactively.

FIGURE 21.7

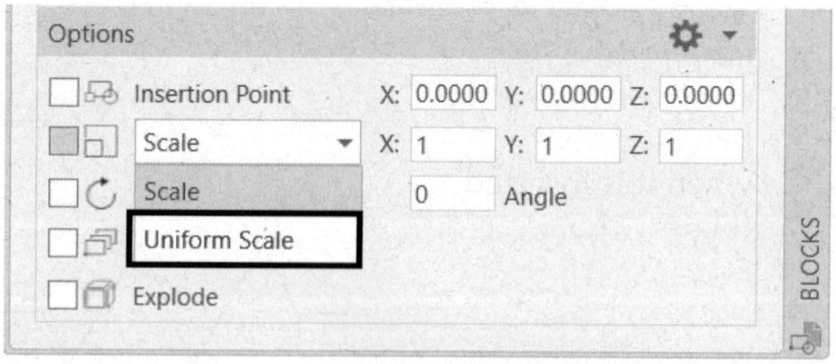


Select *Uniform Scale* from the drop down to ensure the X, Y, and Z values are scaled proportionally (Figure 21.7).

FIGURE 21.8

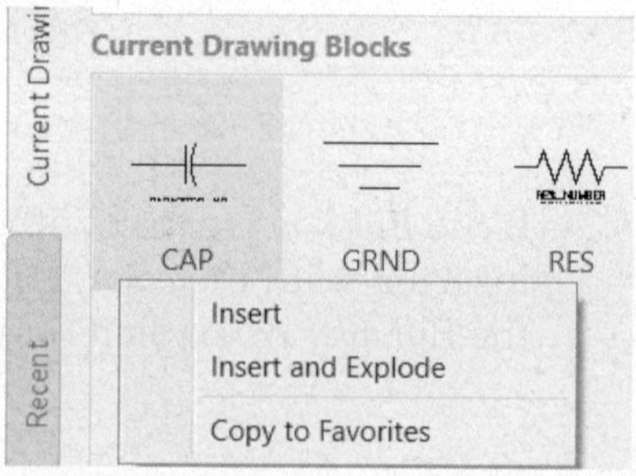


Once you have the Insertion Options set to the desired values, PICK the block from the upper portion of the Blocks palette and drag it into your drawing, or simply move your cursor into the drawing area. The block will appear attached to the crosshairs if you selected to specify the insertion point on screen or if you drag the block into the drawing area.

FIGURE 21.9

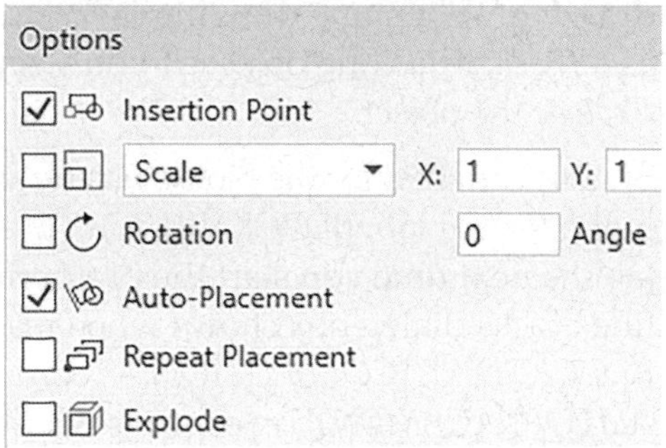


The *Auto-Placement* selection (Figure 21.9) from the Option area of the Blocks palette uses data from blocks in your drawings and libraries, along with machine learning algorithms to automatically scale, rotate, snap to a location, and other options as you place the block. The *BLOCKSDATACOLLECTION* system variable controls whether your data is collected and used to make more relevant suggestions. Setting *BLOCKSDATACOLLECTION* to 1 or *On* allows data to be collected (0 or *Off* to disallow). This feature can make it quick and easy to place blocks, and it can become more useful due to the learning algorithm.

FIGURE 21.10

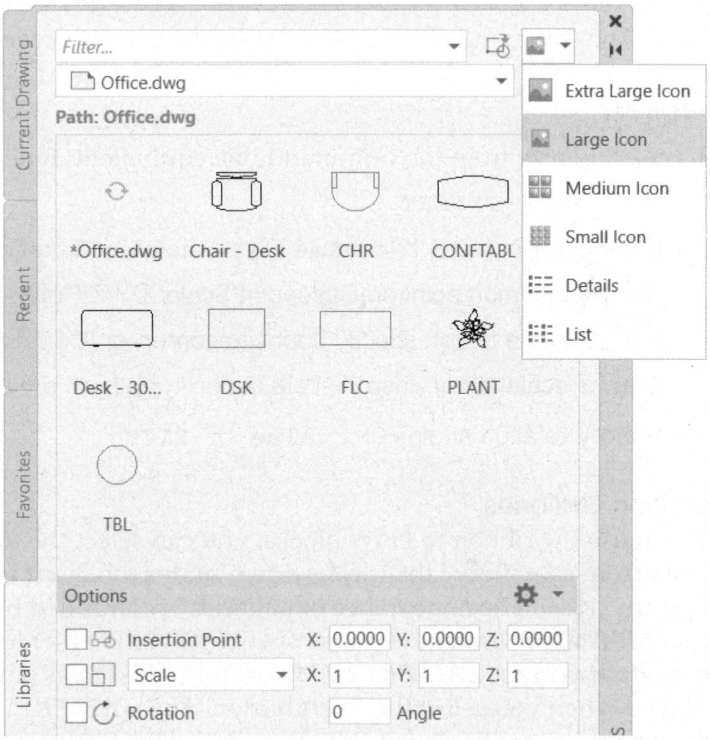


An example is, placing a block of a sink next to a wall. When you insert another instance of the block, it is automatically positioned similarly as you move near a wall feature. If the block was scaled the first time, as you get close to a likely placement feature, the new block will resize as your cursor nears the wall feature. As you make changes, the position, rotation, and scale of the block instances are adjusted to match the other block instance. You have the option to click and accept the suggestion, press Ctrl to move to the next suggestion, or toggle the suggestions off temporarily when placing a block by hold Shift+W simultaneously, or Shift+[ during block insertion.

TIP

The *Repeat Placement* selection allows you to insert the block multiple times. With the box checked, after the first block is placed, placement continues so that you can pick locations for additional block placements.

Right-clicking on a block from the panel shows options for Insert, Insert and Explode, (Fig. 21.8) and others depending on the Insertion Option selections. Pick Insert to place the block at the location entered in the input boxes.

FIGURE 21.11

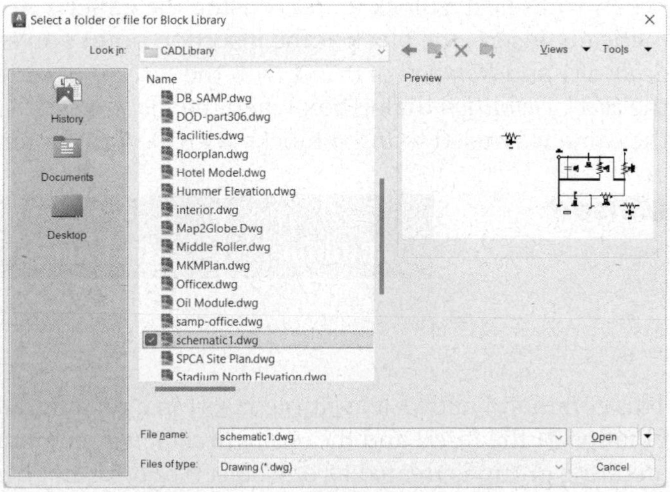


## Inserting *Blocks from Libraries*

Selecting the *Libraries* tab from the left side of the *Blocks* palette shows the drawing files in the current path (Figure 21.10). (You can also select Insert, Blocks from Libraries directly from the Blocks palette.)

The small image icon at the upper right of the *Blocks* palette has selections to view the blocks and drawings as Extra Large, Large, Medium, Small icons, or as Details, or List view of the files.

Use the lower icon from the upper right of the *Blocks* palette to open the *Select a Folder or a File* dialog box (Figure 21.11).

Here you can select <u>any</u> drawing (.DWG file) for insertion or see the block contained within the selected drawing. When one drawing is *Inserted* into another, the entire drawing comes into the current drawing as a *Block*, or as <u>one object</u>. If you want to edit individual objects in the inserted drawing, you must *Explode* the object.

System variables for the Blocks palette are: *BLOCKMRULIST* — controls the number of blocks displayed in the Recent tab, *BLOCKNAVIGATE* — controls the file and blocks that show in the Other Drawing tab the next time you start the program, *BLOCKREDEFINEMODE* — controls whether the redefine blocks dialog box shows when inserting a block with the same name as currently defined block *BLOCKSDATACOLLECTION* — controls whether your block data is collected for auto-placement use., *AUTOPLACEMENT* — controls whether placement suggestions are displayed, *PLACEMENTSWITCH* — controls whether placement suggestions are displayed by default.

The dialog box provides an option to rename the incoming block.

Note: The command *Classicinsert* opens the previous version of the *Insert dialog* box.

## -Insert

To insert blocks from the Command line equivalent, type *-Insert*.

Command: **-INSERT**

Enter block name or [?] : **name** (Type the name of an existing block or .DWG file to insert.)

Specify insertion point or [Basepoint/Scale/X/Y/Z/Rotate]: **PICK** or **option**

Enter X scale factor, specify opposite corner, or [Corner/XYZ] <1>: **value**, **PICK** or **option**

Enter Y scale factor <use X scale factor>: **value** or **PICK**

Specify rotation angle <0>: **value** or **PICK**

### Ribbon Galleries

If you use the ribbon to *Insert* blocks, you can select the desired *Block* from a "gallery" that drops down under the *Insert* button (Figure 21.12). The appearance of galleries is controlled by the *GALLERYVIEW* system variable. *GALLERYVIEW* is set to 1 (on) by default. If you set *GALLERYVIEW* to 0 (off), the *Blocks* palette is displayed when you select the *Insert* button. The *GALLERYVIEW* variable controls the display of Ribbon galleries for Blocks, Dimension Styles, Text Styles, and Table Styles.

When you select a block to *Insert* using the gallery, AutoCAD automatically inserts the block at the insertion point you select, with a *Scale* of 1 and *Rotation* of 0 and with the *Block unit* setting defined in the *Block Definition* dialog box when the *Block* was created. This is the same way used to *Insert* blocks with DesignCenter.

**FIGURE 21.12**

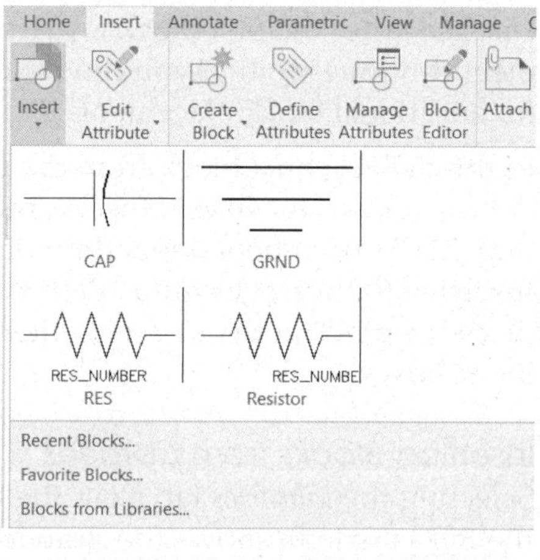


## Minsert

| Ribbon | Menu Bar | Command (Type) | Alias (Type) | Shortcut |
|---|---|---|---|---|
| ... | ... | *Minsert* | ... | ... |

This command allows a <u>multiple insert</u> in a rectangular pattern (Figure 21.13). *Minsert* is actually a combination of the *Insert* and the *Array Rectangular* commands. The *Blocks* inserted with *Minsert* are associated (the group is treated as one object) and cannot be edited independently (unless *Exploded*).

The *Minsert* command syntax yields results similar to that of a *Rectangular Array.*

Command: ***MINSERT***

Enter block name [or ?] <current>: **name**

Specify insertion point or [Basepoint/Scale/X/Y/Z/Rotate]: (**value**), **PICK** or **option**

Enter X scale factor, specify opposite corner, or [Corner/XYZ] <1>: (**value**) or **Enter**

Enter Y scale factor <use X scale factor>: (**value**) or **Enter**

Specify rotation angle <0>: (**value**) or **Enter**

Enter number of rows (—-) <1>: (**value**)

Enter number of columns (|||) <1>: (**value**)

Enter distance between rows or specify unit cell (—-): (**value**) or **PICK** (Value specifies Y distance from *Block* corner to *Block* corner; PICK allows drawing a unit cell rectangle.)

Distance between columns: (**value**) or **PICK** (Specifies X distance between *Block* corners.)

Command:

FIGURE 21.13

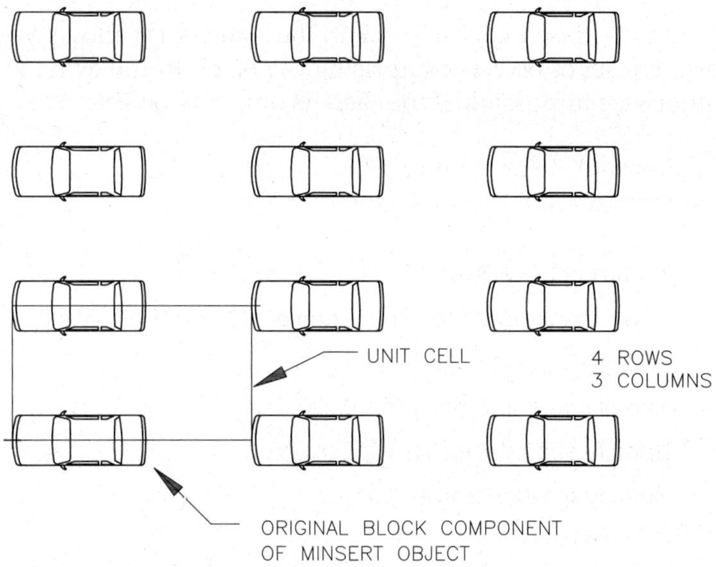


## Explode

| Ribbon | Menu Bar | Command (Type) | Alias (Type) | Shortcut |
|---|---|---|---|---|
| *Home*<br>*Modify*<br>(icon) | *Modify*<br>*Explode* | *Explode* | *X* | ... |

*Explode* breaks a previously inserted *Block* back into its original set of objects (Figure 21.14), which allows you to edit individual objects comprising the shape. If *Allow exploding* was unchecked in the *Block Definition* dialog box when the *Block* was created, the *Block* cannot be *Exploded*. *Blocks* that have been *Minserted* cannot be *Exploded*.

Command: ***EXPLODE***

Select objects: **PICK**

Select objects: **Enter**

Command:

FIGURE 21.14

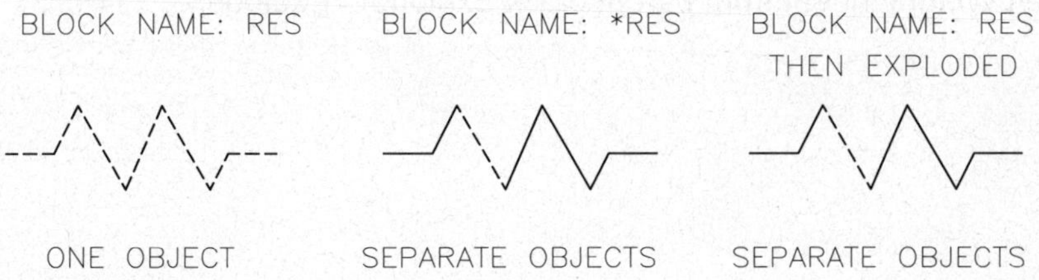

### *-Insert* with *

Using the *-Insert* command with the asterisk (*) allows you to insert a *Block,* not as one object, but as the original set of objects comprising the *Block*. In this way, you can edit individual objects in the *Block*, otherwise impossible if the *Block* is only one object.

The normal *-Insert* command is used (type *-Insert* with a hyphen); however, when the desired Block name is entered, it is prefaced by the asterisk (*) symbol:

Command: ***-INSERT***

Enter Block name (or ?): ***** (**name**) (Type the * symbol, then the name of an existing block or .DWG file to insert.)

Specify insertion point for block:

Specify scale factor for XYZ axes:

Specify rotation angle <0>:

Command:

This action accomplishes the same goal as using the *Insert* dialog box; then *Explode*.

TIP

When you *Insert* a *Block* into a drawing, all layers, linetypes, and colors contained in the *Block* are also inserted into the parent drawing if they do not already exist. If you use *Explode* to break down the *Block* into its component entities, those entities retain their native properties of layer, linetype, and color. For example, you may insert a *Block* that also inserts its layer named FLOORPLAN. If that *Block* is *Exploded*, its objects remain on layer FLOORPLAN. You may instead want to *Explode* the *Block* and have the objects reside on a layer existing in the current drawing, AR-WALL, for example.

### *Xplode*

| Ribbon | Menu Bar | Command (Type) | Alias (Type) | Shortcut |
|---|---|---|---|---|
| ... | ... | *Xplode* | *XP* | ... |

The *Xplode* command is an expanded version of *Explode*. The *Xplode* command allows you to specify new properties for the objects that are exploded. When you use *Xplode*, you can also specify the new *Layer*, *Linetype*, *Color*, *Lineweight*, or choose to *Explode* the *Block* normally.

Command: ***XPLODE***

Select objects to XPlode.

Select objects: **PICK**

Select objects: **Enter**

Enter an option [All/Color/LAyer/LType/LWeight/Inherit from parent block/Explode] <Explode>:

The options are as follows.

### Color

Use this option to specify a color for the exploded objects. Entering *ByLayer* causes the component objects to inherit the color of the exploded object's layer.

```
Enter an option
[All/Color/LAyer/LType/LWeight/Inherit from parent block/Explode] <Explode>: c
New color [Truecolor/COlorbook] <BYLAYER>:
```

### Layer

With this option you can specify the layer of the component objects after you explode them. The default option is to inherit the current layer rather than the layer of the exploded object.

```
Enter an option [All/Color/LAyer/LType/LWeight/Inherit from parent block/Explode] <Explode>: la
Enter new layer name for exploded objects <0>:
```

### Linetype

You can enter the name of any linetype that is loaded in the drawing. The exploded objects assume the specified linetype. Entering *ByLayer* causes the component objects to inherit the linetype of the exploded object's layer. Entering *ByBlock* causes the component objects to inherit the object-specific linetype of the exploded object.

```
Enter an option [All/Color/LAyer/LType/LWeight/Inherit from parent block/Explode] <Explode>: lt
Enter new linetype name for exploded objects <BYLAYER>:
```

### Lineweight

Use this option to specify the lineweight of the objects when they are exploded. The value you enter specifies the lineweight in inches or millimeters, whichever is set in the *Lineweight Settings* dialog box.

```
Enter an option
[All/Color/LAyer/LType/LWeight/Inherit from parent block/Explode] <Explode>: lw
Enter new lineweight <ByLayer>:
```

### Inherit from parent block

This option sets the color, linetype, lineweight, and layer of the exploded objects to that of the *Block* if the *Block* was created using *ByBlock* color, linetype, and lineweight and the objects were drawn on layer 0.

### All

Use this option to specify a layer, color, linetype, and lineweight for the exploded objects.

## *Wblock*

| Ribbon | Menu Bar | Command (Type) | Alias (Type) | Shortcut |
|---|---|---|---|---|
| *Insert*<br>*Block Definition*<br>*Write Block* | *File*<br>*Export...* | *Wblock* | *W* | ... |

The *Wblock* command writes a *Block* out as a separate and complete drawing (.DWG) file. The *Block* used for writing the new file can already exist in the current drawing's *Block* definition table or can be created during the *Wblock* command. Remember that the *Insert* command inserts *Blocks* (from the current drawing's block definition table) or finds and accepts .DWG files and treats them as *Blocks* upon insertion.

There are two ways to create a *Wblock* from the current drawing, (1) using an existing *Block* and (2) using a set of objects not previously defined in the current drawing as a *Block*. If you are using an existing *Block*, a copy of the *Block* is essentially transformed by the *Wblock* command to create a complete AutoCAD drawing (.DWG) file. The original block definition remains in the current drawing's block definition table. In this way, *Blocks* that were originally intended for insertion into the current drawing can be easily inserted into other drawings.

Figure 21.15 illustrates the relationship among a *Block*, the current drawing, and a *WBlock*. In the figure, SCHEM1.DWG contains several *Blocks*. The RES block is written out to a .DWG file using *Wblock* and named RESISTOR. RESISTOR is then *Inserted* into the SCHEM2 drawing.

**FIGURE 21.15**

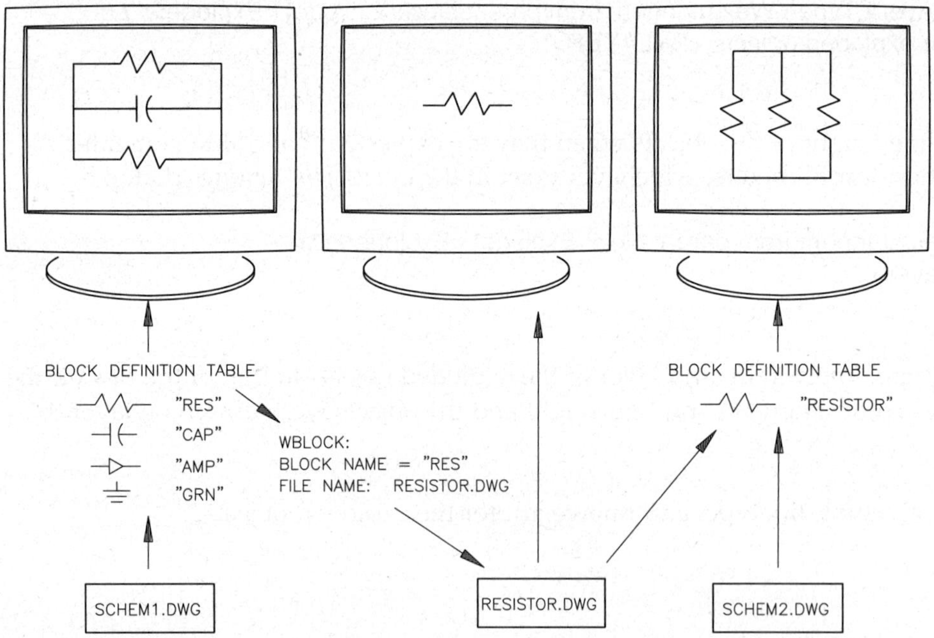


**TIP**

If you want to transform a set of objects into a *Block* to be used in other drawings but not in the current one, you can use *Wblock* to transform (a copy of) the objects in the current drawing into a separate .DWG file. This action does not create a *Block* in the current drawing. As an alternative, if you want to create symbols specifically to be inserted into other drawings, each symbol could be created initially as a separate .DWG file.

The *Wblock* command produces the *Write Block* dialog box (Figure 21.16). You should notice similarities to the *Block Definition* dialog box. Under *Source*, select *Block* if you want to write out an existing *Block* and select the *Block* name from the list, or select *Objects* if you want to transform a set of objects (not a previously defined *Block*) into a separate .DWG file.

**FIGURE 21.16**

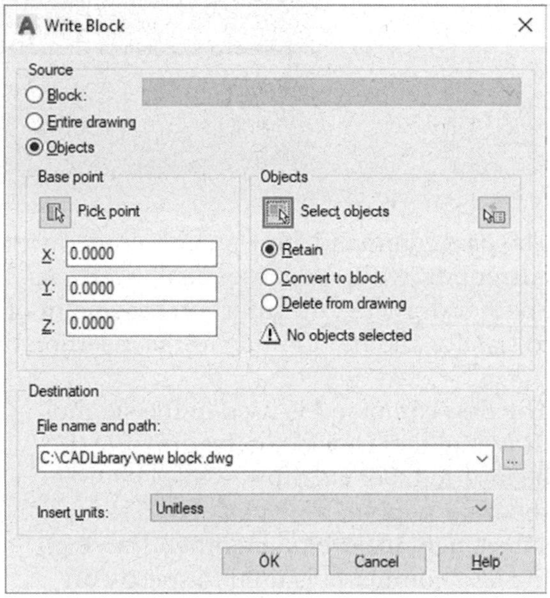


The *Base point* section allows you to specify a base point to use upon insertion of the *Block*. Enter coordinate values or use the *Pick Insertion Base Point* button to pick a location in the drawing. The *Objects* section allows you to specify how you want to treat selected objects if you create a new .DWG file from objects (not from an existing *Block*). You can *Retain* the objects in their current state, *Convert to block*, or *Delete from drawing*.

The *Destination* section defines the desired *File name and path,* and *Insert units*. Your choice for *Insert units* is applicable only when you drag and drop a *Block*, as with Ribbon galleries or DesignCenter. See "DesignCenter" later in this chapter for an explanation of this subject.

If you prefer the Command line equivalent to the *Write Block* dialog box, type *-Wblock* and follow the prompt sequence shown here to create *Wblocks* (.DWG files) from existing *Blocks*.

> Command: ***-WBLOCK***
> (At this point, the *Create Drawing File* dialog box appears, prompting you to supply a name for the .DWG file to be created. Typically, a new descriptive name would be typed in the edit box rather than selecting from the existing names.)
> Enter name of existing block or [= (block=output file)/* (whole drawing)] <define new drawing>:
> (Enter the name of the desired existing *Block*. If the file name given in the previous step is the same as the existing *Block* name, an "=" symbol can be entered, or enter an asterisk to write out the entire drawing.)
> Command:

A copy of the existing *Block* is then created in the selected directory as a *Wblock* (.DWG file).

An alternative method for creating a *Wblock* is using the *Export Data* dialog box accessed from the Application Menu or *File* pull-down menu. Make sure you select *Block (*.DWG)* as the type of file to export (bottom of the dialog box). You can create a *Wblock* (.DWG file) from an existing *Block* or from objects that you select from the screen. After specifying the .DWG name you want to create and the dialog box disappears, you are presented with the same command syntax as shown previously for the *-Wblock* command.

When *Wblocks* are *Inserted*, the *Color, Linetype,* and *Lineweight* settings of the *Wblock* are determined by the settings current when the original objects comprising the *Wblock* were created. The three possible settings are the same as those for *Blocks* (see "*Block*," *Color, Linetype,* and *Lineweight* Settings).

**TIP**

When a *Wblock* is *Inserted*, its parent (original) layer is also inserted into the current drawing. *Freezing* either the parent layer or the layer that was current during the insertion causes the *Wblock* to be frozen.

## Base

| Ribbon | Menu Bar | Command (Type) | Alias (Type) | Shortcut |
|---|---|---|---|---|
| *Home Block Base* | *Draw Block > Base* | *Base* | ... | ... |

The *Base* command allows you to specify an "insertion base point" (see the *Block* command) in the current drawing for subsequent insertions. If the *Insert* command is used to bring a .DWG file into another drawing, the insertion base point of the .DWG is 0,0 by default. The *Base* command permits you to specify another location as the insertion base point.

The *Base* command is used in the symbol drawing, that is, used in the drawing to be inserted. For example, while creating separate symbol drawings (.DWGs) for subsequent insertion into other drawings, the *Base* command is used to specify an appropriate point on the symbol geometry for the *Insert* command to use as a "handle" other than point 0,0 (Figure 21.17).

**FIGURE 21.17**

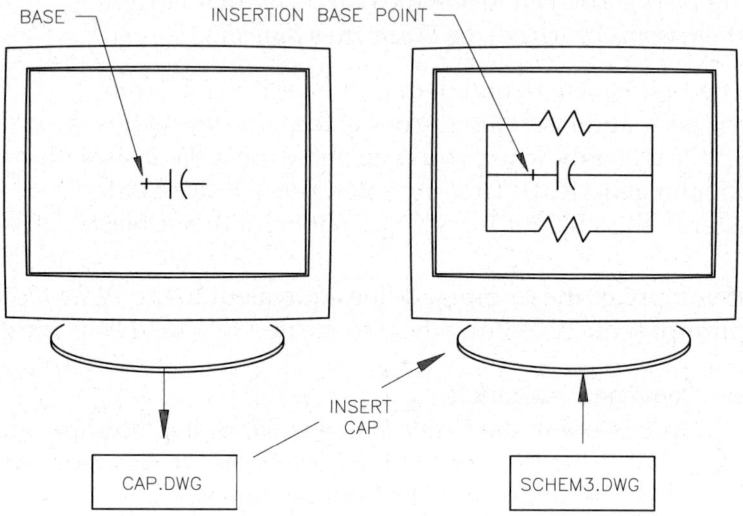


## Redefining Blocks and the Block Editor

If you want to change the configuration of a *Block*, even after it has been inserted, it can be accomplished by redefining the *Block*. When you redefine a block (change a block and save it), all of the previous *Block* insertions in the drawing are automatically and globally updated (Figure 21.18). For each *Block* insertion in the drawing, AutoCAD stores two fundamental pieces of information—the insertion point and the *Block* name. The actual block definition is stored in the block definition table. Redefining the *Block* involves changing that definition.

You can use three methods to redefine a *Block*: 1) use *Refedit* to change and save the block, 2) redraw the block geometry, then use the *Block* command to save the new geometry under the old block name, and 3) use the Block Editor to change the block, then *Save* the block from the editor.

**FIGURE 21.18**

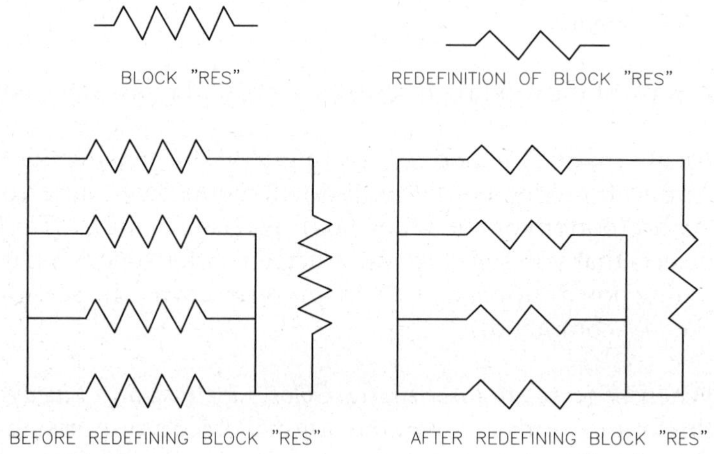

### To Redefine a Block Using *Refedit*

The *Refedit* command "opens" the *Block* for editing, then you make the necessary changes, and finally use *Refclose* to "close" the editing session and save the *Block*. This process has the same result as redefining the *Block*. This method is probably the least efficient method. See "*Refedit*" in Chapter 30, Xreferences.

### To Redefine a Block Using the *Block* Command

To redefine a *Block* by this method, you must first draw the new geometry or change the original "template" set of objects. Alternately, you can *Explode* one insertion of the block in the Drawing Editor, then make the desired changes. (The original unexploded block cannot be included in the new *Block* because a block definition cannot reference itself.) Next, use the *Block* command and select the new or changed geometry. The old block is redefined with the new geometry as long as the original block name is used.

### To Redefine a Block Using the Block Editor

To use the Block Editor, double-click on a block or use the *Bedit* command. See "*Bedit*" next.

## ***Bedit***

| Ribbon | Menu Bar | Command (Type) | Alias (Type) | Shortcut |
|---|---|---|---|---|
| *Home Block* (icon) | *Tools Block Editor* | *Bedit* | *BE* | ... |

To edit and redefine existing blocks, or to create dynamic blocks, use the Block Editor. Produce the Block Editor by using the *Bedit* command by any method shown in the command table above. Alternately, double-click on any block (without attributes) in the drawing (assuming *Dblclkedit* is on) to automatically open the Block Editor.

**FIGURE 21.19**

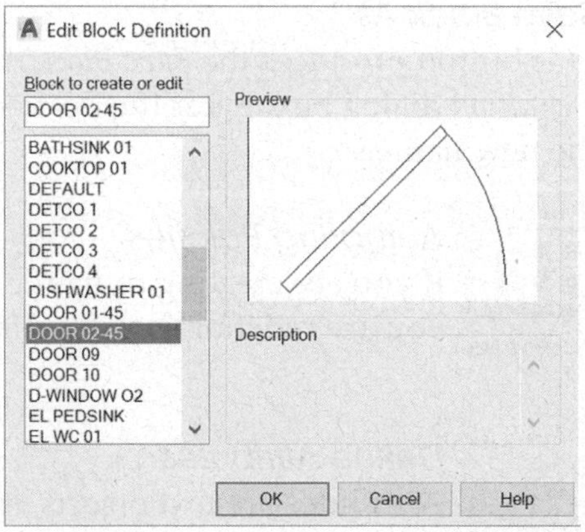


Before the Block Editor appears, the *Edit Block Definition* dialog box appears for you to select the desired block name from the *Block to create or edit* list (Figure 21.19).

Once the desired block is selected in the *Edit Block Definition* dialog box, the Block Editor appears in the drawing area and the *Block Editor* tab appears on the Ribbon (Figure 21.20). Although the *Block Editor* tab and the *Block Authoring* palettes that appear by default offer options for creating dynamic blocks, these options are not commonly used for block editing.

Use "normal" drawing and editing commands for editing block geometry. The Block Editor is similar to the Drawing Editor in that you can draw and edit geometry as you would normally. Most AutoCAD commands operate in the Block Editor, although a few commands, such as those related to plotting, publishing, layouts, views, and viewports, are not allowed.

The selected block geometry also appears in the drawing editor. However, instead of the block appearing as one object, it appears in its original "exploded" state—that is, the individual *Lines, Arcs, Circles,* or other objects that make up the block can be changed. The UCS icon appears to indicate the insertion base point for the block.

NOTE: The *Block Authoring Palettes* appear in the Block Editor by default. If you are not creating dynamic blocks, close the *Block Authoring Palettes.*

**FIGURE 21.20**

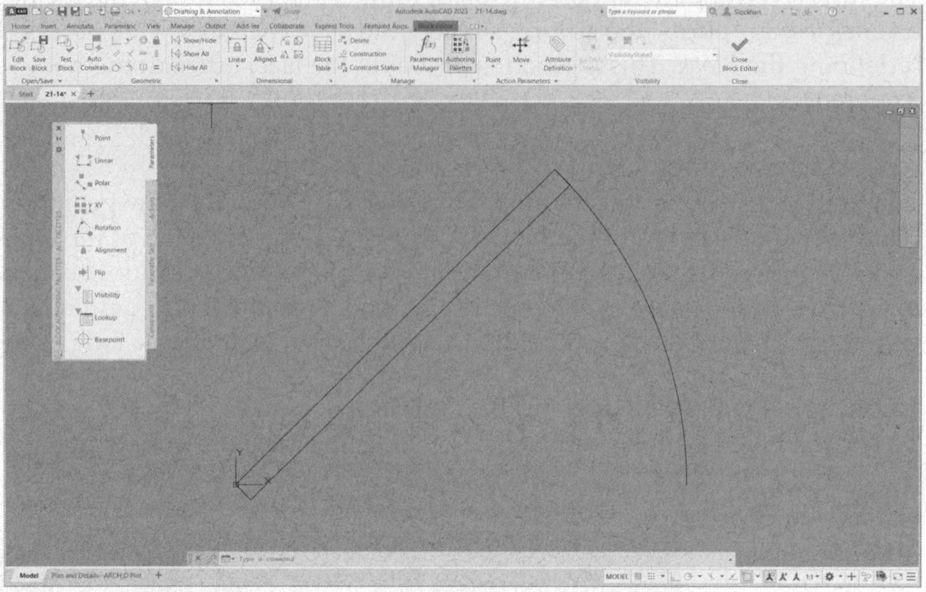

### *Edit or Create Block Definition*

Use this button (far left) to work with a new block in the Block Editor. The *Edit Block Definition* dialog box appears for you to select the desired block name you want to edit.

### *Save Block Definition*

This button activates the *Bsave* command. This command saves the block in the editor to the drawing's block definition table. Any previous insertions of the block in the drawing are automatically changed to the new block definition.

### *Save Block As*

This button produces the *Save Block As* dialog box, which is essentially the same as the *Edit Block Definition* dialog box. Enter the desired name in the edit box and select *OK* to save the current block with the new name.

### *Authoring Palettes*

If you are creating dynamic blocks, use this button to produce the *Block Authoring Palettes*. If you are redefining an existing "normal" block, close the palettes.

### *Define Attributes*

Attributes are text objects attached to a block. See Chapter 22.

### *Close Block Editor*

Close the Block Editor when you have completed the process of creating a new block or redefining an existing block definition and you are ready to return to the Drawing Editor. If you forget to use *Save Block Definition* first, a warning appears and allows you to save your changes.

The other commands that appear in the *Block Editor* tab and the *Block Authoring* palettes are used for creating dynamic blocks and parametric blocks but are not covered in this chapter.

TIP

The *Breplace* command will let you replace a block with another block. Check your system to see if the command is authorized.

## *Blocks* Search Path

You can use the *Options* dialog box to specify the search path used when the *Insert* command attempts to locate .DWG files. As you remember, when *Insert* is used and a *Block* name is entered, AutoCAD searches the drawing's block definition table and, if no *Block* by the specified name is found, searches the specified path for a .DWG file by the name. If the paths of the intended files are specified in the "Support File Search Path," you can type *Insert* and the name of the file, and AutoCAD will locate and insert the file. Otherwise, you have to locate the file using the *Browse* button each time you *Insert* a new file.

The *Files* tab of the *Options* dialog box allows you to list and modify the path that AutoCAD searches for drawings to *Insert* (Figure 21.21). The search path is also used for finding font files, menus, plug-ins, linetypes, and hatch patterns.

**FIGURE 21.21**

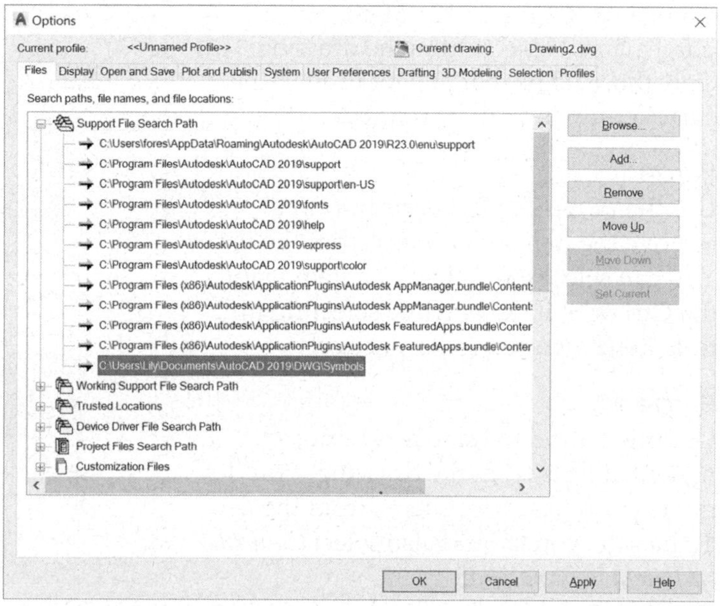


Suppose all of your symbols were stored in a folder called C:\DWG\Symbols. Access the *Options* dialog box by typing *Options* or selecting from the *Tools* pull-down menu. In the *Files* tab, open the *Support File Search Path* section.

Select *Add...* or *Browse...* to enter the location of your symbols as shown in the figure. Use *Move Up* or *Move Down* to specify an order (priority) for searching. Then each time you type "*Insert*," you could enter the name of a file without the path and AutoCAD can locate it.

TIP

AutoCAD can use special Connected support files. Using this AutoCAD finds the necessary support files for drawings in a *Forma* project without requiring configuration for each AutoCAD installation. Connected support files can only be set up by Project Administrator level users in *Forma Data Management*, which has replaced AutoCAD docs. *Forma Data Management* is a separate subscription service, not covered in this text.

## *Purge*

| Ribbon | Menu Bar | Command (Type) | Alias (Type) | Shortcut |
|---|---|---|---|---|
| *Manage*<br>*Cleanup panel*<br>*Purge icon* | *File*<br>*Drawing Utilities >*<br>*Purge...* | *Purge* | *PU* | ... |

*Purge* allows you to selectively delete a *Block* that is not referenced in the drawing. In other words, if the drawing has a *Block* defined but not appearing in the drawing, it can be deleted with *Purge*. In fact, *Purge* can selectively delete <u>any named object</u> that is not referenced in the drawing.

Examples of objects that are named but not referenced are *Blocks* that have been defined but not *Inserted*, *Dimstyles* that have been defined yet no dimensions are created in the style, or *Layers* that exist without objects residing on those layers. The possible unreferenced named objects are:

| | | |
|---|---|---|
| *Blocks* | *Linetypes* | *Section View Styles* |
| *Detail View Styles* | *Materials* | *Shapes* |
| *Dimension Styles* | *Mline Styles* | *Table Styles* |
| *Groups* | *Multileader Styles* | *Text Styles* |
| *Layers* | *Plot Styles* | *Visual Styles* |

Since these named objects exist in the drawing, using *Purge* to delete unused named objects can reduce the file size. This is helpful when drawings are created using template drawings that contain many unused named objects or when drawings are *Inserted* that contain many unused *Layers* or *Blocks*, etc.

*Purge* is especially useful for getting rid of unused *Blocks* because unused *Blocks* can occupy a huge amount of file space compared to other named objects (*Blocks* are the only geometry-based named objects). Although some other named objects can be deleted by other methods (such as selecting *Delete* in the *Text Style* dialog box), *Purge* is the only method for deleting unused *Blocks*.

### Purgeable Items tab

The *Purge* command produces the *Purge* dialog box (Figure 21.22). Here you can purge named objects from the drawing. Only those items that are not used in the current drawing can be purged.

**FIGURE 21.22**

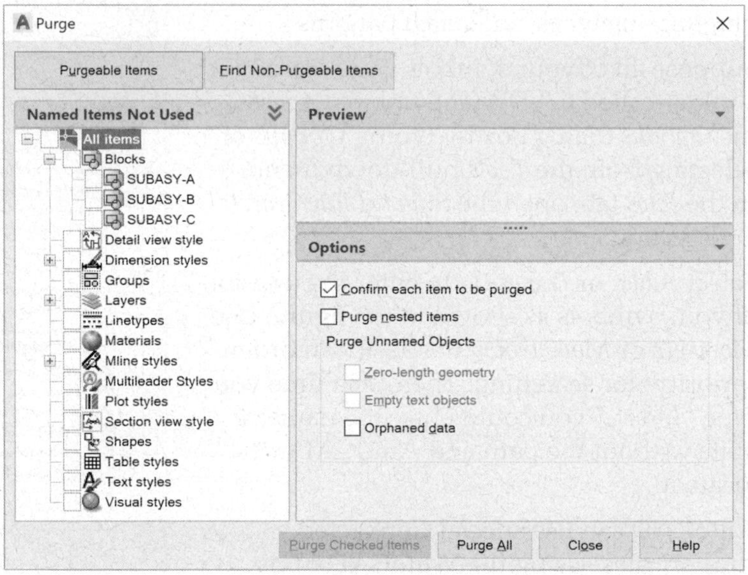


### Purge

Use the *Purge Checked Items* button to remove the items you select from the list. You can select more than one item using the Ctrl or Shift keys while selecting multiple items or a range, respectively.

### Purge All

Use this button to remove all unused named objects from the drawing. You do not have to first select items from the list. To be safe, you should also select *Confirm each item to be purged.*

### Confirm each item to be purged

If the *Confirm each item to be purged* box is checked, AutoCAD asks for confirmation before removing each object. If this box is not checked and you *Purge All*, all unused named objects in the drawing are removed at once without further notification.

### Purge nested items

You can remove unused nested *Blocks* and other nested named objects with this option. This option is particularly helpful if you have *Inserted* other drawings or *Xrefs* into the current drawing since all the named items contained in the inserted drawings are also inserted.

### Purge Unnamed Objects

If this checkbox is enabled, your drawing contains objects that are not visible, such as zero-length geometry or empty text objects that can be purged. Zero-length objects can be accidentally created by grip-snapping one endpoint of a *Line* to the other endpoint, for example. Empty text objects are *Text* lines or *Mtext* objects that contain only spaces but no letters.

### Purge Orphaned Data

A check here performs a drawing scan and removes obsolete DGN linestyle data (from inserted MicroStation drawings).

### Find Non-Purgeable Items tab

Select this tab to view all the named objects contained in the drawing. An explanation as to why a selected item cannot be removed appears in the Possible Reasons area at the middle right of the dialog box. This is helpful in many cases. For objects that can't be purged, the enhanced information includes the number of objects on each layer and their impact on file size. The Preview panel shows the selected named object. You can drag the border of the Preview panel to resize it.

## Rename

| Ribbon | Menu Bar | Command (Type) | Alias (Type) | Shortcut |
|---|---|---|---|---|
| ... | *Format* *Rename...* | *Rename* or *-Rename* | *REN* or *-REN* | ... |

This utility command allows you to rename a *Block* or any named object that is part of the current drawing. *Rename* allows you to rename the named objects listed here:

*Blocks, Detail View Styles, Dimension Styles, Layers, Linetypes, Materials, Plot Styles, Section View Styles, Table Styles, Text Styles, User Coordinate Systems, Views,* and *Viewport* configurations.

If you prefer to use dialog boxes, you can type *Rename* or select *Rename...* from the *Format* pull-down menu to access the dialog box shown in Figure 21.23.

**FIGURE 21.23**

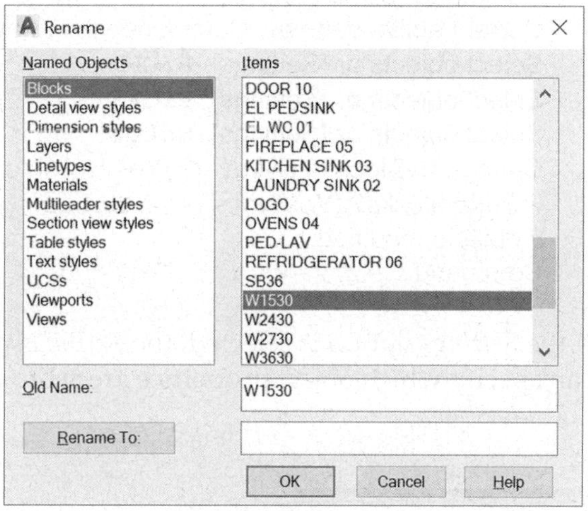


You can select from the *Named Objects* list to display the related objects existing in the drawing. Then select or type the old name so it appears in the *Old Name:* edit box. Specify the new name in the *Rename To:* edit box. You must then PICK *the Rename To:* tile and the new names will appear in the list. Then select the *OK* tile to confirm.

The *Rename* dialog box can be used with wildcard characters to specify a list of named objects for renaming. For example, if you desired to rename all the "DIM-*" layers so that the letters "DIM" were replaced with only the letter "D", the following sequence would be used.

1. In the *Old Name* edit box, enter "DIM-*" and press Enter. All of the layers beginning with "DIM-" are highlighted.
2. In the *Rename To:* edit box, enter "D-*." Next, the *Rename To:* tile must be PICKed. Finally, PICK the *OK* tile to confirm the change.

The *-Rename* command (with a hyphen prefix) can be used to rename objects one at a time. For example, the command sequence might be as follows:

Command: ***-RENAME***
Enter object type to rename [Block/Dimstyle/LAyer/LType/Material/Plotstyle/textStyle/Tablestyle/Ucs/VIew/VPort]: ***b***
Enter old block name: **chair7**
Enter new block name: **desk-chair**
Command:

### Setbylayer

| Ribbon | Menu Bar | Command (Type) | Alias (Type) | Shortcut |
|---|---|---|---|---|
| *Home Modify* (icon) | *Modify Change to Bylayer* | *Setbylayer* | ... | ... |

The *Setbylayer* command allows you to change the *ByBlock* property settings (*color*, *linetype*, *lineweight*, etc.) for selected objects to a *ByLayer* setting. This feature can be very helpful in collaborative environments where drawings and *Blocks* are shared. For example, you may insert *Blocks* from a library or from other drawings that have *ByBlock* settings. Use the *Setbylayer* command to convert the *Blocks* to assume the properties of the layers where the *Blocks* reside in the new drawing.

Command: ***SETBYLAYER***
Current active settings: Color Linetype Lineweight Material PlotStyle
Select objects or [Settings]: **PICK**
Select objects or [Settings]: **PICK**
Select objects or [Settings]: **Enter**
Change ByBlock to ByLayer? [Yes/No] <Yes>: **Enter**
Include blocks? [Yes/No] <Yes>: **Enter**
2 objects modified.
Command:

If the *Settings* option is selected, the *SetByLayer Settings* dialog box is displayed (not shown). Here, you can specify which object properties are set to *ByLayer*: *Color*, *Linetype*, *Lineweight*, *Material*, *Transparency*, and *Plotstyle*.

## DESIGNCENTER

The DesignCenter window (Figure 21.24) allows you to navigate, find, and preview a variety of content, including *Blocks*, located anywhere accessible to your workstation, then allows you to open or insert the content using drag-and-drop.

**FIGURE 21.24**

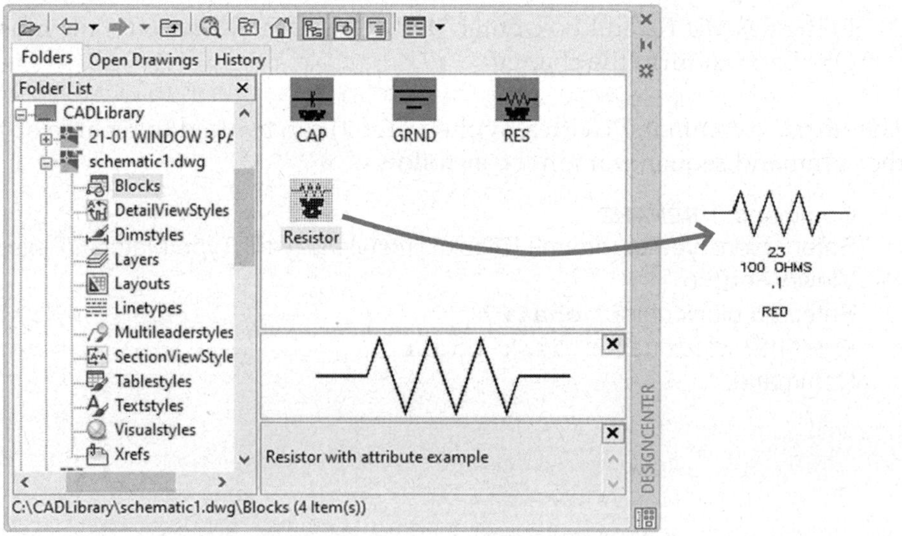


"Content" that can be viewed and managed includes other drawings, *Blocks*, *Dimstyles*, *Layers*, *Layouts*, *Linetypes*, *Table Styles*, *Textstyles*, *Xrefs*, raster images, and URLs (Web site addresses). In addition, if you have multiple drawings open, you can streamline your drawing process by copying and pasting content, such as layer definitions, between drawings. (Dimension Styles and Xrefs are discussed in Chapters 29 and 30, respectively.)

## Adcenter

| Ribbon | Menu Bar | Command (Type) | Alias (Type) | Shortcut |
|---|---|---|---|---|
| *View Palettes* (icon) | *Tools Palettes > DesignCenter* | *Adcenter* | *ADC* | *Ctrl+2* |

Accessing *Adcenter* by any method produces the DesignCenter palette. There are two sections to the window. The left side is called the Tree View and displays a Windows File Explorer-type hierarchical directory (folder) structure of the local system (Figure 21.25). The right side is called the Content Area and displays lists, icons, or thumbnail sketches of the content selected in the Tree View. The Content Area can display *Blocks*, *Dimstyles*, *Layers*, *Layouts*, *Linetypes*, *Table Styles*, *Textstyles*, *Xrefs*, and a variety of other content.

**FIGURE 21.25**

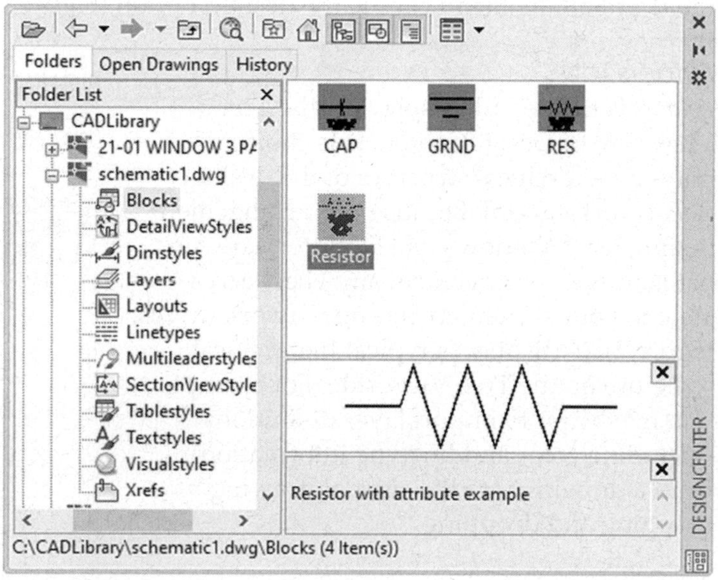


Generally, the Content Area is used to drag and drop the icons or thumbnails from the Content Area into the current drawing. However, you can also streamline a variety of tasks such as those listed here:

- Browse sources of drawing content including open drawings, other drawings, raster images, content within the drawings (*Blocks*, *Dimstyles*, *Layers*, and so on), content on network drives, or content on a Web page.
- Insert, attach, or copy and paste the content (drawings, images, *Blocks*, *Layers*, etc.) into the current drawing.
- Create shortcuts to drawings, folders, and Internet locations that you access frequently.
- Use a special search engine to find drawing content on your computer or network drives. You can specify criteria for the search based on key words, names of *Blocks*, *Dimstyles*, *Layers*, etc., or the date a drawing was last saved. Once you have found the content, you can load it into DesignCenter or drag it into the current drawing.
- Open drawings by dragging a drawing (.DWG) file from the Content Area into the drawing area.

These features of DesignCenter are explained in the following descriptions of the DesignCenter toolbar icons.

### Resizing, Docking, and Hiding the DesignCenter Palette

You can change the width and height of DesignCenter by resting your pointer on one of the borders (not the title bar) until a double arrow appears, then dragging to the desired size. Alternately, rest your pointer on the lower corner (where the bevel appears) to resize both height and width. You can also move the bar between the Content Area and the Tree View area. Dock DesignCenter by clicking the title bar, then dragging it to either edge of the drawing window.

You can hide the DesignCenter palette by using the *Auto-hide* option (click on the *Properties* button at the bottom of the title bar to produce the *Properties* menu as shown in Fig. 21-22).

When *Auto-hide* is on, the palette is normally hidden (only the title bar is visible) and the palette appears when you bring the pointer to the title bar.

## DesignCenter Options

In addition to selecting the icons from the toolbar, you can right-click the Palette background to produce the shortcut menu and choose the desired option (Figure 21.26).

### *Folders* Tab

*Folders* is the default display for the Tree View side of DesignCenter. This choice displays a hierarchical structure of the desktop (local workstation). Because this arrangement is similar to Windows File Explorer, you can navigate and locate content anywhere accessible to your system, including network drives. Figure 21.26 displays a typical hierarchical structure on the Tree View side. For example, you may want to import layer definitions (including color and linetype information) from a drawing into the current drawing by dragging and dropping.

FIGURE 21.26

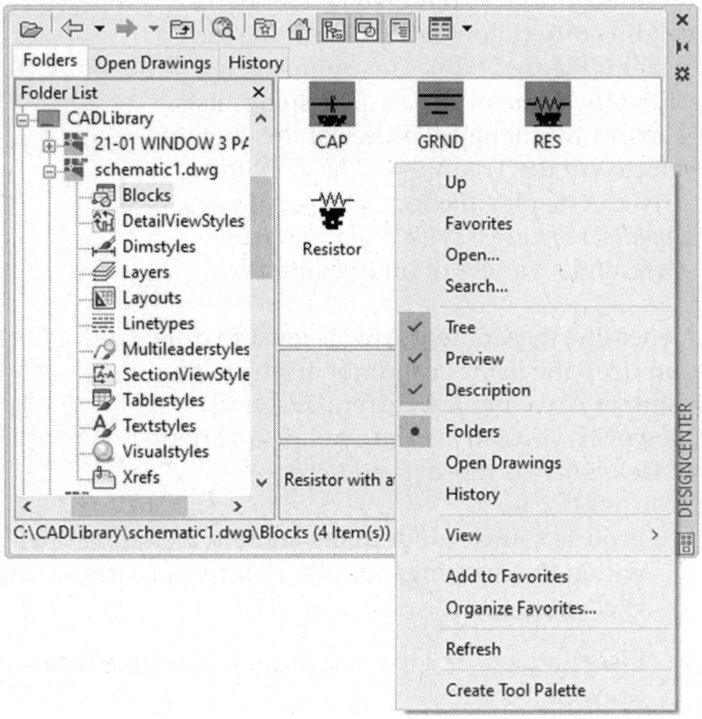


### *Open Drawings* Tab

This option changes the Tree View to display all open drawings (Figure 21.27). This feature is helpful when you have several drawings open and want to locate content from one drawing and import it into the current drawing. As shown in Figure 21.27, you may want to locate *Block* definitions from one drawing and *Insert* them into <u>another</u> drawing. Ensure you make the desired "target" drawing current in the drawing area before you drag and drop content from the Content Area.

FIGURE 21.27

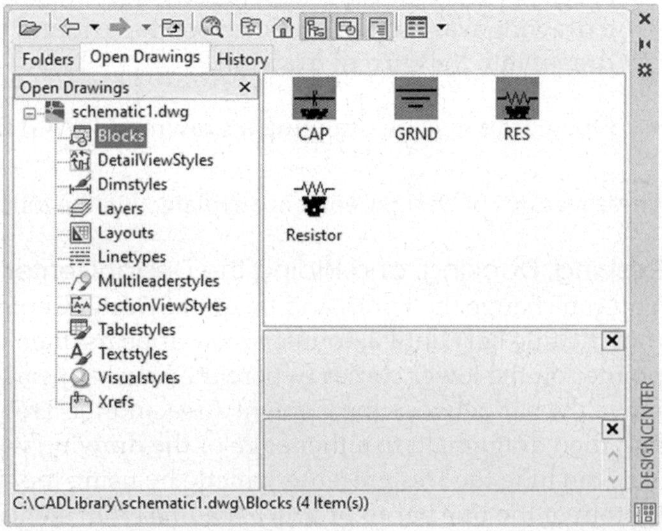

### *History* Tab

This option displays a history (chronological list) of the last 20 file locations accessed through DesignCenter (Figure 21.28). The purpose of this feature is simply to locate the file and load it into the Content Area. Load the file into the Content Area by double-clicking on it.

**FIGURE 21.28**

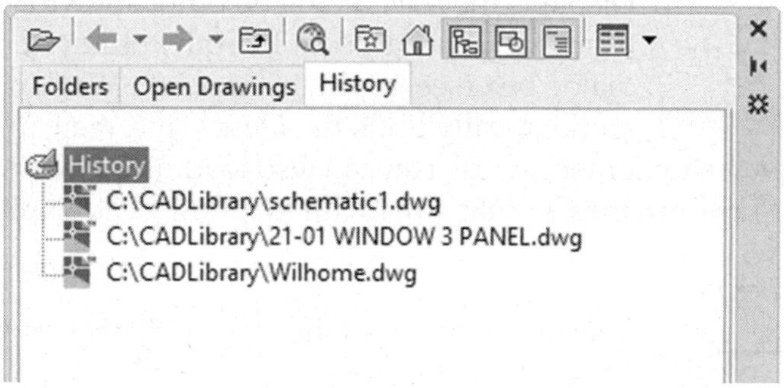


### *Tree View* Toggle

Tree View is helpful for navigating your system for content. Once the desired folder or drawing is found and highlighted in Tree View, you may want to toggle Tree View off so only the Content Area is displayed with the desired content. The desired content may be drawings, *Blocks*, images, or a variety of other content. For example, consider previous Figure 21.28 which displays Tree View on. Figure 21.29 illustrates Tree View toggled off and only the Content Area displayed. The resulting configuration displays only the *Blocks* contained in the selected drawing. Keep in mind that you can also change the Views of the Content Area to display *Large Icons*, *Small Icons*, a *List*, or *Details* (see "Views").

**FIGURE 21.29**

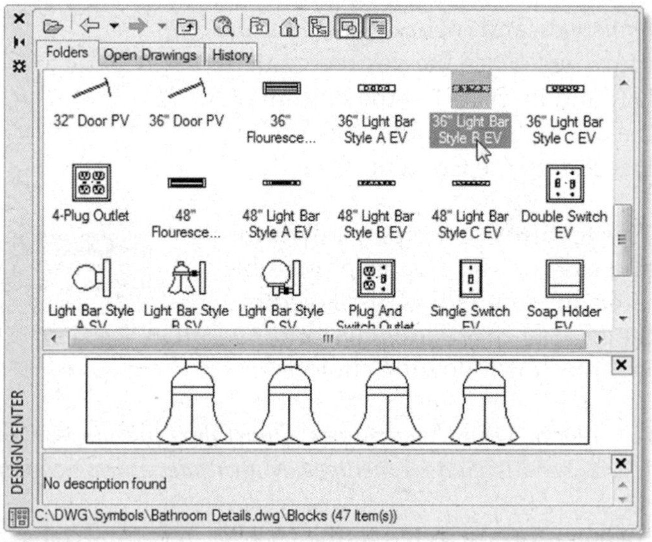


### *Favorites*

This button displays the contents of the AutoCAD Favorites folder in the Content Area. The Tree View section displays the highlighted folder in the Desktop view. You can add folders and files to *Favorites* by highlighting an item in Tree View or the Content Area, right-clicking on it, and selecting *Add to Favorites* from the shortcut menu (Figure 21.30).

**FIGURE 21.30**

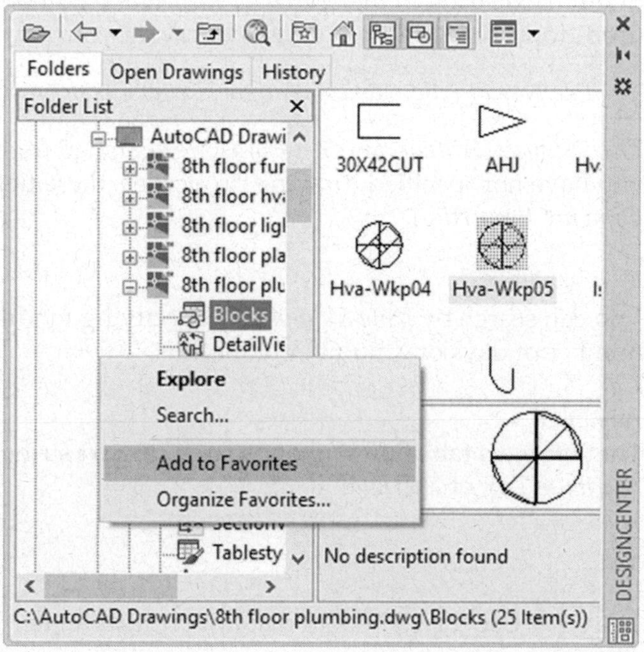

### Load

Displays the *Load* dialog box (not shown), in which you can load the Content Area with content from anywhere accessible from your system. The *Load* dialog box is identical to the *Select File* dialog box (see "*Open*," Chapter 2, Working with Files). After selecting a file, DesignCenter automatically finds the file in Tree View and loads its content (*Blocks*, *Layers*, *Dimstyles*, etc.) into the Content Area. You can also load files into the Content Area using Windows File Explorer (see "Loading the Content Area with Windows File Explorer" at the end of this section).

### Search

This button invokes the *Search* dialog box (Figure 21.31), in which you can specify search criteria to locate drawing files, *Blocks*, *Layers*, *Dimstyles*, and other content within drawings. Once the desired content is found in the list at the bottom of the dialog box, double-click on it to load it into the Content Area.

**FIGURE 21.31**

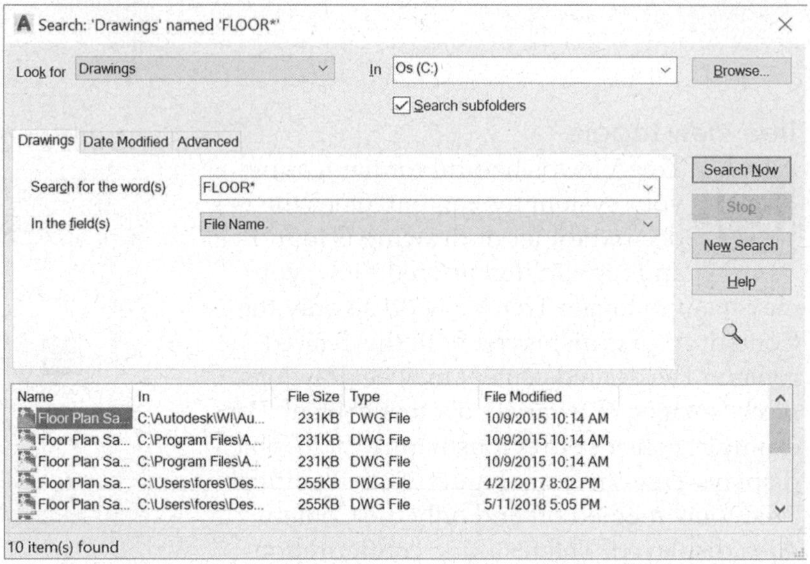


This feature is extremely powerful and easy to use. The list of possible items to search for is displayed in the *Look for* drop-down list and includes the following choices:

*Blocks, Detail View Styles, Dimstyles, Drawings, Drawings and Blocks, Hatch Pattern Files, Hatch Patterns, Layers, Layouts, Linetypes, Multileaderstyles, Section View Styles, Tablestyles, Textstyles, Visualstyles, Xrefs*

If *Drawings* is selected in the *Look for:* list, three tabs (shown in Fig. 21-27) appear to allow you to refine the search criteria.

*Drawings*
Enter a text string in the *Search for the word(s)* edit box. Wildcards can be used. From the *In the field(s)* drop-down list, select what field you want your text string to apply to:

*File Name, Title, Subject, Author, Keywords*

*Title*, *Subject*, *Author*, and *Keywords* are searched for only in the *Drawing Properties* (description), so if you have not specified drawing properties, these fields are not useful (see Chapter 2 for information on *Drawing Properties*).

*Date Modified*
You can search by *Date Modified* by specifying modification between two dates or during the previous X number of days or X number of months.

*Advanced*
The *Advanced* tab allows you to search for text strings in the *Block name, Block and drawing description, Attribute Tag*, or *Attribute Value*.

### Up

Use the *Up* tool to display the next higher level in the Tree View hierarchy.

### *Preview*

This button causes a preview image (thumbnail sketch) of the selected item to appear at the bottom of the Content Area. If there is no preview image saved with the selected item, the *Preview* area is empty. You can resize the preview image by dragging the bar between the Content Area and the preview pane.

To make a preview image appear, you must select an item (drawing, *Block*, or image file) from the Content Area. Selecting items in the Tree View does not cause a preview to appear. Figure 21.32 displays a *Block* preview.

### *Description*

This button displays a text description of the selected item at the bottom of the Content Area (Figure 21.32). The description is displayed for *Blocks* (from the *Description* field when the *Block* is created) and for drawings (from *Drawing Properties,* see *DWGPROPS,* Chapter 2).

The description area can be resized. If you display both *Preview* and *Description* panes, the *Description* pane is displayed below the *Preview* pane, separated from it by a bar.

**FIGURE 21.32**

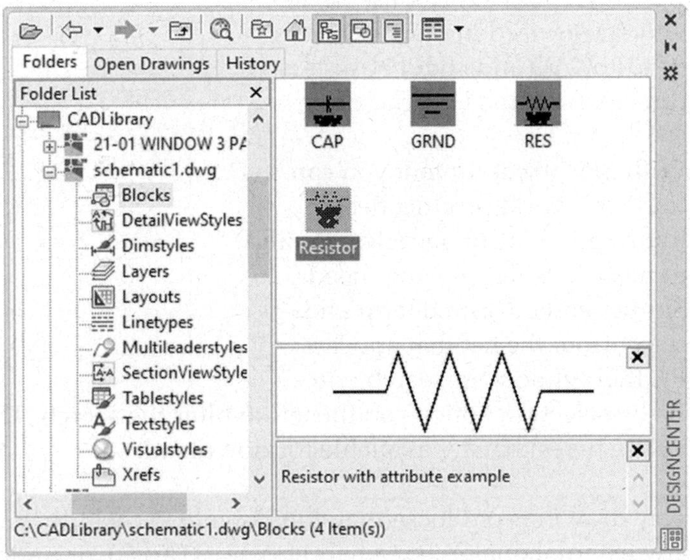


You cannot edit the text description in DesignCenter; however, you can copy it to the Clipboard. Select the text you want to copy, right-click inside the pane, and then select *Copy* from the shortcut menu.

### *Views*

Four possible *Views* of the Content Area are possible. Make the choice by selecting the down arrow to display the options.

*Large icons, Small icons, List, Details*

It is helpful to reset the configuration of the Content Area depending on the job you are performing. For example, if the *Preview* pane is displayed, you can set the Content Area to display only a *List* or *Details*. Or if you prefer to "see" all the drawing files or *Blocks* in the Content Area, it would be better to set the View to *Large icons* and toggle *Preview* off.

Generally, items are sorted alphabetically by name in the Content Area. If you change the *View* to *Details*, you can sort files (not *Blocks*) by name, size, type, and other properties, depending on the type of content displayed in the Content Area. Click on the column header for the column to sort by (click for ascending order, click again for descending order).

### *Back/Forward*

The *Back* button contains a drop-down list displaying content that was previously loaded into the

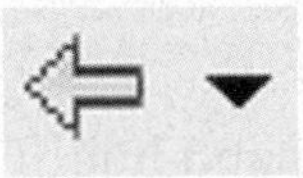

Content Area. Use the drop-down list to locate and load the desired item directly into the Content Area. Once you use this feature (similar to the *Redo* drop-down list), the *Forward* drop-down list holds the list of items you previously used before going *Back*.

**TIP**

### Loading the Content Area with Windows File Explorer

You can use Windows File Explorer to load content into the Content Area. For example, if you are browsing a network drive in Windows File Explorer and locate a drawing file, you can drag the selected file directly into the Content Area. The selected file must be dropped into the Content Area, not the Tree View, Preview, or description areas. If you drop the file into any area other than the Content Area, the drawing is opened into the AutoCAD session.

### *BIMobject Design Content*

The BIMobject website (Figure 21.33) at https://bimobject.com allows you to browse for and download free manufacturer-provided drawings used in a variety of industries. You can download AutoCAD drawing files and blocks under the categories listed. This website replaces Autodesk Seek®, which was used in older versions of AutoCAD and could be accessed directly from the DesignCenter.

FIGURE 21.33

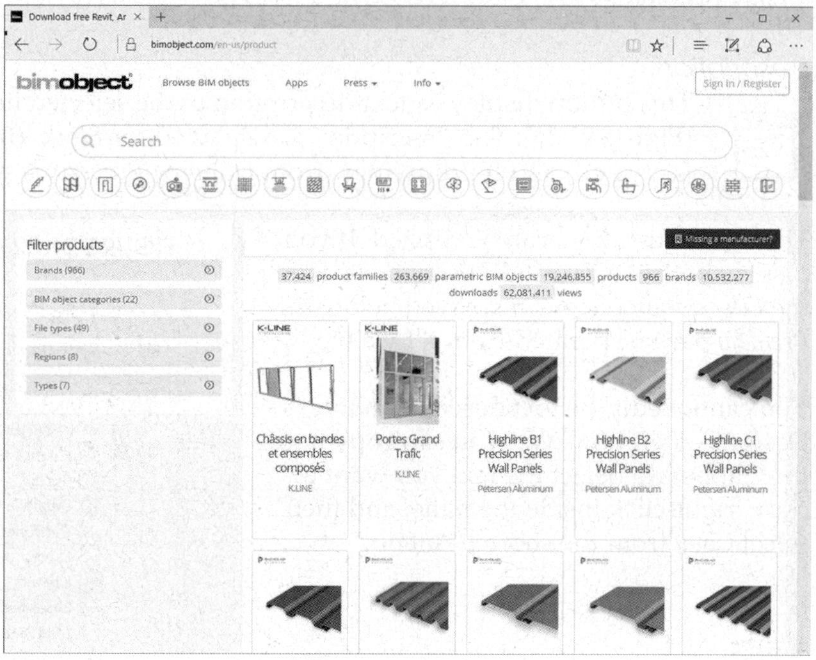


With BIMobject content you can find real-world product design information to enhance designs and to meet specific customer needs. Simply enter a search term and select from the list that appears. Further refine your search criteria by selecting options on the left to filter the search. The left side of the window lists the drawing and image types that are available for download.

Any drawings or blocks that you download include their attribute values. Drawing files do not include any external images, external references (*Xrefs*), or underlay file references (see Chapter 30 for information on *Xrefs*).

## To Open a Drawing from DesignCenter

In addition to using DesignCenter to view and import content contained within drawing files, it is possible to use DesignCenter to *Open* drawings. Drag the icon of the drawing file you want to open from the Content Area and drop it in the drawing area. Remember, you must drag it from the Content Area, not from the Tree View area.

Usually, you would want to drop the selected drawing into a blank (*New*) drawing. Unless the selected drawing is a titleblock and border, drop it into model space, not into a layout (paper space). Make sure the background (drawing area) is visible. You may need to resize the windows displaying any currently open drawings.

A drawing file that is dropped into AutoCAD is actually *Inserted* as a *Block*. The typical Command line prompts appear for insertion point, scale factors, and rotation angle. Generally, enter 0,0 as the insertion point and accept the defaults for X and Y scale factors and rotation angle. If you want to edit the geometry, you will have to *Explode* the drawing.

## To *Insert* a *Block* Using DesignCenter

One of the primary functions of DesignCenter is to insert *Block* definitions into a drawing. When you insert a *Block* into a drawing, the block definition is copied into the drawing database. Any instance of that *Block* that you *Insert* into the drawing from that time on references the original *Block* definition.

You cannot add *Blocks* to a drawing while another command is active. If you try to drop a *Block* into AutoCAD while a command is active at the Command line, the icon changes to a slash circle indicating the action is invalid.

There are two methods for inserting *Blocks* into a drawing using DesignCenter:
1) using drag-and-drop with Autoscaling, and
2) using the *Insert* dialog box with explicit insertion point, scale, and rotation value entry.

### *Block* Insertion with Drag-and-Drop

When you drag-and-drop a *Block* from DesignCenter into a drawing, you cannot key in an insertion point, X and Y scale factors, or a rotation angle. Although you specify an insertion point interactively when you "drop" the *Block* icon, AutoCAD uses Autoscaling to determine the scaling parameters. Autoscaling is a process of comparing the specified units of the *Block* definition (when the *Block* was created) and the *Insertion Scale* set in the *Drawing Units* dialog box of the target drawing. The value options are *Unitless, Inches, Feet, Miles, Millimeters, Centimeters, Kilometers*, and many other choices.

The *Units to scale inserted content* set in the *Drawing Units* dialog box (Figure 21.34) should be set in the target drawing and controls how the *Block* units are scaled when the *Block* is dropped (into the target drawing). It is helpful that the setting here can be changed immediately before dropping a *Block* into the drawing. (Also see "*Units*" in Chapter 6.)

FIGURE 21.34

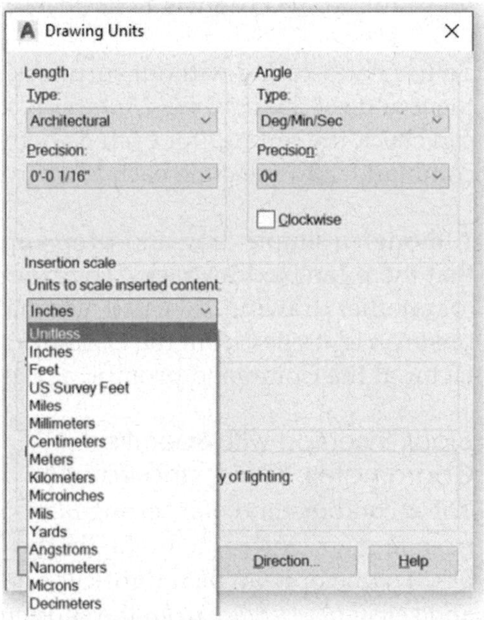


FIGURE 21.35

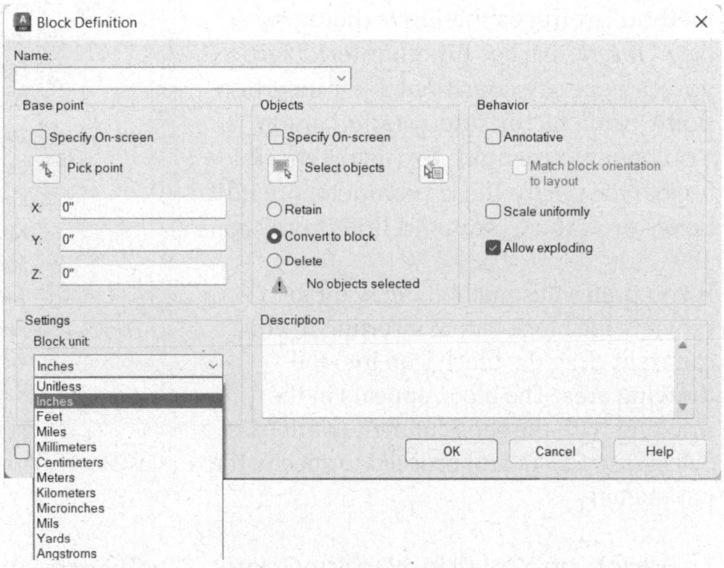


In addition, you can specify the desired *Block* units in the *Block Definition* dialog box (Figure 21.35). This setting is generally assigned when each *Block* is created. (See "*Block*" earlier in this chapter.)

### Resulting Scale Factors with Autoscale (Drag-and-Drop)

This table gives sample scale factors that Autoscaling uses when *Blocks* are inserted into other drawings by the drag-and-drop method. The first row (top) of the table lists the *Block* settings (made in the *Block Definition* dialog box) and the first column (left) lists the drawing units settings (made in the *Drawing Units* dialog box of the target drawing).

| Block Units / Drawing Units | Unitless | Inches | Feet | Millimeters |
|---|---|---|---|---|
| Unitless | 1.000 | 1.000 | 1.000 | 1.000 |
| Inches | 1.000 | 1.000 | 12.000 | 0.039 |
| Feet | 1.000 | 0.083 | 1.000 | 0.003 |
| Millimeters | 1.000 | 25.400 | 304.800 | 1.000 |

Note that when the units in either case (defined in the *Block* or in the target *Drawing Units*) are set to *Unitless*, the scale factor is always 1.00. Also, when the two settings match (*Inches* and *Inches*, or *Feet* and *Feet*), the scale factor is always 1.00. The other cells of the table list a proportion of one unit to the other; for example, .0833 equals 1/12 (feet/inches).

To use Autoscaling without surprises, you should check the units set for the *Blocks* before you insert them and set the *Drawing Units for Block Inserts* in the target drawing according to your needs. You can check the settings for each *Block* by opening the drawing containing the *Blocks*, invoking the *Block* command, and selecting each *Block* from the *Name* drop-down list.

TIP

Although a simple drag-and-drop appears to be a sufficient method for *Block* insertion, be advised that there can be drawbacks. For example, dimension values inside blocks may not be true if a *Block* (or another drawing) is scaled automatically when you drag it into the drawing from DesignCenter. If unsure, right-click from the DesignCenter Content Area and select *Insert Block*, then specify the scale factor at the Command prompt, as described next.

### *Block* Insertion with Specified Coordinates, Scale, and Rotation

The second method of inserting *Blocks* from DesignCenter is to highlight the block name or icon, then right-click on it and select *Insert Block* from the shortcut menu (Figure 21.36). Alternately, double-click on the block name or icon. This method produces the *Insert* dialog box (see "*Insert*" earlier this chapter). Here you can specify coordinates for insertion point, scale factor, and rotation angle. You have your choice (by using check-boxes) to specify these parameters in edit boxes or at the Command line prompts.

If you prefer this method but want to preview the block before inserting it, you can right-drag the block icon into the drawing area. The block appears in the drawing with the shortcut menu options. Select *Insert Block* and proceed to specify the parameters.

**FIGURE 21.36**

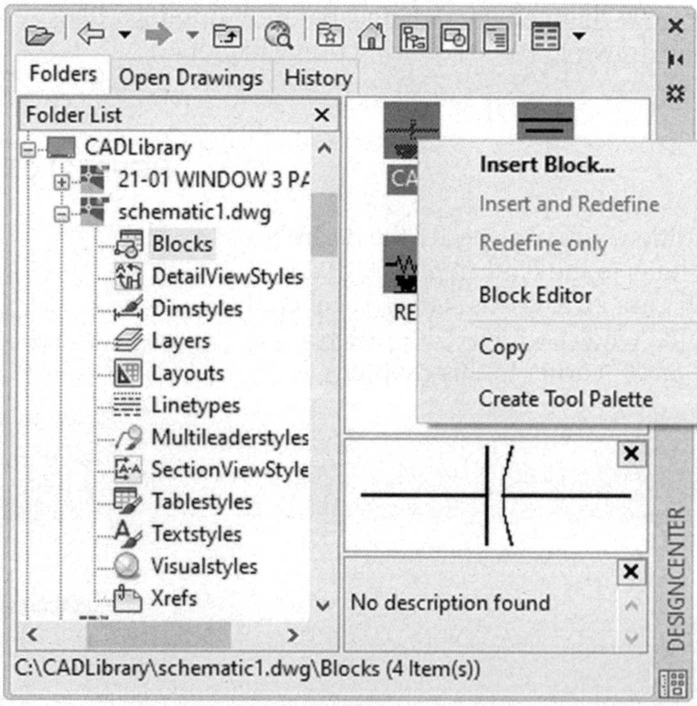


**To Attach an Xref Using DesignCenter** See Chapter 30, Xreferences.

# BLOCK DETECTION AND CONVERSION

AutoCAD machine learning service can analyze your drawing content to recognize objects which you may want to convert into blocks. Select the *Detect and Convert* command (formerly *Detect*) from the *Block palette* or type *Bdetect* at the command prompt. Use the dialog box that appears to select the drawing units. After a wait, the detection results are displayed in the *Detect palette* similar to Figure 21.37. Sometimes no results are detected and the *Search and Convert* selection (*Bsearch* command) may provide better results. *Bsearch* prompts you to select objects to find matching instances to convert to blocks. Detected objects in your drawing become highlighted (Figure 21-38) and the *Detection minitoolbar* (Figure 21-39) appears on your screen. Use it to step through the instances. Select *Convert* from the minitoolbar to show the dialog box (Figure 21-40).

FIGURE 21.37

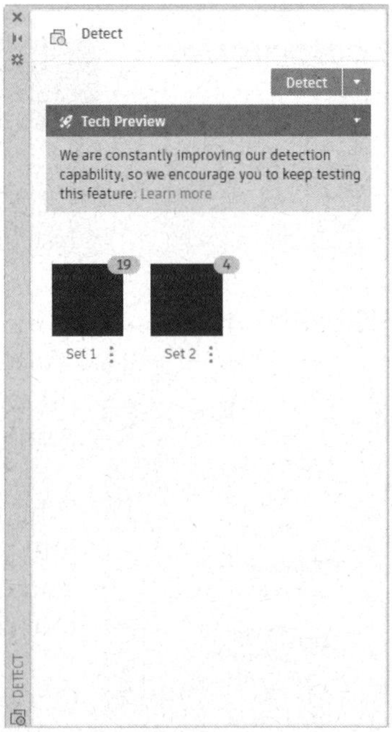


FIGURE 21.38

## Convert to a Block

Using the Convert dialog box to convert geometry or its instances, you can select a block from the drawing, list, a recently used block, or a machine learning-suggested block. Use the *Bconvert* or *Bsearch* command directly to convert selected geometry and identical instances into blocks.

FIGURE 21.40

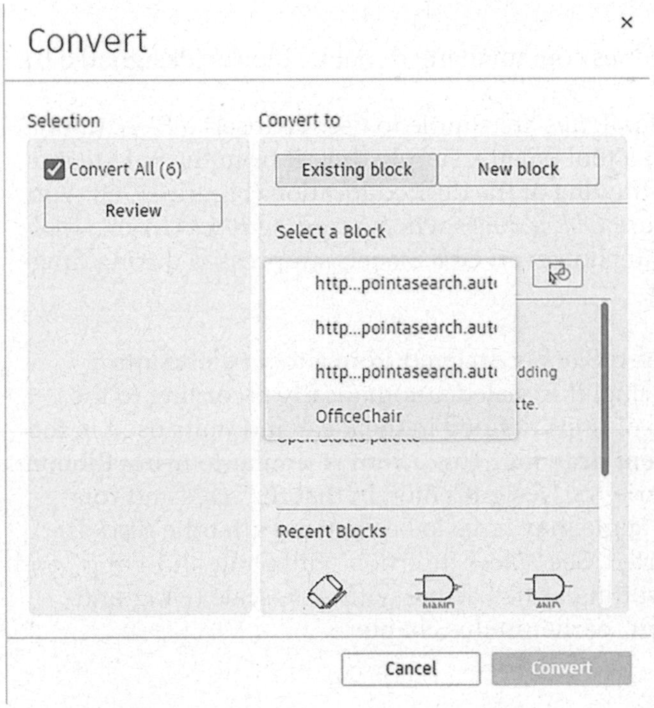


FIGURE 21.39

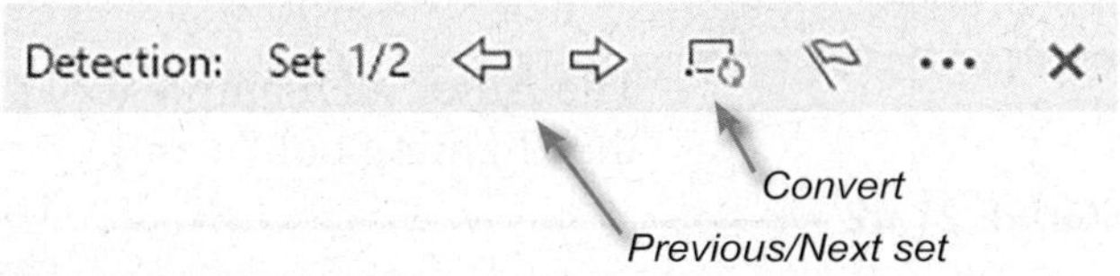


Relevant commands are:

*Bdetect*: shows the Detection palette.
*Bsearch*: shows selected entities and identical instances to convert into blocks.
*Detectclose*: ends the detection review.
*Detectconvert*: shows the Convert dialog box.
*Detectpalette*: opens the Detection palette.
*Detectpaletteclose*: closes the Detection palette.
*Detectprimary*: Specifies the detected instance from which a new block definition is created.
*Detectremove*: Removes selected instances.
*Detectreviewnext* / *Detectreviewprev*: Displays the next / previous set of detected objects.
*Bsearchincludeexistingblocks:* controls default for whether existing block instances are selected.

## TOOL PALETTES

Tool palettes offer a faster and more visible process of inserting *Blocks* and *Hatch* patterns into a drawing. You can use tool palettes to insert a *Block* instead of using the *Insert* command or DesignCenter. You can also use tool palettes to insert hatch patterns (section lines or fill patterns) into a drawing. *Blocks* and hatches that reside on a tool palette are called tools, and several tool properties including scale, rotation, and layer can be set for each tool individually. The features of tool palettes, using tool palettes to insert *Blocks*, and creating tool palettes are discussed in this chapter. Using tool palettes to insert hatch patterns is discussed in Chapter 26, Section Views.

### *Toolpalettes*

| Ribbon | Menu Bar | Command (Type) | Alias (Type) | Shortcut |
|---|---|---|---|---|
| *View Palettes Tool Palettes* | *Tools Palettes > Tool Palettes* | *Toolpalettes* | *TP* | Ctrl+3 |

**FIGURE 21.41**

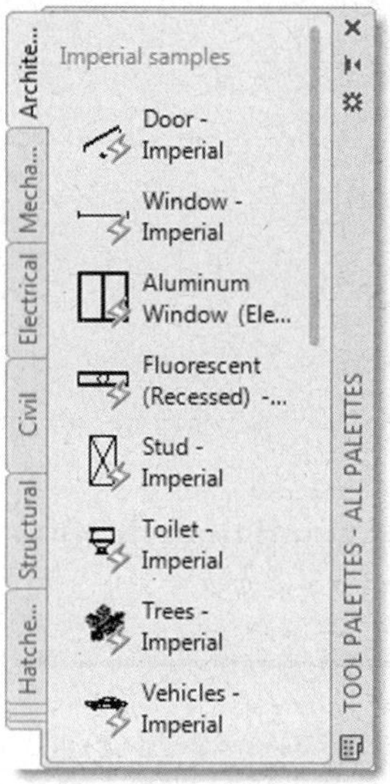


Any of the methods shown in the Command Table above produce the *Tool Palettes* window (Figure 21.41). Tool palettes are the individual tabbed areas (like pages) of the *Tool Palettes* window. The Current Project tab is used with Autodesk's Forma Data Management subscription service.

When you need to add a *Block* to a drawing, you can drag it from the tool palette into your drawing instead of using the *Insert* command or using DesignCenter. You can create your own tool palettes by placing the *Blocks* that you use often on a tool palette.

Once you add blocks from a drawing into the tool palettes, the blocks are "saved" in the tool palettes and can then be dragged and dropped into any other drawing. Therefore, blocks in the tool palettes are not usually located in the current drawing's block definition table, but are automatically located from their source drawing or file. Each block in the tool palettes contains the needed information about its source file.

The AutoCAD sample tool palettes contain many dynamic blocks (designated by the "lighting bolt" icon).

**FIGURE 21.42**

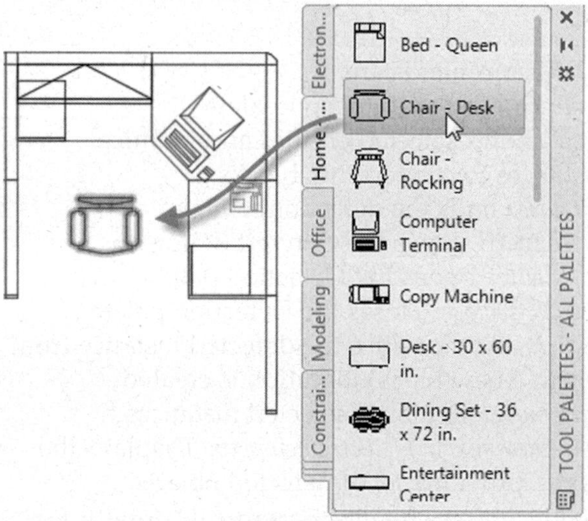


Tool palettes are simple to use. To insert a *Block* (tool) from a tool palette, simply drag it from the palette into the drawing at the desired location (Figure 21.42). You can use *Object Snaps* when dragging *Blocks* from a tool palette; however, *Grid Snap* is suppressed during dragging.

When a *Block* is dragged from a tool palette into a drawing, it is scaled automatically according to the ratio of units defined in the *Block* and units used in the current drawing. This action is similar to using Ribbon galleries or DesignCenter, in that the scale and rotation angle may have to be changed after the *Block* is inserted. See "*Block* Insertion with Drag-and-Drop" and "Resulting Scale Factors with Autoscale (Drag-and-Drop)" earlier in this chapter.

Using the *Tool Palette* window, you can preset the *Block* insertion properties or hatch pattern properties of any tool on a tool palette. For example, you can change the insertion scale of a *Block* or the angle of a hatch pattern. Do this by right-clicking on a tool and selecting *Properties...* (see "Tool *Properties*").

You can also change the scale and rotation angle after inserting the *Block* using grips (see Chapter 19, Grip Editing).

FIGURE 21.43

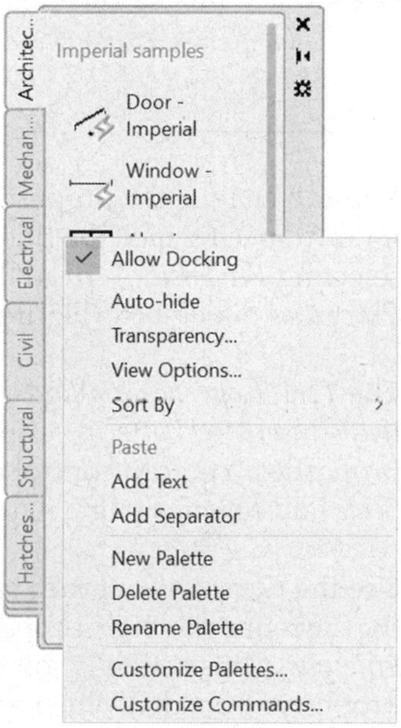


### Palette *Properties*

The options and settings for tool palettes are accessible from shortcut menus in different areas on the *Tool Palettes* window. Right-clicking <u>inside</u> the palette produces the menu shown in Figure 21.43 You can also click on the *Properties* button (on the vertical title bar) to produce a similar menu which adds some options, such as *Move*, *Size*, and *Close*, and removes others, such as *View Options*, *Sort By*, *Add Text*, and *Add Separator*. Palette properties are saved with your AutoCAD profile.

*Move, Size, Close, Allow Docking, Auto-hide*

These options are identical to the same features in DesignCenter. See "Resizing, Docking, and Hiding the DesignCenter Palette" earlier in this chapter.

FIGURE 21.44

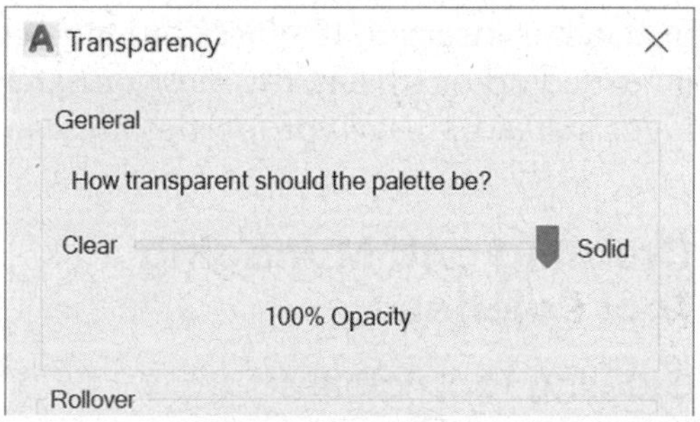


*Transparency...*

Use this option to make the *Tool Palettes* window transparent so you can "see through" the window to the drawing underneath. Set the desired level in the *Transparency* dialog box that appears (Figure 21.44).

FIGURE 21.45

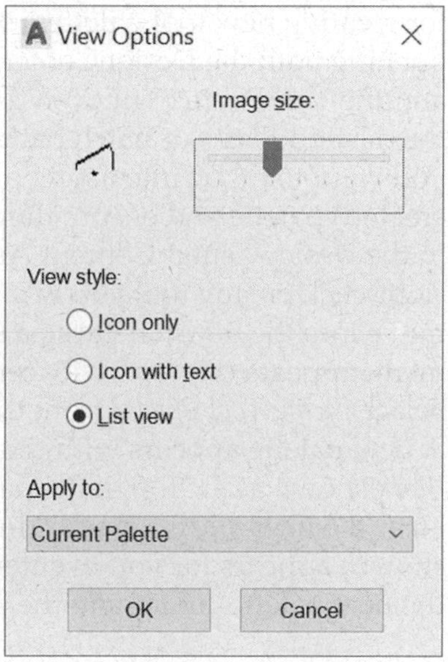


*View Options...*

This option produces the *View Options* dialog box (Figure 21.45). You can specify the *Image size* and the *View style*. The *List View* places the tools with text in a vertical column regardless of the palette width, whereas the *Icon with text* option is best to use when you increase the window width to create multiple columns.

*Add Text, Add Separator, New Palette, Delete Palette, Rename Palette,* and *Customize*

See "Creating and Managing Tool Palettes" later in this section.

### Tool *Properties*

FIGURE 21.46

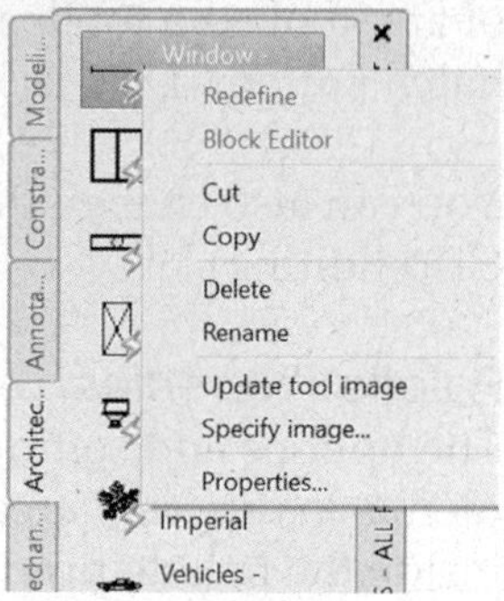


If you right-click on a tool (*Block* or hatch pattern), a shortcut menu appears (Figure 21.46) allowing you to change properties for that specific tool. You can *Cut* any tool and *Paste* it to another palette or *Copy* a tool to another palette. Selecting *Delete* removes it from that palette only. If the source block file has been changed, you can *Update tool image*. Select *Block Editor* to edit the current block in the Block Editor.

You can change the properties (such as scale and rotation angle) for any specific tool. Do this by right-clicking on the tool and selecting *Properties...* from the shortcut menu to produce the *Tool Properties* dialog box (Figure 21.47).

FIGURE 21.47

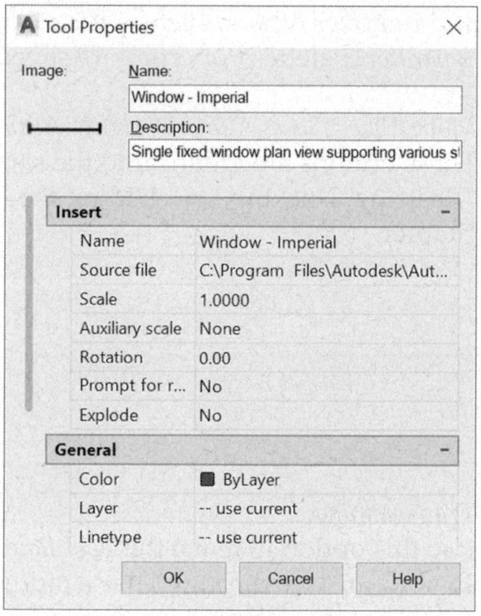


The *Tool Properties* dialog box can have three categories of properties—*Insert* (or *Pattern*) properties, *General* properties, and *Custom* properties. You can control object-specific properties such as *Scale* and *Rotation* angle in the *Insert* list.

Use the *General* properties section only if you want to override the current drawing's property settings such as *Layer*, *Color*, and *Linetype*. (This action is sometimes referred to as setting a tool property override.) When a property override is set, special conditions may exist. For example, if a specified layer does not exist in the drawing, that layer is created automatically when the *Block* or hatch is inserted. If a *Block* or hatch is inserted on a layer that is turned off or frozen, the *Block* or hatch is created on the current layer instead. *Custom* properties are used for dynamic blocks.

## Creating and Managing Tool Palettes

### Creating Tool Palettes using DesignCenter

FIGURE 21.48

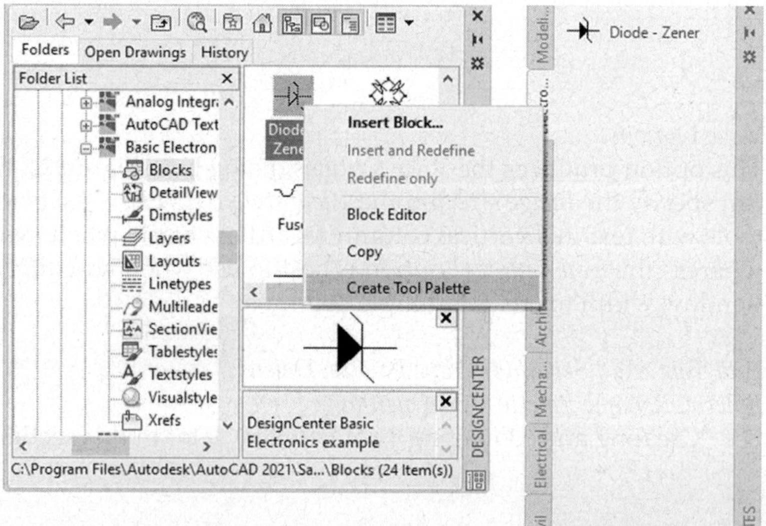


DesignCenter provides a simple method for creating new tool palettes. Begin by opening both the DesignCenter window and the *Tool Palettes* window. Locate the desired *Blocks* or hatch patterns you want to insert into a new (not yet created) palette and ensure they appear in the DesignCenter Content Area. Next, right-click on any tool you want to move into the new palette so a shortcut menu appears (Figure 21.48, center). Select *Create Tool Palette* from the menu. A new palette appears with the selected *Block* shown at the top of the palette. Near the new *Block*, a small edit box (not shown) appears for you to enter in the desired tool palette name. After doing so, you are presented with the new palette containing the selected *Block* (as shown in Figure 21.48, right side).

To add additional *Blocks* or hatch patterns into your new palette, drag and drop them from the DesignCenter Content Area into the new palette (Figure 21.49). Next, use the tool *Properties* shortcut menu, if needed, to preset the scale and rotation angle for the tool, as described earlier.

**FIGURE 21.49**

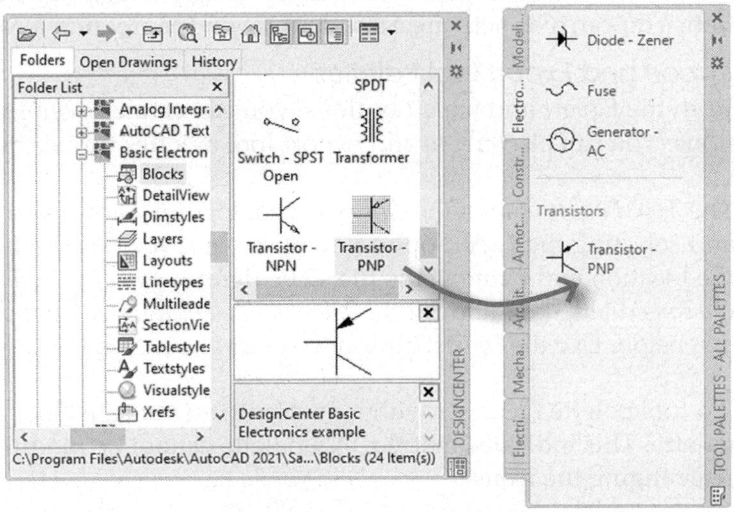


To help organize your new palette, you can add text or add a separator line between sections of the palette. Do this by right-clicking in the blank palette area and select *Add Text* or *Add Separator* from the shortcut menu.

You can also create entire tool palettes fully populated with *Blocks* from all the *Blocks* contained in any drawing using DesignCenter. Do this by locating the drawing file in the Tree View area of DesignCenter, then right-clicking on the drawing file so the shortcut menu appears. Select *Create Tool Palette* (Figure 21.50). A new tool palette is automatically added to the *Tool Palettes* window containing all the *Blocks* from the selected drawing. The new tool palette has the same name as the drawing.

**FIGURE 21.50**

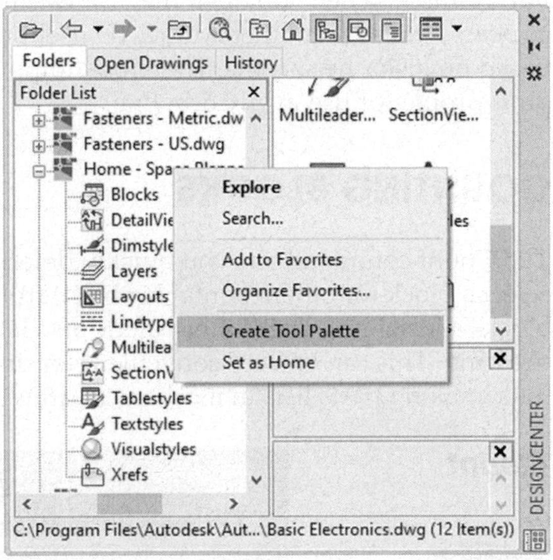


## Creating Tool Palettes using Palette Properties

Selecting any blank area inside a tool palette and right-clicking or selecting the *Properties* button on the *Tool Palette* window produces the palette *Properties* menu. Selecting *New Palette* from this menu prompts you for a name and creates a new palette. You may populate the palette with the desired *Blocks* or hatch patterns using DesignCenter as described previously.

## *Customize*, *Delete*, and *Rename* Tool Palettes

Use the *Properties* button on the *Tool Palette* window or right-click inside a tool palette to produce the palette *Properties* menu. The *Rename* option produces an edit box for you to specify a new name for the current palette. Select *Delete Palette* from the menu to delete an entire palette and all of its contents from the *Tool Palette* window. The *Confirm Palette Deletion* dialog box (not shown) appears for you to *OK* or *Cancel* the deletion.

**FIGURE 21.51**

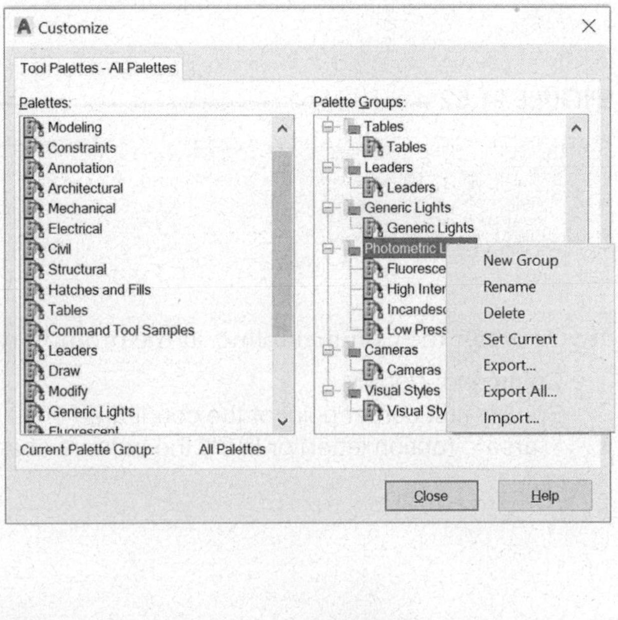

The *Customize* option from the shortcut menu invokes the *Customize* dialog box (Figure 21.51). Here you can change the order of the palettes by simply dragging and dropping the palette names to a new location. You can also activate a shortcut menu for creating a *New Group*.

### *Import* and *Export* Tool Palettes

With the *Export* and *Import* options, you can create several tool palettes to use, each specific applications. You can also create and export tool palettes for use by others in your school or office.

The *Tool Palettes* tab of the *Customize* dialog box allows you to export tool palettes to a file. Right-clicking and selecting the *Export* option invokes the *Export Palette* dialog box (not shown) where you can specify the location and name of the file. (The default path for tool palette files is set in the *Files* tab of the *Options* dialog box under *Tool Palettes File Locations*.) Tool palettes are automatically assigned a .XTP file extension. Use the *Import* button to select and activate a specific tool palette file.

If a tool palette file is set with a read-only attribute, a "lock" icon displays in a lower corner of the tool palette. This indicates that the tool palette cannot be modified beyond changing its display settings and rearranging the icons.

NOTE: Tool palettes are not saved with the drawing file but are saved as part of your AutoCAD profile. For example, when you open the *Tool Palettes* window from any drawing, your specific set of palettes appears, whereas another AutoCAD user on the same computer can have a different set of palettes based on his or her profile. Therefore, importing a tool palette file (*.XTP) adds the specific palettes to your profile for use in any drawing.

## COUNTING BLOCKS

The Count command lets you quickly determine the number of instances of *objects* in a drawing. You select a block (or other countable object) from model space. The toolbar shows the count of the selected blocks's instances. The Count palette also lets you insert a table with the count data into the current drawing. This can be very convenient for drawing take-offs such as window tables in architectural drawings, making parts lists in mechanical drawings, and many more.

### *Count*

| Ribbon | Menu Bar | Command (Type) | Alias (Type) | Shortcut |
|---|---|---|---|---|
| *View*<br>*Palettes*<br>*Count* | | *Count* | | |

**FIGURE 21.52**

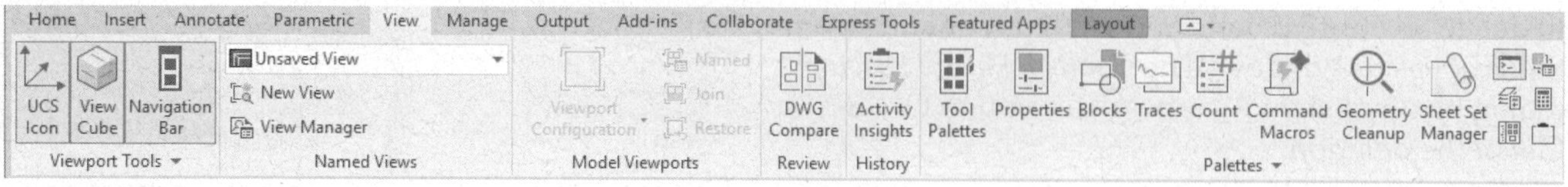


If you prefer the Command line, at the prompt, type *Count* and follow the sequence shown here.

Command: ***COUNT***

Specify first corner point of the count area or [Current area/Entire model space/Object/Polygonal] <Current area>: (option letter) or PICK the corners of a window to specify the area in which to count

The *Count palette* opens (see Figure 21.53) displaying the counted number of instances of the selected block or object. In this case the Chair7 block. Use the green checkmark to close the Count toolbar when finished.

**FIGURE 21.53**

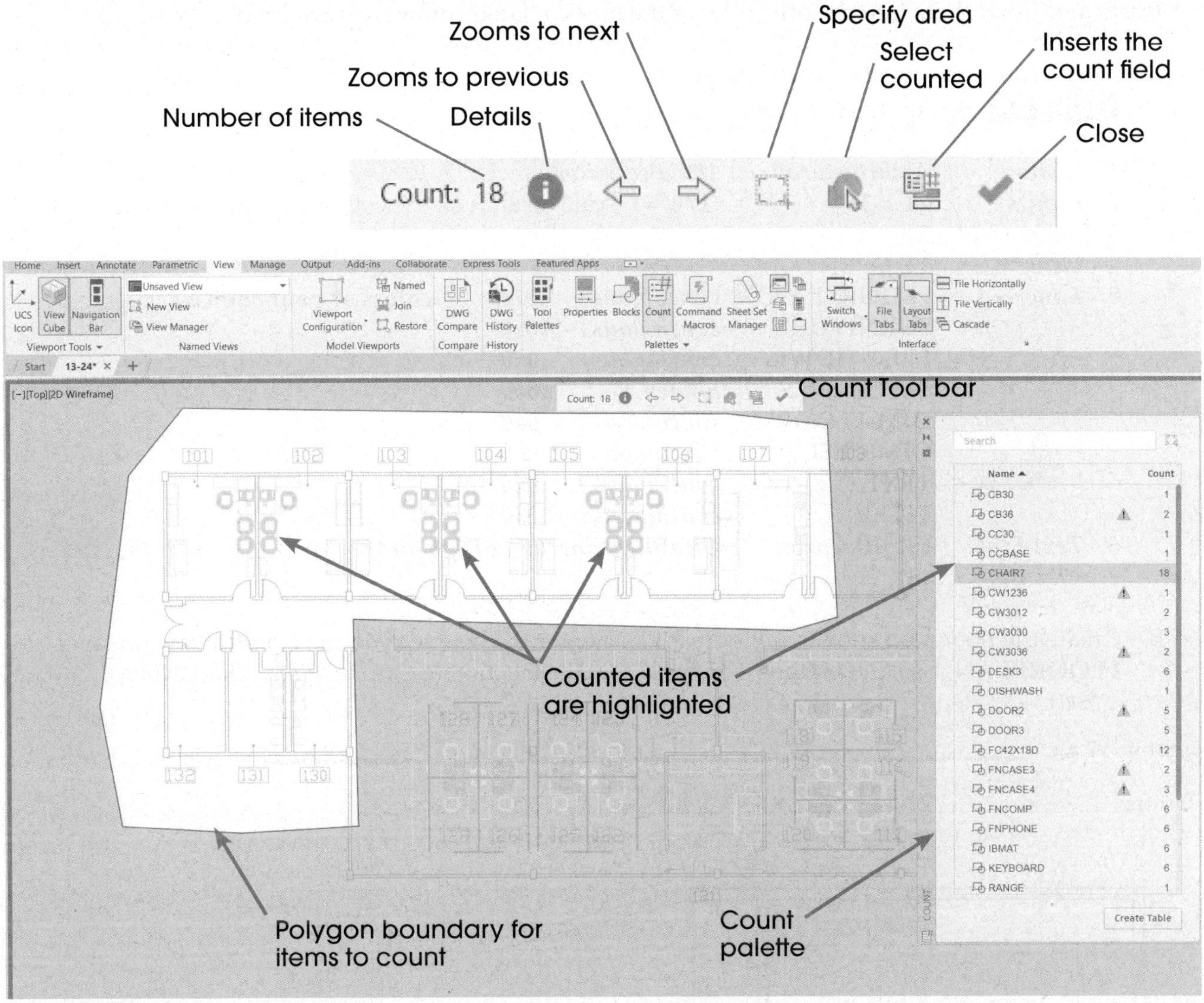


The following system variables related to Count allow you to control the appearance and function of the command results:

COUNTAREA - define area to count the instances of an object or block.

COUNTAREACLOSE - cancels the existing count selection area.

COUNTCHECK - controls the types errors checking in the count result.

COUNTCOLOR - sets the color used to highlight the counted drawing objects.

COUNTERRORCOLOR - sets the color used to highlight counted objects that are potential errors.

COUNTERRORNUM - shows the number of errors in the current count.

COUNTNUMBER - shows the number of the current count.

COUNTPALETTESTATE - reports if the Count palette is open or closed.

COUNTSERVICE - controls the background indexing of the count.

SELECTCOUNT - finds all currently counted objects with properties that match the selected objects; adds them to the selection set.

## CHAPTER EXERCISES

1. ***Block, Insert***

In the next several exercises, you will create an office floor plan, then create pieces of furniture as *Blocks* and *Insert* them into the office. All of the block-related commands are used.

A. Start a *New* drawing. Select the **ACAD.DWT** template. Use *Save* and assign the name **OFF-ATT.** Set up the drawing as follows:

| | | | | |
|---|---|---|---|---|
| 1. *Units* | *Architectural* | *1/2" Precision* | | |
| 2. *Limits* | **48' x 36'** | (1/4"=1' scale on an A size sheet), drawing scale factor = 48 | | |
| 3. *Snap, Grid* | **3** | | | |
| 4. *Grid* | **12** | | | |
| 5. *Layers* | **FLOORPLAN** | *continuous* | *.014* | **colors of your choice** |
| | **FURNITURE** | *continuous* | *.060* | |
| | **ELEC-HDWR** | *continuous* | *.060* | |
| | **ELEC-LINES** | *hidden 2* | *.060* | |
| | **DIM-FLOOR** | *continuous* | *.060* | |
| | **DIM-ELEC** | *continuous* | *.060* | |
| | **TEXT** | *continuous* | *.060* | |
| | **TITLE** | *continuous* | *.060* | |
| 6. *Text Style* | *CityBlueprint* | *CityBlueprint (TrueType font)* | | |
| 7. *Ltscale* | **48** | | | |

B. Create the floor plan shown in Figure 21.54. Center the geometry in the *Limits*. Draw on layer **FLOORPLAN**. Use any method you want for construction (e.g., *Line, Pline, Xline, Mline, Offset*).

FIGURE 21.54

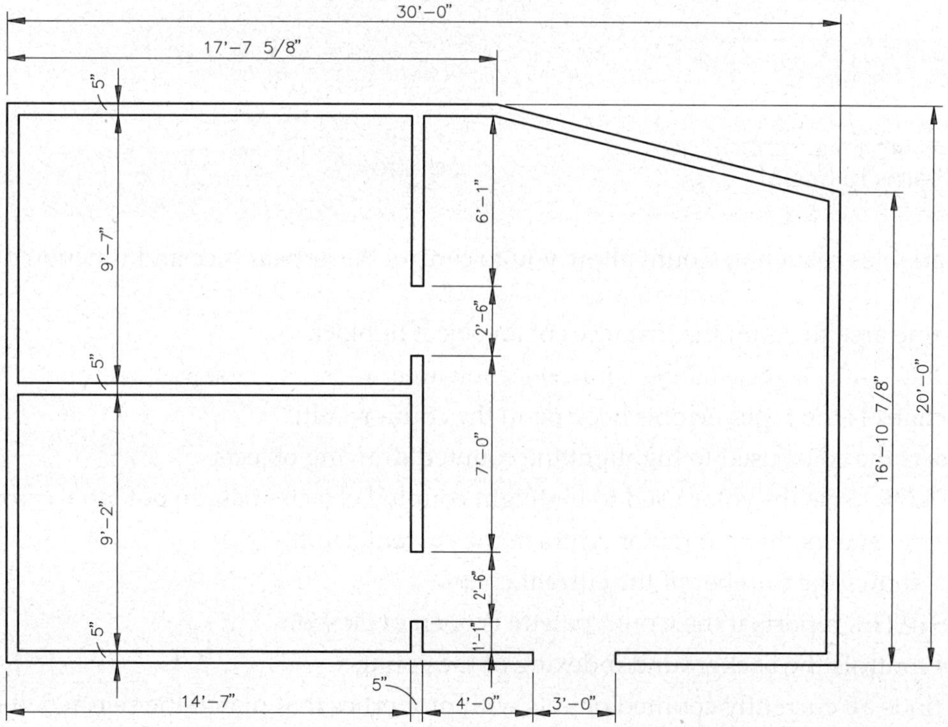

C. Create the furniture shown in Figure 21.55. Draw on layer **FURNITURE**. Locate the pieces anywhere for now. Do not make each piece a *Block*. ***Save*** the drawing as **OFF-ATT**.

Now make each piece a ***Block***. Use the ***name*** as indicated and the ***insertion base point*** as shown by the "blip." Next, use the ***Block*** command again but only to check the list of *Blocks*. Use ***SaveAs*** and rename the drawing **OFFICE.**

**FIGURE 21.55**

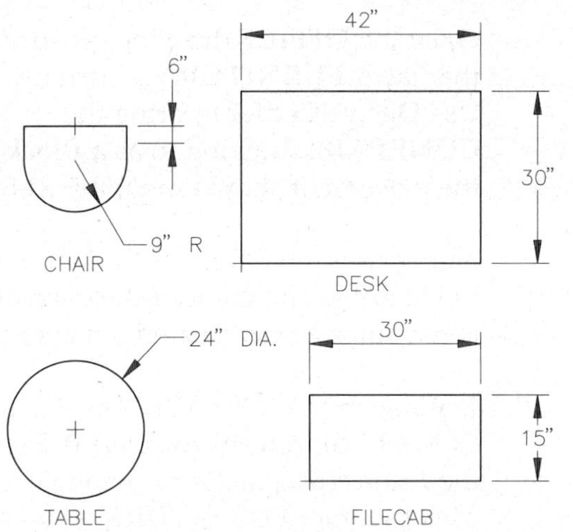


D. Use ***Insert*** to insert the furniture into the drawing, as shown in Figure 21.56. You may use your own arrangement for the furniture, but *Insert* the same number of each piece as shown. ***Save*** the drawing.

**FIGURE 21.56**

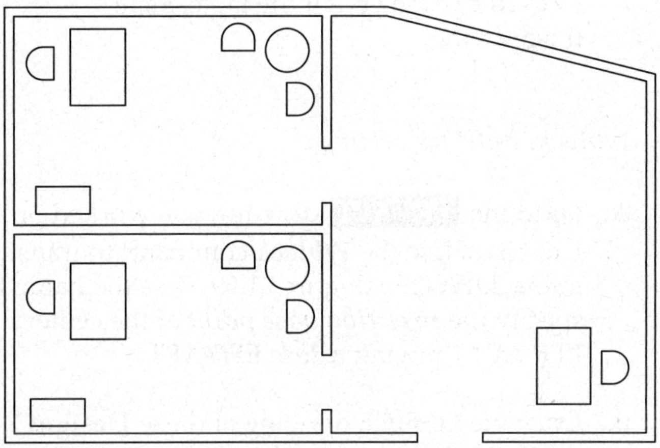

2. **Creating a .DWG file for *Insertion, Base***

Begin a ***New*** drawing. Assign the name **CONFTABL**. Create the table as shown in Figure 21.57 on ***Layer* 0**. Since this drawing is intended for insertion into the **OFFICE** drawing, use the ***Base*** command to assign an insertion base point at the lower-left corner of the table.

When you are finished, ***Save*** the drawing.

**FIGURE 21.57**

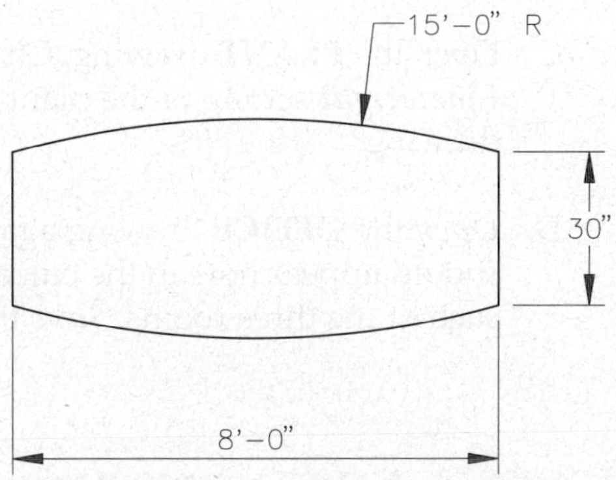

3. *Insert, Explode, Divide*

FIGURE 21.58

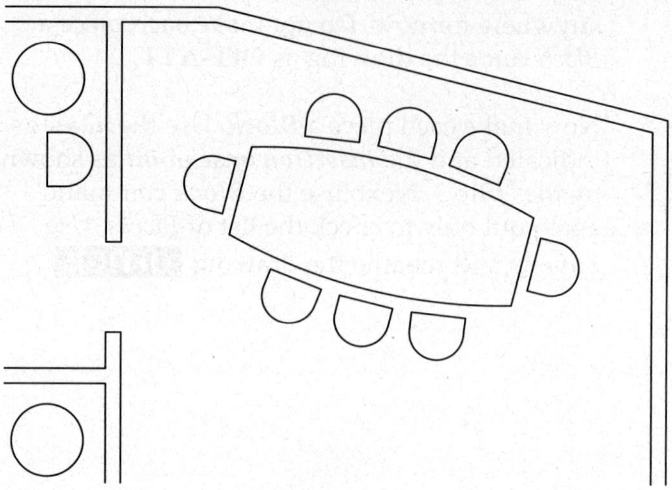

A. ***Open*** the **OFFICE** drawing. Ensure that layer **FURNITURE** is current. Use **DesignCenter** to bring the **CONFTABL** drawing in as a ***Block*** in the placement shown in Figure 21.58.

Notice that the CONFTABL assumes the linetype and color of the current layer, since it was created on layer **0**.

B. ***Explode*** the CONFTABL. The *Exploded* CONFTABL returns to *Layer* **0.** Use the ***Properties*** palette to change it back to ***Layer*** **FURNITURE**. Then use the ***Divide*** command (with the ***Block*** option) to insert the **CHAIR** block as shown in Figure 21.58. Also ***Insert*** a **CHAIR** at each end of the table. ***Save*** the drawing.

4. *Wblock, ByBlock setting*

A. ***Open*** the **EFF-APT** drawing you worked on in Chapter 15 Exercises. Use the ***Wblock*** command to transform the plant into a .DWG file (Figure 21.59). Use the name **PLANT** and specify the ***Insertion base point*** at the center. Do not save the EFF-APT drawing. ***Close*** **EFF-APT**.

FIGURE 21.59

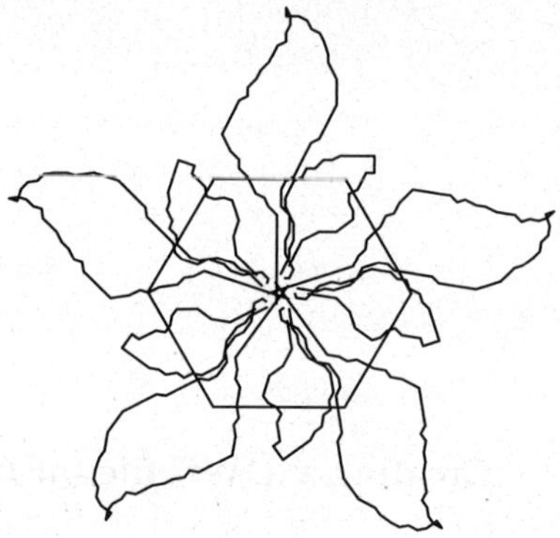

PLANT

B. ***Open*** the **OFFICE** drawing and use **DesignCenter** to ***Insert*** the **PLANT** into one of the three rooms. The plant probably appears in a different color than the current layer. Why? Check the ***Layer*** listing to see if any new layers came in with the PLANT block. ***Erase*** the PLANT block.

C. ***Open*** the **PLANT** drawing. Change the ***Color, Linetype,*** and ***Lineweight*** setting of the plant objects to ***ByBlock***. ***Save*** the drawing.

D. ***Open*** the **OFFICE** drawing again and ***Insert*** the **PLANT** onto the **FURNITURE** layer. It should appear now in the current layer's *color, linetype,* and *lineweight*. ***Insert*** a **PLANT** into each of the three rooms. ***Save*** the drawing.

5. **Redefining a *Block***

After a successful meeting, the client accepts the proposed office design with one small change. The client requests a slightly larger chair than that specified. Open the **Block Editor** and select the **CHAIR** block to edit. Redesign the chair to your specifications. **Save** the CHAIR block and ***Close*** the Block Editor. All previous insertions of the CHAIR should reflect the design change. ***Save*** the drawing. Your design should look similar to that shown in Figure 21.60. ***Plot*** to a standard scale based on your plotting capabilities.

**FIGURE 21.60**

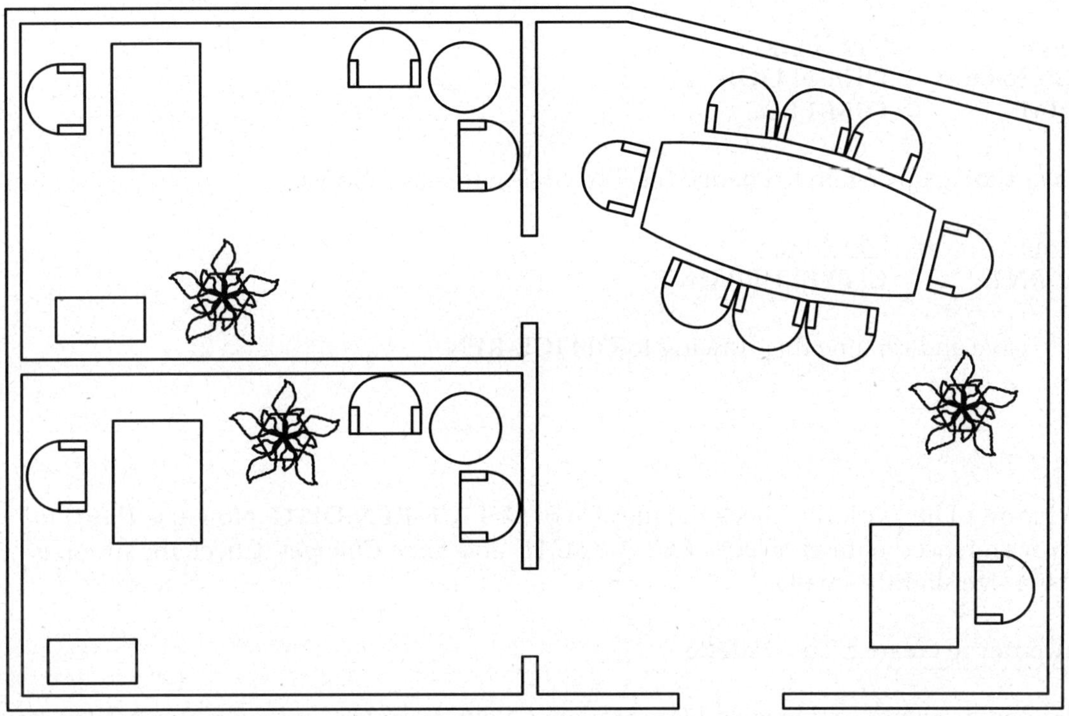

6. ***Rename***

Open the **OFFICE** drawing. Enter *Rename* (or select from the *Format* pull-down menu) to produce the *Rename* dialog box. Rename the *Blocks* as follows:

| New Name | Old Name |
|---|---|
| **DSK** | **DESK** |
| **CHR** | **CHAIR** |
| **TBL** | **TABLE** |
| **FLC** | **FILECAB** |

Next, use the *Rename* dialog box again to rename the *Layers* as indicated below:

| New Name | Old Name |
|---|---|
| **FLOORPLN-DIM** | **DIM-FLOOR** |
| **ELEC-DIM** | **DIM-ELEC** |

Use the *Rename* dialog box again to change the *Text Style* name as follows:

| New Name | Old Name |
|---|---|
| **ARCH-FONT** | **CITYBLUEPRINT** |

Use ***SaveAs*** to save and rename the drawing to **OFFICE-REN**.

7. ***Purge***

Using the Windows File Explorer, check the file size of **OFFICE-REN.DWG.** Now use ***Purge*** to remove any unreferenced named objects. ***Exit AutoCAD*** and ***Save Changes***. Check the file size again. Is the file size slightly smaller?

8. **Use DesignCenter to create a Tool Palette**

A. ***Close*** all open drawings and begin a ***New*** drawing. Next, open ***DesignCenter*** and use the ***Folders*** tab to locate the **OFFICE** drawing in the Tree View area. Expand the plus (+) symbol next to OFFICE.DWG, then click on the word ***Blocks*** just below the drawing name. All of the *Blocks* contained in the drawing should appear in the Content Area.

B. Use the ***Toolpalettes*** command or press **Ctrl+3** to produce the *Tool Palettes* window. Now, back in DesignCenter, locate the OFFICE drawing in the Tree View area and right-click on the drawing name. From the shortcut menu select ***Create Tool Palette***.

C. This action should have automatically created a new tool palette named OFFICE and populated it with all the *Blocks* contained in the drawing. Examine the new tool palette. Right-click on any tool and examine its ***Properties***. Now you have a new tool palette ready to use for inserting the office furniture *Blocks* into any other drawings.

D. Use ***DesignCenter*** again to locate the **HOME - SPACE PLANNER** drawing in the Program Files/Autodesk/AutoCAD 2026/Sample/en-us/DesignCenter folder. In the Tree View area, right-click on the drawing name, then select ***Create Tool Palette*** from the shortcut menu. A new palette should appear in the *Tool Palettes* window fully populated with furniture.

9. **Using Tool Palettes**

In this exercise you will change some of the blocks in the OFFICE-REN drawing using blocks from the *Tool Palettes* window.

A. ***Open*** the **OFFICE-REN** drawing again if not already open. Assume as a result of the meeting with the client, new desks and chairs are requested for the receptionist and the two offices. If not already available, produce the *Tool Palettes* window using the ***Toolpalettes*** command or pressing **Ctrl+3**. Select the new **Home - Space Planner** palette (showing the home furniture).

B. Right-click on the **Chair - Desk** tool and select ***Properties***. In the *Tool Properties* dialog box that appears, note that the scale is **1.00** and the layer is set to **use current**. Check the settings for the **Desk - 30 x 60 in**. With these settings, the new furniture should be usable without changes.

C. Make **FURNITURE** the current layer. ***Erase*** the three desks and the three desk chairs. Drag and drop three chairs and three desks from the palette window and place them near the desired locations in the office. Use ***Rotate*** and ***Move*** to position the new *Blocks* into suitable locations and orientations. Your drawing should look similar to that in Figure 21.61. Use ***SaveAs*** and name the drawing **OFFICE-REV.**

D. Use the Count command to determine how many chairs you will need to order if you use this furniture arrangement.

**FIGURE 21.61**

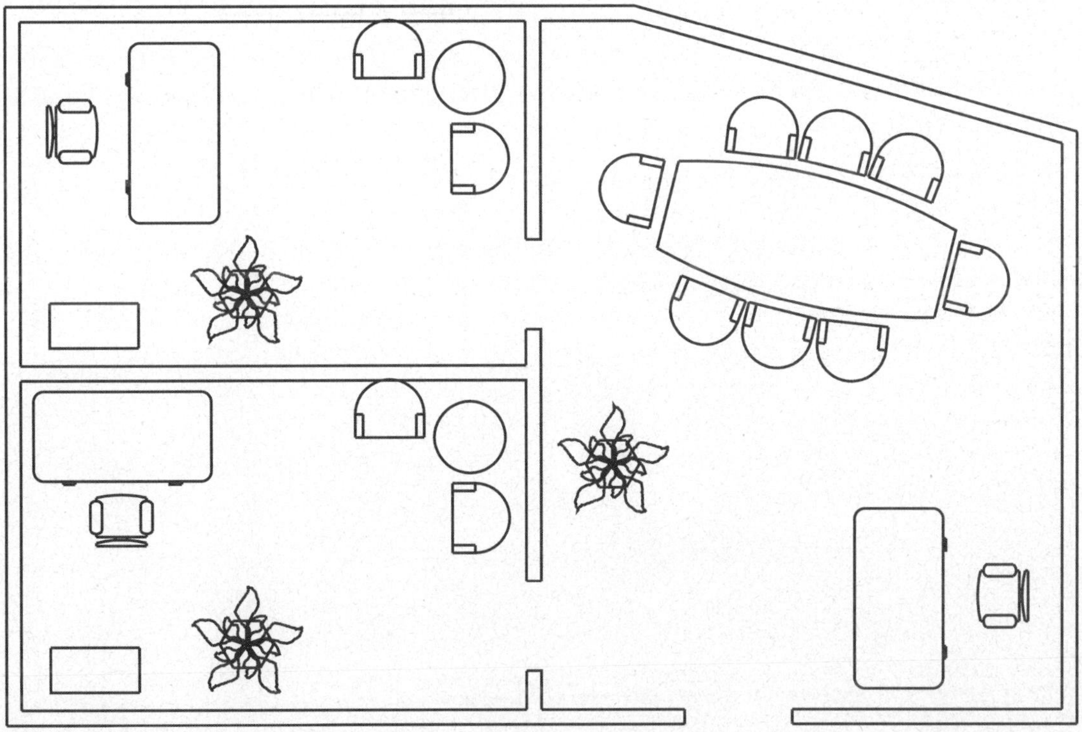

10. Create the process flow diagram shown in Figure 21.62. Create symbols (***Block***) for each of the valves and gates. Use the names indicated (for the *Blocks*) and include the text in your drawing. ***Save*** the drawing as **PFD**.

**FIGURE 21.62**

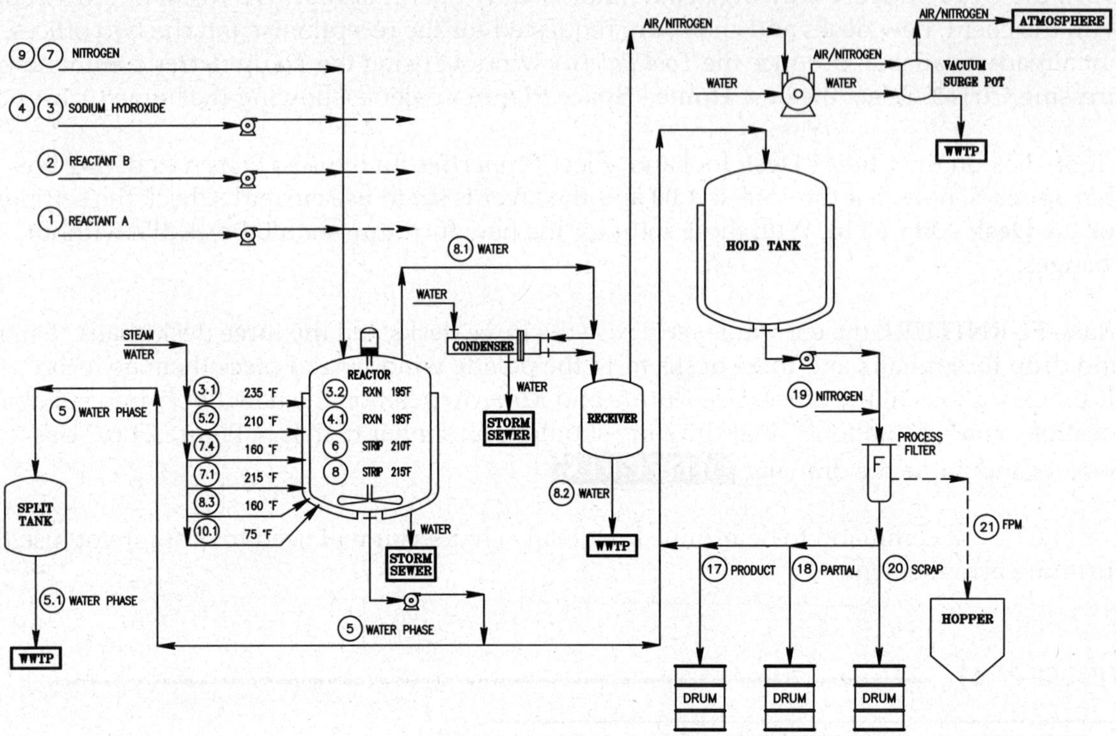

CHAPTER 22

# Block Attributes and Data Links

## Chapter Objectives

After completing this chapter you should:

1. be able to define block attributes using the *Attribute Definition* dialog box and the *-Attdef* command;

2. be able to control the display of attributes in the drawing with *Attdisp*;

3. know how to edit existing attributes with *Attedit*, the *Enhanced Attribute Editor*, and the *Find and Replace* dialog box;

4. be able to edit attributes with *Attedit*;

5. be able to redefine attributes for existing blocks using the *Block Attribute Manager* and the *Attredef* command;

6. be able to use the *Data Extraction* wizard to create extract files of the attribute information contained in one or more drawings;

7. know how to link data in an Excel spreadsheet to an AutoCAD table.

## CONCEPTS

A block attribute is a line of text (numbers or letters) associated with a block. An attribute can be thought of as a label or description for a block. A block can have multiple attributes. The attributes are included with the drawing objects when the block is defined (using the *Block* command). When the block is *Inserted*, its attributes are also *Inserted*.

Since *Blocks* are typically used as symbols in a drawing, attributes are text strings that label or describe each symbol. For example, you can have a series of symbols such as transistors, resistors, and capacitors prepared for creating electrical schematics. The associated attributes can give the related description of each block, such as ohms or wattage values, model number, part number, and cost. If your symbols are doors, windows, and fixtures for architectural applications, attached attributes can include size, cost, and manufacturer. A mechanical engineer can have a series of blocks representing fasteners with attached attributes to specify fastener type, major diameter, pitch, length, etc. Attributes could even be used to automate the process of entering text into a title block, assuming the title block was *Inserted* as a block or separate .DWG file.

Attributes can add another level of significance to an AutoCAD drawing. Not only can attributes automate the process of placing the text attached to a block, but the inserted attribute information can be extracted from a drawing to form a bill of materials or used for cost analysis, for example. Extracted text can be imported to a database or spreadsheet program for further processing and analysis. If desired, extracted text can be inserted directly into a drawing as an AutoCAD *Table*.

Attributes are created with the *Attdef* (attribute define) command. *Attdef* operates similarly to *Text*, prompting you for text height, placement, and justification. Parameters can be adjusted that determine how the attributes will appear when they are later inserted. Similar to annotative text objects (see Chapter 18), attributes can be assigned an *Annotative* parameter so that they change size when the viewport scale (*Viewport Scale*) is changed.

You can link attribute data, or any other data, from an Excel spreadsheet to an AutoCAD table with the *Datalink* command. Data can be changed in either the drawing or the Excel file and then updated to the other using the *Datalink* command.

## CREATING ATTRIBUTES

### Steps for Creating and Inserting Block Attributes

TIP

1. Create the objects that will comprise the block. Do not use the *Block* command yet; only draw the geometry. You can create blocks containing attributes in the Drawing Editor or in the Block Editor. The process is similar using either method, except saving the block is simpler in the Block Editor.

2. Use the *Attdef* command to create and place the desired text strings associated with the geometry comprising the proposed block.

3. Use the *Block* command to convert the drawing geometry (objects) and the attributes (text) into a named block. When prompted to "Select objects:" for the block, select both the drawing objects and the text (attributes).

4. Use *Insert* to insert the attributed block into the drawing. Edit the text attributes as necessary (if parameters were set as such) during the insertion process.

## Attdef

| Ribbon | Menu Bar | Command (Type) | Alias (Type) | Shortcut |
|---|---|---|---|---|
| *Home* *Block Panel* *Define Attributes* (icon) | *Draw* *Block >* *Define Attributes...* | *Attdef or* *-Attdef* | *ATT or* *-ATT* | ... |

*Attdef* (attribute define) allows you to define attributes for future combinations with a block. If you intend to associate the text attributes with drawing objects (*Line, Arc, Circle,* etc.) for the block, it is usually preferred to draw the objects before using *Attdef*. In this way, the text can be located in reference to the drawing objects. *Attdef* can be used in the Drawing Editor or in the Block Editor.

*Attdef* allows you to create the text and provides justification, height, and rotation options similar to the *Text* command. You can also define parameters, called Attribute Modes, specifying how the text will be inserted—*Invisible, Constant, Verify, Preset, Lock position,* and *Multiple lines*.

The *Attdef* command differs depending on the method used to invoke it. If you select from any menus or the icon or type *Attdef*, the *Attribute Definition* dialog box appears (Fig. 22-1). If you type *-Attdef* (with a hyphen prefix), the command line format is used. The Command line format is presented below:

**FIGURE 22-1**

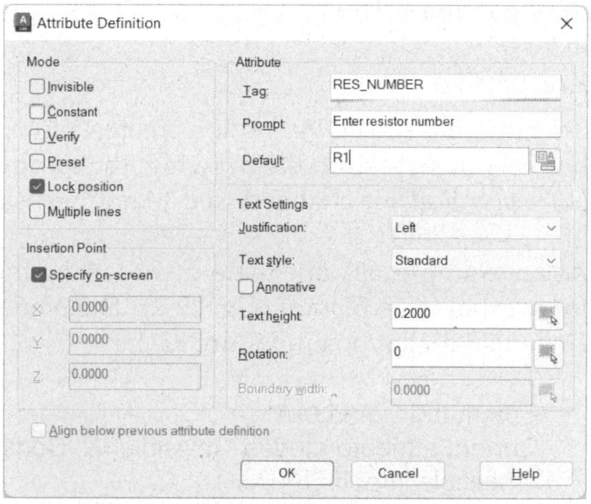


Command: ***-ATTDEF***
Current attribute modes: Invisible=N Constant=N
Verify=N Preset=N Lock position=Y Annotative=N Multiple line=N
Enter an option to change [Invisible/Constant/Verify/Preset/Lock position/Annotative/Multiple lines]
<done>: **Enter** or (**option**)
Enter attribute tag name: (Enter tag, no spaces)
Enter attribute prompt: (Enter desired prompt, spaces allowed)
Enter default attribute value: (Enter desired default text)
Current text style: "Standard" Text height: 0.2000
Specify start point of text or [Justify/Style]: **Enter** or (**option**)
Specify height <0.2000>: **Enter** or (**value**)
Specify rotation angle of text <0>: **Enter** or (**value**)
Command:

The first option, "attribute modes," is described later. For this example, the default attribute modes have been accepted.

The *Attribute Definition* dialog box has identical options that are found in the Command line format, including justification options. The same entries have been made here to define the first attribute example shown in Figure 22-2.

**FIGURE 22-2**

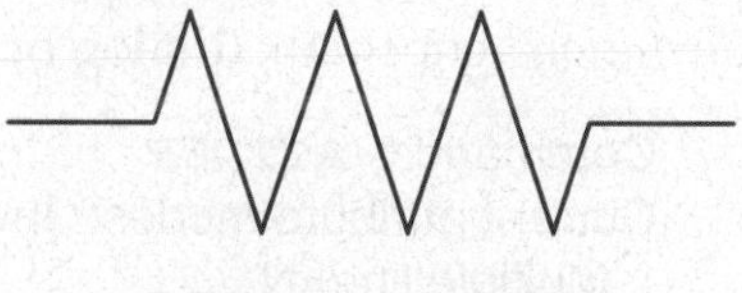


### Attribute Tag

This is the descriptor for the <u>type of text</u> to be entered, such as MODEL_NO, PART_NO, or NAME. <u>Spaces cannot be used</u> in the Attribute Tag.

### *Attribute Prompt*

The prompt is what words (<u>prompt) you want to appear</u> when the block is *Inserted* and when the actual text (values) must be entered. Spaces and punctuation can be included.

### *Default Attribute Value*

This is the <u>default text</u> that appears with the block when it is *Inserted*. Supply the typical or expected words or numbers. If you are using the *Attribute Definition* dialog box, you can enter a text *Field*. In this way, the attribute can automatically change based on changes in the *Field*. (See Chapter 18.)

### *Justify*

Any of the typical text justification methods can be selected using this option.

### *Style*

This option allows you to select from existing text *Styles* in the drawing. Otherwise, the attribute is drawn in the current *Style*.

### *Annotative*

Selecting the *Annotative* option creates annotative attributes. After a block with annotative attributes has been *Inserted* into the drawing, the attributes can automatically change size when the viewport scale (*Viewport Scale*) is changed (see "Annotative Attributes").

As an example, assume that you are using the *Attdef* command to make attributes attached to a resistor symbol to be *Blocked* (Fig. 22-2). The scenario may appear as follows or as presented in the *Attribute Definition* dialog box in Figure 22-1:

Command: ***-ATTDEF***
Current attribute modes: Invisible=N Constant=N Verify=N Preset=N Lock position=Y Annotative=N Multiple line=N
Enter an option to change [Invisible/Constant/Verify/Preset/Lock position/Annotative/Multiple lines] <done>: **Enter**
Enter attribute tag name: **RES_NUMBER**
Enter attribute prompt: **Enter Resistor Number:**
Enter default attribute value: **R1**
Current text style: "Standard"' Text height: 0.2000
Specify start point of text or [Justify/Style]: ***J***
Enter an option [Align/Fit/Center/Middle/Right/TL/TC/TR/ML/MC/MR/BL/BC/BR]: ***C***
Specify center point of text: **PICK**
Specify height <0.2000>: **.125**
Specify rotation angle of text <0>: **Enter**

The attribute would then appear at the selected location, as shown in Figure 22-2. Keep in mind that the *Lines* comprising the resistor were created before creating attributes. Also remember that the resistor and the attribute are not yet *Blocked*. The *Attribute Tag* <u>only</u> is displayed until the *Block* command is used.

Two other attributes could be created and positioned beneath the first one by pressing Enter when prompted for the "<Start point>:" (Command line format) or selecting the "*Align below previous attribute definition*" checkbox (dialog box format). The command syntax may read like this:

Command: ***-ATTDEF***
Current attribute modes: Invisible=N Constant=N Verify=N Preset=N Lock position=Y Annotative=N Multiple line=N
Enter an option to change [Invisible/Constant/Verify/Preset/Lock position/Annotative/Multiple lines] <done>: **Enter**
Enter attribute tag name: **RESISTANCE**

Enter attribute prompt: **Enter Resistor Number:**
Enter default attribute value: **0 ohms**
Current text style: "Standard" Text height: 0.1250
Specify start point of text or [Justify/Style]: **Enter**

Command: ***-ATTDEF***
Current attribute modes: Invisible=N Constant=N Verify=N Preset=N Lock position=Y Annotative=N Multiple line=N
Enter an option to change [Invisible/Constant/Verify/Preset/Lock position/Annotative/Multiple lines] <done>: **Enter**
Enter attribute tag name: **PART_NO**
Enter attribute prompt: **Enter Part Number:**
Enter default attribute value: **0-0000**
Current text style: "Standard"' Text height: 0.1250
Specify start point of text or [Justify/Style]: **Enter**

The resulting unblocked symbol and three attributes appear as shown in Figure 22-3. Only the <u>*Attribute Tags*</u> are displayed at this point. The *Block* command has not yet been used.

**FIGURE 22-3**

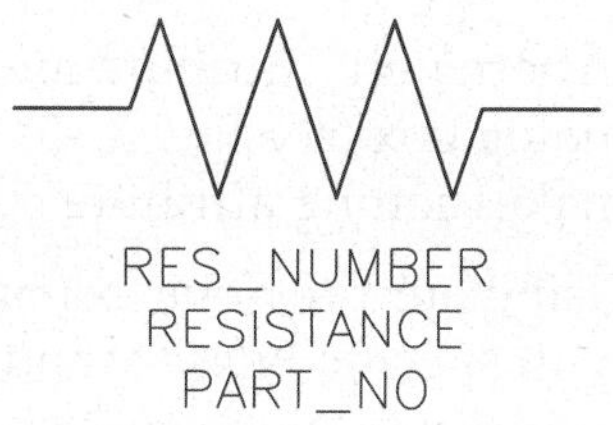


The *Attributes Modes* define the appearance or the action required when you *Insert* the attributes. The options are as follows.

### *Invisible*

An *Invisible* attribute is not displayed in the drawing after insertion. This option can be used to prevent unnecessary information from cluttering the drawing and slowing regeneration time. The *Attdisp* command can be used later to make these attributes visible.

### *Constant*

This option gives the attribute a fixed value for all insertions of the block. In other words, the attribute always has the same text and cannot be changed. You are not prompted for the value upon insertion.

### *Verify*

When attributes are entered in Command line format, this option forces you to verify that the attribute is correct when the block is inserted. This option has no effect when attributes are entered in dialog box format (when *ATTDIA*=1).

### *Preset*

This option prevents you from having to enter a value for the attribute during insertion if you are using the Command line format. In this case, you are not prompted for the attribute value during insertion. Instead, the default value is used for the block, although it can be changed later with the *Attedit* command.

### *Lock Position*

This option locks the position of the attribute within the block definition. If not locked, the attribute can be moved with normal grips once the block is inserted into the drawing. In a dynamic block, an attribute's position must be locked if you intend to include it in an action's selection set.

### *Multiple Lines*

This option allows you to create an attribute as an *Mtext* object. When the attribute value is entered during the *Insert* command, a small *Text Formatting* editor can be used to enter multiple lines of text.

The *Attribute Modes* are toggles (Yes or No). In Command line format, the options are toggled by entering the appropriate letter, which reverses its current position (Y or N). For example, to create an *Invisible* attribute, type **"I"** at the prompt:

Command: ***-ATTDEF***
Current attribute modes: Invisible=N Constant=N Verify=N Preset=N Lock position=Y Annotative=N Multiple line=N
Enter an option to change [Invisible/Constant/Verify/Preset/Lock position/Annotative/Multiple lines] <done>: ***i***
Current attribute modes: Invisible=Y Constant=N Verify=N Preset=N Lock position=Y Annotative=N Multiple line=N
Enter an option to change [Invisible/Constant/Verify/Preset/Lock position/Annotative/Multiple lines] <done>: **Enter**

Note that the second prompt (fourth line) reflects the new option for the previous request.

If *Dynamic Input* is on and set to the default options, select the desired option from the *Dynamic Input* menu (Fig. 22-4). If *Dynamic Input* is off, you can right-click to produce a shortcut menu with the same options.

FIGURE 22-4

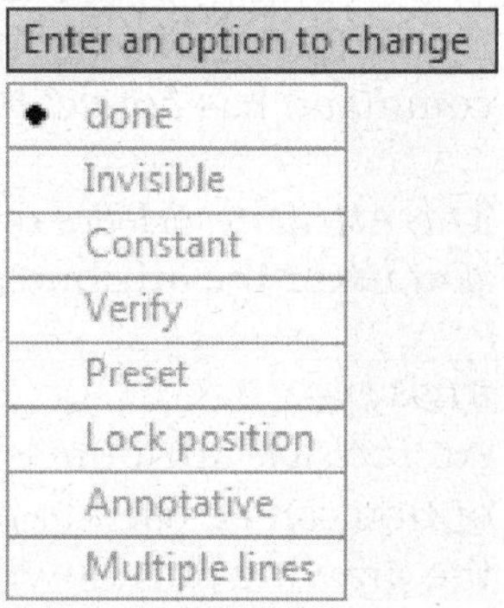


Alternately, attribute modes could be defined using the *Attribute Definition* dialog box (see Fig. 22-5). The *Mode* cluster of checkboxes (upper-left corner) makes setting attribute modes simpler than in Command line format.

Using our previous example, two additional attributes can be created with specific *Invisible* and *Verify* attribute modes. Either the dialog box or Command line format can be used. The following command syntax shows the creation of an *Invisible* attribute:

Command: ***-ATTDEF***
Current attribute modes: Invisible=N Constant=N Verify=N Preset=N Lock position=Y Annotative=N Multiple line=N
Enter an option to change [Invisible/Constant/Verify/Preset/Lock position/Annotative/Multiple lines] <done>: ***i***
Current attribute modes: Invisible=Y Constant=N Verify=N Preset=N Lock position=Y Annotative=N Multiple line=N
Enter an option to change [Invisible/Constant/Verify/Preset/Lock position/Annotative/Multiple lines] <done>: **Enter**
Enter attribute tag name: **MANUFACTURER**
Enter attribute prompt: **Enter Manufacturer:**
Enter default attribute value: **Enter**
Current text style: "Standard" Text height: 0.125000
Specify start point of text or [Justify/Style]: **Enter**
Command:

FIGURE 22-5

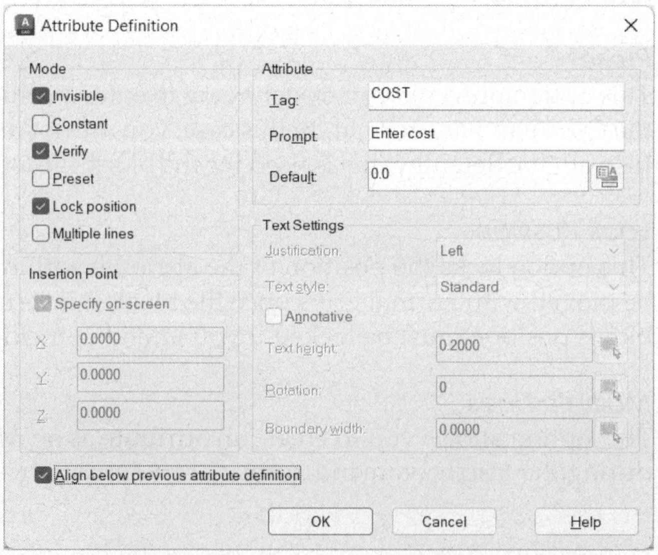


The first attribute has been defined. Next, a second attribute is defined having *Invisible* and *Verify* parameters. Figure 22-5 shows the correct entries in the *Attribute Definition* dialog box for creation of this attribute. *Tags* for *Invisible* attributes appear as any other attribute tags. The *Values*, however, are invisible when the block is inserted.

## Saving the Attributed Block

As indicated earlier, you can create the block geometry and the attributes using either the Drawing Editor or the Block Editor. The next step in the process of creating an attributed block is to save the *Block*. This step is different depending on the method used to create the geometry and attributes.

### Using the Block Editor

If you used the Block Editor for the previous steps, saving the attributed block is simple. Select the *Save Block Definition* button or use the *Bsave* command to complete the process of creating the attributed *Block*. However, using the Block Editor you have no control of the order in which the attributes appear in the *Enter Attributes* dialog box or at the command prompt—you are prompted to enter the attribute values in the order they were created. In other words, when you use the *Insert* command to insert the attributed block into the drawing, prompts to enter the attribute text appear in the order the attributes were created.

### Using the Drawing Editor

If you used the Drawing Editor for the previous steps, you must then use the *Block* or *-Block* command to transform the drawing objects and the attributes into a *Block*. When you are prompted to "Select objects:" during the *Block* command, select the attributes in the order you want their prompts to appear during insertion. In other words, the first attribute selected during the *Block* command is the first attribute prompted for editing during the *Insert* command. PICK the objects with the pickbox one at a time in the desired order. Figure 22-6 shows this sequence. If you select the *Convert to block* option in the *Block Definition* dialog box, then press the *OK* button, you are prompted to enter the attribute text immediately. These attributes are for the original template objects that are being converted to a *Block*.

**FIGURE 22-6**

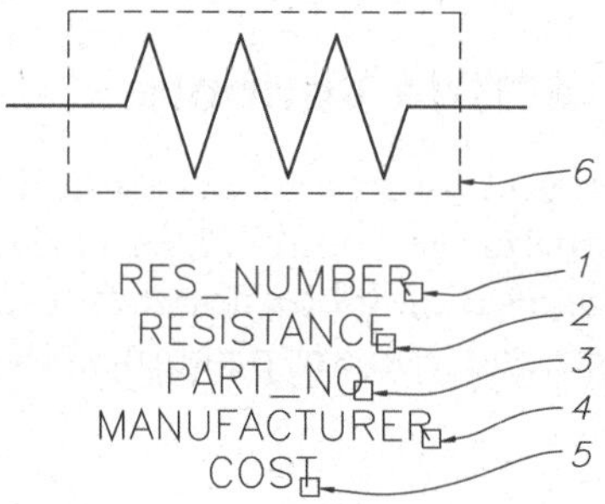


## *Inserting* Attributed Blocks

Use the *Insert* command or Ribbon gallery to bring attributed blocks into your drawing. When the *Insert* dialog box appears (not shown), make the desired settings and select *OK*. Next, specify the insertion point (where you want to place the block in your drawing). Next, the *Enter Attributes* dialog box appears (Fig. 22-7). Enter the desired attribute values for each block insertion.

Notice the prompts that appear are those that you specified as the "attribute prompt" with the *Attdef* command. The order of the prompts matches the order of selection during the *Block* command.

The appearance of the *Enter Attributes* dialog box is controlled by the *ATTDIA* system variable (see "The *ATTDIA* Variable"). If *ATTDIA* is set to 0, the dialog box does not appear and you are prompted to enter the attribute text at the Command line. For example, if *ATTDIA* is set to 0, the following prompt appears during the *Insert* command. Note the repeated prompt (*Verify*) for the COST attribute.

**FIGURE 22-7**

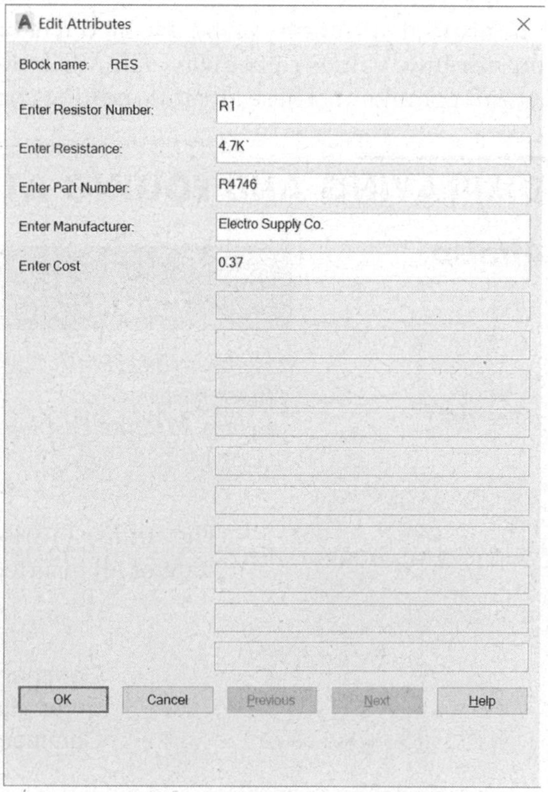

Command: ***INSERT***
Specify insertion point or [Basepoint/Scale/Rotate]: **PICK**
Enter attribute values
Enter Resistor Number: <0>: **R1**
Enter Resistance: <0 ohms>: **4.7K**
Enter Part Number: <0-0000>: **R-4746**
Enter Manufacturer: **Electro Supply Co.**
Enter Cost <0.00>: **.37**
Verify attribute values
Enter Cost <.37>: **Enter**
Command:

Entering the attribute values as indicated by either method above yields the block insertion shown in Figure 22-8. Notice the absence of the last two attribute values since they are *Invisible*.

**FIGURE 22-8**

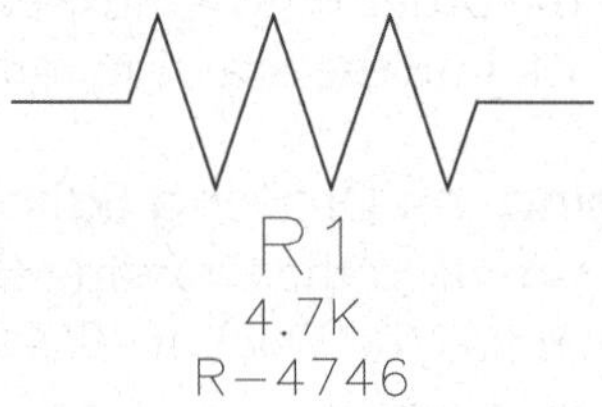


## The *ATTDIA* Variable

**TIP**

When you insert a *Block* containing attributes, you are normally prompted to enter the desired text for the attributes. The *ATTDIA* variable controls how you are prompted. If *ATTDIA* is set to 1, the *Enter Attributes* dialog box appears (see previous Figure 22-7). If *ATTDIA* is set to 0, the dialog box does not appear and you are prompted to enter the attribute text at the Command line.

## The *ATTREQ* Variable

When *Inserting* blocks with attributes, you can force the attribute requests to be suppressed. In other words, you can disable the prompts asking for attribute values when the blocks are *Inserted*. To do this, use the *ATTREQ* (attribute request) system variable. A setting of **1** (the default) turns attribute requesting on, and a value of **0** disables the attribute value prompts.

**TIP**

The attribute prompts can be disabled when you want to *Insert* several blocks but do not want to enter the attribute values right away. The attribute values for each block can be entered at a later time using the *Attedit* command. These attribute editing commands are discussed on the following pages.

# DISPLAYING AND EDITING ATTRIBUTES

## *Attdisp*

| Ribbon | Menu Bar | Command (Type) | Alias (Type) | Shortcut |
|---|---|---|---|---|
| *Home or Insert tab*<br>*Block panel*<br>*Retain Attibute Display*<br>(icon) | *View*<br>*Display >*<br>*Attribute Display* | *Attdisp* | ... | ... |

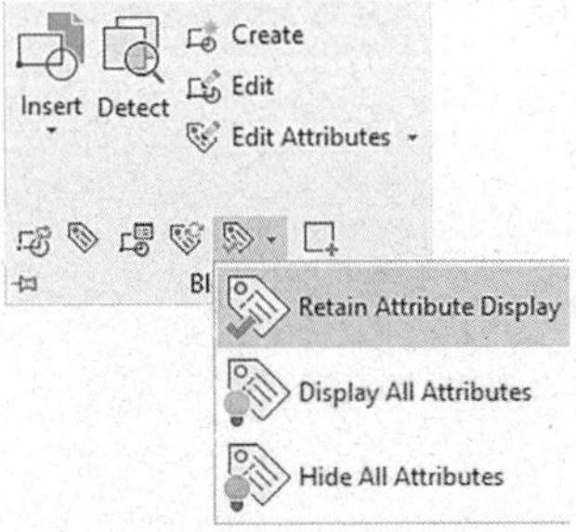


The *Attdisp* (attribute display) command allows you to control the visibility state of all inserted attributes contained in the drawing:

Command: ***ATTDISP***
Enter attribute visibility setting [Normal/ON/OFF] <Normal>: ***n***
Command:

*Attdisp* has three positions. Changing the state forces a regeneration.

*On*
All attributes (normal and *Invisible*) in the drawing are displayed.

*Off*
No attributes in the drawing are displayed.

*Normal*
Normal attributes are displayed; *Invisible* attributes are not displayed.

**FIGURE 22-9**

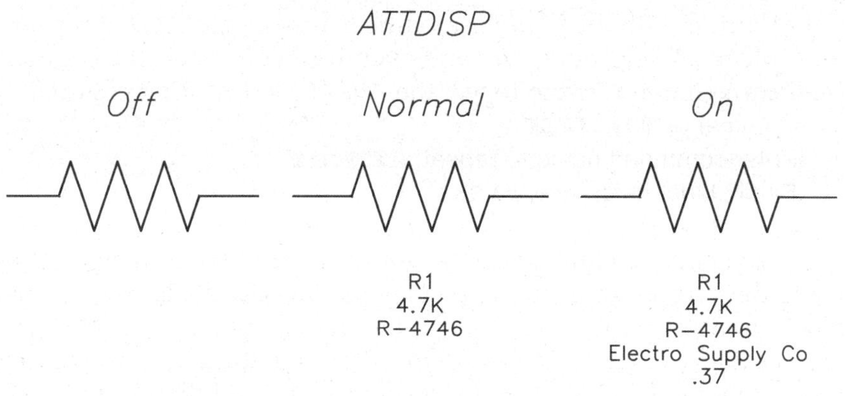


Figure 22-9 illustrates the RES block example with each of the three *Attdisp* options. The last two attributes were defined with *Invisible* modes but are displayed with the *On* state of *Attdisp*. (Since *Attdisp* affects attributes globally, it is not possible to display the three options at one time in a drawing.)

When attributes are defined using the *Invisible* mode of the *Attdef* command, the attribute values are normally not displayed with the block after *Insertion*. However, the *Attdisp On* option allows you to override the *Invisible* mode. Turning all attributes *Off* with *Attdisp* can be useful for many applications when it is desirable to display or plot only the drawing geometry. The *Attdisp* command changes the *ATTMODE* system variable.

## *Attedit*

| Ribbon | Menu Bar | Command (Type) | Alias (Type) | Shortcut |
|---|---|---|---|---|
| .. | ... | *Attedit* | *ATE* | ... |

Typing the *Attedit* command invokes the *Edit Attributes* dialog box (Fig. 22-10). This dialog box allows you to edit only attribute values of existing blocks in the drawing. The configuration and operation of the box are identical to the *Edit Attributes* dialog box (see Fig. 22-11).

When the *Attedit* command is entered, AutoCAD requests that you select a block. Any existing attributed block can be selected for attribute editing.

Command: **ATTEDIT**
Select block reference: **PICK**

At this point, the dialog box appears. After the desired changes have been made, selecting the *OK* tile updates the selected block with the indicated changes.

**FIGURE 22-10**

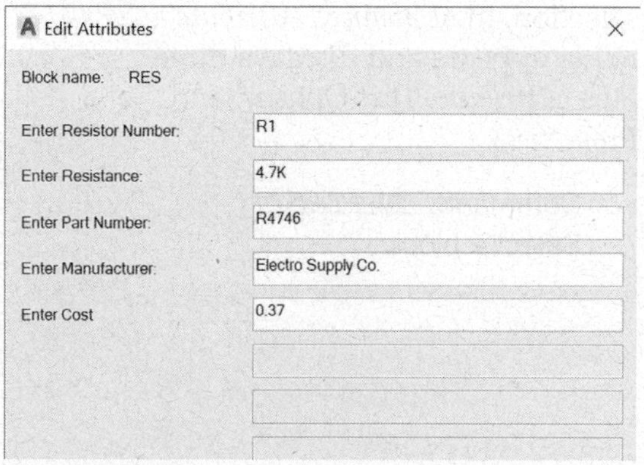

### *Multiple Attedit*

Using *Attedit* with the *Multiple* modifier causes AutoCAD to repeat the *Attedit* command requesting multiple block selections over and over until you cancel the command with Escape. You can use the *Multiple* command first to repeat the *Attedit* (or any other command).

Command: ***MULTIPLE***
Enter command name to repeat: ***ATTEDIT***
Select block reference: **PICK**

TIP

This form of the command can be used to query blocks in an existing drawing. For example, instead of using the *Attdisp* command to display the *Invisible* attributes, *Multiple Attedit* can be used on selected blocks to display the *Invisible* attributes in dialog box form. Another example is the use of *Attdisp* to turn visibility of all attributes *Off* in a drawing, thus simplifying the drawing appearance and speeding regenerations. The *Multiple Attedit* command could then be used to query the blocks to display the attributes in dialog box form.

### *Eattedit*

| Ribbon | Menu Bar | Command (Type) | Alias (Type) | Shortcut |
|---|---|---|---|---|
| *Insert tab Block panel Edit Attribute, Single* | *Modify Object > Attribute > Single...* | *Eattedit* | ... | ... |

The *Eattedit* command produces the *Enhanced Attribute Editor* (Fig. 22-11). This tool is intended as an enhanced *Attedit* tool for existing block insertions. The *Enhanced Attribute Editor* allows you to modify attribute values similar to using *Attedit*; however, you can also modify attribute text settings and properties such as layer, linetype, and color. You can change multiple attributes, one at a time, in one session of the *Enhanced Attribute Editor* by using the *Select block* button in the editor's upper-right corner. Changes you make to the attributes are immediately visible in the drawing. If *Dblclkedit* is *On*, double-clicking on a *Block* with attributes automatically invokes the *Enhanced Attribute Editor*.

TIP

When you invoke the command by any method (shown in the previous command table), AutoCAD prompts you to select a block. You can select the *Block* geometry or any attribute of any *Block* in the drawing. After block selection, the *Enhanced Attribute Editor* appears and displays three tabs (*Attribute*, *Text Options*, and *Properties*).

Command: ***EATTEDIT***
Select a block:

**FIGURE 22-11**

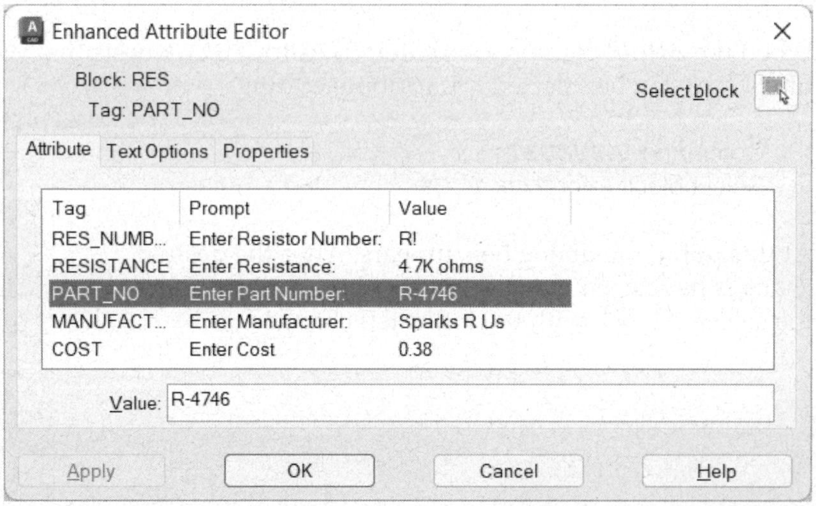

### *Attribute*

When the *Enhanced Attribute Editor* appears, the *Attribute* tab is active by default (see Fig. 22-11). If you PICKed an attribute during block selection, that specific attribute is highlighted in the editor and its text is placed in the *Value* edit box. You can change any attribute value of the *Block*, just as you can using the *Attedit* command.

### *Text Options*

In this tab (Fig. 22-12), you can change any of the text properties for an attribute, such as the *Text Style*, *Justification*, *Height*, *Rotation*, *Width Factor*, *Oblique Angle*, *Annotative* and other properties, depending on the type of text font used (.SHX or .TTF). The changes apply only to the one attribute value you selected in the *Attribute* tab, so before using the *Text Options* tab, select the desired attribute in the *Attribute* tab.

**FIGURE 22-12**

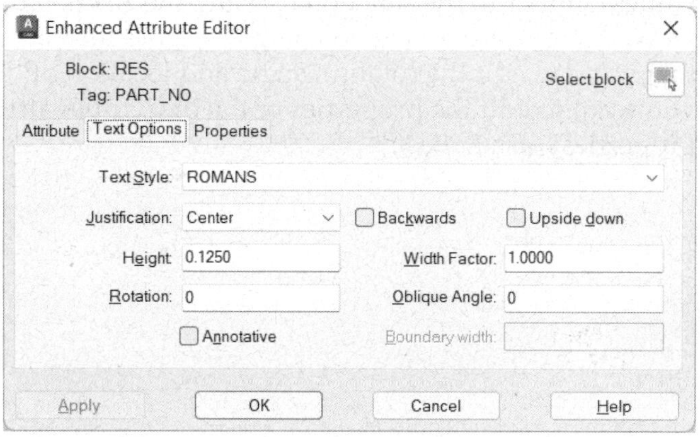


### *Properties*

This tab gives you the ability to change the selected attribute's *Layer*, *Color*, *Linetype*, *Lineweight*, and *Plot style* properties (Fig. 22-13). Since the changes are applied to the highlighted attribute only, remember to select the desired attribute in the *Attribute* tab before using this tab.

**FIGURE 22-13**

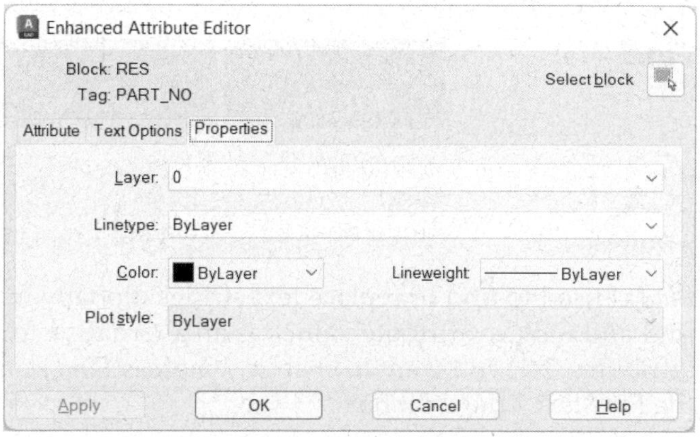


### *Apply*, *OK*, and *Cancel*

Note that the *Enhanced Attribute Editor* has an *Apply* button in addition to the *OK* and *Cancel* buttons. Changes made in any tab of the *Enhanced Attribute Editor* are automatically and immediately displayed in the drawing temporarily until *Apply*, *OK*, or *Cancel* is used. The *Apply* button forces the change permanently and immediately to the drawing. Beware: once you use *Apply*, the changes to the drawing are not canceled if you close the dialog box using the *Cancel* button. Using the *OK* button will also apply all changes; however, unlike using the *Apply* button, you still have the option to use *Cancel* before ending the session. If you are making changes to only one block, use *OK* to apply the changes and close the *Enhanced Attribute Editor*. If you want to change attributes for several blocks, use the *Apply* button between block selection. If you want to cancel applied changes, you must use the *Undo* command immediately after using the *Enhanced Attribute Editor*.

TIP

## Properties

| Ribbon | Menu Bar | Command (Type) | Alias (Type) | Shortcut |
|---|---|---|---|---|
| *View* *Palettes* *Properties* | *Modify* *Properties* | *Properties* | *CH* or *PR* | Ctrl+1 |

You can edit attribute values directly from the *Properties* palette as shown next.

Simply open the *Properties* palette, select the desired attributed *Block*, and all of the attributes appear at the bottom of the palette (Fig. 22-14). You may have to scroll down to view the attributes section in the palette.

Remember, this method is useful only for editing the attribute values, similar to the function of using the *Attedit* command. However, one advantage to using *Properties* instead of *Attedit* is that you can edit the rotation angle and location of the *Block*. If you want to edit the properties of the individual attributes (text properties), use *Eattedit*.

**FIGURE 22-14**

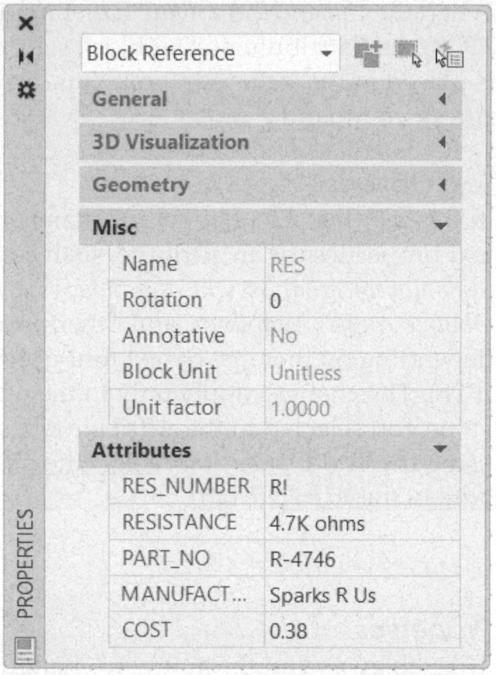


## *Find*

| Ribbon | Menu Bar | Command (Type) | Alias (Type) | Shortcut |
|---|---|---|---|---|
| *Annotate Text* (edit box) | *Edit Find...* | *Find* | ... | (Default Menu) *Find...* |

*Find* is used to find or replace text strings globally in a drawing (see Chapter 18). *Find* can also be used to find and replace attribute values as an alternative to *Attedit*. *Attedit* requires that you actually PICK the attributes that you want to change, whereas *Find* searches the drawing based on the value you enter in the *Find text string* edit box.

TIP

Use the *Find* command to show the *Find and Replace* dialog box (Fig. 22-15). To use *Find* to edit attributes in a drawing, you must ensure that *Find* looks for block attributes when the search is made. To do this, use the *More Options* button in the *Find and Replace* dialog box. Select *Block Attribute Value.*

Specify the attribute value to find in the *Find What* edit box (Fig. 22-16). Enter the replacement attribute value in the *Replace with* edit box. Select the *Find* button to begin the search, then use *Replace* and *Find Next*, or *Replace All* as needed. Refer to Chapter 18 for details on these and other features of the *Find and Replace* dialog box.

**FIGURE 22-15**

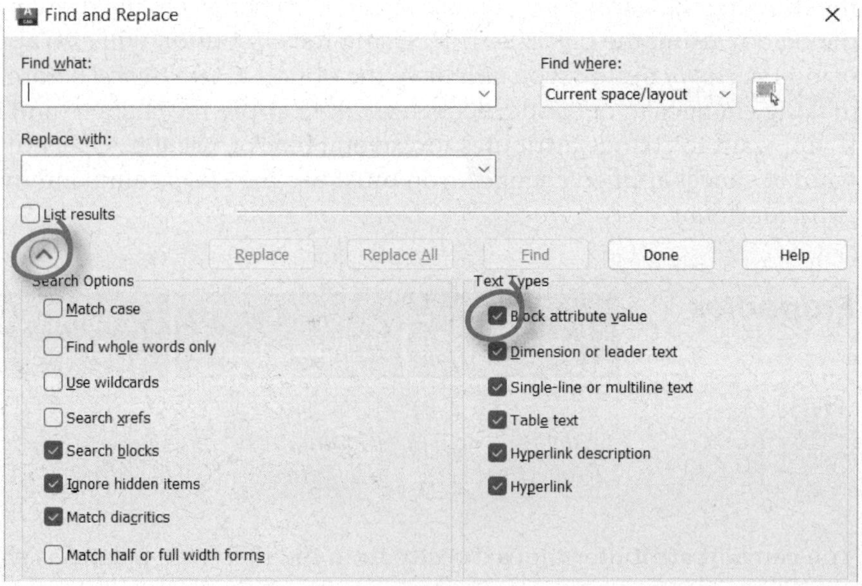

**FIGURE 22-16**

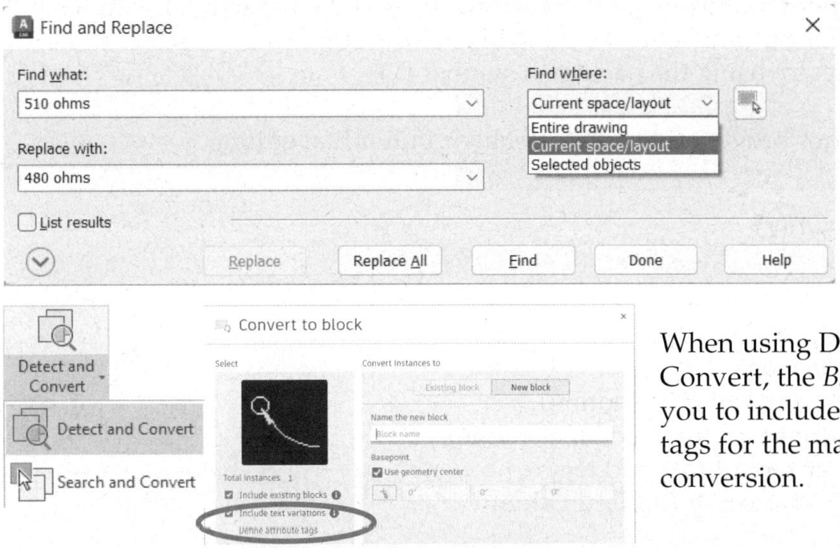


**TIP**

When using Detect and Convert or Search and Convert, the *Bsearch* and *Bdetect* commands allow you to include text variations and define attribute tags for the matching items during the block conversion.

## *-Attedit*

| Ribbon | Menu Bar | Command (Type) | Alias (Type) | Shortcut |
|---|---|---|---|---|
| *Insert Block Multiple* | *Modify Object > Attribute > Global* | *-Attedit* | *-ATE* | ... |

*-Attedit* (Attribute Edit) is an older Command line attribute editing tool that has several features similar to those included in the newer dialog box-based commands. *-Attedit* is similar to *Attedit* and *Eattedit* in that you can edit the values of selected attributes in a drawing. *-Attedit* also has options similar to *Eattedit* that allow you to change the attribute text options and some properties. You can also use *-Attedit* in a manner similar to *Find* to globally edit attribute values in a drawing.

Command: ***-ATTEDIT***
Edit attributes one at a time? [Yes/No] <Y>:

The following prompts depend on your response to the first prompt:

### *N (No)*

Indicates that you want global editing. (In other words, all attributes are selected.) Attributes can then be further selected by block name, tag, or value. Only attribute values can be edited using this global mode.

### *Y (Yes)*

Allows you to PICK each attribute you want to edit. You can further filter (restrict) the selected set by block name, tag, and value. You can then edit any property or placement of the attribute.

No matter which option you choose, the following prompts appear, which allow you to filter, or further restrict, the set of attributes for editing. You can specify the selection set to include only specific block names, tags, or values:

Enter block name specification <*>: Enter desired block name(s)
Enter attribute tag specification <*>: Enter desired tag(s)
Enter attribute value specification <*>: Enter desired value(s)

During this portion of the selection set specification, sets of names, tags, or values can be entered only to filter the set you choose. Commas and wildcard characters can be used. Pressing Enter accepts the default option of all (*) selected.

Attribute values are case sensitive (upper- and lowercase letters must be specified exactly). If you have entered null values (for example, if *ATTREQ* was set to 0 during block insertion), you can select all null values to be included in the selection set by using the backslash symbol (\).

The subsequent prompts depend on your previous choice of global or individual editing.

### *No* (Global Editing—Editing Values Only)

Command: ***-ATTEDIT***
Edit attributes one at a time? [Yes/No] <Y>: ***n***
Performing global editing of attribute values.
Edit only attributes visible on screen? [Yes/No] <Y>: (option)
Enter block name specification <*>: **Enter** or (desired block name)
Enter attribute tag specification <*>: **Enter** or (desired tag)
Enter attribute value specification <*>: **Enter** or (desired value)
Select Attributes: **PICK**
Select Attributes: **PICK**
Select Attributes: **Enter**
2 attributes selected.
Enter string to change: (Enter any existing string)
Enter new string: (Enter any new string)
Command:

The "only attributes visible on the screen" refers to those visible (Y) or those through the entire drawing (N). Answering "N" to this prompt and Enter to the next three would find all attributes in the drawing.

AutoCAD searches for the "string to change" and replaces all occurrences of the string with the new string. No change is made if the string is not found. Remember that only attribute values can be edited with the global mode. This feature has the same capability as using *Find*, described previously.

### *Yes* (Individual Editing—Editing Any Property)

After filtering for the block name, attribute tag, and value, AutoCAD prompts:

Select Attributes: **PICK**
Select Attributes: **PICK** (continue selection)
Select Attributes: **Enter**
*n* attributes selected.
Enter an option [Value/Position/Height/Angle/Style/Layer/Color/Next] <N>:

A marker appears at one of the attributes. Use the *Next* option to move the marker to the attribute you wish to edit. You can then select any other option. After editing the marked attribute, the change is displayed immediately and you can position the marker at another attribute to edit.

You can change the attribute's *Value, Position, Height, Angle, Style, Layer*, or *Color* by selecting the appropriate option. These options are Command line versions of the *Text Options* and *Properties* tabs of the *Enhanced Attribute Editor* (see "*Eattedit*").

## Global and Individual Editing Example

FIGURE 22-17

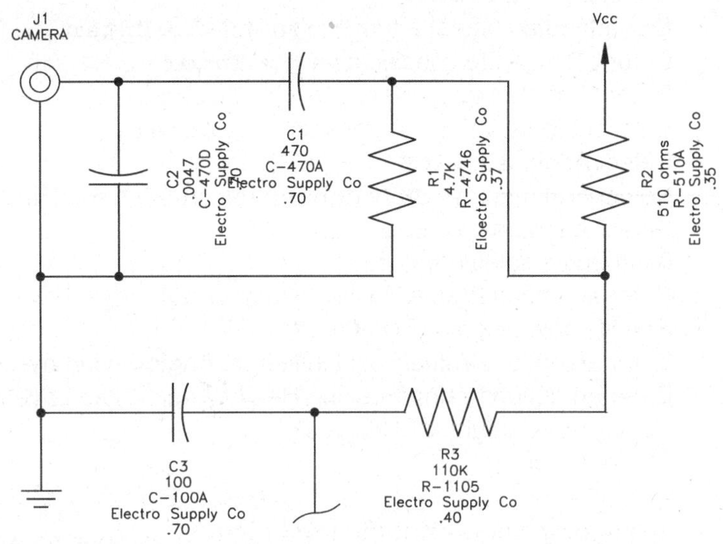


Assume that the RES block and a CAP block were inserted several times to create the partial schematic shown in Figure 22-17. The *Attdisp* command is set to *On* to display all the attributes, including the last two for each block, which are normally *Invisible*.

*-Attedit* is used to edit the attributes of RES and CAP. If you wish to change the value of one or several attributes, the global editing mode of *-Attedit* would generally be used, since values only can be edited in global mode. For this example, assuming that the supplier changed, the MANUFACTURER attribute for all blocks in the drawing is edited. The command syntax is as follows:

Command: ***-ATTEDIT***
Edit attributes one at a time? [Yes/No] <Y>: ***n***
Performing global editing of attribute values.
Edit only attributes visible on screen? [Yes/No] <Y>: **Enter**
Enter block name specification <*>: **Enter**
Enter attribute tag specification <*>: **Manufacturer**
Enter attribute value specification <*>: **Enter**
Select Attributes: **PICK** (Select specifically the 6 MANUFACTURER attributes)
Select Attributes: **Enter**
6 attributes selected.
Enter string to change: **Electro Supply Co**
Enter new string: **Sparks R Us**
Command:

FIGURE 22-18

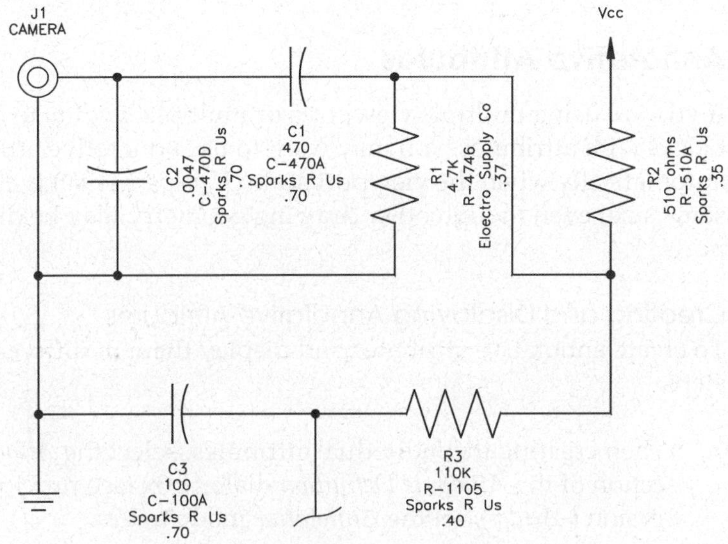


All instances of the "Electro Supply Co" redisplay with the new string. Note that resistor R1 did not change because the string did not match (due to the misspelling, Fig. 22-18). *Attedit* or *Eattedit* can be used to edit the value of that attribute individually.

The individual editing mode of *-Attedit* or *Eattedit* can be used to change the height of text for the RES_NUMBER and CAP_NUMBER attribute for both the RES and CAP blocks. The command syntax is as follows:

Command: **-ATTEDIT**
Edit attributes one at a time? [Yes/No] <Y>: **Enter**
Enter block name specification <*>: **Enter**
Enter attribute tag specification <*>: **Enter**
Enter attribute value specification <*>: **Enter**
Select Attributes: **PICK**
Select Attributes: **PICK** (Continue selecting all 6 attributes)
Select Attributes: **Enter**
6 attributes selected.
Enter an option [Value/Position/Height/Angle/Style/Layer/Color/Next] <N>: ***H***
Specify new height <0.1250>: **.2**
Enter an option [Value/Position/Height/Angle/Style/Layer/Color/Next] <N>: ***N***
Enter an option [Value/Position/Height/Angle/Style/Layer/Color/Next] <N>: ***H***
Specify new height <0.1250>: **.2**
etc.

The result of the new attribute text height for CAP and RES blocks is shown in Figure 22-19. *Attdisp* has been changed to *Normal* (*Invisible* attributes do not display).

**FIGURE 22-19**

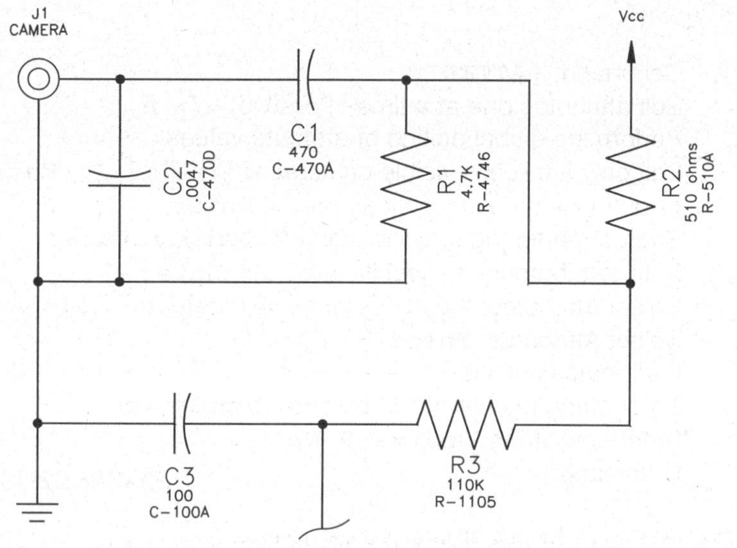


## Annotative Attributes

If you are using multiple viewports or multiple layouts to display the same (model) geometry including blocks with attributes, you may want to use annotative attributes. Annotative attributes can change size automatically when the viewport scale (*Viewport Scale*) is changed, making the text readable (appear the same size) even though other drawing geometry may be displayed larger or smaller in different viewports.

### Creating and Displaying Annotative Attributes

To create annotative attributes and display them in different scales in different viewports, follow these steps:

1. When creating the individual attributes, select the *Annotative* option (checkbox) in the *Text Settings* section of the *Attribute Definition* dialog box (see previous Figure 22-1). If you use the Command line version (*-Attdef*), set the *Annotative* mode to *Yes*.

2. Follow this procedure for <u>every attribute</u> you want to make annotative.
3. Use the *Block* command or the *Block Editor* to create the blocks including the attributes.
4. Create the drawing geometry and *Insert* the blocks. The attributes should appear in model space with the text size that was originally assigned for the attributes using the *Attdef* or *-Attdef* command.
5. Set up multiple viewports in one or more layouts.
6. Make sure the *Add scales to annotative objects when the annotation scale changes* toggle is set (lower-right corner of the Drawing Editor).
7. Set the scale for each viewport by first activating the viewport (double-click inside the viewport or set the *PAPER* toggle) and then selecting the desired viewport scale using the *Viewport Scale* pop-up list or *Viewports* toolbar.

If you have already created attributes and want to change them to annotative at a later time, follow these steps:

1. Use the *Eattedit* command or double-click on an existing attribute to produce the *Enhanced Attribute Editor*.
2. For each attribute in the selected block, activate the *Text options* tab and check the *Annotative* box (see Fig. 22-20).
3. Follow this procedure for every attribute in every block you want to change.
4. Follow steps 5 through 7 from the previous list.

**FIGURE 22-20**

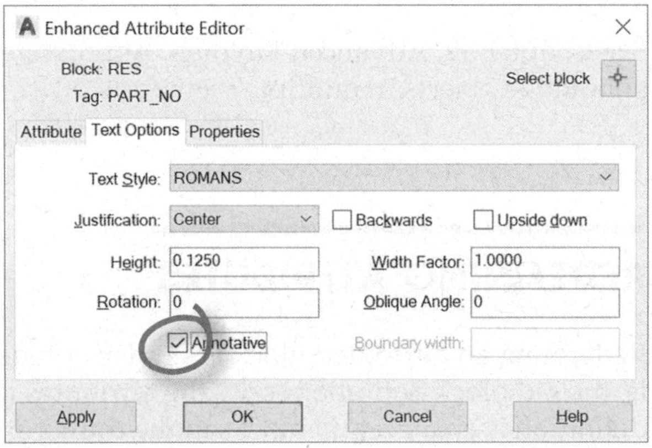


### Annotative Attributes Example

Using our electrical schematic example from the previous pages, assume we changed the existing attributes using the *Enhanced Attribute Editor* as shown in previous Figure 22-20. (Alternately, we could have created annotative attributes originally using the *Attribute Definition* dialog box.) For each block in this drawing, most attributes were changed to *Annotative*; however, the <u>first attribute in each block was not changed</u>.

Assume then two viewports were created using *Vports* to display the schematic in both viewports as shown in Figure 22-21. The viewport on the left is set to display the schematic at 1:1 scale. The viewport on the right is set to display the same geometry at 2:1 scale (note the highlighted viewport border and *Viewports Scale* list). Viewport scale is set by activating each viewport, then using the *Viewport Scale* pop-up list and selecting the appropriate scale.

**FIGURE 22-21**

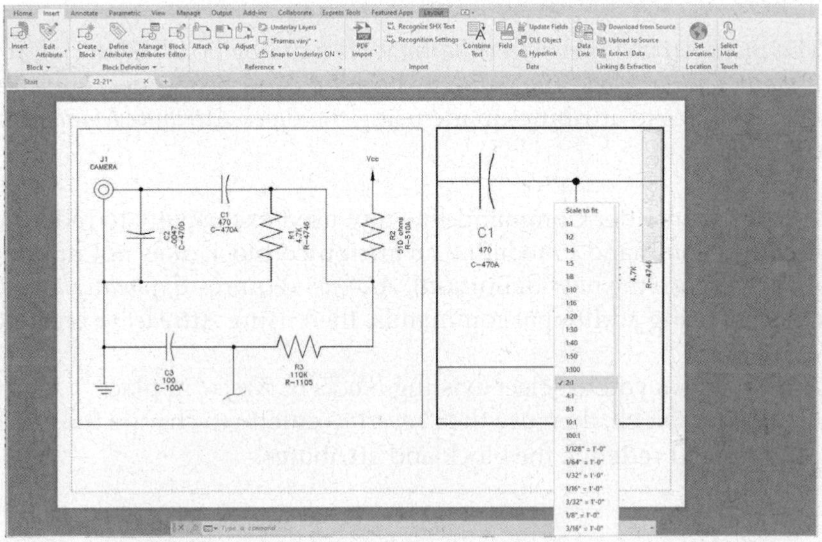

In the left viewport, the blocks and the attributes appear 1:1, just as they were created in model space. In the right viewport, only the "C1" and "R1" blocks appear. Note that the block geometry for both the capacitor and resistor appears twice as large in the right viewport since the geometry is twice the size. Note also that since the first attribute for each block ("C1" and "R1" label) is not annotative, it appears larger than in the left viewport but the same size relative to the blocks. The other attributes, however, are automatically scaled to appear the same size in both viewports or adjusted for the viewport scale.

To make annotative attributes appear at a different scale in a particular viewport, first ensure that the *Add scales to annotative objects when the annotation scale changes* toggle is set. When you use the *Viewports* toolbar or *Viewport Scale* pop-up list to assign a scale for the viewport, AutoCAD automatically creates a new annotative object to appear at the appropriate size based on the viewport scale, then displays only that object. Therefore, multiple objects can exist for each annotative object; however, only the appropriate one is displayed in the related viewport.

See Chapter 32, Advanced Layouts, Annotative Objects, and Plotting for more information on using annotative objects in multiple viewports.

## REDEFINING ATTRIBUTES

Redefining an attributed block is similar to redefining a block (discussed previously in Chapter 21); that is, the old block definition (only the attributes in this case) is replaced with a newer one. When a block definition is changed, all blocks in the drawing that reference that definition are updated and display the new changes; therefore, redefining a block is essentially a global update for all block insertions.

Using one of the tools described in this section to redefine an attributed block also results in a global update—all blocks in the drawing referencing the new block definition reflect the changed attributes. Therefore, the tools explained in this section are used to globally change attribute features such as the *Tag*, *Prompt*, *Default Value*, the order of the attributes that appear upon insertion, or the text properties of the values. If you want to change the attribute values for individual blocks, use *Eattedit*, *Attedit*, *Properties*, or *Find* (see "Displaying and Editing Attributes").

There are several methods you can use to redefine attributes: *Battman* (the *Block Attribute Manager*), *Attredef*, *Refedit*, and the Block Editor.

The *Block Attribute Manager* (*Battman*) is a dialog box-based tool. It allows you to change the *Tag*, *Prompt*, *Default Values*, *Mode*, the value text options and properties, and the order of the attribute prompting. Changes to attributes made using the *Block Attribute Manager* can be immediately reflected in the drawing.

*Attredef* is an older Command-line tool used exclusively to redefine blocks containing attributes (using the *Block* command to redefine an attributed block does not update existing attributes in the drawing that reference the new definition). *Attredef* requires *Exploding* an existing block, making the desired changes using additional commands, then using *Attredef* to create the new block definition.

*Refedit* allows you to select existing *Blocks* or *Xrefs* "in place." With this method, use *Refedit* to select the desired block, then use the *Properties* palette to change the attributes, then use *Refclose* to save the changes and redefine the block and attributes.

## *Battman*

| Ribbon | Menu Bar | Command (Type) | Alias (Type) | Shortcut |
|---|---|---|---|---|
| *Insert*<br>*Block Definition*<br>*Manage Attributes*<br>(icon) | *Modify*<br>*Object >*<br>*Attribute >*<br>*Block Attribute Manager...* | *Battman* | ... | ... |

The *Battman* command invokes the *Block Attribute Manager* (Fig. 22-22). With the *Block Attribute Manager,* you can modify attributes within a *Block* without having to *Explode* or redefine the entire *Block*. You can change almost any property of a *Block's* attributes such as the tag, prompt, default value, mode, text settings, and the order of attribute prompting. All changes made to an attributed *Block* are "pushed" out to all references of the *Block* in the drawing.

**FIGURE 22-22**

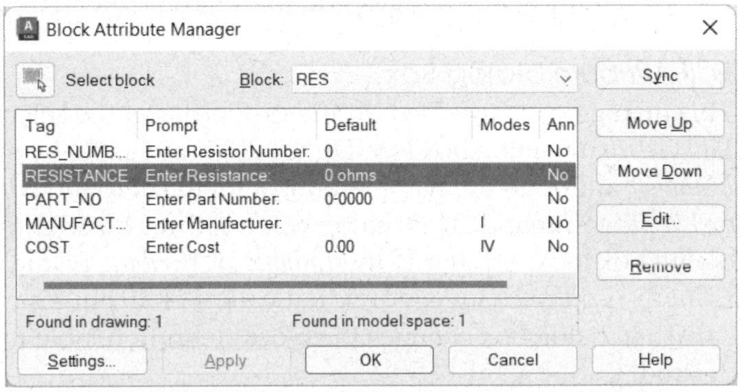


After producing the *Block Attribute Manager,* use the *Select block* button or the *Block*: drop-down list (containing all blocks defined in the drawing) to select the block you want to redefine. Once selected, all the attributes defined for the block appear in the attribute list in the central area. Next, select any attribute from the list you want to redefine. The buttons on the right and bottom of the *Block Attribute Manager* offer the options explained next. Most changes you make to attributes using the *Block Attribute Manager* can be immediately reflected in the drawing if the *Auto preview changes* box is checked in the *Edit Attribute* dialog box (see "*Edit Attribute* Dialog Box").

### *Sync*

The *Sync* button synchronizes the highlighted block to change according to the definition as it exists in the block definition table before using the *Block Attribute Manager*. This action may be necessary only if attributes were redefined using methods other than the *Block Attribute Manager*. Using this button is essentially the same as using the *Attsync* command (see "*ATTSYNC*"). *Sync* does not update the existing attributes in the drawing to new changes you make in the *Block Attribute Manager*—use *Apply* or *OK* to accomplish that action.

### *Move Up, Move Down*

These options change the order in which you are prompted to enter the attribute values upon insertion of the block. Changes to the order made by *Move Up* and *Move Down* <u>do not affect the order that the attributes appear in the inserted block</u>—only the prompting order in the *Edit Attributes* dialog box or at the Command line when using *Insert* or *Ddatte*.

### *Edit*

This button produces the *Edit Attribute* dialog box (see Figure 22-23 and description under "*Edit Attribute* Dialog Box").

### *Remove*

Highlight any attribute in the list, then select *Remove* to delete the attribute from the block definition and from existing insertions of the block if desired. If only one attribute exists, this option is disabled.

### *Settings*

This button produces the *Settings* dialog box (see Figure 22-26 and description).

### *Apply*

Use this button to immediately apply any changes you made in the *Block Attribute Manager* to the block definition table. The changes will also be pushed out to existing insertions of the block unless *Apply changes to existing references* is unchecked in the *Settings* dialog box. You cannot cancel the changes using *Cancel*. Using *Apply* is essentially the same as using *OK*, except the *Block Attribute Manager* does not close. If you want to use the *Block Attribute Manager* to change attributes from more than one block, use *Apply* between block selection.

### *OK*

Use *OK* to apply all changes you made to the block definitions and close the *Block Attribute Manager*.

### *Edit Attribute* Dialog Box

This dialog box is used to edit the properties of the selected attribute in the *Block Attribute Manager*. The *Edit Attribute* dialog box (see Fig. 22-23) is available by selecting the *Edit* button from the *Block Attribute Manager* although the name is almost identical to the *Edit Attributes* dialog box that appears using the *Insert* or *Ddatte* command. This dialog box has three tabs: *Attribute*, *Text Options*, and *Properties*. Although this dialog box looks similar to the *Enhanced Attribute Editor*, their functions are different. The *Enhanced Attribute Editor* is used to change individual values of attributes in existing block insertions, whereas the *Edit Attribute* dialog box changes the block definition table for attributed blocks as well as existing insertions if desired.

*Auto preview changes*

If this box is checked (and *Apply changes to existing references* in the *Settings* dialog box is checked), changes that you make in any of the tabs here and in the *Block Attribute Manager* are displayed in existing block insertions in the drawing. If *Auto preview changes* is not checked, changes are not visible in the drawing until you use *Apply* or *OK*. This option is disabled if *Apply changes to existing references* is unchecked in the *Settings* dialog box.

*Attribute*

The *Attribute* tab (Fig. 22-23) allows you to change the attribute *Tag*, *Prompt*, *Default* value, and *Mode* of the attribute highlighted in the *Block Attribute Manager*.

FIGURE 22-23

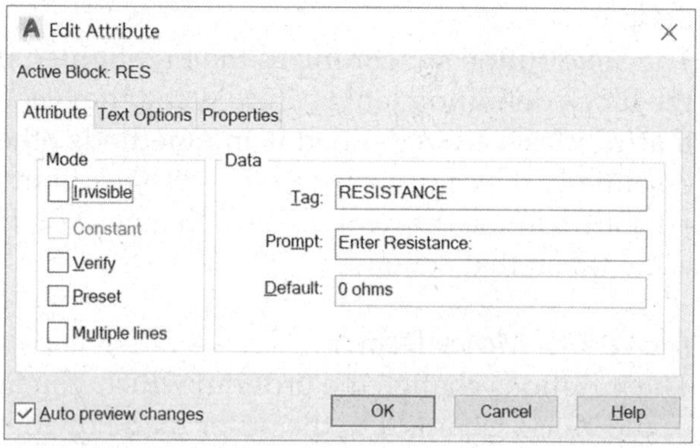


*Text Options*

In this tab (Fig. 22-24), you can change the *Text Style*, *Justification*, *Height*, *Rotation*, *Width Factor*, *Oblique Angle*, *Annotative*, and other options depending on the text font used (.SHX or .TTF). Despite the similarity to the tab of the same name in the *Enhanced Attribute Editor*, this tab changes the block definition, not individual attribute values.

FIGURE 22-24

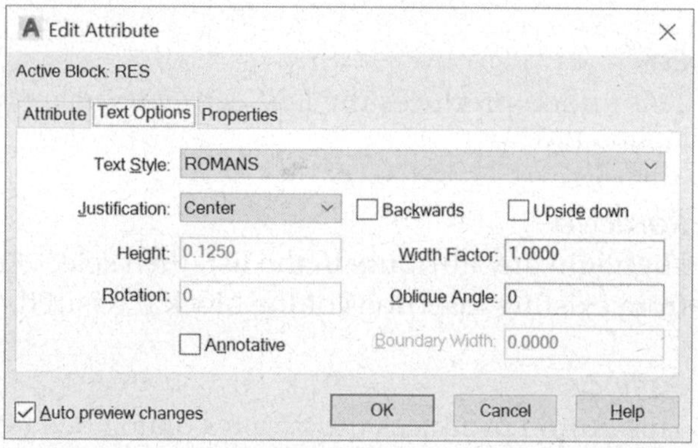

*Properties*

Use the *Properties* tab (Fig. 22-25) to change any attribute values for the block definition such as *Layer*, *Linetype*, *Color*, *Lineweight*, and *Plot style*. This tab is also similar to the tab of the same name in the *Enhanced Attribute Editor*; however, this tab changes the block definition, not individual attribute values.

**FIGURE 22-25**

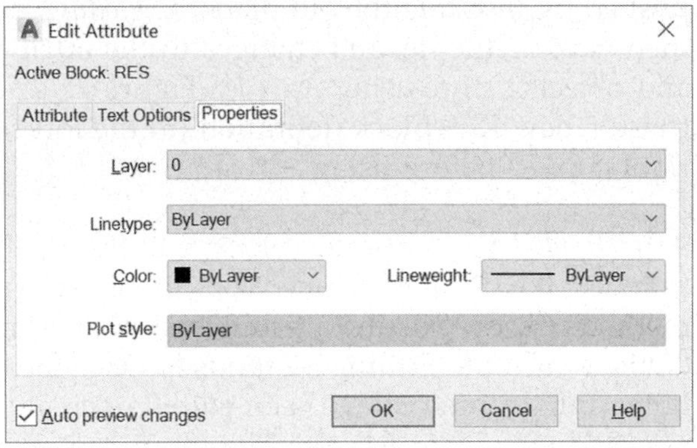


## *Settings* Dialog Box

The *Settings* dialog box (Fig. 22-26) is available through the *Settings* button in the *Block Attribute Manager*.

**FIGURE 22-26**

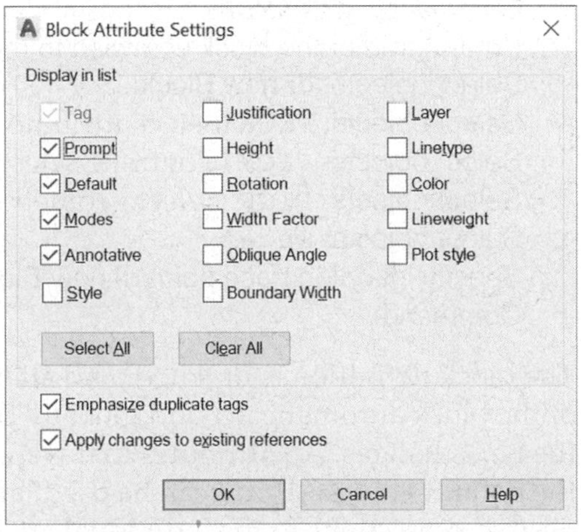


*Display in list*

The *Display in list* section controls which properties of the attributes appear in the attribute list of the *Block Attribute Manager*. Note the options selected in Figure 22-26 and compare the list in Figure 22-22. Selecting multiple properties is helpful if you need to compare attributes in the list based on specific properties.

*Emphasize duplicate tags*

It is possible for one block to contain two or more attributes with the same tag. Such a practice should be avoided to prevent confusion. You can use the *Emphasize duplicate tags* option to force attributes with the same tag to be displayed in the list in red rather than in black. This action makes it easier to locate and change tags that are duplicate.

*Apply changes to existing references*

With this box checked, all changes that are made to attributes in the *Block Attribute Manager* (using *Apply* or *OK*) are applied to the existing block insertions in the drawing. Otherwise, the changes made redefine the block attributes and affect new block insertions but not existing block insertions. This option must be checked to use the *Auto preview changes* feature.

## *Attredef*

| Ribbon | Menu Bar | Command (Type) | Alias (Type) | Shortcut |
|---|---|---|---|---|
| ... | ... | *Attredef* | ... | ... |

The *Attredef* command (Attribute redefine) enables you to redefine an attributed block. To use the *Attredef* feature, the objects (to be *Blocked*) with new attributes defined must be in place before invoking the command. This method is older and tedious compared to redefining attributes using *Battman*.

Creating the new objects and attributes is accomplished by *Inserting* the old block, *Exploding* it, then making the desired changes to the attributes and objects before using *Attredef*. Figure 22-27 illustrates a new RES block definition (in the lower-right corner) before using *Attredef*.

FIGURE 22-27

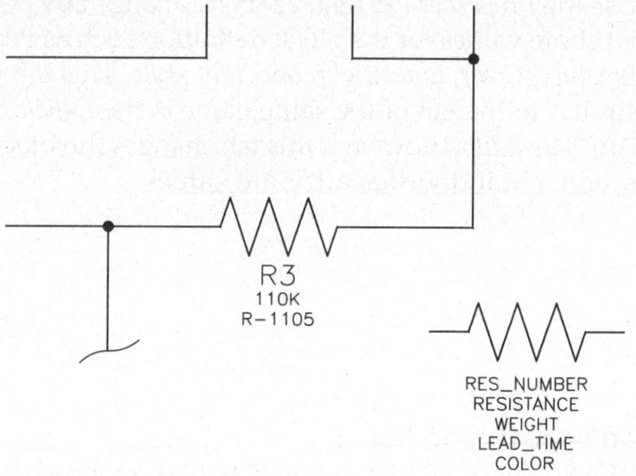


After *Exploding* an inserted block, edit the attributes to reflect the desired change. You can use several methods to edit existing attributes (explained in "Other Attribute Editing Methods"). Once the attributes and geometry have been edited or created to your liking, invoke *Attredef*. The command operates only in Command line format as follows:

Command: ***ATTREDEF***
Enter name of the block you wish to redefine: **res**
Select objects for new Block...
Select objects: **PICK** (Select attributes in order that you want them to appear; see Fig. 22-5.)
Select objects: **PICK** (Continue selecting all attributes.) etc.
Select objects: **PICK** (Select geometry last.)
Select objects: **Enter**
Specify insertion base point of new Block: **PICK**
Command:

The block definition is updated, and all instances of the block automatically and globally change to display the new changes. All attributes that were changed reflect the default values. *Attedit* can be used later to make the desired corrections to the individual attributes.

FIGURE 22-28

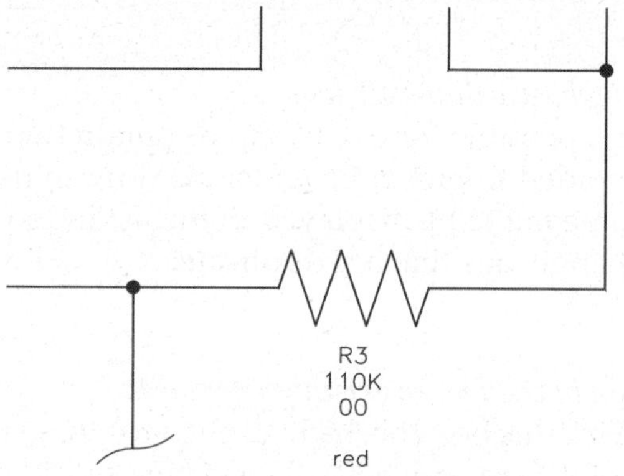


Figure 22-28 displays one of the RES blocks after redefining. Notice that the fourth attribute is invisible as it was before it was *Exploded*. For this attribute, the mode was not changed. The new last attribute, however, was changed to a normal attribute mode and therefore appears after using *Attredef*.

## *Refedit, Properties,* and *Refclose*

TIP

Another method of redefining blocks is to use *Refedit*, the *Properties* palette, and *Refclose*. This technique is also known as "in-place editing" because the references (*Blocks* or *Xrefs*) can be edited after being inserted or attached. However, the *Block Attribute Manager* makes the *Refedit* method seem relatively involved and limited; therefore, this method for the purpose of redefining block attributes is explained briefly in this chapter only for reference. Refer to Chapter 30, Xreferences, for a full explanation of *Refedit* and other related commands and variables for use with *Blocks* and *Xrefs*.

To redefine one block attribute at a time using *Refedit*, *Properties*, and *Refclose*, follow these steps.

1. Invoke the *Properties* palette.
2. Invoke *Refedit*. Select the <u>*Block*</u> (not the attributes) containing the attributes you want to redefine.
3. In the *Reference Edit* dialog box that appears, select the *Display attribute definitions for editing* checkbox. Select *OK*.

4. At the "Select nested objects" prompt, select the *Block* again (not the attributes).
5. Use the *Properties* palette to change the *Tag* or *Prompt* property. When finished, press Escape twice to deselect the *Block*.
6. Use the *Refclose* command and *Save* option to save the changes and redefine the attributes. All insertions of the *Block* are redefined.

When *Blocks* are redefined using this method, the block definition is changed in the current drawing. However, when *Xrefs* are changed using *Refedit*, the original externally referenced drawing can be changed from within the current drawing (see Chapter 30, Xreferences).

## The Block Editor

You can use the Block Editor to change attributed blocks; however, changes to attributes are not automatically updated for existing blocks in the drawing until you use the *Attsync* command. You can use three commands in the Block Editor to redefine blocks: *Ddedit*, *Properties*, or *Attdef*. First open the Block Editor and select the desired block to edit. Next use one of the three commands to edit the attributes. When you are finished in the Block Editor, save the block when you close the editor. The three options are explained next.

FIGURE 22-29

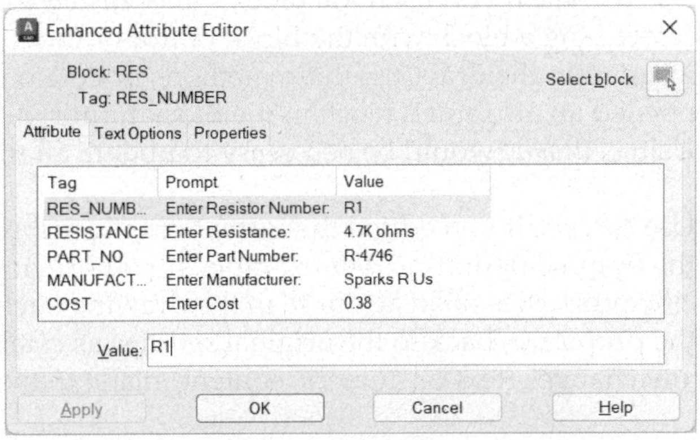


### Ddedit

*Ddedit*, the dialog box for editing text, recognizes the text as attributes and allows you to change the attribute *Tags*, *Prompts*, and *Default Values*. Figure 22-29 illustrates using the *Enhanced Attribute Editor* dialog box for changing the attribute *Prompt*.

FIGURE 22-30

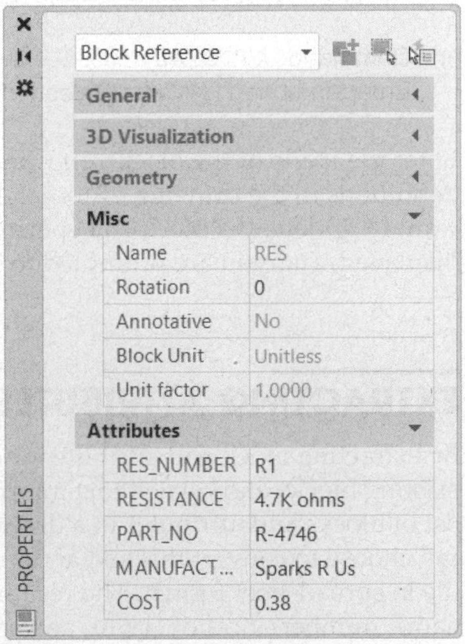


### Properties

Optionally, the *Properties* palette can be invoked. Not only does *Properties* recognize the text as attributes, but the *Tag*, *Prompt*, default *Value*, and attribute modes can be modified. Figure 22-30 displays this dialog box for the same operation. Notice all of the possible changes that can be made.

### Attdef

Instead of using *Ddedit* or *Properties* to change existing attributes, new attributes can be created "from scratch" with *Attdef*.

### Grip Editing Attribute Location

TIP

Grips can be used to "move" the location of the attributes after an attributed *Block* has been inserted. To do this (with Grips on), PICK the *Block* so grips appear at each attribute and at the block insertion point. Next, PICK the attribute you want to move so its grip becomes "hot." The attribute can be moved to any location and still retain its association with the block. (See Chapter 19, Grip Editing.)

### *Attsync*

| Ribbon | Menu Bar | Command (Type) | Alias (Type) | Shortcut |
|---|---|---|---|---|
| *Home* *Block* (icon) | ... | *Attsync* | ... | ... |

The *Attsync* command synchronizes the selected block attributes in the drawing to change according to the definition as it exists in the block definition table. *Attsync* does not affect individual attribute values, only changes made to the properties of the inserted attributes that are different from what is established in the definition table for the block. *Attsync* is usually not needed if you used the *Block Attribute Manager* or *Attredef* to redefine block attributes. This command may be helpful in the following cases.

TIP

Use *Attsync* if you used the Block Editor or *Block* command to redefine a block containing attributes. Redefining a block with the Block Editor or the *Block* command does not update attributes in blocks inserted in the drawing; these methods update only the drawing geometry. For example, you may have opened an attributed block in the Block Editor, deleted an attribute, then saved the *Block* to redefine it. Using *Attsync* would be necessary to update all the existing blocks in the drawing to the new definition.

Use *Attsync* if you edited the format of the attributes by some method and want to change them back to the original definition. For example, if you previously used the *Enhanced Attribute Editor* to change specific properties of an attribute in the drawing such as text height or color, using *Attsync* would change the properties back to the original settings as established in the block's definition. *Attsync* does not affect any changes (text or numeric content) made to the attribute values themselves.

When *Attsync* is invoked, you are prompted for the names of blocks you want to update with the current attributes defined for the blocks.

```
Command: ATTSYNC
Enter an option [?/Name/Select] <Select>:
```

Enter the name of the block you want to update. Entering a question mark (?) displays a list of all block definitions in the drawing. Alternately, you can press Enter to select the block whose attributes you want to update. If a block you specify does not contain attributes or does not exist, an error message is displayed, and you are prompted to specify another block.

## EXTRACTING ATTRIBUTES

By extracting block and attribute information from AutoCAD drawings, you can create a report of the existing blocks and related text information contained in the drawings. In other words, you can select a list of blocks and attributes in a drawing, then extract all the contained information into a table. You can choose to convert this text and/or numeric information into an AutoCAD *Table* or to an external file in spreadsheet format. The resulting table can be used as a report listing a variety of data such as the name, number, location, layer, and scale factor of the blocks, and attribute information such as prices, manufacturers, sizes, dates, or any other attribute data that may be defined in the block insertions.

Because of this attribute extract feature, AutoCAD drawings contain the intelligence to compile or calculate bills of materials, supplier information, employee records, quotations, billing totals, maintenance schedules, catalogs, and so on. Since an AutoCAD *Table* generated from attribute data is linked to the attributes, the data content of the table is updated when changes are made to the attributes. You can use two methods to extract attributes: the *Data Extraction* wizard and the *Attext* command.

The *Eattext* command produces the *Data Extraction* wizard. This wizard makes the process of extracting attributes far simpler and more visual than the earlier methods in AutoCAD. The wizard leads you through the steps of selecting drawings, selecting *Blocks* and/or *Xrefs*, selecting the desired attribute information, selecting the output format, and saving templates for future use. The *Data Extraction* wizard provides you with capabilities not available by other means.

TIP

Alternately, you can use the *Attext* command to produce the older *Attribute Extraction* dialog box or type *-Attext* to invoke the Command line equivalent. Using this method is far more complex than the newer *Data Extraction* wizard because template files must be created externally to AutoCAD and the chances of error are great.

## *Eattext*

| Ribbon | Menu Bar | Command (Type) | Alias (Type) | Shortcut |
|---|---|---|---|---|
| *Insert*<br>*Linking & Extraction*<br>*Extract Data*<br>(icon) | *Tools*<br>*Data Extraction...* | *Eattext* or *DataExtraction* | ... | ... |

The *Data Extraction* wizard leads you through the eight steps to extract block attribute (text and numeric) information from a drawing into a table. The resulting table can be inserted into the drawing as an AutoCAD *Table* or generated as an external spreadsheet format file.

The *Data Extraction* wizard offers a number of features that are not available by other means. With the *Data Extraction* wizard, you can extract attribute data from the current drawing or from multiple drawings and sheet sets. You can also extract attribute data from nested *Blocks* and *Xrefs*. In addition to the ability to choose the type of table to create, you can view and edit the organization of the data before creating the table.

The eight "pages" (steps) required in the *Data Extraction* wizard are explained here. Use the *Next* button to proceed after each page and the *Back* button to go back at any time to change an option.

### *Begin (Page 1 of 8)*

In the first page (Fig. 22-31), decide if you want to create a new data extraction or edit an existing one. If this is your first time to use the *Data Extraction* wizard, select *Create a new data extraction*. If you have used the *Data Extraction* wizard previously and have saved a template file in .DXE or .BLK format, or if a template file exists for your project or company, you can select *Use previous extraction as a template*, then use the browse button to locate and select the file. Template files contain specifications indicating which blocks and attributes are to be extracted and the format of the output table.

FIGURE 22-31

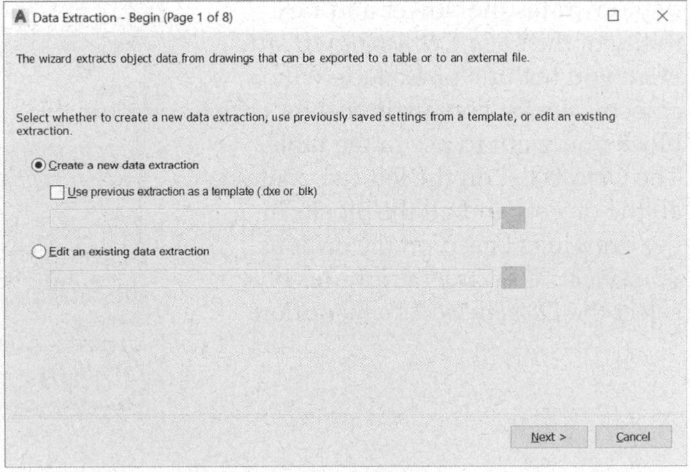

If you are creating a new data extraction, when you select the *Next* button a file navigation dialog box appears for you to save a .DXE template file for future use. (The *Data Extraction* wizard cannot use template files in .TXT format created for early versions of AutoCAD.)

### *Define Data Source (Page 2 of 8)*

In this second page you are prompted to specify the "data source" you want to extract attribute and block information from (Fig. 22-32). Use the *Drawing/Sheet set* button with the *Include current drawing* option to have all blocks in the current drawing considered for extraction. The *Select objects in the current drawing* option allows you to select specific *Blocks* from the current drawing. Use the *Add Folder* and *Add Drawings* options to locate and specify multiple drawings or sheet sets containing the blocks you want to extract information from. The list of drawings you selected appears in the *Drawing files and folders* list. In the (next) *Select Objects* page, you can exclude any of the blocks and other objects you specified in this page if desired.

**FIGURE 22-32**

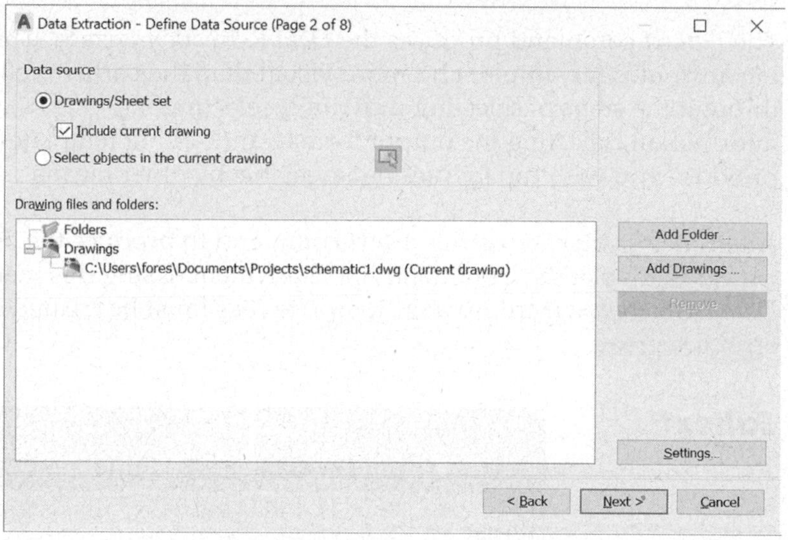


*Settings*

Selecting this button in the second page produces the dialog box shown in Figure 22-33. The data sources you selected previously may contain nested blocks (blocks within blocks) or attached *Xref* drawings that contain blocks. Your choices here determine if you want to include blocks from these sources in the table. For most applications, select *Extract objects from blocks* and *Objects in model space* unless you want to include any attached *Xrefs* or any blocks in layouts (such as title blocks).

**FIGURE 22-33**

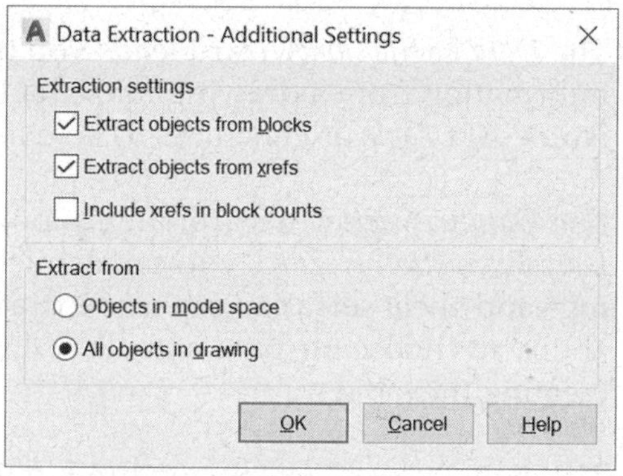


### *Select Objects (Page 3 of 8)*

The *Select Objects* page (Fig. 22-34) demonstrates the power and flexibility of the *Data Extraction* wizard. Here you include or exclude which objects, blocks, and attributes for each block you want to use in the table. The *Objects* list on the left can contain all the objects and all the blocks in the drawings based on the options you select. To extract attributes only, select the *Display blocks only* option.

**FIGURE 22-34**

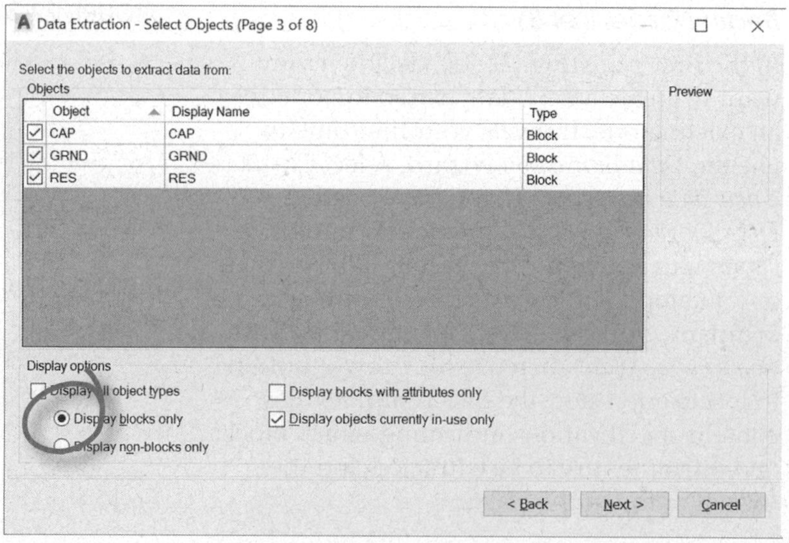

### *Select Properties (Page 4 of 8)*

The *Select Properties* page (Fig. 22-35) allows you to further refine your data search by specifying the type of data in the *Category filter* options on the right. To extract attribute data only, uncheck all options except *Attribute*.

**FIGURE 22-35**

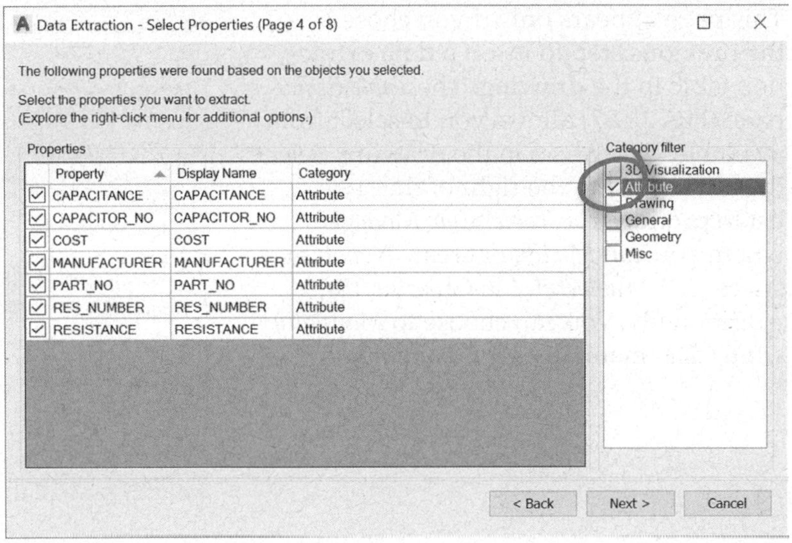


### *Refine Data (Page 5 of 8)*

This page (Fig. 22-36) allows you to set up the format for the table. Click on any column head to sort the entire table by items in that column. You can arrange the table columns by dragging and dropping the column heads to any position. The right-click shortcut menu provides options for sorting data, hiding columns, and renaming columns. You can also use the options below (left) to hide or show the name column, count column, and to combine identical rows.

**FIGURE 22-36**

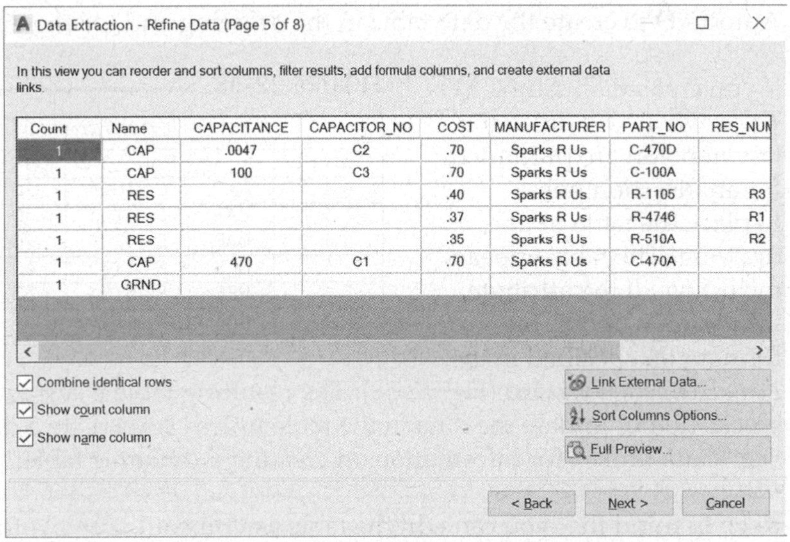


The *Sort Columns Options* button allows you to specify primary and secondary (and more) columns to use for sorting. Select the *Full Preview* button to see the entire table in the current format with the selected options. The preview simulates the appearance of an external file opened in a spreadsheet program.

If you want to include data from an external spreadsheet in the same table as the attribute data, select the *Link External Data* button. Using the *Data Link Manager* that appears, you can specify columns from an external Excel file to include as additional appended columns in the data extraction table (see "Data Links" for more information).

### *Choose Output (Page 6 of 8)*

In this page (not shown), you must specify if you want to create an AutoCAD table in the current drawing (*Insert data extraction table into drawing*), an external file (*Output data into external file .xls, .csv., .mdb, .txt*), or both. The output file formats are Microsoft Excel (.XLS), comma separated format (.CSV), Microsoft Access (.MDB), and ASCII format (.TXT).

### Table Style (Page 7 of 8)

This page appears only if you chose in the previous step to insert a data extraction table in the drawing. The *Table Style* page (Fig. 22-37) allows you to select from available *Table Styles* in the drawing. Select the button just to the right of the style list to produce the *Table Style Manager* where you can *Modify* or create *New* table styles (see "*Tablestyle*" in Chapter 18). Additionally, you can choose to *Manually setup table* (manually set up a table).

**FIGURE 22-37**

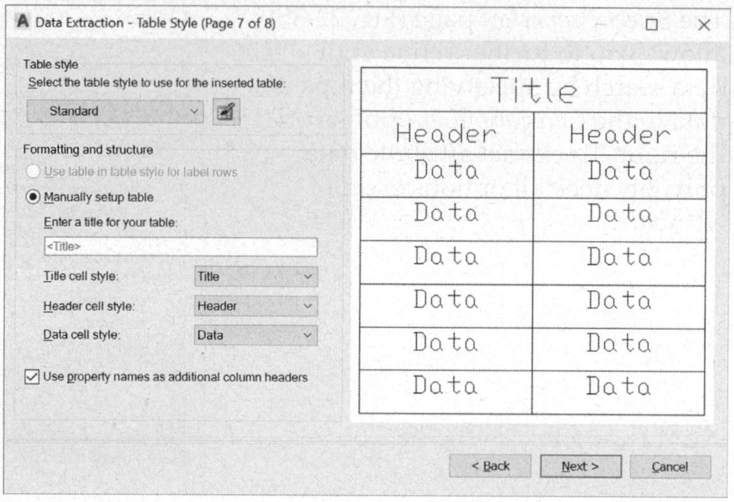


### Finish (Page 8 of 8)

Page 8 (not shown) appears as the last step of either type data extraction you select (table in the drawing or external file). There are no options here other than to select *Finish*. Pressing the *Finish* button causes AutoCAD to create the data table in the drawing and/or the external file.

If you created an AutoCAD *Table*, when you finish the last step you are prompted for an insertion point to place the table in the drawing. The table appears including all the attribute information and in the format you specified in the *Data Extraction* wizard (Fig. 22-38). The resulting table is a static table of text. Changes to the drawing blocks do not update the data in the table unless you create a data link to a Microsoft Excel spreadsheet. See "Data Links" for information on creating a dynamic table.

**FIGURE 22-38**

| CAPACITOR AND RESISTOR DATA | | | | | | |
|---|---|---|---|---|---|---|
| Quantity | Name | PART_NO | COST | MANUFACTURER | CAPACITANCE | RESISTANCE |
| 1 | CAP | C-100A | .70 | Sparks R Us | 100 | |
| 1 | CAP | C-470D | .70 | Sparks R Us | .0047 | |
| 1 | CAP | C-470A | .70 | Sparks R Us | 470 | |
| 1 | RES | R-1105 | .40 | Sparks R Us | | 110K |
| 1 | RES | R-510A | .35 | Sparks R Us | | 510 ohms |
| 1 | RES | R-4746 | .37 | Sparks R Us | | 4.7K |

Keep in mind that you can edit the table as you could any AutoCAD *Table* (see "*Tabledit*" in Chapter 18). You could, say, create additional rows or columns in the table to make calculations in a table. For example, a row could be added at the bottom of the table in Figure 22-38 and a *Formula Field* could be inserted to calculate the *Sum* of the COST attribute column data. Also remember that attributes can contain *Fields* (see "*Attdef*").

If you created an external file (in .CSV, .XLS, .MDB, or .TXT format), the resulting table can be opened in a spreadsheet, database, or other program. An example external file is displayed in Figure 22-39 as it appears in Microsoft Excel. Such a file is not dynamically linked to the attribute information, as has been the case in older releases of AutoCAD.

**FIGURE 22-39**

| | A | B | C | D | E | F | G |
|---|---|---|---|---|---|---|---|
| 1 | Quantity | Name | PART_NO | MANUFACTUR | COST | CAPACITANCE | RESISTANCE |
| 2 | 1 | CAP | C-100A | Sparks R Us | 0.7 | 100 | |
| 3 | 1 | CAP | C-470D | Sparks R Us | 0.7 | 0.0047 | |
| 4 | 1 | CAP | C-470A | Sparks R Us | 0.7 | 470 | |
| 5 | 1 | RES | R-1105 | Sparks R Us | 0.4 | | 110K |
| 6 | 1 | RES | R-510A | Sparks R Us | 0.35 | | 510 ohms |
| 7 | 1 | RES | R-4746 | Sparks R Us | 0.37 | | 4.7K |
| 8 | | | | | | | |

## Updating Attribute Data

In AutoCAD 2006 and AutoCAD 2007, tables created from the *Attribute Extraction* wizard were dynamically linked to the attributes, so changes made to attributes in the drawing could be reflected in the related tables by simply updating the tables. Since AutoCAD 2008, this feature is not available since attributes are not directly linked to AutoCAD tables.

Using AutoCAD's *Data Link Manager*, you can link an AutoCAD table to data in an Excel spreadsheet (see "Data Links"). Although the ability to create a direct link between an AutoCAD table and an Excel file has many useful applications, the process of linking attribute data to a table in the same drawing requires additional steps in newer AutoCAD releases, including maintaining an external data file.

To manage a table of attribute data in current releases of AutoCAD, you must create and maintain an external Excel file to hold the data, then create the attribute table data from the Excel file. When changes are made to attributes, the *Data Extraction* wizard must be used to update the Excel file, then the linked AutoCAD table must be updated from the Excel file.

This process is as follows:

1. Create an external file using the *Output data into external file .xls, .csv., .mdb, .txt* option of the *Data Extraction* wizard.
2. Next, use the *Table* command with the *From a data link* option to create an AutoCAD table from the Excel file. Alternately, you can populate an existing table with data from the Excel file using the table's right-click shortcut menu options.
3. If any attribute data then changes in the drawing, use the *Data Extraction* wizard again with the *Edit an existing data extraction* option to update the existing Excel file.
4. Next, update the AutoCAD table. Normally, you will be prompted to update the table ("data link") when the related external file has changed.

# DATA LINKS

Data in an AutoCAD table can be linked to data in an existing Microsoft Excel (.XLS) spreadsheet. The AutoCAD table can be created directly from the Excel spreadsheet or the table can exist in AutoCAD first, then the link created. Nevertheless, the Excel file must first exist before a data link can be made. You can link data in a drawing table to an entire spreadsheet or you can link only single cells or a range of cells in the spreadsheet to an AutoCAD table and vice versa. Once the dynamic link between the AutoCAD table and the Excel spreadsheet is established, changes made to the data in either one (the AutoCAD table or the Excel spreadsheet) can be updated to reflect those changes in the other (see "*Datalinkupdate*").

The <u>Excel spreadsheet must exist in .XLS or .CSV format</u> before you can create a link. The Excel file can be created three ways:

- Create the spreadsheet from scratch in Excel;
- Use AutoCAD's *Data Extraction* wizard to create an external (.XLS) file; and
- Use an existing AutoCAD table and *Export* it to a .CSV file, then open it in Excel and use *Save As* to create the .XLS file.

You can fill an AutoCAD table with data from an external Excel file by two methods:

- If no table in AutoCAD exists, use the *Table* command in AutoCAD to create the table by selecting the *From a data link* option in the *Insert Table* dialog box.
- If you have an existing AutoCAD table, highlight the cells you want to populate with data and select *Data Link* from the shortcut menu to use AutoCAD's *Data Link Manager* to create the link and to specify the range of data from the spreadsheet to use in the table.

## Datalink

| Ribbon | Menu Bar | Command (Type) | Alias (Type) | Shortcut |
|---|---|---|---|---|
| *Insert Linking & Extraction Data Link* | *Tools Data Links > Data Link Manager* | *Datalink* | *DL* | ... |

The *Datalink* command produces the *Data Link Manager* (Fig. 22-40). This tool is used primarily to create a link between the current AutoCAD drawing and an external Excel (.XLS format) file. The manager does not create tables or update data between the drawing and an external file. In addition to creating a link, you can use a right-click shortcut menu in the *Data Link Manager* to open the Excel file, edit the link (change the file or range of data), rename the link, and delete the link.

FIGURE 22-40

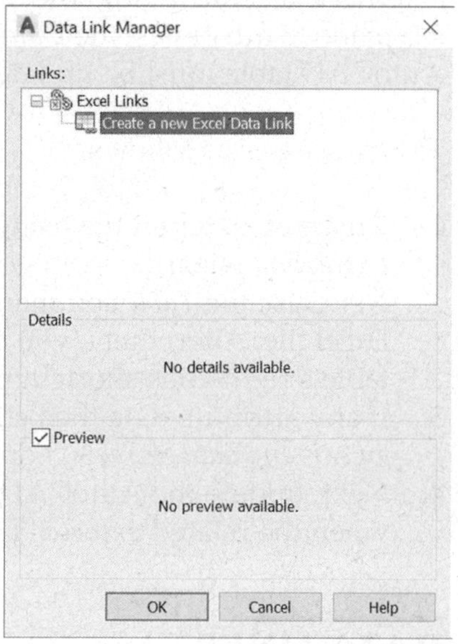


To create a link, select the *Create a new link* line as shown in Figure 22-40. The *Enter Data Link Name* dialog box (not shown) appears for you to assign a name for the link you are creating. This name is internal to AutoCAD and is not the name of the Excel file or any external file.

Next, the *New Excel Data Link* dialog box appears (not shown). Use the *Browse* button to locate and select an existing Excel .XLS file.

Once an Excel file is located, the *New Excel Data Link* dialog box appears (Fig. 22-41). Note that you can link to the entire sheet or link to a range of cells within the spreadsheet. A range is a specific number of connected rows and columns in a spreadsheet or table.

FIGURE 22-41

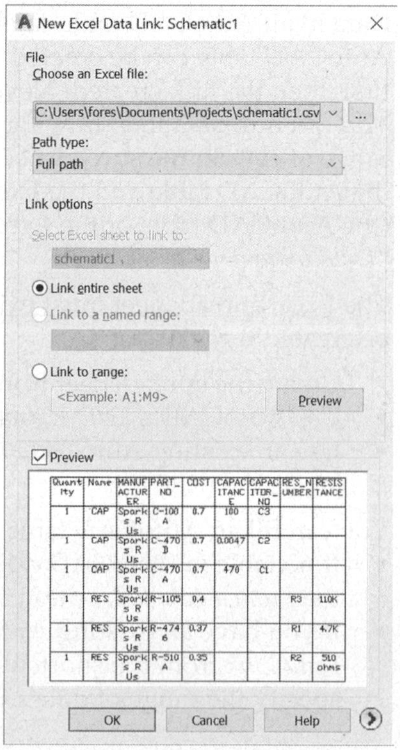


### *Link entire sheet*

Select this option to link the entire Excel spreadsheet (all rows and columns that contain data).

### *Link to a named range*

A "named" range is one that is named and saved in Excel. Any saved named ranges appear in the drop-down list.

### *Link to a range*

Use this option to specify a range of cells in the spreadsheet that have not previously been named in Excel. To define a new range within an Excel spreadsheet, key in the columns (letters) and rows (numbers). For example, a range of the first 2 columns and 8 rows would be specified as A1:B8.

Once the link has been created, the *Data Link Manager* appears again, but with the linked data shown in the preview tile.
Completing this procedure only creates a link. No linked data appears in the drawing until you create a table and populate it with the data.

Therefore, a data link in AutoCAD is a name you assign to a selected range of cells in an external file. If no range is specified, the entire range of cells that contain data in the spreadsheet is included in the link. Therefore, an AutoCAD drawing could contain two data links from the same Excel file. For example, you could have a data link named "Schematic Attributes" for an entire Excel spreadsheet and another data link named "Schematic Attributes Range AB" for a specific range of cells within the spreadsheet.

NOTE: Keep in mind that using the *Datalink* command as described above simply creates a link between the drawing and the external file. It is important to note that you do not have to first create a data link using the *Data Link Manager* to create a table or populate cells in the table. If you want to create an AutoCAD table from scratch using an external file, use the *Table* command with the *From a data link* option to establish the link. If you have an existing table and want to populate all or some of the cells with data from a linked Excel file, highlight the desired cells in the table and select *Data Links* from the right-click shortcut menu. Both of these actions produce the *Data Link Manager*. See the following examples.

TIP

Assuming a link has been created, the right-click shortcut menu in the *Data Link Manager* provides options for managing the links. You can choose to *Open the Excel file* and edit the data directly in Excel. You can also *Edit* the link (change the file or range of data that is linked), *Rename* the link, and *Delete* the link.

**FIGURE 22-42**

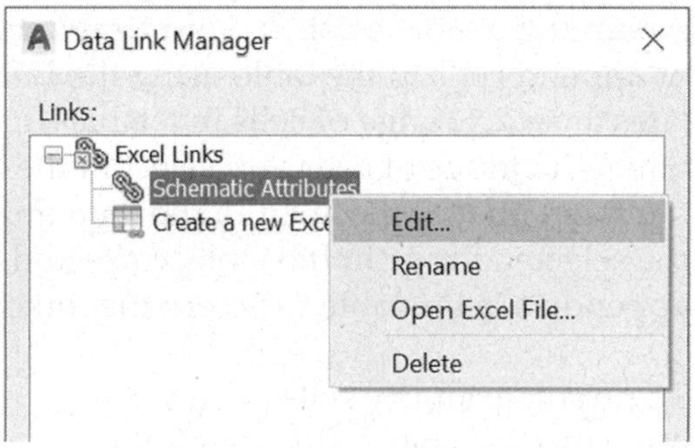


## Data Link Examples

Assuming an external Excel Spreadsheet exists, you can create an AutoCAD table containing data from the table. As previously stated, you can use the *Table* command with the *From a data link* option to create the table from scratch, or you can add data from an external file to an existing table.

### Creating a Table from an Excel Spreadsheet

This procedure is relatively simple. First, use the *Table* command to produce the *Insert Table* dialog box. Select the *From a data link* option. Use the button just to the right of the edit box to produce the *Data Link Manager* (Fig. 22-43).

**FIGURE 22-43**

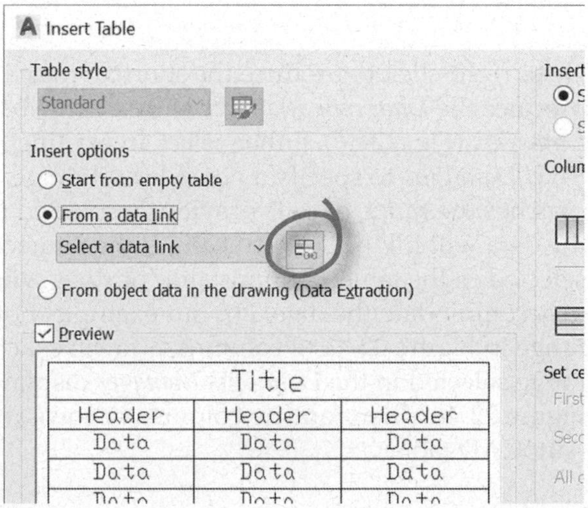

In the *Data Link Manager*, select an existing link (if one has previously been established for the drawing) or select *Create a new Excel Data Link* (see previous Fig. 22-40). Once the link for the table is established, the preview tile displays the new data in the *Data Link Manager* (Fig. 22-44). Select *OK*.

FIGURE 22-44

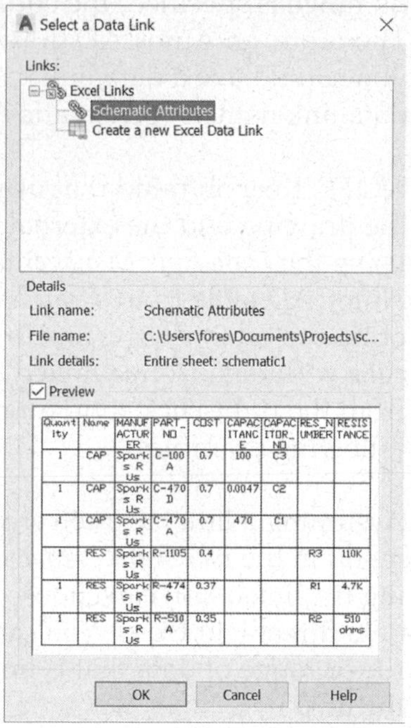


Next, the *Insert Table* dialog box appears again and displays the new data in the preview tile (not shown). Note that options for specifying the table style are disabled. When you select *OK*, you are prompted for an insertion point for inserting the table into the drawing. The table can be edited for format, structure, or contents by double-clicking on a cell (to produce the *Table* editor) or on the border (to produce the table grips), then right-clicking to produce related shortcut menus for editing the table. See Chapter 18 for more information on creating and editing tables.

## Populating Cells of an Existing Table from a Data Link

Assuming a table exists in your drawing, you can populate empty cells or append cells to the table using the *Data Link Manager*. AutoCAD populates an exact range of cells in a table (number of rows and columns) only when the range of cells you select in the AutoCAD table exactly matches the range that is specified in the data link. If the data link range exceeds the selected range in the table, rows and/or columns are automatically appended to the table to accept the linked data.

To populate empty cells in an existing AutoCAD table, first highlight the desired cells using a crossing window. The *Table* editor appears in the Ribbon with the highlighted cells border. The cell grips also appear if Grips are turned on. For example, Figure 22-45 displays an existing table with two columns and eight rows selected.

FIGURE 22-45

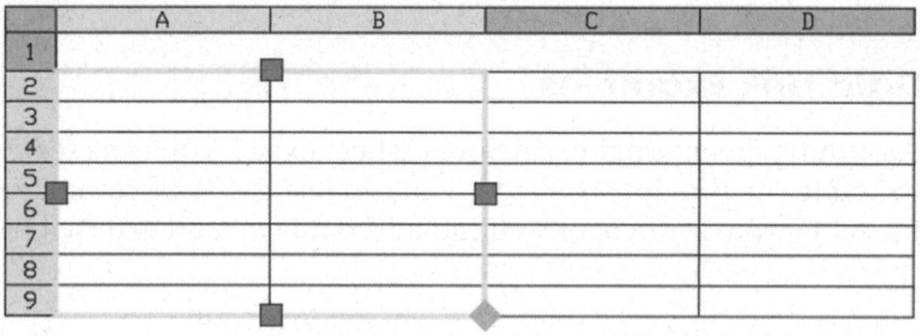


FIGURE 22-46

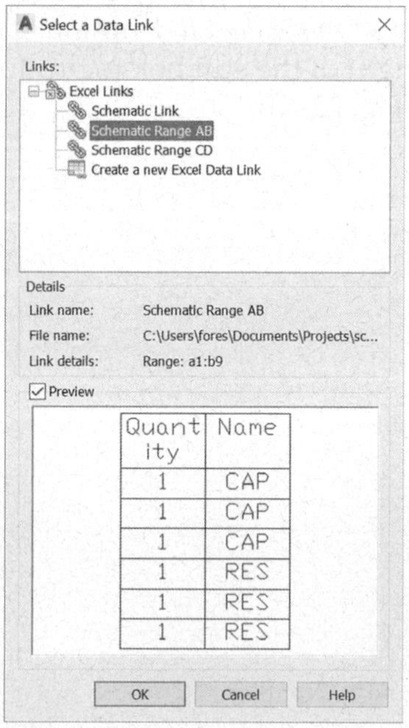


Next, right-click to produce the shortcut menu and select *Data Link...* to produce the *Data Link Manager* with the title bar changed to read *Select a Data Link* (Fig. 22-46). Either select an existing link or select *Create a new Excel Data Link* to specify a new file and range. If the named link exactly matches the range of cells previously selected (in the AutoCAD table), the data will fill the selected cells. If the named link exceeds the range selected in the table, columns and/or rows will be appended to the table to accommodate the data. For our example, note that the selected table range in Figure 22-45 (2 columns, 8 rows) exactly matches the named range selected in the *Data Link Manager* displayed in the preview tile in Figure 22-46. Therefore, no columns or rows will be appended to the AutoCAD table.

### *Datalinkupdate*

| Ribbon | Menu Bar | Command (Type) | Alias (Type) | Shortcut |
|---|---|---|---|---|
| *Insert*<br>*Linking & Extraction*<br>(icon) | *Tools*<br>*Data Links >*<br>*Update Data Links* | *Datalinkupdate* | *DLU* | ... |

Data links can be updated in both directions—that is, changes to the data (contents of the cells) can be made in either the AutoCAD table or the Excel spreadsheet, then automatically updated to the other. If the changes are made first in a linked Excel spreadsheet, the changes can then be "downloaded" or "updated" to the established data link table. If changes are made first to a linked table in your drawing, then you can "upload" or "write out" those changes to your external spreadsheet. There are several ways this can be accomplished. The formal command syntax is shown below.

Command: ***DATALINKUPDATE***
Select an option [Update data link/Write data link] <Update data link>: **Enter**
Select objects or [Data link/all data linKs]: **PICK**
Select objects or [Data link/all data linKs]: **Enter**
1 object(s) found.
1 object(s) updated successfully.
Command:

You can also activate the options of this command by several other methods. Right-click on the *Data Link* icon at the right end of the Status bar, select an AutoCAD table (on its border) and right-click, highlight a cell in the table and right-click, or select the option on the *Data Links Have Changed* notification bubble. However, typical to AutoCAD, the options are confusing since AutoCAD uses different terminology depending on how you invoke this command. To simplify, the following list is given:

<u>To update the AutoCAD table from the Excel file</u>, first change the contents of the desired cells in Excel, then *Save* and *Close* the file. In AutoCAD, use <u>any one</u> of the following options:

1. *Update data link* option, *Datalinkupdate* command
2. *Download Changes from Source File*, right-click menu, highlighted cells in table
3. *Update Table Data Links*, right-click menu, highlighted table
4. *Update All Data Links*, right-click menu, Data Links icon in Status bar
5. *Update all tables using these data links*, notification bubble

<u>To change the AutoCAD table, then update the Excel file</u>, first ensure the Excel file is closed. Next, select the desired cell to change, right-click and select *Locking*, *Unlocked*, from the shortcut menu. Make the changes in the cell, then use one of these options:

1. *Write data link* option, *Datalinkupdate* command
2. *Upload User Changes to Source File*, right-click menu, highlighted cells in table
3. *Write Data Links to External Source*, right-click menu, highlighted table

Note that the AutoCAD table cells must be unlocked to change the contents. Once unlocked, double-click in the cells to make the changes. On the other hand, if changes are first made in the Excel spreadsheet, then updated (or downloaded) to the AutoCAD table, the table cells automatically become locked.

## CHAPTER EXERCISES

1. **Create an Attributed Title Block**

   A. Open the **TBLOCK** drawing that you created in Chapter 18, Exercise 8. Because we need to create block attributes, ***Erase*** the existing text so only the lines remain. (Alternately, you can begin a ***New*** drawing and follow the instructions for Chapter 18, Exercise 8A. Do not complete 8B.)

   B. ***SaveAs*** **TBLOCKAT.** Check the drawing to ensure that 2 text ***Styles*** exist using ***romans*** and ***romanc*** font files (if not, create the two *Styles*). Next, define attributes similar to those described below using ***Attdef***. Substitute your personalized *Tags*, *Prompts*, and *Values* where appropriate or as assigned (such as your name in place of B. R. SMITH):

| Tag | Prompt | Value | Mode |
|---|---|---|---|
| COMPANY | | CADD Design Co. | const |
| PROJ_TITLE | Enter Project Title | PROJ- | |
| SCALE | Enter Scale | 1″=1″ | |
| DES_NAME | | DES.-B.R.SMITH | const |
| CHK_NAME | Enter Checker Name | CHK.- | |
| COMP_DATE | Enter Completion Date | 1/1/17 | |
| CHK_DATE | Enter Check Date | 1/1/17 | |
| PROJ_NO | Enter Project Number | PROJ | verify |

   The completed attributes should appear similar to those shown in Figure 22-47. ***Save*** the drawing.

   C. Use the ***Base*** command to assign the ***Insertion base point***. **PICK** the lower-right corner (shown in Figure 22-47 by the "blip") as the *Insertion base point*.

   Since the entire TBLOCKAT drawing (.DWG file) will be inserted as a block, you do not have to use the *Block* command to define the attributed block. ***Save*** the drawing (as **TBLOCKAT**).

**FIGURE 22-47**

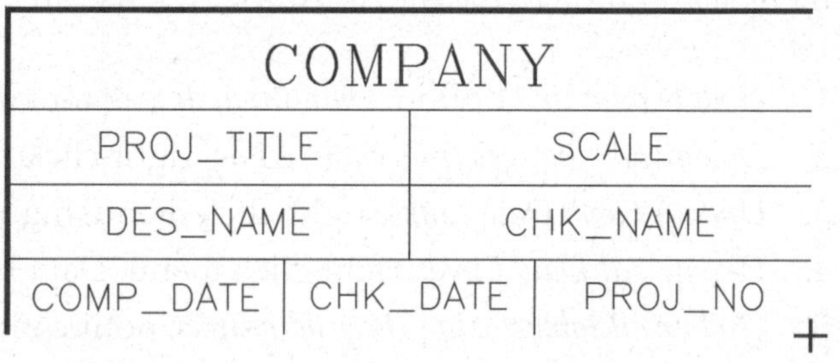


   D. Begin a ***New*** drawing and use the **ASHEET** drawing as a ***template***. Set the **TITLE** layer ***current*** and draw a border with a ***Pline*** of **.02 *width***. Set the ***ATTDIA*** variable to **1** (***On***). Use the ***Insert*** command to bring the **TBLOCKAT** drawing in as a block. The *Enter Attributes* dialog box should appear. Enter the attributes to complete the title block similar to that shown in Figure 22-48. Do not *Save* the drawing.

**FIGURE 22-48**

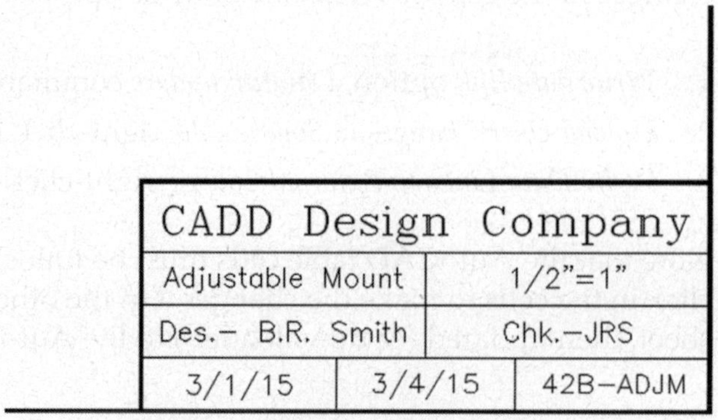

2. **Create Attributed Blocks for the OFF-ATT Drawing**

A. ***Open*** the **OFF-ATT** drawing (not OFFICE-REN) that you prepared in Chapter 21 Exercises. The drawing should have the office floor plan completed and the geometry drawn for the furniture blocks. The furniture has not yet been transformed to *Blocks*. You are ready to create attributes for the furniture, as shown in Figure 22-49.

FIGURE 22-49

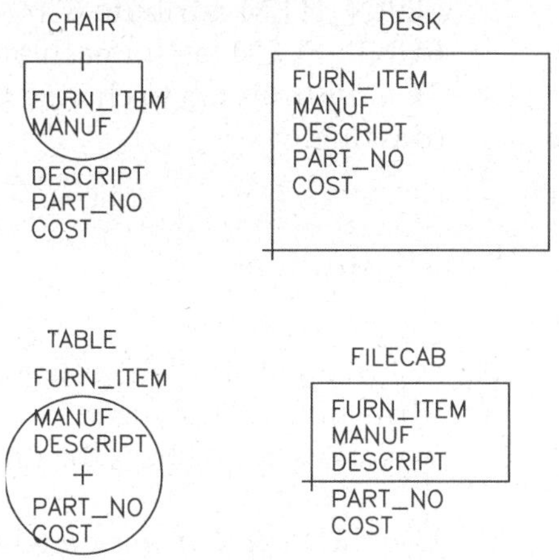


Include the information below for the proposed blocks. Use the same *Tag*, *Prompt*, and *Mode* for each proposed block. The *Values* are different, depending on the furniture item.

| Tag | Prompt | Mode |
|---|---|---|
| FURN_ITEM | Enter furniture item | |
| MANUF | Enter manufacturer | |
| DESCRIPT | Enter model or size | |
| PART_NO | Enter part number | Invisible |
| COST | Enter dealer cost | Invisible |

Values

| | | | |
|---|---|---|---|
| CHAIR | DESK | TABLE | FILE CABINET |
| WELLTON | KERSEY | KERSEY | STEELMAN |
| SWIVEL | 50″ x 30″ | ROUND 2′ | 28 x 3 |
| C-143 | KER-29 | KER-13 | 3-28-L |
| 139.99 | 449.99 | 142.50 | 229.99 |

B. Next, use the ***Block*** command to make each attributed block. Use the block ***names*** shown (Fig. 22-49). Select the attributes in the order you want them to appear upon insertion; then select the geometry. Use the ***Insertion base points*** as shown. ***Save*** the drawing.

C. Set the ***ATTDIA*** variable to **0** (***Off***). ***Insert*** the blocks into the office and accept the default values. Create an arrangement similar to that shown in Figure 22-50.

FIGURE 22-50

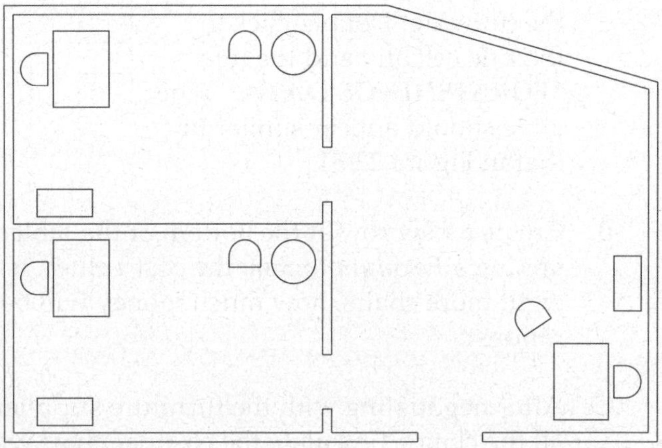

D. Use the ***ATTDISP*** variable and change the setting to ***Off***. Do the attributes disappear?

E. A price change has been reported for all KERSEY furniture items. Your buyer has negotiated an additional $20 off the current dealer cost. Type in the ***Multiple Attedit*** command and select each block one at a time to view the attributes. Make the COST changes as necessary. Use ***SaveAs*** and assign the new name **OFF-ATT2.**

F. Change the ***ATTDISP*** setting back to ***Normal***. It would be helpful to increase the ***Height*** of the FURN_ITEM attribute (CHAIR, TABLE, etc.). Use the ***Enhanced Attribute Editor*** to change the **FUNR_ITEM** tag for each block. You will have to select each block individually, then use the ***Text Options*** tab to change the height to **4"** for each block. If everything looks correct, ***Save*** the drawing.

3. **Extract Attributes**

A. A cost analysis of the office design is required to check against the $9,000 budget allocated for furnishings. This analysis can be accomplished by extracting the attribute data into an AutoCAD *Table*. When finished, the table should appear similar to that in Figure 22-51.

Use the ***Data Extraction*** wizard to extract the block attributes. In the *Begin (Page 1 of 8)* step, ***Create a new data extraction***. Save the data extraction template as **OFF-ATT2.DXE** in your working directory. In the *Define Data Source (Page 2 of 8)* step, select ***Include current drawing***. In the *Select Objects (page 3 of 8)* step, select ***Display blocks only***. In the *Select Properties (page 4 of 8)* step, check only the ***Attribute*** *Category filter*. Next, in the *Refine Data (Page 5 of 8)* step, to force the table to display each insertion of each furniture item, remove the check for ***Combine identical rows***. Remove the check for ***Show name column***. Also in this step, arrange the columns as shown in Figure 22-51 by dragging and dropping the column headers. Sort the list by **FURN-ITEM**. Check the ***Full Preview***, then proceed to the next step.

In the *Choose Output (Page 6 of 8)* step, select ***Insert the extraction data table into drawing***. In the *Table Style (Page 7 of 8)* step, produce the ***Table Style*** manager and specify ***Romans*** font for the data, header, and title. Set the ***Text height*** for the data, header, and title to **4"**, **5"**, and **6"**, respectively. Set the ***Margins*** to **2"**. ***OK*** these dialog boxes, select ***Next*** and ***Finish*** and proceed to insert the table into the drawing. Double-click in the title cell and add the title "**FURNISHINGS DATA**." The table should appear similar to that in Figure 22-51.

**FIGURE 22-51**

| FURNISHINGS DATA | | | | | |
|---|---|---|---|---|---|
| Count | FURN_ITEM | DESCRIPT | MANUF | PART_NO | COST |
| 1 | CHAIR | SWIVEL | WELLTON | C–143 | 139.99 |
| 1 | CHAIR | SWIVEL | WELLTON | C–143 | 139.99 |
| 1 | CHAIR | SWIVEL | WELLTON | C–143 | 139.99 |
| 1 | CHAIR | SWIVEL | WELLTON | C–143 | 139.99 |
| 1 | CHAIR | SWIVEL | WELLTON | C–143 | 139.99 |
| 1 | CHAIR | SWIVEL | WELLTON | C–143 | 139.99 |
| 1 | DESK | 50"x30" | KERSEY | KER–29 | 429.95 |
| 1 | DESK | 50"x30" | KERSEY | KER–29 | 429.95 |
| 1 | DESK | 50"x30" | KERSEY | KER–29 | 429.95 |
| 1 | FILE CABINET | 28x3 | STEELMAN | 3–28–L | 229.99 |
| 1 | FILE CABINET | 28x3 | STEELMAN | 3–28–L | 229.99 |
| 1 | FILE CABINET | 28x3 | STEELMAN | 3–28–L | 229.99 |
| 1 | TABLE | ROUND 2' | KERSEY | KER–13 | 122.50 |
| 1 | TABLE | ROUND 2' | KERSEY | KER–13 | 122.50 |

B. Create a new row at the bottom of the table for totaling the COST column, then add a ***Field*** containing a ***Formula*** to ***Sum*** the cost values. If you plan to purchase a $2800 conference table and eight more chairs, how much money will be left in the budget to purchase plants and wall decorations?

C. After negotiating with the furniture supplier for additional chairs, a 10% discount is offered on all the chairs. Calculate the cost per chair with the discount. Next, use the ***Multiple*** command modifier with the ***Attedit*** command to change the **COST** attribute for the previously inserted chairs to the discounted value. (Alternately, double-click on each chair to change its COST attribute value.)

D. With the savings, you decide to purchase a more expensive desk chair for the two offices. Open the **Block Editor** with the **CHAIR** block. Add arms to the chair. Also, redefine the following attributes using the ***Properties*** palette in the Block Editor to the following:

| | |
|---|---|
| FURN_ITEM | CHAIR-A |
| DESCRIPT | ARM CHAIR |
| PART_NO | C-242 |
| COST | 159.99 |

In the Block Editor, use the ***Save Block As*** option to assign the name **CHAIR-A**. Close the editor. Finally, ***Erase*** the two office desk chairs and ***Insert*** two new arm chairs.

E. Since the original table setup did not include the new CHAIR-A block, you must *Erase* the table and extract the data again. ***Erase*** the existing table of furnishings data. Next, use the ***Data Extraction*** wizard again. In the *Begin (Page 1 of 8)* step, select the ***Use a template*** option and locate the template you saved for the original data extraction. Next, save the new template as **OFF-ATT2-B.DXE**. Accept all the previous settings in page 2. Note that the *Select Objects (Page 3 of 8)* lists the new CHAIR-A block, but you must check it to include its attributes in the new table. In the following steps, use the same options you did to create the first table. Proceed to insert the new table into the drawing. The table should appear with the new CHAIR-A blocks and new cost data. Double-click in the title cell and add the title "**FURNISHINGS DATA**." Finally, create a new row at the bottom of the table for totaling the COST column, then add a ***Field*** containing a ***Formula*** to ***Sum*** the values. ***Save*** the drawing. What is the new total for the new furnishings?

4. **Extract Attributes**

A. Create the electrical schematic illustrated in Figure 22-52. Use *Blocks* for the symbols. The text associated with the symbols should be created as attributes. Use the following block names and tags:

**FIGURE 22-52**

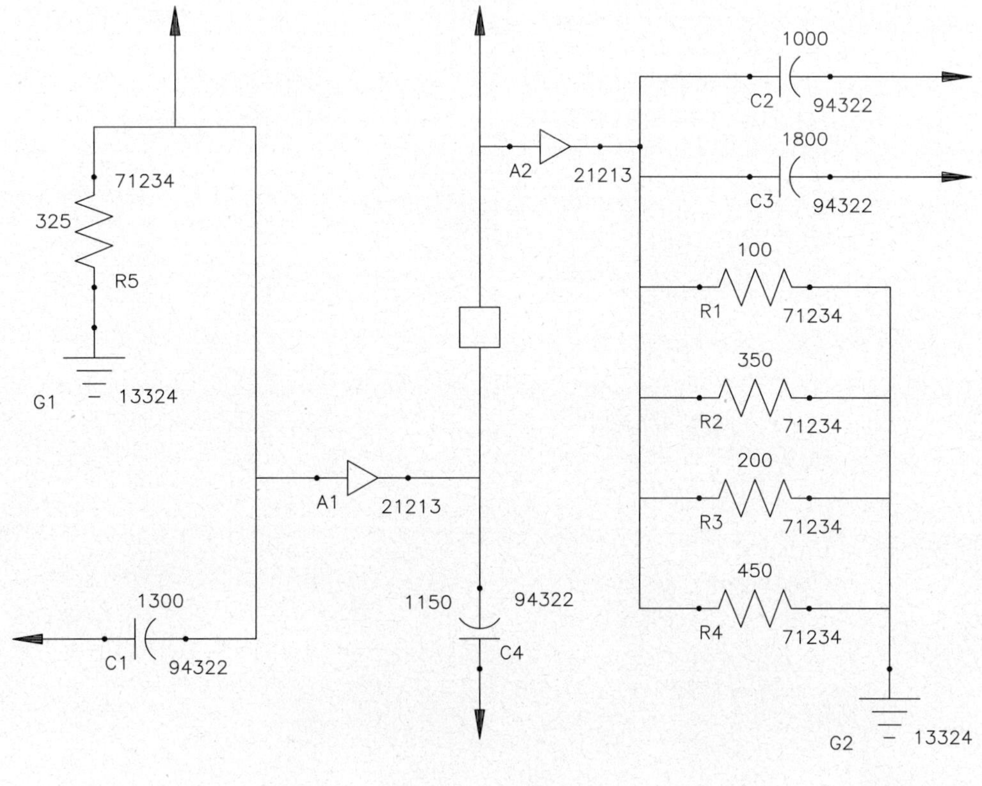

Block Names: RES
CAP
GRD
AMP

Attribute Tags: PART_NUM
MANUF_NUM
RESISTANCE
CAPACITANCE

(Only the RES and CAP blocks have values for resistance or capacitance.)

B. When you are finished with the drawing, create an external .XLS file reporting information for all attribute tags (other data is not needed). Save the drawing as **SCHEM2**.

CHAPTER

# 23

# Internet Tools and Collaboration

## Chapter Objectives

After completing this chapter you should:

1. be able to use the *Browser* command to launch your Web browser from within AutoCAD;
2. be able to share and collaborate using drawings and drawing views;
3. be able to use *Hyperlink* to link your drawings to other drawings or documents on your computer, your network, or the Internet;
4. know how to attach drawings to your email messages using *Etransmit*;
5. know how to use the plot device drivers to create .DWF and .PDF files;
6. be able to *Publish* drawing sets in .DWF and .PDF format;
7. be able to import .PDF files directly into AutoCAD.
8. be able to create *Shared Drawings* that can be viewed on the web in the Autodesk Viewer;

8. be able to mark up AutoCAD drawings;

10. be able to use *Drawing Compare;*

11. be able to use *Activity Insights.*

## CONCEPTS

Through recent years Autodesk created features that allow AutoCAD users to use the latest Internet-related technological advances to enhance their productivity. Multiple tools have been added to enable you to collaborate and share your designs with colleagues and clients using Internet technologies.

AutoCAD includes features that allow you to connect to Web technology, such as publishing Web pages that include drawings, emailing drawings, linking other remotely located drawings and documents to your drawing, creating drawing views or complete drawing sets that are viewable in standard Web browsers, and allowing automatic updating to new features through the Autodesk Web site.

The commands explained in this chapter are listed below.

| | |
|---|---|
| *Browser* | Launch your Web browser from within AutoCAD with the *Browser* command so you can have instant access to the Internet while in your drawing environment. |
| *Push to Forma Data Management* | Sharing your drawings helps teams collaborate to upload, view, and comment using only a Web browser. Forma Data Management lets you upload layouts from multiple drawings as PDFs to a selected project folder on Forma.<br>Note: This feature requires a subscription. |
| *Hyperlink* | Use the *Hyperlink* command to create pointers in your drawings that jump to other files that contain additional graphical, numerical, or textual information to associate to the current drawing. The other files can on your local computer or network or can link to Web pages on the Internet. |
| *eTransmit* | *eTransmit* opens your email system and allows you to include an AutoCAD drawing (and any associated files) as an email attachment. |
| *Exportdwf* | You can create either a DWF (Drawing Web Format) or PDF (Portable Document Format) file of the current drawing by selecting an export command or a plot device ePlot from the *Plot* dialog box. These image formats are compact, vector-based files that are viewable without AutoCAD and easily transported over the Internet. Use Autodesk Design Review to view .DWF files and use Adobe Reader® to view .PDF files. Both viewers are free. |
| *Publish* | This feature allows you to create drawing sets readable with Autodesk Design Review or Adobe Reader®. The drawings are created in .DWF and .PDF format. |
| *Pdfimport* | You can import .PDF files directly into AutoCAD and choose how you want the objects, such as vector geometry, text, and raster images, to be converted. |
| Autodesk Design | This program can be used to view .DWF files. The Autodesk Design Review is free Review and downloadable and can be installed by anyone—without a license for AutoCAD. |
| *Dwfattach* | Use these commands to view .DWF and .PDF images in AutoCAD. You cannot *Pdfattach* alter the images in AutoCAD, but you can *Object Snap* to the geometry and trace over it as an "underlay." |

*Markup* — This command produces the *Markup Set Manager*. This utility allows you to read .DWF files that have been "marked up." Autodesk Design Review is a free program that reviewers can used to mark design revisions needed in the original drawings. Markup Assist can aid in automatically identify markups to create text, leaders, and revision clouds in the original drawing.

*Trace* — Creates a "*trace*" (like an overlay layer) for the AutoCAD web or mobile app used to provide feedback, comments, markups, and design exploration of the drawing without altering drawing content. The Trace palette lets you manage the traces available in the current drawing.

*Share* — Shares a link to a copy of the current drawing to anyone to open in the AutoCAD web app. The link expires after 7 days. Choose between two permission levels: View only, and Edit and save a copy.

## *Browser*

| Ribbon | Menu Bar | Command (Type) | Alias (Type) | Shortcut |
|---|---|---|---|---|
| ... | ... | *Browser* | ... | ... |

AutoCAD includes this command specifically to launch your Web browser from within AutoCAD. An icon button on the *Web* toolbar (Fig. 23-1) and in the file navigation dialog boxes is available to invoke the *Browser* command.

**FIGURE 23-1**

The information about your browser (location and name of the executable file) is stored in the system registry. Therefore, when you use *Browser*, AutoCAD automatically locates and launches your current (default) Web browser from within AutoCAD. This action makes it possible to view any Website or select any viewable files located on your computer. Using Autodesk Design Review, you can view .DWF

**FIGURE 23-2**

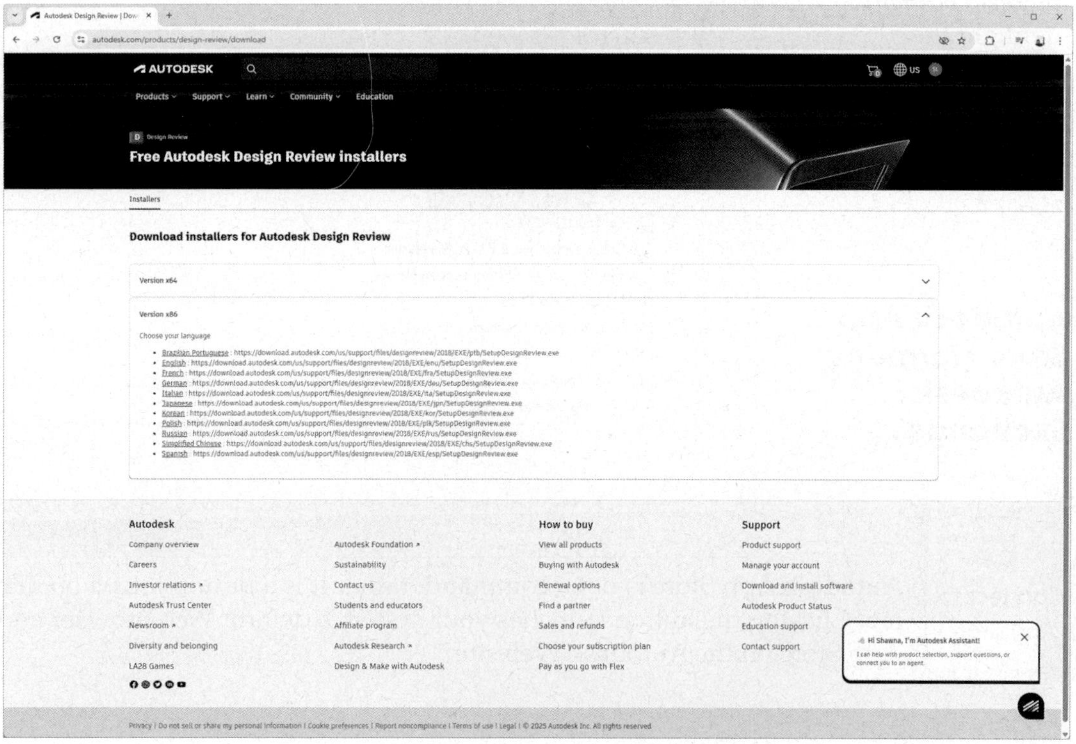

files after you have created them, view other .DWF files on the Internet, and drag-and-drop related .DWG files from the Internet directly into your AutoCAD session.

When you use *Browser*, your registered Web browser is launched. By default (if you have not specified a different file or URL), the Autodesk home page is displayed. You can locate another file or URL by entering it in the *Address* edit box of the browser (Fig. 23-2).

## Inetlocation

| Ribbon | Menu Bar | Command (Type) | Alias (Type) | Shortcut |
|---|---|---|---|---|
| ... | *Tool Options... Files* | *Inetlocation* | ... | ... |

The initial Web address that is located when *Browser* is used can be specified by using the *INETLOCATION* system variable or by using the *Options* dialog box. The default *INETLOCATION* is:

www.autodesk.com

You can specify any other viewable file or Internet address. Using the Command line method, the following prompt is issued:

Command: ***INETLOCATION***
Enter new value for INETLOCATION <"http://www.autodesk.com">: (Enter desired URL)
Command:

If you prefer to use the *Options* dialog box (Fig. 23-3), expand the *Help and Miscellaneous File Names* section. Expand *Default Internet Location*, highlight the address, and select the *Remove* button. Then enter the desired new address.

**FIGURE 23-3**

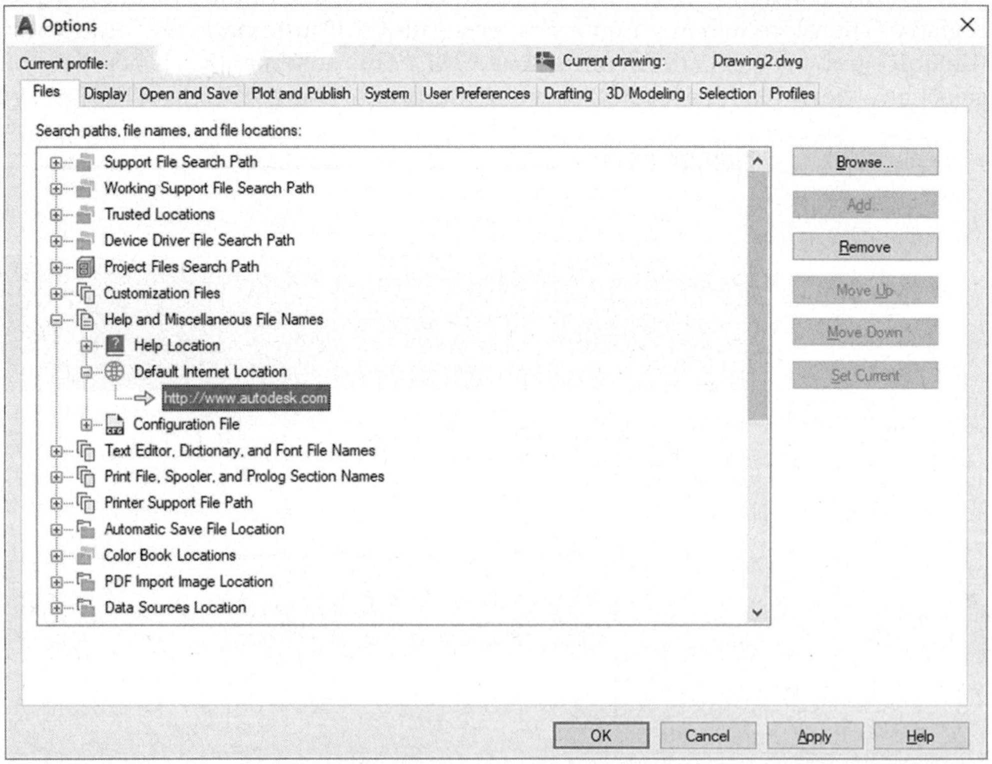


## Autodesk App Store (formerly Autodesk Exchange)

Autodesk App Store is not a command, rather it is a button found on the Featured Apps tab. Clicking the button launches your system's default Web browser and takes you to the App Store on the Autodesk Web site.

The App Store is an online destination for publishing and downloading applications that run with AutoCAD (Fig. 23-4). You can find apps that simplify, automate, and assist you with creating and editing AutoCAD drawings. For example, you can download applications for industries such as architecture, building construction, home design, electrical and electronics, mechanical design, surveying, and many others. In addition, block libraries are available containing blocks for specialized applications. Many of the applications are free or are available for just a few dollars. You can use the selection at the left to display applications for various Autodesk products, such as AutoCAD, Autodesk 3ds Max®, Autodesk Inventor®, Autodesk Revit®, and so on.

**FIGURE 23-4**

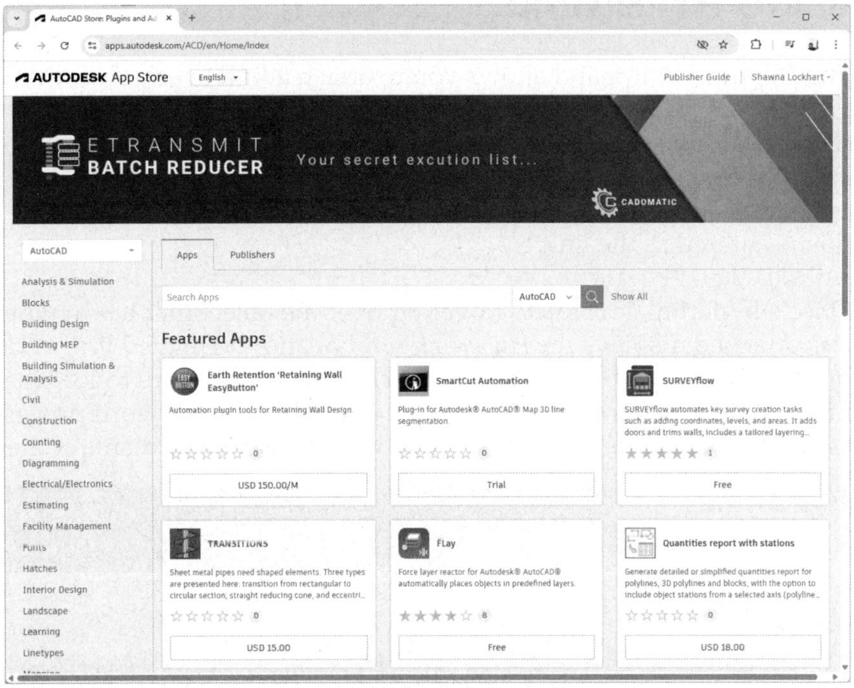


## USE FORMA® TO TRANSFER AutoCAD FILES OVER THE INTERNET

### A360®, Shared Views, and Save to Web and Model are Retired

The A360 Drive, Shared Views, and Save to Web and Mobile features have been replaced. You can still save your drawing in a variety of formats as shown in Figure 23-5, but Save to Web and Mobile is no longer featured.

The new Forma Data Management system lets you make AutoCAD drawing views available to collaborators online. Depending on permissions set in Forma, collaborators are able to mark up or even make changes directly to the drawings.

The *Onlineshare, Onlinesync, Onlinedocs, Olineopenfolder, Onlinesyncsettings, Onlinedesignshare, Openfromcloud, Savetocloud, Publishtoweb, Openfromwebmobile, Psout, Pushtodocsclose, Pushtodocsopen, Savetowebmobile, Sharedviews, Sharedviewsclose, Shareview* commands are no longer in use, as well as the related system variables such as *ONLINESYNCTIME, ONLINEDOCMODE, PUSHTODOCSSTATE, SHAREDVIEWSTATE, SHAREVIEWPROPERTIES and SHAREVIEWTYPE.*

**FIGURE 23-5**

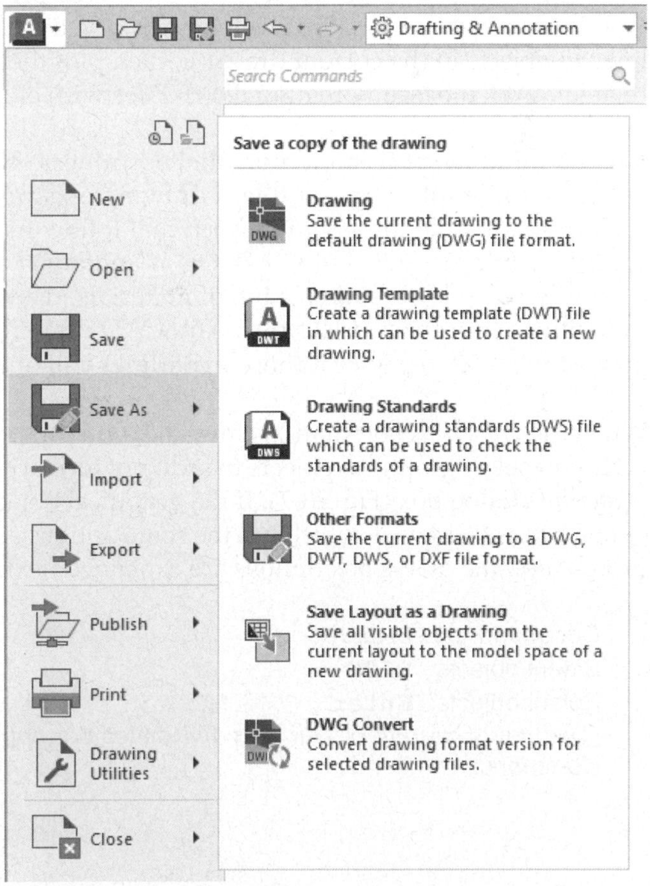

## USING HYPERLINKS

The *Hyperlink* command allows you to attach a link to any AutoCAD graphical object. The link is actually a file name or Web address (URL) associated with the object. Therefore, *Hyperlink* can be used to link to files on the local computer or network (without Internet access) or to connect to files over the Internet, depending on what information is entered as the link.

FIGURE 23-6

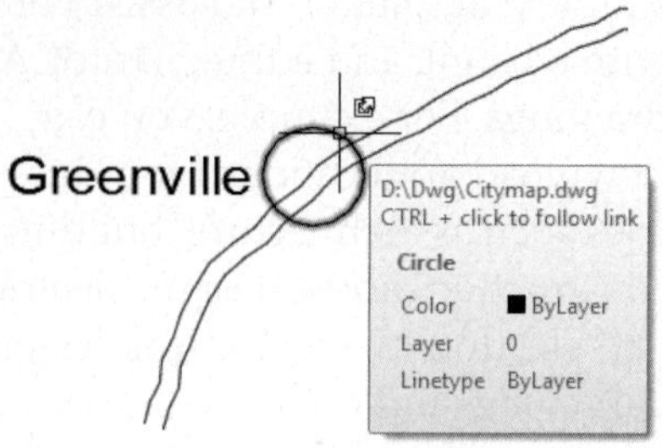


Once a hyperlink is created, hovering over the object that has a hyperlink attached displays the *Hyperlink* symbol and the related link information (Fig. 23-6). Press Ctrl and left-click on the object to follow the link. The link automatically opens the related program (word processor, spreadsheet, or CAD program) and displays the file, or it opens your Web browser and displays the page or image.

***Hyperlink***

| Ribbon | Menu Bar | Command (Type) | Alias (Type) | Shortcut |
|---|---|---|---|---|
| *Insert*<br>*Data*<br>Hyperlink (icon) | *Insert*<br>*Hyperlink...* | *Hyperlink...* | ... | Ctrl+K |

A *Hyperlink* can be used to create links within two types of AutoCAD files: (1) to create links in the current AutoCAD drawing (.DWG) and (2) to create links for a .DWF image that is created from the current drawing but is to be viewed from a Web browser.
To create links to be used in .DWF images, use the *Hyperlink* command as you would normally to attach links to objects in the current drawing, then plot the drawing using the *DWF ePlot* device to create the related .DWF file (see "Creating .DWF Files with ePlot").

Keep in mind the power and potential usefulness of hyperlinks in AutoCAD drawings. Any hyperlink can call another file. For example, a drawing of a mechanical assembly could contain hyperlinks for each component of the assembly such that following the link for each component could open the related component drawing in AutoCAD. Or, as another example, opening another hyperlink for the entire assembly could in turn open a bill of materials in a word processing or spreadsheet program. The same capability hyperlinks added to an AutoCAD drawing can be contained in a .DWF image of the drawing viewed in a Web browser if those hyperlinks call other images normally viewable in a browser.

When the *Hyperlink* command is invoked, you must first select the object you want to attach the link to. If you select graphical objects that do not already contain hyperlinks, AutoCAD displays the *Insert Hyperlink* dialog box (Fig. 23-7). If the graphical objects already contain hyperlinks, the *Edit Hyperlink* dialog box is displayed, which has the same options as the *Insert Hyperlink* dialog box. The Command prompt and the dialog box options are explained next.

Command: ***HYPERLINK***
Select objects: **PICK**
Select objects: **Enter**
(The *Insert Hyperlink* or *Edit Hyperlink* dialog box appears.)
Command:

When the *Insert Hyperlink* dialog box appears, first select (from the buttons on the left side of the dialog box) the kind of link you want to attach to the selected object—an *Existing File or Web Page*, a view of the current drawing (by selecting *View of This Drawing*), or an *Email Address*.

FIGURE 23-7

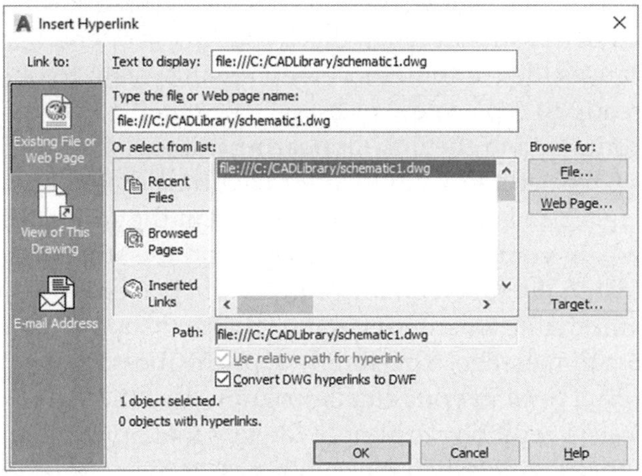


### *Existing File or Web Page*

Selecting the *Existing File or Web Page* button allows you to type the information or select from *Recent Files*, *Browsed Pages*, or *Inserted Links*. In each case, files, pages, and URLs you recently used are displayed. There are also buttons to select another *File* or *Web Page* that is not displayed in the list. Selecting the *Target* button opens the *Select Place in Document* dialog box where you can select a named *View* or layout in a drawing to link to (not the current drawing). The text you enter in the *Text to display:* edit box is the text that appears in the drawing when the cursor passes over the object containing the hyperlink (see Fig. 23-6).

The *Use Relative Path for Hyperlink* checkbox toggles the use of a relative path for the current drawing. If this option is selected, the full path to the linked file is not stored in the drawing with the hyperlink. AutoCAD sets the relative path to the value specified by the *HYPERLINKBASE* system variable or, if this variable is not set, the current drawing path. If the *Convert DWG hyperlinks to DWF* option is checked and the hyperlink is another drawing, the hyperlinked drawing will be converted to a DWF file hyperlink if you publish or plot the current drawing to a DWF file (see "Creating .DWF Files").

The *Remove Link* button appears only in the *Edit Hyperlink* dialog box. You can use this option to remove a previously attached hyperlink from the selected object.

### *View of This Drawing*

If you want a specific view or layout of the current drawing to be associated to the selected object as a hyperlink, you can use this button to select from a list of *Views* contained in the current drawing (Fig. 23-8). This feature allows you to select an object in the current drawing and attach a hyperlink to jump to a different layout or saved view of the drawing. AutoCAD displays that portion of the drawing (view or layout) when the hyperlink is opened.

FIGURE 23-8

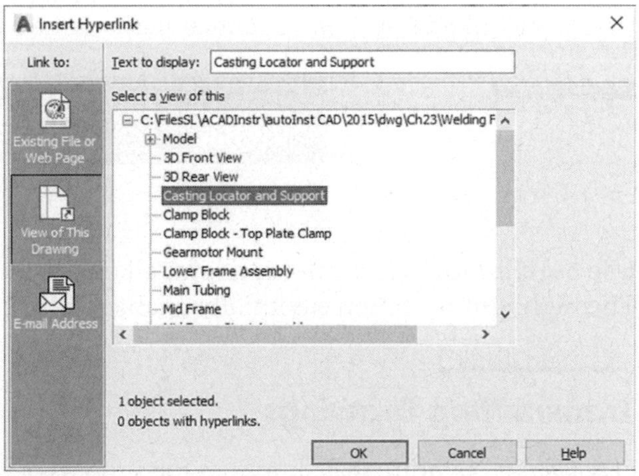

### Email Address

This feature allows you to attach a hyperlink to an object so that when this hyperlink is opened, it causes the computer's default email system to open ready to type a new message. The email address contained in the link (appearing in the *E-mail address:* edit box of the *Insert Hyperlink* dialog box, Fig. 23-9) is automatically entered in the "Send to:" box of your new email message and the subject text in the link (contained in the *Subject:* edit box) is automatically entered in the "Subject:" box of your email message. You simply type the message and select *Send* in your email program. Use the *Text to display:* edit box to supply the text that appears in the drawing when the cursor passes over the object containing the hyperlink.

FIGURE 23-9

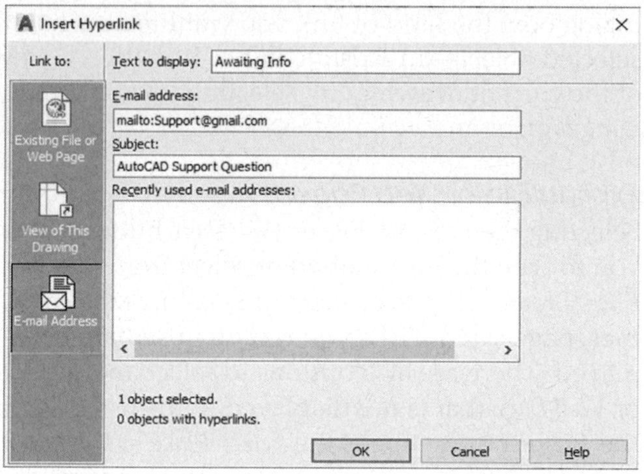


## HYPERLINKBASE

Use the *HYPERLINKBASE* system variable to specify the path you want to use for all relative hyperlinks in the drawing. If no value is specified, the current drawing path is used for all relative hyperlinks. Relative paths are used primarily for file-related hyperlinks rather than Web addresses.

Command: ***HYPERLINKBASE***
Enter new value for HYPERLINKBASE, or . for none <"">:

### Relative and Absolute Paths

With *HYPERLINKBASE* you can specify a relative path that is used by the hyperlinks you create in a drawing. Relative paths afford you greater flexibility and are easier to edit than absolute hyperlinks because with relative hyperlinks, you can update the relative path for all the hyperlinks in your drawing at the same time, rather than editing each hyperlink individually. Absolute hyperlinks work well in situations where the linked files always remain in the same location, which is more common with Web addresses.

### Selecturl

| Ribbon | Menu Bar | Command (Type) | Alias (Type) | Shortcut |
|---|---|---|---|---|
| ... | ... | *Selecturl* | ... | ... |

Use *Selecturl* to "select," or highlight, all objects in the drawing that have hyperlinks (URLs) attached. The highlighting action essentially enables grips on the objects. Press Escape to cancel the grips.

## Transmitting Drawings

*eTransmit* is a feature that simplifies the process of sending AutoCAD drawings by email. The process is straightforward: *Open* the drawing you want to send, use *eTransmit*, enter any notes and select your preferences in the *Create Transmittal* dialog box. AutoCAD then creates a compressed .EXE or .ZIP file of the drawing package and opens your email software with the transmittal file attached. Just write your email note and send it!

## *Etransmit*

| Ribbon | Menu Bar | Command (Type) | Alias (Type) | Shortcut |
|---|---|---|---|---|
| *Application Publish eTransmit* | *File eTransmit* | *Etransmit* | ... | ... |

*eTransmit* operates on the current drawing. First *Open* the drawing you want to transmit. If the drawing contains Xrefs, fonts, plot style tables, or compiled shapes, you can specify each item you want to include in the transmittal. The *eTransmit* command produces the *Create Transmittal* dialog box (Fig. 23-10).

### *Files Tree* Tab

The *Files Tree* tab lists the files to be included in the transmittal set (Fig. 23-10). When you first access the tab, all files associated with the current drawing (such as related Xrefs, drawing standards files, plot styles, plotter configurations, and fonts) are listed in a hierarchical format and have a check appearing in the checkbox before each file name in the list. Remove a check if you do not want the related file included in the set to transmit. For example, if you want to send a drawing that contains Xrefs, but you do not want to include the Xrefs in the transmittal set, remove the checks from all the Xrefed drawings. You can also right-click in the file list to display a shortcut menu from which you can *Check All* or *Uncheck All* and *Expand All* or *Collapse All*. Use the *Add File* button to add additional files to the set. Files that are referenced by URLs (hyperlinks) do not appear in the list and are not included in the transmittal set.

### *Files Table* Tab

The list displays the same files as the Files Tree tab, but in a different format. In this list, files are given in alphabetical order by default but can be sorted by *Filename, Path, Type, Size, Version,* or *Date* in normal or reverse order by clicking on the desired column heading.

### *Add File*

Use this button to open the standard file selection dialog box where you can select additional files to include in the transmittal set.

### *Transmittal Setups*

See "*Transmittal Setups* Dialog Box."

**FIGURE 23-10**

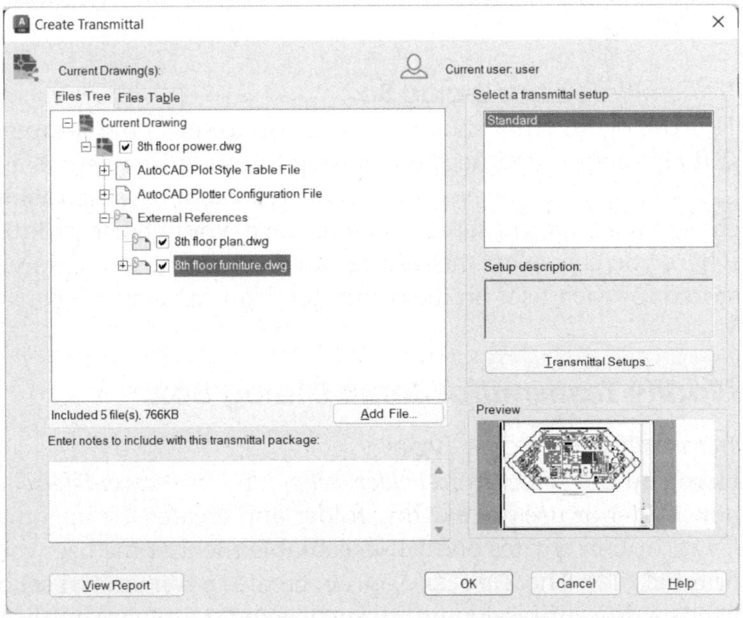

### *View Report*

This button displays the report (Fig. 23-11) that is automatically generated and included with the transmittal set. You can add additional notes in the report by entering the information in the edit box at the bottom of the *Create Transmittal* dialog box. The information automatically generated by AutoCAD explains what steps must be taken by the recipient for the transmittal set to work properly. For example, if AutoCAD detects .PC3 (printer configuration) files or .CTB/.STB (plot style table) files in one of the transmittal drawings, the report instructs the recipient where to copy these files so AutoCAD can detect the files when the transmittal drawing is opened.

FIGURE 23-11

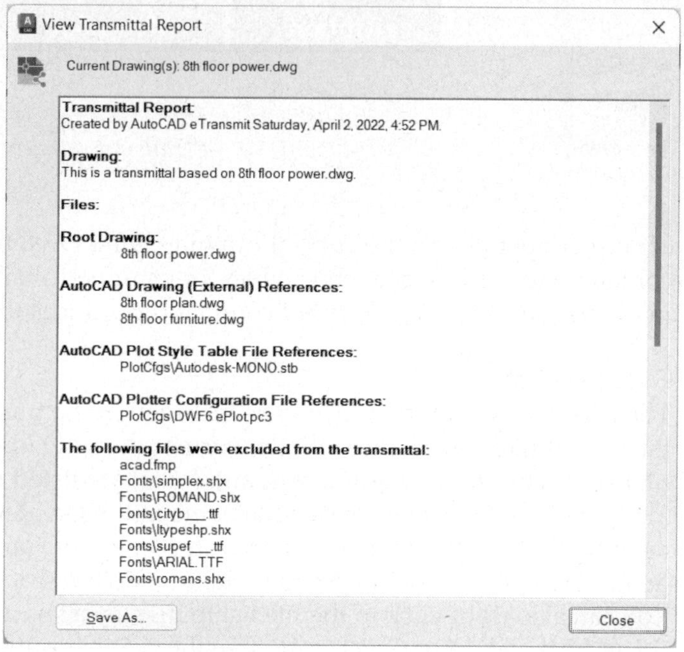


### *Save As*

This button (in the *View Transmittal Report* dialog box) opens the *Save report file* as dialog box where you can specify a location to save a report (.TXT) file. A report file is always generated and included with all transmittal sets; however, you can save an additional copy of the report file for archival purposes using this option.

## Transmittal Setups

A transmittal setup is a saved set of parameters that determine how the transmittal package is created. Selecting the *Transmittal Setups* button in the *Create Transmittal* dialog box produces the *Transmittal Setups* dialog box (not shown) as a "front end" for the *Modify Transmittal Setup* dialog box (see Fig. 23-12).

### *Transmittal Setups* Dialog Box

This dialog box (not shown) allows you to select the setup name you want to use for the transmittal and also acts as an intermediate step to create *New* setups and *Modify* existing setups. If this is your first setup, select the *New* button to create a new setup (using *Standard* as a template) or select *Modify* to change the *Standard* setup. In either case, your choice produces the *Modify Transmittal Setup* dialog box where you can select the settings for the current setup. Any new setups are saved for future use and stored for each user on the computer. You can also select existing setups from the list to *Rename* or *Delete.*

## *Modify Transmittal Setup* Dialog Box

### *Transmittal Package Type*

Here you can select from *Folder (set of files), Self-executable (*.exe), or Zip (*.zip).* The *Folder* option creates a new folder or uses an existing folder and creates the uncompressed transmittal files. The *Self-executable (*.exe)* option creates one self-executable file that the user can double-click to decompress and restore the original files. The *Zip (*.zip)* option creates a transmittal set of files as a compressed .ZIP file. The user needs a decompression utility such as the shareware application WinZip to decompress and restore the files.

### *File format*

You can *Keep the existing file formats* or select from the drop-down list to convert all the drawings to *AutoCAD 2019/LT 2019 drawing format* or earlier formats.

### *Transmittal file folder*

Use this list to select the location on your computer or network where the transmittal set is to be created. You can use the *Browse* button to specify a new location.

FIGURE 23-12

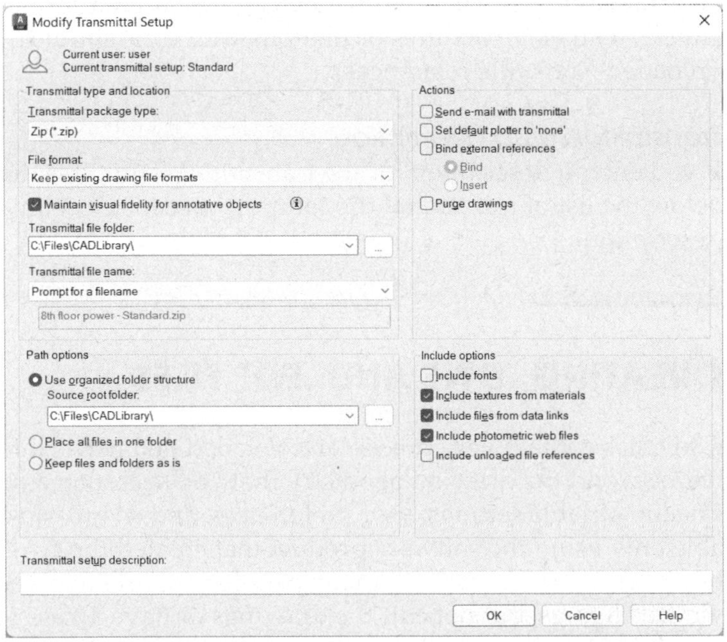


### *Transmittal file name*

Here you specify how the file name is specified when you create the transmittal package. Selecting *Prompt for a file name* allows you to enter the name of the transmittal package. *Increment file name if necessary* uses a logical default file name, and if the file name already exists, a number is added to the end. *Overwrite if necessary* uses a logical default file name, and if the file name already exists, the existing file is automatically overwritten.

### *Use organized folder structure*

When the transmittal package is created, this option duplicates the folder structure for the files being transmitted. The root folder is the top level folder within a hierarchical folder tree. Relative paths remain unchanged and absolute paths within the root folder tree are converted to relative paths. Other folders for fonts, plot configuration files, and sheet sets are created if necessary.

### *Place all files in one folder*

When the transmittal package is created, all files are installed to a single, specified target folder.

### *Keep files and folders as is*

Use this option to preserve the folder structure of all files in the transmittal package. This option is good for decompression and installation on another similar system.

### *Send e-mail with transmittal*

Check this box to launch your default email program when the transmittal set is created. When the program opens, AutoCAD automatically attaches the transmittal set to the email message and enters the drawing set title in the "Subject:" line of your note.

### *Set default plotter to 'none'*

This option changes the printer/plotter setting in the transmittal package to *None* since your local printer/plotter settings are usually not relevant to the recipient.

### *Bind external references*

Binds all Xrefs to the files to which they were attached (see Chapter 30 for information on binding Xrefs).

### *Purge drawings*

Performs a *Purge* on each of the specified drawings to reduce the file size, if possible.

### *Remove Design Feed*

Removes the association to any Design Feed data for the specified drawings.

### *Include fonts, textures from materials, files from data links, photometric web files, Include unloaded file references*

These check boxes specify any additional files that are included with the related drawings. Font files are not generally included since most AutoCAD users already have the associated AutoCAD-supplied

.TTF and .SHX fonts on their systems. Files from data links should be included if the drawings include tables with linked data. If the drawings contain renderings, any custom files (not available with default AutoCAD files) of textures or photometric lights should be included. Typically you would not include unloaded Xrefs (file references).

#### *Transmittal setup description*

If you enter a description for the transmittal setup, it is displayed in the *Transmittal Setups* dialog box below the list of transmittal file setups. You can select any transmittal setup in the list to display its description.

## CREATING .DWF AND .PDF FILES

CAD files contain very precise and detailed geometric information as well as an enormous amount of data beyond the drawing geometry that you see, such as coordinate data, named dependent objects, system variable settings, user preferences, and so on. Most CAD files are also proprietary—viewable only using the software product that created the CAD file. In today's collaborative environment, however, it is necessary for many people who are involved with various aspects of the design process to view drawings, but not edit the drawings or have access to the other data contained in the drawing file. An example of implementing this idea is AutoCAD's *Publish to Web* capability, which facilitates posting drawing images on a Web page in .JPG, .DWF, or .PNG format, so anyone can view the drawings.

### .DWF Files

Autodesk's .DWF file format is also intended for this purpose—viewing a detailed drawing image without AutoCAD and without the ability to access unneeded data. Since the data is compressed, the file is easily transportable over the Internet. Although .JPG and .PNG files can be viewed easily over the Internet, they are raster files and cannot contain much detail. Therefore, Autodesk developed the Design Web Format, or .DWF, which is a vector-based file that can contain incredible geometric detail and additional information, and it is highly compressed for easy transport over the Internet. The .DWF files cannot be viewed using a standard Web browser—they require AutoCAD or Autodesk Design Review.

.DWF files are a good way to share AutoCAD drawing files with colleagues who don't have AutoCAD. Because Autodesk Design Review is free (downloadable from www.autodesk.com) and the interface is easy to use, even individuals with no CAD knowledge can easily view and navigate a .DWF file. You cannot change the .DWF image from within the viewer although you can pan, zoom, print, manipulate layers and views, and activate hyperlinks. (See "Viewing .DWF and .PDF Files.")

### .PDF Files

You can also create .PDF files from AutoCAD drawings. The PDF format (Portable Document Format) was invented by Adobe® and is an industry standard for all types of documents that are exchanged electronically. PDF files can be easily distributed for viewing and printing in the Adobe Reader available from the Adobe website without cost. In addition, .PDF files can be imported directly into AutoCAD (see "Importing .PDF Files"). Using .PDF files, you can share drawings electronically with anyone. Like .DWF files, .PDF files are generated in a vector-based format for maintaining precision.

## Raster Files and Vector Files

The common World Wide Web graphic standard often uses a raster image format such as .GIF or .JPG. These files can be highly compressed, making the information displayed in the images quickly transportable over data transmission lines and devices used for the Internet. However, because raster images are pixel-based, or composed of tiny "dots," viewing detail is not possible—zooming in only enlarges the "dots." Figure 23-13 displays a raster image which has been enlarged. Examples of raster image file formats (file name extensions) are .BMP, .CLP, .DIB, .GIF, .ICO, .IFF, .JPG, .MAC, .MSP, .NIF, .PBM, .PCX, .PSD, .RAS, .SGI, .TGA, .TIF, .XBM, .XPM, .XWD.

**FIGURE 23-13**

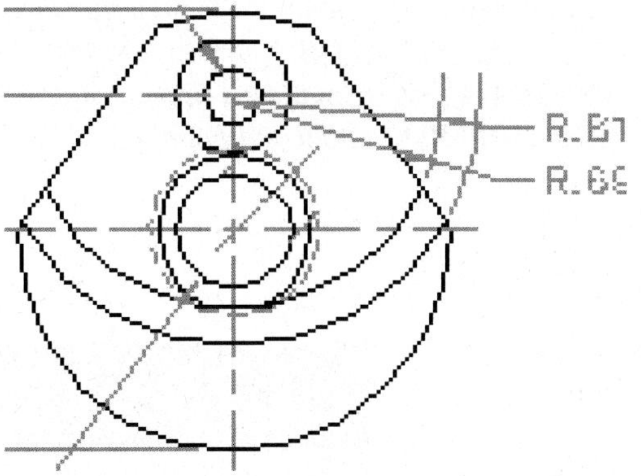

Raster images are of little use to professionals who require drawings with detailed information. Instead, the engineering, architecture, design, and construction industries use vector images to retain data with great precision and detail. CAD programs, which are the primary graphics creation and storage media for these industries, generate vector data. Figure 23-14 illustrates the same vector image (a CAD drawing) in a "zoomed in" display.

**FIGURE 23-14**

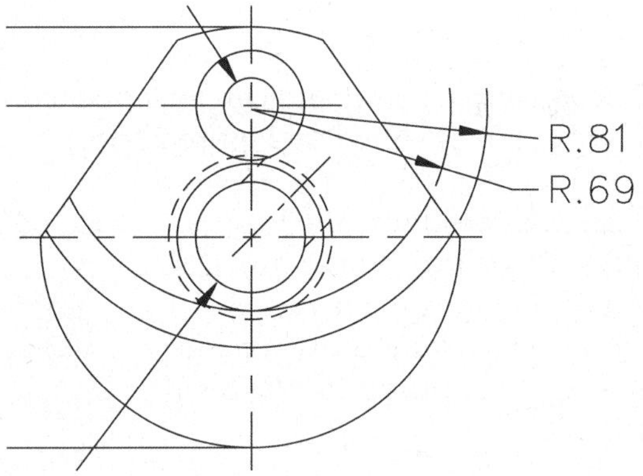


Because a vector file defines a line, for example, as a vector between two endpoints, only one dimension is retained—length. Normally, "zooming in" to a detailed section of the drawing does not create a second dimension (thickness or width) for the line. The thickness of the displayed line on a computer monitor is always one pixel, the size of which is determined by the display device. Examples of vector file formats include .CDR, .CGM, .CMX, .DGN, .DWG, .DWF, .DXF, .GEM, .IGS, .MCS, .P10, .PCL, .PDF, .PGL, .PIC, .PLT, .PRT, .WRL.

NOTE: When creating a .DWF or .PDF file, you can choose to display the drawings with line weight to give a representation of the printed image. In this case, zooming does change line thickness.

## Creating .DWF and .PDF Files with ePlot and Export

You can create .DWF/.DWFx files and .PDF files from AutoCAD drawings. Rather than creating a hard copy print or plot of the current drawing, use the DWF and PDF options to create digital files that can be electronically transferred and read by different programs and devices.

There are two duplicate methods for creating .DWF and .PDF files: 1) you can use the *Plot* command and select the desired "ePlot" or PDF device from the *Printer/Plotter* list in the *Plot* dialog box, or 2) you can use the *Exportdwf* or *Exportpdf* commands available on the Ribbon from the *Output* tab, *Export to DWF/PDF* panel.

The following device options are available to select from in both the *Plot* dialog box and the *Export to DWF/PDF* Ribbon panel.

*AutoCAD PDF (General Documentation)*
*AutoCAD PDF (High Quality Print)*
*AutoCAD PDF (Smallest File)*
*AutoCAD PDF (Web and Mobile)*
*DWF6 ePlot*
*DWFx ePlot (XPS Compatible)*
*DWG To PDF*

### *Exportpdf*

| Ribbon | Menu Bar | Command (Type) | Alias (Type) | Shortcut |
|---|---|---|---|---|
| *Output Export PDF* | ... | *Exportpdf* | ... | ... |

The *Exportpdf* command can be launched using Application icon, Export, Exportpdf, or the ribbon Output tab as shown in Figure 23-15.

FIGURE 23-15

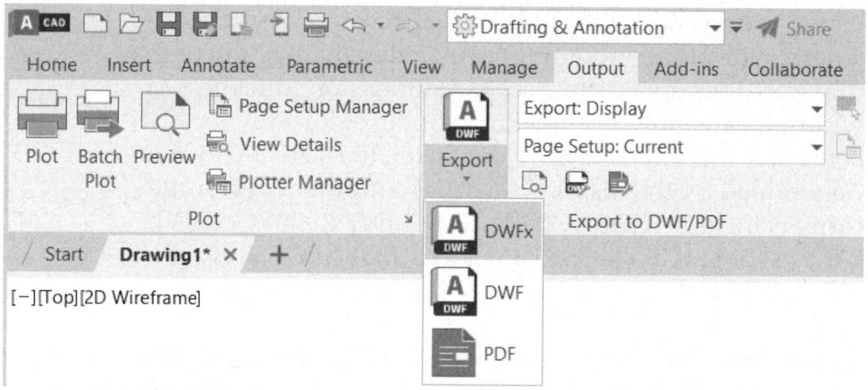


*Exportpdf* produces the *Save as PDF* dialog box (Fig 23-16). Use this dialog box to create a .PDF file of the current drawing and specify the options for the resulting file.

#### *PDF Preset*

Select from *AutoCAD PDF* options or the *DWG To PDF* option. The *AutoCAD PDF* presets provide for four levels of precision (in vector quality) resulting in smaller or larger file sizes.

FIGURE 23-16

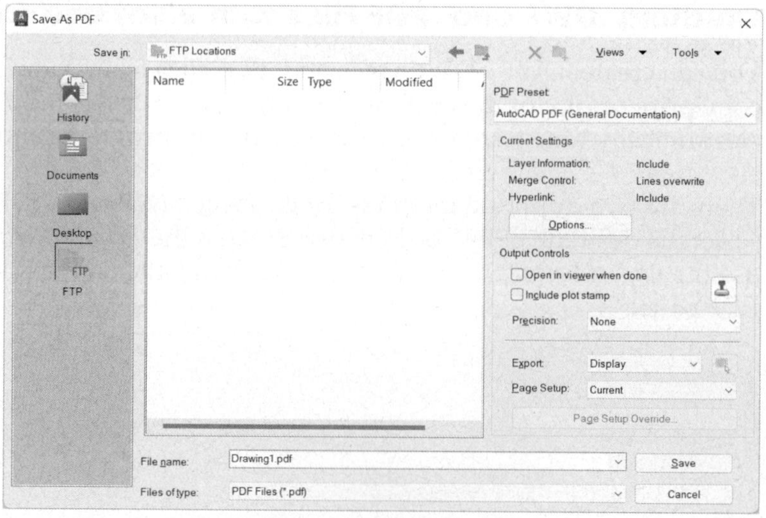


#### *Current Settings*

This section lists the settings as specified in the *Export to PDF Options* dialog box (see *Options*).

#### *Output Controls*

You can choose to automatically *Open in viewer when done* which opens the default PDF viewer on your system when the file is created. You can also *Include plot stamp* (see Chapter 32 for information on plot stamps). The *Export* option specifies the area of the drawing to display (*Display*, *Extents*, or *Window*). You can also include a *Page Setup* if desired.

### *Options*

This button produces the *Export to PDF Options* dialog box (Fig. 23-17). Here you can specify *Quality* and *Data* output for the file. Although the *Vector quality* and *Raster image quality* values can be changed, the default values are predetermined by the particular *AutoCAD PDF* option selected.

You can *Include layer information* and *Include hyperlinks* in the .PDF file. If you *Create bookmarks*, any named *Views* in the drawing become bookmarks in the .PDF file. You can *Capture fonts used in the drawing* (this includes AutoCAD .SHX fonts and Unicode characters) so text in the .PDF file can be accessed, highlighted, copied and searched.

**FIGURE 23-17**

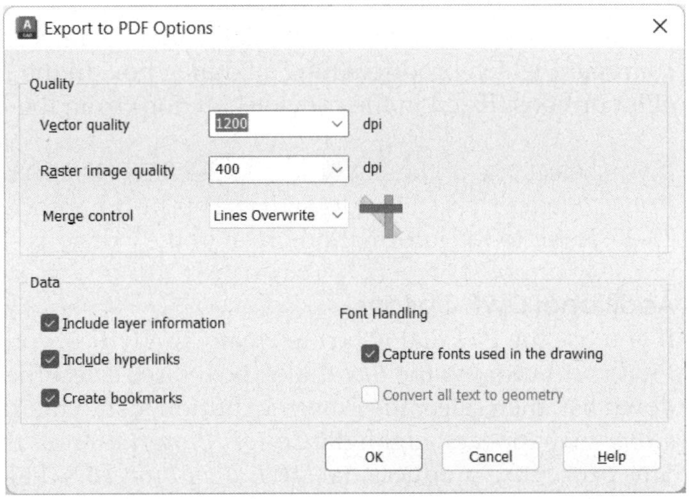


## *Exportdwf*

| Ribbon | Menu Bar | Command (Type) | Alias (Type) | Shortcut |
|---|---|---|---|---|
| *Output Export DWF* | ... | *Exportdwf* | ... | ... |

The *Exportdwf* command produces the *Save as DWF* dialog box (not shown) which is almost the same as the *Save as PDF* dialog box (see Fig. 23-16 and related information). Use this dialog box to create a .DWF file of the current drawing and specify the options for the resulting file.

### *Current Settings*

This section lists the settings as specified in the *Export to DWF Options* dialog box (see *Options*).

### *Options*

This button produces the *Export to DWF Options* dialog box (Fig. 23-18).

*General DWF/PDF options*

The *Location* setting determines the default path for the .DWF file. The *Type* option is enabled only for publishing layouts with the *Publish* command. With *Publish* you can produce one .DWF or .PDF file for each layout or one file containing all layouts (see *Publish*). You can choose to be prompted for a name when each file is created or you can specify a name in the *Name* edit box below. If *Layer information* is set to *Include*, you can control the *On/Off* visibility of the layers in the .DWF image when the resulting .DWF is viewed in the viewer, similar to controlling layer visibility in AutoCAD (see "Autodesk Design Review").

**FIGURE 23-18**

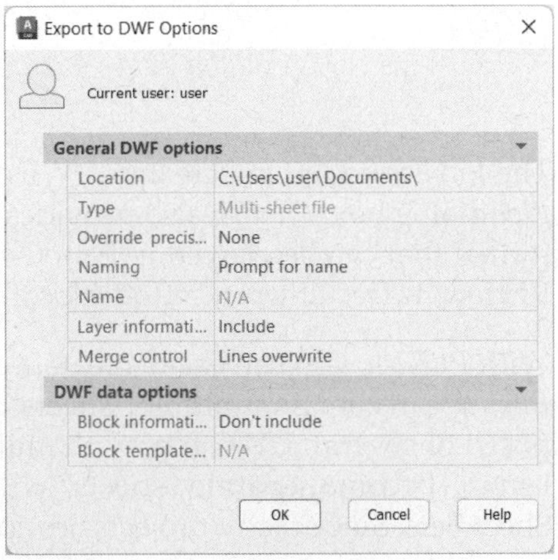


*DWF data options*

If you choose to include *Block information* in the .DWF file, the block property and attribute information will be visible in the .DWF files (see Chapters 21 and 22).

*DWFx* is a newer format similar to DWF, but based on the XML Paper Specification (XPS) that works with the free Microsoft XPS Viewer. The command options are very similar to exporting DXF.

## ePlot Options

There are many ePlot devices available that create .PDF and .DWF files, all available through the *Plot* command which produces the *Plot* dialog box. In the *Plot* dialog box, select from the .PDF and .DWF ePlot options (listed in the previous section) from the *Printer/Plotter Name* drop-down list.

If you select one of the PDF choices, the *PDF Options* button appears in the *Plot* dialog box below the *Printer/Plotter Name* drop-down list. This button produces the *PDF Options* dialog box shown in Figure 23-25. Refer to the information given in the previous section.

### Additional DWF Options

If you use the *Plot* dialog box to create .DWF files, you have access to additional options through the *Properties* button in the *Plot* dialog box. First, select the desired DWF device in the *Plotter/Printer* drop-down list, then select the *Properties* button just to the right of the list. Next, in the *Plotter Configuration Editor* that appears, highlight *Custom Properties* from the list and select the *Custom Properties* button. This procedure produces the *DWF ePlot Properties* dialog box where you can change *Vector Resolution* and *Raster Resolution*, options for *Font Handling*, *Layer information*, *DWF format*, and *Background color*. If a PDF device was selected in the *Printer/Plotter Name* list, selecting the *Properties* button and then *Custom Properties* produces the *PDF Options* dialog box discussed earlier.

### Lineweights

You can create .DWF or .PDF files with or without lineweights. By default, the images are plotted with lineweights. This can cause areas of your .DWF or .PDF files (when zoomed in the viewer) to look significantly different from how they appear as an AutoCAD drawing. If you have not specified lineweight values in the *Layer Properties Manager*, a default lineweight is applied to all graphical objects by Export or ePlot. If you do not want lineweights to appear in the .DWF or .PDF files when viewed in your browser, clear the *Plot object lineweights* option in the *Plot options* section of the *Plot* dialog box expanded section.

## Creating .DWF and .PDF Files with *Publish*

*Publish* automates the process of creating .DWF and .PDF files. *Publish* allows you to combine several drawings and/or layouts into one .DWF or .PDF file.

### *Publish*

| Ribbon | Menu Bar | Command (Type) | Alias (Type) | Shortcut |
|---|---|---|---|---|
| *Output*<br>*Plot*<br>*Batch Plot* | *File*<br>*Publish...* | *Publish* | ... | ... |

The *Publish* command creates .DWF or .PDF files automatically, rather than using a plot device from the *Plot* dialog box. *Publish* also provides you with a utility to create drawing "sets" in either .DWF or .PDF format that can be viewed in AutoCAD or the appropriate viewer. The DWF viewer, Autodesk Design Review, is free at www.autodesk.com, and the PDF viewer, Adobe Reader, is free at www.adobe.com.

With *Publish*, you can easily create a drawing "set" in .DWF or .PDF format. A drawing set created by *Publish* can contain multiple "sheets" within one file. Therefore, you can publish a complete project composed of several .DWG files with multiple layouts. Each drawing's model space layout and paper space layouts become separate "sheets" within the .DWF or .PDF file. You can also use *Publish* effectively with Sheet Sets. Since many projects are composed of several drawings and/or layouts, *Publish* is especially helpful for creating one file that is easily transportable and viewable by other colleagues or clients—particularly when the clients or colleagues do not have AutoCAD.

When you view a multiple sheet .DWF file created by *Publish*, the sheets can be viewed in Autodesk Design Review by selecting the "thumbnails" (Fig. 23-19, left) or by selecting from a list. Each sheet can be from a different layout of the same .DWG file, each can be from a separate .DWG, or the list can contain a combination of both. The sheets you want to appear are specified when you use *Publish* to create the DWF.

FIGURE 23-19

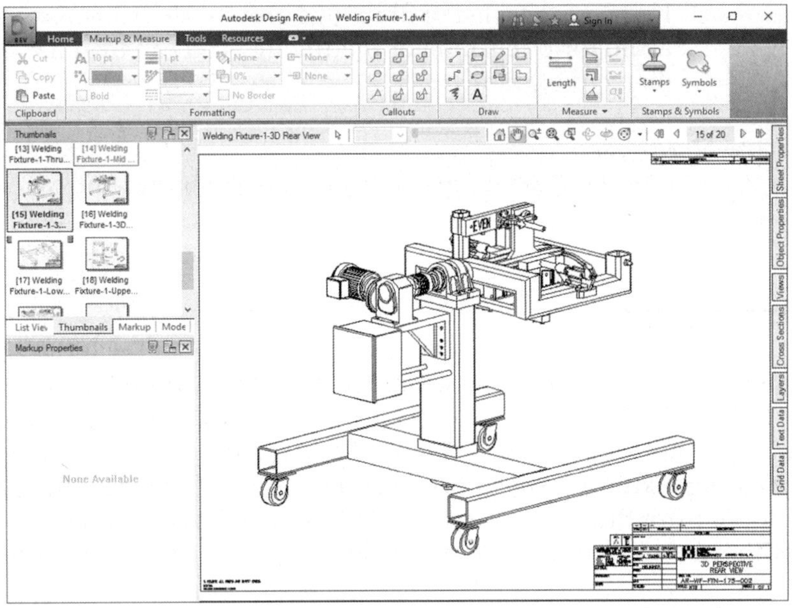


Using *Publish* by any method produces the *Publish* dialog box (Fig. 23-20). Normally, the steps for using this utility are:

FIGURE 23-20

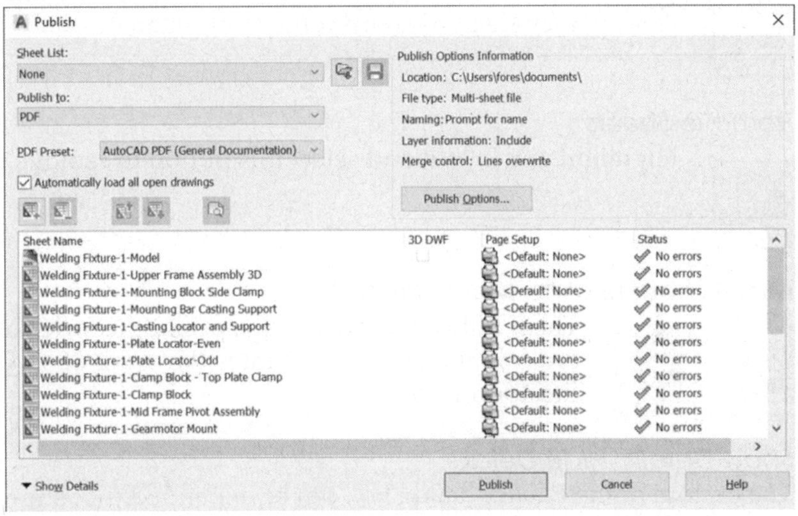


1. Use *Add Sheets* to select the drawings and layouts you want to include in the output file.
2. Select the desired format (*DWF, DWFx*, or *PDF*) from the *Publish to:* drop-down list.
3. If you selected *PDF* from the *Publish to:* list, select the desired *PDF Preset* from the list that appears just below the *Publish to:* list.
4. Change the order that the sheets appear when viewed using *Move Sheet Up* or *Move Sheet Down*.
5. Use *Save Sheet List* if you intend to publish this or a similar group of sheets again.
6. Specify the name and location of the .DWF or .PDF file you want to create.
7. Select *Publish* to create the .DWF or .PDF file.

The options from the buttons or the right-click short-cut menu (not shown) are described here.

### *Sheet List*

If a selection of sheets has been saved and loaded, the sheet list name appears here.

### *Publish to*

Select the desired output format such as *DWF, DWFx*, or *PDF* from this drop-down list. If you select *Plotter named in page setup, Publish* does not create a .DWF or .PDF file. Instead, each set is printed or plotted to a file as specified by the device named in its *Page Setup* column.

### *PDF Preset*

If this section appears, select the desired preset, such as *AutoCAD PDF (General Documentation)*, or *AutoCAD PDF (High Quality Print)*.

### *Load Sheet List*

If you saved a list of drawings previously using *Save Sheet List*, you can load the same list again for publishing or modifying. You can load .DSD (drawing sheet description) or .BP3 (batch plot) files.

### *Save Sheet List*

This choice saves the current list of drawings and layouts in the list of drawing sheets as a .DSD file.

### *Sheet Name*

The *Sheet Name* list displays by default every layout (including the *Model* tab) for each drawing selected with *Add Sheets*. Therefore, selection of one drawing in the *Select Drawings* dialog box usually produces multiple entries in the list of drawing sheets since most drawings contain one model space layout and at least one other layout. In Figure 23-28 for example, all entries are different layouts of one drawing, Welding Fixture-1. The information in the *Page Setup* column is applicable only if you intend to actually plot or print the layouts instead of creating a .DWF or .PDF file (see *Publish to*). The *Status* field indicates whether the page setups were loaded correctly with the layouts.

### *Add Sheets*

This button produces the standard *Select Drawings* dialog box. Select the drawing files you want to include in the output file to be created.

### *Remove Sheets*

Highlight any layout and select this button to remove the layout from the *Sheet List*.

### *Move Sheet Up* and *Move Sheet Down*

These buttons allow you to specify the order of the drawing sheets in the list. The order in the *Sheet List* dictates the order of sheets in the list when they are viewed in the appropriate viewer.

### *Show Details, Hide Details*

The bottom of the *Publish* dialog box can be expanded (not shown) to display information about the sheets and other options using the options in the lower-left corner. In the details section you can display *Selected Sheet Details* simply by highlighting any sheet from the *Sheet List*. You can select multiple copies, specify the precision, include a plot stamp (see "Plot Stamping" in Chapter 32), specify to *Publish* in background, and force the viewer to open when the .DWF or .PDF file is created.

### *Publish*

Select the *Publish* button only when you are ready to begin the publishing operation (all other options are specified). *Publish* creates a .DWF file, .PDF file, or plots to a device or file depending on your choice of *Plotter named in page setup*. If you are publishing to a .DWF or .PDF file, specify the desired file name and location in the dialog box that subsequently appears.

### *Publish Options*

Select this button to produce the *Publish Options* dialog box (see Figures 23-25 and 23-26). Here you specify options when .DWF or .PDF files are created using *Publish*. For information on these options, see the previous section "Creating .DWF and .PDF Files with *ePlot* and Export."

## IMPORTING .PDF FILES

Since AutoCAD 2017, you can import .PDF files directly into AutoCAD. Because .PDF files do not store data such as geometry and text in the same way as .DWG files, there are many options available for how you can convert the data. During the importation process, you can decide how you want to treat the components of the .PDF file (vector geometry, text, raster images, layers, etc.). Depending on the objects in the .PDF, you can decide if vector geometry is converted to AutoCAD objects such as *Plines*, *Blocks*, or *Mtext*. You can create new AutoCAD layers to contain each of the data types (text, vector geometry, raster images), or if the .PDF file contains layers, those layers can be converted to AutoCAD layers. You can also import .PDF files that are currently being viewed in AutoCAD as *PDF Underlays* (see "Viewing .DWF and .PDF Files" next in this chapter). To import .PDF files directly into AutoCAD use the *Import* or *Pdfimport* commands. To import a *PDF Underlay* use the *Pdfimport* command.

### *Import*

| Ribbon | Menu Bar | Command (Type) | Alias (Type) | Shortcut |
|---|---|---|---|---|
| *Insert Import* (icon) | *File Import* | *Import* | ... | ... |

The *Import* command produces the *Import File* dialog box (Figure 23-21). Locate the desired .PDF file on your system and select it from the list. Next, double-click or select *Open*.

NOTE: The *Import File* dialog box can be used to import a variety of files. To import .PDF files, make sure *PDF Files (*.pdf)* is selected in the *Files of type:* drop-down list at the bottom of the dialog box.

Once the desired .PDF file is selected, the *Import PDF* dialog box appears (Figure 23-22, on the next page). There you can specify the location, scale, and rotation for the .PDF file, select what PDF geometry to import and how it is to be converted, specify how PDF layers are treated in AutoCAD, and select from other import options. The specific features of this dialog box are explained below.

**FIGURE 23-21**

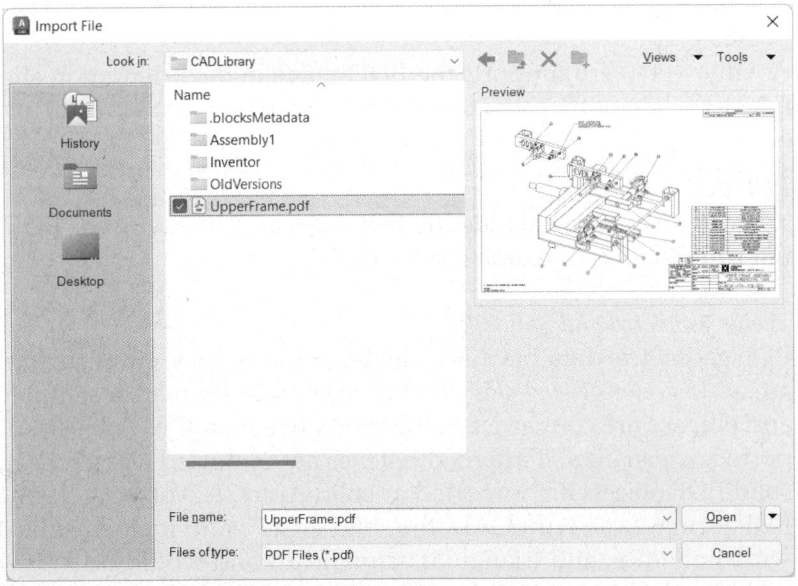


#### *Page to Import*

If the .PDF file contains multiple pages, you can import only a single page at a time. Choose the page to import by entering the page number or by clicking the thumbnail image. Use the button above the upper-right corner of the preview image to toggle between a full-size view and thumbnail views.

### Page size and PDF scale

The preview displays a standard page size with imperial or metric units. The *PDF scale* is the scale of the .PDF file if it were imported and displayed on the indicated sheet. The actual scale for imported PDF image is determined by the entry in the *Scale* edit box, regardless of the display and *PDF scale* of the preview image. See *Location* below.

FIGURE 23-22

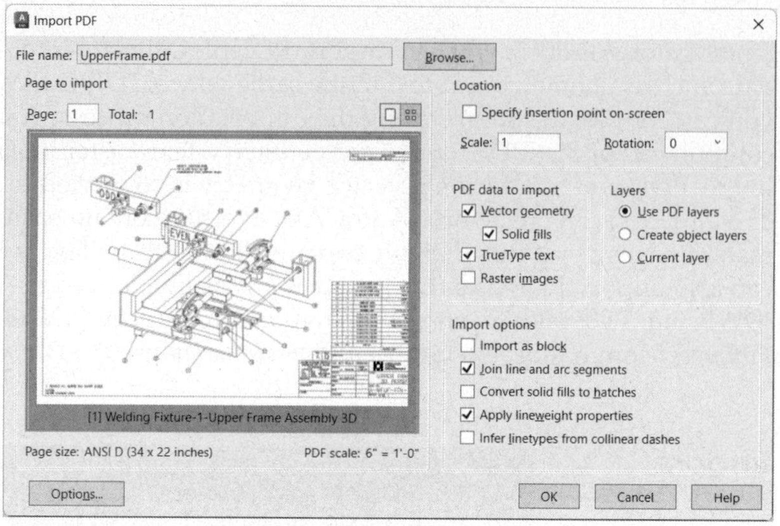


### Browse...

Use the *Browse* button to select from other files on your system to import.

### Location

In this section you can specify the insertion point, scale, and rotation angle for the PDF images to be imported.

*Specify insertion point on-screen*

If this box is checked, you can specify the insertion point with the pointing device or by entering coordinates. If this box is cleared, the PDF objects are inserted at 0,0.

*Scale:*

Enter a value for the scale for the imported PDF image. If the .PDF file was created in AutoCAD, you would typically enter a scale of 1. When importing other PDF images, it may be difficult to determine the scale of the image relative to AutoCAD units. See "Suggested Steps for Scaling and Importing .PDF Files" later in this section.

*Rotation:*

A value of 0 (zero) imports the PDF image in the orientation shown in the preview. Enter a positive or negative value other than zero to rotate the image.

### PDF data to import

You can include or exclude the PDF objects by the type of "PDF data" as listed. PDF files may or may not contain all of the data types listed.

*Vector geometry and Solid fills*

PDF geometric data may include linear paths, arcs, circles, ellipses, Beziér curves, and solid-filled areas. If *Vector geometry* is checked, curves (within an acceptable tolerance) that resemble arcs, circles, and ellipses are converted as the respective AutoCAD objects. Linear paths and other curves are imported as *Polylines*. Patterned hatches are imported as many separate objects. If *Solid fills* is checked, the solid PDF objects are imported as solid AutoCAD objects. If filled areas were originally exported into PDF format from AutoCAD, the solid areas could include solid-filled hatches, 2D solids, wipeout objects, wide polylines, and triangular arrowheads and would be converted back to the same objects. Solid-filled hatches are assigned a 50% transparency so that objects on top or underneath can be seen easily. You can change the color and transparency of these objects manually, or you can select them all by filtering for transparency with the *Qselect* command.

*TrueType text*

PDF files recognize only TrueType text objects. If this box is checked, the PDF TrueType fonts are either matched or substituted with similar fonts available on your system. The imported text is assigned to an

AutoCAD text style that begins with the characters "PDF_" and the TrueType font name. Other fonts (such as SHX fonts) or other text objects are treated as geometric objects and imported as *Polylines*.

*Raster Images*
With this box checked, any raster images included in the .PDF file are imported and converted to .PNG files and attached to the current drawing. The path to each raster image is controlled by the *PDFIMPORTIMAGEPATH* system variable.

### Layers

When PDF data is imported, it is assigned to one or more AutoCAD layers. This section allows you to specify what method is used for assigning imported objects to layers.

*Use PDF layers*
This option creates AutoCAD layers from the layers stored in the PDF file and applies them to the imported objects. The converted layer names have a "PDF" prefix. If no layers are present in the PDF file, object layers are created instead as described next.

*Create object layers*
AutoCAD layers are created for each of the following general types of objects imported from the .PDF file and named as such: *PDF_Geometry*, *PDF_Solid Fills*, *PDF_Images*, and *PDF_Text*.

*Current layer*
This option imports all specified PDF objects to the current layer.

### Import Options

Several other options are available for controlling how PDF objects are processed after being imported.

*Import as block*
This check imports the entire .PDF file as an AutoCAD *Block* rather than as separate objects. This feature is handy if you want to manipulate the PDF image in AutoCAD as a whole rather than individual elements.

*Join line and arc segments*
Here contiguous vector segments (that appear to touch within an acceptable tolerance) are joined into one *Polyline* where possible, rather than being converted to separate *Polylines*. (PDF data is not as precise as AutoCAD; therefore, some PDF vector segments may not actually touch.) See "Considerations for Importing and Exporting .PDF Files."

*Convert solid fills to hatches*
If 2D solid objects exist in the PDF data, this option converts them into AutoCAD solid-filled hatches. 2D solids that can be inferred as arrowheads are excluded and converted to solids. You can later change their properties in AutoCAD if desired.

*Apply lineweight properties*
The lineweight properties of the imported objects can be retained or ignored using this option.

*Infer linetypes from collinear dashes*
If the .PDF file contains sets of short collinear segments, this checkbox causes them to be imported as single *Polyline* segments. These polylines are assigned a dashed linetype named *PDF_Import* with an assigned linetype scale. If desired, you can later reassign different linetypes or appropriate linetype scales to these objects.

## Pdfimport

| Ribbon | Menu Bar | Command (Type) | Alias (Type) | Shortcut |
|---|---|---|---|---|
| *Insert* *Import* (icon) | ... | *Pdfimport* | ... | ... |

*Pdfimport* is used to import a .PDF file (in which case it is equivalent to *Import*), or *PDF Underlay*. A *PDF Underlay* is a .PDF file that must first be attached to the .DWG file through *Pdfattach* before it can be imported. A *PDF Underlay* can only be viewed in AutoCAD, but the image is still an unconverted .PDF file, therefore the PDF objects cannot be individually edited. See the section later in this chapter titled "Viewing .DWF and .PDF Files" for information to *Attach* a *PDF Underlay*.

Command: **`Pdfimport`**
Select PDF underlay or [File] <File>: **`Enter`**
Specify first corner of area to import or [Polygonal/All/Settings] <All>:

### *File*

Use the *File* option to select a .PDF file to import directly into AutoCAD (not an underlay). This option has the same options and results as using the *Import* command to import a .PDF file, as explained in the previous section. Also, see the *Settings* option below.

### *Specify first corner/All/Polygonal*

You can select the entire *PDF Underlay* to import using the *All* option. Alternately, you can specify only a specific area of the underlay to import by selecting two corners with the pointing device or using the *Polygonal* option to draw a polygonal boundary.

### *Settings*

The *Settings* option produces the *PDF Import Settings* dialog box (Fig. 23-23). You will notice the options in this dialog box are identical to those used for importing .PDF files using the *Import* command. Please refer to the previous section describing the *Import* command for an explanation of these options.

**FIGURE 23-23**

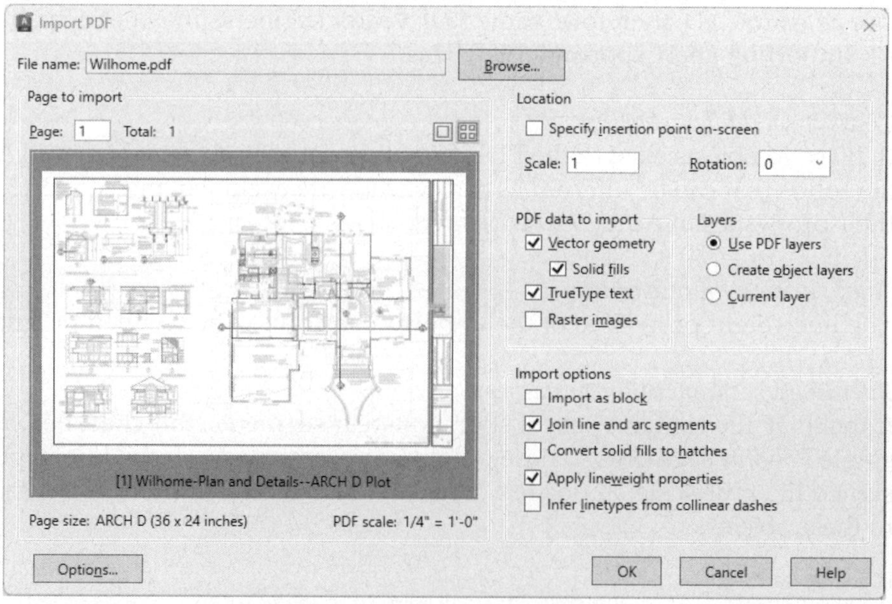

## Importing SHX Fonts from .PDF Files

A .PDF file only recognizes TrueType fonts. When an AutoCAD drawing that uses Autodesk's SHX fonts is exported as a .PDF, SHX fonts in that drawing are converted to geometry instead of text: lines and shapes. When that .PDF file is imported into AutoCAD, text that originally used SHX fonts is imported as Polylines. AutoCAD 2018 adds the Recognize SHX Text utility to compare those shapes against known SHX font files and convert them back into Mtext.

### *PDFSHXTEXT*

| Ribbon | Menu Bar | Command (Type) | Alias (Type) | Shortcut |
|---|---|---|---|---|
| *Insert Import Recognize SHX Text* | ... | *Pdfshxtext* | ... | ... |

Use PDFSHXTEXT after importing a .PDF file that was made from an AutoCAD drawing that used SHX fonts. The text will be presented as geometry.

Command: ***PDFSHXTEXT***
Select objects of [SEttings]: ***PICK***
Select objects or [SEttings]: ***Enter***
Command:

Select the text using one of the standard methods (window, lasso, fence, etc.). You can select multiple lines of text. If the text matches an SHX font specified in the settings and within the given tolerance, it is converted to one or more Mtext objects using that font. Be sure to verify the converted text.

Depending on the amount of text that is converted, you might want to use the Combine Text tool (TXT2MTXT) to unify multiple Mtext objects that result from the conversion into a single Mtext object.

Choosing the *Settings* option launches the PDF Text Recognition Settings dialog box (Figure 23-24), which can also be opened by choosing Recognition Settings in the Import section of the Insert tab.

**FIGURE 23-24**

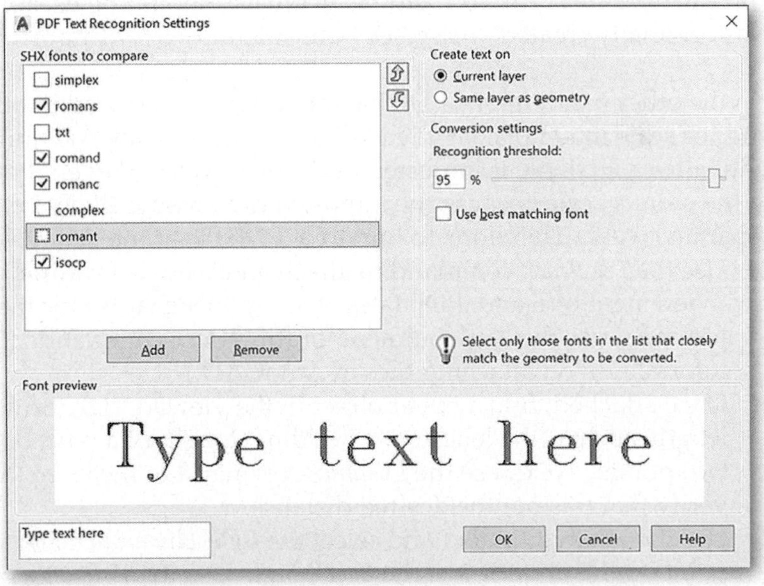


#### *SHX fonts to compare*

This box lists the fonts against which the imported "text" is compared. Don't be tempted to select all fonts in hope of increasing the chances of success. That can result in a mix of fonts. Use the Up and Down ordering buttons to the right of the box to order the fonts. Unless Use best matching font is selected, the first font found that meets the threshold is used, so order can matter.

### Add/Remove

These options allow you to add other SHX fonts to the comparison list, or remove others. If you know the specific font used in the original drawing and it is not listed under SHX fonts to compare, add it here and select it as the only font to compare against.

### Font Preview

By typing text into the smaller "Type text here" box in the lower corner, you can preview individual fonts to get an idea of which are better matches. This can help you narrow your font choices in the SHX fonts to compare box.

### Create text on

Select Current layer to add the Mtext objects to the currently active layer. Select Same layer as geometry to leave the text on the layer that it was stored in the PDF.

### Recognition threshold

Move the slider to specify nearly exact or more loosely matched font results. Generally, it is best to set a high threshold.

### Use best matching font

When selected, this option uses the font with the best match, regardless of its order in the SHX fonts to compare list. When not selected, the first font in the SHX fonts to compare box that meets the threshold is used. Be aware that selecting this option can add to the processing time.

## Suggested Steps for Scaling and Importing .PDF Files

When .PDF files are created by exporting an AutoCAD drawing to a .PDF file, the scale of the geometry is retained in the .PDF file when it is imported back into AutoCAD. Therefore, if you import such a file, you can specify a *Scale* of 1 and the geometry should be useable in AutoCAD again at the same size it was originally created. (Also see "Considerations for Importing and Exporting .PDF Files" below.)

On the other hand, many .PDF files are originally created in other applications and therefore the units and/or scale may not coincide with the desired AutoCAD units or scale. This fact may not be apparent until after a .PDF file is imported into AutoCAD, and even then it may be difficult to adjust the PDF data to the correct scale because the imported data would likely be composed of separate objects possibly on separate layers. Therefore, to import a .PDF file at the desired scale, follow the steps listed next.

1. Use the *Pdfattach* command to attach and view the desired .PDF file in AutoCAD (see "Viewing .DWF and .PDF Files" later in this chapter for information on *Pdfattach*). Alternately, use the *Xref* command or the *Attach* command. (See Chapter 30 for additional information on attaching files in AutoCAD.)
2. Once attached, find a linear object in the viewed PDF for which you know the desired length, such as the length of a building, length of a part, or any known distance between two points. Next, use the *Distance* command to measure that distance on the underlay and verify that it is not the desired distance.
3. Use the *Scale* command and select the *PDF Underlay* (select the PDF border). Select any point on the underlay as the *Base* point. Use the *Reference* option, and then at the "Specify reference length prompt," pick one of the two previously measured distance points. At the "Specify second point" prompt, specify the second known distance point. Finally, at the "Specify new length" prompt, enter the desired value (the correct desired length in AutoCAD units). The *PDF underlay* should then be scaled correctly.
4. Use the *Pdfimport* command to import the *PDF Underlay* and convert the file to AutoCAD objects. Use the *Settings* options to specify the import options. (Complete step 5 before exiting the *Pdfimport* command.)
5. During the *Pdfimport* command you are prompted to keep, detach, or unload the *PDF Underlay*. Select *Detach* to detach the underlay. You no longer need the *PDF Underlay* since the desired PDF objects have been imported.

## Considerations for Importing and Exporting .PDF Files

.PDF files do not maintain the precision that AutoCAD .DWG files do. AutoCAD drawings store data as double-precision floating-point numbers, whereas .PDF files store data as single-precision numbers. Also, .PDF files store vector geometry and some text differently so conversion is not direct or complete. Therefore, the following precautions should be noted.

When Exporting .PDF Files from AutoCAD:

- Coordinates may be rounded off in the PDF such as endpoints, center points, tangent points, arcs and circle centers, etc.
- The dynamic range of a .PDF file is smaller. Therefore, errors of coordinate precision may occur especially for very small objects and for very large objects and locations such as maps or roadways.
- AutoCAD objects (*Lines*, *Arcs*, *Circles*, *Polylines*, etc.) are converted in a .PDF file to line segments and Beziér curves that may or may not be connected segments.
- Only TrueType text is converted properly, whereas .SHX fonts become line segments.
- Many other aspects of the AutoCAD drawings may be lost, such as some object properties, associativity of dimension objects and text, etc.

When Importing .PDF files into AutoCAD:

- Raster objects cannot be easily imported to become drawing objects and text.
- The Markup Import command lets you import markups. See The Markup Tools on page 628.
- The shapes of PDF Beziér curves may be converted to *Circles*, *Arcs*, *Splines*, and *Ellipses* if the precision is within acceptable tolerance. Otherwise, Beziér curves are converted to *Polylines*.
- Non-continuous linetypes are converted to independent, not-connected line segments.

The AutoCAD PDF export and import features should not be used to transfer AutoCAD drawings from one AutoCAD user to another because so much precision and AutoCAD information is lost in the conversion process. Instead, transfer .DWG files between AutoCAD users whenever possible.

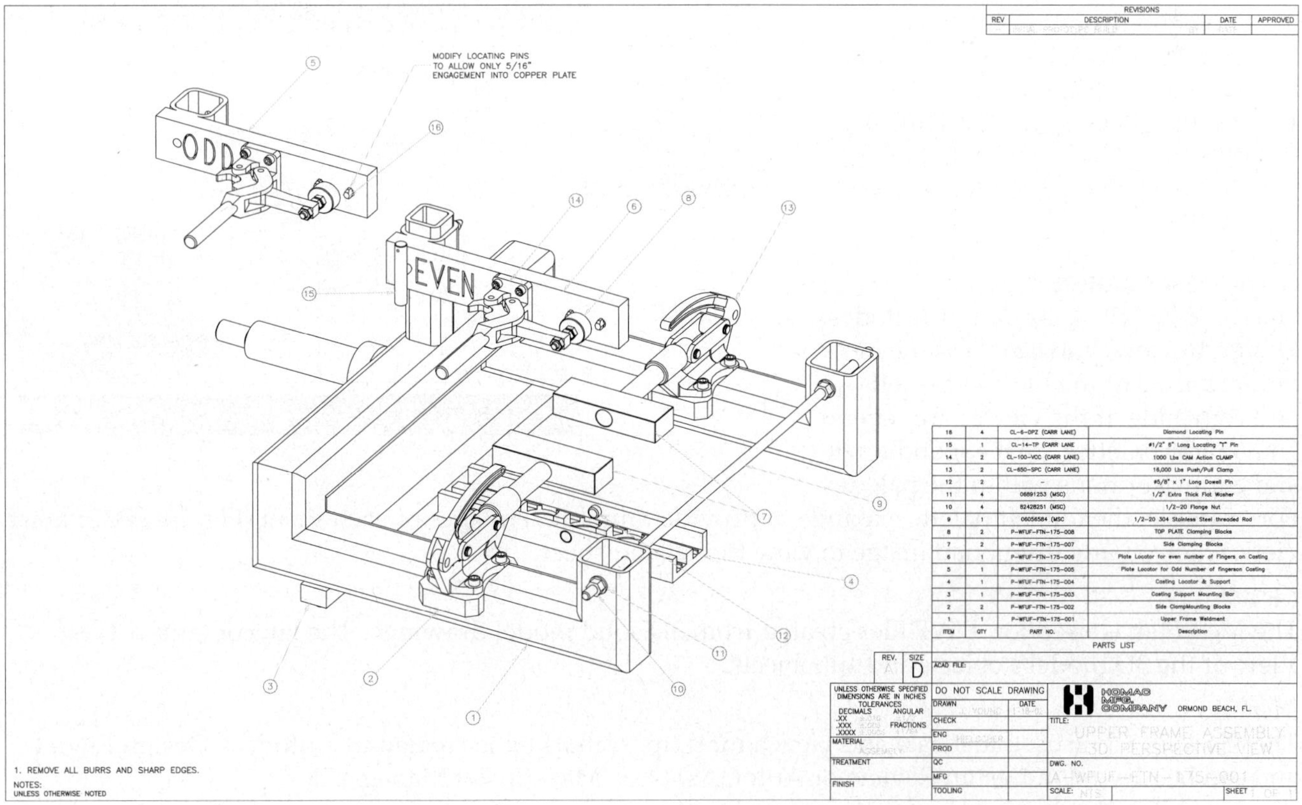

## VIEWING .DWF AND .PDF FILES

Although you can *Import* .PDF files directly into AutoCAD as described in the previous section, there are applications when only viewing the file is needed. Viewing .DWF files is also desireable for some cases. You can view .DWF and .PDF files two ways:

- You can use Autodesk Design Review to view single- or multi-sheet .DWF files. Autodesk Design Review can be downloaded free from www.autodesk.com. View .PDF files using the free Adobe Reader® available at www.adobe.com.
- You can also view .DWF or .PDF files in AutoCAD. This is accomplished by inserting an individual sheet as a *DWF Underlay* or *PDF Underlay*. or by importing the files to convert them as discussed earlier in this chapter.

### Autodesk Design Review

Autodesk Design Review is free, so everyone can view DWF images (not PDF) directly with Autodesk Design Review. Autodesk Design Review can be downloaded free from www.autodesk.com and installed by anyone—without AutoCAD. Keep in mind that AutoCAD Design Review can be used to view and mark up .DWF files.

Tip: You can also download the free DWG TrueView to view, make measurements, and markup .dwg files, and other CAD formats such as Fusion 360, Revit, 3ds Max, BIM 360, and Civil 3D.

In Autodesk Design Review there are several methods that can be used to open and view files:

- Use the *File* pull-down menu in Autodesk Design Review, then use *Open* to locate and open a .DWF file. You can also open DWG and other formats if you have DWG TrueView properly installed.
- Drag and drop the .DWF file from Windows File Explorer into Autodesk Design Review.
- Double-click on the .DWF file in Windows File Explorer to launch Autodesk Design Review to view the file.

FIGURE 23-25

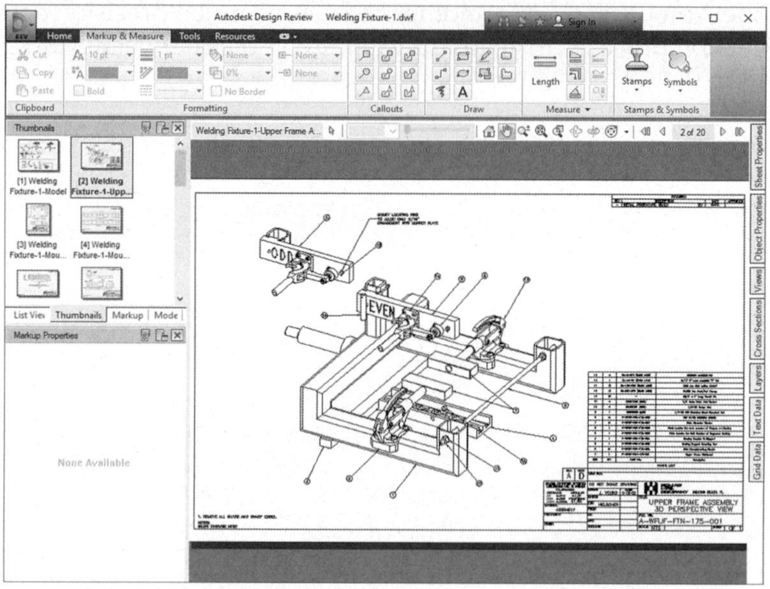


#### Navigator Palettes

Once a .DWF file is loaded in Autodesk Design Review, you can view the individual sheets of the file. On the left and right side of the viewer are several "navigator palettes" that expand to view and control options within the palettes. For example, the top left palette expands to provide thumbnail images of each sheet (Fig. 23-25), or a list view. Click on any thumbnail image to view the related sheet.

*Model*

The *Model* tab is used for .DWF files created from 3D solid model drawings. The tab contains a Tree View of the 3D model's objects and subobjects.

*Markups*

The *Markups* tab is used to locate and view a markup. A markup is created in Autodesk Design Review and can be viewed in Design Review or AutoCAD (see "Markup Set Manager").

*Properties*

The *Properties* palettes allow you to view the properties of a selected sheet, markup, or object. The properties for the selected item display in their individual *Properties* palette.

*Layers*

The *Layers* palette lists all layers for each sheet. You can control which layers are visible in the DWF sheet by toggling the "light bulb" icon on or off. For example, Figure 23-26 displays a sheet with the DIM layer off, so the sheet's dimensions are no longer visible.

*Views*

Similar to an AutoCAD drawing, the related .DWF file can contain the same views. Use the *Views* palette to list views contained in the file. For 3D models, the standard views (top, front, right, etc.) are listed, which you can click to display.

**FIGURE 23-26**

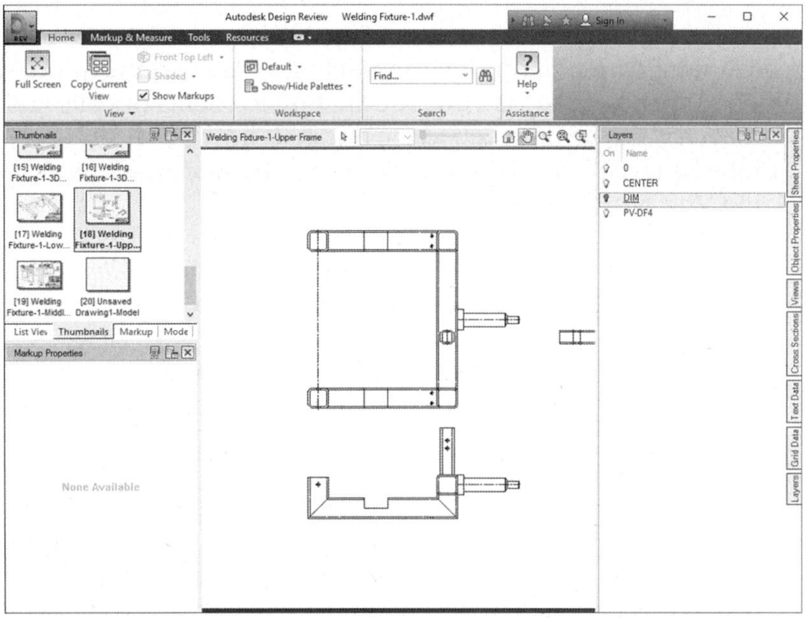


**FIGURE 23-27**

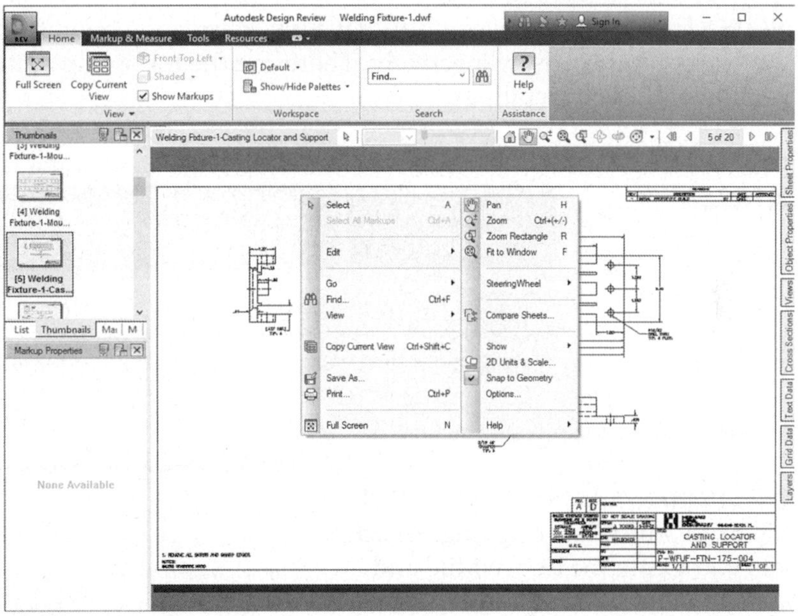


### Shortcut Menu, Pull-down Menus

Right-clicking in the viewing area produces a shortcut menu (Fig. 23-27). This menu combines many options available from both the *View* panel on the *Home* tab and navigation items in the *Canvas* toolbar in the upper right of the viewer.

*Pan*

This option puts you in real-time *Pan* mode and the pointer becomes a "hand" icon. Press the left mouse button to drag the image around on the screen in a manner identical to the *Realtime* option of *Pan* in AutoCAD.

*Zoom* Options

With *Zoom* capabilities of a DWF image, genuine detail can be viewed, similar to capabilities in the original CAD program with the associated CAD drawing. By default, you can *Zoom* at any time (without selecting a menu option) by turning the mouse wheel if you have a wheel mouse. The *Zoom Rectangle* and *Fit in Window* options operate identically to the *Window* and *All* options, respectively, in AutoCAD (see Chapter 10, Viewing Commands).

*Show*

A cascading menu appears with toggles for you to show *Hyperlinks* (if hyperlinks were included in the drawing), *Markups* (if the .DWF files were marked using AutoDesk Design Review), and to show the *Canvas Background* in the viewer.

## *DWF Underlay* and *PDF Underlay* in AutoCAD

One of the primary purposes of using .DWF and .PDF files is to collaborate with clients or colleagues concerning particular designs and details contained in AutoCAD drawings. Using a .DWF or .PDF image, you can pass drawing information to clients or colleagues electronically without risking accidental changes being made to the original drawing file. Using Autodesk Design Review for .DWFs or Adobe Reader® for .PDFs, the files can only be viewed. However, you can view a .DWF file directly in AutoCAD as a *DWF Underlay* and view a .PDF file in AutoCAD as a *PDF Underlay*. These features allow you not only to "see" the images, but the images can be treated as an "underlay" for tracing. Therefore, you can use the images as a basis for drawing and to *Object Snap* directly to objects in the image. You can also control layer visibility of the images.

.DWF and .PDF images react identically in AutoCAD, so the *Attach* command can be used for either. Additional commands and options are used to view and manipulate the two image types. The related commands are:

| | |
|---|---|
| *Attach* | attaches an image as an underlay |
| *Clip* | allows you to clip the image boundary |
| *Adjust* | allows you to fade, change the contrast, or make it monochrome |
| *Dwflayers / Pdflayers* | allows you to control layer visibility of the image |

NOTE: The *Pdfattach, Pdfclip,* and *Pdflayers* commands operate identically to the *Dwfattach, Dwfclip,* and *Dwflayers* commands, respectively. Therefore the following section (three pages) explains only the .DWF version of these capabilities.

### *Attach*

| Ribbon | Menu Bar | Command (Type) | Alias (Type) | Shortcut |
|---|---|---|---|---|
| *Insert Reference Attach* | *Insert DWF Underlay* | *Dwfattach* | ... | ... |

Use *Attach* to attach a .DWF or PDF file to an AutoCAD drawing. *Attach* lets you locate the desired file and specify the insertion point, scale and rotation for the underlay. Keep in mind that one .DWF or PDF file can include multiple sheets, so each sheet that you select is considered a separate underlay.

When you invoke *Attach*, the standard *Select File* navigation dialog box appears for you to select the desired parent file (not shown). Once selected, the *Attach DWF Underlay* dialog appears (Fig. 23-28).

**FIGURE 23-28**

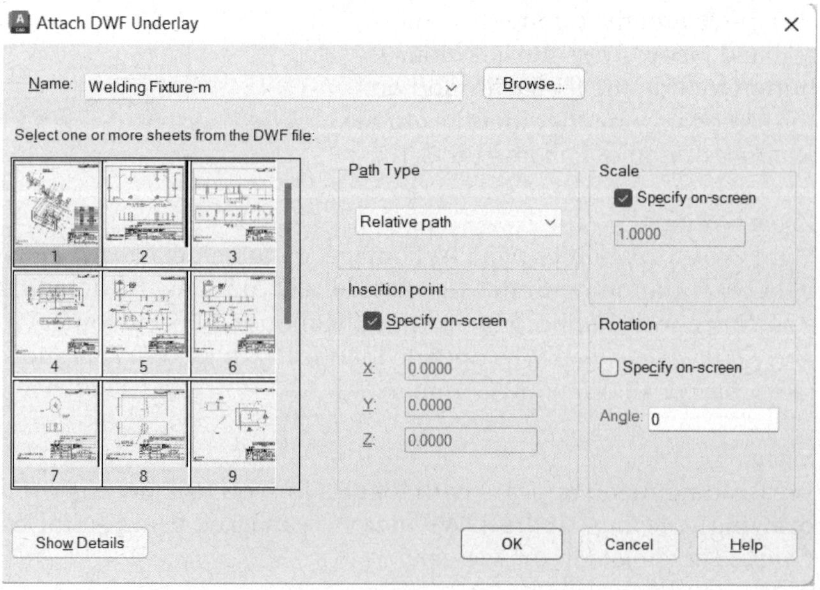

### *Name*

This section identifies the .DWF file you previously selected to attach. The drop-down lists previously attached DWF underlays. To add another copy of a DWF underlay that is already attached, select the DWF image name from the list. Use the *Browse* button to open the *Select DWF File* dialog box.

### *Select a sheet from the DWF file*

This list displays all of the sheets that are found in the .DWF file. Select the desired sheet from the list to attach to the drawing as an underlay. If the .DWF file contains only a single sheet, that sheet is listed. If the .DWF file contains multiple sheets, only a single sheet at a time can be selected for attachment. 3D sheets are not listed.

### *Path Type*

This list allows you to specify one of three types of folder path information to save with an attached DWF underlay: an absolute path, a relative path, or no path. Similar to attaching *Xrefs* or other images, these options are important depending on whether the drawing file or attached image file is later relocated to another folder or computer.

The *Full Path* option saves the entire (absolute) path with the .DWF file and is used when the location of the related files is not expected to change. Use the *Relative Path* option when you expect to relocate the related files together to another drive but using the same directory structure. The *No Path* option saves only the .DWF file name; therefore, this option should be used only when the .DWF file is located in the same folder as the current drawing file.

### *Insertion Point*

Identical to inserting *Blocks*, this section specifies the point in the drawing you want to insert the selected .DWF file. Select *Specify on-screen* to pick a point in the drawing, or remove the check and input coordinates.

### *Scale*

Specifies the scale factor for the selected DWF underlay. Select *Specify On-Screen* to pick a point in the drawing, or clear the checkbox to enter a value for the scale factor.

### *Rotation*

This section specifies the rotation angle of the selected DWF underlay using the same methods as the *Scale* and *Insertion point* sections.

Once the image is inserted into the drawing, you can use it as an "underlay" by tracing directly over the objects or simply using the image as a reference for checking an associated drawing. If you want to *Object Snap* to objects in the image, *DWF Object Snap* must be turned on. Check this using the right-click shortcut menu or click the underlay object to show the DXF Underlay ribbon tab (PDF is similar).

### Underlay Ribbon Tab / Shortcut Menu

To show the menu (Fig. 23-29), first select the DWF image so it becomes highlighted and the insertion point grip appears (assuming grips are on) and the Underlay ribbon tab appears. Deselect the image object (using Esc or clicking off the object) to return to the normal ribbon.

You can also select the image and then right-click to show the shortcut menu.

**FIGURE 23-29**

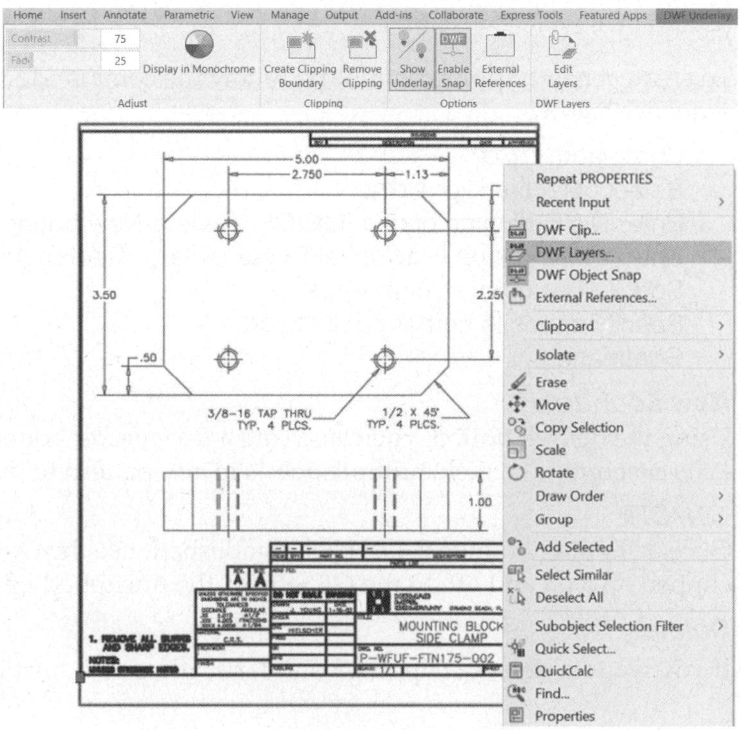

*Enable Snap / DWF Object Snap:* This option must be checked for you to be able to *Object Snap* to objects within the image. This setting is saved in the *DWFOSNAP* system variable (*PDFOSNAP* for PDF underlays).

*Create Clipping Boundary / DWF Clip:* This option invokes the *Clip* command in AutoCAD.

*Edit Layers / DWF Layers:* Use this option to invoke the *Dwflayers* command in AutoCAD (see "*Dwflayers*").

*Display in Monochrome / Contrast / Fade:* Toggle button to make monochrome or restore original colors. Use the sliders to change the contrast or fade the image. You can also use the *Adjust* command options.

*Hide/Show Underlay:* Toggle button to change visibility of the entire image. See "*Dwflayers*" for turning on and off individual layers.

*Shortcut Menu Erase, Move, Copy, Scale, Rotate*

These options affect the entire DWF image—you cannot affect individual objects within the image. Effectively, these options allow you to change the original insertion point, scale, and rotation angle.

### Dwfclip

| Ribbon | Menu Bar | Command (Type) | Alias (Type) | Shortcut |
|---|---|---|---|---|
| *Insert Reference Clip* | ... | *Dwfclip* | ... | ... |

For some applications it is not desirable to show the entire DWF image in the drawing. For example, an image may display a title block and border included on the sheet, but you may want to display only the design geometry rather than the entire sheet. Use the *Clip* command to create a new, smaller, boundary.

For example, examine previous Figure 23-29, which displays the Mounting Block Side Clamp sheet. Using *Clip*, a smaller boundary could be created to display only the profile view, clipping out the title block, border, and flat view (Fig. 23-30).

**FIGURE 23-30**

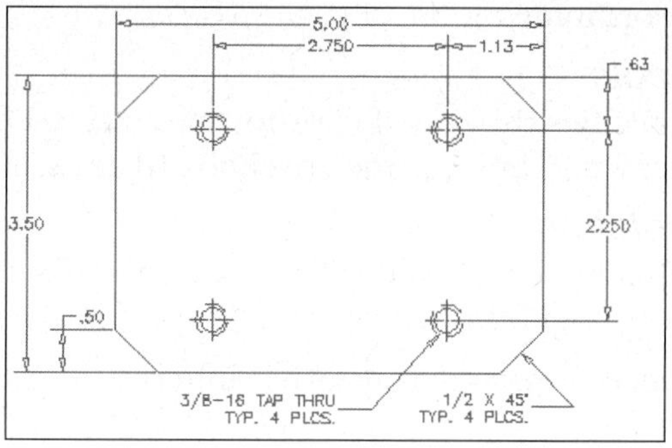


To create a new clipping boundary, follow the command sequence below.

Command: ***CLIP***
Select object to clip: **PICK**
Enter DWF clipping option [ON/OFF/Delete/New boundary] <New boundary>: **Enter**
Enter clipping type [Polygonal/Rectangular] <Rectangular>: **Enter**
Specify first corner point: **PICK**
Specify opposite corner point: **PICK**
Command:

#### *New Boundary*

Using the default option, you can create a *Rectangular* boundary by picking two points or use the *Polygonal* option to pick multiple points in any pattern to define the clipping boundary.

#### *ON/OFF*

Once a clipping boundary has been established, use the *On* and *Off* options to turn the display of the clipped area on and off. In the *Off* setting, the unclipped image is displayed.

#### *Delete*

If you want to delete a clipping boundary rather than turn it off, use this option.

### *Dwflayers*

| Ribbon | Menu Bar | Command (Type) | Alias (Type) | Shortcut |
|---|---|---|---|---|
| *Insert* *Reference* *Underlay Layers* | ... | *Dwflayers* | ... | ... |

The *Dwflayers* command allows you to turn layers in a DWF underlay *On* and *Off*.

Command: ***DWFLAYERS***
Select DWF underlay: **PICK**

Selecting the boundary of a DWF underlay produces the *DWF Underlay* menu. You can use this menu to choose Edit Layers to show a dialog box with the layers and provides the standard "light bulb" icon for controlling the layer on/off status. This is essentially the same tool used in the Autodesk Design Review *Layers* palette (see Figure 23-33).

## To Detach a DWF or PDF Underlay

To sever the link between the AutoCAD drawing and the DWF or PDF image, follow the steps below. For more information on external references, see Chapter 30, Xreferences.

1. From the *View* tab, *Palettes* panel, select *External References* palette.
2. In the *External References* palette that appears, locate the *File References* section and select the DWF or PDF underlay you want to detach.
3. Right-click on the DWF or PDF underlay name and select *Detach* from the shortcut menu.

The underlay is no longer linked to the drawing file. All instances of the underlay are removed from the drawing.

TIP

You can also select the underlay object and use the DWF Underlay ribbon tab that appears to choose External References.

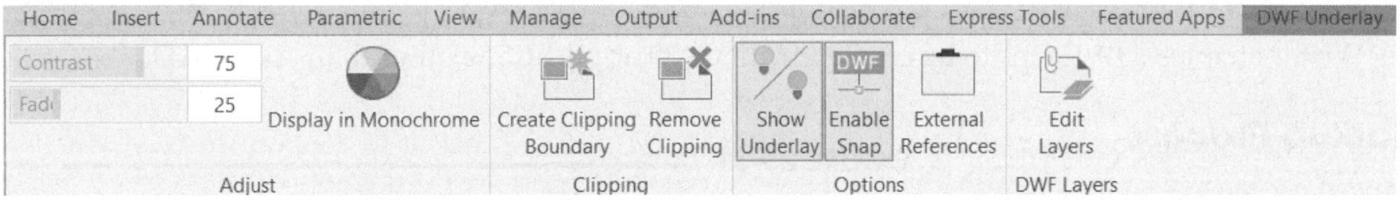


## Viewing URLs (Hyperlinks) in a DWF Image

When viewing a DWF image in Autodesk Design Review or in AutoCAD, passing the mouse pointer over a URL-linked object displays a small hand image. Clicking on the linked object automatically links to the new URL, which causes the text and/or images to display in your default web browser or Autodesk Design Review, depending on the file format (.HTM, .DWF, .JPG, etc.). For example, clicking on a door, window, or fixture in an architectural layout could, in turn, display a table of text information listing price, manufacturer, specifications, etc. Or, clicking on one component of a mechanical assembly could, in turn, display a detailed drawing of the selected part. It would also be possible to view a map, select a building, and see written information about the business or link to the Web site of the business.

NOTE: Depending on the file format, clicking on a URL-linked object in Autodesk Design Review may display the linked object in your system's default web browser, not in Autodesk Design Review.

To create URLs in DWF images, *Open* the original drawing in AutoCAD and use the *Hyperlink* command to specify a URL. Next, use the *Plot* command and select an ePlot device to create the related .DWF file with hyperlinks included (see "Using Hyperlinks" and "Creating .DWF Files with ePlot").

# COLLABORATING AND MARKUP TOOLS

Autodesk provides a variety of tools that can be used by engineers, designers, and architects to collaborate with other professionals. Using AutoCAD and other Autodesk products, the drawing authors can *publish* single- or multi-sheet files for others to view. Autodesk Docs also allows markups on a PDF to be detected by Markup Assist. Using a markup process, clients and colleagues can suggest design changes within the files made available to them for viewing. Finally, the markups can be viewed by the original authors in AutoCAD and changes to the original set of drawings (.DWG files) can be made.

## The Markup Tools

The following tools from Autodesk can be used in the markup process.

1. *Forma Data Management*: Use this subscription cloud-based data management system to let non-AutoCAD users mark up and give feedback on drawings. The markups can be added directly to the drawing. It has many other features to make it easier to collaborate and manage drawings for multiple users.
2. AutoCAD *Markup Set Manager*: Create the original drawings and publish .DWF files to share with your collaborators. Marked up .DWF files are viewed using the *Markup Set Manager* so changes to the original drawings can be made.
3. AutoCAD *Markup Import* and *Markup Assist* commands can be used to view or even convert information in a marked-up .DWF file to make it easier to change to the original drawing content. These commands are only available for AutoCAD subscribers. AutoCAD Docs subscribers can also convert markups from a PDF file.
4. Autodesk Design Review: This tool allows .DWF files to be read and marked up. This is a separate tool that does not run inside AutoCAD that allows project team members to view .DWF files, measure distances, control layers and views, and "red line" drawings by creating marked up drawings with comments, sketches, dimensions, and text. Autodesk Design Review also allows you to reorganize sheets in a sheet set and republish the DWFs with markups.
5. Autodesk DWF Writer: For those without AutoCAD, this free printer driver can be downloaded from the Autodesk site, installed on any Windows system, and set as a system printer. Therefore, DWF Writer enables .DWF files to be created from CAD applications other than AutoCAD.

## The Markup Process

The markup process lets collaborators help make improvements to the design. First, the designer creates drawings in AutoCAD and publishes them as a multi-sheet .DWF file or pushes the drawings to Forma Data Management. Next, the project coordinator, client, or other member of the project team checks the drawings and creates markups. "Markups" areas designate corrections and suggested changes to the drawings in the set. The markups, also called "redlines," generally include a callout, such as a *Revcloud*, and are accompanied by a note indicating the nature of the needed change (Fig. 23-31).

FIGURE 23-31

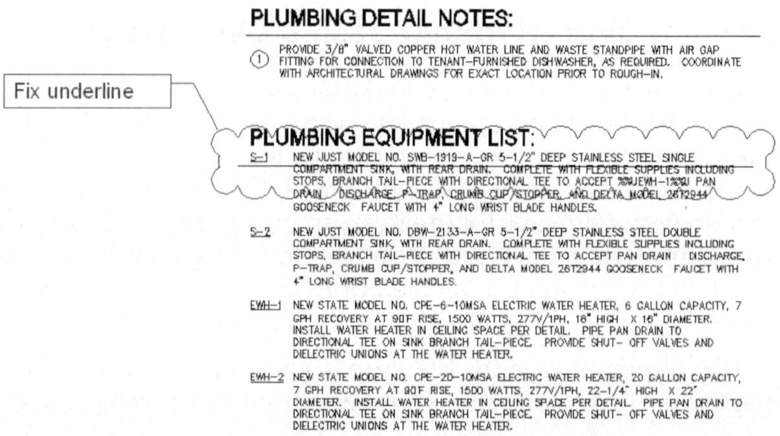

The advantage is revision control for the drawing set is maintained through proper channels and the .DWG files are not edited by every person involved, particularly those team members who may not be skilled at using AutoCAD to create and manage drawings.

Finally, changes to the actual drawings (.DWG) files in AutoCAD are integrated by the drawing author(s). This may be done using tools available in the *Markup Set Manager*. Although .DWF files can be viewed and "redlined" in Autodesk Design Review, changes to the original .DWG files are made in AutoCAD.

The *Collaborate* tab (Figure 23-32) on the ribbon provides tools for sharing your designs with clients, and others, inside and outside your company without the recipient having to have AutoCAD or another drawing viewer. This can save you time in exporting a PDF, DXF or other file format and provide the client the ability to measure, comment and mark up the design from within their web browser.

**FIGURE 23-32**

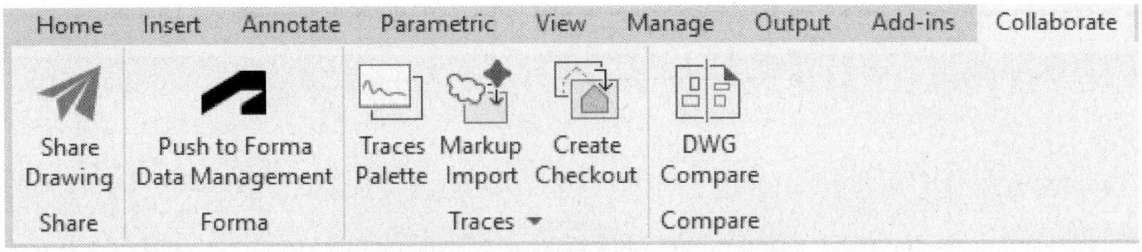


### *Share Drawing (Share URL)*

| Ribbon | Menu Bar | Command (Type) | Alias (Type) | Shortcut |
|---|---|---|---|---|
| *Collaborate* *Share Drawing* | ... | *share* | ... | ... |

You can share a *link* to a copy of the current drawing for someone to view or edit in AutoCAD via the web (*web.autocad.com*) or using AutoCAD mobile with the Share command. All of the related DWG xrefs and images are included. Share works similar to ETRANSMIT in the desktop version of AutoCAD. Anyone can use the link to access the drawing in a web browser or using the mobile app.

Keep the following in mind when using *Share*:

- The link expires in seven days.
- The original drawing or any single referenced drawing can not be larger than 50MB
- The total uploaded drawings cannot exceed 200MB.
- You choose for recipients to *View only* or be able to *Edit and save a copy*.
- Xrefs are saved to the same folder location as the parent file when you save a shared drawing.

**FIGURE 23-33**

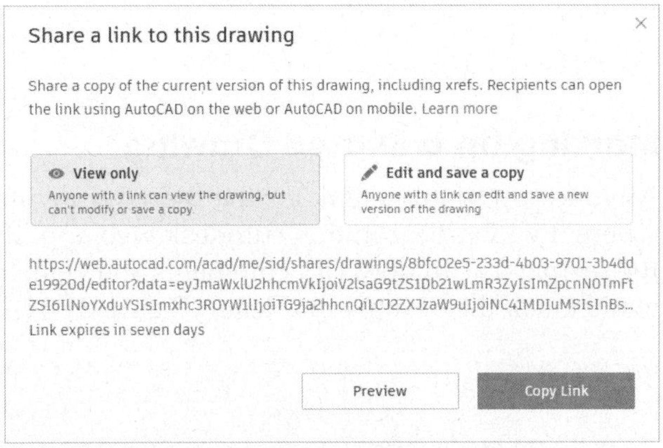


***View only:*** Link lets anyone view the drawing, including all xrefs, but not make changes to the drawing, or save/download a copy.

***Edit and save a copy:*** Link lets anyone edit the drawing and save the drawing and its xrefs as a copy, but not make changes to the original drawing file from the sharer.

To send a link to your drawing, use the *Share Drawing* selection from the ribbon Collaborate tab (see Figure 23-32) or type the *Share* command. You can also select Publish, Share Drawing from the Application icon. The command opens the Share Drawing panel (Figure 23-33)

From the panel, pick *View only* or *Edit and save a copy.*

The link shows in the central part of the panel. Use the Preview button at the lower right to see how the drawing will appear in the AutoCAD web app. The Copy button copies the link to your clipboard, where you can paste it into an email or other document to share with the client, internal commenters, or other shareholders. The linked drawing expires after 7 days, which is helpful to prevent old versions of the drawing haunting you.

Notice the many tools are available in the web app that let the client comment, adjust their view, make measurements, draw revision clouds, show the layer list and other information (Figure 23-34). At the top of the view window are the Draw, Annotate, and Modify tool palettes.

**FIGURE 23-34**

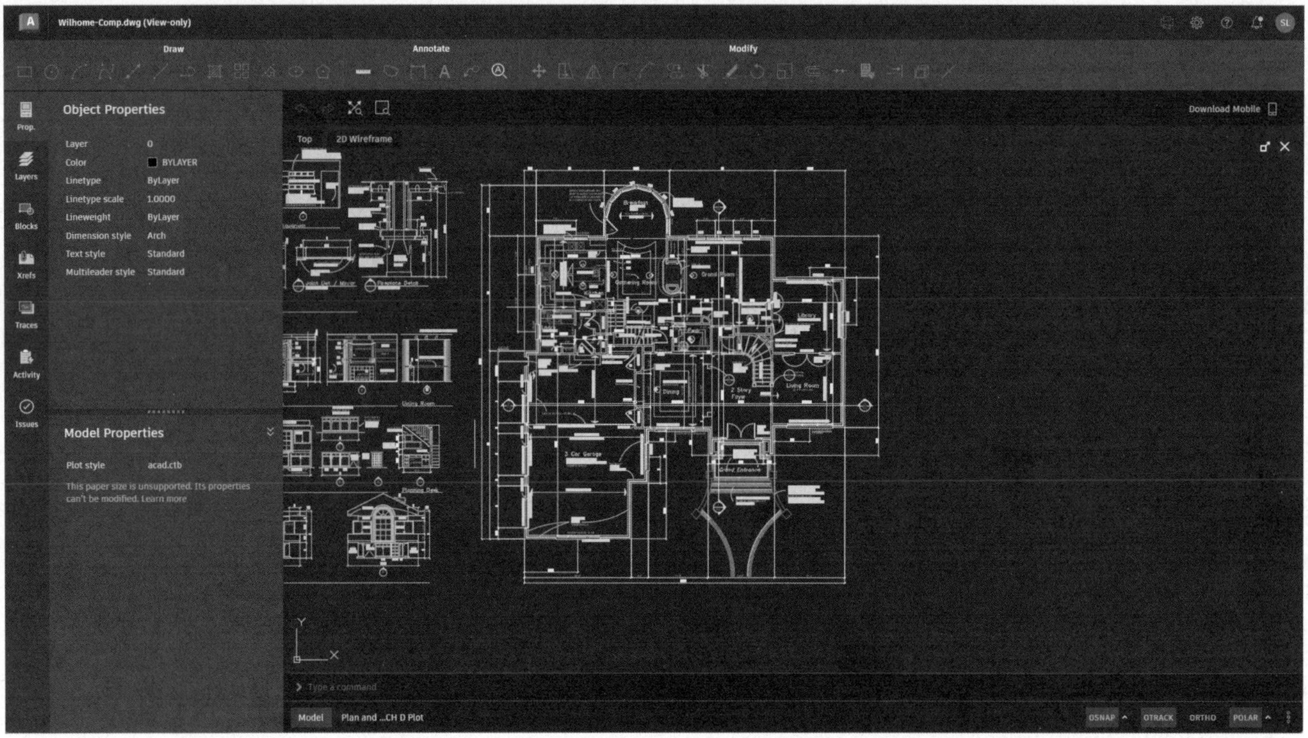


## Marking Up a Shared Drawing

Anyone you send the link for the shared drawing will be able to open the link in their browser window, where it will appear in the Autodesk web app. If you chose *Edit and save a copy* as the permission option, the recipients can now add comments and changes to a copy of the drawing they can share with other members of the design build team (Figure 23-34).

## Forma Data Management

Forma Data Management allows non-AutoCAD users to mark up and provide feedback on drawings via PDFs. The PDF markups can be brought directly in (synchronized) to the DWG file. Forma Data Management also allows project members to share Connected Support Files to locate fonts, plot styles, drawing templates, and so on without added time configuring path locations. With Forma Data Management, External References automatically attempt to locate missing information that may have been renamed or moved. Additionally a user can *checkout* a drawing to edit and propose changes to a portion of a drawing while someone else has the drawing open.

NOTE: You must have a subscription and be logged in to use Forma Data Management.

### *Push to Forma Data Management*

| Ribbon | Menu Bar | Command (Type) | Alias (Type) | Shortcut |
|---|---|---|---|---|
| *Collaborate Push to Forma Data Management* | ... | *pushtoformaopen* | ... | ... |

This command opens the Push to Forma Data Management palette where you can select layouts to upload as PDFs so another person can view, comment on, and mark up the drawing information.

## Making Markups in Autodesk Design Review

.DWF files and sheet sets can be marked up using the Autodesk Design Review. Use the Open file icon from the upper left of Design Review and select your file using the typical file selector. Sheet sets with multiple items will show similar to Figure 23-35 with thumbnails.

Click the thumbnail of the sheet to mark up. Design Review uses ribbon tabs for various tools. Choose the Markup & Measure tab and then use the tools, such as Callouts, Freehand Highlighter, Polycloud and others to add your annotations.

Click the Save icon from the Design Review Quick Access Toolbar to store your annotations.

**FIGURE 23-35**

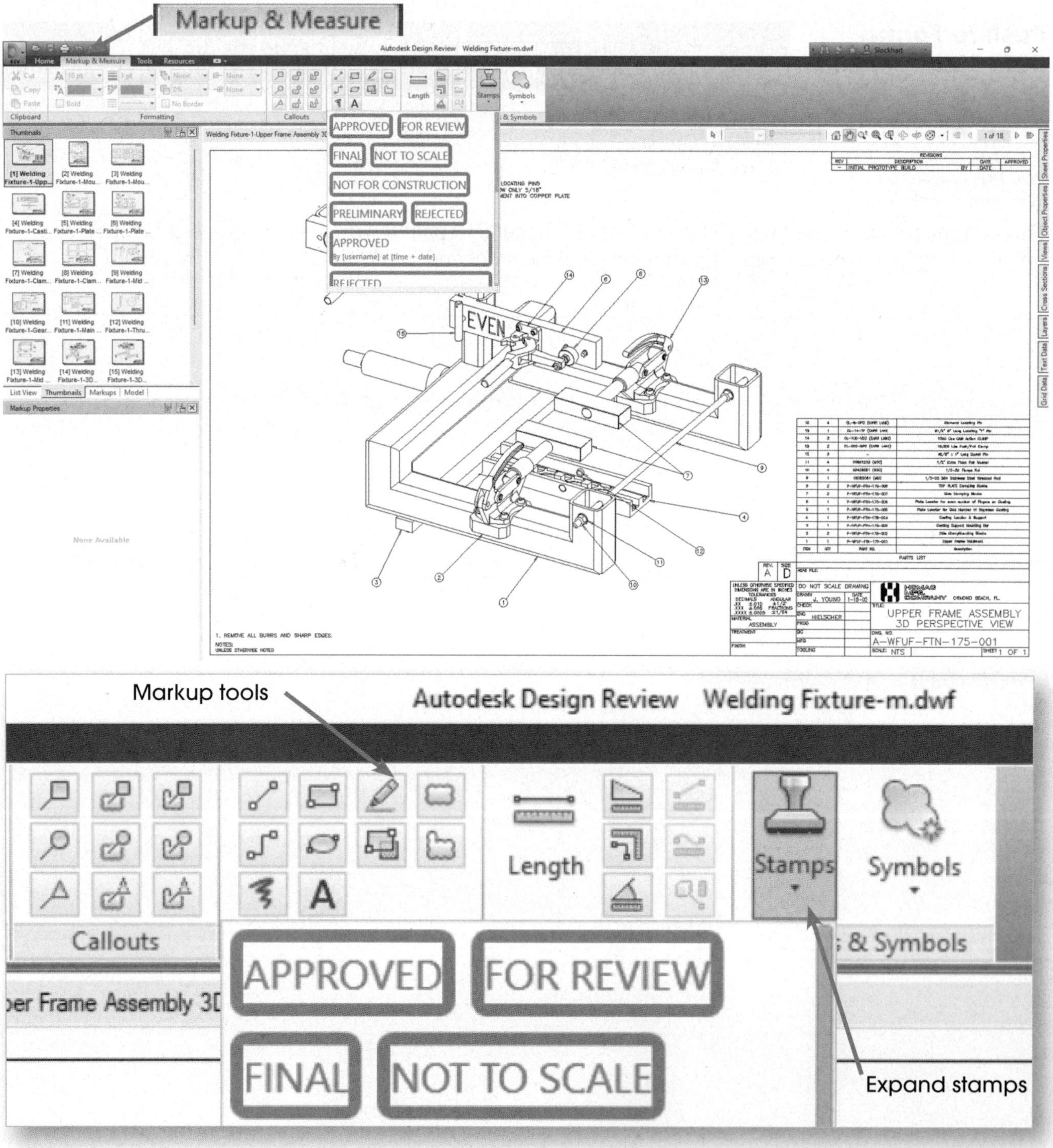

## Markups from Autodesk Design Review

When markups have been made in Autodesk Design Review, they can be viewed in AutoCAD, and the changes can be made to the corresponding .DWG files. The *Markup Set Manager* in AutoCAD facilitates this process. You can open a markup set using the *Opendwfmarkup* command or by opening through the *Markup Set Manager*.

### *Opendwfmarkup*

| Ribbon | Menu Bar | Command (Type) | Alias (Type) | Shortcut |
|---|---|---|---|---|
| ... | *File*<br>*Load Markup Set...* | *Opendwf-markup* | ... | ... |

The *Opendwfmarkup* command simply produces a standard file navigation dialog box. Use it to locate and open any DWF markup set (.DWF) file. Opening the .DWF file automatically invokes the *Markup Set Manager* displaying the contents of the opened markup set.

### *Markup*

| Ribbon | Menu Bar | Command (Type) | Alias (Type) | Shortcut |
|---|---|---|---|---|
| *View*<br>*Palettes*<br>(icon) | *Tools*<br>*Palettes >*<br>*Markup Set*<br>*Manager* | *Markup* | *MSM* | Ctrl+7 |

Use any of the options shown in the previous command table (or use the *Markup* command) to produce the *Markup Set Manager*. The *Markup Set Manager* appears in a palette format (Fig. 23-36). Use the drop-down list options at the top of the *Markup Set Manager* to open a markup set (.DWF file). Doing so displays the markup set list in the *Markups* area of the manager. Each sheet is shown in a hierarchical fashion under the DWF set name. Any markups created for the sheets are displayed by expanding each sheet in the list (clicking the plus symbol next to the sheet name). The *Details* section below gives information about each markup. The *Preview* pane (not shown) displays only the entire sheet, not the marked up area.

FIGURE 23-36

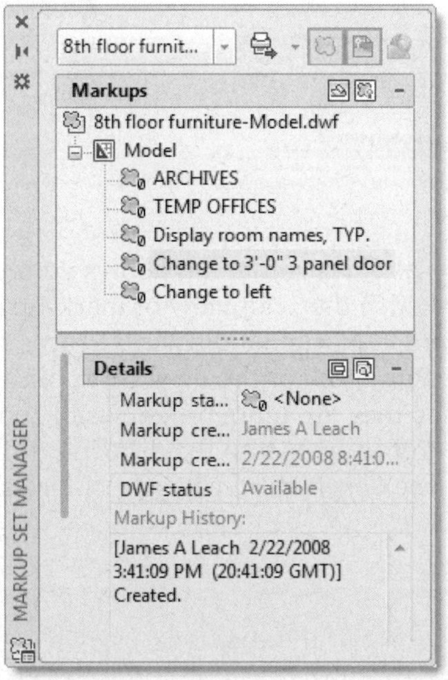

FIGURE 23-37

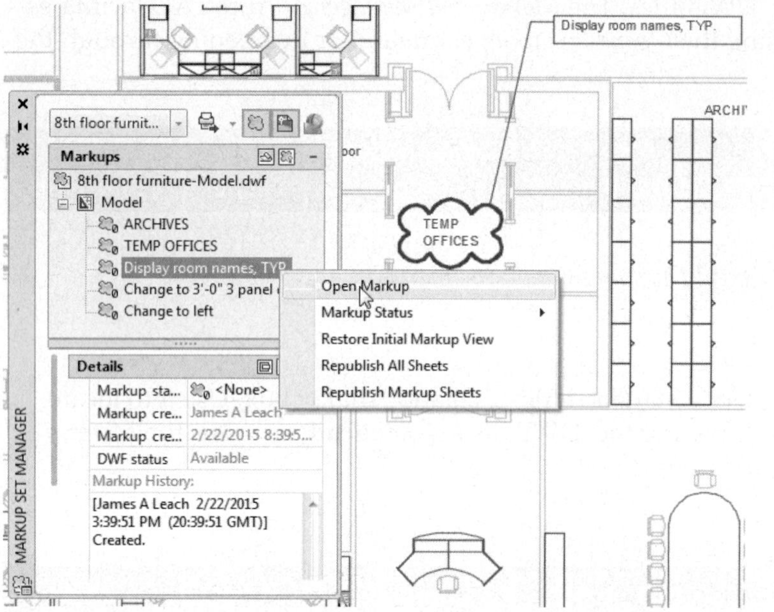


You can view the individual markups by double-clicking on the markup name in the list (Fig. 23-37). This is an extremely useful feature. This action opens the indicated .DWF file and automatically zooms to the area of the markup. Additionally, the corresponding .DWG file is opened. You can toggle between viewing the .DWF and the .DWG files using the two buttons in the upper-right corner of the manager titled *View DWG Geometry (Alt+3)* and *View DWF Geometry (Alt+4)*. In this way, you can see the section of the .DWF marked for change, then toggle to the drawing to actually make the change. You can also use the *View Redline Geometry (Alt+2)* button to toggle the display of the markup in both the .DWF and in the actual corresponding drawing file.

When changes to the original drawing files are made, use the *Markup Sheet Manager* to republish the .DWF files. You can choose to publish only the drawings that were changed or the entire set.

## *Trace*

| Ribbon | Menu Bar | Command (Type) |
|---|---|---|
| *Collaborate Traces Palette* | | *Trace* |

The *Trace* command allows your collaborators to view and comment in a safe overlay of the drawing which they can view or mark-up in the AutoCAD web, desktop, and mobile apps, without risk of changing the original drawing. Traces can only be created in the web and mobile apps, but can be marked up with those or the desktop application. The *Fademarkup* command allows you to fade and unfade markups so they are more or less visible on a drawing. The *Copyfromtrace* command allows you to copy objects from a trace into the drawing. With the trace open and *Traceback* on so the parent drawing can be edited, use *Copyfromtrace* to select objects in the trace, and press Enter to add the trace objects to the drawing.

## DRAWING COMPARE

Small but important changes may exist between drawing revisions. While good file management and drawing check out tools are the best practice, it can be a challenge to keep track of every item, especially in a shared drawing environment. The DWG Compare tool in AutoCAD helps you identify graphical differences between drawings. In a complex drawing, this can be very difficult to do by eye. Missing a small but important revision can be critical.

### *Compare*

| Ribbon | Menu Bar | Command (Type) | Alias (Type) | Shortcut |
|---|---|---|---|---|
| *Collaborate DWG Compare* | *Application (Drawing Utilies) DWG Compare* | *_Compare* | ... | ... |

To use *DWG Compare*, from your current drawing, click the icon from the ribbon Collaborate panel. The file selection dialog box as shown in Figure 23-38 appears. You do not have to have both drawings to compare open. Use the ellipsis button at the right of the file name input box to open the file browser and select any drawing file. The differences between the drawings are highlighted.

**FIGURE 23-38**

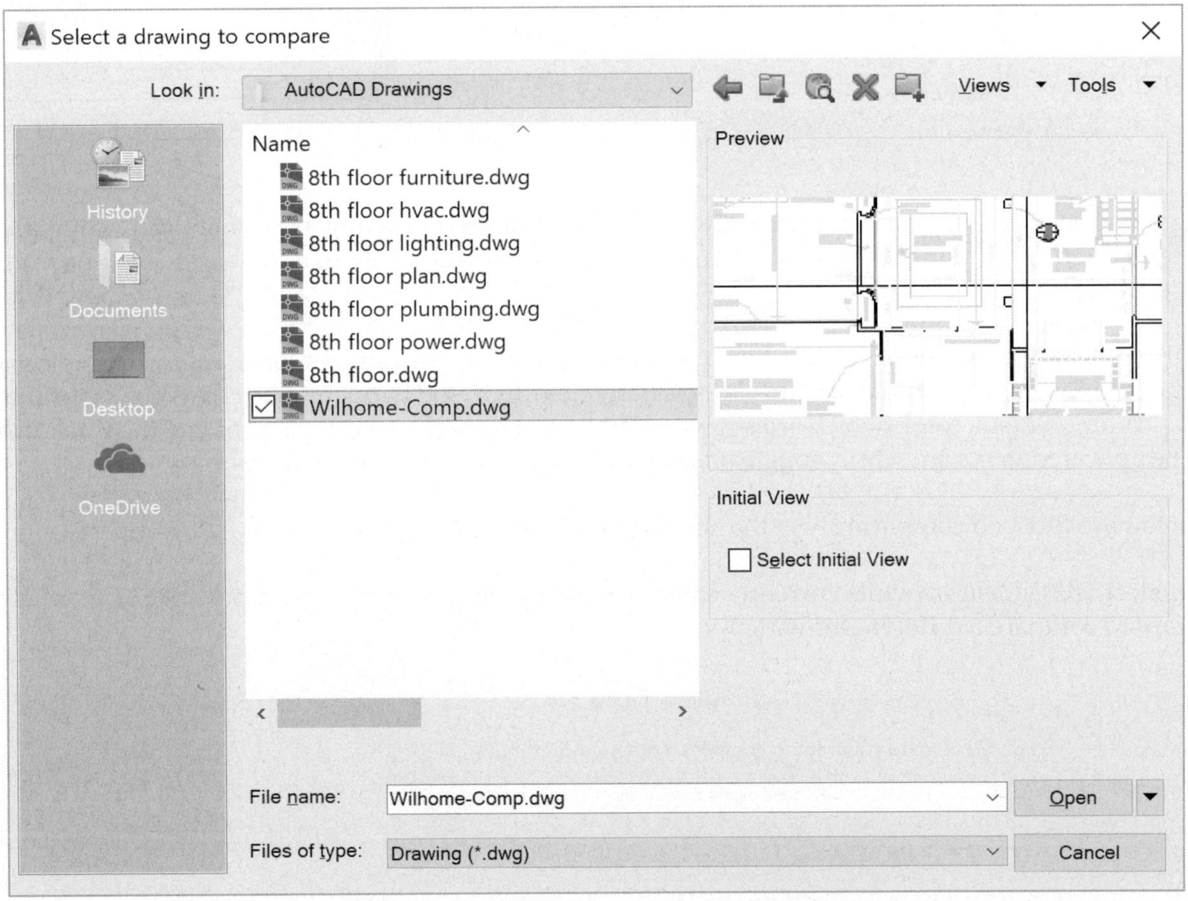

**FIGURE 23-39**

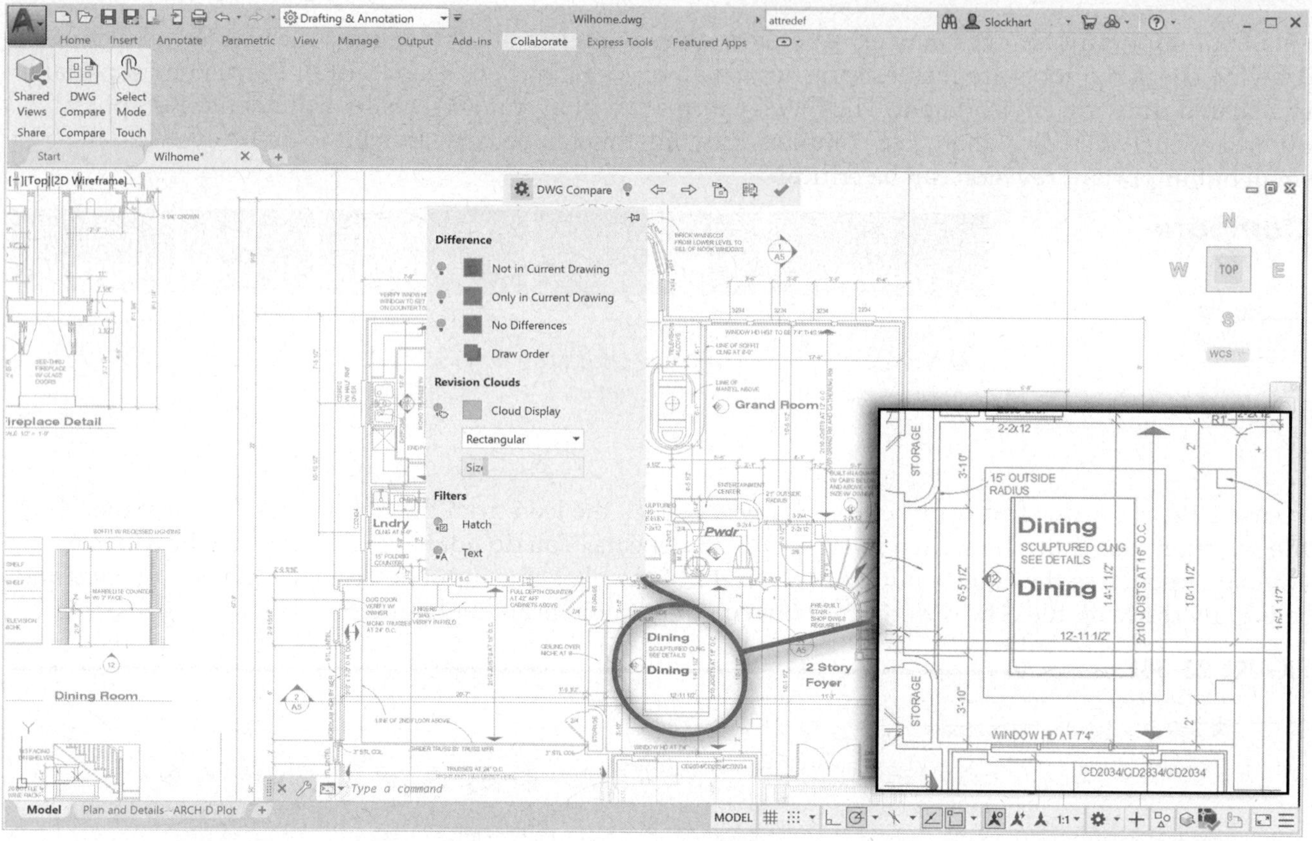


Changes you make in either the current drawing or the compared drawing are dynamically compared and highlighted. The options and controls for drawing comparison show in a toolbar docked near the top of the drawing area (Figure 23-39).

Click the Settings control (which looks like a small gear) to show the options for changing the colors for how the comparisons are displayed (this is especially helpful if you happen to be red-green color blind) and the common items color (which are shown in gray by default). You can also change the draw order for the duplicated items and easily toggle on and off hatching and revision clouds.

When you are finished comparing, use the check icon at the right of the toolbar to exit comparison mode.

If you select *DWG Compare* while currently comparing a drawing, you may see a message as to whether you want to compare a different drawing.

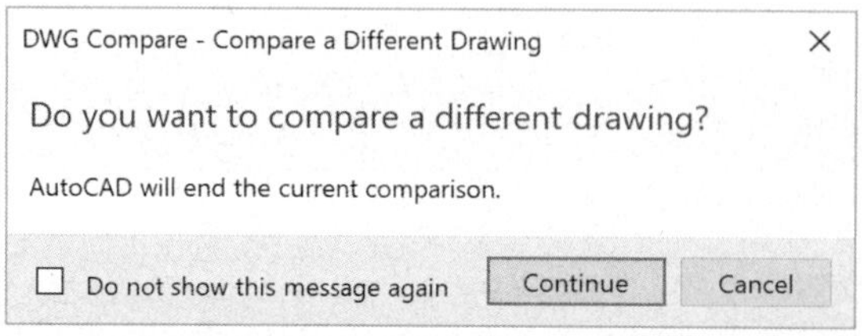

## ACTIVITY INSIGHTS

Another useful tool for tracking drawing revisions is Activity Insights. This tool can be used with files located on Autodesk Docs, Google Drive, OneDrive, Box, or Dropbox. When saving and working on drawings from the cloud along with desktop versions, it can be especially important to track changes easily.

***ACTIVITYINSIGHTSOPEN***

| Ribbon | Menu Bar | Command (Type) | Alias (Type) | Shortcut |
|---|---|---|---|---|
| *View*<br>*History panel*<br>*Activity Insights* | | *Activityinsightsopen* | ... | ... |

To use *Activity Insights*, from your current drawing which was opened from OneDrive or a similar supported service, click the icon from the ribbon View panel. Use the dialog box (Fig 23-40) that appears to filter the versions by date, user, or minimum time increment.
For example, click the small filter icon at the right of the dialog box and select Custom, use the calendar that appears to select a range of two or more dates. Highlight to compare with the current drawing and click Compare.

If you don't see any activities, it can be that your events are being written to a local location such as C:\Users\{username}\AppData\Local\Autodesk\ActivityInsights\Common
instead of a shared location so that drawing activity is stored regardless of who works on a drawing. To set the location, click Application button, Options and use the Files tab, selection to the left of (+) Activity Insights Event Location to the desired file path.

**FIGURE 23-40**

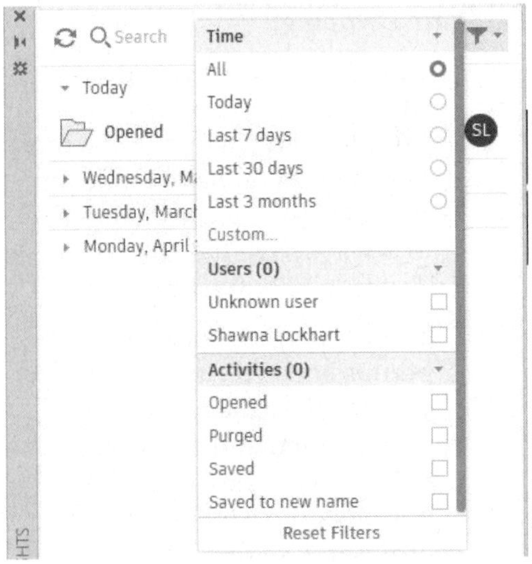

## CHAPTER EXERCISES

1. ***Browser, Inetlocation,* Autodesk Design Review**

   A. Start AutoCAD if not already running. Type in the ***Browser*** command. Note that the default URL is http://www.autodesk.com. Press **Enter** to display your system's default browser and web site.

   B. On the Autodesk site, scroll down to **All Products**, then click to follow the link. On the **Products** page, locate **Design Review**. On the **DWF Viewers** page, select **Learn More About Autodesk Design Review**. If Autodesk Design Review is not installed on your computer, select the **Download Now** button and follow the instructions.

   C. Once the Autodesk Design Review is installed, start AutoCAD again if not already started. Enter ***Inetlocation*** at the command prompt. Change the default location to your favorite news source such as **www.usatoday.com**. Next, use ***Browser*** again, press **Enter** to accept the new URL, and validate that the new site appears in your Internet browser.

2. ***Share Drawing***

   A. Start AutoCAD if not already running. Open one of your completed drawings. ***Save*** the drawing if you have made any changes. (It is recommended to always save the current drawing before using *Share*.)

   B. Invoke ***Share***. Choose the permissions *Edit and save a copy* to allow the result to be downloaded and saved via the link. Use the Copy Link button and paste the link into a browser. Use the editing tools to add a Revision cloud to your drawing. Save a copy of the drawing adding -REV1 to the file name.

**FIGURE 23-41**

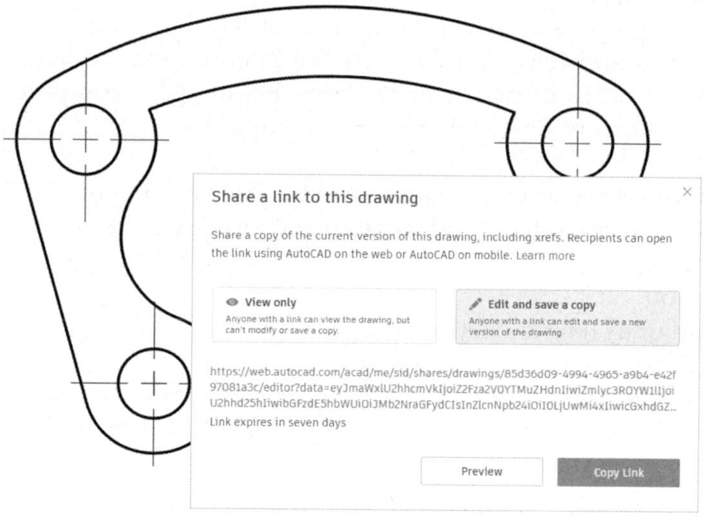


3. ***eTransmit***

   A. Start AutoCAD if not already running. ***Open*** a drawing that you want to transmit to a friend. If you make any changes, ***Save*** the current drawing.

   B. Invoke ***eTransmit***. The *Create Transmittal* dialog box should appear. In the notes section, enter "Here is the drawing that I would like for you to review," or similar message. Select *Transmittal Setups,* then *Modify* the *Standard* setup. Specify a ***Self-extracting executable (*exe)*** as the *Transmittal package type*. Use the ***Browse*** button to specify your working directory as the *Transmittal file folder*. Select *Place all files in one folder*, then clear the other options below. Do not select *OK* yet.

   C. If you have an email system available on the system, select ***Send e-mail with transmittal***. If not, clear the check for this option. Select ***OK***.

D. Depending on your selection in step C, your email system may open with the .EXE file as an attachment. If you chose not to have AutoCAD open your email program, copy the .EXE file to disk, and attach it to an email message at a later time. In either case, send an email to a friend and ask him or her to extract the file and view the drawing in AutoCAD. Request a confirmation and a reply email reporting on the success of the transmittal.

4. **Use *ePlot* to create .DWF files**

A. In AutoCAD, ***Open*** the **T-PLATE** drawing from Chapter 9 Exercises. Activate the ***Model*** tab. Use the ***Plot*** dialog box to select a ***DWF ePlot.pc3*** device. Select ***OK*** to make the .DWF file. Accept the default name (T-PLATE-MODEL) and ensure your working directory is specified as the destination for the file.

B. Start Autodesk Design Review. Use the ***File*** pull-down menu, then the ***Open*** command, and locate the **T-PLATE-MODEL.DWF** to verify the .DWF was created correctly.

C. In AutoCAD, ***Open*** the **GASKETA** drawing from Chapter 9 Exercises. Follow step 1. to make a **.DWF** file. Use Autodesk Design Review to view it.

D. Create another **.DWF** file similar to the previous two, but this time activate a ***Layout*** tab that was set up previously with a viewport. If no layouts were set up, create one viewport but make sure the viewport border is slightly smaller than the printable area (dashed line), then make the .DWF file of the layout.

E. View the new .DWF image with Autodesk Design Review to verify that it was created correctly. The image should appear similar to the Model tab .DWFs, except the viewport border should be visible.

5. **Use *Publish* to create a .DWF file**

A. In AutoCAD, use ***Publish***. When the *Publish* dialog box appears, select ***Add Sheets***. Select four drawings of your choice (for example, **WRENCH**, **GASKETA**, **HAMMER**, and **T-PLATE**).

In the *Sheets to publish* list, specify the layouts you want to publish. Select ***DWF file*** in the ***Publish to*** section below. Invoke the ***Publish Options*** and specify a ***Multi-sheet DWF***, then enter the desired ***DWF file name*** and location, such as **Exercise Drawings**.

B. Select ***Publish*** in the *Publish* dialog box. When publishing is complete, open Autodesk Design Review and view the multi-sheet .DWF file.

C. Your new multi-sheet .DWF file should appear in Autodesk Design Review. Use the *Thumbnail* or *List* area on the left of the viewer to inspect each of the drawings, similar to that in Figure 23-42. When you are finished, close the viewer.

6. **Insert *Hyperlinks***

   A. In AutoCAD, ***Open*** the **GASKETA** drawing. Activate the ***Model*** tab. Use ***SaveAs*** to save the drawing as **GASKETA2**. Close the drawing.

   B. Open the **T-PLATE** drawing. Activate the Model tab. Use SaveAs to save the drawing as **T-PLATE2**. ***Close*** the drawing.

   C. ***Open*** the **GASKET2** drawing. Use ***Text*** to place text at the bottom of the drawing. Enter "**Click here to view the T-PLATE2 drawing.**" The text should appear similar to that shown in Figure 23-43. Create a ***Hyperlink*** for the entire line of text. Associate the **T-PLATE2.DWG** as the file to link. ***Save*** and ***Close*** the drawing.

**FIGURE 23-42**

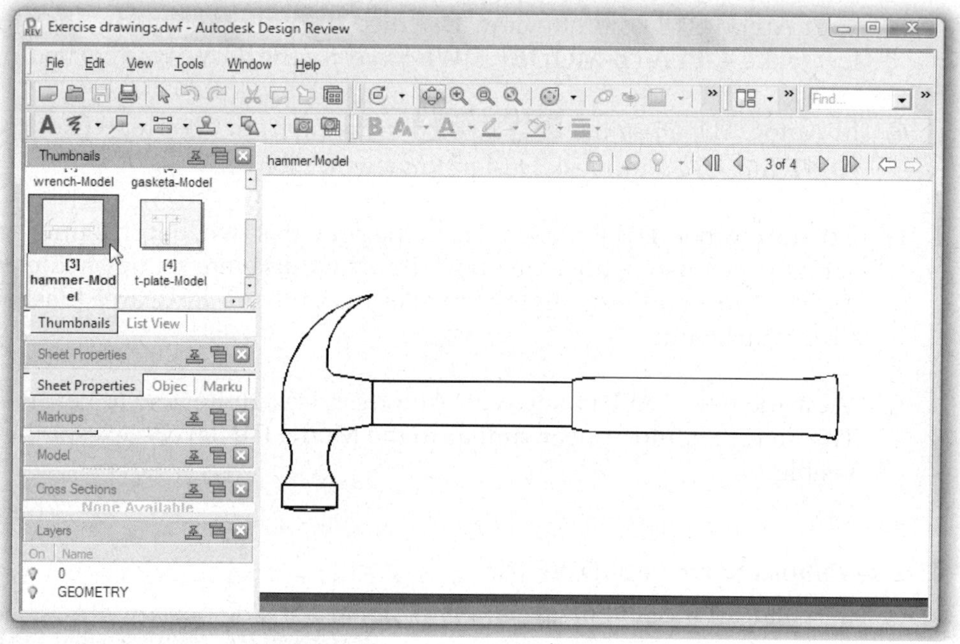


   D. ***Open*** the **T-PLATE2** drawing. Use ***Text*** to place text at the bottom of the drawing, stating "**Click here to view the GASKETA2 drawing**." Create a ***Hyperlink*** for the text that links to the **GASKETA2.DWG**. Save and ***Close*** the drawing.

   E. ***Open*** either drawing. Hover over the text until the link information appears, then hold down the Ctrl key and left-click to follow the link. Do the same in both drawings to "toggle" between the two drawings.

**FIGURE 23-43**

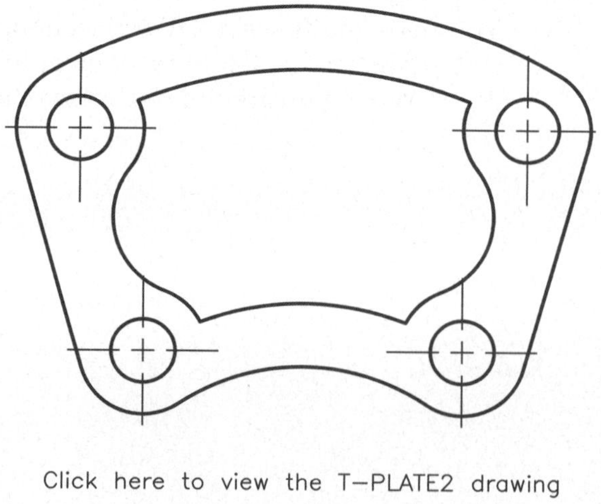


7. **Create .DWF files with *Hyperlinks***

   A. ***Open*** the **GASKETA2** drawing. Activate the ***Model*** tab. Select the hyperlink, right-click to produce the shortcut menu, and select ***Edit Hyperlink***. Edit the hyperlink to link to a .DWF, not the .DWG; for example, change C:\DWG\GASKETA2.DWG to **C:\DWG\GASKETA2.DWF.** Open the ***Plot*** dialog box to create a **.DWF** file. Use the ***DWF6 ePlot.pc3*** device. Make sure you specify **GASKETA2** for the name of the .DWF file, not GASKETA2-Model. Select ***OK*** to produce the .DWF file. ***Save*** and ***Close*** the drawing.

B. ***Open*** the **T-PLATE2** drawing. Follow the same procedure as in step A to create a .DWF file for the T-PLATE2 drawing. Make sure you convert the hyperlink extension to .DWF and save the .DWG with the correct name. ***Save*** and ***Close*** the drawing.

C. Use Autodesk Design Review to view the .DWF images and test the hyperlinks. Make any necessary corrections.

8. **Create a *Shared Drawing***

A. In AutoCAD, ***Open*** the **GASKETA** drawing. Activate the ***Model*** tab.

Use ***SaveAs*** to save the drawing as **GASKETA3**.
Use ***Share Drawing*** to publish the drawing on-line. Give permission to *Edit and save a copy*.

B. Open the shared drawing from step A. Add a comment.

C. Show the shared drawing of **GASKETA3** in the Autodesk web app. Use the markup tools to add a revision cloud around both of the lower circles. Add a note stating: "Change diameter to .44."

9. **Use *Drawing Compare***

A. In AutoCAD, ***Open*** the **b_samp** drawing. Activate the ***Model*** tab. Use Compare from the Collaborate tab to find the differences between the **Floor Plan Sample** drawing and b_sample.

B. Add a revision cloud around the lower rooms. Save the compared file.

**FIGURE 23-44**

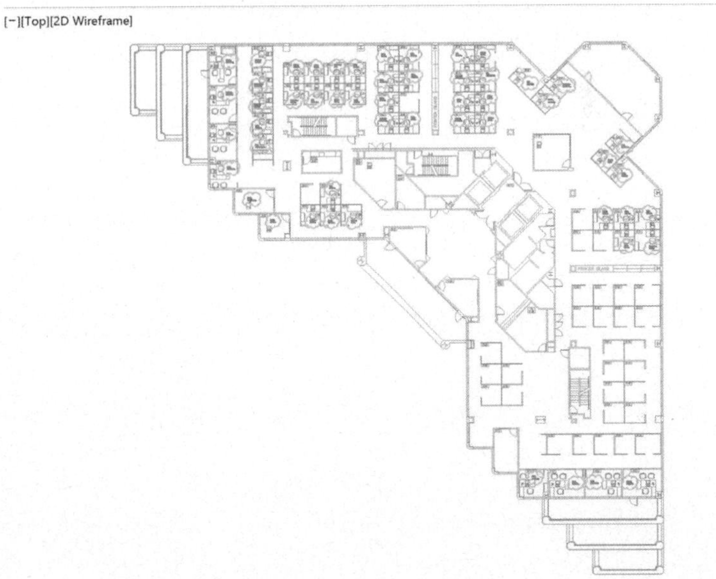

CHAPTER 24

# Multiview Drawing

## Chapter Objectives

After completing this chapter you should:

1. be able to draw projection lines using *Ortho* and *Snap*, and *Polar Tracking*;

2. be able to use *Xline, Ray,* and *Offset* to create construction lines and views;

3. be able to use *Object Snap Tracking* for alignment of lines and views;

4. be able to use construction layers for managing construction lines and notes;

5. be able to use linetypes, lineweights, and layers to draw and manage ANSI standards;

6. be able to create associative center lines using the *Centermark* and *Centerline* commands and to adjust the appearance and properties of existing centerlines;

7. know how to create fillets, rounds, and runouts;

8. know the typical guidelines for creating a three-view multiview drawing.

# CONCEPTS

Multiview drawings are used to represent 3D objects on 2D media. The standards and conventions related to multiview drawings have been developed over years of using and optimizing a system of representing real objects on paper. Now that our technology has developed to a point that we can create 3D models, some of the methods we use to generate multiview drawings have changed, but the standards and conventions have been retained so that we can continue to have a universally understood method of communication.

This chapter illustrates methods of creating 2D multiview drawings with AutoCAD (without a 3D model) while complying with industry standards. Many techniques can be used to construct multiview drawings with AutoCAD because of its versatility. The methods shown in this chapter are the common methods derived from traditional manual techniques. Many other methods are possible, including using object snap tracking, and creating 2D drawings from 3D models.

# PROJECTION AND ALIGNMENT OF VIEWS

Projection theory and the conventions of multiview drawing dictate that the views be aligned with each other and oriented in a particular relationship. AutoCAD has features, such as *Snap, Ortho,* construction lines (*Xline, Ray*), *Object Snap, Polar Tracking* and *Object Snap Tracking*, that can be used effectively for projecting and aligning views.

## Using *Ortho* and *Object Snap* to Draw Projection Lines

*Ortho* (F8) can be used with *Object Snap* to draw projection lines to construct multiview drawings. For example, drawing a Line interactively with *Ortho ON* forces the *Line* to be drawn in either a horizontal or vertical direction.

Figure 24-1 illustrates this feature while drawing projection *Lines* from the top view over to a 45 degree miter line (intended for transfer of dimensions to the side view). The "first point:" of the *Line* originated from the *Endpoint* of the *Line* on the top view.

FIGURE 24-1

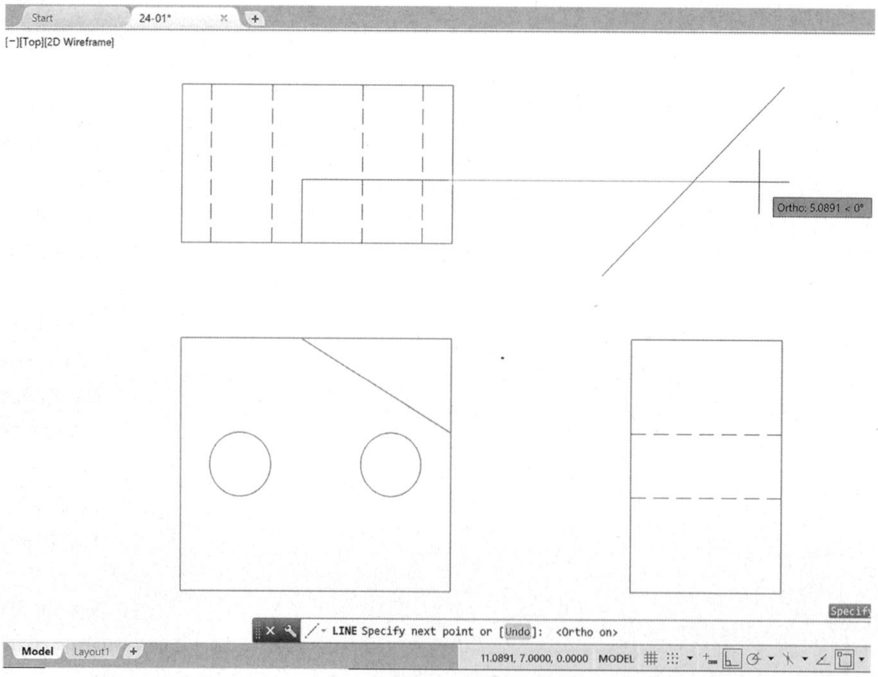

Figure 24-2 illustrates the next step. The vertical projection *Line* is drawn from the *Intersection* of the 45 degree line and the last projection line. *Ortho* forces the *Line* to the correct vertical alignment with the side view.

FIGURE 24-2

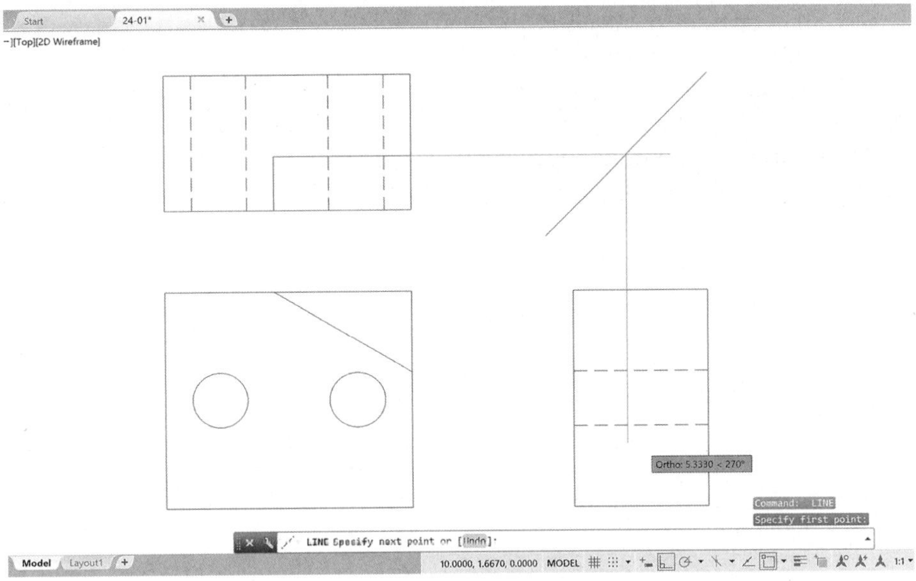


Remember that any draw or edit command that requires PICKing is a candidate for *Ortho* and/or *Object Snap*.

NOTE: *Object Snap* overrides *Ortho*. If *Ortho* is *ON* and you are using an *Object Snap* mode to PICK the "next point:" of a *Line*, the *Object Snap* mode has priority; therefore, the construction may not result in an orthogonal *Line*, unless you make sure to select away from any object.

## Using *Polar Tracking* to Draw Projection Lines

Similar to using *Ortho* and *Object Snap*, you can use *Polar Tracking* and *Object Snap* options to draw projection lines at 90-degree increments. Using the previous example, *Polar Tracking* is used to project dimensions between the top and side views through the 45-degree miter line.

To use *Polar Tracking*, set the desired *Polar Angle Settings* in the *Polar Tracking* tab of the *Drafting Settings* dialog box. Ensure the *Polar* icon appears blue on the Status Bar. You can also set a *Polar Snap* increment or a *Grid Snap* increment to use with *Polar Tracking* (see Chapter 3 for more information on these settings).

FIGURE 24-3

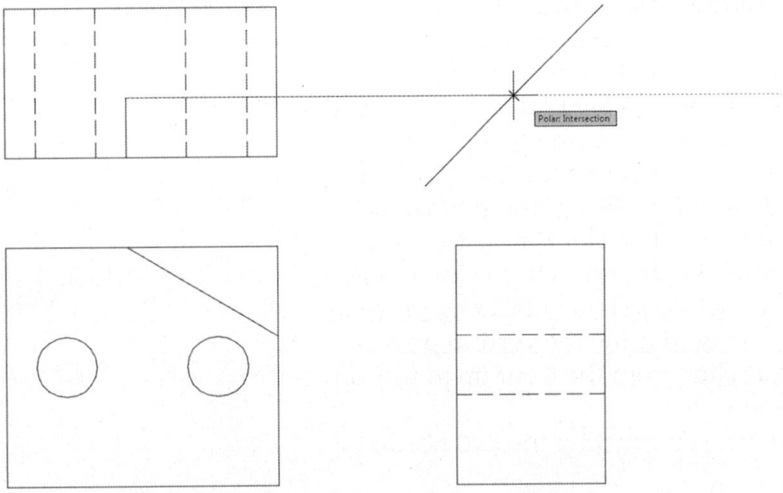


For example, you could use the *Endpoint Object Snap* option to snap to the *Line* in the top view, then *Polar Tracking* forces the line to a previously set Polar increment (0 degrees in this case).

*Object Snaps* work well in conjunction with *Polar Tracking*, as shown in Figure 24-3, such that the current horizontal line *Object Snaps* to its *Intersection* with the 45-degree miter line. (*Object Snaps* override *Ortho*, so you must keep this in mind when using them.)

In the following step, use *Intersection Object Snap* option to snap to the intersection of the previous *Line* and the 45-degree miter line (Fig. 24-4). Again *Polar Tracking* forces the *Line* to a vertical position (270 degrees).

FIGURE 24-4

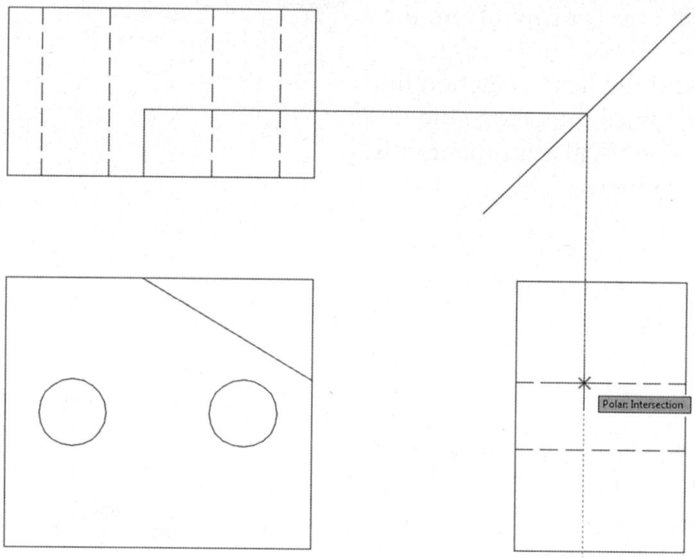


## Using *Object Snap Tracking* to Draw Projection Lines Aligned with *Object Snap* Points

This feature, although the most complex, provides the greatest amount of assistance in constructing multiview drawings. The advantage is the availability of the features described in the previous method (*Polar Tracking* and *Object Snap* for alignment of vertical and horizontal *Lines*) in addition to the creation of *Lines* and other objects that align with *Object Snap* points (*Endpoint*, *Intersection*, *Midpoint*, etc.) of other objects.

To use *Object Snap Tracking*, first set the desired running *Object Snap* options (*Endpoint*, *Intersection*, etc.). Next, toggle on *Object Snap Tracking* at the Status Bar. Use *Polar Tracking* in conjunction with *Object Snap Tracking* by setting polar angle increments (see previous discussion) and toggling on *Polar* on the Status Bar.

For example, to begin the *Line* in Figure 24-5, first "acquire" the *Endpoint* of the *Line* shown in the front view. Note that the "first point" of the new *Line* (in the top view) aligns vertically with the acquired *Endpoint*. At this point in time, the short vertical *Line* for the top view could be drawn from the intersection shown. *Object Snap Tracking* prevents having to draw the vertical projection line from the front up to the top view.

TIP

After constructing the two needed

FIGURE 24-5

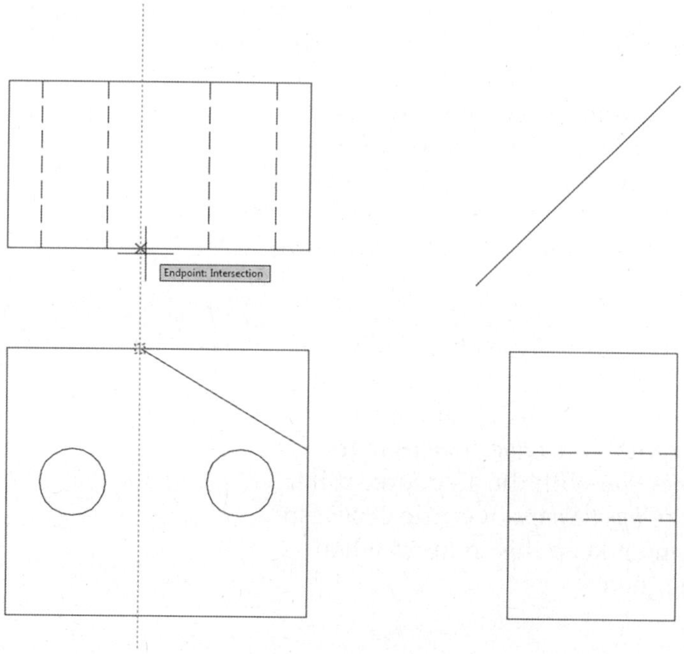

*Lines* for the top view, a tracking vector aligns with the acquired *Endpoint* of the indicated line in the top view and the *Intersection* of the 45-degree miter line (Fig. 24-6). Note that a horizontal projection line is not needed since *Object Snap Tracking* ensures the "first point" aligns with the appropriate point from the top view.

**FIGURE 24-6**

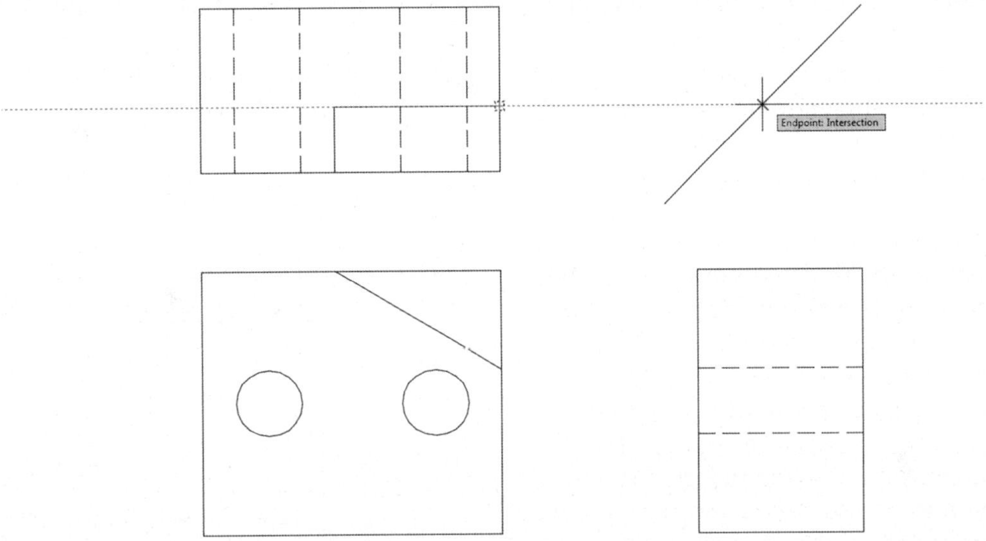


The next step is to draw the short horizontal visible line in the side view (Fig. 24-7). Coming from the "first point" (on the 45-degree miter line found in the previous step), draw the vertical projection line down to the appropriate point in the side view that "tracks" with the related *Endpoint* in the front view (see horizontal tracking vector). Since this *Line* is constructed to the correct endpoint, *Trimming* is unnecessary here, but it would be needed in the previous two methods.

**FIGURE 24-7**

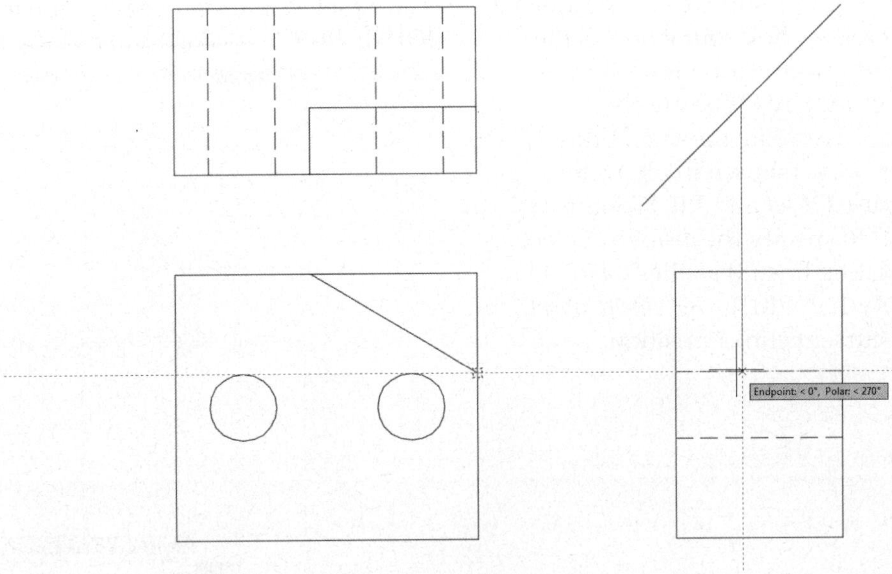


These AutoCAD features (*Object Snap Tracking* in conjunction with *Polar Tracking*) are very helpful features for construction of multiview drawing.

## Using *Xline* and *Ray* for Construction Lines

Another strategy for constructing multiview drawings is to make use of the AutoCAD construction line commands *Xline* and *Ray*.

*Xlines* can be created to "box in" the views and ensure proper alignment (Fig. 24-8). *Ray* is suited for creating the 45-degree miter line for projection between the top and side views. The advantage to using this method is that horizontal and vertical *Xlines* can be created quickly.

**FIGURE 24-8**

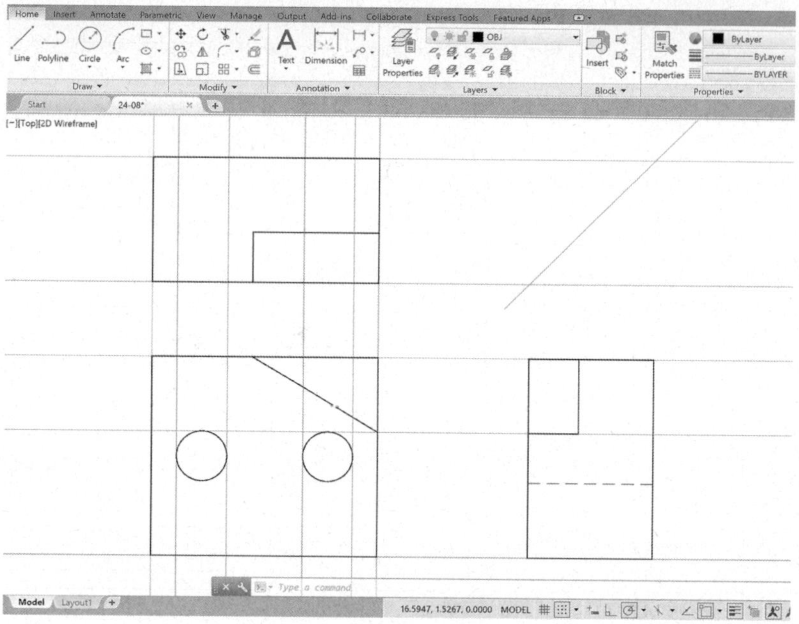


These lines can be *Trimmed* to become part of the finished view. Alternately, other lines could be drawn "on top of" the construction lines to create the final lines of the views. In this case, construction lines should be drawn on a separate layer so that the layer can be frozen before plotting. If you intend to *Trim* the construction lines so that they become part of the final geometry, draw them originally on the view layers.

**TIP**

## Using *Offset* for Construction of Views

An alternative to using the traditional miter line method for construction of a view by projection, the *Offset* command can be used to transfer distances from one view and to construct another. The *Distance* option of *Offset* provides this alternative.

For example, assume that the side view was completed and you need to construct a top view (Fig. 24-9). First, create a horizontal line as the inner edge of the top view (shown highlighted) by *Offset* or other method. To create the outer edge of the top view (shown in phantom linetype), use *Offset* and PICK points (1) and (2) to specify the *distance*. Select the existing line (3) as the "*Object to Offset:*", then PICK the "*side to offset*" at the current cursor position.

**FIGURE 24-9**

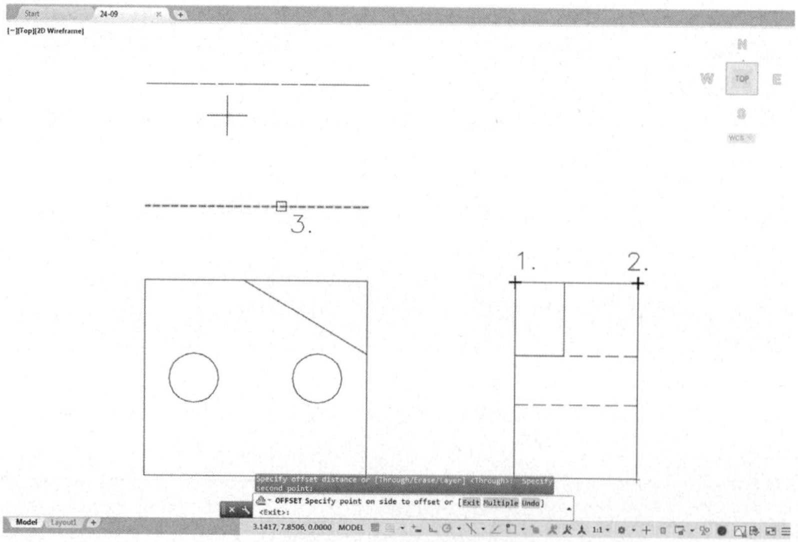

### Realignment of Views Using *Polar Snap* and *Polar Tracking*

Another application of *Polar Snap* and *Polar Tracking* is the use of *Move* to change the location of an entire view while retaining its orthogonal alignment.

For example, assume that the views of a multiview (Fig. 24-10) are complete and ready for dimensioning; however, there is not enough room between the front and side views. You can invoke the *Move* command, select the entire view with a window or other option, and "slide" the entire view outward. *Polar Tracking* ensures proper orthogonal alignment. *Polar Snap* forces the movement to a regular increment so the coordinate points of the geometry retain a relationship to the original points (for example, moving exactly 1 unit).

**FIGURE 24-10**

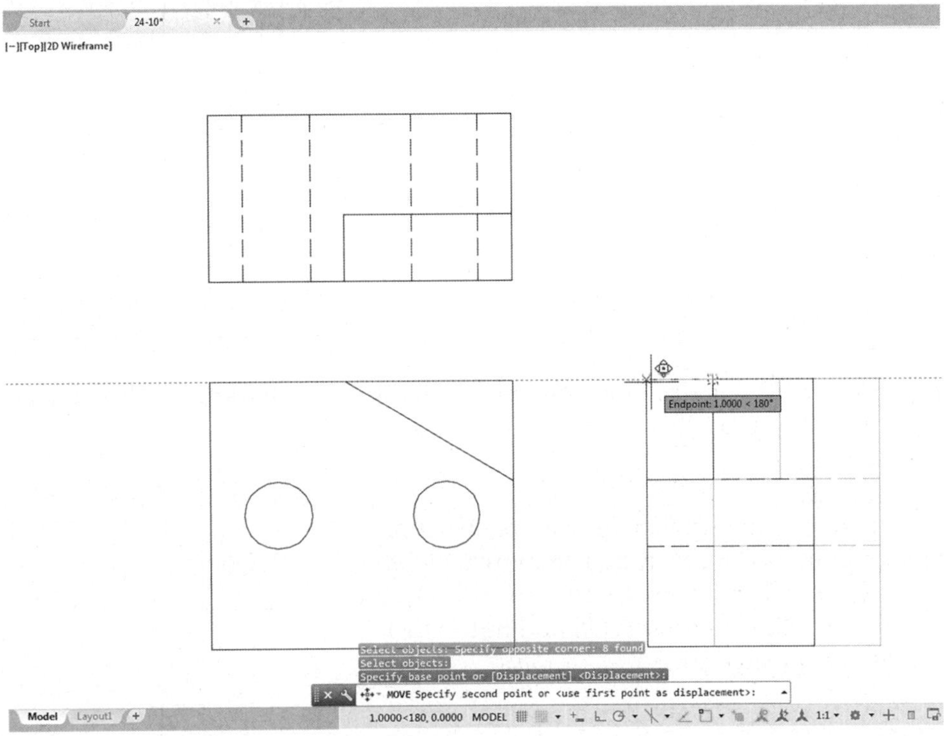


An alternative to moving a view using mouse input is using *Dynamic Input*. Turn *Dynamic Input* on, then key in the desired distance value to move in the distance edit box, and enter an angular value in the angle edit box. Another alternative is moving the view using Direct Distance Entry. With this method, indicate the direction to move with the rubberband line (assuming *Polar* is on), then key in the desired distance at the Command prompt.

## USING CONSTRUCTION LAYERS

The use of layers for isolating construction lines can make drawing and editing faster and easier. The creation of multiview drawings can involve construction lines, reference points, or notes that are not intended for the final plot. Rather than *Erasing* these construction lines, points, or notes before plotting, they can be created on a separate layer and turned *Off* or made *Frozen* before running the final plot. If design changes are required, as they often are, the construction layers can be turned *On*, rather than having to recreate the construction.

There are two strategies for creating construction objects on separate layers:

1. Use Layer 0 for construction lines, reference points, and notes. This method can be used for fast, simple drawings.
2. Create a new layer for construction lines, reference points, and notes. Use this method for more complex drawings or drawings involving use of *Blocks* on Layer 0.

For example, consider the drawing during construction in Figure 24-11. A separate layer has been created for the construction lines, notes, and reference points.

FIGURE 24-11

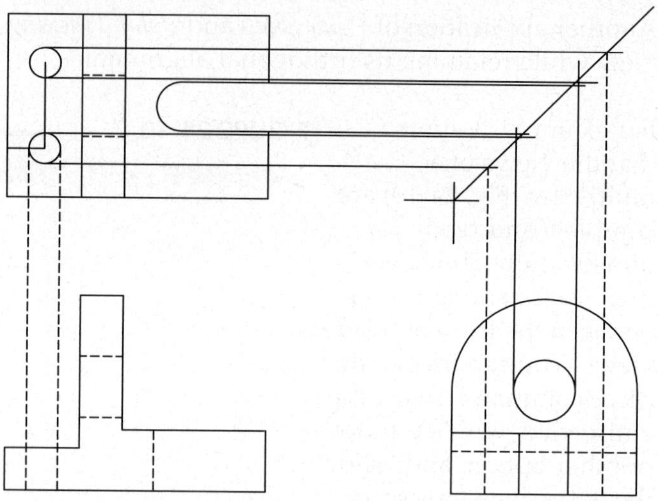

In Figure 24-12, the drawing is ready for making a print. Notice that the construction layer is *Frozen*.

FIGURE 24-12

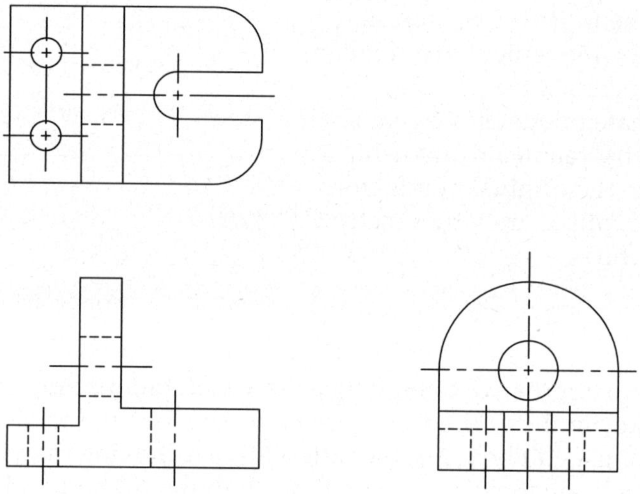

TIP

If you are printing the *Limits*, the construction layer should be *Frozen*, rather than turned *Off*, since *Zoom All* and *Zoom Extents* are affected by geometry on layers turned *Off* and not *Frozen* (unless the objects are *Xlines* or *Rays*).

## USING LINETYPES

Different types of lines are used to represent different features of a multiview drawing. Linetypes in AutoCAD are accessed by the *Linetype* command or through the *Layer Properties Manager*. *Linetypes* can be changed retroactively by the *Properties* palette. *Linetypes* can be assigned to individual objects specifically or to layers (*ByLayer*). See Chapter 11 for a full discussion on this topic.

AutoCAD complies with the ANSI and ISO standards for linetypes. The principal AutoCAD linetypes used in multiview drawings and the associated names are shown in Figure 24-13.

Many other linetypes are provided in AutoCAD. Refer to Chapter 11 for the full list and illustration of the linetypes.

FIGURE 24-13

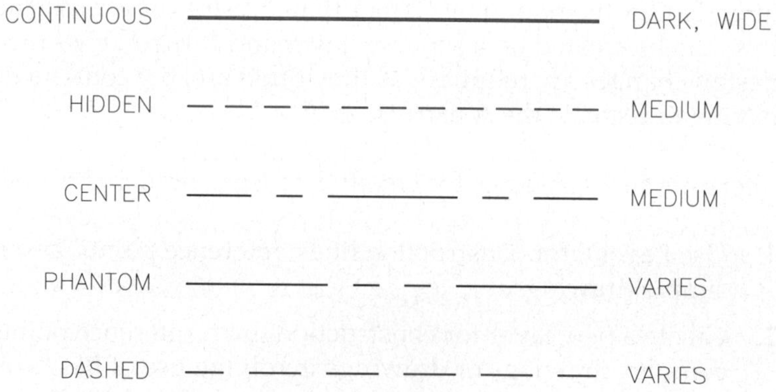

Other standard lines are created by AutoCAD automatically. For example, dimension lines can be automatically drawn when using dimensioning commands (Chapter 28), and section lines can be automatically drawn when using the *Hatch* command (Chapter 26). With AutoCAD 2017, you can create automatic "associative" center lines using the *Centerline* and *Centermark* commands (see "Associative Centerlines and Centermarks" in this chapter). This newer method of drawing center lines and center marks has many advantages over the older method of drawing the *Lines* "manually" and applying center linetypes as described in the next section "Drawing With Hidden and Center Linetypes."

2017

Objects in AutoCAD can have lineweight. This is accomplished by using the *Lineweight* command or by assigning *Lineweight* in the *Layer Properties Manager* (see Chapter 11). Additionally, lineweights can be assigned by using plot styles or assigning plot device lineweights or pen thickness.

## Drawing With Hidden and Center Linetypes

Figure 24-14 illustrates a typical application of AutoCAD *Hidden* and *Center* linetypes. Notice that the horizontal center line in the front view does not automatically locate the short dashes correctly, and the hidden lines in the right side view incorrectly intersect the center vertical line.

Although AutoCAD supplies ANSI standard linetypes, the application of those linetypes does not always follow ANSI standards. For example, you do not have control over the placement of the individual dashes of center lines and hidden lines. (You have control of only the endpoints of the lines and the *Ltscale*.) Therefore, if you draw the *Lines* "manually" (not using the *Centerline* or *Centermark* commands), the short dashes of center lines may not cross exactly at the circle centers, or the dashes of hidden lines may not always intersect as desired.

**FIGURE 24-14**

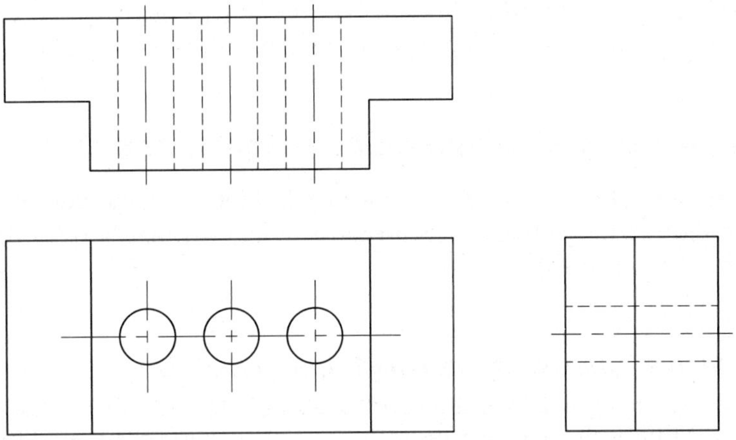

**FIGURE 24-15**

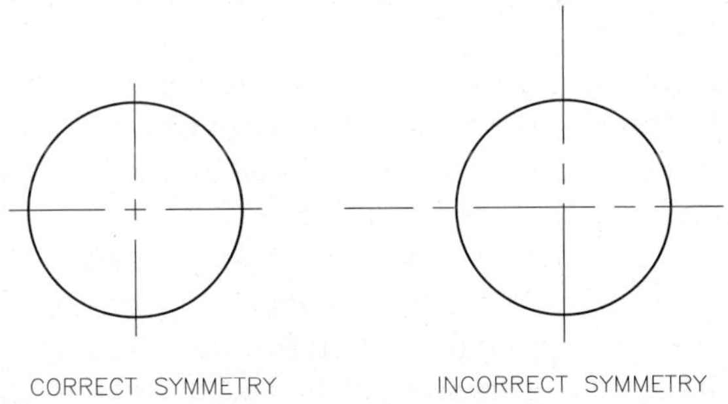


TIP

You do, however, have control of the endpoints of the lines. Draw *Lines* with the *Center* linetype such that the endpoints are symmetric about the circle or group of circles. This action assures that the short dash occurs at the center of the circle (if an odd number of dashes are generated). Figure 24-15 illustrates correct and incorrect technique.

You can also control the relative size of non-continuous linetypes with the *Ltscale* variable. In some cases, the variable can be adjusted to achieve the desired results.

For example, Figure 24-16 demonstrates the use of *Ltscale* to adjust the center line dashes to the correct spacing. Remember that *Ltscale* adjusts linetypes globally (all linetypes across the drawing).

**FIGURE 24-16**

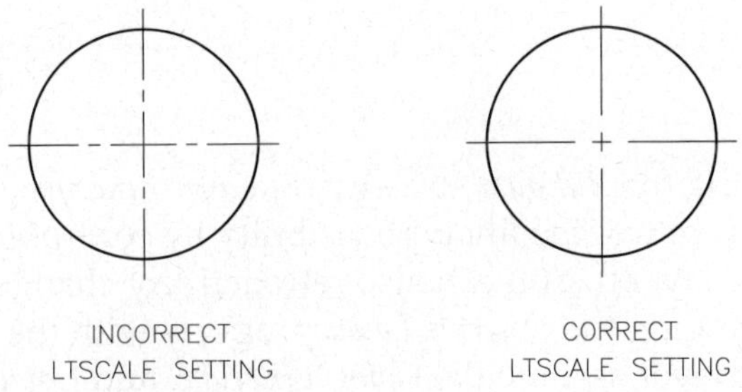


TIP

TIP

When the *Ltscale* has been optimally adjusted for the drawing globally, use the *Properties* palette to adjust the linetype scale of individual objects. In this way, the drawing lines can be created originally with an appropriate global linetype scale without regard to the *Celtscale*. The finished drawing linetype scale can be adjusted if needed with *Ltscale* globally; then only those objects that need further adjusting can be fine-tuned retroactively with *Properties*.

ANSI Standards require that multiview drawings are created with object lines having a dark lineweight, and hidden, center, dimension, and other reference lines created in a medium lineweight (see Figure 24-13).

You can assign *Lineweight* to objects using the *Lineweight* command or assign *Lineweight* to layers in the *Layer Properties Manager*. As described previously in Chapter 11, *Lineweight* can be assigned to individual objects or to layers (*ByLayer*). Use the *Lineweight Settings* dialog box to assign *Lineweight* property to objects. Use the *Layer Properties Manager* to assign *Lineweight* to layers. Generally, the *ByLayer Lineweight* assignment is preferred, similar to the *ByLayer* method of assigning *Color* and *Linetype*.

Additionally, lineweights can be assigned by using plot styles or assigning plot device lineweights or pen thickness (see Chapter 32 for information on plot styles).

## Managing Linetypes, Lineweights, and Colors

There are two strategies for assigning linetypes, lineweights, and colors: *ByLayer* and object-specific assignment. In either case, thoughtful layer utilization for linetypes will make your drawings more flexible and efficient.

### *ByLayer Linetypes, Lineweights,* and *Colors*

The *ByLayer* linetype, lineweight, and color settings are recommended when you want the most control over linetype visibility and plotting. This is accomplished by creating layers with the *Layer Properties Manager* and assigning *Linetype, Lineweight,* and *Color* for each layer. After you assign *Linetypes* to specific layers, you simply set the layer (with the desired linetype) as the *Current* layer and draw on that layer in order to draw objects in a specific linetype.

A template drawing for creating typical multiview drawings could be set up with the layer and linetype assignments similar to that shown in Figure 24-17. Each layer has its own *Color*, *Linetype*, and *Lineweight* setting, and the layer name indicates the type of lines used (hidden, center, object, etc.). This strategy is useful for simple, generic applications.

**FIGURE 24-17**

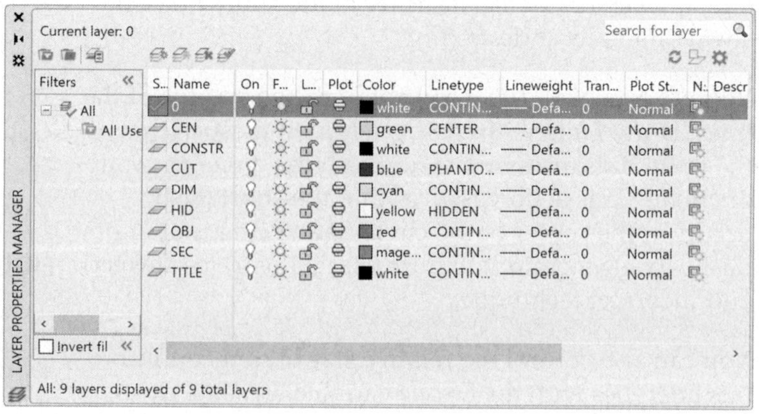


TIP

Using the *ByLayer* strategy (*ByLayer Linetype, Lineweight*, and *Color* assignment) gives you flexibility. You can control the linetype visibility by controlling the layer visibility (show only object layers or hidden line layers). You can also retroactively change the *Linetype, Lineweight,* and *Color* of an existing object by changing the object's *Layer* property with the *Properties* palette or *Matchprop*. Objects changed to a different layer assume the *Color, Linetype,* and *Lineweight* of the new layer.

TIP

Alternately, you can change object-specific (*ByBlock*) property settings for selected objects to *ByLayer* settings using the *Setbylayer* command. *Setbylayer* prompts you to select objects. Properties for the selected objects are automatically changed to the properties of the layer that the objects are on. Properties that can be changed are *Color*, *Linetype*, *Lineweight*, *Material*, and *Plot Style*. See Chapter 11 for more information on *Setbylayer*.

TIP

Another strategy for multiview drawings involving several parts, such as an assembly, is to create layers for each linetype specific to each part, as shown in Figure 24-18. With this strategy, each part has the complete set of linetypes, but only one color per part in order to distinguish the part from others in the display.

Related layer groups could be organized and managed in the *Layer Properties Manager* by creating *Group* filters and *Property* filters (Fig. 24-18). For example, *Group* filters could be created for the "Base," "Fasten," and "Mount" groups. *Property* filters could be created based on *Linetypes* for "Hidden," "Center," and "Object" lines.

FIGURE 24-18

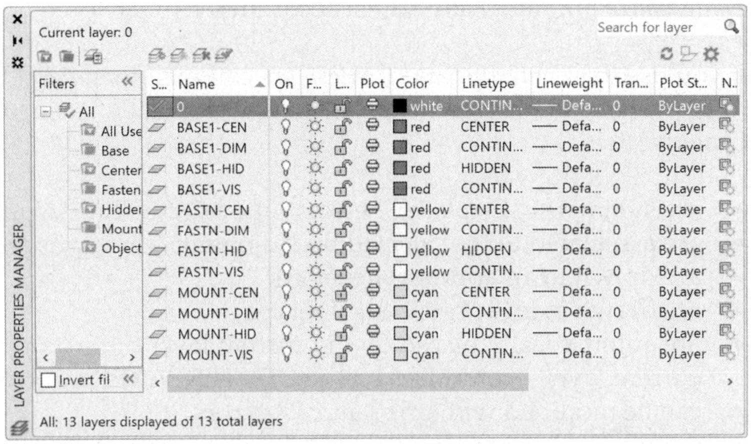


## Object *Linetypes*, *Lineweights*, and *Colors*

Although this method can be complex, object-specific *Linetype*, *Lineweight*, and *Color* assignment can also be managed by skillful utilization of layers. One method is to create one layer for each part or each group of related geometry (Fig. 24-19). The *Color*, *Lineweight*, and *Linetype* settings should be left to the defaults. Then object-specific linetype settings can be assigned (for hidden, center, visible, etc.) using the *Linetype*, *Lineweight*, and *Color* commands.

FIGURE 24-19

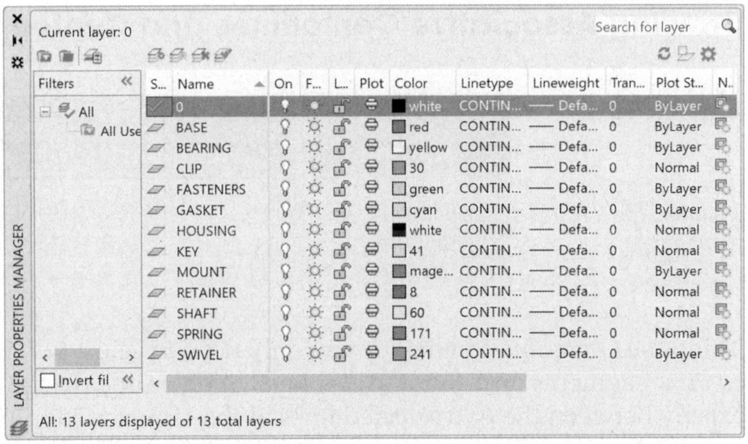


For assemblies, all lines related to one part would be drawn on one layer. Remember that you can draw everything with one linetype and color setting, then use *Properties* to retroactively set the desired color and linetype for each set of objects. Visibility of parts can be controlled by *Freezing* or *Thawing* part layers. You cannot isolate and control visibility of linetypes or colors by this method.

# ASSOCIATIVE CENTERLINES AND CENTERMARKS

AutoCAD 2017 introduced associative center lines and center marks. The *Centerline* command is used to draw one center line symmetrically between two existing *Lines* or *Pline* segments. The *Centermark* command is used to draw a horizontal and a vertical center line through a *Circle* or *Arc*. *Centerlines* and *Centermarks* are associated, or linked, to the geometry for which they designate centers. That is, if the associated *Lines*, *Plines*, *Circles*, or *Arcs* change length, orientation, or position, the *Centerlines* and *Centermarks* automatically adjust accordingly.

If you use the *Centerline* and *Centermark* commands, several things happen automatically (as opposed to drawing center lines and center marks "manually" using the *Line* command). The *Centerlines* and *Centermarks*:

- are symmetrically placed between the selected line segments, circles, or arcs
- are associated to the selected line segments, circles, or arcs
- are drawn to an appropriate length
- are drawn in an appropriate linetype
- are automatically placed on the current layer or on a previously assigned layer regardless of the current layer—controlled by the CENTERLAYER system variable
- can be modified with grips and *Properties*
- maintain short crossing dashes at the center of circles and arcs
- can have other special properties controlled by system variables

The best features of the *Centerline* and *Centermark* commands are that they are very fast to use and relatively simple to control and modify.

## Drawing Associative Centerlines and Centermarks

### *Centerline*

| Ribbon | Menu Bar | Command (Type) | Alias (Type) | Shortcut |
|---|---|---|---|---|
| *Annotate*<br>*Centerlines*<br>*Centerline* | ... | *Centerline* | *CL* | ... |

Using the *Centerline* command, you only need to select two *Lines* or *Pline* segments and AutoCAD creates a center line placed exactly between the two selected lines at the same length plus a small extension called an "overshoot" (Fig. 24-20). The *Centerline* is associative (associated to the selected *Lines* or *Plines*) in that, if the length or orientation of either or both of the selected *Lines* change, the *Centerline* automatically adjusts to the new length and/or orientation. The *Centerline* command **L**ayer option lets you set the layer to be used for the centerlines.

FIGURE 24-20

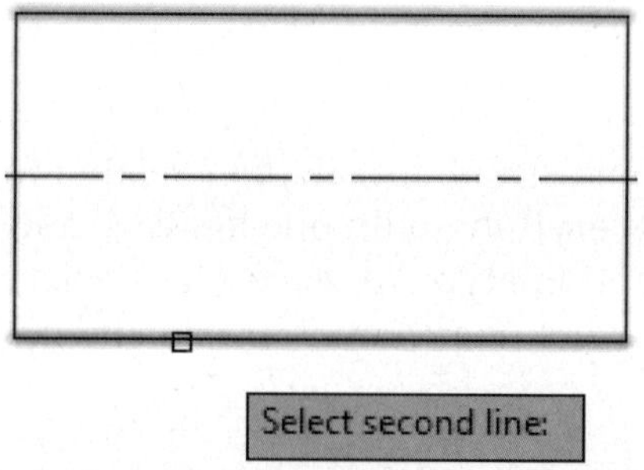


The *Centerline* command is simple to use as shown by the following command prompt.

Command: ***centerline***
Select first line or [Layer]: **PICK**
Select second line: **PICK**
Command:

The selected *Lines* do not have to be the same length or angle. If you select nonparallel lines, the *Centerline* is drawn symmetrically so it bisects the angle between the two line segments. If the two selected lines are of different lengths, the *Centerline* is automatically drawn to match the lengths between the selected two lines plus an overshoot distance (Fig. 24-21). The *Centerline* overshoots the distance between two imaginary lines connecting the endpoints of the selected lines. Several components of the *Centerlines* can be controlled and modified using grips, *Properties*, and system variables (see "Modifying *Centerlines* and *Centermarks*").

FIGURE 24-21

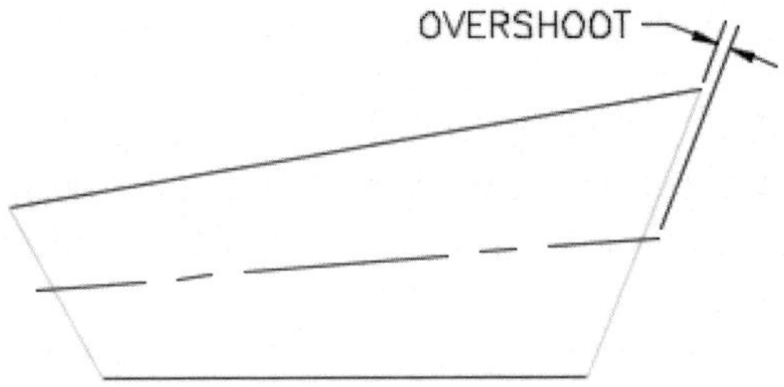


## Centermark

| Ribbon | Menu Bar | Command (Type) | Alias (Type) | Shortcut |
|---|---|---|---|---|
| *Annotate Centerlines Centermark* | ... | *Centermark* | *CM* | ... |

Using the *Centermark* command, you can select a *Circle* or an *Arc* and AutoCAD automatically draws a center mark (one horizontal and one vertical center line crossing at the center). The *Centermark* (with the default settings) is composed of two short dashes that cross at the center and center lines that extend just past the *Circle* or *Arc*. *Centermarks* are associative because they automatically adjust to changes in the selected *Circle* or *Arc* size or position. The **L**ayer option allows you to set the layer used for the centermark.

Command: ***centermark***
Select circle or arc to add center mark or [Layer]: **PICK**
Select circle or arc to add center mark or [Layer]: **Enter**
Command:

Creating a *Centermark* is simple. You only need to select the *Circle* or *Arc* and AutoCAD draws two crossing (vertical and horizontal) center lines (Fig. 24-22).

FIGURE 24-22

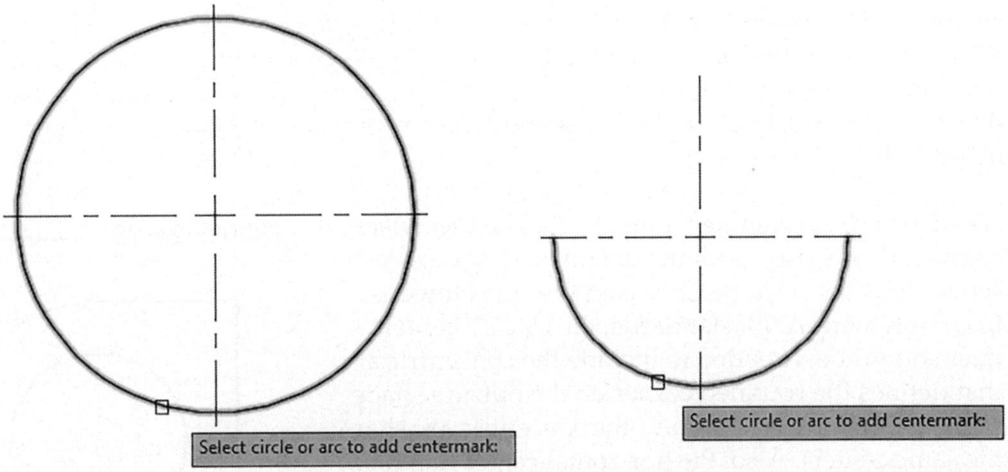


The resulting *Centermark* has two elements: the center cross and the extension lines. The center cross is composed of the two short dashes at the center of the center mark. The extension lines are the lines continuing out from the center cross that extend past the arc or circle. Several features of the *Centermarks* can be controlled using grips, *Properties*, and system variables (see "Modifying *Centerlines* and *Centermarks*" next).

## Modifying *Centerlines* and *Centermarks*

Several individual components of the *Centerlines* and *Centermarks* can be controlled either before or after creating the *Centerlines* and *Centermarks*. Figure 24-23 shows the application of a *Centermark* to a *Circle* with the related components of the *Centermark* labeled. *Centerlines* have some of the same components but do not have a center cross; therefore, *CENTERCROSSGAP* and *CENTERCROSSSIZE* system variables do not apply. *Centerlines* and *Centermarks* can be modified using grips, the *Properties* palette, and many system variables as described in the following sections.

FIGURE 24-23

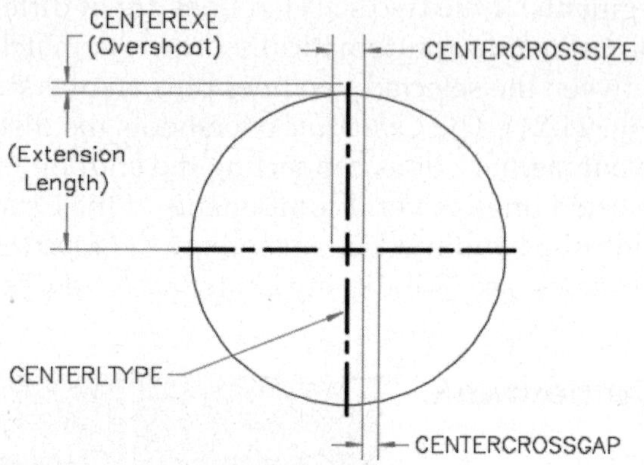


## Grips

Using grips is a visual and intuitive method for editing *Centerlines* and *Centermarks* after they have been drawn. Using grips, you can change the "overshoot" and the "extension length" and move the center line away from the *Circle*, *Arc*, or *Line*. *Centerlines* and *Centermarks* both display the same grips. You cannot use grips to change the center cross size or center cross gap for *Centermarks*.

Figure 24-24 shows *Centermark* grips. The four square grips at the quadrants can be manipulated individually to stretch the center line extension length. The arrow grips change the overshoot. Hovering over any of these two types of grips displays the dynamic dimensions of the extension line or overshoot.

FIGURE 24-24

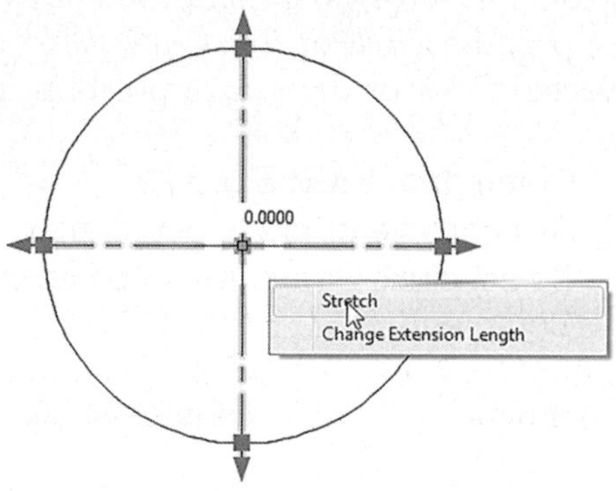


NOTE: When you stretch the center line using the extension length (square) grip, the overshoot moves along with it. However, dragging the overshoot (arrow) grip extends only the overshoot length.

Hovering over the multi-functional center grip displays a menu (see Fig. 24-24) where you can choose to uniformly change all four extension lines (*Change Extension Length*) or move the center lines to another location (*Stretch*). When you use the *Stretch* option to move the center lines but not the object, the center lines for a *Centerline* or *Centermark* become disassociated with the object (*Circle*, *Arc*, *Line*, or *Pline*). Use the *Centerreassociate* command to "reattach" the center lines to the object (see *Centerreassociate*).

As an example, examine Figure 24-25. The *Centermark* command was used with the default settings to create center lines for both the *Circle* and the *Arc*. However, to comply with ANSI standards, the *Circle*'s center lines should be extended to include the concentric arc that defines the rounded corner on the outside shape of the part (since the hole and the concentric arc share the same center). Also, the horizontal center line for the *Arc* on the right side of the part should be shortened on each end to display a gap between the center line and the object lines. Additionally, the top of the vertical center line for that *Arc* is not needed and should be reduced to the center cross.

FIGURE 24-25

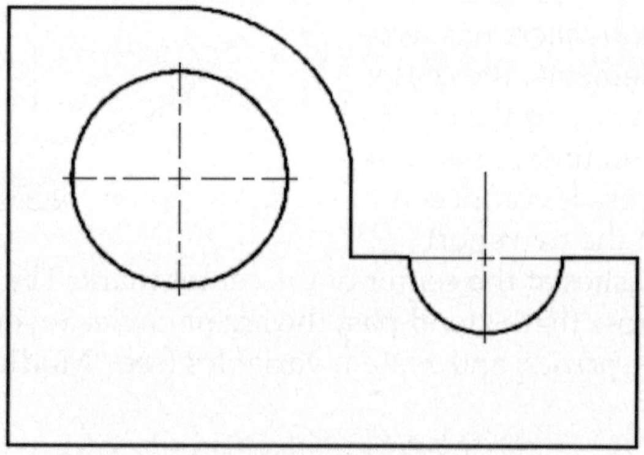

In Figure 24-26 two of the center lines for the *Circle* are being extended past the rounded arc in the outside shape using the overshoot grips. Notice that the overshoot (arrow) grips are used in this case rather than the extension length (square) grips (see the NOTE below). For the arc on the right side of the part, the extension length grips were used to shorten both ends of the horizontal center line to expose a gap between the center lines and object lines that define the part. Also for the arc on the right side of the part, the top of the vertical center line was almost eliminated by dragging the extension length grip downward to the center cross.

FIGURE 24-26

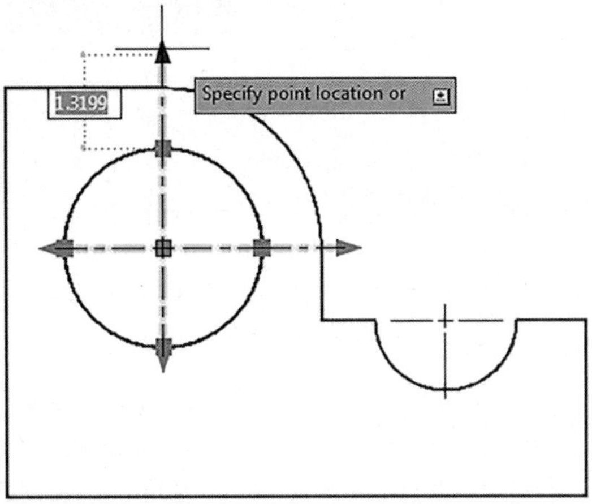


NOTE: If you use grips to stretch a *Centermark* extension length, that individual extension line becomes a fixed length (only on the grip-edited end) and does not change length if the *Circle* is resized. However, if you grip-stretch the center line's overshoot length of a *Circle* or *Arc*, all center lines change length accordingly if the object is subsequently resized. When moving the object, such as a *Circle* or *Arc*, use the object's grip (not the center line grips) to ensure the center lines retain associativity.

TIP

## *Properties*

| Ribbon | Menu Bar | Command (Type) | Alias (Type) | Shortcut |
|---|---|---|---|---|
| *View Palettes Properties* | *Modify Properties* | *Properties* | *PR* | Ctrl+1 |

Instead of using grips to change the overshoot length, you can use the *Properties* palette. After selecting a *Centerline* or *Centermark*, the *Properties* palette displays values for some of the lengths (Fig. 24-27). Enter new values to change lengths for the center line features.

NOTE: The values in the *Properties* palette labeled "extension" actually refer to the overshoot value, not the extension length. You cannot use the *Properties* palette to change the extension length.

TIP

Although there are no grips that appear for the cross size and the cross gap, you can use the *Properties* palette to change the *Cross size* and *Cross gap* values. A value with an "x" suffix makes the value relative to the *Circle's* diameter. See information on the *CENTERCROSSSIZE* and *CENTERCROSSGAP* system variables later in this chapter as well as in previous Figure 24-23.

FIGURE 24-27

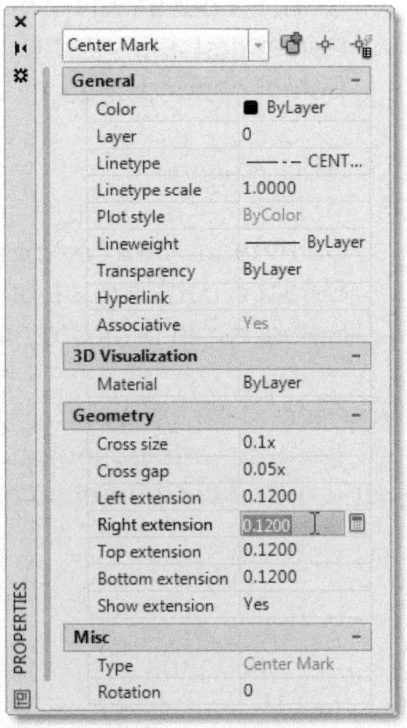

### *Centerdisassociate*

| Ribbon | Menu Bar | Command (Type) | Alias (Type) | Shortcut |
|---|---|---|---|---|
| ... | ... | *Center-disassociate* | ... | ... |

Normally, *Centerlines* and *Centermarks* are associated to the related objects (*Lines*, *Circles*, *Arcs*) such that, if the related objects change size or orientation, the center lines change accordingly. You can use the *Centerdisassociate* command to remove the associativity of *Centermarks* or *Centerlines* from the objects for which they define center lines.

Command: ***Centerdiasassociate***
Select center objects to disassociate: **PICK**
Select center objects to disassociate: **Enter**
1 center object(s) disassociated.
Centerline disassociated..
Command:

### *Centerreassociate*

| Ribbon | Menu Bar | Command (Type) | Alias (Type) | Shortcut |
|---|---|---|---|---|
| ... | ... | *Center-reassociate* | ... | ... |

If a center line for either a *Centermark* or a *Centerline* has become disassociated and you want to "reattach" it, you can use *Centerreassociate* to make the reattachment. For example, a *Centermark* may become disassociated from a *Circle* by accidentally grip-stretching the *Centermark* grip instead of the *Circle* center grip. Use *Centerreassociate* to move and reconnect the center lines to the current location of the *Circle*. The following prompts are displayed depending on which object (*Centerline* or *Centermark*) is selected.
For a *Centerline*:

Command: ***Centerreassociate***
Select a center mark or centerline to reassociate: **PICK** (the *Centerline*)
Select first line: **PICK**
Select second line: **PICK**

For a *Centermark*:

Command: ***Centerreassociate***
Select a center mark or centerline to reassociate: **PICK** (the *Centermark*)
Select circle or arc to associate center mark: **PICK**

TIP

Note that if an extension length (not the overshoot) of a *Centermark* is stretched using grips, the extension line becomes a fixed length. Even though the *Centermark* is technically still associated, it does not readjust if the *Circle* diameter changes. *Centerreassociate* will not remedy this situation.

### *Centerreset*

| Ribbon | Menu Bar | Command (Type) | Alias (Type) | Shortcut |
|---|---|---|---|---|
| ... | ... | *Centerreset* | ... | ... |

The *Centerreset* command can be used to reset the overshoot lengths of a *Centermark* or *Centerline* to the current value specified in the *CENTEREXE* system variable. *Centerreset* does not necessarily restore the *Centermark* or *Centerline* back to the originally created size if the *CENTEREXE* system variable was later changed. The following prompt is displayed.

Command: ***Centerreset***
Select center mark or centerline to reset: **PICK**
Select center mark or centerline to reset: **Enter**
1 center object(s) reset.
Command:

## *Centermark* and *Centerline* System Variables

#### *CENTERLAYER*

This system variable specifies the layer on which center marks and center lines are created. The *CENTERLAYER* system variable applies only to center marks and center lines created with the *Centermark* and *Centerline* commands.

This variable is similar to the *DIMLAYER* and *HPLAYER* system variables for dimensions created with the *Dim* command and hatch patterns created with the *Hatch* command, respectively. The initial value for *CENTERLAYER* is "use current." Valid values include "." (use current), "ByLayer," or any layer name. The setting is saved with the drawing.

With the initial value ("use current"), all newly created *Centerline*s or *Centermark*s are drawn on the current layer. If you have already created a layer for center lines and specified properties for the layer, you can specify that layer for all new *Centermark*s and *Centerline*s. If the desired layer does not yet exist, AutoCAD will automatically create the layer with the name you specify but with default properties of linetype, lineweight, color, etc. However, AutoCAD does not create the new layer until a *Centerline* or *Centermark* is created.

TIP

This feature is very useful and productive when using the *Centerline* and *Centermark* commands to draw center lines. The advantage is that *Centerline*s and *Centermark*s are always drawn on the layer specified by this system variable, even though another layer can be current at the time you create the center lines! Therefore, you do not need to make the center line layer current in order to draw center lines on it.

#### *CENTERLTSCALE*

*CENTERLTSCALE* specifies the linetype scale used by center lines created with the *Centermark* and *Centerline* commands. The initial value is 1.0000 and accepts any real number except zero.

*CENTERLTSCALE* is similar to the *CELTSCALE* variable which holds the linetype scale value for individual objects independent of the drawing's global *LTSCALE*. Likewise, each *Centerline* and *Centermark* object has its own *CENTERSCALE* value.

TIP

In the *Properties* palette, this variable appears in the *Linetype scale* edit box when a *Centerline* or *Centermark* is selected. Rather than changing this variable by command line, the suggested strategy is to first set an appropriate *LTSCALE* globally for all non-continuous lines in the drawing, then create all the objects for the drawing including center lines, and then use the *Properties* palette to adjust the "Linetype scale" value for individual *Centerlines* and *Centermarks* as needed.

### *CENTERLTYPE*

*CENTERLTYPE* specifies the linetype used by center marks and center lines. The initial value is "CENTER2". The *CENTERLTYPE* system variable applies only to the linetype used on the center marks and center lines created with the *Centermark* and *Centerline* commands. This system variable accepts the following values: "." (use current), "ByLayer", or any valid linetype.

The default linetype (CENTER2) is best to use for most cases because with an *Ltscale* setting of 1, and when a drawing is printed 1:1, the short dashes print at the ANSI standard length, 1/8".

### *CENTERLTYPEFILE*

This variable specifies the loaded linetype library file used to create *Centermarks* and *Centerlines*. The default value is "acad.lin." Because the ACAD.LIN file is the only linetype file supplied with AutoCAD, there is no need to change this variable unless you have created or acquired a custom linetype file.

### *CENTEREXE*

The *CENTEREXE* system variable controls the length of the overshoot of center marks and center lines created with the *Centermark* and *Centerline* commands (see Fig. 24-23). The initial value is 0.1200 (in imperial drawing templates) or 3.5000 (in metric drawing templates). *CENTEREXE* accepts only positive real numbers.

This variable sets the length of the overshoot only for new *Centerlines* or *Centermarks*. Once created, the overshoot length can be changed by stretching with grips or by changing the values that appear in the *Properties* palette when selected.

### *CENTERCROSSSIZE*

The *CENTERCROSSSIZE* specifies the length of the short dashes that cross at the center of a *Circle* or *Arc* for center marks created using the *Centermark* command (see Fig. 24-23). The initial value is "0.1x". The value can be absolute, relative, or "ByLineType".

An absolute value is any positive real number. An absolute value (without an "x" suffix) specifies a set length for the line segment of the center cross in units, regardless of the diameter of the *Circle* or *Arc*. A relative value is any positive real number specified with an "x" suffix. A relative value specifies a length relative to the diameter of a *Circle* or an *Arc*. For example, 0.1x is 1/10th of a circle's diameter.

The "ByLineType" setting specifies that the size of the center cross is derived from the linetype assigned to the center mark. With this setting, the short cross for the *Centermark* is equal in length to the other short dashes in the linetype and changes proportionally with the *Ltscale* setting.

TIP

NOTE: With an absolute or a relative *CENTERCROSSSIZE* setting, changes in the *Ltscale* setting have no effect on the size of the center crosses. In these cases, you must use the *Properties* palette to alter the length of center crosses for existing *Centermarks* by changing the values in the *Cross size* edit box.

### *CENTERCROSSGAP*

This setting determines the gap between the center cross and the extension lines of a center mark created with the *Centermark* command (see Fig. 24-23). The initial value is "0.05x" (equal to .05 times the diameter of the associated *Circle* or *Arc*). Similar to the *CENTERCROSSSIZE* variable, the setting can be an absolute value (any positive real number), a relative value (any positive real number specified with an "x" suffix), or "ByLineType". See *CENTERCROSSSIZE* (previous page).

NOTE: The *CENTERCROSSSIZE* and *CENTERCROSSGAP* system variable settings are related to one another. It is recommended that both system variables use the same value type (absolute, relative, or *ByLineType*).

The *CENTERCROSSGAP* system variable setting is ignored (no gap is drawn) when the radius of the *Circle* or *Arc* is too small—less than the sum of half the center cross size and the gap distance.

NOTE: With an absolute or a relative *CENTERCROSSGAP* setting, changes in the *Ltscale* setting have no effect on the size of the center gaps. Therefore, change the values in the *Cross gap* edit box in the *Properties* palette to alter the length of center gaps for existing *Centermarks*.

### *CENTERMARKEXE*

Using this system variable, you can turn on or off the automatic creation of the extension lines' center marks created using the *Centermark* command. The initial value is *On*. When *CENTERMARKEXE* is *On*, *Centermarks* include both the short crossing dashes at the center of the *Circle* or *Arc* and the extension lines and overshoot lengths. When *Off*, the short crossing dashes are created but no other center lines or overshoots are created.

To change the display of extension lines and overshoots for existing *Centermarks*, you can use the *Properties* palette and edit the setting for *Show extension* (*Yes* or *No*). See previous Figure 24-27 at the bottom of the *Geometry* category.

## CREATING FILLETS, ROUNDS, AND RUNOUTS

Many mechanical metal or plastic parts manufactured from a molding process have slightly rounded corners. The otherwise sharp corners are rounded because of limitations in the molding process or for safety. A convex corner is called a round and a concave corner is called a fillet. These fillets and rounds are created easily in AutoCAD by using the *Fillet* command.

The example in Figure 24-28 shows a multiview drawing of a part with sharp corners before the fillets and rounds are drawn.

**FIGURE 24-28**

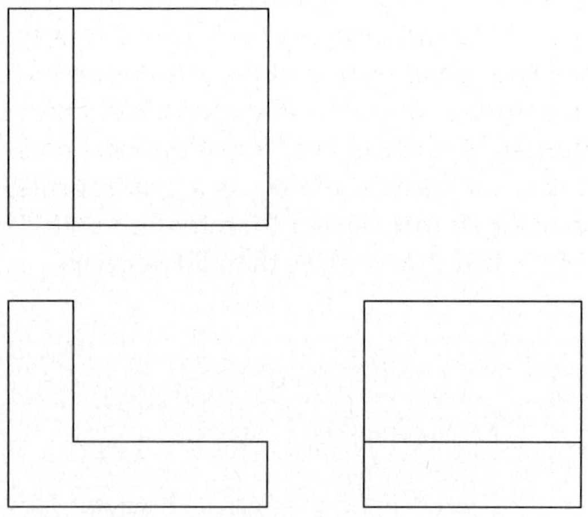

The corners are rounded using the *Fillet* command (Fig. 24-29). First, use *Fillet* to specify the *Radius*. Once the *Radius* is specified, just select the desired lines to *Fillet* near the end to round.

If the *Fillet* is in the middle portion of a *Line* instead of the end, *Extend* can be used to reconnect the part of the *Line* automatically trimmed by *Fillet*, or *Fillet* can be used in the *Notrim* mode.

**FIGURE 24-29**

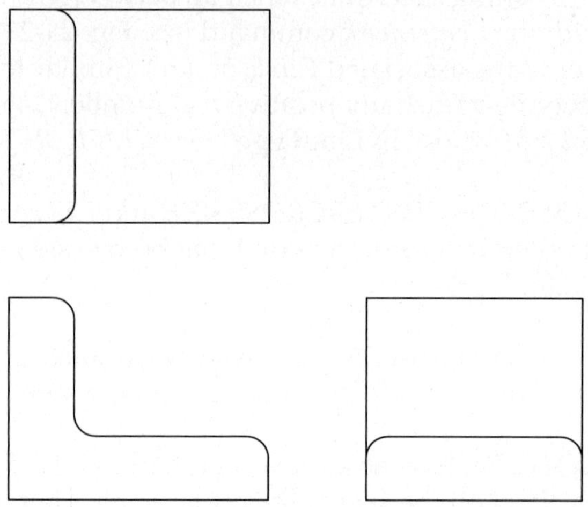

A runout is a visual representation of a complex fillet or round. For example, when two filleted edges intersect at less than a 90-degree angle, a runout should be drawn as shown in the top view of the multiview drawing (Fig. 24-30).

The finish marks (V-shaped symbols) indicate machined surfaces. Finished surfaces have sharp corners.

**FIGURE 24-30**

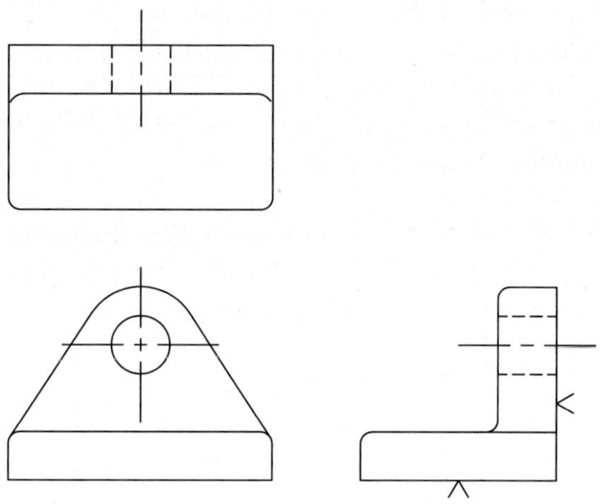

A close-up of the runouts is shown in Figure 24-31. No AutoCAD command is provided for this specific function. The *3point* option of the *Arc* command can be used to create the runouts, although other options can be used. Alternately, the *Circle TTR* option can be used with *Trim* to achieve the desired effect. As a general rule, use the same radius or slightly larger than that given for the fillets and rounds, but draw it less than 90 degrees.

**FIGURE 24-31**

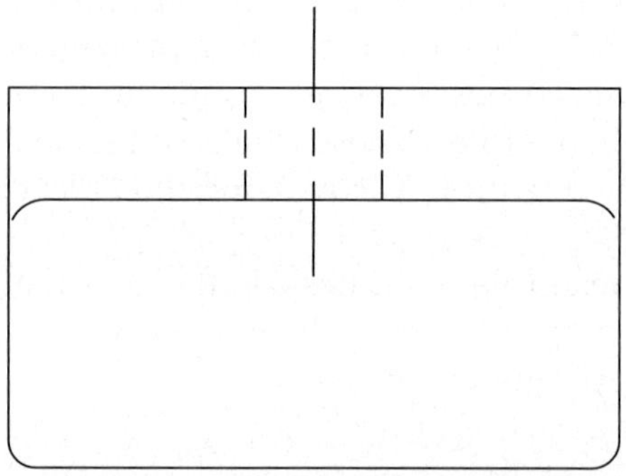

## GUIDELINES FOR CREATING A TYPICAL THREE-VIEW DRAWING

Following are some guidelines for creating the three-view drawing in Figure 24-32. This object is used only as an example. The steps or particular construction procedure may vary, depending on the specific object drawn. Dimensions are shown in the figure so you can create the multiview drawing as an exercise.

**FIGURE 24-32**

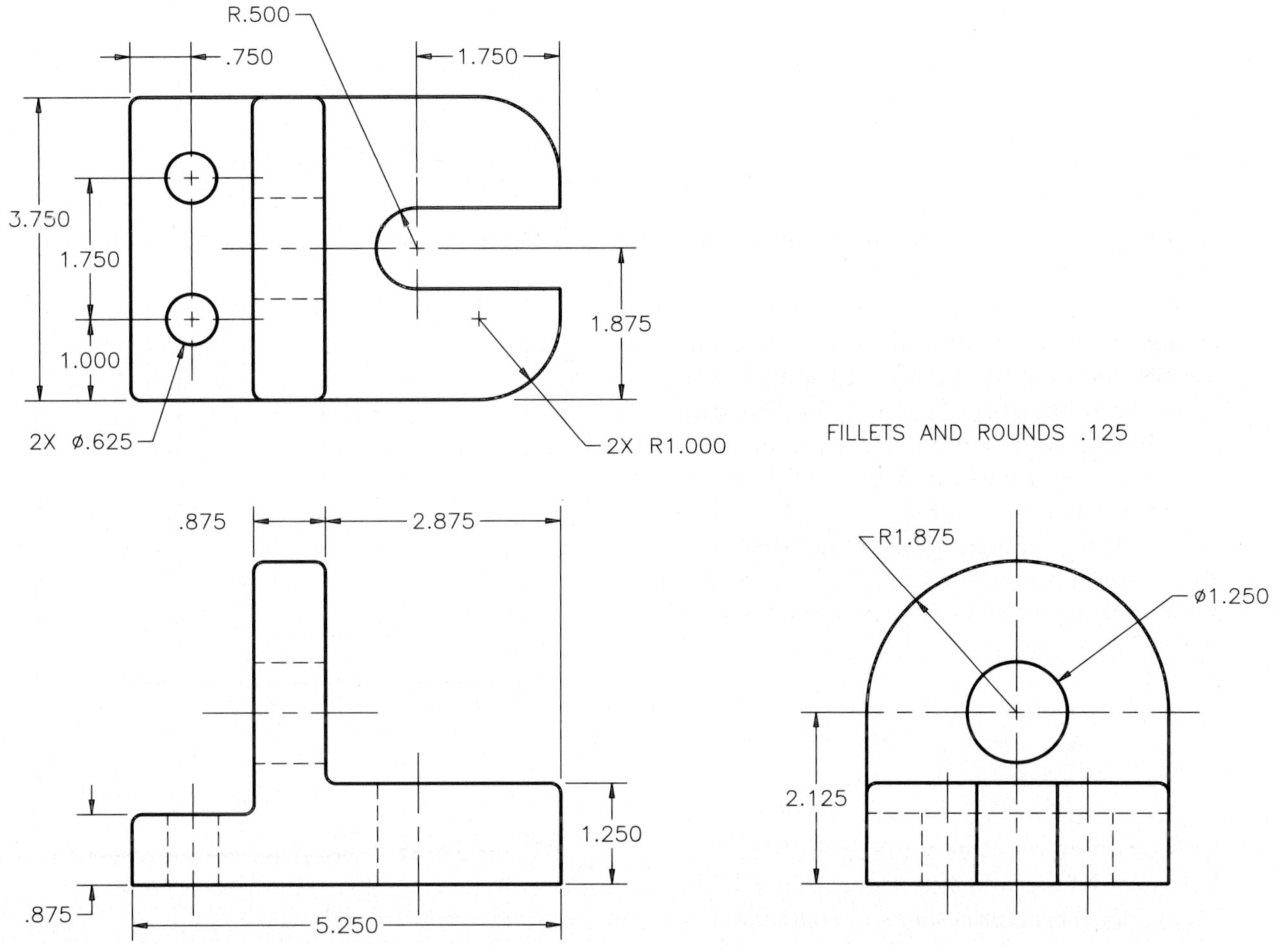


1. Drawing Setup

   *Units* are set to *Decimal* with 3 places of *Precision*. *Limits* of 22 x 17 are set to allow enough drawing space for both views. The finished drawing can be plotted on a B size sheet at a scale of 1″=1″ or on an A size sheet at a scale of 1/2″=1″. *Snap* is set to an increment of .125. *Grid* is set to an increment of .5. Although it is recommended that you begin drawing the views using *Grid Snap*, a *Polar Snap* increment of .125 is set, and *Polar Tracking* angles are set. Turn *Polar* on. Set the desired running *Object Snaps*, such as *Endpoint*, *Midpoint*, *Intersection*, *Quadrant*, and *Center*. Turn on *Object Snap Tracking*. *Ltscale* is not changed from the default of 1. Layers are created (OBJ, HID, CEN, DIM, BORDER, and CONSTR) with appropriate *Linetypes*, *Lineweights*, and *Colors* assigned. Set the *CENTERLAYER* system variable to "CEN".

2. An outline of each view is "blocked in" by drawing the appropriate *Lines* and *Circles* on the OBJ layer similar to that shown in Figure 24-33. Ensure that *Snap* is *ON*. *Ortho* or *Polar Tracking* should be turned *ON* when appropriate. Use the cursor to ensure that the views align horizontally and vertically. Note that the top edge of the front view was determined by projecting from the *Circle* in the right side view.

**FIGURE 24-33**

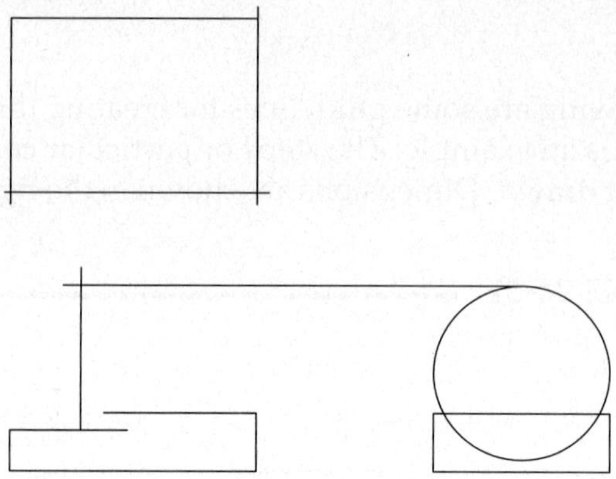

Another method for construction of a multiview drawing is shown in Figure 24-34. This method uses the *Xline* and *Ray* commands to create construction and projection lines. Here all the views are blocked in and some of the object lines have been formed. The construction lines should be kept on a separate layer, except in the case where *Xlines* and *Rays* can be trimmed and converted to the object lines. (The following illustrations do not display this method because of the difficulty in seeing which are object and construction lines.)

**FIGURE 24-34**

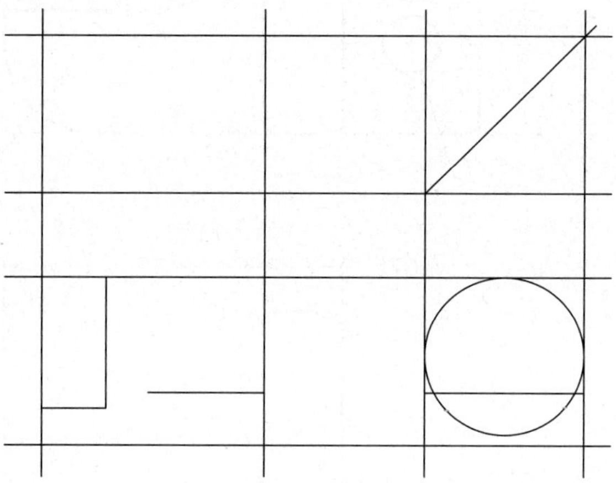

3. This drawing requires some projection between the top and side views (Fig. 24-35). The CONSTR layer is set as *Current*. Two *Lines* are drawn from the inside edges of the two views (using *Object Snap* and *Ortho* for alignment). A 45-degree miter line is constructed for the projection lines to "make the turn." A *Ray* is suited for this purpose.

**FIGURE 24-35**

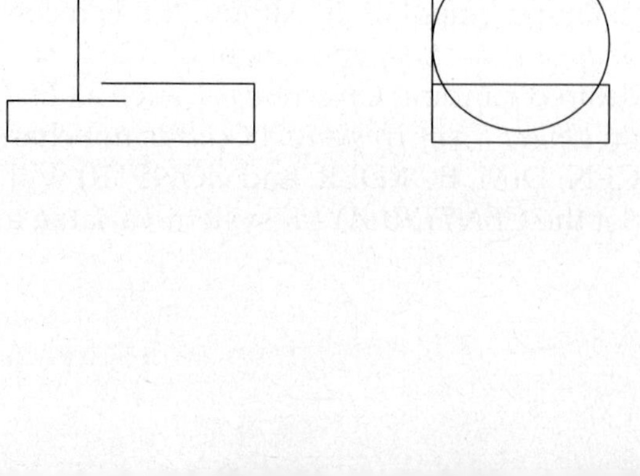

4. Details are added to the front and top views (Fig. 24-36). The projection line from the side view to the front view (previous figure) is *Trimmed*. A *Circle* representing the hole is drawn in the side view and projected up and over to the top view and to the front view. The object lines are drawn on layer OBJ and some projection lines are drawn on layer CONSTR. The horizontal projection lines from the 45-degree miter line are drawn on layer HID awaiting *Trimming*. Alternately, those two projection lines could be drawn on layer CONSTR and changed to the appropriate layer with *Properties* after *Trimming*. Keep in mind that *Object Snap Tracking* can be used here to ensure proper alignment with object features and to prevent having to actually draw some construction lines.

**FIGURE 24-36**

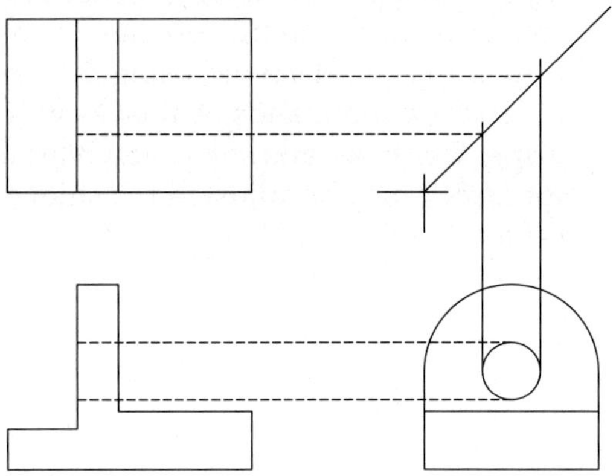

5. The hidden lines used for projection to the top view and front view (previous figure) are *Trimmed* (Fig. 24-37). The slot is created in the top view with a *Circle* and projected to the side and front views. It is usually faster and easier to draw round object features in their circular view first, then project to the other views. Make sure you use the correct layers (OBJ, CONSTR, HID) for the appropriate features. If you do not, *Properties* or *Matchprop* can be used retroactively.

**FIGURE 24-37**

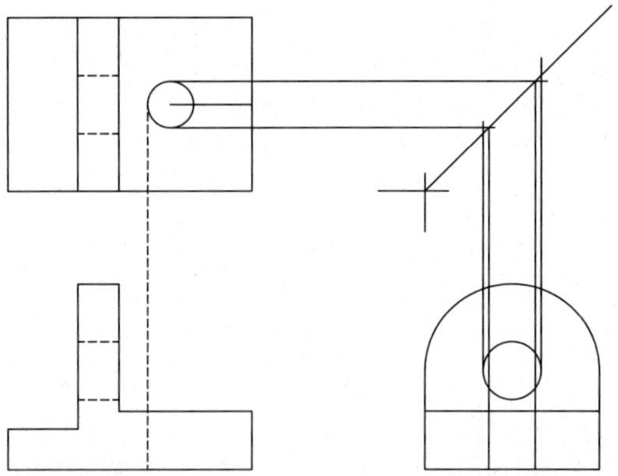

6. The lines shown in the previous figure as projection lines or construction lines for the slot are *Trimmed* or *Erased* (Fig. 24-38). The holes in the top view are drawn on layer OBJ and projected to the other views. The projection lines and hidden lines are drawn on their respective layers.

**FIGURE 24-38**

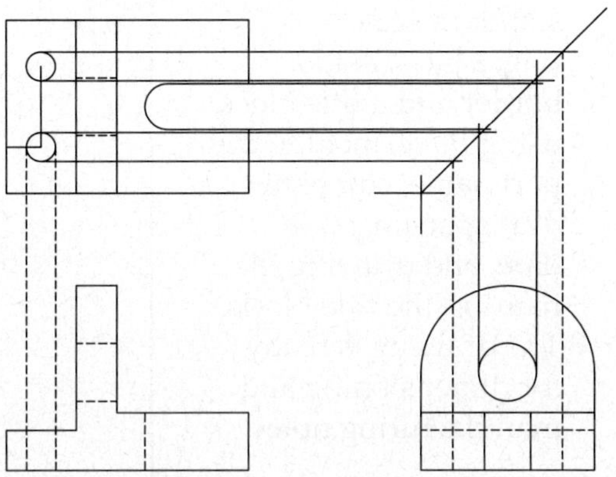

7. *Trim* the appropriate hidden lines (Fig. 24-39). *Freeze* layer CONSTR. On layer OBJ, use *Fillet* to create the rounded corners in the top view. Draw the correct center lines for the holes using the *Centerline* and *Centermark* commands. Adjust as neccessary with grips, *Properties*, and/or system variables. The value for *Ltscale* may be adjusted to achieve the optimum center line spacing. The *Properties* palette can be used to adjust individual object linetype scale.

**FIGURE 24-39**

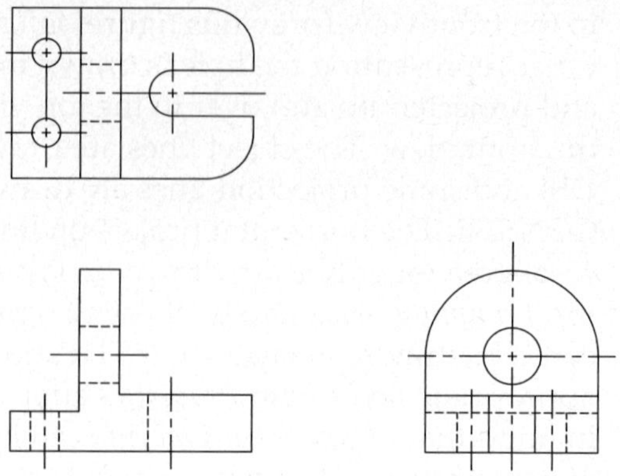

8. Fillets and rounds are added using the *Fillet* command (Fig. 24-40). The runouts are created by drawing a *3point Arc* and *Trimming* or *Extending* the *Line* ends as necessary. Use *Zoom* for this detail work.

**FIGURE 24-40**

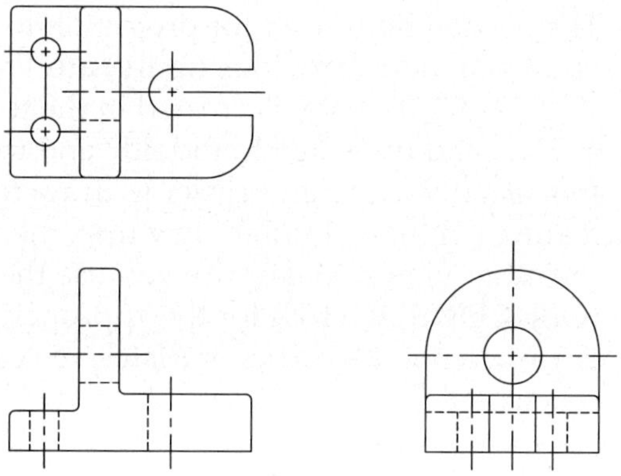

9. Activate a *Layout* tab, configure a print or plot device using *Pagesetup*, make one *Viewport*, and set the *Viewport scale* to a standard scale (Fig. 24-41). Add a border and a title block using *Pline*. Include the part name, company, draftsperson, scale, date, and drawing file name in the title block. The drawing is ready for dimensioning and manufacturing notes.

**FIGURE 24-41**

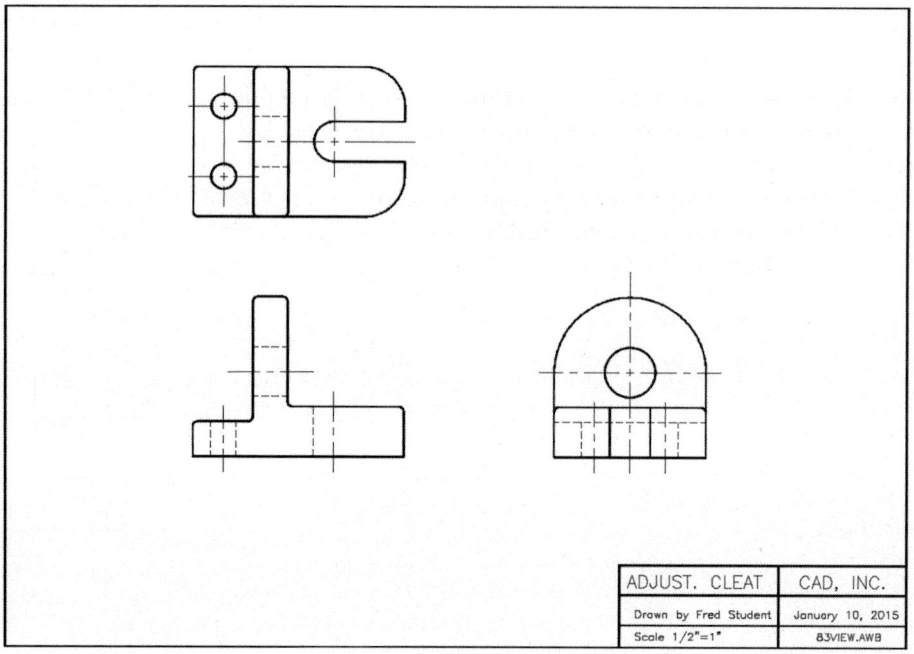

## CHAPTER EXERCISES

1. ***Open*** the **PIVOTARM CH16** drawing.

   A. Create the right side view. Use ***Object Snap*** and ***Ortho*** or *Polar Tracking* to create *Lines* or *Rays* to the miter line and down to the right side view as shown in Figure 24-42. ***Offset*** may be used effectively for this purpose instead. Use ***Extend***, ***Offset***, or ***Ray*** to create the projection lines from the front view to the right side view. Use *Object Snap Tracking* when appropriate.

FIGURE 24-42

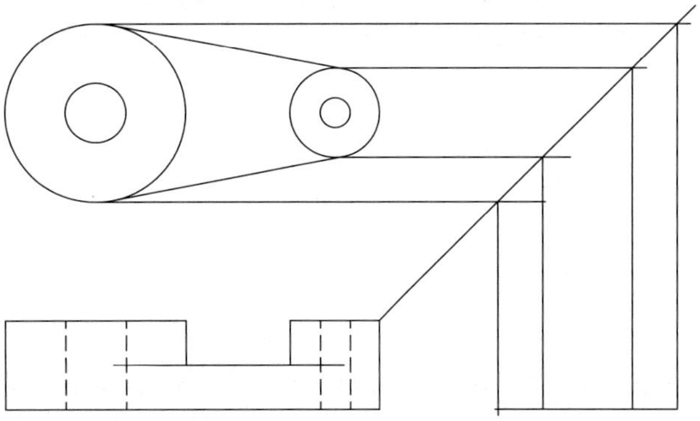

   B. ***Trim*** or ***Erase*** the unwanted projection lines, as shown in Figure 24-43. Draw a ***Line*** or ***Ray*** from the ***Endpoint*** of the diagonal *Line* in the top view down to the front to supply the boundary edge for ***Trimming*** the horizontal *Line* in the front view as shown.

FIGURE 24-43

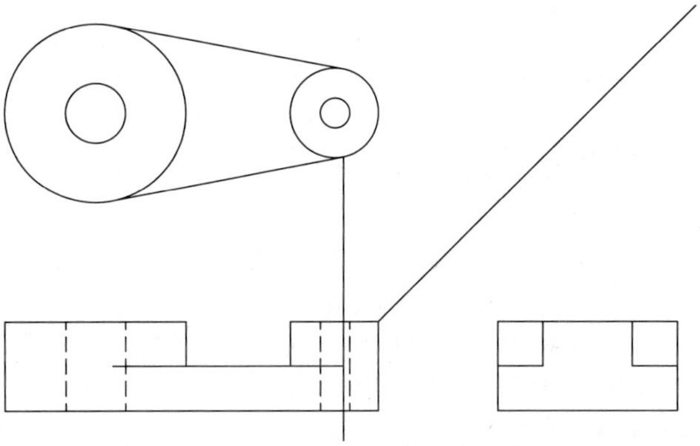

   C. Next, create the hidden lines for the holes by the same fashion as before (Fig. 24-44). Use previously created ***Layers*** to achieve the desired ***Linetypes***. Complete the side view by adding the horizontal hidden ***Line*** in the center of the view.

FIGURE 24-44

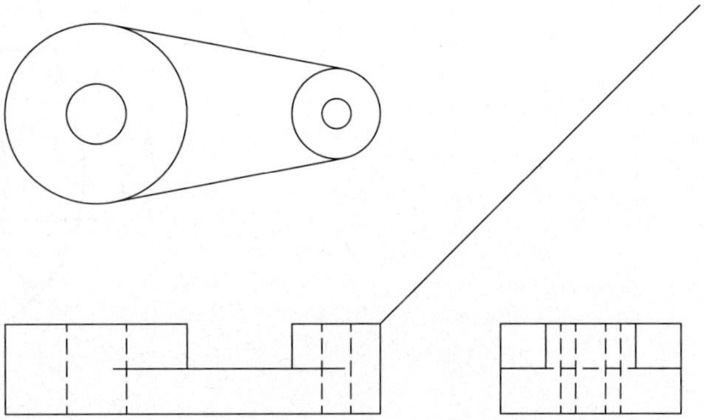

D. Another hole for a set screw must be added to the small end of the Pivot Arm. Construct a ***Circle*** of **4**mm diameter with its center located **8**mm from the top edge in the side view as shown in Figure 24-45. Project the set screw hole to the other views.

**FIGURE 24-45**

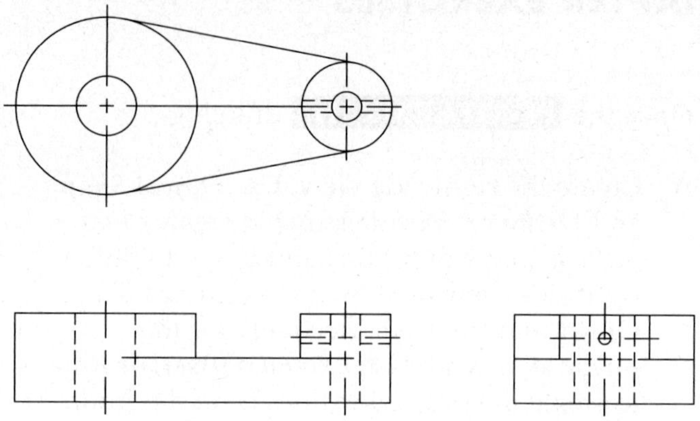

E. Make new layers **CONSTR**, **OBJ**, and **TITLE** and change objects to the appropriate layers with ***Properties***. ***Freeze*** layer **CONSTR**. Use ***Centerline*** and ***Centermark*** to add center lines on the **CEN** layer as shown in Figure 24-45. Change the ***Ltscale*** to **18**. Activate a ***Layout*** tab, configure a print or plot device using ***Pagesetup***, make one ***Viewport***, and set the ***Viewport scale*** to a standard scale. To complete the PIVOTARM drawing, draw a ***Pline*** border (***width*** **.02** x scale factor) in the layout and ***Insert*** the **TBLOCKAT** drawing that you created in Chapter 22 Exercises. ***SaveAs*** **PIVOTARM CH24**.

For Exercises 2 through 5, construct and plot the multiview drawings as instructed. Use an appropriate ***template*** drawing for each exercise unless instructed otherwise. Use conventional practices for ***layers*** and ***linetypes***. Draw a ***Pline*** border with the correct ***width*** and ***Insert*** your **TBLOCK** or **TBLOCKAT** drawing.

2. Make a two-view multiview drawing of the Clip (Fig. 24-46). ***Plot*** the drawing full size (**1=1**). Use the **ASHEET** template drawing to achieve the desired plot scale. ***Save*** the drawing as **CLIP**.

**FIGURE 24-46**

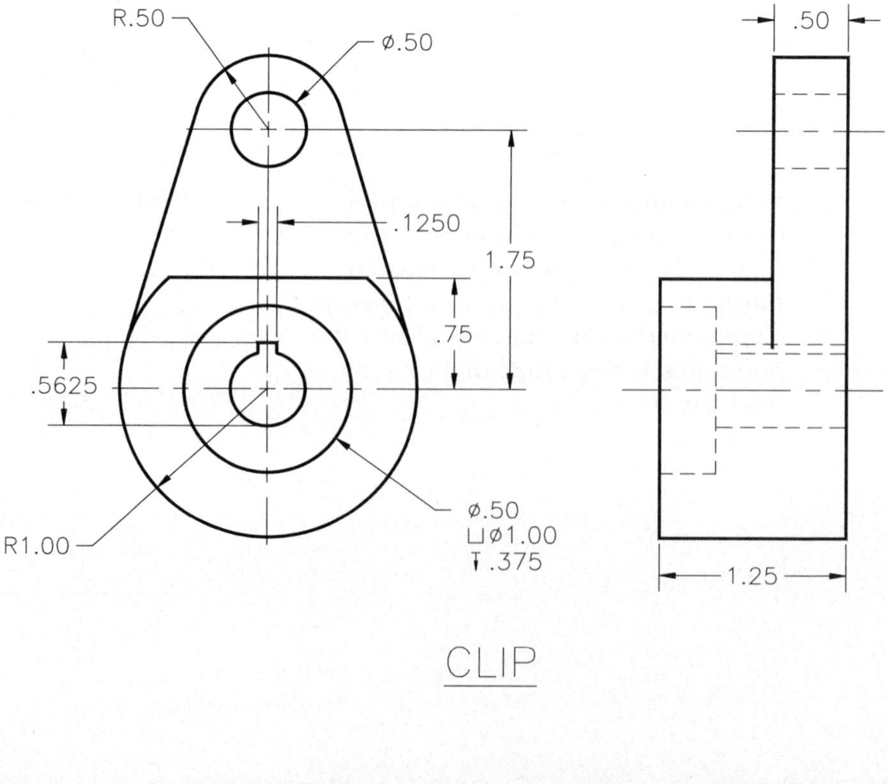

3. Make a three-view multiview drawing of the Bar Guide and ***Plot*** it **1=1** (Fig. 24-47). Use the **BARGUIDE** drawing you set up in Chapter 12 Exercises. Note that a partial ***Ellipse*** will appear in one of the views.

**FIGURE 24-47**

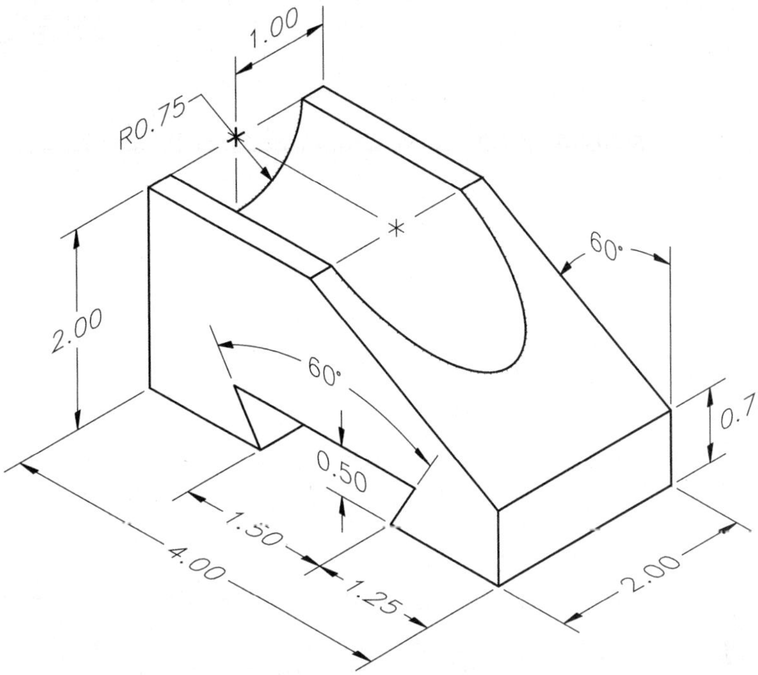


4. Construct a multiview drawing of the Saddle (Fig. 24-48). Three views are needed. The channel along the bottom of the part intersects with the Saddle on top to create a slotted hole visible in the top view. ***Plot*** the drawing at **1=1**. ***Save*** as **SADDLE.**

**FIGURE 24-48**

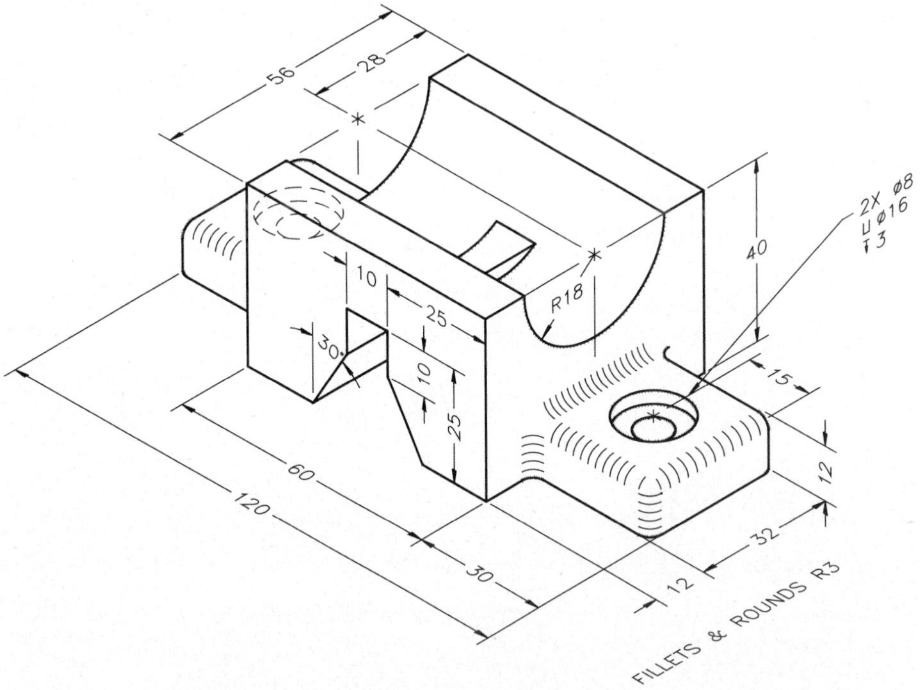

REUSE

5. Draw a multiview of the Adjustable Mount shown in Figure 24-49. Determine an appropriate template drawing to use and scale for plotting based on your plotter capabilities. ***Plot*** the drawing to an accepted scale. ***Save*** the drawing as **ADJMOUNT.**

**FIGURE 24-49**

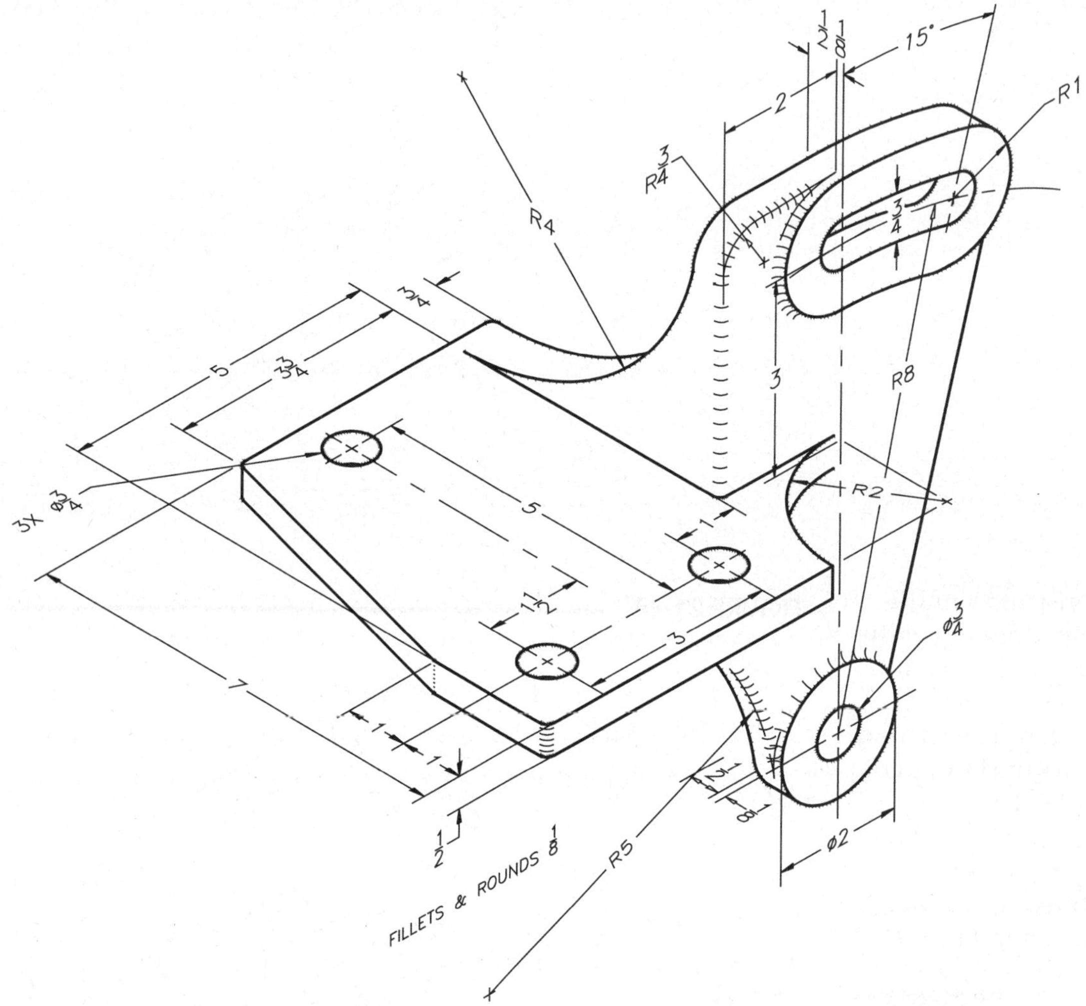

CHAPTER

# 25

# Pictorial Drawings

## Chapter Objectives

After completing this chapter you should:

1. be able to activate the *Isometric Style* of *Snap* for creating isometric drawings;
2. know how to draw on the three isometric planes by using Ctrl+E or toggling the *Isoplane* options on the Status Bar;
3. be able to create isometric ellipses with the *Isocircle* option of *Ellipse*;
4. be able to construct an isometric drawing in AutoCAD;
5. be able to create Oblique Cavalier and Cabinet drawings in AutoCAD.

## CONCEPTS

FIGURE 25-1

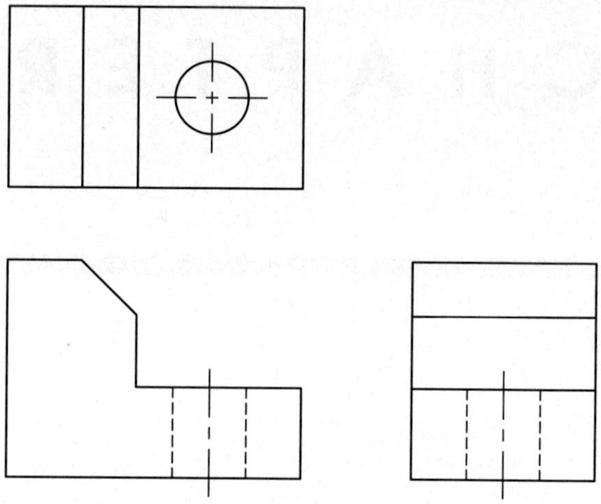

Isometric drawings and oblique drawings are pictorial drawings. Pictorial drawings show three principal faces of the object in one view. A pictorial drawing is a drawing of a 3D object as if you were positioned to see (typically) some of the front, some of the top, and some of the side of the object. All three dimensions of the object (width, height, and depth) are visible in a pictorial drawing.

Multiview drawings differ from pictorial drawings because a multiview only shows two dimensions in each view, so two or more views are needed to see all three dimensions of the object. A pictorial drawing shows all dimensions in the one view. Pictorial drawings depict the object similar to the way you are accustomed to viewing objects in everyday life, that is, seeing all three dimensions. Figure 25-1 and Figure 25-2 show the same object in multiview and in pictorial representation, respectively. Notice that multiview drawings use hidden lines to indicate features that are normally obstructed from view, whereas hidden lines are normally omitted in isometric drawings (unless certain hidden features must be indicated for a particular function or purpose).

FIGURE 25-2

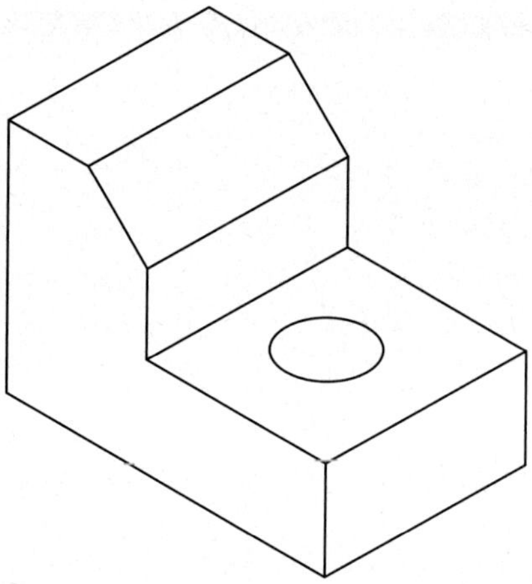

### Types of Pictorial Drawings

Pictorial drawings are classified as follows:

1. Axonometric drawings
   a. Isometric drawings
   b. Dimetric drawings
   c. Trimetric drawings
2. Oblique drawings

**Axonometric** drawings are characterized by how the angle of the edges or axes (axon-) are measured (-metric) with respect to each other.

FIGURE 25-3

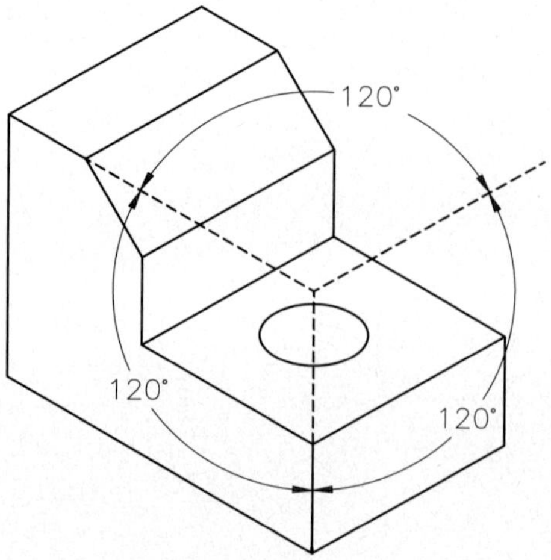


**Isometric** drawings are drawn so that each of the axes have equal angular measurement. ("Isometric" means equal measurement.) The isometric axes are always drawn at 120 degree increments (Fig. 25-3). All rectilinear lines on the object (representing horizontal and vertical edges—not inclined or oblique) are drawn on the isometric axes.

A 3D object seen "in isometric" is thought of as being oriented so that each of three perpendicular faces (such as the top, front, and side) are seen equally. In other words, the angles formed between the line of sight and each of the principal faces are equal.

**Dimetric** drawings are constructed so that the angle between any two of the three axes is equal. There are many possibilities for dimetric axes. A common orientation for dimetric drawings is shown in Figure 25-4. For 3D objects seen from a dimetric viewpoint, the angles formed between the line of sight and each of two principal faces are equal.

FIGURE 25-4

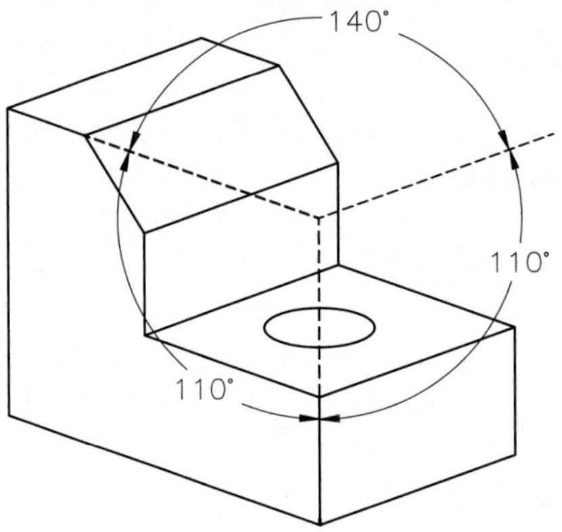


**Trimetric** drawings have three unequal angles between the axes. Numerous possibilities exist. A common orientation for trimetric drawings is shown in Figure 25-5.

FIGURE 25-5

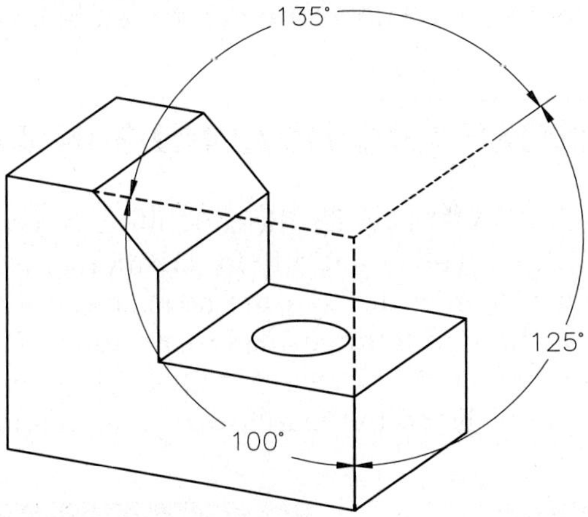


**Oblique** drawings are characterized by a vertical axis and horizontal axis for the two dimensions of the front face and a third (receding) axis of either 30, 45, or 60 degrees (Fig. 25-6). Oblique drawings depict the true size and shape of the front face, but add the depth to what would otherwise be a typical 2D view. This technique simplifies construction of drawings for objects that have contours in the profile view (front face) but relatively few features along the depth. Viewing a 3D object from an oblique viewpoint is not possible.

FIGURE 25-6

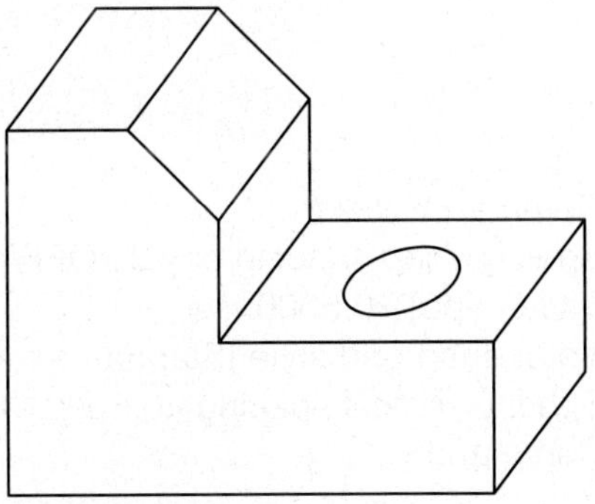

This chapter will explain the construction of isometric and oblique drawings in AutoCAD.

## Pictorial Drawings Are 2D Drawings

Isometric, dimetric, trimetric, and oblique drawings are 2D drawings, whether created with AutoCAD or otherwise. Pictorial drawing was invented before the existence of CAD and therefore was intended to simulate a 3D object on a 2D plane (the plane of the paper). If AutoCAD is used to create the pictorial, the geometry lies on a 2D plane—the XY plane. All coordinates defining objects have X and Y values with a Z value of 0. When the *Isometric* style of *Snap* is activated, an isometrically structured *Snap* and *Grid* appear on the XY plane.

Figure 25-7 illustrates the 2D nature of an isometric drawing created in AutoCAD. Isometric lines are created on the XY plane. The *Isometric Snap* and *Grid* are also on the 2D plane. (The *3Dorbit* command was used to give other than a *Plan* view of the drawing in this figure.)

FIGURE 25-7

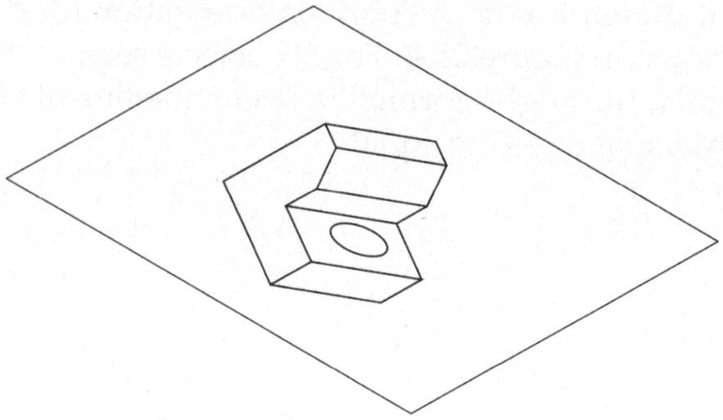

Although 2D pictorial drawings are based on the theory of projecting 3D objects onto 2D planes, they only appear to be 3 dimensional. It is impossible to change the viewpoint once the 2D pictorial drawing is created as you can with a 3D object using a 3D CAD system. In AutoCAD, the *Vpoint* and *3Dorbit* commands can be used to specify the observer's position in 3D space with respect to a 3D model. Chapter 33 discusses the specific commands and values needed to attain axonometric viewpoints of a 3D model.

## ISOMETRIC DRAWING IN AutoCAD

AutoCAD provides the capability to construct isometric drawings. An isometric *Snap* and *Grid* are available, as well as a utility for creation of isometrically correct ellipses. Isometric lines are created with the *Line* command. There are no special options of *Line* for isometric drawing, but isometric *Snap* and *Grid* can be used to force *Lines* to an isometric orientation. Begin creating an isometric drawing in AutoCAD by activating the *Isometric Style* option of the *Snap* command. This action can be done using any of the options listed in the following Command table.

### *Snap*

| Ribbon | Menu Bar | Command (Type) | Alias (Type) | Shortcut |
|---|---|---|---|---|
| ... | *Tools*<br>*Drafting Settings...*<br>*Snap and Grid* | *Snap* | *SN* | F9 or Ctrl+B |

Command: ***SNAP***
Specify snap spacing or [ON/OFF/Aspect/Rotate/Style/Type] <0.5000>: ***s***
Enter snap grid style [Standard/Isometric] <S>: ***i***
Specify vertical spacing <0.5000>: **Enter**
Command:

FIGURE 25-8

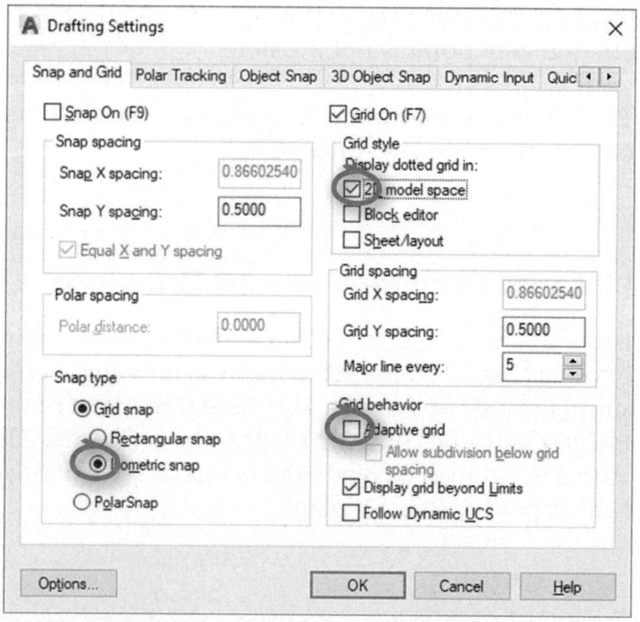


Alternately, toggling the indicated checkbox in the lower-left corner of the *Drafting Settings* dialog box activates the *Isometric snap* and *Grid snap* (Fig. 25-8).

TIP

NOTE: You may want to set the *Grid Style* to display a dotted grid in *2D model space* (Fig. 25-8, upper right). This action makes it easier to see the cursor and the *Lines* as you draw. Also, turning off *Adaptive grid* allows you to see the actual snap increment (Fig. 25-8, lower right).

Note that setting the *Isometric* style of *Snap* changes the *Snap* and *Grid* to align at 30-degree angles from horizontal. Also, the cursor changes to display at an angle so as to be aligned with one of the three "isoplanes."

### *Isodraft*

TIP

Current versions of AutoCAD include a feature for quickly changing the style of the *Snap* and *Grid* to an isometric style called *Isodraft*. The icon for *Isodraft* appears on the Status Bar as shown in Fig. 25-9. Toggling this icon on changes the *Snap* setting to *Isometric* and changes the *Grid* style to dotted. Toggling the icon off returns the *Snap* style to *Standard* and the *Grid* style to lines.

**FIGURE 25-9**

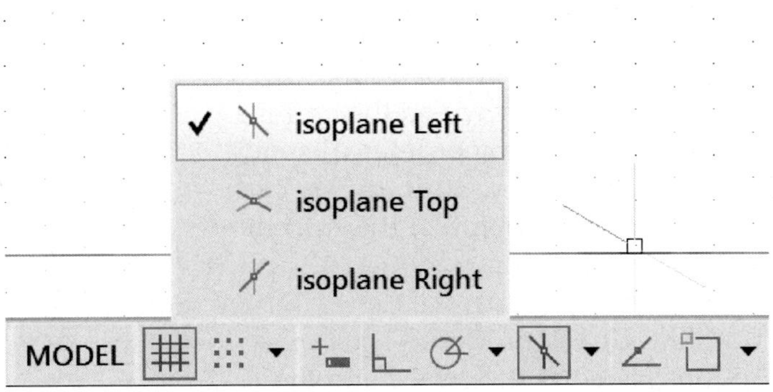


Using Ctrl+E (pressing the Ctrl key and the letter "E" simultaneously) or using the *Isoplane* options next to the *Isodraft* icon on the Status Bar (down arrow, see Figure 25-9) toggles the cursor to one of three possible *Isoplanes* (AutoCAD's term for the three faces of the isometric pictorial). If *Ortho* is *ON*, only isometric lines are drawn; that is, you can only draw *Lines* aligned with the isometric axes. *Lines* can be drawn on only two axes for each isoplane. Toggling Ctrl+E or the *Isoplane* options allows drawing on the two axes aligned with another face of the object. *Ortho* is *OFF* in order to draw inclined or oblique lines (not on the isometric axis). The functions of *Grid* (F7) and *Snap* (F9) remain unchanged.

With *Snap ON*, toggle the coordinate display (F6) several times and examine the read-out as you move the cursor. The Cartesian coordinate format is of no particular assistance while drawing in isometric because of the configuration of the *Grid*.

The relative polar format, however, is very helpful. Use relative polar format for the coordinate display while drawing in isometric.

**FIGURE 25-10**

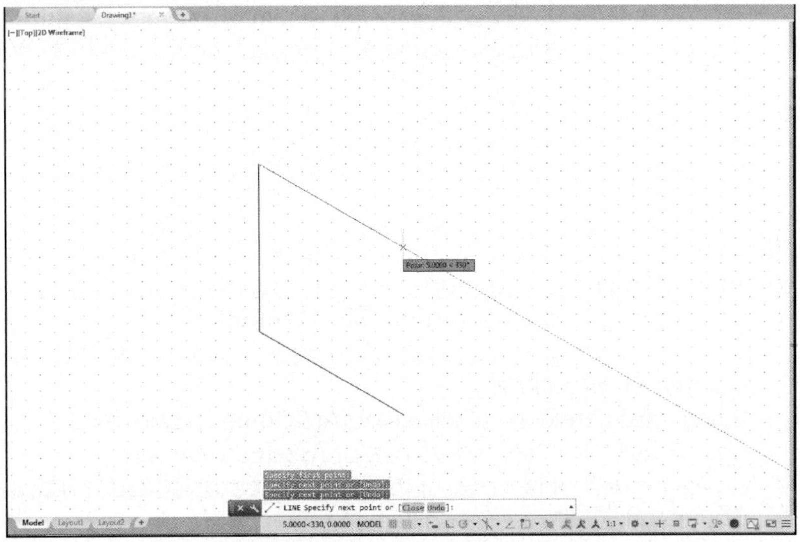

Alternately, use *Polar Tracking* instead of *Ortho*. Set the *Polar Angle Settings* to 30 degrees. The advantage of using *Polar Tracking* is that the current line length is given on the polar tracking tip (see Fig. 25-10). The disadvantage is that it is possible to draw non-isoplane lines accidentally. Only one setting at a time can be used in AutoCAD—*Polar Tracking* or *Ortho*. (The remainder of the figures illustrate the use of *Polar*.)

The effects of changing the *Isoplane* are shown in the following figures. Press Ctrl+E or use the *Isoplane* toggles to change *Isoplane*.

With *Ortho ON*, drawing a *Line* is limited to the two axes of the current *Isoplane*. Only one side of a cube, for example, can be drawn on the current *Isoplane*. Watch the coordinate display (in a polar format) or turn on *Dynamic Input* to give the length of the current *Line* as you draw.

Toggling Ctrl+E or the *Isoplane* options switches the cursor and the effect of *Ortho* and *Polar* to another *Isoplane*. One other side of a cube can be constructed on this *Isoplane* (Fig. 25-11).

**FIGURE 25-11**

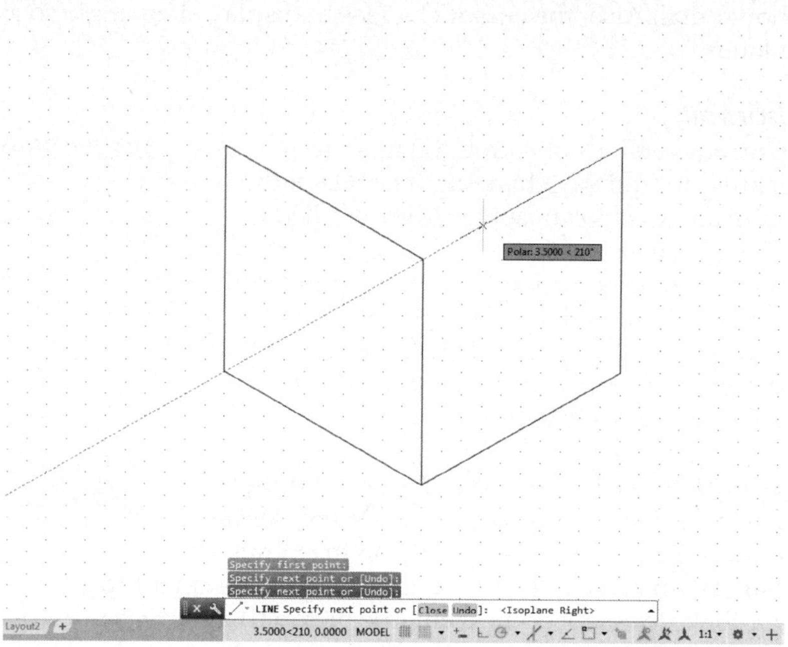


Direct Distance Entry can be of great help when drawing isometric lines. Use Ctrl+E and *Ortho* to force the *Line* to the correct orientation, then enter the desired distance value at the Command line.

## Isometric Ellipses

Isometric ellipses are easily drawn in AutoCAD by using the *Isocircle* option of the *Ellipse* command. This option appears only when the isometric *Snap* is *ON*.

### *Ellipse*

| Ribbon | Menu Bar | Command (Type) | Alias (Type) | Shortcut |
|---|---|---|---|---|
| *Home Draw* (icon) | *Draw Ellipse* | *Ellipse* | *EL* | ... |

**TIP** Although the *Isocircle* option does not appear in the menus, it can be invoked as an option of the *Ellipse* command. The *Isocircle* option of *Ellipse* appears only when the *Snap Type* is set to *Isometric*. You must type "I" to use the *Isocircle* option. The command syntax is as follows:

Command: ***ELLIPSE***
Specify axis endpoint of ellipse or [Arc/Center/Isocircle]: ***i***
Specify center of isocircle: **PICK** or (**coordinates**)
Specify radius of isocircle or [Diameter]: **PICK** or (**coordinates**)
Command:

After selecting the center point of the *Isocircle*, the isometrically correct ellipse appears on the screen on the current *Isoplane*. Use Ctrl+E or *Isoplane* to toggle the ellipse to the correct orientation. When defining the radius interactively, use *Ortho* to force the rubberband line to an isometric axis (Fig. 25-12, next page).

Since isometric angles are equal, all isometric ellipses have the same proportion (major to minor axis). The only differences in isometric ellipses are the size and the orientation (*Isoplane*).

**FIGURE 25-12**

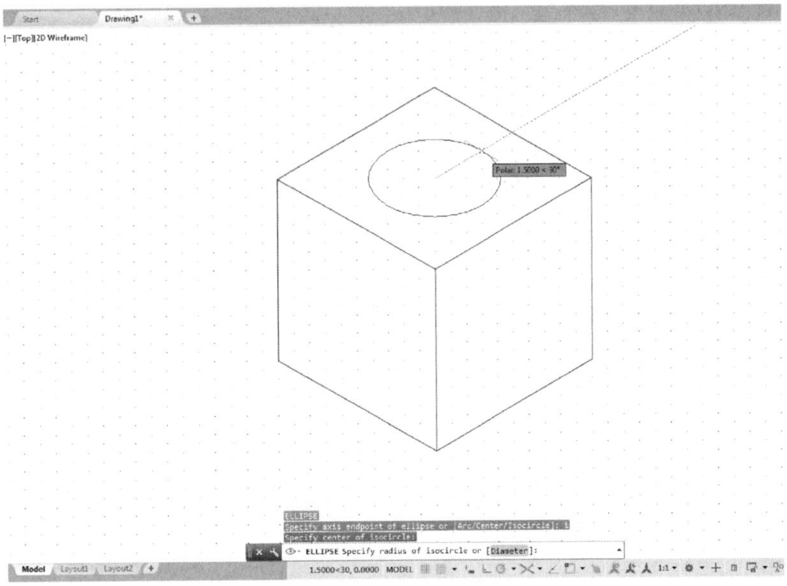


Figure 25-13 shows three ellipses correctly oriented on their respective faces. Use Ctrl+E or *Isoplane* to toggle the correct *Isoplane* orientation: *Isoplane Top*, *Isoplane Left*, or *Isoplane Right*.

**FIGURE 25-13**

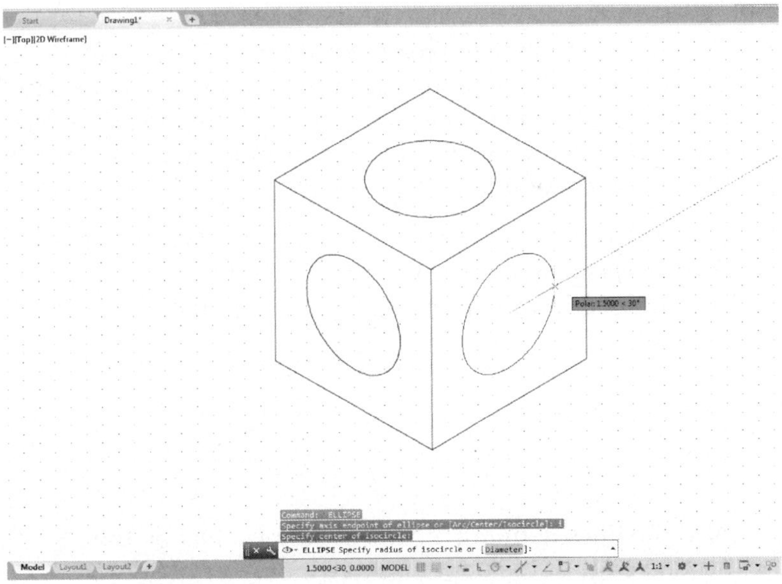


When defining the radius or diameter of an ellipse, it should always be measured in an isometric direction. In other words, an isometric ellipse is always measured on the two isometric axes (or center lines) parallel with the plane of the ellipse.

If you define the radius or diameter interactively, use *Ortho ON*. If you enter a value, AutoCAD automatically applies the value to the correct isometric axes.

## Creating an Isometric Drawing

In this exercise, the object in Figure 25-14 is to be drawn as an isometric.

FIGURE 25-14

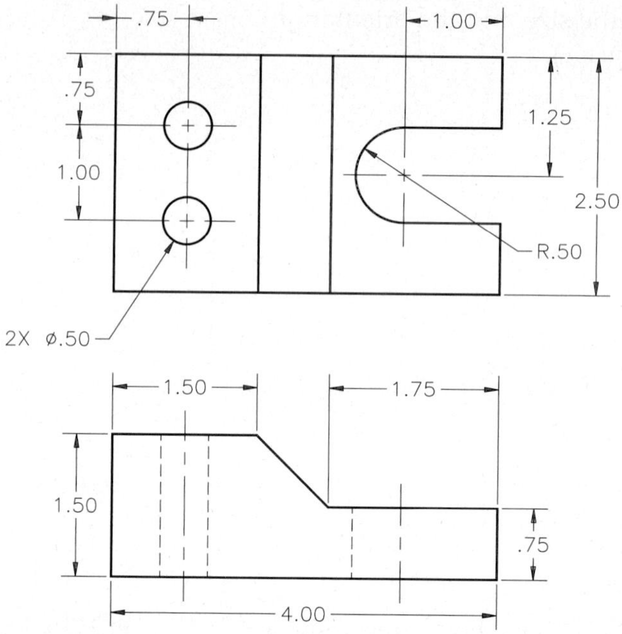


The initial steps to create an isometric drawing begin with the typical setup (see Chapter 6, Drawing Setup):

1. Set the desired *Units*.

2. Set appropriate *Limits*.

3. Set the *Isometric Style* of *Snap* and specify an appropriate value for spacing. Toggle *Adaptive grid* off.

4. The next step involves creating an isometric framework of the desired object. In other words, draw an isometric box equal to the overall dimensions of the object. Using the dimensions given in Figure 25-14, create the encompassing isometric box with the *Line* command (Fig. 25-15).

   Use *Ortho* or *Polar* to force isometric *Lines*. Watch the *coordinate display* display (in a relative polar format) to give the current lengths as you draw or use direct distance entry.

FIGURE 25-15

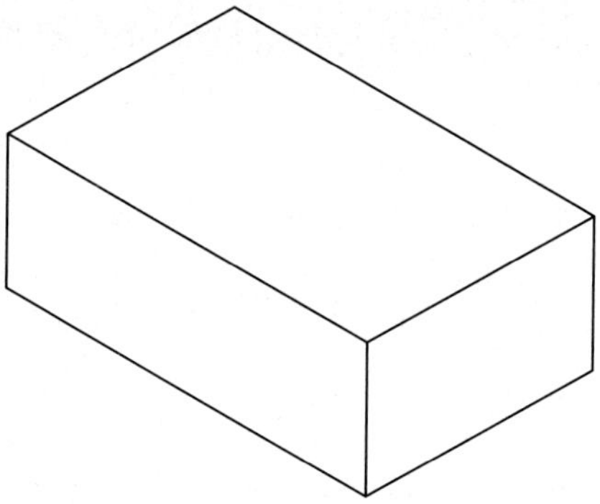

5. Add the lines defining the lower surface. Define the needed edge of the upper isometric surface as shown (Fig. 25-16).

FIGURE 25-16

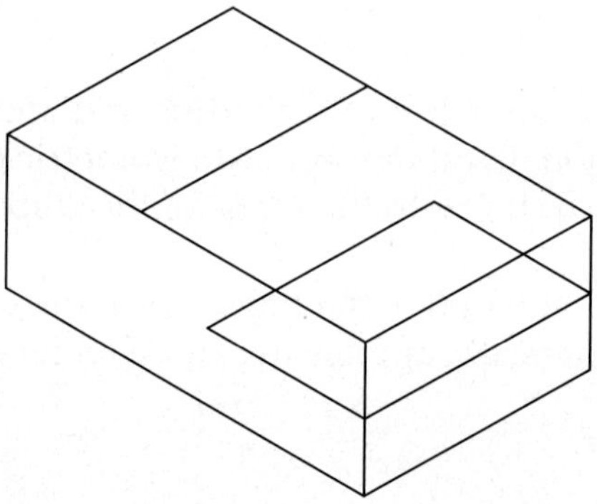

6. The <u>inclined</u> edges of the inclined surface can be drawn (with *Line*) only when *Ortho* is *OFF*. <u>Inclined</u> lines in isometric cannot be drawn by transferring the lengths of the lines, but only by defining the <u>ends</u> of the inclined lines on <u>isometric</u> lines, then connecting the endpoints. Next, *Trim* or *Erase* the necessary *Lines* (Fig. 25-17).

FIGURE 25-17

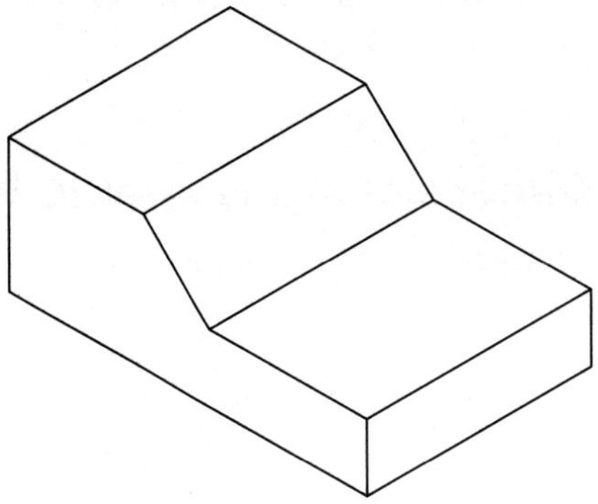

7. Draw the slot by constructing an *Ellipse* with the *Isocircle* option. Draw the two *Lines* connecting the circle to the right edge. *Trim* the unwanted part of the *Ellipse* (highlighted) using the *Lines* as cutting edges (Fig. 25-18).

FIGURE 25-18

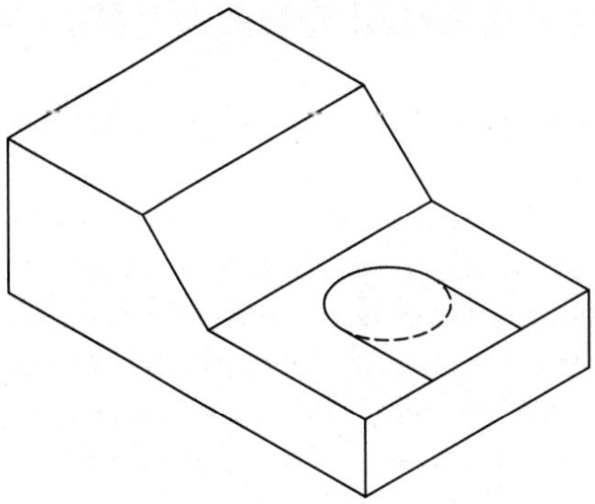

8. *Copy* the far *Line* and the *Ellipse* down to the bottom surface. Add two vertical *Lines* at the end of the slot (Fig. 25-19).

FIGURE 25-19

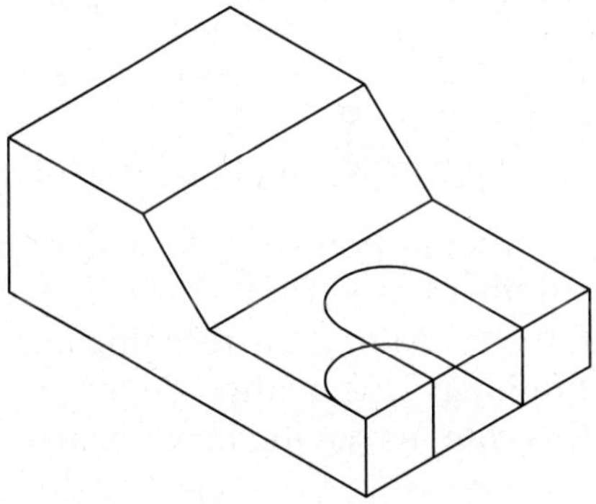

9. Use *Trim* to remove the part of the *Ellipse* that would normally be hidden from view. *Trim* the *Lines* along the right edge at the opening of the slot (Fig. 25-20).

FIGURE 25-20

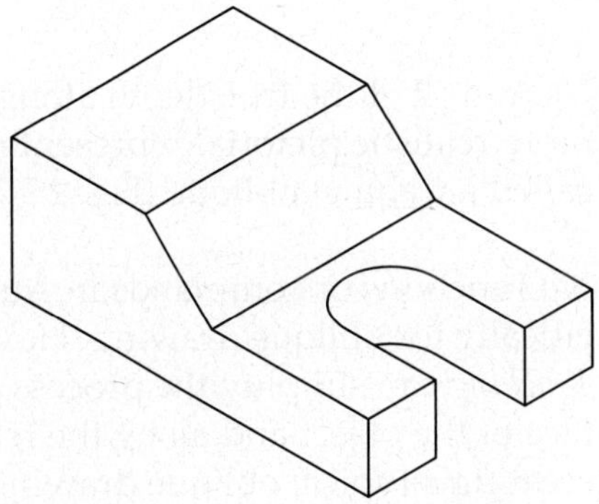

10. Add the two holes on the top with *Ellipse, Isocircle* option (Fig. 25-21). Use *Ortho ON* when defining the radius. *Copy* can also be used to create the second *Ellipse* from the first.

FIGURE 25-21

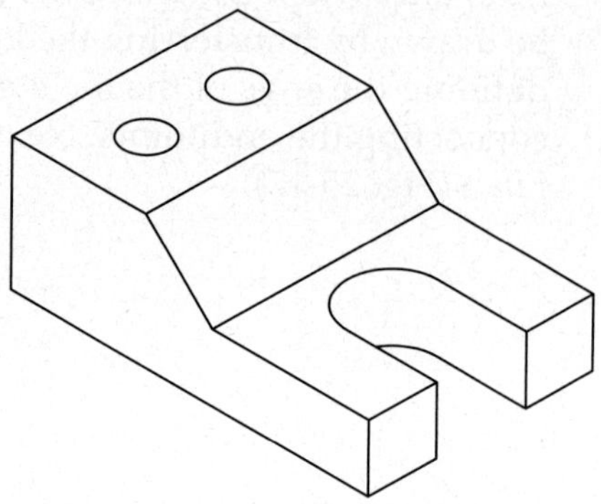

### Dimensioning Isometric Drawings in AutoCAD

Refer to Chapter 29, Dimension Styles and Variables, for details on how to dimension isometric drawings.

## OBLIQUE DRAWING IN AutoCAD

FIGURE 25-22

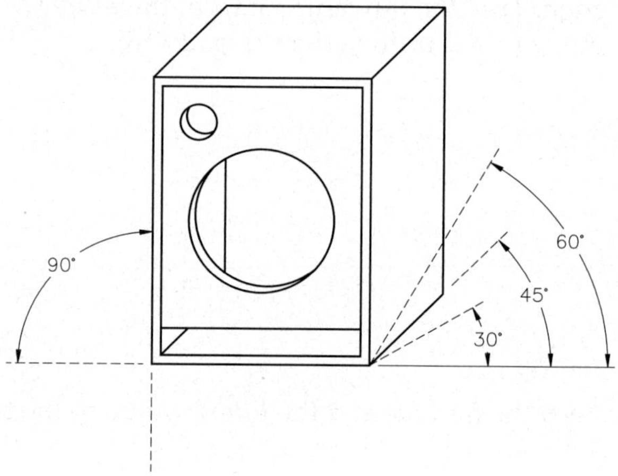


Oblique drawings are characterized by having two axes at a 90 degree orientation. Typically, you should locate the front face of the object along these two axes. Since the object's characteristic shape is seen in the front view, an oblique drawing allows you to create all shapes parallel to the front face true size and shape as you would in a multiview drawing. Circles on or parallel to the front face can be drawn as circles. The third axis, the receding axis, can be drawn at a choice of angles, 30, 45, or 60 degrees, depending on whether you want to show more of the top or the side of the object.

Figure 25-22 illustrates the axes orientation of an oblique
drawing, including the choice of angles for the receding axis.

Another option allowed with oblique drawings is the measurement used for the receding axis. Using the full depth of the object along the receding axis is called Cavalier oblique drawing. This method depicts the object (a cube with a hole in this case) as having an elongated depth (Fig. 25-23).

FIGURE 25-23

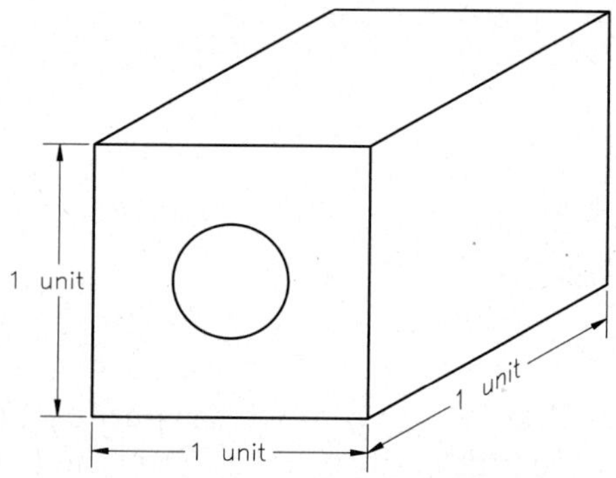


Using 1/2 of the true depth along the receding axis gives a more realistic pictorial representation of the object. This is called a Cabinet oblique (Fig. 25-24).

FIGURE 25-24

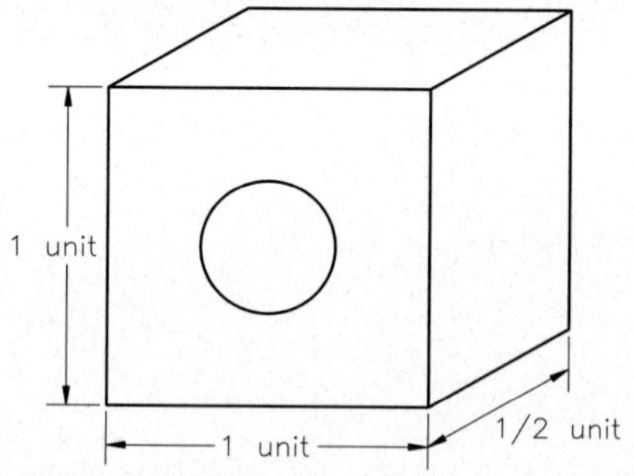


No functions or commands in AutoCAD are intended specifically for oblique drawing. However, *Polar Snap* and *Polar Tracking* can simplify the process of drawing lines on the front face of the object and along the receding axis. The steps for creating a typical oblique drawing are given next.

The object in Figure 25-25 is used for the example. From the dimensions given in the multiview, create a cabinet oblique with the receding axis at 45 degrees.

**FIGURE 25-25**

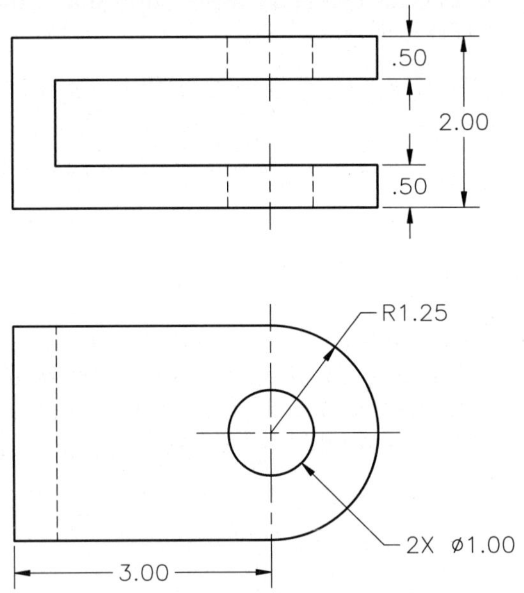


1. Create the characteristic shape of the front face of the object as shown in the front view (Fig. 25-26).

**FIGURE 25-26**

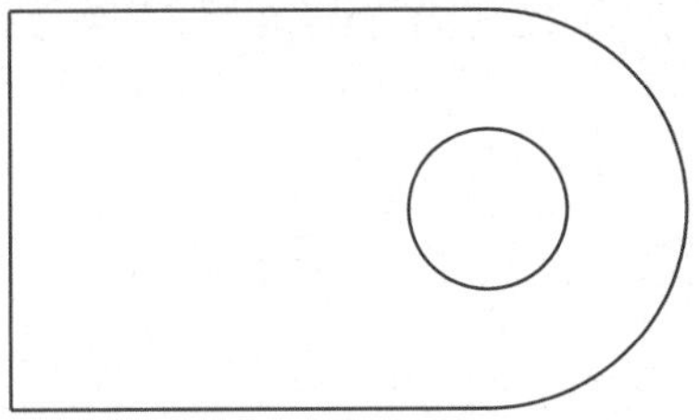

2. Use *Copy* with the multiple option to copy the front face back on the receding axis (Fig. 25-27). *Polar Snap* and *Polar Tracking* can be used to specify the "second point of displacement" as shown in Figure 25-27. Notice that a distance of .25 along the receding axis is used (1/2 of the actual depth) for this cabinet oblique.

**FIGURE 25-27**

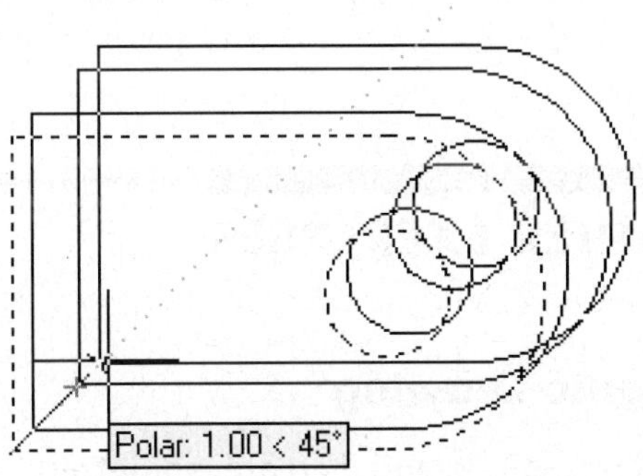


3. Draw the *Line* representing the edge on the upper-left of the object along the receding axis. Use *Endpoint Object Snap* to connect the *Lines*. Make a *Copy* of the *Line* or draw another *Line* .5 units to the right. Drop a vertical *Line* from the *Intersection* as shown (Fig. 25-28).

**FIGURE 25-28**

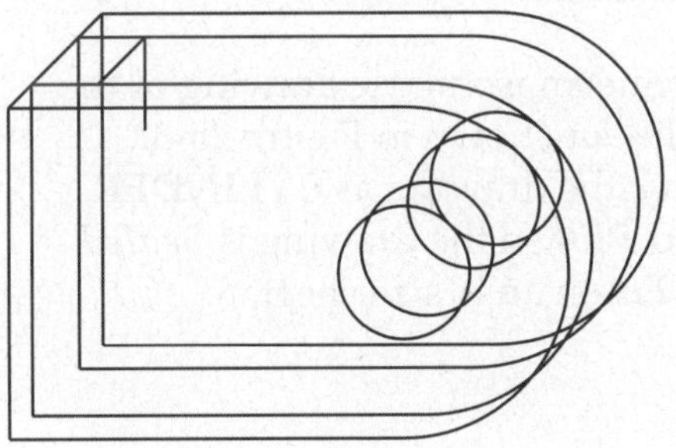

4. Use *Trim* and *Erase* to remove the unwanted parts of the *Lines* and *Circles* (those edges that are normally obscured) (Fig. 25-29).

FIGURE 25-29

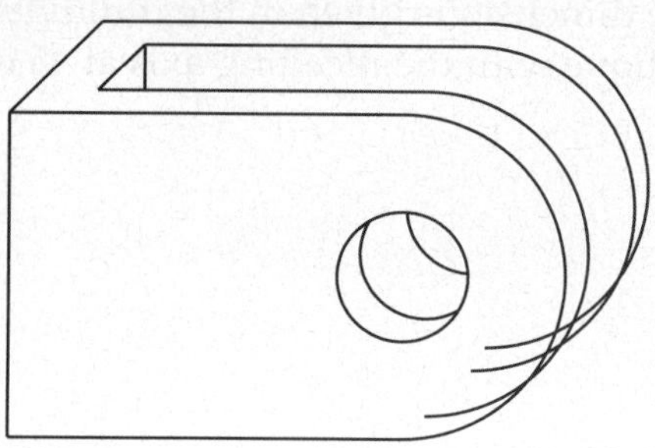

5. *Zoom* with a *window* to the lower-right corner of the drawing. Draw a *Line Tangent* to the edges of the arcs to define the limiting elements along the receding axis. *Trim* the unwanted segments of the arcs (Fig. 25-30).

FIGURE 25-30

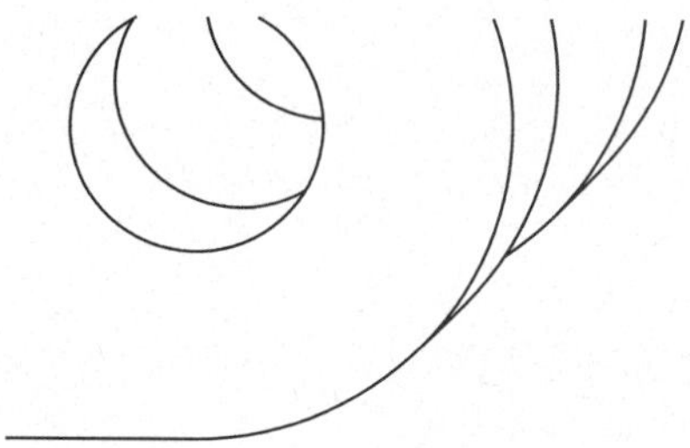

The resulting cabinet oblique drawing should appear like that in Figure 25-31.

FIGURE 25-31

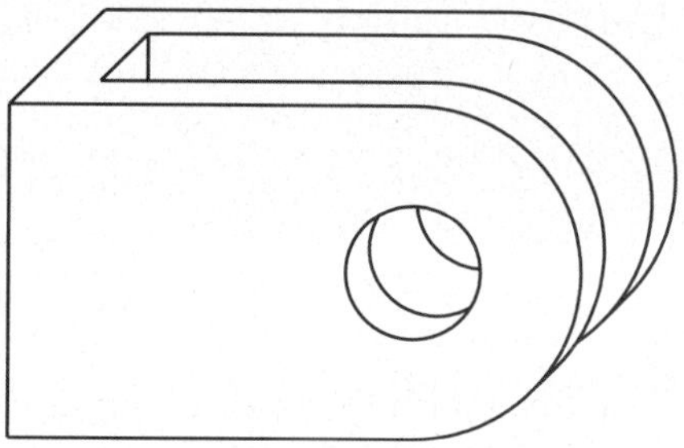

## CHAPTER EXERCISES

### Isometric Drawing

REUSE

For Exercises 1, 2, and 3 create isometric drawings as instructed. To begin, use an appropriate template drawing and draw a ***Pline*** border and insert the **TBLOCKAT.**

1. Create an isometric drawing of the cylinder shown in Figure 25-32. ***Save*** the drawing as **CYLINDER** and ***Plot*** so the drawing is ***Scaled to Fit*** on an A size sheet.

FIGURE 25-32

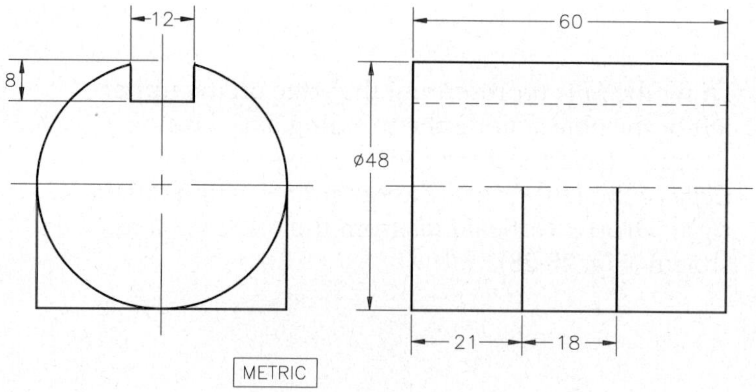

2. Make an isometric drawing of the Corner Brace shown in Figure 25-33. ***Save*** the drawing as **CRNBRACE**. ***Plot*** at **1=1** scale on an A size sheet.

**FIGURE 25-33**

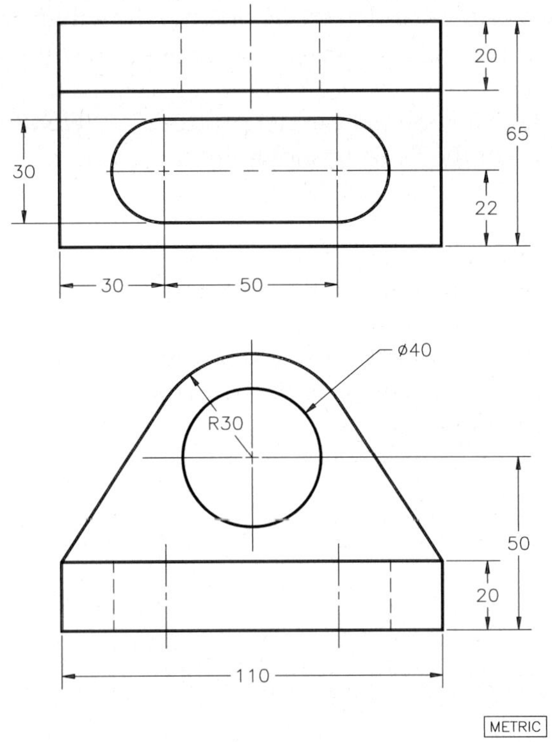


3. Draw the Support Bracket (Fig. 25-34) in isometric. The drawing can be ***Plotted*** at **1=1** scale on an A size sheet. ***Save*** the drawing and assign the name **SBRACKET.**

**FIGURE 25-34**

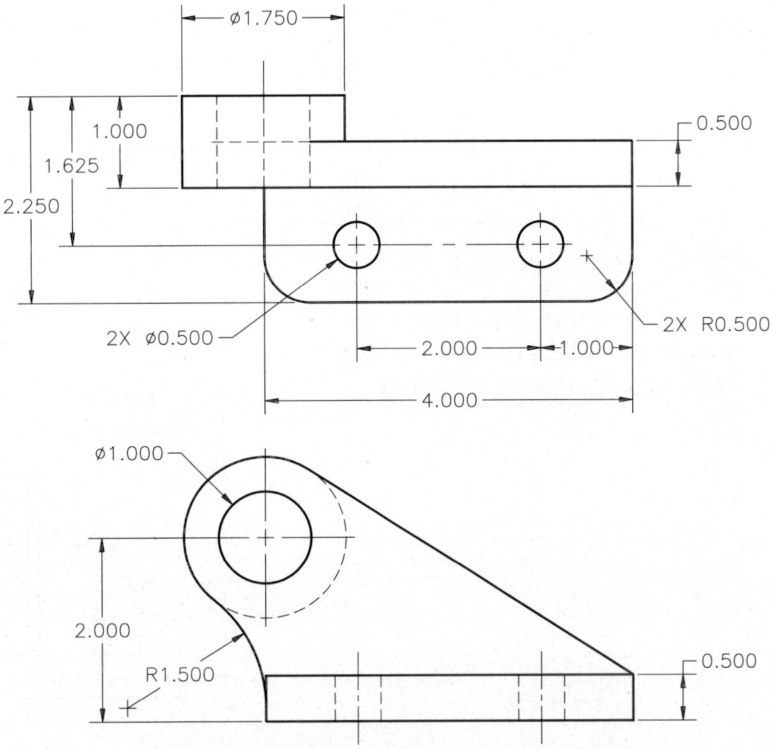


REUSE

REUSE

## Oblique Drawing

For Exercises 4 and 5, create oblique drawings as instructed. To begin, use an appropriate template drawing and draw a *Pline* border and *Insert* the **TBLOCKAT.**

4. Make an oblique cabinet projection of the Bearing shown in Figure 25-35. Construct all dimensions on the receding axis 1/2 of the actual length. Select the optimum angle for the receding axis to be able to view the 15 x 15 slot. ***Plot*** at **1=1** scale on an A size sheet. ***Save*** the drawing as **BEARING**.

FIGURE 25-35

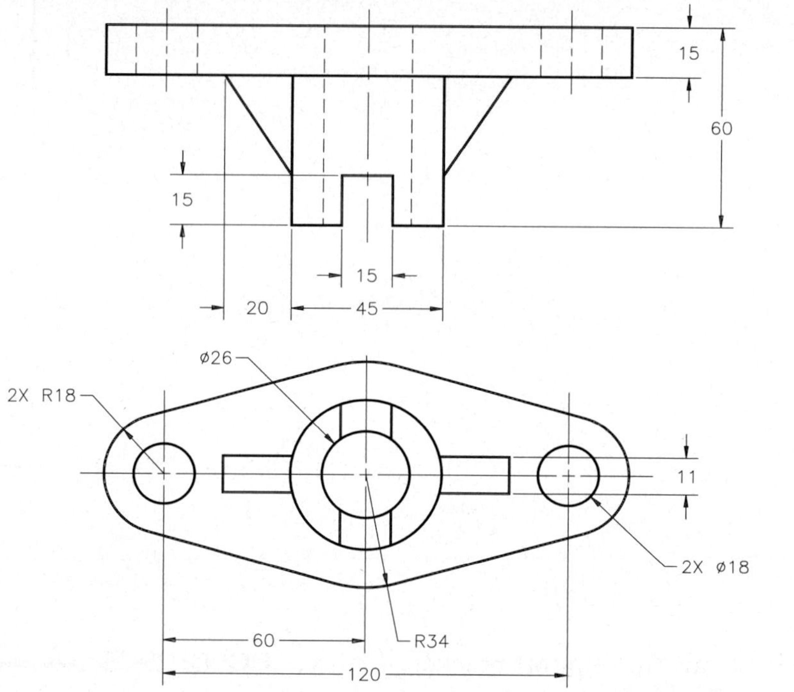


5. Construct a cavalier oblique drawing of the Pulley showing the circular view true size and shape. The illustration in Figure 25-36 gives only the side view. All vertical dimensions in the figure are diameters. ***Save*** the drawing as **PULLEY** and make a ***Plot*** on an A size sheet at **1=1**.

FIGURE 25-36

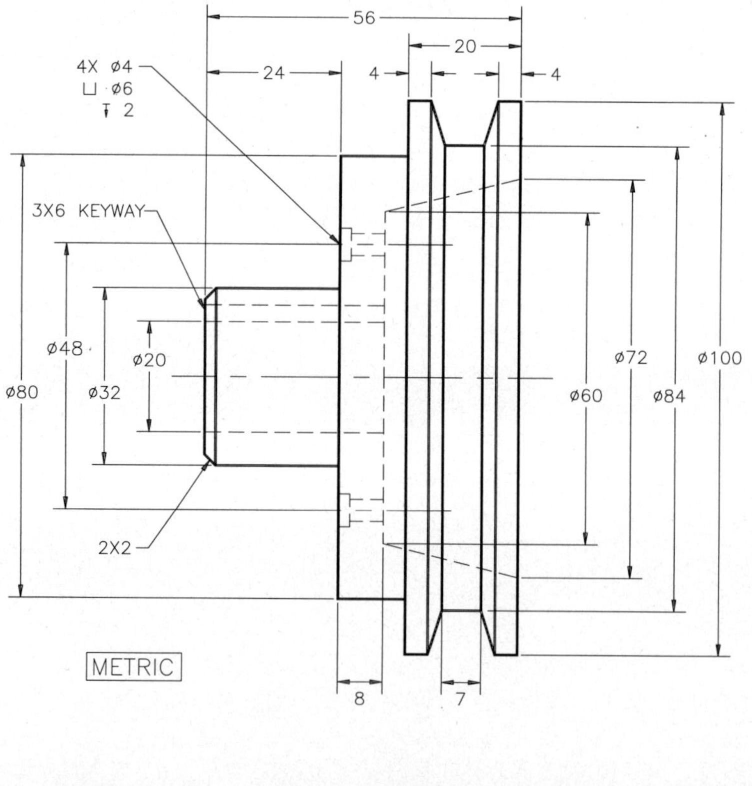

CHAPTER 26

# Section Views

## Chapter Objectives

After completing this chapter you should:

1. be able to use the *Hatch* command to apply associative hatch patterns;

2. know how to define a boundary for hatching using the *Pick Points* and *Select Objects* methods;

3. know how to create *Annotative* hatch patterns for use in multiple viewports;

4. be able to use *-Hatch* to create hatch lines and discard the boundary;

5. be able to drag and drop hatch patterns from a Tool Palette and be able to create your own Tool Palettes;

6. know how to use *Hatchedit* to modify parameters of existing hatch patterns in the drawing;

7. be able to edit hatched areas using *Trim*, *Grips*, selection options, and *Draworder*;

8. know how to draw cutting plane lines for section views.

## CONCEPTS

A section view is a view of the interior of an object after it has been imaginarily cut open to reveal the object's inner details. A section view is typically one of two or more views of a multiview drawing describing the object. For example, a multiview drawing of a machine part may contain three views, one of which is a section view.

Hatch lines (also known as section lines) are drawn in the section view to indicate the solid material that has been cut through. Each combination of section lines is called a hatch pattern, and each pattern is used to represent a specific material. In full and half section views, hidden lines are omitted since the inside of the object is visible.

Hatch lines are used also for many other applications. For example, architectural elevation drawings often include wall sections and use hatch lines to indicate walls that have been "cut through" to reveal the structural components. In addition, floor plans often contain a solid hatch pattern to indicate wall sections, as if the walls were cut through to reveal the "floor."

For mechanical drawings, a cutting plane line is drawn in an adjacent view to the section view to indicate the plane that imaginarily cuts through the object. Arrows on each end of the cutting plane line indicate the line of sight for the section view. A thick dashed or phantom line should be used for a cutting plane line.

This chapter discusses the AutoCAD methods used to draw hatch lines for section views and related cutting plane lines. The *Hatch* command allows you to select an enclosed area and select the hatch pattern and the parameters for the appearance of the hatch pattern, then AutoCAD automatically draws the hatch (section) lines. Existing hatch lines in the drawing can be modified using *Hatchedit*.

## DEFINING HATCH PATTERNS AND HATCH BOUNDARIES

A hatch pattern is composed of many lines that have a particular linetype, spacing, and angle. Many standard hatch patterns are provided by AutoCAD for your selection. Rather than having to draw each section line individually, you are required only to specify the area to be hatched and AutoCAD fills the designated area with the selected hatch pattern. A hatch pattern is inserted as one object. For example, you can *Erase* the inserted hatch pattern by selecting only one line in the pattern, and the entire pattern in the area is *Erased*.

FIGURE 26-1

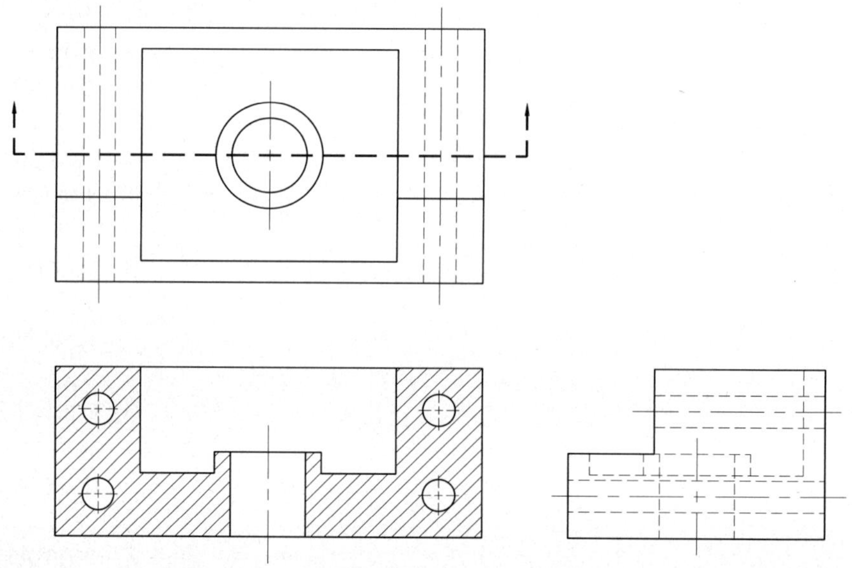

In a typical section view (Fig. 26-1), the hatch pattern completely fills the area representing the material that has been cut through. With the *Hatch* command you can define the boundary of an area to be hatched simply by pointing inside of an enclosed area.

The *Hatch* command is used to create hatch lines in the area that you specify. The *Hatch* command produces the *Hatch Creation* tab on the Ribbon. You can select the desired hatch pattern, then pick inside the area you want to fill, and *Hatch* automatically finds the boundary and fills it with the pattern. You can also drag and drop hatch patterns into a closed area using the Tool Palettes.

Hatch patterns created with *Hatch* are associative. Associative hatch patterns are associated to the boundary geometry such that when the shape of the boundary changes (by *Stretch, Scale, Rotate, Move, Properties*, grips, etc.), the hatch pattern automatically reforms itself to conform to the new shape (Fig. 26-2). For example, if a design change required a larger diameter for a hole, *Properties* could be used to change the diameter of the hole, and the surrounding section lines would automatically adapt to the new diameter.

**FIGURE 26-2**

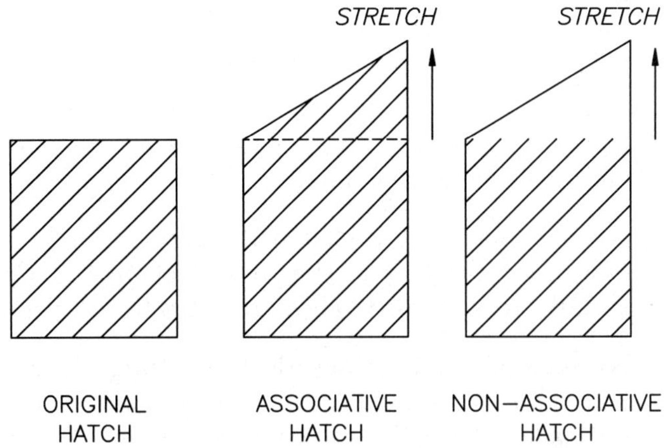


Once the hatch patterns have been drawn, any feature of the existing hatch pattern (created with *Hatch*) can be changed retroactively using *Hatchedit*. The *Hatchedit* dialog box gives access to the same options that were used to create the hatch (in the *Hatch Creation* tab of the Ribbon). Changing the scale, angle, or pattern of any existing section view in the drawing is a simple process.

Hatch patterns can also be assigned an *Annotative* property. Creating annotative hatch patterns is useful if you want to display the same hatched areas at different scales in multiple viewports.

***HPLAYER***

You can use the *HPLAYER* system variable to specify the layer on which hatches are created when the *Hatch* command is used. If you have not previously created a layer for hatches, specify a name for *HPLAYER* at the Command line and AutoCAD will automatically create the layer when the first *Hatch* is created. If you have previously created a layer for hatches, you can select that layer in the *Hatch Layer Override* section of the *Hatch Creation* tab on the Ribbon (see *Hatch*). The advantage of using this feature is that new hatches are always "placed" on the layer specified by this system variable, even though another layer can be current at the time you create the hatches.

The initial value for *HPLAYER* is "use current." Valid values include "." (use current), "ByLayer," or any layer name. This system variable is similar to the *CENTERLAYER* and *DIMLAYER* system variables (see *CENTERLAYER* in Chapter 24 and *DIMLAYER* in Chapter 28).

## Steps for Creating a Section View Using the *Hatch* Command

1. Create the view that contains the area to be hatched using typical draw commands such as *Line, Arc, Circle,* or *Pline*. If you intend to have text or dimensions inside the area to be hatched, add them before hatching.

2. Invoke the *Hatch* command. The *Hatch Creation* tab appears on the Ribbon (see Fig. 26-3).

3. Select the desired pattern from the *Pattern* drop-down list.

4. Specify the *Properties* for the hatch pattern such as the *Scale, Angle,* or *Transparency*. Select a layer for hatches using the *Hatch Layer Override* drop-down list.

5. Define the area to be hatched by PICKing an internal point (*Pick Points* button) or by individually selecting the objects (*Select* button).

6. If needed, specify any other parameters, such as *Origin, Hatch Color, Background Color, Associative,* etc.

7. Preview the hatch to make sure everything is as expected. Adjust hatching parameters as necessary.

8. Apply the hatch by selecting *Close Hatch Creation*. The hatch pattern is automatically drawn and becomes an object in the drawing.

9. If other areas are to be hatched, additional internal points or objects can be selected to define the new area for hatching. The parameters used previously appear again in the *Hatch Creation* tab by default. You can also *Match Properties* from a previously applied hatch.

10. For mechanical drawings, draw a cutting plane line in a view adjacent to the section view. A *Lineweight* is assigned or a *Pline* with a *Dashed* or *Phantom* linetype is used. Arrows at the ends of the cutting plane line indicate the line of sight for the section view.

11. If any aspect of the hatch lines needs to be edited at a later time, *Hatchedit* can be used to change those properties. If the hatch boundary is changed by *Stretch, Rotate, Scale, Move, Properties,* etc., the hatched area will conform to the new boundary.

## Hatch

| Ribbon | Menu Bar | Command (Type) | Alias (Type) | Shortcut |
|---|---|---|---|---|
| *Home Draw* (icon) | *Draw Hatch...* | *Hatch* or *-Hatch* | *BH* or *H* | ... |

*Hatch* allows you to create hatch lines for a section view (or for other purposes) by simply PICKing inside a closed boundary. A closed boundary refers to an area completely enclosed by objects. *Hatch* locates the closed boundary automatically by creating a temporary *Pline* that follows the outline of the hatch area, fills the area with hatch lines, and then deletes the boundary (default option) after hatching is completed. *Hatch* ignores all objects or parts of objects that are not part of the boundary.

Invoking the *Hatch* command produces the *Hatch Creation* tab on the Ribbon as shown in Fig. 26-3.

**FIGURE 26-3**

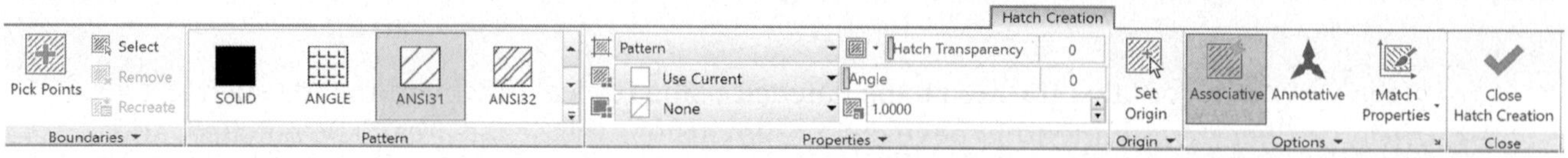


Typically, the first step is to select the desired *Pattern*. After you select the *Pattern* and other parameters for the hatch, you select *Pick points* or *Select objects* to return to the drawing and indicate the area to fill with the pattern.

Hovering over the closed boundary provides a preview of the hatch (Fig. 26-4).

FIGURE 26-4

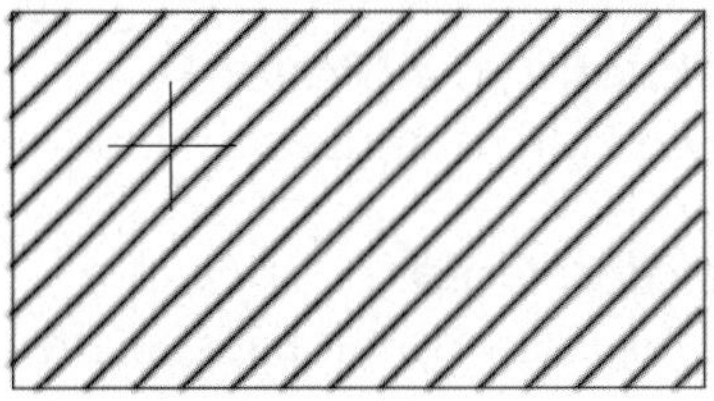

## *Pattern* Panel

### *Pattern*

Selecting the *Pattern* drop-down list (Fig. 26-5) displays the name of each predefined pattern. Use the scroll bar on the right side of the list to display all the available standard patterns that AutoCAD provides (in the ACAD.PAT file). Making a selection dictates the current pattern and causes the pattern to display when a boundary is previewed in the drawing area.

FIGURE 26-5

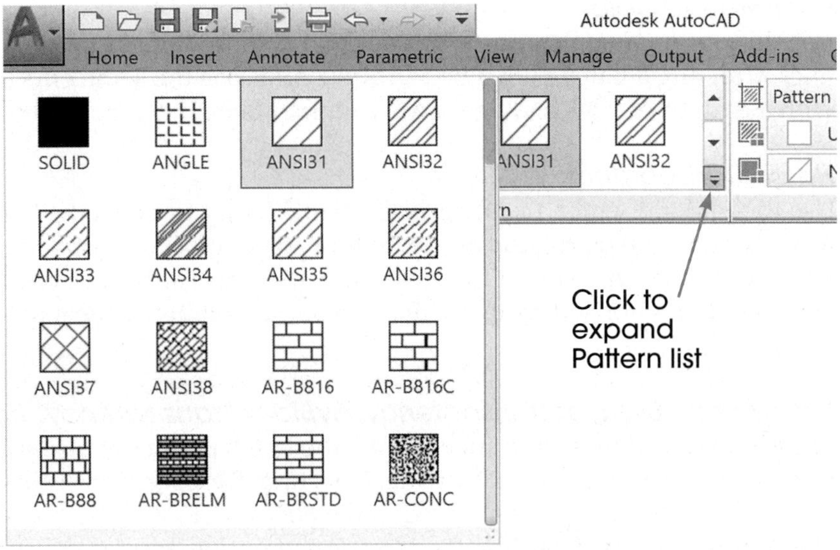


If you use the *Hatch Type* drop-down list in the *Properties* panel, you can select from *Pattern* (appears by default), *Solid*, *Gradient*, or *User-defined* hatch pattern types. Your choice here automatically scrolls to the appropriate section in the *Pattern* drop-down list.

*Pattern*
These patterns are the standard ANSI and architectural hatch patterns typically comprised of lines.

*Solid*
The *Solid* pattern fills the hatch boundary with a solid color. You can use the current *Bylayer* color, *Byblock* color or select from specific colors to override the layer color.

*Gradient*
The *Gradient* hatch types are composed of two colors that blend with a smooth transition. See "*Gradient* Hatches" later in this chapter.

TIP

*User-defined*
To define a simple hatch pattern "on the fly," select *User-defined*. A *User-defined* pattern is a simple pattern composed of horizontal (default angle) lines. You can also create *Double* (perpendicular) lines. All *User-defined* patterns have continuous lines. You can specify the *Spacing* between lines (similar to *Scale*), *Angle, Transparency, Color,* and *Double* options for the pattern.

NOTE: All hatch patterns are created using the current linetype. Therefore, the *Continuous* linetype should be set current when hatching to ensure the selected area is filled with the pattern as it appears in the image tile.

After selecting a pattern, specify the desired *Properties* such as *Scale* and *Angle* of the pattern.

## *Properties* Panel

### *Hatch Type (Pattern)*

See *Pattern* in the previous section.

### *Hatch Color*

This drop-down list specifies a color to use for the pattern. You can use the current layer color (*ByLayer*) or current object-specific color property (*Use Current*). Selecting a color from the color swatches uses that color for the hatch pattern and overrides the current layer and/or object-specific color.

### *Background Color*

Use this option if you want a separate solid color to appear in the boundary area behind the pattern lines. Creating a hatch with a background color has the same effect as hatching the boundary area once with a pattern of lines and again with another *Solid* hatch behind the lines.

### *Hatch Transparency*

Use this slider or enter a value greater than 0 (zero) to cause the hatch pattern to be transparent. A transparency of 0 (zero) causes the lines or solid color to be opaque, whereas a value of 90 (maximum allowed) causes the pattern to be very transparent. The value specified here is saved in the system variable based on your setting using the options shown in the tile just to the left of the slider (explained next).

### *Use Current, ByLayer Transparency, ByBlock Transparency, Transparency Value*

These options allow you to specify how the hatch pattern transparency setting is saved relative to other transparency settings in the drawing. Selecting *Transparency Value* sets the transparency level specifically for new hatches or fills (saved in the *HPTRANSPARENCY* system variable), overriding the current object transparency. Selecting the *Use Current* option applies the current object-specific transparency setting specified for the drawing (*CETRANSPARENCY*) to newly created hatches. Selecting *ByLayer Transparency* or *ByBlock Transparency* sets the current transparency value specified for the current layer or block, respectively, to the newly created hatches.

### *Angle*

The *Angle* specification determines the angle (slant) of the hatch pattern. The default angle of 0 represents whatever angle is displayed in the pattern's image tile. Any value entered deflects the existing pattern (as it appears in the swatch) by the specified value (in degrees). The value entered in this box is held in the *HPANG* system variable.

### *Scale*

The value entered in this edit box is a scale factor that is applied to the existing selected pattern. Normally, this scale factor should be changed proportionally with changes in the drawing *Limits*. Like many other scale factors (*LTSCALE, DIMSCALE*), AutoCAD defaults are set to a value of 1, which is appropriate for the default *Limits* of 12 x 9. If you have calculated the drawing scale factor, enter that value in the *Scale* edit box (see Chapter 12 for information on the "Drawing Scale Factor"). The *Scale* value is stored in the *HPSCALE* system variable.

The scale you specify in this edit box is always relative to the model space geometry. If you are creating *Annotative* hatch patterns to display in multiple viewports at different scales, the hatch pattern scale can be automatically adjusted for the viewport scale (see "Annotative Hatch Patterns").

ISO hatch patterns are intended for use with metric drawings; therefore, the scale (spacing between hatch lines) is much greater than for inch drawings. Because *Limits* values for metric sheet sizes are greater than for inch-based drawings (25.4 times greater than comparable inch drawings), the ISO hatch pattern scales are automatically compensated. If you want to use an ISO pattern with inch-based drawings, calculate a hatch pattern scale based on the drawing scale factor and multiply by .039 (1/25.4).

### *Spacing*

This option is enabled if *User-defined* pattern is specified. Enter a value for the distance between lines.

### *Hatch Layer Override*

Normally the hatch pattern is created on the current layer. If you want to place the hatch on another layer, use this drop-down list to select an existing layer in the drawing on which the new hatch pattern will be created. This action sets the *HPLAYER* system variable.

### *Double*

Only for a *User-defined* pattern, check this box to have a second set of lines drawn at 90 degrees to the original set.

### *Relative to Paper Space*

This option is useful when you want to change the hatch pattern scale relative to paper space units for the current viewport. The viewport must be active, then use *Hatch* or *Hatchedit* to select this option. All other viewports display the hatch pattern scale relative to the one active viewport; therefore, this option is of limited use when you have more than one viewport in the drawing. If you plan to display the hatch pattern in multiple viewports at multiple scales, you may want to use the *Annotative* property (see "Annotative Hatch Patterns") or create multiple hatch patterns on different layers and display each layer only in the appropriate viewport.

### *ISO Pen Width*

You must select an ISO hatch pattern for this tile to be enabled. Selecting an *ISO Pen Width* from the drop-down list automatically sets the scale and enters the value in the *Scale* edit box. See *"Scale."*

## ***Boundaries* Panel (Selecting the Hatch Area)**

Once the hatch pattern and properties have been selected, you must indicate to AutoCAD what area(s) should be hatched. Either the *Pick points* method, *Select boundary objects* method, *Draw* method or a combination can be used to accomplish this.

### *Pick Points*

This tile should be selected if you want AutoCAD to automatically determine the boundaries for hatching. You only need to select a point inside the area you want to hatch. The point selected must be inside a closed shape. When the *Pick points* tile is selected, AutoCAD gives the following prompts:

```
Specify start point or [picK internal point/Select objects/Rectangle/Circle/Mode/seTtings]: K
Pick internal point or [Select objects/Draw/Undo/seTtings]:
```

When an internal point is PICKed, AutoCAD traces and highlights the boundary (Fig. 26-6). The interior area is then analyzed for islands to be included in the hatch boundary. Multiple boundaries can be designated by selecting multiple internal points.

The location of the point selected is usually not critical. However, the point must be PICKed inside the expected boundary. If there are any large gaps in the area, a complete boundary cannot be formed and a boundary error message appears (see *"Gap Tolerance"*).

**FIGURE 26-6**

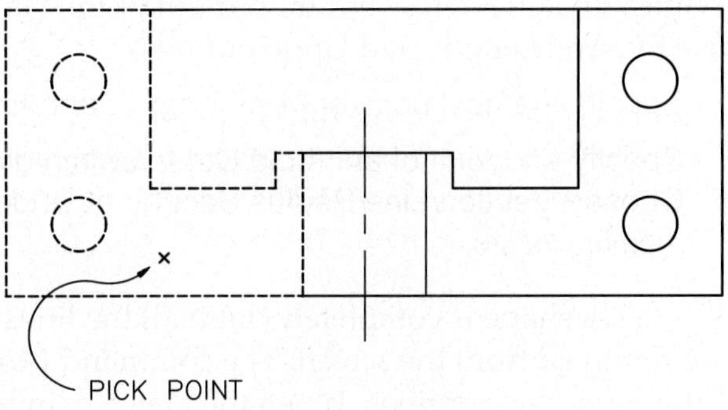

### *Select Boundary Objects*

FIGURE 26-7

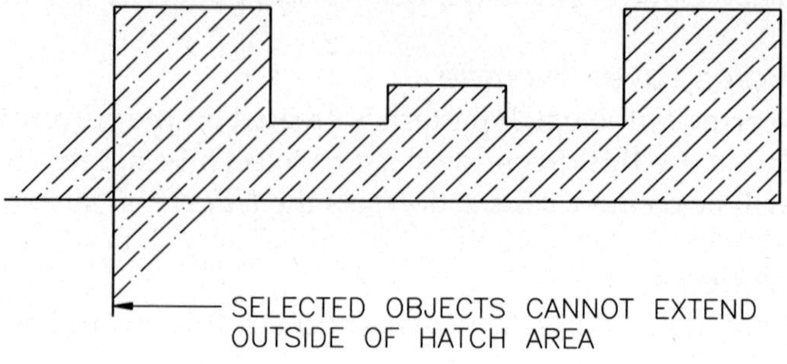


Alternately, you can designate the boundary with the *Select objects* method. Using the *Select Objects* method, you specify the boundary objects rather than let AutoCAD locate a boundary. With the *Select objects* method, no temporary *Pline* boundary is created as with the *Pick points* method. Therefore, the selected objects must form a closed shape with no large gaps or overlaps. If gaps exist or if the objects extend past the desired hatch area (Fig. 26-7), AutoCAD cannot interpret the intended hatch area correctly, and problems may result.

The *Select objects* option can be used after the *Pick points* method to select specific objects for *Hatch* to consider before drawing the hatch pattern lines. For example, if you had created text objects or dimensions within the boundary found by the *Pick points* method, you may then have to use the *Select objects* option to select the text and dimensions (if *Island detection* was off). Using this procedure, the hatch lines are automatically "trimmed" around the text and dimensions.

### *Draw*

FIGURE 26-8

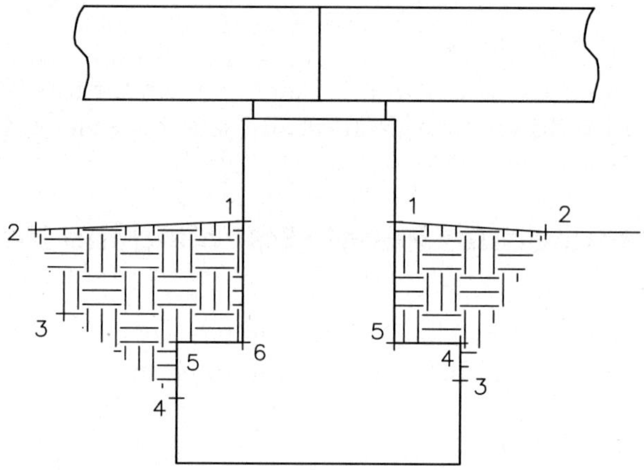


The *Draw* option for hatch boundary is useful when you want to hatch a shape that is not inside a boundary, or is not bordered by other shapes (Fig. 26-8). It makes it quick to put some hatching in a freeform area.

To use the *Draw* option, at the Hatch command prompt,

Pick internal point or [Select objects/Draw/Undo/seTtings]: **D**

Specify start point or [picK internal point/Select objects/Rectangle/Circle/Mode/seTtings]: *Click a location in your drawing to start a boundary* (Fig. 26-9).

Specify the next point or [Arc/Length/Undo/Exit] <Exit>: *click the next location for the boundary*

Specify the next point or [Arc/Close/Length/Undo/Exit] <Exit>: *click the next location for the boundary*

FIGURE 26-9

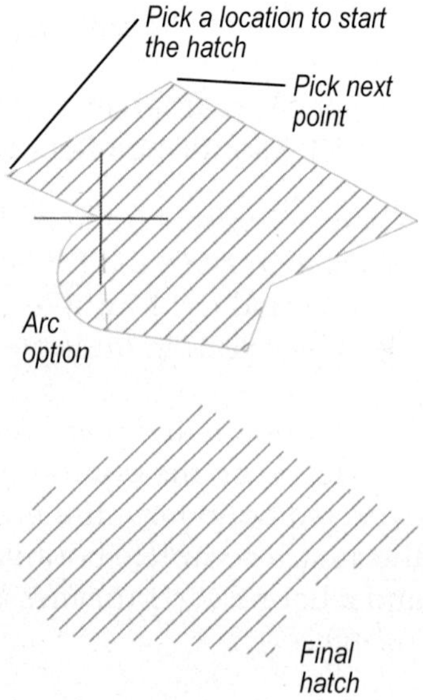


Once a third point is selected you see the hatch appearing in the shape you defined. Use the Close option to join back to the starting point. It is important to use Close and not to click the first point again to prevent creating an overlapping shape.

While still at the Draw option prompt of the Hatch command, you can also use the Arc, Length, and Undo options.

Specify the next point or [Arc/Length/Undo/Exit] <Exit>: **A** [Enter]

Specify endpoint of arc (hold Ctrl to switch direction) or [Angle/CEnter/CLose/Direction/Line/Radius/Second pt/Undo]: *Click a location to begin drawing an arc.*

When the shape is completely defined, the lines drawing the perimeter are removed from the screen. The command prompt still shows the Hatch command *Draw* options. When the Hatch command *Draw* option starts the command line shows important information including the Mode which allows you to choose to define an area for the hatch, or to define a path along which the hatch is created.

Current hatch mode: Area

Specify start point or [picK internal point/Select objects/Rectangle/Circle/Mode/seTtings]: **M**

Specify draw hatch mode [Area/Path] <Area>: **P**

Current hatch mode: Path, Width = 0.2500, Alignment = Center

Specify start point or [picK internal point/Select objects/Rectangle/Circle/ALignment/Width/Mode/seTtings]: *Click a location from your screen. As you click more locations the hatch is drawn along the path you define using the Width and Alignment settings* (Fig. 26-10).

**FIGURE 26-10**

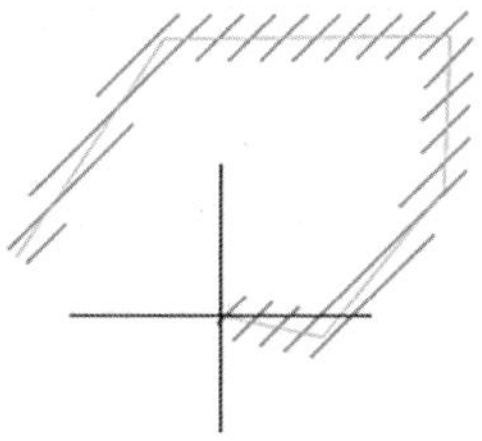

To return to picking internal points inside a hatch boundary, use the K option.

Specify start point or [picK internal point/Select objects/Rectangle/Circle/ALignment/Width/Mode/seTtings]: **K**

### *Remove Boundary Objects*

Clicking this tile allows you to remove previously selected boundaries or to select specific islands (internal objects) to remove from those AutoCAD has found within the outer boundary. If hatch lines have been drawn, be careful to *Zoom* in close enough to select the desired boundary and not the hatch pattern.

### *Recreate Boundary*

This option is always disabled in the *Hatch Creation* tab, but is available for editing hatches. See "*Hatchedit*."

### *Display Boundary Objects*

This option is disabled in the *Hatch Creation* tab but is available for editing hatches. See "*Hatchedit*."

### *Retain Boundaries, Don't Retain Boundaries*

When AutoCAD uses the *Pick points* method to locate a boundary for hatching, a temporary boundary object is created for hatching, then discarded after the hatching process. Choosing a *Retain Boundaries* option forces AutoCAD to keep the boundary. When the box is checked, *Hatch* creates two objects—the hatch pattern and the boundary object. You can specify whether you want to create a *Pline* or a *Region* boundary. You can also elect to discard the boundaries (*Don't Retain*).

### *Specify Boundary Set*

By default, AutoCAD examines all objects in the viewport when determining the boundaries by the *Pick points* method. (*Current Viewport* is selected by default when you begin the *Hatch* command.) For complex drawings, examining all objects can take some time. In that case, you may want to specify a smaller boundary set for AutoCAD to consider. Using the *Select New Boundary Set* option permits you to select objects or select a window to define the new set.

## ***Origin* Panel**

Although you can specify the options, patterns, origin, and hatch boundaries in any order, you typically first select the pattern and specify the boundary (area to be hatched), and then apply the other options. In this way you can preview in the drawing how the optional parameters affect the hatch when you apply them.

Some hatch patterns, such as brick patterns, need to be aligned with a point on the hatch boundary. Figure 26-11 illustrates a brick pattern using the default (current origin) and a *Bottom left* origin. Use the options in this panel to control the starting location of hatch pattern generation.

FIGURE 26-11

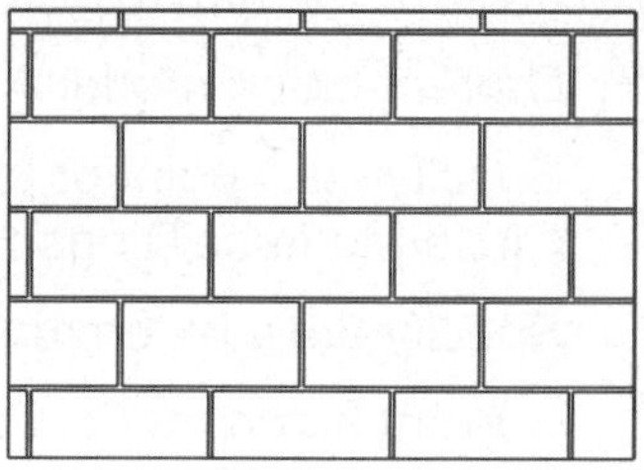

CURRENT ORIGIN

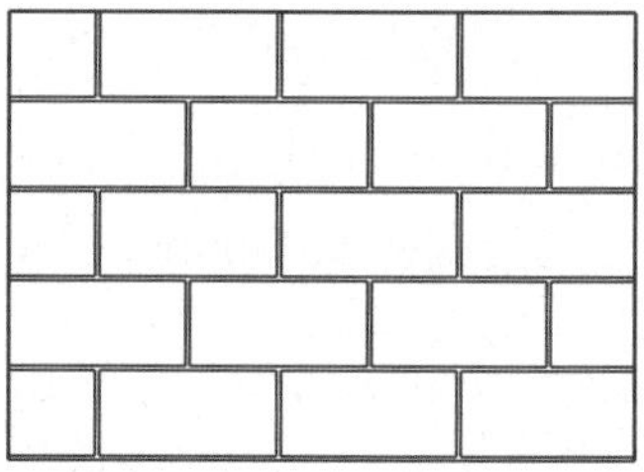

BOTTOM LEFT

### *Set Origin*

This option allows you to select any point in the drawing to specify the new hatch origin point. Typically use an *Object Snap* to specify an exact location.

### *Bottom Left, Bottom Right, Top Left, Top Right, Center*

When you select an area to hatch, AutoCAD automatically locates a closed boundary outside of the point you select. Use one of these options to specify the desired boundary corner or center to use as the origin.

### *Use Current Origin*

This option sets 0,0 of the current coordinate system as the default origin for newly created hatches.

### *Store as Default Origin*

If you use a specified origin, select this button to store the value of the new hatch origin in the *HPORIGIN* system variable for future hatches in the drawing.

## ***Options* Panel**

This panel is a collection of options that are unrelated. The *Gap Tolerance* and *Island Detection* options affect how boundaries are created.

### *Associative*

This checkbox toggles on or off the associative property for hatch patterns. Associative hatch patterns automatically update by conforming to the new boundary shape if the boundary is changed (see Fig. 26-2). A non-associative hatch pattern is static even when the boundary changes or is erased.

### *Annotative*

See "Annotative Hatch Patterns" later in this chapter.

### *Match Properties (Inherit Properties in Hatch and Gradient* dialog box)

Use this section if you want to copy an existing hatch pattern in the drawing, including all the properties, to a new boundary area. This feature works similarly to the *Match Properties* command. You are prompted to select an existing hatch pattern from the drawing and then pick an internal point to specify a new boundary area (where you want to copy the hatch to).

### *Use Current Origin, Use Source Hatch Origin (Match Properties)*

When you copy an existing hatch with *Match Properties*, you have the choice to *Use the Current Origin* (by default, 0,0 of the current coordinate system) or *Use the Source Hatch Origin*. No choice must be made here unless a specific origin was previously specified for the source hatch object.

FIGURE 26-12

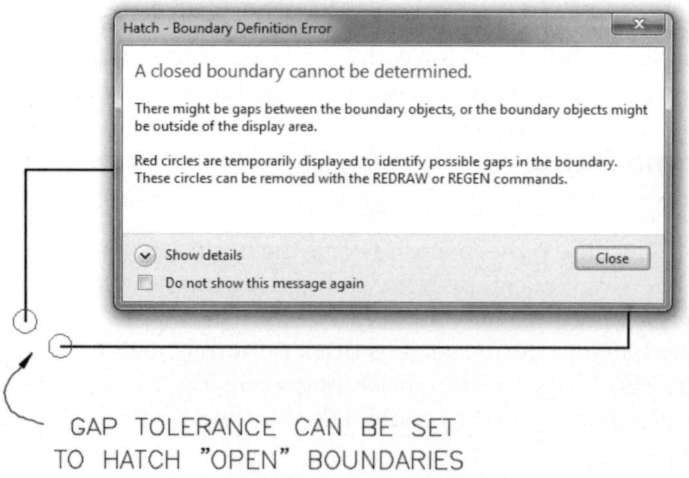


### *Gap Tolerance*

*Tolerance* is set to 0 by default. In this case, all boundaries selected to hatch with the *Hatch* command must be completely closed—that is, no hatch boundaries can contain a "gap." However, if gaps exist in the selected boundary, you can set a *Tolerance* value to compensate for any boundary gaps (Fig. 26-12). You can set the *Gap tolerance* in the *Options* panel of the *Hatch Creation* tab or set the *HPGAPTOL* system variable. The *Gap tolerance* must be greater than the size of the gap in drawing units.

If you attempt to hatch an open area when no *Gap tolerance* is set or a gap exists that is greater than the setting for the *Gap tolerance*, a *Hatch - Boundary Definition Error* warning box appears (see Figure 26-9). However, if a gap exists that is within the *Gap tolerance*, the object will hatch correctly after selecting *Continue hatching this area* in the *Open Boundary Warning* box that appears.

### *Create Separate Hatches*

By default, multiple hatches created with one use of the *Hatch* command are treated as one AutoCAD object, even when the hatches are within separate boundaries. In such a case, multiple hatch areas could be *Erased* by selecting only one of the hatched areas. Check *Create Separate Hatches* to ensure each boundary area forms a unique hatch object.

### *Island Detection*

Check this box to cause AutoCAD to automatically detect any internal areas (islands) inside the selected boundary, such as holes (*Circles*), text, or other closed shapes. *Island detection* is performed according to the following options.

**FIGURE 26-13**

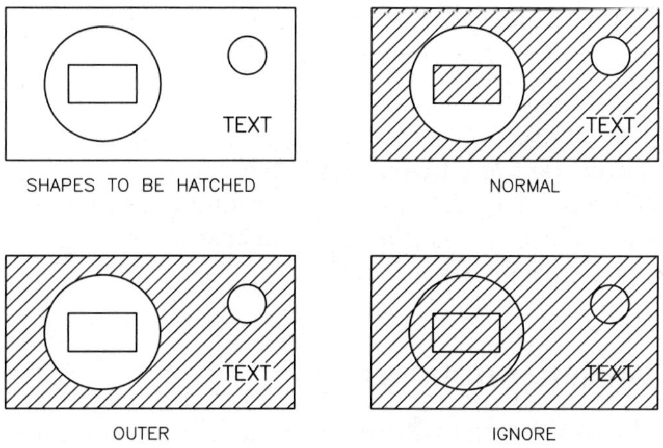


*Normal*

This option should be used for most applications of *Hatch*. Text or closed shapes within the outer border are considered in such cases. Hatching begins at the outer boundary and moves inward, alternating between applying and not applying the pattern as interior shapes or text are encountered (Fig. 26-13).

*Outer*

This option causes AutoCAD to hatch only the outer closed shape. Hatching is turned off for all interior closed shapes (Fig. 26-12).

*Ignore*

*Ignore* draws the hatch pattern from the outer boundary inward, ignoring any interior shapes. The resulting hatch pattern is drawn through the interior shapes (Fig. 26-13).

### Draw Order

Although the draw order does not affect most hatching applications, this option is helpful when you assign solid or gradient hatch patterns and when hatch patterns (containing solid or gradient patterns) overlap each other or other objects (see "*Gradient* Hatches" later in this chapter). For example, a gradient pattern could be created behind or in front of the boundary (Fig. 26-14). Your choice of *Send to back, Bring to front, Send behind boundary, Bring in front of boundary,* or *Do Not Assign* affects the current hatch pattern. You can also use the *Draworder* command to control the display order after creating the hatches (see "*Draworder*" later in this chapter).

**FIGURE 26-14**

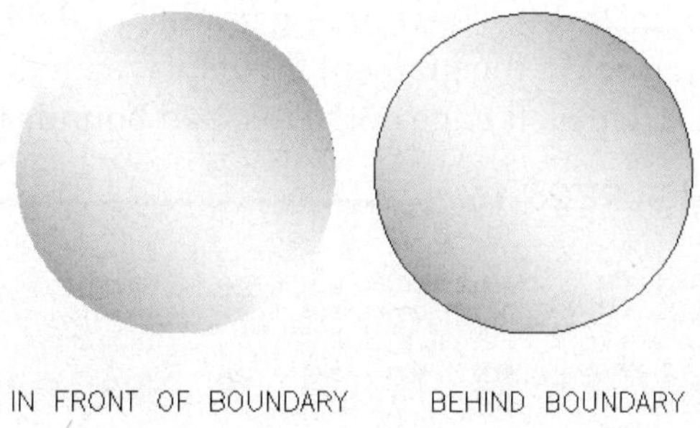

## *Hatch and Gradient* Dialog Box

The *Hatch and Gradient* dialog box (Fig. 26-15) was used in older versions of AutoCAD for creating and editing hatches in contrast to the current AutoCAD's *Hatch Creation* tab on the Ribbon. This dialog box is still available using the *setTings* option of the *Hatch* command (in Command line format). You can also invoke this dialog box using the small arrow in the lower-right corner of the *Options* panel of the *Hatch Creation* tab.

The features of this dialog box are essentially the same as those of the *Hatch Creation* tab; however, some of the wording is different between the two tools. For example, *Match Properties* in the *Hatch Creation* tab has the same function as the *Inherit Properties* in the *Hatch and Gradient* dialog box.

**FIGURE 26-15**

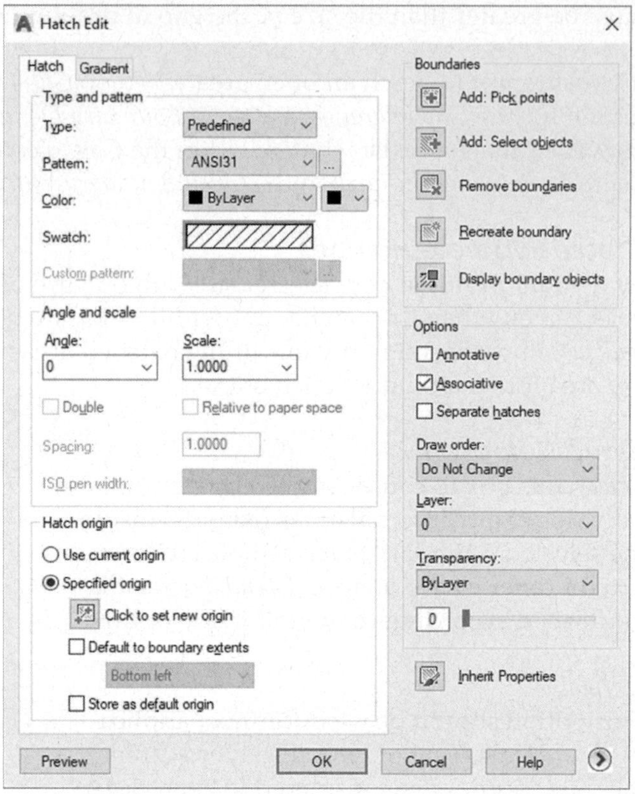


**TIP**

If you want to edit existing hatches, use *Hatchedit*. The *Hatchedit* command invokes the *Hatch and Gradient* dialog box instead of the *Hatch Creation* tab.

## *Gradient* Hatches

The *Gradient* hatches allow you to create a solid fill in a specified closed area (similar to the *Solid* pattern). However, rather than filling the area with one solid color, applying a gradient fill creates a gradual transition between one color and white or black (*One color*) or a transition between two colors (*Two color*).

### *Gradient* Image Tiles

If you use the *Hatch Creation* tab on the Ribbon to create gradient hatches, you can select from nine images in the *Pattern* drop-down list that spread gradient fill in horizontal, vertical, or spherical patterns (Fig. 26-16).

If you prefer to use the *Hatch and Gradient* dialog box, select the *Gradient* tab to produce the image tiles and related options for gradient hatches
(Fig. 26-17). The resulting gradient fill will be spread across the boundary(s) you select. For example, if you select the gradient fill displayed in the first tile (and apply *One color*), the pure color will appear at the extreme left edge of the selected boundary and no color will appear at the extreme right.

**FIGURE 26-16**

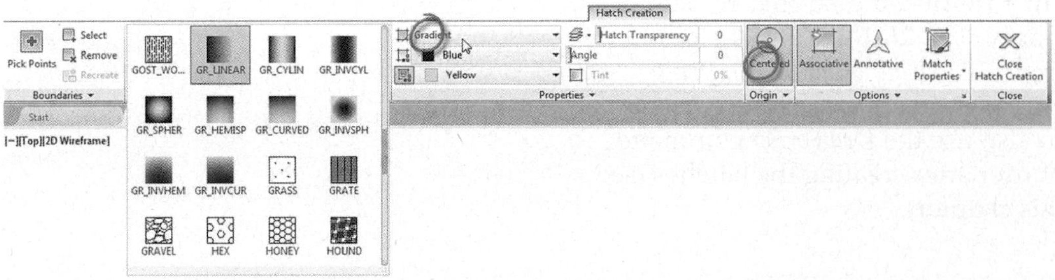

### *One Color*

To create a gradient fill mix of one color with black or white, toggle off the second color button (in the *Hatch Creation* tab) or select the *One Color* button (in the *Hatch and Gradient* dialog box). Select your desired color by using the color swatch to produce the *Select Color* dialog box where you can select from the *Index Color* (ACI), *True Color*, or *Color Books* tabs (see "*Color*" in Chapter 11 for information on the *Select Color* dialog box). Use the *Shade* to *Tint* slider to adjust the amount of white or black to mix with your color selection (*Tint* means white and *Shade* means black). Positioning the slider in the center produces no gradient, only pure hue.

**FIGURE 26-17**

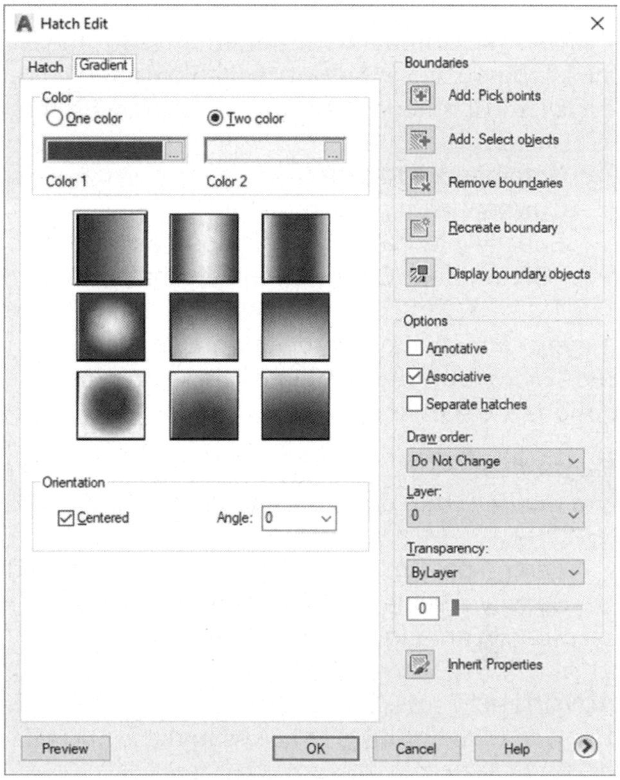


### *Two Color*

To create a gradient fill mix of two colors, make your selection for each of the two colors by using the color swatches to produce the *Select Color* dialog box. The resulting gradient fill is a transition between the two selected colors.

### *Centered*

Checking the *Centered* button ensures that the gradient fill is centered within (or among) the selected boundary(s). For example, Figure 26-18 displays two circles, each with a circular gradient fill applied. The left circle was created with *Centered* checked, the right with *Centered* unchecked.

**FIGURE 26-18**

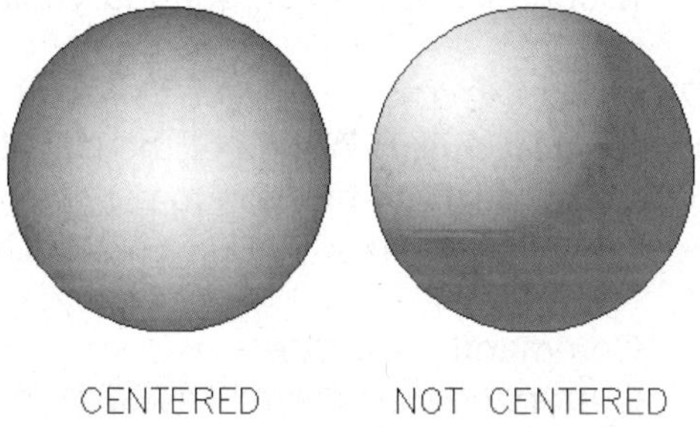


### *Angle*

Normally, the gradient fills are horizontally or vertically oriented (with the default setting of 0). You can produce gradient fills at 15-degree increments by setting an *Angle* value. The nine pattern tiles display the fills at the selected angle. The *Angle* setting can be affected also by the *Centered* option. Figure 26-19 illustrates the left circle filled at 0 degrees and the right circle filled at 45 degrees.

**FIGURE 26-19**

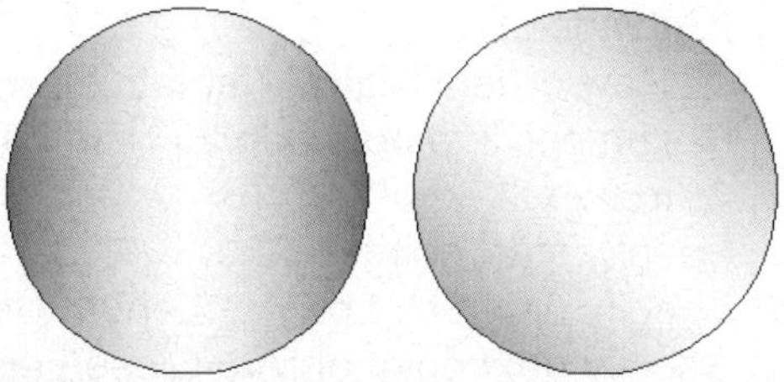

Keep in mind that the resulting gradient fills are determined by the boundary(s) selected. For example, in Figure 26-20 both circles were selected for hatching with one gradient fill (one use of the *Hatch* command), whereas the objects in Figures 26-18 and 26-19 were each hatched individually (*Create Separate Hatches* option).

**FIGURE 26-20**

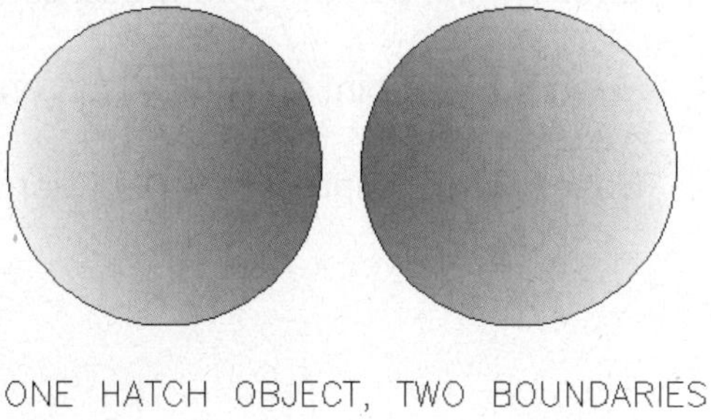

## -*Hatch* Command Line Options

*-Hatch* is the Command line alternative for the *Hatch and Gradient* dialog box. *-Hatch* must be typed at the keyboard since it is not available from the menus. Although you cannot create gradient hatches with *-Hatch*, most other options are available with this method.

> Command: ***-HATCH***
> Current hatch pattern: ANSI31
> Specify internal point or [Properties/Select objects/draW boundary/remove Boundaries/Advanced/ DRaw order/Origin/ANnotative/hatch COlor/LAyer/Transparency]:

As you can see, most of the options in the dialog box version of this command are also available here. See "*Hatch*" for information on these options. The prompts for the *Properties* options and the *Advanced* options are shown next.

### *Properties*

Use the *Properties* option to change the *Pattern, Scale,* and *Angle*.

> Enter a pattern name or [?/Solid/User defined/Gradient] <ANSI31>:
> Specify a scale for the pattern <1.0000>:
> Specify an angle for the pattern <0>:

### *Advanced*

The *Advanced* option give access to these miscellaneous options:

> Enter an option
> [Boundary set/Retain boundary/Island detection/Style/Associativity/Gap tolerance/separate Hatches]:

### *Draw Boundary*

The *Draw boundary* option of the *-Hatch* command serves the same function as the Hatch command Draw option mode. It allows you to pick points to specify an area to hatch, and you can opt to discard the boundary after hatching (Fig. 26-21). The command prompt sequence is as follows.

> Command: ***-HATCH***
> Current hatch pattern: ANSI31
> Specify internal point or [Properties/Select objects/draW boundary/remove Boundaries/ Advanced/ DRaw order/Origin/ANnotative]: ***w***
> Retain polyline boundary? [Yes/No] <N>: **Enter**
> Specify start point: **PICK**
> Specify next point or [Arc/Length/Undo]: **PICK**
> Specify next point or [Arc/Close/Length/Undo]: **PICK**
> Specify next point or [Arc/Close/Length/Undo]: **PICK**
> Specify next point or [Arc/Close/Length/Undo]: ***close***
> Specify start point for new boundary or <Accept>:

**FIGURE 26-21**

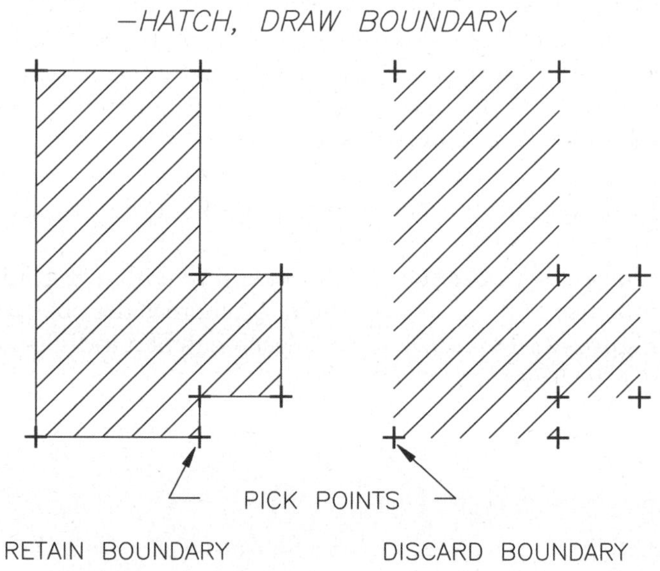

## Annotative Hatch Patterns

TIP

For complex drawings that include multiple viewports that display the same geometry at different scales, it may be desirable to use annotative hatch patterns. Annotative hatch patterns automatically adjust for size when the *Annotative Scale* for each viewport is changed. An annotative hatch pattern can appear the same scale when displayed in multiple viewports that are shown at different viewport scales. As an overview for creating and viewing annotative hatch patterns, follow these steps.

### Steps for Creating and Scaling Annotative Hatch Patterns

1. Create the desired hatch pattern in model space using *Hatch*. Set the desired *Scale* as you would normally, using the "drawing scale factor" (DSF) as the hatch pattern *Scale* value.
2. Select the *Annotative* box to make the hatch pattern associative.
3. Create the desired viewports and display the desired geometry in each viewport.
4. In a layout tab when *PAPER* space is active (not in a viewport), enable the *Add scales to annotative objects...* toggle in the lower-right corner of the Drawing Editor or set the *ANNOAUTOSCALE* variable to a positive value (1, 2, 3, or 4).
5. Set the desired scale for each viewport by double-clicking inside the viewport and then using the *Viewport Scale* pop-up list or *Viewports* toolbar. Alternately, select the viewport object (border) and use the *Properties* palette for the viewport. If the *Annotation Scale* is locked to the *Viewport Scale* (as it is by default), the annotation scale changes automatically to match the *Viewport Scale*. Setting the *Viewport Scale* automatically adjusts the annotation scale for all annotative objects for the viewport.

For example, assume you are drawing a section of a mechanical assembly and plan to print on an ANSI "A" size sheet (11" x 8.5"). Two viewports will be created to display the sectioned assembly in two different scales.

To draw the section in the *Model* tab full size, you change the *Limits* to match the sheet size (11,8.5), resulting in a drawing scale factor = 1. After creating the assembly views, you are ready to create the hatch patterns. Following steps 1 and 2, use *Hatch* to select the desired hatch patterns. Set the hatch *Scale* to 1 (scale = DSF) and select the *Annotative* checkbox. Apply the desired hatch patterns to the sectioned parts.

Next, following steps 4 through 6, switch to a *Layout* tab, set it up for the desired plot or print device using the *Page Setup Manager*, and create the two viewports. Ensure the *Add scales to annotative objects...* is toggled on. Set the viewport scale for each viewport using the *Viewport Scale* pop-up list as shown in Figure 26-22. The *Annotation Scale* is locked to the *Viewport Scale* (by default) so its setting changes accordingly as the *Viewport Scale* changes. The hatch patterns should then appear in each viewport at the appropriate scale since the hatch scale in each viewport is automatically adjusted for the viewport scale. Note that the hatch patterns for the two viewports appear the same size in Figure 26-22.

**FIGURE 26-22**

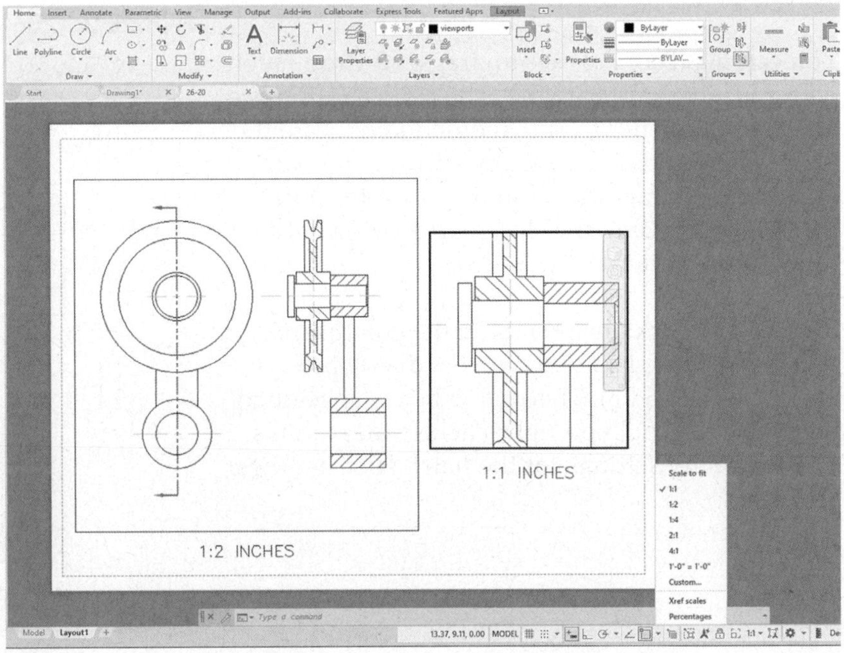

When using annotative hatches, it is simplest to use the drawing scale factor directly as the hatch *Scale*. Since the "scale" for hatch patterns is not based on inches or mm, the annotative feature is valid only for use with multiple viewports. On the other hand, when using annotative text, you can set the text height equal to the paper text height, and then create the text in model space at that scale. Next, set the annotative scale for the *Model* tab using the *Annotation Scale* pop-up list. This procedure changes the text to be properly sized for the model space.

Also different than annotative text and dimension objects, new annotative hatch objects are not created when different *Viewport Scale*s are selected. The existing hatch is simply adjusted for the *Viewport Scale*. For annotative text and dimension objects, when the *Add scales to annotative objects...* toggle is on and a new scale is selected from the *Viewport Scale* list, new annotative objects are created as copies of the originals, but in the new viewport scale.

An alternate method for creating hatch patterns to appear in multiple viewports at different scales is to create hatches of different scales on different layers, then control the layer visibility for the appropriate viewports to display only the correctly scaled hatch pattern.

### Assigning the *Annotative* Property Retroactively

You can change non-annotative hatches in the drawing to annotative by two methods. First, you can double-click on the hatch pattern to produce the *Hatchedit* dialog box and check the *Annotative* box as described previously. You can also invoke the *Properties* palette for a hatched area in a drawing. The *Properties* palette displays an entry for the *Annotative* property for selected hatches. You can use this section of the *Properties* palette to convert non-annotative hatches to annotative by changing the *No* value to *Yes*. In this case, a new entry appears for *Annotative Scale*.

See Chapter 32 for more information about using associative objects.

## *Toolpalettes*

| Ribbon | Menu Bar | Command (Type) | Alias (Type) | Shortcut |
|---|---|---|---|---|
| *View Palettes Tool Palettes* | *Tools Palettes Tool Palettes* | *Toolpalettes* | *TP* | Ctrl+3 |

You can use Tool Palettes to drag-and-drop hatch patterns into your drawing. The process is simple. Open Tool Palettes by any method and locate the palette that contains the hatch pattern you want to use. Next, drag-and-drop the hatch from the palette directly into the closed area you want to fill with the pattern (Fig. 26-23).

If you right-click on any tool in the palette and select *Properties*, its *Tool Properties* dialog box appears (Fig. 26-24). This dialog box is specific to the particular tool you right-clicked, and allows you to specify settings for the hatch such as *Angle* and *Scale*.

FIGURE 26-23

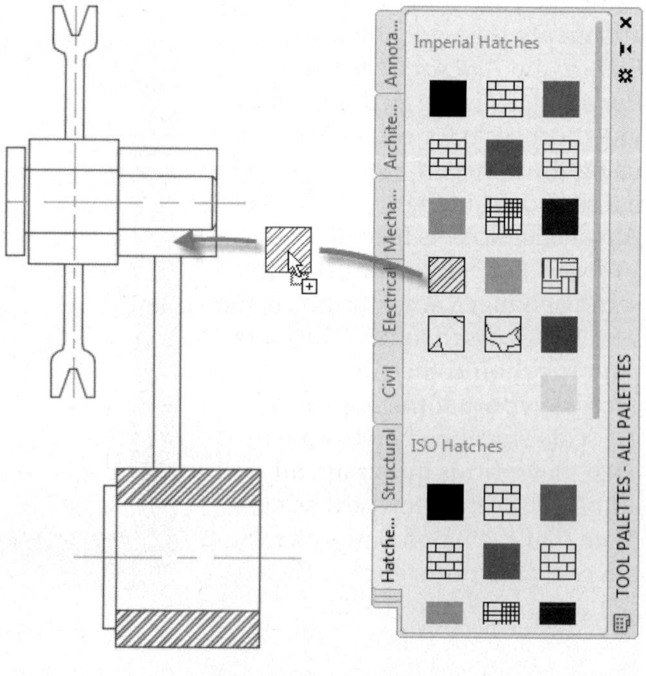

## Creating Tool Palettes

AutoCAD provides only one palette with hatch patterns. Therefore, you will most likely want to create new palettes containing the hatch patterns you use often. To do this, follow these steps:

1. Open Tool Palettes by any method. Right-click in the palette area to produce the shortcut menu (Fig. 26-25). Select *New Palette.*

2. A new palette appears with an edit box (not shown). Enter the title you want to appear on the tab of the new palette, for example, "ANSI Hatches."

3. Open DesignCenter. Use the *Search* button. In the *Search* dialog box that appears (not shown), select *Hatch Pattern Files* in the *Look for* drop-down list and enter ACAD.PAT (or another .PAT file) in the *Search for the name* edit box. Select *Search Now.* When ACAD.PAT appears in the list at the bottom of the *Search* dialog box, double-click on ACAD. PAT. This action loads the hatch patterns into the DesignCenter Content Area.

4. From the DesignCenter Content Area, drag-and-drop the desired hatch patterns individually into your new palette (Fig. 26-26). If you want to specify other than the default *Scale* and *Angle*, right-click on any tool to produce its *Tool Properties* dialog box (see Fig. 26-24).

FIGURE 26-24

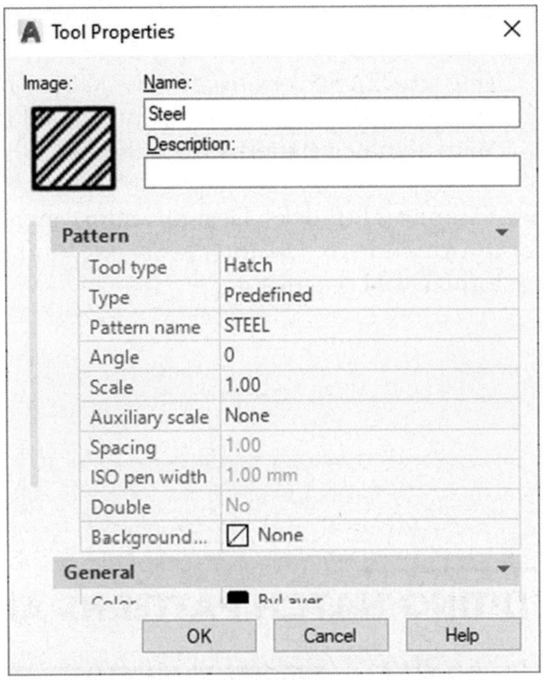


FIGURE 26-25

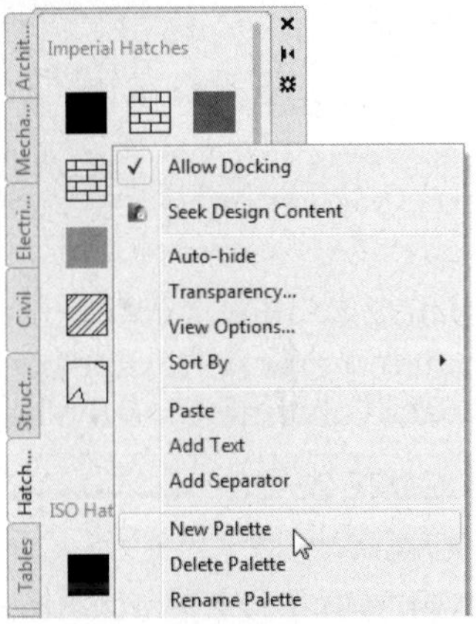


TIP

Alternately, you can create a new tool palette fully populated with all the hatch patterns from the ACAD.PAT file directly from DesignCenter. This is a quicker and simpler method; however, all the patterns from the file are included in your new palette. Follow these steps:

1. Open Tool Palettes by any method.

2. Open DesignCenter. Use the *Search* button. In the *Search* dialog box that appears (not shown), select *Hatch Pattern Files* in the *Look for* drop-down list and enter ACAD.PAT (or another .PAT file) in the *Search for the name* edit box. Select *Search Now.* When ACAD.PAT appears in the list at the bottom of the *Search* dialog box, double-click on ACAD.PAT.

FIGURE 26-26

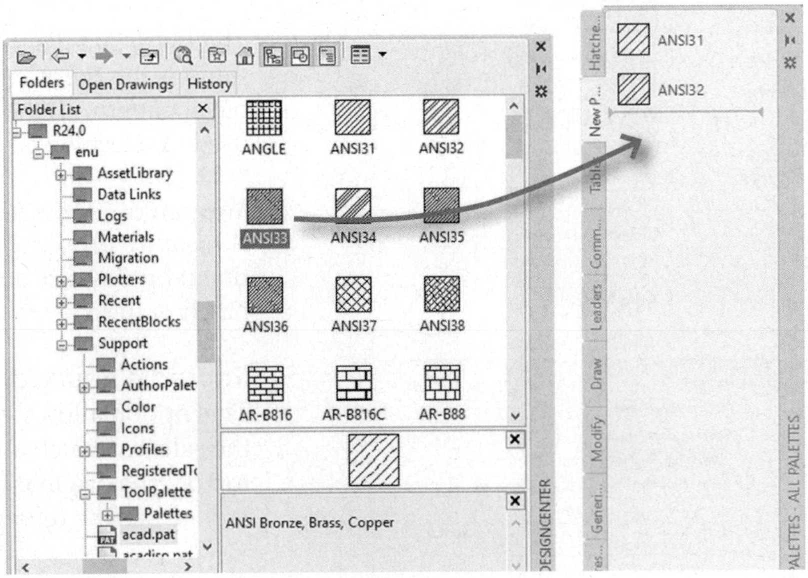


TIP

3. In the DesignCenter Tree View area, right-click on the ACAD.PAT file name to produce a shortcut menu (Fig. 26-27). Select *Create Tool Palette of Hatch Patterns*. A new palette is automatically created fully populated with all hatch patterns from the file.

See Chapter 21, Blocks, DesignCenter, and Tool Palettes, for information on *Deleting*, *Renaming*, and other ways to customize tool palettes.

**FIGURE 26-27**

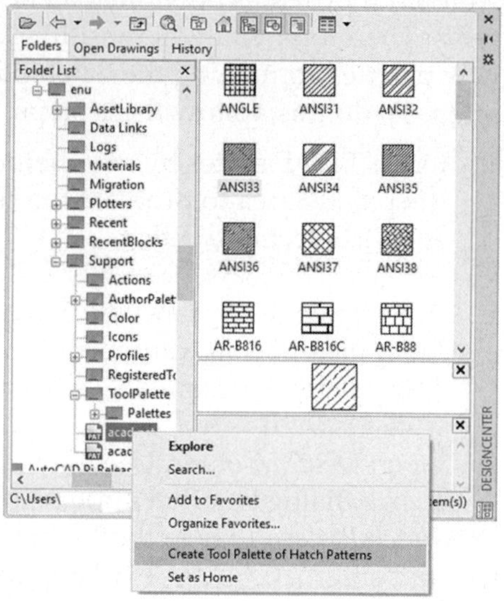


## EDITING HATCH PATTERNS AND BOUNDARIES

### *Hatchedit*

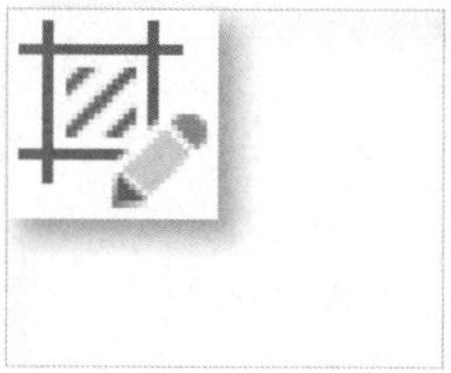

| Ribbon | Menu Bar | Command (Type) | Alias (Type) | Shortcut |
|---|---|---|---|---|
| *Home Modify* (icon) | *Modify Object > Hatch...* | *Hatchedit* or *-Hatchedit* | *HE* | ... |

*Hatchedit* allows you to modify an existing hatch pattern in the drawing. This AutoCADD feature makes hatching more flexible because you can hatch several areas to quickly create a "rough" drawing, then retroactively fine-tune the hatching parameters when the drawing nears completion with *Hatchedit*.

**FIGURE 26-28**

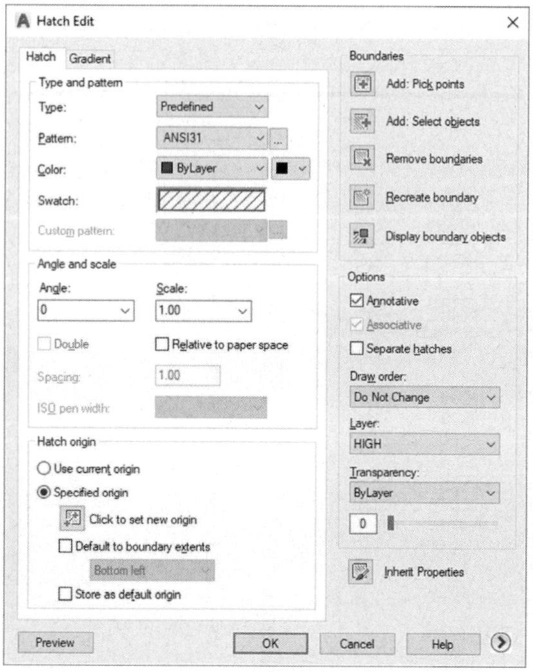


You can show the *Hatch Edit* dialog box (Fig. 26-28) using any method shown in the Command table. The dialog box provides options for changing the *Pattern, Scale, Angle,* and *Type* properties of the existing hatch.

The *Hatch Edit* dialog box (Figure 26-28) is essentially the same as the *Hatch and Gradient* dialog box. When you edit a hatch pattern, all the options are available that were originally used to select the boundary and create the hatch pattern.

**TIP**

You can use *Hatchedit* to retroactively assign the *Annotative* property to hatch patterns. Two additional options are available only when editing a hatch pattern.

#### *Recreate Boundary*

This option allows you to create a duplicate boundary of the selected hatch area. The new boundary is created "on top of" the original boundary, so you have to use *Move, Last,* to see the recreated boundary (Fig. 26-29).

When you select the hatch area to edit, then pick the *Recreate boundary* option, the following prompt appears:

> Enter type of boundary object [Region/Polyline] <Polyline>: **Enter** or (**option**)
> Reassociate hatch with new boundary? [Yes/No] <N>: **Enter** or (**option**)

FIGURE 26-29

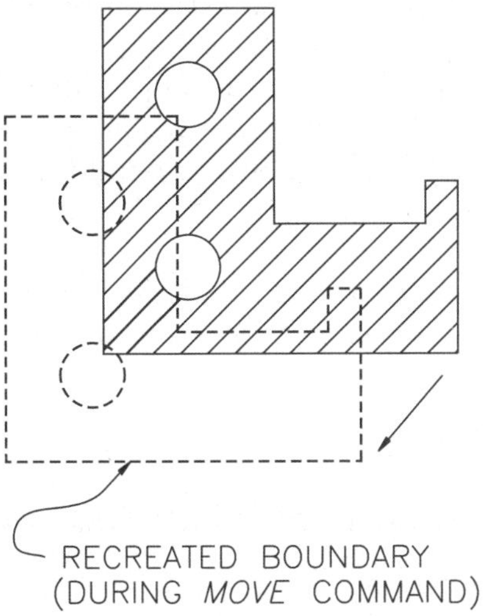


### *Display Boundary Objects*

When you use the *Hatchedit* command and select the desired hatch, you actually select only the hatch object itself, not the boundary. Selecting this button automatically selects the boundary objects that are associated with the hatch. This feature allows you to grip-edit the boundary and hatch object, then return to the *Hatch Edit* dialog box. This option works with associative hatch objects only.

## Using Grips with Hatch Patterns

Grips can be used effectively to edit hatch pattern boundaries of associative hatches. Do this by selecting the boundary with the pickbox or automatic window/crossing window (as you would normally activate an object's grips). Associative hatches retain the association with the boundary after the boundary has been changed using grips (see Fig. 26-2, earlier in this chapter).

FIGURE 26-30

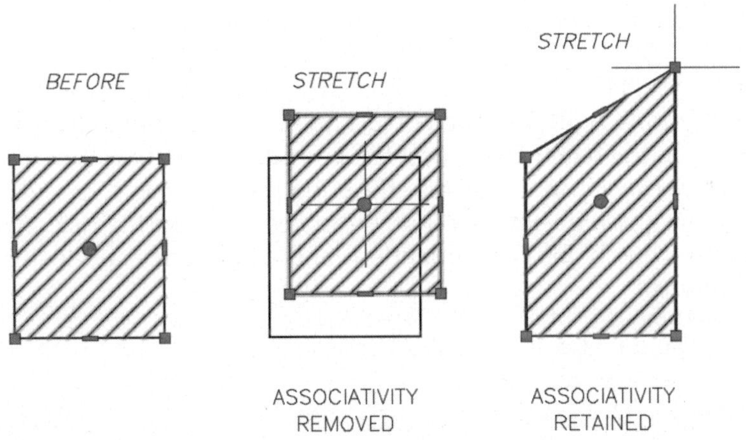


**TIP**

Beware: when selecting the hatch boundary, ensure you edit the boundary and not the hatch object (round grip). If you edit the hatch object only, the associative feature is lost. This can happen if you select the hatch object (round) grip (Fig. 26-30). In this case, AutoCAD issues the following prompt:

> Command: (select grip on hatch object)
> **STRETCH**
> Specify stretch point or [Base point/Copy/Undo/eXit]:
> Hatch boundary associativity removed.
> Command:

If you activate grips on both the hatch object and the boundary and make a boundary grip hot, the boundary and related hatch retain associativity (see Fig. 26-30). You can also use *Multifunctional Grips* to edit hatch boundaries. After selecting the grip, use the Ctrl key to cycle between the options. The hatch pattern will retain associativity.

## Using *Trim* with Hatch Patterns

FIGURE 26-31

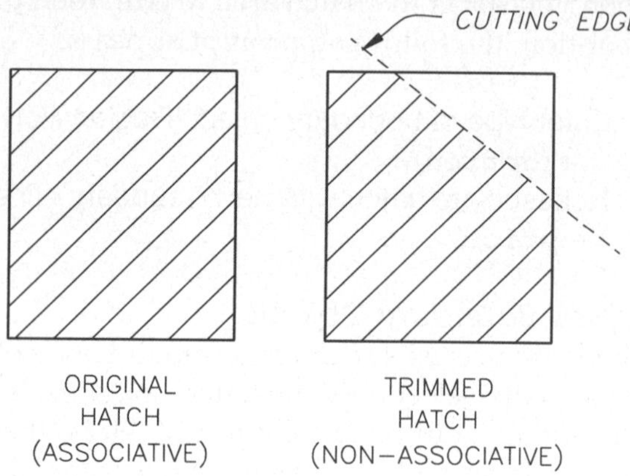


You can use the *Trim* command to trim a hatch object created with *Hatch*. Simply create or use an existing object as a "cutting edge," then invoke *Trim* and trim the hatch object as you would any other *Line, Circle, Arc,* or other object that you would normally trim (Fig. 26-31). Beware, however, because trimming a hatch pattern converts the hatch to a non-associative hatch object. If you then *Move, Scale, Stretch,* or change the boundary in some other way, the hatch pattern will not reform to match the boundary.

## Object Selection Features of Hatch Patterns

When you select a hatch object for editing (PICK only one line of an entire hatch pattern), you actually select the entire hatch object. By default, when you PICK any element of the hatch pattern, the hatch object can be *Moved, Copied,* grip edited, etc. However, when you edit the hatch object but not the boundary, associativity is removed (see "Using Grips with Hatch Patterns").

FIGURE 26-32

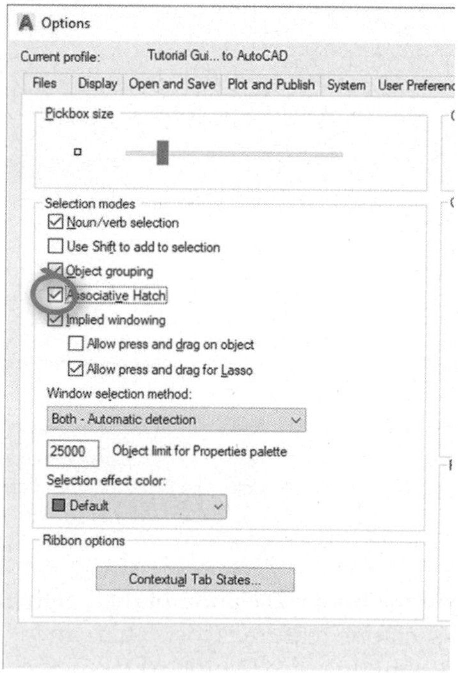


You can easily control whether just the hatch object or the hatch object and boundary together are selected when you PICK the hatch object. Do this by selecting *Associative Hatch* in the *Selection* tab of the *Options* dialog box (Fig. 26-32, lower left). Checking the *Associative Hatch* option sets the *PICKSTYLE* variable to a value of 3.

With *Associative Hatch* selected, PICKing the hatch object results in highlighting both the hatch object and boundary; therefore, you always edit the hatch object and the boundary together. When *Associative Hatch* is not selected, PICKing the hatch object results in only the hatch object selection; therefore, you run the risk of editing the hatch object independent of the boundary and losing associativity.

## Solid *Hatch* Fills

A *Solid* hatch pattern is available. This feature is welcomed by professionals who create hatch objects such as walls for architectural drawings and require the solid filled areas. There are many applications for this feature.

*Solid*-filled hatch areas have versatility because you can control the *Color* of the solid fill. By default, the color of the *Solid* hatch pattern (just like any other hatch pattern) assumes the current *Color* setting. In most cases, the current *Color* setting is *ByLayer*; therefore, the *Solid* hatch pattern assumes the current layer color. You can also assign the *Solid* hatches object-specific *Color*, in which case it is possible to have multiple *Solid* filled areas in different colors all on the same layer.

## FILLMODE

You can use the *Fill* command
(or *FILLMODE* system variable) to control the visibility of the hatched areas. *Fill* also controls the display of solid filled TrueType fonts (see Chapter 18).

If you want to display *Hatch* objects, set *Fill* to *On* (Fig. 26-33). Setting *Fill* to *Off* causes AutoCAD not to display the hatched areas. *Regen* must be used after *Fill* to display the new visibility state. This fill control can be helpful for saving toner or ink during test prints or speeding up plots when much solid fill is used in a drawing.

**FIGURE 26-33**

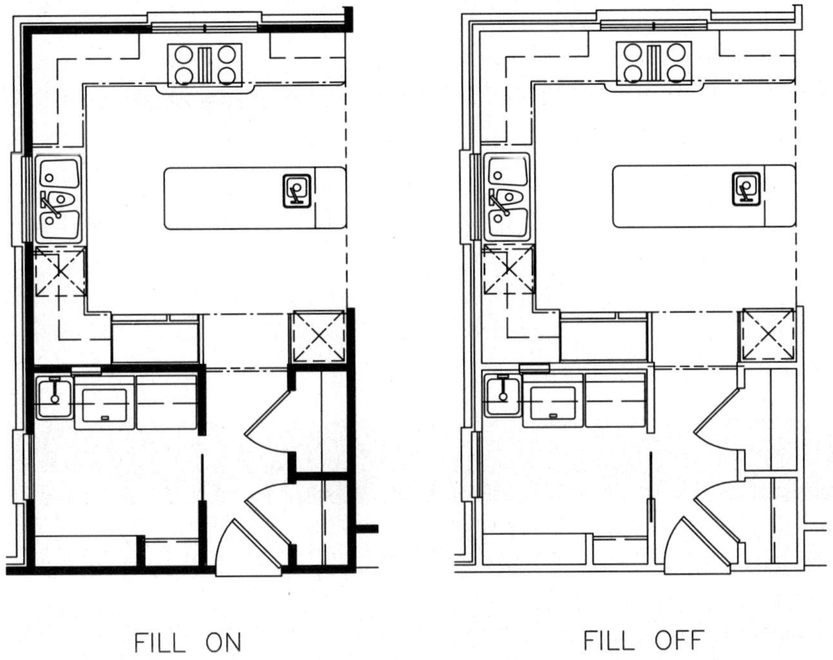


## Draworder

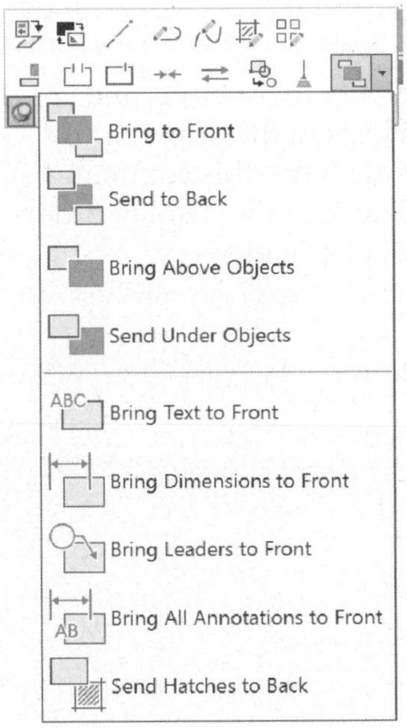


| Ribbon | Menu Bar | Command (Type) | Alias (Type) | Shortcut |
|---|---|---|---|---|
| *Home Modify* (icon) | *Tools Draw Order >* | *Draworder* | *DR* | ... |

It is possible to use solid hatch patterns and raster images in combination with filled text, other hatch patterns and solid images, etc. With these solid areas and images, some control of which objects are in "front" and "back or "above" and "under" must be provided. This control is provided by the *Draworder* command.

For example, if a company logo were created using filled *Text*, a *Sand* hatch pattern, and a *Solid* hatched circle, the object that was created last would appear "in front" (Fig. 26-34, top logo).

The *Draworder* command is used to control which objects appear in front and back or above and under. This capability is necessary for creating prints and plots in black and white and in color.

The *Draworder* command is simple to use. Simply select objects and indicate *Front, Back, Above,* or *Under*.

Command: ***DRAWORDER***
Select objects: **PICK**
Select objects: **Enter**
Enter object ordering option [Above objects/Under objects/Front/Back] <Back>: (**option**)
Regenerating drawing.
Command:

In Figure 26-34, three possibilities are shown for the three objects—*Text, Solid* hatch and boundary, and *Sand* hatch and boundary.

Although you can use *Draworder* to control the order of display for text as well as hatches (as shown in Fig. 26-34), the *Texttofront* command operates for text objects only. Objects that are affected by *Texttofront* are *Text, Mtext,* and dimensions and leaders. The command operates in Command line format.

**FIGURE 26-34**

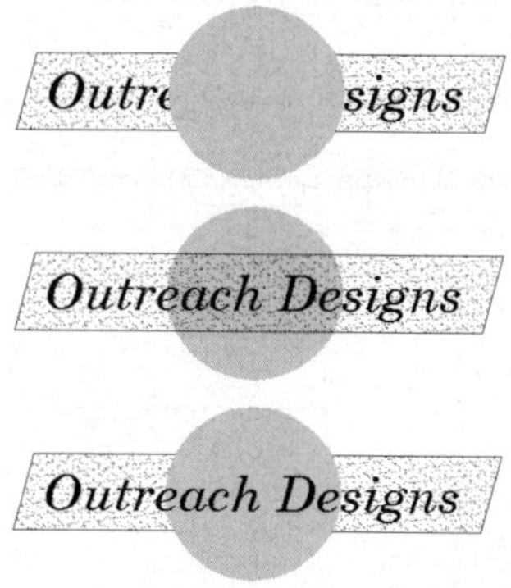


### *Texttofront*

| Ribbon | Menu Bar | Command (Type) | Alias (Type) | Shortcut |
|---|---|---|---|---|
| *Home Modify* (option) | *Tools Draw Order > Bring Text and Dimensions to Front* | *Texttofront* | ... | ... |

Command: ***TEXTTOFRONT***
Bring to front [Text/Dimensions/Leaders/All] <All>: **Enter** or (**option**)
*nn* object(s) brought to front.
Command:

Use the *Text* option to bring all *Text* and *Mtext* objects in the drawing in front of all other objects and the *Dimensions* and *Leaders* options to bring those objects in front of all other objects in the drawing. *All* brings all text, dimensions, and leaders in front of all other objects in the drawing. Since this command operates globally for text, dimension, and leader objects only, use *Draworder* to specify the display order of hatches, then use *Texttofront* to specify the display order for text, dimensions, and leaders.

## DRAWING CUTTING PLANE LINES

In mechanical drawings, most section views (full, half, and offset sections) require a cutting plane line to indicate the plane on which the object is cut. The cutting plane line is drawn in a view adjacent to the section view because the line indicates the plane of the cut from its edge view. (In the section view, the cutting plane is perpendicular to the line of sight, therefore, not visible as a line.)

Standards provide two optional line types for cutting plane lines. In AutoCAD, the two linetypes are *Dashed* and *Phantom,* as shown in Figure 26-35. Arrows at the ends of the cutting plane line indicate the line-of-sight for the section view.

**FIGURE 26-35**

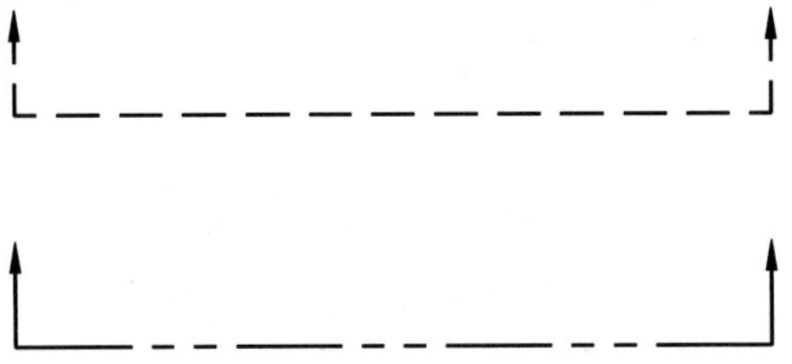

Cutting plane lines should be drawn or plotted in a heavy lineweight. Two methods can be used: use a *Pline* with *Width* or assign a *Lineweight* to the line or layer. If you use a *Pline*, set the *Width* to .02 or .03 times the drawing scale factor. Assigning a *Lineweight* to the line or layer is simpler. A *Lineweight* of .8mm or .031" is appropriate for cutting plane lines.

For the example section view, a cutting plane line is created in the top view to indicate the plane on which the object is cut and the line of sight for the section view (Fig. 26-36). (The cutting plane could be drawn in the side view, but the top is clearer in this case.) First, a *New* layer named CUT is created and the *Dashed linetype* is assigned to the layer. The layer is *Set* as the *Current* layer. Next, construct a *Pline* with the desired *Width* to represent the cutting plane line.

The resulting cutting plane line appears as shown in Figure 26-36 but without the arrowheads. The horizontal center line for the hole (top view) was *Erased* before creating the cutting plane line.

**FIGURE 26-36**

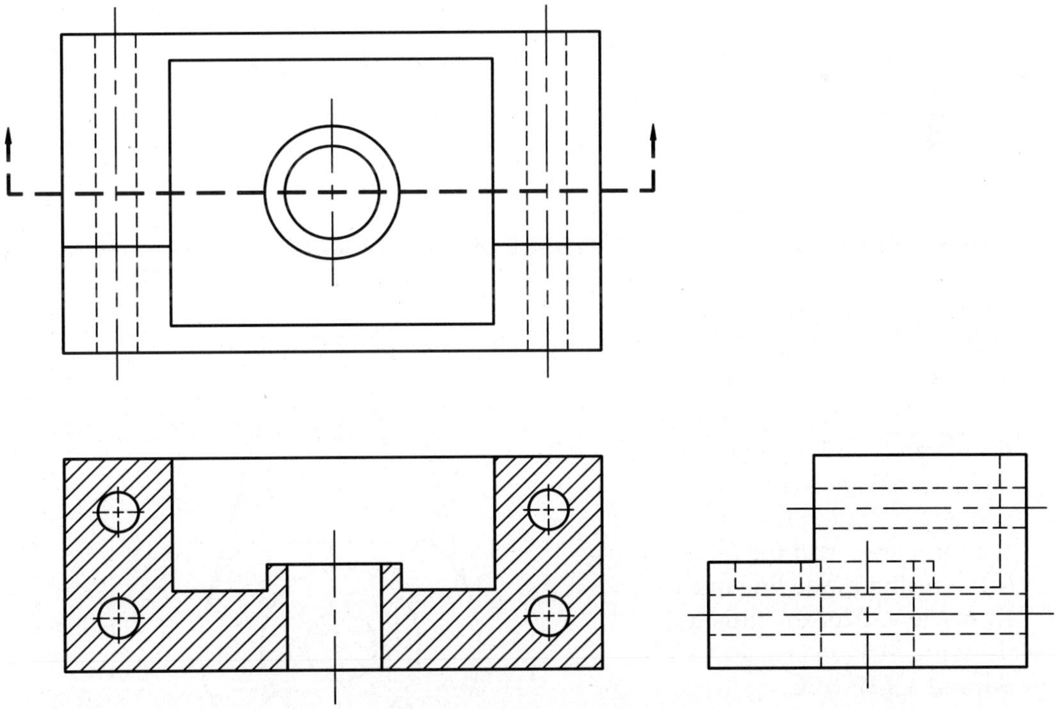

The last step is to add arrowheads to the ends of the cutting plane line.

Arrowheads can be drawn by three methods: (1) use the *Solid* command to create a three-sided solid area; (2) use a tapered *Pline* with beginning width of 0 and ending width of .08, for example; or (3) use the *Leader* command to create the arrowhead (see Chapter 28, Dimensioning, for details on *Leader*). The arrowhead you create can be *Scaled*, *Rotated*, and *Copied* if needed to achieve the desired size, orientation, and locations.

## CHAPTER EXERCISES

For the following exercises, create the section views as instructed. Use an appropriate template drawing for each unless instructed otherwise. Include a border and title block in the layout for each drawing.

1. **Full Section View**

   ***Open*** the **SADDLE** drawing that you created in Chapter 24. Convert the front view to a full section. ***Save*** the drawing as **SADL-SEC.** ***Plot*** the drawing at **1=1** scale.

2. **Full Section View**

   Make a multiview drawing of the Bearing shown in Figure 26-37. Convert the front view to a full section view. Add the necessary cutting plane line in the top view. ***Save*** the drawing as **BEAR-SEC.** Make a ***Plot*** at full size.

**FIGURE 26-37**

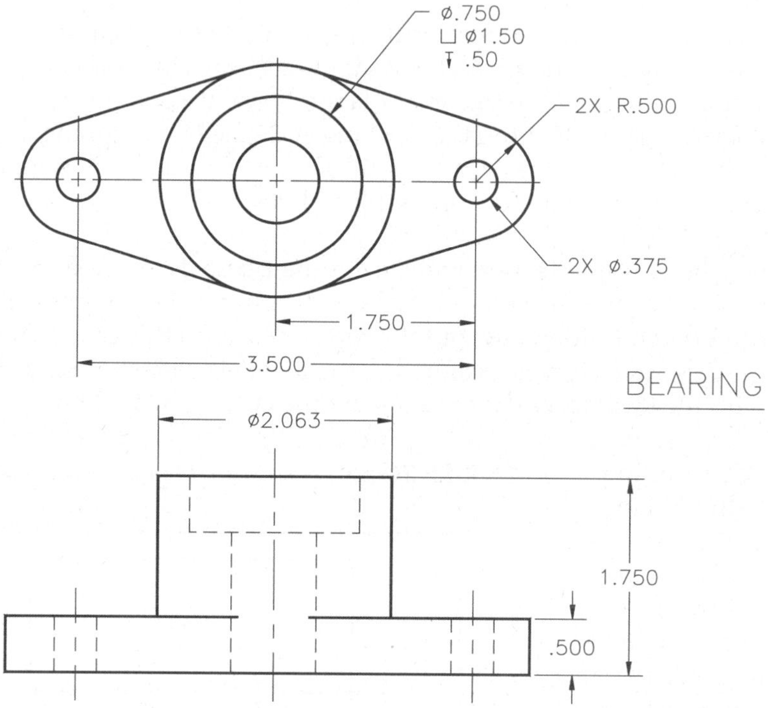


3. **Full Section View**

   Create a multiview drawing of the Clip shown in Figure 26-38. Include a side view as a full section view. You can use the **CLIP** drawing you created in Chapter 24 and convert the side view to a section view. Add the necessary cutting plane line in the front view. ***Plot*** the finished drawing at **1=1** scale and ***SaveAs*** **CLIP-SEC**.

**FIGURE 26-38**

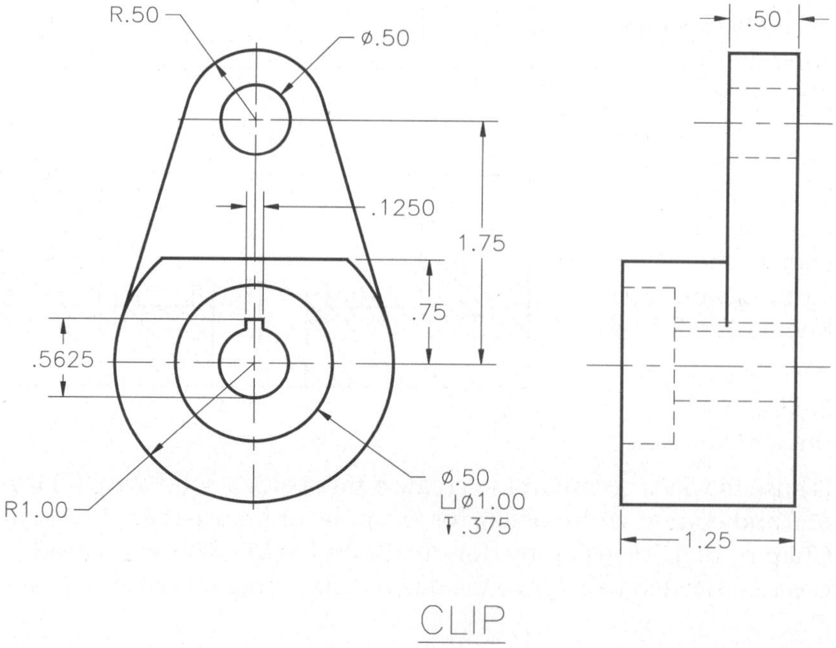

4. **Two Full Section Views**

Create a multiview drawing, including two full sections of the Stop Block as shown in Figure 26-39. Section B–B′ should replace the side view shown in the figure. ***Save*** the drawing as **SPBK-SEC**. ***Plot*** the drawing at **1=2** scale.

**FIGURE 26-39**

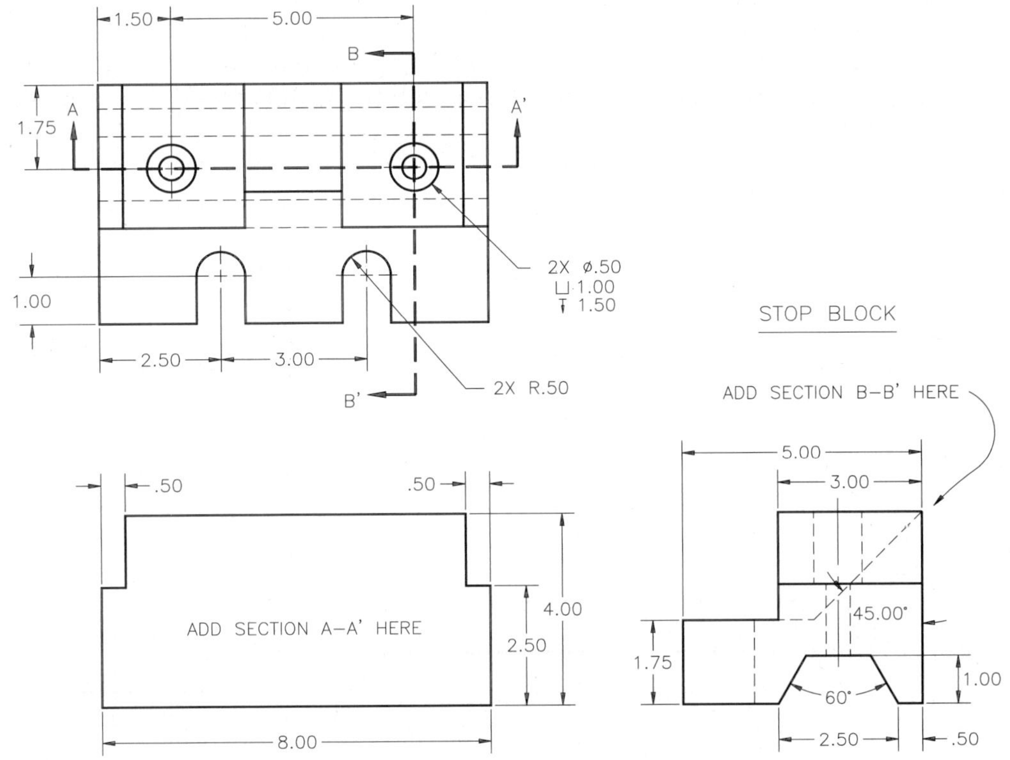


5. **Half Section View**

Make a multiview drawing, including a half section of the Pulley (Fig. 26-40). <u>All vertical dimensions are diameters</u>. Two views (including the half section) are sufficient to describe the part. Add the necessary cutting plane line. ***Save*** the drawing as **PUL-SEC** and make a ***Plot*** at **1:1** scale.

**FIGURE 26-40**

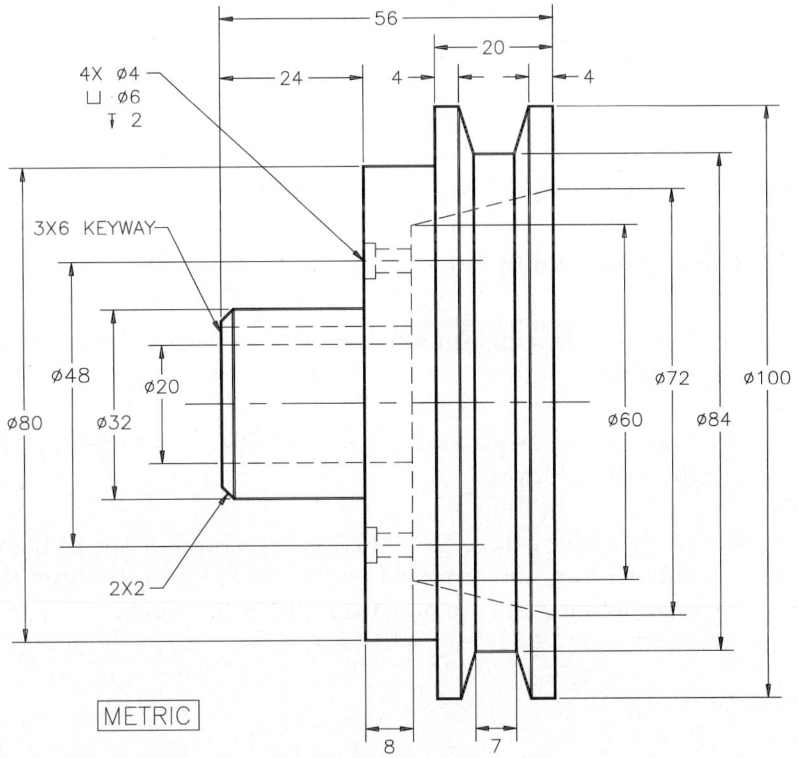

6. **Grade Beam**

Draw the Grade Beam foundation detail in Figure 26-41. Do not include the dimensions in your drawing. Use the ***Hatch*** command to hatch the concrete slab with **AR-CONC** hatch pattern. Use ***Sketch*** to draw the grade line and ***Hatch*** with the **EARTH** hatch pattern. ***Save*** the drawing as **GRADBEAM**.

**FIGURE 26-41**

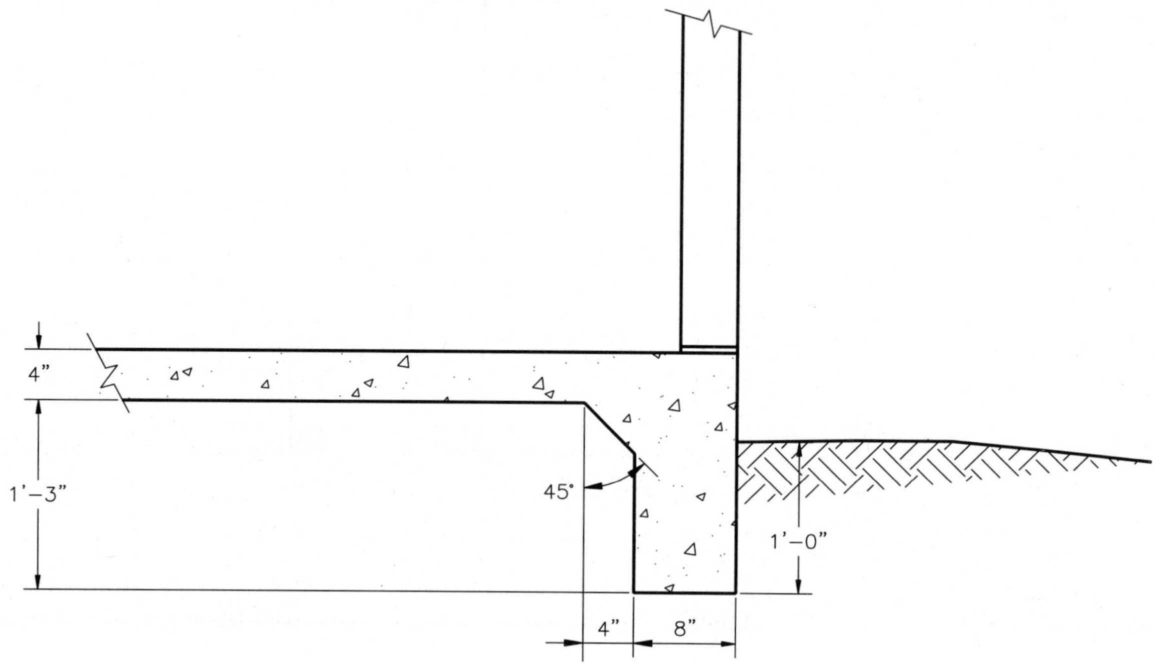


REUSE

7. **Tool Palettes, Solid Fill**

A. ***Open*** the **OFFICE-REV** drawing that you last worked on in Chapter 21. Next, open the **Tool Palettes**.

B. Make a new ***Layer*** and name it **CARPET**. Accept the default color, linetype, etc. Make **CARPET** the ***Current*** layer.

C. Draw a ***Line*** across the doorway threshold of one of the two offices. Next, drag-and-drop a ***Solid*** pattern from the standard *Hatches and Fills* palette into the office. If problems occur, check to ensure that there are no gaps in the office walls.

D. When the solid pattern appears in the office, notice that the pattern may cover the outlines of the walls and furniture items. Use the ***Draworder*** command to move the new hatch pattern to the ***Back***. Your drawing should look similar to that in Figure 26-42.

E. Using the same technique, fill the other two office spaces with carpet from the tool palette or use another wood floor pattern such as ***HOUND***. (Depending on the selected pattern, you may have to select ***Properties*** from the shortcut menu and change the ***Scale***.)

F. Use ***SaveAs*** to save and rename the drawing to **OFFICE-REV-2**.

**FIGURE 26-42**

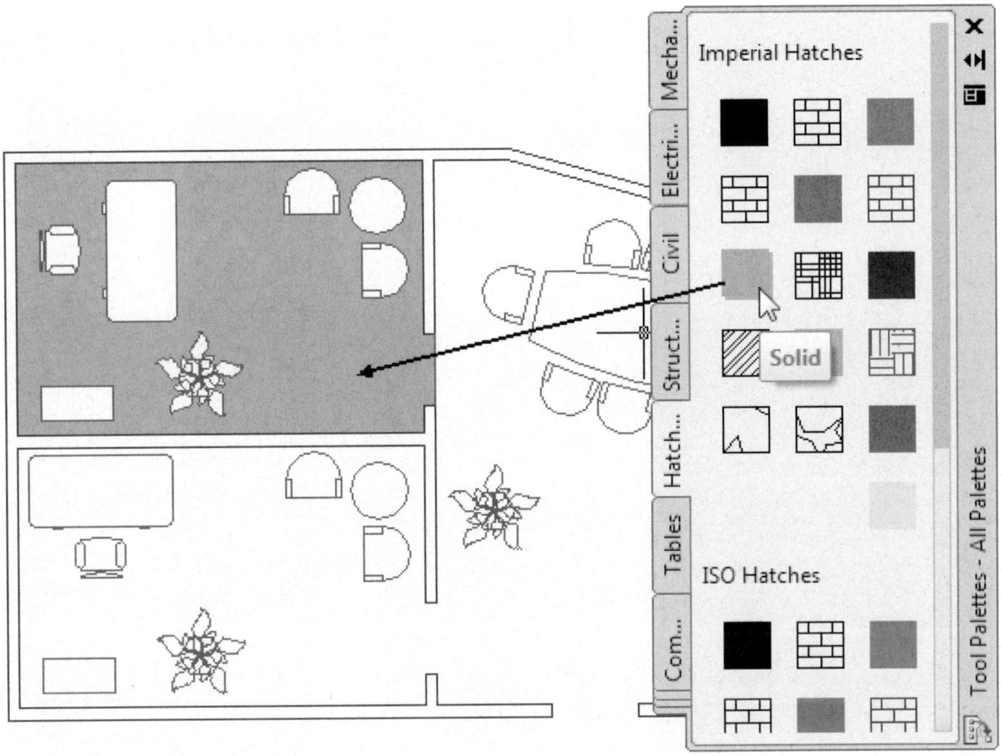


8. **Solid Fill, *Fill* Command**

A. ***Open*** the **OFFICE-REV-2** drawing if not already open. Make a new ***Layer*** and assign the name **WALLS**. Accept the default color, linetype, etc. Make the **WALLS** layer ***Current***.

B. Use the ***Hatch and Gradient*** dialog box to apply the ***Solid*** hatch pattern to the walls of the office.

C. Use the ***Fill*** command and turn ***Off*** the solid fill. Use the ***Regen*** command to apply the new *Fill* setting. Does *Fill* affect only the walls, or does it affect the other carpet solid fill patterns?

D. Turn ***Fill On*** and ***Regen***. ***Save*** the drawing.

CHAPTER

# 27

# Auxiliary Views

## Chapter Objectives

After completing this chapter you should:

1. be able to use the *Rotate* option of *Snap* to change the angle of the *Snap*, *Grid*, and *Ortho*;

2. be able to set the *Increment angle* and *Additional angles* for creating auxiliary views using *Polar Tracking*;

3. know how to use the *Offset* command to create parallel line copies;

4. be able to use *Xline* and *Ray* to create construction lines for auxiliary views.

## CONCEPTS

AutoCAD provides no commands explicitly for the creation of auxiliary views in a mechanical drawing or for angular sections of architectural drawings. However, four particular features that have been discussed previously can assist you in the construction of these drawings. Those features are *Snap* rotation, *Polar Tracking*, the *Offset* command, and the *Xline* and *Ray* commands.

In a mechanical drawing, an auxiliary view may be needed in addition to the typical views (top, front, side). An auxiliary view is one that is normal (the line-of-sight is perpendicular) to an inclined surface of the object. Therefore, the entire view is constructed at an angle by projecting in a 90-degree direction from the edge view of an inclined surface in order to show the true size and shape of the inclined surface. Architectural drawings often include a portion of a plan drawn at some angle, such as a ranch house with one wing at a 45-degree angle to the main body of the house. In either case and depending on the object, the view or rooms could be drawn at any angle, so lines are typically drawn parallel and perpendicular relative to that section of the drawing. The *Snap* rotation feature, *Polar Tracking*, the *Offset* command, and the *Xline* and *Ray* commands help in this task.

An example mechanical part typically requiring auxiliary view construction is shown in Figure 27-1. As you can see, there is an inclined surface that contains two drilled holes. To describe this object adequately, an auxiliary view should be created to show the true size and shape of the inclined surface. Although this chapter uses only this mechanical example, these same techniques would be used for architectural drawings or any other drawings that include views or entire sections of a drawing constructed at some angle.

FIGURE 27-1

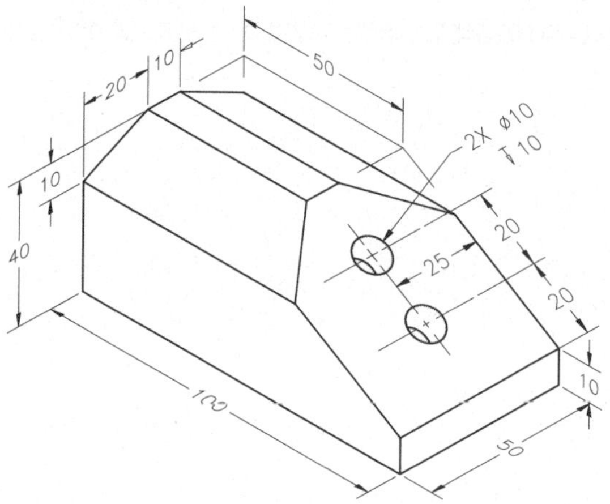


## CONSTRUCTING AN AUXILIARY VIEW

### Setting Up the Principal Views

To begin this drawing, the typical steps are followed for drawing setup (Chapter 12). Because the dimensions are in millimeters, *Limits* should be set accordingly. For example, to provide enough space to draw the views full size and to plot full size on an A sheet, *Limits* of 279 x 216 are specified.

FIGURE 27-2

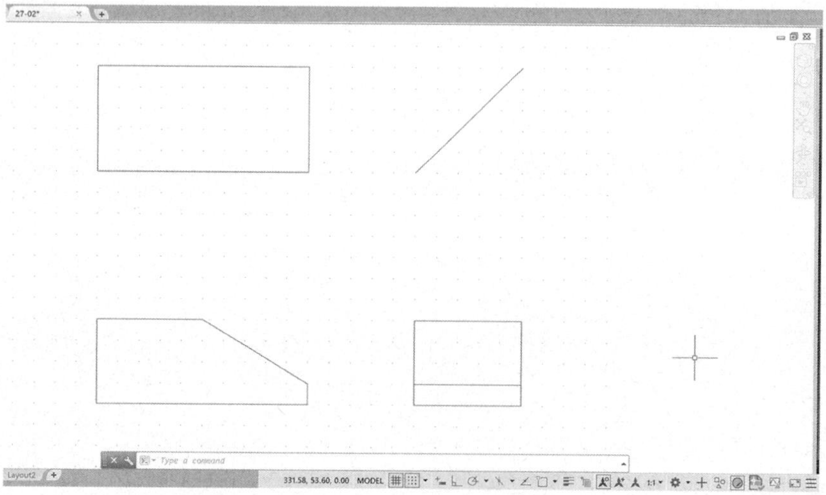

In preparation for the auxiliary view, the principal views are "blocked in," as shown in Figure 27-2. The purpose of this step is to ensure that the desired views fit and are optimally spaced within the allocated *Limits*. If there is too little or too much room, adjustments can be made to the *Limits*. Notice that space has been allotted between the views for a partial auxiliary view to be projected from the front view.

Before additional construction on the principal views is undertaken, initial steps in the construction of the partial auxiliary view should be performed. The projection of the auxiliary view requires drawing lines perpendicular and parallel to the inclined surface. One or more of the three alternatives (explained next) can be used.

## Using *Snap Rotate* and *Ortho*

One possibility to construct an auxiliary view is to use the *Snap* command with the *Rotate* option. This action permits you to rotate the *Snap* to any angle about a specified base point. The *Grid* automatically follows the *Snap*. Turning *Ortho ON* forces *Lines* to be drawn orthogonally with respect to the rotated *Snap* and *Grid*.

In Figure 27-3, the cursor size is changed from the default 5% of screen size to 100% of screen size to help illustrate the orientation of the *Snap, Grid,* and cursor when *Snap* is *Rotated*. You can change the cursor size in the *Display* tab of the *Options* dialog box.

**FIGURE 27-3**

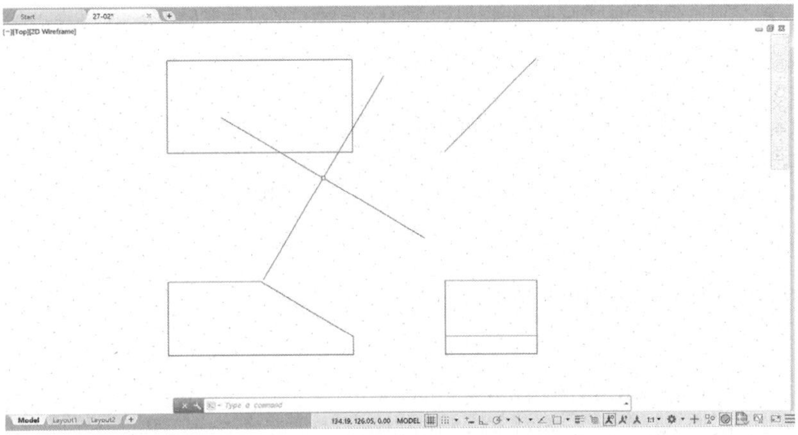

NOTE: You must set the *Grid Style* to *Display a Dotted Grid* for *2D Model Space* in the *Drafting Settings* dialog box. Grid lines do not display once the *Snap* is rotated.

NOTE: The *Rotate* option of *Snap* does not appear as an option in the *Drafting Settings* dialog box or on the Command line. However, the *Rotate* option still functions by typing the letter ***R*** to activate it.

Command: ***SNAP***
Specify snap spacing or [ON/OFF/Aspect/Legacy/Style/Type] <0.5000>: ***r***
Specify base point <0.0000,0.0000>: **PICK** or (**coordinates**) (PICK starts a rubberband line.)
Specify rotation angle <0>: **PICK** or (**value**) (PICK to specify second point to define the angle. See Fig. 27-3.)
Command:

PICK (or specify coordinates for) the endpoint of the *Line* representing the inclined surface as the "base point." At the "rotation angle:" prompt, a value can be entered or another point (the other end of the inclined *Line*) can be PICKed. Use *Object Snap* when PICKing the *Endpoints*. If you want to enter a value but don't know what angle to rotate to, use *List* to display the angle of the inclined *Line*. The *Grid, Snap,* and crosshairs should align with the inclined plane as shown in Figure 27-3.

(An option for simplifying construction of the auxiliary view is to create a new *UCS* [User Coordinate System] with the origin at the new base point. Use the *3Point* option and turn on *Ortho* to select the three points. See Chapter 34, User Coordinate Systems, for more information on this procedure.)

After rotating the *Snap* and *Grid*, the partial auxiliary view can be "blocked in," as displayed in Figure 27-4. Begin by projecting *Lines* up from and perpendicular to the inclined surface. (Make sure *Ortho* is *ON*.) Next, two *Lines* representing the depth of the view should be constructed parallel to the inclined surface and perpendicular to the previous two projection lines. The depth dimension of the object in the auxiliary view is equal to the depth dimension in the top or right view. *Trim* as necessary.

**FIGURE 27-4**

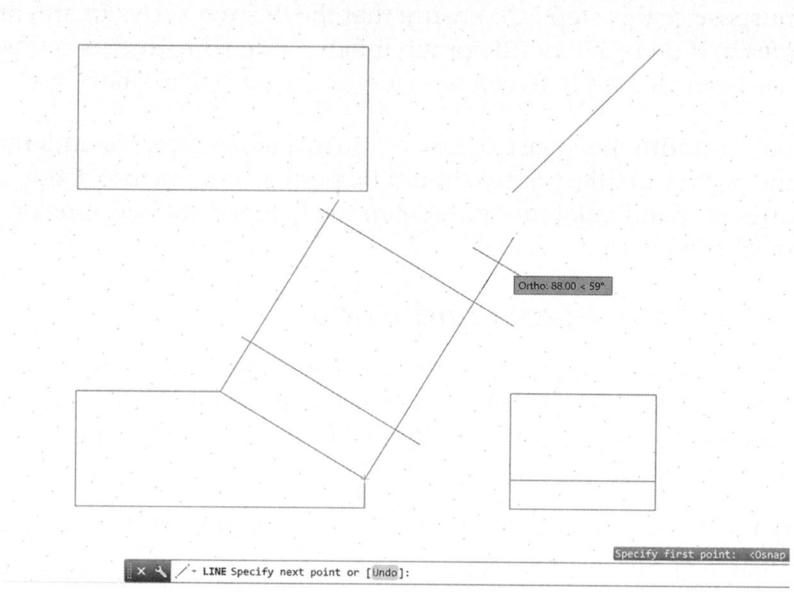


Locate the centers of the holes in the auxiliary view and construct two *Circles*. It is generally preferred to construct circular shapes in the view in which they appear as circles, then project to the other views. That is particularly true for this type of auxiliary since the other views contain ellipses. The centers can be located by projection from the front view or by *Offsetting Lines* from the view outline.

Next, project lines from the *Circles* and their centers back to the inclined surface (Fig. 27-5). Use of a hidden line layer can be helpful here. While the *Snap* and *Grid* are rotated, construct the *Lines* representing the bottom of the holes in the front view. (Alternately, *Offset* could be used to copy the inclined edge down to the hole bottoms, then *Trim* the unwanted portions of the *Lines*.)

**FIGURE 27-5**

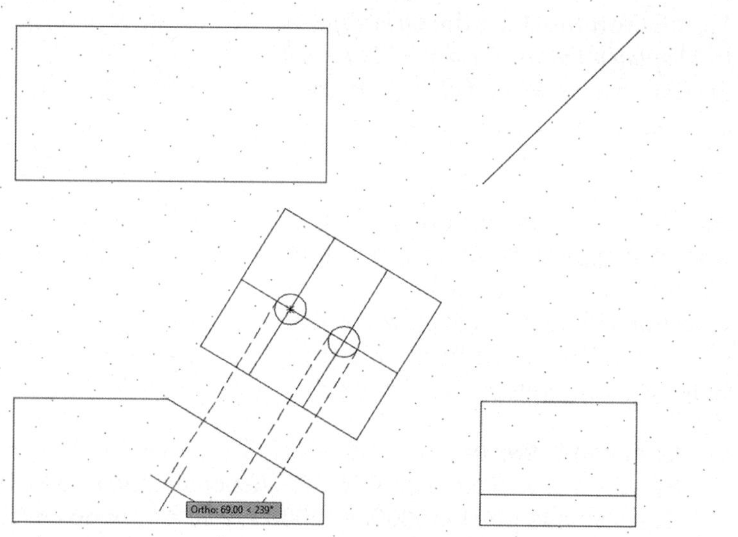


## Rotating *Snap* Back to the Original Position

Before details can be added to the other views, the *Snap* and *Grid* should be rotated back to the original position. It is very important to rotate back using the same base point. Fortunately, AutoCAD remembers the original base point so you can accept the default for the prompt.

TIP

Next, enter a value of **0** when rotating back to the original position. (When using the *Snap Rotate* option, the value entered for the angle of rotation is absolute, not relative to the current position. For example, if the *Snap* was rotated to 45 degrees, rotate back to 0 degrees, not -45.)

Command: ***SNAP***
Specify snap spacing or [ON/OFF/Aspect/Legacy/Style/Type] <2.00>: ***r***
Specify base point <150.00,40.00>: **Enter** (AutoCAD remembers the previous base point.)
Specify rotation angle <329>: **0**
Command:

Construction of multiview drawings with auxiliaries typically involves repeated rotation of the *Snap* and *Grid* to the angle of the inclined surface and back again as needed.

With the *Snap* and *Grid* in the original position, details can be added to the other views as shown in Figure 27-6. Since the two circles appear as ellipses in the top and right side views, project lines from the circles' centers and limiting elements on the inclined surface. Locate centers for the two *Ellipses* to be drawn in the top and right side views.

**FIGURE 27-6**

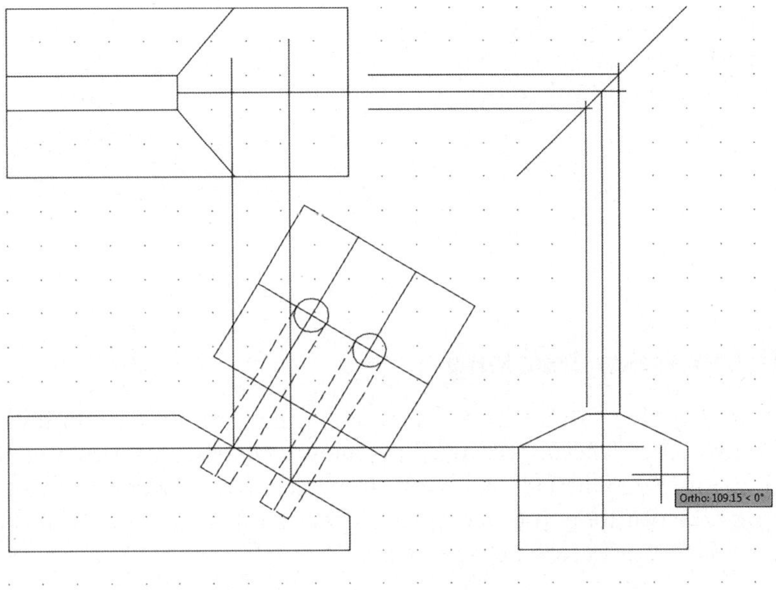


Use the *Ellipse* command to construct the ellipses in the top and right side views. Using the *Center* option of *Ellipse*, specify the center by PICKing with the *Intersection Object Snap* mode. *Object Snap* to the appropriate construction line *Intersection* for the first axis endpoint. For the second axis endpoint (Fig. 27-7), key in the actual circle diameter, since that dimension is not foreshortened.

**FIGURE 27-7**

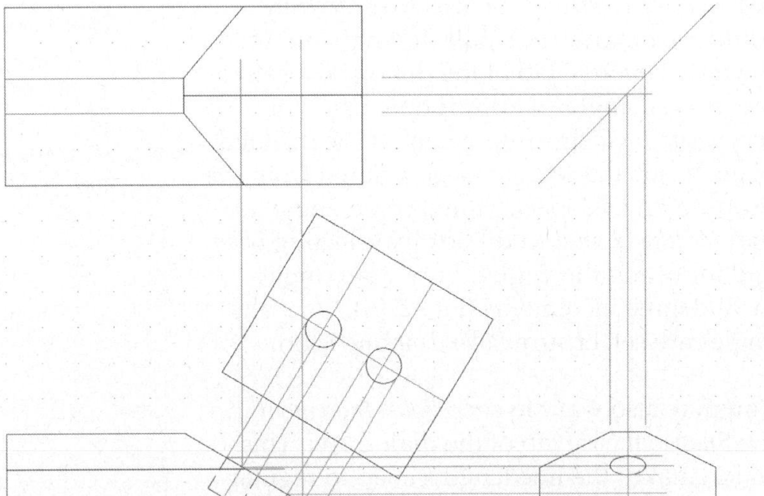

The remaining steps for completing the drawing involve finalizing the perimeter shape of the partial auxiliary view and *Copying* the *Ellipses* to the bottom of the hole positions. The *Snap* and *Grid* should be rotated back to project the new edges found in the front view (Fig. 27-8).

At this point, the multiview drawing with auxiliary view is ready for centerlines, dimensioning, setting up a layout, and construction or insertion of a border and title block.

**FIGURE 27-8**

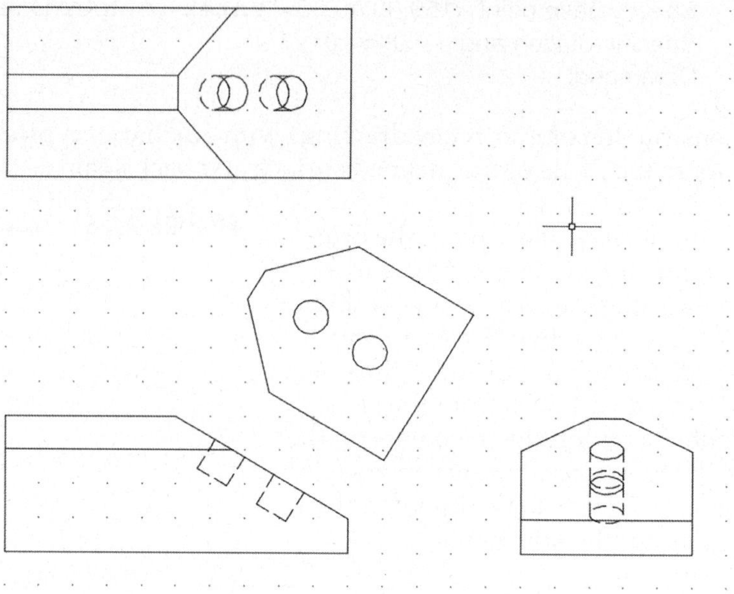

## Using *Polar Tracking*

*Polar Tracking* can be used to create auxiliary views by facilitating construction of *Lines* at specific angles. To use *Polar Tracking* for auxiliary view construction, first use the *List* command to determine the angle of the inclined surface you want to project from. (Keep in mind that, by default, AutoCAD reports angles in whole numbers [no decimals or fractions], so use the *Units* command to increase the *Precision* of *Angular* units before using *List*.)

Once the desired angles are determined, specify the *Polar Angle Settings* in the *Drafting Settings* dialog box. Access the dialog box by right-clicking on the *Polar* icon at the Status Bar, typing *Dsettings*, or selecting *Drafting Settings* from the *Tools* pull-down menu. In the *Polar Tracking* tab of the dialog box, select the desired angles if appropriate from the *Increment angle* drop-down list. If the inclined plane is not at a regular angle offered from the drop-down list, specify the desired angle in the *Additional angles* edit box by selecting *New* and inputting the angles. Enter four angles in 90-degree increments (Fig. 27-9). Once the angles are set, ensure *Polar Tracking* is on.

You may also want to set a *Polar Increment* in the *Snap and Grid* tab of the dialog box. This action makes the line lengths snap to regular intervals. (See Chapter 3 for more information on *Polar Snap* and *Polar Tracking*.)

**FIGURE 27-9**

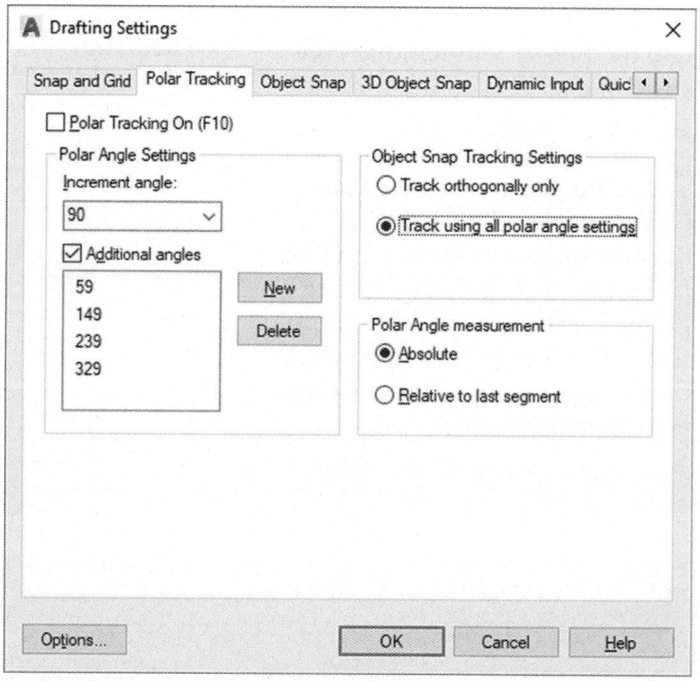

Using *Object Snap*, begin constructing *Lines* perpendicular to the inclined plane. *Polar Tracking* should facilitate the line construction at the appropriate angles (Fig. 27-10).

**FIGURE 27-10**

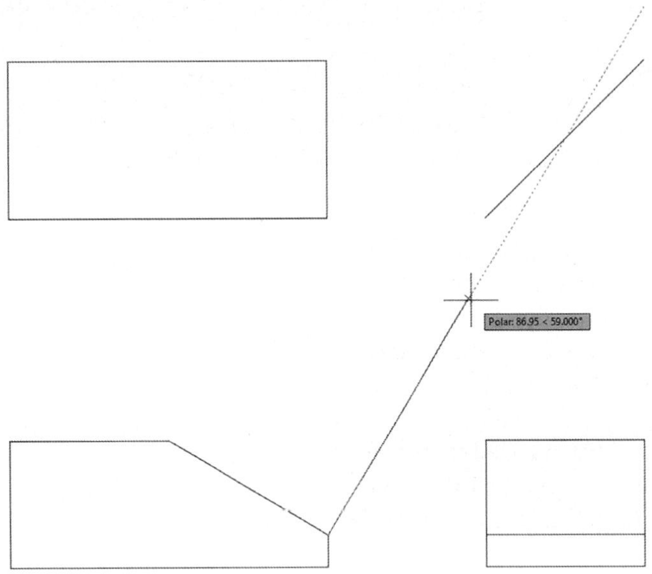


Perpendicular *Line* construction is also assisted by *Polar Tracking* (Fig. 27-11). Continue with this process, constructing necessary lines for the auxiliary view. The remainder of the auxiliary drawing, as described on previous pages (see Fig. 27-5 through Fig. 27-8), could be constructed using *Polar Tracking*. The advantage of this method is that drawing horizontal and vertical *Lines* is also possible while *Polar Tracking* is on.

**FIGURE 27-11**

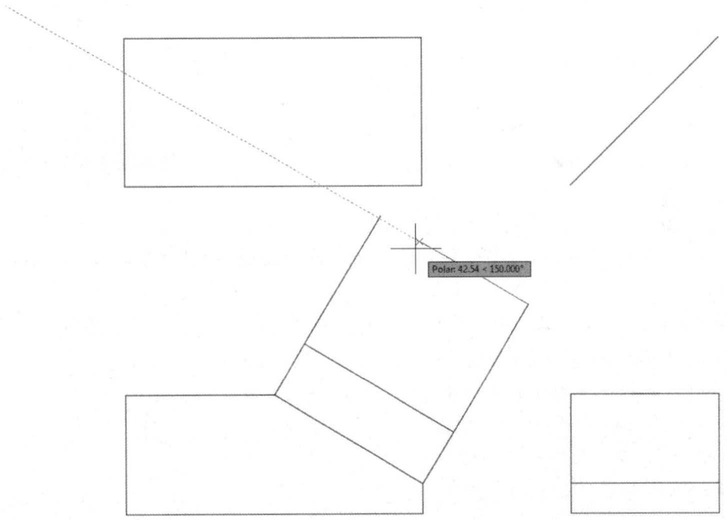


*Object Snap Tracking* can also be employed to construct objects aligned (at the specified Polar angles) with object snap locations (*Endpoint*, *Midpoint*, *Intersection*, *Extension*, etc.). See Chapter 7 for more information on *Object Snap Tracking*.

## Using the *Offset* Command

Another possibility, and an alternative to the *Snap* rotation, is to use *Offset* to make parallel *Lines*. This command can be particularly useful for construction of the "blocked in" partial auxiliary view because it is not necessary to rotate the *Snap* and *Grid*.

### *Offset*

| Ribbon | Menu Bar | Command (Type) | Alias (Type) | Shortcut |
|---|---|---|---|---|
| *Home Modify* (icon) | *Modify Offset* | *Offset* | *O* | ... |

Invoke the *Offset* command and specify a distance. The first distance is arbitrary. Specify an appropriate value between the front view inclined plane and the nearest edge of the auxiliary view (20 for the example). *Offset* the new *Line* at a distance of 50 (for the example) or PICK two points (equal to the depth of the view).

**FIGURE 27-12**

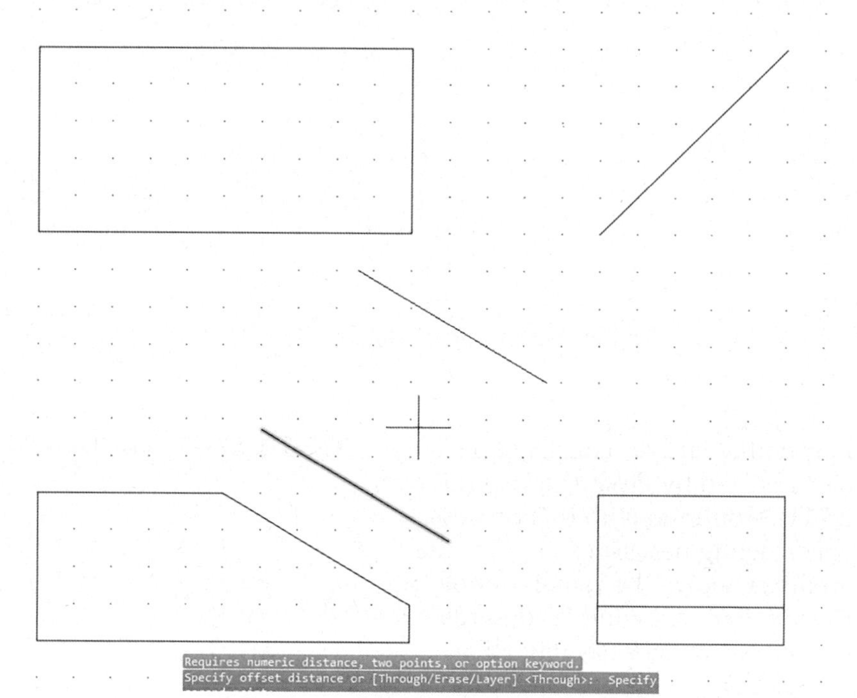


TIP

Note that the *Offset* lines have lengths equal to the original and therefore require no additional editing (Fig. 27-12).

Next, two *Lines* would be drawn between *Endpoints* of the existing offset lines to complete the rectangle. *Offset* could be used again to construct additional lines to facilitate the construction of the two circles in the partial auxiliary view (Fig 27-13).

**FIGURE 27-13**

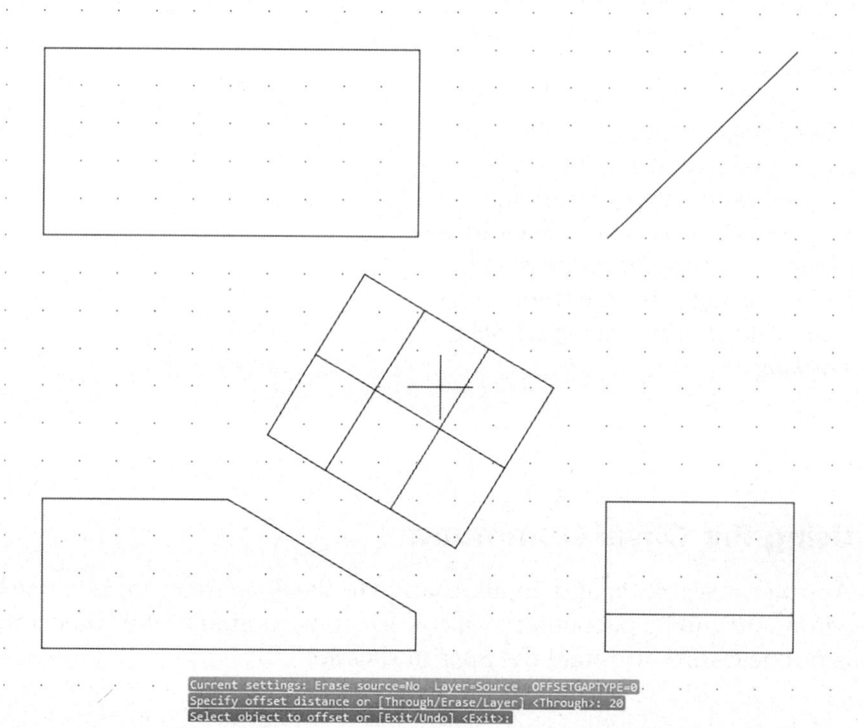


From this point forward, the construction process would be similar to the example given previously (Figs. 27-5 through 27-8). Even though *Offset* does not require that the *Snap* be *Rotated*, the complete construction of the auxiliary view could be simplified by using the rotated *Snap* and *Grid* in conjunction with *Offset*.

## Using the *Xline* and *Ray* Commands

As a fourth alternative for construction of auxiliary views, the *Xline* and *Ray* commands could be used to create construction lines.

### *Xline*

| Ribbon | Menu Bar | Command (Type) | Alias (Type) | Shortcut |
|---|---|---|---|---|
| *Home* *Draw* (icon) | *Draw* *Construction Line* | *Xline* | XL | ... |

### *Ray*

| Ribbon | Menu Bar | Command (Type) | Alias (Type) | Shortcut |
|---|---|---|---|---|
| *Home* *Draw* (icon) | *Draw* *Ray* | *Ray* | ... | ... |

The *Xline* command offers several options shown below:

Command: **XLINE**
Specify a point or [Hor/Ver/Ang/Bisect/Offset]:

The *Ang* option can be used to create a construction line at a specified angle. In this case, the angle specified would be that of the inclined plane or perpendicular to the inclined plane. The *Offset* option works well for drawing construction lines parallel to the inclined plane, especially in the case where the angle of the plane is not known.

Figure 27-14 illustrates the use of *Xline Offset* to create construction lines for the partial auxiliary view. The *Offset* option operates similarly to the *Offset* command described previously. Remember that an *Xline* extends to infinity but can be *Trimmed*, in which case it is converted to *Ray* (*Trim* once) or to a *Line* (*Trim* twice). See Chapter 15 for more information on the *Xline* command.

**FIGURE 27-14**

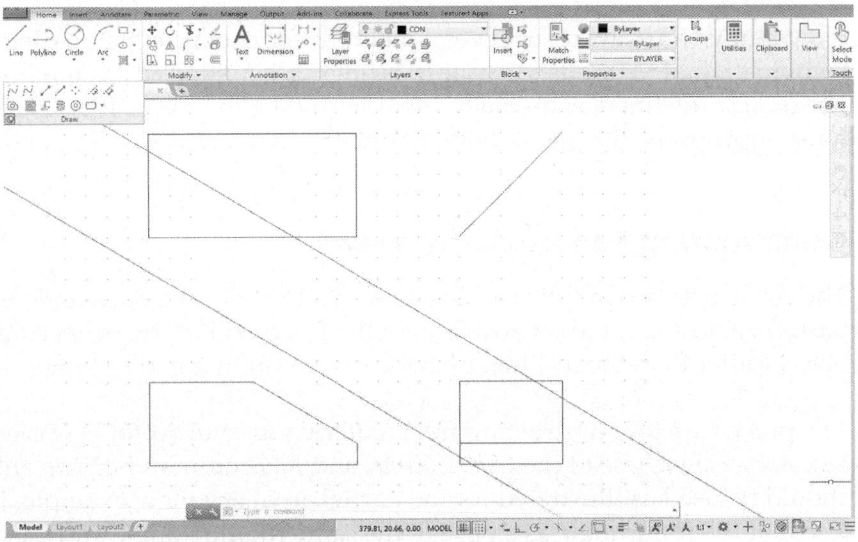

The *Ray* command also creates construction lines; however, the *Ray* has one anchored point and the other end extends to infinity.

Command: ***RAY***
Specify start point: **PICK** or (**coordinates**)
Specify through point: **PICK** or (**coordinates**)

FIGURE 27-15

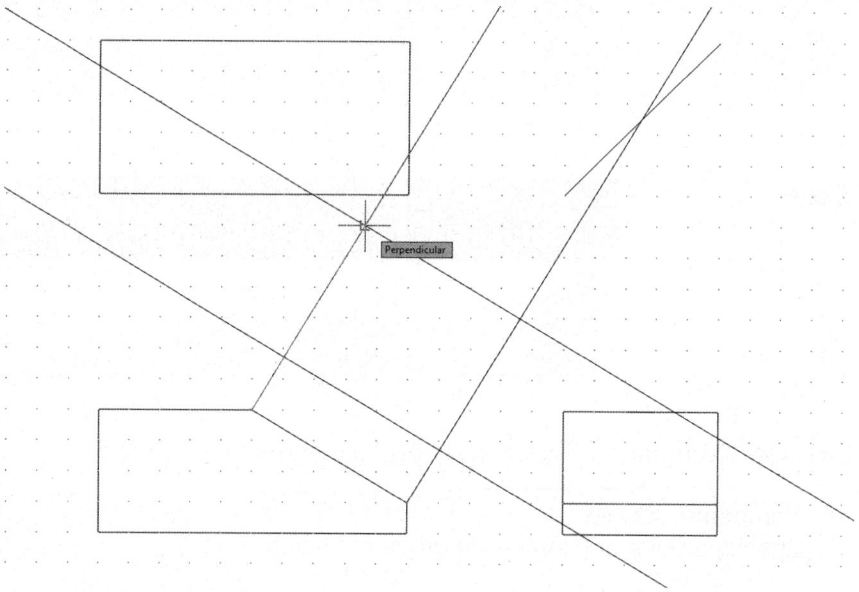


*Rays* are helpful for auxiliary view construction when you want to create projection lines perpendicular to the inclined plane. In Figure 27-15, two *Rays* are constructed from the *Endpoints* of the inclined plane and *Perpendicular* to the existing *Xlines*. Using *Xlines* and *Rays* in conjunction is an excellent method for "blocking in" the view.

There are two strategies for creating drawings using *Xlines* and *Rays*. First, these construction lines can be created on a separate layer and set up as a framework for the object lines. The object lines would then be drawn on top of the construction lines using *Object Snaps* but would be drawn on the object layer. The construction layer would be *Frozen* for plotting. The other strategy is to create the construction lines on the object layer. Through a series of *Trims* and other modifications, the *Xlines* and *Rays* are transformed to the finished object lines.

Now that you are aware of several methods for constructing auxiliary views, use any one method or a combination of methods for your drawings. No matter which methods are used, the final lines that are needed for the finished auxiliary view should be the same. It is up to you to use the methods that are the most appropriate for the particular application or are the easiest and quickest for you personally.

## Constructing Full Auxiliary Views

The construction of a full auxiliary view begins with the partial view. After initial construction of the partial view, begin the construction of the full auxiliary by projecting the other edges and features of the object (other than the inclined plane) to the existing auxiliary view.

The procedure for constructing full auxiliary views in AutoCAD is essentially the same as that for partial auxiliary views. Use of the *Offset, Xline,* and *Ray* commands, *Snap* and *Grid* rotation, and *Polar Tracking* should be used as illustrated for the partial auxiliary view example. Because a full auxiliary view is projected at the same angle as a partial, the same rotation angle and basepoint would be used for the *Snap,* or the same *Increment angle* or *Additional angles* should be used for *Polar Tracking*.

## Reorienting the User Coordinate System

Another way to construct auxiliary views is to align the *UCS* with the inclined surface. A quick way to do this is to drag the UCS icon to a new location and orientation. The *Grid* and *Snap* automatically follow the orientation of the new UCS.

If necessary, turn on display of the UCS icon using the *UCSicon* command with the *ON* option. Click on the UCS icon to show its grips as shown in Figure 27-16.

Hover over the square grip at the origin of the UCS icon to make it "*hot.*" The context menu appears (Figure 27-17) as you hover allowing you to choose *Move and Align*. With *Osnap Intersection* on, drag the UCS icon to the corner of the inclined surface as shown in Figure 27-18.

Click the UCS icon to activate the grips again. Hover over the Y Axis grip as shown in Figure 27-19. The context menu appears. Use it to select Y Axis Direction. Using Osnaps to select accurately, reorient the Y Axis so that it is parallel to the inclined surface.

Now you are ready to draw the auxiliary view.

To restore the UCS to the original appearance, use the UCS command as follows.

Command: ***UCS***
Specify origin of UCS or [Face/NAmed/OBject/Previous/View/World/X/Y/Z/ZAxis] <World>: ***W***

**FIGURE 27-16**

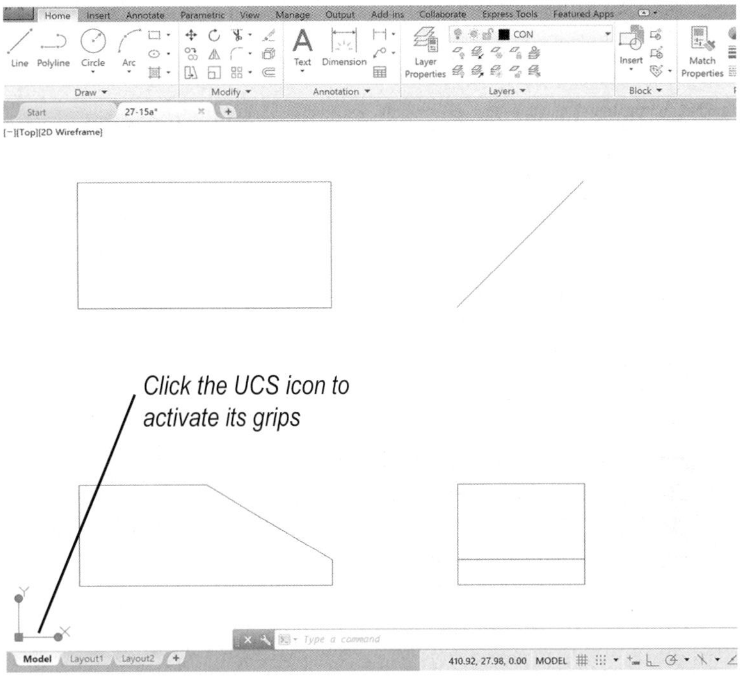


**FIGURE 27-17**

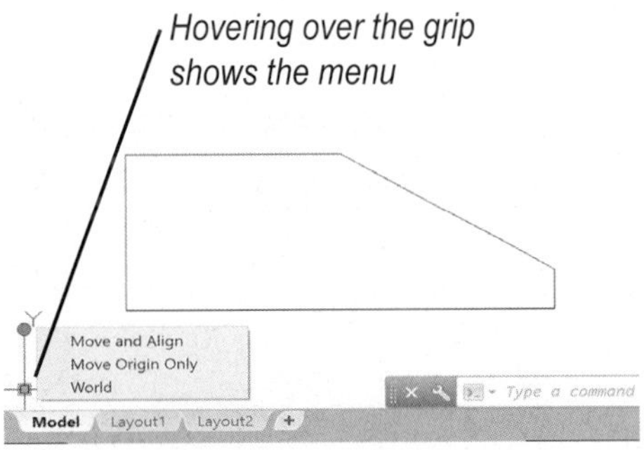


**FIGURE 27-18**

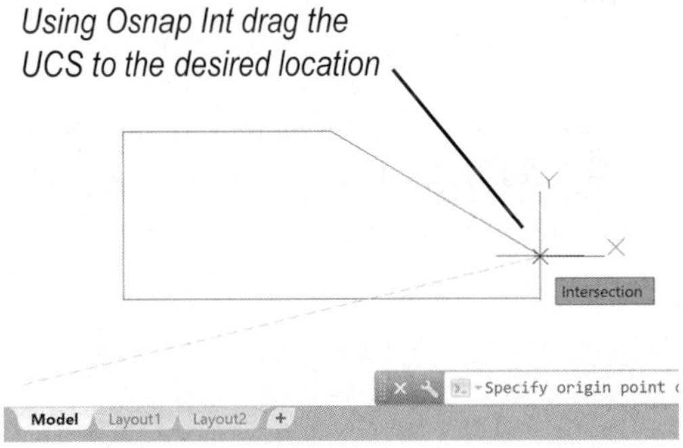


**FIGURE 27-19**

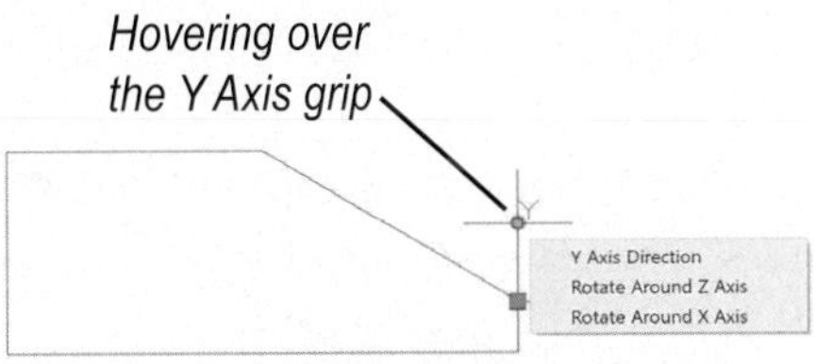


**FIGURE 27-20**

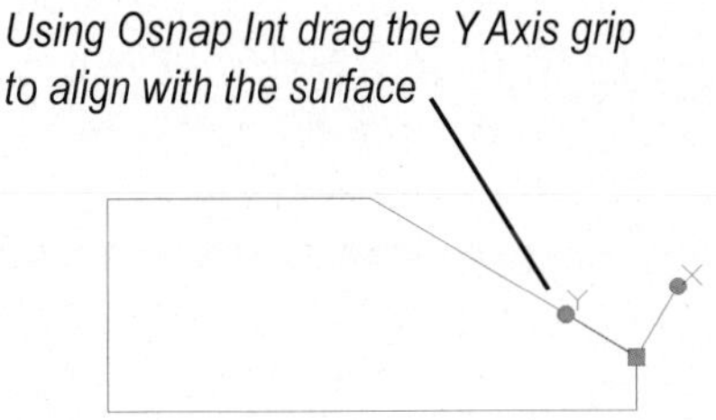

## CHAPTER EXERCISES

For the following exercises, create the multiview drawing, including the partial or full auxiliary view as indicated. Use the appropriate template drawing based on the given dimensions and indicated plot scale.

1. **Partial Auxiliary View**

   Make a multiview drawing with a partial auxiliary view of the example used in this chapter. Refer to Figure 27-1 for dimensions. ***Save*** the drawing as **CH27EX1**. ***Plot*** on an A size sheet at **1=1** scale.

2. **Partial Auxiliary View**

   Recreate the views given in Figure 27-21 and add a partial auxiliary view. ***Save*** the drawing as **CH27EX2** and ***Plot*** on an A size sheet at **2=1** scale.

**FIGURE 27-21**

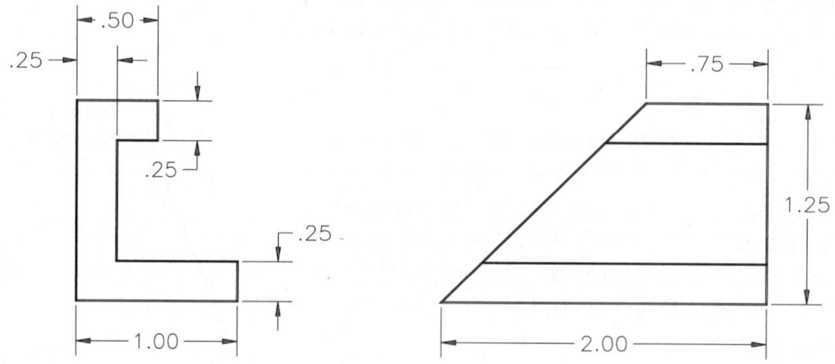


3. **Partial Auxiliary View**

   Recreate the views shown in Figure 27-22 and add a partial auxiliary view. ***Save*** the drawing as **CH27EX3.** Make a ***Plot*** on an A size sheet at **1=1** scale.

**FIGURE 27-22**

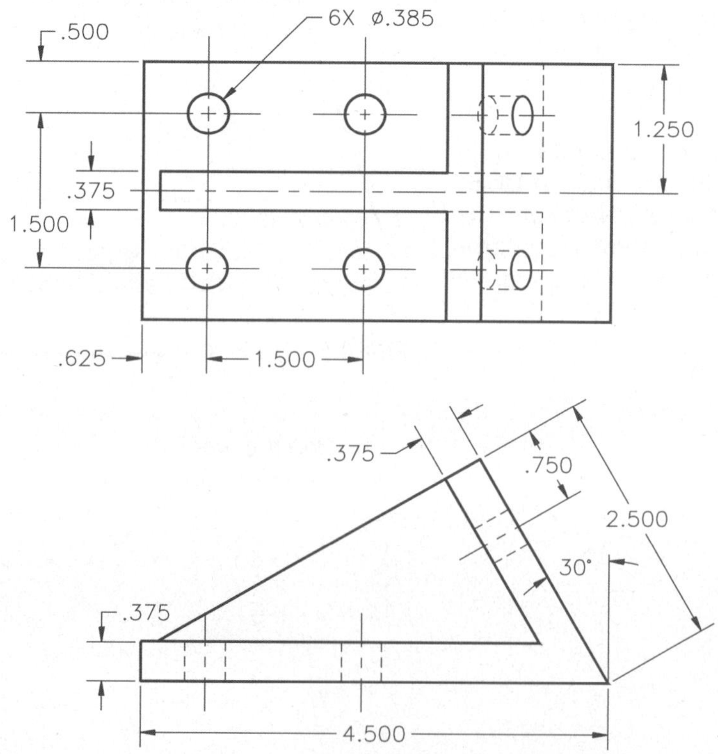

4. **Full Auxiliary View**

Make a multiview drawing of the given views in Figure 27-23. Add a full auxiliary view. ***Save*** the drawing as **CH27EX4**. ***Plot*** the drawing at an appropriate scale on an A size sheet.

**FIGURE 27-23**

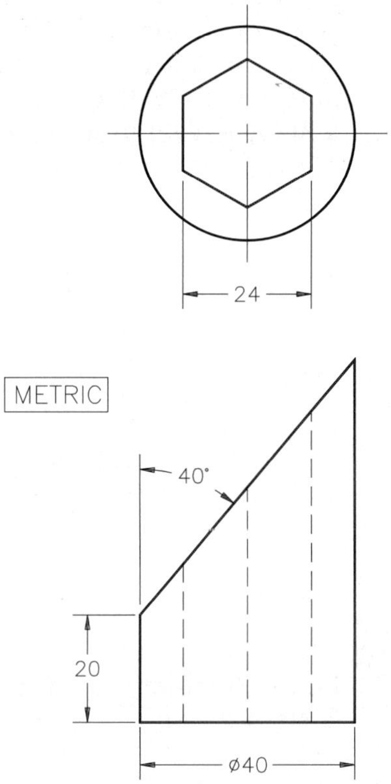


5. **Partial Auxiliary View**

Draw the front, top, and a partial auxiliary view of the holder. During construction, note that the back corner indicated with hidden lines in Figure 27-24 is not a "clean" intersection, but is actually composed of two separate vertical edges. Make a ***Plot*** full size. Save the drawing as **HOLDER**.

**FIGURE 27-24**

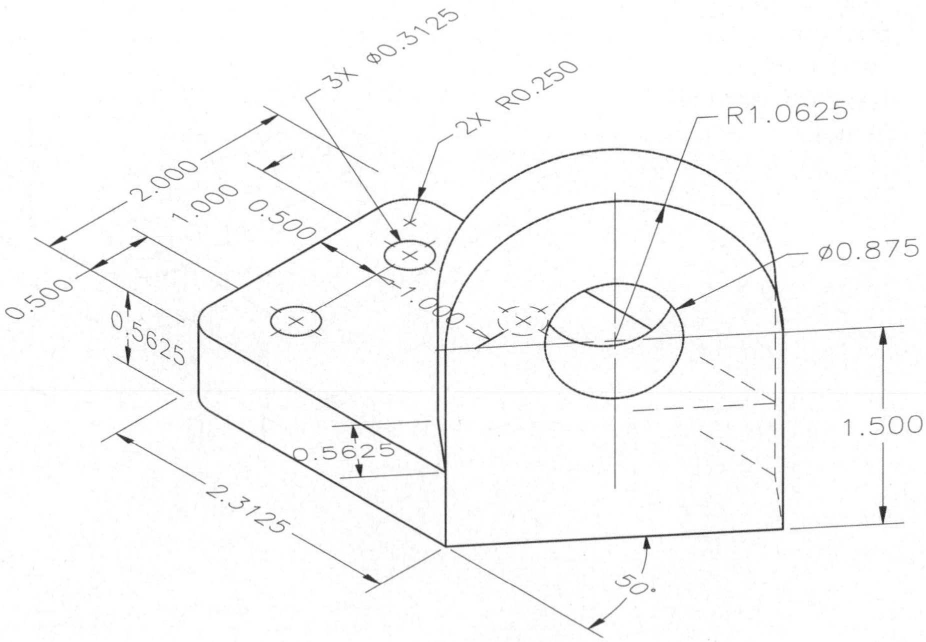

6. **Full Auxiliary View**

Draw three principal views and a full auxiliary view of the V-block shown in Figure 27-25. *Save* the drawing as **VBLOCK**. *Plot* to an accepted scale.

FIGURE 27-25

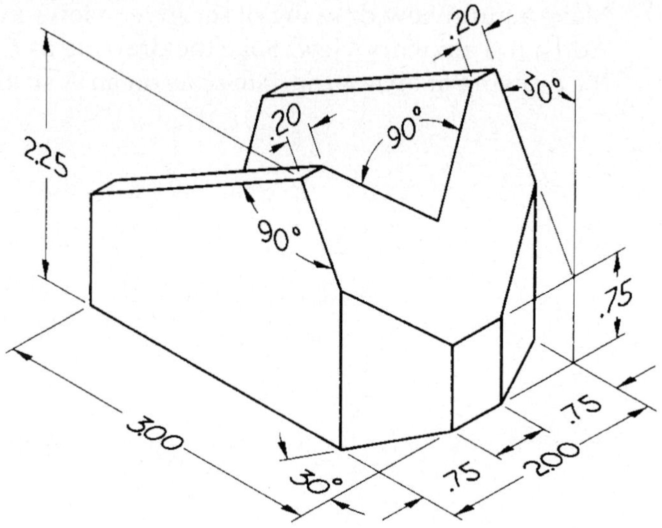


7. **Partial Auxiliary View**

Draw two principal views and a partial auxiliary of the angle brace (Fig. 27-26). *Save* as **ANGLBRAC**.

Make a plot to an accepted scale on an A or B size sheet. Convert the fractions to the current ANSI standard, decimal inches.

FIGURE 27-26

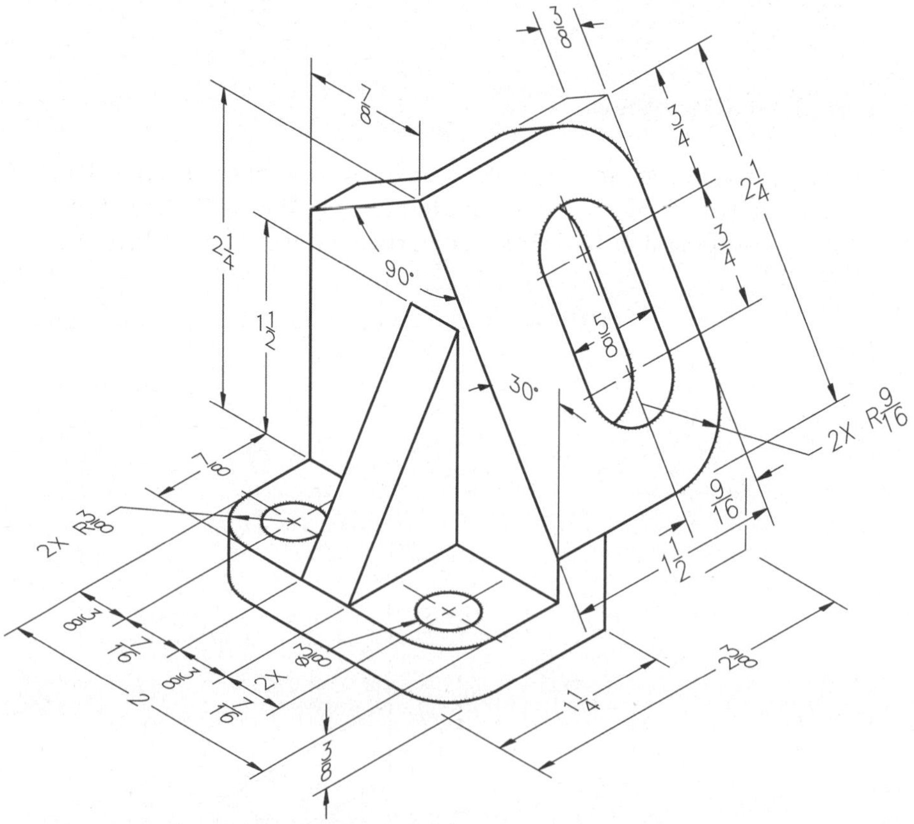

CHAPTER

# 28

# Dimensioning

## Chapter Objectives

After completing this chapter you should:

1. be able to create a variety of dimensions with the *Dim* command;
2. be able to create linear dimensions with *Dimlinear*;
3. be able to append *Dimcontinue* and *Dimbaseline* dimensions to existing dimensions;
4. be able to create *Angular*, *Diameter*, and *Radius* dimensions;
5. know how to affix notes to drawings with *Leaders* and *Qleaders*;
6. be able to use the *Qdim* command to quickly create a variety of associative dimensions;
7. be able to create and apply geometric dimensioning and tolerancing symbols using the *Tolerance* command and dialog boxes;
8. know the possible methods for editing dimensions and dimensioning text.

# CONCEPTS

As you know, drawings created with CAD systems should be constructed with the same size and units as the real-world objects they represent. In this way, the features of the object that you apply dimensions to (lengths, diameters, angles, etc.) are automatically measured by AutoCAD and the correct values are displayed in the dimension text. So if the object is drawn accurately, the dimension values will be created correctly and automatically. Generally, dimensions should be created on a separate layer named DIMENSIONS, or similar, and dimensions should be created in model space (see Chapter 29 for information on dimensioning in paper space).

## Dimension Components

The main components of a dimension are:

1. Dimension line
2. Extension lines
3. Dimension text (usually a numeric value)
4. Arrowheads or tick marks

AutoCAD dimensioning is semi-automatic. When you invoke a command to create a linear dimension, AutoCAD only requires that you PICK an object or specify the extension line origins (where you want the extension lines to begin) and PICK the location of the dimension line (distance from the object). AutoCAD then measures the feature and draws the extension lines, dimension line, arrowheads, and dimension text (Fig. 28-1).

**FIGURE 28-1**

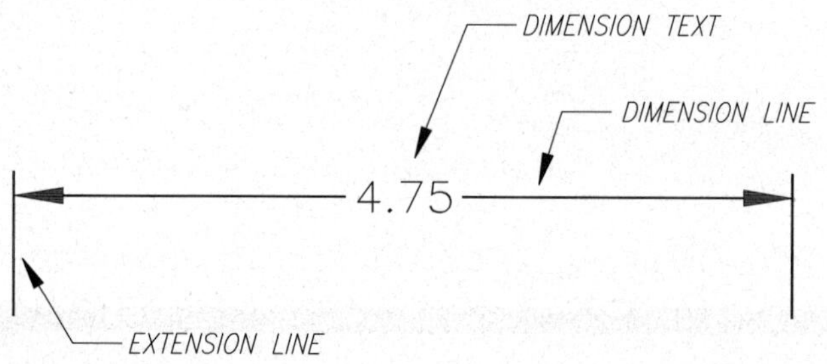


## Dimension Placement

TIP

For linear dimensioning commands, there are two ways to specify placement for a dimension in AutoCAD: you can PICK the object to be dimensioned or you can PICK the two extension line origins. The simplest method is to select the object because it requires only one PICK (Fig. 28-2):

Command: ***DIMLINEAR***
Specify first extension line origin or <select object>: **Enter**
Select object to dimension: **PICK**

**FIGURE 28-2**

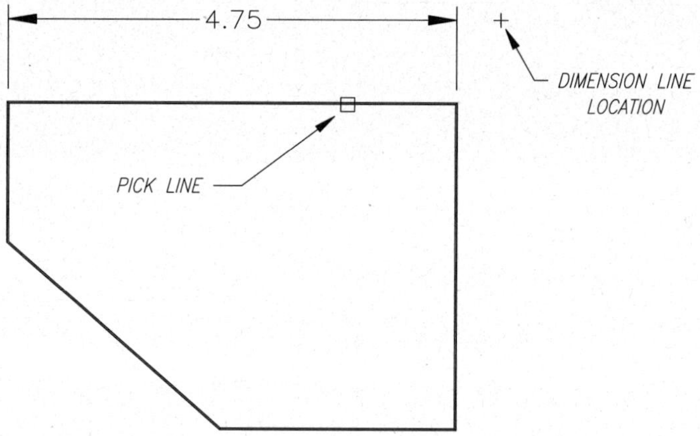


The other method is to PICK the extension line origins (Fig. 28-3). Object Snaps must be used to PICK the object (endpoints in this case) so that the dimension is associated with the object.

Command: ***DIMLINEAR***
Specify first extension line origin or <select object>: **PICK**
Specify second extension line origin: **PICK**

**FIGURE 28-3**

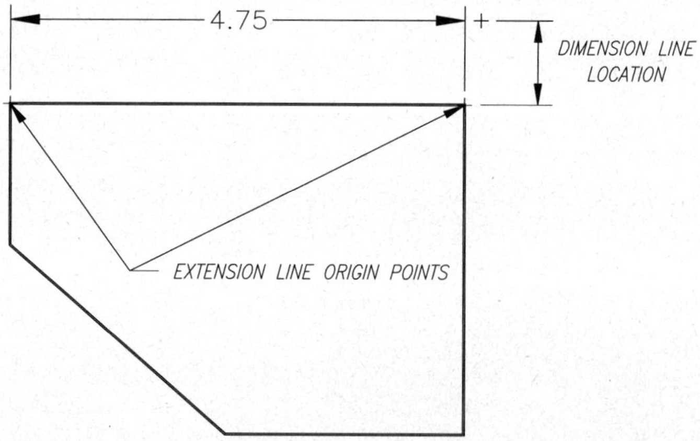

Once the dimension is attached to the object, you specify how far you want the dimension to be placed from the object (called the "dimension line location").

NOTE: Remember that Object Snaps are required if you want to select the "extension line origins." The *Dimlinear* command's default option is to "Specify first extension line origin" so use Object Snaps to select the points. If you instead prefer to simply pick the line to dimension and have AutoCAD automatically select the extension line origins, press Enter at the first prompt (note the previous Command line prompts).

AutoCAD's *Dim* command also allows you to either pick the object to dimension or specify the extension line origins. However, with the *Dim* command, in order to specify the first extension line origin, simply turn on Object Snap and select.

Command: ***DIM***
Select objects or specify first extension line origin:

## Associativity

Dimensioning in AutoCAD is associative (by default). Because the extension line origins are "associated" with the geometry, the dimension text automatically updates if the geometry is *Moved*, *Stretched*, *Rotated*, *Scaled*, or edited using grips.

Because dimensioning is semi-automatic, dimensioning variables are used to control the way dimensions are created. Dimensioning variables can be used to control features such as text or arrow size, direction of the leader arrow for radial or diametrical dimensions, format of the text, and many other possible options. Groups of variable settings can be named and saved as Dimension Styles. Dimensioning variables and dimension styles are discussed in Chapter 29.

**FIGURE 28-4**

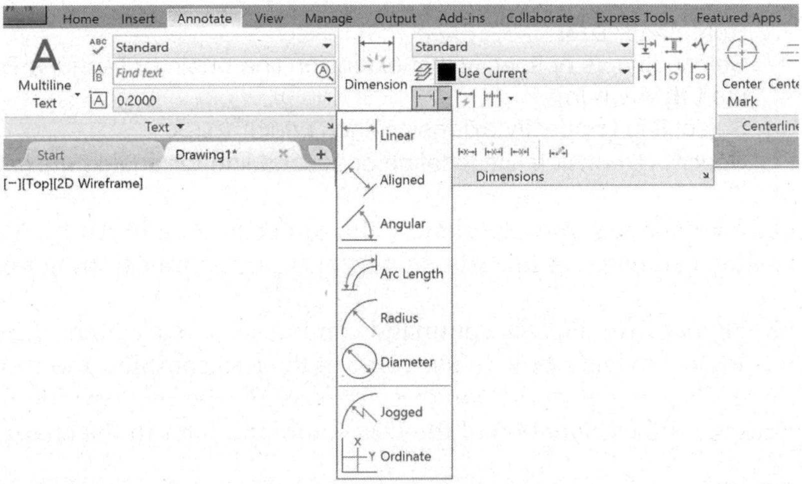


**FIGURE 28-5**

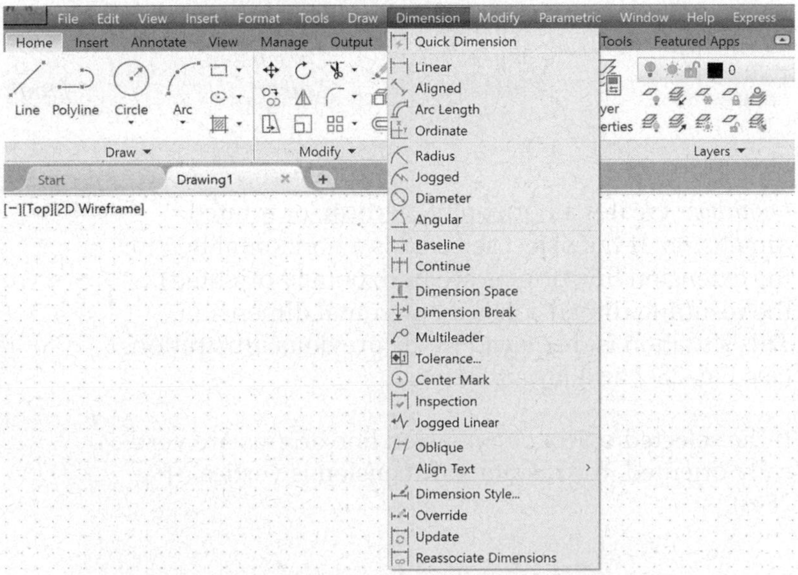


## Accessing Dimension Commands

Dimension commands can be invoked by the typical methods. A *Dimensions* panel is available on both the *Home* tab and the *Annotate* tab of the Ribbon (Fig. 28-4). You can also use the *Dimension* pull-down menu if the Menu Bar is activated (Fig. 28-5). Or, you can activate a *Dimension* toolbar using the toolbar list (*Tools* pull-down menu).

# DIMENSION DRAWING COMMANDS

## *Dim*

The *Dim* command is a newer dimensioning command that combines several older commands. *Dim* allows you to create several types of dimensions and offers options for editing the text objects (dimensional values) and arrangement and spacing of the dimension lines. Prior to 2016, specific commands were needed for each type of dimension (such as *Dimlinear*, *Dimangular*, *Dimaligned*, *Dimdiameter*, or *Dimradius*). Since then the single *Dim* command is used for most functions.

You can use the *Dim* command to create many types of dimensions. The *Dim* command allows you to create the following types of dimensions:

linear, angular, aligned, diametrical, radial, baseline, continuous, ordinate

The Command line prompts are shown below, although not all options appear in the prompts.

Command: **`DIM`**
Select objects or specify first extension line origin or [Angular/Baseline/Continue/Ordinate/aliGn/ Distribute/Layer/Undo]:
Select line to specify extension lines origin: **`PICK`**
Specify dimension line location or second line for angle [Mtext/Text/text aNgle/Undo]:

Once you have some experience with dimensioning in AutoCAD, the *Dim* command may be the primary dimensioning command you use for typical dimensioning tasks.

TIP

NOTE: Because the *Dim* command combines so many other dimension creation and editing commands, it is logical to learn how to use some of the less complex and more specialized commands such as *Dimlinear, Dimangular,* and *Dimradius*, for example, before learning the *Dim* command and all of its options. Find the full explanation of the *Dim* command later in the chapter (after the *Dimradius* command).

## *Dimlinear*

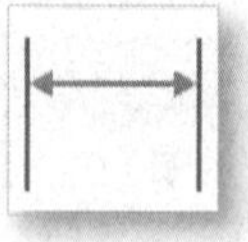

| Ribbon | Menu Bar | Command (Type) | Alias (Type) | Shortcut |
|---|---|---|---|---|
| *Annotate Dimensions Linear* | *Dimension Linear* | *Dimlinear* | DIMLIN or DLI | ... |

*Dimlinear* creates a horizontal, vertical, or rotated dimension. If the object selected is a horizontal line (or the extension line origins are horizontally oriented), the resulting dimension is a horizontal dimension. This situation is displayed in the previous illustrations (see Fig. 28-2 and Fig. 28-3).

If the selected object or extension line origins are vertically oriented, the resulting dimension is vertical (Fig. 28-6).

FIGURE 28-6

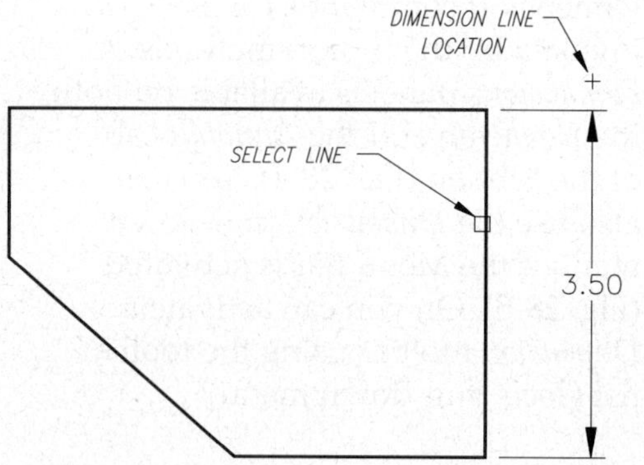

When you dimension an inclined object (or if the selected extension line origins are diagonally oriented), a vertical <u>or</u> horizontal dimension can be made, depending on where you drag the dimension line in relation to the object. If the dimension line location is more to the side, a vertical dimension is created (Fig. 28-7), or if you drag farther up or down, a horizontal dimension results (Fig. 28-8).

**FIGURE 28-7**

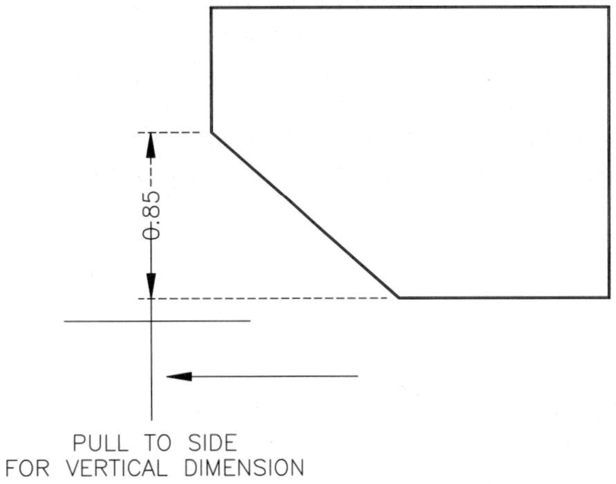


**FIGURE 28-8**

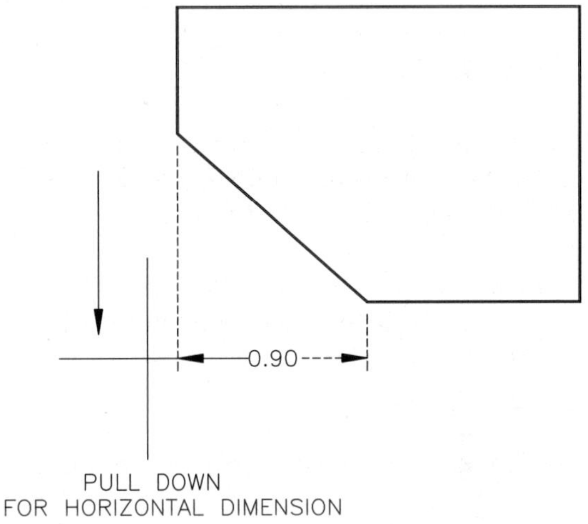


TIP

If you select the extension line origins, it is very important to PICK the <u>object's endpoints</u> if the dimensions are to be truly associative (associated with the geometry). Object Snap should be used to find the object's *Endpoint*, *Intersection*, etc.

Command: ***DIMLINEAR***
Specify first extension line origin or <select object>:
**PICK** (use Object Snaps)
Specify second extension line origin: **PICK** (use Object Snaps)
Specify dimension line location or [Mtext/Text/Angle/Horizontal/Vertical/Rotated]: **PICK** (where you want the dimension line to be placed)
Dimension text = *n.nnnn*
Command:

When you pick the location for the dimension line, AutoCAD automatically measures the object and inserts the correct numerical value. The other options are explained next.

### *Rotated*

If you want the dimension line to be drawn at an angle instead of vertical or horizontal, use this option. Selecting an inclined line, as in the previous two illustrations, would normally create a horizontal or vertical dimension. The *Rotated* option allows you to enter an <u>angular value</u> for the dimension line to be drawn. For example, selecting the diagonal line and specifying the appropriate angle would create the dimension shown in Figure 28-9. This object, however, could be more easily dimensioned with the *Dimaligned* command (see *Dimaligned*) or *Dim* command (see *Dim*).

**FIGURE 28-9**

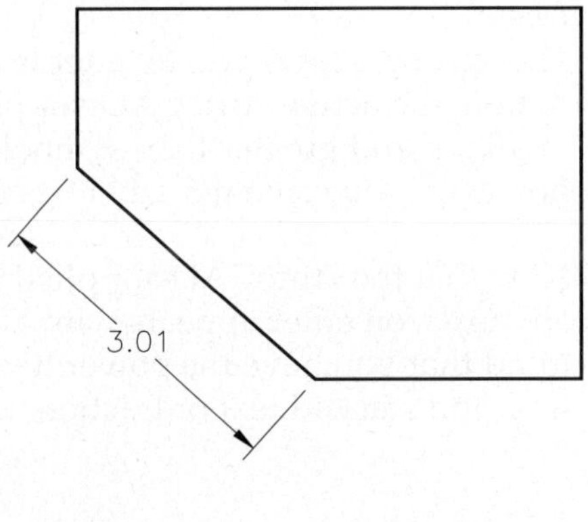

A *Rotated* dimension should be used when the geometry has "steps" or any time the desired dimension line angle is different than the dimensioned feature (when you need extension lines of different lengths). Figure 28-10 illustrates the result of using a *Rotated* dimension to give the correct dimension line angle and extension line origins for the given object. In this case, the extension line origins were explicitly PICKed. The feature of *Rotated* that makes it unique is that you specify the <u>angle</u> that the dimension line will be drawn.

**FIGURE 28-10**

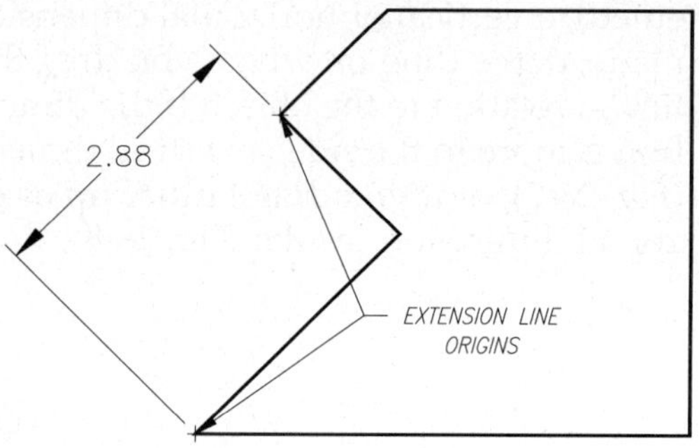


### *Text*

Using the *Text* option allows you to enter any value or characters in place of the AutoCAD-measured text. The measured value is given as a reference at the command prompt.

Command: ***DIMLINEAR***
Specify first extension line origin or <select object>: **PICK**
Specify second extension line origin: **PICK**
Specify dimension line location or [Mtext/Text/Angle/Horizontal/Vertical/Rotated]: ***T***
Enter dimension text <*n.nnnn*>: (**text**) or (**value**)

Entering a value or text at the prompt (above) causes AutoCAD to display that value or text instead of the AutoCAD-measured value. If you want to keep the AutoCAD-measured value but add a prefix or suffix, use the less-than, greater-than symbols (< >) to represent the actual (AutoCAD) value:

Enter dimension text <0.25>: **2X <>**
Specify dimension line location or [Mtext/Text/Angle/Horizontal/Vertical/Rotated]: **PICK**
Dimension text = 0.25
Command:

The "2X <>" response produces the dimension text shown in Figure 28-11.

**FIGURE 28-11**

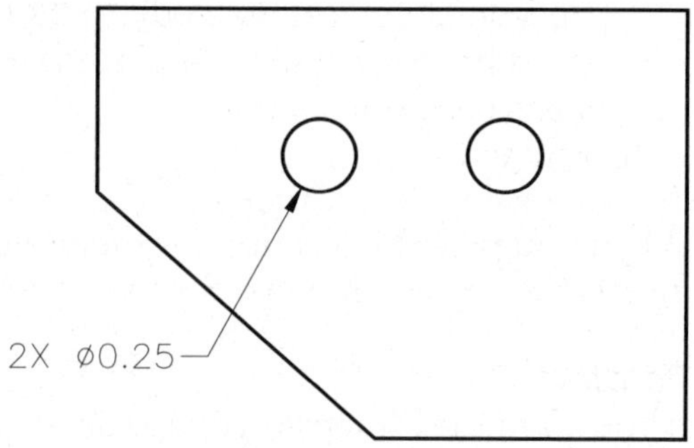


NOTE: Changing the AutoCAD-measured value should be discouraged. If the geometry is drawn accurately, the dimensional value is correct. If you specify other dimension text, the text value is <u>not</u> updated in the event of *Stretching, Rotating,* or otherwise editing the associative dimension.

### *Mtext*

This option allows you to enter text using the *Text Editor* that appears on the Ribbon. With the *Mtext* option, the actual AutoCAD-supplied text appears in the edit box instead of being represented with less-than and greater-than symbols (<>). It is not recommended to change the AutoCAD-supplied text; however, you can enter any text before or after.

Note that the AutoCAD-supplied text appears in a dark background, whereas any text you enter appears with a lighter background (Fig. 28-12). Keep in mind that you have the power to change fonts, text height, bold, italic, underline, and stacked text or fractions using this editor.

**FIGURE 28-12**

### *Angle*

This creates text drawn at the angle you specify. Use this for special cases when the text must be drawn to a specific angle other than horizontal. (It is also possible to make the text automatically align with the angle of the dimension line using the *Dimension Style Manager*. See Chapter 29.)

### *Horizontal*

Use the *Horizontal* option when you want to force a horizontal dimension for an inclined line and the desired placement of the dimension line would otherwise cause a vertical dimension.

### *Vertical*

This option forces a *Vertical* dimension for any case.

## ***Dimaligned***

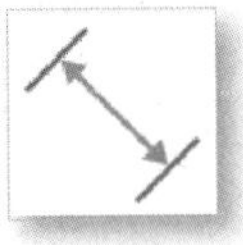

| Ribbon | Menu Bar | Command (Type) | Alias (Type) | Shortcut |
|---|---|---|---|---|
| *Annotate Dimensions Aligned* | *Dimension Aligned* | *Dimaligned* | *DIMALI* or *DAL* | ... |

An *Aligned* dimension is aligned with (at the same angle as) the selected object or the extension line origins. For example, when *Aligned* is used to dimension the angled object shown in Figure 28-13, the resulting dimension aligns with the *Line*. This holds true for either option—PICKing the object or the extension line origins. If a *Circle* is PICKed, the dimension line is aligned with the selected point on the *Circle* and its center. The command syntax for the *Dimaligned* command accepting the defaults is:

**FIGURE 28-13**

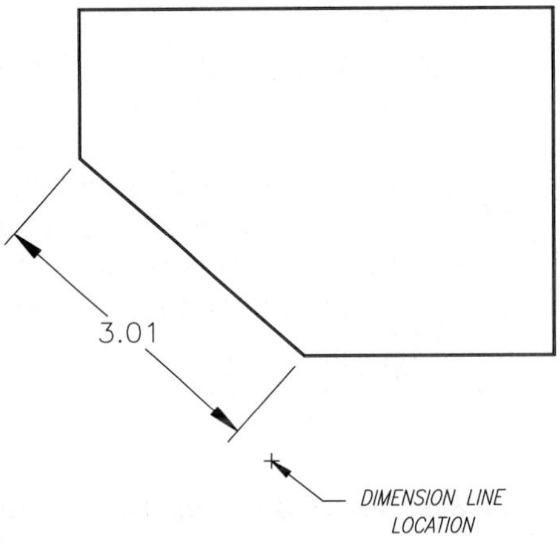


Command: ***DIMALIGNED***
Specify first extension line origin or <select object>: **PICK**
Specify second extension line origin: **PICK**
Specify dimension line location or [Mtext/Text/Angle]: **PICK**
Dimension text = *n.nnnn*
Command:

The three options (*Mtext/Text/Angle*) operate similar to those for *Dimlinear*.

### *Mtext*

The *Mtext* option calls the *Text Editor*. You can alter the AutoCAD-supplied text value or modify other visible features of the dimension text such as fonts, text height, bold, italic, underline, stacked text or fractions, and text style (see "*Dimlinear*," *Mtext*).

### *Text*

You can change the AutoCAD-supplied numerical value or add other annotation to the value in command line format (see "*Dimlinear*," *Text*).

### *Angle*

Enter a value for the angle that the text will be drawn.

**TIP**

The typical application for *Dimaligned* is for dimensioning an angled but <u>straight</u> feature of an object, as shown in Figure 28-13. *Dimaligned* should not be used to dimension an object feature that contains "steps," as shown in Figure 28-10. *Dimaligned* always draws <u>extension lines of equal length</u>.

## *Dimbaseline*

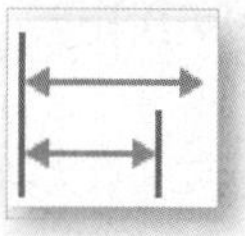

| Ribbon | Menu Bar | Command (Type) | Alias (Type) | Shortcut |
|---|---|---|---|---|
| *Annotate Dimensions Baseline* | *Dimension Baseline* | *Dimbaseline* | *DIMBASE* or *DBA* | ... |

*Dimbaseline* allows you to create a dimension that uses an extension line origin from a previously created dimension. Successive *Dimbaseline* dimensions can be used to create the style of dimensioning shown in Figure 28-14.

**FIGURE 28-14**

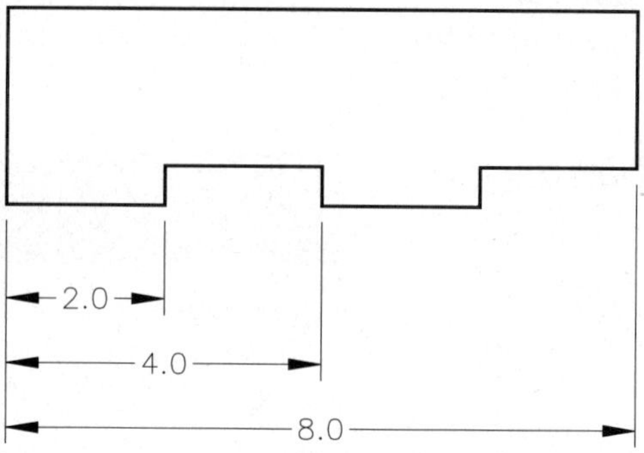


A baseline dimension must be connected to an existing dimension. If *Dimbaseline* is invoked immediately after another dimensioning command, you are required only to specify the second extension line origin since AutoCAD knows to use the previous dimension's first extension line origin:

Command: ***DIMBASELINE***
Specify a second extension line origin or [Select/Undo] <Select>: **PICK**
Dimension text = *n.nnnn*

**FIGURE 28-15**

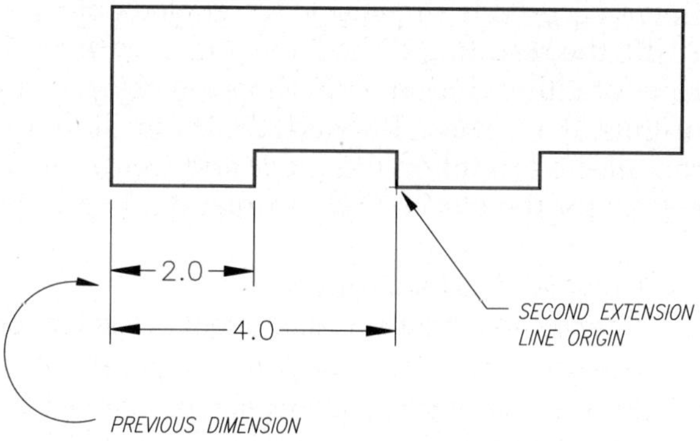


The previous dimension's first extension line is used also for the baseline dimension's first point (Fig. 28-15). Therefore, you specify only the second extension line origin. Note that you are not required to specify the dimension line location. AutoCAD spaces the new dimension line automatically, based on the setting of the dimension line increment variable (Chapter 29).

If you wish to create a *Dimbaseline* dimension using a dimension other than the one just created, use the *Select* option (Fig. 28-16):

Command: ***DIMBASELINE***
Specify a second extension line origin or [Undo/Select] <Select>: ***S*** or **Enter**
Select base dimension: **PICK**
Specify a second extension line origin or [Undo/Select] <Select>: **Enter**
Dimension text = *n.nnnn*

**FIGURE 28-16**

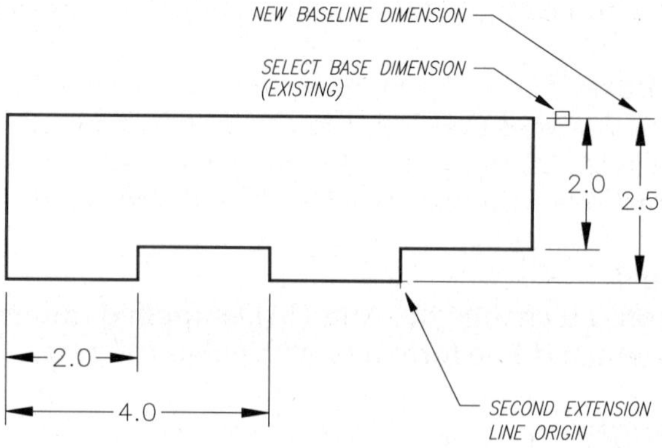


The extension line selected as the base dimension becomes the first extension line for the new *Dimbaseline* dimension.

The *Undo* option can be used to undo the last baseline dimension created in the current command sequence.

*Dimbaseline* can be used with rotated, aligned, angular, and ordinate dimensions.

## *Dimcontinue*

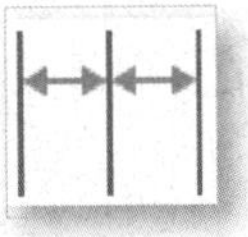

| Ribbon | Menu Bar | Command (Type) | Alias (Type) | Shortcut |
|---|---|---|---|---|
| *Annotate Dimensions Continue* | *Dimension Continue* | *Dimcontinue* | DIMCONT or DCO | ... |

*Dimcontinue* dimensions continue in a line from a previously created dimension. *Dimcontinue* dimension lines are attached to, and drawn the same distance from, the object as an existing dimension.

*Dimcontinue* is similar to *Dimbaseline* except that an existing dimension's second extension line is used to begin the new dimension. In other words, the new dimension is connected to the second extension line, rather than to the first, as with a *Dimbaseline* dimension (Fig. 28-17).

**FIGURE 28-17**

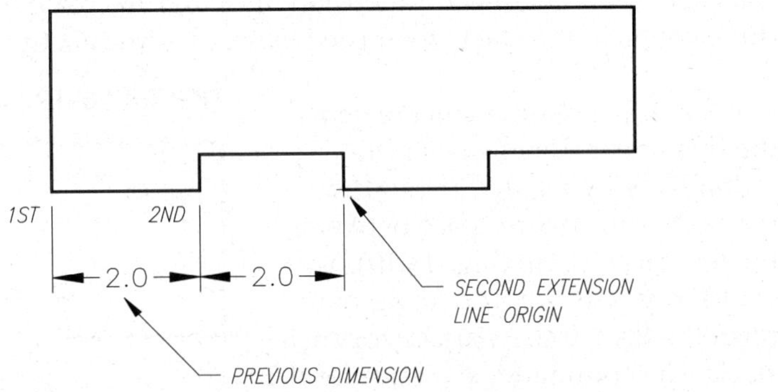


The command syntax is as follows:

Command: ***DIMCONTINUE***
Specify a second extension line origin or
[Undo/Select] <Select>: **PICK**
Dimension text = *n.nnnn*

Assuming a dimension was just drawn, *Dimcontinue* could be used to place the next dimension, as shown in Figure 28-17.

TIP

If you want to create a continued dimension and attach it to an extension line other than the previous dimension's second extension line, you can use the *Select* option to pick an extension line of any other dimension. Then, use *Dimcontinue* to create a continued dimension from the selected extension line (Fig. 28-18).

**FIGURE 28-18**

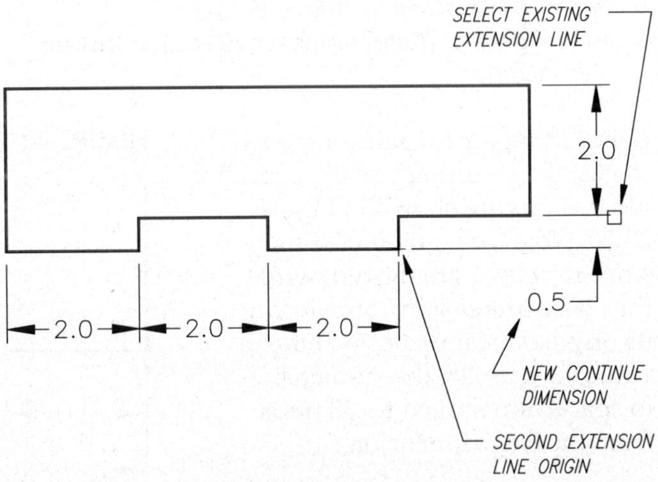


The *Undo* option can be used to undo the last continued dimension created in the current command sequence. *Dimcontinue* can be used with rotated, aligned, angular, and ordinate dimensions.

NOTE: Keep in mind that the new *Dim* command can create continuous and baseline dimensions. The *Baseline* and *Continue* options of *Dim* operate similarly to the *Dimbaseline* and *Dimcontinue* commands.

## Dimspace

| Ribbon | Menu Bar | Command (Type) | Alias (Type) | Shortcut |
|---|---|---|---|---|
| *Annotate Dimensions* (icon) | *Dimension Dimension Space* | *Dimspace* | ... | ... |

The *Dimspace* command automatically spaces parallel linear or angular dimension lines equally. The dimensions can be "stacked" (similar to baseline dimensions) or end to end (similar to continuous dimensions). You can specify a space value or use the *Auto* option to have AutoCAD space the lines automatically. For example, instead of using *Snap* or another method to specify a space interval between "stacked" parallel linear dimension lines as you create each one, you can quickly create all the needed dimensions in the stack, then use *Dimspace* to retroactively space the lines equally.

For example, assume you created the five linear dimensions individually using *Dimlinear*, but did not bother to ensure space between the lines was equal (Fig. 28-19). You could then use *Dimspace* to accomplish the task. Use *Dimspace* for each stack. The command sequence for the three horizontal dimensions in the example is shown below.

**FIGURE 28-19**

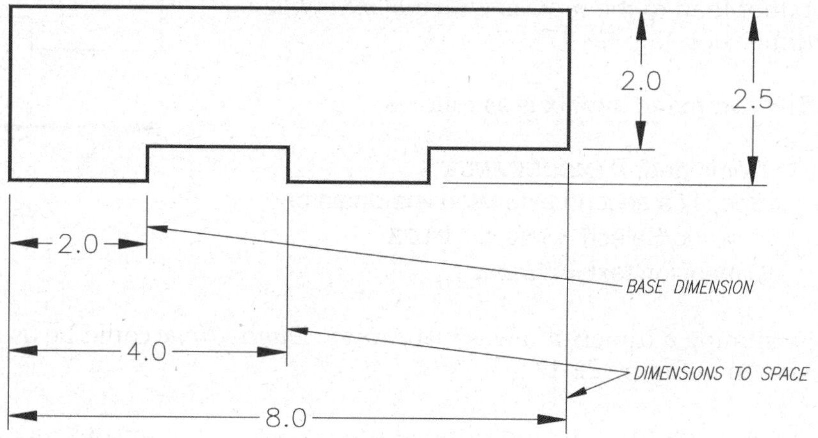


Command: ***DIMSPACE***
Select base dimension: **PICK**
Select dimensions to space: **PICK**
Select dimensions to space: **PICK**
Select dimensions to space: **Enter**
Enter value or [Auto] <Auto>: (**value**) or **Enter**
Command:

Using *Dimspace* for each of the two stacks, the resulting equally spaced dimensions are shown in Figure 28-20. Note that only the "dimensions to space" are moved, while the "base dimension" remains in its original location. If you enter a value such as .38, the "dimensions to space" are moved to .38 units from the base dimension.

**FIGURE 28-20**

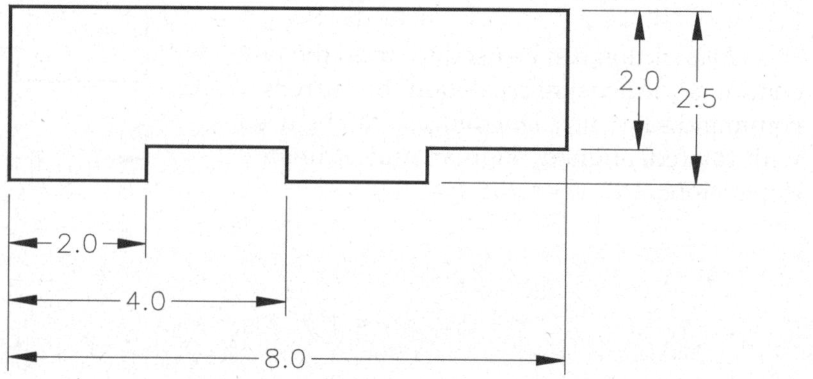


### *Auto*

The *Auto* option calculates the spacing distance automatically based on the text height value specified in the dimension style of the selected base dimension. The resulting space between dimension lines is twice the height of the dimension text.

### End-to-End Alignment

If the dimensions are placed end to end, similar to continuous dimensions, but not perfectly aligned, you can use a spacing value of 0 (zero) to align the selected linear or angular dimensions exactly end to end. For example, if successive dimensions were created individually with *Dimlinear* (not *Dimcontinue*), but not correctly aligned, use *Dimspace* with a value of 0 to achieve perfect end-to-end alignment such as that shown in Figure 28-18. See also the *aliGn* option of *Dim* which can be used to space dimension lines.

## *Dimangular*

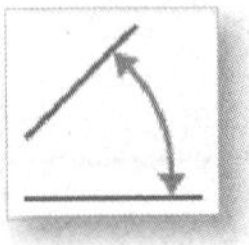

| Ribbon | Menu Bar | Command (Type) | Alias (Type) | Shortcut |
|---|---|---|---|---|
| *Annotate Dimensions Angular* | *Dimension Angular* | *Dimangular* | *DIMANG* or *DAN* | ... |

The *Dimangular* command provides many possible methods of creating an angular dimension.

A typical angular dimension is created between two *Lines* that form an angle (of other than 90 degrees). The dimension line for an angular dimension is radiused with its center at the vertex of the angle (Fig. 28-21). A *Dimangular* dimension automatically adds the degree symbol (°) to the dimension text. The dimension text format is controlled by the current settings for *Units* in the *Dimension Style* dialog box.

**FIGURE 28-21**

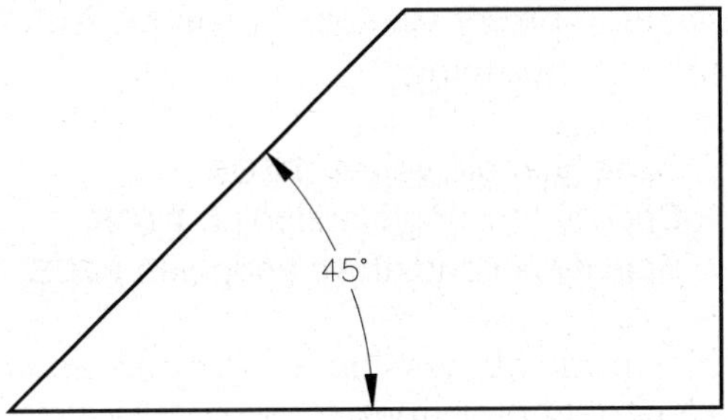


AutoCAD automates the process of creating this type of dimension by offering options within the command syntax. The default options create a dimension, as shown here:

Command: ***DIMANGULAR***
Select arc, circle, line, or <specify vertex>: **PICK**
Select second line: **PICK**
Specify dimension arc line location or [Mtext/Text/Angle/Quadrant]: **PICK**
Dimension text = *nn*
Command:

*Dimangular* dimensioning offers some very useful and easy-to-use options for placing the desired dimension line and text location.

At the "Specify dimension arc line location or [Mtext/Text/Angle/Quadrant]:" prompt, you can move the cursor around the vertex to dynamically display possible placements available for the dimension. The dimension can be placed in any of four positions as well as any distance from the vertex (Fig. 28-22). Extension lines are automatically created as needed.

On the other hand, you can use the *Quadrant* option to lock the dimension lines to one of the four "quadrants" while you drag only the text outside the quadrant (see the four "quadrants" in Fig. 28-22). Extension lines are added if needed.

**FIGURE 28-22**

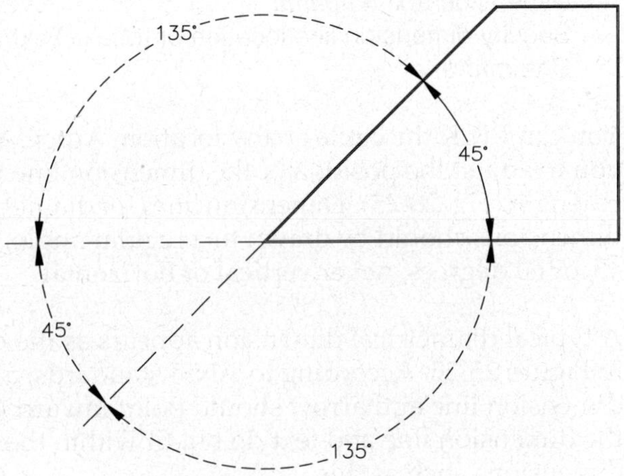

The *Dimangular* command offers other options, including dimensioning angles for *Arcs*, *Circles*, or allowing selection of any three points. If you select an *Arc* in response to the "Select arc, circle, line, or <specify vertex>:" prompt, AutoCAD uses the *Arc's* center as the vertex and the *Arc's* endpoints to generate the extension lines. You can select either angle of the *Arc* to dimension (Fig. 28-23).

**FIGURE 28-23**

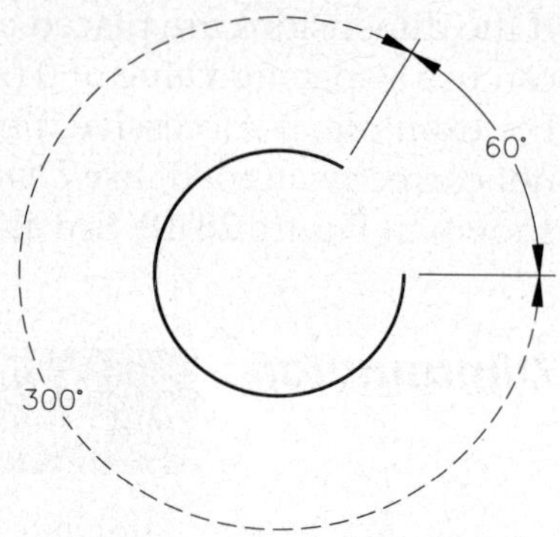


If you select a *Circle*, AutoCAD uses the PICK point as the first extension line origin. The second extension line origin does not have to be on the *Circle*, as shown in Figure 28-24.

**FIGURE 28-24**

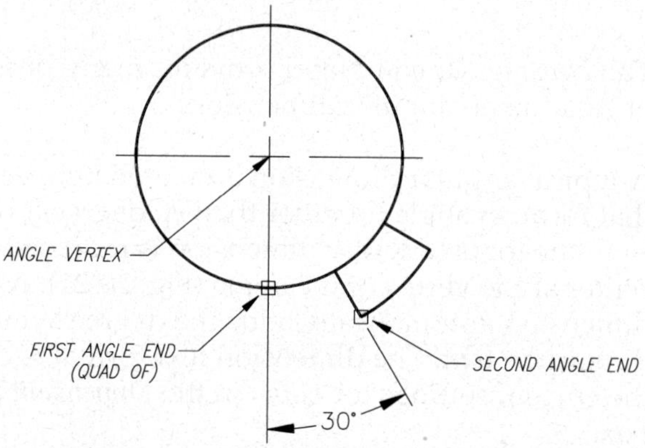


If you press Enter in response to the "Select arc, circle, line, or <specify vertex>:" prompt, AutoCAD responds with the following:

Specify angle vertex: **PICK**
Specify first angle endpoint: **PICK**
Specify second angle endpoint: **PICK**

This option allows you to apply an *Angular* dimension to a variety of shapes.

## *Dimdiameter*

| Ribbon | Menu Bar | Command (Type) | Alias (Type) | Shortcut |
|---|---|---|---|---|
| *Annotate Dimensions Diameter* | *Dimension Diameter* | *Dimdiameter* | DIMDIA or DDI | ... |

The *Dimdiameter* command creates a diametrical dimension by selecting any *Circle*. Diametrical dimensions should be used for full 360 degree *Circles* and can be used for *Arcs* of more than 180 degrees.

Command: ***DIMDIAMETER***
Select arc or circle: **PICK**
Dimension text = *n.nnnn*
Specify dimension line location or [Mtext/Text/Angle]: **PICK**
Command:

**FIGURE 28-25**

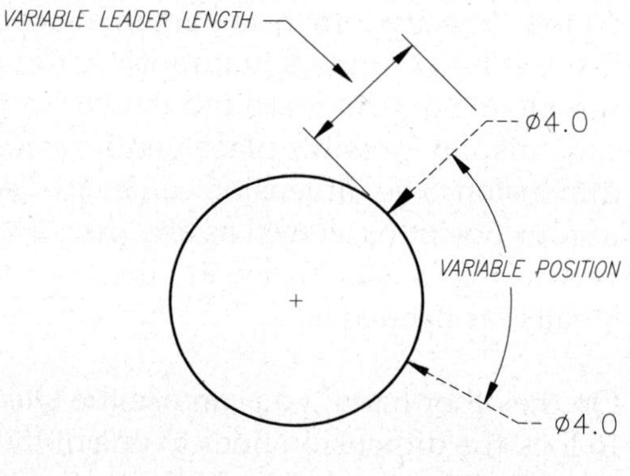


You can PICK the circle at any location. AutoCAD allows you to adjust the position of the dimension line to any angle or length (Fig. 28-25). Dimension lines for diametrical or radial dimensions should be drawn to a regular angle, such as 30, 45, or 60 degrees, never vertical or horizontal.

TIP

A typical diametrical dimension appears as the example in Figure 28-25. According to ANSI standards, a diameter dimension line and arrow should point inward (toward the center) for holes and small circles where the dimension line and text do not fit within the circle. Use the default variable settings for *Dimdiameter* dimensions such as this.

For dimensioning large circles, ANSI standards suggest an alternate method for diameter dimensions where sufficient room exists for text and arrows inside the circle (Fig. 28-26). To create this style of dimensioning, set the variables to *Text* and *Place text manually when dimensioning* in the *Fit* tab of the *Dimension Style Manager* (see Chapter 29).

FIGURE 28-26

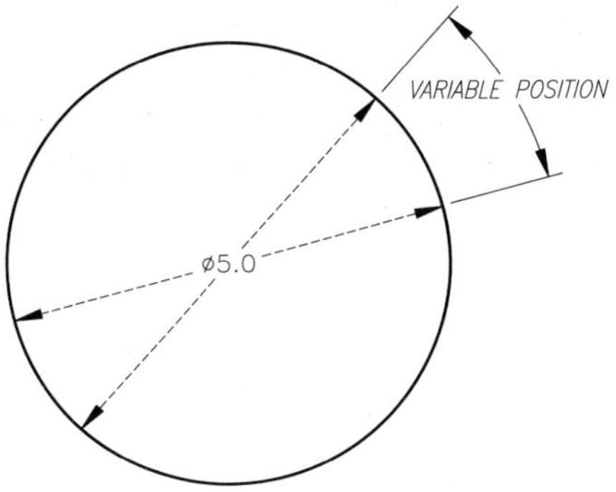


Notice that the *Diameter* command creates center marks at the *Circle's* center. Center marks can also be drawn by using the *Centermark* command (see Chapter 24) or *Dimcenter* command (discussed later). AutoCAD uses the center and the point selected on the *Circle* to maintain its associativity.

### *Mtext/Text*

The *Mtext* or *Text* options can be used to modify or add annotation to the default value. The *Mtext* option summons the *Text Editor* and the *Text* option uses Command line format. Both options operate similar to the other dimensioning commands (see *Dimlinear, Mtext,* and *Text*).

Notice that with diameter dimensions AutoCAD automatically creates the Ø (phi) symbol before the dimensional value. This designation satisfies the ANSI and ISO standards for representing diameters. If you prefer to use a prefix before or suffix after the dimension value, it can be accomplished with the *Mtext* or *Text* option. Remember that the < > symbols represent the AutoCAD-measured value so the additional text should be inserted on <u>either side</u> of the symbols. Inserting a prefix by this method <u>does not override</u> the Ø (phi) symbol (Fig. 28-27).

FIGURE 28-27

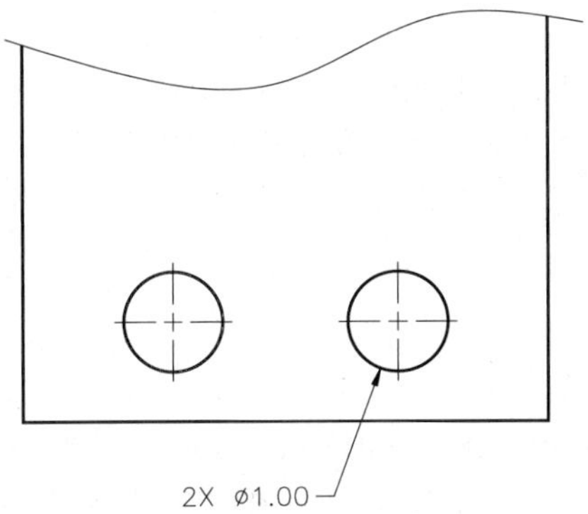


TIP

A prefix or suffix can alternately be added to the measured value by using the *Dimension Style Manager* and entering text or values in the *Prefix* or *Suffix* edit boxes. Using this method, however, <u>overrides</u> the Ø symbol (see Chapter 29).

### *Angle*

With this option, you can specify an angle (other than the default) for the text to be drawn by entering a value.

## Dimradius

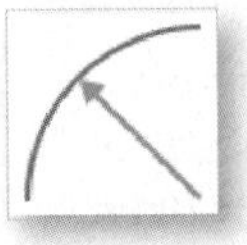

| Ribbon | Menu Bar | Command (Type) | Alias (Type) | Shortcut |
|---|---|---|---|---|
| *Annotate Dimensions Radius* | *Dimension Radius* | *Dimradius* | *DIMRAD* or *DRA* | ... |

*Dimradius* is used to create a dimension for an arc of anything less than half of a circle. ANSI standards dictate that a *Radius* dimension line should point outward (from the arc's center), unless there is insufficient room, in which case the line can be drawn on the outside pointing inward, as with a leader. The text can be located inside an arc (if sufficient room exists) or is forced outside of small *Arcs* on a leader.

Command: ***DIMRADIUS***
Select arc or circle: **PICK**
Dimension text = *n.nnnn*
Specify dimension line location or [Mtext/Text/Angle]: **PICK**
Command:

Assuming the defaults, a *Dimradius* dimension can appear on either side of an arc, as shown in Figure 28-28. Placement of the dimension line is variable. Dimension lines for arcs and circles should be positioned at a regular angle such as 30, 45, or 60 degrees, never vertical or horizontal.

**FIGURE 28-28**

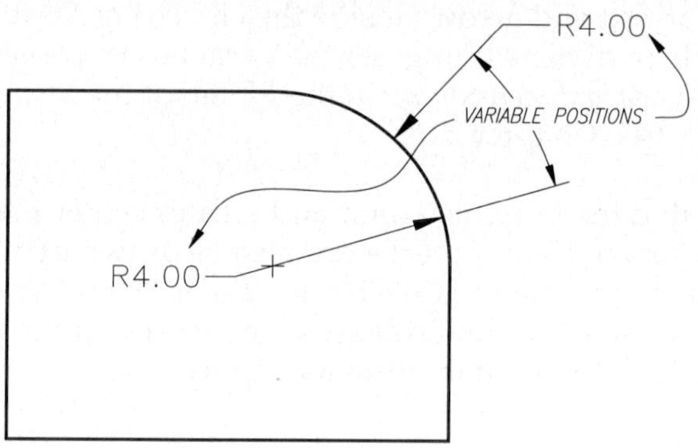


When the radius dimension is dragged outside of the arc, a center mark is automatically created (Fig. 28-28). When the radius dimension is dragged inside the arc, no center mark is created. Center marks can be created using the *Centermark* command (see Chapter 24) or the *Dimcenter* command (discussed later.)

TIP

According to ANSI standards, the dimension line and arrow should point outward from the center for radial dimensions (Fig. 28-29). The text can be placed inside or outside of the *Arc*, depending on how much room exists. To create a *Dimradius* dimension to comply with this standard, dimension variables must be changed from the defaults. To achieve *Dimradius* dimensions as shown in Figure 28-29, set the variables to *Text* and *Place text manually when dimensioning* in the *Fit* tab of the *Dimension Style Manager* (see Chapter 29).

**FIGURE 28-29**

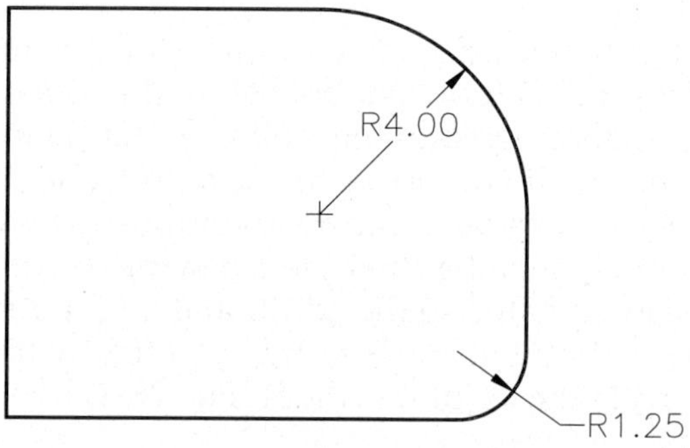


For very small radii, such as that shown in Figure 28-30, there is insufficient room for the text and arrow to fit inside the *Arc*. In this case, AutoCAD automatically forces the text outside with the leader pointing inward toward the center. No changes have to be made to the default settings for this to occur.

**FIGURE 28-30**

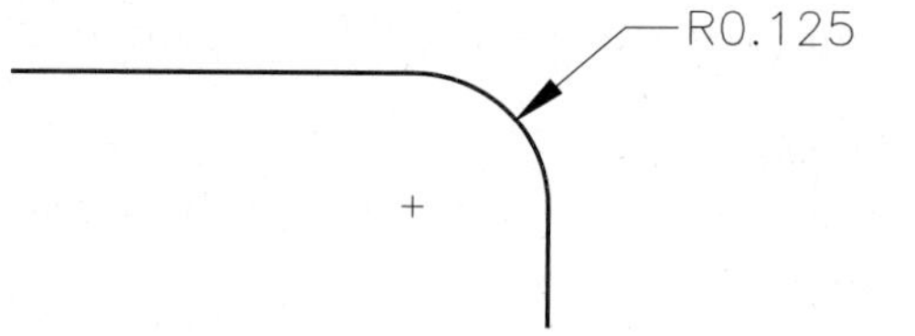


### *Mtext/Text*

These options can be used to modify or add annotation to the default value. AutoCAD automatically inserts the letter "R" before the numerical text whenever a *Dimradius* dimension is created. This is the correct notation for radius dimensions. The *Mtext* option calls the *Text Editor*, and the *Text* option uses the Command line format for entering text. With the *Text* option AutoCAD uses the < > symbols to represent the AutoCAD-supplied value. Entering text inside the < > symbols overrides the measured value. Entering text before the symbols adds a prefix without overriding the "R" designation. Alternately, text can be added by using the *Prefix* and *Suffix* options of the *Dimension Style Manager* series; however, a *Prefix* entered in the edit box replaces the letter "R." (See Chapter 29.)

### *Angle*

With this option, you can specify an angle (other than the default) for the text to be drawn by entering a value.

### *Dim*

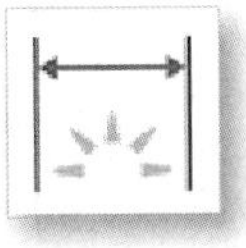

| Ribbon | Menu Bar | Command (Type) | Alias (Type) | Shortcut |
|---|---|---|---|---|
| *Annotate Dimensions Dimension* | ... | *Dim* | ... | ... |

NOTE: The *Dim* command appears on the Ribbon, but does not appear in either the *Dimension* pull-down menu or in the *Dimension* toolbar. You can, however, customize those menus to include the *Dim* command using the *CUI* command. See AutoCAD *Help* for *CUI* (Customize User Interface).

The *Dim* command is a new dimensioning command that combines several older dimensioning commands and offers options for editing the dimensioning text and for organizing the dimension lines. As a matter of fact, the *Dim* command combines all the commands discussed in this chapter previous to this point. That is, *Dim* allows you to accomplish almost everything you can with *Dimlinear, Dimaligned, Dimbaseline, Dimcontinue, Dimspace, Dimangular, Dimdiameter,* and *Dimradius* (previously discussed) as well as *Dimordinate* (discussed near the end of this chapter).

The Command line prompts are shown below, although not all options appear in the prompts.

Command: **DIM**
Select objects or specify first extension line origin or [Angular/Baseline/Continue/Ordinate/aliGn/Distribute/Layer/Undo]: **HOVER** on *Line*
Select line to specify extension lines origin: **PICK**
Specify dimension line location or second line for angle [Mtext/Text/text aNgle/Undo]:

TIP

There are a few differences in the *Dim* command with the operation of command prompts from other dimensioning commands.

1. Although the linear, aligned, diameter and radius options do not appear at the *Dim* command prompt, you can simply select the type of object you want to dimension such as a *Line*, *Circle*, or *Arc*.

2. Unlike other commands, the *Dim* command prompts change as you hover over different object types. The prompts are different depending on the *On/Off* status of *Object Snap*.

3. The *On/Off* status of *Object Snap* determines if you select extension line origins or if you select the dimension object to dimension—there is no option at the Command line to specify which method you want to use. To select objects to dimension (*Line, Arc,* etc.), turn *Object Snap* off. To select extension line origins, simply turn on *Object Snap*.

#### Horizontal, Vertical, Aligned

There are no Command line options for horizontal, vertical, or aligned dimensions. To create a horizontal, vertical, or aligned dimension with *Dim*, select the *Line* to dimension. The resulting dimension will initially be parallel to the selected *Line* as you select the *Line*, as shown in Figure 28-31.

TIP

When dimensioning an angled *Line*, you can create an aligned dimension as shown or drag the dimension to a horizontal or vertical orientation. Also, turning on *Ortho* forces the dimension to a horizontal or vertical orientation.

**FIGURE 28-31**

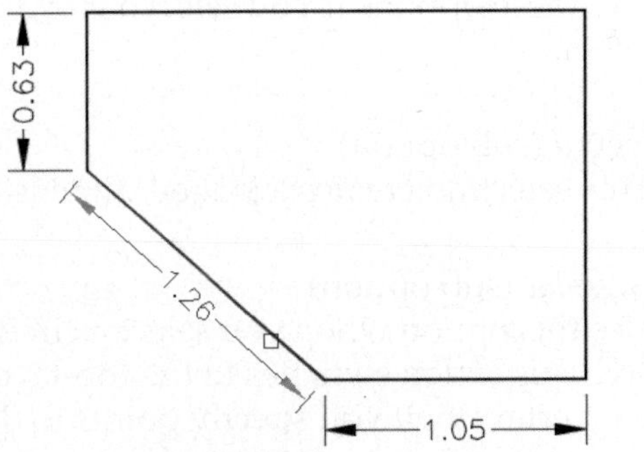


SELECT OBJECT METHOD:
DIMENSION IS INITIALLY PARALLEL TO LINE

If you use the "specify extension line origin" method by turning *Object Snap* on, the resulting dimension is also normally parallel with a selected horizontal or vertical object (technically, parallel with the extension line origins). However, if you are dimensioning an angled line you can drag the resulting dimension to the desired orientation—horizontal, vertical, or aligned as shown in Figure 28-32.

**FIGURE 28-32**

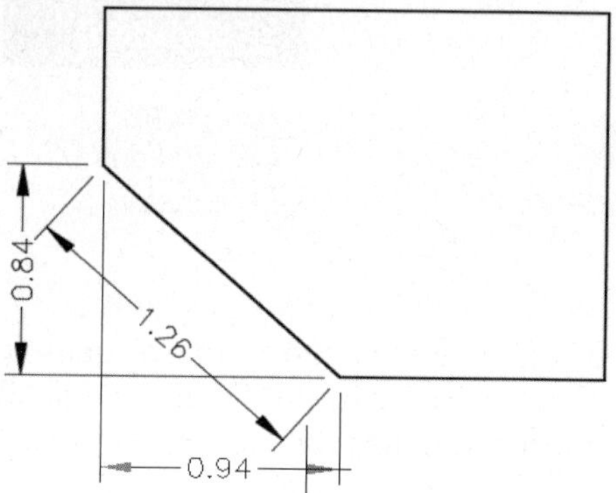


SELECT EXTENSION LINE ORIGIN METHOD:
DIMENSION CAN BE DRAGGED TO
HORIZONTAL, VERTICAL OR ALIGNED

**TIP**

If you are dimensioning "steps," do not use the "select objects" method since the extension lines extend to the object endpoints and therefore would not leave the required gap between extension line and object. Instead, use the "specify extension line origin" method with *Object Snap* on. Remember that the resulting dimension is normally parallel with extension line origins (Fig 28-33, right). You can use *Ortho*, however, to force the dimension to a horizontal or vertical orientation (Fig. 28-33, left).

**FIGURE 28-33**

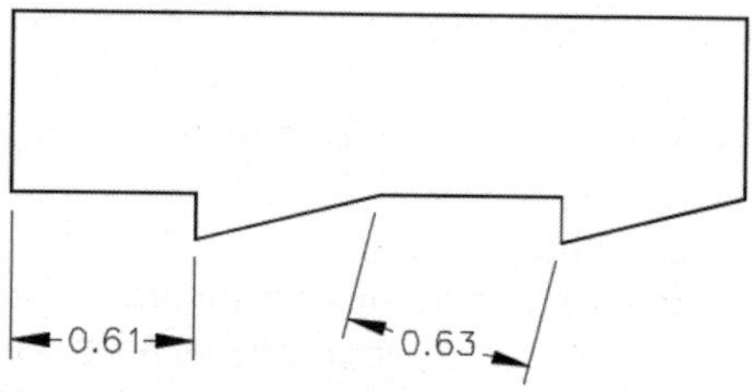


WITH ORTHO WITHOUT ORTHO

## Diameter, Radius

Note that there are no options in the *Dim* command for a diameter or radius dimension. The *Dim* command will automatically create the proper diameter or radius dimension if you select a circle or an arc (Fig. 28-34). (Your dimension line and arrow direction may appear differently than the figure based on your dimension variables settings. See Chapter 29, Dimension Styles and Variables.) As you hover on the circle or arc (before you pick), another set of prompts appears.

**FIGURE 28-34**

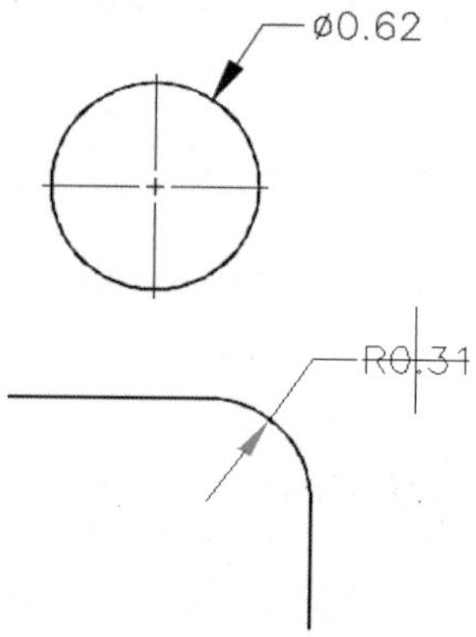


> Command: **DIM**
> Select objects or specify first extension line origin or [Angular/Baseline/Continue/Ordinate/aliGn/ Distribute/Layer/Undo]: **PICK** (*Arc* or *Circle*)
> Select circle to specify diameter or [Radius/Jogged/Angular]:

*Radius/Diameter* (sub option)
Use these options if you select a circle, for example, but want to produce a *Radius* dimension, and vice versa.

*Jogged* (sub option)
This selection creates a jogged dimension line. See *Dimjogged* later in this chapter.

*Angular* (sub option)
Use this option when you select a circle or arc but want to create an angular dimension on the circle or arc. This action switches to the top-level *Angular* option (see next). The center of the circle or arc becomes the vertex, then you specify points to define the angle.

### *Angular*

You can create an angular dimension between two lines with or without the *Angular* option. Using the "select objects" method, you can create an angular dimension by selecting two lines (without specifying *Angular*). Although a temporary linear dimension appears after selecting the first line, continue to select the second line forming the angle (Fig. 28-35). The following prompts appear as you proceed through the steps (note the last prompt).

Select objects or specify first extension line origin or [Angular/
Baseline/Continue/Ordinate/aliGn/ Distribute/Layer/Undo]: **HOVER**
Select line to specify extension lines origin: **PICK**
Select line to specify second side of angle: **PICK**

**FIGURE 28-35**

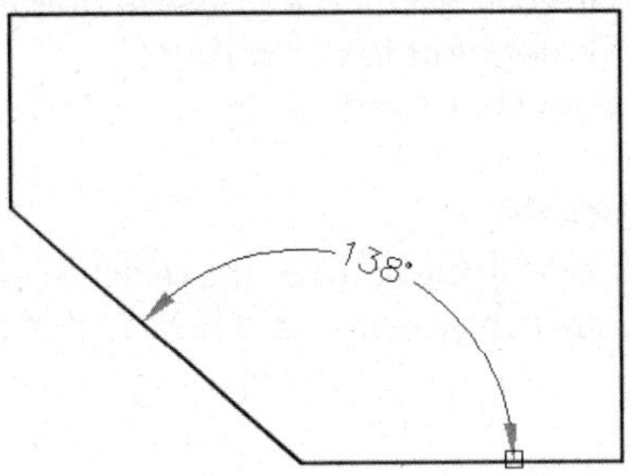


SELECT OBJECT METHOD:
PICK TWO LINES TO SPECIFY AN ANGLE

If you specify the *Angular* option at the first *Dim* command prompt or use the *Angular* option after hovering on a circle or arc, the following message appears.

Select arc, circle, line or [Vertex]: **PICK** (*Line*)
Specify first side of angle or [Undo]: **PICK**
Specify second side of angle or [Mtext/Text/text aNgle/Undo]: **PICK**
Specify angular dimension location or [Mtext/Text/text aNgle/Undo]: **PICK**

Selecting an arc or circle designates the center as the vertex of the angle. Then you are prompted to pick two points to specify the angle. The *Vertex* option allows you to select any point in the drawing as the vertex for the angular dimension, and then specify two points to designate the angle. Remember that you can select two lines to form an angular dimension without using the *Angular* option.

### *Baseline*

The *Baseline* option of *Dim* operates essentially the same as the *Dimbaseline* command (see *Dimbaseline* earlier in this chapter). You must have an existing dimension to use as the first baseline dimension. First select an existing extension line to use as the baseline, then create new dimensions by specifying the "second extension line origins" for the new dimensions. Because the *Baseline* option requires specification of extension line origins, turn on *Object Snap*.

Command: ***DIM***
Select objects or specify first extension line origin or [Angular/Baseline/Continue/Ordinate/ aliGn/
Distribute/Layer/Undo]: ***b***
Current settings: Offset (DIMDLI) = 0.380000
Specify first extension line origin as baseline or [Offset]: **PICK**
Specify second extension line origin or [Select/Offset/Undo] <Select>: **PICK**
Dimension text = *n.nnnn*
Specify second extension line origin or [Select/Offset/Undo] <Select>:

*Offset* (sub option)
The *Offset* is the automatic spacing distance between the "stacked" dimension lines as specified in the *Dimension Styles Manager* (*DIMDLI* system variable). Use the *Offset* option to reset this value.

*Select* (sub option)
You can specify a new first extension line origin as a baseline with the *Select* option.

### *Continue*

This feature creates continuous dimensions just like using the *Dimcontinue* command (see *Dimcontinue* earlier in this chapter). Again, you must have an existing dimension to use as the dimension to continue from. Select an existing extension line to use as the first extension line for the new dimension, then create new dimensions by specifying the "second extension line origins" for the new dimensions. Make sure to turn on *Object Snap*. The newly created dimensions automatically align with the dimension line of selected existing dimension (there is no offset since continuous dimensions are not "stacked").

Specify first extension line origin to continue: **PICK**
Specify second extension line origin or [Select/Undo] <Select>: **PICK**
Dimension text = *n.nnnn*
Specify second extension line origin or [Select/Undo] <Select>:

### *Ordinate*

*Ordinate* dimensions are used in some industries when it is important to give all dimensions relative to one datum point in both X and Y directions. The *Ordinate* option of *Dim* allows you to create ordinate dimensions without having to use the *Dimordinate* command, although both methods operate essentially the same. See *Dimordinate* later in this chapter for more information.

### *aliGn*

The *aliGn* option does not create *Aligned* dimensions as you would logically deduce. Instead, this option moves existing dimension lines so they line up end to end like continuous dimensions. For example, assume you created dimensions in the fashion shown (Fig. 28-36, faded dimensions). Use the *aliGn* option to select the dimension line to keep, then select the dimensions to move into alignment.

TIP

**FIGURE 28-36**

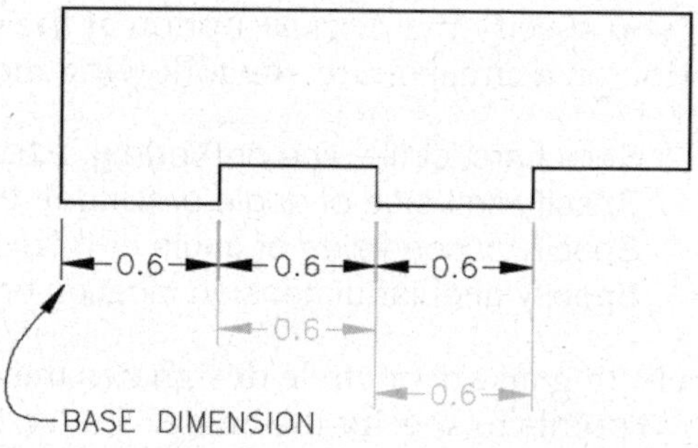


Select base dimension: **PICK**
Select dimensions to align: **PICK**
Select dimensions to align: **PICK**
Select dimensions to align: **Enter**

Note that you want to use this option only with continuous-type dimensions that were not created using the *Continue* option or *Dimcontinue* command since those methods automatically align. Although you can use this option with baseline-type dimensions (even those created with the *Baseline* option or the *Dimbaseline* command) it serves no use since those dimensions must be "stacked" to read the dimensions clearly. Use *Distribute* for baseline-type dimensions (see *Distribute* next).

### *Distribute*

*Distribute* provides a means for spacing stacked baseline dimension lines equally or at a specified distance, similar to that which would automatically be created by the *Dimbaseline* command or the *Baseline* option of *Dim*. Using this option is similar to using the *Dimspace* command discussed earlier but offers more flexibility. This option requires that the selected dimensions have a shared extension line origin; therefore, *Baseline* does not operate for continuous-style dimensions.

Current settings: Offset (DIMDLI) = 0.200000
Specify method to distribute dimensions [Equal/Offset] <Offset>: **Enter**
Select base dimension or [Offset]: **PICK**
Select dimensions to distribute or [Offset]: **PICK**
Select dimensions to distribute or [Offset]: **Enter**
Select dimensions to distribute:

#### *Offset*

This option distributes the dimension lines at the distance specified by the *Offset* setting (*DIMDLI*). The dimension you select as the "base dimension" remains fixed and the "dimensions to distribute" are then spaced according to the *Offset* value to organize the stack (Fig. 28-37). You can select any of the dimensions as the "base dimension." Change the offset value by selecting *Offset* at the appropriate prompt or by using the *Dimension Style Manager* (Chapter 29).

**FIGURE 28-37**

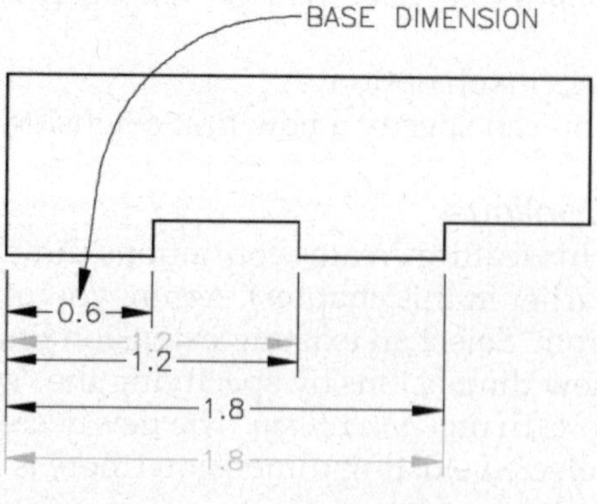

*Equal*
The *Equal* option automatically spaces the dimension lines at equal distances (independent of the *Offset* value) between the inner and outer-most dimension lines (those two dimension lines remain fixed).

### *Layer*
By default, new dimension objects are created on the current layer. You can override this setting and specify on which layer you want new dimensions to reside using the *Layer* option.

Enter layer name or select object to specify layer to place dimensions or . for use current [?/eXit] <"Use Current">:

You can specify the desired layer for new dimensions by two methods. First, you can simply select existing objects on the desired layer. Second, you can type the name of the desired layer. You can also create a new layer for dimensions by typing a new layer name at the prompt. The new layer is then created when you draw the next dimension. The assigned layer name is stored in the *DIMLAYER* system variable. The *?* (question mark) option lists the current layers.

### Text Options
Once the dimensions are placed the command prompt offers these text options.

Specify dimension line location or second line for angle [Mtext/Text/text aNgle/Undo]:

These options are identical to the *Mtext*, *Text*, and *Angle* options of other dimensioning commands.

## *Dimjogline*

| Ribbon | Menu Bar | Command (Type) | Alias (Type) | Shortcut |
|---|---|---|---|---|
| *Annotate Dimensions* (icon) | *Dimension Jogged Linear* | *Dimjogline* | *DJL* | ... |

*Dimjogline* creates a "jog line" in an existing linear dimension. This technique is typically used in cases where it is not practical to draw the entire length of the object and a break line is created to indicate the object is actually longer than depicted (Fig. 28-38).

FIGURE 28-38

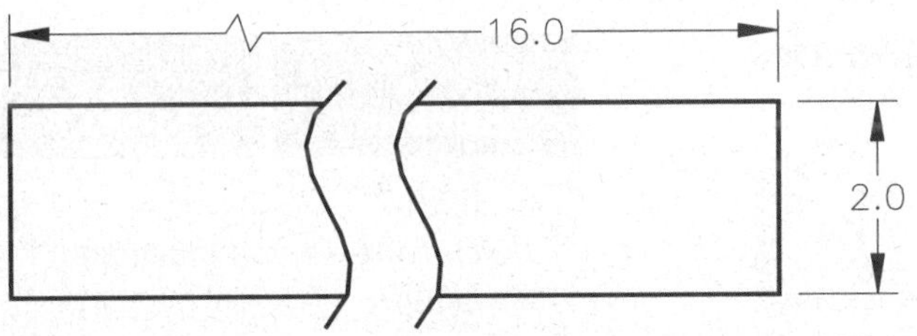


To create the "jog," first create the linear dimension, then use *Dimjogline* to select the dimension and indicate the desired location for the "jog."

Command: ***DIMJOGLINE***
Select dimension to add jog or [Remove]: **PICK**
Specify jog location (or press ENTER): **PICK** or **Enter**
Command:

### *Remove*
This option allows you to remove a jog line previously created with this command.

### *Enter*
Press Enter to place the jog at the midpoint between the dimension text and the first extension line or the midpoint of the dimension line based on the location of the dimension text.

## Dimjogged

| Ribbon | Menu Bar | Command (Type) | Alias (Type) | Shortcut |
|---|---|---|---|---|
| *Annotate Dimensions Jogged* | *Dimension Jogged* | *Dimjogged* | *DJO* | ... |

For an extremely large radius it may not be possible or practical to create a dimension line originating from the arc's actual center, as is traditionally done (see "*Dimradius*"). In such cases, a "jogged" radius dimension line is created. With this type of dimension, the dimension line is drawn from a hypothetical arc center point to the arc; however, the dimension line contains a "jog" to indicate that the origin of the dimension line is not the actual center.

**FIGURE 28-39**

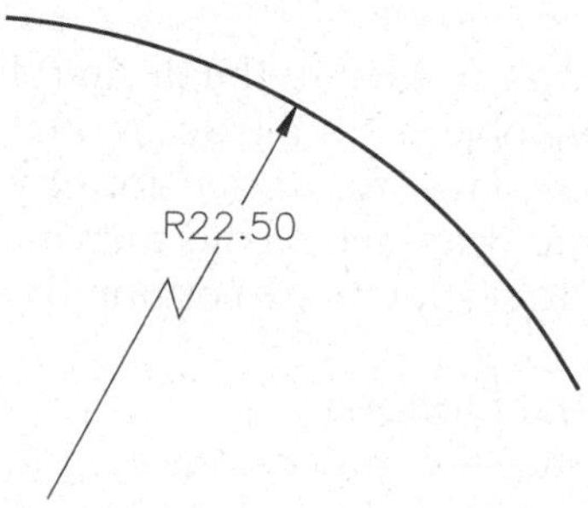


The *Dimjogged* command creates a jogged radius dimension for extremely large arcs (Fig. 28-39). Since the actual center of the arc is not shown by the dimension, you must select the hypothetical center point ("center location override") to use for the origin of the dimension line. You can also specify where you want to place the jog along the dimension line. The *Mtext/Text/Angle* options operate similarly to the *Dimradius* command. See "*Dimradius*."

Command: ***DIMJOGGED***
Select arc or circle: **PICK**
Specify center location override: **PICK** (pick the desired hypothetical arc center)
Dimension text = nn
Specify dimension line location or [Mtext/Text/Angle]: **PICK** (place the location of the dimension text and dimension line)
Specify jog location: **PICK** (pick the location for the "jog")
Command:

## Dimarc

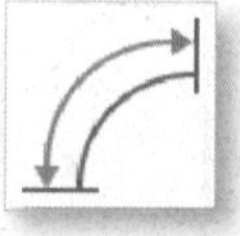

| Ribbon | Menu Bar | Command (Type) | Alias (Type) | Shortcut |
|---|---|---|---|---|
| *Annotate Dimensions Arc Length* | *Dimension Arc Length* | *Dimarc* | *DAR* | ... |

In some cases it may be necessary to dimension the length of an arc along its curve, similar to a partial circumference. The *Dimarc* command automatically measures this length and draws a dimension line concentric to the arc (Fig. 28-40). You select the arc to dimension and place the location of the dimension line.

**FIGURE 28-40**

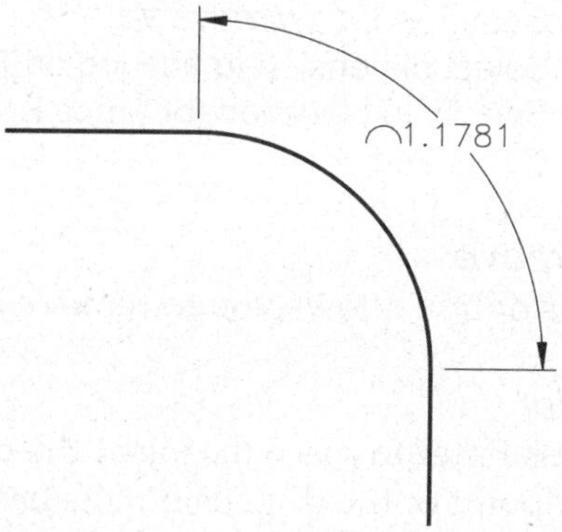


Command: ***DIMARC***
Select arc or polyline arc segment: **PICK**
Specify arc length dimension location, or
[Mtext/Text/Angle/Partial/Leader]: **PICK**
Dimension text = *nn*
Command:

The *Mtext/Text/Angle* options operate similarly to the *Dimradius* command (see "*Dimradius*"). Use the *Leader* option to draw an additional leader from the dimension text to the arc.

Although the *Dimarc* command by default measures the entire length of the selected arc, you can use the *Partial* option to dimension only part of an arc. You are prompted to select two points on the arc to measure. You can use *Object Snaps* to select these two points. In this case, the extension lines for a *Partial* arc length are not perpendicular to the arc as they are with the default (full length) option.

Specify arc length dimension location, or [Mtext/Text/Angle/Partial/Leader]: ***p***
Specify first point for arc length dimension: **PICK**
Specify second point for arc length dimension: **PICK**
Specify arc length dimension location, or [Mtext/Text/Angle/Partial]: **PICK**
Dimension text = *nn*
Command:

## *Dimcenter*

| Ribbon | Menu Bar | Command (Type) | Alias (Type) | Shortcut |
|---|---|---|---|---|
| *Annotate Dimensions (icon)* | *Dimension Center Mark* | *Dimcenter* | *DCE* | ... |

TIP

*Dimcenter* is an older command for drawing center lines for *Arcs* and *Circles*. Since AutoCAD 2017, you can use the *Centermark* command for this purpose with greater ease and more control of the center marks and center lines (see Chapter 26). *Dimcenter*, *Dimdiameter* and *Dimradius* create center marks and center lines, but the resulting lines are much more difficult to adjust than those created with *Centermark*.

Command: ***DIMCENTER***
Select arc or circle: **PICK**
Command:

No matter if the center mark is created by the *Dimcenter* command or by the *Diameter* or *Radius* commands, the center mark can be either a small cross or complete center lines extending past the *Circle* or *Arc* (Fig. 28-41). The type of center mark drawn is controlled by the *Mark* or *Line* setting in the *Lines and Arrows* tab of the *Dimension Style Manager*. It is suggested that a *Dimension Style* be adjusted with these settings just for drawing center marks. (See Chapter 29.)

**FIGURE 28-41**

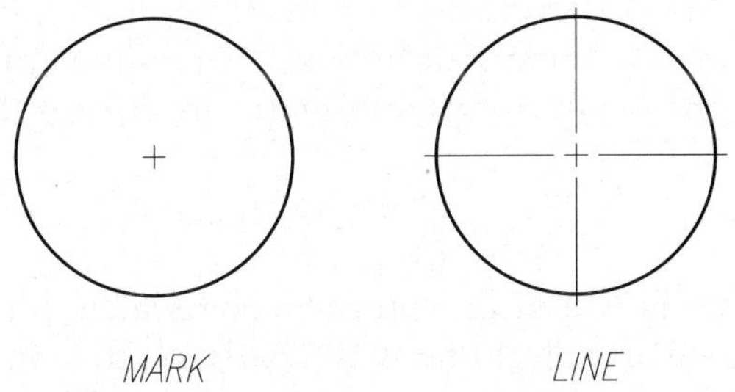


When you are dimensioning, short center marks should be used for *Arcs* of less than 180 degrees, and full center lines should be drawn for *Circles* and for *Arcs* of 180 degrees or more (Fig. 28-42).

**FIGURE 28-42**

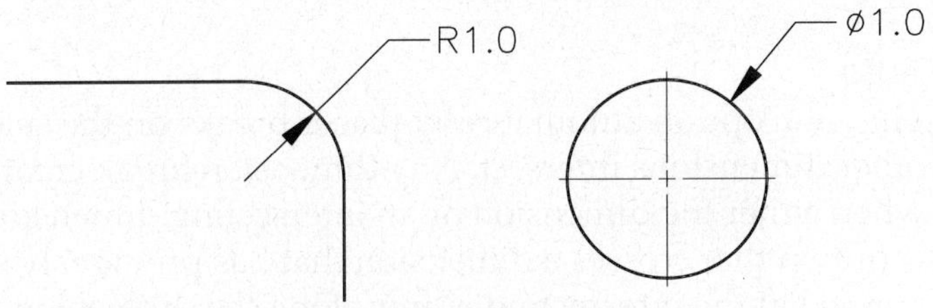


NOTE: Since the center marks created with the *Dimcenter* command are <u>not</u> associative, they may be *Trimmed*, *Erased*, or otherwise edited. The center marks created with the *Dimradius* or *Dimdiameter* commands <u>are</u> associative and cannot be edited.

### Dimbreak

| Ribbon | Menu Bar | Command (Type) | Alias (Type) | Shortcut |
|---|---|---|---|---|
| *Annotate Dimensions* (icon) | *Dimension Dimension Break* | *Dimbreak* | ... | ... |

TIP

According to industry standards for architecture and engineering, you should not cross a dimension line with another dimension line or extension line (although extension lines can cross each other and can cross object lines). However, there are some cases where you cannot avoid crossing a dimension line with another dimension or extension line. For example in Figure 28-43, the "0.40" dimension is crossed by the extension line of the other two dimensions, but even with other arrangements of these dimensions there is no way to avoid crossing a dimension line.

**FIGURE 28-43**

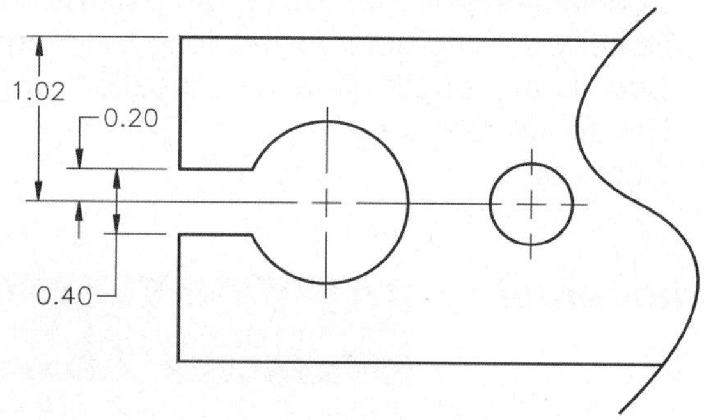


In cases such as that shown, you can use the *Dimbreak* command to create a space in the extension line that causes the infraction. The Command line reports the following:

Command: ***DIMBREAK***
Select a dimension to add/remove break or [Multiple]: **PICK** (select the line on you want to break)
Select object to break dimension or [Auto/Manual/Remove] <Auto>: **PICK** or **Enter**
Command:

When dealing with two dimensions, at the "Select a dimension to add/remove break or [Multiple]:" prompt, select the line to be broken. For the example in Figure 28-44, select the horizontal center/extension line that crosses the "0.40" dimension. At the "Select object to break dimension" prompt, select the dimension line that is to remain unbroken. Alternately, pressing Enter at the next prompt automatically creates the break.

**FIGURE 28-44**

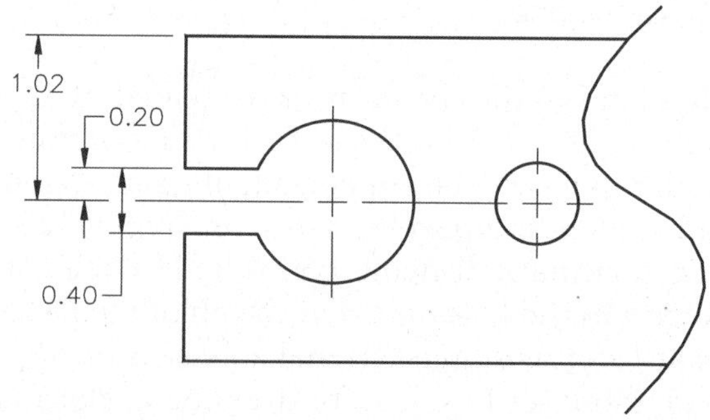


TIP

NOTE: Although you can also break a dimension line using an object line with *Dimbreak*, this practice should be avoided since it violates industry standards. Always try to break extension lines. The options are described next.

#### Multiple

Use this option to specify multiple dimensions to add breaks to or remove breaks from.

#### Auto

The *Auto* option automatically places breaks on the selected dimension or extension line where any other dimensions intersect. Any dimension break created using this option is automatically updated when either the dimension or an intersecting dimension is modified. However, when a new dimension is drawn that crosses a dimension that has previously created dimension breaks, no new breaks are applied at the intersecting points along the dimension object. To add new dimension breaks, you must use *Dimbreak* again.

### *Manual*

This option works similarly to the *Break* command where you can manually select the two points for the break. Dimension breaks can be added to dimensions even when objects do not intersect the dimension or extension lines. You can create only one manual dimension break at a time.

### *Remove*

Use this option to remove previously created breaks from the selected dimensions.

## ***Leader***

| Ribbon | Menu Bar | Command (Type) | Alias (Type) | Shortcut |
|---|---|---|---|---|
| ... | ... | *Leader* | *LEAD* | ... |

The *Leader* command (not *Qleader* ot *Mleader*) allows you to create an associative leader similar to that created with the *Diamcter* command. The *Leader* command is intended to give dimensional notes such as the manufacturing or construction specifications shown next.

Command: ***LEADER***
Specify leader start point: **PICK**
Specify next point: **PICK**
Specify next point or [Annotation/Format/Undo] <Annotation>: **Enter**
Enter first line of annotation text or <options>: **CASE HARDEN**
Enter next line of annotation text: **Enter**
Command:

At the "start point:" prompt, select the desired location for the arrow. You should use an *Object Snap* option (such as *Nearest*, in this case) to ensure the arrow touches the desired object.

A short horizontal line segment called the "landing" is automatically added to the last line segment drawn if the leader line is 15 degrees or more from horizontal. Note that command syntax for the *Leader* in Figure 28-45 indicates only one line segment was PICKed. A *Leader* can have as many segments as you desire.

**FIGURE 28-45**

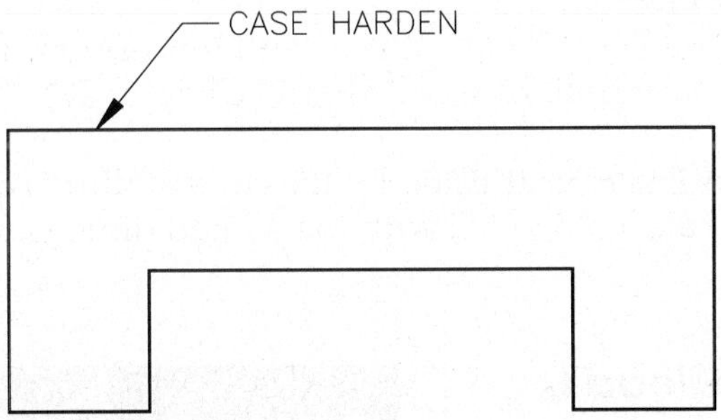


If you do not enter text at the "Annotation:" prompt, another series of options are available:

Command: ***LEADER***
Specify leader start point: **PICK**
Specify next point: **PICK**
Specify next point or [Annotation/Format/Undo] <Annotation>: **Enter**
Enter first line of annotation text or <options>: **Enter**
Enter an annotation option [Tolerance/Copy/Block/None/Mtext] <Mtext>:

### *Format*

This option produces another list of choices:

Enter leader format option [Spline/STraight/Arrow/None] <Exit>:

FIGURE 28-46

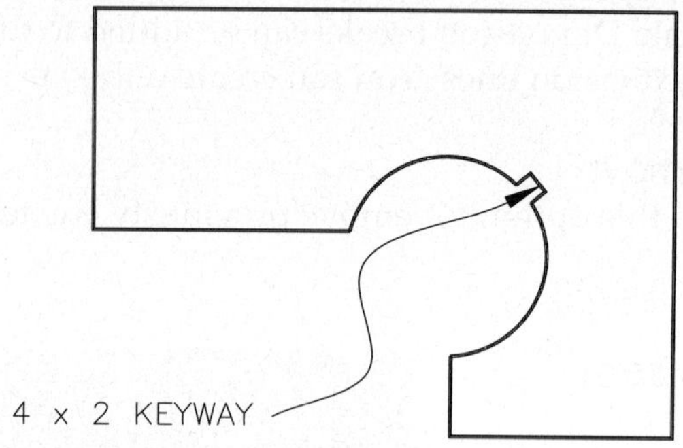


### *Spline/STraight*

You can draw either a *Spline* or straight version of the leader line with these options. The resulting *Spline* leader line has the characteristics of a normal *Spline*. An example of a *Splined* leader is shown in Figure 28-46.

### *Arrow/None*

This option draws the leader line with or without an arrowhead at the start point.

### *Annotation*

This option prompts for text to insert at the end of the *Leader* line.

### *Tolerance*

This option produces a feature control frame using the *Geometric Tolerances* dialog boxes (see "*Tolerance*" later in this chapter).

### *Copy*

You can copy existing *Text* or *Mtext* objects from the drawing to be placed at the end of the *Leader* line. The copied object is associated with the *Leader* line.

### *Block*

An existing *Block* of your selection can be placed at the end of the *Leader* line. The same prompts as the *Insert* command are used.

### *None*

Using this option draws the *Leader* line with no annotation.

### *Undo*

This option undoes the last vertex point of the *Leader* line.

### *Mtext*

The *Text Editor* appears with this option. Text can be entered into paragraph form and you can use the *Mtext* options (see "*Mtext*," Chapter 18).

A *Leader* is affected by the current dimension style settings. You can control the *Leader's* arrowhead type, scale, color, etc., with the related dimensioning variables (see Chapter 29).

## *Qleader*

| Ribbon | Menu Bar | Command (Type) | Alias (Type) | Shortcut |
|---|---|---|---|---|
| ... | ... | *Qleader* | *LE* | ... |

*Qleader* (quick leader) creates a leader with or without text similar in appearance to a leader created with *Leader*. However, *Qleader* allows you to specify preset parameters for the configuration of the text and the leader. These presets prevent you from having to specify the same parameters repeatedly when you create several leaders in a similar fashion for the current drawing. For example, you can specify what type of text object (*Mtext*, *Block*, *Tolerance*, etc.), how many line segments for the leader, what angle to draw the leader lines, what type of arrowheads, and other parameters for the leaders you create.

Command: ***QLEADER***
Specify first leader point, or [Settings]: **PICK** (arrow location)
Specify next point: **PICK** (end of leader)
Specify next point: **Enter**
Specify text width <0.0000>: **Enter**
Enter first line of annotation text <Mtext>: Enter text or press **Enter**
Enter next line of annotation text: Enter text or press **Enter**
Command:

The command sequence above is used to create a one-segment leader with one string of text using the default *Qleader* settings. If you enter a value in response to the "Specify text width <0.0000>:" prompt, the value determines the width of the *Mtext* paragraph. As an alternative, you can press Enter to produce the *Text Editor*.

With *Qleader*, you specify the preset configurations by pressing Enter or entering the letter "s" at the first prompt: "Specify first leader point, or [Settings]<Settings>:." This action produces the *Leader Settings* dialog box. Settings you make in this dialog box control the appearance of subsequent leaders you create in the drawing until a setting is changed. The settings are saved in the current drawing and are not registered in your system. There are three tabs in the dialog box. The *Annotation* tab is described next.

TIP

### *Annotation* Tab

This tab allows you to control presets for the appearance of the leader text as further explained below (Fig. 28-47).

**FIGURE 28-47**

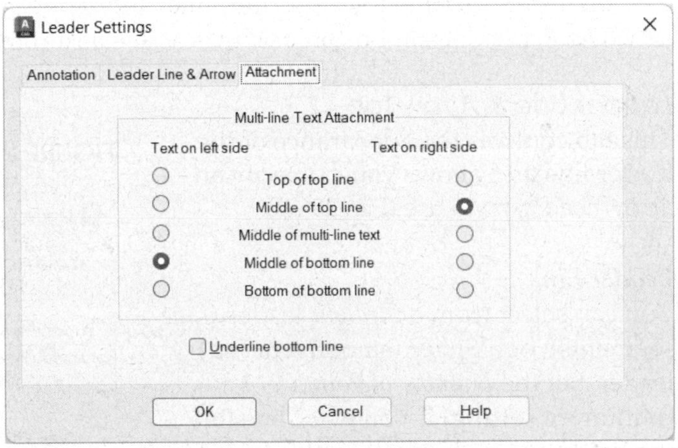


#### *Annotation Type*

This section specifies the type of text objects created (*Mtext*, *Copy an Object*, *Tolerance*, *Block Reference*, or *None*) and the related prompts when you use *Qleader*.

The *Mtext* option creates an *Mtext* object. Depending on your settings in the *Mtext Options* section, you can specify the width of each paragraph you create by responding to the "Specify text width <0.0000>:" prompt, using the *Text Editor*, or pressing Enter to input several individual lines of text.

Specifying *Copy an Object* causes *Qleader* to prompt you to select an existing text object to copy. The text object is then automatically attached to the leader end.

Use the *Tolerance* button to cause *Qleader* to produce the *Geometric Tolerance* dialog box. Here you specify text and symbol content for feature control frames that are attached to the leader end (see "*Tolerance*" for details on geometric tolerancing and this dialog box).

Use the *Block Reference* option if you want to attach a *Block* to the leader end. *Qleader* prompts you to specify an existing *Block* reference defined in the current drawing by producing the "Enter block name or [?]:" prompt. Enter a question mark (?) to list existing *Blocks* in the current drawing.

To create leaders without text attached, select *None* as the *Annotation Type*.

*Mtext Options*
This area is enabled only when *Mtext* is the specified *Annotation Type*. You can select none, one or two options in the section.

When *Prompt for width* is checked, this setting produces the "Specify text width <0.0000>:" prompt at the command line. Enter a value for the paragraph width or press Enter to produce the *Text Editor*. When this setting is not checked, the width prompt does not appear, but you can still produce the *Text Editor* by pressing Enter at the "Enter first line of annotation text:" prompt.

When a leader is drawn from right to left so it is attached to the right side of the text paragraph (Fig. 28-48), AutoCAD normally right-justifies the text. Checking the *Always Left Justify* option always draws the text left justified.

FIGURE 28-48

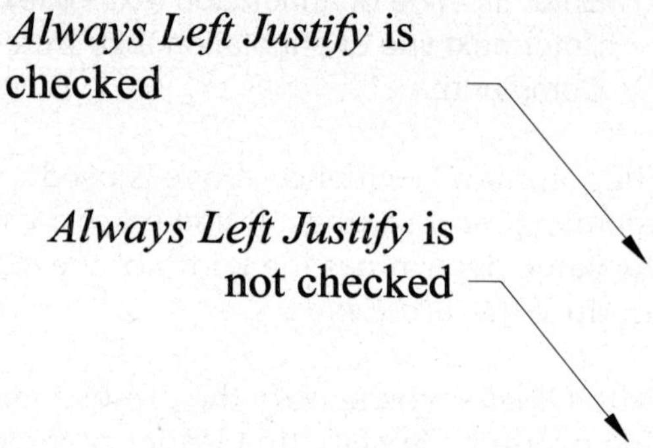


The *Frame Text* option creates a frame around the text (Fig. 28-49).

FIGURE 28-49

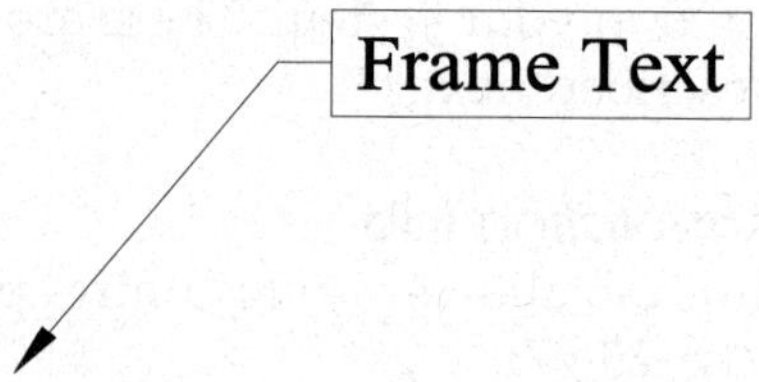


*Annotation Reuse*
When you want to add the same specification note or manufacturing note at several locations in a drawing, the *Annotation Reuse* option can save considerable time. To use the same annotation repeatedly, check *Reuse Next*. After setting this option, create one leader with the desired text. Subsequent leaders automatically appear with the same text. The *Reuse Current* button becomes active as a reminder.

### *Leader Line & Arrow* Tab
This tab controls the appearance of the leader line and allows you to specify an arrowhead type to use (Fig. 28-50).

FIGURE 28-50

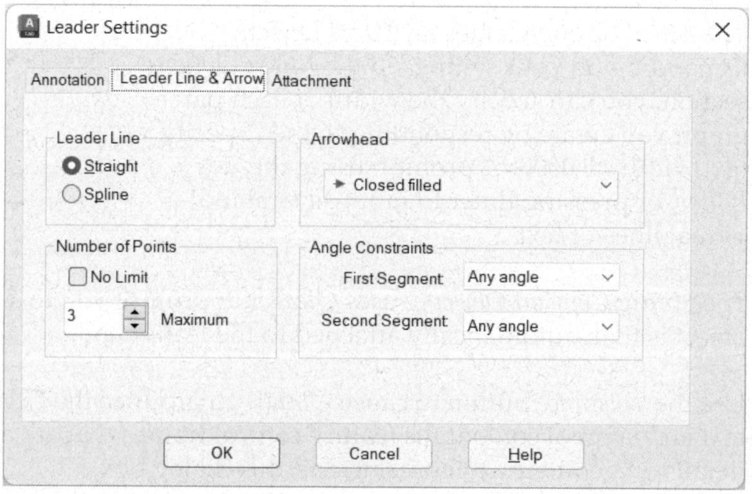


*Leader Line*
You can select from a *Straight* leader (*Line* segments) or a *Spline* (smooth curved) leader. Set the *Number of Points* to 2 (minimum setting) if you want one line segment (two endpoints) or check *No Limit* if you are creating *Spline* leaders, since you need several points to control the shape of a curved leader.

*Arrowhead*
In the *Arrowhead* section, use the drop-down list to select from many different possibilities for the leaders.

*Angle Constraints*
This section allows you to specify (force) an angle for the *Qleader* segments to be drawn. For example, selecting 45 causes the *Qleader* to always be drawn at 45 degrees.

### *Attachment* Tab

The *Attachment* tab (Fig. 28-51) is used for setting the attachment point when *Mtext* is the selected annotation type. This tab is disabled unless *Mtext* is specified in the *Annotation* tab.

FIGURE 28-51

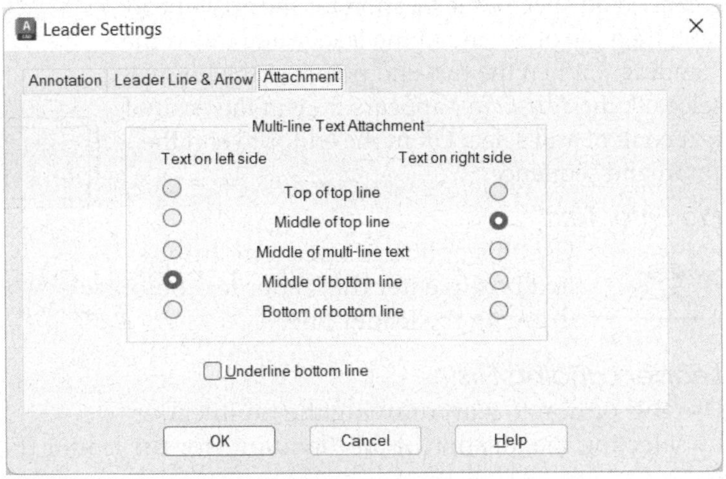


The *Text Formatting Attachment* options can be specified for cases when the text is located on the right or left side of the leader (text is located on the right when a leader is drawn from left to right and vice versa). The default position for text attachment (Fig. 28-52) is *Middle of top line* for text on the right side and *Middle of bottom line* for text on the left side.

FIGURE 28-52

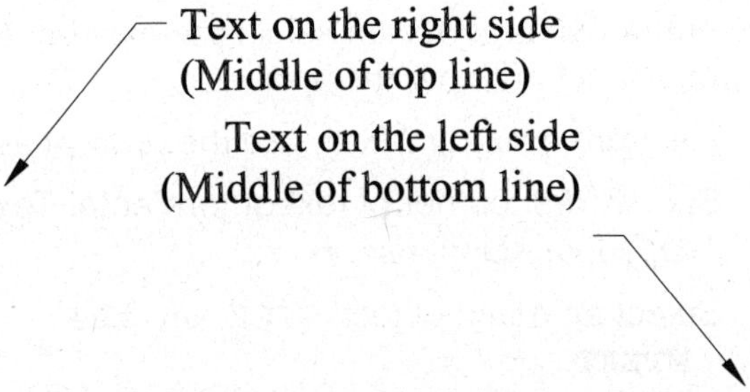


### *Mleader*

| Ribbon | Menu Bar | Command (Type) | Alias (Type) | Shortcut |
|---|---|---|---|---|
| *Annotate Leaders Multileader* | *Dimension Multileader* | *Mleader* | *MLD* | ... |

The *Mleader* command creates a "multileader" object. A multileader is similar to a *Qleader*; however, you have more options and flexibility for annotating drawings using *Mleader.* For example, with a multileader, you can place the arrowhead first, the tail first, or the text content first. The MLEADERLAYER System Variable lets you specify a default layer for new multileaders. You can set it quickly using the Leaders panel option from the ribbon Annotate tab.

A *Multileader* toolbar is available (Fig. 28-53 upper). This toolbar contains buttons to access all related multileader commands. The ribbon Annotate tab, Leaders panel (Fig. 28-53 lower) shows the same options. Included are the *Mleaderstyle* command which allows you to specify the parameters of the leader and text content, and the style drop-down list that allows you to select from existing styles.

FIGURE 28-53

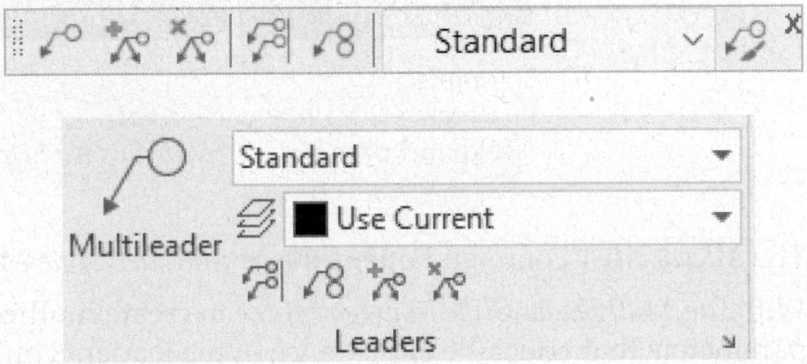


The default *Mleader* command prompt is below:

Command: **MLEADER**
Specify leader arrowhead location or [pre enter Text/leader Landing first/Content first/ Options] <Options>: **PICK**
Specify leader landing location: **PICK**
Command:

Using the default prompts, similar to *Leader* and *Qleader*, you specify the location for the arrowhead first, then the other end of the leader line called the "landing." When the two end points of the line are selected, the *Text Editor* appears. Key in the desired text content and select *OK* in the editor to end the command sequence.

**FIGURE 28-54**

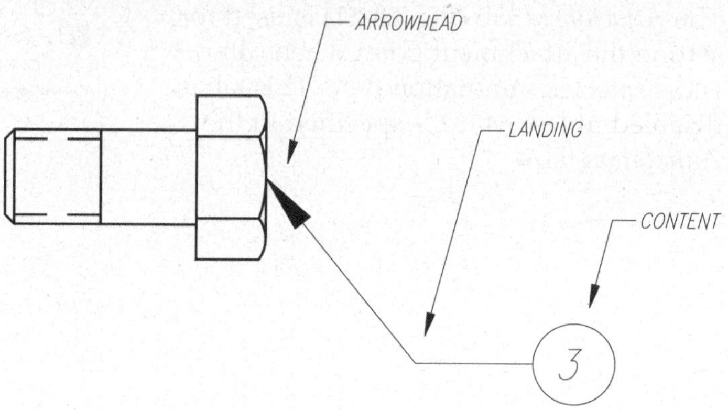


### Pre Enter Text

Use this option to open a multiline text box to enter the leader text before drawing the leader line.

### Leader Landing First

Use this option to start drawing the multileader by selecting the location of the "landing" for the leader (Fig. 28-54).

### Content First

*Content First* allows you to place the *location* of the text first, then pick the location for the arrowhead; to pre-enter text as above; or select pre-existing *Mtext* objects to connect a leader to; enter arrowhead first; landing first; or show the options.

To select pre-existing *Mtext*, use the prompt as below:

Specify first corner of text or [pre enter Text/select Mtext/leader arrowHead first/leader Landing first/Options] <Options>: **`m`**

Select an mtext object: **`PICK on the MTEXT`**

Specify leader arrowhead location: **`PICK`**

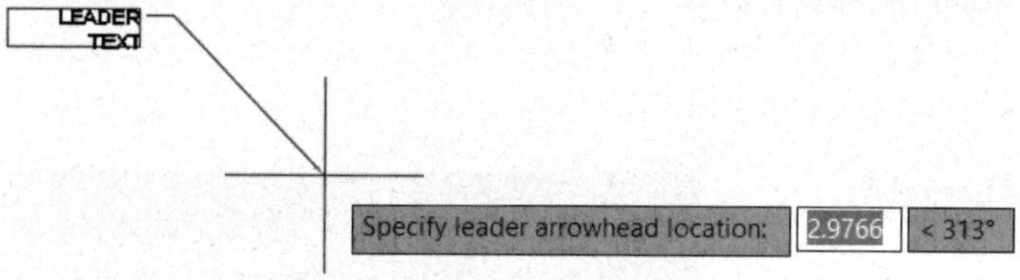


### Options

The format of the leader and content is controlled by the *Multileader Style* and/or *Options* section of the *Mleader* command. Entering "O" at the "Specify leader arrowhead location or [leader Landing first/Content first/Options] <Options>:" prompt provides the following prompt:

Enter an option [Leader type/leader lAnding/Content type/Maxpoints/First angle/Second angle/eXit options] <eXit options>:

These parameters are the similar to those in the *Multileader Style Manager* (*Mleaderstyle* command), described next.

## Mleaderstyle

| Ribbon | Menu Bar | Command (Type) | Alias (Type) | Shortcut |
|---|---|---|---|---|
| *Annotate*<br>*Leaders*<br>(expand panel) | *Format*<br>*Multileader Style* | *Mleaderstyle* | *MLS* | ... |

The *Mleaderstyle* command opens the *Multileader Style Manager* (Fig. 28-55).

With the *Multileader Style Manager* you can create multiple multileader styles, each style with different parameters that affect the appearance of the leaders you then create. You can change the parameters by selecting the *New* or *Modify* buttons which lead to options for specifying the *Leader Format*, *Leader Structure*, and *Content*. You can also use the *Options* section of the *Mleader* command to change some of the parameters for the current multileader.

### *Styles* List

This list displays multileader styles based on your selection under *List*. Select *All Styles* to display all multileader styles available in the drawing. Select *Styles In Use* to display only the multileader styles that are referenced by multileaders in the current drawing. *Standard* and *Annotative* are included in the ACAD.DWT and ACADISO.DWT template drawings. The current style is highlighted.

### *Set Current*

This button sets the selected style from the *Styles* list as the current style. Subsequently created multileaders are created using this multileader style.

### *New*

Selecting this button produces the *Create New Multileader Style* dialog box (not shown) in which you can specify new style names and proceed to define new multileader styles.

### *Delete*

*Delete* the multileader style selected in the *Styles* list with this option. A style cannot be deleted if multileaders using it have been created in the drawing.

**FIGURE 28-55**

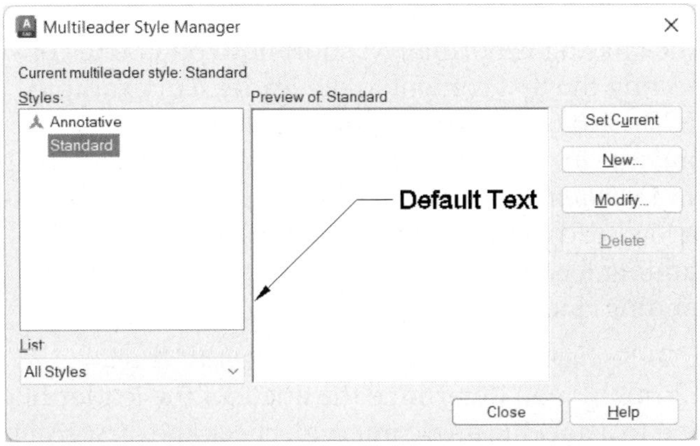


### *Modify*

Use this button to display the *Modify Multileader Style* dialog box to modify the selected multileader style (Fig. 28-56). The appearance of the multileaders you create in the current style can be altered by specifying the parameters in this dialog box, described next. The three tabs of this dialog box control the *Leader Format*, *Leader Structure*, and text *Content*.

### *Leader Format* Tab

Use this tab to specify the parameters for the leader lines and arrowheads (Fig. 28-56).

*Type*
You can choose a straight leader, a spline, or no leader line (only content).

*Color, Linetype, Lineweight*
These options affect the appearance of the leader line arrowhead but not the text content.

*Arrowhead Symbol* and *Size*
These options control the shape (symbol) and the size of the multileader arrowheads.

*Break Size*
If a multileader is selected with the *Dimbreak* command (to break the multi-leader), this setting determines the size of the break.

**FIGURE 28-56**

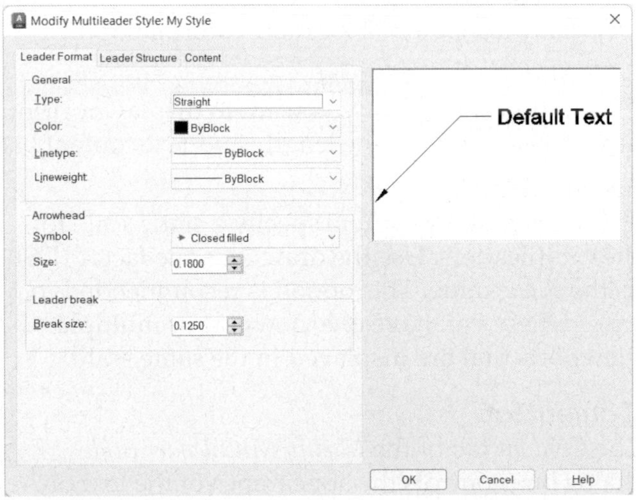

### *Leader Structure* Tab

Set parameters for the leader line here, including the scale and landing specifications (Fig. 28-57).

FIGURE 28-57

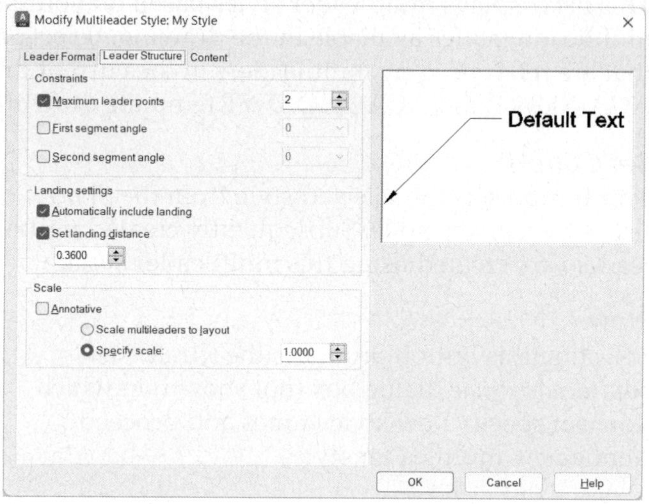


*Maximum Leader Points*

This determines the maximum number of points for the leader line. You always pick at least two points: the arrowhead location and landing location (start of the landing line).

The landing is normally a short horizontal line connecting the text content to the leader. For example, if 2 leader points are specified, you would pick only the arrowhead location and landing location. If *Maximum leader points* is set to 3, you select the arrowhead location, an additional point, and a landing location, resulting in two sections plus the landing line.

*First Segment Angle* and *Second Segment Angle*

Normally you determine the angle of the leader line by picking points. If you prefer to force the leader line to a specific angle interval, check the *First segment angle* box and select an angle value. For example, specify 45 to draw the leader line at any 45-degree increment. Use the *Second segment angle* only if you have 3 or more leader points.

*Automatically Include Landing:* Checking this box (checked by default) attaches a short horizontal landing line to the multileader.

*Set Landing Distance:* This value determines the fixed length of the multileader landing line.

*Annotative*

You can create annotative multileaders using this checkbox. Annotative objects are useful when you want to show the same objects in multiple viewports at different scales. When multileaders are annotative, a *Scale* is not specified here to determine the size of the arrowhead, landing line length, and text content. Instead, the *Annotation Scale* or *VP Scale* for the viewport determines the size of these features. (See "Annotative Text" in Chapter 18, "Annotative Dimensions" in Chapter 29, and "Annotative Objects" in Chapter 32 for more information.)

*Scale Multileaders to Layout*

This option is useful if you want to display or print only one viewport at a time and want to scale the arrowhead, landing line length, and text content based on the scaling in a selected viewport.

*Specify Scale*

This option allows you to specify a static scale for the multileaders. Use the drawing scale factor (DSF) as the *Scale* value. This option is recommended in cases where you have one viewport or multiple viewports that are displayed in the same scale.

FIGURE 28-58

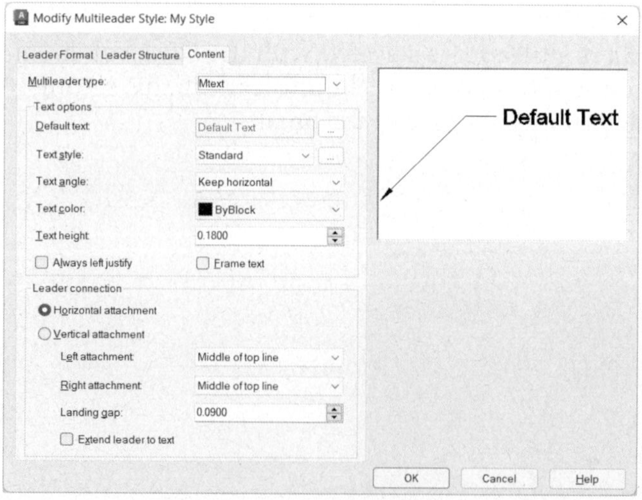


### *Content* Tab

The *Content* tab of the *Modify Multileader Style* dialog box controls the appearance of the text or other content for the leader (Fig. 28-58).

*Multileader Type*

This shows different options for a multileader that contains text, a block, or no content.

***MText Options***

*Default Text*

Use this section if you want the same default text for all multileaders. Use the browse button to produce the *Text Editor*, then enter the desired text.

*Text Style, Text Angle, Text Color, Text Height*

These options specify a predefined text style, angle, color, and height for the text. All text styles included in the drawing are displayed in the *Text Style* list.

*Always Left Justify*

Since you can create multiple lines of text for a multileader, the text lines must justify at the right or left side of the "paragraph." Normally, text is automatically justified depending on the direction of the leader. Use this option to force text to be always left justified independent of the direction of the leader.

*Frame Text*

This option creates a rectangular box around the multileader text content.

*Horizontal Attachment, Left Attachment, Right Attachment*

These options determine the connection point between the "paragraph" of text and the landing line. The choices include options for underlining. The *Left attachment* and *Right attachment* control the connection when the text is to the left or right of the leader, respectively. The *Landing Gap* value specifies the distance between the landing line end and the multileader text.

*Vertical Attachment, Top Attachment, Bottom Attachment*

A vertical attachment does not include a landing line between the text and the leader. The leader can be attached to the top center or bottom center of the text content. You can opt to use an overline or underline.

*Extend Leader to Text*

This option extends the leader line to the text rather than ending at the text bounding box (Fig. 28-59).

**FIGURE 28-59**

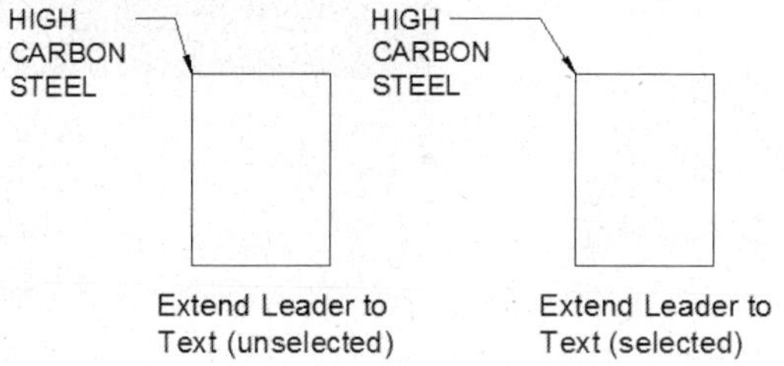


***Block Options***

*Source Block*

Specifies the block used for multileader content. Predefined shapes are available (*Circle, Slot, Box,* etc.) when you want "bubbles" to surround the text. Text is entered as attributes for these blocks. *User blocks* contained in the drawing are also available.

*Attachment*

You can attach the block to the leader by specifying the *Center extents* or the normal *Insertion point* specified for the block.

TIP

*Color*

Specifies the color of the multileader block content. The block color control has effect only if the object color included in the block is set to *ByBlock*.

*Scale*

Allows you to enter a scale for the block content inserted as part of the leader.

| **Mleaderedit** | Ribbon | Menu Bar | Command (Type) | Alias (Type) | Shortcut |
|---|---|---|---|---|---|
|  | *Annotate Leaders* (icons) | *Modify Object > Multileader > Add, Remove* | *Mleaderedit* | *MLE* | ... |

Use *Mleaderedit* to add leaders to or remove leaders from an existing multileader.

If you add leaders, the new leaders automatically connect to an end of the existing landing line or text content depending on the direction of the new leader (Fig. 28-60). For example, to add a new leader line, use the following prompts:

Command: ***MLEADEREDIT***
Select a multileader: **PICK**
Specify leader arrowhead location or [Remove leaders]: **PICK**
Specify leader arrowhead location or [Remove leaders]: **Enter**
Command:

FIGURE 28-60

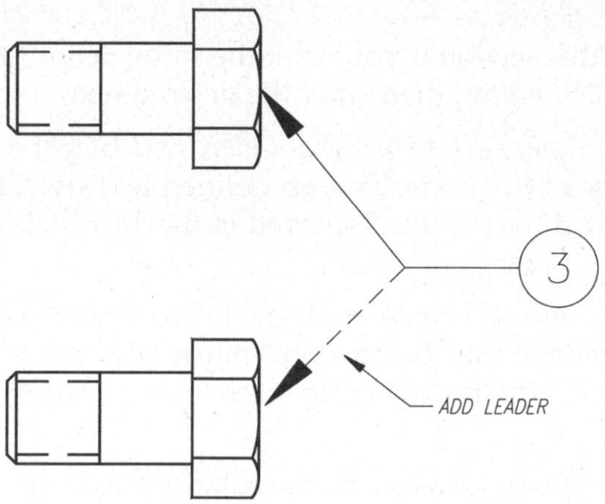


Use the default option to add a leader line to the selected multileader object. The new leader line is added to the left or right of the selected multileader, depending on the location of the cursor when you pick (Fig. 28-60). In the case that more than two leader points are defined in the current multileader style, you are prompted to specify additional points.

### *Remove Leader*

Simply select a leader line from the selected multileader object to remove.

## *Mleaderalign*

| Ribbon | Menu Bar | Command (Type) | Alias (Type) | Shortcut |
|---|---|---|---|---|
| *Annotate Leaders* (icon) | *Modify Object > Multileader > Align* | *Mleaderalign* | *MLA* | ... |

*Mleaderalign* is used to align the text content for a group of multileaders. For example, assume your drawing contained two multileaders as shown in Figure 28-61. In this case, the text content does not align vertically.

To remedy this inconsistency, use *Mleaderalign*. Follow the prompts shown next.

Command: ***MLEADERALIGN***
Select multileaders: **PICK**
Select multileaders: **PICK**
Select multileaders: **Enter**
Current mode: Use current spacing
Select multileader to align to or [Options]: **PICK**
Specify direction: **PICK**
Command:

FIGURE 28-61

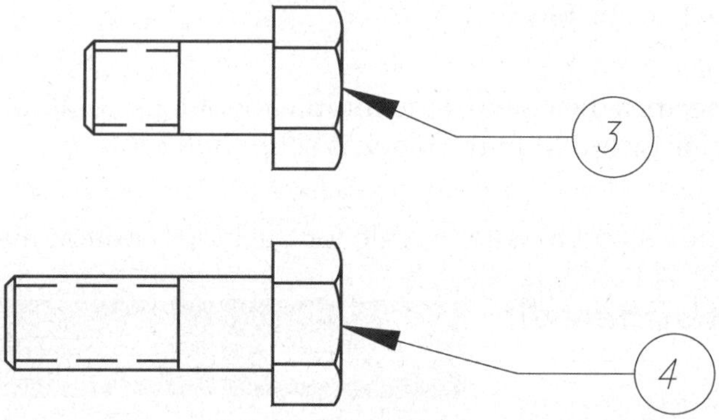

After selecting the two leaders, select one of the leaders to use as the reference at the "Select multileader to align to" prompt. At the "Specify direction" prompt, an alignment vector appears (attached to the cursor) for you to use as a guide. Swing this guide to the desired orientation, click, and the text content automatically aligns (Fig. 28-62).

FIGURE 28-62

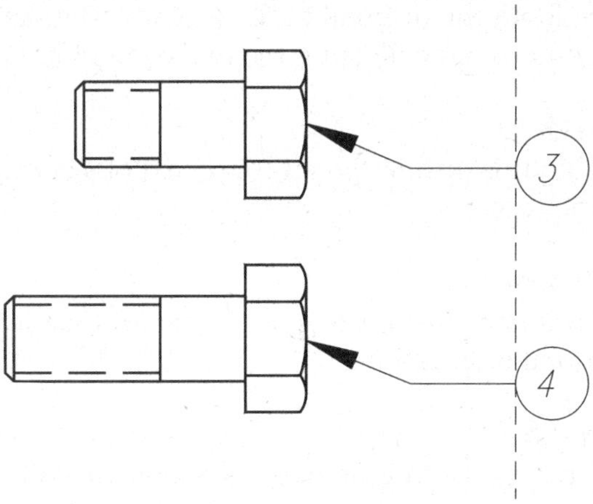


The following options are available:

Select multileader to align to or [Options]: *o*
Enter an option [Distribute/make leader segments
  Parallel/specify Spacing/Use current spacing]
<Use current spacing>:

### *Distribute*

This option causes the content to be spaced evenly between two selected points that you specify.

### *Make Leader Segments Parallel*

This option makes the selected multileader lines parallel. This option does not align the text content. You may want to first use the *Use current spacing* option to align the text, then use *Make leader lines parallel*.

### *Specify Spacing*

Enter a value to specify the spacing between the landing lines of the selected multileaders. You then indicate the desired direction between the landing lines in the "Specify direction" step.

### *Use Current*

This option aligns the text content (vertically or horizontally) based on the multileader you select at the "Select multileader to align to" prompt.

## *Mleadercollect*

| Ribbon | Menu Bar | Command (Type) | Alias (Type) | Shortcut |
|---|---|---|---|---|
| *Annotate Leaders* (icon) | *Modify Object > Multileader > Collect* | *Mleadercollect* | MLC | ... |

*Mleadercollect* simply combines multiple leader lines into one, then stacks the text content based on your specification. For example, assume you have an assembly with two multileaders indicating the components with part number bubbles (Fig. 28-63). You can use *Mleadercollect* to combine the leaders into one leader containing both text bubbles.

FIGURE 28-63

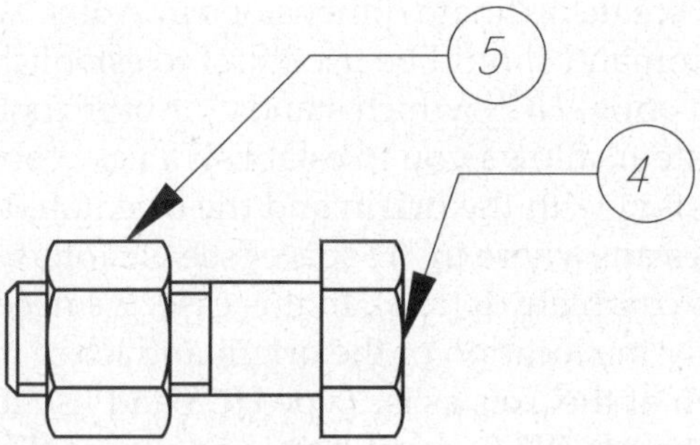


Command: ***MLEADERCOLLECT***
Select multileaders: **PICK**
Select multileaders: **PICK**
Select multileaders: **Enter**
Specify collected multileader location or [Vertical/
  Horizontal/Wrap] <Horizontal>: **PICK**
Command:

The resulting drawing would display only one leader containing both the original bubbles of text (Fig. 28-64). The possible methods for collecting the text content are given next.

FIGURE 28-64

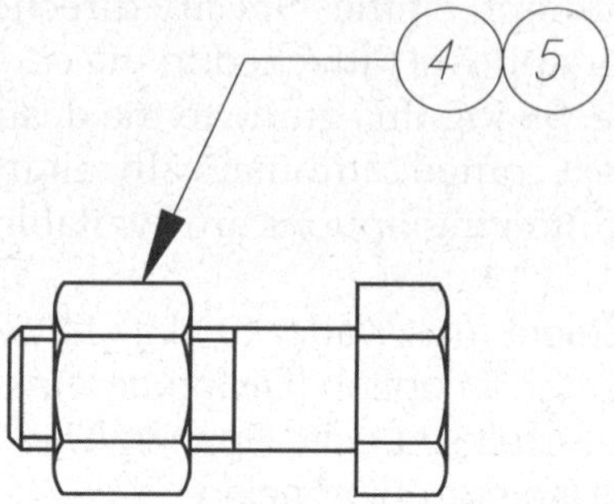


### *Vertical*

With this option the multiple text objects are placed in a vertical orientation.

### *Horizontal*

Similarly, this option places the multileader collection in a horizontal orientation.

### *Wrap*

When you combine many leaders, the collection of text contents can create a long string. In such a case, use *Wrap* to specify a width for a wrapped multileader collection. Specify either a *Width* or a maximum *Number* of blocks per row in the multileader collection.

## Dimordinate

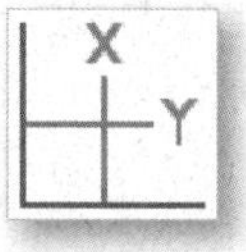


| Ribbon | Menu Bar | Command (Type) | Alias (Type) | Shortcut |
|---|---|---|---|---|
| *Annotate Dimensions Ordinate* | *Dimension Ordinate* | *Dimordinate* | *DIMORD* or *DOR* | ... |

Ordinate dimensioning is a specialized method of dimensioning used in the manufacturing of flat components such as those in the sheet metal industry. Because the thickness (depth) of the parts is uniform, only the width and height dimensions are specified as Xdatum and Ydatum dimensions.

TIP

*Dimordinate* dimensions give an Xdatum or a Ydatum distance between object "features" and a reference point on the geometry treated as the origin, usually the lower-left corner of the part. This method of dimensioning is relatively simple to create and easy to understand. Each dimension is composed only of one leader line and the aligned numerical value (Fig. 28-65).

FIGURE 28-65

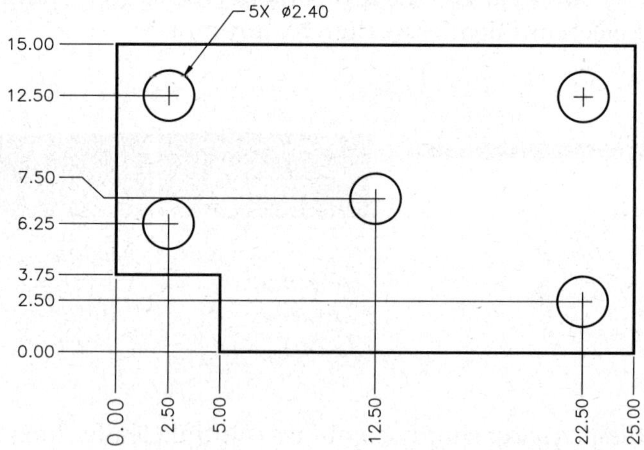


To create ordinate dimensions in AutoCAD, the *UCS* command should be used first to establish a new 0,0 point. *UCS*, which stands for user coordinate system, allows you to establish a new coordinate system with the origin and the orientation of the axes anywhere in 3D space (see Chapters 35 and 36 for complete details). In this case, we need to change only the location of the origin and leave the orientation of the axes as is. Type ***UCS*** and use the ***Origin*** option to PICK a new origin as shown (Fig. 28-66).

FIGURE 28-66

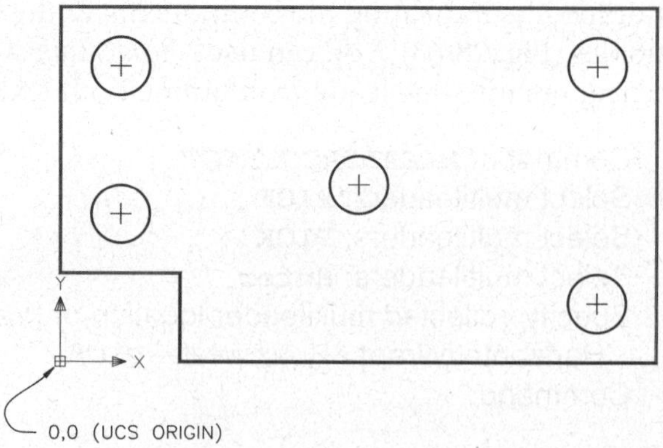

When you create a *Dimordinate* dimension, AutoCAD only requires you to (1) PICK the object "Feature" and then (2) specify the "Leader endpoint:". The dimension text is automatically aligned with the leader line.

It is not necessary in most cases to indicate whether you are creating an Xdatum or a Ydatum. Using the default option of *Dimordinate*, AutoCAD makes the determination based on the direction of the leader you specify (step 2). If the leader is perpendicular (or almost perpendicular) to the X axis, an Xdatum is created. If the leader is (almost) perpendicular to the Y axis, a Ydatum is created.

The command syntax for a *Dimordinate* dimension is this:

Command: ***DIMORDINATE***
Specify feature location: **PICK**
Specify leader endpoint or
[Xdatum/Ydatum/Mtext/Text/Angle]: **PICK**
Dimension text = *n.nnnn*
Command:

NOTE: You can use either the *Dimordinate* command or the *Ordinate* option of the *Dim* command to create ordinate dimensions. The only difference is that the *Dim* command gives a preview of the dimension before selecting the "feature location."

A *Dimordinate* dimension is created in Figure 28-67 by PICKing the object feature and the leader endpoint. That's all there is to it. The dimension is a Ydatum, yet AutoCAD automatically makes that determination, since the leader is perpendicular to the Y axis. It is a good practice to turn *Ortho* or *Polar Tracking On* in order to ensure the leader lines are drawn horizontally or vertically.

**FIGURE 28-67**

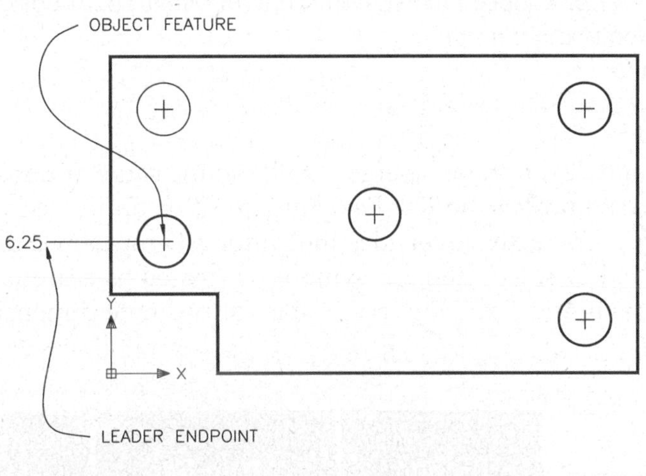


An Xdatum *Dimordinate* dimension is created in the same manner. Just PICK the object feature and the other end of the leader line (Fig. 28-68). The leader is perpendicular to the X axis; therefore, an Xdatum dimension is created.

**FIGURE 28-68**

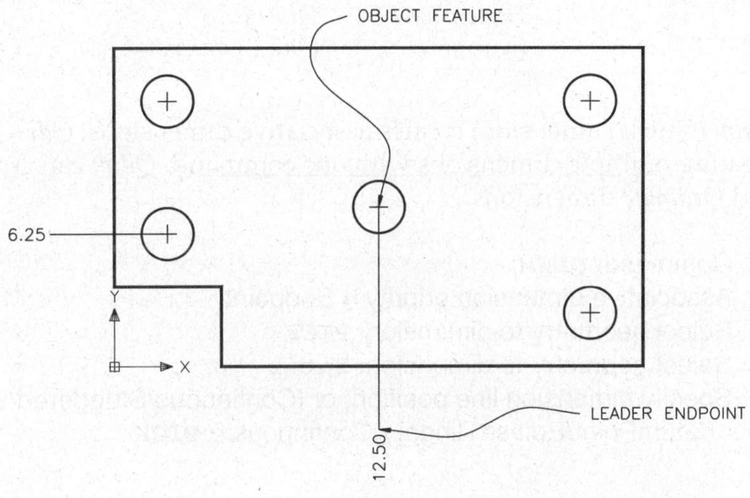

The leader line does not have to be purely horizontal or vertical. In some cases, where the dimension text is crowded, it is desirable to place the end of the leader so that sufficient room is provided for the dimension text. In other words, draw the leader line at an angle. (*Ortho* must be turned *Off* to specify an offset leader line.) AutoCAD automatically creates an offset in the leader as shown in the 7.50 Ydatum dimension in Figure 28-69. As long as the leader is more perpendicular to the X axis, an Xdatum is drawn and vice versa.

**FIGURE 28-69**

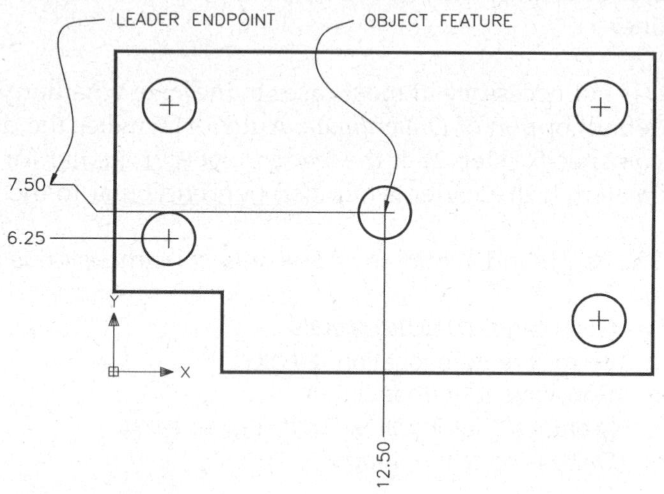


### *Xdatum/Ydatum*

The *Xdatum* and *Ydatum* options are used to specify one of these dimensions explicitly. This is necessary in case the leader line that you specify is more perpendicular to the other axis that you want to measure along. The command syntax would be as follows:

Command: ***DIMORDINATE***
Specify feature location: **PICK**
Specify leader endpoint or [Xdatum/Ydatum/Mtext/Text/Angle]: ***X***
Specify leader endpoint or [Xdatum/Ydatum/Mtext/Text/Angle]: **PICK**
Dimension text = *n.nnnn*
Command:

### *Mtext/Text*

The *Mtext* and *Text* options operate similar to the same options of other dimensioning commands. The *Mtext* option summons the *Text Editor* and the *Text* option uses Command line format. Any text placed inside the < > characters overrides the AutoCAD-measured dimensional value. Any additional text entered on either side of the < > symbols is treated as a prefix or suffix to the measured value. All other options in the *Text Editor* are usable for ordinate dimensions.

## *Qdim*

| Ribbon | Menu Bar | Command (Type) | Alias (Type) | Shortcut |
|---|---|---|---|---|
| *Annotate Dimensions* (icon) | *Dimension Quick Dimension* | *Qdim* | ... | ... |

*Qdim* (Quick Dimension) creates associative dimensions. *Qdim* simplifies the task of dimensioning by creating multiple dimensions with one command. *Qdim* can create *Continuous, Baseline, Radius, Diameter*, and *Ordinate* dimensions.

Command: ***QDIM***
Associative dimension priority = Endpoint
Select geometry to dimension: **PICK**
Select geometry to dimension: **Enter**
Specify dimension line position, or [Continuous/Staggered/Baseline/Ordinate/Radius/Diameter/datumPoint/Edit/seTtings] <Continuous>: **PICK**

When the *Edit* option is invoked, small markers appear at the extension line origins called dimension points (Fig. 28-78). You can *Add* or *Remove* dimensions by adding or removing dimension points. For example, using the *Remove* option, three dimensions are reduced to one by removing the indicated dimension points.

**FIGURE 28-78**

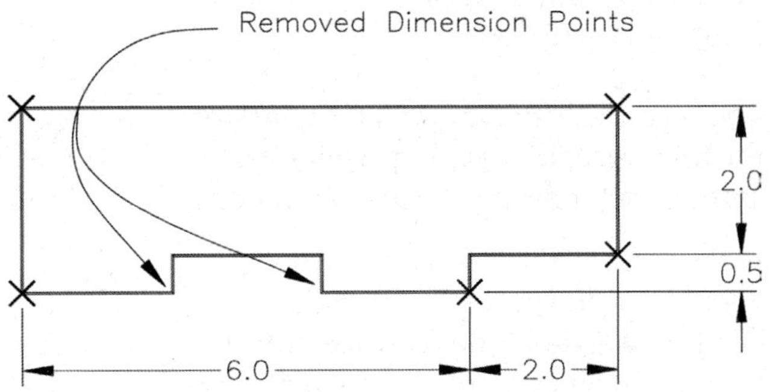


### *Tolerance*

| Ribbon | Menu Bar | Command (Type) | Alias (Type) | Shortcut |
|---|---|---|---|---|
| *Annotate Dimensions* (icon) | *Dimension Tolerance...* | *Tolerance* | *TOL* | ... |

Geometric Dimensioning and Tolerancing (GDT) has become an essential component of detail drawings of manufactured parts. Standard symbols are used to define the geometric aspects of the parts and how the parts function in relation to other parts of an assembly. The *Tolerance* command in AutoCAD produces a series of dialog boxes for you to create the symbols, values, and feature control frames needed to dimension a drawing with GDT. Invoking the *Tolerance* command produces the *Symbol* dialog box and the *Geometric Tolerance* dialog box (see Fig. 28-81 and Fig. 28-82 on the next page). These dialog boxes can also be accessed by the leader commands. Unlike general dimensioning in AutoCAD, geometric dimensioning is <u>not associative</u>; however, you can use *Textedit* to edit the components of an existing feature control frame.

Some common examples of geometric dimensioning may be a control of the flatness of a surface, the parallelism of one surface relative to another, or the positioning of a group of holes to the outside surfaces of the part. Most of the conditions are controls that cannot be stated with the other AutoCAD dimensioning commands.

This book does not provide instruction on how to use geometric dimensioning; rather, it explains how to use AutoCAD to apply the symbology. The ASME Y14.5M - 1994 Dimensioning and Tolerancing standard is the authority on the topic in the United States. The symbol application presented in AutoCAD is actually based on the 1982 release of the Y14.5 standard. Some of the symbols in the 1994 version are modified from the 1982 standard.

The dimensioning symbols are placed in a feature control frame (FCF). This frame is composed of a minimum of two sections and a maximum of three different sections (Fig. 28-79).

**FIGURE 28-79**

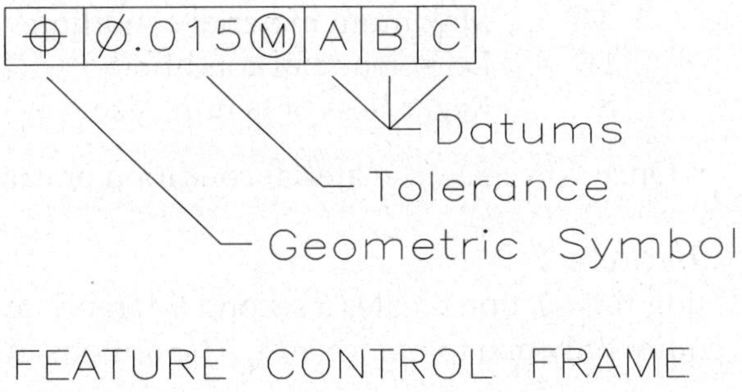

The first section of the FCF houses one of 14 possible geometric characteristic symbols (Fig. 28-80).

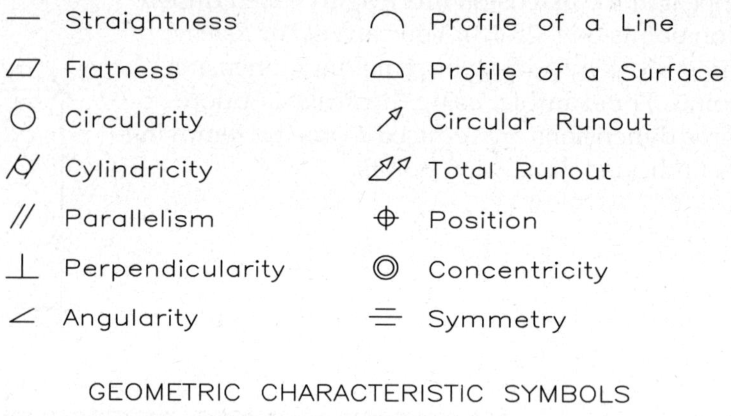


FIGURE 28-80

The second section of the FCF contains the tolerance information. The third section includes any datum references.

In AutoCAD, the *Tolerance* command allows you to specify the values and symbols needed for a feature control frame. The lines comprising the frame itself are automatically generated. Invoking *Tolerance* by any method produces the *Geometric Tolerance* dialog box (Fig. 28-81).

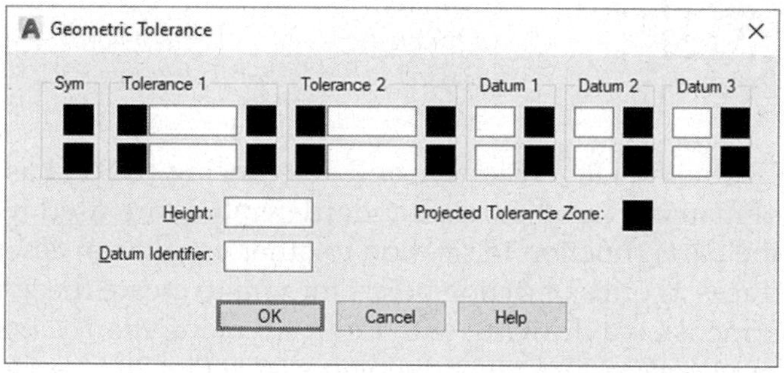


FIGURE 28-81

The *Geometric Tolerance* dialog box (Fig. 28-81) appears for you to specify the tolerance values and datum. The areas of the *Geometric Tolerance* dialog box are explained below.

### Sym

Click in this empty tile to produce the *Symbol* dialog box (Fig. 28-82). Here you select the geometric characteristic symbol to use for the new feature control frame. The selected symbol then appears in the *Sym* tile in the *Geometric Tolerance* dialog box.

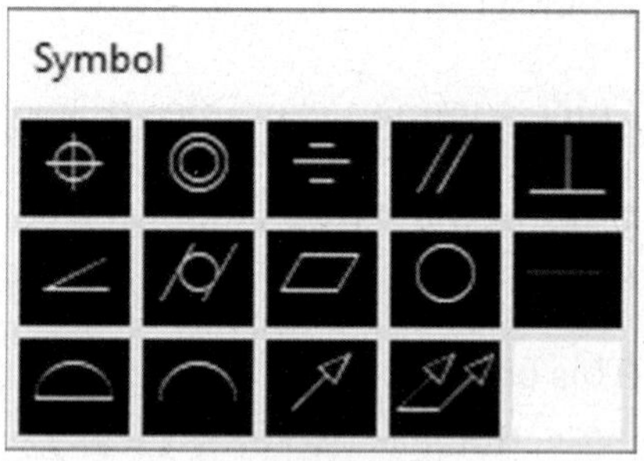


FIGURE 28-82

### Tolerance 1

This area is used to specify the first tolerance value in the feature control frame. This value specifies the amount of allowable deviation for the geometric feature. The three sections in this cluster are:

*Dia* (first box)
If a cylindrical tolerance zone is specified, a diameter symbol can be placed before the value by clicking in this area.

*Value* (second box)
Enter the tolerance value in this edit box.

*MC* (third box)
A material condition symbol can be placed after the value by choosing this tile. The *Material Condition* dialog box appears for your selection (see Fig. 28-83):

| | |
|---|---|
| *M* | Maximum material condition |
| *L* | Least material condition |
| *S* | Regardless of feature size |

Once you select a material condition or cancel, the *Geometric Tolerance* dialog box reappears.

### Tolerance 2

Using this section creates a second tolerance area in the feature control frame. GDT standards, however, require only one tolerance area. The options in this section are identical to *Tolerance 1*.

### *Datum 1*

This section allows you to specify the primary datum reference for the current feature control frame. A datum is the exact (theoretical) geometric feature that the current feature references to establish its tolerance zone.

*Datum* (first box)
Enter the primary datum letter (A, B, C, etc.) for the FCF in this edit box.

*MC* (second box)
Choosing the *MC* tile allows you to specify a material condition modifier for the datum using the *Material Condition* dialog box (Fig. 28-83).

**FIGURE 28-83**

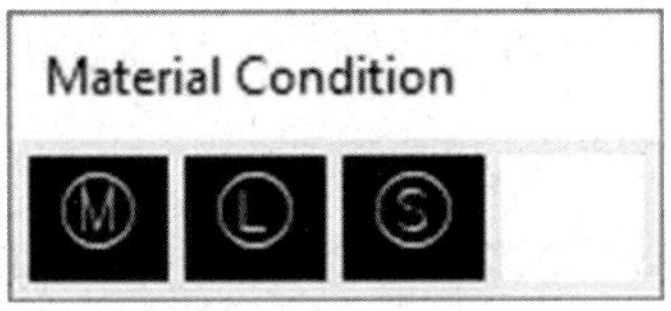


### *Datum 2*

This section is used if you need to create a second datum reference in the feature control frame. The options are identical to those for *Datum 1*.

### *Datum 3*

Use this section if you need to specify a third datum reference for the current feature control frame.

### *Height*

The edit box allows entry of a value for a projected tolerance zone in the feature control frame. A projected tolerance zone specifies a permissible cylindrical tolerance that is projected above the surface of the part. The axis of the control feature must remain within the stated tolerance.

### *Projected Tolerance Zone*

Click this section to insert a projected tolerance zone symbol (a circled "P") after the value.

### *Datum Identifier*

This option creates a datum feature symbol. The edit box allows entry of the value (letter) indicating the datum. A hyphen should be included on each side of the datum letter, such as "-A-".

After using the dialog boxes and specifying the necessary values and symbols, AutoCAD prompts for you to PICK a location on the drawing for the feature control symbol or datum identifier:

Enter tolerance location: **`PICK`**

The feature control symbol or datum identifier is drawn as specified.

## Editing Feature Control Frames

The AutoCAD-generated feature control frames are not associative and can be *Erased*, *Moved*, or otherwise edited without consequence to the related geometry objects. Feature control frames created with *Tolerance* are treated as one object; therefore, editing an individual component of a feature control frame is not possible unless *Textedit* is used. If *Dblclkedit* is *On*, you can double-click on the FCF.

*Textedit* is normally used to edit text, but if an existing feature control frame is selected in response to the "Select an annotation object" prompt, the *Geometric Tolerance* dialog box appears. The edit boxes in this dialog box contain the values and symbols from the selected FCF. Making the appropriate changes updates the selected FCF. (See Chapter 18 for detailed information on *Textedit*.)

## Basic Dimensions

Basic dimensions are often required in drawings using GDT. Basic dimensions in AutoCAD are created by using the *Dimension Style Manager* (or by entering a negative value in the *DIMGAP* variable). The procedure is explained briefly in the following example (application 5) and discussed further in Chapter 29, Dimension Styles and Variables.

## GDT-Related Dimension Variables

Some aspects of how AutoCAD draws the GDT symbols can be controlled using dimension variables and dimension styles (discussed fully in Chapter 29). These variables are:

| | |
|---|---|
| *DIMCLRE* | Controls the color of the FCF |
| *DIMCLRT* | Controls the color of the tolerance text |
| *DIMGAP* | Controls the gap between the FCF and the text, and controls the existence of a basic dimension box |
| *DIMTXT* | Controls the size of the tolerance text |
| *DIMTSTSTY* | Controls the style of the tolerance text |

## Geometric Dimensioning and Tolerancing Example

Six different geometric dimensioning and tolerancing examples are shown on the SPACER drawing in Figure 28-84. Each example is indicated on the drawing and in the following text by a number, 1 through 6. The purpose of the examples is to explain how to use the geometric dimensioning features of AutoCAD. In each example, any method shown in the *Tolerance* command table can be used to invoke the command and dialog boxes.

**FIGURE 28-84**

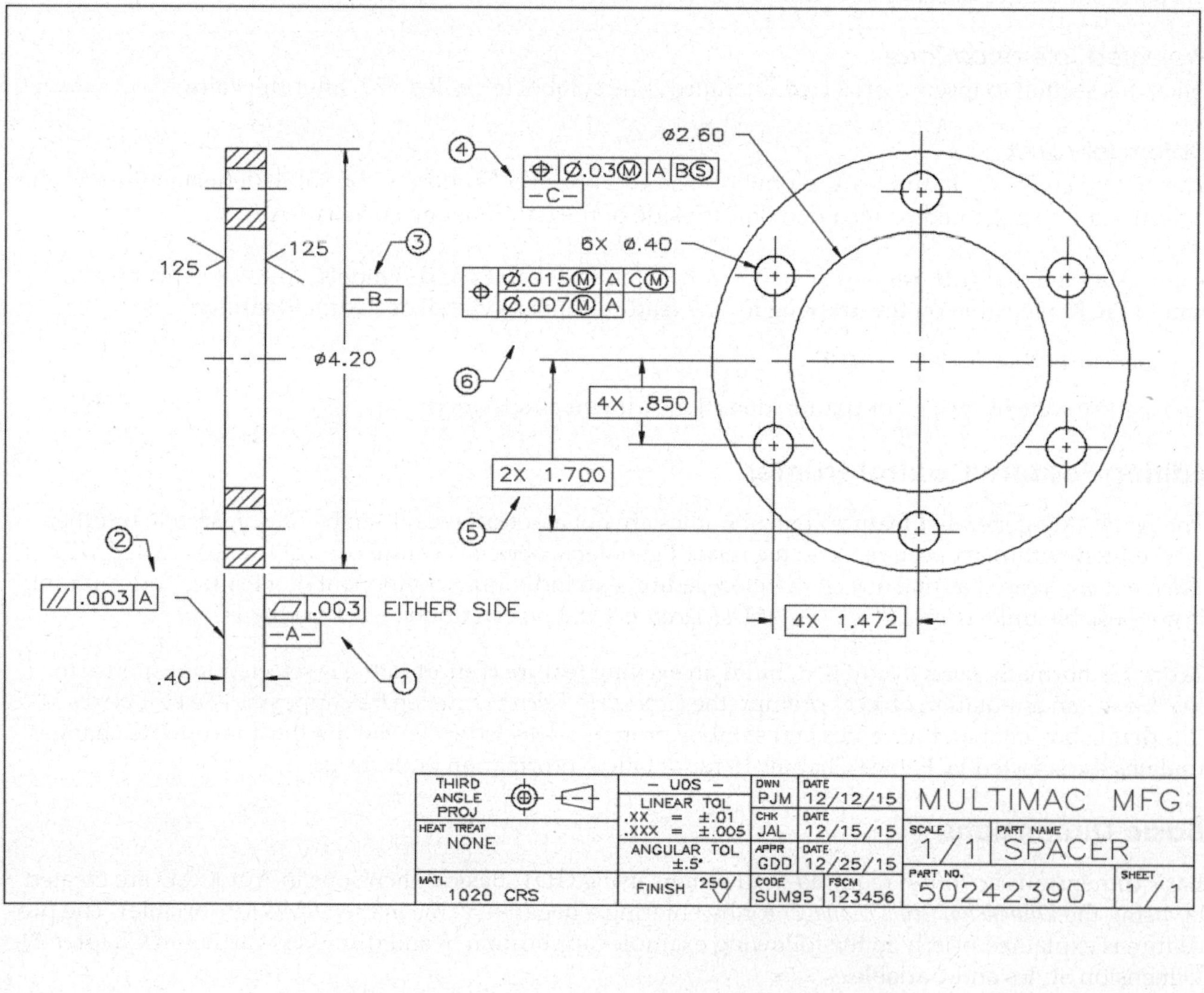

1. **Flatness Application**
The first specification applied is flatness. In addition to one of the surfaces being controlled for flatness, it is also identified as a datum surface. Both conditions are applied in the same application (Fig. 28-85).

Command: ***TOLERANCE*** (The *Geometric Tolerance* dialog box appears.)
PICK the ***Sym*** box. When the *Symbol* dialog box appears, select the ***flatness*** symbol. The *Geometric Tolerance* dialog box reappears.
PICK the ***Tolerance 1*** edit box and type "**.003**" as the tolerance.
PICK the ***Datum Identifier*** box and type "**-A-**".
PICK the ***OK*** button. (The *Geometric Tolerance* dialog box disappears.)
Enter tolerance location: **PICK** (PICK a point on the right extension line where you want the FCF to be attached, see Fig. 28-86.)
Command:

**FIGURE 28-85**

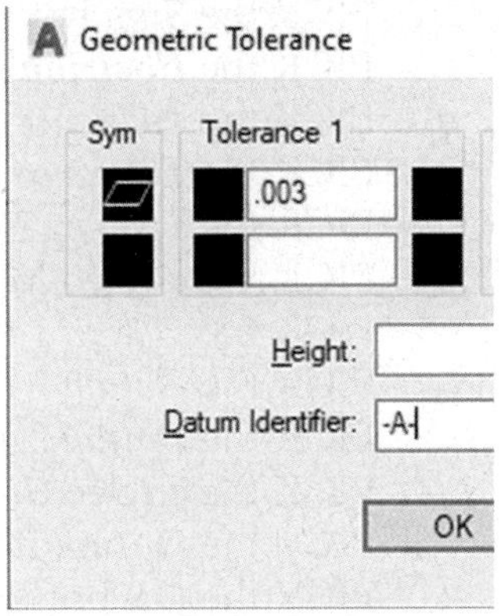


Because either side of the SPACER can be chosen for the flatness specification, the note "EITHER SIDE" is entered next to the FCF in previous Figure 28-84.

**FIGURE 28-86**

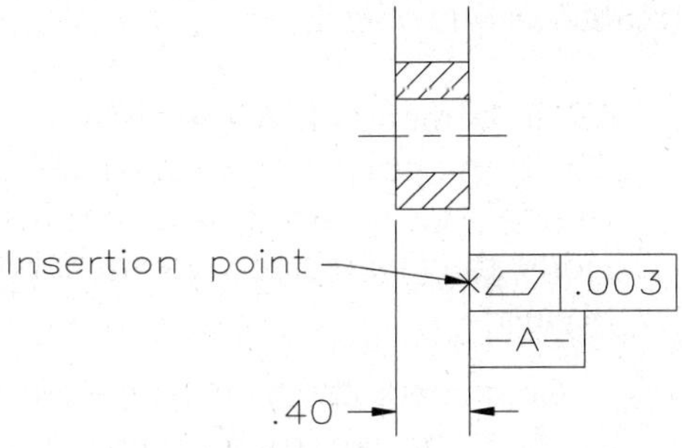


2. **Parallelism Application**
The second application is a parallelism specification to the opposite side of the part. This control is applied using a *Leader*:

Command: ***LEADER***
Specify leader start point: **PICK** (PICK a point on the left extension line, Fig. 28-84.)
Specify next point: **PICK** (PICK a location slightly up and to the left.)
Specify next point or [Annotation/Format/Undo] <Annotation>: **Enter**
Enter first line of annotation text or <options>: **Enter**
Enter an annotation option [Tolerance/Copy/Block/None/Mtext] <Mtext>: **t** (The *Geometric Tolerance* dialog box appears.) PICK the ***Sym*** box. When the *Symbol* dialog box appears, select the ***parallelism*** symbol. The *Geometric Tolerance* dialog box reappears.
PICK the ***Tolerance 1*** edit box and type "**.003**" as the tolerance.
PICK the ***Datum 1*** edit box and type "**A**".
PICK the ***OK*** button. (The *Geometric Tolerance* dialog box clears and the parallelism FCF is placed to the left of the leader.)
Command:

If the leader had projected to the right of the controlled surface, the FCF would be placed on the right of the leader.

3. **Datum Application**
The third application is the B datum on the 4.20 diameter. This datum is used later in the position specification.

Command: ***TOLERANCE*** (The *Geometric Tolerance* dialog box appears.)
PICK the ***Datum Identifier*** edit box and type "**-B-**". PICK the ***OK*** button.
Enter tolerance location: **PICK** (PICK a point on the dimension line of the 4.20 diameter.)
Command:

4. **Position Application**
The fourth application is a position specification of the 2.60 diameter hole. It is identified as a C datum because it will be used in the position specification of the six mounting holes. This specification creates a relationship between inside and outside diameters and a perpendicularity requirement to datum A.

Command: ***TOLERANCE*** (The *Geometric Tolerance* dialog box appears.)
PICK the ***Sym*** box so the *Symbol* dialog box appears.
PICK the ***Position*** symbol. (The *Geometric Tolerance* dialog box reappears.)
PICK the *Tolerance 1* ***Dia*** box. (This places a diameter symbol in front of the tolerance.)
PICK the *Tolerance 1* ***Value*** box and type "**.03**" as a tolerance.
PICK the *Tolerance 1* ***MC*** box. (The *Material Condition* dialog box appears.)
PICK the circled ***M***.
PICK the *Datum 1* ***Datum*** box and type "**A**".
PICK the *Datum 2* ***Datum*** box and type "**B**".
PICK the *Datum 2* ***MC*** box to produce the *Material Condition* dialog box.
PICK the circled ***S***.
PICK the ***Datum Identifier*** box and type "**-C-**".
PICK the ***OK*** button to return to the drawing.
Enter tolerance location: **PICK** (PICK a point that places the FCF below the 2.60 diameter [Fig. 28-84].)

If you need to move the FCF with *Move* or grips, select any part of the FCF or its contents since it is treated as one object.

5. **Basic Dimension Application**
The fifth application concerns basic dimensions. Position uses basic (theoretically exact) location dimensions to locate holes because the tolerances are stated in the FCF. Before applying these dimensions, the *Dimension Style* must be changed. (See Chapter 29 for a complete discussion of Dimension Styles.)

Command: ***DDIM*** (The *Dimension Style Manager* appears.)
PICK the ***Modify*** button.
PICK the ***Tolerances*** tab.
PICK the ***Method*** drop-down list and select ***Basic***.
PICK the ***OK*** button, then ***Close*** the *Dimension Style Manager*.
Command: **dimlinear** (Use the *dimlinear* command to apply each of the three linear basic dimensions shown in Fig. 28-84.)

6. **Composite Position Application**
The sixth application is the composite position specification. A composite specification consists of two separate lines but only one geometric symbol. Achieving this result requires entering data on the top and bottom lines in the *Geometric Tolerance* dialog box.

Command: ***TOLERANCE*** (The *Geometric Tolerance* dialog box appears.)
PICK the ***Sym*** box so the *Symbol* dialog box appears.
PICK the ***position*** symbol. (The *Geometric Tolerance* dialog box reappears.)
PICK the *Tolerance 1* ***Dia*** box in the *Geometric Tolerance* dialog box.
PICK the *Tolerance 1* ***Value*** box and enter "**.015**" as a tolerance.
PICK the *Tolerance 1* ***MC*** box. (The *Material Condition* dialog box appears.)
PICK the circled ***M***.
PICK the *Datum 1* ***Datum*** box and type "**A**".
PICK the *Datum 2* ***Datum*** box and type "**C**".
PICK the *Datum 2* ***MC*** box. (The *Material Condition* dialog box appears.)
PICK the circled ***M***.
PICK the ***Sym*** box under the position symbol to produce the *Symbol* dialog box.
PICK the ***position*** symbol.
PICK the *Tolerance 1* ***Dia*** box on the second line.
PICK the *Tolerance 1* ***Value*** box on the second line and type "**.007**" as a tolerance.
PICK the *Tolerance 1* ***MC*** box on the second line. (The *Material Condition* dialog box appears.)
PICK the circled ***M***.

PICK the *Datum 1* **Datum** box and type "**A**".
PICK the ***OK*** button. (The *Geometric Tolerance* dialog box clears.)
Enter tolerance location: **`PICK`** (PICK a point that places the FCF below the .40 diameter, Fig. 28-84.)
Command:

### Datum Targets

The COVER drawing (Fig. 28-87) uses datum targets to locate the part in 3D space. AutoCAD provides no commands to apply datum targets. The best way to apply these symbols is to use *Blocks* and attributes. The target circle diameter is 3.5 times the letter height. The dividing line is always drawn horizontally through the center of the circle.

**FIGURE 28-87**

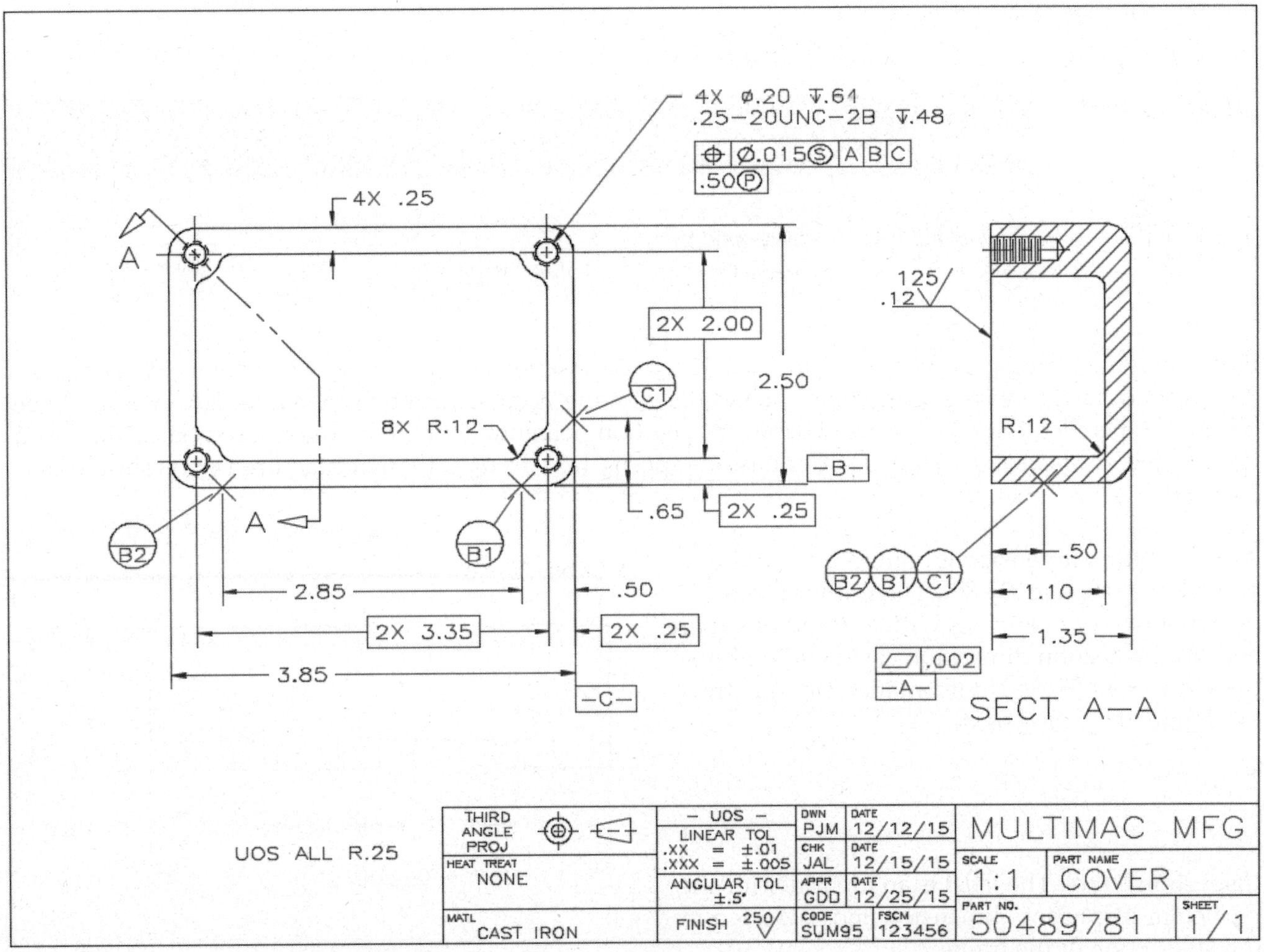


### Projected Tolerance Zone Application

The COVER drawing also uses a projected tolerance with the position specification. A projected tolerance zone is used primarily with internally threaded holes and dowel holes. In this case, the concern is not the position and orientation of the threaded hole, but rather the position and orientation of the fastener inserted into the hole, specifically the shank of the fastener; therefore, the tolerance zone projects above the surface of the part and not within the part. This condition is especially important when the hole allowance is small.

Command: ***TOLERANCE*** (The *Geometric Tolerance* dialog box appears.)
PICK the ***Sym*** box so the *Symbol* dialog box appears.
PICK the ***Position*** symbol. (The *Geometric Tolerance* dialog box reappears.)
PICK the ***Height*** box in the *Geometric Tolerance* dialog box and enter "**.50**".
PICK the ***Projected Tolerance Zone*** box. A circled *P* appears.
PICK the ***OK*** button in the *Geometric Tolerance* box.
Enter tolerance location: **PICK** (PICK a point that places the FCF below the .20 diameter, Fig. 28-87.)
Command:

This section has presented only the mechanics of GDT symbol application in AutoCAD. Geometric dimensioning and tolerancing can be a complicated and detailed subject to learn. Many different combination possibilities may appear in a feature control frame. However, for any one feature, there are very few possibilities. Knowing which of the possibilities is best comes from a fundamental knowledge of GDT. Knowledge of GDT increases your understanding of design, tooling, manufacturing, and inspection concepts and processes.

## *Diminspect*

| Ribbon | Menu Bar | Command (Type) | Alias (Type) | Shortcut |
|---|---|---|---|---|
| *Annotate Dimensions* (icon) | *Dimension Inspection* | *Diminspect* | ... | ... |

In the manufacturing process, tools, molds, and related machinery will slowly wear down. This causes the finished parts to vary slightly over time as the wearing occurs, such that the parts may eventually be out of the specified tolerance range. Therefore, a certain percentage of the manufactured parts may need to be tested or inspected. Inspection dimensions specify how frequently manufactured parts should be checked.

An inspection dimension is composed of a frame and text values (Fig. 28-88). The frame is composed of two parallel lines with a round or square end. An inspection dimension can contain up to three different fields of information (text) within the frame, described next.

**FIGURE 28-88**

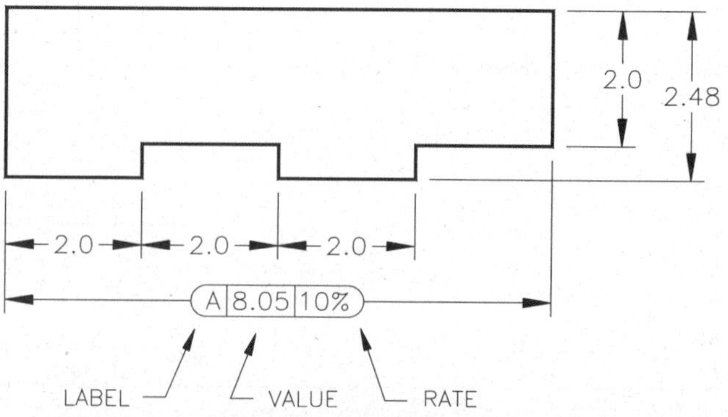


Inspection Label: The label is an identifier for the particular inspection dimension and is located in the left section of the frame.

Dimension Value: Located in the center section of the inspection frame, the dimension value is the original dimension value that is supplied by the previous dimension command (linear, angular, continue, etc.) before the inspection dimension is added. The dimension value can contain tolerances, text (both prefix and suffix), and the measured value.

Inspection Rate: This text, expressed as a percentage, indicates the frequency that the dimension value should be inspected. The rate is located in the right section of the inspection frame.

The *Diminspect* command uses the *Inspection Dimension* dialog box to provide a means to add or remove inspection dimensions from an existing dimension in the drawing (Fig. 28-89). You can add an inspection dimension to any type of dimension object. Alternately, you can enter *-Diminspect* at the command prompt and access the options in Command line mode.

FIGURE 28-89

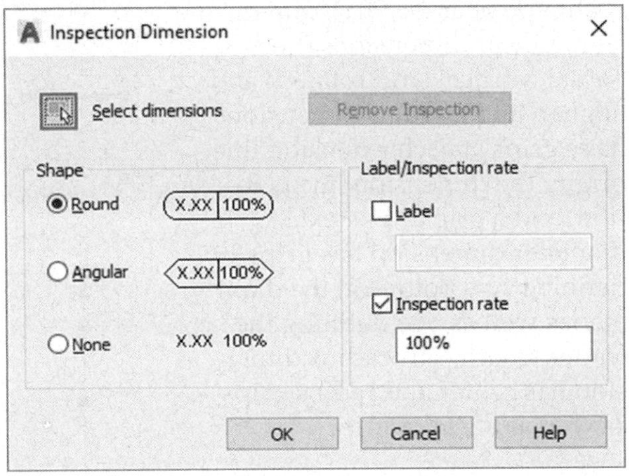


### *Select Dimensions*

First, use this button to select an existing dimension from the drawing to apply the inspection frame, then press enter to return to the *Inspection Dimension* dialog box. Alternately, select dimensions before activating the *Diminspect* command.

### *Remove Inspection*

This option allows you to remove a previously created inspection dimension from a dimension object in the drawing.

### *Shape*

Select from *Round*, *Angular*, or *None* to specify the shape of the inspection dimension frame.

### *Label*

If you want to create a label, use the checkbox to turn on the display of the *Label* field, then enter the desired label for the dimension. Enter any text as the label.

### *Inspection Rate*

If you want to include an inspection rate section in the frame, use this checkbox to turn on the display of the *Inspection Rate* field, then enter the desired rate value. The value is expressed as a percentage, and the valid range is 0 to 100.

# EDITING DIMENSIONS

## Grip Editing Dimensions

Grips can be used effectively for editing dimensions. Any of the grip options (STRETCH, MOVE, ROTATE, SCALE, and MIRROR) is applicable. Depending on the type of dimension (linear, radial, angular, etc.), grips appear at several locations on the dimension when you activate the grips by selecting the dimension at the Command: prompt. Associative dimensions offer the most powerful editing possibilities, although nonassociative dimension components can also be edited with grips. There are many ways in which grips can be used to alter the measured value and configuration of dimensions.

Figure 28-90 shows the grips for each type of associative dimension. Linear dimensions (horizontal, vertical, and rotated) and aligned and angular dimensions have grips at each extension line origin, the dimension line position, and a grip at the text. A diameter dimension has grips defining two points on the diameter as well as one defining the leader length. The radius dimension has center and radius grips as well as a leader grip.

**FIGURE 28-90**

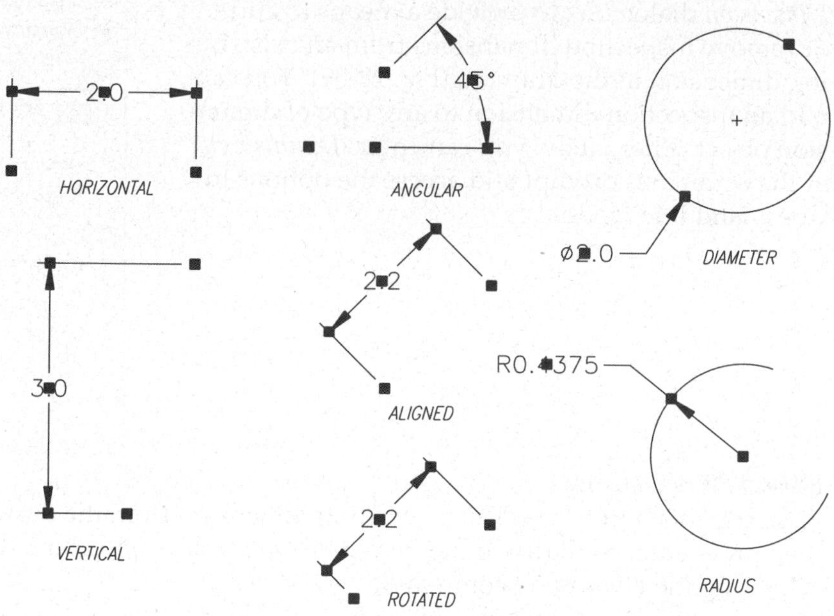


With dimension grips, a wide variety of editing options are possible. Any of the grips can be PICKed to make them **hot** grips. All grip options are valid methods for editing dimensions.

For example (Fig. 28-91), a horizontal dimension value can be increased by stretching an extension line origin grip in a horizontal direction. A vertical direction movement changes the length of the extension line. The dimension line placement is changed by stretching its grips. The dimension text can be stretched to any position by manipulating its grip.

**FIGURE 28-91**

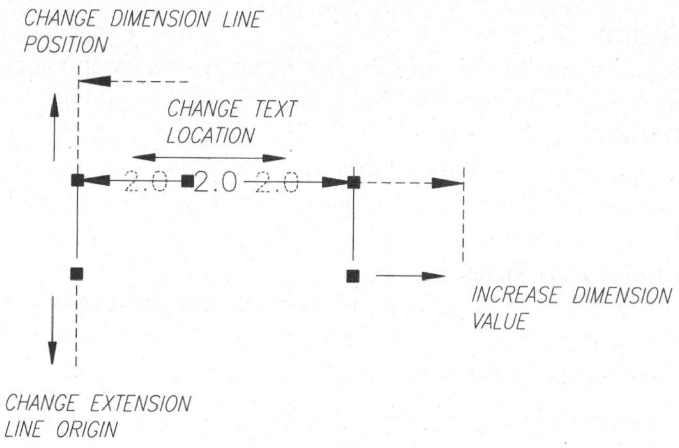


An angular dimension can be increased by stretching the extension line origin grip. The numerical value automatically updates. Stretching a rotated dimension's extension line origin allows changing the length of the dimension as well as the length of the extension line. An aligned dimension's extension line origin grip also allows you to change the aligned angle of the dimension (Fig. 28-92).

**FIGURE 28-92**

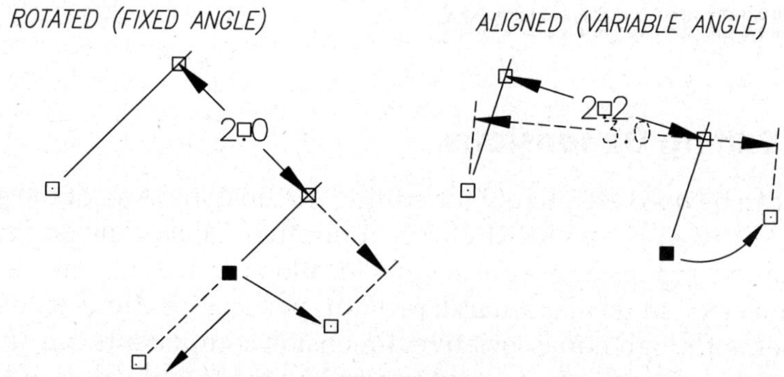

Rotating the <u>center</u> grip of a radius dimension (with the ROTATE option) allows you to reposition the location of the dimension around the *Arc*. Note that the text remains in its original horizontal orientation (Fig. 28-93).

**FIGURE 28-93**

*RADIUS*

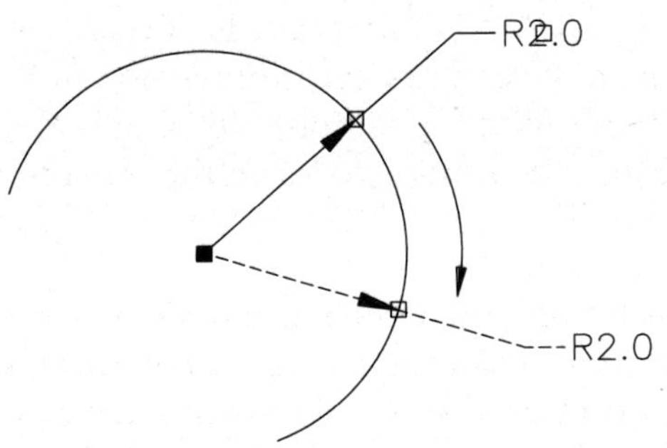


TIP

You can move the dimension <u>text</u> independent of the dimension line with grips for dimensions that have a *DIMTMOVE* variable setting of 2. Use an existing dimension, and change the *DIMTMOVE* setting to 2 (or change the *Fit* setting in the *Dimension Style Manager* to *Over the dimension line without a leader*). The dimension text can be moved to any location without losing its associativity. See *"Fit (DIMTMOVE),"* Chapter 29.

Many other possibilities exist for editing dimensions using grips. Experiment on your own to discover some of the possibilities that are not shown here.

**FIGURE 28-94**

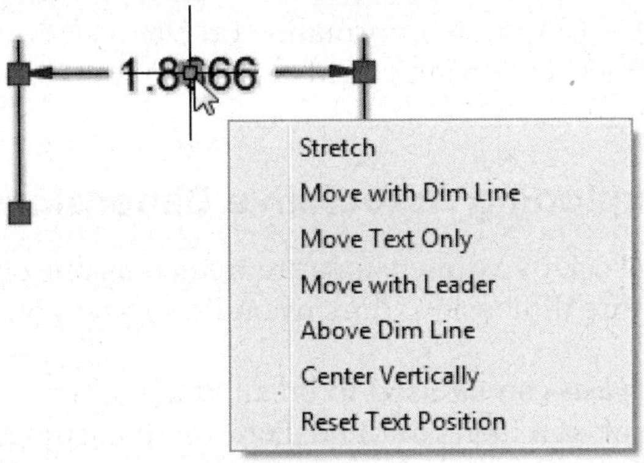


## Multifunctional Grips

TIP

*Multifunctional Grips* provide the same options as *Grips*. However, they increase efficiency by reducing the number of clicks. Figure 28-94 shows the multi-function grips active on a horizontal dimension. Hover over the desired grip to produce the grips menu.

### *Dim Text Position*

Several options are available using the multi-functional grips. The *Center Vertically* and *Reset Text Position* options are duplicates of the *Dimtedit* command options of the same names (see *"Dimtedit"*). The *Move text only*, *Move with leader*, and *Move with dim line* options change the settings of the *DIMTMOVE* system variable for the dimension. The *Move* options operate best with grips on. For example, you can select the *Move text only* option, make the text grip hot, then drag the text to any position in the drawing while still retaining the associative quality of the dimension (Fig. 28-95). (For more information on these options, see *"DIMTMOVE"* in Chapter 29.)

**FIGURE 28-95**

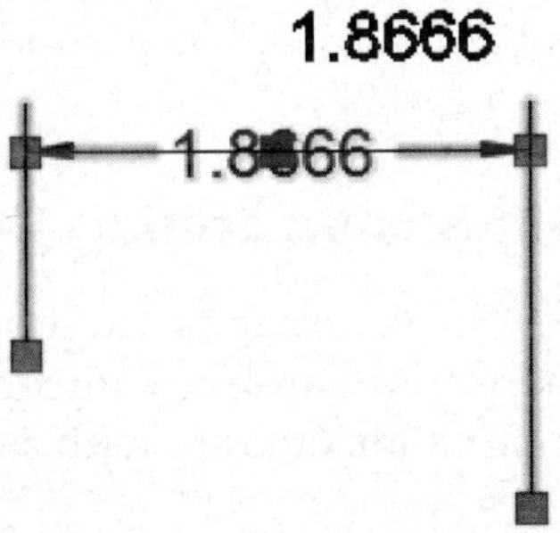


### *Precision*

Use the dimension right-click menu to access the *Precision* option. Clicking the cascading menu will allow you to select from a list of decimal or fractional precision values. For example, you could change a dimension text value of "5.0000" to "5.00" with this menu. Your choice changes the *DIMDEC* system variable for the selected dimension (Fig. 28-96). (For more information, see *"DIMDEC"* in Chapter 29.)

**FIGURE 28-96**

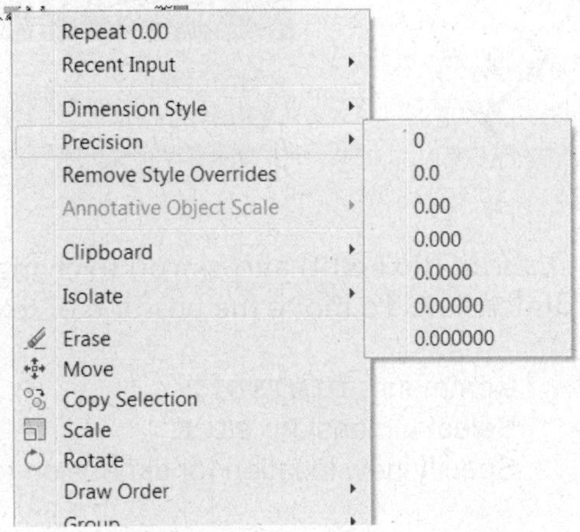


### *Dim Style*

Use this option from the dimension right-click menu to change the selected dimension from one *Dimension Style* to another. Changing a dimension's style affects the appearance characteristics of the dimension (Fig. 28-96). (For more information on *Dimension Styles*, see Chapter 29.)

*Flip Arrow*

Use the multi-function grip to flip the arrow nearest to the point you right-click to produce the menu (see Fig. 28-97). In other words, if both arrows pointed from the center of the dimension out to the extension lines (default position), flipping both arrows would cause the arrows to be on the outside of the extension lines pointing inward (Fig. 28-98).

**FIGURE 28-97**

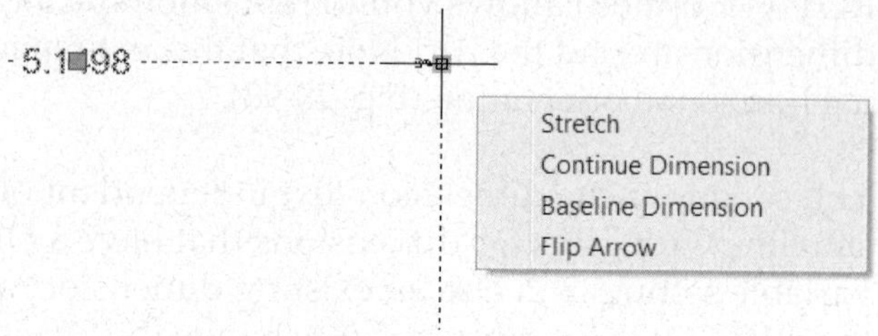


Although the *Centered* and *Home Text* options are duplicates of the *Dimtedit* command, the other options in this shortcut menu actually change system variables for the individual dimensions. These options do not change the variable settings for the *Dimension Style* of the selected dimension, but create dimension style overrides for the individual dimension. (For more information on *Dimension Styles* and dimension style overrides, see Chapter 29.)

**FIGURE 28-98**

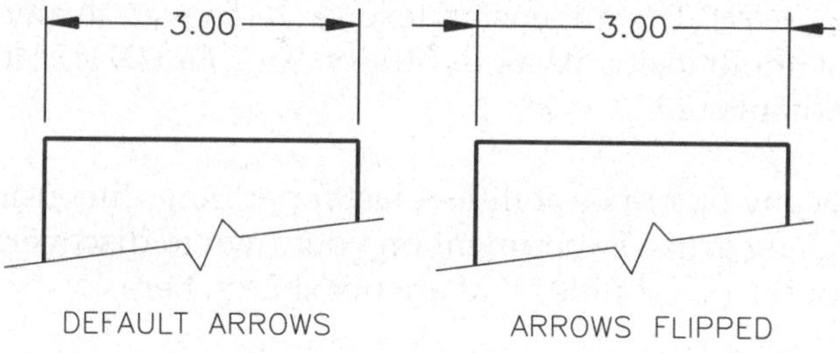


## Exploding Associative Dimensions

Associative dimensions are treated as one object. If *Erase* is used with an associative dimension, the entire dimension (dimension line, extension lines, arrows, and text) is selected and *Erased*.

*Explode* can be used to break an associative dimension into its component parts. The individual components can then be edited. For example, after *Exploding*, an extension line can be *Erased* or text can be *Moved*.

There are two main drawbacks to *Exploding* associative dimensions. First, the associative property is lost. Editing commands or *Grips* cannot be used to change the entire dimension but affect only the component objects. More important, Dimension Styles cannot be used with unassociative dimensions.

## Dimension Editing Commands

Several commands are provided to facilitate easy editing of existing dimensions in a drawing. Most of these commands are intended to allow variations in the appearance of dimension text. These editing commands operate with associative and nonassociative dimensions but not with exploded dimensions.

### *Dimtedit*

| Ribbon | Menu Bar | Command (Type) | Alias (Type) | Shortcut |
|---|---|---|---|---|
| *Annotate Dimensions* (icon) | *Dimension Align Text >* | *Dimtedit* | *DIMTED* | ... |

*Dimtedit* (text edit) allows you to change the position or orientation of the text for a single associative dimension. To move the position of text, this command syntax is used:

Command: ***DIMTEDIT***
Select dimension: **PICK**
Specify new location for dimension text or [Left/Right/Center/Home/Angle]: **PICK**

At the "Specify new location for dimension text" prompt, drag the text to the desired location. The selected text and dimension line can be changed to any position, while the text and dimension line retain their associativity (Fig. 28-99).

FIGURE 28-99

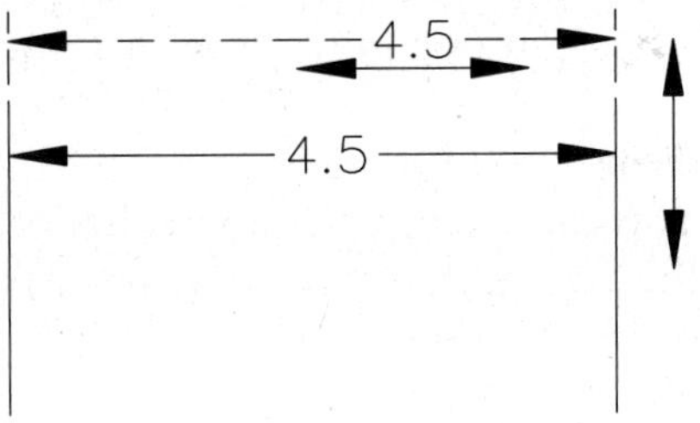


### Angle

The *Angle* option works with any *Horizontal, Vertical, Aligned, Rotated, Radius*, or *Diameter* dimensions. You are prompted for the new text angle (Fig. 28-100):

Command: ***DIMTEDIT***
Select dimension: **PICK**
Specify new location for dimension text or [Left/Right/Center/Home/Angle]: ***a***
Specify angle for dimension text: **45**
Command:

FIGURE 28-100

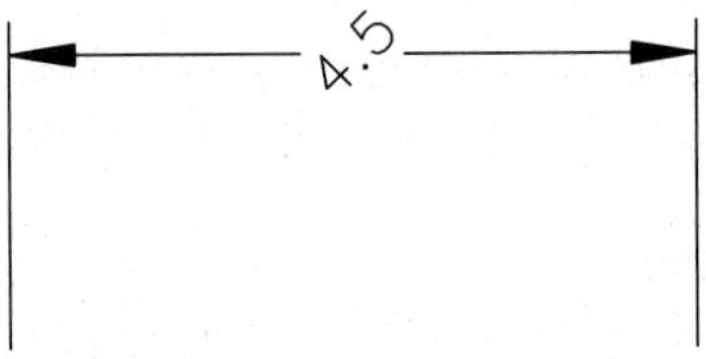


### Home

The text can be restored to its original (default) rotation angle with the *Home* option. The text retains its right/left position.

### Right/Left

The *Right* and *Left* options automatically justify the dimension text at the extreme right and left ends, respectively, of the dimension line. The arrow and a short section of the dimension line, however, remain between the text and closest extension line (Fig. 28-101).

### Center

The *Center* option brings the text to the center of the dimension line and sets the rotation angle to 0 (if previously assigned; Fig. 28-101).

FIGURE 28-101

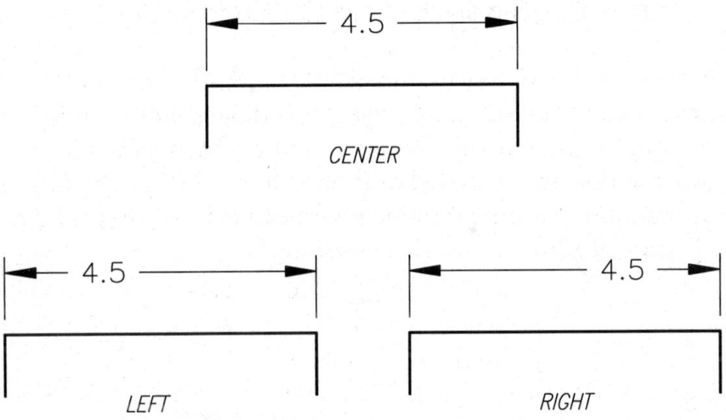


## Dimedit

| Ribbon | Menu Bar | Command (Type) | Alias (Type) | Shortcut |
|---|---|---|---|---|
| ... | ... | *Dimedit* | *DIMED* or *DED* | ... |

The *Dimedit* command allows you to change the angle of the extension lines to an obliquing angle and provides several ways to edit dimension text. Two of the text editing options (*Home, Rotate*) duplicate those of the *Dimtedit* command. Another feature (*New*) allows you to change the text value and annotation.

Command: ***DIMEDIT***
Enter type of dimension editing [Home/New/Rotate/Oblique] <Home>:

### Home

This option moves the dimension text back to its original angular orientation angle without changing the left/right position. *Home* is a duplicate of the *Dimtedit* option of the same name.

### New

You can change the text value of any existing dimension with this option. When you invoke this option, the *Text Editor* appears with "0.0000" as the string of text in the dark background. Enter the desired text, press *OK*, then you are prompted to select a dimension. The text you entered is applied to the selected dimension.

**TIP** The key feature of the *New* option is that the original AutoCAD-measured value can be restored in the case that it was "manually" changed for some reason. Simply invoke *Dimedit* and the *New* option, then do not enter any new text in the *Text Editor*, and press the *OK* button. As prompted next, select the desired dimension, and the original AutoCAD-measured value is automatically restored.

### Rotate

AutoCAD prompts for an angle to rotate the text. Enter an absolute angle (relative to angle 0). This option is identical to the *Angle* option of *Dimtedit*.

### Oblique

This option is unique to *Dimedit*. Entering an angle at the prompt affects the extension lines. Enter an absolute angle:

```
Enter obliquing angle (press ENTER for none):
```

Normally, the extension lines are perpendicular to the dimension lines. In some cases, it is desirable to set an obliquing angle for the extension lines, such as when dimensions are crowded and hard to read (Fig. 28-102) or for dimensioning isometric drawings (see Chapter 29, "Dimensioning Isometric Drawings").

**FIGURE 28-102**

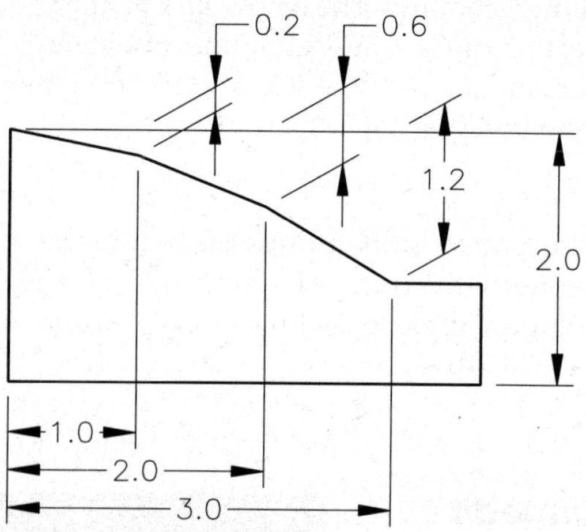


## Properties

| Ribbon | Menu Bar | Command (Type) | Alias (Type) | Shortcut |
|---|---|---|---|---|
| *View Palettes Properties* | *Modify Properties* | *Properties* | *MO* or *PR* | (Edit Mode) *Properties* |

The *Properties* command (discussed in Chapter 16) can be used effectively for a wide range of editing purposes. Right-clicking on any dimension and selecting *Properties* will open the *Properties* palette. The palette gives a list of all properties of the dimension including dimension variable settings.

The dimension variable settings for the selected dimension are listed in the *Lines & Arrows*, *Text*, *Fit*, *Primary Units*, *Alternate Units*, and *Tolerances* sections. Each section is expandable, so you can change any property (dimension variable setting) in each section for the dimension (Fig. 28-103). Each entry in the left column represents a dimension variable; changing the entries in the right column changes the variable's setting.

**FIGURE 28-103**

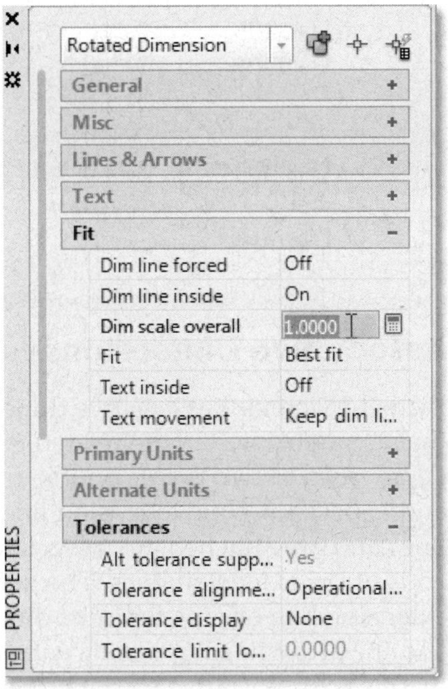


Each of the six categories (*Lines & Arrows*, *Text*, *Fit*, *Primary Units*, *Alternate Units*, and *Tolerances*) corresponds to the tabs in the *Dimension Style Manager*. Typically, you create dimension styles in the *Dimension Style Manager* by selecting settings for the dimensioning variables, then draw the dimensions using one of the styles. Each style has different settings for dimension variables, so the resulting dimensions for each style appear differently.

Using the *Properties* palette, you can change the dimension variable settings for any one or more dimensions retroactively. Making a change to a dimension variable with this method does not change the previously created dimension style, but it does create an override for that particular dimension. For more information on dimension variables, dimension styles, and dimension style overrides, see Chapter 29, Dimension Styles and Variables.

TIP

## Customizing Dimensioning Text

As discussed earlier, you can specify dimensioning text other than what AutoCAD measures and supplies as the default text using the *Text* or *Mtext* option of the individual dimension creation commands. In addition, the text can be modified at a later time using the *Properties* command or the *New* option of *Dimedit*.

The *Text* option and *Properties* palette use the less-than and greater-than symbols (<>) to represent the AutoCAD-measured dimensional value, whereas the *Mtext* option displays the AutoCAD-measured value with a dark background. Either way, placing text or numbers inside the symbols or inside the dark background overrides the correct measurement (not advised) and creates static text. Any override text is not associative and is not updated in the event of *Stretching*, *Rotating*, or otherwise editing the associative dimension. The original text can only be retrieved using the *New* option of *Dimedit*.

Text placed outside of the <> symbols or outside of the dark background, however, acts only as a prefix/suffix to the AutoCAD-measured value. To insert a prefix or suffix using the *Text* option, *Mtext* option, or the *Properties* palette, enter a text string before or after the AutoCAD-measured value (such as "2X <>") to retain the AutoCAD-measured value including any symbols such as a diameter symbol (Ø) or a radius designator (R). You can add a *Prefix* or *Suffix* in the *Dimension Style Manager*; however, this method overrides any symbols included with the measured value.

TIP

This feature is important in correct ANSI standard dimensioning for entering the number of times a feature occurs. For example, one of two holes having the same diameter would be dimensioned "2X Ø1.00" (see Fig. 28-27).

You can also enter special characters by using the Unicode values or "%%" symbols. The following codes can be entered outside the AutoCAD-measured value for *Dimedit New*, *Mtext*, the *Text* option of dimensioning commands, or *Properties* palette:

| Enter: | Enter: | Result | Description |
|---|---|---|---|
| \U+2205 | %%c | ø | diameter (metric) |
| \U+00b0 | %%d | ° | degrees |
| | %%o | ‾‾‾‾ | overscored text |
| | %%u | ____ | underscored text |
| \U+00b1 | %%p | ± | plus or minus |
| | %%*nnn* | varies | ASCII text character number |
| \U+*nnnn* | | varies | Unicode text hexadecimal value |

## Associative Dimensions

AutoCAD creates associative dimensions by default. Associative dimensions contain "definition points" that are associated to the related geometry. For example, when you use the *Dimlinear* command and then *Object Snap* to two points on the drawing geometry in response to the "Specify first extension line origin" and "Specify second extension line origin," two definition points are created at the ends of the extension lines and are attached, or "associated," to the geometry. You may have noticed these small dots at the ends of the extension lines. The definition points are generally located where you PICK, such as at the extension line origins for most dimensioning commands, or the arrowhead tips for the leader commands and the arrowhead tips and centers for diameter and radius commands. All of the dimensioning commands are associative by default except *Mleader, Qleader, Leader, Dimcenter*, and *Tolerance*.

When you create the first associative dimension, AutoCAD automatically creates a new layer called DEFPOINTS. The layer is set to not plot and that property cannot be changed. AutoCAD manages this layer automatically, so do not attempt to alter this layer.

TIP

There are two advantages to associative dimensions. First, if the associated geometry is modified by typical editing methods, the associated definition points automatically change; therefore, the dimension line components (dimension lines and extension lines) automatically change and the numerical value automatically updates to reflect the geometry's new length, angle, radius, or diameter. Second, existing associative dimensions in a drawing automatically update when their dimension style is changed (see Chapter 29 for information on dimension styles).

Examples of typical editing methods that can affect associative dimensions are *Chamfer, Extend, Fillet, Mirror, Move, Rotate, Scale, Stretch, Trim* (linear dimensions only), *Array* (if rotated in a *Polar* array), and grip editing options.

Dimensions can have three possible associativity settings controlled by the *DIMASSOC* system variable.

Associative dimensions — (*DIMASSOC* = 2) Associative dimensions automatically adjust their locations, orientations, and measurement values when the geometric objects associated with them are modified. Associative dimensions in a drawing will update when the dimension style is modified. These fully associative dimensions are used in AutoCAD 2002 through the current release.

Nonassociative dimensions — (*DIMASSOC* = 1) Nonassociative dimensions do not change automatically when the geometric objects they measure are modified. These dimensions have definition points, but the definition points must be included (selected) if you want the dimensions to change when the geometry they measure is modified. The existing dimensions in a drawing will update when the dimension style is modified. In AutoCAD 2000 and previous releases, these dimensions were called associative.

Exploded dimensions — (*DIMASSOC* = 0) This type of dimension is actually a group of separate objects (lines, arrows, text) rather than a single dimension object. These dimensions do not have definition points and do not change when the geometry is modified. Exploded dimensions do not update when the dimension style is changed. These dimensions can also be created using *Explode* on existing associative or nonassociative dimensions.

Characteristics of the three types of dimensions (associative, nonassociative, and exploded) are explained in the three illustrations here.

An associative dimension is fully associative; that is, if the geometry is changed (in this case by the *Scale* command) the related dimensions components (dimension lines, extension lines, and measured value) automatically update. The shape in Figure 28-104 is a *Pline*, so the *Scale* command affects the entire object. Note that only the single *Pline* object is selected to scale, <u>not the dimensions</u>.

**FIGURE 28-104**

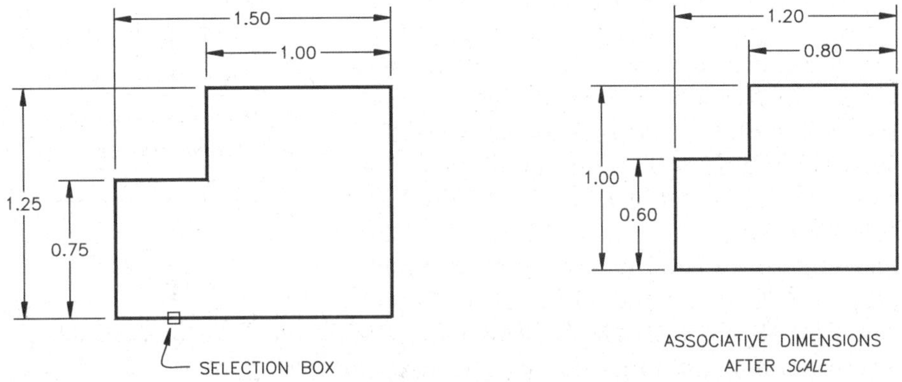


Figure 28-105 displays a similar action—using *Scale* to change an object—but with nonassociative dimensions. Nonassociative dimensions do not automatically change when the related objects are modified unless the dimension definition points are included in the selection set. Note in this case the *Pline* and only two vertical dimensions (1.25 and 0.75) are selected and are therefore scaled, while the horizontal dimensions (1.50 and 1.00) are not selected and therefore are not scaled.

**FIGURE 28-105**

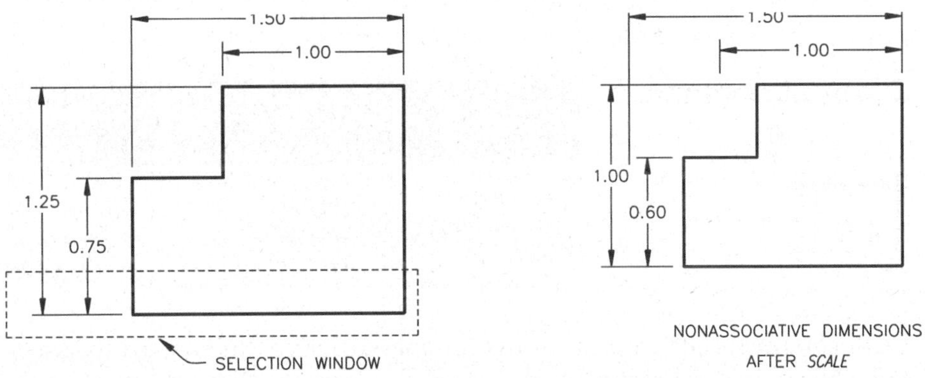


The characteristics of exploded dimensions are explained in Figure 28-106. *Exploded* dimensions are composed of several individual and unrelated components (arrows, text, lines).

**FIGURE 28-106**

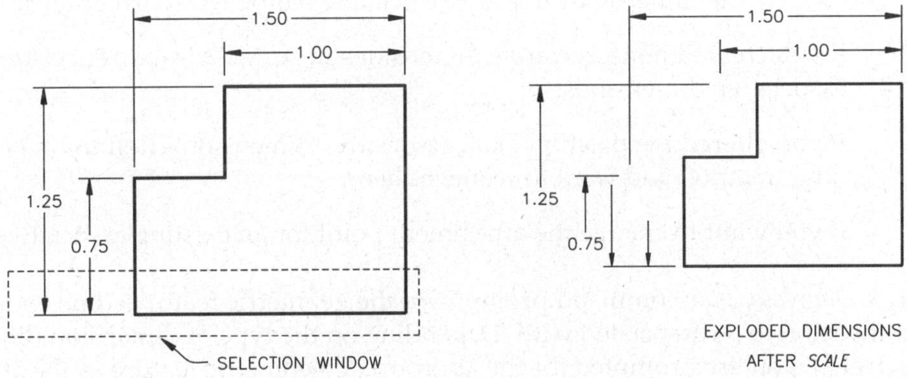

When *Scale* is used, only the selected individual dimension components are affected, so the resulting dimensions are unusable—essentially, the dimensions fall apart. In addition, exploded dimensions do not update if their original dimension style is modified.

If you want to determine whether a dimension is associative or nonassociative, you can use the *Properties* palette or the *List* command. You can also use the *Quick Select* dialog box to filter the selection of associative or nonassociative dimensions. A dimension is considered associative even if only one end of the dimension is associated with a geometric object. Associativity is *not* maintained between a dimension and a block reference if the block is redefined. Associativity is not possible when dimensioning *Multiline* objects.

Remember that the associativity status for newly created dimensions (associative, nonassociative, or exploded) is controlled by the *DIMASSOC* system variable (2, 1, 0, respectively). You can set the *DIMASSOC* system variable at the Command line or use the *Options* dialog box (Fig. 28-107). In the *User Preferences* tab, a check in the *Make new dimensions associative* box (right center) sets *DIMASSOC* to 2, while no check in this box sets *DIMASSOC* to 1. The *DIMASSOC* setting does not affect existing dimensions, only subsequently created ones. The *DIMASSOC* variable is not saved with a dimension style.

**FIGURE 28-107**

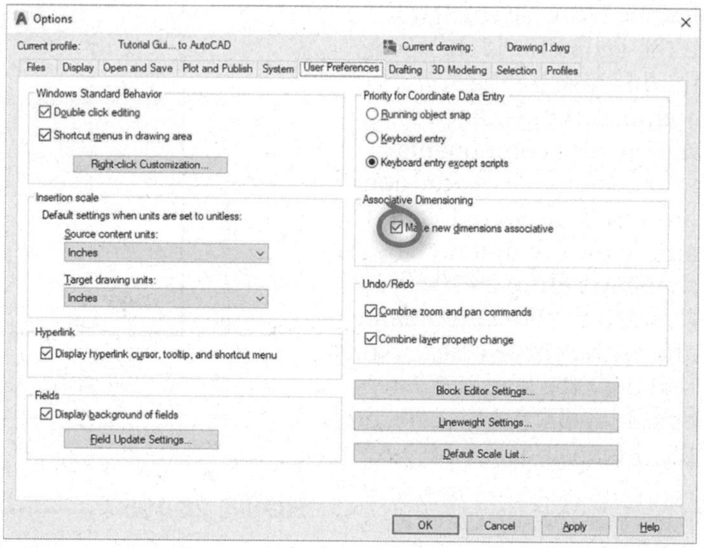


### *Dimreassociate*

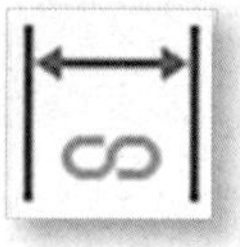

| Ribbon | Menu Bar | Command (Type) | Alias (Type) | Shortcut |
|---|---|---|---|---|
| *Annotate Dimensions (icon)* | *Dimension Reassociate Dimensions* | *Dimreassociate* | *DRE* | ... |

With *Dimreassociate*, you can convert nonassociative dimensions to associative dimensions. In addition, you can change the feature of the drawing geometry (line endpoint, etc.) that an existing definition point is associated with. This command is especially helpful in several situations, such as:

if you are updating a drawing completed in AutoCAD 2000 or a previous release and want to convert the dimensions to fully associative AutoCAD (current release) dimensions;

if you created nonassociative dimensions in AutoCAD (current release) but want to convert them to associative dimensions;

TIP

if you altered the position of an associative dimension's definition points accidentally with grips or other method and want to reattach them;

if you want to change the attachment point for an existing associative dimension.

The *Dimreassociate* command prompts for the geometric features (line endpoints, etc.) that you want the dimension to be associated with. Depending on the type of dimension (linear, radial, angular, diametrical, etc.), you are prompted for the appropriate geometric feature as the attachment point.

For example, if a linear dimension is selected to reassociate, AutoCAD prompts you to specify (select attachment points for) the extension line origins.

Command: ***DIMREASSOCIATE***
Select dimensions to reassociate ...
Select objects or [Disassociated]: **PICK** (linear dimension selected)
Select objects or [Disassociated]: **Enter**
Specify first extension line origin or [Select object] <next>: **PICK**
Specify second extension line origin <next>: **PICK**
Command:

When you select the "dimensions to reassociate," the associative and nonassociative elements of the selected dimension(s) are displayed one at a time. A marker is displayed for each extension line origin (or other definition point) for each dimension selected (Fig. 28-108). If the definition point is not associated, an "X" marker appears, and if the definition point is associated, an "X" within a box appears. To reassociate a definition point, simply pick the new object feature you want the dimension to be attached to. If an extension line is already associated and you do not want to make a change, press Enter when its definition point marker is displayed. Figure 28-108 displays a partly associated linear dimension with its extension line origin markers during the *Dimreassociate* command.

**FIGURE 28-108**

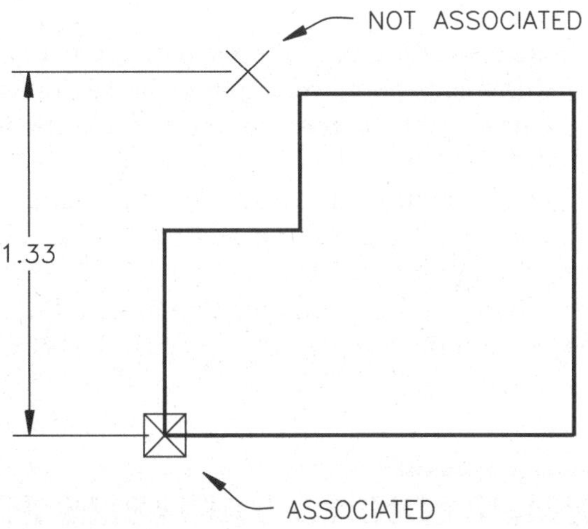


### *Disassociated*

Use this option to have automatically AutoCAD locate disassociated dimensions. AutoCAD finds and highlights disassociated dimensions in the drawing. Reassociate these dimensions by selecting the "first extension line origin" and "second extension line origin" as before. Use the *next* option to cause AutoCAD to cycle to the next disassociated extension line.

## *Dimdisassociate*

| Ribbon | Menu Bar | Command (Type) | Alias (Type) | Shortcut |
|---|---|---|---|---|
| ... | ... | *Dimdisassociate* | *DDA* | ... |

The *Dimdisassociate* command automatically converts selected associative dimensions to nonassociative. The command prompts for you to select only the dimensions you want to disassociate, then AutoCAD makes the conversion and reports the number of dimensions disassociated.

Command: ***DIMDISASSOCIATE***
Select dimensions to disassociate ...
Select objects: **PICK**
Select objects: **Enter**
*nn* disassociated.
Command:

You can use any selection method including *Qselect* to "Select objects." *Dimdisassociate* filters the selection set to include only associative dimensions that are in the current space (model space or paper space layout) and not on locked layers.

### *Dimregen*

| Ribbon | Menu Bar | Command (Type) | Alias (Type) | Shortcut |
|---|---|---|---|---|
| ... | ... | *Dimregen* | ... | ... |

*Dimregen* updates the locations of all associative dimensions in the current drawing. *Dimregen* does not alter the associativity features of dimensions; it affects only the display of associative dimensions. *Dimregen* may be needed to regenerate the display of associative dimensions in three cases:

- after panning or zooming with a wheel mouse within a viewport in a paper space layout when dimensions created in paper space are associated with drawing objects in model space (see Chapter 29, "Dimensioning in Paper Space Layouts");
- after opening an AutoCAD 2002 to 2017 drawing that has been modified with a previous version of AutoCAD and the dimensioned objects have been modified;
- after opening a drawing containing *Xrefs* if the external reference geometry is dimensioned in the current drawing and the *Xrefs* have been modified.

## DIMENSIONING VARIABLES INTRODUCTION

Now that you know the commands used to create dimensions, you need to know how to control the appearance of dimensions. For example, after you create dimensions in a drawing, you may need to adjust the size of the dimension text and arrowheads, or change the text style, or control the number of decimal places appearing on the dimension text.

The appearance of dimensions is controlled by setting the dimensioning variables. You can ensure an entire group of dimensions appear the same by using one dimension style for the dimensions. A dimension style is the set of variables that all the dimension objects in the group reference.

TIP

Although you can change many settings that control the appearance of dimensions, there is one dimensioning setting that is the most universal, that is, the *Scale for dimension features*. The *Scale for dimension features* controls all of the size-related features of dimension objects, such as the arrowhead size, the text size, the gap between extension lines and the object, and so on. Even though each of those features can be changed individually, changing the *Scale for dimension features* controls all of the size-related features of dimensions together.

There are three methods for controlling the scale for dimensioning features:

1. set an *Overall Scale* value;
2. make the dimensions *Annotative*;
3. set the dimension scale based on a (viewport) layout.

Using the *Overall Scale* method is the simplest and most widely used method. With this method, dimensions are displayed at a static scale—the same scale (in relation to the geometry) in all viewports and layouts. The *Overall Scale* value should be set based on the drawing scale factor. The *Overall Scale* value is stored in the *DIMSCALE* variable.

Annotative dimensions can be made to automatically change size when the *Annotation Scale* or *VP scale* is changed. This method is recommended when the drawing will be displayed in multiple viewports at different scales. This setting changes the *DIMANNO* variable (0 = off, 1 = on). This option works similarly to annotative text (Chapter 18).

The *Scale dimensions to a layout* (more specifically, scale to a viewport) method is useful when you want to set the size of dimension features based on one viewport at a time. This is an older method of automatically adjusting the dimension scale based on the viewport scale, but it is limited in that only one scale can be displayed at a time. This setting changes the *DIMSCALE* variable to 0.

**TIP**

It is highly recommended that you use the *Overall Scale* method while you learn the dimensioning basics. Since you will at least initially be creating simpler drawings to be displayed in one scale, setting the *Overall Scale* will avoid having to deal with complexities that are unnecessary as you learn the fundamentals.

To change the *Overall Scale*, open the *Dimension Style Manager* (by selecting it from the *Dimension* pull-down menu, *Dimension* toolbar, or by typing *D*). Select the *Modify* button, access the *Fit* tab, and enter the desired value in the *Use overall scale of* edit box (Fig. 28-109). This action changes the overall size for all dimensions created using that particular dimension style.

**FIGURE 28-109**

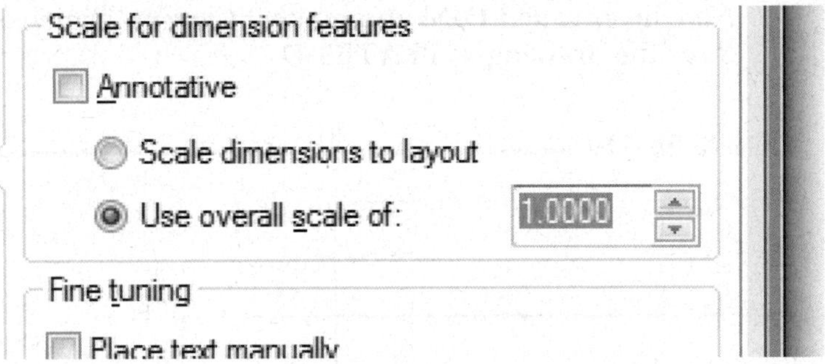


For example, assume you created a drawing and applied the first few dimensions, but noticed that the dimension text was so small that it was barely readable. Instead of changing the dimension variable that controls the size of just the dimension text, change the *Overall Scale*.

This practice ensures that the text, arrowheads, gaps, extension, and all size-related features of the dimension are changed together and are therefore correctly proportioned. Using the *Dimension Style Manager* to change this variable also ensures that all the dimensions created in that dimension style will have the same size.

Chapter 29 explains all the dimensioning variables, including *Overall Scale* and annotative dimensions, as well as dimension styles. While working on the exercises in this chapter, you will be able to create dimensions without having to change any dimension variables or dimension styles, although you may want to experiment with the *Overall Scale*. Knowledge and experience with both the dimensioning commands and dimensioning variables is essential for AutoCAD users in a real-world situation. Make sure you read Chapter 29 before attempting to dimension drawings other than those in these exercises.

## CHAPTER EXERCISES

Only four exercises are offered in this chapter to give you a start with the dimensioning commands. Many other dimensioning exercises are given at the end of Chapter 29, Dimension Styles and Variables. Since most dimensioning practices in AutoCAD require the use of dimensioning variables and dimension styles, that information should be discussed before you can begin dimensioning effectively.

The units that AutoCAD uses for the dimensioning values are based on the *Unit format* and the *Precision* settings in the *Primary Units* tab of the *Dimension Style Manager*. You can dimension these drawings using the default settings, or you can change the *Units* and *Precision* settings for each drawing to match the dimensions shown in the figures. (Type *D*, select *Modify*, then the *Primary Units* tab.)

Other dimensioning features may appear different than those in the figures because of the variables set in the AutoCAD default STANDARD dimension style. For example, the default settings for diameter and radius dimensions may draw the text and dimension lines differently than you desire. After reading Chapter 29, those features in your exercises can be changed retroactively by changing the dimension style.

REUSE

1. ***Open*** the PLATES drawing that you created in Chapter 9 Exercises. ***Erase*** the plate on the right. Use ***Move*** to move the remaining two plates apart, allowing 5 units between. Create a ***New*** layer called **DIM** and make it ***Current***. Dimension the two plates, as shown in Figure 28-110. ***Save*** the drawing as **PLATES-D**.

**FIGURE 28-110**

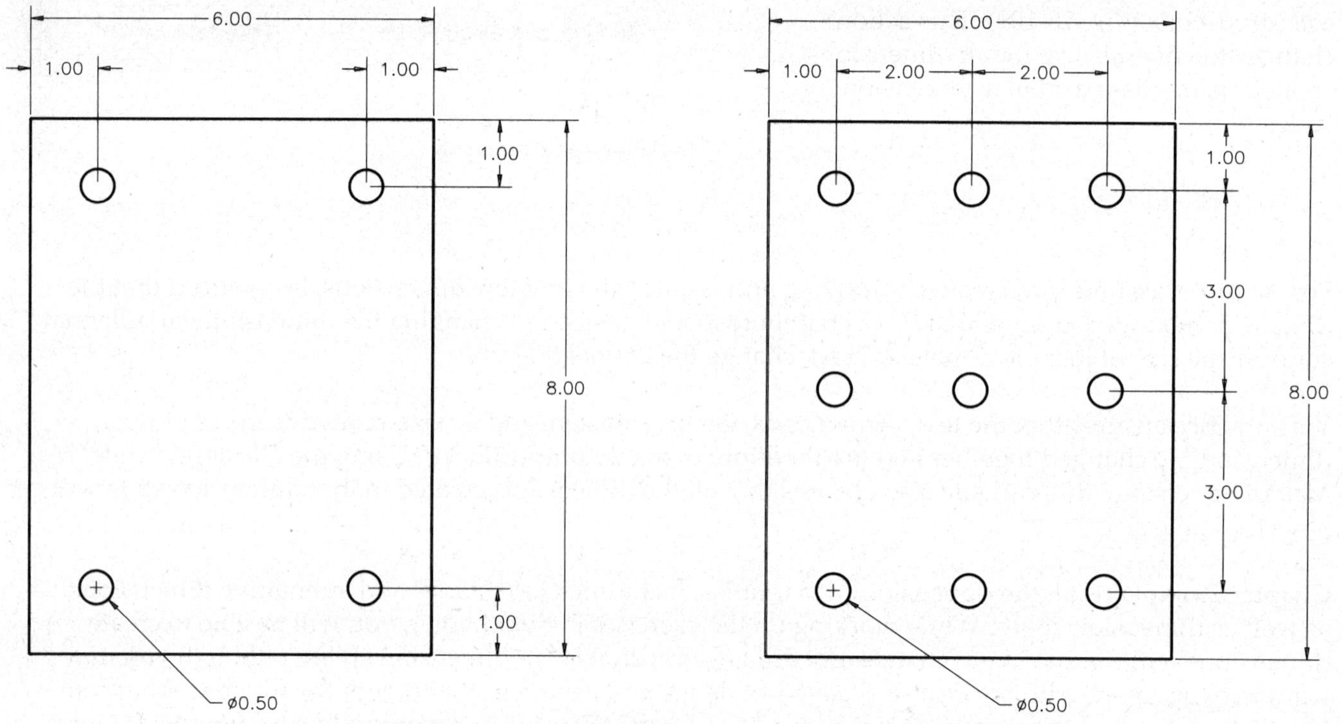

2. *Open* the **PEDIT1** drawing. Create a *New* layer called **DIM** and make it *Current*. Dimension the part as shown in Figure 28-111. *Save* the drawing as **PEDIT1-DIM**. Draw a *Pline* border of **.02** *width* and *Insert* **TBLOCK** or **TBLOCKAT**. *Plot* on an A size sheet using *Scale to Fit*.

FIGURE 28-111

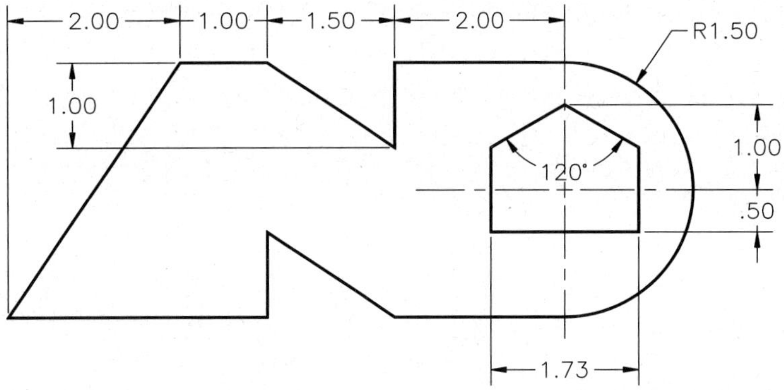


3. *Open* the **GASKETA** drawing that you created in Chapter 9 Exercises. Create a *New* layer called **DIM** and make it *Current*. Dimension the part as shown in Figure 28-112. *Save* the drawing as **GASKETD.**

Draw a *Pline* border with **.02** *width* and *Insert* **TBLOCK** or **TBLOCKAT** with an **8/11** scale factor. *Plot* the drawing *Scaled to Fit* on an A size sheet.

FIGURE 28-112

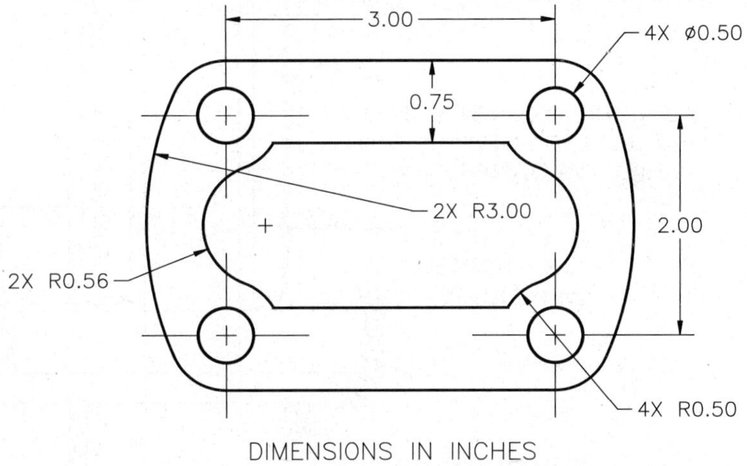


4. *Open* the **BARGUIDE** multiview drawing that you created in Chapter 24 Exercises. Create the dimensions on the **DIM** layer. Keep in mind that you have more possibilities for placement of dimensions than are shown in Figure 28-113.

FIGURE 28-113

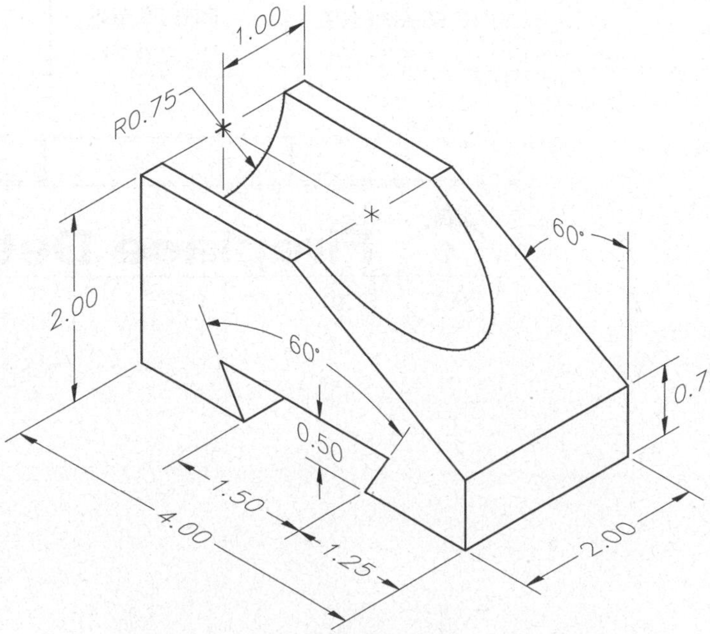

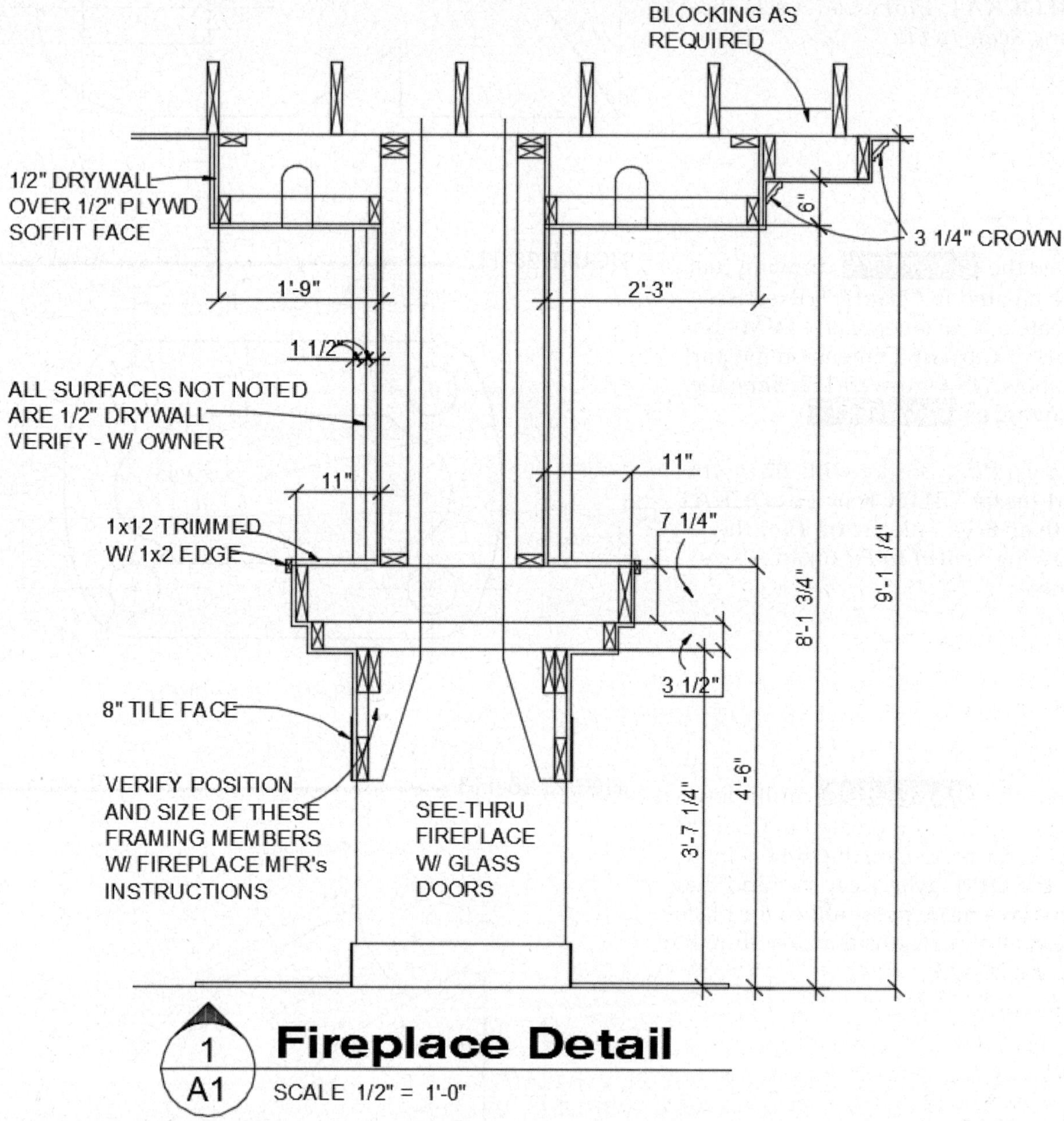


**Fireplace Detail**

SCALE 1/2" = 1'-0"

**WILHOME.DWG Courtesy of AutoDesk, Inc.**

CHAPTER

# 29

# Dimension Styles and Variables

## Chapter Objectives

After completing this chapter you should:

1. be able to control dimension variable settings using the *Dimension Style Manager*;
2. be able to save and restore dimension styles with the *Dimension Style Manager*;
3. know how to create dimension style families and specify variables for each child;
4. know how to create and apply dimension style overrides;
5. be able to modify dimensions using *Update*, *Dimoverride*, *Properties*, and *Matchprop*;
6. know the guidelines for dimensioning;
7. be able to create annotative dimensions;
8. know how to dimension isometric drawings.

# CONCEPTS

## Dimension Variables

Since a large part of AutoCAD's dimensioning capabilities are automatic, some method must be provided for you to control the way dimensions are drawn. A set of 80 dimension variables allows you to affect the way dimensions are drawn by controlling sizes, distances, appearance of extension and dimension lines, and dimension text formats.

An example of a dimension variable is *DIMSCALE* (if typed) or *Overall Scale* (if selected from a dialog box, Fig. 29-1). This variable controls the overall size of the dimension features (such as text, arrowheads, and gaps). Changing the value from 1.00 to 1.50, for example, makes all of the size-related features of the drawn dimension 1.5 times as large as the default size of 1.00 (Fig. 29-2). Other examples of features controlled by dimension variables are arrowhead type, orientation of the text, text style, units and precision, suppression of extension lines, fit of text and arrows (inside or outside of extension lines), and direction of leaders for radii and diameters (pointing in or out). The dimension variable changes that you make affect the appearance of the dimensions that you create.

**FIGURE 29-1**

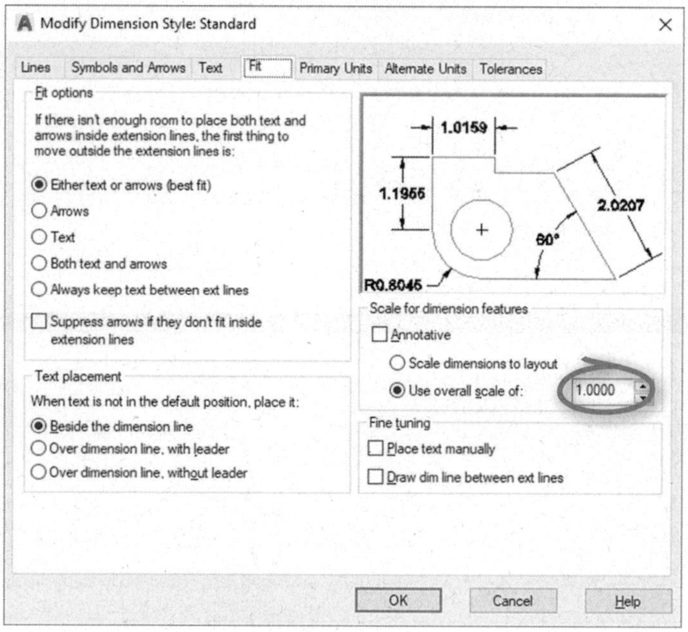


There are two basic ways to control dimensioning variables:

1. Use the *Dimension Style Manager* (Fig. 29-3).
2. Type the dimension variable name in Command line format.

The dialog boxes employ "user-friendly" terminology and selection, while the Command line format uses the formal dimension variable names but cannot be used for all settings.

**FIGURE 29-2**

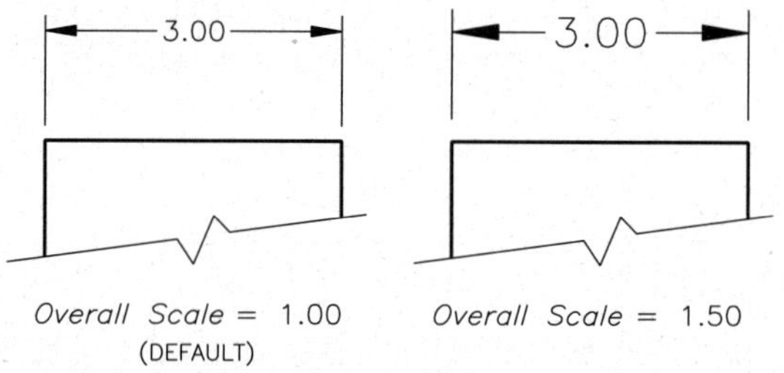


Changes to dimension variables are usually made before you create the affected dimensions. Dimension variable changes are not always retroactive, in contrast to *LTSCALE*, for example, which can be continually modified to adjust the spacing of existing non-continuous lines. Changes to dimensioning variables affect existing dimensions only when those changes are *Saved* to an existing *Dimension Style* that was in effect when previous dimensions were created. Generally, dimensioning variables should be set before creating the desired dimensions although it is possible to modify dimensions and *Dimension Styles* retroactively.

**FIGURE 29-3**

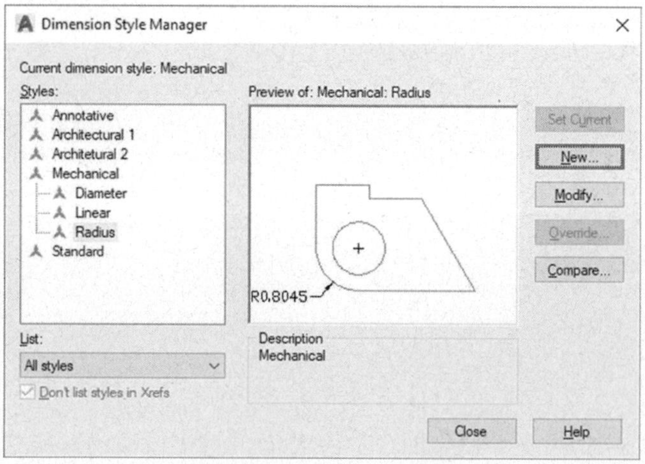

## Dimension Styles

All dimensions are part of a dimension style. The default dimension style is called *Standard* (the *Annotative* style is also supplied). The *Standard* dimension style has all the default dimension variable settings for creating dimensions with a typical size and appearance. Similar to layers, you can create, name, and specify settings for any number of dimension styles. Each dimension style contains the dimension variable settings that you select. A dimension style is a group of dimension variable settings that has been saved under a name you assign. When you set a dimension style *Current*, AutoCAD remembers and resets that particular combination of dimension variable settings.

You can create a dimension style for each particular "style" of dimension (Fig. 29-4). For example, you may want to create one dimension style for most of your normal dimensions, another dimension style for limit dimensions (as shown in the diameter dimension), and another style for dimensions without extension lines (as shown in the interior slot). Each time you want to draw a particular style of dimension, select the style name from the list, make it the "current" style, then begin drawing dimensions.

**FIGURE 29-4**

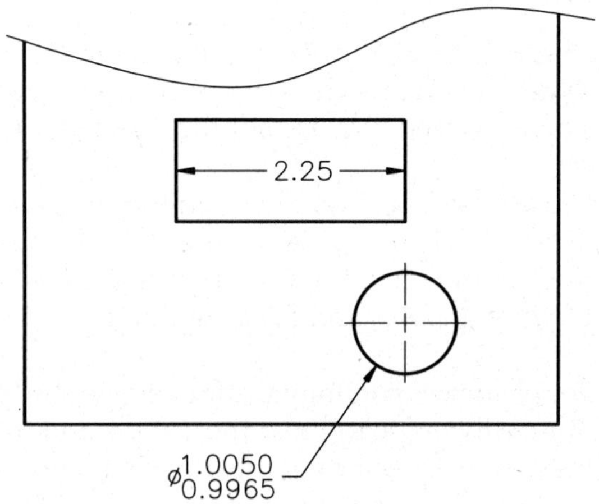


To create a new dimension style, select the *New* button in the *Dimension Style Manager* (see Fig. 29-3), then use the tabs that appear in the dialog box to specify the appearance of the dimensions. Selecting options in these tabs (see Fig. 29-14) sets the values for related dimension variables that are saved in the drawing. When a dimension style is *Set Current*, the dimensions you create with that style appear with the dimension variable settings you specified. Creating dimensions with the current *Dimension Style* is similar to creating text with the current *Text Style*—that is, the objects created take on the current settings assigned to the style.

TIP

Another advantage of using dimension styles is that existing dimensions in the drawing can be globally modified by making a change and saving it to the dimension style(s). To do this, select *Modify* in the *Dimension Style Manager* or use the *Save* option of *-Dimstyle* after changing a dimension variable in Command line format. This action saves the changes for newly created dimensions as well as automatically updating existing dimensions in the style.

## Dimension Style Families

In the previous chapter, you learned about the various types of dimensions, such as linear, angular, radial, diameter, ordinate, and leader. You may want the appearance of each of these types of dimensions to vary, for example, all radius dimensions to appear with the dimension line arrow pointing out and all diameter dimensions to appear with the dimension line arrow inside pointing in. It is logical to assume that a new dimension style would have to be created for each variation. However, a new dimension style name is not necessary for each type of dimension. AutoCAD provides dimension style families for the purpose of providing variations within each named dimension style.

The dimension style names that you create are assigned to a dimension style family. Each dimension style family can have six children. The children are linear, angular, diameter, radial, ordinate, and leader. Although the children take on the dimension variable settings assigned to the family, you can assign special settings for one or more children. For example, you can make a radius dimension appear slightly different from a linear dimension in that family. Do this by selecting the *New* button in the *Dimension Style Manager*, then selecting the type of dimension (*linear, radius, diameter*, etc.) you want to change from the *Create New Dimension Style* dialog box that appears (Fig. 29-5).

**FIGURE 29-5**

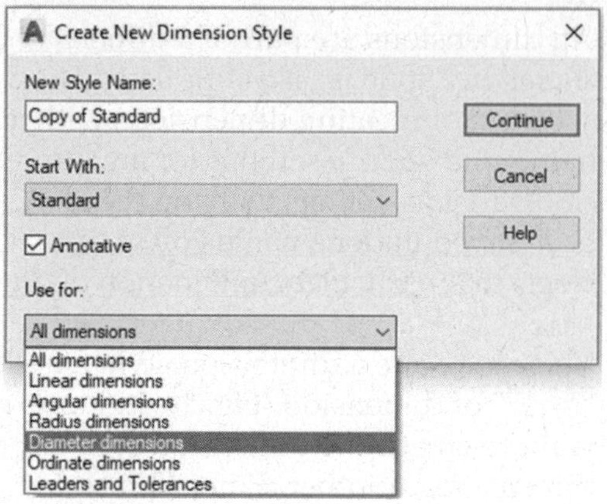


Select *Continue* to specify settings for the child. When a dimension is drawn, AutoCAD knows what type of dimension is created and applies the selected settings to that child.

Dimension styles that have children (special variable settings set for the *linear*, *radial*, *diameter*, etc. dimensions) appear in the list of styles in the *Dimension Style Manager* as sub-styles branching from the parent name. For example, notice the "Mechanical" style in Figure 29-3 has special settings for its *Diameter*, *Linear*, and *Radial* children.

For example, you may want to create the "Mechanical" dimension style to draw the arrows and dimension lines inside the arc but drag the dimension text outside of the arc for *Radial* dimensions (as shown in Fig. 29-6). To do this, create a new child for the "Mechanical" family as previously described, and set "Mechanical" as the current style. When a *Radius* dimension is drawn, the dimension line, text, and arrow should appear as shown in Figure 29-6.

**FIGURE 29-6**

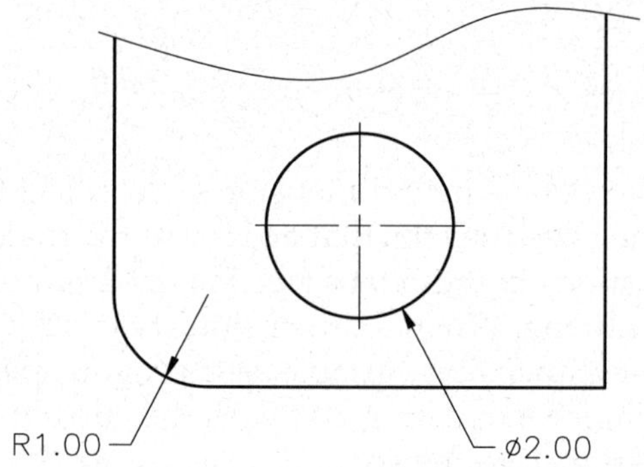


In summary, a dimension style family is simply a set of dimension variables related by name. Variations within the family are allowed, in that each child (type of dimension) can possess its own subset of variables. Therefore, each child inherits all the variables of the family in addition to any others that may be assigned individually.

## Dimension Style Overrides

If you want only a few dimensions in a dimension style to have a special appearance, you can create a dimension style override. An override does not normally affect the parent style (unless you save the override setting to the style). Therefore, existing dimensions in the drawing are not affected by a dimension style override. You can create a dimension style override two ways.

The more practical method of creating a dimension style override is to select an existing dimension, then use the *Properties* palette to change the desired variable setting. This method (explained in "Modifying Existing Dimensions") creates a dimension style override for the selected dimension only and does not affect any other dimensions in the style.

Alternately, you can create a dimension style override in the *Dimension Style Manager*, then create the dimensions. First, make the desired style *Current*, then select the *Override* button. Proceed to specify the desired variable settings from the seven tabs that appear. A new branch under the family name appears in the *Dimension Style Manager* styles list named "<style overrides>" (see Fig. 29-8 under "Architectural 2"). Finally, create the new dimensions. To clear the overrides, simply make another style current. (Alternately, type any dimension variable name at the Command line and change the value. The override is applied only to the newly created dimensions in the current style and clears when another style is made current.)

## DIMENSION STYLES

### *Dimstyle or Ddim*

| Ribbon | Menu Bar | Command (Type) | Alias (Type) | Shortcut |
|---|---|---|---|---|
| *Annotate Dimensions* (arrow) | *Dimension Dimension Style...* | *Dimstyle or Ddim* | *DIMSTY, DST*, or *D* | ... |

The *Dimstyle* or *Ddim* command produces the *Dimension Style Manager* (Fig. 29-7). This is the primary interface for creating new dimension styles and making existing dimension styles current. This dialog box also gives access to seven tabs that allow you to change dimension variables. Features of the *Dimension Style Manager* that appear on the opening dialog box (shown in Fig. 29-7) are described in this section. The tabs that allow you to change dimension variables are discussed in detail later in this chapter in the section "Dimension Variables."

**FIGURE 29-7**

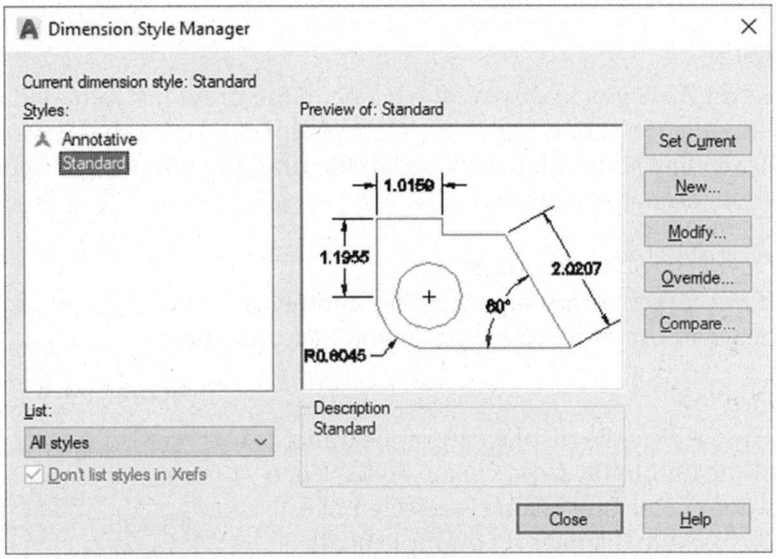


The *Dimension Style Manager* makes the process of creating and using dimension styles easier and more visual than using Command line format. Using the *Dimension Style Manager*, you can *Set Current*, create *New* dimension styles, *Modify* existing dimension styles, and create an *Override*. You also have the ability to view a list of existing styles including children and overrides; see a preview of the selected style, child, or override; examine a description of a dimension style as it relates to any other style; and make an in-depth comparison between styles. You can also control the entries in the list to include or not include Xreferenced dimension styles. The options are explained next.

#### *Styles*

This list displays all existing dimension styles in the drawing, including *Xrefs*, depending on your selection in the *List* section below. The list includes the parent dimension styles, the children that are shown on a branch below the parent style, and any overrides also branching from family styles.

The buttons on the right side of the *Dimension Style Manager* affect the highlighted style in the list. For example, to modify an existing style, select the style name from the list and select the *Modify* button. If you want to create a *New* style based on an existing style, select the style name you want to use as the "template," then select *New*.

Select (highlight) any dimension style from the *Styles* list to display the appearance of dimensions for that style in the *Preview* tile. Selecting a style from the list also makes the *Description* area list the dimension variables settings compared to the current dimension style.

**FIGURE 29-8**

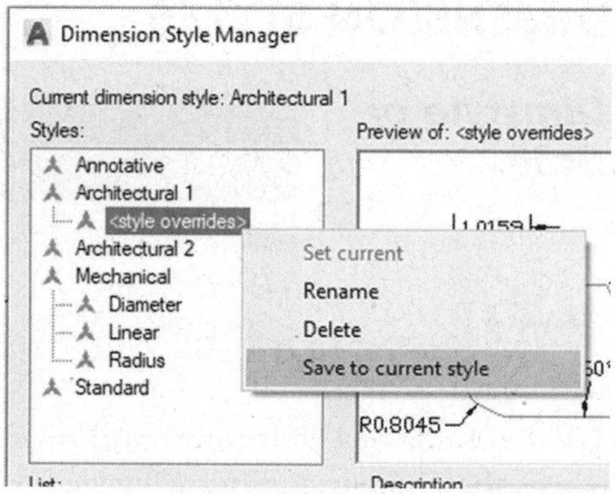


TIP

You can right-click to use a shortcut menu for the selected dimension style (Fig. 29-8). The shortcut menu allows you to *Set Current*, *Rename*, or *Delete* the selected dimension style. Only dimension styles that are unreferenced (no dimensions have been created using the style) can be deleted.

If you have created a style override (by pressing the *Override* button, then setting variables), you can save the overrides to the dimension style using the right-click shortcut menu. Normally, overrides are automatically discarded when another dimension style is made current. This shortcut menu is the only method available in the *Dimension Style Manager* to save overrides. (You can also use the *-Dimstyle* command to save overrides.)

### *List*

Select *All Styles* to display the list of all the previously created or imported dimension styles in the drawing. Selecting *Styles In Use* displays only styles that have been used to create dimensions in the drawing—styles that are saved in the drawing but are not referenced (no dimensions created in the style) do not appear in the list.

### *Don't List Styles in Xrefs*

If the current drawing references another drawing (has an *Xref* attached), the *Xref's* dimension styles can be listed or not listed based on your selection here.

### *Preview*

The *Preview* tile displays the appearance of the highlighted dimension style. When a dimension style name is selected, the *Preview* tile displays an example of all dimension types (linear, radial, etc.) and how the current variable settings affect the appearance of those dimensions. If a child is selected, the *Preview* tile displays only that dimension type and its related appearance. For example, Figure 29-9 displays the appearance of the selected child, *Mechanical: Radial*.

**FIGURE 29-9**

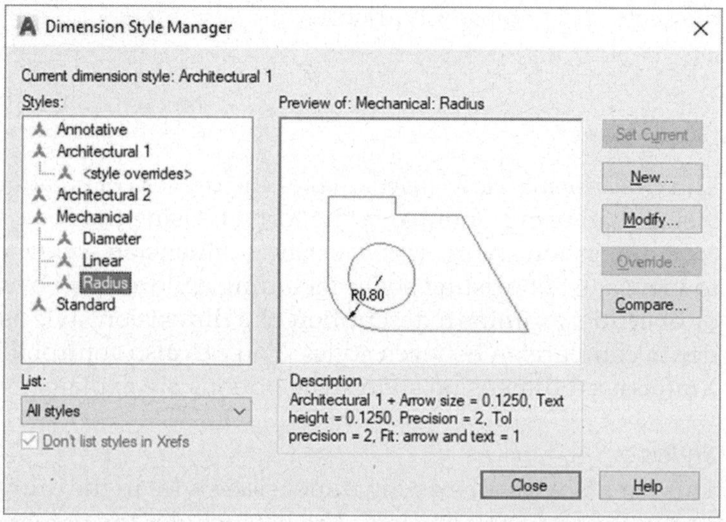


### *Description*

The description area is a valuable tool for determining the variable settings for a dimension style. When a dimension style from the list is selected, the *Description* area lists the differences in variable settings from the current style.

For example, assume the "Architectural 2" dimension style was created using "Standard" as the template. To display the dimension variable settings that have been changed (the differences between Standard and Architectural 2), make Standard the current style, then select Architectural 2 and examine the description area, as shown in Figure 29-10. You can use the same procedure to display the variable settings assigned to a child (the differences between the family style and the child), as shown in the *Description* area in Figure 29-9.

**FIGURE 29-10**

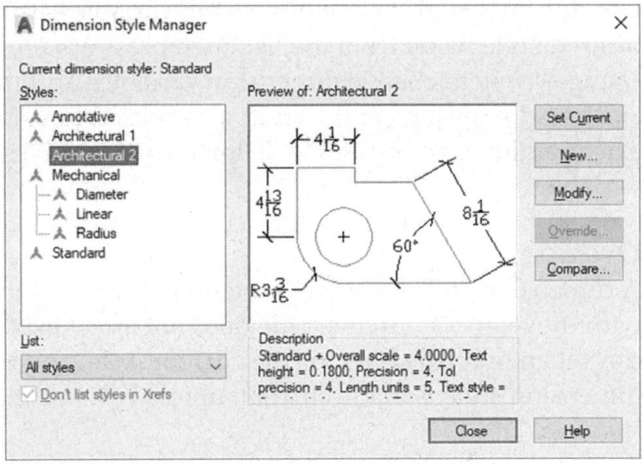


### *Set Current*

Select the desired existing style from the list on the left, then select *Set Current*. The current style is listed at the top of the *Dimension Style Manager*. When you *Close* the dialog box, the current style is used for drawing new dimensions. If you want to set an *Override* to an existing style, you must first make it the current style.

### *New*

Use the *New* button to create a new dimension style. The *New* button invokes the *Create New Dimension Style* dialog box (Fig. 29-11). The options are discussed next.

**FIGURE 29-11**

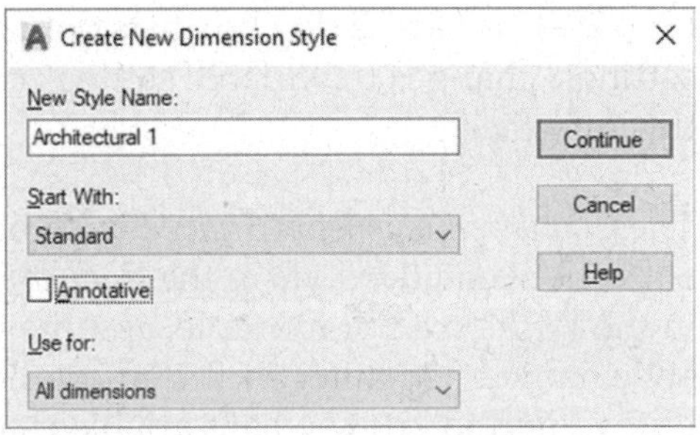


*New Style Name*

Enter the desired new name in the *New Style Name* edit box. Initially the name that appears is "Copy of (current style)."

*Start With*

After assigning a new name, select any existing dimension style to use as a "template" from the *Start With* drop-down list. Initially (until dimension variable changes are made for the new style), the new dimension style is actually a copy of the *Start With* style (has identical variable settings). By default, the current style appears in the *Start With* edit box.

*Use For*

If you want to create a dimension style family, select *All Dimensions*. In this way, all dimension types (*linear, radial, diameter*, etc.) take on the new variable settings. If you want to specify special dimension variable settings for a child, select the desired dimension type from the drop-down list (see Fig. 29-5).

*Continue*

When the new name and other options have been selected, press the *Continue* button to invoke the *New Dimension Style* dialog box. Here you select from the seven tabs to specify settings for any dimension variable to apply to the new style. See the "Dimension Variables" section for information on setting dimension variables.

NOTE: Generally you should not change the Standard dimension style. Rather than change the Standard style, create *New* dimension styles using Standard as the base style. In this way, it is easy to restore and compare to the default settings by making the Standard style current. If you make changes to the Standard style, it can be difficult to restore the original settings.

TIP

### *Modify*

Use this button to change dimension variable settings for an existing dimension style. First, select the desired style name from the list, then press *Modify* to produce the *Modify Dimension Style* dialog box. Using *Modify* to change dimension variables automatically saves the changes to the style and updates existing dimensions in the style, as opposed to using *Override*, which does not change the style or update existing dimensions. See the "Dimension Variables" section for information on setting dimension variables.

### *Override*

A dimension style override is a temporary variable setting that affects only new dimensions created with the override. An override does not affect existing dimensions previously created with the style. You can set an override only for the current style, and the overrides are automatically cleared when another dimension style is made current, unless you use the right-click shortcut menu and select *Save to current style* (see Fig. 29-8).

To create a dimension style override, you must first select an existing style from the list and make it the current style. Next, select *Override* to produce the *Override Current Style* dialog box. See the "Dimension Variables" section for information on setting dimension variables to act as overrides.

### *Compare*

TIP

This feature is very useful for examining variable settings for any dimension style. Press the *Compare* button to produce the *Compare Dimension Styles* dialog box (Fig. 29-12). This dialog box lists only dimension variable settings; changes to variable settings cannot be made from this interface.

The *Compare Dimension Styles* dialog box lists the differences between the style in the *Compare* box and the style in the *With* box. For example, assume you created a new style named "Architectural 2" from the "Standard" style, but wanted to know what variable changes you had made to the new style. Select Architectural 2 in the *Compare* box and Standard in the *With* box. The variable changes made to Architectural 2 are displayed in the central area of the box. Also note that the settings for each variable are listed for each of the two styles.

A useful feature of this dialog box is that the formal variable names are listed in the *Variable* column. Many experienced users of AutoCAD prefer to use the formal variable names since they do not change from release to release.

TIP

You can also use the *Compare Dimension Styles* dialog box to give the entire list of variables and related settings for one dimension style. Do this by selecting the desired dimension style name from the *Compare* drop-down list and select <*none*> from the *With* list (Fig. 29-13).

**FIGURE 29-12**

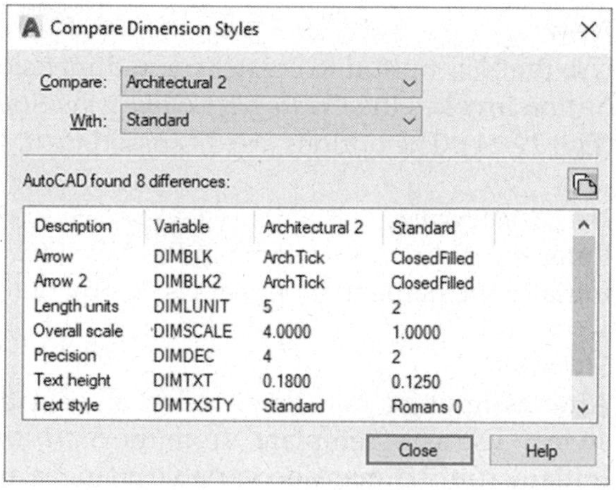


**FIGURE 29-13**

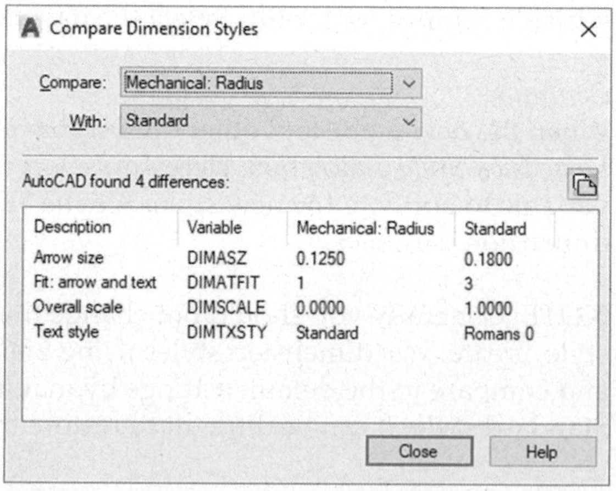

## -Dimstyle

| Ribbon | Menu Bar | Command (Type) | Alias (Type) | Shortcut |
|---|---|---|---|---|
| ... | ... | *-Dimstyle* | ... | ... |

The *-Dimstyle* command is the Command line equivalent to the *Dimension Style Manager*. Operations that are performed by the dialog box can also be accomplished with the *-Dimstyle* command. The *Apply* option offers one other important feature that is not available in the dialog box:

Command: ***-DIMSTYLE***
Current dimension style: Standard
Enter a dimension style option [ANnotative/Save/Restore/STatus/Variables/Apply/?] <Restore>:

### *Save*

Use this option to create a new dimension style. The new style assumes all of the current dimension variable settings comprised of those from the current style plus any other variables changed by Command line format (overrides). The new dimension style becomes the current style.

Enter a dimension style option [ANnotative/Save/Restore/STatus/Variables/Apply/?] <Restore>: ***s***
Enter name for new dimension style or [?]:

### *Restore*

This option prompts for an existing dimension style name to be restored. The restored style becomes the current style. You can use the "select dimension" option to PICK a dimension object that references the style you want to restore.

### *STatus*

*Status* gives the current settings for all dimension variables. The displayed list comprises the settings of the current dimension style and any overrides. All dimension variables that are saved in dimension styles are listed:

Enter a dimension style option [ANnotative/Save/Restore/STatus/Variables/Apply/?] <Restore>: ***st***

| | | |
|---|---|---|
| DIMASO | Off | Create dimension objects |
| DIMSTYLE | Standard | Current dimension style (read-only) |
| DIMADEC | 0 | Angular decimal places |
| DIMALT | Off | Alternate units selected |

etc.

### *Variables*

You can use this option to list the dimension variable settings for any existing dimension style. You cannot modify the settings.

Enter a dimension style option [ANnotative/Save/Restore/STatus/Variables/Apply/?] <Restore>: ***v***
Enter a dimension style name, [?] or <select dimension>:

Enter the name of any style or use "select dimension" to PICK a dimension object that references the desired style. A list of all variables and settings appears.

### *~stylename*

TIP

This variation can be entered in response to the *Variables* option. Entering a dimension style name preceded by the tilde symbol (~) displays the differences between the current style and the (~) named style.

For example, assume that you wanted to display the differences between Standard and the current dimension style (Mechanical, for example). Use *Variables* with the ~ symbol. (The ~ symbol is used as a wildcard to mean "all but.") This option is very useful for keeping track of your dimension styles and their variable settings.

Command: ***-DIMSTYLE***
Current dimension style: Mechanical
Enter a dimension style option [ANnotative/Save/Restore/STatus/Variables/Apply/?] <Restore>: ***v***
Enter a dimension style name, [?] or <select dimension>: ***~standard***
Differences between STANDARD and current settings:

| | STANDARD | Current Setting |
|---|---|---|
| DIMSCALE | 1.0000 | 0.7500 |
| DIMUPT | Off | On |
| DIMTXSTY | Standard | Mech1 |

### *Apply*

Use *Apply* to update dimension objects that you PICK with the current variable settings. The current variable settings can contain those of the current dimension style plus any overrides. The selected object loses its reference to its original style and is "adopted" by the applied dimension style family.

Enter a dimension style option [ANnotative/Save/Restore/STatus/Variables/Apply/?] <Restore>: ***a***
Select objects:

TIP

This is a useful tool for changing an existing dimension object from one style to another. Simply make the desired dimension style current, then use *Apply* and select the dimension objects to change to that style.

TIP

Using the *Apply* option of *-Dimstyle* is identical to using the *Update* button available on the *Dimensions* panel, *Annotate* tab of the Ribbon. See "*-Dimstyle, Apply*" in the "Modifying Existing Dimensions" section later in this chapter.

### *ANnotative*

This option creates annotative dimensions. Annotative dimensions, like annotative text and other annotative objects, can change size when the *Annotative Scale* or the *VP Scale* is changed. Annotative dimensions are useful when you want to show the drawing geometry (including dimensions) in multiple viewports at different scales. See "Annotative Dimensions."

Enter a dimension style option [ANnotative/Save/Restore/STatus/Variables/Apply/?] <Restore>: ***an***
Create annotative dimension style [Yes/No] <Yes>: **Enter**
Enter name for new dimension style or [?]: (enter name)
Command:

## DIMENSION VARIABLES

Now that you understand how to create and use dimension styles and dimension style families, let's explore the dimension variables. Two primary methods can be used to set dimension variables: the *Dimension Styles Manager* and Command line format. First, we will examine the dialog box (accessible through the *Dimension Styles Manager*) that allows you to set dimension variables. Dimension variables can alternately be set by typing the formal name of the variable (such as *DIMSCALE*) and changing the desired value (discussed after this section on dialog boxes). The dialog box offers a more "user-friendly" terminology, edit boxes, checkboxes, drop-down lists, and a preview image tile that automatically reflects the variables changes you make.

## Changing Dimension Variables Using the Dialog Box Method

In this section, the dialog box that contains the seven tabs for changing variables is explained (see Fig. 29-14). The seven tabs indicate different groups of dimension variables. The tab names are:

*Lines, Symbols and Arrows, Text, Fit, Primary Units, Alternate Units, Tolerances*

Access to this dialog box from the *Dimension Style Manager* is accomplished by selecting *New, Modify,* or *Override*. Practically, there is only one dialog box that allows you to change dimension variables; however, you might say there are three dialog boxes since the title of the box changes based on your selection of *New, Modify,* or *Override*. Depending on your selection, you can invoke the *Create Dimension Style, Modify Dimension Style,* or *Override Current Style* dialog box. The options in the boxes are identical and only the titles are different; therefore, we will examine only one dialog box.

In the following pages, the heading for the paragraphs below include the dialog box option and the related formal dimension variable name in parentheses. The following figures typically display the AutoCAD default setting for a particular variable and an example of changing the setting to another value. These AutoCAD default settings (in the "Standard" dimension style) are for the ACAD.DWT template drawing and for *Start from Scratch, Imperial Default Settings*. If you use the *Metric Default Settings* or other template drawings, dimension variable settings may be different based on dimension styles that exist in the drawing other than Standard. See "Using *Setup Wizards* and Template Drawings" later in this chapter.

### *Lines* Tab

The options in this tab change the appearance of dimension lines and extension lines (Fig. 29-14). The preview image automatically changes based on the settings you select in the *Dimension Lines* and *Extension Lines* sections.

FIGURE 29-14

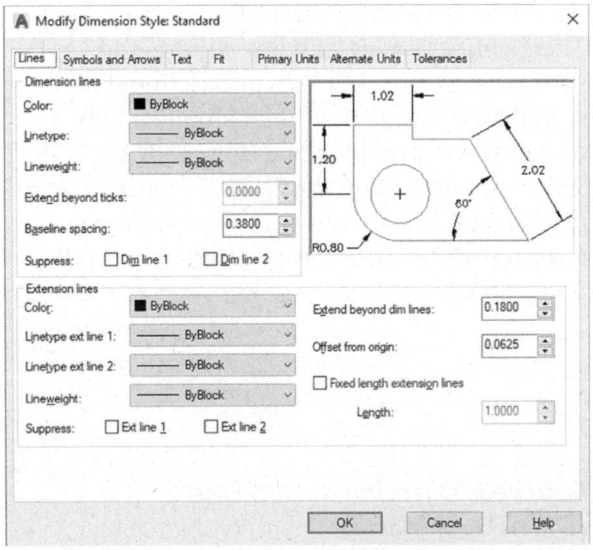


### *Dimension Lines* Section (*Lines* Tab)

#### Color (DIMCLRD)

The *Color* drop-down list allows you to choose the color for the dimension line. Assigning a specific color to the dimension lines, extension lines, and dimension text gives you more control when color-dependent plot styles are used because you can print or plot these features with different line widths, colors, etc. This feature corresponds to the *DIMCLRD* variable (dim color dimension line).

FIGURE 29-15

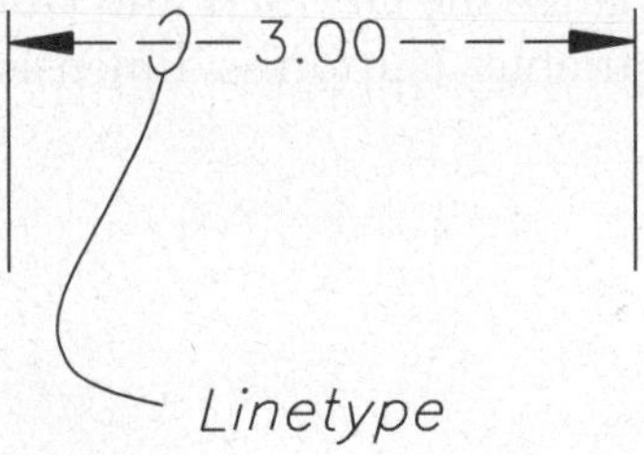


#### Linetype (DIMLTYPE)

Generally, linetypes for dimensions are *Continuous*; however, for special applications you may want to change this setting (Fig. 29-15). Select from the drop-down list. This setting is stored in the *DIMLTYPE* variable.

*Lineweight (DIMLWD)*

Use this option to assign lineweight to dimension lines (Fig. 29-16). Select any lineweight from the drop-down list or enter values in the *DIMLWD* variable (dim lineweight dimension line). Values entered in the variable can be -1 (*ByLayer*), -2 (*ByBlock*), 25 (*Default*), or any integer representing 100th of mm, such as 9 for .09mm.

**FIGURE 29-16**

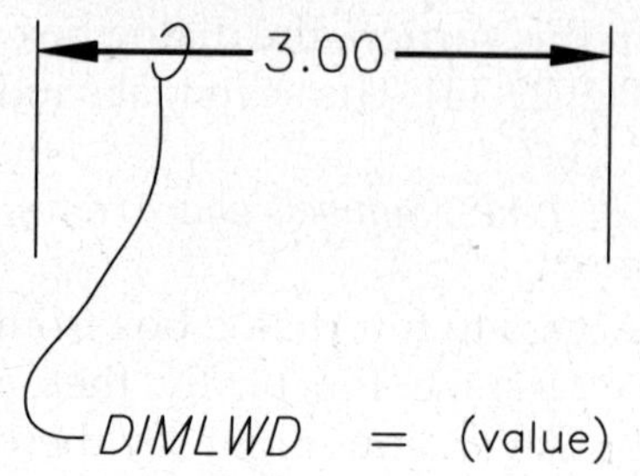


*Extend beyond ticks (DIMDLE)*

The *Extend beyond ticks* edit box is disabled unless the *Oblique, Integral,* or *Architectural Tick* arrowhead type is selected (in the *Arrowheads* section). The *Extend* value controls the length of dimension line that extends past the dimension line (Fig. 29-17). The value is stored in the *DIMDLE* variable (dimension line extension). Generally, this value does not require changing since it is automatically multiplied by the *Overall Scale* (*DIMSCALE*).

**FIGURE 29-17**

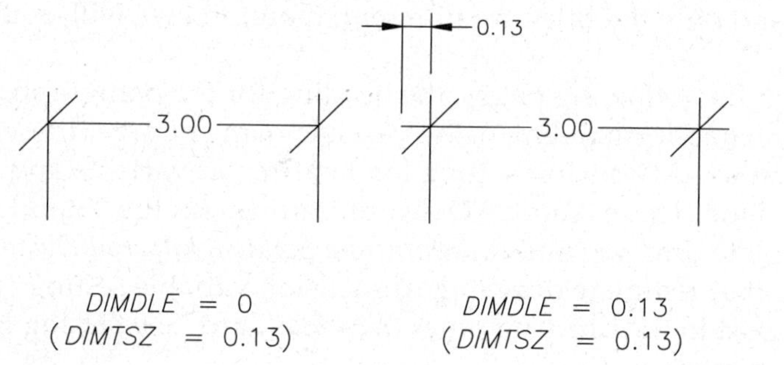


*Baseline spacing (DIMDLI)*

The *Baseline spacing* edit box reflects the value that AutoCAD uses in baseline dimensioning to "stack" the dimension line above or below the previous one (Fig. 29-18). This value is held in the *DIMDLI* variable (dimension line increment). This value rarely requires input since it is affected by *Overall Scale* (*DIMSCALE*).

**FIGURE 29-18**

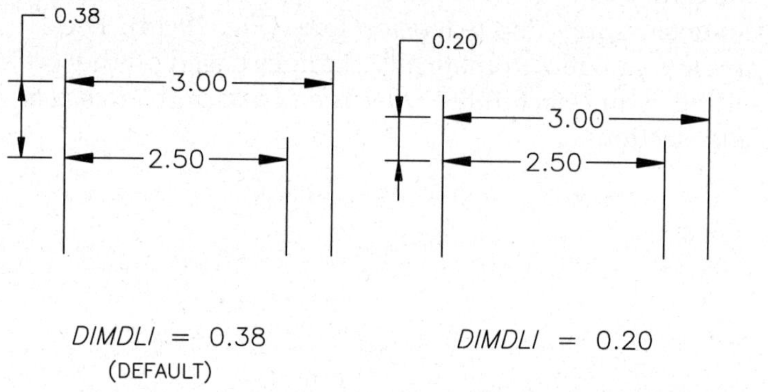


*Suppress Dim Line 1, Dim Line 2 (DIMSD1, DIMSD2)*

This area allows you to suppress (not draw) the *1st* or *2nd* dimension line or both (Fig. 29-19). The first dimension line would be on the "First extension line origin" side or nearest the end of object PICKed in response to "Select object to dimension." These toggles change the *DIMSD1* and *DIMSD2* dimension variables (suppress dimension line 1, 2).

**FIGURE 29-19**

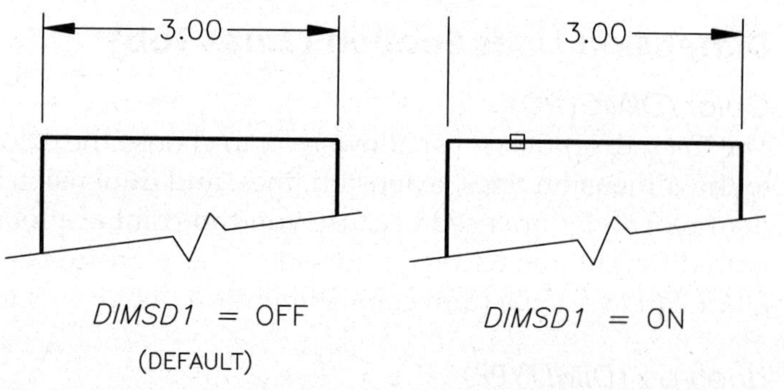

## *Extension Line* Section (*Lines* Tab)

### *Color (DIMCLRE)*

The *Color* drop-down list allows you to choose the color for the extension lines. You can also activate the standard *Select Color* dialog box. The color assignment for extension lines is stored in the *DIMCLRE* variable (dim color extension line).

### *Linetype Ext Line 1, Ext Line 2 (DIMLTEX1, DIMLTEX2)*

These drop-down lists allow you to set the linetype for the first and second extension lines (Fig. 29-20). The linetype settings for extension line 1 and extension line 2 are stored in the *DIMLTEX1* and *DIMLTEX2* variables, respectively.

**FIGURE 29-20**

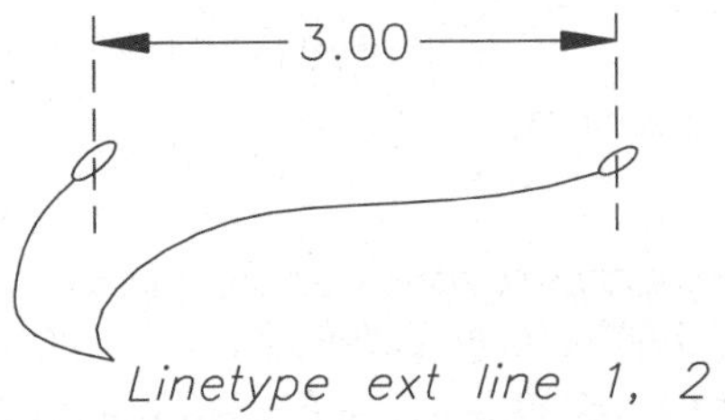


### *Lineweight (DIMLWE)*

This option is similar to that for dimension lines, only the lineweight is assigned to extension lines (Fig. 29-21). Select any lineweight from the drop-down list or enter values in the *DIMLWE* (dim lineweight extension line) variable (*ByLayer* = -1, *ByBlock* = -2, *Default* = 25, or any integer representing 100th of mm).

**FIGURE 29-21**

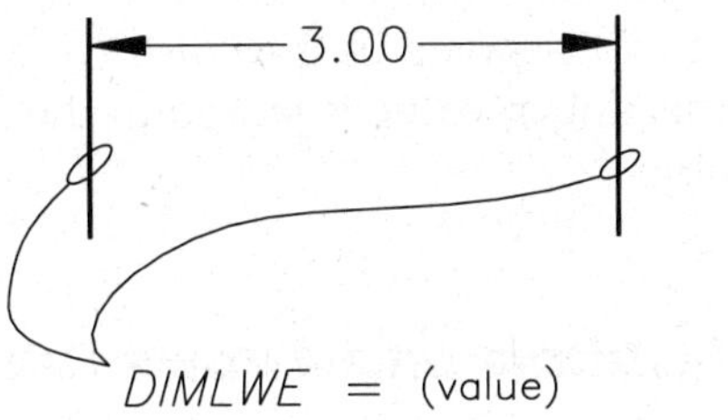


### *Suppress Ext Line 1, Ext Line 2 (DIMSE1, DIMSE2)*

This area is similar to the *Dimension Line* area that controls the creation of dimension lines but is applied to extension lines. This area allows you to suppress the *1st* or *2nd* extension line or both (Fig. 29-22). These options correspond to the *DIMSE1* and *DIMSE2* dimension variables (suppress extension line 1, 2).

**FIGURE 29-22**

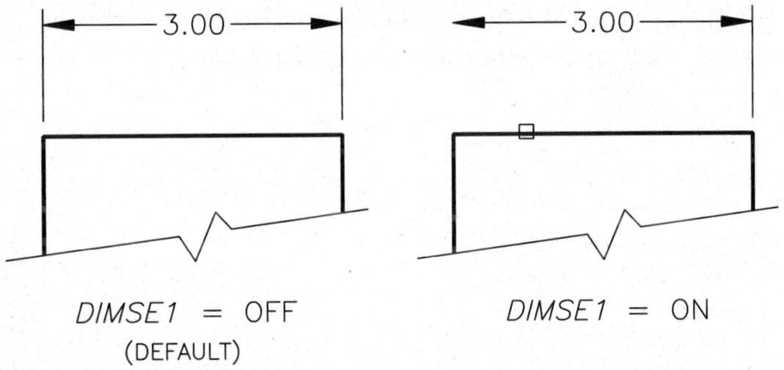


### *Extend beyond dim lines (DIMEXE)*

The *Extend beyond dim lines* edit box reflects the value that AutoCAD uses to set the distance for the extension line to extend beyond the dimension line (Fig. 29-23). This value is held in the *DIMEXE* variable (extension line extension). Generally, this value does not require changing since it is automatically multiplied by the *Overall Scale* (*DIMSCALE*).

**FIGURE 29-23**

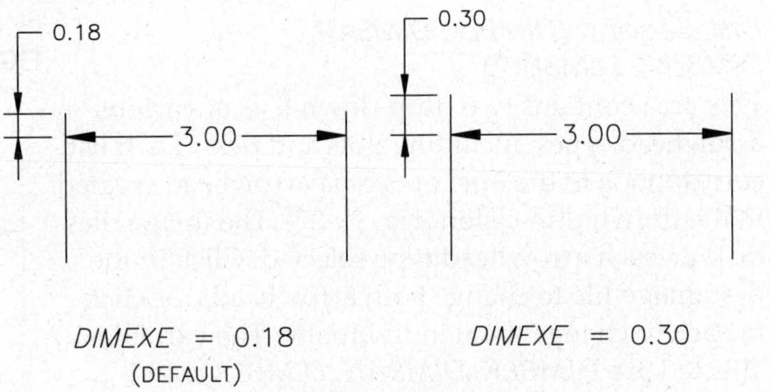

### *Offset from origin (DIMEXO)*

The *Offset from origin* value specifies the distance between the origin points and the extension lines (Fig. 29-24). This offset distance allows you to PICK the object corners, yet the extension lines maintain the required gap from the object. This value rarely requires input since it is affected by *Overall Scale* (*DIMSCALE*). The value is stored in the *DIMEXO* variable (extension line offset).

**FIGURE 29-24**

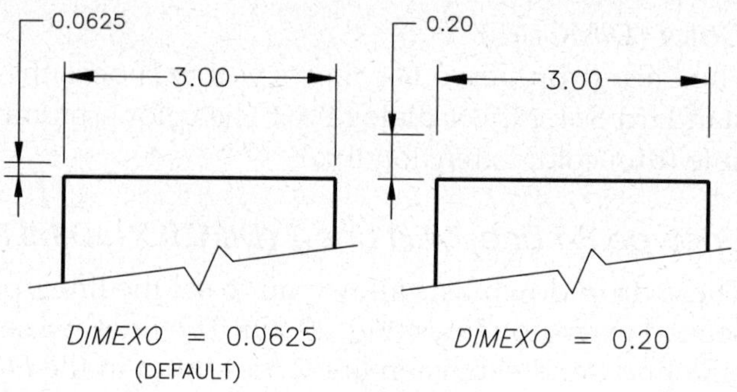


### *Fixed Length Extension Lines (DIMFXLON, DIMFXL)*

This checkbox creates dimensions with extension lines all the same length (Fig. 29-25). The *Length* value determines the length of the extension lines measured from the dimension line toward the "extension line origin." The checkbox setting (on or off) is saved in the *DIMFXLON* variable, and the length value is stored in the *DIMFXL* variable.

**FIGURE 29-25**

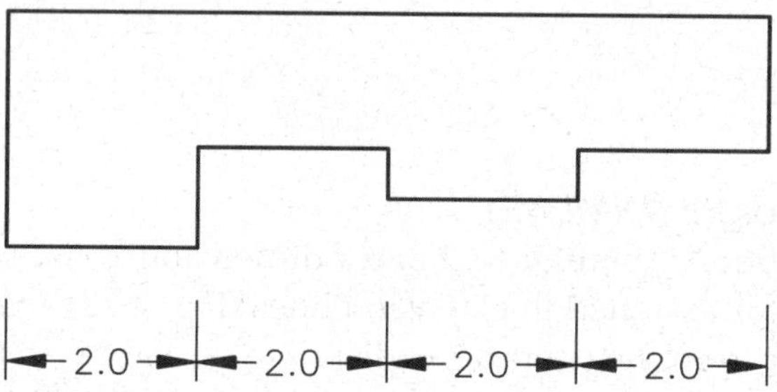


## *Symbols and Arrows* Tab

Use this tab to set your preferences for the appearance of arrowheads, center marks, and special arc and radius symbols (Fig. 29-26). The image tile updates to display your choices for the *Arrowheads* and *Center Marks* sections only.

**FIGURE 29-26**

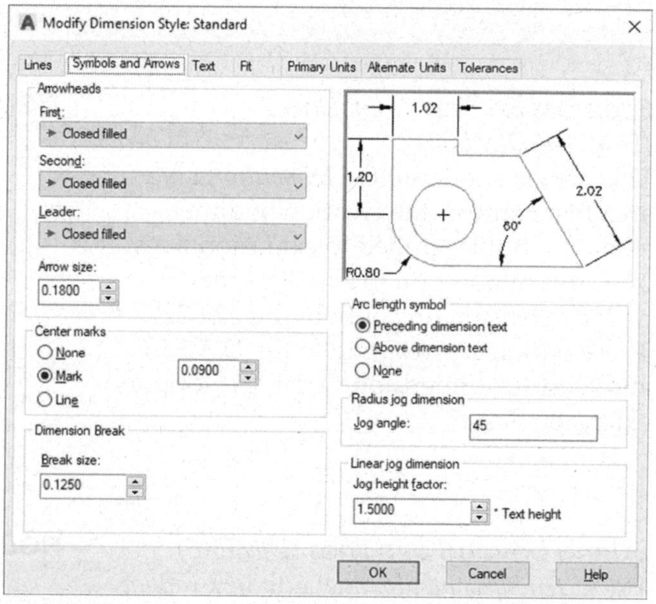


## *Arrowheads* Section (*Symbols and Arrows* Tab)

### *First, Second (DIMBLK, DIMSAH, DIMBLK1, DIMBLK2)*

This area contains two drop-down lists of various arrowhead types, including dots and ticks. Each list corresponds to the *First* or *Second* arrowhead created in the drawn dimension (Fig. 29-27). The image tiles display each arrowhead type selected. Click in the first image tile to change both arrowheads, or click in each to change them individually. The variables affected are *DIMBLK, DIMSAH, DIMBLK1,* and *DIMBLK2*.

**FIGURE 29-27**

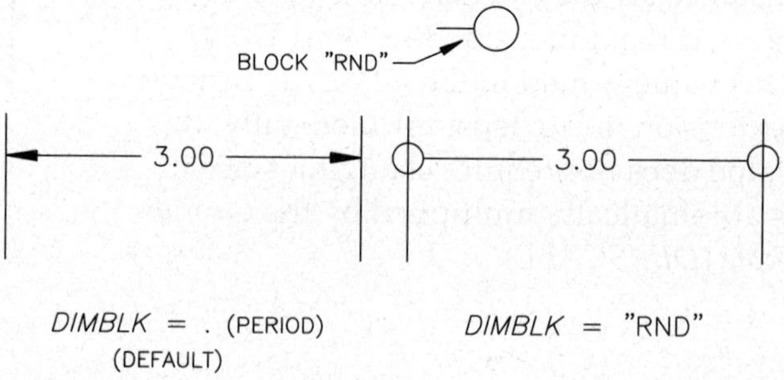

*DIMBLK* specifies the *Block* to use for arrowheads if both are the same (Fig. 29-27). When *DIMSAH* is on (separate arrowheads), separate arrowheads are allowed for each end defined by *DIMBLK1* and *DIMBLK2*. You can enter any existing *Block* name in the *DIMBLK* variables.

### *Leader (DIMLDRBLK)*

This drop-down list specifies the arrow type for leaders (Fig. 29-28). This setting does not affect the arrowhead types for dimension lines. Select any option from the drop-down list or enter a value in the *DIMLDRBLK* variable (dim leader block). (For a list of arrowhead entries, see "Dimension Variables Table.")

**FIGURE 29-28**

### *Arrow size (DIMASZ)*

The size of the arrow can be specified in the *Arrow size* edit box. The *DIMASZ* variable (dim arrow size) holds the value (Fig. 29-29). Remember that this value is multiplied by *Overall Scale* (*DIMSCALE*).

**FIGURE 29-29**

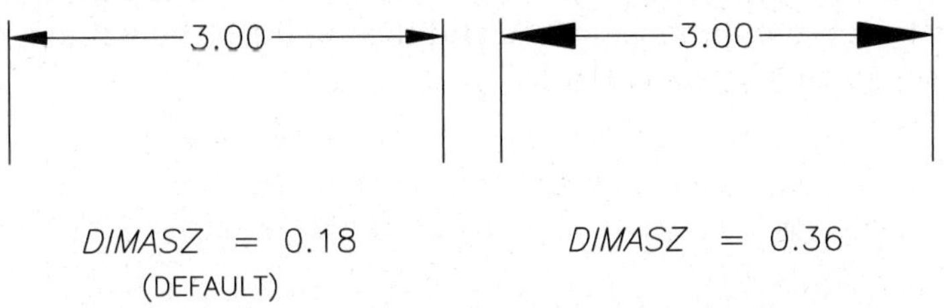


## Other Sections (*Symbols and Arrows* Tab)

### *Center Marks (DIMCEN)*

The list here determines how center marks are drawn when the dimension commands *Dimcenter*, *Dimdiameter*, or *Dimradius* are used (not *Centerline* or *Centermark*). The image tile displays the *Mark*, *Line*, or *None* feature specified. This area actually controls the value of one dimension variable, *DIMCEN* by using a 0, positive, or negative value (Fig. 29-30). The *None* option enters a *DIMCEN* value of 0.

**FIGURE 29-30**

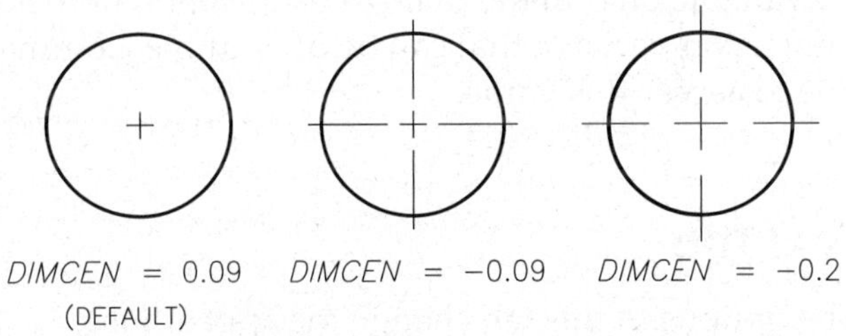


### *Size (DIMCEN)*

The *Size* edit box controls the size of the short dashes and extensions past the arc or circle. The value is stored in the *DIMCEN* variable (Fig. 29-30). Only positive values can be entered in this edit box.

### *Break Size*

This value determines the size of the break when the *Dimbreak* command is used to create a space in a dimension line. The *Size* represents the length of line that is broken (Fig. 29-31). This setting is available only in the dialog box format. There is no dimension variable that can used at the Command line to access this value.

**FIGURE 29-31**

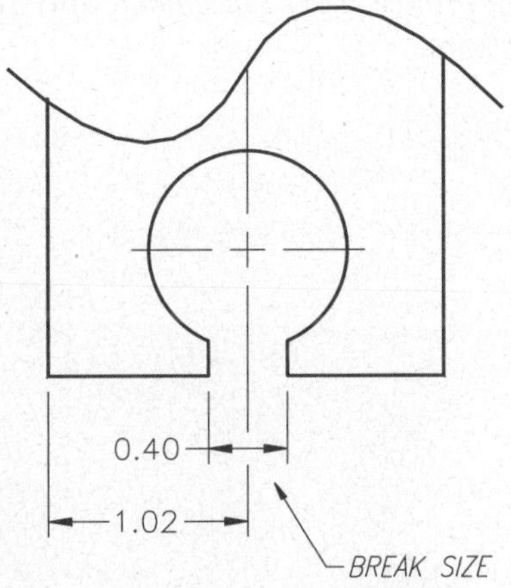

### *Arc Length Symbol (DIMARCSYM)*

Use this option to determine the presence and location of the symbol that appears for an *Arc Length* dimension. Specify the placement of the arc length symbol by selecting *Preceding Dimension Text* or *Above Dimension Text* (Fig. 29-32). Select *None*
to suppress the display of the arc length symbol. The dialog box settings correspond to *DIMARCSYM* variable settings of 0 (preceding), 1 (above), and 2 (off).

**FIGURE 29-32**

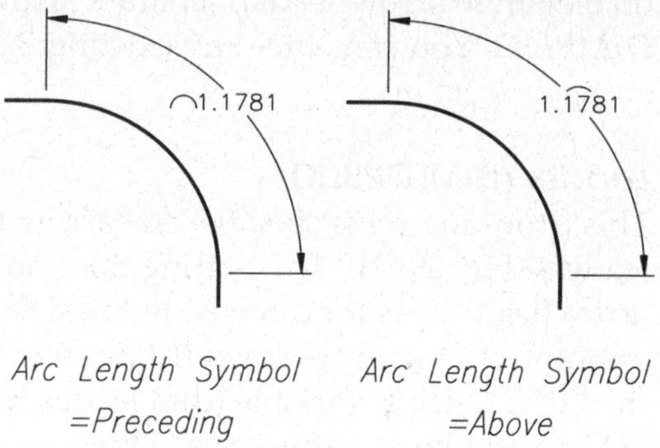


### *Jog Angle (DIMJOGANG)*

This setting specifies the angle of the "jog" line for *Jogged* radius dimensions (Fig. 29-33). The *Jog Angle* value determines the angle between the dimension line and the jog line. The value is saved in the *DIMJOGANG* variable.

**FIGURE 29-33**

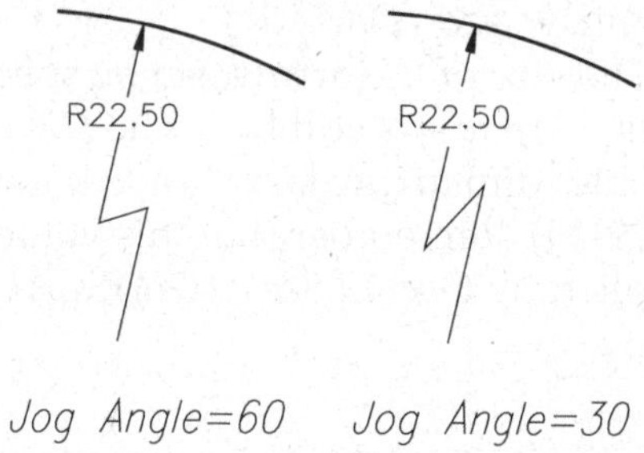


### *Jog Height Factor*

This setting affects dimensions that have a linear jog applied by the *Dimjogline* command. The value in the edit box specifies a multiplier for the text height to dictate the height for jog lines (Fig. 29-34). This setting is available only in the dialog box format. There is no dimension variable that can be used at the Command line to access this value.

**FIGURE 29-34**

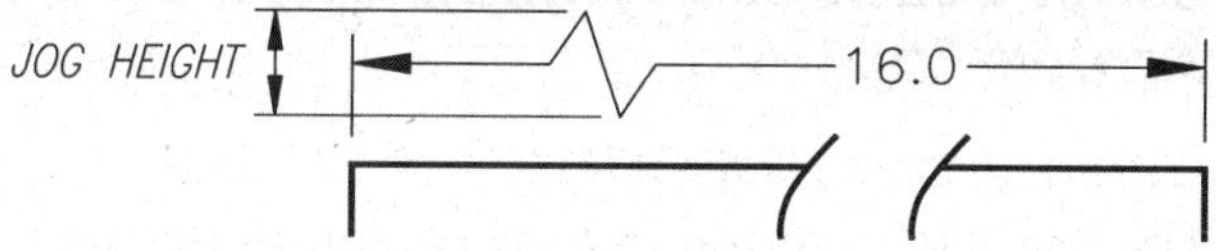


## ***Text* Tab**

The options in this tab change the appearance, placement, and alignment of the dimension text (Fig. 29-35). The preview image automatically reflects changes you make in the *Text Appearance*, *Text Placement*, and *Text Alignment* sections.

**FIGURE 29-35**

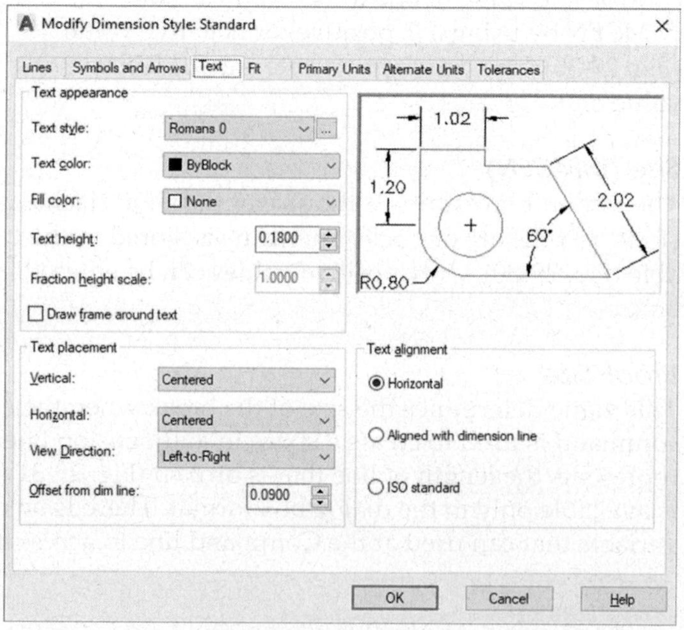

## *Text Appearance* Section (*Text* Tab)

### *Text style (DIMTXSTY)*

This feature of AutoCAD allows you to have different text styles for different dimension styles. The text styles are chosen from a drop-down list of existing styles in the drawing. You can also create a text style "on the fly" by picking the button just to the right of the *Text style* drop-down list, which produces the standard *Text Style* dialog box. The text style used for dimensions remains constant (as defined by the dimension style) and does not change when other text styles in the drawing are made current (as defined by *Style, Text,* or *Mtext*). The *DIMTXSTY* variable (dim text style) holds the text style name for the dimension style.

### *Text color (DIMCLRT)*

Select this drop-down list to select a color or to activate the standard *Select Color* dialog box. The color choice is assigned to the dimension text only (Fig. 29-36). This is useful for controlling the text appearance for printing or plotting when color-dependent plot styles are used. The setting is stored in the *DIMCLRT* variable (dim color text).

**FIGURE 29-36**

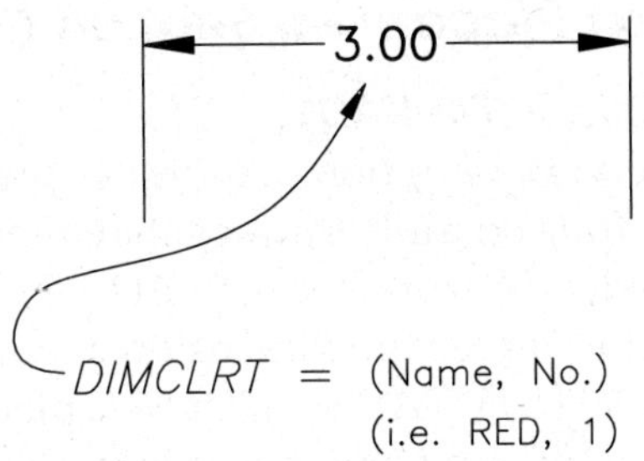


### *Fill Color (DIMTFILL, DIMTFILLCLR)*

You can create a colored background for your dimensions using this option (Fig 29-37). Use the drop-down list to select a color or to activate the *Select Color* dialog box. You can also enter color name or number. The setting for type of background (none, drawing, or color) is saved in the *DIMTFILL* variable and the color value is saved in the *DIMTFILLCLR* variable.

**FIGURE 29-37**

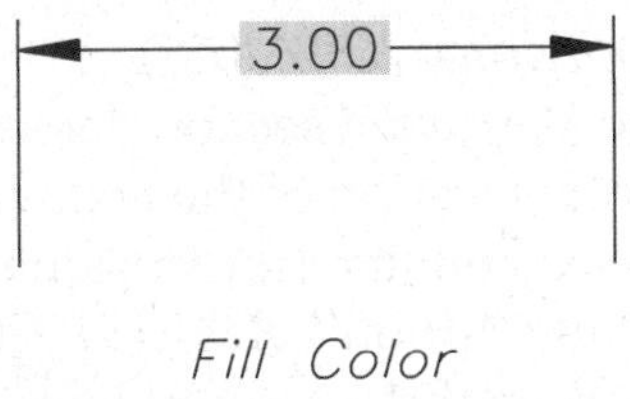


### *Text height (DIMTXT)*

This value specifies the primary text height; however, this value is multiplied by the *Overall Scale* (*DIMSCALE*) to determine the actual drawn text height. Change *Text height* only if you want to increase or decrease the text height in relation to the other dimension components (Fig. 29-38). Normally, change the *Overall Scale* (*DIMSCALE*) to change all size-related features (arrows, gaps, text, etc.) proportionally. The text height value is stored in *DIMTXT*.

**FIGURE 29-38**

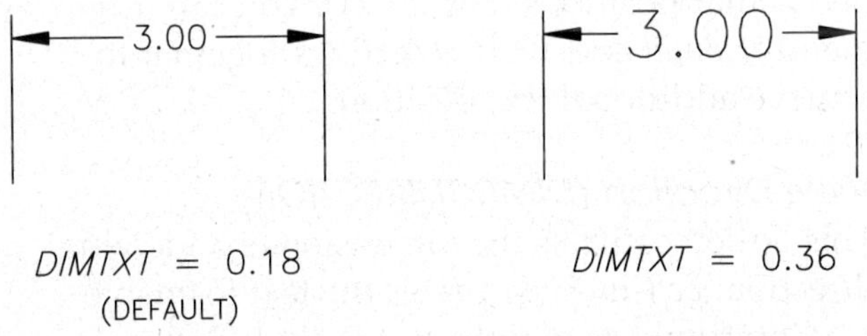


### *Fraction height scale (DIMTFAC)*

When fractions or tolerances are used, the height of the fractional or tolerance text can be set to a proportion of the primary text height. For example, 1.0000 creates fractions and tolerances the same height as the primary text; .5000 represents fractions or tolerances at one-half the primary text height (Fig. 29-39). The value is stored in the *DIMTFAC* variable (dim tolerance factor).

**FIGURE 29-39**

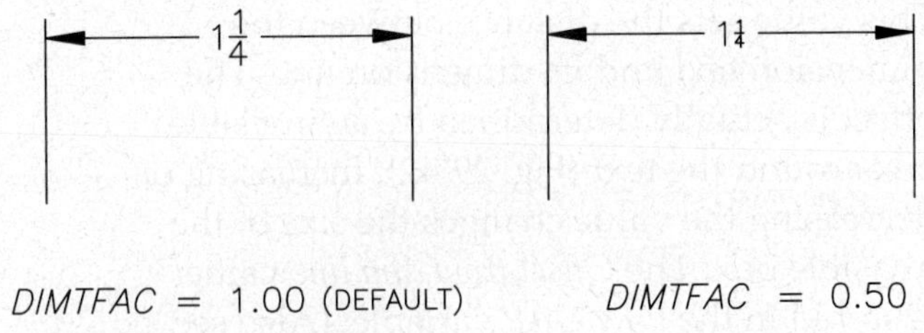

*Draw frame around text (DIMGAP)*

The "frame" or "gap" is actually an invisible box around the text that determines the offset from text to the dimension line. This option sets the *DIMGAP* variable to a negative value which makes the box visible (Fig. 29-40). This practice is standard for displaying a basic dimension. See *"Tolerance"* (geometric dimensioning and tolerancing) in Chapter 28. See also *Offset from dimension line* in the *"Text Placement"* section.

**FIGURE 29-40**

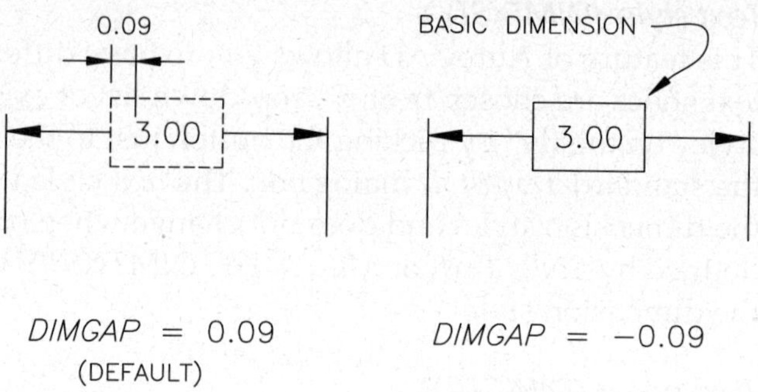


## ***Text Placement* Section (*Text* Tab)**

*Vertical (DIMTAD)*

The *Vertical* option determines the vertical location of the text with respect to the dimension line. There are four possible settings that affect the *DIMTAD* variable (dim text above dimension line). *Centered,* the default option (*DIMTAD* = 0), centers the dimension text between the extension lines. The *Above* option places the dimension text above the dimension line except when the dimension line is not horizontal (*DIMTAD* = 1). *Outside* places the dimension text on the side of the dimension line farthest away from the extension line origin points—away from the dimensioned object (*DIMTAD* = 2). The *JIS* option places the dimension text to conform to the Japanese Industrial Standard (*DIMTAD* = 3).

*Horizontal (DIMJUST)*

The *Horizontal* section determines the horizontal location of the text with respect to the dimension line (dimension justification). The default option (*DIMJUST* = 0) centers the text between the extension lines. The other four choices (*DIMJUST* = 1-3) place the text at either end of the dimension line in parallel and perpendicular positions (Fig. 29-41). You can use the *Horizontal* and *Vertical* settings together to achieve additional text positions.

**FIGURE 29-41**

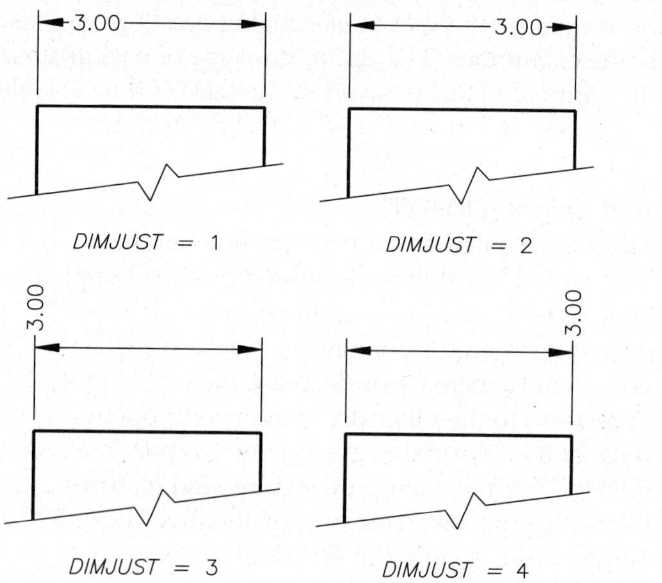


*View Direction (DIMTXTDIRECTION)*

This option controls the dimension text viewing direction. *Left-to-Right* places the text to enable reading from left to right and *Right-to-Left* places the text to enable reading from right to left (upside-down for horizontal text).

*Offset from dim line (DIMGAP)*

This value sets the distance between the dimension text and its dimension line. The offset is actually determined by an invisible box around the text (Fig. 29-42). Increasing or decreasing the value changes the size of the invisible box. The *Offset from dim line* value is stored in the *DIMGAP* variable. (Also see *"Draw frame around text."*)

**FIGURE 29-42**

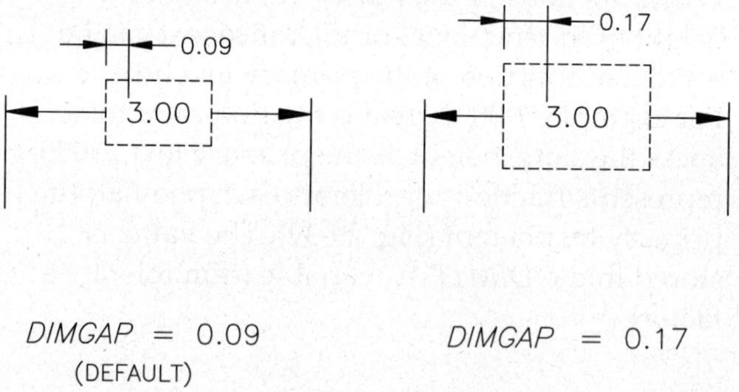

## *Text Alignment* Section (*Text* Tab)

The *Text Alignment* settings can be used in conjunction with the *Text Placement* settings to achieve a wide variety of dimension text placement options.

### *Horizontal (DIMTIH, DIMTOH)*

This radio button turns on the *DIMTIH* (dim text inside horizontal) and *DIMTOH* (dim text outside horizontal) variables. The text remains horizontal even for vertical or angled dimension lines (see Fig. 29-43 and Fig. 29-44). This is the correct setting for mechanical drawings (other than ordinate dimensions) according to the ASME Y14.5M-1994 standard, section 1.7.5, Reading Direction.

**FIGURE 29-43**

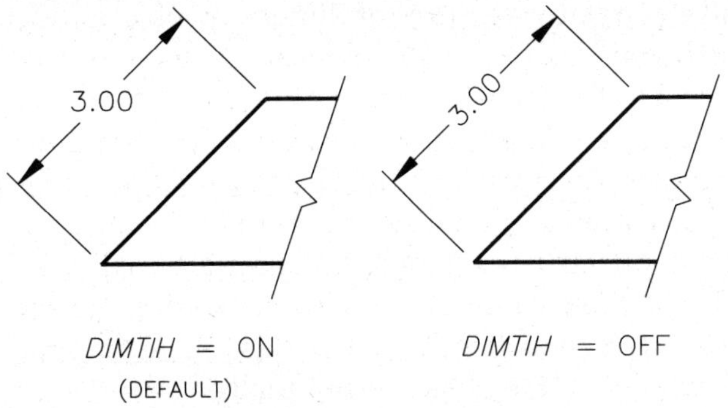


### *Aligned with dimension line (DIMTIH, DIMTOH)*

Pressing this radio button turns off the *DIMTIH* and *DIMTOH* variables so the text aligns with the angle of the dimension line (see Fig. 29-43 and Fig. 29-44).

### *ISO Standard (DIMTIH, DIMTOH)*

This option forces the text inside the dimension line to align with the angle of the dimension line (*DIMTIH* = off), but text outside the dimension line is horizontal (*DIMTOH* = on).

**FIGURE 29-44**

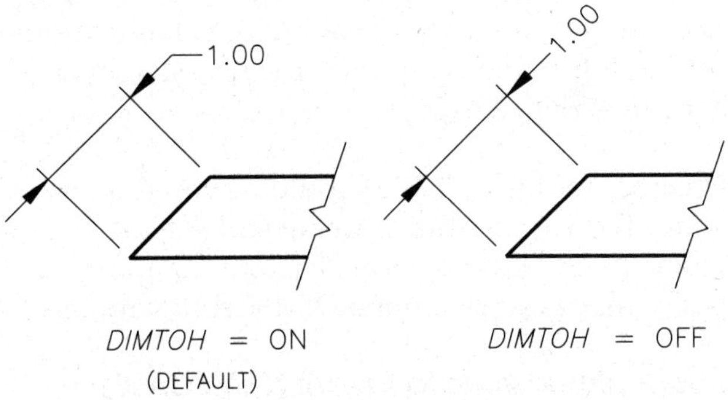


## *Fit* Tab

The *Fit* tab allows you to determine the *Overall Scale* for dimensioning components, how the text, arrows, and dimension lines fit between extension lines, and how the text appears when it is moved (Fig. 29-45).

**FIGURE 29-45**

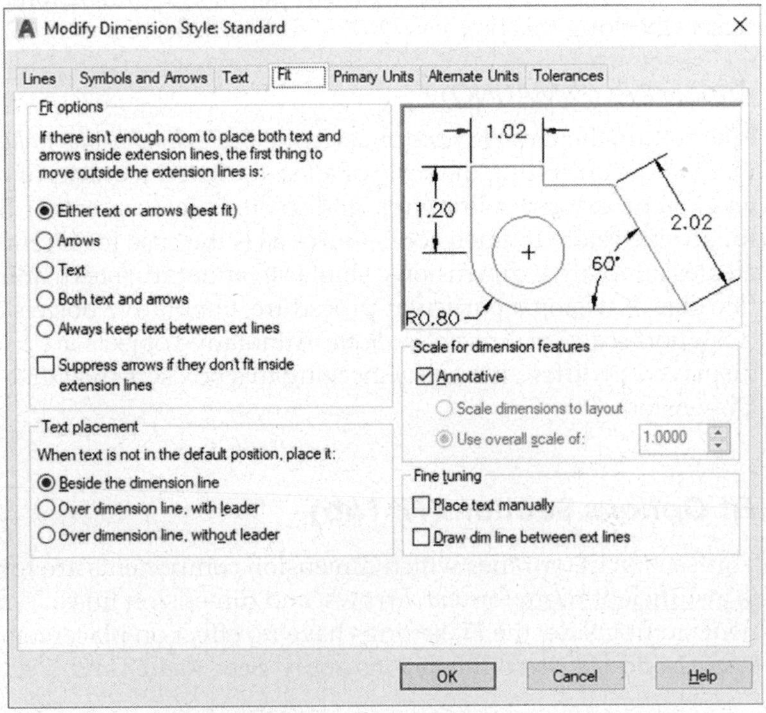

## ***Scale for Dimension Features*** **Section (*Fit* Tab)**

Although this is not the first section in the dialog box, it is presented first because of its importance.

### Overall Scale (DIMSCALE)

The *Overall Scale* value globally affects the scale of all size-related features of dimension objects, such as arrowheads, text height, extension line gaps (from the object), extensions (past dimension lines), etc. All other size-related (variable) values appearing in the dialog box series are multiplied by the *Overall Scale*. Notice how all dimensioning features (text, arrows, gaps, offsets) are all increased proportionally with the *Overall Scale* value (Fig. 29-46). Therefore, to keep all features proportional, change this one setting rather than each of the others individually.

**FIGURE 29-46**

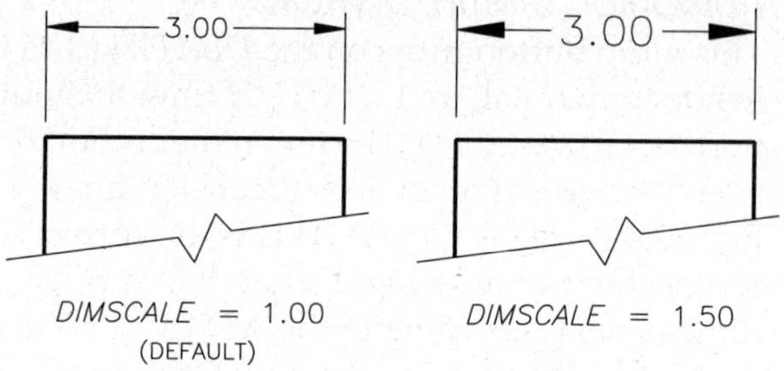


Although this area is located on the right side of the box, it is probably the most important option in the entire series of tabs. Because the *Overall Scale* should be set as a family-wide variable, setting this value is typically the first step in creating a dimension style (Fig. 29-47).

**FIGURE 29-47**

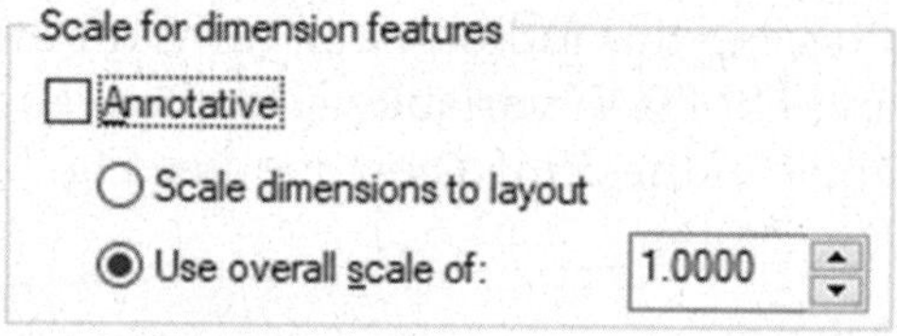


Changes in this variable should be based on the *Limits* and plot scale. You can use the drawing scale factor to determine this value. (See "Drawing Scale Factor," Chapter 12.) The *Overall Scale* value is stored in the *DIMSCALE* variable.

### Scale dimensions to layout (DIMSCALE)

Checking this box allows you to retroactively change the size of the dimensions in a viewport based on the viewport scale by selecting all the dimensions (in the viewport), then using the *Update* button from the Ribbon's *Annotate* tab, *Dimensions* panel. The dimensions display in a new size based on the viewport scale. Checking this box sets *DIMSCALE* to 0. See Chapter 32 for applications for this setting.

### Annotative (DIMANNO)

Like setting the *Overall Scale* or selecting *Scale dimensions to layout*, the *Annotative* option is an alternate method of controlling the scale of all size-related features of dimension objects, such as arrowheads, text height, extension line gaps, and so on. However, rather than setting a static value as a multiplier for all size-related dimension features as is the case for *Overall Scale*, this option is dynamic since it creates annotative dimensions, similar to annotative text, annotative hatch patterns, and other annotative objects. Using a particular procedure, annotative objects can change scale based on the *VP Scale* (viewport scale) or *Annotative Scale*. Annotative objects are useful when you have multiple viewports displayed at different scales. Checking this box sets the *DIMANNO* variable to 1 (on). See "Annotative Dimensions."

## ***Fit Options*** **Section (*Fit* Tab)**

This section determines which dimension components are forced outside the extension lines only if there is insufficient room for text, arrows, and dimension lines. In most cases where space permits all components to fit inside, the *Fit* settings have no effect on placement. *Linear*, *Aligned*, *Angular*, *Baseline*, *Continue*, *Radius*, and *Diameter* dimensions apply. See "*Radius* and *Diameter* Variable Settings."

*Either text or arrows (DIMATFIT)*
This is the default setting for the STANDARD dimension style. AutoCAD makes the determination of whether the text or the arrows are forced outside based on the size of arrows and the length of the text string (Fig. 29-48). This option often behaves similarly to *Arrows*, except that if the text cannot fit, the text is placed outside the extension lines and the arrows are placed inside. However, if the arrows cannot fit either, both text and arrows are placed outside the extension lines. The *DIMATFIT* (dimension arrows/text fit) setting is 3.

**FIGURE 29-48**

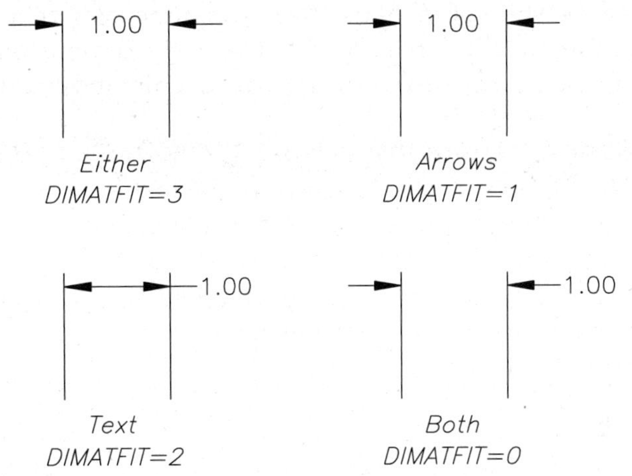


*Arrows (DIMATFIT)*
The *Arrows* option forces the arrows on the outside of the extension lines and keeps the text inside. If the text absolutely cannot fit, it is also placed outside the extension lines (see Fig. 29-48). This option sets *DIMATFIT* to 1.

*Text (DIMATFIT)*
The *Text* option places the text on the outside and keeps the arrows on the inside unless the arrows cannot fit, in which case they are placed on the outside as well (see Fig. 29-48). For this option, *DIMATFIT* = 2.

*Both text and arrows (DIMATFIT)*
The *Both text and arrows* option keeps the text and arrows together always. If space does not permit both features to fit between the extension lines, it places the text and arrows outside the extension lines (see Fig. 29-48). You can set *DIMATFIT* to 0 to achieve this placement.

*Always keep text between ext lines (DIMTIX)*
If you want the text to be forced between the extension line no matter how much room there is, use this option (Fig. 29-49). Pressing this radio button turns *DIMTIX* on (text inside extensions).

**FIGURE 29-49**

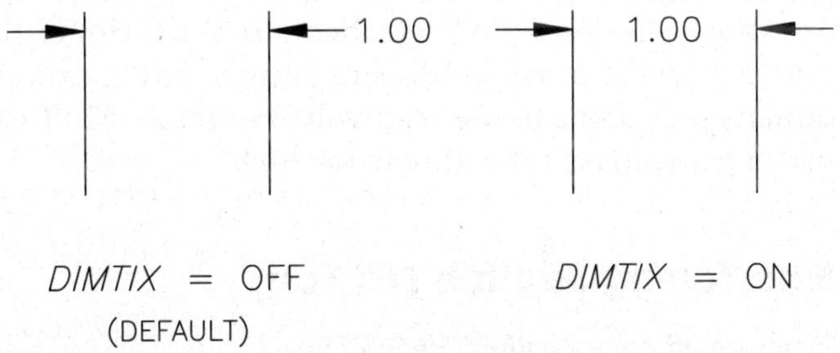


*Suppress arrows if they don't fit (DIMSOXD)*
When dimension components are forced outside the extension lines and there are many small dimensions aligned in a row (such as with *Continue* dimensions), the text, arrows, or dimension lines may overlap. In this case, you can prevent the arrows and the dimension lines from being drawn entirely with this option. This option suppresses the arrows and dimension lines only when they are forced outside (Fig. 29-50). The setting is stored as *DIMSOXD* = on (suppress dimension lines outside extensions).

**FIGURE 29-50**

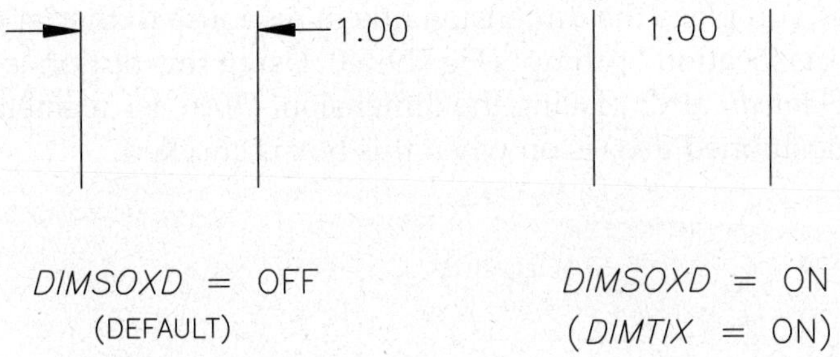

## *Text Placement* Section (*Fit* Tab)

This section of the dialog box sets dimension text movement rules. When text is moved either by being automatically forced from between the dimension lines based on the *DIMATFIT* setting (the *Fit Options* above this section) or when you actually move the text with grips or by *Dimtedit*, these rules apply.

### *Beside the dimension line (DIMTMOVE)*

This is the normal placement of the text—aligned with and beside the dimension line (Fig. 29-51). The text always moves when the dimension line is moved and vice versa. *DIMTMOVE* (dimension text move) = 0.

**FIGURE 29-51**

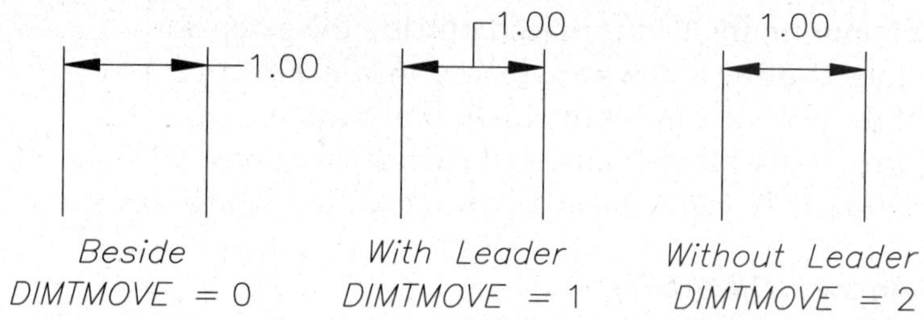


### *Over the dimension line, with leader (DIMTMOVE)*

This option creates a leader between the text and the center of the dimension line whenever the text cannot fit between the extension lines or is moved using grips (see Fig. 29-51). *DIMTMOVE* = 1.

### *Over the dimension line, without leader (DIMTMOVE)*

Use this setting to have the text appear above the dimension line, similar to *DIMTMOVE* = 1, but without a leader. This occurs only when there is insufficient room for the text between the extension lines or when you move the text with grips. *DIMTMOVE* = 2.

**FIGURE 29-52**

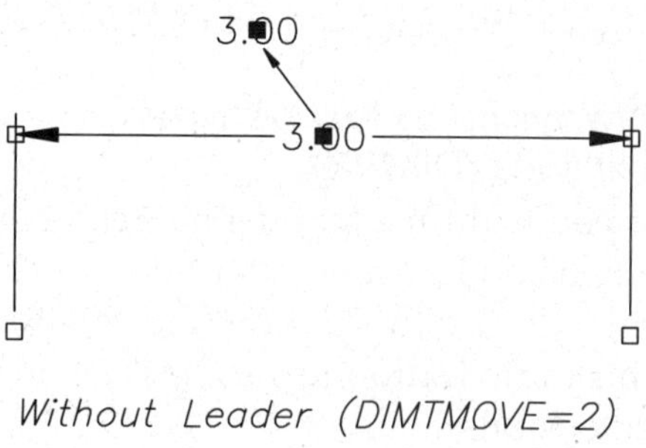


**TIP**

There is an important benefit to this setting (*DIMTMOVE* = 2). When there is sufficient room for the text and arrows between extension lines, this setting has no effect on the placement of the text. When there is insufficient room, the text moves above without a leader. In either case, if you prefer to move the text to another location with grips or using *Dimtedit*, the text moves as if it were "detached" from the dimension line. The text can be moved independently to any location and the dimension retains its associatively. Figure 29-52 illustrates the use of grips to edit the dimension text.

## *Fine Tuning* Section (*Fit* Tab)

### *Place text manually (DIMUPT)*

When you press this radio button, you can create dimensions and move the text independently in relation to the dimension and extension lines as you place the dimension line in response to the "specify dimension line location" prompt (Fig. 29-53). Using this option is similar to using *Dimtedit* after placing the dimension. *DIMUPT* (dimension user-positioned text) is on when this box is checked.

**FIGURE 29-53**

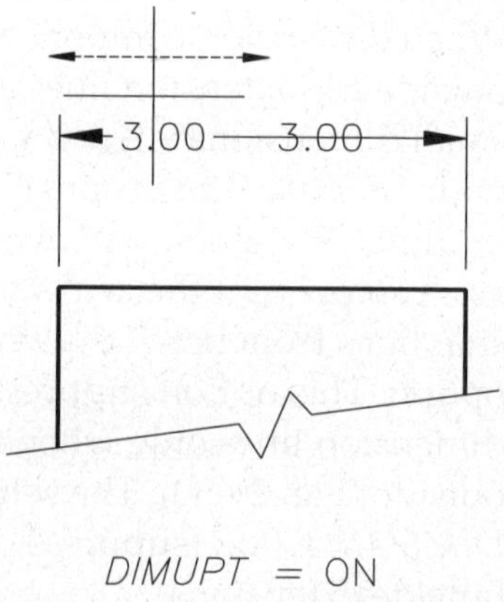

*Draw dim line between ext lines (DIMTOFL)*
Occasionally, you may want the dimension line to be drawn inside the extension lines even when the text and arrows are forced outside. You can force a line inside with this option (Fig. 29-54). A check in this box turns *DIMTOFL* on (text outside, force line inside).

**FIGURE 29-54**

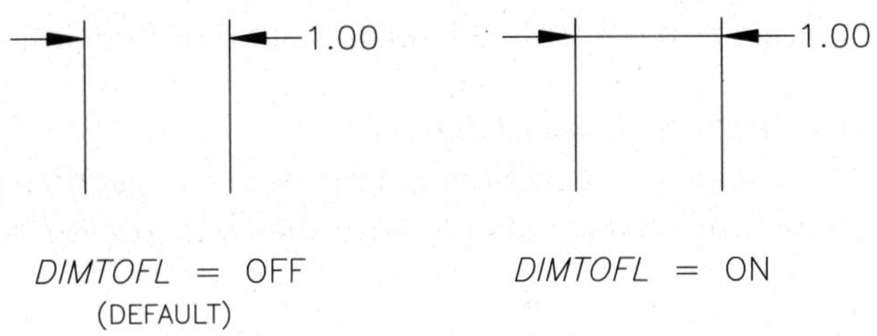


### Radius and Diameter Variable Settings

For creating mechanical drawing dimensions according to ANSI standards, the default settings in AutoCAD are correct for creating *Diameter* dimensions but not for *Radius* dimensions. The following variable settings are recommended for creating *Radius* and *Diameter* dimensions for mechanical applications.

For *Diameter* dimensions, the default settings produce ANSI-compliant dimensions that suit most applications—that is, text and arrows are on the outside of the circle or arc pointing inward toward the center. For situations where large circles are dimensioned, *Fit Options* (*DIMATFIT*) and *Place text manually* (*DIMUPT*) can be changed to force the dimension inside (Fig. 29-55).

**FIGURE 29-55**

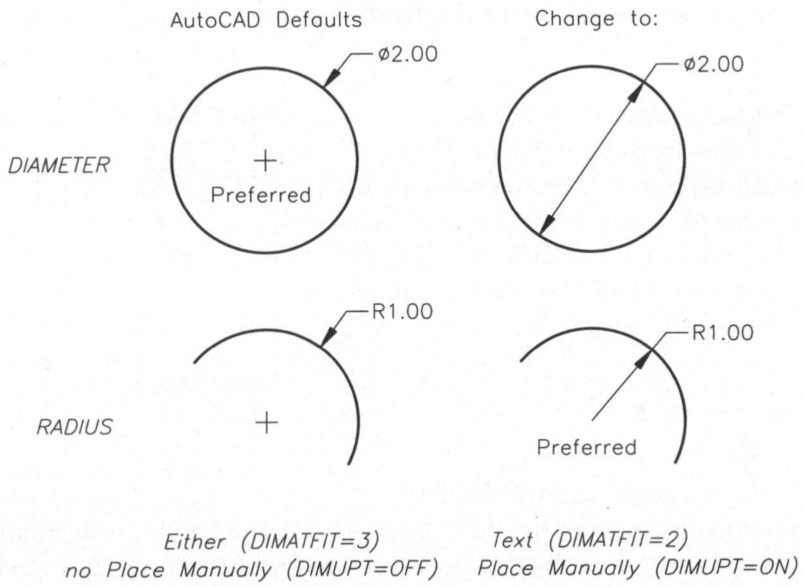


TIP

For *Radius* dimensions the default settings produce incorrect dimensioning practices. Normally (when space permits) you want the dimension line and arrow to be inside the arc, while the text can be inside or outside. To produce ANSI-compliant *Radius* dimensions, set *Fit Options* (*DIMATFIT*) and *Place text manually* (*DIMUPT*) as shown in Figure 29-55. Radius dimensions can be outside the arc in cases where there is insufficient room inside.

## *Primary Units* Tab

The *Primary Units* tab controls the format of the AutoCAD-measured numerical value that appears with a dimension (Fig. 29-56). You can vary the numerical value in several ways such as specifying the units format, precision of decimal or fraction, prefix and/or suffix, zero suppression, and so on. These units are called primary units because you can also cause AutoCAD to draw additional or secondary units called *Alternate Units* (for inch and metric notation, for example).

**FIGURE 29-56**

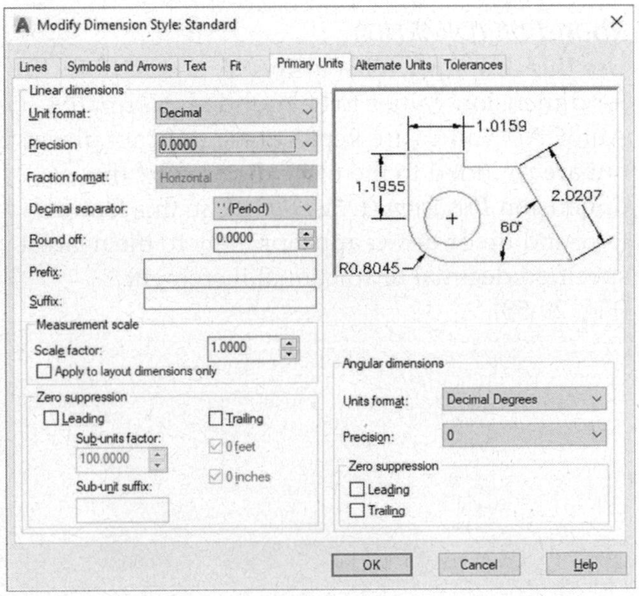

### ***Linear Dimensions* Section (*Primary Units* Tab)**

This section controls the format of all dimension types except *Angular* dimensions.

#### *Unit format (DIMLUNIT)*

The *Units format* section drop-down list specifies the type of units used for dimensioning. These are the same unit types available with the *Units* dialog box (*Decimal, Scientific, Engineering, Architectural,* and *Fractional*) with the addition of *Windows Desktop*. The *Windows Desktop* option displays AutoCAD units based on the settings made for units display in Windows Control Panel (settings for decimal separator and number grouping symbols). Remember that your selection affects the units drawn in dimension objects, not the global drawing units. The choice for *Units format* is stored in the *DIMLUNIT* variable (dimension linear unit). This drop-down list is disabled for an *Angular* family member.

#### *Precision (DIMDEC)*

The *Precision* drop-down list in the *Dimension* section specifies the number of places for decimal dimensions or denominator for fractional dimensions. This setting does not alter the drawing units precision. This value is stored in the *DIMDEC* variable (dim decimal).

#### *Fraction format (DIMFRAC)*

Use this drop-down list to set the fractional format. The choices are displayed in Figure 29-57. This option is enabled only when *DIMLUNIT* (*Unit format*) is set to 4 (*Architectural*) or 5 (*Fractional*).

**FIGURE 29-57**

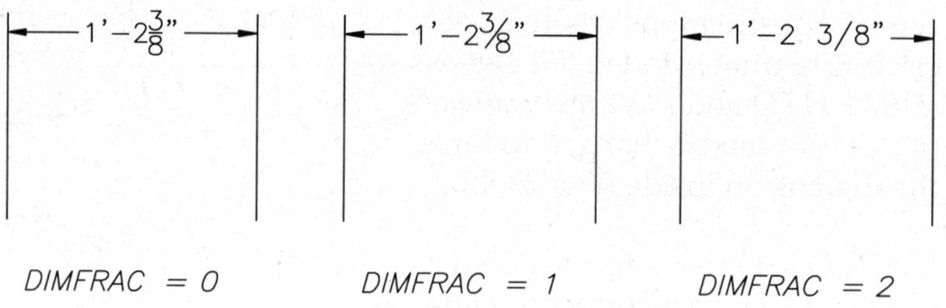


#### *Decimal separator (DIMDSEP)*

When you are creating dimensions whose unit format is decimal, you can specify a single-character decimal separator. Normally a decimal (period) is used; however, you can also use a comma (,) or a space. The character is stored in the *DIMDSEP* (dimension decimal separator) variable.

#### *Round off (DIMRND)*

Use this drop-down list to specify a precision for dimension values to be rounded. Normally, AutoCAD values are kept to 14 significant places but are rounded to the place dictated by the dimension *Precision* (*DIMDEC*). Use this feature to round up or down appropriately to the nearest specified decimal or fractional increment (Fig. 29-58).

**FIGURE 29-58**

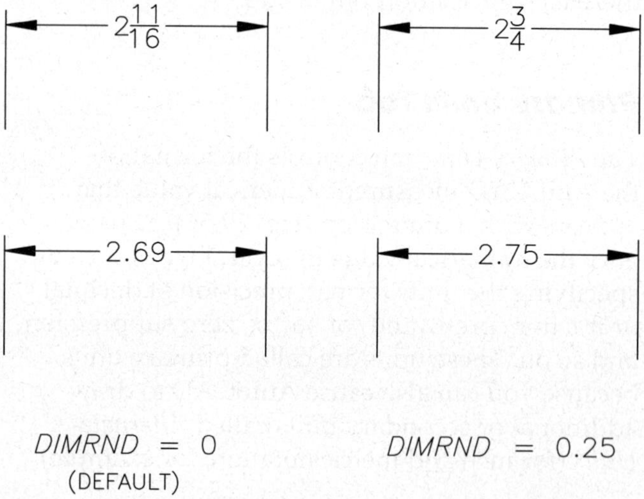

### *Prefix/Suffix (DIMPOST)*

The *Prefix* and *Suffix* edit boxes hold any text that you want to add to the AutoCAD-supplied dimensional value. A text string entered in the *Prefix* edit box appears before the AutoCAD-measured numerical value and a text string entered in the *Suffix* edit box appears after the AutoCAD-measured numerical value. For example, entering the string " mm" or a " TYP." in the *Suffix* edit box would produce text as shown in Figure 29-59. (In such a case, don't forget the space between the numerical value and the suffix.) The string is stored in the *DIMPOST* variable.

**FIGURE 29-59**

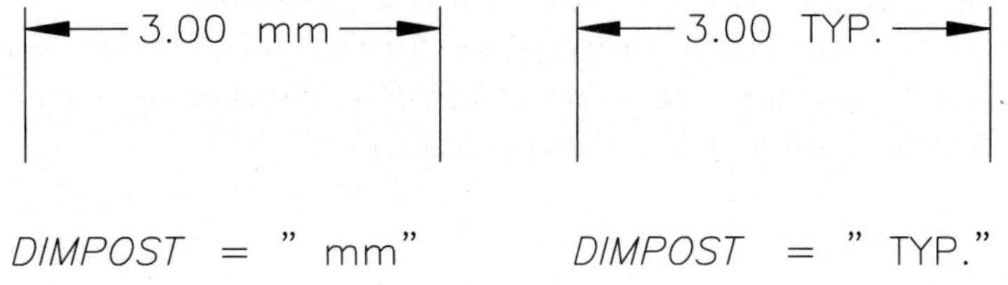


If you use the *Prefix* box to enter letters or values, any AutoCAD-supplied symbols (for radius and diameter dimensions) are overridden (not drawn). For example, if you want to specify that a specific hole appears twice, you should indicate by designating a "2X" before the diameter dimension. However, doing so by this method overrides the phi (Ø) symbol that AutoCAD inserts before the value. Instead, use the *Mtext/Text* options within the dimensioning command or use *Dimedit* or *Textedit* to add a prefix to an existing dimension having an AutoCAD-supplied symbol. Remember that AutoCAD uses a dark background in the *Text Editor* to represent the AutoCAD-measured value, so place a prefix in front of the dark area.

## ***Measurement Scale* Section (*Primary Units* Tab)**

### *Scale factor (DIMLFAC)*

Any value placed in the *Scale factor* edit box is a <u>multiplier</u> for the AutoCAD-measured numerical value. The default is 1. Entering a 2 would cause AutoCAD to draw a value two times the actual measured value (Fig. 29-60). This feature might be used when a drawing is created in some scale other than the actual size, such as an enlarged detail view in the same drawing as the full view. Changing this setting is <u>unnecessary</u> when you create associative dimensions in paper space attached to objects in model space (see "Dimensioning in Paper Space").

**FIGURE 29-60**

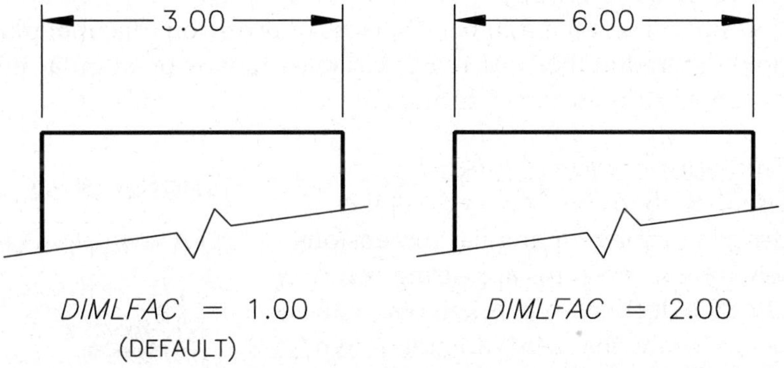


However, if you are using versions of AutoCAD previous to 2002, or if you are using nonassociative dimensions in paper space, change this setting to create dimensions that display other than the actual measured value. This setting is stored in the *DIMLFAC* variable (dimension length factor).

### *Apply to layout dimensions only (DIMLFAC)*

Use this checkbox to apply the *Scale factor* to dimensions placed in a layout for <u>nonassociative</u> dimensions or for drawings in versions earlier than AutoCAD 2002. For example, assume you have a detail view displayed at 2:1 in a viewport and want to place nonassociative dimensions in paper space, set the *Scale factor* to .5 and check *Apply to layout dimensions only* so the measured values adjust for the viewport scale. A check in this box sets the *DIMLFAC* variable to the negative of the *Scale factor* value. Associative dimensions drawn in paper space automatically adjust for the viewport scale, so changing this setting is unnecessary (see "Dimensioning in Paper Space").

## *Zero Suppression* Section (*Primary Units* Tab)

### *Leading/Trailing/0 Feet/0 Inches (DIMZIN)*

*Leading* and *Trailing* are enabled for *Scientific, Decimal, Engineering* and *Fractional* units. The *0 Feet* and *0 Inches* checkboxes are enabled for *Architectural* and *Engineering* units.

For example, assume primary *Units format* was set to *Architectural* and *Zero Suppression* was checked for *0 Inches* only. Therefore, when a measurement displays feet and no inches, the 0 inch value is suppressed (Fig. 29-61, top dimension). On the other hand, since *0 Feet* is not checked, a measurement of less than 1 foot would report 0 feet (Fig. 29-61, bottom dimension).

**FIGURE 29-61**

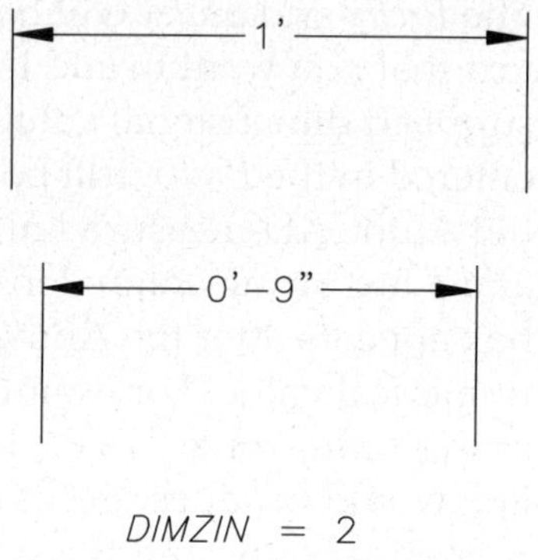


### *Sub-units factor, Sub-unit suffix*

Use this feature to calculate and display the distance in a sub-unit when the measured distance is less than one unit. For example, to force .96m to display as 96cm, enter "100" as the *Sub-units factor* and "cm" as the *Sub-units suffix*.

## *Angular Dimensions* Section (*Primary Units* Tab)

### *Units format (DIMAUNIT)*

The *Units format* drop-down list sets the unit type for angular dimensions, including *Decimal degrees, Degrees Minutes Seconds, Gradians,* and *Radians*. This list is enabled only for parent dimension style and *Angular* family members. The variable used for the angular units is *DIMAUNIT* (dimension angular units).

### *Precision (DIMADEC)*

This option sets the number of places of precision (decimal places) for angular dimension text. The selection is stored in the *DIMADEC* variable (dimension angular decimals). This option is enabled only for the parent dimension style and *Angular* family member.

### *Zero Suppression (DIMAZIN)*

Use these two checkboxes to set the desired display for angular dimensions when there are zeros appearing before or after the decimal. A check in one of these boxes means that zeros are not drawn for that case.

## *Alternate Units* Tab

This tab allows you to display alternate, or secondary, dimensioning values along with the primary AutoCAD-measured value when a dimension is created (Fig. 29-62). Typically, alternate units consist of millimeter values given in addition to the decimal inch values. This practice is often called "dual dimensioning." The options in this tab control the display and format of alternate units. Notice that most of these options are equivalent to those found in the *Primary Units* tab.

**FIGURE 29-62**

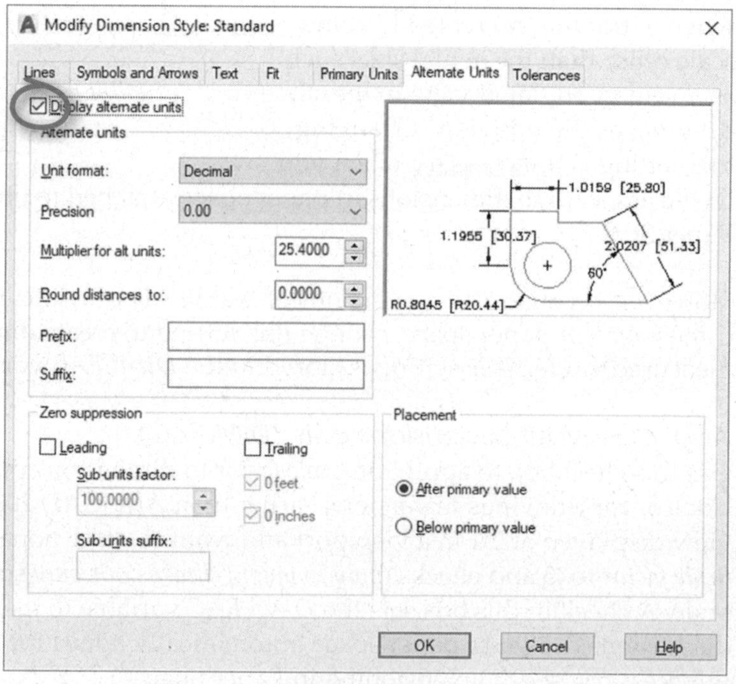

*Display alternate units (DIMALT)*
By default, AutoCAD displays only one value for a dimension—the primary unit. If you want to have AutoCAD measure and create an additional value for each dimension, check this box (Fig. 29-63). All options in this tab are disabled unless *Display alternate units* is checked. The presence of alternate units is controlled by setting the *DIMALT* variable on (dimension alternate).

**FIGURE 29-63**

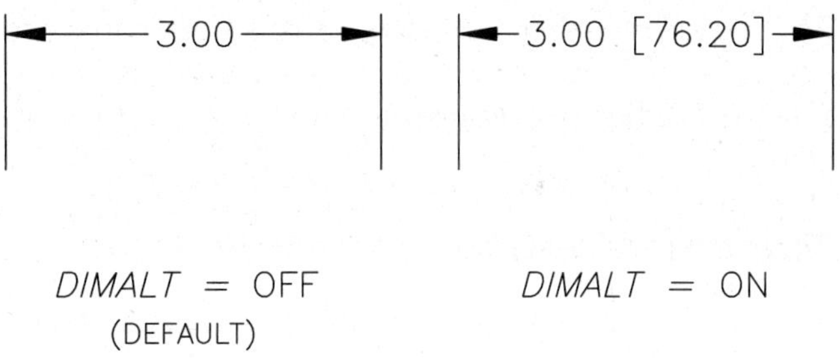


## ***Alternate Unit* Section (*Alternate Units* Tab)**

*Unit format (DIMALTU)*
This drop-down list sets the format for alternate units. The setting is stored in the *DIMALTU* variable (dimension alternate units). This is the alternate units equivalent to *Units format* for *Primary Units.*

*Precision (DIMALTD)*
Use this drop-down list to set the decimal precision for alternate units. Note that the alternate units precision is controlled independently of the primary units precision (Fig. 29-64). The value is stored in the *DIMALTD* variable (dimension alternate decimals).

**FIGURE 29-64**

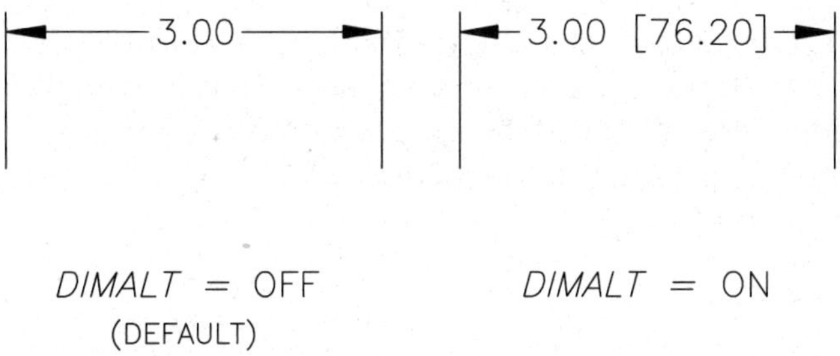


*Multiplier for alt units (DIMALTF)*
This is the alternate units equivalent for the primary units scale factor (*DIMLFAC*). In other words, the AutoCAD-measured primary units value is multiplied by this factor to determine the displayed value for alternate units (Fig. 29-65). You can also enter the desired multiplier in the *DIMALTF* (dimension alternate factor) variable.

**FIGURE 29-65**

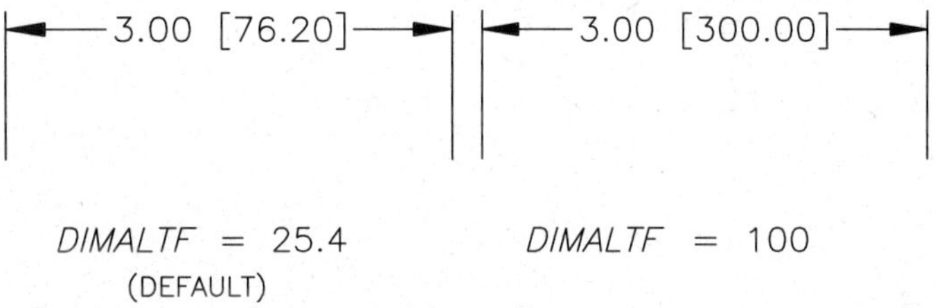


*Round distances to (DIMALTRND)*
If you do not want the alternate units value to be displayed as the actual measurement, but to be rounded to nearest regular increment, enter the desired increment in this edit box. For example, you may want alternate units to be displayed with two places of precision (to the right of the decimal) but to round to the nearest millimeter. Alternately, set the increment using the *DIMALTRND* variable (alternate rounding). This is the alternate units equivalent to the *Round off* option in the *Primary Units* tab (see Fig. 29-58).

*Prefix/Suffix (DIMAPOST)*
This section allows you to include a prefix and/or suffix with the alternate units measured value. For example, you may want to display "mm" after the alternate units value (Fig. 29-66). The prefix and/or suffix is stored in the *DIMAPOST* variable. (See also *Prefix/Suffix* in the *Primary Units* tab for more information.)

**FIGURE 29-66**

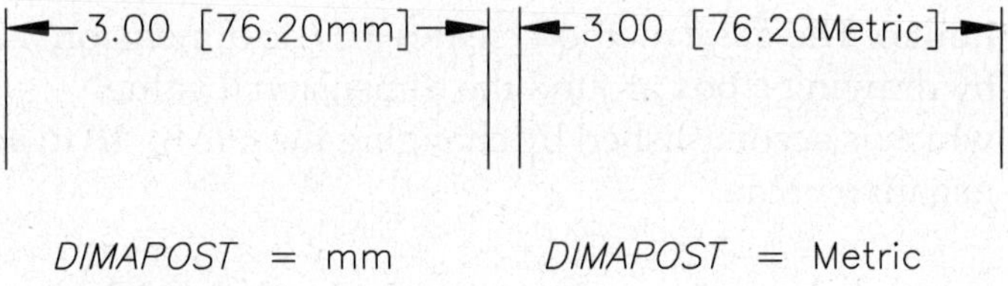

## ***Zero Suppression*** **Section (*Alternate Units* Tab)**

The *Zero Suppression* section controls how zeros are drawn in alternate units values when they occur. A check in one of these boxes means that zeros are not drawn for that case. Your selection sets the value for *DIMALTZ*. See the description for *Zero Suppression*, *Primary Units* Tab for more information and illustration.

## ***Placement*** **Section (*Alternate Units* Tab)**

This section only has two options; both are related to the *DIMPOST* variable. Normally, the alternate units values are placed *After primary value*. You can instead toggle *Below primary value* to yield a display as shown in Figure 29-67, right dimension. Selecting *Below primary value* sets the *DIMPOST* variable to "\x."

**FIGURE 29-67**

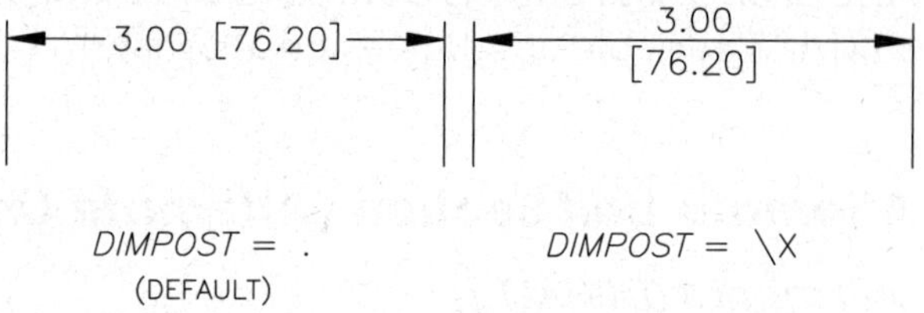


## ***Tolerances*** **Tab**

The *Tolerances* tab allows you to create several formats of tolerance dimensions such as limits, two forms of plus/minus dimensions, and basic dimensions (Fig. 29-68). Most options in this tab are disabled until you select a *Method*.

**FIGURE 29-68**

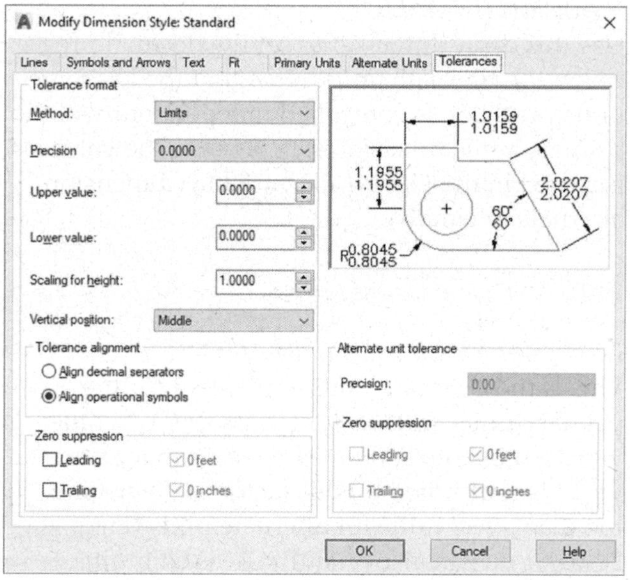


## ***Tolerance Format*** **Section (*Tolerances* Tab)**

### Method (DIMTOL, DIMLIM, DIMGAP)

The *Method* option displays a drop-down list with five types: *None, Symmetrical, Deviation, Limits,* and *Basic*. The four possibilities (other than *None*) are illustrated in Figure 29-69. The *Symmetrical* and *Deviation* methods create plus/minus dimensions and turn on the *DIMTOL* variable (dimension tolerance). The *Limits* method creates limit dimensions and turns the *DIMLIM* variable on (dimension limits). The *Basic* method creates a basic dimension by drawing a box around the dimensional value, which is accomplished by changing the *DIMGAP* to a negative value.

**FIGURE 29-69**

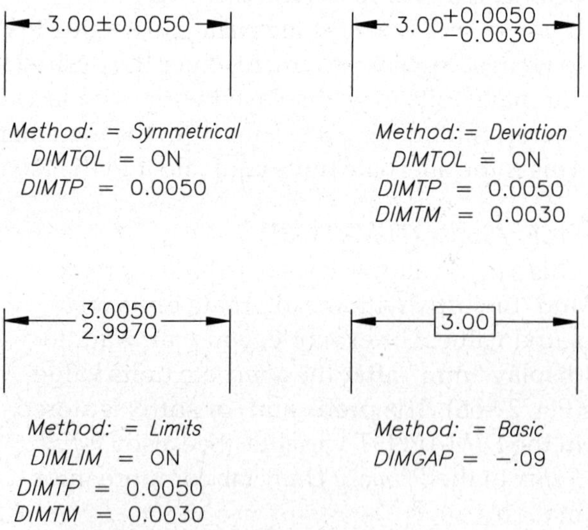

### *Precision (DIMTDEC)*

Use the *Precision* drop-down list to set the precision (number of places to the right of the decimal place) for values when drawing *Symmetrical, Deviation,* and *Limits* dimensions. The precision is stored in the *DIMTDEC* variable (dimension tolerance decimals).

### *Upper Value/Lower Value (DIMTP, DIMTM)*

An *Upper Value* and *Lower Value* can be entered in the edit boxes for the *Deviation* and *Limits* method types. In these cases, the *Upper Value* (*DIMTP*—dimension tolerance plus) is added to the measured dimension and the *Lower Value* (*DIMTP*—dimension tolerance minus) is subtracted (Fig. 29-69). An *Upper Value* only is needed for *Symmetrical* and is applied as both the plus and minus value.

### *Scaling for height (DIMTFAC)*

The height of the tolerance text is controlled with the *Scaling for height* edit box value. The entered value is a proportion of the primary dimension value height. For example, a value of .50 would draw the tolerance text at 50% of the primary text height (Fig. 29-70). The setting affects the *Symmetrical, Deviation,* and *Limits* tolerance methods. The value is stored in the *DIMTFAC* (dimension tolerance factor) variable. Note that this is the same variable that controls the *Fraction height scale* for architectural and fractional values.

**FIGURE 29-70**

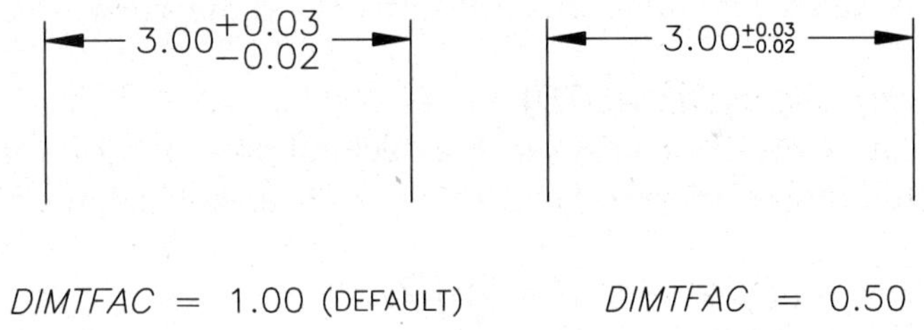


### *Vertical position (DIMTOLJ)*

For *Deviation, Symmetrical,* and *Limits* tolerance methods, you can control the placement of the tolerance values in relation to the primary units values (Fig. 29-71). The choices are *Top, Middle,* and *Bottom* and set the *DIMTOLJ* variable (dimension tolerance justification) to 2, 1, and 0, respectively. The *Top* option aligns the primary text and the top tolerance value, and the *Bottom* option aligns the primary text and the bottom tolerance value.

**FIGURE 29-71**

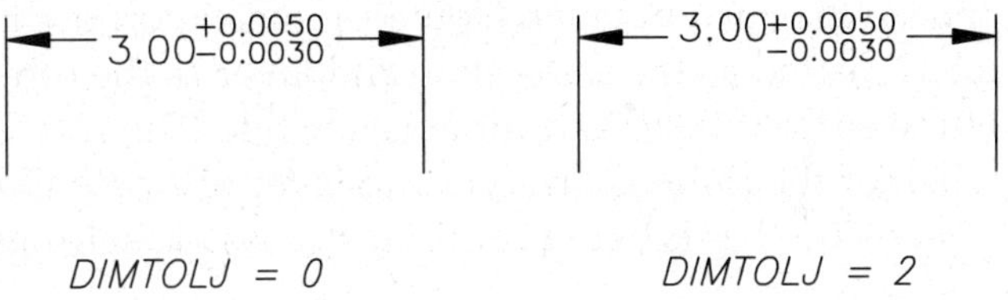


### *Tolerance Alignment*

This section controls the alignment of upper and lower tolerance values only when *Tolerance Method* is set to *Symmetrical, Deviation,* or *Limits* and proportional TrueType fonts are used for the tolerance values. There are no system variables that can be accessed to set these options.

### *Align Decimal Separators*

The tolerance values are vertically aligned by their decimal separators (Fig. 29-72).

### *Align Operational Symbols*

The tolerance values are vertically aligned by their operational symbols (Fig. 29-72).

**FIGURE 29-72**

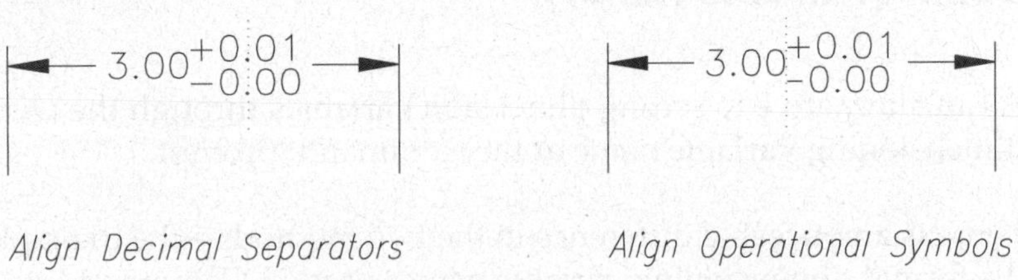

### *Zero Suppression (DIMTZIN) (Tolerance Format Section)*

The *Zero Suppression* section controls how zeros are drawn in *Symmetrical, Deviation,* and *Limits* tolerance dimensions when they occur. A check in one of these boxes means that zeros are not drawn for that case. The *Zero Suppression* section here operates identically to the *Zero Suppression* section of the *Primary Units* tab, except that it is applied to tolerance dimensions (see *Zero Suppression, Primary Units* Tab for more information). The affected variable is *DIMTZIN* (dimension tolerance zero indicator).

## ***Alternate Unit Tolerance* Section (*Tolerances* Tab)**

When you have specified that alternate units are to be drawn (in the *Alternate Units* tab) and you have turned on some form of tolerance *Method* (*Symmetrical, Limits,* etc.), the alternate units will automatically display as a tolerance along with the primary units. In other words, when alternate units are on, both primary and alternate units are drawn the same—with or without tolerances.

### *Precision (DIMALTTD)*

This drop-down list sets the decimal precision for the alternate units tolerances (when alternate units and tolerances are on). The *DIMALTTD* variable (dimension alternate tolerance decimal) holds the setting.

### *Zero Suppression (DIMALTTZ)*

When alternate units and tolerances are on, use these check boxes to determine how leading and trailing zeros are treated. The setting is stored in the *DIMALTTZ* variable (dimension alternate tolerance zeros). See *Zero Suppression* in the *Primary Units* Tab section for more information on zero suppression.

## ***DIMCONTINUEMODE***

TIP

Normally when you create a new dimension, that dimension is created with the current dimension style and on the current layer. However, when you use a *Continue* or *Baseline* dimension, you typically want it to be created as the same style and layer as the other dimensions you select to continue from, regardless of the current layer and dimension style. The *DIMCONTINUEMODE* variable allows you to determine whether the dimension style and layer of a new *Continue* or *Baseline* dimension are inherited from the dimension that is being continued. Possible settings are:

0 Use the current dimension style and layer.

1 Use the dimension style and layer of the dimension that is being continued or used as a baseline (default setting).

*DIMCONTINUEMODE* must be typed at the command prompt. This variable is not accessible through the *Dimension Style Manager*.

## **Changing Dimension Variables Using the Command Line Format**

## ***Dim...* (variable name)**

As an alternative to setting dimension variables through the *Dimension Style Manager*, you can type the dimensioning variable name at the Command: prompt.

There is a noticeable difference in the two methods—the dialog boxes use different nomenclature than the formal dimensioning variable names used in Command line format; that is, the dialog boxes use descriptive terms that, if selected, make the appropriate change to the dimensioning variable. The formal dimensioning variable names accessed by Command line format, however, all begin with the letters *DIM* and are accessible only by typing.

Another important but subtle difference in the two methods is the act of saving dimension variable settings to a dimension style. Remember that all drawn dimensions are part of a dimension style, whether it is Standard or some user-created style. When you change dimension variables by the Command line format, the changes become overrides until you use the *Save* option of the *-Dimstyle* command or the *Dimension Style Manager*. When a variable change is made, it becomes an override that is applied to the current dimension style and affects only the newly drawn dimensions. Variable changes must be *Saved* to become a permanent part of the style and to retroactively affect all dimensions created with that style.

TIP

All other dimensioning controls are accessible using the dimension variables explained in the following Dimensioning Variables Table. All these dimension variables except *DIMASSOC* are included in every dimension style.

In order to access and change a variable's setting by name, simply type the variable name at the Command: prompt. For dimension variables that require distances, you can enter the distance (in any format accepted by the current *Units* settings) or you can designate by (PICKing) two points.

For example, to change the value of the *DIMSCALE* to **.5**, this command syntax is used:

```
Command: DIMSCALE
Enter new value for dimscale <1.0000>: .5
```

## *DIMASSOC* (Associative Dimensions)

Since AutoCAD 2002, dimensions have full associativity. The *DIMASSOC* variable controls the associative feature. The *DIMASSOC* variable setting cannot be saved in a dimension style; therefore, one dimension style can contain associative and nonassociative dimensions but not exploded dimensions (exploded dimensions can be created from a dimension style but cannot be updated since they do not reference the style). The *Dimension Style Manager* does not provide access to the *DIMASSOC* variable. *DIMASSOC* must be changed by Command line format or by using the *Options* dialog box (see "Associative Dimensions" in Chapter 28).

## Dimension Styles in the Template Drawings

### ACAD.DWT Template Drawing

The STANDARD dimension style is the default dimension style used in all the ACAD.DWT template drawings (acad.dwt, acad3D.DWT, acad–Named Plot Styles.dwt, and acad–Named Plot Styles3D.dwt). The ACAD.DWT drawing is the template or basis used when you begin AutoCAD with the *Start Drawing* tile in the *Create* screen, use the *New* command or otherwise select the ACAD.DWT as the template drawing, or choose all the inch defaults in the *Setup Wizards*.

### ACADISO.DWT Template Drawing

ISO is the acronym for International Organization for Standardization and is the authority on metric standards. If you select the ACADISO template drawing, either of the ACADISO plot styles template drawings, or if you select *Metric* in *Start from Scratch* in the *Create New Drawing* dialog box, the ISO-25 dimension style is the default dimension style.

The following table (on the next page) lists the differences between the dimension variable settings in the STANDARD (ACAD.DWT) dimension style and the ISO-25 (ACADISO.DWT) dimension style. All dimension variables not listed in the table have the same setting.

Dimension Variable Differences between STANDARD and ISO-25 Dimension Styles

| Variable | Description | ANSI | ISO-25 |
|---|---|---|---|
| *DIMALTD* | Alternate unit decimal places | 2 | 3 |
| *DIMALTF* | Alternate unit scale factor | 25.4000 | 0.0394 |
| *DIMASZ* | Arrow size | 0.1800 | 2.5000 |
| *DIMCEN* | Center mark size | 0.0900 | 2.500 |
| *DIMDEC* | Decimal places for dimensions | 4 | 2 |
| *DIMDLI* | Dimension line spacing | 0.3800 | 3.7500 |
| *DIMDSEP* | Decimal separator | . | , |
| *DIMEXE* | Extension above dimension line | 0.1800 | 1.2500 |
| *DIMEXO* | Extension line origin offset | 0.0625 | 0.625 |
| *DIMGAP* | Gap from dimension line to text | 0.0900 | 0.6250 |
| *DIMTAD* | Place text above the dimension line | 0 | 1 |
| *DIMTDEC* | Tolerance decimal places | 4 | 2 |
| *DIMTIH* | Text inside extensions is horizontal | On | Off |
| *DIMTOFL* | Force line inside extension lines | Off | On |
| *DIMTOH* | Text outside horizontal | On | Off |
| *DIMTOLJ* | Tolerance vertical justification | 1 | 0 |
| *DIMTXT* | Text height | 0.1800 | 2.5000 |
| *DIMTZIN* | Tolerance zero suppression | 0 | 8 |
| *DIMZIN* | Zero suppression | 0 | 8 |

### ANNOTATIVE Dimension Style

Although it is not set as the current dimension style, the ANNOTATIVE dimension style is provided in all template drawings supplied with current releases of AutoCAD. This dimension style differs from the other style that appears in a selected template drawing only with respect to one dimension variable setting—that is, *DIMANNO* (*Annotative*) is set to 1 (on) in the ANNOTATIVE style, whereas *DIMANNO* is set to 0 (off) in the other dimension styles. In other words, when the ANNOTATIVE style appears with the STANDARD style, it is identical to STANDARD except for the *DIMANNO* setting, and when it appears with the ISO-25 style, it is identical to ISO-25 except for the *DIMANNO* setting. Therefore, the ANNOTATIVE dimension style is different in the ACAD and the ACAD-ISO templates. In addition, using the ANNOTATIVE dimension style accomplishes the same action as using either the STANDARD or ISO-25 dimension style and changing the setting for *Annotative*.

## MODIFYING EXISTING DIMENSIONS

Even with the best planning, a full understanding of dimension variables, and the correct use of *Dimension Styles*, it is probable that changes will have to be made to existing dimensions in the drawing due to design changes, new plot scales, or new industry/company standards. There are several ways that changes can be made to existing dimensions while retaining associativity and membership to a dimension style. The possible methods are discussed in this section.

### Modifying a Dimension Style

Existing dimensions in a drawing can be modified by making one or more variable changes to the dimension style family or child, then *Saving* those changes to the dimension style.

This process can be accomplished using either the *Dimension Style Manager* or Command line format. When you use the *Modify* option in the *Dimension Style Manager* or the *Save* option of the *-Dimstyle* command, the existing dimensions in the drawing automatically update to display the new variable settings.

### Properties

| Ribbon | Menu Bar | Command (Type) | Alias (Type) | Shortcut |
| --- | --- | --- | --- | --- |
| *View Palettes Properties* | *Modify Properties* | *Properties* | *PR* or *CH* | (Edit Mode) *Properties or Ctrl+1* |

Remember that *Properties* can be used to edit existing dimensions. Using *Properties* and selecting one dimension displays the *Properties* palette with all of the selected dimension's properties and dimension variables.

Using *Properties* you can modify any aspect of one or more dimensions, including text, and access is given to the *Lines and Arrows, Text, Fit, Primary Units, Alternate Units,* and *Tolerances* categories. Any changes to dimension variables through *Properties* result in overrides to the selected dimension objects only but do not affect the dimension style. *Properties* has essentially the same result as *Dimoverride* (see "*Dimoverride*"). *Properties* can also be used to modify the dimension text value. *Properties* of dimensions are also discussed in Chapter 28.

TIP

## Creating Dimension Style Overrides and Using *-Dimstyle*, *Apply* (*Update*)

You can modify existing dimensions in a drawing without making permanent changes to the dimension style by creating a dimension style override, then using *-Dimstyle*, *Apply* (*Update*) to apply the new setting to an existing dimension. This method is preferred if you wish to modify one or two dimensions without modifying all dimensions referencing (created in) the style.

TIP

To do this, use either the Command line format or *Dimension Style Manager* to set the new variable. In the *Dimension Style Manager*, select *Override* and make the desired variable settings in the *Override Current Style* dialog box. This creates an override to the current style. In Command line format, simply enter the formal dimension variable name and make the change to create an override to the current style, then use the *Update* button on the Ribbon to apply the current style settings plus the override settings to existing dimensions that you select. The overrides remain in effect for the current style unless the variables are reset to the original values or until the overrides are cleared. You can clear overrides for a dimension style by making another dimension style current in the *Dimension Style Manager*.

### -Dimstyle Apply

| Ribbon | Menu Bar | Command (Type) | Alias (Type) | Shortcut |
| --- | --- | --- | --- | --- |
| *Annotate Dimensions* (icon) | *Dimension Update* | *-Dimstyle Apply* | ... | ... |

*-Dimstyle, Apply (Update)* can be used to update existing dimensions in the drawing to the current settings. The current settings are determined by the current dimension style and any dimension variable overrides that are in effect (see previous explanation above).

This is an excellent method of modifying one or more existing dimensions without making permanent changes to the dimension style that the selected dimensions reference. *Update* is actually the *-Dimstyle* command (with the hyphen) and the *Apply* option.

For example, if you wanted to change the *DIMSCALE* of several existing dimensions to a value of 2, change the variable by typing *DIMSCALE*. Next use *Update* (*-Dimstyle*, *Apply*) and select the desired dimension. That dimension is updated to the new setting. The command syntax is the following:

Command: ***DIMSCALE***<br>
New value for DIMSCALE <1.0000>: **2**<br>
Command:

Command: ***-DIMSTYLE***<br>
Current dimension style: Standard Annotative: No<br>
Enter a dimension style option<br>
[ANnotative/Save/Restore/STatus/Variables/Apply/?] <Restore>: ***apply***<br>
Select objects: **PICK** (select a dimension object to update)<br>
Select objects: **PICK** (select a dimension object to update)<br>
Select objects: **Enter**<br>
Command:

TIP

Beware, *Update* creates an override to the current dimension style. You should reset the variable to its original value unless you want to keep the override for creating other new dimensions. You can clear overrides for a dimension style by making another dimension style current in the *Dimension Style Manager*. Keep in mind that you can save the override to the dimension style by using the *-Dimstyle* command with the *Save* option.

Also, see the explanation near the beginning of this chapter for "*-Dimstyle*."

## *Dimoverride*

| Ribbon | Menu Bar | Command (Type) | Alias (Type) | Shortcut |
|---|---|---|---|---|
| *Annotate Dimensions* (icon) | *Dimension Override* | *Dimoverride* | *DIMOVER* or *DOV* | ... |

TIP

*Dimoverrride* grants you a great deal of control to edit existing dimensions. The abilities enabled by *Dimoverride* are similar to the effect of using the *Properties* palette. *Dimoverride* enables you to make variable changes to dimension objects that exist in your drawing without creating an override to the dimension style that the dimension references (was created under). For example, using *Dimoverride,* you can make a variable change and select existing dimension objects to apply the change. The existing dimension does not lose its reference to the parent dimension style nor is the dimension style changed in any way. In effect, you can override the dimension styles for selected dimension objects. There are two steps: set the desired variable and select dimension objects to alter.

Command: ***DIMOVERRIDE***<br>
Enter dimension variable name to override or [Clear overrides]: (**variable name**)<br>
Enter new value for dimension variable <current value>: (**value**)<br>
Enter dimension variable name to override: **Enter**<br>
Select objects: **PICK**<br>
Select objects: **PICK** or **Enter**<br>
Command:

The *Dimoverride* feature differs from creating dimension style overrides in two ways: (1) *Dimoverride* applies the changes to the selected dimension objects only, so the overrides are not appended to the parent dimension styles, but only the selected objects; and (2) *Dimoverride* can be used once to change dimension objects referencing multiple dimension styles, whereas to make such changes to dimensions by creating dimension style overrides requires changing all the dimension styles, one at a time. The effect of using *Dimoverride* is essentially the same as using the *Properties* palette.

TIP

*Dimoverride* is useful as a "backdoor" approach to dimensioning. Once dimensions have been created, you may want to make a few modifications, but you do not want the changes to affect dimension styles (resulting in an update to all existing dimensions that reference the dimension styles). *Dimoverride* offers that capability. You can even make one variable change to affect all dimensions globally without having to change multiple dimension styles. For example, you may be required to make a test print of the drawing in a different scale than originally intended, necessitating a new global *Dimscale*. Use *Dimoverride* to make the change for the plot:

```
Command: DIMOVERRIDE
Enter dimension variable name to override or [Clear overrides]: DIMSCALE
Enter new value for dimension variable <1.0000>: .5
Enter dimension variable name to override: Enter
Select objects: (window entire drawing) Other corner: 128 found
Select objects: Enter
Command:
```

This action results in having all the existing dimensions reflect the new *DIMSCALE*. No other dimension variables or any dimension styles are affected. Only the selected objects contain the overrides. *Dimoverride* does not append changes (overrides) to the original dimension styles, so no action must be taken to clear the overrides from the styles. After making the plot, *Dimoverride* can be used with the *Clear* option to change the dimensions (by object selection) back to their original appearance. The *Clear* option is used to clear overrides from dimension objects, not from dimension styles.

### *Clear*

The *Clear* option removes the overrides from the selected dimension objects. It does not remove overrides from the current dimension style:

```
Command: DIMOVERRIDE
Enter dimension variable name to override or [Clear overrides]: c
Select objects: PICK
Select objects: Enter
Command:
```

The dimension then displays the variable settings as specified by the dimension style it references without any overrides (as if the dimension were originally created without the overrides). Using *Clear* does not remove any overrides that are appended to the dimension style so that if another dimension is drawn, the dimension style overrides apply.

## *Matchprop*

| Ribbon | Menu Bar | Command (Type) | Alias (Type) | Shortcut |
|---|---|---|---|---|
| *Home Properties Match Properties* | *Modify Match Properties* | *Matchprop* | *MA* | ... |

*Matchprop* can be used to "convert" an existing dimension to the style (including overrides) of another dimension in the drawing.

For example, if you have two linear dimensions, one has *Oblique* arrows and *Romans* text font (*Dimension Style* = "Oblique") and one is a typical linear dimension (*Dimension Style* = "Standard") as in Figure 29-73, "before."

You can convert the typical dimension to the "Oblique" style by using *Matchprop*, selecting the "Oblique" dimension as the "source object" (to match), then selecting the typical dimension as the "destination object" (to convert). The typical dimension then references the "Oblique" dimension style and changes appearance accordingly (Fig. 29-73, "after"). Note that *Matchprop* does not alter the dimension text value.

**FIGURE 29-73**

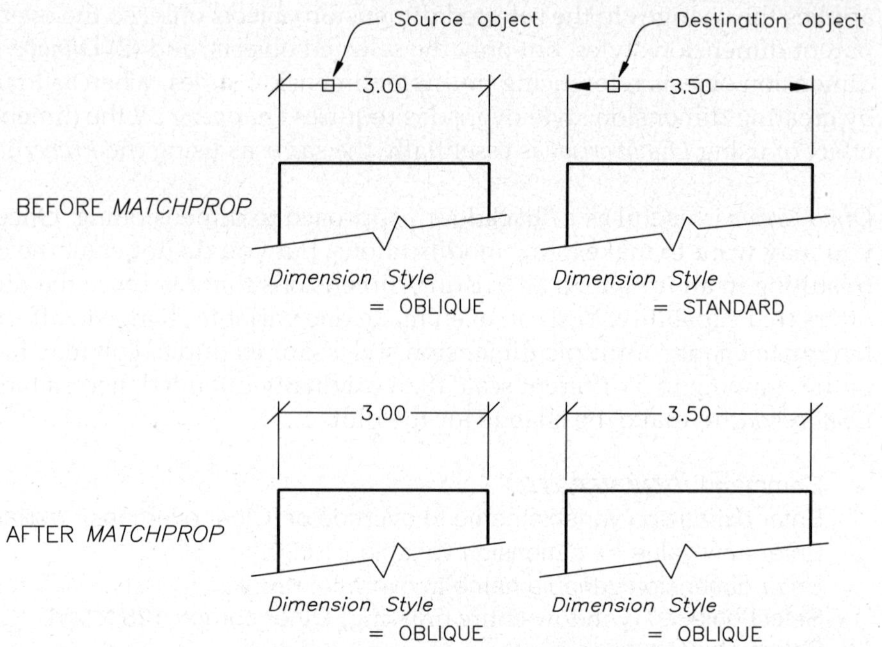


**TIP** Using the *Settings* option of the *Matchprop* command, you can display the *Property Settings* dialog box (Fig. 29-74). The *Dimension* box under *Special Properties* must be checked to "convert" existing dimensions as illustrated above.

**FIGURE 29-74**

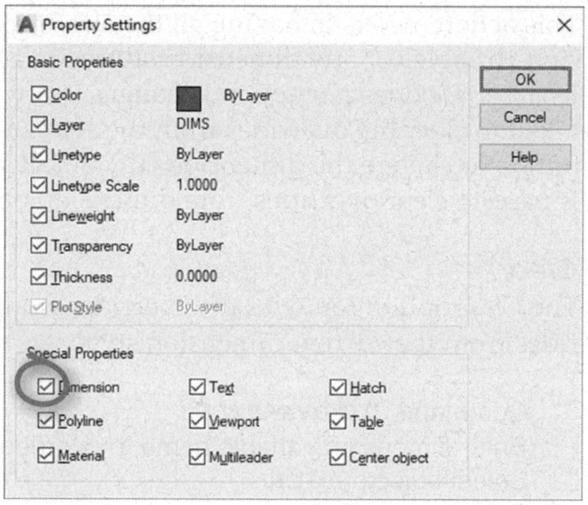


Using *Matchprop* is a fast and easy method for modifying dimensions from one style to another. However, this method is applicable only if you have existing dimensions in the drawing with the desired appearance that you want others to match.

## Dimension Right-Click Menu

If you select any dimension, you can right-click to produce a shortcut menu (Fig. 29-75). This menu contains options that allow you to change the *Dimension Style*, the *Precision*, and other options explained next.

**FIGURE 29-75**

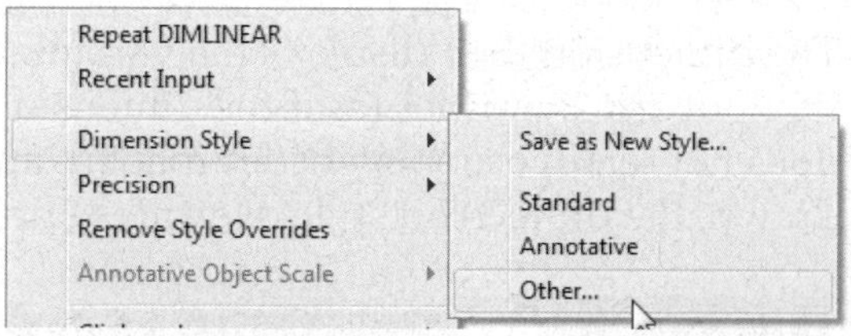


### *Dimension Style*

Use this option to change the selected dimension from one dimension style to another. You can easily create a new dimension style or change the dimension to another style, similar to using *Match Properties*. Select *Other* to produce a list of all the dimension styles contained in the drawing.

### *Precision*

Use this cascading menu to select from a list of decimal or fractional precision values. For example, you could change a dimension text value of "5.0000" to "5.00" with this menu. Your choice creates a DIMDEC dimension style override for the selected dimension.

### *Remove Style Overrides*

Use this option to clear all dimension style overrides for the selected dimension only. This option has the same results as using the *Clear* option of the *Dimoverride* command.

### *Annotative Object Scale*

Several options are available from this menu item when an annotative dimension object is highlighted. You can add annotative scales to or delete scales from the highlighted object. When a highlighted annotative dimension has several scales, you can synchronize their positions. See "Annotative Dimensions" and see Chapter 32 for more information on annotative objects.

## GUIDELINES FOR NON-ANNOTATIVE DIMENSIONING IN AutoCAD

Listed in this section are some guidelines to use for dimensioning a drawing using dimensioning variables and dimension styles. Although there are other strategies for dimensioning, two strategies are offered here as a framework so you can develop an organized approach to dimensioning.

In almost every case, dimensioning is one of the last steps in creating a drawing, since the geometry must exist in order to dimension it. You may need to review the steps for drawing setup, including the concept of drawing scale factor (Chapter 12).

### Strategy 1. Dimensioning a Single Drawing

This method assumes that the fundamental steps have been taken to set up the drawing and create the geometry. Assume this has been accomplished:

> Drawing setup is completed: *Units, Limits, Snap, Grid, Ltscale, Layers,* border, and titleblock.
> The drawing geometry (objects comprising the subject of the drawing) has been created.

Now you are ready to dimension the drawing subject (of the multiview drawing, pictorial drawing, floor plan, or whatever type of drawing).

1. Create a *Layer* (named DIM, or similar) for dimensioning if one has not already been created. Set *Continuous* linetype and appropriate color. Make it the *current* layer.

2. Set the *Overall Scale (DIMSCALE)* based on drawing *Limits* and expected plotting size (Fig. 29-76).

**FIGURE 29-76**

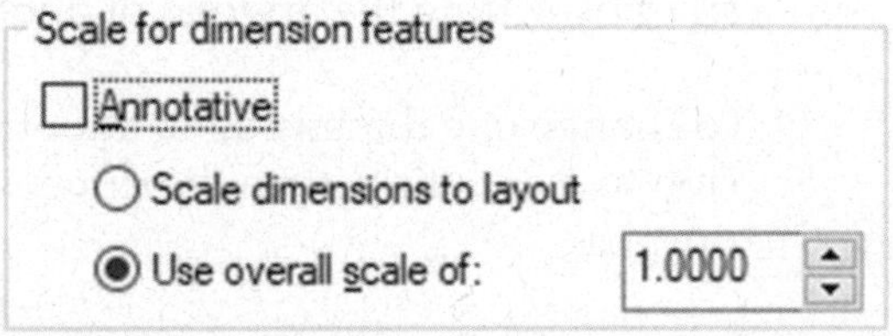


For plotted dimension text of 3/16":

Multiply *Overall Scale (DIMSCALE)* times the drawing scale factor. The default *Overall Scale* is set to 1, which creates dimensioning text of approximately 3/16" (default *Text Height:* or *DIMTXT* =.18) when plotted full size. All other size-related dimensioning variables' defaults are set appropriately.

For plotted dimension text of 1/8":

Multiply *Overall Scale* times the drawing scale factor, times .7. Since the *Overall Scale* times the scale factor produces dimensioning text of .18, then .18 x .7 = .126 or approximately 1/8". (See "Optional Method for Fixed Dimension Text Height.") (*Overall Scale* [*DIMSCALE*] is one variable that usually remains constant throughout the drawing and therefore should generally have the same value for every dimension style created.)

3. Make the other dimension variable changes you expect you will need to create most dimensions in the drawing with the appearance you desire. When you have the basic dimension variables set as you want, *Save* this new dimension style family as TEMPLATE or COMPANY_STD (or other descriptive name) style. If you need to make special settings for types of dimensions (*Linear, Diameter,* etc.), create "children" at this stage and save the changes to the style. This style is the fundamental style to use for creating most dimensions and should be used as a template for creating other dimension styles. If you need to reset or list the original (default) settings, the STANDARD style can be restored.

4. Create all the relatively simple dimensions first. These are dimensions that are easy and fast and require no other dimension variable changes. Begin with linear dimensions, then progress to the other types of dimensions.

5. Create the special dimensions next. These are dimensions that require variable changes. Create appropriate dimension styles by changing the necessary variables, then save each set of variables (relating to a particular style of dimension) to an appropriate dimension style name. Specify dimension variables for the classification of dimension (children) in each dimension style when appropriate. The dimension styles can be created "on the fly" or as a group before dimensioning. Use TEMPLATE or COMPANY_STD as your base dimension style when appropriate.

   For the few dimensions that require variable settings unique to that style, you can create dimension variable overrides. Change the desired variable(s), but do not save to the style so the other dimensions in the style are not affected.

6. When all of the dimensions are in place, make the final adjustments. Several methods can be used:

   A. If modifications need to be made familywide, change the appropriate variables and save the changes to the dimension style. This action automatically updates existing dimensions that reference that style.

   B. To modify the appearance of selected dimensions, use the *Properties* palette or the dimension right-click shortcut menu. Generally, these methods affect only the selected dimensions by creating dimension style overrides, so all dimensions in the style are not changed. You could also use *Dimoverride* for individual dimensions.

   C. Alternately, change the appearance of individual dimensions by creating dimension style overrides, then use *Update* or *-Dimstyle, Apply* to update selected dimensions. Clear the overrides by setting the original or another dimension style current.

   D. To change one dimension to adopt the appearance of another, use *Matchprop*. Select the dimension to match first, then the dimension(s) to convert. *-Dimstyle, Apply* can be used for this same purpose.

   E. Use *Dimoverride* to make changes to selected dimensions or to all dimensions globally by windowing the entire drawing. *Dimoverride* has the advantage of allowing you to assign the variables to change and selecting the objects to change all in one command. What is more, the changes are applied as overrides only to the selected dimensions but do not alter the original dimension styles in any way.

   F. If you want to change only the dimension text value, use *Properties* or *Dimedit, New*. *Dimedit, New* can also be used to reapply the AutoCAD-measured value if text was previously changed. The location of the text can be changed with *Dimtedit* or grips.

G. Grips can be used effectively to change the location of the dimension text or to move the dimension line closer or farther from the object. Other adjustments are possible, such as rotating a *Radius* dimension text around the arc or flipping arrows.

## Strategy 2. Creating Dimension Styles as Part of Template Drawings

1. Begin a *New* drawing or *Open* an existing *Template*. Assign a descriptive name.

2. Create a DIM *Layer* for dimensioning with *continuous* linetype and appropriate color and lineweight (if one has not already been created).

TIP

3. Set the *Overall Scale* accounting for the drawing *Limits* and expected plotting size. Use the guidelines given in Strategy 1, step 2. Make any other dimension variable changes needed for general dimensioning or required for industry or company standards.

4. Next, *Save* a dimension style named TEMPLATE or COMPANY_STD. This should be used as a template when you create most new dimension styles. The *Overall Scale* is already set appropriately for new dimension styles in the drawing.

5. Create the appropriate dimension styles for expected drawing geometry. Use TEMPLATE or COMPANY_STD dimension style as a base style when appropriate.

6. *Save* and *Exit* the newly created template drawing.

7. Use this template in the future for creating new drawings. Restore the desired dimension styles to create the appropriate dimensions.

Using a template drawing with prepared dimension styles is a preferred alternative to repeatedly creating the same dimension styles for each new drawing.

## Optional Method for Fixed Dimension Text Height in Template Drawings

To summarize Strategy 1, step 2., the default *Overall Scale* (*DIMSCALE* =1) times the default *Text Height* (*DIMTXT* =.18) produces dimensioning text of approximately 3/16″ when plotted to 1=1. To create 1/8″ text, multiply *Overall Scale* times .7 (.18 x .7 = .126). As an alternative to this method, try the following.

For 1/8″ dimensions, for example, multiply the initial values of the size-related variables by .7; namely

| | | |
|---|---|---|
| *Text Height* | (*DIMTXT*) | .18 x .7 = .126 |
| *Arrow Size* | (*DIMASZ*) | .18 x .7 = .126 |
| *Extension Line Extension* | (*DIMEXE*) | .18 x .7 = .126 |
| *Dimension Line Spacing* | (*DIMDLI*) | .38 x .7 = .266 |
| *Text Gap* | (*DIMGAP*) | .09 x .7 = .063 |

Save these settings in your template drawing(s). When you are ready to dimension, simply multiply *Overall Scale* (1) times the drawing scale factor.

Although this method may seem complex, the drawing setup for individual drawings is simplified. For example, assume your template drawing contained preset *Limits* to the paper size, say 11 x 8.5. In addition, the previously mentioned dimension variables were set to produce 1/8″ (or whatever) dimensions when plotted full size. Other variables such as *LTSCALE* could be appropriately set. Then, when you wish to plot a drawing to 1=1, everything is preset. If you wish to plot to 1=2, simply multiply all the size-related variables (*Limits*, *Overall Scale*, *Ltscale*, etc.) times 2!

## DIMENSIONING IN PAPER SPACE LAYOUTS

Generally, dimensions are created in model space and are attached to model space geometry. In this way, you can make one or more layouts, each with one or more viewports that display the model space geometry, and the dimensions are visible by default in each of the layouts and viewports. Assuming the dimensions are created on a dimensioning layer or layers, you can control the display of the dimensions in each viewport using viewport-specific layer visibility controls in the *Layer Manager*. This strategy is used for most AutoCAD drawings except for certain cases.

It is possible to create dimensions in a paper space layout associated with (attached to) model space geometry inside a viewport. To explain, a typical drawing includes geometry in model space (such as mechanical part views, a floor plan, or a diagram) and at least one paper space layout used to display and print the model geometry. You can create the dimensions in a layout (in paper space) and *Object Snap* to objects in model space (inside a viewport). These new dimensions are fully associative and display the actual measurement value of the drawing objects in model space units.

This example drawing (Fig. 29-77) has some dimensions in model space (left viewport) and some dimensions in paper space (right viewport). Note that the dimensions for the small cutout have been omitted in model space (left viewport). All of the dimensions for the detail (right viewport) are created in paper space. Some of the dimensions for the detail are actually outside of the viewport border, but the two short horizontal dimensions (.5 and .25) appear to be in model space even though they were actually created in paper space. Note that all dimensions associated to model space geometry, whether created in model space or paper space, display the correct value of the model feature they are measuring.

**FIGURE 29-77**

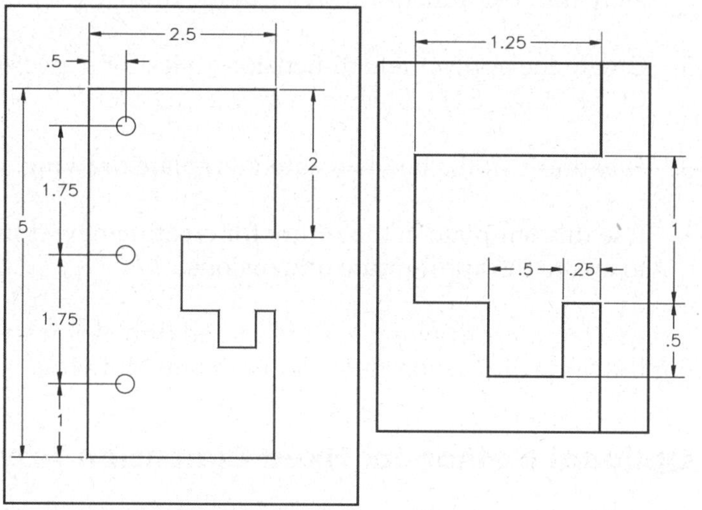


Figure 29-78 displays the completed drawing after adding some annotation and freezing the viewport border layer. Note that you cannot detect which dimensions are in model space and which are in paper space.

**FIGURE 29-78**

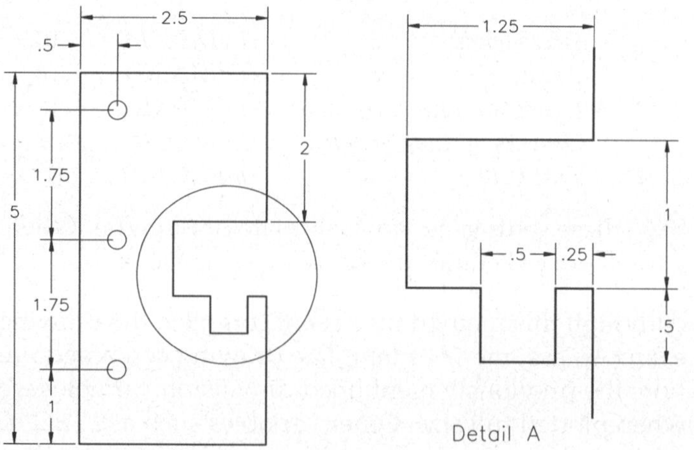

## When to Dimension in Paper Space

Although it is still recommended to create dimensions in model space for most situations, there are some cases where creating dimensions in paper space could be used. Generally, whenever you want specific dimensions to appear for only one of several viewports, it may be useful to create those dimensions in paper space. Here are two specific examples.

If you have created a 3D solid model, then generated 2D views (a multiview layout) from the model with each view in a separate paper space viewport, consider creating the dimensions for each view in paper space. (See Chapter 37, 2D Drawings from 3D Models.)

If you want one or more detail views (small, enlarged sections of a larger drawing) each in a separate viewport or layout and at different scales than the full drawing display, consider creating dimensions for the detail views in paper space. Dimensioning in paper space for this application is an alternative to using annotative dimensions (see "Annotative Dimensions").

For the situations listed above, the advantages of creating dimensions in paper space attached to objects in model space (inside a viewport) are these:

Setting *DIMSCALE* or *Overall Dimension Scale* (the size of the dimension text, arrowheads, etc.) for paper space dimensions is simplified since you are concerned only with paper space units and not with the viewport scale (the scale of the geometry that appears in the viewports). This is especially important when you have several detail views of a drawing and each detail view is at a different scale than the full drawing view. Paper space dimensions appear in one size, whereas model space dimensions appear in different sizes when the drawing is scaled differently in each viewport.

Creating dimensions in paper space ensures that those dimensions appear only for that viewport but not for other displays of the model geometry appearing in other viewports and layouts. Therefore, you do not have to use different layers for different sets of dimensions and the *Layer Properties Manager* to set viewport-specific visibility (which dimension layers you want to appear in which viewports).

Dimensioning a 3D model requires that dimensions for each side or view of the model appear on different planes (created by using construction planes called User Coordinate Systems or UCSs in AutoCAD). For a multiview-type setup of a 3D model, dimensioning in paper space prevents having to use different UCSs as well as different layers for the dimensions that should appear in each view (on each plane of the 3D model). Also, you do not have to control viewport-specific layer visibility to ensure that dimensions for the top view do not appear in the front view, and so on.

TIP

Although it appears that these advantages might outweigh the practice of dimensioning in model space, most drawings require that the dimensions appear in multiple viewports or layouts; therefore, dimensioning in model space is the more common practice. Dimensioning in paper space eliminates the possibility of showing the same dimensions in multiple viewports or layouts. See Chapters 32 and 37 for more information and applications of dimensioning in paper space and using annotative dimensions.

## Procedure for Dimensioning in Paper Space

Dimensioning in paper space differs from dimensioning in model space only in a few respects such as determining the *DIMSCALE* (*Overall Scale*) and the need to use *Dimregen* when zooming or panning in the viewport; therefore, the procedure for dimensioning in paper space is similar to that for dimensioning in model space. Here is a typical procedure to use when you need paper space dimensions.

1. Complete the model space geometry.

2. (Optional, depending on the situation and desired result.) Create a layer named DIM or DIMENSIONS and set it *Current*. Use the *Dimension Style Manager* to create a dimension style for the model space dimensions. Set the *Overall Scale* and other variables and save the style. Create the dimensions you need in model space. These dimensions are those that you want to appear in multiple viewports.

3. Set up the desired layouts and viewports and set the desired viewport scale for each viewport (so the model geometry appears in the desired scale in each viewport).

4. Create a layer named DIM-PS or DIMENSIONS-PAPERSPACE and set the layer *Current*.

5. Use the *Dimension Style Manager* to create a dimension style for the paper space dimensions by copying the style for model space dimensions and rename the style DIM-PS or similar. Set the *Overall Scale* so the dimensions print in paper space units. (Normally, an *Overall Scale* of 1 is used for paper space dimensions since the drawing is printed from the layout at 1:1.) The other dimension variables normally would not change (from the model space settings).

6. Activate the desired layout and ensure that paper space is active (the cursor appears in paper space, not in the viewport). With *Object Snap* on, create the dimensions in paper space but *Object Snap* the extension line origins, etc. to the objects in model space. The location of the dimension text can be placed outside or inside the viewport borders but the actual dimension objects are in paper space. Figure 29-79 displays two paper space dimensions, one has text inside the viewport border.

**FIGURE 29-79**

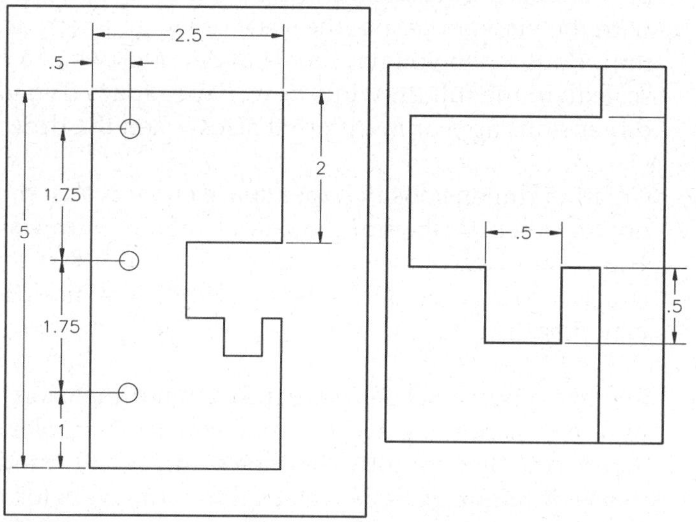


7. Complete the paper space dimensions. If you need to *Zoom* or *Pan* inside the viewport or change the viewport scale (using *Zoom XP* or the *Viewport Scale* list), you should use *Dimregen* to force AutoCAD to redisplay the location of the paper space dimension objects to align correctly with the new location of the model space geometry. Figure 29-80 illustrates the drawing after using *Pan* but before using *Dimregen*.

**FIGURE 29-80**

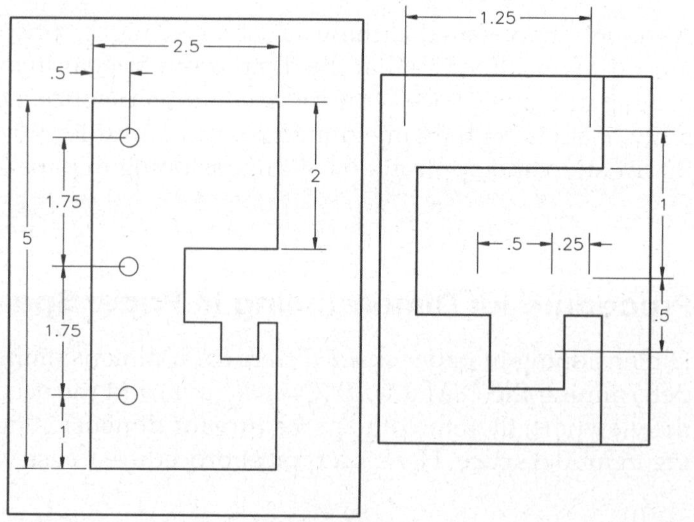


8. Complete any additional notation or objects in paper space. In the *Layer Properties Manager*, *Freeze* or disable plotting for the viewport border layer (see Fig. 29-78).

## ANNOTATIVE DIMENSIONS

For complex drawings including multiple layouts and viewports, it may be desirable to create annotative objects. Annotative objects, such as annotative text, annotative dimensions, and annotative hatch patterns, automatically adjust for size when the *Annotative Scale* for each viewport is changed. That is, annotative objects can be automatically resized to appear "readable" in different viewports at different scales. The following section gives a quick overview of how to create annotative dimensions for use in multiple viewports at different scales.

### Steps for Creating and Viewing Basic Annotative Dimensions

These steps assume the drawing geometry is created but dimensions have not been created.

1. Produce the Dimension Style Manager by any method.
2. In the Dimension Style Manager, create a New dimension style or Modify either the Standard or Annotative style. In the Fit tab, check the Annotative box if not already checked. Set the Text Height for dimensions to the actual paper space height you want to use. Set any other desired variable settings (Text Style, Units Format and Precision, etc.). Select the OK button.
3. In the Model tab, set the Annotation Scale for model space. You can use the Annotation Scale pop-up list or set the CANNOSCALE variable. Use the reciprocal of the "drawing scale factor" (DSF) as the annotation scale (AS) value, or AS = 1/DSF.
4. Create the dimensions in the Model tab. The dimensions should appear in the appropriate size. If you plan to create a viewport for a detailed view, you can wait to apply these dimensions later.
5. Create the desired viewports if not already created. Typically, one viewport is intended to show the entire model space geometry, and would therefore be displayed at about the same scale as that in model space. Other viewports may be created to display the geometry in other scales, such as for a detail view. Specify the desired plot device for the layouts using the Page Setup Manager.
6. In a layout tab when paper space is active (not in a viewport), enable the Add scales to annotative objects when the annotative scale changes toggle in the lower-right corner of the Drawing Editor or set the ANNOAUTOSCALE variable to a positive value (1, 2, 3, or 4). This setting remains active for all layouts.
7. Next, activate the layout in which you intend to show the entire model space geometry. Set the desired scale for the viewport by double-clicking inside the viewport and using the Viewport Scale pop-up list or Viewports toolbar. (Alternately, select the viewport object [border] and use the Properties palette for the viewport.) Setting the Viewport Scale automatically adjusts the annotation scale for the viewport. The dimensions should appear in the appropriate size for the viewport. The dimensions appearing in this viewport do not change size if the Viewport Scale is the same scale that you selected for the Annotation Scale in model space.
8. Once the viewport scale is set, select the Lock/Unlock Viewport toggle in the lower-right corner of the Drawing Editor to lock the viewport. If you Zoom in an unlocked viewport, the scale can get out of synchronization with the selected scale.
9. Activate another viewport by double-clicking inside the viewport. Set the viewport scale for the viewport using the Viewport Scale pop-up list, Viewports toolbar, or Properties palette. Since the Annotation Scale is locked to the Viewport Scale, the annotation scale changes automatically to match the Viewport Scale.
10. Repeat step 9 for other viewports.
11. Create any other dimensions in detail views.
12. Make the necessary adjustments to the dimensions.

This procedure can become complex depending on the number of viewports and annotative scales. See Chapter 32 for more information on creating annotative objects in multiple viewports, creating detail views, and adjusting annotative dimension objects.

### Annotative Dimension Example

For example, assume you are drawing a stamped mechanical part. In order to draw the stamping in the *Model* tab full size, you change the default *Limits* (12, 9) by a factor of 4 to arrive at *Limits* of 48,36 (drawing scale factor = 4). After creating the part outline and mounting holes, you are ready to begin dimensioning.

Following steps 1 through 3, create a new dimension style and check the *Annotative* box in the *Fit* tab. Set any other desired variables, such as *Text Height*, *Text Style*, *Units Format* and *Precision*. Next, set the annotative scale for the *Model* tab using the *Annotation Scale* pop-up list in the lower-right corner of the Drawing Editor (Fig. 29-81). Using the drawing scale factor (DSF) of 4, set the *Annotation Scale* to 1:4 (or AS = 1/DSF). (Also see "Steps for Drawing Setup Using Annotative Objects" in Chapter 12.) Next (step 4), create the overall dimensions in the *Model* tab (Fig. 29-81).

FIGURE 29-81

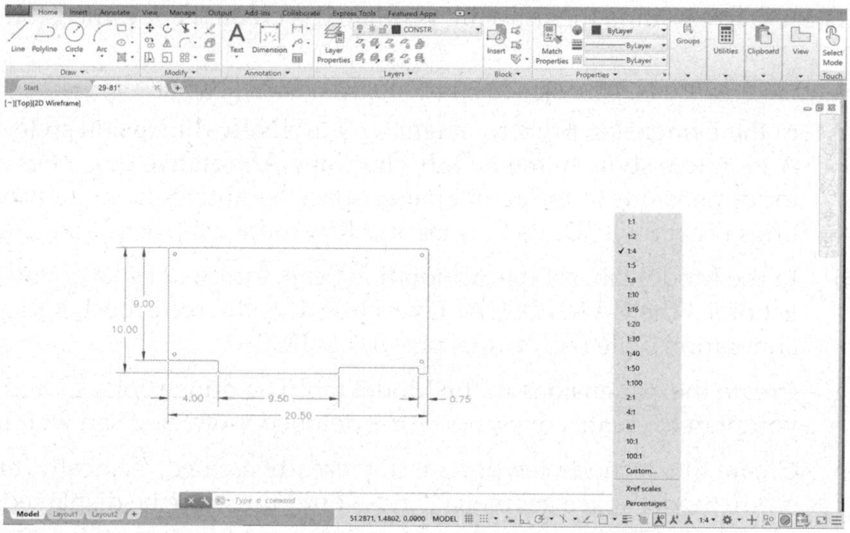

Next, following steps 5 and 6, switch to a *Layout* tab, set it up for the desired plot or print device using the *Page Setup Manager*, and create a viewport to display the entire model space geometry. Set the viewport scale for the viewport using the *Viewport Scale* pop-up list as shown in Figure 29-82. Note that the *Annotation Scale* is locked to the *Viewport Scale* (by default) so its setting changes accordingly as the *Viewport Scale* changes. The dimensions may not appear to be changed in this viewport if the *Viewport Scale* is the same as that selected in model space for the *Annotation Scale*.

FIGURE 29-82

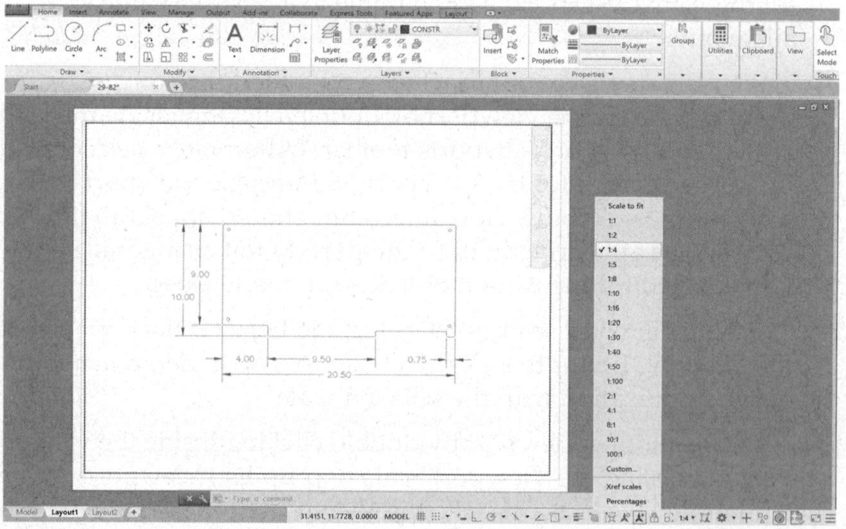

Note that at this point, using only one viewport, there is no real advantage to using annotative dimensions as opposed to non-annotative dimensions. The dimensions appear in the *Model* tab and *Layout* tab at the correct size, just as if non-annotative dimensions were used.

To illustrate setting the annotative scale for other viewports, create another layout with two viewports, one to display the entire stamping at 1:4 scale and the second viewport to display only a detail of one corner at 1:2 scale. Examining the display of the drawing in the two viewports at this point, the dimensions would be displayed at the same size relative to the geometry.

To change the display of the annotative text in the left viewport, make the viewport active and select 1:2 from the *Viewport Scale* pop-up list (Fig. 29-83). Once this is completed, the size of the geometry will be changed to exactly 1:2, and since the *Annotation Scale* is locked to the *Viewport Scale*, the annotative dimensions will be displayed at the correct size. Simply stated, the dimensions will appear at the same (paper space) size as those in the other viewport.

**FIGURE 29-83**

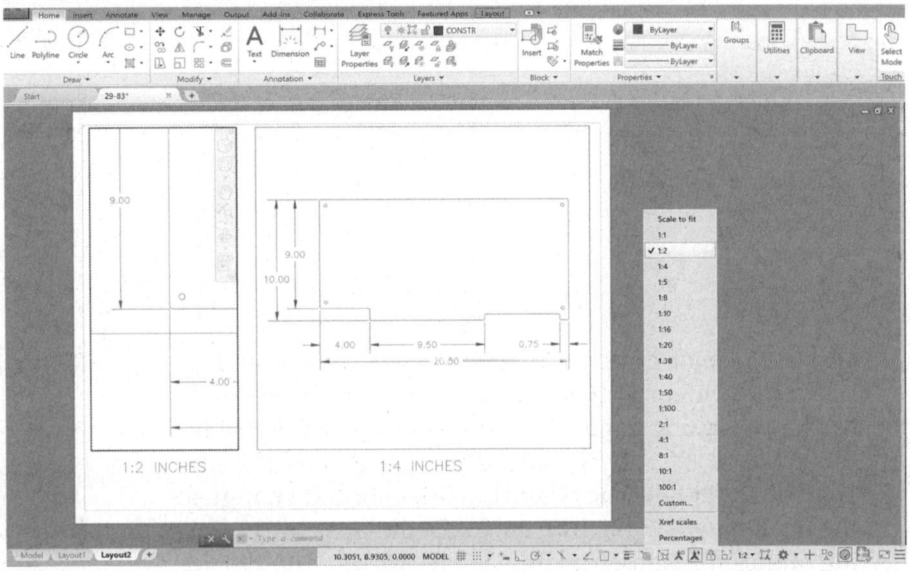


## Multiple Annotative Dimension Objects

When a different *Viewport Scale* is selected and the annotative scale for the dimensions changes, the original dimension objects do not actually change size. Instead, a new annotative dimension object is created as a copy of the original, but in the new scale (Fig. 29-84). When the *Add scales to annotative objects...* toggle (the *ANNOAUTOSCALE* variable) is on, new dimension objects are created for each new scale selected from the *Viewport Scale* list. Therefore, note that a single annotative object can have multiple *Annotative Scales* (or *Object Scales*), one for each time the object is displayed in a different scale.

**FIGURE 29-84**

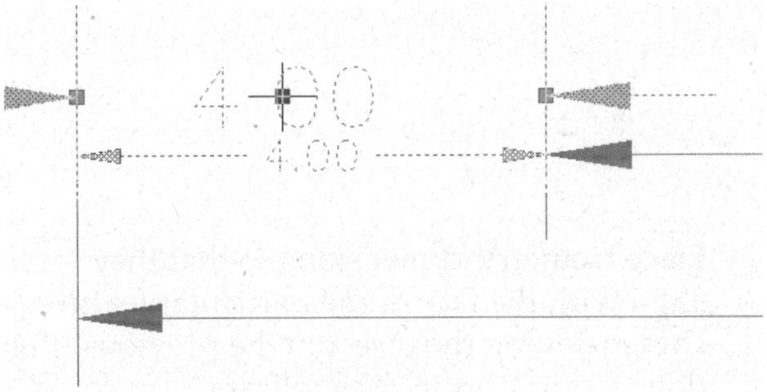


TIP

The *SELECTIONANNODISPLAY* system variable, when set to 1, allows you to see each dimension object when selected, as shown in Figure 29-84. The dimensions can be manipulated individually using grips.

See Chapter 32 for more information on using annotative objects in multiple viewports, creating dimensions for detail views, and using grips to manipulate annotative dimensions.

# DIMENSIONING ISOMETRIC DRAWINGS

Dimensioning isometric drawings in AutoCAD is accomplished using *Dimaligned* dimensions and then adjusting the angle of the extension lines with the *Oblique* option of *Dimedit*. The technique follows two basic steps:

1. Use *Dimaligned* or the *Rotated* option of *Dimlinear* to place a dimension along one edge of the isometric face. Isometric dimensions should be drawn on the isometric axes lines (vertical or at a 30° rotation from horizontal; Fig. 29-85).

FIGURE 29-85

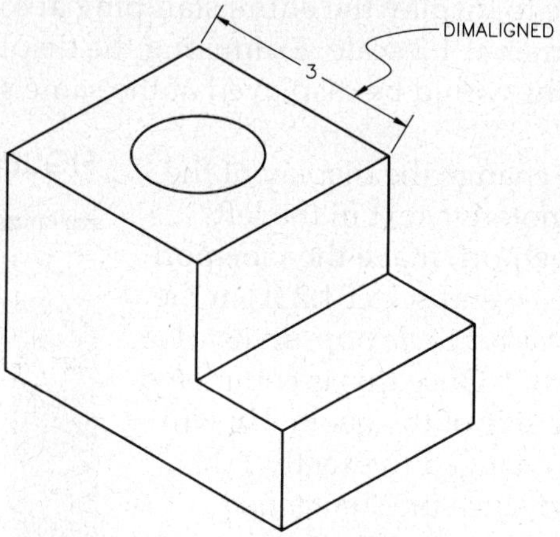


2. Use the *Oblique* dimensioning option of *Dimedit* and select the dimension just created. When prompted to "Enter obliquing angle," enter the desired value (**30** in this case) or PICK two points designating the desired angle (Fig. 29-86). The extension lines should change to the designated angle. In AutoCAD, the possible obliquing angles for isometric dimensions are **30**, **150**, **210**, or **330**.

FIGURE 29-86

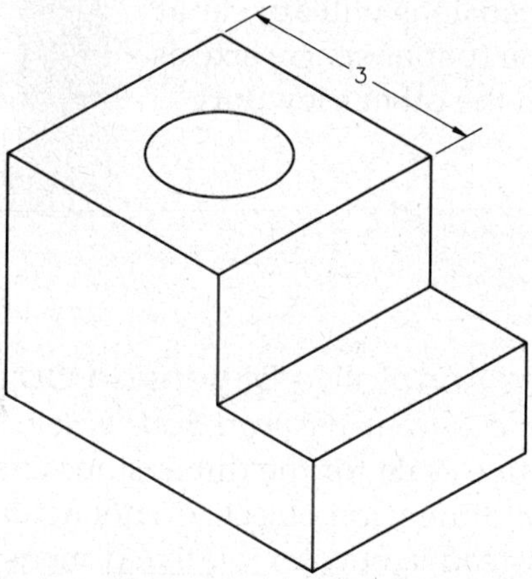


Place isometric dimensions so that they align with the face of the particular feature. Dimensioning the object in the previous illustrations would continue as follows.

Create a *Dimlinear, vertical* dimension along a vertical edge (Fig. 29-87).

FIGURE 29-87

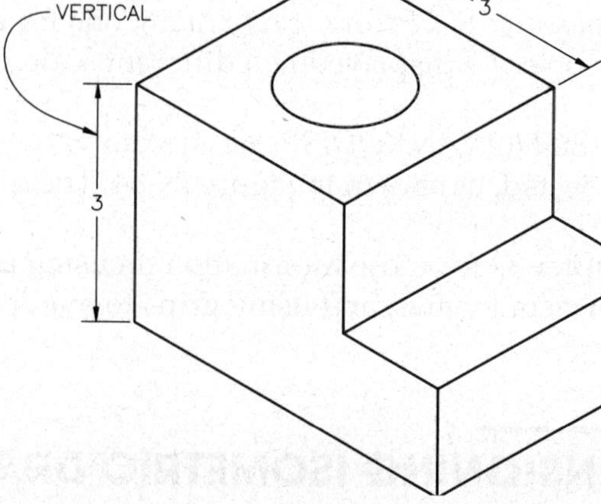

Use *Dimedit, Oblique* to force the dimension to an isometric axis orientation. Enter a value of **150** or PICK two points in response to the "Enter obliquing angle:" prompt (Fig 29-88).

**FIGURE 29-88**

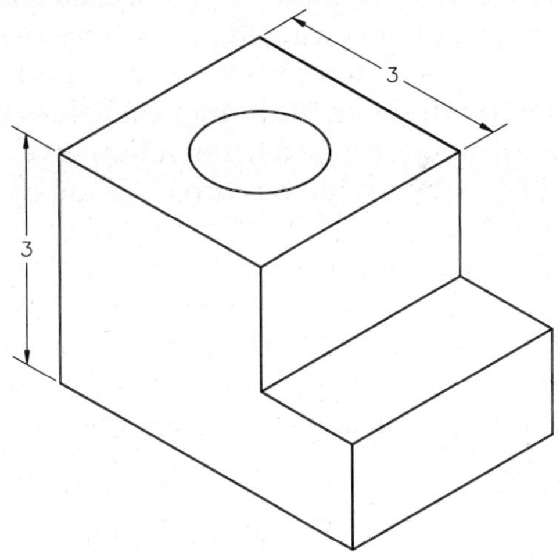


For isometric dimensioning, the extension line origin points <u>must</u> be aligned with the isometric axes. If not, the dimension is not properly oriented. This is important when dimensioning an isometric ellipse (such as this case) or an inclined or oblique edge. Construct centerlines for the ellipse on the isometric axes. Next, construct an *Aligned* dimension and *Object Snap* the extension line origins to the centerlines. Finally, use *Dimedit Oblique* to reorient the angle of the extension lines (Fig 29-89).

**FIGURE 29-89**

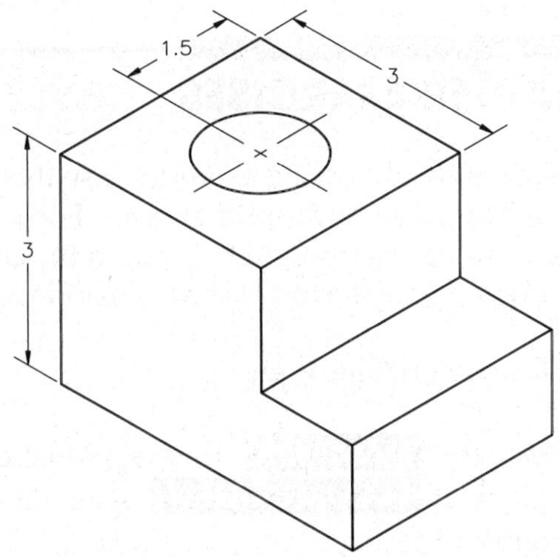


Using the same technique, other appropriate *Dimlinear vertical* or *Aligned* dimensions are placed and reoriented with *Oblique*. Use a *Leader* to dimension a diameter of an *Isocircle,* since *Dimdiameter* cannot be used for an ellipse (Fig. 29-90).

**FIGURE 29-90**

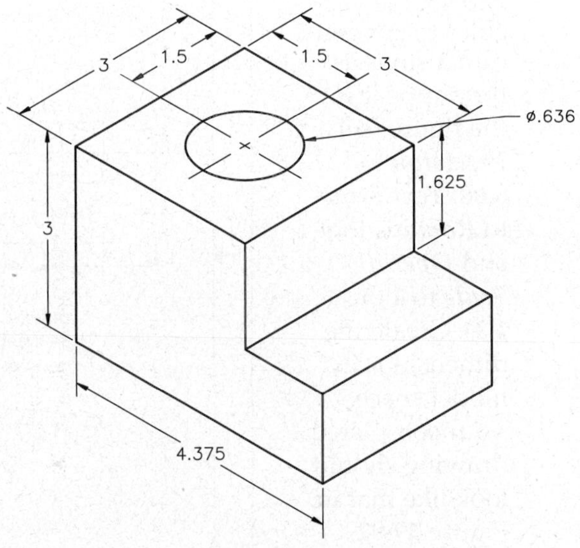

The dimensioning text can be treated two ways. (1) For quick and simple isometric dimensioning, unidirectional dimensioning (all values read from the bottom) is preferred since there is no automatic method for drawing the numerical values in an isometric plane. (2) As a better alternative, text *Styles* could be created with the correct obliquing angles (30 and -30 degrees). The text must also be rotated to the correct angle using the *Rotate* option of the dimension commands or by using *Dimedit Rotate* (Fig. 29-91). Optionally, *Dimension Styles* could be created with the correct variables set for text style and rotation angle for dimensioning on each isoplane.

**FIGURE 29-91**

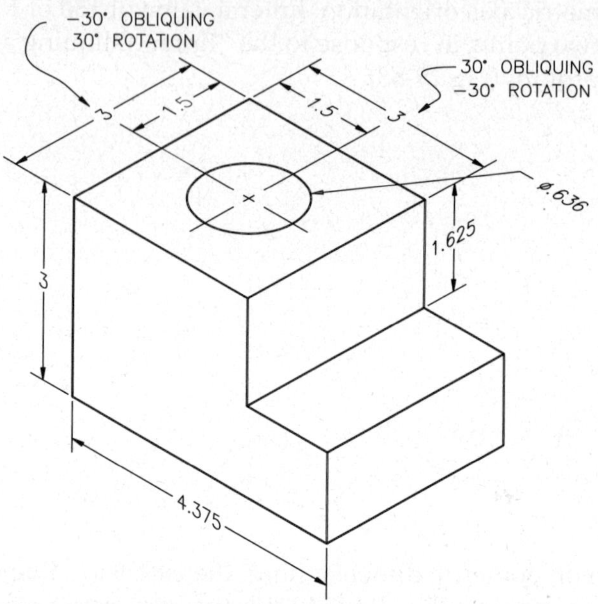


## CHAPTER EXERCISES

For each of the following exercises, use the existing drawings, as instructed. Create dimensions on the DIM (or other appropriate) layer. For 1. through 8., follow the "Guidelines for Non-annotative Dimensioning in AutoCAD" given in the chapter, including setting an appropriate *Overall Scale* based on the drawing scale factor. Use dimension variables and create and use dimension styles when needed.

1. **Dimension One View**

   ***Open*** the **HAMMER** drawing you created in Chapter 13 Exercises and use the ***Saveas*** command to rename it **HAMMER-DIM.** Add all necessary dimensions as shown in Chapter 13 Exercises, Figure 13-39.

   Create a ***New*** dimension style and set the variables to generate dimensions as they appear in the figure (set ***Precision*** to **0.00**, ***Text Style*** to ***Romans*** font, and ***Overall Scale*** to **1** or **1.3**). Create the dimensions in model space. Your completed drawing should look like that in Figure 29-92.

**FIGURE 29-92**

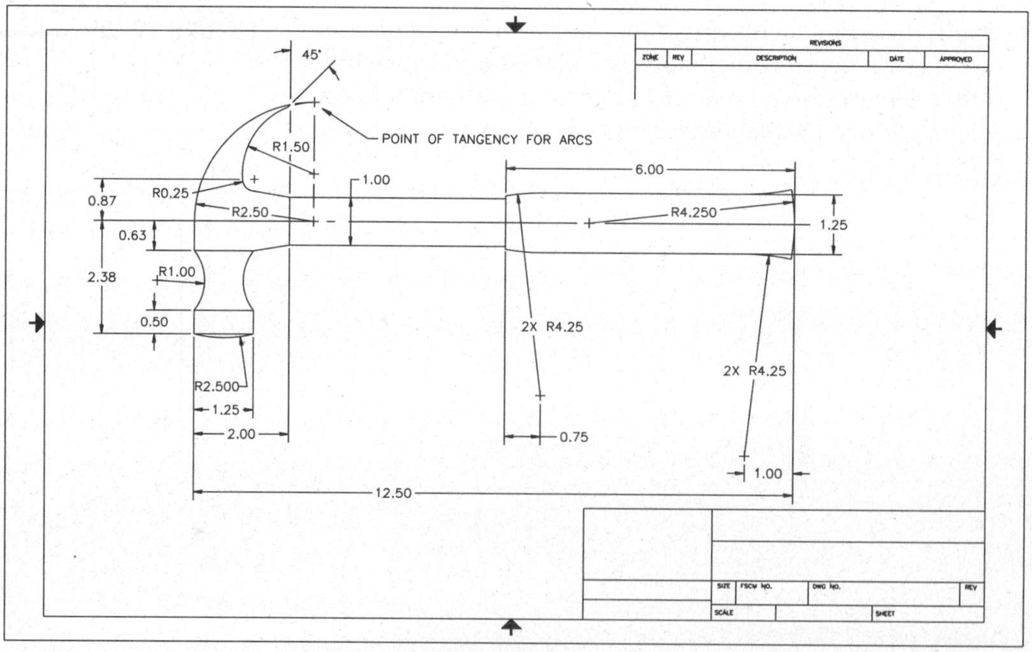

2. **Dimensioning a Multiview**

*Open* the **SADDLE** drawing that you created in Chapter 24 Exercises. Set the appropriate dimensional ***Units*** and ***Precision***. Add the dimensions as shown. Because the illustration in Figure 29-93 is in isometric, placement of the dimensions can be improved for your multiview. Use optimum placement for the dimensions. ***Save*** the drawing as **SADDL-DM** and make a ***Plot*** to scale.

**FIGURE 29-93**

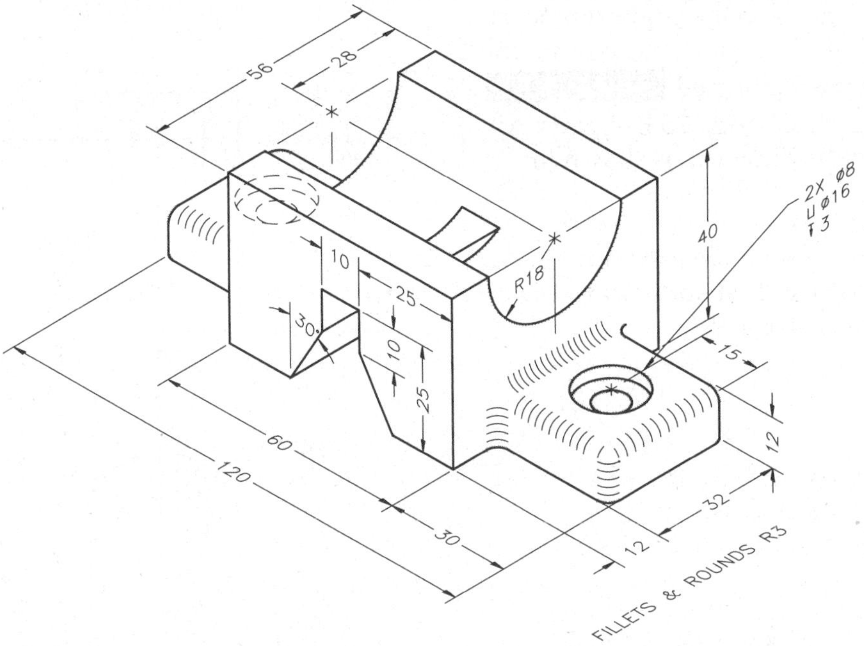


3. **Architectural Dimensioning**

*Open* the **OFFICE** drawing that you completed in Chapter 21. Dimension the floor plan as shown in Figure 21-87. Add ***Text*** to name the rooms. ***Save*** the drawing as **OFF-DIM** and make a ***Plot*** to an accepted scale and sheet size based on your plotter capabilities.

4. **Dimensioning an Auxiliary**

*Open* the **ANGLBRAC** drawing that you created in Chapter 27. Dimension as shown in Figure 29-94, but convert the dimensions to *Decimal* with **Precision** of **.000**. Use the "Guidelines for Dimensioning." Dimension the slot width as a ***Limit*** dimension—**.6248/.6255**. ***Save*** the drawing as **ANGL-DIM** and ***Plot*** to an accepted scale.

**FIGURE 29-94**

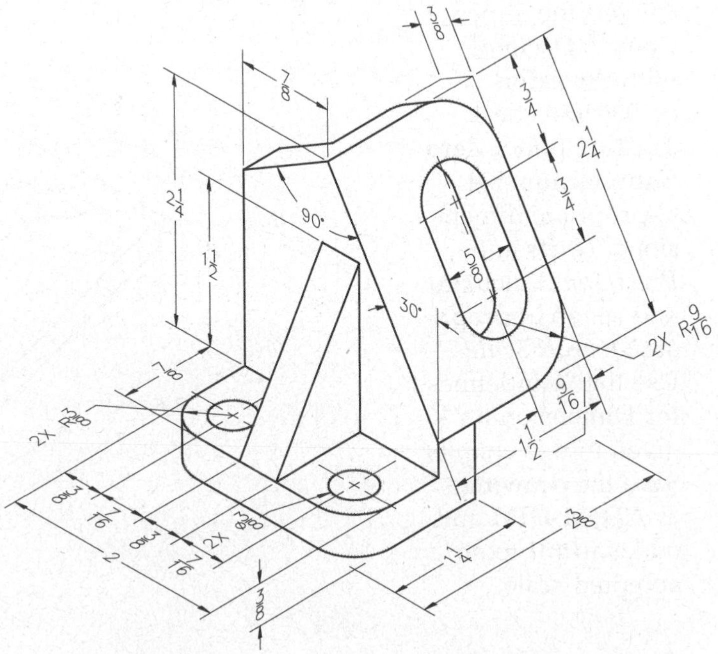

5. **Isometric Dimensioning**

Dimension the support bracket that you created as an isometric drawing named **SBRACKET** in the Chapter 25 Exercises. All of the dimensions shown in Figure 29-95 should appear on your drawing in the optimum placement. *Save* the drawing as **SBRCK-DM** and *Plot* to **1=1** on an A size sheet.

**FIGURE 29-95**

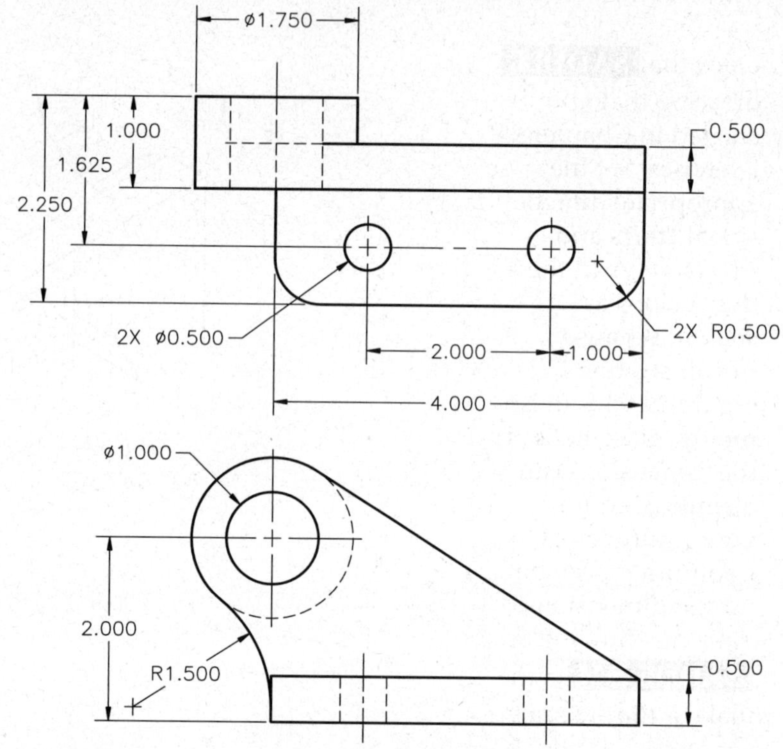


6. **Dimensioning a Multiview**

*Open* the **ADJMOUNT** drawing that you completed in Chapter 24. Add the dimensions shown in Figure 29-96 but convert the dimensions to *Decimal* with **Precision** of **.000** and check **Trailing** under **Zero Suppression**. Set appropriate dimensional *Units* and *Precision*. Calculate and set an appropriate *Overall Scale*. Use the "Guidelines for Dimensioning" given in this chapter. Save the drawing as **ADJM-DIM** and make a *Plot* to an accepted scale.

**FIGURE 29-96**

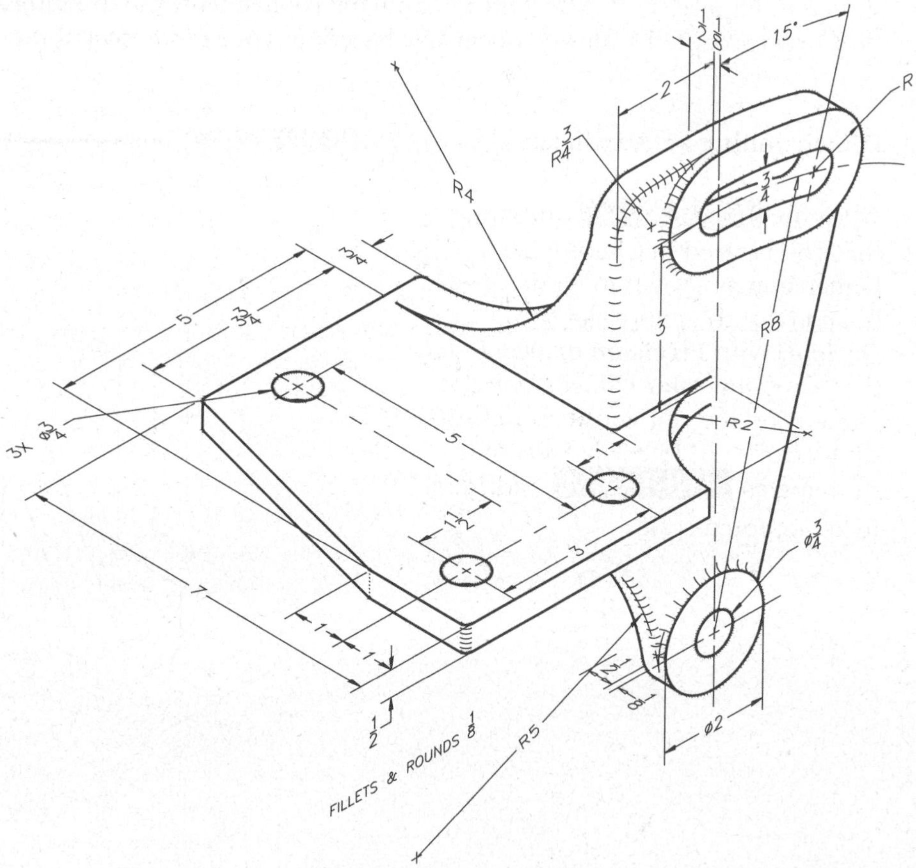

7. **Geometric Dimensioning and Tolerancing**

This exercise involves Flatness, Profile of a Surface, and Position applications. ***Open*** the dimensioned **ANGL-DIM** drawing you worked on in Exercise 4 and add the following geometric dimensions:

A. **Flatness** specification of **.002** to the bottom surface.

B. **Profile of a Surface** specification of **.005** to the angled surface relative to the bottom surface and the right side surface of the 1/4 dimension. The 30° angle must be ***Basic***.

C. **Position** specification of **.01** for the two mounting holes relative to the bottom surface and two other surfaces that are perpendicular to the bottom. Remember, the location dimensions must be ***Basic*** dimensions.

Use ***SaveAs*** to save and name the drawing **ANGL-TOL**.

8. **Geometric Dimensioning and Tolerancing**

This exercise involves a cylindricity and runout application. Draw the idler shown in Figure 29-97. Apply a **Cylindricity** specification of **.002** to the small diameter and a **Circular Runout** specification of **.007** to the large diameter relative to the small diameter as shown. The P730 neck has a **.07** radius and **30°** angle. ***Save*** the drawing as **IDLERDIM**.

FIGURE 29-97

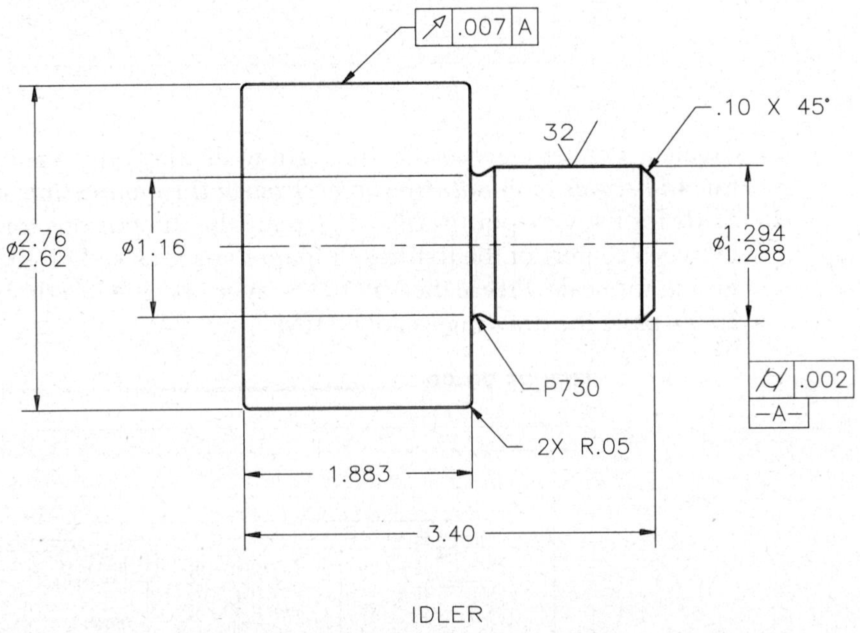


9. **Annotative Dimensions**

A. ***Open*** the **2BR-APT** drawing you worked with in Chapter 20 Exercises. Make a ***New*** dimension style and set the variables to generate dimensions as they appear in Chapter 15 Exercises. In the ***Fit*** tab, check the ***Annotative*** box, then select ***OK*** to save the style. Create the dimensions in model space on a layer called **DIM**.

B. Create a ***New Layout***. Use ***Pagesetup*** for the layout and select an appropriate ***Plot Device*** that can use a "**B**" size sheet, such as an HP 7475 plotter. Set the layout to a "B" size sheet. Next, use DesignCenter to locate and insert the **ANSI-B title block** from the Template folder. If you do not have the ANSI-B title block on your computer, you can download it from the website listed on the back cover of this book.

Make a new layer named **VPORTS** and create one viewport as shown in Figure 29-98. Set the viewport scale to display the apartment floor plan at **1/4″=1′** scale.

**FIGURE 29-98**

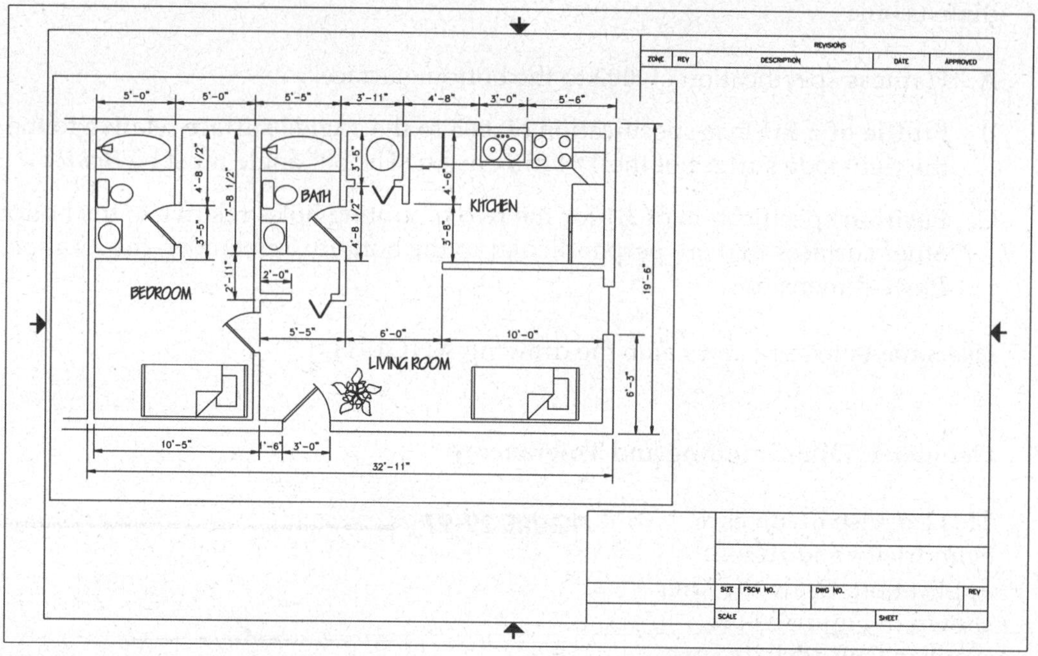


C. Create a smaller viewport on the right to display only one of the bathrooms. Turn off the toggle for ***Add scales to annotative objects when the annotation scale changes***. Set the ***Viewport Scale*** for the viewport to **1/2″=1′**. Create the dimensions for the plumbing wall (distance between centers of the fixtures) in paper space as shown in Figure 29-99. Label each viewport giving the scale. ***Freeze*** the **VPORTS** layer to achieve a drawing like that shown in Figure 29-99. Save the drawing as **APT-DIM**.

**FIGURE 29-99**

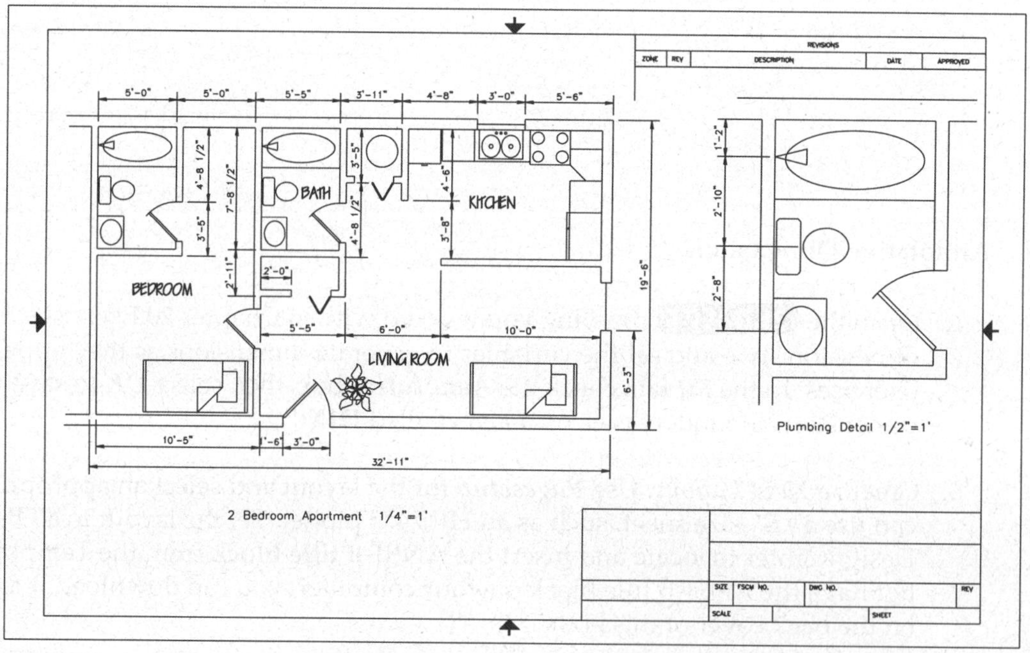

CHAPTER 30

# Xreferences

## Chapter Objectives

After completing this chapter you should:

1. know the differences and similarities between an *Xrefed* drawing and an *Inserted* drawing;
2. be able to *Attach* an externally referenced drawing to the current drawing;
3. understand that an *Overlay* drawing cannot be nested;
4. be able to *Reload*, *Unload*, or *Detach* an externally referenced drawing;
5. be able to *Bind* an entire *Xrefed* drawing and *Xbind* individual named objects;
6. be able to *Xclip* an externally referenced drawing so only a portion of the *Xref* appears;
7. be able to control demand loading with *INDEXCTL* and *XLOADCTL*;
8. be able to use *Xopen* to open an *Xref* drawing for editing in a separate window;
9. know how to edit *Blocks* and *Xrefs* "in place" using *Refedit* and how to control in-place editing with *XEDIT*.

## CONCEPTS

When you *Insert* a drawing as a *Block*, the inserted drawing becomes part of the current drawing—it is considered an "internal" reference. An *Inserted Block* (internal reference) is static—it does not change unless you *Explode*, edit, and redefine it. In contrast, an *Xref* (external reference) is a reference to another drawing. An *Xrefed* drawing is viewable from within the current drawing, but it is not *Inserted*. The *Xref* geometry and associated "named objects" do not become a part of the current drawing by default, although the *Xrefed* layer visibility can be controlled. When the original *Xref* drawing is changed, the changes are visible from within the current drawing.

| *Insert* | *Xref* |
|---|---|
| Any drawing can be *Inserted* as a *Block*. | Any drawing can be *Xrefed*. |
| The drawing "comes in" as a *Block*. | The drawing is visible as an *Xref*. |
| The *Block* drawing is a permanent part of the current drawing. | The *Xref* is not permanent, only "attached" or "overlayed." |
| The current drawing file size increases approximately equal to the *Block* drawing size. | The current drawing increases only by a small amount (enough to store information about loading the *Xref* ). |
| The *Inserted Block* drawing is static. It does not change. | Each time the current drawing is opened, it loads the most current version of the *Xref* drawing. |
| If the original drawing that was used as the *Block* is changed, the *Inserted Block* does not change because it is not linked to the original drawing. | If the original *Xref* drawing is changed, the changes are automatically reflected when the current drawing is *Opened*. |
| Objects of the *Inserted* drawing cannot be changed in the current drawing unless *Refedit*, the Block Editor, or *Explode* is used. | The *Xref* can be changed from the current drawing by using *Refedit* or use *Xopen* to edit the *Xref* in a separate window. |
| The current drawing can contain multiple *Block*s. | The current drawing can contain multiple *Xrefs*. |
| A *Block* drawing cannot be converted to an *Xref*. | An *Xref* can be converted to a *Block*. The *Bind* option makes it a *Block*—a permanent part of the current drawing. |
| A *Block* is *Inserted* on the current layer. | An *Xref* drawing's layers are also *Xrefed* and visibility of its layers can be controlled independently. |
| Any associated "named objects" of a *Block* can be used in the current drawing. | Any "named objects" of an *Xref* can be used in the current drawing only if *Xbind* is used. |
| *Blocks* can be nested. | An *Attached Xref* can be nested. |
| Any *Block* that is referenced by another *Block* is "nested." | An *Overlay Xref* cannot be nested. |
| A "circular" reference is not allowed with *Blocks* (X references Y and Y references X) because nested *Blocks* are also referenced. | Circular references are automatically detected (when the drawing you are *Xrefing* contains a nested *Xref* of the current drawing). Any nested *Xrefs* will load, but circular *Xrefs* are terminated. |

| *Insert* | *Xref* |
|---|---|
| When a *Block* is inserted, you can control the properties of the block's layers. | When a drawing is *Attached* or *Overlayed*, you can control the properties of the *Xref's* layers. |
| You can *Xclip* the *Block* so that only portions of the entire *Block* drawing are visible. | You can *Xclip* the *Xref* so that only portions of the entire *Xref* drawing are visible. |

As you can see, an *Xrefed* drawing has similarities and differences to an *Inserted* drawing. The following figures illustrate both the differences and similarities between an *Xrefed* drawing and an *Inserted* one.

The relationship between the current (parent) drawing and a drawing that has been *Inserted* and one that has been *Xreferenced* is illustrated here (Fig. 30-1). The *Inserted* drawing becomes a permanent part of the current drawing. No link exists between the parent drawing and the original *Inserted* drawing. In contrast, the *Xrefed* drawing is not a permanent part of the parent drawing but is a dynamic link between the two drawings. In this way, when the original *Xrefed* drawing is edited, the changes are reflected in the parent drawing (if a *Reload* is invoked or when the parent drawing is *Opened* next). The file size of the parent drawing for the *Inserted* case is near the sum of both the drawings whereas the file size of the parent drawing for the *Xref* case is only the original size plus the link information.

**FIGURE 30-1**

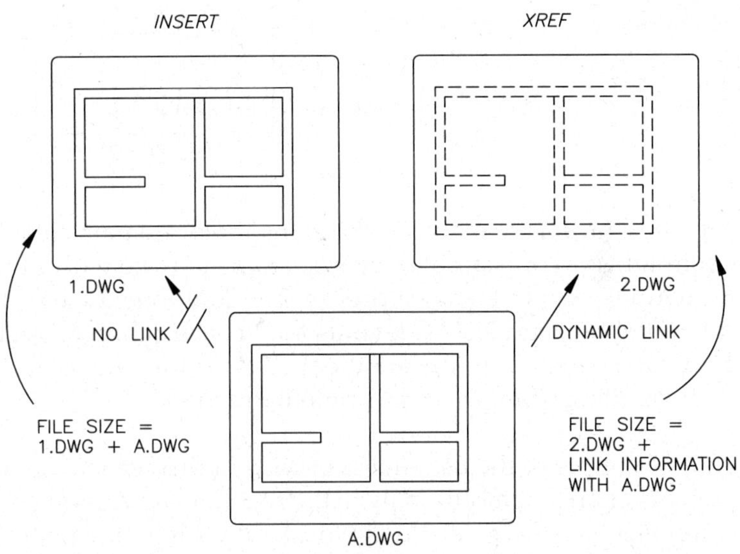


Some of the similarities between *Xreferenced* drawings and *Inserted* drawings are shown in the following two figures.

Any number of drawings can be *Xrefed* or *Inserted* into the current drawing (Fig. 30-2). For example, component parts can be *Xrefed* to compile an assembly drawing.

**FIGURE 30-2**

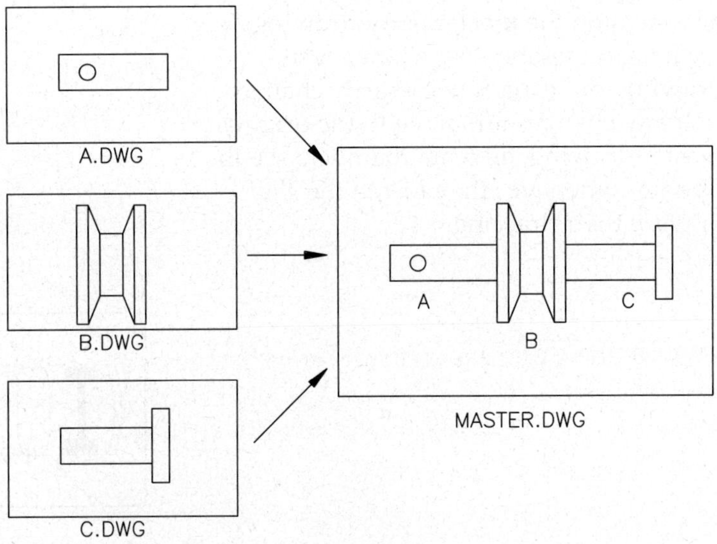

The current drawing can contain nested *Xrefs*. This feature is also similar to *Blocks*. For example, the OFFICE drawing can *Xref* a TABLE drawing and the TABLE drawing can *Xref* a CHAIR drawing. Therefore, the CHAIR drawing is considered a "nested" *Xref* in the OFFICE drawing (Fig. 30-3).

**FIGURE 30-3**

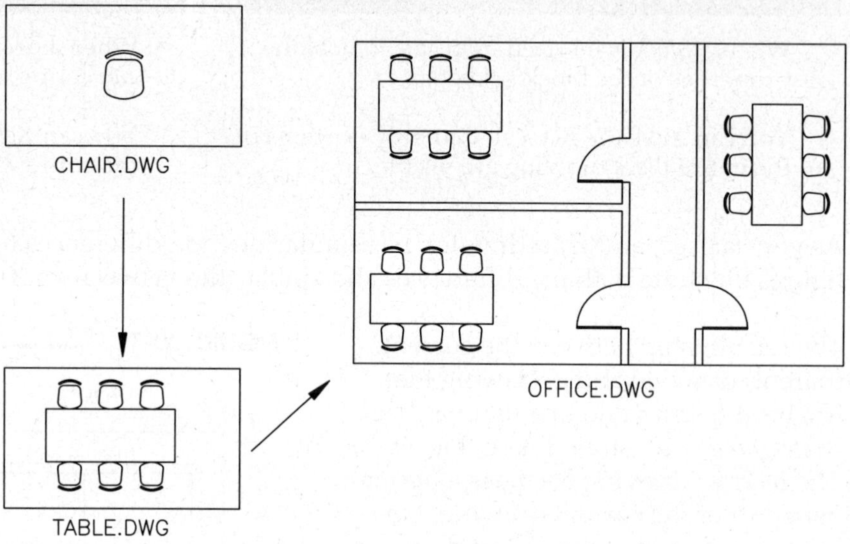


*Xref* drawings have many applications. *Xrefs* are particularly useful in a networked office or laboratory environment. For example, several people may be working on one project, each person constructing individual components of a project. The components may be mechanical parts of an assembly; or electrical, plumbing, and HVAC layouts for a construction project; or several areas of a plant layout. In any case, each person can *Xref* another person's drawing as an external reference without fear of inadvertently making changes to the original drawing.

As an example of the usefulness of *Xrefs*, a project coordinator could *Open* a new drawing, *Xref* all components of an assembly, analyze the relationships among components, and plot the compilation of the components (see Fig. 30-2). The master drawing may not even contain any objects other than *Xrefs*, yet each time it is *Opened*, it would contain the most up-to-date component drawings.

In another application, an entire team can access (*Xref, Attach*) the same master layout, such as a floor plan, assembly drawing, or topographic map. Figure 30-4 represents a mechanical design team working on an automobile assembly and all accessing the master body drawing. Each team member "sees" the master drawing, but cannot necessarily change it. If any changes are made to the original master drawing, all team members see the updates whenever they *Reload* the *Xref* or *Open* their drawing.

**FIGURE 30-4**

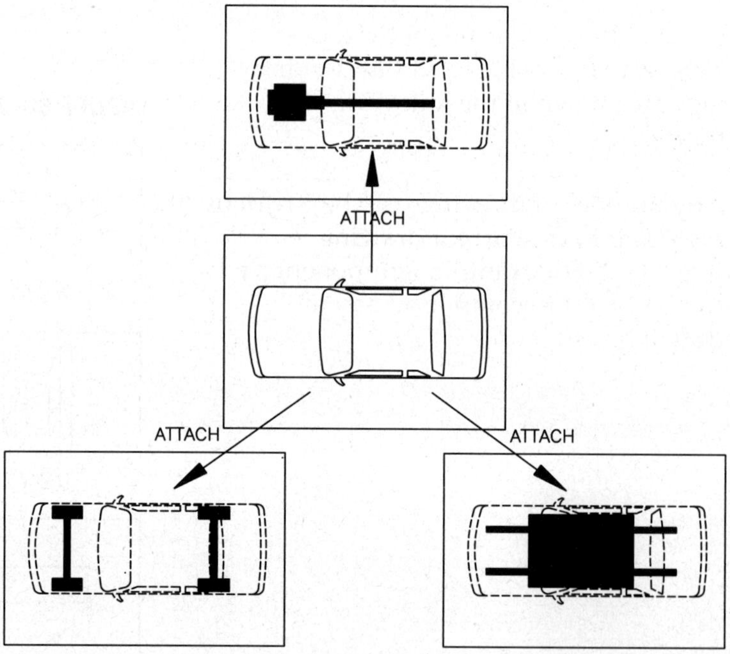

You also have the capability to clip portions of the *Xref* drawing that you do not want to see. Once you attach an *Xref*, you can use the *Xclip* command to draw a rectangular or polygonal (any shaped) boundary to determine what area of the *Xref* to view—all areas outside the boundary become invisible. For example, assume a site plane is *Xrefed* in order to calculate drainage around a house (Fig. 30-5).

**FIGURE 30-5**

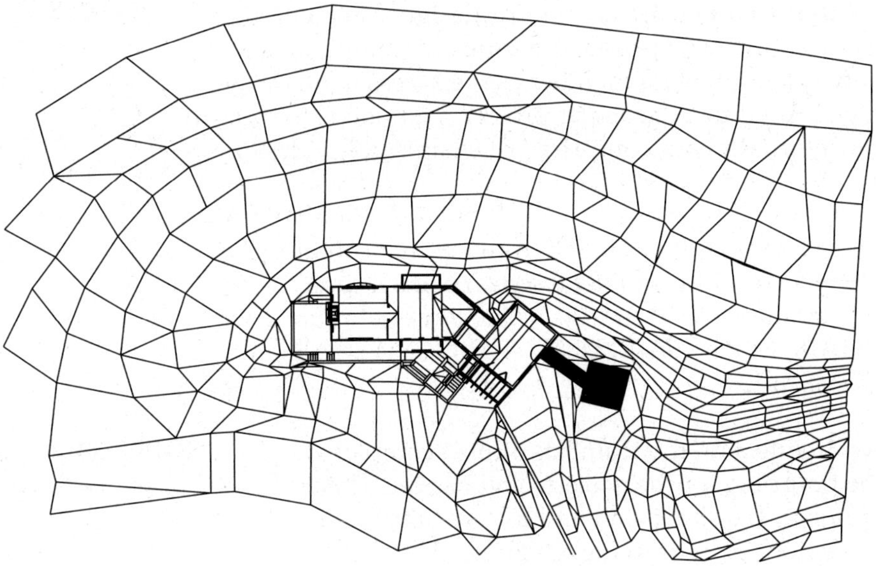

*Xclip* can be used to display only a portion of the Xref drawing needed for the calculations (Fig. 30-6).

**FIGURE 30-6**

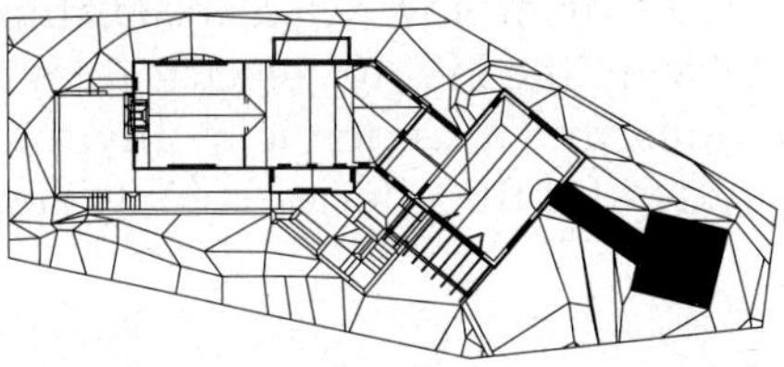

Internal features known as "demand loading," "spatial indexing," and "layer indexing" allow AutoCAD to load only the portion of the drawing that appears in the clip boundary into the current drawing. Demand loading and spatial and layer indexing can save a considerable amount of time and memory when Xreferencing large drawings such as maps or large floor plans, because only the small portion of the drawing that is needed (in the clip boundary) is actually loaded.

The named objects (*Layers*, *Text Styles*, *Blocks*, *Views*, etc.) that may be part of an *Xref* drawing become dependent objects when they are *Xrefed* to the parent drawing. These dependent objects cannot be renamed, changed, or used in the parent drawing. They can, however, be converted individually (permanently attached) to the parent drawing with the *Xbind* command.

Even though an *Xrefed* drawing's named objects (layers, linetypes, text styles, etc.) are available, you cannot normally draw on or edit any of its layers. You do, however, have complete control over the visibility of the *Xref* drawing's layers. The *State* of the layers (*On*, *Off*, *Freeze*, and *Thaw*) of the *Xref* drawing can be controlled like any layers in the parent drawing.

**FIGURE 30-7**

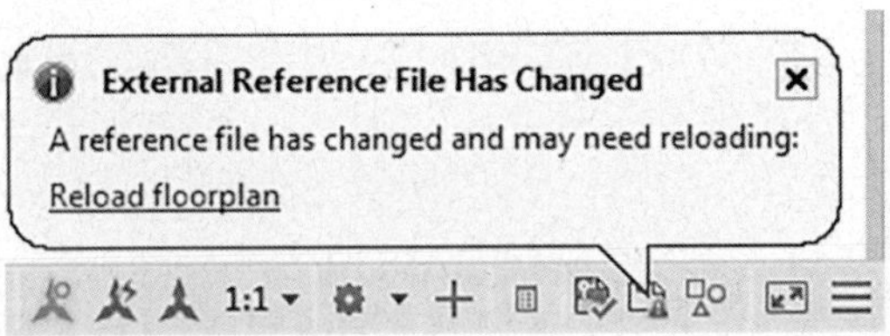


TIP

If a drawing used as an *Xref* needs editing, three methods can be used. First, you can close the parent drawing, then *Open* the original external drawing, make the edits and *Save* the drawing. When the parent drawing is opened again, the new *Xref* is loaded. Second, while in the parent drawing, you can use the *Xopen* command to open the original (used as an *Xref*) to make the changes. In this case, when the *Xref* drawing is changed and saved, AutoCAD notifies you that a change has been made so you can reload the *Xref* in the parent drawing (Fig. 30-7). The parent drawing then reflects the new changes in the *Xref*. Third, you can use the *Refedit* command to edit the *Xref* from within the parent drawing, but only if the *Xedit* variable is set to 1 in the *Xrefed* drawing. Any changes made are "sent back" to the original *Xref* drawing.

## An *Xref* Example

Assume that you are an interior designer in an architectural firm. The project drawings are stored on a network so all team members have access to all drawings related to a particular project. Your job is to design the interior layout comprised of chairs, tables, desks, and file cabinets for an office complex. You use a prototype drawing named INTERIOR.DWG for all of your office interiors. It is a blank drawing that contains only block definitions (not yet *Inserted*) of each chair, desk, etc., as shown in Figure 30-8. The *Block* definitions are CHAIR, CONCHAIR, DESK, TABLE, FILECAB, and CONTABLE.

FIGURE 30-8

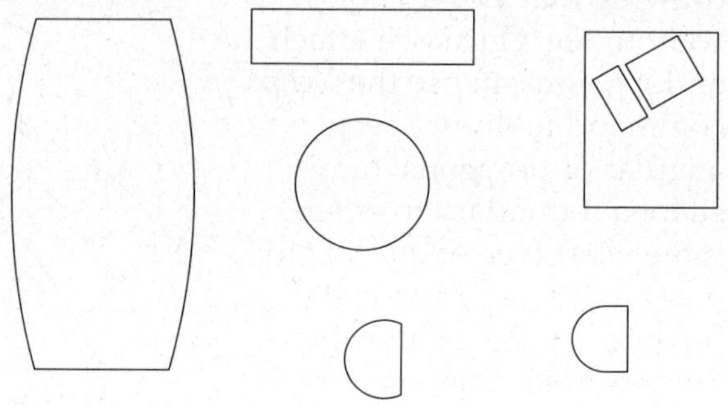

An architect on the team has almost completed the floorplan for the project. You use the *Attach* option of the *Xref* command to "see" the architect's floor plan drawing, OFFICEX.DWG. The visibility of the individual dependent layers can be controlled with the *Layer* command or *Layer Control* drop-down list. For example, layers showing the HVAC, electrical layout, and other details are *Frozen* to yield only the floor plan layer needed for the interior layout (Fig. 30-9).

FIGURE 30-9

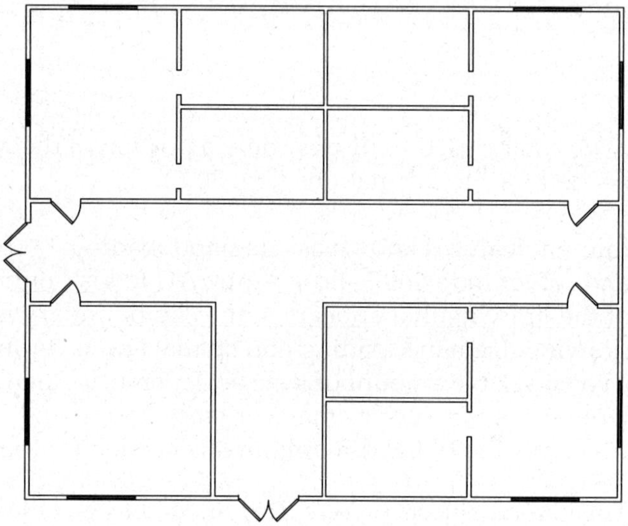

Next, you draw objects in the current drawing or, in this case, *Insert* the office furniture *Block*s that are defined in the current drawing (INTERIOR.DWG). The resulting drawing would display a complete layout of the office floor plan plus the *Inserted* furniture—as if it were one drawing (Fig. 30-10). You *Save* your drawing and go home for the evening.

FIGURE 30-10

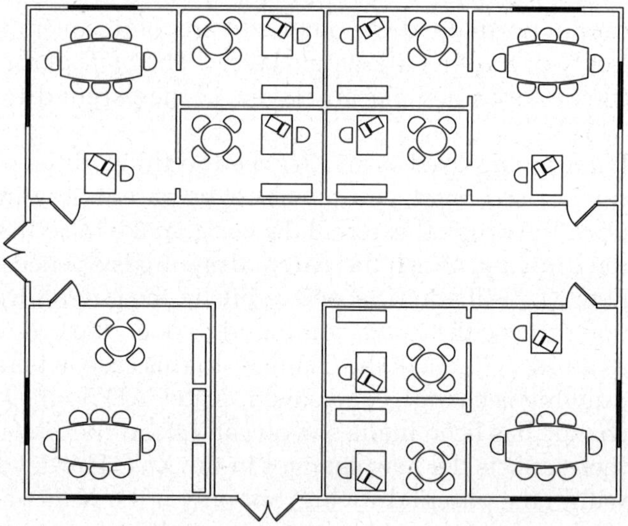

That evening, the architect makes a change to the floor plan. He works on the original OFFICEX drawing that you *Xrefed*. The individual office entry doors in the halls are moved to be more centrally located. The new layout appears in Figure 30-11.

**FIGURE 30-11**

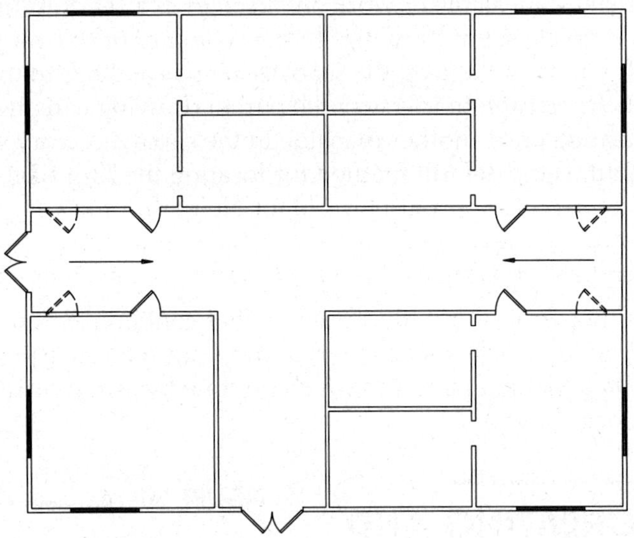

The next day, you *Open* the INTERIOR drawing. The automatic loading of the OFFICEX drawing displays the latest version of the office floor plan with the design changes. The appropriate changes to the interior drawing must be made, such as relocating the furniture by the hall doors (Fig. 30-12). When all parts of the office complex are completed, the INTERIOR (parent) drawing could be plotted to show the office floor plan and the interior layout. The resulting plot would include objects in the parent drawing and the visible *Xref* drawing layers.

**FIGURE 30-12**

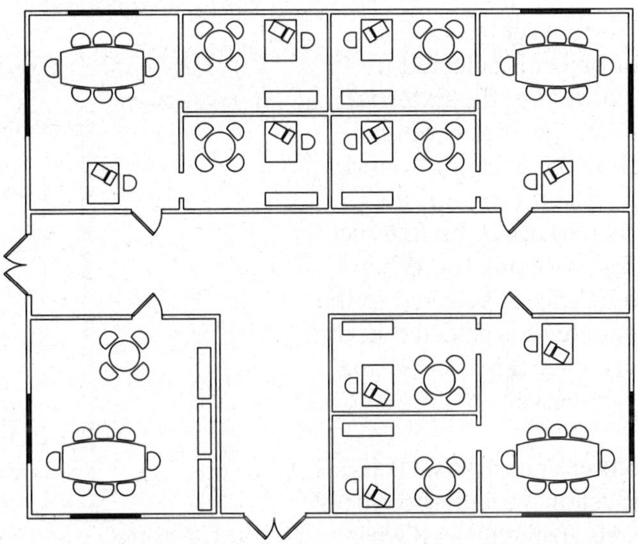

AutoCAD's ability to externally reference drawings makes team projects much more flexible and efficient. If the *Xref* capability did not exist, the OFFICEX drawing would have to be *Inserted* as a *Block*, and design changes made later to the OFFICEX drawing would not be apparent. The original OFFICEX *Block* would have to be *Erased*, *Purged*, and the new drawing *Inserted*.

The *Xclip* feature gives you the capability to isolate only a portion of an *Xrefed* drawing. For example, assume you wanted to discuss a particular room with the architect, so you open a *New* drawing and *Xref* the INTERIOR drawing. The OFFICEX drawing also is *Xrefed* because it is nested (previously *Xrefed* into the INTERIOR drawing) so the new drawing appears the same as the original INTERIOR drawing (see Fig. 30-12). To isolate only the room in discussion, *Xclip* is used to draw a rectangular boundary around the desired area (Fig. 30-13).

**FIGURE 30-13**

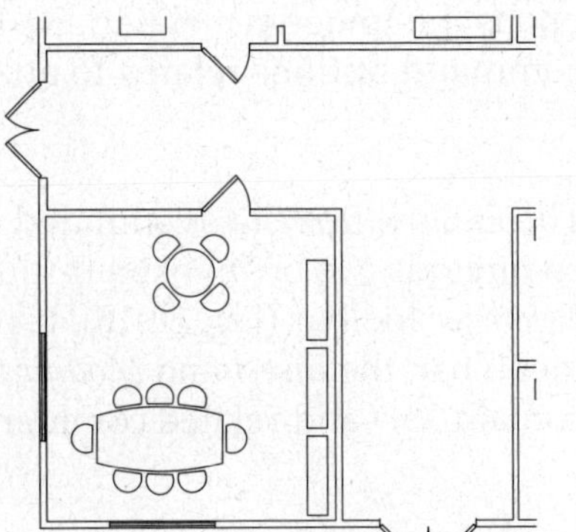

Often at the end of a project, the final set of drawings is sent to the client. Because the *Xref* capability makes it feasible to work on a design as a set of individual component drawings, it is necessary to send the complete set of drawings to the customer or to combine the set into one drawing to submit.
If you want to keep the set of drawings as *Xrefs* to the parent drawing, the *Xrefs* can be saved with a *Relative Path*. In this way the parent drawing can always locate and load the *Xrefs* as long as they were loaded onto another machine in the same directory structure relative to the location of the parent. Relative path is the default method for locating the Xref files. The REFPATHTYPE system variable lets you set the default Xref path type (0 for No path, 1 for Relative path, or 2 for Full path). Another strategy when sending drawings is to combine them into one drawing using *Bind*. *Bind* brings an *Xrefed* drawing into the current drawing as if it were *Inserted*, so it becomes a permanent part of the parent drawing (as a *Block*).

**TIP** When working with drawings that contain references, the File References pane's Missing toggle allows you to view missing external drawing reference (xref) files. Connected Support Files used within Forma Data Management can save you time by automatically resolving paths of missing external drawing references (xrefs).

## CREATING AND USING *XREFS*

When a reference file or object is selected in a drawing, the External Reference panel opens showing xref command icons, where you can click to open the *External Reference* palette. When a reference is selected in the palette, it is selected in the drawing (Fig. 30-14) and vice-versa.

Note: The ERHIGHLIGHT (System Variable) controls xrefs are highlighted when selected in the External References palette or the drawing window. Set to 1 highlighting is on, set to 0 then xrefs are not highlighted.

FIGURE 30-14

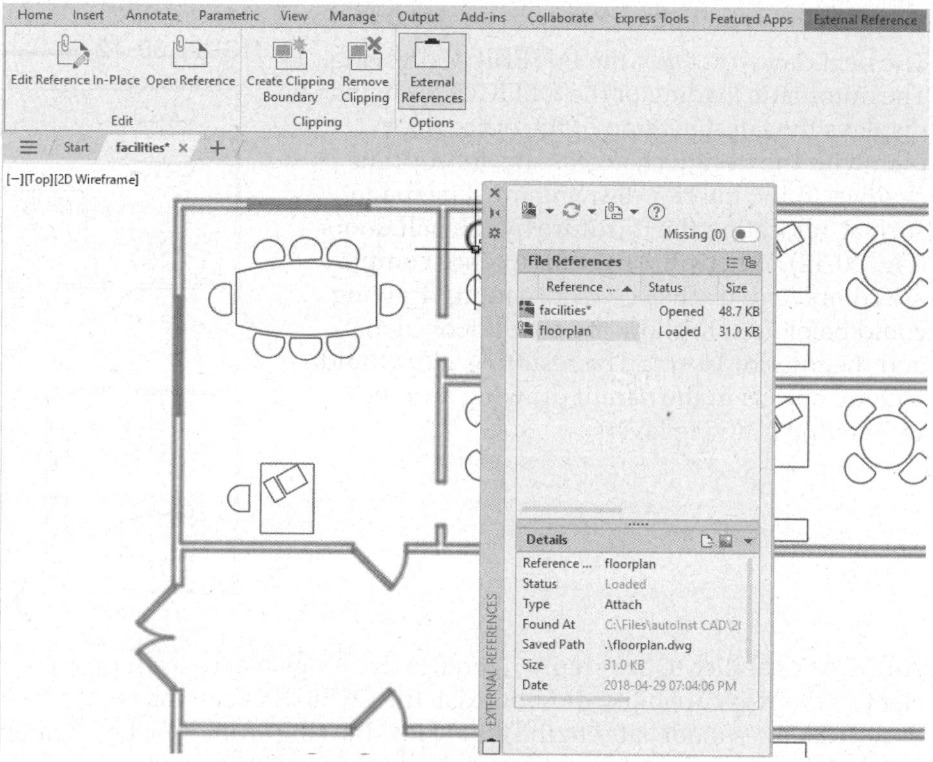


Similarly, selecting an attached image file opens the Image panel (Fig. 30-15) showing command options related to attached images..

FIGURE 30-15

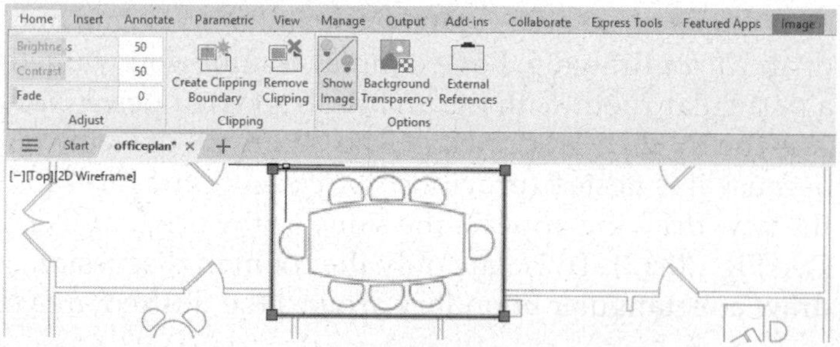


The *Externalreferences* command and related commands can be invoked through the *Reference* toolbar (Fig. 30-16). If you prefer the menu bar, the *Insert* and *Modify* pull-downs contain *Xref* and related commands.

FIGURE 30-16

### *Xref* or *External-references*

| Ribbon | Menu Bar | Command (Type) | Alias (Type) | Shortcut |
|---|---|---|---|---|
| *View Palettes* (icon) | *Insert External References* | *Xref* or *Externalreferences* | *XR* or *ER* | ... |

Both the *Externalreferences* command and the *Xref* command open the *External References* palette (Fig. 30-17). This palette allows you to create and manage external references (*Xrefs*). The central area of the palette lists drawings that are currently *Attached*. The options for managing the external references are available using a right-click shortcut menu, as shown in Figure 30-17 or through the menu choices.

If you prefer to type, you can use the *-Xref* command (notice the hyphen prefix). Using *-Xref* displays the following prompt:

Command: ***-XREF***
[?/Bind/Detach/Path/pathType/
Unload/Reload/Overlay/Attach/Show]
<Attach>:

Whether you use *Externalreferences*, *Xref*, or *-Xref*, the *Attach*, *Overlay*, *Detach*, *Reload*, *Unload*, *Bind*, *Path*, *pathType* and *Show* functions are similar, except that -Xref produces the command line version.

**FIGURE 30-17**

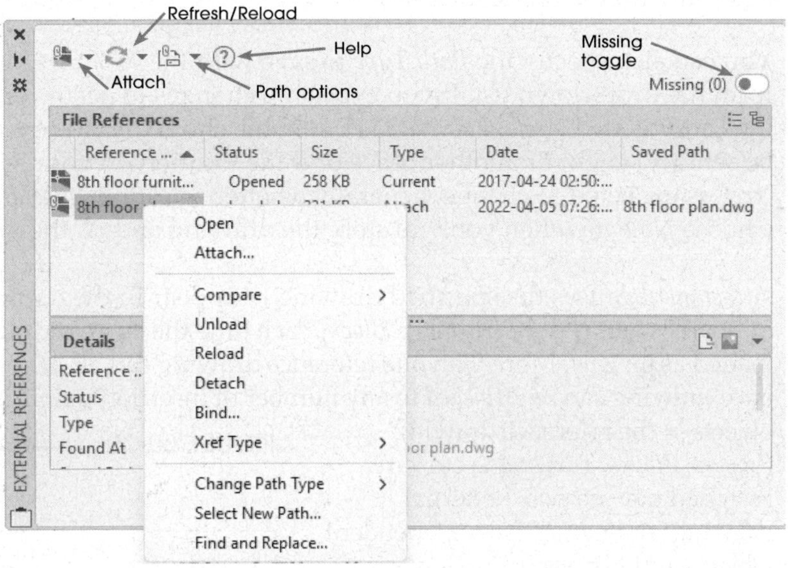


**FIGURE 30-18**

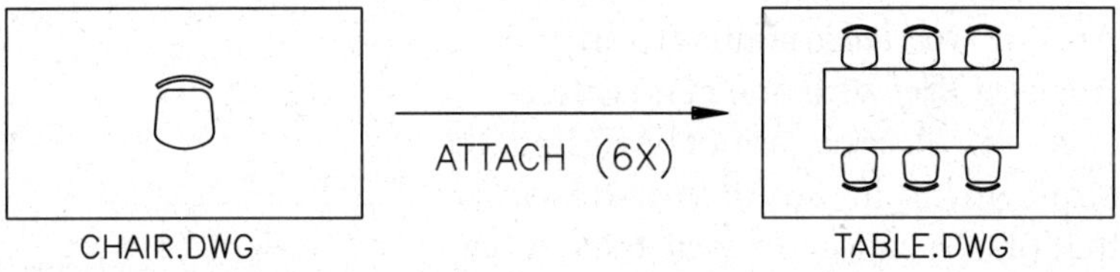


### *Attach*

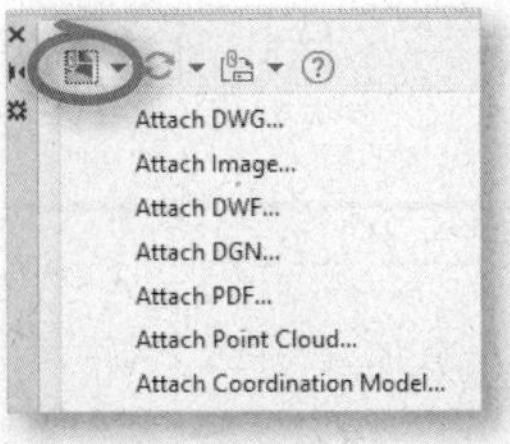


This option attaches one drawing to another by making the *Xref* drawing visible in the "parent" drawing. Even though the *Xref* drawing is visible, it cannot be edited from within the parent drawing unless *Refedit* is used. *Attach* creates a link between two drawings (Fig. 30-18).

Selecting the *Attach* option produces the standard *Select File* dialog box where you locate and select the .DWG file to attach (not shown). When you have selected the desired file to *Xref*, select *OK* to dismiss the *Select File* dialog box and produce the *Attach External Reference* dialog box (see Fig. 30-19).

**FIGURE 30-19**

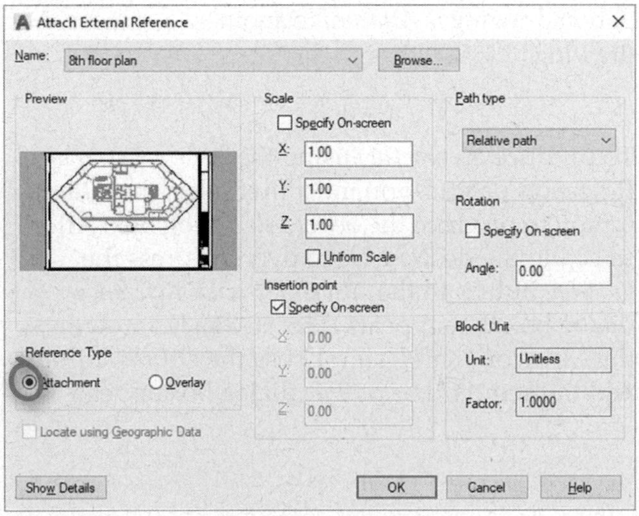

When the *Attach External Reference* dialog box appears (Fig. 30-19), specify the *Reference Type—Attachment* or *Overlay*. Ensure the *Attachment* button is pressed. (See "*Overlay*" next.)
(Typing the *-Xref* command and using the *Attach* option produces the *Select File* dialog box directly—the *External References* palette does not appear.) Once the desired file is specified, the *Attach External Reference* dialog box appears again for further action.

Notice the similarity of the edit boxes (Fig. 30-19) to the *Insert* command. The *Insertion Point, Scale, Rotation,* and *Block Unit* options operate identically to the *Insert* command. Similar to the action of *Insert*, the base point of the *Xref* drawing is its 0,0 point unless a different base point has been previously defined with the *Base* command.

You can also specify the *Path Type* to save with the *Xref* by selecting *Full Path, Relative Path* or *No Path* from the drop-down list. If you expect no changes to occur for the directory system used for the drawings and related *Xrefs*, *Full Path* is a suitable choice. However, if the drawing and related *Xrefs* might be sent to a client or another office or if the location on your system for the parent drawing and related *Xrefs* is expected to change, generally a better selection is *Relative Path* (see "Managing Xref Drawings"). Choose *No Path* when you will store the drawing and all the xref'd files in the same folder.

*Attachment* causes the specified drawing to appear in the current drawing similar to the way a drawing appears when it is *Inserted* as a *Block*. Each time the current drawing is *Opened*, the *Attached* drawing is loaded as an *Xref*. More than one reference drawing can be *Attached* to the current drawing. A single reference drawing can be *Attached* to any number of insertion points in the current drawing. The dependent objects in the reference drawing (*Layers, Blocks, Text styles*, etc.) are assigned new names, "external drawing | old name" (see "Dependent Objects and Names").

**FIGURE 30-20**

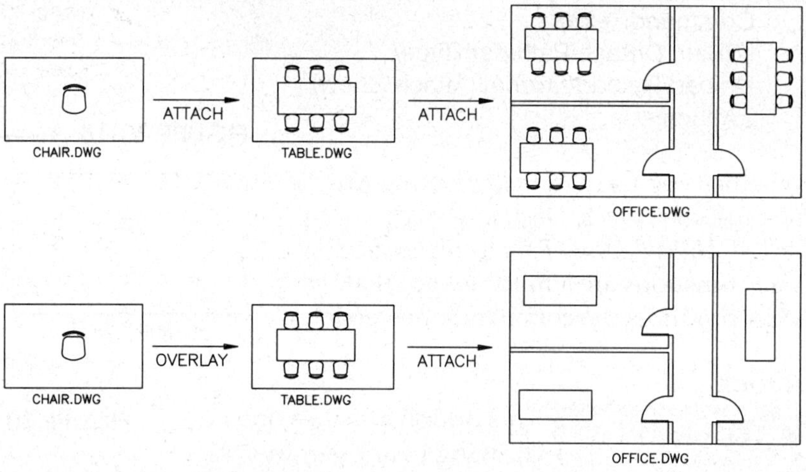


### Overlay

An *Xref Overlay* is similar to an *Attached Xref* with one main difference—an *Overlay* cannot be nested. In other words, if you *Attach* a drawing that (itself) has an *Overlay*, the *Overlay* does not appear in your drawing. On the other hand, if the first drawing is *Attached*, it appears when the parent drawing is *Attached* to another drawing (Fig. 30-20).

**FIGURE 30-21**

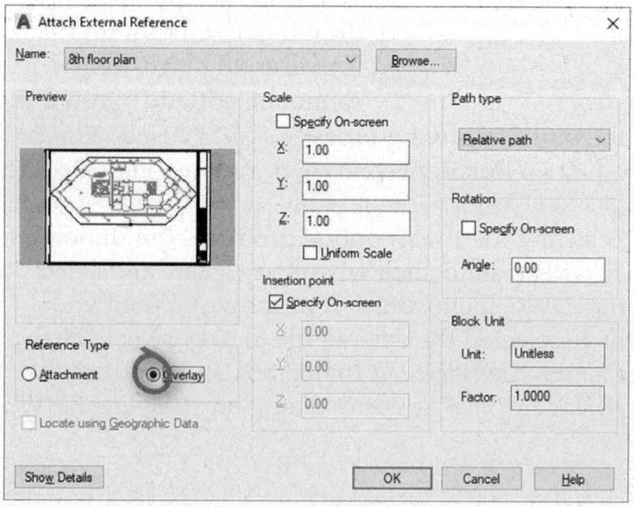


To produce an *Overlay* using the *Attach External References* palette, you must first select the *Attach* option to produce the *Select File* dialog box. After selecting the desired file to overlay, press the *Overlay* button in the *Attach External Reference* dialog box that appears (Fig. 30-21). If you type the *-Xref* command, simply use the *Overlay* option and the standard *Select File* dialog box appears.

The *Overlay* option prevents "circular" *Xrefs* from appearing by preventing unwanted nested *Xrefs*. This is helpful in a networking environment where many drawings *Xref* other drawings. For example, assume drawing B has drawing A as an *Overlay*. As you work on drawing C, you *Xref* and view only drawing B without drawing B's overlays—namely drawing A. This occurs because drawing A is an *Overlay* to B, not *Attached*. If all drawings are *Overlays*, no nesting occurs (Fig. 30-22).

**FIGURE 30-22**

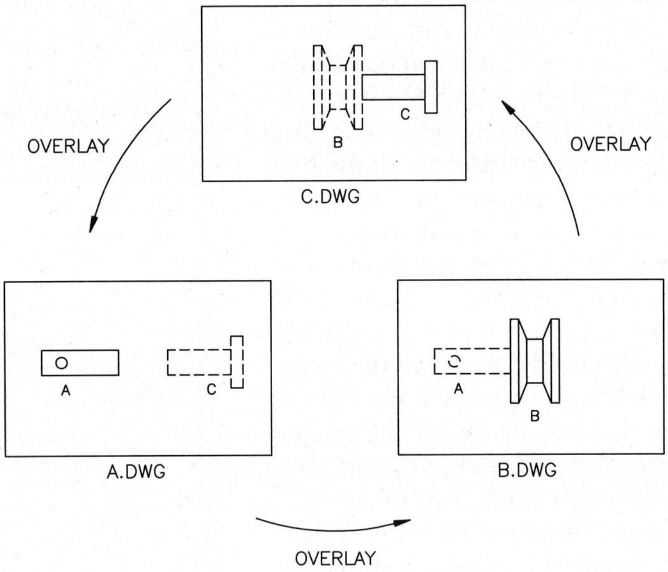


**FIGURE 30-23**

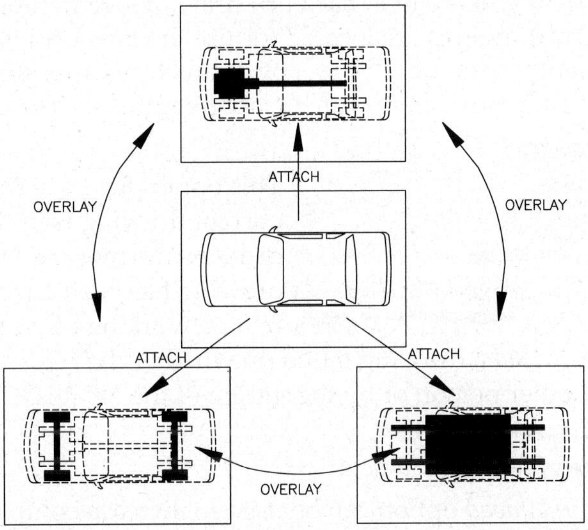


The overlay concept is particularly useful in a concurrent engineering project, where all team members must access each other's work simultaneously. In the case of a design team working on an automobile assembly, each team member can *Attach* the master body drawing and also *Overlay* the individual subassembly drawings (Fig. 30-23). *Overlay* enables each member to *Xref* all drawings without bringing in each drawing's nested Xreferences.

There are times when you cannot foresee the potential use of a drawing and therefore do not *Overlay* rather than *Attach*. For example, in a large design office it is quite possible to *Xref* a drawing on the network that includes your current drawing as an *Attached Xref*, thus causing a circular reference. If this occurs, an *AutoCAD Alert* dialog box appears giving notification of the circular reference and presents the option to continue or not (Fig. 30-24). If you respond with *Yes*, AutoCAD reads in the *Xref* and any nested *Xrefs* to the point where it detects the circularity, then terminates. Answering *No* cancels the *Attach*.

**FIGURE 30-24**

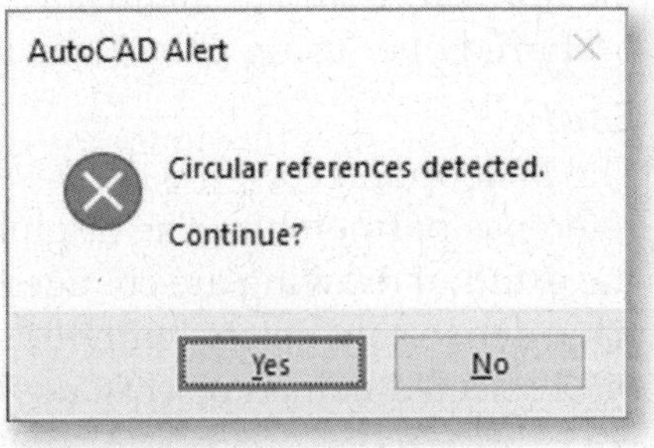


### *Detach*

*Detach* breaks the link between the parent drawing and the *Xref*. Using this option causes the *Xrefed* drawing to "drop" from view immediately. When the parent drawing is *Opened*, the previously *Attached* drawing is no longer loaded as an *Xref*. Nested *Xrefs* cannot be *Detached* or selected to *Bind* since they are actually part of another drawing. You can *Detach* the drawing that contains the nested *Xref*.

### Details/Preview

The lower section of the *External References* palette can display *Details* or a *Preview* depending on your selection (Fig. 30-25, right-center buttons). The *Details* option gives information about the highlighted *Xref* from the list above. The *Status*, *Size*, and *Date* fields are read-only; however, you can change the *Reference Name*, *Type* (*Attach* or *Overly*), and *Found At* (path) fields. For example, use the browse button at the right of the *Found At* field to locate a referenced drawing that has been moved to a different folder or drive. In such a case, the *Status* of the *Xref* is listed above as *Not Found*, as shown in this figure for the "8th floor lighting" drawing.

**FIGURE 30-25**

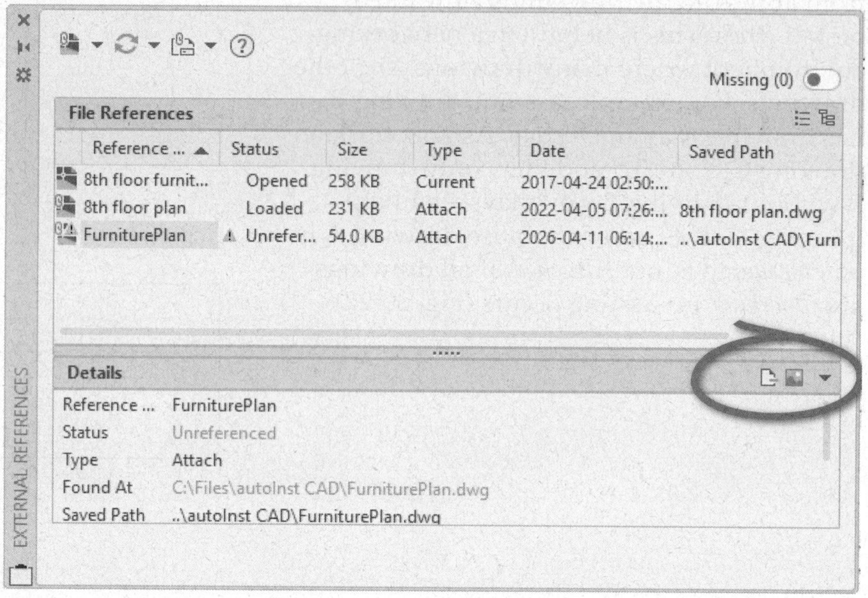


TIP

When you archive (back up) drawings, remember also to back up the drives and directories where the *Xref* drawings are located (or use the *Bind* option when the project is finished). One possible file management technique is to specify a directory especially for *Xrefs* (see "Managing Xref Drawings"). In a network environment, access rights to *Xref* directories must be granted.

### Reload

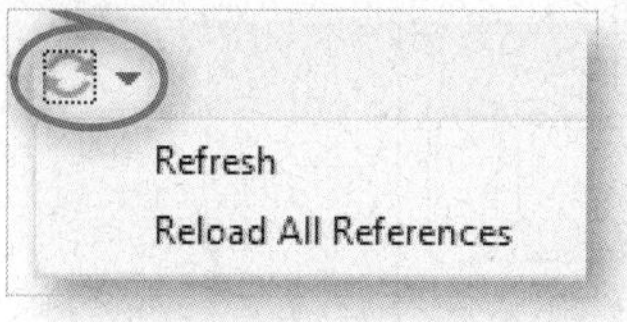


This option forces a *Reload* of the external drawing at any time while the current drawing is in the Drawing Editor and ensures that the current drawing contains the most recent version of the external drawing. You are notified if a drawing has been changed and requires *Reloading*. (See also "*XREFNOTIFY*.")

In a networking environment, you may not be able to *Attach*, *Overlay*, *Bind*, or *Reload* a drawing based on whether the referenced drawing is currently being edited (open) by another person or by the setting of the *XLOADCTL* system variable in the referenced drawing (see "*XLOADCTL*").

TIP

### Unload

The *Unload* option can be used to increase computing speed in a drawing. In cases when you need to view (load) drawings only on an as-need basis, *Unload* them. This causes the current (parent) drawing to open more quickly, speeds regenerations, and uses less memory. The unloaded *Xref* is not visible, but the link information remains in the current drawing. To view the *Xref* again, use *Reload*. This method is preferred over using *Detach* when you expect to use the *Xref* again.

### Bind

The *Bind* option converts the *Xref* drawing to a *Block* in the current drawing, then terminates the external reference partnership. The original *Xrefed* drawing is not affected. The names of the dependent objects in the external drawing are changed to avoid possible conflicts in the case that the parent and *Xref* drawing have dependent objects with the same name (see "Dependent Objects and Names"). If you want to only bind selected named objects, use the *Xbind* command.

When the *Bind* button is selected (in the *External References* palette), the *Bind Xrefs* dialog box appears for you to select the *Bind Type* (see Fig. 30-29). This option determines how the names of dependent objects (such as layers and blocks) appear in the drawing after they become a permanent part of the parent drawing (see "Dependent Objects and Names").

### *Open*

Using this option, you can open an externally referenced drawing in a separate window (of the same AutoCAD session), make any needed changes, then *Save* the drawing. When the opened drawing has been changed and saved, AutoCAD notifies you (in the system tray) that the drawing has been changed and needs reloading (see Fig. 30-7). (See also "*XREFNOTIFY*.") This method is sometimes easier than using *Refedit* for editing an *Xref*. *Opening* a drawing through the *External References* palette (*Xopen*) overrides the setting of the *XLOADCTL* and *XEDIT* variables (see "Demand Loading" and "Editing *Xrefs* and *Blocks*").

### *List View* and *Tree View*

The central area of the *External References* palette is used to list and give the status of drawings that have been *Xrefed*. Use the two buttons in the upper-right corner of the palette to display the information in a list that can be ordered alphabetically by column (*List View*) or in a hierarchical tree structure (*Tree View*).

### *List View*

This option (default) displays the *Xrefs* alphabetically by name. You can sort the list by column in forward or reverse order (*Reference Name*, *Size*, *Type*, *Date*, etc.) by selecting the column heading and clicking once or twice. You can resize the column width by dragging the column separator left or right. The *Status* column displays the state of each reference. The following states are possible:

| *Status* | Description |
|---|---|
| *Loaded* | *Xref* was found when drawing was opened or reloaded |
| *Unload* | *Xref* will be unloaded when dialog box is closed |
| *Unloaded* | *Xref* is currently unloaded |
| *Needs Reloading* | *Xref* has changed, use *Reload* to force reload when dialog box is closed |
| *Reload* | *Xref* will be reloaded when dialog box is closed |
| *Loaded, recent changes* | *Xref* was loaded but has been changed recently |
| *Not found* | *Xref* was not found in search paths when drawing was opened or reloaded |
| *Unresolved* | *Xref* was found but could not be read by AutoCAD |
| *Unreferenced* | *Xref* was attached but is erased |
| *Orphaned* | Nested *Xref* is attached to an *Xref* that is not loaded |
| *Open* | *Xref* will be opened for editing in another window when dialog is closed |

### *Tree View*

The *Tree View* shows the *Xrefs* in a hierarchical tree structure. This option is particularly useful if you have nested *Xrefs* (Fig. 30-26). Note that the icons also indicate the type and status of each referenced drawing (*Overlay*, *Attach*, *Reload*, *Unload*, *Not found*, etc.).

Orphaned xrefs are identified with a yellow warning symbol. You can right-click these and use the shortcut menu to choose Attach to quickly find and add the parent file.

**FIGURE 30-26**

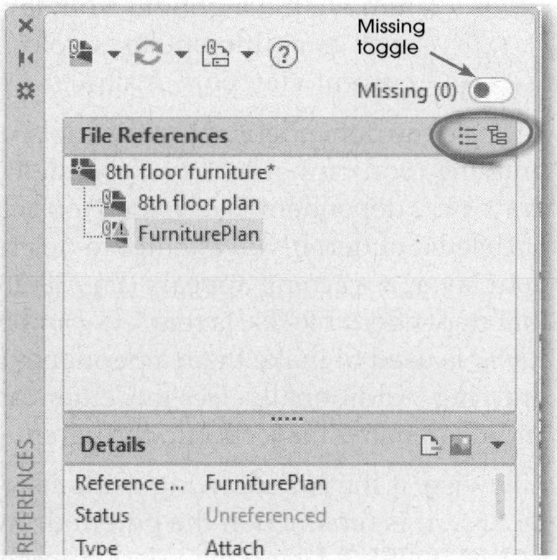


TIP

### *Missing Toggle*

Turn the Missing toggle on to display all of the missing references in the File References pane. From there you can select the xref and update its path. Projects using Forma Data Management can automate resolving missing references.

## Dependent Objects and Names

The named objects that have been created as part of a drawing become dependent objects when that drawing is *Xrefed*. The named objects in an *Xref* drawing can be any of the following:

*Blocks, Layers, Linetypes, Dimension Styles, Text Styles*

These dependent objects <u>cannot</u> be renamed or changed in the parent drawing. Dependent text or dimension styles cannot be used in the parent drawing. You cannot draw on dependent layers from the parent drawing; you can only control the visibility of the dependent layers. You <u>can</u>, however, *Bind* the entire *Xref* and the dependent named objects, or you can bind individual dependent objects with the *Xbind* command. *Xbind* converts an individual dependent object into a permanent part of the parent drawing. (See "*Xbind*.")

**FIGURE 30-27**

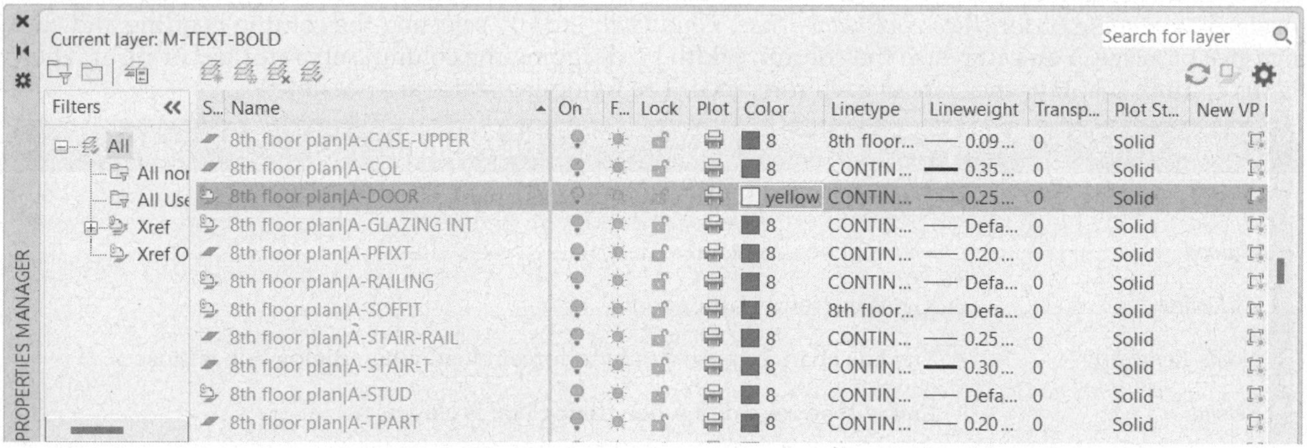


When an *Xref* contains named objects, the names of the dependent objects are changed (in the parent drawing). A new naming scheme is necessary because conflicts would occur if the *Xref* drawing and the parent drawing both contained *Layers* or *Block*s, etc., with the same names, as would happen if both drawings used the same template drawing. When a drawing is *Attached*, its original object names are prefixed by the drawing name and separated by a pipe ( | ) symbol.

For example, if the *8th floor plan* drawing is *Xrefed* and it contains a layer named A-DOOR, it is listed (in the parent drawing) as 8th floor plan | A-DOOR (Fig. 30-27).

Also note that *Xref* layers are shown in a faded grey color in the *Layer Properties Manager* and in the Ribbon drop-down layers list in order to distinguish them from other layers in the drawing (Fig. 30-27).

A toggle at the upper right of the Layer Properties Manager makes it easy to highlight and remove the highlight from layers which have overrides. This includes both Xref overrides, such as a different layer color, and Viewport overrides, such as a different lineweight in the current viewport. A similar icon identifies the overridden xref'd layer.

Because the dependent objects cannot be used in the parent drawing (you cannot insert a dependent block nor can you draw on a dependent layer), these objects cannot be renamed or deleted either. If you attempt to delete or rename a dependent layer, a warning appears (Fig. 30-28). (Keep in mind that dependent blocks, layers, etc. can be used if *Bind* or *Xbind* is used to make them a permanent part of the parent drawing. Additionally, DesignCenter can be used to drag-and-drop named objects into the parent drawing.)

**FIGURE 30-28**

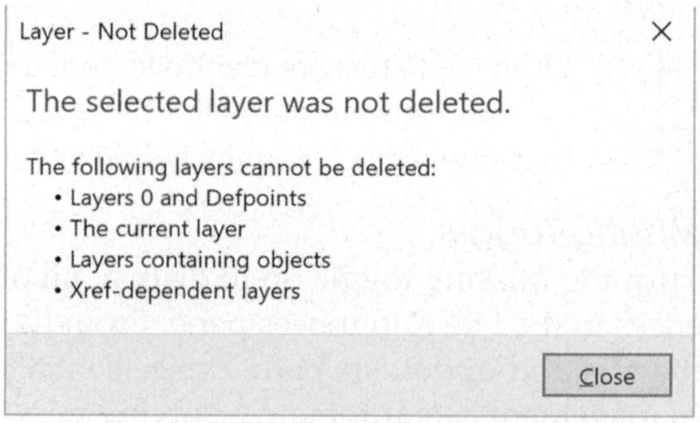


Likewise, if the *Xref* drawing contains a block named DESK2, it is renamed in the parent drawing to DB_SAMP | DESK2. Using the *-Block* command to reveal a listing of blocks in the current drawing yields the following display. Note the tally below of "User Blocks," "External References," and "Dependent Blocks":

```
Command: -block
Enter block name or [?]: ?
Enter block(s) to list <*>:
Defined blocks.
 "CENOS"                                Xref: resolved
 "DB_SAMP"                      Xref: resolved
 "DB_SAMP|CHAIR7"                       Xdep: "DB_SAMP"
  "DB_SAMP|COMPUTER"            Xdep: "DB_SAMP"
  "DB_SAMP|DESK2"               Xdep: "DB_SAMP"
  "DB_SAMP|DESK3"                       Xdep: "DB_SAMP"
  "DB_SAMP|DOOR"                Xdep: "DB_SAMP"
  "DB_SAMP|DR-36"               Xdep: "DB_SAMP"
  "DB_SAMP|DR-69P"                      Xdep: "DB_SAMP"
  "DOD-part306"                 Xref: resolved
  "SUBASY-A"                    Xref: resolved
  "SUBASY-B"                    Xref: resolved
  "SUBASY-C"                    Xref: resolved
User        External            Dependent       Unnamed
Blocks      References          Blocks          Blocks
0           6                   7               0
Command:
```

The same naming scheme operates with all dependent objects—*Dimension Styles*, *Text Styles*, *Views*, etc.

### Dependent Object Names with *Bind* and *Xbind*

If you want to *Bind* the *Xref* drawing, the link is severed and the *Xref* drawing becomes a *Block* in the parent drawing. Or, if you want to bind only individual named objects, the *Xbind* command can be used to bring in specific *Blocks* or *Layer* names. The names are converted from the previous dependent naming scheme (using the *Xref* drawing name as a prefix with a | [pipe] separator) to another name depending on the command and option you use.

**FIGURE 30-29**

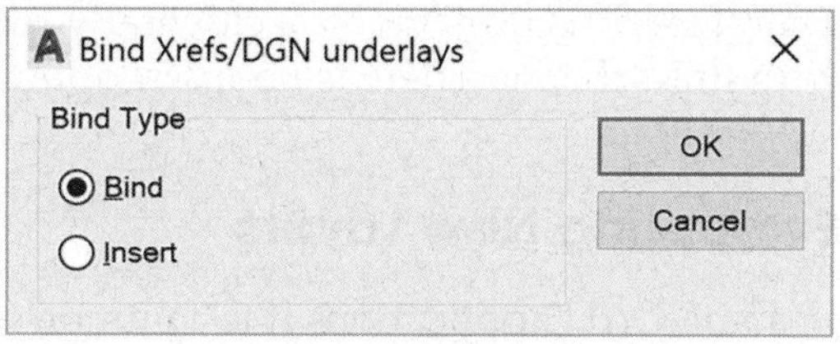


If you use *Xbind* or use *Bind* with the default *Bind* option, the names of dependent objects keep the *Xref* drawing name, but the | (pipe) symbol is changed to a \$*n*\$ separator, where *n* represents a number. For example, assume you use the *Bind* button (in the *External References* palette) with the *8th floor plan* drawing from the previous example. The *Bind Xrefs* dialog box appears (Fig. 30-29). The default (*Bind*) option is the traditional naming scheme. That is, the *Xref* drawing name prefix stays with the object name.

**FIGURE 30-30**

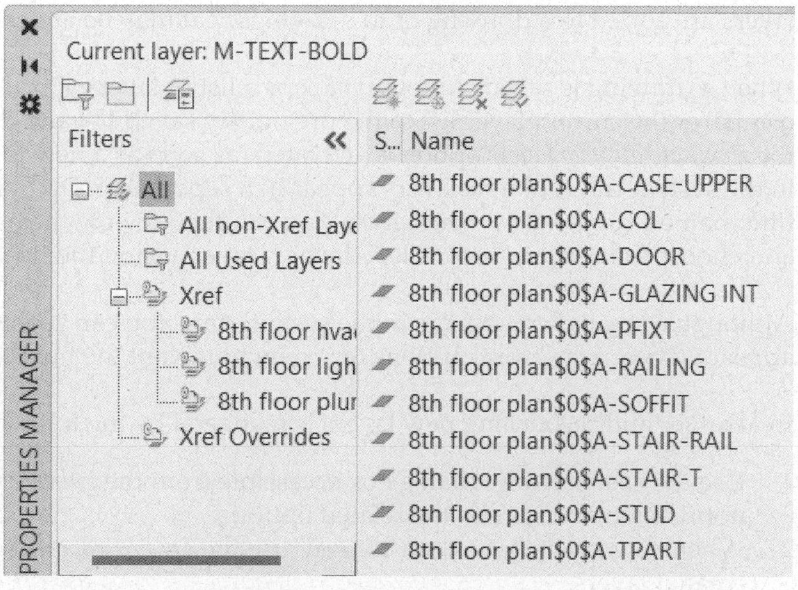


Using the *Bind* option, the resulting layer names would appear in the *Layer Properties Manager* listing (Fig. 30-30). Notice the dependent layer previously named ***8th floor plan | XGRIDS*** is changed to ***8th floor plan\$0\$XGRIDS***. These layers can be renamed at this point because they are now a permanent part of the current drawing. Using the *Xbind* command to "import" selected named objects results in the same layering scheme illustrated here.

Also note that the layer names are no longer displayed as faded grey text since they are now a permanent part of the drawing.

On the other hand, if you select the *Insert* option in the *Bind Type* dialog box, the *Xref* drawing name prefix is dropped from the object names when they are "imported." Therefore, the named objects assume their original name, assigned when they were created in the original drawing. Using the previous *Xref* example (*8th floor plan*), the layer names would appear in the *Layer Properties Manager* listing as XGRID, WALLTYPE12, etc. (Fig. 30-31).

**FIGURE 30-31**

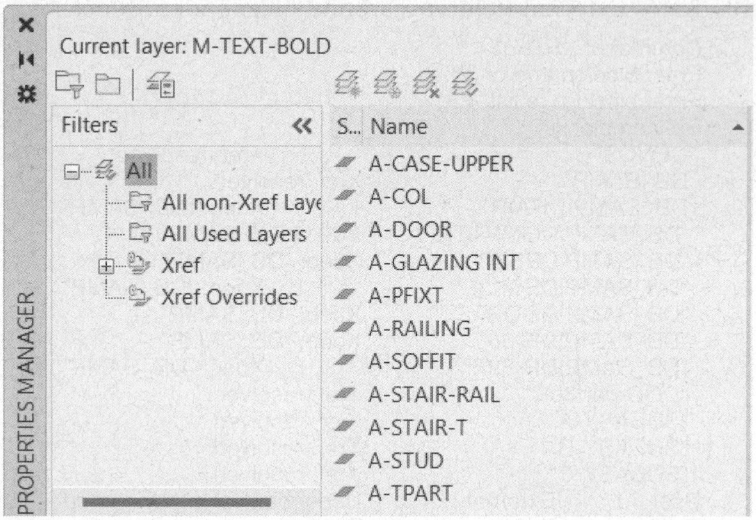


TIP

This method (*Bind*, *Insert*) is preferred if you want to *Bind* an *Xrefed* drawing that was created from the same template drawing as the current (parent) drawing. In this way, the same layers, dimension styles, text styles, etc. are not duplicated when you *Bind* the *Xref*. However, this method should not be used when two different *Blocks* exist with the same name or two layers exist with different linetype and color settings that should be maintained. When a dependent *Block* or layer having the same name as that in the parent drawing is "imported" with *Bind*, *Insert*, the *Block* definition and the layer properties of the parent drawing take precedence. Therefore, it is possible to accidentally redefine a *Block* or lose linetype and color properties of a layer with this scheme. There is no *Insert* option for *Xbind*.

## Evaluating New Layers

In some CAD applications it is common for drawings to be accessible to multiple users, such as an engineering or architectural project where several people collaborate on a drawing or set of drawings. In these cases, it is important to maintain levels of security with respect to changes that may occur to the drawings. The layer evaluation feature in AutoCAD saves all newly created layers in a special group filter named *Unreconciled New Layers*. You can also specify when a notification bubble appears if new layers are added to a drawing or to drawings that may be attached as Xrefs.

When a drawing is saved, AutoCAD keeps a list of layers. Once layer evaluation is enabled, AutoCAD compares the current layer list to the previously saved list and detects the creation of any new layers. If the *Evaluate all new layers* option is activated, as soon as a new layer is created, a small indicator appears in the system tray. All new layers appear in a separate group in the *Layer Properties Manager* as a group filter named *Unreconciled New Layers*. You can then review the list, manually select and view the layers, and choose to keep, or "reconcile," the new layers using the *Reconcile* option from the right-click menu.

Although a small alert appears in the system tray, you can specify that an additional notification bubble appears. There are several options for when you want the bubble to appear.

Evaluating and reconciling new layers is managed by the following three steps:

1. Use the *Layer Settings* dialog box accessible from the *Layer Properties Manager* to enable new layer notification and specify the related options.
2. When new layers have been created, use the *Layer Properties Manager* to view, or "evaluate," newly created layers.
3. Decide whether or not to reconcile (permanently add) the new layers. Reconcile new layers using the right-click menu in the *Layer Properties Manager*.

### *Layer Settings*

Use the *Settings* button (gear icon) in the upper-right corner of the *Layer Properties Manager* to show the *Layer Settings* dialog box (Fig. 30-32).

**FIGURE 30-32**

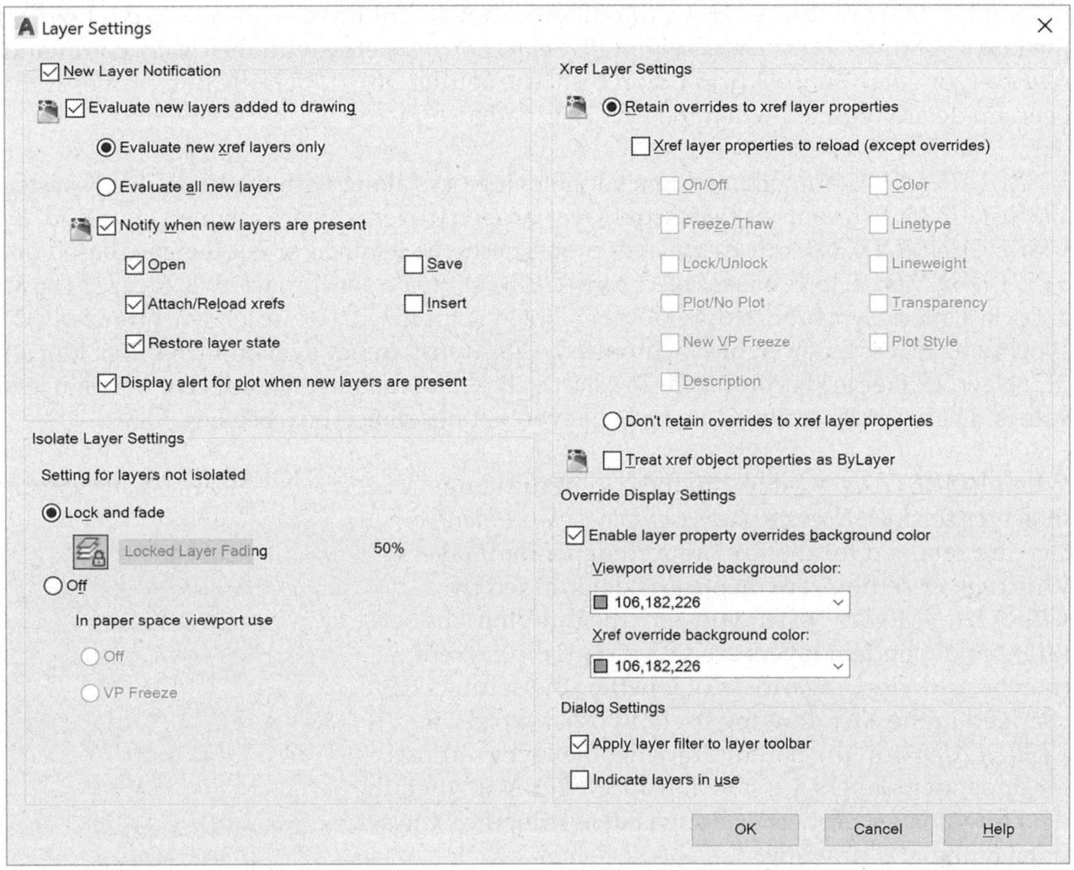


*Evaluate new layers added to drawing*
Select *Evaluate new layers added to drawing* to turn on evaluation and reconciliation function. You have the choice to either *Evaluate all new layers* or *Evaluate new xref layers* only. If *Evaluate all new layers* is checked and a new layer is created in the drawing any time after this function is enabled, the new layer(s) are saved in a group filter named *Unreconciled New Layers*. If the *Evaluate new xref layers* option is selected, AutoCAD checks the externally referenced drawings when the current drawing is opened or plotted.

*Notify when new layers are present*
This section sets when a notification bubble appears, not when AutoCAD makes the check for newly created layers. Checking the *Open* option causes the notification bubble to appear when you open the drawing, and the *Save* option warns you when you use the *Save* command. You can also have the notification bubble appear when you *Attach* or *Reload Xrefs*, or *Insert* other drawings and newly created layers exist in those drawings.

*Display alert for plot when new layers are present*
This check causes the notification bubble to appear when you use the *Plot* command and new layers exist.

*Xref Layer Settings*
These options let you choose the Xref layer properties you want to reload, making it easier to control Xref overrides. These options are stored using the VISRETAINMODE system variable.

*Treat Xref object properties as ByLayer* lets you force the properties of Xref objects to use the ByLayer property.

*Dialog Settings*
Use this section to specify a list of new layers is created as a filter in the Layer drop-down lists.

*Override Display Settings*
Use this to specify the *Xref* and/or *Viewport override background color* (see Chapter 32 for more information on viewport overrides).

## *VISRETAIN/VISRETAINMODE/XREFOVERRIDE*

*VISRETAIN* is a system variable, not a command; therefore, you must type *VISRETAIN* at the command prompt or use the *Layer Settings* dialog box *Retain overrides to xref layer properties* button to change this setting. This variable controls the dependent (*Xref* drawing) layer settings in the parent drawing. Remember that you cannot draw or edit on a dependent layer without *Refedit*; however, you can change the layer's settings (*Freeze/Thaw, On/Off, Color, Linetype*, etc.) with the *-Layer* command, *Layer Properties Manager*, or *Layer Control* drop-down box. The setting of *VISRETAIN* determines if the settings that have been made are retained when the *Xref* is reloaded.

*VISRETAINMODE* is another system variable. It works along with the VISRETAIN system variable. *VISRETAINMODE* controls ***which*** xref layer property overrides are retained on reload. Setting *VISRETAINMODE to* 0 causes ***all*** layer properties to be retained or not retained based on *VISMODE*. Setting *VISRETAINMODE* to 1 causes layers turned On/Off to be synchronized; 2: Freeze/Thaw synchronized; 4: Lock/Unlock synchronized; 8: Plot/No Plot synchronized.; 16: Color synchronized; 32: Linetype synchronized; 64: Lineweight synchronized; 128: Transparency synchronized; 256: Plot style synchronized; 512: New VP Freeze synchronized. To selectively set more than one property to retain, enter the sum of the values. These can be easily set using the Layer Settings dialog box. See Fig. 30-38.

**TIP**

When *VISRETAIN* is set to 1 in the parent drawing, the layer settings (*Freeze/Thaw, On/Off, Color, Linetype*, etc.) are retained for the Xreferenced dependent layers. Which layer settings are retained is determined by *VISRETAINMODE*. When you *Save* the drawing, the settings for dependent layers are saved with the current (parent) drawing, regardless of whether the settings have changed in the *Xref* drawing itself. In other words, if *VISRETAIN*=1 in the parent drawing, the layer settings for dependent layers are saved. If *VISRETAIN*=0, then the dependent layer property setting is determined by the settings in the *Xref* drawing.

*VISRETAIN* also determines whether or not changes to the path of nested *Xrefs* are saved. You can change the path of an *Xref* or of a nested *Xref* (location AutoCAD searches for *Xrefs*) with the *Path* option of *-Xref* or the *Path type* drop-down list in the *External References* palette.

The *XREFOVERRIDE* system variable can be used to set objects on Xref layers that have their properties set as if to ByLayer (setting 1). When XREFOVERRIDE is set to 0, objects on xref layers use their assigned object properties. This can be set in the Layer Settings dialog box by checking *Treat xref object properties as ByLayer.*

**FIGURE 30-38**

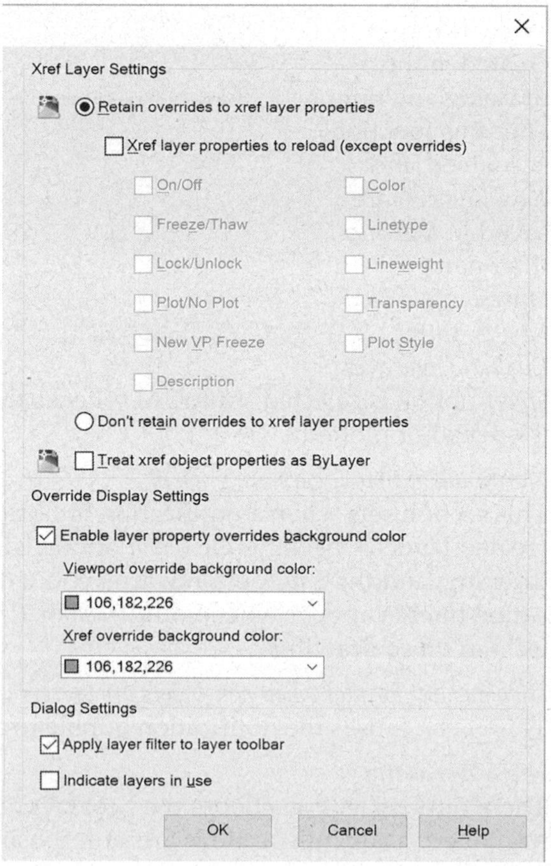

## Evaluating and Reconciling Layers

FIGURE 30-33

Whenever the small warning appears in the system tray, newly created layers are detected. A notification bubble may appear depending on your settings. Open the *Layer Properties Manager* to reveal the new layer(s) located in the *Unreconciled New Layers* group filter (Fig. 30-33). If you want to keep any new layers, select the layer(s), right-click and select *Reconcile Layer* from the shortcut menu. This action moves the selected layer(s) from the group filter into the *All Used* list. You cannot delete an unreconciled layer unless all objects on the layer are *Erased*.

FIGURE 30-34

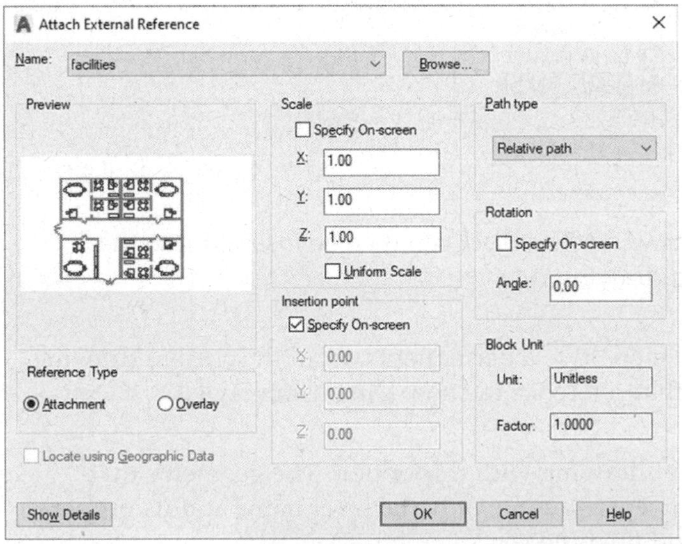


### *Xattach*

| Ribbon | Menu Bar | Command (Type) | Alias (Type) | Shortcut |
|---|---|---|---|---|
| *Insert Reference Attach* | *Insert DWG Reference* | *Xattach* | *XA* | ... |

*Xattach* is a separate command from *Xref* but accomplishes the same action as *Xref, Attach.* If you use *Xref*, the *External References* palette appears, then you must select the *Attach…* tile to invoke the *Select File* and *Attach External Reference* dialog boxes. However, you can use *Xattach* to invoke the *Select File* and *Attach External Reference* dialog boxes directly (Fig. 30-34). Therefore, use the *Xattach* command when you want to Attach or Overlay an *Xref* but do not want to view or manage the list of current Xrefs.

NOTE: *Attach* can be used to attach images, drawings, and other file formats. *Attach* can be used instead of *Xattach*; however, the file type (.dwg) must be specified.

### *Xbind*

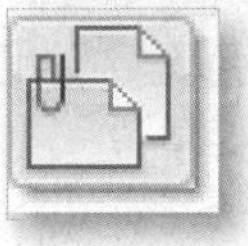

| Ribbon | Menu Bar | Command (Type) | Alias (Type) | Shortcut |
|---|---|---|---|---|
| ... | *Modify Object > External Reference > Bind...* | *Xbind* or *-Xbind* | *XB* or *-XB* | ... |

**TIP**

*Xbind* is a separate command, not an option of the *Xref* command as are most other Xreference controls. *Xbind* is similar to *Bind*, except that it binds an <u>individual</u> dependent object, whereas *Bind* converts the <u>entire</u> *Xref* drawing to a *Block*. Any of the listed dependent named objects (see "Dependent Objects and Names") in a drawing can be converted to a permanent usable part of the current drawing with *Xbind*. In effect, *Xbind* makes a copy of the named object since the original named object is not removed from the dependent drawing.

The *Xbind* command produces the *Xbind* dialog box (Fig. 30-35). All *Attached* and *Overlayed Xrefs* are listed. Expand any *Xref* to reveal the dependent object types (*Block*, *Dimstyle*, etc.) that can be imported. Expand each dependent object type to list the named objects of the type for that particular drawing. Select the desired object, then pick *Add ->* to add the object to the current (parent) drawing. The following object types are valid.

**FIGURE 30-35**

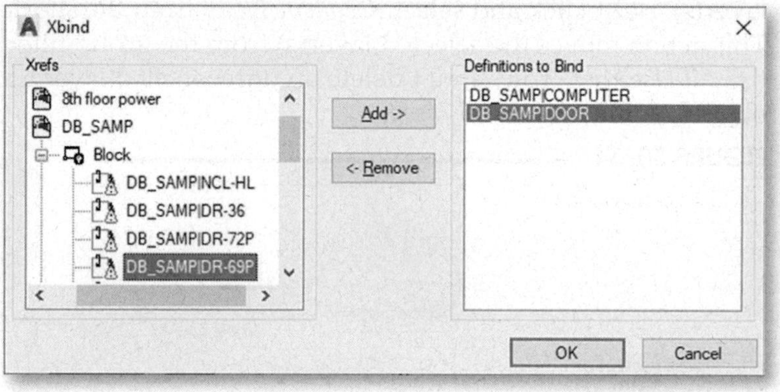


### *Block*

Select this option to *Xbind* a *Block* from the *Xrefed* drawing. The *Block* can then be inserted into the current drawing and has no link to the original *Xref* drawing.

### *Dimstyle*

This option allows you to make a dependent Dimension Style a permanent part of the current drawing. Dimensions can be created in the current drawing that reference the new Dimension Style.

### *Layer*

Any dependent layer can be brought into the current drawing with this option. The geometry that resides on that layer does <u>not</u> become part of the current drawing, only the layer name and its properties. *Xbinding* a layer also *Xbinds* the layer's color and linetype.

### *Linetype*

A linetype from an *Xrefed* drawing can be bound to the current drawing using this option. This action has the same effect as loading the linetype.

### *Textstyle*

If a text style exists in an *Xrefed* drawing that you want to use in the current drawing, use this option to bring it into the current drawing. New text can be created or existing text can be modified to reference the new text style.

If you use (type) the *-Xbind* command, the Command line version of the command is invoked and individual dependent objects must be typed:

Command: **-xbind**
Enter symbol type to bind [Block/Dimstyle/LAyer/LType/Style]:

You must know and type the exact name of the dependent object name, including the dependent drawing name prefix and pipe character; therefore, this method is cumbersome compared to using the *Xbind* dialog box.

## *Adcenter*

| Ribbon | Menu Bar | Command (Type) | Alias (Type) | Shortcut |
|---|---|---|---|---|
| *View Palettes (icon)* | *Tools Palettes > DesignCenter* | *Adcenter* | *DC* or *ADC* | Ctrl+2 |

As an alternative to *Xref*, *Xattach*, and *Xbind*, you can use DesignCenter to drag-and-drop drawings or named objects into a drawing.

FIGURE 30-36

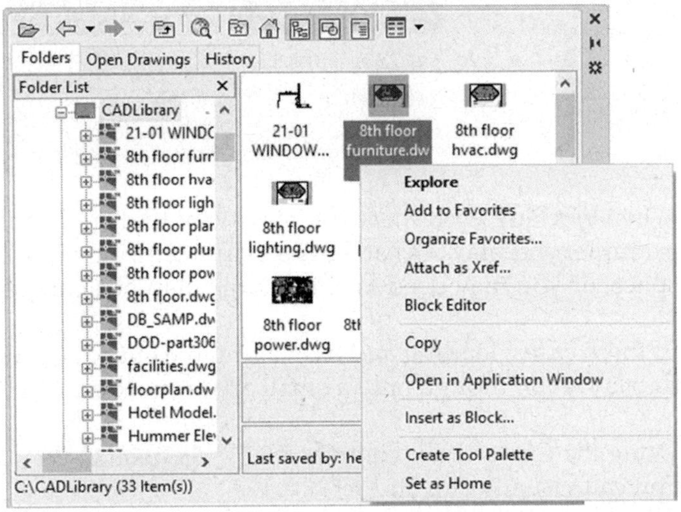


### Using DesignCenter to *Attach* or *Overlay Xrefs*

Normally when you drag-and-drop a .DWG file from the DesignCenter Content Area (window on the right) directly into the current drawing, it is *Inserted* as a *Block*. You can, however, use DesignCenter to attach *Xrefs*.

To *Attach* or *Overlay* an *Xref*, use the Tree View window (left window) to locate the folder that contains the drawing file you want to *Xref*, so all drawings contained in the folder appear in the Content Area (right window). Next, right-click on the desired drawing in the Content Area. Alternately right-drag (drag-and-drop, but holding down the right button) the desired drawing. This causes a shortcut menu to appear (Fig. 30-36). Select *Attach as Xref* from the menu. This causes the *External Reference* dialog box to appear (see Fig. 30-34). Note that you can specify *Attach* or *Overlay* in the *External Reference* dialog box. Proceed to specify the desired options in the dialog box and bring in the new *Xref*. The result is identical to using *Xattach* or the *External References* palette to *Attach* or *Overlay* the *Xref*.

### Using DesignCenter as an Alternative to *Xbind*

DesignCenter can also be used to drag-and-drop named objects from any drawing into the current drawing. *Xbind* allows you to bring named objects into the current drawing but only from Xrefed drawings. With DesignCenter, you can drag-and-drop from any drawing. Named objects that can be imported with DesignCenter are *Blocks*, *Dimstyles*, *Layers*, *Layouts*, *Linetypes*, *Textstyles* and *Xrefs*.

TIP

To bring named objects into the current drawing, use the Tree View window (Fig. 30-37, left window) to locate the drawing file that contains the named objects you want to import, then expand the drawing (plus sign) so all named objects contained in the drawing appear in the Content Area (right window). Next, drag-and-drop the desired named object from the Content Area into the drawing area. This causes the selected named object to become part of the current drawing.

See Chapter 21 for more detailed information on DesignCenter.

FIGURE 30-37

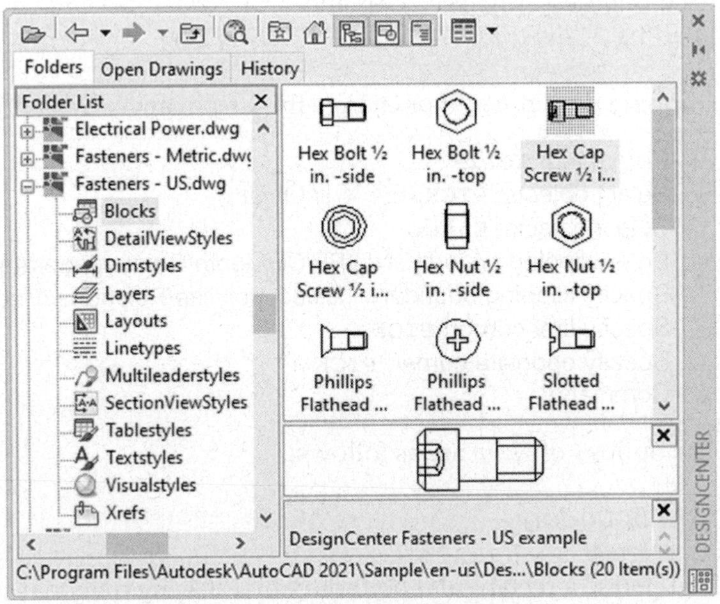

### Xclip

| Ribbon | Menu Bar | Command (Type) | Alias (Type) | Shortcut |
|---|---|---|---|---|
| *Insert Reference Clip* | *Modify Clip > Xref* | *Xclip or Clip* | *XC* | (Edit Mode) *Clip Xref* |

When working with *Xrefs* you may want to display only a specific part of the attached drawing. For example, you may *Xref* an entire manufacturing plant layout (floor plan) in order to redesign an office space, or you may need to *Xref* a large map of a city just to design a new interchange at one location.

In these cases, loading and displaying the entire *Xref* would occupy most of the system resources just for display, even though only a small area is needed.

Using the *Clip* or *Xclip* command, you can visually clip an Xref drawing or a *Block* to display only the portion inside the area defined by the clip boundary (Fig. 30-39). You can define the boundary as a rectangle or as a many-sided polygon of virtually any shape. All geometry inside the boundary is displayed normally. The portions of the *Xrefed* drawing (or *Block*) that fall outside the boundary are not only restricted from the display, but a feature called "demand loading" can be used to prevent the unwanted portions of the *Xref* from loading so system resources are not used unnecessarily (see "Demand Loading," "*INDEXCTL*" and "*XLOADCTL*").

**FIGURE 30-39**

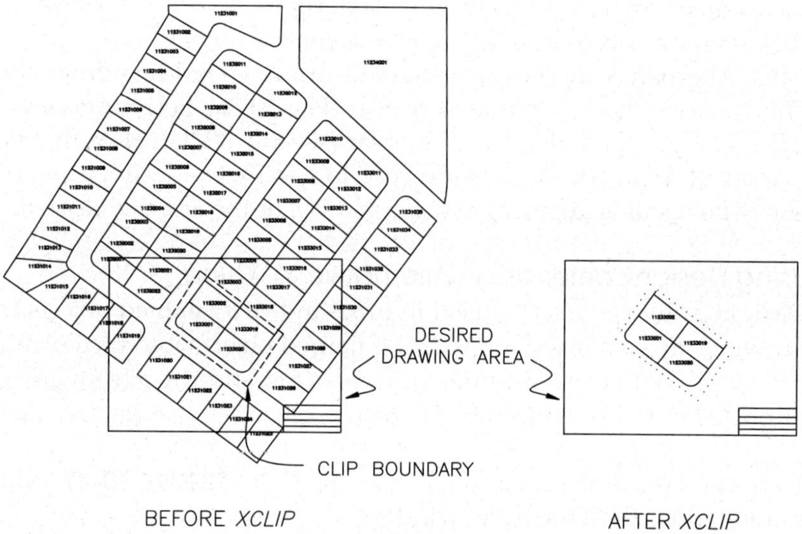


Clipping does not edit or change the *Xref* in any way, it only prevents portions from being displayed:

Command: ***XCLIP***
Select objects: **PICK** (the *Xref* Object)
Select objects: **Enter**
Enter clipping option [ON/OFF/Clipdepth/Delete/generate Polyline/New boundary] <New>: ***n***
Specify clipping boundary: [Select polyline/Polygonal/Rectangular/Invert clip] <Rectangular>: ***r***
Specify first corner: **PICK**
Specify opposite corner: **PICK**
Command:

The options of *Xclip* are as follows.

#### *New Boundary*

*Rectangular*

You define a rectangular boundary by PICKing diagonal corners. The clipping boundary is on the XY plane of the current UCS.

*Polygonal*

Select points as the vertices of a polygon. The polygon can have any shape and number of sides.

*Select Polyline*
Select an existing polyline (*Pline*) to use for the boundary. The existing polyline must consist of straight segments but can be open.

### ON/OFF
*ON* displays only the *Xref* geometry inside the clipping boundary. *OFF* displays the entire *Xref*, ignoring the clipping boundary.

### Clipdepth
Use this option to set front and back clipping planes on the *Xref* object. This option is useful for 3D geometry when you want to clip portions of the object behind a back clipping plane and in front of a front clipping plane. You must specify a clip distance from and parallel to the clipping boundary.

### Delete
You can delete an existing boundary for the selected *Xref* or *Block* with this option. *Erase* cannot be used to delete a clipping boundary.

### Generate Polyline
Automatically draws a *Pline* along an existing clipping boundary. The new *Pline* assumes the current layer, linetype, and color settings. This option is helpful if you want to change the clipping boundary using *Pedit* with the new *Pline*, then use the *Polyline* option to establish the new boundary.

### Invert clip
This option inverts the current mode of the clipping boundary. The two modes are *Inside mode* (objects inside the boundary are hidden) and *Outside mode* (objects outside the boundary are hidden).

## XCLIPFRAME

| Ribbon | Menu Bar | Command (Type) | Alias (Type) | Short-cut |
|---|---|---|---|---|
| ... | *Modify*<br>*Object >*<br>*External Reference >*<br>*Frame* | XCLIPFRAME | ... | ... |

*XCLIPFRAME* is a system variable that can be used to control the display of the clipping boundary created by the *Xclip* command. For example, in some cases it may be desirable to display the boundary (Fig. 30-40), while in other cases no boundary may be preferred (Fig. 30-41).

The possible *XCLIPFRAME* settings are:

0 The frame is not visible nor is it plotted. The frame reappears temporarily if you select the clipped object.
1 The clipped *Xref* frame is both displayed and plotted.
2 The clipped *Xref* frame is displayed but not plotted (default setting).

NOTE: The *FRAME* system variable (accessible from the Ribbon, *Insert* tab, *Reference* panel) can be used to control *XCLIPFRAME* since *FRAME* controls clipping frames for all *Xrefs* and other images.

**FIGURE 30-40**

*XCLIPFRAME* = 1

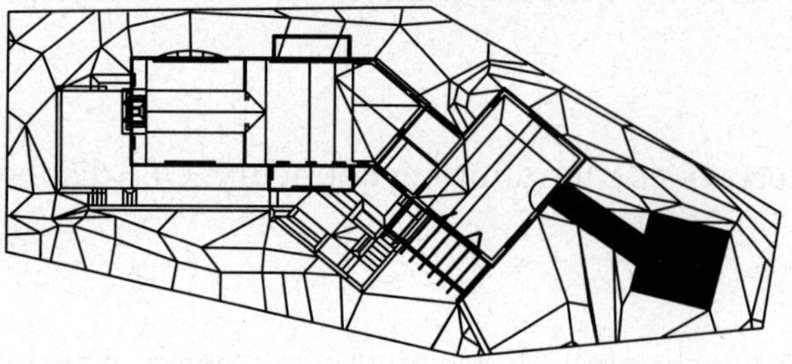

**FIGURE 30-41**

*XCLIPFRAME* = 0

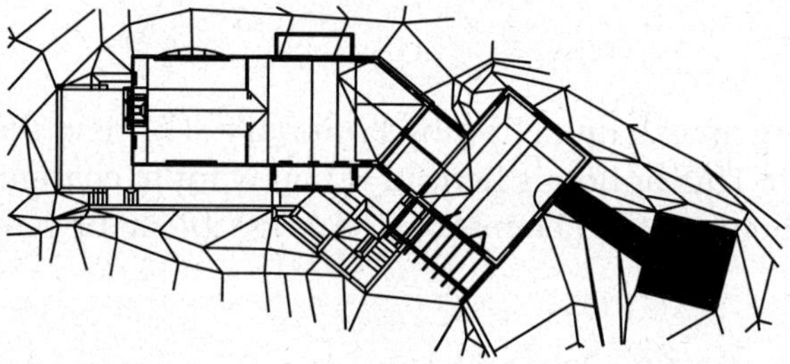

## Demand Loading

When *Xref* clipping is used to display only a selected area of an *Xrefed* drawing or block, it logically follows that the area outside the clipping boundary is of little or no interest in the current drawing. In order to conserve system resources and save time loading and regenerating these unwanted drawing areas, AutoCAD provides a feature called "demand loading."

The *Partialopen* and *Partiaload* commands allow you to open only part of a drawing based on *Layers, Views,* or by selecting a particular area of a drawing to load by specifying a boundary (see "*Partialopen*" and "*Partiaload*" in Chapter 2, Working with Files, for more information). These features are similar to the *Xclip* feature for *Xrefs* and can also take advantage of demand loading.

Demand loading allows AutoCAD to load only the information (about the *Xref* or partially opened drawing) that is needed to display the geometry in the clipping boundary. To take maximum advantage of demand loading, the demand loading feature must be enabled in the parent or partially opened drawing and the *Xref* drawings must be saved with layer and spatial indexes enabled. When a clipping boundary is formed, AutoCAD uses the layer index and spatial index in the *Xrefed* or partially opened drawing to categorize which objects are fully within, partially within, and not within the clipping boundary and which layers are visible and not visible.

| | |
|---|---|
| Spatial Index | Updated in the *Xref* or partially opened drawing when a clipping boundary is formed to display a portion of the drawing's objects |
| Layer Index | Updated in the *Xref* or partially opened drawing when layer visibility is changed (*Freeze, Thaw, On, Off*) |

Using the layer index and spatial index, AutoCAD determines which portions of the *Xrefed* or partially opened drawing file to load. This process provides performance advantages; however, certain demand loading controls must be used. The controls for demand loading are the *INDEXCTL* and *XLOADCTL* system variables.

TIP

*INDEXCTL* enables spatial indexing and layer indexing and should be set in the *Xref* drawings and in drawings intended to be used with *Partialopen* and *Partiaload*. *XLOADCTL* is intended to be set in the parent drawing for an *Xref*, but has no usefulness for the *Partialopen* and *Partiaload* features.

## *INDEXCTL*

*INDEXCTL* is a system variable, not a command; therefore, you must type *INDEXCTL* at the command prompt or use the *Save As* dialog box to change this setting.

TIP

When a drawing is used as an *Xref*, and particularly a clipped *Xref*, demand loading can increase performance in the parent drawing. Similarly, when *Partialopen* or *Partiaload* is used to display a drawing, demand loading can also increase performance. In order for this to occur, spatial indexing and/or layer indexing must be enabled in the *Xref* or partially opened drawing(s). *INDEXCTL* controls how a drawing is loaded when it is used as an *Xref*. In other words, *INDEXCTL* is set in the *Xref* drawing (the drawing to be referenced), not in the parent drawing (the current drawing when using the *Xref* command).

Considering this in another way, assume you *Open* a drawing (parent drawing) and *Xref* two other drawings. Next, you create a clipping boundary around each *Xref*, so only a portion of each *Xref* is visible. Then you *Save* the parent drawing. In order to take advantage of demand loading and realize a performance increase when you *Open* the parent drawing again, layer and spatial indexing must first be enabled in each of the two *Xrefed* drawings for demand loading to occur. *Open* each (original) *Xref* drawing, and set *INDEXCTL* to 1, 2, or 3, then *Save*.

The *INDEXCTL* variable can be set by Command line format or can be set by using *SaveAs*, selecting *Tools*, then selecting the *Options...* button. The *Saveas Options* dialog box appears (Fig. 30-42) with four options for *Index Type*.

**FIGURE 30-42**

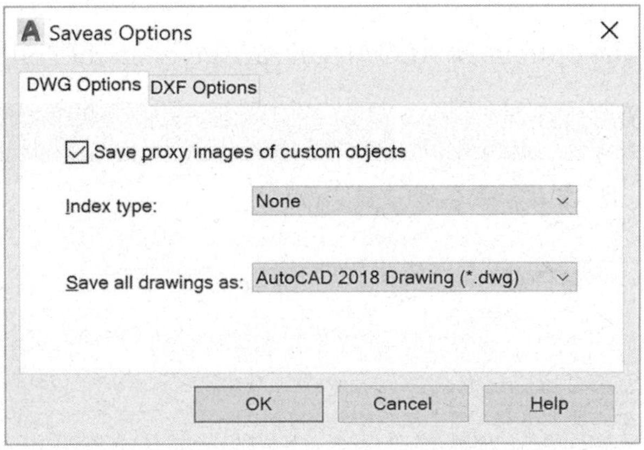


The *INDEXCTL* system variable is set to 0 by default, in which no spatial or layer indexes are created. The settings for *INDEXCTL* are as follows.

| Setting | Export Options Setting | Effect |
|---|---|---|
| 0 | *None* | no indexes created |
| 1 | *Layer* | layer indexes created |
| 2 | *Spatial* | spatial indexes created |
| 3 | *Layer & Spatial* | both layer and spatial indexes created |

To take full advantage of demand loading, set *INDEXCTL* to 3 in the drawing you anticipate using as an *Xrefed* or partially opened drawing.

## XLOADCTL

TIP

*XLOADCTL* is a system variable, so you must type *XLOADCTL* at the command prompt or use the *Options* dialog box to change this setting.

The *XLOADCTL* variable set in the parent drawing (1) turns on or off demand loading for the current drawing and (2) controls what access other users have to *Xrefed* drawings. *XLOADCTL* is intended to be set in the parent drawing for an *Xref* but has no usefulness for the *Partialopen* and *Partiaload* features.

Demand loading is dynamic. If the clipping boundary is changed, the spatial index in the *Xrefed* drawing is automatically updated and if an *Xref* layer is frozen or thawed, the layer index is updated. This unique feature allows information from the *Xrefed* drawing to be loaded "on demand."

However, there is one more aspect to demand loading for *Xrefs*—that is, when demand loading is enabled, AutoCAD places a lock on all *Xrefed* drawings so that it can read and modify the information it needs (indexes) on demand. Therefore, other users can *Open* the *Xrefed* drawings as "read only" but cannot *Save* changes to them. Alternately, you can use the *Copy* option of demand loading, which creates a temporary copy of the *Xrefed* drawing(s) for use by the parent drawing and leaves the original *Xref* drawing available for use by others to edit and *Save*.

NOTE: *XLOADCTL* has no effect on <u>your</u> ability to use *Xopen* (or the *Open* option in the *External References* palette). In other words, if *XLOADCTL* is set to 1 (others cannot edit), you can still open and edit a currently *Xrefed* drawing using *Xopen*; however, other users can only open the original *Xref* drawing as "read-only."

The options of *XLOADCTL* are given in the following table.

| Setting | *Options* Setting | Effect |
|---|---|---|
| 0 | *Disabled* | demand loading is disabled<br>*Xref* drawing is loaded in its entirety<br>others can access, edit, and save the *Xref* file |
| 1 | *Enabled* | demand loading is enabled<br>locks original *Xref* drawing for use by parent drawing only |
| 2 | *Enabled with copy* | demand loading is enabled<br>creates a temporary copy of the *Xref* for use by the parent drawing<br>others can access, edit, and save original *Xref* file |

TIP

The default setting for *XLOADCTL* is 2. If you are in a networking environment, a setting of 1 ensures others cannot alter the *Xref*, while a setting of 2 allows other users involved in a particular project (related to the *Xrefed* drawing) to continue working on the original *Xref* files.

*XLOADCTL* can also be set in the *Open and Save* tab of the *Options* dialog box (Fig. 30-43, right center). The three options (*Disabled*, *Enabled*, and *Enabled with copy*) are explained in the previous table.

## XLOADPATH

This system variable is used when *XLOADCTL* is set to 2 (*Enabled with copy*). Normally, when AutoCAD creates the copy of *Xref* drawings, the temporary copy is placed in the directory specified as *Temporary External Reference File Location* in the *Files* tab of the *Options* dialog box. You can use the *XLOADPATH*

FIGURE 30-43

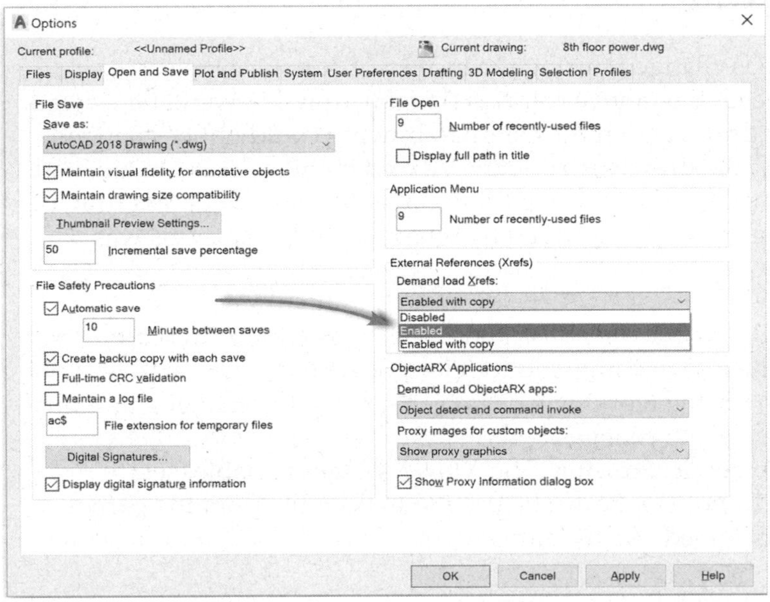


variable directly (in Command line format) or the *Options* dialog box to specify another location for temporary files. In the *Files* tab of the *Options* dialog box, expand *Temporary External Reference File Location*, highlight the previously specified location, then select the *Browse...* button to specify a new location for temporary files.

As a summary for setting demand loading and related controls, Figure 30-44 illustrates the appropriate drawing files (parent drawing and related *Xref* drawings) for setting the *XLOADCTL*, *XLOADPATH*, and *INDEXCTL* variables.

FIGURE 30-44

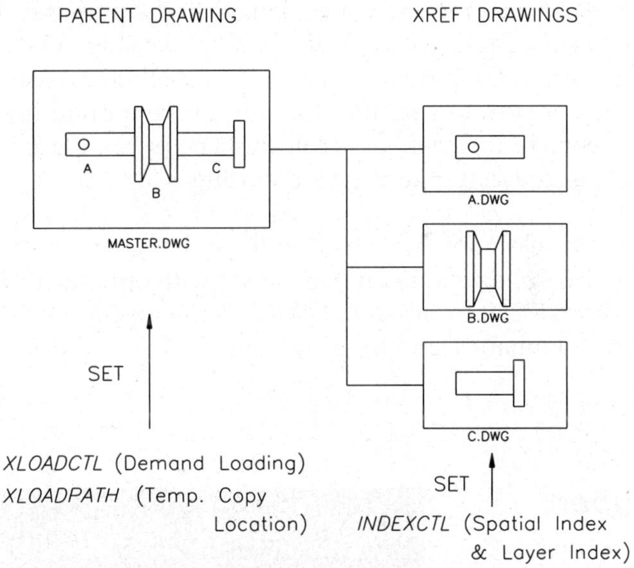


TIP

## *XREFOVERRIDE*

If you *Xref* drawings that contain objects whose visual properties (such as color, linetype, lineweight, etc.) are not set to *ByLayer*, you have no control over these visibility settings for those objects. For example, when *Xrefed* objects have color set to a *ByBlock* setting, you cannot control the appearance of those objects using the *Layer Properties Manager*. Therefore, you can use the *XREFOVERRIDE* variable to force non-*ByLayer* objects on *Xref* layers to be treated as if their properties are set to *ByLayer* and each external reference layer can have its own set of layer overrides. The settings are as follows, but only apply when the visual properties of the objects on the external reference drawing are not set to *ByLayer*.

0 Visibility settings for (non-*ByLayer*) objects on *Xref* layers cannot be controlled (legacy behavior).

1 Non-*ByLayer* objects on *Xref* layers are treated as if their properties are set to *ByLayer* and each *Xref* layer can have its own set of layer overrides.

## EDITING FOR *XREFS* AND *BLOCKS*

The strategy for editing external drawings has changed during the history of AutoCAD releases. When external references were first introduced, the innovation was that any drawing could be viewed (externally referenced) from within another drawing, but the external drawing could not be changed from within the parent drawing. In this way, the drawings could be viewed by many users to facilitate working in collaborative design environments, yet the lack of editing capabilities in the parent drawing prevented the externally referenced drawings from being edited by unauthorized users. This was AutoCAD's built-in protection for *Xrefs*.

In AutoCAD 2000, a concept was introduced that allowed *Xrefs* and *Blocks* to be edited "in place," or from within the parent drawing. First, use the *Refedit* command to select individual *Xrefs* or *Blocks* to be edited, then use *Refset* to add to or remove from the working set of objects, make the changes, and finally use *Refclose* to close the editing session and save the changes.
The changes are "sent back" to the original external drawing. The *XEDIT* system variable (set in the original *Xref* drawing) controls whether or not *Refedit* can be used to make edits from the parent drawing, so security for the *Xrefs* can be maintained when desired.

Since AutoCAD 2004, a simpler and more direct method of editing *Xrefs* is available—the *Xopen* command (also see "*Xref*" and the *Open* option in the *External References* palette). Using *Xopen* from within the parent drawing, you can open any external reference drawing in a separate window, make changes, *Save*, and view the changes in the parent drawing immediately upon reloading the *Xref*. The *Xopen* command overrides the *XEDIT* variable, which in previous versions disabled editing capabilities from within the parent drawing. This method is clean and simple; however, there are no AutoCAD commands or system variables that can prevent unauthorized users from changing the original externally referenced drawings. Controls must be put in place by the CAD managers using network security measures to prevent unauthorized editing.

You can select any *Xref* object in the drawing and right-click to produce a shortcut menu (Fig. 30-45) with options for *Edit Xref in-place* (*Refedit* command), *Open Xref* (*Xopen* command), *Clip Xref* (*Xclip* command), and for producing the *External References* palette.

**FIGURE 30-45**

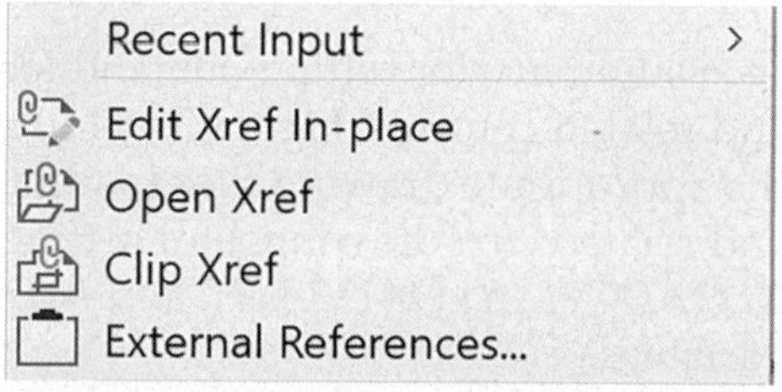


### *Xopen*

| Ribbon | Menu Bar | Command (Type) | Alias (Type) | Shortcut |
|---|---|---|---|---|
| *External Reference Open Reference* | *Tools Xref and Block In-place Open Reference* | *Xopen* | ... | (Edit Mode) *Open Xref* |

The *Xopen* command allows a drawing currently being viewed as an *Xref* from within the parent drawing to be opened and edited in a separate window. In older versions of AutoCAD, only the *Refedit* command could be used to edit an *Xref* drawing from within the parent drawing.

If you use *Xopen* in Command line format, you are prompted to select the desired *Xref* you want to edit. After selection, you are presented with the new drawing window and the original externally referenced drawing.

Command: ***XOPEN***
Select Xref: **PICK**
Command:

Alternately, use the *Open* option in the *External References* palette (Fig. 30-46). Selecting the shortcut menu option (*Open*) immediately opens the new window with the selected *Xref* drawing. You have full editing capabilities when you invoke *Xopen* by any method to open an *Xref* drawing.

When the changes are made and *Saved*, you are normally notified in the parent drawing that the *Xref* has changed. A notification icon or bubble appears at the system tray (depending on the setting of *XREFNOTIFY* system variable) indicating the name of the drawing that has changed and that it requires reloading (Fig. 30-47). To reload the drawing, use the *Reload* button in the *External References* palette (see Fig. 30-46). Also see "*XREFNOTIFY*."

FIGURE 30-46

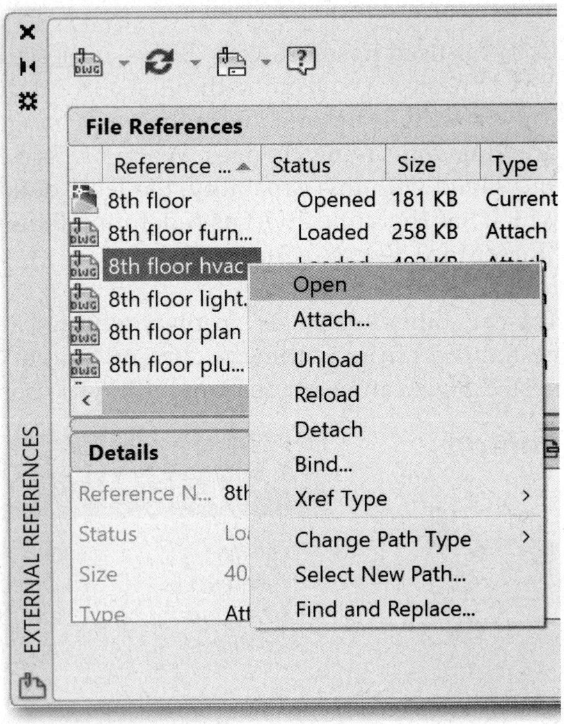


TIP

NOTE: Opening a drawing using <u>*Xopen*</u> (by any method) <u>overrides</u> the setting of both the *XLOADCTL* and *XEDIT* variables. When *XEDIT* is set to 0 (in the original external drawing), the drawing currently being viewed as an *Xref* cannot be edited using <u>*Refedit*</u> in the parent drawing. When *XLOADCTL* is set to 1 (in the parent drawing), the drawing currently being viewed as an *Xref* cannot be edited by another user (technically, by another AutoCAD session, even if it is the same user as the person viewing the *Xref*). In other words, when viewing an *Xref* (in which its *XEDIT* variable is set to 0 and the parent's *XLOADCTL* variable is set to 1), you can use the *Xopen* command to edit the original drawing, but you cannot use the normal *Open* command or the *Refedit* command to edit the drawing.

FIGURE 30-47

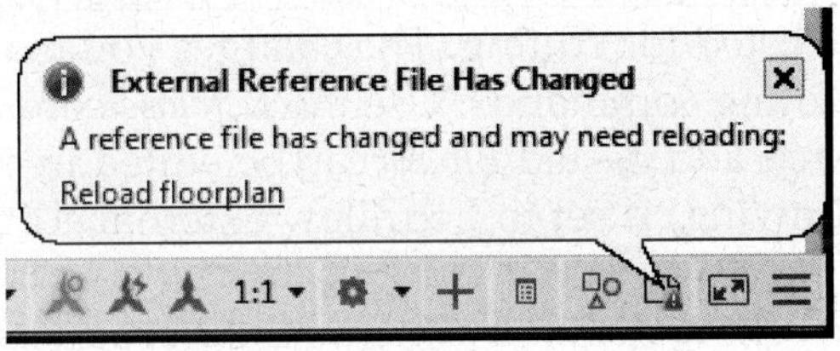


## In-Place *Xref* and *Block* Editing

You can also edit *Xrefs* and *Blocks* "in place" or from within the parent drawing using the *Refedit* command and a series of other commands and variables described next. Although these features can be used in current releases, the *Xopen* command is a simpler method for editing *Xrefs*; however, using *Refedit* to change *Blocks* from within the parent drawing is still a viable method since *Blocks* cannot be edited "in place" using *Xopen*. The commands and variables for editing *Blocks* or *Xrefs* "in place" are:

| | | |
|---|---|---|
| *Refedit* | (command) | Allows you to select individual objects in an *Xref* or *Block* and edit that geometry from within the current "parent" drawing |
| *Refset* | (command) | Allows you to add or remove objects to and from a working set while editing *Blocks* and *Xrefs* from the current drawing |
| *Refclose* | (command) | Closes the current reference editing session and saves or discards changes made to the *Xref* or *Block* geometry back to the original *Xref* drawing file or *Block* definition |
| *XEDIT* | (variable) | Controls whether a drawing file can be edited in place when referenced by another drawing |
| *XFADECTL* | (variable) | Controls the percentage of fading that occurs to objects not in the working set to edit |

With the "in-place *Xref* and *Block* editing" capability, anyone who can *Xref* a drawing can also change that drawing and save it back to the original file unless *XEDIT* is set to 0 in the original *Xref* drawing. This situation presents a potential danger since the default setting for *XEDIT* is 1, which allows in-place editing. Therefore, it is suggested that the *XEDIT* system variable be changed to 0 for all template drawings in multi-user project environments.

*Refedit* is used to select *Xrefs*, *Blocks*, or nested *Xrefs* and *Blocks* to edit. When *Refedit* is used to change an *Xref* or nested *Xref* from within the current drawing and *Refclose* is used to save the changes, the original *Xref* drawing is updated with the new changes, which has the same effect as opening the original *Xref* drawing and saving changes. When *Blocks* defined in the current drawing or nested *Blocks* are edited and saved with this capability, the *Block* definition is saved in the original drawing containing the *Block*, which has the same effect as redefining the *Block* in its original drawing by the traditional method (see "Redefining *Blocks*," Chapter 21).

The capability known as "in-place editing" allows you to change geometry in an *Xrefed* drawing or nested *Block* from within the current "parent" drawing. With in-place editing, changes to an *Xref* or nested *Block* can be made from within the parent drawing, then saved back to the original file.

## Refedit

| Ribbon | Menu Bar | Command (Type) | Alias (Type) | Shortcut |
|---|---|---|---|---|
| *External Reference Edit Reference In-Place* | *Tools In-place Xref and Block Edit > Edit Reference* | *Refedit* | ... | (Edit Mode) *Edit Xref In-place* |

The *Refedit* command simply allows you to select which *Xref* or *Block* you want to edit and to pick the specific geometry you want to change. Only one *Xref* or *Block* at a time (and its nested objects) can be selected for editing. The changes you make to the selected *Xref* or *Block* are accomplished using typical editing commands. Use the *Refclose* command to save the changes back to the original drawing file. *Xrefs* and nested *Blocks* can be edited in place only if their *XEDIT* system variable (in the original *Xrefed* drawing) is set to 1 to allow external editing.

Invoking *Refedit* by any method produces the "Select reference" prompt. Select any *Block* or *Xref* from the drawing area using the pickbox or window.

Command: ***REFEDIT***
Select reference: **PICK** (*Xref* or *Block*)
(*Reference Edit* dialog box appears)

After selecting an *Xref* or *Block*, the *Reference Edit* dialog box appears (Fig. 30-48). The name of the selected reference appears under *Reference name:* along with its preview image on the right.

**FIGURE 30-48**

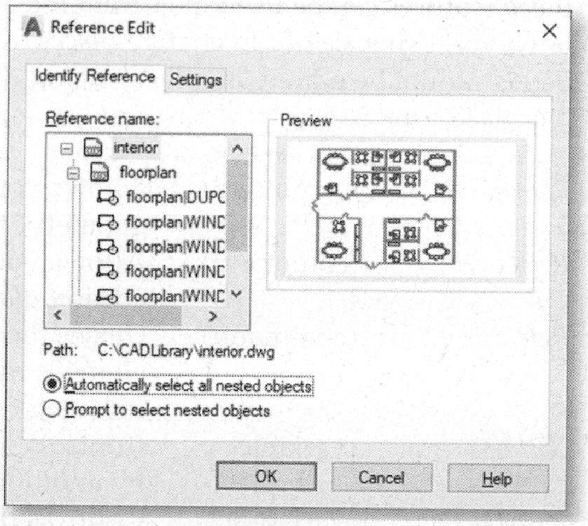


You can choose to *Automatically select all nested objects*, or you can choose that AutoCAD *Prompt to select nested objects*. If you want to edit only a few *Lines* or *Circles* in your *Xref* or *Block*, you should choose to be prompted. Select *OK* to return to the drawing editor.

The "Select nested objects" prompt requests that you select geometry contained in the selected reference object such as a *Line*, *Circle*, or *Block* that you want to edit. This becomes your working set. After selection, the working set becomes visually distinct because all other objects not in the working set become faded. For example, Figure 30-49 displays geometry comprising the table and chairs as the "nested objects," or working set, and the remainder of the drawing is faded. The amount of fading is controlled by the *XFADECTL* system variable (see "*XFADECTL*" later in this section).

FIGURE 30-49

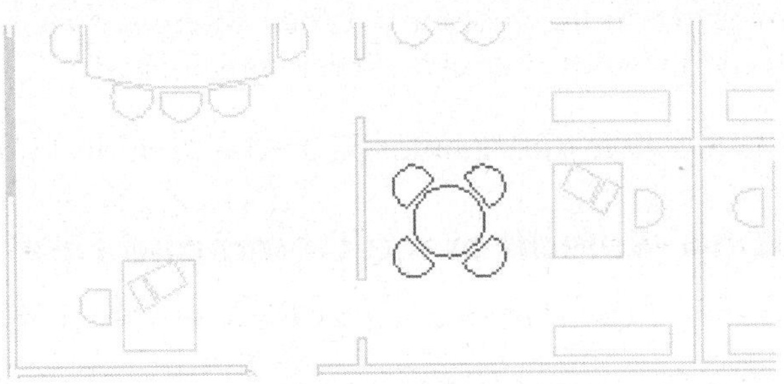

The working set can then be edited and changed in any way using normal editing commands and grips. To save changes made to the working set, use *Refclose* with the *Save* option. The changes are saved back to the original *Xref* drawing file or to the *Block* definition in the drawing that contains the *Block*. If a warning is issued when you select "nested objects" (Fig. 30-50), editing is not allowed because the *XEDIT* system variable is set to 0 in the original *Xrefed* drawing or original drawing containing the nested *Block*.

FIGURE 30-50

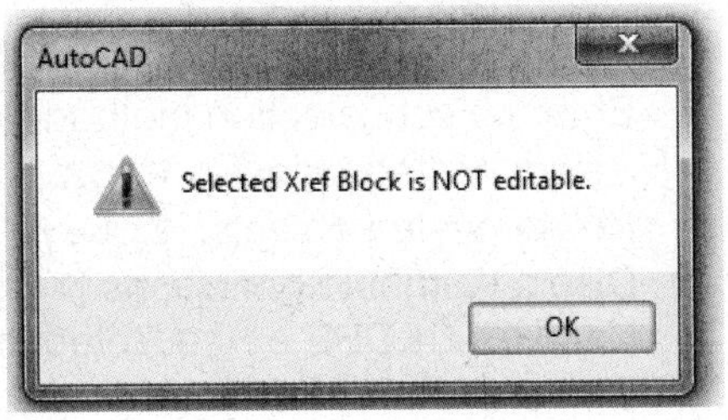


*Refedit* allows you to edit nested *Xrefs* and nested *Blocks*. When you select a reference object in response to the first *Refedit* prompt, "Select reference," the *Reference Edit* dialog appears and a hierarchical nesting structure is displayed for the reference objects (Fig. 30-51). Select the desired nesting level from the list to cause AutoCAD to highlight the selected reference in the drawing.

FIGURE 30-51

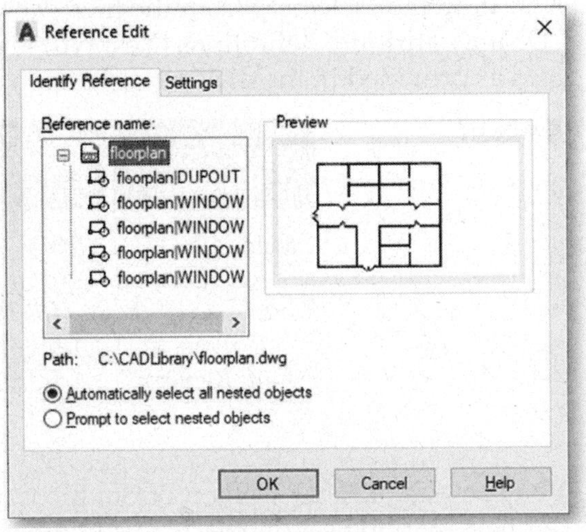


### *Settings* Tab

The *Settings* tab (Fig. 30-52) provides the following options.

#### *Create Unique Layer, Style, and Block Names*

When this box is checked (default position), the layer and dependent object names of selected *Xrefs* are prefixed with the *Xref* drawing name and $x$, similar to the naming convention used when you *Bind* an *Xref*. If the box is not checked, dependent layer and object names appear in the current drawing as they would in the original *Xref* drawing.

#### *Display Attribute Definitions for Editing*

This feature lets you edit attribute definitions (tags) in *Block* references from within the current drawing. When this box is checked, the attribute tags in *Blocks* (in the current drawing) or in nested *Blocks* are displayed instead of the attribute values themselves. You can select the attribute tags for editing along with the *Block* geometry. When changes are saved back to the *Block* reference, only new insertions of the *Block* are affected. Attributes of the previous *Block* insertions are not changed. See Chapter 22, Block Attributes and Data Links, for more details on this subject.

FIGURE 30-52

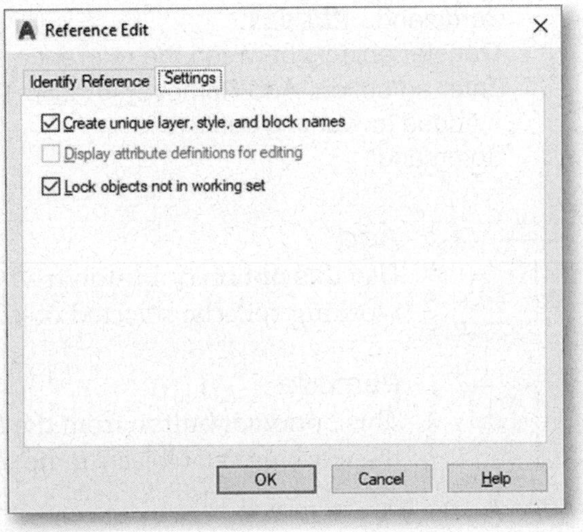


#### *Lock Objects Not in Working Set*

This option automatically locks all objects not in the working set including objects in the parent drawing (see *Refset*). Any selected locked objects behave as if they are on a locked layer. This feature prevents you from accidentally selecting and editing objects in the parent drawing.

If you select objects in an *Xref* as the working set, but they are on a locked layer, you must *Unlock* the layer to edit the geometry. When changes are saved back to the *Xref*, the layer state of the original *Xref* drawing is not changed—it remains locked.

The *Refedit* feature cannot be used with individual *Blocks* that have been inserted using *Minsert*.

## Using *-Refedit* at the Command Line

Typing *-Refedit* produces the Command line version shown below.

```
Command: -REFEDIT
Select reference: PICK
Select nesting level [Ok/Next] <Next>: o
Enter object selection method [All/Nested] <All>: n
Select nested objects: PICK
Select nested objects: Enter
Display attribute definitions [Yes/No] <No>: Enter
Use REFCLOSE or the Refedit toolbar to end reference editing session.
Command:
```

The Command line prompt to "Select nesting level [Ok/Next] <Next>:" serves the same function as the *Next* button or selecting from the hierarchical list in the *Reference Edit* dialog box. Note the last prompt, "Display attribute definitions [Yes/No] <No>:", provides the same level of access for editing attribute tags as provided in the dialog box version of the command.

### *Refset*

| Ribbon | Menu Bar | Command (Type) | Alias (Type) | Short-cut |
|---|---|---|---|---|
| *External Reference Edit Reference Add, Remove* | *Tools Xref and Block In-place Edit > Add to Workset or Remove from Workset* | *Refset* | ... | ... |

You can *Add* or *Remove* objects from the working set using *Refset* while editing in place. You initially define a working set of objects to edit when you select objects at the "Select nested objects:" prompt during the *Refedit* command. The working set is visually distinct while the other objects in the drawing become faded. You can add or remove to or from this working set using *Refset*.

```
Command: REFSET
Transfer objects between the reference set and drawing...
Enter an option [Add/Remove] <Add>: PICK
1 Added to working set
Command:
```

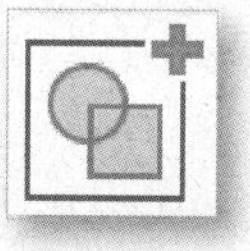

#### *Add*

Use this option or button from the *Edit Reference* panel to select objects to add to the working set. The selected objects change from faded to distinct and can then be edited.

#### *Remove*

This option or button from the *Edit Reference* panel allows you to select objects to remove from the working set. Objects removed from the working set become faded and cannot be edited.

## Refclose

| Ribbon | Menu Bar | Command (Type) | Alias (Type) | Shortcut |
|---|---|---|---|---|
| *External Reference Edit Reference Save, Discard* | *Tools In-place Xref and Block Edit > Save Reference Edits* | *Refclose* | ... | ... |

Use the *Refclose* command to *Save* back or *Discard* changes made to objects during an in-place editing session. In other words, use *Refedit* to select *Xrefs* and *Blocks* and objects within those references to edit, modify the geometry, then use *Refclose* to close the editing session and save or discard the modifications to the working set.

Command: ***REFCLOSE***
Enter option [Save/Discard reference changes] <Save>: **Enter**
Regenerating model.
1 block instances updated
(Block name) redefined.
Command:

You can also *Save* or *Discard* edits by using the options on the *Tools* pull-down menu, *Edit Reference* panel, or the *Refedit* toolbar. The *Edit Reference* panel appears after a reference is selected for editing and disappears after changes to the reference are saved or discarded.

FIGURE 30-53

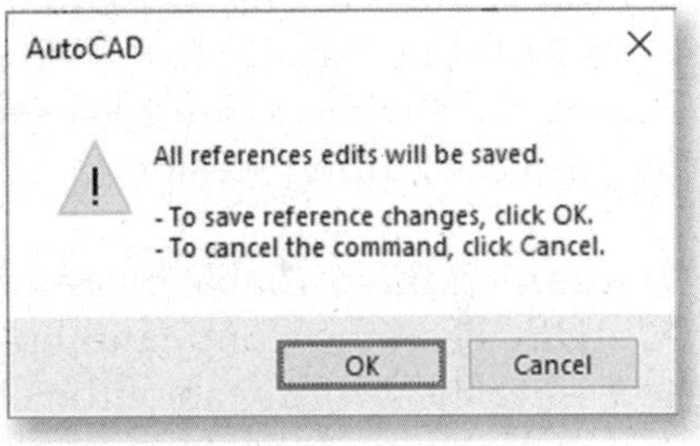

### *Save*

Saving changes saves the modifications made to the working set back to the original reference (*Xref* or *Block*) drawing file. The change is made without actually opening the reference drawing or recreating the block. If you decide to *Save* changes, you must confirm the dialog box that appears to complete the desired action (Fig. 30-53).

Using the *Save* option of *Refclose* does not save changes made to other objects (not in the in-place editing working set)—you must use the full *Save* command to save those changes. If you use *Refedit* and then *Refclose Save* to make edits to *Blocks* contained only in the current drawing, you must also *Save* the drawing.

### *Discard Reference Changes*

Use this option if you want to close the in-place editing session and discard all changes made to the working set. A dialog box appears similar to that shown in Figure 30-53. Any changes you have made to other objects in the current drawing (not in the working set) are not discarded.

## XEDIT

*XEDIT* is a system variable, so you can type *XEDIT* at the command prompt or use the *Options* dialog box to change this setting.

The value of this system variable determines whether or not a drawing can be edited in place when it is used as an *Xref*. <u>*XEDIT* should be set in the original *Xref* drawing</u>, not in the "parent" drawing. The possible settings are:

0 The drawing cannot be edited while *Xrefed* by another drawing.
1 The drawing can be edited in place when used as an *Xref* in another drawing.

You can change this system variable two ways: enter *XEDIT* at the Command line and change the value, or use the *Open and Save* tab of the *Options* dialog box (see Fig. 30-38). If you use the *Options* dialog box, placing a check in *Allow others to Refedit current drawing* sets the *XEDIT* variable to 1. The variable is saved in the drawing.

**TIP**

The default setting of *XEDIT* in AutoCAD template drawings is 1, so in-place editing can occur without changing this variable. Keep in mind that in this case anyone who can *Xref* a drawing can also make an accidental or unauthorized change to that drawing and save it back to the original file. This situation presents a potential danger in multi-user project environments. Therefore, it is suggested that the *XEDIT* system variable be changed to 0 for all template drawings in multi-user project environments.

## XFADECTL

*XFADECTL* is a system variable. Change this setting at the command prompt or use the *Options* dialog box.

This system variable controls the percentage of fading during an in-place editing session. When *Refedit* is used and reference geometry is selected for editing, it becomes the working set and remains distinct (not faded). All other objects in the current drawing become faded (see Fig. 30-49). The default value is 0. The *XFADECTL* range is from 0 (0%, no fading) to 90 (90% fading). Fading ends when *Refclose* is used to close the in-place editing session.

FIGURE 30-54

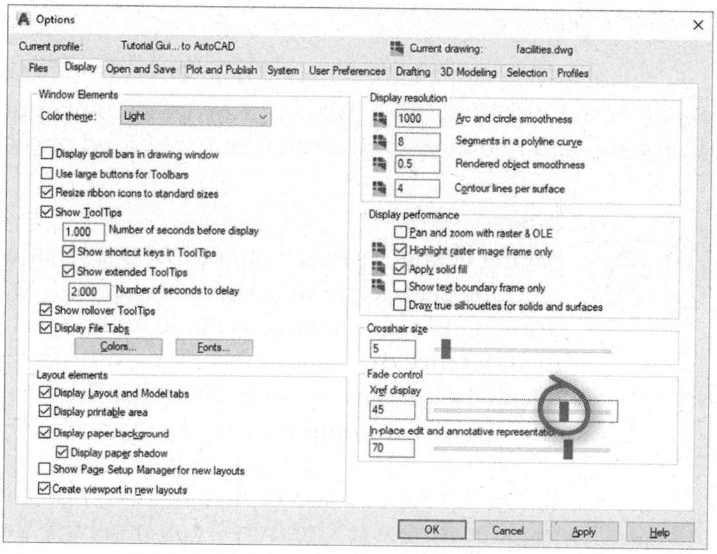


You can set this variable by entering *XFADECTL* at the Command line or by accessing the *Display* tab of the *Options* dialog box (Fig. 30-54, lower right). Use the *In-place edit and annotative representations* slider to control the *XFADECTL* variable. *XFADECTL* is saved in the system registry, so once the variable is set, you do not have to reset it whenever new *Xrefs* are edited.

## XREFNOTIFY

*XREFNOTIFY* is a system variable. You can change this setting at the command prompt.

*XREFNOTIFY* is a system variable that controls the display of the notification icon and balloon message that appear at the lower-right corner of the Status bar when *Xrefs* have been updated or are not found. The settings are described below. This variable is saved in the system registry.

0 *Xref* notification is disabled.

1 *Xref* icon is enabled. This setting turns on the *Xref* icon in the Status bar when *Xrefs* are attached to the current drawing. If *Xrefs* are not found, the icon is displayed with a yellow alert symbol (!).

2 *Xref* notification and balloon messages are enabled. This setting turns on the *Xref* icon, as in a setting of 1, but also displays balloon messages in the same area when *Xrefs* are modified.

## MANAGING *XREF* DRAWINGS

TIP

Because of the external reference feature, the contents of a drawing can be stored in multiple drawing files and directories. In other words, if a drawing has *Xrefs*, the drawing is actually composed of several drawings, each possibly located in a different directory. This means that special procedures to handle drawings linked in external reference partnerships should be considered when drawings are to be backed up or sent to clients. Several considerations are listed here:

1. Create a project name for the drawing with the *PROJECTNAME* variable. The project name is stored in the drawing file. Notify the client so the client can create the same project name on his or her system (with the *Files* tab of the *Options* dialog box) to store the drawings and *Xrefs* (see "*PROJECTNAME*").
2. When *Attaching Xrefs* to a drawing intended to be used by clients, use the *No path* or *Relative path* options (in the *External Reference* dialog box). If the *No path* option is used, you can copy the *Xref* drawings on another system into the same directory (folder) as the parent drawing. If the *Relative path* option is used, back up and copy the same directory structure (from the parent drawing down only) on the new system as on the original system.
3. Modify the current drawing's path to the external reference drawing so both are stored in the same directory, then archive them together.
4. Archive the directory structure of the external reference drawing with the drawing which references it.
5. Make the external reference drawing a permanent part of the current drawing with the *Bind* option of the *Xref* command prior to archiving. This option is preferred when a finished drawing set is sent to a client.
6. Use the *Reference Manager* to manage the paths of drawings and the related *Xrefs*. For example, if you copied the parent drawing and related *Xrefs* onto another system with a different directory structure than that saved in the *Full path* on the original system when the *Xrefs* were *Attached*, launch the *Reference Manager* on the new system. Using *Reference Manager*, you can change the path for a group of *Xrefs* globally. See "Reference Manager."
7. Use Autodesk's Forma Data Management cloud-based system which groups and stores project files for shared access, version control, and easy collaboration including using connected support files.

### *PROJECTNAME*

Project names make it easier for AutoCAD users to exchange drawings that contain *Xrefs*. The *PROJECTNAME* variable allows you to specify a project name that AutoCAD uses to point to an alternate search path when trying to locate *Xrefs*. *PROJECTNAME* is a system variable, not a command. The *PROJECTNAME* value must first be set in the system registry using the *Options* dialog box, and then be set in the drawing using command line format.

If you load a drawing containing several *Xrefs* given to you by a colleague from another company, AutoCAD first attempts to locate the *Xrefs* by searching the hard-coded paths. (The path of an *Xref* is "hard coded" when the drawing is originally *Attached* or *Overlayed* or if the *Path type* option is used in the *External References* palette). The *Xrefs* are not found initially unless you have an identical directory structure on your system or network, and the *Xrefs* have been copied into the directories accordingly. If an *Xref* cannot be found because the drawing is not located in the hard-coded path, AutoCAD then uses the drawing's *PROJECTNAME* to search for the *Xref* in the location specified by your system registry.

There are two components to this scheme: (1) the *PROJECTNAME* variable is saved in the parent drawing file, and (2) the directory location for that project name is saved in your system registry. This scheme makes it possible for collaborating designers to exchange drawings containing *Xrefs* when the designers have different directory structures. The drawing that is exchanged contains the project name, but the location (path) for that project name can be different for each system (saved in the system registries). Each designer's system contains the same project name but has a different drive and/or directory mapping. In other words, if an *Xref* cannot be found in the hard-coded path, AutoCAD then searches for the *Xref* by using the *PROJECTNAME* variable in the drawing file, but the location for the project name is saved in your system registry.

The value for the drawing's *PROJECTNAME* variable (*Project1*, for example) can only be set using the *PROJECTNAME* variable in command line format. The *PROJECTNAME* variable is saved in the parent drawing file. However, you must use the *Files* tab of the *Options* dialog box (but not the *PROJECTNAME* variable in command line format) to specify the directory mapping (location), which is then stored in your system registry.

TIP

### System Registry Project Name

To assign a search path to be saved in your system registry, you must use the *Files* tab of the *Options* dialog box (Fig. 30-55). You cannot use the *PROJECTNAME* variable or any other method of Command line entry to add, remove, or modify the directory location for a project name in the system registry. Using the dialog box, follow these steps.

1. Double-click *Project Files Search Path*, then select *Add*. A project folder named *Project1* appears (or *Projectx*, where x is the next available number).
2. Enter a new name or press Enter to accept *Project1*.
3. Double-click to expand *Project1* (or other name you entered).
4. Select *Browse*. Use the *Browse for Folder* dialog box that appears (not shown) to locate the desired folder on your system where you want to store *Xref* drawings for the project.
5. Select *OK* in both dialog boxes to save the project name location to your system registry.

The action above assigns a project name to the system registry.

**FIGURE 30-55**

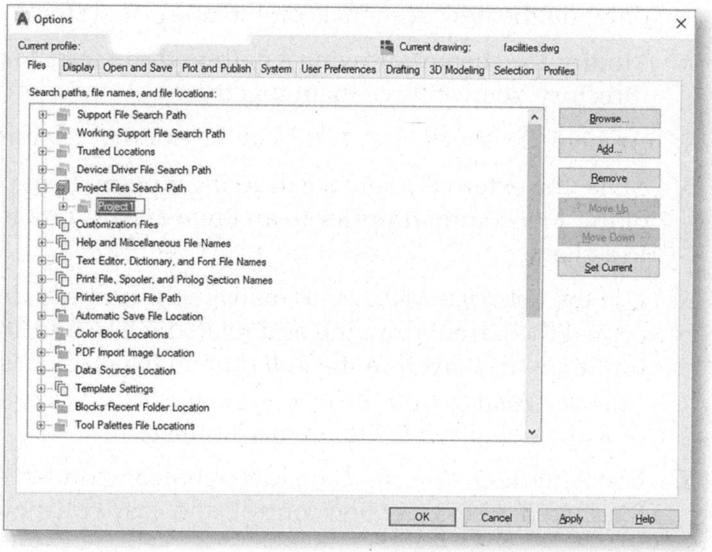


### Saving the *PROJECTNAME* in the Drawing File

Use the *PROJECTNAME* variable at the command line to save the drawing's component of the project name. Do this after registering the project name in your system using the *Options* dialog box. The Command prompt appears as follows:

Command: ***PROJECTNAME***
Enter new value for PROJECTNAME, or . for none <" ">: **project1**
Command:

The action on the previous page assigns a project name to the drawing file.
If you try this step before using the *Options* dialog box to register the project name on your system, the following prompt appears:

Command: ***PROJECTNAME***
Enter new value for PROJECTNAME, or . for none <" ">: **project1**
"project1" not found in the registry. Use the Options dialog Files tab to create the project name and set the project search paths.
Command:

## *Project Aware Connected Support Files*

Connected support files are only available for *Forma Project Data Management* (formerly a feature of *Autodesk Docs*) users. After creating the *Forma Project Data Management project* and uploading project drawings, someone with Project Administrator level access can configure the project to use *Connected support files.* This can be very useful in a large distributed organization, as it makes it easier to manage shared AutoLISP files, drawing files standards, templates, custom hatch patterns (.pat), fonts, linetypes (.lin), plot configurations, and more.

The PROJECTAWARE System Variable Informs you of the project path state of the current drawing. 0=Path settings are defined by an AutoCAD user profile; 1=Path settings are defined by a *project.*

## Reference Manager

Reference Manager, a separate application from AutoCAD, allows you to manage all references to a drawing contained in external files. The paths to all referenced files are saved in each AutoCAD drawing. However, Reference Manager allows you to list referenced files in selected drawings and to modify the saved reference paths without opening the drawing files in AutoCAD because Reference Manager is a stand-alone application that you run outside of AutoCAD. Use the Start Menu on your computer to launch Reference Manager from the Autodesk group.

References to a drawing contained in external files can be text fonts, attached images, plot configurations, and externally referenced drawings (*Xrefs*). The Reference Manager helps you manage the paths (directory locations) for all references to a drawing (Fig. 30-56). This is especially helpful when drawings or the files that they reference (such as *Xrefs*) are moved to different folders or to different disk drives, such as when drawings are used by clients. Also use Reference Manager to fix groups of unresolved references.

**FIGURE 30-56**

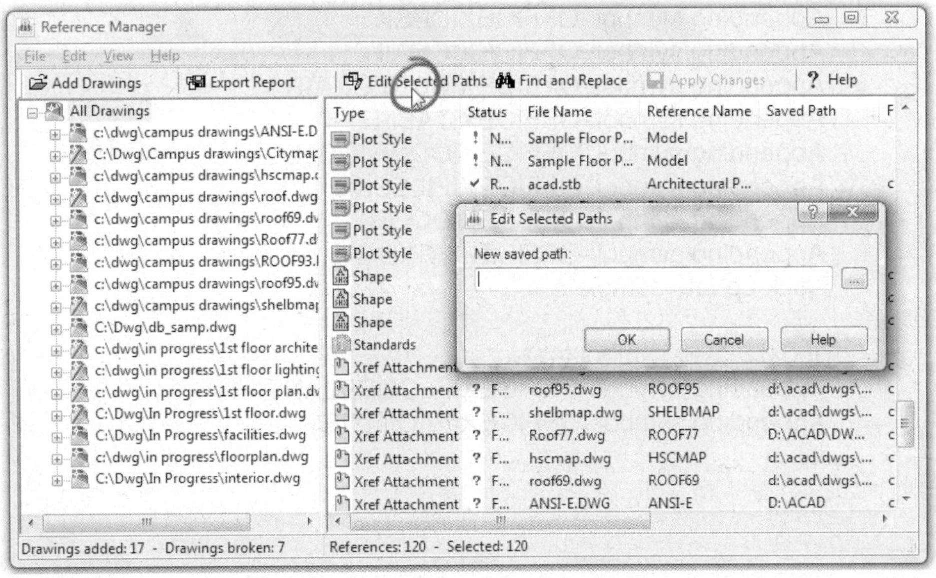


In Reference Manager, use the *Edit Selected Paths* option to specify new paths for *Xref* drawings (see Fig. 30-56). You can select multiple *Xrefs* and change the paths for all of them together.

## *XREFCTL*

AutoCAD has a tracking mechanism for the *Xref* activity in drawings. An external ASCII log file can be kept for all drawings that contain information on external references. The *XREFCTL* variable controls the creation of the *Xref* log files. *XREFCTL* is a system variable that can be changed at the command prompt.

If *XREFCTL* is set to 1, a log file registers each *Attach*, *Overlay*, *Bind*, *Detach*, *Reload*, and *Unload* of each external reference for every drawing. These log files have the same name as the related drawing with a file extension ".XLG." The log files are placed in the same directory as the related drawing. If a drawing is *Opened* again and the *Xref* command is used, the *Xref* activity (*Attach*, *Overlay*, *Bind*, etc.) is appended to the existing log file. The .XLG files have no direct connection to the drawing, so they can be deleted if desired without consequence to the related drawing file. If *XREFCTL* is set to 0 (default setting), no log is created. The *XREFCTL* variable is saved in the system registry, so once it is set, it affects all drawings. An example .XLG file is shown here. This file lists the activity from an earlier example in this chapter when the OFFICEX drawing was *Xrefed* to the INTERIOR drawing.

```
================================

Drawing: D:\Dwg\Interior.dwg
Date/Time: 03/24/15 08:55:16
Operation: Attach Xref

================================

Attach Xref OFFICEX: D:\Dwg\Officex.dwg
    Searching in ACAD search path

    Update block symbol table:
     Appending symbol: OFFICEX|DUPOUT
     Appending symbol: OFFICEX|L1-TBLCK
     Appending symbol: OFFICEX|DESK
     Appending symbol: OFFICEX|CHAIR
     Appending symbol: OFFICEX|CONFCHAIR
     Appending symbol: OFFICEX|TABLE
     Appending symbol: OFFICEX|CONFTABLE
     Appending symbol: OFFICEX|FILECAB
     Appending symbol: OFFICEX|DOOR
     Appending symbol: OFFICEX|WINDOW
    Block update complete.

    Update layer symbol table:
     Appending symbol: OFFICEX|FLOORPLAN
     Appending symbol: OFFICEX|FACILITIES
    etc.
```

## CHAPTER EXERCISES

1. ***Xref Attach***

Use the **SLOTPLATE2** drawing that you modified last (with *Stretch*) in Chapter 9 Exercises. The slot plate is to be manufactured by a stamping process. Your job is to nest as many pieces as possible within the largest stock sheet size of 30" x 20" that the press will handle.

A. ***Open*** the **SLOTPLATE2** drawing and use the ***Base*** command to specify a basepoint at the plate's lower-left corner. ***Save*** and ***Close*** the drawing.

B. Begin a *New* drawing using the **C-D-SHEET** prototype and assign the name **SLOTNEST**. Set ***Limits*** to **34** x **22**. Draw the boundary of the stock sheet (30" x 20"). ***Xref Attach*** the **SLOTPLATE2** into the sheet drawing multiple times as shown in Figure 30-57 to determine the optimum nesting pattern for the slot plate and to minimize wasted material. The slot plate can be rotated to any angle but cannot be scaled. Can you fit 12 pieces within the sheet stock? ***Save*** and ***Close*** the drawing.

**FIGURE 30-57**

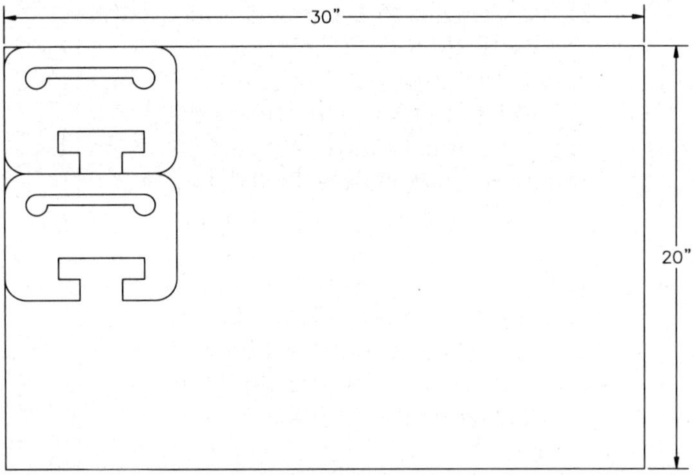


C. In the final test of the piece, it was determined that a symmetrical orientation of the "T" slot in the bottom would allow for a simplified assembly. ***Open*** the **SLOTPLATE2** drawing and use ***Stretch*** to center the "T" slot about the vertical axis as shown in Figure 30-58. ***Save*** the new change and ***Close*** the drawing.

**FIGURE 30-58**

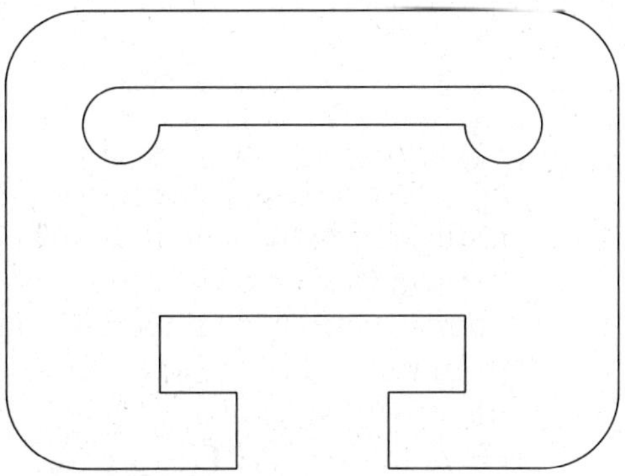

D. ***Open*** the **SLOTNEST** drawing. Is the design change reflected in the nested pieces?

2. ***Attach, Detach, Xbind, VISRETAIN***

As the interior design consultant for an architectural firm, you are required to place furniture in an efficiency apartment. You can use some of the furniture drawings (*Blocks*) that you specified for a previous job, the office drawing, and *Insert* them into the apartment along with some new *Blocks* that you create.

A. Use the **C-D-SHEET** as a ***template*** and set ***Architectural Units*** and set ***Limits*** of **48'** x **36'**. (Scale factor is 24 and plot scale is 1/2"=1' or 1=24). Assign the name **INTERIOR**. ***Xref*** the **OFFICE** drawing that you created in Chapter 21 Exercises. Examine the layers (using the ***Layer Control*** drop-down list) and list (**?**) the ***Blocks***. ***Freeze*** the **TEXT** layer. Use ***Xbind*** to bring the **CHAIR**, **DESK**, and **TABLE** ***Blocks*** and the **FURNITURE** ***Layer*** into the INTERIOR drawing. ***Detach*** the **OFFICE** drawing. Then ***Rename*** the new ***Blocks*** and the new ***Layer*** to the original names (without the **OFFICE$0$** prefix). In the INTERIOR drawing, create a new ***Block*** called **BED**. Use measurements for a queen size (60" x 80"). ***Save*** the INTERIOR drawing.

REUSE

B. *Xref* the **EFF-APT2** drawing (from Chapter 18 Exercises) into the INTERIOR.DWG. Change the ***Layer*** visibility to turn ***Off*** the **EFF-APT2 | TEXT** layer. ***Insert*** the furniture ***Blocks*** into the apartment on ***Layer*** **FURNITURE**. Lay out the apartment as you choose; however, your design should include at least one insertion of each of the furniture *Blocks* as shown in Figure 30-59. When you are finished with the design, ***Save*** and ***Close*** the INTERIOR drawing.

**FIGURE 30-59**

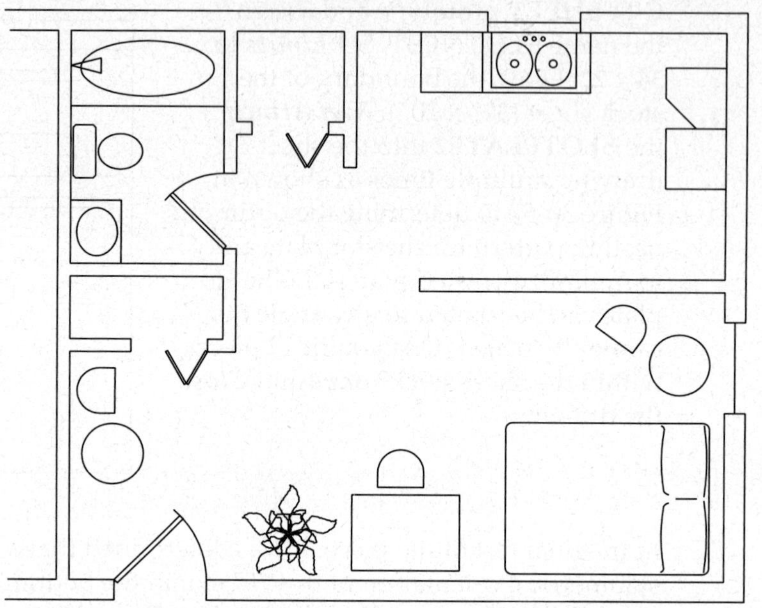

C. Assuming you are now the architect, a design change is requested by the client. In order to meet the fire code, the entry doorway must be moved farther from the end of the hall within the row of several apartments. ***Open*** the **EFF-APT2** drawing and use ***Stretch*** to relocate the entrance **8′** to the right as shown in Figure 30-60. ***Save*** and ***Close*** the EFF-APT2 drawing.

**FIGURE 30-60**

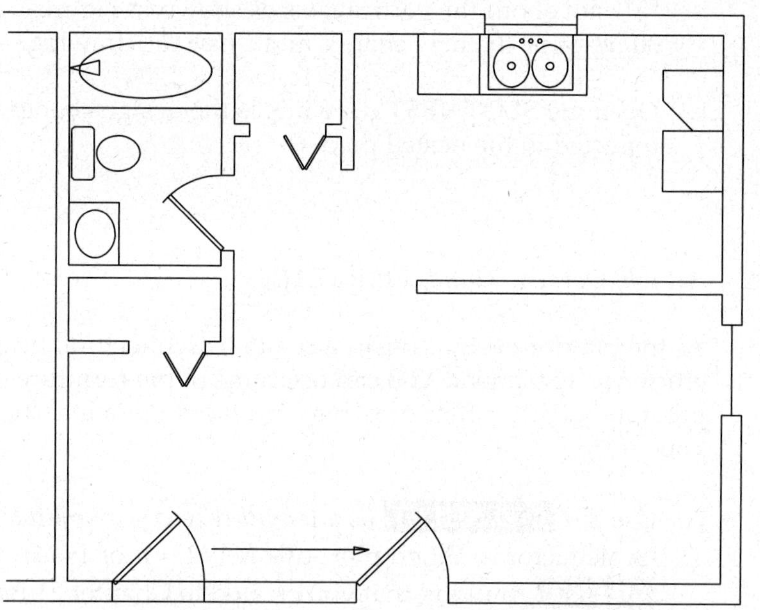

D. Next (as the interior design consultant), ***Open*** the **INTERIOR** drawing. Notice that the design change made by the architect is reflected in your INTERIOR drawing. Does the text appear in the drawing? ***Freeze*** the **TEXT** layer again, but this time change the ***VISRETAIN*** variable to **1**. Make the necessary alterations to the interior layout to accommodate the new entry location. ***Save*** the INTERIOR drawing.

3. ***Xref Attach***

As project manager for a mechanical engineering team, you are responsible for coordinating an assembly composed of 5 parts. The parts are being designed by several people on the team. You are to create a new drawing and *Xref* each part, check for correct construction and assembly of the parts, and make the final plot.

A. The first step is to create two new parts (as <u>separate</u> drawings) named **SLEEVE** and **SHAFT**, as shown in Figure 30-61. Use appropriate drawing setup and layering techniques. In each case, draw the two views on <u>separate layers</u> (so that appearance of each view can be controlled by layer visibility). Use the ***Base*** command to specify an insertion point at the right end of the rectangular views at the centerline.

**FIGURE 30-61**

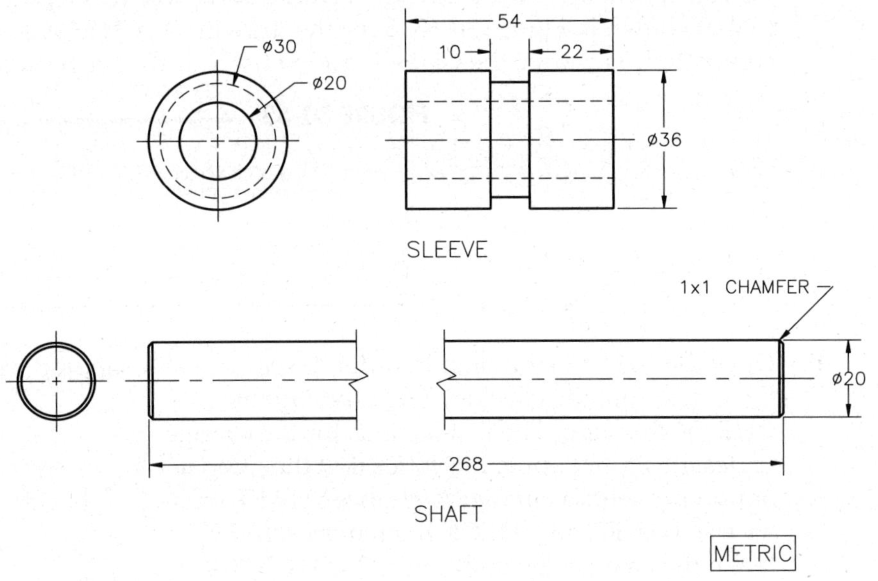


B. You will eventually ***Xref*** the drawings **SHAFT**, **SLEEVE**, **PUL-SEC** (from Chapter 26 Exercises), **SADDLE** (from Chapter 24 Exercises), and **BOLT** (from Chapter 9 Exercises) to achieve the assembly as shown in Figure 30-62. Before creating the assembly, ***Open*** each of the drawings and use ***Properties*** to move the desired view (with all related geometry, no dimensions) to a new layer. You may also want to set a ***Base*** point for each drawing.

REUSE

**FIGURE 30-62**

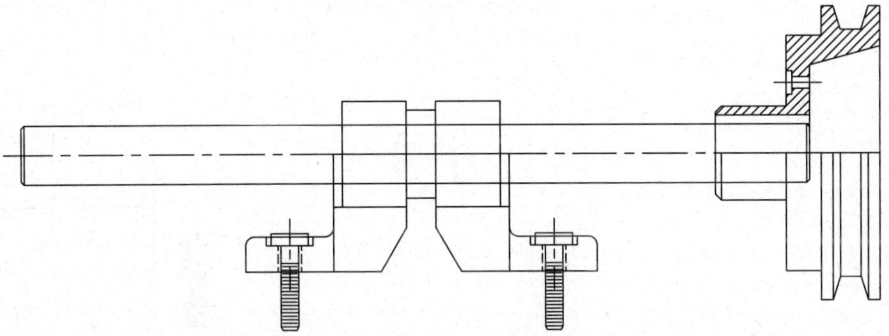

C. Finally, begin a ***New*** drawing (or use a ***Template***) and assign the name **ASSY**. Set appropriate ***Limits***. Make a ***Layer*** called **Xref** and set it ***Current***. ***Xref*** the drawings to complete the assembly. The **BOLT** drawing must be scaled to acquire a diameter of **6**mm. Use the ***Layer Control*** drop-down list to achieve the desired visibility for the *Xrefs*. Set ***VISRETAIN*** to **1**. Make a ***Plot*** to scale. ***Save*** and ***Close*** the **ASSY** drawing.

4. ***Xref Overlay***

A. Assume you are now the project coordinator/checker for the assembly. (If it is possible in your lab or office, exchange the SHAFT drawings with another person so that you can perform this step for each other.)

Begin a *New* drawing and make a *Layer* called **CHECK**. Next, ***Xref Overlay*** the **SHAFT** drawing (of your partner). On the CHECK layer, insert some ***text*** by any method and a dimension similar to that shown in Figure 30-63. ***Save*** the drawing as **CHECK** and then ***Close*** the drawing. (When completed, exchange the CHECK and SHAFT drawings back.)

**FIGURE 30-63**

B. ***Open*** the **SHAFT** drawing. In order to see the notes from the checker, ***Xref Overlay*** the **CHECK** drawing. The instructions for the change in design should appear. (Notice that the *Overlay* option prevents a circular *Xref* since SHAFT references CHECK and CHECK references SHAFT.) Make the changes by adding the 3 x 3 NECK to the geometry (Fig. 30-64). ***Save*** and ***Close*** the **SHAFT** drawing (with the *Overlay*).

**FIGURE 30-64**

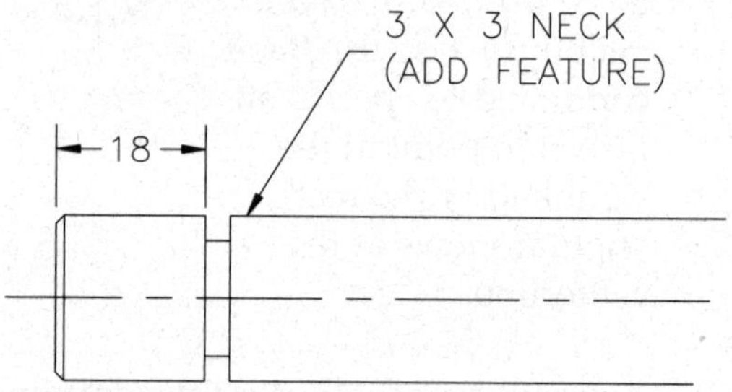


C. Now ***Open*** the **ASSY** drawing. The most recent *Xrefs* are automatically loaded, so the design change should appear (Fig. 30-65). Why don't the checker notes (from CHECK drawing) appear? Finally, ***Save*** the drawing and make another plot.

**FIGURE 30-65**

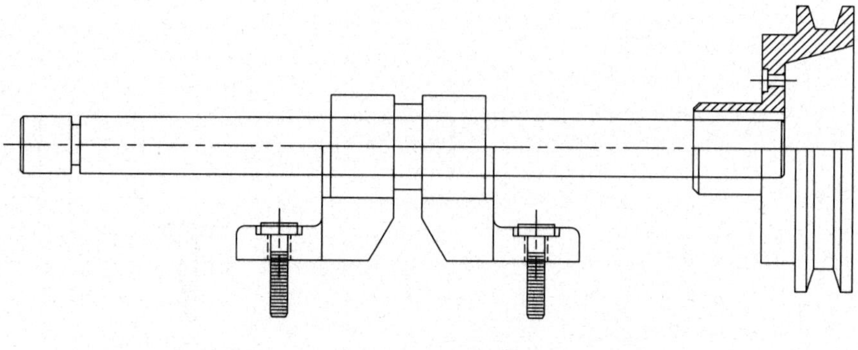

5. ***Layouts, Vports***

This exercise serves as a review of the basic procedures and commands for using an *Xref* and creating layouts and viewports to display and print an existing drawing.

A. ***Open*** the **ASSY** drawing that you created in previous exercises (make sure all of the *Xrefed* drawings are also located in your working directory). Use ***Xref*** with the ***Bind*** option to bind all of the *Xrefed* drawings. Use ***SaveAs*** to save and name the drawing as **ASSY-PS**.

B. Activate a ***Layout*** tab. Use ***Pagesetup*** to select the paper size and to set the plotting options for the device you use to plot with. Make a ***New Layer*** named **TITLE** and set it as the ***Current*** layer. Draw a border and draw, *Insert*, or *Xref* a title block (in paper space).

C. Make a ***New Layer*** called **VIEWPORT** or **VPORT** and set it as ***Current***. Next, use ***Vports*** to create one viewport. Make the viewport as large as possible while staying within the confines of the title block and border. Double-click inside the viewport to activate model space in the viewport, and use ***Zoom XP*** or the ***Viewport scale*** drop-down list (in the *Viewports* toolbar) to scale the model geometry times paper space (remember that the geometry is in millimeter units). Enter the scale in the title block (scale = XP factor or viewport scale).

D. Turn ***Off*** the **VIEWPORT** layer and ***Save*** the drawing (as ASSY-PS). ***Plot*** the drawing from the layout at **1=1**. Your drawing should appear similar to Figure 30-66.

**FIGURE 30-66**

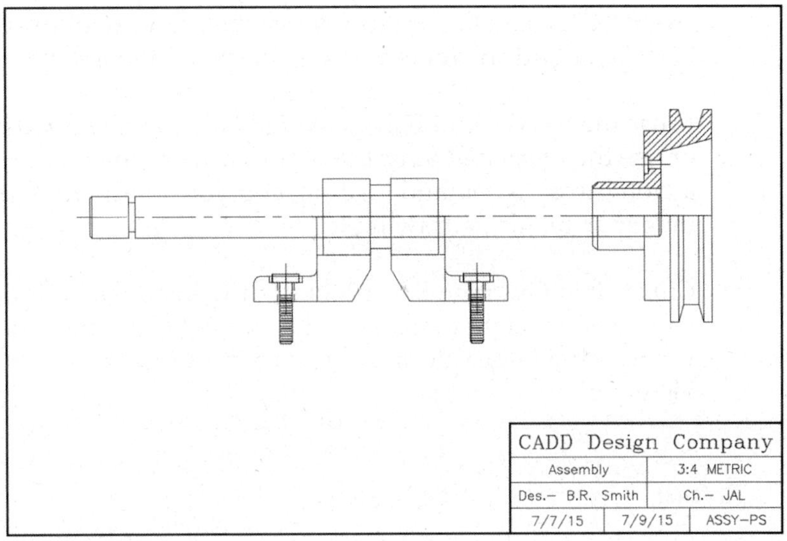


6. ***DesignCenter, XCLIP, XLOADCTL, INDEXCTL***

A. ***Open*** the **FLOOR PLAN SAMPLE.DWG** provided with AutoCAD. In the AutoCAD opening ***Create*** page, select ***Explore Sample Drawings***. Find the ***Floor Plan Sample.dwg*** in the ***Database Connectivity*** folder. Use ***SaveAs*** to change the name to **SAMP-OFFICE** and change the location of the drawing (using *SaveAs*) to your working directory.

B. Begin a ***New*** drawing. Use ***Save*** and assign the name **SAMP-CLIP**. Next, use ***DesignCenter*** to ***Attach*** as an ***Xref*** the **SAMP-OFFICE** drawing. Use an ***Insertion point*** of **0,0**, accept the defaults for scale factors and rotation angle, then ***Zoom All***. The entire SAMP-OFFICE drawing should appear.

C. Create a clipping boundary using the ***Xclip*** command. Make a ***Rectangular*** boundary around room 6150 (lower right corner office). ***Zoom*** in to display only the displayed office, 6150. Now, use the ***Layer*** command or *Layer Control* drop-down list to ***Freeze*** the **SAMP-OFFICE | EMPLOYEE** layer. Your drawing should look like Figure 30-67.

**FIGURE 30-67**

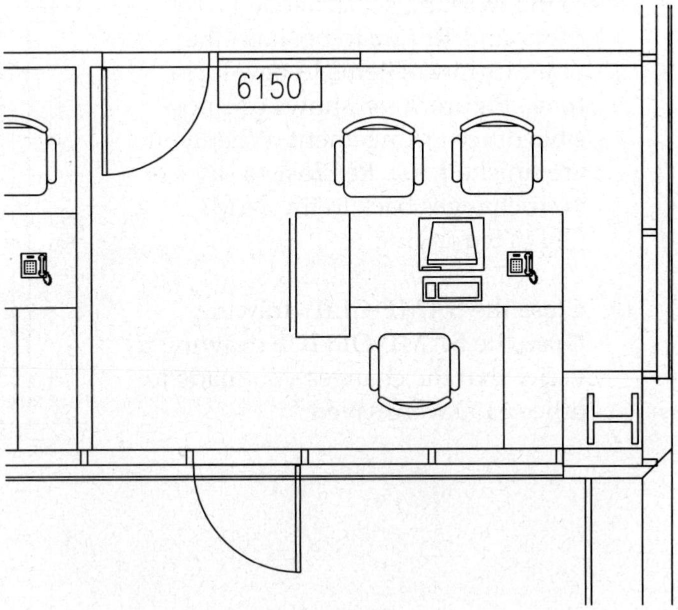

D. Use ***XLOADCTL*** or use the ***Open and Save*** tab of the ***Options*** dialog box to set the variable to **0** or ***Disabled***. ***Save*** and ***Close*** the drawing (SAMP-CLIP).

E. Now ***Open*** **SAMP-CLIP** again and use your watch or count seconds to time how long it takes to load the drawing and Xref drawing. At this point, demand loading is disabled and layer and spatial indexing are not being utilized so the drawing and Xref loading time should be at its maximum for your system.

F. Next, ***Close*** the **SAMP-CLIP** drawing and ***Open*** the **SAMP-OFFICE** drawing. Use the ***INDEXCTL*** variable or the ***SaveAs*** command with the ***Options...*** button to set indexing on for both layer and spatial indexes (a setting of 3). ***Save*** the SAMP-OFFICE drawing.

G. ***Close*** the **SAMP-OFFICE** drawing and ***Open*** the **SAMP-CLIP** drawing again. Use ***XLOADCTL*** or use the ***Open and Save*** tab of the ***Options*** dialog box to set the variable to **1** or ***Enabled***. This action turns on demand loading, and you should now take advantage of the spatial and layer indexing. ***Save*** the drawing.

H. ***Close,*** then ***Open*** the **SAMP-CLIP** drawing again and time the loading process. You should experience a slight increase in the loading time speed now that demand loading is enabled. When using large (file size) Xrefs or multiple Xrefs, this feature can offer a big performance increase.

7. ***Refedit***

A. ***Open*** the **SAMP-CLIP** drawing. Assuming you are planning to move into office 6150, your employer gives you the responsibility to arrange the furniture in your new office to your liking.

B. Use ***Refedit*** and select any wall at the "Select reference:" prompt. The *Reference Edit* dialog box should appear with the SAMP-OFFICE drawing listed as the *Reference name*. Select ***OK*** to dismiss the dialog box. At the "Select nested reference:" prompt, select two chairs. You should see all but the two chairs fade to a lighter color shade. Move the chairs to new locations in the office.

C Use ***Refset*** to add other items, such as the desk, computer, and phone, to the working set as needed. Use ***Move*** and ***Rotate*** to position the other furniture items to new locations. Figure 30-68 shows one possible office arrangement. When you are finished, use ***Refclose*** to save your changes back to the SAMP-OFFICE drawing.

D. ***Close*** the **SAMP-CLIP** drawing. ***Open*** the **SAMP-OFFICE** drawing to verify that the changes you made to office 6150 were saved.

**FIGURE 30-68**

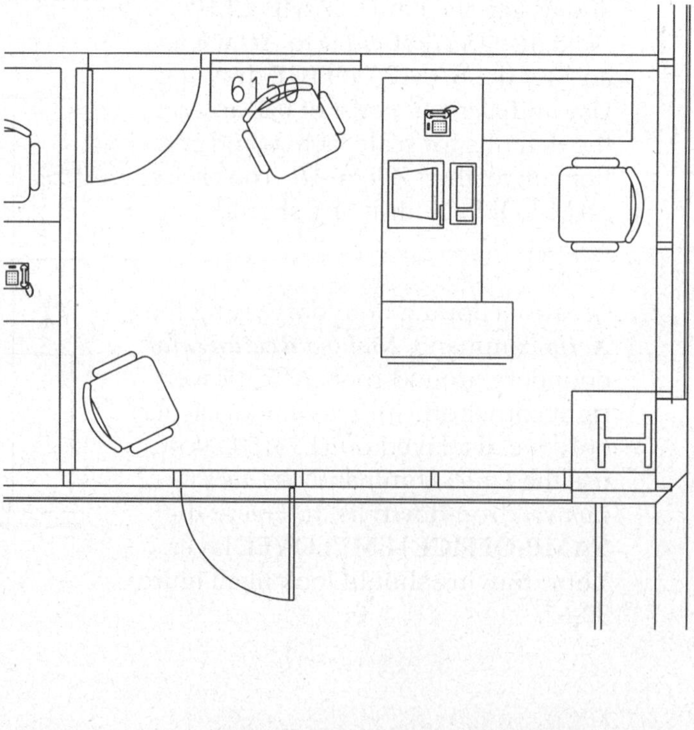

CHAPTER 31

# Object Linking and Embedding (OLE)

## Chapter Objectives

After completing this chapter you should:

1. be able to copy and paste objects from one open AutoCAD drawing to another open AutoCAD drawing;
2. be able to use AutoCAD's Object Linking and Embedding (OLE) feature to copy information from a document (in a software program such as AutoCAD, Microsoft Word, or Microsoft Excel) and temporarily store it in the Clipboard;
3. be able to paste information from the Clipboard into a drawing;
4. be able to embed information from another software program into an AutoCAD drawing;
5. be able to link information from another software program into an AutoCAD drawing;
6. be able to manage OLE information in a drawing.

# CONCEPTS

The Microsoft® Windows operating systems enable AutoCAD users to extend their graphic communication abilities within, and beyond, the Drawing Editor. Microsoft Windows introduced the capability to "copy" and "paste" objects. As an AutoCAD user, there are two basic ways in which you can copy and paste objects. First, it is possible to copy and paste AutoCAD objects between multiple open drawings. Secondly, you can copy and paste between AutoCAD and other Windows software applications.

## Cut and Paste Between AutoCAD Drawings

AutoCAD allows you to open multiple drawings at the same time in one AutoCAD session. One of the greatest advantages for doing this is the ability to copy objects from one drawing and paste them into another drawing. This feature can save hours of redrawing the same or similar geometry and increase your ability to collaborate with others on design ideas.

When objects are copied from one AutoCAD drawing, the pasted objects in the receiving drawing are still normal AutoCAD objects. Once the objects have been pasted, there are no special functions or commands that you need to manipulate and edit them. There are also commands in AutoCAD specifically intended for you to handle the objects during the cut and paste process, such as *Copy with Base Point* and *Copy to Original Coordinates*. For example, Figure 31-1 displays the AutoCAD Drawing Editor with two drawings open. Objects from the drawing on the left are copied with the *Copyclip* command and placed into the drawing on the right with the *Pasteclip* command.

You can have multiple drawings open in one AutoCAD session, and copying and pasting between drawings is transparent—that is, even though the basic Windows Object Linking and Embedding (OLE) mechanism is being utilized, there are no special functions you need to use.

The following commands can be used to cut (erase), copy, and paste objects between AutoCAD drawings:

**FIGURE 31-1**

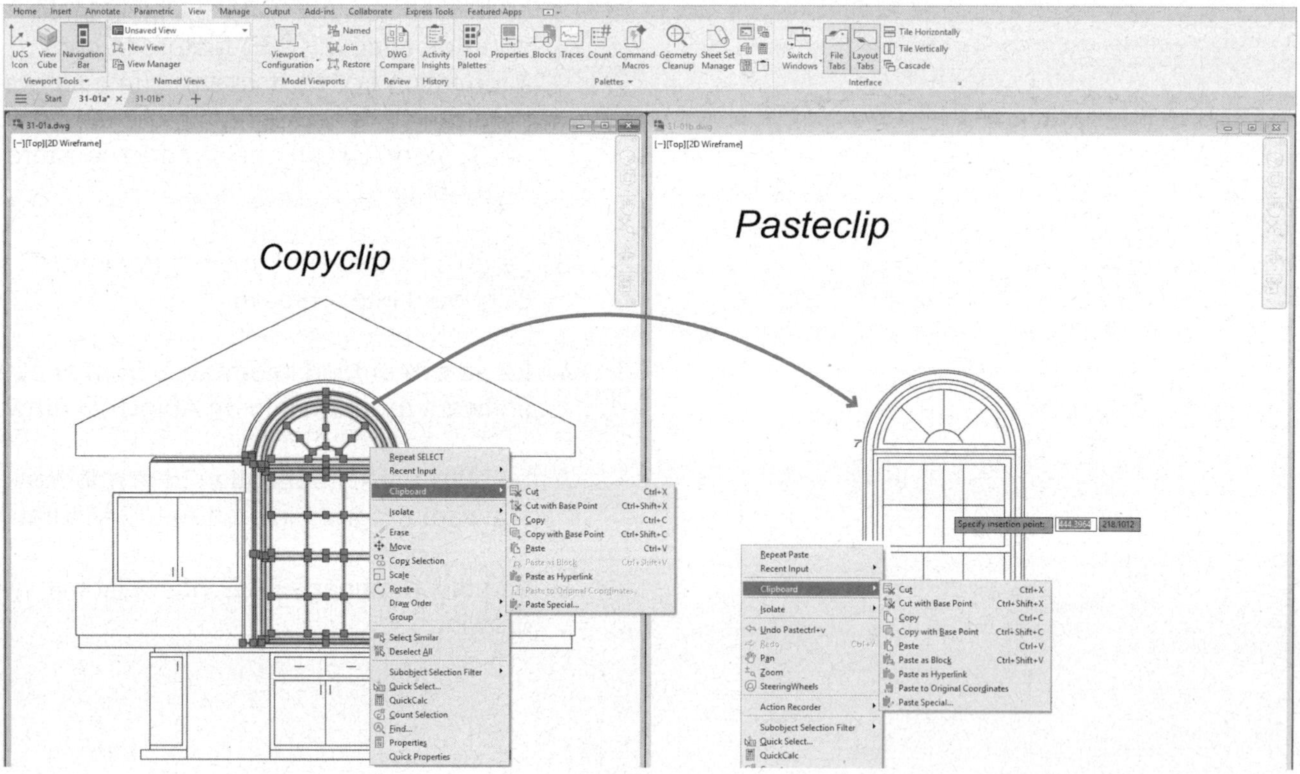

| Menu Command | Formal Command | Accelerator | Result |
|---|---|---|---|
| *Copy* | *Copyclip* | *Ctrl + C* | copies selected objects to the Clipboard |
| *Cut* | *Cutclip* | *Ctrl + X* | copies objects to the Clipboard and cuts (erases) them from the original drawing |
| *Copy with Base Point* | *Cutbase* | *Ctrl + Shift + X* | copies objects to the Clipboard, with a specified base point, and removes them from the drawing |
| *Copy with Base Point* | *Copybase* | *Ctrl + Shift + C* | copies objects to the Clipboard and asks for a base point |
| *Copy Link* | *Copylink* | | copies all objects to the Clipboard |
| *Paste* | *Pasteclip* | *Ctrl + V* | pastes Clipboard objects into the drawing |
| *Paste as Block* | *Pasteblock* | *Ctrl + Shift + V* | pastes Clipboard objects into the drawing as a *Block* |
| *Paste to Original Coordinates* | *Pasteorig* | | pastes Clipboard objects into the new drawing at the same location as the original drawing |

## Cut and Paste Between AutoCAD and Other Windows Applications (OLE)

You can also easily communicate 2D and 3D AutoCAD designs with other Windows applications and benefit from the Windows Object Linking and Embedding (OLE) feature. Unlike using only one software application (AutoCAD) to copy and paste, this concept combines objects created in two different software applications but displayed in only one. Microsoft developed the method called object linking and embedding, or OLE, to handle dissimilar objects. The standard OLE features of linking and embedding operate as smoothly and easily from within AutoCAD as they do in other Windows software such as Microsoft® Word, Excel, and others.

AutoCAD's support of the OLE features allows you to include data and information created in an application, such as Word, within your AutoCAD drawing. You can also take AutoCAD drawing information and use it within Word documents, Excel spreadsheets, and so on. This feature works best if all the applications using or sharing the common information support the OLE feature. But it is not absolutely necessary for a software application to support the OLE feature for that application to display AutoCAD drawing information by way of the copy and paste functions available within Windows.

AutoCAD's OLE features utilize the standard Windows *Copy, Cut,* and *Paste* commands common to all Windows programs. OLE, however, extends these Windows commands so information copied from one application (such as a Word document) and pasted into another application (like an AutoCAD drawing) can be *linked* back to the original source document, also known as the server. The source document is a separate document; it is only displayed within the client document (the AutoCAD drawing in this example).

OLE objects are treated by AutoCAD as any other object—that is, you can use the normal selection methods (pickbox, window, crossing window, etc.) to *Move, Erase, Copy, Scale,* or *Array* the OLE object.

Select the OLE object by selecting the "border" around the object. The OLE object borders appear in the Drawing Editor, but do not appear in a plot or print.

Object Linking and Embedding enables you to create a bill of materials within, for example, Word, which is designed specifically for word processing, then paste it into an AutoCAD drawing. First, create and format the bill of materials within Word (Fig. 31-2). Next, *Copy* the formatted text onto the Clipboard and then *Paste* it into the AutoCAD drawing. In AutoCAD (the client), it will have the same appearance that it has in the Word (the source) document.

FIGURE 31-2

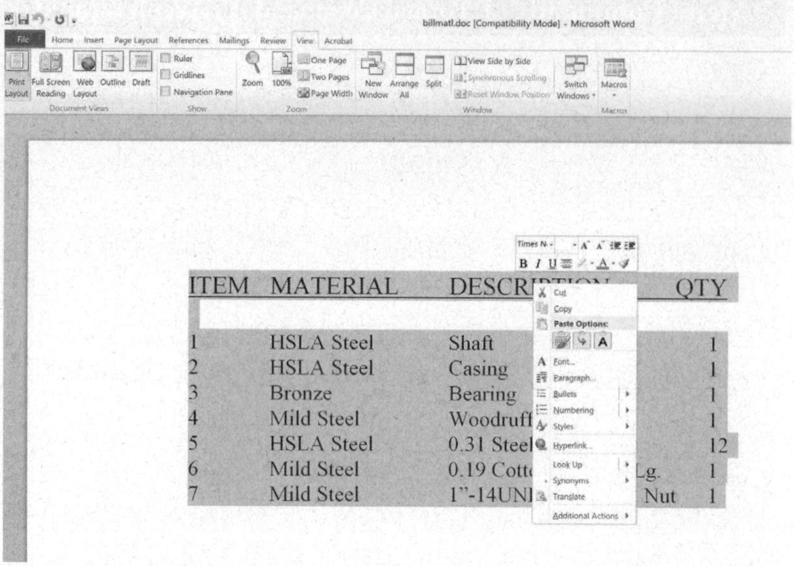


When a non-AutoCAD OLE object is pasted into AutoCAD (Fig. 31-3), you have the choice to paste it as an embedded or a linked object. A linked object is simply a picture of the original source document, whereas an embedded object is wholly contained in AutoCAD. Once the OLE object is pasted into AutoCAD as either an embedded object or linked object, you can double-click on the object in the drawing to cause the source application to open (in this case, Word) with the document loaded and ready for editing. Changing and saving the document causes the object in AutoCAD to change. If the object is embedded, only the AutoCAD object is changed—not the original source document. If the object is linked, the original source document is changed and is reflected in AutoCAD.

FIGURE 31-3

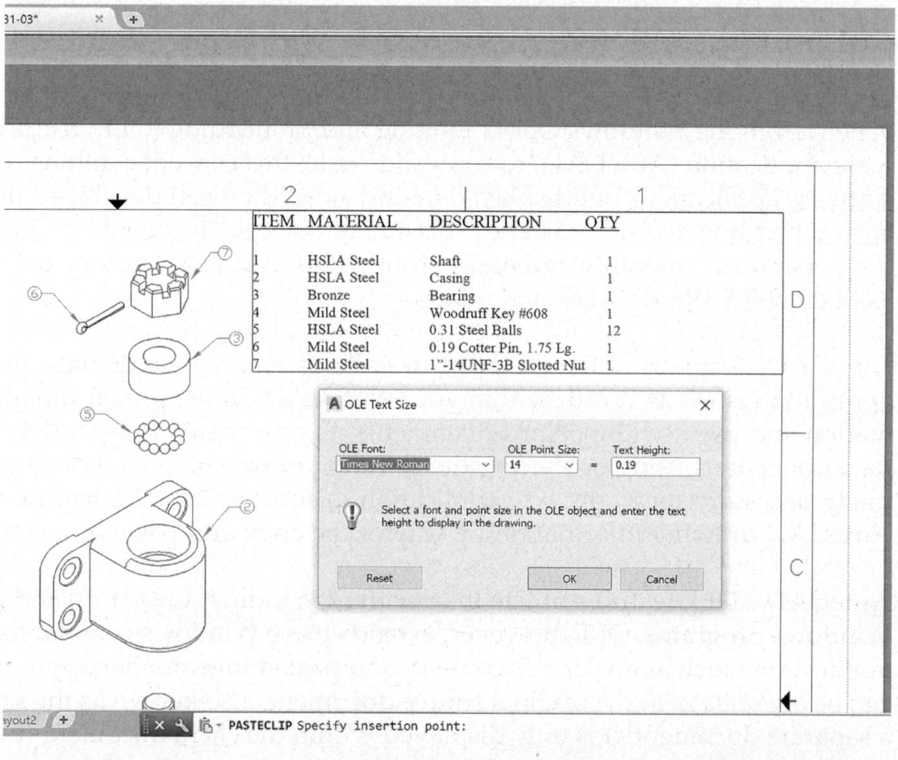


There are two strategies for pasting information from other applications into your AutoCAD drawing:

1. Embed the object when pasting
   Embedding is similar to using the AutoCAD *Insert* command. When embedded, the object becomes a permanent part of the drawing. If the source application that created the object supports OLE, you can double-click on the embedded object and the source application is started so you can edit the contents of the object for the insertion of the (embedded) objects in AutoCAD only. The original source document remains unchanged.

2. Link the object when pasting
   Linking is similar to the AutoCAD *Xref* command. When the object is linked, it is not part of the drawing; it is only displayed in the drawing. When you double-click on the linked object, the source application is started with the original document loaded and ready for editing. Any changes made to the original source document are reflected in the drawing automatically or upon a manual update, depending on your update setting in the OLE *Links* dialog box. Similar to an *Xref*, when the original document is edited, the updated version appears in the client application.

These AutoCAD commands are used to copy, cut, link, or embed objects between AutoCAD and other applications:

| Menu Command | Formal Command | OLE Result |
|---|---|---|
| *Copy* | *Copyclip* | Embedded or linked |
| *Cut* | *Cutclip* | Embedded |
| *Paste* | *Pasteclip* | Embedded |
| *Paste Special* | *Pastespec* | Embedded or linked |
| *Paste as Hyperlink* | *Pasteashyperlink* | Hyperlink |

The main difference between linked and embedded is how the information is stored—with the client (application that "displays" the document) or server (source application that created the document).

## The Windows Clipboard

Windows manages the OLE features. The Windows Clipboard is a key mechanism used throughout the process of cutting and pasting within AutoCAD and linking and embedding other information in an AutoCAD drawing. The Clipboard is a Windows feature which stores information temporarily in memory so that it can be utilized (or pasted) into another application. This enables you to create the information only once and then share it with other documents where it is needed.

For example, assume the bill of materials mentioned earlier was created, spell checked, and formatted in Word, then selected and copied to the common memory area called the Clipboard. Then the information was pasted (similar to the AutoCAD *Insert* command) into an AutoCAD drawing with the same content and appearance it had in the original document. The same information could then be pasted (inserted) into a spreadsheet with the identical content and appearance it has in the AutoCAD drawing and in the original Word document that created it.

The Clipboard retains the most recent information that was cut or copied from an application currently running in the Windows environment. When another piece of information (e.g., graphical objects or text) is cut or copied, it replaces the last group of information stored within the Clipboard's memory area.

The options encountered when performing a paste vary depending upon the AutoCAD paste command being used (*Pasteclip, Pastespec*) and the type of object information being pasted from the Clipboard. The following table lists some common object types and options that are available when pasting objects into an AutoCAD drawing.

| Object Type | Paste Object As: | | | | | |
|---|---|---|---|---|---|---|
| | AutoCAD Vectors | AutoCAD Block | AutoCAD Text | Bitmap | Picture | Image |
| **AutoCAD Vector Objects** (*Lines, Arcs, Circles, Blocks, Plines,* etc.) | ✓ | ✓ | | | | |
| **Other Vector Objects** (primarily Windows metafile) | | ✓ | | | | |
| **AutoCAD Text Objects** (*Mtext, Text*) | | ✓ | ✓ | | ✓ | ✓ |
| **Formatted Text Objects** (such as Word text or Excel spreadsheet) | | ✓ | ✓ | ✓ | ✓ | ✓ |

To summarize, first create the document (to be copied) in its source application (Word, for example). While the document is open, use *Copy* or similar OLE command in that application to copy the document contents (called an OLE object) to the Clipboard. Next, open the client application (AutoCAD, for example) and use *Paste, Paste Special,* or a similar OLE command to paste the OLE object. The OLE object appears in the client application just as it did in its original format (in the source application). Later, you can double-click on the OLE object to open the original (source) application and modify the object. If the object is <u>linked</u>, the original source document is updated; if the object is <u>embedded</u>, only the OLE object is changed, but the source document (file) is not changed.

## OLE RELATED COMMANDS

AutoCAD OLE related commands covered in this chapter are:

*Cutclip, Cutbase, Copyclip, Copylink, Copybase, Pasteclip, Pasteblock, Pasteorig, Pastespec, Insertobj, Pasteashyperlink, Olelinks, Oleopen,* and *Olescale*

Most OLE related commands are available from the *Edit* pull-down menu (Fig. 31-4). The *Insertobj* command is accessible from the *Insert* pull-down menu.

Most OLE commands are also available on the Ribbon in the *Clipboard* panel of the *Home* tab.

FIGURE 31-4

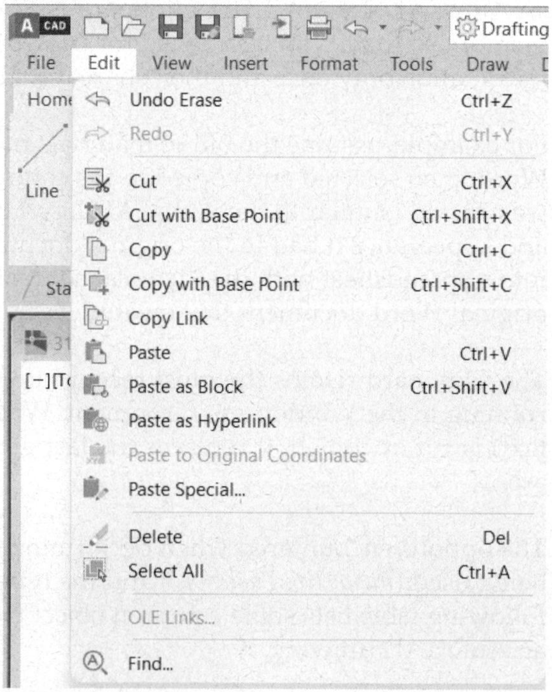

In addition, right-clicking in the drawing area when no commands are in use produces the default shortcut menu displaying the OLE commands (Fig. 31-5).

FIGURE 31-5

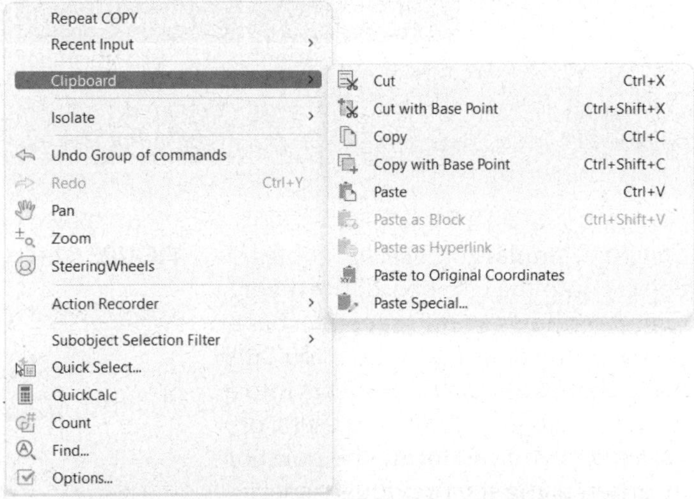


### *Copyclip*

| Ribbon | Menu Bar | Command (Type) | Alias (Type) | Shortcut |
|---|---|---|---|---|
| *Home Clipboard* (icon) | *Edit Copy* | *Copyclip* | ... | Ctrl+C or (Default Menu) *Copy* |

You can use the *Copyclip* command to select the objects from the current drawing to be placed on the Clipboard. *Copyclip* prompts you to "Select objects:". You can use any of the normal AutoCAD selection options (like *Window, Crossing, Fence, Group*, etc.). You can also pre-select the objects (known as Noun-Verb selection) you want placed on the Clipboard, and then issue the *Copyclip* command to avoid the "Select Objects:" prompt:

Command: ***COPYCLIP***
Select objects: **PICK** (Select objects to be copied from the drawing and placed on the Clipboard)
Select objects: **Enter**
Command:

AutoCAD objects that are selected with *Copyclip* remain in the drawing. Copies of the objects are temporarily stored on the Clipboard.

The *Copyclip* command can be used when you plan to paste the objects back into AutoCAD (using *Pasteclip, Pastespec*, or *Pasteblock*) or when you intend to link or embed them within other applications. When the objects are pasted into another application, you may have little control over where they are placed, depending on the host application.

When you paste the objects into the same or other AutoCAD drawing, you can drag the objects into the desired location. The *Copyclip* command does not prompt for a base point; therefore, pasting the objects involves using a base point at the lower-left corner of the set of objects (Fig. 31-6).

FIGURE 31-6

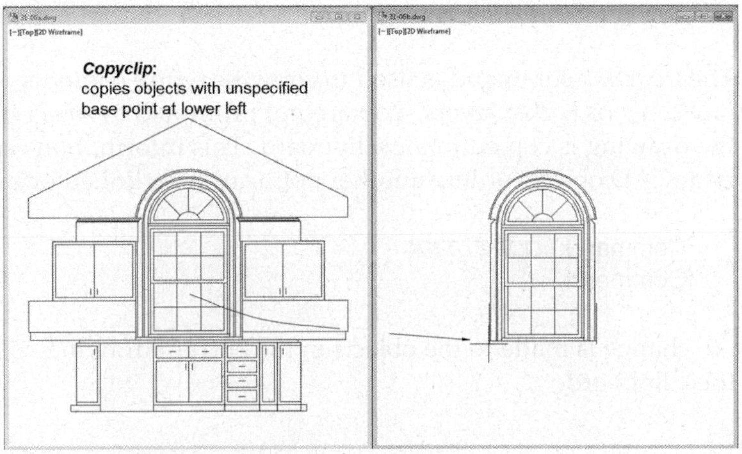

## Copybase

| Ribbon | Menu Bar | Command (Type) | Alias (Type) | Shortcut |
|---|---|---|---|---|
| ... | *Edit*<br>*Copy with Base Point* | *Copybase* | ... | (Default Menu)<br>*Copy with Base Point* |

*Copybase*, similar to *Copyclip*, copies objects onto the Clipboard; however, *Copybase* allows you to specify a base point for the object(s). When you later paste the set of *Copybase* objects into a drawing using *Pasteclip*, *Pastespec*, or *PasteBlock*, you can locate the point of insertion using the previously specified base point (Fig. 31-7).

FIGURE 31-7

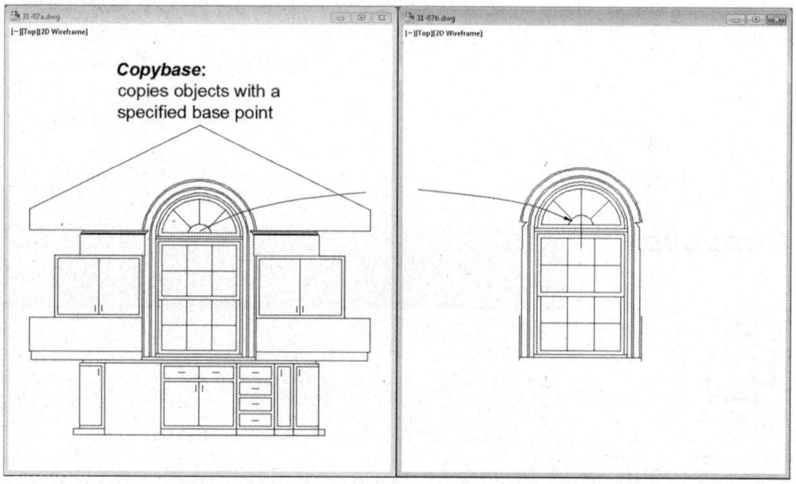


In contrast, *Copyclip* does not allow you to specify a base point but automatically determines a base point, usually in the lower left corner of the object set (see Fig. 31-6). Therefore, *Copybase* is generally preferred to *Copyclip* for typical operations within AutoCAD because you have more direct control of the insertion point.

If you intend to use *Pasteorig* to paste the object set, it does not matter if you use *Copyclip* or *Copybase* since the object set is pasted to the original coordinates regardless of the base point. (See "*Pasteorig*," later in this chapter.)

## Copylink

| Ribbon | Menu Bar | Command (Type) | Alias (Type) | Shortcut |
|---|---|---|---|---|
| ... | *Edit*<br>*Copy Link* | *Copylink* | ... | ... |

The *Copylink* command is used to copy the entire contents of the current drawing, including objects on *Frozen* or *Locked* layers. You are not prompted to select individual objects. Whatever is contained in the drawing is copied to the Clipboard. This information can then be pasted into AutoCAD as normal AutoCAD objects or into another application as linked or embedded objects.

```
Command: COPYLINK
Command:
```

No change is made to the objects in the current drawing. Copies of the objects are temporarily stored in the Clipboard.

### Cutclip

| Ribbon | Menu Bar | Command (Type) | Alias (Type) | Shortcut |
|---|---|---|---|---|
| *Home Clipboard* (icon) | *Edit Cut* | *Cutclip* | ... | Ctrl+X or (Default Menu) *Cut* |

The *Cutclip* command places selected objects onto the Clipboard. However, unlike *Copyclip*, it removes the selected objects from the drawing. If desired, you can use the AutoCAD *Oops* command to "unerase" the objects deleted from the drawing by *Cutclip*. If you use *Oops*, you must use it before you delete any other objects from the drawing since *Oops* brings back only the last set of *Erased* objects.

```
Command: CUTCLIP
Select objects: PICK (Select objects to be removed from the drawing and placed on the Clipboard)
Select objects: Enter
Command:
```

Use this command when you do not need the selected items in the current drawing any longer but wish to use them in another drawing or another application. In this case, *Cutclip* allows you to remove them from the current drawing and paste them into another application. For example, you may want to open a second AutoCAD drawing, then cut objects from the first drawing and paste them into the second drawing. This helps to reduce the current drawing file size and also eliminates the need to draw the items again. Since the objects temporarily remain on the Clipboard, the deleted items can be pasted into other drawings or documents where they are needed.

### Cutbase

| Ribbon | Menu Bar | Command (Type) | Alias (Type) | Shortcut |
|---|---|---|---|---|
| *Home Clipboard* (icon) | *Edit Cut with Basepoint* | *Cutbase* | ... | Ctrl+Shirt+X or (Default Menu) *Cutbase* |

The *Cutbase* command places selected objects onto the Clipboard. However, unlike *Cutclip*, it lets you specify a basepoint. When you paste the objects from the Clipboard into another location using *Pasteclip*, *Pastespec*, or *PasteBlock*, the previously specified base point is used when pasting.

### Pasteclip

| Ribbon | Menu Bar | Command (Type) | Alias (Type) | Shortcut |
|---|---|---|---|---|
| *Home Clipboard* (icon) | *Edit Paste* | *Pasteclip* | ... | Ctrl+V or (Default Menu) *Paste* |

You can copy information from AutoCAD or another application and paste it into your drawing with AutoCAD's *Pasteclip* command. If you copy AutoCAD objects and paste them into the same or another AutoCAD drawing, the objects are inserted as the original AutoCAD objects having the same properties. If you *Pasteclip* text from a word processing application or data from a spreadsheet application, the pasted object is embedded within the AutoCAD drawing.

### Pasting AutoCAD Objects

First, you must use *Copyclip*, *Copybase*, *Copylink*, or *Cutclip* to copy AutoCAD objects to the Clipboard. Next, use *Pasteclip* to insert the objects into the same or another open AutoCAD drawing. AutoCAD prompts for the base point.

Command: ***PASTECLIP***
Specify insertion point: **PICK**
Command:

TIP

The objects are inserted into the new location or new drawing. AutoCAD objects are pasted as the original objects and retain their original properties such as layer, color, linetype, text style, dimension style, and so on. Therefore, layers on which the pasted objects reside and other named objects associated with the pasted objects (linetypes, dimension styles, etc.) are added to the new drawing if they did not previously exist in the host drawing.

### Pasting Non-AutoCAD Objects

If you have used Copy or similar OLE command from a word processing or spreadsheet program, for example, those objects are copied to the Clipboard. In AutoCAD, use *Pasteclip* to bring the objects into AutoCAD as embedded objects.

Command: ***PASTECLIP***

When you copy objects that contain text from another application into an AutoCAD drawing using *Pasteclip*, the point size of the text in the source application is automatically converted to an approximate equivalent size in AutoCAD units. AutoCAD uses the appropriate units depending on whether the OLE object is inserted into a layout or into model space. If you *Paste* the object into a layout, the object is automatically scaled based on the layout's sheet size and units. If you *Paste* the OLE object into model space, the scale is based on the current page setup (if one has been set) and the *MSOLESCALE* system variable (see "*MSOLESCALE*").

When *Pasteclip* is used with non-AutoCAD objects containing text, the *OLE Text Size* dialog box appears (Fig. 31-8). This dialog box gives you additional control of the scale of the OLE object by manipulating the *Text Height* for a particular *OLE Font* or the *OLE Point Size*. The *OLE Text Size* dialog box can be produced at a later time by highlighting an existing non-AutoCAD OLE object and using the *Olescale* command (see "*Olescale*").

**FIGURE 31-8**

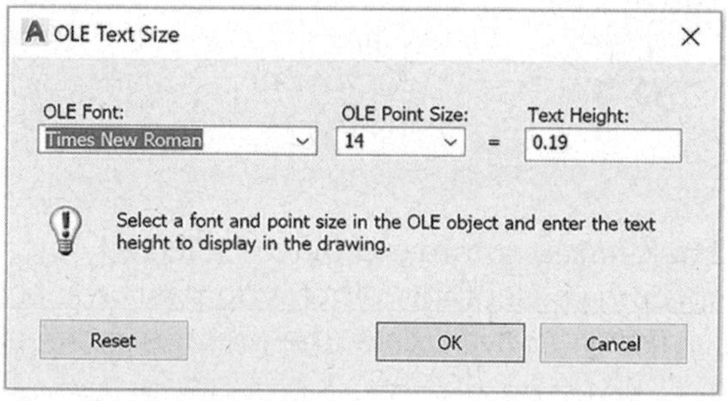


The *OLE Text Size* dialog box allows you to set the scale for the OLE object. Although it appears as if you can change the font and the point size, that is not the case—the *OLE Font* and *OLE Point Size* drop-down lists display only the fonts and sizes already contained in the OLE object. This tool is used to set the scale of the OLE object by specifying a *Text Height* value. Do this by selecting any combination of *OLE Font* and *OLE Point Size*, then entering the desired *Text Height* value. The *Text Height* value determines the scale for the entire OLE object.

In other words, assume the *Text Height* displayed a default value of 0.1389 for the 10 point Times New Roman font. Changing that value to 0.2778 would double the scale of the entire OLE object (0.2778 = 0.1389 x 2). Additionally, the *Text Height* for all other point sizes for all other fonts would automatically double. If you want to change the scale of the OLE object back to the original size, select *Reset*.

Remember that OLE objects in AutoCAD are treated as any other AutoCAD objects—that is, you can *Move, Erase, Copy,* or *Array* them. For example, instead of using the *OLE Text Size* dialog box, you could use the *Scale* command to scale OLE objects.

Once the non-AutoCAD object is inserted, it becomes an embedded object. If the embedded object's source application supports OLE, the object is displayed in the original application's format. The embedded object can be edited using the original application that created it by double-clicking on the object.

Embedded objects are copies of the original information. They have no connection, or link, to the original source document in which they were created. Therefore, editing the original objects using the source application does not update the embedded (copied) object. Alternately, if you use the *Pastespec* command instead of *Pasteclip*, you have the choice of embedding or linking the objects.

Keep in mind that using *Pasteclip* to paste an AutoCAD object into another AutoCAD drawing does not create an embedded object. Pasted AutoCAD objects are inserted as exact copies of the original object set, complete with all associated properties and named objects, unless *Pasteblock* is used.

## Pasteorig

| Ribbon | Menu Bar | Command (Type) | Alias (Type) | Shortcut |
|---|---|---|---|---|
| *Home Clipboard Paste to Original Coordinates* | *Edit Paste to Original Coordinates* | *Pasteorig* | ... | (Default Menu) *Paste to Original Coordinates* |

The *Pasteorig* command is designed specifically to be used for copying AutoCAD objects into other AutoCAD drawings. *Pasteorig* inserts the object(s) into the new drawing at the same coordinate position as the original drawing. Therefore, you are not prompted for an insertion point.

**TIP**

To illustrate, open two AutoCAD drawings and use *Cutclip*, *Copyclip*, *Copylink*, or *Copybase* to copy objects from one drawing onto the Clipboard. Next, make the other AutoCAD drawing current by clicking in it and then use *Pasteorig*. The object(s) are pasted to the same coordinate location as in the original drawing (Fig. 31-9). This feature makes it simple to create or edit a second drawing based on geometry from a previously created drawing.

**FIGURE 31-9**

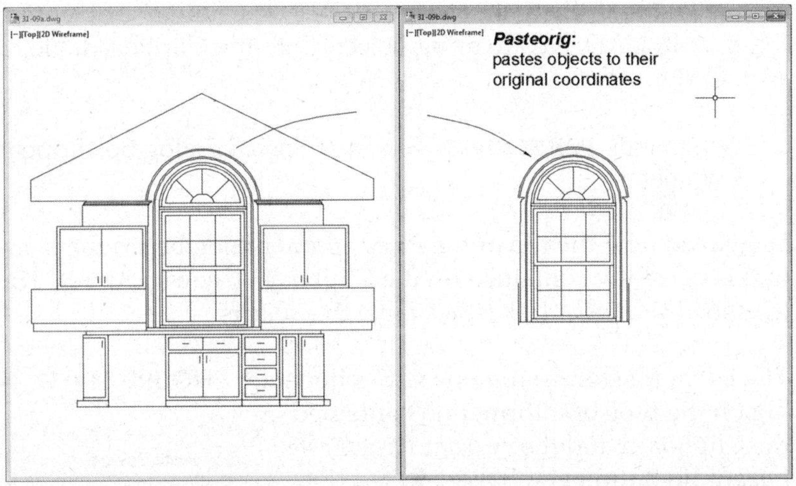


Note that there is no added control by using *Copybase* to specify a base point when you capture the original objects since *Pasteorig* automatically locates the new objects at their original coordinates regardless of a specified base point. *Pasteorig* does not operate for copying objects back to the original drawing. *Pasteorig* does not function for copying non-AutoCAD objects from other applications into AutoCAD.

## Pasteblock

| Ribbon | Menu Bar | Command (Type) | Alias (Type) | Shortcut |
|---|---|---|---|---|
| *Home Clipboard Paste as Block* | *Edit Paste as Block* | *Pasteblock* | ... | (Default Menu) *Paste as Block* |

*Pasteblock* is used for copying AutoCAD objects into the same or other AutoCAD drawings. *Pasteblock* operates similar to *Pasteclip* except that *Pasteblock* inserts AutoCAD objects into AutoCAD as *Blocks*. This can be useful if you want to manipulate all objects from one paste operation as one object—a *Block*.

## *Pastespec*

| Ribbon | Menu Bar | Command (Type) | Alias (Type) | Shortcut |
|---|---|---|---|---|
| *Home Clipboard Paste Special* | *Edit Paste Special...* | *Pastespec* | *PA* | ... |

*Pastespec* is intended to be used primarily to paste objects from other applications into AutoCAD.

The *Pastespec* command gives you the choice of embedding or linking objects that are pasted into the drawing. *Pastespec* produces the *Paste Special* dialog box (Fig. 31-10). Your selection of *Paste* or *Paste Link* determines if the objects from the Clipboard are embedded or linked, respectively.

**FIGURE 31-10**

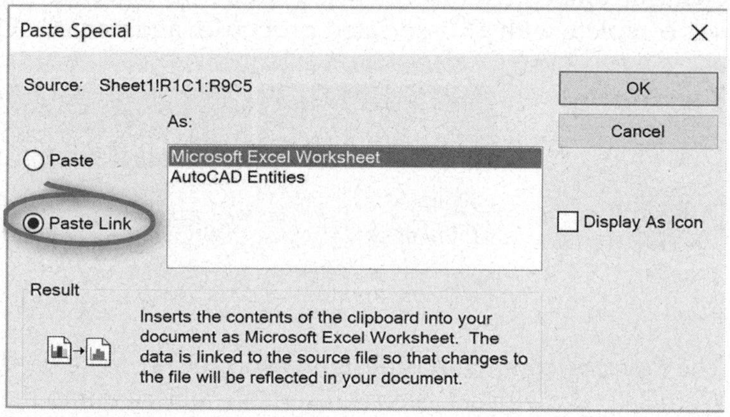


If the information copied from an application that supports OLE is linked within your AutoCAD drawing, a "picture" of the original document is displayed that links (or references) the original source document that was used to create it.

To use *Pastespec,* first you must use *Copy* or a similar OLE command from another application (Word, Excel, AutoCAD, etc.) to bring objects onto the Clipboard. Next, use the *Pastespec* command in AutoCAD:

Command: ***PASTESPEC*** (The *Paste Special* dialog box appears)
Command:

Displayed near the top of the *Paste Special* dialog box is the *Source:* application which created the object that is currently contained on the Clipboard (see Fig. 31-10). The two radio buttons on the left allow you to embed (*Paste*) or link (*Paste Link*) the objects.

The list of (Paste) *As:* options varies depending on the type of information contained on the Clipboard and the current *Paste/Paste Link* radio button you have selected (Fig. 31-11).
If the current information on the Clipboard is from an AutoCAD drawing, the *Paste Link* option is not available (grayed out). All of the options in this dialog box are described in detail next.

**FIGURE 31-11**

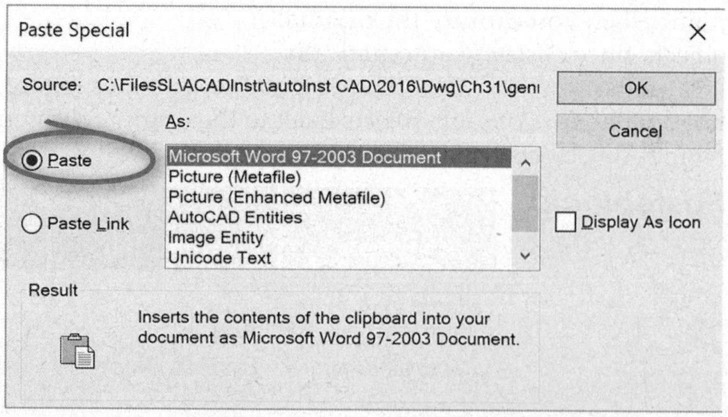


### *Paste*

When the *Paste* radio button is selected, the object becomes embedded within the drawing. You can select the type of object to be inserted in the drawing by making a selection from the (Paste) *As:* list (see Fig. 31-11). The list of object types varies depending on the source of the objects on the Clipboard.

### *Paste Link*

Selecting the *Paste Link* button causes AutoCAD to display a "picture" of the object in the drawing. The information displayed is not actually copied into the drawing; it is linked back to the source document from which it was originally copied. This option is available only when the information contained on the Clipboard was created with an application that supports OLE.

TIP

If the copied content is from a Microsoft Excel file, you can use the *AutoCAD Entities* option of *Paste Link* to copy the Excel spreadsheet into AutoCAD as an AutoCAD *Table*. This procedure has the same results as creating a "data link" by other methods. Creating a linked AutoCAD table offers many advantages over creating a normal OLE object. Since the linked data is contained in an AutoCAD table, you can alter the format (style) of the table as well as update the table data directly in AutoCAD and then send it back to the Excel document. (See "Creating and Editing Tables" in Chapter 18 and "Data Links" in Chapter 22.)

### *Result*

The *Result* box at the bottom of the dialog box describes the current object on the Clipboard. It also indicates how the pasted objects are managed in the drawing based on your selections within the dialog box (*Paste/Paste Link* and *Paste As:* list). This is a very helpful feature that dynamically explains your options as they are selected.

### *Display As Icon*

You can display the source application's icon instead of displaying the actual information pasted from the Clipboard by choosing *Display As Icon* (Fig. 31-12). Since only the application's icon is visible in the drawing, you must double-click on it to view the linked or embedded information. This icon is a shortcut that points to the location of the information pasted from the Clipboard. This can be useful in a situation where the linked information is relevant to the project depicted in your drawing, but it is not necessary to display that information directly in the drawing.

**FIGURE 31-12**

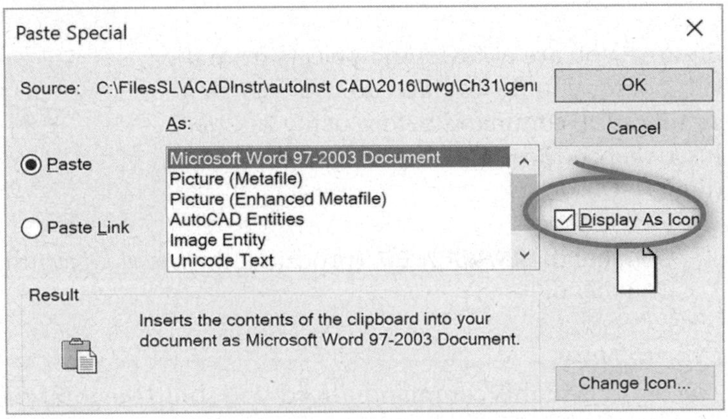


TIP

For example, you may be displaying a *Block* of a pump in the drawing. Specific text information such as the pump's specifications, manufacturer's model number, cost, and ordering information could be included within a Word document which is pasted into your drawing and displayed as an icon. The information is important for reference, but it is not necessary to display the information directly in your drawing.

TIP

If the information represented by the icon is embedded, double-clicking on the icon activates the application used to create it and displays the contents of the information that was pasted into the drawing in its original format. Changes made to the information as a result of this action are changed for the embedded information only and are not reflected in the source (original) document.

If the *Icon* information is linked, double-clicking on the icon activates the application used to create it and in turn displays the original file that contains the information. The information referenced by the pasted icon becomes highlighted in the source document. Once the source document is saved and closed, the updated version is linked to the icon displayed within your drawing. In this case, the source document itself has actually been changed.

### *Change Icon*

Whenever you have selected the *Display As Icon* checkbox, a preview of the application's default icon is displayed directly beneath the checkbox area. A new button also appears labeled *Change Icon*. This allows you to change the icon if you so desire.

## Insertobj

| Ribbon | Menu Bar | Command (Type) | Alias (Type) | Shortcut |
|---|---|---|---|---|
| *Insert Data (icon)* | *Insert OLE Object...* | *Insertobj* | *IO* | ... |

TIP

*Insertobj* allows you to paste specific information from another application into your drawing as either a linked or embedded object. It functions in much the same way as the *Pastespec* command. The key difference between *Insertobj* and *Pastespec* is that *Insertobj* allows you to paste a file without having the application or document currently open and a particular portion of it selected.

**FIGURE 31-13**

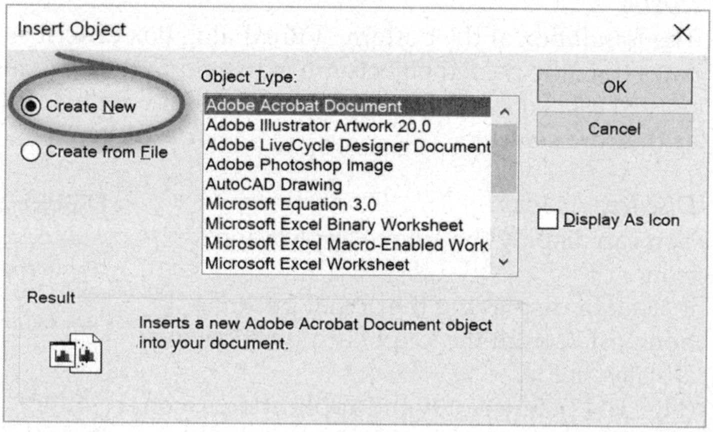


Because you are not copying objects from the Clipboard, you do not have to use *Copy* or a similar OLE command before using *Insertobj*. *Insertobj* produces the *Insert Object* dialog box (Fig. 31-13).

Command: ***INSERTOBJ*** (produces the *Insert Object* dialog box)
Command:

TIP

### *Create New*

The options in this command afford you some flexibility. If the file doesn't exist, you can create a new file by selecting *Create New*, then the application you want to use to create it. The applications available to you (registered on your system) are listed in the *Object Type* listing. The selected application is opened when you select *OK* so that you can create the new document. When you are finished creating the new document and you close the application, the new document (object) is embedded into your drawing.

### *Create from File*

When the *Create from File* radio button is selected (Fig. 31-14), you can select the *Browse* button to look for files available in your available drives and folders. After selecting a specific file, you can paste it into your drawing as an embedded or linked object. If the object is to be linked, the display of the linked object may be limited to an icon based on the file type and the OLE capabilities of the source application.

**FIGURE 31-14**

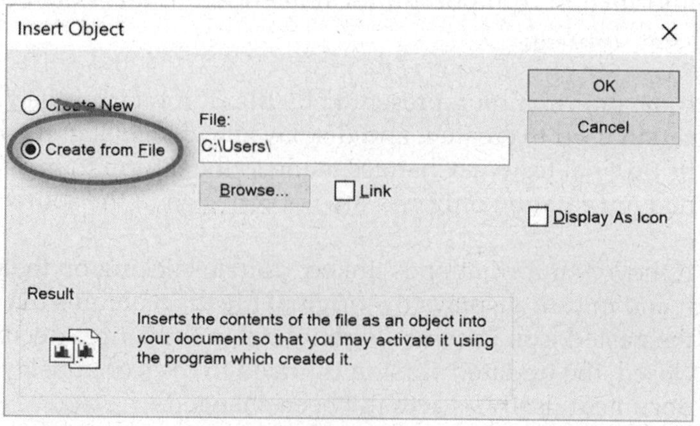

## Pasteashyperlink

| Ribbon | Menu Bar | Command (Type) | Alias (Type) | Short-cut |
|---|---|---|---|---|
| *Home Clipboard Paste as Hyperlink* | *Edit Paste as Hyperlink* | *Pasteashyperlink* | ... | ... |

The *Hyperlink* command (not *Pasteashyperlink*) produces the *Insert Hyperlink* dialog box. That dialog box is used to create a link between an AutoCAD object and a Web page or another document. A hyperlink has a similar function to an embedded OLE object, in that opening the hyperlink opens the related document (usually your Web browser) and displays the Web page or other document. A hyperlink, however, is attached to an AutoCAD object, so passing the cursor over the object displays the hyperlink symbol and right-clicking on the object allows you to open the hyperlink. (See Chapter 23, Internet Tools and Collaboration, for more information on Hyperlinks.)

The *Pasteashyperlink* command or *Paste as Hyperlink* option from the Menu bar or Ribbon is another method of creating a hyperlink as an alternative to the *Insert Hyperlink* dialog box; however, you can use *Pasteashyperlink* to link only to documents, not to Web addresses. The resulting hyperlink is attached to an AutoCAD object. The main difference is that with the *Pasteashyperlink* command, you first open the document you want to link to, use *Copy*, then open AutoCAD and use *Paste as Hyperlink*. AutoCAD simply asks you to select an object to attach the link to.

```
Command: PASTEASHYPERLINK
Select objects: PICK
Select objects: Enter
Command:
```

The resulting hyperlink (to the remote document) is an embedded object in the drawing attached to a specific AutoCAD object. The procedure is as follows:

1. open the source program, then open the desired document you want to link to;
2. highlight some text, such as the title or first line of the document;
3. select *Copy* from the program's *Edit* pull-down menu while leaving the document open;
4. toggle to AutoCAD and select *Paste as Hyperlink* from the *Edit* menu;
5. select the AutoCAD object you want to attach the document to.

TIP

Two conditions must exist for *Pasteashyperlink* to operate: 1) the document that you *Copy* from must be a previously saved document (not a Web page), and 2) the document that you copy from must be open at the time that you *Paste*.

The text that you select in step 2 appears at the cursor when you pass the cursor over the AutoCAD object. Opening the link produces the source program and displays the entire document that contains the text that you selected to *Copy*. Therefore, the selected text has significance only in that it appears in AutoCAD at the cursor, so you should select text that indicates the name or purpose of the linked document.

*Pasteashyperlink* is essentially another way to hyperlink a document to an object, but with this method you select the link from within the linked document rather than browsing for the document (file name) when establishing a *Hyperlink* from AutoCAD in the *Insert Hyperlink* dialog box.

## MANAGING OLE OBJECTS

### *Olelinks*

| Ribbon | Menu Bar | Command (Type) | Alias (Type) | Shortcut |
|---|---|---|---|---|
| ... | *Edit*<br>*OLE Links* | *Olelinks* | ... | ... |

When you have *Linked* an object within an AutoCAD drawing, you can control if it is automatically or manually updated whenever changes are made to the original (source) document. The *Olelinks* command activates the *Links* dialog box with options for the maintenance of object links. If no objects are currently *Linked* within the drawing, the *OLE Links...* option in the *Edit* menu is grayed out and typing *Olelinks* invokes no response from AutoCAD.

Command: **`OLELINKS`** (activates the *Links* dialog box if links exist in the current drawing)
Command:

The following are features included in the *Links* dialog box (Fig. 31-15).

**FIGURE 31-15**

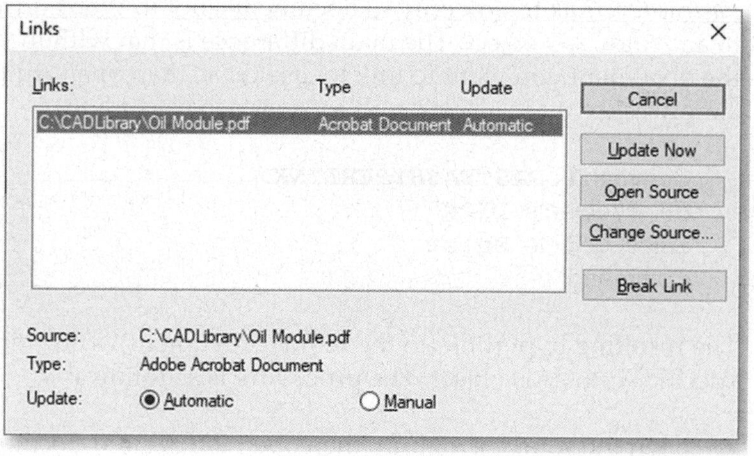


#### *Links*, *Type*, and *Update* Columns

The list box in the main section of the dialog box provides a listing of all linked objects in the drawing. The *Links* column displays the source information (file name and path). The *Type* is the source application type. The *Update* column lists the current status of the update option.

#### *Source*

Found in the lower-left corner of the dialog box, this description specifies the drive, path, application, and additional information such as object type or selection area, depending on the source application.

#### *Type*

Also listed in the lower-left corner of the dialog box is the *Type* of link for the highlighted item. This is the application-specific name (source application) of the currently selected linked object. The application-specific name is sometimes followed by an OLE link number assigned to the object by AutoCAD.

#### *Update*

Two *Update* radio buttons are located at the bottom of the dialog box. The *Automatic* option causes updating of the linked object whenever its source document is modified while the drawing is open. Setting the *Update* function to *Manual* prevents the link from updating when revisions are made to the source. When *Manual* is selected, use *Update Now* to update the link.

#### *Update Now*

This button forces the links selected from the list to be updated. When you close the dialog box, listed data reflects the newly updated status of the source object.

#### *Open Source*

This option opens the currently selected link with the source application so it can be modified. The linked information is highlighted when the source application is opened.

### *Change Source*

This option produces the *Change Source* dialog box where you can select a different source file to link.

### *Break Link*

Use this option to sever the link of the object to its source file and automatically convert it to a "Static OLE object" or Windows metafile object. It can no longer be updated if the source is changed. The object can no longer be edited with the source application or with AutoCAD commands.

For example, if a spreadsheet displaying a bill of materials for the drawing is created in Excel and then linked inside the drawing, it can be modified at any time with Excel simply by double-clicking on the spreadsheet picture displayed on the drawing. If, however, the link is broken, the spreadsheet picture is converted to a metafile picture object and can no longer be edited using Excel. Static OLE objects can be edited or removed with normal AutoCAD commands.

## Olescale

| Ribbon | Menu Bar | Command (Type) | Alias (Type) | Shortcut |
|---|---|---|---|---|
| ... | ... | *Olescale* | ... | ... |

The *Olescale* command produces the *OLE Text Size* dialog box (see previous Fig. 31-8), which allows you to alter the scale of OLE objects in AutoCAD. This dialog box is the same dialog box that appears when you first paste a non-AutoCAD OLE object into AutoCAD using *Pasteclip* (see "*Pasteclip*").

## Oleopen

| Ribbon | Menu Bar | Command (Type) | Alias (Type) | Shortcut |
|---|---|---|---|---|
| ... | ... | *Oleopen* | ... | ... |

*Oleopen* opens the OLE object's source application so you can edit the document. You must first highlight the OLE object, then use this command. You can also double-click on the OLE object to produce this command. The result of editing the OLE object differs based on whether the object is linked or embedded. If the OLE object is linked, the original source document is edited. If the OLE object is embedded, only the object in AutoCAD is changed, not the source document.

### OLE Right-Click Shortcut Menu

You can highlight an OLE object, then right-click to produce the shortcut menu shown in Figure 31-16. The *OLE* cascading menu has several options.

### *Open*

Use this option to open the OLE object's source application to edit the OLE object. This option invokes the *Oleopen* command.

FIGURE 31-16

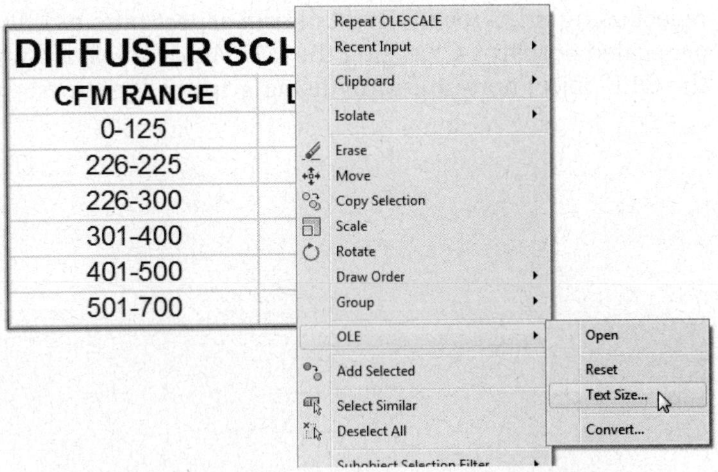

### *Reset*

Use this option to reset the OLE scale to the original scale when the object was inserted.

### *Text Size*

If the OLE object contains text, this option allows you to change the text size using the *OLE Scale* dialog box (see "*Pasteclip*"). When the text size is changed, the table containing the text (if existing) automatically adjusts accordingly.

### *Convert*

This option produces the *Convert* dialog box (not shown) that operates only for embedded objects. The options available in the dialog box differ based on the selected OLE object type. For example, highlighting a Microsoft Excel Worksheet object would allow you to convert the object to a Microsoft Excel Chart.

## Other AutoCAD Commands with OLE Objects

OLE objects in AutoCAD behave similarly to other native AutoCAD objects. For example, you can select an OLE object with the pickbox, window, crossing window, or other object selection method. You can add or remove the OLE objects to or from selection sets. Therefore, you can use standard Modify commands with OLE objects, such as *Erase, Move, Copy, Array,* and *Scale*.

OLE objects are sensitive to Object Snaps, such as *Endpoint, Midpoint*, etc. Therefore, you can draw *Lines*, for example, and *Object Snap* to OLE objects. Or, you can *Scale* an OLE object that Object Snaps to a predrawn rectangular area, or *Move* an OLE object that *Object Snaps* to a title block or border.

You can also use grips to modify the size and location of the OLE object. Simply highlight the OLE object (a table in this case), and drag a grip to resize the table (Fig. 31-17). This technique is an alternate to using the *OLE Scale* dialog box.

**FIGURE 31-17**

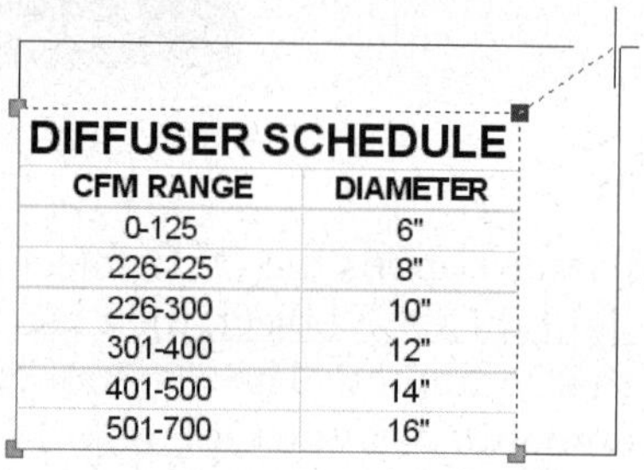


**TIP**

Additionally, you can use the *Properties* palette to manipulate OLE features such as *Width, Height, Scale %,* and *Position* of the OLE object (Fig. 31-18). Normally, the *Lock Aspect* is set to *Yes* so scaling the OLE object using grips results in a uniform aspect ratio (width and height are scaled equally). Changing the *Lock Aspect* to *No* allows you to scale the OLE object non-uniformly using grips or *Properties*.

**FIGURE 31-18**

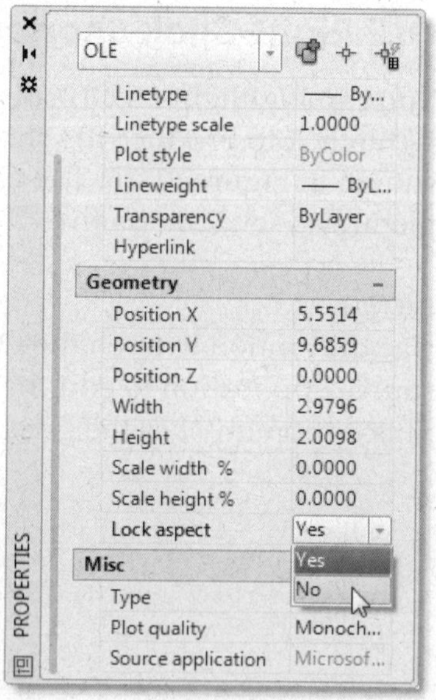

## OLE-RELATED SYSTEM VARIABLES

Four system variables affect the appearance of OLE objects in the Drawing Editor and in prints and plots:

*MSOLESCALE, OLEHIDE, OLEQUALITY,* and *OLESTARTUP.*

### MSOLESCALE

When pasting an OLE object containing text into model space, the *MSOLESCALE* variable controls the initial size of the object. *MSOLESCALE* affects only current insertions, not previously pasted OLE objects. The initial value is 1.000. A value of 0.000 scales the OLE object by the *DIMSCALE* value. (If you *Paste* the object into a layout, the object is automatically scaled based on the layout's sheet size and units.)

### OLEHIDE

*OLEHIDE* controls the visibility of OLE objects in AutoCAD. Your choice for this variable setting affects the AutoCAD Drawing Editor screen as well as printing and plotting. The possible settings are:

0 All OLE objects are visible (initial setting)
1 OLE objects are visible in paper space only
2 OLE objects are visible in model space only
3 No OLE objects are visible

The *OLEHIDE* system variable setting is saved in the system registry.

### OLEQUALITY

*OLEQUALITY* sets the default quality level for plotting and printing OLE objects. This setting corresponds to the *OLE plot quality* option selected in the *Plot and Publish* tab of the *Options* dialog box. When *OLEQUALITY* is set to 3 (default setting), the quality level is assigned automatically based on the type of object. For example, spreadsheets and tables are set to 0, color text and pie charts are set to 1, and photographs are set to 2. The four options are:

0 Monochrome
1 Low graphics
2 High graphics
3 Automatically Select

### OLESTARTUP

*OLESTARTUP* determines whether the source application of an embedded OLE object loads when plotting. Loading the OLE source application may improve the plot quality, depending on the source application and type of OLE object. The possible settings are:

0 Does not load the OLE source application (initial setting)
1 Loads the OLE source application when plotting

The *OLESTARTUP* system variable setting is saved in the drawing file.

## CHAPTER EXERCISES

1. **AutoCAD Objects:** ***Copylink, Copyclip, Copybase, Pasteclip, Pasteorig***

In this exercise you will gain experience copying and pasting between two open AutoCAD drawings. You will use existing drawings to copy most of the needed geometry to develop a half section view.

A. ***Open*** the **SADDL-DM** drawing you created in the Chapter 29 exercises. Activate the ***Model*** tab, or if you created a titleblock in model space, ***Freeze*** its layer so you see only the geometry (including center lines and hidden lines) and dimensions. Also, ***Freeze*** layer **DIM**.

B. With SADDL-DM open, begin a ***New*** drawing based on the **ACAD.DWT** template. Use the ***Window*** pull-down menu to ***Tile Vertically***. ***Zoom Extents*** in the SADDL-DM window.

C. Use the ***Copylink*** command to automatically copy all objects from SADDL-DM. Use ***Pasteorig*** to bring the objects into the new drawing. This should create a display similar to that shown in Figure 31-19. Notice that all layers come into the new drawing when *Copylink* is used, including layers that are frozen. Since you do not want all these layers, ***Exit*** the new drawing without saving changes and begin a ***New*** drawing again using the **ACAD.DWT** template drawing.

**FIGURE 31-19**

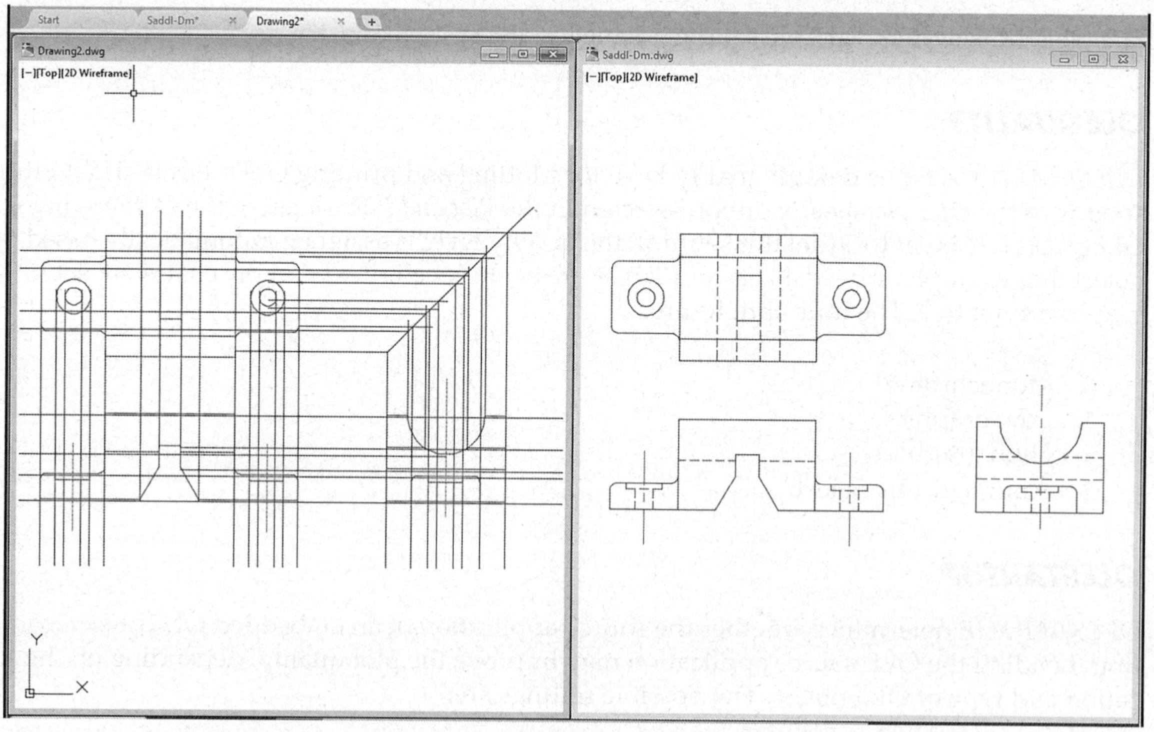


D. Now use ***Copyclip*** and select only the visible geometry from the SADDL-DM drawing. Activate the new drawing and use ***Pasteorig***. Perform a ***Zoom Extents***. The two drawings should appear identical, except note that the new drawing's linetypes appear to be continuous. Change the ***Ltscale*** to **15**. ***Save*** the new drawing as **SADDLE-HALF SECTION**.

FIGURE 31-20

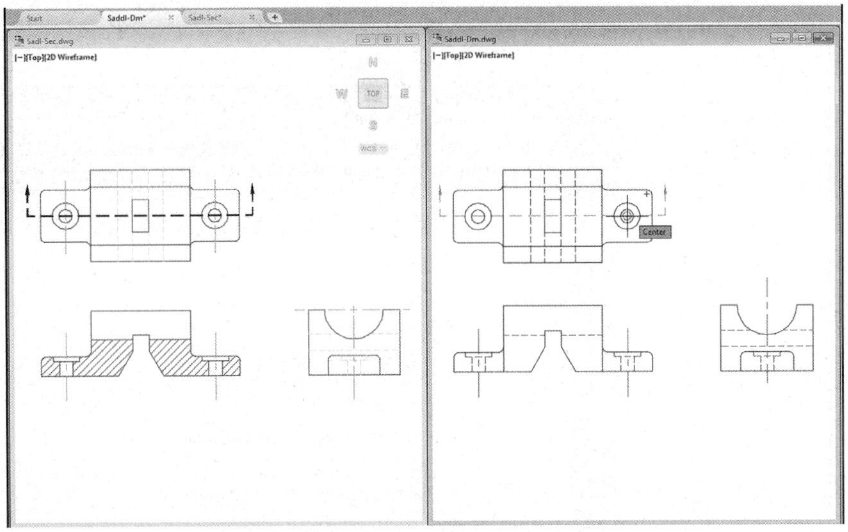


E. Next, ***Close*** SADDL-DM and do not save changes. ***Open*** the **SADL-SEC** drawing from Chapter 26 exercises, and ***Tile Vertically***. In order to copy the cutting plane line from SADL-SEC, use ***Copybase*** and specify a base point at the ***Center*** of one of the two holes in the top view. Since the two drawings (SADDL-DM and SADL-SEC) may have different coordinate locations for the geometry, do not use *Pasteorig*, but use ***Pasteclip*** to place the cutting plane line in the SADDLE-HALF SECTION drawing at the ***Center*** of the appropriate hole. Your display should appear similar to that shown in Figure 31-20.

F. ***Close*** the SADL-SEC drawing and do not save changes. Maximize the new drawing and continue to generate a half section view. Your results should appear similar to that shown in Figure 31-21.

FIGURE 31-21

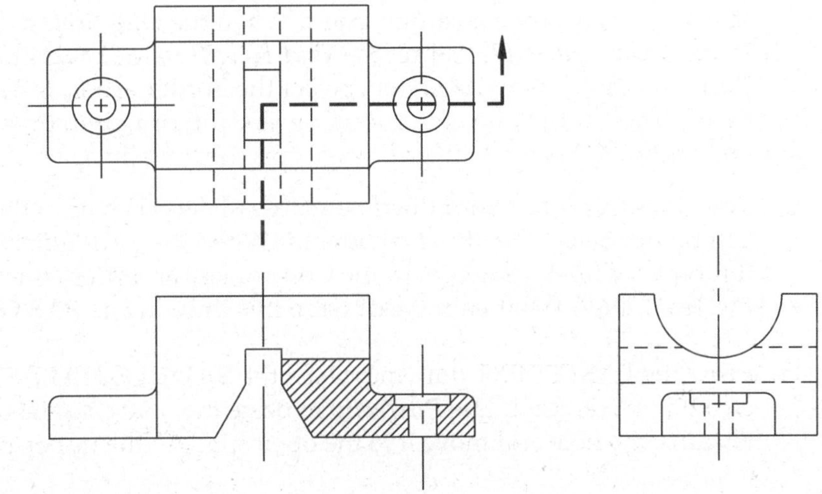

G. This technique (using copy and paste between AutoCAD drawings) is useful especially in cases such as this when you need geometry from two or more drawings. What is a better method to use when you need to create a drawing similar to only one other drawing?

2. ***Copy, Pasteclip***

For this exercise, you will write a short paragraph of General Notes to be used in an AutoCAD drawing. Since the notes are text objects, it is easier to use a word processor to write and format the notes, *Copy* them to the Clipboard, and then use *Pasteclip* in AutoCAD to insert the notes into your drawing.

A. Begin a ***New*** drawing (using the **ACAD.DWT** template). Also, while AutoCAD is running, start **Word** and compose a short paragraph of General Notes similar to the note shown in Figure 31-22 (the word "copywrighted" is intentionally misspelled). After typing the notes, highlight the entire block of text and the select ***Copy*** from the ***Edit*** pull-down menu. This action copies the Word document onto the Clipboard.

**FIGURE 31-22**

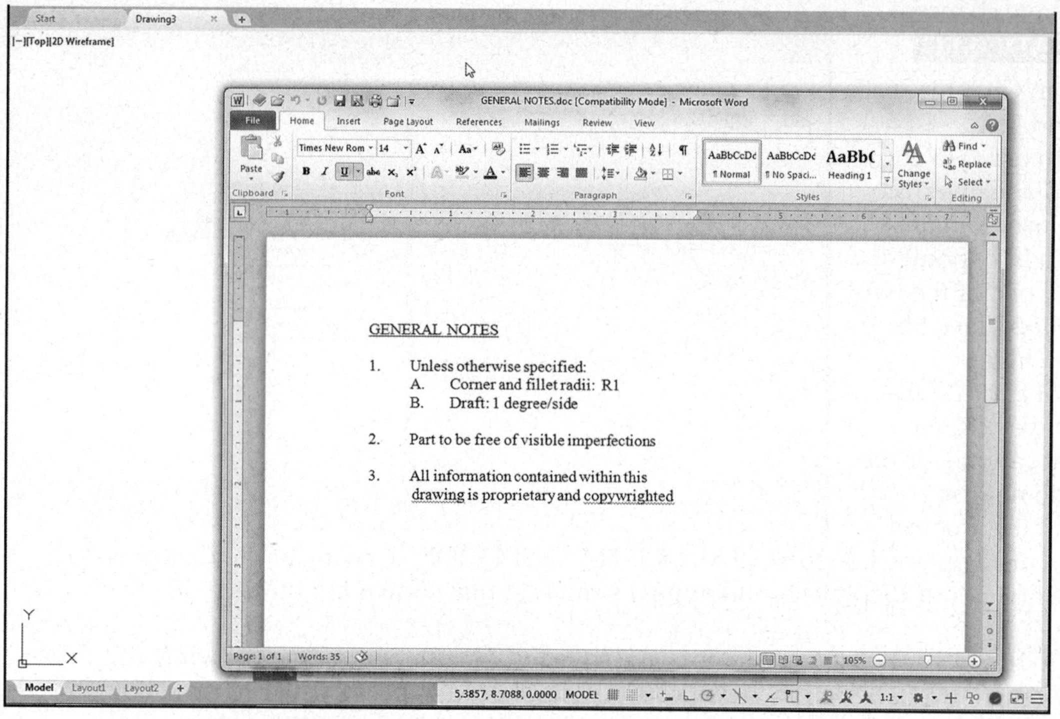


B. Now return to the new drawing you started in step A (you can use the **Alt+Tab** key sequence to switch to AutoCAD). In AutoCAD, use ***Pasteclip***. When the OLE object "box" appears, select an appropriate insertion point. When the *OLE Scale* dialog box appears, select the ***OLE Point Size*** of **10** or **12**, and set the ***Text Height*** value to **.2**. ***Zoom*** if necessary to see the notes. Turn on Grips if not already on. Select the border around the object to display the grips at each corner. Select one of the corner grips and drag it in or out to resize the note. Notice the text within the pasted image changes proportionally as you change the box size.

C. The notes from the Word document are still stored within the current memory of the Clipboard. Select the ***Mtext*** command. When the *Text Editor* appears, do not enter in any text, but type a Ctrl+V (hold down the Control key and press the "v" key). Select the ***OK*** button in the *Text Editor*. What happened? Save this drawing as **PASTETXT**.

D. ***Close*** the PASTE TEXT drawing. ***Open*** the **SADDLE-HALF SECTION** drawing that you created in Exercise 1. Use ***Pasteclip*** to paste the note contained on the Clipboard. Use the grips to resize the note and move it to the open space in the upper-right corner. ***Save*** the drawing.

3. **Double-clicking on an Embedded Object**

Assume that you printed the **SADDLE-HALF SECTION** drawing and noticed the misspelled word. Follow this procedure to change the spelling of the word "copywrighted" to "copyrighted."

A. *Open* the **SADDLE-HALF SECTION** drawing if not already opened. Point to the pasted (embedded) note object and then double-click on it using the mouse. What happens?

B. Now correct the misspelled word in **Word**. Then select ***Exit*** from the ***File*** pull-down menu in Word. Next, view the notes in AutoCAD. What happened to the notes? ***Save*** the drawing, but do not *Exit*.

CHAPTER

# 32

# Advanced Layouts, Annotative Objects, and Plotting

## Chapter Objectives

After completing this chapter you should:

1. be able to use the *Layer Properties Manager* to control layer visibility and layer properties in viewports;
2. be able to set *PSLTSCALE* for multiple viewports and layouts;
3. know how to dimension in paper space;
4. know how to create annotative dimensions and text for display at different scales in multiple viewports;
5. know how to edit annotative dimensions using *Objectscale, Annoupdate, Annoreset,* and related system variables;
6. be able to create a new Plot Style Table using *Add-A-Plot Style Table Wizard*;
7. be able to edit a Plot Style Table using the Plot Style Table Editor;
8. be able to assign named plot styles to layers and objects;
9. know how to apply plot stamps to your prints and plots.

## CONCEPTS

This chapter is concerned with advanced concepts of using layouts and plotting those layouts—specifically creating multiple viewports per layout, creating multiple layouts, using annotative objects in multiple viewports, attaching Plot Style Tables to layouts, and assigning plot styles to layers and objects. To isolate these complex ideas as an attempt to simplify their explanation, the chapter is divided into three sections: Advanced Layouts, Annotative Objects, and Advanced Plotting.

Topics discussed in this chapter are arranged as follows:

Advanced Layouts

- Layer Visibility for Viewports (*Layer, Current VP Freeze, New VP Freeze, Vplayer*)
- Linetype Scale in Viewports (*Psltscale*)
- Dimensioning in Paper Space
- The "Reverse Method" for Calculating Drawing Scale Factor
- Inserting Named Views as Viewports

Annotative Objects

- Setting Annotation Scale and Viewport Scale
- Creating and Displaying Annotative Objects in Multiple Viewports
- Adding and Deleting Scale Representations for Annotative Objects
- Adjusting Positions for Individual Annotative Object Scale Representations
- Linetype Scales for Annotative Drawings
- Applications for Multiple Viewports

Advanced Plotting

- Plot Style Tables and Plot Styles
- Plot Stamping

Shared Views

- Using Shared Views to Produce a Shareable Link Viewable in a Browser

## ADVANCED LAYOUTS

From previous chapters, you are familiar with creating and setting up drawings, creating layouts and viewports, and plotting layouts. These concepts are fundamental for setting up and creating drawings in AutoCAD; however, they were based on using only one viewport per layout. The first half of this chapter deals with setting up multiple viewports and multiple layouts.

To this point, you have learned the following. Objects that represent the subject of the drawing (model geometry) are normally drawn in model space. Dimensioning and related annotation (text) are traditionally performed in model space because they are directly associated to the model geometry. A layout simulates the sheet of paper you will print or plot on, so it is used to prepare the desired display of the model geometry for plotting. A viewport is created in a layout with the *Vports* command in order to "see" the model geometry. The display of the geometry in the viewport should be scaled to achieve an appropriate scale for the drawing. Since a layout represents the plotted sheet, you normally plot the layout at 1:1 scale. The geometry displayed in the viewports is scaled to the desired plot scale by using the *Viewport Scale* list, *Viewports* toolbar, the *Properties* palette for the viewport, or a *Zoom XP* factor. With the layout, you can use the *Page Setup* and *Plot* dialog boxes to specify and save the print or plot parameters such as plot device, paper size, and orientation, etc.

There are many uses for creating multiple viewports per layout and/or creating multiple layouts in one drawing, each displaying different drawing geometry or the same geometry at different scales. For example, you may want to create one plot to show several views of one drawing, such as an overall view of an assembly and several detail views of individual components. Or you may want to show a floor plan and several views of individual rooms. You may have the need to show several separate drawings on one plot, which can be accomplished by *Xrefing* several drawings, creating several viewports, and controlling each viewport to display only the layers associated with one *Xref*. In another application, you may need to set up multiple layouts for creating multiple sets of plots, each layout displaying a different combination of layers. Using this scheme, for example, you can create and save a layout to plot the floor plan and the plumbing details for the plumbing contractor, a layout to plot the floor plan and the electrical layout for the electrical contractor, a layout for the floor plan and the HVAC layout, and so on.

When using multiple viewports and multiple layouts, the setup becomes more complex but offers more alternatives. Remember, you must scale the display of the geometry in each viewport and you must control the layer visibility in each viewport (which layers are visible in which viewports). When you have multiple viewports, you cannot always effectively use the "drawing scale factor" to determine the viewport scale since that idea is based on preparing the drawing for only one large viewport. In addition, you cannot effectively control which layers are visible in which viewports using the global *Freeze/Thaw* or *On/Off* controls.

This section explains how you can control the viewport-specific layer visibility using the *VP Freeze* button and *New VP Freeze* button in the *Layer Properties Manager* or *Layer* drop-down list. The *Layer Properties Manager* is also explained for controlling "viewport layer property overrides" (properties of *Color*, *Linetype*, and *Lineweight*) for individual viewports. In addition, another strategy for calculating the drawing scale factor for viewports is discussed. The "reverse method" is used to help determine appropriate values for *Ltscale*, *Dimscale*, text height, etc. based on the viewport scale.

This section also presents a totally different strategy you may want to consider when using multiple viewports and layouts, that is, creating dimensions in paper space, creating text in paper space, and setting *Psltscale* to paper space. This strategy prevents you from having to calculate and set the text height, *Dimscale*, and *Ltscale* for each viewport and having to control the layer visibility specific to each viewport.

Historically, the use of paper space, viewports, and layouts has changed from early AutoCAD releases having only a few capabilities (and therefore few applications) to current releases offering advanced capabilities and applications. In contrast to early AutoCAD releases, the common use of paper space, viewports, and layouts represents a major paradigm shift in the way drawings are prepared and plotted. Although this technology continues to advance and become more "user-friendly," these concepts are not simple to understand nor easy to carry out. Because there are currently different applications, strategies, and personal preferences for creating and setting up multiple viewports and layouts, this chapter presents the alternatives and discusses the advantages and disadvantages of each. In this way, you can make the appropriate decisions based on your specific applications, preferences, and level of technology.

## Layer Visibility for Viewports

AutoCAD provides you with the capability to create multiple layouts. In each layout, you can create one or several viewports using *Vports* or *-Vports*. Even though you can have multiple viewports in a drawing, each viewport displays the same model space geometry, right? Well, yes and no. That is, there is only one model space, and by default the same model geometry is displayed in every viewport.

You can control what parts of the model space geometry are displayed in each viewport by
(1) using *Zoom*, *Pan*, and other viewing commands to display different areas of model space and
(2) using layer visibility to control what layers appear in which viewports.

**Layer**

| Ribbon | Menu Bar | Command (Type) | Alias (Type) | Shortcut |
|---|---|---|---|---|
| *Home* *Layers* *Layer Properties* | *Format* *Layer...* | *Layer or -Layer* | *LA or -LA* | ... |

The *Layer* command produces the *Layer Properties Manager* dialog box. This dialog box can be used to control which layers are visible in which viewports. Two small icon buttons in the *Layer Properties Manager* dialog box are used for viewport-specific layer visibility: *VP Freeze* and *New VP Freeze*.

Figure 32-1 displays the *Layer Properties Manager* showing the *VP Freeze* and *New VP Freeze* columns (last two visible columns). Some of the icons do not appear when the *Model* tab is current. If a layout tab is selected and a viewport is active (the cursor appears in a viewport), use these icons to freeze or thaw layers in the current viewport or to freeze or thaw layers in new viewports (the selected layer will be frozen or thawed for any new viewports that are created).

**FIGURE 32-1**

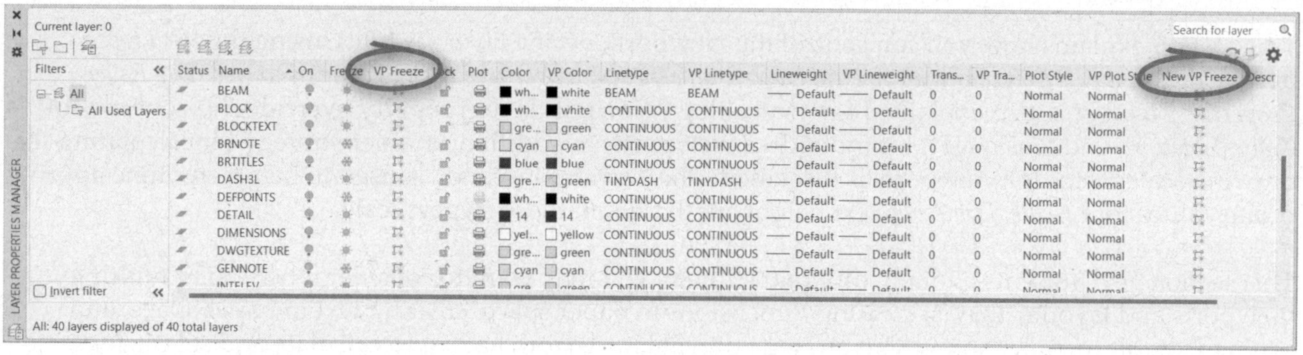


### *VP Freeze*

For example, assume you have two viewports created in a layout and each viewport displays the same model space geometry (Fig. 32-2, on the next page). Notice that in the left viewport all layers are visible compared to the right viewport, in which the dimension layer is not visible. To freeze the "Dimensions" layer in the right viewport only, first make the right viewport current, then use *Layer* to produce the *Layer Properties Manager* and select the *VP Freeze* icon to change this layer to a frozen state for that viewport (snowflake; see Fig. 32-1).

If the left viewport were made active and then the *Layer Properties Manager* invoked, these icons for all layers would appear thawed (sunshine) for the current (left) viewport.

When a viewport is current, the *VP Freeze* icons that appear give the visibility state of layers specific to the viewport that is current when the list is displayed. In other words, the icons (*VP Freeze*) may display different information depending on which viewport is current when the list is viewed. Changing the *Freeze/Thaw* icon affects the layer's state globally (for all viewports in the drawing), while changing the *VP Freeze* icon affects the layer's state only for the current viewport.

TIP

A *Layout* tab itself is considered a paper space viewport when dealing with layer visibility. That is, if you are in paper space (*Pspace* is active, not in a viewport) and you PICK the *VP Freeze* icon, the highlighted layer becomes frozen for the entire layout (the current viewport) but not in the individual viewports or in the *Model* tab.

FIGURE 32-2

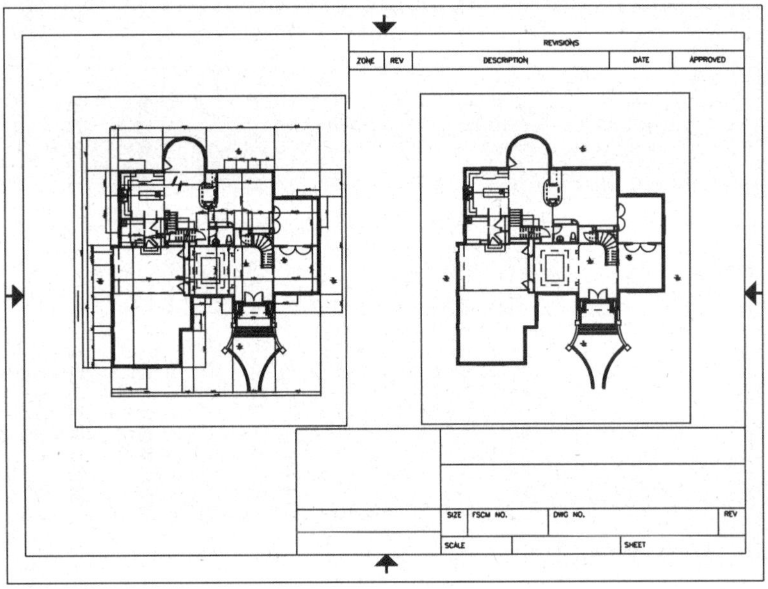


TIP

A *Freeze or Thaw in Current Viewport* icon appears in the *Layer* drop-down list (Fig. 32-3). This icon has the identical function as the *VP Freeze* icon in the *Layer Properties Manager*. The *New VP Freeze* button is not available in the drop-down list, only in the *Layer Properties Manager*.

### New VP Freeze

This icon prevents the selected layer from appearing in any viewports that are created from that time on. Any layers selected with the *New VP Freeze* attribute do not appear in subsequently created viewports.

FIGURE 32-3

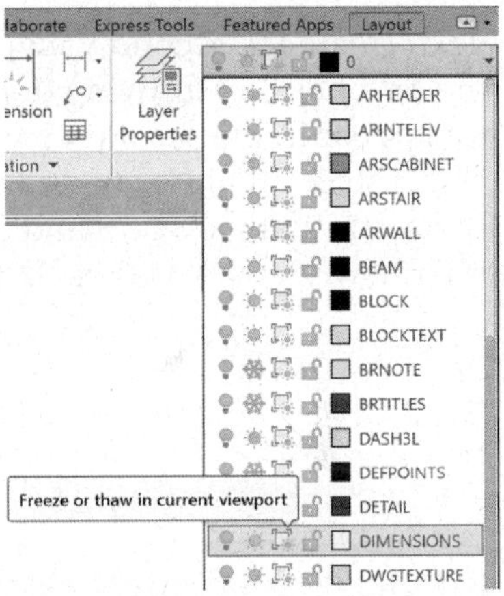


This feature is helpful when you need several viewports, each to display a separate drawing. For example, assume you want to *Xref* several drawings, but only want one drawing to appear in each viewport.

Assume you created one viewport and *Xrefed* a drawing into model space so it (all its layers) appeared in the viewport (Fig. 32-4). If you were to create a second viewport, the same drawing would normally appear in the second viewport just as all model space geometry normally appears in all viewports. If you were to *Xref* a second drawing into model space, it would also normally appear in <u>both</u> viewports. (See the Xreferences chapter for details.)

FIGURE 32-4

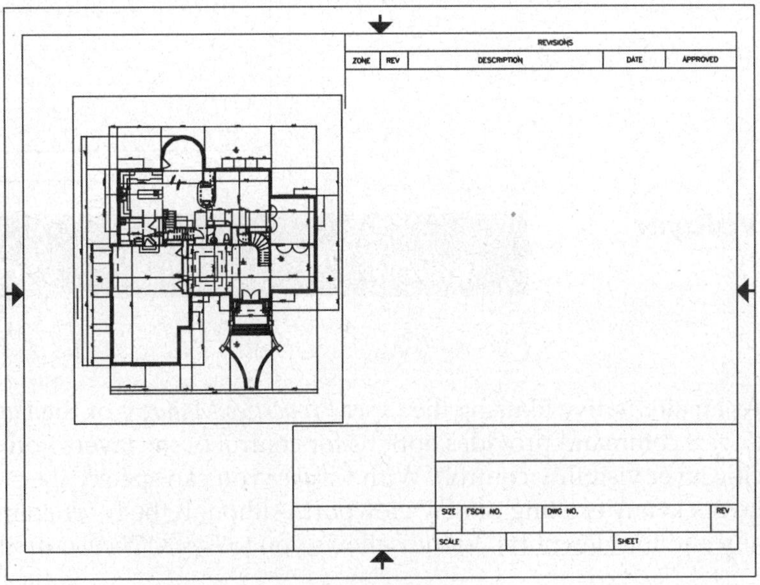


TIP

To avoid this problem, after you create one viewport and *Xref* the first drawing but before you create the second viewport, use *New VP Freeze* for all layers of the *Xref* drawing (Fig. 32-5, all "Xref1 | *" layers). (Unlike the *VP Freeze* icon, the *New VP Freeze* icon displays the same information for a specific layer no matter which viewport is current when the list is displayed.)

**FIGURE 32-5**

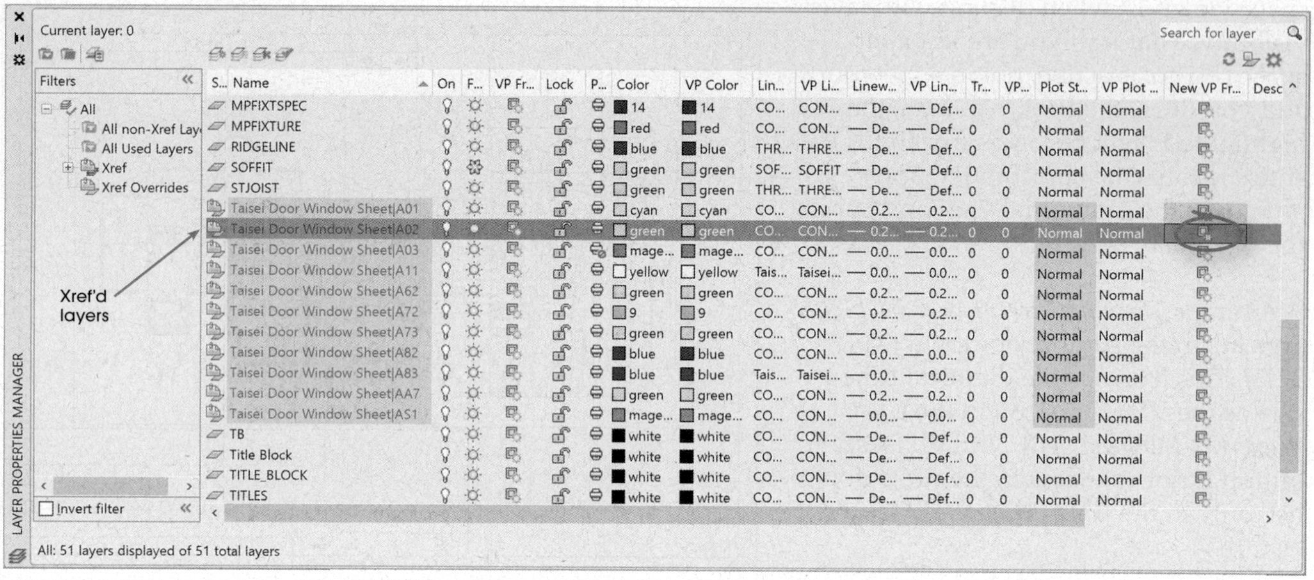


Note the *Layer Properties Manager* includes filters for *Xref* layers (for any drawing containing *Xrefs*) to assist you in controlling properties of those layers.

Next, create the second viewport. The original *Xref* drawing does not appear in the second viewport (right) since all of its layers are frozen for that viewport and all other new viewports (Fig. 32-6).

**FIGURE 32-6**

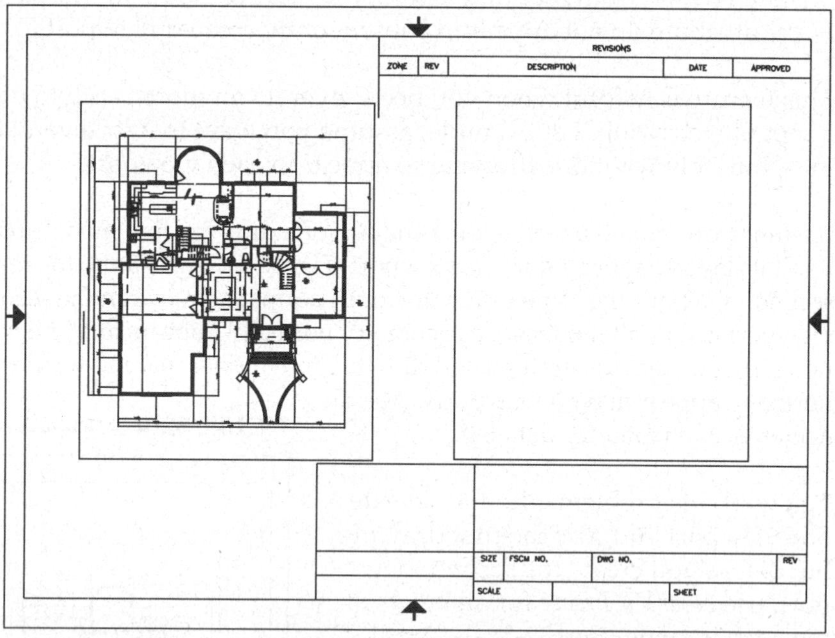


## *Vplayer*

| Ribbon | Menu Bar | Command (Type) | Alias (Type) | Shortcut |
|---|---|---|---|---|
| ... | ... | *Vplayer* | ... | ... |

As an alternative to using the *Layer Properties Manager* or the *Layer* drop-down list, the *Vplayer* (Viewport Layer) command provides options for control of the layers you want to see in each viewport (viewport-specific layer visibility control). With *Vplayer* you can specify the state (*Freeze/Thaw* and *Off/On*) of the drawing layers in any existing or new viewport. Although the *Layer* command allows you to *Freeze/Thaw* layers globally and for viewports, *Vplayer* allows you to *Freeze/Thaw* only the viewport-specific layer setting.

A *Layout* tab must be set current for the *Vplayer* command to operate fully, and a layer's global (*Layer*) settings must be *On* and *Thawed* to be affected by the *Vplayer* settings. You should normally use *Vplayer* while you are in paper space or in any viewport. The simplest method, however, is to make the desired viewport current, then use *Vplayer* or the *Layer Manager*.

For example, you may have two viewports set up in a layout, such as in Figure 32-7. Assume you want to *Freeze* the DIM (dimensioning) layer for the right viewport, so you make that the current viewport (your cursor is in the right viewport). In order to control the visibility of the DIM layer for the current viewport, the command syntax shown below would be used.

**FIGURE 32-7**

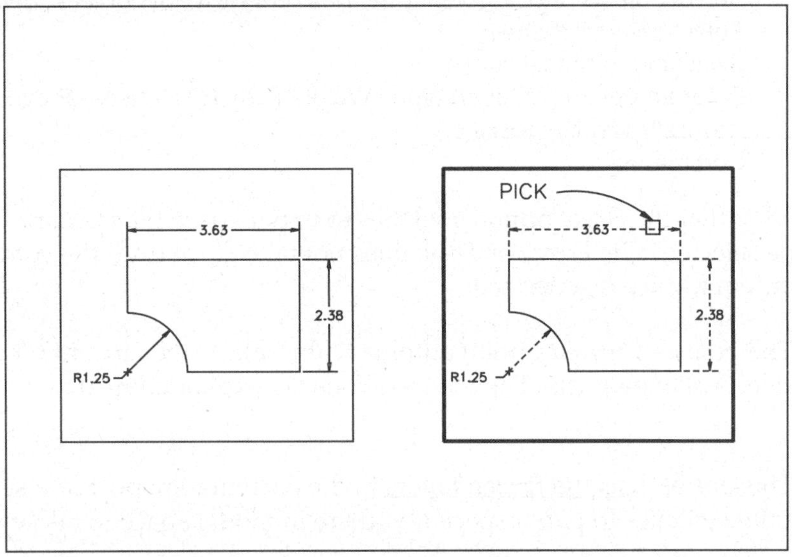


Command: ***VPLAYER***
Enter an option [?/Color/Ltype/LWeight/PStyle/TRansparency/Freeze/Thaw/Reset/reMoveoverrides/Newfrz/Vpvisdflt]: ***f***
Enter layer name(s) to freeze or <select objects>: **Enter**
Select objects: (Select an object on the desired layer. See Fig. 32-7.)
Select objects: **Enter**
Enter an option [All/Select/Current/eXcept current] <Current>: **Enter**
Enter an option [?/Color/Ltype/LWeight/PStyle/TRansparency/Freeze/Thaw/Reset/reMoveoverrides/Newfrz/Vpvisdflt]: **Enter**
Command:

This action results in *Freezing* the DIM layer (containing the highlighted objects) in the right viewport.

You do not have to make the desired viewport current before using *Vplayer*—your cursor can be in paper space or in another viewport. Invoke *Vplayer* at any time. Assume that you are in the left viewport, but want to *Freeze* the DIM layer for the right viewport. In the command sequence on the next page, notice that AutoCAD allows you to switch to paper space to PICK the desired viewport object (viewport borders can be PICKed only from paper space). This sequence results in *Freezing* the DIM layer for the right viewport (Fig. 32-8).

**FIGURE 32-8**

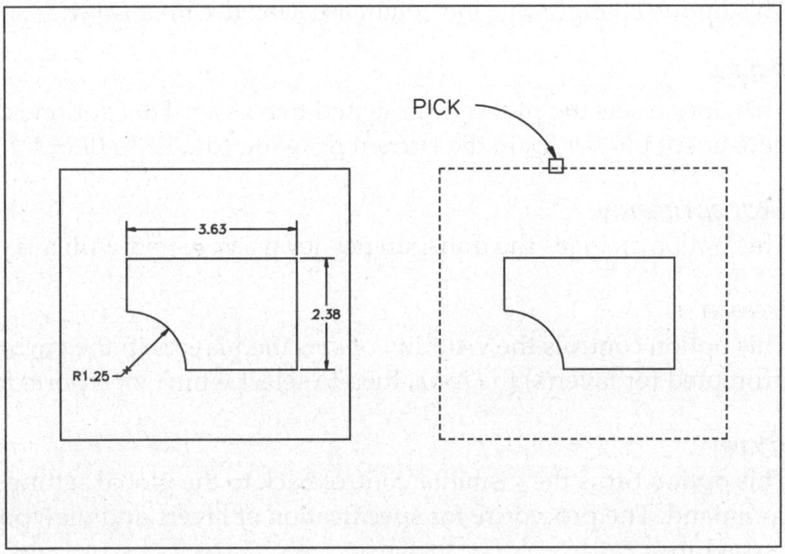

Command: ***VPLAYER***
Enter an option[?/Color/Ltype/LWeight/PStyle/TRansparency/Freeze/Thaw/Reset/reMoveoverrides/
Newfrz/Vpvisdflt] : ***f***
Enter layer name(s) to freeze or <specify layers by object selection>: **dim**
Specify viewport(s) [All/Select/Current/eXcept current] <Current>: **s**
Switching to Paper space.
Select objects: **PICK** (Select the desired viewport object. See Fig. 32-8.)
Select objects: **Enter**
Switching to Model space.
Enter an option [?/Color/Ltype/LWeight/PStyle/TRansparency/Freeze/Thaw/Reset/reMoveoverrides/
Newfrz/Vpvisdflt]: **Enter**
Command:

Note that the *Select* option switches to paper space if necessary so that the viewport objects (borders) can be selected. The *Freeze* or *Thaw* does not take place until the command is completed and the drawing is automatically regenerated.

The *Vplayer* ***Current*** option automatically selects the current viewport. The ***All*** option automatically selects all viewports. The other options are explained next.

### *?*

This option lists the frozen layers in the current viewport or a selected viewport. AutoCAD automatically switches to paper space if you are in model space to allow you to select the viewport object.

### *Color*

This option changes the color associated with a layer. You can use True Color, which allows you to specify color settings using the Red, Green, and Blue (RGB) color model. Over sixteen million colors are available when using True Color functionality or Color Book, which allows you to specify colors using third-party color books or user-defined color books.

### *Ltype*

This option changes the linetype associated with a *Layer*.

### *LWeight*

This option changes the lineweight associated with a *Layer*.

### *PStyle*

This option sets the plot style assigned to a *Layer*. This option is not available if you are using color-dependent plot styles in the current drawing (the *PSTYLEPOLICY* system variable is set to 1).

### *TRansparency*

This option changes the transparency level associated with a layer.

### *Freeze*

This option controls the visibility of specified layers in the current or selected viewports. You are prompted for layer(s) to *Freeze*, then to select which viewports for the *Freeze* to affect.

### *Thaw*

This option turns the visibility control back to the global settings of *On/Off/Freeze/Thaw* in the *Layer* command. The procedure for specification of layers and viewports is identical to the *Freeze* option procedure.

### *Reset*

*Reset* returns the layer visibility status to the default if one was specified with *Vpvisdflt*. You are prompted for "Layer(s) to Reset" and viewports to select by "All/Select/Current/eXcept current".

### Removeoverrides

Removes specific viewport layer property overrides for selected layers, or all overrides. You are prompted for "Layer(s) to Reset" and viewports to select by "All/Select/Current/eXcept current".

TIP

### *Newfrz*

This option creates new layers. The new layers are frozen in all viewports. This is a shortcut when you want to create a new layer that is visible only in the current viewport. Use *Newfrz* to create the new layer, then use *Vplayer Thaw* to make it visible only in that (or other selected) viewport(s). You are prompted for "New viewport frozen layer name(s):". You may enter one name or several separated by commas.

### *Vpvisdflt*

This option allows you to set up layer visibility for new viewports. This is handy when you want to create new viewports but do not want any of the existing layers to be visible in the new viewports. This option is particularly helpful for *Xrefs*. For example, suppose you created a viewport and *Xrefed* a drawing "into" the viewport. Before creating new viewports and *Xrefing* other drawings "into" them, use *Vpvisdflt* to set the default layer visibility to *Off* for the existing *Xrefed* layers in new viewports. (See the example for *Layer, New VP Freeze*.) You are prompted for a list of layers and whether they should be *Thawed* or *Frozen*.

## Layer Property Overrides

You can also control viewport-specific visibility (*VP Freeze* and *New VP Freeze*), and the individual layer properties of *Color*, *Linetype*, *Lineweight*, *Transparency*, and *Plotstyle* for individual viewports with the *Layer Properties Manager*.

When the *Color*, *Linetype*, *Lineweight*, *Transparency*, and *Plotstyle* properties of a layer are set for a specific viewport, they are called viewport layer property overrides. This is because each layer has global (drawing-wide) *Color*, *Linetype*, *Lineweight*, *Transparency*, and *Plotstyle* properties. Therefore, setting these properties for individual viewports overrides the global layer properties.

Viewport property overrides are controlled in the *Layer Properties Manager*. Columns display the *VP Color*, *VP Linetype*, *VP Lineweight*, and *VP Plotstyle* (Fig. 32-9). Use the slider at the bottom of the layer listing or expand the *Layer Properties Manager* to make the last columns visible, then select the desired color, linetype, lineweight, transparency or plotstyle.

FIGURE 32-9

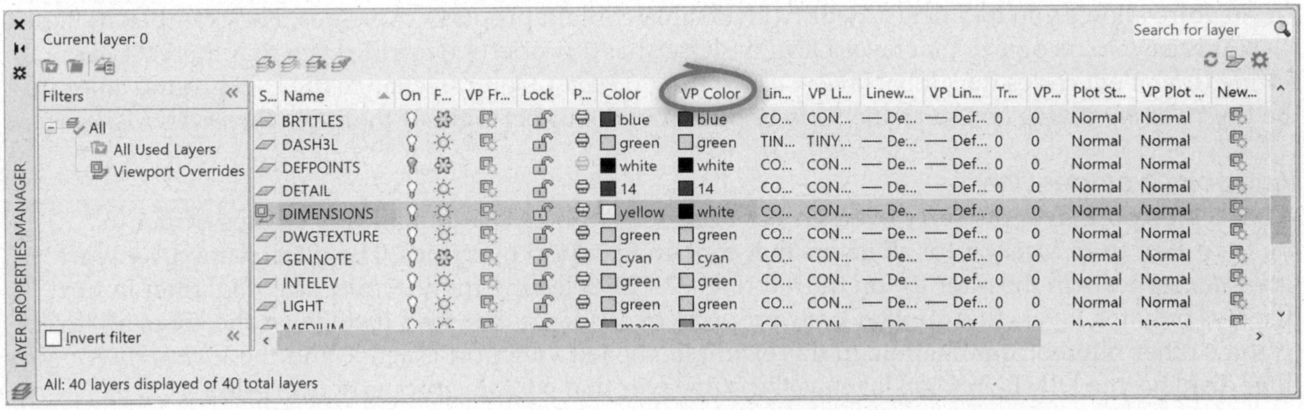

FIGURE 32-10

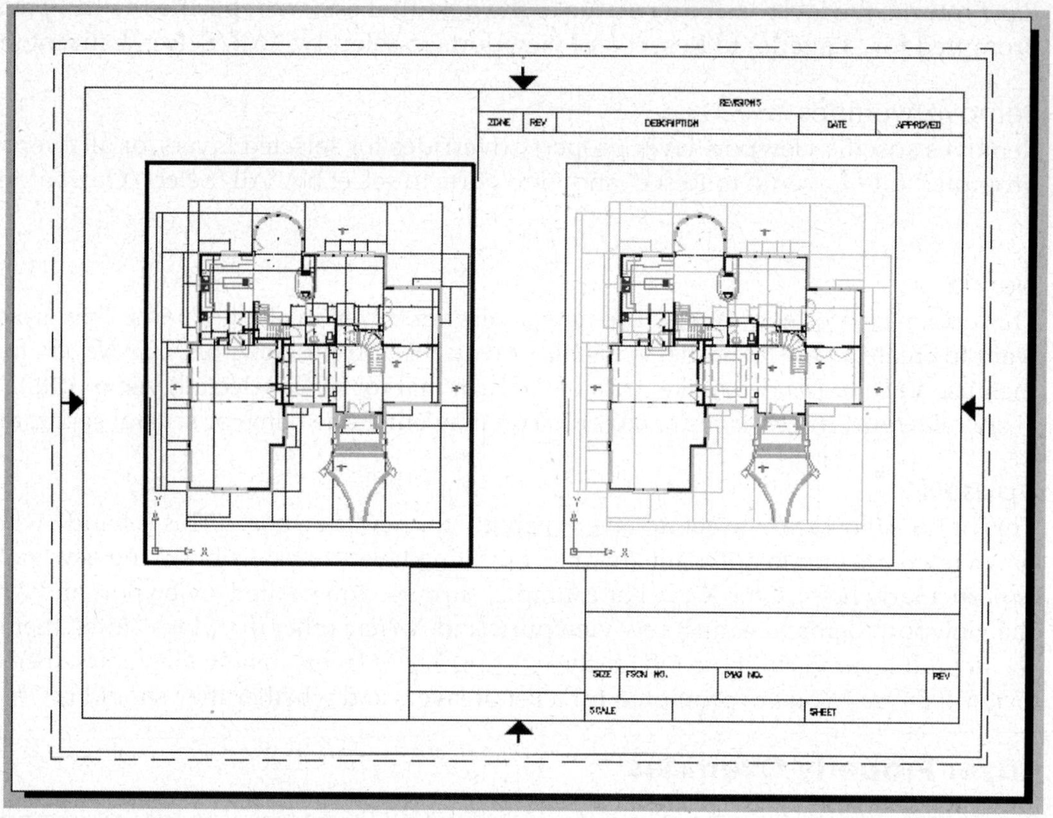


The viewport layer property overrides are powerful because they allow the same objects to be displayed in different viewports with different properties. This means that you can create different viewports and display the same model geometry but in different colors, linetypes, lineweights, or plotstyles. What is more, it is easy to remove viewport layer property overrides for one or more layers or viewports with a menu item selection.

For example, suppose you want the DIMENSIONS layer to appear in a darker color in the left viewport than in the right viewport. Do this by making the left viewport active, invoking the *Layer Properties Manager*, then selecting the color swatch in the *VP Color* column for the layer to change the color (see Fig. 32-9). The resulting drawing might appear as that in Figure 32-10.

### Property Overrides Highlight

When a viewport layer property override is set, the layer name and overridden properties become highlighted. That is, the background color in the *Layer Property Manager* is displayed in a light blue color. This feature allows you to quickly identify layers that contain property overrides. For example, in Figure 32-11 it is easy to recognize that layer DIMENSIONS has a property override such that the *VP Color* setting (*white*) overrides the global layer *Color* setting (*yellow*). You can change the background color in the *Layer Setting* dialog box (pick the *Settings* button in the upper right of the *Layer Properties Manager*).

### *Viewport Overrides* Filter

When a viewport layer property override has been set, AutoCAD automatically creates a filter in the *Layer Properties Manager* for all layers that contain viewport overrides. The filter, named *Viewport Overrides*, appears in the filter list on the left (Fig. 32-12). Selecting the *Viewport Overrides* filter in turn displays only the layers that contain layer property overrides rather than displaying the list of all layers or some other filtered combination. In this example, the left viewport is active and the *Viewport Overrides* filter displays the DIMENSIONS layer as the only layer that contains property overrides.

**FIGURE 32-11**

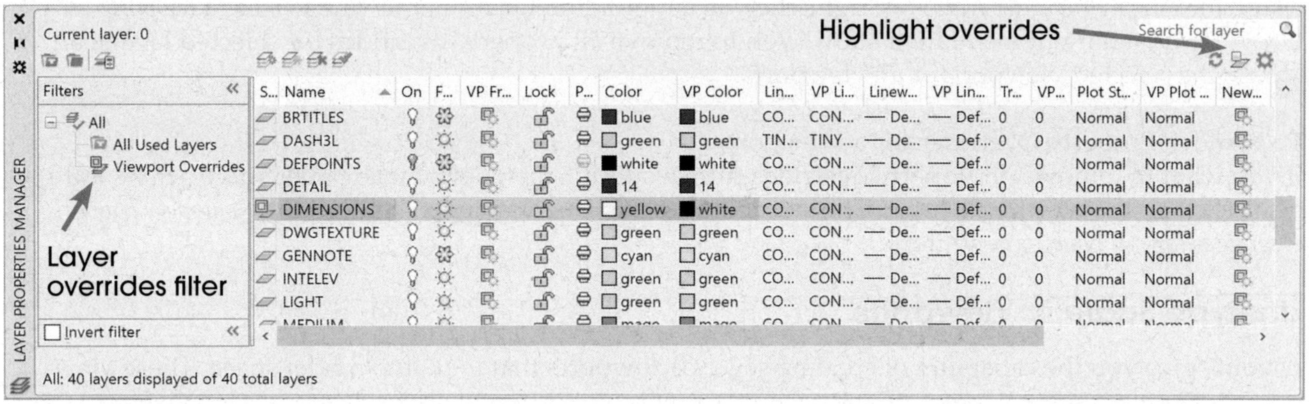


NOTE: Keep in mind that the *Layer Properties Manager* displays viewport layer property overrides (as highlighted) and *Viewport Overrides* filter only for the active viewport. Even though many viewport layer property overrides may exist in the drawing, you must make an individual viewport active to see the layer property overrides and *Viewport Overrides* filter for the specific viewport.

TIP

## Removing Layer Property Overrides in Viewports

Rather than having to select the individual viewports, then select the layers in the *Layer Properties Manager* and reset the color, linetype, and so on, there are several easy ways to remove the layer property overrides. You can remove individual property overrides, overrides for selected viewports, overrides for selected layers, or overrides for all layers or viewports.

**FIGURE 32-12**

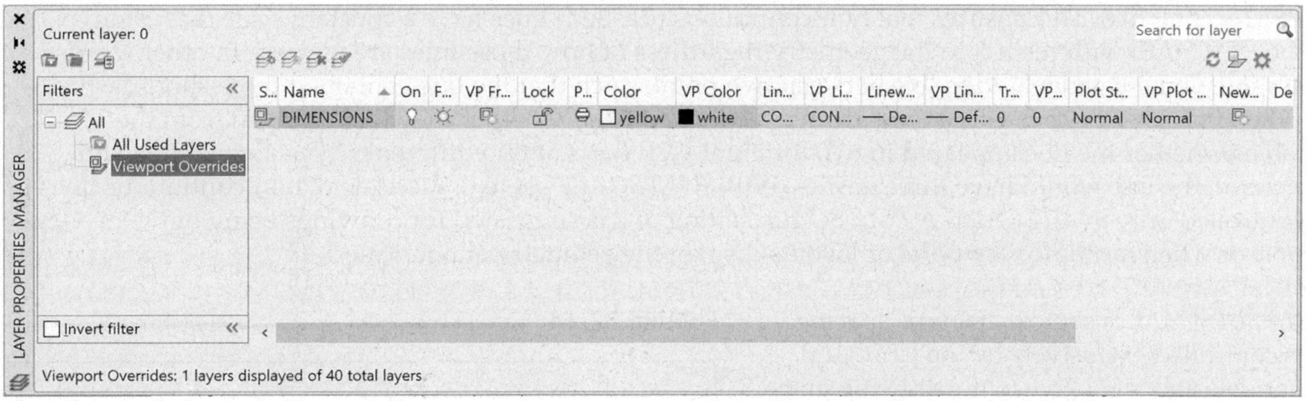


To Remove Individual Property Overrides: From the *Layer Properties Manager*, right-click on a highlighted viewport layer override property, such as *VP Color*. The *Remove Viewport Overrides for>* shortcut menu item allows you to remove that individual property override or remove all property overrides for selected layers, all layers, the selected viewport, or all viewports (Fig. 32-13). You can also use the *VPLayer* command *Removeoverrides* option as you learned in the last section.

**FIGURE 32-13**

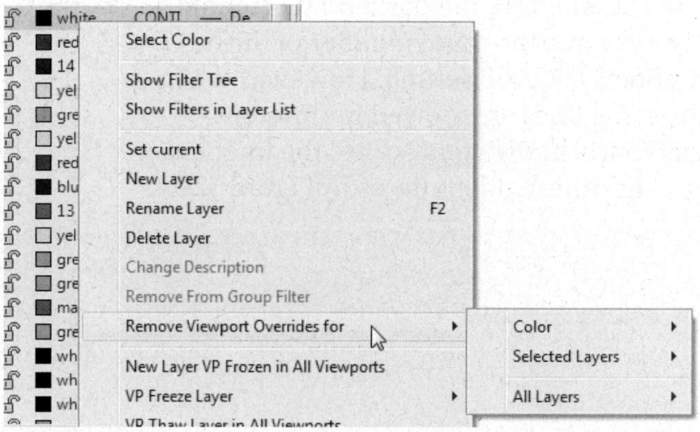

To Remove Property Overrides for Selected Layers:
Using the *Layer Properties Manager*, right-click on a highlighted layer name. The *Remove Viewport Overrides for>* shortcut menu item allows you to remove all property overrides for selected layers, all layers, the selected viewport, or all viewports.

To Remove Property Overrides for a Viewport:
If you want to remove all property overrides from a specific viewport, make paper space active and right-click on the viewport border. The shortcut menu for the viewport includes the *Remove Viewport Overrides for All Layers* menu item.

## Linetype Scale in Viewports

Layouts give you the capability of creating several viewports that look into model space. These viewports can contain several views of one or more drawings at different scales. Because each viewport can display linetype scaling differently based on viewport scale, the *PSLTSCALE* variable is needed to control how multiple linetype scales appear in one layout. Another variable, *MSLTSCALE,* is used to control linetype scale in the *Model* tab when multiple annotative scales are used (see "Annotative Objects" later in this chapter).

### *PSLTSCALE*

*LTSCALE* controls linetype scaling globally for the drawing, whereas the *PSLTSCALE* variable controls the linetype scaling for non-continuous lines for viewing in paper space. *LTSCALE* is a variable that accepts any value other than 0, whereas *PSLTSCALE* is a variable that accepts either a 0 (*Off*) or 1 (*On*). *LTSCALE* can still be used to change linetype scaling globally regardless of the *PSLTSCALE* setting.

If you want the linetype scale to always appear the same with respect to the drawing geometry, set *PSLTSCALE* to 0. This ensures that non-continuous (dashed) lines have a constant scale (based on the *LTSCALE*) with respect to the geometry regardless of how those lines are viewed. In other words, assume you created a hidden line between two points such that the line contained three short dashes. When *PSLTSCALE* is 0, you would always see three dashes whether they were viewed from the *Model* tab or whether the line appeared in two different viewports at two different scales. Even when you *Zoomed,* the line would have three dashes. With *PSLTSCALE* set to 0, the scale of non-continuous lines is controlled only by *LTSCALE.* A *PSLTSCALE* setting of 0 is suggested for drawings using only one viewport or when multiple viewports or layouts display the geometry at one scale.

If *PSLTSCALE* is set to 0, linetype spacing is controlled exclusively by the *LTSCALE* variable and is relative to the drawing units in model space (or in paper space when lines are created there). In other words, if *PSLTSCALE* is 0, the dashed lines would always have the same number of dashes for one *LTSCALE* setting. However, when the same lines are viewed in different viewports in different scales, the linetypes appear different lengths as in Figure 32-14.

**FIGURE 32-14**

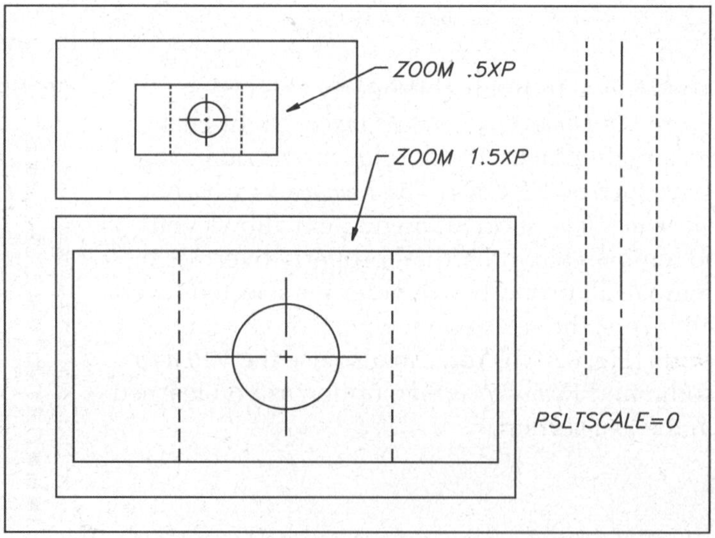

**TIP**

A *PSLTSCALE* setting of 1 (the default setting) is recommended when you are viewing the same geometry in multiple viewports at different scales. A *PSLTSCALE* setting of 1 makes all non-continuous lines appear to have the same linetype scaling, regardless of viewport scale (or *Zoom XP*) settings. Technically, *PSLTSCALE* set to 1 displays all non-continuous lines relative to paper space units so that all lines in paper space and in different viewports appear the same. Therefore, the model geometry in viewports can have different viewport scales (or *Zoom XP* factors), yet the linetypes display with the same scale (Fig. 32-15).

**FIGURE 32-15**

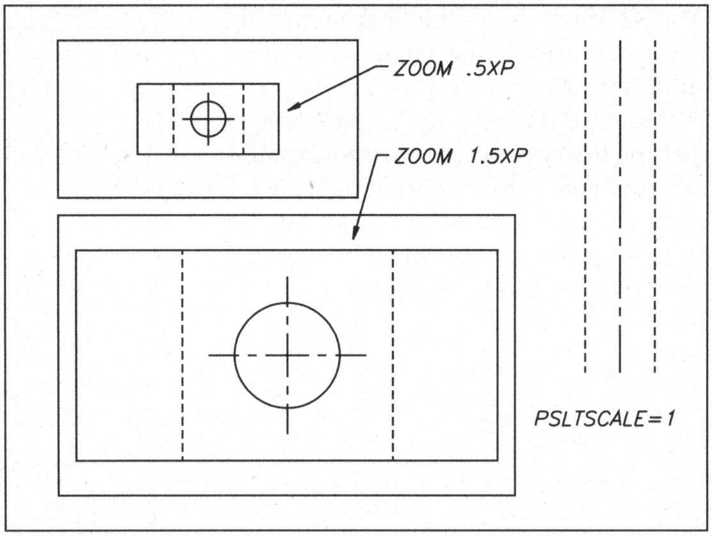


Set *PSLTSCALE* at the Command prompt by entering *Psltscale* and typing 1 or 0. *PSLTSCALE* can also be set using the *Linetype* command which produces the *Linetype Manager* (Fig. 32-16, lower left). You must use the *Show Details* button for this feature to appear. A check appearing in the *Use paper space units for scaling* checkbox sets *PSLTSCALE* to 1. The *Global scale factor* represents the *LTSCALE* value and the *Current object scale* represents the *CELTSCALE* value (see Chapter 12 for more information). If *PSLTSCALE* is changed, a *Regenall* must be used to display the effects of the new setting.

**FIGURE 32-16**

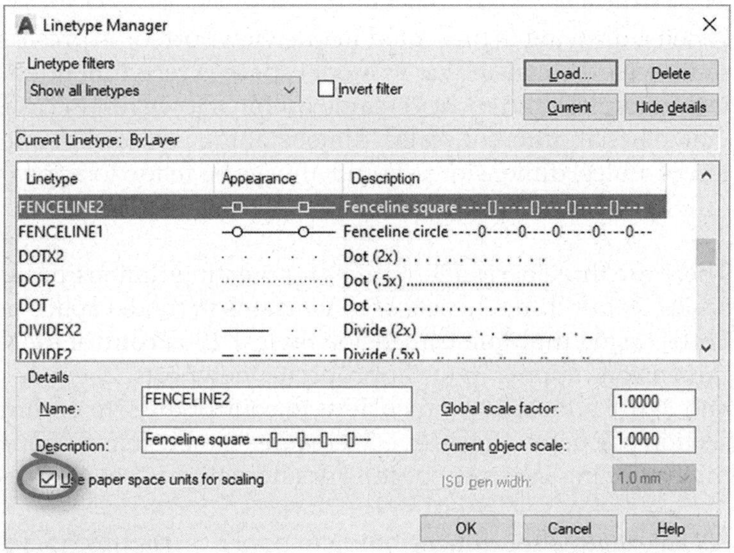


As an example, consider the crankshaft drawing (Fig. 32-17) that appears in its entirety in the top viewport at 1:1 and as a detail in the lower viewport at 2:1. Notice that the hidden line dashes are almost imperceptible in the top viewport, whereas the hidden lines in the lower viewport are much clearer. In this case, *PSLTSCALE* = 0, so the linetype scale is proportional to the model space geometry, no matter at what scale it is viewed.

**FIGURE 32-17**

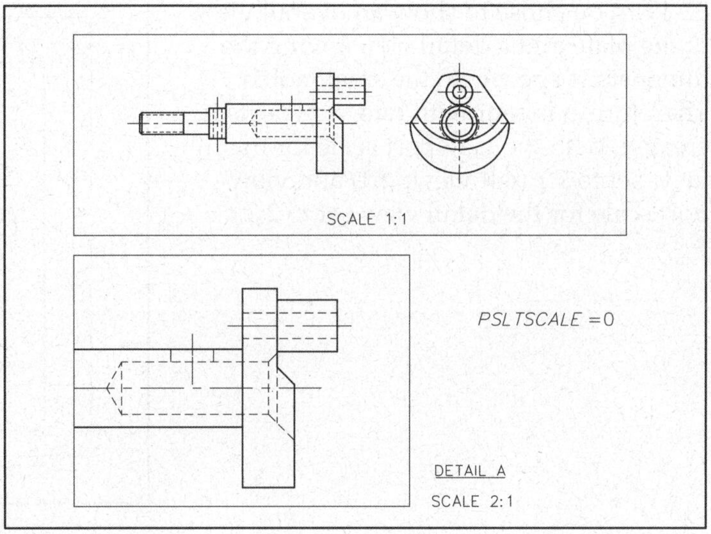

In contrast, when *PSLTSCALE* is set to 1 (Fig. 32-18), hidden line dashes in both viewports appear the same—they are automatically adjusted for the viewport scale. Another way to grasp this is that the linetype scale is proportional to paper space units, not model space units. In most cases where a layout contains multiple viewports, a *PSLTSCALE* setting of 1 is more practical.

**FIGURE 32-18**

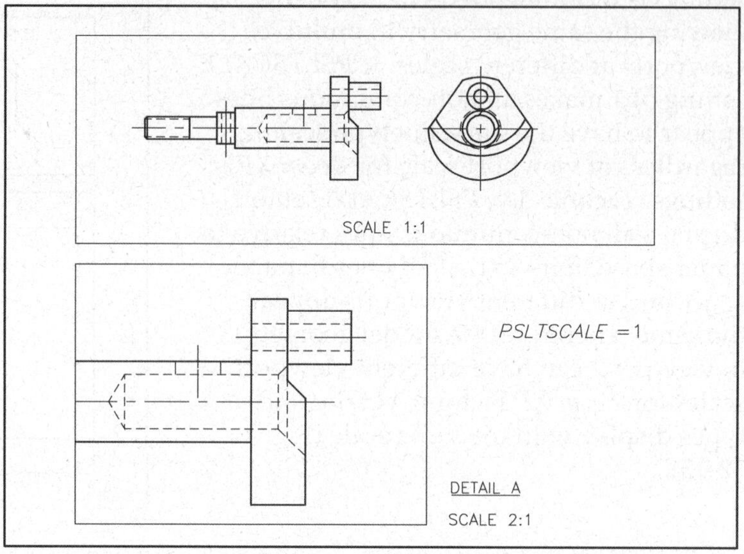


## Dimensioning in Paper Space

When a drawing is presented in one viewport, or is presented in more than one viewport but at the same scale, dimensioning in model space is recommended. In these cases, dimensioning in model space is the simplest and most versatile option. However, in cases where a drawing is displayed in multiple viewports at different scales, dimensioning is more complex. Even though the geometry appears in different scales, dimensions should always be made to appear at the same size (dimension text should typically be .125" or 6 mm).

There are three basic alternatives for creating dimensions displayed in multiple viewports at different scales: 1) put different sets of dimensions in model space on different layers and with different *Overall Scales* (using multiple Dimension Styles), then control the viewport-specific layer visibility so the correct dimensions appear in the appropriate viewports, 2) create dimensions in paper space for each viewport, and 3) use annotative dimensions to control the size of dimensions that appear in each viewport. This section discusses alternative 2, creating dimensions in paper space (see Chapter 29 for basic concepts on dimensioning in paper space). For alternative 3, see "Annotative Objects" later in this chapter.

As an example for dimensioning in paper space, assume you created a mechanical drawing of a plate to be stamped (Fig. 32-19). You chose to show an overall view of the plate and a detail view giving the dimensions specific to the small notch. Therefore, a layout with two viewports is created, with the viewport scale for the full view set to 1:1 (left viewport) and viewport scale for the detail view set to 2:1.

**FIGURE 32-19**

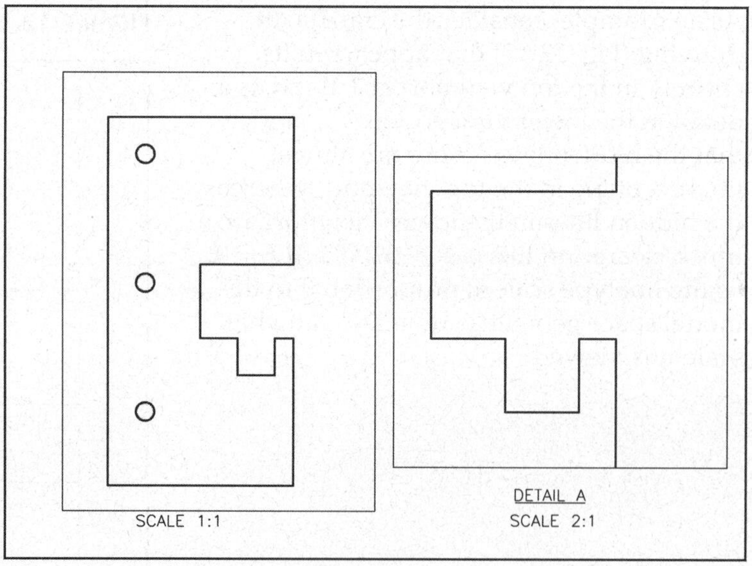

Assume you then activate the *Model* tab and begin to dimension the object as you would normally. Set the *Overall Scale* for dimensions based on your drawing scale factor so the dimensions are correctly sized for the viewport displaying the entire drawing. Checking the layout to see how the first few dimensions look, you notice that the dimensions appear in both viewports, and those in the detail view are also scaled 2:1 with the geometry (Fig. 32-20).

**FIGURE 32-20**

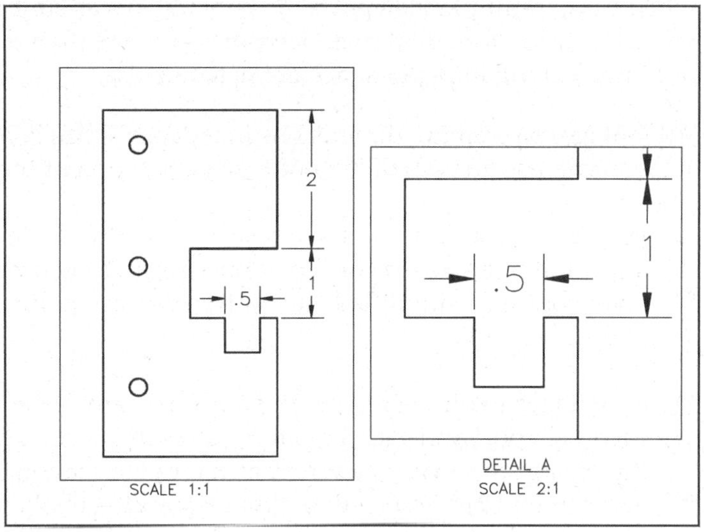


As one solution to the problem, erase all the model space dimensions. Then, in the layout, begin dimensioning in paper space with the dimensions associated with (attached to) the model space geometry (Fig. 32-21). You can use the same dimension style as before with an *Overall scale* of 1, since the layout will be printed 1:1. All dimensions have the same size, since they are all in the same space. Note that it appears that some dimensions are actually inside the viewport while others fall outside the viewport (when the viewport layers are frozen, this is not noticeable). In addition, dimensions attached to the geometry in the detail view reflect the actual model space measurement even though the geometry is scaled differently. Dimensions for the notch and for the overall part are placed in the appropriate viewport.

**FIGURE 32-21**

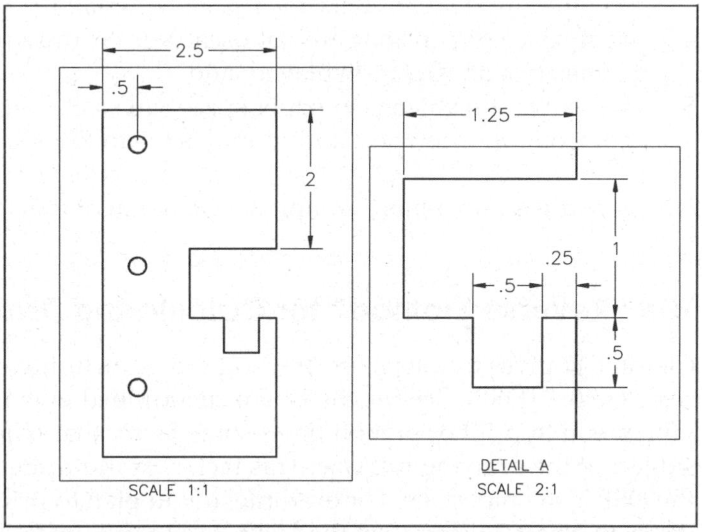


As the last step, the viewport borders layer is *Frozen* and detail area is indicated in the full view. Dimensioning in paper space allows you to dimension a drawing shown in different viewports at different scales (Fig. 32-22).

**FIGURE 32-22**

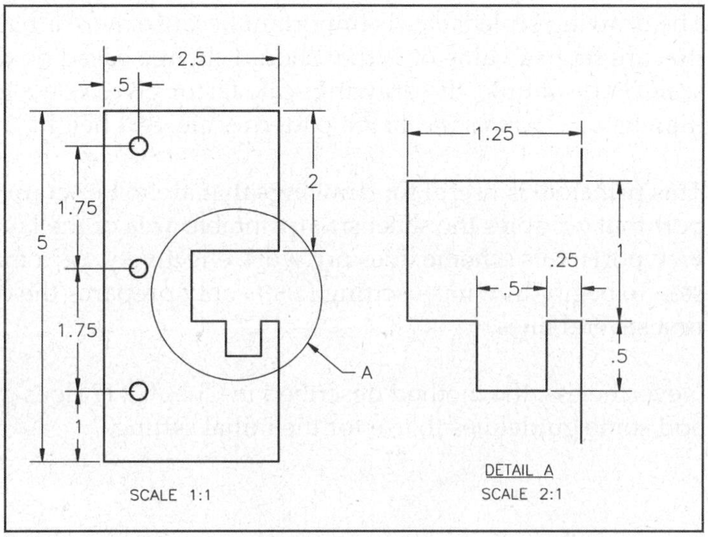

Although dimensioning in paper space appears to be a simple solution to the complexities that arise when using multiple viewports, it is not always as simple as the previous example. Keep in mind that you may need more than two viewports or more than one layout. Some advantages and disadvantages of dimensioning in paper space are listed below.

TIP

Advantages to creating dimensions in a paper space layout include:

1. you can use an *Overall scale* of 1, so calculation of the drawing scale factor (for dimensions, at least) is simpler;
2. you can attach dimensions to objects in multiple viewports at different scales and all dimensions appear at one size, so you do not necessarily need more than one dimension style for one layout; and
3. layer control is simplified, since all paper space dimensions appear in the layout and do not appear in other layouts.

TIP

Disadvantages to dimensioning in paper space include:

1. each viewport and each layout requires its own set of dimensions, so you may have to draw all dimensions for every viewport and for every layout, even when you show the same geometry;
2. you cannot *Copy* associative dimensions effectively from one viewport to another since copies are not associative;
3. you cannot copy associative dimensions effectively from one layout to another layout unless you copy the entire layout with the *Layout* command;
4. paper space dimensions are not useable if the drawing is used as an *Xref*, since only the model geometry of an *Xref* is displayed; and
5. associative dimensions in paper space can easily become disassociated if the geometry is moved or zoomed in the viewport so you may have to *Reassociate* the dimensions.

For more ideas and a more complete discussion of this topic, see "Applications for Multiple Viewports."

## The "Reverse Method" for Calculating Drawing Scale Factor

Chapter 12 gives the steps for drawing setup, including an explanation for calculating the "drawing scale factor" (DSF). Briefly, the DSF is determined when you set the *Limits*. Set the *Limits* to the sheet size you intend to print or plot on times some factor that will provide adequate space for you to draw the subject of the drawing full size. This factor, or multiplier, is the drawing scale factor. The reciprocal of the DSF is the plot scale. For example, if you plan to print on an 11 x 8.5 sheet and need enough space to draw an object 36" long, set the *Limits* to 11 x 8.5 times 4, or 44 x 34. So, the drawing scale factor is 4. The plot scale is then 1:4.

The drawing scale factor is important because there are many size-related variables and features (many that are set to a value of 1) that should change based on the size of the drawing and the intended plot scale. For example, the drawing scale factor gives you a guideline for the initial settings for *Ltscale*, dimension *Overall scale*, hatch pattern scale, text height, and so on.

This principle is useful for drawings that are to be set up in the model tab or in a layout with one viewport that occupies the sheet size (printable area of the layout). However, when you plan to have multiple viewports, this scheme does not work effectively. As a matter of fact, setting *Limits* is not a necessary step to begin drawing—setting *Limits* only prepares the drawing area with enough space to "see" those first several lines.

Nevertheless, the method described in Chapter 12 does give you an understanding of the relationships and some guidelines to use for the initial settings.

Regardless of whether you plan to use only one viewport or multiple viewports, you need some method to help you determine how to set sizes of such things as text, dimension *Overall scale* (assuming you intend to create text and dimensions in model space), hatch pattern scale, linetype scale, etc. A different method for calculating drawing scale factor, called the "reverse method," is given here. The reverse method works well when:

- you plan to display the drawing in multiple viewports at different scales;
- you plan on using one viewport to fill the layout but have not yet decided on a plot device or sheet size; or
- you are ready to begin drawing but do not yet know some information such as what plot devices or sheet sizes you will use, what sizes you will create the viewports, or what scales you will display the geometry.

## Steps for Calculating the Drawing Scale Factor by the "Reverse Method"

TIP

This method works by finding the viewport scale by trial-and-error using the *Viewport Scale* pop-up or *Viewport Scale* drop-down list. The viewport drawing scale factor is the reciprocal of the viewport scale.

1. Determine the approximate space you will need to create the subject of the drawing full size in the actual units. Set the *Limits* accordingly but only to provide enough space. *Limits* do not have to be in any proportion or size. *Zoom All.*

   (As an alternative, you do <u>not</u> have to set the *Limits*. You can instead, *Zoom* out, estimating about how much area you will need to "see" the entire drawing.)

2. Begin creating the model geometry. Draw at least an outline of the area you'll need, "box in" the views, or draw the parts of the geometry that occupy the majority of the drawing area. You can continue to draw the geometry, until completion if desired, or at least until you are ready to dimension, hatch, add text, or for some reason require a drawing scale factor.

3. Activate a *Layout* tab. Use *Pagesetup* to assign a plot device and a sheet size. Activate the *Viewports* toolbar (including the *Viewport Scale* drop-down list). Use *Vports* to create the viewport or viewports that you need.

4. Activate the viewport for which you want to display the full view of the geometry. *Zoom Extents*, or use real-time *Zoom*, to display the approximate area of the drawing you want in the viewport. Use one of the layer tools to set the desired viewport-specific layer visibility to ensure all objects that you intend to show appear in the viewport.

5. Use the *Viewport Scale* drop-down list or *Viewport Scale* pop-up to set (by trial and error) an appropriate scale for the viewport. You may need to *Pan* occasionally to "see" the desired area of the drawing. (If you are planning to create dimensions in model space or paper space, ensure you allow enough room either inside or outside the viewport.) Once the viewport scale is set, the drawing scale factor for that viewport is the reciprocal of the scale appearing in the *Viewport Scale* box. (Optional—once you have settled on the desired scale for the viewport, activate paper space, highlight the viewport border, and *Lock* the viewport using the *Properties* palette or *Lock/Unlock Viewport* toggle.)

6. To determine the <u>drawing scale factor for the drawing (or viewport)</u>, convert the viewport scale (calculate the reciprocal). Proportional scales (such as 1:2, 1:4, 1:8) can be converted directly by using the second number as the DSF (2, 4, 8, respectively). For feet and inch scales (for example, 1/4"=1'), multiply the denominator (bottom number of the fraction) by 12 (such as 4 x 12 = 48).

As an alternate method, activate paper space if not already active. Highlight the viewport border and use the *List* command to give the "Scale relative to Paper space: *n.nnnn* xp" (where *n.nnnn* represents a value). This value is the viewport scale as a decimal value. To find the drawing scale factor for the viewport, calculate 1/*n.nnnn*. This is the drawing scale factor for that viewport.

7. Use the drawing scale factor to set the size-related scales such as dimension *Overall scale* (if you are dimensioning in model space), *Ltscale* (if this is the main viewport), hatch pattern scale, text size (for model space text), and so on. For non-annotative dimensions and text you create for the viewport in model space, use layers named DIM-1, TEXT-1, or similar, so visibility can be controlled for this and other viewports.

This method can be used for any case when you plot from a layout, including when you have only one viewport that occupies the entire layout.

## ANNOTATIVE OBJECTS

As you already know, many objects in AutoCAD such as text, dimensions, hatch patterns, and attributes can have an annotative property. Unlike the normal drawing geometry (lines, arcs, circles, etc.), objects such as text, dimensions, hatch patterns, and block attributes are likely to require modifications to size when displayed at multiple scales in separate viewports. For example, when a floor plan or a mechanical part drawing includes one viewport displaying the entire drawing and other viewports displaying detail views, the text and dimensions, in order to be readable, would have to be displayed at different sizes in each viewport based on the viewport scale. Annotative text, dimensions, hatch patterns, and block attributes are designed specifically for this application.

This section discusses applications for displaying annotative objects in multiple viewports at different scales. Next, other applications that require multiple viewports are discussed including alternative solutions (using annotative and non-annotative objects) and comparisons between the alternatives. Previous chapters in this book have given you a foundation on using the individual annotative objects. If you need information on individual annotative objects, refer to the following chapters:

Chapter 12, Advanced Drawing Setup
Chapter 18, Text and Tables
Chapter 22, Block Attributes
Chapter 26, Section Views
Chapters 28 and 29 for information on annotative dimensions

### Setting Annotation Scale and Viewport Scale

#### Model Space Annotation Scale

Model space (the *Model* tab) has only an annotation scale. The model space annotation scale can be set using the *Annotation Scale* pop-up list in the *Model* tab (lower-right corner of the Drawing Editor) or by entering the *CANNOSCALE* system variable at the Command line.

Generally, you would begin drawing by setting the *Annotation Scale* (*CANNOSCALE* variable) to the scale you intend to print the drawing in the primary viewport (full view). In most cases, you would set the *Annotation Scale* to the reciprocal of the Drawing Scale Factor. Use the following formula, where AS is the *Annotation Scale* and DSF is the Drawing Scale Factor:

AS = 1/DSF

In this way, when you create annotative objects, they appear in the correct size based on the current annotation scale. For example, if you want to create annotative text that appears at .125" in the printed drawing, first create the text *Style*, check the *Annotative* property, and set the *Paper Height* to .125. If you then create text objects in model space, the text objects are created in the correct size based on the *Annotation Scale*. The text's paper height is multiplied by the annotation scale, or:

Paper Height x Annotation Scale = Model Space Text Height

If the text style *Paper Height* is set to .125, and model space *Annotation Scale* is set to 1:2, the resulting text objects would appear at .250 (twice as large when displayed or printed at 1:2).

Annotative dimension objects work similarly. If you want dimension text values to be printed at .125", set the dimension style to be *Annotative* (*Fit* tab) and the *Text Height* (*Text* tab) to be .125. The resulting dimensions' text in model space would appear at .250 for a model space *Annotation Scale* of 1:2.

NOTE: If a dimension style references a text style that has a fixed paper height, the dimension *Text Height* (*Text* tab) is also fixed to that height.

### Paper Space Viewport Scale and Annotation Scale

Paper space viewports have both a viewport scale and an annotation scale. There are controls for both scales, labeled *Viewport Scale* and *Annotation Scale*. Each viewport has its own *Viewport Scale* and *Annotation Scale*. When paper space is current, the annotation scale is always 1:1 and cannot be changed.

Set the *Viewport Scale* for a viewport to display the model space geometry at the desired scale. Since you plot from paper space at 1:1, the *Viewport Scale* for the viewport is normally the same as the plot scale for the geometry. In most cases, both the *Viewport Scale* and the *Annotation Scale* should be set to the same value, so setting the *Viewport Scale* from the status bar synchronizes both scales to the *Viewport Scale*.

#### *Viewport Scale Sync Button*

Typically, you want the *Viewport Scale* and the *Annotation Scale* to be the same, or synchronized. AutoCAD provides a *Viewport Scale Sync* button just to the right of the *Viewport Scale* pop-up list. Once this button is selected, any change made to the *Viewport Scale* using the pop-up list on the Status bar automatically changes the *Annotation Scale* to match. The icon displays the synchronization symbol and the tool tip reports "Viewport scale is equal to Annotation Scale."

NOTE: Initially (when the button is a light blue color and displays the tool tip "Viewport scale is not equal to Annotation Scale – Click to synchronize"), selecting the *Viewport Scale Sync* button changes the *Viewport Scale* to match the *Annotation Scale*. Next, select an appropriate scale from the *Viewport Scale* pop-up list.

You can also change the *CANNOSCALE* value of the current viewport "manually" at the Command line. In this case, the *Annotation Scale* changes but the *Viewport Scale* remains unchanged. The *CANNOSCALE* value can only be set to a scale that appears on the pop-up list and must be typed exactly as shown.

If a paper space viewport is unlocked and you *Zoom* in or out, the *Viewport Scale* changes but the *Annotation Scale* remains unchanged. In most cases, you should then reset the *Viewport Scale* after making any edits to synchronize both values. You can also lock viewports using the *Lock/Unlock Viewport* toggle located to the left of the *Viewport Scale* control. Locking prevents accidentally changing the viewport scale.

## Creating and Displaying Annotative Objects in Multiple Viewports

As a simple example for displaying annotative objects in multiple viewports at different scales, the previous mechanical stamped plate example is used. You may want to review the "Steps for Creating and Viewing Basic Annotative Dimensions" in Chapter 29 since this example generally follows those steps.

Assume that most of the drawing geometry is already created in model space but text and dimensions have not yet been created. Using the *Style* command, a new text style is created with the *Annotative* property and the *Paper Text Height* value is set to the desired value. Then using the *Dimension Style Manager*, a *New* annotative dimension style is created by checking the *Annotative* box in the *Fit* tab. Additional settings are made to the dimension style, such as setting the *Text Height* for dimensions to the actual paper space height (*Text Height* is already set if the current *Text Style* is used for the dimension style).

**FIGURE 32-23**

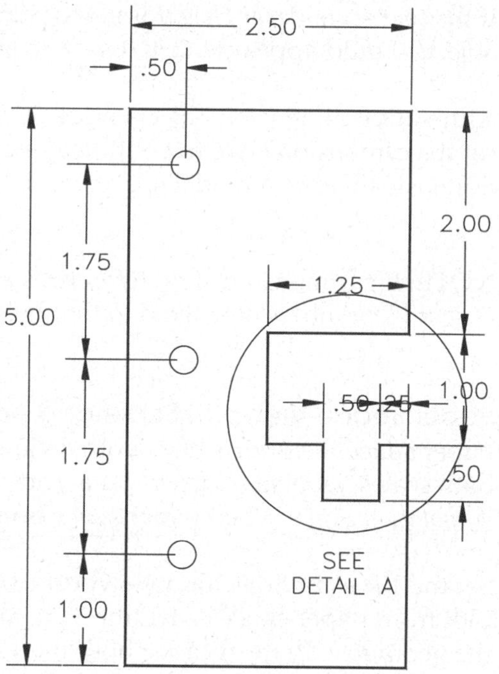


In the *Model* tab, the *Annotation Scale* for model space is set using either the *Annotation Scale* pop-up list (lower-right) or using the *CANNOSCALE* variable. The reciprocal of the "drawing scale factor" (DSF) is used as the annotation scale. Next, all the needed dimensions are created, including those that eventually appear only in the detailed view (Fig. 32-23). As an alternative, dimensions for detailed views could be applied at a later time.

The *Text* command is used to create two lines of annotative text ("SEE DETAIL A") to reference the detail view. To designate the detail area, a circle is drawn on a separate layer that can be controlled for viewport-specific visibility at a later time. (NOTE: The standard method for designating a detail view is to use only a single letter label and leader; however, text is used in this case to explain the annotative features.)

In the layout tab, the desired viewports are created. In this case, one viewport is intended to show the entire model space geometry and a second viewport is created to display a detail view. Next, the *Autoscale* button is enabled (*Add scales to annotative objects when scale changes – On*) located on the Status bar just to the left of the *Viewport Scale* and *Viewport lock* buttons. The desired scale for viewports (*1:1* in this case) is set by double-clicking inside a viewport and using the *Viewport Scale* pop-up list or *Viewports* toolbar. In our example, both viewports have been set to 1:1 and, therefore, display the geometry, text, and dimensions at the same size in both viewports (Fig. 32-24).

**FIGURE 32-24**

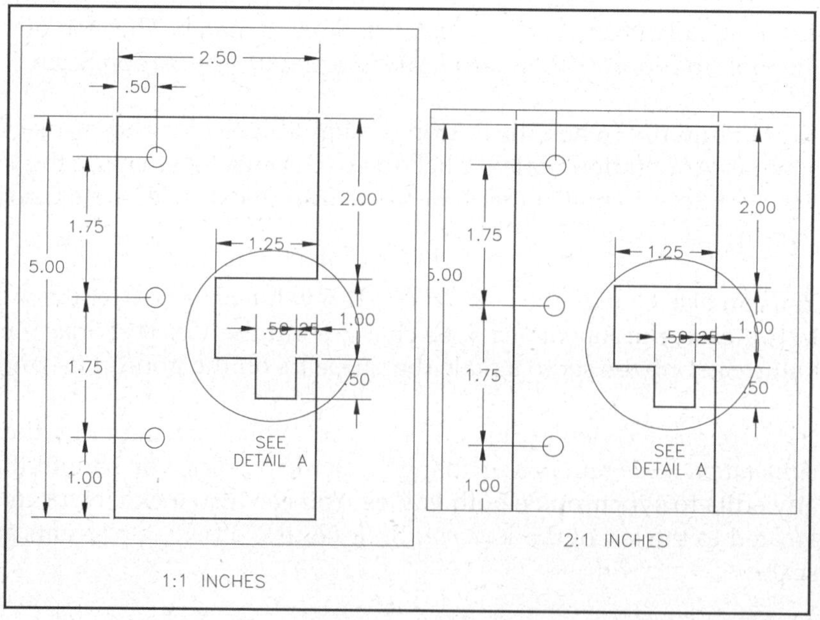

Next, the scale for the detail view (right viewport) is set to 2:1 by the same method (Fig. 32-25). Note that the geometry is now displayed at 2:1 scale. The dimensions and text, however, have automatically changed to display at the appropriate size based on the viewport scale. Because the *Add scales to annotative objects when the annotative scale changes* toggle was set, the dimensions were adjusted for the viewport scale.

FIGURE 32-25

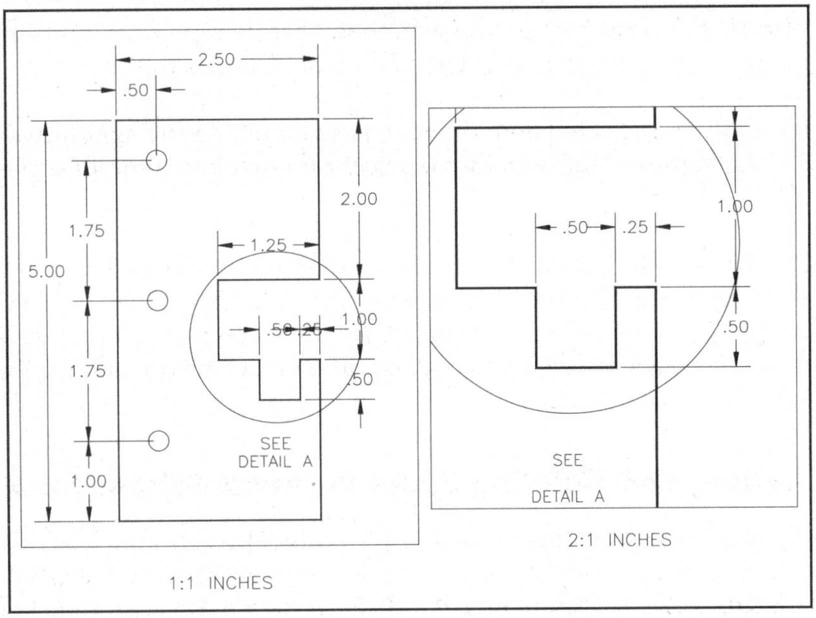


Note also that a few other changes have occurred to the annotative objects based on the new scale. The location of the dimensional values and arrowheads for the ".50" and ".25" dimensions have been automatically adjusted correctly, but the spacing between the two lines of *Text* ("SEE" and "DETAIL A") have not been adjusted for scale (an annotative *Mtext* object would adjust).

### Scale Representations—Multiple Annotative Object Copies

Technically, the original annotative text and dimension objects have not actually changed size. Instead, new annotative objects have been created as copies of the originals but in the new scale. AutoCAD uses the term "scale representations" to describe the multiple annotative object copies since each copy represents a different scale.

In our example, the geometry was set to be displayed at two different scales; therefore, two sets of annotative objects (or two scale representations) were created, one for each scale. AutoCAD automatically displays only the appropriate set of annotative objects based on the scale specified for the viewport.

For example, Figure 32-26 displays both of the ".50" scale representations created for the notch height. The larger dimension object was created to display at 1:1 scale and the smaller dimension object was created to display at 2:1 scale (since the objects are displayed twice as large in the right viewport, the dimensions should be half as large).

FIGURE 32-26

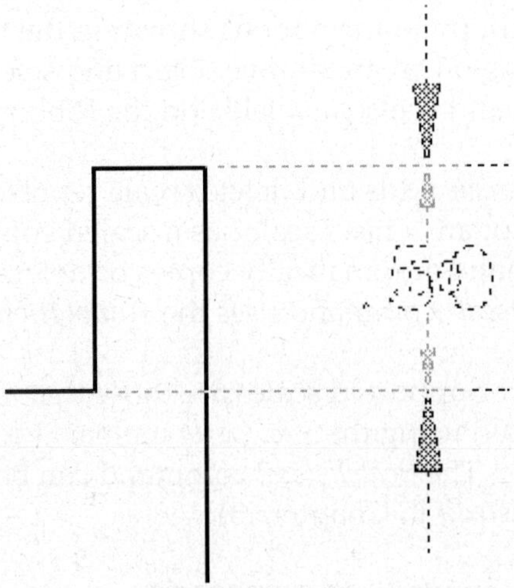


You can actually see all scale representations of annotative objects that have been created by selecting the objects. If the *SELECTIONANNODISPLAY* system variable is set to 1, you can highlight an annotative dimension object and display all copies of the annotative object. See "*SELECTIONANNODISPLAY*" next.

These multiple annotative objects, or multiple scale representations, are also known as "object scales." You can add or delete these annotative objects for particular scales using the *Objectscale* command (see "*Objectscale*").

### *SELECTIONANNODISPLAY*

This system variable controls whether all scale representations of annotative objects are displayed when an annotative object is selected. The settings are as follows:

0 Only the selected annotative objects matching the annotative scale for the viewport are displayed.
1 All copies of the annotative objects are displayed for all scales.

When *SELECTIONANNODISPLAY* is set to 1, the annotative objects in the current annotative scale are displayed in full intensity while the other annotative object copies are displayed also, but in a "dimmed" state (note the "dimmed" small arrowheads in Fig. 32-26). The intensity of the faded display is set by the *XFADECTL* variable. The *XFADECTL* value can range from 0 to 90 and represents the fading percentage (0 is not faded, or full intensity, and 90 represents 90% faded). The default *XFADECTL* setting is 50.

## Adding and Deleting Scale Representations for Annotative Objects

Annotative objects can have multiple scale representations as displayed in Figure 32-26. A scale representation is one copy, or scaled version, of a multi-scaled annotative object. Once the original dimension (original scale representation) is created using a dimensioning command, additional scale representations (copies of the dimension for new scales) can be created automatically by two methods:

1. toggle on the *Add scales to annotative objects when the annotative scale changes* (or set the *ANNOAUTOSCALE* variable) and select a new *Viewport Scale*; or
2. use the *Objectscale* command.

The *Objectscale* command can be used to create new scale representations or delete existing ones.

### *Objectscale*

| Ribbon | Menu Bar | Command (Type) | Alias (Type) | Shortcut |
|---|---|---|---|---|
| *Annotate*<br>*Annotation Scaling*<br>(icons) | *Modify*<br>*Annotative*<br>*Object Scale* | *Objectscale* | ... | ... |

In addition to the options shown in the table above, the *Objectscale* command can be invoked by right-clicking on an annotative object and selecting *Annotative Object Scale* from the shortcut menu. The command is also available on the Ribbon.

*Objectscale* adds and deletes scale representations for existing annotative objects. For this command, think of an "object scale" as a scaled copy, or scale representation, of an annotative object. The *Objectscale* command automatically copies or deletes selected annotative objects for a particular scale. The *Objectscale* command uses the *Annotation Object Scale* dialog box to add and delete supported scales.

NOTE: Supported scales are those that are listed in the scale list for the drawing. The scale list can be accessed using the *Viewports* toolbar, *Viewport Scale* list, *Annotation Scale* list, plot scale list, and *Properties* palette. The *Scalelistedit* command can be used to add or delete items in the scale list for the drawing (see *Scalelistedit* in Chapter 13).

Command: ***OBJECTSCALE***
Select annotative objects: **PICK**
Select annotative objects: **Enter**
Command:

At the "Select annotative objects:" prompt, select the dimension objects you want to add or delete. Once annotative objects are selected, the *Annotation Object Scale* dialog box appears (Fig. 32-27).

To delete a scale representation for the selected objects, simply highlight the scale from the list and select the *Delete* button. This procedure is confusing because scales are not actually deleted. Instead, the selected annotative objects that appear in the scale are deleted.

To add a scale, select the *Add* button to produce the *Add Scales to Object* dialog box (not shown). This dialog box simply contains all the supported scales for the drawing (see previous NOTE on supported scales). Adding a scale actually makes a copy, or scale representation, for each selected annotative object at the selected scale.

**FIGURE 32-27**

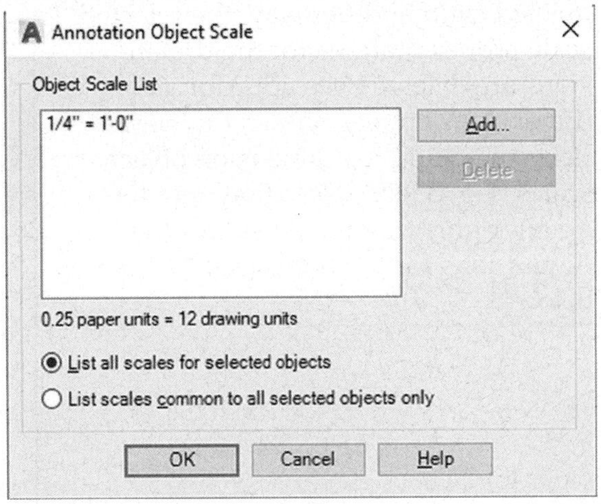

As an alternative to using the *Objectscale* command or button, you can use the *Add Current Scale* and *Delete Current Scale* buttons available on the Ribbon. These buttons are handy options of the *Objectscale* command. *Add Current Object Scale* adds a scale representation of a selected object at the current annotation scale (*CANNOSCALE*) value. *Delete Current Object Scale* allows you to select a scale representation object to delete.

An alternative to using the *Objectscale* command is to produce the *Properties* palette for selected annotative objects (Fig. 32-28). Selecting the browse ("...") button just to the right of the current *Annotative* scale edit box produces the *Annotative Object Scale* dialog box shown in Figure 32-27.

**FIGURE 32-28**

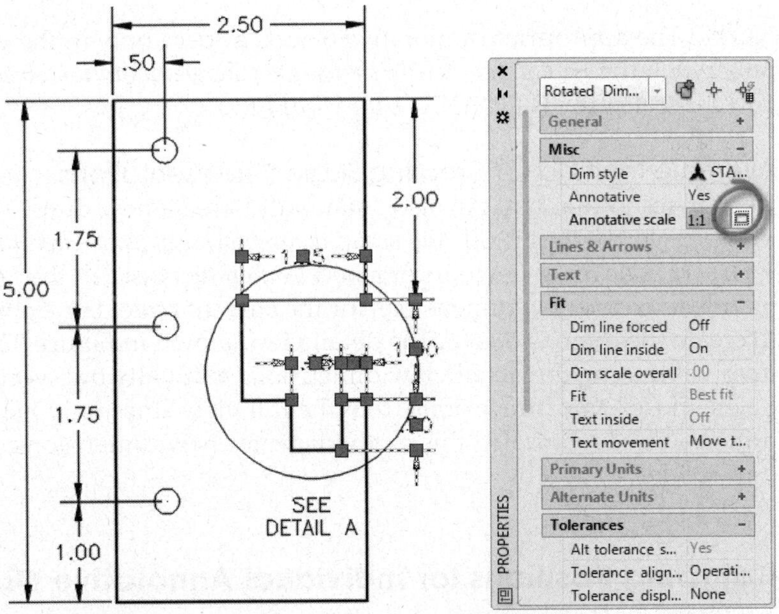


### Example for Deleting Object Scales

As an example for deleting object scales, examine our previous example of the stamped plate. As evident by examining previous Figure 32-25, it is not necessary to display all the dimension and text objects in both viewports. That is, the dimensions for the notch detail can be deleted from the left viewport displaying the full plate. In addition, the "SEE" text object can be deleted from the detail view. Since a single annotative object can have multiple copies (or scale representations), one for each time the object is displayed in a different scale, it is possible to delete annotative text and dimension objects for specific viewports.

The dimensions that are needed only for the detail view can be deleted for the full view. Either the *Objectscale* command or *Properties* palette is used to delete the needed dimension objects (object scales). Note the selection of dimensions (shown with grips) in Figure 32-28.

Once the selected dimensions are deleted (said another way, the selected scale representations for the dimensions are deleted), the appropriate dimensions appear only in the viewports matching the dimension object scales (Fig. 32-29). Note that only the detail (right) viewport displays the dimensions for the notch (see NOTE below).

**FIGURE 32-29**

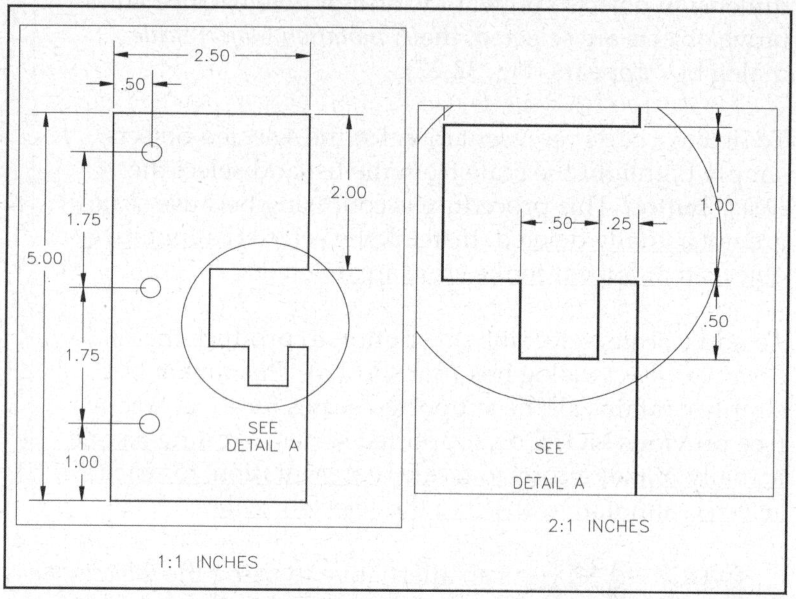


Next, text objects for the detail view should be adjusted by deleting only the "SEE" text object's *object scale.* (The line "DETAIL A" can be used to label the view.) The *Objectscale* command is used again to select the "SEE" text object and delete the 2:1 text object scale representation. Using the same procedure, the vertical "2.00" dimension is removed from the detail view by deleting the dimension's 2:1 scale representation. Both of these annotative objects should then appear only in the full view (1:1 object scales).

NOTE: The appropriate annotative objects appear only in the viewports matching the object scales, but only when the *Annotation Visibility* toggle (*Show annotation objects*) is set to show annotative objects for current scale only (see "*ANNOALLVISIBLE*").

### Alternate Method for Creating Scale Representations

Keep in mind that you can also "manually" create new dimensions for a particular annotation scale. In other words, toggle off *Add scales to annotative objects when the annotative scale changes*, then set a new *Viewport Scale* for a viewport or a new *Annotation Scale* in the *Model* tab. Next, use dimensioning commands to create new dimensions for the current scale. For example, this method could have been used to create the dimensions for the detail view shown in Figure 32-31 (right viewport). This method would involve creating the detail view dimensions manually but would avoid having to use *Objectscale* to delete those same dimensions from the full view since only one scale representation exists for the new dimensions. Technically, this method creates new dimensions, not scale representations of existing dimensions.

## Adjusting Positions for Individual Annotative Object Scale Representations

A problem with annotative objects is that copies of the original objects (scale representations) that are created for different scales can appear crowded or spaced too far apart. This is common with annotative dimensions that are created to match a wide range of scales. Although the dimension text, arrowheads, and gaps are adjusted automatically, the distance between successive stacked dimension lines is not adjusted. Also, when the dimension text and arrowheads are created at a larger size (to display at a smaller scale) than the original annotative dimension, they can be forced outside of the extension lines and interfere with other dimensions. In addition, as is the case in our example, multiple single-line text objects are not adjusted for line spacing.

Since annotative object scale representations are actually copies of the original object, one for each object scale created for the drawing, it is possible to adjust the position of individual copies (each scale representation). This is accomplished by using the object grips to move the objects. However, only the one annotative object copy (scale representation) that matches the current *Annotative Scale* is affected by the object's grips. Therefore, whether you adjust the annotative objects in the *Model* tab or in a paper space viewport, the correct *Annotation Scale* must be set to match the target object.

TIP

### Adjusting Annotative Objects in the *Model* Tab

Using the *Model* tab to select and reposition individual annotative object scale representations allows you to zoom as needed without having to zoom in a paper space viewport and change the viewport scale. Activate the *Model* tab, then set the *Annotation Scale* to match the desired scale representation of the object you want to adjust.

### Adjusting Annotative Objects in Paper Space Viewports

Adjusting annotative objects in paper space viewports avoids having to change the *Annotative Scale*; however, zooming changes the viewport scale. Activate the viewport that matches the scale representation of the individual object you want to adjust. If you must *Zoom*, you should then reset the *Viewport Scale* after adjustment.

### Example for Adjusting Locations for Individual Annotative Objects

In our example, the "DETAIL A" text object can be used to label the view; however, it must be moved upward so it will be visible when the drawing is panned in the viewport. In this case, the *Model* tab is made active and the *Annotation Scale* is set to *2:1* to match the desired text object's scale. Zooming in, activating the grips, and dragging reveals the small copy of the annotative text object (the 2:1 scale representation) as it is adjusted to the desired position (Fig. 32-30).

**FIGURE 32-30**

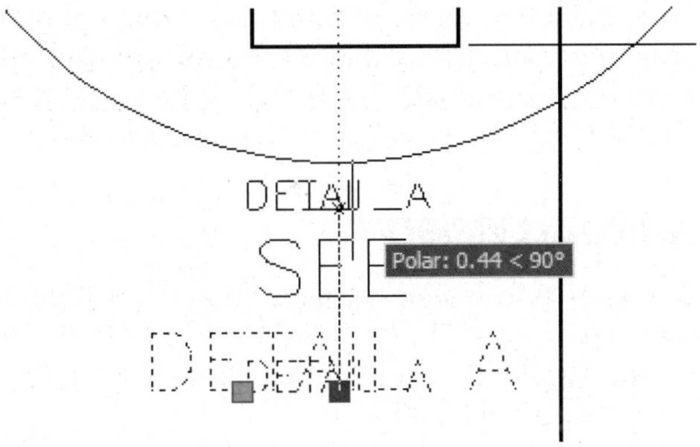


The stamping plate drawing is shown in the final form in Figure 32-31. Note the adjustment to the "DETAIL A" text object so it appears as a label for the detail view (right viewport) while the "SEE" text object is deleted for the 2:1 scale. Also note that only the appropriate dimension objects appear in the appropriate viewports based on viewport scale. The layer for the circle around the detail area in the left viewport is frozen for the right viewport and the viewport layer is frozen.

**FIGURE 32-31**

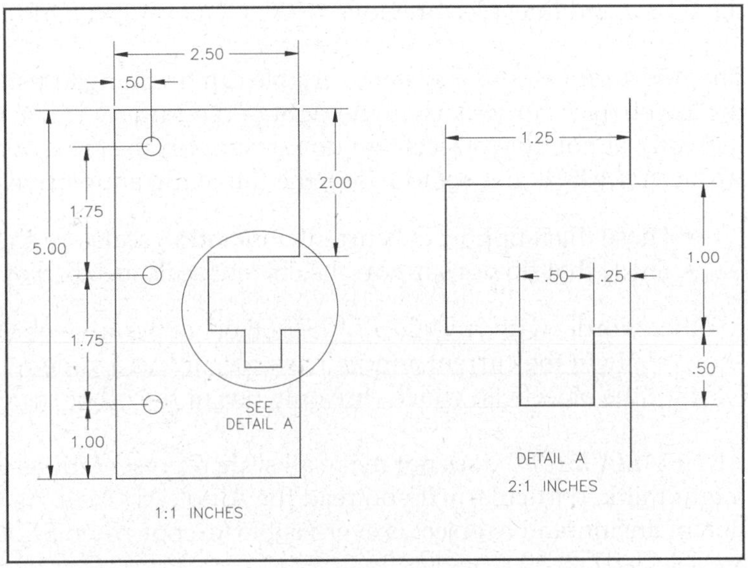

### *Annoreset*

| Ribbon | Menu Bar | Command (Type) | Alias (Type) | Shortcut |
|---|---|---|---|---|
| *Annotate Annotation Scaling Sync Scale Positions* | *Modify Annotative Object Scale > Synchronize Multiple-scale Positions* | *Annoreset* | ... | ... |

If you have adjusted the position of individual annotative objects, such as annotative dimensions or annotative text objects, but want to reposition them to the same location, you can use *Annoreset*. Use the *Annoreset* command to reset the location of all scale representations for an annotative object to that of the current scale representation.

Command: ***ANNORESET***
Reset alternate scale representations to current position
Select objects: **PICK**
Select objects: **Enter**
Command:

Note that all scale representations for an annotative object are repositioned to the location of the current scale representation. In other words, multiple copies of the annotative object are moved to the current location of the object that matches the current *Annotative Scale*. Therefore, before using *Annoreset*, set the *Annotative Scale* to match the target object's scale.

## *ANNOALLVISIBLE*

There are two system variables that control the visibility of annotative objects that have multiple scale representations: *SELECTIONANNODISPLAY* and *ANNOALLVISIBLE*. When these two variables are off (set to 0), only the annotative objects with scale representations matching the current *Annotative Scale* for the viewport are visible.

*SELECTIONANNODISPLAY*, when set to 1 (on), allows you to see all scale representations for annotative objects when the objects are selected. This is the only way to see all scale representations for a selected annotative object, but the object must be selected. *SELECTIONANNODISPLAY* must be typed at the Command line. (See previous "*SELECTIONANNODISPLAY*.")

The *ANNOALLVISIBLE* system variable can be changed using the *Annotation Visibility* toggle in the lower-right corner, just to the right of the *Viewport Scale* list. When *ANNOALLVISIBLE* is set to 0 (off), annotative objects that do not support the current annotation scale are hidden. When *ANNOALLVISIBLE* is set to 1 (on), the following annotative objects are shown:

- Those that support the current annotation scale are shown in the current scale only.
- Those that do not support the current scale are displayed in one other scale only.

In other words, with *ANNOALLVISIBLE* on, annotative objects having multiple scale representations are shown only in the current annotative scale, unless a scale representation for the current scale does not exist for the objects, in which case only one of the other scale representations is shown.

*ANNOALLVISIBLE* does not make all scale representations for all annotative objects visible, as one might think, particularly if you read the AutoCAD *Help*. As a matter of fact, only one scale representation of an annotative object is ever visible (except when *SELECTIONANNODISPLAY* is on and objects are selected).

For example, *Annotation Visibility* (*ANNOALLVISIBLE*) is turned on for the stamped plate drawing in Figure 32-32. Compare this figure with previous Figure 32-31 where *Annotation Visibility* is turned off. Note that in this figure the left viewport (current scale 1:1) displays the dimensions for the detail view (2:1 scale representations), and in the right viewport (current scale 2:1) the arrowhead for the vertical "2.00" dimension is displayed (1:1 scale representation).

The *ANNOALLVISIBLE* setting is saved individually for model space and each layout. An annotative object's layer must be *On* and *Thawed* for the objects to display.

**FIGURE 32-32**

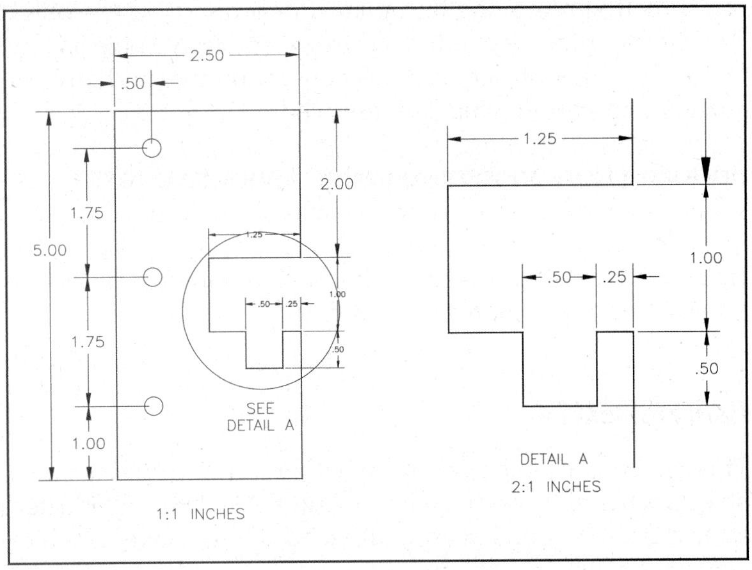


In general, you should turn off annotation visibility, especially when the drawing is complete or ready to print or plot. However, when you want to add or delete scales to existing annotative objects or inspect drawings created by others, it is helpful to see other scale representations for the annotative objects.

## Annoupdate

| Ribbon | Menu Bar | Command (Type) | Alias (Type) | Shortcut |
|---|---|---|---|---|
| ... | ... | *Annoupdate* | ... | ... |

*Annoupdate* can be used to update non-annotative objects, such as dimensions and text created in older releases of AutoCAD, to annotative objects. Technically, *Annoupdate* causes dimension, text, attribute, and table objects to match the annotative property of the object's style, whether the style is annotative or non-annotative.

Command: ***ANNOUPDATE***
Select objects: **PICK**
Select objects: **Enter**
*n* found *n* was updated
Command:

Before existing objects can be updated to annotative objects using *Annoupdate*, the objects' style must first be set to *Annotative*, then the objects can be updated to the new style. For example, to update non-annotative dimensions to annotative dimensions, see the following steps.

### Updating Non-annotative Dimensions to Annotative Dimensions

This process is useful especially for updating older AutoCAD drawings to use the annotative features.

1. Produce the *Dimension Style Manager*. Select the dimension style from the *Styles* list that was used to create the existing dimensions.
2. In the *Fit* tab, change the *Scale for dimension features* to *Annotative*. Select *OK*, make the style *Current*, then *Close* the *Dimension Style Manager*.
3. In model space, invoke *Annoupdate*. Using a window or crossing window, select all existing dimensions.

On the other hand, annotative dimensions and other annotative objects can be converted to non-annotative by first removing the *Annotative* property of the objects' style, then using *Annoupdate*. When the *Annotative* property is removed from annotative objects (annotative objects are updated to a non-annotative style), the object becomes non-annotative and any alternate scale representations, other than the current annotation scale, are removed.

#### Updating Non-annotative text to Annotative Text

When an existing non-annotative text object in the drawing is updated to an annotative text style, the text becomes annotative and supports the current annotative scale. If the text style has a fixed *Paper Height*, the text object is set to that height. However, if the text style's *Paper Height* is set to 0, the size of the existing text remains the same.

### *SAVEFIDELITY*

This system variable specifies whether or not drawings are saved with visual fidelity for annotative objects when they are viewed in AutoCAD 2007 and earlier releases. Visual fidelity affects how multiple scale representations of annotative objects are saved. You can set this variable at the Command line using the following values.

1 Visual fidelity is maintained.
0 Visual fidelity is not maintained.

Alternately, you can use the *Options* dialog box, *Open and Save* tab, *Maintain Visual Fidelity for Annotative Objects* check box to turn visual fidelity on or off.

If you work primarily in model space, do not use annotative objects, or do not exchange drawings with others, it is recommended that you turn off visual fidelity (set *SAVEFIDELITY* to 0). However, if you use annotative objects and need to exchange drawings with others, then visual fidelity should be turned on (*SAVEFIDELITY* set to 1).

If visual fidelity is on and the drawing is saved, annotative objects are separated and each set of scale representations is saved on a separate layer in unnamed blocks. The layers are named based on the objects' original layer name but appended with a number. Therefore, visibility for annotative objects opened in AutoCAD 2007 and earlier releases can be controlled using layer visibility controls.

If visual fidelity is off, the drawing is saved, and then opened in AutoCAD 2007 or earlier, only one set of objects is displayed in model space based on the current annotation scale when the drawing was saved. More annotation objects at different scales may be displayed in model space and in paper space viewports if *ANNOALLVISIBLE* was on when the drawing was saved.

## Linetype Scales for Annotative Drawings

There are three linetype scaling variables that can affect non-continuous lines in AutoCAD drawings: *LTSCALE*, *PSLTSCALE*, and *MSLTSCALE*.

*LTSCALE* — Linetype Scale. The global size-related variable that affects all lines and affects the other two variables. *LTSCALE* can be set to any value and is retroactive when changed. The *LTSCALE* value is never changed by the setting of the other two variables.

*PSLTSCALE* — Paper Space Linetype Scale. This variable determines if non-continuous lines displayed in paper space are automatically scaled based on the viewport scale (*Viewport Scale*). *PSLTSCALE* can be set to 1 (on) or 0 (off). (See previous discussion this chapter, "*PSLTSCALE*.")

*MSLTSCALE* Model Space Linetype Scale. This variable determines if non-continuous lines displayed in model space are automatically scaled based on the *Annotation Scale* in the *Model* tab. *MSLTSCALE* can be set to 1 (on) or 0 (off). (See "*MSLTSCALE*" next.)

## MSLTSCALE

This variable controls the scaling of non-continuous lines only in the *Model* tab display based on the *Annotation Scale* in model space and the *LTSCALE* setting.

0 Linetypes displayed in the *Model* tab are not scaled by the model space annotation scale.
1 Linetypes displayed in the *Model* tab are scaled by the model space annotation scale.

The default *MSLTSCALE* setting is 1 (on) for current AutoCAD template drawings. *MSLTSCALE* is set to 0 when you open drawings created in AutoCAD 2007 and earlier.

Basically, this variable (when on) causes the *Model* tab display of non-continuous lines to change based on the *Annotation Scale* in model space. The *LTSCALE* value is not changed by the *Annotation Scale*. Instead, the line spacing is displayed at the *LTSCALE* value divided by the *Annotation Scale* in model space. For example, with *MSLTSCALE* set to 1 (on), an *LTSCALE* setting of 1.00, and *Annotation Scale* in model space set to 1:2, linetypes would be scaled to 2 (twice as large). In other words, with *MSLTSCALE* on, the display in the *Model* tab would be:

Linetype scale = *LTSCALE* / *Annotation Scale*

Using the *MSLTSCALE* variable in this manner accomplishes the same results in model space as setting the *LTSCALE* x drawing scale factor (the recommended method), but with one major shortcoming. That is, *MSLTSCALE* has no effect on the display of non-continuous lines in paper space viewports. Therefore, even if *MSLTSCALE* is on, you must still calculate and set the *LTSCALE* setting as you would normally to control the linetype spacing in paper space. Additionally, when *MSLTSCALE* is on and you change the *Annotation Scale* in the *Model* tab, you must then *Regen* to see the new linetype setting for existing lines.

TIP

Using the *MSLTSCALE* feature is not a substitute for calculating and setting a linetype scale as previously recommended (*LTSCALE* = Drawing Scale Factor). Turning off *MSLTSCALE* and using *LTSCALE* only to control scaling of linetypes in model space is a more efficient and predictable strategy. In addition, when *MSLTSCALE* and *PSLTSCALE* are both on, it can be confusing since you see the actual *LTSCALE* only when *Annotation Scale* and *Viewport Scale* are set to 1:1.

A *MSLTSCALE* setting of 0 is recommended for most applications. However, setting *MSLTSCALE* to 1 (on) might be useful for applications in which you must use a very large zoom ratio and seeing linetypes is important. For example, if you want to work in model space to create and edit annotative objects for detail views to be displayed in paper space viewports in scales much different than the full view (such as 10:1 or greater) and seeing linetypes is important when zoomed in, *MSLTSCALE* may be useful.

### Recommended Procedure for Setting Linetype Scale for Multiple Viewport Applications

For most applications involving multiple viewports, the following process is recommended:

1. Set *LTSCALE* to the drawing scale factor.
2. Set *PSLTSCALE* to 1 (on).
3. Set *MSLTSCALE* to 0 (off).
4. Create the drawing including all non-continuous lines.
5. Display geometry in the paper space viewports using the *Viewport Scale*.
6. If any individual non-continuous lines need to be adjusted for scaling, use the *Properties* palette to change the individual lines' *Linetype Scale*.

## Applications for Multiple Viewports

The capability of AutoCAD to create multiple viewports in one layout and multiple layouts in one drawing offers many possibilities for displaying a drawing. Because of features such as viewport-specific layer visibility, layer property overrides, dimensioning in paper space, and annotative objects, different features of the same geometry can be shown using multiple viewports. In addition, one drawing can be printed and plotted in multiple ways using multiple layouts and viewports. It is even possible to print or plot several drawings in one layout using *Xrefs*.

A few examples of typical applications that make use of multiple viewports and/or layouts are:

- architectural drawings that display several aspects of one drawing, such as an overall view of a floor plan and enlarged views of particular rooms or areas
- mechanical drawings that display one view of the entire part or assembly and additional detail views of particular aspects of the part or individual components of the assembly
- civil engineering or construction drawings that show an overall plan and one or more detailed views of a specific aspect of the site or structure
- 3D mechanical drawings that show several views of one part or assembly, each view showing a different side or "view" of the 3D model, as in a multiview drawing (see Chapter 37).
- architectural or engineering drawings that display identical geometry in multiple viewports or layouts, but each display emphasizing specific functions or components by color, lineweight, screening, etc.
- multiple architectural, engineering, construction, or design drawings displayed on one sheet, accomplished by *Xrefing* the drawings and displaying one drawing per viewport

In many cases in industry, drawings created in older releases of AutoCAD require updating, possibly to utilize layouts and viewports to print the geometry with different devices or display the geometry in different scales. Three examples of using multiple viewports or layouts are given next: 1) using two layouts to plot the same geometry at different scales, 2) using one layout to show the same geometry in two viewports at different scales, and 3) using one layout to display *Xref* drawings in multiple viewports at different scales.

## Using Two Layouts to Plot the Same Geometry at Different Scales

For the first problem, consider the drawing shown in Figure 32-33. This mechanical part was originally created in an older release of AutoCAD and set up to plot only on a "C" size engineering sheet (22" x 17"). *Limits* were set to 22 x 17 so the geometry could be drawn full size, then the layout was set up with an ANSI C title block. Since the *Limits* size equaled the sheet size, viewport scale was set to 1:1 and the layout was plotted at 1:1. Dimension *Overall scale*, *Ltscale*, and other size-related variables were set to 1. Dimensions were created in model space. The resulting plot is a full size drawing of the part. The dimension text and other text as well as hidden line dashes are plotted at the ANSI standard .125" (1/8").

**FIGURE 32-33**

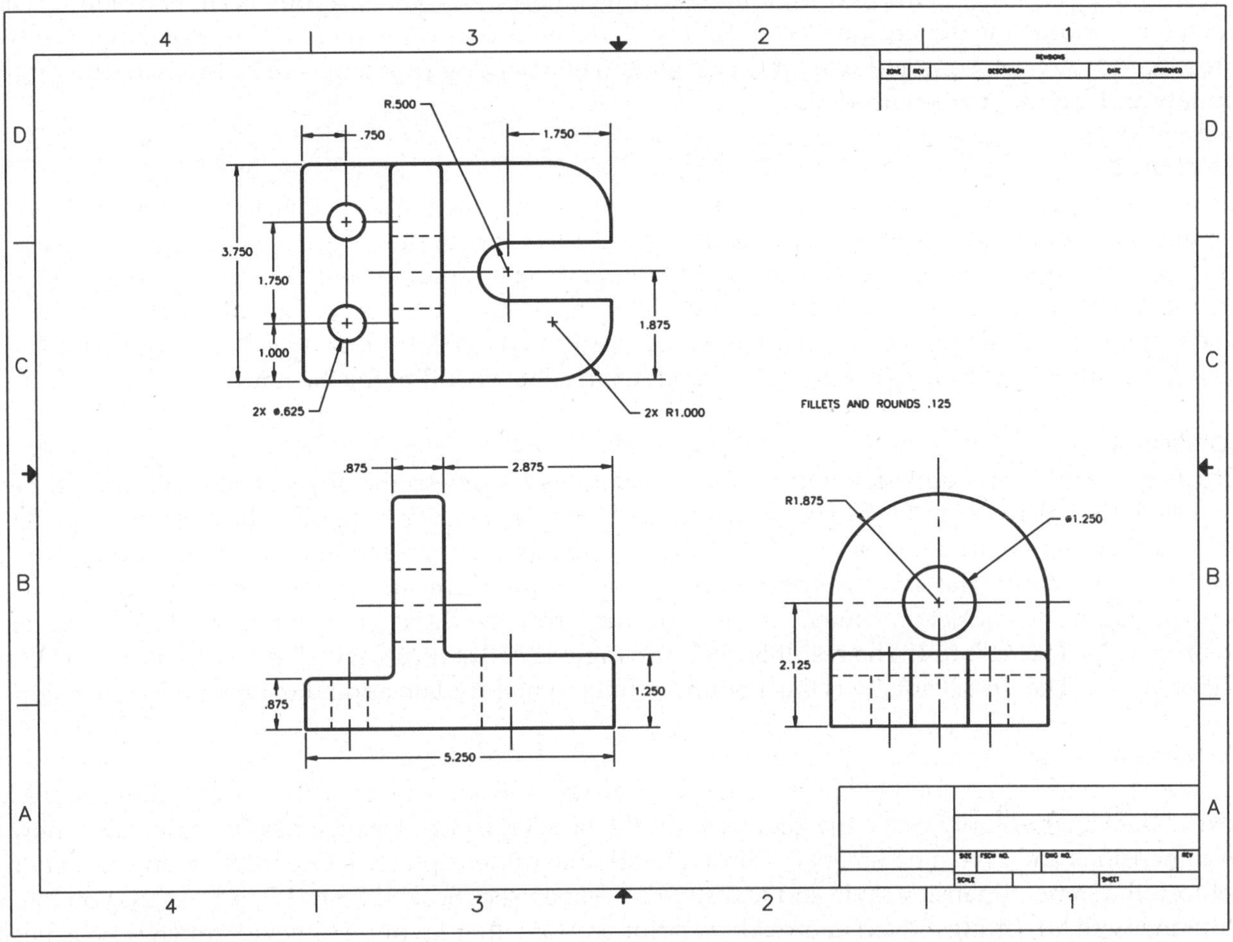


Assume the client calls and requests a fax of the drawing or an electronic copy of the drawing that he can print on his laser printer (ANSI "A" size 11 x 8.5 sheet). A new layout and viewport could be created with an ANSI A title block and the geometry scaled to fit the viewport at .375, or 3/8"=1" scale. However, making a print at this scale generates dimension text, other text, and hidden line dashes at about .047" (3/64")—much too small to be read easily. Here are several possible solutions.

### Solution 1

Create a second layout and viewport for an "A" size sheet. Scale the geometry for the new viewport. Display the same model space non-annotative dimensions in the new layout at a new *Overall Scale*. Change the text height to make the text readable. Change the *Ltscale* to 2.666 (the reciprocal of .375) in order to make hidden lines recognizable. In addition, the placement of several dimensions must be adjusted to accommodate the (relatively) increased text size (Fig. 32-34). One problem with this solution now is that the drawing is prepared and saved only for A size prints since the *Ltscale* and dimension scale have changed.

**FIGURE 32-34**

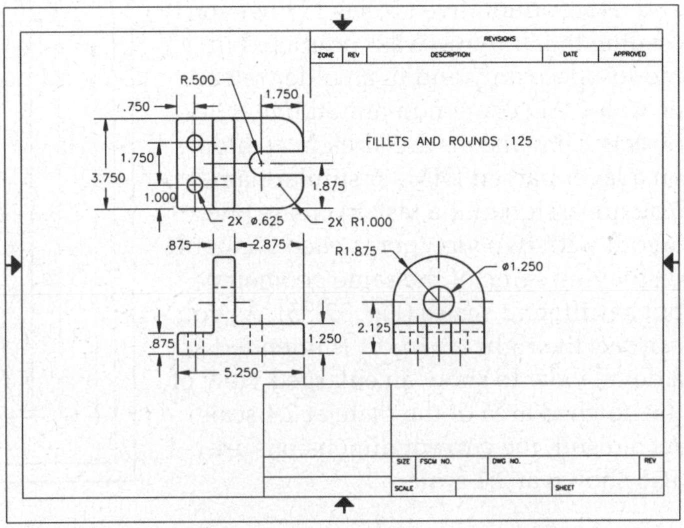

### Solution 2

Create a duplicate file. In the new separate drawing, create the ANSI A layout. As in Solution 1, scale the geometry, change the dimension *Overall Scale* and make needed adjustments. Now you have a drawing prepared for A size prints and one for C size plots. The resulting problem is there are two drawings to update and maintain when needed.

### Solution 3

Create a second layout and viewport for an "A" size sheet. Scale the geometry for the new viewport. Create a new dimension style and change the *Overall scale* appropriately for the new layout. Create a separate set of the model space dimensions and place them on a separate layer. Use viewport-specific layer visibility to display the correct dimension set in the appropriate viewports (layouts). Set *PSLTSCALE* to 1. The result is one drawing with all settings saved, ready to plot or print using either layout. The disadvantage is the amount of time to create, adjust, and manipulate visibility for the new dimensions.

### Solution 4

Create a second layout and viewport for an "A" size sheet. Convert the original text style and dimension style to *Annotative* and make the styles current. Use *Annoupdate* to update the existing dimensions and text objects. Create a new set of dimensions and text (scale representations) automatically by setting the *Viewport Scale* for the new viewport. Make needed adjustments to the new dimension and text scale representations to correct for dimension line spacing, crowded dimension text, flipped arrowheads, and so on. Set *PSLTSCALE* to 1. The result is one drawing with all settings saved, ready to plot or print using either layout. The disadvantage is the amount of time to manipulate and adjust the new dimensions.

### Solution 5

Create a second layout and viewport for an "A" size sheet. *Freeze* the layer for existing dimensions and text in the viewport only. Scale the geometry for the new viewport. Using a new layer, create a new set of dimensions and text in paper space. Since the original dimension style *Overall Scale* and text style was set to 1, the same dimension style and text style could be used. Set *PSLTSCALE* to 1. Again you have one drawing with all settings saved and ready to print or plot either layout. However, much time is needed to create the new set of dimensions and these dimensions are not easily adjusted for future changes.

## Using Two Viewports to Display the Same Geometry at Different Scales

For the second application, consider a drawing of the stamped plate. (This is the same drawing used previously to discuss both creating dimensions in paper space and using annotative objects.) However, assume the drawing was completed in model space only and in an older release of AutoCAD using non-annotative objects. Dimensions are already created on a layer named DIM, or similar name. You are assigned the task to create one layout with two viewports, each viewport displaying some of the same geometry but at different scales (Fig. 32-35). As you can see, the right viewport is intended as a detail view to show an enlarged view of the notched area of the plate at 2:1 scale. As a result, the current dimensions are also shown at 2:1 scale.

**FIGURE 32-35**

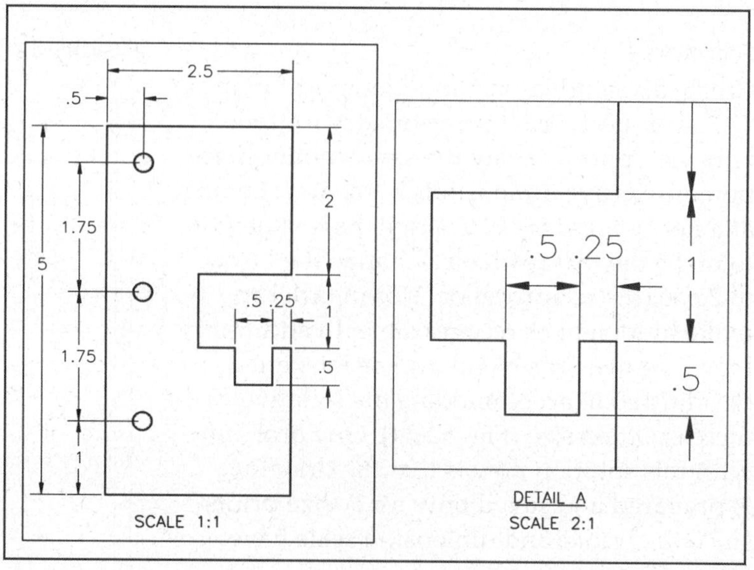

### Solution 1

Create a new layer named DIM-2 and move only the notch dimensions (at 1:1) to the new layer. Create another layer named DIM-3 and re-create the notch dimensions on the new layer and in a new dimension style with the *Overall scale* set to .5, since these dimensions will be displayed a 2:1. (You can make the new dimension style by copying the existing one, then changing the *Overall scale*.) Now there are three sets of dimensions: primary dimensions for the full view (at 1:1), dimensions for the notch for full view (at 1:1), and dimensions for the notch for the detail (at 2:1). Use viewport-specific layer visibility control to display the appropriate sets of dimensions in the appropriate views. This solution provides flexibility for future use because you have a full set of model space dimensions for the full view (layer DIM plus DIM-2) and one set for the detail view (DIM-3).

### Solution 2

Convert the original dimension style to *Annotative* and make the style current. Use *Annoupdate* to update the existing dimensions to annotative dimensions. Create a new set of dimensions (scale representations) for the detail viewport automatically by setting the *Viewport Scale* for the viewport (similar to previous Figure 32-25). Make needed adjustments to the new dimension scale representations to correct for dimension line spacing, crowded dimension text, flipped arrowheads, and so on. Next, use *Objectscale*, select all the dimensions that are not used for the detail view and delete the 2:1 object scale. Use *Objectscale* again, select all the dimensions used only for the detail view and delete the 1:1 object scale. The result is one drawing ready to plot or print displaying dimensions correctly for the two layouts (similar to previous Figure 32-31). The disadvantage is the amount of time needed to convert to annotative dimensions, create new scale representations, and delete objects scales. If in the future you wanted to show all dimensions in one view, a new layout with new dimensions could be created with relative ease.

### Solution 3

As shown in previous Figures 32-21 and 32-22, *Erase* all of the model space dimensions and create new dimensions in paper space at 1:1. See the example in "Dimensioning in Paper Space." Although this solution is relatively simple, the main drawback is the limitation of having dimensions in paper space—these dimensions are useable only for this layout. A view with all dimensions is not possible unless a new layout and an entire new set of dimensions is created manually.

### Solution 4

*Erase* only the model space dimensions for the notch appearing in the detail view, and then re-create them in paper space (as in Fig. 32-36, right viewport). Create these dimensions on a separate layer, for example, DIM-2. Note that part of one model space dimension (large arrowhead) appears in the detail view. To correct for this, the DIM layer is frozen for the right viewport only (*VP Freeze*). The DIM-2 layer does not have to be frozen for the left viewport since these dimensions appear only in the layout. This is probably the simplest and quickest solution. However, as with Solution 3, there are limitations such as not being able to display any view of the drawing with all dimensions.

**FIGURE 32-36**

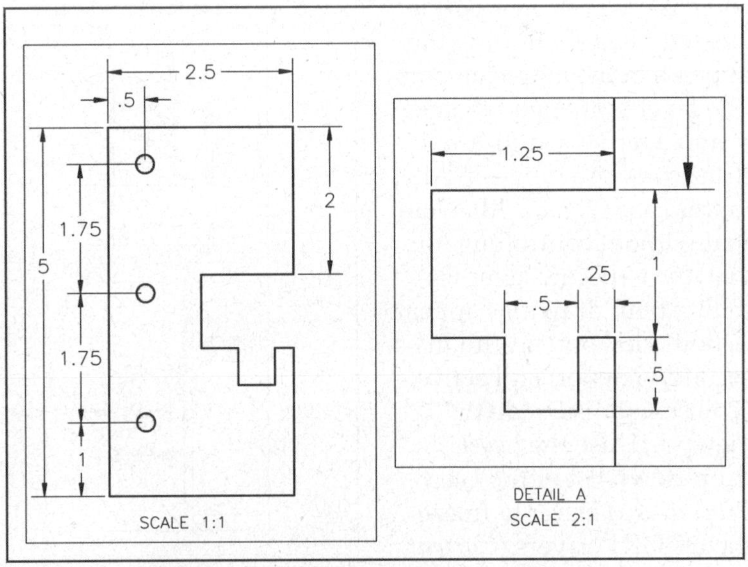

## Using One Layout to Plot *Xref* Drawings in Multiple Viewports at Different Scales

In this problem, one layout is created with three viewports to display and plot three *Xrefed* drawings at different scales. Although this sounds complex, it is a fairly straightforward procedure. The drawing typically contains no geometry (other than the *Xrefs*) and little or no text other than simple notes, titleblock and border.

Begin a *New* drawing, immediately activate a *Layout* tab and use *Pagesetup*. Set up the layout for the desired plot device and sheet size. In most cases, a large sheet size is required since multiple drawings will be displayed and plotted. *Insert* the desired titleblock and border. Create the first viewport with the *Vports* command on layer VIEWPORTS. *Save* the drawing.

Before *Xrefing* the first drawing, it is important to set a common layer current, such as layer 0. (If layer VIEWPORTS is current when the drawings are *Xrefed* and that layer is later *Frozen*, the *Xrefs* will not appear.) Also, double-click inside the viewport since you want to *Xref* the drawing into model space, not paper space. Use *Xref* and *Attach* the desired drawing. In this example, the DOOR drawing is referenced (Fig. 32-37). Use the *Viewports* toolbar or other method to set the viewport scale.

**FIGURE 32-37**

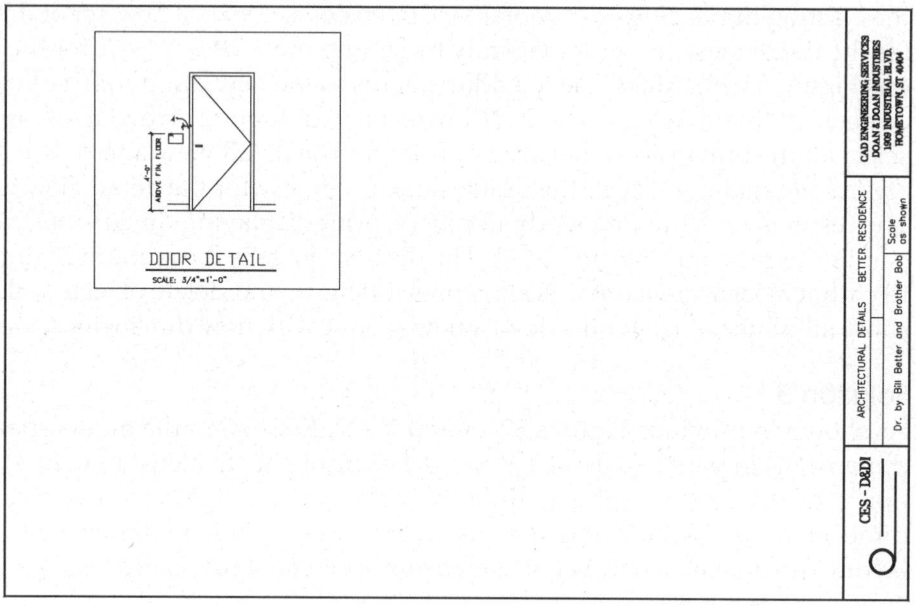


A second viewport is created on layer VIEWPORTS. As soon as the new viewport is created, the DOOR drawing appears in the new viewport. The layer is changed to layer 0, and *Xref* is used to *Attach* the second drawing—WALL, in this case (Fig. 32-38). Note that without controlling for viewport-specific layer visibility, both drawings appear in both viewports. With the second viewport still active (your cursor is inside the viewport), use the *Layer* drop-down list or the *Layer Properties Manager* to freeze the DOOR | * layers (*Current VP Freeze*).

**FIGURE 32-38**

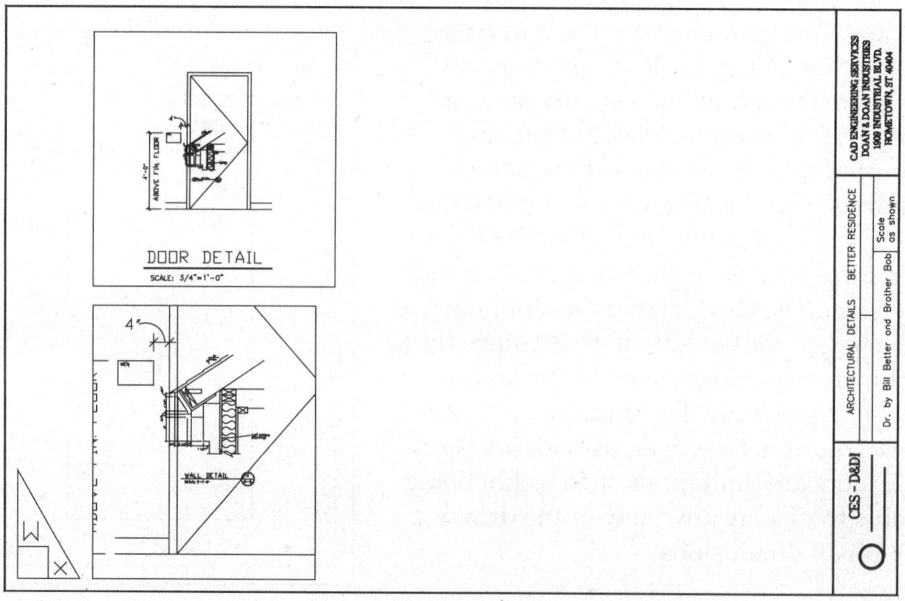

Next, click in the DOOR viewport and freeze the WALL | * layers. Additionally, to prevent the *Xref* drawings from appearing in any new viewports you create, select *New VP Freeze* for all the DOOR | * and WALL | * layers. Use the *Viewports* toolbar or other method to set the viewport scale in the new viewport.

FIGURE 32-39

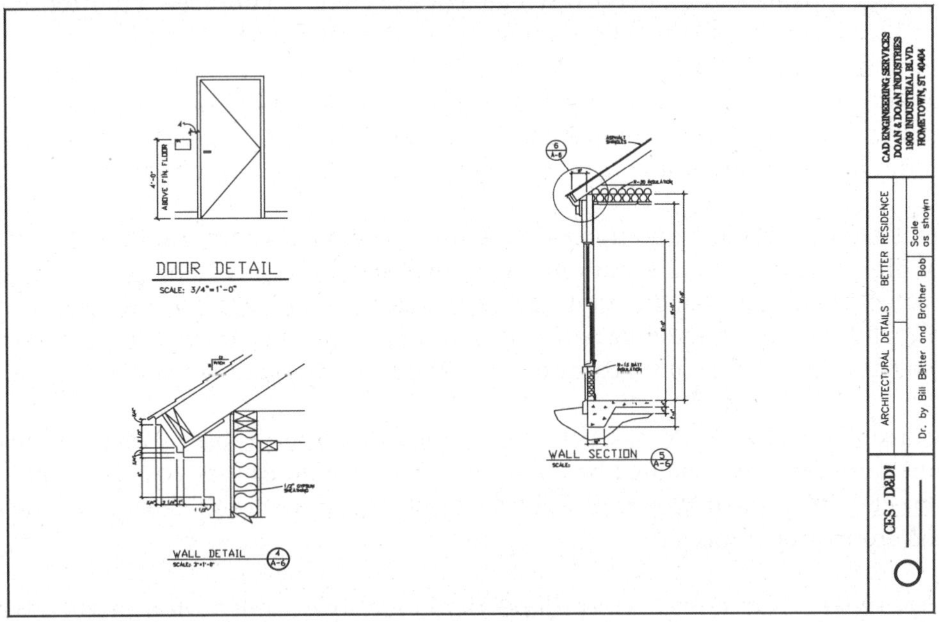


As in the previous step, create a new viewport on the appropriate layer, reset the current layer, then *Xref* the third drawing (SECTION) into model space. In this case, the previous two drawings do not appear in the new viewport; however, the third drawing appears on all three viewports (not shown). Use *VP Freeze* to freeze the SECTION drawing in the first two viewports. Set the viewport scale for the last viewport. *Freeze* the VIEWPORTS layer (globally) to complete the drawing (Fig. 32-39). Remember to set the *VISRETAIN* variable to 1; otherwise, none of the viewports' layer visibility settings will be saved with the drawing.

Since, in this case, it was known for each of the drawings at what scales they were to be plotted, the scale for each could be set independently using the viewport scale. If dimensions were created for these drawings in model space and the recommended methods for annotative or non-annotative dimensions were used, the dimensions could be displayed effectively in this parent drawing and the dimension layers could be controlled for visibility. If the dimensions for the *Xrefed* drawings were created in paper space, they would be unusable in the parent drawing since you cannot display any layouts of an externally referenced drawing.

## PLOT STYLE TABLES AND PLOT STYLES

A plot style is an object property. A plot style controls how an object looks on a plotted drawing. By modifying an object's plot style, you can override that object's drawing color, linetype, and lineweight for the plot. You can also specify additional features such as line end styles, join styles, fill styles, dithering, gray scale, pen assignment, and screening. You can use plot styles if you need to plot the same drawing with different "looks."

When a drawing is plotted, three main components are referenced to develop the finished plot or print:

1. layout settings you set up in the *Plot* dialog box and *Page Setup* dialog box (saved in the .DWG file);
2. plotter configuration information you specify for the plot device (saved in the device's .PC3 file); and
3. settings that are assigned in a Plot Style Table that you attach (saved in .CTB and .STB files).

The third component, a Plot Style Table, is optional for plotting. This section discusses Plot Style Tables and plot styles.

Before explaining plot styles, it would help to consider how the appearance of objects is controlled in plotted drawings in pre-AutoCAD 2000 releases. In these releases, the color of an object determined how objects appeared in plotted drawings. For example, if you used a plotter with colored pens, it was normally configured so the object's color would plot with a pen of the same color. Alternately, you could configure the print or plot device so colors in the drawing would correspond to certain lineweights on the finished page. Normally, the objects' color is set to *ByLayer*, so you can consider that the layer's color rather than the object's color determined the plotted appearance. The practice of associating colors with certain plotter pen numbers is known as making "pen assignments."

This strategy is still valid in current AutoCAD releases and will most likely be used for the majority of plots and prints. As a matter of fact, several color-dependent Plot Style Tables are supplied with AutoCAD that utilize this same concept; that is, an object's color determines how the object appears in the plot. There is one limitation to this system, however. With a color-dependent Plot Style Table, all objects in the same color generally have the same appearance on the plotted sheet.

Current AutoCAD releases also offer named Plot Style Tables, in which the contained plot styles are not automatically assigned by color but can have names that you designate and can be assigned to any object. With named Plot Style Tables, objects that appear in one color in the drawing can be plotted with different appearances.

Plot styles, even the new color-dependent Plot Style Tables, offer many advantages. With plot styles you can assign line end styles, joint styles, fill patterns, gray scales, screen pattern percentages, and pen assignments. Therefore, a color-dependent Plot Style Table is simply a list of colors, and for each color in the drawing you can assign a specific lineweight, linetype, line end style, joint style, fill pattern, gray scale, screen percentage, color, and pen assignment for the plot or print.

Unlike some other object properties, plot style is optional. Objects will print or plot based on color even if no plot style is assigned to it or if no Plot Style Table is attached. If you assign a plot style to an object, and then detach or delete the Plot Style Table that defines that plot style, the plot style will have no effect on the object.

## What is the Difference between a Plot Style and a Plot Style Table?

It is important to make the distinction between a plot style and a Plot Style Table because of how they are assigned or what they are attached to. This book uses lowercase letters for the terms "plot style" and upper- and lowercase for "Plot Style Table" to help make that distinction.

TIP

plot style — A property assigned to objects that determines the way the object appears in a plot or print. A plot style contains several appearance characteristics such as color, linetype, lineweight, line end style, joint style, fill pattern, gray scale, screen percentage, and pen assignment.

Plot Style Table — A Plot Style Table is a collection of many plot styles. A color-dependent Plot Style Table contains 255 plot styles; each plot style is a color. A Plot Style Table is "attached" to the *Model* tab or a *Layout* tab.

TIP

The Plot Style Table is attached to a layout. You can attach the same or a different Plot Style Table to each layout (the *Model* tab and all *Layout* tabs). You attach a Plot Style Table to a layout using the *Page Setup* dialog box. Since different layouts can have different Plot Style Tables attached, two similar layouts could be plotted with two different "looks." Alternately, one layout could be plotted with one Plot Style Table and later plotted with another Plot Style Table to achieve a different "look."

For example, Figure 32-40 displays the AutoCAD-supplied ACAD.CTB color-dependent Plot Style Table. The colors listed in the table (Color 1, Color 2, Color 3, etc.) are plot styles. Each color in the drawing takes on the appearance defined by its plot style, so everything drawn in Color 1 can have a particular plotted color, grayscale, pen number, screening, linetype, and so on, whereas everything drawn in Color 2 can take on a different appearance.

**FIGURE 32-40**

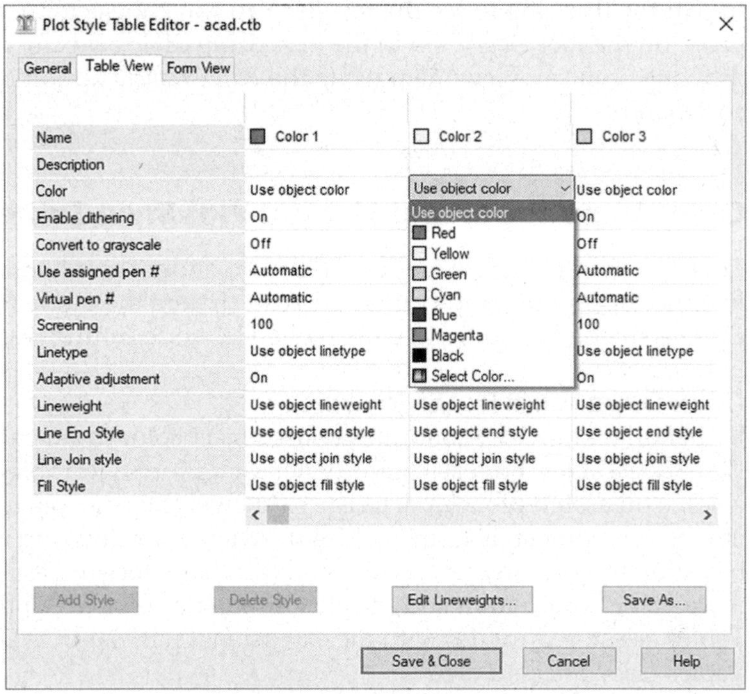


## Color Plot Style Table Example

As an example, consider a construction drawing set up in one *Layout*, but the project has two phases. Using two Plot Style Tables, the layout could be plotted for the demolition phase showing one appearance, then later plotted during the construction phase with a different appearance. The floor plan in Figure 32-41 illustrates the demolition phase. Walls to be demolished are most apparent, shown in 100% screen (dark), while unchanged walls are in 50% screen (medium), and new construction is barely noticeable in 30% screen (light). Note the Plot Style Table in Figure 32-40 is set up to plot Color 3 (demolition) in 100%, Color 2 (unchanged) in 50%, and Color 1 (new construction) in 30% screen.

**FIGURE 32-41**

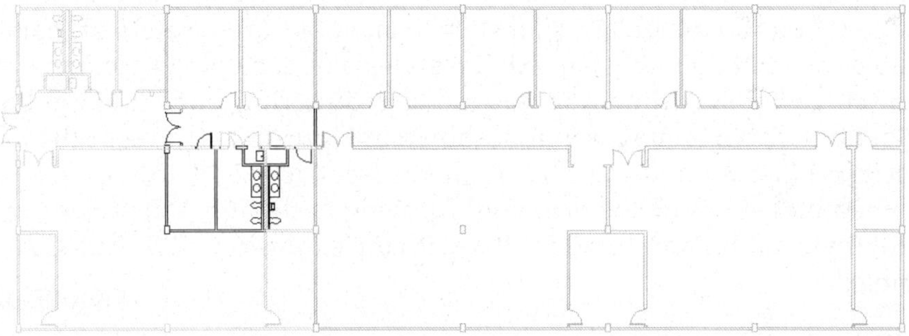

Figure 32-42 shows the same layout but for the construction phase. In this case, a second Plot Style Table is attached to the layout that applies the opposite screen percentages to Color 1 and Color 3, thereby emphasizing the walls needed for new construction (100% screen, dark) and fading the demolished walls (30% screen, light).

**FIGURE 32-42**

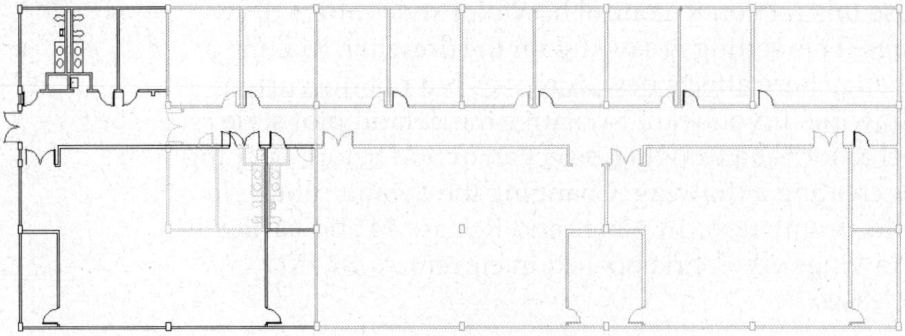

When you use plot styles you have the option to display the plot styles in the layout so you don't have to plot the drawing to see the results. You can choose to display plot styles by selecting *Display Plot Styles* on the *Plot Device* tab in the *Page Setup* dialog box. If you choose not to display plot styles in the drawing, you can view them using the *Full Preview* option in the *Plot* dialog box or by using the *Preview* command.

## Color-Dependent and Named Plot Style Tables

There are two plot style modes: Color-Dependent and Named. Each drawing you open in AutoCAD is in either one mode or the other. You can change the mode for new drawings using the *Plotting* tab in the *Options* dialog box to change the plot style mode setting (see Fig. 32-43 and following explanation).

### Color-Dependent Plot Style Tables

By default, AutoCAD continues to use object color to control output effects by creating color-dependent Plot Style Tables. For each color-dependent Plot Style Table there are 255 plot styles; each plot style is a color. You assign plot characteristics (lineweight, screening, dithering, end and joint types, fill patterns, etc.) to each plot style (see Fig. 32-40). When your drawing is plotted, the appearance of each line is based on its color in the drawing and the characteristics assigned to that color. In a color-dependent Plot Style Table, you cannot add, delete, or rename color-dependent plot styles. Color-dependent Plot Style Tables are stored in files with the extension .CTB. AutoCAD supplies several color-dependent Plot Style Tables (see "*Stylesmanager*").

### Named Plot Style Tables

A named Plot Style Table is not based on a list of colors but is based on a list of plot style names that you define. You can assign any plot style to any object regardless of that object's color. Each named Plot Style Table can contain many named plot styles. For each plot style name you define, you can assign specific line characteristics (lineweight, screening, dithering, end and joint types, fill patterns, etc.). A named Plot Style Table is similar to that shown in Figure 32-40, except the individual plot style names can be assigned instead of using color numbers (see Fig. 32-49, top row). Normally, you would then assign the named plot styles to layers in your drawing (although plot styles can be assigned to any object). Named Plot Style Tables are stored in files with the extension .STB. AutoCAD supplies several named Plot Style Tables.

**FIGURE 32-43**

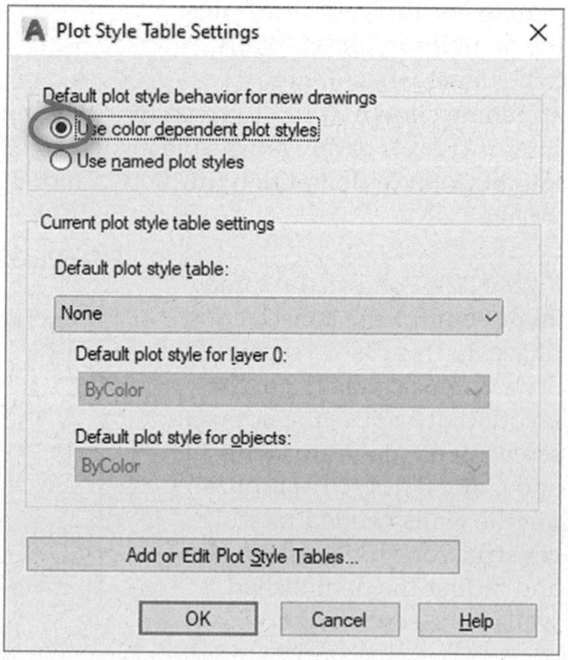


Since AutoCAD can use only one of the two types of Plot Style Tables at a time (named or color-dependent), make the desired choice by using the *Plot and Publish* tab of the *Options* dialog box, then selecting the *Plot Style Table Settings* button on the bottom-right. This action produces the *Plot Style Table Settings* dialog box (Fig. 32-43).

#### *Default Plot Style Behavior for New Drawings*

Use this section to control how plot styles affect drawings. This setting is saved with the drawing. Making a change here affects new drawings but not the current drawing. If you want to change the default plot style behavior for a drawing, select an option before opening or creating a drawing. Changing the default plot style affects only new drawings and Release 14 and earlier drawings when first opened in current AutoCAD releases.

*Use Color-Dependent Plot Styles*
By default, AutoCAD uses color-dependent plot styles for new drawings and pre-AutoCAD 2000 drawings. You can also create and edit color-dependent plot styles by setting the *PSTYLEPOLICY* system variable to 1. If a drawing is saved with *Use Color-Dependent Plot Styles* as the default, you can change it to *Use Named Plot Styles* later. However, once a drawing is saved with *Use Named Plot Styles* as the default, you can change it to use color dependent plot styles with *Convertpstyles*.

*Use Named Plot Styles*
This check causes AutoCAD to use named plot styles as the default in new drawings. This action allows you to use the AutoCAD-supplied named Plot Style Tables or create new named Plot Style Tables. You can also set the *PSTYLEPOLICY* system variable to 0 to enable named Plot Style Tables. Once a drawing is saved with *Use Named Plot Styles* as the default, you can change it to use color dependent plot styles using *Convertpstyles*.

*Default Plot Style Table*
From this drop-down list, select the default Plot Style Table to attach to new drawings. If you are using color-dependent plot styles (*Use Color Dependent Plot Styles* is checked), this option lists all color-dependent Plot Style Tables (.CTB files) found in the search path as well as the value of *None*. If you are using named plot styles (*Use Named Plot Styles* is checked), this option lists all named Plot Styles Tables (.STB files).

*Default Plot Style for Layer 0*
*Use Named Plot Styles* must be checked to enable this drop-down list. Your choice sets the default plot style for Layer 0 for new drawings. The list displays the default value *Normal* and any plot styles defined in the currently loaded Plot Style Table. You can also control the default plot style for Layer 0 using the *DEFLPLSTYLE* system variable (saved in the drawing).

*Default Plot Style for Objects*
*Use Named Plot Styles* must be checked to enable this drop-down list. Here you set the default plot style that is assigned when you create new objects. The list displays a *ByLayer*, *ByBlock*, and *Normal* style, and lists any plot styles defined in the currently loaded Plot Style Table. Alternately, you can set the *DEFPLSTYLE* system variable.

*Add or Edit Plot Style Tables*
Selecting this button invokes the *Plot Style Table Manager*, which is a separate system window. See "*Stylesmanager*" later in this chapter.

## Attaching a Plot Style Table to a Layout

To attach a Plot Style Table to a *Layout* or *Model* tab, first select the desired *Model* tab or *Layout* tab. Invoke the *Page Setup Manager*, then select *New* or *Modify* to produce the *Page Setup* dialog box. In the upper-right corner of the *Page Setup* dialog box (Fig. 32-44), use the *Plot style table (pen assignments)* drop-down list to select the desired table (.CTB or .STB file).

Use the *Edit...* button to invoke the *Plot Style Table Editor*. (See the section titled "Editing Plot Styles Assigned to Objects.") Selecting the *New...* button invokes the *Add-A-Plot Style Table Wizard*. (See "Creating a Plot Style Table.") You can choose to *Display Plot Styles* if you want the layout to reflect the appearance of the plot styles in the layout (so the layout looks like a *Full Preview*).

FIGURE 32-44

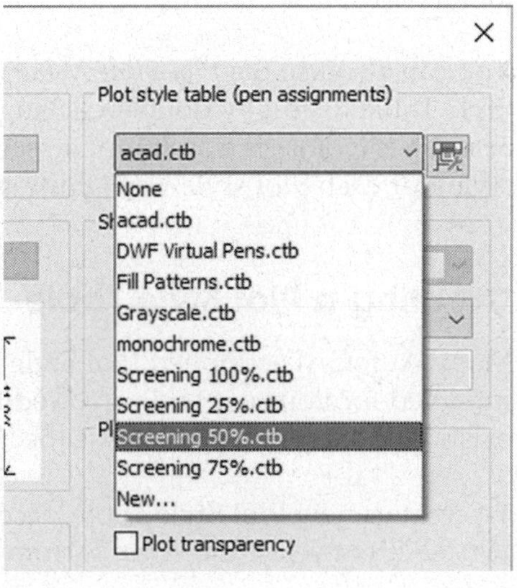

**TIP** If you want to change the name of a named Plot Style Table from Style 1, Style 2, etc. to something more descriptive (such as walls, electrical, plumbing), it should be done before opening a drawing that references the table. If the names are changed after the Plot Style Tables have been attached to a drawing, the plot style names in the drawing will not match the names of the styles in the attached Plot Style Table.

### *Stylesmanager*

| Ribbon | Menu Bar | Command (Type) | Alias (Type) | Shortcut |
|---|---|---|---|---|
| ... | *File Plot Style Manager...* | *Stylesmanager* | ... | ... |

The *Stylesmanager* command produces the *Plot Style Manager*. You create Plot Style Tables and edit Plot Style Tables with the *Plot Style Manager*. Selecting the *Plot Styles Manager* produces a window displaying the *Add-A-Plot Style Table Wizard* and the list of available Plot Style Tables (.CTB and .STB files). This separate application window runs independently of AutoCAD (Fig. 32-45).

**FIGURE 32-45**

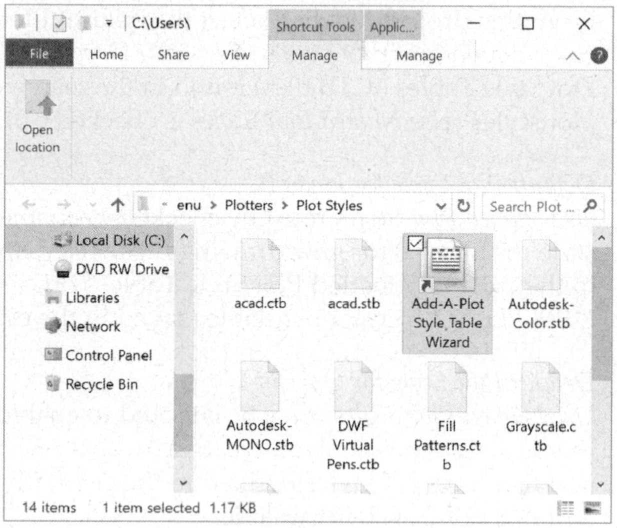


Color-dependent Plot Style Tables are stored as .CTB files, and named Plot Style Tables are stored as .STB files. To find the location of the .CTB and .STB files on your system, use the drop-down list in the *Address* section at the top of the window, or use *Options*, *Files* tab, and expand the *Printer Support File Path*, then *Plot Style Table Search Path* sections.

To create a new Plot Style Table using the *Plot Style Manager*, select the *Add-A-Plot Style Table Wizard* (see Fig. 32-45). After creating new Plot Style Tables, they automatically appear in the window. See "Creating a Plot Style Table."

You can also use the *Plot Style Manager* to edit existing Plot Style Tables. Using the manager, select a Plot Style Table to edit by double-clicking its icon. Editing a table involves assigning the appearance characteristics (color, pen assignment, screening, linetype, lineweight, line end style, line join style, and fill style) for each plot style. See "Editing Plot Styles."

## Creating a Plot Style Table

Many AutoCAD-supplied Plot Style Tables are available and have "generic" plot styles. These tables are intended for your use and can be edited for your needs using the *Plot Style Table Editor*. You can use the existing Plot Style Tables or use the *Add-A-Plot Style Table Wizard* to create other tables.

To create a new Plot Style Table, access the *Add-A-Plot Style Table Wizard* from the *Plot Style Manager* (see Fig. 32-45) or select the *Wizards* from the *Tools* pull-down menu.

In the *Add-A-Plot Style Table Wizard*, you can *Start from scratch*, *Use an existing plot style table*, *Use My R14 Plotter Configuration*, or *Use a PCP or PC2 file*. Selecting the first choice, *Start from Scratch*, you can create a color-dependent Plot Style Table or a named Plot Style Table (Fig. 32-46).

**FIGURE 32-46**

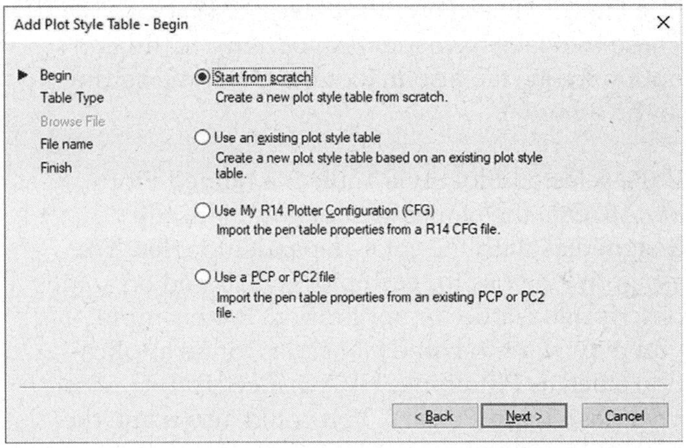


The *Add-A-Plot Style Table Wizard* guides you through the steps to easily create the new table. There are only a few steps involved with creating a new Plot Style Table. Once the table is created, your specific preferences on how objects appear in plots are achieved by editing the table and creating the individual plot styles using the *Plot Styles Table Editor*. The wizard contains explanatory text on each page, so no additional explanation is required here. If you feel that you do need help, look in the *AutoCAD Help*. Once the table is created, edit it using the *Plot Style Table Editor* as explained in the next section.

Plot Style Table Conversion from Previous Releases
In releases of AutoCAD previous to 2000, the components needed for plotting were saved in the ACAD. CFG and the .PCP or .PC2 files. If you want, you can use the *Add-A-Plot Style Table Wizard* to convert your older Release 14 pen assignments to Plot Style Tables.

These new Plot Style Tables will be available in the list in addition to the supplied Plot Style Tables. See Figure 32-46, last two options (*Use My R14 Plotter Configuration (CFG)* and *Use a PCP or PC2 File*). See also "*Pcinwizard*" in Chapter 14.

**FIGURE 32-47**

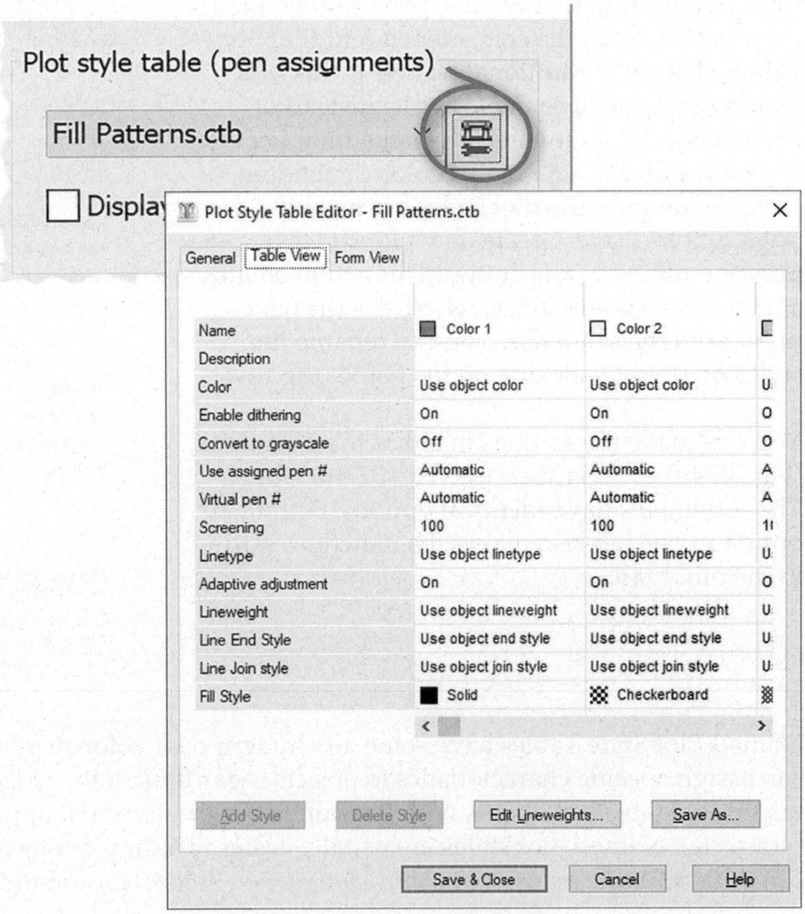


## Editing Plot Styles

To edit plot styles, use the *Edit...* button in the *Page Setup* dialog box. This action invokes the *Plot Style Table Editor* (Fig. 32-47).

If the selected Plot Style Table is color-dependent, the assignment to objects is automatic; that is, each of the 255 plot styles in the table is already assigned to an object color. However, you can set the *Fill Style*, *Line Join Style*, *Line End Style*, *Lineweight*, *Line Type*, *Screening*, and so on for each color as shown. Therefore, each drawing color takes on the appearance of the selected options. You cannot change the plot style names (Color 1, Color 2, etc.) for a color-dependent Plot Style Table.

You can make the settings in either the *Table View* (see Fig. 32-47) or *Form View* (Fig. 32-48) tabs. These two tabs have identical options. Setting an option in one tab also makes the matching setting in the other tab.

**FIGURE 32-48**

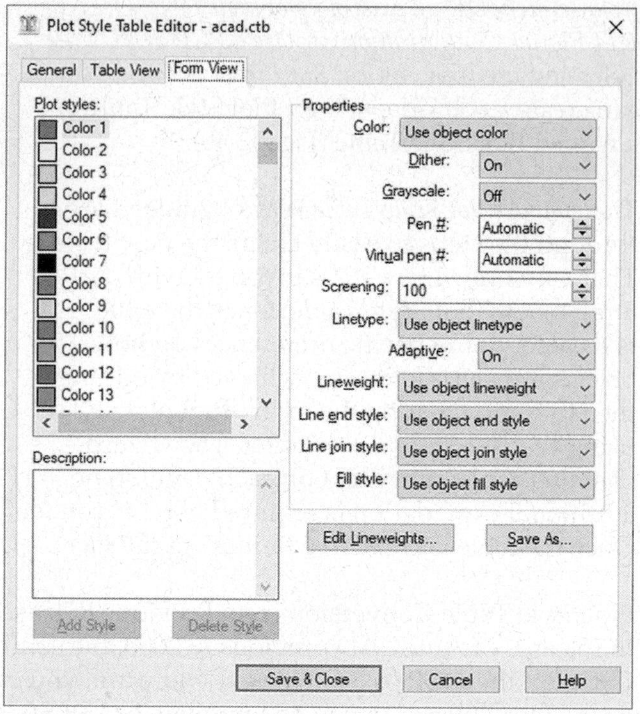


If the selected Plot Style Table is a named Plot Style Table, the names are not automatically assigned as with the color-dependent tables. You assign the names for each plot style based on some descriptive feature or application. For example, you may want to name plot styles for an application, such as Plumbing, HVAC, Electrical, or possibly Phase 1 and Phase 2. You could also name the plot styles according to a feature you assign; for example, you may use the names Wide, Medium, and Thin to describe lineweights that are defined in the styles.

Figure 32-49 displays a named Plot Style Table in the *Plot Style Table Editor*. All of the same features (color, dither, grayscale, pen number, etc.) are available as in the color-dependent Plot Style Tables. The only additional feature is that you assign the name of each plot style, instead of the styles being automatically named for a color (Color 1, Color 2, etc.) as in a color-dependent table. By default, the plot styles are named Style 1, Style 2, etc. Create new plot styles by pressing the *Add Style* button. This action creates a new style named Style#, where # is the next number in the sequence. You can rename the styles by double-clicking on the name (top row).

**FIGURE 32-49**

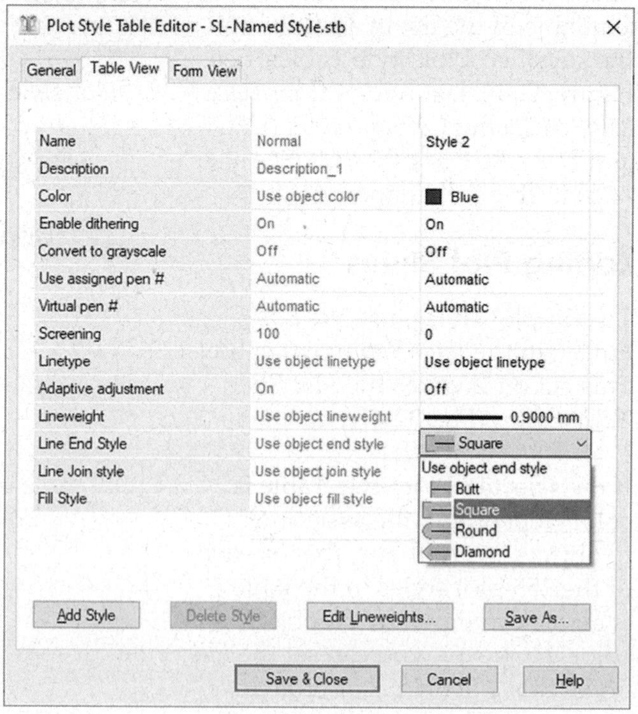


You can make the settings in either the *Table View* (Fig. 32-49) or *Form View* (see Fig. 32-48) tabs. These two tabs have identical options. Setting an option in one tab also makes the matching setting in the other tab.

**TIP**

Named Plot Style Tables have some advantages over color-dependent tables. With named tables, you can assign specific characteristics to objects regardless of the objects' screen color, whereas with color-dependent tables all objects with the same screen color must appear the same in the plot (have the same plot style). Named plot styles are usually easier to assign to objects in complex drawings (when the drawing contains many colors and layers). See the discussion in the next section.

## Assigning Named Plot Styles to Objects

Keep in mind that each drawing can have only one of the two types of Plot Style Tables attached—color-dependent or named Plot Style Tables. The desired table type should be set in the *Options* dialog box before creating a drawing (see "Color-Dependent Plot Style Tables" and "Named Plot Style Tables").

### Color-Dependent Plot Style Tables

The process of assigning plot styles to objects is necessary only for named Plot Style Tables. This is based on the fact that color-dependent plot styles are already assigned to colors, not objects. In other words, each color-dependent plot style (named Color 1, Color 2, etc.) is automatically assigned to its matching color. Figure 32-50 explains this idea. Notice that in the figure displaying the *Layer Properties Manager*, the *Plot Styles* column is disabled since assignment is automatic. Drawings that have color-dependent Plot Style Tables attached display a disabled *Plot Style* drop-down list on the *Object Properties* toolbar.

**FIGURE 32-50**

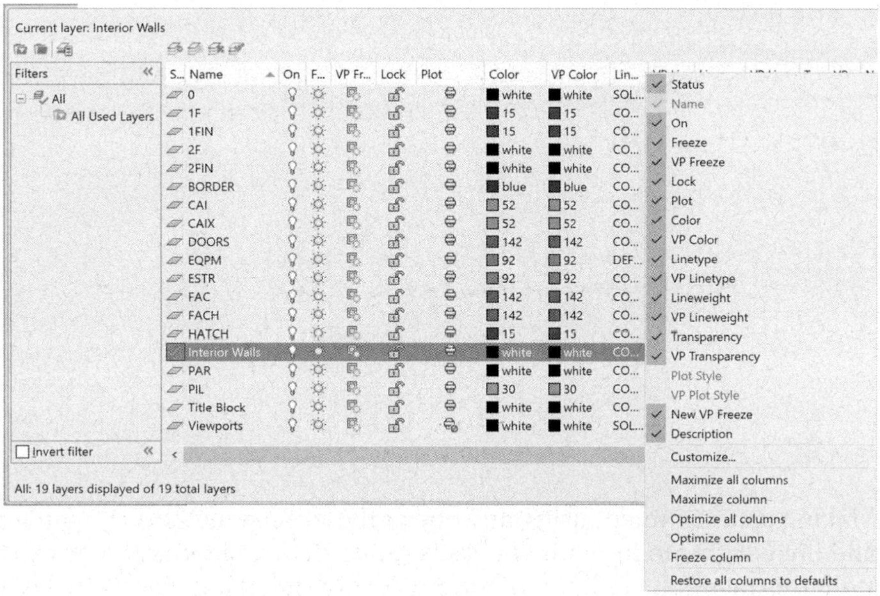


### Named Plot Style Tables

Plot styles in a named Plot Style Table can be attached to any object. Individual plot styles within a named table are often assigned to layers. Assigning plot styles to individual objects offers much more flexibility but can be a tedious job and create a fairly complex layout. If plot styles are assigned both to layers and objects, the plot style assigned to an object overrides the plot style assigned to the layer.

TIP

If you want to assign plot styles to individual objects (rather than to layers), three methods are available: the *Plot Style* drop-down list on the *Properties* panel, the *Properties* palette, or the *Plotstyle* command.

**FIGURE 32-51**

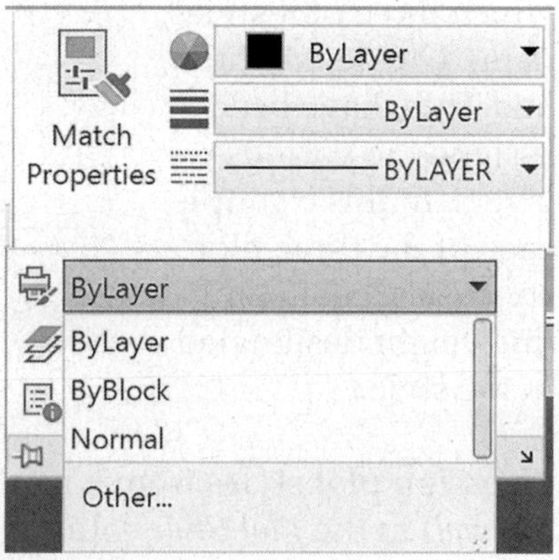


Figure 32-51 displays the *Plot Style* drop-down list on the *Properties* panel. Using this method, you would select the desired plot style from the list. The selected style becomes the current object-specific plot style—that is, any object created after the setting is made will have the selected plot style. This method is similar to specifying an object-specific color, linetype, or lineweight (see Chapter 11). The *Plot Style* drop-down list on the *Properties* panel cannot be used to assign color-dependent plot styles.

You can also use the *Plotstyle* command to assign object-specific plot styles (see "*Plotstyle*" next). Using *Plotstyle* has the same result as using the *Plot Style* drop-down list. The *Plotstyle* command can only be used to assign named plot styles.

**TIP**

Figure 32-52 displays the *Properties* palette. To use this method, select the desired object(s), then invoke the *Properties* palette. Select the desired plot style from the drop-down list. This method differs from the two previously described methods in that using the *Properties* palette is retroactive for the selected objects only, whereas the *Plotstyle* command and the *Object Properties* toolbar *Plot Style* drop-down list set plot styles for newly created objects.

**FIGURE 32-52**

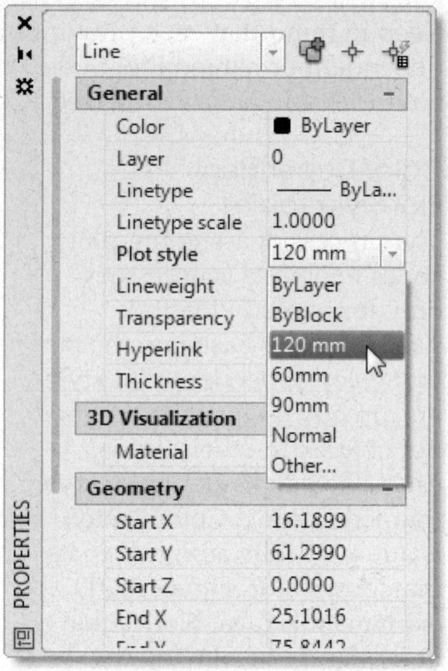


Making an assignment using any one of these three methods is similar to assigning colors, linetypes, and lineweights to individual objects rather than assigning the properties by the object's layer (*ByLayer*). With this method, you lose the simplicity of organizing objects by layer and color; however, that is one of the underlying reasons for using named plot styles—so objects of different colors can be plotted in different ways.

Alternately, named plot styles can be assigned to layers. This is generally a simpler method, particularly for complex drawings with many objects. Figure 32-53 displays *the Layer Properties Manager* in which three plot styles, named Wide, Medium, and Thin, have been assigned to various layers. In this example, each of the three plot styles has a matching lineweight designated in the style.

**FIGURE 32-53**

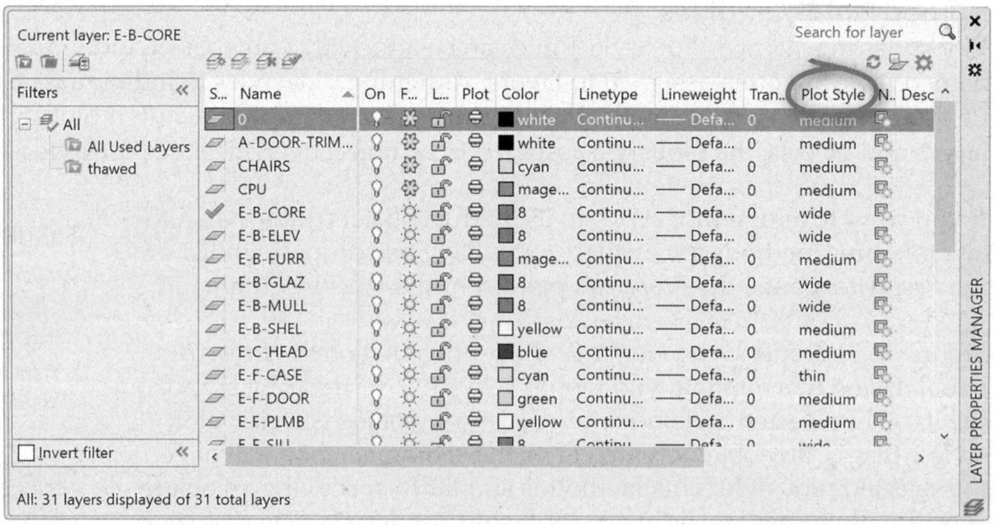


To assign plot styles from a named Plot Style Table to layers, click on the plot style name (usually *Normal*) in the *Plot Style* column for a specific layer. The *Select Plot Style* dialog box appears (Fig. 32-54 on the next page) providing the list of plot styles to select from. All objects on the selected layer are then plotted with the parameters (screening, pen number, lineweight, line end joints, etc.) as previously designated in the plot style.

Note that the information given at the bottom of the dialog box indicates the *Active Plot Style Table* and the layout that it is *Attached to*. You can select a different Plot Style Table from this list to attach to the layout; however, doing so changes the setting previously made in the *Page Setup* dialog box. In other words, you can assign plot styles to objects and layers, but you cannot attach Plot Style Tables to individual objects or layers, only to layouts.

**FIGURE 32-54**

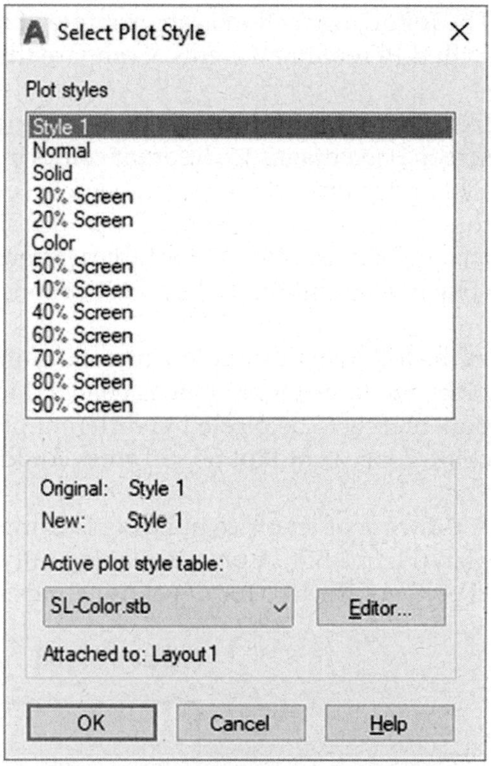


## *Plotstyle*

| Ribbon | Menu Bar | Command (Type) | Alias (Type) | Shortcut |
|---|---|---|---|---|
| ... | *Format Plot Style...* | *Plotstyle* or *-Plotstyle* | ... | ... |

As previously described briefly, the *Plotstyle* command is used to set the current named plot style for newly created objects. It has the same function as selecting a current plot style from the *Plot Style* drop-down list on the *Properties* panel. That is, the selected named plot style becomes the plot style assigned to all newly created objects.

The *Plotstyle* command produces the *Current Plot Style* dialog box (not shown). This dialog box is essentially the same as the *Select Plot Style* dialog box that appears when assigning plot styles to layers (see Fig. 32-54), except with *Plotstyle* the styles are assigned to objects. The *Plotstyle* command operates only for drawings that were created in named Plot Style Table mode. See the previous discussion, "Assigning Named Plot Styles to Objects."

To use the Command line version, type *-Plotstyle*. The command prompts are shown here.

```
Command: -PLOTSTYLE
Current plot style is "ByLayer"
Enter an option [?/Current] :
```

Enter the name of the desired plot style. The question (?) option lists all plot styles in the currently attached table. The current drawing must be in named Plot Style Table mode and you must have a named Plot Style Table attached to the layout or viewport for this command to operate.

## Named Plot Style Table Example

The following section is an example of using one named Plot Style Table to plot a drawing two times with different lineweights. Compare this example with the example given earlier in the chapter.

In the earlier example, two similar color-dependent Plot Style Tables were used to assign different screen percentages to different colors in the drawing. The two plots were created by attaching a different Plot Style Table to the layout before creating each plot. Using several color-dependent Plot Style Tables for one layout is a reasonable practice since the individual plot style names (color numbers) cannot change between Plot Style Tables. Therefore, all color-dependent Plot Style Tables have the same plot style names, although the specifications for each style (screen percentage in this case) can be different.

In the following example, only one named Plot Style Table is attached to a layout. To achieve different plots, the lineweight value is changed for individual plot styles to achieve the desired effects. In addition, objects to be plotted in different plot styles (lineweights) are drawn in the same color; therefore, color-dependent Plot Style Tables could not be used.

A drawing of an office building is completed and a layout is created to display and plot only the HVAC plan (Fig. 32-55). A plot displaying only the building core and the HVAC information is sent to the HVAC contractors for developing bids.

**FIGURE 32-55**

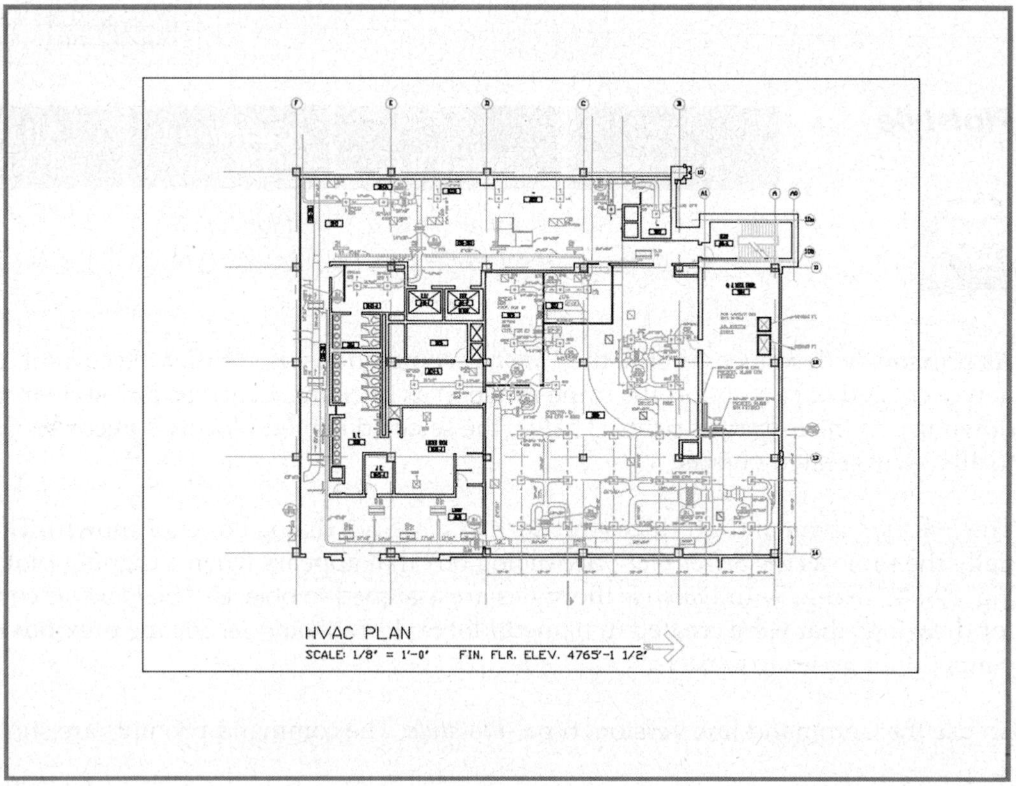


After a contractor is retained, it is discovered that one large duct unit must be specially manufactured. In order to make the unit more apparent for the manufacturer, a special plot must be made by applying a heavier lineweight to the unit.

Since all of the HVAC objects are drawn in one color (see Fig. 32-55, light colored objects), a color-dependent Plot Style Table cannot be used. Instead, a named Plot Style Table is created called "LW mm" (Fig. 32-56). Several plot styles are created with different lineweight settings; each plot style is named according to weight (1.20mm, .90mm, .60mm, etc.). The Plot Style Table is then attached to the layout using the *Page Setup* dialog box.

**FIGURE 32-56**

Next, objects representing the duct unit are selected and the *Properties* palette is invoked (Fig. 32-57). In the *Plot Style* row for all 25 objects [note the "Group (25)" displayed at the top of the box], the desired plot style (named "90mm") is selected and assigned to the objects.

**FIGURE 32-57**

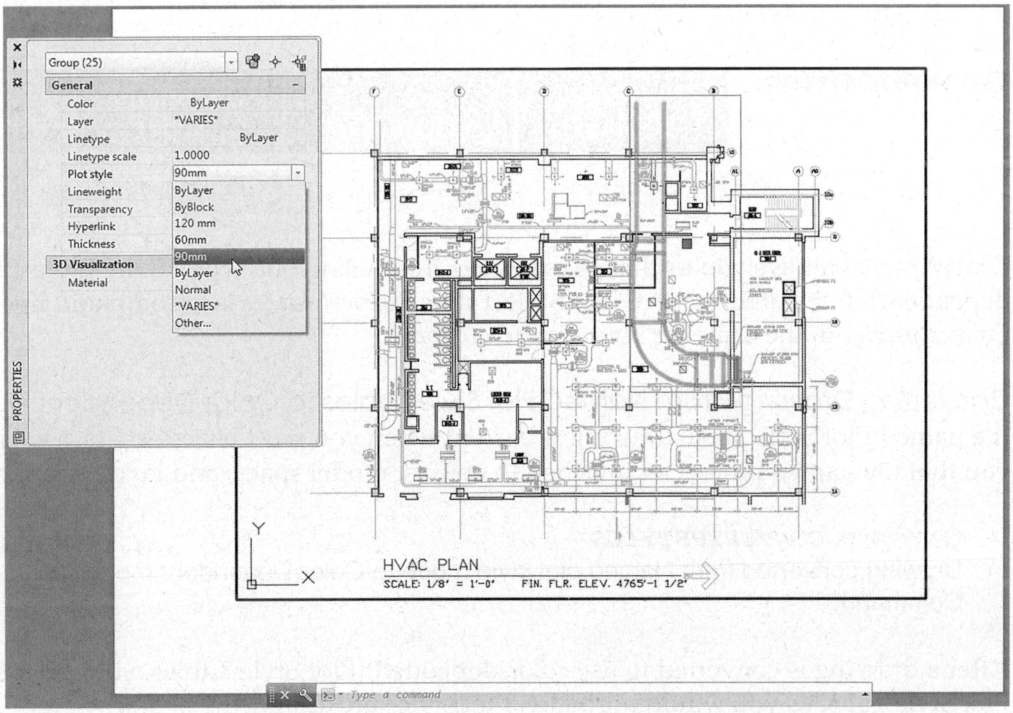

The finished plot reveals the special duct unit plotted in the wide lines (Fig. 32-58). Note that these objects are plotted differently even though they share the layer and color of the other HVAC components.

FIGURE 32-58

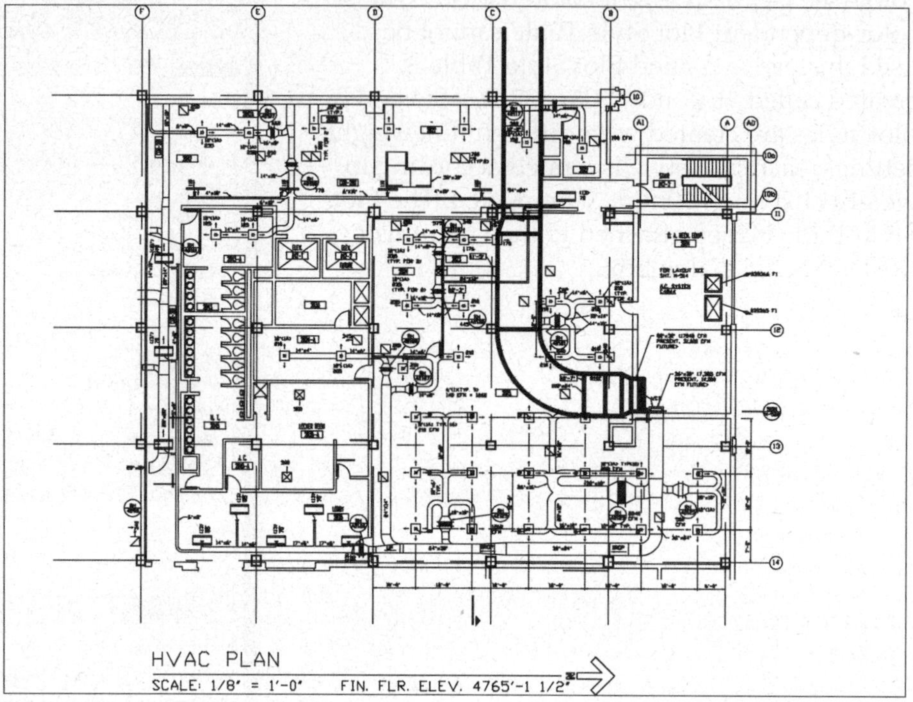


If another plot must be made to reveal other duct units in the layout, the process of applying the lineweight-based plot styles to particular objects within one layer can be used again. If you wanted to prepare a plot with uniform lineweights for all objects, it would be simple to select all objects and then use the *Properties* palette to assign the same plot style to all objects.

## *Convertpstyles*

| Ribbon | Menu Bar | Command (Type) | Alias (Type) | Shortcut |
|---|---|---|---|---|
| ... | ... | *Convertpstyles* | ... | ... |

*Convertpstyles* allows you to change drawings from using one type of Plot Style Table (named or color-dependent) to the other type of Plot Style Table. *Convertpstyles* is a Command line-driven utility. Type *Convertpstyles* in the drawing you want to convert.

### Converting Drawings from Named Plot Style Tables to Color-Dependent Plot Style Tables

If a named Plot Style Table drawing is current when you use *Convertpstyles*, a small dialog box advises you that the named plot styles attached to objects, model space, and layouts will be detached.

Command: ***CONVERTPSTYLES***
Drawing converted from Named plot style mode to Color Dependent mode.
Command:

After a drawing is converted to use color-dependent Plot Style Tables, you can assign a color-dependent Plot Style Table as you would normally. Plot styles are assigned by color.

### Converting Drawings from Color-Dependent Plot Style Tables to Named Plot Style Tables

Before you use *Convertpstyles* in a color-dependent drawing, you should use *Convertctb* to convert the attached color-dependent tables to named tables. You can convert the color-dependent Plot Style Tables assigned to the drawing to named Plot Style Tables using *Convertctb* (see *Convertctb* next). After converting the tables, use *Convertpstyles* to reattach the newly converted named Plot Style Tables.

TIP

Assuming *Convertctb* has been used first in the drawing, type *Convertpstyles*. AutoCAD displays the standard file navigation dialog box for you to select the named Plot Style Table file to attach to the converted drawing. You must select a named Plot Style Table that was converted using *Convertctb* or created from a PC2 or PCP file. Normally you should select the named Plot Style Table that was converted from the color-dependent Plot Style Table that was assigned to the same drawing. The selected named Plot Style Table is assigned to model space and to all layouts. Drawing layers are each assigned a named plot style (from the converted Plot Style Table) that has the same plot properties that their color-dependent plot style had. After going through this conversion, you can change the named Plot Style Table assignment or assign other named Plot Style Tables to model space or layouts.

### *Convertctb*

| Ribbon | Menu Bar | Command (Type) | Alias (Type) | Shortcut |
|---|---|---|---|---|
| ... | ... | *Convertctb* | ... | ... |

*Convertctb* is intended to be used in a drawing that you want to convert from using color-dependent Plot Style Tables to use named Plot Style Tables. First use *Convertctb* to convert the attached color-dependent tables to named tables, then use *Convertpstyles* in the drawing to convert the drawing (see *Convertpstyles*).

When you type *Convertctb* at the Command prompt in the drawing, the standard file navigation dialog box appears for you to select the desired color-dependent Plot Style Table, then reappears for you to specify a name (you should use the same name as the .CTB) and location for the new (now named) Plot Style Table. You should use *Convertctb* to convert all attached tables. The original color-dependent tables are not changed.

*Convertctb* creates one named plot style for each color that has unique plot properties, one named plot style for each group of colors that are assigned the same plot properties, and a default named plot style called NORMAL.

## PLOT STAMPING

Plot stamping provides options for you to have an informational text "stamp" plotted on each drawing. For example, the drawing name, date, and time of plot can be added to each plot for tracking and verification purposes.

### *Plotstamp*

| Ribbon | Menu Bar | Command (Type) | Alias (Type) | Shortcut |
|---|---|---|---|---|
| ... | ... | *Plotstamp* | ... | ... |

To enable plot stamping, you must use the expanded section of the *Plot* dialog box, then check *Plot Stamp On* (not shown).

To specify the text you want to appear on the plot, select the *Plot Stamp Settings* button that appears next to *Plot Stamp On* when you check it, or enter *Plotstamp* at the Command line. This action produces the *Plot Stamp* dialog box (Fig. 32-59).

**FIGURE 32-59**

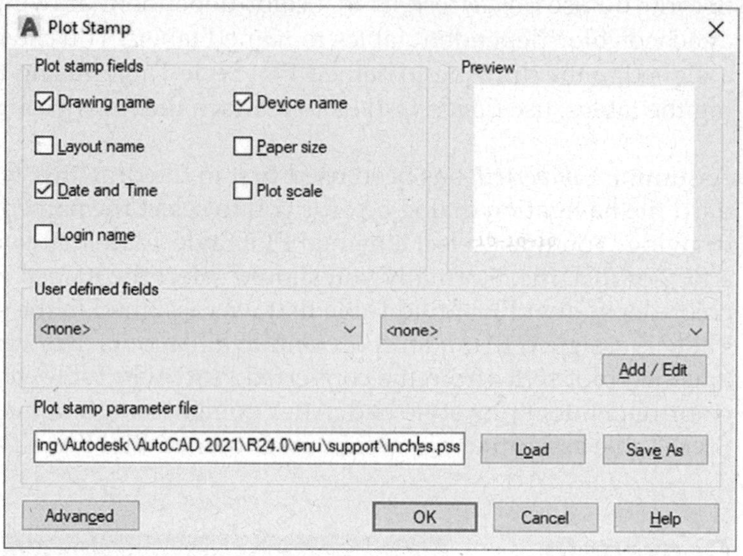


**TIP**

Saving plot stamp parameters involves changing a plot stamp parameter (.PSS) file. Two .PSS files are supplied by AutoCAD, INCHES.PSS and MM.PSS. You can change either of these files or create new files. A simple method for creating your own .PSS files is to use either of the AutoCAD-supplied files as a template, make the desired changes in the *Plot Stamp* dialog box, then use the *Save As* button to save the changes to a new file name. (See "*Plot stamp parameter file*.")

The .PSS files are stored (by default for stand-alone installations) under your user name in the Plot Styles folder. To find the location of the .PSS files on your system, see the edit box in the *Plot stamp parameter file* section (lower-left corner), or use *Options*, *Files* tab, and expand the *Working Support File Search Path* section.

### *Plot stamp fields*

In this area, check any items you want to appear in the plotted drawing. Each piece of text is added to the plot stamp separated by a comma. Content for *Drawing name*, *Layout name*, *Device name*, *Paper size*, and *Plot scale* is automatically obtained from the AutoCAD drawing. *Plot scale* may not be an appropriate item to select when you plot a layout since this text is actually the plot scale that appears in the *Plot* dialog box (normally set to 1:1), not the viewport scale (the actual plot scale of the geometry). Content for *Date and Time* and *Login name* are obtained from the computer's system data.

### *Preview*

This image tile indicates only the location and orientation of the plot stamp, not the text content. You cannot see a preview of the actual plot stamp using the *Full Preview* button or the *Preview* command; you must make a plot.

### *User defined fields*

You can specify additional text to include in the plot stamp by selecting the *Add/Edit* button. The *User Defined Fields* dialog box appears (not shown) for you to enter new text or edit existing user text fields. Once specified, this personalized text appears in the *User defined fields* drop-down lists (see Fig. 32-59). Two user defined fields can be added to a plot stamp.

### *Plot stamp parameter file*

You can store the current plot stamp information in a file with a .PSS extension. For example, if you wanted to create a standard plot stamp for your company or school, first specify the plot stamp fields and advanced options you want, then save them to a .PSS file. Other users can then access the saved file and stamp their plots based on those standard settings.

To create a plot stamp parameter file, use either the AutoCAD-provided MM.PSS or INCHES.PSS file as a template, then save your changes to a new name using the *Save As* button. The *Load* button displays the standard file selection dialog box in which you can specify the location of the parameter file you want to use. The MM.PSS or INCHES.PSS files differ only in millimeter or inch units used for the default text height and offset.

### *Advanced Options*

This button produces the *Advanced Options* dialog box (Fig. 32-60) where you can specify the *Location*, *Orientation*, and *Text properties* of the plot stamp. The *Location* and *Orientation* are reflected in the *Preview* tile of the *Plot Stamp* dialog box. The *X offset* and *Y offset* are the distances from the edge of the printable area or paper border you want the stamp to appear.

**FIGURE 32-60**

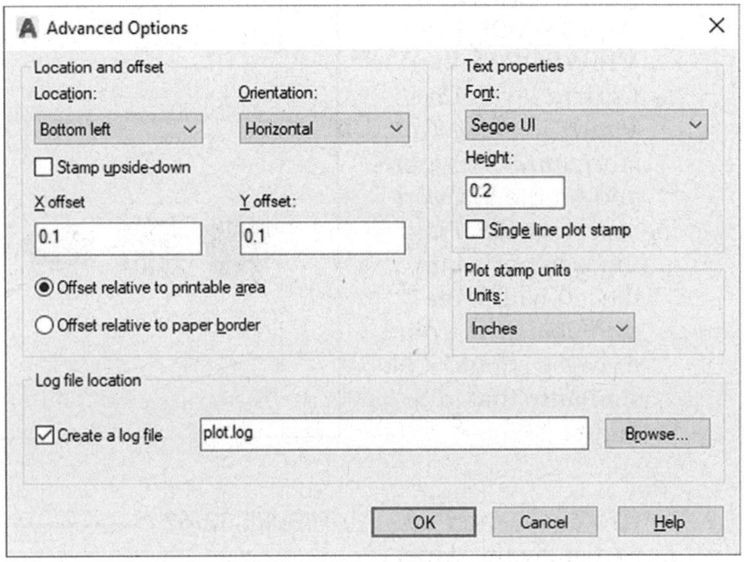


In the *Log file location* area you can indicate if you want to *Create a log file* and specify the name and location of the file. A log file is an ASCII text file that contains a record for each plot created, and each record lists the same items you specified for the plot stamp such as the date and time, drawing name, device name, and so on.

TIP

## CHAPTER EXERCISES

1. **Multiple Viewports and Viewport-Specific Layer Visibility**

   A. ***Open*** the **Single Cavity Mold** drawing. The **Single Cavity Mold** drawing is available from the website for this text (the website address is listed on the back cover and in the Preface under "Online Resources"). Use ***SaveAs*** to rename the drawing **Ch32 SCM** and locate it in your working directory. Activate each of the layout tabs and the model tab to familiarize yourself with the drawing.

   B. Activate the **Plan of Ejector** tab. Right-click on the tab and select ***Move or Copy*** from the shortcut menu. In the *Move or Copy* dialog box, select ***Plan of Ejector*** and ***Create a copy***, then ***OK***. Now activate the new layout, right-click and ***Rename*** the tab **Ejector Details**.

   C. With the new layout activated, use ***Pagesetup*** to select an available printer as the *Printer/Plotter*. Select ***Letter*** as the *Paper Size*. Select ***OK*** to save the settings for the layout. Next, use the ***Layer*** drop-down list and select ***Freeze or thaw in current viewport*** to freeze layer **Bourder**. Make a test print of the layout. Note in the print that the dimensions are barely readable at that size, and the linetype scale is too large.

D. Using **grips**, select the existing viewport and **STRETCH** it to modify the viewport size to occupy only the right half of the printable area as shown in Figure 32-61. Next, make layer **LAYOUT-VIEWPORTS** the ***Current*** layer. Use ***Vports*** to make ***Two: Horizontal*** viewports and set the *Viewport Spacing* to **.25**. ***Fit*** the viewports into the left half of the printable area. Your drawing should look similar to that in Figure 32-61.

**FIGURE 32-61**

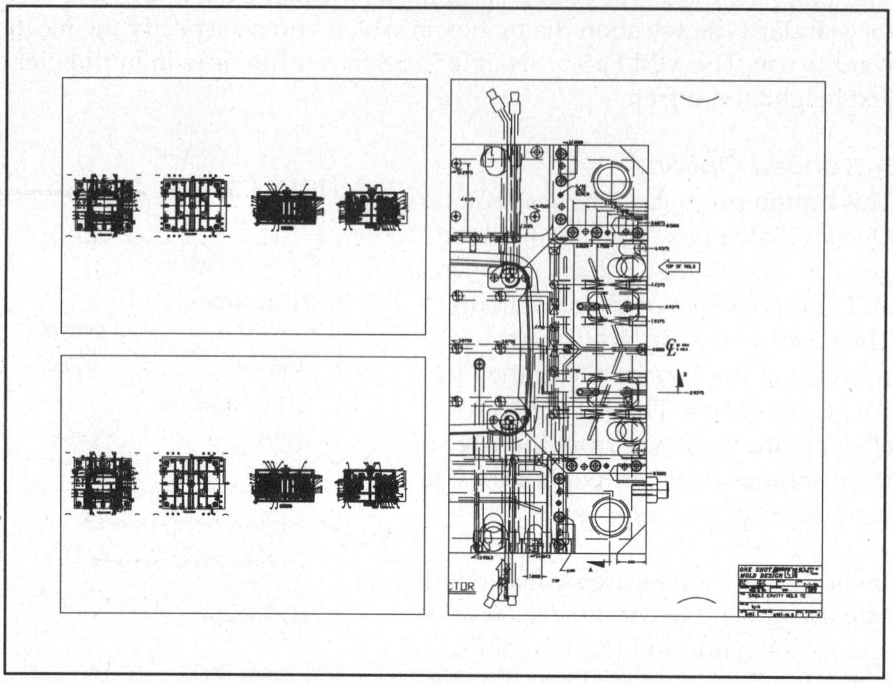


E. Use ***Zoom*** and ***Pan*** to display the upper-right corner of the mold in the small top viewport and the lower-right corner of the mold in the bottom viewport.

**FIGURE 32-62**

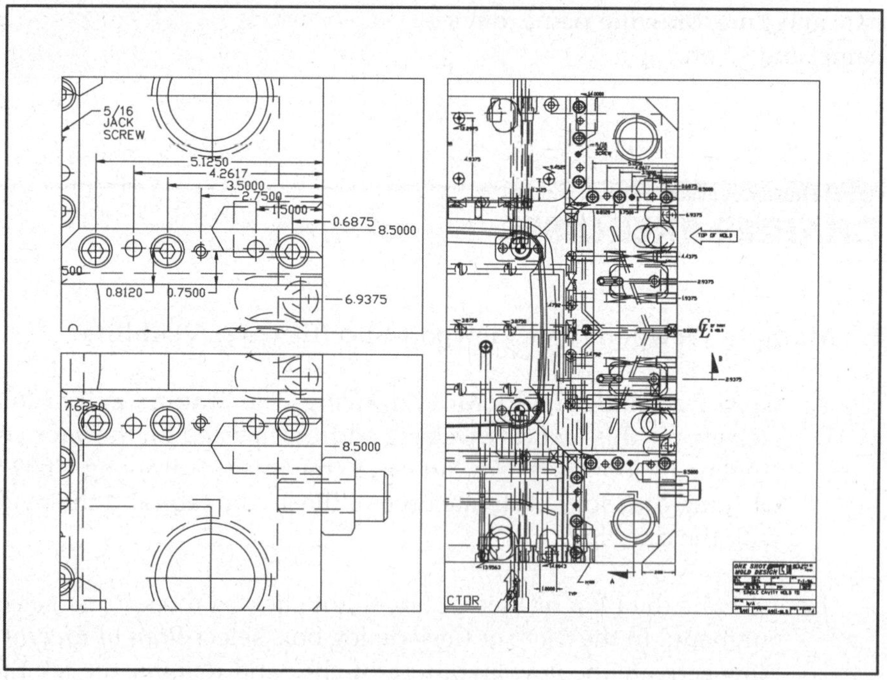


Use the ***Viewports*** pop-up list and set the viewport scales as follows:

| | |
|---|---|
| top left viewport | **1:2** |
| bottom left viewport | **1:2** |
| right viewport | **3:16** (.1875) |

The resulting display should look like Figure 32-62.

F. Activate the right viewport and use the ***Layer Properties Manager*** and select ***VP Freeze*** to freeze these layers for the right viewport only:

**A-H20**
**AIR**
**B-H20**
**DIMF**
**SLID1H20**
**SLID2H20**

G. Activate paper space. On layer **NOTES**, add two circles and leaders to indicate Detail A and B. Make the text size **.12** and the text font ***Romans***. Add the labels and scales for the two small viewports as shown in Figure 32-63. Set ***LTSCALE*** to **.5**. Finally, ensure ***PSLTSCALE*** is set to **1** so the linetype dashes appear equal in all viewports. ***Save*** the drawing and make a print of the layout.

**FIGURE 32-63**

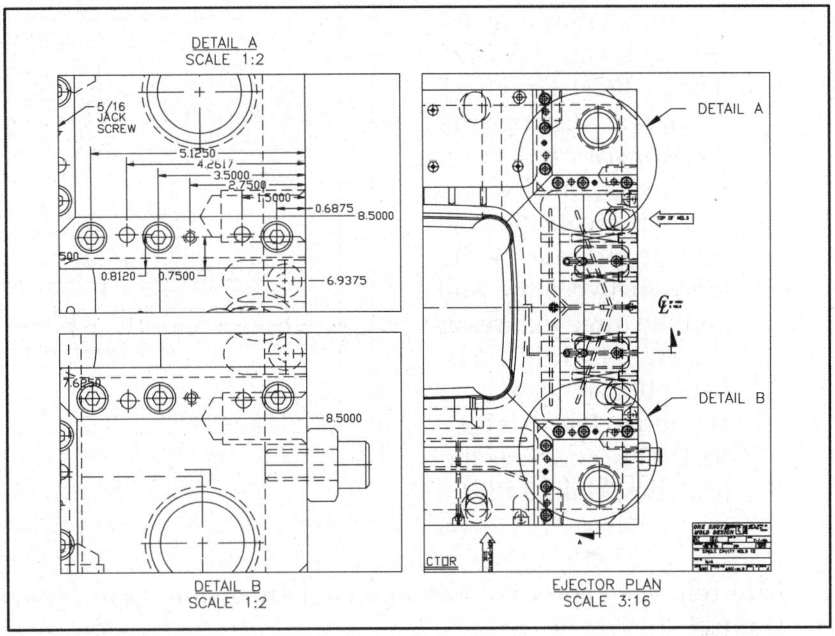


2. **Multiple Viewports and Dimensioning in Paper Space**

In this exercise, you will create a layout with two viewports to show an overall view of a drawing and a detail in a different scale. You will dimension the drawing both in model space and in paper space.

A. ***Open*** the **2BR-APT** you worked on last in Chapter 20 Exercises. Create a text ***Style*** using ***Roman Simplex*** font. Create a new dimension style named **ARCH-1** using the new font and with other settings to create architectural style dimensions as shown in Chapter 15, Figure 15-45. The drawing scale factor is **48**, so set the appropriate ***Overall scale***. In model space, create dimensions for the apartment on the **DIM** layer. Make sure you add dimensions also to the new bedroom in your drawing. If you need assistance placing the dimensions, refer to Figure 32-64 on the next page. Use ***SaveAs*** and name the drawing **2BR-APT-DIM**.

B. Create a new layout using the ***Template*** option. Select the ***ANSI B template*** with an ***ANSI B title block***. These templates are available from the website for this text (the website address is listed on the back cover and in the Preface under "Online Resources"). Copy the templates into your working folder or into the folder with the other standard AutoCAD templates. (If you cannot find the folder for the AutoCAD templates folder, use the *Options* dialog box *Files* tab to locate the path.) Next, use ***Pagesetup*** to specify a plot device and paper size for an ***ANSI B*** size sheet.

C. In the layout, create one viewport on the **VPORTS** layer that occupies about 2/3 of the layout area. ***Zoom*** and ***Pan*** as needed to show the entire floor plan. Set the viewport scale to **1/4″=1′**. If the entire floor plan does not appear in the viewport, make the necessary adjustments to the viewport border using **grips**.

D. Next, make a second smaller viewport on the right side of the layout. ***Zoom*** and ***Pan*** to display only the bathroom at the far left of the floor plan. Set the viewport scale to 1/2″=1′.

E. Make a new layer named **DIM-PS** and set the properties the same as layer DIM. Make a ***New*** non-annotative dimension style called **ARCH-2** using ARCH-1 as a "template." Change the ***Overall scale*** to **1**. In paper space and on layer DIM-PS, create dimensions to indicate the distances between the wall and the three fixtures as shown in Figure 32-64. Label the viewports as shown, ***Freeze*** layer **VPORTS**, and ***Save*** the drawing. Make a ***Plot***.

**FIGURE 32-64**

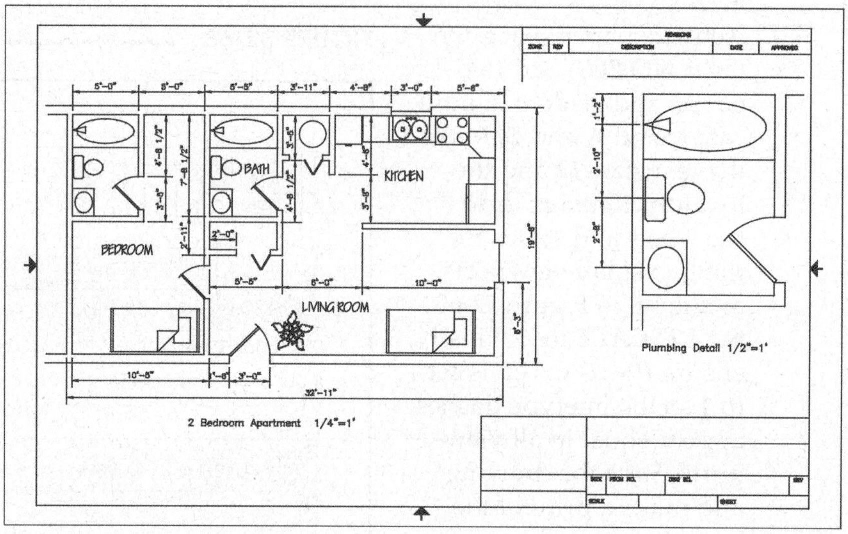


3. **Multiple Viewports, Non-annotative Dimensioning in Model Space, and Viewport-Specific Layer Control**

In this exercise, you will create a layout with three viewports to show an overall view of a drawing and two details in a different scale. You will dimension the drawing in model space and control the visibility for each viewport.

A. ***Open*** the **EFF-APT2** drawing that you worked on in Chapter 30 Exercises (with the door *Stretched* to the center of the front wall). ***Erase*** the plant. Create a text ***Style*** using ***Roman Simplex*** font. In model space, create non-annotative dimensions for the interior of each room on the **DIM** layer, similar to those dimensions you created in Exercise 2. (You may consider also opening the **2BR-APT-DIM** drawing from the last exercise and using copy and paste to copy the needed dimensions to the new drawing. In this case, however, the dimensions will have to be reassociated.) ***Zoom*** in, and on another ***New Layer*** named **DIM2** give the dimensions for the wash basin in the bath and the counter for the kitchen sink.

B. To set up the layout, you have two choices: (1) use a ***Layout Template,*** or (2) create and set up the layout from scratch. In either case, use the template, plot device, and paper size of your choice.

C. Make a ***New Layer*** or use the existing layer named **VIEWPORT** and set it as the ***Current*** layer. Use ***Vports*** to make one viewport on the right side of the layout, occupying approximately 2/3 of the page. Use ***Vports*** again to make two smaller viewports in the remaining space on the left.

D. Use *SaveAs* and name the drawing **EFF-APT-3**. Next, use *SaveAs* again and name the new drawing **EFF-APT-ANNO** (this copy of the drawing will be used for another exercise). *Close* the drawing, then *Open* drawing **EFF-APT-3**.

E. Activate the large viewport (in *Mspace*) and use *Viewport Scale* or other viewport scale method to achieve the largest possible display of the apartment to an accepted scale. In the top-left viewport, use display controls to produce a scaled detail of the bath. Produce a scaled detail of the kitchen sink in the lower-left viewport, similar to that shown in Figure 32-65.

**FIGURE 32-65**

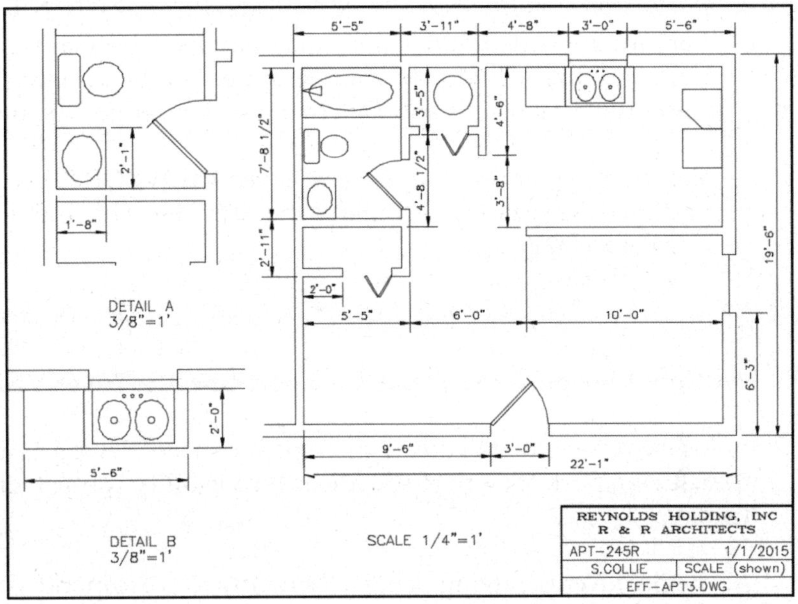


F. Use the *Layer Properties Manager* to *VP Freeze* layer **DIM2** in the large viewport and *VP Freeze* layer **DIM** in the two small viewports. Use *Properties* and select the dimensions that appear in the two detailed views and set a new *Dim scale overall* (in the *Fit* category) so the dimensions appear at the appropriate size for the viewport scale.

G. Return to paper space and *Freeze* layer **VIEWPORT**. Use *Text* (in paper space) to label the detail views and give the scale for each. *Save* the drawing as **EFF-APT3** and *Plot* from paper space at **1:1**. The final plot should look similar to Figure 32-65.

4. **Multiple Viewports, Annotative Dimensioning**

A. Open drawing **EFF-APT-ANNO** that you created in the last exercise. Open the *Dimension Style Manager* and change the current dimension style (used to create the room dimensions) to *Annotative*. In the *Model* tab, use the *Annoupdate* command to update all the existing dimensions.

B. Return to the layout you set up previously with the three viewports. For the large viewport, set the appropriate *Viewport Scale* to display the entire apartment as large as possible. The dimensions should automatically adjust to display correctly based on the new viewport scale. (If the original dimensions were not created in the correct *Overall scale*, correct scale representations of the dimensions should be automatically created.)

C. In the top-left viewport, use *Viewport Scale* to create a detail view of the bath to an accepted scale. *Pan* as needed. In the bottom-left viewport, use *Viewport Scale* to create a detail view to display the kitchen sink to an accepted scale. *Pan* as needed.

D. To add dimensions to each of these viewports, you will create new annotative dimensions directly in each viewport in the layout, rather than in the *Model* tab. Using the current annotative dimension style, the dimensions should be correctly sized based on the viewport scale. Create the new dimensions for the two sinks (as shown in previous Figure 32-65) using the ***Linear*** (*Dimlinear*) command. Make sure the new dimensions do not appear in the main viewport by using the ***Annotation Visibility*** toggle.

E. For any unwanted dimensions that may have been automatically created in the two small viewports, but that you do not want to appear in those viewports, use the ***Objectscale*** command to select the dimensions and ***Delete*** the scale used for the main viewport.

F. Return to paper space and ***Freeze*** layer **VIEWPORT**. Use ***Text*** in paper space to label the views and give the scale for each. ***Save*** the drawing. The drawing should look similar to that in previous Figure 32-65.

5. **Multiple Viewports, *Xref*, and Viewport-Specific Layer Visibility**

REUSE

In this exercise, you will create three paper space viewports and display three *Xrefed* drawings, one in each viewport. Viewport-specific layer visibility control will be exercised to produce the desired display.

A. Begin a ***New*** drawing using a ***Template*** to correspond to the sheet size you intend to plot on. Activate a ***Layout*** tab. Use ***Pagesetup*** to select the paper size and to set the plotting options for the device you use to plot with. Create ***New Layers*** named **TITLE**, **XREF**, and **VIEWPORT**. On the **VIEWPORT** layer, use ***Vports*** to create a viewport occupying about 1/2 of the page on the right side.

B. Make the **XREF** layer ***Current***. Double-click inside the viewport to activate model space in the viewport. In the viewport, ***Xref*** the GASKETD drawing from Chapter 28 Exercises. Use ***Viewport Scale*** or use the ***Viewport scale*** drop-down list (in the *Viewports* toolbar) to scale the gasket appropriately to paper space. Use the ***Layer Properties Manager*** to ***Freeze*** (globally) the **GASKETD | DIM** layer, then use ***New VP Freeze*** for all **GASKETD*** layers.

C. Option 1. Change to ***PAPER*** (or double-click in paper space) and create two more viewports on the left side on layer **VIEWPORT**, each equal in size to about half of the viewport on the right. Set layer **XREF** ***Current*** and ***Xref*** the GASKETB (Chapter 9) and the GASKETC (Chapter 16) drawings. Change to ***MODEL*** and use the ***Layer Properties Manager*** or ***Vplayer*** to constrain only one gasket to appear in each viewport.

Option 2. Change to ***PAPER*** (or double-click in paper space) and create one more viewport on layer **VIEWPORT** equal to one half the size of the viewport on the right. Set layer **XREF** ***Current*** and ***Xref*** only the GASKETB drawing. Change to ***MODEL*** and use the ***Layer Properties Manager*** or ***Vplayer*** to set the **GASKETB** visibility for the existing and new viewports. Repeat these steps for GASKETC.

D. Use ***Viewport Scale*** or the *Viewport Scale* drop-down list to produce a scaled view of each new gasket. Change to ***PS*** and ***Freeze*** layer **VIEWPORT**. Use ***Text*** to label the gaskets. ***Plot*** the drawing to scale. Your plot should look similar to Figure 32-66. ***Save*** the drawing as **GASKETS**.

**FIGURE 32-66**

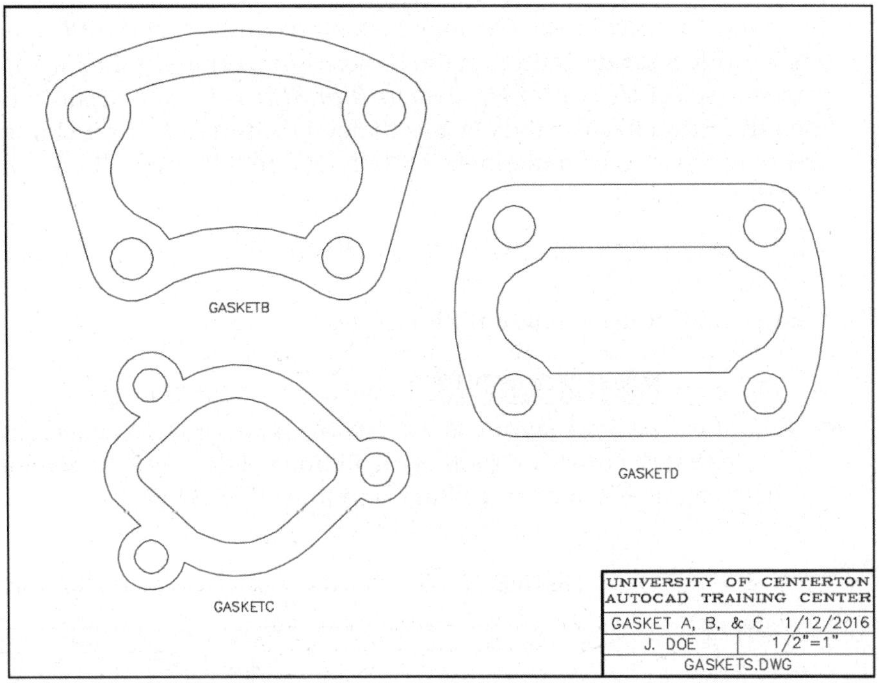


6. **Create a Color-Dependent Plot Style Table**

Start AutoCAD and begin a ***New*** drawing by any method. Open the ***Plot Style Manager***. Use the ***Add-A-Plot Style Table Wizard*** to create a new Plot Style Table. In the ***Begin*** step, select ***Start from Scratch***. In the next step, select ***Color-Dependent Plot Style Table***, then proceed to assign the name **SCREEN PERCENTS**. When you are finished, check to make sure the new Plot Style Table appears in the *Plot Style Manager*.

7. **Edit a Color-Dependent Plot Style Table**

Open the ***Plot Style Manager***. Double-click on the **SCREEN PERCENTS** Plot Style Table to cause the *Plot Style Table Editor* to appear. Use the *Table View* or *Form View* to assign ***Screening*** values of **100**, **75**, **50**, and **25** to plot styles **Color 1**, **Color 2**, **Color 3**, and **Color 4**, respectively. Select ***Save & Close*** to save the new settings. Reopen the new Plot Style Table to ensure your settings were saved.

8. **Create and Edit a Named Plot Style Table**

Open the ***Plot Style Manager*** if not yet open. Use the ***Add-A-Plot Style Table Wizard*** to create a new Plot Style Table. In the ***Begin*** step, select ***Start from Scratch***. In the next step, select ***Named Plot Style Table***, then proceed to assign the name **LINEWEIGHTS**. In the *Finish* step, select the ***Plot Style Table Editor*** button.

In the *Plot Style Table Editor*, create five plot styles named **1.20mm**, **.90mm**, **.60mm**, **.30mm**, and **.15mm**. Assign the respective settings from the ***Lineweight*** drop-down list. Select the ***Save & Close*** button. Finally, select the ***Finish*** button to close the wizard. In the *Plot Style Table Manager*, double-click on the new **LINEWEIGHTS.STB** file and check your settings.

9. **Set the Default Plot Style Behavior**

In AutoCAD, invoke the ***Options*** dialog box and select the ***Plot and Publish*** tab. Select the ***Plot Style Table Settings*** button in the lower-right corner. In the *Plot Style Table Settings* dialog box that appears, select ***Use Color Dependent Plot Styles***. Under *Default Plot Style Table*, select **None**. Select the ***OK*** button to close the *Options* dialog box and save your changes. Now, new drawings that you create will use color-dependent Plot Style Tables by default but can later be changed to use named Plot Style Tables.

REUSE

10. **Assigning Color-Dependent Plot Styles**

A. ***Open*** the **HAMMER-DIM** drawing you created in Chapter 29 Exercises. Make a plot from the *ANSI B Title Block* layout as it exists from the previous exercise (before assigning plot styles). Note that the drawing plots with all lines in the same lineweight, making it difficult to discern between object lines and dimension lines (Fig. 32-67).

**FIGURE 32-67**

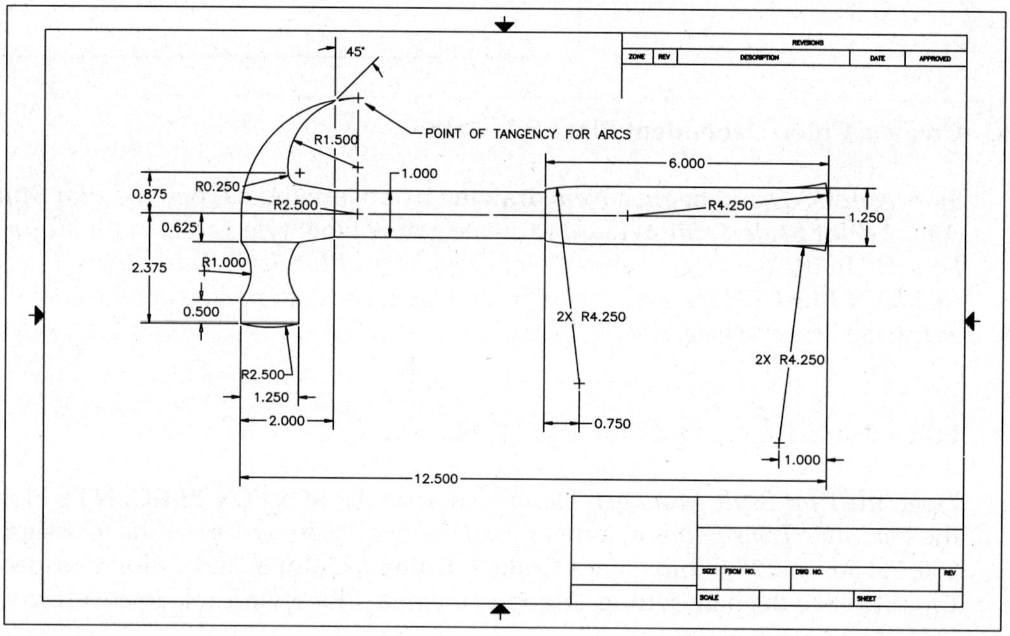


B. From the layout tab, invoke the ***Page Setup*** dialog box and attach the **SCREEN PERCENTS** Plot Style Table you created and edited in Exercises 6 and 7. Select the ***Edit*** button to view the plot styles settings (Color 1=100, Color 2=75, etc.). Use ***SaveAs*** and assign the name **HAMMER-PSTYLES**.

C. Use the ***Layer Properties Manager*** to see which plot styles are assigned to which layers. Since the plot styles are assigned by color, you will have to change the layer colors to achieve the desired screen percentages. Make the **GEOMETRY** layer color 1 (red) and the **DIMENSIONS** layer color 3 (green).

D. Use the ***Page Setup*** dialog box again and check ***Display Plot Styles*** in the expanded section. Check the layout to ensure the plot styles are assigned correctly. Your display should look similar to Figure 32-68. Finally, make a ***Plot***. The plot should display the geometry at a 100 percent screen and the dimensions in a lighter screening. Save the drawing.

**FIGURE 32-68**

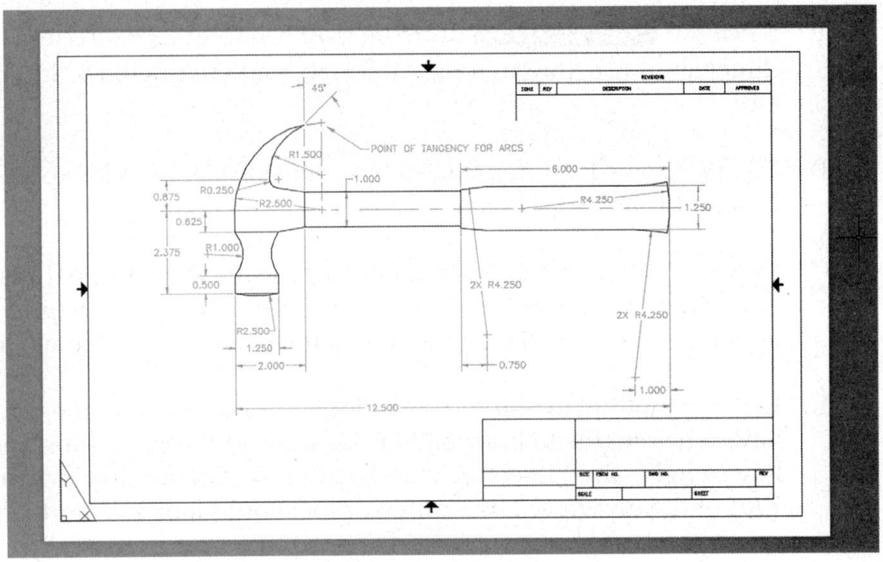


E. Experiment by changing the **DIMENSIONS** layer to color 3 or 4. Make plots to see the results.

11. **Assign Named Plot Styles**

A. ***Open*** the **FLOOR PLAN SAMPLE** drawing from the Program Files/AutoCAD 2017/Samples/Database Connectivity folder. This drawing was created to use a named Plot Style Table. Assuming you are the general contractor, you are required to create a plot that displays the entire plan but highlights the building's concrete core.

B. From the layout tab, invoke the ***Page Setup*** dialog box and attach the **LINEWEIGHTS.STB** Plot Style Table you created and edited in Exercise 7. Select the ***Edit*** button to view the plot styles settings and style names. Also check ***Display Plot Styles***. Use ***SaveAs*** and assign the name **VP_SAMP-PSTYLES** and save the file to your working directory.

C. Open the ***Layer Properties Manager***. ***Freeze*** all the ***Cyan*** layers except **E-F-TERR**. Also ***Freeze*** layers **FILE_CABINETS** and **FURNITURE**.

D. Assign the **.15mm** plot style to all layers. Then assign the **.90mm** plot style to layer **E_B_CORE**. Your display should look like Figure 32-69 (this figure does not display the viewport). ***Save*** the drawing. Make a ***Plot*** and check your results.

**FIGURE 32-69**

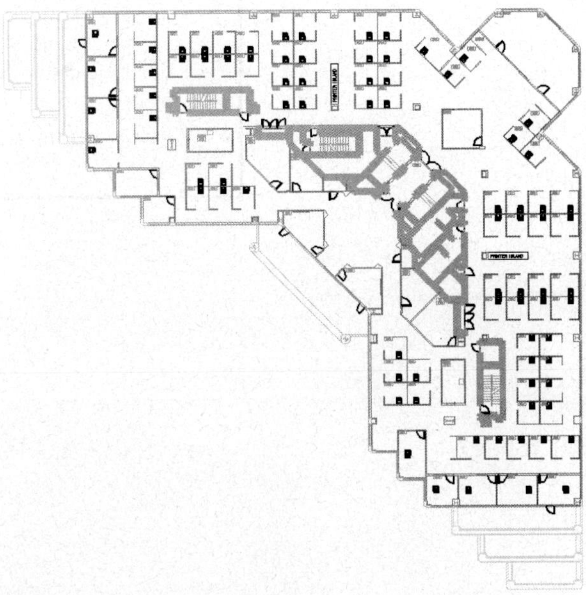

REUSE

12. **Assign Named Plot Styles**

A. ***Open*** the **SADDL-DM** drawing from Chapter 29 Exercises. The model geometry and dimensions are drawn in the *Model* tab, and the titleblock and border are inserted into a layout tab.

B. Use Windows Explorer to search for the **ACAD.CTB**. Make a ***Copy*** of the file and ***Rename*** it **LINEWEIGHTS.CTB**. ***Exit*** Explorer.

C. In AutoCAD, select the layout tab and use the ***Page Setup*** dialog box to attach the new Plot Style Table that you created in the previous step. Also select ***Display Plot Styles***. There should be no apparent change in the layout. Use ***SaveAs*** and assign the name **SADDL-DIM-PLOTSTYLES**.

D. Use any method to edit the new Plot Style Table so that the color of the geometry object lines have a heavy (thick) lineweight assigned and the dimension, hidden line, and center line colors have a light (thin) lineweight assigned. If the layout displays the geometry as you intend, make a ***Plot*** of the drawing. The resulting plot should look similar to Figure 32-70. ***Save*** the drawing.

**FIGURE 32-70**

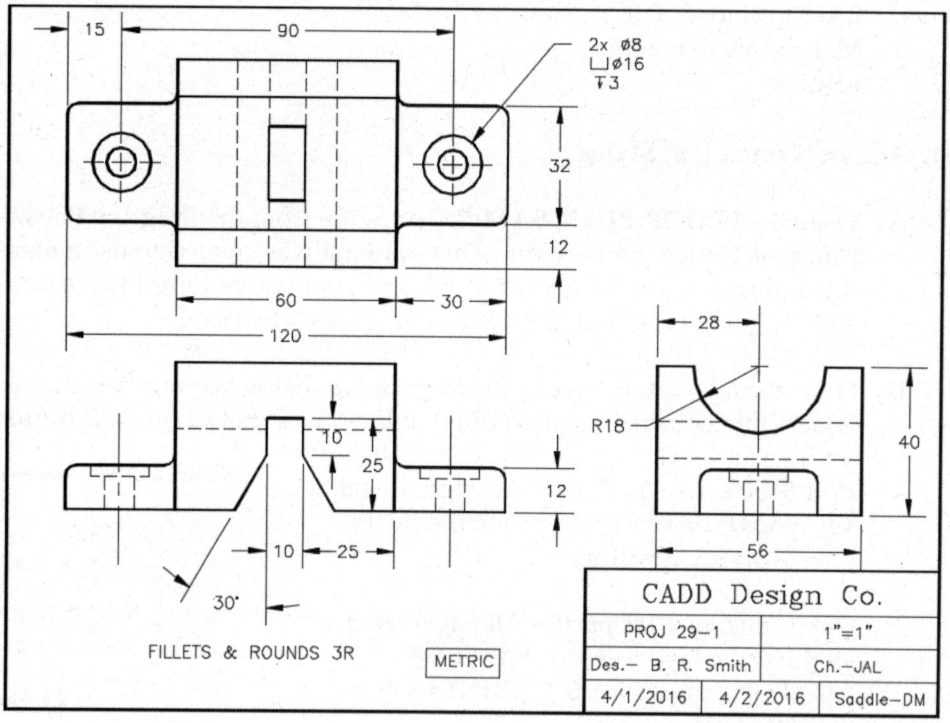

CHAPTER

# 33

# 3D Basics, Navigation, and Visual Styles

## Chapter Objectives

After completing this chapter you should:

1. be able to set up AutoCAD for 3D drawings;
2. know the characteristics of wireframe, surface, and solid models;
3. know the formats for 3D coordinate entry;
4. understand the orientation of the World Coordinate System, the purpose of User Coordinate Systems, and the use of the *Ucsicon*;
5. be able to use 3D navigation and view commands to generate different views of 3D models;
6. be able to create named views and use them in layouts;
7. be able to use the *Visual Styles* control panel and the *Visual Styles Manager* to specify variations of wireframe, hidden, and shaded displays of 3D models.

## SETTING UP AUTOCAD TO DRAW IN 3D

In AutoCAD often you will want to begin drawing in 3D by 1) setting up the workspace and 2) activating one of the 3D drawing templates.

FIGURE 33-1

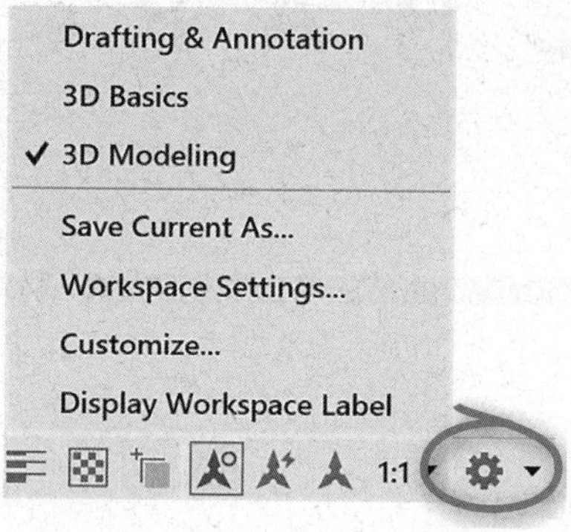


### Setting Up a Workspace

Besides the *Drafting & Annotation* workspace, two other workspaces are available in AutoCAD for drawing in 3 dimensions (3D): the *3D Basics* and the *3D Modeling* workspace. The *3D Basics* workspace is used for creating simple 3D shapes, while the *3D Modeling* workspace provides a more robust selection of options for creating and editing surfaces and solids.

You can set the workspace by selecting the desired option from the *Workspaces* pop-up list from the status bar area (Fig. 33-1). If the *Workspace Switching* icon doesn't show on the status bar, turn it on using the *Customization* selection from the status bar.

#### The *3D Basics* Workspace

The *3D Basics* workspace is structured in the "Ribbon" format and contains basic 3D drawing and editing commands (Fig. 33-2).

FIGURE 33-2

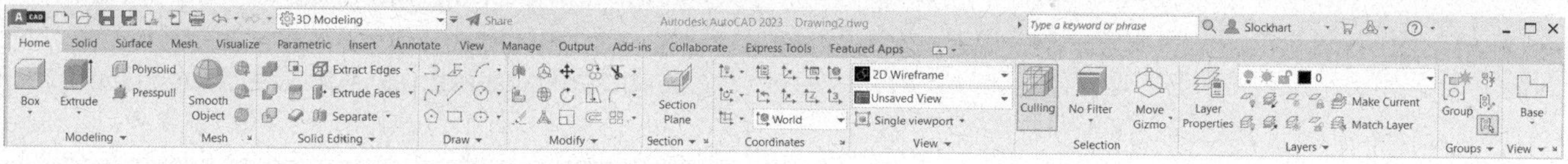


#### The *3D Modeling* Workspace

The *3D Modeling* workspace is also structured in the "Ribbon" format similar to the *3D Basics* workspace and contains panels for more advanced 3D modeling and editing (Fig. 33-3).

FIGURE 33-3

NOTE: For the purpose of locating commands on the Ribbon, the command tables in this book refer to the *3D Modeling* workspace. Also, this book displays all workspaces with a light background, whereas the AutoCAD default background color is dark for the *3D Basics* and *3D Modeling* workspaces.

FIGURE 33-4

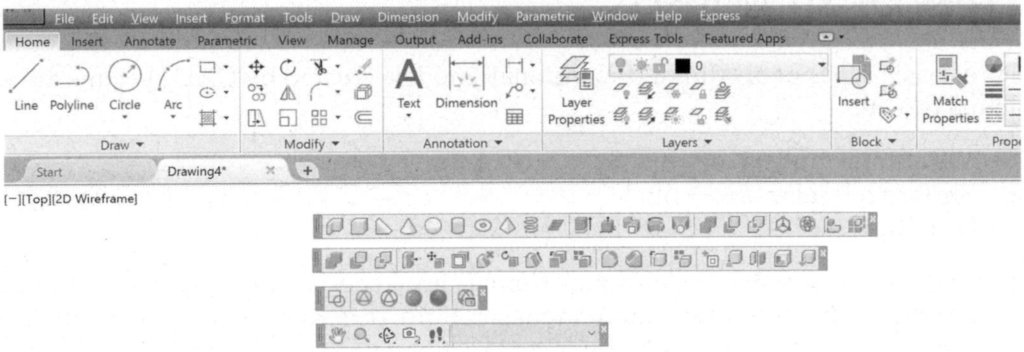


The *Drafting & Annotation* Workspace for 3D Modeling

To use the *Drafting & Annotation* workspace for 3D modeling, you can activate several toolbars to provide access to the commands useful for 3D modeling, editing, and navigating such as:

*Modeling*
*Solid Editing*
*Visual Styles*
*3D Navigation*

In addition, the *Orbit*, *Camera Adjustment*, *Smooth Mesh*, and *Smooth Mesh Primitives* toolbars are available for specific tasks.

## Activating the ACAD3D or ACADISO3D Drawing Template

For drawing in 3D, it is highly recommended that you use one of the provided 3D drawing templates rather than one of the 2D templates such as ACAD.DWT. The 3D templates are ACAD3D.DWT (for inches) and ACADISO3D (for metric). These drawings begin in a 3D perspective mode and have default shades of color assigned to 3D object faces and to other areas of the drawing, such as the ground and sky. Figure 33-3 displays a 3D template drawing. Note the 3D coordinate icon and the perspective grid.

To begin with a 3D template, use the *New* command to begin a new drawing. When the *Select Template* dialog box appears, select the **ACAD3D.DWT** or **ACADISO3D.DWT** template (Fig. 33-5).

If you intend to create 3D drawings continually, you might want to use the *Options* dialog box and change the *Default Template File Name for QNEW* to one of these 3D templates.

FIGURE 33-5

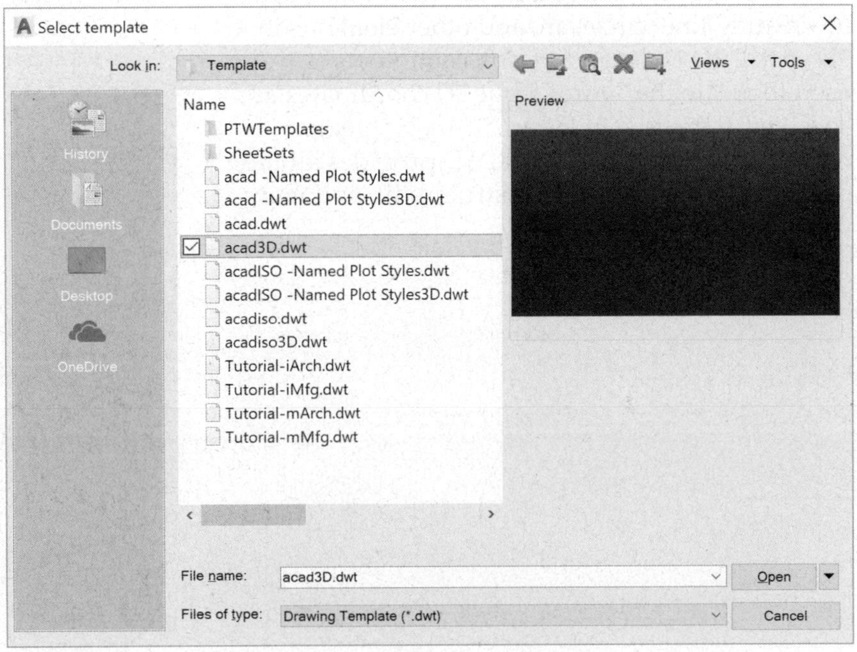


TIP

It is possible to use any other 2D drawing template, such as ACAD.DWT, to create 3D models. This might be useful, for example, if you decide to convert portions of a previously created 2D drawing to a 3D model. However, if you are starting from scratch to create a 3D model, the 3D templates should be used. The 3D templates have several system variable settings optimized for 3D geometry, particularly those variables that enhance visualization of a 3D object.

# TYPES OF 3D MODELS

Three basic types of 3D (three-dimensional) models created by CAD systems are used to represent actual objects.

1. Wireframe models
2. Surface models
3. Solid models

These three types of 3D models range from a simple description to a very complete description of an actual object. The different types of models require different construction techniques, although many concepts of 3D modeling are the same for creating any type of model on any type of CAD system.

## Wireframe Models

"Wireframe" is a good descriptor of this type of modeling. A wireframe model of a cube is like a model constructed of 12 coat-hanger wires. Each wire represents an edge of the actual object. The surfaces of the object are not defined; only the boundaries of surfaces are represented by edges. No wires exist where edges do not exist. The model is see-through since it has no surfaces to obscure the back edges. A wireframe model has complete dimensional information but contains no volume. Examples of wireframe models are shown in Figures 33-6 and 33-7.

FIGURE 33-6

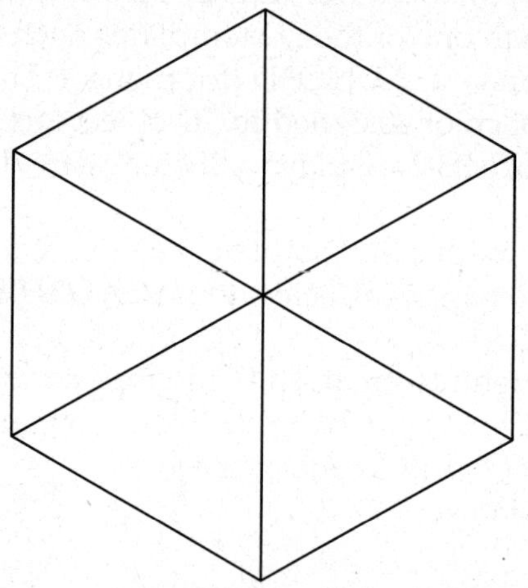

Wireframe models are relatively easy and quick to construct; however, they are not very useful for visualization purposes because of their "transparency." For example, does Figure 33-6 display the cube as if you are looking toward a top-front edge or looking toward a bottom-back edge? Wireframe models tend to have an optical illusion effect, allowing you to visualize the object from two opposite directions unless another visual clue, such as perspective, is given.

With AutoCAD, a wireframe model is constructed by creating line, circle, arc and other elements in 3D space. 2D *Draw* and *Modify* commands are used to create the "wires," but 3D coordinates are specified. The cube in Figure 33-6 was created with 12 *Line* segments. AutoCAD provides all the necessary tools to easily construct, edit, and view wireframe models.

FIGURE 33-7

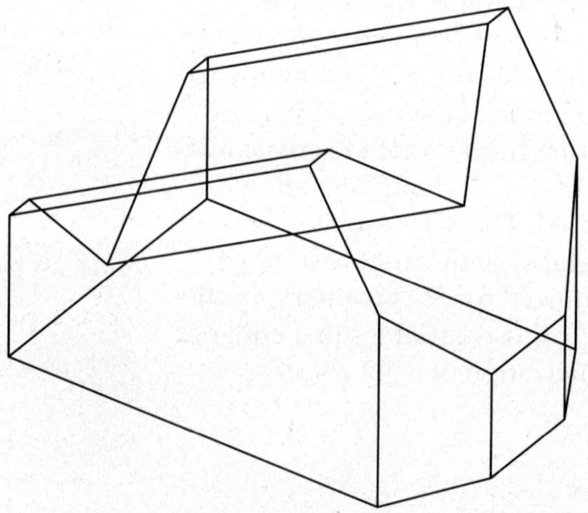

A wireframe model offers many advantages over a 2D engineering drawing. Wireframe models are useful in industry for providing computerized replicas of actual objects. A wireframe model is dimensionally complete and accurate for all three dimensions. Visualization of a wireframe is generally better than a 2D drawing because the model can be viewed from any position or a perspective can be easily attained. The 3D database can be used to test and analyze the object three dimensionally. The sheet metal industry, for example, uses wireframe models to calculate flat patterns complete with bending allowances. A wireframe model can also be used as a foundation for construction of a surface model. Because a wireframe describes edges but not surfaces, wireframe modeling is appropriate to describe objects with planar or single-curved surfaces, but not compound curved surfaces.

**FIGURE 33-8**

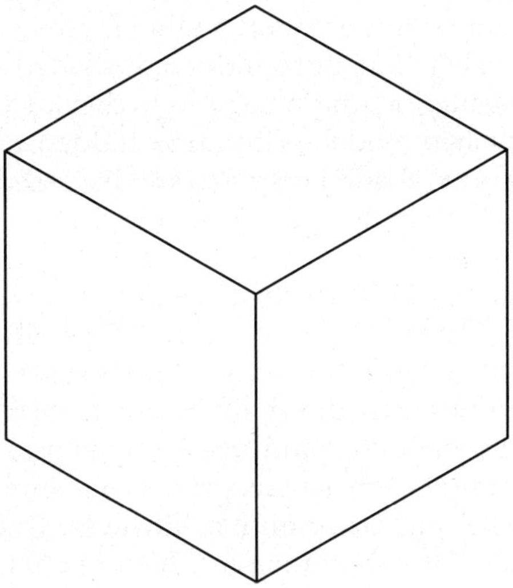

## Surface Models

Surface models provide a better description of an object than a wireframe, principally because the surfaces as well as the edges are defined. A surface model of a cube is like a cardboard box—all the surfaces and edges are defined, but there is nothing inside. Therefore, a surface model has volume but no mass. A surface model provides a better visual representation of an actual 3D object because the front surfaces obscure the back surfaces and edges from view. Figure 33-8 shows a surface model of a cube, and Figure 33-9 displays a surface model of a somewhat more complex shape. Notice that a surface model leaves no question as to which side of the object you are viewing.

Surface models require a relatively tedious construction process. Each surface must be constructed individually. Each surface must be created in, or moved to, the correct orientation with respect to the other surfaces of the object. In AutoCAD, a surface can be constructed in a number of ways. Often, wireframe elements are used as a framework to build and attach surfaces. The complexity of the construction process of a surface is related to the number and shapes of its edges.

Because there are several generations of surfacing capabilities in AutoCAD, surface modeling in AutoCAD is somewhat ambiguous. Traditionally, surface modeling in AutoCAD was accomplished by creating polygon mesh surfaces such as a *Rulesurf, Tabsurf, Edgesurf,* or *Revsurf.* These older legacy surfaces can be used to construct surface models as shown in Figure 33-9, but cannot be used in conjunction with solid models. In addition, a region (see Chapter 15) can be used as a surface, but a region has different properties than a mesh and the *Region* command does not appear in any of the 3D modeling menus.

AutoCAD offers a number of newer surfaces much different than polygon meshes or regions. The *3D Modeling* workspace provides several commands that can be used to create surfaces as well as solid models. For example, *Extrude, Revolve, Sweep,* and *Loft* create a surface if applied to an "open" 2D object such as a *Line, Spline, Arc,* or *Pline.* However, *Extrude, Revolve, Sweep,* and *Loft* create a solid if applied to a "closed" 2D shape such as a *Circle, Ellipse, Region,* or closed *Pline.* In addition, only these newer surfaces (not regions or polygon meshes) can be used to "slice" solid models.

Surface models are capable of wireframe, hidden, or various shaded displays. A wireframe representation might be used during construction, but a hidden or shaded representation might be used to display the finished model. Wireframe, hidden, and various shaded displays are discussed later in this chapter (see "Visual Styles").

FIGURE 33-9

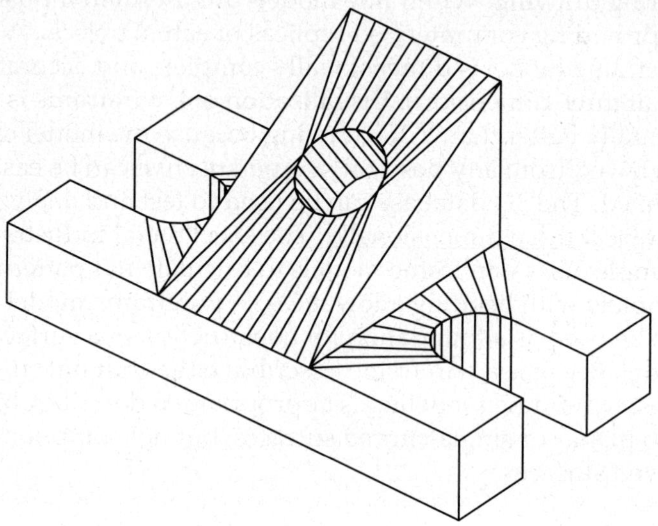

AutoCAD 2010 introduced mesh modeling. These new meshes are technically surfaces but can be converted to 3D solids. Meshes offer the ability to create incrementally smoothed and free-form shapes. For example, a mesh box primitive is similar to a solid box primitive; however, you can adjust the smoothness (which rounds the corners and edges) as well as alter the shape using "gizmo" tools. The finished shape can then be converted to a solid.

## Solid Models

Solid modeling is the most complete and descriptive type of 3D modeling. A solid model is a complete computerized definition of the actual object. A solid model contains the complete surface and edge description, as well as the interior features of the object. If a solid model is cut in half (sectioned), the interior features become visible. Since a solid model is "solid," it can be assigned material characteristics to define the mass properties of the object. Because solid models define a volume, most solid modeling systems include capabilities to automatically calculate volumetric and mass properties.

FIGURE 33-10

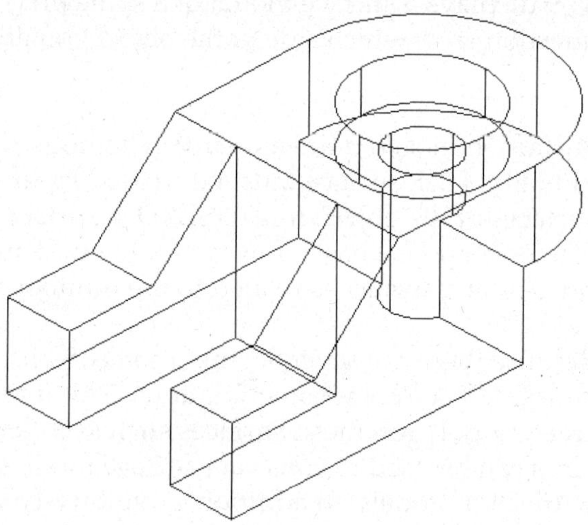

Techniques for constructing solid models are often simpler than those for constructing surface models. AutoCAD uses a process known as Constructive Solid Geometry (CSG) modeling. CSG is characterized by its simple and straightforward construction techniques of combining "primitive" shapes (boxes, cylinders, wedges, etc.) utilizing Boolean operations (*Union*, *Subtract*, and *Intersect*). CSG models store the "history" (record of primitives and Boolean operations) used to construct the models.

The CSG construction typically begins by specifying dimensions for simple primitive shapes such as boxes or cylinders, then combining the primitives using Boolean operations to create a "composite" solid. Other primitives and/or composite solids can be combined by the same process. Several repetitions of this process can be continued until the desired solid model is finally achieved.

Figure 33-10 displays a solid model constructed of simple primitive shapes combined by Boolean operations. CSG construction and editing techniques are discussed in detail in Chapters 35 and 36.

Technically, AutoCAD is a hybrid modeler. That is, the solid modeling kernel is used as a combination Boundary Representation (B-Rep) and CSG modeler. Boundary representation modeling defines a model in terms of its edges and surfaces (rather than by primitives and Booleans) and determines the solid model based on which side of the surfaces the model lies. The B-Rep capabilities (which are automatic and not apparent to the user) are used by AutoCAD for the display of the models (such as when a "mesh" representation is shown). However, the *Brep* command can be used to remove a composite solid's CSG history and reduce the editing capabilities of the model.

**FIGURE 33-11**

Solid models, like surface models, are capable of wireframe, hidden, or various shaded displays. For example, a wireframe display might be used during construction (see Fig. 33-10) and hidden or shaded representation used to display the finished model (Fig. 33-11). See "Visual Styles" in this chapter for more information.

The newer AutoCAD surfaces can be used to *Slice* solids. For example, a *Lofted* surface can be used to *Slice* through a box solid to create a complex-curved face on a solid. To do this operation, only the new surface types (those created by *Extrude*, *Revolve*, *Sweep*, and *Loft* applied to an "open" 2D object such as a *Line*, *Spline*, *Arc*, or *Pline*) can be used. The older, pre-AutoCAD 2007 surfaces, such as polygon meshes and regions, cannot be used for this purpose.

## 3D COORDINATE ENTRY

When creating a model in three-dimensional drawing space, the concept of the X and Y coordinate system, which is used for two-dimensional drawing, must be expanded to include the third dimension, Z, which is measured from the origin in a direction perpendicular to the plane defined by X and Y. Remember that two-dimensional CAD systems use X and Y coordinate values to define and store the location of drawing elements such as *Lines* and *Circles*. Likewise, a three-dimensional CAD system keeps a database of X, Y, and Z coordinate values to define locations and sizes of two- and three-dimensional elements.

For example, a *Line* is a two-dimensional object, yet the location of its endpoints in three-dimensional space must be specified and stored in the database using X, Y, and Z coordinates (Fig. 33-12). The X, Y, and Z coordinates are always defined in that order, delineated by commas. The AutoCAD Coordinate Display displays X, Y, and Z values.

FIGURE 33-12

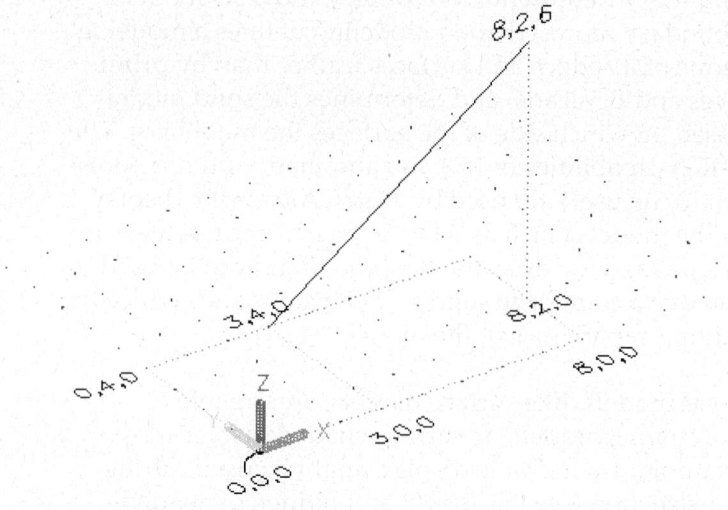


The icon that appears (usually in the lower-left corner of the AutoCAD Drawing Editor) is the Coordinate System icon (sometimes called the UCS icon). When you draw in the 2D mode the icon indicates only the X and Y axis directions (Fig. 33-13).

FIGURE 33-13

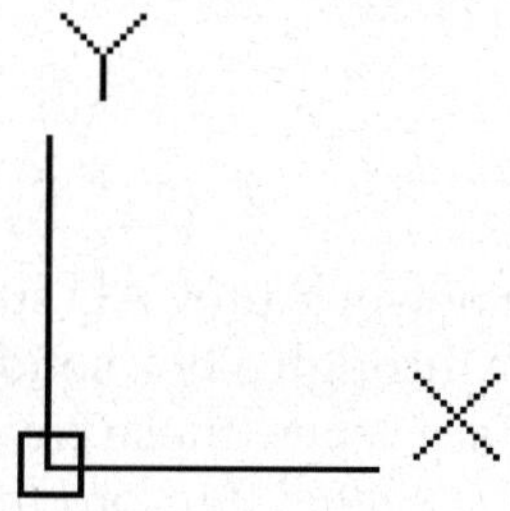


When you change your display of an AutoCAD drawing to one of the 3D *Visual Styles* or begin a drawing using the ACAD3D.DWT or ACADISO3D.DWT drawing template, the icon changes to a colored, 3-pole icon (Fig. 33-14). This UCS icon displays the directions for the X (red), Y (green), and the Z (blue) axes and is located by default at the origin, 0,0. Technically, the setting for *Visual Styles* controls which icon appears—the *2D wireframe* option displays the simple "wire" icon, whereas any 3D option displays the icon shown in Figure 33-14.

FIGURE 33-14

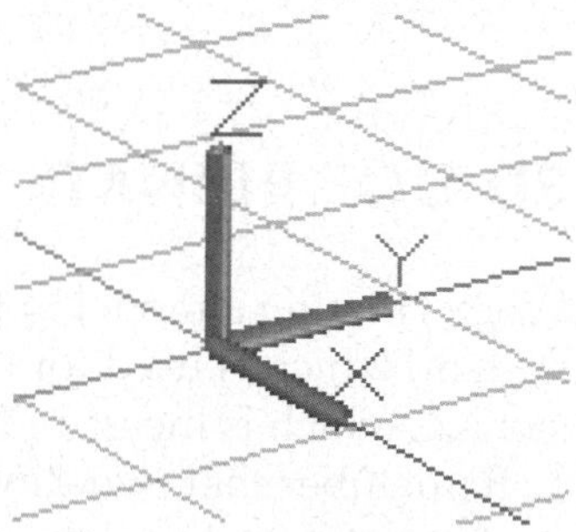


## 3D Coordinate Entry Formats

Because construction in three dimensions requires the definition of X, Y, and Z values, the methods of coordinate entry used for 2D construction must be expanded to include the Z value. The methods of command entry used for 2D construction are valid for 3D coordinates with the addition of a Z value specification. Relative Polar coordinate specification (@dist<angle) is expanded to form two other coordinate entry methods available explicitly for 3D coordinate entry.

The methods of coordinate entry for 3D construction follow:

| | | |
|---|---|---|
| 1. **Mouse Input** | **PICK** | Use the cursor to select points on the screen. *Object Snap,* Object Snap Tracking, or point filters must be used to select a point in 3D space; otherwise, points selected are on the XY plane. |
| 2. **Absolute Cartesian coordinates** | **X,Y,Z** | Enter explicit X, Y, and Z values relative to point 0,0,0. |
| 3. **Relative Cartesian coordinates** | **@X,Y,Z** | Enter explicit X, Y, and Z values relative to the last point. |
| 4. **Cylindrical coordinates (relative)** | **@dist<angle,Z** | Enter a distance value, an angle in the XY plane value, and a Z value, all relative to the last point. |
| 5. **Spherical coordinates (relative)** | **@dist<angle<angle** | Enter a distance value, an angle in the XY plane value, and an angle from the XY plane value, all relative to the last point. |
| 6. **Dynamic Input** | **dist,angle** or **X,Y,Z** | Enter a distance value in the first edit box, press Tab, enter an angular value in the second box. You can also enter X,Y,Z coordinates separated by commas to establish relative Cartesian coordinates. |
| 7. **Direct distance entry** | **dist,direction** | Enter (type) a value, and move the cursor in the desired direction. |

Cylindrical and spherical coordinates can be given without the @ symbol, in which case the location specified is relative to point 0,0,0 (the origin). This method is useful if you are creating geometry centered around the origin. Otherwise, the @ symbol is used to establish points in space relative to the last point.

Examples of each of the 3D coordinate entry methods are illustrated in the following section. In the illustrations, the orientation of the observer has been changed from the default plan view in order to enable the visibility of the three dimensions.

### Mouse Input

Figure 33-15 illustrates using the interactive method to PICK a location in 3D space. *Object Snap* must be used to PICK in 3D space. Any point PICKed with the input device without *Object Snap* will result in a location on the XY plane (unless you are establishing a Dynamic UCS). In this example, the *Endpoint Object Snap* mode is used to establish the "Specify next point:" of a second *Line* by snapping to the end of an existing vertical *Line* at 8,2,6:

**FIGURE 33-15**

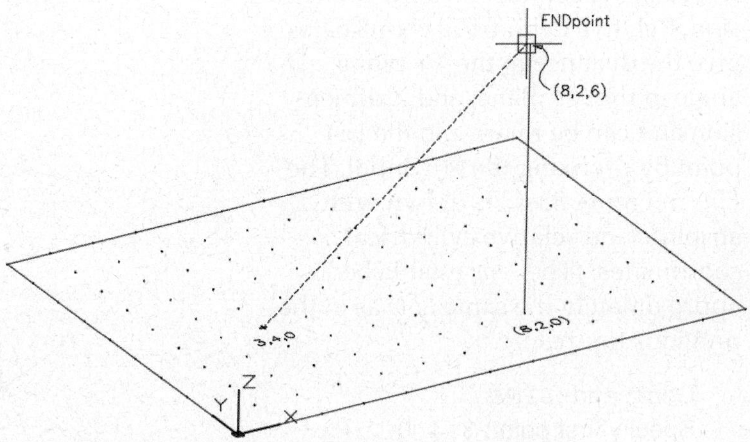


Command: ***LINE***
Specify first point: **3,4,0**
Specify next point or [Undo]: ***endpoint*** of **PICK**

### Absolute Cartesian Coordinates

Figure 33-16 illustrates the absolute coordinate entry to draw the *Line*. The endpoints of the *Line* are given as explicit X,Y,Z coordinates:

Command: ***LINE***
Specify first point: **3,4,0**
Specify next point or [Undo]: **8,2,6**

**FIGURE 33-16**

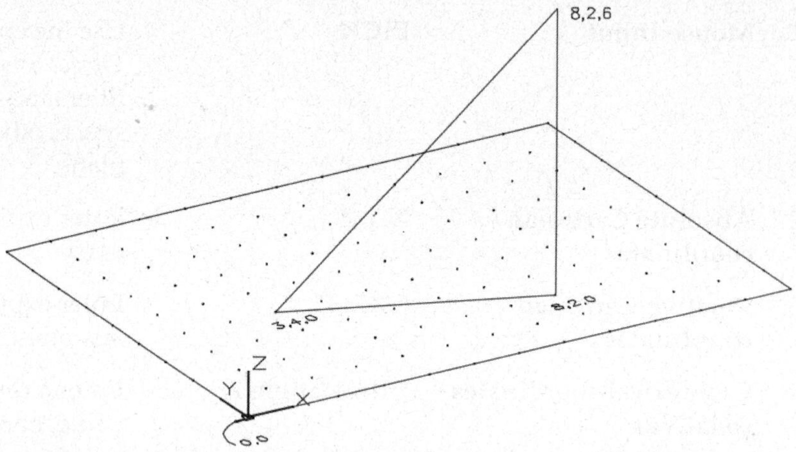


### Relative Cartesian Coordinates

Relative Cartesian coordinate entry is displayed in Figure 33-17. The "Specify first point:" of the ***Line*** is given in absolute coordinates, and the "Specify next point:" end of the *Line* is given as X,Y,Z values relative to the last point:

Command: ***LINE***
Specify first point: **3,4,0**
Specify next point or [Undo]: **@5,-2,6**

**FIGURE 33-17**

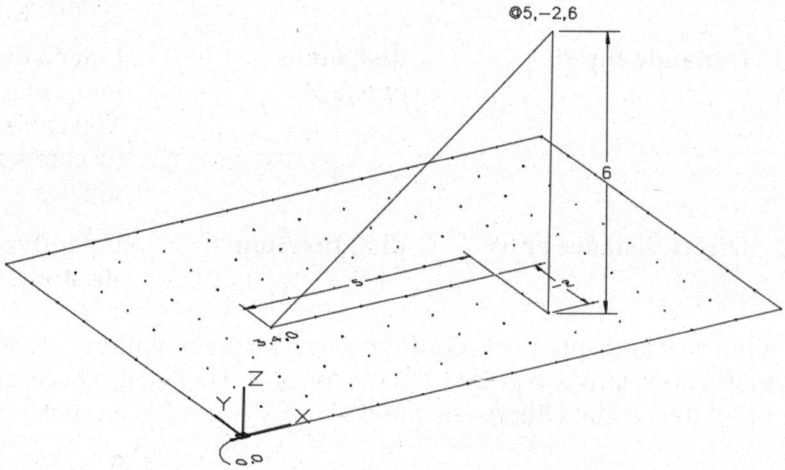


### Cylindrical Coordinates (Relative)

Cylindrical and spherical coordinates are an extension of Polar coordinates with a provision for the third dimension. Relative cylindrical coordinates give the distance in the XY plane, angle in the XY plane, and Z dimension and can be relative to the last point by prefixing the @ symbol. The *Line* in Figure 33-18 is drawn with absolute and relative cylindrical coordinates. (The *Line* established is approximately the same *Line* as in the previous figures.)

Command: ***LINE***
Specify first point: **3,4,0**
Specify next point or [Undo]: **@5<-22,6**

**FIGURE 33-18**

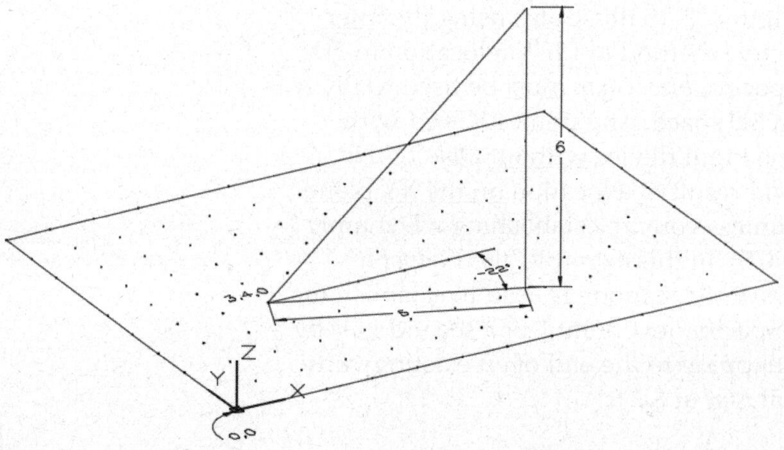

### Spherical Coordinates (Relative)

Spherical coordinates are also an extension of Polar coordinates with a provision for specifying the third dimension in angular format. Spherical coordinates specify a distance, an angle <u>in</u> the XY plane, and an angle <u>from</u> the XY plane and can be relative to the last point by prefixing the @ symbol. The distance specified is a <u>3D distance</u>, not a distance in the XY plane. Figure 33-19 illustrates the creation of approximately the same line as in the previous figures using absolute and relative spherical coordinates.

**FIGURE 33-19**

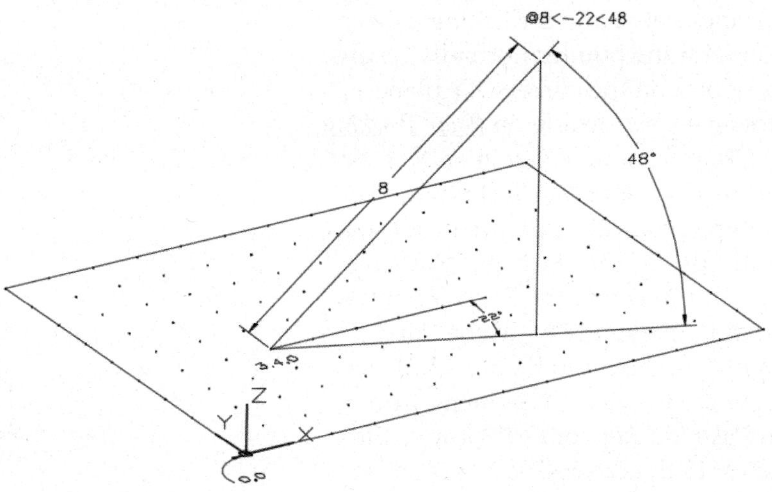


Command: ***LINE***
Specify first point: **3,4,0**
Specify next point or [Undo]: **@8<-22<48**

### *Polar Tracking*

*Polar Tracking* can be used when constructing and editing 3D models. When *Polar Tracking* is turned on in the 3D drawing mode, the *Polar Tracking* vector appears in the current XY plane <u>or in a plane parallel to the XY plane</u> when designating points. If the "first point" or "base point" is located in 3D space, the resulting tracking vectors appear parallel to the coordinate system XY plane (in the XY plane of the point). By default, the tracking vectors appear for 90-degree increments but can be set to any increment using the *Polar Tracking* tab of the *Drafting Settings* dialog box (see Chapter 3).

In 3D mode the *Polar Tracking* vector capability is extended to operate for the <u>Z axis</u>. For example, assume you used the *Move* command to move a small *Box* from atop a larger *Box* (Figure 33-20). With *Polar Tracking* on, you could move the small box from the "base point" in a +Z or –Z direction as well as in the parallel XY plane Polar tracking angles.

**FIGURE 33-20**

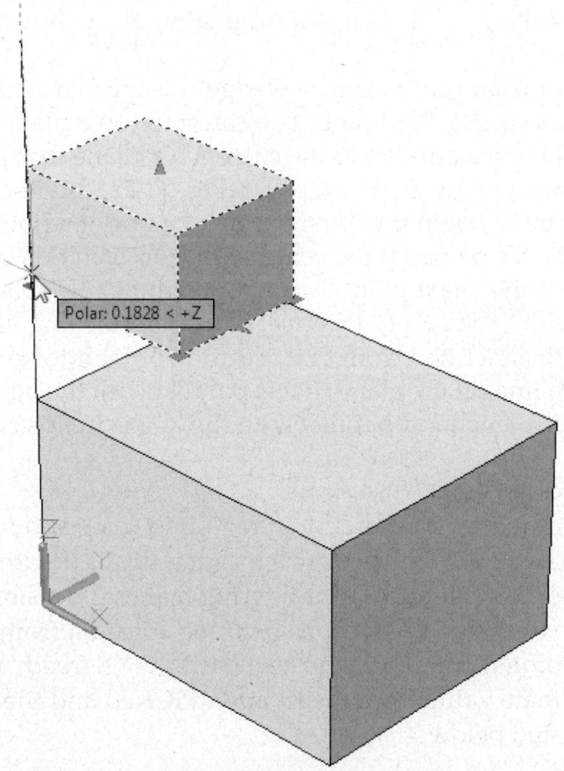


### *Ortho*

The *Ortho* feature can also be used for drawing or editing in 3D space. When *Ortho* is on, the "rubber band line" (from the last point to the current cursor position) is forced to an <u>X, Y, or Z direction</u>. This feature is similar to using *Polar Tracking* in 90-degree increments.

### Direct Distance Entry

FIGURE 33-21

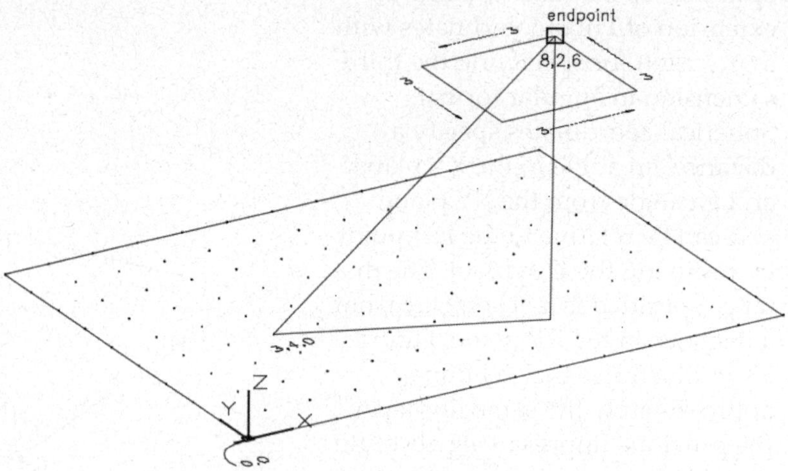


With Direct Distance Entry, entering a distance value and indicating a direction with the pointer normally results in a point on the current XY plane. However, by turning on Polar Tracking or Ortho, you can draw in an X, Y, or Z direction from the "last point." For example, to draw a square in 3D space with sides of 3 units, begin by using *Line* and selecting the *Endpoint* of the existing diagonal *Line* as the "first point:" as shown in Figure 33-21. To draw the first 3-unit segment, turn on *Polar Tracking* or *Ortho*, move the cursor in the desired X, Y, or Z direction, and enter "3". Next, move the cursor in the desired (90 degree) direction and enter "3," and so on.

### *Object Snap Tracking*

*Object Snap Tracking* functions in 3D similar to the way that *Object Snap* and Direct Distance entry operate—that is, when a point is "acquired" by one of the *Object Snap* modes, tracking occurs in the plane of the acquired point. To be more specific, when you acquire a point in 3D space (not on the XY plane), tracking vectors appear originating from the acquired object in a plane parallel to the current XY plane.

FIGURE 33-22

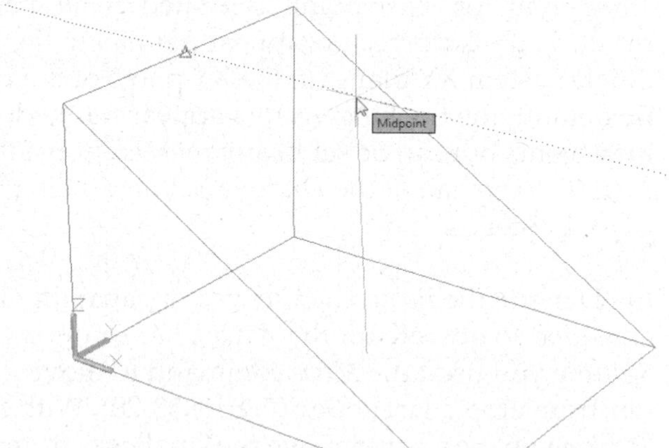


For example, assume a wedge was created with its base on the XY plane. You can track on a plane in 3D space parallel to the current XY plane that passes through the acquired point (Fig. 33-22). In other words, begin the "first point" of a line at a point on the XY plane (at the center of the base, in this case). For the "next point," acquire a point in 3D space with *Object Snap Tracking*. Note that tracking vectors are generated from that acquired point into space on an imaginary plane that is parallel with the current XY plane (and parallel with the X axis in this case).

### Point Filters

Point filters are used to filter X and/or Y and/or Z coordinate values from a location PICKed with the pointing device. Point filtering makes it possible to build an X,Y,Z coordinate specification from a combination of point(s) selected on the screen and point(s) entered at the keyboard. A .XY (read "point XY") filter would extract, or filter, the X and Y coordinate value from the location PICKed and then prompt you to enter a Z value. Valid point filters are listed below.

| | |
|---|---|
| .X | Filters (finds) the X component of the location PICKed with the pointing device. |
| .Y | Filters the Y component of the location PICKed. |
| .Z | Filters the Z component of the location PICKed. |
| .XY | Filters the X and Y components of the location PICKed. |
| .XZ | Filters the X and Z components of the location PICKed. |
| .YZ | Filters the Y and Z components of the location PICKed. |

The .XY filter is the most commonly used point filter for 3D construction and editing. Because 3D construction often begins on the XY plane of the current coordinate system, elements in Z space are easily constructed by selecting existing points on the XY plane using .XY filters and then entering the Z component of the desired 3D coordinate specification by keyboard. For example, in order to draw a line in Z space two units above an existing line on the XY plane, the .XY filter can be used in combination with *Endpoint Object Snap* to supply the XY component for the new line. See Figure 33-23 for an illustration of the following command sequence:

**FIGURE 33-23**

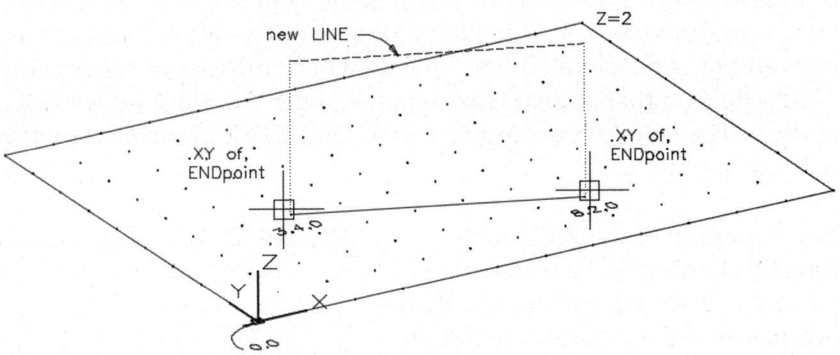


Command: **LINE**
Specify first point: **.XY** of
**endpoint of PICK** (Select one *Endpoint* of the existing line on the XY plane.)
(need Z) **2**
Specify next point or [Undo]: **.XY** of
**endpoint of PICK** (Select the other *Endpoint* of the existing line.)
(need Z) **2**

Typing **.XY** and pressing **Enter** cause AutoCAD to respond with "of" similar to the way "of" appears after typing an *Object Snap* mode. When a location is specified using an .XY filter, AutoCAD responds with "(need Z)" and, likewise, when other filters are used, AutoCAD prompts for the missing component(s).

### *Dynamic Input*

In the normal Polar format (distance and angle edit boxes), *Dynamic Input* operates only in the current XY plane. You can, however, create User Coordinate Systems in 3D space and use *Dynamic Input* on the XY plane of these coordinate systems. See "Coordinate Systems" next. With the default settings, if you instead enter an X, Y, and Z coordinate set separated by commas, *Dynamic Input* accepts the coordinates as relative Cartesian coordinates (from the last point).

NOTE: *Dynamic Input* settings are made in the *Dynamic Input* tab of the *Drafting Settings* dialog box where you can change *Dynamic Input* to absolute coordinate mode for second points. You can also provide an additional edit box for a Z coordinate component if you change the *Show Z Field for Pointer Input* checkbox in the *Options* dialog box, *3D Modeling* tab.

## COORDINATE SYSTEMS

In AutoCAD two kinds of coordinate systems can exist, the World Coordinate System (WCS) and one or more User Coordinate Systems (UCS). The World Coordinate System always exists in any drawing and cannot be deleted. The user can also create and save multiple User Coordinate Systems to make construction of a particular 3D geometry easier. Only one coordinate system can be active at any one time in any one view or viewport, either the WCS or one of the user-created UCSs. You can have more than one UCS active at any one time if you have several viewports active (model space or layout viewports).

## The World Coordinate System (WCS) and WCS Icon

The World Coordinate System (WCS) is the default coordinate system in AutoCAD for defining the position of drawing objects in 2D or 3D space. The WCS is always available and cannot be erased or removed but is deactivated temporarily when utilizing another coordinate system created by the user (UCS). The icon that appears (by default) in the Drawing Editor (Fig. 33-14) indicates the orientation of the WCS. The icon, whose appearance (*ON, OFF*) is controlled by the *Ucsicon* command, appears for the WCS and for any UCS.

The orientation of the WCS with respect to Earth may be different among CAD systems. In AutoCAD the WCS has an architectural orientation such that the XY plane is a horizontal plane with respect to Earth, making Z the height dimension. A 2D drawing (X and Y coordinates only) is thought of as being viewed from above, sometimes called a plan view. This default 2D orientation is like viewing a floor plan—from above (Fig. 33-24). Therefore, in a 3D AutoCAD drawing, imagine this same orientation of objects in space; that is, X is the width dimension, Y is the depth dimension, and Z is height.

**FIGURE 33-24**

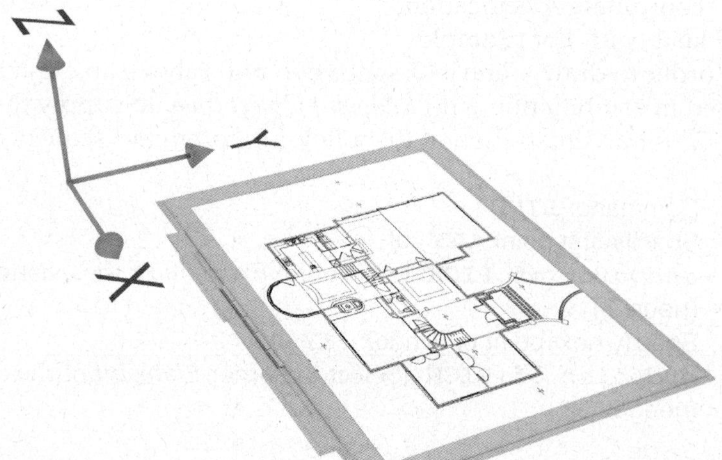

## User Coordinate Systems (UCS) and Icons

There are no User Coordinate Systems that exist as part of the AutoCAD template drawings as they come "out of the box." UCSs are created to suit the 3D model when and where they are needed.

Creating geometry is relatively simple when having to deal only with X and Y coordinates, such as in creating a 2D drawing or when creating simple 3D geometry with uniform Z dimensions. However, 3D models containing complex shapes on planes not parallel with the XY plane are good candidates for UCSs. A User Coordinate System is thought of as a construction plane created to simplify creation of geometry on a specific plane or surface of the object. The user creates the UCS, aligning its XY plane with a surface of the object, such as along an inclined plane, with the UCS origin typically at a corner or center of the surface (Fig. 33-25).

**FIGURE 33-25**

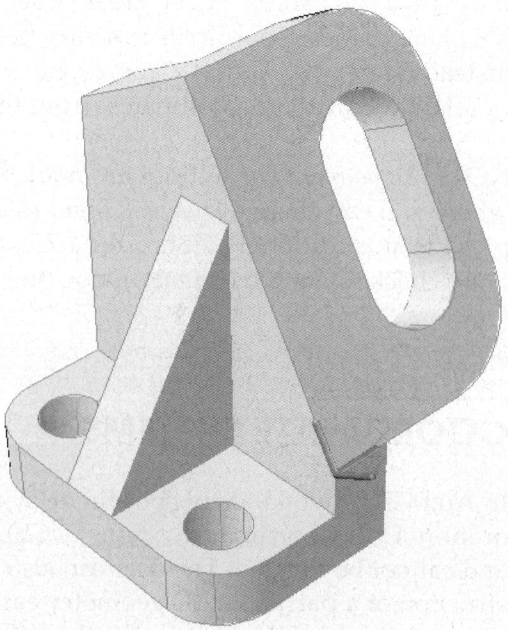

The user can then create geometry aligned with that plane by defining only X and Y coordinate values of the current UCS. The Coordinate display in the Status Bar always displays the X, Y and Z values of the current coordinate system, whether it is WCS or UCS. The *Snap, Grid,* and *Polar Tracking* automatically align with the current coordinate system, providing *Snap* points, tracking vectors, and enhanced visualization of the current XY "construction" plane (Fig. 33-26). Practically speaking, it is easier in most cases to specify only X and Y coordinates with respect to a specific plane on the object rather than calculating X, Y, and Z values with respect to the World Coordinate System. For example, a UCS would have been created on the inclined surface to construct the slotted hole shown in Figure 33-26.

**FIGURE 33-26**

User Coordinate Systems can be created by any of several options of the *UCS* command. Once a UCS has been created, it becomes the current coordinate system and remains current until another UCS is activated. Since it is likely you would alternate between different coordinate systems during construction and editing of the 3D model, it is suggested that UCSs be saved (by using the *UCS Manager*) for possible future geometry creation or editing.

### Dynamic UCSs

In addition to static UCSs created with the *UCS* command, you can use temporary coordinate systems called Dynamic User Coordinate Systems; however, Dynamic User Coordinate Systems operate only with solid models. A Dynamic User Coordinate System, or Dynamic UCS, appears automatically only during the creation of a primitive such as a *Box*, *Wedge*, or *Cylinder*. Use the *Dynamic UCS* toggle on the Status line to activate the Dynamic UCS feature.

When the *Dynamic UCS* toggle is on, a Dynamic UCS appears on the surface you select to establish the base of a primitive. You can select which surface you want by hovering your cursor over any surface of an existing solid during a primitive command. For example, Figure 33-27 displays a Dynamic UCS appearing on the inclined surface while constructing a *Box* primitive. The Dynamic UCS disappears after the primitive has been created.

**FIGURE 33-27**

## The Right-Hand Rule

AutoCAD complies with the right-hand rule for defining the orientation of the X, Y, and Z axes. The right-hand rule states that if your right hand is held partially open, the thumb, first, and middle fingers define positive X, Y, and Z directions, respectively.

More precisely, if the thumb and first two fingers are held out to be mutually perpendicular, the thumb points in the positive X direction, the first finger points in the positive Y direction, and the middle fingers point in the positive Z direction (Fig. 33-28).

FIGURE 33-28

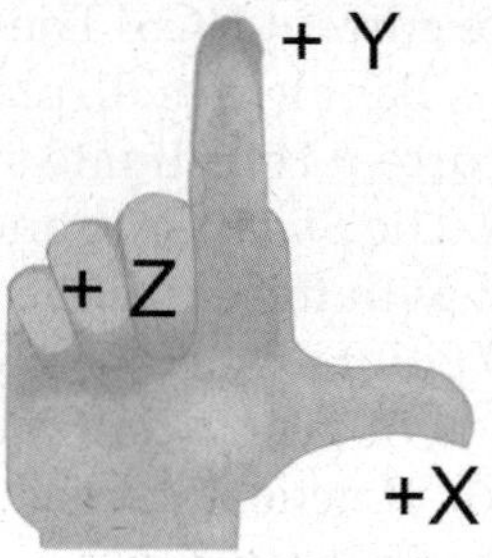


In this position, looking toward your hand from the tip of your middle finger is like the default AutoCAD viewing orientation in 2D mode—positive X is to the right, positive Y is up, and positive Z is toward you.

When you rotate geometry in 3D space or create a new User Coordinate System, it is possible that you want to rotate about the existing X, Y, or Z axis. In doing this you must indicate whether you want to rotate in a positive or negative direction. The right-hand rule is used to indicate the direction of positive rotation about any axis. Imagine holding your right hand closed around any axis but with the thumb extended and pointing in the positive direction of the axis (Fig. 33-29). The direction your fingers naturally curl indicates the positive rotation direction about that axis. This principle applies to all three axes.

FIGURE 33-29

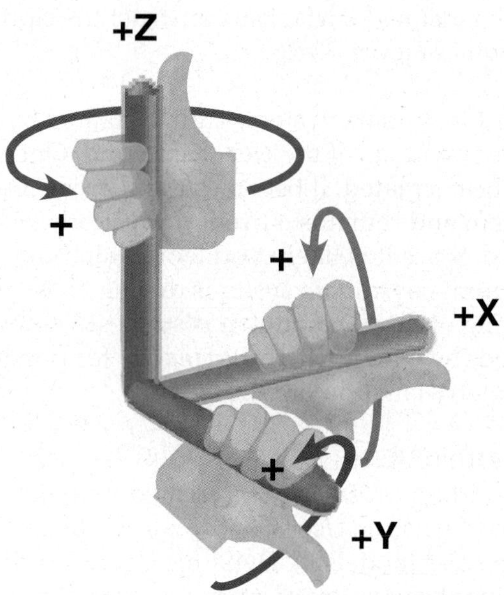


### *Ucsicon*

| Ribbon | Menu Bar | Command (Type) | Alias (Type) | Shortcut |
|---|---|---|---|---|
| *View*<br>*Viewport Tools*<br>*UCS Icon* | *View*<br>*Display >*<br>*USC Icon >* | *Ucsicon* | ... | ... |

The *Ucsicon* command controls the appearance and positioning of the Coordinate System icon. In order to aid your visualization of the current UCS or the WCS, it is highly recommended that the Coordinate System icon be turned *ON* and positioned at the *ORigin*.

Command: ***UCSICON***
Enter an option [ON/OFF/All/Noorigin/ORigin/Properties] <ON>: ***on***
Command:

The *On* option causes the Coordinate System icon to appear (Fig. 33-30, lower left).

**FIGURE 33-30**

The *Origin* setting causes the icon to move to the origin of the current coordinate system (Fig. 33-31).

The icon may not always appear at the origin after using some viewing commands like *3Dorbit*. This is because the origin (0,0) may be located outside the border of the drawing screen (or viewport) area, which causes the icon to appear in the lower-left corner. In order to cause the icon to appear at the origin, try zooming out until the icon aligns itself with the origin.

**FIGURE 33-31**

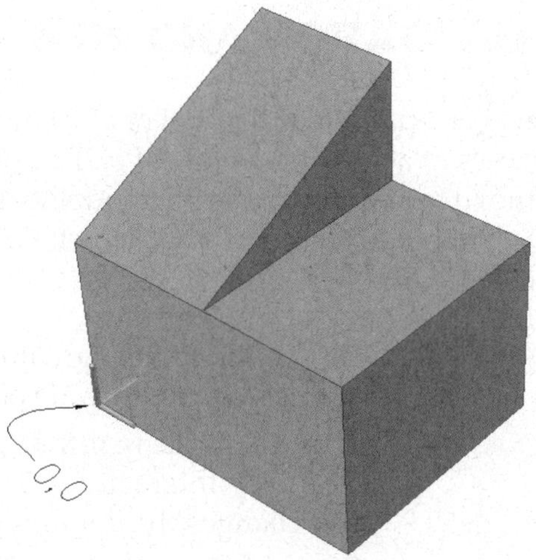


Options of the *Ucsicon* command are:

| | |
|---|---|
| ***ON*** | Turns the Coordinate System icon on. |
| ***OFF*** | Turns the Coordinate System icon off. |
| ***All*** | Causes the *Ucsicon* settings to be effective for all viewports. |
| ***Noorigin*** | Causes the icon to appear always in the lower-left corner of the screen, not at the origin. |
| ***ORigin*** | Forces the placement and orientation of the icon to align with the origin of the current coordinate system. |
| ***Selectable*** | Turns On/Off the the ability to select the icon and relocate it using grips. |
| ***Properties*** | Shows the UCS Icon dialog box. |

### Properties

The *Properties* option produces the *UCS Icon* dialog box (Fig. 33-32). Ironically, the *UCS Icon Style* and the *UCS Icon Size* have no effect for the 3-pole colored icon that appears in 3D mode. It is actually the *Visual Styles* option that determines which icon is displayed: the *2D wireframe* option displays the simple "wire" icon, whereas any 3D option displays the 3-pole colored icon. You can change the *UCS icon color* for model space and paper space. The default option for *Model space icon color* is either *Black* or *White*—the opposite of whatever you set for the *Model* tab background color in the *Options* dialog box.

**FIGURE 33-32**

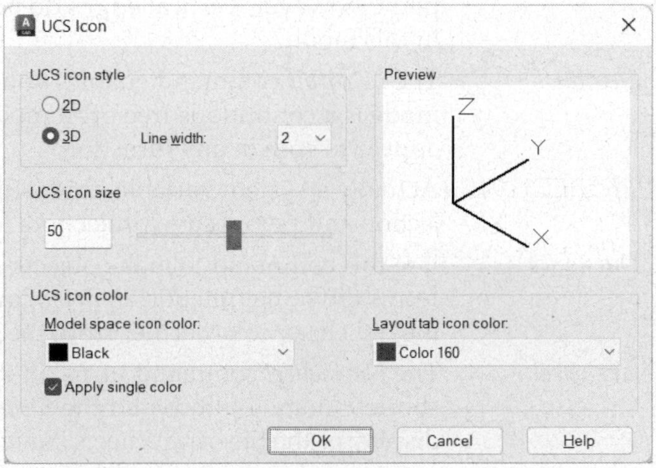

**TIP** The *UCS* icon can also be manipulated using multifunction *Grips*. You can easily move the origin (Fig. 33-33) or any of the 3-dimensions (Fig. 33-34), and automatically align the *UCS* with objects and curved surfaces and solids.

FIGURE 33-33

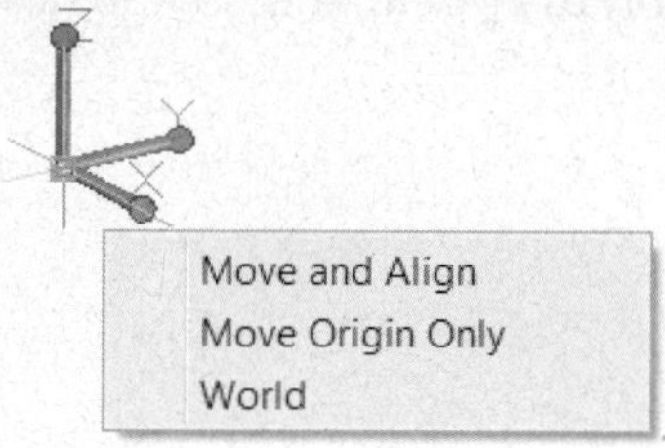


FIGURE 33-34

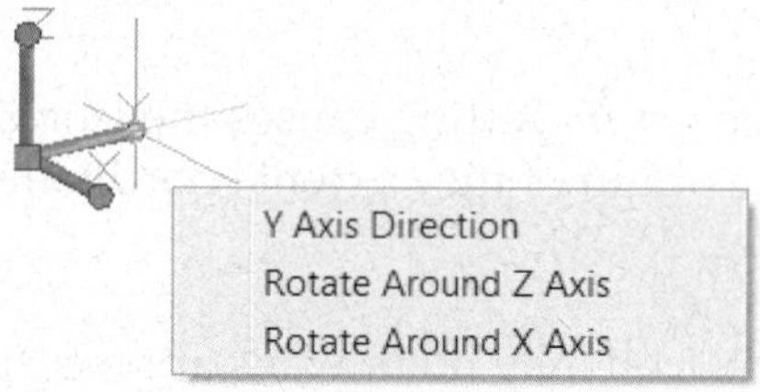


## 3D NAVIGATION AND VIEW COMMANDS

A "view" is a particular display of a 2D drawing or a 3D model. The term "navigate" is descriptive of the process of achieving a view of a 3D model—that is, you as the observer move around, move closer to, or move farther from, the 3D model to arrive at a particular view. The commands listed below allow you to change the direction from which you view a 3D model or otherwise affect the view of the 3D model.

| | |
|---|---|
| *Pan* | This command is the same as the real-time *Pan* used for 2D drawings. This command enables you to drag 3D objects about the view interactively. |
| *Zoom* | Identical to the real-time *Zoom* used for 2D drawings, you can enhance your view of 3D objects by interactively zooming in and out. For a view in perspective projection, *Zoom* also changes the amount of perspective as the objects appear nearer or farther. |
| *Navvcube* | *Navvcube* turns the ViewCube on or off and provides access to the *ViewCube Settings* dialog box. The ViewCube is an excellent tool to use to provide normal views of a 3D model such as top, front, right, and standard isometric views. |
| *3Dorbit* | *3Dorbit* controls interactive viewing of objects in 3D. You can dynamically view the object from any point, even while the object retains its visual style. The constrained orbit (default) mode locks the vertical edges to a vertical position. |
| *3Dforbit* | The free orbit mode of *3Dorbit* allows you to dynamically view the object from any point; however, vertical edges do not remain vertical. You can "spin" the view around in this mode. |
| *3Dcorbit* | The *3Dcorbit* command enables you to set the 3D objects into continuous motion. This mode is a continuous free orbit mode since the vertical axis is not locked and objects appear to spin in any direction. |
| *PERSPECTIVE* | Actually a system variable, this setting toggles the current view between parallel projection and perspective projection. |
| *3Ddistance* | Use this command to make objects appear closer or farther away. This command performs the same function as *Zoom*, in that it changes the amount of perspective as if you moved closer to or farther from the 3D objects. |
| *Navswheel* | The *Navswheel* command turns on the current SteeringWheel and provides access to shortcut menu options and the *SteeringWheels Settings* dialog box where you control the display of the SteeringWheels. Several SteeringWheels are available. SteeringWheels incorporate many navigation commands into one tool. |

*View* — The *View* command produces the *View Manager*. Here you can create new views and manage the specifications for the *Current View, Model Views*, and *Layout Views*. You can also select from preset orthographic views and isometric viewpoints. This dialog box provides access to view settings such as camera and target locations and lens length.

*3Dclip* — *3Dclip* allows you to establish front and/or back clipping planes in the *Adjust Clipping Planes* window.

*Camera* — You can create one or more cameras where each camera represents a view of the 3D model. You can specify a X,Y, and Z camera position, target position, and lens length for each camera. You can edit a camera's settings using grips or the *Properties* palette.

*3Dswivel* — This command simulates the effect of turning the current camera or view.

*-Vpoint* — *-Vpoint* is an old AutoCAD command that allows you to change your viewpoint of a 3D model. Three options are provided: *Vector, Tripod,* and *Rotate.*

*Vpoint* — Also a legacy command, this tool allows you to select a viewing angle in degrees.

*Plan* — The *Plan* command automatically gives the observer a plan (top) view of the object. The plan view can be with respect to the WCS (World Coordinate System) or an existing UCS (User Coordinate System).

*Grid* — The *Grid* command offers several settings that affect the appearance of the grid in 3D mode and control how the grid lines react to zooming.

*Vports* — You can use *Vports* effectively during construction of 3D objects to divide the screen into several sections, each displaying a different standard 3D view.

TIP

Many of the 3D navigation commands discussed in this section can be accessed using the *View* tab, *Navigate* panel of the Ribbon (*3D Modeling* workspace) or using the *3D Navigation* toolbar. You must activate the *Navigate* panel by right-clicking on the *View* tab and selecting *Show Panels.* These and other navigation and viewing commands can be accessed using the *View* pull-down menus.

When using 3D navigation tools such as *Pan, Zoom,* ViewCube, *3Dorbit,* etc., it is important to imagine that the observer moves about the object rather than imagining that the object rotates. The object and the coordinate system (WCS and icon) always remain stationary and always keep the same orientation with respect to Earth. Since the observer moves and not the geometry, the objects' coordinate values retain their integrity, whereas if the geometry rotated within the coordinate system, all coordinate values of the objects would change as the object rotated. The viewing commands change only the viewpoint of the observer.

## Pan

| Ribbon | Menu Bar | Command (Type) | Alias (Type) | Shortcut |
|---|---|---|---|---|
| *View*<br>*Navigate*<br>*Pan* | *View*<br>*Pan >*<br>*Realtime* | *Pan* | *P* | (Default Menu)<br>*Pan* |

The cursor changes to a hand cursor when this command is active. Identical to the real-time *Pan* command, you can click and drag the view of the 3D objects to any location in your drawing area. However, when any 3D visual style is displayed and *Perspective* is toggled on, *Pan* displays a true 3D effect. That is, panning slightly changes the direction from which the 3D objects are viewed. As an alternative to invoking this command, you can press and hold the mouse wheel to activate *Pan*. To end this command, press the Esc key or select *Exit* from the right-click shortcut menu.

## Zoom

| Ribbon | Menu Bar | Command (Type) | Alias (Type) | Shortcut |
|---|---|---|---|---|
| *View*<br>*Navigate*<br>*Realtime* | *View*<br>*Zoom >*<br>*Realtime* | *Zoom* | *Z* | (Default Menu)<br>*Zoom* |

This *Zoom* is identical to the real-time version of the *Zoom* command used in a 2D drawing. However, when viewing 3D objects in any 3D visual style and with *Perspective* toggled on, *Zoom* performs an actual 3D zoom; that is, zooming changes the distance the object appears from the observer. Technically, zooming changes the distance between the camera (observer) and the target (objects). Therefore, *Zoom* changes the size of the objects relative to screen size as well as the amount of perspective.

The *Zoom* cursor is a magnifying glass icon. Press and hold the left mouse button, then move up to zoom in (enlarge image) and down to zoom out (reduce image). As an alternative to invoking this command, you can turn the mouse wheel forward and backward to zoom in and out, respectively. To end this command, press the Esc key or select *Exit* from the right-click shortcut menu.

NOTE: In releases of AutoCAD previous to AutoCAD 2007, the *Zoom* command and the *3Dzoom* command changed only the size of the 3D objects relative to the screen but did not change the amount of perspective. In other words, the distance between the camera and target, and therefore the amount of perspective, did not change with *Zoom* or *3Dzoom*. The *3Ddistance* command was used to change the camera-target distance or the amount of perspective.

Since AutoCAD 2007, *Zoom* and *3Ddistance* perform the same action—change the size of the objects on the screen and change the camera-target distance. Also see NOTE in "*3Ddistance*."

## Navvcube

| Ribbon | Menu Bar | Command (Type) | Alias (Type) | Shortcut |
|---|---|---|---|---|
| *View*<br>*Viewport Tools*<br>*View Cube* | *View*<br>*Display >*<br>*ViewCube* | *Navvcube* | *Cube* | ... |

The ViewCube (Fig. 33-35) is an excellent tool to use to provide normal views of a 3D model such as top, front, right, and standard isometric views, as well as other views. The ViewCube is interactive, simple and fast to use, and its functions are intuitive. The ViewCube is not a command-driven tool. The *Navvcube* command only turns the ViewCube on or off and provides access to the *ViewCube Settings* dialog box.

**FIGURE 33-35**

```
Command: NAVVCUBE
Enter an option [ON/OFF/Settings] <ON>:
```

### Face, Corner, and Edge Options

To change your view to a standard view using the ViewCube, select any face, corner, or edge of the cube. For example, selecting the corner of the cube connecting the front, right, and top faces (as shown) produces a view of the forklift as shown in Figure 33-36, on the next page. This is a standard isometric view—that is, you see an equal amount of the three faces of the object.

The view shown in Figure 33-36 is a "southeast" isometric view since it displays the object as if the observer was looking from a southeast position (note the position of the north, south, east, and west direction labels). For a southeast isometric, the front, top, and right faces are visible.

FIGURE 33-36

Figure 33-37 displays a "southwest" view of the forklift; that is, imagine looking at the 3D model from a southwest position. Note the position of the north, south, east, and west direction labels. This view shows the front, left, and top faces of the model equally.

FIGURE 33-37

To view a face of the 3D model such as shown in a multiview drawing, you can select any face of the ViewCube labeled *Front*, *Right*, *Left*, *Back*, *Top*, or *Bottom*. Alternately, you can select the letter *S* for the front view, *E* for the right view, *N* for back view, and *W* for the left view (see Fig. 33-35). The circle below the cube with the *N*, *S*, *E*, and *W* indicators is called the ViewCube "compass." Figure 33-38 displays the *Front* face of the forklift model.

FIGURE 33-38

NOTE: For any viewing or navigation command, selecting an option such as the *Front* or *Top* produces the expected view of the 3D model only if the model is oriented in space correctly. Technically, the *Front* option shows a viewing direction normal to (perpendicular to) the XZ plane from a negative Y position (note the coordinate system icon in Figures 33-36 and 33-37). The *Top* option would look down on the XY plane from a positive Z position. Therefore, to achieve the expected viewing results, the 3D model must be placed in space such that its base is oriented parallel to the XY plane and the front face is oriented parallel to the XZ plane of the World Coordinate System.

Figure 33-39 displays the forklift looking at the right side. Note that in this and the other views (Figs. 33-36 through 33-40), the views are shown in perspective projection. In contrast, you can display the views in parallel projection, similar to a normal view in a multiview drawing. (See "*PERSPECTIVE*" for a technical description of these projection modes.) You can choose to display the faces (*Top*, *Front*, *Right*, etc.) in either parallel or perspective projection (see "*ViewCube* Shortcut Menu").

When the ViewCube is displayed such that only one of the faces appears (*Front*, *Top*, *Right*, etc.), you can select one of the four small arrowheads around the cube to display the adjacent faces (Fig. 33-39). In addition, you can rotate the model in 90-degree increments using the curved arrows just above the ViewCube. Note that none of these arrows appears until the cursor is near the ViewCube.

You can also choose to display a 3D model showing two faces by selecting any edge of the ViewCube. For example, display a view of both the front and top by selecting the edge of the ViewCube between *Front* and *Top* as shown in Figure 33-40.

**FIGURE 33-39**

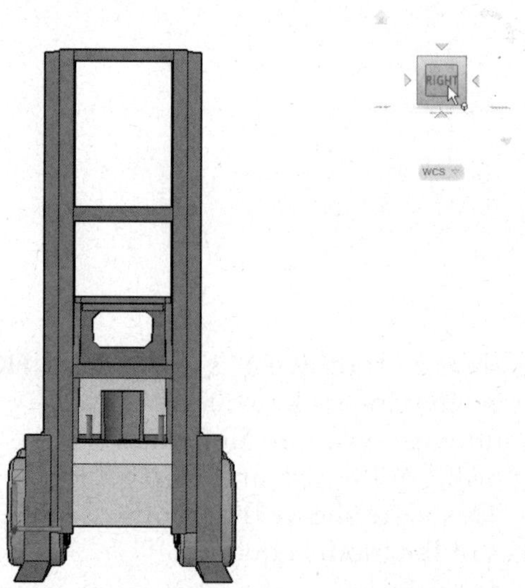


**FIGURE 33-40**

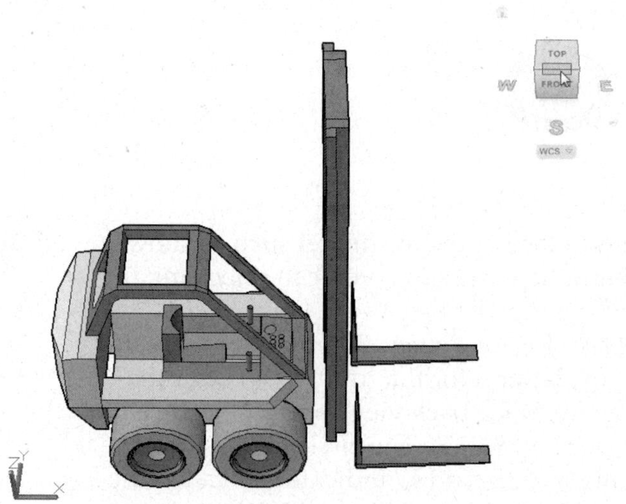


## Dragging the ViewCube

In addition to the face, corner, and edge options, which produce standard views of the model, you can dynamically "drag" the ViewCube to achieve a display of the model in any position. Do this by positioning the cursor on the ViewCube and holding down the button, then dragging the cube around its center. The model display dynamically follows the position of the ViewCube. If you rest near a standard position, the ViewCube and model will snap to the nearest standard view by default (unless turned off in the *ViewCube Settings* dialog box).

### Home

Selecting the "home" icon above the ViewCube displays the model in its home position. You can specify the home position for each drawing using the *Set Current View as Home* option in the ViewCube shortcut menu (see "ViewCube Shortcut Menu"). The default home position is a southwest isometric view.

### *WCS* (UCS Setting)

You can use the indicator just beneath the ViewCube to specify the activate coordinate system. Use the drop-down list to select the World Coordinate System (*WCS*), to select a named User Coordinate System (UCS), or to create a *New UCS*. Ironically, this indicator has no effect on specified views unless the *Orient ViewCube to Current UCS* option is checked in the *ViewCube Settings* dialog box, in which case the views are based on the selected coordinate system.

### ViewCube Shortcut Menu

Right-clicking on the ViewCube produces the shortcut menu shown in Figure 33-41. The options are as follows:

**FIGURE 33-41**

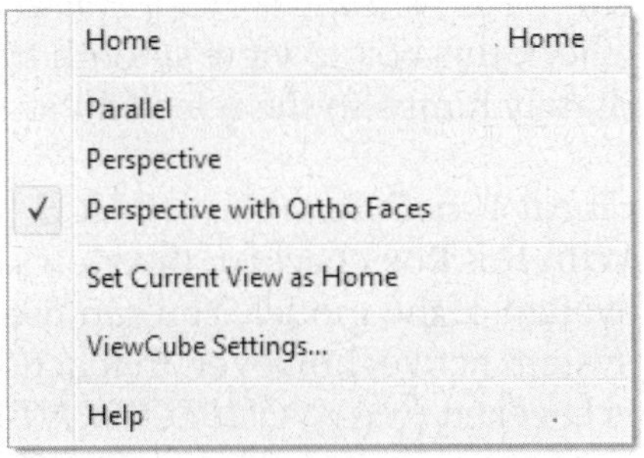


*Home*

This option places the model in its home position.

*Parallel, Perspective*

Select one of these options to display the model in the desired mode for every view. See "*PERSPECTIVE*" for a description of these types of projections.

*Perspective with Ortho Faces*

This option produces the isometric views in perspective projection and the orthographic views (*Front*, *Top*, *Right*, etc.) in parallel projection. You cannot achieve a parallel projection on orthographic views by dragging the ViewCube—you must select a face. See "*PERSPECTIVE*" as well as the description adjacent to Figure 33-39.

*Set Current View as Home*

Select this option to set the current view as the home position. The home position is displayed by selecting the *Home* option or the "home" icon above the ViewCube. This setting is stored with the drawing. Southwest isometric is the default home position.

*ViewCube Settings*

This selection produces the *ViewCube Settings* dialog box.

### *ViewCube Settings* Dialog Box

The dialog box displayed in Figure 33-42 allows you to specify how the ViewCube appears and reacts to selections.

**FIGURE 33-42**

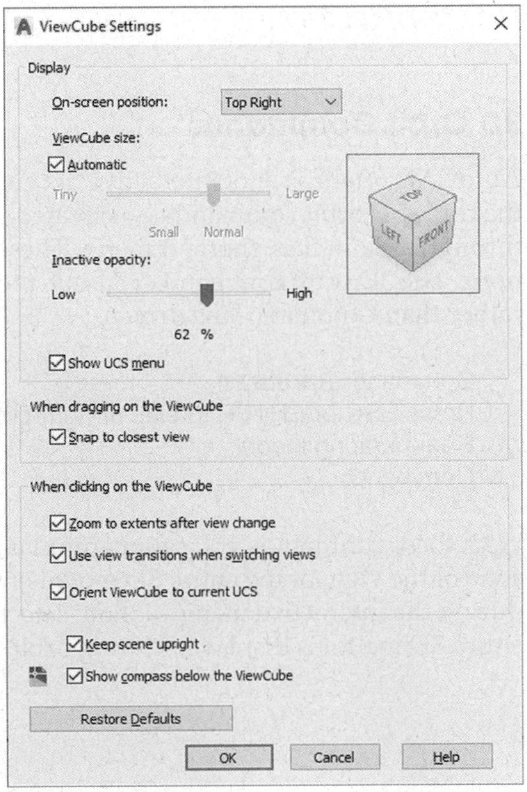


*On-screen Position*

Select the desired location where the ViewCube appears (*NAVVCUBELOCATION* system variable).

*ViewCube Size*

When *Automatic* is unchecked, this slider controls the display size of the ViewCube (*NAVVCUBESIZE* system variable).

*Inactive Opacity*
Use this slider to control the opacity level of the ViewCube when it is inactive (*NAVVCUBEOPACITY* system variable).

*Show UCS Menu*
This check controls the display of the UCS drop-down menu below the ViewCube.

*Snap to Closest View*
When you drag the ViewCube to achieve a view, the display is adjusted to the nearest preset view position if this box is checked.

*Zoom to Extents After View Change*
When this box is checked, the model is forced to fit the current viewport after a view change.

*Use View Transitions When Switching Views*
Check this box to view smooth transitions when switching between views. Otherwise, the model immediately jumps to the selected views.

*Orient ViewCube to Current UCS*
With this box checked, the views selected using the ViewCube are based on the current coordinate system of the model. You can use the indicator just beneath the ViewCube to make a specific coordinate system active; however, that action has no effect unless the *Orient ViewCube to Current UCS* checkbox is checked (*NAVVCUBEORIENT* system variable). Generally, views should be produced relative to the WCS. See Chapter 34, User Coordinate Systems.

*Keep Scene Upright*
Specifies whether the viewpoint of the model can be turned upside-down or not.

*Show Compass Below the ViewCube*
This check controls whether the compass (the circle with the *N*, *S*, *E*, and *W* indicators) is displayed below the ViewCube. The north direction indicated on the compass is the value defined by the *NORTHDIRECTION* system variable.

## 3D Orbit Commands

AutoCAD offers a set of interactive 3D viewing commands, *3Dorbit*, *3Dforbit*, and *3Dcorbit*. When one of the three 3D orbit commands is activated, all commands and related options in the group are available through a right-click shortcut menu. These commands and options are discussed in the following sections. The 3D orbit commands all issue the same simple command prompt because they are interactive rather than Command-line driven.

```
Command: 3DORBIT
Press ESC or ENTER to exit, or right-click to display shortcut-menu.
Regenerating model.
Command:
```

With these commands you can manipulate the view of 3D models by dragging your cursor. You can control the view of the entire 3D model or of selected 3D objects. If you want to see only a few objects during the interactive manipulation, select those objects before you invoke the command, otherwise the entire 3D model is displayed during orbit.

### 3Dorbit

| Ribbon | Menu Bar | Command (Type) | Alias (Type) | Shortcut |
|---|---|---|---|---|
| *View Navigate Orbit* | *View Orbit > Constrained Orbit* | *3Dorbit* | *3DO* | ... |

*3Dorbit* controls viewing of 3D objects interactively. After activating the command, simply hold down the left mouse button and drag the pointer around the screen to change the view of the selected 3D objects. Drag up and down to view the objects from above or below, and drag left and right to view the objects from around the sides. All objects in the drawing are displayed in the interactive display unless individual objects are selected before invoking the command. As stated in the Command prompt, press Esc, Enter, or select *Exit* from the right-click shortcut menu to end the command.

Using the *3Dorbit* command forces the interactive movement to a "constrained orbit." That is, the Z axis is constrained from changing so it always remains in a vertical orientation. This limitation keeps all vertical edges of the 3D objects in a vertical position, just as real objects are oriented (Fig. 33-43). Use *3Dforbit* (free orbit) to change the orientation of the vertical axis.

NOTE: Remember that you are changing the view. When moving the cursor about the screen, it appears that the 3D objects rotate; however, in theory the target remains stationary while the observer, or camera, moves around the objects.

**FIGURE 33-43**

#### *3Dorbit* Shortcut Menu Options

During the *3Dorbit*, *3Dforbit*, or *3Dcorbit* commands, you can use the shortcut menu to access other 3D viewing commands and options (Fig. 33-44). Using commands from the shortcut menu does not exit the *3Dorbit* session (unless, of course, you select *Exit*).

Most of the options from this menu are discussed immediately following and in other sections of this chapter.

**FIGURE 33-44**

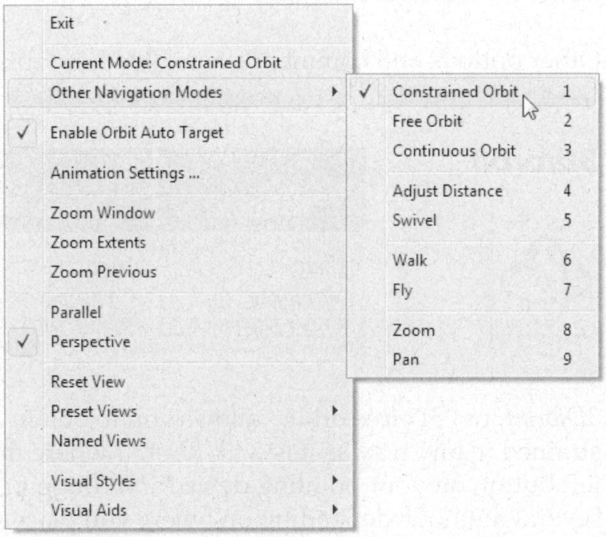

*Enable Orbit Auto Target*
This option forces the center of the orbit (target) to the center of the objects. The resulting effect is that the observer is moving around the objects—the objects are the center of rotation. This setting is recommended for most applications. When this option is not checked, the center of the viewport becomes the target (center of rotation), disregarding the location of the objects.

*Zoom Window*
Select *Zoom Window* to zoom into a smaller area by defining a window, similar to the normal *Zoom* command with the *Window* option. With the *3Dorbit Zoom Window*, you must click and drag to define two diagonal corners of the window rather than click at each corner. The display remains only during the *3Dorbit* command and reverts to the previous display when you exit *3Dorbit*.

*Zoom Extents*
Select *More, Zoom Extents* from the shortcut menu to perform a typical *Zoom Extents*, similar to the normal *Zoom* command with the *Extents* option. The view is centered and sized so all selected objects are displayed in the 3D view.

*Zoom Previous*
This option displays the last view during the current *3Dorbit* session. Only one previous view is saved.

*Reset View*
You can use *Reset View* from the shortcut menu to undo all changes to the view during the *3Dorbit* session and reset to the view before using *3Dorbit* and still remain in the *3Dorbit* session.

*Visual Aids, Compass*
Selecting the *Compass* toggle from the shortcut menu produces a sphere composed of three circles in 3D space. The three circles represent rotation about each of the three axes, X, Y, and Z.

*Visual Aids, Grid*
This option toggles the grid lines that appear on the XY plane of the current UCS (see Fig. 33-3). By default, the number of grid lines changes as you zoom in or out. The number of grid lines and the adaptive behavior of the grid is controlled by the *Grid* command or the related system variables. See *Grid* later in this chapter. The *Grid* command cannot be activated during a *3Dorbit* session.

*Visual Aids, UCS Icon*
Use this option to toggle the UCS icon on and off. The UCS icon can enhance your orientation of the view because it changes as the view changes, even during perspective projection. Alternately, use the *Ucsicon* command to control the icon visibility.

Other options and commands available from this menu are discussed on the following pages.

## 3Dforbit

| Ribbon | Menu Bar | Command (Type) | Alias (Type) | Shortcut |
|---|---|---|---|---|
| *View Navigate Free Orbit* | *View Orbit > Free Orbit* | *3Dforbit* | ... | ... |

*3Dforbit*, or "3D free orbit," allows you to obtain a wide variety of views. The orbit movement is not constrained in any way as it is with *3Dorbit* where the Z axis always remains vertical. By holding down the left button on your pointing device and dragging your cursor, you can control the view of 3D objects by several methods depending on where you place your cursor with respect to the "arc ball."

When you invoke *3Dforbit*, the current viewport displays the selected objects and the "arc ball" (Fig. 33-45). The arc ball is a circle with each quadrant denoted by a smaller circle. Moving the cursor over different areas of the arc ball changes the direction in which the view rotates. The center of the arc ball is the target.

There are four possible methods of rotating your view depending on where you place the cursor and the direction you move it. The cursor icon changes to indicate one of these four methods. The four positions are as follows.

**FIGURE 33-45**

*Inside the Arc Ball*

This icon is displayed when you move the cursor inside the arc ball. Imagine a sphere around the model. Click and drag the pointing device (hold down the left button) to rotate the sphere (and 3D model) around the target point (center of the sphere) in any direction. Note that the movement is not confined to the "constrained orbit" action (as with *3Dorbit*) and that the coordinate system Z axis (and therefore, vertical edges of objects) can be reoriented (see Fig. 33-45).

*Outside the Arc Ball*

Move the cursor outside the arc ball, then click and drag to "roll" the display. This action rotates the view around the center of the arc ball. Imagine rotating the view about a Z axis protruding perpendicular from the screen.

*Left and Right Quadrants*

This icon is displayed when you move the cursor over either of the small circles on the left or right quadrants of the arc ball, then click and drag to the left or right. This action allows you to rotate the view about a vertical axis that passes through the center of the arc ball or about an imaginary Y axis on the screen.

*Top and Bottom Quadrants*

To rotate the view about a horizontal axis passing through the center of the arc ball, place the cursor over either of the small circles on the top or bottom quadrants of the arc ball, then click and drag up or down. This action represents rotation about an imaginary X screen axis.

Remember that the center of the arc ball is the target and remains stationary. The arc ball always appears in the center of your screen or current viewport. To center your objects in the arc ball, use the *Pan* (*3Dpan*) option or *Enable Orbit AutoTarget* option from the shortcut menu.

### 3Dcorbit

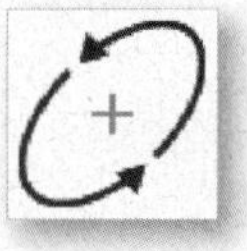

| Ribbon | Menu Bar | Command (Type) | Alias (Type) | Shortcut |
|---|---|---|---|---|
| *View Navigate Continuous Orbit* | *View Orbit > Continuous Orbit* | *3Dcorbit* | ... | ... |

The *3Dcorbit* command allows you to set the selected 3D objects into continuous motion. To set the 3D objects in motion, click in the drawing area and hold down the left button, drag the cursor in any direction, then release the button while the cursor is in motion. The objects continue to spin in the direction of your cursor motion. The speed of the cursor movement determines the speed of the spin. Change the direction and speed of the object movement by clicking and dragging the cursor again. You can stop the motion at any position and discontinue *3Dcorbit* by pressing Enter or the Esc key. Alternately, right-click and select *Exit* or another option.

## PERSPECTIVE

*PERSPECTIVE* is actually a system variable, not a command. The *PERSPECTIVE* variable can be changed by command line or by selecting the *Perspective* and *Parallel* options in the various shortcut menus. A setting of 1 turns on the perspective projection and a setting of 0 displays parallel projection.

```
Command: PERSPECTIVE
Enter new value for PERSPECTIVE <1>:
```

NOTE: You can adjust the amount of perspective using the *View Manager* (see "Adjusting the Amount of Perspective").

Parallel projection is used for 2D drawing and it is automatically turned on when the *2D Wireframe* visual style is set. However, parallel projection displays 3D objects unrealistically because parallel edges of an object remain parallel in the display. A good example of this theory is shown in Figure 33-46 where the grid lines remain parallel to each other, no matter how far they extend away from the observer. Parallel projection does not take into account the distance the observer is from the object.

**FIGURE 33-46**

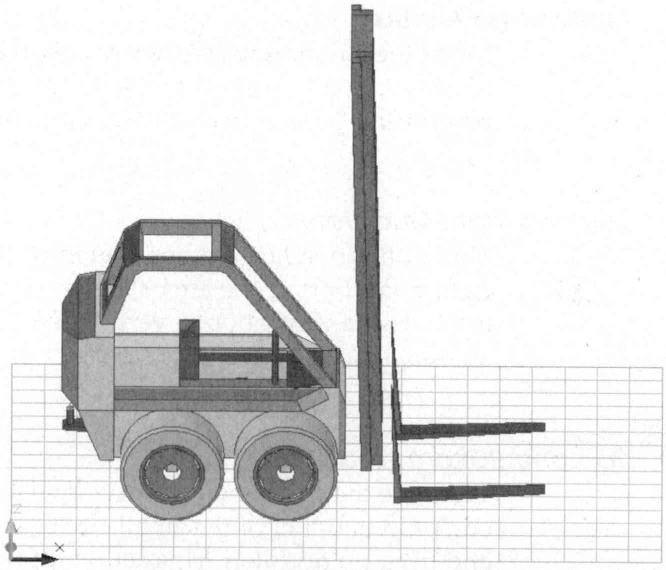

Perspective projection simulates the way we actually see 3D objects because it takes into account the distance the observer is from the object. In AutoCAD, this distance is determined by the distance from our eyes (camera) to the objects (target). Therefore, since objects that are farther away appear smaller, parallel edges on a 3D object appear to converge toward a point as they increase in distance. Figure 33-47 shows the same objects as the previous figure but in perspective projection. Perspective projection can be displayed (*PERSPECTIVE* is set to 1) for any 3D objects shown in any of the 3D visual styles.

**FIGURE 33-47**

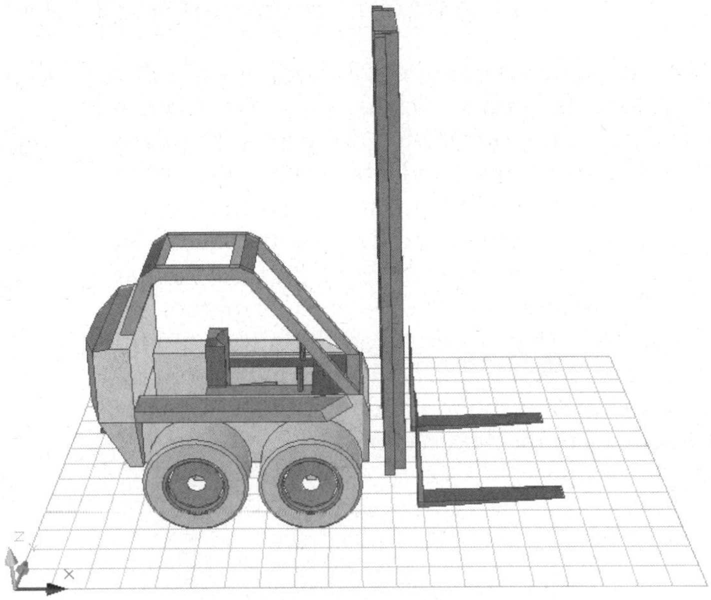

### *3Ddistance*

| Ribbon | Menu Bar | Command (Type) | Alias (Type) | Shortcut |
|---|---|---|---|---|
| ... | *View*<br>*Camera >*<br>*Adjust Distance* | *3Ddistance* | ... | ... |

The *3Ddistance* command changes the distance between the camera (observer) and the target (objects). When perspective projection is on, the amount of perspective changes as a function of camera-target distance. When you invoke *3Ddistance*, the icon changes to a double-sided arrow shown in perspective. Hold the left mouse button down and move the cursor upward to bring the objects closer (and to achieve more perspective), or move the cursor down to move the objects farther away (and achieve less perspective).

Since AutoCAD 2007, the *Zoom* command serves the same function as *3Ddistance*; therefore, you can use *Zoom* instead of *3Ddistance* to accomplish the same display. Normally, you think of using *Zoom* to change the size of objects on the screen; however, object size, camera-target distance, and amount of perspective are now interrelated in AutoCAD.

NOTE: Historically (in versions previous to AutoCAD 2007), neither *Zoom* nor *3Dzoom* changed the camera-target distance but changed only the object size on screen. *3Ddistance* was the intended method for changing the camera-target distance. Therefore, you used *3Ddistance* to change the amount of perspective for the objects, then *3Dzoom* to adjust the size of the objects on the screen keeping the same perspective. The advantage of this older method was that you could adjust amount of perspective and object size independently. Since AutoCAD 2007, you can adjust the amount of perspective independent of screen size by changing a camera's setting for *Lens length* or for *Field of View*. These settings can be changed in either the *View* panel of the *Home* tab or in the *View Manager* (see "Adjusting the Amount of Perspective").

## View Presets

FIGURE 33-48

The View Presets provide standard viewing directions for a 3D model, such as *Top*, *Front*, *Right*, and standard isometric viewing directions. The View Presets are not commands that can be entered at the Command line but are selections you can make from any of several methods described here. The View Presets are available from the *View* toolbar (Fig. 33-48) and from the *Views* panel on the *Visualize* tab (Fig. 33-49). In addition, you can find the View Presets in the *3D Views* option from the *View* menu (not shown) and in the *View Manager* (described in the next section).

FIGURE 33-49

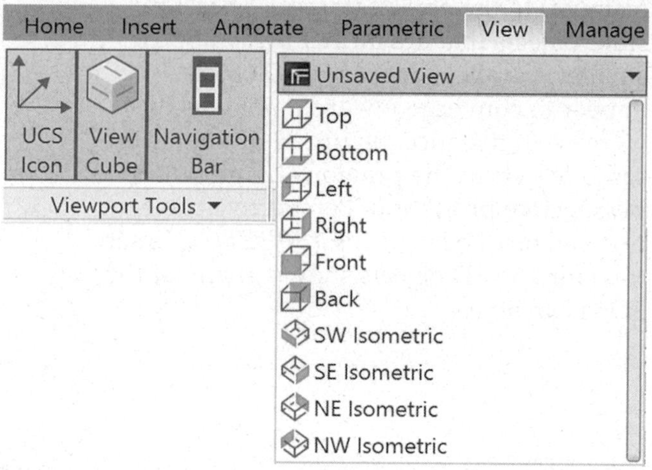


Each of the 3D View options is described next. The FORKLIFT drawing is displayed in these figures in parallel projection for several View Presets. It is important to remember that for each view, imagine you are viewing the object from the indicated position in space. Imagine the observer changes position while the object remains stationary.

It is important to note that the View Presets tool is an older feature of AutoCAD than the ViewCube and that the View Presets are a duplication of similar options in the ViewCube. However, there is one distinction of the View Presets. That is, using the orthographic options of the View Presets (*Top*, *Bottom*, *Left*, *Right*, *Front*, *Back*) can actually create new User Coordinate Systems if the *UCSOrtho* system variable is set to 1. See NOTEs below.

NOTE: View Presets display the object (or orients the observer) with respect to the World Coordinate System by default. For example, the *Top* option always shows the plan view of the WCS XY plane. Even if another coordinate system (UCS) is active, AutoCAD temporarily switches back to the WCS to attain the selected viewpoint.

TIP

NOTE: It is also important to note that if you use the *Top*, *Bottom*, *Front*, *Back*, *Right*, or *Left* option, and the *UCSOrtho* system variable is set to 1 (default setting), the UCS automatically changes to align with the view. Set *UCSOrtho* to 0 if you do not want to create a new UCS for each of these views (for more information on this setting, see Chapter 34, User Coordinate Systems).

### *Top*

The object is viewed from the top (Fig. 33-50). Selecting this option shows the XY plane from above (the default orientation when you begin a *New* drawing). Notice the position of the WCS icon. This orientation should be used periodically during construction of a 3D model to check for proper alignment of parts.

FIGURE 33-50

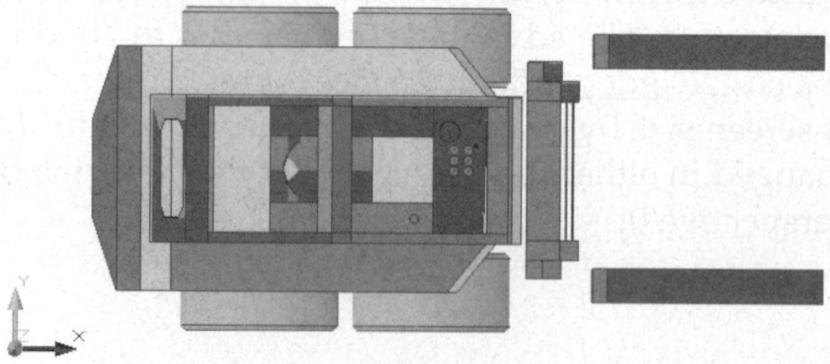

### *Bottom*

This option displays the object as if you are looking up at it from the bottom. The Coordinate System icon appears backward when viewed from the bottom.

### *Left*

This option looks at the object from the left side.

### *Right*

Imagine looking at the object from the right side (Fig. 33-51). This view is used often as one of the primary views for mechanical part drawing.

**FIGURE 33-51**

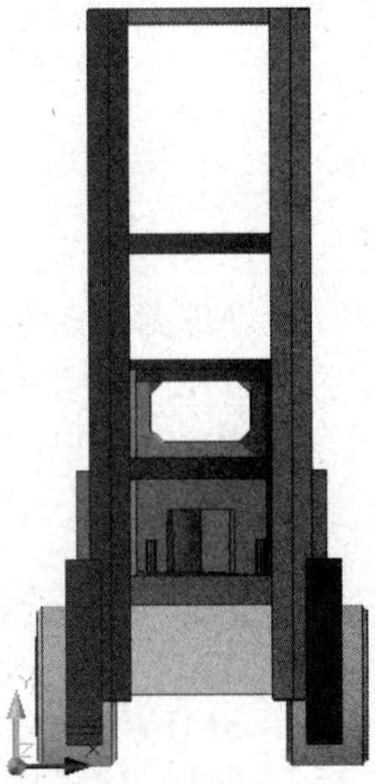

### *Front*

Selecting this option displays the object from the front view. This common view is used often during the construction process. The front view is usually the "profile" view for mechanical parts (Fig. 33-52).

**FIGURE 33-52**

### *Back*

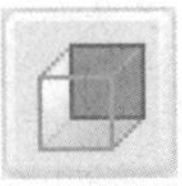

*Back* displays the object as if the observer is behind the object. Remember that the object does not rotate; the observer moves.

### *SW Isometric*

A *SW Isometric* is a view of the object as if the viewer were looking from a southwest position. Isometric views are used more than the "orthographic" views (front, top, right, etc.) for constructing a 3D model. Isometric viewpoints give more information about the model because you can see three dimensions instead of only two dimensions (as in the previously discussed views).

### *SE Isometric*

The *Southeast Isometric* view is generally the first choice for displaying 3D geometry. This view shows the object as if the observer was located southeast in relation to the 3D objects. If the object is constructed with its base on the XY plane so that X is length, Y is depth, and Z represents height, this orientation shows the front, top, and right sides of the object. Try to use this viewpoint as your principal mode of viewing during construction. Note the orientation of the WCS icon (Fig. 33-53).

**FIGURE 33-53**

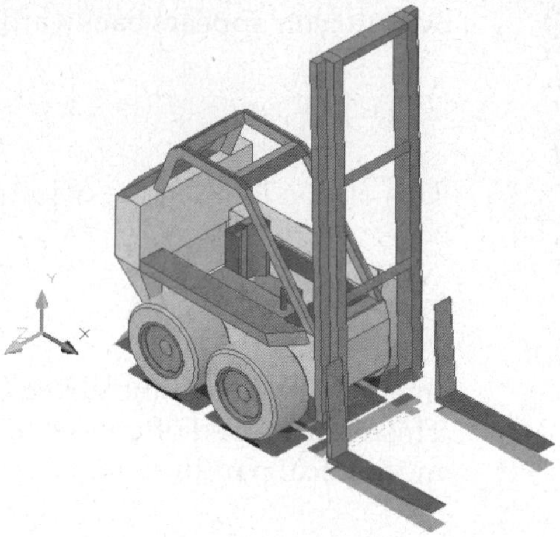

### *NE Isometric*

The *Northeast Isometric* view shows the right side, top, and back (if the object is oriented in the manner described earlier).

### *NW Isometric*

This viewpoint allows the observer to look at the left side, top, and back of the 3D object (Fig. 33-54).

**FIGURE 33-54**

## *View*

| Ribbon | Menu Bar | Command (Type) | Alias (Type) | Shortcut |
|---|---|---|---|---|
| *Visualize*<br>*Named Views*<br>*View Manager* | *View*<br>*Named Views...* | *View* | *V* | ... |

The *View* command produces the *View Manager*, which provides several functions useful for viewing 3D models. You can assign a name, save, and restore viewed areas of a drawing. You can access information and controls to the current view, any camera, and any saved views. You can also use this manager to adjust the amount of perspective displayed for selected views (see "Adjusting the Amount of Perspective").

For basic information on the *Set Current*, *New*, *Update Layers*, *Edit Boundaries*, and *Delete* functions of the *View Manager* see "*View*" in Chapter 10. The *View Manager* also provides access to the *View Presets* (described in the previous section) as shown in Figure 33-55.

**FIGURE 33-55**

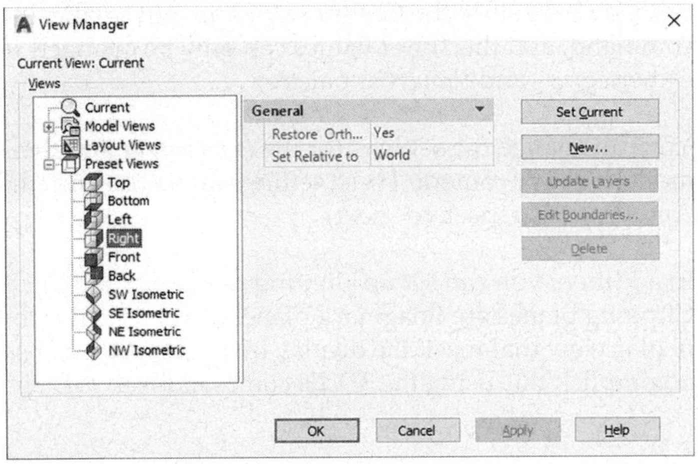


### *New*

When you create a *New* view of a 3D model, AutoCAD actually creates a camera and inserts a camera icon into the drawing. The view settings and camera are saved under the view name. Therefore, you can access information about the view in the *View Manager* or by selecting the view's camera icon and using the *Properties* palette.

### *Preset Views*

You can change your display to any of the *Preset Views* which are identical to those available from the *View* toolbar, the *View* panel drop-down list, or the *Views* menu, as explained earlier (see "View Presets"). To generate a 3D view using the *View Manager*, several steps are required. First, select the desired view preset from the list. Next, make the desired view current by selecting the *Set Current* button. Alternately, you can double-click on the desired view to make it current. Finally, select the *OK* button to produce the view.

**TIP**

It is strongly suggested that you ensure the *Set Relative to:* option is set to *World* (see Fig. 33-55). This setting causes the 3D view that is generated to be relative to the World Coordinate System. Using another option can make the resulting view difficult to predict. (See Chapter 34 for more information on this option and the *UCSBASE* system variable.)

Also keep in mind that if the *Restore Orthographic UCS* option is set to *Yes* when you use the *Top*, *Bottom*, *Front*, *Back*, *Right*, or *Left* option, the UCS automatically changes to align with the view (so the new XY plane is parallel with the screen). Change this setting to *No* if you do not want to create a new UCS with each orthographic view you select. (See Chapter 34 for information concerning the *Restore Orthographic UCS* option and the *UCSOrtho* system variable.)

### *Current and Model Views*

The *View Manager* not only allows you to set views current, but provides access to information and controls for the current view and for any saved views. Note that both saved views and any cameras that have been created are listed under *Model Views*. You can control such settings as *Name*, *Category*, and *Visual Style*, as shown in Figure 33-56.

**FIGURE 33-56**

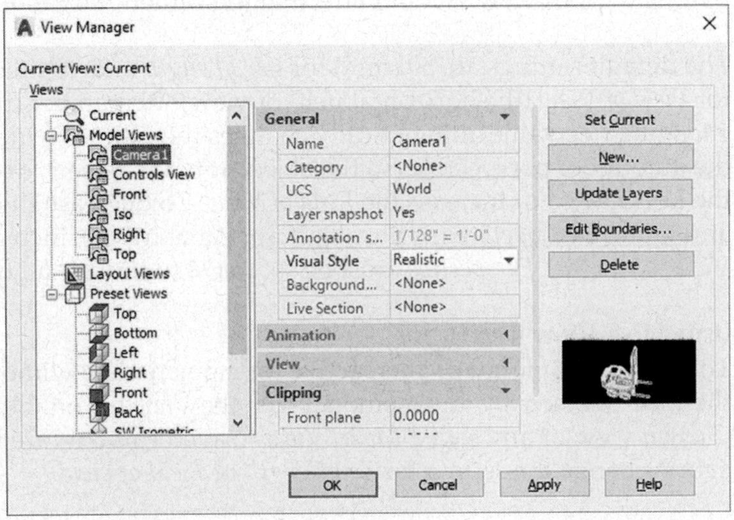

Note that the settings for *Camera* and *Target* are read-only in the *View Manager*. These values cannot be changed here since the *Camera* values are automatically generated using *3Dorbit* or similar navigation command, and the *Target* values can only be changed using the *3Dswivel* or *Camera* commands (and camera grips and *Properties* palette).

You can change the settings for the *Lens length* and the *Field of View* for the *Current* view or any saved model view or camera. These settings affect the amount of perspective for the view (see "Adjusting the Amount of Perspective" next).

In addition, you can set up clipping planes for the *Current* view or any saved model view or camera. Clipping planes are imaginary planes perpendicular to the line of sight (parallel to the screen) for a particular view that mask the display of objects. You can set a front or back clipping plane. You can also accomplish this using the *3Dclip* command (see "*3Dclip*").

## Adjusting the Amount of Perspective

Regardless of the size objects appear on the screen or plot, it is helpful to make some objects appear very large (such as a building) by increasing the amount of perspective, or to make some objects appear very small (such as a nut or bolt) by reducing the amount of perspective. To accomplish this, you must be able to adjust the amount of perspective that appears in a view independent of the size of the objects on the screen. However, the *Zoom* and the *3Ddistance* commands change both the screen size and the appearance of perspective (camera-target distance) together. (See previous discussion in the "*3Ddistance*" section.)

You can adjust the amount of perspective independent of the object's size on screen by changing a camera's setting for *Lens length* or the setting for *Field of View*. You can use either the *View* panel or the *View Manager* to change these settings for the *Current* view (if no named views or cameras have been saved) and for named views and cameras. If you created a *Camera* or saved a *Model View*, you can also select the camera icon and use the *Properties* palette to change the settings.

### Using the *View* Panel

In the expanded section of the *View* panel on the *Home* tab of the Ribbon, (Fig. 33-57), the slider and two edit boxes are used to change the settings for *Lens length* or *Field of view*. You can use either the slider or one of the two edit boxes to change the values. These two settings are codependent, so changing one automatically changes the other. This action adjusts the amount of perspective for a given view without affecting the camera-target distance.

**FIGURE 33-57**

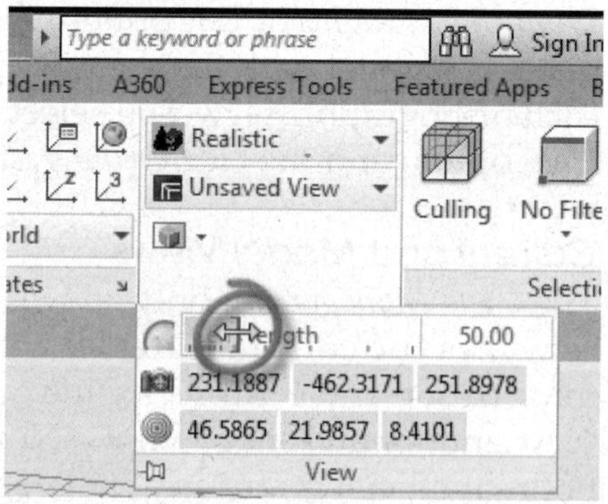


The default settings are 50 (mm) for *Lens Length* and 40 (degrees) for *Field of View* (the horizontal field of view). Remember that changing one value automatically changes the other. To increase the amount of perspective (so the object appears closer), decrease the *Lens Length* or increase the *Field of View*. To decrease the amount of perspective (so the object appears farther), increase the *Lens Length* or decrease the *Field of View*. Note that *Perspective* must be set to *On* to immediately view these changes.

TIP

### Using the *View Manager*

To adjust the amount of perspective for a given view without affecting the camera-target distance, open the *View Manager* by any method. From the *View* list on the left side of the *View Manager*, select the *Current* view or any saved *Model Views* (saved cameras are listed under *Model Views*). In the *View* category, change the setting for *Lens length* or *Field of view*.

### 3Dclip

| Ribbon | Menu Bar | Command (Type) | Alias (Type) | Shortcut |
|---|---|---|---|---|
| ... | ... | *3Dclip* | ... | ... |

Clipping planes are invisible planes that can pass through the view. Anything in front of the front clipping plane or in back of the back clipping plane becomes invisible when the planes are toggled on. The clipping planes are always parallel with the screen and perpendicular to the viewing direction (the screen Z axis).

Using *3Dclip* produces the *Adjust Clip Planes* window (Fig. 33-58). The window displays the current view from above so the clip planes can be seen on edge (appear as horizontal lines). The white (or black) line on the bottom of the window represents the front clipping plane and the green line on the top (if visible) represents the back clipping plane. As you move the clip planes through the 3D objects, the results are displayed in the drawing area. To see the results of adjusting clipping planes, the clipping planes must be toggled on. The controls for the *Adjust Clip Planes* window are available if you select the icon buttons or right-click in the window to produce the shortcut menu.

**FIGURE 33-58**

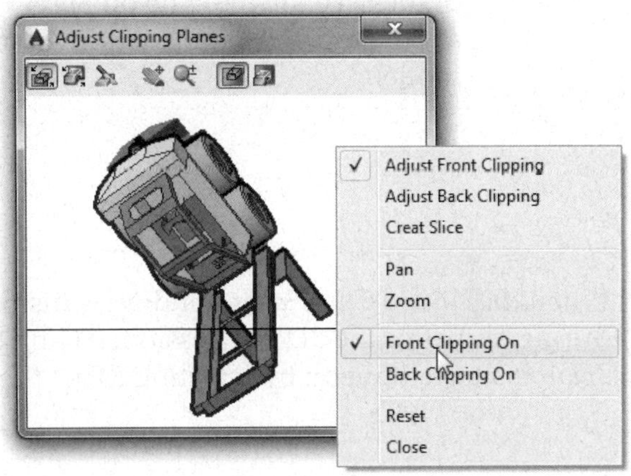


#### Front Clipping On

You must toggle the front clipping plane visibility on (depress the icon button or check the shortcut menu) if

you want to see the results of adjusting the front clipping plane in the drawing area. Toggle this option on before using *Adjust Front Clipping*. Once the desired clipping is set, you can toggle this on or off for the display in the drawing area.

#### Back Clipping On

Toggle this option on before using *Adjust Back Clipping*. Like *Front Clipping On*, you must toggle the back

clipping plane on to see the results of adjusting the back clipping plane in the drawing area. Once the desired clipping is set, you can toggle this on or off for the display in the drawing area.

#### Adjust Front Clipping

Move the horizontal black line near the bottom of the

window upward to adjust the front clipping plane into the 3D objects. Any objects or parts of objects in front of the front clipping plane disappear. Notice how most of the fork and lift sections of the forklift are removed by the front clipping plane in Figure 33-59.

**FIGURE 33-59**

### *Adjust Back Clipping*

Move the horizontal green line near the top of the window to adjust the back clipping plane. Generally, move the plane downward to pass through the 3D objects. Any objects or parts of objects in back of the back clipping plane disappear as shown in Figure 33-60.

**FIGURE 33-60**

### *Slice*

Make this selection if you want to move both front and back clipping planes together. First, adjust either clipping plane to determine the thickness of the slice, then toggle on *Slice*. You can achieve a view to display only a thin section of your 3D model.

When the *Adjust Clip Planes* window is dismissed, the clipping planes remain active (control the display) during and after the *3Dorbit* session. At any time, control the visibility of clipping planes using the *3Dorbit* shortcut menu by selecting *More*, then the *Front Clipping On* toggle or *Back Clipping On* toggle.

## Camera

| Ribbon | Menu Bar | Command (Type) | Alias (Type) | Shortcut |
|---|---|---|---|---|
| *Visualize Camera* (icon) | *View Create Camera* | *Camera* | *CAM* | ... |

For any 3D view in AutoCAD, the display is determined by the location of the observer, known as the camera, and the target, which is generally in the center of the 3D objects. For most views, such as those achieved through the use of the ViewCube or *3Dorbit*, you determine the position of the camera by the particular observation point you generate. In this case, an imaginary camera is positioned at the eye of the observer. Also, AutoCAD automatically calculates the position of the target at the centroid (geometrical center) of the 3D objects. For a view created using the ViewCube, *3Dorbit*, or similar navigation command, AutoCAD creates only an imaginary camera.

You can instead create a physical camera in AutoCAD, including a camera icon in the 3D model, using the *Camera* command. You can alternately create a camera by saving a named *View* (see "*View*" discussed earlier). Using the *Camera* command, you can create a camera and position a target. In addition to the camera's location in 3D space, you can specify a camera name, a target location, the lens length, field of view, and clipping planes. Since a camera created with the *Camera* command appears in the view list in the *View* panel and in the *View Manager*, you can restore the camera view at any time. You can *Erase* a camera by selecting its icon.

Camera allows you to set different camera and target locations by specifying 3D coordinate values or by picking points in 3D space. A camera icon appears when you select the camera location and a box representing the field of view appears when you select the target (Fig. 33-61). If you PICK points to define locations in 3D space, use *Object Snap* modes. Alternately, use *List* to locate a desired geometry location before entering specific coordinate values.

**FIGURE 33-61**

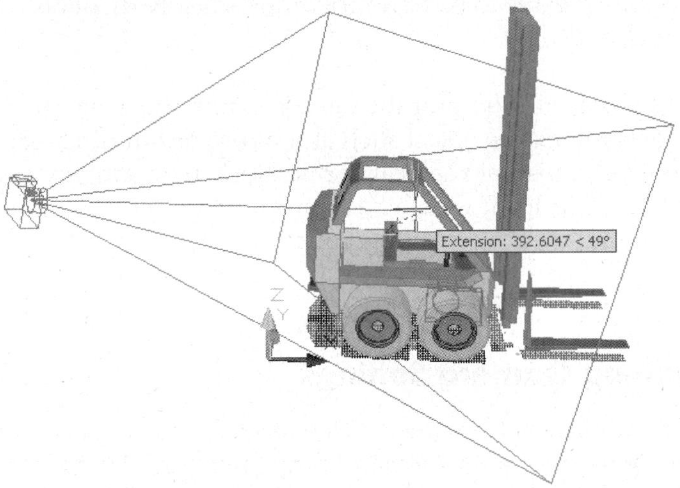


Command: ***CAMERA***
Current camera settings: Height=20.0000 Lens Length=50.0000 mm
Specify camera location: **PICK** or **coordinates**
Specify target location: **PICK** or **coordinates**
Enter an option [?/Name/LOcation/Height/Target/LEns/Clipping/View/eXit]<eXit>:

The options are briefly described below.

### *Name*

Assign a name for the camera. Otherwise AutoCAD assigns a name such as "Camera1." The camera name appears in any list of views.

### *LOcation*

Use this option to change the location of the camera. Enter coordinate values or PICK a location using *Object Snaps*.

### *Height*

This option keeps the camera's current X,Y location, but changes the *Camera Z* value for the current camera. This also resets the default *Height* value when creating subsequent cameras.

### *Target*

Use this option to change the location of the target, which determines the direction the camera points. Enter coordinate values or PICK a location using *Object Snaps*.

### *LEns*

You can specify a particular *Lens length* in millimeters using this option. A camera's lens length affects the distance perception of the view, similar to changing the strength of a telescopic lens. Therefore, changing the *Lens length* effectively changes the amount of perspective for the view (see "Adjusting the Amount of Perspective"). Changing the *Lens length* value also automatically changes the camera's *Field of View* setting.

### *Clipping*

You can establish *Front* and *Back* clipping planes for the camera. You are prompted for an "offset from target plane" in units. Clipping planes are invisible planes that can pass through the view. Any parts of the 3D model in front of the front clipping plane or in back of the back clipping plane become invisible when the planes are toggled on. (See "*3Dclip*" for more information.)

### *View*

Answering *Yes* to this prompt automatically displays a view as seen by the camera and ends the *Camera* command.

A good application for the *Camera* command is an architectural 3D model. In this case, the model may represent a large object such as a house or building—an object that you might want to see from inside. In contrast, most mechanical applications are relatively small and only viewed from outside. For an architectural model, it would be appropriate to set the camera location inside a room and set different target locations at each end of the room.

## Editing Camera Settings

Once a *Camera* (or named *View*) has been established, you can edit the settings such as lens length, field of view, camera position and target position. Two methods can be used to edit the settings: the camera's grips and the *Camera Preview*, or the camera's *Properties* palette.

### Camera Display

Before you can access the camera's grips or *Properties* palette, you must turn on the camera's icon so it appears in the drawing, if it is not already visible. Control the camera display using the *CAMERADISPLAY* system variable (0 = off, 1 = on) or the *3D Modeling* workspace, *Visualize* tab, *Camera* panel.

### Camera Grips and *Camera Preview*

If you click on a camera icon in the 3D model, several grips appear and the *Camera Preview* window appears (Fig. 33-62). You can select one of the following grips and drag it to establish a new location or setting: *Target location* grip, *Camera location* grip, *Camera & target location* grip and one of the *Lens length/FOV* grips. In addition, if you select an axis or plane of the Dynamic UCS that appears, you could change the *Target distance* or one of the previously mentioned grips along the selected axis or plane. (See Chapter 34, User Coordinate Systems, for more information on the Dynamic UCS.)

**FIGURE 33-62**

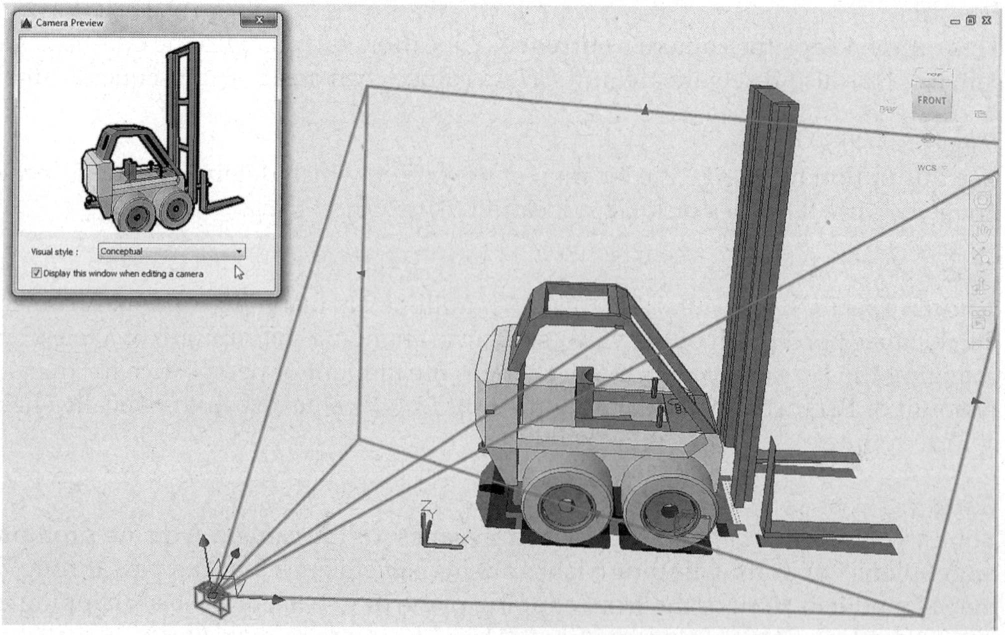


When you select a camera icon, the *Camera Preview* window also appears displaying the camera's current view (Fig. 33-62, upper-left corner). The view is dynamic so any changes to the view affected by dragging the grips are immediately displayed in the window. Note that you can specify a *Visual style* for the *Camera Preview*.

### Camera *Properties* palette

As an alternative, you can edit the values for the camera settings using the *Properties* palette. Activate the *Properties* palette, then select the camera icon in the 3D model. Note that all the fields can be edited, including camera and target position edit boxes (Fig. 33-63).

FIGURE 33-63

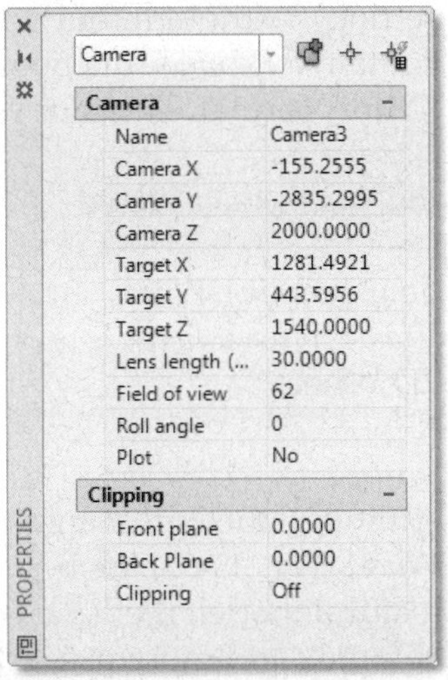


## *3Dswivel*

| Ribbon | Menu Bar | Command (Type) | Alias (Type) | Shortcut |
|---|---|---|---|---|
| ... | *View Camera > Swivel* | *3Dswivel* | ... | ... |

The 3D view is determined by the location of the observer, known in AutoCAD as the camera, and the target, which is generally in the center of the 3D objects. The *3Dswivel* command changes the target for the current view. Using *3Dswivel* simulates the effect of turning a camera on a tripod to change the viewing area. For example, if you moved the cursor (arched arrow) to tilt the camera upward, the objects in the viewfinder would move down and vice versa. Likewise, if you turned the camera to the right, objects in the viewfinder would move to the left.

If you use the *Camera* command to create a camera, you specify the location of the *Target* in the process. However, if a camera is created by saving a *View*, the target is automatically calculated as the center of the 3D objects. In either case, you can later change a camera's target setting using *3Dswivel*, dragging the camera's *Target* grip, or using the camera's *Properties* palette. You cannot change the target for a view that has not been saved as a named view.

## *Navswheel*

| Ribbon | Menu Bar | Command (Type) | Alias (Type) | Shortcut |
|---|---|---|---|---|
| *View Navigate* (icons) | *View SteeringWheels* | *Navswheel* | *Wheel* | ... |

SteeringWheels provide several tools for navigating about a model and are very similar to using the ViewCube, *3Dorbit*, *Pan*, *Zoom*, and *3Dswivel* tools. Although the SteeringWheels accomplish the same functions as those commands discussed previously, the SteeringWheels are somewhat more interactive and dynamic. The *Rewind* function allows you to rewind dynamically through the previous views; however, the rewind contents are valid only for the current session and cannot be saved. Although the SteeringWheels are useful for navigating "inside" an architectural model and can be used for demonstration purposes, they are not as efficient as using the basic navigation commands (ViewCube, *3Dorbit*, *Pan*, *Zoom*) for non-architectural models. The *Full Navigation Wheel* is displayed in Figure 33-64.

FIGURE 33-64

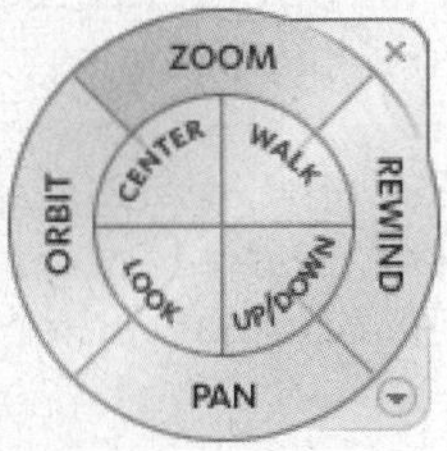

The *Navswheel* command produces the current SteeringWheel (based on the *NAVSWHEELMODE* setting). Although four wheels are available, three are used for navigating 3D models. All the wheels contain a *Rewind* function but differ in the other tools offered. All of the wheels are also available in a "Mini" format. The four wheels are:

| | |
|---|---|
| *Full Navigation Wheel* | Contains eight of the nine SteeringWheel functions (*NAVSWHEEL-MODE* 2) |
| *Tour Building Wheel* | Contains *Rewind, Look, Up/Down,* and *Forward* (*NAVSWHEELMODE* 1) |
| *View Object Wheel* | Contains *Rewind, Zoom, Center,* and *Orbit* (*NAVSWHEELMODE* 0) |
| 2D Wheel | Contains only *Zoom, Pan,* and *Rewind* tools and is useful for 2D viewing (*NAVSWHEELMODE* 3) (See Chapter 10, Viewing Commands.) |

Although the *Full Navigation Wheel* and the *Tour Building Wheel* (Fig. 33-65) are best for navigating "inside" an architectural model, the *View Object Wheel* (Fig. 33-66) is less complex and more useful for examining "outside" a model, such as a mechanical part or assembly.

**FIGURE 33-65**

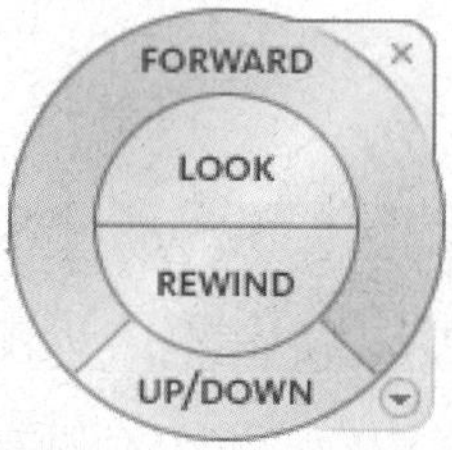


**FIGURE 33-66**

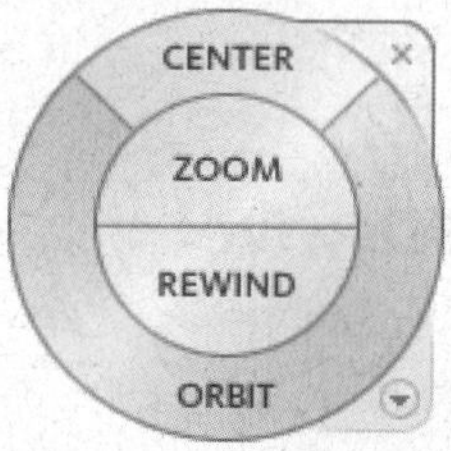


The options on the wheels are totally interactive. The wheels follow the cursor about the drawing so you can continually *Zoom, Pan, Orbit,* etc. The wheels appear on the screen until dismissed or another command is used. Therefore, the wheels are not typically used during the process of creating and editing a model, but are more suited to examining a model.

## SteeringWheel Shortcut Menu

Right-clicking anywhere on the wheel or on the down arrow (lower-right corner) produces the shortcut menu shown in Figure 33-67. Here you can activate any of the three 3D wheels in full-size or "mini" format. The 2D wheel is available only by setting *NAVSWHEELMODE* to 3. In addition, you can select from the following options.

**FIGURE 33-67**

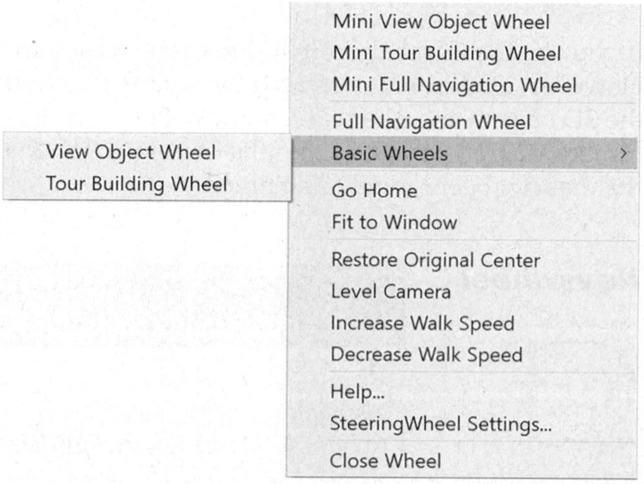


*Go Home*
This option returns the model to its home position. You can specify the home position for the model using the right-click shortcut menu of the ViewCube (see "*Navvcube*"). The default home position is a southwest isometric view.

*Fit to Window*
This option fits the model to the window, similar to *Zoom Extents*.

*Restore Original Center*
The *Center* tool on the wheel allows you to specify a center for *Orbit* and *Zoom*. Use this option to restore the original center location.

*Level Camera*
This option rotates the current view to align with the XY ground plane.

*Increase/Decrease Walk Speed*
Clicking this option increases/decreases the walk speed used for the Walk tool by a factor of two.

*SteeringWheel Settings*
This option produces the *SteeringWheel Settings* dialog box. See "*SteeringWheel Settings*."

**FIGURE 33-68**

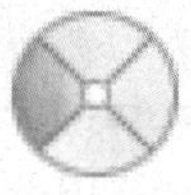


The three 3D wheels are available in a "mini" format. The mini wheels contain the same tools as the larger version, but without text on the wheel. To operate a mini wheel, move the cursor around the wheel near the segments to cause the segments to change color and text descriptors to appear. Figure 33-68 displays the *Mini View Object Wheel* with the *Orbit* tool activated.

## SteeringWheel Navigation Tools

The *Full Navigation Wheel* contains all the tools except *Forward*, whereas the other wheels contain a smaller set of tools. The tools operate as described in the following section. For many of the tools, first move the SteeringWheel to the desired location to set the "center" or "pivot" for the action, then select the option and hold down the button.

### *Zoom*

Locate the cursor and SteeringWheel at the desired center, then select the *Zoom* tool and hold down the button. When the "magnifying glass" icon appears, move up or down to zoom in or out, as usual. This tool serves the same function as normal *Zoom* command or using the mouse wheel to zoom, but is more complex.

Rather than holding down the button to activate *Zoom*, you can single-click the *Zoom* button to automatically zoom the current view about two-times size. To do this, check *Enable Single Click Incremental Zoom* in the *SteeringWheel Settings* dialog box.

### *Orbit*

This option is similar to using *3Dorbit* in that the position of the observer (camera) changes while the focal point (target or "center") is constant. Locate the cursor (and wheel) in the desired center, and then select *Orbit*. It is recommended to use the *Center* option first to locate a "Pivot" point before using *Orbit*, otherwise a central point in the view (or coordinate 0,0 if in the current view) is used as the pivot point if a *Center* is not designated (see "*Center*"). Alternately (if checked in the *SteeringWheel Settings* dialog box), you can cancel the SteeringWheel, then select an object to use as the center, then invoke the SteeringWheel again, then use the *Orbit* tool (see "*SteeringWheel Settings*").

### *Pan*

This action operates identically to the *Pan* command or using the mouse wheel to pan realtime. No *Center* is needed.

### *Center*

The *Center* option allows you to select a focal point on the model to use during *Orbit* or *Zoom*. (This option is named *Center*, but the designated point on the model is labeled "Pivot"). You can select any point in the current view as the pivot. AutoCAD attaches the pivot to visible surfaces or edges on objects in the view. Once the pivot point is designated, the current view automatically pans to locate the pivot point in the center of the drawing window.

### *Rewind*

Rewind allows you to wind backwards through all of the views generated in the drawing session. This is the primary feature that makes the SteeringWheels unique. Each time you use the *Zoom*, *Pan*, *Orbit*, and other options, including commands such as *Zoom*, *Pan*, and *3Dorbit* outside of the SteeringWheels, the sequence of views that you generate is temporarily saved so that you can *Rewind* back through them.

When you select *Rewind*, a set of thumbnails appears where each thumbnail represents each generated view in reverse chronological order from right to left (Fig. 33-69). Use the slider to dynamically rewind the set of views in the drawing window. Arrows may appear in place of thumbnails based on your selection in the *SteeringWheel Settings* dialog box for when to generate thumbnails.

**FIGURE 33-69**

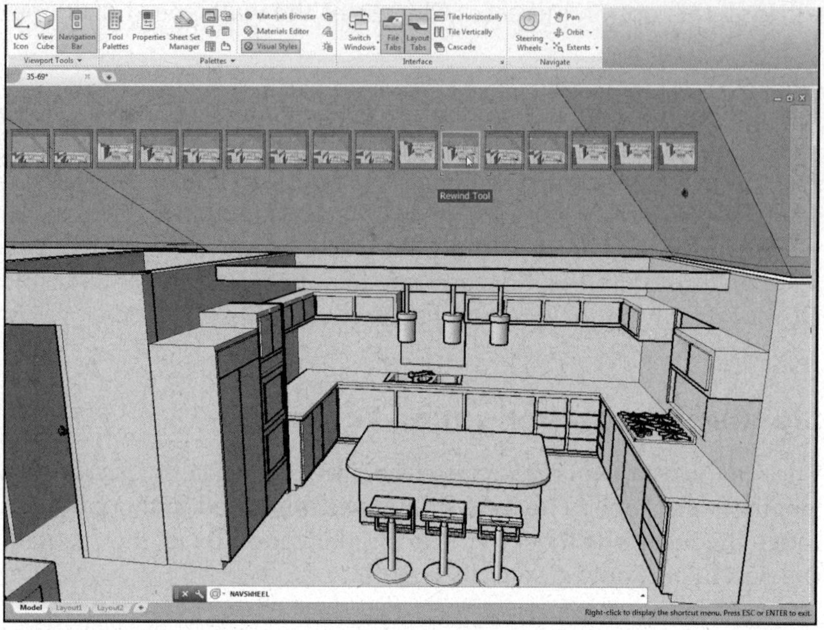


*Rewind* is useful for demonstration purposes but is effective only for the current drawing session. Because the thumbnails and sequence of views are not saved when the drawing is saved, the usefulness of *Rewind* is limited to "realtime" demonstrations.

### *Look*

The *Look* tool is similar to *3Dswivel*; that is, the observer (camera) remains stationary while the focal point of the view (target) changes. This tool is excellent for navigating "inside" an architectural model. Using *Look* you can pan left and right as well as up and down while standing in the same position—to look around a room, for example.

### Up/Down

The *Up/Down* tool changes the vertical (Z axis) position of both the observer (camera) and focal point (target) using a vertical slider. For example, in an architectural model you could use this tool to move from a position on the first floor directly above to the second floor.

### *Walk*

The *Walk* tool is unique in that it incorporates several capabilities. Combining the features of *Look*, *Zoom*, and *Up/Down*, you can virtually walk through an architectural model. When *Walk* is activated, move the mouse to direct the small arrow in the direction you want to walk (Fig. 33-70, circled arrow). Moving the arrow around the center dot indicates the direction (like *Look*), and moving the arrow farther or nearer the center dot adjusts the speed of the walk (like *Zoom*). You can hold down the Shift key to switch to *Up/Down* mode.

**FIGURE 33-70**

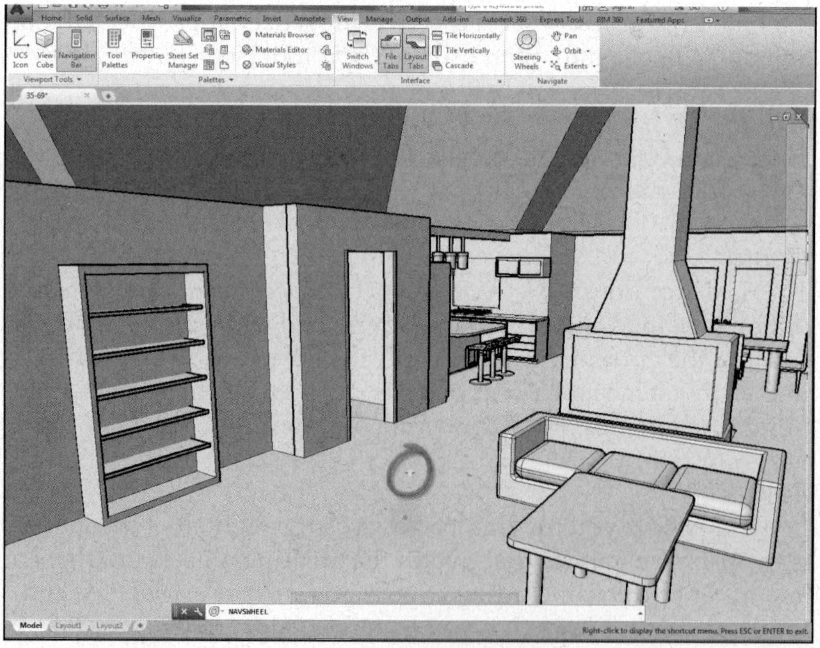

Another unique capability of *Walk* is that it automatically adds incremental views (and thumbnails) to the *Rewind* sequence. The other SteeringWheel tools create a *Rewind* view only at the end of the motion. The *Walk Speed* increment can be set in the *SteeringWheel Settings* dialog box.

#### *Forward*

The *Forward* tool is a slider that moves in and out toward the designated surface. This tool operates just like the normal *Zoom* command. First, point the wheel at the desired surface you want to zoom to, and then select the *Forward* tool to cause the slider to appear. The *Forward* tool is available only on the *Tour Building Wheel*.

### *SteeringWheel Settings*

Select the down arrow or right-click on any SteeringWheel and select *SteeringWheel Settings* to produce the dialog box shown in Figure 33-71. Here you can specify the SteeringWheels' appearance and functions.

**FIGURE 33-71**

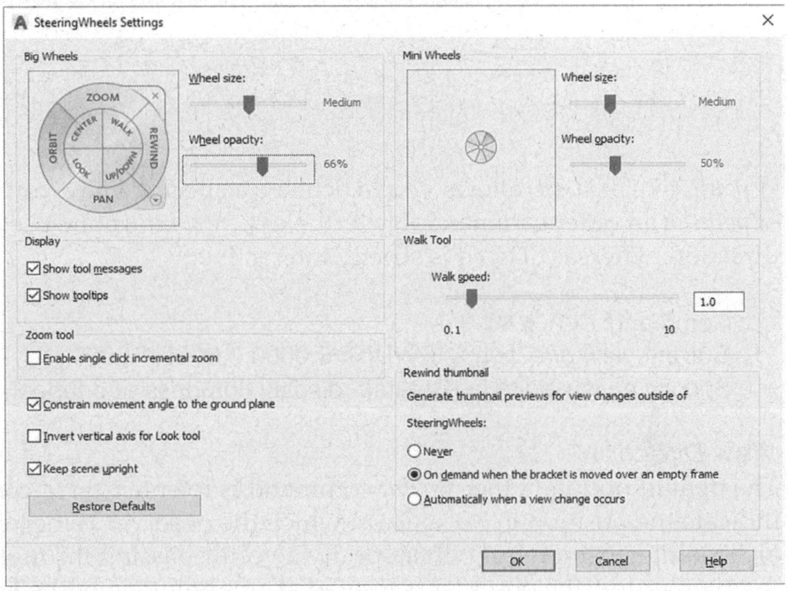


#### *Big Wheels, Mini Wheels*

These sliders allow you to adjust the size and opacity level for the wheels.

#### *Display*

Use the checkboxes to specify whether you want tooltips and tool messages to appear. If you select *Show the pinned wheel at startup* and then activate the SteeringWheel, the *Full Navigation SteeringWheel* appears in the lower-left corner of the AutoCAD window rather than the current wheel appearing at the cursor. In addition, you cannot immediately use the wheel until you cycle through six "mini tutorial" options to select the desired wheel.

#### *Zoom Tool*

*Enable Single Click Incremental Zoom*

This option allows you to zoom in at about two-times size when you single-click the *Zoom* tool.

#### *Invert Vertical Axis for Look Tool*

When not checked (default setting), you move the cursor up to *Look* up. When checked, move the cursor down to *Look* up.

#### *Maintain Up Direction for Orbit Tool*

When checked, the *Orbit* tool operates in constrained orbit mode (*3Dcorbit*). If not checked, the *Orbit* tool operates in free orbit mode (*3Dforbit*).

#### *Use Selection Sensitivity for Orbit Tool*

If checked, you can select objects before a wheel is displayed to define the pivot point for the *Orbit* tool (see previous discussion in "SteeringWheel Navigation Tools, *Orbit*").

### Walk Tool

*Constrain Walk Angle to Ground Plane*
This option controls whether you can use the Shift key to adjust the current view along the Z direction when using the *Walk* tool.

*Walk Speed*
Use this slider to set the movement speed for the *Walk* tool.

### Rewind Thumbnail

When you use the *Rewind* tool, thumbnail previews of previous views appear; however, if thumbnails do not exist, a "rewind" arrow symbol appears instead. Thumbnail previews are always generated for view changes that occur using the SteeringWheels. This section determines if and when thumbnail previews are generated if you use commands outside of the SteeringWheels such as *3Dorbit*, *Pan*, *Zoom*.

## -Vpoint

| Ribbon | Menu Bar | Command (Type) | Alias (Type) | Shortcut |
|---|---|---|---|---|
| ... | *View*<br>*3D Views >*<br>*Viewpoint* | *-Vpoint* | *-VP* | ... |

*-Vpoint*, like *3Dorbit*, allows you to achieve a specific view (or viewpoint) of a 3D object; however, *-Vpoint* is an older command. Nevertheless, *-Vpoint* can be used to specify an exact viewing angle or direction, whereas *3Dorbit* is strictly interactive.

```
Command: -VPOINT
Current view direction: VIEWDIR=0.0000,0.0000,1.0000
Specify a view point or [Rotate] <display compass and tripod>:
```

### View Direction

The default option of the *-Vpoint* command is to enter X,Y,Z coordinate values. The coordinate values indicate the position in 3D space at which the observer is located. The coordinate values do not specify an absolute position but rather specify a vector passing through the coordinate position and the origin. In other words, the observer is located at any point along the vector looking toward the origin.

Even though the *-Vpoint* command generates either a perspective or a parallel projection (whichever setting is current), *-Vpoint* does not compute a true distance. Therefore, the magnitude of the coordinate values is of no importance, only the relationship among the values. Values of 1,-1,1 would generate the same display as 2,-2,2.

### Rotate

The name of this option is somewhat misleading because the object is not rotated. The *Rotate* option prompts for two angles in the WCS (by default), which specify a vector indicating the direction of viewing. The two angles are (1) the angle in the XY plane and (2) the angle from the XY plane. The observer is positioned along the vector looking toward the origin.

```
Command: -VPOINT
Current view direction: VIEWDIR=0.0000,0.0000,1.0000
Specify a view point or [Rotate] <display compass and tripod>: r
Enter angle in XY plane from X axis <270>: 315
Enter angle from XY plane <90>: 35.3
Command:
```

The first angle is the angle in the XY plane at which the observer is positioned looking toward the origin. This angle is just like specifying an angle in 2D. The second angle is the angle that the observer is positioned up or down from the XY plane. The two angles are given with respect to the WCS.

Angles of **315** and **35** specified in response to the *Rotate* option display an almost perfect isometric viewing angle. An isometric drawing often displays some of the top, front, and right sides of the object. For some regularly proportioned objects, perfect isometric viewing angles can cause visualization difficulties, while a slightly different angle can display the object more clearly.

### *Display compass and tripod*

Similar to *3Dorbit*, the *Display compass and tripod* method of *-Vpoint* is a (somewhat) interactive method for adjusting your view; however, this option does not actually display the drawing geometry. Therefore, it is generally preferred to use the ViewCube or *3Dorbit* instead of this option of *-Vpoint*.

This option displays a three-pole axis system which rotates dynamically on the screen as the cursor is moved within a "globe." When you PICK the desired viewing position, the geometry is displayed in that orientation. Because the viewing direction is specified by PICKing a point without actually seeing the 3D objects, it is difficult to specify the desired viewpoint.

When the *Tripod* method is invoked, the current drawing temporarily disappears and a three-pole axis system appears at the center of the screen indicating the orientation of the X, Y, and Z axes for the new *-Vpoint*. The axes are dynamically rotated by moving the cursor (a small cross) in a small "globe" at the upper-right of the screen. The center of the "globe" represents the North Pole, so moving the cursor to that location generates a plan, or top, view. The small circle of the globe represents the Equator, so moving the cursor to any location on the Equator generates an elevation view (front, side, back, etc.). The outside circle represents the South Pole, so locating the cursor there shows a bottom view. When you PICK, the axes disappear and the current drawing is displayed from the new viewpoint.

## *Vpoint*

| Ribbon | Menu Bar | Command (Type) | Alias (Type) | Shortcut |
|---|---|---|---|---|
| ... | *View<br>3D Views ><br>Viewpoint Presets...* | *Vpoint* | *VP* | ... |

The *Vpoint* command is an older command used mostly before the introduction of *3Dorbit*. *Vpoint* produces the *Viewpoint Presets* dialog box (Fig. 33-72). This tool is an interface for attaining viewpoints that could otherwise be attained with the ViewCube, *3Dorbit*, or the *-Vpoint* command.

*Vpoint* serves the same function as the *Rotate* option of *-Vpoint*. You can specify angles *From: X Axis* and *XY Plane*. Angular values can be entered in the edit boxes, or you can PICK anywhere in the image tiles to specify the angles. PICKing in the enclosed boxes results in a regular angle (e.g., 45, 90, or 10, 30) while PICKing near the sundial-like "hands" results in irregular angles (Fig. 33-72). The *Set to Plan View* tile produces a plan view.

**FIGURE 33-72**

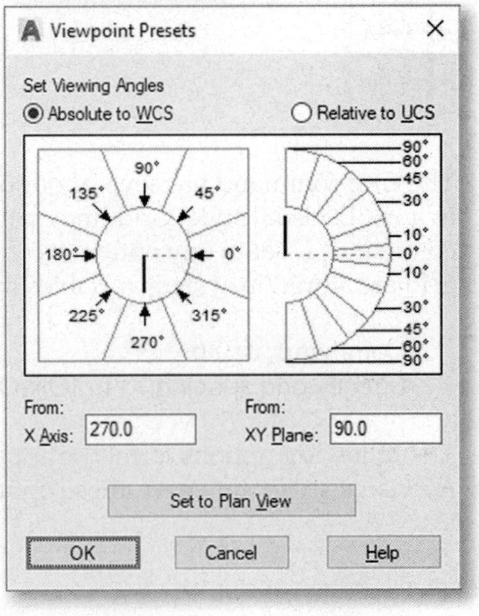

The *Relative to UCS* radio button calculates the viewing angles with respect to the current UCS rather than the WCS. Normally, the viewing angles should be absolute to WCS, but certain situations may require this alternative. Viewing angles relative to the current UCS can produce some surprising viewpoints if you are not completely secure with the model and observer orientation in 3D space.

NOTE: When you use this tool, ensure that the *Absolute to WCS* radio button is checked unless you are sure the specified angles should be applied relative to the current UCS.

### *Plan*

| Ribbon | Menu Bar | Command (Type) | Alias (Type) | Shortcut |
|---|---|---|---|---|
| ... | *View*<br>*3D Views >*<br>*Plan View >* | *Plan* | ... | ... |

This command is useful for quickly displaying a plan view of any UCS:

Command: ***PLAN***
Enter an option [Current ucs/Ucs/World] <Current>: (**option**) or **Enter**
Regenerating model.
Command:

Responding by pressing Enter causes AutoCAD to display the plan (top) view of the current UCS. Typing *W* causes the display to show the plan view of the World Coordinate System. The *World* option does not cause the WCS to become the active coordinate system but only displays its plan view. Invoking the *UCS* option displays the following prompt:

Enter name of UCS or [?]:

The *?* displays a list of existing named UCSs. Entering the name of an existing UCS displays a plan view of that UCS.

### *Grid*

| Ribbon | Menu Bar | Command (Type) | Alias (Type) | Shortcut |
|---|---|---|---|---|
| ... | *Tools*<br>*Drafting Settings...*<br>*Snap and Grid* | *Grid* | ... | F7 or Ctrl+G |

The *Grid* command has several options applicable to control the appearance of the grid that appears in any 3D visual style. Grid lines rather than grid dots are displayed when a *Visual Style* (*Vscurrent* command) is set to any option except *2D Wireframe* (see "Visual Styles"). The values that control the grid are stored in system variables that are saved with the drawing file. The prompt is as follows.

Command: ***GRID***
Specify grid spacing(X) or [ON/OFF/Snap/Major/aDaptive/Limits/Follow/Aspect] <0.5000>:

The following options affect the appearance of both the grid dots (*2D Wireframe*) and the grid lines (any 3D visual style); however, these options are especially applicable for 3D visual styles.

### *Major*

The 3D grid contains major and minor grid lines. This value (5 is the default) specifies the frequency of major grid lines compared to minor grid lines. This setting is also controlled by the *GRIDMAJOR* system variable.

### *aDaptive*

Use this option to control the density of grid lines when zoomed in or out.

Turn adaptive behavior on [Yes/No] <Yes>:

When *aDaptive* is turned on, the density of grid lines or dots is limited when you zoom out, which prevents the grid pattern from becoming too dense. This setting is also controlled by the *GRIDDISPLAY* system variable.

Allow subdivision below grid spacing [Yes/No] <No>:

If this subdivision option is turned on, additional, more closely spaced grid lines or dots are generated when you zoom in. The frequency of these grid lines is determined by the frequency of the major grid lines.

### *Limits*

This option determines if the grid is displayed beyond the area specified by the *Limits* command.

Display grid beyond Limits [Yes/No] <Yes>:

Since *Limits* are generally not used for 3D drawings, the default option for the ACAD3D.DWT is *Yes*.

### *Follow*

This option controls if the grid plane changes to follow the XY plane of the Dynamic UCS. The default setting is *No*, in which case the grid plane remains with the current UCS XY plane. This setting is also controlled by the *GRIDDISPLAY* system variable.

## *Vports*

| Ribbon | Menu Bar | Command (Type) | Alias (Type) | Shortcut |
|---|---|---|---|---|
| *Visualize* *Model Viewports* *Named* (icon) | *View* *Viewports* > | *Vports* | ... | ... |

The *Vports* command creates viewports. Either model space viewports or paper space (layout) viewports are created, depending on which space is current when you activate *Vports.* See Chapters 10 and 12 for basic information about *Vports* and using viewports for 2D drawings.

This section discusses the use of viewports to enhance the construction of 3D geometry. Options of *Vports* are available with AutoCAD specifically for this application.

When viewports are created in the *Model* tab (model space), the screen is divided into sections like tiles, unlike paper space viewports. Tiled viewports (viewports in model space) affect only the screen display. The viewport configuration cannot be plotted. If the *Plot* command is used, only the current viewport is plotted.

To use viewports for construction of 3D geometry, use the *Vports* command to produce the *Viewports* dialog box (Fig. 33-73). Make sure you invoke *Vports* in the *Model* tab (model space). To generate a typical orthographic view arrangement in your drawing, first ensure that the current display is set to *Parallel projection*. Next, invoke the *Viewports* dialog box and select three or four viewports from the list in the *Standard viewports* list, then select the *3D* option in the *Setup:* drop-down list (see Fig. 33-73, bottom center). This action produces a typical top, front, southeast isometric, and side view, depending on whether you selected three or four viewports.

**FIGURE 33-73**

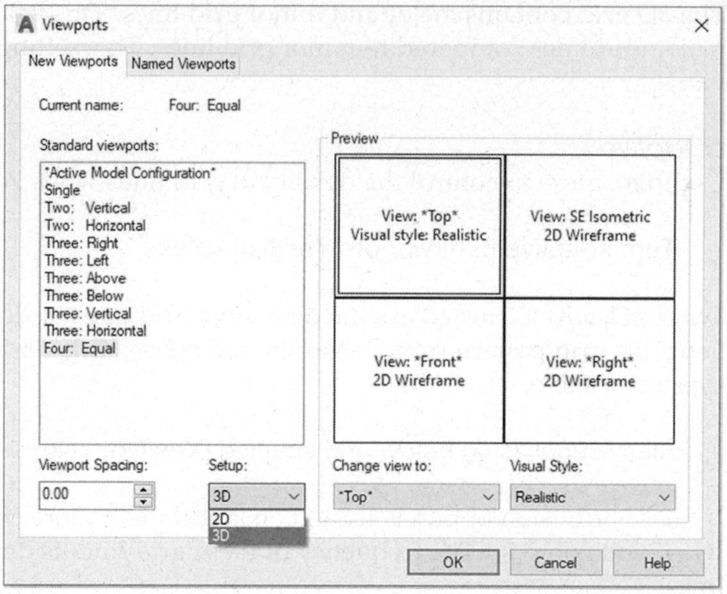


Assuming the options shown in Figure 33-73 were used with the FORKLIFT drawing, the configuration shown in Figure 33-74 would be generated automatically. Note, however, that the objects are not sized proportionally in the viewports. Since a *Zoom Extents* is automatically performed, each view is sized in relation to its viewport, not in relation to the other views.

**FIGURE 33-74**

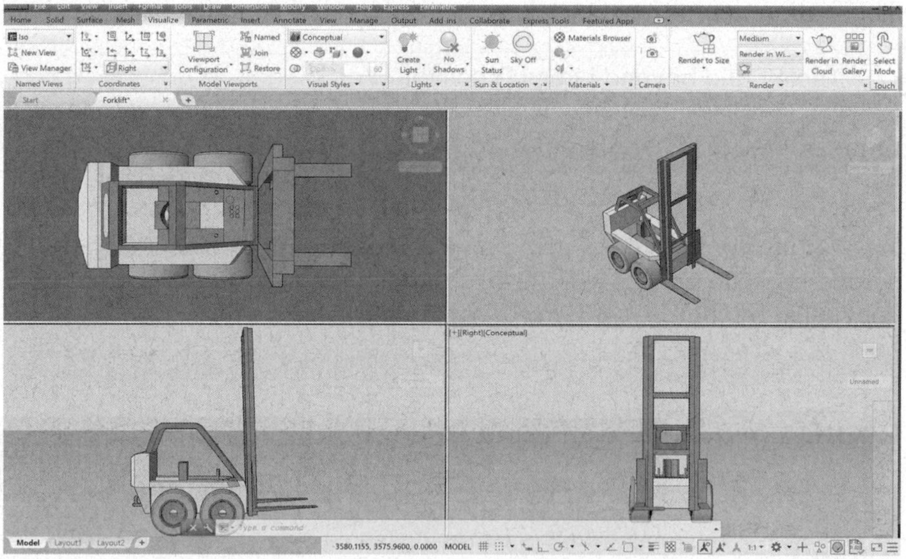

Next, if you desire all views to display the object proportionally, use *Zoom* with a constant scale factor (such as 1, not 1x) in each viewport to generate a display similar to that shown in Figure 33-75 (on the next page). You can then change other settings as desired for each viewport, such as *Visual style*, *Perspective*, etc.

NOTE: If *Perspective Projection* is turned on before using the *3D* option in the *Setup:* drop-down list, the objects are sized equally in the respective viewports.

NOTE: The setting of *UCSOrtho* has an effect on the resulting UCSs when the *3D Setup* option of the *Viewport* dialog box is used. If *UCSOrtho* is set to 0, no new UCSs are created when *Vports* is used. However, if *UCSOrtho* is set to 1, a new UCS is generated for each new orthographic view created so that the XY plane of each new UCS is parallel with the view. In other words, each orthographic view (not the isometric) now shows a UCS with the XY plane parallel to the screen. See Chapter 34 for more information on the *UCSOrtho* system variable.

FIGURE 33-75

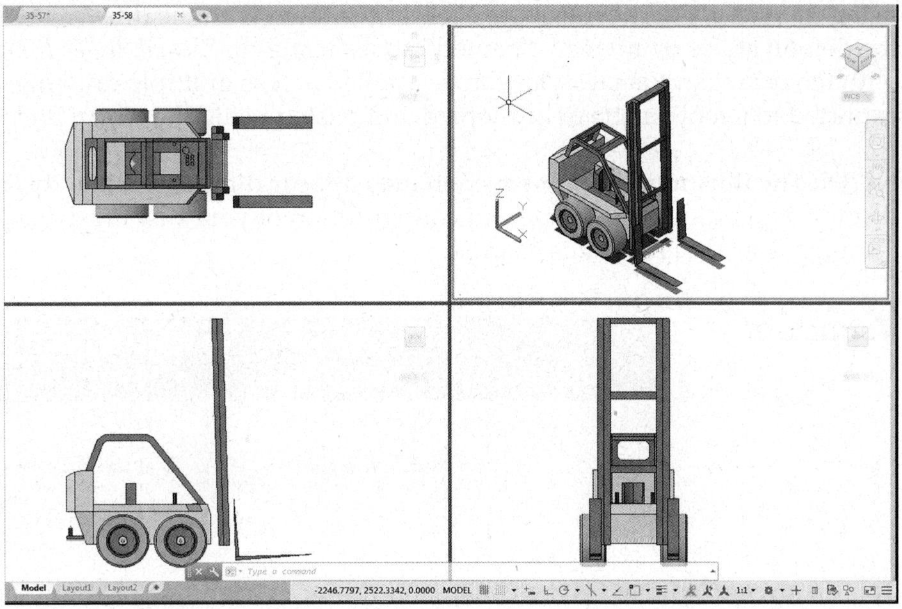


## VISUAL STYLES

A "visual style" is a group of settings that specify the display of 3D objects. The display, or appearance, of 3D objects includes such characteristics as wireframe, hidden, or shaded displays, treatment of object edges, and application of shadows. Visual styles that display shaded surfaces on an object or that generate shadows generally increase the realistic appearance of 3D objects and can greatly enhance your visibility and understanding of the model, especially for complex geometry. Many preset "default" visual styles are available and can be applied to surface and solid models, whereas only wireframe displays are appropriate for wireframe models since those objects do not contain surfaces. You can specify a visual style for each viewport. When a visual style is applied, the model retains its appearance until you change it.

Applying visual styles replaces commands used in older versions of AutoCAD such as *Shademode*, *Hide*, and *Hlsettings*. Instead of using many individual commands and typing system variables, you apply a visual style to the model and change the properties of the current visual style using the *Visual Styles* panel or the *Vscurrent* command. The *Visual Styles* panel is located in the *Visualize* tab of the Ribbon (Fig. 33-76). As soon as you apply a visual style or change its settings, you can see the effect in the viewport. Using the *Visual Styles* control panel allows you to select from many default visual styles and modify the current style to your needs. Any changes are applied to a visual style called "*Current*."

FIGURE 33-76

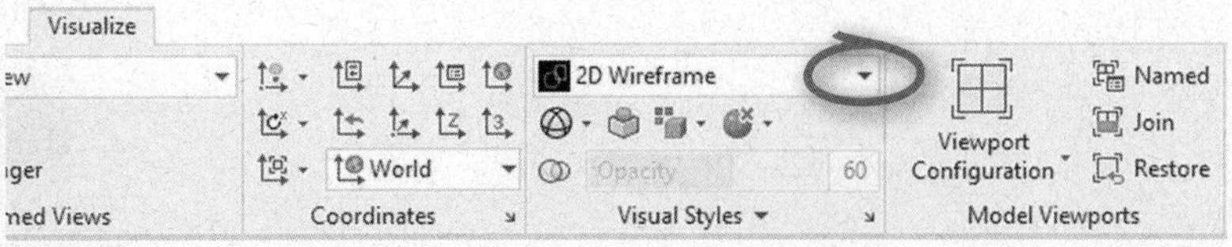


Changes you make to the *Current* style are saved with the drawing only if *Current* is the applied style when the drawing is saved. If you want to create new "named" visual styles, save them with the drawing, and apply them to any viewport or other drawing, use the *Visual Styles Manager* (see "*Visual Styles Manager*").

NOTE: Visual styles are saved in the drawing file. Whether you make changes to the *Current* visual style using the *Visual Styles* control panel and save them with the drawing or whether you make changes to the default styles or any new "named" styles using the *Visual Styles Manager*, visual styles are saved with the drawing. Visual styles can be applied across multiple drawings only if named styles are exported to a tool palette or are copied and pasted using the *Visual Styles Manager*.

NOTE: The illustrations in this section may appear differently than those shown in your computer display depending on the type and configuration of your display driver. See "Graphics Performance Tuning" near the end of this chapter.

### *Vscurrent*

| Ribbon | Menu Bar | Command (Type) | Alias (Type) | Shortcut |
|---|---|---|---|---|
| ... | *View*<br>*Visual Styles >* | *Vscurrent* | *VS* | ... |

The *Vscurrent* command is the Command-line equivalent to using the same options in the *Visual Styles* control panel. Note that the Command line prompt provides the same options as using the control panel drop-down list.

Command: ***VSCURRENT***
Enter an option
[2dwireframe/Wireframe/Hidden/Realistic/Conceptual/Shaded/shaded with Edges/shades of Gray/SKetchy/X-ray/Other] <2dwireframe>:

The options are described in the following section "Default Visual Styles."

## Default Visual Styles

The drop-down list in the *Visual Styles* panel provides access to the ten default visual styles: *2D wireframe, Wireframe, Hidden, Conceptual, Realistic, Shaded, Shaded with Edges, Shades of Grey, Sketchy,* and *X-ray* (Fig. 33-77). Although each of these styles can be modified to your liking and other new styles created, a description and illustration for each of these styles is given next.

FIGURE 33-77

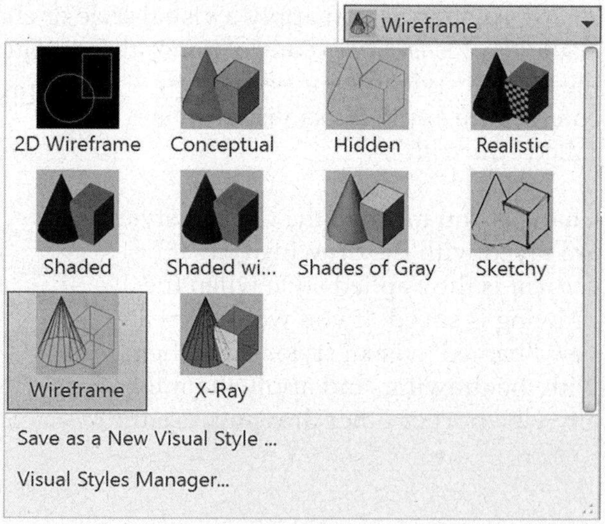

### *2D Wireframe*

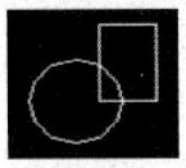

This option displays only lines and curves. The 3D geometry is defined only by the edges representing the surface boundaries. The model is displayed in parallel projection—perspective projection mode is disabled. This display uses the black model space background typically used for 2D drawing. The UCS icon is displayed as a "wireframe" icon. Also, linetypes and lineweights are visible. Any raster and OLE objects are also visible in the display. Generally used for 2D geometry, this option can be used for 3D geometry when lineweights and linetypes are important (Fig. 33-78).

**FIGURE 33-78**

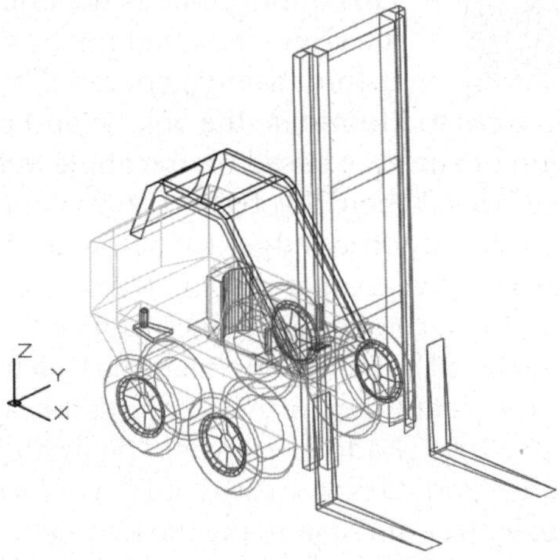


### *Wireframe*

Generally used to display 3D geometry when a wireframe representation is needed, this mode displays the objects using lines and curves to represent the surface edges (Fig. 33-79). Since all edges are displayed, this option is useful for 3D construction and editing. The shaded 3D UCS icon appears in this mode. Linetypes and lineweights are not displayed and raster and OLE objects are not visible. The *Compass* is visible if turned on. The lines are displayed in the material colors if you have applied *Materials*, otherwise they appear in the defined line or layer color.

**FIGURE 33-79**

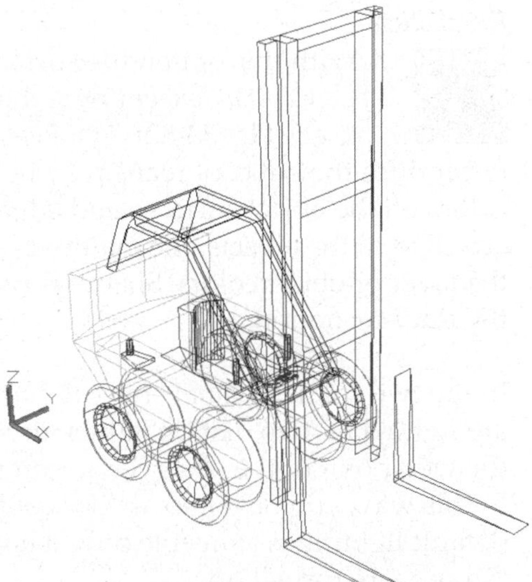


### *Hidden*

Use this mode to display 3D objects similar to the *Wireframe* representation but with hidden lines suppressed. In other words, edges that would normally be obscured from view by opaque surfaces become hidden (Fig. 33-80).

NOTE: The *Hidden* visual style replaces the older *Hide* command. *Hide* can still be used with the *2D Wireframe* style to remove obscured (hidden) lines from the display; however, *Hide* in this mode is not persistent—using *Regen* or *3Dorbit* cancels the display generated by *Hide*. You can use *Hide* to display meshed surfaces for curved edges. *Hide* is also affected by the setting of the *DISPSILH* variable; that is, if *DISPSILH* is set to 1, no mesh lines appear on curved surfaces.

**FIGURE 33-80**

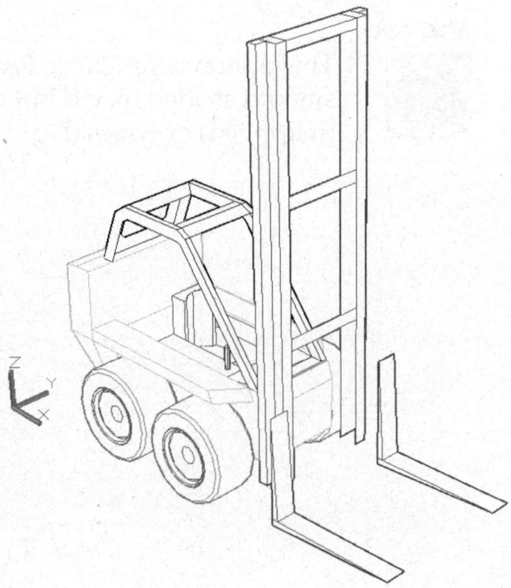

### *Conceptual*

This option shades the objects and smoothes the edges between polygon faces. Shading is automatically generated by using default light sources to illuminate the objects and adjusting the illumination of surfaces based on the angle between the lights and surfaces. When 3D objects contain curved geometry, this option smoothes edges (which would normally appear for curved surfaces) to display true curved geometry by eliminating the facet, or mesh, lines (notice the forklift wheels in Fig. 33-81). This option uses the *Gooch* face style which can show details better by softening the contrast between lighted areas and shadowed areas. The lighter areas display warm tones and darker areas display cool tones. The effect is less realistic, but it can make the details of the model easier to see by using color contrast instead of light/dark contrast. Shadows can be applied with this option.

**FIGURE 33-81**

### *Realistic*

With this option the surfaces are also shaded like *Conceptual*, except here the *Realistic* face style is used (Fig. 33-82). The *Realistic* face style is meant to produce the effect of realism by using the object or layer colors on the object surfaces and edges. If *Materials* have been attached to the objects, the materials are displayed rather than the layer or object colors. Shadows can be applied also with the *Realistic* option.

In the two shaded visual styles (*Conceptual* and *Realistic*), faces are lighted by two distant light sources that move to follow the viewpoint of the observer as you move around the model. In this way, you never see a "dark side" of the model. The default lighting is available only when other lights, including the sun, are turned off.

**FIGURE 33-82**

### *Shaded*

This option is similar to *Realistic*. It produces a smooth shaded model but does not show textures (if applied) or edges (Fig. 33-83).

**FIGURE 33-83**

### *Shaded with Edges*

This style is the same as *Shaded*. It produces a smooth shaded model with visible edges (Fig. 33-84).

**FIGURE 33-84**

### *Shades of Gray*

This option produces a gray color effect by using monocolor face color mode. Facet and silhouette edges are turned on and materials and textures are turned off (Fig. 33-85).

**FIGURE 33-85**

### *Sketchy*

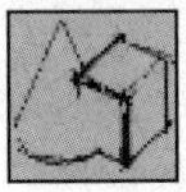

*Sketchy* produces a hand-sketched effect by using the overhang and jitter edges (Fig. 33-86).

**FIGURE 33-86**

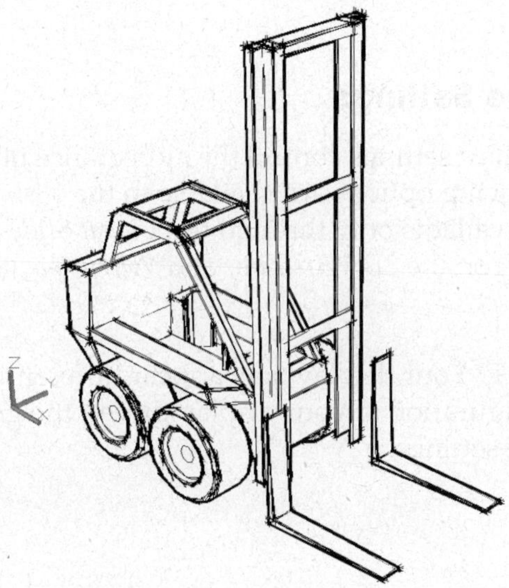

### *X-ray*

*X-ray* changes the opacity of faces to make the whole scene partially transparent. This style is similar to *Shaded with Edges*, except the opacity property is set to 50% (Fig. 33-87).

FIGURE 33-87

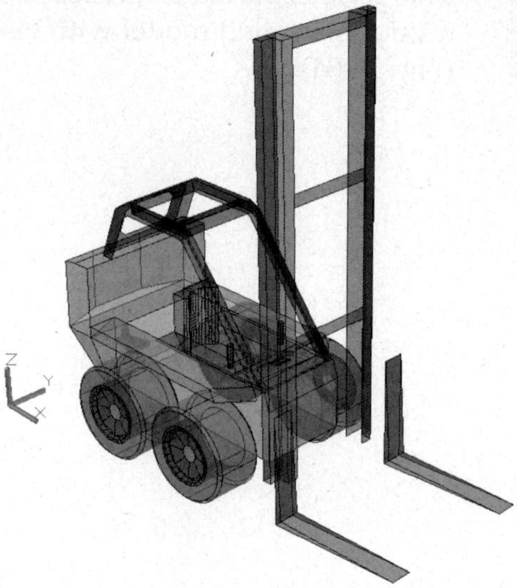

## Visual Style Properties

You can select a visual style and change its properties settings at any time using the options described next. These options are available in the *Visual Styles* and panels of the *Visualize* tab on the Ribbon (Fig. 33-88). The changes you make are reflected immediately in the viewport where the visual style is applied. You can change settings for faces, edges, and the environment. Any changes you make to a visual style are saved in a temporary visual style named "*Current*." You can save the *Current* style only by saving the drawing while the *Current* style is active. If another default visual style is selected from the drop-down list, the previous settings are lost. Although no commands exist for these options, system variables are used to store the settings. You can also make and save custom visual styles using the *Visual Styles Manager* (see "*Visual Styles Manager*").

FIGURE 33-88

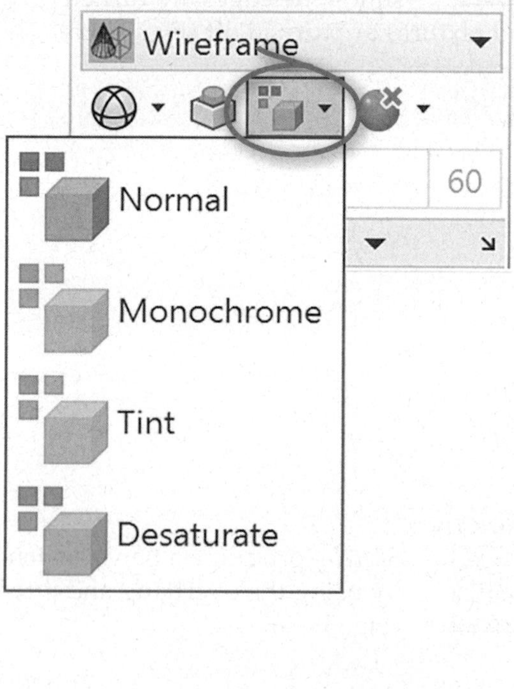


## Face Settings

The face settings control the appearance of the surfaces on all 3D objects in the viewport. Most of the following options are available in the *Visual Styles* control panel. A few of these options for face settings are available only through the *Visual Styles Manager* (see "*Visual Styles Manager*"). Face settings have no effect on the *2D Wireframe* and *Wireframe* visual styles.

NOTE: Your display may appear differently than that shown in the following figures depending on the configuration of your display driver, the graphics performance settings, and the combination of face and edge settings.

### *X-ray Effect*

The *X-ray Effect* setting makes the 3D objects appear transparent as shown in Figure 33-89. If the edge mode is set to *Isolines* or *Facet Edges*, the normally obscured edges of the objects would also become visible in *X-ray* mode. If multiple objects are included in the model, the objects located behind become visible through the foremost objects.

**FIGURE 33-89**

### *Face Color Mode*

The face color options control the display of colors on surfaces of a solid or surface model. This group of four options is available using the flyout on the *Visual Styles* panel. The options below correspond to the order in which the options appear on the flyout. The settings are saved in the *VSFACECOLORMODE* system variable.

*Regular*
This option does not apply a face color modifier. Therefore, the faces appear in the color or material as assigned to the object or layer.

*Monochrome*
This option displays the model in a default gray or blue-gray color, depending on the *Face Style*. If multiple colors or materials are assigned to objects or faces, use this option to display the model in all one color. The default monochrome color setting can be changed in the *Visual Styles Manager*.

*Tint*
This setting changes the hue and saturation value of face colors to a default gray or yellow-gray, depending on the *Face Style*. If multiple colors or materials are assigned to objects or faces, the model is displayed in multiple values of gray when the *Face Style* is *Realistic*.

*Desaturate*
Use this option to reduce the saturation component of the object's assigned colors by 30 percent. In other words, the *Desaturate* option makes the hues less intense. Therefore, the objects appear in a softer variation of the original colors.

### *Face Style*

The options on this flyout group define the type of shading used on faces. The setting is stored in the *VSFACESTYLE* system variable. The order of the options below correspond to the order of the buttons on the flyout group.

*No Face Style*
This option does not apply a face style but presents the objects in a 3D hidden style display. The other face settings are disabled.

*Realistic Face Style*
This is the default setting and is intended to display the objects as close as possible to how the faces would appear in the real world. For example, if *Face Color Mode* is set to *Realistic Face Colors*, the objects appear in the actual colors or materials assigned.

*Warm-Cool Face Style*
In the *Realistic* face style, shading is accomplished by changing the actual face colors to lighter and darker values of the assigned colors. The *Warm-Cool* face style instead uses cool and warm hues of the assigned colors for shading. The advantage is that darker faces are generally more visible in a *Warm-Cool* style than in a *Realistic* style.

## Shadow Settings

The shadow settings that you can control are the display of shadows and the background display. Shadows can be controlled in the *Lights* panel; however, you must use the *Visual Styles Manager* to set a background (see *"Visual Styles Manager"*).

### *No Shadows*
Turn shadows off to increase performance. When shadows are off, real-time navigation (such as with *3Dorbit*) becomes faster since shadows are not calculated with each change of display. This option sets the *VSSHADOWS* system variable to a value of 0.

### *Ground Shadows*
This option calculates ground shadows only. These shadows appear when parts of the objects cast a shadow on the ground, such as shown by the over-hanging surface in Figure 33-90. This option sets the *VSSHADOWS* system variable to a value of 1.

### *Full Shadows*
Using this setting all shadows appear, which includes ground shadows as well as shadows on objects that would be caused by surfaces that lie between the light sources and other object surfaces. Displaying full shadows sets the *VSSHADOWS* system variable to a value of 2.

**FIGURE 33-90**

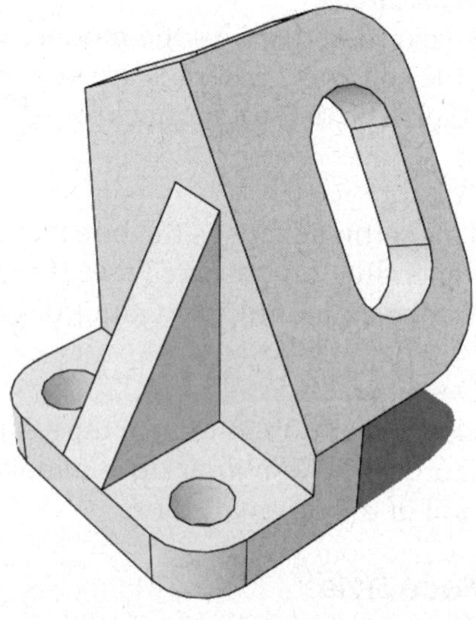

TIP

NOTE: To display full shadows, hardware acceleration is required. To access these settings, enter *Graphicsconfig* at the Command prompt. In the *Graphics Performance* dialog box, turn *Hardware Acceleration On*. When *Hardware Acceleration* is *Off*, full shadows cannot be displayed. See *"Graphics Performance"* near the end of this chapter.

## Edge Effects

There are many options that determine how edges are displayed in the model. These options are all available in the *Edge Effects* panel.

### *No Edges, Isolines, Facet Edges*

These settings control how the edges of the model are displayed. Your choice changes the *VSEDGES* system variable setting to 0, 1, or 2, respectively.

*No Edges*

When *No Edges* is selected, the surfaces of the object are differentiated only by the shading that displays surfaces lighter or darker. The edges between surfaces are not highlighted in any way (Fig. 33-91).

**FIGURE 33-91**

*Isolines*

In this case, the edges of the objects are displayed as well as the isolines that appear on curved surfaces. You can change the number of isolines using the *ISOLINES* system variable. The default *ISOLINES* setting is 4, so each of the holes in Figure 33-92 displays two visible lines (and two lines are not visible).

**FIGURE 33-92**

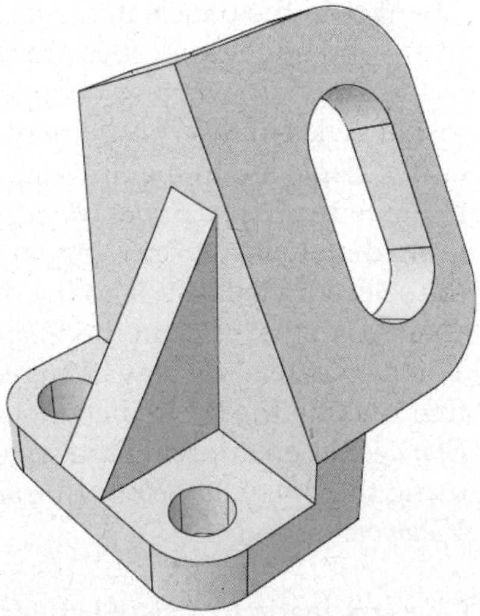

*Facet Edges*

This setting causes edge lines to appear at all facet edges as well as normal (non-curved) edges. The display is similar to that shown in Figure 33-92 (*Isolines* setting), except that additional lines would be displayed on all facet edges of curved surfaces (in the holes, for example). *Facet Edges* must be on to display *Obscured Edges* and *Intersection Edges*.

*Edge Color*

If the edge mode is set to either *Isolines* or *Facet Edges*, you can use the drop-down list to set the color for the edges. The setting is stored in the *VSEDGECOLOR* system variable.

### *Intersection Edges*

This display option is useful only when your 3D model contains multiple solids that intersect or overlap. In this case, turning on *Intersection Edges* displays lines of intersection wherever solids connect. The edge mode must be set to *Facet Edges* to use this option. The on/off setting is stored in the *VSINTERSECTIONEDGES* system variable. You can use the color drop-down list to specify the color for intersection edges (*VSINTERSECTIONCOLOR* system variable). If you prefer a non-continuous linetype for the edge lines, use the *VSINTERSECTIONLTYPE* system variable. To increase performance, turn off the display of intersection edges.

## *Visualstyles*

| Ribbon | Menu Bar | Command (Type) | Alias (Type) | Shortcut |
|---|---|---|---|---|
| *Visualize*<br>*Visual Styles*<br>(arrow) | *View*<br>*Visual Styles >*<br>*Visual Styles Manager* | *Visualstyles* | *VSM* | ... |

The *Visual Styles Manager* (Fig. 33-93) gives you complete control of accessing, modifying, and creating visual styles. You can modify any style by selecting its image in the *Available Visual Styles in Drawing* section, then making changes to the face, environment, and edge properties below. Changes you make to a style are saved with the drawing.

**FIGURE 33-93**

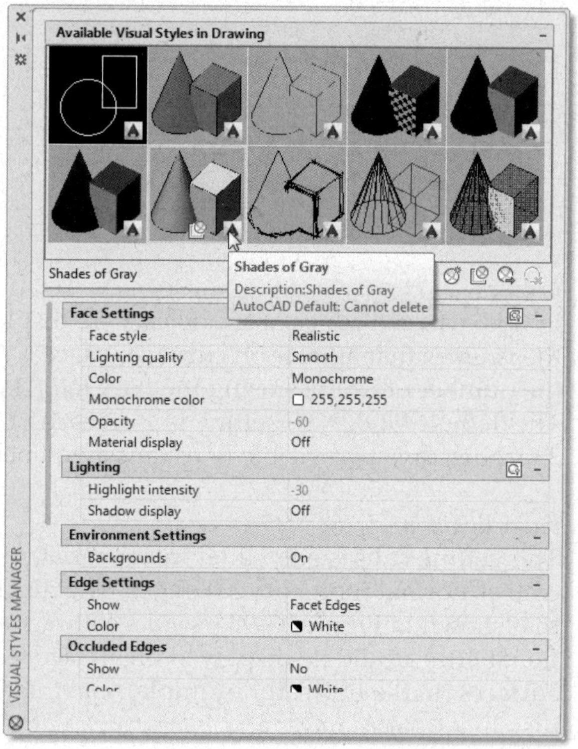


NOTE: The *Visual Styles* and *Edge Effects* panels (on the Ribbon) and the *Visual Styles Manager* operate differently in these respects:

1. Changes in the panels made to the face, environment, and edge properties are immediately reflected in the drawing, whereas changes made to one of the visual styles in the *Visual Styles Manager* are not generally displayed in the drawing. However, changes made in the *Visual Styles Manager* can be displayed immediately in the drawing only if you first apply the selected visual style to the drawing, then make changes to that style in the *Visual Styles Manager*. You can instead display changes retroactively by first making the desired changes in the *Visual Styles Manager*, then applying the style to the drawing using the *Apply the Selected Visual Style to Current Viewport* button.

2. Changes made to a visual style using the panels are applied to a temporary style named *Current* but are not saved unless *Current* is the applied style when you save the drawing. However, using the *Visual Styles Manager*, changes to any style, including the default styles, are automatically saved with that drawing only. Therefore changes made to any of the default styles do not affect the default styles of other drawings. Named styles can be exported to a tool palette or copied and pasted for use in other drawings.

### *Available Visual Styles in Drawing*

At the top of the palette-style interface, the *Available Visual Styles in Drawing* section displays image tiles of every visual style that is available in the drawing (the ten default styles and any named styles). The selected visual style displays a yellow border and the name of the current style is displayed below the image tiles. The *Face Settings*, *Environment Settings*, and *Edge Settings* of the selected visual style are displayed in the panels below. The AutoCAD drawing icon may appear on an image tile. The default visual styles that are shipped with the product display the product icon in the lower-right corner of the image tile. If one of the styles is applied to the drawing, an icon is displayed in that image tile (see Fig. 33-93, where the *Shades of Gray* style is applied to the viewport).

### Shortcut Menu and Tool Strip Buttons

You can right-click in the image tile area to display the shortcut menu (Fig. 33-94). In addition, several buttons are available just below the image tiles.

**FIGURE 33-94**

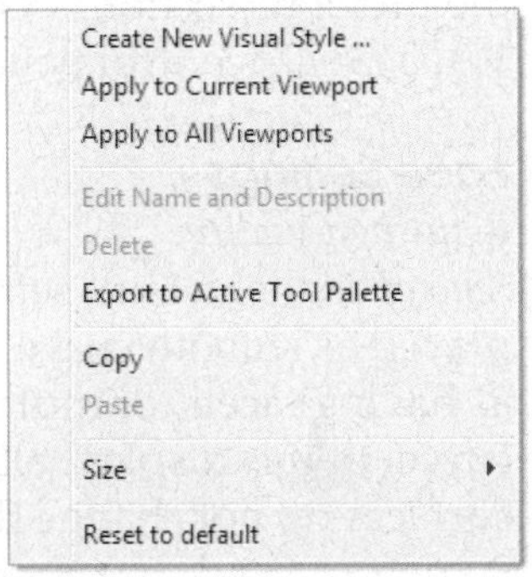


*Create New Visual Style*

Use this option to create and save a "named" visual style. The *Create New Visual Style* dialog box appears (not shown) where you enter a name and an optional description. A new sample image is placed at the end of the panel and becomes the selected style. The initial settings for a new style are based on the *Realistic* style.

*Apply Selected Visual Style to Current Viewport*

This option applies the selected visual style to the current viewport. This button is useful if you want to immediately see how changes you make to the visual style affect the drawing.

*Export the Selected Visual Style to the Tool Palette*

Use this option if you want to use the selected visual style in other drawings. A tool (button) is created for the selected visual style and placed on the active tool palette. Therefore, open the desired tool palette before using this option. If the *Tool Palettes* window is closed, it is automatically opened and the tool is placed on the top palette.

*Delete*

Use this option to remove the selected visual style from the drawing. A default visual style or one that is in use (currently applied) cannot be deleted.

*Apply to All Viewports*

This option applies the selected visual style to all viewports in the drawing.

*Edit Name and Description*

The *Edit Name and Description* dialog box is displayed so you can change the name, add a description, or change an existing description. The description is displayed in a tooltip when the cursor hovers over the sample image.

*Copy*

The current visual style and image is copied to the Windows Clipboard. You can then paste it into the *Available Visual Styles* panel or paste it into the *Tool Palettes* window to create a copy of the visual style.

*Paste*

After using *Copy*, you can paste a visual style into the *Available Visual Styles* panel or paste it into the *Tool Palettes* window. Since you can *Copy* and *Paste* between multiple open drawings in AutoCAD, this option provides another method for exporting and using a visual style in another drawing.

*Size*

Use this setting to specify the size of the image tiles displayed in the *Available Visual Styles* panel. *Medium* is the default size. The *Full* option fills the panel with one image.

*Reset to Default*

If you change the properties of one of the default visual styles, you can use this button to restore the original settings.

### *Face Settings*, *Environment Settings*, and *Edge Settings*

Properties of the faces, environment, and edges for the selected visual style can be changed to your preferences using this section of the *Visual Styles Manager*. Changes are automatically applied and saved to the selected visual style. These property options are displayed in the *Visual Styles* control panel (refer to the previous section, "Visual Style Properties"). However, the following options are not available in the *Visual Styles* control panel.

### *Face Settings*

*Lighting Quality*

*Smooth* is the default setting in which all curved surfaces are displayed as smooth curves. *Faceted* lighting displays curved surfaces as having facets, or multiple flat surfaces, around a normally smooth curve. In this display, all curved surfaces display facets but planar surfaces do not change (Fig. 33-95).

FIGURE 33-95

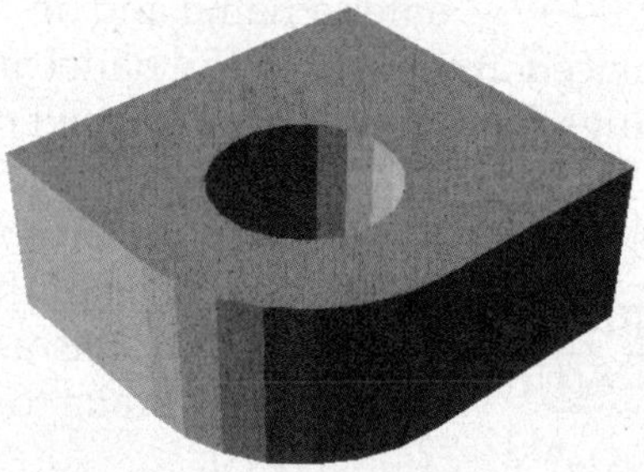

With *Smoothest* lighting quality, the colors are computed for individual pixels (Fig. 33-96). To use the *Smoothest* option, the *Per-Pixel Lighting* must be turned on in the *Graphics Performance* dialog box.

FIGURE 33-96

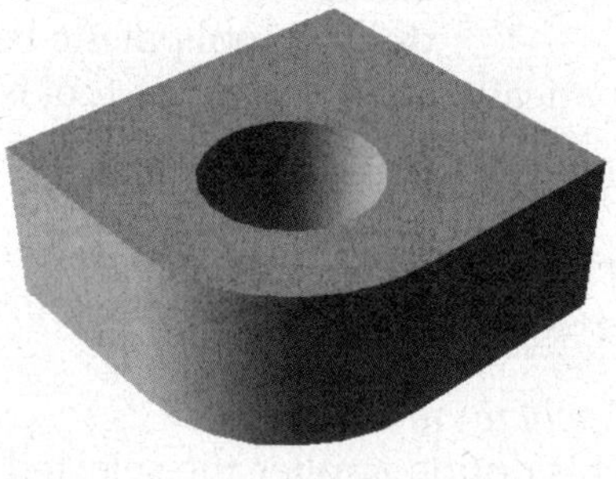

*Opacity*

This value controls the opacity, or the level of transparency, of objects. Greater values display more opacity while smaller values make the objects more transparent. The *Opacity* toggle button and value input here serve the same function as the *X-ray* mode button and slider on the Ribbon.

*Material Display*

These settings, *Materials and Textures, Materials,* and *Off*, control the display of materials on objects. These properties affect the display only if materials and/or textures have been applied to the objects. These settings are available only when the *Realistic* or *Conceptual* visual style, or a visual style based on them, is selected.

### *Lighting*

The *Highlight Intensity* property controls the size of highlights on faces without materials. Highlights are the shiny spots that appear on smooth, hard surfaces. Higher (positive) values produce greater highlights and lower (negative) values produce a dull surface. The *Highlight Intensity* toggle button turns this feature on or off.

### *Occluded Edges*

You can display the occluded edges (hidden edges) by setting *Show* to *Yes* (Fig. 33-97). To use this display, edge mode must be set to *Facet Edges* or *Isolines*. When *Occluded Edges* is on, you can set the desired color for those edges using the drop-down list. Use the *Color* option to display occluded edges in virtually any color.

**FIGURE 33-97**

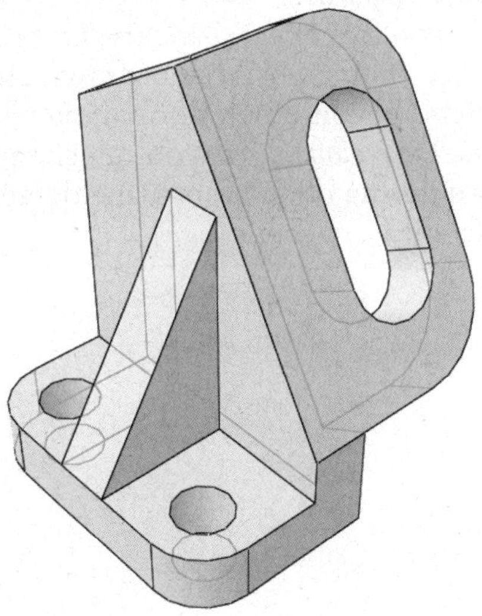

You can create a display that shows the occluded lines as hidden lines (or other non-continuous) similar to that in Figure 33-98. To do this, first set the *Face Style* to *None* and set the *Edge Settings* to *Facet Edges* or *Isolines*. Next, set the *Linetype* to a *Dashed* option.

**FIGURE 33-98**

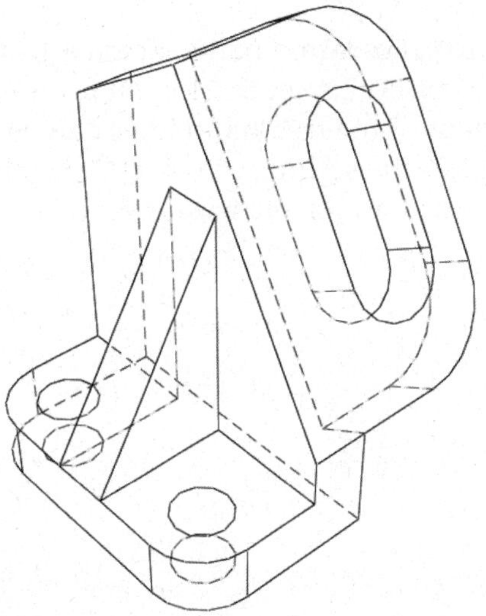

### *Intersection Edges*

This display option is useful only when your 3D model contains multiple solids that intersect or overlap. In this case, turning on *Intersection Edges* displays lines of intersection wherever solids connect. The edge mode must be set to *Facet Edges* or *Isolines* to use this option. You can use the drop-down lists to specify the *Color* and *Linetype* for the intersection edges. To increase performance, turn off the display of intersection edges.

### *Silhouette Edges*

If you want only the outline edges to appear around the object, turn *Silhouette Edges* on and set the *Edge Settings* to *None*, as shown in Figure 33-99. You can, however, turn edges to *Isolines* or *Facet Edges*, in which case the silhouette edges still appear heavier than the other edges. You can control the width of silhouette edges by changing the *Width* value. Silhouette edges are not displayed on wireframe or transparent objects.

**FIGURE 33-99**

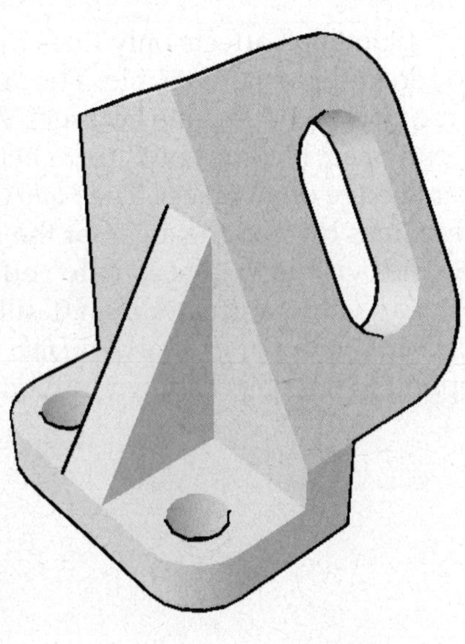

### Edge Modifiers

*Line Extensions*

Turn on this effect using the *Line Extensions edges* toggle button. This option gives a hand-drawn effect because it makes lines extend beyond their normal intersection (Fig. 33-100). When *Line Extensions* is on, you can change the length of the overhang by entering other values (the default setting varies based on the current visual style).

**FIGURE 33-100**

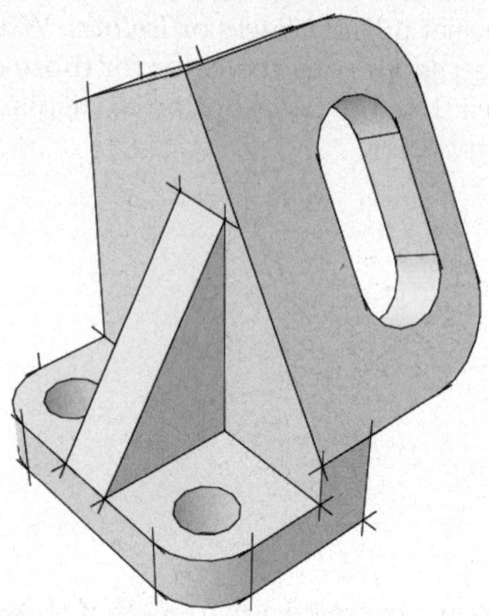

*Jitter*

*Jitter* gives a free-hand sketch effect as shown in Figure 33-101. This option is set to *Medium* for the *Sketchy* visual style. The extent of the free-hand lines can be set to *High, Medium* or *Low*—the higher the setting, the more "sketch" lines that appear. For a realistic free hand-sketched effect, turn on both *Jitter* and *Line Extensions*.

**FIGURE 33-101**

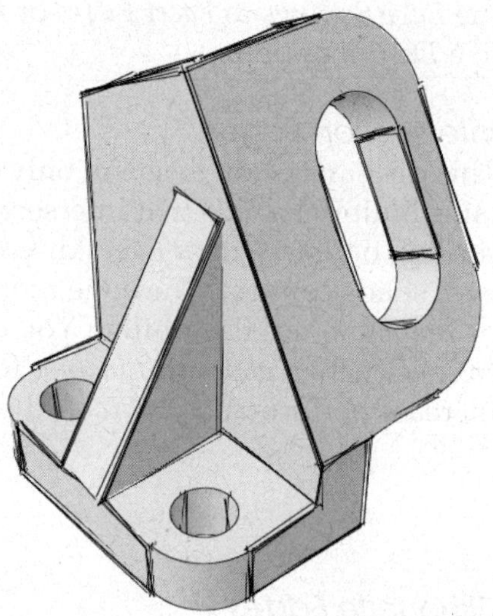

*Halo Gap %*

The "halo gap" affects only lines that are partially hidden and partially visible. The "gap" begins at the point where a line becomes visible and extends past ("comes out" from behind) another visible edge of an object. The *Halo Gap %* shortens these lines by creating a gap at the point where it becomes visible to give a "halo" effect. When the halo gap value is greater than 0, silhouette edges are not displayed. For example, a *Halo Gap %* of *10* is applied to the object shown in Figure 33-102.

**FIGURE 33-102**

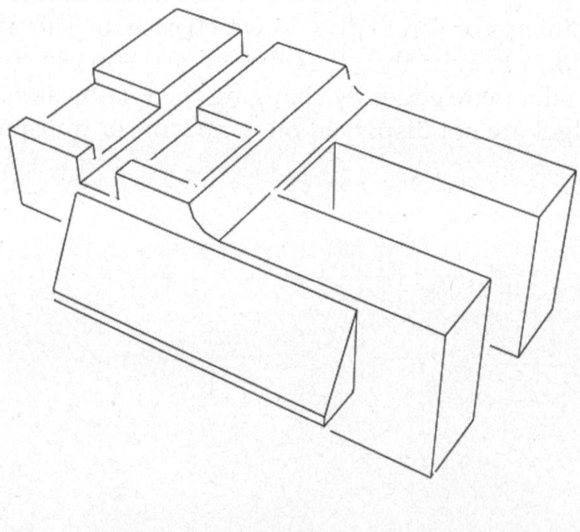

## Named Views and Viewports

The *NAmed* option of the *MIEW* command can be used to directly insert a previously-saved model-space named view into the current layout, automatically creating the layout viewport and placing the view. The newly created layout viewport is set to a default scale or the current annotative scale, and starts out locked. You can unlock the layout viewport and make changes as needed. The layout viewports are easily moved, resized, and scaled.

This feature can make it easy to create the layout you desire, but requires just a bit of setup. First of all you must create the named view. Recall the *View Manager* selection available on the ribbon *Visualize* tab. This is a convenient way to make new named model space views.

**FIGURE 33-103**

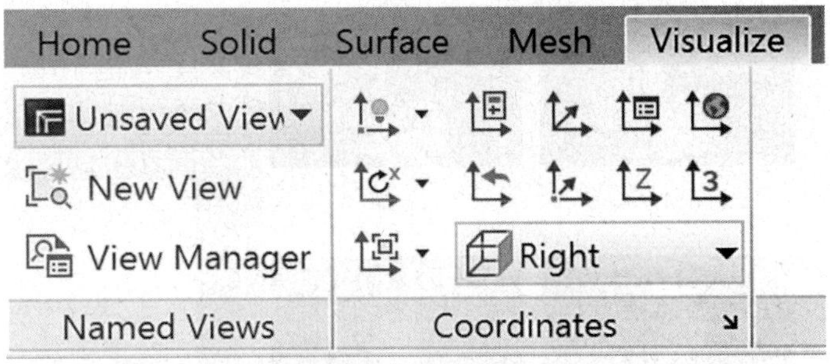


### *Creating a Named View*

Establish an onscreen view of your model space drawing using the 3D Navigation and View Commands presented earlier in this chapter.

Select *New View* (see Figure 33-103) to show the New View/Shot Properties dialog box (Figure 33-104). Notice that you can expand the options using the arrow symbol in the dialog box.

Use the View name input box to enter the name for your view. With the Current display button selected. You can use the expanded panel options to set the view properties, such as the visual style. You will learn more about USCs in the next chapter. Select OK to save the named view.

Next, activate a *Layout*. See Figure 33-105. The contextual Layout tab appears on the ribbon as shown in Figure 33-106.

**FIGURE 33-104**

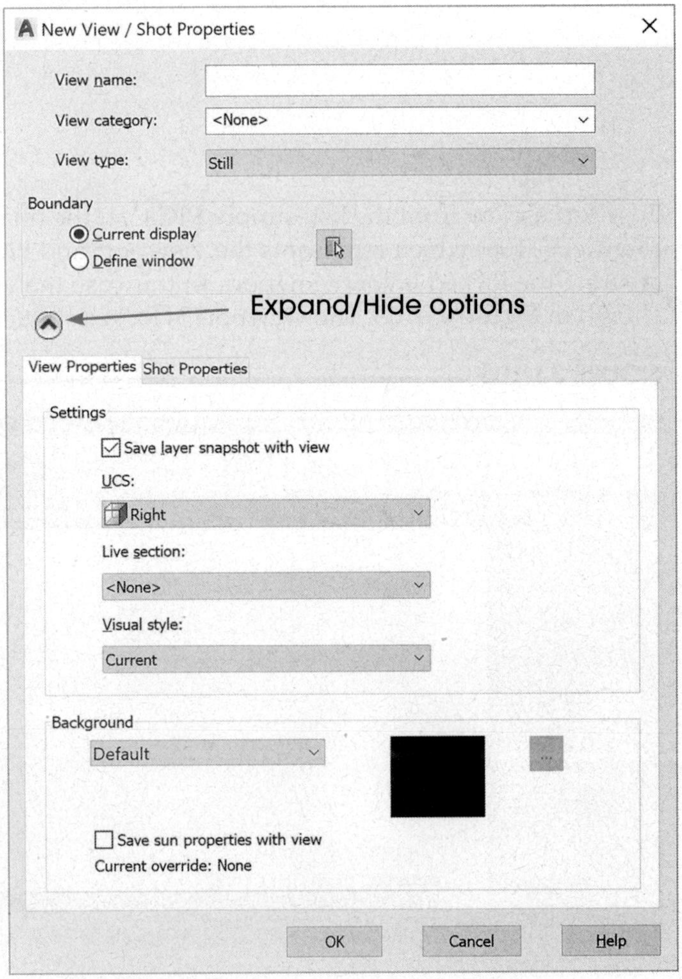


**FIGURE 33-105**

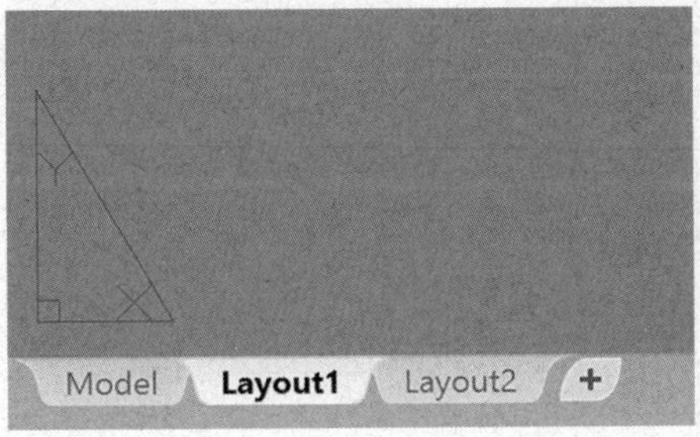

FIGURE 33-106

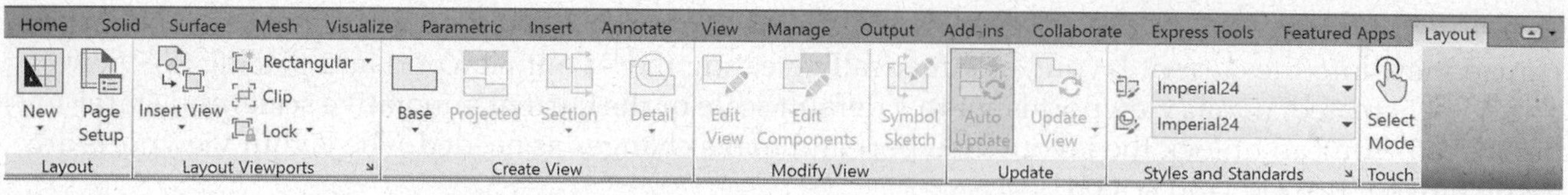


FIGURE 33-107

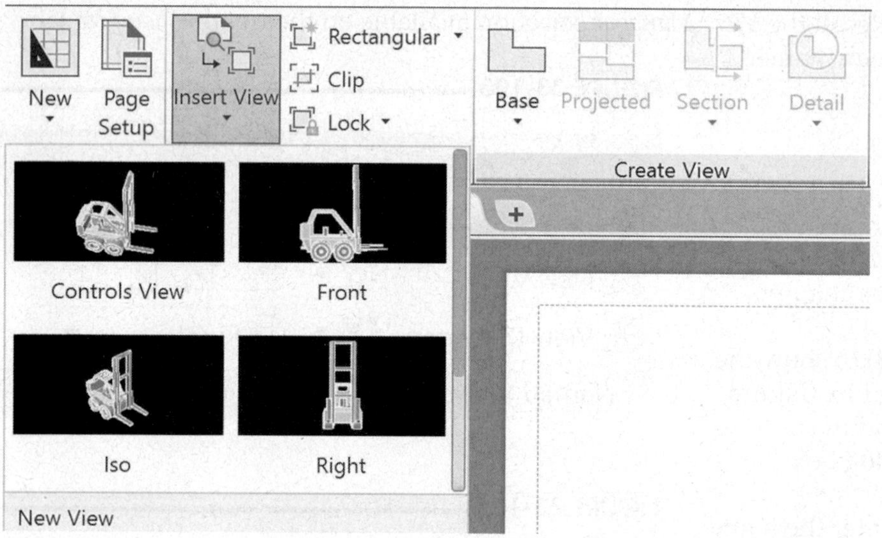


**TIP**

The *GALLERYVIEW* system variable controls whether you see images in the Insert View drop down list or a simple list. You can set it at the Command prompt; a value of 1 shows images, 0 shows the simple list.

Figure 33-107 shows named views in the Forklift drawing example. These were saved previously using the methods described earlier.

To insert a view from the list, simply PICK on the image or list item. A box appears on the crosshairs (Figure 33-108) which represents the viewport and view which will be created. PICK to place the view, creating the locked layout viewport. In this case the view Controls View was selected to create the view shown in Figure 33-108. The viewport is locked, but can easily be unlocked if you wish to make changes.

FIGURE 33-108

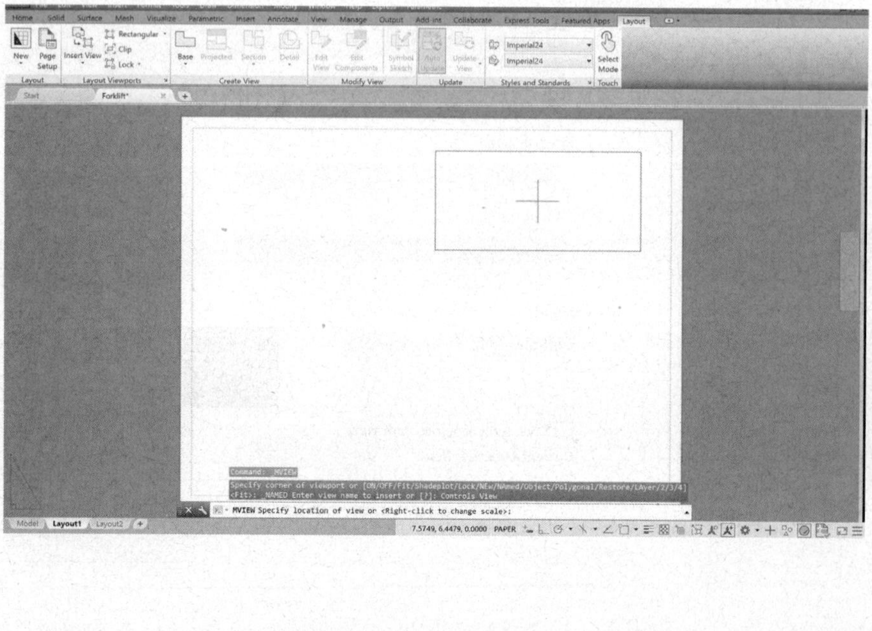

**FIGURE 33-109**

Any named view can be inserted into the layout. It does not have to be a view from a 3D model. Zooming in on areas and creating named views for details is a quick way to place details in a 2D layout as well as a 3D layout. There are many uses for this method to quickly create layout views.

TIP

## Graphics Performance Tuning

3D Modeling in combination with visual styles, rendering, and animation can create heavy demands on your computer's graphics display capabilities. Therefore, AutoCAD provides a tool for tuning your graphics display for optimum performance. The "performance tuner" automatically tunes your system for you when you install AutoCAD. You can view the *Tuner Log* to examine the automatic settings, and you can access and change these controls to your own preferences using the dialog box.

The *Graphicsconfig* command produces the *Graphics Performance* dialog box (Fig. 33-110). Alternately, the dialog box can be accessed from the *Graphics Performance* button on the *System* tab of the *Options* dialog box. This dialog box allows you to view and control the 3D display performance. Using this tool you can view AutoCAD's automatic tuning of your system or change the settings to your needs.

When you work with 3D display effects such as shading, shadows, or backgrounds, the performance of your system slows down. If you are using many of these effects and your computer system's graphics display cannot handle the load, your system may become too slow to display everything in a reasonable amount of time. Therefore, these effects can be set to optimize the performance of the display based on effects you feel are important for the current situation.

The following options are displayed in the *Graphics Performance* dialog box.

### Hardware Setup

This section at the top of the dialog box displays the graphics hardware information such as the video card, driver version, and the current virtual device on your computer system.

### Effects Settings - Hardware Acceleration

You can use this section to turn on or off hardware acceleration. It is recommended to leave this setting *On* unless you are experiencing graphics-related problems with AutoCAD. The *On* setting enables the use of hardware acceleration and provides control to all effects settings available for the current driver. Turning *Hardware Acceleration Off* disables most or all of those options listed below.

FIGURE 33-110

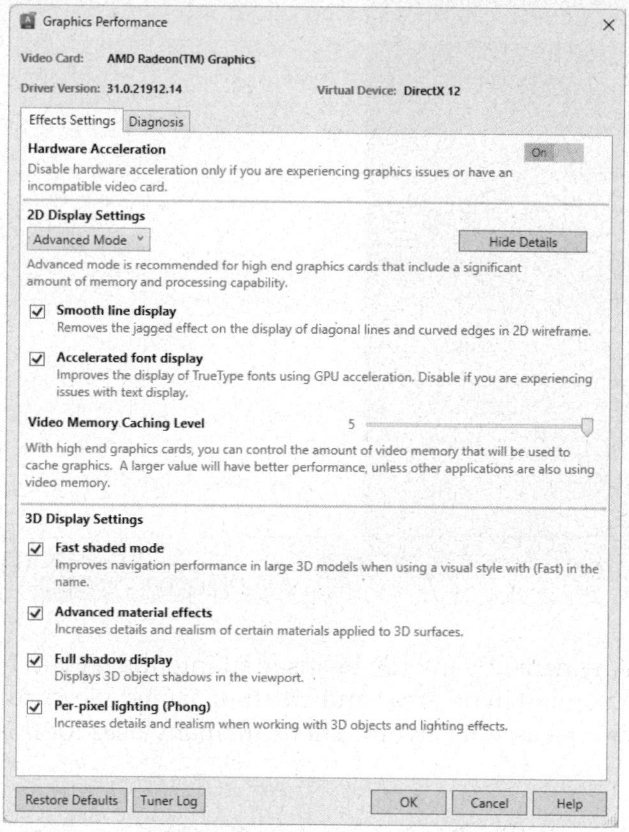


### Smooth Line Display

When object edges are displayed in a line close to, but not exactly aligned with, the physical orientation of pixels on the screen display, a jagged edge or abrupt change in the line can result. Anti-aliasing is the term used to describe the effect of smoothing pixels on the display to avoid jagged edges. Turn this option on to reduce the jagged effect on the display of diagonal lines and curved edges.

### Accelerated Font Display

Selecting this uses GPU acceleration to enhance the visual quality of TrueType fonts. Deselect this if you encounter problems with text display.

### Fast Shaded Mode

Specifies whether the new cross platform 3D graphics system is turned on or off. In the Shaded or Shaded with Edges, Hidden, and Conceptual visual styles, (Fast) is shown in the viewport control to indicate that modern 3D graphics are being used.

### Advanced Material Effects

When materials are applied to objects, this feature enhances the visual effects of the materials and provides a more realistic display of the objects.

### Full Shadow Display

This item must be checked for *Full Shadows* to be displayed in the viewport. Only *Ground Shadows* are available with this setting unchecked.

### Per-Pixel Lighting (Phong)

Phong lighting enables the computation of colors for individual pixels in the display. With this option turned on, the 3D objects and lighting effects on object surfaces in the viewport appear smoother.

### Uncompressed Textures

When textures are applied to objects, a great increase in the demand of your graphics capability is required to display the textures realistically. Checking this option increases the amount of video memory available to display better quality textures in drawings that contain materials with images or that have attached images.

#### *Restore Defaults*

If you change the automatically tuned *Graphics Performance* settings, you can later use this option to reset the values to the defaults based on AutoCAD's original evaluation of your graphics card.

#### *Tuner Log*

When AutoCAD initially evaluates your system's graphics capabilities, it decides whether to use software or hardware implementation for particular display features and sets the *Graphic Performance* accordingly. Features that work with your system are turned on, and features that might not work with your system are turned off. In this process, a tuner log is created. Display the *Tuner Log* to display the result of the evaluation, including other information about your graphics card.

NOTE: Since hardware acceleration can affect other applications you use on your computer while you are running AutoCAD, it is recommended that you do not use any type of remote access application or windows emulating software (such as NetMeeting or Remote Desktop) in conjunction with hardware acceleration. Such applications may not support hardware acceleration and, as a result, can cause general display failure and instability. You can disable hardware acceleration prior to starting the remote access software or use the */NOHARDWARE* command line switch to start an AutoCAD-based product in Software mode.

## *Shade, SHADEDIF, SHADEDGE*

The command *Shade* and the system variables *SHADEDIF* and *SHADEDGE* are older shading techniques that still appear in AutoCAD, most likely for legacy scripts and customized menus. Although these older shading methods are generally more difficult to use, they do not provide any additional benefits over using visual styles. It is recommended that you use visual styles, the newer method of 3D object display.

## *Render*

The *Render* command is used to create a realistic rendering of the 3D model (solid or surface) model. AutoCAD's rendering capabilities provide you with an amazing amount of control for generating photo-realistic renderings, including lighting and controls, backgrounds, shadows, reflections, texture mapping, and landscaping effects.

## Printing 3D Models

You can print 3D models in any visual style. If you have a laserjet, inkjet, or electrostatic type printer configured as the plot device (not a pen plotter), you can select from several options from *Shadeplot* section of the (expanded) *Plot* dialog box. You could also select these options or the *Rendered* option from the *Properties* palette for the viewport.

### Printing a Shaded Image From the *Model* Tab

If you want to print a shaded 3D drawing from the *Model* tab, use the *Shaded Viewport Options* section of the *Plot* dialog box, expanded section (Fig. 33-111). Use the *Shadeplot* drop-down list to select from all of the default visual styles, *Legacy wireframe*, *Legacy hidden*, *3D Hidden*, *3dWireframe*, and *Rendered* options (Fig. 33-104). Alternately, you can apply a visual style to the drawing display, then select *As Displayed* from this list to print the drawing as it appears on screen. You can also select from several resolution qualities for the print (see the "Shaded Viewport Options" section in Chapter 14).

NOTE: The *Legacy Wireframe* option in the drop-down list is essentially the same as the *3D Wireframe* visual style. However, the *Legacy Hidden* option differs from the *3D Hidden* option based on your setting of the *DISPSILH* system variable. If *DISPSILH* is set to 0 (the default), a *Hidden* display generates mesh lines on curved surfaces. The display of meshed lines is also affected by the *FACETRES* system variable setting.

### Printing a Shaded Image From a *Layout* Tab

If you are printing from a layout and want to specify a shaded image for one or more viewports, use the *Shadeplot* option of the viewport's *Properties* palette. Here you can select from the same options (Fig. 33-112) as in the *Plot* dialog box.

FIGURE 33-111

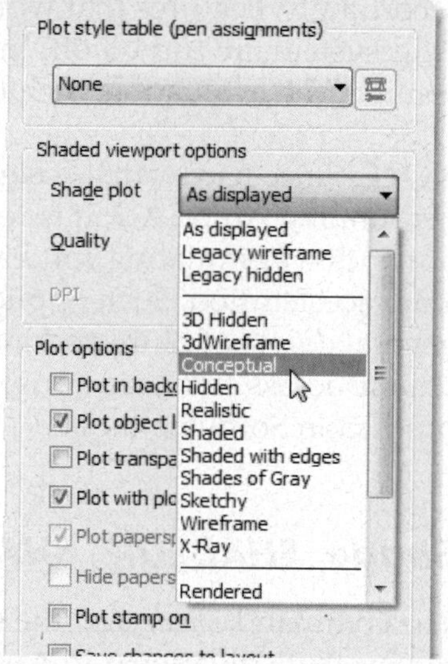


Alternately, you can type the *-Vports* command to assign the *Shadeplot* option for one or more viewports. <u>You must be in a *Layout* for the *Shadeplot* options to appear in the Command prompt.</u>

```
Command: -VPORTS
Specify corner of viewport or [ON/OFF/Fit/Shadeplot/Lock/Object/
Polygonal/Restore/2/3/4] <Fit>: s
Shade plot? [As displayed/Wireframe/Hidden/Visual Styles/
Rendered] <As displayed>: v
[3dwireframe/3dHidden/Realistic/Conceptual/Other] <Realistic>:
```

The *Rendered* option prints the drawing according to the settings made in the *Render* dialog box, including any specified lights, materials, and background. If no rendering settings have been specified, the *Rendered* option produces a rendered image similar to the *Realistic* visual style.

FIGURE 33-112

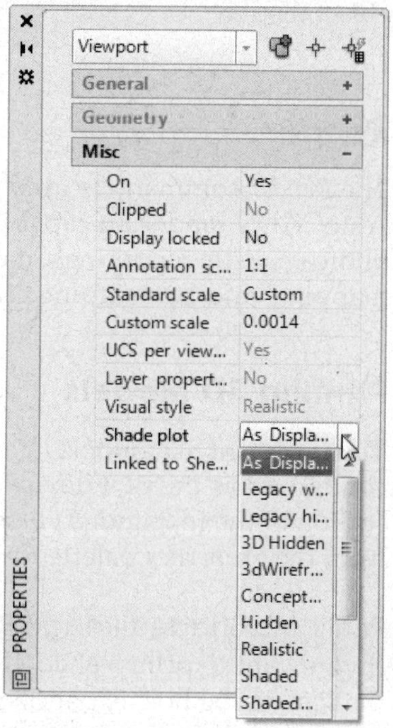

## CHAPTER EXERCISES

1. What are the three types of 3D models?

2. What characterizes each of the three types of 3D models?

3. What kind of modeling techniques does AutoCAD's solid modeling system use?

4. What are the seven formats for 3D coordinate specification?

5. Examine the 3D geometry shown in Figure 33-113. Specify the designated coordinates in the specified formats below.

   A. Give the coordinate of corner D in absolute Cartesian format.

   B. Give the coordinate of corner F in absolute Cartesian format.

   C. Give the coordinate of corner H in absolute Cartesian format.

   D. Give the coordinate of corner J in absolute Cartesian format.

   E. What are the coordinates of the line that define edge F-I?

   F. What are the coordinates of the line that define edge J-I?

   G. Assume point E is the "last point." Give the coordinates of point I in relative Cartesian format.

   H. Point I is now the last point. Give the coordinates of point J in relative Cartesian format.

   I. Point J is now the last point. Give the coordinates of point A in relative Cartesian format.

   J. What are the coordinates of corner G in cylindrical format (from the origin)?

   K. If E is the last point, what are the coordinates of corner C in relative cylindrical format?

**FIGURE 33-113**

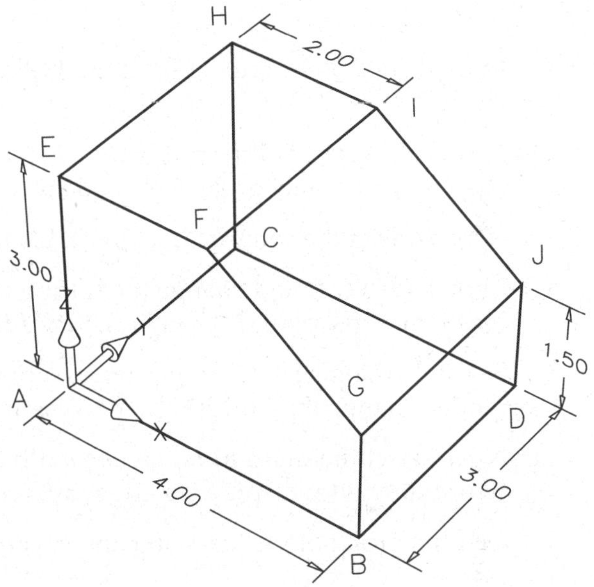


NOTE: For the following exercises you will use several AutoCAD sample drawings that are provided on the website for this book (the website address is listed on the back cover and in the Preface under "Online Resources"). Four of the drawings are used in these exercises: TRUCK MODEL, OPERA, WATCH, and R300-20. These drawings are previous release AutoCAD sample drawings and are provided courtesy of Autodesk.

6. In this exercise you will use a sample drawing and use the *Visual Styles* panel to practice with five of the default visual styles.

   A. ***Open*** the **TRUCK MODEL** drawing. Select the *Model* tab if not already current.

B. In the *Visual Styles* panel, select the ***Hidden*** visual style. Your display should look similar to that in Figure 33-114. Note that perspective projection should be on. If perspective projection is not on, change the *PERSPECTIVE* system variable to **1**.

**FIGURE 33-114**

C. In the *Visual Styles* panel, select the ***Wireframe*** visual style. Do you see all the edges of the model?

D. Next, select the ***2D Wireframe*** visual style. Note that the display has been automatically changed to parallel projection mode for this visual style.

E. Type ***Hide*** at the command prompt. Do mesh lines occur on the truck?

F. Type ***DISPSILH*** and change the setting to **1.** Type ***Hide*** at the command prompt again. Did the mesh lines disappear? Change ***DISPSILH*** back to **0** and try the ***Hide*** command again.

G. Select the ***Hidden*** visual style again. Note that the truck is displayed in the parallel projection mode. Change the *PERSPECTIVE* system variable to **1**.

H. Next, select the ***Conceptual*** style. Finally, change the visual style to ***Realistic***. Which of these two styles better display the edges and contrast between surfaces?

I. ***Exit*** the Truck Model drawing and do not save the changes.

7. In this exercise, you open an AutoCAD sample drawing and use the *Visual Styles* panel on the Ribbon and the *View Manager* to practice with the preset views.

A. ***Open*** the **OPERA** drawing. Select the ***Model*** tab. ***Freeze*** layers **E-AGUAS**, **E-PREDI**, **E-TERRA**, **E-VIDRO**, and **O-PAREDE**. Use the *Visual Styles* panel to generate a ***Conceptual*** display.

B. Select a ***Top*** view from the *Views* panel. ***Zoom*** in if necessary to fully display the objects on the screen. Examine the view and notice the orientation of the coordinate system icon.

Next, type ***UCSORTHO*** at the command line. Change the setting to ***0*** (zero). This action prevents the User Coordinate System from changing when the view changes.

C. Next, produce the *View Manager*. From the ***Preset Views***, select a ***SE Isometric*** view, select ***Set Current***, then ***OK***. ***Zoom*** if necessary. Examine the icon and find north (Y axis). Consider how you (the observer) are positioned as if viewing from the southeast.

D. Produce a ***Front*** view. ***Zoom*** if needed. This is a view looking north. Notice the icon. The Y axis should be pointing away from you.

E. Generate a ***Right*** view. Produce a ***Top*** view again. ***Zoom*** whenever needed for a full-screen display.

F. Next, view the OPERA from a ***SE Isometric*** viewpoint once again. Generate a ***Hidden*** visual style. Your view should appear like that in Figure 33-115. Notice the orientation of the WCS icon on your screen.

**FIGURE 33-115**

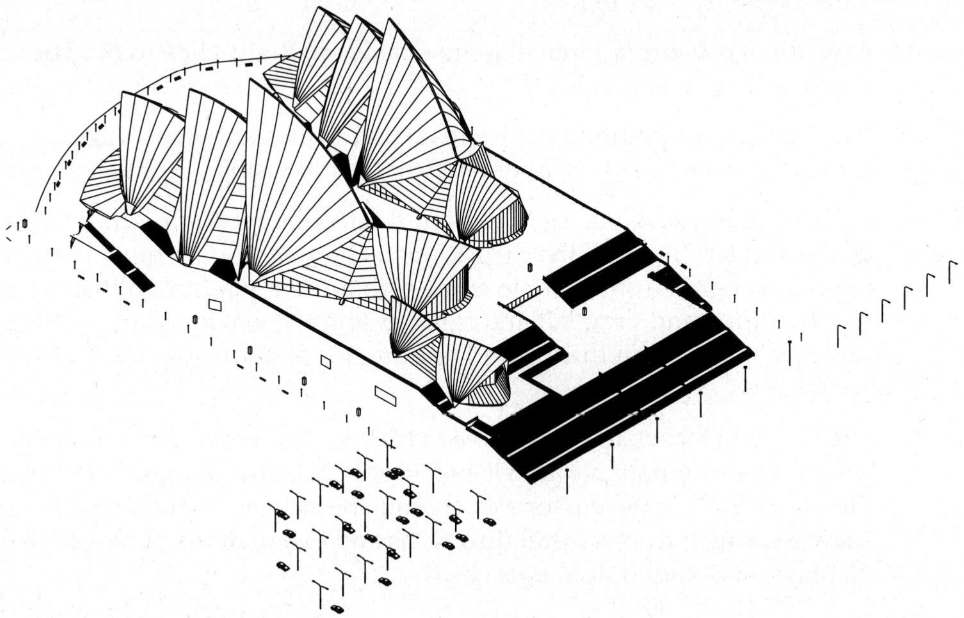

G. Finally, generate a ***NW Isometric*** and ***NE Isometric***. In each case, examine the WCS icon to orient your viewing direction.

H. Experiment and practice more with *Views* panel and the *View Manager*. Do not save the drawing.

8. In this exercise you will practice using the *Views* panel and *Zoom*.

A. ***Open*** the **TRUCK MODEL** again that you used in Exercise 6. Do not change the current visual style throughout this exercise. Type "***UCS***," then use the ***World*** option. This action makes the World Coordinate System the current coordinate system. Use the ***Ucsicon*** command and ensure the icon is at the ***Origin*** (near the center of the truck). Notice the orientation of the truck with respect to the WCS. The X axis points outward from the right side of the truck, Y positive is to the rear of the truck, and Z positive is up.

B. Use the *Views* panel to generate a ***Top*** view. Note the position of the X, Y, and Z axes. You should be looking down on the XY plane and Z points up toward you.

C. Generate a ***SE Isometric*** view. Note the position of the X, Y, and Z axes.

D. Ensure perspective is turned off. Generate a ***Right*** view. Notice that AutoCAD automatically performs a ***Zoom Extents*** so all of the drawing appears on the screen. ***Freeze*** layer **BASE**. Now use ***Zoom Extents*** to display the truck at the maximum screen size.

E. Finally, use each of the following preset views in the order given and notice how the truck appears to change its position for each view. Remember that you should imagine that the viewer moves around the object for each view, rather than imagining that the object rotates.

***SE Isometric, Top, SE Isometric, Bottom, SE Isometric, Right, SW Isometric, Left, SE Isometric, Front, SE Isometric***

F. Close the drawing but do not save the changes.

9. Now you will get some practice with ***3Dorbit*** and the shortcut menu options.

   A. ***Open*** the **WATCH** drawing. Activate the ***Model*** tab. Now generate a ***SE Isometric*** view to get a good look at the watch. Notice that the components are disassembled, similar to an exploded view assembly drawing.

   B. Now invoke ***3Dorbit***. Note that the *Grid* is on. Right-click to produce the shortcut menu, then under ***Visual Aids*** toggle off the ***Grid***. In the shortcut menu, select ***Visual Styles***, then ***Realistic***.

   C. Now drag your pointing device to change your viewpoint. Generate your choice for a view that shows the watch so all components are mostly visible. Right click and ***Exit*** *3Dorbit*.

   D. Activate ***3Dorbit*** again. Right click and select ***Other Navigation Modes***, then ***Free Orbit***. With this mode, try "rolling" the display by using your cursor outside the arc ball. Next, try placing your cursor on the small circle at the right or left quadrant of the arc ball. Hold down the left mouse button and drag left and right to spin the watch about an imaginary vertical axis. Try the same technique from the top and bottom quadrant to spin the watch about an imaginary horizontal axis.

   E. Finally, switch back to ***Constrained Orbit*** and generate your choice of views that shows the watch so all components are visible. Right click and change to ***Perspective*** projection (you may then have to drag the cursor and change the display slightly to activate the perspective display). View the watch from several directions and examine the perspective effect. Arrive at a view that displays all of the components clearly.

   F. Right click to produce the shortcut menu and select ***Other Navigation Modes***, then ***Continuous Orbit***. Hold down the left mouse button, drag the mouse, and release the button to put the watch in a continuous spin.

   G. Finally, ***Exit*** *3Dorbit*. ***Close*** the drawing and do not save changes.

10. In this exercise you will create four model space viewports with the *3D* option.

   A. ***Open*** the **WATCH** drawing again. Activate the ***Model*** tab. Turn off ***Grid***. Change the *Visual Style* to ***Realistic***. Check the setting of the ***UCSOrtho*** variable by typing it at the command line and ensuring it is set to **0**. A setting of 0 means no new UCSs will be created when new views are generated.

   B. Invoke the ***Vports*** command (make sure you are in model space). In the *New Viewports* tab of the *Viewports* dialog box, select ***Four: equal***. At the bottom of the dialog box in the *Setup* drop-down list, select ***3D***. Select the ***OK*** button.

   C. Your display should show a top, front, right side, and southeast isometric view. Notice that the watch is not sized proportionally with respect to the other views. Activate the front viewport (bottom right). Type ***Zoom*** at the command line and enter a value of **1.5**. Do the same for the other two orthographic views. Your display should look like that shown in Figure 33-116 (on the next page).

   D. Activate the isometric viewport. Change the *Visual Style* to ***Conceptual***. This is a good viewport arrangement to use when you are constructing 3D models because you can see several views of the object on your screen at once. Remember this arrangement when you begin working with the exercises in the next several chapters.

   E. ***Close*** the drawing but do not save the changes.

**FIGURE 33-116**

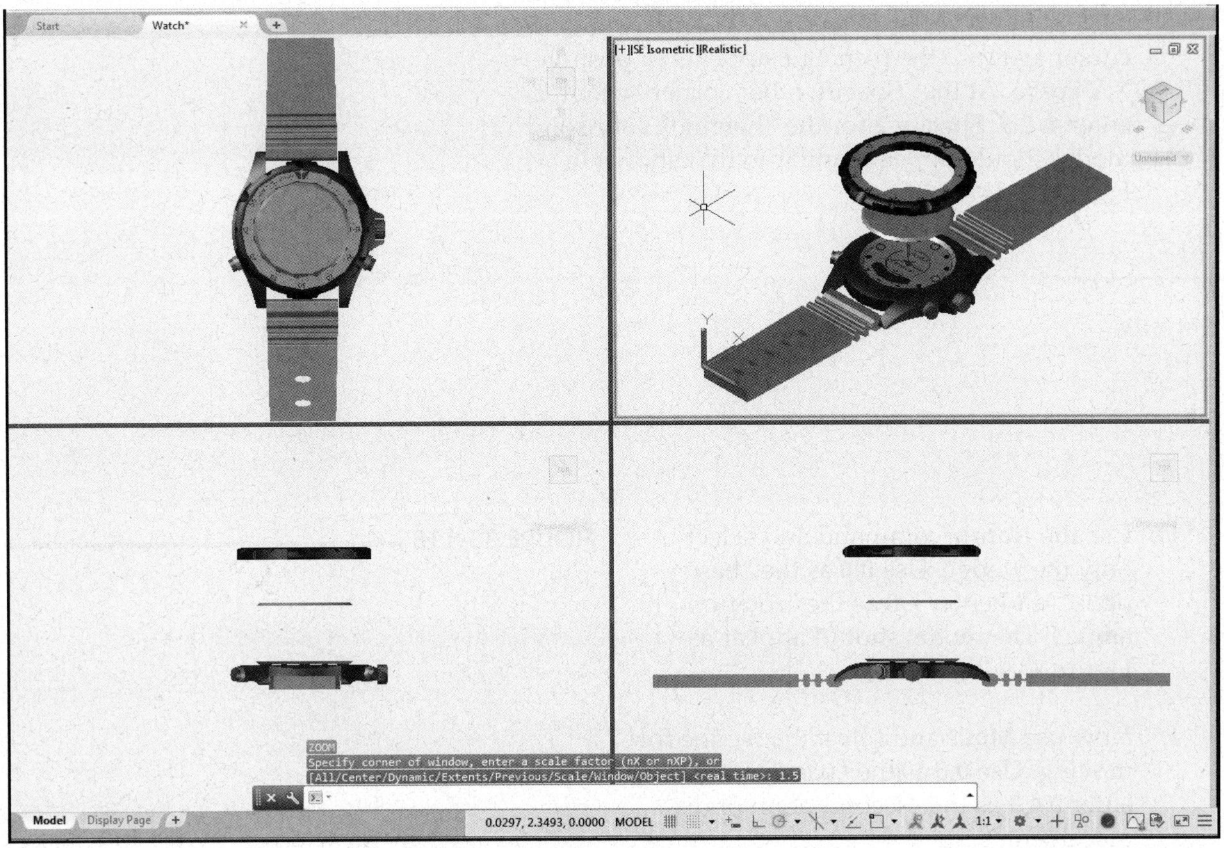


11. In this exercise, you will create a simple solid model to get an introduction to some of the solid modeling construction techniques discussed in Chapter 35. You will use this model again in these chapter exercises for practicing visual style edge and face settings. You will also use this model for creation of User Coordinate Systems in Chapter 34.

    A. Start AutoCAD in the 3D environment (ensure you are using the ***3D Modeling*** workspace). Begin a ***New*** drawing using the ACAD3D.DWT template drawing. Use ***Save*** and name the drawing **SOLID1.** Also make sure the ***Ucsicon*** is ***On*** and set to appear at the ***ORigin***. Use the ***Layer Properties Manager*** to create a new layer named **MODEL** and set the layer's color (in the *Index Color* tab) to **253**. Make **MODEL** the ***Current*** layer.

    B. Type or select ***Box*** to create a solid box. When the "Specify first corner" prompt appears, enter **0,0,0**. Then move your cursor so the rectangle appears in positive X,Y space. At the "Specify other corner" prompt, enter **5,4**. Enter a value of **3** for the "height." Change your viewpoint to ***SE Isometric***, then ***Zoom*** if necessary to view the box. Use ***3Dorbit*** to change the view slightly.

C. Change the visual style to ***Wireframe***. Type or select the ***Wedge***. When the prompts appear, use **0,0,0** as the "Corner of wedge." Then move your cursor again so the rectangle appears in positive X,Y space. At the "Specify other corner" prompt, enter **4, 2.5**. Specify **2** for the "height." Your solid model should appear similar to that shown in Figure 33-117.

**FIGURE 33-117**

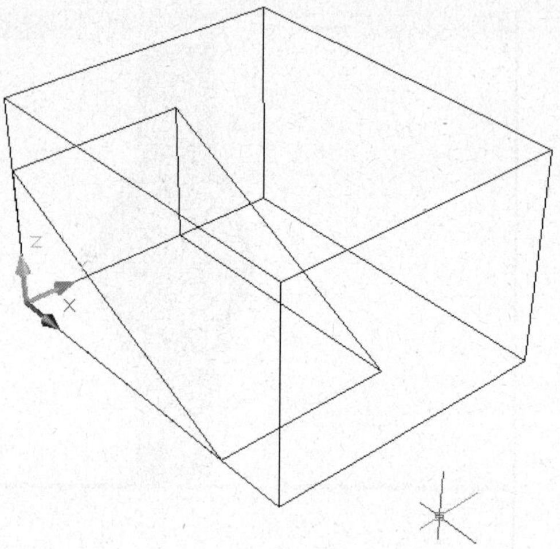

D. Use the ***Rotate*** command and select only the wedge. Use **0,0** as the "base point" and enter **-90** as the "rotation angle." The model should appear as Figure 33-118.

E. Now use ***Move*** and select the wedge for moving. Use **0,0** as the "base point" and enter **0,4,3** as the "second point of displacement."

**FIGURE 33-118**

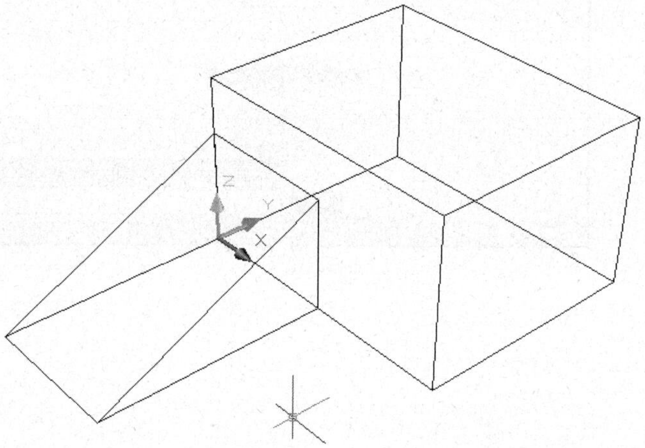

F. Finally, type or select ***Union***. When prompted to "Select objects:", PICK both the wedge and the box. Change the visual style to ***Conceptual***. The finished solid model should look like Figure 33-119. ***Save*** the drawing.

**FIGURE 33-119**

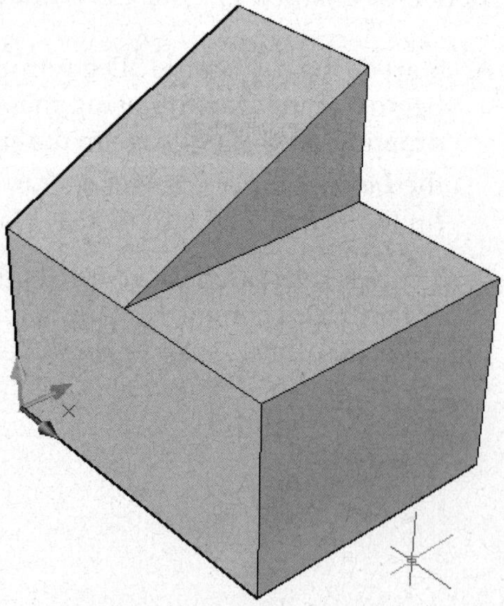

12. In this exercise you will use the SOLID1 drawing from the previous exercise and practice with the face and edge settings in the *Visual Styles* panel and the *Visual Styles Manager*.

TIP

A. If not already open, ***Open*** the **SOLID1** drawing. It should be set to the default ***Conceptual*** visual style. Use the ***X-Ray Effect*** toggle in the *Visual Styles* panel. Your display should now appear as "see-through." Depending on your system's display driver and settings in the *Graphics Performance* dialog box, the display may have a "screen door" effect.

B. Make sure the default ***Conceptual*** visual style is set again. Near the top center of the *Visual Styles* panel, find the flyout button for the four face color modes. Try each of the following face color settings and examine the resulting display: ***Normal, Monochrome, Tint,*** and ***Desaturate***. Next, use the ***Layer Properties Manager*** to change the MODEL layer color to **132**. Finally, experiment with the four face color settings again. Change the visual style to ***Realistic***. Experiment with the four face color settings again and note the differences that occur while the ***Realistic*** style is current.

C. Change the visual style several times between ***Conceptual*** and ***Realistic***. Notice that the face style button automatically changes between *Realistic face style* and *Warm-Cool face style*.

D. Next, activate the ***Visual Style Manager*** (not the *View Manager*). Arrange your drawing (with ***Zoom*** and ***Pan***) to see both the 3D model and the *Visual Style Manager*. Select the ***Conceptual*** visual style from the images at the top. Double-click on the *Conceptual* image to set it as the current visual style for the drawing (a small icon should appear at the bottom-center of the image tile). Next, under *Silhouette Edges*, set *Show* to **No**. In the *Edge Modifiers* section, use the small toggle buttons to turn on ***Line Extension edges*** and ***Jitter edges***. Set the *Line Extensions* value to **12** to achieve a display similar to that shown in Figure 33-120.

**FIGURE 33-120**

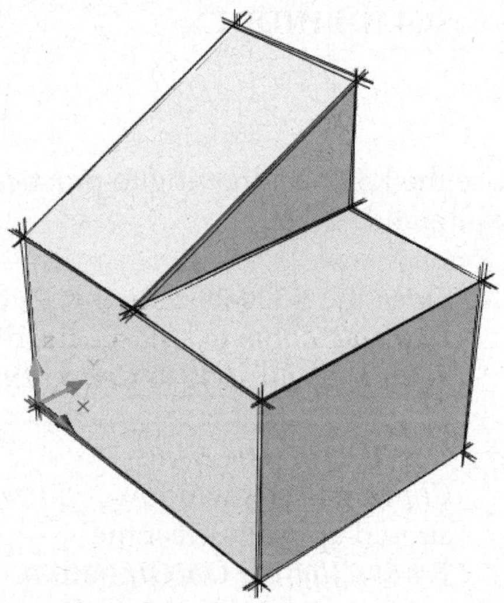

E. Next, select the ***Realistic*** visual style and right-click to produce the shortcut menu. Select ***Apply to Current Viewport***. Examine the change to the model. Then, select the ***Conceptual*** visual style again, right-click to produce the shortcut menu and select ***Apply to Current Viewport***. Examine the change to the model (similar to that in Figure 33-113). Note that all the options that you set previously for this visual style are saved in the drawing.

F. Use any method to select a ***SW Isometric*** display. ***Zoom*** and/or ***Pan*** as necessary to see the object clearly again. Produce a display similar to that shown in Figure 33-121 by applying the ***Hidden*** visual styles and using the *Edge Modifiers* section of the *Visual Styles Manager* to set the *Halo Gap %* to **10**.

**FIGURE 33-121**

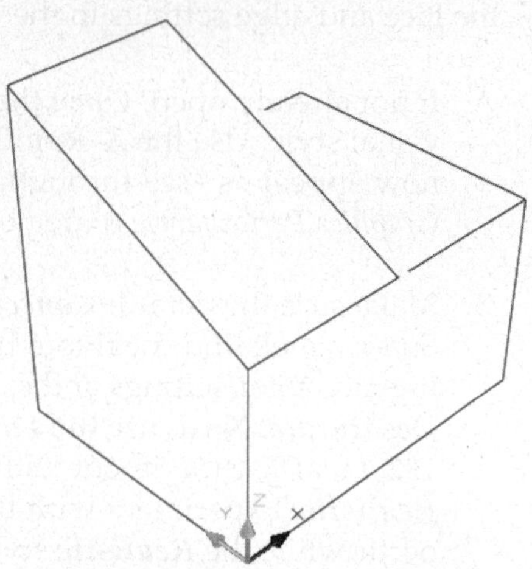

G. Finally, create a display similar to that in Figure 33-122 by changing the settings in the ***Occluded Edges*** section to *Show* **Yes** and *Linetype* ***Dashed***.

**FIGURE 33-122**

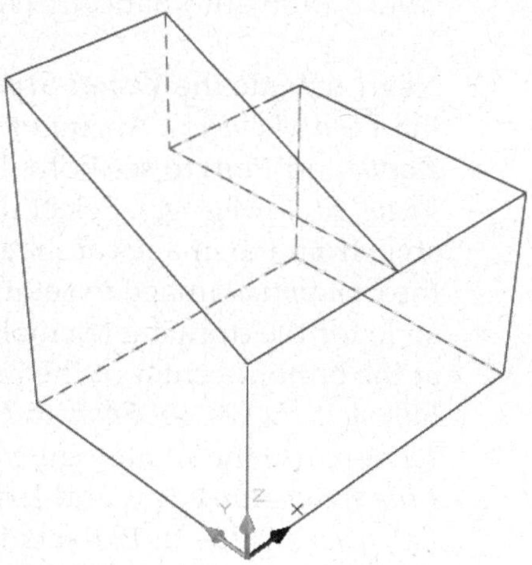

H. Use ***Saveas*** and change the name of the drawing to **SOLID1 HIDDEN**.

13. Use the R300-20 drawing to practice with the ***3Dclip*** command.

A. ***Open*** the **R300-20** drawing. Activate the ***Model*** tab. Use ***Pan*** and ***Zoom*** to enlarge the display to fill the screen. Set the visual style to ***Conceptual***.

B. Use ***3Dclip***. The *Adjust Clipping Planes* window should appear. Select the ***Front Clipping On/Off*** button (or you can right-click inside the window to select from a shortcut menu). Next, select ***Adjust Front Clipping Plane***. Adjust the plane so it passes approximately one-half of the way through the assembly in the lower-left corner of the drawing. Your display should appear similar to that shown in Figure 33-123. Notice how the interior features of the assembly are clearly shown. Finally, close the *Adjust Clipping Planes* window.

**FIGURE 33-123**

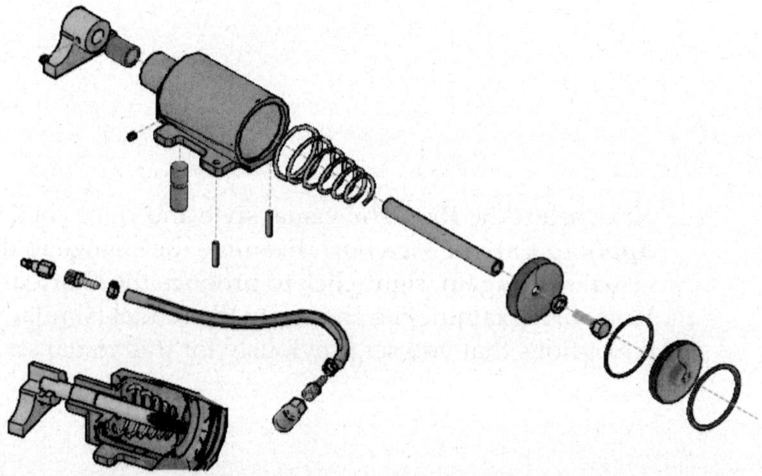

C. ***Close*** the drawing and <u>do not save</u> the changes.

# CHAPTER 34

# User Coordinate Systems

## Chapter Objectives

After completing this chapter you should:

1. know how to create, *Save*, *Restore*, and *Delete* User Coordinate Systems using the *UCS* command;

2. know how to use the *UCS Manager* to create *Save*, *Restore*, and *Delete* User Coordinate Systems;

3. be able to create UCSs by the *Origin, Zaxis, 3point, Object, Face, View, X, Y,* and *Z* methods;

4. be able to create *Orthographic* UCSs using the UCS Manager;

5. know the function of the *UCSVIEW, UCSORTHO, UCSFOLLOW,* and *UCSVP* system variables.

## CONCEPTS

No User Coordinate Systems exist as part of any of the AutoCAD default template drawings as they come "out of the box." UCSs are created to simplify construction of the 3D model when and where they are needed. UCSs are applicable to all types of 3D models—wireframe models, surface models, and solid models.

When you create a multiview or other 2D drawing, creating geometry is relatively simple since you only have to deal with X and Y coordinates. However, when you create 3D models, you usually have to consider the Z coordinates, which makes the construction process more complex. Constructing some geometries in 3D can be challenging when the objects have shapes on planes not parallel with, or perpendicular to, the XY plane or not aligned with the WCS (World Coordinate System).

FIGURE 34-1

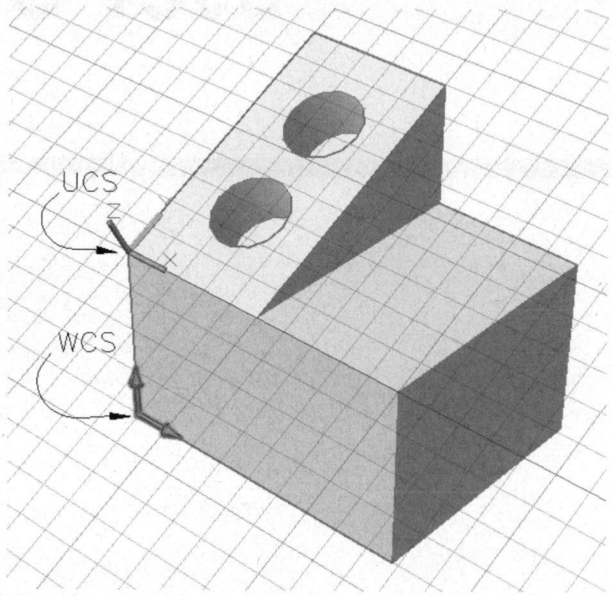


UCSs are created when needed to aid in constructing a 3D object. For example, imagine specifying the coordinates for the centers of the cylindrical shapes in Figure 34-1, using only world coordinates. Instead, creating a UCS on the slanted face of the object containing the cylinders (the inclined plane), makes the construction process much simpler. To draw on the new UCS, only the X and Y coordinates (of the UCS) would be needed to specify the centers, since anything drawn on that plane has a Z value of 0. The Coordinates display in the Status Bar always displays the X, Y, and Z values of the current coordinate system, whether it is the WCS or a UCS.

Generally, you would create a UCS aligning its XY plane with a surface of the object, such as along an inclined plane, with the UCS origin typically at a corner or center of the surface. Any coordinates that you specify (in any format) are assumed to be user coordinates (coordinates that lie in or align with the current UCS). Even when you PICK points interactively, they fall on the XY plane of the current UCS. The *Snap, Polar Snap,* and *Grid* automatically align with the current coordinate system XY plane, enhancing your usefulness of the construction plane.

## CREATING AND MANAGING USER COORDINATE SYSTEMS

You can create User Coordinate Systems several ways. The *Ucs* command is the traditional method for creating new coordinate systems and provides several options for orienting the new UCSs. The UCS Manager (*Ucsman* command) allows you to create orthographically positioned UCSs as well as name, save, and restore UCSs. You can also have UCSs created for you automatically when you create or restore an orthographic view.

You can create and save multiple UCSs in one drawing. Rather than recreating the same UCSs multiple times, you can assign a name and save the UCSs that you create so they can be restored at a later time. Although you can have multiple UCSs, only one UCS per viewport can be active at any one time.

You often use multiple UCSs in a 3D drawing. This chapter explains the features for creating and managing UCSs wisely. First the chapter discusses how to create UCSs, then naming, saving, and restoring UCSs are explained. At the end of the chapter, the system variables that control UCSs are discussed.

If you are constructing solid models, you can create Dynamic UCSs. Dynamic UCSs are temporary coordinate systems used for constructing a solid, but they cannot be saved or named. Dynamic UCSs are discussed in Chapter 35.

TIP

NOTE: When creating 3D models, it is recommended that you use the *Ucsicon* command with the *On* option to make the icon visible and the *Origin* option to force the icon to the origin of the current coordinate system.

### *Ucs*

| Ribbon | Menu Bar | Command (Type) | Alias (Type) | Shortcut |
|---|---|---|---|---|
| *Visualize Coordinates* (icon) | *Tools New UCS >* | *Ucs* | ... | ... |

The *UCS* command allows you to create, save, restore, and delete User Coordinate Systems. There are many options available for creating UCSs. The *UCS* command always operates only in Command line mode.

Command: ***UCS***
Current ucs name: *WORLD*
Specify origin of UCS or [Face/NAmed/OBject/Previous/View/World/X/Y/Z/ZAxis] <World>:

FIGURE 34-2

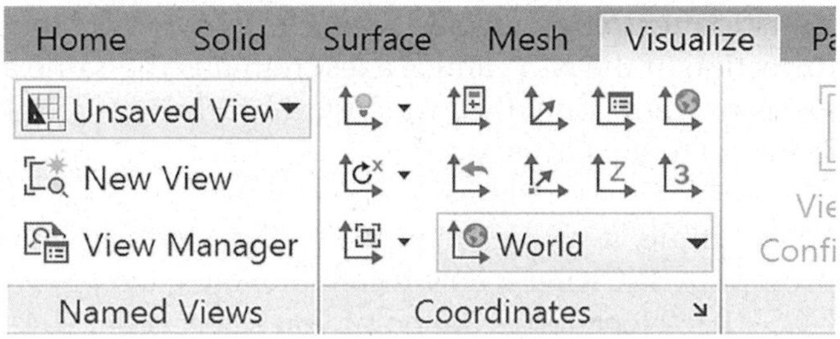


Although the *UCS* command operates in Command line mode, you can access the command tool buttons in the *3D Modeling* workspace in the *Visualize* tab, *Coordinates* panel (Fig. 34-2).

If you prefer the Menu Bar, the UCS commands can be accessed from the *Tools* menu and from two available toolbars (Fig. 34-3).

FIGURE 34-3

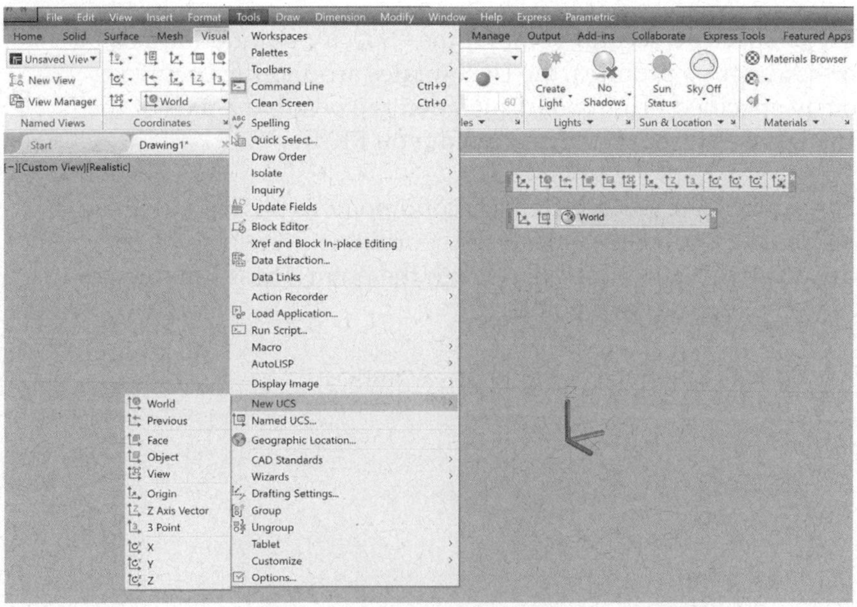

**TIP**

When you create UCSs, AutoCAD prompts for points. These points can be entered as coordinate values at the keyboard or you can PICK points on existing 3D objects. *Object Snap* modes should be used to PICK points in 3D space (not on the current XY plane). You can also use multifunctional grips to change the UCS. In addition, understanding the right-hand rule is imperative when creating some UCSs (see Chapter 33).

NOTE: While learning these options, it is suggested that you turn off Dynamic UCS (*Dynamic UCS* on the Status bar). Dynamic UCS affects the orientation of the UCS as you select points in 3D space. Dynamic UCS is explained in Chapter 35.

The options of the UCS command are explained next.

The default option of the *Ucs* command defines a new UCS using one, two, or three points.

```
Command: UCS
Current ucs name: *WORLD*
Specify origin of UCS or [Face/NAmed/OBject/
  Previous/View/World/X/Y/Z/ZAxis] <World>:
```

If you specify a single point, the origin of the current UCS shifts to the point you specify. The orientation of the new UCS (direction of the X, Y, and Z axes) remains the same as its previous position, only the origin location changes. This option is identical to the *Origin* option.

For example, if you respond to the "Specify origin of UCS" prompt by PICKing a new location using *Object Snap*, the new UCS's 0,0,0 location is the point you select (Fig. 34-4). Note that the X,Y,Z orientation is the same as the previous UCS position. At the following prompt, press Enter to accept the first point.

FIGURE 34-4

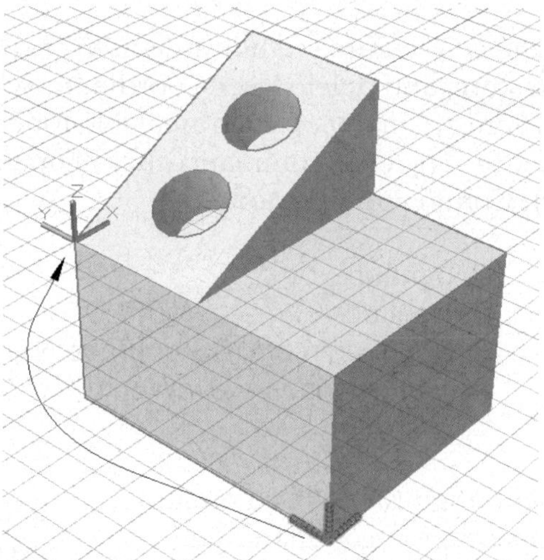

```
Specify point on X-axis or <Accept>:
```

If you specify a second point (at the "Specify point on X-axis or <Accept>:" prompt), the UCS rotates around the previously specified origin point such that the positive X axis of the UCS passes through the new point you PICK.

For example, if you selected a second point as shown in Figure 34-5 as the "point on X axis," the coordinate system spins so the positive X direction aligns with the point. Press Enter to limit the input to two points.

FIGURE 34-5

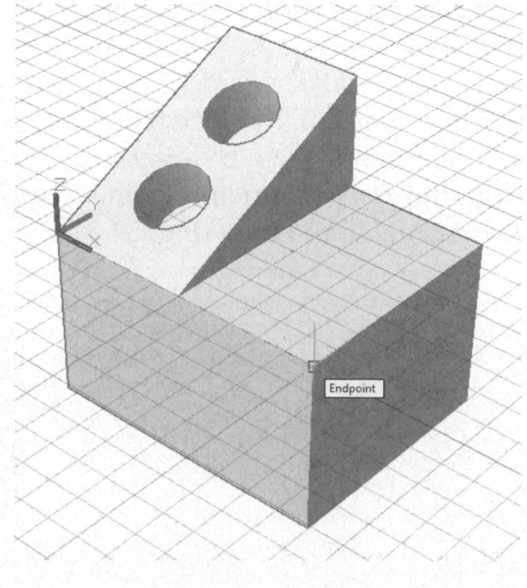


```
Specify point on XY plane or <Accept>:
```

You can specify a third point at the prompt shown above. If you specify a third point, the UCS rotates around the X axis such that the positive Y half of the XY plane of the UCS contains the point. Remember that the first two points lock the origin and the direction for positive X, so the third point only specifies a location in the XY plane. This option (specifying three points at the default prompt) is identical to the separate *3Point* option.

For example, if the third point was specified on the inclined plane as shown in Figure 34-6, the XY plane of the resulting UCS would align itself with the inclined plane.

**FIGURE 34-6**

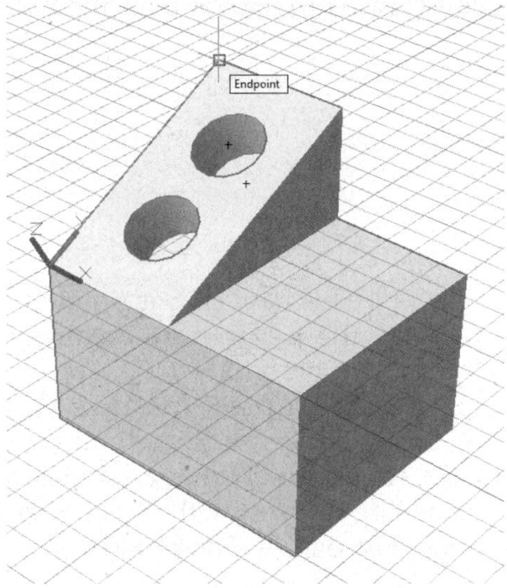


### *ORIGIN*

The *Origin* option defines a new UCS by allowing you to specify a point for the new 0,0,0 location. The orientation of the X,Y, and Z axes remains the same as the previous UCS.

Command: ***UCS***
Current ucs name: *WORLD*
Specify origin of UCS or
[Face/NAmed/OBject/Previous/View/World/X/Y/Z/ZAxis] <World>: ***o***
Specify new origin point <0,0,0>:

This option is identical to using the default option to PICK one point, then pressing Enter to accept that point as the new origin.

As an option to using the UCS command, you can use multifunction grips on the UCS icon to manipulate it. For example, you can hover over the origin grip to produce the shortcut menu shown in Figure 34-7. The *Move Origin Only* option is similar to the *Origin* option of *UCS*.

**FIGURE 34-7**

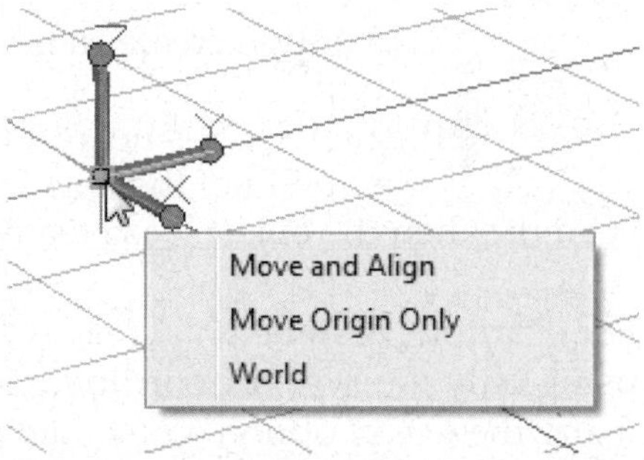


NOTE: The *UCSSELECTMODE* variable must be set to 1 (on) to use the UCS icon multifunctional grips.

### *ZAxis*

You can define a new UCS by specifying an origin and a direction for the Z axis. Only the two points are needed. The X or Y axis generally remains parallel with the current UCS XY plane, depending on how the Z axis is tilted (Fig. 34-8). AutoCAD prompts:

Specify new origin point or [Object] <0,0,0>:
 **PICK**
Specify point on positive portion of Z-axis
<0.0000,0.0000,0.0000>: **PICK**

**FIGURE 34-8**

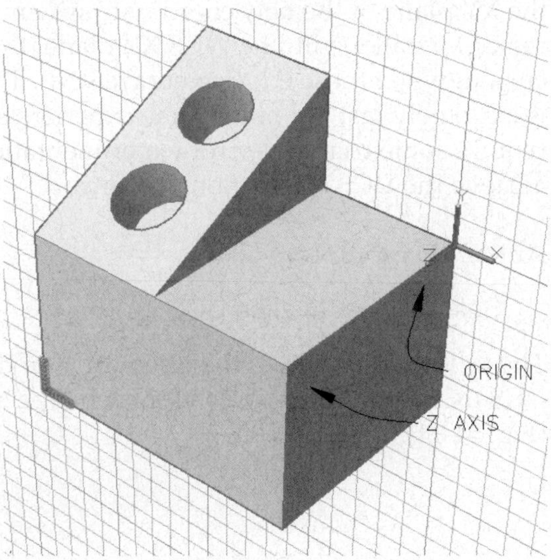

**TIP**

The *Z Axis Direction* option can be selected using multifunctional grips (Fig. 34-9). Hover over the Z-axis grip to produce the menu.

**FIGURE 34-9**

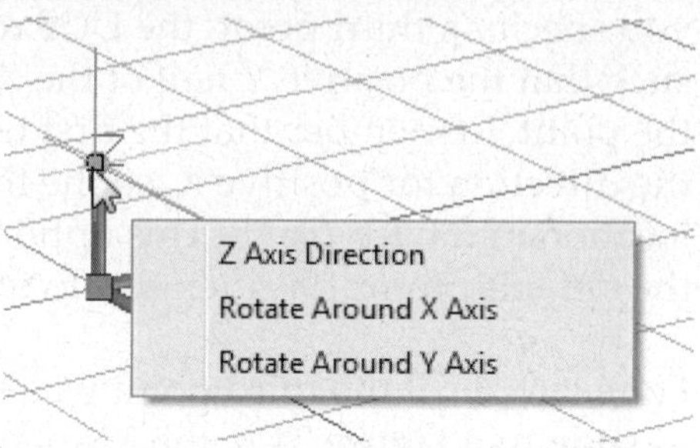


*3point*

**TIP**

This new UCS is defined by (1) the <u>origin</u>, (2) a point on the <u>X axis</u> (positive direction), and (3) a point on the <u>Y axis</u> (positive direction) or <u>XY plane</u> (positive Y). This is the most universal of all the UCS options (works for most cases). It helps if you have geometry established that can be PICKed with *Object Snap* to establish the points (Fig. 34-10). The prompts are:

**FIGURE 34-10**

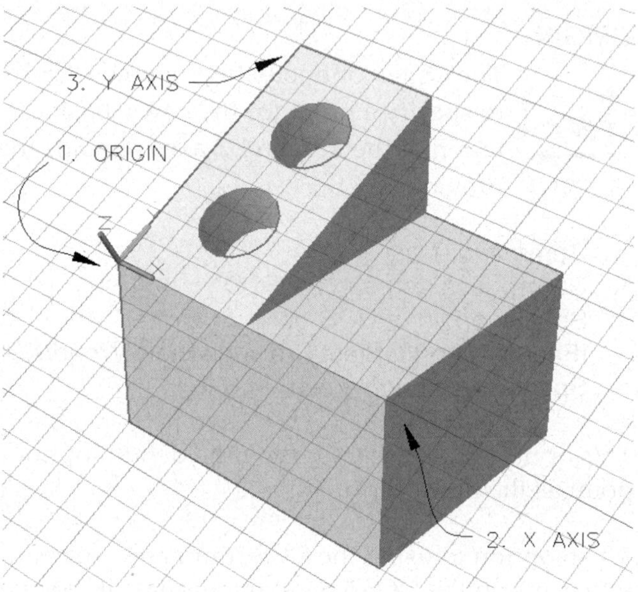


Command: ***UCS***
Current ucs name: *WORLD*
Specify origin of UCS or [Face/NAmed/OBject/ Previous/View/ World/X/Y/Z/ZAxis] <World>: *3*
Specify new origin point <0,0,0>: **PICK**
Specify point on positive portion of X-axis <0.000,0.000,0.000>: **PICK**
Specify point on positive-Y portion of the UCS XY plane <0.000,0.000,0.000>: **PICK**

*Object*

This option creates a new UCS aligned with the selected object. You need to PICK only one point to designate the object. The origin of the new UCS is nearest the endpoint of the edge (or center of circular shape) you select. The orientation of the new UCS is based on the type of object selected and the XY plane <u>that was current when the object was created</u>.

For example, assume you created a solid model using only the World Coordinate System. Then, using the *Object* option, you could quickly create new UCSs aligned with edges of the model; however, the XY plane of the new coordinate systems would remain parallel with the WCS XY plane (that was current when those edges were created). This option is extremely fast but takes some experience and may require some trial-and-error experimentation to achieve the UCS orientation you intend.

**FIGURE 34-11**

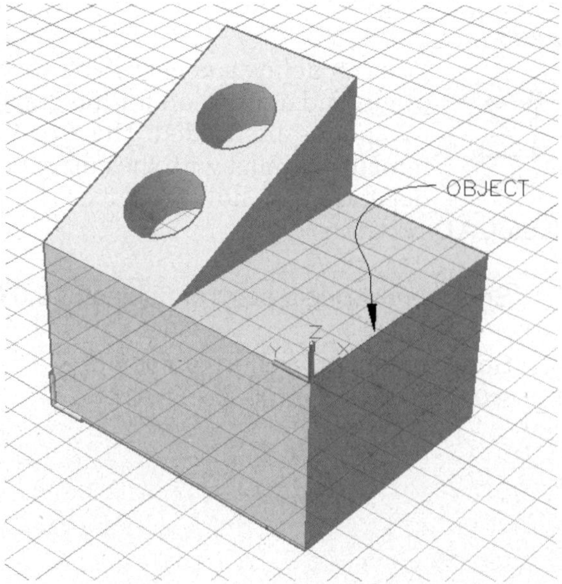


AutoCAD prompts:

Select object to align UCS:

The following list gives the orientation of the UCS using the *Object* option for each type of object (Fig. 34-11).

| Object | Orientation of New UCS |
|---|---|
| *Line* | The end nearest the point PICKed becomes the new UCS origin. The new X axis aligns with the *Line*. The XY plane keeps the same orientation as the previous UCS. |
| *Circle* | The center becomes the new UCS origin with the X axis passing through the PICK point. |
| *Arc* | The center becomes the new UCS origin. The X axis passes through the endpoint of the *Arc* that is closest to the pick point. |
| *Point* | The new UCS origin is at the *Point*. The X axis is derived by an arbitrary but consistent "arbitrary axis algorithm." (See the *AutoCAD Help*.) |
| *Pline (2D)* | The *Pline* start point is the new UCS origin with the X axis extending from the start point to the next vertex. |
| *Solid* | The first point of the *Solid* determines the new UCS origin. The new X axis lies along the line between the first two points. |
| *Dimension* | The new UCS origin is the middle point of the dimension text. The direction of the new X axis is parallel to the X axis of the UCS that was current when the dimension was drawn. |
| *3Dface* | The new UCS origin is the first point of the *3Dface*, the X axis aligns with the first two points, and the Y positive side is on that of the first and fourth points. |
| *Text, Insertion, Attribute* | The new UCS origin is the insertion point of the object, while the new X axis is defined by the rotation of the object around its extrusion direction. Thus, the object you PICK to establish a new UCS will have a rotation angle of 0 in the new UCS. |

### Face

The *Face* option operates with solids, surfaces, or meshes. A *Face* is a planar or curved surface on an object, as opposed to an edge. When selecting a *Face*, you can place the cursor directly on the desired face when you PICK, or select an edge of the desired face.

TIP

Specify origin of new UCS or [Face/NAmed/OBject/Previous/View/World/X/Y/Z/ZAxis] <World>: **f**
Select face of solid, surface, or mesh: **PICK**
(Select desired face of 3D solid to attach new UCS. Edges or faces can be selected.)
Enter an option [Next/Xflip/Yflip] <accept>:

*Next*
Since you cannot actually PICK a surface, or if you PICK an edge, AutoCAD highlights a surface near where you picked or adjacent to a selected edge (Fig. 34-12). Since there are always two surfaces joined by one edge, the *Next* option causes AutoCAD to highlight the other of the two possible faces. Press Enter to accept the highlighted face.

*Xflip*
Use this option to flip the UCS icon 180 degrees about the X axis. This action causes the Y axis to point in the opposite direction (see Fig. 34-12).

**FIGURE 34-12**

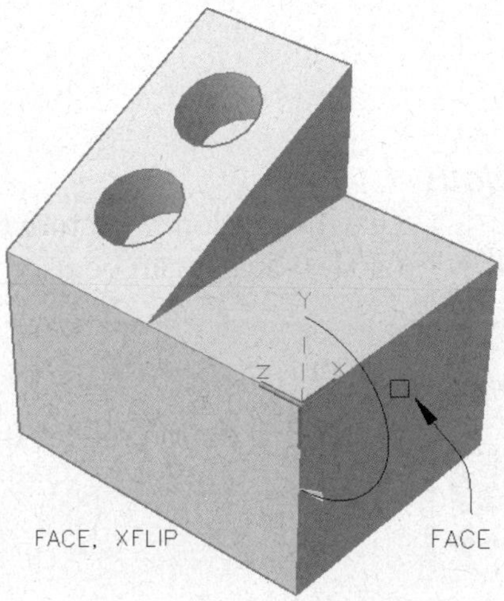

*Yflip*

This option causes the UCS icon to flip 180 degrees about the Y axis. The X axis then points in the opposite direction.

### *View*

This *UCS* option creates a UCS parallel with the screen (perpendicular to the viewing angle) (Fig. 34-13). The UCS <u>origin</u> remains unchanged. This option is handy if you wish to use the current viewpoint and include a border, title, or other annotation. There are no options or prompts.

**FIGURE 34-13**

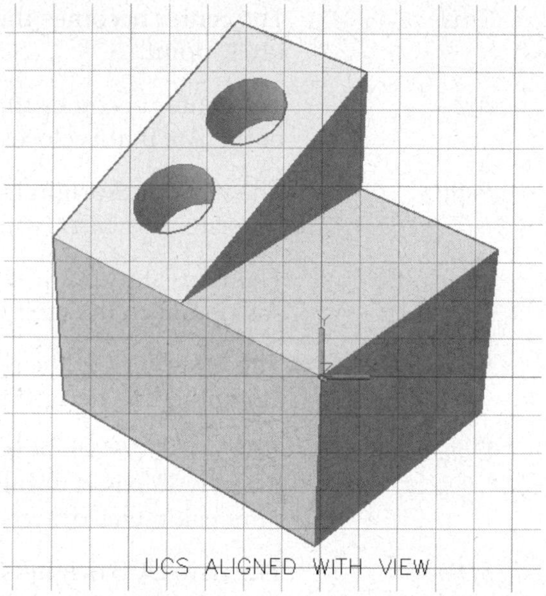


### *X Y Z*

Each of these options rotates the UCS about the indicated axis according to the right-hand rule.

The Command prompt is:

Specify rotation angle about *n* axis <90>:

The angle can be entered by PICKing two points or by entering a value. This option can be repeated or combined with other options to achieve the desired location of the UCS. It is imperative that the right-hand rule is followed when rotating the UCS about an axis (see Chapter 33).

**FIGURE 34-14**

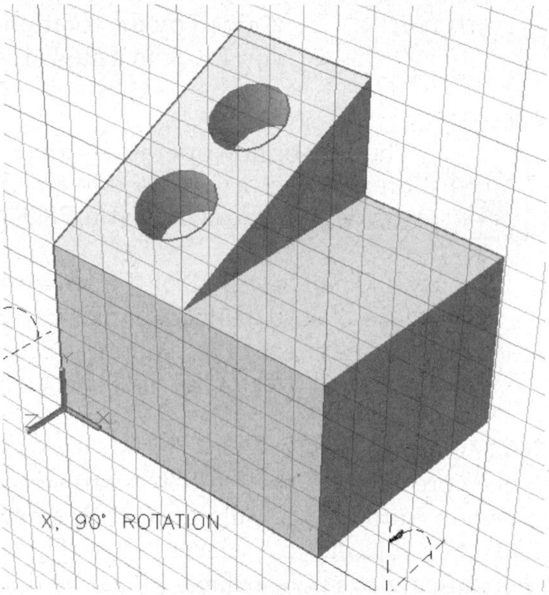


Notice that in the prompt for the *X*, *Y*, or *Z* axis rotate options, the default rotation angle is 90 degrees ("<90>" as shown above). If you press Enter at the prompt, the axis rotates by the displayed amount (90 degrees, in this case, Fig. 34-14). You can set the *UCSAXISANG* system variable to another value which will be displayed in brackets when you next use the *X*, *Y*, or *Z*, axis rotate options. Valid values are 5, 10, 15, 18, 22.5, 30, 45, 90, 180. See *"UCSAXISANG"* in the System Variables section of this chapter.

### *Previous*

Use this option to restore the previous UCS. AutoCAD remembers the ten previous UCSs used. *Previous* can be used repeatedly to step back through the UCSs.

### *World*

Using this option makes the WCS (World Coordinate System) the current coordinate system.

### Named

The *Named* option allows you to *Save* the current UCS, *Restore* previously named UCSs, and *Delete* named UCSs.

```
Command: UCS
Current ucs name: *WORLD*
Specify origin of UCS or [Face/NAmed/OBject/Previous/View/World/X/Y/Z/ZAxis] <World>: na
Enter an option [Restore/Save/Delete/?]:
```

*Save*

Invoking this option prompts for the name you want to assign for saving the current UCS. Up to 256 characters can be used in the name. Entering a name causes AutoCAD to save the current UCS. The *?* option and the UCS Manager list all previously saved UCSs.

*Restore*

Any *Save*d UCS can be restored with this option. The *Restored* UCS becomes the current UCS. The ? option and the UCS Manager list the previously saved UCSs.

*Delete*

You can remove a *Save*d UCS with this option. Entering the name of an existing UCS causes AutoCAD to delete it.

*?*

This option lists the named UCSs (UCSs that were previously *Saved*). Also listed are the origin coordinates and directions for the X, Y, and Z axes.

NOTE: It is important to remember that changing a UCS or creating a new UCS does not change the display (unless the *UCSFOLLOW* system variable is set to 1). Only one coordinate system can be active in a view. If you are using viewports, each viewport can have its own UCS if desired.

## Dducsp

| Ribbon | Menu Bar | Command (Type) | Alias (Type) | Shortcut |
|---|---|---|---|---|
| *Visualize Coordinates* (arrow) | ... | *Dducsp* | ... | ... |

The *Dducsp* command invokes the UCS Manager. It can be used to create new UCSs or to restore existing UCSs. The UCS Manager can also be used to set specific UCS-related system variables, described later. Activating the UCS manager from the ribbon causes the *Settings* tab to be uppermost.

Typing the *Dducsp* command activates specifically the *Orthographic UCSs* tab of the UCS Manager (Fig. 34-15). Here you can create new UCSs. *Orthographic UCSs* are not automatically saved as named UCSs but are created as you select them; that is, you can create, save, and restore additional but separate UCSs with the names Front, Top, Right, etc.

**FIGURE 34-15**

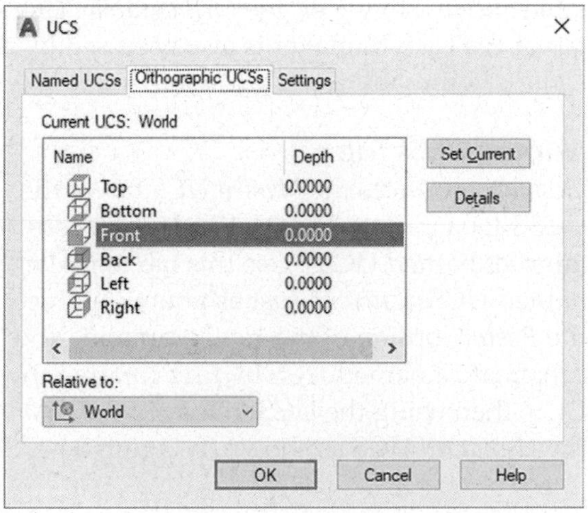

To create one of these *Orthographic UCSs*, you can double-click the desired UCS name, or highlight the desired UCS name (one click) and select the *Set Current* button. Press *OK* to return to the drawing to see the new UCS. Keep in mind that although the new UCSs are normal to the front, top, right, etc. orthographic views, they are not necessarily attached to the front, top, or right faces of the objects. The new UCSs typically have the same origin as the WCS, unless a new *Depth* is specified or a named UCS is selected in the *Relative to:* drop-down list.

### Depth

Notice in the *Orthographic UCS* tab that each UCS can have a specified depth. *Depth* is a Z dimension or distance perpendicular to the XY plane to locate the new UCS origin. Select the *Depth* value of any UCS to produce the *Orthographic UCS Depth* dialog box (Fig. 34-16).

FIGURE 34-16

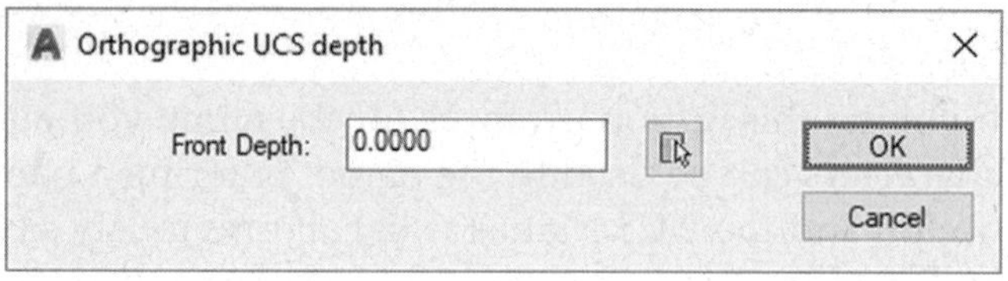


### Details

*Details* shows the coordinates for the origin of the highlighted UCS. The direction vector coordinates for the X, Y, and Z axes are also listed. This has the same function as the *?* option of the *UCS* command.

### Relative to:

TIP

The *Relative to:* drop-down list allows you to select any named UCS as the base UCS. When the *Front*, *Top*, *Right*, or other UCS option is selected, it is created in an orthographic orientation with respect to the origin and orientation of the selected named UCS. Typically, you want to ensure *World* is selected as the base UCS unless you have some other specific application. The setting in the *Relative to:* list is stored in the *UCSBASE* system variable.

## Ucsman

| Ribbon | Menu Bar | Command (Type) | Alias (Type) | Shortcut |
|---|---|---|---|---|
| *Visualize Coordinates (icon)* | *Tools Named UCS...* | *Ucsman* | *UC* | ... |

The *Ucsman* command invokes the UCS Manager (Fig. 34-17). The UCS Manager has three tabs: *Named UCSs*, *Orthographic UCSs* and *Settings*. The UCS Manager can be accessed by several methods shown in the table above. However, the *Orthographic UCSs* tab of the UCS Manager is also accessed by the *Dducsp* command (see *Dducsp*).

### Named UCSs Tab

*Ucsman* produces the *Named UCS* tab of the UCS Manager (see Fig. 34-17). The list contains all named UCSs. Use this tab to make named UCSs *Current*, just as you would use the *Restore* option of the *Ucs* command. You can create a named UCS by first creating the UCS, then using the UCS Manager to *Rename* the *Unnamed* UCS as shown in Figure 34-17.

FIGURE 34-17

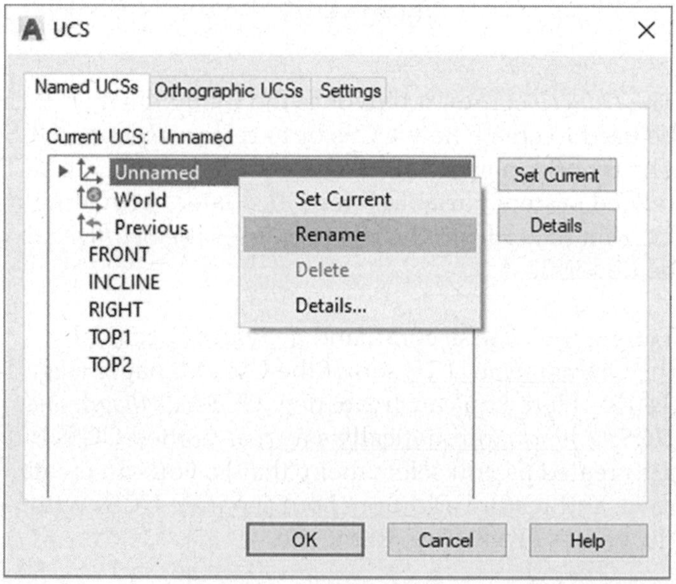

You can also create a named UCS by using the *Save* option of the *UCS* command. In this tab you can double-click the desired UCS name or highlight (single-click) the desired name and then select *Set Current*. Either way, you must then select *OK* to restore the desired UCS. The *Details* button lists the origin coordinates and X, Y, and Z direction vectors of the highlighted UCS.

### *Settings* Tab

The *Settings* tab can be accessed directly only by invoking the UCS Manager. However, you can use the *Dducsp* or *Dynamic UCS* commands or buttons explained earlier to produce the other tabs, then change tabs. The *Settings* tab (Fig. 34-18) allows you to control the UCS icon and two system variables.

FIGURE 34-18

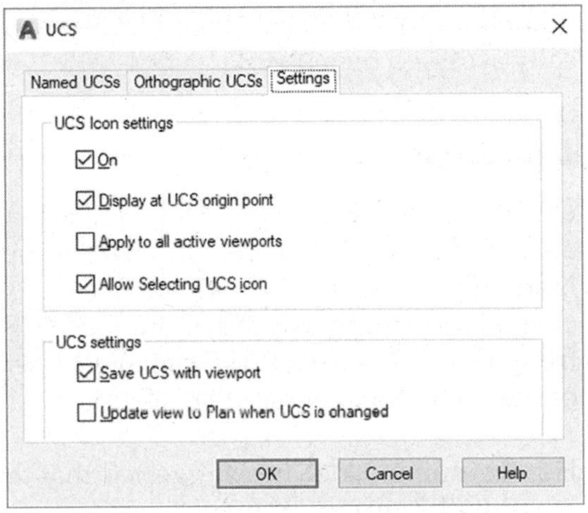


*UCS Icon settings*

Settings in this section are identical to controls of the *Ucsicon* command (see the *Ucsicon* command in Chapter 33). The *On* checkbox toggles the UCS icon on or off. *Display at UCS origin point* moves the icon to the coordinate system origin (otherwise it is located in the lower left corner of the screen) and has the identical functions as the *Origin* and *Noorigin* options of the *Ucsicon* command. *Apply to all viewports* is the same as the *All* option of the *Ucsicon* command; that is, the settings made in this section apply to the icons in all active viewports when multiple viewports are used. With the *Allow Selecting UCS Icon* box checked, the UCS icon is highlighted when the cursor moves over it, and you can click to access the UCS icon grips (*UCSSELECTMODE* variable).

*UCS Settings*

These two settings control two UCS-related system variables that affect only the currently active viewport when multiple viewports are used. *Save UCS with viewport* controls the *UCSVP* system variable and affects the UCS of the current viewport. When you are using multiple viewports, you have the option of locking the orientation of a UCS with a specific viewport so that when a new UCS is created or an existing UCS is restored, the locked UCS remains unchanged (that is, the UCS of the viewport that is active when the box is checked remains unchanged). *Update view to Plan when UCS is changed* controls the *UCSFOLLOW* system variable which affects the View of the active viewport. The viewport that is active when this box is checked will automatically change to a plan view of a new UCS whenever a new UCS is created or an existing UCS restored. When both boxes are checked, *Update view to Plan when UCS is changed* overrides *Save UCS with viewport.* (See "UCS System Variables" next in this chapter.)

## ViewCube *UCS* Drop-Down List

TIP

Although the ViewCube provides a method for selecting from named UCSs, this feature should be used with caution. The ViewCube drop-down list displays the current UCS (Fig. 34-19). Selecting a named UCS from the list makes it the current UCS—this is a very handy method of changing UCSs. However, if you then use the ViewCube to change views, the new view that appears is relative to the named UCS. This can be very confusing. For most cases, if you want to use the ViewCube to change views, first select *WCS* from the drop-down list.

FIGURE 34-19

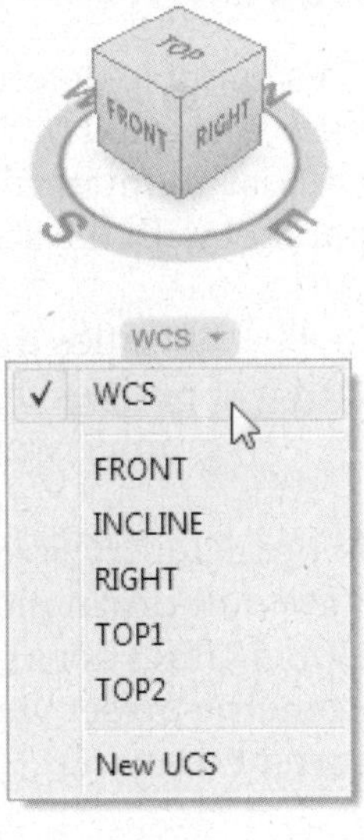

## UCS SYSTEM VARIABLES

AutoCAD offers a great amount of control and flexibility for UCSs, such as the ability to automatically create UCSs when creating views, saving UCSs with views, and having multiple UCSs current when using multiple viewports. Because of these options, it is easy to be overwhelmed if you are not aware of the system variables that control UCSs. This section is provided to help avoid confusion and to offer guidelines for UCS use. Not all UCS system variables are explained here, but only those that are typically accessed by a user.

### UCSBASE

*UCSBASE* stores the name of the UCS that defines the origin and orientation when *Orthographic* UCSs are created using the *UCS, Orthographic* option or the *Orthographic UCSs* tab of the UCS Manager. Normally the World Coordinate System is used as the base UCS, so when a new orthographic UCS is created it is simply rotated about the WCS origin to be normal to the selected orthographic view (top, front, right, etc.). *UCSBASE* can also be set in the *Relative to:* drop-down list of the *Orthographic UCSs* tab of the UCS Manager (see Fig. 34-15).

TIP

For beginning users, it is suggested that this variable be left at its initial value, *World*. This variable is saved in the current drawing.

### UCSVIEW

*UCSVIEW* determines whether the current UCS is saved with a named view. For example, suppose you create a new 3D viewpoint, such as a southeast isometric, to enhance your visualization while you construct one area of a model. You also create a UCS on the desired construction plane. With *UCSVIEW* set to 1, you then *Save* that view and assign a name such as RIGHT. Whenever the RIGHT view is later restored, the UCS is also restored, no matter what UCS is current before restoring the view. Without this control, the current UCS would remain active when the RIGHT view is restored. *UCSVIEW* has two possible settings.

0 The current UCS is not saved with a named view.
1 The current UCS is saved whenever a named view is saved (initial setting).

TIP

*UCSVIEW* is easily remembered because it saves the *UCS* with the *VIEW*. Remember that the view must be named to save its UCS.

### UCSORTHO

This variable determines whether an orthographic UCS is automatically created when an orthographic view is created or restored. For example, if *UCSORTHO* is set to 1 and you select or create a front view, a UCS is automatically set up to be parallel with the view. In this way, the XY plane of the UCS is always parallel with the screen when an orthographic view is restored. The settings are as follows.

0 Specifies that the UCS setting remains unchanged when an orthographic view is restored.
1 Specifies that the related orthographic UCS is automatically created or restored when an Orthographic view is created or restored (initial setting).

TIP

In order for *UCSORTHO* to operate as expected, you must select a preset orthographic view from the *View* pull-down menu, *Orthographic* option of the *-View* command, or an orthographic view icon button. *UCSORTHO* does not operate with views created using *Vpoint*, *Plan*, the *3Dorbit* commands, or for any isometric preset views. *UCSORTHO* operates for views as well as viewports. *UCSORTHO* is saved in the current drawing.

A good exercise you can use to understand *UCSORTHO* is to use *Vports* to create four tiled viewports in model space and select the *3D Setup*. With *UCSORTHO* set to 1, the arrangement shown in Figure 34-20 would result. Here, each orthographic view has its own orthographically aligned UCS. Notice the grid appears in all orthographic viewports because each viewport has its own coordinate system with the XY plane parallel to the view. The UCS icon, however, displays only in the current viewport.

**FIGURE 34-20**

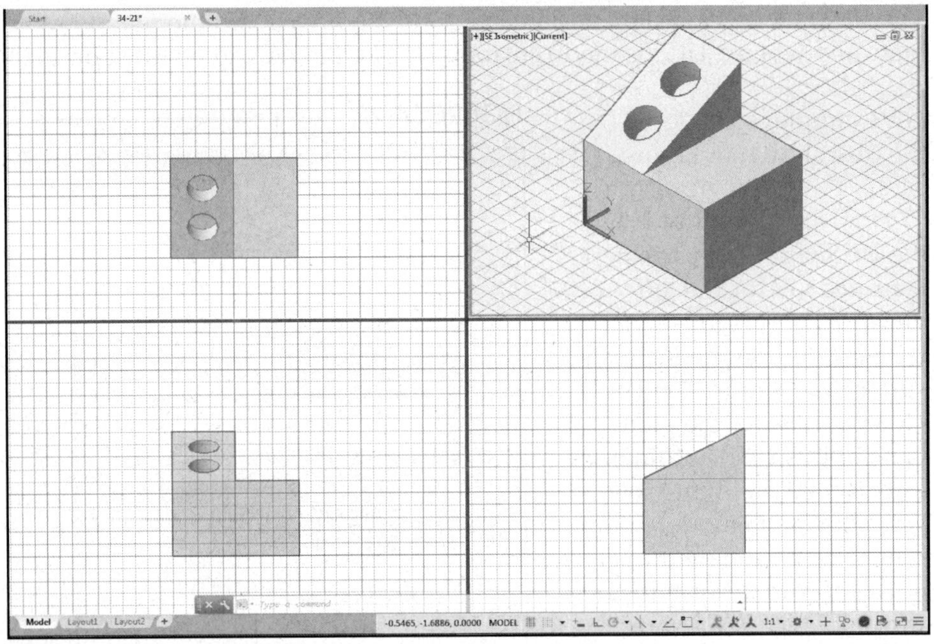

If *UCSORTHO* is set to 0 and the *Vports* command is used to create a *3D Setup*, the arrangement shown in Figure 34-21 would be created. Notice the grid appears only in the top two viewports where the XY plane is visible. Since each view displays the same coordinate system, the grid is not displayed in the bottom two viewports because the XY plane is perpendicular to the two views. The UCS icon displays only in the current viewport.

**FIGURE 34-21**

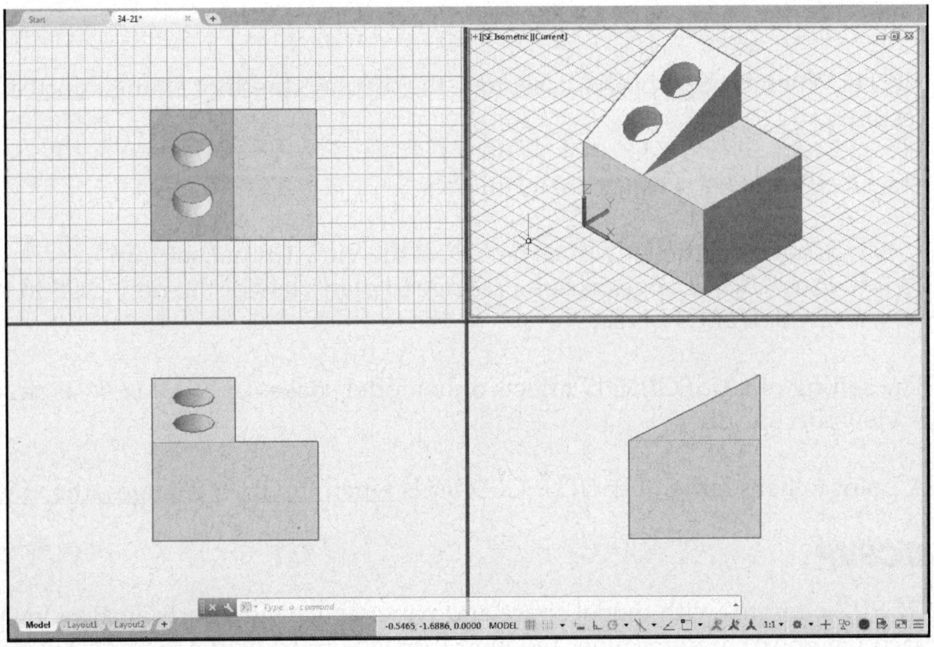

A point to help remember *UCSORTHO* is when the view changes to an orthographic view, the UCS changes.

TIP

## UCSFOLLOW

The *UCSFOLLOW* system variable, if set to 1 (*On*), causes the plan view of a UCS to be displayed automatically when a UCS is made current. *UCSFOLLOW* can be set separately for each viewport.

For example, consider the case shown in Figure 34-22. Two tiled viewports (*Vports*) are being used, and the *UCSFOLLOW* variable is turned *On* for the left viewport only. Notice that a UCS created on the inclined surface is current. The left viewport shows the plan view of this UCS automatically, while the right viewport shows the model in the same orientation no matter what coordinate system is current. If the WCS were made active, the right viewport would keep the same viewpoint, while the left would automatically show a plan view of the new UCS when the change was made.

**FIGURE 34-22**

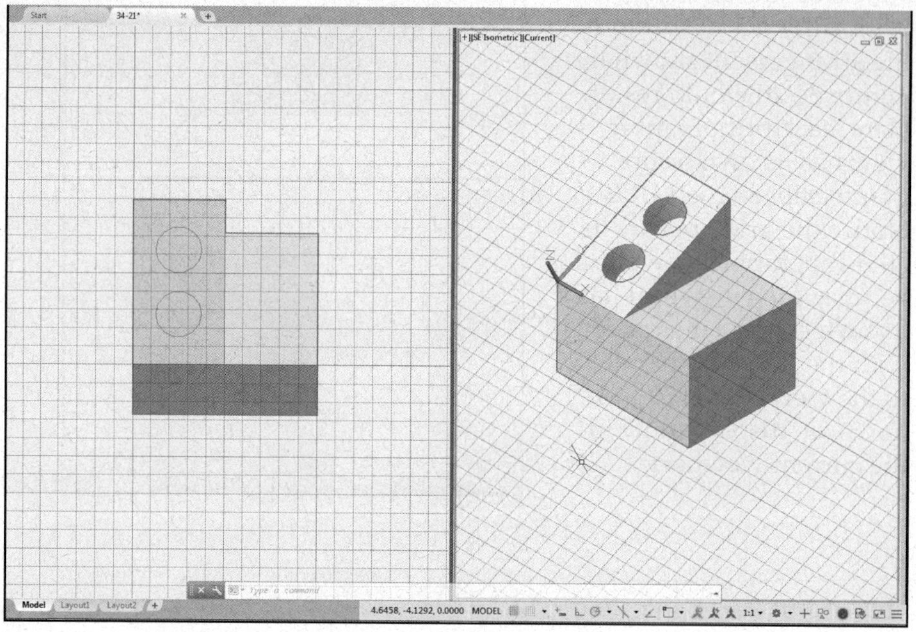


NOTE: When *UCSFOLLOW* is set to 1, the display does not change until a new UCS is created or restored.

The *UCSFOLLOW* settings are as follows.

0 Changing the UCS does not affect the view (initial setting).
1 Any UCS change causes a change to plan view of the new UCS in the viewport current when *UCSFOLLOW* is set.

**TIP**

The setting of *UCSFOLLOW* affects only model space. *UCSFOLLOW* is saved in the current drawing and is viewport specific.

A point to help remember *UCSFOLLOW* is when the UCS changes, the view follows.

## UCSVP

*UCSVP* operates with model space and paper space viewports and is viewport specific (can be set for each viewport). It determines whether the UCS in the active viewport remains locked or changes when another UCS is made current in another viewport. *UCSVP* is similar to *UCSVIEW* where the UCS is saved with the *View*, only with *UCSVP* the UCS is saved with the viewport. The possible *UCSVP* settings are shown below.

0 The UCS setting is not locked to the viewport so the UCS reflected becomes that of any other viewport that is made active.
1 The viewport and UCS are locked. Therefore, the UCS is saved in that viewport and does not change when another viewport displaying another UCS becomes active (initial setting).

### *UCS3DPERPDISPLAYSETTING*

Normally (with the initial setting), the UCS icon is displayed when perspective view is on (*PERSPECTIVE*=1) and a 3D visual style is current. If you want the UCS icon to disappear when you apply a visual style to a 3D model in perspective, set *UCS3DPERPDISPLAYSETTING* to 0 (off). The settings are as follows:

0 Off—the UCS icon is not displayed when perspective is turned on and a 3D visual style is current.
1 On—the UCS icon is displayed when perspective is turned on and a 3D visual style is current.

NOTE: The *UCSICON* command must also be set to *ON* to display the UCS icon.

### *UCS3DPARADISPLAYSETTING*

The UCS icon normally displays when a 3D model is shown in parallel projection (*PERSPECTIVE*=0) and a 3D visual style is current. Set *UCS3DPARADISPLAYSETTING* to 0 (off) if you do not want the UCS icon to appear when you apply a visual style to a 3D model when perspective mode is off.

0 Off—the UCS icon is not displayed when perspective is turned off and a 3D visual style is current.
1 On—the UCS icon is displayed when perspective is turned off and a 3D visual style is current.

### *UCSSELECTMODE*

This system variable determines whether the UCS icon can be selected and manipulated with multifunction grips. Normally (with the initial setting), you can hover over the UCS icon and access grips for the origin and each axis to provide an alternative to using the *UCS* command. Set *UCSSELECTMODE* to 0 to turn off this feature.

0 The UCS icon is not selectable.
1 The UCS icon is selectable for use with multifunction grips.

### *UCSAXISANG*

This variable stores the default angle when rotating the UCS around one of its axes using the *X*, *Y*, or *Z* options of the *UCS* command. When one of these options is used, the following prompt is displayed.

```
Specify rotation angle about n axis <90>:
```

The initial value is 90, so when you use one of these options to rotate the UCS, you can press Enter to rotate 90 degrees instead of entering a value. If you want another angle to appear as the default (in the < > brackets), enter the value in the *UCSAXISANG* variable. Valid values are 5, 10, 15, 18, 22.5, 30, 45, 90, 180. The variable setting is saved in the system registry.

## USER COORDINATE SYSTEMS AND 3D MODELING

Now that you understand the basics of 3D modeling discussed in Chapters 33 and 34, you are ready to progress to constructing and editing solid models. The use of User Coordinate Systems as discussed in Chapter 34 is applicable to all types of 3D modeling—wireframe, surface, and solid models.

### Dynamic User Coordinate Systems

You can create Dynamic UCSs. Dynamic UCSs, however, can be created only with solid models. Dynamic UCSs are similar to other UCSs, but they are created "on the fly" during a primitive creation command such as *Box*, *Wedge*, or *Cylinder*. Therefore, Dynamic UCSs are temporary coordinate systems that cannot be saved or named. Dynamic UCSs are discussed in Chapter 35.

## CHAPTER EXERCISES

1. Using the model in Figure 34-23, assume a UCS was created by rotating about the X axis 90 degrees from the existing orientation (World Coordinate System).

   A. What are the absolute coordinates of corner F?

   B. What are the absolute coordinates of corner H?

   C. What are the absolute coordinates of corner J?

2. Using the model in Figure 34-23, assume a UCS was created by rotating about the X axis 90 degrees from the existing orientation (WCS) and then rotating 90 degrees about the (new) Y.

   A. What are the absolute coordinates of corner F?

   B. What are the absolute coordinates of corner H?

   C. What are the absolute coordinates of corner J?

FIGURE 34-23

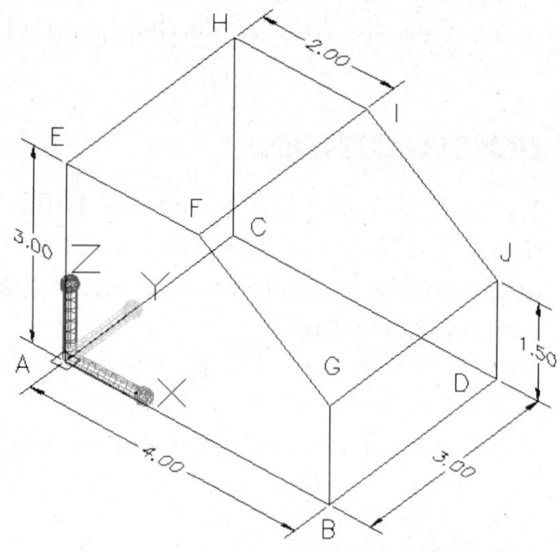

3. *Open* the **SOLID1** drawing that you created in Chapter 33, Exercise 11. View the object from a *SE Isometric* view. Make sure that the *UCSicon* is *On* and set to the *ORigin*.

   A. Create a *UCS* with a vertical XY plane on the front surface of the model as shown in Figure 34-24. *Save* the UCS as **FRONT**.

**FIGURE 34-24**

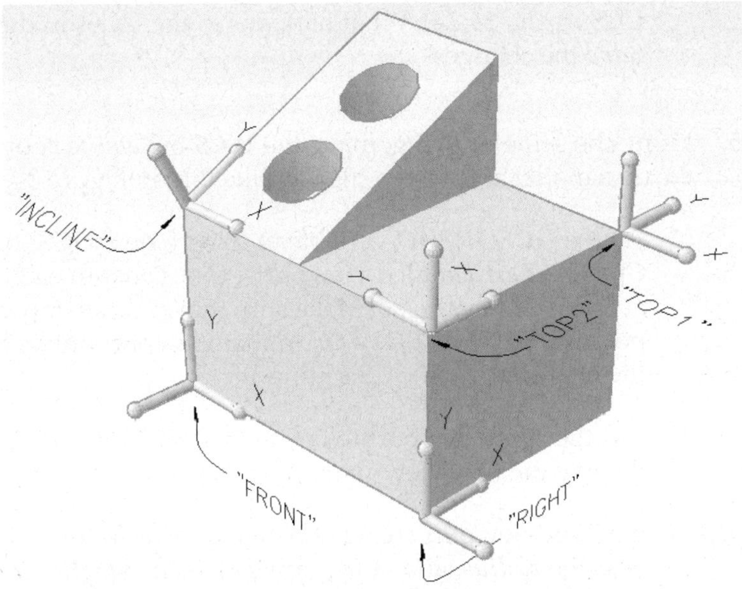


   B. Change the coordinate system back to the *World*. Now, create a *UCS* with the XY plane on the top horizontal surface and with the orientation as shown in Figure 34-24 as TOP1. *Save* the UCS as **TOP1**.

   C. Change the coordinate system back to the *World*. Next, create the *UCS* shown in the figure as TOP2. *Save* the UCS under the name **TOP2**.

   D. Activate the *WCS* again. Create and *Save* the **RIGHT** UCS.

   E. Use *SaveAs* and change the drawing name to **SOLIDUCS**.

4. In this exercise, you will create two viewport configurations, each with a different setting for *UCSORTHO*. In the first configuration, only one UCS exists, and in the second configuration, a new UCS is created for each viewport.

   A. *Open* the **SOLID1** drawing again. Use *SaveAs* and change the name to **VPUCS1**. Set the *UCSORTHO* variable to **0** so no new UCSs are created when you set up viewports. Ensure you are in model space and use the *Vports* command to create *Four Equal* viewports. Also in the *Viewports* dialog box, select the *3D Setup*. The resulting drawing should look like that in Figure 34-21. Notice that only one UCS appears in all viewports. Next, use any viewport and create a new UCS in the same orientation as the "Front" UCS shown in Figure 34-24. What happens in the other viewports? *Close* **VPUCS1** and *Save* the changes.

B. ***Open*** the **SOLID1** drawing again. Use ***SaveAs*** and change the name to **VPUCS2**. This time, set the ***UCSORTHO*** variable to **1** so when you set up viewports new orthographically oriented UCSs are created. Set up the same viewport configuration as in the previous step. Your new viewport configuration and the resulting new orthographic UCSs should look like that in Figure 34-20. In the isometric view create a new UCS on the inclined plane (like the "Incline" UCS in Fig. 34-24). What happens to the UCSs in the other viewports? ***Close*** **VPUCS2** and ***Save*** the changes.

5. Using the same drawing, make the ***WCS*** the active coordinate system. In this exercise, you will create a model and screen display like that in Figure 34-22.

A. Set the ***UCSORTHO*** variable to **0** so no new UCSs are created when you set up viewports. Create **2** ***Vertical*** tiled viewports. Use ***Zoom*** in each viewport if needed to display the model at an appropriate size. Make sure the *UCSicon* appears at the origin. Activate the left viewport. Set ***UCSFOLLOW On*** in that viewport only. (The change in display will not occur until the next change in UCS setting.)

B. Use the right viewport to create the **INCLINE** ***UCS***, as shown in Figure 34-24. ***Regenall*** to display the new viewpoint.

C. In either viewport create two cylinders as follows: Type ***cylinder***; specify **1.25,1** as the ***center***, **.5** as the ***radius***, and **-1** (negative 1) as the ***height***. The new cylinder should appear in both viewports.

D. Use ***Copy*** to create the second cylinder. Specify the existing center as the "Basepoint, of displacement." To specify the basepoint, you can enter coordinates of **1.25,1** or **PICK** the center with ***Object Snap***. Make the copy **1.5** units above the original (in a Y direction). Again, you can PICK interactively or use relative coordinates to specify the second point of displacement.

E. To create two holes, ***Subtract*** the two cylinders from the main solid. The resulting solid should look like that in Figure 34-22. Use ***SaveAs*** and assign the name **SOLID2**.

CHAPTER 35

# Solid Modeling Construction

## Chapter Objectives

After completing this chapter you should:

1. be able to create solid model primitives using the following commands: *Box, Wedge, Cone, Cylinder, Torus, Pyramid, Polysolid,* and *Sphere*;

2. be able to create swept solids from existing 2D shapes using the *Extrude, Loft, Revolve,* and *Sweep* commands;

3. be able to create Dynamic User Coordinate Systems;

4. be able to move solids in 3D space using *3Dmove, 3Dalign,* and *3Drotate*;

5. be able to create new solids in specific relation to other solids using *Mirror3D* and *3DArray*;

6. be able to combine multiple primitives into one composite solid using *Union, Subtract,* and *Intersect*;

7. know how to create beveled edges and rounded corners using *Chamfer* and *Fillet*.

## CONCEPTS

A solid model is a virtually complete representation of the 3D shape of a physical object. Solid modeling differs from wireframe or surface modeling in two fundamental ways: (1) the information is more complete in a solid model and (2) the method of construction of the model itself is relatively easy.

The basic solid modeling construction process is to use simple 3D solids, called "primitives," then combine them using "Boolean operations" to create a complex 3D model called a "composite solid." This modeling technique is known as Constructive Solid Geometry, or CSG, modeling. Although the process is relatively simple, you have tremendous flexibility during construction using tools such as UCSs and Dynamic UCSs, commands for moving and rotating, grip editing, and many other editing commands and techniques. AutoCAD uses a solid modeling engine, or "kernel," called ACIS.

### Solid Model Construction Process

The process used to construct most composite solid models follows four general steps:

1. Construct simple 3D primitive solids such as *Box*, *Cylinder*, or *Wedge*, or create 2D shapes and convert them to 3D solids by *Extrude*, *Revolve*, *Sweep*, or *Loft*.

2. Create the primitives in location relative to the associated primitives using UCSs or move, rotate, or mirror the primitives into the desired location relative to the associated primitives.

3. Use Boolean operations (*Union*, *Subtract*, and *Intersect*) to combine the primitives to form a composite solid.

4. Make necessary design changes to features of a composite solid using a variety of editing tools such as grips, *Solidedit*, and using surfaces to *Slice* solids.

This chapter discusses the basic construction process organized into the following sections: 1) Solid Primitives Commands, 2) Dynamic User Coordinate Systems, 3) Commands to Move, Rotate, and Copy Solids, and 4) Boolean Operations. Each section is briefly described next. Chapter 36 discusses advanced editing techniques, such as grip editing, editing faces and edges, sectioning, and display variables.

### Solid Primitives Commands

Solid primitives are the basic building blocks that make up more complex solid models. The commands that create 3D primitives "from scratch" are listed here:

| | |
|---|---|
| *Box* | Creates a solid box or cube |
| *Cone* | Creates a solid cone with a circular or elliptical base |
| *Cylinder* | Creates a solid cylinder with a circular or elliptical base |
| *Polysolid* | Creates a series of connected solid segments having a specified height and width |
| *Pyramid* | Creates a solid right pyramid with any number of sides |
| *Sphere* | Creates a solid sphere |
| *Torus* | Creates a solid torus |
| *Wedge* | Creates a solid wedge |

Commands that create solid primitives from existing 2D shapes are listed next.

| | |
|---|---|
| *Extrude* | Converts a closed 2D object such as a *Pline, Circle, Region,* or *Planar Surface* to a solid by adding a Z dimension |
| *Loft* | Creates a solid using a series of 2D cross sections (curves or lines) that define the shape |
| *Revolve* | Converts a closed 2D object such as a *Pline, Circle,* or *Region* to a solid by revolving the shape about an axis |
| *Sweep* | Creates a solid by sweeping a closed planar profile along an open or closed 2D or 3D path |

## Dynamic User Coordinate Systems

As you learned in Chapter 34, the *UCS* command creates a User Coordinate System that is static—that is, it remains active until you change to another coordinate system. This chapter discusses how to create and control Dynamic Coordinate Systems.

A Dynamic User Coordinate System (DUCS) is a temporary UCS that appears automatically when you use a primitive command such as *Box, Cylinder,* or *Pyramid* and disappears when you complete the command. You can use the *Dynamic UCS* button on the Status bar to toggle this feature on or off. You cannot use the *Ucs* command to create a Dynamic User Coordinate System. A Dynamic UCS appears only with some primitive solid creation commands but not with surface or wireframe models.

For example, suppose you wanted to create a *Box* on the surface of an existing solid. When you use the *Box* command, you are prompted for the "first corner." With *Dynamic UCS* on, you can create a temporary UCS on an existing solid to begin drawing the *Box*. You determine the location of the XY plane of the Dynamic UCS by "hovering" over any planar face of an existing solid. You can also specify the direction of the X and Y axes based on the location of the cursor in relation to the highlighted plane. Once the Dynamic UCS icon appears in the desired orientation, PICK to specify the "first corner" for the *Box*. The Dynamic UCS disappears once the new *Box* is complete.

## Commands to Move, Rotate, Mirror, and Scale Solids

It is not always efficient or practical to create every primitive solid initially in the desired location and/or orientation. In many cases primitives are created in another position, then they are moved, rotated, or otherwise aligned into the desired position before combining them into composite solids. Also, some primitives can be created from others by mirroring or arraying existing solids. This section of the chapter discusses the following commands:

| | |
|---|---|
| *Move* | The command used for moving objects in 2D can also be used in 3D space |
| *3Darray* | Makes a polar array about any axis or a rectangular array with rows, columns, and levels |
| *3Dalign* | Connects two solids together by matching three alignment points on each solid |
| *3Dmove* | A temporary grip tool allows you to restrict the movement to an axis or a plane |
| *3Drotate* | A temporary grip tool allows you to specify an axis about which to rotate |
| *3Dscale* | A temporary grip tool allows you to scale an object along a plane or axis |
| *Mirror3D* | Creates a mirror copy of solids by specifying a mirror plane |

## Boolean Operations

Primitives are combined to create complex solids by using Boolean operations. The Boolean operators are listed below. An illustration and detailed description are given for each of the commands.

| | |
|---|---|
| *Union* | Unions (joins) selected solids. |
| *Subtract* | Subtracts one set of solids from another. |
| *Intersect* | Creates a solid of intersection (common volume) from the selected solids. |

Primitives are created at the desired location or are moved into the desired location before using a Boolean operator. In other words, two or more primitives can occupy the same space (or share some common space) yet are separate solids. When a Boolean operation is performed, the solids are combined or altered in some way to create one solid. AutoCAD takes care of deleting or adding the necessary geometry and displays the new composite solid complete with the correct configuration and lines of intersection.

FIGURE 35-1

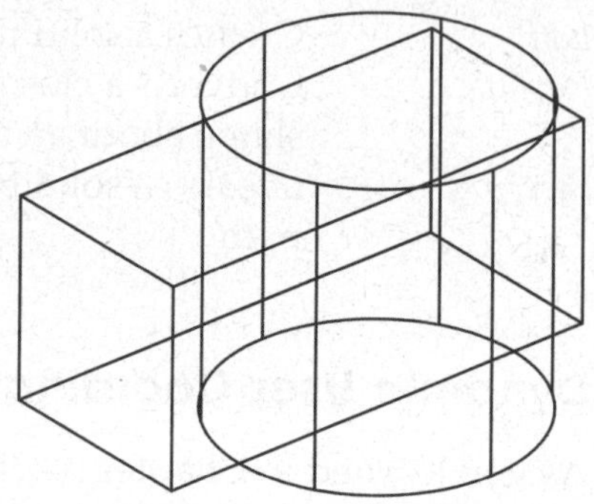

Consider the two solids shown in Figure 35-1. When solids are created, they can occupy the same physical space. A Boolean operation is used to combine the solids into one composite solid and it interprets the resulting utilization of space.

### *Union*

*Union* creates a union of the two solids into one composite solid (Fig. 35-2). The lines of intersection between the two shapes are calculated and displayed by AutoCAD.

FIGURE 35-2

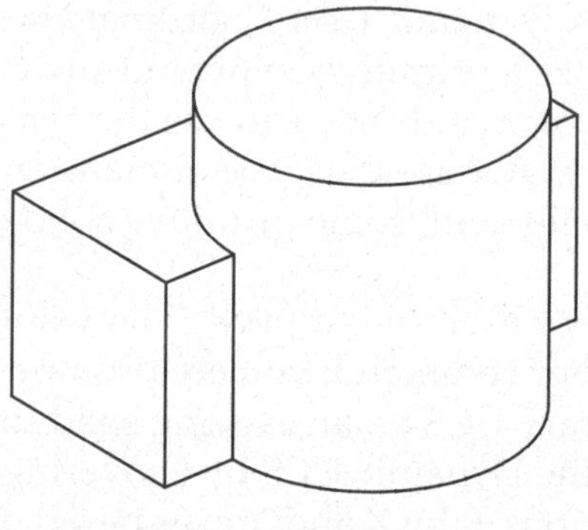

### *Subtract*

*Subtract* removes one or more solids from another solid. ACIS calculates the resulting composite solid. The term "difference" is sometimes used rather than "subtract." In Figure 35-3, the cylinder has been subtracted from the box.

FIGURE 35-3

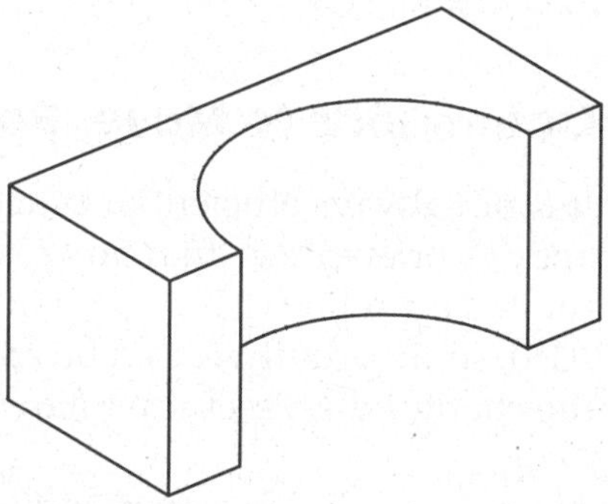

### *Intersect*

*Intersect* calculates the intersection between two or more solids. When *Intersect* is used with *Regions* (2D surfaces), it determines the shared area. Used with solids, as in Figure 35-4, *Intersect* creates a solid composed of the shared volume of the cylinder and the box. In other words, the result of *Intersect* is a solid that has only the volume which is part of both (or all) of the selected solids.

FIGURE 35-4

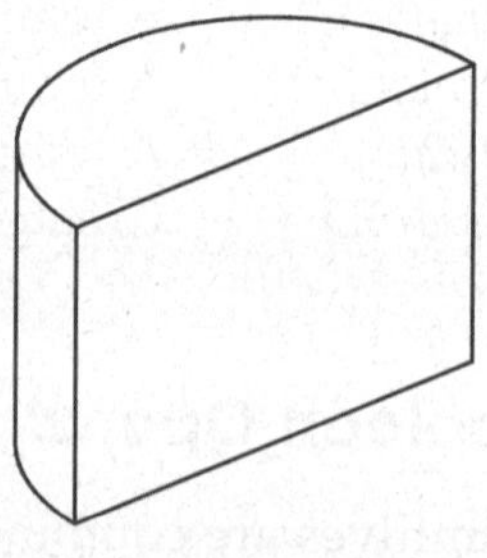

### *Chamfer* and *Fillet*

The *Chamfer* command creates an angled edge between two surfaces, and a *Fillet* creates a rounded edge between two surfaces. Although the *Chamfer* command and the *Fillet* command are not Boolean operators, they are included in this section. With each of these commands, you specify dimensions to create a wedge primitive (with *Chamfer*) or a rounded primitive (with *Fillet*) and AutoCAD automatically performs the Boolean needed to add or subtract the primitive from the selected solids.

## SOLID PRIMITIVES COMMANDS

This section explains the commands that allow you to create primitives used for construction of composite solid models. The commands allow you to specify the dimensions and the orientation of the solids. Once primitives are created, they are combined with other solids using Boolean operations to form composite solids.

NOTE: If you use a pointing device to PICK points, use *Object Snap* when possible to PICK points in 3D space. If you do not use *Object Snap*, the selected points are located on the current XY construction plane, so the true points may not be obvious. It is recommended that you use *Object Snap* or enter values.

The commands for creating solid primitives and modifying solids can be assessed several ways. If you are using the *3D Modeling* workspace, the *Home* tab provides the *Modeling*, *Solid Editing*, and *Modify* panels (Fig. 35-5).

**FIGURE 35-5**

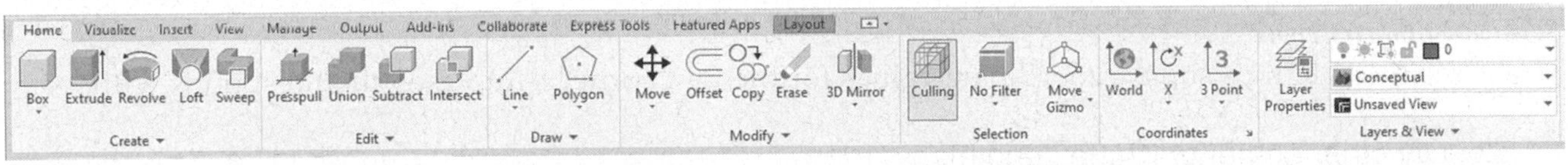


In the Menu Bar, the solid primitives commands are available from the *Draw* menu, then *Modeling* menu (Fig. 35-6). Commands for moving and editing solids are located in the *Modify* menu (not shown).

Two toolbars are available titled *Modeling* and *Solid Editing* that contain almost all the commands you need for solid modeling (Fig. 35-7).

In addition to these methods, each of the solid modeling commands are accessible by typing in the command name at the Command prompt, such as *Box* or *Wedge*.

**FIGURE 35-6**

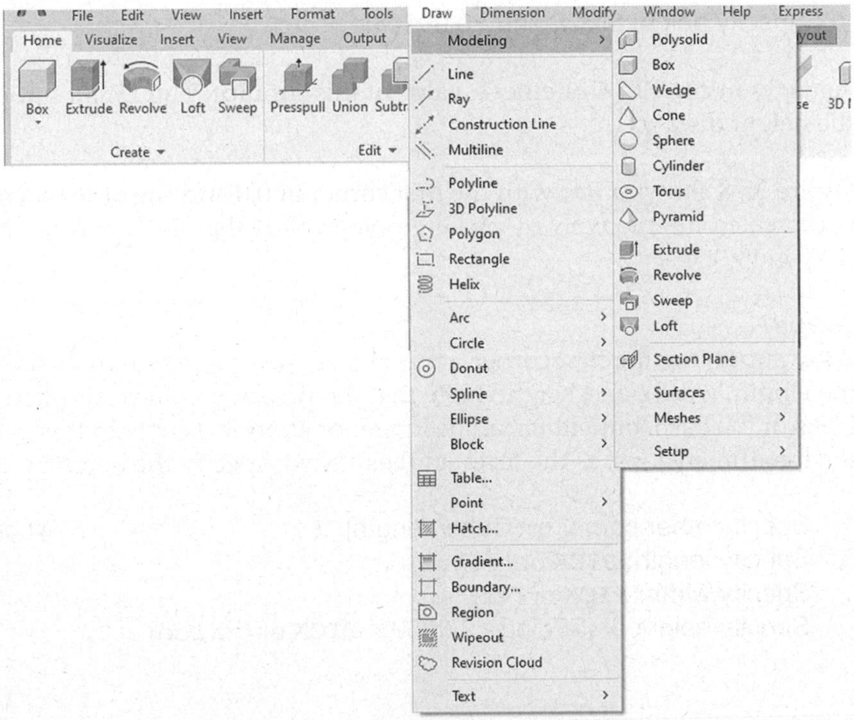


**FIGURE 35-7**

## Box

| Ribbon | Menu Bar | Command (Type) | Alias (Type) | Shortcut |
|---|---|---|---|---|
| *Home Modeling Box* | *Draw Modeling > Box* | *Box* | ... | ... |

*Box* creates a solid box primitive to your dimensional specifications. The base of the box is always oriented parallel to the current XY plane (WCS, UCS, or Dynamic UCS ). You can specify dimensions of the box by PICKing or by entering values. The box can be defined by (1) giving the corners of the base, then height, (2) by locating the center, then the corners and height, or (3) by giving each of the three dimensions.

Command: ***BOX***
Specify first corner or [Center]: **PICK** or (**coordinates**) or **C**

### *Specify first corner*

You can either PICK or supply coordinate values for the first corner. AutoCAD then responds with:

Specify other corner or [Cube/Length]:

If you PICK the other corner interactively, the length and width of the base are parallel with the X and Y axes. AutoCAD then requests the height.

Specify height or [2Point] <0.0000>:

Again, you can PICK or enter a value at this prompt. The *2Point* option allows you to PICK two points as the height distance.

Figure 35-8 shows a *Box* with the first corner at 0,0 and the other corner at 5,4. The height is 3. This box was created interactively by picking points. Note that the box is oriented with the edges aligned with the X,Y, and Z axes.

### *Length*

After supplying the first corner, you can use the *Length* option to specify three dimensions for the box: the length, width, and height. Note that the first two values supplied (*Length* and *Width*) are the dimensions of the base, but either can be longer or shorter (*Length* in this case is not necessarily the longer of the two dimensions, as the term implies). Next, specify the height.

Specify other corner or [Cube/Length]: ***l***
Specify length: **PICK** or (**value**)
Specify width: **PICK** or (**value**)
Specify height or [2Point] <0.0000>: **PICK** or (**value**)

**FIGURE 35-8**

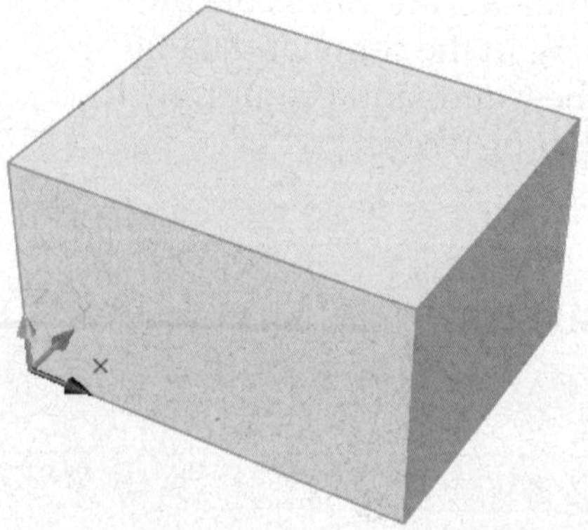

NOTE: Using the *Length* option, the direction of the length and width is determined by the current location of the crosshairs; therefore, the edges of the box's base are not aligned with the X and Y axes unless *Polar Tracking* or *Ortho* is on. In addition, the box can be created in negative X,Y, and Z directions.

### Cube

The *Cube* option requires only one dimension to create the *Box*. You can PICK a distance or enter a value. In either case, the direction of the edges of the base is determined by the current location of the crosshairs.

### Center

With this option you first locate the center of the box, then supply the dimensions of the box by PICKing or entering values.

Command: ***BOX***
Specify first corner or [Center]: ***c***
Specify center: **PICK** or (**coordinates**)
Specify corner or [Cube/Length]:

FIGURE 35-9

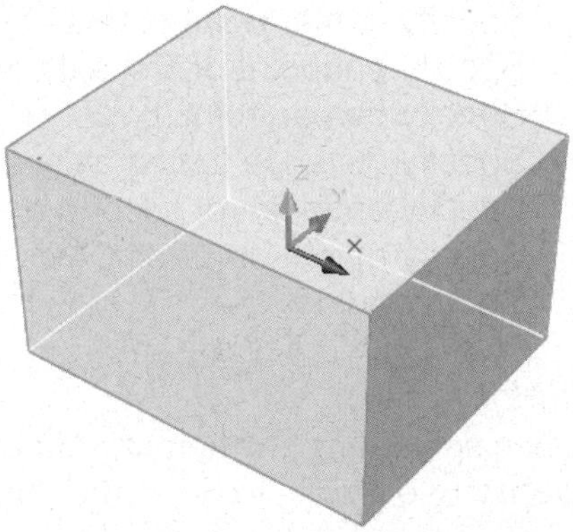

Figure 35-9 displays a *Box* created using the *Center* method. Note that the center is the volumetric center, not the center of the base.

## Cone

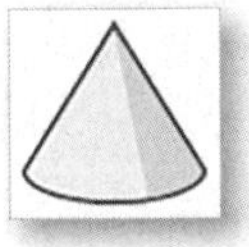

| Ribbon | Menu Bar | Command (Type) | Alias (Type) | Shortcut |
| --- | --- | --- | --- | --- |
| *Home Modeling Cone* | *Draw Modeling > Cone* | *Cone* | ... | ... |

*Cone* creates a right circular or elliptical solid cone ("right" means the axis forms a right angle with the base). By the default method, you specify the center location, radius (or diameter), and height. With this option, the orientation of the cylinder is determined by the current UCS so that the base lies on the XY plane and height is perpendicular (in a Z direction). Alternately, a different orientation can be defined by using the *Axis endpoint* or *3Ppoint* option.

### Center Point

Using the defaults (PICK or supply values for the center, radius, and height), the cone is generated in the orientation shown in Figure 35-10.

FIGURE 35-10

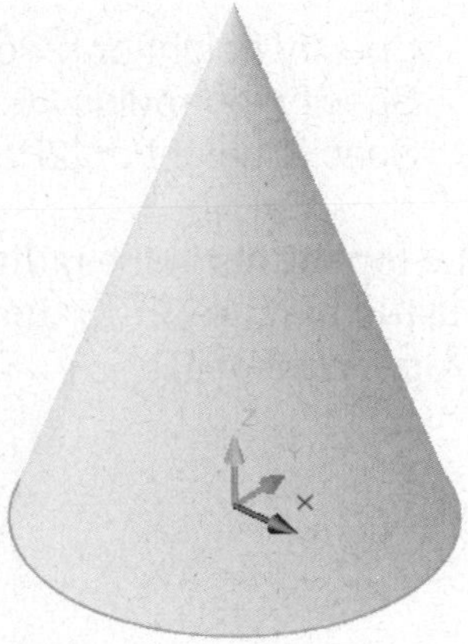

The default prompts are shown as follows:

Command: ***CONE***
Specify center point of base or [3P/2P/Ttr/Elliptical]: **PICK** or (**coordinates**)
Specify base radius or [Diameter]: **PICK** or (**coordinates**)
Specify height or [2Point/Axis endpoint/Top radius]: **PICK** or (**value**)

### *3P/2P/Ttr*

Note that you can define the location and diameter of the base using the *3Point*, *2Point*, or *Ttr* (tangent, tangent, radius) methods. These options work similar to creating a *Circle* by the same methods. Using the *2Point* and *Ttr* methods, the *Cone* base always lies in a plane parallel with the UCS, no matter which points in 3D space are selected. However, the *3Point* option allows a different orientation such that the base of the *Cone* lies in the plane defined by the three points. These options and principles operate similarly for creating a *Cylinder* or *Torus*.

### *Elliptical*

This option draws a cone with an elliptical base. You specify two axis endpoints to define the elliptical base. The elliptical cone in Figure 35-11 was created using the following specifications :

Command: **CONE**
Specify center point of base or [3P/2P/Ttr/Elliptical]: **e**
Specify endpoint of first axis or [Center]: **c**
Specify center point: **0,0**
Specify distance to first axis: **2**
Specify endpoint of second axis: **1**
Specify height or [2Point/Axis endpoint/Top radius]: **3**
Command:

**FIGURE 35-11**

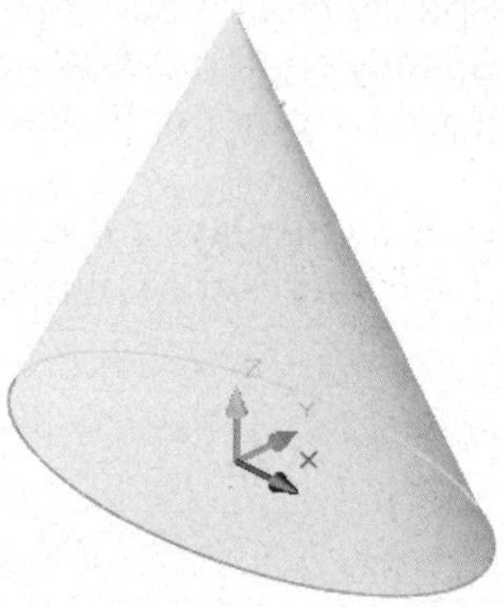

### *2Point*

When specifying the height, you can use this option to interactively specify two points to determine the height distance.

### *Axis endpoint*

Invoking the *Axis endpoint* option (after the base has been established) displays the following prompt:

Specify apex point: **PICK** or (**coordinates**)

Locating a point for the axis endpoint defines the height and orientation of the cone. The axis of the cone is aligned with the line between the specified center point of the base and the apex, and the height is equal to the distance between the two points. Figure 35-12 shows a *Cone* with the base at 0,0 and the *Axis endpoint* located on the X axis.

**FIGURE 35-12**

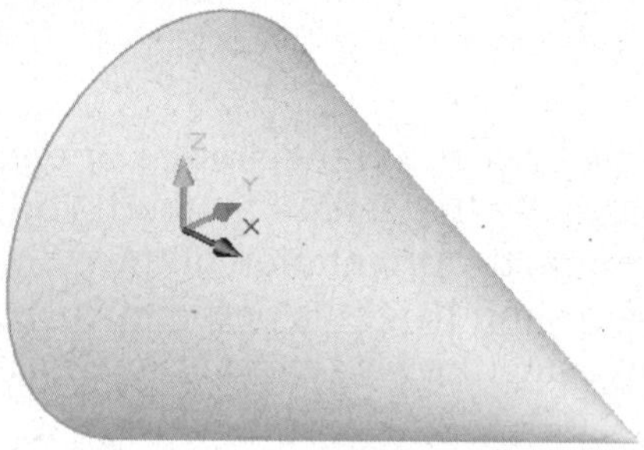

### *Top radius*

The *Top radius* option allows you to create a frustum (a truncated right cone with the top plane parallel to the base). The following prompt is displayed:

Specify height or [2Point/Axis endpoint/Top radius]: **t**
Specify top radius <0.0000>: **PICK** or (**value**)
Specify height or [2Point/Axis endpoint]: **PICK** or (**value**)

The top radius is the radius for the circle at the top of the frustum. Figure 35-13 shows a *Cone* with a base radius of 2 and a top radius of 1.

**FIGURE 35-13**

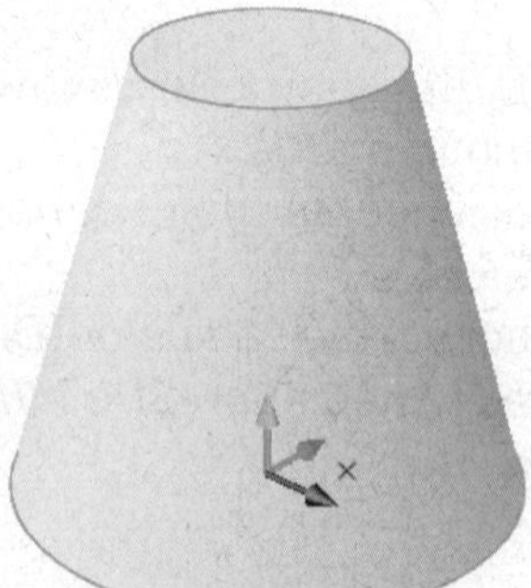

## *Cylinder*

| Ribbon | Menu Bar | Command (Type) | Alias (Type) | Shortcut |
|---|---|---|---|---|
| *Home Modeling Cylinder* | *Draw Modeling > Cylinder* | *Cylinder* | ... | .. |

*Cylinder* creates a cylinder with an elliptical or circular base with a center location, diameter, and height you specify. Default orientation of the cylinder is determined by the current UCS, such that the circular plane is coplanar with the XY plane and height is in a Z direction. However, the orientation can be defined otherwise by the *Axis endpoint* or *3Point* option.

**FIGURE 35-14**

### *Center Point*

The default options create a cylinder in the orientation shown in Figure 35-14:

Command: ***CYLINDER***
Specify center point of base or [3P/2P/Ttr/Elliptical]: **PICK** or (**coordinates**)
Specify base radius or [Diameter] <0.0000>: **PICK** or (**value**)
Specify height or [2Point/Axis endpoint] <0.0000>: **PICK** or (**value**)
Command:

### *3P/2P/Ttr*

You can define the location and diameter of the base using the *3Point*, *2Point*, or *Ttr* (tangent, tangent, radius) methods. These options work similar to creating a *Circle* by the same methods. If you use the *2Point* and *Ttr* methods, you can select any points in 3D space to define the base of the *Cylinder*; however, the base always lies in a plane parallel with the UCS. On the other hand, if you use the *3Point* option, the base lies in the same plane as the three specified points. These options and principles are the same for creating a *Cone* base diameter or *Torus* diameter.

### *Elliptical*

This option draws a cylinder with an elliptical base. The elliptical cone in Figure 35-15 was created using the following specifications.

**FIGURE 35-15**

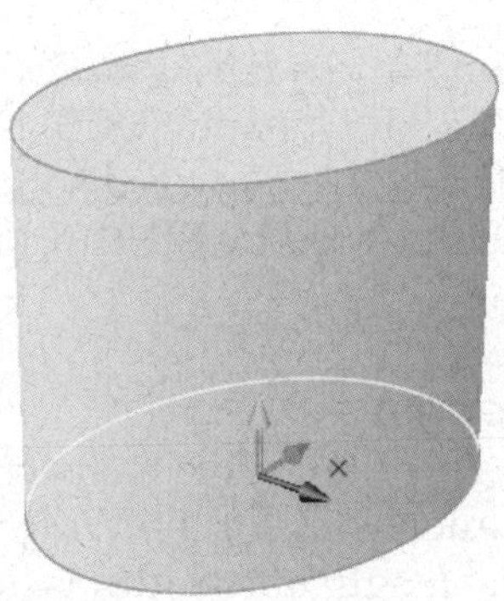

Command: ***CYLINDER***
Specify center point of base or [3P/2P/Ttr/Elliptical]: ***e***
Specify endpoint of first axis or [Center]: ***c***
Specify center point: **0,0**
Specify distance to first axis <0.0000>: **2**
Specify endpoint of second axis: **1.5**
Specify height or [2Point/Axis endpoint] <0.0000>: **3**

### *2Point*

When specifying the height, you can use this option to interactively specify two points to determine the height distance.

### *Axis endpoint*

FIGURE 35-16

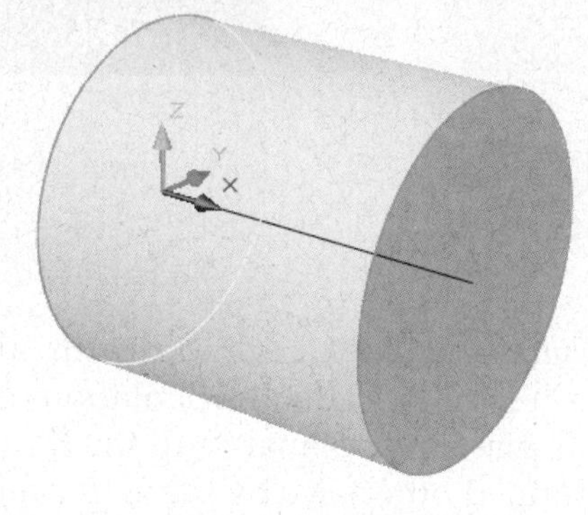

Invoking the *Axis endpoint* option (after the base has been established) allows you to determine both the height and orientation for the cylinder.

> Command: ***CYLINDER***
> Specify center point of base or [3P/2P/Ttr/Elliptical]: **PICK** ***(Endpoint)***
> Specify base radius or [Diameter] <0.0000>: **1.5**
> Specify height or [2Point/Axis endpoint] <0.0000>: ***a***
> Specify axis endpoint: **PICK** ***(Endpoint)***

The cylinder in Figure 35-16 was created using the *Axis endpoint* method. Here, the *Line* along the X axis was previously established. The *Cylinder* was created by using *Endpoint* to snap to the two endpoints for the "center point of base" and "axis endpoint" (see the prompts listed above).

## *Wedge*

| Ribbon | Menu Bar | Command (Type) | Alias (Type) | Shortcut |
|---|---|---|---|---|
| *Home Modeling Wedge* | *Draw Modeling > Wedge* | *Wedge* | *WE* | ... |

FIGURE 35-17

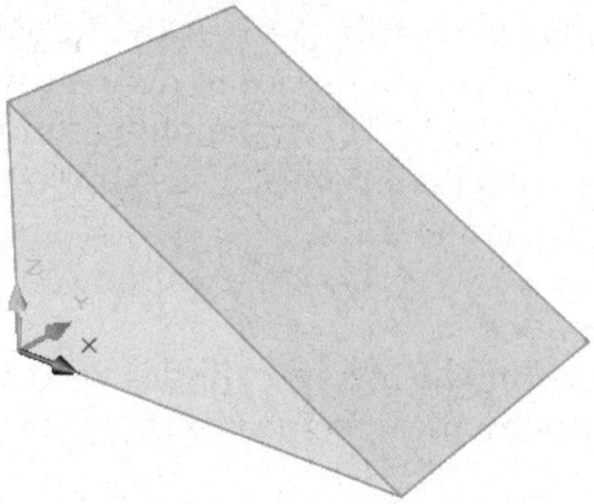

Creating a *Wedge* solid primitive is similar to creating a *Box* (see "*Box*"). Note that the prompts are the same, and therefore the methods for creating a *Wedge* are identical to those for creating a *Box*. The difference between a *Wedge* and a *Box* with the same dimensions is the *Wedge* contains half the volume of the *Box*—a *Wedge* is like a *Box* cut in half diagonally.

> Command: ***WEDGE***
> Specify first corner or [Center]: **PICK** or (**coordinates**)
> Specify other corner or [Cube/Length]: **PICK** or (**coordinates**)
> Specify height or [2Point] <0.0000>: **PICK** or (**value**)
> Command:

Similar to a *Box*, a *Wedge's* base is always parallel to the XY plane of the current coordinate system (WCS, UCS, or Dynamic UCS). Additionally, using the default option ("Specify first corner"), the *Wedge* always slopes down toward second ("other") corner. Notice the orientation of the *Wedge* in Figure 35-17, which was created by PICKing the "first corner" at the origin and the "other corner" in an XY-positive direction.

### *Length*

As an alternative to specifying two corners and a height, you can use the *Length* option to specify three dimensions for the *Wedge*: the length, width, and height. NOTE: Using the *Length* option, the direction of the length and width is determined by the current location of the crosshairs; therefore, the edges of the *Wedge's* base are not aligned with the X and Y axes unless *Polar Tracking* or *Ortho* is on. In addition, the *Wedge* always slopes down toward the direction indicated for the *Length*.

### *Cube*

The *Cube* option requires only one dimension to create the *Wedge*. You can PICK a distance or enter a value. In either case, with this option the *Wedge* always slopes down toward the direction indicated by the current location of the crosshairs.

### *Center*

With this option you first locate a "center" for the *Wedge*, then supply the dimensions by PICKing or entering values.

Command: ***WEDGE***
Specify first corner or [Center]: ***c***
Specify center: **0,0**
Specify corner or [Cube/Length]:

Figure 35-18 displays a *Wedge* created by specifying 0,0 as the *Center* method (see prompts above). Note that the designated center is not the volumetric center but the center of the sloping side of the *Wedge*. If you imagine the *Wedge* as one half of a *Box*, the center would be the volumetric center for the imaginary *Box* (see "*Box*").

**FIGURE 35-18**

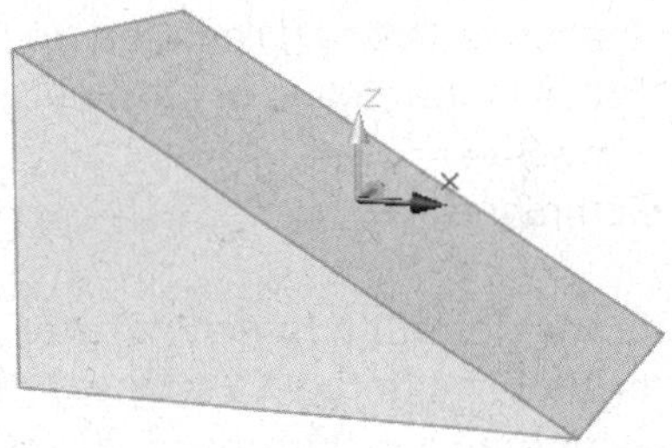

*Sphere* allows you to create a sphere by defining its center point and radius or diameter.

## *Sphere*

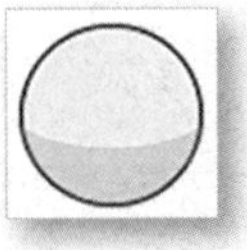

| Ribbon | Menu Bar | Command (Type) | Alias (Type) | Shortcut |
|---|---|---|---|---|
| *Home Modeling Sphere* | *Draw Modeling > Sphere* | *Sphere* | ... | ... |

Command: ***SPHERE***
Specify center point or [3P/2P/Ttr]: **PICK** or (**coordinates**)
Specify radius or [Diameter] <0.0000>: **PICK** or (**value**)

Creating a *Sphere* with the default options and using 0,0 as the center would create the shape shown in Figure 35-19.

**FIGURE 35-19**

### *3P/2P/Ttr*

As options, you can define the location and diameter of the *Sphere* using the *3Point*, *2Point*, or *Ttr* (tangent, tangent, radius) methods. For example, you could use the *3Point* option and *Object Snap* to create a *Sphere* location and diameter based on three points on existing solids. These options are similar to the same options for creating a *Cone* base, *Cylinder* base, or *Torus*.

### *Torus*

| Ribbon | Menu Bar | Command (Type) | Alias (Type) | Shortcut |
|---|---|---|---|---|
| *Home* *Modeling* *Torus* | *Draw* *Modeling >* *Torus* | *Torus* | *TOR* | ... |

*Torus* creates a torus (donut shaped) solid primitive. Three specifications are needed: (1) the center location, (2) the radius or diameter of the torus (from the center of the *Torus* to the centerline of the tube), and (3) the radius or diameter of the tube. AutoCAD prompts:

Command: ***TORUS***
Specify center point or [3P/2P/Ttr]: **PICK** or (**coordinates**)
Specify radius or [Diameter] <0.0000>: **PICK** or (**value**)
Specify tube radius or [2Point/Diameter] <0.0000>: **PICK** or (**value**)
Command:

**FIGURE 35-20**

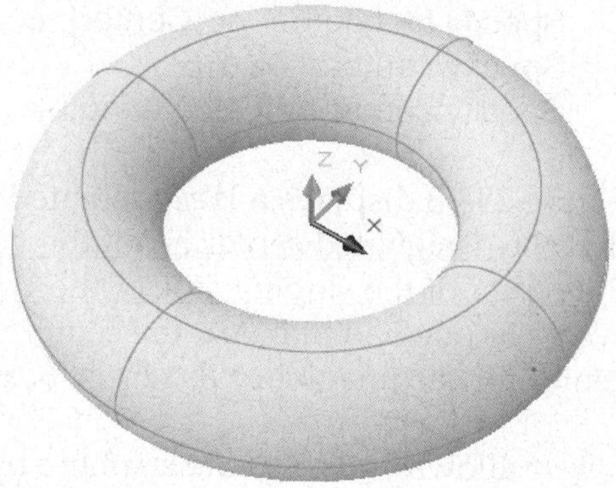

Using the default ("Specify center point") option, the *Torus* is oriented such that the *Torus* diameter (tube centerline) lies in the XY plane of the current UCS and the rotational axis of the tube is parallel with the Z axis of the UCS. Figure 35-20 shows a *Torus* created using the default orientation and the center of the torus at 0,0,0. The *Torus* has a torus radius of 3 and a tube radius of 1. (To improve the visibility of this figure, the visual style edge setting is *Isolines*.)

A self-intersecting torus is allowed with the *Torus* command. A self-intersecting torus is created by specifying a tube radius equal to or greater than the torus radius. Figure 35-21 illustrates a torus with a torus radius of 3 and a tube radius of 3.

**FIGURE 35-21**

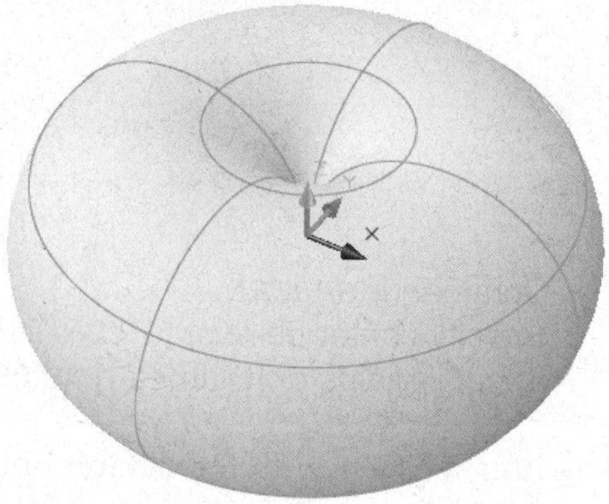

#### *3P/2P/Ttr*

Similar to the same options for drawing a *Circle*, you can define the location and diameter of the *Torus* using the *3Point*, *2Point*, or *Ttr* (tangent, tangent, radius) methods. For example, you could use the *3Point* option and *Object Snap* to create a *Torus* location and diameter based on three points on existing solids. Using the *2Point* and *Ttr* methods, the *Torus* always lies in a plane parallel with the UCS, whereas the *3Point* option allows a different orientation. These options are similar to the same options for creating *Cone* base or *Cylinder* base.

### *Pyramid*

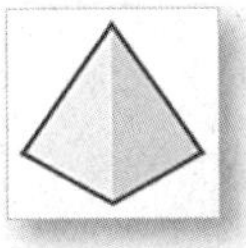

| Ribbon | Menu Bar | Command (Type) | Alias (Type) | Shortcut |
|---|---|---|---|---|
| *Home Modeling Pyramid* | *Draw Modeling > Pyramid* | *Pyramid* | *PYR* | ... |

The *Pyramid* command creates a right pyramid with any number of sides. With the default options, you specify the center of the base, the base radius, and the height. The resulting *Pyramid* has its base parallel to the XY plane of the current UCS, unless the *Axis endpoint* method is used.

Command: ***PYRAMID***
4 sides Circumscribed
Specify center point of base or [Edge/Sides]: **PICK** or (**coordinates**)
Specify base radius or [Inscribed] <0.0000>: **PICK** or (**value**)
Specify height or [2Point/Axis endpoint/Top radius] <0.0000>: **PICK** or (**value**)
Command:

FIGURE 35-22

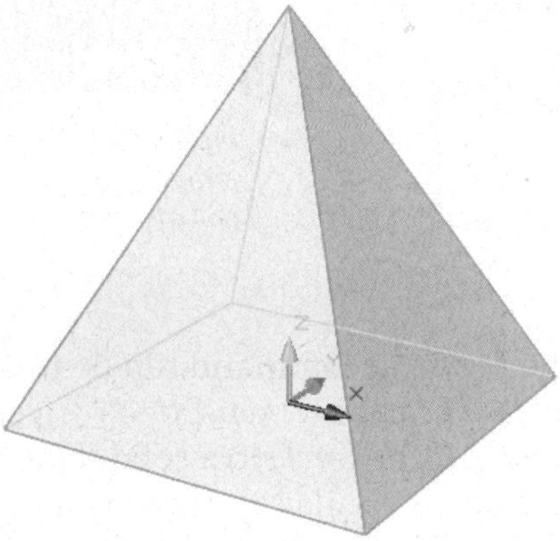

Figure 35-22 shows a four-sided pyramid with the base located at 0,0, a base radius (inscribed) of 2, and a height of 4.

#### *Edge*

Use the *Edge* option to PICK two points in space to define the length and orientation of one edge. The *Pyramid* base is generated on the left side of the defined line going from the first to the second point.

Specify center point of base or [Edge/Sides]: ***e***
Specify first endpoint of edge: **PICK** or (**coordinates**)
Specify second endpoint of edge: **PICK** or (**coordinates**)

#### *Sides*

Use this option to specify the number of sides (not including the base) for the *Pyramid*. Any practical number of sides can be specified with 3 as the minimum.

#### *Inscribed/Circumscribed*

This feature specifies how the base radius is drawn—either inscribed within or circumscribed about an imaginary circle. This method is similar to defining the size of a *Polygon*.

FIGURE 35-23

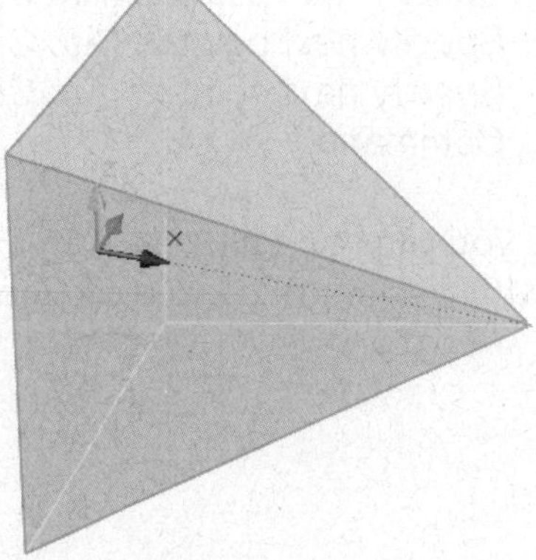

#### *2Point*

This option is the same as for a *Cone* or *Cylinder*. You can PICK two points to specify the height of the *Pyramid* such that the resulting height is equal to the indicated distance; however, the *Pyramid* does not change its orientation from that set by the base radius.

#### *Axis endpoint*

This option sets the height as well as the orientation for *Pyramid*. The height is equal to the distance between the center of the base and the axis endpoint selected. For example, Figure 35-23 shows a *Pyramid* similar to that in the previous figure but with the *Axis endpoint* selected at 4,0. This option is the same as for a *Cone* or *Cylinder*.

### *Top radius*

Specifying a value for the *Top radius* creates a truncated pyramid such that the top surface is parallel with the base. Figure 35-24 displays a *Pyramid* with a base radius (inscribed) of 2, a *Top radius* of 1, and a height of 3.

FIGURE 35-24

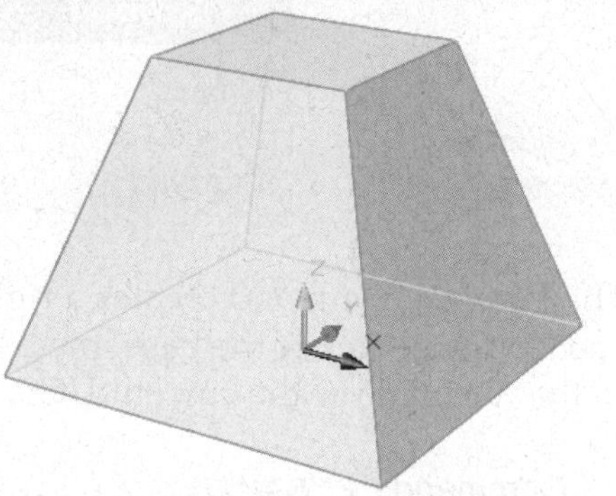

## *Polysolid*

| Ribbon | Menu Bar | Command (Type) | Alias (Type) | Shortcut |
|---|---|---|---|---|
| *Home Modeling* (icon) | *Draw Modeling > Polysolid* | *Polysolid* | *PSOLID* | ... |

The *Polysolid* command can be used to draw a solid. *Polysolid* can also be used to convert a *Line*, 2D *Pline*, *Arc*, or *Circle* to a solid. Drawing a solid with *Polysolid* is similar to drawing a *Polyline*. However, a *Polysolid* is a rectangular shape that has a prescribed height as well as width. You interactively select points to determine the lengths of the sections as you draw. A *Polysolid* can have curved segments, but the profile (cross section) is always rectangular by default. For example, Figure 35-25 displays a *Polysolid* created by selecting five points for the segment endpoints and using the default *Height* and *Width* settings. See the following prompts.

FIGURE 35-25

```
Command: POLYSOLID
Specify start point or [Object/Height/Width/Justify] <Object>: PICK
Specify next point or [Arc/Undo]: PICK
Specify next point or [Arc/Undo]: PICK
Specify next point or [Arc/Close/Undo]: PICK
Specify next point or [Arc/Close/Undo]: Enter
Command:
```

As you can see, *Polysolid* is an excellent command for drawing walls—having a prescribed height and width (rectangular cross section) but having different lengths that you PICK.

### *Height/Width*

Use these options to set the height and width for subsequent *Polysolids* that you draw or convert.

Command: ***POLYSOLID***
Specify start point or [Object/Height/Width/Justify] <Object>: ***h***
Specify height <4.0000>: (**value**)

Alternately, use the *PSOLHEIGHT* and *PSOLWIDTH* system variables to set the default height and width for subsequent solids. The settings are not retroactive for previously drawn or converted segments.

### *Object*

With the *Polysolid* command, you can use the *Object* option to convert an existing object into a *Polysolid*. An existing *Line*, 2D *Pline*, *Arc*, or *Circle* can be converted to a *Polysolid*. When the 2D object is converted, it assumes the currently specified height and width.

Specify start point or [Object/Height/Width/Justify] <Object>: ***o***
Select object: **PICK**

### *Justify*

Use this option to set the solid to be generated on the left, right, or center of the points you PICK when defining the segment endpoints. The justification is based on the direction from the first to the second point you PICK when drawing the segments.

Command: ***POLYSOLID***
Specify start point or [Object/Height/Width/Justify] <Object>: ***j***
Enter justification [Left/Center/Right] <Center>:

### Drawing Arcs

You can draw arc segments as you create a *Polysolid* using the *Arc* option when prompted for the "next point." The *Arc* option adds an arc segment to the solid such that the default starting direction of the arc is tangent to the last drawn segment as shown in Figure 35-26. The process of drawing arcs and the command line options are similar to drawing a *Pline*.

**FIGURE 35-26**

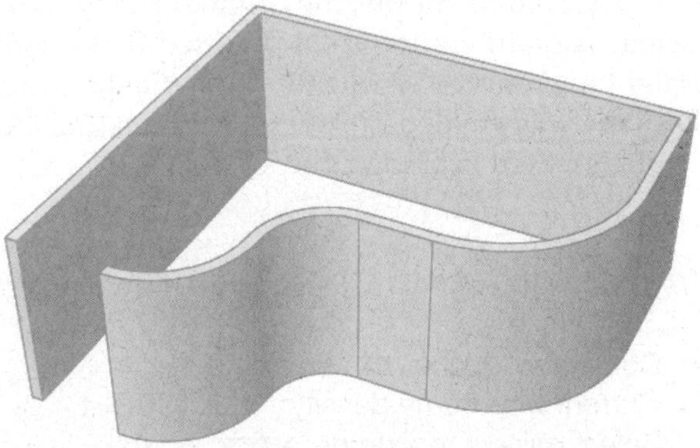

Specify the next point or [Arc/Close/Undo]: ***a***
Specify endpoint of arc or [Close/Direction/Line/Second point/Undo]:

For a full explanation of these options, see "*Pline*" in Chapter 8.

### Extrude

| Ribbon | Menu Bar | Command (Type) | Alias (Type) | Shortcut |
|---|---|---|---|---|
| *Home Modeling Extrude* | *Draw Modeling > Extrude* | *Extrude* | *EXT* | ... |

*Extrude* means to add a third dimension (height) to an existing 2D shape. For example, if you *Extrude* a *Circle*, the resulting shape is a cylinder. If you *Extrude* a *Line*, the resulting shape is a planar surface. Therefore, *Extrude* can create a surface or a solid, depending on the 2D shape used. If you *Extrude* an open 2D shape (one having endpoints), such as a *Line*, *Arc*, or open *Pline*, a surface is created. However, if you *Extrude* an existing closed 2D shape such as a *Circle, Polygon, Rectangle, Ellipse, Pline, Spline, Region*, or planar surface, the resulting extrusion is a solid. For example, Figure 35-27 displays a closed *Pline* (before) to create a solid (after) using *Extrude*. Extruding solids is discussed in this section.

**FIGURE 35-27**

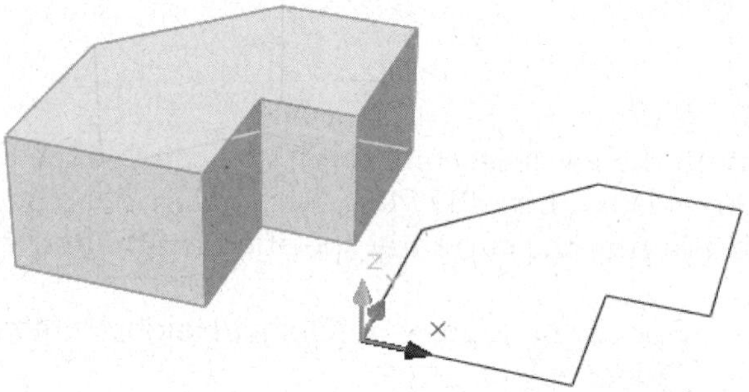

*Extrude* can simplify the creation of many solids that may otherwise take much more time and effort using typical primitives and Boolean operations. The versatility of this command lies in the fact that any closed shape that can be created by or converted to a *Pline, Spline, Region*, surface, etc., no matter how complex, can be transformed into a solid by *Extrude*. The closed 2D shape cannot be self-intersecting (crossing over itself). For example, the original shape used for the extrusion in Figure 35-28 (a closed *Pline*) contains straight and curved segments. Note that the original 2D shape was created on a vertical plane (the UCS XY plane), then extruded perpendicularly.

**FIGURE 35-28**

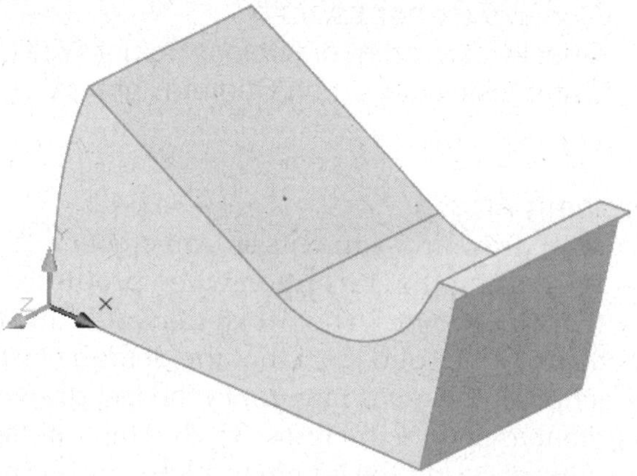

First, create the closed 2D shape, then invoke *Extrude*.

```
Command: EXTRUDE
Current wire frame density: ISOLINES=4
Select objects to extrude: PICK
Select objects to extrude: Enter
Specify height of extrusion or [Direction/Path/Taper angle]: PICK or (value)
Command:
```

By default the 2D shape is consumed (deleted) by the *Extrude* command. The *DELOBJ* system variable controls if the 2D shape is deleted by *Extrude*.

### *Taper angle*

A taper angle can be specified for the extrusion. Entering a positive taper angle causes the sides of the extrusion to slope inward (Fig. 35-29, right), while a negative taper angle causes the sides to slope outward (Fig. 35-29, left). This is helpful, for example, for developing molded parts that require a slight draft angle to facilitate easy removal from the mold.

**FIGURE 35-29**

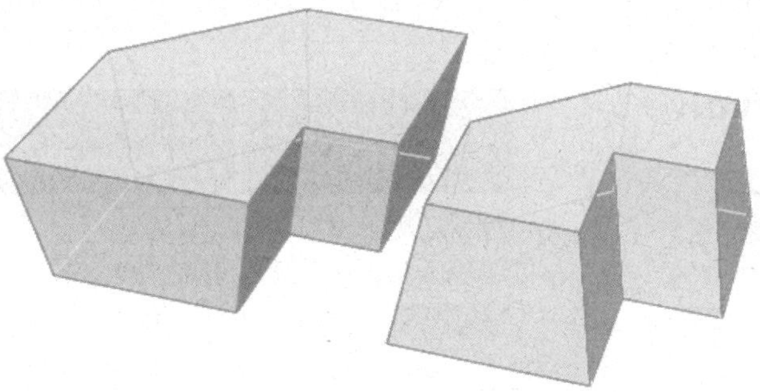

### *Direction*

Normally, with the default method of *Extrude*, the selected 2D shape is extruded perpendicular to the plane of the shape regardless of the current UCS orientation, as shown in the previous three figures. However, you can use the *Direction* option to specify an extrusion direction other than perpendicular. Select two points to determine the direction.

**FIGURE 35-30**

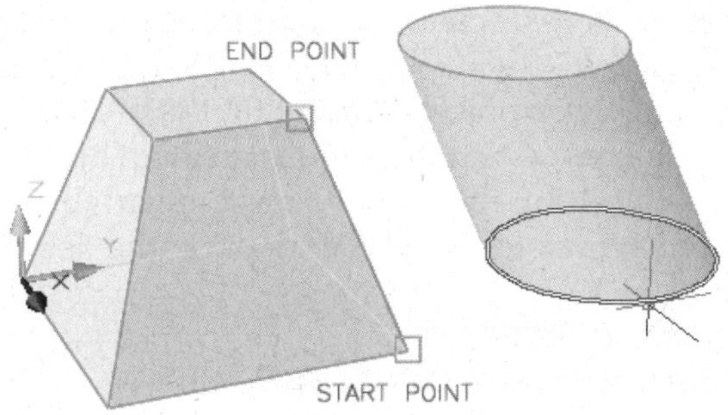


Specify height of extrusion or [Direction/Path/
  Taper angle]: ***d***
Specify start point of direction: **PICK** or (**coordinates**)
Specify end point of direction: **PICK** or (**coordinates**)

The two points of direction can be anywhere in 3D space and can be determined by selecting points on existing objects. The two points only indicate direction and do not change the orientation or position of the original 2D shape. For example, Figure 35-30 extrudes a *Circle* using the *Direction* option to *Object Snap* to the *Endpoints* along the edge of a *Pyramid*.

### *Path*

The *Path* option allows you to sweep the 2D shape called a "profile" along an existing line or curve called a "path." The path can be composed of a *Line, Arc, Ellipse, Pline, Spline,* or *Helix* (or can be different shapes converted to a *Pline*). The path does not have to be in a plane (2D) but can be in 3D space, such as in the form of a helix.

**FIGURE 35-31**

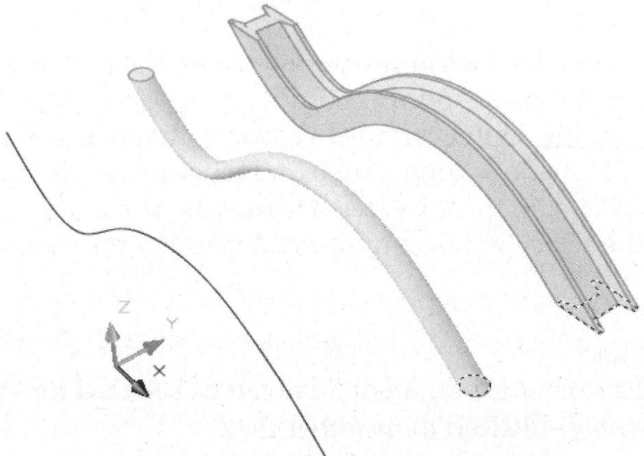

Specify height of extrusion or [Direction/Path/
  Taper angle]: ***p***
Select extrusion path or [Taper angle]: **PICK**

Figure 35-31 shows a *Circle* and a closed *Pline* (on the XY plane) extruded along a *Spline* path (that lies in the XZ plane). In this case, the same *Spline* path was selected for each extrusion. The path temporarily moves to the location of the profile for the extrusion. The 2D shapes that are extruded (shown highlighted) are consumed by default, but the path is not deleted. Notice how the original 2D shapes that were extruded change orientation as they extrude along the path.

The *Sweep* command can also be used to create solids similar to *Extrude* with the *Path* option. See "*Sweep*."

### *Revolve*

| Ribbon | Menu Bar | Command (Type) | Alias (Type) | Shortcut |
|---|---|---|---|---|
| *Home* *Modeling* *Revolve* | *Draw* *Modeling >* *Revolve* | *Revolve* | *REV* | ... |

*Revolve* allows you to revolve an open or closed 2D shape (profile) about a selected axis. Like *Extrude*, the resulting shape can be a surface or solid depending on the original shape used. *Revolve* creates a solid if the original profile used is "closed," such as a *Pline, Polygon, Circle, Ellipse, Spline*, or a *Region* object. By default, the profile is consumed (deleted) when the solid is created, but it can be retained by setting the *DELOBJ* system variable to 0. The command syntax for *Revolve* (accepting the defaults) is:

FIGURE 35-32

Command: ***REVOLVE***
Current wire frame density: ISOLINES=4
Select objects to revolve: **PICK**
Select objects to revolve: **Enter**
Specify axis start point or define axis by [Object/X/Y/Z] <Object>: **PICK**
Specify axis endpoint: **PICK**
Specify angle of revolution or [STart angle] <360>: **Enter** or (**value**)
Command:

Using the default option of *Revolve*, you must specify two points to define the axis to revolve about. You can select two points on the 2D object as the axis. The profile in Figure 35-32 (right) was used to create the 3D object (left) by selecting two points on the vertical edge.

To define the axis of revolution using the default option or the *Object* option, you can select points on other objects or specify coordinates. The profile is always revolved about an axis in the same plane as the profile. However, the two points that define the axis for revolving do not have to be coplanar with the 2D shape as long as one of the points lies in the same plane and the axis is not perpendicular to the 2D shape. In this case, *Revolve* uses an axis in the plane of the revolved shape aligned with the direction of the selected points or object. The other options for selecting an axis of revolution are described next.

#### *Object*

A *Line* or single segment *Pline* can be selected for an axis. The positive axis direction is from the closest endpoint PICKed to the other end.

#### *X/Y/Z*

Uses the positive X, Y, or Z axis of the current UCS as the axis of rotation. However, keep in mind that an axis perpendicular to the plane of the profile cannot be used.

### *Start angle*

After defining the axis of revolution, *Revolve* requests the number of degrees for the object to be revolved. Any angle can be entered. For example, one possibility for revolving a profile is shown in Figure 35-33 where the profile is generated through 270 degrees.

FIGURE 35-33

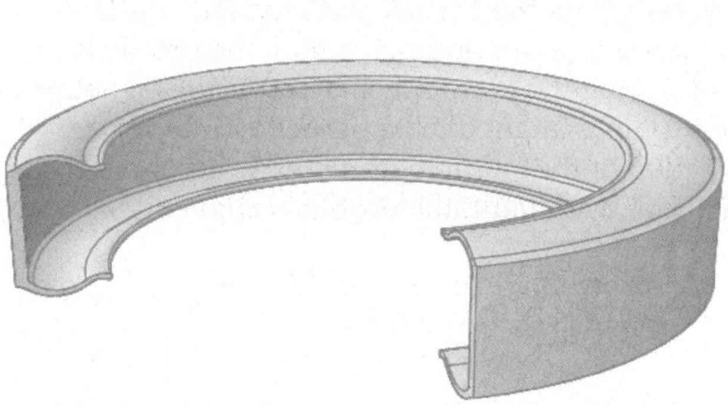

By default, the object is revolved from its current plane in a counter-clockwise direction. If you revolve the object less than 360 degrees, you can use the *Start angle* option to specify an angular value from the current 2D shape position to begin the *Revolve*.

## Sweep

| Ribbon | Menu Bar | Command (Type) | Alias (Type) | Shortcut |
|---|---|---|---|---|
| *Home Modeling Sweep* | *Draw Modeling > Sweep* | *Sweep* | ... | ... |

The *Sweep* command allows you to create a new solid or surface by sweeping an open or closed planar curve (profile) along an open or closed 2D or 3D path. *Sweep* is similar to using *Extrude* with the *Path* option; however, *Sweep* has more versatility. Like *Extrude* and *Revolve*, you create a solid from a "closed" 2D shape, and you create a surface from an "open" 2D shape. To *Sweep*, simply select an existing profile and a path.

Command: ***SWEEP***
Current wire frame density: ISOLINES=4
Select objects to sweep: **PICK**
Select objects to sweep: **Enter**
Select sweep path or [Alignment/Base point/Scale/Twist]: **PICK**
Command:

With *Sweep*, you can use a *Circle*, *Ellipse*, *Region*, closed *Pline*, or a surface as the profile to create a swept solid. The path can be any one of the following: *Line*, *Arc*, *Pline*, *Circle*, *Ellipse*, *Spline*, *Helix*, or edges of solids. The path can be closed and does not have to be planar. By default settings, the profile is deleted when swept, but the path is not deleted. The *DELOBJ* system variable setting determines if the profile, path, neither, or both are deleted.

The *Sweep* command automatically moves the profile into position at the end of the path. By default, the orientation of the profile is adjusted to sweep in a direction perpendicular to the path along its entire length. For example, Figure 35-34 displays the before and after stages of sweeping a *Circle* profile along a *Helix* path. Note that the *Circle* was created in the same plane as the base of the *Helix* but was automatically moved into position and realigned perpendicularly to the path as it was swept.

FIGURE 35-34

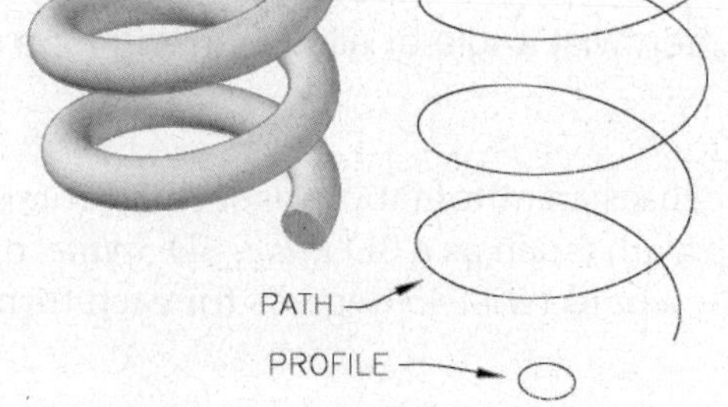

*Alignment*

Different from the *Extrude* with the *Path* option, *Sweep* provides an option for how the profile is swept along the path. *Alignment* specifies whether the profile is aligned to be perpendicular to the direction of the sweep path or left in its current orientation. By default, the profile is aligned.

FIGURE 35-35

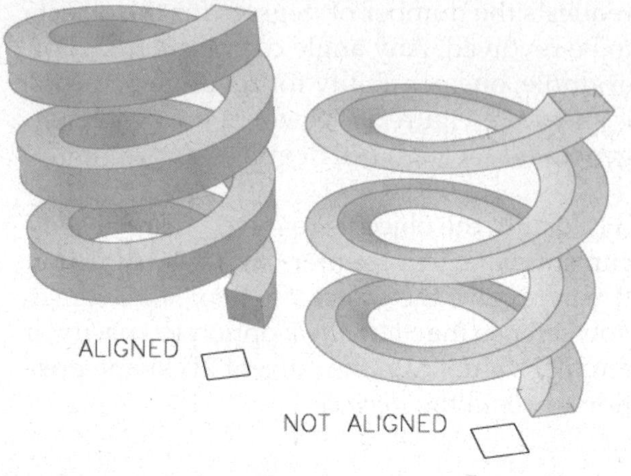


Align sweep object perpendicular to path before sweep [Yes/No] <Yes>:

If you answer *No* to this option, the profile is left in its original orientation with respect to the first point along the sweep path. From that point, the profile is realigned along the length of the path to keep the same orientation with respect to the path as the path turns. For example, Figure 35-35 displays a comparison between the two options for *Alignment*. Notice the "flat" appearance of the helix on the left (not aligned), yet the profile turns along the length of the path.

*Base Point*

By default, the center of the profile is used as the base point when it is swept along the path. For example, in Figure 35-34, the center of the *Circle* becomes the base point. Use the *Base point* option to specify a new base point for the objects to be swept. If the specified point does not lie on the plane of the selected objects, it is projected onto the plane.

*Scale*

You can use this option to scale the profile for a sweep operation. Applying a scale factor causes the profile to be tapered uniformly as it is swept from the start to the end of the sweep path (Fig. 35-36, right). You can specify a scale factor or enter *R* for the reference option.

FIGURE 35-36

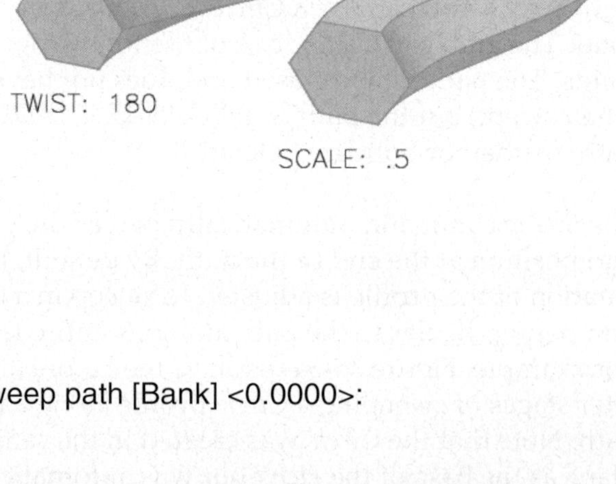


Enter scale factor or [Reference] <1.0000>:

The *Reference* option of *Scale* allows you to specify the scale by picking two points or entering values.

*Twist*

Using this option, you can twist the profile as it is swept along the length of the path. You specify a value for the twist angle. The twist angle specifies the amount of rotation along the entire length of the sweep path (Fig. 35-36, left). Specify an angle value less than 360.

Enter twist angle or allow banking for a non-planar sweep path [Bank] <0.0000>:

*Bank*

*Bank* causes an automatic, unspecified rotation (twist) of the profile being swept along a non-planar sweep path (such as a 3D *Pline*, 3D *Spline*, or *Helix* sweep path). For example, using the *Bank* option, a profile would twist 45 degrees for each turn of a *Helix*.

### *Loft*

| Ribbon | Menu Bar | Command (Type) | Alias (Type) | Shortcut |
|---|---|---|---|---|
| *Home Modeling Loft* | *Draw Modeling > Loft* | *Loft* | ... | ... |

With the *Loft* command, you can create a new solid or surface by specifying a series of cross sections. Similar to *Extrude*, *Revolve*, and *Sweep*, the resulting shape (surface or solid) depends on if the 2D shape(s) are "open" (having endpoints) or "closed" (one connected loop). *Loft* creates a solid when the cross-section shapes are closed.

Command: **LOFT**
Current wire frame density: ISOLINES=4, Closed profiles creation mode = Solid
Select cross sections in lofting order or [POint/Join multiple edges/MOde]: **PICK**
Select cross sections in lofting order or [POint/Join multiple edges/MOde]: **PICK**
Select cross sections in lofting order or [POint/Join multiple edges/MOde]: **PICK**
Select cross sections in lofting order or [POint/Join multiple edges/MOde]: **Enter**
Enter an option [Guides/Path/Cross sections only/Settings] <Cross sections only>: **Enter**
Command:

The cross sections define the shape of the resulting solid. You must specify at least two cross sections when you use the *Loft* command. *Loft* then draws a solid in the space between (connecting) the closed cross sections. *Loft* will "morph" between the shapes to create a smooth transition. For example, Figure 35-37 displays a lofted solid between three cross sections: a square *Pline* at the base, a *Circle* in the middle, and a 6-sided *Polygon* at the top. The solid was created using the above prompts.

**FIGURE 35-37**

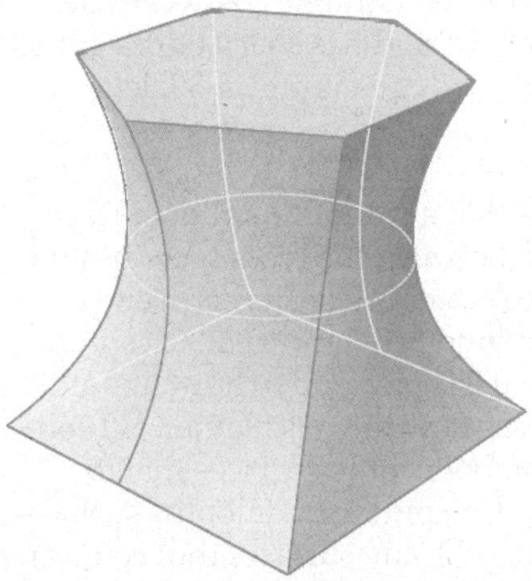

Closed cross sections that can be used with *Loft* to create solids are *Circle*, closed 2D *Pline*, closed 2D *Spline*, and *Ellipse*. (Keep in mind that *Plines* include *Rectangles* and *Polygons*.) The *DELOBJ* system variable setting determines whether the cross sections, guides, and path are automatically deleted when the solid is created.

#### *MOde*

This option determines whether the resulting lofted object is a *Surface* or *Solid*.

#### *POint*

Instead of using a closed cross section, you can loft to a point. The selected point can be the start or end point of the loft. Use the *POint* option immediately before selecting any point as the start or end shape.

#### *Join multiple edges*

This option allows you to select multiple connected (end-to-end) edges of a solid or surface to operate as a closed 2D cross section.

### *Settings*

After you have selected cross sections, you can use the *Settings* option to produce the *Loft Settings* dialog box (Fig. 35-38). Here you can select from several methods for controlling the contour of the lofted shape at its cross sections.

**FIGURE 35-38**

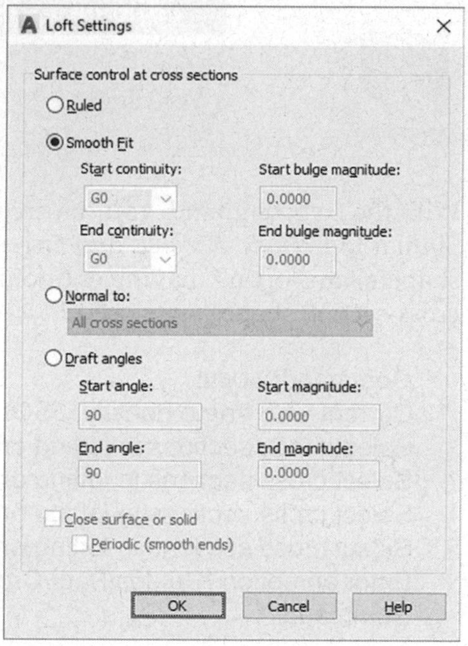


**FIGURE 35-39**

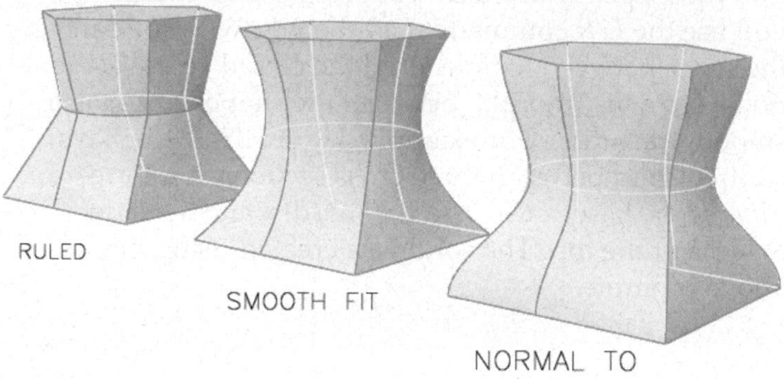


*Ruled*

This option specifies that the solid or surface is ruled (straight) between the cross sections and has sharp edges at the cross sections. See Figure 35-39, left.

*Smooth Fit*

Specifies that a smooth solid or surface is drawn between the cross sections and has sharp edges at the start and end cross sections. See Figure 35-39, center. *Continuity* is a measure of how smoothly two curves flow into each other. G0 continuity smooths only surfaces that are collinear. G1 continuity smooths surfaces where the end tangents match at the common edges. G2 continuity smooths any surfaces that connect. *Bulge magnitude* is a measure of how much a surface curves or "bulges" as it flows into another surface. Magnitude can be between 0 and 1 where 0 is flat and 1 curves the most. You can also modify the continuity and bulge magnitude with special grips on the lofted shape.

*Normal to*

Controls the surface normal (perpendicular) of the solid where it passes through the cross sections. You can determine that surfaces are normal to *All cross sections* or specific cross-sections, such as *Start Cross Section*, *End Cross Section*, and *Start and End Cross Sections*. See Figure 35-39, right.

*Draft Angles*

This section controls the draft angle and magnitude of the first and last cross sections of the lofted solid or surface. The draft angle is the beginning direction of the surface. The 90 (degree) setting is defined as 90 degrees from the cross section and 0 is defined as outward from the plane of the curve. The *Magnitude* controls the relative distance of the surface from the start cross section in the direction of the draft angle before the surface starts to bend toward the next cross section.

Figure 35-40 displays two lofted solids, each created from the same two cross sections. As you can see, the left solid's surfaces protrude outward applying the angle of greater than 90 for both the start and end profiles (120 in this case), whereas the solid on the right curves inward with angles of less than 90 (60 in this case).

**FIGURE 35-40**

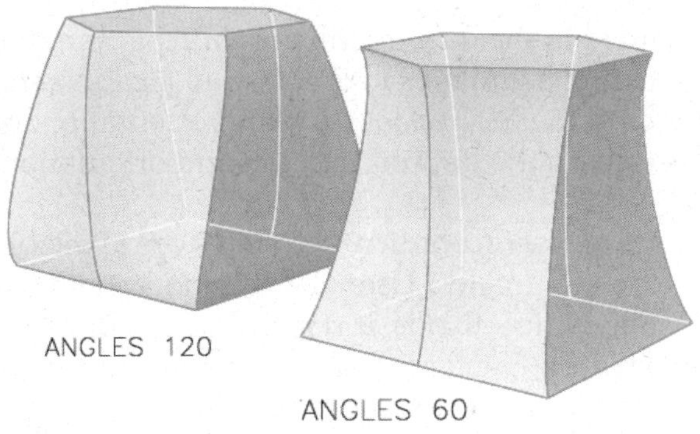


*Guides*

With the *Guides* option, you can select multiple curves to define the contours of the solid or surface. Guide curves are lines or curves that define the form of the solid by adding wireframe information to the object. Objects that can be used as guides are *Splines* (2D or 3D) and *Plines* (2D or 3D). You can use guide curves to control exactly how points are matched up on corresponding cross sections to produce predictable results. Figure 35-41 displays a shape that might be created with guides that might be difficult to attain through other *Loft* options.

**FIGURE 35-41**

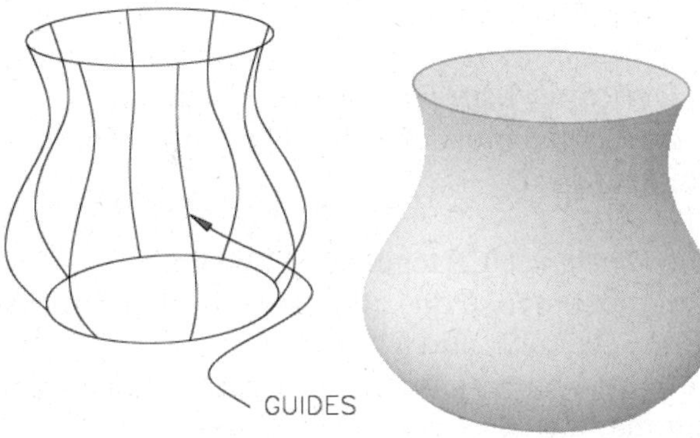


In order for guides to work correctly, each guide curve must intersect each cross section, start on the first cross section, and end on the last cross section. You can select any number of guide curves for the lofted surface or solid.

*Path*

With the *Path* option, you can select a single path curve to define the shape of the solid. This method is similar to *Sweep*, except here you use multiple profiles (cross sections) to determine the shape of the solid along the path. The path curve must intersect the planes of all the profiles. Objects that can be used as a *Loft Path* include *Spline* (2D or 3D), *Helix*, *Circle*, *Ellipse*, *Pline* (2D or 3D).

For example, the solid in Figure 35-42 was created using the *Path* option of *Loft*. Note that the path curve is a *Spline*. Also note that although the profiles are similar, they are of varying sizes and orientations.

**FIGURE 35-42**

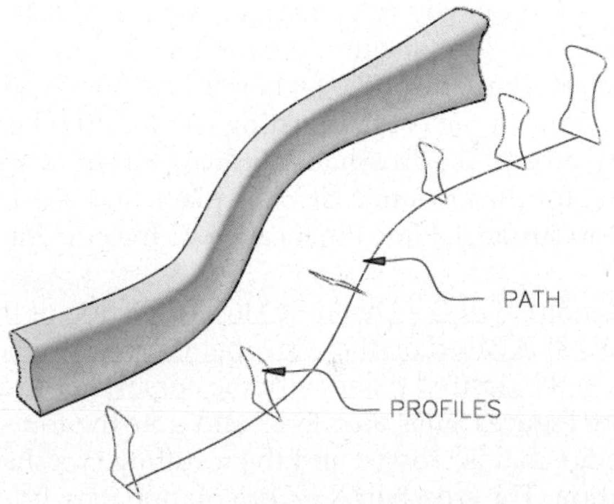

## DYNAMIC USER COORDINATE SYSTEMS

As explained in the introduction of this chapter, the process of solid modeling construction involves either creating primitives in the desired location and orientation or moving, rotating, or copying the primitives to the desired location and orientation. Normally, the desired approach is to create the primitives directly in the desired location and orientation. To do this, User Coordinate Systems must be used.

Two kinds of User Coordinate Systems are available during the solid modeling construction process. Firstly, you can create a User Coordinate System using the *UCS* command, as you learned in Chapter 34. This UCS is static—that is, it remains active until you change to another coordinate system. Secondly, you can create a Dynamic User Coordinate System (DUCS). A Dynamic UCS is a temporary UCS that remains active only during a solid primitive creation command such as *Box* or *Wedge*, then it disappears.

Use the *Dynamic UCS* button on the Status bar to toggle this feature on or off (Fig. 35-43). If *Dynamic UCS* is toggled on, a Dynamic User Coordinate System appears automatically during primitives commands.

**FIGURE 35-43**

### Using Dynamic User Coordinate Systems

Three conditions must exist for a Dynamic Coordinate System to appear:

1. The *Dynamic UCS* toggle must be on.
2. You must invoke one of these primitive commands: *Box*, *Wedge*, *Cone*, *Sphere*, *Cylinder*, *Pyramid*, *Torus*, *Polysolid*, and *Mesh*.
3. You must hover your cursor over a planar surface of an existing solid during the first prompt of the command.

### Dynamic Coordinate System Example

With the *Dynamic UCS* feature, you can temporarily align the XY plane of the UCS with a plane on a solid model. For example, assume you wanted to create a *Wedge* on the top surface of an existing *Box* and orient the *Wedge* so it sloped down toward the left vertical face. When you use the *Wedge* command, you are prompted for the "first corner." With *Dynamic UCS* on, you can create a temporary UCS on the top face of the *Box* to begin drawing the *Wedge*. You specify the desired XY plane of the Dynamic UCS by "hovering" over any planar face of the existing solid until it becomes highlighted and the crosshair labels appear indicating the direction for the Dynamic UCS XY plane and Z axes (Fig. 35-44, top surface). Note the location of the current UCS.

**FIGURE 35-44**

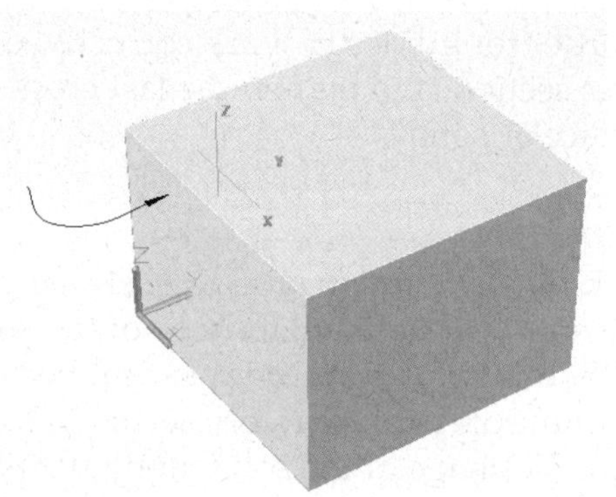

**TIP**

The orientation of the Dynamic UCS (direction of the X, Y, and Z axes) is based on the edge that is crossed when you approach the desired plane with the cursor. For example, compare Figures 35-44 and 35-45 and note the top surface's edge that is crossed and the resulting crosshair XYZ orientation. The crosshair XYZ orientation may be critical for creating some primitives such as a *Wedge*. Once the *Dynamic UCS* icon appears in the desired orientation, PICK to specify the "first corner" for the *Box*.

**FIGURE 35-45**

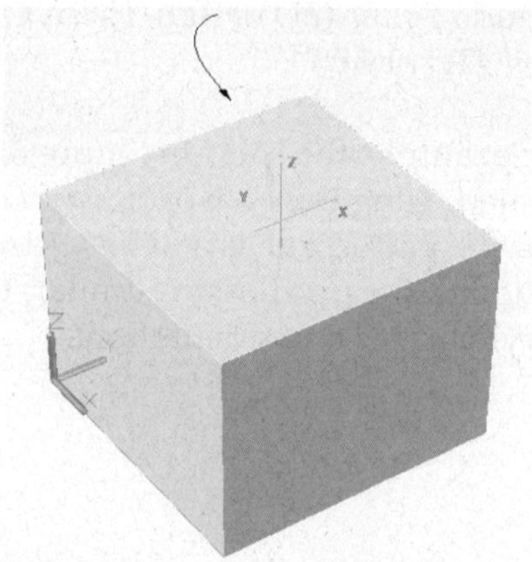

In Figure 35-46, the *Midpoint* of the right edge is selected as the "first corner." Note that once the first corner is established, the UCS icon appears in that location and orientation. Next, the *Endpoint* at the far left corner is selected as the "other corner" for the base of the *Wedge*.

**FIGURE 35-46**

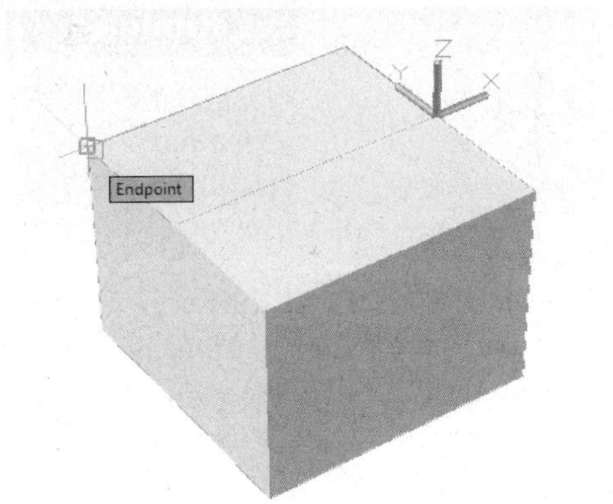


Proceeding with the *Wedge* command, the "height" is specified next to complete the primitive. Figure 35-47 displays the wedge while the height is being PICKed. Note the UCS icon still located in the selected "first corner" location.

NOTE: Notice in Figure 35-47 that the *Wedge* slopes down toward the –X axis. Remember that a *Wedge* always slopes down in the direction of the second corner.

Once the primitive command is complete, the *Dynamic UCS* disappears. The normal (non-dynamic) UCS becomes active again and the UCS icon returns to its previous location and orientation (see Fig. 35-44).

**FIGURE 35-47**

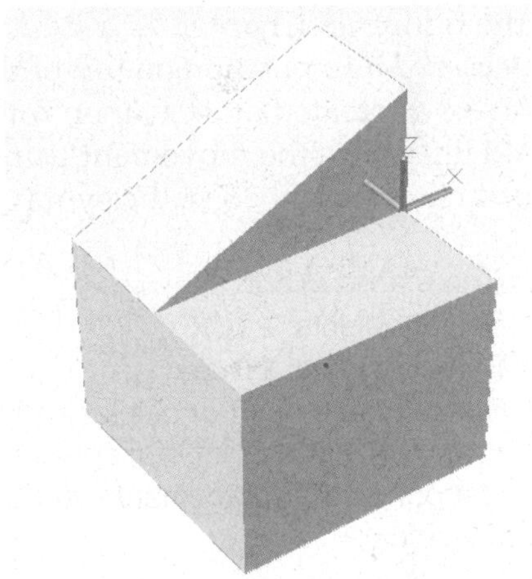

### Dynamic UCS Principles

Several principles of Dynamic UCSs are followed:

TIP

A. The X axis of the Dynamic UCS is always aligned with an edge of the face and the positive direction of the X axis always points toward the right half of the screen.
B. Only the front faces of a solid are detected by the Dynamic UCS .
C. If *Snap* mode is turned on, it aligns temporarily to the Dynamic UCS .
D. You can temporarily turn off the Dynamic UCS by pressing F6 or Shift+Z while moving the crosshairs over a face.

## COMMANDS TO MOVE, ROTATE, MIRROR, AND SCALE SOLIDS

When you create the desired primitives, Boolean operations are used to construct the composite solids. However, the primitives must be in the correct position and orientation with respect to each other before Boolean operations can be performed. You can either create the primitives in the desired position during construction (by using UCSs) or move the primitives into position after their creation. Several methods that allow you to move solid primitives are explained in this section.

## Move

| Ribbon | Menu Bar | Command (Type) | Alias (Type) | Shortcut |
|---|---|---|---|---|
| *Home* *Modify* (icon) | *Modify* *Move* | *Move* | *M* | (Edit Mode) *Move* |

The *Move* command that you use for moving 2D objects in 2D drawings can also be used to move 3D primitives. Generally, *Move* is used to change the position of an object in one plane (translation), which is typical of 2D drawings. *Move* can also be used to move an ACIS primitive in 3D space if *Object Snaps* or 3D coordinates are used.

*Move* operates in 3D just as you used it in 2D. Previously, you used *Move* only for repositioning objects in the XY plane, so it was only necessary to PICK or use X and Y coordinates. Using *Move* in 3D space requires entering X, Y, and Z values or using *Object Snaps*.

For example, to create the composite solid used in the figures in Chapter 34, a *Wedge* primitive was *Moved* into position on top of the *Box*. The *Wedge* was created at 0,0,0, then rotated. Figure 35-48 illustrates the movement using absolute coordinates described in the syntax below:

Command: ***MOVE***
Select objects: **PICK**
Select objects: **Enter**
Specify base point or displacement: **0,0**
Specify second point of displacement or <use first point as displacement>: **0,4,3**
Command:

**FIGURE 35-48**

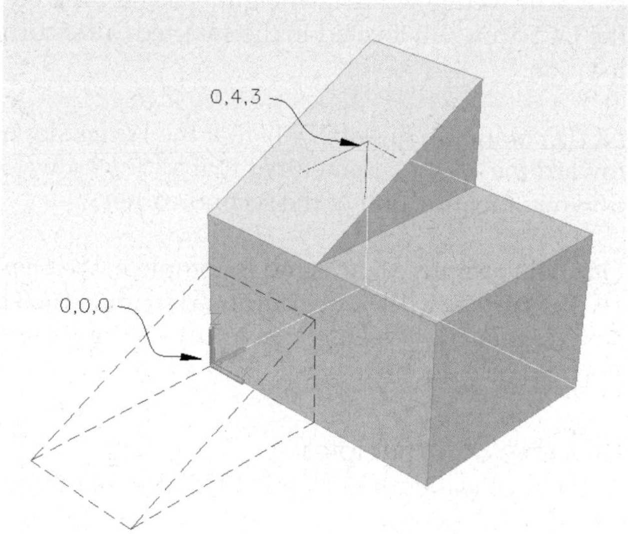


Alternately, you can use *Object Snaps* to select geometry in 3D space. Figure 35-49 illustrates the same *Move* operation using *Endpoint Object Snaps* instead of entering coordinates.

**FIGURE 35-49**

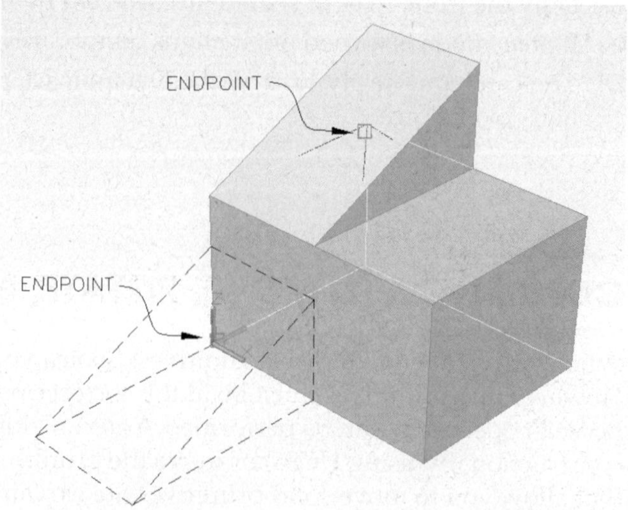

### *3Dalign*

| Ribbon | Menu Bar | Command (Type) | Alias (Type) | Shortcut |
|---|---|---|---|---|
| *Home Modify* (icon) | *Modify 3D Operations > 3D Align* | *3Dalign* | *3AL* | ... |

*3Dalign* provides a means of aligning one 3D solid with another. *3Dalign* is very powerful because it automatically performs 3D translation and rotation if needed. All you have to do is select the points on two 3D objects that you want to align (connect).

Aligning one solid with another can be accomplished by connecting source points (on the solid to be moved) to destination points (on the stationary solid). Alternately, you can short cut this procedure by indicating a source plane or destination plane instead of selecting 3 points. You should always use *Object Snap* modes to select the source and destination points to assure accurate alignment.

*3Dalign* performs a translation (like *Move*) and two rotations (like *Rotate*), each in separate planes to align the points as designated. The motion performed by *3Dalign* is actually done in three steps. For example, assume you wanted to move the *Wedge* on top of the *Box* and also reorient the *Wedge*. The command syntax for 3D alignment is as follows:

Command: ***3DALIGN***
Select objects: **PICK**
Select objects: **Enter**
Specify source plane and orientation ...
Specify base point or [Copy]: **PICK** (S1)
Specify second point or [Continue] <C>: **PICK** (S2)
Specify third point or [Continue] <C>: **PICK** (S3)
Specify destination plane and orientation ...
Specify first destination point: **PICK** (D1)
Specify second destination point or [eXit] <X>: **PICK** (D2)
Specify third destination point or [eXit] <X>: **PICK** (D3)

In this case, as shown in Figure 35-50, the "base point" (S1) is connected to the first destination point (D1). This first motion is a translation. The base point and the first destination point always physically touch.

Next, the vector defined by the first and second source points (S1 and S2) is aligned with the vector defined by the first and second destination points (D1 and D2). The length of the segments between the first and second points on each object is not important because AutoCAD only considers the vector direction. This second motion is a rotation along one axis.

Finally, the third set of points are aligned similarly (S2-S3 with D2-D3). This third motion is a rotation along the other axis, completing the alignment.

**FIGURE 35-50**

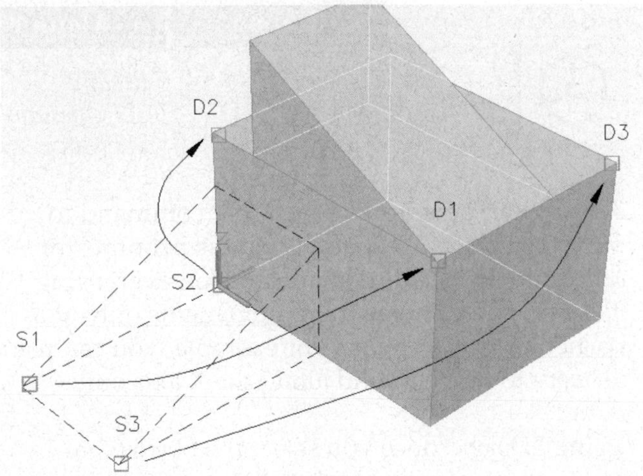

In some cases, such as when cylindrical objects are aligned, only two sets of points have to be specified. For example, if aligning a shaft with a hole, the first set of points (source and destination) specify the attachment of the base of the shaft with the bottom of the hole. The second set of points specify the alignment of the axes of the two cylindrical shapes. A third set of points is not required because the radial alignment between the two objects is not important. When only two sets of points are specified, AutoCAD asks if you want to "continue" based on only two sets of alignment points.

You can also use the "source plane" method to align the *Wedge* with the *Box* as shown in Figure 35-51. This method uses the first source point (S1), or "base point," to indicate the XY plane of the source object. Once this base point is selected, press Enter to begin specifying the destination points, as shown in the following prompts.

Command: ***3DALIGN***
Select objects: **PICK**
Select objects: **Enter**
Specify source plane and orientation ...
Specify base point or [Copy]: **PICK** (S1)
Specify second point or [Continue] <C>:
**Enter**
Specify destination plane and orientation ...
Specify first destination point: **PICK** (D1)
Specify second destination point or [eXit] <X>: **PICK** (D2)
Specify third destination point or [eXit] <X>: **Enter**

**FIGURE 35-51**

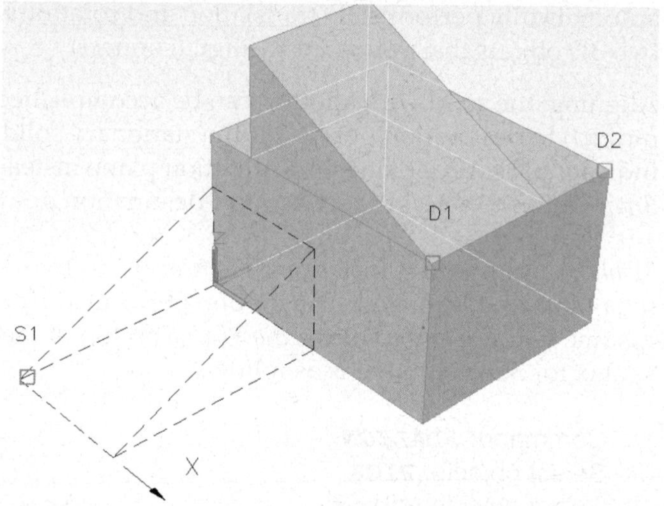


Only the first two destination points need to be specified. *3Dalign* uses the positive X direction of the source plane to make the alignment between the first two destination points (D1 and D2). Since the planar orientation of the source and destination planes is the same, no other rotation is needed, so pressing Enter completes the alignment.

## 3Dmove

| Ribbon | Menu Bar | Command (Type) | Alias (Type) | Shortcut |
|---|---|---|---|---|
| *Home Modify* (icon) | *Modify 3D Operations > 3D Move* | *3DMove* | *3M* | ... |

*3Dmove* can be used like the *Move* command to move objects in 3D space. Even the prompts for *Move* and *3Dmove* are identical. However, using *3Dmove*, you can constrain movements only to a particular axis or plane. For example, you can use *3Dmove* to move a solid along the X axis only.

Using *3Dmove*, once you select the object(s) to move, the move grip tool, or "gizmo," is displayed. Next, select the desired base point, just as you would with the *Move* command. The move gizmo locks to the selected "base point." For example, in Figure 35-52, the move gizmo appears during the "Specify base point" prompt.

**FIGURE 35-52**

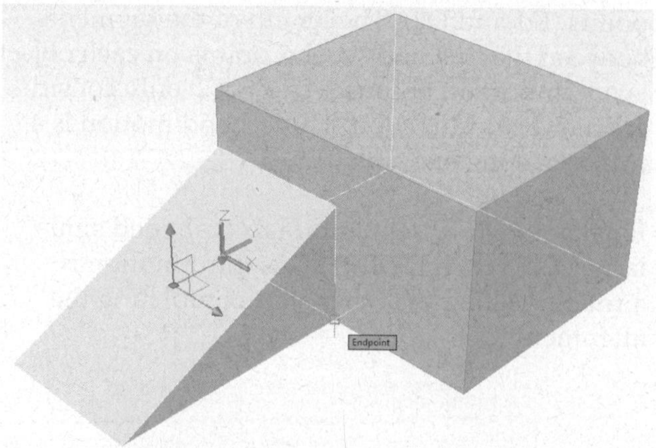

In this case, one *Endpoint* of the *Wedge* is at the base point.

Command: ***3DMOVE***
Select objects: **PICK**
Select objects: **Enter**
Specify base point or [Displacement]<Displacement>: **PICK**
Specify second point or <use first point as displacement>: **PICK**

When the move gizmo appears, select any axis to constrain the movement of the selected objects to the indicated gizmo axis. Once selected, the axis changes to a yellow-gold color. For example, selecting the X axis of the gizmo displays a vector along the axis (Fig. 35-53, left). Likewise, selecting the Y or Z axis highlights the appropriate move vector (Fig. 35-53, center and right).

**FIGURE 35-53**

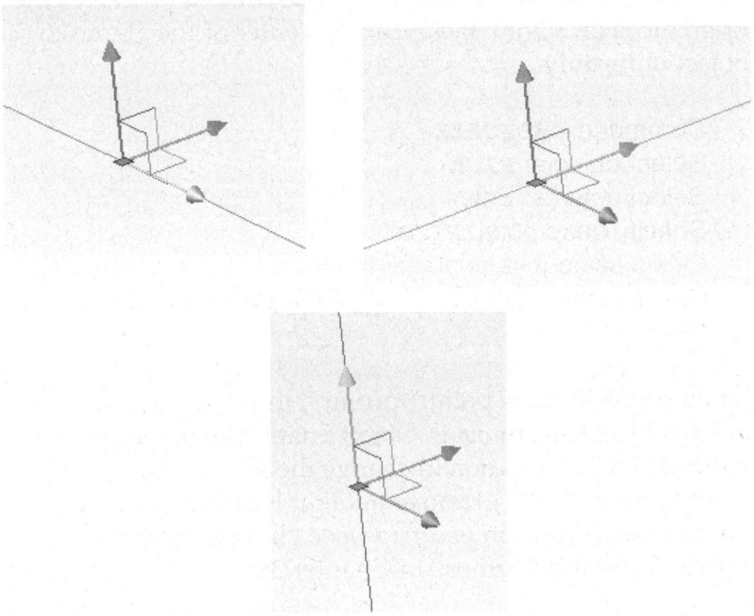

You can also constrain the movement to a plane. Do this by selecting one of the small squares on the move gizmo. For example, if you want to move the selected object within the XZ plane, select the small square connecting the X and Z axes (Fig. 35-54).

**FIGURE 35-54**

From the previous *Box* and *Wedge* example, selecting the Z axis on the move gizmo would constrain movement of the *Wedge* to a positive or negative Z direction, as shown in Figure 35-55. Next, move the *Wedge* to the desired location on the Z axis, then PICK or enter a value as the "second point." If you want to relocate the gizmo at another location or plane on the object, you can use grips (see "Editing Solids Using Grips and Gizmos" in Chapter 36).

**FIGURE 35-55**

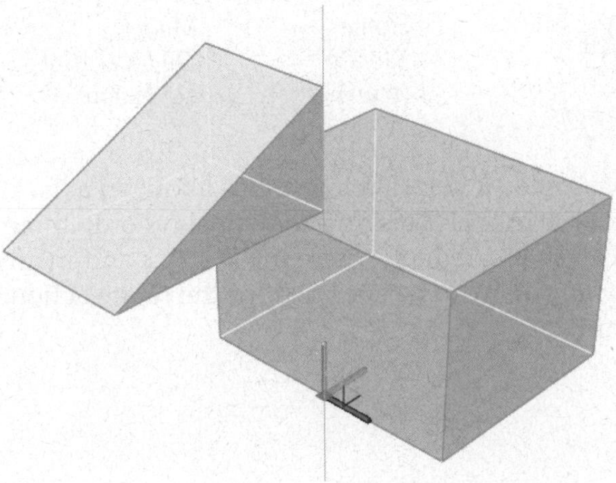

## 3Dscale

| Ribbon | Menu Bar | Command (Type) | Alias (Type) | Shortcut |
|---|---|---|---|---|
| *Home* *Modify* (icon) | ... | *3Dscale* | *3S* | ... |

The *3Dscale* command allows you to scale the selected objects uniformly based on a scale factor. The *3Dscale* command produces the scale grip tool, or gizmo (Fig. 35-56). Pick near the center of the gizmo to scale the selected object uniformly.

FIGURE 35-56

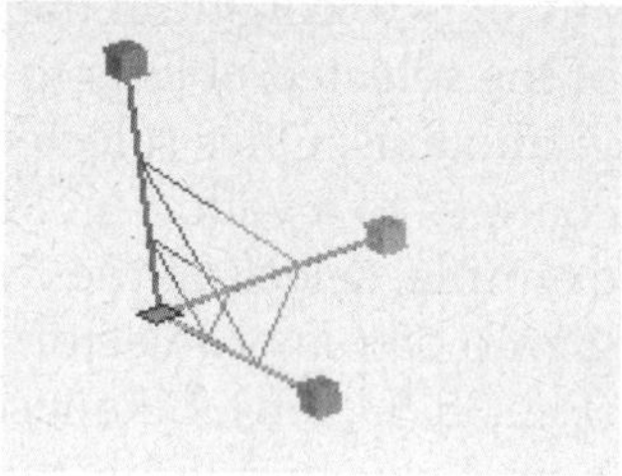

Command: ***3DSCALE***
Select objects: **PICK**
Select objects: Enter
Specify base point: **PICK**
Pick a scale axis or plane: PICK (select the center area of the gizmo)
Specify scale factor or [Copy/Reference] <1.0000>: **PICK** (or option)
Command:

FIGURE 35-57

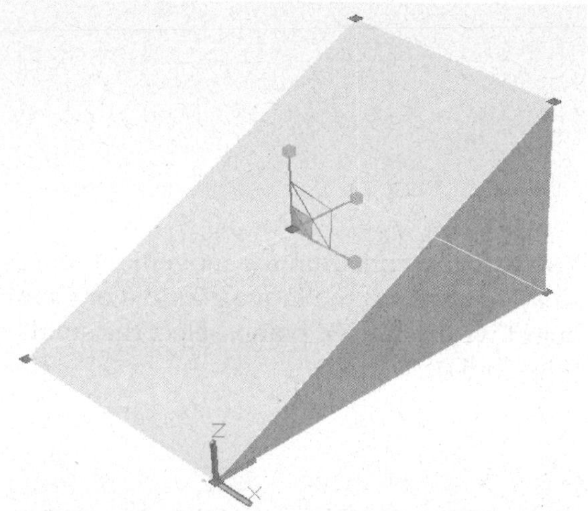

At the "Specify base point:" prompt, locate the gizmo at any object feature using Object Snaps. Enter a scale value or PICK a location to change the scale. If you want to relocate the gizmo at another location or plane on the object, you can use grips (see "Editing Solids Using Grips and Gizmos" in Chapter 36).

## 3Drotate

| Ribbon | Menu Bar | Command (Type) | Alias (Type) | Shortcut |
|---|---|---|---|---|
| *Home* *Modify* (icon) | *Modify* *3D Operations >* *3D Rotate* | *3Drotate* | *3R* | ... |

*3Drotate* allows you to rotate solids about an axis. You specify the axis to rotate about by using the *3Drotate* grip tool. This grip tool displays only three possible rotational axes. However, since the grip tool can be located anywhere in 3D space and in any orientation using the *Dynamic UCS* feature, virtually any axis can be specified for the rotate action. The grip tool appears at the "Specify base point" prompt.

Command: ***3DROTATE***
Current positive angle in UCS: ANGDIR=counterclockwise ANGBASE=0
Select objects: **PICK**
Select objects: **Enter**
Specify base point: **PICK**
Pick a rotation axis or type an angle: **PICK** (Select an axis handle to specify the rotation axis)
Specify angle start point: **PICK**
Specify angle end point: **PICK**
Command:

The rotate grip tool appears at the base point you specify. At the "Pick a rotation axis" prompt, select one of the circular "handles." Once a circular handle is selected, the related axis is displayed. For example, Figure 35-58 (left) displays selection of the X axis handle. Note the cursor location in the figure. When the handle is selected, the related axis vector remains on the screen until the rotate action is complete. Figure 35-58 (right) displays selection of the Z axis handle.

**FIGURE 35-58**

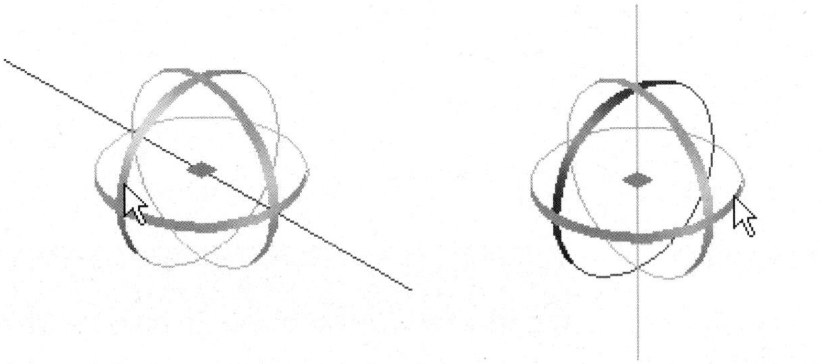

For example, suppose you wanted to rotate the wedge shown in Figure 35-59 about the Z axis at its nearest corner. Using *3Drotate*, select the wedge at the "Select objects" prompt, then select the nearest corner of the wedge at the "Specify base point" prompt to locate the *3Drotate* grip tool. At the "Pick a rotation axis" prompt, select the Z axis handle (note the cursor location).

**FIGURE 35-59**

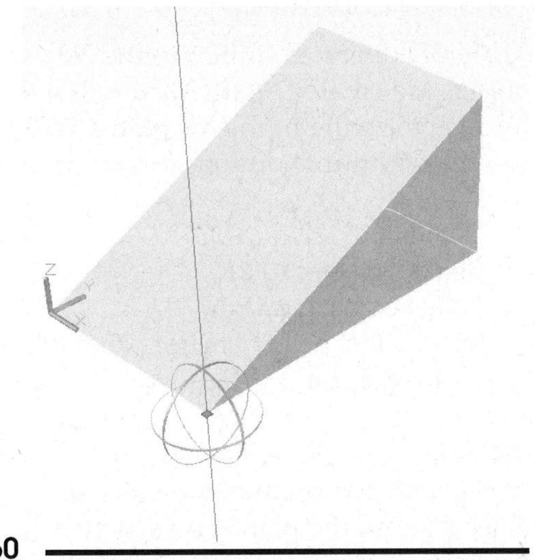

Next, you must specify an "angle start point." In our example, the *Endpoint* at the far right corner on the base is selected as the "angle start point." Finally, you can rotate the object about the selected point in real time at the "Specify angle endpoint" prompt, as shown in Figure 35-60.

**FIGURE 35-60**

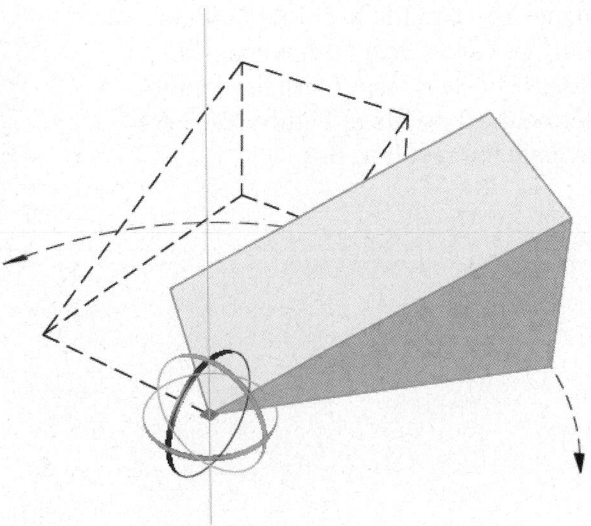

Using the *3Dmove*, *3Dscale*, or *3Drotate* commands, you are limited with respect to the manipulation of the gizmos, which in turn limit your ability to edit the object. For example, you may want to relocate the gizmo at another location or plane on the object, such as shown in Figure 35-61, where the gizmo is aligned with the inclined top face of the solid. Instead of using the *3Dmove*, *3Dscale*, or *3Drotate* commands to do this, you can use grips (see "Editing Solids Using Grips and Gizmos" in Chapter 36).

**FIGURE 35-61**

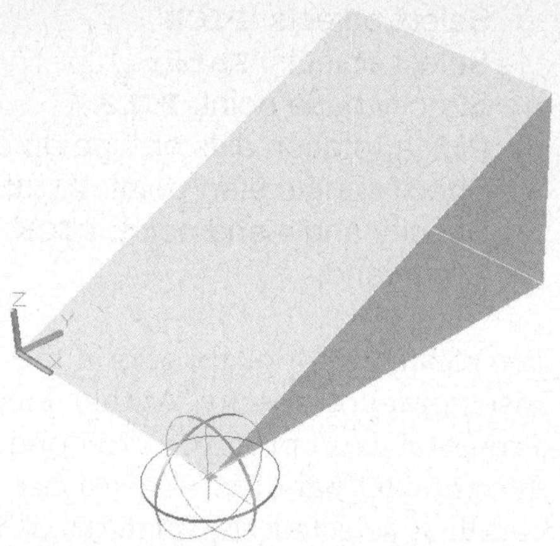

## *Mirror3D*

| Ribbon | Menu Bar | Command (Type) | Alias (Type) | Shortcut |
|---|---|---|---|---|
| *Home Modify* (icon) | *Modify 3D Operations > 3D Mirror* | *Mirror3D* | ... | ... |

*Mirror3D* operates similar to the 2D version of the command *Mirror* in that mirrored replicas of selected objects are created. With *Mirror* (2D) the selected objects are mirrored about an axis. The axis is defined by a vector lying in the XY plane. With *Mirror3D*, selected objects are mirrored about a plane. *Mirror3D* provides multiple options for specifying the plane to mirror about:

Command: ***MIRROR3D***
Select objects: **PICK**
Select objects: **Enter**
Specify first point of mirror plane (3 points) or [Object/Last/Zaxis/View/XY/YZ/ZX/3points] <3points>: **PICK** or (**option**)

### *3points*

The *3points* option mirrors selected objects about the plane you specify by selecting three points to define the plane. You can PICK points (with or without *Object Snap*) or give coordinates. *Midpoint Object Snap* is used to define the 3 points in Figure 35-62 A to achieve the result in B.

**FIGURE 35-62**

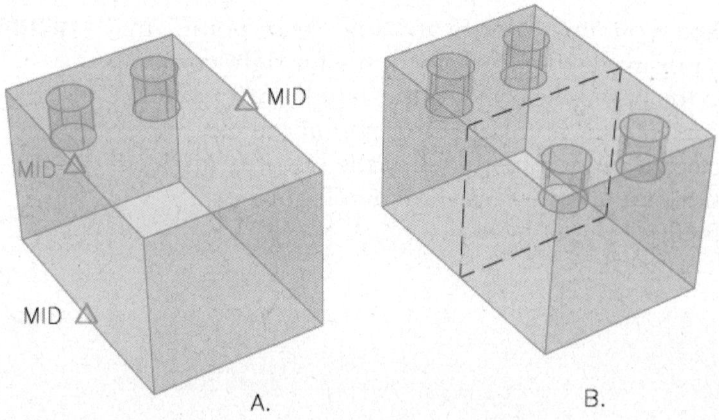

### *Object*

Using this option establishes a mirroring plane with the plane of a 2D object. Selecting an *Arc* or *Circle* automatically mirrors selected objects using the plane in which the *Arc* or *Circle* lies. The plane defined by a *Pline* segment is the XY plane of the *Pline* when the *Pline* segment was created. Using a *Line* object or edge of a solid is not allowed because neither defines a plane. Figure 35-63 shows a box mirrored about the plane defined by the *Circle* object. Using *Subtract* produces the result shown in B.

**FIGURE 35-63**

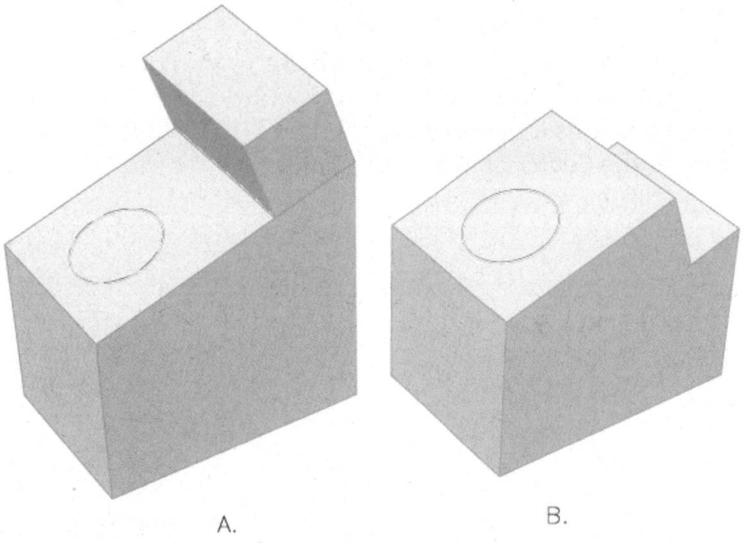


### *Last*

Selecting this option uses the plane that was last used for mirroring.

### *Zaxis*

With this option, the mirror plane is the XY plane perpendicular to a Z vector you specify. The first point you specify on the Z axis establishes the location of the XY plane origin (a point through which the plane passes). The second point establishes the Z axis and the orientation of the XY plane (perpendicular to the Z axis). Figure 35-64 illustrates this concept. Note that this option requires only two PICK points.

**FIGURE 35-64**

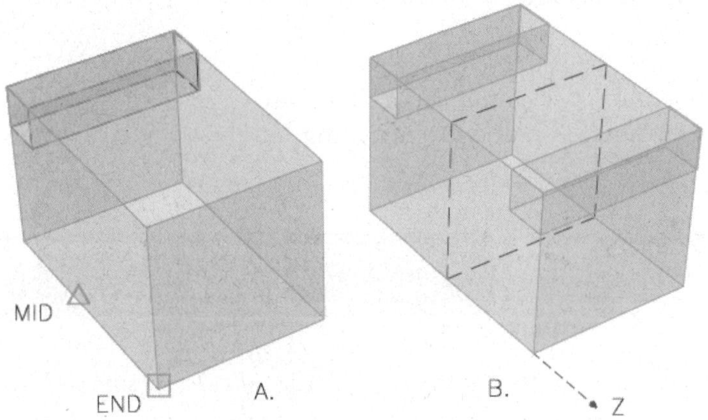


### *View*

The *View* option of *Rotate3D* uses a mirroring plane parallel with the screen and perpendicular to your line of sight based on your current viewpoint. You are required to select a point on the plane. Accepting the default (0,0,0) establishes the mirroring plane passing through the current origin. Any other point can be selected. You must change your viewpoint to "see" the mirrored objects (Fig. 35-65).

**FIGURE 35-65**

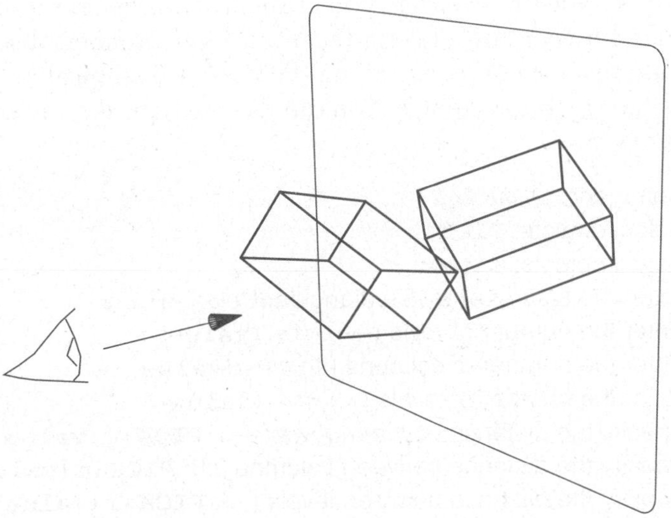

### *XY*

This option situates a mirroring plane parallel with the current XY plane. You can specify a point through which the mirroring plane passes. Figure 35-66 represents a plane established by selecting the *Center* of an existing solid.

FIGURE 35-66

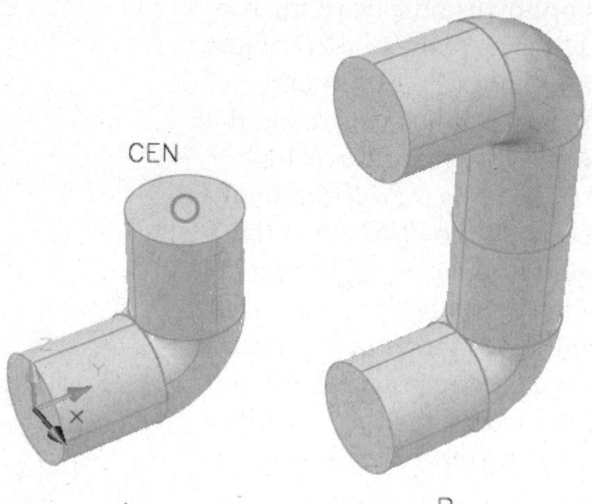


### *YZ*

Using the *YZ* option constructs a plane to mirror about that is parallel with the current YZ plane. Any point can be selected through which the plane will pass (Fig. 35-67).

FIGURE 35-67

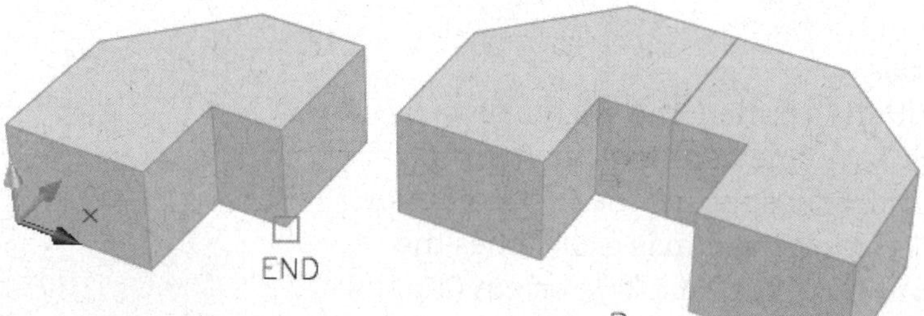


### *ZX*

This option uses a plane parallel with the current ZX plane for mirroring.

## 3Darray

| Ribbon | Menu Bar | Command (Type) | Alias (Type) | Shortcut |
|---|---|---|---|---|
| ... | *Modify 3D Operations > 3D Array* | *3Darray* | *3A* | ... |

### *Rectangular*

With this option of *3Darray*, you create a 3D array specifying three dimensions—the number of and distance between rows (along the Y axis), the number/distance of columns (along the X axis), and the number/distance of levels (along the Z axis). Technically, the result is an array in a prism configuration rather than a rectangle. You can use the typical array commands to set the *Levels* feature to create 3D arrays, also.

Command: ***3DARRAY***
Select objects: **PICK**
Select objects: **Enter**
Enter the type of array [Rectangular/Polar] <R>: ***r***
Enter the number of rows (—-) <1>: (**value**)
Enter the number of columns (|||) <1>: (**value**)
Enter the number of levels (...) <1>: (**value**)
Specify the distance between rows (—-): **PICK** or (**value**)
Specify the distance between columns (|||): **PICK** or (**value**)
Specify the distance between levels (...): **PICK** or (**value**)
Command:

The selection set can be one or more objects. The entire set is treated as one object for arraying. All values entered must be positive.

Figures 35-68 and 35-69 illustrate creating a *Rectangular 3Darray* of a cylinder with 3 rows, 4 columns, and 2 levels.

**FIGURE 35-68**

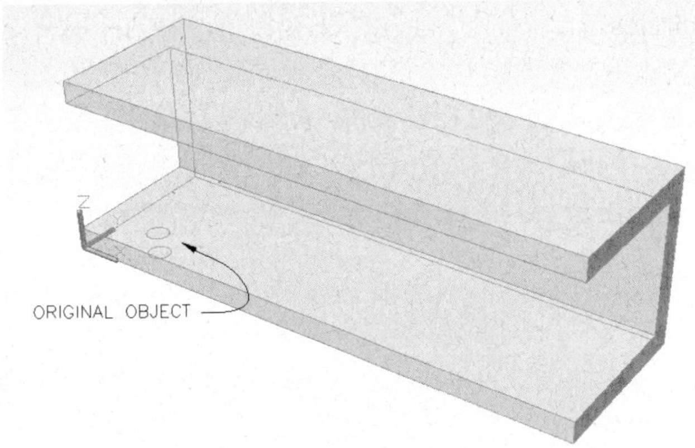


The cylinders are *Subtracted* from the extrusion to form the finished part (Fig. 35-69).

**FIGURE 35-69**

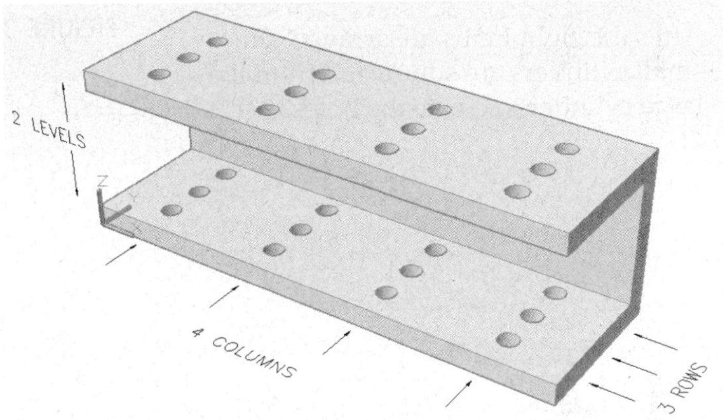


### *Polar*

Similar to a *Polar Array* (2D), this option creates an array of selected objects in a <u>circular</u> fashion. The only difference in the 3D version is that an array is created about an <u>axis of rotation</u> (3D) rather than a point (2D). Specification of an axis of rotation requires two points in 3D space:

Command: ***3DARRAY***
Select objects: **PICK**
Select objects: **Enter**
Enter the type of array [Rectangular/Polar] <R>: ***p***
Enter the number of items in the array: (**value**)
Specify the angle to fill (+=ccw, -=cw) <360>: **PICK** or (**value**)
Rotate arrayed objects? [Yes/No] <Y>: ***y*** or ***n***
Specify center point of array: **PICK** or (**coordinates**)
Specify second point on axis of rotation: **PICK** or (**coordinates**)
Command:

In Figures 35-70 and 35-71, a *3Darray* is created to form a series of holes from a cylinder. The axis of rotation is the center axis of the large cylinder specified by PICKing the *Center* of the top and bottom circles.

**FIGURE 35-70**

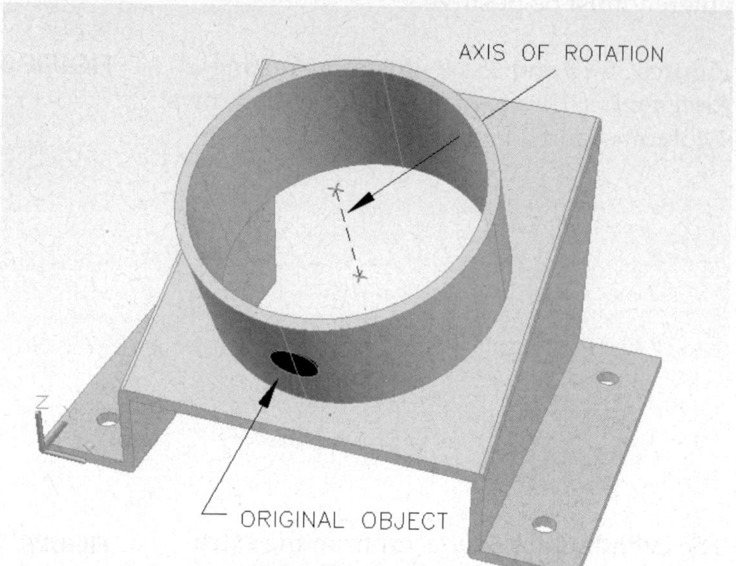


After the eight items are arrayed, the small cylinders are subtracted from the large cylinder to create the holes.

**FIGURE 35-71**

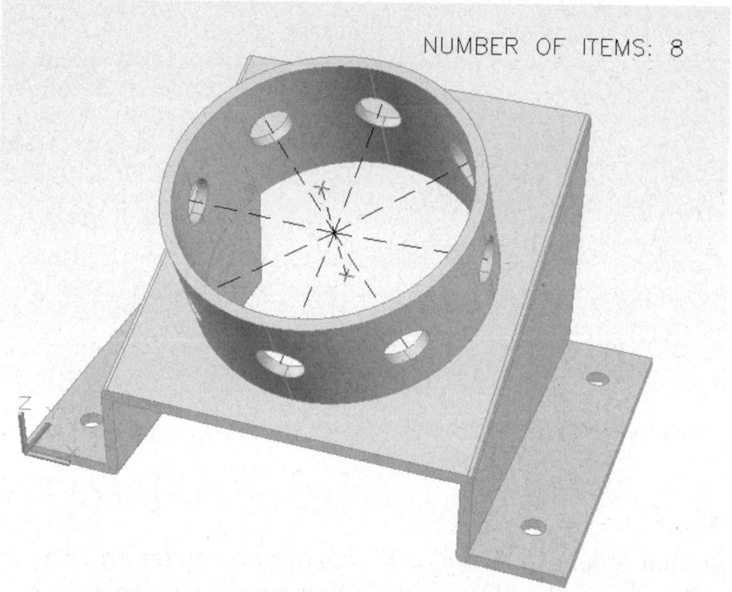


You can also use the *Path* option of *Array* (or *Arraypath*) to create 3D geometry (Fig. 35-72).

**FIGURE 35-72**

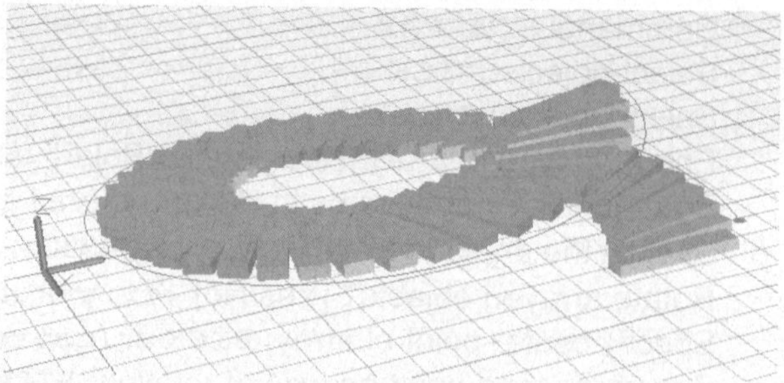

TIP

NOTE: As an alternative to *3Darray*, you can use the *Array* command (or *Arrayrect*, *Arraypolar*, and *Arraypath*) to create 3D arrays. The *Array* command (and each of the options) has a *Levels* option which allows you to specify a third dimension for the array. The *Array* (and *Arrayrect*, *Arraypolar*, and *Arraypath*) commands have the advantage of creating associative arrays that can be edited with grips.

## BOOLEAN OPERATION COMMANDS

Once the individual 3D primitives have been created and moved into place, you are ready to put together the parts. The primitives can be "assembled" or combined by Boolean operations to create composite solids. The Boolean operations found in AutoCAD are listed in this section: *Union*, *Subtract*, and *Intersect*.

### Union

| Ribbon | Menu Bar | Command (Type) | Alias (Type) | Shortcut |
|---|---|---|---|---|
| *Home Solid Editing* (icon) | *Modify Solid Editing > Union* | *Union* | *UNI* | ... |

*Union* joins selected primitives or composite solids to form one composite solid. Usually, the selected solids occupy portions of the same space, yet are separate solids. *Union* creates one solid composed of the total encompassing volume of the selected solids. (You can union solids even if the solids do not overlap.) All lines of intersections (surface boundaries) are calculated and displayed by AutoCAD. Multiple solid objects can be unioned with one *Union* command:

Command: ***UNION***
Select objects: **PICK** (Select two or more solids.)
Select objects: **Enter** (Indicate completion of the selection process.)
Command:

Two solid boxes are combined into one composite solid with *Union* (Fig. 35-73). The original two solids (A) share the same physical space. The resulting union (B) consists of the total contained volume. The new lines of intersection are automatically calculated and displayed. *Obscured edges* was used to enhance visualization.

**FIGURE 35-73**

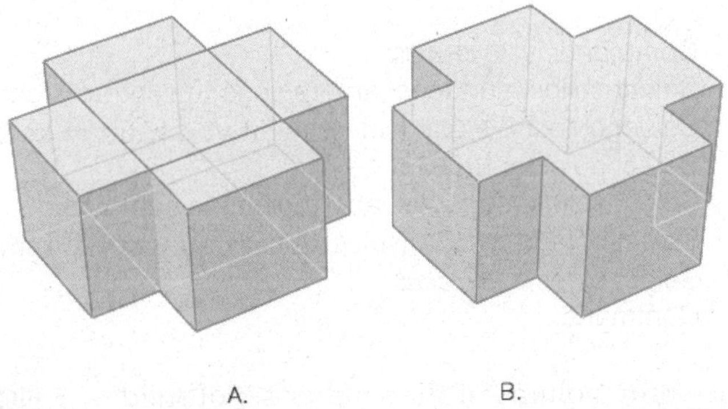


Because the volume occupied by any one of the primitives is included in the resulting composite solid, any redundant volumes are immaterial. The two primitives in Figure 35-74 A yield the same enclosed volume as the composite solid B. (*X-ray mode* has been used for this figure.)

**FIGURE 35-74**

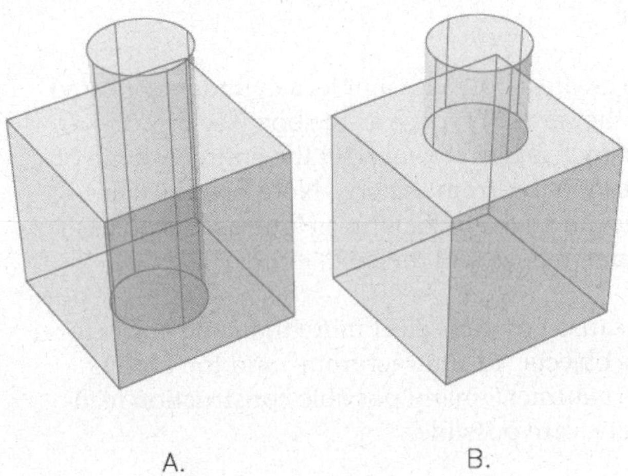


Multiple objects can be selected in response to the *Union* "Select objects:" prompt. It is not necessary, nor is it efficient, to use several successive Boolean operations if one or two can accomplish the same result. Two primitives that have coincident faces (touching sides) can be joined with *Union*. Several "blocks" can be put together to form a composite solid.

Figure 35-75 illustrates how several primitives having coincident faces (A) can be combined into a composite solid (B). Only one *Union* is required to yield the composite solid.

FIGURE 35-75

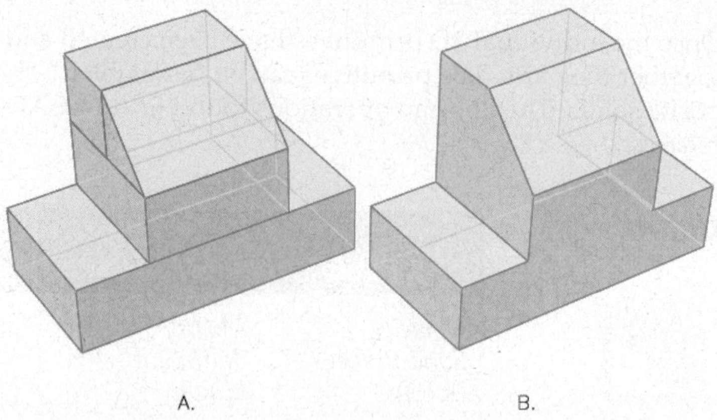


## Subtract

| Ribbon | Menu Bar | Command (Type) | Alias (Type) | Shortcut |
|---|---|---|---|---|
| *Home Solid Editing* (icon) | *Modify Solid Editing > Subtract* | *Subtract* | *SU* | ... |

*Subtract* takes the difference of one set of solids from another. *Subtract* operates with *Regions* as well as solids. When using solids, *Subtract* subtracts the volume of one set of solids from another set of solids. Either set can contain only one or several solids. *Subtract* requires that you first select the set of solids that will remain (the "source objects"), then select the set you want to subtract from the first:

Command: ***SUBTRACT***
Select solids, surfaces, and regions to subtract from...
Select objects: **PICK** (select what you want to keep)
Select objects: **Enter**
Select solids, surfaces, and regions to subtract...
Select objects: **PICK** (select what you want to remove)
Select objects: **Enter**
Command:

The entire volume of the solid or set of solids that is subtracted is completely removed, leaving the remaining volume of the source set.

To create a box with a hole, a cylinder is located in the same 3D space as the box (see Fig. 35-76). *Subtract* is used to subtract the entire volume of the cylinder from the box. Note that the cylinder can have any height, as long as it is at least equal in height to the box.

Because you can select more than one object for the objects "to subtract from" and the objects "to subtract," many possible construction techniques are possible.

FIGURE 35-76

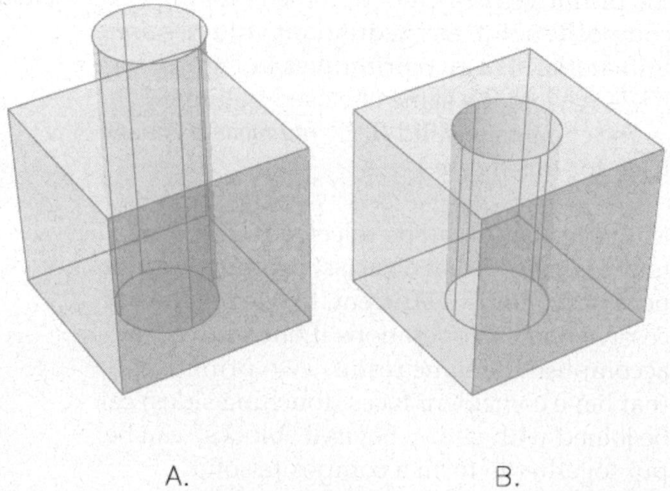

If you select multiple solids in response to the select objects "to subtract from" prompt, they are automatically unioned. This is known as an *n*-way Boolean operation. Using *Subtract* in this manner is very efficient and fast.

TIP

Figure 35-77 illustrates an *n*-way Boolean. The two boxes (A) are selected in response to the objects "to subtract from" prompt. The cylinder is selected as the objects "to subtract...". *Subtract* joins the source objects (identical to a *Union*) and subtracts the cylinder. The resulting composite solid is shown in B.

**FIGURE 35-77**

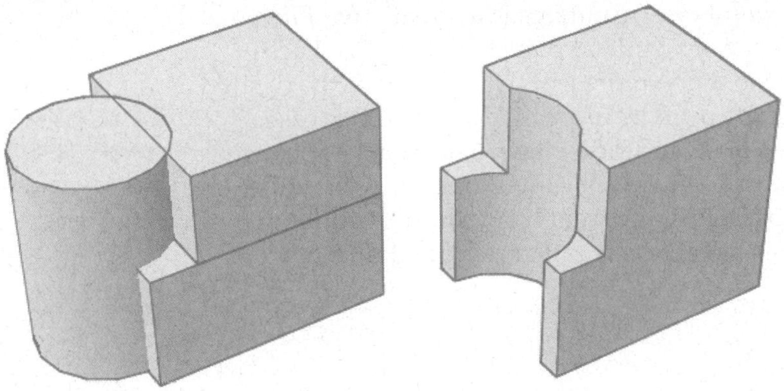

A. B.

### *Intersect*

| Ribbon | Menu Bar | Command (Type) | Alias (Type) | Shortcut |
|---|---|---|---|---|
| *Home*<br>*Solid Editing*<br>(icon) | *Modify*<br>*Solid Editing* ><br>*Intersect* | *Intersect* | *IN* | ... |

*Intersect* creates composite solids by calculating the intersection of two or more solids. The intersection is the common volume shared by the selected objects. Only the 3D space that is part of all of the selected objects is included in the resulting composite solid. *Intersect* requires only that you select the solids from which the intersection is to be calculated.

Command: ***INTERSECT***
Select objects: **PICK** (Select all desired solids.)
Select objects: **Enter** (Indicates completion of the selection process.)
Command:

An example of *Intersect* is shown in Figure 35-78. The cylinder and the wedge share common 3D space (A). The result of the *Intersect* is a composite solid that represents that common space (B). (*X-ray mode* has been used for this figure.)

**FIGURE 35-78**

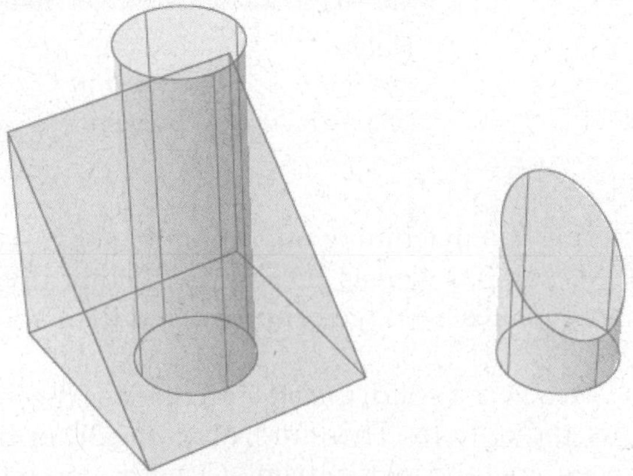

A. B.

*Intersect* can be very effective when used in conjunction with *Extrude*. A technique known as reverse drafting can be used to create composite solids that may otherwise require several primitives and several Boolean operations. Consider the composite solid shown in Figure 38-78A. Using *Union*, the composite shape requires four primitives.

A more efficient technique than unioning several box primitives is to create two *Pline* shapes on vertical planes (Fig. 35-79). Each *Pline* shape represents the outline of the desired shape from its respective view: in this case, the front and side views. The *Pline* shapes are intended to be extruded to occupy the same space. It is apparent from this illustration why this technique is called reverse drafting.

**FIGURE 35-79**

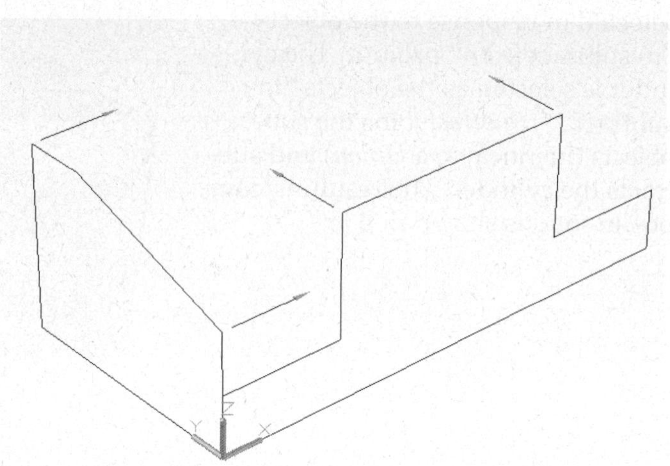

The two *Pline* "views" are extruded with *Extrude* to comprise the total volume of the desired solid (Fig. 35-80 A). Finally, *Intersect* is used to calculate the common volume and create the composite solid (B).

**FIGURE 35-80**

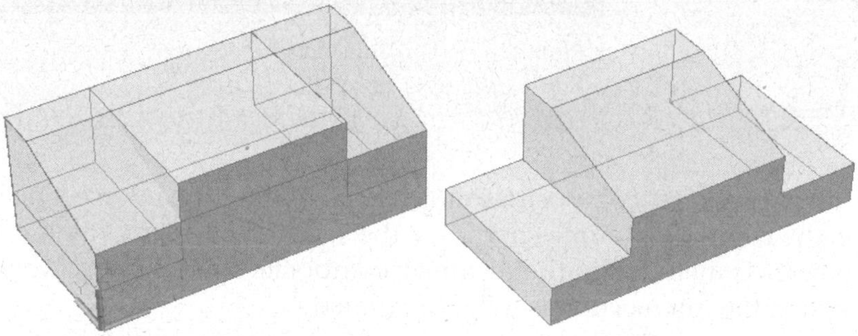

A. B.

### *Chamfer*

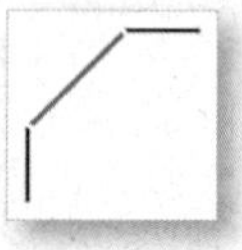

| Ribbon | Menu Bar | Command (Type) | Alias (Type) | Shortcut |
|---|---|---|---|---|
| *Home*<br>*Modify*<br>*Chamfer* | *Modify*<br>*Chamfer* | *Chamfer* | *CHA* | ... |

Chamfering is a machining operation that bevels a sharp corner. *Chamfer* chamfers selected edges of an AutoCAD solid as well as 2D objects. Technically, *Chamfer* (used with a solid) is a Boolean operation because it creates a wedge primitive and then adds to or subtracts from the selected solid.

When you select a solid, *Chamfer* recognizes the object as a solid and switches to the solid version of prompts and options. Therefore, all of the 2D options are not available for use with a solid, only the "distances" method. When using *Chamfer*, you must both select the "base surface" and indicate which edge(s) on that surface you wish to chamfer:

Command: ***CHAMFER***
(TRIM mode) Current chamfer Dist1 = 0.5000, Dist2 = 0.5000
Select first line or [Polyline/Distance/Angle/Trim/Method/mUltiple]: **PICK** (Select solid at desired edge)
Base surface selection...
Enter surface selection option [Next/OK (current)] <OK>: ***N*** or **Enter**
Specify base surface chamfer distance <0.5000>: (**value**)
Specify other surface chamfer distance <0.5000>: (**value**)
Select an edge or [Loop]: **PICK** (Select edges to be chamfered)
Select an edge or [Loop]: **Enter**
Command:

When AutoCAD prompts to select the "base surface," only an edge can be selected since the solids are displayed in wireframe. When you select an edge, AutoCAD highlights one of the two surfaces connected to the selected edge. Therefore, you must use the "Next/<OK>:" option to indicate which of the two surfaces you want to chamfer (Fig. 35-81 A). The two distances are applied to the object, as shown in Figure 35-81 B.

**FIGURE 35-81**

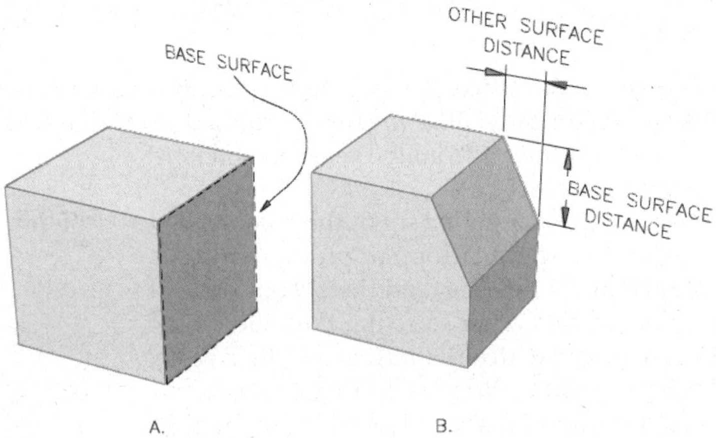


You can chamfer multiple edges of the selected "base surface" simply by PICKing them at the "Select an edge or [Loop]:" prompt (Fig. 35-82). If the base surface is adjacent to cylindrical edges, the bevel follows the curved shape.

**FIGURE 35-82**

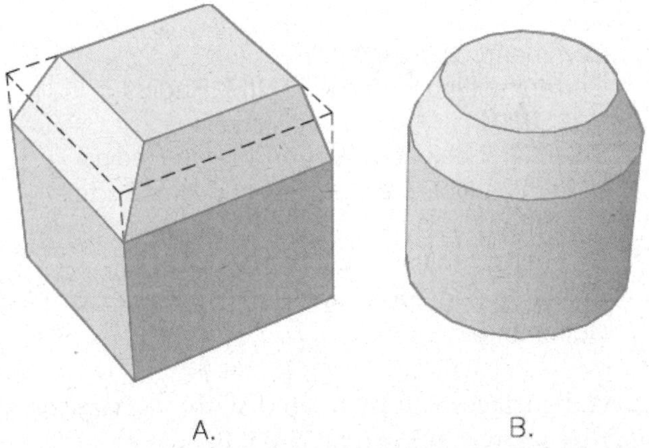


### *Loop*

The *Loop* option chamfers the entire perimeter of the base surface. Simply PICK any edge on the base surface (Fig. 35-83).

Select an edge or [Loop]: ***l***
Select an edge loop or [Edge]: **PICK** (Select edges to form loop)
Select an edge or [Loop]: **Enter**
Command:

**FIGURE 35-83**

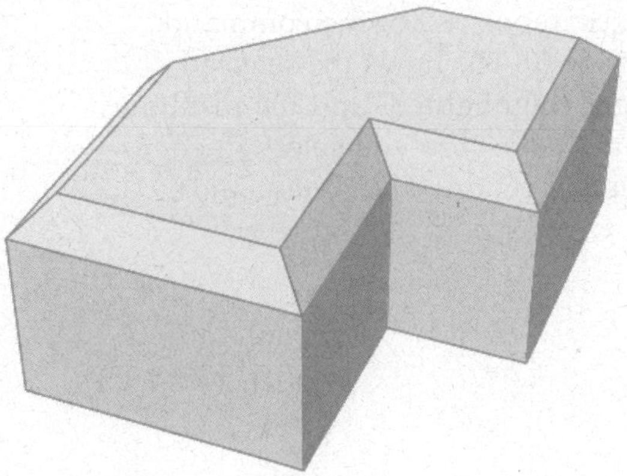

### Edge

The *Edge* option switches back to the "Select edge" method.

### Multiple

The *Multiple* option simply repeats the *Chamfer* command.

## Fillet

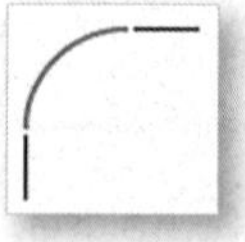

| Ribbon | Menu Bar | Command (Type) | Alias (Type) | Shortcut |
|---|---|---|---|---|
| *Home Modify Fillet* | *Modify Fillet* | *Fillet* | *F* | ... |

*Fillet* creates fillets (concave corners) or rounds (convex corners) on selected solids, just as with 2D objects. Technically, *Fillet* creates a rounded primitive and automatically performs the Boolean needed to add or subtract it from the selected solids.

When using *Fillet* with a solid, the command switches to a special group of prompts and options for 3D filleting, and the 2D options become invalid. After selecting the solid, you must specify the desired radius and then select the edges to fillet. When selecting edges to fillet, the edges must be PICKed individually. Figure 35-84 depicts concave and convex fillets created with *Fillet*.

**FIGURE 35-84**

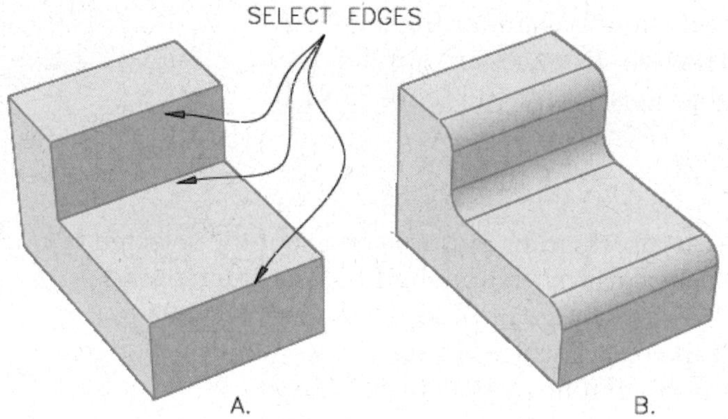


Command: **FILLET**
Current settings: Mode = TRIM, Radius = 0.5000
Select first object or [Undo/Polyline/Radius/Trim/Multiple]: **PICK** (Select desired edge to fillet)
Enter fillet radius <0.5000>: (**value**)
Select an edge or [Chain/Radius]: **Enter** or **PICK** (Select additional edges to fillet)
Command:

Curved surfaces can be treated with *Fillet,* as shown in Figure 35-85. If you want to fillet intersecting concave or convex edges, *Fillet* handles your request, provided you specify all edges in one use of the command. Figure 38-85 shows the selected edges (highlighted) and the resulting solid. Make sure you select all edges together (in one *Fillet* command).

**FIGURE 35-85**

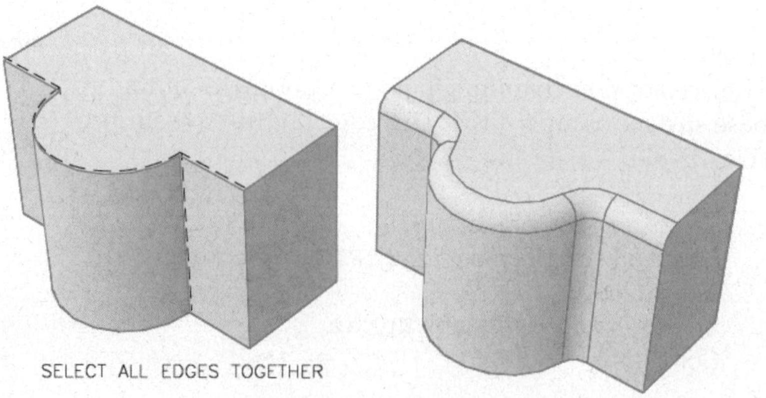

### *Chain*

The *Chain* option allows you to fillet a series of connecting edges. Select the edges to form the chain (Fig. 35-86). If the chain is obvious (only one direct path), you can PICK only the ending edges, and AutoCAD will find the most direct path (series of connected edges):

> Select an edge or [Chain/Radius]: ***c***
> Select an edge chain or [Edge/Radius]: **PICK**

**FIGURE 35-86**

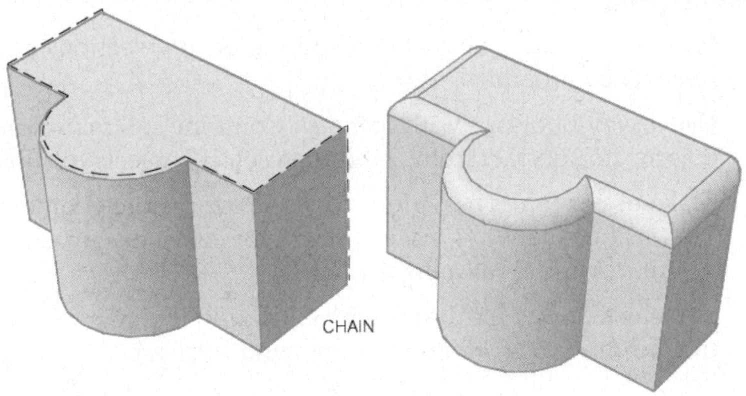


### *Edge*

This option cycles back to the "<Select edge>:" prompt.

### *Radius*

This method returns to the "Enter radius:" prompt.

### *Multiple*

The *Multiple* option causes the *Fillet* command to repeat.

## DESIGN EFFICIENCY

Now that you know the complete sequence for creating composite solid models, you can work toward improving design efficiency. The typical construction sequence is (1) create primitives, (2) ensure the primitives are in place by using UCSs or any of several move and rotate options, (3) combine the primitives into a composite solid using Boolean operations, and (4) make necessary design changes to composite solids using a variety of editing tools (see Chapter 36). The typical step-by-step, "building-block" strategy, however, may not lead to the most efficient design. In order to minimize computation and construction time, you should minimize the number of Boolean operations and, if possible, the number of primitives you use.

For any composite solid, usually several strategies could be used to construct the geometry. You should plan your designs ahead of time, striving to minimize primitives and Boolean operations.

For example, consider the procedure shown in Figure 35-75. As discussed, it is more efficient to accomplish all unions with one *Union*, rather than each union as a separate step. Even better, create a closed *Pline* shape of the profile, then use *Extrude*. Figure 35-77 is another example of design efficiency based on using an *n*-way Boolean. Multiple solids can be unioned automatically by selecting them at the select objects "to subtract from" prompt of *Subtract*. Also consider the strategy of reverse drafting, as shown in Figure 35-79. Using *Extrude* in concert with *Intersect* can minimize design complexity and time.

TIP

In order to create efficient designs and minimize Boolean operations and primitives, keep these strategies in mind:

- Execute as many subtractions, unions, or intersections as possible within one *Subtract*, *Union*, or *Intersect* command.
- Use *n*-way Booleans with *Subtract*. Combine solids (union) automatically by selecting multiple objects "to subtract from," and then select "objects to subtract."
- Make use of *Plines* or regions for complex profile geometry, then *Extrude* the profile shape. This is almost always more efficient for complex curved profile creation than using multiple primitives and Boolean operations.
- Make use of reverse drafting by extruding the "view" profiles (*Plines* or *Regions*) with *Extrude*, then finding the common volume with *Intersect*.

## CHAPTER EXERCISES

1. What are the typical four steps for creating composite solids?

2. Consider the two solids in Figure 35-87. They are two extruded hexagons that overlap (occupy the same 3D space).

   A. Sketch the resulting composite solid if you performed a *Union* on the two solids.

   B. Sketch the resulting composite solid if you performed an *Intersect* on the two solids.

   C. Sketch the resulting composite solid if you performed a *Subtract* on the two solids.

**FIGURE 35-87**

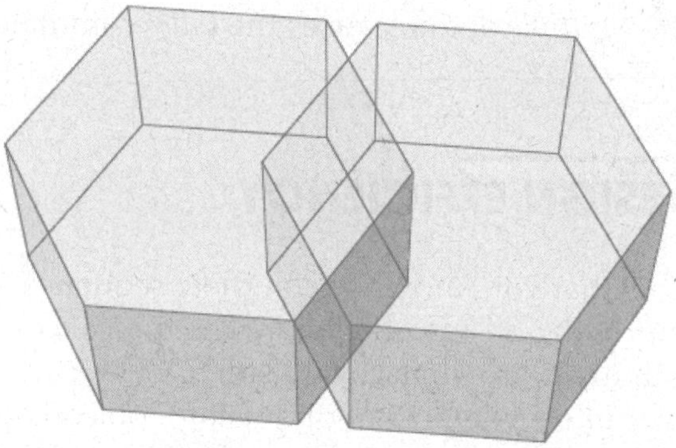

**FIGURE 35-88**

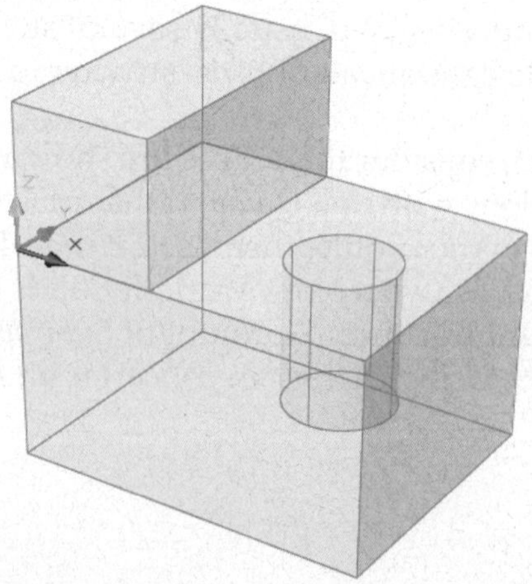

REUSE

3. Begin a drawing and assign the name **CH35EX3.**

   A. Create a ***box*** with the lower-left corner at **0,0,0**. The ***Lengths*** are **5**, **4**, and **3**.

   B. Create a second ***box*** using a Dynamic UCS as shown in Figure 35-88 (use the ***ORigin*** option). The *box* dimensions are **2** x **4** x **2**.

   C. Create a ***Cylinder***. Use a Dynamic UCS as in the previous step. The *cylinder* ***Center*** is at **3.5,2** (of the *UCS*), the ***Diameter*** is **1.5**, and the ***Height*** is **-2**.

   D. ***Save*** the drawing.

E. Perform a ***Union*** to combine the two boxes. Next, use ***Subtract*** to subtract the cylinder to create a hole. The resulting composite solid should look like that in Figure 35-89. ***Save*** the drawing.

**FIGURE 35-89**

4. Begin a drawing and assign the name **CH35EX4.**

REUSE

A. Create a ***Wedge*** at point **0,0,0** with the ***Lengths*** of **5, 4, 3**.

B. Create a ***3point UCS*** option with an orientation indicated in Figure 35-90. Create a *Cone* with the ***Center*** at **2,3** (of the *UCS*) and a ***Diameter*** of **2** and a ***Height*** of **-4**.

**FIGURE 35-90**

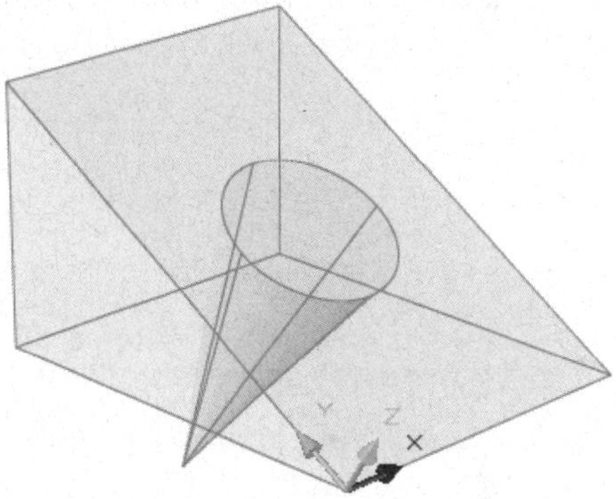


C. ***Subtract*** the cone from the wedge. The resulting composite solid should resemble Figure 35-91. ***Save*** the drawing.

**FIGURE 35-91**

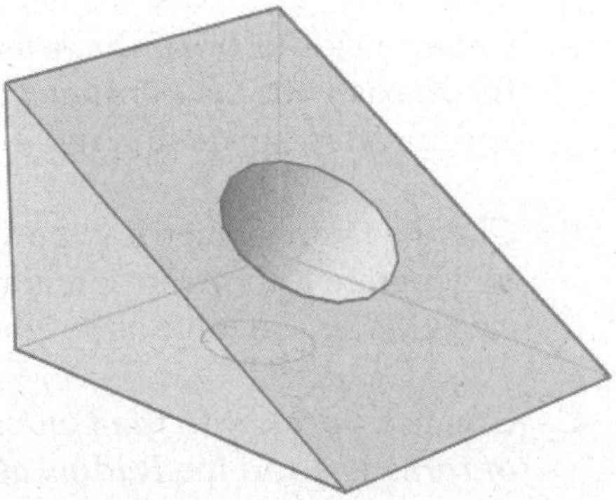

5. Begin a drawing and assign the name **CH35EX5.** Display a ***Plan*** view.

   A. Create 2 ***Circles*** as shown in Figure 35-92, with dimensions and locations as specified. Use ***Pline*** to construct the rectangular shape. Combine the 3 shapes into a *Region* by using the ***Region*** and ***Union*** commands or converting the outside shape into a *Pline* using *Trim* and *Pedit*.

**FIGURE 35-92**

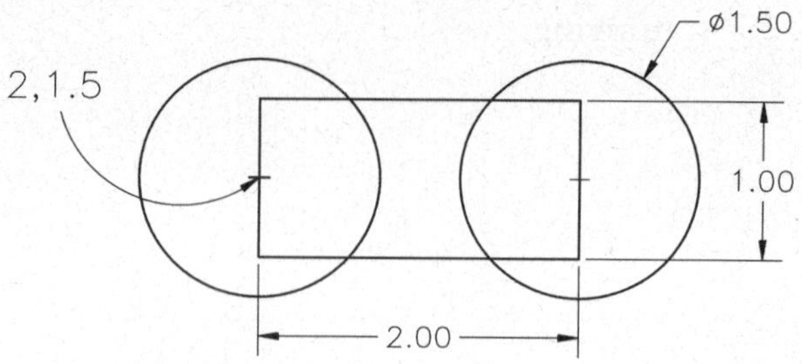


   B. Change the display to an isometric-type view. ***Extrude*** the *Region* or *Pline* with a ***Height*** of **3** (no *taper angle*).

   C. Create a ***Box*** with the lower-left corner at **0,0**. The ***Lengths*** of the box are **6, 3, 3**.

   D. ***Subtract*** the extruded shape from the box. Your composite solid should look like that in Figure 35-93. ***Save*** the drawing.

**FIGURE 35-93**

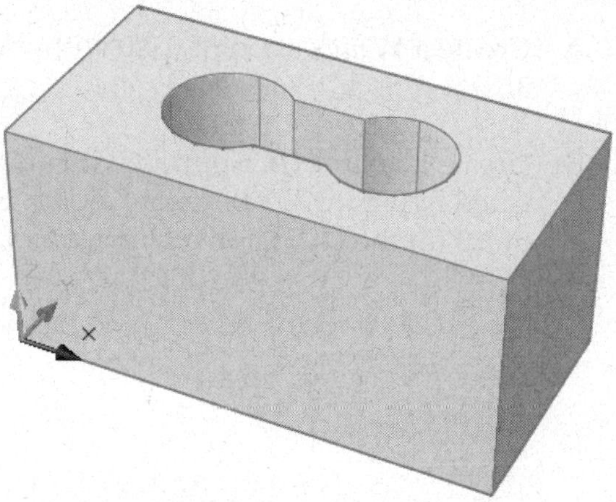

6. Begin a drawing and assign the name **CH35EX6.** Display a ***Plan*** view.

   A. Create a closed ***Pline*** shape symmetrical about the X axis with the locational and dimensional specifications given in Figure 35-94.

   B. Change to an isometric-type view. Use ***Revolve*** to generate a complete circular shape from the closed *Pline*. Revolve about the **Y** axis.

   C. Create a ***Torus*** with the ***Center*** at **0,0**. The ***Radius of torus*** is **3** and the ***Radius of tube*** is **.5**. The two shapes should intersect.

**FIGURE 35-94**

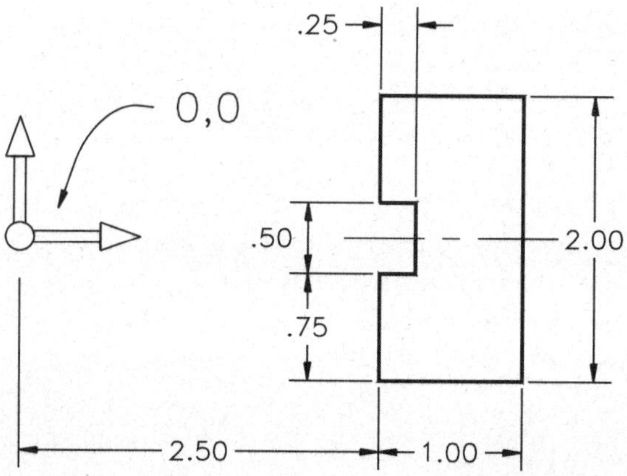

D. Generate a display like Figure 35-95.

E. Create a ***Cylinder*** with the ***Center*** at **0,0,0**, a ***Radius*** of **3**, and a ***Height*** of **8**.

F. Use ***3DRotate*** to rotate the revolved *Pline* shape **90** degrees about the X axis (the *Pline* shape that was previously converted to a solid—not the torus). Next, move the shape up (positive Z) **6** units with ***3DMove***.

G. Move the torus up **4** units with ***3DMove***.

**FIGURE 35-95**

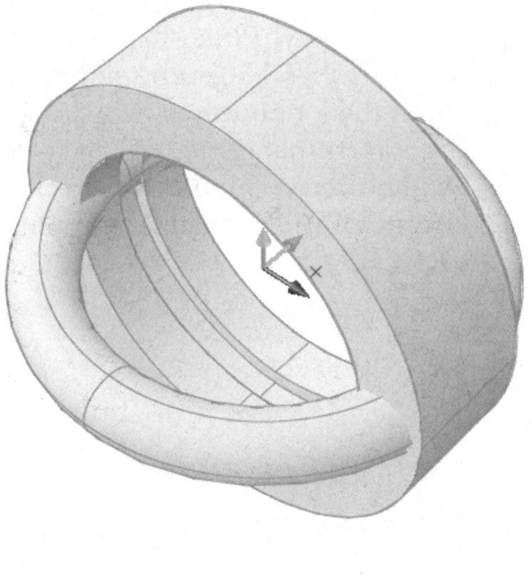

H. The solid primitives should appear as those in Figure 35-96.

**FIGURE 35-96**

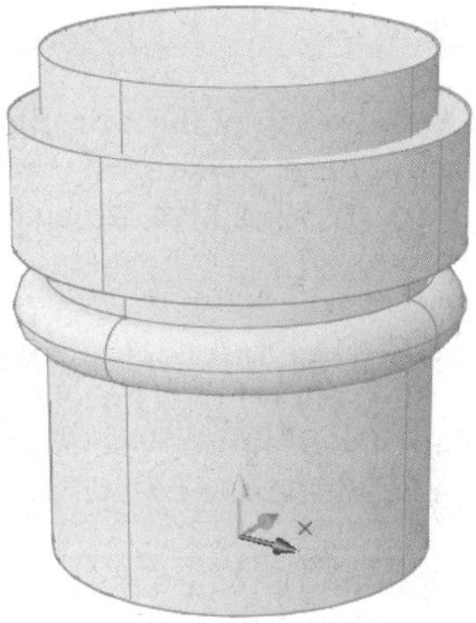

I. Use ***Subtract*** to subtract both revolved shapes from the cylinder. The solid should resemble that in Figure 35-97. ***Save*** the drawing.

**FIGURE 35-97**

7. Begin a ***New*** drawing or use a ***Template.*** Assign the name **FAUCET1.**

A. Draw 3 closed ***Pline*** shapes, as shown in Figure 35-98. Assume symmetry about the longitudinal axis. Use the WCS and create 2 new ***UCS***s for the geometry. Use ***3point Arcs*** for the "front" profile. ***Save*** the drawing as **FAUCET2**.

**FIGURE 35-98**

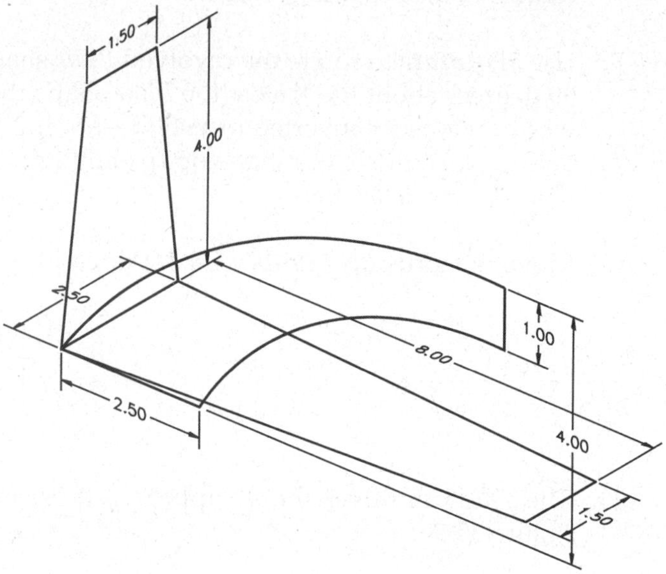


B. ***Extrude*** each of the 3 profiles into the same space. Make sure you specify the correct positive or negative ***Height*** value.

C. Finally, use ***Intersect*** to create the composite solid of the Faucet (Fig. 35-99). ***Save*** the drawing. Use ***Saveas*** and assign the name **FAUCET1**.

D. (Optional) Create a nozzle extending down from the small end. Then create a channel for the water to flow (through the inside) and subtract it from the faucet.

**FIGURE 35-99**

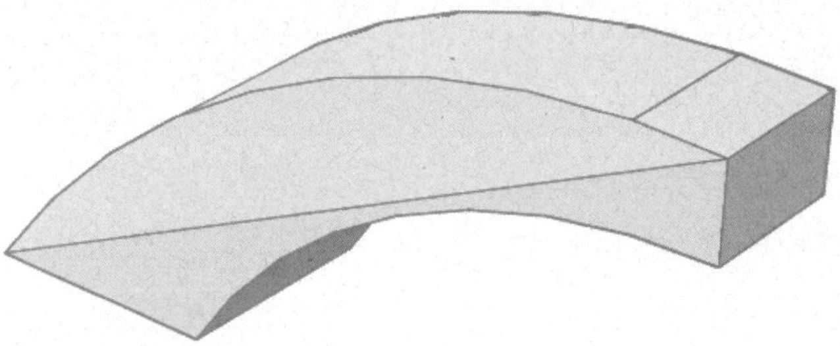

8. ***Open*** the **FAUCET2** drawing you saved in the last exercise. From this drawing you will create an improved faucet using the *Loft* command.

A. ***Erase*** the bottom profile (on the XY plane). Create a ***Rectangle*** on the XY plane as shown in Figure 35-100 using the *Endpoints* of the other profiles on the XY plane for the "first corner" and "other corner."

**FIGURE 35-100**

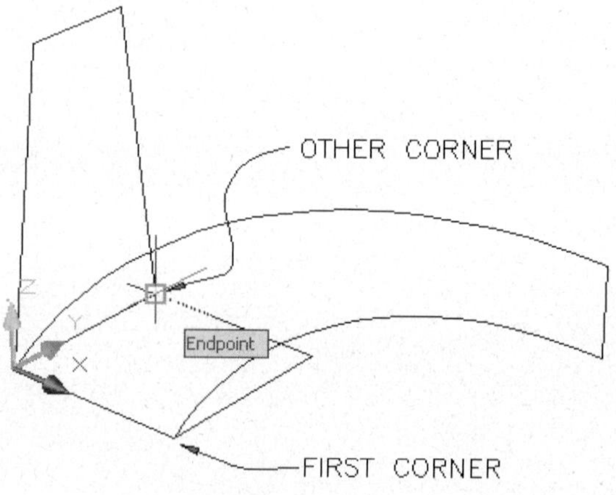

B. Create a new ***UCS*** at the spout end of the faucet as shown in Figure 35-101. Create a ***Rectangle*** as shown with ***Dimensions*** of **1.5** and **1.0**. Next, use ***3Dmove*** to move the new *Rectangle* **.5** units in a positive X direction relative to the current UCS.

FIGURE 35-101

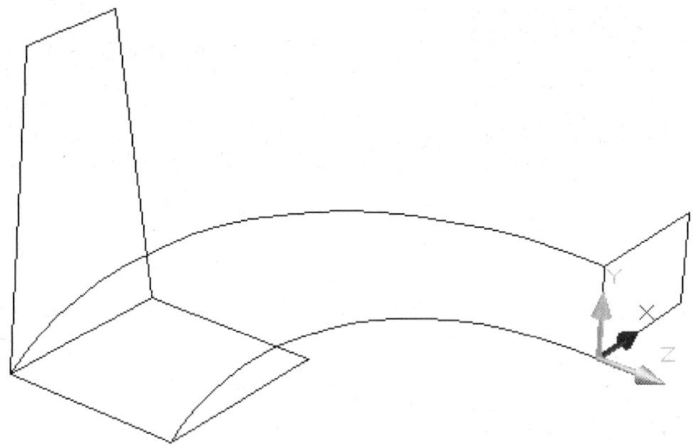

C. Change the ***UCS*** back to ***World***. ***Erase*** the remaining two original profiles and keep only the two new *Rectangles*. Draw a diagonal ***Line*** between the opposite corners (*Endpoints*) for each *Rectangle*, as shown in Figure 35-102. Next, create a ***Spline*** by defining three points—the two ends are at the ***Midpoints*** of the diagonal *Lines* and the middle point is at 4.15, 1.25, 2.66. Finally, ***Erase*** the two diagonal *Lines*.

FIGURE 35-102

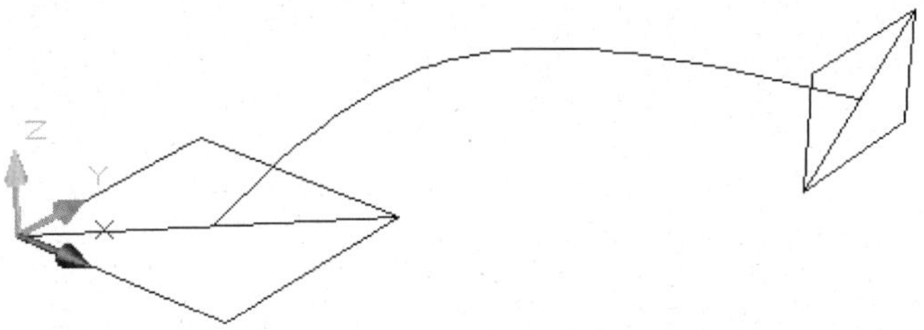

D. Use the ***Loft*** command. At the "Select cross-sections in lofting order:" prompt, select the two *Rectangles*. Use the ***Path*** option. At the "Select path curve:" prompt, select the *Spline*. Your completed faucet should look like that in Figure 35-103.

FIGURE 35-103

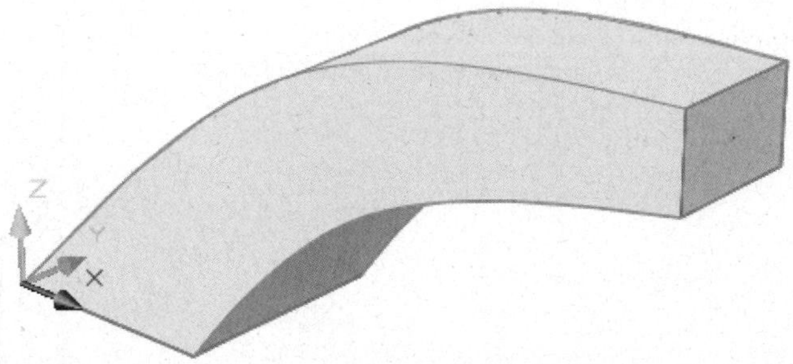

REUSE

9. Construct a solid model of the bar guide in Figure 35-104. Strive for the most efficient design. It is possible to construct this object with one ***Extrude*** and one ***Subtract***. Save the model as **BGUID-SL.**

**FIGURE 35-104**

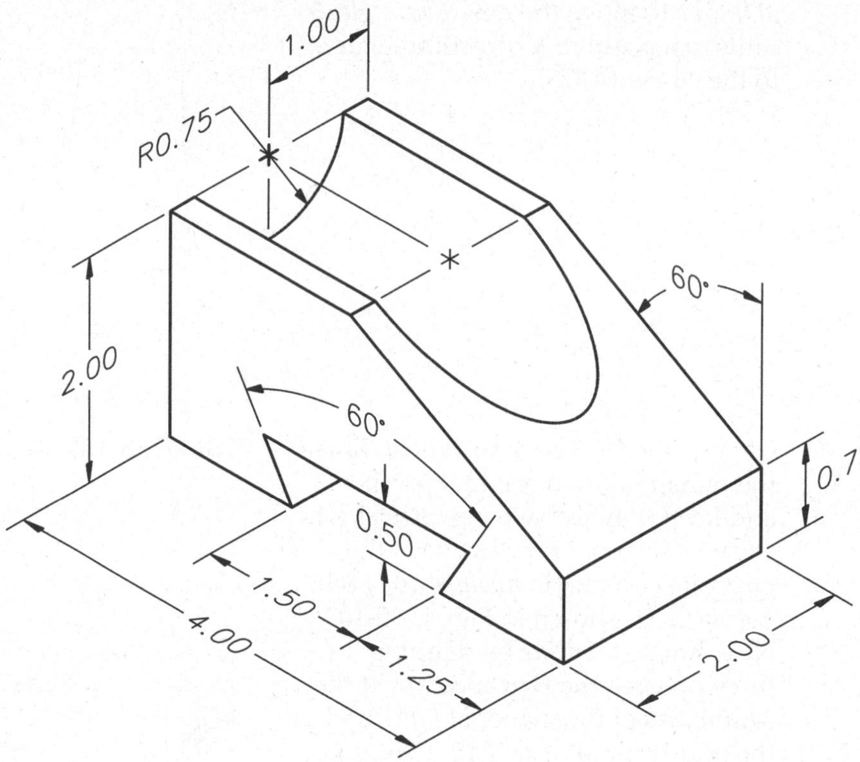


REUSE

10. Make a solid model of the V-block shown in Figure 35-105. Several strategies could be used for construction of this object. Strive for the most efficient design. Plan your approach by sketching a few possibilities. Save the model as **VBLOK-SL.**

**FIGURE 35-105**

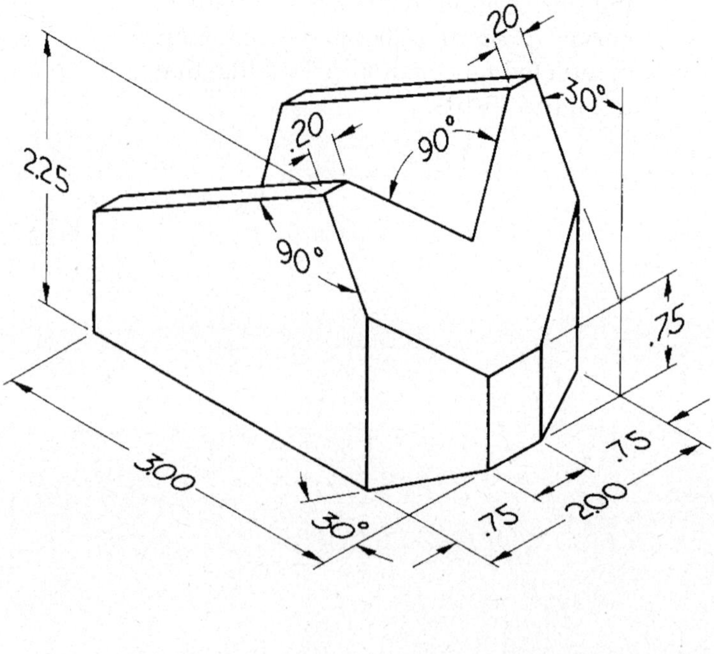

11. Construct a composite solid model of the support bracket (Fig. 35-106) using efficient techniques. Save the model as **SUPBK-SL**.

**FIGURE 35-106**

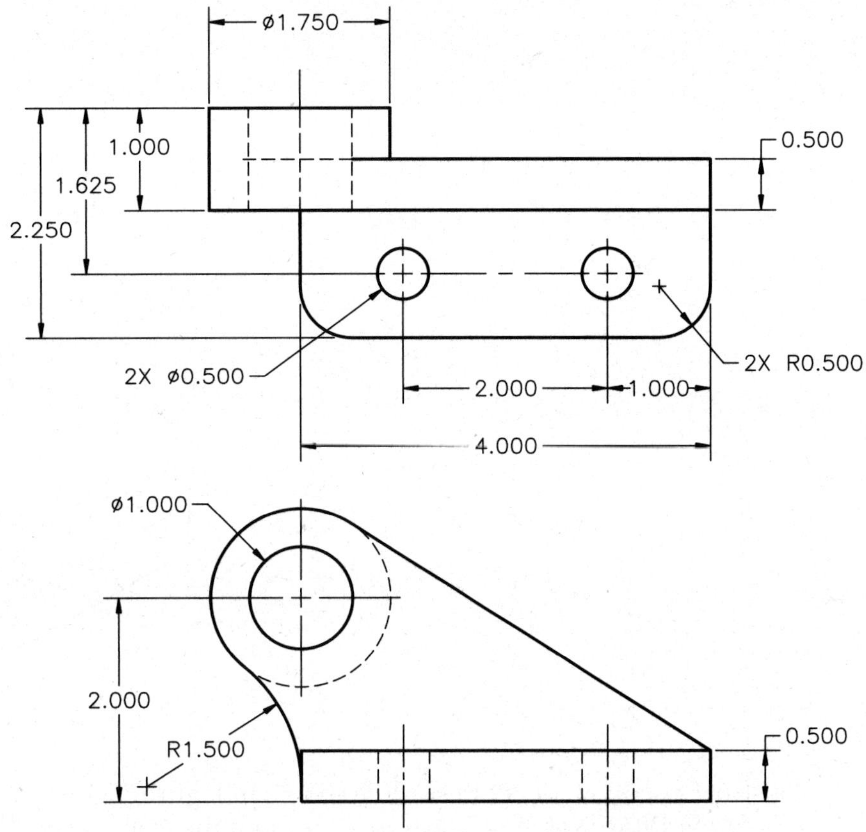


12. Construct the swivel shown in Figure 35-107. The center arm requires ***Sweeping*** the 1.00 x 0.50 rectangular shape along an arc path through 45 degrees. Save the drawing as **SWIVEL.**

REUSE

**FIGURE 35-107**

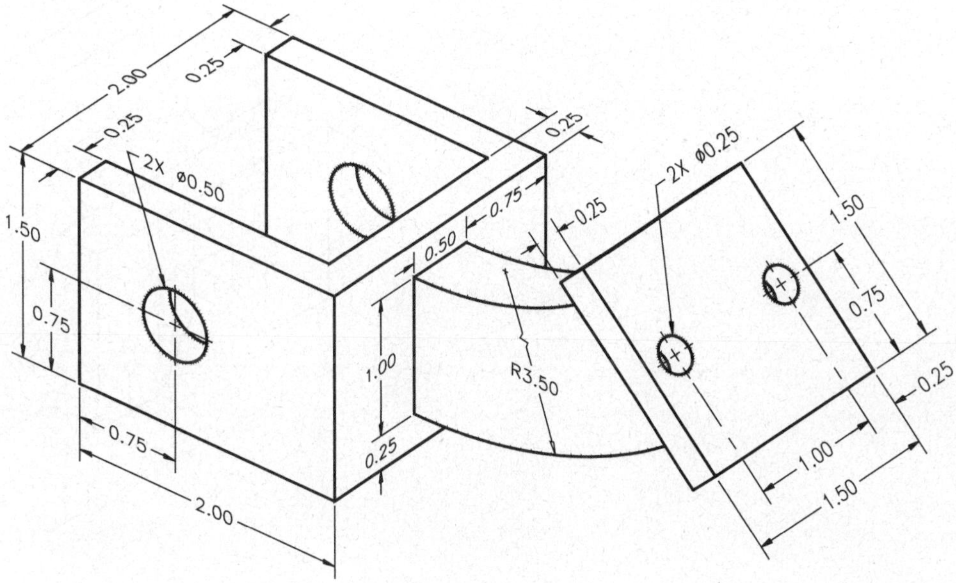

13. Construct a solid model of the angle brace shown in Figure 35-108. Use efficient design techniques. Save the drawing as **AGLBR-SL.**

**FIGURE 35-108**

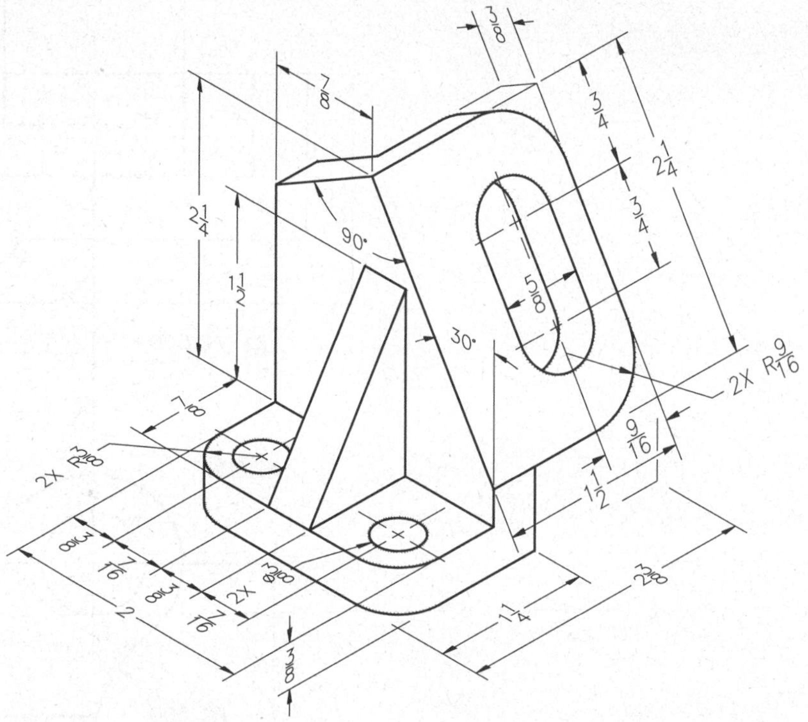


14. Construct a solid model of the saddle shown in Figure 35-109. An efficient design can be utilized by creating *Pline* profiles of the top "view" and the front "view," as shown in Figure 38-110, on the next page. Use *Extrude* and *Intersect* to produce a composite solid. Additional Boolean operations are required to complete the part. The finished model should look similar to Figure 35-111, on the next page. Save the drawing as **SADL-SL.**

**FIGURE 35-109**

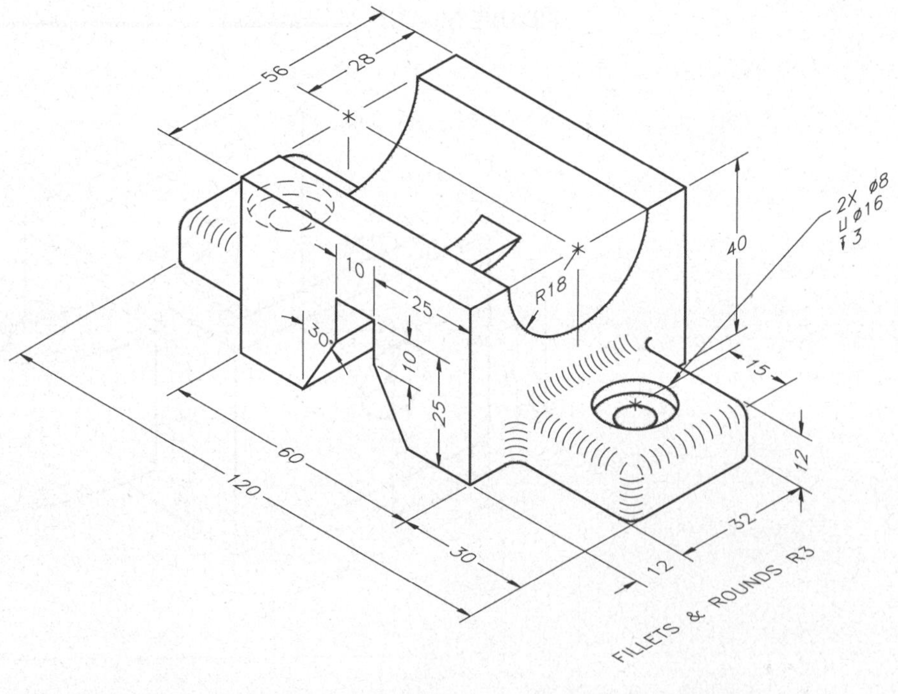

**FIGURE 35-110**

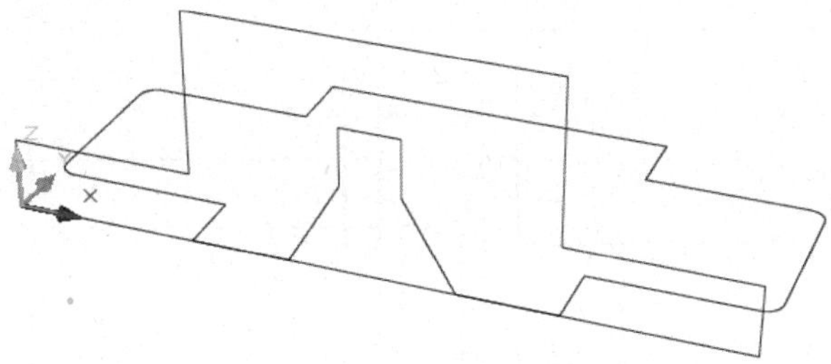

**FIGURE 35-111**

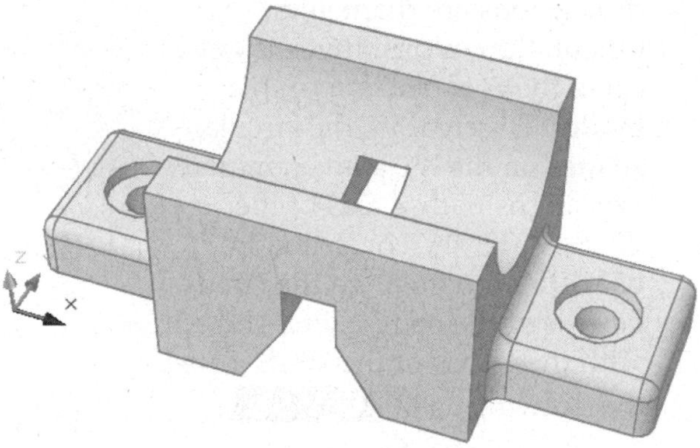

REUSE

15. Construct a solid model of a bicycle handle bar using the *Sweep* command. First, using the World Coordinate System, create the top section of the bars using a ***Pline*** having a straight section 370 units in length and a 60 radius arc on each end as shown in Figure 35-112. Next, create a new ***UCS*** as shown in Figure 35-112 as SIDE. On the XY plane of the new UCS, create one "drop" section of the bars using another ***Pline*** with a straight section of 97 units and an arc section with a 60 unit radius. ***Copy*** this drop section to the other end of the bars. ***Save*** the drawing as **DROPBAR.**

**FIGURE 35-112**

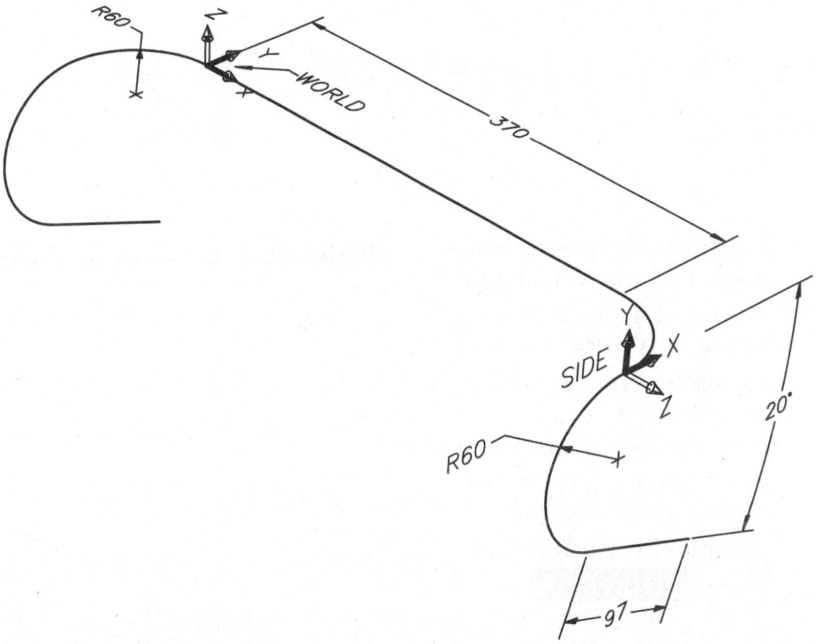


Next, create a ***Circle*** with 28 unit diameter at any location. Finally, use the ***Sweep*** command three times, one for each section of the bars. In each case, select the ***Circle*** as the "object to sweep" and select the ***Pline*** section as the "sweep path." Finally, ***Union*** the three sections into one composite solid. ***Erase*** the *Circle*. ***Save*** the drawing. The completed handle bar should look like that in Figure 35-113.

**FIGURE 35-113**

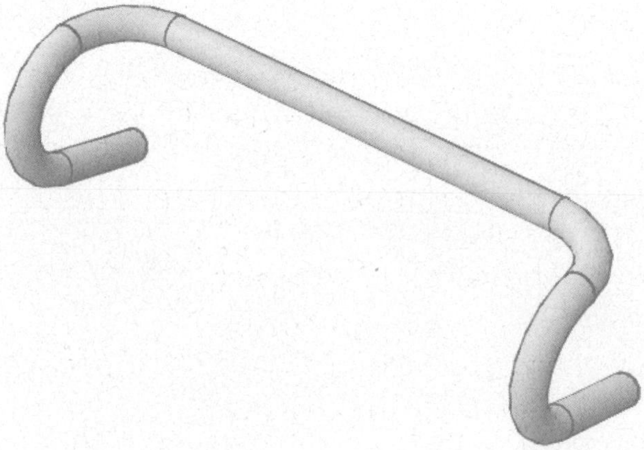

16. Create a solid model of the pulley (Fig. 35-114). All vertical dimensions are diameters. Orientation of primitives is critical in the construction of this model. Try creating the circular shapes on the XY plane (circular axis aligns with Z axis of the WCS). After the construction, use ***3DRotate*** to align the circular axis of the composite solid with the Y axis of the WCS. ***Save*** the drawing as **PULLY-SL.**

**FIGURE 35-114**

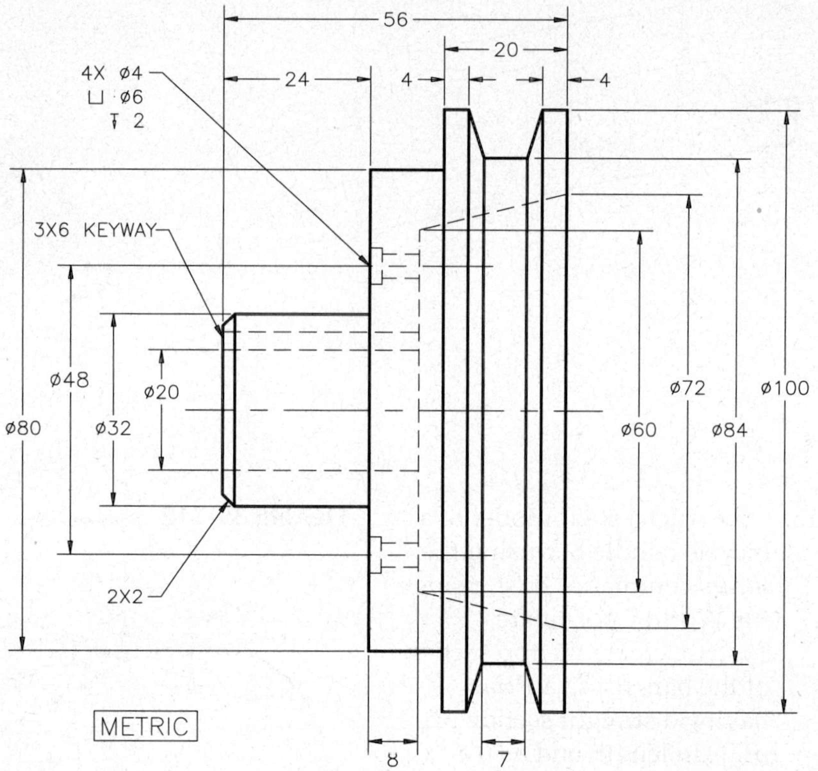


17. Create a composite solid model of the adjustable mount (Fig. 35-115). Use ***Move*** and other methods to move and align the primitives. Use of efficient design techniques is extremely important with a model of this complexity. Assign the name **ADJMT-SL.**

**FIGURE 35-115**

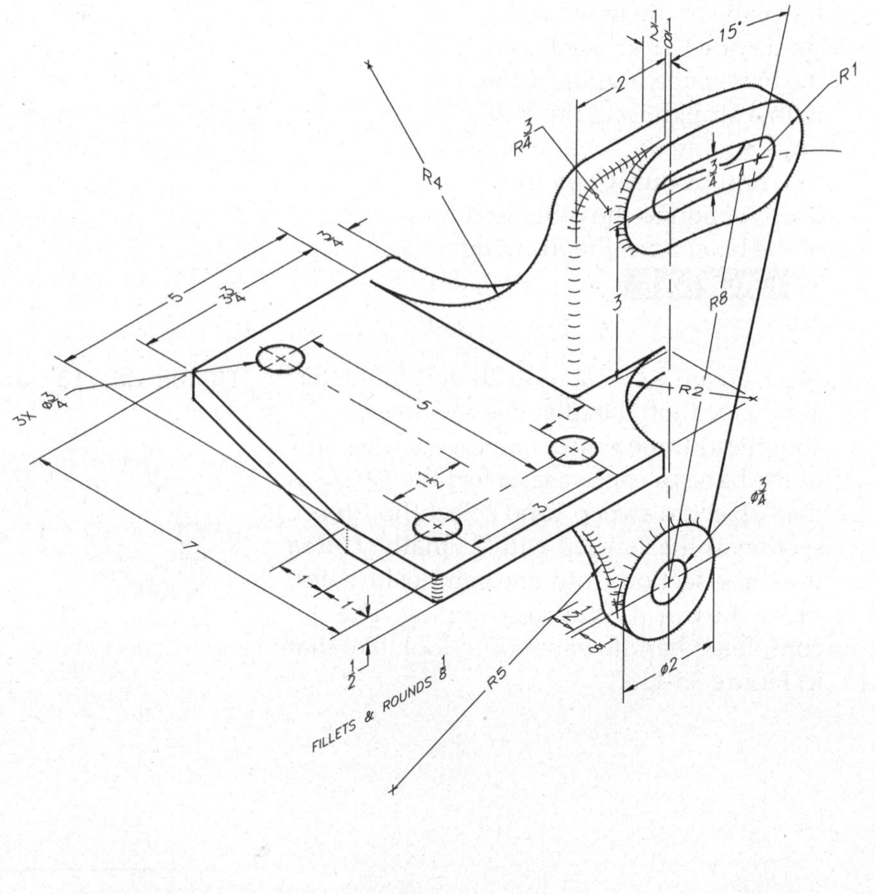

CHAPTER 36

# Solid Model Editing

## Chapter Objectives

After completing this chapter you should:

1. be able to edit the properties of solid models using the *Properties* palette;
2. be able to edit primitive solids and composite solids using grips and gizmos;
3. be able to use subobject selection to edit faces, edges, and vertices of solids;
4. know how to use *Solidedit* to accomplish a variety of solid editing tasks;
5. be able to use the utility solid editing commands: *Presspull*, *Interfere*, and *Imprint*;
6. be able to calculate solids data with *Massprop* and be able to export solids data with *Stlout*, *3Dprint*, and *Acisout*.

## CONCEPTS

The basic solid modeling construction process is to create simple 3D solids, called "primitives," then combine them using "Boolean operations" to create a complex 3D model called a "composite solid." A number of primitive types and construction techniques were discussed as well as strategies to augment the construction process, including using Dynamic User Coordinate Systems.

This chapter discusses several options you can use to edit composite solids, including editing individual primitives, faces, edges, and vertices. In addition, advanced solid editing techniques are discussed involving creating planar and complex-curved surfaces, then shaping solids by "slicing" them using these surfaces.

The topics in this chapter are arranged as follows:

Editing Primitives using the *Properties* Palette
Editing Solids Using Grips and Gizmos
Solids Editing Commands: *Solidedit, Slice, Presspull, Sectionplane, Interfere,* and *Imprint*
Using Solids Data

Using the information from Chapter 35 and the solid editing techniques discussed in this chapter, you should be able to create practically any shape or configuration of solid model.

## EDITING PRIMITIVES USING THE *PROPERTIES* PALETTE

You can use the *Properties* palette to edit basic dimensional characteristics of solid primitives. Even if a primitive has been combined into a composite solid using a Boolean operation, it can be edited by holding down the Ctrl key and selecting the desired primitive as a "subobject." Editing primitives using the *Properties* palette is fairly simple and straightforward. This section explains the following procedures:

Editing *Properties* of individual primitives
Composite solid subobject selection
Editing *Properties* of composite solid subobjects (primitives)

The illustrations in this section may display grips or gizmos on the selected primitives only to indicate which primitives are selected for editing. Grips and gizmos are not discussed, however, until the next section of this chapter.

NOTE: Grips and gizmos can be on or off when using the *Properties* palette to edit solids. Grips and gizmos have no effect on editing capabilities using the *Properties* palette.

### Editing *Properties* of Individual Primitives

The basic dimensional and positional properties of individual solids are available for editing using the *Properties* palette. The procedure is simple: with the *Properties* palette displayed on the screen, select the desired primitive, then change the desired values in the *Geometry* section of the *Properties* palette.

For example, assume you want to change the height of a *Box* primitive. Selecting the box displays the primitive's properties in the *Properties* palette (Fig. 36-1). Find the *Height* edit box in the *Geometry* section of the palette and enter the desired value.

Notice that the grips or the current gizmo may appear on the primitive when selected based on the *GRIPS* and *DEFAULTGIZMO* setting. The gizmo and grips are not needed to access the *Properties* palette. See "Editing Solids Using Grips and Gizmos" for more information.

**FIGURE 36-1**

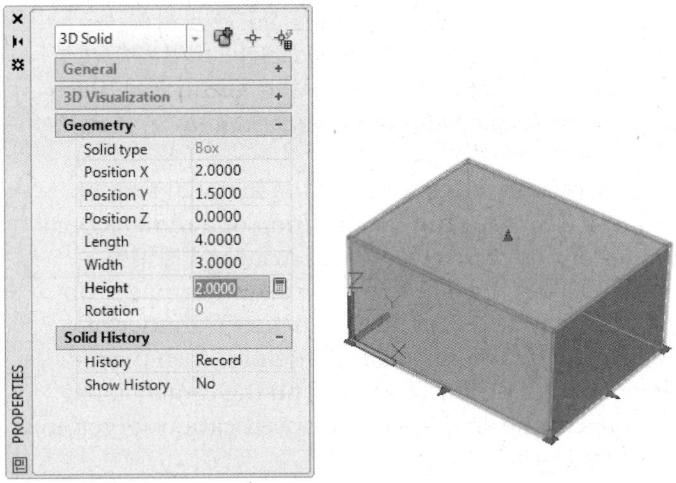


Notice that changing a dimensional or positional value in the *Properties* palette immediately changes the primitive. In this case, the *Height* of the box was changed from 2.0000 to 4.0000.

**FIGURE 36-2**

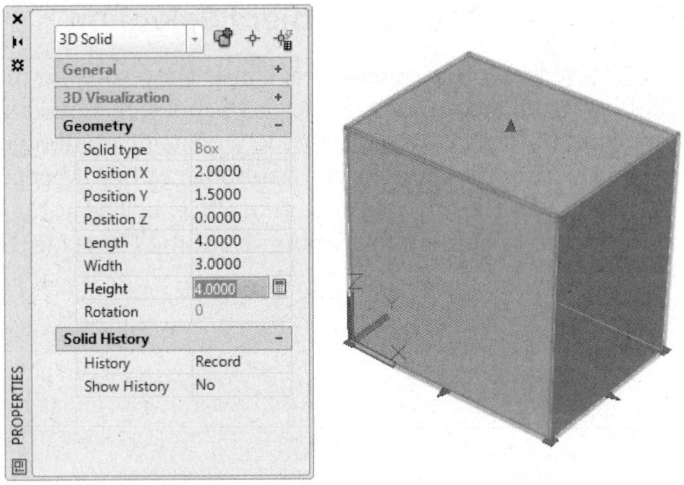


The positional values of primitives can be changed by the same method. Values that appear in the *Position X*, *Position Y*, and *Position Z* edit boxes represent the center of the base. This location is also indicated by a square grip at the base's center. Note that in our example the UCS icon appears at its original location. In Figure 36-3 the *Position Y* value was changed from 1.5000 to 4.0000.

**FIGURE 36-3**

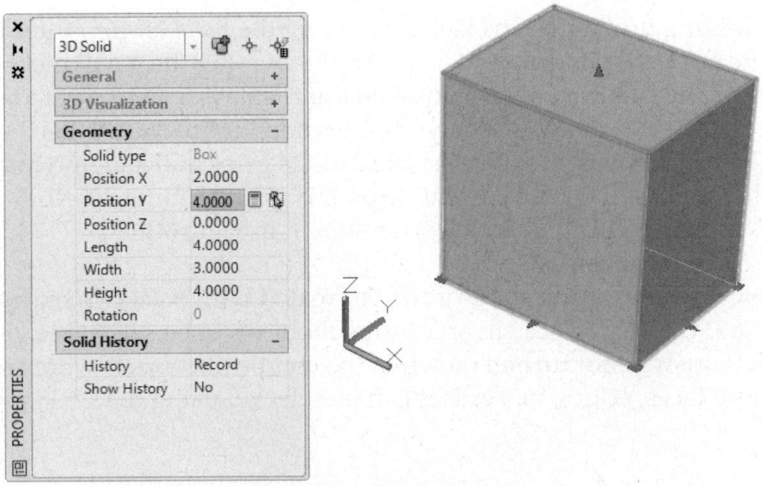

Although all primitives have *Position X*, *Position Y*, and *Position Z* properties, each primitive type has unique dimensional values. For example, a *Cylinder* has a field for *Radius*, *Height*, and *Elliptical* (Fig. 36-4). If *Elliptical* is set to *Yes*, additional fields appear for *Major radius* and *Minor radius*.

A *Cone* primitive has similar dimensional properties to a *Cylinder* with the addition of a dimensional value for *Top radius* (see "Grip Editing Primitive Solids"). A *Pyramid*, as another example, has property fields for *Base radius*, *Top radius*, (number of) *Sides*, *Height*, and *Rotation*. Therefore, each primitive contains properties based on the dimensional features of the shape and the specifications given to create the solid.

FIGURE 36-4

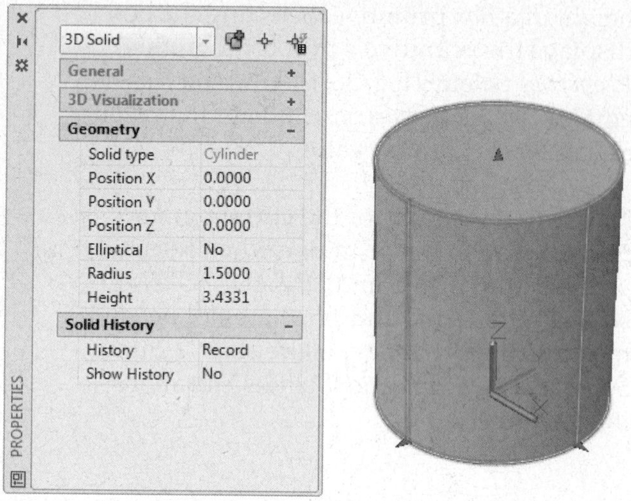


## Composite Solid Subobject Selection

TIP

A subobject is a component of a solid. Therefore, an individual primitive that is part of a composite solid is a subobject. You can select any subobject by holding down the Ctrl key or using a subobject selection filter, then hovering the cursor over a subobject until it becomes highlighted, then using the left mouse button to PICK it. The subobject selection filters are available on the Ribbon from the *Home* tab, *Selection* panel as shown in Figure 36-5.

FIGURE 36-5

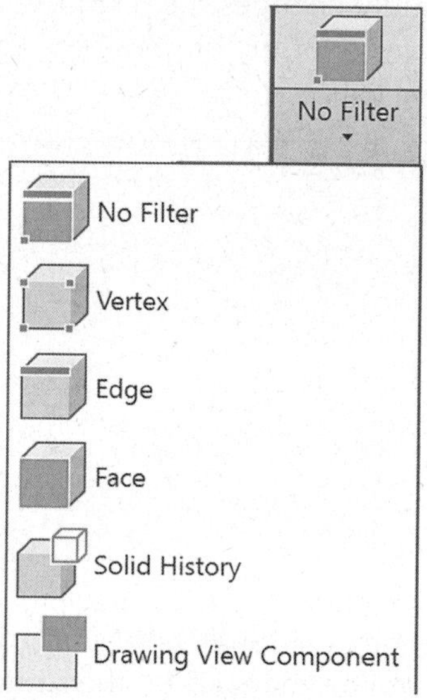


Holding down the Ctrl key or selecting the *Solid History* filter from the *Selection* panel (see Fig. 36-5) and hovering the small pickbox over the smaller *Box* primitive on top displays a "highlighted" effect for the small *Box* only, as shown in Figure 36-6. When the highlighted subobject is selected, its grips and a gizmo may appear based on the current grips and gizmo settings (*GRIPS* and *DEFAULTGIZMO* variables), as shown in Figure 36-7.

TIP

NOTE: Composite solids must be created with *Solid History* turned *On* (*SOLIDHIST* = 1) in order for subobjects to be selectable. If *Solid History* was not turned on when the composite objects were created, only faces, edges, or vertices can be selected using the Ctrl key.

FIGURE 36-6

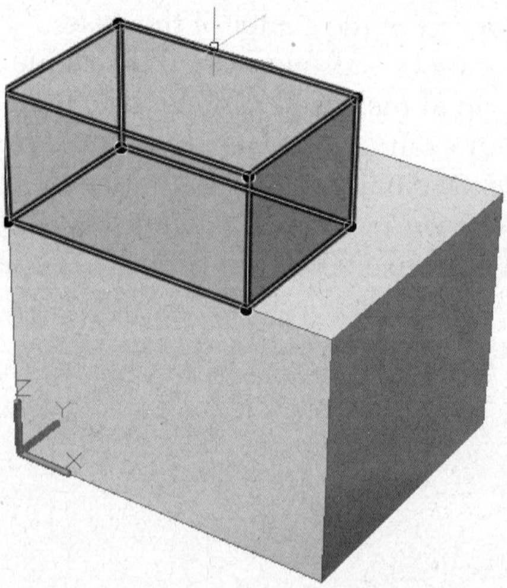

### Editing *Properties* of Composite Solids

To change the properties of a composite solid subobject using the *Properties* palette, hold down the Ctrl key, hover over the desired primitive until it becomes highlighted, then PICK (see previous NOTE). The *Properties* palette should display the locational and dimensional properties of the primitive solid only, not the properties of the composite solid (Fig. 36-7). Grips do not have to be turned on to edit the primitive using *Properties*.

FIGURE 36-7

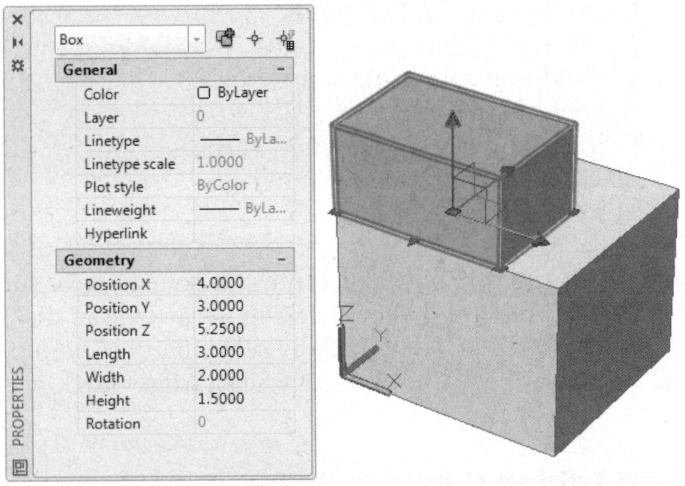


For example, assume you want to change both the location and dimensions of the subobject (*Box* primitive). Once the subobject is selected, change the desired values in the *Geometry* section of the *Properties* palette (Fig. 36-8). Note that in our example, the *Position Y* was changed from 3.0000 to 3.5000, so the *Box* moved in from the left edge of the base feature. In addition, the *Length* value was changed from 3.0000 to 2.0000. Note that the *Length*, *Width*, and *Height* are relative to the center of the base of the primitive; therefore, the length was reduced by 0.5000 on each end of the subobject.

FIGURE 36-8

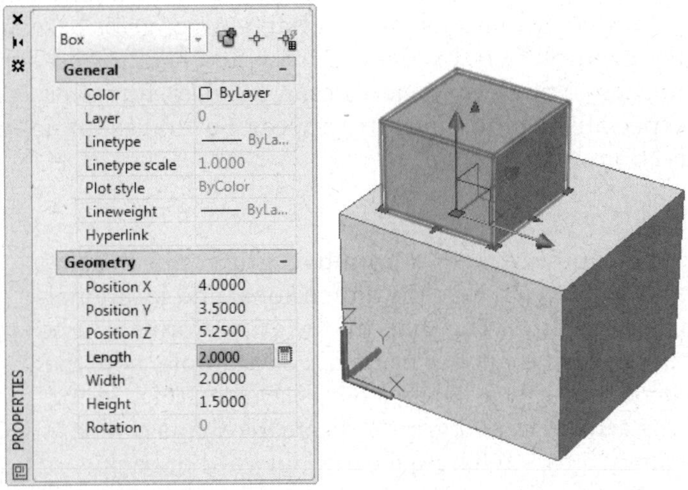


The important concept to note is that when you edit individual primitives (subobjects) of a composite solid using *Properties* or grips, the composite solid remains one unit. In other words, although the small box in our example changed dimensions and location, it is still *Unioned* to the large box.

## EDITING SOLIDS USING GRIPS AND GIZMOS

The most intuitive method for editing solids is using the grips and gizmos that are available. When grips and gizmos are turned on, they appear on the object when selected. The grips and gizmos allow you to change the solids in different ways as described in this section. Turn grips on and off using the *GRIPS* variable (0 = off, 1 or 2 = on), and control the current gizmo using the *Selection* panel in the *Home* tab of the *3D Modeling* workspace (see Fig. 36-5) or changing the *DEFAULTGIZMO* variable setting (0= move, 1 = rotate, 2 = scale, 3 = no gizmo).

You can also edit individual subobjects within solids. "Subobject" is the term AutoCAD uses to describe a component of a solid. You can select composite solids to edit, individual primitives contained in composite solids, and even individual faces, edges, and vertices.

There are numerous possibilities for editing solids using grips since this editing method is so dynamic and flexible. Every situation cannot be described in this section; therefore, only the basic principles and most common possibilities are discussed.

This section will explain the following concepts:

- Grip editing primitive solids
- Move, rotate, and scale gizmos
- Grip editing composite solids
- Subobjects and selection
- Grip editing faces, edges, and vertices
- Selecting overlapping primitives
- History settings

TIP

NOTE: To use grip editing features for solids, make sure grips are turned on. Do this by setting the *GRIPS* variable to 1 or 2 or checking *Show Grips* in the *Selection* tab of the *Options* dialog box. Also turn *Culling Off* (*CULLINGOBJ* = 0) by selecting the *Culling* toggle (see Fig. 36-5) if you want to see grips that are behind or beneath solids (obscured from view).

## Grip Editing Primitive Solids

Assuming grips are turned on, when a solid is selected, one or more grips appear. Each type of solid primitive displays its own special types of grips based on the geometric characteristics of the primitive. For example, a *Box* is defined by width, height, and depth, so appropriate grips appear allowing you to change those characteristics. A *Cylinder*, however, is defined by a base diameter and height, so grips especially for those features appear. Several examples of solids and the related grips are described in this section.

### *Box*

When you select a *Box* primitive, grips appear that allow you to change the dimensional and locational characteristics. The shape of the grip informs you how the primitive can be changed. For example, note the grips that appear when a *Box* is selected (Fig. 36-9). The arrow-shaped grips indicate the shape can be changed by stretching the grip forward or backward in the specified direction. Square grips indicate motion in any direction.

**FIGURE 36-9**

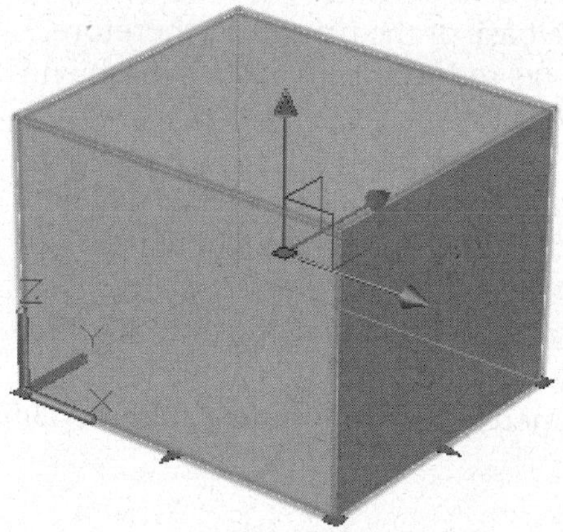

Change the dimensions of the *Box* using any of the arrow-shaped grips. For example, select a grip to make it hot, then stretch the face of the solid by moving the grip inward or outward as illustrated in Figure 36-10. The arrow grips stretch <u>only</u> in the indicated direction. *Polar Tracking* does not have to be turned on, but may help indicate the current length and can be used in conjunction with *Polar Snap*. You can also enter a value at the command prompt to specify a distance to stretch the solid.

Specify point location or [Base point/ Undo/eXit]:
`PICK` or (`value`)

**FIGURE 36-10**

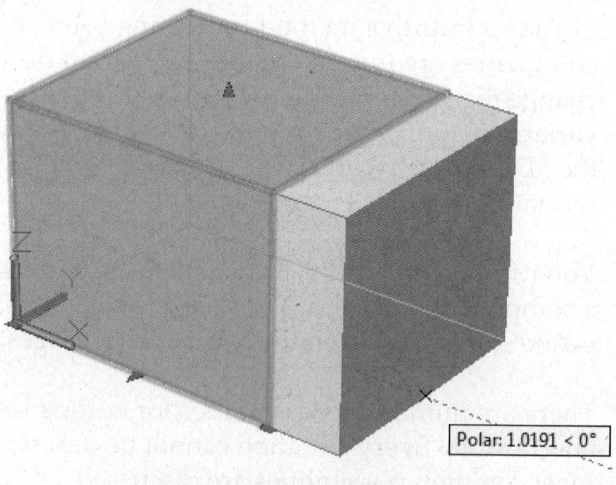

Stretch any other face inward or outward using the appropriate arrow grip. For example, change the height of the *Box* by stretching the arrow grip on top or bottom (Fig. 36-11).

**FIGURE 36-11**

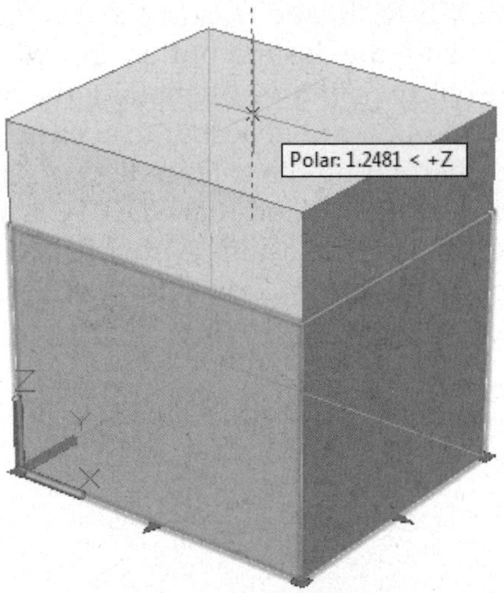


The square grips indicate that you can stretch the corner of the solid in any direction. For example, you can select a corner (square) grip to make it hot, then shorten one dimension while lengthening another (Fig. 36-12). The same prompt appears at the Command line where you can enter coordinates for the new location of the corner.

Specify point location or [Base point/Undo/eXit]:
`PICK` or (`coordinates`)

**FIGURE 36-12**

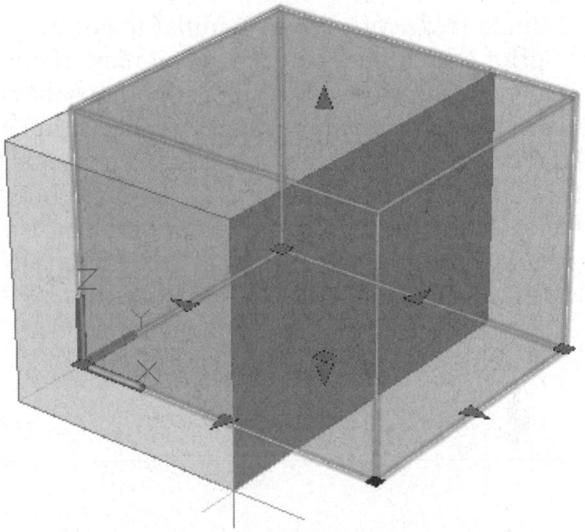

If you select the square grip at the center of the base, the location of the entire primitive can be changed as shown in Figure 36-13. *Polar Tracking* can be used to ensure movement along a specified angle.

When the solid's "base" grip is selected (the square grip at the center of the base), the Command prompt changes to the standard grip options.

** STRETCH **
Specify stretch point or [Base point/ Copy/Undo/ eXit]:

NOTE: You must turn *Culling Off* (*CULLINGOBJ* = 0) if you want to see grips that are behind or beneath solids (obscured from view).

**FIGURE 36-13**

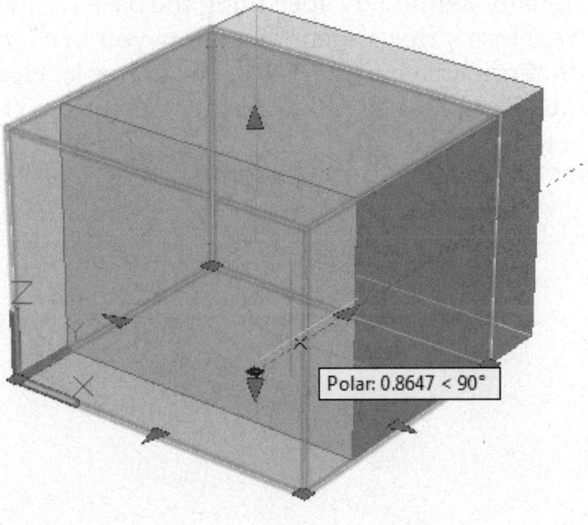

Cycle through the five options (STRETCH, MOVE, ROTATE, MIRROR, and SCALE) to change the entire primitive using the desired option. For example, you can rotate the entire primitive by making the base grip hot, then pressing Enter or spacebar to cycle to ROTATE, then dynamically rotating the solid in the current XY plane (Fig. 36-14). Keep in mind that STRETCH and MOVE accomplish the same results. Also, MIRROR has an effect on the solid only if the Copy suboption is used.

**FIGURE 36-14**

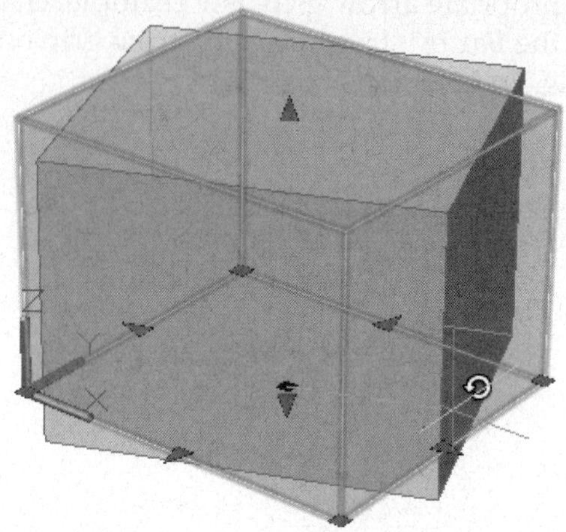

### *Wedge*

Editing a *Wedge* primitive is similar to editing a *Box* in that the grips that appear on each are similar. The height grip on a *Wedge* appears at the highest point of the solid and can be used to stretch the height as shown in Figure 36-15. All other grip options are the same as for a *Box*, including the prompt that appears on the Command line. Remember that selecting the grip at the center of the base allows you to cycle through the five normal grip options (STRETCH, MOVE, ROTATE, MIRROR, and SCALE).

**FIGURE 36-15**

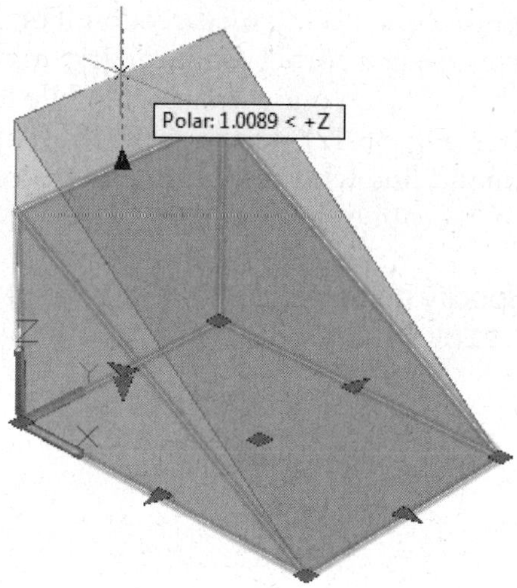


### *Cone*

Grips that appear on a *Cone* primitive when selected are displayed in Figure 36-16. Remember that a *Cone* is dimensionally defined by specifying the base radius and height; therefore, grips appear that allow you to change those dimensional characteristics. For example, Figure 36-16 illustrates dragging the top grip upward to change the height of the solid.

**FIGURE 36-16**

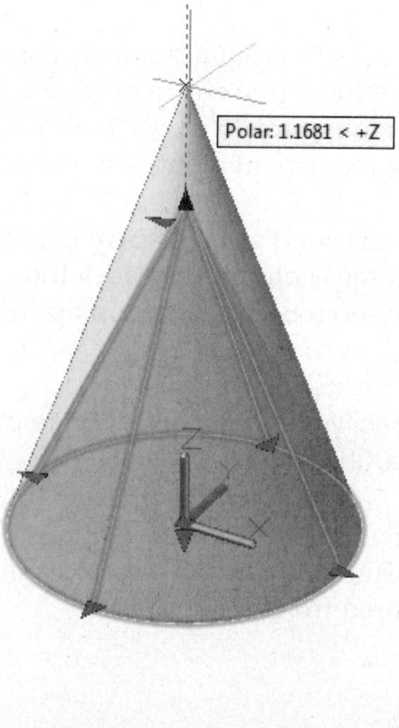

FIGURE 36-17

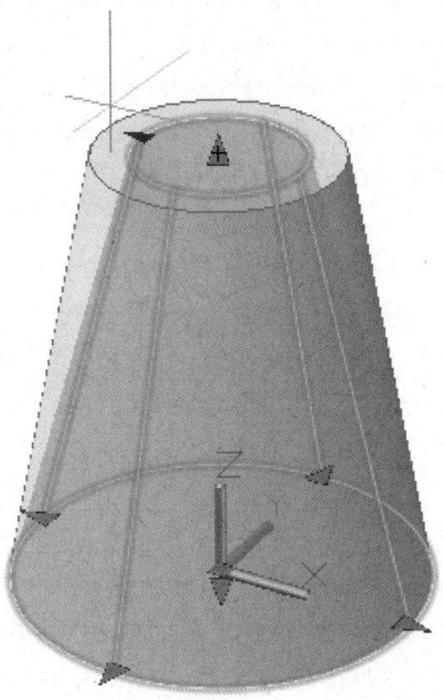

Also remember that a *Cone* can have a top radius. In such a case, the cone is a frustum (truncated cone). Even if a top radius was not defined initially and a normal cone was created, a top radius grip appears (see previous Fig. 36-16). Figure 36-17 displays the process of dragging the top radius grip to create a frustum.

### Sphere

In the case of a *Sphere,* four arrow-shaped grips appear (not shown). Therefore, you can change only the diameter dimension. In addition, a center grip appears for changing location.

### Cylinder

FIGURE 36-18

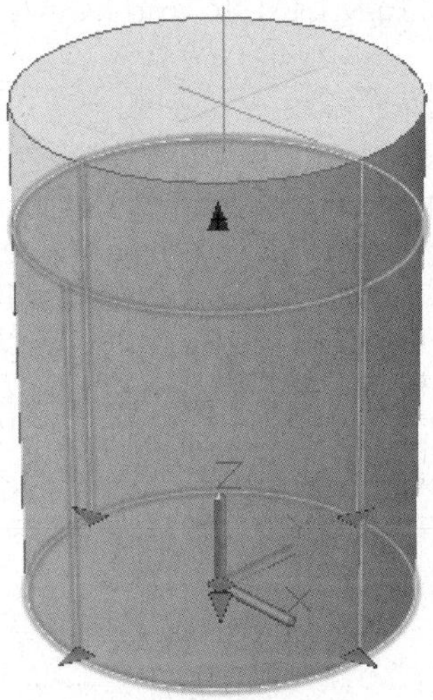

A *Cylinder* is defined by the base radius and height, so the related grips appear when the primitive is selected as shown in Figure 36-18. You can change the height at the top or bottom grip or change the radius using one of the four arrow grips. Move the solid or access the five normal grip options by selecting the grip at the center of the base.

### Pyramid

FIGURE 36-19

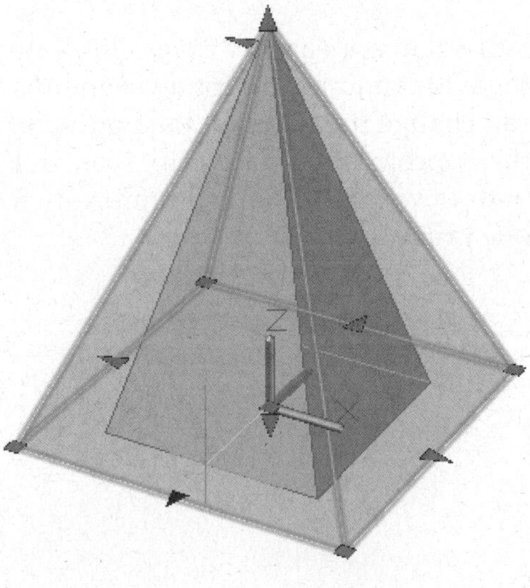

A *Pyramid* is also defined by the base radius and height. As you can see in Figure 36-19, selecting a *Pyramid* primitive reveals grips at each corner and along each side of the base; however, you can change only the base diameter with these eight grips. Selecting any one of the corner grips or side grips changes all corners and sides uniformly. One side or corner cannot be changed individually because a *Pyramid* primitive is a right pyramid (the apex is always directly above the center of the base).

Note that a *Pyramid* can have a top radius, similar to a *Cone*. Therefore, a related top radius grip can be used to create a top radius for a *Pyramid* originally defined with no top radius, as shown in Figure 36-20. Likewise, you can change a previously defined top radius to a *Pyramid* with an apex (top radius of 0).

**FIGURE 36-20**

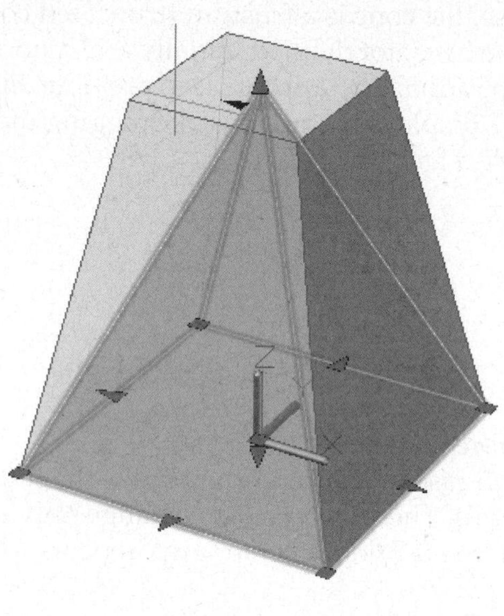

NOTE: Although a *Pyramid* can be defined initially with as many sides as you want, you cannot change the original number of sides using grips. You can, however, use the *Properties* palette to reset the original number of sides to any practical number.

### *Torus*

When a *Torus* primitive is selected for grip editing (not shown), four arrow grips appear allowing you to change the tube radius. In addition, one arrow grip allows you to change the torus radius. You can also use the grip at the center to change the location of the primitive or to access the five normal grip editing options (STRETCH, MOVE, ROTATE, MIRROR, and SCALE).

### *Polysolid*

A *Polysolid* can be altered in a number of ways using grips. Remember that a *Polysolid* is a multi-segmented shape and has a predefined width (between "walls") and a predefined height. Like a *Pline*, a *Polysolid* is created by specifying points for each segment. When selected, a *Polysolid's* corner grips allow you to change the location of the segment vertices. For example, Figure 36-21 illustrates stretching a corner grip to change the location of one vertex.

**FIGURE 36-21**

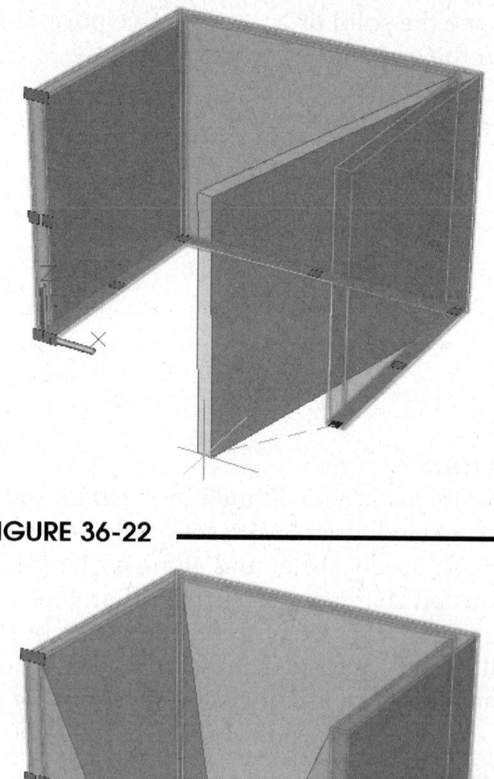

Grips that appear on a *Polysolid* allow you to change both the width and height for all segments. For example, you can change the width at the bottom of the *Polysolid* using the appropriate square grip. Note in Figure 36-22 that the bottom width is changed uniformly for all segments with one grip.

**FIGURE 36-22**

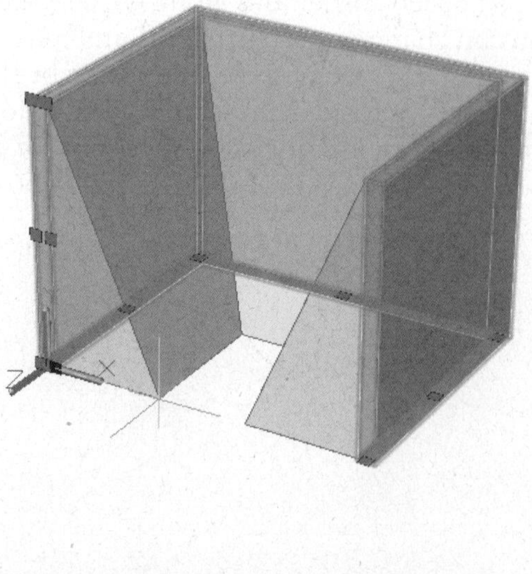

Similarly, dragging one of the top width grips changes the width for all segments uniformly. In this case as shown in Figure 36-23, the top width is increased uniformly as well as the height for all outside faces.

**FIGURE 36-23**

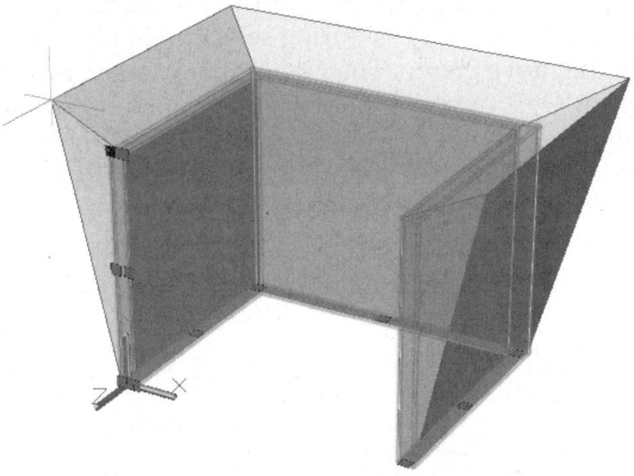

## Move, Rotate, and Scale Gizmos

There are three ways to edit solids using grips and/or gizmos:

A. Use the *3Dmove, 3Drotate,* or *3Dscale* commands. This allows use of the gizmos without grips. (See Chapter 35 for information on these commands.)

B. Use the normal grip options without the gizmos (as described in the previous section). To do this, simply select any grip on the solid, face, edge, without selecting a gizmo if one appears (*GRIPS* must be on). The STRETCH option appears at the command prompt. You can cycle through all five options by pressing Enter.

C. Use the move, rotate, and scale gizmos with grips or with *Object Snaps*. Either of these methods works basically the same. To do this, select the object (*DEFAULTGIZMO* must be set to anything but 3, or off) so a gizmo appears. Next move the gizmo to the desired location. When you select one of the gizmo planes or axes, the MOVE, ROTATE, SCALE, or STRETCH option appears at the command prompt. Use that option to change the object, or press enter to switch to another option (MOVE, ROTATE, SCALE, or STRETCH) and related gizmo. This method is described next.

TIP

NOTE: You can use the *Selection* panel of the *Solid* tab of the Ribbon (see Fig. 36-5) to select the default gizmo that appears when the object is selected (*DEFAULTGIZMO* system variable). No matter which gizmo appears first, you can select the gizmo and then cycle through to the other gizmos by pressing the spacebar.

### Using Gizmos with Grips or *Object Snaps*

When a solid is selected and assuming gizmos are on (*DEFAULTGIZMO* is set to anything but 3, or off), a gizmo appears at the center of the object (Fig. 36-24). The move gizmo appears by default unless this setting is changed. You can then activate the gizmo by selecting a plane or axis, or you can move the gizmo to a different location on the object. To move the gizmo to the desired location, select the gizmo grip to make it hot, and then relocate the gizmo. To move the gizmo effectively, use grips or *Object Snaps*.

**FIGURE 36-24**

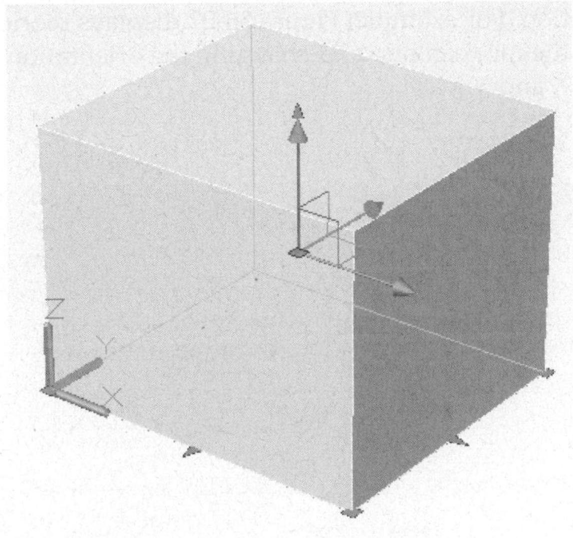

When the gizmo is at the desired location, select the desired axis or plane on the grip tool. Figure 36-25 displays this process during selection of the X axis of the move gizmo. The gizmo is now activated and the MOVE option appears at the Command prompt allowing you to move the solid along the gizmo's X axis. Pressing the Esc key allows you to select a different axis or plane on the move gizmo.

FIGURE 36-25

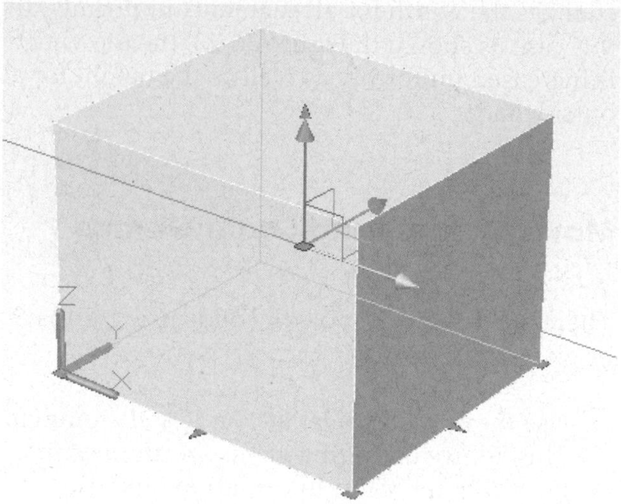


When any gizmo is used, only the MOVE, ROTATE, SCALE, and STRETCH options appear on the Command line when you press Enter to cycle through the options. For example, if you cycle to ROTATE, the move gizmo is automatically replaced by the rotate gizmo. Figure 36-26 illustrates rotating the *Box* about the gizmo's X axis as previously selected. You can select a different axis to rotate about by first pressing the Esc key, then selecting a different axis on the rotate gizmo.

FIGURE 36-26

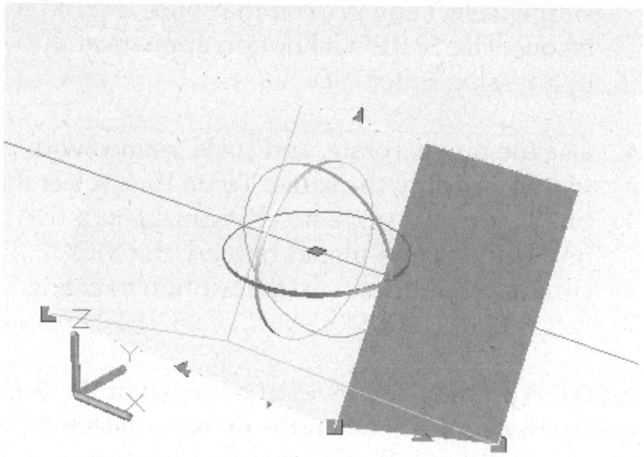


If you relocate a gizmo, you can use *Dynamic UCS* to assist in moving the gizmo to the desired orientation. For example, Figure 36-27 displays reorienting the move gizmo and changing the orientation of the X and Y axes.

FIGURE 36-27

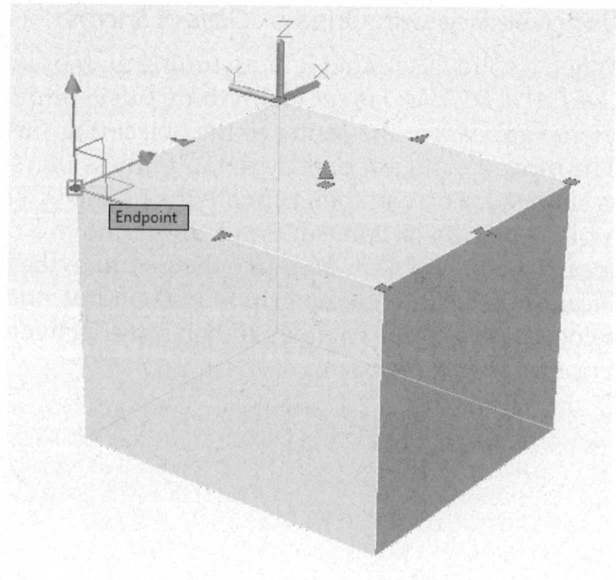

## Grip Editing Composite Solids

When you select a composite solid for editing with grips, only the one grip appears at the center of the base as shown in Figure 36-28. The other grips that appear on individual primitives do not appear by default, but can be selected by using a filter or holding down the Ctrl key and hovering the cursor over a primitive, then selecting it (see "Subobjects and Selection" next).

NOTE: You must turn *Culling Off* (*CULLINGOBJ* = 0) if you want to see grips that are behind or beneath solids (obscured from view).

**FIGURE 36-28**

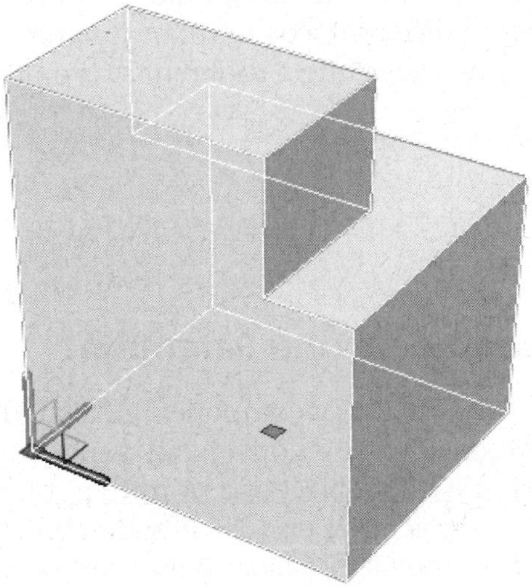

The single base grip that appears when a composite solid is selected, however, allows you to edit the composite solid in the same way as described in the previous section. That is, if you select the base grip so it becomes hot, you can cycle through all five normal grip options: STRETCH, MOVE, ROTATE, SCALE, and MIRROR. For example, using the normal grip options, once the base grip is hot, you cycle to the ROTATE option to rotate the composite solid in the XY plane of the grip (Fig. 36-29).

**FIGURE 36-29**

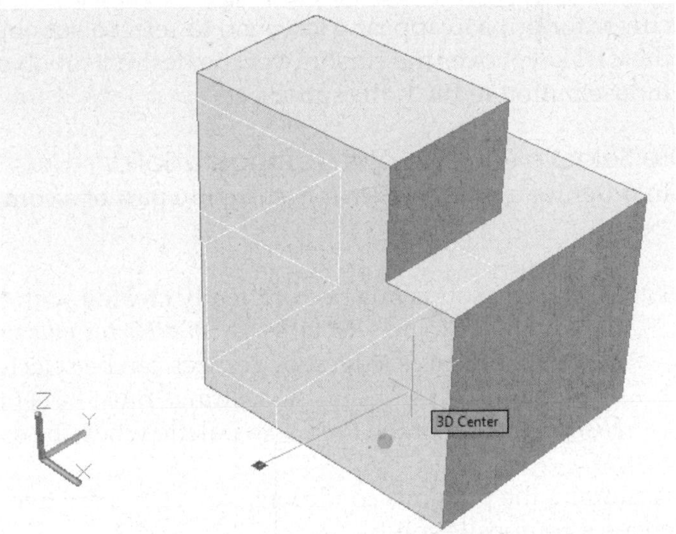


If gizmos are turned on, hover over the grip until the gizmo automatically relocates to the location of the base grip. At this point, select the desired axis or plane on the gizmo (Fig. 36-30). You can press the Esc key to select a different axis or plane on the gizmo. Remember that only the STRETCH, MOVE, SCALE, and ROTATE options appear on the Command line when you use the gizmo.

**FIGURE 36-30**

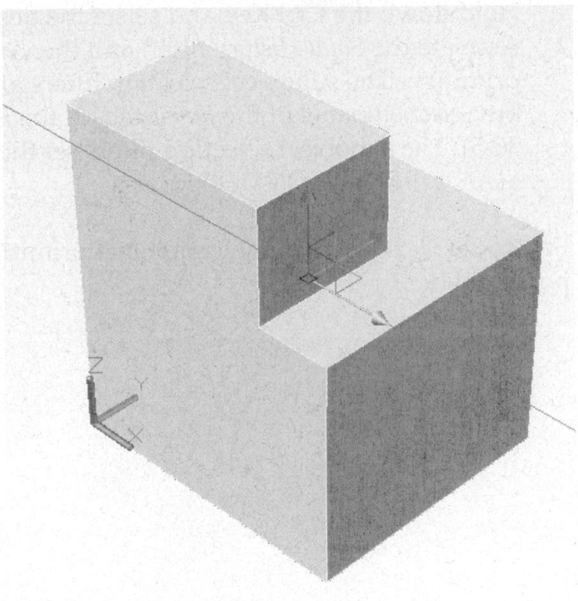

For example, if you cycle to ROTATE, the move gizmo is automatically replaced by the rotate gizmo. By default the composite solid rotates about the selected axis (Fig. 36-31). You can select a different axis to rotate about by first pressing the Esc key, then selecting a different axis on the rotate gizmo.

FIGURE 36-31

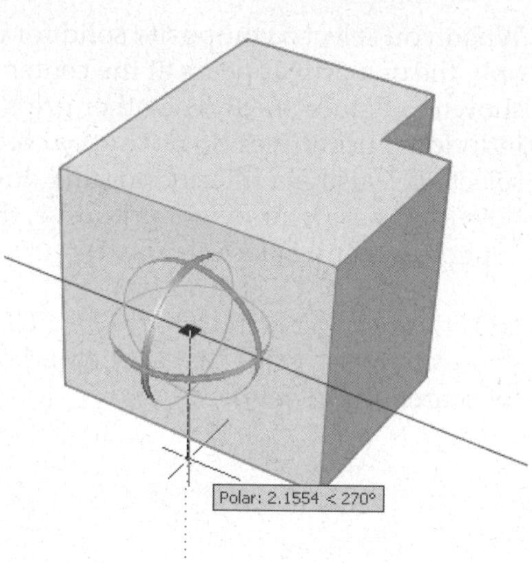


## Subobjects and Selection

A "subobject" is a component of a solid. A primitive within a composite solid is considered a subobject. Also, a face, edge, or vertex is a subobject of a primitive solid. The grips that appear on subobjects are exclusive to that type of subobject depending on the type of subobject: edge, vertex, face, or type of primitive.

As you already know, to edit a solid (not a subobject) using grips, you simply select the solid with the cursor for grips to appear. However, to select a subobject, you can use a selection filter or hold down the Ctrl key, hover the cursor over the desired subobject until it becomes highlighted, then use the left mouse button to PICK the subobject.

### To Select a Primitive of a Composite Solid

In order for a primitive (that is currently part of a composite solid) to be selectable, two conditions must exist:

1. Composite solids must be originally created with *Solid History* turned *On* (*SOLIDHIST* = 1) in order for primitives to be selectable. If *Solid History* was not turned on when the composite objects were created, only faces, edges, or vertices can be selected.
2. The composite solid's history settings must be set to *Record*. This can be accomplished in the *Solid History* section of the *Properties* palette when the solid is selected.

TIP

Assuming the previous conditions exist, there are two ways to select one primitive that was used to create a composite solid:

1. Hold down the Ctrl Key and select the desired primitive.
2. Activate the *Solid History* filter, and then select the desired primitive. The subobject selection filters are located on the *Selection* panel of the *Home* tab on the Ribbon (see Fig. 36-5). The subobject selection filters set the appropriate *SUBOBJSELECTIONMODE* value.

Figure 36-32 displays the selection of a primitive (*Box*) in a composite solid.

FIGURE 36-32

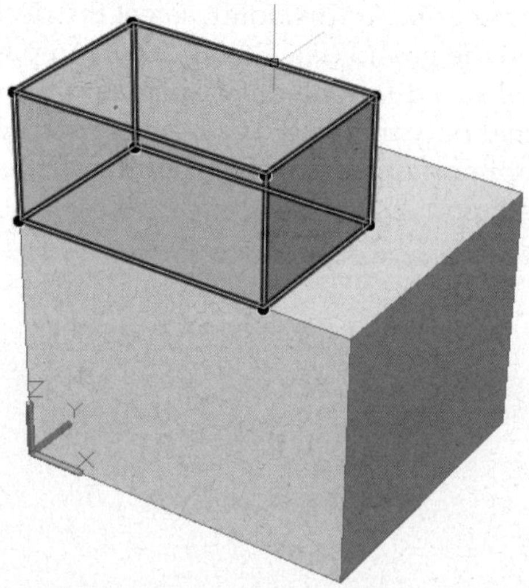

### To Select an Edge, Face, or Vertex

Edges, faces, and vertices of any solid can be selected. There are two ways to select these subobjects:

1. Activate the appropriate subobject selection filter (*Edge*, *Face*, or *Vertex*), and then select the desired subobject(s). The subobject selection filters are located on the *Selection* panel of the *Home* tab on the Ribbon. The subobject selection filters set the appropriate *SUBOBJSELECTIONMODE* value.
2. Hold down the Ctrl Key and select the desired subobject. Using the Ctrl key allows you to select any edge, face, or vertex since this method overrides any subobject selection filters.

FIGURE 36-33

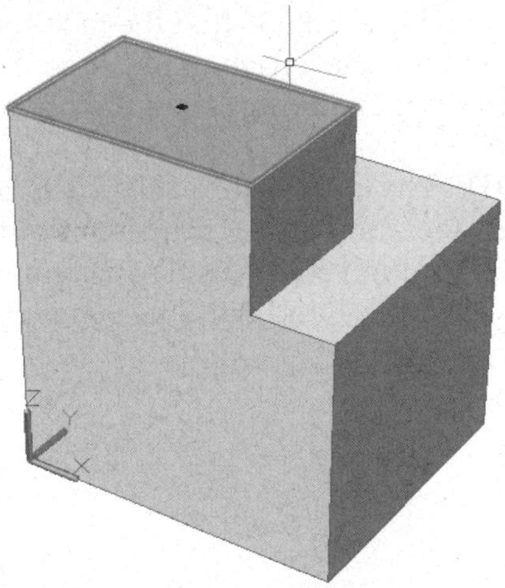

Selected edge, face, and vertex subobjects display different types of grips, depending on the subobject type. The composite solid in Figure 36-33 displays the top face selected. One grip appears at the center of a face.

FIGURE 36-34

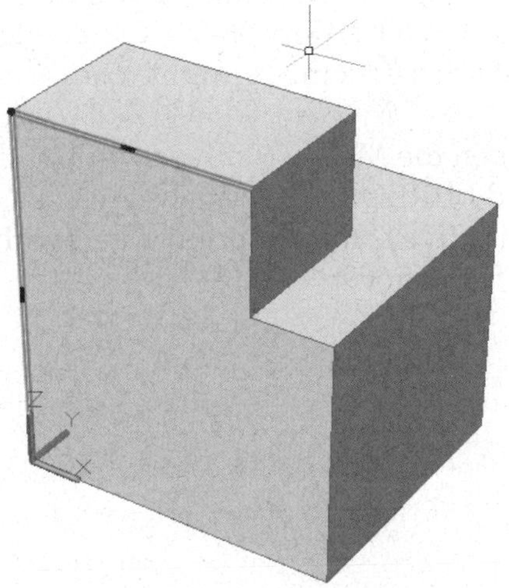

TIP

NOTE: If you cannot select the desired face or edge of a composite solid because it resides on an original primitive, the composite solid's history settings must be set to *None* in the *Solid History* section of the *Properties* palette.

You can select more than one subobject on any 3D object. Also, the selection set can include more than one type of subobject. For example, Figure 36-34 displays the selection of two edges and one vertex. The best way to select multiple types of subobjects is to press and hold Ctrl rather than changing selection filters. You can remove an item from the selection set by pressing and holding Shift and selecting it again.

## Grip Editing Faces, Edges, and Vertices

FIGURE 36-35

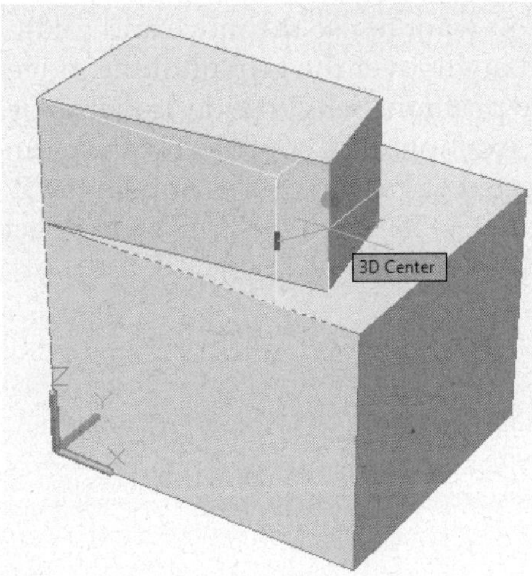


An individual face, edge, and vertex on a solid has a distinct grip shape. Each type of grip can be used to edit the object feature in predefined ways. Similar to using grips to edit primitive solids and composite solids, you can choose to edit with or without the move and rotate gizmos. The methods for activating the move and rotate tools for faces, edges, and vertices are the same as previously described for primitives and composite solids.

### Edge Grips

Several examples for using edge grips are described and illustrated in this section; however, many more possibilities exist. Using our composite solid example, assume the top right edge grip was selected for editing. Without using the move and rotate gizmos, you can cycle to all five grip options. Using the STRETCH option would allow you to stretch the selected edge in any direction as shown in Figure 36-35. Note that other edges and faces are also affected; however, only the primitive containing the selected edge is changed.

Cycling to the ROTATE option allows you to rotate the edge. Note in Figure 36-36 that the face containing the selected edge is rotated but the length of the edge itself does not change. Other primitives in a composite solid are not affected.

NOTE: The composite's *Solid History* setting may affect how edge, face, and vertex grips manipulate the objects. Figures 36-35 through 36-39 display a composite solid created with *SOLIDHIST*=1 and with the *Solid History* set to *Record*.

**FIGURE 36-36**

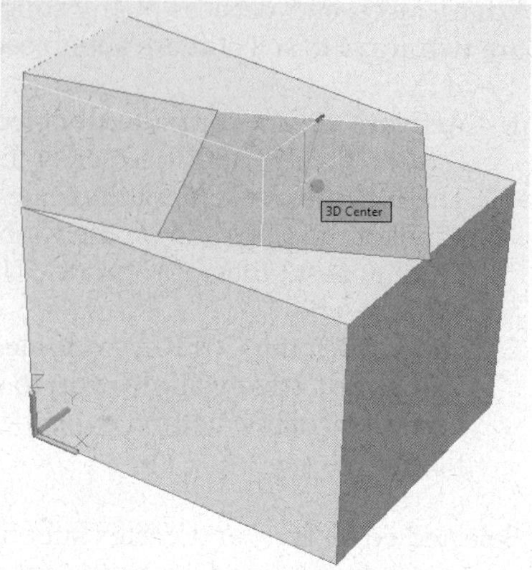


The SCALE option allows you to change the length of the selected edge (Fig. 36-37). Other edges and faces are affected but only those on the primitive containing the edge.

When the MOVE (with or without the move and rotate gizmos) and MIRROR grip options are used with an edge grip, the entire primitive is moved or mirrored without any other changes or deformations.

**FIGURE 36-37**

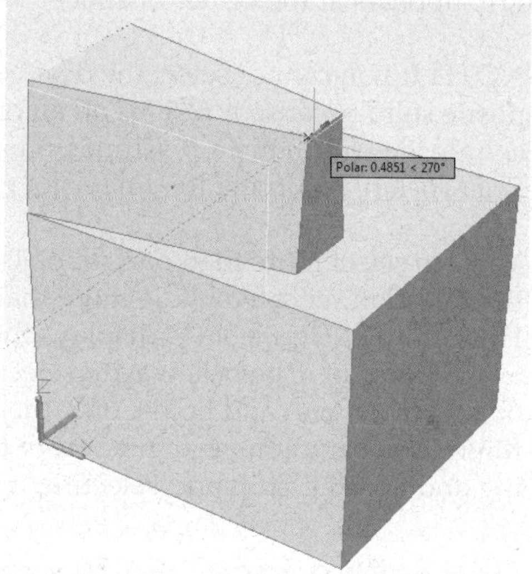


If you want to use the move and rotate gizmos, press Ctrl and hover over the grip until the move tool relocates to the grip position. Select the desired axis or plane on the gizmo. For example, you could select the gizmo's Z axis to ensure the selected edge grip moves only in the Z direction, as indicated in Figure 36-38. Press the Esc key to select another axis or plane.

**FIGURE 36-38**

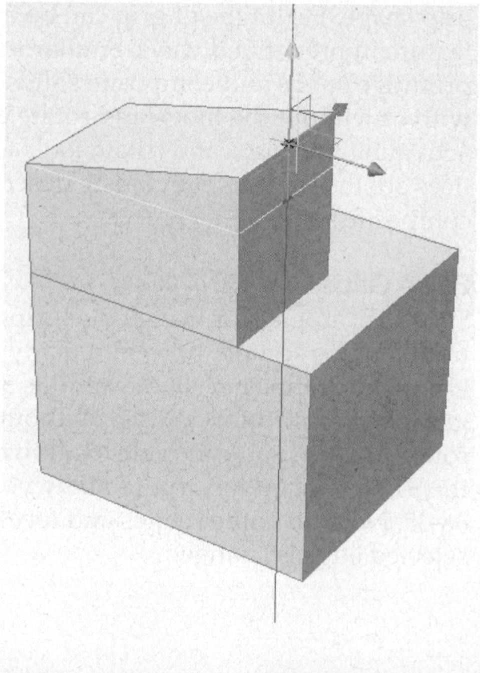

When you cycle to the ROTATE option, the rotate gizmo appears. By default, rotation occurs about the previously selected axis. Press Esc to select another axis on the rotate gizmo. In Figure 36-39, the rotate gizmo's X axis is used. Beware since the selected edge does not change length using any option other than SCALE; therefore, the primitive may become deformed depending on the axis of rotation as shown in Figure 36-39.

**FIGURE 36-39**

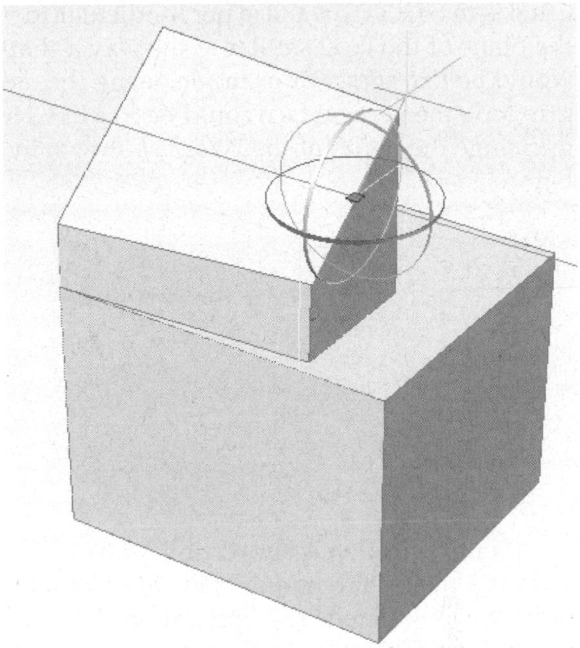

Another possibility for using an edge grip is shown in this example of a *Polysolid* primitive (Fig. 36-40). Note that one edge grip has been selected for editing. Using the STRETCH mode with the move gizmo allows a change to be made to the width of a single segment of the *Polysolid*. This is in contrast to changing the width of all segments using the primitive's grips normal grips (not the subobject edge grip) as shown previously in Figures 36-22 and 36-23. Because there are so many types of primitive and composite solid combinations, the possibilities for editing with grips are practically endless.

**FIGURE 36-40**

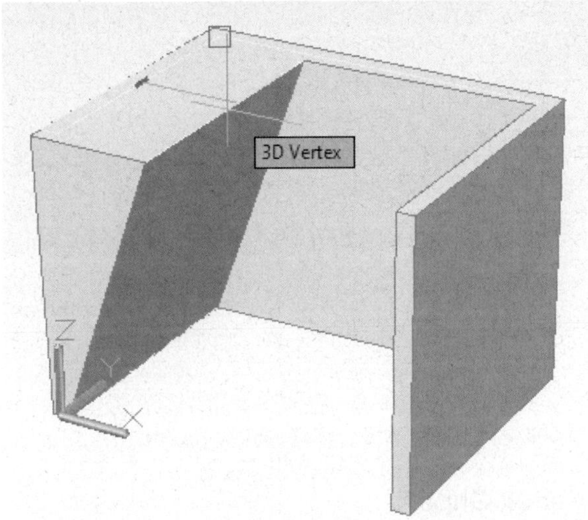


## Face Grips

A face grip of a primitive solid or of a composite solid can be selected using the same procedure as explained earlier (using the Ctrl key). Faces of solids are usually planar but can include some curved faces. Grips that appear on faces of solids appear as a circle on the plane of the face as shown on the vertical front face of the *Wedge* primitive in Figure 36-41.

**FIGURE 36-41**

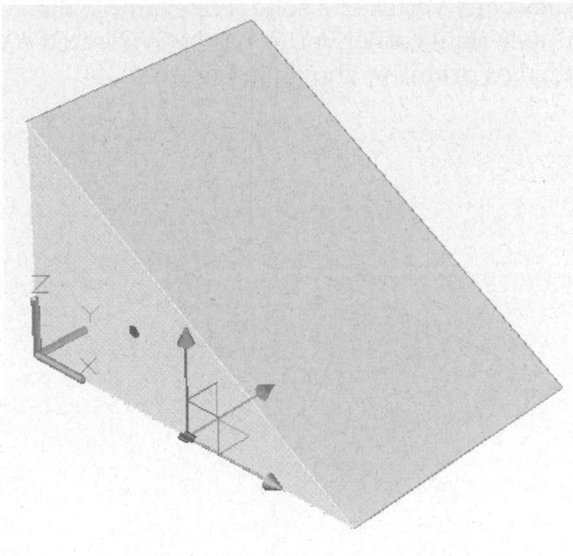

One of the most useful applications of using face grips is to STRETCH a face perpendicular to the plane of the face, similar to the way a shape would be *Extruded*. For example, using the move grip tool, the vertical face could be STRETCHed to change the depth of the *Wedge* as shown in Figure 36-42.

**FIGURE 36-42**

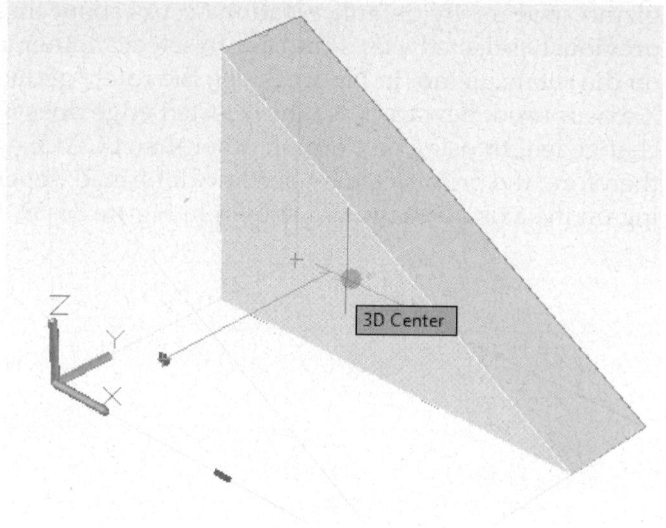


Another application is shown in Figure 36-43. In this case, the right face of a *Pyramid* (with a top radius) was selected. The illustration shows the face being STRETCHed using the move gizmo.

**FIGURE 36-43**

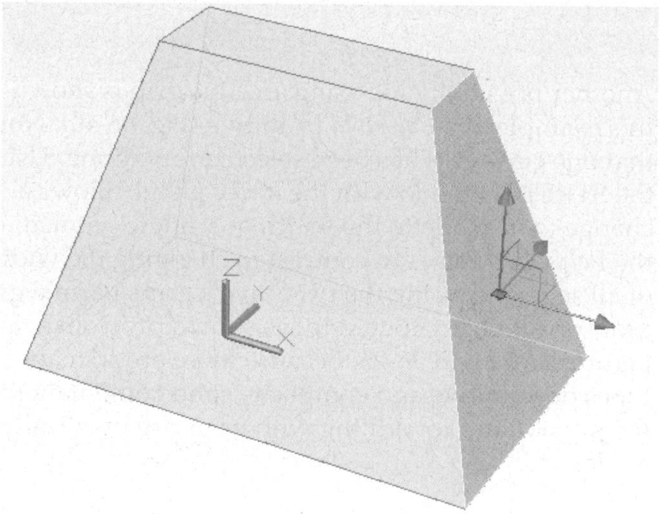

### Vertex Grips

A vertex grip can be selected by using Ctrl to select any vertex of a solid. For example the upper-right corner vertex has been selected on the *Box* primitive shown in Figure 36-44.

**FIGURE 36-44**

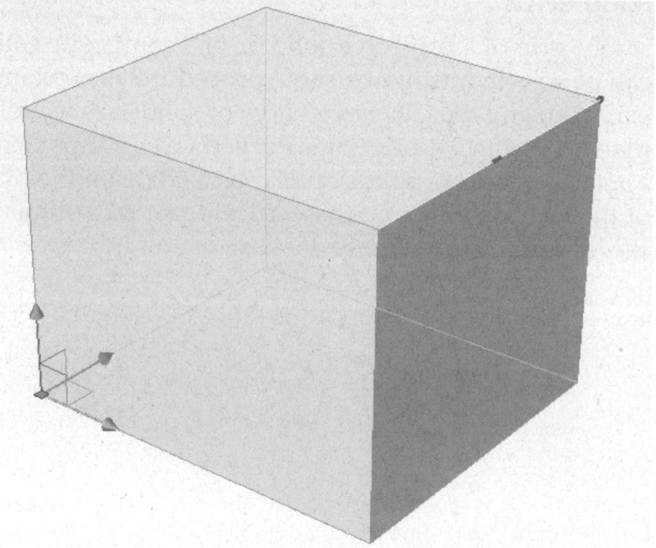

A vertex grip can be STRETCHed as shown in Figure 36-45. Only the selected vertex is affected along with any connected faces and edges. Interestingly, only the STRETCH can be used effectively with vertex grips. Since a vertex represents one point (with no length or dimension), it cannot be scaled or rotated. If the MOVE or MIRROR option is used, the entire solid is affected.

NOTE: Keep in mind that multiple grips can be selected. For example, you may want to ROTATE two faces together. Accomplish this by holding the Ctrl key down while selecting each face. Additionally, any combination of grip types can be selected. For example, you could select one face grip, one edge grip, and one vertex grip.

**FIGURE 36-45**

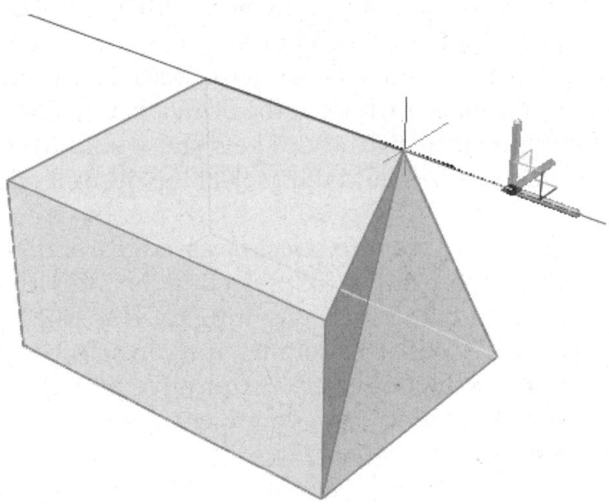

## Selecting Overlapping Subobjects

In some cases the subobject you want to select may be behind, or overlapping, another subobject, and therefore, difficult to PICK. When you "hover" over composite solids while holding Ctrl, the subobject in the foreground is detected first.

### *CULLINGOBJ*

By default, the *Culling* filter is on (*CULLINGOBJ* system variable=1). This prevents you from selecting subobjects that are normally obscured from view. If you want to select subobjects that are behind solids, turn the *Culling* filter off (*CULLINGOBJ*=0). This allows you to select any face, edge, or vertex.

For example, Figure 36-46 displays a composite solid with the three back edges selected. This is accomplished by turning *Culling Off,* then using the *Edge* filter to select the desired edges.

**FIGURE 36-46**

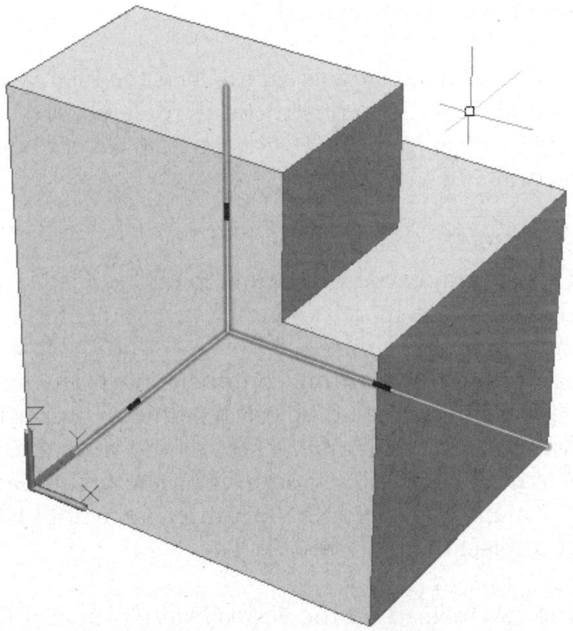

TIP

## History Settings

You can control the history settings for solids globally (for the entire drawing) and individually (for individual composite solids). Use the *SOLIDHIST* and *SHOWHIST* system variables to control the global history settings and the Properties palette to control history settings for individual solids.

### *SOLIDHIST*

TIP

This system variable controls whether new composite solids retain a history of their original components. When set to 1, all subsequently created composite solids retain a history of the original primitive objects. Therefore the primitives of a composite solid can be selected to edit. The possible settings are:

0 Sets the *History* property to *None* for new solids, and no history is retained (initial setting).

1 Sets the *History* property to *Record* for new solids, and solids retain a history of their original objects.

### Changing *Solid History* for Individual Primitives with *Properties*

If a composite solid was created with *SOLIDHIST On* (set to 1), the properties palette would display *History* as *Record*. With this setting, you can select individual edges, faces, and vertices only using the subject selection filters in the *Selection* panel of the Ribbon. However, if you prefer using the Ctrl key, you would first have to select the individual primitive, and then select the desired edges, faces, or vertices.

If you always want to select faces, edges, and vertices within composite solids without first selecting the individual primitive, you can change the solid's *History* setting to *None* using the *Properties* palette (Fig. 36-47). With this setting, you can select any face, edge, or vertex without having to select the primitive first. Remember that using the *Properties* palette allows you to change properties for the selected object(s) only.

FIGURE 36-47

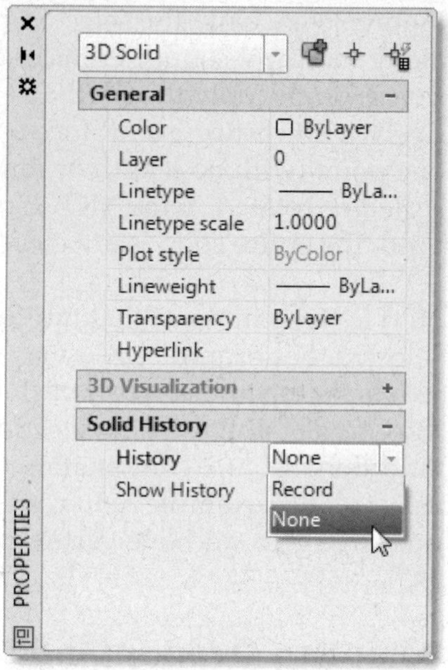


For example, assume the *History* setting for the composite solid shown in Figure 36-48 was changed to *None*. Selection of the L-shaped face would be possible with this setting. Otherwise, with a *History* setting of *Record*, the entire top face of the original *Box* primitive would be selected and highlighted.

FIGURE 36-48

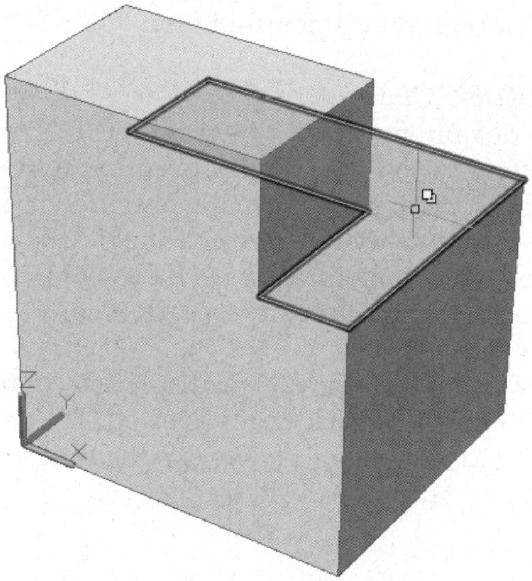

NOTE: Once *History* is set to *None*, the history of the composite solid's construction is deleted and cannot be retrieved. Setting the *History* back to *Record* begins a new record.

### *SHOWHIST*

This system variable controls the *Show History* property for solids in a drawing. When you combine two or more primitives into a composite solid using a Boolean operation, the resulting composite normally appears to be one singular solid object. Assume a composite solid was created using *Subtract* to remove a small *Box* primitive from a large *Box* primitive as shown in Figure 36-49. Normally, the individual primitives are not visible unless you select them as subobjects.

FIGURE 36-49

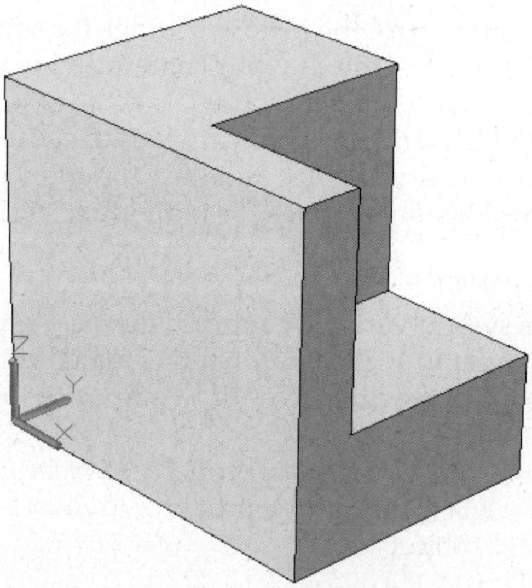

You can instead set the *Show History* property to *Yes* so the individual primitives comprising a composite solid are visible. Use the *SHOWHIST* system variable to control the global (drawing-wide) setting or the *Properties* palette to control the setting for individual composite solids.

The possible *SHOWHIST* system variable settings are:

0 Sets the *Show History* property to *No* (read-only) for all solids. Overrides the individual *Show History* property settings for solids. You cannot view the original primitives that were used to create the solid.
1 Initial setting. Does not override the individual *Show History* property settings for solids.
2 Displays the history of all solids by overriding the individual *Show History* property settings for solids. You can view the original primitives that were used to create the solid.

FIGURE 36-50

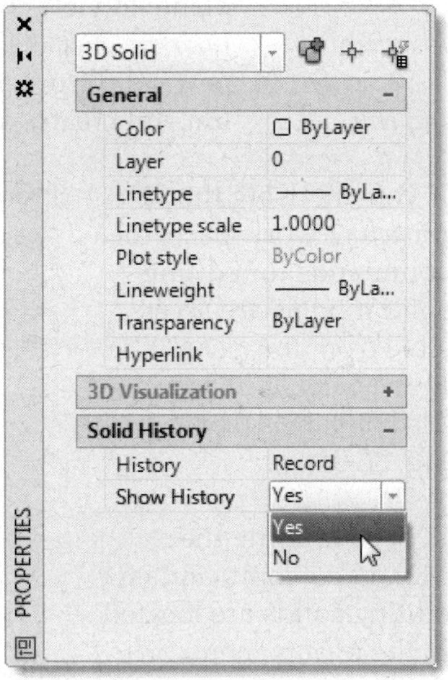


### Changing *Show History* for Individual Solids with *Properties*

As shown in previous Figure 36-49, with the default *SHOWHIST* settings, the individual primitives of a composite solid are not visible unless you select them as subobjects. If a composite solid's *Show History* property is set to *Yes* in the *Properties* palette, the individual primitives that make up the composite solid can be viewed. To do this, select the composite solid and change its *Show History* setting to *Yes* as shown in Figure 36-50.

When a composite solid's *Show History* setting is set to *Yes*, the individual primitives that were used to create the solid are visible at all times as shown in Figure 36-51.

FIGURE 36-51

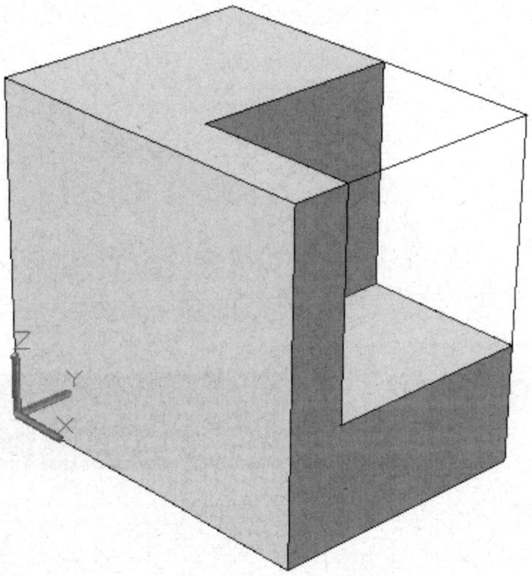

## SOLIDS EDITING COMMANDS

In addition to using grips to edit composite solids, AutoCAD offers several commands that allow you to change the geometry.

*Solidedit* This command has options for extruding, moving, rotating, offsetting, tapering, copying, coloring, separating, shelling, cleaning, checking, and deleting faces and edges of 3D solids.

*Slice* Use this command to cut through a solid using a plane or a surface. You can keep both parts of the solid or delete part.

*Presspull* This unique command combines creating an extrusion and performing a Boolean all in one command. The extrusion is created on a solid surface and then pushed into the solid's face (*Subtract*) or pulled out from the face (*Union*).

| | |
|---|---|
| *Sectionplane* | Use this command to create an imaginary variable cutting plane, similar to a clipping plane, to view the inside of a solid. |
| *Interfere* | *Interfere* checks two or more solids to see if they overlap in 3D space. You can create a new solid from an overlapping volume. |
| *Imprint* | Imprint creates an "edge" element on a solid that can be used for editing with grips. |

If you are using the *3D Modeling* workspace, the commands for editing solid models are accessible from the *Home* tab in the *Solid Editing* and *Modify* panels (see Fig. 36-52).

If you are using the Menu Bar, commands for editing solids are located in the *Modify* menu on the *Solids Editing* and *3D Operations* cascading menus (Fig. 36-52).

**FIGURE 36-52**

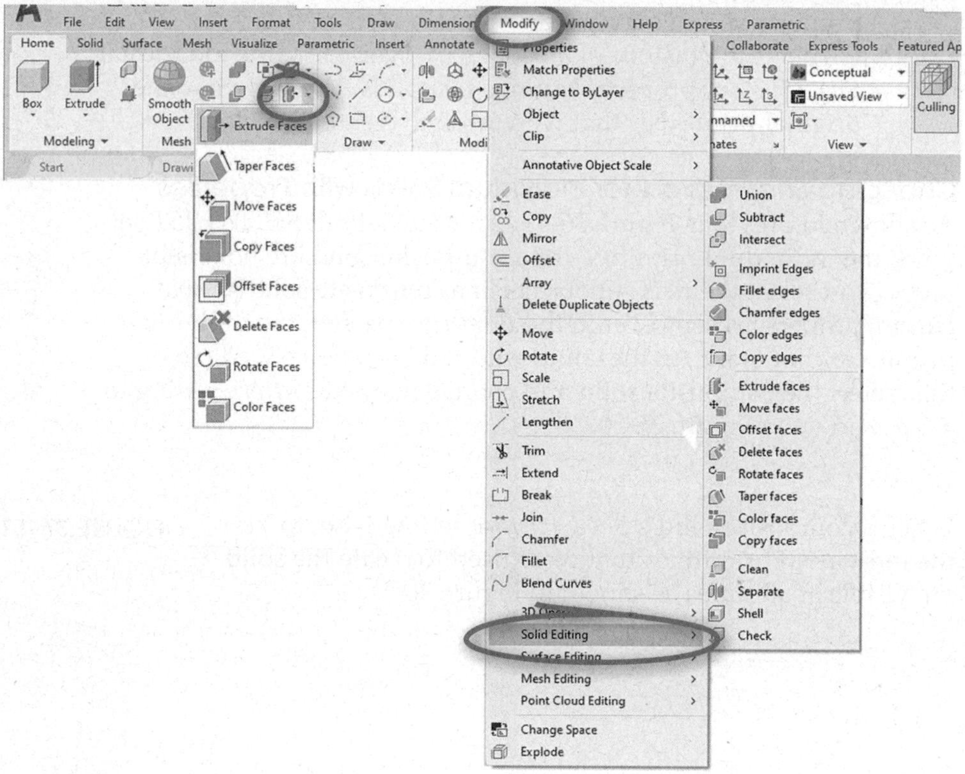


## *Solidedit*

| Ribbon | Menu Bar | Command (Type) | Alias (Type) | Shortcut |
|---|---|---|---|---|
| *Home*<br>*Solid Editing*<br>(icon) | *Modify*<br>*Solid Editing >* | *Solidedit* | ... | ... |

The *Solidedit* command has several "levels" of options. The first level prompts you to specify the type of geometry you want to edit—*Face*, *Edge*, or *Body*. The second level of options depends on your first level response as shown below.

Command: ***SOLIDEDIT***
Solids editing automatic checking: SOLIDCHECK=1
Enter a solids editing option [Face/Edge/Body/Undo/eXit] <eXit>: ***f***
Enter a face editing option
[Extrude/Move/Rotate/Offset/Taper/Delete/Copy/coLor/mAterial/Undo/eXit] <eXit>:

TIP

NOTE: The subobject selection filters can override the selection of a *Face*, *Edge*, or *Body*. Therefore, set the selection filter to *No Filter* in the *Selection* panel of the *Home* tab on the Ribbon (see Fig. 36-5).

*Solidedit* operates on three types of geometry: *Edges, Faces* and *Bodies*. An *Edge* is defined as the common edge between two surfaces that has the appearance of a line, arc, circle, or spline (Fig. 36-53 A). *Faces* are planar or curved surfaces of a 3D object (Fig. 36-53 B). A *Body* is defined as an existing 3D solid or a non-solid shape created with a *Solidedit* option.

**FIGURE 36-53**

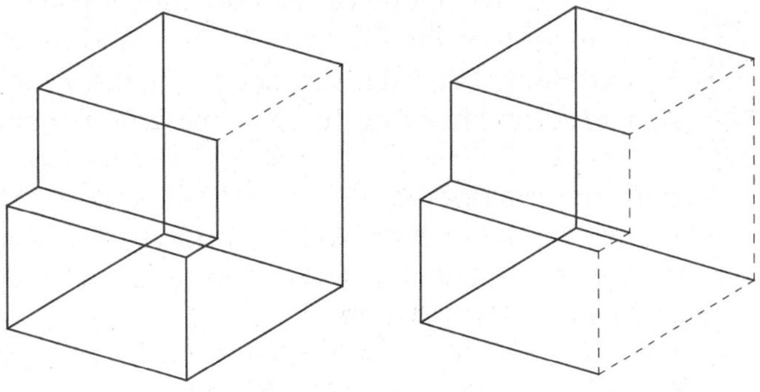

A. B.

TIP

Object selection is an integral part of using *Solidedit*. For example, once you have specified the type of geometry and the particular editing option, the *Solidedit* command prompts to select *Edges, Faces,* or a *Body*.

Select faces or [Undo/Remove]: Select a face or enter an option
Select faces or [Undo/Remove/ALL]: Select a face or enter an option

When you are prompted to select *Faces* or *Edges*, select the desired face or edge directly with the pickbox. You will most likely go through a series of adding and removing geometry until the desired set of lines is highlighted. Remember that you can hold down the Shift key and select objects to *Remove* them from the selection set.

If you use this command often, it may help to change the composite solid's *History* setting to *None* using the *Properties* palette (see previous discussion, "History Settings"). Setting the *History* to *None* allows you to use the Ctrl key while hovering over solids to select individual faces and edges rather than the entire primitives that make up the solid.

TIP

Ensure that you highlight exactly the intended geometry before you proceed with editing. If an incorrect set is selected, you can get unexpected results or an error message can appear such as that below.

Modeling Operation Error:
  No solution for an edge.

For example, many of the *Face* options allow selection of one or multiple faces. In Figure 36-54 both selection sets are valid faces or face combinations, but each will yield different results. Figure 36-54 A indicates selection of one face and B indicates selection of several faces defining an entire primitive. Add and remove faces or edges until you achieve the desired geometry.

**FIGURE 36-54**

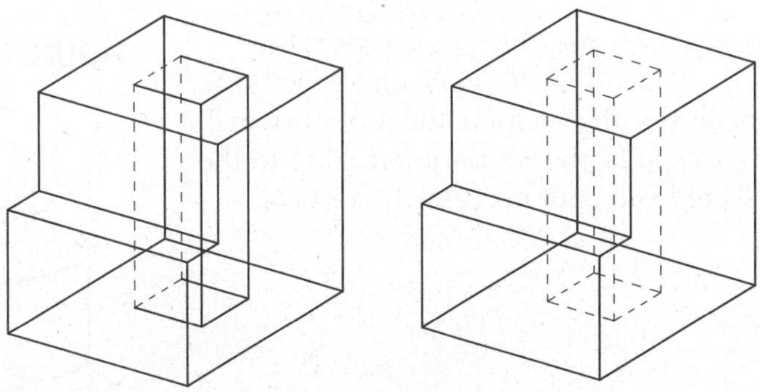

A. B.

## ***Face* Options**

The *Face* options edit existing surfaces and create new surfaces. A *Face* is a planar or curved surface. *Faces* that are part of existing 3D solids can be altered to change the configuration of the 3D composite solid. Individual *Faces* can be edited and multiple *Faces* comprising a primitive can be edited. New surfaces can be created from existing *Faces*, but entirely new independent solids cannot be created using these editing tools.

### Extrude

The *Extrude* option of *Solidedit* allows you to extrude any *Face* of a 3D object in a similar manner to using the *Extrude* command to create a 3D solid from a *Pline*. This capability is extremely helpful if you need to make a surface on a 3D solid taller, shorter, or longer. You can select one or more faces to extrude at one time.

Command: ***SOLIDEDIT***
Solids editing automatic checking: SOLIDCHECK=1
Enter a solids editing option [Face/Edge/Body/Undo/eXit] <eXit>: ***f***
Enter a face editing option
[Extrude/Move/Rotate/Offset/Taper/Delete/Copy/coLor/Undo/eXit] <eXit>: **e**
Select faces or [Undo/Remove]: **PICK**
Select faces or [Undo/Remove/ALL]: **PICK** or remove
Select faces or [Undo/Remove/ALL]: **Enter**
Specify height of extrusion or [Path]: (**value**)
Specify angle of taper for extrusion <0>: **Enter** or (**value**)
Solid validation started.
Solid validation completed.

For example, Figure 36-55 illustrates the extrusion of a face to make the 3D object taller, where A indicates the selected (highlighted) face and B shows the result. This extrusion has a 0 degree taper angle.

**FIGURE 36-55**

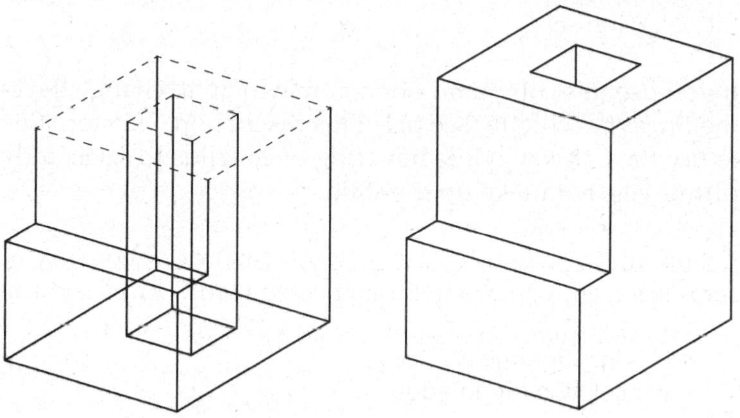

A. B.

Specifying a positive angle tapers the face inward (Fig. 36-56 A) and specifying a negative angle tapers the face outward (B). *Height* is always perpendicular to the selected *Face*, not necessarily vertical.

**FIGURE 36-56**

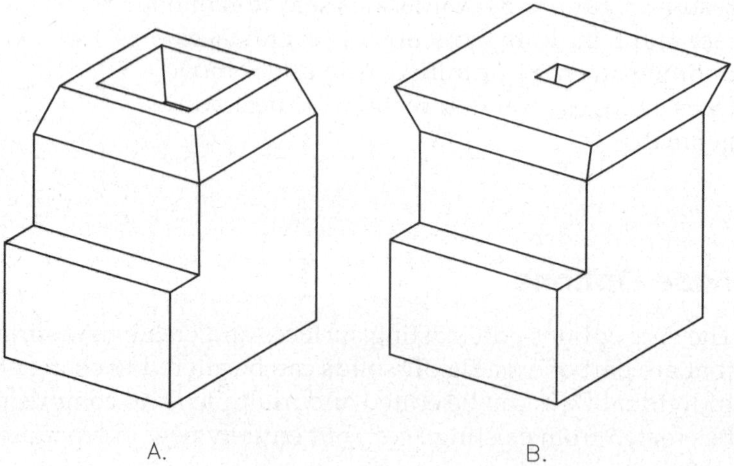

A. B.

An internal face can be selected for extrusion. In Figure 36-57, a single face is selected (A) and the *Height* value specified is greater than the distance to the outer face of the solid, creating an open side on the object (B). *Extrude* allows an internal face to pass through an external face as shown in Figure 36-57 B. Several other options, such as *Move, Rotate,* and *Offset* also allow the interior primitives to pass through the bounding box.

**FIGURE 36-57**

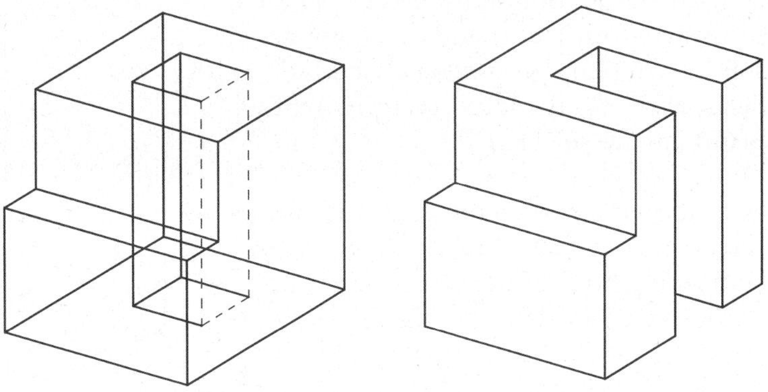


Faces can be extruded along a *Path*. The *Path* object can be a *Line, Circle, Arc, Ellipse, Pline,* or *Spline*. For example, Figure 36-58 illustrates extrusion of a face along a *Line* path (A) to yield the result shown on the right (B).

**FIGURE 36-58**

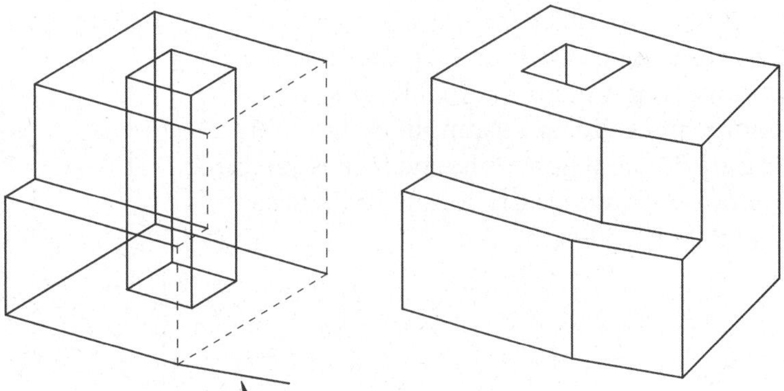


NOTE: Entering a positive value at the "Specify height of extrusion or [Path]:" prompt increases the volume of the solid and entering a negative value decreases the volume of the solid. For example, entering a positive value for extruding the face in Figure 36-57 A would result in a smaller hole.

TIP

### *Move*

With the *Move* option, you can move a single face of a 3D solid or move an entire primitive within a 3D solid. This capability is particularly helpful during geometry construction or when there is a design change and it is required to alter the location of a hole or other feature within the confines of the composite 3D model.

[Extrude/Move/Rotate/Offset/Taper/Delete/Copy/coLor/Undo/eXit] <eXit>: ***m***
Select faces or [Undo/Remove]: **PICK**
Select faces or [Undo/Remove/ALL]: **PICK** or remove
Select faces or [Undo/Remove/ALL]: **Enter**
Specify a base point or displacement: **PICK**
Specify a second point of displacement: **PICK**

For example, Figure 36-59 illustrates using the *Move* option to relocate a hole primitive (four faces) within a composite solid. In this case you must ensure all faces, and only the faces, comprising the primitive are selected.

With the *Move* option the internal features can be moved to extend into or past the "bounding box" of the composite solid, as they can with *Extrude* (see Fig. 36-57).

**FIGURE 36-59**

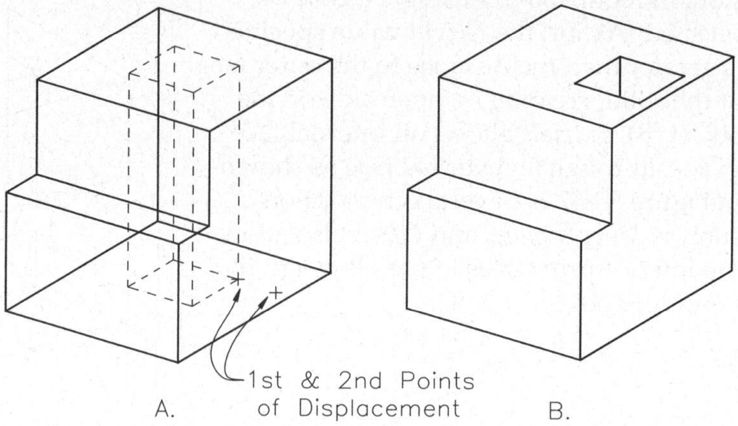


Another possibility for *Move* is to move only one or selected faces to alter the configuration of an internal feature of a composite solid. An example is shown in Figure 36-60, where only one face is selected to move. Compare the results in Figures 36-59 B and 36-60 B.

**FIGURE 36-60**

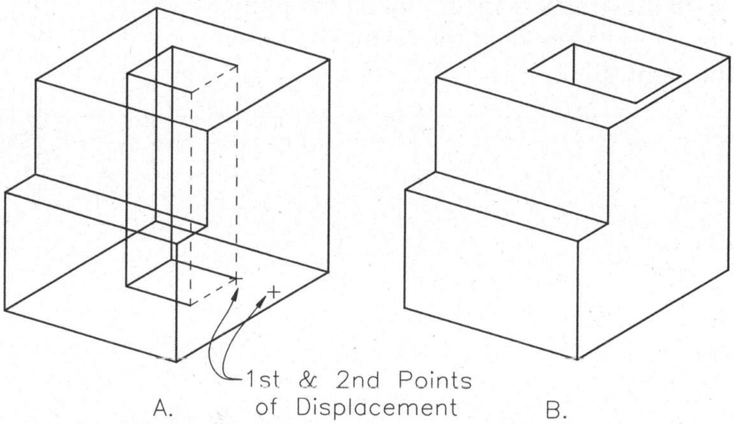


Keep in mind that *Move* can be used to move a singular external face, achieving the same result as *Extrude* (see previous Fig. 36-55).

### *Offset*

The 2D *Offset* command makes a parallel copy of a *Line*, *Arc*, *Circle*, *Pline*, etc. The *Offset* option of *Solidedit* makes a 3D offset. A simple application is making a hole larger or smaller.

[Extrude/Move/Rotate/Offset/Taper/Delete/Copy/coLor/Undo/eXit] <eXit>: *o*
Select faces or [Undo/Remove]: **PICK**
Select faces or [Undo/Remove/ALL]: **PICK** or remove
Select faces or [Undo/Remove/ALL]: **Enter**
Specify the offset distance: **PICK** or (**value**)

You can offset through a point or specify an offset distance. Specify a positive value to increase the volume of the solid or a negative value to decrease the volume of the solid.

TIP

Figure 36-61 demonstrates how an internal feature, such as a hole, can be *Offset* to effectively change the volume of the hole. Since positive values increase the size of the solid, a negative value was used here. In other words, holes inside a solid become smaller when the solid is *Offset* larger.

**FIGURE 36-61**

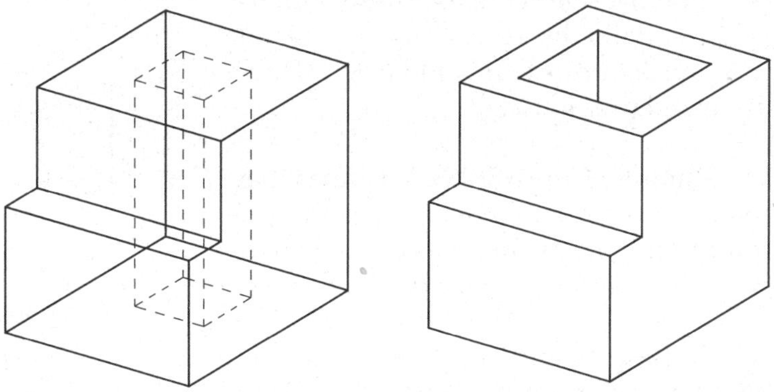

A. B.

### *Delete*

This option of *Solidedit* deletes faces from composite solids. Although you cannot delete one planar face from a solid, you can delete one curved face, such as a cylinder, that comprises an entire primitive. You can also delete multiple faces that comprise a primitive or entire feature.

[Extrude/Move/Rotate/Offset/Taper/Delete/Copy/coLor/Undo/eXit] <eXit>: ***d***
Select faces or [Undo/Remove]: **PICK**
Select faces or [Undo/Remove/ALL]: **PICK** or remove
Select faces or [Undo/Remove/ALL]: **Enter**
Solid validation started.
Solid validation completed.

As an example, Figure 36-62 A displays two selected faces to be deleted from a composite solid. The result is displayed in B. In this case, both faces must be highlighted for the deletion to operate.

For the same composite solid, if all four internal faces comprising the hole are selected, as shown in previous Figure 36-61 A, the hole could be deleted.

**FIGURE 36-62**

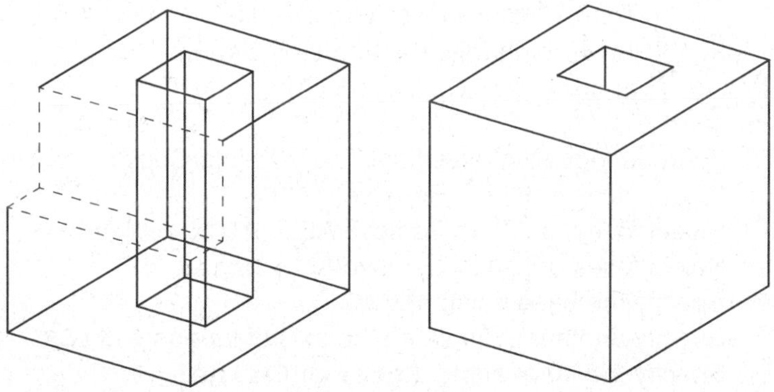

A. B.

### *Rotate*

*Rotate* is helpful during geometry construction and for design changes when components within a composite solid must be rotated in some way.

[Extrude/Move/Rotate/Offset/Taper/Delete/Copy/coLor/Undo/eXit] <eXit>: ***r***
Select faces or [Undo/Remove]: **PICK**
Select faces or [Undo/Remove/ALL]: **PICK** or remove
Select faces or [Undo/Remove/ALL]: **Enter**
Specify an axis point or [Axis by object/View/Xaxis/Yaxis/Zaxis] <2points>: **PICK**
Specify the second point on the rotation axis: **PICK**
Specify a rotation angle or [Reference]: **PICK** or **(value)**

Several methods of rotation are possible based on the selected axis. These options are essentially the same as those available with the *3DRotate* command (see "*3DRotate*" discussed previously).

For example, Figure 36-63 illustrates the rotation of a primitive within a composite solid. Here, the primitive is rotated about two points defining one edge of the hole.

NOTE: Positive and negative rotation complies with the right-hand rule. Also keep in mind that when PICKing two points to define the axis of rotation, positive direction on the axis is determined from the first to the second point picked.

One face of a composite solid can be selected for rotation within a composite solid (Fig. 36-64 A). The selected face is rotated about two points defining one edge of the face.

**FIGURE 36-63**

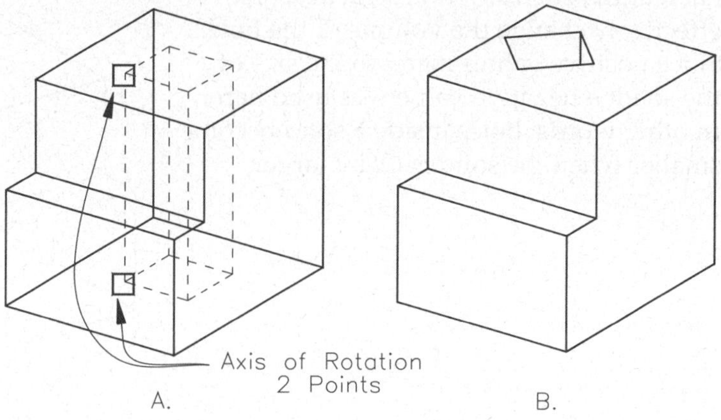


**FIGURE 36-64**

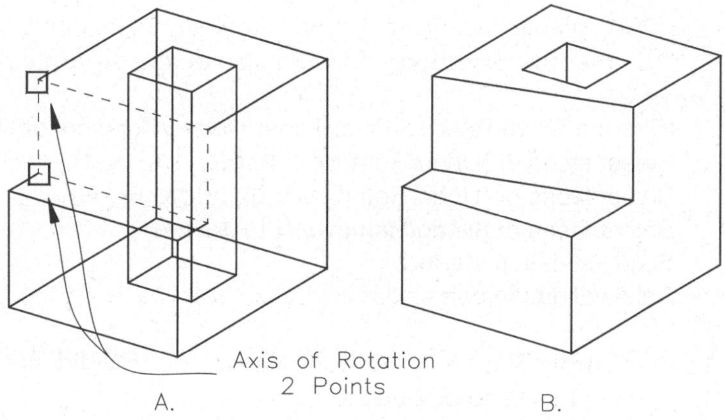


### *Taper*

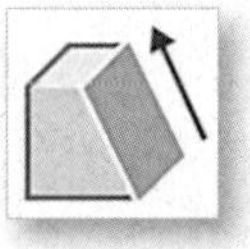

*Taper* angles a face. You can use *Taper* to change the angle of planar or curved faces (Fig. 36-65).

[Extrude/Move/Rotate/Offset/Taper/Delete/Copy/coLor/Undo/eXit] <eXit>: **t**
Select faces or [Undo/Remove]: **PICK**
Select faces or [Undo/Remove/ALL]: **PICK** or remove
Select faces or [Undo/Remove/ALL]: **Enter**
Specify the base point: **PICK**
Specify another point along the axis of tapering: **PICK**
Specify the taper angle: **Enter** or (**value**)

TIP

The rotation of the taper angle is determined by the order of selection of the base point and second point. Although AutoCAD prompts for the "axis of tapering," the two points actually determine a reference line from which the taper angle is applied. The taper starts at the first base point and tapers away from the second base point. The rotational axis passes through the first base point and is perpendicular to the line between the base points. The line between the base points does not specify the rotational axis, as implied.

**FIGURE 36-65**

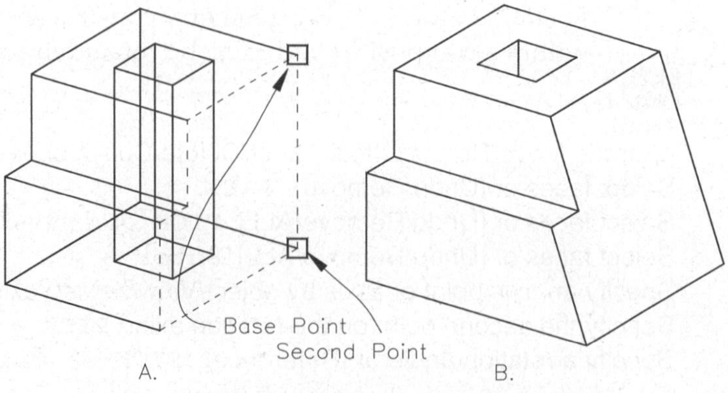

To explain this idea, Figure 36-65 A (previous page) illustrates the selection of a planar face, the base point, and the second point. An angle of –20 degrees (negative value) is entered to achieve the results shown in B. Positive angles taper the face inward (toward the solid) and negative values taper the face outward (away from the solid).

For some cases, either *Taper* or *Rotate* could be used to achieve the same results. For example, both Figure 36-65 B and previous Figure 36-64 B could be attained using *Taper* or *Rotate* (although different points must be selected depending on the option used).

Curved faces can also be selected for applying a *Taper*. Figure 36-66 A illustrates a cylindrical hole primitive. The entire cylinder primitive is selected as one *Face* object. Note the selection of the base point and second point. *Taper* converts the cylindrical hole into a conical hole (B).

**FIGURE 36-66**

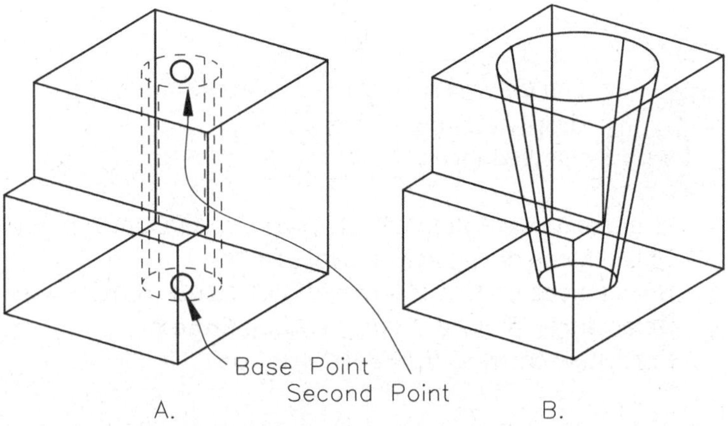


### *Copy*

As you would expect, the *Copy* option copies *Faces*. However, the resulting objects are *Regions* or *Bodies*.

[Extrude/Move/Rotate/Offset/Taper/Delete/Copy/coLor/Undo/eXit] <eXit>: **c**
Select faces or [Undo/Remove]: **PICK**
Select faces or [Undo/Remove/ALL]: **PICK** or remove
Select faces or [Undo/Remove/ALL]: **Enter**
Specify the base point or displacement: **PICK** or (**value**)
Specify a second point of displacement: **PICK**

Although any *Face* can be selected (planar or curved), Figure 36-67 shows how *Copy* can be used to create a *Region* from a planar face. Typically, the *Copy* option would be used to create a new *Region* from existing 3D solid geometry. The *Extrude* command could then be used on the *Region* to form a new 3D solid from the original face.

If a curved *Face* is selected to *Copy*, the resulting object is a *Body*. This type of body is a curved surface, not a solid.

**FIGURE 36-67**

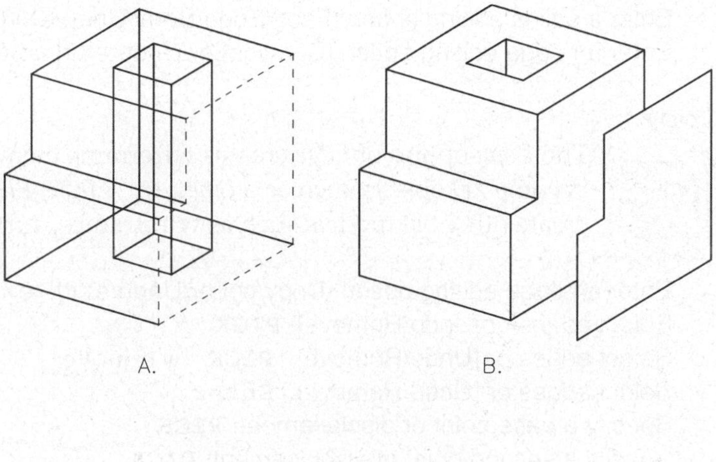

Using the same composite model as in previous figures, assume you were to *Copy* only the five vertical faces indicated in Figure 36-68 A. After selecting the second point of displacement (B), notice only those five vertical surfaces result. Not included are the faces defining the hole or any horizontal faces. Therefore, you can select *All* faces comprising a composite solid to *Copy*, and the resulting model is a surface model composed of *Regions*.

**FIGURE 36-68**

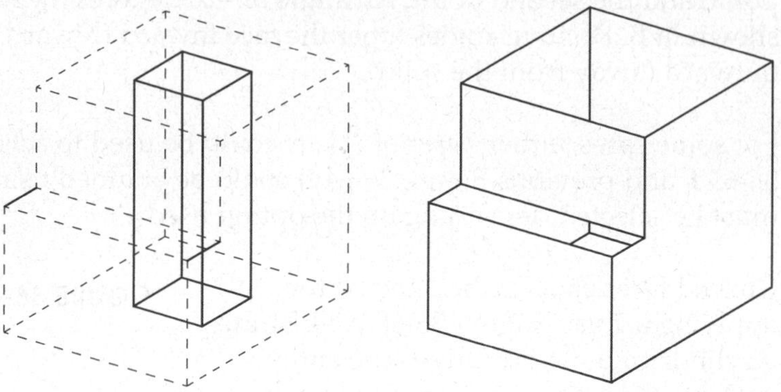

A. B.

### *Color*

The *Color Face* option of *Solidedit* simply changes the color of selected faces.

[Extrude/Move/Rotate/Offset/Taper/Delete/Copy/coLor/Undo/eXit] <eXit>: ***l***
Select faces or [Undo/Remove]: **PICK**
Select faces or [Undo/Remove/ALL]: **PICK** or remove
Select faces or [Undo/Remove/ALL]: **Enter**
Enter new color <BYLAYER>: (**color**)

The resulting display of the 3D solid depends on the *Visual Style* setting. Only the object edges appear in the designated color when *2D wireframe*, *Wireframe*, or *Hidden* settings are used. For the *Conceptual* and *Realistic* options, all surfaces are displayed in shaded intensities of the designated color. The edges are displayed in a lighter or darker intensity of the colors (unless the *Color Edge* option is used).

## ***Edge*** **Options**

*Edges* are lines or curves that define the boundary between *Faces*. Copying an *Edge* would result in a 2D object. The selection process for *Edges* is critical, but not as involved as selecting *Faces*. You can use multiple selection methods (the same options as with *Edges*), and selection may involve a process of adding and removing objects. As with *Face* selection, make sure you have the exact desired set highlighted before continuing with the procedure. Remember to set the selection filter on the Ribbon to *No Filter*. The command syntax is as follows.

Command: ***SOLIDEDIT***
Solids editing automatic checking: SOLIDCHECK=1
Enter a solids editing option [Face/Edge/Body/Undo/eXit] <eXit>: ***edge***
Enter an edge editing option [Copy/coLor/Undo/eXit] <eXit>:

### *Copy*

The *Copy* option of *Edge* creates wireframe elements, not surfaces or solids. Copied edges become 2D objects such as a *Line*, *Arc*, *Circle*, *Ellipse*, or *Spline*. These elements could be used to create other 3D models such as wireframes, surfaces, or solids.

Enter an edge editing option [Copy/coLor/Undo/eXit] <eXit>: ***c***
Select edges or [Undo/Remove]: **PICK**
Select edges or [Undo/Remove]: **PICK** or remove
Select edges or [Undo/Remove]: **Enter**
Specify a base point or displacement: **PICK**
Specify a second point of displacement: **PICK**

One or multiple *Edges* can be selected. After indicating the second point of displacement, the selected edges are extracted from the solid and copied to the new location (Fig. 36-69).

Keep in mind that the *Copied Edges* are wireframe elements. These new elements can be used for construction of other 3D geometry, such as with the *Body, Imprint* option (see "*Body* Options" later in this section).

**FIGURE 36-69**

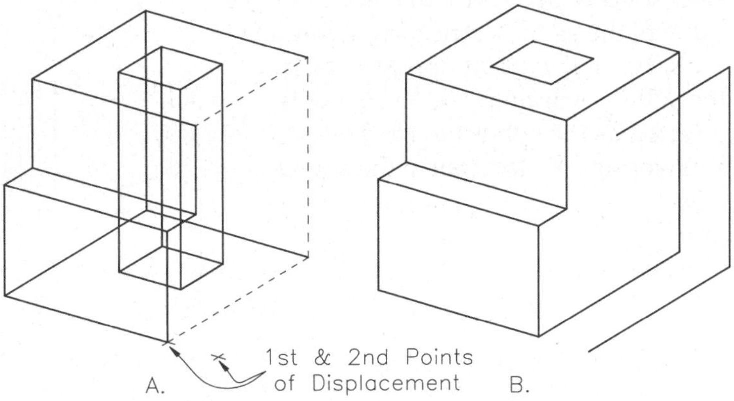


### Color

The *Color* option for *Edges* operates similarly to the *Color* option for *Faces* except that only the *Edges* are affected. This option is useful in cases where you need the edges to be more or less visible to bring attention to certain features or components of a solid, or in cases when you want the surfaces of a solid to be shaded in one color and the edges to appear in another.

```
Enter an edge editing option [Copy/coLor/Undo/eXit] <eXit>: l
Select edges or [Undo/Remove]: PICK
Select edges or [Undo/Remove]: PICK or remove
Select edges or [Undo/Remove]: Enter
Enter new color <BYLAYER>: (color)
```

The selected edges appear in the new color. The *Visual Style* setting also affects the appearance of the surfaces and edges. When *Visual Style* is set to any option that causes the edges to appear (*2D wireframe, Wireframe*, or *Hidden*), the edges appear in the selected color. In the *Conceptual* and *Realistic* modes, only the surfaces appear without edges.

## *Body* Options

A *Body* is typically a 3D solid. Any primitive or composite solid can be selected as a *Body*. However, some *Solidedit* options can create a *Body* that is a non-solid. For example, selecting a curved *Face* and using *Copy* creates a curved surface that AutoCAD lists as a *Body*. Several options are offered here to alter the configuration of a *Body* or to create or edit 2D geometry used to interact with a *Body*.

### Imprint

*Imprint* is used to attach a 2D object on an existing face of a solid (*Body*). The new *Imprint* can then be used with *Extrusion* to create a new 3D solid.

Objects that can be used to make the *Imprint* can be an *Arc, Circle, Line, Pline, Ellipse, Spline, Region*, or *Solid*. The selected object to *Imprint* must touch or intersect the solid in some way, such as when a *Circle* lies partially on a *Face*, or when two solid volumes overlap. When *Imprint* is used, one or more 2D components becomes attached, or imprinted, to an existing 3D solid face. The resulting *Imprint* has little usefulness of itself, but is an intermediate step to creating a new 3D solid shape on the existing solid.

```
[Imprint/seParate solids/Shell/cLean/Check/Undo/eXit] <eXit>: i
Select a 3D solid: PICK
Select an object to imprint: PICK
Delete the source object <N>: y
Select an object to imprint: Enter
```

For example, Figure 36-70 A displays a 3D solid with a *Circle* that is on the same plane as the vertical right face of the solid. The *Imprint* option is used, the solid is selected, and the *Circle* is selected as the "object to imprint." Answering "Y" to "Delete the source object," the resulting *Imprint* is shown in B.

**FIGURE 36-70**

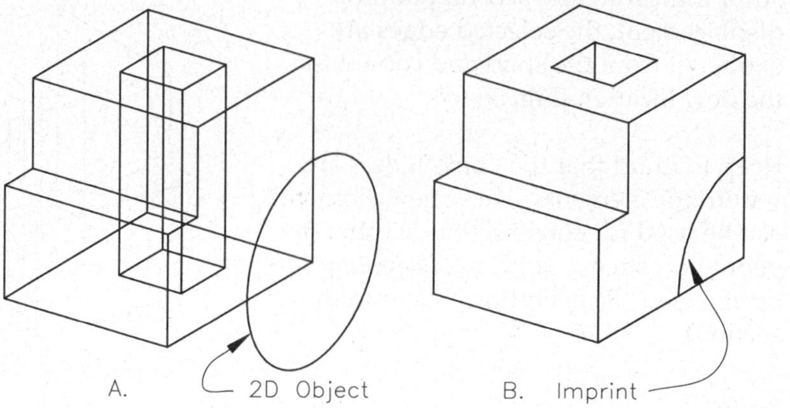


The *Imprint* creates a separate *Face* on the object—in this case, a total of two coplanar *Faces* are on the same vertical surface of the object. The new *Face* can be treated independently for use with other *Solidedit* options. For example, *Extrude* could be used to create a new solid feature on the composite solid (Fig. 36-71).

**FIGURE 36-71**

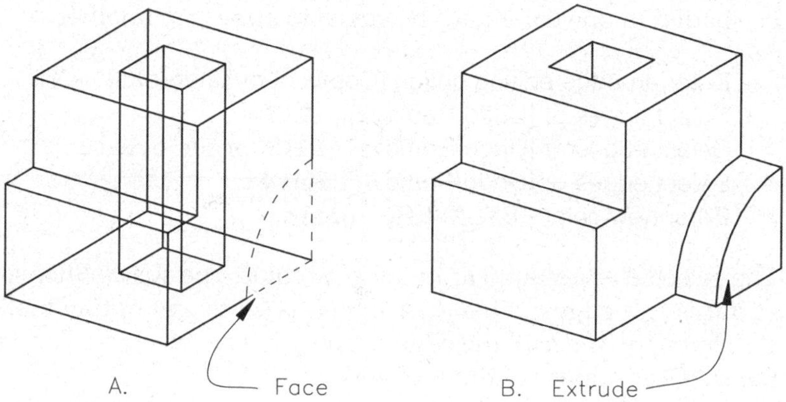


### *Clean*

The *Clean* option deletes any 2D geometry on the solid. For example, you may use *Imprint* to "attach" 2D geometry to a *Body* (solid) for the intention of using *Extrude*. *Imprinted* objects become permanently attached to the *Body* and cannot be removed by *Erase*. If you need to remove an *Imprint*, use the *Clean* option to do so. One use of *Clean* removes all *Imprints*.

[Imprint/seParate solids/Shell/cLean/Check/Undo/eXit] <eXit>: ***l***
Select a 3D solid: **PICK**

### *Separate*

It is possible in AutoCAD to *Union* two solids that do not occupy the same physical space and have two distinct volumes. In that case, the two shapes appear to be separate, but are treated by AutoCAD as one object (if you select one, both become highlighted).

**TIP** Occasionally you may intentionally or inadvertently use *Union* to combine two or more separate (not physically touching) objects. *Shell* and *Offset*, when used with solids containing holes, can also create two or more volumes from one solid. *Separate* can then be used to disconnect these into discrete independent solids. *Separate* cannot be used to disconnect or "break down" *Unioned*, *Subtracted*, or *Intersected* solids that form one volume.

[Imprint/seParate solids/Shell/cLean/Check/Undo/eXit] <eXit>: ***p***
Select a 3D solid: **PICK**

### Shell

*Shell* converts a solid into a thin-walled "shell." You first select a solid to shell, then specify faces to remove, and enter an offset distance. An example would be to convert a solid cube into a hollow box—the thickness of the walls equal the "shell offset distance." If no faces are removed from the selection set, the box would have no openings. Faces that are removed become open sides of the box.

[Imprint/seParate solids/Shell/cLean/Check/Undo/eXit] <eXit>: ***s***
Select a 3D solid: **PICK**
Remove faces or [Undo/Add/ALL]: **PICK**
Remove faces or [Undo/Add/ALL]: **PICK**
Remove faces or [Undo/Add/ALL]: **Enter**
Enter the shell offset distance: (**value**)
Solid validation started.
Solid validation completed.

A positive value entered at the "shell offset distance" prompt creates the wall thickness outside of the original solid, whereas a negative value creates the wall thickness inside the existing boundary.

An example is shown in Figure 36-72. The left object (A) indicates the solid with all the selected faces highlighted and the faces removed from the selection set not highlighted. A negative offset distance is used to yield the shape shown on the right (B).

Don't expect to achieve the desired results the first time. As with most *Solidedit* options, the process of adding and removing faces is an inexact procedure since any edge you select can highlight either one of two faces.

**FIGURE 36-72**

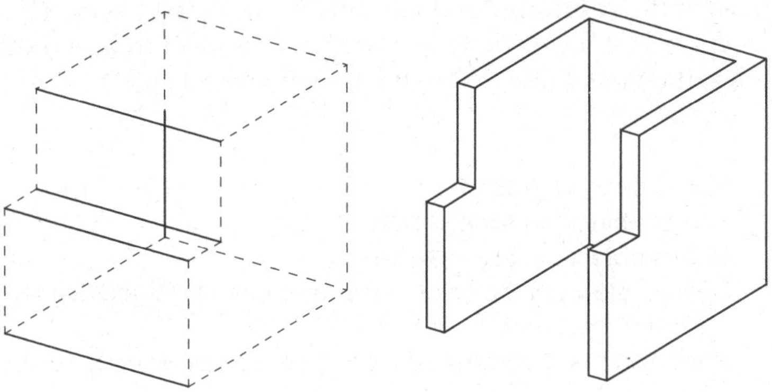

A. B.

### Check

Use this option to validate the 3D solid object as a valid solid, independent of the *SOLIDCHECK* system variable setting.

[Imprint/seParate solids/Shell/cLean/Check/Undo/eXit] <eXit>: ***c***
Select a 3D solid: **PICK**
This object is a valid Shape Manager solid.

The *SOLIDCHECK* variable turns the automatic solid validation on and off for the current AutoCAD session. By default, *SOLIDCHECK* is set to 1, or on, to validate the solid. The *Solidedit* command displays the current status of the variable as shown in the command syntax below.

Command: ***SOLIDEDIT***
Solids editing automatic checking: SOLIDCHECK=0
Enter a solids editing option [Face/Edge/Body/Undo/eXit] <eXit>

## *Slice*

| Ribbon | Menu Bar | Command (Type) | Alias (Type) | Shortcut |
|---|---|---|---|---|
| *Home Solid Editing* (icon) | *Modify 3D Operations > Slice* | *Slice* | *SL* | ... |

The *Slice* command is a separate command (not part of the *Solidedit* options). However, *Slice* is a useful command for solids construction and editing for special cases and is therefore discussed in this chapter for those purposes.

**TIP** A solid that contains oblique surfaces is a likely candidate for using *Slice*. For example, the simplest method of construction for the solid shown in Figure 36-73 is to create the rectilinear shapes (using *Box*), then use *Slice* to create the oblique surfaces.

**FIGURE 36-73**

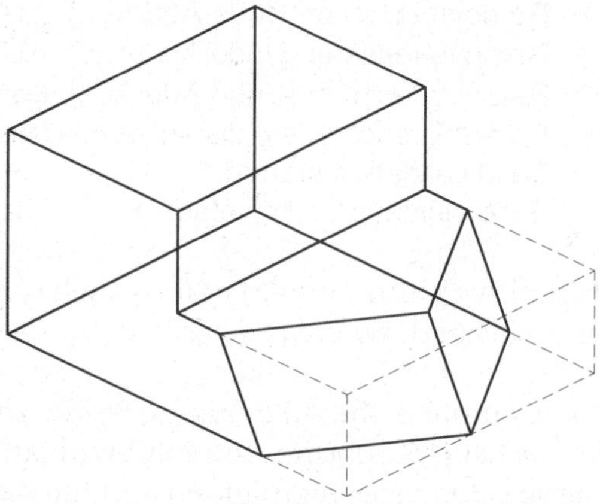

With the *Slice* command you can cut a solid into two "halves" by defining a plane in any orientation. You can keep both sides of the resulting solid(s) or PICK one side to keep. (If in doubt as to which side to keep or how to select it, keep both sides, then use *Erase* to remove the unwanted side.)

Command: ***SLICE***
Select objects to slice: **PICK**
Select objects to slice: **Enter**
Specify start point of slicing plane or [planar Object/Surface/Zaxis/View/XY/YZ/ZX/3points] <3points>:
(option specific prompts)
Specify a point on desired side or [keep Both sides] <Both>: **PICK** or **Enter**
Command:

One oblique surface for the solid (Fig. 36-74) is created by first drawing a 2D *Line* on the solid, then defining a *3point* slicing plane using *Object Snaps* (*Endpoint*, *Endpoint* of *Line*, then *Midpoint* of solid edge). This figure shows the resulting solids after using *Slice* and keeping both sides. Duplicating this procedure for the other oblique surface creates the solid shown previously in Figure 36-73.

**FIGURE 36-74**

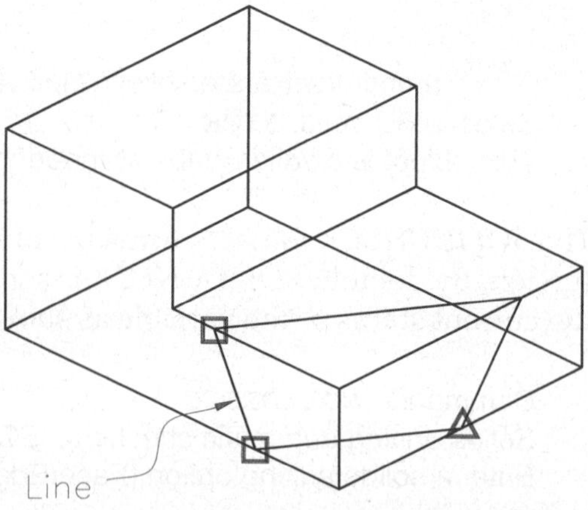


**TIP** *Slice* can be used for many applications, such as an alternative to using *Wedge*, *Chamfer*, or *Solidedit*, *Taper*. Additionally, if you wanted to cut a section of a solid away, such as creating a notch or a half section (cutting away one quarter of an object), consider making multiple slicing planes with *Slice*, keeping all sides, *Erase* the unwanted section, then using *Union* to join the sections back into one solid.

*Slice* can also be used to create a true solid section. *Slice* cuts a solid or set of solids on a specified cutting plane. The original solid is converted to two solids. You have the option to keep both halves or only the half that you specify. Examine the object in Figure 36-75.

**FIGURE 36-75**

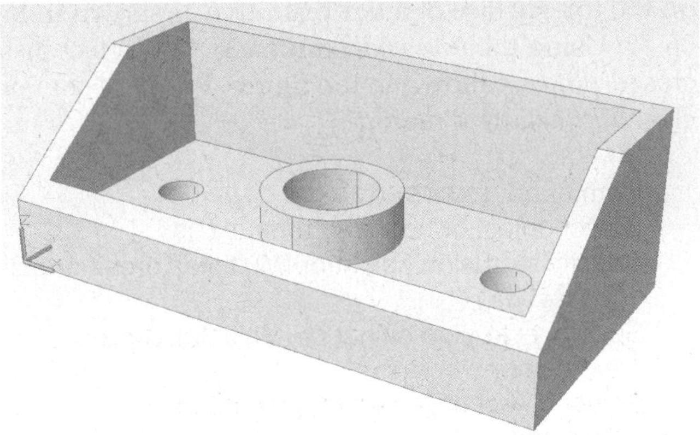

To create a true section of the example, use *Slice*. The *ZX* method is used to define the cutting plane midway through this solid. Next, the new solid to retain was specified by PICKing a point on that geometry. The resulting sectioned solid is shown in Figure 36-76. Note that the solid on the near side of the cutting plane was not retained.

**FIGURE 36-76**

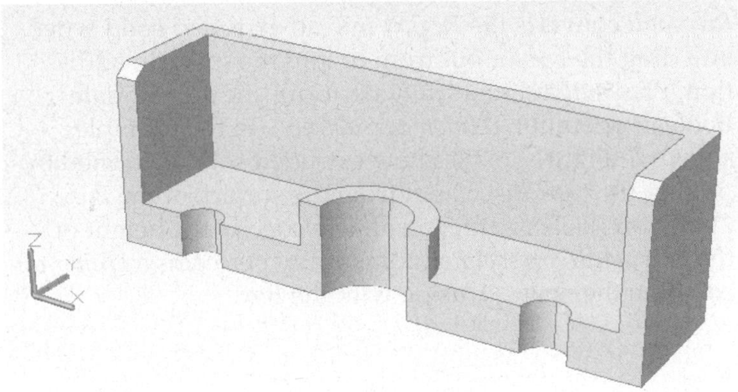

Entering ***B*** at the "keep both sides" prompt retains the solids on both sides of the cutting plane. Otherwise, you can pick a point on either side of the plane to specify which half to keep.

### *Surface* option

Another very important and useful application for the *Slice* command is using surfaces to slice solids. That is, instead of slicing a solid using an imaginary plane (*XY*, *YZ*, *ZX*, or *3points* options), you can select an existing surface as the slicing plane using the *Surface* option. The existing surface can be one created by commands such as *Extrude*, *Sweep*, *Revolve*, or *Loft*. Using this method to slice a solid allows you to create solids containing a variety of complex curved shapes.

## ***Presspull***

| Ribbon | Menu Bar | Command (Type) | Alias (Type) | Shortcut |
|---|---|---|---|---|
| *Home Modeling* (icon) | ... | *Presspull* | ... | ... |

*Presspull* is an unusual command because it accomplishes two tasks: 1) it converts a closed 2D shape such as a *Circle*, *Pline*, *Polygon*, etc. into an extruded solid, and 2) it performs a Boolean *Union* or *Subtract* based on whether you "pull" or "push" the extrusion. The closed 2D shape to extrude must lie in a plane. *Presspull* is simple to use since there are only two steps. The command prompts you only during the first step.

For example, assume that you create a closed 2D *Pline* on the top surface of a *Box* primitive as shown in Figure 36-77. Using the *Presspull* command, you select inside the closed *Pline* as shown in the figure. *Presspull* converts the closed *Pline* into a region.

FIGURE 36-77

Command: ***PRESSPULL***
Select object or bounded area: **PICK**
Specify extrusion height or [Multiple]: press or pull
  or **(value)**
Specify extrusion height or [Multiple]: **Enter**
1 extrusion(s) created
Select object or bounded area: **Enter**
Command:

Next, you simply "pull" or "press" to create a new extruded solid from the 2D shape. In other words, *Presspull* converts the *Region* into an extruded solid when you drag the cursor out from or into the solid. In addition, *Presspull* automatically performs the appropriate Boolean operation (*Union* or *Subtract*). In the example shown in Figure 36-78, a new extruded solid is formed by pulling the new *Region* out from the surface of the *Box*. Once you PICK or enter a value to specify the height of the extrusion, *Presspull* automatically performs a *Union* to combine the new extrusion with the *Box*.

FIGURE 36-78

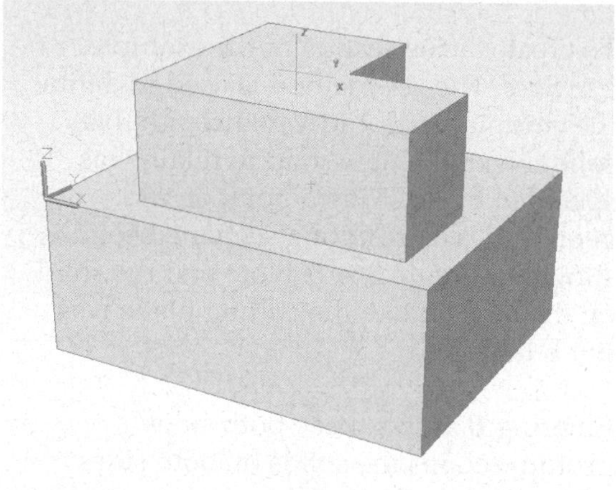

On the other hand, if you "press" the solid into the *Box*, *Presspull* performs a *Subtract*. In Figure 36-79, the new solid is formed by pulling the *Region* down into and through the *Box*. At this point the height has not been specified so no Boolean operation has been performed. Two solids exist at this point.

FIGURE 36-79

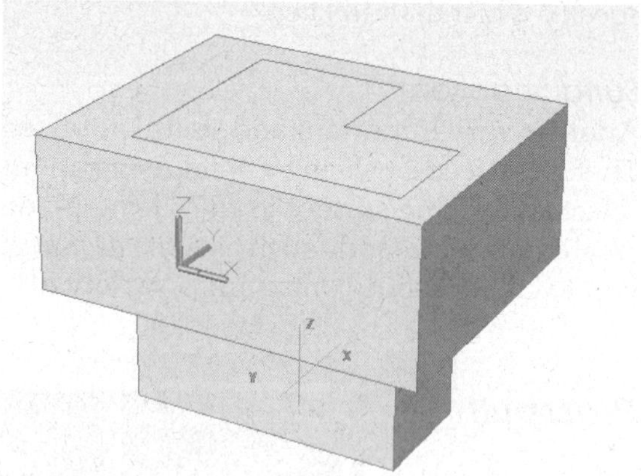

When you PICK or enter a value to define the height of the extrusion, the Boolean is automatically performed. In this case (Fig. 36-80), the new extruded solid is *Subtracted* from the *Box* to form a new composite solid.

FIGURE 36-80

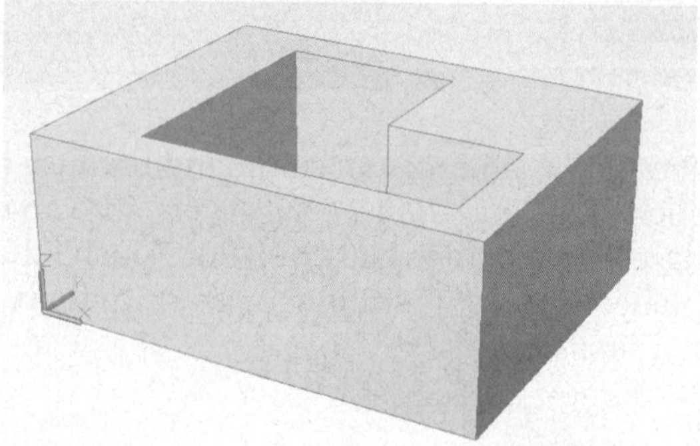

The 2D shapes that can be used with *Presspull* do not have to be single objects. *Presspull* works like the *Boundary* command so any areas bounded by other edges can be selected. You can press or pull any of the following types of bounded areas:

- Areas that could be selected using *Hatch* or *Boundary* by picking a point (with zero gap tolerance)
- Areas enclosed by crossing coplanar, linear geometry, including edges and geometry in blocks
- Closed *Plines*, *Regions*, 3D faces, and 2D solids that consist of coplanar vertices
- Areas created by geometry (including edges on faces) drawn coplanar to any face of a 3D solid

As another example, *Presspull* can be used to create an extruded solid not connected to another solid. In addition, the boundary does not have to be an object such as a closed *Pline* or *Region* since the needed *Region* is automatically created by *Presspull*.

For example, examine the 2D shapes (*Circle*, *Line*, and *Pline*) shown in Figure 36-81. Although two closed areas exist within these shapes, the shapes are not all closed and do not lie on the plane of an existing solid. Nevertheless, *Presspull* can be used to create a new extruded solid from one of the closed areas.

**FIGURE 36-81**

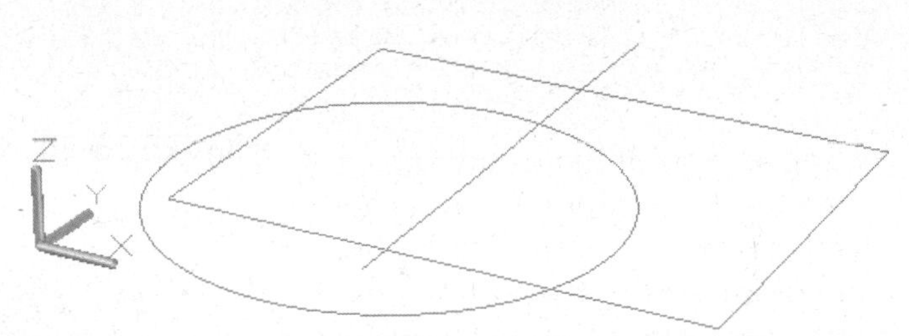


Invoking *Presspull*, then selecting inside one of the closed areas automatically creates a *Region* that can be pulled to form a new solid as shown in Figure 36-82. No Boolean operation is performed by *Presspull* since the new solid was not created from the plane of an existing solid.

**FIGURE 36-82**

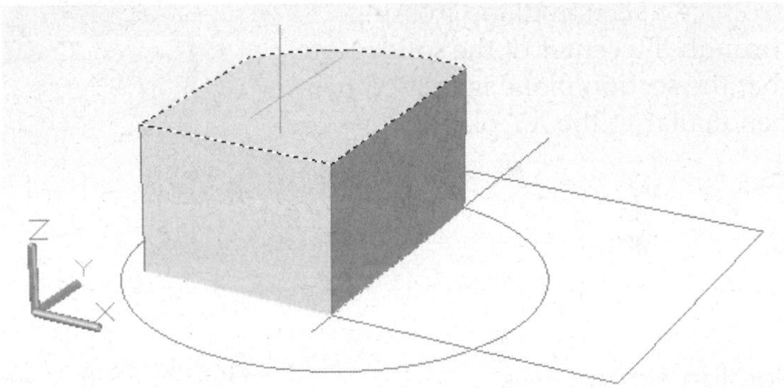


## *Sectionplane*

| Ribbon | Menu Bar | Command (Type) | Alias (Type) | Shortcut |
|---|---|---|---|---|
| *Home Section Section Plane* | *Draw Modeling > Section Plane* | *Sectionplane* | *SPANE* | ... |

*Sectionplane* creates a translucent clipping plane that can pass through a solid to reveal the inside features of the solid. The section plane is actually a section <u>object</u> that can be moved or revolved to dynamically display the solid's interior features. The section plane can contain "jogs" to create an offset cutting plane. The section object is manipulated using grips that appear on the section plane object.

Command: ***SECTIONPLANE***
Select a face or any point to locate section line or [Draw section/Orthographic]: **PICK**
Specify through point: **PICK**

You can select a face to create a section plane or specify two points to create the ends of the section plane. Selecting a face on a solid aligns the section object parallel to that face. For example, selecting the right face at the first prompt would create the section plane shown in Figure 36-83.

FIGURE 36-83

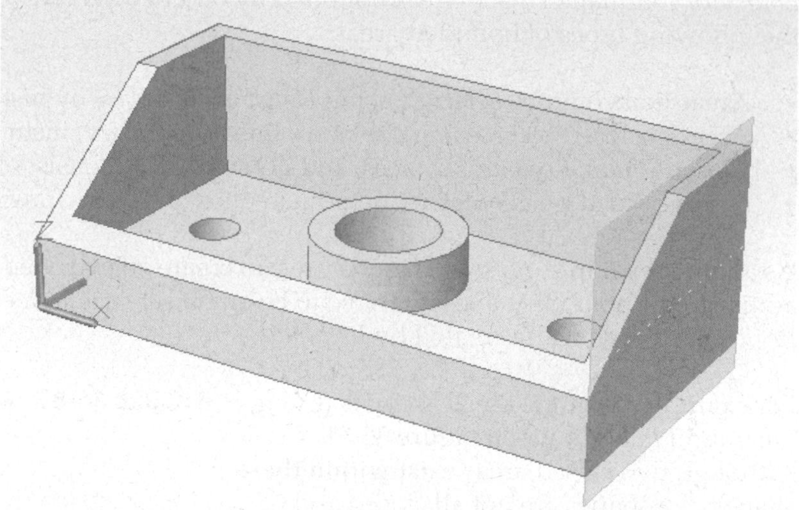

Selecting any point that is not on a face creates the first point of a section object independent of the solid. The second point creates the section object. For example, selecting two points just to the right and left of the composite solid shown in Figure 36-84 would produce a section plane crossing through the center of the solid. Note that the section plane is created perpendicular to the XY plane.

FIGURE 36-84

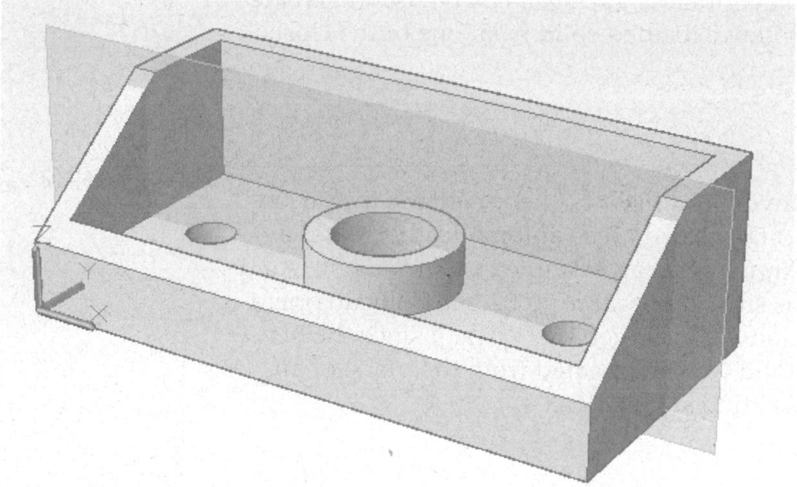

## Section Plane Grips

Note that selecting the section plane object makes its grips appear (Fig. 36-85). Grips can be used to rotate the section plane to a different orientation, change the length of the section plane, flip the section plane, and drag the section plane into a new position. For example, Figure 36-85 displays the direction grip being dragged through the solid. PICKing the point at the center of the hole would move the section plane to that location, and the object would appear cut in half as shown in the figure.

FIGURE 36-85

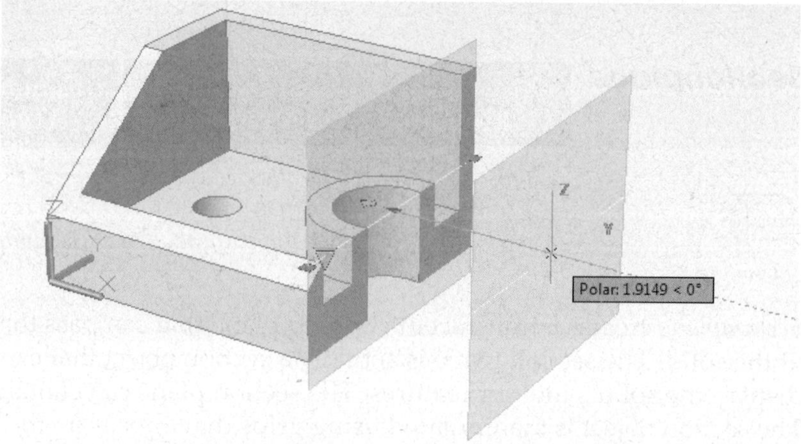

Once one of the square grips at the end of the section plane object is made hot, the normal five grip options appear (STRETCH, MOVE, ROTATE, SCALE, and MIRROR). For example, you can rotate the section plane by making a square grip hot, then cycling to the ROTATE option as shown in Figure 36-86.

**FIGURE 36-86**

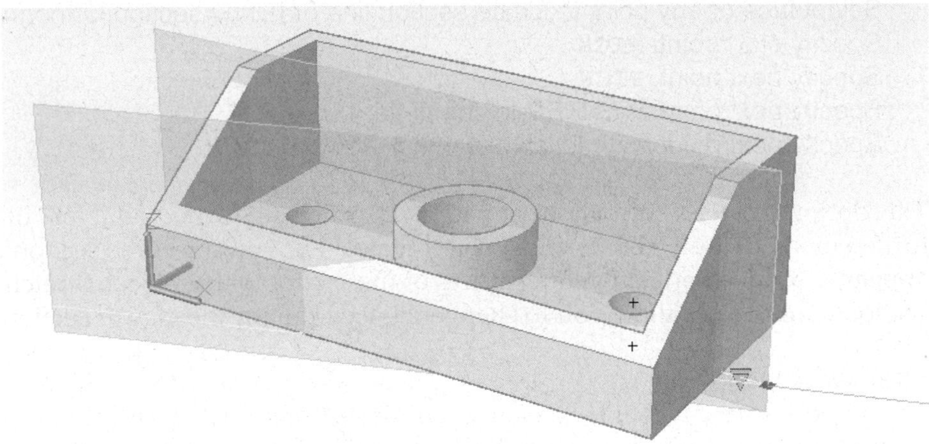

### Live Sectioning

To make solid geometry on one side of the section plane appear invisible, the *Live Section* property for the section plane object must be set to *Yes* (Fig. 36-87). When a section plane is created using the "face" option, *Live Section* property is set to *Yes* by default as shown previously in Figure 36-85. However, when you create a section plane by selecting points, *Live Section* is set to *No*. For example, note that in Figure 36-84 the section plane passes through the center of the solid, yet both sides of the plane are visible.

**FIGURE 36-87**

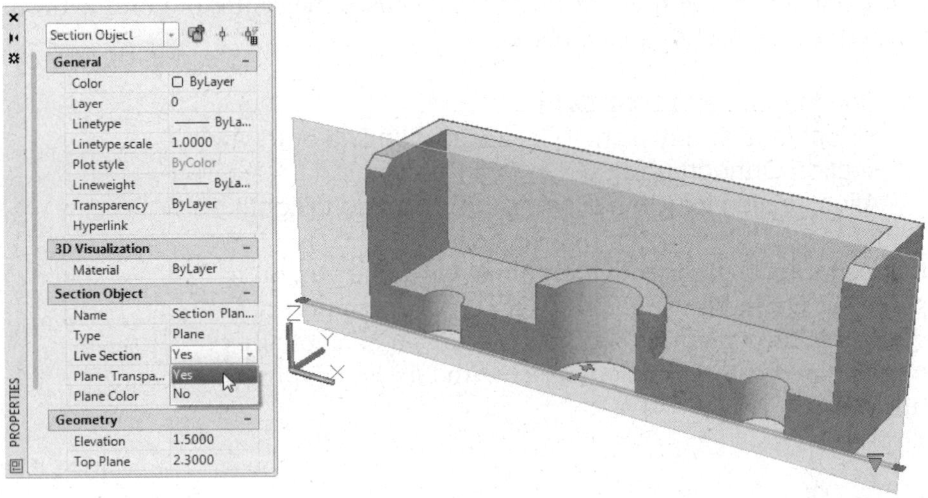


Using the *Properties* palette to set the *Live Section* to *Yes* makes the solid geometry on one side of the plane invisible as shown in Figure 36-87. The "flip" arrow grip determines which side of the section plane is invisible.

**FIGURE 36-88**

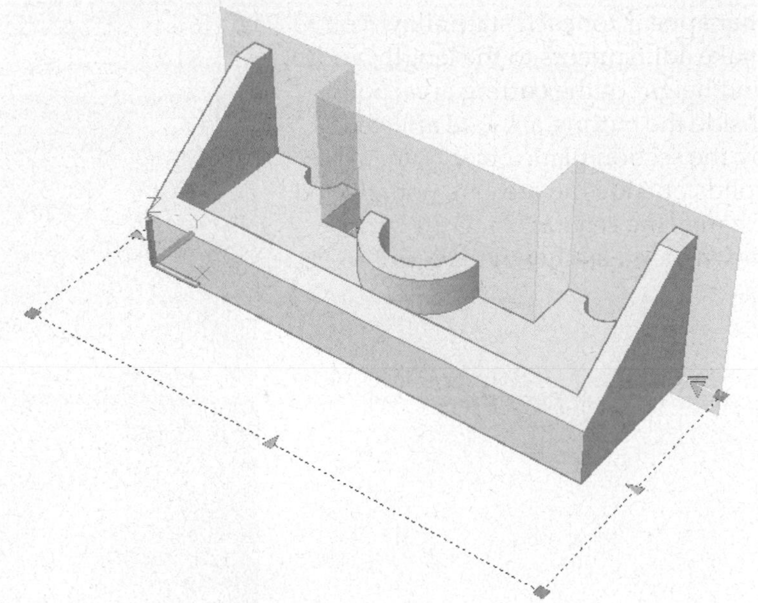

#### *Draw Section*

The *Draw section* option allows you to create an offset section plane. Offset section planes are used when "jogs" in the section line are needed to pass through multiple features of the solid as shown in Figure 36-88.

```
Command: SECTIONPLANE
Select face or any point to locate section line or [Draw section/Orthographic]: d
Specify start point: PICK
Specify next point: PICK
Specify next point or ENTER to complete: Enter
Specify next point in direction of section view: PICK
```

This option creates a section object with live sectioning turned off. Note that the section boundary is turned on by default (see "Section Object States"). If you select the section plane object so the grips appear, you can change several features of the section plane object. Stretch grips for each section of the section plane object allow you to change the location for the jogged sections (Fig. 36-88).

### *Orthographic*

This option is very useful for creating typical sections. *Orthographic* aligns the section object to a specific orthographic orientation relative to the UCS. The resulting section contains all 3D objects in the model. If a single composite solid is located correctly with respect to the USC (lower-left corner at 0,0,0 and front side along the X axis), AutoCAD uses the center of volume of the solid (or multiple solids) to create the new section plane object.

```
Command: SECTIONPLANE
Select face or any point to locate section line or [Draw section/Orthographic]: o
Align section to: [Front/Back/Top/Bottom/Left/Right]:
```

For example, using the *Right* option creates the section plane shown in Figure 36-89. Using the *Front* option produces a section plane as shown in Figure 36-87. Note that this option creates a section object with *Live Sectioning* turned on.

**FIGURE 36-89**

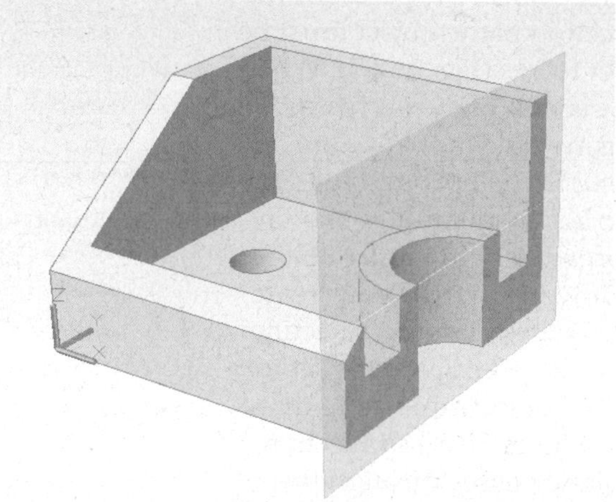

### Section Object States

Section objects have four states: *Plane* (a 2D plane), *Slice*, *Boundary* (a 2D box), and *Volume* (a 3D box). Grips that appear for each state allow you to make adjustments to the length, width, and height of the cutting area. Solids inside the cutting area are affected by the section plane object, whereas solids outside the area are not affected. Change the section using the drop-down menu as shown in Figure 36-90.

**FIGURE 36-90**

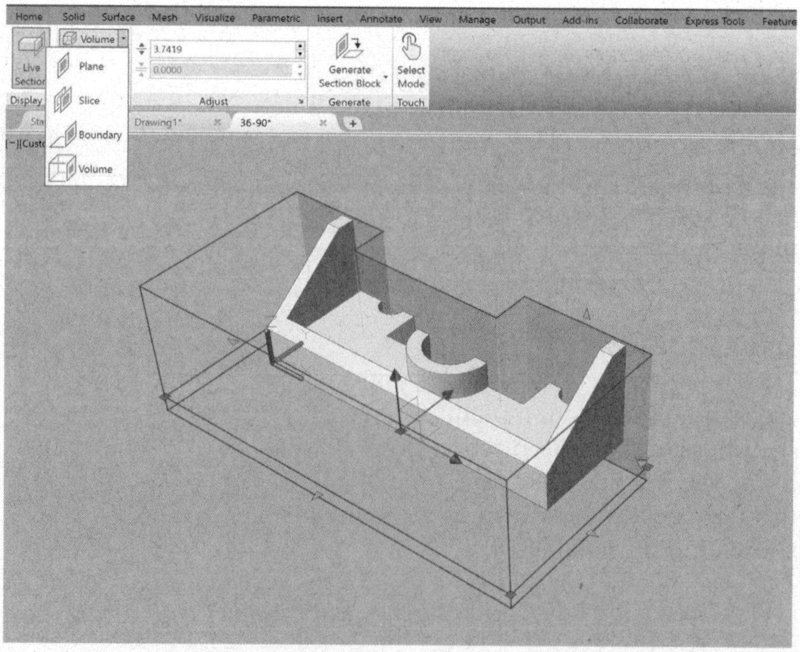

*Plane* — The cutting plane extends infinitely in all directions. Only the section line and transparent section plane indicator are displayed (Fig. 36-89).

*Slice* — A *Slice* is a thin slice of the object with a thickness value (distance perpendicular to the section plane). A 2D box shows the thickness along the direction of the cutting plane.

*Boundary* — The cutting plane along the Z axis extends infinitely. A 2D box shows the XY extents of the cutting plane (Fig. 36-88).

*Volume* — A 3D box shows the extents of the cutting plane (Fig. 36-90).

## *Sectionplane* Contextual Tab

The *Sectionplane* command has several options that appear on the *Section Plane* contextual tab. It automatically appears when you select a section plane (Fig. 36-91). Some of these options also appear on the *Home* tab's *Section* drop-down panel. The options are explained next.

**FIGURE 36-91**

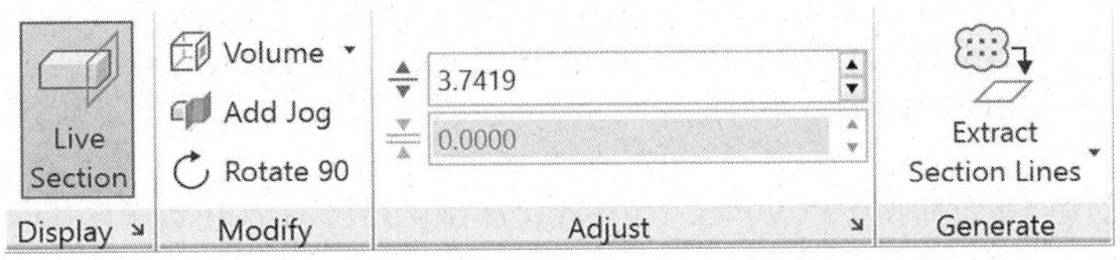


### *Live Section* (*Display* panel)

This toggle turns on or off live sectioning for a selected section object. This toggle has the same result as using the *Properties* palette to turn live sectioning on or off (see Fig. 36-87).

### *Plane, Slice, Volume, Boundary* (*Modify* panel)

Your choice from this drop-down list sets the display option for the selected section plane. These display options are the same as those available on the section object states grip. See "Section Object States" and Figure 36-90.

### *Add Jog* (*Modify* panel)

You can use this option to create a jogged segment to the selected section object. This button actually invokes the *Sectionplanejog* command.

```
Command: SECTIONPLANEJOG
Specify a point on the section line to add jog:
```

Selecting a point on the highlighted cutting plane line adds a new grip at the selected location. Use this grip to alter the section plane as explained earlier (see "Section Plane Grips").

### *Rotate 90* (*Modify* Panel)

Each time you select this button the section object is rotated around the section line 90 degrees. Keep in mind this may be helpful to create multiple slices using the *Generate Section Block* option.

TIP

### *Section Offset* (*Adjust* Panel)

The value that appears in this edit box is the current distance that the selected section plane is from the WCS origin measured in the direction perpendicular to the section plane. You can change this value to move the section plane towards or away from the WCS origin (the plane remains parallel to its current position).

### *Slice Thickness* (*Adjust* Panel)

If the section plane state is set to *Slice* (see "Section Object States"), the value listed in this edit box is the thickness of the slice. Change this value to increase or decrease the slice thickness of the section object.

### *Generate Section Block* (*Generate* Panel)

Using this tool you can create a 2D or 3D block from the selected section plane. The new *Block* can be a *2D* or *3D* block. The new *Block* can be brought back into the drawing on the current XY plane or exported to a file based on the specifications in the *Create Section/Elevation* dialog box that appears. This button actually produces the *Sectionplanetoblock* command.

FIGURE 36-92

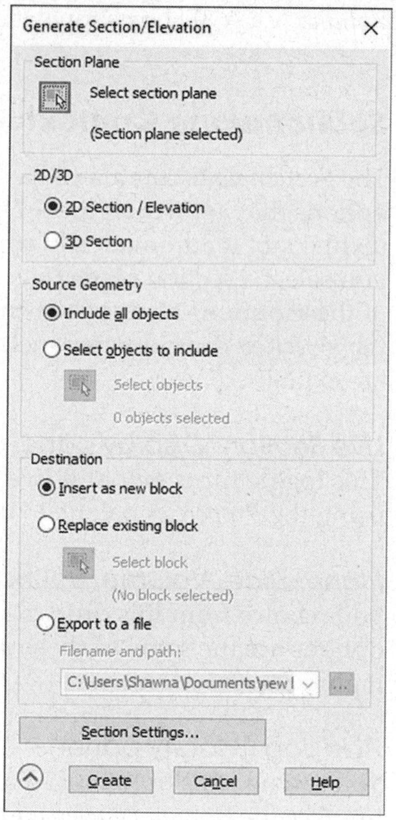


Command: ***SECTIONPLANETOBLOCK***
Units: Inches Conversion: 1.0000
Specify insertion point or [Basepoint/Scale/X/Y/Z/Rotate]: PICK or (coordinates)
Enter X scale factor, specify opposite corner, or [Corner/XYZ] <1>: **Enter** or (**value**)
Enter Y scale factor <use X scale factor>: **Enter** or (**value**)
Specify rotation angle <0>: **Enter** or (**value**)
Command:

The *Sectionplanetoblock* command initially produces the *Create Section/ Elevation* dialog box (Fig. 36-92). Here you can specify the options for the *Block* that will be created.

#### *Section Plane*

If not already selected, specify the desired section plane to use.

#### *2D/3D*

Select if you want to create a *2D* block or *3D* block. A *2D* block is a zero-thickness slice of the model that would appear on the section plane. A *3D* block is a copy of the 3D model exactly as it would appear with *Live Sectioning* on.

#### *Source Geometry*

You can *Include all objects* in the drawing as they would be cut from the section plane (as an infinite plane) or *Select objects to include* from the drawing. To create a block of only the portion of the model containing the section plane, select only that object using the *Select objects* button to select from the drawing.

#### *Destination*

If you *Insert as new block*, the block will come into the drawing on the current XY plane, even though (for a 2D block) this may be a different orientation from the current section plane slice. You can also *Replace existing block* in the drawing or save the new block to a .DWG file by selecting *Export to file.*

#### *Section Settings*

This button produces the *Section Settings* dialog box (see "*Section Settings* Dialog Box" next).

#### *Extract Section Lines*

This option invokes the *Pcextractsection* command which is intended only for creating sections from point clouds.

#### *Create*

When all settings are specified to your liking, use the *Create* button to create the new *Block*. If your choice was to *Insert as new block*, at this point you will be prompted for the *Insertion point*, *X scale factor*, *Y scale factor*, and *Rotation* angle. See the Command prompt above.

## *Section Settings* Dialog Box

Produce the *Section Settings* dialog box by using the arrow at the bottom of the *Section* panel in the *Home* tab or by using the *Section Settings* button in the *Create Section Elevation* dialog box. This dialog box (Fig. 36-93) allows you to specify the advanced settings for using the *Sectionplane* command's options for creating 2D and 3D blocks as well as settings to be used for *Live Sectioning*. The advantage of this dialog box is that 1) you can have preselected options for using *Section Plane*, and 2) you can specify that, for each drawing, all section objects have similar settings or different section objects have different settings.

FIGURE 36-93

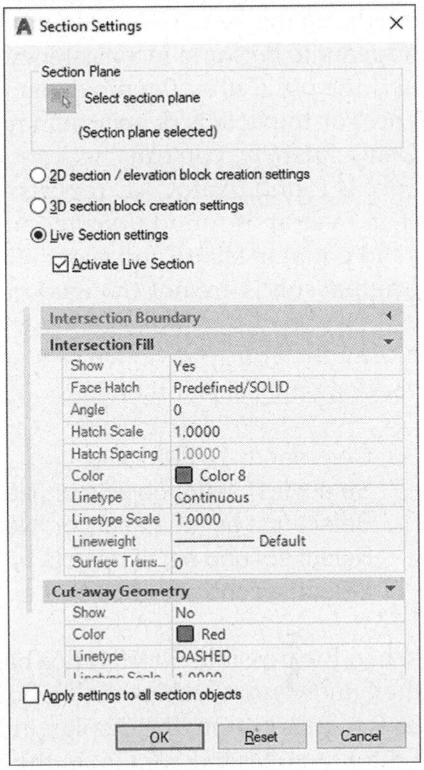


### *2D Section/Elevation Block Creation Settings*
### *3D Section Block Creation Settings*
### *Live Section Settings*

Select one of the three radio buttons near the top of the dialog box to set the specifications for that type of sectioning. For each of the three options here (*2D*, *3D*, or *Live Sectioning*), a different set of parameters appears in the list below applicable only for that type of section. You could think of this dialog box as a *Properties* palette for section objects.

If you select the *Live Section Settings* radio button, options appear below for the *Intersection Boundary*, *Intersection Fill*, and *Cut-away Geometry*. Note that, for example, you can specify settings for the hatch pattern, angle, scale, spacing, color, linetype, linetype scale, lineweight, and transparency. The *2D* settings and *3D* settings are similar to *Live Sectioning* with only a few exceptions or additions.

### *Apply settings to all section objects*

Check this box to apply the settings to all section objects in the drawing. If this box is not checked, settings apply only to the selected object. In this way you can have different settings (such as hatch pattern, hatch scale, color, etc.) for different section objects.

## *Interfere*

| Ribbon | Menu Bar | Command (Type) | Alias (Type) | Shortcut |
|---|---|---|---|---|
| *Home*<br>*Solid Editing*<br>(icon) | *Modify*<br>*3D Operations >*<br>*Interference Checking* | *Interfere* | *INF* | ... |

In AutoCAD, unlike real life, it is possible to create two solids that occupy the same physical space. *Interfere* checks solids to determine whether or not they interfere (occupy the same space). If there is interference, *Interfere* reports the overlap and allows you to create a new solid from the interfering volume, if you desire. For example, consider the two solids shown in Figure 36-94. The two shapes fit together as an assembly. The locating pin on the part on the right should fit in the hole in the left part.

FIGURE 36-94

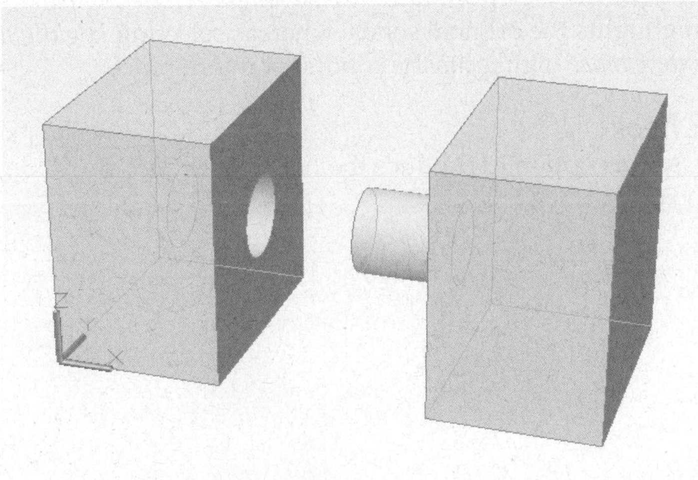

Sliding the parts together until the two vertical faces meet produces the assembly shown in Figure 36-95. There appears to be some inconsistency in the assembly of the hole and the pin. Either the pin extends beyond the hole (interference) or the hole is deeper than necessary (no interference). Using *Interfere*, you can check for an overlap. If no interference is found, AutoCAD reports "Objects do not interfere." If an overlap is found, *Interfere* automatically creates a new solid equal in shape and size to the interfering volume. The original solids are not changed in any way.

**FIGURE 36-95**

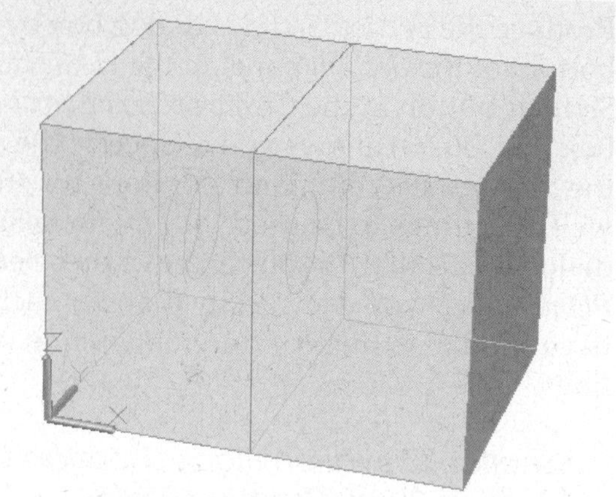

Normally, you specify two sets of solids for AutoCAD to check against each other:

Command: **INTERFERE**
Select first set of objects or [Nested selection/Settings]: **PICK**
Select first set of objects or [Nested selection/Settings]: **Enter**
Select second set of objects or [Nested selection/checK first set] <checK>: **PICK**
Select second set of objects or [Nested selection/checK first set] <checK>: **Enter**

When the possible interference has been checked, the *Interference Checking* dialog box appears (Fig. 36-96). In addition, the display in the Drawing Editor changes to zoom in on the solid(s) of interference. If multiple interferences occur, you can use the *Previous* and *Next* options to highlight each area. Remember that new solids are created equal to the interfering volume. Use these new solids to change and correct your solid model or delete the solids by checking *Delete interference objects created on Close.*

**FIGURE 36-96**

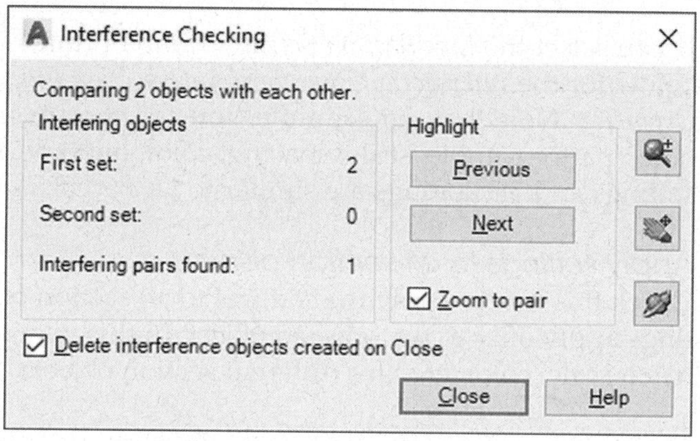


### *Nested Selection*

Using this option during the prompts allows you to select individual solid objects that are nested in Blocks and Xrefs.

### *Settings*

If you use the *Settings* option during the first prompt, the *Interference Settings* dialog box appears (Fig. 36-97). Here you can change the *Visual Style* for the interference objects and for the viewport. *Selecting Highlight interfering pair* highlights the original solids, whereas selecting *Highlight interference* highlights the solid(s) of interference.

### *Check*

Use this option to produce the *Interference Checking* dialog box.

**FIGURE 36-97**

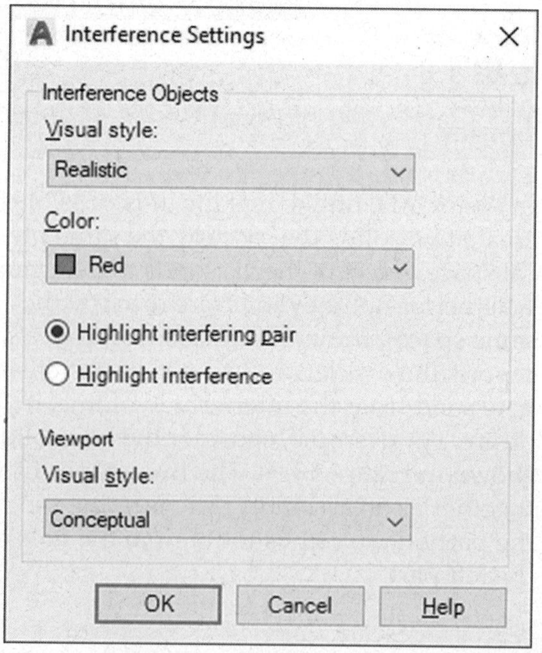

### *Imprint*

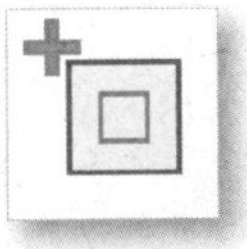

| Ribbon | Menu Bar | Command (Type) | Alias (Type) | Shortcut |
|---|---|---|---|---|
| *Solid*<br>*Solid Editing*<br>*Imprint* | *Modify*<br>*Solid Editing >*<br>*Imprint Edges* | *Imprint* | ... | ... |

The *Imprint* command can be used to create new edges on the solids that can be edited with grips. As you remember, *Imprint* is an option of the *Solidedit* command; however, *Imprint* is also a stand-alone command and has direct access through the *Imprint* button on the *Solid Editing* panel. The *Imprint* command operates exactly like the *Imprint* option of *Solidedit* (see previous coverage of *Solidedit*). That is, you can make an imprint from any *Arc, Circle, Line, Pline, Ellipse, Spline, Region*, or solid that lies in the plane of a solid. See also page 1145.

The advantage of using *Imprint* to create new edges on a solid is that the new edges can be used to change the solid using grip editing methods. For example, assume you added *Line* to the top of an existing box, then used *Imprint* to "convert" the line to a new edge on the solid as shown in Figure 36-98.

Command: ***IMPRINT***
Select a 3D solid: **PICK**
Select an object to imprint: **PICK**
Delete the source object [Yes/No] <N>: ***y***
Select an object to imprint: **Enter**
Command:

**FIGURE 36-98**

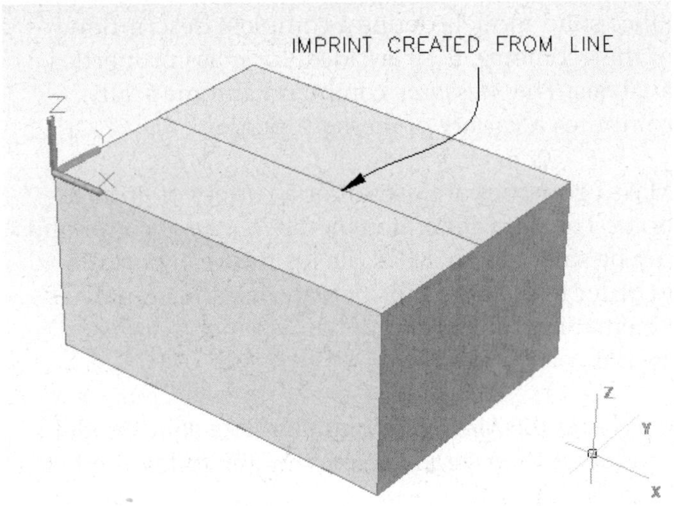


Once the new edge is created, use the Ctrl key to hover the new edge, then PICK to select the edge only. Next, hover over the edge grip until the gizmo moves to the edge grip location, then select the desired axis on the gizmo. The solid shape can be changed using the gizmo to STRETCH the new edge. For example, the gizmo's Z axis is selected in Figure 36-99 and the new edge is moved down to create the desired solid shape.

NOTE: For best results, use the move and rotate gizmo when editing a solid using an added edge created by *Imprint*.

**FIGURE 36-99**

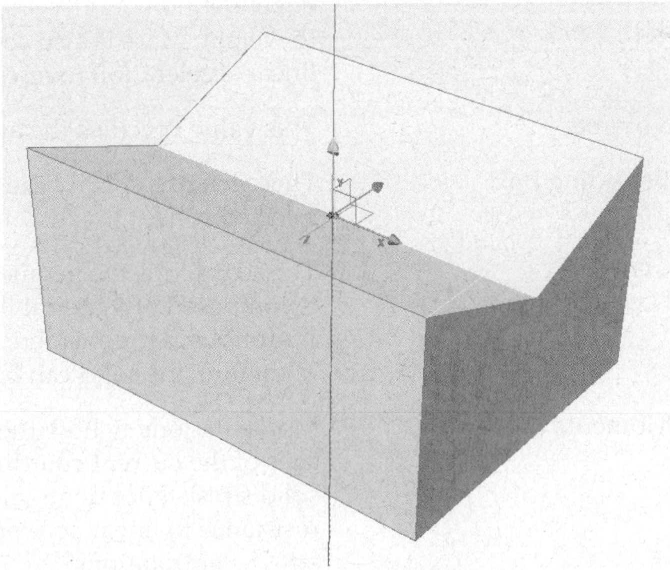

## USING SOLIDS DATA

AutoCAD offers a few commands that allow you to extract, import, and export data from solid model geometry. The *Massprop* command calculates a variety of properties for selected solids. Commands available for exporting and importing solid model geometric data for use with other programs are *Stlout*, *3Dprint*, *Acisin*, and *Acisout*.

### Massprop

| Ribbon | Menu Bar | Command (Type) | Alias (Type) | Shortcut |
|---|---|---|---|---|
| ... | *Tools Inquiry > Region/Mass Properties* | *Massprop* | ... | ... |

Since solid models define a complete description of the geometry, they are ideal for mass properties analysis. The *Massprop* command automatically computes a variety of mass properties.

Mass properties are useful for a variety of applications. The data generated by the *Massprop* command can be saved to an .MPR file for future exportation in order to develop bills of material, stress analysis, kinematics studies, and dynamics analysis (Fig. 36-100).

Applying the *Massprop* command to a solid model produces a text screen displaying the following list of calculations:

FIGURE 36-100

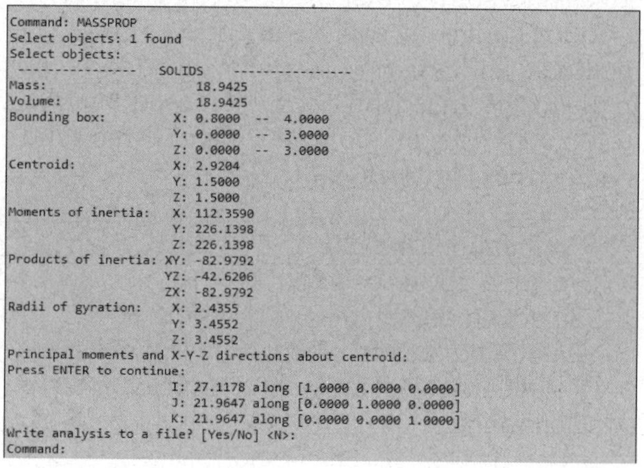

```
Command: MASSPROP
Select objects: 1 found
Select objects:
 ----------------   SOLIDS    ----------------
Mass:                    18.9425
Volume:                  18.9425
Bounding box:         X: 0.8000  --  4.0000
                      Y: 0.0000  --  3.0000
                      Z: 0.0000  --  3.0000
Centroid:             X: 2.9204
                      Y: 1.5000
                      Z: 1.5000
Moments of inertia:   X: 112.3590
                      Y: 226.1398
                      Z: 226.1398
Products of inertia: XY: -82.9792
                     YZ: -42.6206
                     ZX: -82.9792
Radii of gyration:    X: 2.4355
                      Y: 3.4552
                      Z: 3.4552
Principal moments and X-Y-Z directions about centroid:
Press ENTER to continue:
                      I: 27.1178 along [1.0000 0.0000 0.0000]
                      J: 21.9647 along [0.0000 1.0000 0.0000]
                      K: 21.9647 along [0.0000 0.0000 1.0000]
Write analysis to a file? [Yes/No] <N>:
Command:
```

Mass — Mass is the quantity of matter that a solid contains. Mass is determined by density of the material and volume of the solid. Mass is not dependent on gravity and, therefore, different from but proportional to weight. Mass is also considered a measure of a solid's resistance to linear acceleration (overcoming inertia).

Volume — This value specifies the amount of space occupied by the solid.

Bounding Box — These lengths specify the extreme width, depth, and height of the selected solid.

Centroid — The centroid is the geometrical center of the solid. Assuming the solid is composed of material that is homogeneous (uniform density), the centroid is also considered the center of mass and center of gravity. Therefore, the solid can be balanced when supported only at this point.

Moments of Inertia — Moments convey how the mass is distributed around the X, Y, and Z axes of the current coordinate system. These values are a measure of a solid's resistance to <u>angular</u> acceleration (mass is a measure of a solid's resistance to <u>linear</u> acceleration). Moments of inertia are helpful for stress computations.

Products of Inertia — These values specify the solid's resistance to angular acceleration with respect to two axes at a time (XY, YZ, or ZX). Products of inertia are also useful for stress analysis.

Radii of Gyration — If the object were a concentrated solid mass without holes or other features, the radii of gyration represent these theoretical dimensions (radius about each axis) such that the same moments of inertia would be computed.

Principal Moments and X, Y, Z Directions — In structural mechanics, it is sometimes important to determine the orientation of the axes about which the moments of inertia are at a maximum. When the moments of inertia about centroidal axes reach a maximum, the products of inertia become zero. These particular axes are called the principal axes, and the corresponding moments of inertia with respect to these axes are the principal moments (about the centroid).

TIP

Notice the "Mass:" is reported as having the same value as "Volume." This is because AutoCAD solids do not have material properties assigned to them, so AutoCAD assumes a density value of 1. To calculate the mass of a selected solid in AutoCAD, use a reference guide (such as a machinist's handbook) to find the material density and multiply the value times the reported volume (mass = volume x density).

### *Stlout*

| Ribbon | Menu Bar | Command (Type) | Alias (Type) | Shortcut |
|---|---|---|---|---|
| ... | *File*<br>*Export...*<br>**.stl* | *Stlout* | ... | ... |

The *Stlout* command converts an AutoCAD (ACIS) solid model or assembly to a file suitable for use with additive manufacturing technology (such as 3D printing, laser sintering, or stereo lithography). Use *Stlout* if you want to use an ACIS solid file to create a prototype part or finished part. This technology reads the CAD data and creates a part by using a laser to join or solidify microthin layers of plastic or wax polymer or powdered metal. A complex 3D prototype can be created from a CAD drawing in a matter of hours.

*Stlout* writes an ASCII or binary .STL file from an AutoCAD ACIS solid model. The model must reside entirely within the positive X,Y,Z octant of the WCS (all part geometry coordinates must be positive):

Command: ***STLOUT***
Select solids or watertight meshes: **PICK**
Select solids or watertight meshes: **Enter**
Create a binary STL file ? [Yes/No] <Y>: Enter **N** to create an ASCII rather than binary file.
Command:

AutoCAD displays the *Create .STL File* dialog box for you to designate a name and path for the file to create.

When you design parts with AutoCAD to be generated by stereolithography apparatus, it is a good idea to create the profile view (that which contains the most complex geometry) parallel to the XY plane. In this way, the aliasing (stair-step effect) on angled or curved surfaces, caused by incremental passes of the laser, is minimized.

### 3Dprint

| Ribbon | Menu Bar | Command (Type) | Alias (Type) | Shortcut |
|---|---|---|---|---|
| *Output*<br>*3D Print*<br>*Send to 3D Print Service* | ... | *3dprint* | *3DP* | ... |

As an alternative to the *Stlout* command, you can use the *3Dprint* command. This accomplishes the same goal as *Stlout* by generating an .STL file of your model. However, *3Dprint* provides a dialog box interface (Fig. 36-101). At the "Select solids or watertight meshes:" prompt, you can select multiple objects; however, only those object types are included in the STL file—all other geometry is discarded.

**FIGURE 36-101**

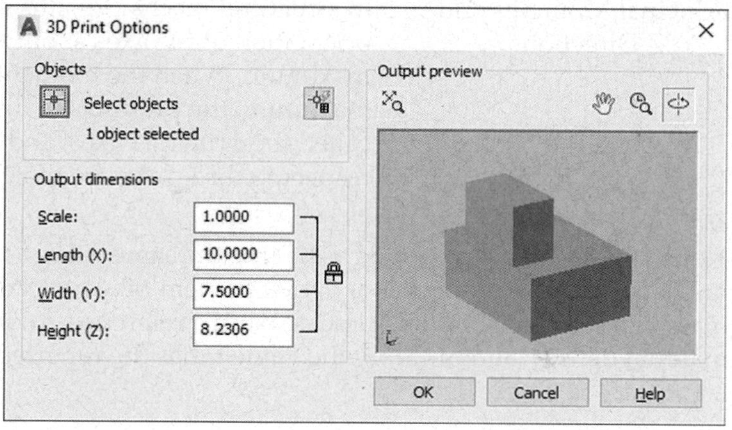


*Objects*
The dialog box provides *Select Objects* and a *Quick Select* button so you can add or remove objects from the selection set.

*Output Dimensions*
You can modify the dimensions of the bounding box and specify the scale of 3D solids and watertight meshes within the bounding box. These dimensions are locked proportionally so that if you modify any one of the values, the others are automatically adjusted.

*Output Preview*
You can use the *Zoom Extents, Pan, Zoom,* and *Orbit* buttons to change your view of the selected objects in the preview window. If you modify the preview, the output dimensions remain unaffected, and vice versa. The dialog box is resizable, so you can enlarge the preview if desired.

### Acisin

| Ribbon | Menu Bar | Command (Type) | Alias (Type) | Shortcut |
|---|---|---|---|---|
| *Insert*<br>*Import* | *Insert*<br>*ACIS File...* | *Acisin* | ... | ... |

The *Acisin* command imports an ACIS solid model stored in a .SAT (ASCII) file format. This utility can be used to import .SAT files describing a solid model created by AutoCAD or other CAD systems that create ACIS solid models. The *Select ACIS File* dialog box appears, allowing you to select the desired file. AutoCAD reads the file and builds the model in the current drawing. You can also use the more general Import command to import a variety of formats into the current drawing.

### Acisout

| Ribbon | Menu Bar | Command (Type) | Alias (Type) | Shortcut |
|---|---|---|---|---|
| ... | *File*<br>*Export... *.sat* | *Acisout* | ... | ... |

This utility is used to export an ACIS solid model to a .SAT (ASCII) file format that can later be read by AutoCAD or other CAD systems utilizing the ACIS modeler. The *Create ACIS File* dialog box is used to define the desired name and path for the file to be created.

## CHAPTER EXERCISES

1. **Editing Primitives using *Properties***

*Open* the CH35EX3 drawing that you created in Chapter 35 Exercises. Invoke the ***Properties*** palette. Hold down the **Ctrl** key while hovering and selecting the hole (*Cylinder* primitive) so properties for the primitive appear in the palette (Fig. 36-102). Edit the hole to have a new *Radius* of **1.0000** and a *Height* of **1.0000**. (Note that the "height" is the dimension in the negative Z direction in this case.) Use ***SaveAs*** and name the drawing **CH36EX1**.

FIGURE 36-102

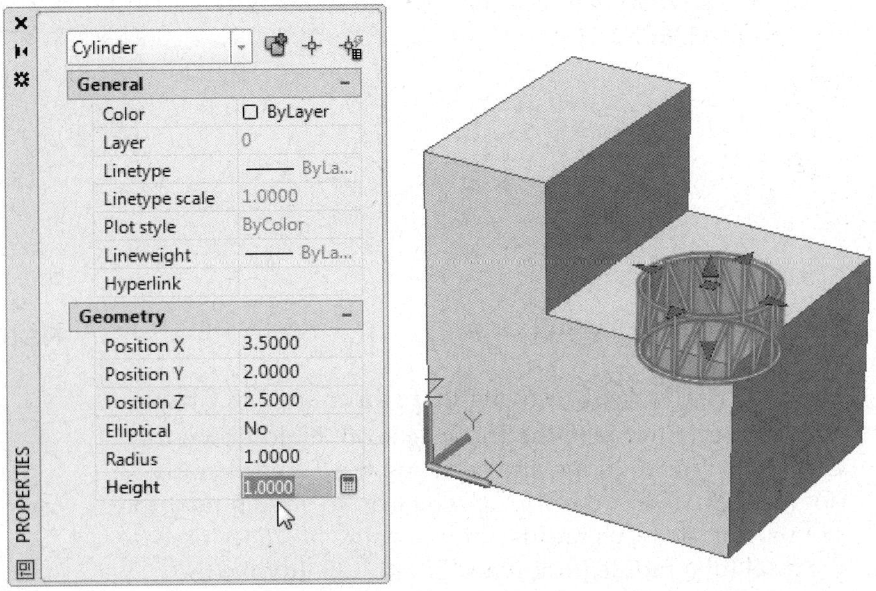


2. **Editing Primitives using *Properties***

If not already open, *Open* the **CH36EX1** drawing that you just created in the previous exercise. Invoke the ***Properties*** palette again. This time, hover and select the *Box* primitive so its properties appear in the palette (Fig. 36-103). Edit the ***Width*** dimension of the *Box* to a value of **3.0000**. Use ***SaveAs*** and name the drawing **CH36EX2**.

FIGURE 36-103

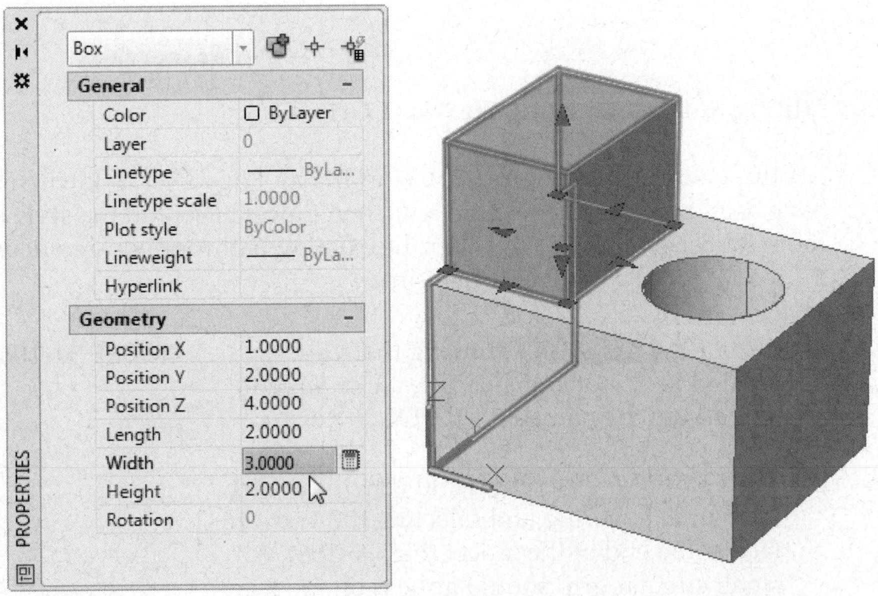

3. **Editing Primitives using Grips**

*Open* the **CH35EX5** drawing that you created in Chapter 35 Exercises. Turn on ***Polar Tracking***. Hold down the **Ctrl** key while hovering and selecting the *Box* primitive. Select the grip in the center of the base on the right side to make it hot, then stretch the grip 3 units toward the center of the solid (Fig. 36-104). This action effectively "cuts" the solid in half. Use ***SaveAs*** and name the drawing **CH36EX3**.

FIGURE 36-104

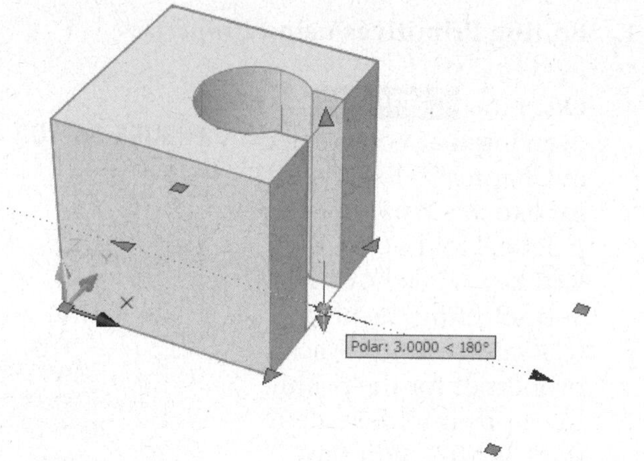


4. **Editing Primitives using Grips**

*Open* the **CH35EX6** drawing that you created in Chapter 35 Exercises. Ensure ***Polar Tracking*** is on. Hold down the **Ctrl** key while hovering and selecting the *Torus* primitive (Fig. 36-105). Select the appropriate grip to stretch the tube radius, not the torus radius. Enter a value of **.75** to increase the total tube radius (originally .50) to 1.25. Invoke the ***Properties*** palette to ensure the new *Tube Radius* is a value of **1.2500**. Use ***SaveAs*** and name the drawing **CH36EX4**.

FIGURE 36-105

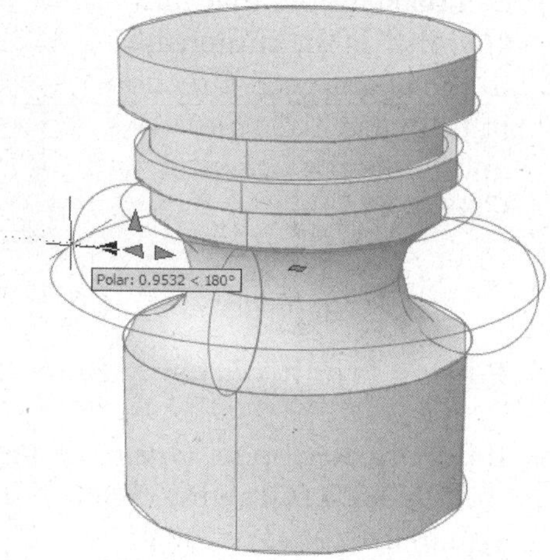


5. **Editing Subobjects using the Move Grip Tool**

For this exercise, assume you are working as a designer of kitchen and bathroom fixtures. You have been contracted to design a kitchen faucet based on the style of a recently designed bathroom faucet. You can easily alter the existing bathroom design using the move grip tool to extend the faucet for use with a kitchen sink.

A. *Open* the **FAUCET1** drawing that you created in Chapter 35 Exercises. Use ***SaveAs*** and assign the name **FAUCETK**.

B. Turn on ***Polar Tracking***. Hold down the **Ctrl** key while hovering and selecting the vertical face at the end of the faucet (Fig. 36-106). A small circular grip should appear on the face. Hover the cursor over the grip until the move grip tool relocates to the grip (Fig. 36-106). Select the X axis of the move grip tool.

FIGURE 36-106

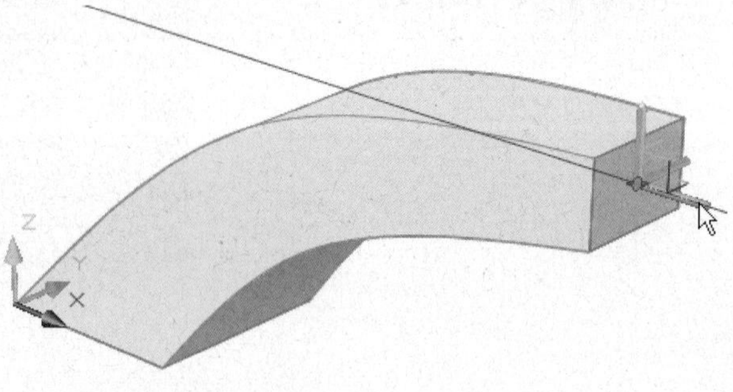

C. Stretch the end face of the faucet in a positive X direction using ***Polar Tracking***. Enter a value of **3** to make the faucet 3 units longer.

D. Depending on your setting of the *DELOBJ* system variable, some of the original geometry used to create the faucet may still remain. If the original *Rectangle* is visible as shown in Figure 36-107, select the object so it becomes highlighted, then use ***Erase*** to delete it.

**FIGURE 36-107**

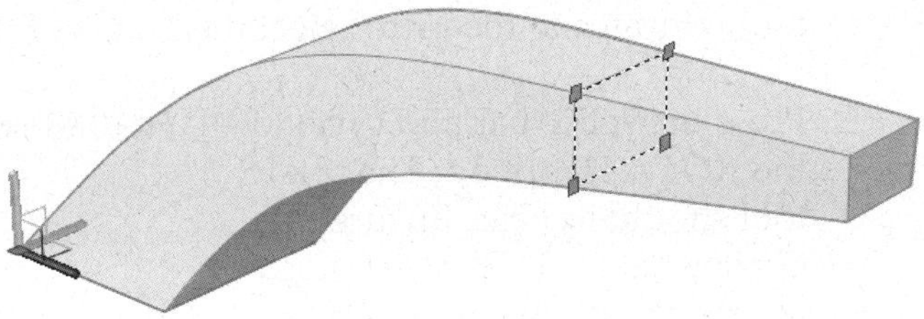

6. **Editing Subobjects using the Move Grip Tool**

As the designer for the National Bicycle Racing Team, a request has been made to add 20mm to the grip portion of the drop bars used on the team bikes. This change can be made using the move grip tool to extend the length of the current design.

**FIGURE 36-108**

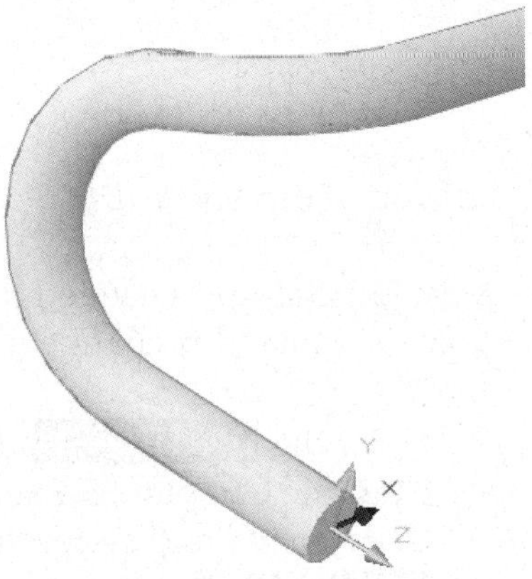

REUSE

A. ***Open*** the DROPBAR drawing that you created in Chapter 35 Exercises. Use ***SaveAs*** and assign the name **DROPBAR-L**.

B. Create a new ***UCS*** using the ***3 Point*** method. Use the ***Center*** *Object Snap* for the ***Origin*** and ***Quadrant*** *Object Snaps* for the ***X*** and ***Y*** directions. Your new UCS should look like that in Figure 36-108.

C. Turn on ***Polar Tracking***. Hold down the **Ctrl** key while hovering or use the ***Face*** filter, then select the face at the end cap of the bars (at the new UCS origin) so the circular grip appears on the face. Hover the cursor over the grip until the move grip tool relocates to the grip. Select the X axis of the move grip tool and stretch the end cap out **20** mm.

D. Perform the same action to extend the length of the other end of the drop bar. ***Save*** the drawing.

7. **Editing Subobjects using the Rotate Grip Tool**

A design change has been requested to rotate the mounting surface on the Swivel by 15 degrees. This change can be made using the rotate grip tool.

A. ***Open*** the SWIVEL drawing that you created in Chapter 35 Exercises. Use ***SaveAs*** and assign the name **SWIVEL2**.

**FIGURE 36-109**

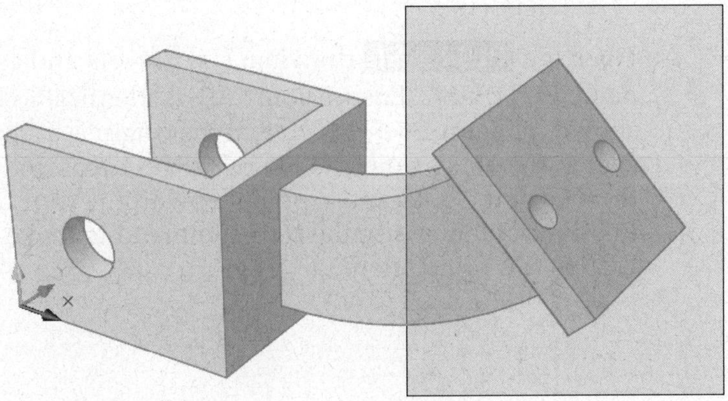

REUSE

B. Hold down the **Ctrl** key and use a window to select the entire mounting plate as shown in Figure 36-109. Hover the cursor over the face grip in the center of the top surface until the move grip tool relocates to the grip. Select the X axis of the move grip tool.

C. Press the space bar and cycle to the **ROTATE** option. The rotate tool should appear. Ensure the appropriate axis is selected on the tool to allow you to rotate the entire mounting plate as shown in Figure 36-110. Enter a value of **15** for the rotation. ***Save*** the drawing.

**FIGURE 36-110**

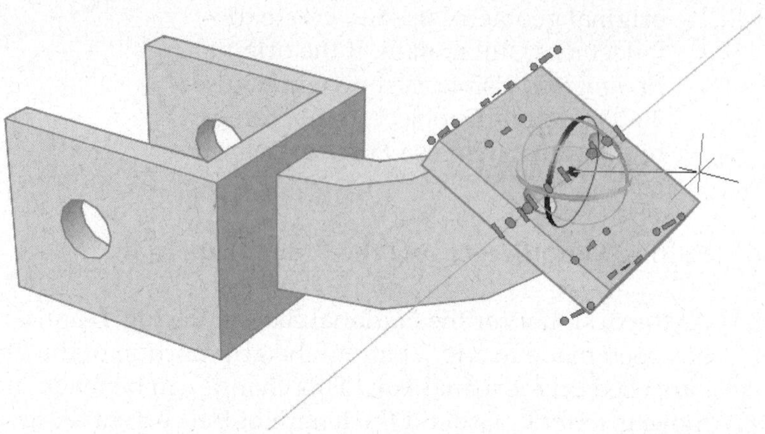

8. **Editing using *Solidedit***

A design change is needed to change the angle of the inclined surface on the Bar Guide by 10 degrees. Make this change using the *Rotate Faces* option of *Solidedit*.

A. ***Open*** the **BGUID-SL** drawing from the Chapter 35 Exercises. Use ***SaveAs*** and assign the name **BGUID-SL-2**.

B. Invoke the ***Solidedit*** command and the ***Rotate Faces*** option. Select the inclined plane as the face to rotate. Use ***Endpoint*** *Object Snap* to specify 2 points to define the axis of rotation. Select the endpoints of the bottom edge of the surface as shown in Figure 36-111. Rotate the surface **10** degrees. ***Save*** the drawing.

**FIGURE 36-111**

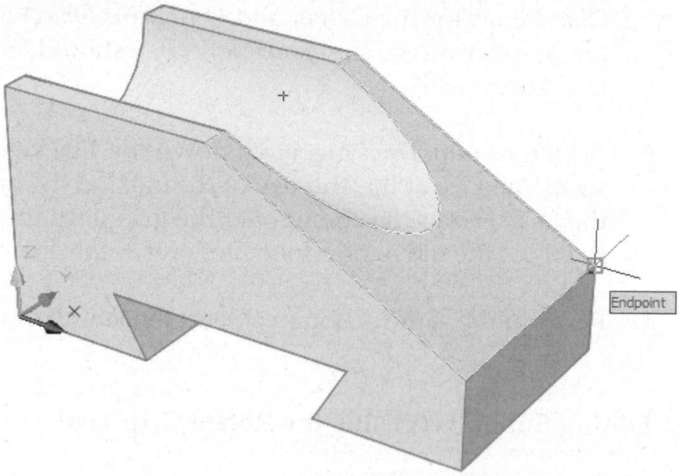


9. **Using *Sectionplane***

A. ***Open*** the **PULLY-SL** drawing. Use ***SaveAs*** and assign the name **PULLY-SC**. Ensure your UCS is identical to that shown in Figure 36-112 such that the origin is at the front center, the X axis points to the right, the Y axis goes through the axis of the hole, and Z points upward. If your UCS is not in this orientation, use the ***UCS*** command to make a new UCS in this orientation.

**FIGURE 36-112**

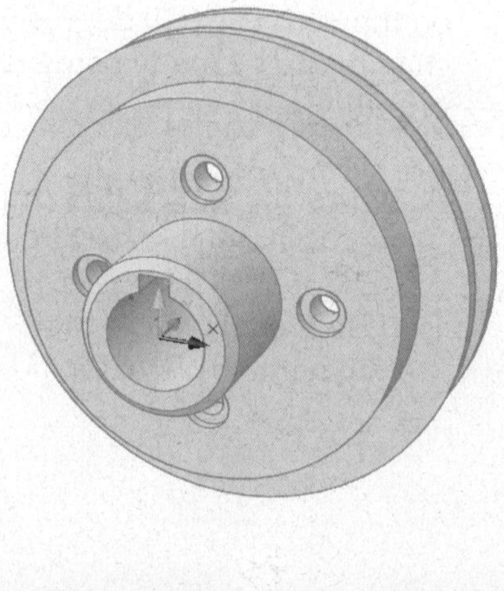

B. Invoke ***Sectionplane***. Use the ***Orthographic*** option and align the section to the ***Right*** side. A new section plane should appear as shown in Figure 36-113.

C. Experiment with the section plane by selecting it to activate the grips, then using the move grip tool to pass the section plane through the pulley to reveal its inner features.

D. Invoke the ***Properties*** palette, select the section plane object, and set the ***Live Section*** property to ***No***. The section plane should appear in the drawing, but the pulley should not appear cut. ***Save*** the drawing.

**FIGURE 36-113**

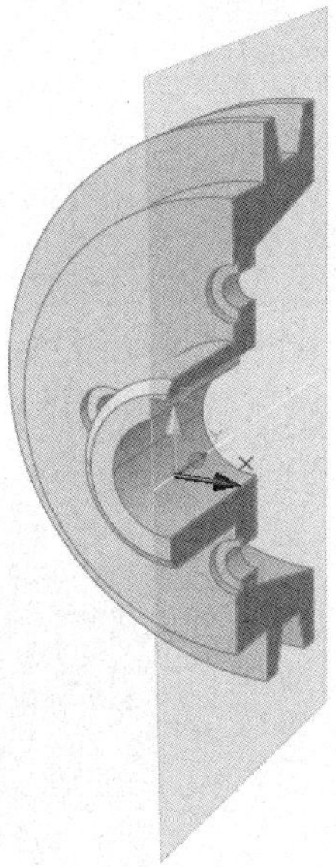

10. **Using *Sectionplane***

A. ***Open*** the **SADL-SL** drawing from the Chapter 35 Exercises. Use ***SaveAs*** and assign the name **SADL-SC**.

B. Invoke ***Sectionplane*** to create a new section plane like that shown in Figure 36-114.

C. If you have problems, ensure your UCS is identical to that shown in Figure 36-114. Also check to make sure *Live Sectioning* is on.

**FIGURE 36-114**

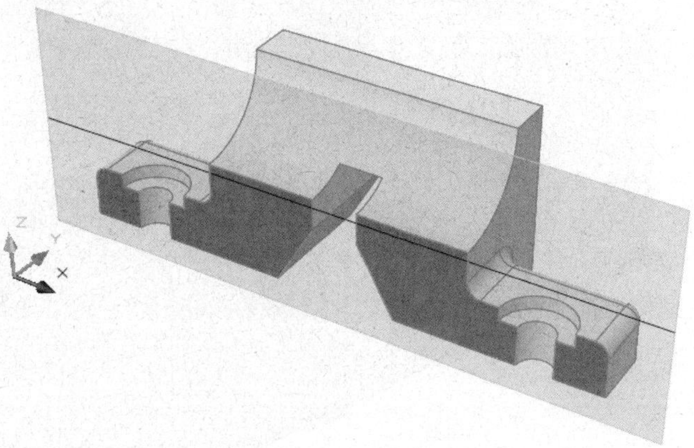

D. Experiment with the section plane by selecting it to activate the grips, then using the move grip tool to pass the section plane through the pulley to reveal its inner features. ***Save*** the drawing.

11. **Calculating *Mass Properties***

***Open*** the **SADL-SL** drawing that you created in Chapter 35 Exercises. Calculate ***Mass Properties*** for the saddle. Write the report out to a file named **SADL-SL.MPR**. Use a text editor to examine the file.

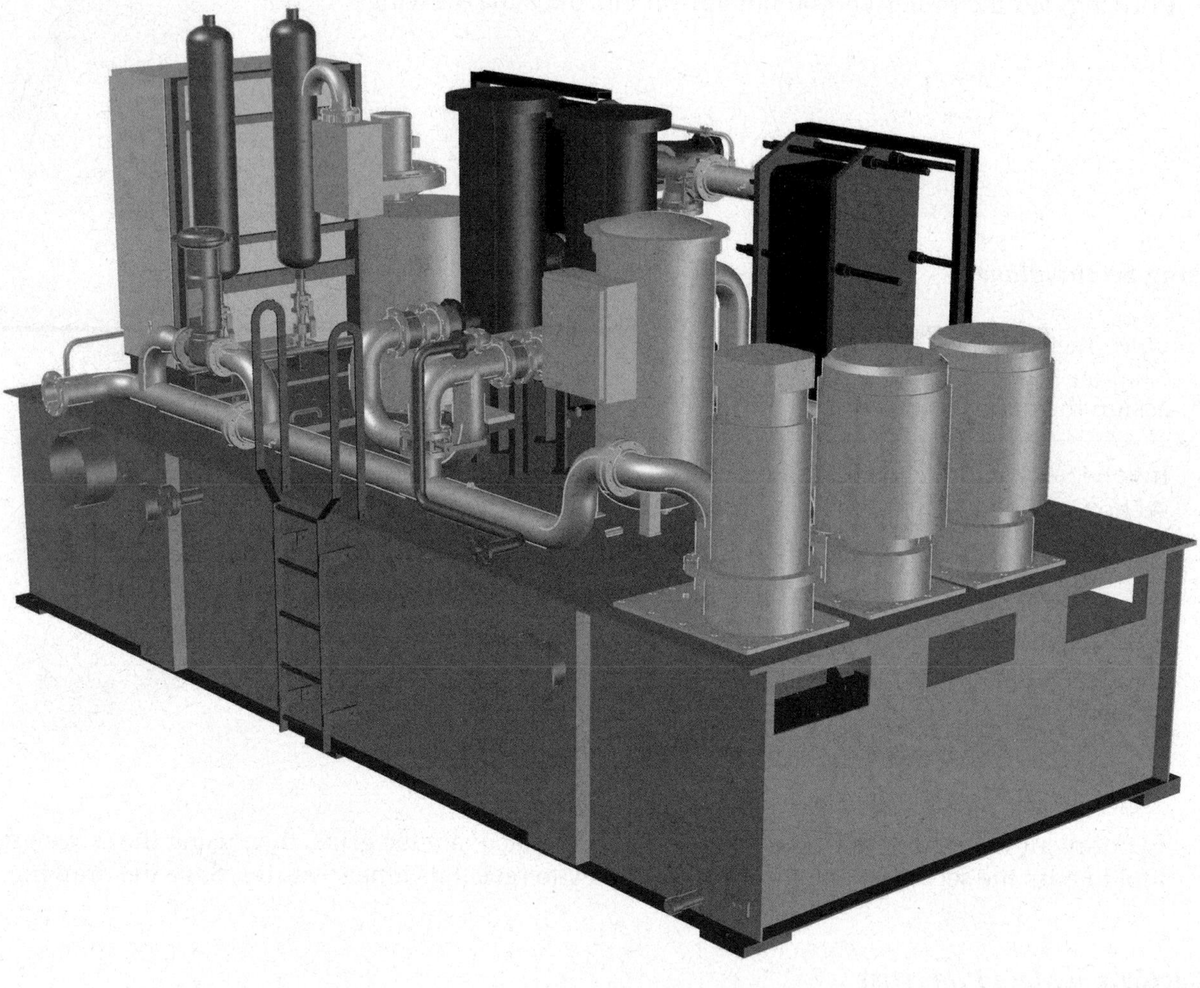

**OIL MODULE.DWG Courtesy of Autodesk, Inc.**

CHAPTER 37

# 2D Drawings from 3D Models

## Chapter Objectives

After completing this chapter you should:

1. be able to create a base (primary) view;
2. be able to create projected (secondary) views;
3. be able to create an isometric pictorial view;
4. be able to create section views;
5. be able to create detail views;
6. understand how to control section view styles and detail view styles; and
7. be able to update views when the model changes.

# CONCEPTS

Orthographic drawings have been used as the industry standard method to describe and document 3-dimensional objects. Orthographic drawings include multiview drawings and isometric drawings. Isometric drawings describe a 3-dimensional object in one "pictorial" view showing 3 dimensions, but drawn on a 2-dimensional plane. Multiview drawings describe a 3-dimensional object by showing multiple 2-dimensional views, such as a front view, top view, and side view. These views have a specific arrangement such that the front and side views align horizontally and the front and top views align vertically. Therefore, we consider the views to be "projected." Typically, multiview drawings include the dimensional specifications for the objects. Therefore, a complete multiview drawing is a 2-dimensional drawing used to describe the 3-dimensional object in preparation for manufacturing and to document the design for future use and possible revisions.

Chapter 24, Multiview Drawing, discusses commands and methods for creating 2D views of 3D objects. The process there is to draw each view "from scratch" using the *Line*, *Circle*, and other commands and using *Polar Tracking* and other tools to project views. This process has been used for decades—that is, first create a 2D drawing, then use the drawing to manufacture the 3D part.

With the development of 3D CAD systems the process has changed. With our current technology, it is possible and preferable to first create the desired 3D model, and then use that model to somewhat automatically create the needed 2D multiview drawings and isometric views. Since we can manufacture the finished parts directly from the 3D model, the 2D multiview drawings are used more for documentation—a chronological record of the design and revisions to the design.

**FIGURE 37-1**

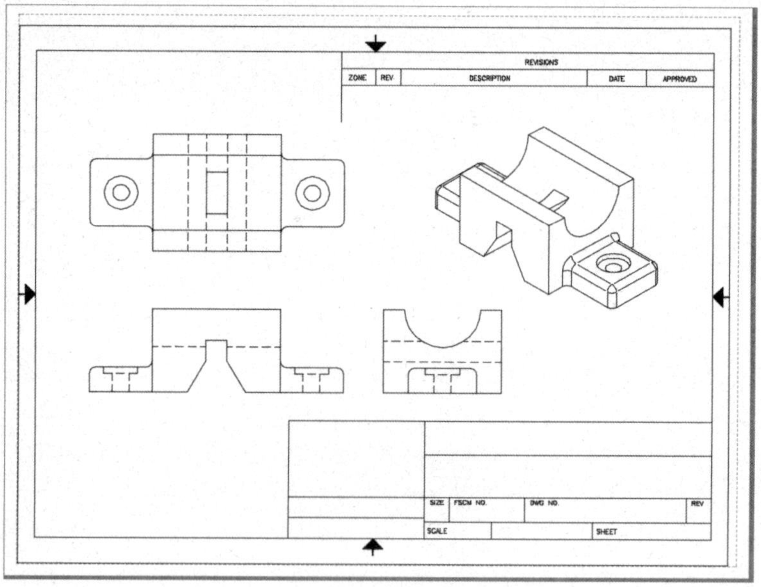


This chapter explains how to use AutoCAD to create 2D documentation drawings. Although the commands can be used to create a 2D drawing from an Autodesk Inventor model as well, only the options used for AutoCAD 3D models are discussed here. Figure 37-1 displays an AutoCAD drawing created from a 3D solid model. The commands needed to create, edit, and update such a drawing are explained. The commands that are discussed in this chapter are listed below.

| | |
|---|---|
| *Viewbase* | creates the base (primary) view |
| *Viewedit* | use to edit parameters for an existing view |
| *Viewproj* | creates a view projected from an existing view |
| *Viewstd* | opens the *Drawing Standard* dialog box used for setting drafting standards |
| *Viewsection* | creates a projected section view from any existing view |
| *Viewsectionstyle* | specifies the parameters for creating a section view |
| *Viewcomponent* | changes display of a section view to a slice, outline, or normal view |
| *Viewdetail* | creates a projected detail view from any existing view |
| *Viewdetailstyle* | specifies the parameters for creating a detail view |
| *Viewupdate* | updates the existing views to any changes made to the model |
| *Viewsymbolsketch* | applies constraints between the model and the cutting plane or detail boundary |

The commands listed here are available in the *3D Modeling* workspace. The *Viewbase* command is the first command needed to begin the process. *Viewbase* (*Base*) is located in the *View* panel of the *Home* tab of the Ribbon. *Viewbase* requires you to select the 3D object or component (of an assembly) and creates the first view. Additional projected views can be created with *Viewbase* or *Viewproj*.

However, the full array of commands needed to create a documentation drawing is not available until you first use the *Viewbase* command. Once you create a base view, use the *Layout* tab from the Ribbon (Figure 37-2) to access commands needed to create and edit a 2D documentation drawing including section views and detail views.

**FIGURE 37-2**

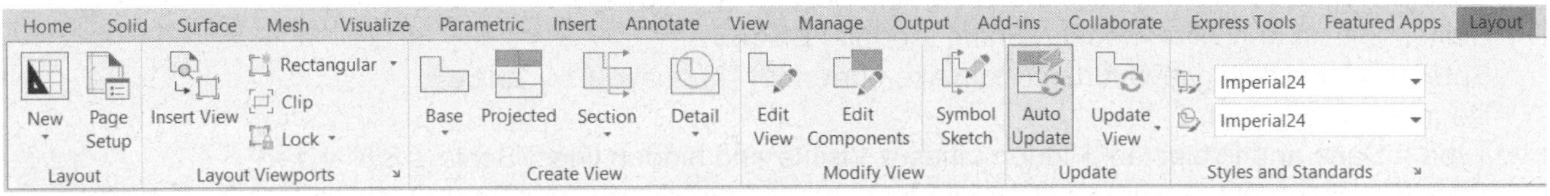


## COMMANDS

The following commands explain the process of creating a 2D drawing in a layout from a 3D model in model space. The first several commands are explained here in a sequential order similar to the typical process you would use to create a documentation drawing. Before using any of the following commands, you must first have a 3D model that you intend to show in a multiview drawing arrangement. Figure 37-3 shows a solid model of the Saddle that will be used in the next several pages to describe the process.

**FIGURE 37-3**

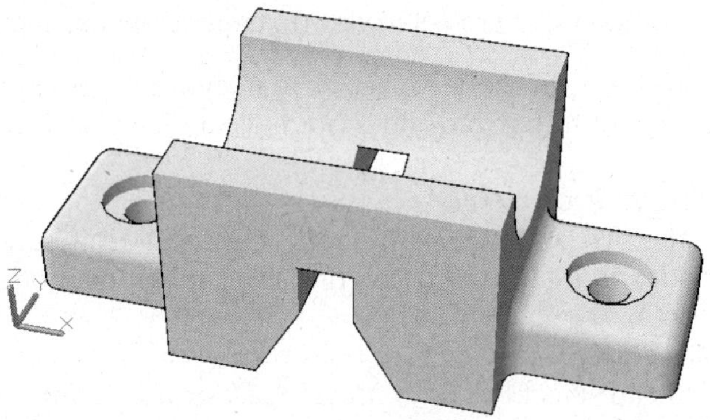

NOTE: *Viewbase* and the other commands in this chapter can be used to create a 2D drawing from an AutoCAD 3D model or from an Autodesk Inventor model. This chapter discusses only the options used for AutoCAD 3D models.

### *Viewbase*

| Ribbon | Menu Bar | Command (Type) | Alias (Type) | Shortcut |
|---|---|---|---|---|
| *Home*<br>*View*<br>*Base* | ... | *Viewbase* | ... | ... |

Use *Viewbase* to create the first (base) view in a drawing. Although all other views are derived from the base view, you can create additional projected views when you first use the *Viewbase* command. If more than one solid or surface object exists in model space, *Viewbase* allows you to select individual objects but can include all visible objects within model space.

The *Viewbase* command displays the *Drawing View Creation* tab on the Ribbon (Figure 37-4). (If the Ribbon is not active, you can use Command line prompts of the *Viewedit* command to change the options for the base view and projected views.) See "*Viewbase* Options."

The Command line displays the following prompts. In this case *Viewbase* was used to create a base (front) view, top view, side view, and isometric view as described in the following paragraphs and figures.

Command: ***VIEWBASE***

**FIGURE 37-4**

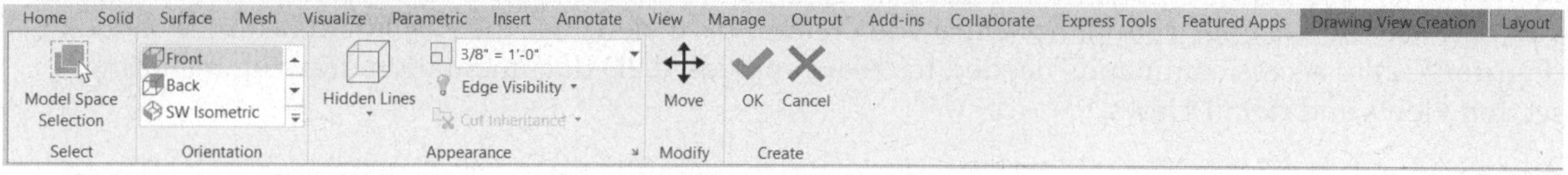


Specify model source [Model space/File] <Model space>: **Enter**
Select objects or [Entire model] <Entire model>: **Enter**
Enter new or existing layout name to make current or [?] <Layout1>: **Enter**
Regenerating layout.
Type = Base and Projected Hidden Lines = Visible and hidden lines Scale = 3/8" = 1'-0"
Specify location of base view or [Type/sElect/Orientation/Hidden lines/Scale/Visibility] <Type>: **PICK**
Select option [sElect/Orientation/Hidden lines/Scale/Visibility/Move/eXit] <eXit>: **Enter**
Specify location of projected view or <eXit>: **PICK**
Specify location of projected view or [Undo/eXit] <eXit>: **PICK**
Specify location of projected view or [Undo/eXit] <eXit>: **PICK**
Specify location of projected view or [Undo/eXit] <eXit>: **Enter**
Base and 3 projected view(s) created successfully.

You can be in model space or in a layout when invoking this command. Make sure that the layout you use is not initialized (does not have a viewport already created).

### *Model Space/File*

In the first step ("Specify model source") designate if you want to use a model existing in AutoCAD model space (*Model Space*) or use an Inventor model (*File*).

#### *Model Space*

You can select the *From Model Space* option from the Ribbon or press Enter at the Command line prompt. If you are in a layout, *Viewbase* automatically uses the existing model in model space. If you are in model space, the following prompt is displayed. Select the entire model or select a single component (if the drawing contains multiple objects).

Select objects or [Entire model] <Entire model>: **PICK**

#### *File*

If you use the *File* option or if model space does not contain any visible solids or surfaces, the *Select File* dialog box is displayed to enable you to select an Inventor model.

Once the model is selected, AutoCAD either uses the current layout and prompts you to specify the location for the base view or prompts for the layout name to use if you are in model space. You can enter a new layout name to create.

Enter new or existing layout name to make current or [?] <Layout1>: **Enter** or (**name**)
Regenerating layout.
Type = Base and Projected Hidden Lines = Visible and hidden lines Scale = 3/8" = 1'-0"
Specify location of base view or [Type/sElect/Orientation/Hidden lines/Scale/Visibility] <Type>: **PICK**

Once AutoCAD opens the layout (unless already open), pick the location for the base view. *Viewbase* places the base view in the location you specify, but it is temporarily displayed shaded as shown in Figure 37-5. AutoCAD uses the other settings as listed in the prompt and automatically calculates a scale based on the model dimensions and layout sheet size (See "*Viewbase* Options" for changing the settings).

**FIGURE 37-5**

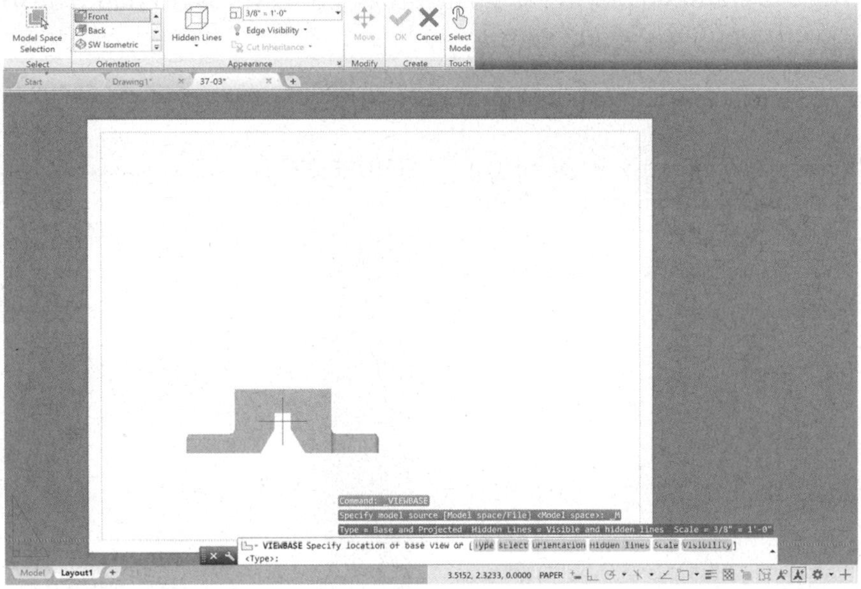


TIP

NOTE: AutoCAD considers the default base view as the front view. AutoCAD uses the object's face parallel to the World Coordinate System XZ plane as the front view for the model. In other words, the viewing direction for the base view is normal to (perpendicular to) the WCS XZ plane (see Figure 37-3). Therefore, before using *Viewbase*, it is best to orient the model so the desired front surface is parallel with the WCS XZ plane. Also see the *Orientation* option.

To place a second (projected) view without specifying other options, press Enter at the next prompt and pick a location for the next view.

Select option [sElect/Orientation/Hidden lines/Scale/Visibility/Move/eXit] <eXit>: **`Enter`**
Specify location of projected view or <eXit>: **`PICK`**

For example, Figure 37-6 illustrates locating a projected top view. AutoCAD automatically aligns the view with the base view. For special cases such as removed sections, you can break the orthogonal alignment while placing the view by tapping (press and release) the Shift key. Also see "Drawing View Editing, Moving Views" for breaking the alignment after placing the view.

**FIGURE 37-6**

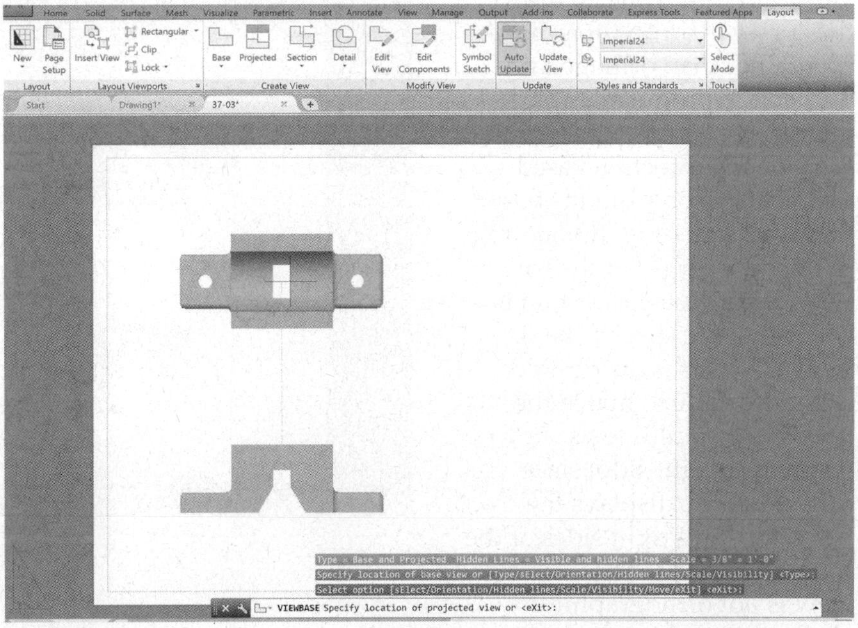

You can continue placing as many projected views as needed. The following prompt is displayed as long as you continue placing views.

Specify location of projected view or [Undo/eXit] <eXit>: **PICK**

Figure 37-7 illustrates picking a location for the side view. Note that alignment with the front view is automatically maintained.

**FIGURE 37-7**

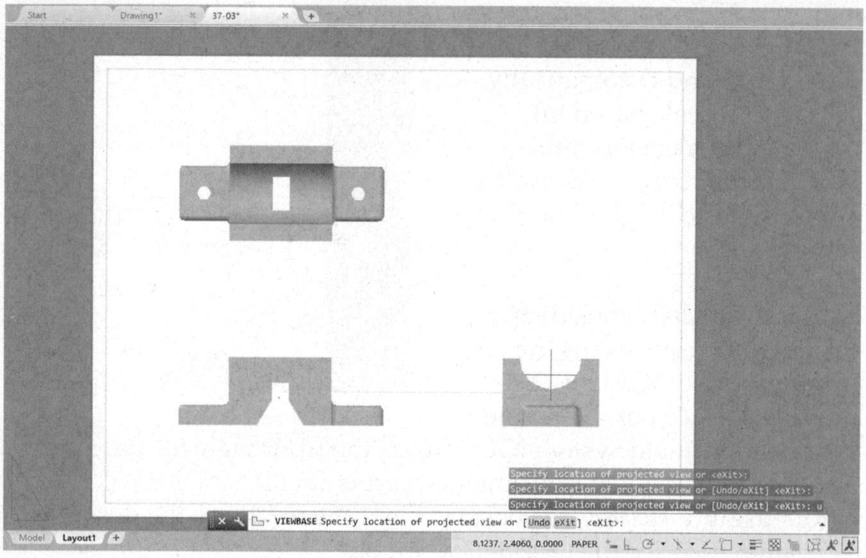


When placing views at the "Specify location of projected view" prompt, you can drag the cursor in an angular direction rather than orthogonally. This action produces an isometric view of the model with the viewing direction based on the direction from the base view that you drag. If you drag to the upper-right of the base view, the viewing direction is a SE isometric (Figure 37-8). Typically, this is the desired isometric view to match the selection of multiviews shown (front, top, right side) since a SE isometric displays the front, top, and right sides of the model. Note that an isometric view is not orthographically aligned or constrained by the location of the base view.

**FIGURE 37-8**

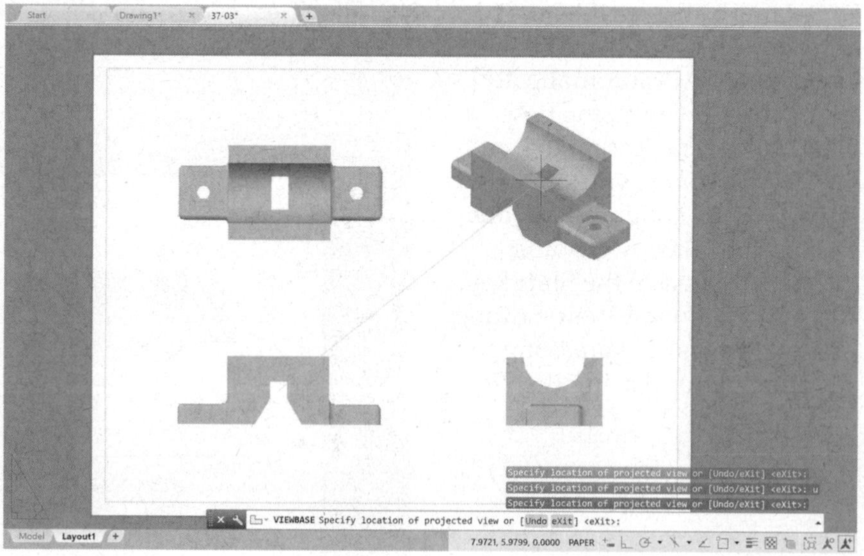

When you have placed all the needed views, press Enter to end the command.

Specify location of projected view or [Undo/eXit] <eXit>: **Enter**
Base and 3 projected view(s) created successfully.

When the *Viewbase* command ends, the model views are converted to 2D lines representing a correct multiview drawing. That is, the model, previously displayed as shaded (or as other visual style), is now displayed with visible and hidden lines (Figure 37-9). The resulting display is based on the options previously set or set before placing the first view (see "*Viewbase* Options").

**FIGURE 37-9**

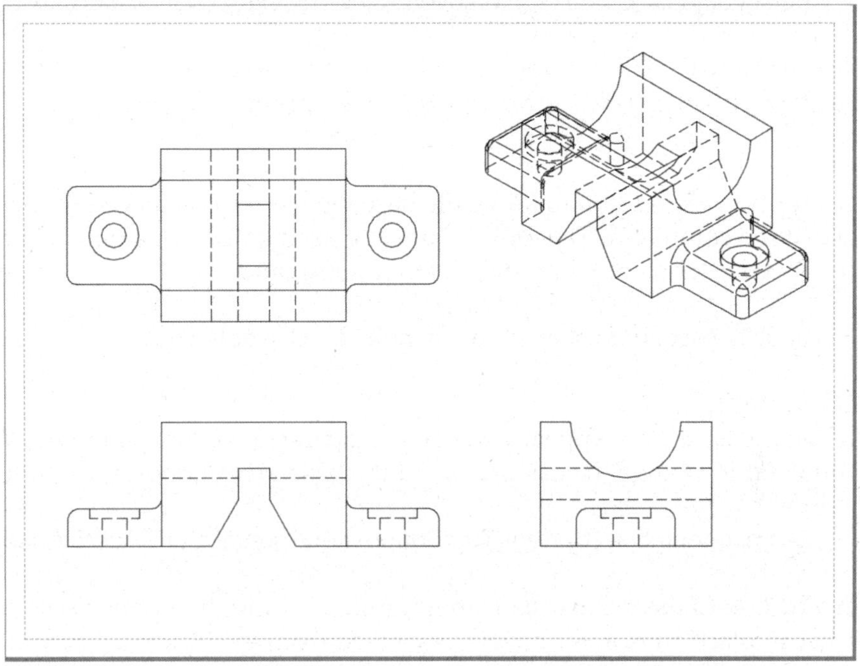

## *Viewbase* Options

When you initiate the *Viewbase* command, the first step is to select the model and specify the desired layout to use. Next, the Ribbon displays the *Drawing Creation* tab providing options for the new views (Figure 37-10). In addition, you can use the Command line to change options.

**FIGURE 37-10**

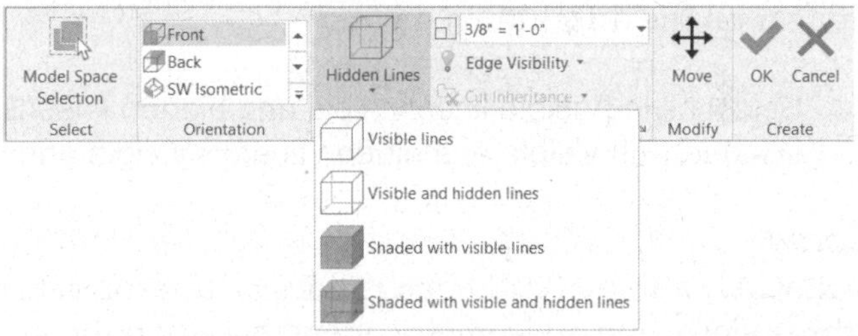


NOTE: These next options can be used only while placing the base view. When you place additional projected views the *Drawing Creation* tab becomes inactive and the Command line no longer offers the prompts. Additional views automatically inherit properties (such as scale, hidden lines, orientation, and visibility options) from the base view.

Type = Base and Projected  Hidden Lines = Visible and hidden lines Scale = 3/8" = 1'-0"
Specify location of base view or [Type/sElect/Orientation/Hidden lines/Scale/Visibility] <Type>:

### *Type*

Use this option to specify if the command continues after creating the base view (single view) or goes on to create projected views.

Enter a view creation option [Base only/base and Projected] <Base and Projected>:

### Select

If you are in a layout when you use the *Viewbase* command, AutoCAD automatically selects all available solids and surfaces in model space. When more than one object exists in model space, you can exclude or include additional solids and surfaces from the base view using the *Select* option to specify objects to add or remove.

Select objects to add or [Remove/Entire model/LAYout] <return to layout>:

*Remove* allows you to remove objects from the selection set in model space and then reopens the previous layout. The *Layout* option returns to the previous layout.

### Move

This option allows you to move the base view after it is placed in the drawing area without forcing you to exit the command. This option must be used before placing additional views although views can also be moved after completing the *Viewbase* command.

Specify second point or <use first point as displacement>:

### Orientation

When the base view appears, the viewing direction is normal to the WCS XZ plane by default (*Front*). If you want to change the orientation of the base view, choose from the following preset orientations:

Select orientation [Current/Top/Bottom/Left/Right/Front/BAck/SW iso/SE iso/NE iso/NW iso] <Front>:

In addition to the normal six orthographic viewing directions (*Top, Bottom, Left, Right, Front,* and *Back*), you can place an isometric view as viewed from the SW, SE, NE, and NW.

### Hidden Lines

Use this option to specify the display style to use for the base view. See Figure 37-10 for the options that appear on the drop-down menu.

Select style [Visible lines/vIsible and hidden lines/Shaded with visible lines/
sHaded with visible and hidden lines] <Visible and hidden lines>:

### Scale

AutoCAD automatically determines an appropriate scale to use for a typical multiview arrangement (three views and an isometric view) based on the model dimensions and the paper size assigned to the layout.

Enter scale <0.0417>:

You can change the scale to any other value, particularly if you need only one or two views or need a large number of views. The scale value affects the base view. Projected views derived from the base view automatically inherit the same scale.

### Visibility

Many visibility options are available for the display of the base view and projected views; however, some or all of these options may not be available for the selected model depending on the geometric configuration of the model.

Select type [Interference edges/Tangent edges/Bend extents/Thread features/Presentation trails/eXit]
<eXit>:

*Interference edges*
If the base view includes an assembly of components and an interference condition exists, this option turns visibility of interference edges on or off. When turned on, the base view displays both hidden and visible edges that are otherwise excluded due to an interference condition.

FIGURE 37-11

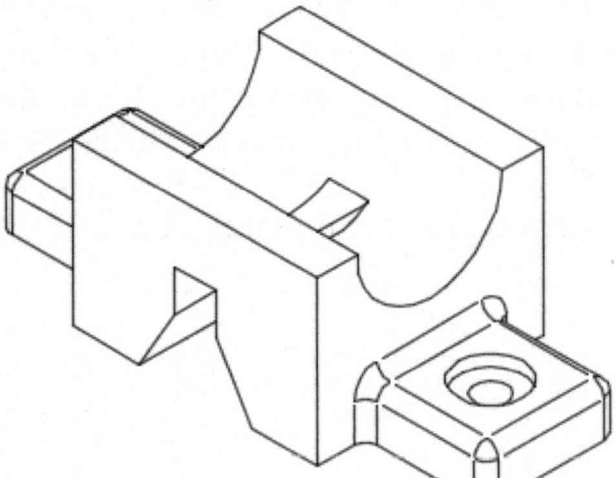

*Tangent edges*
This option affects models that contain rounded surfaces that are tangent to other surfaces such as when fillets exist. When the *Tangent edges* option is turned on, the selected view displays a line to show the intersection of surfaces meeting tangentially, as shown on the filleted edges in Figure 37-11. Without this option, no edges would be visible where the fillets exist—only the outlines of the surfaces. This setting does not affect completely round surfaces such as holes.

*Tangent edges foreshortened*
If you choose to turn on *Tangent edges*, you can choose to slightly shorten the length of tangential edges to differentiate them from visible edges. Turning on *Tangent edges* and *Tangent edges foreshortened* is recommended for isometric views of objects that contain filleted edges such as that in Figure 37-11. Note the small gaps at the corners of the fillets.

*Thread features, Presentation trails, Bend features*
These options affect Autodesk Inventor models.

### Viewedit

| Ribbon | Menu Bar | Command (Type) | Alias (Type) | Shortcut |
|---|---|---|---|---|
| *Layout*<br>*Modify View*<br>*Edit View* | ... | *Viewedit* | ... | ... |

*Viewedit* allows you to edit existing views including base views, projected views, section views, and detail views. You can invoke *Viewedit* by two methods: 1) type or select the *Viewedit* command and 2) double-click on a view to edit.

Command: **VIEWEDIT**
Select view: **PICK**
Select option [sElect/Hidden lines/Scale/Visibility/eXit] <eXit>:

*Viewedit* produces the *Drawing View Editor* tab (Figure 37-12). This tab and the command prompt options offer essentially the same options as the *Drawing View Creation* tab shown in Figure 37-4 and the *Viewbase* Command line prompts, with a few exceptions.

FIGURE 37-12

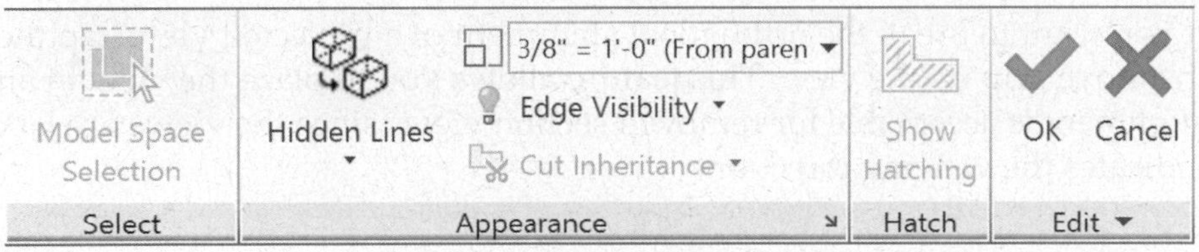

Please refer to the previous section "*Viewbase* Options" for an explanation of the *Select, Hidden lines, Scale,* and *Visibility* options. The exceptions are 1) you can choose additional options that appear on the *Drawing View Editor* when a section view or detail view is selected, and 2) you cannot move a view with *Viewedit*, you must use a different method described in the "*Drawing View* Editing" section (next).

## Drawing View Editing

There is an additional method for editing views that allows you to interactively move views and to change the scale of the views. This method does not involve the use of a command but it does activate two grips in the view and produces the *Drawing View* tab on the Ribbon (Fig. 37-13). To access this method, select any existing view (single-click, not double-click). Once selected, the view and its viewport border become highlighted. The available options are as follows.

**FIGURE 37-13**

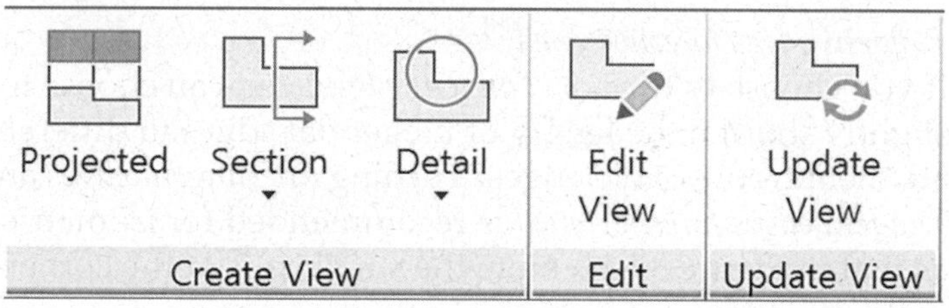


TIP

### Changing View Scale

Click the arrow grip to cause the scale list to pop up (Figure 37-14). Select any scale from the list. If you change the scale for the base view, all projected views also change to the new scale because projected views always inherit properties from the "parent." However, you can change the scale of projected views independently. For example, you may want to change the scale of an isometric view since it is not truly a projected (orthogonal) view and therefore could be displayed at a different scale than the orthogonally aligned views.

**FIGURE 37-14**

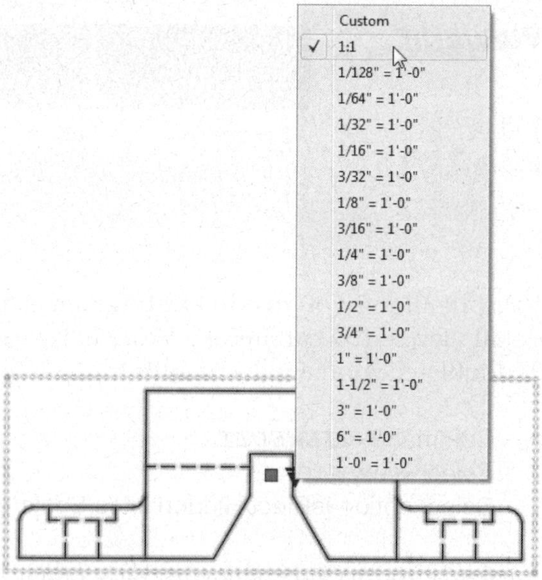


### Moving Views

Select the center square grip that appears in any view to move the view to a new location (Figure 37-15). Any orthogonally projected views automatically move so as to retain the orthogonal alignment. Isometric views do not move when the parent view is moved.

TIP

If you want to break the orthogonal alignment of a projected view, tap the Shift key once while dragging the move grip on the view. This feature allows you to place the view in any position. A non-orthogonal alignment is acceptable for removed section views since the views are labeled and the cutting plane line indicates the viewing direction.

### Deleting Views

To delete a view, select the view so it is highlighted, then use the *Erase* command or press the Delete key. Deleting the base (parent) view also deletes projected views in some, but not all cases.

FIGURE 37-15

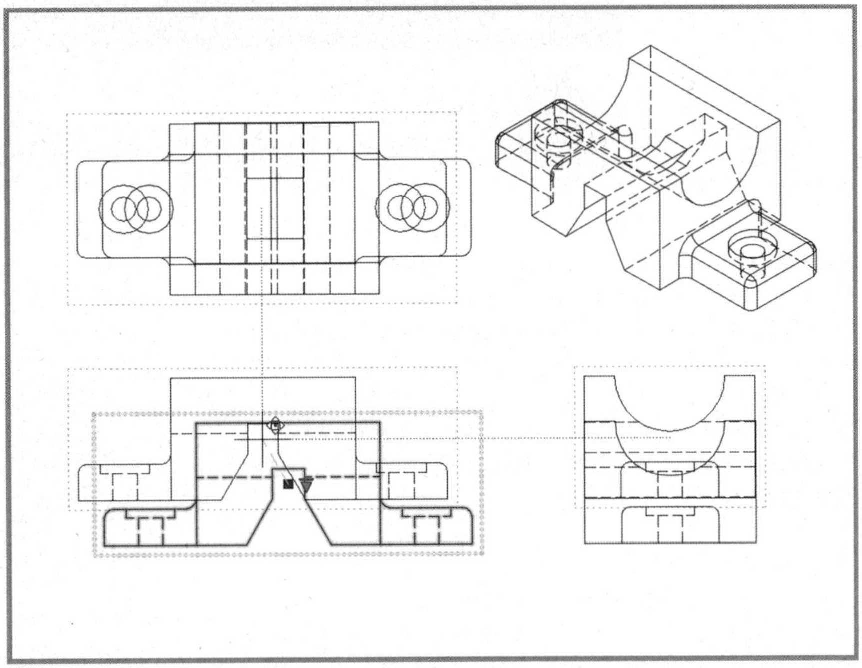

### *Edit View*

This option produces the *Viewedit* command.

### *Projected View*

Selecting this button invokes the *Viewproj* command (see "*Viewproj*").

### *Section*

Use this button to access the *Viewsection* command (see "*Viewsection*").

### *Detail*

Select this button to produce the *Viewdetail* command (see "*Viewdetail*").

### *Update View*

This button invokes the *Viewupdate* command (see "*Viewupdate*").

## *Viewproj*

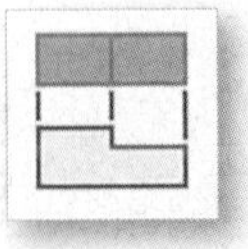

| Ribbon | Menu Bar | Command (Type) | Alias (Type) | Shortcut |
|---|---|---|---|---|
| *Layout*<br>*Create View*<br>*Projected* | ... | *Viewproj* | ... | ... |

*Viewproj* creates one projected view from any existing view. The views that can be created include orthogonal views (views that are projected perpendicularly to the viewport's vertical and horizontal boundary edges) and isometric views (views projected at any other angle than vertical and horizontal). *Viewproj* is simple to use and operates identically to *Viewbase* when creating projected views.

Command: ***VIEWPROJ***
Select parent view: **PICK**
Specify location of projected view or <eXit>: **PICK**
Specify location of projected view or [Undo/eXit] <eXit>: **Enter**
1 projected view(s) created successfully.

NOTE: Projected views inherit properties from the parent view, such as visibility options, scale, and hidden lines.

TIP

You cannot create a view projected from an inclined surface of the object so views such as auxiliary views are not possible with *Viewproj* (see "Creating Auxiliary Views with *Viewsection*").

### Viewstd

| Ribbon | Menu Bar | Command (Type) | Alias (Type) | Shortcut |
|---|---|---|---|---|
| *Layout* *Styles and Standards* (arrow) | ... | *Viewstd* | ... | ... |

*Viewstd* produces the *Drafting Standard* dialog box (Figure 37-16). Here you can select the standards that are used when you create new views using *Viewbase, Viewproj,* etc. These options are explained below.

FIGURE 37-16

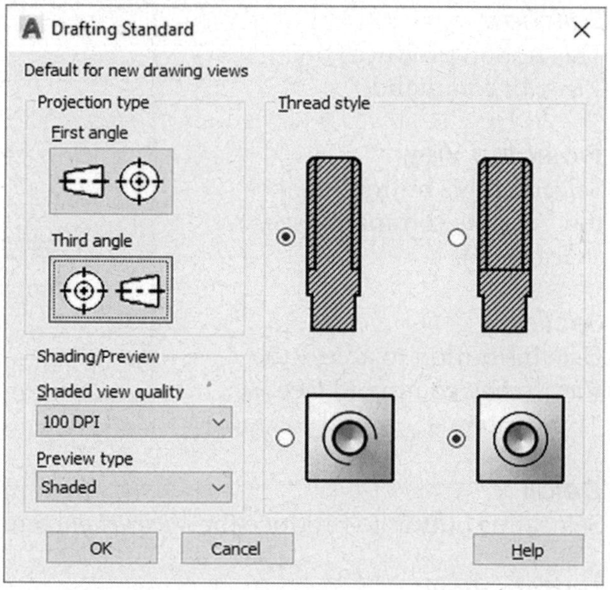


NOTE: These options are used for subsequently created (new) views only. The options do not affect existing views.

#### Projection Type

*Third angle* projection dictates that the top view is projected above the front view, the bottom view is projected below the front view, the right side view of the object is projected to the right of the front view, and so on as shown previously. Third angle projection is typically used for drawings in the United States and therefore is the ANSI (American National Standards Institute) standard for mechanical drawings.

*First angle* projection, however, places the top view below the bottom view, the bottom view above the front view, and the right side view of the object on the left side of the front view, and so on as shown in Figure 37-17. Note that in this figure the front (base) view is placed first in the upper-left so the top view can be correctly projected beneath the front view. The left side view of the object is placed to the right of the front (since the object is symmetrical in this case, the right and left sides of the object appear the same).

FIGURE 37-17

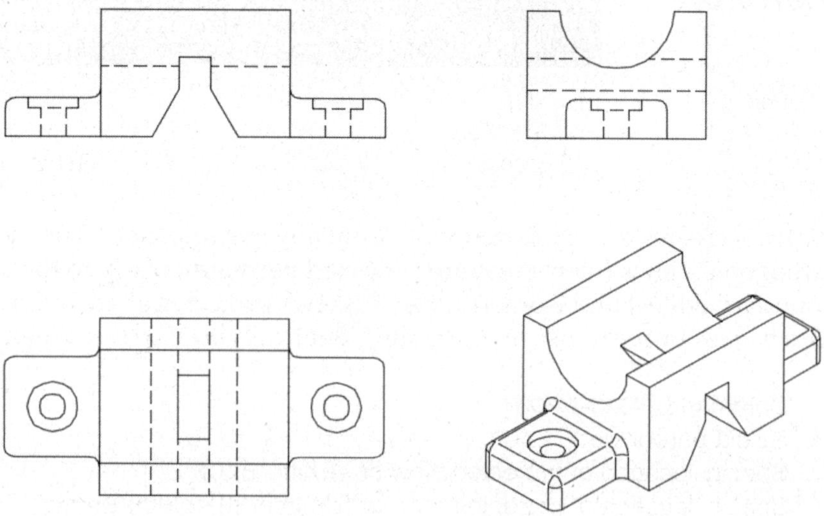

First angle projection is used in European countries and in many other countries outside the United States. Medical imaging devices such as MRI machines and other devices typically use the first angle standard.

NOTE: Since the *Drafting Standard* options do not affect existing views, you cannot retroactively change a first angle projection to a third angle projection and vice versa. However, it is possible to have one drawing file with a third angle projection in one layout and a first angle projection in another. That scenario is possible by setting the *Drafting Standard* before creating each set of views.

#### *Thread Style*
This option is used for Inventor models only.

#### *Shaded view quality*
This setting determines the default resolution for shaded drawing views. Resolutions above 150 DPI may not be achievable for large models. In such a case, the resolution adjusts automatically.

#### *Preview type*
The preview is the temporary graphic of the object shown during view creation. Options are a shaded preview or a bounding (rectangular) box. Use a bounding box if the model is complex and a shaded preview takes too much time to generate.

### *Viewsection*

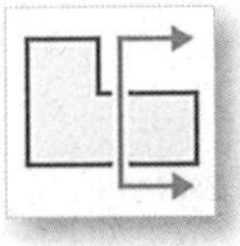

| Ribbon | Menu Bar | Command (Type) | Alias (Type) | Shortcut |
|---|---|---|---|---|
| *Layout*<br>*Create View*<br>*Section* | ... | *Viewsection* | ... | ... |

Use the *Viewsection* command to create a projected section view from a base view or any existing view. Several standard types of section views can be created such as full, half, offset, and aligned sections.

Command: ***VIEWSECTION***
Select parent view: **PICK**
Hidden Lines = Visible lines Scale = .75 (From parent)
Specify start point or [Type/Hidden lines/Scale/Visibility/Annotation/hatCh] <Type>: **PICK**
Specify next point or [Undo]: **PICK**
Specify next point or [Undo/Done] <Done>: **Enter**
Specify location of section view or: **PICK**
Select option [Hidden lines/Scale/Visibility/Projection/Depth/Annotation/hatCh/Move/eXit] <eXit>: **Enter**
Section view created successfully.

A section view is specified by creating a cutting plane line at the "Specify start point" and "Specify next point" prompts. Once the cutting line is specified, the section preview is attached to the cursor for placement, along with the alphabetical section label identifier. An alignment vector from the center of the parent view and perpendicular to the cutting line is displayed to force an orthogonal alignment. (Although the view is projected orthogonally, the alignment can be broken using the Shift key to position the view anywhere.) When the section view is created, *Viewsection* creates a label identifier for the cutting plane line and section view such as A-A or B-B.

The *Viewsection* options are available on the Command line and on the *Section View Creation* tab that appears on the Ribbon when the *Viewsection* command is active (Figure 37-18).

**FIGURE 37-18**

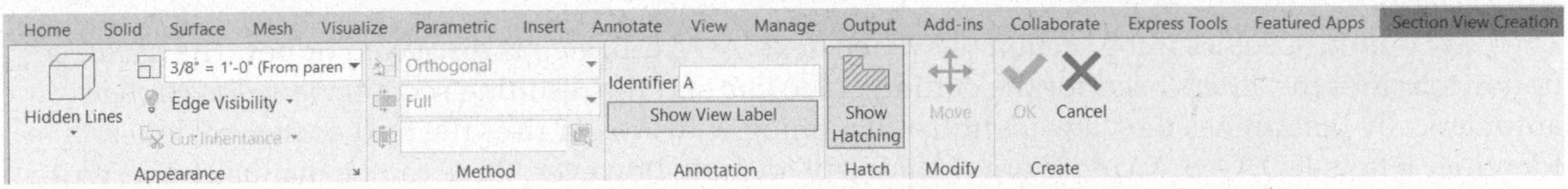

### Type

You can select from several *Types* of section views. The presented options depend on the type you choose.

*Full*

Use this option to create a full section of the complete base view.

*Half*

Use *Half* if you want to project a view but section only part of the object. Create a cutting plane line that cuts through part of the object then turns (typically at 90 degrees toward the view projection) and ends outside the view. Therefore, you must select three points minimum ("Start point," "next point," and "end point"). Once the end point of the section line is specified, the section preview is attached to the cursor, along with the alphabetical section label identifier. (See "Section View Examples.")

*Offset*

An offset section is created by specifying a "jogged" cutting plane line. Since offset sections can have any number of "jogs" or offsets, you are prompted for the "next point" until you select *Done*. The section view preview is attached to the cursor along with the alphabetical section label identifier. (See "Section View Examples.")

*Aligned*

An aligned section is similar to an offset section in that the cutting plane line can contain several sections. Those sections are then revolved to be projected onto one plane. Therefore, you are repeatedly prompted for the "next point" until you select *Done*. (See "Section View Examples.")

*Object*

You can use some existing part geometry to designate the cutting plane with this option. Once the existing geometry is selected and the section view is created, the cutting line created from that geometry is associated to the parent view but is not constrained to the view geometry (see "*Viewsymbolsketch*").

TIP

NOTE: You can draw *Lines* or other objects in a layout (in paper space) to define the orientation and location of the intended cutting plane line. Use *Object Snap* and other drawing tools to create the geometry. Use the *Object* option of *Viewsection* and select the paper space geometry to create the cutting plane line.

### Hidden Lines

These settings determine the linetype display for the section view. The options (*Visible, Visible and hidden, Shaded with visible lines,* and *Shaded with visible and hidden lines*) are identical to those for *Viewbase* and *Viewedit*. (See "*Viewbase* Options" and "Drawing View Editing.")

### Scale

Use this option to specify a scale for the section view if you want to change the presented scale. By default, the scale of the section view is inherited from the parent view. (See "*Viewbase* Options" and "Drawing View Editing.")

### Visibility

You can specify the visibility options to set for the section view. Object visibility options are model specific and some options may not be available in the selected model. (See "*Viewbase* Options" and "Drawing View Editing.")

### Annotation

The *Label* option specifies if the section view label text (A-A, B-B, etc.) is displayed or not. The *Identifier* option specifies the letters used for the cutting plane line and the resulting section view. *Viewsection* automatically determines the view identifier beginning with A-A or uses the next sequential letter. Identifier letters I, O, Q, S, X, and Z are excluded by default; however, these can be manually overwritten using the *Section View Style Manager* (see "*Viewsectionstyle*").

### Hatch

This option (*Yes* or *No*) determines whether or not hatch is displayed in the section view. The hatch parameters can be changed after view creation by clicking on the hatch pattern to produce the *Drawing View Hatch Editor* (similar to the *Hatch Editor* explained in Chapter 26).

### Projection

The *Projection* options are *Normal* and *Orthogonal*. Using either of these options during *Viewsection* apparently has no effect during the placement of a section view. In addition, no documentation on the *Projection* option exists in the current AutoCAD *Help*. Nevertheless, it is possible to create a non-orthogonal placement of the section view by tapping the Shift key to break the alignment.

### Depth

Use this option to specify the depth of the section view—that is, the depth of the object that is displayed beyond cutting plane. When you initiate the *Depth* option, a sliding plane appears in the parent view so you can interactively select a depth to display in the section view. As you move the slider, the section view displays the depth in the section view in real time. You can alternately indicate a depth by entering a positive numeric value. The *Full* option creates a typical section, that is, one that displays the entire object that would normally be visible beyond the cutting plane. Use the *Slice* option to display only the slice of the solid cut by the cutting plane.

## Section View Examples

### Full Section

Assume a documentation drawing was required for the model shown in Figure 37-19. Because this object has a cavity in the bottom surface, it is decided to show a section view rather than relying completely on hidden lines in a normal front view to describe that feature. Therefore, the drawing will contain a section view as the front view, a top view, a side view, and an isometric view.

FIGURE 37-19

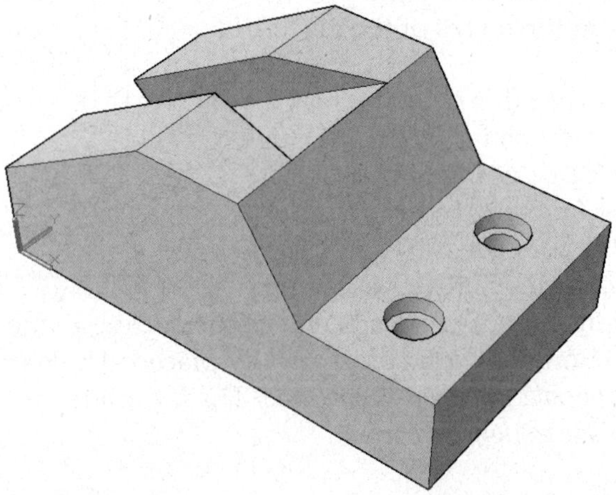

Begin by creating a base view with the *Viewbase* command (Figure 37-20). Note in this case the base view is actually the top view. This change is accomplished using the *Orientation* option when the base view is created and specifying *Top*.

FIGURE 37-20

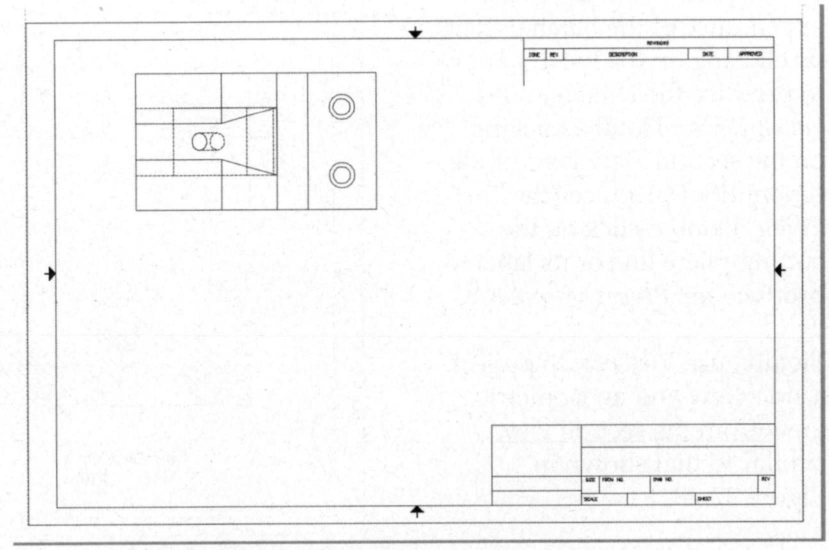

TIP

NOTE: When drawing cutting plane lines, it is suggested to use *Object Snap* and *Object Snap Tracking* to ensure the first point of the cutting plane line aligns orthogonally with the longitudinal axis of the object or with the feature you want to cut through (using the *midpoint* of the hole in this case).

Figure 37-21 displays the action during the "Specify start point" prompt.

Also note that the "Start point" determines the location for the directional arrows on the cutting plane. Therefore, pick a point just off the object (not on an object line). You can *Stretch* the cutting plane line ends later if needed using grips.

**FIGURE 37-21**

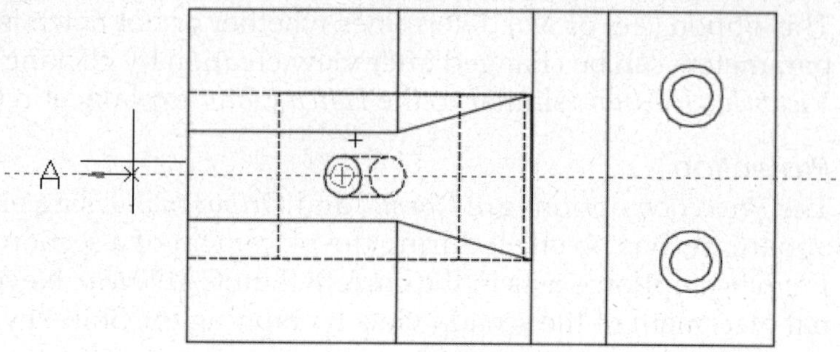


Next, place the section view beneath and orthogonally aligned with the top view (Figure 37-22). Note that the interior features of the cavity on the bottom surface are clearly shown. Also note that careful selection of the "Start point" and "next point" produces a cutting plane line with arrows in clear sight away from the edges of the object.

**FIGURE 37-22**

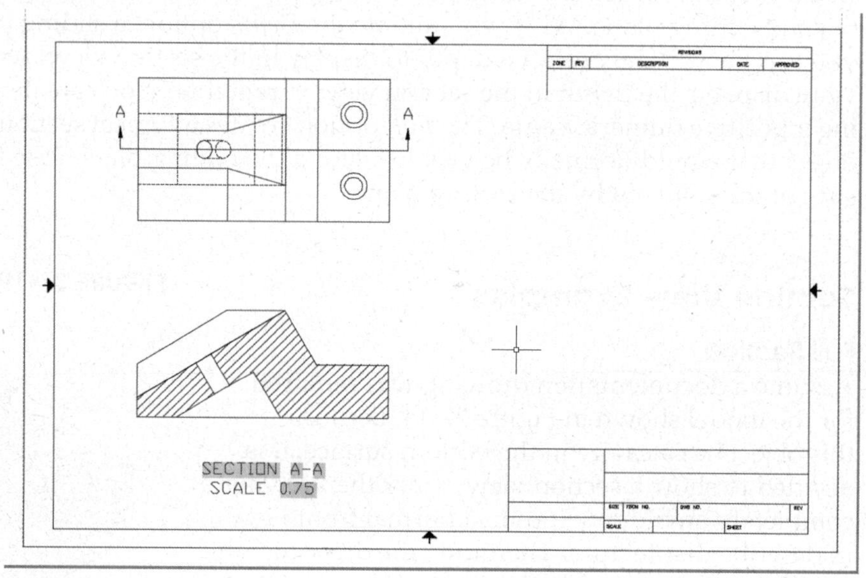


Because this drawing was created using the ACAD3D. DWT drawing template, the *Imperial24 Section View Style* was used. Note that *Viewsection* creates a cutting plane label (A) at each end of the cutting plane line. Also a label for the section view and the scale of the section view are automatically placed. The parameters for these items are determined by settings in the *Imperial24 Section View Style*. These parameters can be changed using the *Section View Style Manager*. See "*Viewsectionstyle*" next.

Once the section view has been created, you can edit the appearance of the hatch pattern by clicking on the hatch pattern to produce the *Drawing View Hatch Editor*. Double-clicking on the section view label, scale, or identifier produces the *Text Editor*. Double-click on the cutting plane line or its label to produce the *Properties* palette.

Finally, use *Viewproj* to project a side view and an isometric view <u>from the section view</u> similar to that shown in Figure 37-23.

**FIGURE 37-23**

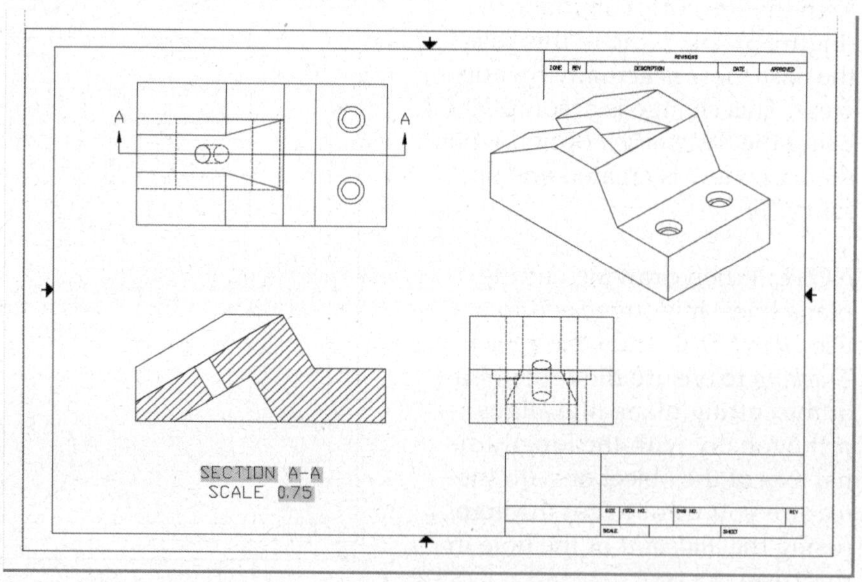

TIP

Keep in mind that the projected views inherit properties from the parent view, so the side view and isometric view do not include hidden lines by default. Justify this by double-clicking on the side view to produce the *Drawing View Editor* and changing the display to *Visible and hidden lines*. Also, the default projected isometric view includes the section. Therefore, double-click on this view to produce the *Drawing View Editor* and select *Cut Inheritance* and remove the check for *Section cut*. These actions are needed to create the drawing as shown in Figure 37-23.

### Half Section

The half section option allows you to create a section through only a portion of the object. Technically, half sections are used for objects that are symmetrical. In that way, the half section view shows one half of the view "in section" and the other half of the view as a normal view except without hidden lines. Since the object is symmetrical about the central axis, the "normal" half of the view does not need hidden lines because the sectioned half describes the interior features.

Figure 37-24 shows a documentation drawing of the Saddle with a half section. Note that the front view is comprised of a section on the right half of the view showing the internal features while the left half is a normal front view but without hidden lines. The hidden lines are not needed since the object is symmetrical and the sectioned half represents internal features for both halves. Also note that because the isometric view is projected from the section view, the offset section is inherited and is not changed for this drawing using *Viewedit*. Since this object is metric, the default *Metric50 Section View Style* is used.

**FIGURE 37-24**

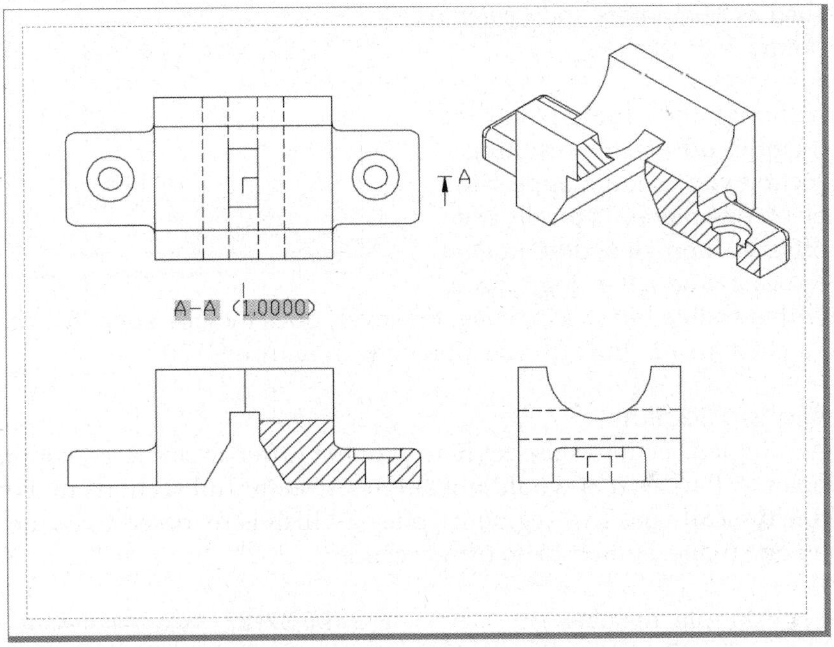


This drawing is most easily created by placing the top view as the base view and using the *Orientation* option to specify *Top*. Since the cutting plane line in the top view extends horizontally only half way through the part and then cuts vertically down through the front of the view, three points must be specified. Ensure you use *Object Snap* and *Object Snap Tracking* when selecting the "Start point," next point," and "end point." Figure 37-25 shows how the "next point" is tracked from the *Midpoints* of the adjacent vertical and horizontal edges.

**FIGURE 37-25**

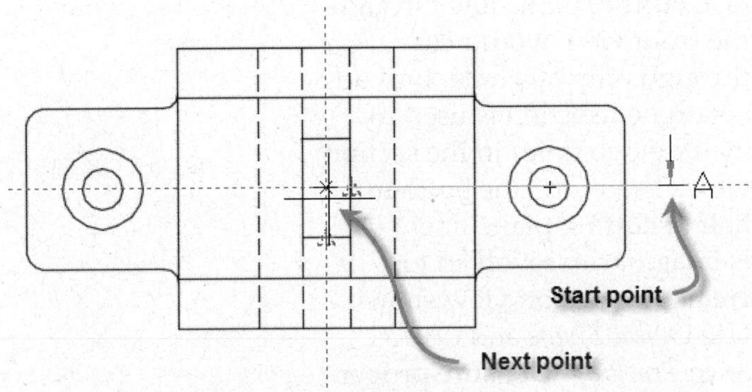


TIP

Since the right side view and isometric view were projected from the half section view (see previous Fig. 37-24), <u>*Viewedit* was used to change the inherited properties</u> to *Visible and hidden lines* for the side view.

### Offset Section

An offset section is created by specifying a "jogged" cutting plane line. Therefore the cutting plane line typically has multiple segments or "jogs." In this way you can include several features such as non-aligned holes in one section view.

Notice in Figure 37-26 the cutting plane line in the top view jumps from one horizontal plane to another in order to cut through two holes. Again, use *Object Snap* and *Object Snap Tracking* when prompted for the points to ensure the cutting plane line passes precisely through the desired points such as *Mid* points and *Center* points.

FIGURE 37-26

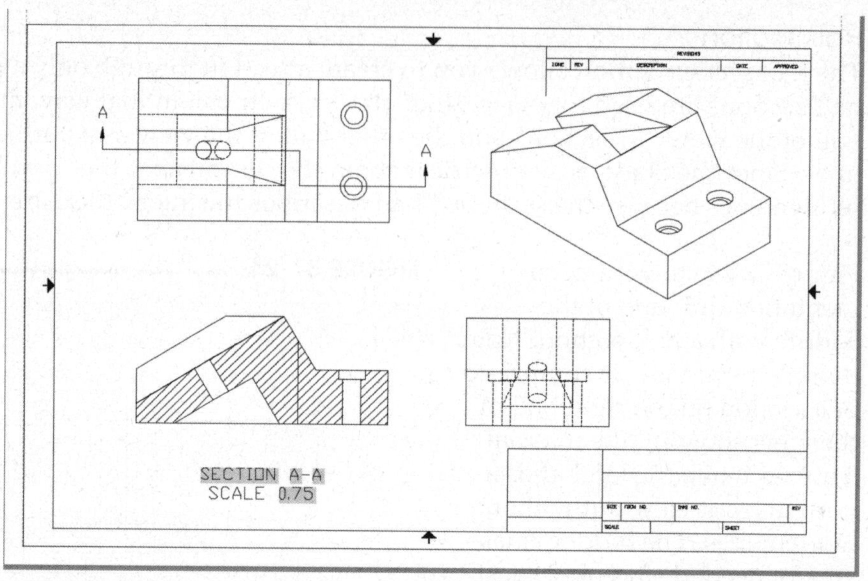


Although the cutting plane line includes offsets, the resulting section view should appear to be on one plane. Typically the offset section view does not indicate where the "jogs" occur with a visible line. *Viewsection*, however, does include such lines as shown by the vertical line inside the hatched area in the offset section view in Figure 37-26.

### Aligned Section

An aligned section is used when an object (often cylindrical) contains equally spaced features such as holes or flanges that would not fall in a straight full section cut. For an aligned section, the cutting plane line typically has two segments, one of which is revolved to create a section view that appears to be all on one plane, similar to an offset section.

For example, the object in Figure 37-27 contains three equally spaced holes. In a typical full section view, a vertical cutting plane line through the front view would cut through only one hole. Instead, an aligned section is used to include two holes in the section view. Therefore, the bottom half of cutting plane line A-A is imaginarily revolved to create the section view shown. Use *Object Snaps* and *Object Snap Tracking* to ensure proper alignment when creating the cutting plane line.

FIGURE 37-27

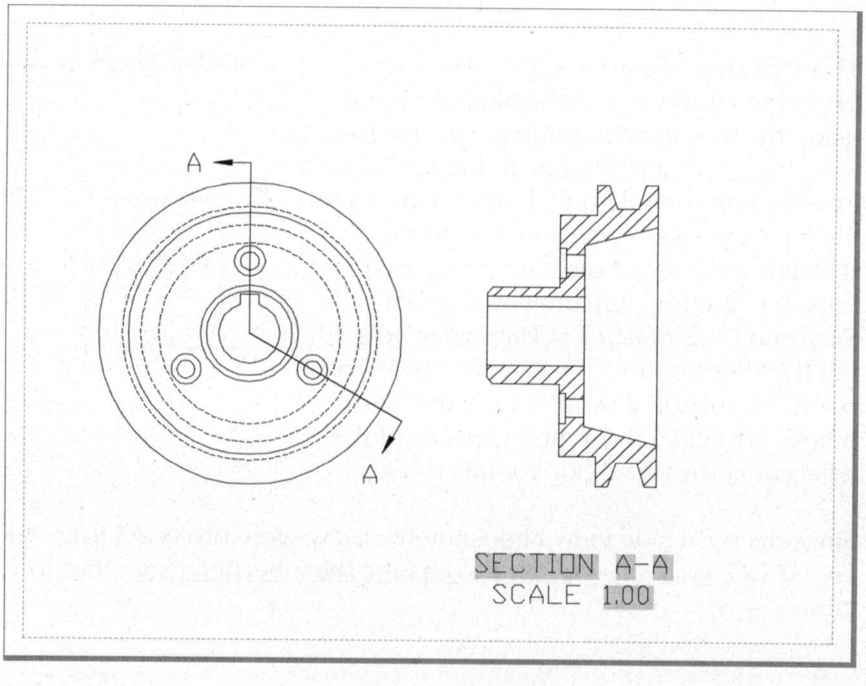

### Isometric Section

TIP

It is easy to create an isometric section view; however, an isometric section view is created using *Viewproj*, not *Viewsection*. Since projected views inherit properties from the parent view, first create a normal (orthogonal) section view using *Viewsection*, and then use *Viewproj* to create a sectioned isometric view projected from the section view. You can later use *Viewedit* to remove the inherited section cut from the isometric view if desired.

## *Viewsectionstyle*

| Ribbon | Menu Bar | Command (Type) | Alias (Type) | Shortcut |
|---|---|---|---|---|
| *Layout*<br>*Styles and Standards*<br>(icon) | ... | *Viewsectionstyle* | ... | ... |

When you create a section view using *Viewsection*, a few components of the section view are automatically created and added to the drawing such as the section (hatch) lines, cutting plane line arrows and identifiers, a section view label, and the section view scale. The *Section View Style* dictates the appearance of these components.

The *Viewsectionstyle* command produces the *Section View Style Manager* (Figure 37-28). This manager allows you to control the appearance of the cutting plane line, identifiers, view label, scale text, and section view hatching.

FIGURE 37-28

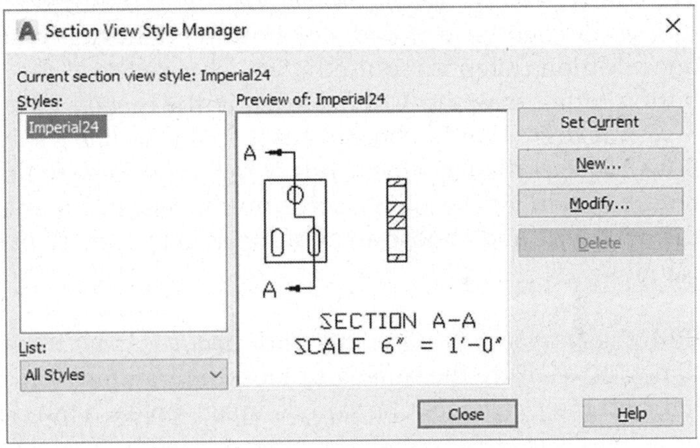


Using the *Section View Style Manager* to specify the format of section view objects is similar to using the *Dimension Style Manager* to specify the format of dimensions. In fact, the *Section View Style Manager* operates similarly to the *Dimension Style Manager*. You can have multiple styles in one drawing, create new styles, and modify existing styles using the managers. However, unlike settings that control dimension styles, settings that control section view styles cannot be changed directly using system variables at the Command line. No system variables related to section view styles are listed in the AutoCAD documentation.

Currently, AutoCAD supplies only two section view styles—*Imperial24* and *Metric50*. The *Imperial24* style is included in all template drawing files (acad.dwt and related) intended to be used for inch units and the *Metric50* style is included in all template drawing files (acadISO.dwt and related) intended to be used for metric units.

Options in the opening dialog box of the *Section View Style Manager* are basically the same as those of the *Dimension Style Manager*. *Current section view style* displays the name of the section view style that is applied to section views you subsequently create. The *Styles* section lists section view styles available in the current drawing file. Use the *List* drop-down to filter the list (select *All Styles* to display all section view styles available in the drawing file and select *Styles in use* to display only the section view styles that are referenced by section views in the current drawing). The *Preview* panel displays a sample image depicting the settings of the style that is selected in the *Styles* list. The buttons on the right have the following functions.

### *Set Current*

This button sets the section view style selected in the *Styles* list as the current style. All new section views are created using this section view style.

### *New*

Using *New* produces the *Create New Section View Style* dialog box, in which you can define new section view styles.

### *Modify*

This option produces the *Modify Section View Style* dialog box, in which you can modify section view styles.

### *Delete*

Use this choice to delete the section view style selected in the *Styles* list. This button is disabled if the style selected in the *Styles* list is currently in use by an existing section view.

NOTE: You cannot delete or rename the *Imperial24* and *Metric50* section view styles.

## *Modify Section View Style* and *New Section View Style* Dialog Boxes

The contents of the *Modify Section View Style* dialog box and the *New Section View Style* dialog box are identical except one additional step is required if you choose the *New* button rather than the *Modify* button in the *Section View Style Manager*. That is, choosing *New* in the *Section View Style Manager* displays the *Create New Section Style* dialog box shown in Figure 37-29. Here you can assign a name for the new style and choose an existing style to *Start With* or "clone."

**FIGURE 37-29**

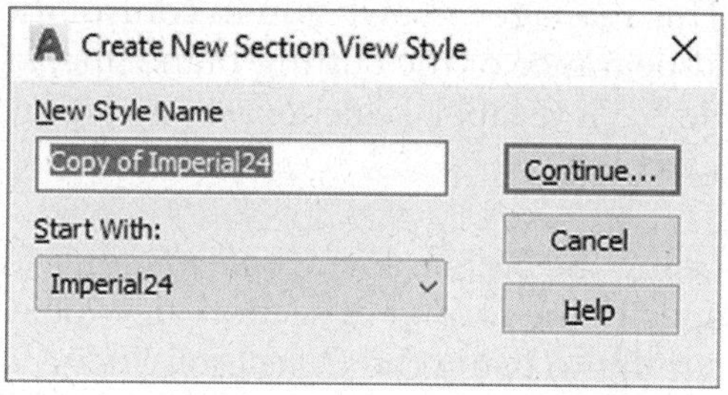


Both the *New Section View Style* and *Modify Section View Style* dialog boxes have the same four tabs: the *Identifier and Arrows* tab, the *Cutting Plane* tab, the *View Label* tab, and the *Hatch* tab. The sample image on each tab displays the effects of changing each option.

### *Identifier* and *Arrows* Tab

The *Identifier and Arrows* tab provides options for how the cutting plane line arrows and labels (identifiers) appear in the cutting plane line view (not the section view). Changing these options is easy and clearly visible since the image window displays each choice as you make it (Figure 37-30).

TIP

**FIGURE 37-30**

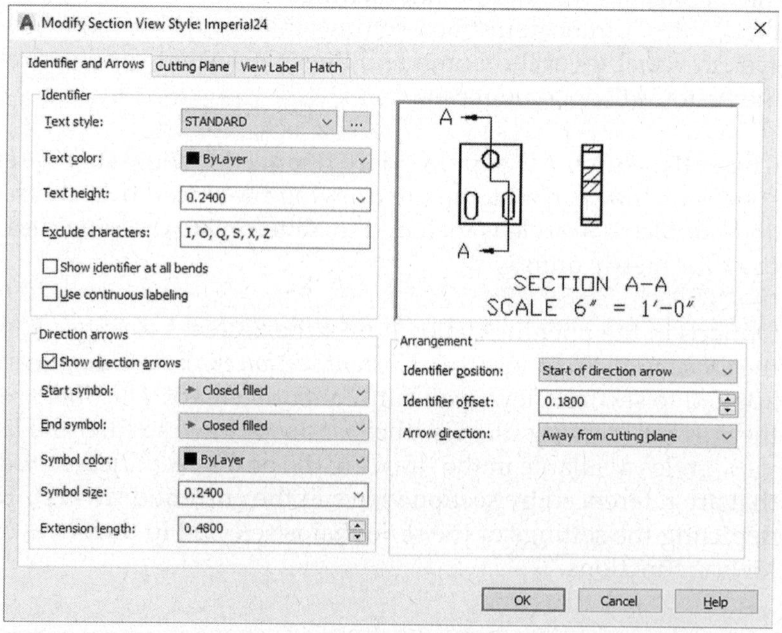

*Identifier*
Identifiers are the text labels that appear at each arrow of the cutting plane line. You can control the *Text style* and font, the *Text color*, and the *Text height*. The characters used are automatic beginning with the letter A and are sequential if multiple sections are created. All letters are used except those appearing in the *Exclude characters* edit box. Identifiers normally appear only at each arrowhead unless *Show identifier at all bends* is used. *Use continuous labeling* causes sequential lettering (A, B, C, etc.) to be used at each arrow rather than the same letter at each arrow.

*Direction arrows*
This section specifies the parameters for the arrowheads. You can specify a symbol for the *Start symbol* and the *End symbol*, as well as the *Symbol color* and *Symbol size*. The *Extension length* is the length of the extension line perpendicular to the cutting plane (distance between the cutting plane line and base of the arrowhead).

*Arrangement*
This section controls the placement of cutting plane line identifiers and direction arrows in relation to the start and end point of the section line. The *Identifier* position is the location of the labels in relation to the arrowheads (five options). The distance between the identifier and arrowhead is the *Identifier offset*. The *Arrow direction* is not the direction the arrows point, but is where the arrows are attached to the cutting plane line, *Away from cutting plane* (ANSI standard) or *Towards cutting plane* (ISO standard).

NOTE: The actual direction of the arrows indicates the direction you view the object in the section view. The direction that the arrows point is not changeable and is automatically determined by 1) the placement of the view in relation to the parent view and 2) the drafting standard (first angle or third angle).

TIP

### *Cutting Plane* Tab

This tab provides options for controlling the appearance of the cutting plane line (Figure 37-31). To understand what the different line segments are, apply a *Line color* to the lines and examine the image window.

TIP

NOTE: The *Show end and bend lines* check box and the *Show cutting plane lines* check box must be checked for any of the options in this section to be applied.

**FIGURE 37-31**

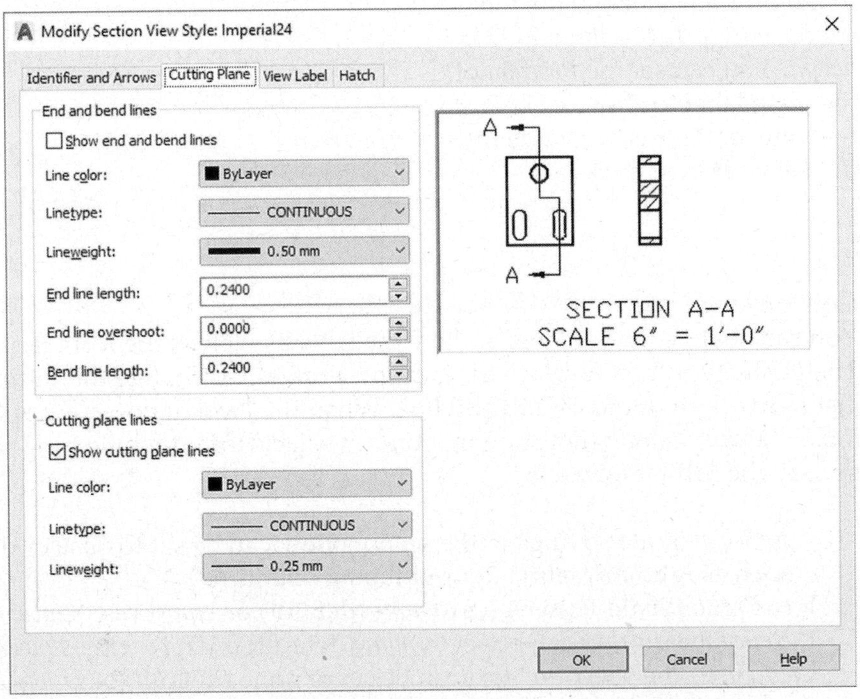


*End and bend lines*
These options specify the segments of the cutting plane line that extend from any bends or "jogs" (bend lines) and the extensions of the lines past the object. The end line is the extension of the cutting plane line that extends past the object on the cutting plane. This distance is also controlled by the original placement of the "start point" and "next points" during creation. Bend lines are the segments of the cutting plane line just past the bends.

You can specify *Line color, Linetype,* and *Lineweight*. Changing the value in the *End line length* or *Bend line length* causes that segment of the cutting plane line to shorten or extend. *End line overshoot* is a segment of cutting plane line that can extend past the bend at the line parallel to the arrowhead.

NOTE: You can also turn on multifunctional grips on the cutting plane line in the drawing to display at different line segments. Some of these cutting plane line segments can also be stretched using grips.

*Cutting plane lines*
This section controls the properties for the other line segments on the cutting plane. You can specify the *Line color, Linetype,* and *Lineweight*. Turn off the display of the cutting plane line completely by removing the check for *Show cutting plane lines*.

### *View Label* Tab

Use this tab to specify the appearance of the label text that appears near the section view (Figure 37-32). Keep in mind that the preview window can be used to see any changes made to the options.

**FIGURE 37-32**

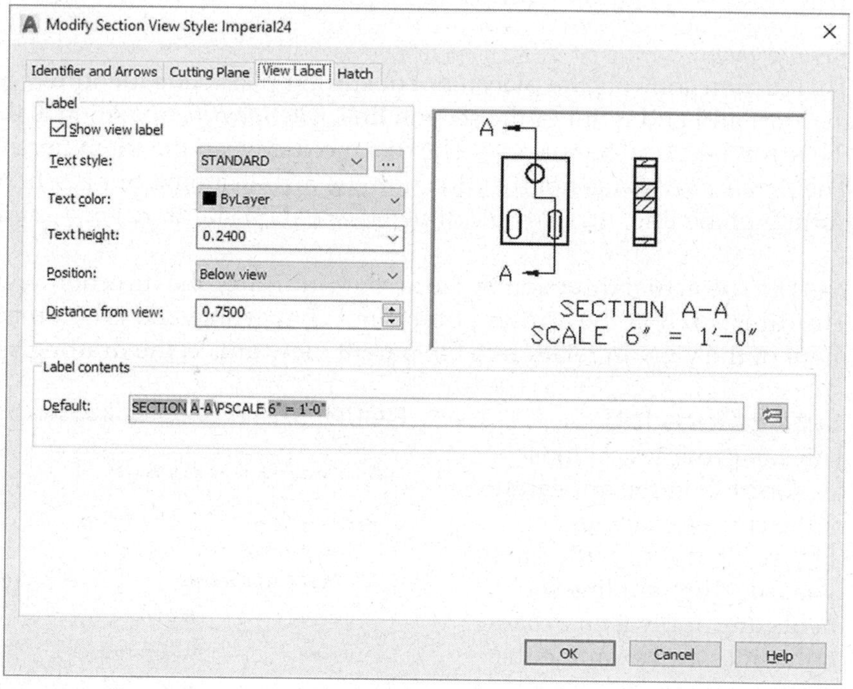


*Label*
The *Show view label* box must be checked for any of the options to be applied. You can change the *Text style* to any style existing in the drawing, change the font used, change *Text color*, and *Text height*. You can set the *Position* of the view label to above or below the view and specify a value for the *Distance from view*.

*Label contents*
You can change the contents of the view label as well as the format of the individual text information. The *Default* edit box displays the current settings. To change the settings, select the *Field dialog box* button just to the right of the *Default* edit box. When the *Field* dialog box appears with the field category set to *Model Documentation* (not shown), you can select from the following fields to appear in the label and specify the format for each:

*ViewScale* - scale settings of the section view can be set to *none* or displayed in one of several formats such as ratio formats (1:2) or unit equivalents (6"=1').
*ViewSectionEndId* - can be set to *none* (default) or *Upper case, Lower case, First capital,* or *Title case.*
*ViewSectionStartId* - can be set to *none* (default) or *Upper case, Lower case, First capital,* or *Title case.*
*ViewType* - the "Section" text can be set to none (default) or *Upper case, Lower case, First capital,* or *Title case.*

NOTE: Place the cursor in the *Default* edit box to insert or edit the existing contents. When you exit the *Field* dialog box, the selected field is inserted at the current cursor position within the *Default* edit box. If the cursor is not within the *Default* box, the entire contents of the *Default* box are replaced.

### *Hatch* Tab

This tab controls the appearance of the hatch lines (Figure 37-33). The *Show hatching* box must be checked for hatch lines to appear in the preview window and in the drawing section view.

*Hatch*

You can change the *Pattern* to any type typically available in AutoCAD. You can specify the *Hatch color*, *Background color*, *Hatch scale*, and *Transparency* settings.

*Hatch angle*

Select an angle from the list or select *New* to specify an angle not displayed in the list. You can highlight and *Delete* any angle from the list.

**FIGURE 37-33**

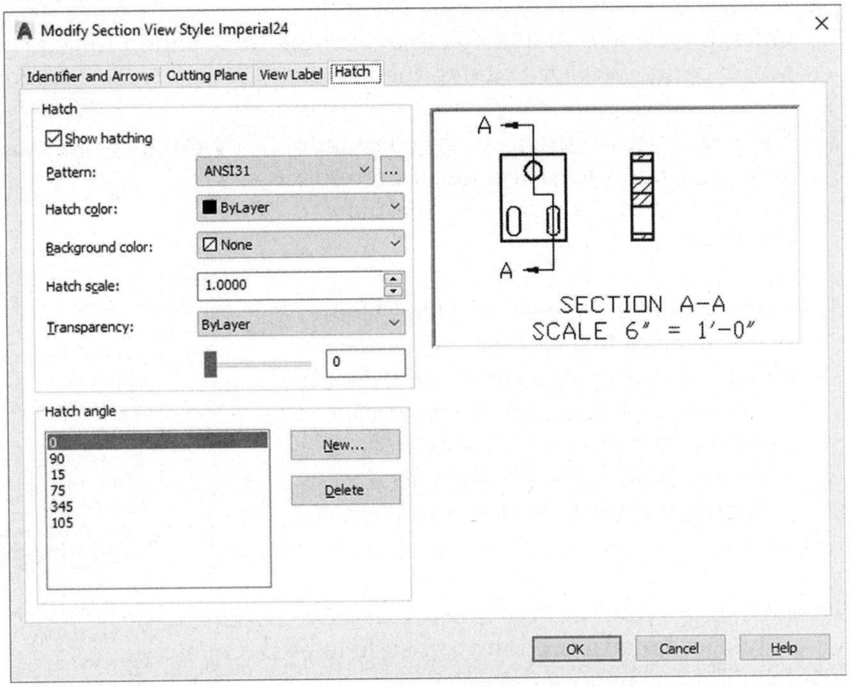


## *Section View Style* and *Detail View Style* Issues

What if you want to use one basic style for both metric and inch drawings, or if you create or convert a metric drawing using an inch template or vice versa? There are two issues in such a case with respect to using *Section View Styles* and *Detail View Styles*. First, AutoCAD does not provide a method to globally change the scale of all the size-related features of a section or detail view style such as an overall scale factor similar to the *Overall dimension scale* factor in the *Dimension Style Manager*. Second, only one section style and one detail style is provided in any one template drawing. The ACADISO.DWT templates contain only the *Metric50* styles and the ACAD.DWT templates contain only the *Imperial24* styles.

Assume you created an inch drawing using one of the inch templates such as acad.dwt. Then you were required to make a metric version of the part. This would be relatively easy because you could *Scale* the part by a factor of 25.4 so all the previous units (inches) would now correspond to millimeters (metric). Dimensions could also be easily scaled to match the new drawing by setting the *Overall dimension* scale factor also to 25.4. *Section View Styles* and *Detail View Styles*, however, cannot be easily scaled without multiplying every value appearing in the appropriate style manager by 25.4. One solution to this problem is explained next.

### Importing The Metric *Section View Style* and *Detail View Style* in a Non-Metric Drawing and Vice Versa

If you use a non-ISO template (not one of the ACADISO.DWT templates) for creating section views or detail views with the *Viewsection* or *Viewdetail* commands, the metric section style (*Metric50*) and metric detail style (*Metric50*) are not available. The reverse of this scenario is also true for the *Imperial24* styles. Therefore, it is recommended that you use DesignCenter to import those styles from the appropriate AutoCAD-provided templates into the current drawing when needed.

TIP

For example, to import the *Metric50 Section View Style* and *Detail View Style*, follow these steps.

1. Open the *Options* dialog box to determine the location for your template drawings. In the *Files* tab, expand *Template Settings*, then *Drawing Template File Location*.

2. Open Autodesk *DesignCenter* (command alias *ADC*). In the *Folders* tab, search for the location of your drawing template files (Figure 37-34).

3. Expand one of the two ACADISO templates (*acadISO-Named Plot Styles.dwt* or *acadISO-Named Plot Styles3D.dwt*). Select *Section View Styles*. Drag the *Metric50* icon from the preview panel into your drawing. Also expand *DetailViewStyles* and drag the *Metric50* icon into your drawing.

This action provides both the metric detail view style and metric section view style to be available for use in your drawing.

FIGURE 37-34

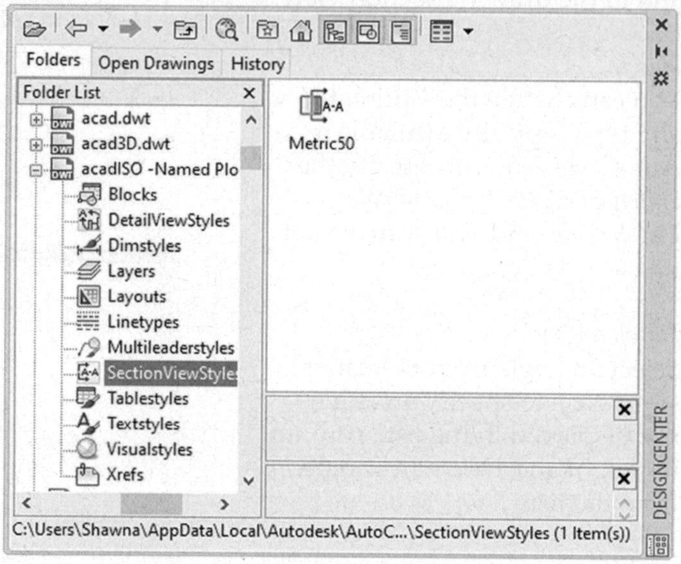


## Viewcomponent

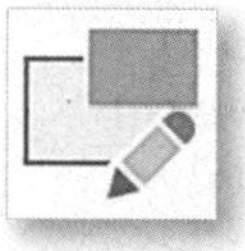

| Ribbon | Menu Bar | Command (Type) | Alias (Type) | Shortcut |
|---|---|---|---|---|
| *Layout*<br>*Modify View*<br>*Edit Components* | ... | *Viewcomponent* | ... | ... |

The *Viewcomponent* command allows you to select components from a drawing view for editing. According to the AutoCAD documentation, currently the only components that can be selected and edited are section views.

Command: ***VIEWCOMPONENT***
Select component: **PICK**
Select component: **Enter**
Select section participation [None/Section/sLice] <None>: (option)
Section participation changed for 1 component(s).

As an example, assume we apply the *Viewcomponent* command to the section view that was created in a previous example (shown earlier as a complete drawing in Figure 37-35).

*Section*
Use this option to display the normal section as shown in Figure 37-35.

FIGURE 37-35

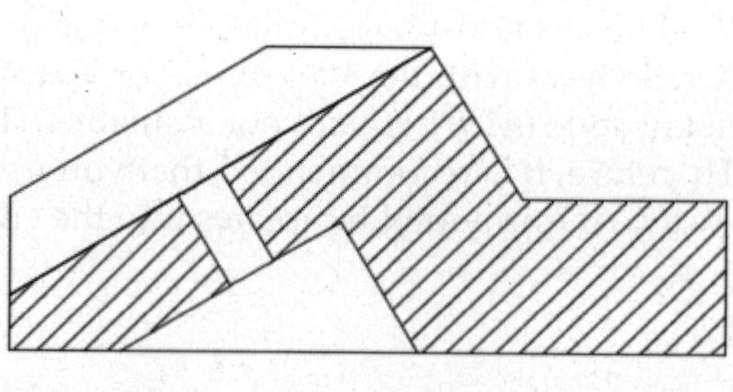

This option is typically useful only if one of the other two options has been used or if the *Slice* or *Depth* option of *Viewsection* or *Viewedit* has been used and you want to restore the view to the normal section view.

### None

You can use the *None* option to remove the sectioned portion of the view. The result is an outline of the view without hidden lines and without the sectioned components (Figure 37-36).

**FIGURE 37-36**

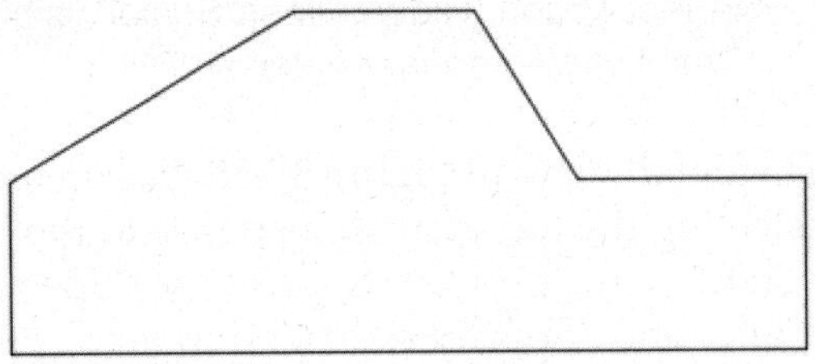

### Slice

This option specifies that zero-depth geometry is created when you section the component (Figure 37-37). In other words, only the slice of the object that is on the cutting plane appears. All other parts of the object normally visible behind the cutting plane are not shown. This option has the same result as the *Slice* option of *Viewsection* or *Viewedit*.

**FIGURE 37-37**

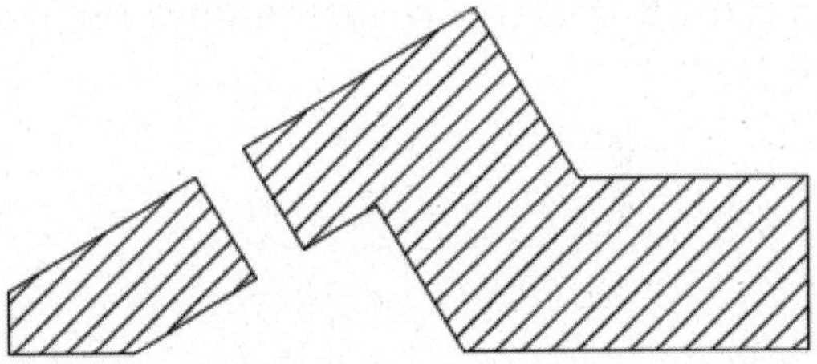

## Viewdetail

| Ribbon | Menu Bar | Command (Type) | Alias (Type) | Shortcut |
|---|---|---|---|---|
| *Layout*<br>*Create View*<br>*Detail* | ... | *Viewdetail* | ... | ... |

The *Viewdetail* command allows you to create a detail view in a documentation drawing. First create the needed base view and projected views, and then use *Viewdetail* to "capture" a segment of any existing view to create a detail view from it. For example, Figure 37-38 shows a documentation drawing including a detail view. This drawing was produced by first creating the orthogonal multiviews and the isometric view with *Viewbase*. Next, *Viewdetail* was used to create the detail view (left side) of one of the three lugs from the top view.

**FIGURE 37-38**

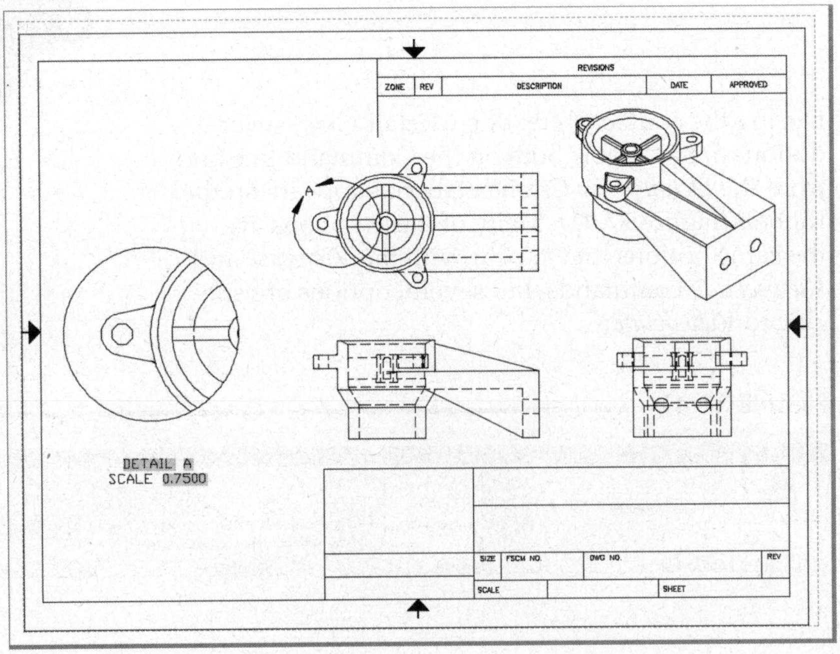

Command: ***VIEWDETAIL***
Select parent view: **PICK**
Boundary = Circular  Model edge = Smooth  Scale = 1:1
Specify center point or [Hidden lines/Scale/Visibility/Boundary/model Edge/Annotation] <Boundary>: **PICK**
Specify size of boundary or [Rectangular/Undo]: **PICK**
Specify location of detail view: **PICK**
Select option [Hidden lines/Scale/Visibility/Boundary/model Edge/Annotation/Move/eXit] <eXit>: **Enter**
Detail view created successfully.

The detail view in Figure 37-38 was created by first selecting the top view in response to the "Select parent view" prompt. Next, at the "Specify center point" and "Specify size of boundary" prompts, a location in the existing top view was picked using *Quadrant Osnap* and the boundary size was specified by dragging the cursor (Figure 37-39). The "center point" represents the center for the new detail view and the boundary size determines the area you want to capture.

**FIGURE 37-39**

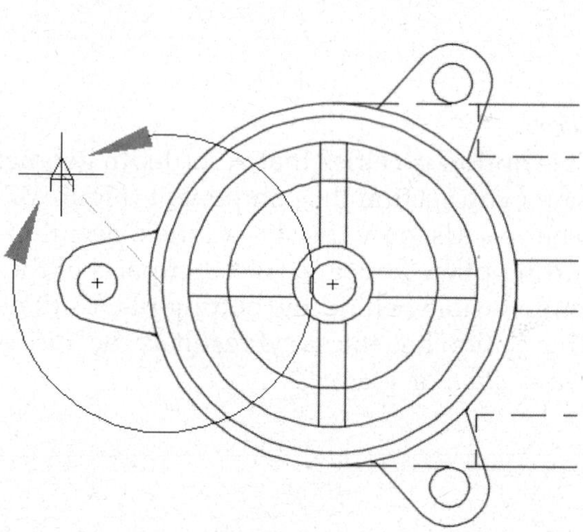

At the "Specify location of detail view" prompt, drag the captured segment of the object to the desired location for the new detail view (Figure 37-40). In this example, the *Scale* option was used to enlarge the detail view.

**FIGURE 37-40**

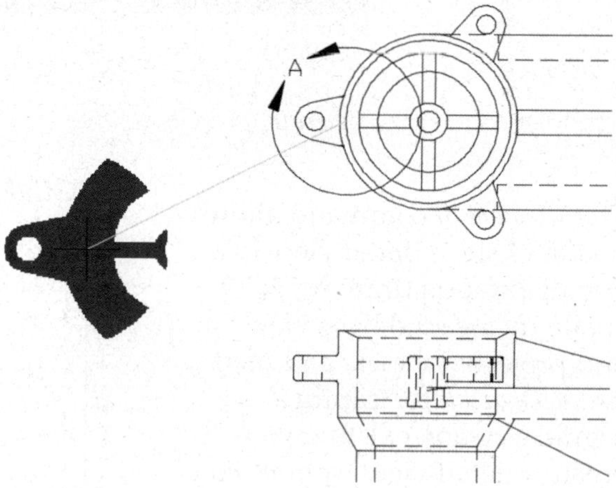


During the course of creating a detail view, several options are available both on the Command line and from the *Detail View Creation* tab that appears on the Ribbon (Figure 37-41). Many of these options are the same options that appear with the *Viewbase* and *Viewsection* commands, but several options are specific to *Viewdetail*.

**FIGURE 37-41**

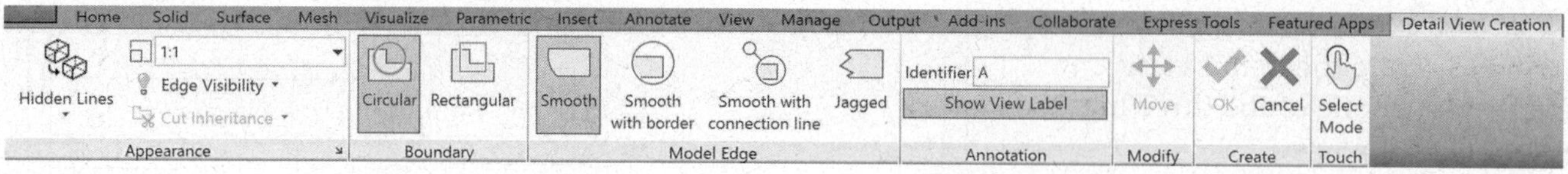

### Hidden Lines

These settings determine the linetype display for the detail view. The options (*Visible, Visible and hidden, Shaded with visible lines*, and *Shaded with visible and hidden lines*) are identical to those for *Viewsection*, *Viewbase* and *Viewedit*. (See "*Viewbase* Options" and "Drawing View Editing.")

### Scale

Use this option to specify a scale for the detail view if different from the scale presented at the Command line prompt when invoking the command. By default, the scale of the parent view is inherited. (See "*Viewbase* Options" and "Drawing View Editing.")

### Visibility

You can specify the visibility options to set for the detail view. Object visibility options are model specific and some options may not be available in the selected model. (See "*Viewbase* Options" and "Drawing View Editing.")

### Annotation

The *Label* option specifies if the detail view label text (A, B, etc.) is displayed or not. The *Identifier* option specifies the label for the circular or rectangular boundary. *Viewdetail* automatically determines the view identifier beginning with the letter A or uses the next sequential letter for the identifier.

### Boundary

You can specify the boundary as being *Circular* or *Rectangular*.

### Model Edge

This option specifies how the edge of the object lines appear in the detail view and if a border appears or not. The *Smooth* option causes the object edges to appear as if they were cut cleanly and no border is drawn around the detail view. *Smooth with Border* draws the circular or rectangular boundary and cuts the object lines within cleanly (see Figure 37-38). The *Smooth with Connection line* option includes the border and adds a leader line connecting the detail view with the parent view. The *Jagged* option has no border and makes the object edges appear "broken off."

TIP

NOTE: Keep in mind that you can use *Viewedit* to produce the *Detail View Editor* tab on the Ribbon and change the options for the detail view. Options on the *Detail View Editor* tab are the same as those on the *Detail View Creation* tab.

## Viewdetailstyle

| Ribbon | Menu Bar | Command (Type) | Alias (Type) | Shortcut |
|---|---|---|---|---|
| *Layout*<br>*Styles and Standards*<br>(icon) | ... | *Viewdetailstyle* | ... | ... |

Similar to a *Section View Style*, the appearance of detail views created with *Viewdetail* is controlled by the settings in the *Detail View Style*. Also similar to a *Section View Style*, there are two existing styles that AutoCAD provides—*Imperial24* for inch drawings and *Metric50* for metric drawings. (See *Viewsectionstyle*.)

The *Viewdetailstyle* command produces the *Detail View Style Manager* (Figure 37-42). This is the front end dialog box for detail styles and allows you to preview and *Set Current* detail view styles as well as provides access for you to create *New* and *Modify* existing detail view styles. Functions of this dialog box are described next and are similar to the *Section View Style Manager* and the *Dimension Style Manager*.

**FIGURE 37-42**

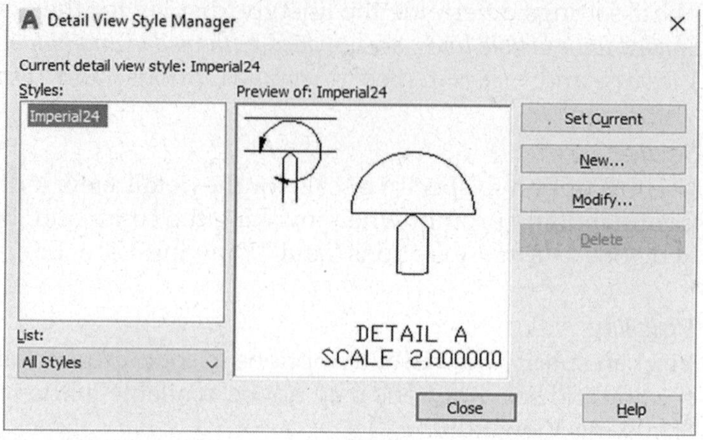


Remember that you can have multiple styles in one drawing, create new styles, and modify existing styles using this manager. However, unlike settings that control dimension styles, settings that control detail view styles cannot be changed directly using system variables at the Command line. No system variables related to detail view styles are listed in the AutoCAD documentation.

### *Set Current*

This button sets the detail view style selected in the *Styles* list as the current style. All new detail views are created using this detail view style.

### *New*

Using *New* produces the *Create New Detail View Style* dialog box, in which you can define new detail view styles.

### *Modify*

This option produces the *Modify Detail View Style* dialog box, in which you can modify detail view styles.

### *Delete*

Use this choice to delete the detail view style selected in the *Styles* list. This button is disabled if the style selected in the *Styles* list is currently in use by an existing detail view.

NOTE: You cannot delete or rename the *Imperial24* and *Metric50* detail view styles.

## *Modify Detail View Style* and *New Detail View Style* Dialog Boxes

The contents of the *Modify Detail View Style* dialog box and the *New Detail View Style* dialog box are identical except one additional step is needed if you choose the *New* button rather than the *Modify* button in the *Detail View Style Manager*. Choosing *New* in the *Detail View Style Manager* displays the *Create New Detail Style* dialog box (not shown, refer to previous Figure 37-29). Here you can assign a name for the new style and choose an existing style to clone.

Both the *New Detail View Style* and *Modify Detail View Style* dialog boxes have the same three tabs: *Identifier* tab, *Detail Boundary* tab, and *View Label* tab. The sample image on each tab displays the effects of changing each option.

### *Identifier* Tab

This tab controls the appearance of the identifier and boundary appearing <u>in the parent view</u> (Figure 37-43). The *Identifier* consists only of the letter used in the parent view to identify the related detail view. You can change the *Text style* and font, the *Text color*, and the *Text height*. The characters used are automatic beginning with the letter "A" and are sequential if multiple detail views are created.

**FIGURE 37-43**

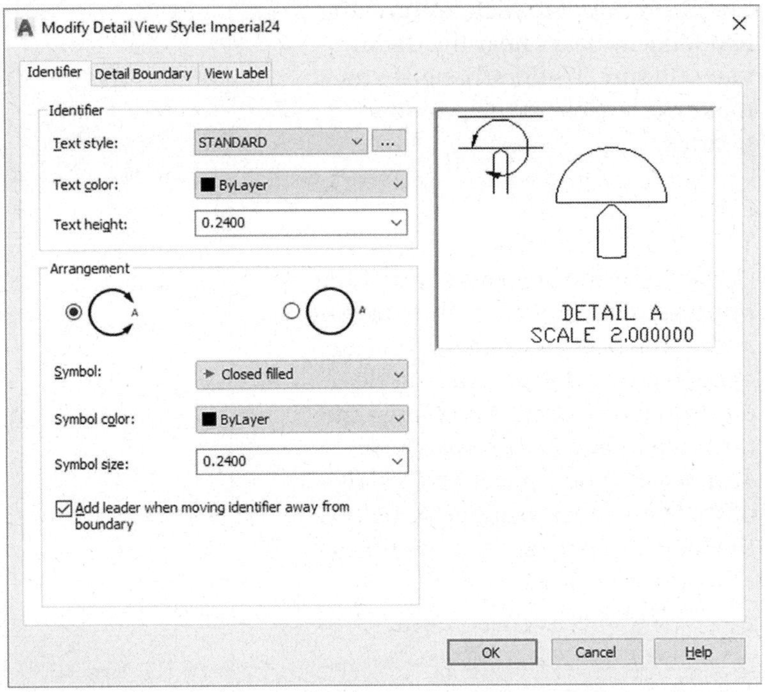


The *Arrangement* section dictates the appearance of the boundary and arrow symbols in the parent view (if used) to designate which area of the view is "captured" for the detail view. Choose either an open or closed boundary, the *Symbol* (arrowhead type) used, the *Symbol color*, and *Symbol size*. When the closed boundary is selected, no symbols appear and the identifier is connected to the boundary by a short leader. The *Add leader when moving identifier away from boundary* refers to a temporary dashed line that appears only while dragging to specify the detail view location.

### *Detail Boundary* Tab

This tab affects the appearance of the boundary lines <u>in the detail view</u> (Figure 37-44). The *Boundary line* section controls the *Line color*, *Linetype*, and *Lineweight* of the boundary line in the parent view. The *Model Edge* section controls how the boundary is drawn in the detail view. For example, the *Smooth* and *Jagged* options draw only the segment of boundary that intersects the object, whereas *Smooth with border* and *Smooth with connection line* draw a complete boundary designating the area of the detail view. You can also control the *Line color*, *Linetype*, and *Lineweight* of the boundary line in the detail view. The *Connection line* refers to the leader appearing only when *Smooth with connection line* is selected. You can select the leader *Line color*, *Linetype*, and *Lineweight*.

**FIGURE 37-44**

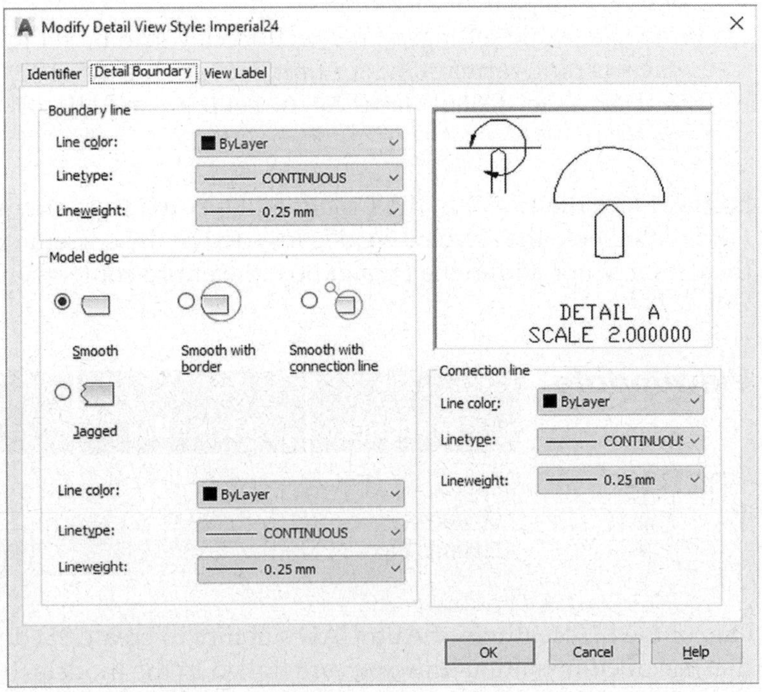

### *View Label* Tab

The *View Label* tab dictates how the text label appears near the detail view (Figure 37-45). Changes you make are visible in the preview window.

**FIGURE 37-45**

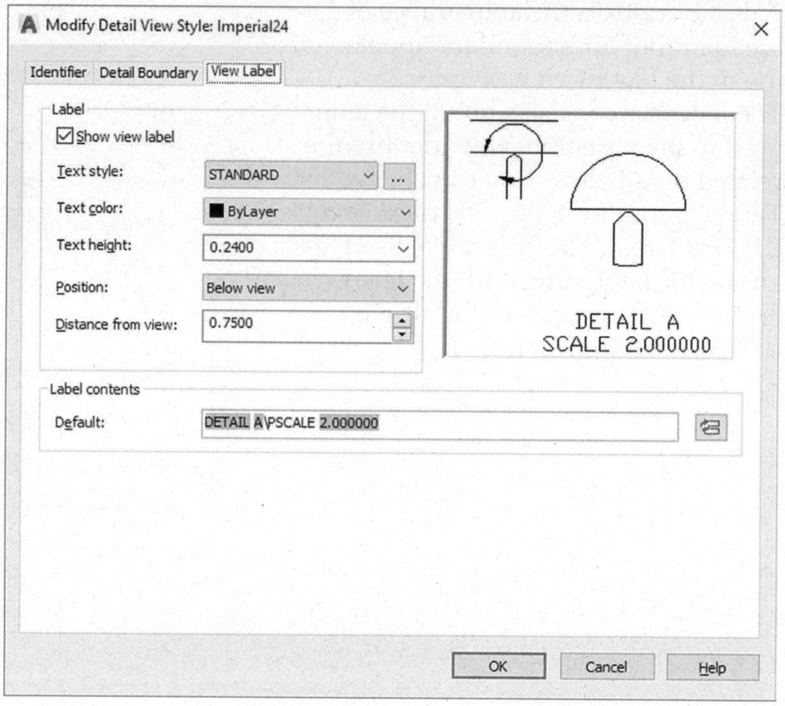


*Label*

The *Show view label* box must be checked for the label and any of the options to be applied in the drawing and in the preview panel. You can change the *Text style* to any style existing in the drawing, change the font used, change *Text color*, and *Text height*. You can set the *Position* of the view label to above or below the view and specify a value for the *Distance from view.*

*Label contents*

Use this edit box to change the format and the contents of the view label. The *Default* edit box displays the current settings. To change the settings, select the *Field dialog box* button just to the right of the *Default* edit box. When the *Field* dialog box appears with the field category set to *Model Documentation* (not shown) you can select from the following fields to appear in the label and the format for each:

*ViewDetailId* – the one letter identifier for the detail can be set to none (default) or *Upper case, Lower case, First capital,* or *Title case.*

*ViewScale* - scale settings of the section view can be set to none or displayed in one of several formats such as ratio formats (1:2) or unit equivalents (6″=1′).

*ViewType* – the "Detail" label can be set to none (default) or *Upper case, Lower case, First capital,* or *Title case.*

NOTE: Place the cursor in the *Default* edit box to insert or edit the existing contents. When you exit the *Field* dialog box, the selected field is inserted at the current cursor position within the *Default* edit box. If the cursor is not within the *Default* box, the entire contents of the *Default* box are replaced.

## *Viewupdate*

| Ribbon | Menu Bar | Command (Type) | Alias (Type) | Shortcut |
|---|---|---|---|---|
| *Layout*<br>*Update*<br>*Update View* | ... | *Viewupdate* | ... | ... |

One of the best features of AutoCAD's ability to create 2D documentation drawings from 3D models is that the documentation drawings are linked to the models. In that way, if the model is changed in any way, the drawing can be updated to display those changes.

For example, suppose you created a documentation drawing of the Saddle as shown in Figure 37-46. The drawing consists of three views including a section view plus an isometric view.

Next, suppose you are required to change the model. For example, assume the holes need to be relocated slightly outwards and the top of the inverted cut through the bottom of the front view is lowered a bit. Those design changes are made by editing the solid model in the *Model* tab (model space).

**FIGURE 37-46**

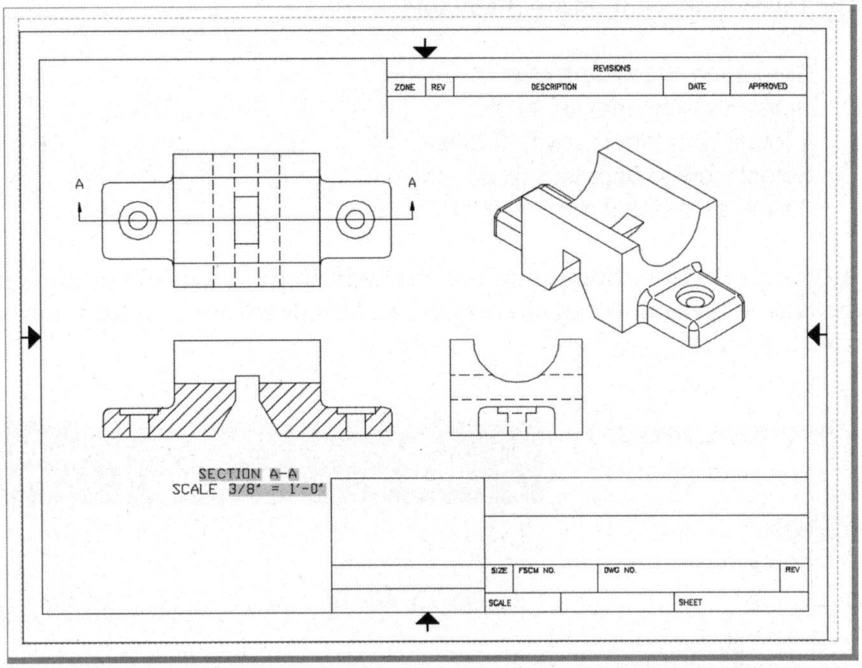


When you switch back to the layout containing the documentation drawing, AutoCAD notifies you that the views need updating (Figure 37-47). Each drawing view that has become out-of-date because the source model has changed is displayed with red corners and a balloon message appears at the lower-right corner of the screen notifying you to make the updates.

**FIGURE 37-47**

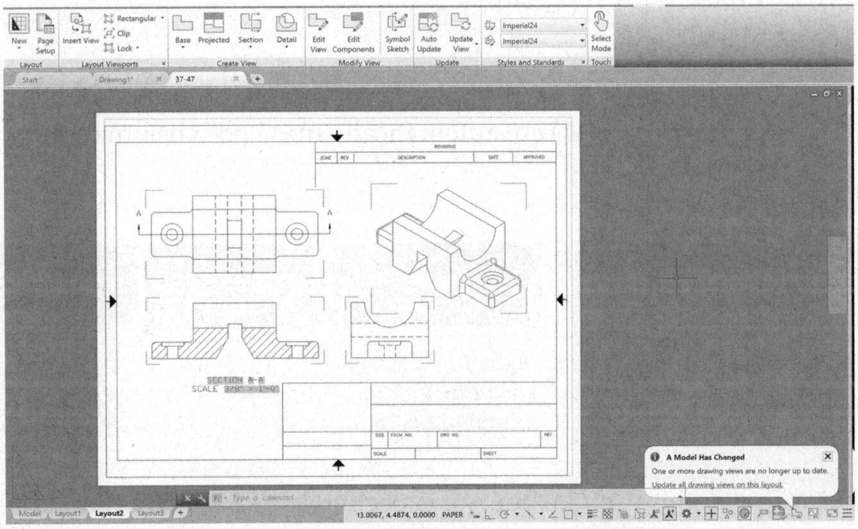


TIP

NOTE: If *VIEWUPDATEAUTO* is on, the updates are automatically made when changes to the model occur. See *VIEWAUTOUPDATE* next.

There are several ways you can make the updates to the views.

1. You can use the *Viewupdate* command,
2. Select <u>*Update all drawing views on this layout*</u> in the balloon message below the layout,
3. Left-click on the *Drawing Views* icon in the system tray and select *Update All Views*, or
4. Turn on *Autoupdate* (*VIEWUPDATEAUTO* system variable).

With most of these methods and with the *Viewupdate* command, you can also update views individually. The following prompts are displayed.

Command: ***VIEWUPDATE***
Select view to update: **PICK**
1 found (2 related views), 3 total
Select view to update: **Enter**
3 view(s) updated successfully.

Although an *All* option is not included within the prompt, you can enter "all" to select all views within the current layout. As an alternative, an *Update All Views* button is available as a drop-down in the *Update* panel.

## *Viewupdateauto*

| Ribbon | Menu Bar | Command (Type) | Alias (Type) | Shortcut |
|---|---|---|---|---|
| *Layout*<br>*Update*<br>*Update All Views* | ... | *Viewupdateauto* | ... | ... |

*VIEWUPDATEAUTO* is actually a system variable that turns automatic updating on or off. However, there is a button on the Ribbon, *Layout* tab, *Update* panel you can use to set *VIEWUPDATEAUTO* to 0 (off) or 1 (on). The settings are:

0 No drawing views are updated automatically when the source model changes (button is not highlighted).
1 All drawing views are automatically updated when the source model changes (button is highlighted).

## *Viewsymbolsketch*

| Ribbon | Menu Bar | Command (Type) | Alias (Type) | Shortcut |
|---|---|---|---|---|
| *Layout*<br>*Modify View*<br>*Symbol Sketch* | ... | *Viewsymbolsketch* | ... | ... |

This command is useful if you are expecting to make many changes to a model that has an associated documentation drawing containing a section or detail view. Using dimensions or geometric constraints, you can constrain the cutting plane line or the detail view boundary to model geometry. Then if the model is changed, the cutting plane line or detail view boundary should maintain its location with the desired geometry.

If *Infer constraints* are turned on at the time you create section views or detail views, AutoCAD automatically infers constraints with the detail boundaries and/or cutting plane lines. Therefore, if view updates take place (the model changes), the constraints ensure that the section lines and detail boundaries retain their positions in relation to the features they document. In some cases, you may require constraints that are more complex than the automatically applied constraints. Use *Viewsymbolsketch* to select the parent view of a section or detail view and add or delete constraints. *Viewsymbolsketch* opens an editing environment so you can add or delete constraints between section lines or detail boundaries and drawing view geometry.

## Creating Auxiliary Views Using *Viewsection*

TIP

Although there is nothing like a "Viewauxiliary" command in AutoCAD, you can use the *Viewsection* command to create an auxiliary view in a documentation drawing. Since auxiliary views typically do not include direction arrows, identifiers, or labels, you must suppress those features using the *Section View Style Manager*.

For example, assume you created the solid model shown in Figure 37-48 and now you want to create a documentation drawing including a front view, top view, and side view. Since these normal orthogonal views do not show the true size of the inclined plane, an auxiliary view is required.

Begin by creating the front, top, and side views as you would normally using the *Viewbase* command. You can create all views with *Viewbase*, or create only one view and then create projected views with *Viewproj*.

FIGURE 37-48

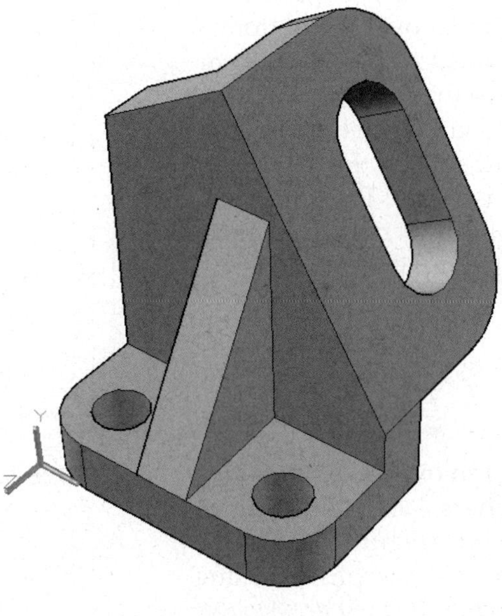

To create the auxiliary view, use *Viewsection*. When prompted to "Select parent view," select the view from which you want to project the auxiliary view (this is normally the view that shows the edge view of the inclined plane—the front view in this case). When prompted to "Specify start point" and "specify next point" to define the cutting plane line, select the two endpoints of the inclined surface. Ensure you use *Object Snap* when selecting points. Next, drag the new view to the desired location at the "Specify location of section view" as shown in Figure 37-49.

FIGURE 37-49

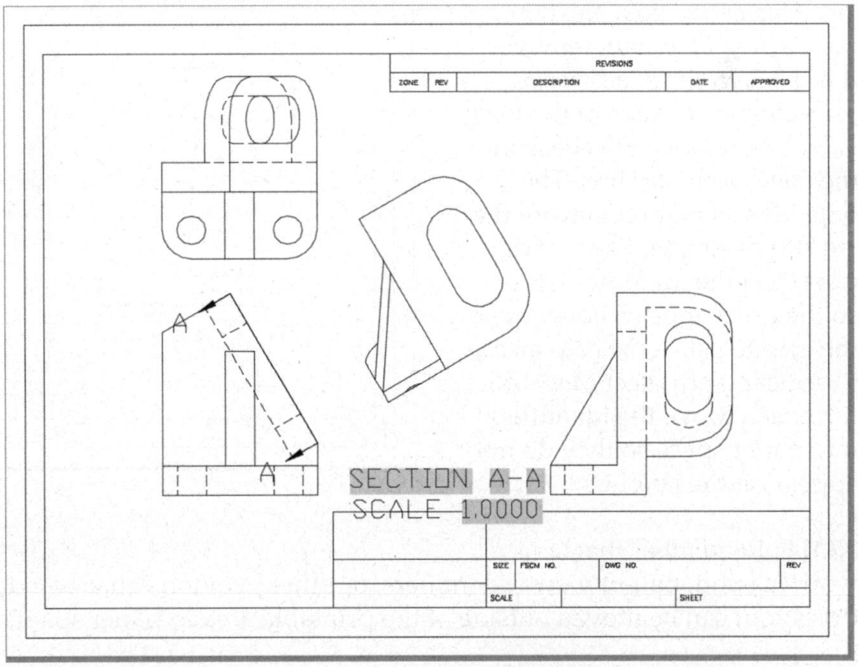

TIP

Note that since the cutting plane line stops at the bottom of the inclined surface the entire model is not fully displayed in the resulting auxiliary view. This can be rectified by extending the cutting plane line. You can edit the cutting plane line using grips. In this case, hover over the bottom grip, select *Add Segment* from the grip menu, then drag the line down to extend past the bottom edge of the model (Figure 37-50). The entire model is now displayed in the updated auxiliary view.

**FIGURE 37-50**

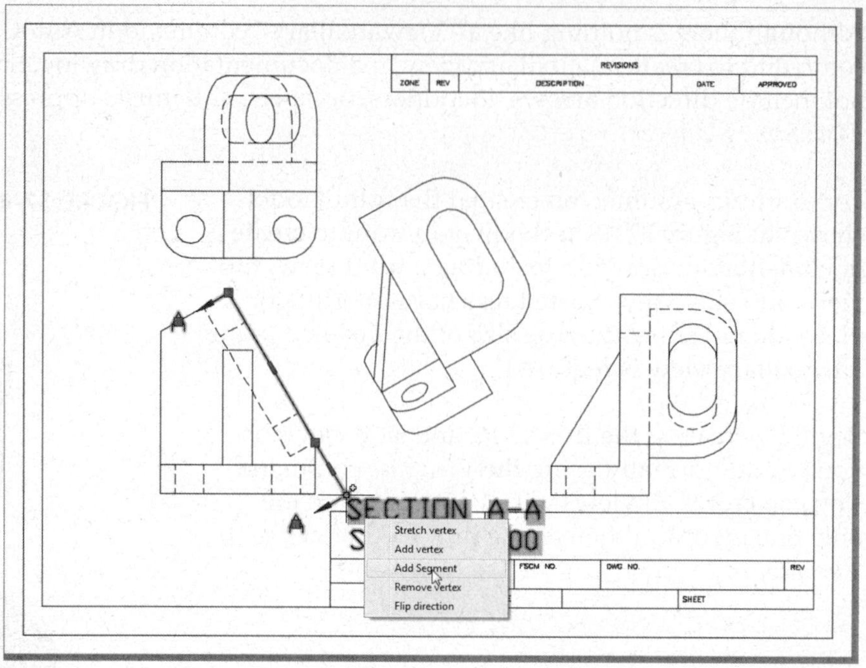


Next, remove the section view identifiers, label, and cutting plane line (Figure 37-51). This change is accomplished using the *Section View Style Manager.* In the *Identifier and Arrows* tab, remove the check in the *Show direction arrows* box. Next, in the *Cutting Plane* tab, remove the check in the *Show cutting plane lines* box. Also, in the *View Label* tab, remove the check in the *Show view label* box. The only objects that remain are the cutting plane identifiers. You can select the identifier letters so the grips appear, hover over the grip to cause the grip menu to appear, and select *Move identifier only*. Move the identifiers into paper space so they do not appear on the layout.

**FIGURE 37-51**

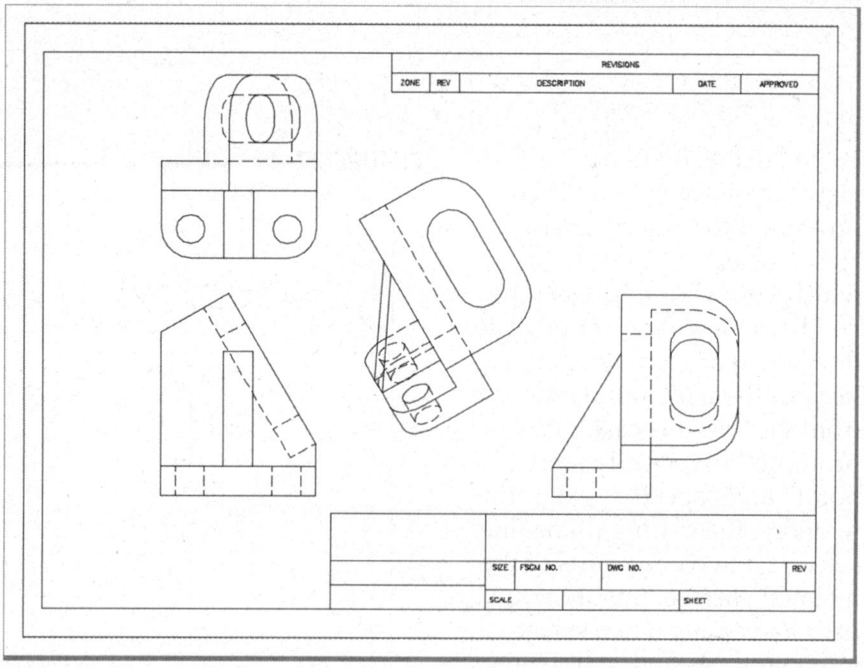


TIP

NOTE: Remember that, as a work-around, parent views, identifiers, or other geometry that you do not want to appear in the print of the layout can be moved outside of the printable area in paper space.

NOTE: You cannot delete the cutting plane line since that action also deletes the entire section view.

If only a partial auxiliary view is required (only the inclined surface is displayed rather than the entire object), use the *Slice* option of *Viewedit*. A partial auxiliary view results since only geometry that appears on the cutting plane line is displayed (Figure 37-52).

**FIGURE 37-52**

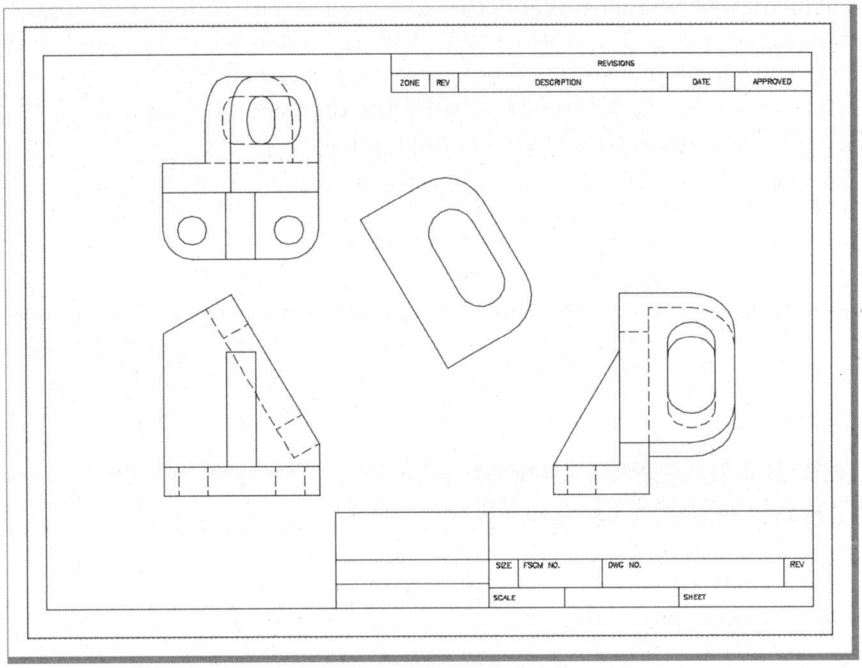


## Dimensioning a Documentation Drawing

Dimensioning a documentation drawing is relatively straightforward. Since you can dimension in paper space, you have the ability to dimension the documentation drawing (in paper space) rather than dimension the 3D model in model space.

Dimension commands operate properly in paper space because you can *Object Snap* directly to the model geometry in a drawing view, even though the dimensions exist in paper space. Therefore, you can add dimensions in paper space that appear around and on top of drawing views (Figure 37-53).

You can normally set the *Overall scale* for dimensions in the *Dimension Style Manager* to 1 since the layout is normally printed at a scale of 1:1. Even though the drawing views may be scaled to the actual geometry size, the dimension objects measure the model geometry. Therefore, the dimensional measurements represent the true model dimensions.

**FIGURE 37-53**

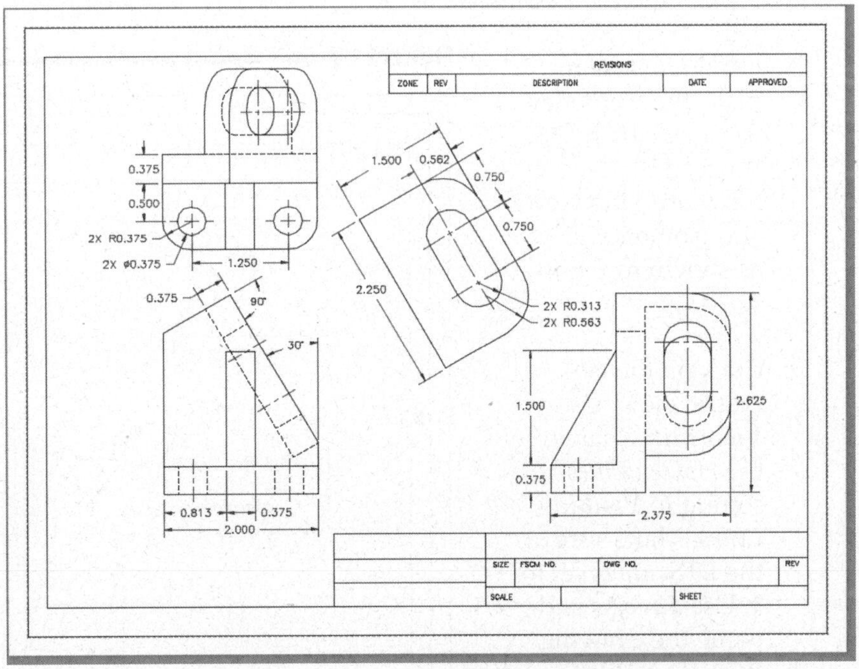

### Helpful Hints

There may be some problems that occur when the dimension objects become disassociated from the model geometry. The Annotation Monitor, when turned on, will report these problems and allow you to make the reassociations. This issue may occur especially when the model changes, then the views are *Updated*. Annotation Monitor reports the problems by producing a yellow badge with an exclamation mark (!) at those disassociated connections.

In some cases, the Annotation Monitor reports problems when, in fact, problems do not exist. You can turn off the Annotation Monitor in such a situation.

NOTE: Some *Object Snaps* may not work properly in some situations. For example, you cannot use the *Intersection* mode to locate the intersection between a paper space object and drawing view object.

## CHAPTER EXERCISES

1. ***Viewbase, Viewedit***

A. ***Open*** the **BGUID-SL.DWG** you created in the Chapter 35 Exercises. Ensure that the model is oriented so the front face (looking through the dovetail notch on the bottom) is parallel with the XZ plane of the WCS.

B. Activate an unused layout. Ensure that no viewports exist in the layout (***Erase*** any existing viewports). Use ***Pagesetup***, select ***Modify***, and ensure the ***Paper size*** is set to ***Letter (8.50 X 11.00 Inches)***. ***Insert*** the **ANSI A TITLE BLOCK.DWG** into the layout. (If you have not already downloaded this title block, go to the website address listed on the back cover and in the Preface under "Online Resources.")

C. Use ***Viewbase*** to place the front view. Set the ***Scale*** to **.75**. Proceed to place a top view, side view, and isometric view as shown in Figure 37-54.

D. Click on the isometric view. Use ***Viewedit***. Change the *Hidden Lines* setting to ***Visible Lines***. Make sure the ***LTScale*** is set to **1**. Use ***SaveAs*** and name the drawing **BGUID-VIEWS. DWG**.

**FIGURE 37-54**

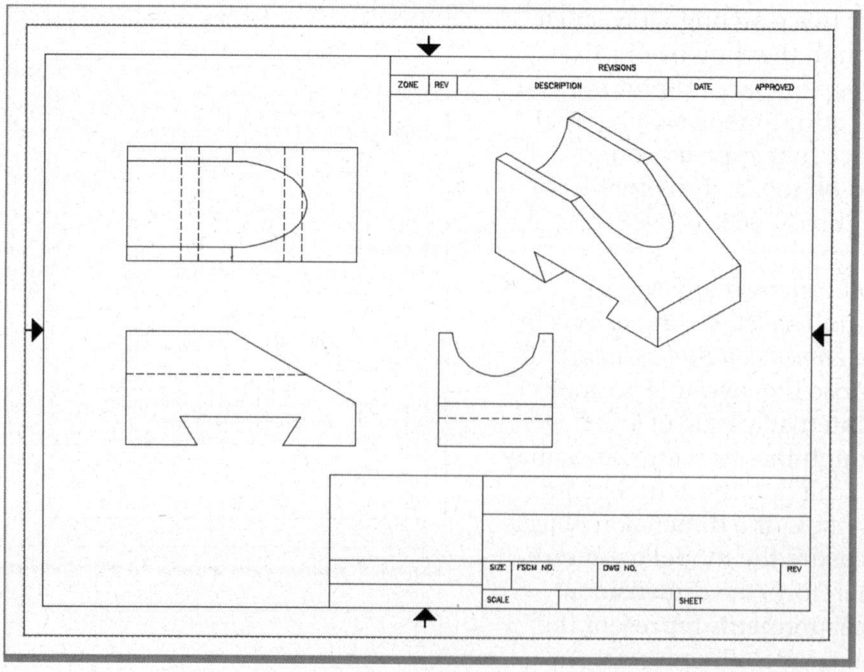

2. *Viewbase, Viewedit*

A. ***Open*** the **SADDLE.DWG** you created in the Chapter 35 Exercises. Use ***Saveas*** and name the drawing **SADDLE-VIEWS.DWG**. Ensure that the model is oriented so the front face (looking through the inverted notch on the bottom) is parallel with the XZ plane of the WCS.

B. Activate an unused layout. Ensure that no viewports exist in the layout (***Erase*** any existing viewports). Use ***Pagesetup***, select ***Modify***, and ensure the ***Paper size*** is set to ***Letter (8.50 X 11.00 Inches)***. Next, in the *Plot Scale* section under *Scale*, set *1 inches* = **25.4** *units* so the sheet is converted to mm.

C. ***Insert*** the **ANSI A TITLE BLOCK.DWG** into the layout. (If not already downloaded, go to the website address listed on the back cover and in the Preface under "Online Resources.") Before insertion, set the *Scale* (in the *Insert* dialog box) to **25.4**. Use **0,0** as the ***Insertion Point***.

D. At the Command line, set the ***LTScale*** to **25**. Next, use ***Viewbase*** to place the front view. Set the ***Scale*** to **3:4**. Proceed to place a top view, side view, and isometric view as shown in Figure 37-1 (near the beginning of this chapter).

E. Click on the isometric view. Use ***Viewedit***. Change the *Hidden Lines* setting to ***Visible Lines***. Also, set the ***Edge Visibility*** to ***Tangent Edges*** and ***Tangent Edges Foreshortened***. If needed, move the views using grips to achieve the best placement on the sheet. ***Save*** the drawing as **SADDLE-VIEWS.DWG**.

3. *Viewsection, Viewproj, Viewsectionstyle*

A. ***Open*** the **SADDLE-VIEWS.DWG** from the previous exercise. Use ***Saveas*** and name the new file **SADDLE-VIEWS-SECTION.DWG**. Use *DesignCenter* (command alias *ADC*) to import the ***Metric50*** *Section View Style* into the drawing. If you need help, follow the steps given in this chapter in the section "Importing the Metric Section View Style and Detail View Style in a Non-Metric Drawing and Vice Versa." Use the *Layout* tab to set ***Metric50*** as the current *Section View Style*.

**FIGURE 37-55**

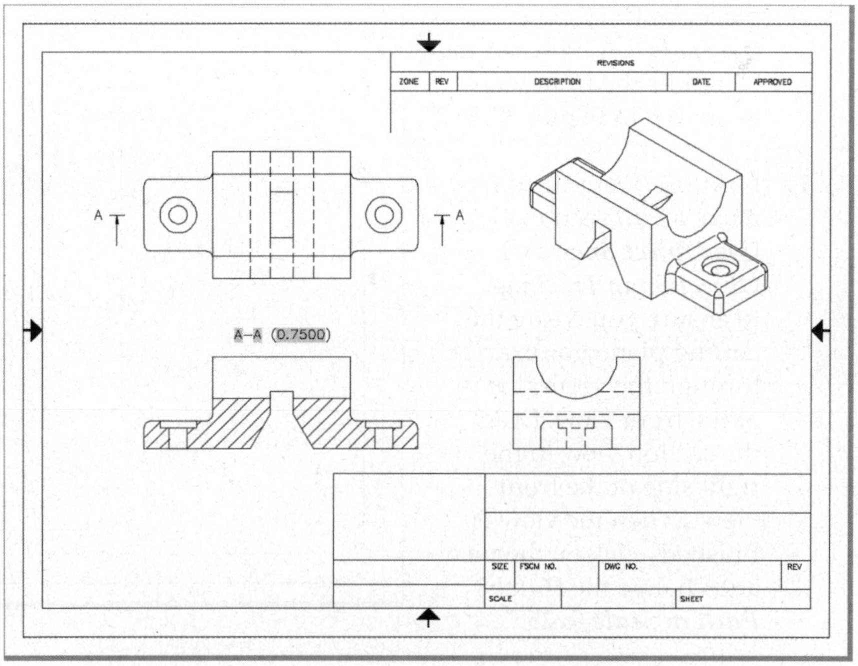


B. ***Erase*** the front and side views. Next, use ***Viewsection***, select the top view as the *Parent View*, place a full section below the top view (in place of the front view). Use ***Object Snap*** and ***Object Snap Tracking*** to ensure you create the cutting plane line exactly through the center axis of the top view. When the view is finished, click on the view and change the ***Hatch Pattern Scale*** to **25**.

C. Use ***Viewproj*** to place a right side view projected from the front (section) view. Since the new view inherits properties from the parent, you must make the hidden lines visible by using ***Viewedit*** and selecting ***Visible and Hidden Lines***.

D. Open the ***Section View Style Manager*** and ***Modify*** the ***Metric50*** style. In the ***View Label*** tab, change the *Text Height* to **3**. In the ***Identifier and Arrows*** tab change the *Text Height* to **3** and the *Symbol Size* to **3**. Select the ***OK*** button, and then ***Close*** the *Section View Style Manager*. Use grips to move the view label closer to the view. Compare your drawing to Figure 37-55. Make any needed adjustments. ***Save*** the drawing.

**4.** ***Viewbase, Viewsection, Viewsectionstyle***

A. ***Open*** the **PULLY-SL.DWG** from the Chapter 35 Exercises. Use ***Saveas*** and name the new file **PULLEY-VIEWS-SECTION.DWG**. Use *DesignCenter* (command alias ***ADC***) to import the ***Metric50*** *Section View Style* and the ***Metric50*** *Detail View Style* into the drawing. If you need help, follow the steps given in this chapter in the section "Importing the Metric Section View Style and Detail View Style in a Non-Metric Drawing and Vice Versa." Use the *Layout* tab to set ***Metric50*** as the current *Section View Style*.

B. Ensure that the model is oriented so the front face (the cylindrical shape) is parallel with the XZ plane of the WCS.

C. Activate an unused layout. Ensure that no viewports exist in the layout (***Erase*** any existing viewports). Use ***Pagesetup***, select ***Modify***, and ensure the ***Paper size*** is set to ***Letter (8.50 X 11.00 Inches)***. Next, in the *Plot Scale* section under *Scale*, set *1 inches* = **25.4** *units* so the sheet is converted to mm.

D. ***Insert*** the **ANSI A TITLE BLOCK.DWG** into the layout. (If not already downloaded, go to the website address listed on the back cover and in the Preface under "Online Resources.") Before insertion, set the *Scale* (in the *Insert* dialog box) to **25.4**. Use **0,0** as the ***Insertion Point***.

E. At the Command line, set the ***LTScale*** to **25**. Next, use ***Viewbase*** to place the front view. Set the ***Scale*** to **1**. Proceed to complete the front view as shown in Figure 37-56.

F. Next use ***Viewsection*** to make a ***Full*** section view. Use ***Object Snap*** and ***Object Snap Tracking*** to ensure you create the cutting plane line exactly through the vertical axis of the front view. Drag the section view to the right side of the front view. When the view is finished, click on the view and change the ***Hatch Pattern Scale*** to **25**.

**FIGURE 37-56**

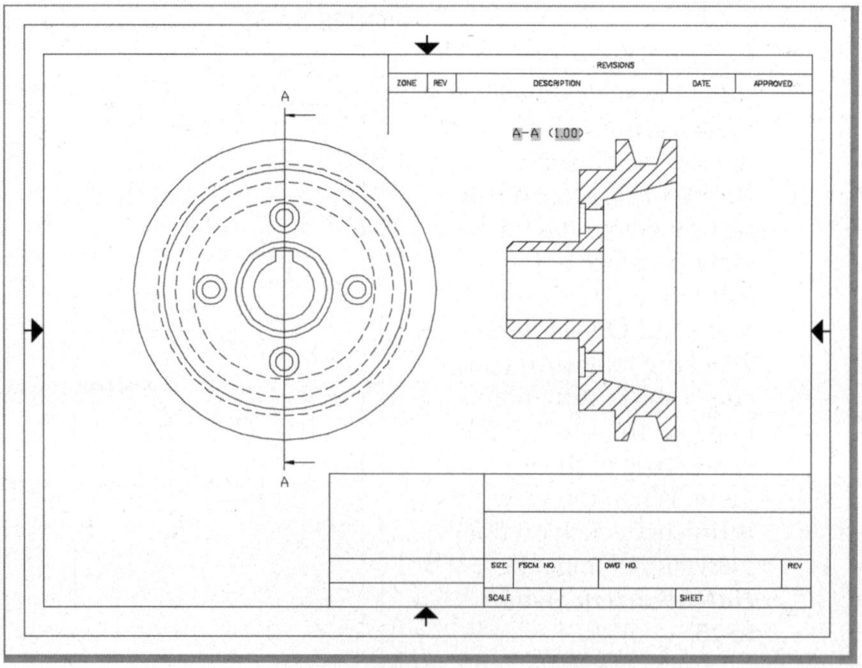

G. Open the ***Section View Style Manager*** and ***Modify*** the ***Metric50*** style. In the ***View Label*** tab, change the *Text Height* to **3**. In the ***Identifier and Arrows*** tab change the *Text Height* to **3** and the *Symbol Size* to **3**. In the ***Cutting Plane*** tab, check ***Show cutting plane lines***. Select the ***OK*** button, and then ***Close*** the *Section View Style Manager*. Finally, use grips to move the view label to an appropriate location.

H. Compare your results to Figure 37-56. Make any needed adjustments. ***Save*** the drawing.

**5.** ***Viewbase, Viewsection, Viewsectionstyle***

A. ***Open*** the **VBLOK-SL.DWG** you created in the Chapter 35 Exercises. Use ***SaveAs*** and name the new file **VBLOK-VIEWS-AUX.DWG**. Ensure that the model is oriented so the front face (the 3.00" x 2.25" face) is parallel with the XZ plane of the WCS.

B. Activate an unused layout. Ensure that no viewports exist in the layout (***Erase*** any existing viewports). Use ***Pagesetup***, select ***Modify***, and ensure the ***Paper size*** is set to ***Letter (8.50 X 11.00 Inches)***. ***Insert*** the **ANSI A TITLE BLOCK.DWG** into the layout. (If not already downloaded, go to the website address listed on the back cover and in the Preface under "Online Resources.") Use **0,0** as the ***Insertion Point***.

C. Use ***Viewbase*** to place the front view. Set the ***Scale*** to **.75**. Proceed to place a top view, right side view, and isometric view as shown in Figure 37-57. Ensure there is enough room to project an auxiliary view between the front and side views.

**FIGURE 37-57**

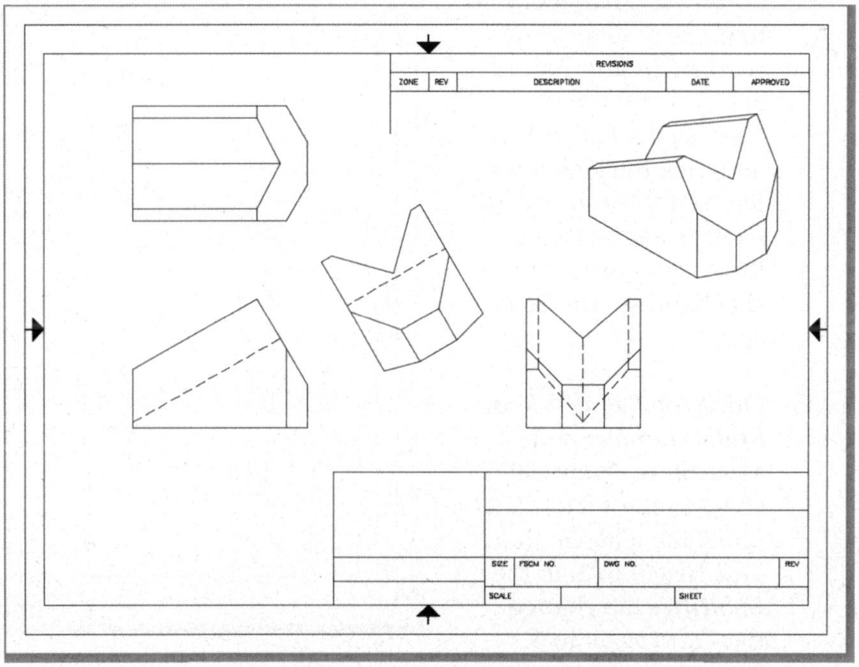


D. In order to make a full auxiliary view, you must use *Viewsection* since there is no auxiliary view creation command. Create a ***Full Section*** view using ***Viewsection***. Select the front view as the *parent view*. Create the cutting plane line from the two ***Endpoints*** of the inclined surface and drag the view towards the center of the drawing.

E. Note that since the cutting plane line ends at the bottom of the inclined surface, the entire object is not projected to the auxiliary view. To correct this, select the cutting plane line so the ***grips*** appear, hover over the bottom grip, and select ***Add Segment***. Now stretch the bottom grip to extend the line down, making sure that the extension is aligned with the existing segment. Make sure you turn off *Osnap* during this operation. This action should cause the entire object to display properly in the auxiliary view.

F. Next, click on the auxiliary view and use ***Viewedit*** to set the line display to ***Visible and hidden lines***. Using the same method, change the setting in the auxiliary to ***Visible lines***.

G. Finally, open the ***Section View Style Manager*** for the ***Imperial24*** style and ***Modify*** the settings. In the ***Identifier and Arrows*** tab, remove the check for *Show direction arrows*. In the ***Cutting Plane*** tab, remove the check for *Show cutting plane lines*. In the ***View Label*** tab, remove the check for *Show view label*. Select the ***OK*** button, and then ***Close*** the *Section View Style Manager*. Finally, click on the identifier letters for the cutting plane line, hover over an identifier grip, and select ***Move identifier only***. Move the identifier into paper space outside of the sheet. Do this for both identifiers.

H. Compare your results to Figure 37-57. Make any needed adjustments. ***Save*** the drawing.

6. ***Viewdetail, Viewdetailstyle***

A. ***Open*** the **PULLEY-VIEWS-SECTION.DWG** drawing from the exercise earlier in this chapter. Use ***Saveas*** and name the new file **PULLEY-VIEWS-DETAIL.DWG**. Use any method and move the two views to the right to provide space for a detail view as shown in Figure 37-58.

**FIGURE 37-58**

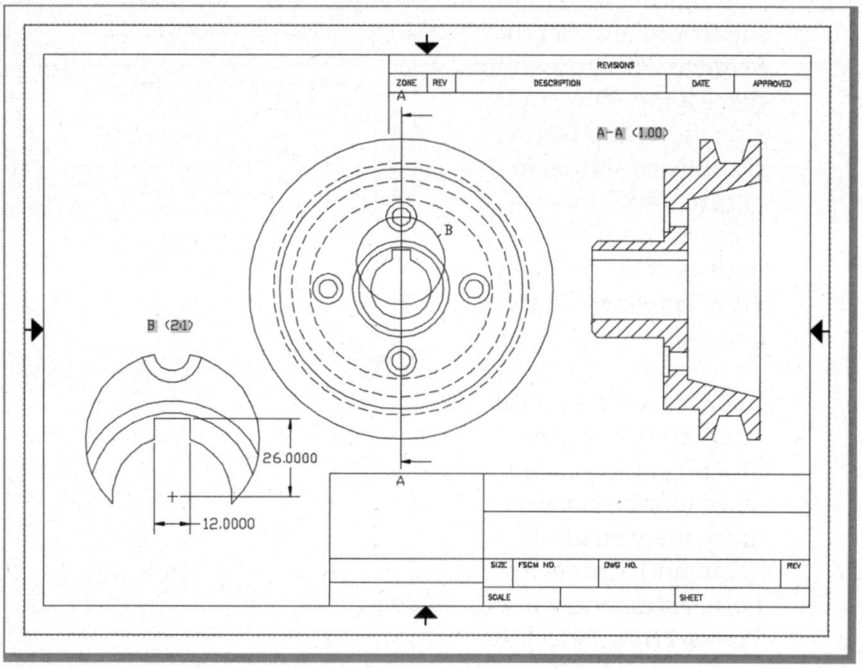


B. Make ***Metric50*** the ***current*** *Detail View Style*. Use ***Viewdetail*** to make a detail view with a ***Circular*** boundary. Select the front view as the *Parent View* and pick the keyway as the center for the detail. Drag the view to the lower-left corner of the sheet and set the ***Scale*** as **2**.

C. Open the ***Detail View Style Manager*** and ***Modify*** the **Metric50** style. In the ***View Label*** tab, change the *Text Height* to **3**. In the ***Identifier*** tab change the *Text Height* to **3**. Then use grips to move the view label to an appropriate location closer to the view. Adjust the location of the new view as needed.

D. Use the ***Dimension Style Manager*** to set the dimension ***Overall Scale*** to **18**. Create a ***Center Mark*** and two ***Linear*** dimensions for the keyway as shown.

E. Compare your results to Figure 37-58. Make any needed adjustments. ***Save*** the drawing.

**7.** ***Viewbase, Viewsection, Viewedit***

***Open*** the **AGLBR-SL.DWG** you created in the Chapter 35 Exercises. Use ***Saveas*** and name the new file **AGLBR-AUX-DIM.DWG**. Create a documentation drawing of the Angle Brace as shown earlier in the chapter in Figure 37-53. Using the skills you learned from completing the previous exercises, use ***Viewbase*** to project a front view, top view, and right side view at a scale of 1, as shown in Figure 37-53. Use ***Viewsection*** to create an auxiliary view and ***Viewedit*** to show only a ***Slice***. Finally, create the needed dimensions as shown.

REUSE

Helpful Hints:

A. You can *Insert* the ANSI A Title Block, but you will have to *Explode* it and make alterations to the lower-right corner to provide more room for the views.

B. You should use the *Imperial24 Section View Style*. Make changes as needed to remove the display of the cutting plane line, identifiers, and view label.

C. Make sure you move the views to allow plenty of space for dimensioning. Avoid moving views after placing dimensions because this may cause dimensions to become disassociated. Turn off the *Annotation Monitor* if needed.

D. Make any needed changes to the *Dimension Style* before placing dimensions. Dimensions existing in the drawing may not update properly when the *Dimension Style* is modified because they may be disassociated.

CHAPTER 38

# Parametrics

## Chapter Objectives

After completing this chapter you should:

1. know the two types of parameters that provide parametric control of drawing geometry;
2. know 12 geometric constraints to apply to drawing geometry;
3. know how to add dimensional constraints;
4. know the difference between fully constrained, and underconstrained conditions;
5. apply and configure implied constraints;
6. apply and configure automatic constraints;
7. use the parameters manager to create equations and set values for dimension parameters.

## PARAMETRIC CONSTRAINT AND DIMENSION CONCEPTS

AutoCAD's parametric commands let you control the relationships, size, and proportions of 2D drawing objects to maintain specific geometric properties that adjust as your design changes. You can use formulas to control feature sizes and create relationships between features in your design. You can also make "families" of parts in a similar design in a variety of sizes. Parametric or as it is often called, constraint-based modeling, makes it easy to experiment with different design ideas without spending a lot of time redrawing. You can incorporate design requirements into the drawing geometry, which is useful in both 2D and 3D design.

The two main parameter types are geometric constraints, and dimensional constraints. *Geometric constraints* like parallelism, perpendicularity, tangency, and concentricity let you control the 2D drawing geometry to quickly create complex geometry and create relationships between drawing objects. *Dimensional constraints* control feature sizes through defined values or by using equations. It is typically best to apply geometric constraints first, to control the feature relationships, and then control the sizes.

AutoCAD's parametric tools are available from the ribbon Parametric tab (Figure 38.1) from the *Drafting and Annotation* workspace or from the *3D Modeling* Workspace and also from Parametric on the Menu bar (Figure 38.2).

**FIGURE 38.1**

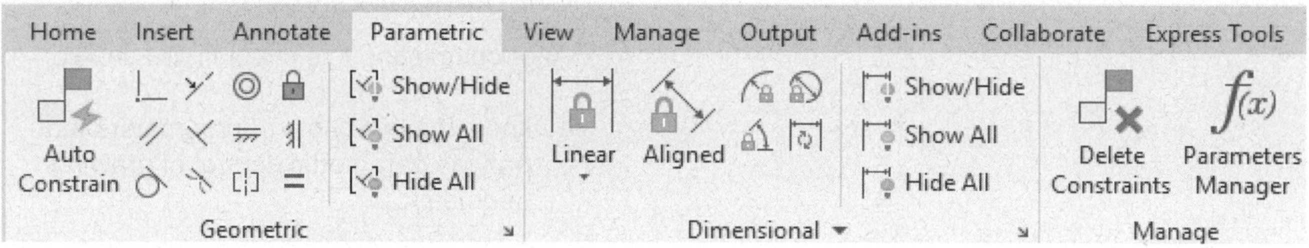


## GEOMETRIC CONSTRAINTS

A geometric constraint lets you define a relationship between 2D objects or points on objects. When you edit the drawing, the constraint conditions are maintained. The available constraints are: coincident, perpendicular, parallel, tangent, horizontal, vertical, collinear, concentric, smooth, symmetric, equal, and fix.

Each object type has only certain points to which the constraint can be applied. The order of selection and the location selected on the drawing object affect how the constraint applies to that geometry. A good understanding of the basics of drawing with AutoCAD is helpful for getting the desired results.

**FIGURE 38.2**

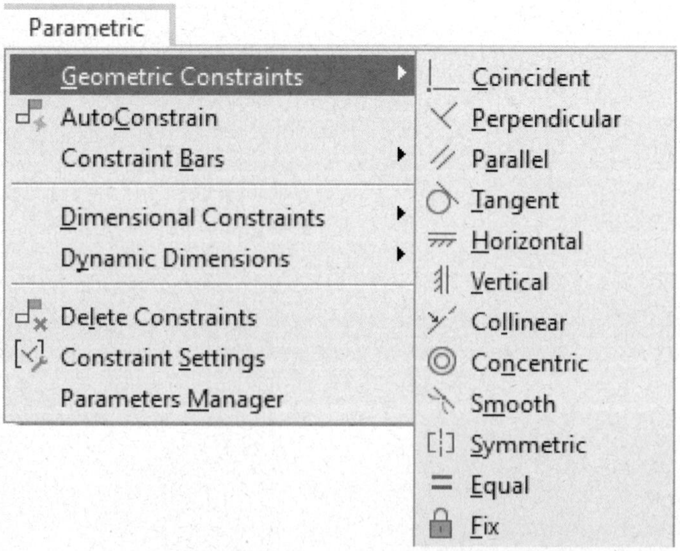

The following table outlines the valid constraint points for objects.

| *OBJECT TYPE* | *VALID CONSTRAINT SELECTIONS* |
|---|---|
| ***Line*** | Endpoints, midpoint |
| ***Arc*** | Center, endpoints, midpoint |
| ***Spline*** | Endpoints |
| ***Circle*** | Center |
| ***Ellipse*** | Center, major and minor axes |
| ***Polyline*** | Endpoints, midpoints of line and arc subobjects, center of arc subobjects; Segments within a polyline can be constrained individually |
| ***Xref, attribute, table*** | Insertion point |
| ***Block*** | Insertion point, nested entities |
| ***Text, multiline text*** | Insertion point, alignment point |

When a geometric constraint is applied to drawing objects, a small symbol called a constraint icon (Figure 38.3) appears near the items to indicate their relationship. The constraint icon symbols are similar to those shown at right in Figure 38.2.

The coincident constraint appears as a small blue box, but the larger symbol appears if you hover your mouse over the item. Complex geometry can get quite cluttered so you can use the Show/Hide options from the Parametric tab Geometric panel.

**FIGURE 38.3**

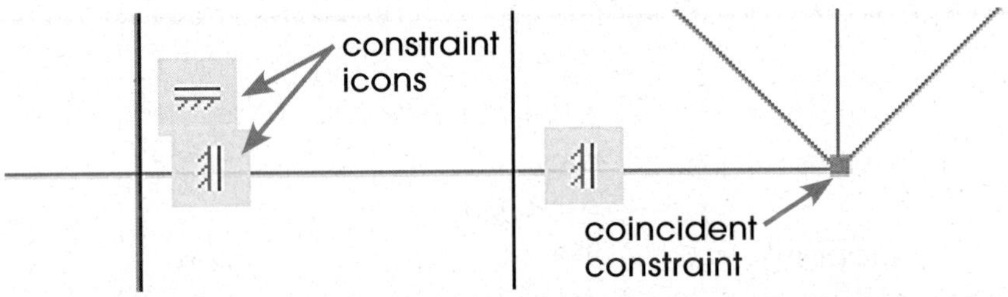


### GEOMCONSTRAINT

| Ribbon | Menu Bar | Command (Type) | Alias (Type) | Shortcut |
|---|---|---|---|---|
| | | GEOMCONSTRAINT | *GCON* | |

A single command, GEOMCONSTRAINT, can be used to apply any of the geometry constraints. Each constraint also has an individual command, which will be described next. GEOMCONSTRAINT applies a geometric relationship between objects or points on objects. It has options for each of the geometric constraint types. As the options of this command are the same as in the individual commands, they are not described here, but in the individual commands.

To use GEOMCONSTRAINT, enter the command name, or the alias GCON at the command prompt and then select the constraint type. The prompts you see following depend on the constraint selected.

```
Command: GEOMCONSTRAINT
Enter constraint type [Horizontal/Vertical/Perpendicular/PArallel/Tangent/SMooth/Coincident/CONcentric/
COLlinear/Symmetric/Equal/Fix] <Coincident>:
```

## GCCOINCIDENT

| Ribbon | Menu Bar | Command (Type) | Alias (Type) | Short-cut |
|---|---|---|---|---|
| *Parametric Geometric Coincident* | *Parametric Geometric Constraints Coincident* | GCCOINCIDENT | | |

*Coincident constraint*: Two selected points must stay connected or a point must stay connected to a curve or line (or its extension). The second selected point moves to be connected to the first point chosen. You cannot preselect the item and then select the command (noun-verb selection) when adding contraints.

Command: GCCOINCIDENT
Select first point or [Object/Autoconstrain] <Object>: **PICK** (Select near the endpoint on the first object)
Select second point or [Object] <Object>: **PICK** (Select near the endpoint on the second object)
Command:

Figure 38.4 shows an example of adding a coincident constraint between a line and an arc. When selecting during GCCOINCIDENT (and the other geometric constraint commands), you do not have to use osnaps to find the endpoint. Just click near the desired endpoint, the endpoint will be selected automatically as long as you click nearer to that endpoint than the other endpoint. A special marker appears to show you what is being selected. Coincident constraints display as blue squares, smaller than the grips in the image.

**TIP** It is better to select a bit away from where objects intersect, as you may not be sure which one is being constrained. It can be helpful to select a bit away from the endpoint on the object you desire as that makes it easier to see the endpoint selection icon. Notice that when the arc is selected first, the line moves into position with the arc. When the line is selected first, the arc is moved.

**FIGURE 38.4**

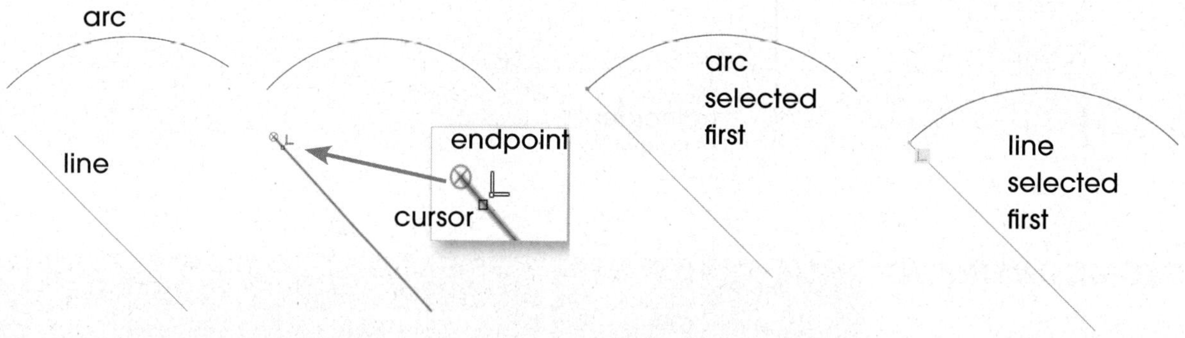


**FIGURE 38.5**

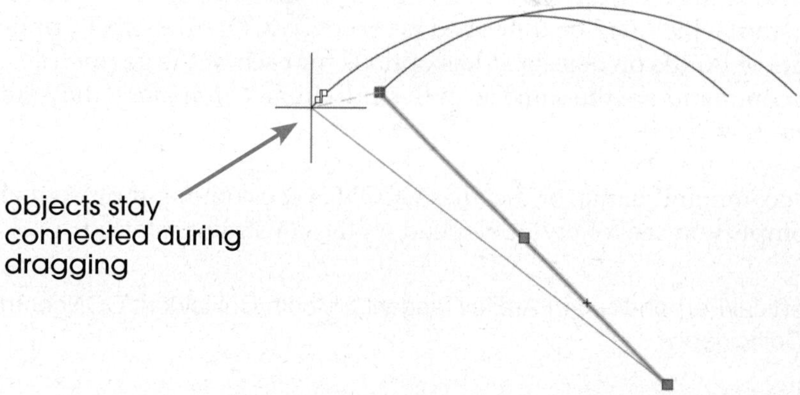


Once the coincident constraint is applied, moving the endpoint will cause the other object to move to remain coincident, even when it is not selected (Figure 38.5). Using the grips and dragging the item is an easy way to test if the constraints are applied as you intended.

### GCPERPENDICULAR

| Ribbon | Menu Bar | Command (Type) | Alias (Type) |
|---|---|---|---|
| *Parametric Geometric Perpendicular* | *Parametric Geometric Constraints Perpendicular* | GCPERPENDICULAR | |

*Perpendicular constraint*: Two selected lines (or polyline segments) must maintain a 90 degree angle to one another. The one selected second is constrained perpendicular to the first selected. The lines do not necessarily intersect, but are perpendicular along their infinite paths (Figure 38.6).

Command: GCPERPENDICULAR
Select first object: **PICK** (select a line)
Select second object: **PICK** (select a line; the second line adjusts to be perpendicular to the first)

**FIGURE 38.6**

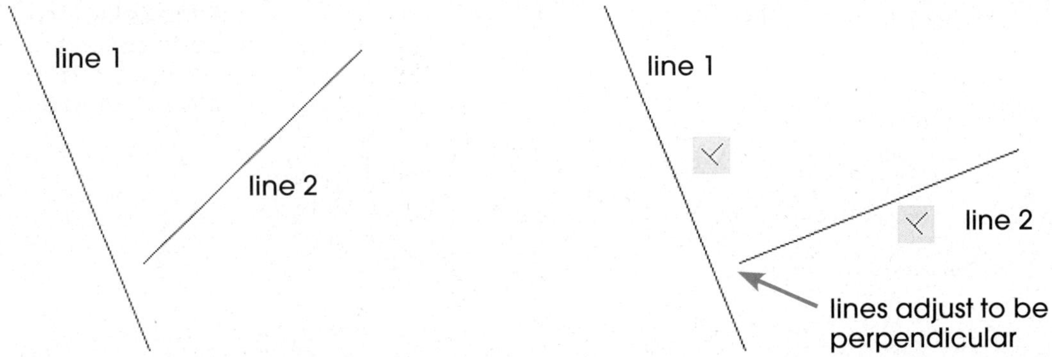


### GCPARALLEL

| Ribbon | Menu Bar | Command (Type) | Alias (Type) |
|---|---|---|---|
| *Parametric Geometric Parallel* | *Parametric Geometric Constraints Parallel* | GCPARALLEL | |

*Parallel constraint*: Two selected lines must remain parallel. The one selected second is constrained to maintain the same angle parallel to the first selected (Figure 38.7).

Command: GCPARALLEL
Select first object: **PICK** (select a line)
Select second object: **PICK** (select a line; the second line adjusts to be parallel to the first)

**FIGURE 38.7**

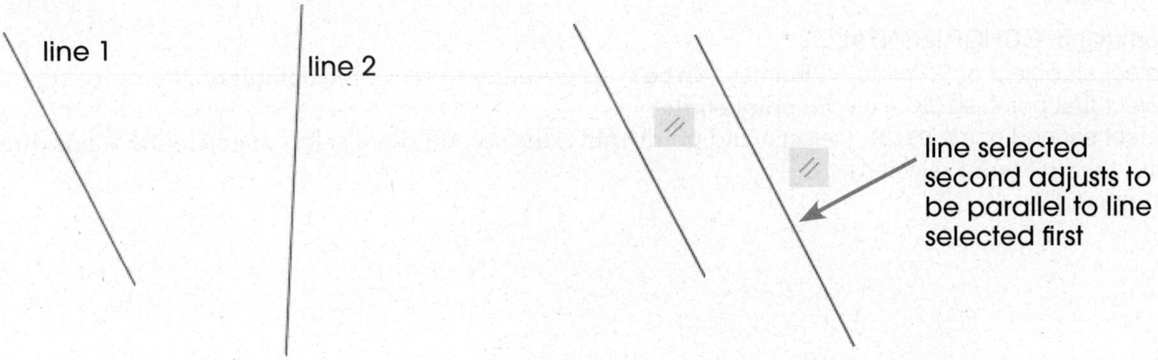

### GCTANGENT

| Ribbon | Menu Bar | Command (Type) | Alias (Type) |
|---|---|---|---|
| *Parametric Geometric Tangent* | *Parametric Geometric Constraints Tangent* | GCTANGENT | |

*Tangent constraint*: Two curves (one can be a line) must maintain tangency to each other or their extensions. The one selected second is constrained to maintain tangency to the first selected.

Command: GCTANGENT
Select first object: **PICK**
Select second object: **PICK**
Command:

**FIGURE 38.8**

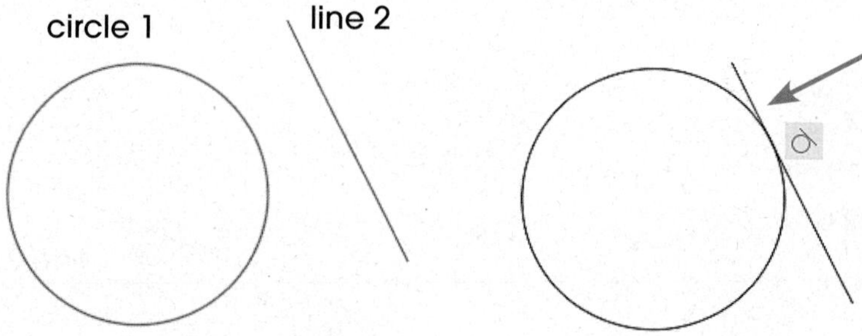


### GCHORIZONTAL

| Ribbon | Menu Bar | Command (Type) | Alias (Type) |
|---|---|---|---|
| *Parametric Geometric Horizontal* | *Parametric Geometric Constraints Horizontal* | GCHORIZONTAL | |

*Horizontal constraint*: A line or a pair of points must remain parallel to the X axis of the current coordinate system.

Command: GCHORIZONTAL
Select an object or [2Points] <2Points>: **PICK** (select the line to become horizontal)
Command:

#### *2Points Option*

The prompts for aligning a pair of points horizontally (parallel to the X axis of the current coordinate system) are shown next.

Command: GCHORIZONTAL
Select an object or [2Points] <2Points>: **Enter** (press enter to select the default of 2Points or type 2P)
Select first point: **PICK** (select a point entity)
Select second point: **PICK** (select a 2nd point entity; the second point select aligns in the x-axis direction with the first point selected)
Command:

## GCVERTICAL

| Ribbon | Menu Bar | Command (Type) | Alias (Type) |
|---|---|---|---|
| *Parametric Geometric Vertical* | *Parametric Geometric Constraints Vertical* | GCVERTICAL | |

*Vertical constraint*: Lines or a pairs of points on an object must remain parallel to the Y axis of the current coordinate system.

Command: GCVERTICAL
Select an object or [2Points] <2Points>: **PICK** (select the line to become vertical)
Command:

The prompts for aligning a pair of points vertically (parallel to the Y axis of the current coordinate system) are similar to those for horizontal.

## GCCOLLINEAR

| Ribbon | Menu Bar | Command (Type) | Alias (Type) | Short-cut |
|---|---|---|---|---|
| *Parametric Geometric Colinear* | *Parametric Geometric Constraints Colinear* | GCCOLINEAR | | |

*Collinear constraint*: Two or more line segments must remain along the same line, or its extension. The one selected second maintains alignment with the one selected first (Figure 38.9).

Command: GCCOLLINEAR
Select first object or [Multiple]: **PICK** (select a line)
Select second object: **PICK** (select a line, the second line adjust to become colinear with the first)
Command:

**FIGURE 38.9**

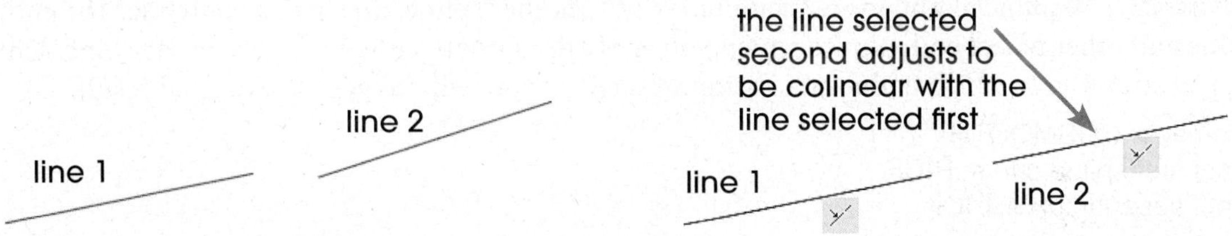


### *Multiple*

Using this option returns the prompts shown next. You can use this method to pick additional lines to make colinear to the first one selected.

Select first object or [Multiple]: **M Enter**
Select first object: PICK (select a line)
Select object to make collinear to first: **PICK** (select a line)
Select object to make collinear to first: **PICK** (select a line)
Select object to make collinear to first: **Enter**
Command:

## GCCONCENTRIC

| Ribbon | Menu Bar | Command (Type) | Alias (Type) | Short-cut |
|---|---|---|---|---|
| *Parametric Geometric Concentric* | *Parametric Geometric Constraints Concentric* | GCCONCENTRIC | | |

*Concentric constraint*: Two arcs, circles, or ellipses must maintain the same center point. The one selected second keeps concentric to the one selected first. Figure 38.10 shows an example using circles, but arcs or ellipses can also be constrained to be concentric.

Command: GCCONCENTRIC
Select first object: PICK
Select second object: PICK

**FIGURE 38.10**

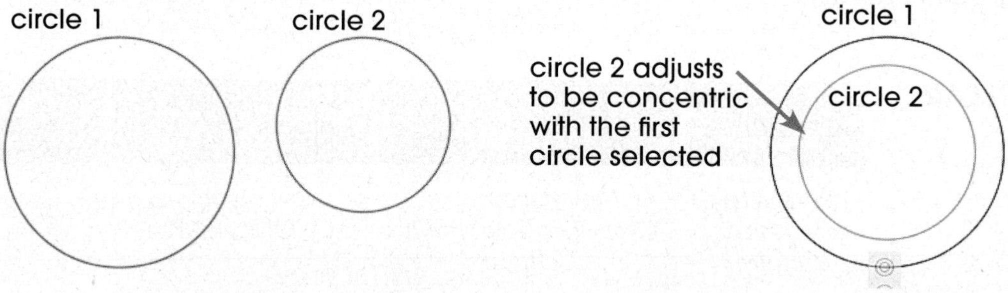


## GCSMOOTH

| Ribbon | Menu Bar | Command (Type) | Alias (Type) | Short-cut |
|---|---|---|---|---|
| *Parametric Geometric Smooth* | *Parametric Geometric Constraints Smooth* | GCSMOOTH | | |

*Smooth constraint*: A spline maintains *G2 continuity* with another spline, line, arc, or polyline. The endpoints of the spline and other object endpoint to which you apply the smooth constraints are made coincident. When selecting, pay attention to which endpoint is being selected or you will not get the expected result.

Command: GCSMOOTH
Select first spline curve: PICK
Select second curve: PICK

**FIGURE 38.11**

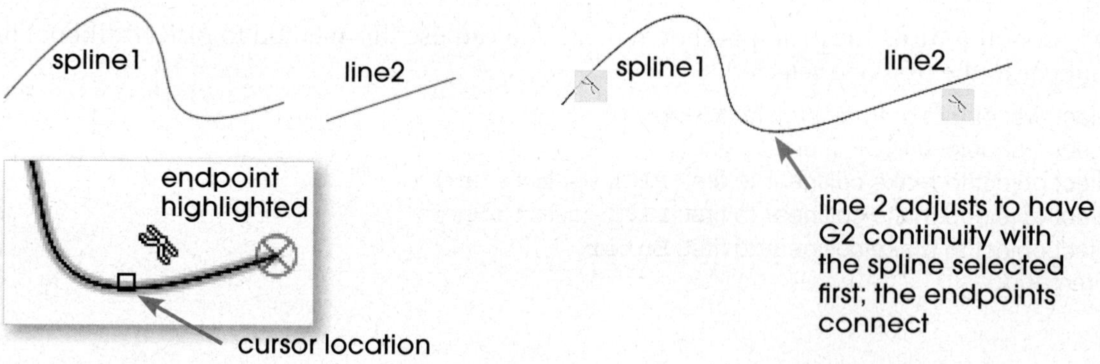

## GCSYMMETRIC

| Ribbon | Menu Bar | Command (Type) | Alias (Type) | Short-cut |
|---|---|---|---|---|
| *Parametric Geometric Symmetric* | *Parametric Geometric Constraints Symmetric* | GCSYMMETRIC | | |

*Symmetric constraint*: Two selected objects or points must remain symmetric about the selected symmetry line.

Command: GCSYMMETRIC
Select first object or [2Points] <2Points>: PICK (select an object)
Select second object: PICK (select an object)
Select symmetry line: PICK (select a line for the objects to be symmetric about)
Command:

### *2Points*

The prompts for selecting points to be symmetric about a line are shown next. The points can be endpoints of a line, arc, pline, spline, point objects, and so on.

Select first object or [2Points] <2Points>: **Enter**
Select first point: PICK (select a point)
Select second point: PICK (select a point)
Select symmetry line: PICK (select a line for the objects to be symmetric about)
Command:

**FIGURE 38.12**

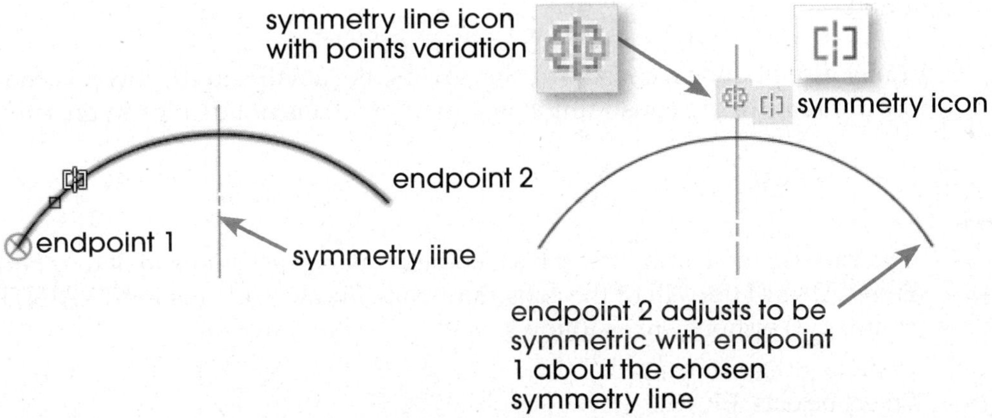


## GCEQUAL

| Ribbon | Menu Bar | Command (Type) | Alias (Type) | Short-cut |
|---|---|---|---|---|
| *Parametric Geometric Equal* | *Parametric Geometric Constraints Equal* | GCEQUAL | | |

*Equal constraint*: Selected arcs and circles must maintain the same radius, or selected lines must maintain the same length. Choosing the Multiple option allows you to continue selecting objects to make equal size.

Command: GCEQUAL
Select first object or [Multiple]: PICK
Select second object: PICK

## GCFIX

| Ribbon | Menu Bar | Command (Type) | Alias (Type) | Short-cut |
|---|---|---|---|---|
| *Parametric Geometric Fix* | *Parametric Geometric Constraints Fix* | GCFIX | | |

*Fix constraint*: Points, line endpoints or curve points must stay in fixed position on the world coordinate system. You will find this a very useful constraint. Without fixing an endpoint on the coordinate system, the geometry may update or move in ways you did not expect. Fixing an axis line on the coordinate system and referencing drawing objects to it, makes it easier for you to predict how the drawing geometry will update. Consider fixing a horizontal and vertical line with their centers at 0,0, or some other geometry (like the center of a circle) to which other objects will be constrained.

TIP

Command: GCFIX
Select point or [Object] <Object>: PICK
Command:

# VIEWING CONSTRAINTS

**FIGURE 38.13**

constraint bar

Constraint icons give you visual cues to the objects that are geometrically constrained. When an object has several constraints applied, the icons are grouped into a constraint bar as shown in Figure 38.13. You can drag icons and constraint bars to move them on screen and you can control whether they show or are hidden.

A quick way to hide a constraint bar is using the small x in the lower corner. This does not remove the constraint, it just hides it so it is not displayed on screen.

**FIGURE 38.14**

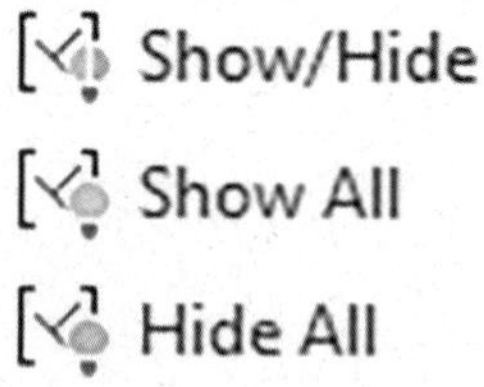


You can use the Parametric tab's Geometric panel selections to Show/Hide, Show All, or Hide All of the constraint bars. Show/Hide (CONSTRAINTBAR command) options are as follows:

Command: ConstraintBar
Select objects: PICK
Select objects: **Enter**
Enter an option [Show/Hide/Reset]<Show>: (enter the option letter)

### *Constraint Icons Appearances*

Constraint icons have variations to provide information on how the constraint is applied.

| Variation | Description |
|---|---|
|  | Small circle indicates constraint locks a point rather than the object |
|  | Small circle indicates two points are constrained vertical rather than an object. The horizontal constraint has a similar variation. |

| Variation | Description |
|---|---|
|  | Small blue axis symbol on a horizontal or vertical constraint icon indicates it is not parallel to the current UCS X or Y axis. The added blue symbol is removed when the UCS is aligned with the constraint X or Y axis again. |
|  | The variations can combine as shown in this example which includes the point symbol and the UCS axis added. |
|  Point Line Object | Icon variations of the symmetry line for constraint defined by a point, line, or an object. |

### *Cursor Badge for Constraint*

A special cursor badge appears (Figure 38.15) when you hover the cursor over a constrained drawing object. This is a useful visual cue to remind you that you can only make changes to the geometry that will not violate the constraint that is applied.

**FIGURE 38.15**

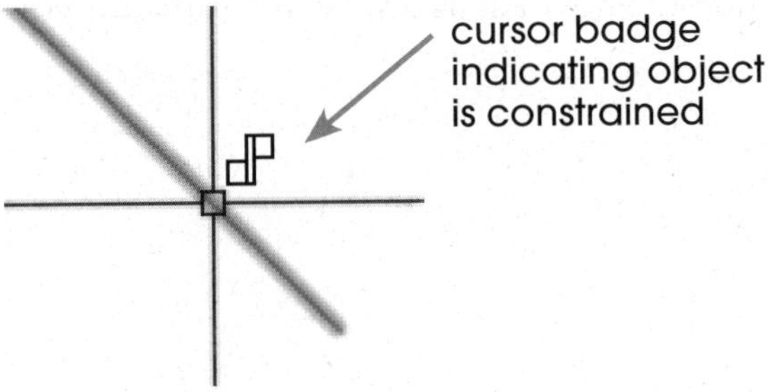


### *Highlighting Constraints*

It can be difficult at times to know which constraint bar or icon relates to which feature. You can determine which object the icon applies to in two ways (Figure 38.16):

- Click an object with a constraint applied and the associated constraint bar becomes highlighted.
- Click a constraint icon, the objects associated with that geometric constraint become highlighted.

**FIGURE 38.16**

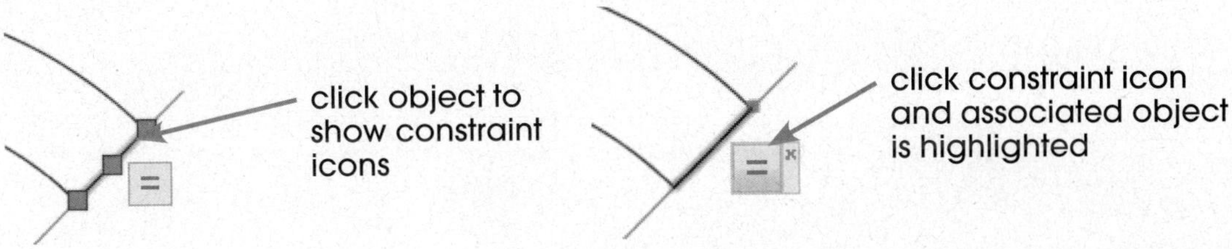

## DELETING CONSTRAINTS

Often you will need to delete a constraint to allow more freedom for the geometry to reposition, or to apply a different one. You can delete constraints as a group, or individually.

### *DELCONSTRAINT*

| Ribbon | Menu Bar | Command (Type) | Alias (Type) | Short-cut |
|---|---|---|---|---|
| *Parametric Geometric Fix* | *Parametric Geometric Constraints Fix* | GCFIX | | |

The DELCONSTRAINT command lets you delete all constraints from a selection set.

Command: DELCONSTRAINT
All constraints will be removed from selected objects...
Select objects: PICK
Select objects: **`Enter`**
1 constraint(s) removed

You can also delete individual constraints by right-clicking on the constraint icon to show the menu and choosing Delete as shown in Figure 38.18. Keep in mind that a quick way to hide a constraint icon is using the small x in the upper corner. This does not remove the constraint, it just hides it so it is not displayed on screen. You can use the PGeometric panel selection to Show All as a way to verify that you deleted the constraint.

**FIGURE 38.17**

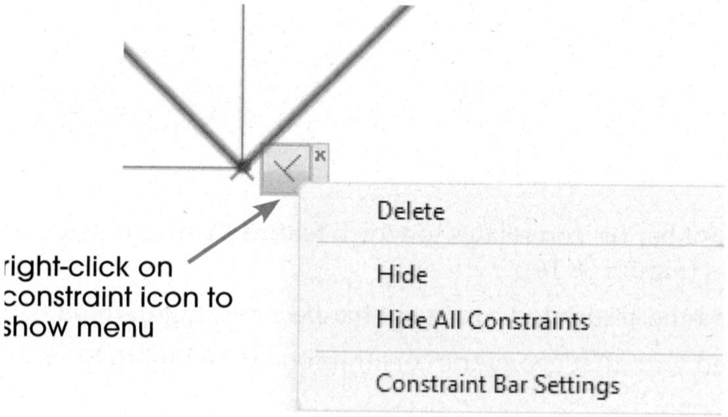

## CONSTRAINT SETTINGS

The Constraint Settings dialog box gives you control over the types of geometric constraints displayed on the constraint bars. Options make constraint bars appear automatically or only show temporarily when a constrained object is selected or hovered over. You can also choose to display only certain icon types—for example only concentric constraints or only perpendicular constraints. This is useful when the screen is crowded, or when you are troubleshooting a complicated constraint set.

### *CONSTRAINTSETTINGS*

| Ribbon | Menu Bar | Command (Type) | Alias (Type) |
|---|---|---|---|
| *Parametric Geometric arrow* | *Parametric Constraint Settings* | CONSTRAINTSETTINGS | *CSETTINGS* |

You can launch the Constraint Settings dialog box by entering the command name or alias, or by selecting the tiny arrow at the right of the Parametric tab Geometric panel as shown in Figure 38.18.

**FIGURE 38.18**

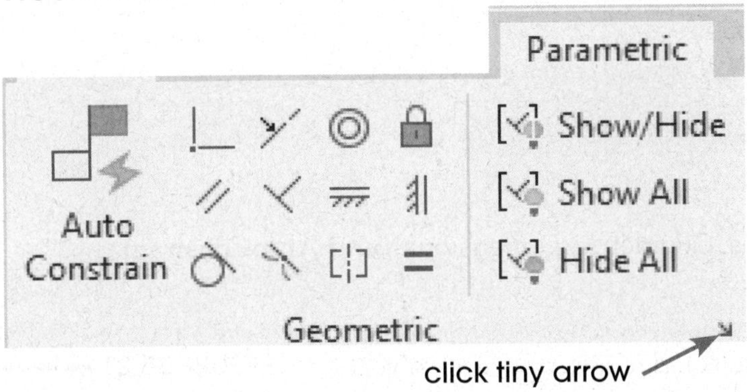


**FIGURE 38.19**

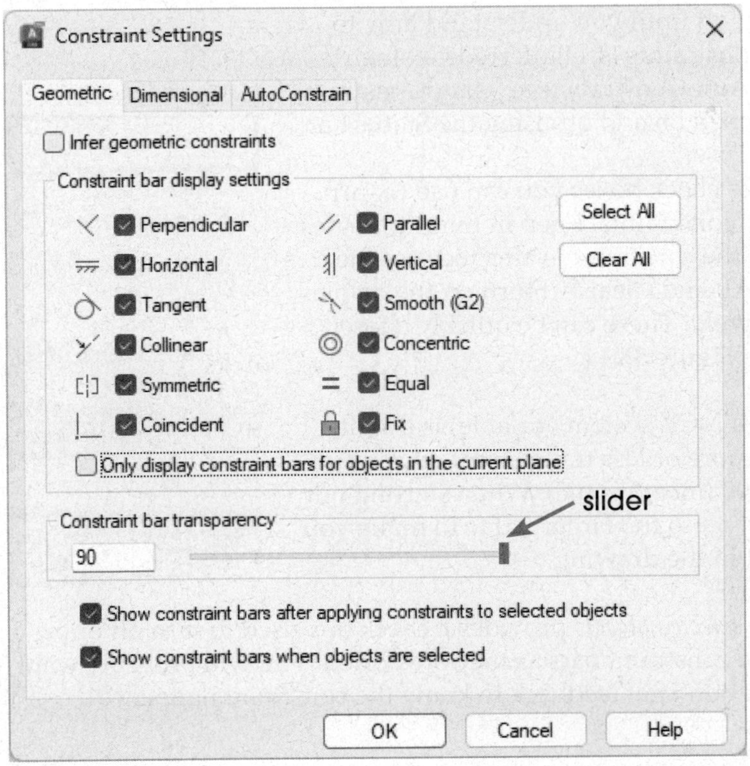

**FIGURE 38.20**

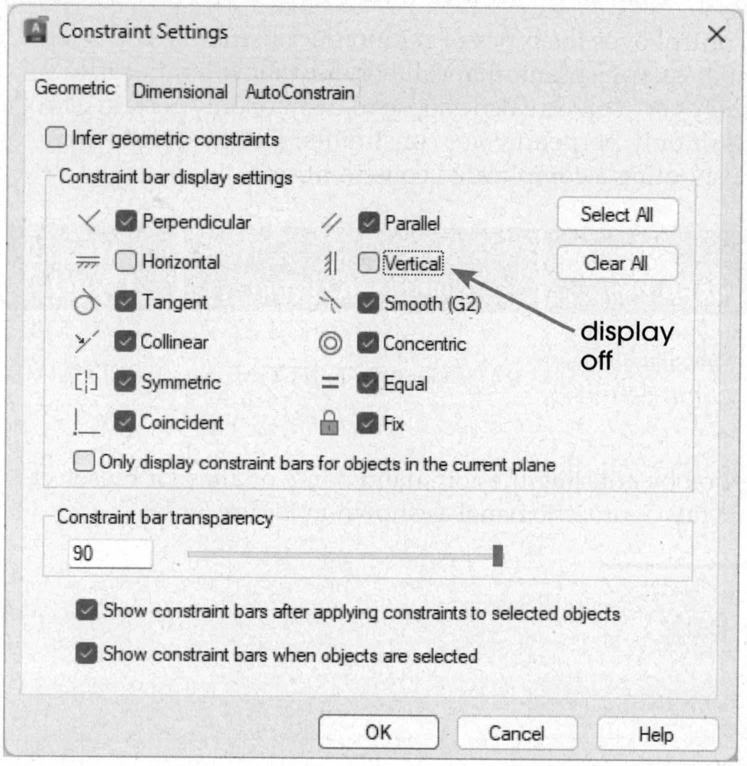


The Constraints Settings dialog box has three tabs: Geometric, Dimensional, and AutoConstrain

### *Geometric Tab*

**FIGURE 38.21**

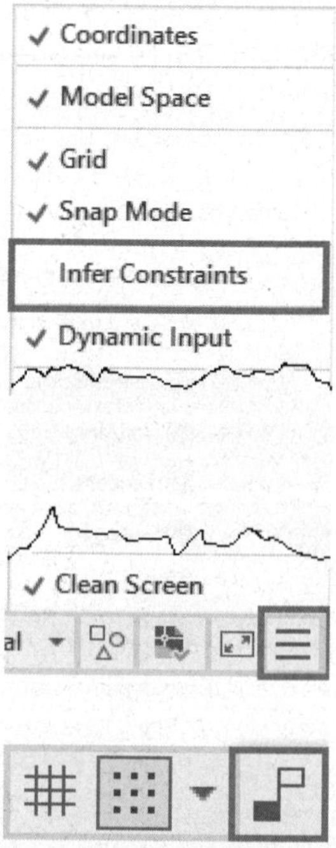


With the check box for *Infer geometric constraints* selected (check appears), object snaps that you use to create geometry generate geometric constraints when the selected objects or points can meet the constraint conditions. This sets the CONSTRAINTINFER system variable to 0 (off) or 1 (on). Inferred constraints can be handy, but it is probably best to leave it off until you understand how to apply constraints manually. Infer geometric constraints (see"Inferred Geometric Constraints" on page 1228) is not identical to AutoConstrain (see "AutoConstrain" on page 1227). Infer constraints can also be turned on and off using the Status bar.

The *Constraint bar display settings* area provides check boxes you can use to turn off/on the display of each of the 12 geometric constraints. Keep in mind that these constraints are still active, only the display of the constraint is affected, not the constraint condition itself. Buttons for Select All and Clear All turn on the display of all constraints, or turn them all off, respectively. These can be quickly chosen from the Parametric tab, Geometric panel; see Figure 38.14.

*Constraint bar transparency* (CBARTRANSPARENCY system variable) is a slider that lets you make the constraint icons show more or less transparent on screen. The setting ranges from 10 (barely visible) to 90 (most visible). When starting out with geometric constraints you may want this set to the larger value to make you more aware of the conditions that are applied in the drawing.

*Show constraint bars after applying constraints to selected objects* provides a check box used to turn off or on the automatic update of the screen to show the constraint bars as they are added. You will probably want to leave this turned on initially, otherwise you don't get feedback to know the constraint happened.

*Show constraint bars when objects are selected* is also useful to leave on. It can help you verify which drawing objects have constraints applied (see Figure 38.16).

### *Dimensional Tab*

The Dimensional tab of the Constraints Setttings dialog box is used to set the display format options for dimensional constraints. See "Overconstrained Vs UnderConstrained" on page 1231 for more information on applying dimensional constraints.

**FIGURE 38.22**

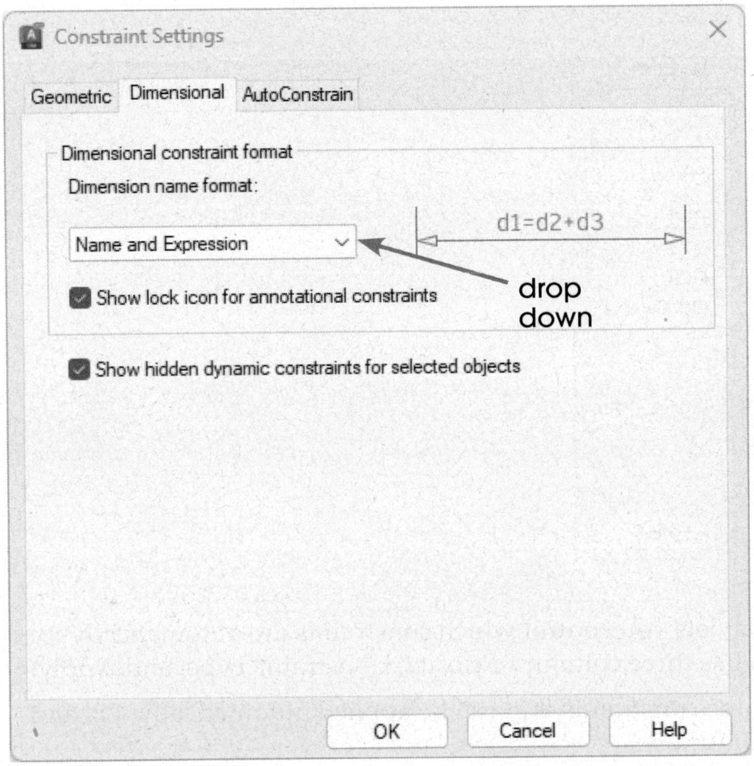


**FIGURE 38.23**

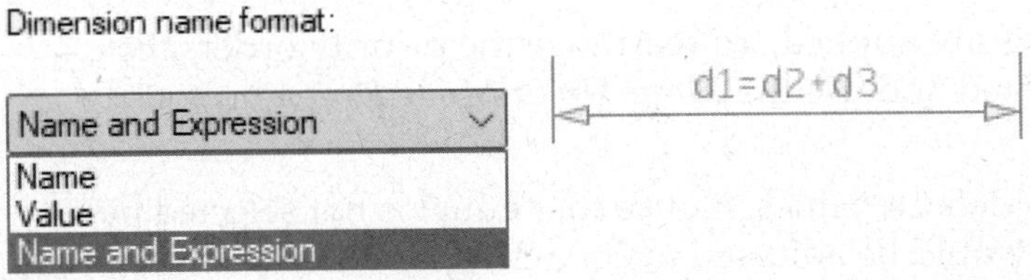


The *Dimension name format* area has a drop-down you use to choose the format for the display of dimensional constraints. This can be the dimension name (like a variable name), the dimension value only, or the dimension name and expression. Using equations to set dimension values is covered later in this chapter. The image tile shows an example of the selected format. See Figure 38.23.

The check box for *Show lock icon for annotational constraints* lets you choose to show lock icon when an object has an annotational constraint applied (DIMCONSTRAINTICON system variable).

The check box for *Show hidden dynamic constraints for selected objects* sets whether or not to display selected dynamic constraints that have been set to hide.

TIP

### AutoConstrain Tab

**FIGURE 38.24**

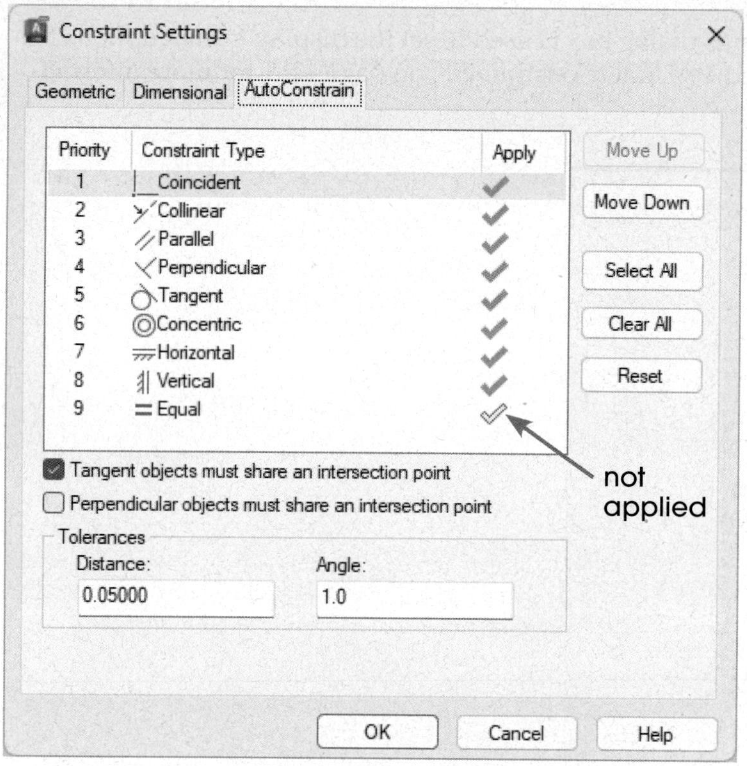


The upper center area of the AutoConstrain tab lets you control which constraints are automatically applied and their application order. This area has three columns: Priority, Constraint Type, and Apply.

- ***Constraint Type:*** lists the names of the nine constraints that can be applied automatically. Fix and Smooth constraints are not applied automatically.
- ***Apply:*** click the check mark to control whether that constraint is applied by the AutoConstrain command. The check mark appears filled in when that constraint will be applied. The ***Select All*** and ***Clear All*** buttons let you quickly change to all or no constraints applied during automatic constraining.
- ***Priority:*** lists the order in which constraints are applied. To rearrange the priority order, click a constraint type so that it appears highlighted and use the ***Move Up*** or ***Move Down*** buttons to change the order of the item
- ***Reset:*** resets AutoConstrain settings to the default values. Notice that Equal is not selected by default as lines within the stated tolerance would be adjusted to equal lengths automatically—which value to be equal to depends on a number of factors.
- ***Tangent Objects Must Share an Intersection Point:*** checking this box lets you require two curves to share a common point within the set distance tolerance before the tangent constraint is automatically applied.
- ***Perpendicular Objects Must Share an Intersection Point:*** checking this box requires lines either intersect or the endpoint of one line is coincident with the other line within the set distance tolerance.
- ***Tolerances:*** these entry boxes let you set the Distance applied to coincident, concentric, tangent, and collinear constraints and the Angle applied to horizontal, vertical, parallel, perpendicular, tangent, and collinear constraints. For example, if the distance between the endpoints of two line segments is greater than 0.05000 (the default) the lines will not be set to Coincident; an angle at 88.5 degrees from horizontal will not have a Vertical constraint set when the angle tolerance is set to 1.

## *AutoConstrain*

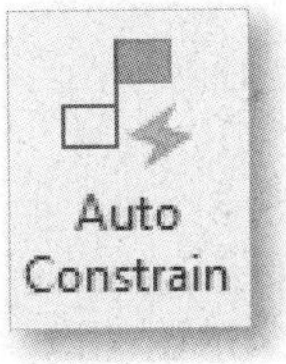


| Ribbon | Menu Bar | Command (Type) | Alias (Type) |
|---|---|---|---|
| *Parametric Geometric Auto Constrain* | *Parametric AutoConstrain* | AUTOCONSTRAIN | |

The AUTOCONSTRAIN command lets you apply the currently defined constraint set to selected objects. AutoConstrain uses the constraint settings from the dialog box (Figure 38.24) to generate geometric constraints for you. Constraints are applied in priority order if the object meets the constraint conditions. During this process the software checks to see if the selected objects are within the tolerances you specified in the AutoConstrain tab.

For example, two different lines will have their endpoints set coincident when that constraint is active and the points are within the set distance tolerance. Tangent and Perpendicular have special selection boxes for requiring an intersection point. This means that before the tangent constraint is automatically applied, the software checks that the objects are within the set distance tolerance of each other; within the tolerance distances they are assumed to be intersecting even if they do not share an exact point. When those constraints do *not* require intersection, the objects can be an even larger distance apart.

The priority of the constraints is important. Applied constraints prohibit the geometry from re-adjusting to satisfy a different constraint requirement with lower priority. For example, if two lines are first set to be perpendicular, they can't also be set to be parallel.

Command: AUTOCONSTRAIN
Select objects or [Settings]: PICK (select all of the objects you want included)
Select objects or [Settings]: **Enter** (to end selection)
X constraint(s) applied to Y object(s)
Command:

The Settings option opens the Constraint Settings dialog box as shown in Figure 38.24. If you have cleared all of the constraint types to be applied, you will see the message "Geometric constraints will not be applied" when you use the AUTOCONSTRAIN command. You must have at least one constraint selected.

**FIGURE 38.25**

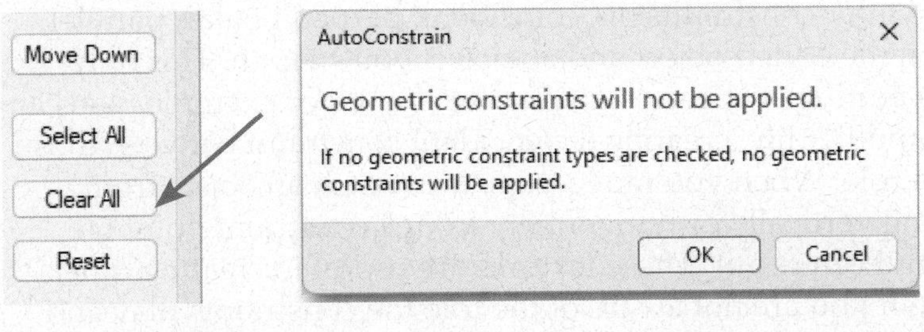


CONSTRAINTSOLVEMODE is a system variable that controls methods of constraint solving. It can be set to 0 or 1. A setting of 0 allows drawing objects to be resized when a constraint is applied or modified; setting the value to 1 retains the size of the geometry. When CONSTRAINTSOLVEMODE is set to 0 you may get odd results. To satisfy a set of constraint conditions, one object may suddenly be made very small or large compared to another.

## INFERRED GEOMETRIC CONSTRAINTS

Enabling Infer Constraints mode automatically applies constraints between the objects you are creating or editing. In this mode, some of the object snaps that you use when creating geometry also apply geometric constraints. Similar to AUTOCONSTRAIN (page 1227), constraints are applied only if the objects meet the required conditions. Objects do not reposition when inferred constraints are added. This can be a bit confusing if you have it turned on inadvertently as you may not notice a change in behavior until you try to edit the objects. See "Infer geometric constraints" on page 1224 for additional settings information.

Not all object snaps that you use will add constraints even with this mode on. Intersection, Apparent Intersection, Extension, and Quadrant object snap modes do not have a related constraint that would make sense to infer, thus they do not infer a constraint.

Not all constraint types can be inferred by using an object snap. The Fix, Smooth, Symmetric, Concentric, Equal and Collinear constraints ***do not*** get applied. The Horizontal, Vertical, Coincident, Perpendicular, Parallel, Concentric, and Colinear constraints are constraints that can be inferred.

**FIGURE 38.26**

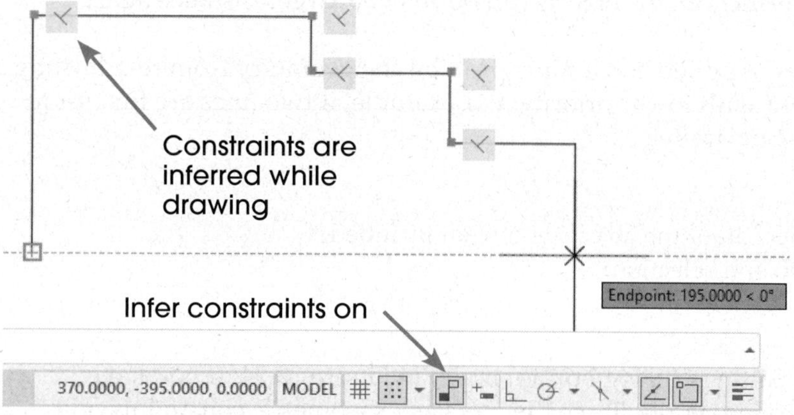


Some drawing and editing commands infer constraints even when an object snap is not used. Many of these make sense when you think about the geometry you intend to create using them. The LINE and PLINE commands infer coincident constraints on connected endpoints; when within the specified angle tolerance, a Perpendicular constraint may be added (see for an example). Using the Close option infers a coincident constraint between the start point of the first line and the endpoint of the last line. RECTANG applies parallel constraints and a perpendicular constraint to the closed polyline it creates. The FILLET command applies tangent and coincident constraints to the arc and the pair of lines. CHAMFER applies coincident constraints between the new line and selected pair of lines.

When moving, copying, or stretching with the Infer constraints on, coincident, perpendicular, parallel, or tangent constraints may be applied between the edited object and an object you snap to. This only happens if the base point of the object you are editing is already a constraint point. For example, you can stretch a line and snap its endpoint to that of another line creating a coincident constraint between those two endpoints. Object tracking can also play a role. When you move, copy, or stretch an object from a constrained base-point and are object-tracking vertically or horizontally from a constraint point on another object, a vertical or horizontal constraint can result. These features are useful in creating quick accurate geometry, but may be very confusing if you are not aware of the role the constraints play and how to edit the constraints when necessary.

## DIMENSIONAL CONSTRAINTS

Geometric constraints allow you to control the relationships of drawing objects and points in the drawing. ***Dimensional constraints*** let you control the distance between drawing objects (including polyline segments), between points on the objects, the angles between objects or points, and the size of arcs and circles. Dimensional constraints appear similar to dimensions that you add in the drawing during documentation, but they are *not* the same. When you add a dimensional constraint to an object, that dimension "drives" the size or angle of that feature. A dimension constraint *variable* is automatically created for storing the value. These automatically get names such as *d1* or *dia1*, but you can and should rename them to relate clearly to your drawing. If you change the dimensional constraint value, all of the constraints on that object are evaluated and the objects update automatically.

Drawing dimensions you used previously while working through this text are made of regular drawing objects, such as lines and text. These dimensions get their value by calculating from the drawing object(s) they are attached to using the special Defpoints layer. Unlike typical drawing dimensions, dimensional constraints ***control*** the size of the feature.

**FIGURE 38.27**

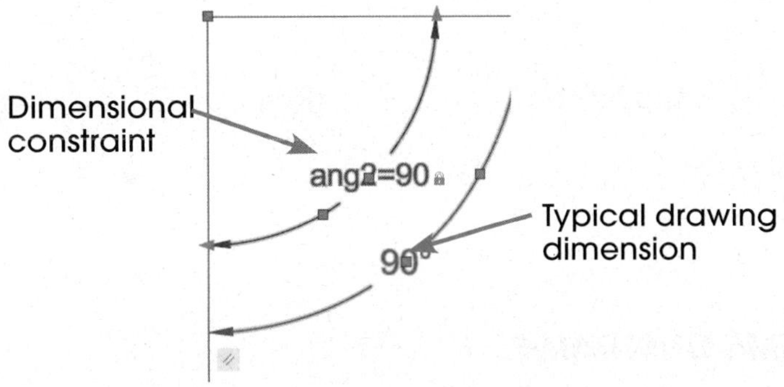


AutoCAD's dimensional constraints can be either dynamic or annotational. Dynamic dimensional constraints are the default and are typical in parametric drawings. Annotational constraints are provided to let you control the dimension style and plot dimensional constraints. You can use the Properties palette to change dynamic dimensional constraints to annotational constraints and vice-versa. Generally you will use dynamic dimensions. Annotational constraints will not be covered in any detail, but you can use your knowledge of dimension styles to understand their features.

***Dynamic dimensional constraints***:

- stay the same size when zoomed
- can all be turned on or off as a group
- using a preset dimension style
- automatically position the dimension text
- have special triangle grips you use to change the dimensional constraint value
- don't display when the drawing is plotted

***Annotational dimensional constraints***:

- change size when zoomed
- can be on different layers
- using the dimension styles
- have grips similar to regular drawing dimensions
- display when the drawing is plotted

The CONSTRAINTNAMEFORMAT system variable controls the display of dimensional constraints. It can be set to 0, 1, or 2. Setting to the default value of 2 shows the variable name and value; 1 shows just the value; and 0 shows just the variable name. See Figure 38.28.

The LUPREC and AUPREC system variables control the number of decimal places displayed in dimensional constraints.

**FIGURE 38.28**

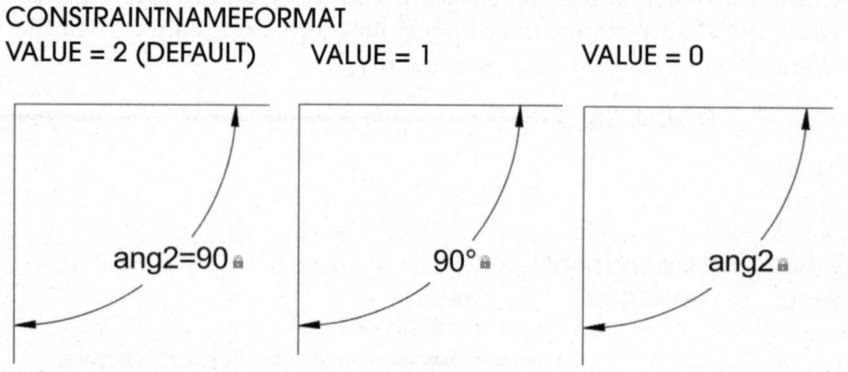


## *DIMCONSTRAINT*

The DIMCONSTRAINT command provides command line options to add dimensional constraints to objects or points on objects. You can also use it to convert associative dimensional constraints to dynamic dimensional constraints. Each of the dimensional constraint options also has a related command that you can use directly. They are not covered separately, but the options and command names are listed below. Their icons are available from the Parametric tab, Dimensional panel as shown in Figure 38.29.

See "Valid Constraint Selections" on page 1213 for a list of the valid constraint points for an object.

**FIGURE 38.29**

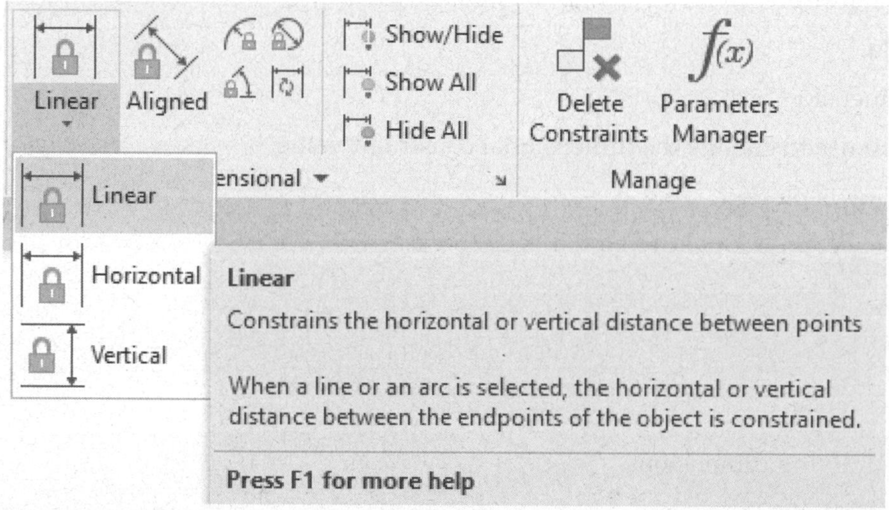

The DIMCONSTRAINT command options function the same as the commands.

| *OPTION NAME* | *COMMAND NAME* | *DESCRIPTION* |
|---|---|---|
| Linear | DCLINEAR | Creates a horizontal, vertical, or rotated constraint based on the locations of the extension line origins and the dimension line. |
| Horizontal | DCHORIZONTAL | Constrains the X distance between points on an object, or between two points on different objects. |
| Vertical | DCVERTICAL | Constrains the Y distance between points on an object, or between two points on different objects. |
| Aligned | DCALIGNED | Constrains the distance between two points on different objects. |
| Angular | DCANGULAR | Constrains the angle between line or polyline segments, the angle swept out by an arc or a polyline arc segment, or the angle between three points on objects. |
| Radius | DCRADIUS | Constrains the radius of a circle or an arc. |
| Diameter | DCDIAMETER | Constrains the diameter of a circle or an arc. |
| Form | DCFORM | Specifies whether the dimensional constraint being created is dynamic or annotational. |
| Convert | DCCONVERT | Converts associative dimensional constraints to dynamic dimensional constraints. |

TIP

Note: The L option (last object drawn) is not allowed in the DIMCONSTRAINT command as the constraint behavior is dependent on where you pick the object.

## OVERCONSTRAINED VS UNDERCONSTRAINED

When you apply constraints to drawing objects, the software must solve for the resulting positions of the drawing geometry. As you are adding constraints, you may see a message telling you the geometry would be *overconstrained* (see Figure 38.30) and therefore the constraint cannot be applied. The software does checking that will not allow you to overconstrain the drawing. You cannot add a condition that would potentially contradict an existing constraint or dimensions.

**FIGURE 38.30**

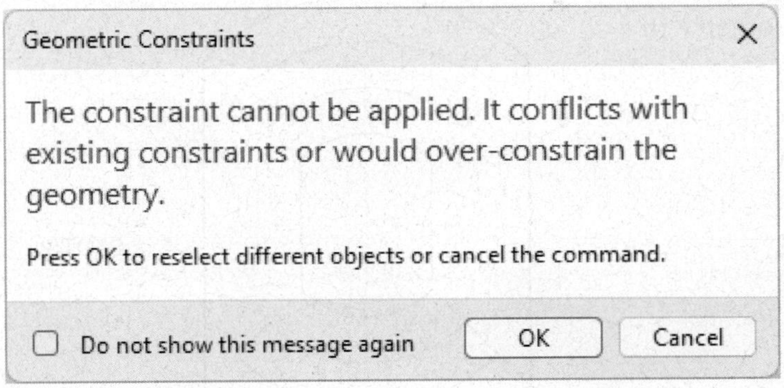

A simple example is shown in Figure 38.31. Using the RECTANGLE command with inferred constraints on results in two sets of parallel constraints and one perpendicular constraint. If you try to add a dimensional constraint for the angle, you will see a message saying that would overconstrain the geometry. The dimension constraint is a variable that drives the angle value and is allowed to be something other than 90°, but can not do so and maintain the parallel and perpendicular constraints that already exist.

To fix this type of issue, delete the constraint(s) that are restricting the geometry changing in the way you would like. In this example, deleting the perpendicular constraint would allow the angle to change. The sides would still remain parallel, but would be allowed to be an angle other than 90°.

**FIGURE 38.31**

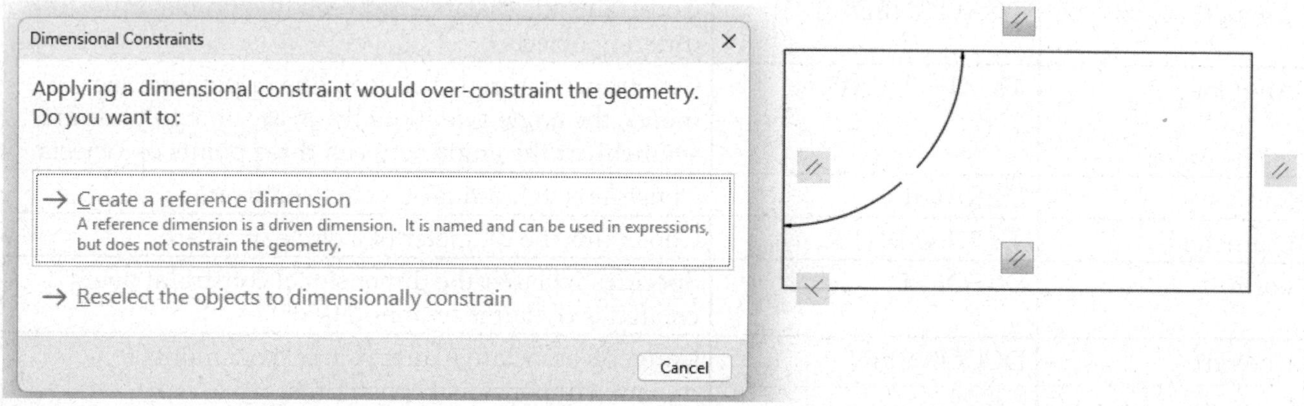


*Underconstrained* geometry just means that portions of the shape are free to move. When you use MOVE, ROTATE, SCALE, or STRETCH drawing geometry that has some constraints, the constraints are maintained during editing. The ability to modify underconstrained objects depends on the object type and the constraints you have applied to it. You might think of the constraints as restricting the degrees of freedom of motion for the object. Fully constrained geometry cannot be dragged, stretched, etc. except by changing dimensional constraint values, as all of the positions are locked down.

## CONSTRAINTS EXAMPLE

Here's a step by step example for constraining a simple gasket such as that shown in Figure 38.32. One issue when starting out is that there are multiple correct ways you can constrain the geometry. Keep in mind the ways in which the design may change and create geometric and dimensional constraints that will allow the drawing geometry to update to reflect that design intent.

**FIGURE 38.32**

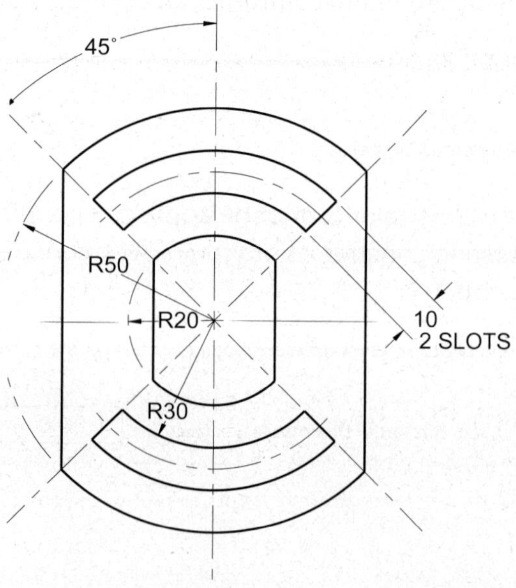

## Tips for Getting Started Constraining Geometry

FIGURE 38.33

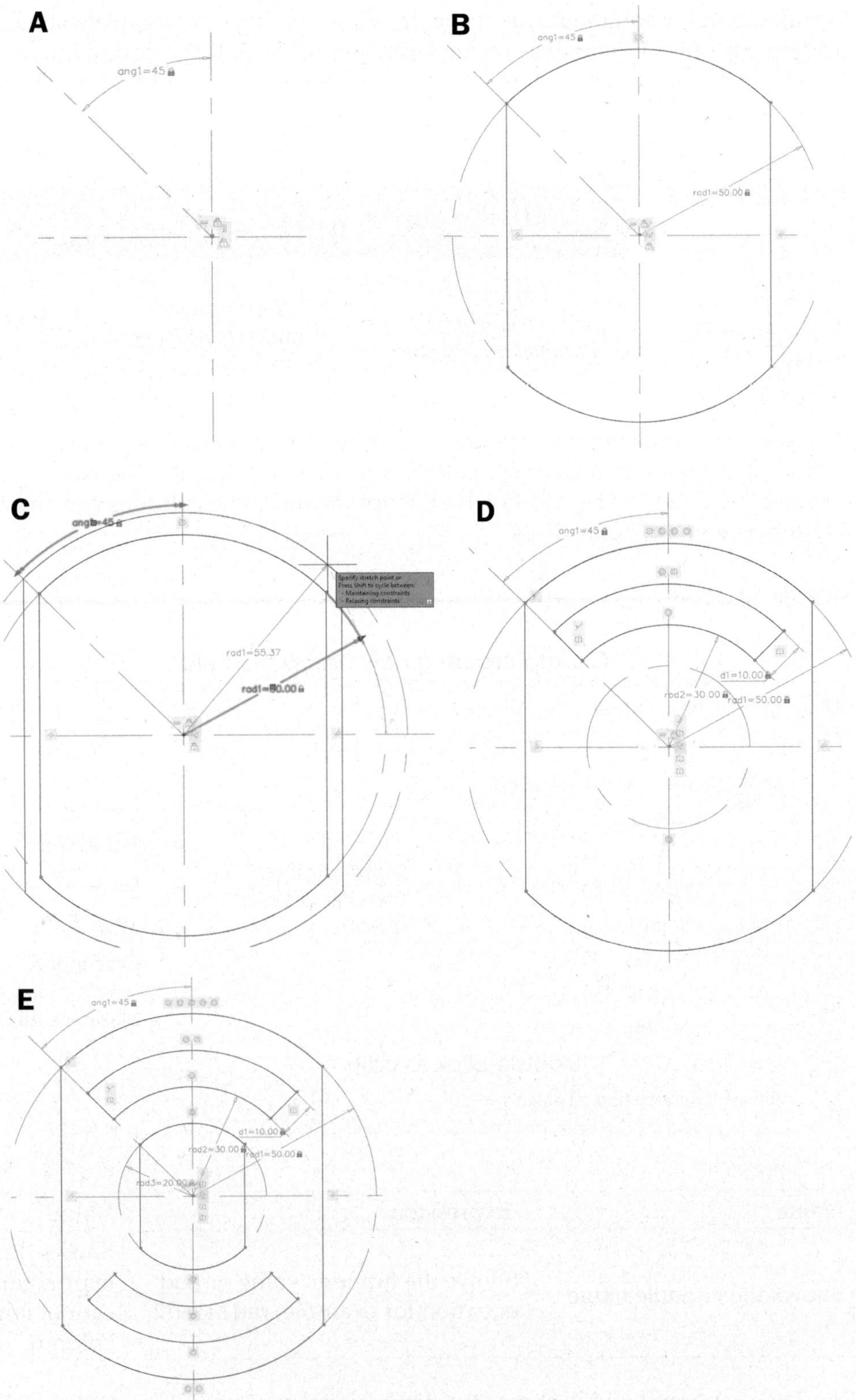


- Start with features that are unlikely to change, such as the main centerlines as shown in Figure 38.33 A. Consider adding a lock constraint to fix them on the coordinate system.
- Instead of drawing all of the geometry and then trying to constraint it all, use smaller increments. Start with just a main feature.
- Keep order in mind when selecting. For example, when adding a coincident constraint the second object selected moves to be coincident to the first item selected. To keep the arc endpoint on the angled line in this example, the angled line is the first object selected (Figure 38.33 B).
- Test your constraint set by dragging geometry or a dimensional constraint to verify that the constraints function as intended (Figure 38.33 C). If things "fall apart" you may need to add coincident constraints.
- Add additional features that relate to the main ones you have already constrained (Figure 38.33 D & E).

# PARAMETER VARIABLES AND EQUATIONS

You can use the Parameters Manager to add mathematical expressions that define dimensional constraints. You can also type values and have the drawing geometry update to match the value. The Parameters Manager lets you define custom user variables to use within dimensional constraints and to make complex equations that drive drawing geometry. You can also include predefined functions. Adding custom user variables make it easy to understand the design intent when looking at the dimensional constraint expressions. It also makes it easy to edit a constant or value and have it update the geometry.

## *PARAMETERS*

| Ribbon | Menu Bar | Command (Type) | Alias (Type) |
|---|---|---|---|
| *Parametric Manage Parameters Manager* | *Parametric Parameters Manager* | PARAMETERS | *PAR* |

Entering the Parameters command or selecting it from the ribbon, shows the Parameters Manager palette. The Parameters Manager palette is made of a grid showing Name, Expression, and Value columns, by default. The additional Description and Type columns can be added using the shortcut menu as shown in Figure 38.34.

**FIGURE 38.34**

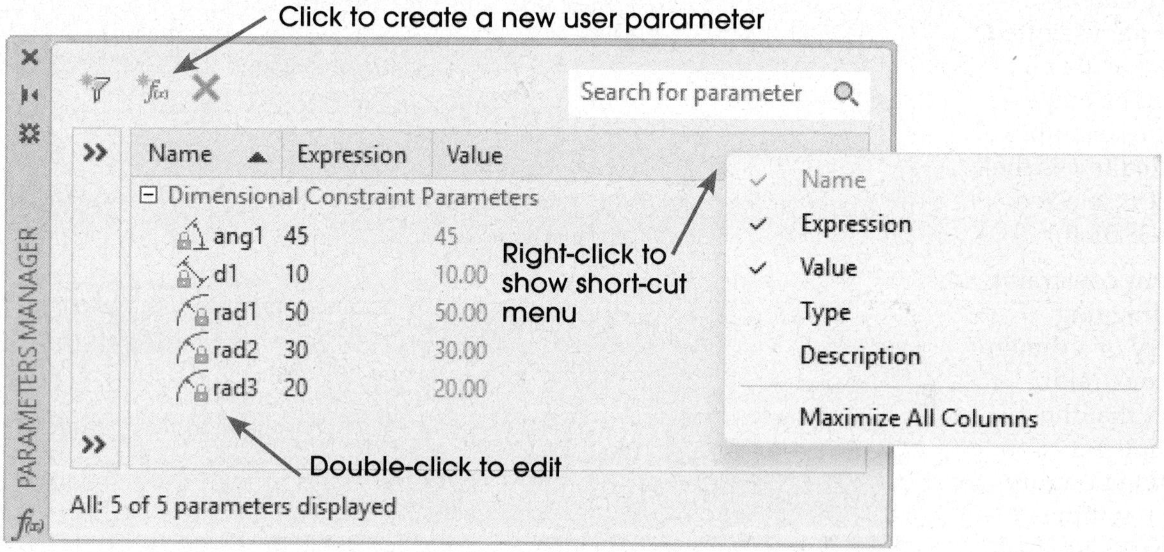


| Name | Expression | Value |
|---|---|---|
| Shows the variable name | Shows the numeric value or and equation for example, rad1+rad2 | Shows the evaluated expression value. You cannot edit the Value column. |

The Type column if added shows the dimensional constraint type or the variable value. The Description column shows comments or notes you have added to that variable.

## Creating Dimensional Variables

Whenever you apply a dimensional constraint to an object in the drawing, a new create a dimensional variable is automatically created. Figure 38.34 shows an example of those created making the gasket from Figure 38.33.

FIGURE 38.35

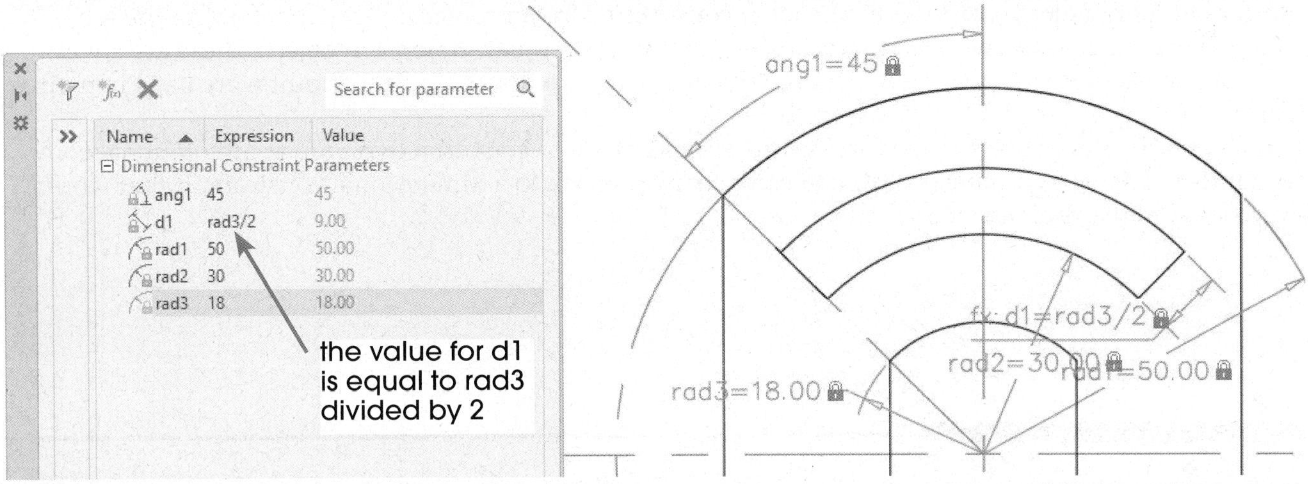


FIGURE 38.36

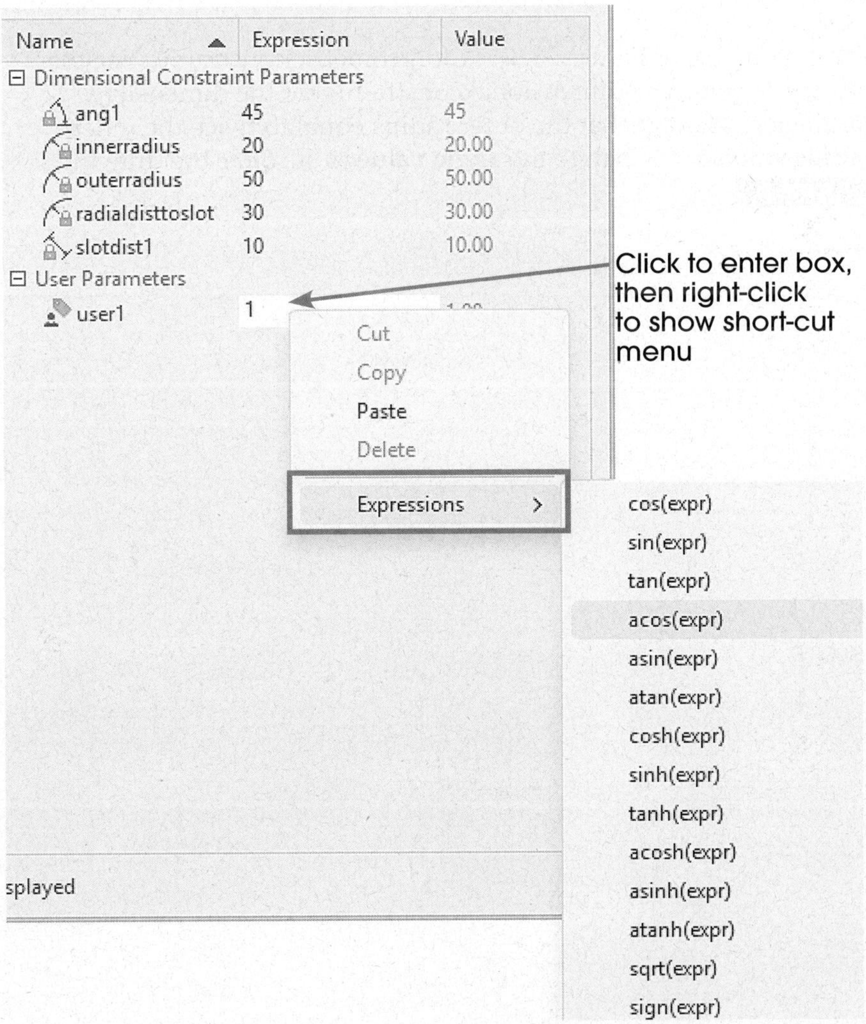


Click the New User Parameter icon from near the top of the Parameters Manager palette to create a new user variable or as a short-cut, double-click an empty area at the bottom of the Name list. The names you give variables must be alphanumeric and cannot begin with a number, contain spaces, or exceed 256 characters. Values can range within -1e100 and 1e100.

You can create expressions for the variables. The Expressions shortcut menu contains common mathematical functions. Invalid expressions display in red text as you are editing.

When you rename a dimensional variable, all of the instances of that variable automatically update in the drawing and in the Parameters Manager palette.

**TIP**

Using the shortcut menu from the Name and Expression cells lets you copy and paste the values.

When you delete a dimension variable from the list, the dimensional constraint in the drawing is removed. Other variables that referenced the deleted one use its previously evaluated value as a constant to keep the geometry stable.

### Reference Parameters

A reference parameter is a *driven* dimensional constraint. Driven constraints do not control the associated geometry. They receive a measured value similar to a regular dimension object. These are a way to display measurements that overconstrain the geometry if a dimension constraint were used in their place.

You can set the Reference property in the Properties palette to convert a dynamic or annotational constraint to a reference parameter. But, you can't change it back to a dimensional constraint if that would overconstrain the geometry.

## CHAPTER EXERCISES

1. Angle *Gasket*

   Begin a *New* drawing and create the geometry in Figure 38.37. Add geometric constraints. See the example in the chapter to help you get started. Use dimensional constraints for the dimensions shown in the figure. Using the Parameters Manager set the outer radius equal to twice the inner radius value. Change the inner radius value to 8. Change the angle value to 40. *Save* the drawing and assign the name **ANGLEGASKET.**

**FIGURE 38.37**

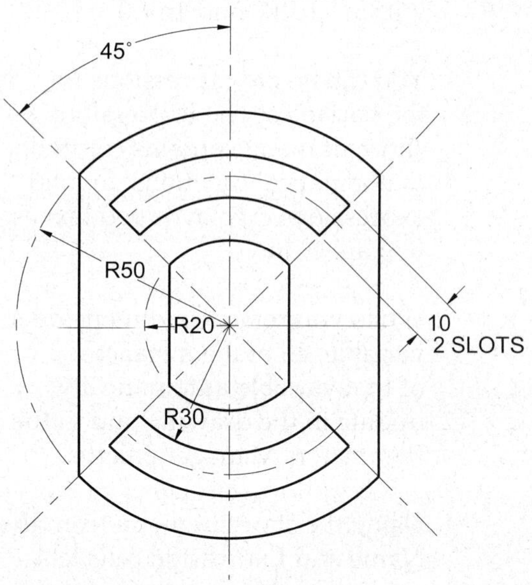

**FIGURE 38.38**

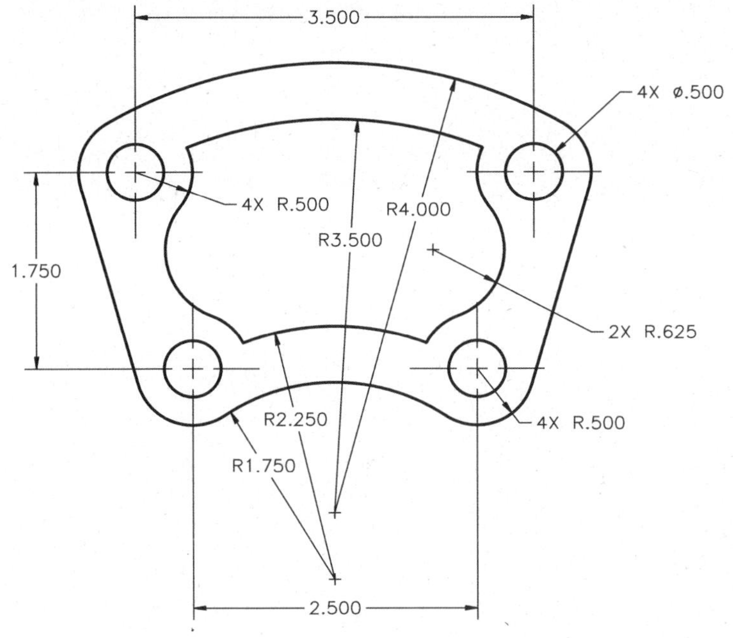


2. Create the Gasket as shown in Figure 38.38 using geometric constraints. Locate and draw the four .5″ diameter ***Circles***, then create the concentric .5″ radius arcs as full ***Circles***, then ***Trim***. Add construction geometry on a separate layer to use to aid in constraining features. Make use of ***Object Snaps*** and ***Trim*** whenever applicable. ***Save*** the drawing and assign the name **GASKETA2.**

**FIGURE 38.39**

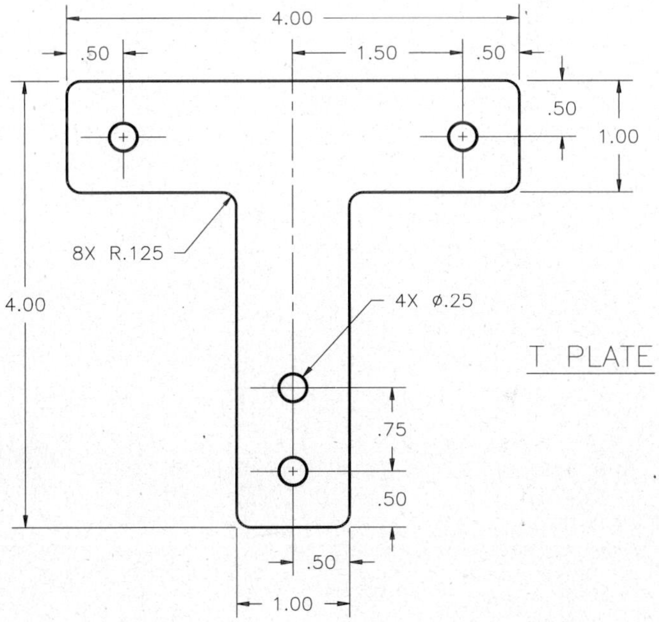


3. Create the "T" Plate shown in Figure 38.39 using inferred constraints. Use ***Fillet*** to create all the fillets and rounds as the last step. Set a constraint so that the hole diameters are equal. Simplify the dimensional constraints by creating user variables for dimensions which have the same value, for example, the 1.00 width. ***Save*** the drawing as **T-PLATE-INFER**.

# A

# Appendix A

AutoCAD 2027 Command Alias List
Sorted by Command

# APPENDIX A

## AutoCAD Command Alias List Sorted by Command, See AutoCAD Help

You can show the AutoCAD Alias Editor by typing ***Alias*** at the command prompt. The dialog box appears on your screen. Click the heading to sort by either Alias or AutoCAD Command. Use Add to create your own command aliases. You can also remove aliases. This is handy when you want to use those letters for a different command.

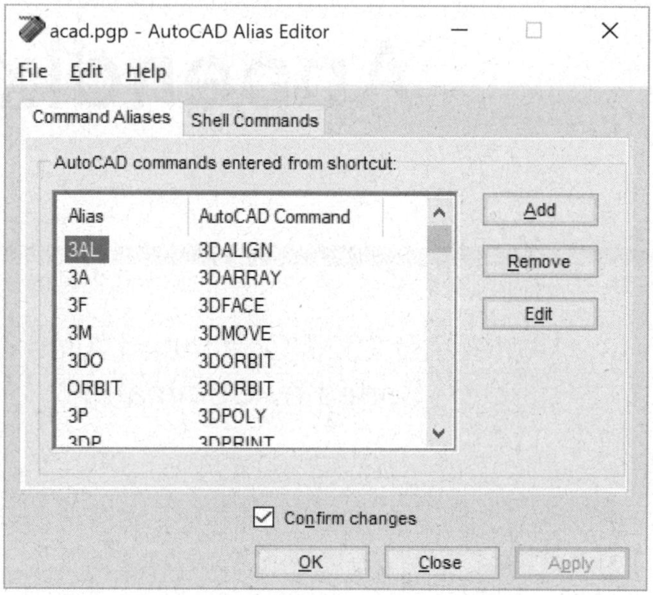


| *Command* | *Alias* |
|---|---|
| *-ACTSTOP* | *-ARS* |
| *-ACTUSERMESSAGE* | *-ARM* |
| *-ARRAY* | *-AR* |
| *-ATTDEF* | *-ATT* |
| *-ATTEDIT* | *-ATE* |
| *-ATTEDIT* | *ATTE* |
| *-BLOCK* | *-B* |
| *-BOUNDARY* | *-BO* |
| *-COORDINATIONMODELATTACH* | *-CMATTACH* |
| *-ExportToAutoCAD* | *AECTOACAD* |
| *-GROUP* | *-G* |
| *-HATCH* | *-H* |
| *-IMAGE* | *-IM* |
| *-INSERT* | *-I* |
| *-LAYER* | *-LA* |
| *-LAYOUT* | *LO* |
| *-LINETYPE* | *-LT* |
| *-LINETYPE* | *-LTYPE* |
| *-MTEXT* | *-T* |
| *-OSNAP* | *-OS* |
| *-PAN* | *-P* |
| *-PARAMETERS* | *-PAR* |
| *-PARTIALOPEN* | *PARTIALOPEN* |
| *-PURGE* | *-PU* |
| *-RENAME* | *-REN* |
| *-UNITS* | *-UN* |
| *-VIEW* | *-V* |
| *-VISUALSTYLES* | *-VSM* |
| *-VPOINT* | *-VP* |
| *-WBLOCK* | *-W* |
| *-XBIND* | *-XB* |
| *-XREF* | *-XR* |
| *3DALIGN* | *3AL* |
| *3DARRAY* | *3A* |
| *3DFACE* | *3F* |
| *3DMOVE* | *3M* |
| *3DORBIT* | *3DO* |
| *3DORBIT* | *ORBIT* |
| *3DPOLY* | *3P* |
| *3DPRINT* | *3DP* |
| *3DROTATE* | *3R* |
| *3DSCALE* | *3S* |
| *3DWALK* | *3DNavigate* |
| *3DWALK* | *3DW* |
| *ACTRECORD* | *ARR* |
| *ACTSTOP* | *ARS* |
| *ACTUSERINPUT* | *ARU* |
| *ACTUSERMESSAGE* | *ARM* |
| *ADCENTER* | *ADC* |
| *ADCENTER* | *DC* |
| *ALIGN* | *AL* |
| *ALLPLAY* | *APLAY* |
| *ANALYSISZEBRA* | *ZEBRA* |
| *APPLOAD* | *AP* |
| *ARC* | *A* |
| *AREA* | *AA* |
| *ARRAY* | *AR* |

| | |
|---|---|
| ATTDEF | ATT |
| ATTEDIT | ATE |
| ATTIPEDIT | ATI |
| BACTION | AC |
| BCLOSE | BC |
| BCPARAMETER | CPARAM |
| BEDIT | BE |
| BLOCK | B |
| BOUNDARY | BO |
| BPARAMETER | PARAM |
| BREAK | BR |
| BSAVE | BS |
| BVSTATE | BVS |
| CAMERA | CAM |
| CENTERLINE | CL |
| CENTERMARK | CM |
| CHAMFER | CHA |
| CHANGE | -CH |
| CHECKSTANDARDS | CHK |
| CIRCLE | C |
| COLOR | COL |
| COLOR | COLOUR |
| COMMANDLINE | CLI |
| CONSTRAINTBAR | CBAR |
| CONSTRAINTSETTINGS | CSETTINGS |
| COORDINATIONMODELATTACH | CMATTACH |
| COPY | CO |
| COPY | CP |
| CTABLESTYLE | CT |
| CVHIDE | POINTOFF |
| CVREBUILD | REBUILD |
| CVSHOW | POINTON |
| CYLINDER | CYL |
| DATAEXTRACTION | DX |
| DATALINK | DL |
| DATALINKUPDATE | DLU |
| DBCONNECT | DBC |
| DDGRIPS | GR |
| DELCONSTRAINT | DELCON |
| DIMALIGNED | DAL |
| DIMANGULAR | DAN |
| DIMARC | DAR |
| DIMBASELINE | DBA |

| | |
|---|---|
| DIMCENTER | DCE |
| DIMCONSTRAINT | DCON |
| DIMCONTINUE | DCO |
| DIMDIAMETER | DDI |
| DIMDISASSOCIATE | DDA |
| DIMEDIT | DED |
| DIMJOGGED | JOG |
| DIMJOGGED | DJO |
| DIMJOGLINE | DJL |
| DIMLINEAR | DLI |
| DIMORDINATE | DOR |
| DIMOVERRIDE | DOV |
| DIMRADIUS | DRA |
| DIMREASSOCIATE | DRE |
| DIMSTYLE | D |
| DIMSTYLE | DST |
| DIST | DI |
| DIVIDE | DIV |
| DONUT | DO |
| DRAWINGRECOVERY | DRM |
| DRAWORDER | DR |
| DSETTINGS | DS |
| DSETTINGS | SE |
| DVIEW | DV |
| EDITSHOT | ESHOT |
| ELLIPSE | EL |
| ERASE | DELETE |
| ERASE | E |
| EXPLODE | X |
| EXPORT | EXP |
| EXTEND | EX |
| EXTERNALREFERENCES | ER |
| EXTRUDE | EXT |
| FILLET | F |
| FILTER | FI |
| FLATSHOT | FSHOT |
| GEOGRAPHICLOCATION | GEO |
| GEOGRAPHICLOCATION | NORTH |
| GEOGRAPHICLOCATION | NORTHDIR |
| GEOMCONSTRAINT | GCON |
| GRADIENT | GD |
| GROUP | G |
| HATCH | BH |

| | |
|---|---|
| HATCH | H |
| HATCHEDIT | HE |
| HATCHTOBACK | HB |
| HIDE | HI |
| HIDEPALETTES | POFF |
| IMAGE | IM |
| IMAGEADJUST | IAD |
| IMAGEATTACH | IAT |
| IMAGECLIP | ICL |
| IMPORT | IMP |
| INSERT | I |
| INSERTOBJ | IO |
| INTERFERE | INF |
| INTERSECT | IN |
| ISOLATEOBJECTS | ISOLATE |
| JOIN | J |
| LAYER | LA |
| LAYERSTATE | LAS |
| LAYERSTATE | LMAN |
| LENGTHEN | LEN |
| LINE | L |
| LINETYPE | LT |
| LINETYPE | LTYPE |
| LIST | LI |
| LIST | LS |
| LTSCALE | LTS |
| LWEIGHT | LINEWEIGHT |
| LWEIGHT | LW |
| MARKUP | MSM |
| MATBROWSEROPEN | MAT |
| MATCHPROP | MA |
| MEASURE | ME |
| MEASUREGEOM | MEA |
| MESHCREASE | CREASE |
| MESHREFINE | REFINE |
| MESHSMOOTH | CONVTOMESH |
| MESHSMOOTH | SMOOTH |
| MESHSMOOTHLESS | LESS |
| MESHSMOOTHMORE | MORE |
| MESHSPLIT | SPLIT |
| MESHUNCREASE | UNCREASE |
| MIRROR | MI |
| MIRROR3D | 3DMIRROR |

| | |
|---|---|
| MLEADER | MLD |
| MLEADERALIGN | MLA |
| MLEADERCOLLECT | MLC |
| MLEADEREDIT | MLE |
| MLEADERSTYLE | MLS |
| MLINE | ML |
| MOVE | M |
| MSPACE | MS |
| MTEXT | MT |
| MTEXT | T |
| MVIEW | MV |
| NAVSMOTION | MOTION |
| NAVSMOTIONCLOSE | MOTIONCLS |
| NAVSWHEEL | WHEEL |
| NAVVCUBE | CUBE |
| NEWSHOT | NSHOT |
| NEWVIEW | NVIEW |
| OFFSET | O |
| OPTIONS | OP |
| OSNAP | OS |
| PAN | P |
| PARAMETERS | PAR |
| PASTESPEC | PA |
| PDFSHX | EPDFSHX |
| PEDIT | PE |
| PLINE | PL |
| PLOT | PRINT |
| POINT | PO |
| POINTCLOUDATTACH | PCATTACH |
| POINTLIGHT | FREEPOINT |
| POLYGON | POL |
| POLYSOLID | PSOLID |
| PREVIEW | PRE |
| PROPERTIES | CH |
| PROPERTIES | MO |
| PROPERTIES | PR |
| PROPERTIES | PROPS |
| PROPERTIESCLOSE | PRCLOSE |
| PSPACE | PS |
| PTYPE | DDPTYPE |
| PURGE | PU |
| PUSHTODOCSCLOSE | P2DCLOSE |
| PUSHTODOCSOPEN | P2D |

| | |
|---|---|
| PYRAMID | PYR |
| QLEADER | LE |
| QUICKCALC | QC |
| QUICKCUI | QCUI |
| QUICKPROPERTIES | QP |
| QUIT | EXIT |
| QVDRAWING | QVD |
| QVDRAWINGCLOSE | QVDC |
| QVLAYOUT | QVL |
| QVLAYOUTCLOSE | QVLC |
| RECTANG | REC |
| REDRAW | R |
| REDRAWALL | RA |
| REGEN | RE |
| REGENALL | REA |
| REGION | REG |
| RENAME | REN |
| RENDER | RR |
| RENDERCROP | RC |
| RENDERPRESETS | RP |
| RENDERWINDOW | RW |
| REVOLVE | REV |
| ROTATE | RO |
| RPREF | RPR |
| SCALE | SC |
| SCRIPT | SCR |
| SECTION | SEC |
| SECTIONPLANE | SPLANE |
| SECTIONPLANEJOG | JOGSECTION |
| SEQUENCEPLAY | SPLAY |
| SETVAR | SET |
| SHADEMODE | SHA |
| SHEETSET | SSM |
| SHOWPALETTES | PON |
| SLICE | SL |
| SNAP | SN |
| SOLID | SO |
| SPELL | SP |
| SPLINE | SPL |
| SPLINEDIT | SPE |
| STANDARDS | STA |
| STRETCH | S |
| STYLE | ST |

| | |
|---|---|
| SUBTRACT | SU |
| SURFBLEND | BLENDSRF |
| SURFEXTEND | EXTENDSRF |
| SURFFILLET | FILLETSRF |
| SURFNETWORK | NETWORKSRF |
| SURFOFFSET | OFFSETSRF |
| SURFPATCH | PATCH |
| SURFSCULPT | CREATESOLID |
| TABLE | TB |
| TABLESTYLE | TS |
| TEXT | DT |
| TEXTALIGN | TA |
| TEXTEDIT | DDEDIT |
| TEXTEDIT | ED |
| TEXTEDIT | TEDIT |
| THICKNESS | TH |
| TILEMODE | TI |
| TOLERANCE | TOL |
| TOOLBAR | TO |
| TOOLPALETTES | TP |
| TORUS | TOR |
| TRIM | TR |
| UCSMAN | UC |
| UNION | UNI |
| UNISOLATEOBJECTS | UNHIDE |
| UNISOLATEOBJECTS | UNISOLATE |
| UNITS | UN |
| VIEW | V |
| VIEWGO | VGO |
| VIEWPLAY | VPLAY |
| VISUALSTYLES | VSM |
| VPOINT | DDVPOINT |
| VPOINT | VP |
| VSCURRENT | VSM |
| WBLOCK | W |
| WEDGE | WE |
| XATTACH | XA |
| XBIND | XB |
| XCLIP | XC |
| XLINE | XL |
| XREF | XR |
| ZOOM | Z |

## AutoCAD 2027 New Commands

| | |
|---|---|
| *CHECKOUT* | *Check out and modify selected geometry in the Trace workspace while someone else has that drawing open.* |
| *GEOMETRYCLEANUPCLOSE* | *Closes the Geometry Cleanup palette.* |
| *GEOMETRYCLEANUPOPEN* | *Opens the Geometry Cleanup palette which displays potential gap, overshoot, undershoot, and angular errors for the current drawing.* |
| *GEOMETRYERRORREVIEWCLOSE* | *Closes the Geometry Cleanup review window.* |
| *PUSHTOFORMACLOSE* | *Closes the Push to Forma Data Management palette.* |
| *PUSHTOFORMAOPEN* | *Opens the Push to Forma Data Management palette.* |

## AutoCAD 2027 New Variables

| | |
|---|---|
| *CHECKOUTDISPLAYMODE* | *Controls the visibility of checked out objects in the trace and the drawing.* |
| *CHECKOUTFADECTL* | *Controls the amount of fading when TRACEMODE is inactive.* |
| *GEOMETRYCLEANUPSTATE* | *Indicates whether the Geometry Cleanup palette is open or closed.* |
| *GEOMETRYCLEANUPZOOMPAN* | *When set to 1, clicking a node in the tree view will auto-zoom directly to the highlighted instance(s).* |
| *PUSHTOFORMASTATE* | *Indicates whether the Push to Forma Data Management palette is open or closed.* |

## AutoCAD 2027 Discontinued Commands/Variables

**Discontinued Command/Variable**
*OPENFROMWEBMOBILE, PSOUT, PUSHTODOCSCLOSE, PUSHTODOCSOPEN, SAVETOWEBMOBILE, SHAREDVIEWS, SHAREDVIEWSCLOSE, -SHAREVIEW, SHAREVIEW, PUSHTODOCSSTATE, SHAREDVIEWSTATE, SHAREVIEWPROPERTIES, SHAREVIEWTYPE*

## AutoCAD 2026 New Commands

| | |
|---|---|
| *BDETECT* | *Displays the Detect palette where you can start object detection and manage the objects that are eligible for conversion.* |
| *BSEARCH* | *Displays the Convert dialog box, which provides options to convert selected entities and identical instances into blocks.* |
| *IMAGEASYNCWAIT* | *Ensures that raster images being loaded in the background during file opening are complete before proceeding.* |

## AutoCAD 2026 New Variables

| | |
|---|---|
| *BSEARCHINCLUDEEXISTINGBLOCKS* | *Controls whether existing block instances are selected by BSEARCH command by default.* |
| *IMAGEASYNC* | *Controls whether images are asynchronous loaded in the background during the opening of a drawing file.* |
| *MAXINTERSECTIONCURVEPOINTS* | *Controls the upper limit of the number of control points or vertices that splines and polylines can include in the intersection calculations.* |
| *PROJECTAWARE* | *Informs user/developer of the project path state of the current drawing.* |
| *TEXTTOATTRIBUTE* | *In the BSEARCH command, if text objects are selected, controls whether text variations will be searched for conversion into block attributes.* |

## AutoCAD 2025 New Commands/Variables

*ACTIVITYINSIGHTSCLOSE, ACTIVITYINSIGHTSOPEN, BCONVERT, -BCONVERT, DETECT, DETECTCLOSE , DETECTCONVERT, DETECTREMOVE , DETECTREVIEWNEXT, DETECTREVIEWPREV, ACTIVITYINSIGHTSSTATE, ACTIVITYINSIGHTSSUPPORT, BLOCKEXCLUDECOLOR, BLOCKINCLUDECOLOR, BLOCKTARGETCOLOR, ENABLESYNCPDF, DX12FRAMERATEUNLIMITED, HPDRAWMODE, HPPATHALIGNMENT, HPPATHWIDTH,*

# B

# Appendix B

Buttons and Special Keys

# APPENDIX B

## BUTTONS AND SPECIAL KEYS

### Mouse and Digitizing Puck Buttons

Depending on the type of mouse or digitizing puck used for cursor control, a different number of buttons are available. The *User Preferences* tab of the *Options* dialog box can be used to control the appearance of shortcut menus and to customize the actions of a right-click. In any case, the buttons have the following default settings.

| | | |
|---|---|---|
| #1 (left mouse) | **PICK** | Used to select commands or point to locations on screen. |
| #2 (right mouse) | **Shortcut menu or Enter** | Generally, activates a shortcut menu (see Chapter 1, "Shortcut Menus" and Chapter 2, "Windows Right-Click Shortcut Menus"). Otherwise, performs the same action as the Enter key on the keyboard. |
| press (center wheel) | ***Pan*** | Activates the realtime *Pan* command. |
| turn (center wheel) | ***Zoom*** | Activates the realtime *Zoom* command. |

### Function (F) Keys

Function keys in AutoCAD offer a quick method of turning on or off (toggling) drawing aids.

| | | |
|---|---|---|
| **F1** | ***Help*** | Opens a help window providing written explanations on commands and variables (see Chapter 5). |
| **F2** | ***Flipscreen*** | Activates a text window showing the previous command line activity (see Chapter 1). |
| **F3** | ***Object Snap Toggle*** | If Running Object Snaps are set, toggling this key temporarily turns the Running Object Snaps off so that a point can be picked without using Object Snaps. If no Running Object Snaps are set, F3 produces the *Object Snap* tab of the *Drafting Settings* dialog box (see Chapter 7). |
| **F4** | ***3DOsnap*** | Toggles *3D Object Snap* on or off. You can also set the *3D Object Snap* settings with the *3DOSMODE* system variable. |
| **F5** | ***Isoplane*** | When using an *Isometric* style *Snap* and *Grid* setting, toggles the crosshairs (with *Ortho* on) to draw on one of three isometric planes (see Chapter 25). |
| **F6** | ***Dynamic UCS*** | Toggles *Dynamic UCS* on and off (see Chapter 35). |
| **F7** | ***Grid*** | Turns the *Grid* on or off (see Chapter 1, "Drawing Aids"). |
| **F8** | ***Ortho*** | Turns *Ortho* on or off (see Chapter 1, "Drawing Aids"). |
| **F9** | ***Snap*** | Turns *Snap* (*Grid Snap* or *Polar Snap*) on or off (see Chapter 1, "Drawing Aids" and Chapter 3, "*Polar Tracking* and *Polar Snap*"). |
| **F10** | ***Polar Tracking*** | Turns *Polar Tracking* on or off (see Chapter 1, "Drawing Aids" and Chapter 3, "*Polar Tracking* and *Polar Snap*"). |
| **F11** | ***Object Snap Tracking*** | Turns *Object Snap Tracking* on or off (see Chapter 7). |
| **F12** | ***Dynamic Input*** | Toggles *Dynamic Input* (see Chapter 1, "Drawing Aids" and Chapter 3, "*Dynamic Input*"). |

## Control Key Sequences (Accelerator Keys)

Accelerator keys (holding down the Ctrl key and pressing another key simultaneously) invoke regular AutoCAD commands or produce special functions. Several have the same duties as F3 through F11.

| | | |
|---|---|---|
| **Ctrl+A** | *Select* | Selects all objects. |
| **Crtl+B** (F9) | *Snap* | Turns *Snap* on or off. |
| **Ctrl+C** | *Copyclip* | Copies the highlighted objects to the Windows clipboard. |
| **Ctrl+D** (F6) | *Dynamic UCS* | Toggles *Dynamic UCS* on and off (see Chapter 35). |
| **Ctrl+E** (F5) | *Isoplane* | When using an *Isometric* style *Snap* and *Grid* setting, toggles the crosshairs (with *Ortho* on) to draw on one of three isometric planes. |
| **Ctrl+F** (F3) | *Object Snap Toggle* | If Running *Object Snaps* are set, pressing Ctrl+F temporarily turns off the Running *Object Snaps* so that a point can be picked without using *Object Snaps*. If there are no Running *Object Snaps* set, Ctrl+F produces the *Object Snap Settings* dialog box. This dialog box is used to turn on and off Running *Object Snaps* (discussed in Chapter 7). |
| **Ctrl+G** (F7) | *Grid* | Turns the *Grid* on or off. |
| **Ctrl+H** | *PICKSTYLE* | Toggles *PICKSTYLE* On (1) or Off (0) and toggles selectable *Groups* on or off. |
| **Ctrl+J** | *Enter* | Executes the last command. |
| **Ctrl+K** | *Hyperlink* | Activates the *Hyperlink* command. |
| **Ctrl+L** (F8) | *Ortho* | Turns *Ortho* on or off. |
| **Ctrl+N** | *New* | Invokes the *New* command to start a new drawing. |
| **Ctrl+O** | *Open* | Invokes the *Open* command to open an existing drawing. |
| **Ctrl+P** | *Plot* | Produces the *Plot* dialog box for creating and controlling prints and plots. |
| **Ctrl+Q** | *Exit* | Exits AutoCAD. |
| **Ctrl+S** | *Qsave* | Performs a quick save or produces the *Saveas* dialog box if the file is not yet named. |
| **Ctrl+T** | *Tablet* | If a digitizing tablet is configured, Ctrl+T toggles the tablet on or off. |
| **Ctrl+U** (F10) | *Polar Tracking* | Turns *Polar Tracking* on or off. |
| **Ctrl+V** | *Pasteclip* | Pastes the clipboard contents into the current AutoCAD drawing. |
| **Ctrl+W** | *Selection Cycling* | Toggles *Selection Cycling* on or off (discussed in Chapter 4). |
| **Ctrl+X** | *Cutclip* | Cuts the highlighted objects from the drawing and copies them to the Windows clipboard. |
| **Ctrl+Y** | *Redo* | Invokes the *Redo* command. |
| **Ctrl+Z** | *Undo* | Undoes the last command. |
| **Ctrl+0** | *Clean Screen* | Toggles *Clean Screen* on and off. |
| **Ctrl+1** | *Properties* | Toggles the *Properties* palette. |
| **Ctrl+2** | *DesignCenter* | Toggles DesignCenter. |
| **Ctrl+3** | *Toolpalettes* | Toggles the *Tool Palettes* window. |
| **Ctrl+4** | *Sheet Set Manager* | Toggles the *Sheet Set Manager.* |
| **Ctrl+6** | *dbConnect* | Toggles the *DBConnect Manager.* |
| **Ctrl+7** | *Markup Set Manager* | Toggles the *Markup Set Manager.* |
| **Ctrl+8** | *QuickCalc* | Opens the *Sheet Set Manager.* |
| **Ctrl+9** | *Command Line* | Opens or closes the Command Line. |

## Special Key Functions

| | |
|---|---|
| **Esc** | The Escape key cancels a command, menu, or dialog box or interrupts processing of plotting or hatching. |
| **Space bar** | In AutoCAD, the space bar performs the same action as the Enter key or #2 button. Only when you are entering text into a drawing does the space bar create a space. |
| **Enter** | If Enter or Space bar is pressed when no command is in use (the open Command: prompt is visible), the last command used is invoked again. |

# Index

## Symbols



## A

## D

## W



## X



## Y



## Z

# Chapter Exercise Index

## CHAPTER EXERCISE INDEX

List of Drawing Exercise File Names (when files were created or changed)